2025 IEEE Applied Power Electronics Conference and Exposition (APEC 2025)

Atlanta, Georgia, USA
16-20 March 2025

Pages 586-1273

IEEE Catalog Number: CFP25APE-POD
ISBN: 979-8-3315-1612-3

**Copyright © 2025 by the Institute of Electrical and Electronics Engineers, Inc.
All Rights Reserved**

Copyright and Reprint Permissions: Abstracting is permitted with credit to the source. Libraries are permitted to photocopy beyond the limit of U.S. copyright law for private use of patrons those articles in this volume that carry a code at the bottom of the first page, provided the per-copy fee indicated in the code is paid through Copyright Clearance Center, 222 Rosewood Drive, Danvers, MA 01923.

For other copying, reprint or republication permission, write to IEEE Copyrights Manager, IEEE Service Center, 445 Hoes Lane, Piscataway, NJ 08854. All rights reserved.

*** *This is a print representation of what appears in the IEEE Digital Library. Some format issues inherent in the e-media version may also appear in this print version.*

IEEE Catalog Number: CFP25APE-POD
ISBN (Print-On-Demand):   979-8-3315-1612-3
ISBN (Online):   979-8-3315-1611-6
ISSN: 1048-2334

Additional Copies of This Publication Are Available From:

Curran Associates, Inc
57 Morehouse Lane
Red Hook, NY 12571 USA
Phone: (845) 758-0400
Fax: (845) 758-2633
E-mail: curran@proceedings.com
Web: www.proceedings.com

TABLE OF CONTENTS

Versatile Controller Architecture for a Universal DC Fast Charging Front-End 1
  Anurag Singh, Sayan Paul, Tejas Bhuse, Trent Martin, Hien Nguyen, Inder Vedula, Nikola Milivojeviœ, Dragan Maksimoviœ, Luca Corradini

A 10 kV SiC MOSFET based Three-Phase Single-Stage Isolated MVAC/LVDC Converter for Solid State Transformer Applications 9
  Anup Anurag, Chi Zhang, Rudy Wang, Peter Barbosa

Direct Digital Control Applied to T-Type Vienna Rectifiers for Power Factor Correction 16
  Jun-Yang Chang, Tsai-Fu Wu, Chien-Chih Hung, Jui-Yang Chiu

Active Power Decoupling Method based on Dual Active Bridge Converter without Additional Components 21
  Kosuke Takeuchi, Takashi Ohno, Hiroki Watanabe, Yuki Nakata, Jun-Ichi Itoh

An ANPC-Based Building Block for Medium-Voltage Applications 27
  Ahmed Rahouma, Hui Cao, David A. Porras, Zhuxuan Ma, Yue Zhao, Juan C. Balda

Analog Control of a 2.5 kW GaN based CRM PFC with Input Filter Optimization 34
  Naveed Ishraq, Ayan Mallik

An iTHD and Efficiency Optimized Control Method for Triangular Conduction Mode Totem Pole Bridgeless PFC with Zero Current Detection 42
  Brent McDonald, Sheng-Yang Yu

Resonance Current Suppression for AC-DC Active-Clamp Flyback Converter by Triangular Current Mode 48
  Yasuo Uchida, Hiroki Watanabe, Jun-Ichi Itoh

A Universal DC Fast Charging Front-End with Optimized Film Capacitor Design 54
  Sayan Paul, Anurag Singh, Tejas Bhuse, Trent Martin, Hien Nguyen, Inder Vedula, Nikola Milivojeviœ, Dragan Maksimoviœ, Luca Corradini

Power Characterization of a 1200-V/800-V 22-kVA 30-kHz Unity-Gain Dual-Active-Bridge Converter Prototype 62
  Radhika Sarda, Abishek Sethupandi, Madasamy Palavesha Thevar, Howe Li Yeo, Praveenkumar Palani, Vaisambhayana B. Sriram, Anshuman Tripathi

Design of Fully Soft-Switched Semi-Dual Active DC-DC Converter for Battery Charging Application 69
  Siva Prabhakar, Shiladri Chakraborty, Sandeep Anand

A ZCS-ZVS Strategy for Low Impedance Dual Active Bridges in MHz Range 77
  Pushkar Saraf, Michael Solomentsev, Alex Hanson

A 6.6 kW Highly Efficient Reconfigurable Dual Active Bridge Converter Designed using Planar Transformer, SiC-Fets and Monolithic Bidirectional Devices 90
  Reza Barzegarkhoo, Fabian Groon, Arkadeb Sengupta, Marco Liserre

Interleaved Switched-Inductor-Based SIPO Partial Power Converter Module for Battery Management Systems 98
  Fengwang Lu, Henry Shu-Hung Chung

Single Sensor-Based Fault Localization and Detection in GaN Three-Phase Dual Active Bridge Converters 103

Satyam Sa, Yi Han, Cheng Feng Wang, Olivier Trescases

Enhanced Cocharge Operation Scheme in Bidirectional PhaseShift Full-Bridge Converters with Eliminated Voltage Overshoot and Reduced Freewheeling Current.................. 111

Tien-Sheng Li, Minh Ngo, Rolando Burgos, Dong Dong

DC Bias Elimination in Isolated DC-DC Converters using Fundamental-Frequency Ripple 118

Arkadeb Sengupta, Thiago Antonio Pereira, Marco Liserre

Tunable Matching Network with Dual Phase-Switched Impedance Modulation Actuators.................. 124

Alexander Jurkov, David Perreault

Soft-Switched Pulsed Bias Plasma Supply System 132

Julia Estrin, Alexander Jurkov, David J. Perreault

Analysis and Design of a Cyclo-Active-Bridge Inverter for Single-Stage Three-Phase Grid Interface.................. 139

Mian Liao, Tanuj Sen, Yang Wu, Minjie Chen

Modular Nanosecond Pulse Generator Leveraging GaN and SiC for Versatility and Performance 147

P. Briz, H. Sarnago, O. Lucía

A Variable Frequency Technique for EMI and Efficiency Improvements in High-Level Count Flying Capacitor Multilevel Converters 151

Francesca Giardine, Sahana Krishnan, Logan Horowitz, Robert C. N. Pilawa-Podgurski

Analysis and Implementation of Minimum-Sensor Capacitor Voltage Estimators for Flying Capacitor Multilevel Converters 157

S. Tahmid Mahbub, Rahul K. Iyer, Ivan Z. Petriœ, Robert C. N. Pilawa-Podgurski

Single-Stage Bidirectional High-Frequency Link DC to Three-Phase AC (4-Wire) Grid-Tied Microinverter.................. 164

Aniruddh Marellapudi, Satish Belkhode, Joseph Benzaquen Sune, Deepak Divan

Analysis and Design of a Constant Current LCC Class-E Inverter 171

Ju Gao, Ziheng Liu, Jiayin He, Hongjie Peng, Chengkang Ao, Jinyan Wang

Series Connected Class-E Push-Pull Converters using GaN HEMT for High-Efficiency RF Generators in Float Zone Silicon Production 175

Faheem Ahmad, Thore Stig Aunsborg, Jannick Kjær Jørgensen, Stig Munk-Nielsen

State of the Art 1.7kV Lateral GaN HEMTs, an Alternative to SiC 180

Karthick Murukesan, Robert Yang, Kamal Varadarajan, Sorin Georgescu, Doug Kang

Modeling and Characterization of Current and Future 1.2 kV Wide Bandgap Semiconductor-Based MOSFETs.................. 185

Sushanta Gautam, Austin M. Szczublewski, Samuel K. Atwimah, Aidan P. Fox, William M. Collings, Tolen Nelson, Daniel G. Georgiev, Raghav Khanna, Andrew D. Koehler, Karl D. Hobart

2.5-kV 6.4-ns 100-kHz Repetitive GaN Marx Generator.................. 192

Ruize Sun, Ci Pan, Wanjun Chen, Bo Zhang

Novel Dual Output LDO Architecture in 650-V GaN Technology for Power ICs 195

Plinio Bau, Thanh Hai Phung, Deniz Aygun, Bart Coomans, Mike Wens

Impact of Substrate Bias on the Stability of Bidirectional GaN HEMT in Hard- and Soft-Switching 202
Qihao Song, Hongchang Cui, Qiang Li, Yuhao Zhang

Characterization of LED Driven GaN-Based Photoconductive Switches 207
Samuel K. Atwimah, Tolen M. Nelson, Geoffrey M. Foster, Daniel G. Georgiev, Andrew D.
Koehler, Alan G. Jacobs, Karl D. Hobart, Micheal R. Hontz, Raghav Khanna

Development and Validation of Repetitive Transient Gate Overvoltage Rating for GaN HEMTs 214
Ricardo Garcia, Angel Espinoza, Siddhesh Gajare, Shengke Zhang

Junction Temperature Monitoring of GaN HEMT by using On-Resistance with Voltage Clamp and
Current Shunt 219
Xiao Wang, Mingrui Zou, Jiakun Gong, Yulei Wang, Zheng Zeng

False Turn-On Failure and Protection of p-Gate GaN HEMT in MHz Class-E Resonant Inverter 225
Ziheng Liu, Ju Gao, Hongjie Peng, Jiayin He, Jinyan Wang, Maojun Wang

Heat Extraction from Ferrite Cores using Metallic Laminations 231
Alyssa Brown, Duy T. Nguyen, Alex J. Hanson

Folded Flex-PCB Winding Planar Transformer for High-Frequency Isolated DC-DC Converters 238
Soundhariya G. Soundararajan, Hans Wouters, Wout Vanderwegen, Wilmar Martinez

Winding Strategy Analysis and Optimization for High-Current Matrix Transformer 246
Bima Nugraha Sanusi, Pinhe Wang, Michael A. E. Andersen, Ziwei Ouyang

Investigation on Impact of Transformer Parasitic Capacitance on Standby Power Consumption in
Power Converters 252
Kamran Kamran, Andrea Russo, Federica Cammarata, Claudia Malannino, S. Yuri Ciardo,
Ziwei Ouyang

PCB-Winding Integrated Transformer for 800-V Dual Active Bridge Converter using 1.2-kV GaN
Devices 258
Hans Wouters, Wei-Ren Lin, Nicolas Pirson, Thomas Jochmans, Yu Zuo, Wilmar Martinez

Comparative Assessment of Inductance Modeling for PCB-Based Circular Spiral Coils in Inductive
Power Transfer Systems 266
Gaia Petrillo, Drazen Dujic

Compact Air-Core Inductors for Variable Frequency Soft-Switching in 3 Phase Inverters 272
Youssef A. Fahmy, Matthias Preindl

Simulation and Experimental Research on Cooling Performance of Fully-Immersed Evaporative
Cooling High-Frequency Transformer 278
Zhanlei Liu, Lingyu Zhu, Yuntian Gao, Yongliang Dang, Cao Zhan, Shengchang Ji

High-Efficiency PCB-Embeddable Inductor for Vertical Power IVR Applications 285
Youssef Kandeel, Liang Ye, John Flannery, Cian Ó Mathúna, Ranajit Sai, Seamus O'Driscoll,
Takayuki Tsuchida, Naoya Terauchi, Sumiaki Kishimoto, Toshio Hiraoka, Masanori Nagano

An Adaptive Zero Current Switching Control Technique for Multi-Resonant Switched-Capacitor
Converters 291
Haifah B. Sambo, Rose A. Abramson, Sahana Krishnan, Robert C. N. Pilawa-Podgurski

Small-Signal Analysis and External Ramp Design for Multiphase Current-Mode Constant On-Time
Control with Phase Overlapping 299
Sundaramoorthy Sridhar, Qiang Li

Multiphase Constant-on-Time Minimum-Deviation Controller for Modern Processors 307
  Duo Li, Gianluca Roberts, Aleksandar Prodić, Alan Wu

Closed-Loop Control of a Dual-Side Series/Parallel Piezoelectric-Resonator-Based DC-DC
Converter 315
  Wen-Chin B. Liu, Gaël Pillonnet, Patrick P. Mercier

High-Bandwidth Embedded Rogowski Coil on Multilayer Substrate with Minimal Contribution to
Power Loop Inductance 321
  Takahiro Okamoto, Masataka Ishihara, Kazuhiro Umetani, Eiji Hiraki

Operating and Switching Frequency Circulating Current Control in Paralleled High Power
Adjustable Speed Drives with Common DC Link 327
  Kevin Lee, Zhihao Song, Wenxi Yao, Bo Wei

Mixed-Signal Sliding Mode Controller for Non-Inverting Buck-Boost Photovoltaic DC Optimizers 334
  Anurag Singh, Sayan Paul, Dragan Maksimović, Luca Corradini

A Current Sensorless Output Voltage Tracking Controller-Observer for a Boost Inverter using
Feedback Linearization 342
  Ion Leandro Dos Santos, Tailan Orlando, Yohannes Amilcar Tekle Scherer, Telles Brunelli
  Lazzarin, Hector Bessa Silveira

Modeling and Control of a Cyclo-Active-Bridge Inverter for Single-Stage Three-Phase Grid
Interface 349
  Tanuj Sen, Mian Liao, Yang Wu, Minjie Chen

Turn-On Transient Modeling of 10 kV SiC MOSFET Half-Bridge Power Module in LTspice 357
  Nianzun Qi, Jannick Kjær Jørgensen, Gao Liu, Zhixing Yan, Morten Rahr Nielsen, Asger
  Bjørn Jørgensen, Hongbo Zhao, Stig Munk-Nielsen

A Compact, Automated Sawyer-Tower System for Characterization of the High-Frequency, Soft-
Switching C_{oss} Loss of Wide Bandgap Devices 363
  Katherine Liang, Malachi Hornbuckle, Juan Rivas-Davila

Enhancing Behind-the-Meter Visibility of Grid Edge PV Systems and Electric Vehicle Charging
Loads Through Integration of Compact Low-Cost Sensors 370
  Mehrnaz Madadi, Paul Ohodnicki, Subhashish Bhattacharya

Supercapacitor based TMS Pulse Generator Design-Experimental Results Versus MATLAB
MOSFET Simulation Model 378
  Soniya Raju, Nihal Kularatna, Marcus Wilson, Alistair Steyn-Ross

Application of Artificial Intelligence for Modeling SiC Power MOSFETs 385
  Fredo Chavez, Danial Bavi, Sourabh Khandelwal

Multi-Objective Design Automation in Power Electronics using Bayesian Optimization Techniques 389
  Tung-Tan Nguyen, Man-Hay Pong, Huang-Jen Chiu

Reduced Order Thermal Modelling of Multi-Chip Silicon Carbide Power Modules 395
  Aamir Rafiq, Blake Nelson, Marshal Olimmah

Design and Evaluation of Dual-Resolver Emulation for Control System Verification in Aerospace
Actuation Applications 401
  Tomas Sadilek, Julian Opificius, Jason Wright, Alec Leslie, Jeremie Tuzizila, Cesar Alzate,
  Hunter Burnett, Joshua Atkinson, Justin Stricula

Un-Terminated Blackbox Modeling for Electric Machines 409
Xinliang Yang, Vladimir Mitrovic, Qing Lin, Rolando Burgos

7.2 kW GaN-Based DAB Converter with 37 kW/L Power Density and High Efficiency 416
Esmaeil Jalalabadi, Xiaoyu Wang, Jaksa Rubinic, Yang Jiao, Lucas Lu

A Novel Interleaving Method for High Power Integrated Electric Vehicle Charger with Three-Phase
Permanent Magnet Synchronous Motor 423
Ryota Tanaka, Toshihiro Kai, Kenta Takishima, Yoshiyuki Nagai, Tetsuya Hayashi, Kantaro Yoshimoto

A Three-Phase CLLC Resonant Converter with Integrated Planar Magnetics for 22-kW On-Board
Chargers 429
Tianlong Yuan, Zhangwei Xiang, Abdelrahman Ali, Feng Jin, Qiang Li, Wendell Da-Cunha-Alves, Xiaoshan Liu

Reconfigurable LLC Resonant Converter for Wide Voltage Range and Reduced Voltage Stress in
DC-Connected EV Charging Stations 436
Yu Zuo, Xiaobing Shen, Bangli Du, Qingcheng Sui, Tim Geboers, Wilmar Martinez

Design and Control of GaN based Three-Phase / Single-Phase Combo Three-Level Flying
Capacitor PFC for OBC Applications 442
Nidhi Haryani, Laszlo Huber, Anup Anurag, Juan Ruiz, Peter Barbosa

Optimization Strategy for Battery Electric Vehicle (BEV) DC Fast Charging (FC) in Cold
Environments 449
Seif Sarofim, Cheng Feng Wang, Satyam Sa, Avram Kachura, Isaac Muscat, Olivier Trescases

DC-Link Voltage Reduction with Synergetic Common-Mode Voltage Control of Single-Phase Two-
Stage Non-Isolated EV Chargers 457
Dongsu Lee, Juwon Lee, Jung-Ik Ha

DC-DC Converter Architecture for Fast Electric Vehicle (EV) Battery Charging Applications 464
Shibaji Basu, Arjun Ivimey, Praveen Jain

Fast Simulator for the Estimation of Inverter DC-Link Temperature in e-Drives Subjected to Highly
Variable Working Cycles 472
Simone Giuffrida, Fabio Mandrile, Radu Bojoi

A Monolithic Regulated 160 MHz Resonant DC-DC Converter 479
Giacomo Ripamonti, Stefano Michelis, Georgios Bantemits, Pablo Daniel Antoszczuk, Khalil Khalife, Nils Hans Van Der Blij, Sokratis Koseoglou, Mattia Balutto, Francesco Driussi, Stefano Saggini

Reconfigurable Trans-Inductor Voltage Regulator with Improved Light Load Efficiency in Data
Center Applications 485
Ziyao Wang, Zehui Li, Haoyu Wang

Fully Integrated Voltage Regulators (FIVRs) with Package In-Situ Coupled CoaxMIL Inductor for
High Power Density Microprocessor Applications 491
Jaeil Baek, Beomseok Choi, Siddharth Kulasekaran, Huong Do, Brandon Marin, Jose Chavarria, Leigh Wojewoda, Kaladhar Radhakrishnan

Multiphase Lateral Flux Indirect Coupled Inductor for Vertical Power Delivery Voltage Regulator
Module 498
Adhistira M. Naradhipa, Qiong Wang, Qiang Li

A High Density Three-Level Quadratic Buck Hybrid Converter for 48V-to-PoL Conversion..........................505
    Kejia Wang, Si Yuan Sim, Yin Quen Choong, Xin Zhang, Sriharsh Pakala, Cheng Huang

Air-LEGO: A Magnetic-Free Ultra-Thin 24V-to-1V 120A VRM with Air-Coupled Inductors........................510
    Haoran Li, Wenliang Zeng, Youssef Elasser, Minjie Chen

A 15A 48V-Input Dual-Path Hybrid Dickson Converter with 6 mm³ Low Saturation Current
Inductors for Point-of-Load Conversion 518
    Hua Chen, Young-Seok Noh, Minxiang Gong, Vivek De, Arijit Raychowdhury

An Ultra-Fast Control Strategy and Pre-Current-Balancing Measures Prepared for Rapid Transients
in Constant On-Time Controllers 524
    Yijie Qian, Yuan Gao, Wenze Shu, Lingyun Li, Shen Xu, Weifeng Sun

Loosely Coupled Trans-Inductor Voltage Regulator (LC-TLVR) Inductor as Compensation Inductor
(Lc)...530
    Pavan Kumar, Arturo Sanchez Hernandez

Novel Complex Permeability Model of Powder Magnetic Materials..538
    Lukas Mueller, James Cox, Jun Wang, Enrique Garcia

Design Study Evaluating Impact of Gap Loss on Nanocrystalline Inductor Cores with Experimental
Validation 544
    Maurice Sturdivant, Brandon Grainger, Christopher Bracken, Paul R. Ohodnicki

A Permanent Magnet Variable Inductor for DC Fault Current Limiting Applications 552
    Mark Nations, Subhashish Bhattacharya

Design-Oriented Modeling and Multi-Objective Optimization of Two-Phase Coupled Inductors in
Multiphase PWM Converters 558
    Yicheng Zhu, Jiarui Zou, Robert C. N. Pilawa-Podgurski

MagNetX: Extending the Magnet Database for Modeling Power Magnetics in Transient 566
    Hyukjae Kwon, Shukai Wang, Haoran Li, Youssef Elasser, Gyeong-Gu Kang, Daniel Zhou,
    Davit Grigoryan, Minjie Chen

Non-Monotonic Influence of DC Bias on Ferrite Core Loss Up to 10 MHz with Sine Wave
Excitation 573
    Bohua Zhang, Martin Pfost

Comprehensive SPICE Model for Inductors Considering Magnetic Losses Under DC Bias Current 579
    Yuki Sato, Hirokazu Matsumoto, Junichi Kotani, Shohei Tomioka, Kenichiro Tanaka

Indented Core to Reduce and Desensitize Inductor's Fringing Losses without Increasing Volume................586
    Rajaie Nassar, Promit Datta, Guo-Quan Lu, Christina DiMarino, Khai Ngo

Coupled Inductor Analysis and Finite Element Modeling Assisted Design for Boost Extender
Topology...594
    Vikas Kumar Rathore, Michael Evzelman, Mor Mordechai Peretz

Stability Analysis of Current-Limited Grid-Forming Inverters with Frequency Stabilization: An
Equivalent Impedance Approach..602
    Bowen Yang, Gab-Su Seo

Revisit Active Power Oscillation in Multi-Virtual Synchronous Generators Gride 609
    Junjie Xiao, Pavol Bauer, Zian Qin

A Novel Current Control Technique for Off-Grid Single-Phase Inverters 616
Arpan Laha, Abirami Kalathy, Praveen Jain, Majid Pahlevani

Intelligent Low-Bandwidth Frequency Controller for VSGs at Economic Dispatch in Islanded
Microgrid 622
Shraf Eldin Sati, Ahmed Al-Durra, Hatem H. Zeineldin, Tarek H.M. El-Fouly, Ehab F. El-Saadany

Hardware-in-the-Loop of a Grid Forming Control Strategy Applied to a DC Off-Grid Green
Hydrogen Production System 629
Diego Montoya-Acevedo, René Contreras-Barrios, Ángel Maureira-Riquelme, Esteban Ibáñez-Muñoz, Catalina Gonzalez-Castaño, Carlos Restrepo

Experimental Validation of a 40kW, 480V Point-to-Point DC Interlinks for Controller-Agnostic,
Interoperable Networked Microgrids 637
Maximiliano Ferrari, Michael Starke, John Smith, Joao Pereira, Misael Montejano

Andronov-Hopf Oscillator-Based Grid-Forming Converters with Embedded Disturbance Rejection
for Non-Ideal Loading Condition 645
Vikram Roy Chowdhury, Gab-Su Seo, Barry Mather

Estimation of Rectifier Output Current of the LLC Converter 651
Xin Wu, Yi Zhou, Haihong Long, Dehong Xu

A 100kHz Digitally Controlled 10kW, 2-Channel Solar MPPT Converter using 3-Level Topology
with >75W/in³ Power Density and >98.5% Peak Efficiency 658
Ranajay Mallik, Akshat Jain

A Bootstrapless KY-S-Hybrid Buck-Boost Converter with Full Range iLs Reduction and 400%
Line Transient Response Acceleration for AI-Mobile Application 664
Chuan-En Chang, Cheng-Ta Chuang, Hao-Ran Huang, Chieh-Ju Tsai, Ching-Jan Chen

Digital Control of a 600-V to 28-V 20-kW Two-Stage DC-DC Converter 670
Shreyas B. Shah, Rachit Pradhan, Jiaqi Yuan, Mohamed Ibrahim, Ahmed Elezab, Samuel Hemming, Giorgio Pietrini, Piranavan Suntharalingam, Mario F. Cruz, Ali Emadi

Self-Calibrated Digital Current Emulation for High-Frequency Hysteretic Current-Mode Control in
GaN PFC Converters 676
Mohammad Shawkat Zaman, Olivier Trescases

High-Frequency Flying Capacitor Four-Level Drain Supply Modulator 682
Audrey Cheshire, Paul Flaten, Zoya Popoviœ, Dragan Maksimoviœ

Discontinuous Modulation Strategy for Voltage and Temperature Balancing of MMCs 689
Davide D'Amato, Stayner Nóbrega Barros, Jun-Hyung Jung, Marco Liserre

Damping Control and Improvement of Grid-Forming Inverter from a Wideband Stability
Perspective 696
Rui Kong, Subham Sahoo, Yubo Song, Frede Blaabjerg

A Grid-Forming Split-Phase Three-Leg Inverter with Unbalanced Loading and Active Power
Decoupling 703
Namwon Kim, Renata Kimpara, Michael Starke

Completely Decentralized Active and Reactive Power Control of Grid-Connected Cascaded H-
Bridge Inverters with Integrated Battery Storage 711
Soham Dutta, Brian Johnson

Small-Signal Modeling and Damping Design of Unfolding-Based Single Stage AC-DC Converter using the Extra Element Theorem 719
Dakota Goodrich, Aditya Zade, Shubhangi Gurudiwan, Mahmoud Mansour, Regan Zane, Hongjie Wang

Methods to Enhance Cybersecurity of Multiple Inverters in Large Grid Connected PV / Battery Energy Storage Systems 727
Hasan Ibrahim, Jaewon Kim, Peng-Hao Huang, Vishwam Raval, Prasad Enjeti

Optimal DC-DC Converter Topology and Control Algorithm for Fuel Cell Electric Vehicle with Series-Connected Supercapacitor 733
Hyeon Soo Kim, Yun Seong Hwang, Seung Hyun Kang, Man Jae Kwon, Byoung Kuk Lee

Reliability-Constrained Design of a High-Gain Power Optimizer based on a Real Mission Profile 738
Stefano Cerutti, Francesco Iannuzzo, Ariya Sangwongwanich, Tamás Kerekes, Mario Giuseppe Pavone, Francesco Gennaro, Natale Aiello, Francesco Musolino, Paolo Stefano Crovetti

Submodule Voltage Balancing Technique of Solar MMC for Firing the Switches using Integrated PWM Modules 746
Ahmed Elsanabary, Saad Mekhilef, Mokhtar Aly, José Rodriguez

Single-Stage High-Frequency-Link Split-Phase Microinverter with High Voltage Gain based on Buck-Boost AC Chopper 751
Xuewen Li, Jia Liu, Jinjun Liu

Fault Diagnosis and Tolerant Strategy for Triple-Port Hydrogen Converter using SSA-Optimized Random Forest Algorithm 757
Shiqi Zhang, Yiyina Teng, Naizhe Diao, Xiaoqiang Guo, Vladimir Terzija, Lichong Wang

Resilient Operation for Grid-Connected Cascaded H-Bridge Multilevel Inverter with Improving PV Source Stress 761
Jinli Zhu, Yuan Li, Hector Akuta, Jeonghun Kim, Uthandi Selvarasu, Shumeng Wang, Vikram Roy Chowdhury, Brad Lehman, Fang Z. Peng

A Medium Voltage Grid-Connected PV Inverter with a New Modular High Voltage Gain Converter Featuring Internal Modified Voltage Doubling Balancers 768
Kajanan Kanathipan, Muhammad Ali Masood Cheema, John Lam

Split-Source Common-Ground Inverter for Photovoltaic Applications 775
Mahmoud A. Gaafar, Mohamed Orabi, Samir Kouro, Ahmed Ibrahim, Eltaib Abdeen D. Ibrahim

Comprehensive Investigation and Proposal of a New Wireless Charging Road Structure using Low-Environmental-Impact Magnetic Concrete 782
Shuntaro Inoue, Yuko Kano, Shin Tajima

Design of a Bidirectional High Power Inductive Power Transfer System with Auxiliary Winding for Automotive Applications 788
Luis Ruiz Chamorro, Nikola Mirkoviœ, Alberto Delgado Expósito, Pedro Alou Cervera, Miroslav Vasiœ

Mutual Inductance and Load Identification Method based on the Voltage Transients of WPT Systems 795
Xiaosheng Wang, C.Q. Jiang, Yibo Wang, Liping Mo

Digitally Controlled Misalignment-Tolerant Inductive Power Transfer System with Adaptive Hybrid Compensation for CC/CV Charging of E-Scooter 801
Niranjan Shrestha, V.S.R.Varaprasad Oruganti, Sheldon Williamson

On/Off Control of Modular Inductive Power Transfer System 809
Kunxiao Zhou, Guangdong Ning, Heyuan Li, Xinlin Wang, Minfan Fu

Receiver Side Regulation of LCC Wireless Power Transfer System with Variable Notch Filter 815
Hsin-Che Hsieh, Jih-Sheng Lai

84.7 Percent Peak Efficiency Stress Tolerant DC DC Buck Converter for Li Ion Battery Driven Standby Circuits in 18nm FDSOI 821
Gautam Dey Kanungo, Pijush Kanti Panja, Vikas Bugade, Kallol Chatterjee

Leveraging Ultrasound and Neural Networks for Non-Invasive Power Converter Efficiency Estimation 828
Youssof Fassi, Vincent Heiries, Jérôme Boutet, Julien Marianne, Sébastien Martin, Mathilde Chareyron, Clément Chambon, Sébastien Boisseau

A Load-Independent Multi-Relays Wireless Power Transfer with Self-Regulation and Single Compensation Network 834
Jong-Hun Kim, Najam Ul Hassan, Seogyong Jeong, Myeong-Ho Kim, Min-Sik Kim, Jee-Hoon Jung, Byunghun Lee, Se-Un Shin

A GaN-Based Single-Stage Solid-State Transformer Replacement for 40 VA Class 2 Line-Frequency Transformers 840
Allen T. Nguyen, Charles R. Sullivan

Survey of Components and Topologies for High-Efficiency and High-Power Density 48V DC-DC Converters 848
Joseph Winkler, Niklas Deneke, Bernhard Wicht

A Novel Solid-State Circuit Breaker using B-TRAN™ 854
Mudit Khanna, Ruiyang Yu, Milad Tayebi, Jiankang Bu, Jeffrey Knapp

Development of a Supercritical Fluid-Insulated Fast Mechanical Switch for MVDC Hybrid Circuit Breakers 860
Zhiyang Jin, Qichen Yang, Alfonso Cruz, Lukas Graber

Dynamic Impedance Matching for a Variable Reluctance Energy Harvesting Application with Constrained Space 868
Fernando Pérez, Alejandro Redondo, Airán Francés, Gabriel Mujica

Renewable Energy-Powered DC-Converted Refrigerator based on a Supercapacitor-Assisted Technique 874
Nirashi Polwaththa Gallage, Nihal Kularatna, Alistair Steyn-Ross, Dulsha Kularatna-Abeywardana

Design and Evaluation of Flexible Inductors for Wearable Power Electronics 880
Sean Logi, F. Selin Bagci, Katherine A. Kim

Design of Boost Power Factor Corrector and Asymmetrical Half-Bridge Flyback Converter for USB-PD Applications 887
Yun-Keng Cheng, Tsorng-Juu Liang, Kai-Hui Chen, Ming-Chang Tsou

Computationally Efficient Current Sensorless Predictive Control for PMSM Drive Fed by a Matrix Converter with CMV-Free Operation 895
Ali Sarajian, Ibrahim Harbi, Quanxue Guan, Davood Arab Khaburi, Ralph Kennel, José Rodriguez, Patrick Wheeler, Mokhtar Aly

PMSM Motor Drive with Current Direct Digital Control and Near 1st-Order Speed Control 900
Po-Chang Lee, Tsai-Fu Wu, Han Ku, Chien-Chih Hung, Jui-Yang Chiu

Fault-Tolerant Multilevel Converter for Multiphase Switched Reluctance Motor Drives based on q+2 Converter 906
Mahmoud A. Gaafar, Mohamed Orabi, Hao Chen, Mostafa Dardeer

Uncertainty-Aware Artificial Intelligence for Gear Fault Diagnosis in Motor Drives 912
Subham Sahoo, Huai Wang, Frede Blaabjerg

Neural Network based Digital Twin Health Monitoring of BLDC Motor Drives for Robots 919
Mohamed Y. Metwly, Benjamin Luckett, Landon Clark, JiangBiao He, Biyun Xie

MTPA Control using Predictive P&O Method for Dual Parallel Surface-Mounted Permanent Magnet Synchronous Motor Drives Fed by a Single Inverter 925
Jae-Seong Kim, Kyo-Beum Lee

A Novel I-f Startup Strategy with Smooth Transition to Sensorless Control for CSI-Fed PMSM Drives used in Submersible Pumps 930
Milad Bahrami-Fard, Majid Ghasemi Korrani, Babak Fahimi

Simulation-Assisted Design and Implementation of an Electrically Excited Synchronous Motor Drive System 938
Shih-Gang Chen, Jun-Ming Hsu, Chun-Yen Chen, Ming-Shi Huang

Implementation and Analysis of Direct Torque Control on High-Speed PMSMs: A Comparative Study of Commercial and Laboratory-Developed Motors 943
Md Moniruzzaman, Kishor Joshi, Md Rashedur Rahman, Md Khurshedul Islam, Seungdeog Choi, Masoud Karimi Ghartemani

A Ferrite based Carbon Reinforced Composite Wrapped IPM Rotor Design for High-Speed Traction Applications 951
Md Rashedur Rahman, Md Khurshedul Islam, Md Moniruzzaman, Seungdeog Choi, Han-Gyu Kim, Andrew Walters

A Novel Phase-Mode Controller for Resonant Converters 958
Claudio Adragna, Daniele Cazzaniga, Stefano Manzoni

A Regulated 36V-60V-Input VIN-Insensitive Resonant Switched-Capacitor Converter with Large Voltage Conversion Ratio 966
Yichao Ji, Jingyi Yuan, Lin Cheng

A Hybrid Switched Capacitor Converter Enabling Capacitive-Based Wireless Power Transfer for Battery Charging Applications 971
Jade Sund, Samantha Coday

A 48V to 50-110V Resonant Power-Bus Charger with Reduced Conduction Loss for MHz-Frequency Long-Range LiDAR Driver 978
Hangxiao Ma, Xuchu Mu, Yang Jiang, Weihang Zhang, Jincheng Zhang, Rui P. Martins, Pui-In Mak

A Trajectory Controlled 48-to-24 V Resonant Switched Capacitor Converter with 98.7% Efficiency and Ultrafast Dynamic Response 983
Hélène T.W. Ma Yang, Liang Wang, Haoyu Wang, Wai Tung Ng

Low Power, Non-Isolated, Extremely-High Step-Up, Quasi-Resonant Hybrid DC–DC Converter 990
Kumar Joy Nag, Aleksandar Prodiœ

Isolated Soft-Switching Flying-Capacitor based Quasi-Resonant Step-Up Converter................................... 997
Kumar Joy Nag, Aleksandar Prodiœ

Accurate Small-Signal Phasor Transformation-Based Modeling of Secondary-Side Diode-Bridge Rectifiers for Battery Charging Applications 1004
Aditya Zade, Regan Zane

High-Efficiency Isolated Piezoelectric Transformers for Magnetic-Less DC-DC Power Conversion........... 1012
Sourav Naval, Wentao Xu, Mustapha Touhami, Jessica D. Boles

First Characterization of GaN Power Device and IC at Deep Cryogenic Temperatures Down to 100 mK.. 1020
Xin Yang, Matthew Porter, Zineng Yang, Zichen Xi, Liyang Jin, Liyan Zhu, Linbo Shao, Yuhao Zhang

Dynamic Environment-Aware Lifetime Prediction of SiC MOSFET Modules Through LSTM 1026
Md Zakir Hasan, Seungdeog Choi, Youssef Aider, Prashant Singh, Chun-Hung Liu

Guarding-Based C-V Characterization of 10 kV SiC MOSFET in Half-Bridge Module Configuration.. 1034
Nianzun Qi, Gao Liu, Zhixing Yan, Shaokang Luan, Pawel Piotr Kubulus, Yuan Gao, Stefan Meyer, Hongbo Zhao, Asger Bjørn Jørgensen, Stig Munk-Nielsen

Automated Characterization Platform for Comprehensive Dynamic R_{dson} Assessment of GaN HEMTs from 50 K to 400 K... 1040
Tian Qiu, Zheyu Zhang, Purushottam Khadka, Ahmed Siraj, Dilip Rana

A Gate Driving Scheme for GaN Git with Enhanced Short Circuit Capability for Motor Drive Application 1047
Zongjie Zhou, Yan Cheng, Kevin J. Chen

Online Detection and Reduction of the Influence of Parameter Tolerance of Paralleled SiC MOSFETs in an EV Inverter Environment.. 1051
Hadiuzzaman Syed, Jochen Streit, Robert Kragl, Muhammad Muneeb Alam, Alberto Martinez-Limia, Karl Oberdieck, Ertuðrul Sönmez

Dynamic Current Sharing Issues with Paralleling SiC Power MOSFETs.. 1058
Ching-Yao Liu, Chen-Chan Lee, Jih-Sheng Lai

Integrated Short-Circuit Protection Design based on Dual-Channel Gate Driver for Series Connected Medium-Voltage SiC MOSFETs 1063
Rui Wang, Drazen Dujic

Long-Term High-Temperature Dynamic Gate Stress Reliability of a Last-Generation, Automotive-Grade, Planar 1200 V SiC MOSFET 1070
Giuseppe Mauromicale, Alessandro Sitta, Michele Fiore, Michele Calabretta, Francesco Iannuzzo

Innovative Gate Driver Structure Achieving Low Time Skew Across Isolation Barrier for Parallel Connected SiC Modules 1076
Louison Gouy, Anne-Sophie Descamps, Nicolas Ginot, Christophe Batard

Fully Integrated Closed-Loop Active Gate Driver IC with Real-Time Control of Gate Current Change Timing by Gate Current Sensing 1084
Yaogan Liang, Katsuhiro Hata, Makoto Takamiya

Analyze and Design of Digitally Load Current Modulated Active Gate Driver for GaN HEMTs based Buck DC-DC 1090
Wentao Liu, Zhina Lian, Taotao Wu, Xiaochuan Peng, Hao Min

Impact of Real-Time Variable Gate-Drive Strength on Drive Cycle Efficiency in SiC Inverter-Fed PMSM Traction Drives 1096
Matteo Pizzuto, Aiswarya Balamurali, Aniket Anand, Narayan C. Kar

Demonstration of Efficiency Increase of 350 V-to-13.3 V Isolated DC-DC Converters for Electric Vehicles by Active Gate Driving 1102
Yohei Sukita, Katsuhiro Hata, Hiroki Kondo, Kenichi Watanabe, Kenichi Nagayoshi, Makoto Takamiya

A Multi-Level Active Gate Driver for Achieving Thermal Balance in Parallel Connected Power MOSFETs 1108
Jingyuan Liang, Lingwei Sun, Wen Tao Cui, Wai Tung Ng, Motomitsu Iwamoto, Haruhiko Nishio

A Fast Short-Circuit Protection Method for Ohmic Gate P-GaN HEMT based on Gate Charge 1114
Yue Wu, Xi Jiang, Song Yuan, Xiaowu Gong, Zhaoheng Yan, Jiahong Chen, Yun Xu, Jinjie Liu

Comparison of Ultrafast-Rise-Time Gate Drivers for Wide-Bandgap Devices in Sub-Microsecond Pulsed Power Applications 1121
Soham Roy, Duy T. Nguyen, Neeraj Anantha, Alex J. Hanson

A Discrete Multilevel Active Gate Driver for GaN HEMTs to Optimize the Switching Behavior 1129
Celine Lawniczak, Martin Pfost

Attenuation of Fundamental Component of Differential Mode Noise using Active EMI Filter 1135
Guru Abhilash Mulumudi, Naveed Ishraq, Ayan Mallik

Graph Neural Network based Performance Modeling for the Dual Active Bridge Converter with Operational Generalization 1143
Weihao Lei, Fanfan Lin, Xinze Li, Xiaokun Bao, Xin Zhang

An Augmented State Space Modelling Approach for DC-DC Converter Start-Up in Closed Loop 1148
Waseah Anjum, Arkadeb Sengupta, Marco Liserre

The Utilization of a Parallel Computing Algorithm for Accelerating Switching-Level Modeling of Power Electronics Simulations in a T-Type PV Inverter 1153
Buck F. Brown III, Liwei Wang, Zheyu Zhang, Johan Enslin, Yi Li

A New Reduced Order Analytical Switching Model for eGaN HEMTs 1159
Ruqi Li, Douglas Arduini, Phen Lumod, Shobhana Punjabi, River Lin, Harold Gutierrez

Proposal of an Alternative Reverse Recovery Calculation Method.. 1167
Brian Deboi, Blake Nelson, Austin Curbow

Improvement of CM EMI Attenuation Ability of Transformer with Negative Capacitor 1173
  Qinghui Huang, Yiming Li, Yirui Yang, Shuo Wang, Yanwen Lai, Zhedong Ma

Damping Factor based PCB Parasitic Inductance Value Optimization to Minimize Voltage
Overshoot and Settling Time of Semiconductors 1179
  Reza Shahbazi, Yunting Liu

Hardware Implementation of Virtual Resistance based FRT Logic in Programmable 3-Level ANPC
Inverters..............1184
  Mohammad Safayet Hossain, Shuvangkar Chandra Das, Paychuda Kritprajun, Amin Banaie,
  Tapas Barik, Deepak Ramasubramanian, Aboutaleb Haddadi, Evangelos Farantatos, Ulrich
  Muenz

Rad-Hard PSFB Controller for High-Voltage Space Applications 1190
  Reynaldo S. Gonzalez, Robert E. Bolaños

Modeling, Control and Digital Implementation of a Buck Converter Operating in Triangular
Current Mode for a Wide Output Voltage Range Space Application..............1197
  Regina Ramos, Sara Pérez, Guillermo Núñez, Pedro Alou, Javier Torres

Thermal Model and Optimization of a Multi-Winding Transformer for Lunar Surface Power
Transmission..............1203
  Zhining Zhang, Yuzhou Yao, Junchong Fan, Juchen Yang, Robert Guenther, Pengyu Fu, Jin
  Wang

Active Gate Driver Power Supply for High-Reliability Applications 1211
  Joseph P. Kozak, Juan Ramirez, Jesse Lin, Allison Orr, Alexander Martin, Hala Tomey

A Hybrid Energy Storage System for eVTOL Unmanned Aerial Vehicles using Supercapacitors..............1217
  Ali Alenezi, PengHao Huang, Prasad Enjeti

Evaluation of Retired Lithium-Ion Batteries for Second-Life Applications Through Electrochemical
Impedance Spectroscopy 1224
  Latha Anekal, Sheldon Williamson

Uninterruptable Non-Isolated Integrated Power Electronics Converter (UNIPEC) for Commercial
Truck Auxiliary Power Unit 1230
  Pouya Zolfi, Ahmad Alzahrani, Ayman El-Refaie

Investigation of Electrical Safety for Non-Isolated Single-Phase On-Board Chargers used in
BEV/PHEV 1237
  Soya Kataoka, Shohei Funatsu, Hiroaki Matsumori, Takashi Kosaka, Keisuke Nakamura,
  Subrata Saha

An 8-Level Flying Capacitor Multilevel Converter for Electric Aircraft Pulse Deicing 1242
  Nicole Stokowski, Andrew Freeman, Aidan Rodgers, Aria Delmar, Jonathan Sengstock, Alex
  Solecki, Andrew Stillwell

Impact of Position Measurement Delay Angle on Performance of PMSM Drives for Electric Power
Steering in a Wide Speed Range..............1248
  Yingzhe Wu, Hengbin Zhang, Yuxiang Xue, Lisheng Wang, Hui Li, Shan Yin

Physical Parameter Estimation for a Two-Level VSI Three-Phase PMSM Electric Drivetrain 1255
  Bernard Steyaert, Ananda Tjakra Adisurja, Matthias Preindl

A Novel Two-Dimensional Random Switching Frequency PWM Method for Variable Frequency Drives 1261

  Mostafa Abarzadeh, Kevin Lee

Optimized Maximum Torque and Minimum Loss Fault-Tolerant Control Schemes for Dual Three-Phase PMSM 1267

  Syed Mohammad Maaz, Dong-Choon Lee

Wireless Actuation of Magnetic Robots with a Modular 60 mT 3-D Helmholtz Coil System 1274

  Konstantinos Manos, Yifan Rao, Tuo Zhao, Kevin Liu, Daniel Zhou, Calvin Nguyen, Eric Chen, Glaucio H. Paulino, Minjie Chen

A Versatile PHIL based Motor Emulator Testbench using a High-Performance Power Amplifier Testbench 1279

  Seyedeh Nazanin Afrasiabi, Rajendra Thike, Mathews Boby, K. S. Amitkumar

A 450V Three Phase GaN IPM Achieving 99.1% Efficiency in Smallest 12mm x12mm Package for 250W Power Delivery without Heatsink 1286

  Maik Kaufmann, Manu Balakrishnan, Stefan Herzer, Anand Chellamuthu, Hely Zhang

FEA based High-Frequency Synthesis for the Design and Optimization of GaN-Based Dual Three Phase Motor Drive System 1294

  Syed Imam Hasan, Alper Uzum, Ashraf Siddiquee, Yilmaz Sozer, Krishna Namburi

Evaluation of Passive Common-Noise Canceller Considering Both of Thermal Equilibrium and Common-Mode Noise Cancellation 1299

  Koji Mitsui, Kenshiro Katsura, Koki Notake, Koji Yamaguchi

Performance Evaluation of Isolated DC/DC Converters in Modularized Bridge Rectifier Solid-State Transformer 1305

  Zhenchao Li, Andrea Cervone, Drazen Dujic

Active and Reactive Power Flow Control of the Dual Active Bridge Converter 1311

  Lauryn Morris, Thomas W. Francois, Jonathan Saelens, Oroghene Oboreh-Snapps, Arnold Fernandes, Praneeth Uddarraju, Sophia A. Strathman, Jonathan W. Kimball

Comparative Analysis of Carbon Footprints and Material Usage of Solid-State Transformers and Low-Frequency-Transformer-Based MVac-LVdc Interfaces for High-Power EV Charging 1318

  Luc Imperiali, Rudy Wang, Anup Anurag, Peter Barbosa, Johann W. Kolar, Jonas Huber

Trade Study of Isolation Requirements and Magnetic Core Selection for Medium Frequency-Medium Voltage Transformers 1326

  Mohendro Kumar Ghosh, Mark A. Juds, Brandon Grainger, Ahmad El Shafei, Bogdan S. Borowy, Paul Ohodnicki

Comparative Evaluation of a Multilevel LLC Resonant Converter for a Modular DC/DC Stage in a Electrolyzer Power Supply 1334

  Samuel S. Queiroz, Levy F. Costa

Cost-Effectiveness Assessment of SiC MOSFET and Si IGBT Semiconductors in a Three-Level Resonant Converter for Solid-State Transformer 1341

  Samuel S. Queiroz, Levy F. Costa

Comparative Performance Analysis of Medium Voltage 3L-ANPC and 3L-DNPC Pole Enabled by Series-Connection of 10kV SiC MOSFETs and 10kV SiC JBS Diodes for Sine Triangle PWM Operation 1347
> Sanket Parashar, Shubham Rawat, Nithin Kolli, Raj Kumar Kokkonda, Subhashish Bhattacharya

A Zero Harmonic Distortion Master Converter for Medium Voltage Microgrids 1355
> Gabriel V. Ramos, Dener A. de L. Brandão, Thiago M. Parreiras, Danilo I. Brandão, Braz de J.C. Filho

An MILP Approach for Modeling and Analyzing the BESS for Smoothing Renewable Fluctuations Considering BESS Capacity Attenuation in the Bulk Power System with High Inverter-Based Resource Penetration 1363
> Hualong Liu, Wenyuan Tang

Thermal and Efficiency Characterization of Immersion Cooled SiC Traction Inverter 1368
> Yiju Wang, Reza Ilka, JiangBiao He

FPGA-Based Hybrid Simulator for Real-Time 3-D Temperature Monitoring of Power Converters 1375
> Xianghao Mo, Daniel Ríos Linares, Regina Ramos, Miroslav Vasiœ

A New Subassembly Concept for Enhanced Heat Dissipation and Reliability of Power Module 1383
> Yosuke Nakata, Yuji Sato, Shin Uegaki, Jun Fujita, Akihiko Furukawa, Masayoshi Tarutani

Stand-Alone R_{DS-ON} Sensor for In-Situ Prognostic, Protection and Reliability Enhancement of Power Converters 1388
> Zaheen Mustakin, Qiang Mu, Lucas Pereira, Jiale Zhou, Tiefu Zhao, Babak Parkhideh

Electrical Evaluation of a Modular High Voltage 3D Power Module using Direct Dielectric Liquid Cooling 1396
> Omar Sanjakdar, Yvan Avenas, Rachelle Hanna, Guillaume Piquet Boisson, Emmanuel Marcault, Antoine Philippe

Board Level Reliability of Gull-Wing, Micro-Leaded and Lead-Less Packaged MOSFETs in Automotive Environments 1403
> Christopher Liu, Vijayakrishna Satyamsetti, Xuanjing Wei, Christian Radici, Peter Vines, Wayne Lawson

Cost Effective and High Noise Immunity Methodology for Aging Evaluation of DC-Link Capacitors in Traction Inverters 1408
> Seyed Hossein Aleyasin, Fausto Stella, Radu Bojoi, Enrico Vico

A 3D Structure of Single-Sided Cooling Power Module with Low Thermal Resistance and Low Inductance 1414
> Hirofumi Hisamochi, Koki Notake, Yoshiaki Takahashi, Koji Yamaguchi

Aging of Y-Capacitor in an EMI Filter and Its Impact on Common-Mode Noises 1420
> Tahmid Ibne Mannan, Seungdeog Choi, Subarto Kumar Ghosh, Md Moniruzzaman

2200A/48V-to-1V Low-Profile Direct Power Converter with Standard PCB Transformer 1427
> Alejandro Figueroa, Pablo Mazariegos, Álvaro Cobos, Javier Goicoechea, Alejandro Castro, José A. Cobos

Single-Stage 48V-to-1V Regulator with a Half-Turn Transformer and Current-Doubler Rectifier 1433
> Xinmiao Xu, Qiang Li

Ultra-Low-Profile Single-Stage Voltage Regulator Module (VRM) for Next-Generation AI Accelerators.. 1439
  Xufu Ren, Jinfeng Zhang, Zhenshuai Rong, Borong Hu, Teng Long

Novel TLVR Operation in Multi-Stage Voltage Regulator Module with Current Multipliers..................... 1444
  Kevin Zufferli, Roberto Rizzolatti, Mario Ursino, Simone Mazzer, Gerald Deboy, Stefano Saggini

Interphase LC-Oscillation Suppression with Fast Line-Transient Response in 48-V Series-Capacitor Buck Converters for Automotive Applications 1451
  W.L. Jiang, Y. Liu, N. Khan, J. Pigott, H.J. Bergveld, V. Chaturvedi, O. Trescases

An Approach to Compensate for Low Frequency DC-Link Voltage Ripple in High Power ANPC Inverter 1459
  Shaozhe Wang, Ankit Vivek Deshpande, Rolando Sandoval, Erick Pool-Mazun, Enrique Garza-Arias, Prasad Enjeti

A Cascaded Multilevel Inverter System with Hot-Swapping and Fault Isolation Capability for Improved Resiliency... 1465
  Uthandi Selvarasu, Vikram Roy Chowdhury, Shumeng Wang, Jinli Zhu, Mahshid Amirabadi, Yuan Li, Brad Lehman

Layout Optimization for Parasitic Inductance Reduction of GaN-Based NPL.X Multilevel Inverter 1473
  Ali Halawa, Jinyeong Moon, Woongkul Lee

Topology Selection and Design Methodology for SiC based Solar Photovoltaic Medium Voltage Direct Grid Connect Inverters 1481
  Jenson Joseph C. Attukadavil, Baylon G. Fernandes

EMI Modeling of PCB-Based Three-Level Active Neutral-Point-Clamped GaN Converter....................... 1489
  Mohammad Hassan Adeli, Necmi Altin, Erkan Deniz, Adel Nasiri

A Novel Layout for Improving Current Sharing of Paralleled SiC MOSFETs with TO-247 Package........... 1495
  Che-Wei Chang, Matthias Spieler, Rolando Burgos, Ayman El-Refaie, Renato Amorim Torres, Dong Dong

A Sensor-Less IGBT On-State Voltage Estimation Method using Inverter Control Variables 1501
  Shuyu Ou, Subham Sahoo, Ariya Sangwongwanich, Yongjie Liu, Frede Blaabjerg

A Novel Non-Intrusive Online Monitoring Method for Diagnosing the Lift-Off of Bonding Wires in SiC MOSFETs.. 1507
  Keqi Song, Henry Shu-Hung Chung, Ho-Tin Tang

Optimizing MOSFET Selection for EMC-Critical Automotive Applications 1512
  Sacha J. Cazzitti, Christian Radici, Andrew J. Forsyth, Cheng Zhang, Peter Vines

Improving Dynamic Current Sharing Between Parallel MOSFETs by Optimizing Device Parameters 1519
  Kunal Jha, Kapil Kelkar, Marina Hedenik, David Penof

A 21.6 kW/L Two-Phase Immersion-Cooled Isolated DC-DC Converter 1529
  Aleksandar Ristic-Smith, Kawsar Ali, Daniel Rogers

Extraction of Common Mode Parasitic Capacitance in Balance Filter for the Prediction of EMI Noise Suppression 1537
  Qiuzhe Yang, Xingyu Chen, Zijian Wang, Qiang Li

A 660W, 96% Efficiency 3D Heterogeneously Integrated Digital DC/DC Power Module for Vertical Power Delivery 1544
Haoyu Wang, Xuliang Wang, Yan Wang, Xiaosen Liu

Planar Rogowski Coil-Based Switch Current Measurement for a 1.2 kV SiC MOSFET Embedded Die PCB... 1551
Matthias Spieler, Che-Wei Chang, Ayman M. El-Refaie, Dong Dong, Rolando Burgos

Effect of Magnetic Couplings on Conducted EMI of GaN-Based PFC Converter 1557
Tyler McGrew, Qiang Li

Optically-Controlled 3.3 kV SiC MOSFET with Fast Switching Speed and Low Optical Power 1564
Xin Yang, Guannan Shi, Liyang Jin, Yuan Qin, Matthew Porter, Che-Wei Chang, Xiaoting Jia, Dong Dong, Linbo Shao, Yuhao Zhang

Optimization Techniques for Parallel-Connected Devices in IPMs for Consumer Use 1569
Keisuke Kawamoto, Haruhiko Murakami, Teruaki Nagahara, Michael Rogers, Akiko Goto, Shoji Saito, Koichiro Noguchi

Investigating the Temperature Dependency and Operating Parameters of a Self-Driving Active Gate Driver 1576
Vin Loong Choo, Martin Pfost

Use of Switched-Capacitor Circuit to Generate Negative Gate-Source Voltage Pulses 1582
Ho-Tin Tang, Henry Shu-Hung Chung

An Optically Isolated Gate Driver with Simultaneous Data and Power Transmission Through a Miniaturized, Efficient Photonic Platform.. 1590
Jiajun Li, Mariia Klymenko, Yanqiao Li, William Scheideler, Jason T. Stauth

Optimal Shared Energy Storage Capacity Configuration in Multi-Energy Microgrids Considering Battery Lifetime Loss based on Relaxation Techniques 1597
Hualong Liu, Wenyuan Tang

Virtual Resistance Control for an Active Battery Management System 1602
Alastair P. Thurlbeck, Ashraf Siddiquee, Mithat John Kisacikoglu, Yilmaz Sozer

Internal Voltage Source Saturation Impact on Stability Limits of Grid Forming Converter 1610
Divyanshu Bansal, Aravind G., L. Umanand

A Zero Harmonic Distortion Grid-Connected Grid-Forming Converter for Battery Energy Storage System Applications 1615
Gabriel V. Ramos, Thiago M. Parreiras, Fangzhou Zhao, Xiongfei Wang, Braz de J.C. Filho

Single Cell Energy Router Justification for Three Phase Near Zero Energy Buildings 1622
Hossein Nourollahi Hokmabad, Tala Hemmati Shahsavar, Oleksandr Matiushkin, Tanel Jalakas, Oleksandr Husev, Juri Belikov

A Multi-UAV Charging Station Enabling Free Landing by Grid Pattern Transmitter..................................... 1629
Jungho Kim, Hyunkyeong Jo, Seoktae Seo, Bonyoung Lee, Hyungki Min, Franklin Bien

Capacitor Design for Self-Resonant Coils for Long-Distance Wireless Power Transfer System................... 1635
Mostak Mohammad, Vandana Rallabandi, Omer C. Onar, Gui-Jia Su

A 10.4-kW High-Power-Transfer-Density Multi-MHz Capacitive Wireless Power Transfer System for EV Charging Utilizing Stacked-Inverter Stacked-Rectifier Architecture 1640

Dheeraj Etta, Miguel Alvarez Dominguez, Sounak Maji, Syed Saeed Rashid, Khurram K. Afridi

Reduced-Fringing-Field Multi-MHz Capacitive Wireless Power Transfer System using Metasurface-Based Couplers with Active Field Cancellation 1646

Syed Saeed Rashid, Dheeraj Etta, Matteo Ciabattoni, Francesco Monticone, Khurram K. Afridi

Living Object Detection in Wireless Power Transfer Systems using Remote Capacitive Bio-Signals Monitoring.. 1653

Bruno M.G. Rosa, Paul D. Mitcheson

Modified N:1 Switched Capacitor Converter with Reduced Capacitor DC Bias Voltage for High Power Density 1659

Taewoo Lee, Dam Yun, Sunghyuk Choi, Jung-Ik Ha

Wide Range Digital Control for Three-Level Buck Converters with Sensorless Flying-Cap Voltage Balancing... 1666

Hossein Hajisadeghian, Giovanni Bonanno

A Comparative Investigation of a New Continuous Voltage Conversion Ratio Approach in a Zero-Inductor Voltage Converter... 1673

Sina Salehi Dobakhshari, Aamna Nasir Hameed, Binghui He, Mojtaba Forouzesh, Yan-Fei Liu

A 96.1% Peak Efficiency, 6.8 kW/in³, 48V-to-6V On-Package Intermediate Bus Converter with LV-GaN Power Transistors... 1681

Mausamjeet Khatua, Nachiket Desai, Harish Krishnamurthy, Sheldon Weng, Jingshu Yu, Huong Do, Samuel Bader, Han Wui Then, Krishnan Ravichandran, James Tschanz, Kaladhar Radhakrishnan, Vivek De

A 48V to 2.4V-5V 95.8%-Peak-Efficiency 869W/in³-Power-Density Fibonacci Dual-Path Hybrid DC-DC Converter with Inductor Current Reduction and Low Output Resistance.................................... 1687

Yichao Ji, Zeguo Liu, Lin Cheng

An Ultra-Fast Very Large Scale Interleaved Li-Fi Transmitter 1693

Daniel H. Zhou, Konstantinos Manos, Minjie Chen

Isolated PWM DC-DC Converter with Single Magnetic Component, ZVS and Self-Balanced Switched-Capacitor Voltage 1701

Pablo M. Gil, Juan Rodríguez, Diego G. Lamar

Analysis and Design of a Low-Complexity ZVS Buck-Boost Converter 1707

Burkhard Ulrich

A High Conversion-Ratio Hybrid Series-Parallel DC-DC Converter with Pseudo-Soft-Charging and Inductor Current Frequency Multiplication... 1715

Avinash Maddela, Kishalay Datta, Jason T. Stauth

A Real-Time Variation Control of Deadtime in GaN-Based Bidirectional Buck-Boost Converter for Lithium-Ion Battery Formation System... 1723

Jong-Hun Lim, Go Woon Heo, Je-Yeong Lim, Dong Hwan Kim, Byoung Kuk Lee

A Space Vector PWM Strategy for Charging of Bootstrap Capacitor in Three-Level Neutral-Point-Clamped Inverter 1728
Anantha Hegde, Asamira Suzuki, Hirokazu Nakamura, Takamune Kabashima, Koji Higashiyama, Keiji Akamatsu

A Complementary Carrier based PWM Strategy for Average Current Sampling of Three-Phase Inverter using Single Current Sensor 1734
Byeong-Il Kim, Joon-Seok Kim, Yeongsu Bak, June-Seok Lee

Short-Circuit Ride-Through for a CRM-Based Soft-Switching Three-Phase Inverter 1741
Xingyu Chen, Gibong Son, Qiang Li

Modified Space Vector Modulation with Low Bandwidth Sensor to Reduce Losses in Soft Switching Three-Phase Inverters 1746
Md Didarul Alam, Nazmul Hassan, Iqbal Husain, Liming Liu, Hongrae Kim

A Feedforward Ripple Reduction Control Strategy based on a Hybrid GaN/Si Interleaved Inverter 1754
Mowei Lu, Jurgis Reinotas, Xiaoyang Tian, Stefan M. Goetz

IGBT Comparison for Optimized Switching Behavior in the SiC/Si-Hybrid Switch 1759
Adrian Amler, Thomas Heckel, Daniel Ruppert, Cornelius Rettner, Martin März

Forward Recovery and its Mitigation in Hybrid Si/SiC-Based DC–AC Converters 1767
Yan Zhou, Thomas Lehmeier, Adrian Amler, Martin März

Real-Time IGBT Module Ageing Characterization Through Temperature Monitoring 1774
Quirc Perez-Farre, Luis F. Gomez-Rivera, Carlos Lopez-Torres, Kai Dannehl, Antoni Garcia-Espinosa, Alejandro Paredes-Camacho

Experimental Validation of Triangular SOA via Infrared Thermography of a MOSFET Die Operating in the Thermally Unstable Linear-Mode for Automotive Applications 1781
Yacine Ayachi Amor, Christian Radici, Kerry J. Abrams, Philip Ellis, Peter Vines, Wayne Lawson

Feasibility Study of the SuperIGBT: A Series-Connected High Voltage IGBT with a Single Gate 1786
Junhong Tong, Alex Q. Huang, Huanghaohe Zou, Zhiyuan Ma

Low Profile, Laminated Nife Transformers for Flyback Converters 1791
Xuan Wang, Reza Mounesi, Matthew Catanoso, Matthew Fox, Adel Nasiri, Mark G. Allen

Comprehensive Demonstration of New Magnetic Designs Utilizing Magnetic Anisotropy of the Cores for Integrated Magnetics 1797
Yota Takamura, Honami Nitta, Tatsuya Miyazaki, Kimito Yamanaka, Ryosuke Ishido, Akira Namba, Keisuke Fujisaki, Shigeki Nakagawa

A Two – Stage Artificial Neural Network (ANN) – based Design and Optimization of High Frequency Transformers for Dual Active Bridge Converter 1803
Lufan Zhou, Alberto Delgado Expósito, Adam Ruszczyk, Simon Round, Miroslav Vasiœ

Modeling and Optimizing Winding Arrangement for Gapped Planar Magnetics based on Artificial Neural Network 1810
Hanqing Cao, Bima Nugraha Sanusi, Ziwei Ouyang

Free-Shape Optimization of VHF Air-Core Inductors using a Constraint-Aware Genetic Algorithm 1816
Thomas Guillod, Charles R. Sullivan

Organic Direct Bonded Copper-Based Rapid Prototyping for Silicon Carbide Power Module Packaging 1824

Shuofeng Zhao, Joshua Major, Douglas DeVoto, Sarwar Islam, Xiaoling Li, Mike Tant, Faisal Khan, Sreekant Narumanchi

Discrete Power Device Packaging with Integrated Direct Two-Phase Cooling 1832

Jinpeng Cheng, Jinxiao Wei, Hao Feng, Li Ran

Investigation of Die Top-Side Re-Metallization for SiC-Based Double-Side Cooled Power Modules 1836

Narayanan Rajagopal, Christina DiMarino

Design of Low Parasitic Inductance GaN HEMT Flip-Chip Power Module 1844

Mohammad Dehan Rahman, Tanzila Akter, Abu Shahir Md Khalid Hasan, H. Alan Mantooth, Xiaoqing Song

A Scalable Dual-Orthogonal-Cooling Packaging Concept for Parallel-Series SiC Chips 1850

Ekaterina Muravleva, Youssef Abotaleb, Blake Anderson, Zichen Zhang, Boyi Zhang, Jerry Hudgins, Jun Wang

Parasitic Impact Analysis and Design of Hybrid EMI Filter for Active Clamp Flyback SMPS 1858

Tahmid Ibne Mannan, Seungdeog Choi, Masoud Karimi-Ghartemani

Overview of Dynamic Characterization of Switches for Three Phase Voltage Source, Current Source, and Matrix Converter Applications 1866

Sneha Narasimhan, Sathya Rupan Thirumoorthi, Subhashish Bhattacharya

Advanced Modeling Technique of Class-E Inverter Considering Low R_{on} of eGaN FETs and Different Design Procedures 1874

Manas Palmal, Jungwon Choi

PiezoNet and Data-Driven Models for Time-Domain Characterization of Piezoelectric Resonators 1882

Davit Grigoryan, Mian Liao, Haoran Li, Shukai Wang, Tanuj Sen, Matthew Tan, Minjie Chen

A New Gate Charge De-Embedding Method for Accurate On-Wafer Characterization of HV MOSFET Devices 1889

João R.R.O. Martins, Rachid Hamani, Vincent Quenette, Joerg Gessner

4 kW Auxiliary Power Module for Electric Vehicles Utilizing a Dual-Phase LLC DC-DC Converter 1892

Mojtaba Forouzesh, Xiang Yu, Yan-Fei Liu, Paresh C. Sen

New Reverse Mode Control Method of Phase-Shift Full-Bridge Converter for Bidirectional Auxiliary Power Module 1899

Jongyoon Chae, Dongmin Kim, Dongmin Choi, Gun-Woo Moon

In-Situ EV EIS with a High-Density Flying Capacitor Multi-Level Converter Supercapacitor System 1905

Avram Kachura, Gaël Vergès, Samantha K. Murray, Olivier Trescases

A Novel 500-kHz LLC-T Resonant Converter with Wide Output Range 1913

Zhengming Hou, Dong Jiao, Jih-Sheng Lai

High Efficiency Traction Drive Operation with a Partial Load Three-Phase Triangular Current Mode Modulation Concept 1919

Bhaskar Chatterjee, Jan Allgeier, Thomas Plum, Marc Hiller

Analysis of Maximum Power Transfer Limit for Linear Operation of Dual-Active-Bridge Converters 1927
Radhika Sarda, Ezequiel Ramos Rodriguez, Gaowen Liang, Glen G. Farivar, Josep Pou, Vaisambhayana B. Sriram, Anshuman Tripathi

Enhanced Control for Integrated Active Power Decoupling in Single-Phase Three-Level Flying Capacitor PFC Converter 1935
Gleisson Balen, Cristian Blanco, Ángel Navarro-Rodríguez, Pablo García, Rafael Peña-Alzola

Improving Transient Stability of PLL-Synchronized Grid-Following Inverters 1940
Surya Prakash, Kalpana Beura, Mohamed Alkhatib, Omar Al Zaabi, Khalifa Al Hosani, Utkal Ranjan Muduli

Online Impedance-Based Analysis for Power System Stability Assessment using Transformer-Less and Filter-Less Switch-Mode Perturbation Generator 1946
Tomoya Ide, Yuko Hirase, Cheng Huang, Takanori Isobe

PIR-R Control for Three-Phase Grid-Connected Inverter with Unbalanced Grid Current Correction 1953
Haneen Ghanayem, Xingyu Yang, Mohammad Alathamneh, R.M. Nelms

Design and Placement of a Passive Clamp Snubber for Isolated SEPIC and Cuk Converters Working as Automatic Power Factor Correctors 1959
Abraham López, Juan Rodríguez, Duberney Murillo-Yarce, Javier Sebastián, Diego G. Lamar

Current Sensorless Control Strategy for Single-Phase T-Type PFC Converter 1967
Che-Yu Lu, Jia-En Zeng

Three-Phase Single-Stage Multiport AC-DC Converter with Integrated DC-DC Conversion Stages 1972
Asad Hameed, Gerry Moschopoulos

High Efficiency AC-Adapter Realized by Voltage-Clamper with Mid-Voltage AHB Converter using Synchronous Rectification 1977
Shuichiro Motoori, Akihiro Kawano, Toshiyuki Zaitsu, Riku Tatetsu, Kohei Sebata, Kazuki Miyanjou, Kimihiro Nishijima

Active Soft Switching Technique for Single Phase Series Capacitive Link Universal Rectifier 1983
Anran Wei, Brad Lehman, Mahshid Amirabadi

A Multi Mode Control Algorithm for Totem-Pole Bridgeless PFC 1990
Bosheng Sun, Sheng-Yang Yu, Amir Hussain

Protection Strategy for Flying Capacitor Totem-Pole PFC Under the AC Drop Transient 1995
Yanqing Wu, Wending Zhao, Zhenhai Zhu, Xinke Wu

Three-Phase with Three Single-Phase Single-Stage Isolated AC-DC Converters for EV Charging Station Applications 2002
Misha Kumar, Peter M. Barbosa, Juan M. Ruiz

400V SiC in Next-Generation 3-Level Flying Capacitor Bridgeless Totem-Pole PFC 2009
Rytis Beinarys, Seamus O'Driscoll

Extended Smart-Link Quasi-Single-Stage 3-Phase AC-DC Power Supply Module for AI-Driving Data Centers 2014
D. Biadene, J. Huber, J.W. Kolar, P. Mattavelli

A New Three-Phase Multi-Mode AC/DC LLC Converter with Output-Controlled Active Rectifier (with V2G and G2V Functions) for Fast DC Charging Application 2022
Xiaoyi Xia, John Lam

Capacitorless Notch Resonant Converters for Miniaturized LLCLC Resonant Converters in Electric Vehicle Charging Applications 2029
Haitham Kanakri, Euzeli Cipriano Dos Santos Jr., Maher Rizkalla

Multiple-Core Transformer Design based on Half-Turn Structure in Two-Stage DC-DC Converter for Battery Storage System 2035
Yilei Li, Bima Nugraha Sanusi, Pinhe Wang, Tianming Luo

Bidirectional DC-DC Converter Utilizing Coupled Inductors for Energy Storage System 2043
Wen-Hsuan Lee, Jiann-Fuh Chen, Hsuan Liao, Kuo Fu Liao

Comparison of 2-Level and Quasi-2-Level Topologies in a Bidirectional Isolated DC-DC Converter for MVDC Networks 2051
José Andrés Aguilar Croston, Jean-Yves Gauthier, Cyril Buttay, Maryam Saeedifard, Besar Asllani, Piotr Dworakowski

Sling Forward Converter for Offline Operation: Achieving High Efficiency and Wide Voltage Range Performance 2059
Nasherul Islam, Guozhu Chen, Honglei Miao, Fuxing Zhang

A Pulse Width Alternating Modulation Strategy for Three-Level Buck-Boost Converter 2066
Xinlong He, Caifeng Liu, Xudong Zou, Jiaao Zou, Tianyi Zhang, Yong Kang

ZVT Circuit Applied for Wide Input Range Isolated Converters 2070
Linguo Wang, Zhongyin Guo, Junjie Zhu, Bing Zhang, Zhiling Zuo, Xiaoguang Gao, Guangji Ma

Impact of Asymmetrical Leakage Inductance on a 380 V-12 V LLC Converter with Synchronous Rectifier for DCX Application 2075
Jinshu Lin, Shan Yin, Chen Song, Honglang Zhang, Minhai Dong, Limei Xu, Hui Li

Start-Up Techniques and Universal Closed Loop Control of Immittance Network based Resonant Converter 2082
Ripun Phukan, Misha Kumar, Randy Beckemeyer, Juan Ruiz, Peter Barbosa

Multi-Objective Efficiency-Oriented Optimization for DAB Converters Minimizing Current Stress and Backflow Power with Soft-Switching Assurance 2088
Kun Wang, Ian Laird, Jun Wang

An ISOP-PSFB PWM Converter based on Coupled Output Inductors and Phase-Shifted Modulation with Full ZVZCS Range 2096
Kang Hong, Guo Xu, Guangfu Ning, Mei Su

Design and Implementation of a GaN-Based Soft-Switched Series-Capacitor Buck Converter Operating at the CCM-DCM Boundary for High-Performance Computing Systems 2101
Ramin Rahimzadeh Khorasani, Kolman Puterman Ghitelman, Madhavan Swaminathan

Intrinsic Feedback Model for Coupled-Damped Self-Balancing of General Multiphase Hybrid Converters 2109
Haoran Xu, Weijia Hao, Desheng Zhang, Run Min, Qiaoling Tong, Xuecheng Zou

A High-Efficiency Switching Oscillation Suppression Strategy based on Damped Oscillation for Synchronous DC-DC Converter 2117
Hao Yuan, Chuan Ni, Zhengyu Ye, Wei Lu, Hui Xue, Ting Qian

Efficient and Streamlined Demodulation Strategy for High-Frequency Talkative DC-DC Converters 2125
Abdelmoumin Allioua, Hendrik Gockel, Gerd Griepentrog

A 90.9% Peak Efficiency KY Single-Inductor Bipolar-Output Converter with Conductance Modulation Controller for Active-Matrix Organic Light Emitting Diode Power Supply 2131
Sheng-Han Yu, Chieh-Ju Tsai, Hao-Ran Huang, Ching-Jan Chen

Constant-on-Time Control for Zero-Bias Trans-Inductor Voltage Regulators 2138
Hank Zeng, Justin Lee, Rixin Lai, Hang Shao

An Improved PFM Control Scheme for Three-Level Buck Converter based on Ton Extension Achieving an 810% Frequency Reduction 2143
Yi-Chun Chang, Chieh-Ju Tsai, Ting-Lun Lee, Ching-Jan Chen

A Concept for Current Ripple or Transient Improvements in Multiphase Converters 2149
Alexandr Ikriannikov, Alex Gao

System Solutions and Design Trade-Offs to the Input Filter Interactions with Battery Chargers 2157
Xigen Zhou, Dan Mavencamp, Kuang-Yao Cheng

Modeling and Implementation of a Zero Bias TLVR 2162
Lei Wang, Travis Guthrie, Peyman Asadi, Mark Alexander, Kunrong Wang, Brandon Howell

cGANET-Enhanced Voltage Gain Modeling: Elevating CLLC Converter Accuracy 2167
Yu Zuo, Xiaobing Shen, Fanghao Tian, Jiaze Kong, Hans Wouters, Wilmar Martinez

Capacitive vs Inductive Coupling based DC-DC Converter Operating in MHz Switching Frequency Range 2173
Saeid Pourjafar, Parham Mohseni, Oleksandr Husev, Ryszard Strzelecki, Oleksandr Matiushkin

LLC Converter Main Transformer Losses: Eliminating Air Gaps and Integrating Parallel External Inductors 2179
Yu-Chen Liu, Shang-Syun Wu

Small-Signal Phasor Modeling of T-Type Bridge-Based Single-Sided and Double-Sided LCC Resonant Converters for WPT Applications 2194
Aditya Zade, Shubhangi Gurudiwan, Regan Zane

A Hybrid Three-Port Topology for Urban Charging Stations 2202
Mohammadreza Khodaparast Klidbari, Naser Souri, Zahra Sadat Habibolahi, Hamid Montazeri Hedeshi

Reconfigurable H5-Bridge based LLC-DAB Sigma Converter for EV Fast Charging Stations 2207
Huangsheng Xu, Mingde Zhou, Qishan Pan, Haoyu Wang

A Resonant Reset Forward Converter with Ultra-High Conversion Gain using Differential Transformation Technique (DTT) 2213
Shubham Srivastava, Mandeep S. Rana, Santanu K. Mishra

Full-Range ZVS Modulation of Switched Capacitor Converter for Sensorless Voltage Balancing 2220
Md Tanvir Ahammed, Wensong Yu

Dimensional Parasitics Absorption in Capacitively-Isolated Ćuk Converter for Medium-Voltage High Step-Down Converters 2228

Aakash Kamalapur, Jung-Soo Bae, Mark Cairnie, Rajaie Nassar, Jack Knoll, Dushan Boroyevich, Guo-Quan Lu, Christina DiMarino, Qiang Li, Khai D. T. Ngo

A 36-to-60V Input Dual-Phase 2MHz 93%-Efficiency ZVT Series-Parallel Hybrid Buck Converter using Single Auxiliary Inductor and Adaptive Time Multiplexing Control 2236

Qi Cheng, Hoi Lee

Improved Efficiency in a 10 W Class-Φ_2 Converter Utilizing a Resonant Gate Drive 2241

Malachi Hornbuckle, Katherine Liang, Juan Rivas-Davila

The Analysis and Design of a Resonant Capacitively-Isolated Cockcroft-Walton Converter 2249

Elizabeth Rabenold, Raiphy Jerez, Samantha Coday

SHSC: Non-Isolated High-Density 4:1 IBC for 48 V Applications 2254

Mario Ursino, Roberto Rizzolatti, Simone Mazzer

High-Performance Current Multiplier: A Hybrid Switched Capacitor Solution for High-Current Applications 2260

Kevin Zufferli, Roberto Rizzolatti, Mario Ursino, Simone Mazzer, Gerald Deboy, Stefano Saggini

Representation and Design Methodology for Generalized Switched-Capacitor Converter Topologies 2268

Seokwon Choi, Dam Yun, Jung-Ik Ha

A 48-V-to-1-V Gallium Nitride Switching Bus Converter for Processor Vertical Power Delivery with 2.7 mm Thickness and 3048 W/in³ Power Density 2276

Jiarui Zou, Yicheng Zhu, Nathan M. Ellis, Logan Horowitz, Robert C. N. Pilawa-Podgurski

Ripple Reduction and Efficiency Improvement of Always-Dual-Path Hybrid DC-DC Converter based on Phase Shift Operation 2284

Katsuhiro Hata, Shinsaku Tanaka, Toru Ashikaga, Yasuhiro Rikiishi

Ultralocal PQ Theory: A New Approach for Model-Free Predictive Direct Power Control of Shunt Active Power Filters 2290

Mahdi S. Mousavi, Abolfazl Nassaji, Ibrahim Harbi, Behnam Nikmaram, S. Alireza Davari, Mokhtar Aly, José Rodriguez

Symmetrical Balanced Circuit for Common-Mode Noise Mitigation in LCL-T Resonant Converter 2296

Ripun Phukan, Boyi Zhang, Juan Ruiz, Peter Barbosa

A Single-Phase Soft-Switching Buck-Boost Inverter 2303

Lukas Wipprecht, Burkhard Ulrich

Low-Complexity Model Predictive Control Method for Driving Dual Induction Motors Fed by Five-Leg Inverter 2311

Jun Young Lee, Eun Woo Lee, Dongho Choi, June-Seok Lee

Overvoltage Mitigation Filter using High-Frequency Cable Modeling in Long Transmission Lines for Silicon Carbide Inverter Systems 2317

Yun-Jin Lee, Kyo-Beum Lee

Power Delivery Network (PDN) Design and Analysis to Achieve Low Impedance in Fast Edge Rate DC-DC Converters for EMI Compliance 2322

Manraj Singh Ladhar, Sheldon Williamson

Enhancing the Performance of Dual Input Split Source Inverters using an Advanced Modulation Strategy 2327

Mustafa Abu-Zaher, Fang Zhuo, Mokhtar Aly, Mahmoud A. Gaafar, Mohamed Orabi, José Rodriguez, Alaaeldien Hassan, Jiachen Tian, Samir Kouro

A Novel GaN-HEMT Single-Phase Single-Stage Buck-Boost Micro-Inverter Topology for PV Applications 2332

Pengwei Li, Uiliam Kutrolli, Ali Bazzi

A Dynamic Current Sharing Method using Novel Clip Considering Mutual Inductance Coupling 2343

Zexiang Zheng, Jianwei Lv, Yiyang Yan, Baihan Liu, Yifan Zhang, Linhao Ren, Jiaxin Liu, Cai Chen, Yong Kang, Xiong Zhang, Hao Yu, Wei Jiang

Application-Oriented Test Setup for Measuring Dynamic Output and Transfer Characteristics of GaN-HEMTs 2348

Philipp Swoboda, Martin Fein, Simon Frank, Andreas Liske, Marc Hiller

Mitigating Gate Voltage Oscillation in Parallel SiC Power Modules for xEV 2356

Hideo Komo, Michael Rogers, Mark Steiner, Eric Motto, Koichi Taguchi, Chihiro Kawahara, Junichi Nakashima, Yasushige Mukunoki, Seiichiro Inokuchi, Rei Yoneyama

Switching Performance Comparison of Low-Voltage GaN and Si Devices 2361

Tianxiao Chen, Haoyang Liu, Pedro A.M. Bezerra, Eckart Hoene, Sibylle Dieckerhoff

Modeling of Switching Transients for Frequency-Domain CM EMI Analysis in Double Sided Cooling Power Modules 2369

Sijia Liu, Liu Yang, Heng Zhang, Yifan Zhang, Zexiang Zheng, Jianwei Lv, Jiaxin Liu, Cai Chen, Yong Kang, Yuebin Zhou, Daming Wang, Shuang Zhao

Leakage Current Detection Scheme for Aging Test of 10kV SiC MOSFET Power Module 2375

Peiyang Ding, Hong Zhang, Tianshu Yuan, Qiling Chen, Jiacheng Guo, Dingkun Ma, Peiyuan Sun, Ting Hou, Laili Wang

Physics-Informed Neural Network Approach for Early Degradation Trajectory Prediction of Power Semiconductor Modules 2380

Jie Kong, Yi Zhang, Yichi Zhang, Lukas Wick, Frederik Lillebæk Hansen, Dao Zhou, Huai Wang

Nonlinear Output Capacitance of Bidirectional Gallium Nitride Power Switches 2387

Michael Bosch, Jeremy Nuzzo, Dominik Koch, Mathias C.J. Weiser, Ingmar Kallfass

Novel Approach of Determining and Predicting SiC MOSFET's on Resistance from Device Case Temperature using Machine Learning 2393

Paul Bradford, Conner Deppe, Hongjie Wang

Comparison of Static Characteristics in GaN HEMTs Across 50K to 400K Considering Diverse Techniques and Statistical Variation 2400

Purushottam Khadka, Saumil Shivdikar, Zheyu Zhang, Tian Qiu, Ahmed Siraj

Compact Model of β-Ga$_2$O$_3$ Schottky Barrier Diode 2407

Abu Shahir Md Khalid Hasan, Mohammad Dehan Rahman, Tanzila Akter, Md Majharul Islam, Md Maksudul Hossain, Xiaoqing Song, H. Alan Mantooth

DC-Link Capacitor Board Design for Low Parasitic Inductance 2413

Mikayla Benson, Lifang Yi, Kangbeen Lee, Jinyeong Moon, Woongkul Lee

First Demonstration of a Gallium Oxide Power Converter 2419
*Joshua J. Piel, Elizabeth A. Sowers, Daniel M. Dryden, Thaddeus J. Asel, Adam T. Neal,
Brenton A. Noesges, Shin Mou, Andrew J. Green*

Optimized Integrated EMI Filter Design in SiC Power Modules with Terminal Inductor for Better
High-Frequency EMI Suppression 2426
*Yifan Zhang, Wenzhe Xu, Jianwei Lv, Yiyang Yan, Baihan Liu, Sijia Liu, Jiaxin Liu, Cai Chen,
Yong Kang, Xiong Zhang, Hao Yu, Wei Jiang*

Balanced Technique using Integrated Winding Coupled Inductor for High-Power Density Two-
Phase Interleaved Boost Converter 2431
Yuta Imaeda, Jun Imaoka, Masayoshi Yamamoto, Hiroyuki Onishi

MagNetX: Foundation Neural Network Models for Simulating Power Magnetics in Transient 2438
*Shukai Wang, Hyukjae Kwon, Haoran Li, Youssef Elasser, Gyeong-Gu Kang, Daniel Zhou,
Davit Grigoryan, Minjie Chen*

Revisiting Models of Common Mode Inductors to Include the Magnetized Capacitance Effect 2446
Rafael Bogo Portal Chagas, Marcelo Lobo Heldwein

A High Frequency Coupled Inductor with Distributed Air Gap for High Power DC-DC Converters 2453
Muhammad Fasih Uddin, Ahmed H. Ismail, Peyman Darvish, Baher Abu Sba, Yue Zhao

High-Power Planar Transformer Design for Four-Port Converters 2461
Arya Sadasivan, Behrooz Mirafzal

Optimal Design of Inductors with Aluminum Litz Wire for Inductive Power Transfer Systems 2468
Jesús Acero, Claudio Carretero, Ignacio Lope, Óscar Lahuerta, José-Miguel Burdío

Analytic Design of Flat-Wire Inductors for High-Current and Compact DC-DC Converters 2474
Sajjad Mohammadi, James L. Kirtley, Alireza Namadmalan

Insulation Dielectric Loss of High-Frequency Transformer Under Square Voltage Excitation with
Edge Oscillation 2482
Zhanlei Liu, Lingyu Zhu, Yuntian Gao, Yongliang Dang, Cao Zhan, Shengchang Ji

Improved High-Speed Thermal Analysis based on Two-Step Simulation for High-Frequency
Transformers 2488
Zheyuan Yi, Kai Sun, Qiang Li, Zengyang Liu

Core Material Characterization Under DC Bias Conditions 2495
Jonas Mühlethaler, Fabrice Locher, Frédéric Mathieu, Edward Herbert

A Low-Cost Setup and Procedure for Measuring Losses in Inductors 2502
Burkhard Ulrich

Effect of Temperature of Additively Manufactured Cores 2510
Ken Johnson, Ali Bazzi

Extreme Temperature Permeability Engineered Soft Magnetics 2516
Tyler W. Paplham, Alex M. Leary, Paul R. Ohodnicki Jr.

An Isolated RF Power Combining Approach with Multiple Decoupled Input Coils 2521
Ziyang Xu, Yifan Zhao, Zhan Liu, Alex J. Hanson, Ming Liu

Simulation of a Custom Core, 15kV Isolated Gap Transformer Optimized for High Power Density 2527
Andrew Galamb, Fei Teng, Srdjan Lukic

Low Interwinding Capacitance Design for PCB-Winding based Transformer in Self-Powered Gate Drive Power Supply for High-Voltage SiC MOSFET 2535
Yuan Zhou, Li Zhang, Yilun Chen, Tianxiang Yin, Lei Lin

Integrated 4-Level Dual-Phase Superimposed Quadratic Power Converter for High-Density Direct 48V/1V Conversion 2541
Prosenjit Ghosh, Jin Woong Kwak, Fei Zhou, D. Brian Ma

Compensation Method for Unbalance of the Multi-Channel Class E Power Amplifier using the Closed Loop Frequency Control 2547
Kyungmin Lee, Sungku Yeo

High Temperature Operation of Digital Gate Driver Integrated Into a Power Module 2551
Kazuma Saiga, Shohei Zaizen, Satoshi Nakano, Shigeru Kusunoki, Kiyoto Watabe, Katsuhiro Hata, Makoto Takamiya, Shin-Ichi Nishizawa, Wataru Saito

Evaluation Index-Based Multiphysics Coupling Model and Analysis Methodology for High-Reliable Power Supply Module 2556
Haoyu Wang, Xuliang Wang, Yan Wang, Xiaosen Liu

Electrical Characterization of Modular 3D Packaging Assembled with Compressed Metal Foams 2562
Paul Bruyere, Alexis Derbey, Betina Zynger-Capaverde, Yvan Avenas, Eric Vagnon, Jean-Luc Schanen, Jean-Michel Guichon, Omar Sanjakdar

Improvement in Short-Circuit Robustness of SiC-MOSFETs based Power Modules using Two-Level Turn-On (2LTO) 2569
Muhammad Muneeb Alam, Saad Khalid, Nisar Ahmed Khan, Ngoc Ho Tran, Sebastian Strache

GaN-Based Two Stage Point-of-Load (PoL) Converter with 2.5D Embedded Substrate Implementation 2576
Samuel Defaz, Yang Li, Fang Luo

Near-Field Coupling Mitigation of the Noise from High Voltage DC-Link Decoupling Capacitors in Voltage Source Converters 2582
Yuxuan Wu, Kushan Choksi, Samuel Defaz, Fang Luo

Advantages of Paralleling SiC MOSFETs in High-Performance Power Modules 2589
Steffen Beushausen, Dominik Alexander Ruoff, Wenqi Zhou, Karl Oberdieck

A SiC Half-Bridge Power Module based on Liquid Metal Packaging for High Performance and Low Thermal Stress 2597
Wei Mu, Ameer Janabi, Luke Shillaber, Borong Hu, Teng Long

Analysis and Modeling of Radiated EMI Considering Coupling Between Power Converter and Power Cable with LC-Type EMI Filter 2603
Qinghui Huang, Yingjie Zhang, Shuo Wang, Yirui Yang, Zhedong Ma, Yanwen Lai

Simple Prediction Method for Impacts of Switching Characteristics on EMI Noise of a Three-Phase PWM Inverter 2610
Shinobu Nagasawa, Toshiya Tadakuma, Keita Takahashi

Coaxially Nested 3.3 kV SiC MOSFET Packages with Uniform Interpackage Electric Field Distribution 2616
Jack Knoll, Mark Cairnie, Christina DiMarino

Thermal Modeling and Performance of a Bare-Die Embedded PCB for High Power Density Converters Design 2624

Shahid Aziz Khan, Feng Zhou, Mengqi Wang, DucDung Le, Shivam Chaturvedi

Research on the Voltage Fluctuation Suppression Strategy in Weak Grid Under Pulsed Power Load Integration 2628

Xi Chen, Jiazheng Zhang, Mingjun Bao

An Optimized Firmware-Based Cycle-by-Cycle Current Limiting Method for Power Electronic Converters in UPS 2634

Teng Wu, Hong Liu

Frequency Stop-Band Management System for DC-DC Converters 2640

Alessandro Bertolini, Alberto Cattani, Claudio Luise, Alessandro Gasparini

Multi-Stage Model Predictive Control with Enhanced Discrete-Time Models for Multilevel Inverters 2647

Hoang Le, Apparao Dekka, Deepak Ronanki, Abdul R. Beig

Direct Effective Power Control (D-EPC) for LLC Resonant Converters Operating in Boost Mode using Event-Driven-Timer based Digital Controller 2654

Yuto Yoshimura, Kenji Funatani, Kazuhiro Umetani, Toshiyuki Zaitsu, Akito Nakagaki, Masataka Ishihara, Eiji Hiraki

Mitigation Method of Resonance Between Paralleled On-Line UPS 2660

Teng Wu, Zhenguo Huo, Shangxian Ning

An Extra-Element Small-Signal Model for a Current-Fed Resonant Dual-Active-Bridge Converter 2667

Paolo Sbabo, Paolo Mattavelli, Giorgio Spiazzi, Andrea Petucco

Concurrent Charge Distribution and Time-Optimal Control for Unordered Single-Inductor Dual-Output Converter 2675

Xuliang Wang, Haoyu Wang, Yang Liu, Yunxin Wang, Boran Zhang, Hongru Liu, Yan Wang, Xiaosen Liu

Circulating Current Control with Loss Reduction for Parallel Connected Inverters 2681

Shun Endo, Takae Shimada, Masato Ando, Yuuichi Mabuchi, Masaki Miyamae, Naoki Takayama, Yohei Matsumoto, Naoto Onuma

Analysis of Power and Power Spectral Density for Quaternary Random Pulse Position Modulation 2687

Hung-Chi Chen, Hsiang-Kai Wu, Chih-Chiang Wu

Bidirectional CLLC Converter using a Hybrid Control Method for Wide Voltage Range Applications 2692

Jhih-Cheng Hu, Hong-Xuan Liao, Chien-Lung Liu, Wei Wang, Ming-Shi Huang

Design and Control of a High-Bandwidth Dual Active Bridge DC-DC Converter 2698

Alper Uzum, Syed Imam Hasan, Yilmaz Sozer, Kenneth A. Loparo

Unified Model Predictive Control for DC-DC Buck Converters: From Start-Up to Steady-State Operation 2703

Zhengchen Guo, R.M. Nelms

A Novel IPPC Method for Precise Overload Protection and Burst Mode Operation in LLC Resonant Converters 2708

Manikanta Pallantla, Ramkumar S

An Improved Current-Sensorless Model Predictive Voltage Control for Four-Leg Voltage Source Inverters.. 2713
Heng Guo, Yuxin Wei, Mengmeng Jing, Wenlong Ding, Bin Duan, Chenghui Zhang

A Highly Integrable, Modular and Multi-Functional Fault Monitoring Active Gate Driver with Parallel Buffers for a Global Enhanced Reliability of Gen. 3 SiC Power MOSFETs 2718
Mathis Picot-Digoix, Léo Seugnet, Frédéric Richardeau, Jean-Marc Blaquière, Sébastien Vinnac, Thanh-Long Le, Stéphane Azzopardi

A 24 – 16 V to 0.8 – 1.2 V Merged 4-Stage Hybrid-SC-SL Converter with 96.5% Peak Efficiency and Larger Than 50% iL Reduction.. 2725
Chien-Hao Tseng, Cheng-Ta Chuang, Chieh-Ju Tsai, Ching-Jan Chen

Innovation Active Gate Drive Method (Named TriC3™) for MOSFET Heat Reduction and EMI 2730
Hisashi Sugie

A KY Buck-Boost Converter with Extended Ramp Control Achieving 1500% Output Variation Reduction for Smooth Mode Transition 2735
Yu-Ting Hung, Chieh-Ju Tsai, Ching-Jan Chen, Chun-Yu Hsieh

An USB Cable based Extended Conversion Range L-First Hybrid-Converter using Valley-Virtual-Inductor-Current-Mode Control with Auto-Tracking Slope Compensation Against ±50% Inductance Variation.. 2741
Chun-I Li, Chieh-Ju Tsai, Ching-Jan Chen

Impact of Gate Resistor Configurations on Current Balancing in Paralleled SiC MOSFETs 2746
Yifu Zhang, Shashank Karanth, Emanuel Eni

Exploring the Potential of FPGA in High-Frequency Switching DC-DC Boost Converters using Model Predictive Control 2752
Qingcheng Sui, Bangli Du, Yu Zuo, Wilmar Martinez

A 7 Bit 5A 6.7 GHz Gate-Shaping Digital Gate Driver with Burst-Sampling ADC for Iterative Switching Optimization of SiC Power MOSFETs 2757
Tobias Zekorn, Kenny Vohl, Erik Wehr, Leon Weihs, Michael Hanhart, Ralf Wunderlich, Stefan Heinen

Decentralized Interleaving of Series-Stacked DC-DC Converters via Extremum-Seeking Control 2764
Ivan Petriœ, Vignesh Iyer, Shoudong Hu, Chirayu Rajpurohit, Bailey Sauter, Milan Iliœ, Luca Corradini, Dragan Maksimoviœ

Online Dead-Time Control for Half Bridges without Preliminary Training based on Switching Transient Steepness 2772
Lukas Knappstein, Niklas Falkenberg, Martin Pfost

Impedance-Based State-of-Health Estimation for Lithium-Ion Battery Management Systems 2779
Mohammad K. Al-Smadi, Jaber A. Abu Qahouq

Stability Analysis and Resonance Damping of LC Filter-Based Voltage Source Converter with Single-Loop Voltage Control.. 2785
Aravind G., Divyanshu Bansal, L. Umanand

Finite Control Set Model Predictive Control Combined with Online Junction Temperature Estimation for Reliability Enhancement of Voltage Source Inverters 2790
Qiang Mu, Jiale Zhou, Zaheen Mustakin, Lucas Pereira, Babak Parkhideh, Tiefu Zhao

Framework for Dynamic Control and Operation of Power Electronics Interfaces 2797
Radha Sree Krishna Moorthy, Steven Campbell

Achieving Soft-Charging and Over 20% Input Current Ripple Reduction in a 48-to-6 V Dickson
Converter using 3-Phase Split-Phase Control 2805
Nagesh Patle, Rose A. Abramson, Sahana Krishnan, Jiarui Zou, Robert C. N. Pilawa-Podgurski

Experimental Verification of Circuit-Losses Analysis-Model of DC-Output Converter Developed
using Approximated Equations from Measurement Data and Datasheet Data 2813
Ryota Kondo, Tsuyoshi Funaki

Scattering Parameter Measurement System using Probes for Surface Mount Devices Operating in
the Frequency Range from 50 kHz to 1 GHz 2821
Ryoko Kishikawa, Masahiro Horibe, Tomokazu Shoji, Shigenori Yabuta, Toshi Ohi, Ryo Takeda, Takamasa Arai

Optical Transformer Design with Additional Common-Mode Noise Reduction Winding for Flyback
DC-DC Converters 2828
Yusuke Irie, Shinichiro Eguchi, Yoichi Ishizuka, Toshiro Takeuchi, Akio Iwabuchi, Takahiro Koga, Toshiyuki Tanaka

Enhanced Bus Voltage Stability Through Digital Twin-Enabled Adaptive Controller Tuning 2833
Matthew Belanger, Andy Wong, Kerry Sado, Enrico Santi

Modeling and Performance Characterization of Lithium-Ion Capacitor at Different Temperature
and Voltage Values 2840
Mohammad K. Al-Smadi, Jaber A. Abu Qahouq, Sajad Saberi

Conveniently Identify Coils in Inductive Power Transfer System using Machine Learning 2846
Yifan Zhao, Mowei Lu, Ting Chen, Heyuan Li, Xiang Gao, Zhenbin Zhang, Minfan Fu, Stefan M. Goetz

Accurate Modeling of LLC Resonant Converters with Enhanced Analytical Approach Considering
of Parasitic Capacitance 2851
Dong Jiao, Zhengming Hou, Jih-Sheng Lai

High-Frequency Conditioning Circuits for Power-Related Information Extraction in Non-
Sinusoidal Power Electronic Systems 2857
Haoyu Wang, Yuanxin Zhang, Di Mou, Alex Hanson, Shiqi Ji

Transconductance Model of the Dual Active Bridge Converter Under Single and Dual Phase Shift
Control 2865
Jared Cronin, Andrew Wunderlich, Enrico Santi

Lumped Parameter Modeling for Real-Time Thermal Regulation of Li-Ion Battery Packs 2871
Utkal Ranjan Muduli, Mohamed Shawky El Moursi, Khalifa Al Hosani, Ahmed Al-Durra

A Physics-Based Temperature Dependent Analytical Model for 2DEG Density in AlGaN/GaN
HEMT Devices 2877
Kashfia Tajmim Nabila, Jerry L. Hudgins

Comparative Analysis of Stator-PM Machines: Design Optimization and Electromagnetic
Performance Evaluation 2883
Maryam Salehi, Madhav Manjrekar

Elimination of Deadtime Effect on Resolver Offset Estimation using the Pulsating Current Command for Electric Vehicle Application 2889
Yingfeng Ji, Nurani Chandrasekhar

A Generic Load Emulator for Testing Motor Drives of E-Mobility 2894
Qingzheng Zhang, Kaiyuan Feng, Changsheng Hu, Dehong Xu

Design and Implementation of Power Assisted Control System for E-Bikes 2900
Che-Yu Lu, Tzu-Ping Cheng

A Hybrid PWM Strategy with Reduced Common-Mode Voltage and Extended Output Voltage Linearity for Adjustable Speed Drives 2907
Zhe Zhang, Kevin Lee

Single-Phase Open-Circuit Fault-Tolerant Control of Three-Phase PMSM Drives 2913
Yuichiro Minato, Yuki Nakata, Jun-Ichi Itoh

Multi-Vendor Encoder Position Sensing Interface using Programmable IP based Solution............................ 2920
Rajul Bhambay, Dhaval Khandla, Pratheesh Gangadhar, Thomas Leyrer, Achala Ram, Manoj Koppolu, Archit Dev

Sensorless Control Method at Low-Speed Range using High-Frequency Voltage Injection for Synchronous Reluctance Motors Considering to Nonlinear Characteristic Due to Magnetic Saturation 2924
Sota Takizawa, Sari Maekawa

Hybrid Control Scheme for Permanent Magnet Gear Motor.. 2932
Bing Li, Takayoshi Matsuo, Ahmed Sayed-Ahmed, Yujia Cui, Jiangang Hu

Cost-Effective Fault Diagnosis for Motor and Inverter using Bootstrap Charging and Single DC Link Current Sensor 2937
Gyu Cheol Lim, Won Hyo Jeong, Kahyun Lee, Jung-Ik Ha

Improved PWM to Suppress Motor Overvoltage Caused by Voltage Reflection 2943
Sung-Oh Kim, Kyo-Beum Lee

Analysis of Double Pulsing Effect in Motor Drives based on Vector Diagram.. 2948
Byeong-Woo Kang, Kyo-Beum Lee

A Novel Speed Sensor-Less Control of a Solar-Powered PMSM Drive 2953
Abirami Kalathy, Arpan Laha, Praveen Jain, Majid Pahlevani

Design of a Compact Low-Loss MMC Double Submodule for MVDC and HVDC Applications 2960
Ali Sharaf Addin, Rainer Marquardt, Thomas Brückner

A Series-Type Dynamic Voltage Restorer Control Strategy to Cope with Voltage Swell............................ 2968
Jiazheng Zhang, Hongyu Chen, Xi Chen, Mingjun Bao

Machine Learning Approach for Accurate Lithium-Ion Battery Temperature Prediction using Electrochemical Features Independent of Battery SOC and SOH.. 2973
Vincent Masabiar Tingbari, Oluwaseun Isaiah Ekuewa, Anshul Nagar, Asad Abbas, Jamil Umar, Yuxin Zhang, Woonki Na, Jonghoon Kim

A Battery Strings Circulating Current Blocking Method for Battery Energy Storage Systems 2981
Haihong Long, Ziang Sun, Yucheng Fan, Xin Wu, Dehong Xu

A Hybrid Multilevel Converter-Based High-Gain Isolated DC/DC Converter for Grid-Tied Energy Storage Applications 2986
Pengyu Fu, Yizhou Cong, Jin Wang, Anant Agarwal

LCL Filter Parameter Selection using Graphical Method for a 13.8 kV ac 1.1 MVA 7-Level Flying Capacitor Grid-Connected Converter Utilizing Variable Switching Frequency 2992
Arthur Mendes, David Nam, Mingze Gao, Thimothy Thacker, Dong Dong, Rolando Burgos

Online Extraction of Electrochemical Impedance Spectroscopy Pattern based on EV Load Profile and Short Time Fourier Transform for Diagnosis of Lithium-Ion Battery Safety 3000
Miyoung Lee, Dongcheol Lee, Youngmin Bae, Jongchan An, Garam Yang, Woonki Na, Jonghoon Kim

Enhanced Incremental Capacity Analysis for Evaluating Battery Degradation Mechanisms of Optimized Fast Charging Methods 3006
Taehyeon Gong, Jaehyeong Lee, Sungjun Lee, Yura Kim, Bomyeong Ko, Woonki Na, Sungjin Choi, Jonghoon Kim

Co-Estimation of SOC and SOT in Lithium-Ion Batteries using an RLS-Based Heat Generation Model 3012
Seongkyu Lee, Eunjin Kang, Minhyeok Kim, Seunghyun Lee, Minwoo Song, Jaea Lee, Woonki Na, Jonghoon Kim

Three-Stage Adaptive Control Strategy for Stability Improvement of Grid-Connected Inverter in Weak Grid 3018
Longxiang You, Sicong Jin, Xin Zhang, Zuoshuai Wang, Sunqing Wang

Degradation Analysis of Offshore Bifacial PV Modules Under Multiple Climatic Stressors 3024
Aidha Muhammad Ajmal, Yongheng Yang

A Flexible Energy Management System for Solar Powered Electric-Bus Charging Stations 3030
Supun Amarathunga, Pasan Gunawardena, Xiaoting Wang, Yunwei Li

A Vienna Rectifier based Grid-Connected Powertrain for Hydrokinetic Turbine Systems 3036
Peidong Li, Md Tariquzzaman, Yue Cao

Condition Monitoring for DC-Link Capacitors and PV Arrays based on the Start-Up Process of the PV System 3042
Yongjie Liu, Ariya Sangwongwanich, Chen Liu, Xing Wei, Shuyu Ou, Tamás Kerekes, Jiahong Liu, Huai Wang

Electrically and Thermally Efficient Reliable Power Converter Design for Micro–Hydrokinetic Turbine 3048
Md Tariquzzaman, Peidong Li, Yue Cao

Comprehensive Evaluation of Cyber Attacks on Grid-Connected Smart Inverters 3054
Rishabh Singla, Vishwam Raval, Hasan Ibrahim, Jaewon Kim, Prasad Enjeti, Narsimha Reddy

Parallel Operation of Grid-Forming Converters based on Kuramoto Oscillators with Virtual Cable Emulation for Improved Power Sharing 3059
Vikram Roy Chowdhury, Gab-Su Seo, Barry Mather

Enhancing Hydrogen Production in Hybrid Standalone Microgrids 3064
Utkal Ranjan Muduli, Mohamed Shawky El Moursi, Khalifa Al Hosani, Ahmed Al-Durra

LSTM-Based Sub-Synchronous Oscillation Detection Scheme for Type 4 Wind Farm Interfaced with Weak AC Grid 3071

Omar Abu-Rub, Muhammad F. Umar, Jana A. Sheikh Ali, Yazan Qiblawey, Abdulrahman Alassi, Maryam Saeedifard, Mohammad B. Shadmand

A Study of Module Design Method to Suppress the Oscillation Occurs Between Parallel-Connected Power Devices 3077

Shinji Yato, Hiroto Sakai, Hideo Araki, Shumei Shimosako

A High-Efficient Hybrid Traction Inverter in Electric Vehicle Applications 3083

Yousefreza Jafarian, Omid Salari, Praveen Jain, Alireza Bakhshai, Mohamed Z. Youssef

Dual-Use of Onboard Chargers to Achieve Controllable DC Bus Voltage for Electric Vehicles 3089

Anuj Maheshwari, Elie Libbos, Arijit Banerjee

Isolated Single-Phase Onboard Chargers for BEV/PHEV using Active Power Decoupling Technology 3096

Yoshiki Amano, Keigo Nishimura, Hiroaki Matsumori, Takashi Kosaka, Kenichi Nagayoshi, Kenichi Watanabe

A Practical Use of xEVCap: The Modular and Standard DC-Link Capacitor Solution for the Main EV Powertrain Inverter 3100

David Olalla, Tomas Wagner, Fernando Rodriguez, Alberto Espinar

Optimized Bidirectional On-Board Charger using a Novel Unfolder-DAB Topology 3109

Héctor Sarnago, Ignacio Álvarez, Pablo Briz, Óscar Lucía

Critical Thermal Characterization of Next-Generation Solid-State Batteries for Automotive Battery Management Systems 3114

Chandan Chetri, Sheldon Williamson

Nanocrystalline CMC Inductors for EV Charging: Trade Studies and Testing Standardization 3119

Christopher Bracken, Mark A. Juds, Paul R. Ohodnicki, Bharadwaj Reddy Andapally, Jose Gato

Predicting Efficiency of On-Board and Off-Board EV Charging Systems using Machine Learning 3124

Mohamed Yasko, Fanghao Tian, Wilmar Martinez, Johan Driesen

High-Power and High-Speed Multi-Channel VCSEL Arrays with GaN Driver for Automotive LiDAR 3129

Yifu Liu, Sichao Li, Junlei He, Changyu Hu, Bill He, Karthik Krishnamurthy, Andy Shen

Double Pulse Test Platform for Hybrid SiC-IGBT Switch Characterization and Optimal Gate Control Strategy for EV Traction Inverters 3133

Rosario Attanasio, Harsha Ademane, Ryan Satterlee, Gianni Vitale

Critical Role of Individual Cell Temperature Monitoring in Mitigating Thermal Runaway and Reducing Accelerated Degradation in Lithium-Ion Batteries 3141

Mohit Sharma, Akash Samanta, William Locke, Sheldon Williamson

Loss-Optimized Design of a Triple Active Bridge DC-DC Converter for an Electric Vehicle Application 3147

Sreejith Chakkalakkal, Kyle Kozielski, Wesam Taha, Yicheng Wang, Aniket Anand, Ali Emadi

A Magnetic-Less DC/DC Converter with Pulse Charging for 800 V Powertrains from 400 V DC Fast Chargers 3155

Duc Dung Le, Shivam Chaturvedi, Shahid Aziz Khan, Mengqi Wang, Mohamed Elshaer

Boosting Charger Efficiency: A GaN-Based Flyback Converter with Energy Recycling 3160
Ahmad Nabizadah, Majid Ghasemi Korrani, Babak Fahimi

A Hybrid Three-Level Buck Converter with Flying Supercapacitor for High Load Current Surge
Capability using Peak Current Mode Control 3167
Finlay Lodge, Rafael Peña-Alzola, Martin MacFadyen, Patrick Norman, Mark Sweet,
Graeme Burt

Supercritical Carbon Dioxide (sCO.)-Cooled Current Source Inverter-based Integrated Motor Drive
for MW-Scale Electric Aviation Applications 3174
Hang Dai, John Yagielski, Thomas Jahns, Kum-Kang Huh, Vandana Rallabandi, Libing
Wang, Tarak Saha, Wenda Feng, Bulent Sarlioglu

The Challenge of Thermal Runaway in Soft Magnetic Materials for Inductive Power Transfer 3181
Yibo Wang, Ben Zhang, Weisheng Guo, Tianlu Ma, Sheng Ren, C.Q. Jiang

A Capacitively Coupled Alternative Electric Field Control for Freeze-Free based High Quality
Food Preservation 3187
Jaeyong Cho, Junhyeong Park, Sung-Bum Park, Daehyun Kim, Jinsoo Choi

The Characteristics of the Long Length Primary Loop and the Power Supply for the SCMaglev's
DWPT System 3194
Keisuke Yamamoto, Jun Enomoto, Shunsaku Koga, Junichi Kitano

A Wireless EV Charging System with a Double-Sided LCC Network using Variable Switching
Frequency and DC-Link Voltage Control 3200
Chae-Lyn Kim, Hyeonu Jo, Ju-A Lee, Dong Hyeon Sim, Byoung Kuk Lee

Class E/EF Inductive Power Transfer to Achieve Stable Output Under Variable Low Coupling 3206
Yifan Zhao, Mowei Lu, Heyuan Li, Zhenbin Zhang, Minfan Fu, Stefan M. Goetz

A Motorized Air-Core Variable Inductance Winding Structure 3212
Xindong Li, Sampath Jayalath, Cheng Zhang

Wireless Power Transfer System with Automatic Tuning Capability in Metallic Environment 3220
Renjie Zhang, Yue Wu, Delin Zhao, Yaohua Li, Yongbin Jiang, Yi Tang, Huan Yuan, Xiaohua
Wang, Mingzhe Rong

Design of Wireless Power Transmitters for Enhanced Transmission Distance and Output Power 3227
Kaiyuan Wang, Shuang Zhao, Shuye Shang, Eric Ka-Wai Cheng, Siew-Chong Tan, Yun Yang

Optimization of Wireless Power Transfer Waveforms and In-Vivo Receivers for Implantable
Medical Devices 3232
Hanbing Liu, Xin Zan

Comparison of Compact Power Amplifier Designs for High Frequency Resonant Wireless Power
Transfer Systems at 6.78 MHz using High-Q Resonators 3241
Manuel Rueß, Kilian Müller, Mathias C.J. Weiser, Ingmar Kallfass

Analysis and Design of Capacitive Coupling Wireless Power Transfer System using Load-
Independent Class-EF Inverter 3248
Takumi Kobayashi, Yutaro Komiyama, Akihiro Konishi, Hiroaki Ota, Yuki Ito, Taichi Mishima,
Takeshi Uematsu, Kien Nguyen, Hiroo Sekiya

Design and Optimization of a 600 W Wireless Drone Charger for High Gravimetric Power Density 3253
Arka Basu, Daniel Costinett

Stabilization Method for DC-Bus Oscillation in Dynamic Wireless Power Transfer Systems........................ 3261
Yuki Ochiai, Keisuke Kusaka

Unveiling Aliasing Effect on Resonant Pole Locations in Wireless Battery Chargers 3267
Anwesha Mukhopadhyay, Daniel Costinett

Integrated Hybrid Inductive and Capacitive Power Transfer System with Asymmetrical PCB Self-Resonator.. 3275
Yao Wang, Zhen Sun, Xiangrong Zhang, Yun Yang, Shu Yuen Ron Hui

High Frequency Noise Reduction Method of the Class E Power Amplifier... 3281
Kyungmin Lee, Sungku Yeo

Single-Stage Three-Phase Buck-Matrix Rectifier with Series-Parallel Connected Transformers for High-Power 48 V Data Center Power Supplies.. 3285
Yuki Ishikura, Chinmay Bhagat

Sector Transition PWM Modulation Scheme for a Three-Phase Isolated Buck-Matrix Rectifier................. 3291
Chinmay Bhagat, Yuki Ishikura

Adaptive Capacitance Circuit for Optimal Dynamic Impedance Matching in Variable Reluctance Energy Harvesting Applications 3298
Alejandro Redondo, Fernando Pérez, Sofía García, Gabriel Mujica, Airán Francés

Gallium Nitride (GaN) based Topology Comparison for Low Power Battery Charging Applications 3304
Jai Aditya Chaudhary, Rosario Attanasio, Gianni Vitale

Server Motherboard Power Performance Study Under Immersion Cooling Environment........................... 3312
Meng Wang, Haiyan Wang, Pavan Kumar, Haijin Zhang, Xiang Li, Fengwei Bian, Jianting Deng, Jiaqi Zhu, Yiming Lei

Practical PCB Design Considerations for GaN HEMTs based Isolated DC-DC Converter 3316
Gaureej Gauttam, Harish S. Krishnamoorthy, Sai Sushma Pasupuleti

Data-Driven Characterization and Forecasting of Metal-Oxide Varistor Degradation in DC Circuit Breakers... 3321
Zhi Jin Zhang, Yang Liu, Lukas Graber, Maryam Saeedifard

A Thyristor-Based Fault Current Bypass Solid-State Circuit Breaker for DC Microgrid Applications 3328
Jiale Zhou, Xiuhu Sun, Qiang Mu, Tiefu Zhao

Single-Stage Three-Phase AC-AC Isolated Inertialess Converter (IIC) for Industrial Drives...................... 3334
Brad Houska, Decheng Yan, Aniruddh Marellapudi, Satish Belkhode, Joseph Benzaquen Sune, Deepak Divan

Author Index

Indented Core to Reduce and Desensitize Inductor's Fringing Losses without Increasing Volume

Rajaie Nassar, Promit Datta, Guo-Quan Lu, Christina DiMarino, and Khai Ngo
Center for Power Electronics Systems (CPES)
The Bradley Department of Electrical and Computer Engineering
Virginia Polytechnic Institute and State University
Blacksburg, VA, USA
Email: {rajaienassar, promitdatta, gqlu, dimaricm, kdtn}@vt.edu

Abstract—There is a constant push towards more compact and efficient power electronic converters. However, the fringing losses of gapped magnetic components inhibit size reduction by requiring extra space to keep the winding away from the gaps. Concurrent high efficiency and high power density cannot thus be achieved with conventional gapping approaches. A core structure that reduces the fringing and overall losses without requiring additional space is presented here, simultaneously enabling higher efficiency and power density. The geometry also reduces the sensitivity of the winding to its position in the winding window compared to the conventional discrete gapping approach. The idea is verified in a planar inductor design, decreasing the power loss of a soft-switched unloaded buck converter by more than 32%.

Keywords—fringing fields, fringing loss, air gaps, ac resistance, compact magnetics, planar magnetics, high-frequency inductors

I. INTRODUCTION

The fringing field caused by gapped magnetic cores can significantly increase their loss by generating excessive eddy currents in the winding [1, 2]. This results in a higher ac resistance, especially at the elevated switching frequencies enabled by wide bandgap semiconductors. Various strategies have been proposed to mitigate these fringing losses [3, 4], with one standard approach being to place the winding farther from the gap because the fringing intensity decays with distance [5, 6]. While an easy solution, this method is sensitive to the winding-to-gap spacing and the manufacturing tolerances that control this distance. More importantly, it inefficiently uses the winding window and demands extra space to position the turns away from the gaps. Because volume is detrimental in size-constrained magnetics, the required additional distance presents a tradeoff between the size of gapped magnetic components and their ac winding loss [1]. Planar magnetics are an example where the corresponding converters must fit in the limited space of their target applications, such as server racks and portable electronics [7-11]. A planar inductor with conventional discrete gaps is shown in Fig. 1(a), illustrating the fringing fields that prevent the simultaneous reduction of the window height and the ac loss.

Introduced here is a core geometry that allows narrow winding windows while minimizing fringing losses typically associated with conventional discrete gaps, aiming to diminish the tradeoff between the energy density and efficiency of compact magnetic components. The planar inductor in Fig. 1(b) is used here to demonstrate the concept. It has a single-layer printed circuit board (PCB) winding to conform to the low-profile, narrow window requirement and reduce turn-to-turn proximity effects. A distributed gapping solution is presented in [12] for inductors with turns on multiple PCB layers.

The geometry and its benefits are first presented in Section II. Design considerations are discussed in Section III along with 3-D Finite-Element Simulations (FES) of the prototypes. Experimental loss results are finally presented in Section IV for the inductor prototypes, confirming the loss reduction of the idea in planar magnetics.

II. INDENTED-CORE INDUCTOR

The motivation behind the introduced geometry is the shape of the magnetic field lines illustrated in Fig. 1(a). The encircled

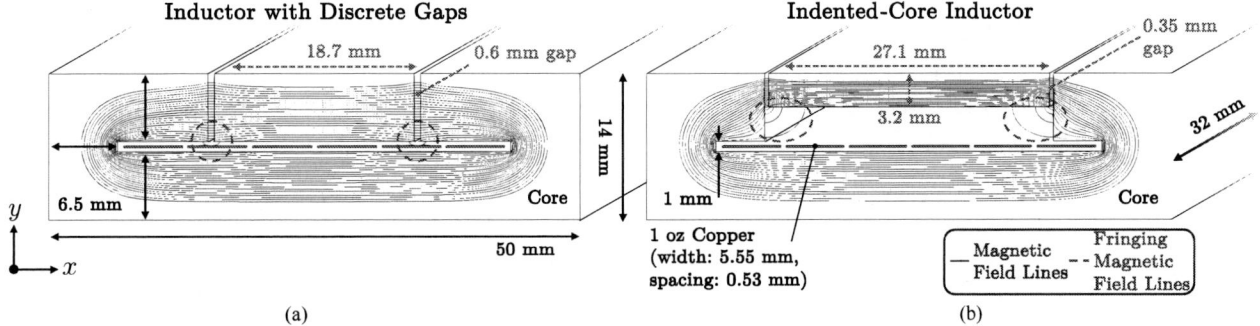

Fig. 1. Planar inductor structures with dimensions like the UI core in [16], illustrating the magnetic lines path and fringing effects: (a) with conventional discrete gaps and (b) with the presented indented-core inductor (ICI). The solid-line dimensions are common to both structures while the dashed-line dimensions are unique.

This work was supported by the U.S. Department of Energy, Advanced Research Project Agency-Energy (ARPA-E) through the DE-AR0001568 award and the High-Density Integration (HDI) mini consortium at the Center for Power Electronics Systems (CPES), Virginia Tech.

979-8-3315-1612-3/25 $31.00 © 2025 IEEE

fringing lines generated by the discrete gaps have a large magnetic field component perpendicular to each turn's surface. This causes significant fringing loss which is proportional to the square of the orthogonal field component [6, 13]:

$$P_{fringe} = \frac{1}{6\rho}(\pi\mu_0 H_y f)^2 w^3 t \qquad (1)$$

where P_{fringe} is the fringing loss; ρ is the resistivity of the turns' material, μ_0 is the permeability of the vacuum; H_y is the y-component magnitude of the magnetic field intensity of the fringing magnetic field lines at the top surface of the turns (perpendicular to the winding's flat surface in Fig. 1); f is the switching frequency; w is the width of the turns; and t is the thickness of the turns. If the dominant component of the fringing magnetic field were parallel to the surface of the winding, the fringing loss could have been minimized according to (1). Thus, the structure in Fig. 1(b) introduces an indentation in the middle segment of the gapped I-core piece to divert the fringing magnetic field lines away from the winding, making them almost parallel to the surface of the turns The simulated H_y component along the top surface of the turns for the structures of Fig. 1 is plotted in Fig. 2 (ANSYS Maxwell 2D EddyCurrent solver). The gaps cause positive and negative peaks at the winding surface (encircled), with the conventional discrete gaps reaching magnitudes 12 and 6 times higher than the Indented-Core Inductor (ICI) for the positive and negative peaks, respectively. This causes more distortion in the current distribution in the winding of the inductor with discrete gaps, resulting in more significant field peaks around the ends of the turns than in the ICI (sharp peaks indicated with arrows). The lower fringing profile of this ICI thus entails the benefits discussed in the subsections hereafter.

A. Lower Loss for the Same Outer Volume

The minimized perpendicular fringing component in ICIs significantly reduces the localized fringing loss as described by (1). This is achieved without changing the size of the magnetic component, as seen in Fig. 1. This improvement manifests as a reduction in the ac winding resistance. For example, 2-D FES simulations predict at least 2.5 times smaller resistance when comparing the discrete-gap inductor best case (minimum) with the ICI worst case (maximum) in Fig. 3. The core loss in the

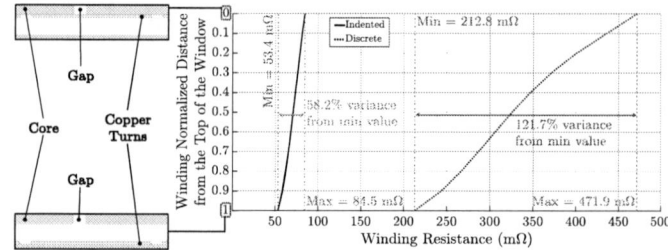

Fig. 3. The simulated winding resistance of the discrete-gap and indented-core inductors of Fig. 1 as a function of the winding position (sinusoidal excitation). The ICI geometry reduces the total winding resistance by at least 2.5 times over the inductor with discrete gaps. It also decreases the variation in resistance from the minimum value by more than two times.

indented segment increases less than the reduction in the winding loss, lowering the overall loss.

B. Desensitization of the Winding Resistance to its Position in the Window

Unlike the conventional discrete gaps approach, the ICI core shape forces the fringing field to be more dominantly parallel to the turns in the winding window, diminishing the change in the perpendicular fringing component along the window's height. As a result, the sensitivity of the winding's ac resistance to its position in the window in an ICI is reduced by two times compared to discrete gaps, as shown in Fig. 3. This means that the turns do not necessarily need to be spaced away from the gaps to reduce their ac resistance, also allowing for less strict manufacturing tolerances during assembly.

C. Smaller Outer Volume for the Same Loss

As previously mentioned, a common approach to reducing the fringing losses of an inductor with discrete gaps is to space the turns away from the gaps. However, even when the turns are at the farthest distance from the gaps in the structure of Fig. 1(a), the ac resistance is still higher than that of an ICI inductor, as Fig. 3 shows. This is due to the low-profile requirement, which forces a small window for the turns (assuming the core dimensions are optimal). Thus, to reduce the ac resistance of the inductor with discrete gaps to a level comparable to an ICI or to meet the loss requirements, the window has to be larger to leave adequate room for spacing the turns from the gaps. A structure that achieves this is shown in Fig. 4, where the window height was increased to 2 mm from the 1 mm of Fig. 1(a), making the total height of the structure 15 mm instead of 14 mm. With the turns furthest away from the gaps, the ac resistance of the discrete gaps with a spaced winding is 104.1 mΩ, much closer

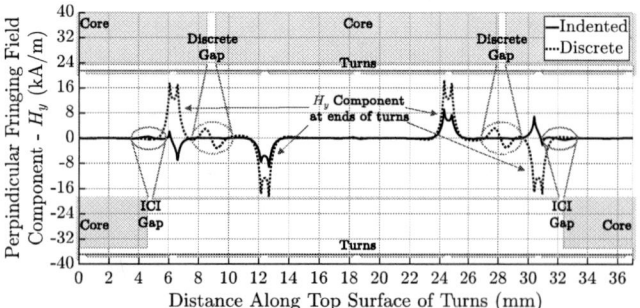

Fig. 2. The simulated magnitude of the perpendicular fringing component along the top surface of the winding (solid blue line) for the discrete gaps and indented-core inductors of Fig. 1. The excitation current is 11.5 A, and the windings are centered in the window (position 0.5 in Fig. 3).

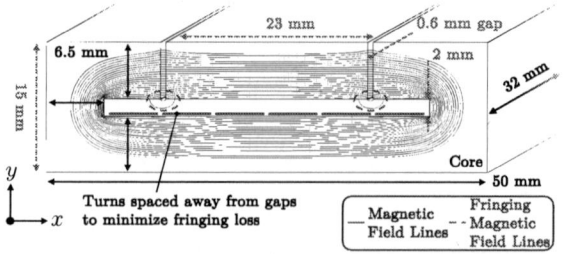

Fig. 4. Planar inductor structure with discrete gaps and similar dimensions of the inductor in Fig. 1(a). The winding window height was increased in this inductor to space the winding away from the gaps and reduce the fringing loss.

to the ICI's max ac resistance of 84.5 mΩ (compared to 212.8 mΩ previously). The minimum and maximum resistance values for the structures presented in Fig. 1 and Fig. 4 are summarized in Table I. The inductors with discrete gaps and spaced turns need to be larger to match the ICI's resistance.

Because an ICI has a smaller volume for the same loss of an inductor with discrete gaps, the total height of low-profile converters where planar magnetics are the tallest component can be reduced, increasing the converter's power density for the same power rating (by 7% between Fig. 4 and Fig. 1(b)). In such systems, reducing the inductor's height also decreases the quantity of required closure materials and their cost.

D. Improved Heat Extraction Capability

Fully utilizing the winding window of planar magnetics (assuming the PCB winding thickness is equal to the window height) confines the winding within the ferrite core (a poor thermal conductor). Even if the loss of the inductor with discrete gaps is acceptable in terms of efficiency, extracting the heat from the structure is challenging, causing a heat management issue. Increasing the window height, as in Fig. 4 can solve this problem by reducing the fringing loss and introducing a space between the turns and the core that could be used for heat extraction (forced air, thermally-conducting fillers, ceramics, etc.) at the cost of increased volume. The ICI geometry offers the same space for heat extraction without increasing the structure's volume. The related tradeoffs on the ICI's core loss and saturation current will be discussed in the next section.

III. CASE STUDY

A. Design Dimensions and Optimization

The structures in Fig. 1 and Fig. 4 are based on the dimensions of a commercial UI50 planar core by DMEGC [16]. The only change is that the thickness of all sides was kept at 6.5 mm to keep the flux density constant in all segments of the discrete gaps structure. The pieces making the gapped I-segment were manufactured separately and not cut from a single piece to avoid increased core losses related to surface stress [14]. A commercial planar core size was chosen to serve as an example of the effectiveness of the discussed structure.

The structures were designed for a target inductance of 10 µH. The six copper turns fill 90% of the winding window width to maximize horizontal space utilization and cause the

copper loss to dominate the total inductor loss (about two-thirds). Maximizing the horizontal dimension also discourages the magnetic field from jumping vertically across the window (the y-direction in Fig. 1) between the turns or at either end of the winding since the window height is comparable to the length of the gaps, which could cause significant eddy current losses.

Finally, the distance between the two gaps on all the structures was optimized to achieve minimum losses for a switching frequency of 400 kHz and 10 A peak sinusoidal current. For the ICI structure of Fig. 1(b), the indentation amount (effectively the width of the indented segment) was included in the optimization. The Adaptive Single-Objective (Gradient) optimization algorithm in ANSYS Maxwell 2D EddyCurrent solver was used to ensure the inductance stays within ±30% of 10 µH and minimize the total loss.

Six-ounce thick copper was assumed in the optimization to minimize the resistance without increasing the losses due to the skin effect (two-skin depths at 400 kHz is ~2×0.103 mm). However, the prototypes were built using one-ounce boards to minimize manufacturing cost and time. Regardless of the thickness, the same relative effects of Fig. 3 are seen in the inductors' windings, which are summarized in Table I. The table indicates that the one-ounce thickness shows the lowest benefit of the ICI structure and the highest variance among the other thicknesses, providing a conservative demonstration.

B. Increased Core Loss in the Indented Segment

Because the total flux remains relatively constant through the magnetic structure, the smaller cross-section of the indented

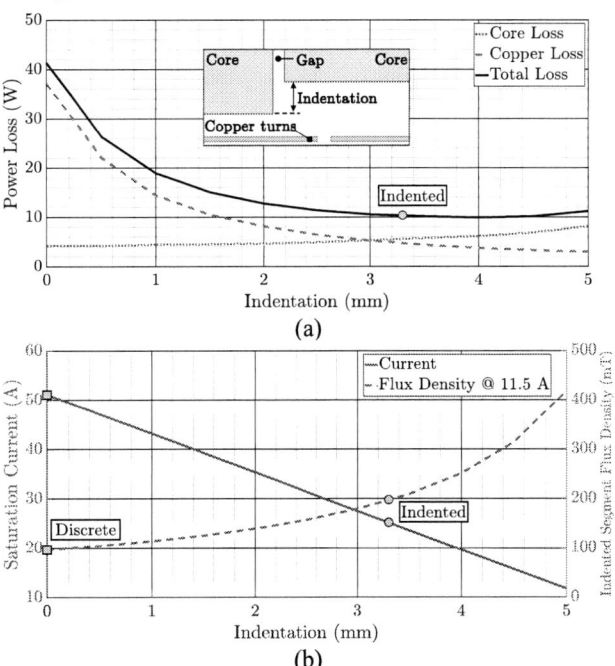

Fig. 5. Design optimization and limiting plots: (a) the simulated ICI losses as a function of the segment indentation (dimensions of Fig. 1(b) but the lengths of the gaps were changed with indentation to maintain 10.2 µH; 11.5 A_{pk} sinusoidal excitation) and (b) the calculated saturation current for a 430 mT saturation flux in the indented segment and its magnetic flux density at 11.5 A (DMEGC DMR96A material limits at 100°C). The square and circular markers correspond to the discrete and ICI designs of Fig. 1.

TABLE I. MINIMUM AND MAXIMUM WINDING RESISTANCE OF THE DESIGNED STRUCTURES WITH VARYING COPPER THICKNESS

Structure	Discrete Gaps (Fig. 1(a))			ICI (Fig. 1(b))			Discrete Spaced (Fig. 4)		
Copper thickness (oz)	1	3	6	1	3	6	1	3	6
R_{min} (mΩ)	213	110	96.3	53.4	25.0	21.8	104	49.3	41.5
R_{max} (mΩ)	472	235	199	84.5	38.0	29.2	382	183	154
Variance of R_{max} from R_{min}	122%	114%	107%	58.2%	52.0%	33.9%	267%	271%	271%
Least ICI Improvement Over Discrete Gaps ($R_{Dis,min}/R_{ICI,max}$)	---			2.52×	2.89×	3.30×	---		

segment causes the magnetic flux density to be higher than the rest of the structure, increasing with the indentation, as shown in Fig. 5. The loss components are plotted in Fig. 5(a) as a function of the indentation. For a fair comparison, the gaps had to be decreased with more indentation to keep the inductance constant at around 10.2 µH because of the increasing reluctance of the indented segment. As expected, the copper loss reduces quickly as more indentation is introduced due to the reduced perpendicular fringing field, dropping by about seven times between the maximum (0 mm) and minimum (4 mm) total loss points. On the other hand, the core loss increases only by 50% between the same two points. This is due to three reasons. First, the volume of the piece decreases with increasing indentation, lightly counteracting the increase in loss due to intensified flux density. Second, the increased flux density happens only in the indented segment, which makes up about 25% of the total core volume. Thus, even though the flux density more than doubles between the two points, as denoted by the markers in Fig. 5(b), the core loss increases more slowly than the copper loss decreases. Third, the loss of the inductor with no indentation is dominated by the copper loss, which comprises about 90% of its total loss due to the large fringing component. Thus, the slight increase in the core loss will not affect the total loss as much as a decrease in the copper loss. As a result of these three reasons, the total loss decreases with more indentation up to an optimal point and then increases again. The ICI design is slightly lower than the optimal point. However, it is still near the optimal minimum loss because of the wide valley and has a lower operating flux density in the indented segment.

The other consideration is the saturation current since the thinner indented segment would be the first piece to saturate in the structure. The saturation current is inversely proportional to the indentation according to the flux linkage equality:

$$I_{sat} = \frac{N B_{sat} A}{L}$$
$$= \frac{6 \cdot 430 \text{ mT} \cdot 32 \cdot (6.5 - \ell_{ind}) \text{ mm}^2}{10.5 \text{ µH}} \quad (2)$$

where I_{sat} is the saturation current; N is the number of turns; B_{sat} is the saturation flux density; A is the cross-sectional area; ℓ_{ind} is the indentation; and L is the inductance. The saturation current curve in Fig. 5(b) was created using (2), and the flux density was calculated by making it the subject of the equation and plugging in 11.5 A for the current (peak value in experiments). The saturation curve in Fig. 5(b) suggests that the ICI can handle only about half of what the discrete gaps structure can. However, one must remember that the inductor with discrete gaps suffers from a significantly higher total loss (four times without optimized gap spacing) and will run into the thermal limit before saturation becomes a consideration. A larger window structure can be utilized to manage heat extraction and enable reaching the saturation limit, but the larger volume becomes the tradeoff. Thus, one can achieve two of the following: adequate heat extraction, higher saturation current, or smaller volume, but not the three at once. Further work that could achieve the three extremes in size-constrained magnetics would be beneficial but is not the focus of this paper.

C. 3-D Finite-Element Simulations (FES)

To confirm the benefits of the ICI structure, a 1-oz, 2-layer Printed Circuit Board (PCB) was designed to implement the six turns according to the dimensions in Fig. 1. Multiple PCBs were prepared with the turns either on the bottom or the top layer to test the sensitivity of the winding's ac resistance to its location in the window, as shown in Fig. 3. Five prototypes were planned in total: two ICIs with turns on the bottom and top, two inductors with discrete gaps and turns on the bottom and top, and an inductor with discrete gaps, a larger window and turns on the bottom, for which the 3-D FES simulation models are shown in Fig. 6. The simulated current and flux densities using ANSYS Maxwell 3D EddyCurrent solver are shown in Fig. 7 and Fig. 8, respectively. The copper turns were kept about 0.1-0.2 mm away from the top or bottom of the winding window to account for the PCB mask and assembly inaccuracies.

As expected, Fig. 7 shows that the current density distribution is more uniform in the ICI structures than in the inductors with discrete gaps, even when a larger window is used. The discrete gaps cause significant current crowding directly under them, making the fringing losses of these inductors higher than ICIs. Within each structure, the fringing effect on the current crowding is worse when the turns are on the top layer because they are closer to the gaps. In the discrete gaps case, the current spreads over a larger area with the bottom turns (Fig. 7(c)) compared to the turns on top (Fig. 7(d)), thus reducing the fringing loss and the corresponding ac resistance. The current distribution on the bottom-layer turns of the inductor with discrete gaps and the larger window in Fig. 7(e) is similar to the distribution of the ICI with the turns on the top layer in Fig. 7(b), which makes their ac resistances closer in value as discussed in Subsection II.C and shown in Fig. 12. The flux density shown in Fig. 8 agrees well with the calculated values marked in Fig. 5,

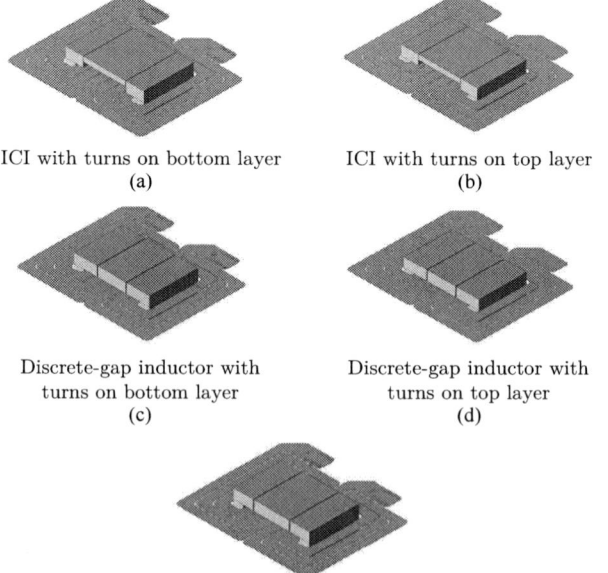

ICI with turns on bottom layer
(a)

ICI with turns on top layer
(b)

Discrete-gap inductor with turns on bottom layer
(c)

Discrete-gap inductor with turns on top layer
(d)

Discrete-gap inductor with a larger window and turns on bottom layer
(e)

Fig. 6. Models for 3-D FES: dimensions of the models in (a) and (b) are from Fig. 1(b); dimensions for (c) and (d) are from Fig. 1(a); and dimensions for (e) are from Fig. 4.

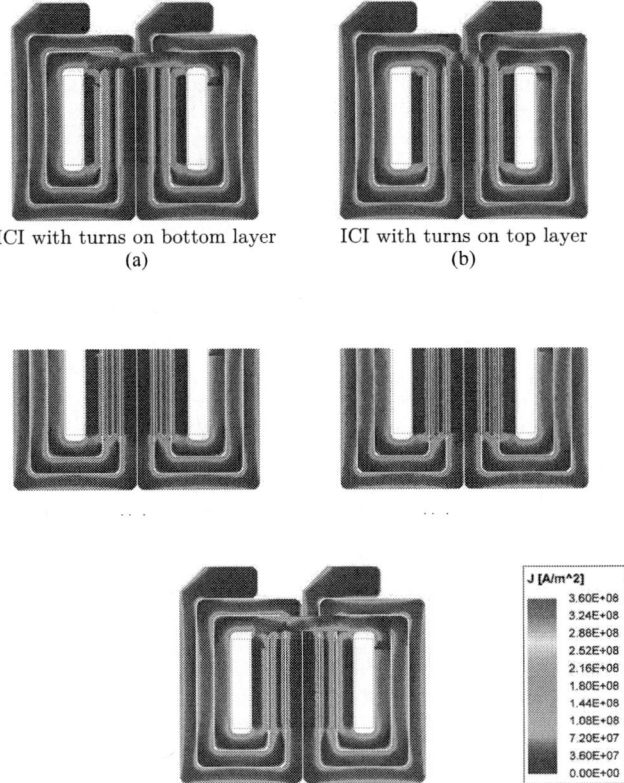

ICI with turns on bottom layer (a)   ICI with turns on top layer (b)

Fig. 7. The simulated current density for 11.5 A in the turns of the corresponding models of Fig. 6.

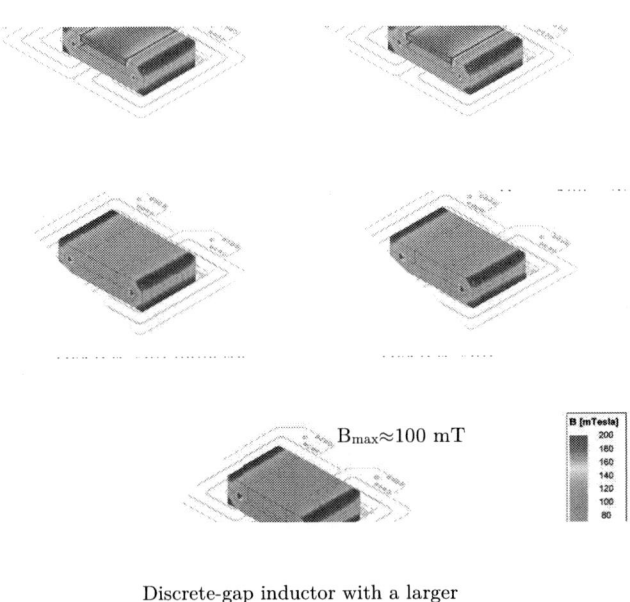

$B_{max} \approx 100$ mT

Discrete-gap inductor with a larger window and turns on bottom layer
(e)

Fig. 8. The simulated magnetic flux density for 11.5 A on the core surface of

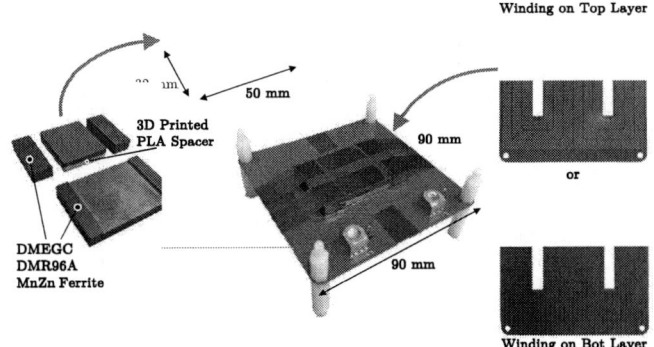

Fig. 9. The assembly process of the inductor prototype is shown with turns on the top layer (closest to the gaps)

showing the tradeoff of a higher flux density in the indented segment and thus slightly higher core loss as discussed in Subsection III.B. Other than in the indented segment, the flux density distribution is the same between the discrete gaps and ICI structures and does not vary with the position of the turns.

IV. PROTOTYPES AND MEASUREMENT RESULTS

Prototypes of the five planar PCB inductors of Fig. 6 were built according to the dimensions in Fig. 1 and Fig. 4. The turns shown in Fig. 7 were routed on the top and bottom layers of a 2-layer, 1-oz PCB with a thickness of 0.8 mm to almost fill the 1.0 mm-high window of the first four prototypes. The same PCB winding was used for the fifth prototype (discrete gaps) with the larger 2 mm-high window and held against the bottom of the winding window to space the turns as far as possible from the gaps. Multiple layers of 0.05 mm Kapton® tape and plastic shims were used to implement the necessary gaps, and a 3D-printed spacer supports the indented piece in the ICI prototypes. Each inductor was assembled using four core pieces (DMEGC DMR96A MnZn ferrite), as in Fig. 9. The completed prototypes are shown in Fig. 10.

A. Impedance Measurements

The impedance of the four prototypes was measured with a Keysight E4990A impedance analyzer using a 42941A probe and summarized in Fig. 11. The impedances and inductances match well, which is essential for the test setup of Fig. 14, staying within ±3% of 10.5 μH. The resistance for the five prototypes was simulated in 3-D FES as a function of frequency using the models of Fig. 6 and compared to the measurement in Fig. 12. The simulated and measured resistances match well within the rough range of 30 kHz – 300 kHz. The low-frequency dc resistance discrepancy could be due to the different measurement contact resistance for the different prototypes, the screw terminals that were not modeled in the 3-D FES models, and the reduced accuracy of the impedance analyzer at lower frequencies for small impedances but is not a concern here. In addition to the terminals, the high permittivity of the core and the different parasitic capacitances (turn-to-turn, turns-to-core, turns-to-connecting trace on opposite layer, etc.) cannot be modeled in FES without adding unnecessary computation

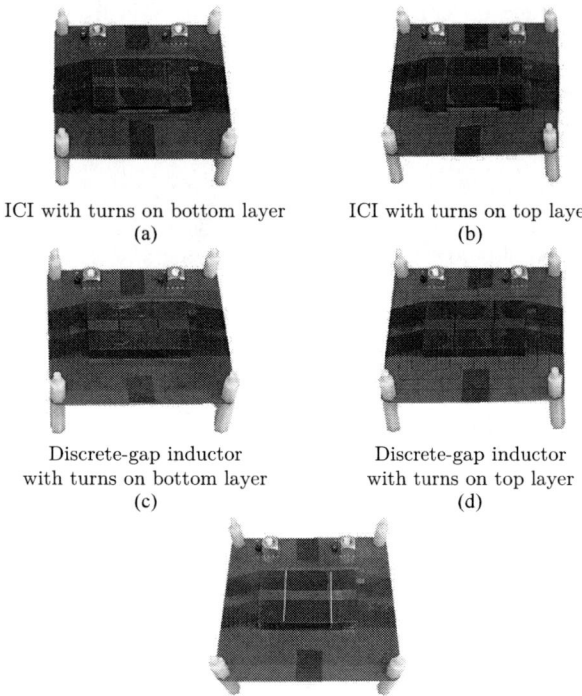

ICI with turns on bottom layer
(a)

ICI with turns on top layer
(b)

Discrete-gap inductor
with turns on bottom layer
(c)

Discrete-gap inductor
with turns on top layer
(d)

Discrete-gap inductor with a larger window and turns on bottom layer
(e)

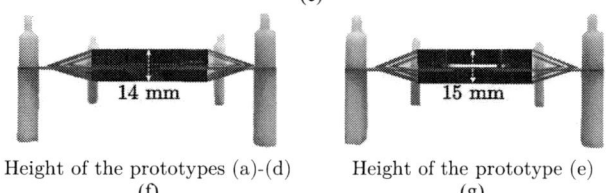

14 mm

15 mm

Height of the prototypes (a)-(d)
(f)

Height of the prototype (e)
(g)

Fig. 10. Five assembled inductors to test the extreme points of Fig. 3 and the design of Fig. 4.

a series inductance-resistance branch (which is not entirely accurate), resulting in disagreement at higher frequencies. The measured resistances at the 400 kHz switching frequency are summarized in Table II.

Within the mentioned range, the resistances show a good agreement, especially in terms of the order that one would expect: the prototype with discrete gaps and turns on top has the

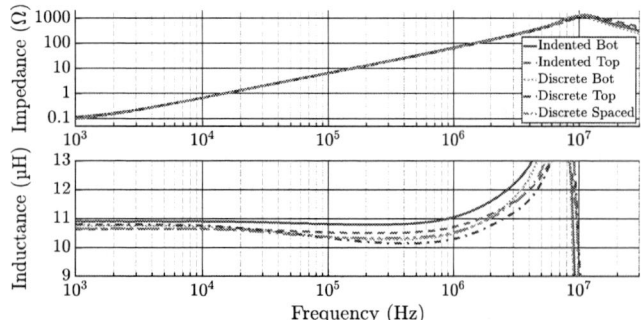

Fig. 11. The measured impedance and inductance of the prototypes of Fig. 10 using a Keysight E4990A impedance analyzer and a 42941A probe. The inductances range from 10.15 μH to 10.82 μH at the intended switching frequency of 400 kHz.

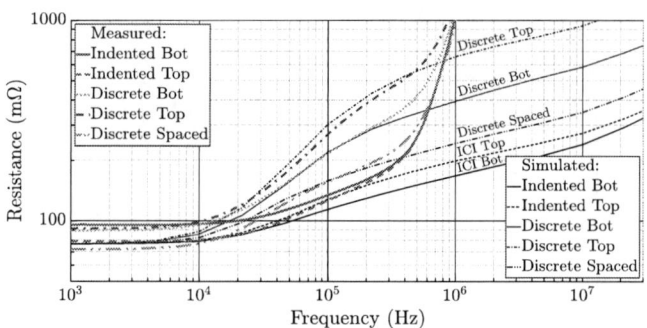

Fig. 12. The simulated resistance (11.5 A_pk current) of the 3-D FES models in Fig. 6 and the measured resistance (500 mV_pk voltage) of the prototypes of Fig. 10 using a Keysight E4990A impedance analyzer and a 42941A probe.

TABLE II.   THE MEASURED RESISTANCES IN FIG. 12 AT 400 KHZ

Prototype	Indented Bot	Indented Top	Discrete Bot	Discrete Top	Discrete Spaced
Measured Resistance (mΩ)	214.0	210.4	367.8	513.4	249.3
Variance of R_{max} from R_{min}	1.7%		39.6%		---

highest resistance, followed by the discrete bot inductor, then the spaced-turns discrete prototype, and finally, the two ICIs. While the ICI with bottom turns was expected to have a lower resistance than the ICI with top turns, its higher inductance seen in Fig. 11 indicates that the implemented gaps might be slightly smaller than intended. This results in a higher fringing loss [3, 6], causing the two ICIs to have similar resistance values at the desired switching frequency. The resistance variation reported in Table II is smaller than that in Table I because the prototypes require additional routing to connect the six turns in series. This adds extra copper outside the winding window that sees the same current distribution in all the prototypes, as seen in Fig. 7. The ac resistance contribution of this part of the winding is thus the same for all the prototypes, diluting the larger variation seen in 2-D simulations. As a result, the ac resistance variation of the prototypes with discrete gaps is underestimated, and the desensitization benefit of the ICIs is overestimated. The ICIs still provide a significant improvement over the prototypes with discrete gaps of at least 1.7 times lower resistance. Implementing the 2-D cross-sections of Fig. 1 in the two windows of an EI core is expected to give more considerable resistance variations as in Table I.

B. Losses Measurements in an Unloaded, Soft-Switched Buck Converter

To assess the benefits of the ICIs, the prototypes were operated in an unloaded buck converter at 50% duty. The expected loss breakdown for the prototypes at 400 V is shown in Fig. 13. It was calculated using the measured resistance values of Table II, the dimensions of the prototypes in Fig. 1 and Fig. 4, and the Improved Generalized Steinmetz Equation (iGSE) as follows [15]:

$$P_{loss} = P_{copper} + P_{core}$$

$$= I_{rms}^2 R_{meas} + k_i V f^\alpha (D^\alpha + (1-D)^{1-\alpha}) B_\Delta^\beta \quad (3)$$

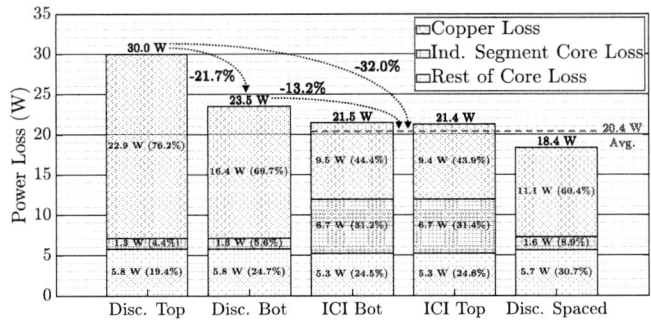

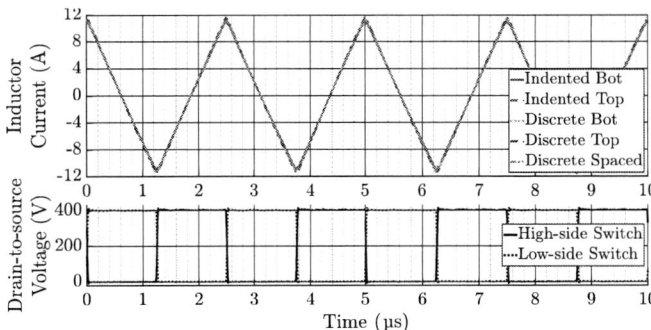

Fig. 13. The calculated loss breakdown of the prototypes of Fig. 10 using the measured resistance values in Table II, the dimensions in Fig. 1 and Fig. 4, and the ferrite material Steinmetz Coefficients according to (3). The ICI prototypes and the spaced-turns inductor are expected to have ~32% lower total losses.

Fig. 15. The resulting waveforms from the setup in Fig. 14 at 400 V input, showing soft switching operation. All the prototypes have a very similar current waveform (~23 A_{pk-pk}) and drain-to-source voltage (only one shown).

where I_{rms} is the average rms current of the five measurements in Fig. 15 (6.675 A); R_{meas} is the measured resistance from Table II; k_i is a factor given by $k_i = \frac{k}{2^{\beta+1} \pi^{\alpha-1} (0.2761 + \frac{1.7061}{\alpha+1.354})}$; V is the ferrite piece volume; f is the switching frequency (400 kHz); D is the duty cycle (0.5); B_Δ is the peak-to-peak flux density in the core piece; and k, α, and β are the Steinmetz coefficients of the core material (0.0108, 1.8524, 2.7039 at 100°C, respectively). The core loss of the indented segment was separated from the rest of the ferrite pieces to show the breakdown within the core. Since the core structure remains the

same for either the prototypes with discrete gaps or the ICIs, the core loss is the same for each structure regardless of the position of the turns in the window. The loss breakdown shows that the ICIs reduce the total loss by up to 32% over inductors with discrete gaps and a small window at the price of an increased core loss due to the increased flux density in the indented segment, as discussed in Subsection III.B. Spacing the gaps achieves the same effect (slightly lower loss) at the cost of an increased volume. Additionally, the plot illustrates how the change in the resistance due to the winding position can significantly increase the total loss of an inductor with discrete gaps (~28%) due to the amplified fringing losses while the ICI prototypes desensitize the winding to the fringing fields and achieve similar losses in both cases.

The prototypes were tested using the setup of Fig. 14 with the operating waveforms at 400 V shown in Fig. 15. The converter uses low ESR capacitors and operates in soft switching (as seen in the drain-to-source voltages in Fig. 15) to minimize and maintain the losses of capacitors and switches at the same level across the different inductors. Because the inductances of the prototypes are almost equal, the converter operating conditions do not vary significantly. The measured input power reduction is thus dominantly due to the change in inductor loss, which is confirmed by the identical inductor currents in Fig. 15. The shown drain-source voltages of the switches are from the ICI bot prototype operation, but all the prototypes have similar soft-switching waveforms. The input

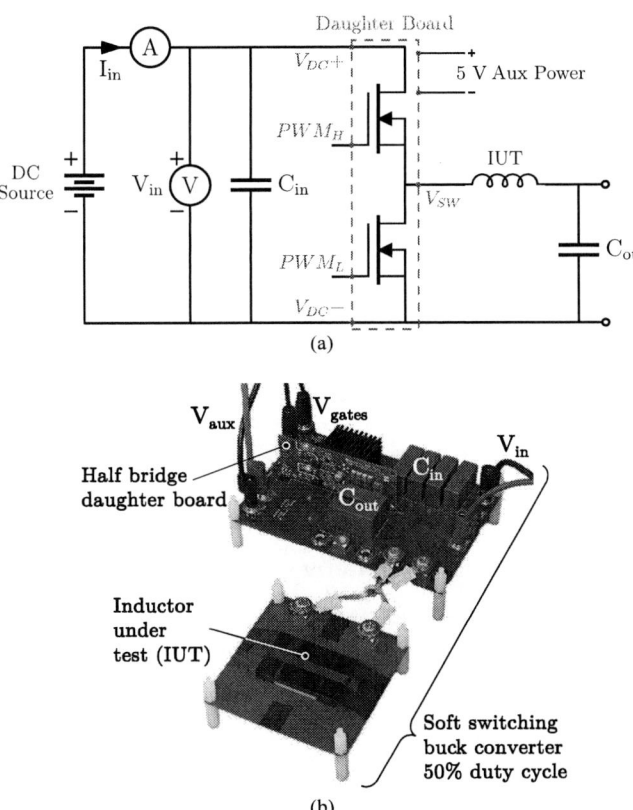

Fig. 14. The test setup implementing the built prototypes of Fig. 10 in an unloaded soft-switched buck converter at 50% duty cycle: (a) circuit schematic and (b) equivalent hardware implementation.

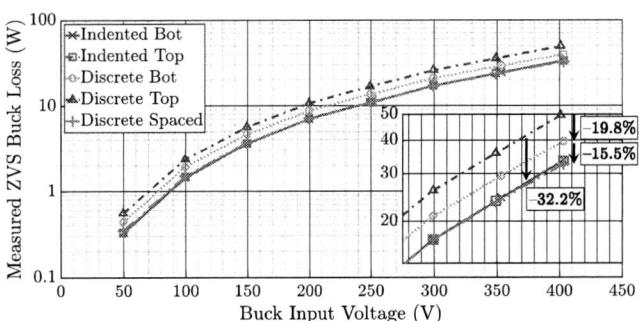

Fig. 16. The measured total input power of the unloaded soft-switched buck converter of Fig. 14 with the prototypes of Fig. 10. The inductors put the converter in the same operating conditions. Thus, the loss reduction is dominantly due to the decrease in the inductor loss.

power was measured using 34405A Agilent multimeters and summarized in Fig. 16 as a function of the varied input voltage.

The ICI prototypes constantly have a lower power loss than the inductors with discrete gaps, decreasing the total converter losses by more than 32% across the tested range. The difference between the losses in Fig. 16 and Fig. 13 is due to the additional switching and capacitors losses included in the measurements (estimated at ~10 W and ~2 W from spice simulations, respectively) and the elevated temperature of the windings, which increases the measured copper loss. The calculations also overestimate the core loss by assuming a constant flux density in all the core pieces other than the indented segment (not entirely accurate, as seen in Fig. 8). However, the same trend is observed with similar loss reductions (although the percentage reduction should be larger in Fig. 13 to get the same percentage in the total loss of Fig. 16). The volume and desensitization benefits previously discussed are also reflected on the total converter loss: the ICI prototypes achieve the same loss as a larger inductor with discrete gaps, and the loss does not change with the location of the winding within the window. The ICI loss is also at least 15% lower than an inductor with discrete gaps of the same size. These results confirm three of the four advantages discussed in Section II.

V. Conclusions and Future Work

This paper presented a core geometry that reduces the fringing losses in gapped magnetic components. Using an indented ferrite segment at the location of the gap diverts the fringing flux away from the winding, decreasing the magnitude of the harmful perpendicular fringing component at the surface of the turns. As a result, the winding's ac resistance is reduced for the same volume of a conventionally gapped inductor, decreasing the inductor's copper and total losses. Alternatively, the size of the inductor with discrete gaps can be reduced without sacrificing efficiency, thus increasing the power density and reducing related material costs. The introduced geometry also decreases the sensitivity of the winding's ac resistance to its position in the window and enables better heat extraction from the structure. Five prototypes were built and tested in an unloaded buck converter, reducing the total converter loss by more than 32% and confirming the structure's potential to simultaneously achieve high efficiencies and power densities using low-profile or size-constrained magnetics.

Further work is needed to confirm the heat extraction advantage discussed in this work. Also, the full benefits of the discussed geometry can be demonstrated by applying it to two-windowed cores (like EI sets) and other core shapes. A more generalized and systematic design approach would also be valuable in extending the geometry beyond the presented case study. The discussion in III.B also highlighted the need for a solution that simultaneously achieves reduced loss, smaller volume, and better heat extraction.

Acknowledgment

The authors would like to thank DMGEC for kindly donating the ferrite core samples used in this study.

References

[1] K. D. T. Ngo and M. H. Kuo, "Effects of air gaps on winding loss in high-frequency planar magnetics," in *PESC '88 Record., 19th Annual IEEE Power Electronics Specialists Conference*, Kyoto, Japan, 1988, vol. 2: IEEE, pp. 1112-1119, doi: 10.1109/pesc.1988.18251.

[2] Z. Ouyang and M. A. E. Andersen, "Overview of Planar Magnetic Technology—Fundamental Properties," *IEEE Transactions on Power Electronics*, vol. 29, no. 9, pp. 4888 - 4900, Sep. 2014, doi: 10.1109/TPEL.2013.2283263.

[3] H. Jiankun and C. R. Sullivan, "AC resistance of planar power inductors and the quasidistributed gap technique," *IEEE Transactions on Power Electronics*, vol. 16, no. 4, pp. 558-567, 2001, doi: 10.1109/63.931082.

[4] P. Ren, W. Chen, X. Huang, Y. Chen, Y. Wang, and X. Yang, "AC Copper Loss Reduction in Planar Inductors With Magnetic Building Blocks-Based Gapless Parallel Symmetrical Magnetoresistance Structure," *IEEE Journal of Emerging and Selected Topics in Power Electronics*, vol. 11, no. 4, pp. 4295-4312, 2023, doi: 10.1109/JESTPE.2023.3278689.

[5] R. A. Jensen and C. R. Sullivan, "Optimal core dimensional ratios for minimizing winding loss in high-frequency gapped-inductor windings," in *Eighteenth Annual IEEE Applied Power Electronics Conference and Exposition, 2003. APEC '03.*, Miami Beach, FL, USA, 2003, vol. 2: IEEE, pp. 1164-1169, doi: 10.1109/APEC.2003.1179363.

[6] W. A. Roshen, "Fringing Field Formulas and Winding Loss Due to an Air Gap," *IEEE Transactions on Magnetics*, vol. 43, no. 8, pp. 3387-3394, 2007, doi: 10.1109/TMAG.2007.898908.

[7] Y. Liu, A. Kumar, S. Pervaiz, D. Maksimovic, and K. K. Afridi, "A high-power-density low-profile DC-DC converter for cellphone battery charging applications," in *2017 IEEE 18th Workshop on Control and Modeling for Power Electronics (COMPEL)*, Stanford, CA, USA, 2017 2017: IEEE, pp. 1-6, doi: 10.1109/COMPEL.2017.8013362.

[8] Y. Dou, Z. Ouyang, P. Thummala, and M. A. E. Andersen, "PCB embedded inductor for high-frequency ZVS SEPIC converter," in *2018 IEEE Applied Power Electronics Conference and Exposition (APEC)*, San Antonio, TX, USA, 2018: IEEE, pp. 98-104, doi: 10.1109/APEC.2018.8340994.

[9] A. Nabih and Q. Li, "Low-Profile and High-Efficiency 3 kW 400 V-48 V LLC Converter with a Matrix of Four Transformers and Inductors for 48V Power Architecture for Data Centers," in *2021 IEEE Energy Conversion Congress and Exposition (ECCE)*, Vancouver, BC, Canada, 2021: IEEE, 2021, pp. 1813-1819, doi: 10.1109/ECCE47101.2021.9595881.

[10] F. Jin, A. Nabih, T. Yuan, and Q. Li, "A High-Efficiency High-Density Three-Phase CLLC Resonant Converter With a Universally Derived Three-Phase Integrated Transformer for On-Board-Charger Application," *IEEE Transactions on Power Electronics*, vol. 39, no. 4, pp. 4350-4366, 2024, doi: 10.1109/TPEL.2024.3354679.

[11] T. Yuan, F. Jin, and Q. Li, "Analysis and Comparison of Integrated Planar Transformers for 22-kW On-Board Chargers," *IEEE Transactions on Power Electronics*, vol. 39, no. 9, pp. 11368-11385, 2024, doi: 10.1109/TPEL.2024.3410878.

[12] S. Mukherjee, Y. Gao, and D. Maksimovic, "Reduction of AC Winding Losses Due to Fringing-Field Effects in High-Frequency Inductors With Orthogonal Air Gaps," *IEEE Transactions on Power Electronics*, vol. 36, no. 1, pp. 815-828, 2021, doi: 10.1109/TPEL.2020.3002507.

[13] E. C. Snelling, *Soft Ferrites: Properties and Applications*, 1st ed. London, U.K.: Iliffe Books Ltd, 1969, p. 27.

[14] J. Knowles, "The origin of the increase in magnetic loss induced by machining ferrites," *IEEE Transactions on Magnetics*, vol. 11, no. 1, pp. 44-50, 1975, doi: 10.1109/tmag.1975.1058549.

[15] K. Venkatachalam, C. R. Sullivan, T. Abdallah, and H. Tacca, "Accurate prediction of ferrite core loss with nonsinusoidal waveforms using only Steinmetz parameters," in *IEEE Workshop on Computers in Power Electronics*, Mayaguez, PR, USA, 2002: IEEE, doi: 10.1109/cipe.2002.1196712.

[16] DMEGC, "UI cores UI50," datasheet. [Online], Available: https://dongyangdongci.oss-cn-hangzhou.aliyuncs.com/uploads/2023032 9/aa74741d54c8772be463ac3e8ffa6d27.pdf. [Accessed Aug. 18, 2024]

Coupled Inductor Analysis and Finite Element Modeling Assisted Design for Boost Extender Topology

Vikas Kumar Rathore, *Student Member, IEEE,* Michael Evzelman,*Member, IEEE* and Mor Mordechai Peretz, *Member, IEEE*

The Center for Power Electronics and Mixed-Signal IC, Department of Electrical and Computer Engineering
Ben-Gurion University of the Negev, P.O. Box 653, Beer-Sheva, 8410501 Israel
rathore@post.bgu.ac.il; evzelman@bgu.ac.il; morp@bgu.ac.il
http://www.pemic.org/

Abstract – **Theoretical and finite element analysis of four coupled inductor structures based on the common cores suitable for Boost Extender topology is presented. The impact of each setup on the boost extender topology is evaluated and a design method that targets optimal operation point of the converter is demonstrated. Coupled inductors built around the most popular EE and UU cores are examined. The impact on the self-inductance of each winding and the coupling coefficient between the windings as a function of the air gap size and location are evaluated. It was found that a UU core structure based coupled inductor is a good candidate to address the challenges of the losses and parasitic ripple generated due to the coupling between the two inductors in the boost extender converter. Theoretical predictions agree well and are complemented by a finite element magnetic simulation software suite ANSYS Maxwell, which is then validated experimentally on a full scale 100W experimental prototype.**

Keywords –Boost extender, Ansys maxwell, High voltage gain, Coupled inductor, Core structure, Effective inductance.

I. INTRODUCTION

In recent years, the popularity of high-density power converters is on the rise. Modern applications such as Electric Vehicles (EVs), cloud computing, and renewable energy harvesting and processing pose a significant challenge in reducing the size, weight and increasing the efficiency of the power converter [1]-[3]. One of the challenges is to carry out high conversion ratios in a single step, to both increase the efficiency and reduce the space. For this purpose, a variety of high conversion gain solutions have been presented in the literature [4]-[6]. High density in power converters is commonly achieved by increasing the switching frequency or reducing the size of passive elements, such as inductors and capacitors [7]. However, since the magnetic materials aren't following the development pace of switching elements, the path of further increasing switching frequency is rather limited. An alternative path is reducing the size of the passive components, which occupy substantial space and weight.

Going with the capacitors, there are several works that demonstrate how a large passive capacitor can be transformed into a small electronic capacitor, reducing the space and increasing power density [8], [9]. The other way to go is

reducing the size of magnetic elements. For this purpose, a popular method is an interleaving technique [10], [11], where the switching frequency is effectively multiplied by the number of paralleled units, and input and or output ripples could be partially cancelled, enabling smaller overall components. Another approach used at certain compatible topologies is inductors coupling [12]-[15], where two or more inductors are coupled on the same magnetic core, reducing overall magnetic element count, and providing some additional benefits such as ripple reduction [16], [17].

The two most common setups to build a coupled inductor are either the EE core, or the toroid core [18], [19]. EE based cores are more flexible in terms of adjustment and changes, since an airgap could be changed at any stage of the design and its overall handling during the development phase. Toroidal cores are used when the relative permeability is already known, and some methods such as distributed air core are used to implement it to gain some higher performance in terms of uniform flux along the core, and physically rigid structure for good manufacturability. However, beyond the points mentioned above, coupled inductor design poses an additional set of challenges, where the selection of coupling coefficient is one of the major ones.

Most of the magnetic design methods are empirical and require some degree of iterations to reach the desired set of parameters.

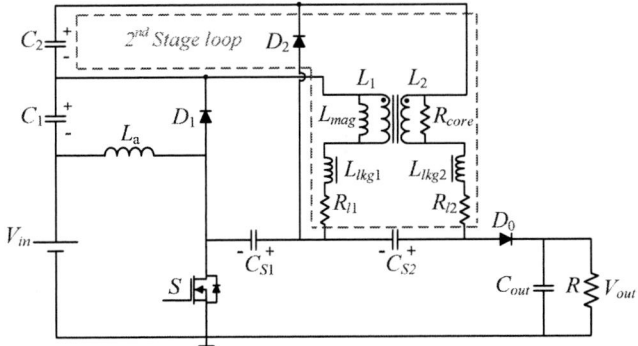

Fig. 1. Boost Extender topology with coupled inductor and parasitic components around it.

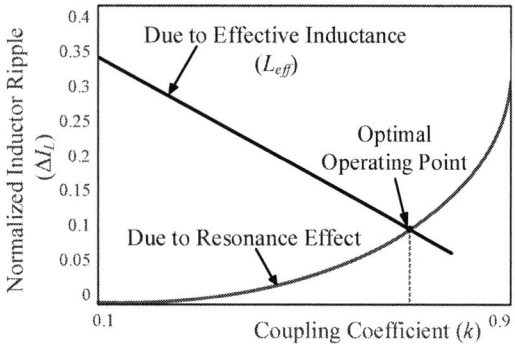

Fig. 2. Combined effect due to effective inductance and resonance effect.

One popular magnetic analysis method is based on creating an equivalent magnetic circuit, and either analyzing it theoretically or in a numerical based circuit simulator. Another method is to employ a dedicated software suite such as FEMM or Ansys Maxwell, which are solving Maxwell equations to evaluate the magnetic and electrical connections between the parameters at play [20]-[22]. There are some mentions in the literature of the methods to select coupling coefficient [23], [24], and there are some circuit level approaches that demonstrate some particular solutions, providing no structural design insight [25].

In this work, the method for selecting coupling coefficient for boost extending topology is presented. The method utilizes theoretical evaluation where possible, and in the cases where a pure theoretical approach is limited provide the way to simulate the core design in Ansys Maxwell magnetic simulation pack, in order to facilitate the desired behaviour of the coupled inductor to achieve a circuit level goals such as ripple reduction and increase in power converter efficiency. The work includes both examination of several magnetic structures of the EE and UU core coupled inductors, and a way to select the optimal operation point for the lowest losses of the boost extending converter.

The rest of the paper is organized as follows: Section II revisits the theoretical evaluation of effective inductance and parasitic resonance effect on coupled inductor ripple in boost extender topology; Section III presents the theoretical analysis of some common coupled inductor structures; Section IV shows Ansys Maxwell simulations and extends the theoretical analysis of section III; Section V presents the experimental results; Section VI draws the conclusion.

II. BOOST EXTENDER TOPOLOGY AND ITS LOSS FEATURES

A modular boost extending topology is shown in Fig. 1. The capacitor stacking approach on back end side of regular boost converter [26] offers several advantages such as low voltage stress across the semiconductor devices, high modularity and high voltage gain. However, each extension stage requires an inductor, increasing the size and weight of the converter. As it turns out, the inductors of each of the stages are compatible to be coupled on a single core, i.e. the voltages across all the inductors are proportional to each other and are aligned in time [27]. Coupling reduces the size and weight of the converter enabling higher power density and maintaining the same voltage stresses across the components and similar efficiency compared to uncoupled boost extender topology. While coupling two or more inductors on a single core offer many advantages, it presents several design challenges. These include inductor core

design and coupling coefficient, k, selection. For the case of coupled inductor-based boost extender topology two major loss effects have been determined in [25] that need to be considered while setting up inductors size and estimating associated converter losses.

It is assumed here that the major loss contribution of the boost extender topology is the ripple amplitude as discussed in [25]. In the same reference, two separate processes contributing to the ripple were delineated. The first, is caused by V/L, where the capacitor voltages are applied to the effective inductance L_{eff} of the coupled inductors and linearly impact the ripple (Fig. 2). The second effect is the parasitic resonance effect that is developed between the leakage inductances and the stacked capacitors next to each side of the coupled inductor. During circuit operation, L_1 and L_2 are coupled and the total uncoupled parts of the loop highlighted in blue in Fig. 1, $l_{kg\_T} = l_{lkg1} + l_{lkg2}$ and total capacitance of the same loop $C_t = C_1 + C_{S2}$ begin to resonate because of the initial current left in the leakages from the previous operation phase. The normalized change of ripple amplitude in the coupled inductor as a function of coupling coefficient is shown in Fig. 2. The black straight trace demonstrates the impact of the first, V/L linear effect, while a curved blue trace represents the impact of the resonance effect on the ripple. The two effects are naturally combined to form a total current ripple. The case demonstrated in Fig. 2 is a unique opportunity for coupled inductor design, such that the traces are monotonic and have an opposite effect on the current ripple versus coupling coefficient. For this case the coupled inductor design needs to comply with self-inductance increase with coupling coefficient increase, while being able to provide sufficiently high coupling coefficient to enjoy the benefits of size reduction stemming from building two inductors on a single core.

III. THEORETICAL ANALYSIS OF POTENTIAL COUPLED INDUCTOR STRUCTURES FOR BOOST EXTENDER TOPOLOGY

Optimal coupled inductor for the boost extender topology should exhibit the characteristics such as the behaviour of the coupling curves outlined above and the resulting losses, self-inductance versus air gap behaviour, and the number of turns. In this section the most common, potential candidate structures for coupled inductor are evaluated. A common magnetic core, the EE core is used and three of the coupling setups are examined (Fig. 3). In addition, a UU core is evaluated as well, since it is practically the EE core without the center leg, and this configuration exhibits certain advantageous features that will be discussed later. Some additional variations of the EE core and two windings such as the case of both inductors winded around the center leg of the EE core were omitted, due to their behavioural similarity to either one of the options discussed below.

The theoretical analysis of the coupled inductor structures is carried out using the equivalent magnetic circuit of the core [28]-[30]. An example of the analysis is done on all gapped EE core shown in Fig. 4a. Each of the core legs are represented by their equivalent reluctance shown as the resistors in the equivalent circuit of Fig. 4b, $R_{l(outer)}$, $R_{l(Center)}$. The reluctances are expressed in (1), where all the parameters are according to notation in Fig. 4, and μ_r and μ_0 are the relative and vacuum permeabilities respectively. The magnetomotive force of each of the windings

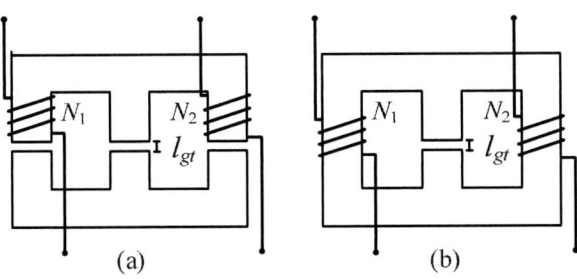 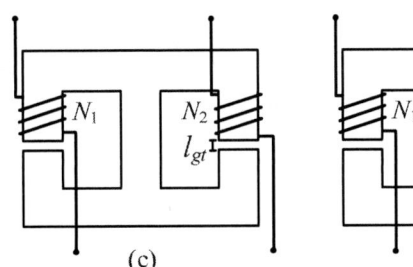 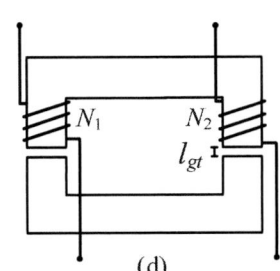

Fig. 3. Coupled inductor design configurations.

(Fig. 4a) is represented by the voltage source NI (Fig. 4b). Next the equivalent circuit of Fig. 4b is analysed and the effective inductance, L_{eff}, of each winding (2) and coupling coefficient, k, between the cores (3) are calculated.

$$R_{l(Center)} = \frac{l_{center} - l_{gt}}{\mu_r \mu_0 (A_c)} + \frac{l_{gt}}{\mu_0 (A_c)} \; ; \; R_{l(Outer)} = \frac{l_e - 2 l_{gt}}{\mu_r \mu_0 (A_o)} + \frac{2 l_{gt}}{\mu_0 A_o} \quad (1)$$

$$L_{eff} = N^2 \frac{R_{l(Center)} + R_{l(Outer)}}{R_{l(Outer)}^2 + 2 R_{l(Center)} R_{l(Outer)}} \quad (2)$$

$$(k) = \frac{R_{l(Center)}}{R_{l(Center)} + R_{l(Outer)}} \quad (3)$$

The UU core equivalent circuit model is shown in Fig. 4c. The effective inductance is summarized in (4), where R_x is the total reluctance that consists of the outer leg reluctance $R_{l(Outer)}$, and the air gap reluctances between both legs R_{lgt1} and R_{lgt2}, assuming equal air gaps at both legs, these reluctances are equal and represented by R_{lgt} (Fig. 4c). The reluctances of core ($R_{l(Outer)}$), air gap reluctances and the total reluctance R_x are summarized in (5).

$$L_{eff} = \frac{2N^2}{R_x} \quad (4)$$

$$R_{l(Outer)} = \frac{l_e - 2 l_{gt}}{\mu_0 \mu_r A_o}; \; R_{lgt} = \frac{l_{gt}}{\mu_0 A_o}; \; R_x = \frac{l_e - 2 l_{gt}}{\mu_0 \mu_r A_o} + \frac{2 l_{gt}}{\mu_0 A_o} \quad (5)$$

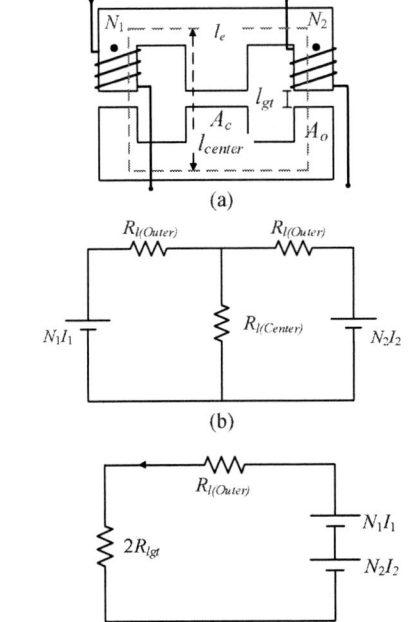

Fig. 4. (a) EE core structure; (b) EE core Equivalent magnetic circuit (Reluctance model); (c) UU core equivalent magnetic circuit.

As the case study an ETD34 core is selected (which is similar to EE core), with core magnetic material N87 by EPCOS -TDK, and its parameters (Table I) are used to evaluate the behavior of each of the coupled inductor structures of Fig. 3. The UU core Fig. 3d, is based on the ETD34 core with the center leg removed for results consistency to the EE core case.

The results of all legs gapped core structure of Fig. 3a are shown in Fig. 5. The structure behavior demonstrates some loose coupling peaking at $k = 0.2$, and unsatisfactory control over the coupling coefficient range. The tendency of the effective inductance to go down with coupling coefficient increase is demonstrated in Fig. 5. This is an opposite to the optimal effective inductance behavior presented in Fig. 2.

The results of middle leg gapped structure of Fig. 3b are shown in Fig. 6. This structure demonstrates a wide range of coupling coefficients (Fig. 6). The coupled inductance L_{mag} in this structure is constant, and independent of the air gap. The minimum effective inductance in this case is L_{mag}, bounded from the bottom, and occurs once the air gap is too large to sustain any significant magnetic flux. The coupling adjustment is carried out by adding leakage to the constant coupled inductance, which results in higher effective inductance, and lower coupling coefficient. Like in the previous case, the tendency of the effective inductance to go down with coupling coefficient increase (Fig. 6) doesn't fit the optimal effective inductance behavior required for the boost extender (Fig. 2).

Next coupled inductor structure candidate is an outer legs gapped EE core of Fig. 3c. This structure behaves according to the requirements of boost extender topology, where the effective inductance of each winding goes up with increase in the coupling coefficient (Fig. 7). This requirement creates a negative trend in current ripple of the boost extender topology with increase in coupling coefficient, assisting in overall converter efficiency gains. However, the range of the coupling this structure can provide in this case is $0 \sim 0.1$ (Fig. 7), which is rather limited, and resembles two uncoupled inductors, lacking the benefits of coupling two inductors on a single core.

TABLE I: ETD34 AND N87 MATERIAL PARAMETERS

Parameters	Values
Center leg area (A_c)	97.1mm^2
Center leg lenght(l_{center})	29.3 mm
Outer leg area (A_o)	48.55 mm^2
Total outer leg lenght (l_e)	115 mm
Ferrite core	(ETD) N87
Core permeability (μ_r)	1670
Maximum flux density (B_{max})	0.3T

The final case is the UU core structure of Fig. 3d. Due to the nature of the UU core, the theoretical analysis based on the magnetic circuit model for this case is limited to effective inductance versus air gap trace (Fig. 8), and this structure is further analyzed in the next chapter using ANSYS Maxwell. No analytical result are available for the coupling coefficient due to the limiting ability to model the leakage flux path and its magnitude relative to the main magnetic flux path.

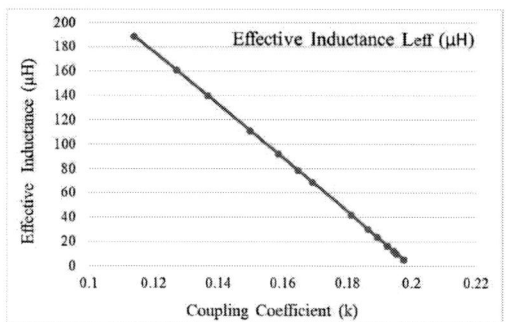

Fig. 5: All leg gapped core structure, Effective Inductance L_{eff} vs Coupling coefficient k.

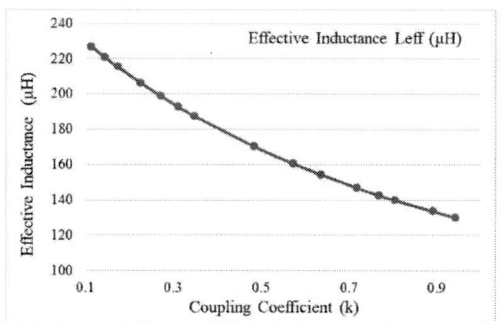

Fig. 6: Middle leg gapped core structure, Effective Inductance L_{eff} vs Coupling coefficient k.

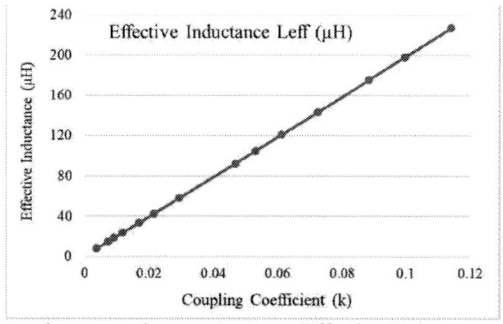

Fig. 7: Outer leg gapped core structure, Effective Inductance L_{eff} vs Coupling coefficient k.

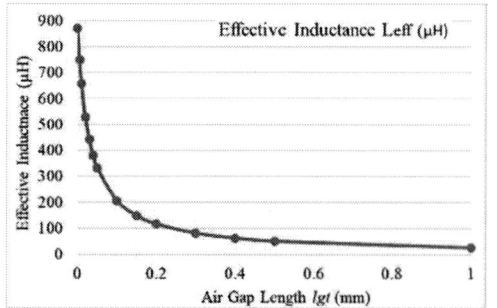

Fig. 8. UU core structure, Coupling coefficient k vs Air gap length l_{gt}.

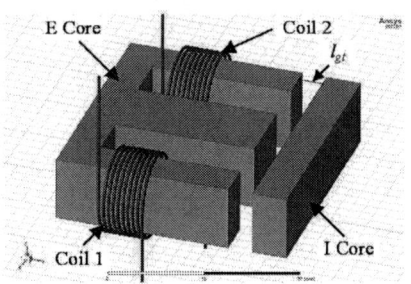

Fig. 9. All leg air gapped EI coupled inductor design for air gap parameter sweep.

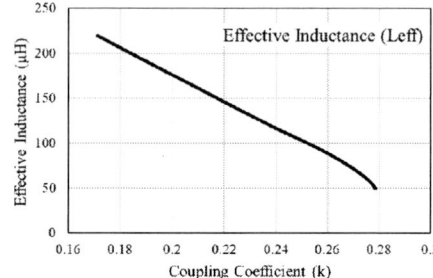

Fig. 10. All leg gapped EI core structure Ansys Maxwell simulation results, Effective Inductance L_{eff} vs Coupling coefficient k.

IV. ANSYS SIMULATION RESULTS

To validate the theoretical analysis and to enable the analysis of the UU core structure, modelling of the four coupled inductor designs of Fig. 3 is carried out with ANSYS Maxwell software suite that enables numerical examination of electromagnetic structures. The methodology consists of building the coupled inductor in 3D space, dividing it into a polygon mesh of finite elements which interact between one another. Finally, Maxwell equations are solved using appropriate magnetostatic solver, resulting in interactions between the mesh elements, which can be summarized to electrical parameters relevant to this study, such as inductance and coupling coefficients.

First, the design of all leg gapped (Fig. 3a) EE core structure is analysed. The 3D structure of the EE core model is created. To assist with parametric sweep of the air gap, a slight component rearrangement was done to the structure, where the two E-cores were substituted by a single E-core with double the leg length (Fig. 9), closed with an I core (Fig. 9). The structure dimensions conform to the parameters of ETD 34 core (Table I), except the length of the side legs, to enable magnetic path closure during the parametric sweep simulation. Assembled 3D structure was assigned to be a ferrite material and the relative permeability of it was set to $\mu_r = 1670$, which is the permeability of the N87 ferrite by TDK EPCOS. Next copper coils for the coupled inductor are created by adjusting parameters of predefined polygon helix. Specifically, a total of 9 turns are selected for the EE core case with a wire diameter of 0.5mm and a pitch of 0.8mm. The starting helix radius is set to be 7mm, ensuring proper winding over both outer legs of the E core. The winded coil configuration is illustrated in Fig. 9. After winding coils over the outer leg of the resulting EI core, an excitation current ($I_{exciation}$) is applied to both coils of the coupled inductor. The containing medium of the inductor was set to be air so that the gap will be air filled. To measure changes in the effective inductance L_{eff} and coupling coefficient k with respect to the air gap length l_{gt}, the parametric sweep analysis was performed on air gap length, moving the I-core away from E-core to create a larger gap (Fig. 9).

The behaviour of effective inductance with respect to coupling coefficient for the case of EE core structure of Fig. 3a, using the simulation model of Fig. 9, is shown in Fig. 10. Based on the results obtained from Ansys Maxwell, it is concluded that the coupling coefficient variation for all leg gapped structure core is limited, remaining around 0.28. This indicates that the coils of the coupled inductor are loosely coupled, with a significant portion of the magnetic flux left uncoupled. Furthermore, as shown in Fig. 10, the variation in effective inductance L_{eff} with respect to coupling coefficient k is minimal, making it unsuitable for boost extender, where higher coupling coefficients are needed to benefit from volume savings due to the two inductors coupling on the same core.

The next coupled inductor configuration that is analysed using Ansys software is the middle leg air-gapped structure, as shown in Fig. 3b. To analyse the 3D structure of EE core, like in the previous case, a single E core is created with double the legs length and is complemented by an I-core, to achieve an air gap between the middle leg and an I-core (Fig. 11). To facilitate air core changes only in the middle leg during the simulation sweep, the outer legs of the core were created just slightly longer. It enables to maintain closed magnetic path through the outer legs, while creating an air gap in the middle leg. The copper coils are winded around the outer legs of the core as explained in the previous design. The overall dimensions of the 3D structure complied to ETD 34 core as summarised in Table I.

After creating the 3D structure, a parametric sweep is performed for the air gap in the middle leg l_{gt} by moving the I-core. The variation of inductance is measured with respect to coupling coefficient k as shown in Fig. 12. From simulation results it has been observed that the coupling coefficient k is varying over the range of 0.1 to 0.87, which could be considered a wide variation.

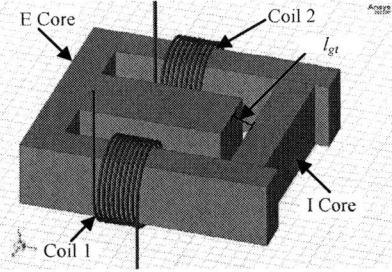

Fig. 11. Middle leg air gapped EI coupled inductor design with copper coils.

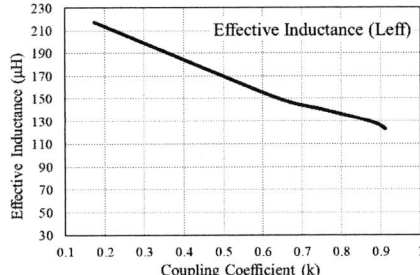

Fig. 12. Middle leg gapped EI core structure Ansys simulation results, Effective Inductance L_{eff} vs Coupling coefficient k.

But similar to previous discussed case the inductance value is inversely proportional to gap length, and going down with respect to coupling coefficient k. This behaviour of inductance align with the theoretical explanation but is not suitable for boost extender topology, as explained in previous theoretical section.

The next coupled inductor design is the outer leg gapped structure of EE core as shown in Fig. 3c. In order to create the 3D geometrical model of outer leg gapped structure, first a T shape core is modeled in the software as shown in Fig. 13. To complement the model to the EE core with outer air gaps, two L shaped core sections are created to be placed on both sides of the T shape (Fig. 13). The length of the L core sections is twice the length of a single E-core leg, like in the previous cases. The length of the T-core is slightly longer than twice the length of the E core legs to allow for seamless air gap sweep during the simulation.

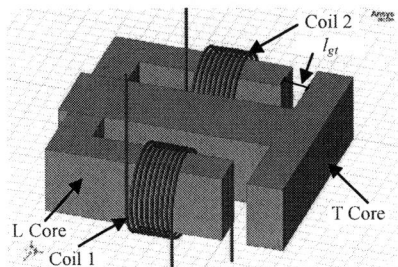

Fig. 13. Complete Outer leg air gapped EE coupled inductor design with copper coils.

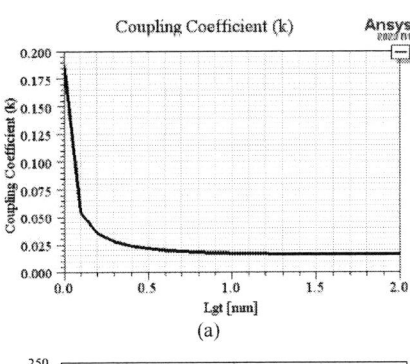

(a)

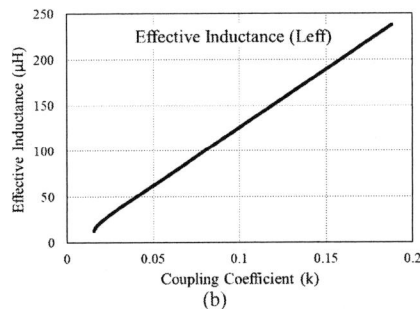

(b)

Fig. 14. Outer leg gapped EE core structure Ansys simulation results, (a) Coupling coefficient k vs Air gap length l_{gt}; (b) Effective Inductance L_{eff} vs Coupling coefficient k.

TABLE II: SIMULATION PARAMETERS

Parameters	Values
Initial Effective Inductance L_{eff}	260 µH
Initial air gap (l_{gt})	0mm
Excitation Current ($I_{L1} = I_{L2}$)	1A
Maximum inductor ripple current (ΔI_L)	2A
No. of turn ($N_1 = N_2$)	9
Current desnity of windings (J)	4.7 A/mm²

The sweep in air gap is carried out by sliding the T shaped core right and away from the L shaped sections (Fig. 13). The copper coils are winded on both sides of the L shaped core sections (Fig. 13). The overall core structure dimensions were maintained according to the ETD 34 core (Table I).

Ansys Maxwell simulation results for outer leg air gapped core are shown in Fig. 14. This core configuration demonstrates a loosely coupled behaviour, spanning a narrow range of coupling coefficients, k, as a function of an air gap length l_{gt} (Fig. 14a). This is unacceptable for boost extender topology, since the benefits of size reduction and power density increase at low coupling coefficients is rather minor. On the positive side, the increase in effective inductance of each winding of this structure comes along with the increase of the coupling coefficient, which is in line with the requirements for the boost extender topology (Fig. 14b). The Ansys Maxwell simulation results agree well with the theoretical core structure analysis presented in the previous section.

The next core structure that based on the theoretical analysis is a good candidate to comply with the requirements of the ripple interactions presented in Fig. 2 is the UU core coupled inductor design. To model the UU core design, an E-core is modified by removing the middle leg, as illustrated in Fig. 15.

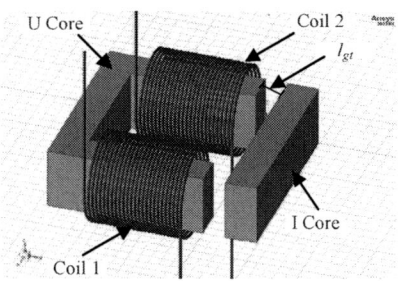

Fig. 15. UU core coupled inductor design, Complete UI structure with copper coils.

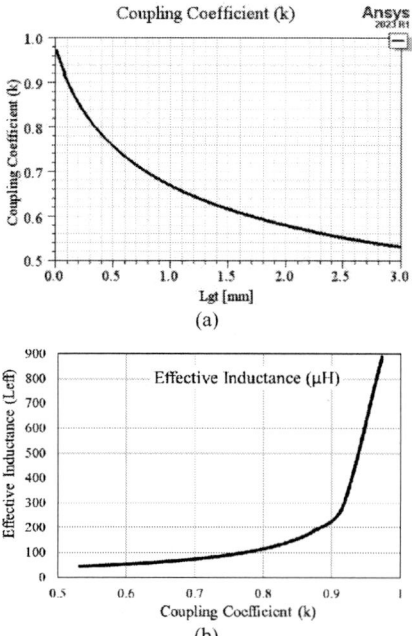

Fig. 16. UU core structure Ansys simulation results; (a) Coupling coefficient k vs Air gap length l_{gt}, (b) Effective Inductance L_{eff} vs Coupling coefficient k.

As in the previous cases and for simulation simplicity, the U core legs length is created double the length of the ETD 34 core (Table I), and an I core is added on top to create the air gaps on both sides (Fig. 15). As a result of the middle leg removal the overall core area is decreased in half, and the magnetic path length is increased to be nearly twice as long. Since effective inductance is proportional to magnetic area and inversely proportional to magnetic path length, the effective inductance in this setup has been reduced nearly four-fold. To keep the comparison consistent with the EE core cases discussed earlier, the number of turns in the UU case is increased from 9 to 25 (Fig. 15). Using the structure of Fig. 15 and following the simulation parameters described for all leg air gap EE core (Table II), the simulation results of the UU core are summarized in Fig. 16. The coupling coefficient spans the range between 0.5 to 1 (Fig. 16a), which is a good range suitable for size and power density benefits due to the windings coupling. The result of Fig. 16b demonstrates an increase in effective inductance with increase in coupling coefficient, which is a desired feature for the boost extender topology as shown in Fig. 2. Fig. 16b complements the missing result from the theoretical analysis, showing a change in effective inductance L_{eff} with respect to coupling coefficient.

V. EXPERIMENTAL RESULTS

To validate the theoretical derivations presented in this study, an experimental prototype has been built and evaluated. From the theoretical results of section III and the complementary Ansys Maxwell results of section IV, the most suitable structure based around UU core is selected, as it demonstrates the most potential to both performance increase in terms of ripple and losses reduction, along with size decrease of the magnetic component. The U cores are prepared from ETD 34 cores, by mechanically removing the center leg of the core. A copper wire with a diameter of $d = 0.5$mm is winded around the outer legs of the UU core as described in Fig. 15. The rest of the experimental parameters of the converter and coupled inductor are summarized in Table III. The variation in coupling coefficient is achieved by changing the air gap as discussed in the previous sections. For the matter of keeping the most efficient operation point of the boost extender converter, the CBCM mode [26], the effective inductance is kept constant around $L_{eff} = 100\mu$H while the increase in air gap is compensated by increasing the number of turns. Starting with 31 turns for the lowest coupling case the number of turns has reached 7 turns for the highest coupling case. However, for the same converter current, increasing the number of turns results in a higher magnetic field in the core and a higher strain on the core. To keep the consistency and for fair comparison between the results, a core utilization factor is introduced, to estimate the amount of stress applied to the core, comparing to its maximum stress specification, i.e. relative utilization. The utilization factor, U_f, is a ratio between the maximum magnetic field specified for the core material $B_{Max(Datasheet)}$, and an actual peak magnetic field developed as a result of the converter operation $B_{Peak(Exp)}$ (6). The value of $B_{Max(Datasheet)}$ is set to 0.3T for all the experiments, and used as the reference for maximum core utilization. The magnetic field developed in the core $B_{Peak(Exp)}$ is estimated based on the peak current flowing through the inductor, I_{Peak}, and is calculated from experimental results using (7):

$$U_f = \frac{B_{Peak(Exp)}}{B_{Max(Datasheet)}} \tag{6}$$

$$B_{Peak(Exp)} = \frac{I_{Peak}L_{eff}}{NA_o} \tag{7}$$

To create a point of reference two uncoupled UU core-based inductors were placed in the second stage of boost extender (Fig. 1) instead of a single coupled inductor. The inductors are built with 17 turns each, and an inductance of 100μH is achieved by adjusting the air gap. The results for inductor currents I_{L1} and I_{L2} are shown in Fig. 17a. The peak-to-peak current was measured to be 1.23A, and the inductor current lines are straight, indicating that there is a single effect that causes the ripple, and it is the linearly increasing ripple V/L, caused by the capacitor's voltages V applied to the inductance L. The core utilization factor calculated for this case is around 44% (Table IV), which means that ETD34 core is significantly underutilized for two coupled inductors case and the setup of Table III. Practically speaking, that means that two uncoupled cores half the size of ETD34 could be a suitable design for the power level of 100W.

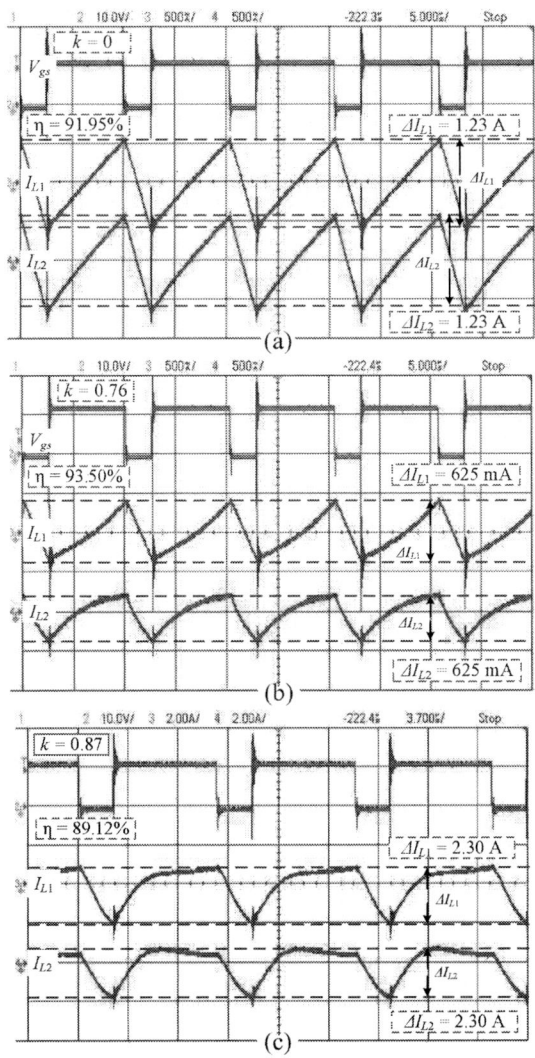

Fig. 17. Experimental waveforms for Coupling coefficients: (a) Coupling coefficient $k = 0$, 500mA/Div., (b) Coupling coefficient $k = 0.76$, 500mA/Div., (c) Tight coupling $k = 0.87$, 2A/Div. Top trace - Gate signal, Middle trace - Inductor Current I_{L1}, Bottom trace - Inductor Current I_{L2}, Time base: 5μs/Div. and 3.70μs/Div.

TABLE III: EXPERIMENTAL PROTOTYPE PARAMETERS

Parameters	Values
Effective Inductance (L_{eff})	100 μH
Wire diamerter (d)	0.5mm
Input Voltage (V_{in})	20V
Ouput Voltage (V_{out})	200V
Duty Cycle (D)	0.75
Switching Frequency (f_s)	100kHz
Output power (P_{out})	100W
Auxillary capacitor ($C_1 = C_2$)	20 μF
Series capacitors ($C_{S1} = C_{S2}$)	6.8 μF
Output Capacitor (C_{out})	50 μF

TABLE IV: SUMMARY OF EXPERIMENTAL RESULTS

(k)	ΔI_L	(η)	($B_{Peak(Exp)}$)	(U_f)
0 (Uncoupled)	1.23 A	91.95 %	0.132 T	44%
0.23	843 mA	92.14 %	0.122 T	41%
0.5	762 mA	92.68 %	0.168 T	56%
0.65	718 mA	92.75%	0.206 T	69%
0.76	625 mA	93.50%	0.276 T	92%
0.81	1.80A	91.86%	0.745 T	248%
0.86	2.15A	89.40%	0.773 T	258%
0.87	2.30 A	89.12%	1.08T	360%

To achieve some lower coupling coefficient, $k = 0.23$, 31 turns are winded on both outer legs of the UU core, and the effective inductance $L_{eff}= 100$μH is obtained by adjusting the air gaps of the two core legs. The resulting peak to peak current in this case is 843mA. The utilization factor, calculated using (6), is approximately 41% along with magnetic flux density of 0.122T (Table IV). The interaction between the two different current ripple effects contributes to lower overall current ripple in each of the windings, and a slight increase in efficiency, while a single core utilization factor remains nearly the same as for the two uncoupled inductors case, demonstrating the advantages of coupled inductor in the boost extender system.

Several more experiments with UU core based coupled inductor and different coupling coefficients are carried out. The number of turns is adjusted to fit the target effective inductance of 100μH, while the coupling coefficient is adjusted using the air gap. The results, including inductor currents ΔI_L, maximum flux density $B_{Peak(Exp)}$, core utilization factors U_f and boost extender conversion efficiency are summarized in Table IV. Waveforms are shown for coupling coefficients of $k = 0$ (Fig. 17a), $k = 0.76$ (Fig. 17b) and $k = 0.87$ (Fig. 17c).

The cases of coupling coefficients of $k = 0.81$ and higher, are exceeding the maximum allowable magnetic flux density of the core $B_{Max (Datasheet)} = 0.3$T, and the core becomes saturated. The utilization factor in these cases exceeds 100% (Table IV) and renders these setups unacceptable for implementation. In Fig. 17c, for $k = 0.87$, core overutilization is demonstrated, were high peak to peak currents begin to develop, and the shape of the current becomes similar to trapezoidal, highlighting the nonlinear effect it experiences.

Following the results of Table IV, the case of coupling coefficient of $k = 0.76$ is the most suitable to address the specifications outlined in Table III. The efficiency in this case peaks at 93.5%, and core utilization factor reaches 92%, which

979-8-3315-1612-3/25 $31.00 © 2025 IEEE

is very close to full core utilization, leaving some space for potential overshoots/overloads. This is in line with the theoretical results presented in Section II, where an optimal coupling coefficient point is found once the coupling inductor behaviour is maintained according to Fig. 2. The UU core is found to be a suitable coupling inductor structure candidate for optimizing the efficiency of the boost extender topology.

VI. CONCLUSION

An analysis of coupled inductor physical structure on the losses and efficiency of the boost extender topology is evaluated. Several common coupled inductor structures based around EE and UU cores are tested. Theoretical and finite element analysis are applied, followed by the experimental validation. The impact of the structural changes on the self-inductance and the coupling coefficient of the coupled inductor are evaluated. It was found that a UU core structure based coupled inductor is a good candidate to address the challenges of the losses and parasitic ripple generated due to the coupling between the two inductors in the boost extender converter. Experimental and Ansys Maxwell simulation results agree well with the theoretical predictions.

ACKNOWLEDGMENT

This research was funded by the Israel Ministry of Energy.

REFERENCES

[1] M. Hirakawa, M. Nagano, Y. Watanabe, K. Andoh, S. Nakatomi and S. Hashino, "High power density DC/DC converter using the close-coupled inductors," 2009 *IEEE Energy Conversion Congress and Exposition, San Jose, CA, USA*, 2009, pp. 1760-1767.

[2] Y. Suh, T. Kang, H. Park, B. Kang and S. Kim, "Bi-directional power flow rapid charging system using coupled inductor for electric vehicle," 2012 *IEEE Energy Conversion Congress and Exposition (ECCE), Raleigh, NC, USA*, 2012, pp. 3387-3394.

[3] C. C. Chan and K. T. Chau, "An overview of power electronics in electric vehicles," *IEEE Trans. on Industrial Electron.* vol.44, no.1, pp.3-13, 1997

[4] R. -J. Wai, C. -Y. Lin, R. -Y. Duan and Y. -R. Chang, "High-Efficiency DC-DC Converter With High Voltage Gain and Reduced Switch Stress," in *IEEE Transactions on Industrial Electronics*, vol. 54, no. 1, pp. 354-364, Feb. 2007.

[5] Qun Zhao and F. C. Lee, "High-efficiency, high step-up DC-DC converters," in *IEEE Transactions on Power Electronics*, vol. 18, no. 1, pp. 65-73, Jan. 2003.

[6] Chung-Wook Roh, Seung-Hoon Han, Sung-Soo Hong, Sug-Chin Sakong and Myung-Joong Youn, "Dual-coupled inductor-fed DC/DC converter for battery drive applications," in *IEEE Transactions on Industrial Electronics*, vol. 51, no. 3, pp. 577-584, June 2004.

[7] J. Zhang, J. -S. Lai, R. -Y. Kim and W. Yu, "High-Power Density Design of a Soft-Switching High-Power Bidirectional dc–dc Converter," in *IEEE Transactions on Power Electronics*, vol. 22, no. 4, pp. 1145-1153, 2007.

[8] S. Chowdhury, E. Gurpinar and B. Ozpineci, "High-Energy Density Capacitors for Electric Vehicle Traction Inverters," 2020 *IEEE Transportation Electrification Conference & Expo (ITEC), Chicago, IL, USA*, 2020, pp. 644-650.

[9] D. Cittanti et al., "Analysis and Design of a High Power Density Full-Ceramic 900 V DC-Link Capacitor for a 550 kVA Electric Vehicle Drive Inverter," 2022 *International Power Electronics Conference (IPEC-Himeji 2022- ECCE Asia), Himeji, Japan*, 2022, pp. 1144-1151.

[10] V. K. Rathore, M. Evzelman and M. M. Peretz, "Interleaving Boost Extender Topology," 2023 IEEE 24th *Workshop on Control and Modeling for Power Electronics, Ann Arbor, MI, USA*, 2023, pp. 1-6.

[11] W. Li, Y. Zhao, Y. Deng and X. He, "Interleaved Converter With Voltage Multiplier Cell for High Step-Up and High-Efficiency Conversion," *IEEE Trans. on Power Electronics*, vol. 25, no. 9, pp. 2397-2408, Sept. 2010.

[12] Pit-Leong Wong, Q. Wu, Peng Xu, Bo Yang and F. C. Lee, "Investigating coupling inductors in the interleaving QSW VRM," *APEC 2000. Fifteenth Annual IEEE Applied Power Electronics Conference and Exposition*, 2000, vol. 2, pp. 973-978, New Orleans, LA, USA.

[13] S. Ćuk and Zhe Zhang, "Coupled-inductor analysis and design," 1986 17th *Annual IEEE Power Electronics Specialists Conference, Vancouver, BC, Canada*, 1986, pp. 655-665.

[14] S. Cuk, "New magnetic structures for switching converters," in *IEEE Transactions on Magnetics*, vol. 19, no. 2, pp. 75-83, March 1983.

[15] Z. Zhang, "Coupled-Inductor Magnetics in Power Electronics," *Ph.D. thesis, California Institute of Technology*, 1986.

[16] S. Cuk and R. D. Middlebrook, "Advances in Switched-Mode Power Conversion Part I," in *IEEE Transactions on Industrial Electronics*, vol. IE-30, no. 1, pp. 10-19, Feb. 1983.

[17] S. Cuk, "A new zero-ripple switching DC-to-DC converter and integrated magnetics," *IEEE Trans. on Magnetics*, vol. 19, no. 2, pp. 57-75, 1983.

[18] Jieli Li, C. R. Sullivan and A. Schultz, "Coupled-inductor design optimization for fast-response low-voltage DC-DC converters," *APEC. Seventeenth Annual IEEE Applied Power Electronics Conference and Exposition (Cat. No.02CH37335), Dallas, TX, USA*, 2002, pp. 817-823.

[19] González-Castaño, C., Restrepo, C., Giral, R., García-Amoros, J., Vidal-Idiarte, E. and Calvente, J. (2020), Coupled inductors design of the bidirectional non-inverting buck–boost converter for high-voltage applications. *IET Power Electronics*, 13: 3188-3198.

[20] L. Duong, C. Somers and K. al-Haddad, "Design of a Four Windings Coupled Inductors for High Frequency Converters Applications," 2023 *IEEE International Conference on Industrial Technology (ICIT), Orlando, FL, USA*, 2023, pp. 1-6.

[21] K. J. Hartnett, J. G. Hayes, M. G. Egan and M. S. Rylko, "CCTT-Core Split-Winding Integrated Magnetic for High-Power DC–DC Converters," *IEEE Trans. on Power Electronics*, vol. 28, no. 11, pp. 4970-4984, 2013.

[22] Z. Dang and J. A. Abu Qahouq, "Modeling and design guidelines of high density power inductor for battery power unit," 2016 *IEEE Applied Power Electronics Conference and Exposition (APEC), Long Beach, CA, USA*, 2016, pp. 2114-2121.

[23] A. De Nardo, N. Femia, F. Forrisi, and M. Granato, "Design of SEPIC with coupled inductors," in *Proc. PCIM Europe, Nurnberg, Germany*, May 2009, pp. 238–239

[24] G. Di Capua and N. Femia, "A Critical Investigation of Coupled Inductors SEPIC Design Issues," in *IEEE Transactions on Industrial Electronics*, vol. 61, no. 6, pp. 2724-2734, June 2014.

[25] V. K. Rathore, M. Evzelman and M. M. Peretz, "Coupled Inductor Design Methodology for Optimization of Boost Extending Topology," 2024 *IEEE Applied Power Electronics Conference and Exposition (APEC), Long Beach, CA, USA*, 2024, pp. 1389-1395.

[26] V. K. Rathore, M. Evzelman and M. M. Peretz, "Non-Isolated High Conversion Ratio Boost Extender Based on Back-end Series Capacitor Stacking," 2022 *IEEE 23rd Workshop on Control and Modeling for Power Electronics (COMPEL), Tel Aviv, Israel*, 2022, pp. 1-7.

[27] V. K. Rathore, M. Evzelman and M. M. Peretz, "Coupled Inductor Based Non-Isolated High Conversion Ratio Boost Extender," 2023 *IEEE Applied Power Electronics Conference and Exposition (APEC), Orlando, FL, USA*, 2023, pp. 22-28.

[28] J. Imaoka et al., "A Magnetic Design Method Considering DC-Biased Magnetization for Integrated Magnetic Components Used in Multiphase Boost Converters," in *IEEE Transactions on Power Electronics*, vol. 33, no. 4, pp. 3346-3362, April 2018.

[29] L. Yang and S. Wang, "A compensation winding structure for balanced three-phase coupled inductor," 2017 *IEEE Applied Power Electronics Conference and Exposition (APEC), Tampa, FL, USA*, 2017, pp. 868-875.

[30] Shogo Aoto, Shota Kimura, Jun Imaoka, Masayoshi Yamamoto, An Investigation of Size-reduction Effects in High Power Density Boost Converters with a Magnetic Coupling, *IEEJ Transactions on Electronics, Information and Systems*, 2015, Volume 135, Issue 7, Pages 776-784.

Stability Analysis of Current-Limited Grid-Forming Inverters with Frequency Stabilization: An Equivalent Impedance Approach

Bowen Yang and Gab-Su Seo

Power Systems Engineering Center, National Renewable Energy Laboratory, Golden, CO 80401, USA

email: bowen.yang@nrel.gov, gabsu.seo@nrel.gov

Abstract—The rapid deployment of inverter-based resources (IBRs) in modern power grids aims to integrate renewable energy, yet the prevalence of grid-following (GFL) inverters raises stability concerns due to their dependency on a stiff grid. This has led to increased attention to grid-forming (GFM) inverters, which generate internal reference signals and can synchronize with the grid, essential for transitioning to IBR-dominated grids. However, challenges remain in understanding GFM inverters' capabilities and limitations in stabilizing grids with fewer synchronous machines, particularly regarding transient stability during disturbances. Inverters' strict overcurrent limits demand effective current limiting for protection, but current limiting risks instability as it prevents GFM inverters from meeting the power injection set by primary control leading to instability. This instability mechanism necessitates frequency stabilization methods to enhance inverter transient behavior, especially during prolonged disturbances. Since frequency stabilization methods rely on the current limiting conditions, understanding the interactions between current limiters and frequency stabilization methods is crucial. To fill the gap, this paper proposes an equivalent impedance-based model to analyze the interactions between current limiters and stabilization methods and their effects on GFM transient stability. The model, validated through numerical models and experiments, provides insights for improving GFM stability and reliability during grid disturbances.

Index Terms—Grid-forming inverter, stability, current limiting, equivalent impedance, frequency stabilization.

I. INTRODUCTION

In recent years, there has been significant deployment of inverter-based resources (IBRs) in modern power grids to integrate more renewable energy resources and address climate change [1]. However, most IBRs currently operate as grid-following (GFL) inverters, which raise critical stability concerns due to their reliance on grid-side voltage to generate inverter reference angles [2], [3]. These concerns have driven a paradigm shift towards grid-forming (GFM) inverters [2], [4], [5], which generate their own internal reference angles and

This work was authored by the National Renewable Energy Laboratory, operated by Alliance for Sustainable Energy, LLC, for the U.S. Department of Energy (DOE) under Contract No. DE-AC36-08GO28308. Funding provided in part by the U.S. Department of Energy Office of Energy Efficiency and Renewable Energy under the Solar Energy Technologies Office Award Number 38637. The views expressed in the article do not necessarily represent the views of the DOE or the U.S. Government. The U.S. Government retains and the publisher, by accepting the article for publication, acknowledges that the U.S. Government retains a nonexclusive, paid-up, irrevocable, worldwide license to publish or reproduce the published form of this work, or allow others to do so, for U.S. Government purposes.

synchronize with the grid using various primary controllers, such as droop control [6], virtual synchronous machines (VSM) [7], and dispatchable virtual oscillators (dVOC) [8]. Despite this shift, transitioning from conventional synchronous generator-based power systems to grids dominated by GFM inverters presents numerous challenges, with many research questions yet to be answered. One of the most critical issues is transient stability during grid disturbances [4], [9]. Power electronics-based inverters have stringent overcurrent limits due to hardware constraints, necessitating effective current-limiting control to quickly curtail overcurrents and prevent irreversible hardware damage [2], [4], [10], [11]. To address this issue, various current limiters have been proposed, including current reference saturation [12]–[14], virtual impedance [15]–[18], and hybrid methods [19]–[21]. However, when inverters enter current-limiting mode, they can no longer supply the required active or reactive power. Consequently, the actual injected power no longer matches the power setpoints or the power needed for synchronization, causing wind-up in the primary controller and leading to deviations in the inverter's internal frequency and reference angle, ultimately resulting in instability [9], [22]–[24]. To address this problem, frequency stabilization methods have been proposed to enhance the stability of primary controllers during inverter current-limiting conditions [24]–[26]. Since most frequency stabilization methods rely on estimations of the current-limiting conditions, understanding the interactions between various current limiters and frequency stabilization methods is crucial for assessing their impact and effectiveness on transient stability. This paper proposes an equivalent impedance-based model to provide an effective, efficient, and intuitive approach for understanding these interactions and their impacts on GFM inverter transient stability. The key contributions of this work are summarized as follows:

1) An intuitive equivalent impedance model is proposed and validated through numerical and experimental results.
2) An analytical form of the GFM inverter power-angle characteristics is derived to support transient stability analysis and control optimization.
3) A practical criterion is developed based on the proposed model to assess GFM inverter transient stability during grid-side voltage disturbances.

979-8-3315-1612-3/25 $31.00 © 2025 IEEE

The rest of the paper is organized as follows: Section II describes the examined current limiter and frequency stabilization method, Section III derives the proposed impedance model, Section IV provides experimental validation results, and Section V provides the final remarks.

II. GFM INVERTERS WITH CURRENT LIMITER AND FREQUENCY STABILIZATION

Since inverters typically have significantly lower overcurrent tolerance (1.2 p.u.–2.0 p.u.) compared to conventional synchronous generators (SGs) (5.0 p.u.–10.0 p.u.), it is crucial for GFM control to implement effective measures, such as current limiters, to swiftly curtail output current and prevent irreversible damage to converter hardware [9]. Depending on specific control strategies and the placement of current limiters, various methods have been proposed to address this issue. However, the core principle of most approaches involves adaptively adjusting the inverter output voltage to suppress overcurrent. Consequently, during the current-limiting mode, the inverter can no longer provide the power necessary to meet the droop law and thus synchronize with the grid. To further explain the instability mechanism, given that most GFM inverter primary controls incorporate droop-like characteristics for instantaneous power sharing and frequency stabilization, the internal reference angle of the GFM inverter will naturally drift from its initial equilibrium point due to an imbalance in the droop equation (1). If the fault is not cleared within the critical clearing time (CCT), the GFM inverter will lose synchronization with the grid, leading to instability. To overcome this challenge, specific countermeasures are necessary for GFM inverters to endure prolonged fault conditions. A common approach, recently explored in the literature, involves introducing compensations to the primary control to balance (1) during the current-limiting mode, thereby enabling the inverter to maintain grid synchronization. Such methods are generally considered frequency stabilization techniques, with compensations typically corresponding to current-limiting conditions. Due to this dependency, understanding the interactions between various current limiters and frequency stabilization methods is not straightforward but crucial for evaluating their impact on transient stability [9].

$$\omega = \omega_0 + k_P(P^\star - P) \tag{1a}$$
$$E^\star = E_0 + k_Q(Q^\star - Q) \tag{1b}$$

Since GFM inverters are programmed to behave as a voltage source behind an impedance and most current limiters introduce a voltage drop to the inverter output voltage, an equivalent impedance model can be naturally introduced, with the impacts of the current limiter represented as an equivalent virtual impedance between the droop-controlled voltage source and the inverter's inherent impedance. In a similar manner, because the compensation introduced by frequency stabilization methods depends on current limiting, it can also be modeled as a quantifying impedance equivalent to this voltage drop. Following this insights and modeling approach, details of the proposed equivalent impedance model are elaborated in Section III. The remainder of this section introduces recently proposed current limiter and frequency stabilization methods, which serve as example cases to demonstrate the proposed model, given the broad applicability of the proposed equivalent impedance approach to various current limiting and frequency stabilization methods.

Fig. 1 presents the examined GFM control system including the current limiting and frequency stabilization methods. The current limiter under study [19] integrates a saturation current limiter with virtual impedance embedded in the voltage controller's anti-windup terms to enhance power and voltage support during grid disturbances. The limiting ratio ρ is defined by (2), where $I_{\max}$ represents the current limit, $Z_g = R_g + j\omega_g L_g = |Z_g|\angle\varphi_g$ is the grid-side line impedance, and v_g is the grid voltage. Subsequently, the virtual impedance $Z_{vi} = R_{vi} + jX_{vi} = |Z_{vi}|\angle\varphi_{vi}$ is introduced into the anti-windup terms of the voltage controller. In steady-state operation, the voltage controller's output can be calculated using (3). Since ρ and $|Z_{vi}|$ together determine the overall anti-windup gain, $|Z_{vi}|$ alone does not affect the resulting equivalent impedance model, as shown in Section III. We consider a common choice for $|Z_{vi}| = (k_p^{vc})^{-1}$ to align with the voltage controller design. This approach leaves the phase angle φ_{vi} as the primary tuning parameter for the current limiter.

$$\rho = \min(1, \frac{I_{\max}}{|i^\star|}) \tag{2}$$

$$E^\star - (\rho i^\star Z_g + v_g) - (1 - \rho)i^\star Z_{vi} = 0 \tag{3}$$

The examined frequency stabilization method [24] employs a fictitious power concept defined by (4) along with compensated droop equations, where $Z_x = R_x + jX_x = |Z_x|\angle\varphi_x$ represents the quantifying impedance for fictitious power, $\kappa = \frac{3}{2}$ is the conversion scaling factor, k_P and k_Q are the droop gains for active and reactive power, and $*$ denotes the complex conjugate.

$$\omega_g = \omega_0 + k_P(P^\star - P - P_f) \tag{4a}$$
$$E^\star = E_0 + k_Q(Q^\star - Q - Q_f) \tag{4b}$$
$$P + jQ = \kappa e \cdot i^* \tag{4c}$$
$$P_f + jQ_f = \kappa e \cdot \left(\frac{E^\star - e}{Z_x}\right)^* \tag{4d}$$

Building on (3) and (4), we derive the overall system dynamic model and numerically compute the power-angle curves to analyze the transient stability of the GFM inverters. These numerical results serve as a benchmark to validate the accuracy of the proposed impedance model. It is important to note that this numerical model assumes an ideal inner current loop. Additionally, in the following analysis, we assume that the reactive power droop effect is negligible, allowing the voltage reference to approximate to the nominal value E_0.

III. PROPOSED EQUIVALENT IMPEDANCE MODEL

To establish an intuitive and interpretable framework for analyzing the complex interactions between various current

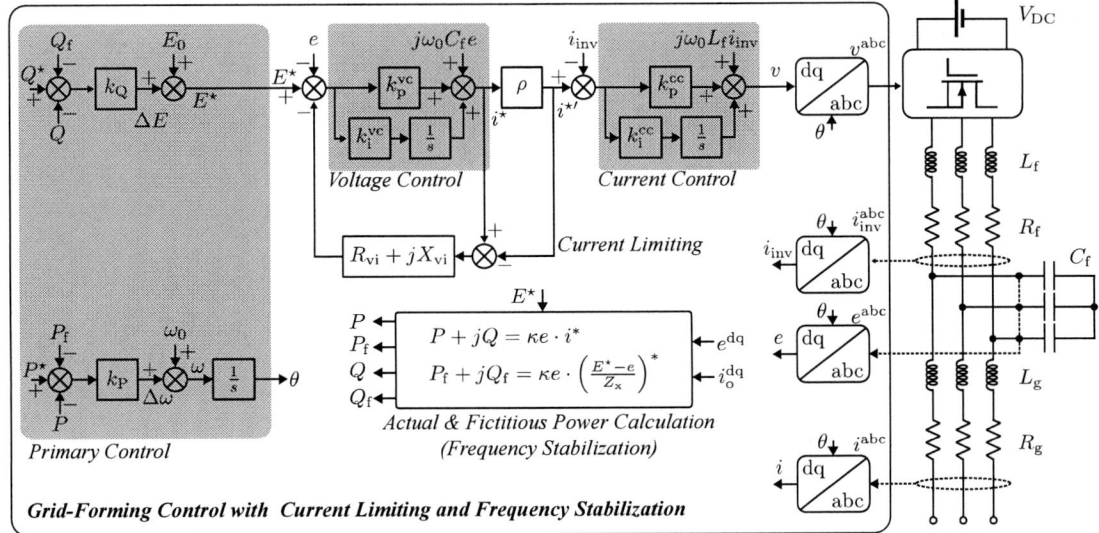

Fig. 1. Control diagram of the GFM inverter with current limiting and frequency stabilization under study.

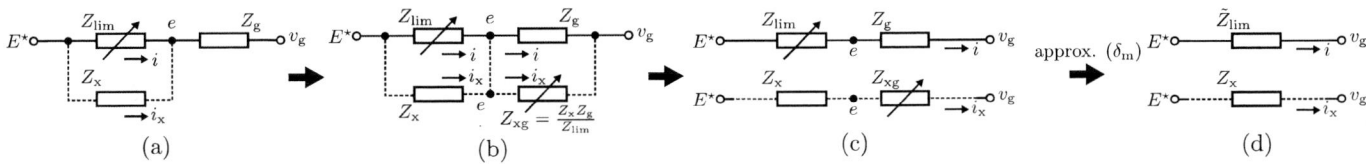

Fig. 2. Proposed equivalent impedance model for GFM inverters with current limiting and frequency stabilization: (a) baseline, (b) modification for "real" and virtual part separation, (c) resultant impedance model after separation, and (d) approximated model to provide further insights. Solid lines represent the real power path, and dotted lines represent the virtual power path.

limiters and frequency stabilization methods, an equivalent impedance model is proposed in this section. As discussed in Section II, the primary function of most current limiters is to adjust the inverter output voltage to suppress overcurrent. This adjustment of the inverter output voltage can be modeled as a voltage drop across an equivalent impedance, such as $Z_{\lim}$ shown in Fig. 2. From (3), it is straightforward to derive $Z_{\lim}$ as given in (5), where the equivalent impedance angle $\varphi_{\lim} = \varphi_{vi}$.

$$Z_{\lim} = \frac{1-\rho}{\rho} Z_{vi} \tag{5}$$

In addition, based on the concept of fictitious power shown in Fig. 1 and (4), the power compensation introduced by the frequency stabilization method is equivalent to the power produced by Z_x in parallel with $Z_{\lim}$, as illustrated in Fig. 2(a). The equivalent compensation current is given by $i_x = \frac{E^\star - e}{Z_x}$.

Since the Z_x branch represents a virtual power injection, we aim to separate this virtual branch from the "real" branch. To achieve this, the circuit manipulation shown in Fig. 2 (b) and (c) is adopted. As illustrated in Fig. 2(c), the actual power injection is determined by the current limiter $Z_{\lim}$, while the virtual power compensation is governed by the quantifying impedance Z_x and a coupling impedance $Z_{xg} = \frac{Z_x Z_g}{Z_{\lim}}$. From this equivalent circuit, an intuitive understanding of the dynamics between current limiters and frequency stabilization can be established:

1) The power-angle characteristics, reflecting the actual power injection of the inverter, are determined solely by the current limiters.
2) The virtual power compensation for frequency stabilization is governed by Z_x and the coupling terms associated with the current limiters, Z_{xg}.
3) Although the virtual power compensation does not directly affect the actual power-angle curves, it influences the newly established equilibrium point during current limiting, which indirectly impacts the actual power injection of the inverter.

With the established equivalent circuit shown in Fig. 2(c), the analytical form of both actual and virtual power under given grid conditions can be derived. It is important to note that, since the ultimate objective of calculating power is to examine the droop equations in (4) and assess the transient stability of the GFM inverters, the power calculation should be performed at the point of the output voltage measurement, e. Keeping this in mind, the analytical expressions for these power values are derived from (6).

$$S = P + jQ = \kappa e \cdot \left(\frac{E^\star - v_g}{Z_{\lim} + Z_g} \right)^* \tag{6a}$$

$$S_f = P_f + jQ_f = \kappa e \cdot \left(\frac{E^\star - v_g}{Z_x + Z_{xg}} \right)^* \tag{6b}$$

$$P = \frac{|E^\star|^2}{|Z_{\lim} + Z_g|^2}\left(|Z_g|(\cos\varphi_g - k_f\cos(\delta - \varphi_g)) - k_f^2|Z_{\lim}|(\cos\varphi_{vi} - \frac{1}{k_f}\cos(\delta + \varphi_{vi}))\right) \tag{7a}$$

$$Q = \frac{|E^\star|^2}{|Z_{\lim} + Z_g|^2}\left(|Z_g|(\sin\varphi_g + k_f\sin(\delta - \varphi_g)) - k_f^2|Z_{\lim}|(\sin\varphi_{vi} - \frac{1}{k_f}\sin(\delta + \varphi_{vi}))\right) \tag{7b}$$

$$P_f = \frac{|E^\star|^2}{|Z_{\lim} + Z_g|^2}\cdot\frac{|Z_{\lim}|}{|Z_x|}\left(|Z_g|(\cos\varphi_{xg} - k_f\cos(\delta - \varphi_{xg})) - k_f^2|Z_{\lim}|(\cos\varphi_x - \frac{1}{k_f}\cos(\delta + \varphi_x))\right) \tag{7c}$$

$$Q_f = \frac{|E^\star|^2}{|Z_{\lim} + Z_g|^2}\cdot\frac{|Z_{\lim}|}{|Z_x|}\left(|Z_g|(\sin\varphi_{xg} + k_f\sin(\delta - \varphi_{xg})) - k_f^2|Z_{\lim}|(\sin\varphi_x - \frac{1}{k_f}\sin(\delta + \varphi_x))\right) \tag{7d}$$

Through mathematical derivation, the analytical power function with respect to the power angle δ is developed in (7). Note that the power angle δ used here is defined as the difference between the inverter's internal reference angle θ and the grid voltage angle θ_g: $\delta = \theta - \theta_g$. To distinguish this from the common convention, we denote $\delta' = \theta_g - \theta$. Additionally, the ratio between the internal voltage reference and the grid voltage is defined as $k_f = \frac{|v_g|}{|E^\star|}$. To solve (7), several additional terms need to be derived.

Firstly, since the inverter output current is limited to I_{max}, the total impedance magnitude can be calculated using (8), where the voltage difference ΔE is determined from (9) using the law of cosines.

$$|Z_{\lim} + Z_g| = \max(|Z_g|, \frac{\Delta E}{I_{max}}) \tag{8}$$

$$\Delta E^2 = |E^\star|^2(1 + k_f^2 - 2k_f\cos\delta) \tag{9}$$

Then, the equivalent impedance for the current limiter, $Z_{\lim}$, can be calculated without explicitly deriving the limiting ratio ρ, as shown in (10).

$$|Z_{\lim}| = -|Z_g|\cos(\varphi_g - \varphi_{vi}) \\ + \sqrt{|Z_{\lim} + Z_g|^2 - |Z_g|^2\sin^2(\varphi_g - \varphi_{vi})} \tag{10}$$

Meanwhile, since $Z_{xg} = \frac{Z_x Z_g}{Z_{\lim}}$, the impedance angle of Z_{xg} can be calculated using (11).

$$\varphi_{xg} = \varphi_x + \varphi_g - \varphi_{vi} \tag{11}$$

By substituting (8)-(11) into (7), we can analytically calculate the perceived power-angle characteristics (actual power plus virtual power as a function of angle) and assess the transient stability of the GFM inverter under given grid conditions. Compared to numerical models, this analytical approach is efficient and practical.

Furthermore, we can take an additional step to establish a practical criterion for determining the transient stability of the GFM inverter during grid-side voltage drops. First, from (5), we know that during a severe voltage drop, the inverter output current will be significantly limited, resulting in $\rho \ll 1$. Under these conditions, $|Z_{\lim}| \gg |Z_g|$, and $Z_{\lim}$ will dominate the "real" branch in Fig. 2(c). Meanwhile, according to (12), the virtual branch impedance can be approximated by Z_x alone.

Therefore, in these conditions, the equivalent circuit model can be further simplified as shown in Fig. 2(d).

$$Z_x + Z_{xg} = \frac{Z_x(Z_{\lim} + Z_g)}{Z_{\lim}} \approx Z_x \tag{12}$$

Based on the equivalent circuit shown in Fig. 2(d), the perceived active power is given by:

$$P_{perc} = P + P_f = \frac{|E^\star||v_g|}{|Z_{\lim}|^2}(|Z_{\lim}|\cos\varphi_{vi}\cos\delta' \\ + |Z_{\lim}|\sin\varphi_{vi}\sin\delta') + \frac{|v_g|^2}{|Z_{\lim}|^2}|Z_{\lim}|\cos\varphi_{vi} \\ + \frac{|E^\star||v_g|}{|Z_x|^2}(|Z_x|\cos\varphi_x\cos\delta' \\ + |Z_x|\sin\varphi_x\sin\delta') + \frac{|v_g|^2}{|Z_x|^2}|Z_x|\cos\varphi_x. \tag{13}$$

Since a grid-side voltage drop decreases the perceived active power, a practical criterion is to ensure that the perceived active power-angle curves intersect the active power setpoints, $P^\star$. To guarantee this, the maximum perceived active power should be at least equal to $P^\star$. Thus, we can first calculate the δ' corresponding to the maximum P_{perc} using (14).

$$\frac{d}{d\delta'}P_{perc} = 0 \tag{14}$$

Note that $|Z_{\lim}|$ is also a function of δ', which significantly complicates the calculation. However, based on the power-angle curves of conventional synchronous generators (SGs), we know that the maximum active power injection typically occurs around $\delta' = 90°$. Thus, we approximate $|Z_{\lim}|$ as $|\tilde{Z}_{\lim}| = |Z_{\lim}|\big|_{\delta'=90°}$ using (8), (9), and (10). Then, by substituting $|\tilde{Z}_{\lim}|$ into (13) and (14), we can determine the corresponding δ'_m using (15).

$$\delta'_m = \arctan\left(\frac{|Z_x|\sin\varphi_{vi} + |\tilde{Z}_{\lim}|\sin\varphi_x}{|Z_x|\cos\varphi_{vi} + |\tilde{Z}_{\lim}|\cos\varphi_x}\right) \tag{15}$$

We can then establish a practical criterion for examining the transient stability of the GFM inverter during grid-side voltage drops, as shown in Proposition 1.

Proposition 1: Consider any GFM inverter connected to the grid under specific grid conditions. The inverter will maintain synchronization with the grid and withstand the given

TABLE I
GFM INVERTER HARDWARE TESTBED PARAMETERS

Parameter	Value	Unit	Parameter	Value	Unit
V_{base}	16	V	L_{f}	$120 \cdot 10^{-6}$	H
S_{base}	32.0	VA	R_{f}	$200 \cdot 10^{-3}$	Ω
I_{base}	1.15	A	C_{f}	$10 \cdot 10^{-6}$	F
k_{p}^{vc}	0.60	pu	L_{g}	$5 \cdot 10^{-3}$	H
k_{i}^{vc}	1010.64	pu	R_{g}	$40 \cdot 10^{-3}$	Ω
k_{p}^{cc}	0.06	pu	$I_{\text{max,rms}}$	1.2	pu
k_{i}^{cc}	94.25	pu	k_{P}	2	%
$P^{\star}$	1.0	pu	k_{Q}	5	%
$Q^{\star}$	0	pu	ω_0	$60 \cdot 2\pi$	rad/s
$|Z_{\text{vi}}|$	1.66	pu	$|Z_{\text{x}}|$	0.10	pu

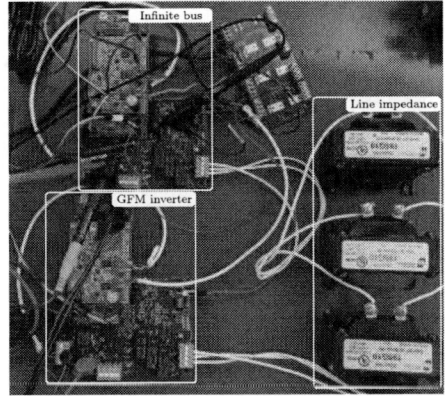

Fig. 3. Picture of the SMIB setup for hardware experiments.

grid conditions if (16) is satisfied, where P_{perc} and δ'_{m} are calculated according to (13) and (15), respectively.

$$P_{\text{perc}}\big|_{\delta'=\delta'_{\text{m}}} \geq P^{\star} \tag{16}$$

In summary, this section proposed an equivalent impedance model to analyze the complex interactions between various current limiters and frequency stabilization methods. Based on the proposed model, an intuitive interpretation can be established to understand how current limiters and virtual power compensations affect the perceived power and transient stability of GFM inverters. Analytical functions are derived from the proposed model to efficiently calculate power-angle curves, allowing for more accurate analysis of GFM inverter transient stability. Finally, a practical criterion is provided to easily test whether the designed current limiters and frequency stabilization methods can maintain stability under given grid conditions. It is also worth noting that while the mathematical models elaborated in this section are based on the GFM control shown in Fig. 1, the proposed method can be readily extended to most current limiters and frequency stabilization techniques. To further validate the accuracy and effectiveness of the proposed methods, numerical and experimental results will be presented in Section IV.

IV. VALIDATION RESULTS

This section provides numerical and experimental results to further validate the accuracy and effectiveness of the proposed

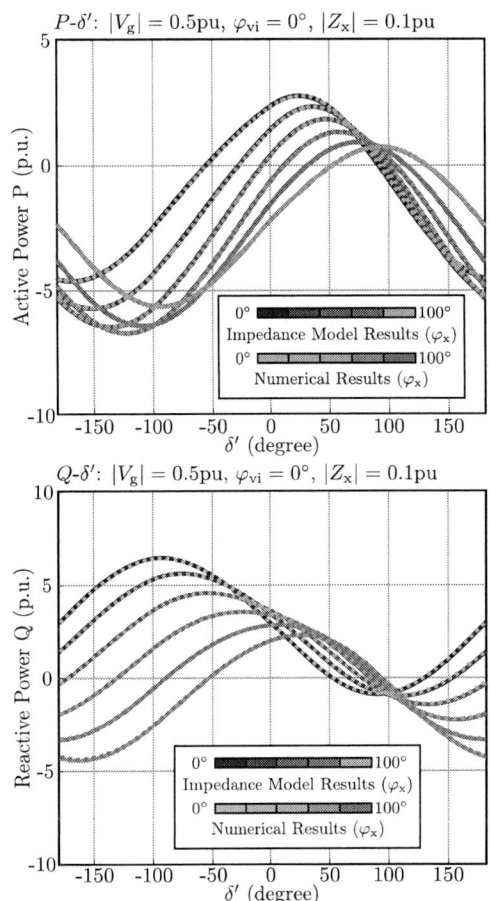

Fig. 4. Comparison between impedance model (red solid lines) and numerical results (blue dashed lines): perceived power-angle curves.

equivalent impedance model. The experimental testbed utilizes a single-machine-infinite-bus (SMIB) setup, with detailed experimental parameters listed in Table I and a picture of the testbed shown in Fig. 3.

First, the numerical results calculated from (2), (3), and (4) are collected and compared with the results obtained from the proposed model. Fig. 4 shows the resulting power-angle curves for a grid-side voltage of $|v_{\text{g}}| = 0.5$ p.u.. As illustrated, the proposed model well matches the numerical results, proving its accuracy.

Secondly, Fig. 5 plots the actual and perceived power-angle curves based on the proposed model with varying φ_{vi} and φ_{x} in a scenario where the grid-side voltage drops from 1.0 p.u. to 0.5 p.u. In Fig. 5, the red dots indicate the original equilibrium point before the fault initiates, while the blue dots show the new equilibrium points established under different current limiter and frequency stabilization settings. From Fig. 5, it is evident that although the actual power-angle curves are governed by the current limiter angle φ_{vi}, the newly established equilibrium point is influenced by both φ_{vi} and φ_{x}, which together determine the inverter's power injections during disturbances. Additionally, it is clear that in two specific cases, $\varphi_{\text{vi}} = 0°$, $\varphi_{\text{x}} = 90°$ and $\varphi_{\text{vi}} = 0°$, $\varphi_{\text{x}} = 110°$,

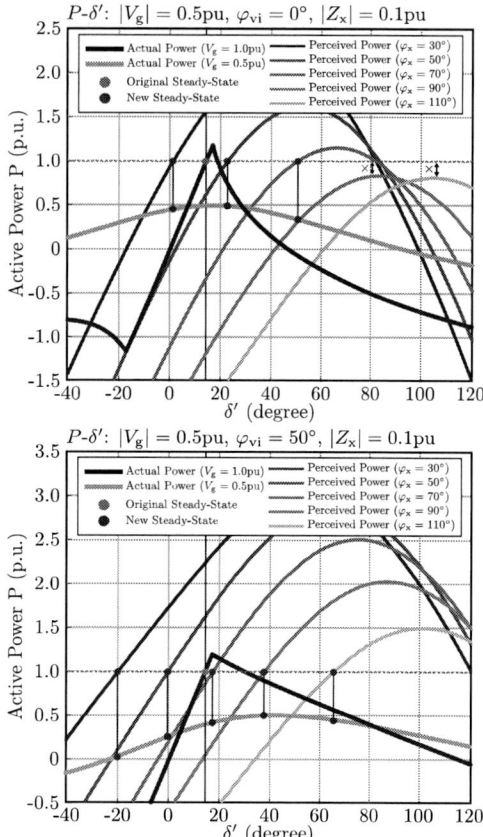

Fig. 5. Power-angle curves with various setups of current limiting and frequency stabilization.

TABLE II
$P_{\text{perc}}|_{\delta'_m}$ (p.u.), WITH $|Z_x| = 0.1$ p.u. AND $|v_g| = 0.5$ p.u..

φ_{vi} \ φ_x	50°	60°	70°	80°	90°	100°	110°
0°	1.57	1.35	1.14	0.95	0.82	0.76	0.80
10°	1.81	1.60	1.37	1.15	0.96	0.83	0.77
20°	2.05	1.85	1.63	1.39	1.17	0.98	0.84
30°	2.29	2.12	1.90	1.66	1.42	1.19	0.99
40°	2.52	2.36	2.18	1.95	1.70	1.45	1.21
50°	2.73	2.63	2.47	2.26	2.01	1.75	1.48
60°	2.94	2.88	2.76	2.58	2.35	2.08	1.80
70°	3.13	3.12	3.05	2.90	2.70	2.45	2.16
80°	3.31	3.36	3.34	3.24	3.07	2.84	2.56

no achievable equilibrium points exist because the maximum perceived power does not reach the active power setpoint of 1.0 p.u.. Based on this observation, we can further validate the practical criterion proposed in Proposition 1. To do so, Table II lists the calculated $P_{\text{perc}}|_{\delta'_m}$, which clearly shows that the results align with the analysis from Fig. 5.

Finally, to further validate the proposed model, the experimental results are shown in Fig. 6, where the active and reactive power profiles are plotted for a case study in which the grid-side voltage drops from 1.0 p.u. to 0.5 p.u. for 5 seconds. In Fig. 6, the red and green dashed lines denote the initiation and clearing times of the voltage drop, respectively.

It is evident from Fig. 6 that the transient stability of the inverter aligns with the predictions from Table II, validating the practicality of Proposition 1. Additionally, the actual active power injections observed in Fig. 6 are consistent with the analysis from Fig. 5, further supporting the accuracy and effectiveness of the proposed model.

V. CONCLUSION

With GFM inverters adopting various current limiting and frequency stabilization methods to enhance stability and ensure sustainable operation during different grid disturbances, the interactions between current limiters and frequency stabilization methods in primary controllers—and their impact on GFM inverter transient stability—require further attention. This paper proposed an equivalent impedance-based approach that provides an effective, practical, and intuitive means to analyze these interactions and their impacts on GFM inverter transient stability, while also aiding in current limiter and frequency stabilization design. In summary, the proposed model achieves the following objectives:

1) It offers an efficient method to accurately estimate the power-angle characteristics of GFM inverters across various current limiting and frequency stabilization configurations;

2) It establishes a practical criterion for assessing GFM inverter transient stability;

3) It provides analytical forms of the GFM power-angle characteristics, which can be leveraged for advanced control design and optimization.

Additionally, the proposed model is validated through both numerical simulations and experimental results. Although this work demonstrates the model using a specific control setup, the model can be readily extended to a wide range of current limiting and frequency stabilization methods.

REFERENCES

[1] B. Kroposki, B. Johnson, Y. Zhang, V. Gevorgian, P. Denholm, B.-M. Hodge, and B. Hannegan, "Achieving a 100% renewable grid: Operating electric power systems with extremely high levels of variable renewable energy," *IEEE Power and Energy Magazine*, vol. 15, no. 2, pp. 61–73, 2017.

[2] B. Kroposki, "UNIFI specifications for grid-forming inverter-based resources (v. 2)," National Renewable Energy Laboratory (NREL), Golden, CO (United States), Tech. Rep., 2024.

[3] NERC, "Grid forming technology: Bulk power system reliability considerations," *NERC, Atlanta*, 2021.

[4] Y. Lin, J. H. Eto, B. B. Johnson, J. D. Flicker, R. H. Lasseter, H. N. Villegas Pico, G.-S. Seo, B. J. Pierre, and A. Ellis, "Research roadmap on grid-forming inverters," National Renewable Energy Lab.(NREL), Golden, CO (United States), Tech. Rep., 2020.

[5] AEMO, "Voluntary specification for grid-forming inverters," *AEMO, Melbourne*, 2023.

[6] M. C. Chandorkar, D. M. Divan, and R. Adapa, "Control of parallel connected inverters in standalone ac supply systems," *IEEE transactions on industry applications*, vol. 29, no. 1, pp. 136–143, 1993.

[7] H.-P. Beck and R. Hesse, "Virtual synchronous machine," in *2007 9th international conference on electrical power quality and utilisation*. IEEE, 2007, pp. 1–6.

[8] G.-S. Seo, M. Colombino, I. Subotic, B. Johnson, D. Groß, and F. Dörfler, "Dispatchable virtual oscillator control for decentralized inverter-dominated power systems: Analysis and experiments," in *2019 IEEE Applied Power Electronics Conference and Exposition (APEC)*. IEEE, 2019, pp. 561–566.

979-8-3315-1612-3/25 $31.00 © 2025 IEEE

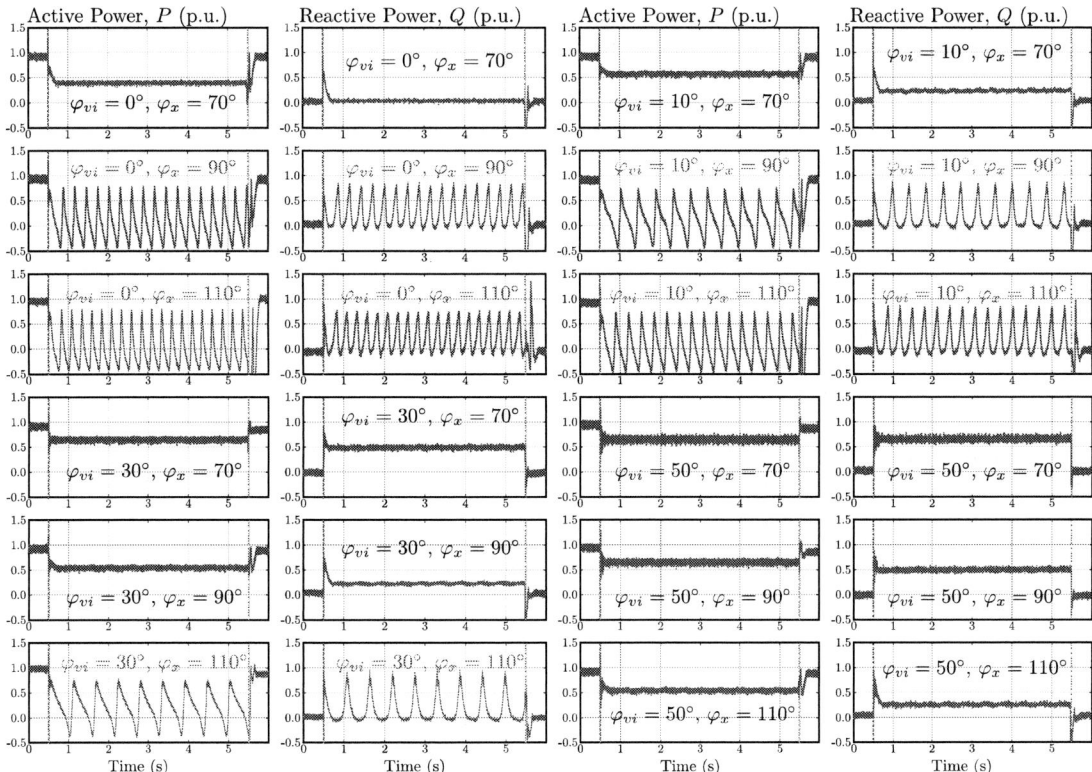

Fig. 6. Experimental results of GFM inverter active and reactive power with various setups of current limiting and frequency stabilizations. Red and green dashed lines indicate the 0.5 pu voltage drop inception and clear time, respectively.

[9] N. Baeckeland, D. Chatterjee, M. Lu, B. Johnson, and G.-S. Seo, "Overcurrent limiting in grid-forming inverters: A comprehensive review and discussion," *IEEE Trans. Power Electron.*, vol. 39, no. 11, pp. 14 493–14 517, Nov. 2024.

[10] "IEEE standard for interconnection and interoperability of distributed energy resources with associated electric power systems interfaces," *IEEE Std 1547-2018*, 2018.

[11] H. Wu, X. Wang, and L. Zhao, "Design considerations of current-limiting control for grid-forming capability enhancement of vscs under large grid disturbances," *IEEE Transactions on Power Electronics*, 2024.

[12] B. Mahamedi, M. Eskandari, J. E. Fletcher, and J. Zhu, "Sequence-based control strategy with current limiting for the fault ride-through of inverter-interfaced distributed generators," *IEEE Transactions on Sustainable Energy*, vol. 11, no. 1, pp. 165–174, 2018.

[13] M. G. Taul, X. Wang, P. Davari, and F. Blaabjerg, "Current limiting control with enhanced dynamics of grid-forming converters during fault conditions," *IEEE Journal of Emerging and Selected Topics in Power Electronics*, vol. 8, no. 2, pp. 1062–1073, 2019.

[14] E. Rokrok, T. Qoria, A. Bruyere, B. Francois, and X. Guillaud, "Transient stability assessment and enhancement of grid-forming converters embedding current reference saturation as current limiting strategy," *IEEE Transactions on Power Systems*, vol. 37, no. 2, pp. 1519–1531, 2021.

[15] A. D. Paquette and D. M. Divan, "Virtual impedance current limiting for inverters in microgrids with synchronous generators," *IEEE Transactions on Industry Applications*, vol. 51, no. 2, pp. 1630–1638, 2014.

[16] X. Wang, Y. W. Li, F. Blaabjerg, and P. C. Loh, "Virtual-impedance-based control for voltage-source and current-source converters," *IEEE Transactions on Power Electronics*, vol. 30, no. 12, pp. 7019–7037, 2014.

[17] X. Lu, J. Wang, J. M. Guerrero, and D. Zhao, "Virtual-impedance-based fault current limiters for inverter dominated ac microgrids," *IEEE Transactions on Smart Grid*, vol. 9, no. 3, pp. 1599–1612, 2016.

[18] M. Miranbeigi, P. M. Gajare, J. Benzaquen, P. Kandula, and D. Divan, "On the passivity of grid-forming converters—role of virtual impedance," in *2022 IEEE Applied Power Electronics Conference and Exposition (APEC)*. IEEE, 2022, pp. 650–656.

[19] N. Baeckeland and G.-S. Seo, "Novel hybrid current limiter for grid-forming inverter control during unbalanced faults," in *2023 11th International Conference on Power Electronics and ECCE Asia (ICPE 2023-ECCE Asia)*. IEEE, 2023, pp. 1517–1522.

[20] W. Du and S. M. Mohiuddin, "A two-stage current limiting control strategy for improved low-voltage ride-through capability of direct-droop-controlled, grid-forming inverters," in *2023 IEEE Energy Conversion Congress and Exposition (ECCE)*. IEEE, 2023, pp. 2886–2890.

[21] L. Huang, C. Wu, D. Zhou, and F. Blaabjerg, "A power-angle-based adaptive overcurrent protection scheme for grid-forming inverter under large grid disturbances," *IEEE Transactions on Industrial Electronics*, vol. 70, no. 6, pp. 5927–5936, 2022.

[22] X. Lyu, W. Du, S. M. Mohiuddin, S. P. Nandanoori, and M. Elizondo, "Improved transient stability analysis of multi-loop droop-controlled grid forming inverters with current limiter," in *2024 IEEE Power & Energy Society Innovative Smart Grid Technologies Conference (ISGT)*. IEEE, 2024, pp. 1–5.

[23] N. Bottrell and T. C. Green, "Comparison of current-limiting strategies during fault ride-through of inverters to prevent latch-up and wind-up," *IEEE Transactions on Power Electronics*, vol. 29, no. 7, pp. 3786–3797, 2013.

[24] N. Baeckeland and G.-S. Seo, "Enhanced large-signal stability method for grid-forming inverters during current limiting," in *2023 IEEE 24th Workshop on Control and Modeling for Power Electronics (COMPEL)*. IEEE, 2023, pp. 1–8.

[25] T. Liu, X. Wang, F. Liu, K. Xin, and Y. Liu, "A current limiting method for single-loop voltage-magnitude controlled grid-forming converters during symmetrical faults," *IEEE Transactions on Power Electronics*, vol. 37, no. 4, pp. 4751–4763, 2021.

[26] S. Cen, M. Awal, and I. Husain, "Dynamic phasor modeling and analysis of sequence decomposed grid-forming control under unbalanced faults," in *2024 IEEE Applied Power Electronics Conference and Exposition (APEC)*. IEEE, 2024, pp. 2730–2736.

979-8-3315-1612-3/25 $31.00 © 2025 IEEE

Revisit Active Power Oscillation in Multi-Virtual Synchronous Generators Grid

Junjie Xiao
Electrical Sustainable Energy
Delft University of Technology
Delft, The Netherlands
J.Xiao-2@tudelft.nl

Pavol Bauer
Electrical Sustainable Energy
Delft University of Technology
Delft, The Netherlands
P.Bauer@tudelft.nl

Zian Qin
Electrical Sustainable Energy
Delft University of Technology
Delft, The Netherlands
Z.Qin-2@tudelft.nl

Abstract—Active power oscillation (APO) issues may arise during the deployment of multiple parallel converters with Virtual Synchronous Generator (VSG) control in Microgrids. To that end, the equivalent circuit models of a converter with VSG control are proposed, which intuitively reveals the root cause of APOs. Accordingly, a graph-theory-based virtual impedance is introduced to harmonize parameters among involved VSGs, effectively eliminating APOs. Simulation and experimental results verify the improvements of the proposed control.

Index Terms—Power oscillation, distributed control, damping control, virtual synchronous generator.

I. INTRODUCTION

THE distributed generation (DG) technology has received widespread attention. The transition from droop control to VSG control for DG control can enhance critical frequency indicators such as the rate of change of frequency (RoCoF), thereby benefiting grid stability [1]. However, introducing oscillatory dynamics complicates the system, potentially leading to significant active power oscillations. These oscillations occur when multiple VSGs operate in stand-alone (SA) mode [2]. The large instantaneous currents associated with these oscillations can trigger overcurrent protection mechanisms, exacerbating system stability issues [3], [4].

Various variants of VSG control have been proposed to suppress oscillations. They can be broadly classified into two main categories: model-free and model-based approaches. For instance, the model-free approach, such as [5], [6], detects frequency deviations from its nominal and assigns different inertia at different phases. Specifically, larger inertia is applied to counteract these deviations when the DG frequency deviates from the common frequency. In contrast, smaller inertia accelerates convergence when DG frequencies align with the common frequency. This method ensures that all DG frequencies promptly synchronize with the common frequency. Furthermore, [7] proposes a self-adaptive inertia and damping combination control method to enhance frequency stability through an interleaving control technique. In short, these adaptive methods introduce a nonlinear element into DG operation, potentially altering the carefully designed inertia.

Another philosophy incorporates an extra feedback loop into the original VSG. For instance, in [8], deviations between a DG's frequency and the system's common frequency are detected, leading to adjustments of the DG. Studies in [9], [10] integrate variations in a DG's power, frequency, or phase during transients to establish feedback loops. These feedback loops fed the oscillation element into the VSG decision process, achieving disturbance compensation. Moreover, by utilizing low-bandwidth communication, the graph theory-based secondary frequency control can achieve the frequency consensus under SA mode [11]–[13]. In [14], [15], frequency disparities with neighboring DGs are employed to develop a mutual damping term. Despite the technical effectiveness of these model-free methods, they become effective only after oscillations occur and have been detected.

In addition to model-free methods, model-based approaches are employed to mitigate APO. For instance, [16] adjusts the damping and inertia coefficients simultaneously to determine and maintain the optimal damping ratio, thereby suppressing power and frequency oscillations throughout the operation. Similarly, [17] proposes an additional damping correction loop that adjusts the system damping ratio without affecting the steady-state frequency droop characteristic. However, this approach changes the preset inertial response of the VSG. In [18], the active power reference is feedforward to compensate for the VSG frequency to enhance damping. However, this feedforward controller is designed explicitly for changes in power reference, so it does not mitigate power allocation in SA mode. Moreover, [19] analyzes the power oscillation mechanism and utilizes virtual impedance to suppress power oscillations caused by line mismatches. In [20], parameter design principles are defined to eliminate all transient circulating power theoretically. [21] utilizes a phase feedforward path to replace the traditional frequency compensation path; this can enhance the damping. Nevertheless, an ideal parameter necessitates full knowledge of the system. Although these model-based methods provide effective damping for oscillation mitigation, they typically require foreknowledge of the system, such as feeder impedance. This cannot be assured in practical scenarios.

Another gap is that the existing models explaining the APO issues induced by the VSG transfer function are overly complicated and absent physical meaning. This article proposes an intuitive modeling perspective for VSG oscillation analysts, which visualizes the VSG control loop as a circuit element

979-8-3315-1612-3/25 $31.00 © 2025 IEEE

with resistance, inductance, and capacitance and considers the load switch as a current source. Moreover, this paper offers suggestions for suppressing the APO in multiple VSG systems.

This paper's main contributions are summarized as follows:

1) The VSG equivalent circuit model is proposed, which provides clear physical interpretations. In this model, inertia, damping, and feeder impedance are analogized to capacitance, resistance, and inductance, respectively. Load switches are excitation sources injecting power into the circuit. The power oscillations are viewed as LC resonance phenomena.

2) A distributed virtual impedance method is proposed to attenuate oscillations in the SA mode. It benefits faster response to load variations and less communication dependence.

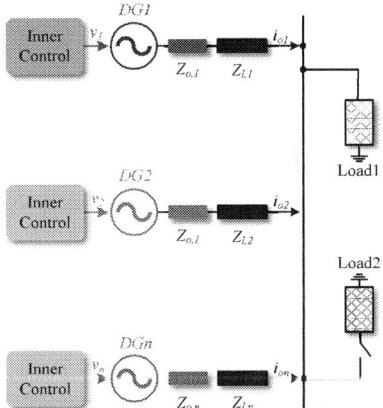

Fig. 1: The microgrid configuration consisting of n DGs.

II. REVISIT VSG CONTROL

This section reviews the two most discussed primary controls among n-distributed generators-tied microgrids, as shown in Fig.1, where the inverter can be modeled as a voltage in series with an output impedance $Z_{o,i}$ and feeder impedance $Z_{l,i}$ of i-th inverter. Herein, $Z_{l,g}$ is the grid impedance. This paper assumes feeder impedance is predominantly inductive.

A. Review on Traditional Droop and VSG Control

The power control loops of droop and VSG control are shown in Fig.2. The difference between VSG and droop control mainly focuses on the active power control loops, and the reactive power loops are ignored.

The active power loop for droop control is shown in (1).

$$\omega = \omega_0 - m_p \frac{\omega_c}{s + \omega_c}(P - P_r) \tag{1}$$

where ω represents the generated angular frequency reference of the inverter output voltage, ω_0 is the nominal value of angular frequency, m_p is the droop coefficient, P represents the inverter output active power, P_r is the nominal value of active power, m_p is the droop coefficient, and ω_c is the cutoff angular frequency of the low-pass filter.

The active power control equation for VSG is shown in (2).

$$P_r - m_g(\omega - \omega_0) - P - k_d(\omega - \omega_0) = M \frac{d(\omega - \omega_0)}{dt} \tag{2}$$

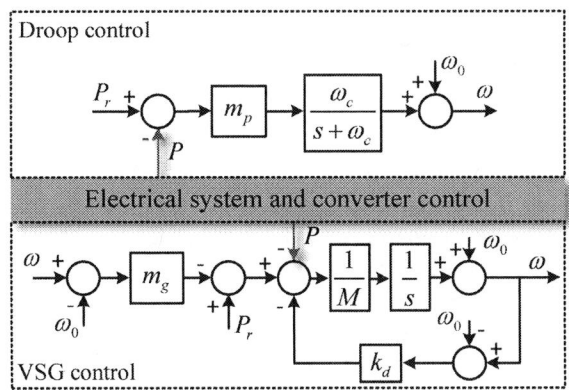

Fig. 2: Active power loop implementation

where m_g is the proportional coefficient of the governor, k_d is the damping factor, and M is the moment of inertia.

Accordingly, the small-signal model of droop control and VSG is simplified, as shown in Fig.3 and Fig.4, respectively.

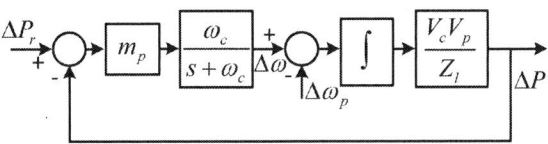

Fig. 3: Small-signal model of droop control.

In Fig.3, ω_p is the angular frequency of the point of common coupling (PCC). V_c represents the unit output voltage, V_p is the PCC voltage, and Z_l denotes the feeder impedance, ω_p is the PCC frequency disturbance. In the following, $K = V_c V_p/Z_l$.

Considering the RoCoF requirements, the dynamic performance of angular frequency when the load changes is the main focus in the SA mode. For droop control, the small-signal transfer function of angular frequency change $\Delta\omega$ over loading transition ΔP is shown in (3).

$$G_{d,sa} = \frac{\Delta\omega}{\Delta P} = -m_p \frac{\omega_c}{s + \omega_c} = -m_p \frac{1}{s/\omega_c + 1} \tag{3}$$

By simplifying the inertial and damping term in Fig.2 with $J = M$ and $D = k_d + m_g$. The small signal model of VSG in Fig.2 can be simplified as shown in Fig.4.

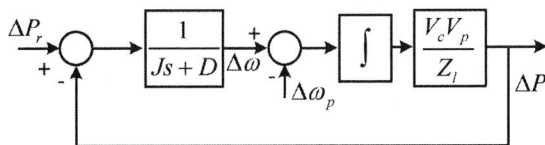

Fig. 4: Small-signal model of VSG control.

The transfer function of VSG control in stand-alone mode is shown in (4), respectively.

$$G_{v,sa} = \frac{\Delta\omega}{\Delta P} = -\frac{1}{Js + D} = -\frac{1}{D} \frac{1}{Js/D + 1} \tag{4}$$

979-8-3315-1612-3/25 $31.00 © 2025 IEEE

Combining (3)-(4), with $m_p = 1/D$, $\omega_c = D/J$, the droop control can be equivalent to VSG control. In this paper, the VSG is adopted for the power converter control verification.

B. Active Power Oscillation with VSG control

As VSG simulates the synchronous generator's inertia and damping characteristics, the oscillation characteristics are inevitably introduced. This subsection investigates the mechanism of active power oscillation in SA mode.

The active power across the feeder is shown in (5).

$$P_i = \frac{V_i V_p}{Z_i} sin\delta_i = K_i \frac{\Delta\omega_i - \Delta\omega_p}{s} \tag{5}$$

where δ_i is the power angle difference of the i-th converter and the PCC power angle. Within a multi-unit system, the coupling of different units is caused by PCC frequency variation.

Based on the small signal diagram of VSG, the transfer function from the PCC voltage fluctuates to i-th VSG output variation characteristics can be expressed as (6):

$$\begin{cases} \Delta P_i = \dfrac{-K_i(J_i s + D_i)}{J_i s^2 + D_i s + K_i} \Delta\omega_p \\ \Delta\omega_i = \dfrac{K_i}{J_i s^2 + D_i s + K_i} \Delta\omega_p \end{cases} \tag{6}$$

The output power of the involved DGs is equal to the load power P_L, which can be represented as in (7).

$$\sum_{i=1}^{n} \Delta P_i = \Delta P_L \tag{7}$$

By combining equations (6)-(7), the transfer function that describes the interaction between load changes and PCC frequency variations for VSGs under SA mode can be derived as shown in (8). This derivation facilitates calculating the PCC frequency responses of VSGs under varying load conditions.

$$\Delta\omega_p = -\frac{1}{\sum\limits_{i=1}^{n} G_i(s)(J_i s + D)} \Delta P_L \tag{8}$$

where $G_i = -K_i/(J_i s^2 + D_i s + K_i)$. For a multi-VSG scenario, the load change causes a variation in PCC frequency. As shown in (6), the different transfer functions from PCC frequency to DG's outputs lead to a DG's dynamic disparity, which contributes to the oscillations. By combining equations (6)-(8), the transfer function describes the dynamics and steady state of the DGs when the load switch is shown as in (9).

$$\begin{cases} \Delta\omega_i = -\dfrac{G_i(s)}{\sum\limits_{k=1}^{n} G_k(s)(J_k s + D)} \Delta P_L \\ \Delta P_i = -\dfrac{G_i(s)(J_i s + D)}{\sum\limits_{k=1}^{n} G_k(s)(J_k s + D)} \Delta P_L \end{cases} \tag{9}$$

Based on the analysis above, increasing the damping coefficient D or decreasing the inertia coefficient J within a certain range can suppress active power and frequency oscillations in multi-VSG systems during load changes. However, since D

is coupled with the droop coefficient, representing the steady-state frequency deviation, modifying D will inevitably change the frequency deviation nadir. Additionally, decreasing J is undesirable for VSGs as it may violate the RoCoF rules. Consequently, a trade-off between the dynamic and steady-state performance of the VSG is unavoidable. Moreover, existing virtual impedance techniques may be ineffective, as exact parameter matching of parallel VSGs may not be fully achievable. While inserting substantial virtual impedance into the control loop can mitigate oscillations, it also leads to considerable and unexpected voltage drops.

III. EQUIVALENT IMPEDANCE CIRCUIT OF VSG

An equivalent circuit model is developed in this section to understand the root cause of APO issues intuitively, which ultimately leads to mitigation measures.

A. Single-VSG Equivalent

From the VSG small signal in Fig.4, the following equation can be rephrased as in (10) and (11):

$$\frac{1}{K_i} \frac{d\Delta P_i}{dt} = \Delta\omega_i - \Delta\omega_p \tag{10}$$

$$\Delta P_{ri} = \Delta P_i + J_i \frac{\Delta\omega_i}{dt} + D_i \Delta\omega_i \tag{11}$$

Tab.I shows the analogy relationships between the control and circuit variables. where subscript i represents the DGi's parameter, while subscript g represents those of the utility grid, and subscript p represents the PCC's parameter.

TABLE I: Correspondence between VSG and circuit variables

circuit	U_i	U_p	I_i	I_{ri}	I_L	R_i	C_i	L_i	L_g
VSG	$\Delta\omega_i$	$\Delta\omega_p$	ΔP_i	ΔP_{ri}	ΔP_L	$1/D_i$	J_i	$1/K_i$	$1/K_g$

With the analogy, ω_i is equivalent to the voltage U_i, representing the frequency change. ΔP_i is the equivalent to the current I_i, representing the active power change. J_i is equivalent to a capacitance C_i, is the inertia coefficient; $1/D_i$ is the resistance R_i, is the damping factor; $1/K_i$ is equivalent to an inductance L_i, representing feeder impedance term. According to Tab.I, (10) and (11), the VSG model can be equivalent to the circuit model, as shown in (12) and (13).

$$L_i \frac{dI_i}{dt} = U_i - U_p \tag{12}$$

$$I_{ri} = I_i + C_i \frac{dU_i}{dt} + \frac{U_i}{R_i} \tag{13}$$

As shown in (7), the current should follow $\sum\limits_{i=1}^{n} I_i = I_L$.

Combining (6)(7) and (12)(13), the VSG model can be analogized to the circuit in Fig.5. Accordingly, the $P - f$ relationship in the VSG is analogous to the $I - V$ relationship in a second-order RLC circuit. The inertia coefficient J_i suppresses frequency changes similarly to how the capacitor stabilizes circuit voltage. The damping coefficient D_i governs

979-8-3315-1612-3/25 $31.00 © 2025 IEEE

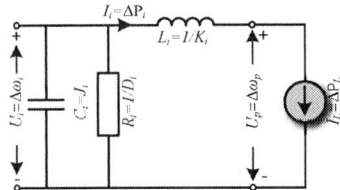

Fig. 5: Single VSG equivalent circuit.

the angular frequency changes in output power, analogous to how resistance determines the voltage change in the circuit.

The current source I_L is enabled when the load switches. Consequently, the current I_i increases to I_L. The capacitance C_i reduces the voltage change rate U_i, analogying that VSG provides inertia and maintains RoCoF. The steady-state U_i is determined by R_i, which acts as the droop coefficient that dictates the frequency deviation. As the comparison in Section II, a key distinction between droop and VSG control is the inclusion of capacitance in the latter.

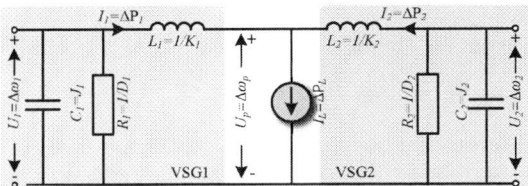

Fig. 6: Multi-VSG's equivalent circuit perspective in SA mode

B. Multi-VSG Equivalent

In this section, the transient of a multi-VSG system under SA mode is intuitively analyzed through the resonance in its equivalent impedance circuits. Subsequently, the resonance in the equivalent circuit can be quantitatively analyzed, providing insights for deriving circuit parameter configuration rules. Similarly, VSG parameters can be configured to eliminate power oscillations during the VSG's transient.

Based on Fig.5, the multi-VSG equivalent circuit can be derived. In this section, a two-VSG system is considered, for example, and can be expanded to a n VSG system. They can be equivalent to the impedance circuit perspective as shown in Fig.6. The interaction $I_1 + I_2 = I_L$ holds throughout the entire operation. This indicates that the circuit operates in parallel, and the current sharing ratio $I_1:I_2$ is determined by the equivalent impedance of each branch, where the impedance Z_{ei} is shown in (14).

$$Z_{ei} = \frac{U_p}{I_i} = -\frac{1}{C_i s + R_i} - s L_i \quad (14)$$

With a given current source, the current I_i is determined by the impedance, where the resonance may occur and the resonance inconsistency leads to disproportional current sharing. Herein, the circuit current I_i sharing analysis includes: 1) steady-state current sharing and 2) dynamic current sharing. The steady-state current sharing is characterized by the proportional setting of the resistors R_i, consistent with the damping

coefficients D_i. Moreover, accurate dynamic current sharing means no current oscillation is within the system. This requires that the circuit model's impedance Z_{ei} remains proportional throughout the dynamic process.

Accordingly, the circuit elements should be tuned proportionally to avoid oscillation. Converting to the VSG control variable in Tab.I, the VSG parameters should be set as in (15). Here, P_{mi} denotes the maximum output active power capacity of i-th converter.

$$\frac{P_{mi}}{P_{mj}} = \frac{J_i}{J_j} = \frac{D_i}{D_j} = \frac{K_i}{K_j} = \frac{Z_j}{Z_i} \quad (15)$$

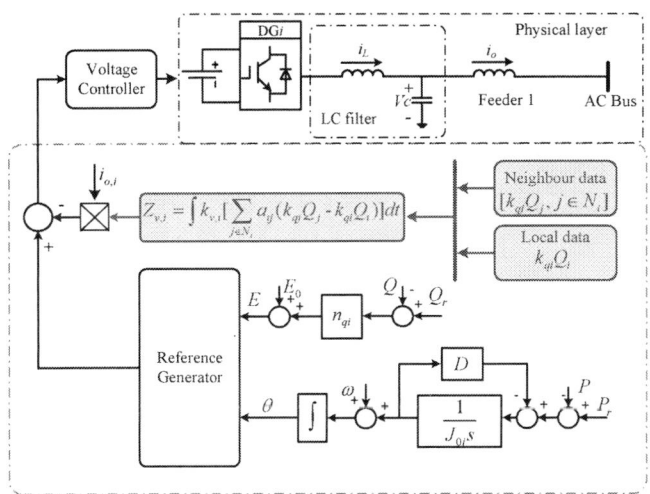

Fig. 7: Control structure of the proposed method.

IV. PROPOSED CONTROL DESIGN

With the parameter design method proposed in Section III, the transient circulating current in a multi-VSG system can be theoretically eliminated. While the inertia and damping settings can be satisfied by configuring J to maintain the same ratio as D across multiple VSGs, achieving the ideal impedance ratio is often impractical due to uncertainties in actual line inductances. To address this limitation, this section derives a new VSG control strategy based on a n-VSG system. The oscillations can be mitigated by suitably harmonizing the virtual impedance, thereby tuning the equivalent impedance. Analogizing to the circuit model in Fig.6, the inductance L_i is tunned, while the capacitance C_i and resistance R_i can be set properly by the DG's output capacity ratio.

Accordingly, mismatched equivalent output impedance can lead to uneven reactive power sharing. This indicates that the equivalent impedance has been well adjusted when the reactive power is proportionally distributed. This adjustment can help mitigate active power oscillations when the load is switched.

As the feeder is assumed to be inductive, the power flowing through the feeder impedance results in a voltage drop ΔV_i, which can be expressed as:

$$\Delta V_i \approx \frac{X_{e,i} Q_i}{V_{c,i}} \quad (16)$$

where $X_{e,i}$ is the equivalent impedance of i-th VSG. In [22]–[24], it is demonstrated that proper design of the virtual impedance enables modification of the equivalent feeder impedance $X_{e,i}$. This adjustment, in turn, facilitates control of the voltage drop among the units, thereby promoting proportional sharing of reactive power.

1) Communication network modelling: The microgrid's communication network is represented using an undirected cyber graph [25], illustrating how converters exchange information with neighboring units. Each VSG communicates with its adjacent VSGs through the communication network. It can be described by the communication adjacency matrix $A = (a_{ij})_{N \times N}$. The entry a_{ij} is set to 1 if units i and j are regularly communicating, and 0 otherwise. The degree of vertex ζ_i is defined as $d_i = \sum_{j=1}^{N} a_{ij}$. The corresponding degree matrix is $D_M = \text{diag}(d_1, \ldots, d_N)$. The Laplacian matrix L of the communication network is defined as $L = D_M - A$.

2) Graph Theory-Based Virtual Impedance implementation: The proposed method is shown in Fig.7. The inverters exchange information related to reactive power $(Q_1/D_1, \ldots, Q_N/D_i)$ with their adjacent units to achieve a reactive power consensus when the impedance has been appropriately adjusted. The reshaped consensus algorithm-based virtual fundamental impedance is expressed as (17).

$$Z_{v,i} = \int k_{v,i} [\sum_{j \in N_i} a_{ij}(Q_j/D_j - Q_i/D_i)] dt \qquad (17)$$

where the parameter $k_{v,i}$ determines the bandwidth of the virtual impedance loop, which should be slower than that of the reactive power loop. The reactive power calculation typically includes a low-pass filter, which sets the speed of the reactive power calculation. The upper bound of $k_{v,i}$ is constrained by the bandwidth of this low-pass filter, while the lower bound is dictated by the minimum required speed for virtual impedance adjustments. The adaptively adjusted impedance $Z_{v,i}$ is influenced by neighboring information and the local unit's state. When reactive power is improperly allocated, the consensus algorithm prompts the controller to adjust the virtual impedance. This modification aims to achieve balanced reactive power distribution, ensuring proportional sharing across equivalent impedance.

V. VERIFICATION

The proposed strategy has been tested in Simulink to validate its effectiveness, where three inverters connected in parallel are considered. In this microgrid system, the output side of the inverters is connected to the AC bus through an LC filter and line impedance. The expected active power-sharing ratio is assumed to be 1:2:3. In this paper, a ROCOF threshold of 1 rad/s^2 is assumed, as it falls within the typical range of minimum and maximum values reported in [26]. However, this threshold may be adjusted based on practical requirements. The verification parameters for both simulation and experiments are shown in Tab.II.

TABLE II: Parameters for Verification

Symbol	Description	Value 1
U_{dc}	DC voltage	300V
f_s	Switch frequency	20kHz
Z_L	Feeder impedance	2.2mH
L_f	Inductor of LC filter	2.2mH
C_f	Capacitor of LC filter	12μF
ω_0	Nominal angular frequency	314rad/s
V_0	Nominal voltage amplitude	150V
$k_{v,i}$	Virtual Impedance loop gain	0.01

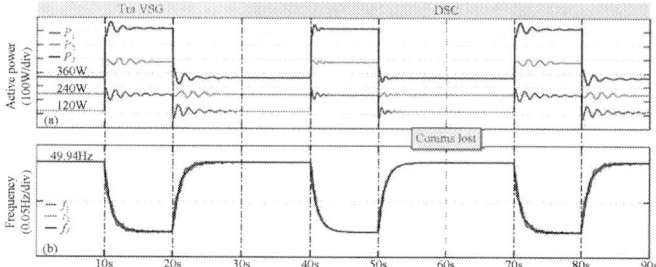

Fig. 8: Dynamics with the distributed secondary control in [12]

A. Simulation Result

Fig.1 shows the simulation structure. Herein, $J_1 = D_1 = 300$, $J_2 = D_2 = 600$, $J_3 = D_3 = 900$. The simulation illustrates the effectiveness of the proposed method over the traditional VDG and the method in [12] under SA mode.

Fig.8 compares traditional VSG control, and the distributed secondary control proposed in [12]. The output active power and frequency are displayed in Figure 8(a) and Figure 8(b), respectively. Initially, the conventional VSG control is used to regulate the microgrids, resulting in a proportional steady-state active power-sharing ratio of 1:2:3. At 10 seconds, a 700W load is added, which leads to decreased frequency. However, the active power and frequency experience oscillations due to the mismatched feeder impedance. When the load is suddenly switched off, the system recovers to its original power level, but oscillations persist in the dynamics. At 30 seconds, the distributed secondary control (DSC) proposed in [12] is activated, providing extra damping for the VSG system. As seen at 40 and 50 seconds, where the load is stepped on and off, the oscillations are relatively smaller than the conventional VSG control. Since DSC necessitates a communication network,

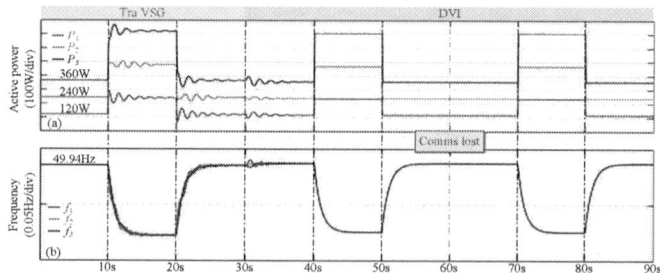

Fig. 9: Dynamics with the proposed distributed virtual impedance

it is reasonable to consider the scenario of communication loss. At 60 seconds, the communication is removed, indicating that the inverters can no longer receive information from each other. In this case, when the load increases and decreases at 70 and 80 seconds, respectively, the active power and frequency oscillations are equivalent to those observed under traditional VSG control, indicating that DSC loses its effectiveness in mitigating oscillations. This demonstrates that DSC is not robust against communication disruptions.

Fig.9 compares the proposed distributed virtual impedance (DVI) control and the conventional VSG control. Similarly, the load increases and decreases at 10 s and 20 s, respectively, leading to active power and frequency oscillations. At 30 seconds, the distributed virtual impedance control is activated. While a slight oscillation occurs due to the tuned impedance affecting the active power slightly, the system demonstrates improved stability. When the load is switched at 40 and 50 seconds, the active power and frequency smoothly transition to their steady state without significant oscillations. This illustrates the effectiveness of the proposed DVI control method. At 60 seconds, the communication is removed. The DVI has been fixed and will not be changed anymore. Therefore, the parameters can remain matched for the rest of the operation. Consequently, even with load changes at 70 and 90 seconds under the no-communication scenario, the active power and frequency do not experience oscillations. This procedure suggests that the proposed DVI control method is more immune to communication delays and interruptions than the DSC.

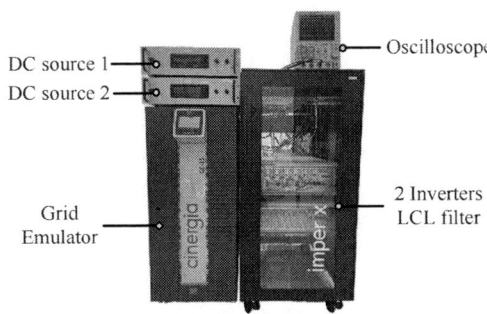

Fig. 10: Experiment setup.

B. Experiment Result

The proposed adaptive control strategy has also been validated through experiments, which involved two VSG-based inverters and an ideal grid emulator. The experimental setup is illustrated in Fig.10. In this setup, the parameters are set as $J_1 = D_1 = 100$ and $J_2 = D_2 = 200$.

This study's experimental results for active power, frequency, and reactive power are depicted in Fig.11. The system initially undergoes a loading phase followed by an unloading phase. The results demonstrate that significant oscillations between all sources are evident in the output active power and frequency when conventional VSG control is employed. This is accompanied by disproportionate sharing of reactive power. Notably, poor frequency dynamic performance, such

as significant frequency overshoot, poses a risk of unexpected load-shedding or extensive blackouts.

Fig.12 illustrates the effectiveness of the proposed DVI method in mitigating active power oscillations. Following system development and subsequent DVI activation, dynamic adjustments of the reactive power are observed. These adjustments result in the reactive power Q_1 and Q_2 approaching their expected values, ultimately achieving a ratio of 1:2. As discussed in the previous section, appropriate reactive power sharing implies that the equivalent impedance of the DGs has been proportionally tunned. When inertia and damping coefficients are proportionally set following the DGs' maximum output capacity and RoCoF requirements, all parameters are harmonized, thereby eliminating the oscillations caused by VSG control. This is evident in Fig.12, where post-DVI activation, load variations do not induce oscillations. Furthermore, with DVI, the frequency change rate remains as expected.

VI. Conclusion

The VSG control can be revisited from an impedance circuit perspective, where VSG oscillations are analogous to LC resonance for an Intuitive understanding of the power oscillations issue. A distributed virtual impedance is proposed to harmonize the parameters and attenuate oscillations in SA mode to address these oscillations. The application of the proposed method yields several benefits: 1) The VSG control and its oscillations can be understood in a more intuitive way. 2) power oscillations can be precisely and quickly attenuated without requiring prior knowledge of the feeder impedance.

Acknowledgement

ECS4DRES is supported by the Chips Joint Undertaking under grant agreement number 101139790 and its members, including the top-up funding by Germany, Italy, Slovakia, Spain, and The Netherlands.

References

[1] X. Meng, J. Liu, and Z. Liu, "A generalized droop control for grid-supporting inverter based on comparison between traditional droop control and virtual synchronous generator control," *IEEE Transactions on Power Electronics*, vol. 34, no. 6, pp. 5416–5438, 2019.

[2] Q.-C. Zhong and G. Weiss, "Synchronverters: Inverters that mimic synchronous generators," *IEEE Transactions on Industrial Electronics*, vol. 58, no. 4, pp. 1259–1267, 2011.

[3] S. Chen, Y. Sun, H. Han, Z. Luo, G. Shi, L. Yuan, and J. M. Guerrero, "Active power oscillation suppression and dynamic performance improvement for multi-vsg grids based on consensus control via coi frequency," *International Journal of Electrical Power & Energy Systems*, vol. 147, p. 108796, 2023.

[4] J. Zhu, C. D. Booth, G. P. Adam, A. J. Roscoe, and C. G. Bright, "Inertia emulation control strategy for vsc-hvdc transmission systems," *IEEE transactions on power systems*, vol. 28, no. 2, pp. 1277–1287, 2012.

[5] X. Hou, Y. Sun, X. Zhang, J. Lu, P. Wang, and J. M. Guerrero, "Improvement of frequency regulation in vsg-based ac microgrid via adaptive virtual inertia," *IEEE Transactions on Power Electronics*, vol. 35, no. 2, pp. 1589–1602, 2020.

[6] M. Li, W. Huang, N. Tai, L. Yang, D. Duan, and Z. Ma, "A dual-adaptivity inertia control strategy for virtual synchronous generator," *IEEE Transactions on Power Systems*, vol. 35, no. 1, pp. 594–604, 2020.

979-8-3315-1612-3/25 $31.00 © 2025 IEEE   614

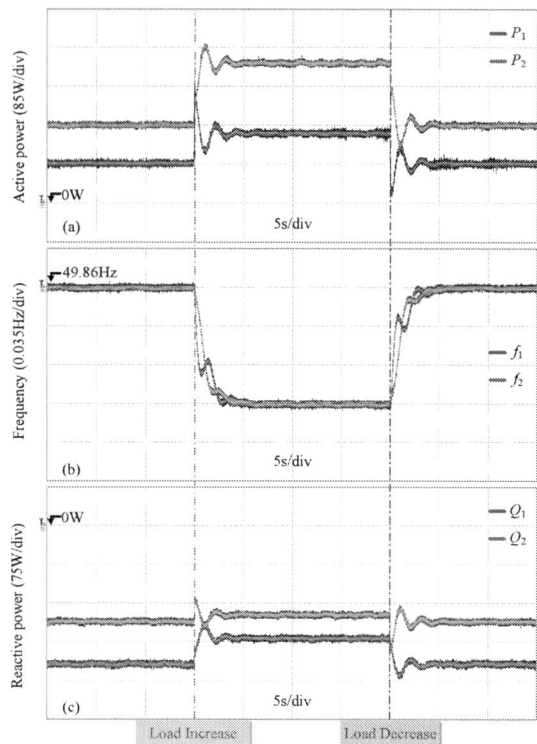

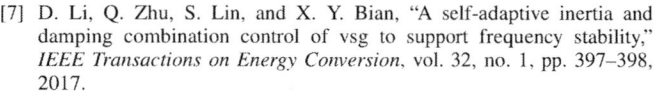

Fig. 11: Dynamic of the VSG control in SA mode

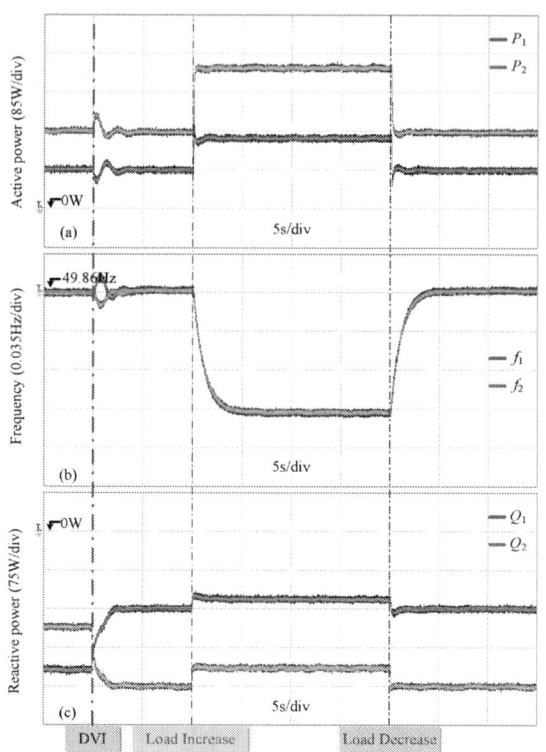

Fig. 12: Dynamic of the proposed control in SA mode.

[7] D. Li, Q. Zhu, S. Lin, and X. Y. Bian, "A self-adaptive inertia and damping combination control of vsg to support frequency stability," *IEEE Transactions on Energy Conversion*, vol. 32, no. 1, pp. 397–398, 2017.

[8] L. Li, Y. Sun, Y. Liu, P. Tian, and S. Shen, "A communication-free adaptive virtual inertia control of cascaded-type vsgs for power oscillation suppression," *International Journal of Electrical Power & Energy Systems*, vol. 149, p. 109034, 2023.

[9] M. Chen, D. Zhou, and F. Blaabjerg, "Active power oscillation damping based on acceleration control in paralleled virtual synchronous generators system," *IEEE Transactions on Power Electronics*, vol. 36, no. 8, pp. 9501–9510, 2021.

[10] Z. Wang, Y. Chen, X. Li, Y. Xu, C. Luo, Q. Li, and Y. He, "Active power oscillation suppression based on decentralized transient damping control for parallel virtual synchronous generators," *IEEE Transactions on Smart Grid*, vol. 14, no. 4, pp. 2582–2592, 2023.

[11] S. Fu, Y. Sun, L. Li, Z. Liu, H. Han, and M. Su, "Power oscillation suppression in multi-vsg grid by adaptive virtual impedance control," *IEEE Systems Journal*, vol. 16, no. 3, pp. 4744–4755, 2022.

[12] M. Shi, X. Chen, J. Zhou, Y. Chen, J. Wen, and H. He, "Frequency restoration and oscillation damping of distributed vsgs in microgrid with low bandwidth communication," *IEEE Transactions on Smart Grid*, vol. 12, no. 2, pp. 1011–1021, 2021.

[13] J. Xiao, L. Wang, Z. Qin, and P. Bauer, "A resilience enhanced secondary control for ac micro-grids," *IEEE Transactions on Smart Grid*, vol. 15, no. 1, pp. 810–820, 2023.

[14] X. Gao, D. Zhou, A. Anvari-Moghaddam, and F. Blaabjerg, "An adaptive control strategy with a mutual damping term for paralleled virtual synchronous generators system," *Sustainable Energy, Grids and Networks*, vol. 38, p. 101308, 2024.

[15] Y. Yu, S. K. Chaudhary, G. D. A. Tinajero, L. Xu, J. C. Vasquez, and J. M. Guerrero, "Active damping for dynamic improvement of multiple grid-tied virtual synchronous generators," *IEEE Transactions on Industrial Electronics*, vol. 71, no. 4, pp. 3673–3683, 2024.

[16] F. Wang, L. Zhang, X. Feng, and H. Guo, "An adaptive control strategy for virtual synchronous generator," *IEEE Transactions on Industry Applications*, vol. 54, no. 5, pp. 5124–5133, 2018.

[17] S. Dong and Y. C. Chen, "Adjusting synchronverter dynamic response speed via damping correction loop," *IEEE Transactions on Energy Conversion*, vol. 32, no. 2, pp. 608–619, 2017.

[18] Y. Yu, S. K. Chaudhary, G. D. A. Tinajero, L. Xu, N. N. B. A. Bakar, J. C. Vasquez, and J. M. Guerrero, "A reference-feedforward-based damping method for virtual synchronous generator control," *IEEE Transactions on Power Electronics*, vol. 37, no. 7, pp. 7566–7571, 2022.

[19] Z. Shuai, W. Huang, Z. J. Shen, A. Luo, and Z. Tian, "Active power oscillation and suppression techniques between two parallel synchronverters during load fluctuations," *IEEE Transactions on Power Electronics*, vol. 35, no. 4, pp. 4127–4142, 2020.

[20] B. Yang, H. Li, S. Xu, H. Liu, and S. Lu, "Systematic methods to eliminate the transient circulating powers in the multi-vsgs system," *IEEE Transactions on Smart Grid*, vol. 15, no. 1, pp. 179–190, 2024.

[21] M. Li, P. Yu, W. Hu, Y. Wang, S. Shu, Z. Zhang, and F. Blaabjerg, "Phase feedforward damping control method for virtual synchronous generators," *IEEE Transactions on Power Electronics*, vol. 37, no. 8, pp. 9790–9806, 2022.

[22] J. Xiao, L. Wang, P. Bauer, and Z. Qin, "Virtual impedance control for load sharing and bus voltage quality improvement in low voltage ac microgrid," *IEEE Transactions on Smart Grid*, 2023.

[23] J. Xiao, L. Wang, Y. Wan, P. Bauer, and Z. Qin, "Distributed model predictive control based secondary control for power regulation in ac microgrids," *IEEE Transactions on Smart Grid*, 2024.

[24] J. Xiao, L. Wang, Z. Qin, and P. Bauer, "An adaptive cyber security scheme for ac micro-grids," in *2022 IEEE Energy Conversion Congress and Exposition (ECCE)*. IEEE, 2022, pp. 1–6.

[25] J. Xiao, L. Wang, P. Bauer, and Z. Qin, "Cyber secure-oriented communication network design for microgrids," in *2024 IEEE 15th International Symposium on Power Electronics for Distributed Generation Systems (PEDG)*. IEEE, 2024, pp. 1–6.

[26] M. H. Bollen and F. Hassan, *Integration of distributed generation in the power system*. John wiley & sons, 2011.

A Novel Current Control Technique for Off-Grid Single-Phase Inverters

Arpan Laha, Abirami Kalathy, Praveen Jain, and Majid Pahlevani

Department of Electrical and Computer Engineering, Queen's University, Kingston, Canada
Emails: arpan.laha@queensu.ca, 18ak67@queensu.ca, praveen.jain@queensu.ca, majid.pahlevani@queensu.ca

Abstract—This paper presents a novel control scheme for the off-grid operation of single-phase inverters. The proposed method generates a sinusoidal output current reference that regulates the voltage at the point of common coupling (PCC) of the inverters. The proposed controller is robust to changes in load and output impedance, ensuring stable operation without the necessity for virtual impedance compensation. As a result, the proposed scheme can effectively regulate the output voltage within a narrow range by eliminating the voltage reference drop typically caused by impedance compensation. Unlike conventional droop laws, the design of this controller does not require adjustments based on output impedance characteristics. Experimental results demonstrate the superior dynamic performance of the proposed control technique.

Index Terms—off-grid inverters, current control techniques, decentralized control, power sharing, distributed energy resources, virtual impedance.

I. INTRODUCTION

The transition to renewable energy sources is a vital step in combating climate change. Distributed energy resources (DERs), including solar photovoltaics (PVs) and wind turbines (WTs), are central to this transition, offering significant benefits such as reduced greenhouse gas emissions, lower transmission losses, and flexible installation options, particularly in remote locations [1], [2]. These advantages have driven the increasing integration of DERs into modern power systems, with inverters serving as a critical component in facilitating this integration. Fig. 1 shows a block diagram of a solar microinverter system. Off-grid systems powered by DERs often face challenges related to limited energy storage capacity, making the reliability of inverters highly dependent on the robustness of their control systems. Advanced control strategies are essential to achieve a rapid dynamic response for managing load transients, stable operation across diverse load types (resistive, inductive, capacitive, and nonlinear), efficient and proportional power sharing during parallel operation, and tight regulation of voltage and frequency within acceptable limits.

In parallel-operated inverter systems, efficient power sharing is crucial, whether achieved with [3], [4] or without communication [5], [6]. However, communication-based methods are often hindered by reliability issues, scalability challenges, and susceptibility to interference, prompting a preference for decentralized control approaches based on the droop concept. Despite its widespread use, the conventional droop technique (CDT) is limited by its sensitivity to output impedance varia-

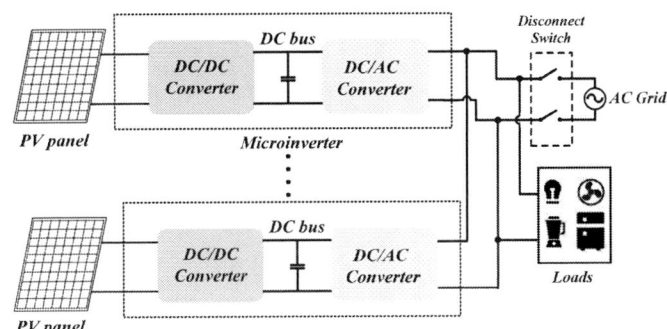

Fig. 1. Representation of a typical microinverter system.

tions, impacting both power-sharing precision and overall system stability [7], [8]. Virtual impedance compensation is often used to improve power distribution among inverters [9] and enhance the stability of the control system [10], [11]. However, a significant drawback of virtual impedance compensation-based techniques is the increased voltage deviation range from the nominal set point, which contradicts existing standards for off-grid systems that require the load terminal voltage to be regulated within a predefined narrow range [12], [13]. Additionally, these techniques can also increase voltage distortions for non-linear loads since the sensed currents with high harmonic content are fed back into the voltage control loop [14]. Furthermore, extracting active and reactive power in single-phase droop control systems poses challenges due to the need to orthogonalize noisy inverter currents, potentially leading to system sluggishness and stability issues.

This paper presents a novel control technique for the off-grid operation of single-phase inverters where the inverter operates as a voltage-controlled current source rather than as a pure voltage source. The proposed control approach ensures stable operation and accurate power sharing, regardless of the nature or magnitude of the output impedance. Furthermore, it eliminates the need for power calculations, thereby avoiding the use of output current filtering, which enhances system stability and dynamic response. By obviating the need for virtual impedance compensation, the proposed technique prevents associated voltage drops. Experimental results validate the robustness of this approach, demonstrating its effectiveness in handling load transients, enabling seamless parallel operation with highly reactive loads, and ensuring stable performance with nonlinear loads while achieving improved power sharing.

979-8-3315-1612-3/25 $31.00 © 2025 IEEE

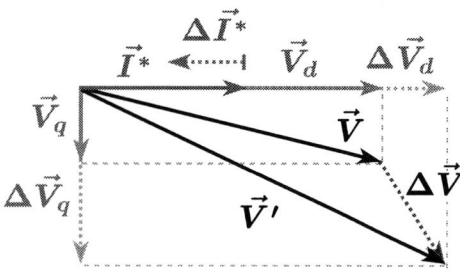

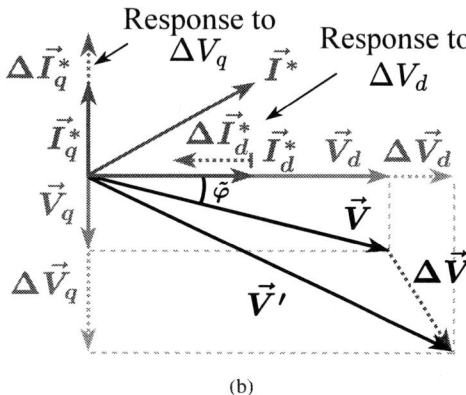

(b)

Fig. 2. Phasor representation of the voltages and currents using (a) the control law in (1) (b) the proposed control law in (3) and (4).

II. CONTROL DESIGN

A. Current Amplitude Reference Control

The objective is to generate a sinusoidal current reference of the form $i_L^* = I^* \sin \varphi_i$, where I^* is the amplitude and φ_i is the phase angle of the current reference. An obvious candidate for I^* is given by

$$I^* = k_I(V_{nom} - V_{amp}) \tag{1}$$

where V_{nom} is the nominal value of the inverter's peak output voltage at no load condition, V_{amp} is the actual output voltage amplitude, and k_I is a positive gain coefficient for the current reference I^*. The rationale behind this control law is the inverse relationship between the current amplitude and the voltage amplitude; as the latter increases, the former should decrease, and vice versa. This control law is particularly effective for resistive loads where voltage and current are in phase. However, if the load is reactive, the output current i_L will be out of phase with the output voltage v_L, preventing i_L from reaching its peak simultaneously with v_L. Due to this phase misalignment, the instantaneous current drawn by the load does not accurately represent the voltage amplitude at that instant. As a result, modifying the current reference amplitude I^* to correct the voltage amplitude V_{amp} may not successfully regulate the PCC voltage v_L.

Fig. 2(a) depicts the phasor diagram for the control law in (1) for a case with the current i_L leading v_L. Here, $\vec{I}^*$ and $\vec{V}$ are used as phasor notations for the current i_L and voltage v_L, respectively. The reference axis is chosen to be

along $\vec{I}^*$, and the voltage $\vec{V}$ can be resolved into direct ($\vec{V}_d$) and quadrature components ($\vec{V}_q$) along this axis. If the phase difference between the voltage and current phasor is $\tilde{\varphi}$, the orthogonal components are $V_d = V_{amp} \cos \tilde{\varphi}$ and $V_q = V_{amp} \sin \tilde{\varphi}$, where $\vec{V}_d$ is aligned along the current phasor $\vec{I}^*$. From Fig. 2(a), it can be observed that a perturbation in the load voltage $\Delta \vec{V}$ has two components - $\Delta \vec{V}_d$ along $\vec{I}^*$ and $\Delta \vec{V}_q$ orthogonal to $\vec{I}^*$. It is evident from Fig. 2(a) that while perturbations in V_d (ΔV_d) can be controlled by I^*, perturbations in V_q (ΔV_q) cannot be effectively addressed using this approach.

Based on this reasoning, it can be inferred that V_q must also be controlled to maintain system stability under diverse load conditions and varying power factors. A possible strategy to achieve this could involve splitting the current reference i_L^* into two constituent components as follows:

$$i_L^* = I_d^* \sin \varphi_i + I_q^* \cos \varphi_i, \tag{2}$$

where I_d^* is the direct component of current that controls V_d, and I_q^* is the quadrature component of current that controls V_q. Fig. 2(b) illustrates the phasor diagram for the proposed control in (2), where $\vec{I}_d^*$ serves as the reference axis, and $\tilde{\varphi}$ is the angle between $\vec{V}$ and $\vec{I}_d^*$.

The control law for I_d^* and I_q^* can be expressed as:

$$I_d^* = k_{I_d}(V_{nom} - V_{amp}), \tag{3}$$
$$I_q^* = -k_{I_q} V_q \tag{4}$$

where k_{I_d} and k_{I_q} are the positive gain coefficients for the current amplitude references. The rate of change of I_d^* with respect to V_d and I_q^* with respect to V_q can be derived as:

$$\frac{\partial V_d}{\partial I_d^*} = -\frac{\sqrt{V_d^2 + V_q^2}}{k_{I_d} V_d} < 0 \tag{5}$$

and

$$\frac{\partial V_q}{\partial I_q^*} = -\frac{1}{k_{I_q}} < 0. \tag{6}$$

Equations (5) and (6) exhibit negative impedance characteristics, with variations in I_d^* and I_q^* opposing changes in V_d and V_q, respectively, as illustrated in Fig. 2(b). This indicates that the control actions remain stable irrespective of the load and output impedance characteristics.

B. Frequency Reference Control

The objective now is to align the $I_{d,ref}$ axis of all inverters operating in parallel. This can be done by controlling the frequency of the inverters. A simple control law can be defined to accomplish this:

$$\omega_i = \omega_{nom} + k_\omega V_q = \omega_{nom} + k_\omega V_{amp} \sin \tilde{\varphi} \tag{7}$$

where k_ω is a positive constant representing the frequency droop gain, ω_{nom} is the nominal frequency, and ω_i is the frequency of the current reference. If the output voltage v_L is defined using the sine reference, $v_L = V_{amp} \sin \varphi_v$, where

979-8-3315-1612-3/25 $31.00 © 2025 IEEE

φ_v is the phase angle of the output voltage, then $\tilde{\varphi} = \varphi_v$
Differentiating this with respect to time, we get:

$$\dot{\tilde{\varphi}} = \dot{\varphi}_v - \dot{\varphi}_i = \omega_v - \omega_i = \omega_v - \omega_{nom} - k_\omega V_{amp} \sin \tilde{\varphi}$$

where ω_v is the frequency of the output voltage at the
It should be noted that ω_v and ω_i may be different d
transients. From (8), the steady-state equilibrium point
satisfies the following equation:

$$\sin \tilde{\varphi}^* = \frac{\omega_v - \omega_{nom}}{k_\omega V_{amp}}$$

Let us define μ as:

$$\mu = \frac{\omega_v - \omega_{nom}}{k_\omega V_{amp}}$$

For real solutions of $\tilde{\varphi}$, μ must lie within the interval $[-$
Hence, for frequency entrainment, the following conc
must be satisfied:

$$\omega_{nom} - k_\omega V_{amp} \leq \omega_v \leq \omega_{nom} + k_\omega V_{amp}$$

Different fixed points emerge depending on whether
greater or less than ω_{nom}. Specifically:

- If $\omega_v > \omega_{nom}$, the two equilibrium points are:

$$\tilde{\varphi}_1^* = \sin^{-1}(\mu), \quad \tilde{\varphi}_2^* = \pi - \sin^{-1}(\mu)$$

- If $\omega_v < \omega_{nom}$, the two equilibrium points are:

$$\tilde{\varphi}_1^* = \sin^{-1}(\mu), \quad \tilde{\varphi}_2^* = -\pi - \sin^{-1}(\mu)$$

The system dynamics can be modeled as:

$$\dot{\tilde{\varphi}} = f(\tilde{\varphi}) = \omega_v - \omega_{nom} - k_\omega V_{amp} \sin(\tilde{\varphi}) \quad (14)$$

The stability of the system shown in (14) can be determined
by analyzing the sign of the derivative of f with respect to $\tilde{\varphi}$:

$$\frac{\partial f}{\partial \tilde{\varphi}} = -k_\omega V_{amp} \cos(\tilde{\varphi}) \quad (15)$$

Evaluating this derivative at each equilibrium point gives:

- At $\tilde{\varphi}_1^* = \sin^{-1}(\mu)$:

$$\left. \frac{\partial f}{\partial \tilde{\varphi}} \right|_{\tilde{\varphi}=\tilde{\varphi}_1^*} = -k_\omega V_{amp} \sqrt{1 - \mu^2} \quad (16)$$

This indicates that the fixed point $\tilde{\varphi}_1^*$ is stable.

- At $\tilde{\varphi}_2^* = \pi - \sin^{-1}(\mu)$ or $\tilde{\varphi}_2^* = -\pi - \sin^{-1}(\mu)$:

$$\left. \frac{\partial f}{\partial \tilde{\varphi}} \right|_{\tilde{\varphi}=\tilde{\varphi}_2^*} = k_\omega V_{amp} \sqrt{1 - \mu^2} \quad (17)$$

This indicates that the fixed point $\tilde{\varphi}_2^*$ is unstable. Fig.
3 presents a phase portrait illustrating the dynamics of the
state variable $\tilde{\varphi}$ and its derivative $\dot{\tilde{\varphi}}$. The curve intersects the
$\tilde{\varphi}$-axis at two equilibrium points, $\tilde{\varphi}_1^*$ and $\tilde{\varphi}_2^*$. The stability
of these points can be inferred from the direction of the red
arrows along the curve. At $\tilde{\varphi}_1^*$, the arrows converge toward the
equilibrium, indicating a stable fixed point. In contrast, at $\tilde{\varphi}_2^*$,
the arrows diverge, indicating an unstable fixed point.

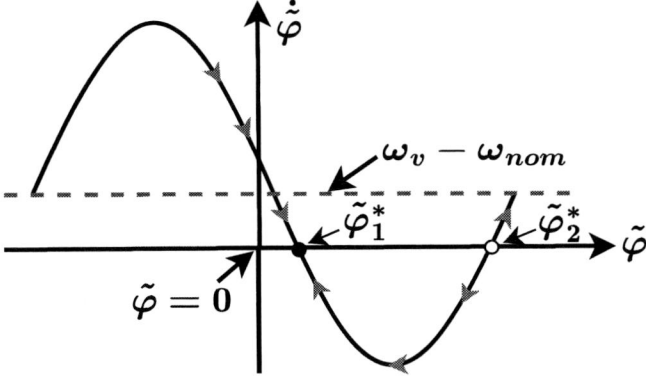

Fig. 3. Phase portrait illustrating the dynamics of $\tilde{\varphi}$, with two equilibrium points ($\tilde{\varphi}_1^*$ and $\tilde{\varphi}_2^*$) and their respective stability characteristics.

If $|\mu| > 1$, the sinusoidal trajectory does not intersect the
$\tilde{\varphi}$-axis, meaning the system has no fixed points and frequency
entrainment is not possible. At $|\mu| = 1$, when $\tilde{\varphi} = \pi/2$,
the dynamical system undergoes a saddle-node bifurcation,
resulting in the emergence of two fixed points for $|\mu| < 1$:
a stable fixed point ($\tilde{\varphi}_1^*$) and an unstable one ($\tilde{\varphi}_2^*$). Hence, the
stable equilibrium point for φ can be expressed as:

$$\tilde{\varphi}^* = \sin^{-1} \left(\frac{\omega_v - \omega_{nom}}{k_\omega V_{amp}} \right) \quad (18)$$

Equation (18) shows that, at steady state, the phase difference between the output voltage and $\vec{I}_d^*$ will be the same for all
parallel-connected inverters, provided that k_ω is the same for
all inverters. With the synchronization of $\vec{I}_d^*$, all the inverters
observe the same value for V_q based on the measured voltage
at the PCC, v_L. This leads to the automatic synchronization
of $\vec{I}_q^*$, thereby aligning the output currents of all the inverters
across the system.

C. Complete Control System Framework

The block diagram of the proposed off-grid control system
is shown in Fig. 4. The proposed control law generates a
sinusoidal current reference signal (i_L^*) according to (2) using
only the sensed voltage at the PCC. The Orthogonal Signal
Generation (OSG) block, implemented using a Second Order
Generalized Integrator (SOGI), extracts the $\alpha\beta$ components
of the voltage v_L. The Park Transformation ($\alpha\beta$-dq) block
calculates the V_q component of the voltage (considering a
sine reference). A Proportional-Resonant (PR) controller is
preferred as the current controller due to its high gain at
the operating frequency, making it more suitable for tracking
a sinusoidal reference current signal. The Sinusoidal Pulse-
Width Modulator (SPWM) block generates the switching
pulses for the inverter switches.

III. PARAMETER SELECTION

The careful selection of the gain coefficients k_{I_d}, k_{I_q}, and
k_ω for the current reference is critical for controlling the
permissible voltage and frequency variations and achieving

979-8-3315-1612-3/25 $31.00 © 2025 IEEE

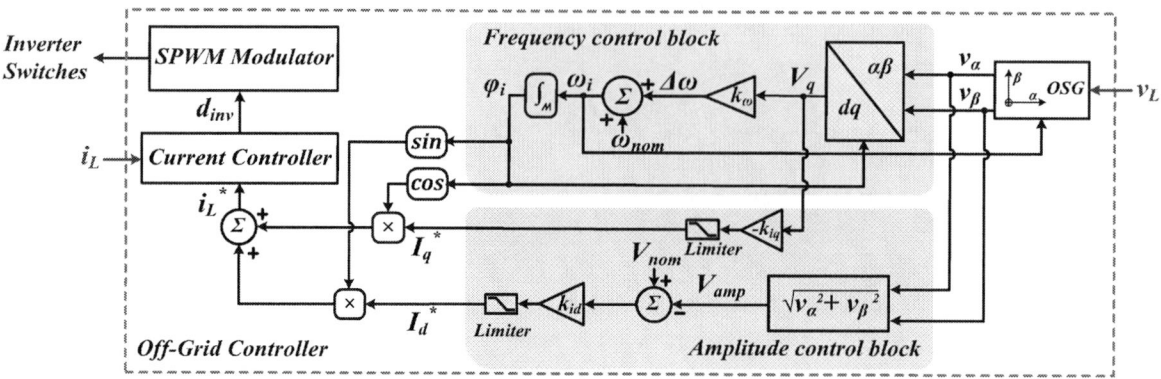

Fig. 4. Proposed control scheme for offgrid operation of solar microinverters.

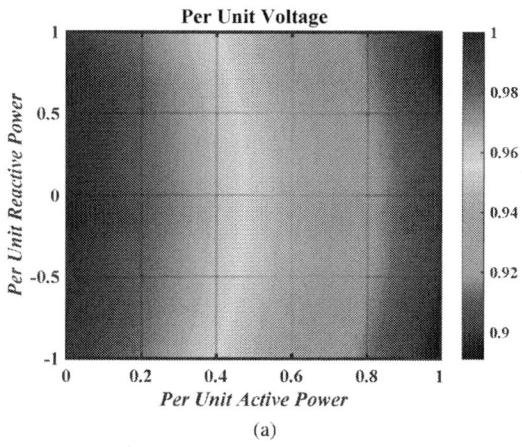

(a)

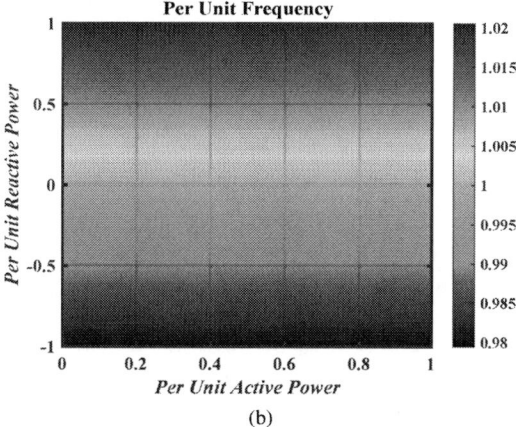

(b)

Fig. 5. Plot of (a) per unit voltage and (b) per unit frequency with variations in per unit values of active and reactive power.

the desired operational behavior. Typically, the permissible voltage droop during normal operation is within 10% below the nominal voltage. According to the desired specifications in Table I and using equation (3), k_{I_d} can be calculated to be $0.42\,\Omega^{-1}$. Using equations (3) and (4), the apparent power output of an inverter using the proposed control strategy can

be expressed as:

$$S_o = \frac{1}{2}V_{amp}\sqrt{k_{I_d}^2(V_{nom}-V_{amp})^2 + k_{I_d}^2 V_{amp}^2 \sin^2(\tilde{\varphi})} \quad (19)$$

From (19), it is evident that multiple solutions of $V_{L,amp}$ could correspond to a given power level, S_o. $\tilde{\varphi}$ also satisfies the following equation:

$$\tilde{\varphi} = \tan^{-1}\left(\frac{Q_o}{P_o}\right) - \tan^{-1}\left(\frac{k_{I_q}V_{amp}\sin(\tilde{\varphi})}{k_{I_d}(V_{nom}-V_{amp})}\right) \quad (20)$$

where P_o and Q_o represent the active and reactive power delivered by the inverter. Using equations (19) and (20), it can be determined that for a rated $2\,\text{kVA}$ inductive load, values of k_{I_q} exceeding $0.55\,\Omega^{-1}$ result in the system exhibiting multiple solutions for V_{amp}. The presence of multiple steady-state solutions can lead to oscillations and potentially destabilize the system. For all other load conditions, the worst-case permissible value of k_{I_q} is higher than $0.55\,\Omega^{-1}$. Therefore, k_{I_q} must be limited to values below $0.55\,\Omega^{-1}$. Although higher k_{I_q} values improve transient response by enabling faster adjustments of the output current to voltage changes, careful tuning is required to balance the benefits of improved response times against the risk of system oscillations.

The permissible frequency range for inverters, as determined by regulatory requirements, typically allows a deviation of $\pm 2\%$ from the nominal frequency. The largest deviation occurs at rated reactive power delivery. Taking into account the previously chosen values of k_{I_d} and k_{I_q}, the value of k_ω is calculated to be 0.0535 Hz/V. Fig. 5(a) and Fig. 5(b) display the plots of the steady-state values of the per unit output voltage amplitude and frequency, respectively, as functions of the per unit active power (P_o) and per unit reactive power (Q_o) delivered.

IV. EXPERIMENTAL RESULTS

Experimental prototypes of PV microinverters were developed according to the specifications outlined in Table I. Each microinverter features a two-stage power conditioning system: a DC/DC stage followed by a DC/AC stage. Four parallel DC/DC converters extract power from individual PV

TABLE I
SYSTEM SPECIFICATIONS

Parameter	Value/Range
Output Voltage Range	207 - 230 V
Maximum RMS Current	9.66 A
Nominal Output Voltage	230 V
Rated Output Power	1.8 kVA
Output Frequency Range	58.8 - 61.2 Hz
Nominal Output Frequency	60 Hz
Switching Frequency of Inverter Bridge	57 kHz
Inverter Inductance	2.5 mH
Inverter Input DC Voltage	420 V

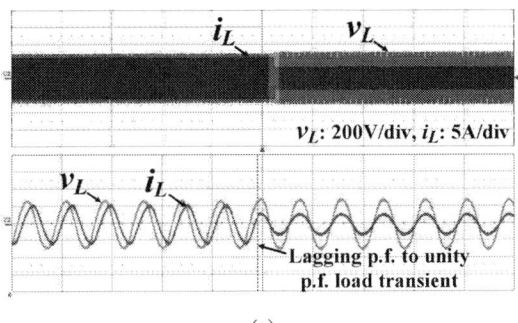

(a)

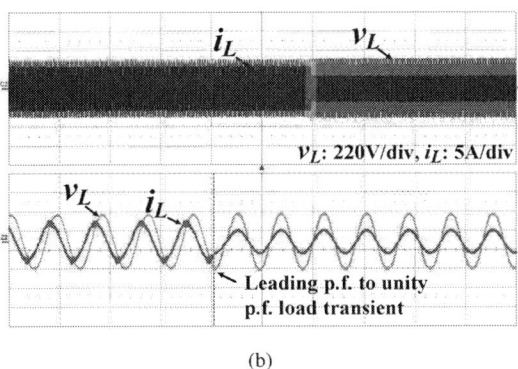

(b)

Fig. 6. Experimental waveforms showing the transient response of the proposed control system during sudden load changes: (a) from a lagging power factor to a unity power factor load, and (b) from a leading power factor to a unity power factor load.

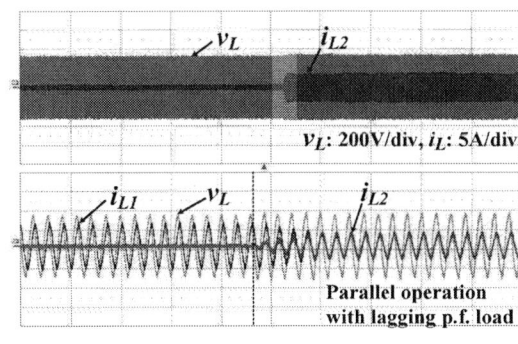

(a)

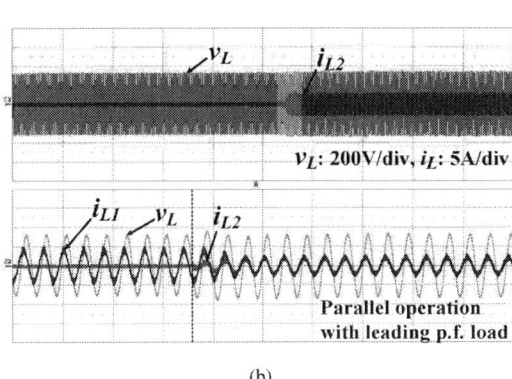

(b)

Fig. 7. Experimental waveforms illustrating the paralleling operation of two inverters under different load conditions: (a) lagging power factor, and (b) leading power factor.

panels, maintaining regulation of the DC bus voltage. The DC/AC converter uses the proposed off-grid control method to generate AC voltage in compliance with the specifications in Table I. The control system is digitally implemented using an Intel Cyclone X 10CL025YE144I7G FPGA.

Fig. 6 illustrates the control system's response to reactive load transients, showing the waveforms for the voltage at the PCC (v_L) and the inverter's output current (i_L). In Fig. 6(a), the system settles within 1.6 ms following a load change from 850 VA at 0.65 pf lagging to 400 W at unity power factor. Similarly, in Fig. 6(b), the system settles within 1.9 ms after a load transition from 800 VA at 0.62 pf leading to 480 W. These results demonstrate the system's rapid transient response to varying load conditions, enabled by precise control of both direct ($I_{d,ref}$) and quadrature ($I_{q,ref}$) current components.

Fig. 7 illustrates the scenario in which an inverter initiates parallel operation with another while connected to: (a) a lagging power factor load (750 VA, 0.7 pf) and (b) a leading power factor load (1.1 kVA, 0.72 pf). The output currents of the inverters are represented as i_{L1} and i_{L2}, respectively. As the proposed controller operates independently of the type of output impedance, the two inverters seamlessly join in parallel within 2–3 line cycles, sharing the load current equally. The waveforms in Fig. 7 not only highlight the robustness of the proposed control system across different types of output impedance but also demonstrate its capability to enable rapid parallel operation of inverters.

Experimental results with a large nonlinear load were conducted to demonstrate the proposed control technique's ability to proportionally share currents between two inverters. The nonlinear load consisted of a combination of domestic appliances, including table fans (70 W), CFL lights (100 W), and a blender (450 W), alongside industrial equipment such as a battery charger (1.6 kW). The total load on the two inverters was 2.2 kW. Fig. 8 shows the time-domain and frequency-domain waveforms of the output voltage and currents for the two inverters operating under the proposed control technique. These results highlight the stable operation of the inverters with a nonlinear load and their precise current-sharing performance.

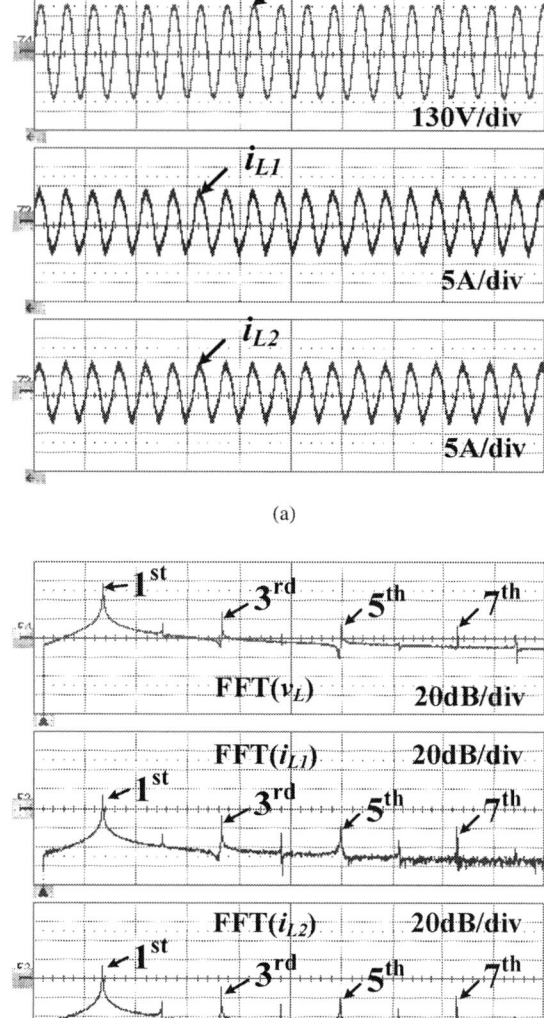

(a)

(b)

Fig. 8. Experimental waveforms depicting parallel inverter operation with the proposed control method under nonlinear loads: (a) time-domain characteristics and (b) frequency-domain characteristics.

V. CONCLUSION

This paper proposes a novel sinusoidal current control technique that ensures fast and robust operation of off-grid single-phase inverters. The technique is designed to handle loads with highly varying power factors and can also provide a very fast dynamic response to load transients. The overall control system is robust to variations in output impedance and does not require structural adjustments when the nature of the impedance changes. Additionally, it eliminates the need for impedance compensation, ensuring compliance with stringent voltage regulation range requirements. Experimental results validate the effectiveness of the proposed control method in managing load transients across varying power factors, enabling seamless parallel operation with highly reactive loads,

and maintaining stability with nonlinear loads.

REFERENCES

[1] M. Liserre, T. Sauter, and J. Y. Hung, "Future Energy Systems: Integrating Renewable Energy Sources into the Smart Power Grid Through Industrial Electronics," *IEEE Ind. Electron. Mag.*, vol. 4, no. 1, pp. 18-37, March 2010.

[2] N. Gu, J. Cui, and C. Wu, "Power-Electronics-Enabled Transactive Energy Market Design for Distribution Networks," *IEEE Trans. Smart Grid*, vol. 13, no. 5, pp. 3968-3983, Sept. 2022.

[3] P. P. Patankar, M. M. Munshi, R. R. Deshmukh, and M. S. Ballal, "A Modified Control Method for Grid Connected Multiple Rooftop Solar Power Plants," *IEEE Trans. Ind. Appl.*, vol. 57, no. 4, pp. 3306-3316, July-Aug. 2021.

[4] W. Huang, Z. Shuai, X. Shen, Y. Li, and Z. J. Shen, "Dynamical Reconfigurable Master–Slave Control Architecture (DRMSCA) for Voltage Regulation in Islanded Microgrids," *IEEE Trans. Power Electron.*, vol. 37, no. 1, pp. 249-263, Jan. 2022.

[5] J. M. Guerrero, L. Hang, and J. Uceda, "Control of Distributed Uninterruptible Power Supply Systems," *IEEE Trans. Ind. Electron.*, vol. 55, no. 8, pp. 2845-2859, Aug. 2008.

[6] H. Safamehr, I. Izadi, and J. Ghaisari, "Robust V-I Droop Control of Grid-Forming Inverters in the Presence of Feeder Impedance Variations and Nonlinear Loads," *IEEE Trans. Ind. Electron.*, vol. 71, no. 1, pp. 504-512, Jan. 2024.

[7] Q.-C. Zhong and T. Hornik, *Control of Power Inverters in Renewable Energy and Smart Grid Integration*. New York, NY, USA: Wiley, 2013.

[8] Q.-C. Zhong, "Robust Droop Controller for Accurate Proportional Load Sharing Among Inverters Operated in Parallel," *IEEE Trans. Ind. Electron.*, vol. 60, no. 4, pp. 1281-1290, April 2013.

[9] A. S. Vijay, N. Parth, S. Doolla, and M. C. Chandorkar, "An adaptive virtual impedance control for improving power sharing among inverters in islanded AC microgrids," *IEEE Trans. Smart Grid*, vol. 12, no. 4, pp. 2991–3003, Jul. 2021.

[10] M. G. Taul, X. Wang, P. Davari, and F. Blaabjerg, "Current limiting control with enhanced dynamics of grid-forming converters during fault conditions," *IEEE J. Emerg. Sel. Topics Power Electron.*, vol. 8, no. 2, pp. 1062–1073, Jun. 2020.

[11] O. M. Anubi and S. Ameli, "Robust Stabilization of Inverter-Based Resources Using Virtual Resistance-Based Control," *IEEE Contr. Syst. Lett.*, vol. 6, pp. 3295-3300, 2022.

[12] Y. Dong, B. Ren, and Q.-C. Zhong, "Bounded Universal Droop Control to Enable the Operation of Power Inverters Under Some Abnormal Conditions and Maintain Voltage and Frequency Within Predetermined Ranges," *IEEE Trans. Ind. Electron.*, vol. 69, no. 11, pp. 11633-11643, Nov. 2022.

[13] American National Standard for Electric Power Systems and Equipment Voltage Ratings (60 Hertz). ANSI C84.1-2016, 2016.

[14] W. Wang, X. Zeng, X. Tang, and C. Tang, "Analysis of microgrid inverter droop controller with virtual output impedance under nonlinear load condition," *IET Proc. Power Electron.*, vol. 7, no. 6, pp. 1547–1556, Jun. 2014.

979-8-3315-1612-3/25 $31.00 © 2025 IEEE

Intelligent Low-Bandwidth Frequency Controller for VSGs at Economic Dispatch in Islanded Microgrid

Shraf Eldin Sati, Ahmed Al-Durra, Hatem H. Zeineldin, Tarek H.M. EL-Fouly and Ehab F. El-Saadany

Department of Electrical and Computer Engineering

Khalifa University of Science and Technology

Abu Dhabi, UAE

{100060832, ahmed.aldurra, hatem.zeineldin, tarek.elfouly, ehab.elsadaany}@ku.ac.ae

Abstract—**This paper proposes a distributed low-bandwidth intelligent controller (DLBIC) to restore islanded microgrid (IMG) frequency and maintain economic dispatch (ED) among inverter-based distributed generators (IBDGs). The DLBIC utilizes local measurements and receives a single variable from the nearest DG, thereby reducing communication link bandwidth and network infrastructure costs. Notably, the DLBIC is model-free, operating without prior knowledge of the IMG model and parameters. It outperforms other approaches by effectively handling the nonlinearity and complexity of IBDGs without extensive training. Leveraging reinforcement learning principles, the DLBIC can learn autonomously and efficiently manage uncertainties and disturbances within the IMG. The effectiveness of the DLBIC is demonstrated through tests against load connection and disconnection scenarios using the IEEE 38-node benchmark. Results indicate that the proposed intelligent controller successfully restores IMG frequency, mitigates frequency nadir/overshoot, eliminates oscillations, and achieves optimal ED.**

Index Terms—**Distributed secondary frequency control, economic dispatch, intelligent controller, inverter-based distributed generators, islanded microgrid, virtual synchronous generator.**

I. INTRODUCTION

Aiming to reduce pollution, enhance efficiency, and minimize transmission losses, the concept of islanded systems such as islanded microgrids (IMGs) has increased. IMG consists of loads and distributed generations (DGs) interfaced with the grid via power inverters [1]. Key challenges include optimizing power flow, enhancing system reliability and robustness, and improving frequency stability. A hierarchical control structure governs IMG dynamics: primary control layer (PCL) stabilizes frequency, voltage, and power sharing between DGs; secondary control layer (SCL) corrects deviations caused by PCL operations; tertiary control layer (TCL) manages power flow at the connection point between the grid and IMGs [2].

In IMG's PCL, $Q - V$ and $P - \omega$ droop control methods enable communication-free power sharing between DGs and enhance voltage and frequency stability. However, steady-state errors remain challenging, addressed by centralized, decentralized, or distributed SCL schemes [3], [4]. Centralized SCL uses a PI regulator to adjust frequency and voltage but is vulnerable to system failures due to communication dependency. Decentralized SCL, based on local DG variables, reduces reliance on central control but requires detailed knowledge of IMG topology. Distributed SCL offers a middle ground, improving reliability without central control.

Distributed approaches use limited communication networks to share information among nearby controllers, enhancing expandability and dependability. [5] focus on removing frequency deviations and achieving accurate active power distribution over a set time frame. [6] propose a fuzzy cooperative SCL framework using model predictive control, targeting similar objectives, while [2] explores a sliding mode regulator. To overcome challenges related to real-time data exchange and minimize communication bandwidth dependency, event-triggered strategies are discussed in [7]–[12]. These strategies aim to proportionally share loads and improve IMG reliability and flexibility. However, their functionality does not account for inertia synthesis [13] or economic dispatch (ED).

To dispatch the IMG economically while maintaining sufficient virtual inertia to retain the rate of change of frequency (RoCoF) among permissible ranges, [14] proposes integrating the incremental cost (IC) formula in the PCL of VSG; however, the model faces persistent frequency deviations. The authors of [15]–[18] present various distributed algorithms ranging from consensus to event-trigger approaches for primary, secondary, and ED functions. While in [19], instead of setting the power references for each DG to zero to form the IMG and generate the operating frequency, the authors manipulate the power references to reduce the total generation cost by forecasting the set points using the Back-propagation neural network model. Nevertheless, these studies [15]–[20], neglect modeling virtual inertia, rendering the system vulnerable to high RoCoF following disturbances.

Therefore, this manuscript proposes a distributed low-bandwidth intelligent controller (DLBIC) to restore IMG frequency while preserving ED across the DGs and maintaining sufficient virtual inertia to slow the RoCoF. The proposed DLBIC significantly outperforms existing SCL methods. Current methods face several challenges: they rely heavily on specific operating conditions, which reduces their robustness and reliability. While online tuning approaches exist, they require a precise and complex mathematical model, making them time-consuming to develop. IMGs' dynamic and varying conditions mean no single mathematical model can apply universally to all scenarios. Variations in the IMG's configuration or parameters can lead to model mismatches, adversely affecting the controller's performance, especially during steady-state operations. In contrast, the DLBIC addresses these issues,

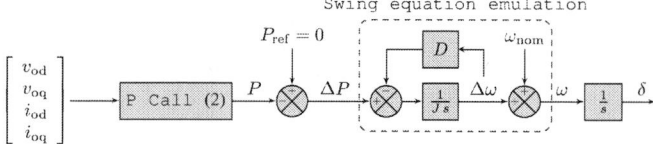

Fig. 1: Power regulation schema of grid-forming VSG in IMG.

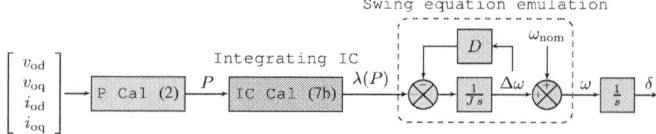

Fig. 2: Integrating IC formula in the power regulation schema of grid-forming VSG to form ED-VSG model.

offering a more adaptable and reliable solution for managing IMG conditions. The proposed DLBIC uses the brain's emotional learning-based intelligence controller (BELBIC). The BELBIC's main characteristic is its model-free structure, allowing it to handle uncertainty, nonlinearity, and complexity in power converters.

The remaining parts of this article are arranged in an ensuing manner: The problem statements are defined in Section II. Section III introduces the proposed intelligent controller and its design. Section IV demonstrates the controller's effectiveness through simulations, and Section V outlines the paper.

II. PROBLEM STATEMENT

The following section reviews VSG implementation in IMGs and details the integration of IC function into the VSG framework. This integration allows IMGs to maintain ED during steady-state operation while supplying virtual inertia during transients through a unified ED-VSG model. Gaps in ED-VSG implementation within the PCL are identified. The section also explores correcting frequency deviation employing the integral controller presented in [3] for VSG without ED.

A. Grid-Forming VSG Formation in IMGs

The VSG enhances the frequency stability of IMGs by emulating the synchronous generator inertia within the inverter. This emulation attenuates the RoCoF, providing a promising solution for securing the operation of low-inertia systems. The swing formula is incorporated in the power regulation loop to simulate rotor inertia, forming the VSG as shown in Fig. 1 and expressed as follows:

$$P_{\text{ref}} - P = \Delta P = J\frac{\text{d}(\omega - \omega_{\text{nom}})}{\text{d}t} + D(\omega - \omega_{\text{nom}}), \quad (1)$$

P_{ref} and P are the VSG's virtual active power reference and output, and ω is the virtual angular frequency. $\omega_{\text{nom}} = 2\pi F_{\text{nom}}$ is the nominal angular frequency, with D and J denoting the virtual damping and inertia coefficient, respectively. Eq. 1 shows a straightforward VSG model considering a pole pair. The input to the VSG is ΔP, where P is calculated utilizing the inverter output current and voltage as detailed below:

$$P = \frac{\omega_{\text{F}}}{s + \omega_{\text{F}}}(v_{\text{od}}i_{\text{od}} + v_{\text{oq}}i_{\text{oq}}) \quad (2)$$

Where i_{od}, i_{oq}, v_{od}, and v_{oq} are the output currents and voltages in the dq-reference frame, with ω_{F} being the power filter cutoff frequency. Eq. (1) simplifies the system by using ω_{nom} instead of the AC-bus frequency, thus removing the need for a phase-locked loop as mentioned in [14]. Moreover,

incorporating ω_{nom} as feedforward before ω eliminates the rotor's initialization delay, guaranteeing that the steady-state response is unaffected.

For reactive power control, $Q - V$ droop control is used:

$$v_{\text{od}}^* = v_{\text{nom}} - m_{\text{q}}(Q - Q^{\text{ref}}), \quad v_{\text{oq}}^* = 0. \quad (3)$$

Where v_{od}^* and v_{oq}^* represent the references voltage amplitude in dq coordinate, v_{nom} is the nominal voltage, m_{q} is the $Q - V$ droop coefficient, Q^{ref} is the reactive reference power and Q is the reactive power output, which calculated as follows:

$$Q = \frac{\omega_{\text{F}}}{s + \omega_{\text{F}}}(v_{\text{oq}}i_{\text{od}} - v_{\text{od}}i_{\text{oq}}) \quad (4)$$

In IMGs, setting P_{ref} and Q_{ref} to zero achieves the desired grid-forming operation, resulting in the subsequent straightforward VSG representation in the s-domain:

$$\Delta\omega = \omega_{\text{nom}} - \omega = \frac{1}{Js + D}P, \quad (5)$$

Where $\Delta\omega = 2\pi\Delta F$ represents the angular frequency deviation caused by the VSG characteristic.

B. Integrating IC Function Into VSG To Form ED-VSG

Minimizing the total generation cost in IMG involves operating the system in ED, where the cost minimization can be illustrated with the constraints as follows:

$$\min \quad \sum_{j=1}^{N} CF_j^{\diamond}(P_j) \quad (6a)$$

$$\text{s.t.} \quad \sum_{i=1}^{N} P_j = P_{\text{Demand}}, \quad (6b)$$

$$P_j^{\text{mx}} \le P_j \le P_j^{\text{mi}}, \quad j = 1, \ldots, N \quad (6c)$$

Where P_{Demand}, P_j^{mx}, and P_j^{mi} denote the IMG total demand, and the upper and lower power thresholds of the jth IBDG, respectively. According to the Lagrangian approach utilized for the cost formula in (6), optimal generation expenses are minimized when all IBDGs possess identical IC values. The IC for each IBDG can be determined from the cost function $(CF^{\diamond})$ as follows.

$$CF_j^{\diamond}(P_j) = \alpha_j^{\diamond} + \beta_j^{\diamond} P_j + \gamma_i^{\diamond} P_j^2 + \rho_j^{\diamond} e^{\epsilon_j^{\diamond} P_j}, \quad (7a)$$

$$IC \equiv \lambda(P) = \frac{\partial CF_j^{\diamond}(P_j)}{\partial P_j} = a_j^{\diamond} P_j + b_j^{\diamond} e^{c_j^{\diamond} P_j} + d_j^{\diamond}, \quad (7b)$$

Where $\lambda(P)$ is a nonlinear IC formula, which is the derivative of power generation cost $(CF_j^{\diamond}(P_j))$ concerning the power

(P). $a_j^\diamond$, $b_j^\diamond$, $c_j^\diamond$, and $d_j^\diamond$ are the IC coefficients detailed in [4].

The ED-VSG model has two main goals: 1) ensuring the cost-effective running of the IMG by allocating the demand among VSGs based on their generation expense attributes, and 2) providing emulated inertia to reduce the RoCoF after disruptions. Using the ED-VSG requires two primary modifications to the VSG power regulation loop: configuring the control framework and adjusting parameters accordingly.

Configuring VSG Power Control Loop Structure:
Firstly, the input variable shifts to the IC ($\lambda(P)$) from active power (P), illustrated in Fig. 2. This transformation significantly influences power distribution strategies and system stability. The revised setup is expressed as follows:

$$\Delta\omega = \frac{1}{Js + D}\lambda(P) \tag{8}$$

where J, $\Delta\omega$, $\lambda(P)$, and D retain their respective meanings as defined in eqs. (1), (5), and (7b).

Adjusting Parameter Settings:
Secondly, regarding parameter adjustments, the D gain for all VSGs must be uniformly configured to equalize the IC during steady-state (ss). This uniform D setting ensures IC balance across all VSGs, considering they operate at a common frequency during ss. Notably, at ss, the influence of the inertia term is minimal, and the model functions akin to an IC-based droop controller for economically distributing load among VSGs based on the principle of equal IC. Therefore, the ED-VSG model can be formulated as:

$$\Delta\omega^{\text{ss}} = \frac{1}{D}\lambda(P), \tag{9}$$

where $\Delta\omega^{\text{ss}}$ represents the angular frequency deviation at ss, and $\frac{1}{D}$ denotes the droop gain constant ($\lambda(P) - \Delta\omega^{\text{ss}}$). To clarify, achieving ED requires equalizing the IC for each DG:

$$\lambda(P)_{\text{DG}_1} = \lambda(P)_{\text{DG}_2} = \cdots = \lambda(P)_{\text{DG}_N} = \lambda(P)^* \tag{10}$$

Where $\lambda(P)^*$ is the optimal IC value when eq. (10) satisfied.

The IC and frequency responses in Figs. 3 and 4 illustrate the behavior of the ED-VSG model during load connection and disconnection scenarios for the benchmark IMG, where the altered load equals 33.33% of the total load. The IMG benchmark details are illustrated in Section IV. Fig. 3 shows the alignment of IC values among the IBDGs, ensuring ED convergence during steady-state. During transients, the inertia term becomes relevant, slowing the RoCoF. As described in (9), the ED-VSG correlates $\lambda(P)$ with $\Delta\omega^{\text{ss}}$ through the droop gain $\frac{1}{D}$. This relationship causes persistent frequency deviation, as depicted in Fig. 4. Notably, the model fails to restore the system frequency, resulting in a steady-state frequency error of 0.23 Hz. The parameter D significantly influences both persistent frequency deviation and ED convergence. A lower D accelerates ED attainment but increases frequency deviation and reduces stability limit. The persistent frequency deviation can be demonstrated by employing the final value

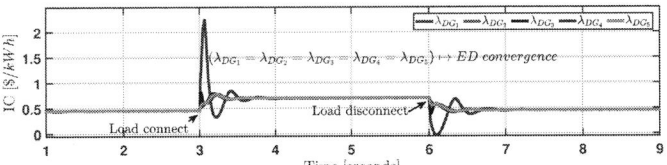

Fig. 3: IC based on the ED-VSG model for load changes.

theorem in response to a step load change, represented as:

$$\Delta\omega_{t\to\infty} = \lim_{s\to 0} s\Delta\omega = \lim_{s\to 0} s\frac{\lambda(\Delta P)/s}{Js + D} = \frac{\lambda(\Delta P)}{D} \tag{11}$$

$\lambda(\Delta P)$ is the change in IC due to the load disturbance. Eq. (11) and Fig. 4 illustrate that the ED-VSG model behaves like an IC-based droop, resulting in persistent frequency deviation, thereby necessitating a SCL to restore system frequency.

C. Investigating Frequency Restoration for the ED-VSG using the Integral Controller

Secondary frequency controllers help return the frequency of IMGs to their nominal values. In VSG applications, various control methods, such as proportional-integral, fuzzy, integral, or adjustable controllers, are employed. While these approaches have demonstrated effectiveness in restoring frequency in IMGs, their application in scenarios where VSGs operate in ED mode remains largely unexplored. This subsection examines the feasibility of using an integral controller in the ED-VSG framework, mimicking previous research in [3], which focused on VSGs operating without ED. Extending the integral approach to the ED-VSG framework yields the following model:

$$\Delta\omega = \frac{1}{Js + D + \frac{k}{s}}\lambda(P), \tag{12}$$

where k represents the integral gain used to compensate $\Delta\omega$. Applying the final value theorem for a step change allows us to determine the steady-state characteristic (final value) of $\Delta\omega$:

$$\Delta\omega_{t\to\infty} = \lim_{s\to 0} s\frac{\lambda(\Delta P)/s}{Js + D + \frac{k}{s}} = \frac{\lambda(\Delta P)}{\infty} = 0. \tag{13}$$

As highlighted in Eq. (13), the integral controller effectively eliminates the steady-state error to $\Delta\omega^{\text{ss}} = 0$. However, this approach introduces a critical issue: the model loses the relationship between $\lambda(P)$ and $\Delta\omega$, resulting in disparate ICs that disrupt ED convergence, as will be demonstrated in the result Section IV. This conflict between restoring frequency and maintaining ED functionality compromises the ED-VSG's primary objective.

III. PROPOSED DISTRIBUTED LOW-BANDWIDTH INTELLIGENT CONTROLLER (DLBIC)

This section aims to clarify the contribution and present the mathematical formulation and design of the proposed controller. In subsection II-C, we demonstrated the failure of the decentralized integral controller to preserve ED. Consequently, the proposed intelligent distributed controller aims to correct

979-8-3315-1612-3/25 $31.00 © 2025 IEEE

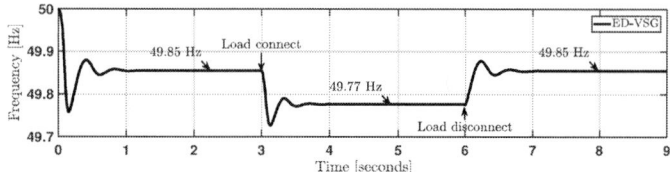

Fig. 4: IMG frequency response for load based on ED-VSG.

frequency deviation and maintain ED by adjusting the PCL of the ED-VSG. Therefore, the final steady-state frequency and IC values of the DGs are as follows:

$$\lim_{t \to \infty} \omega = \omega_{\text{nom}} \tag{14a}$$

$$\lim_{t \to \infty} \lambda_{\text{DG}_{1,2,\dots,N}}(P) = \lambda(P)^* \tag{14b}$$

$\lambda(P)^*$ is the optimal IC value for each DG to ensure ED.

The schematic representation of the DLBIC is displayed in Fig. 5. The DLBIC receives two variables: one from local measurement and another from the nearest IBDG, thereby minimizing communication link bandwidth and network setup costs. Specifically, the controller uses the local frequency deviation ($\Delta\omega$) from the PCL and receives the IC value from the nearest unit ($\lambda(P)_{\text{NU}}$). Fig. 6 illustrates the communication network between units and transmitted data. Then, the controller generates the correction signal that eliminates the $\Delta\omega$ and ($\lambda(P) - \lambda(P)_{\text{NU}} = \Delta\lambda(P)$). Where $\Delta\lambda(P)$ is the mismatch in IC values between the units.

The BELBIC is chosen for the DLBIC due to its unique features. Its model-free nature, fewer adjustable parameters in emotional controllers, and simplistic control framework make it ideal for practical applications. BELBIC operates on a reinforcement learning approach, efficiently addressing uncertainties and disturbances within the system. In contrast to traditional reinforcement learning techniques in machine learning, BELBIC does not require exhaustive training. It excels due to its minimal computational complexity, online learning capability, and independence from prior understanding of IMG dynamics, making it superior to other intelligent controllers like neural networks and fuzzy logic.

A. Mathematical Formulation of the BELBIC

This subsection details the BELBIC controller's components and their mathematical formulation. The controller consists of the Amygdala (Am), responsible for emotional learning, the Thalamus, the sensory cortex, and the Orbitofrontal (Orb) cortex. It receives two inputs: emotional signal (E_m^s) and sensory input (S_e^n). The Thalamus processes the S_e^n signals, including filtering or noise cancellation, and sends the output to the sensory cortex. The sensory cortex then relays this information to the Am and the Orb cortex. The BELBIC output (u) is the difference between the outputs of Am and Orb, as shown in below:

$$u = Am - Orb \tag{15}$$

Network Am receives inputs from both S_e^n and E_m^s. The S_e^n input undergoes multiplication by a pre-defined weight (W) to produce the output of Am, which is given by:

$$Am = S_e^n W \tag{16}$$

where W varies according to the integral formula:

$$W = \int_0^t \delta w + W(0), \tag{17a}$$

$$\delta w = \alpha S_e^n [max(0, E_m^s - Am - A_n] \tag{17b}$$

$$A_n = \max[S_e^n] M_a \tag{17c}$$

α represents the rate of learning, A_n denotes a neuron receiving the highest sensory signals directly from the thalamus, and $\max[S_e^n]$ represents the peak value among all sensory signals. The behavior of M_a are illustrated by

$$M_a = \int_0^t \delta m_a dt + M_a(0). \tag{18}$$

Orb receives inputs from S_e^n and E_m^s, in addition to the previous model output. The output of network Orb is determined by scaling the S_e^n signal with a connection weight Z.

$$Orb = S_e^n Z, \tag{19a}$$

$$Z = \int_0^t \delta z dt + Z(0), \tag{19b}$$

$$\delta z = \beta S_e^n [Am - Orb - E_m^s] \tag{19c}$$

β is the inhibition rate, initializing $W(0) = Z(0) = M_a(0) = 0$; the BELBIC output in eq. (15) can be defined as

$$u = S_e^n [\alpha \int_0^t S_e^n [max(0, E_m^s - Am$$
$$-A_n] dt - \beta \int_0^t S_e^n [Am - Orb - E_m^s] dt] \tag{20}$$

The Am and Orb networks update their adaptive weights using internal rules (Eqs. (17b) and (19c)). The Orb network mirrors the Am network, adjusting weights to achieve desired inhibition. Ensuring the BELBIC model's robustness within the SCL requires identifying conditions for internal stability, where the outputs of the Am and Orb networks asymptotically converge. We will explore these convergence conditions next.

B. Condition of Convergence

Considering the adjustments in network weights as described in eqs. (15) to (19c), there is a configuration of the α and β parameters and S_e^n signal for which

1) $1 > |1 - \alpha S_e^{n2}|$
2) $1 > |1 - \beta S_e^{n2}|$

This ensures that the weights of Am and Orb asymptotically converge. *Proof.* BELBIC's operation can be divided into two stages: transient and steady-state. Firstly, at the transient stage, eq. (17b) can be expressed as:

$$\delta w = \alpha S_e^n [E_m^s - Am - A_n] \tag{21}$$

979-8-3315-1612-3/25 $31.00 © 2025 IEEE   625

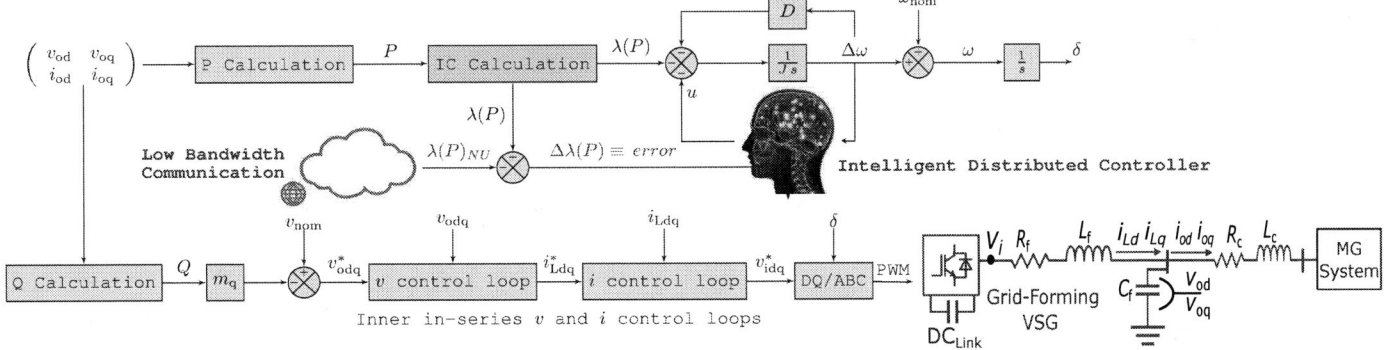

Fig. 5: Proposed distributed low-bandwidth intelligent controller for grid-forming ED-VSG in IMG

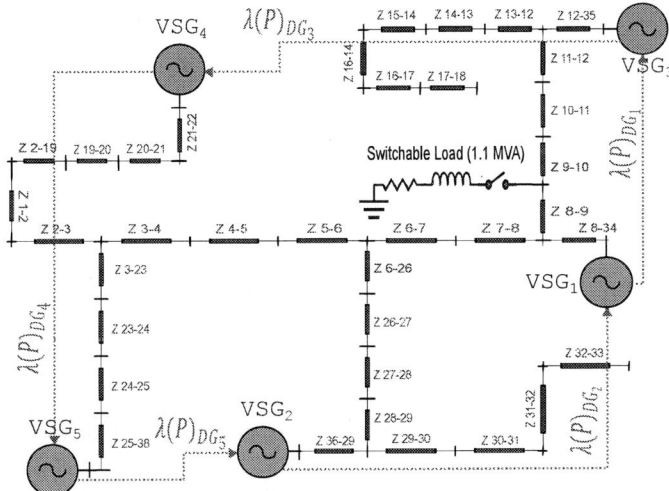

Fig. 6: The IEEE 38-node benchmark system for performance validation; red dotted lines denote the communication graph.

At the steady-state stage, the change in the weights of networks Am and Orb is zero, i.e.,

$$\delta w = \delta m_a = \delta z = 0 \tag{22}$$

By implementing condition (22) to eqs. (21) and (19c), and assuming $S_e^n \neq 0$,

$$E_m^s = A_n = S_e^n M_a = u \tag{23}$$

Assume that m_a and m_a^* are the weights of Am during and after adaptation, respectively. Let $\acute{E}_m^s = S_e^n M_a$ and $\acute{E}_m^s = S_e^n M_a^*$ is the E_m^s signal before and after adaptation. The weight adjustment δm_a can be described as follows:

$$\delta m_a = \alpha S_e^n [max(0, (E_m^s - \acute{E}_m^s)] \tag{24}$$

When $E_m^s - \acute{E}_m^s > 0$, eq. (24) decreases to

$$\mu m_a = \alpha S_e^n (E_m^s - \acute{E}_m^s) \tag{25a}$$
$$= \alpha S_e^n (S_e^n m_a^* - S_e^n M_a) \tag{25b}$$
$$= \alpha S_e^{n2} (M_a^* - M_a) = \alpha S_e^{n2} \tilde{M}_a \tag{25c}$$

where $\tilde{M}_a = M_a^* - M_a$. Over a small period δt, M_a varies as

$$M_a(t+\delta t) = M_a + \delta m_a \tag{26a}$$
$$\tilde{M}_a(t+\delta t) = G_a^*(t+\delta t) - M_a(t+\delta t) \tag{26b}$$
$$= \tilde{M}_a - \alpha S_e^{n2} \tilde{M}_a = (1 - \alpha S_e^{n2}) \tilde{M}_a \tag{26c}$$

Therefore, $\tilde{M}_a(t+\delta t)$ converges to $\tilde{M}_a$ if $|1 - \alpha S_e^{n2}| < 1$. The adaptation in Orb is formulated as follows:

$$\delta z = \beta S_e^n (Am - Orb - E_m^s) \tag{27a}$$
$$= \beta S_e^n (0 - S_e^n Z - S_e^n G_a^*) \tag{27b}$$
$$= -\beta S_e^{n2} (G_a^* + Z) = -\beta S_e^{n2} \tilde{Z} \tag{27c}$$

where $\tilde{Z} = G_a^* + Z$. The term Z changes as

$$Z(t+\delta t) = Z + \delta z \tag{28a}$$
$$\tilde{Z}(t+\delta t) = G_a^*(t+\delta t) + Z(t+\delta t) \tag{28b}$$
$$= G_a^*(t+\delta t) + Z + \delta z \tag{28c}$$
$$= \tilde{Z} - \beta S_e^{n2} \tilde{Z} = (1 - \beta S_e^{n2}) \tilde{Z} \tag{28d}$$

Thus, $\tilde{Z}(t+\delta t)$ converges to $\tilde{Z}$ if $|1 - \beta S_e^{n2}| < 1$.

Remark. When choosing values for α and β, it is important to consider the convergence criteria outlined in III-B.

Effective performance with BELBIC requires establishing an empirical relationship between S_e^n, E_m^s, and u.

C. Design of S_e^n and E_m^s in BELBIC

The inputs S_e^n and E_m^s for BELBIC are chosen as follows

$$S_e^n = \xi_1 (\Delta\omega + \Delta\lambda(P)) + \xi_2 \int (\Delta\omega + \Delta\lambda(P)) dt \tag{29}$$

$$E_m^s = \xi_3 (\Delta\omega + \Delta\lambda(P)) + \xi_4 \int (\Delta\omega + \Delta\lambda(P)) dt + \xi_5 u \tag{30}$$

Where ξ_1, ξ_2, ξ_3, ξ_4, and ξ_5 are weighting coefficients for the S_e^n and E_m^s functions, fine-tuned through trial and error. The S_e^n and E_m^s functions are designed to ensure minimal overshoot, quick response, zero steady-state error, and low oscillation from a specified reference.

979-8-3315-1612-3/25 $31.00 © 2025 IEEE

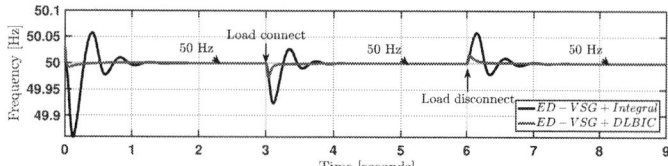

Fig. 7: IMG frequency response for load changes based on ED-VSG+Integral and the proposed ED-VSG+DLBIC.

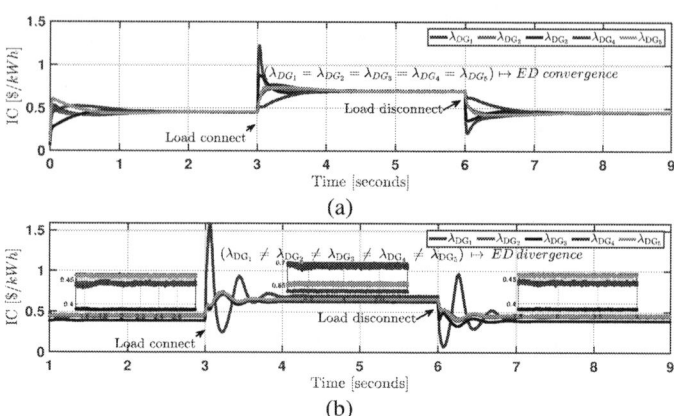

(a)

(b)

Fig. 8: IC response for the five DGs for load changes controlled based on: (a) ED-VSG+DLBIC, (b) ED-VSG+Integral.

IV. PERFORMANCE VALIDATION

The DLBIC's effectiveness in restoring frequency and achieving ED is validated using time-domain simulations and control-in-loop (CIL) OPAL-RT experiments. The IEEE 38-node system (Fig. 6), with five IBDGs totaling 3.5 MVA and a load of 2.78 MW at 0.85 lagging power factor, serves as the test system [21]. A 1.1 MVA switchable bulk load is connected at t=3 seconds and disconnected at t=6 seconds to assess the performance. Fig. 7 shows the DLBIC quickly restoring frequency to its nominal value without overshooting. Unlike the integral approach [3], the DLBIC mitigates frequency nadir, dampens oscillations, and achieves ED (Fig. 8a). During transients, IC value unification is briefly lost but re-equalizes within milliseconds, demonstrating the controller's fast dynamic ED capability. In contrast, the integral controller fails to maintain ED due to disparate IC values (Fig. 8b).

The CIL real-time simulator (RTS) validates the practical execution of the DLBIC in 1) restoring the frequency of the IMG and 2) facilitating ED by unifying IC values. The CIL setup comprises the OPAL-RT OP5707XG RTS, a PC equipped with RT-LAB software (V.2024.3), and a digital oscilloscope for capturing analog output waveforms, as illustrated in Fig. 9. The OP5707XG unit has eight analog output pathways to oversee IMG frequency deviation (ΔF), RoCoF, voltage deviation (ΔV), and IC among the five IBDGs. The proposed model accepts two analog inputs: one from local measurement and one from the nearest IBDG. The controller was assessed on the IEEE 38-node IMG to analyze load connect and disconnect. Fig. 10 demonstrates eliminating ΔF and equalizing IC values for the five IBDGs in real-time, ensuring both ED and frequency regulation.

V. CONCLUSION

This research proposes a distributed low-bandwidth intelligent controller to restore the IMG frequency, preserve the ED, and provide virtual inertia. The proposed distributed controller uses a low-bandwidth communication channel with few communication links, thus reducing communication costs and improving system reliability and stability. The proposed approach demonstrates significant advantages over traditional controllers regarding computational efficiency, practical applicability, and adaptability. Moreover, its model-free structure allows for handling IMG complexity, nonlinearity, and uncertainty.

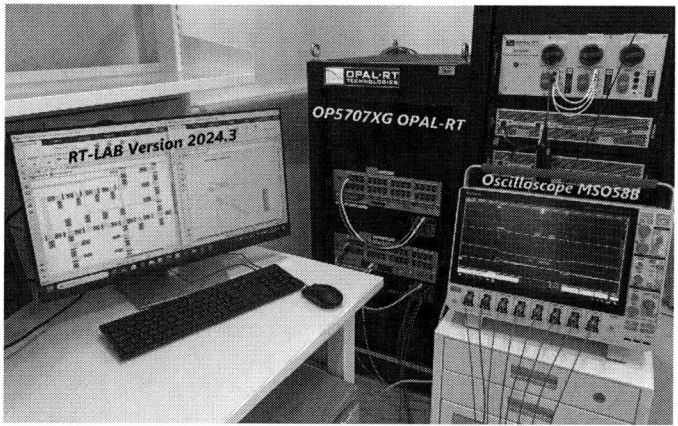

Fig. 9: Real-time execution using the OPAL-RT setup.

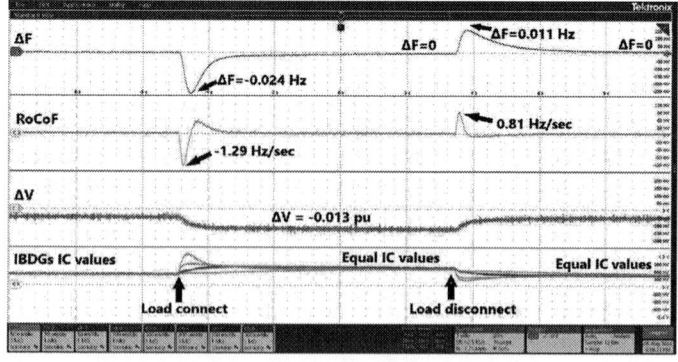

Fig. 10: ΔF, RoCoF, ΔV, and IC values responses for the proposed ED-VSG+DLBIC for load changes.

979-8-3315-1612-3/25 $31.00 © 2025 IEEE   627

REFERENCES

[1] S. E. Sati, A. Al-Durra, H. Zeineldin, T. H. El-Fouly, and E. F. El-Saadany, "Adaptive virtual inertia and damping for frequency stability enhancement using a seamless compensator," in *2023 IEEE PES Conference on Innovative Smart Grid Technologies-Middle East (ISGT Middle East)*. IEEE, 2023, pp. 1–5.

[2] N. Sarrafan, M.-A. Rostami, J. Zarei, R. Razavi-Far, M. Saif, and T. Dragičević, "Improved distributed prescribed finite-time secondary control of inverter-based microgrids: Design and real-time implementation," *IEEE Transactions on Industrial Electronics*, vol. 68, no. 11, pp. 11 135–11 145, 2021.

[3] K. Jiang, H. Su, H. Lin, K. He, H. Zeng, and Y. Che, "A practical secondary frequency control strategy for virtual synchronous generator," *IEEE Trans. Smart Grid*, vol. 11, no. 3, pp. 2734–2736, 2020.

[4] S. E. Sati, A. Al-Durra, H. H. Zeineldin, T. H. EL-Fouly, and E. F. El-Saadany, "Decentralized frequency restoration and stability enhancement for virtual synchronous machines at economic dispatch in islanded microgrid," *Applied Energy*, vol. 377, p. 124544, 2025.

[5] Y. Xu, H. Sun, W. Gu, Y. Xu, and Z. Li, "Optimal distributed control for secondary frequency and voltage regulation in an islanded microgrid," *IEEE Trans. Ind. Informat.*, vol. 15, no. 1, pp. 225–235, 2018.

[6] Y. Shan, J. Hu, K. W. Chan, and S. Islam, "A unified model predictive voltage and current control for microgrids with distributed fuzzy cooperative secondary control," *IEEE Trans. Ind. Informat.*, vol. 17, no. 12, pp. 8024–8034, 2021.

[7] C. Liu, X. Wang, Y. Ren, X. Wang, and J. Zhang, "A novel distributed secondary control of heterogeneous virtual synchronous generators via event-triggered communication," *IEEE Trans. Smart Grid*, vol. 13, no. 6, pp. 4174–4189, 2022.

[8] G. Zhao, L. Jin, H. Cui, and Y. Wang, "Distributed dynamic event-triggered secon- dary control for islanded microgrids with disturbances and communication de- lays: A hybrid systems approach," *IEEE Trans. Ind. Informat.*, vol. 19, no. 8, pp. 8795–8805, 2023.

[9] Y. Chen, K.-W. Lao, D. Qi, H. Hui, S. Yang, Y. Yan, and Y. Zheng, "Distributed self-triggered control for frequency restoration and active power sharing in islanded microgrids," *IEEE Trans. Ind. Informat.*, 2023.

[10] S. Deng, L. Chen, X. Lu, T. Zheng, and S. Mei, "Event-based distributed frequency control in harsh communication conditions," *IEEE Trans. Ind. Informat.*, vol. 18, no. 6, pp. 3777–3786, 2022.

[11] M. Shi, X. Chen, J. Zhou, Y. Chen, J. Wen, and H. He, "Frequency restoration and oscillation damping of distributed vsgs in microgrid with low bandwidth communication," *IEEE Trans. Smart Grid*, vol. 12, no. 2, pp. 1011–1021, 2021.

[12] W. Kang, Y. Guan, YunYu, B. Arbab-Zavar, J. C. V. Q., and J. M. Guerrero, "Distributed event-triggered secondary frequency regulation by sharing hess power in microgrids," *IEEE Trans. Smart Grid*, pp. 1–1, 2024.

[13] S. E. Sati, A. Al-Durra, H. Zeineldin, T. H. EL-Fouly, and E. F. El-Saadany, "A novel virtual inertia-based damping stabilizer for frequency control enhancement for islanded microgrid," *International Journal of Electrical Power & Energy Systems*, vol. 155, p. 109580, 2024.

[14] S. E. Sati, M. B. Abdelghany, B. Hamad, A. Al-Durra, H. Zeineldin, T. H. EL-Fouly, and E. F. El-Saadany, "Economic power-sharing and stability enhancement for virtual synchronous generators in islanded mg," *IEEE Transactions on Power Systems*, pp. 1–17, 2024.

[15] H. Xin, L. Zhang, Z. Wang, D. Gan, and K. P. Wong, "Control of island ac microgrids using a fully distributed approach," *IEEE Trans. Smart Grid*, vol. 6, no. 2, pp. 943–945, 2015.

[16] Z. Li, Z. Cheng, J. Liang, J. Si, L. Dong, and S. Li, "Distributed event-triggered secondary control for economic dispatch and frequency restoration control of droop-controlled ac microgrids," *IEEE Trans. Sustain. Energy*, vol. 11, no. 3, pp. 1938–1950, 2020.

[17] E. Espina, R. J. Cárdenas-Dobson, J. W. Simpson-Porco, M. Kazerani, and D. Sáez, "A consensus-based distributed secondary control optimization strategy for hybrid microgrids," *IEEE Trans. Smart Grid*, vol. 14, no. 6, pp. 4242–4255, 2023.

[18] Z. Li, Z. Cheng, J. Si, and S. Li, "Distributed event-triggered hierarchical control to improve economic operation of hybrid ac/dc microgrids," *IEEE Trans. Power syst.*, vol. 37, no. 5, pp. 3653–3668, 2022.

[19] Y. Shan, J. Hu, and B. Shen, "Distributed secondary frequency control for ac microgrids using load power forecasting based on artificial neural network," *IEEE Trans. Ind. Informat.*, vol. 20, no. 2, pp. 1651–1662, 2024.

[20] H. Xin, R. Zhao, L. Zhang, Z. Wang, K. P. Wong, and W. Wei, "A decentralized hierarchical control structure and self-optimizing control strategy for f-p type dgs in islanded microgrids," *IEEE Trans. Smart Grid*, vol. 7, no. 1, pp. 3–5, 2016.

[21] D. Singh, R. K. Misra, and D. Singh, "Effect of load models in distributed generation planning," *IEEE Trans. Power syst.*, vol. 22, no. 4, pp. 2204–2212, 2007.

Hardware-in-the-Loop of a Grid Forming Control Strategy Applied to a DC Off-Grid Green Hydrogen Production System

Diego Montoya-Acevedo
Facultad de Ingenierías
Universidad de Talca
Curicó, Chile
diego.montoya@utalca.cl

René Contreras-Barrios
Facultad de Ingenierías
Universidad de Talca
Curicó, Chile
rcontreras18@alumnos.utalca.cl

Angel Maureira-Riquelme
Facultad de Ingenierías
Universidad de Talca
Curicó, Chile
amaureira15@alumnos.utalca.cl

Esteban Ibáñez-Muñoz
Facultad de Ingenierías
Universidad de Talca
Curicó, Chile
eibanez18@alumnos.utalca.cl

Catalina Gonzalez-Castaño
Energy Transformation Center
Universidad Andres Bello
Santiago de Chile, Chile
catalina.gonzalez@unab.cl

Carlos Restrepo
Facultad de Ingenierías
Universidad de Talca
Curicó, Chile
crestrepo@utalca.cl

Abstract—Green hydrogen production has become an attractive solution to reduce the effects of climate change using renewable energy systems because it is a decarbonization vector. To produce green hydrogen efficiently from an isolated direct current microgrid, it is necessary to hybridize the system with different energy storage devices like batteries and supercapacitors to ensure continuous operation of the electrolyzer when the Photovoltaic energy source is not available. This paper presents a grid-forming control strategy applied to an isolated DC microgrid. Extensive hardware-in-the-Loop implementations are carried out under different operating conditions, showing a reliable and continuous operation. Results show that it is possible to form a DC grid that maintains the power conditions of the electrolyzer constant to maximize its utilization rate under photovoltaic power variations with the backup of battery and supercapacitor systems.

Index Terms—grid forming, DC microgrids, hydrogen production, energy management, HIL.

I. INTRODUCTION

The energy transition and green hydrogen production are relevant challenges in the global commitment to respecting sustainable development. This transition has brought a paradigm change in the generation energy system, proposing new configurations such as isolated microgrids harnessing renewable resources as solar radiation and wind strength far from the consumption centers. These isolated microgrids allow the diversification of new processes like hydrogen generation from these renewable sources [1]. Hydrogen is an energetic vector that can be produced through water electrolysis without harming the environment [2]. Despite the environmental benefits, one of the challenges is their economic viability; it is estimated that the costs would reduce by around 60% to 80 % as the demand for hydrogen and renewable energy production increases, and it is necessary to explore novelty and scalable configurations to enhance the production efficiency and reduce

the capital costs (CAPEX) and operational costs (OPEX) [3]. In the literature and commercially, different configurations of On-Grid and Off-Grid green hydrogen production exists [4], [5]; however, these approaches present inefficiencies due to multiple conversion stages from direct current (DC) to alternating current (AC), which makes the process redundant, and the power levels implemented are very high [6], [7]. photovoltaic (PV) energy shows a high potential for this application; however, due to its intermittent nature, it is necessary to use different energy storage systems as batteries (B) and supercapacitors (SC) to meet the power demand in slow and fast dynamic response connecting the electrolyzer (Elz) to them forming a hybrid isolated microgrid. These elements operate with different dynamics and multiple frequency rates. When the system is emulated in a typical simulation platform as Simulink, Plecs, or PSim, these dynamics can not be developed accurately, and the time representation is not based in real-time. Another scope is that the development of different control strategies in real equipment can be expensive and risky. Therefore, the hardware-in-the-loop (HIL) is presented as a cost-effective and advantageous simulation platform that guarantees real-time operation, allowing the development of novel control techniques before real equipment implementation. Performing an isolated microgrid simulation for green hydrogen production requires a high level of computing due to the modeling of all the elements that comprise it. Within this scope, various platforms allow HIL to be performed. Some authors developed many microgrid HIL simulations to evaluate the behavior of diverse controllers [8] that evaluate a HIL simulation in an OPAL-RT, with different levels of a hierarchical control based on energy management operation under islanded and grid-connected modes. Other authors use less cost-expensive options, which use a DSP to evaluate the performance of DC-DC and DC-AC converters that demon-

979-8-3315-1612-3/25 $31.00 © 2025 IEEE

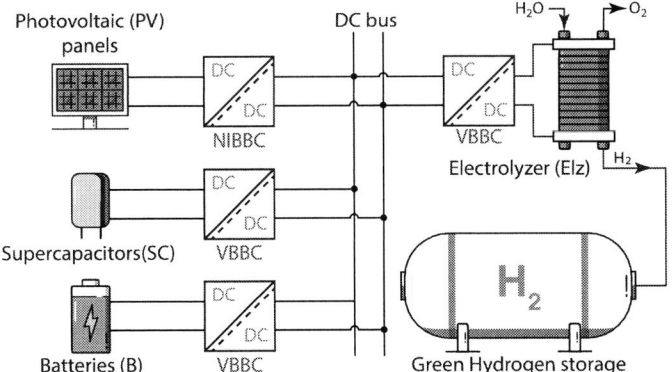

Fig. 1. Hybrid DC Off-grid microgrid to green hydrogen production.

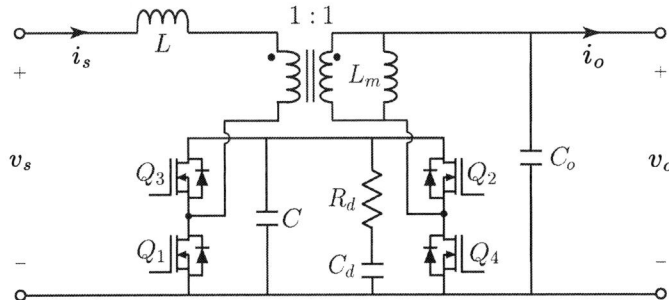

Fig. 2. Dynamic model Versatile Buck-Boost Converter.

strate a high fidelity with real systems [9], [10]. Different control strategies are evaluated in HIL applied to different microgrids [11]–[13]. In [14], the authors propose a household solution implemented in a power-hardware in the loop (PHIL) integrating B, SC, Elz, and fuel cells (FC), with the purpose of better using PV production with a model predictive control (MPC) with classical DC-DC converter topologies. This approach found good performance in energy management to produce and consume hydrogen; however, the battery charge and discharge cycles are affected considerably by the lack of direct control of the energy shared by the SC, reducing the lifetime of these devices. The validation of a DC grid-forming control scheme applied to green hydrogen production has yet to be validated. Elz requires a constant DC voltage supply to produce hydrogen, which is possible by implementing a hierarchical control that forms the grid, ensures the power supply during their operations, and maximizes the utilization rate. An option classified as accessible and efficient to simulate the microgrid using HIL is through an RT Box, which has high computational power and is also affordable. The challenges in the microgrid simulation, as was mentioned before, come from the different dynamics of the devices that compose it because not all respond similarly. For example, the battery has a slower response because it has the function of providing power over a long time, while the supercapacitor is capable of delivering its power in the order of milliseconds, providing support to the microgrid. On the other hand, photovoltaic generation presents intermittent power delivery. The implementation of converters, which have high switching frequency rates, is also considered, which is the reason that the computational cost of the implementation is relatively high.

This paper proposes for the first time a grid-forming control strategy with extensive HIL implementations in two RT Box platforms of an isolated microgrid for green hydrogen production. The aim is to evaluate the performance of the different models developed in real-time, considering the state of charge of the B and SC. The paper is organized as follows: first, a complete description of the microgrid model for green hydrogen production is presented in Section II. Grid-Forming to DC microgrid is described in Section III, then, HIL system description is given in Section IV. Later, HIL implementation and results of the Grid-Forming control over the DC microgrid are shown in Section V. Finally, conclusions and future work are given.

II. SYSTEM DESCRIPTION

The configuration of the DC microgrid is presented in Fig. 1; this configuration comprises a PV system as the primary source of energy, an SC system to attend to the demand in fast dynamics responses from the PV or the load, a B energy system to low dynamics of the system and steady-state operation, and an Elz to produce hydrogen with a hydrogen storage tank, these elements are connected to a DC bus through DC-DC power electronic converters. The PV system is connected by a classical unidirectional non-inverting buck-boost converter (NIBBC), and the other devices are connected with a novelty DC-DC bidirectional topology called the versatile buck-boost converter (VBBC), shown in Fig. 2. This converter has significant advantages over the conventional approach. Their primary characteristics highlight the bandwidth to voltage regulation, scalable power processing, and an improvement in conversion efficiency. In addition, the VBBC offers a smooth transition between the buck and boost operation modes, which makes it versatile in diverse applications. This power converter can implement reconfigurable control methods that allow optimal adaptation under different operating conditions and specific photovoltaic system needs or Elz [14], [15]

A. PV system

The operation and design of PV systems are based on their electrical characteristics, especially the current-voltage relationship of the cells, which varies according to irradiance and temperature. This evaluation facilitates the modeling of PV system behavior in either an idealized or detailed manner. Ideal models simplify physical processes and enable quick performance assessments, making them attractive to engineers and researchers focused on renewable energy and power electronics. On the other hand, detailed models incorporate complex processes such as charge carrier generation and recombination, shading effects, and internal resistance, providing a highly accurate representation of the functioning of photovoltaic cells.

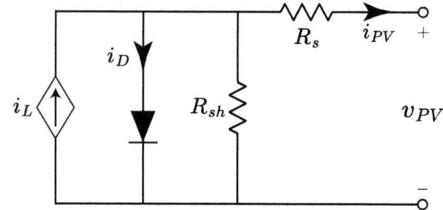

Fig. 3. Five parameters model of PV modules.

For this research, a five-parameter model is used to capture the PV system dynamics as a function of irradiance (R) and temperature (T) [16]. This model includes a current source, a diode, a series resistance, and a shunt resistance, providing an accurate representation of the dependence of photovoltaic modules on irradiance and temperature, as well as their current-voltage behavior. The schematic of this model is presented in Fig. 3. The current (i_{PV}) and voltage (v_{PV}) of the photovoltaic panel are described by the following equations:

$$i_{PV} = i_L - i_D - i_{R_{sh}} - i_{R_s} \quad (1)$$

$$v_{PV} = \left(k\frac{T}{q}\right) nIN_{cell} \quad (2)$$

where i_L is the current generated by the current source as a function of the incident irradiance, and i_D, i_{Rsh}, and i_{Rs} represent the currents through the diode, shunt resistance, and series resistance, respectively, all influenced by voltage and temperature. Here, k denotes the Boltzmann constant, q the electron charge, T the panel temperature in Kelvin, nI the diode ideality factor, and N_{cell} the number of cells in series. The model used for the PV system is idealized, which helps save computational costs associated with more detailed models.

B. Battery and supercapacitor

The energy storage system using batteries in this microgrid is based on a simplified model that represents the battery's behavior as a controlled voltage source, where E_B is in series with constant resistance, R_B Fig. 4. This model emulates the battery's charge and discharge characteristics [17], with V_B defined by the equation:

$$V_B = V_{B,o} - K\left(\frac{Q}{Q - i_B(t)}\right) + Ae^{(-B \int i_B(t)\,dt)} \quad (3)$$

In this model, $V_{B,o}$ represents the open-circuit voltage of the battery, Q denotes the battery capacity in ampere-hours (Ah), and K is the polarization voltage, accounting for internal losses during charging and discharging. The parameter A represents the amplitude associated with the exponential region of the battery's voltage response, while B serves as the time constant for this region, measured in Ah^{-1}. Finally, $i_B(t)$ is the internal current of the battery at a given time t. This formulation captures the essential aspects of the battery's charge and discharge cycle.

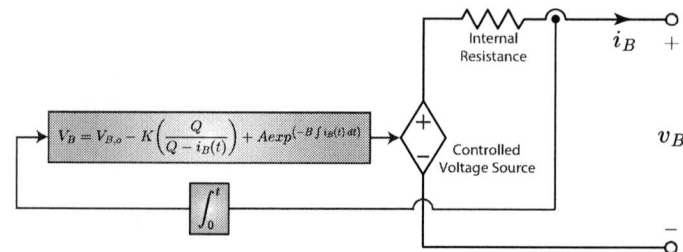

Fig. 4. Model of Battery Energy Storage System.

In a B system, one of the main parameters is the state of charge (SOC_B). Generally, the SOC is defined as the actual energy stored E_B divided by the nominal energy capacity $E_{B,nom}$ as is presented in 4. $E_{B,nom}$ is the maximum amount of energy that is allowable energy batteries that can be stored in the system and is given by the manufacturer.

$$SOC_B = \frac{E_B}{E_{B,nom}} \quad (4)$$

The main function of an SC is to compensate for sudden power fluctuations from PV and load, in this sense represented by the Elz. The SC prevents high charging and discharging currents of B, contributing to a robust and stable operation. The supercapacitor energy storage system is modeled through a resistor-capacitor (RC) network with a voltage-dependent capacitance like Fig. 5. This model specifically focuses on energy storage, allowing transient response to be neglected [18]. The supercapacitor's capacitance is nonlinear and defined as $C = \frac{Q}{V}$. When the voltage across the capacitance is zero, the supercapacitor has an initial capacitance C_0. As the voltage increases, the capacitance grows approximately linearly, modeled by:

$$C(v_{SC}) = C_0 + k_v v_{SC} \quad (5)$$

where k_v represents the slope of the supercapacitor's characteristic curve [19].

This device must operate with limits to the SOC_{SC}; this parameter can be considered as an operation band defined where minimum and maximum values are established to ensure a proper operation. Fig. 6 describes the operation zones

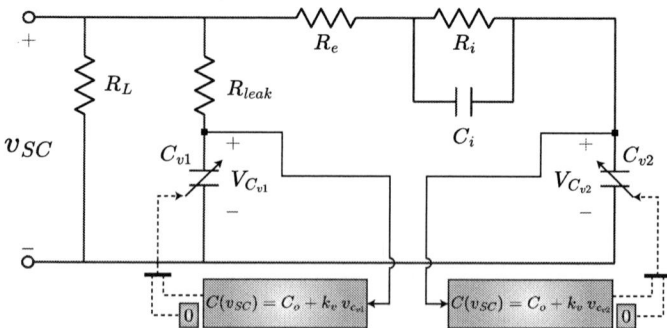

Fig. 5. Supercapacitor Model.

979-8-3315-1612-3/25 $31.00 © 2025 IEEE

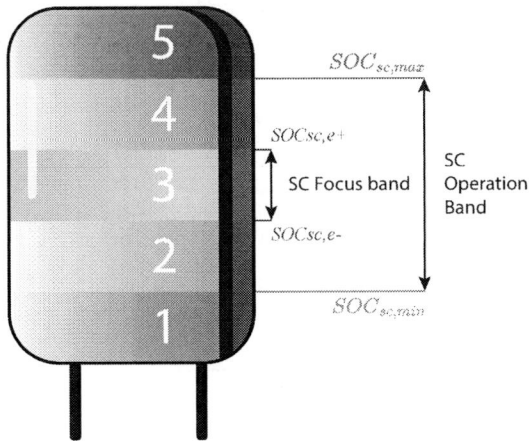

Fig. 6. Operation regions of SOC_{SC}.

of the SOC_{SC}. In region 1, the SC is in an over-discharge condition, where $SOC_{SC} < SOC_{SC,min}$. In this condition, the system cannot supply energy to the grid. The $SOC_{SC,min}$, represents the acceptable minimum charge value of the SC. On the other hand, in Region 5, the SC is over-charged, which means $SOC_{SC} > SOC_{SC,max}$, which implies that the system is not capable of absorbing the energy of the microgrid. In this case, $SOC_{SC,max}$ indicates the maximum charge value of the SC. Regions 2 to 4 correspond to the SC's optimal operation areas, which can absorb and supply energy from and to the microgrid, respectively. These regions are ideal for proper operation and will be used for tertiary control strategies in future works.

C. Electrolyzer

Proton exchange membrane (PEM) electrolyzers are essential for producing green hydrogen from renewable energy sources, using a proton-conducting polymer membrane to enable the electrolysis process, which separates hydrogen and oxygen molecules [20]. For this research, a model was established that incorporates electrochemical conversion, the applied electric current i_{Elz}, and electrolyte concentration to accurately estimate hydrogen production, with system temperature and pressure being critical factors for optimizing electrolyzer performance.

The model is based on Butler-Volmer kinetics at the electrodes and transport resistance within the polymer electrolyte [21]. Designed for temperatures between 20–80°C, it represents PEM electrolyzer operation as a variable load system, where output voltage depends on input current, temperature, and internal factors such as cell dimensions and properties. This simplification facilitates extended simulations. When current is applied to the electrodes of the PEM electrolytic cell, the operating voltage V_{cell} is defined by the reversible voltage U_{rev}:

$$
\begin{aligned}
U_{rev}(T) = 1.1584 &- 1.5421 10^{-3} T \\
&+ 9.523 10^{-5} T \ln(T) \\
&+ 9.84 10^{-8} T^2
\end{aligned} \tag{6}
$$

$$
\begin{aligned}
V_{cell} =& U_{rev} + \frac{RT}{2\alpha_a F} \ln\left(\frac{i}{i_{o,a}}\right) \\
&+ (R_{ele} + R_{ion})\, i + \frac{RT}{2\alpha_c F} \ln\left(1 - \frac{i}{i_{o,c}}\right)
\end{aligned} \tag{7}
$$

where R is the universal gas constant, F is Faraday's constant, and α_a and α_c are the charge transfer coefficients at the anode and cathode, respectively. i represents the electrolyzer current, while i_a and i_c are the currents at the anode and cathode. Due to limited manufacturer data, this semi-empirical model was developed using Multiobjective Particle Swarm Optimization (PSO) for parameter estimation from experimental data, simplifying (8) to:

$$
V_{cell} = a\left(U_{rev} + \frac{RT}{F} \ln\left(\frac{i_{Elz}}{d}\right) + c\frac{i_{Elz}}{d}\right) \tag{8}
$$

and representing the hydrogen flux H_2 as:

$$
H_2 = ae\frac{i_{Elz}}{d}\frac{1}{2F}\exp\left(\frac{f+gT}{i_{Elz}/d}\right) \tag{9}
$$

where a is the number of cells, b corresponds to $i_{o,a}$, c represents the equivalent resistance $R_{eq} = R_{elec} + R_{ion}$, d is the effective electrolyte area, and e, f, and g denote current density and temperature dependencies. Enhanced by real-time optimization, this approach provides accurate estimates of the electrolyzer's terminal voltage and hydrogen flow.

III. Grid-forming Control

The primary objective of the DC microgrid is to maximize the utilization rate of the electrolyzer (Elz) for continuous hydrogen production, despite fluctuations in photovoltaic (PV) power and the limited energy storage capacity of battery (B) and supercapacitor (SC) systems. Since electrolyzers require a constant voltage to maintain the electrochemical reaction and high current levels to optimize hydrogen production, the microgrid must provide high reliability and adaptability in various operational scenarios derived from the energy sources. The Grid-Forming approach consists of three stages: Primary Control, Droop Control, and Secondary Control, interconnected through a centralized communication scheme, as shown in Fig. 7. In this configuration, the battery controller operates in a grid-forming role, while the supercapacitor controller functions in a grid-following role, with the primary control references depending on battery performance. The PV system incorporates Maximum Power Point Tracking (MPPT) control to maximize renewable energy capture.

Primary control oversees all generating units and loads within a centralized system, ensuring rapid responses to changes in demand or generation and adjusting the outputs of the battery and PV voltage bus converter (VBBC) as needed. Algorithms

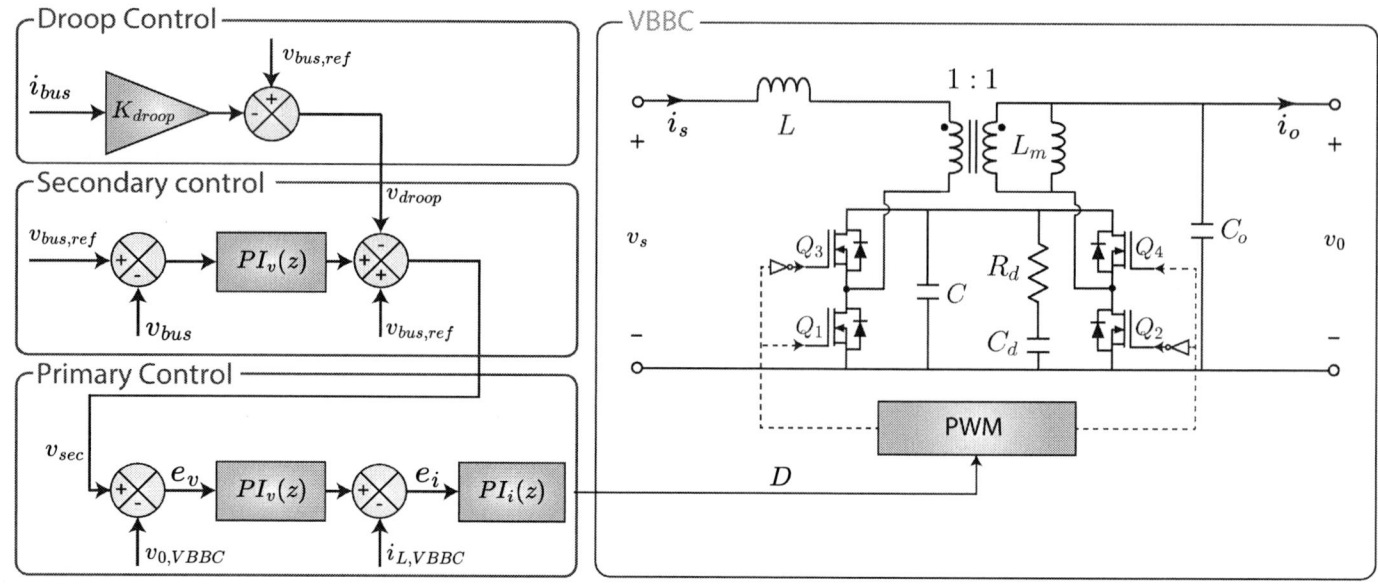

Fig. 7. Grid-Forming control approach to Off-Grid microgrid for green hydrogen production.

optimize system stability and efficiency, while droop control, though decentralized, integrates with the centralized scheme to enhance stability and load distribution by simulating voltage drop characteristics and balancing the electrolyzer load among PV, battery, and SC sources [22]. Secondary control finely tunes the system by making precise adjustments to restore nominal operating values, such as voltage levels, after disturbances [23]. It optimizes load distribution and adjusts the PV and converter references, enhancing microgrid efficiency and prolonging equipment lifespan through real-time monitoring and continuous adjustments.

IV. HIL SYSTEM DESCRIPTION

The different models presented in Section III of the isolated DC microgrid to green hydrogen production are tested under the Grid-Forming control strategy to maximize the utilization rate of the Elz. A HIL simulation was implemented; this approach guarantees a real-time operation with a precise representation of the different dynamics of the devices. Models were developed in PLECS, a specialized power electronics platform that allows the modeling and simulation of entire systems like the microgrid evaluated in this paper [24]. The HIL simulation was implemented in two RT Box connected through a Small Form-Factor Pluggable protocol (SFP) [25]. The measured variables were implemented as scalable analog outputs to be shown in the oscilloscope. Fig. 8, shows the distribution of the DC Off-grid microgrid to green hydrogen production. RT Box 1 contains the B and SC models with the respective VBBC, operating as grid forming and grid following, respectively. Hierarchical control presented in Fig. 7 is also implemented here. The output voltages and currents of B and SC are sent with SFP and established in a coupling circuit in RT Box 2, which contains the PV system with the MPPT algorithm and the Elz.

V. HIL IMPLEMENTATION RESULTS

To validate the Grid-Forming control scheme in this paper, a DC microgrid with a PV rated at 5.4 kW, 53 V on the DC bus, is considered the main parameter of the microgrid, as listed in Table I. The simulation time duration is 2000 s to validate the behavior of the SOC of B and SC; the sample time of the system is $T_S = 5\mu s$. Extensive time-domain simulations and HIL implementation of the Grid-Forming control approach are carried out with an irradiance scenario that changes its value every 500 s, as is presented in Fig. 9. It can be observed that each step of power change presents a fast dynamic of the SC as expected, absorbing high levels of current protecting B while SOC is reached. The tunned parameters of the control stage are presented in [26].
Fig. 10 (a) presents the power and SOC behavior of the battery during the simulation. The charging and discharging cycles of the battery are remarkable when PV power variations are made. Similarly, Fig. 10 (b) depicts the power and SOC behavior. SOC_{SC} accomplished the objective of operating in a focus band lower than the $SOC_{SC,max}$ to be capable of handling the absorption or delivery of high levels of power as was presented in Section II. The formed grid to the Elz is presented in Fig. 11. The control strategy establishes the DC

TABLE I
DC Microgrid Parameters

Device	Ref	Value	Unit	Amount	Power
Panasonic (PV)	EVPV380	380	W	14	5320 W
Batteries (B)	RS PRO	1200	Ah	4	4800 W
Cornell (SC)	DCMC454	0.450	F	1	1200 W
ENAPTER (Elz)	EL4.0 DC	2400	W	1	2400 W

979-8-3315-1612-3/25 $31.00 © 2025 IEEE

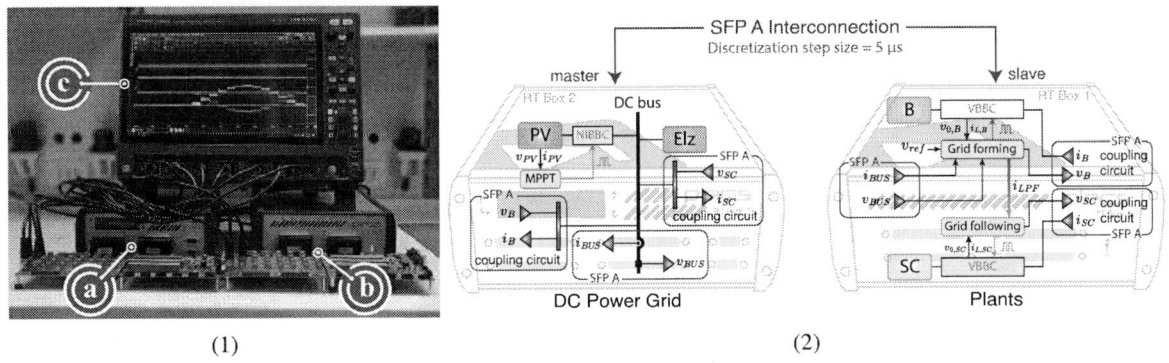

Fig. 8. HIL of DC Off-Grid microgrid to green hydrogen production. (1): HIL setup, (a) RT Box 2 with SFP. (b) RT Box 1 with SFP. (c) Oscilloscope. (2) SFP diagram connection.

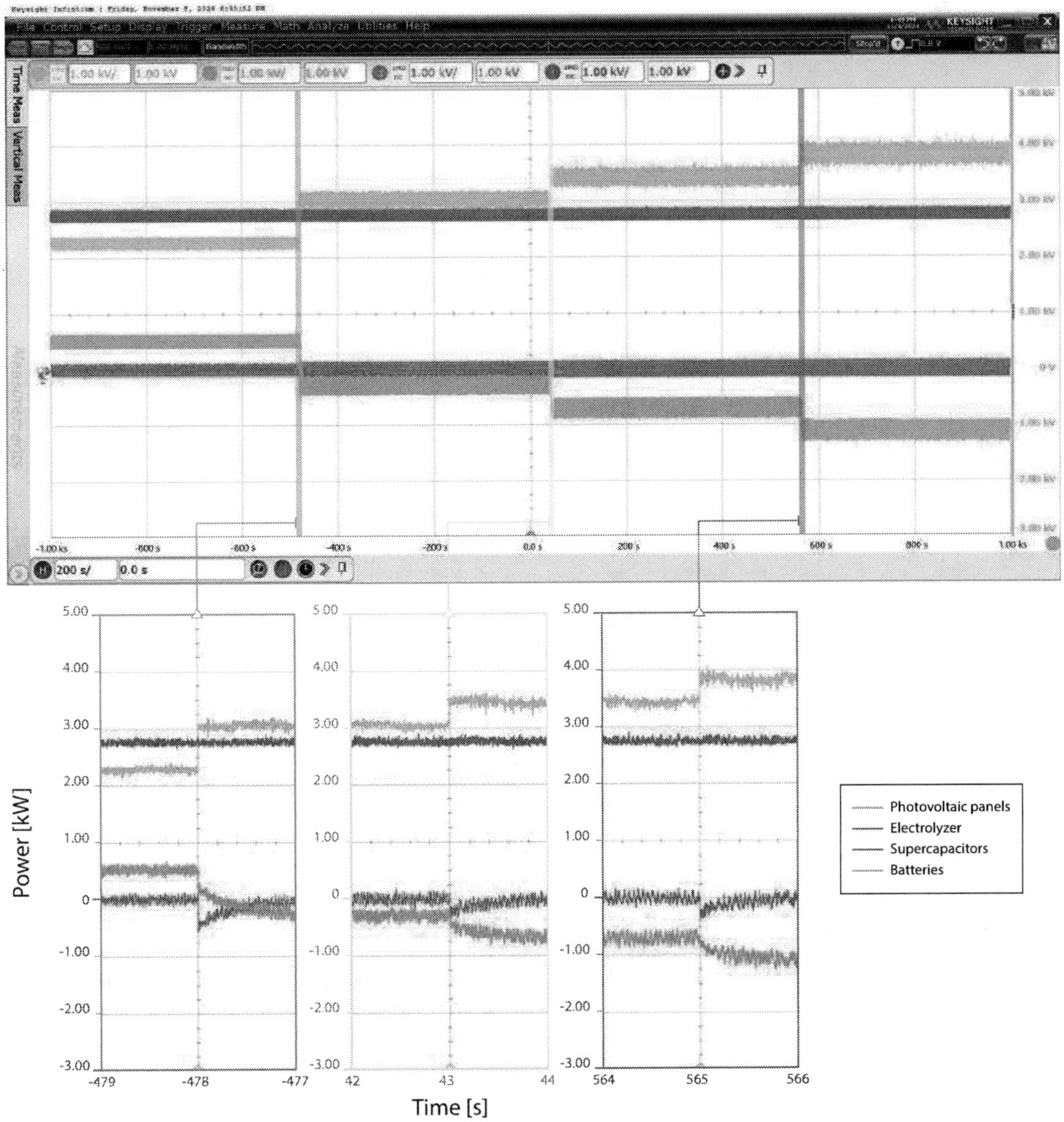

Fig. 9. HIL simulation results of devices power behavior: CH1: PV power (1 kV/div), CH2: B power (1 kV/div), CH3: SC power (1 kV/div), CH4: Elz power (1 kV/div).

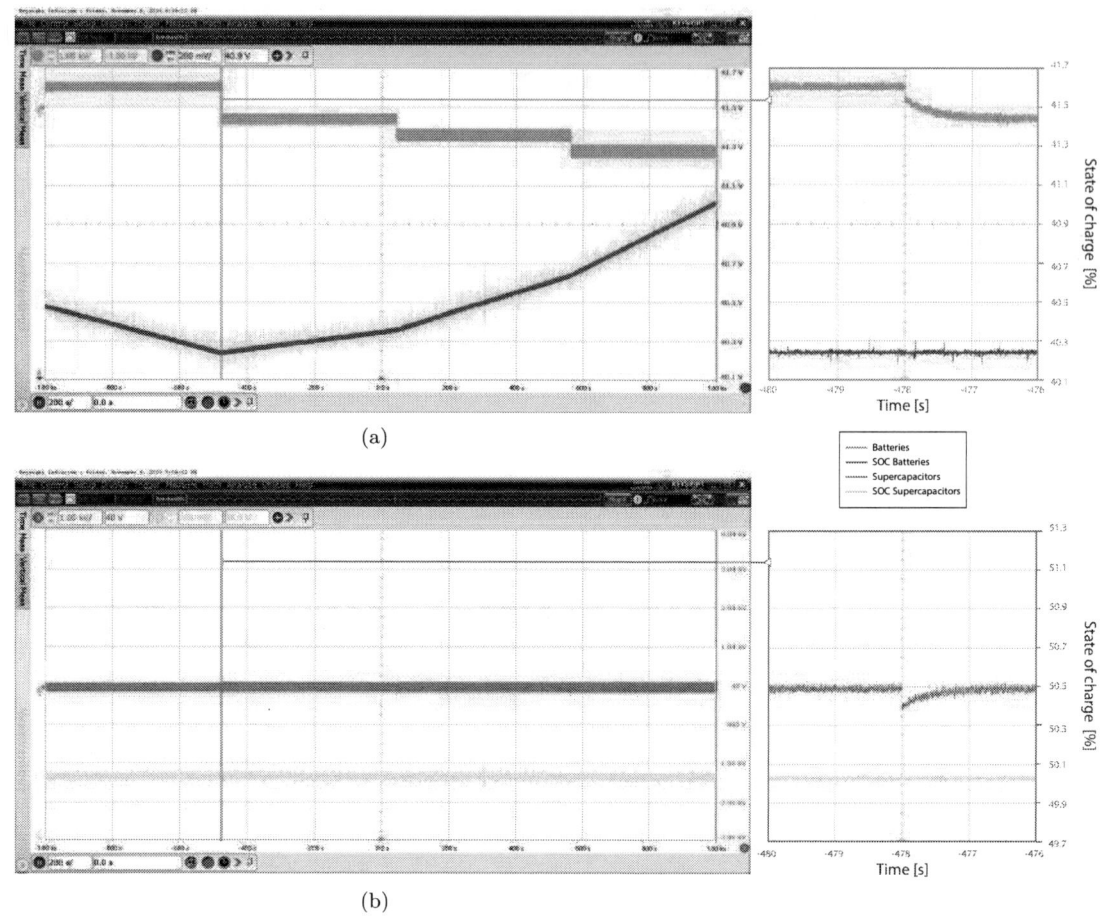

(a)

(b)

Fig. 10. HIL simulation results B and SOC power and state of charge: (a) CH2: B power (1 kV/div), CH7: B SOC (200 mV/div), (b) CH2: SC power (1 kV/div), CH7: SC SOC (200 mV/div).

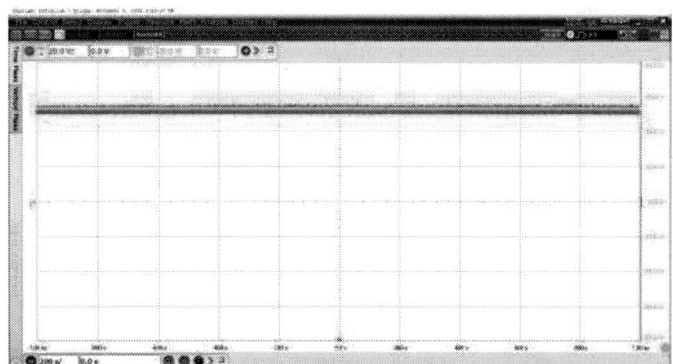

Fig. 11. HIL simulation results of DC bus current and voltage: CH5: DC bus voltage (20 V/div), CH6: Elz current (20 V/div).

bus voltage at 55 V, which is the necessary voltage to feed the line impedances and the Elz nominal voltage. It is important to remark that this value is constant during all simulation times, independently of the above PV power variations. The ELz produces 1287 normalized liters of hydrogen (Nl) that, according to its molar mass at standard temperature and

pressure conditions (1 atm, 0°), which is equivalent to 0.0899 g/L, produce 0.115 kg of hydrogen at a constant rate.

Grid-forming control results show that the control references used in the primary control of the B and SC operations are adjusted according to PV variations. When the PV power changes from 2.2 kW to 3 kW, the stabilization time of B is 2.268 s, and the SC is 616 ms. Finally, the SOC_B discharging rate is 2.2 %, and the charging rate is 1.87 % during the simulation time. These values indicate the proper operation of the models proposed in this paper and validate the real-time execution.

Conclusions and future work

This work proposes a model and a HIL real-time simulation for an off-grid microgrid with green hydrogen production and a Grid Forming Control approach to ensure constant power levels to the Elz to maximize its utilization rate and maximize the green hydrogen production. The microgrid comprises the following devices: PV, B, SC, and Elz, which work as DC-load. This model was developed by analyzing the dynamics of the devices in response to a specific DC load and considering the variability of solar resources, as well as the SOC behavior of B and SC. Initial HIL results show the desired DC Power and de DC bus voltage necessary to operate the Elz and maximize their utilization rate with a reliable operation of all the devices involved in the microgrid when the irradiance profile changes abruptly; it is remarkable the fast response of the SC when fast dynamics of the system are required to protect the lifetime of B and Elz. In future work, a tertiary layer will be included in the hierarchical control of the grid-forming scheme to optimize in an economical way the behavior of the devices, including the variations of the SOC of B and SC. Also, a hydrogen fuel cell will be considered to increase the system's reliability.

ACKNOWLEDGMENT

This work was supported in part by Doctoral Program in Engineering Systems at la Universidad de Talca under the Foreign Students Scholarship N° 202349810, the Thematic Network RIBIERSE-CYTED (723RT0150) and in part by the Chilean Government under projects ANID/FONDEQUIP EQM170054/, ANID/FONDECYT/3220126, ANID/FONDECYT/1231015, SERC Chile ANID/FONDAP/1522A0006, and the Millenium Institute on Green Ammonia as Energy Vector MIGA (ANID/Millennium Science Initiative Program/ICN2021 023.

REFERENCES

[1] International Energy Agency, "Global Energy and Climate Model Documentation – 2023," 2023, international Energy Agency, Paris, France. [Online]. Available: https://www.iea.org/

[2] International Renewable Energy Agency, "Global hydrogen trade to meet the $1.5°C$ climate goal: Part III – Green hydrogen cost and potential," 2022, international Renewable Energy Agency, Abu Dhabi.

[3] World Energy Council, "Working Paper: Hydrogen Demand and Cost Dynamics," 2021, world Energy Council, United Kingdom.

[4] V. Samano-Ortega, H. Rodriguez-Estrada, E. Rodríguez-Segura, J. Padilla-Medina, J. Aguilera-Alvarez, and J. Martinez-Nolasco, "Power sharing control in a grid-tied dc microgrid: Controller hardware in the loop validation," *Applied Sciences*, vol. 11, no. 19, 2021. [Online]. Available: https://www.mdpi.com/2076-3417/11/19/9295

[5] L. Li, Y. Han, Q. Li, Y. Pu, C. Sun, and W. Chen, "Event-triggered decentralized coordinated control method for economic operation of an islanded electric-hydrogen hybrid dc microgrid," *Journal of Energy Storage*, vol. 45, p. 103704, 2022. [Online]. Available: https://www.sciencedirect.com/science/article/pii/S2352152X21013785

[6] M. Chen, S.-F. Chou, F. Blaabjerg, and P. Davari, "Overview of power electronic converter topologies enabling large-scale hydrogen production via water electrolysis," *Applied Sciences*, vol. 12, no. 4, 2022. [Online]. Available: https://www.mdpi.com/2076-3417/12/4/1906

[7] M. Jayachandran, R. K. Gatla, A. Flah, A. H. Milyani, H. M. Milyani, V. Blazek, L. Prokop, and H. Kraiem, "Challenges and opportunities in green hydrogen adoption for decarbonizing hard-to-abate industries: a comprehensive review," *IEEE Access*, 2024.

[8] M. N. Tasnim, T. Ahmed, S. Ahmad, and S. Mekhilef, "Hardware in the loop implementation of the control strategies for the ac-microgrid in opal-rt simulator," in *2023 IEEE IAS Global Conference on Renewable Energy and Hydrogen Technologies (GlobConHT)*. IEEE, 2023, pp. 1–6.

[9] R. F. Bastos, G. H. Fuzato, C. R. Aguiar, R. V. Neves, and R. Q. Machado, "Model, design and implementation of a low-cost hil for power converter and microgrid emulation using dsp," *IET Power Electronics*, vol. 12, no. 14, pp. 3833–3841, 2019.

[10] I. Jayawardana and C. N. M. Ho, "A power electronics-based power hil real time simulation platform for evaluating pv-bes converters on dc microgrids," in *2021 IEEE Energy Conversion Congress and Exposition (ECCE)*. IEEE, 2021, pp. 688–693.

[11] M. A. Aslam, S. A. R. Kashif, M. Adeel, M. U. Shahid, M. Iqbal, and M. A. Riaz, "A controller hardware in loop framework for microgrid control applications," in *2023 6th International Conference on Energy Conservation and Efficiency (ICECE)*. IEEE, 2023, pp. 1–5.

[12] V. Samano-Ortega, H. Rodriguez-Estrada, E. Rodríguez-Segura, J. Padilla-Medina, J. Aguilera-Alvarez, and J. Martinez-Nolasco, "Power sharing control in a grid-tied dc microgrid: Controller hardware in the loop validation," *Applied Sciences*, vol. 11, no. 19, p. 9295, 2021.

[13] J. Wachter, L. Gröll, and V. Hagenmeyer, "Adaptive feedforward control for dc/dc converters in microgrids-a power hardware in the loop study," in *2021 9th International Conference on Smart Grid (icSmartGrid)*. IEEE, 2021, pp. 49–56.

[14] F. K/bidi, C. Damour, D. Grondin, M. Hilairet, and M. Benne, "Power management of a hybrid micro-grid with photovoltaic production and hydrogen storage," *Energies*, vol. 14, no. 6, 2021. [Online]. Available: https://www.mdpi.com/1996-1073/14/6/1628

[15] C. Restrepo, C. González-Castaño, and R. Giral, "The versatile buck-boost converter as power electronics building block: Changes, techniques, and applications," *IEEE Industrial Electronics Magazine*, vol. 17, no. 1, pp. 36–45, 2023.

[16] T. N. Olayiwola, S.-H. Hyun, and S.-J. Choi, "Photovoltaic modeling: A comprehensive analysis of the i–v characteristic curve," *Sustainability*, vol. 16, no. 1, p. 432, 2024.

[17] O. Tremblay, L.-A. Dessaint, and A.-I. Dekkiche, "A generic battery model for the dynamic simulation of hybrid electric vehicles," in *2007 IEEE vehicle power and propulsion conference*. Ieee, 2007, pp. 284–289.

[18] F. Naseri, S. Karimi, E. Farjah, and E. Schaltz, "Supercapacitor management system: A comprehensive review of modeling, estimation, balancing, and protection techniques," *Renewable and Sustainable Energy Reviews*, vol. 155, p. 111913, 2022.

[19] S. Karthikeyan, B. Narenthiran, A. Sivanantham, L. D. Bhatlu, and T. Maridurai, "Supercapacitor: Evolution and review," *Materials Today: Proceedings*, vol. 46, pp. 3984–3988, 2021.

[20] B. Yodwong, D. Guilbert, M. Phattanasak, W. Kaewmanee, M. Hinaje, and G. Vitale, "Proton exchange membrane electrolyzer modeling for power electronics control: A short review," *C*, vol. 6, no. 2, p. 29, 2020.

[21] D. Falcão and A. Pinto, "A review on pem electrolyzer modelling: Guidelines for beginners," *Journal of cleaner production*, vol. 261, p. 121184, 2020.

[22] R. Dadi, K. Meenakshy, and S. Damodaran, "A review on secondary control methods in dc microgrid," *Journal of Operation and Automation in Power Engineering*, vol. 11, no. 2, pp. 105–112, 2023.

[23] K. D. R. Felisberto, P. T. de Godoy, D. Marujo, A. B. de Almeida, and R. de Barros Iscuissati, "Trends in microgrid droop control and the power sharing problem," *Journal of Control, Automation and Electrical Systems*, vol. 33, no. 3, pp. 719–732, 2022.

[24] Plexim, "Innovative tools for power electronics simulation," 2024, accessed: Nov. 14, 2024. [Online]. Available: https://www.plexim.com/

[25] J. F. Patarroyo, J. Pfannschmidt, K. Amitkumar, J.-N. Paquin, and W. Li, "Considerations for digital real-time simulation, control-hil, and power-hil in microgrids/der studies," *Microgrids: Theory and Practice*, pp. 579–614, 2024.

[26] D. Montoya, A. Maureira, "Grid-forming control applied to isolated hybrid dc microgrid for green hydrogen production," 2024, accessed: 2024-09-22. [Online]. Available: https://damawinipeg.wordpress.com/investigacion-2/

Experimental Validation of a 40kW, 480V Point-to-Point DC Interlinks for Controller-Agnostic, Interoperable Networked Microgrids

Maximiliano Ferrari, Michael Starke, John Smith, Joao Pereira, Misael Montejano
Oak Ridge National Laboratory Oak Ridge, TN, US.

,

Abstract— This paper presents the experimental validation of point-to-point dc-interlinks for interconnecting two solar-based, laboratory-scale AC microgrids. DC interlinks provide a solution to numerous technical and operational challenges encountered in networked microgrids, including precise power flow control, stable and fast synchronization, enhanced stability, and improved voltage and frequency regulation. By decoupling microgrids through power converters, dc-interlinks enable power exchange among microgrids that can be owned by different entities (such as communities, utilities, or universities) and managed by diverse microgrid controller vendors. This characteristic makes this dc-interlinks a promising solution for networking real-world microgrids. The presented point-to-point dc interlink utilizes two four-quadrant converters: a 40kW 3-phase ac/dc regulating the dc-link voltage to 800V, and a 3-phase dc/ac controlling power flow. These converters, connected via a 20-foot dc cable, interconnect two ac microgrids operating at 480V, each featuring energy storage, photovoltaic generation, and load emulation. Experimental validation employs commercially available off-the-shelf (COTS) converters and real-world data from solar-powered microgrids in Adjuntas, Puerto Rico. To the authors' knowledge, this work provides the first at-scale experimental validation of dc-interlinks for networked ac microgrids using COTS inverters, demonstrating their practicality and effectiveness in addressing real-world operational challenges.

Keywords—DC interlinks, point-to-point dc interlinks, Networked microgrids, Microgrids testbed, Grid-Forming

I. INTRODUCTION

Microgrid technology has matured to become a versatile solution, a popular means for integrating and managing renewable energy and energy storage systems. Microgrid are being applied across a diverse set of applications including communities, industrial complexes, utility distribution networks, and commercial and unviersity campuses. Microgrid adoptions have observed significant growth which is expected to contiue with a projected Compound Annual Growth Rate (CAGR) of 15%. This is driven by the rising demand for renewable energy integration and enhanced reliability in energy delivery [1].

Geographically collocated microgrids can interconnect to form a networked microgrid (NMG). Multiple traditional microgrids can couple into single unified islanded entities sharing resources to enhance efficiency and resiliency. These NMGs must find a means for the individual microgrids to share in the energy and stability requirements of the overall system, allowing power exchange and coordinated control actions.

Notice: This manuscript has been authored by UT-Battelle, LLC, under contract DE-AC05-00OR22725 with the US Department of Energy (DOE) Office of Electricity. The US government retains and the publisher, by accepting the article for publication, acknowledges that the US government retains a nonexclusive, paid-up, irrevocable, worldwide license to publish or reproduce the published form of this manuscript, or allow others to do so, for US government purposes. DOE will provide public access to these results of federally sponsored research in accordance with the DOE Public Access Plan (https://www.energy.gov/doe-public-access-plan).

The benefits of such an architecture can be categorized into three key dimensions: operational benefits, economic advantages, and benefits for the larger grid. I) Operational benefits: The interconnection of microgrids enhances operational flexibility, facilitates resource sharing, and improves the management of distributed energy resources (DERs). It enables global optimization of energy allocation, surpassing the local optimization possible with standalone microgrids. NMG also boosts reliability and resilience, as microgrids can support one another during outages. II) Economic Benefits: NMG can present themselves as a stronger entity in energy markets, potentially leading to improved Return on Investment (RoI) for their owners, particularty if resiliency gains are included in this calculation as down-time resiliency gains are considered, as reducing downtime for critical facilities—such as hospitals, data centers, or emergency service. Operational Benefits for the Grid: Interconnected microgrids, as more predictable and stable systems, contribute to enhanced overall grid reliability.This improvement benefits regional customers by improving service quality and reducing the likelihood and impact of outages. These benefits have been explored in prior works [2,7], which delve into the various aspects of operational and economic advantages as well as the implications for grid reliability and resilience.

Most of the published literature on NMGs have concentrated on higher-level control strategies for clustering operations, both in centralized and distributed controls [5-7]. However, many real-world application-oriented aspects remain insufficiently explored. For example, today microgrids have predominately been deployed as ac systems as this integrates more easily into traditional distribution network compared to dc architectures. A key consideration in NMG is the type of electrical interconnection between ac microgrids— selection of a ac or dc interlink. This choice has significant implications regarding voltage and frequency regulation, power quality, microgrid-to-microgrid synchronization, microgrid boundary expansion, and

979-8-3315-1612-3/25 $31.00 © 2025 IEEE

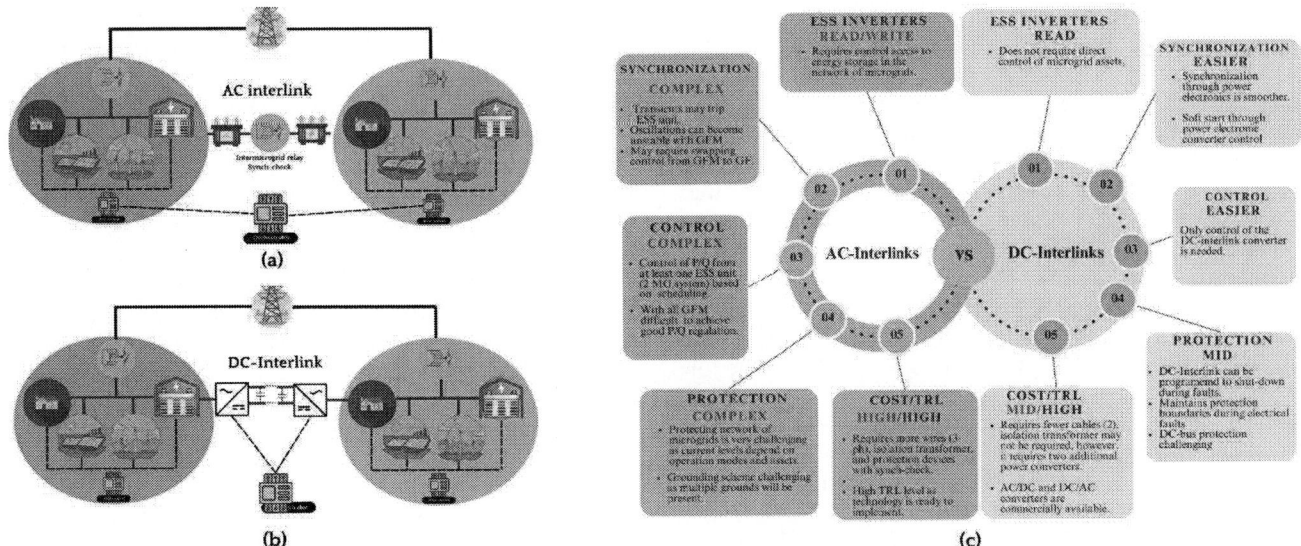

Fig 1: Comparison AC-interlinks and DC-interlinks: (a) Interconnection type (top ac) bottom (dc) interlinks. (b) high-level comparison between AC and DC interlinks

overall stability and protection coordination. The interconnection type also influences overall microgrid cluster management [8]. For instance, ac interlinks may necessitate greater controllability of local microgrid assets to achieve effective power flow control. This can present challenges when integrating microgrids with diverse controllers or when connecting various microgrid types.

This paper addresses the gap in experimental hardware validation of dc interlinks for the interconnection of ac microgrids and contributes to the broader discussion of their advantages over AC interlinks. The novelties of this work are:

Addressing Experimental Validation: This paper fills a gap in experimental hardware validation of dc interlinks for interconnecting ac microgrids, advancing understanding of their advantages over ac interlinks. It provides the detailed experimental validation of dc interlinks for connecting two three-phase AC microgrid systems operating at 480V.

Utilization of Commercially Available Technology: The study employs commercially available off-the-shelf (COTS) converters, offering a practical and replicable approach that simplifies adoption in real-world applications.

Integration of Real-World Data: The paper integrates operational data from a community microgrid in Adjuntas, Puerto Rico, showing the relevance and applicability of dc-links to increase islanding time of microgrids in places vulnerable to natural disasters.

Exploration of Point-to-Point DC Converters: Unlike prior studies that focus on back-to-back converter configurations for similar applications, this work investigates point-to-point dc converters, which present opportunities to reduce interconnection costs and coordination complexities.

The proposed approach demonstrates benefits for geographically co-located microgrids separated by longer distances, addressing a key challenge in microgrid interconnection.

II. BACKGROUND LITERATURE ON DC INTERLINKS

Dc-interlinks (using ac-dc and dc-ac converters) are used in many applications to decouple ac systems. In various electrical systems incorporating motors, dc-interlinks separate the grid frequency from the rotational machine speed (variable speed fans, vehicle propulsion systems, and wind generation technology) [9-11]. These are usually single directional with input composed commonly of rectifier bridges. In transmission networks, dc-interlinks separate the different regional operating systems Electric Reliability Council of Texas (ERCOT), the Eastern Interconnect, and Western Interconnect [12], allowing for independent grid operations (frequency and inertia) while still allowing energy exchange. In a similar concept, these interlinks can be used at a much smaller scale to isolate and separate microgrids. This technology has the potential for enabling interoperability among microgrids with diverse controllers and serves as a critical step toward transitioning NMGs concepts from research to practical, real-world applications.

While dc interlinks have previously been explored in the context of microgrid-to-bulk-grid coupling [13] and microgrid-to-microgrid coupling [14,15], most published research remains focused on model development and simulations rather than hardware implementation. For example, the authors in [13] discuss a Microgrid Building Block (MBB) approach, which replaces the point of common coupling (PCC) with a back-to-back power converter, presenting simulation results to demonstrate advantages for both grid-tied and islanded operations. Similarly, the studies in [14,15] examine a set of AC microgrids interconnected by back-to-back converters, using a generalized small-signal model and Opal-RT simulations to validate feasibility. In [16], a uniform control strategy for bidirectional AC/DC interlinking converters is proposed for linking AC and DC microgrids, with hardware-in-the-loop simulations in Opal-RT for validation. DC interlinks were also considered in the networked operation of the IIT-Bronzeville microgrids [2]; however, the authors stated in this paper that this was a conceptual exploration, and the publication lacks both experimental validation and implementation details. For interconnecting microgrids separated by long distances, medium voltage DC (MVDC) is a more suitable solution than low-voltage DC (LVDC) interconnections due to its lower losses and higher efficiency. However, most existing studies on MV solutions, such as those employing back-to-back converters with multilevel cascaded H-bridge (CHB) and dual active bridge (DAB) topologies, focus on simulations rather than experimental validation [17]. Nevertheless, MVDC demonstrates significant potential as a viable and efficient technology to address the challenges of interconnecting microgrids over long distances. Although these studies published establish a feasibility and theoretical foundation, they lack experimental validation and the associated implementation strategies required to turn the technology into practical applications.

III. AC versus DC

NMGs require a physical electrical connection between the microgrids. This interconnection can be established using either ac or dc interlinks. The choice of interconnection type significantly influences key factors such as control, synchronization, protection, and the cost of implementation and management. Furthermore, the coordination of these microgrids needs a higher order system such as a distribution management system or orchestrator that can play the role of an independent coordinator. Figure 1 provides a high-level comparison of these metrics, highlighting the trade-offs associated with each interconnection type.

a) AC-interlinks

AC interlinks electrically couples the microgrids through ac wiring where the phases at each microgrids are matched and bonded. For this application, an isolation transformer may be needed as each microgrid has its own grounding scheme and would prevent multiple ground loops to be present in the network, which can be a challenge for protection [18]. A protective device with synch-check capability and protection functions is needed at the point of connection between the microgrids for interconnecting and isolating microgrids.

With ac interlinks, determining the mix between grid-forming (GFM) and grid following (GF) controls and the settings determines important factors such as reliability, flexibility of operation, scalability and design of the energy management systems. Operating individually or in separate microgrids, each microgrid must have at least one GFM asset. In a consolidated form as a network, at least one microgrid must have a GFM

asset while other assets can operate in either GF or GFM (depending on the type of GFM). For many microgrids, the GFM asset is represented by the energy storage systems (ESS). The reliability of the network can be increased when choosing to operate all GFM capable assets in GFM [19]. This facilitates power sharing through distributed droop control, which only requires local measurements of voltage, current, and frequency to determine real and reactive power output. However, managing multiple microgrids with GFM inverters also bring several challenges. Standalone island microgrid operations to networked microgrids (by closing the interconnecting switch) can introduce system oscillations and result in tripping of GFM assets. Voltage and frequency regulation are heavily dependent on the accurate tuning of droop gains and virtual impedance. Large load transients can also introduce significant voltage and frequency fluctuations, which might activate protection mechanisms. The interlink protective device should be configured with maximum current thresholds to safeguard against potential wire damage.

Critically, networking ac-interlinks may require extensive data sharing and control access among the local microgrids and the higher-level network microgrids controller, or microgrid orchestrator, which introduces further complexities due to ownership models, data privacy concerns, and technical incompatibilities. The microgrid orchestrator should have control access to some local microgrid assets, typically, the energy storage system (ESS) to control power flow exchange between microgrids (e.g. by writing on power offsets) or to control changes from GFM to GF. This is particularly challenging when dealing with multi-vendor ESS systems, proprietary technology, or microgrid controllers from different vendors. For a real-world application, implementing a microgrid orchestrator controller would require collaboration between local microgrid controller to allow a higher-level controllability of their assets. Liability may become a hurdle when integrating multiple controllers in a multi-owner system. Although some research in distributed optimization [7,20], may ease some of these challenges, this application may be challenging in real deployments as it requires co-optimization with local microgrid controllers, which may not be supported by some microgrid controller vendors. Currently, there are no microgrid controllers that apply distributed optimization to allow interoperable microgrid systems.

These technical and ownership related complexities can make AC coupled microgrids particularly challenging option for networked microgrid deployment.

b) DC-interlinks

DC interlinks ease many of the technical and operational challenges encountered when interconnecting ac microgrids. DC interlinks decouples through power electronics the dynamics of independent microgrids, improving and each microgrid can operate their ESS in Voltage-Frequency (VF) control, ensuring stiff regulation of these magnitudes. The synchronization transients are minimized, as turn on/off the

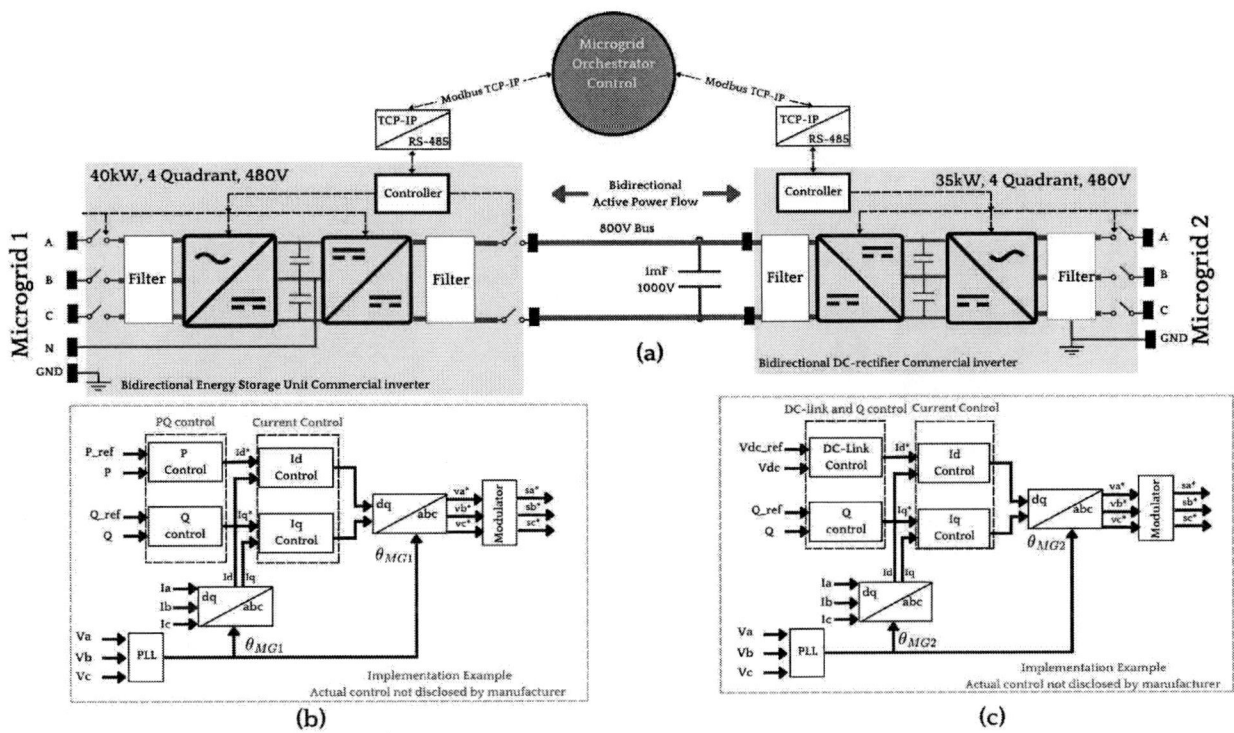

Fig 2: DC-interlinks implemented using COTS inverters. (a) High level block diagram of converter topology. (b) Implementation example of energy storage inverter with PQ controls. (c) Implementation example of dc-link voltage regulator.

power electronics is smooth. The transitions process is notably straightforward. DC-interlinks may allow for a straight-forward solution to make microgrids in a network interoperable. From a controls perspective. Higher level This factor not only simplifies the control strategy but also maintains data privacy as no changes are made to the local controllers to accommodate networked operation capabilities. DC-interlinks is controller agnostic and allows interoperability among microgrid are owned and controlled by different entities. Through dc-interlink, the microgrid orchestrator controls power flow exchange based on an optimization, or logical controller, which only requires limited information from each microgrid (e.g. Stage of Change (SOC)). The local microgrid controllers react to the inter-microgrid power exchange (either surplus or deficit) by controlling their local assets. With this approach, the controller does not have to interact directly with any of the local of the microgrids in the network. This allows for a more straight forward, plug-and play solution for power exchange additional to solving the technical challenges of AC coupling microgrids including synchronization transients.

IV. POINT TO POINT DC INTERLINK

For the experimental validation of DC-interlinks two four-quadrant, commercial-of-the-shelf converters (COTS) were deployed at ORNL networked microgrid testbed. These inverters UL certified and comply with IEEE 1547-2018 standards. The commercially availability and standards compliance increases the technology reediness level TRL level of the proposed DC-interlinks for real-world application of NMGs. Figure 2 shows the block diagram of the power

converter from the converter's spec sheet [21], and an example implementation of its primary-level controller, which are provided as example as details about the actual control is not disclosed by the manufacturer.

Fig 3: DC-interlinks implemented using commercial 3ph, dc-rectifier converter.

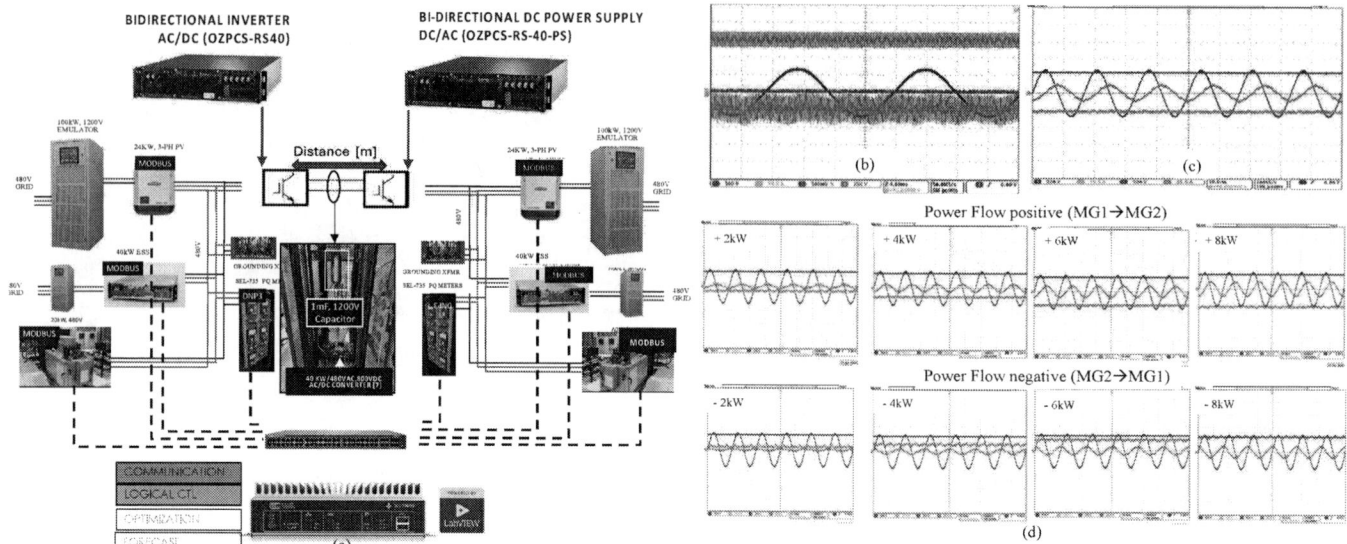

Fig. 4: (a) ORNL networked microgrid testbed. (b) A 100 μF DC interlink capacitor caused instability in the converters as power flow increased. (c) Replacing the capacitor with a 1 mF component significantly reduced DC interlink voltage oscillations, enabling higher power flow. (d) Bidirectional power flow was achieved between microgrids.

A photo of the dc-rectifier using COTS converter is shown in Figure 3. This converter is controlled using RS-485 Modbus RTU. To integrate it with the higher-level microgrid orchestrator, an Advantech Modbus RTU to TCP/IP converter was employed. Additionally, a large 1000 mF capacitor was connected to the DC output of the converter, which links to a second power quality (PQ) converter in the second microgrid. This capacitor has enhanced stability during increased power exchange between the dc interlinks.

a) Bidirectional DC rectifier converter

The DC rectifier is four quadrant, 38kW, 480V, 3-wire power converter. Its main purpose is to maintain stiff regulation of the DC-link voltage regardless of the direction of the power flow. Although the manufacturer does not disclose control of the converter, the most typical controller for this type of converter is Voltage-oriented control (VOC) [22]. The current control loop regulates the exchanged grid current, while the dc-voltage control loop regulates the dc voltage. The implementation of both control loops is well reported in the scientific literature [23], existing a great number of technical solutions (e.g., PI control, proportional-resonant control, repetitive control, etc.).

a) Energy Storage Inverter Control

The PQ converter is a four quadrant, 40kW, 480V, 4-wire power converter. The main purpose of this converter is to regulate active and reactive power regardless of the direction of the dc power. A popular control strategy for this converter is VOC [22]. Where the current control loop regulates the exchanged microgrid current, while the power control loop regulates the active and reactive power independently. The implementation of both control loops is well reported [23].

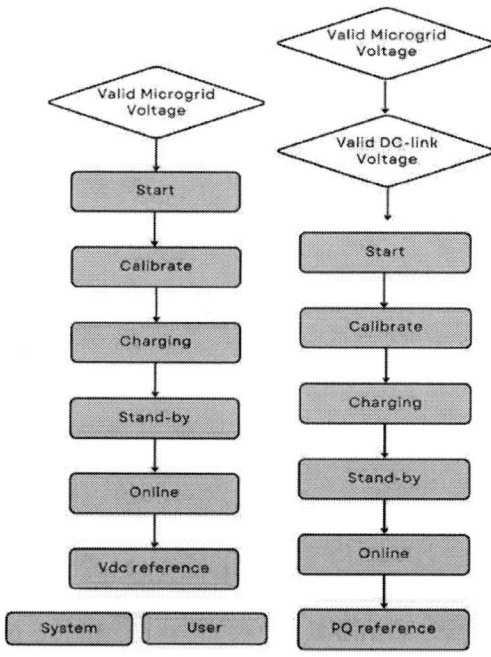

Fig 5: DC-interlink converter startup sequence

a) Start-up Sequence

The start-up sequence of the dc-interlink must follow a sequence, depicted in Fig 5. The first converter to start-up in the DC-rectifier. This converter requires valid ac voltage, which is stablished by the local grid-forming unit in the microgrid (e.g. energy storage system). Once the inverter detects valid voltage, the start-up sequence is provided through Modbus TCP/IP

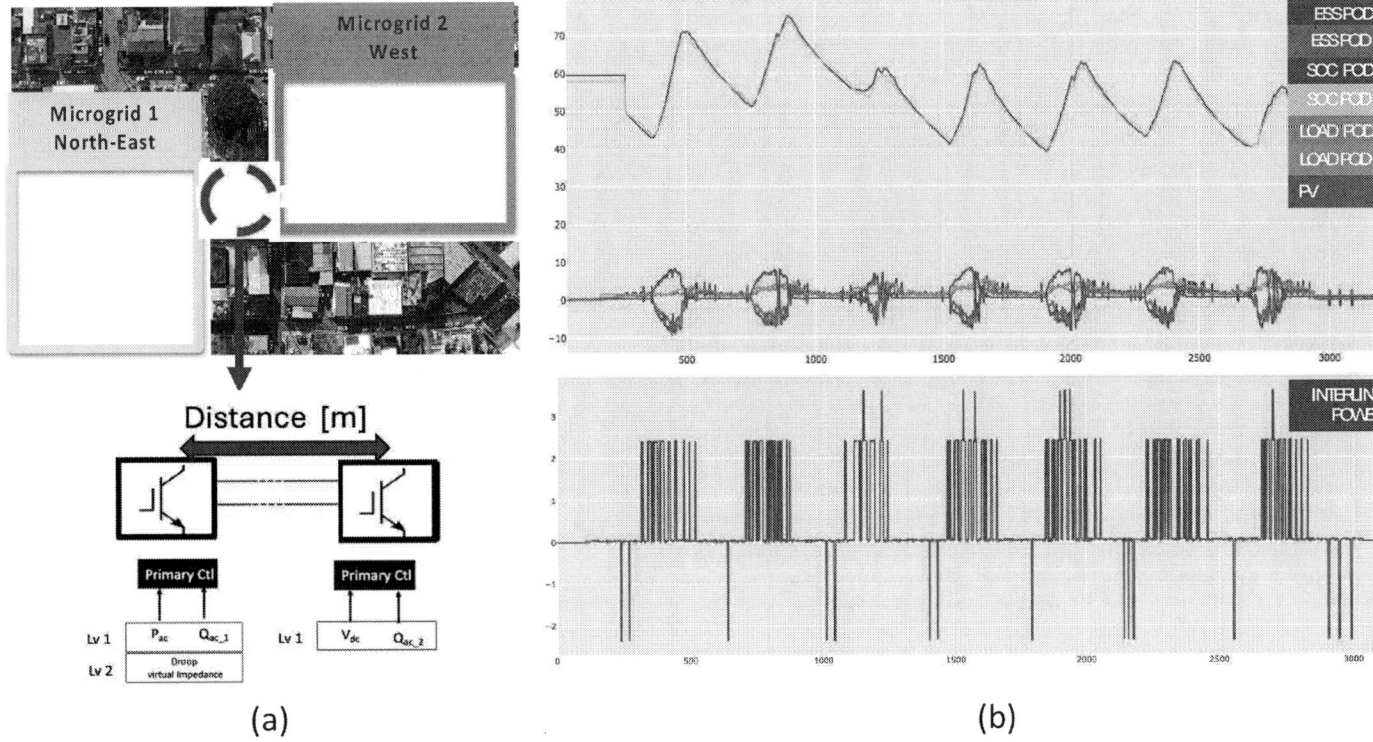

Fig 6: (a) Microgrid system in Adjuntas Puerto Rico. Credit, Casa Pueblo (b) Results from ORNL ac networked microgrid testbed, where the power flow between microgrids is achieved through the point-to-point dc-interlink. Real data from the microgrid used for the experiments.

commands to the dc-rectifier converter. The converter goes to proprietary internal sequences: calibrate, charging, stand-by and online. Once the converter is online the dc-link voltage is being regulated to the user specified voltage. The PQ converter requires valid ac voltage and dc voltage. The ac voltage must be stablished in the second microgrid by the grid-forming unit. The dc voltage is stablished by the dc-rectifier following the steps previously mentioned. Once online, the, the PQ converter receives commands from the orchestrator, the higher-level controller. The PQ values can be determined by an optimization, or can be calculated to achieve SOC balancing, as will be shown in the experimental section.

V. EXPERIMENTAL RESULTS

The dc-interlink was validated on ORNL's networked microgrid testbed, which incorporates distribution grade protection and switchgear, off-the shelf 3-ph PV inverter and energy storage inverter (ESS) that can be configured as 3-wire or 4-wire, supporting resource emulators, and microgrid controllers. The local ESS is represented as a coupling of a DC source emulator and inverter can operate in grid-forming and grid-following control. In grid following, grid assets follow IEEE 1547-2018. While in grid-forming, the inverter has been programmed to operate in droop control with virtual impedance. A diagram of the testbed and the equipment is shown in Figure 4 (a). Figure 4(b) and 4(c) show oscilloscope data of the DC interlink voltage (green), current (cyan), and

Grid Microgrid 1 voltage (blue). When a small capacitor (100 µF/1000 V) was used, oscillations appeared in the DC interlink. To eliminate these oscillations, a larger film capacitor (1000 µF/1000 V) was installed at the DC rectifier terminals. This adjustment effectively suppressed the oscillations, enabling greater power transfer between the microgrids. A detailed analysis of this behavior will be provided in future publications.

Figure 4(c) show oscilloscope waveforms the DC rectifier phase-A voltage (blue), AC current (green), DC-bus voltage (purple), and DC current (cyan) at various power levels. During this test, the active power setpoint of the PQ converter was adjusted within a range of [-20% to 20%] of the rated 40 kW capacity of the converters to evaluate bidirectional power flow between the two AC microgrid systems. Notably, the DC-link voltage remained stable regardless of the power flow direction.

a) Validation with data from Community Microgrid in Adjuntas Puerto Rico

Figure 6 presents the experimental validation of the proposed dc interlink, using as a case study two community-owned microgrids located in Adjuntas, Puerto Rico. These microgrids were designed to provide affordable and reliable electricity to businesses situated in the town square [19]. In their current configuration, these microgrids operate independently of each other. The experimental results shown in Figure 6 shows the

979-8-3315-1612-3/25 $31.00 © 2025 IEEE

potential resiliency benefits of networking these systems using the proposed DC interlink. Data for this experiment were collected directly from the field using dataloggers and PV inverter communication interfaces. This field data was imported into the ORNL network microgrid testbed, where it was scaled down to match the power rating of the laboratory setup.

The use case analyzed in this experiment assumes a scenario where one of the microgrids, identified as Pod 2, experiences a 40% loss in PV generation. To address this, power flow between the two microgrids is coordinated through a microgrid orchestrator implementing a state-of-charge (SOC) balancing algorithm. This algorithm aims to minimize the SOC difference between the two microgrids, thereby extending the islanding duration compared to stand-alone operation. In the absence of this coordination, the affected microgrid would rapidly deplete its battery due to the generation shortfall. Networking the microgrids allows surplus PV generation from the healthy microgrid to be exported to the affected one, maximizing the overall utilization of available renewable energy resources. The results obtained in this figure are accelerated and the time axis is in seconds and 3500 seconds corresponds to a week in the real system. As shown, the interlink power flow is mainly positive, as power from the microgrid with full capacity flow to the second microgrid to maintain the SOC of both systems close to each other. Without interconnecting the microgrids, the PV in the healthy microgrid would have to be curtailed to prevent the batteries to overcharge. By exporting power to the second microgrid the use of renewables is optimized. By networking microgrids with the proposed dc-interlink and control the system can run in island for one week with a minimum SOC of around 40%.

Figure 7 shows the voltage and frequency regulation of both microgrid systems using dc and ac interlinks. A key advantage of interconnecting microgrids through dc-interlinks is that the grid-forming inverters in Microgrid 1 and Microgrid 2 can operate in Voltage Frequency (VF) control mode. This configuration ensures stiff voltage control within the microgrids, thereby improving power quality. In contrast, the voltage and frequency regulation with ac-interlinks depends on the droop curves and virtual impedance configuration of the grid-forming inverters. Therefore, the level of voltage regulation, as demonstrated in the example shown in Fig. 6(b), is not stiff and the voltage and frequency deviations depend on the load and generation.

VI. CONCLUSIONS

This paper presents the benefits of networking ac microgrids via dc-interlinks and bridges a gap in the field by providing an experimental validation of a 40kW dc-interlinks networking two, 480V three-phase ac laboratory-scale microgrid systems. A use-case with two microgrids in Adjuntas Puerto Rico is presented to demonstrate the resiliency gains by networking microgrids through the proposed dc-interlink. DC-interlinks also improve voltage regulation compared with ac coupled microgrids with grid-forming inverters in droop control. This work demonstrates the feasibility of dc interlinks with

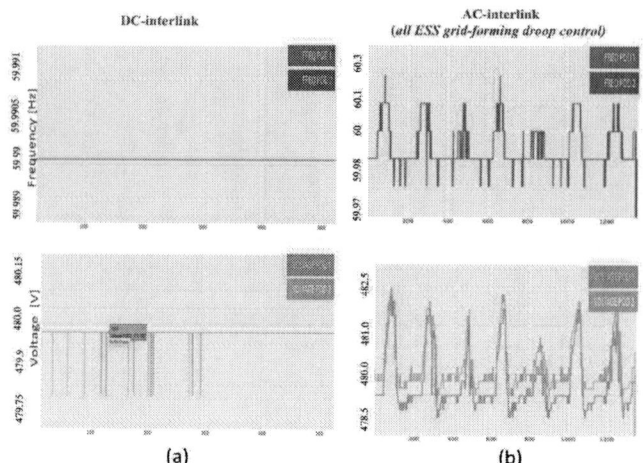

Fig 7: (a) DC-interlinks frequency and voltage regulation. (b) AC-interlinks frequency and voltage regulation with two grid-forming inverters in droop control.

experimental results using commercially available converters. This practical approach facilitates faster adoption in real-world systems.

REFERENCES

[1] Microgrids Market Trends and Analysis by Technology, Installed Capacity, Generation, Key Players and Forecast, 2022.

[2] Chen, Bo, Wang, Jianhui, Lu, Xiaonan, Chen, Chen, and Zhao, Shijia. *Networked Microgrids for Grid Resilience, Robustness, and Efficiency: A Review.* United States: N. p., 2020. Web. doi:10.1109/tsg.2020.3010570.

[3] M. Shahidehpour, Z. Li, S. Bahramirad, Z. Li and W. Tian, "Networked Microgrids: Exploring the Possibilities of the IIT-Bronzeville Grid," in IEEE Power and Energy Magazine, vol. 15, no. 4, pp. 63-71, July-Aug. 2017, doi: 10.1109/MPE.2017.2688599.

[4] L. Che, M. Shahidehpour, A. Alabdulwahab and Y. Al-Turki, "Hierarchical Coordination of a Community Microgrid With AC and DC Microgrids," in IEEE Transactions on Smart Grid, vol. 6, no. 6, pp. 3042-3051, Nov. 2015, doi: 10.1109/TSG.2015.2398853.

[5] F. Feng, P. Zhang, Y. Zhou and L. Wang, "Distributed Networked Microgrids Power Flow," in IEEE Transactions on Power Systems, vol. 38, no. 2, pp. 1405-1419, March 2023, doi: 10.1109/TPWRS.2022.3175933.

[6] Z. Li, M. Shahidehpour, F. Aminifar, A. Alabdulwahab and Y. Al-Turki, "Networked Microgrids for Enhancing the Power System Resilience," in Proceedings of the IEEE, vol. 105, no. 7, pp. 1289-1310, July 2017, doi: 10.1109/JPROC.2017.2685558.

[7] Liu, G., Ferrari, M.F., Chen, Y.: A Mixed integer linear programming-based distributed energy management for networked microgrids considering network operational objectives and constraints. IET Energy Syst. Integr. 5(3), 320–337 (2023). https://doi.org/10.1049/esi2.12103

[8] A. Sundararajan, G. Liu, M. Starke, R. K. Moorthy and C. Irwin, "Networked Microgrid Ownership, Data, and Control Implications: Challenges and Open Questions," 2024 IEEE Power & Energy Society General Meeting (PESGM), Seattle, WA, USA, 2024, pp. 1-5, doi: 10.1109/PESGM51994.2024.10688903.

[9] [Aazmi] M. A. Aazmi, M. I. Fahmi, M. Z. Aihsan, H. F. Liew and M. Saifizi, "A review on VFD Control and Energy Management System of Induction Motor for Electric Vehicle," *2021 IEEE 19th Student Conference on Research and Development (SCOReD)*, Kota Kinabalu, Malaysia, 2021, pp. 36-41.

[10] [Vilathgamuwa] D. M. Vilathgamuwa and S. D. G. Jayasinghe, "Rectifier systems for variable speed wind generation - a review," 2012 IEEE

International Symposium on Industrial Electronics, Hangzhou, China, 2012, pp. 1058-1065

[11] M. Ferrari, M. Orendorff, T. Smith and M. A. Buckner, "OPEN-CODE, REAL-TIME EMULATION TESTBED OF GRID-CONNECTED TYPE-3 WIND TURBINE SYSTEM WITH HARDWARE VALIDATION," 2019 IEEE Applied Power Electronics Conference and Exposition (APEC), Anaheim, CA, USA, 2019, pp. 66-70, doi: 10.1109/APEC.2019.8722049.

[12] U.S. Department of Energy. (n.d.). Connecting the Country with HVDC. Office of Electricity. Retrieved from https://www.energy.gov/oe/articles/connecting-country-hvdc

[13] C. -C. Liu et al., "Microgrid Building Blocks: Concept and Feasibility," in IEEE Open Access Journal of Power and Energy, vol. 10, pp. 463-476, 2023, doi: 10.1109/OAJPE.2023.3282188.

[14] M. Naderi, Y. Khayat, Q. Shafiee, T. Dragičević, H. Bevrani and F. Blaabjerg, "Interconnected Autonomous ac Microgrids via Back-to-Back Converters—Part II: Stability Analysis," in IEEE Transactions on Power Electronics, vol. 35, no. 11, pp. 11801-11812, Nov. 2020, doi: 10.1109/TPEL.2020.2986695.

[15] M. Naderi, Y. Khayat, Q. Shafiee, T. Dragicevic, H. Bevrani and F. Blaabjerg, "Interconnected Autonomous AC Microgrids via Back-to-Back Converters—Part I: Small-Signal Modeling," in IEEE Transactions on Power Electronics, vol. 35, no. 5, pp. 4728-4740, May 2020, doi: 10.1109/TPEL.2019.2943996.

[16] J. Wang, C. Jin and P. Wang, "A Uniform Control Strategy for the Interlinking Converter in Hierarchical Controlled Hybrid AC/DC Microgrids," in IEEE Transactions on Industrial Electronics, vol. 65, no. 8, pp. 6188-6197, Aug. 2018, doi: 10.1109/TIE.2017.2784349.

[17] J. Choi, J. P. Pinto, M. S. Chinthavali and A. Adib, "Medium Voltage Energy Hub Based on Multilevel Cascaded H Bridge-Dual Active Bridge Back-to-Back Converter for Power Distribution Feeders Interconnection and Multiple Simultaneous Grid Services," 2022 IEEE Energy Conversion Congress and Exposition (ECCE), Detroit, MI, USA, 2022

[18] J. Mohammadi, F. Badrkhani Ajaei and G. Stevens, "Grounding the AC Microgrid," in IEEE Transactions on Industry Applications, vol. 55, no. 1, pp. 98-105, Jan.-Feb. 2019, doi: 10.1109/TIA.2018.2864106.

[19] Usman Bashir Tayab, Mohd Azrik Bin Roslan, Leong Jenn Hwai, Muhammad Kashif, A review of droop control techniques for microgrid, Renewable and Sustainable Energy Reviews, Volume 76, 2017, Pages 717-727, ISSN 1364-0321, https://doi.org/10.1016/j.rser.2017.03.028.

[20] Liu, G.; Ferrari, M.F.; Ollis, T.B.; Sundararajan, A.; Olama, M.; Chen, Y. Distributed Energy Management for Networked Microgrids with Hardware-in-the-Loop Validation. Energies 2023, 16, 3014. https://doi.org/10.3390/en16073014

[21] Oztek Corporation, "RS-40 and RS-38" [Online]. Available: https://oztekcorp.com/products/. [Accessed: Nov. 18, 2024].

[22] J. Rocabert, A. Luna, F. Blaabjerg and P. Rodríguez, "Control of Power Converters in AC Microgrids," in IEEE Transactions on Power Electronics, vol. 27, no. 11, pp. 4734-4749, Nov. 2012, doi: 10.1109/TPEL.2012.2199334.

[23] M. P. Kazmierkowski, F. Blaabjerg, and R. Krishnan, Control in Power Electronics—Selected Problems. New York, NY, USA: Academic, 2002.

[24] M. Ferrari, B. Ollis, M. Starke and A. Massol-Deya, "The Well-Connnected Microgried: A Network of Microgrids can Withstand Severestorms," in IEEE Spectrum, vol. 60, no. 10, pp. 36-43, October 2023, doi: 10.1109/MSPEC.2023.10271348.

Andronov-Hopf Oscillator-based Grid-Forming Converters with Embedded Disturbance Rejection for Non-Ideal Loading Condition

Vikram Roy Chowdhury, Gab-Su Seo, and Barry Mather

Power Systems Engineering Center, National Renewable Energy Laboratory, Golden, CO 80401, USA

email: vikram.roychowdhury@nrel.gov, gabsu.seo@nrel.gov, barry.mather@nrel.gov

Abstract—**This paper introduces a modified Andronov-Hopf Oscillator for improved grid-forming inverter operation under balanced, unbalanced, and nonlinear loads. Unlike traditional models, this design adapts the limit cycle to an ellipse for unbalanced loads and a distorted curve for nonlinear loads, eliminating the need for additional control mechanisms and simplifying system architecture. The oscillator enhances voltage quality with reduced harmonic distortion, improving system robustness and adaptability. It offers a scalable solution for integration into existing infrastructures, optimizing energy distribution and reducing operational costs. This novel approach can not only facilitate the transition to inverter-rich grids but also address challenges in integrating renewable energy. Extensive simulations and case studies highlight its potential. The system's effectiveness is validated through implementations in MATLAB/Simulink and PLECS along with real-time controller hardware in the loop, underscoring its ability to enhance power quality under different loading conditions.**

Index Terms—**Andronov-Hopf oscillators, grid-forming converters, Lyapunov energy function, disturbance rejection, controller hardware in the loop, total harmonic distortion.**

I. INTRODUCTION

Integrating multiple energy sources interfaced with power electronics converters to collectively supply a common load necessitates their parallel operation [1]–[6]. This configuration can not only enhance the reliability of energy supply but also improve the overall efficiency of energy distribution systems. However, it is widely recognized that most renewable energy resources, such as photovoltaic (PV) systems, wind energy, and wave energy, are inherently intermittent and unpredictable in nature. This variability underscores the critical need for advanced control architectures to ensure their effective and reliable operation [7]–[10]. These control strategies are fundamental to maintaining the stability of the power grid and delivering an uninterrupted power supply, even in the face of fluctuations in resource availability. Additionally, they play

This work was authored by the National Renewable Energy Laboratory, operated by Alliance for Sustainable Energy, LLC, for the U.S. Department of Energy (DOE) under Contract No. DE-AC36-08GO28308. Funding provided by U.S. Department of Energy Office of Energy Efficiency and Renewable Energy Water Power Technologies Office. The views expressed in the article do not necessarily represent the views of the DOE or the U.S. Government. The U.S. Government retains and the publisher, by accepting the article for publication, acknowledges that the U.S. Government retains a nonexclusive, paid-up, irrevocable, worldwide license to publish or reproduce the published form of this work, or allow others to do so, for U.S. Government purposes.

a pivotal role in enabling the grid to adapt dynamically to varying energy demands, ensuring a seamless and balanced operation under diverse conditions. As renewable energy adoption continues to accelerate globally, the development and refinement of these control mechanisms become increasingly essential to support the transition to a sustainable, resilient, and efficient energy future. These systems not only address the challenges of integration but also position renewable energy as a cornerstone of modern power systems.

In this paper, a novel control methodology is presented that leverages the Andronov-Hopf oscillators (AHO) [11], [12] for grid-forming (GFM) converter applications, with the unique ability to adapt the shape of its limit cycle based on load dynamics. This adaptive capability enables the oscillator to flexibly modify its behavior. For instance, under unbalanced linear loads, the limit cycle of the AHO transitions to an elliptical shape, reverting to a circular form in balanced conditions, while harmonic disturbances induce a distorted closed-loop cycle. These dynamic adjustments, coupled with feed-forward control terms, allow the generation of high-fidelity reference signals for the inverter, ensuring sinusoidal voltages at the point of common coupling (PCC) that are free from distortions across balanced, unbalanced, and nonlinear loading conditions, in alignment with GFM converter standards [13]–[15]. The proposed approach significantly enhances power quality and ensures robust performance under diverse operational scenarios. The flexibility of the AHO framework enables real-time adaptation to fluctuating grid conditions, thereby improving system resilience and reliability. By harnessing the intrinsic stabilizing properties of oscillators, this method inherently maintains voltage stability without relying on complex compensatory algorithms, resulting in improved operational efficiency and potential cost savings. The methodology has been thoroughly validated through simulations conducted in MATLAB/Simulink and PLECS, supported by detailed case studies that highlight its effectiveness in maintaining power stability and quality. The paper is structured as follows: Section II details the step-by-step derivation of the proposed Andronov-Hopf Oscillator with embedded disturbance rejection architecture; Section III presents simulation results and discussion of key case studies; and Section IV concludes the findings.

979-8-3315-1612-3/25 $31.00 © 2025 IEEE

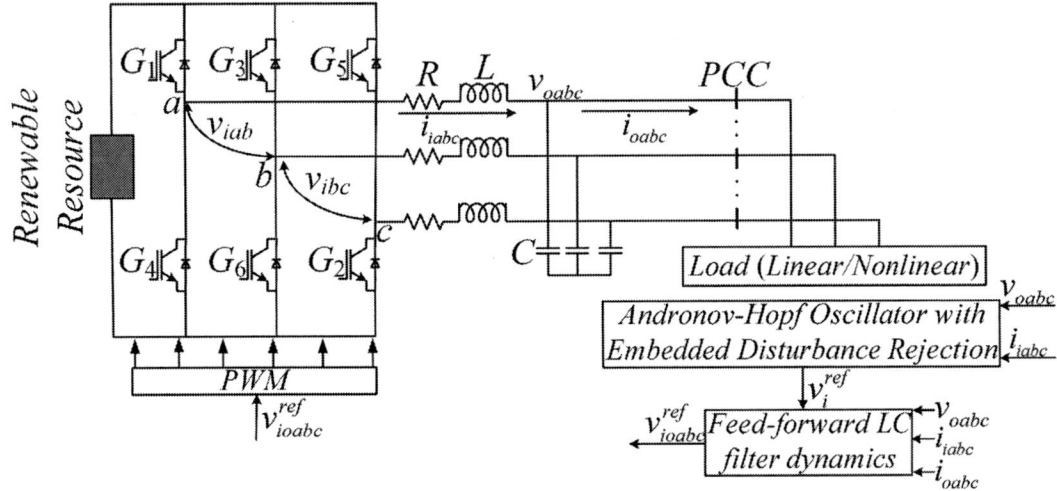

Fig. 1: Simplified block diagram of the proposed system for GFM converters.

II. ANDRONOV-HOPF OSCILLATOR WITH EMBEDDED DISTURBANCE REJECTION FOR GFM CONVERTERS

The simplified block diagram of the proposed system is presented in Fig. 1. In general, the dc bus comprises of the renewable resource, PV, wind, wave, etc. The ac side is connected through an interfacing LC filter to the PCC which can either be connected to a balanced, unbalanced, or nonlinear load. If pursued as a GFM asset, the primary objective of such a converter is to behave as a voltage source, following standards being laid for GFM converters such as UNIFI specifications [13].The general representation of AHO as elaborated in [11], [12] is presented as:

$$\begin{aligned}
\dot{x} &= \mu\left(r^2 - x^2 - y^2\right)x + \omega y, \\
\dot{y} &= \mu\left(r^2 - x^2 - y^2\right)y - \omega x,
\end{aligned} \tag{1}$$

where x and y are the states of the system, μ is a system constant equivalent to the bandwidth of the controller, r is the steady-state output magnitude, and ω is the oscillation frequency. The steady-state solution for the AHO can be: $x = r\sin(\omega t + \theta_o)$ and $x = r\cos(\omega t + \theta_o)$, where θ_o is the initial phase of the system. This paper proposes a modified architecture of the AHO to improve disturbance rejection during unbalanced and nonlinear loading conditions. A cascaded AHO architecture mitigates voltage unbalance and harmonics, ensuring balanced output voltages with low THD at PCC for unbalanced and nonlinear loads. The modified AHO architecture in stationary $\alpha\beta$ domain (chosen to reduce the number of equations) is to cater for balanced or unbalanced loading conditions presented as:

$$\dot{v}_{c\alpha}^{ref} = \begin{bmatrix} \mu\left(|V_{nom}|^2 - \frac{v_{c\alpha}^{ref2}}{a^2} - \frac{v_{c\beta}^{ref2}}{b^2}\right)v_{c\alpha}^{ref} \\ + \left(\omega_{nom}v_{c\beta}^{ref} - k_i i_{i\alpha} + k_v v_{c\alpha}\right) \end{bmatrix},$$

$$\dot{v}_{c\beta}^{ref} = \begin{bmatrix} \mu\left(|V_{nom}|^2 - \frac{v_{c\alpha}^{ref2}}{a^2} - \frac{v_{c\beta}^{ref2}}{b^2}\right)v_{c\beta}^{ref} \\ + \left(-\omega_{nom}v_{c\alpha}^{ref} - k_i i_{i\beta} + k_v v_{c\beta}\right) \end{bmatrix}, \tag{2}$$

where ω_{nom} is the nominal frequency, $|V_{nom}|$ is the magnitude of the nominal voltage, $v_{c\alpha\beta}^{ref}$ are the oscillatory voltages used as references for the PCC capacitor to the inner filter dynamics, $v_{c\beta}$ and $i_{L\alpha\beta}$ are measured capacitor voltages and inductor currents, k_v and k_i are gains for capacitor voltages and inductor currents' feed-forward inside the AHO architecture, a and b are coefficients catering for unbalanced linear loads and are defined by: $a = \frac{|V_{nom}|}{v_{c\alpha}^{peak}}$ $b = \frac{|V_{nom}|}{v_{c\beta}^{peak}}$ respectively. The $v_{c\alpha\beta}^{peak}$ are computed based on quadrature axis component generation utilizing dual second order generalized integrator (DSOGI) architecture [16] therefore, computed peak values can be in generalized format are given by: $v_{c\alpha}^{peak} = \sqrt{v_{c\alpha}^2 + v_{c\beta(computed)}^2}$ and $v_{c\beta}^{peak} = \sqrt{v_{c\alpha(computed)}^2 + v_{c\beta}^2}$. To cater for harmonics, a family of AHO oscillators are implemented for 3rd, 5th, 7th... harmonics and in generalized format is presented as:

$$\dot{v}_{c\alpha}^{ref} = \begin{bmatrix} \frac{\mu}{n}\left(|V_{nom}^n|^2 - \frac{v_{c\alpha}^{nref2}}{a_n^2} - \frac{v_{c\beta}^{nref2}}{b_n^2}\right)v_{c\alpha}^{nref} \\ \left(+n\omega_{nom}v_{c\beta}^{nref} - \frac{k_i}{n}i_{i\alpha}^n + \frac{k_v}{n}v_{c\alpha}^n\right) \end{bmatrix},$$

$$\dot{v}_{c\beta}^{ref} = \begin{bmatrix} \frac{\mu}{n}\left(|V_{nom}^n|^2 - \frac{v_{c\alpha}^{nref2}}{a_n^2} - \frac{v_{c\beta}^{nref2}}{b_n^2}\right)v_{c\beta}^{nref} \\ + \left(-n\omega_{nom}v_{c\alpha}^{nref} - \frac{k_i}{n}i_{i\beta}^n + \frac{k_v}{n}v_{c\beta}^n\right) \end{bmatrix}, \tag{3}$$

where $n = 3, 5, 7...$ is the harmonic order, measured values $v_{c\alpha\beta}^n$ and $i_{i\alpha\beta}^n$ are computed based on DSOGI architecture tuned at the specific frequencies, $|V_{nom}^n|$ is chosen based on application specificity considering the facts of how much harmonic load in general can be demanded for, the coefficients for possible unbalance during nonlinear loading are given by: $a_n = \frac{|V_{nom}^n|}{v_{c\alpha}^{npeak}}$ and $b_n = \frac{|V_{nom}^n|}{v_{c\beta}^{npeak}}$. The composite value of the generated reference voltages for the capacitor on the PCC are then given by: $v_{c\alpha}^{composite} = v_{c\alpha}^{ref} + v_{c\alpha}^{nref}$ and $v_{c\beta}^{composite} = v_{c\beta}^{ref} + v_{c\beta}^{nref}$ for α and β axes respectively.

979-8-3315-1612-3/25 $31.00 © 2025 IEEE

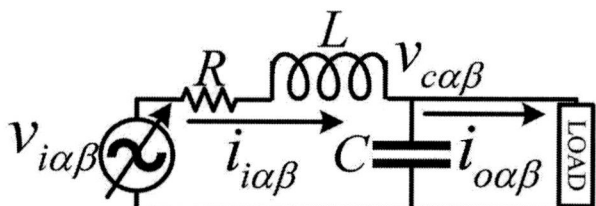

Fig. 2: $\alpha\beta$-axes equivalent circuit of the inverter.

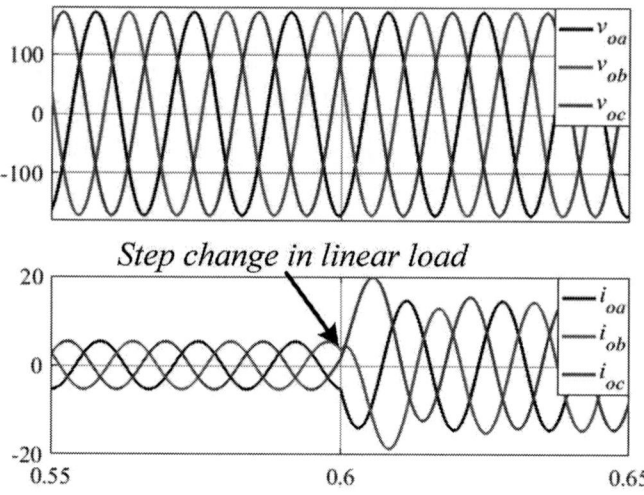

Fig. 3: simulation result of the GFM converter under a step change in linear load at PCC.

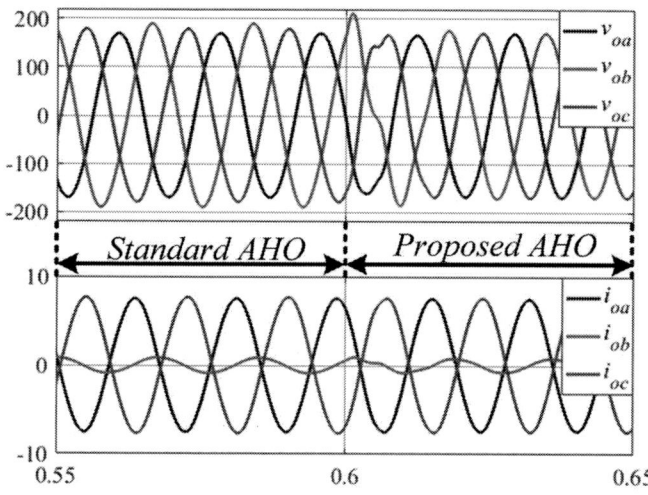

Fig. 4: Simulation result for inverter performance comparison of the proposed AHO to the conventional AHO for unbalanced linear loading.

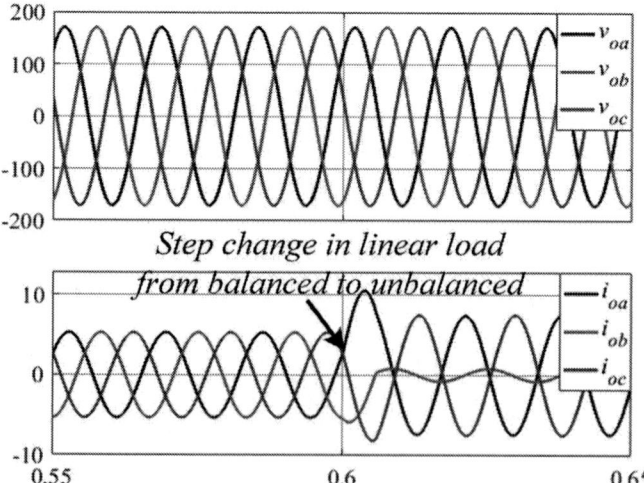

Fig. 5: Simulation result for a step change in linear load from balanced to unbalanced at PCC.

In general the AHO can be simplified and can be shown to embed a nonlinear resistive droop architecture. For (2), define: phasor voltage magnitude $V = \sqrt{v_{i\alpha}^2 + v_{i\beta}^2}$ and the phase angle $\theta = tan^{-1}\left(\frac{v_{i\beta}}{v_{i\alpha}}\right)$. Using these definitions, and (2) gives:

$$\dot{V} = \gamma\left(|V_{nom}|^2 - |V|^2\right)V - \frac{2k_i}{3V}P,$$

$$\dot{\theta} = \omega = \left|\omega_{nom} + \frac{2k_i}{3V^2}Q\right|,$$

(4)

where $P = \frac{3}{2}\left(v_{i\alpha}i_{i\alpha} + v_{i\beta}i_{i\beta}\right)$ and $Q = \frac{3}{2}\left(v_{i\alpha}i_{i\beta} - v_{i\beta}i_{i\alpha}\right)$ are the active and reactive powers supplied by the converter to the PCC. As observed from (4), this AHO architecture embeds a resistive nonlinear droop within itself thereby enhancing the power sharing performance as well as the stability of the overall system [17]. In general, (4) can be separately computed for either negative sequence or harmonics and the overall analyses can be accomplished for the proposed AHO architecture. These generated outputs of the AHO oscillator are used for the feed-forward filter dynamics as briefly presented in the next section to compute the reference voltages for the inverter pulse width modulator (PWM). To ensure better disturbance rejection and dynamic response, feed-forward dynamics of the LC filter is added to the overall control architecture. The equivalent single-phase representation from the inverter to the PCC for $\alpha\beta$ axes of the GFM converter is presented in Fig. 2. The capacitor voltage and the inductor current dynamics are presented as:

$$L\frac{di_{i\alpha\beta}^{ref}}{dt} = -Ri_{i\alpha\beta}^{ref} + v_{i\alpha\beta}^{ref} - v_{c\alpha\beta}^{composite},$$

$$C\frac{dv_{c\alpha\beta}^{composite}}{dt} = i_{i\alpha\beta}^{ref} - i_{o\alpha\beta},$$

(5)

where $i_{o\alpha\beta}$ are the measured load currents, $i_{i\alpha\beta}^{ref}$ are the generated reference currents for the LC filter dynamics, R is the loss resistance for the interfacing reactor, $v_{c\alpha\beta}^{composite}$ are the generated capacitor voltage references from the AHO based oscillator outputs (2) and (3). With these feed-forward terms, in conjunction with the generated capacitor voltage references from the proposed AHO based oscillator, the inverter voltage references $v_{i\alpha\beta}^{ref}$ are computed and used for the PWM. The next section shows the results and discussions for important case studies with the proposed architecture.

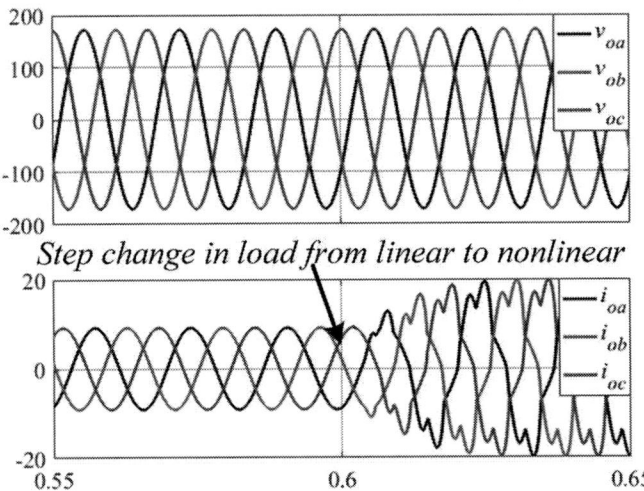

Fig. 6: Simulation result for a step change in load from linear to nonlinear at PCC.

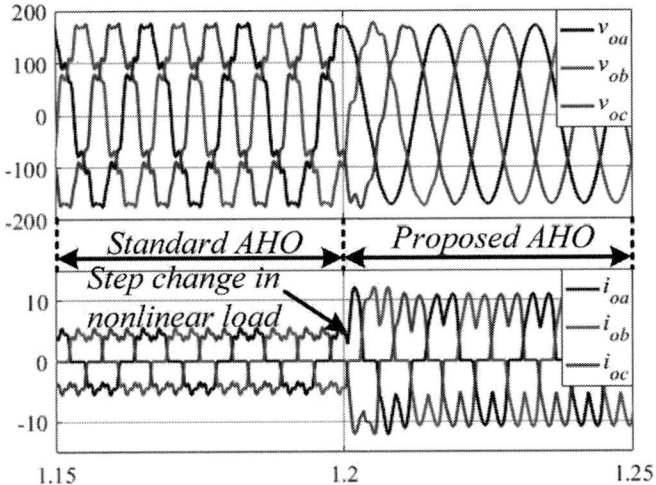

Fig. 7: Simulation result for the performance evaluation during a step change in nonlinear load with standard to proposed AHO.

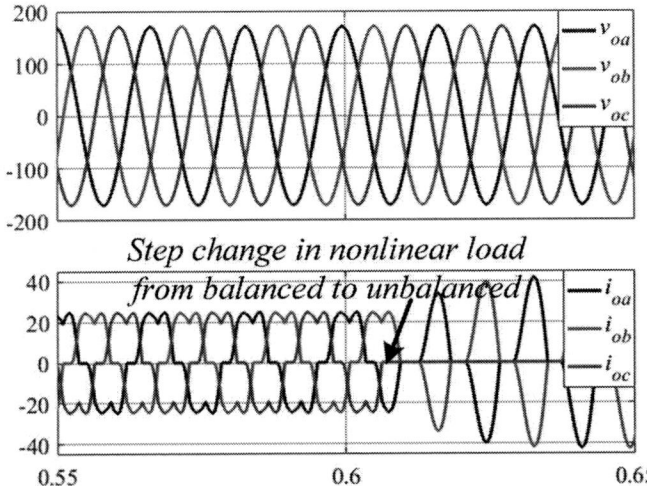

Fig. 8: Simulation result for a step change in nonlinear load from balanced to unbalanced at PCC.

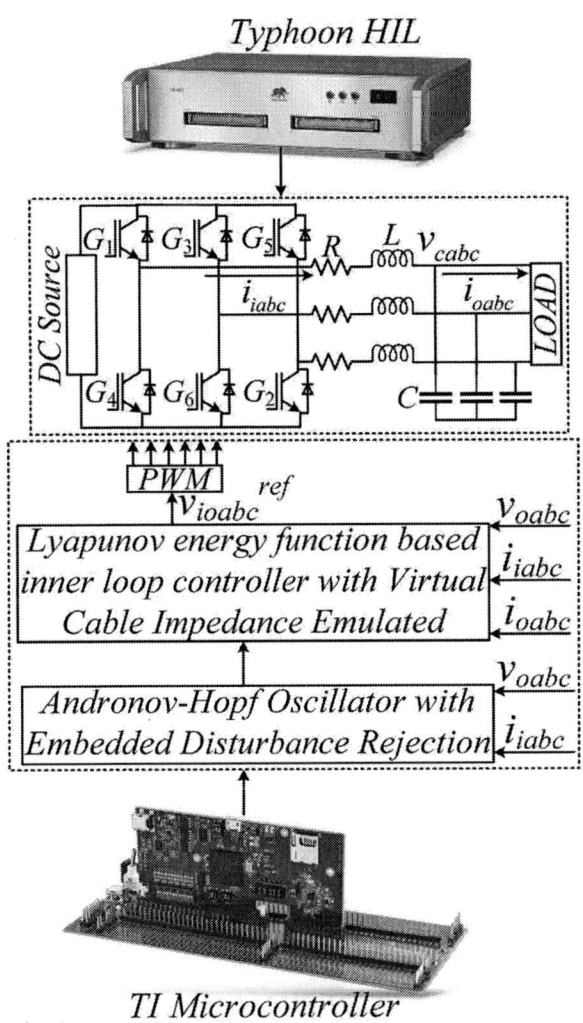

Fig. 9: Diagram for the CHIL-based validation setup.

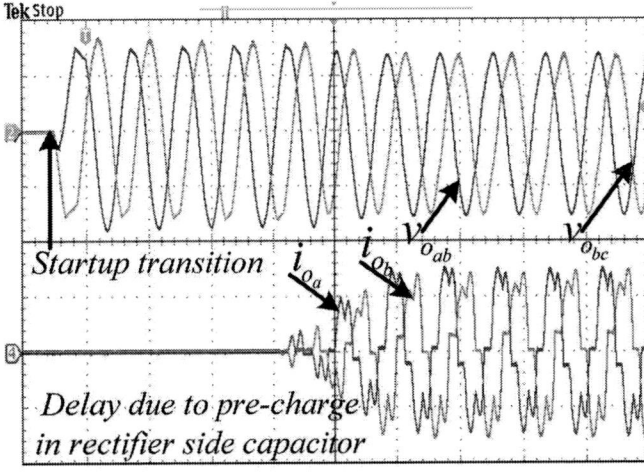

Fig. 10: CHIL results for voltage buildup of the proposed AHO-based GFM converter under nonlinear loading.

979-8-3315-1612-3/25 $31.00 © 2025 IEEE

TABLE I: PLANT AND COMPENSATOR PARAMETERS

Parameter	Value
Power Rating	5 KVA
V_{dc}	400 V
v_{PCC}	120 V(rms)
L	5.0 mH
R	1.0 Ω
C	20.0 μF
ω_{nom}	377.0 rad/s
k_i	200
k_v	25
μ	0.2

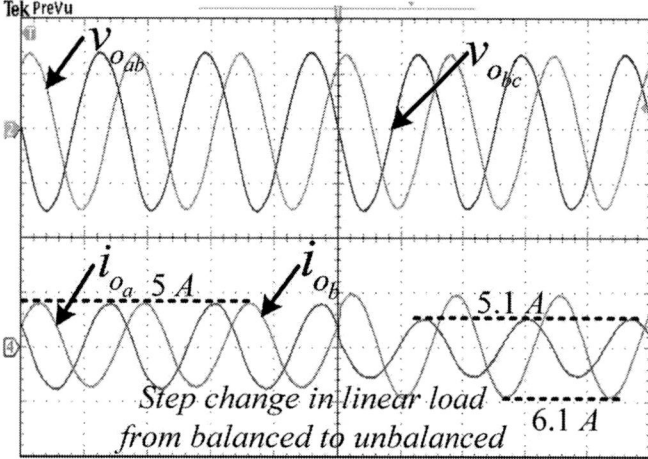

Fig. 11: CHIL results for a linear load step change from balanced to unbalanced with the proposed AHO-based control architecture.

III. RESULTS AND DISCUSSIONS

The overall system is modeled in MATLAB/Simulink and PLECS domain and various case study results are presented. The parameters of simulation are elaborated in Table I. DSOGI filter damping coefficient used to compute unbalance and

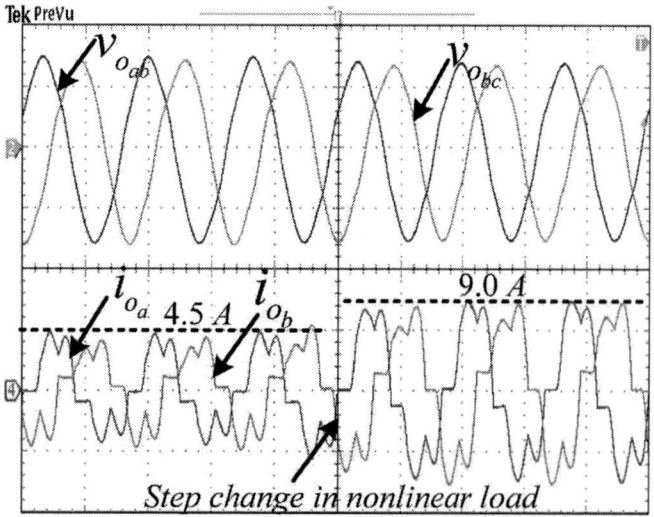

Fig. 12: CHIL results for a step change in nonlinear load with the proposed AHO architecture.

harmonics $k = 0.8$ for negative sequence and $\frac{k}{n}$ for the harmonics where $n = 1, 3, 5, 7...$ are the harmonic order. using these parameters, the result showing the step change in load during balanced linear loading condition is presented in Fig. 3. This result shows the seamless dynamic performance of the proposed architecture during a change in linear balanced loading condition. The next result shows the transition when the elliptical AHO is disabled during an unbalanced loading condition and switched back on as presented in Fig. 4. It is observed from this result that the proposed architecture can successfully mitigate for the unbalances in the PCC voltage waveform during a linear unbalanced loading condition. The next result shows the step change in load from balanced to unbalanced as shown in Fig. 5. It is observed from this result that during such a transient, the PCC voltage is still maintained to be distortion-free with low harmonic content as per requirement proving the effectiveness of the proposed approach. The next result shows the transition from linear loading to nonlinear loading as presented in Fig. 6. As observed from this result , during this transition the PCC voltage can be still maintained within the standards specified in [18] indicating the superior dynamic performance achievable with the proposed architecture. The next result shows the performance during step change of nonlinear loading without enabling the the proposed AHO and operating under a standard AHO architecture as shown in Fig. 7. It can be inferred from this result that with the proposed AHO architecture enabled it is possible to mitigate all the harmonics from the PCC voltage and maintain low total harmonic distortion (THD) voltage on the PCC indicating the efficacy of the porpoised approach. The result showing the transition from balanced nonlinear loading to unbalanced nonlinear loading is presented in Fig. 8. It is observed from this result that the PCC voltage can still be maintained within acceptable limits specified by the requirement. all these results collectively show the superior performance metric achievable by the proposed modified AHO architecture. To further substantiate the overall architecture is implemented in Typhoon-CHIL platform and important case study results are presented showing the effectiveness of the proposed approach. The parameters of the CHIL is the same as those for the offline simulations presented in Table I. The result showing the startup with the proposed architecture with an uncontrolled diode rectifier load is presented in Fig. 10. It is observed that with the proposed architecture, it is possible to accomplish a successful voltage buildup under a highly nonlinear load and still ensure low THD on the load voltage as prescribed in [18]. The result showing step change of the linear loads (with resistive to resistive-inductive) from balanced to unbalanced is presented Fig. 11. Notice that the controlled variable, i.e., the generated oscillator voltages, can remain balanced with low-harmonic-content ($< 5\%$ THD) under this transient, indicating a stable voltage regulation. Finally, the result showing the step change in the diode rectifier load is presented in Fig. 12 which shows it is still possible to maintain low THD load voltage under this transient indicating the effectiveness of the proposed approach.

979-8-3315-1612-3/25 $31.00 © 2025 IEEE

IV. CONCLUSION

This paper presented a novel Andronov-Hopf nonlinear oscillator with an adaptively adjustable limit cycle based on load conditions. The proposed approach modifies the limit cycle shape to an ellipse for unbalanced loading, maintains a circle for balanced conditions, and transitions to a distorted enclosed shape during harmonic disturbances by combining various harmonic waveforms to tightly regulate the voltage with high power quality under nonideal conditions. The major and minor axes of the ellipse are functions of the degree of unbalance at the PCC voltage, allowing for adaptive adjustments to ensure balanced sinusoidal PCC voltage. The feed-forward dynamics effectively mitigate any residual voltage unbalance or distortion, demonstrating the efficacy of the proposed architecture. The comprehensive analyses, case study results, and real-time simulation and experiments validated the effectiveness of the novel solution, highlighting its potential for enhancing power quality and system stability.

REFERENCES

[1] M. P. Kazmierkowski and L. Malesani, "Current control techniques for three-phase voltage-source PWM converters: A survey," *IEEE Transactions on Industrial Electronics*, vol. 45, pp. 691–703, October 1998.

[2] S. B. Kjaer, J. K. Pedersen, and F. Blaabjerg, "A review of single-phase grid-connected inverters for photovoltaic modules," *IEEE Transactions on Industry Application*, vol. 41, pp. 1292–1306, Oct 2005.

[3] M. Karimi-Ghartemani, "Linear and pseudo linear enhanced phased locked loop (ePLL) structures," *IEEE Transactions on Industrial Electronics*, vol. 61, pp. 1464–1474, Mar 2014.

[4] N. Barrera Gallegos, M. Molinas, and V. Gasca Segura, "Synchronization properties of voltage source converters when seen as coupled oscillators based on the kuramoto model," in *MATHMOD 2018 Extended Abstract Volume, ARGESIM Report 55.* Vienna, Austria: ARGESIM, 2018, pp. 89–90.

[5] A. Timbus, M. Liserre, R. Teodorescu, P. Rodriguez, and F. Blaabjerg, "Evaluation of current controllers for distributed power generation systems," *IEEE Transactions on Power Electronics*, vol. 24, pp. 654–664, March 2009.

[6] F. Blaabjerg, R. Teodorescu, M. Liserre, , and A. V. Timbus, "Overview of control and grid synchronization for distributed power generation systems," *IEEE Transactions on Industrial Electronics*, vol. 53, pp. 1398–1409, October 2006.

[7] D. G. Holmes, T. A. Lipo, B. P. McGrath, and W. Y. Kong, "Optimized design of stationary frame three phase ac current regulators," *IEEE Transactions on Power Electronics*, vol. 24, pp. 2417–2426, November 2009.

[8] M. Malinowski, M. P. Kazmierkowski, S. Hansen, F. Blaabjerg, and G. D. Marques, "Virtual-flux-based direct power control of three-phase PWM rectifiers," *IEEE Transactions on Industry Applications*, vol. 37, pp. 1019–1027, Jul./Aug 2001.

[9] P. C. Loh and D. G. Holmes, "Analysis of multiloop strategies for LC/CL/LCL-filtered voltage-source and current-source inverters," *IEEE Transactions on Industry Applications*, vol. 41, pp. 644–654, Apr 2005.

[10] P. Mattavelli, "An improved deadbeat control for ups using disturbance observers," *IEEE Transactions on Industrial Electronics*, vol. 52, pp. 206–212, Feb 2005.

[11] M. Li, Y. Gui, Y. Guan, J. Matas, J. M. Guerrero, and J. C. Vasquez, "Inverter parallelization for an islanded microgrid using the hopf oscillator controller approach with self-synchronization capabilities," *IEEE Transactions on Industrial Electronics*, vol. 68, no. 11, pp. 10 879–10 889, 2021.

[12] S. Luo, W. Chen, X. Li, and Z. Hao, "A new virtual inertial strategy for andronov–hopf oscillator based grid-forming inverters," *IEEE Journal of Emerging and Selected Topics in Power Electronics*, pp. 1–1, 2024.

[13] UNIFI Consortium, "Specifications for grid-forming inverter-based resources: version 2," National Renewable Energy Lab.(NREL), Golden, CO (United States), Tech. Rep., 2024.

[14] B. B. Johnson, S. V. Dhople, A. O. Hamadeh, and P. T. Krein, "Synchronization of parallel single-phase inverters with virtual oscillator control," *IEEE Transactions on Power Electronics*, vol. 29, no. 11, pp. 6124–6138, 2014.

[15] B. Johnson, M. Rodriguez, M. Sinha, and S. Dhople, "Comparison of virtual oscillator and droop control," in *2017 IEEE 18th Workshop on Control and Modeling for Power Electronics (COMPEL), Stanford, CA, USA*, pp. 1–6.

[16] P. Rodriguez, R. Teodorescu, I. Candela, A. V. Timbus, M. Liserre, and F. Blaabjerg., "New positive-sequence voltage detector for grid synchronization of power converters under faulty grid conditions," in *Proc. 37th Annual IEEE Power Electronics Specialist Conf.*, Jeju, South Korea, June 2006.

[17] D. De and V. Ramanarayanan, "A proportional bm + multiresonant controller for three-phase four-wire high-frequency link inverter," *IEEE Transactions on Power Electronics*, vol. 25, p. 899–906, April 2010.

[18] IEEE Standards Association, "IEEE standard for interconnection and interoperability of distributed energy resources with associated electric power systems interfaces," in *IEEE Std 1547-2018 (Revision of IEEE Std 1547-2003) - Redline*, 2018, p. 1–227.

Estimation of Rectifier Output Current of the LLC Converter

Xin Wu, Yi Zhou, Haihong Long and Dehong Xu
College of Electrical Engineering
Zhejiang University
Hangzhou 310027, China
wuxin2020@zju.edu.cn

Abstract— **An estimation method of the rectifier output current of the LLC converter is proposed to avoid a bulk current sensor, which results in parasitic inductor in the rectifier loop and compromises the LLC converter performance. The principle of the estimation method is introduced. Then control of the LLC converter with the estimated rectifier output current is designed. Finally, the proposed rectifier output current estimation method is verified by the experiment.**

Keywords— LLC converter, rectifier output current estimation, current control.

I. Introduction

Solid state transformer (SST) has a big potential to be applied in data centers, fast charging stations and renewable energy power distribution systems etc. SST is typically composed of two stages: a front-end AC/DC converter and isolated DC/DC converter. To improve the efficiency and power-density of SSTs, the LLC converter is commonly used as the isolated DC/DC converter [1] – [4].

For LLC converter, conventional single-voltage-loop control sometimes cannot meet the requirement of fast dynamics. Several control schemes have been investigated by the predecessors to improve the control dynamics [5] – [10]. Charge control or one-cycle control has better dynamics . However it is suitable to be implemented with the analog circuits [6], [7]. The cascade voltage and current control is commonly used control [8] – [10].

In this case, the output current of the rectifier often needs to be sensed as the controlled variable of the inner current loop, which reflects the charging current of the output capacitor. Thus a bulk current sensor is required to measure the current due to the large current. The bulk current sensor not only has a large size, but will also introduce additional parasitic inductance in the secondary side of the LLC converter, which compromises the performance of the converter. In [9], the primary-side current is sensed as the controlled variable of the inner loop. However, it cannot directly reflect the variation of the output power due to the effect of the magnetizing current of the transformer.

In this paper, an estimation method of the rectifier output current of the LLC converter is proposed to avoid a bulk current sensor. Firstly, the principle of the estimation method including the digital implementation is introduced in Section II. In Section III, the design of the control loop based on the estimation method is discussed. The proposed estimation method is verified by the experiment, and the experimental results are shown in Section IV. Finally, the conclusion is

given in Section V.

II. Principle of the Estimation Method

In this section, the control diagram for MV three-level LLC converter is introduced. The principle of rectifier output current estimation is explained.

A. Control scheme

Three-level LLC converter with the cascade voltage and current control is shown in Fig. 1. Instead of measured directly by a sensor, the rectifier output current i_{rec} is estimated according to variables such as the resonant current i_p, output voltage v_o, switching frequency f_s, and the synchronic signal SYN input. The i_{rec} estimation block can calculate the average value of i_{rec} each switching period.

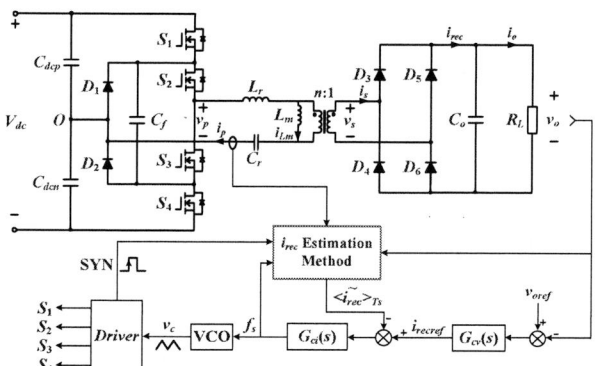

Fig. 1. Control diagram of the three-level LLC coverter based on the i_{rec} estimation method.

B. Estimation of i_{rec} when $f_s \leq f_r$

Typical waveforms of three-level LLC converter in a switching period when $f_s < f_r$ are shown in Fig. 2.

It is assumed that the output voltage v_o keeps constant in a switching period ($v_o = V_o$) and the magnetizing current i_{Lm} is constant at the interval $[T_r / 2, T_s / 2]$. It is derived [11]

$$I_{Lm} = \frac{nV_o}{4L_m f_r} \tag{1}$$

The magnetizing current i_{Lm} in the half switching period is derived [11]

$$i_{Lm}(t) = \begin{cases} -\dfrac{nV_o}{4L_m f_r} + \dfrac{nV_o}{L_m}t & 0 \leq t < T_r / 2 \\[2mm] \dfrac{nV_o}{4L_m f_r} & T_r / 2 \leq t < T_s / 2 \end{cases} \tag{2}$$

---

* This work is supported by National Natural Science Foundation of China under Grant 52037010.

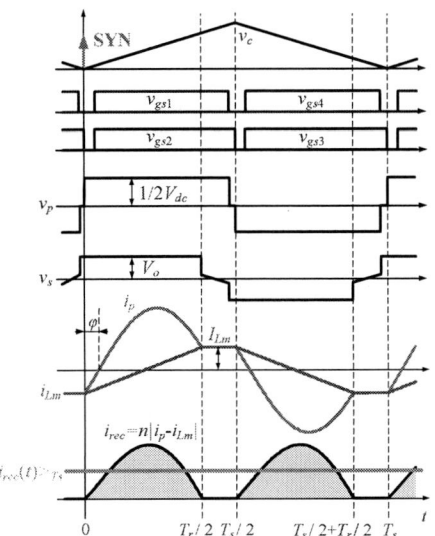

Fig. 2. Typical waveforms of three-level LLC converter in a switching period when $f_s < f_r$.

The integral value of i_{rec} in the half switching period is expressed as

$$\int_0^{T_s/2} i_{rec}(t)dt = \int_0^{T_s/2} n\left(i_p(t) - i_{Lm}(t)\right)dt$$
$$= n\int_0^{T_s/2} i_p(t)dt - n\int_0^{T_s/2} i_{Lm}(t)dt \qquad (3)$$

The integral value of i_{Lm} in the half switching period is derived as follows

$$\int_0^{T_s/2} i_{Lm}(t)dt = \int_0^{T_r/2} i_{Lm}(t)dt + \int_{T_r/2}^{T_s/2} i_{Lm}(t)dt$$
$$= I_{Lm} \cdot \frac{T_s - T_r}{2} = \frac{nV_o}{8L_m f_r}\left(\frac{1}{f_s} - \frac{1}{f_r}\right) \qquad (4)$$

Then equation (3) is rewritten as

$$\int_0^{T_s/2} i_{rec}(t)dt = n\int_0^{T_s/2} i_p(t)dt - \frac{n^2 V_o}{8L_m f_r}\left(\frac{1}{f_s} - \frac{1}{f_r}\right) \qquad (5)$$

The average value of i_{rec} in a switching period is derived as

$$\langle i_{rec}(t)\rangle_{T_s} = 2nf_s\int_0^{T_s/2} i_p(t)dt - \frac{n^2 V_o}{4L_m f_r}\left(1 - \frac{f_s}{f_r}\right) \qquad (6)$$

C. Estimation of i_{rec} when $f_s > f_r$

Typical waveforms of three-level LLC converter in a switching period when $f_s > f_r$ are shown in Fig. 3. Compared with Fig. 2, it can be seen that the interval $[T_r / 2, T_s / 2]$ disappears because the voltage on the magnetizing inductor L_m is clamped to either nV_o or $-nV_o$. Additionally, the intersection instant of the primary side current i_p and the magnetizing current i_{Lm} is t_0 instead of 0 in the first half of the switching period. As a result, the stages in the first half of the switching period are divided into $[0, t_0]$ and $[t_0, T_s / 2]$.

The amplitude of the magnetizing current i_{Lm} is derived as

$$I_{Lm} = \frac{nV_o}{4L_m f_s} \qquad (7)$$

The magnetizing current i_{Lm} in the half switching period is derived

$$i_{Lm}(t) = \begin{cases} -\dfrac{nV_o}{4L_m f_s} - \dfrac{nV_o}{L_m}(t - t_0) & 0 \le t < t_0 \\[3mm] -\dfrac{nV_o}{4L_m f_s} + \dfrac{nV_o}{L_m}(t - t_0) & t_0 \le t < T_s / 2 \end{cases} \qquad (8)$$

The integral value of i_{rec} in the half switching period is expressed as

$$\int_0^{T_s/2} i_{rec}(t)dt = \int_0^{T_s/2} n\left|i_p(t) - i_{Lm}(t)\right|dt$$
$$= n\left(-\int_0^{t_0} i_p(t)dt + \int_{t_0}^{T_s/2} i_p(t)dt\right) + n\left(\int_0^{t_0} i_{Lm}(t)dt - \int_{t_0}^{T_s/2} i_{Lm}(t)dt\right) \qquad (9)$$

According to half-wave symmetry, it is obtained that

$$i_{Lm}(t) = -i_{Lm}(t + T_s / 2) \qquad (10)$$

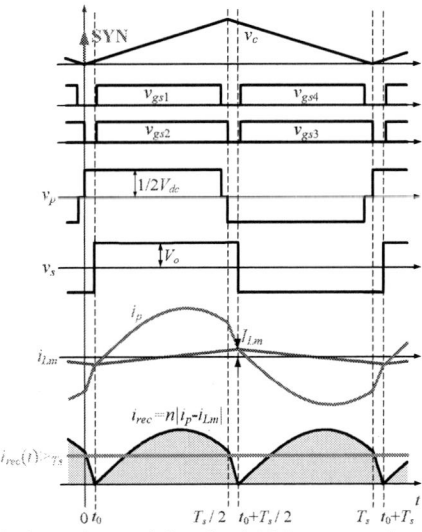

Fig. 3. Typical waveforms of three-level LLC converter in a switching period when $f_s > f_r$.

The second term of equation (9) is simplified as

$$\int_0^{t_0} i_{Lm}(t)dt - \int_{t_0}^{T_s/2} i_{Lm}(t)dt$$
$$= -\int_{T_s/2}^{t_0 + T_s/2} i_{Lm}(t)dt - \int_{t_0}^{T_s/2} i_{Lm}(t)dt \qquad (11)$$
$$= -\int_{t_0}^{t_0 + T_s/2} i_{Lm}(t)dt = 0$$

By substituting equation (11) into equation (9), it is derived

$$\int_0^{T_s/2} i_{rec}(t)dt = n\left(-\int_0^{t_0} i_p(t)dt + \int_{t_0}^{T_s/2} i_p(t)dt\right) \qquad (12)$$

The average value of i_{rec} in a switching period is derived as

$$\langle i_{rec}(t)\rangle_{T_s} = 2nf_s\left(-\int_0^{t_0} i_p(t)dt + \int_{t_0}^{T_s/2} i_p(t)dt\right) \qquad (13)$$

To calculate the above equation, instant t_0, i.e. the intersection instant of i_p and i_{Lm}, needs to be detected. It is realized by comparing measured i_p with $-I_{Lm}$ expressed by equation (7).

D. Digital implementation

For the requirement of MV isolation, two individual digital controllers (TI TMS320F28335) are used in MV side and LV side, and connected by optical fibers, as shown in Fig. 4.

The timing sequence of the control is shown Fig. 5. For the LV-side controller, the output voltage v_o is sampled after receiving the synchronic signal SYN (Step 1) and uploaded via SCI in the first half of T_s (Step 2).

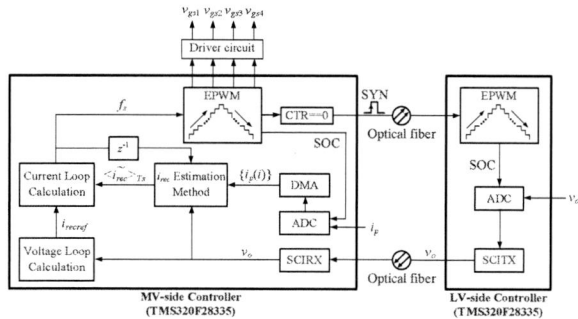

Fig. 4. Hardware architecture of LLC coverter controller.

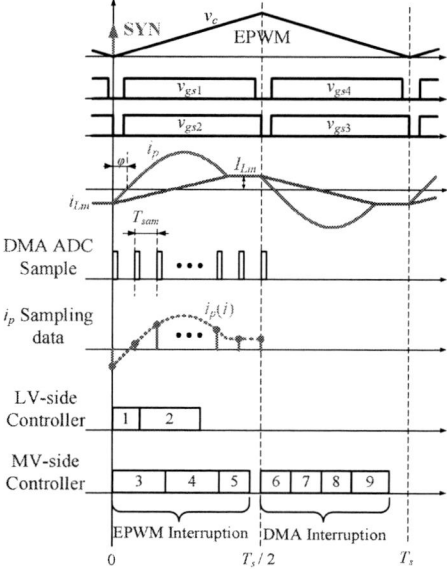

Fig. 5. Timing sequence of the software ($f_s < f_r$). Step decomposition—1: LV-side sampling; 2: Upload of LV-side sampling data via SCI; 3: Protection, mode idenficaition, etc.; 4: Reception of LV-side sampling data via SCI; 5: Voltage loop calculation; 6: i_{rec} estimation; 7: HV-side sampling except i_p; 8: Current loop calculation; 9: PWM calculation.

For the MV-side controller, the resonant current i_p is sampled with the interval T_{sam} via DMA in the first half of T_s. In this case, $T_{sam} = 680$ ns and the number of sampling points for rated operation point is 18. As shown in Fig. 4 and Fig. 5, the MV-side controller receives the uploaded data of v_o from the LV-side controller via SCI (Step 4). Then the voltage loop is calculated and the reference signal of i_{rec} — i_{recref} is obtained (Step 5). In the second half of T_s, the sampling of i_p is finished. $<i_{rec}(t)>_{Ts}$ is estimated with the

sampling data $\{i_p(i)\}$, f_s and v_o using equation (6) or (13) (Step 6). Then the current loop is calculated with i_{recref} and the estimated $<i_{rec}(t)>_{Ts}$, and f_s of next period is obtained (Step 8). Finally, the driving signals of $S_1 \sim S_4$ are generated via EPWM (Step 9). Other functions such as protection are also realized by the MV-side controller (Step 3 and Step 7).

III. DESIGN OF CONTROL LOOP

In this section, the dynamic model of the LLC converter is introduced. According to the model, both single-voltage-loop control and cascaded control are designed. Then these two designs are compared.

A. Dynamic model

To derive the dynamic model, a static operation point is assumed, where the switching frequency F_s is below the resonant point, i.e. $F_s < f_r$.

According to [12], when the LLC converter operates at switching frequency $F_s < f_r$, the transfer function from the switching frequency f_s to the output voltage v_o is expressed as:

$$G_{vf}^0(s) = \frac{\hat{v}_o(s)}{\hat{f}_s(s)} = 2\pi G_{DC} \frac{1}{1 + \dfrac{s}{Q_p \omega_p} + \left(\dfrac{s}{\omega_p}\right)^2} \quad (14)$$

where G_{DC}, Q_p and ω_p are calculated as [12]

$$G_{DC} = \frac{V_{dc}}{2n} \frac{L_n}{\Omega_0} \frac{\left[\left(\frac{1}{\Omega_n^2} - \Omega_n^2\right)\left(\frac{\pi^2}{8}QL_n\right)^2 - \left(L_n + 1 - \frac{1}{\Omega_n^2}\right)\left(\frac{2}{\Omega_n^2}\right)\right]\frac{1}{\Omega_n} \cdot \frac{1}{\sin\left(\frac{\pi}{2}\Omega_n\right)} + \left[\left(L_n + 1 - \frac{1}{\Omega_n^2}\right)^2 + \left(\left(\frac{1}{\Omega_n} - \Omega_n\right)\cdot\frac{\pi^2}{8}QL_n\right)^2\right]\left(-\frac{\pi}{2}\frac{\cos\left(\frac{\pi}{2}\Omega_n\right)}{\sin\left(\frac{\pi}{2}\Omega_n\right)}\right)}{\left[\sqrt{\left(L_n + 1 - \frac{1}{\Omega_n^2}\right)^2 + \left(\left(\frac{1}{\Omega_n} - \Omega_n\right)\cdot\frac{\pi^2}{8}QL_n\right)^2}\right]^3}$$

$$(15)$$

$$Q_p = \frac{8n}{\pi^2} R_L \sqrt{\frac{C_o}{L_e}} \quad (16)$$

$$\omega_p = \sqrt{\frac{1}{L_e \dfrac{\pi^2}{8n^2} C_o}} \quad (17)$$

The intermediate variables in equations (15) ~ (17) are defined as [12]

$$\begin{cases} \Omega_0 = 2\pi f_r \\ \Omega_s = 2\pi F_s \\ \Omega_n = \Omega_s / \Omega_0 \\ L_n = L_m / L_r \\ L_e = \left(1 + \Omega_0^2 / \Omega_s^2\right)L_r + \left(1 - \Omega_s / \Omega_0\right)L_m \\ Q = \sqrt{L_r / C_r} / \left(n^2 R_L\right) \end{cases} \quad (18)$$

According to Fig. 1, the load impedance of the LLC converter is expressed as

979-8-3315-1612-3/25 $31.00 © 2025 IEEE

$$Z_o(s) = \frac{\hat{v}_o(s)}{\hat{i}_{rec}(s)} = \frac{R_L}{R_L C_o s + 1} \quad (19)$$

By combining equations (14) and (19), the transfer function from f_s to the rectifier output current i_{rec} is derived:

$$G^0_{if}(s) = \frac{\hat{i}_{rec}(s)}{\hat{f}_s(s)} = \frac{G^0_{vf}(s)}{Z_o(s)} = \frac{2\pi G_{DC}}{R_L} \frac{R_L C_o s + 1}{1 + \dfrac{s}{Q_p \omega_p} + \left(\dfrac{s}{\omega_p}\right)^2} \quad (20)$$

Since $G^0_{vf}(s)$ and $G^0_{if}(s)$ are negative, for the convenience of loop analysis, we will analysis $G_{vf}(s) = -G^0_{vf}(s)$ and $G_{if}(s) = -G^0_{if}(s)$.

The experimental parameters of the three-level LLC converter are listed in TABLE I. The analysis of frequency characteristics is based on the rated operation point of the low voltage frequency-sweeping test ($V_{dc} = 400$ V) shown in TABLE I.

TABLE I. EXPERIMENTAL PARAMETERS

Parameter	Value	
	High voltage experiment	Low voltage frequency-sweeping test
Rated power P_{orate}	7.42 kW	1.15 kW
DC voltage V_{dc}	1000 V	400 V
Output voltage V_o	77 ~ 115 V (104 V rated)	31 ~ 46 V (41 V rated)
Output filter capacitance C_o	400 µF	
Turn ratio of the transformer n	17:3	
Resonant inductance L_r	80.2 µH	
Resonant capacitance C_r	136.5 nF	
Magnetizing inductance L_m	275.4 µH	
Rated load resistance	1.458 Ω	
Resonant frequency f_r	48.1 kHz	
Switching frequency f_s	35 ~ 80 kHz (41.9 kHz rated)	

The frequency characteristics of $G_{vf}(s)$ and $G_{if}(s)$ are drawn in Fig. 6 and Fig. 7 respectively according to equation (14) and (20). In the two figures, both the PLECS simulation and the experiment results are added. Three curves coincide well in the figures. It is observed that there are resonant peaks around 2.4 kHz in both figures caused by conjugate poles. With regards to the phase-frequency characteristics, both the simulation and experiment characteristics show larger phase delay than the theoretical model in higher disturbance frequency range.

Finally, the frequency characteristics measured by the experiment are used to design the control. The experimental measurement data is fitted with polynomials to obtain smooth frequency characteristics curve. Since $G_{vf}(s)$ and $G_{if}(s)$ are proportional to V_{dc} according to equation (15), the frequency characteristics obtained at $V_{dc} = 400$ V are extended to high voltage as follows:

$$G_{vf}(s) = \frac{V_{dc}}{400} G_{vf\_400V}(s) \quad (21)$$

$$G_{if}(s) = \frac{V_{dc}}{400} G_{if\_400V}(s) \quad (22)$$

The following loop design is based on the rated operation point of the high voltage experiment ($V_{dc} = 1000$ V) listed in TABLE I. $G_{vf}(s)$ and $G_{if}(s)$ are derived according to equation (21) and (22), respectively.

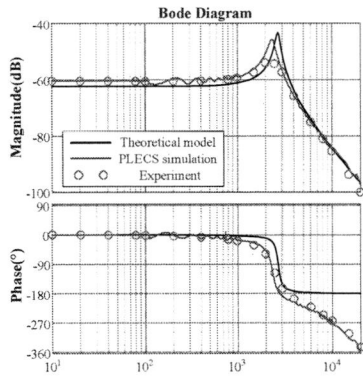

Fig. 6. The frequency characteristic of $G_{vf}(s)$.

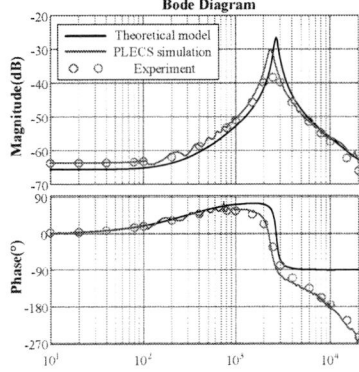

Fig. 7. The frequency characteristic of $G_{if}(s)$.

B. Single-voltage-loop control

The diagram of single-voltage-loop control is shown in Fig. 8, where $G_{cv}(s)$, $G_d(s)$ and $H_v(s)$ are the transfer functions of the voltage controller, digital delay and voltage-sampling feedback part, respectively.

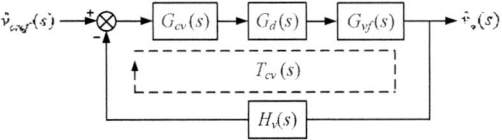

Fig. 8. The diagram of single-volage-loop control.

The open loop gain $T_v(s)$ is

$$T_v(s) = G_d(s) \cdot G_{vf}(s) \cdot H_v(s) \quad (23)$$

The closed loop gain $T_{cv}(s)$ is

$$T_{cv}(s) = G_{cv}(s) \cdot G_d(s) \cdot G_{vf}(s) \cdot H_v(s) \quad (24)$$

Following PI control is used

$$G_{cv}(s) = K_{pv} + \frac{K_{Iv}}{s} \quad (25)$$

The crossing frequency of $T_{cv}(s)$ — f_{cv} is set about 1/10 of the conjugate poles frequency ($\approx$ 2.4 kHz) so the resonant peak caused by the conjugate poles is below 0 dB. The corner frequency of PI controller — f_{zv} is set higher than the conjugate poles frequency to achieve larger gain margin. In this case, $f_{cv} = 200$ Hz and $f_{zv} = 10$ kHz.

The following equations can be obtained:

$$\begin{cases} \left| T_v \left(j 2 \pi f_{cv} \right) \right| = \dfrac{1}{\left| G_{cv} \left(j 2 \pi f_{cv} \right) \right|} \\ \dfrac{K_{iv}}{2 \pi K_{pv}} = f_{zv} \end{cases} \tag{26}$$

By solving the equations (26), it is calculated that K_{pv} = 8.68 and K_{iv} = 545407.4. After compensation, the phase margin is 84.4° and the gain margin is 12.6 dB. The Bode diagrams of $T_v(s)$, $T_{cv}(s)$ and $G_{cv}(s)$ are shown in Fig. 9. It can be observed that the bandwidth of the voltage loop is mainly limited by the resonant peak.

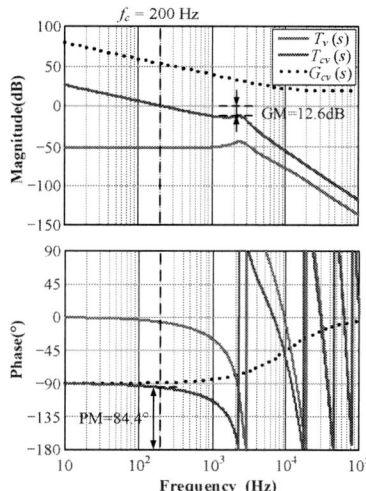

Fig. 9. Bode diagram of $T_v(s)$, $T_{cv}(s)$ and $G_{cv}(s)$.

C. Cascaded control

The diagram of cascaded control is shown in Fig. 10. The rectifier output current i_{rec} is estimated as the inner loop with the proposed method.

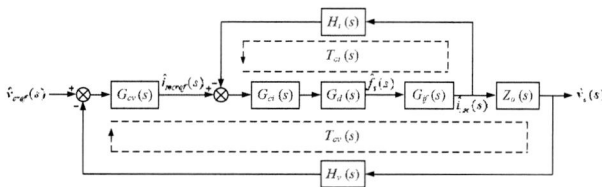

Fig. 10. The diagram of cascaded control.

For the current loop, the open loop gain $T_i(s)$ is

$$T_i(s) = G_d(s) \cdot G_{if}(s) \cdot H_i(s) \tag{27}$$

The closed loop gain $T_{ci}(s)$ is

$$T_{ci}(s) = G_{ci}(s) \cdot G_d(s) \cdot G_{if}(s) \cdot H_i(s) \tag{28}$$

Following P control is used

$$G_{ci}(s) = K_{pi} \tag{29}$$

The crossing frequency of $T_{ci}(s)$ — f_{ci} is set about 1/10 of the switching frequency F_s (41.9 kHz) to suppress the ripple. In this case, f_{ci} = 3000 Hz and K_{pi} is calculated as

$$K_{pi} = \frac{1}{\left| T_i \left(j 2 \pi f_{ci} \right) \right|} = 38.9 \tag{30}$$

After compensation, the phase margin is 74.9° and the gain margin is 10.7 dB. The bode diagrams of $T_i(s)$, $T_{ci}(s)$ and $G_{ci}(s)$ are shown in Fig. 11.

The closed-loop transfer function of current loop is

$$G_{clsi}(s) = \frac{G_{ci}(s) \cdot G_d(s) \cdot G_{if}(s)}{1 + G_{ci}(s) \cdot G_d(s) \cdot G_{if}(s) \cdot H_i(s)} \tag{31}$$

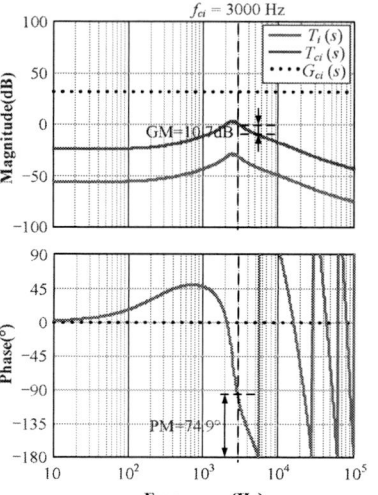

Fig. 11. Bode diagram of $T_i(s)$, $T_{ci}(s)$ and $G_{ci}(s)$.

For the voltage loop, the open loop gain $T_v(s)$ is

$$T_v(s) = G_{clsi}(s) \cdot Z_o(s) \cdot H_v(s) \tag{32}$$

The closed loop gain $T_{cv}(s)$ is

$$T_{cv}(s) = G_{cv}(s) \cdot G_{clsi}(s) \cdot Z_o(s) \cdot H_v(s) \tag{33}$$

Following PI control is used as equation (25) for the voltage loop.

The crossing frequency of $T_{cv}(s)$ — f_{cv} is set about 1/10 ~ 1/5 of f_{ci}. The corner frequency of PI controller — f_{zv} is set higher than the conjugate poles frequency to achieve larger gain margin. In this case, f_{cv} = 500 Hz and f_{zv} = 10 kHz. According to equations (26), K_{pv} = 0.60 and K_{iv} = 37567.1. After compensation, the phase margin is 72.7° and the gain margin is 11.4 dB. The Bode diagrams of $T_v(s)$, $T_{cv}(s)$ and $G_{cv}(s)$ are shown in Fig. 12.

Comparing $T_v(s)$ in Fig. 9 and Fig. 12, it can be seen the resonant peak caused by the conjugate poles is attenuated and moves to higher frequency. The damping of the system increases and the higher bandwidth of voltage loop can be achieved.

IV. EXPERIMENTAL VERIFICATION

In this section, a three-level LLC converter is set up to verify the proposed i_{rec} estimation method and control design. The high voltage experimental parameters are listed in TABLE I.

979-8-3315-1612-3/25 $31.00 © 2025 IEEE

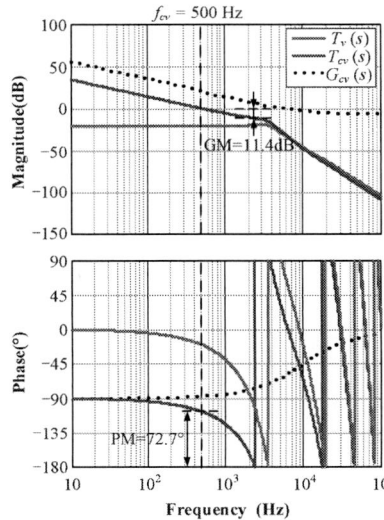

Fig. 12. Bode diagram of $T_v(s)$, $T_{cv}(s)$ and $G_{cv}(s)$ for cascaded control.

A. Verification of i_{rec} estimation method

In order to verify the feasibility of the proposed i_{rec} estimation method, the estimation results are compared with experimental measurement results in Fig. 13. It is observed that the estimated values matches the measurement values well at full load and 50% load. The estimation accuracy decreases at 10% load.

Fortunately, the estimated i_{rec} is used in the inner current loop while the output voltage is regulated by the outer voltage loop. Hence, the accuracy of the output voltage control is not so susceptible to the estimation error.

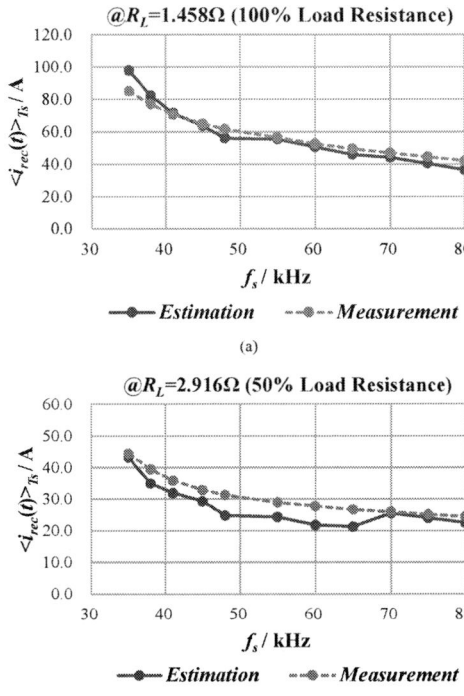

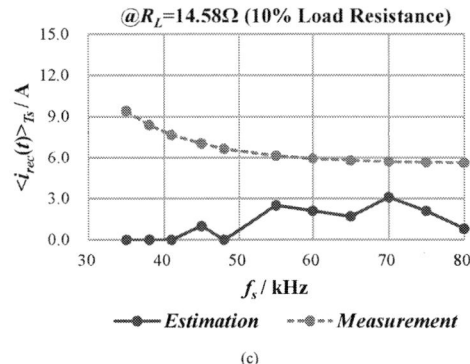

Fig. 13. The estimation and measurement values of $<i_{rec}(t)>_{Ts}$ at different operation points: (a) 100% load; (b) 50% load; (c) 10% load.

B. Dynamic response test

Firstly, the dynamic response of single-voltage-loop control when the load steps up and down is tested in rated condition. The waveforms of v_p, v_o, i_o and i_p are shown in Fig. 14. It is defined the duration when v_o enters within ±0.5% range of the final value as the settling time t_s. A longer time ring is observed during transient. The settling time is longer than 2 ms as shown in Fig. 14.

The dynamic response of the cascaded control based on the proposed i_{rec} estimation method is shown in Fig. 15. Compared with single-voltage-loop control, the settling time is reduced to about 700 μs. Owing to the function of the inner current loop, the damping of the system increases and higher bandwidth can be achieved.

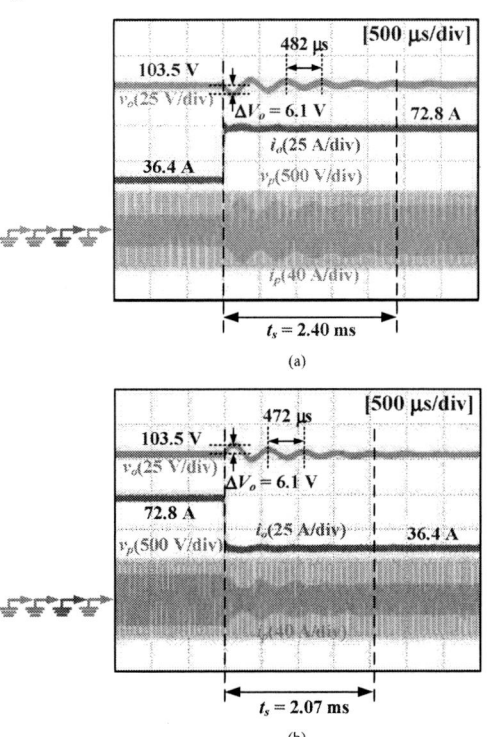

Fig. 14. The transient waveforms of single-voltage-loop control. (a) Half load to full load; (b) Full load to half load.

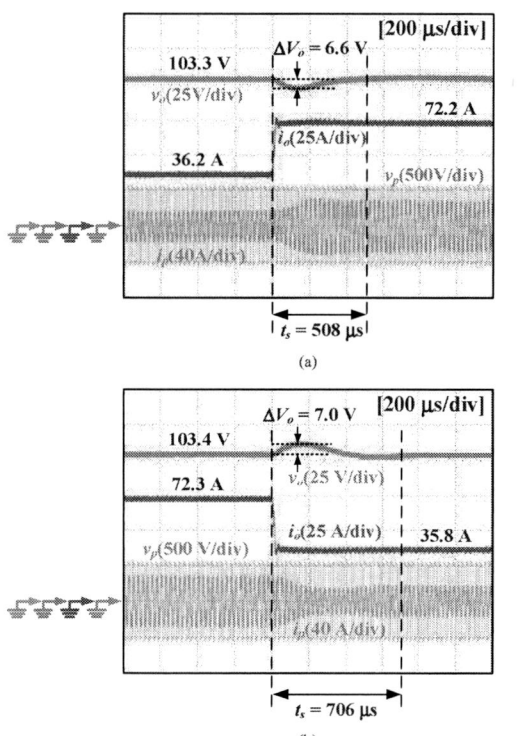

Fig. 15. The transient waveforms of cascaded control with the proposed i_{rec} estimation method. (a) Half load to full load; (b) Full load to half load.

V. CONCLUSION

In this paper, an estimation method of rectifier output current for LLC converter is proposed. The principle of the proposed method and the digital implementation are introduced. The proposed method is verified by the experiment. Further it shows that cascaded control based on the proposed estimation method outperforms the single-voltage-loop control. Higher power SST module test with the three-level LLC converter will be reported in the near future.

REFERENCES

[1] C. Zhao, D. Dujic, A. Mester, J. K. Steinke, M. Weiss, S. L.-Schmid, T. Chaudhuri and P. Stefanutti, "Power Electronic Traction Transformer—Medium Voltage Prototype," *IEEE Trans. Power Electron.*, vol. 61, no. 7, pp. 3257-3268, July 2014.

[2] M. Lee, C. -S. Yeh, O. Yu, J. -W. Kim, J. -M. Choe and J. -S. Lai, "Modeling and Control of Three-Level Boost Rectifier Based Medium-Voltage Solid-State Transformer for DC Fast Charger Application," *IEEE Trans. Transport. Electrific.*, vol. 5, no. 4, pp. 890-902, Dec. 2019.

[3] D. Rothmund, T. Guillod, D. Bortis and J. W. Kolar, "99% Efficient 10 kV SiC-Based 7 kV/400 V DC Transformer for Future Data Centers," *IEEE J. Emerg. Sel. Topics Power Electron.*, vol. 7, no. 2, pp. 753-767, June 2019.

[4] C. Lu, W. Hu and F. C. Lee, "Neutral-Point Voltage Balancing Methods of Series-Half-Bridge LLC Converter for Solid State Transformer," *IEEE Trans. Power Electron.*, vol. 36, no. 6, pp. 7060-7073, June 2021.

[5] W. Feng, F. C. Lee and P. Mattavelli, "Simplified Optimal Trajectory Control (SOTC) for LLC Resonant Converters," *IEEE Trans. Power Electron.*, vol. 28, no. 5, pp. 2415-2426, May 2013.

[6] Q. Xu. R. Zou, T. Wen, S. Du, X. Li, D. Lu and H. Hu, "A Nonlinear Load Current Feedforward Strategy for the Charge-Controlled LLC Converter and Its Digital Implementation to Improve the Dynamic Response," *IEEE Trans. Ind. Electron.*, vol. 70, no. 10, pp. 10195-10203, Oct. 2023.

[7] Z. Hu, Y. -F. Liu and P. C. Sen, "Bang-Bang Charge Control for LLC Resonant Converters," *IEEE Trans. Power Electron.*, vol. 30, no. 2, pp. 1093-1108, Feb. 2015.

[8] F. Degioanni, I. G. Zurbriggen and M. Ordonez, "Dual-Loop Controller for LLC Reson-ant Converters Using an Average Equivalent Model," *IEEE Trans. Power Electron.*, vol. 33, no. 11, pp. 9875-9889, Nov. 2018.

[9] S. Zong, H. Luo, W. Li, X. He and C. Xia, "Theoretical Evaluation of Stability Improvement Brought by Resonant Current Loop for Paralleled LLC Converters," *IEEE Trans. Ind. Electron.*, vol. 62, no. 7, pp. 4170-4180, July 2015.

[10] D. Cittanti, M. Gregorio, E. Vico, F. Mandrile, E. Armando and R. Bojoi, "High-Performance Digital Multiloop Control of LLC Resonant Converters for EV Fast Charging With LUT-Based Feedforward and Adaptive Gain," *IEEE Trans. Ind. Appl.*, vol. 58, no. 5, pp. 6266-6285, Sept.-Oct. 2022.

[11] W. Chen, Y. Gu and Z. Lu. A Novel Three Level Full Bridge Resonant DC-DC Converter Suitable for High Power Wide Range Input Applications[C]. *Proceeding of the 2007 22nd Annual IEEE Applied Power Electronics Conference and Exposition – APEC 2007.* Anaheim: IEEE, 2007: 373-379.

[12] S. Tian, F. C. Lee and Q. Li, "Equivalent Circuit Modeling of LLC Resonant Converter," *IEEE Trans. Power Electron.*, vol. 35, no. 8, pp. 8833-8845, Aug. 2020.

979-8-3315-1612-3/25 $31.00 © 2025 IEEE

A 100kHz digitally controlled 10kW, 2-channel solar MPPT converter using 3-level topology with >75W/in^3 power density and >98.5% peak efficiency

Ranajay Mallik
System Research and Applications Lab
STMicroelectronics
Greater Noida, India
ranajay.mallik@st.com

Akshat Jain
System Research and Applications Lab
STMicroelectronics
Greater Noida, India
akshat.jain@st.com

Abstract— Not much literature is available on the design and implementation of a high power (>5kW) high density solar MPPT converter (either buck or boost) targeting the 200-850V range and with multi-level topology, to target ~850VDC. This paper illustrates a 10kW MPPT controller implemented with SiC 650V MOSFETs but in a three-level flying capacitor configuration and it targets an input voltage of 425-850V with an output of 400V nominal. There are two independent channels with active current sharing on the common output 400V DC bus for feeding the battery bank or an inverter. The peak power conversion efficiency of the system exceeds 98.5% and a standard Perturb and Observe algorithm is used for the MPPT function. One single 32-bit STM32G4 ARM MCU controls all the functions of MPPT, Buck conversion of two channels, flying capacitor balancing of two channels, output regulation and protection. It is targeted at renewables and energy storage applications.

Keywords—multi-level converter, 3-level topology, digitally controlled, two channel MPPT converter, SiC, flying capacitor balancing

I. INTRODUCTION

The system in this paper uses SiC MOSFETs coupled with powerful digital control to reach a 75W/in^3 power density and 98.5% peak conversion efficiency while running at 100kHz. IGBTs cannot operate at very high frequency due to their slow switching, causing excessive dynamic losses and hence need proportionately larger passives. Silicon MOSFETs are suitable at low-medium power but lose out to IGBTs due to the MOSFET's finite on-resistance. The power dissipated in a MOSFET is proportional to the square of the current times its R$_{ds}$ON, but an IGBT has a uniform saturation voltage, V$_{CE}$ sat, and hence the loss is proportional to current. SiC MOSFET is a clear winner at 100kHz levels and targeting the 10-20kW range. It offers a balance of high power density, high efficiency and a higher return on capital for the deployed renewable or energy storage system, in the long run.

II. KEY DESIGN CONSIDERATIONS

Significant analysis and work on three-level converters for use on chip and for telecom applications exist. [1], [2], [3] highlight the converters that are designed for either 100kW or for 3kW smaller systems while useful works, such as in [4], [5] and [6] provide a lot of insight into operation of a three-level converter but are restricted to a single channel and much

lower power, and not so much for high voltage, high power renewable energy applications. The system under discussion targets a) dual channel operation for connecting 2 high voltage photovoltaic strings independently b) MPPT plus Buck conversion c) Flying capacitor balancing d) Very high-power density running at 100kHz and, e) Output power combining into a single DC bus all controlled by a single MCU. The actual system is shown in Fig.1. while in Fig.2 the block diagram of the implementation is highlighted. From the dimensions, the power density is 76.6W/in^3.

Fig.1: Actual System 310 x 115 x 60 mm

III. FLYING CAPACITOR AND INDUCTOR : CRUCIAL ELEMENTS

The three-level flying capacitor buck converter offers higher efficiency and higher power density and aids in restricting the MOSFET withstanding voltage to half the input. [7], [8]. It effectively halves the inductor size because at steady state balance, the effective switching frequency seen by the inductor is 2 x F$_{sw}$. This also means that the flying capacitor must always be kept at half of the input voltage under all operating conditions to ensure system reliability and high performance. Similarly, the inductor must be sized in a way that DC and AC losses, ripple current and size are all minimized. In the system presented here, we have designed the main inductor and flying capacitor under the following constraints:

Table 1: Design Specifications

Parameters	Value
Input Voltage (min, max)	425, 850VDC
Output Voltage	375-400VDC
Iout max	12.5A
Inductor Ripple current	30% max
Flying Cap ripple voltage	10% max

979-8-3315-1612-3/25 $31.00 © 2025 IEEE

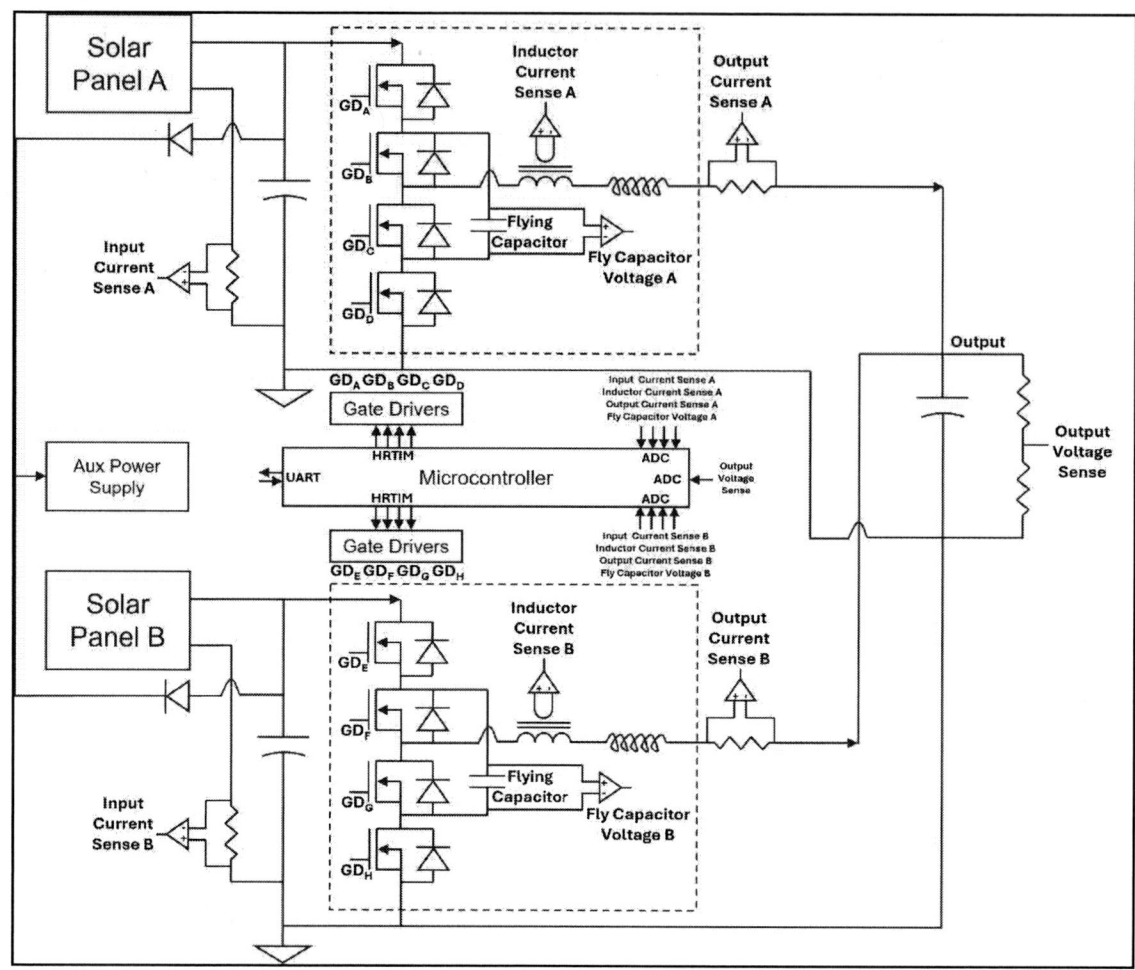

Fig.2: Block diagram of the 2-channel 3-level converter developed

Flying capacitor ripple voltage is given by:

$$\frac{0.5 * I_o * D}{C * f_{sw}}$$

Inductor ripple current can be expressed as:

$$\frac{V_{in} * (1 - D) * (D - 0.5)}{L * f_{sw}} \; for \; D > 0.5$$

From Fig.2, there are two channels working independently and delivering 5kW each into the common output bus. So, each channel delivers 400 x 12.5A = 5000W and that is the basis of the calculations for each channel. From the equations above, and considering practical choice, the flying capacitor is 3uF (3 x 1uF in parallel). The inductance value for a three-level buck converter may be expressed as:

$$L = \frac{V_{in}{}^{max} * 0.25}{8 * I_{ripple}{}^{max} * I_o{}^{max} * f_{sw}}$$

A SENDUST core swinging inductance has a benefit of consistently low ripple current over the entire load and duty cycle range, because the instantaneous inductance is a function of the Ampere-turns in the inductor as governed by

its magnetizing curve in Fig.3, Fig.4 shows the ripple reduction possible [14] and Fig.5 the actual chokes.

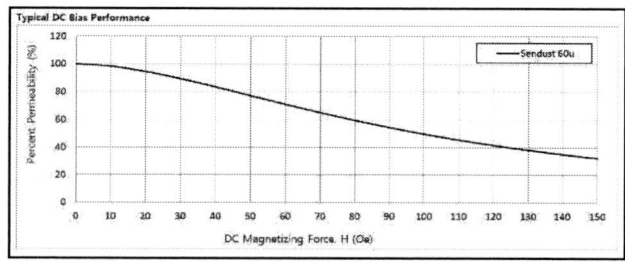

Fig.3: Magnetizing curve of 60μ SENDUST

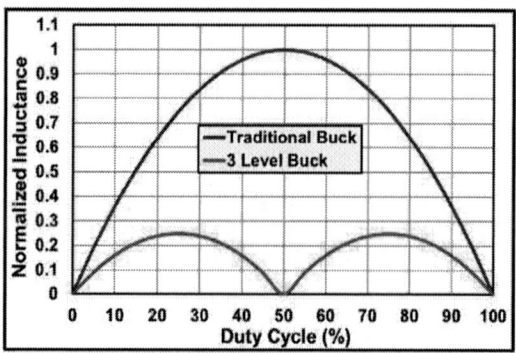

Fig.4: Ripple current : 2-level vs 3-level

979-8-3315-1612-3/25 $31.00 © 2025 IEEE

Fig.5: Two channel inductors

For a maximum load current of 12.5A and Vin(max) of 850V, F_{sw} = 100kHz, L = 70µH with a maximum ripple of 30% while it should be 700µH for 10% of full load. This span is comfortably covered by the inductor over the entire load current range. A ferrite core would be much larger physically, and with a fixed inductance, ripple would have been progressively higher at lighter loads as explained, since the inductance calculation would have been based on the maximum power output. It would have been difficult to achieve this power density or performance with a ferrite core inductor. [10] offers good insight into inductor selection.

IV. CHOICES OF SWITCHING DEVICES AND LOSSES

SiC MOSFET body diodes have much lower stored charge (Q_{rr}) than silicon types, therefore improving conversion efficiency significantly. SiC MOSFET output capacitance and channel on-resistance is also comparatively lower than for silicon types, leading to yet more efficiency savings. So, a trade-off between maximum heatsink temperature, heatsink size and a conservative loss budget was set up during the specifications phase. For high efficiency, the MOSFET losses were targeted at 1% of the total power throughput, given that we can express the four major losses with some approximations, are as follows:

A. Switching loss

$$0.5 * f_{sw} * \left(\frac{V_{dd} * 0.5 * I_L *}{[t_{fall} + t_{rise}]} \right) + \left(\frac{C_{oss} * [V_{dd}^2]}{2} \right)$$

B. Total Gate drive loss

$$V_g * f_{sw} * 4 * Q_g$$

C. Conduction loss

$$I_{rms}^2 * [D * (2 * R_{ds}{}^{on}) + (1 - D) * (2 * R_{ds}{}^{on})]$$

D. Dead time loss

$$V_{DD} * \left[\begin{array}{l} I_L{}^{max} * (t_{DT}(Q1,3) + t_{DT}(Q2,4)) \\ + I_L{}^{min} * (t_{DT}(Q1,3) + t_{DT}(Q2,4)) \end{array} \right] * f_{sw}$$

The gate drive resistors chosen [11], [12] were a tradeoff between switching time, dv/dt constraints and EMI management. With a 6.8ohms ON and a 2.2ohms OFF resistance with a +19/-4V bipolar drive at 100kHz, it has been managed a realistic T_{on} = 55nS and a T_{off} = 60nS with an isolated gate driver. The measured end to end efficiency at 5kW/channel is 97.7% and a peak efficiency of 98.6%, V_{mpp} (max power point) 650V, Vout 400V, Iout 12.5A DC.

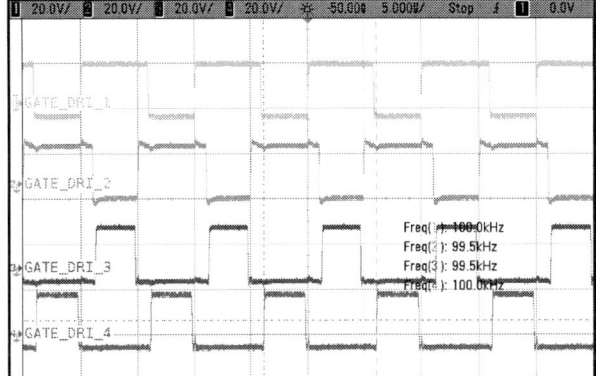

Fig.6: Gate Drive Signals

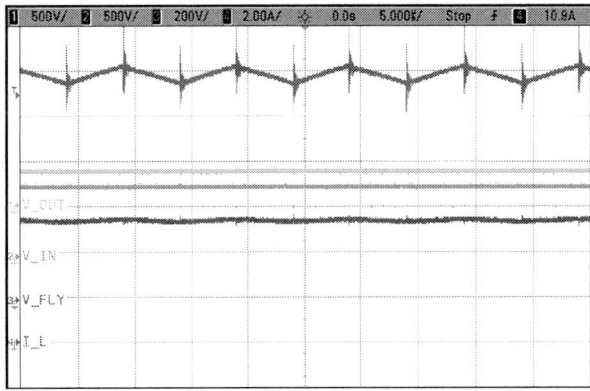

Fig.7: Voltages and current operating at 5kW

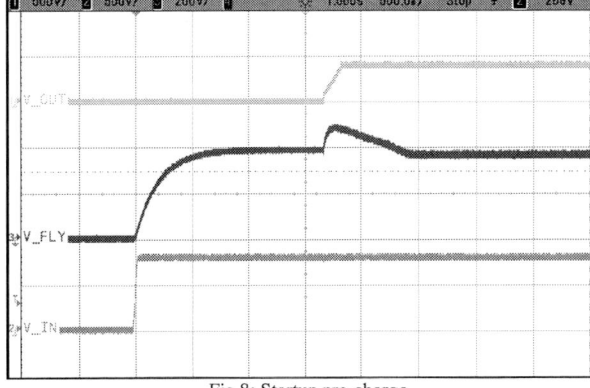

Fig.8: Startup pre-charge

V. CONTROL ALGORITHM OF THE SYSTEM DEVELOPED

The system works in average current mode control implemented in firmware. At very light loads or during start-up the system works like a voltage mode control to reach the regulation set point. After the voltage set point is reached and load current starts flowing the system firmware moves on to average current mode control.

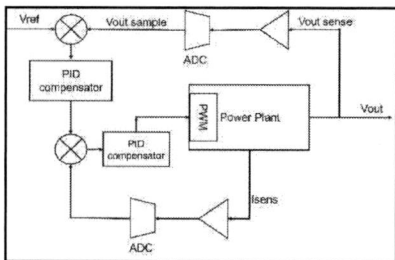

Fig.9: Control scheme

Since this is an MPPT system, the load range can be varied and so can the operating voltage range on the input side. At light loads, or, when the input is very high, but the output current is small, it is difficult to implement peak current mode control due to insufficient current and very high noise due to hard switching. Further, the output must still be regulated to meet the limits of the downstream inverter or load. Hence starting up the system at light, or zero load, with voltage mode control adds to the reliability. As the current demand peaks up, from about 20% of rated full load current, the microcontroller moves on to average current mode control. The scheme also avoids the large peak to average current errors in cases where the inductor goes discontinuous. Overall, we mitigate noise immunity issues, need for slope compensation under a wide operating range and yet achieve the performance of a current mode control.

The block diagram shown in Fig.9 shows the overall control strategy. The inductor current is sensed using a tunnel magneto-resistive device with an isolated output and the output voltage is sampled with standard resistive dividers. These are measured with two ADC channels and depending on the operating state: either voltage mode control or average current mode control is applied.

To manage dual-channel MPPT based on three-level converter topology, the firmware implemented includes two control loops i.e. outer control loop and inner control loop. The outer control loop manages the MPPT routine while the inner control manages to regulate the flying capacitor voltage and output voltage as shown in Fig.10. The system implementation requires a mix of fast computing and high-resolution TIMERs and ADC peripherals.

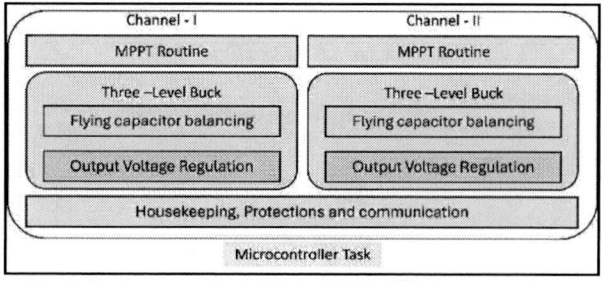

Fig.10: Tasks managed by MCU in 10kW MPPT charger developed

VI. Gate Drive and Flying Capacitor Balancing

The gate drive of channel A switches GD$_A$ and GD$_B$ are phase shifted by 180° while the gate drive GD$_D$ is complementary to gate drive GD$_A$ and the gate drive GD$_C$ is complementary

to gate drive GD$_B$. Fig.6. highlighting the gate drive working at 65% duty cycle.

The flying capacitor voltage is measured at ADC peripheral and is compared with half of the input voltage. The duty cycle is updated based on the PI control loop implementation such that if the flying capacitor voltage is greater than half of the input voltage then the duty cycle for switches QA and QC will decrease and for switches QB and QD will increase in equal steps and vice versa as highlighted in Table 2.

Table 2: Duty cycle change of the gate drive signals to have Flying capacitor voltage balanced at half of input voltage

If flying capacitor voltage is greater or lesser than half of input voltage?	Switch QA Gate Drive GD$_A$	Switch QB Gate Drive GD$_B$	Switch QC Gate Drive GD$_C$	Switch QD Gate Drive GD$_D$
Greater	Decrease	Increase	Decrease	Increase
Lesser	Increase	Decrease	Increase	Decrease

VII. Noiseless ADC Sampling

Any digital control loop takes several input signals, which are read at the ADC peripheral. If ADC sampling coincides with the ON/OFF of any power switch in the converter, then the noise from the hard switching transition will affect the small signal being measured and lead to variation in the measured value. No amount of analog filtering or averaging in the firmware can remove the error without introducing a big delay. For reasonable digital control speed, it is to be ensured that the measurements of the signals are not only clean but acquired in a minimum amount of time. The system developed has a 425 to 850V DC input voltage range, while the output is fixed at 400V DC. So, in normal working conditions the minimum duty cycle will be around 47% while the startup is from 25% duty cycle. Also, the system developed is dual channel so overall there are eight PWM signals that need to be controlled as shown in Fig.11. In the firmware, the PWM signals are synced with the master TIMER peripheral which means the starting/reset of each TIMER peripheral is known. With this the ADC sampling is scheduled at 20% period of the master TIMER where there is no MOSFET switching transition happening at all, ensuring noise-free measurements at the ADC peripheral.

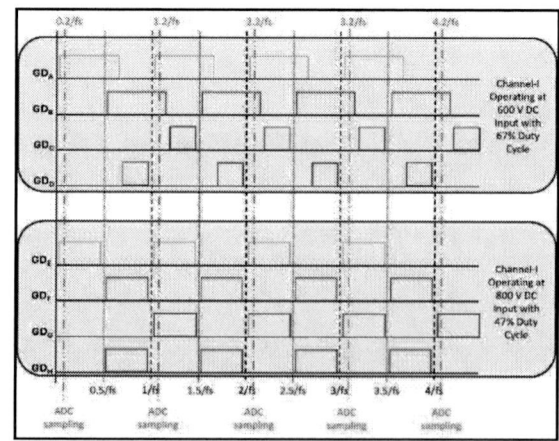

Fig.11: ADC sampling at intervals to avoid switching noise

VIII. Testing Results

The board has been tested under wide input voltage range. Fig.12 highlight the efficiency performance with respect to power output at 600V, 700V and 800V input voltage. The key components used in the 2-channel three level converter design has been highlighted in Table 3.

Table 3: Components used

Device Purpose	Device Part Number
SiC MOSFETs used in three level converter	SCT040W65G3AG
Gate Driver	STGAP2SM
Inductor	Sendust 60µ x 2 : 800µH
Flying capacitor	1uF x 3 R71PI410050H6K
Microcontroller	STM32G474

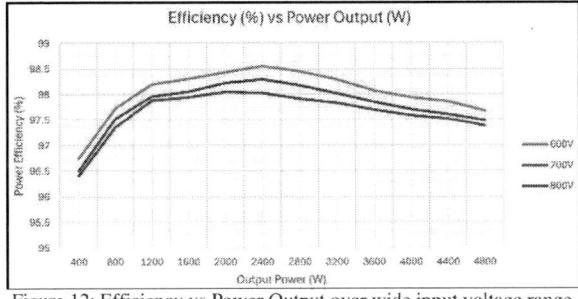

Figure 12: Efficiency vs Power Output over wide input voltage range

Fig.13 and Fig.14 highlight the load transients and Fig.15 highlights the thermal behavior at maximum load.

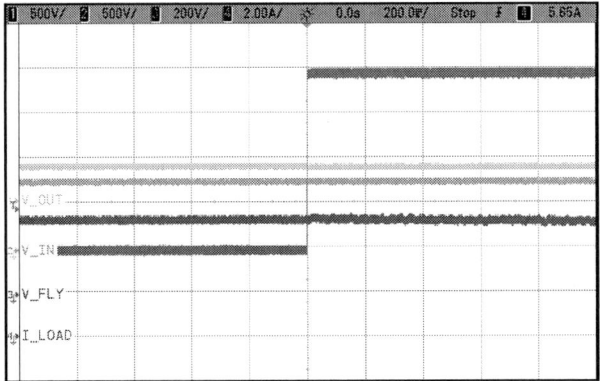

Fig.13: Load Transient from 4A to 12A

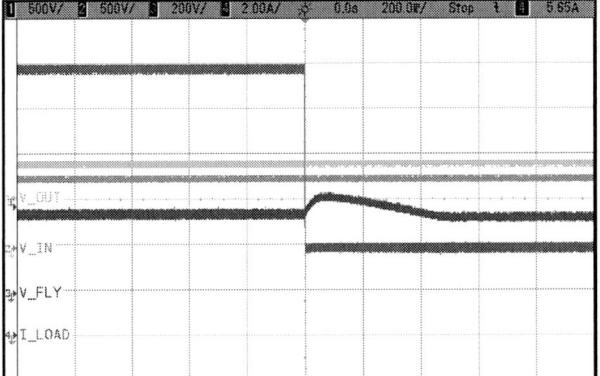

Fig.14: Load Transient from 12A to 4A

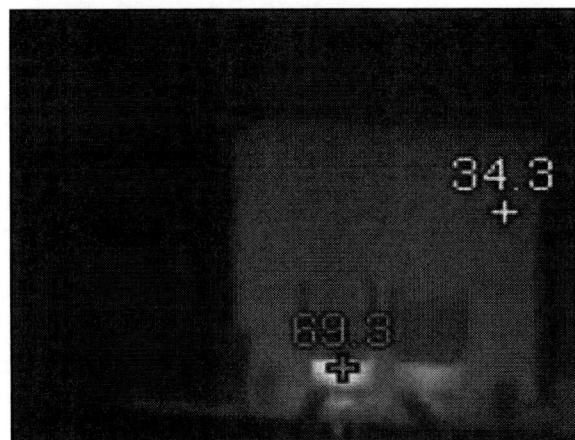

Fig.15: Thermal Image at maximum load

IX. Output Current Sharing of Two Channels

As can be seen from Fig.2, the two channels are independent and run in parallel to each other with the constraint of a minimum current threshold being delivered by each output. The output is not diode ORed, to increase efficiency. Instead, active current balancing has been implemented thanks to the powerful STM32G4 MCU. [12] is a typical simulation-based approach but in this practical case, the current is sensed at 2 locations i) The inductor current for implementing average current mode control ii) The DC output for measuring and summing the output to the load. The output current is continuously checked to also implement the MPPT function. It is also important to sense and keep the inductor current in continuous mode so that the individual converters behave like a current source feeding a common load. The unique inductor design helps to achieve that over a broad load current range.

Since each channel has its own current control loop, during start-up, the MCU assigns one channel as the master and the other as a slave. The converter output voltage is set by the master channel and the slave follows the master channel voltage while maintaining the maximum current possible at that voltage level governed by the MPPT subroutine. Of course there has to be a current demand from the downstream inverter. If the load requirement itself is very low, the controller might choose not to activate both the channels. At full load condition, and identical input, the firmware balances the current from each channel to within 3%.

Conclusions and Future Work

The system has provision in hardware, for both buck and boost configuration. The work on the three-level boost will be completed next. In addition, a mixed control using both voltage mode and average current mode depending on the load condition is discussed. There is also a scope to improve efficiency further by introducing ZVS over some part of the operating range to further flatten the efficiency curve. This will also evolve into a three-level non-inverting buck boost and then to bi-directional application for energy storage. The efficiency will be further optimized with even more advanced SiC MOSFETs.

Acknowledgment

Special thanks to our colleague Fabrizio Di Franco from STMicroelectronics, Catania, Italy for their support throughout the development.

REFERENCES

[1] Y. Liu, "A Full SiC MOSFET DCDC Boost Power Module Using 2 kV SiC MOSFET for 1500V String Solar Inverter Applications," 2024 IEEE Applied Power Electronics Conference and Exposition (APEC), Long Beach, CA, USA, 2024, pp. 188-191, doi: 10.1109/APEC48139.2024.10509478.

[2] . -M. Choe, J. -S. Lai, J. -U. Lim and G. -H. Choe, "Test and analysis of 3kW PV battery energy storage system," 2015 IEEE Energy Conversion Congress and Exposition (ECCE), Montreal, QC, Canada, 2015, pp. 4610-4615, doi: 10.1109/ECCE.2015.7310312.

[3] Y. Zhang, Y. Bezawada, R. Fu, W. Tian and R. M. Winter, "Study of a 3kW high-efficient wide-bandgap DC-DC power converter for solar power integration in 400V DC distribution networks," 2017 IEEE 12th International Conference on Power Electronics and Drive Systems (PEDS), Honolulu, HI, USA, 2017, pp. 680-684, doi: 10.1109/PEDS.2017.8289115.

[4] Yousefzadeh, Vahid & Alarcon, E. & Maksimovic, Dragan. (2005). Three-level buck converter for envelope tracking in RF power amplifiers. Proc. IEEE Applied Power Electronics Conference (APEC'05). 3. 1588 - 1594 Vol. 3. 10.1109/APEC.2005.1453248.

[5] Li, Fang et al. "Non-Inverting Three-Level Buck-Boost Converter for Wide Voltage Range Application." 2018 IEEE Energy Conversion Congress and Exposition (ECCE) (2018): 4870-4875.

[6] Y. Zhang et al., "Multilevel Non-Inverting Buck-Boost Converter with Low-Frequency Ripple-Shaping Based Controller for Operating in Step-down/Step-up Transition Region," 2018 IEEE 19th Workshop on Control and Modeling for Power Electronics (COMPEL), Padua, Italy, 2018, pp. 1-7, doi: 10.1109/COMPEL.2018.8460071.

[7] M. Trabelsi and L. Ben-Brahim, "Experimental photovoltaic power supply based on flying capacitors multilevel inverter," 2011 International Conference on Clean Electrical Power (ICCEP), Ischia, Italy, 2011, pp. 578-583, doi: 10.1109/ICCEP.2011.6036314.

[8] S. F. Graziani, A. Barchowsky and B. M. Grainger, "Design Considerations of a Flying Capacitor Multilevel Flyback Converter for DC-DC and Pulsed Power Applications," IECON 2018 - 44th Annual Conference of the IEEE Industrial Electronics Society, Washington, DC, USA, 2018, pp. 1140-1145, doi: 10.1109/IECON.2018.8591838.

[9] D. Jiao and A. Q. Huang, "Pre-Charge Strategy and Light-Load Voltage Balance Enhancement for DC/AC Flying Capacitor Converter," 2024 IEEE Texas Power and Energy Conference (TPEC), College Station, TX, USA, 2024, pp. 1-6, doi: 10.1109/TPEC60005.2024.10472214.

[10] F. H. A. Gerfer, "7 Design tips for selection of power inductors," 2016 IEEE 2nd Annual Southern Power Electronics Conference (SPEC), Auckland, New Zealand, 2016, pp. 1-4, doi:10.1109/SPEC.2016.7846170.

[11] B. Zhao, H. Qin, X. Nie and Y. Yan, "Evaluation of isolated gate driver for SiC MOSFETs," 2013 IEEE 8th Conference on Industrial Electronics and Applications (ICIEA), Melbourne, VIC, Australia, 2013, pp. 1208-1212, doi: 10.1109/ICIEA.2013.6566550.

[12] M. R. S. To and E. R. Magsino, "Elementary and simplified control of digital parallel buck converters," 2012 IEEE International Conference on Control System, Computing and Engineering, Penang, Malaysia, 2012, pp. 72-76, doi: 10.1109/ICCSCE.2012.6487118.

[13] https://www.changsung.com/file/product/cores/5.%20Sendust%20Cores/TDS_CS400060G.pdf datasheet Chang Sung CS400060

[14] Biswas, Suvankar & Reusch, David. (2018). GaN Based Switched Capacitor Three-Level Buck Converter with Cascaded Synchronous Bootstrap Gate Drive Scheme.

A Bootstrapless KY-S-Hybrid Buck-Boost Converter with Full Range iLs Reduction and 400% Line Transient Response Acceleration for AI-Mobile Application

Chuan-En Chang
Graduate Institute of Electronics Engineering
National Taiwan University
Taipei, Taiwan
r13943025@ntu.edu.tw

Cheng-Ta Chuang
Graduate Institute of Electrical Engineering
National Taiwan University
Taipei, Taiwan
darcyboogi286@gmail.com

Hao-Ran Huang
Graduate Institute of Electrical Engineering
National Taiwan University
Taipei, Taiwan
r13921025@ntu.edu.tw

Chieh-Ju Tsai
Graduate Institute of Electronics Engineering
National Taiwan University
Taipei, Taiwan
f04943123@ntu.edu.tw

Ching-Jan Chen
Graduate Institute of Electrical Engineering
National Taiwan University
Taipei, Taiwan
chenjim@ntu.edu.tw

Abstract—In this paper a novel bootstrapless hybrid buck-boost dc-dc converter for AI-mobile application is proposed. It leverages KY-boost converter combining with S-hybrid converter to achieve full voltage-conversion-ratio (VCR) inductor current stress reduction. As a result, 6.8 uH 0805 small form factor inductor could be employed. Moreover, a bootstrapless driving strategy is introduced to make sure full NMOS power switches adopted with small on-chip area and 0 off-chip bootstrap capacitors. On the other hand, nowadays, the AI-mobile application is facing stringent line transient challenges on the buck-boost converter. This paper leverages a line transient feedforward technique to reduce the line to output perturbation up to 400% compared to the conventional control V-mode algorithm. Simulated in a 180 nm CMOS process, this design achieved 95.5% peak efficiency at V_{IN} = 4.2 V to V_O = 3.3 V.

Keywords—*KY-S hybrid converter, buck-boost converter, boostrapless, line transient feedforward*

I. INTRODUCTION

The integration of AI with mobile computing requires a buck-boost converter to provide a stable 3.3 V output for AI-Mobile-SoCs, powered by a 2.5 V to 4.2 V battery. Ensuring high efficiency and robustness against line transients—caused by the load pulsing currents of AI accelerators—is essential to meet the high-performance demands of modern converters. As shown in Fig. 1, the converter must regulate these dynamic conditions and deliver reliable performance across a wide range of operating scenarios. However, conventional buck-boost converters, with their low efficiency and slow transient response, fall short of these high-demand applications. Consequently, recent work has introduced innovative topologies and control schemes to enhance converter performance.

As proposed in [1], the KY converter provides an effective solution for voltage boosting with low output ripple and ratio of inductor current and output current, making it suitable for high-precision, low-ripple applications. Similarly, the S-Hybrid DC-DC converter described in [2] achieves voltage step-down with a highly integrated architecture by positioning the inductor at the input side. This design significantly reduces size and power loss by avoiding large inductance values that typically come with high DCR and bulky dimensions. However, these approaches are not fully compatible with lithium battery conditions due to conversion ratio limitations inherent to their topologies and operating conditions.

This paper introduces a hybrid buck-boost converter that combines a KY-boost converter with an S-hybrid-buck converter, as shown in Fig. 2. The proposed structure enhances power density by minimizing voltage swings and inductor current ratio, allowing for the selection of smaller inductors. Section II first reviews the advantages of KY-boost and S-hybrid converters, then presents the new KY-S hybrid converter, covering its operating principles and steady-state analysis. The bootstrap capacitor, a critical issue in advanced topologies, is also addressed. Thus, in Section III, we incorporate a bootstrapless driving system with the KY-S hybrid converter to mitigate the need for large bootstrap capacitance. Additionally, in Section IV, we introduce a generalized V_{IN}-feedforward scheme that further accelerates line transient response, achieving up to 400% reduction in output voltage deviation. Simulation results are presented in Section V, and conclusions are drawn in Section VI.

979-8-3315-1612-3/25 $31.00 © 2025 IEEE

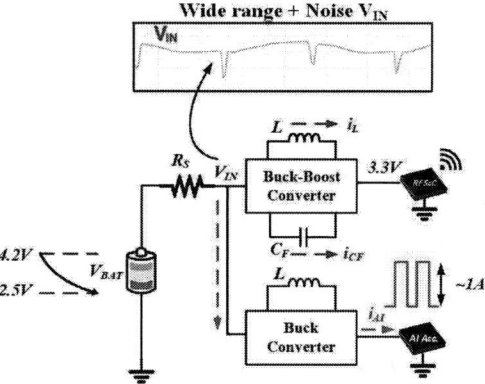

Fig. 1. High-performance converter application example.

II. KY-S Buck-Boost Hybrid Converter Architecture

A. KY-S Hybrid Buck-Boost Converter

In high-efficiency applications like AI, conduction loss is a critical factor that cannot be overlooked. To minimize loss and improve precision, it is essential to reduce both inductor current and output voltage ripple. The proposed KY-S hybrid converter, illustrated in Fig. 2, combines the advantages of the KY-boost and S-hybrid step-down converters. The KY-boost converter decreases the ratio between the inductor current and output current, which is achieved by incorporating a switched capacitor (SC) network at the input side of the topology. Similarly, the S-hybrid step-down converter has an SC network on the output side, effectively reducing output voltage ripple.

As a result, we propose a new topology that merges the KY-boost and S-hybrid converters: the KY-S hybrid buck-boost converter. This design consists of six MOSFET switches (S_1 to S_6), an output capacitor (C_O), an inductor (L), and two energy-transferring capacitors (C_{F1} and C_{F2}). The input and output are assumed to be V_{IN} and V_o, respectively.

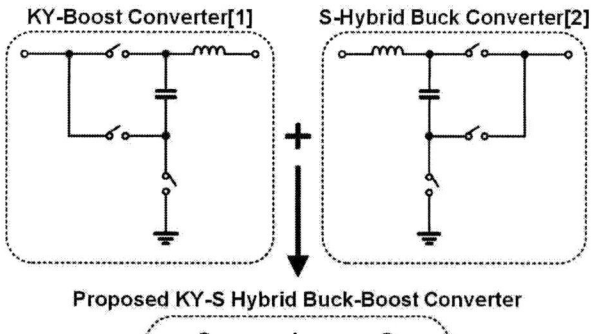

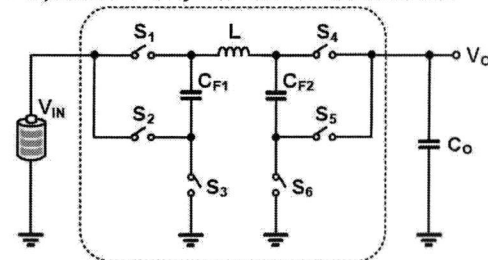

Fig. 2. Proposed KY-S hybrid buck-boost converter

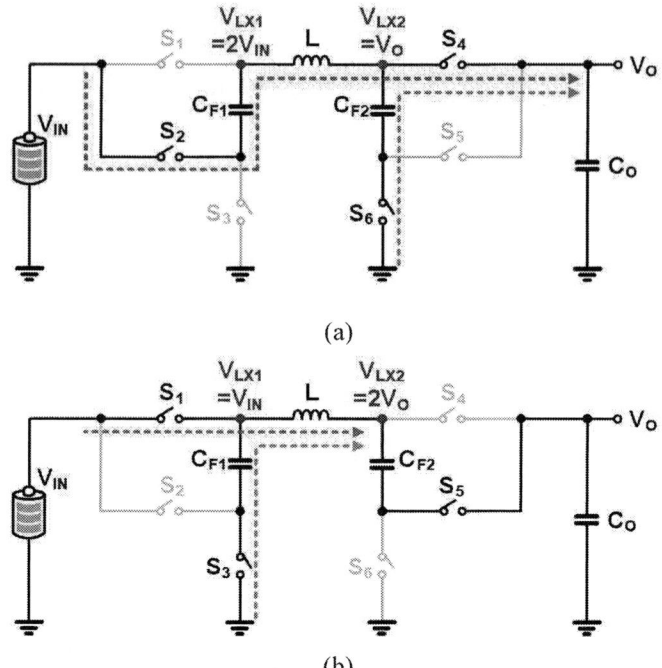

(a)

(b)

Fig. 3. Working principle of KY-S hybrid buck-boost converter: (a) Phase 1 (Φ_1) current path and switches operation (b) Phase 2 (Φ_2) current path and switches operation

B. Operating Principle

The converter operates in two phases (Φ_1 and Φ_2) during a complete switching period T_s. Φ_1 corresponds to the interval from 0 to DT_s, where D is the duty ratio. Φ_2 accounts for the remaining period, which lasts for $(1-D)T_s$. Φ_1 and Φ_2 also represent the charging and discharging periods of the inductor, respectively. In Φ_1, refer to Fig. 3(a), C_{F1} is in series with V_{IN} and C_{F2} is in parallel with V_O by turning on switches S_2, S_4, and S_6 while turning off the other switches. The inductor L is magnetized by a cross voltage of $2V_{IN} - V_O$, since the front-end switching node (V_{LX1}) is charged to V_{IN} in the previous cycle in Φ_2. During this period, C_{F1} and C_{F2} are both discharging to form dual current paths i_L, the current flows through inductor L, and i_{CF2}, the current flows through capacitor C_{F2} from the ground, to support the load current i_O.

In phase Φ_2, refer to Fig. 3(b), C_{F1} is in parallel with V_{IN} and C_{F2} is in series with V_O by turning on switches S_1, S_3, and S_5 while turning off the other switches. During the period, C_{F1} is charged by V_{IN} and C_{F2} in series of V_O is charged by L as it demagnetizes with a cross voltage of $V_{IN}-2V_O$. The back-end switching node (V_{LX2}) equals to $2V_O$ because V_{CF2} is charged to Vo in the previous phase. On the current side, i_L flows through C_{F2} and to the output to support the loading solely.

C. Conversion Ratio and Inductor Current

In this section, we dig deeper to explain the conversion ratio and the reduction of the inductor current in detail. As in Fig. 3, the converter boosts the front-end switching node (V_{LX1}) to $2V_{IN}$ and maintains the back-end switching node (V_{LX2}) at V_O during the on-duty cycle (D) to charge the inductor.

979-8-3315-1612-3/25 $31.00 © 2025 IEEE

Conversely, during the off-duty cycle (1-D), V_{LX1} connects to V_{IN} while V_{LX2} is boosted to $2V_O$ to discharge the inductor during the off-duty cycle (1-D). With the volt-second balance, the conversion ratio (M) can be derived, as shown in (1):

$$M = \frac{V_O}{V_{IN}} = \frac{1+D}{2-D}, (0.5 < M < 2) \tag{1}$$

From equation (1), we realized the converter operates as a buck-boost converter without additional mode-switching and functions continuously by simply controlling duty ratio (D), which enormously reduces the complexity of the control loop.

On the other hand, when the duty cycle is on (Φ_1), there is a power delivery path to the output other than the inductor L, formed by C_{F2}. In order to maintain capacitor charge balance, C_{F2} must output an equivalent charge (1-D)$i_L T_s$ to the output since the inductor charges the capacitor with (1-D)$i_L T_s$ at Φ_2. When the duty cycle is off (Φ_2), the current to output charge is provided by the inductor only via (1-D)$i_L T_s$. Therefore, the overall output charge combines the inductor current charge and the C_{F2} charge, resulting in a reduction in the dc value of the inductor current, as derived in (2):

$$\frac{i_L}{i_O} = \frac{M+1}{3} \tag{2}$$

Fig. 4 compares the inductor current ratio reduction of the prior buck-boost converter [3]-[6] with the proposed converter. The proposed converter reduces inductor current stress by 23% and 40% at V_{IN} = 2.5 V and 4.2 V, respectively. This significant reduction prevents the need for a bulky low DCR inductor. Consequently, a 58 mOhm 0805 tiny SMD inductor can be adopted to maximize power density.

III. BOOSTRAPLESS KY-S HYBRID BUCK-BOOST CONVERTER

Inspired by previous work on monolithic bootstrap capacitor solutions [7], the proposed KY-S Hybrid Buck-Boost Converter employs a bootstrapless design by reusing the flying capacitor as a bootstrap capacitor, rather than relying on a large on/off chip bootstrap capacitor. The schematic of this approach is illustrated in Fig. 5, where all switches are implemented with low-voltage 5 V devices [8], [9].

The operating principles of the bootstrapless driving scheme for the KY-S Hybrid Buck-Boost Converter are as follows. First, for switches S_3 and S_6, traditional drivers switch between 0 and V_{IN}, resulting in a gate-source voltage (V_{GS}) of V_{IN} when S_3/S_6 are on, and a drain-source voltage (V_{DS}) of V_{IN}/V_O when off. Second, for S_2 and S_5, the driver switches between V_{IN}/V_O and $2V_{IN}/2V_O$. When S_2/S_5 conducts, the source terminal of S_2/S_5 equals V_{IN}/V_O, and the switching nodes V_{LX1} and V_{LX2} reach $2V_{IN}/2V_O$, yielding a V_{GS} of V_{IN}/V_O when on. When S_2/S_5 turns off, V_{DS} equals V_{IN}/V_O. Lastly, V_{LX2} is connected to S_1's driver, and V_{LX1} to S_4's driver, ensuring that the V_{GS} of S_1 and S_4 is $(2V_O-V_{IN})/(2V_{IN}-V_O)$ and the V_{DS} of S_1/S_4 is V_{IN}/V_O when off. Table I summarizes the V_{GS} and V_{DS} values for all switches.

By eliminating the need for charge pumps or dedicated bootstrap capacitors, this bootstrapless topology leverages the flying capacitor and inductor terminals V_{LX1} and V_{LX2} to supply the driver voltage. As a result, an all-LV-NMOS-based

bootstrapless topology is achieved, successfully reducing the bootstrap capacitor requirement by four.

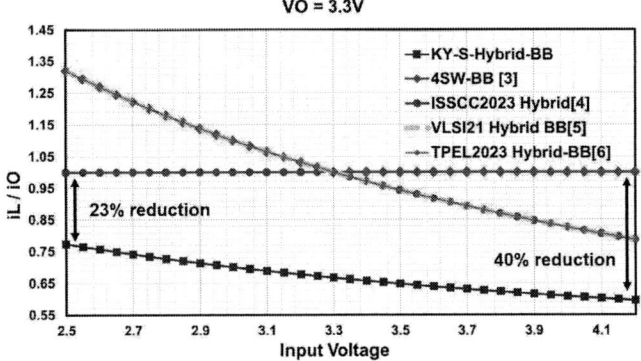

Fig. 4. Inductor current ratio reduction

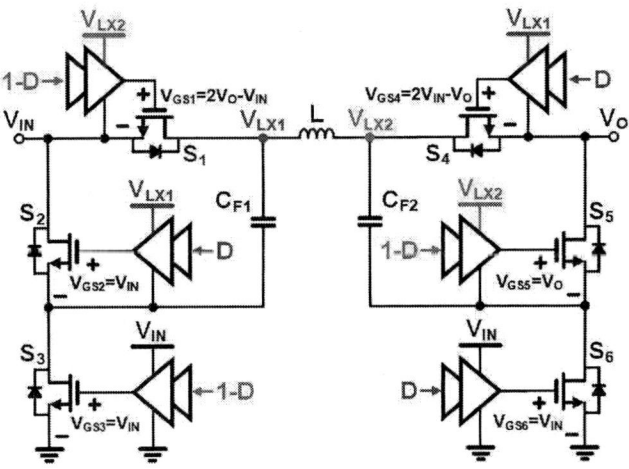

Fig. 5. The bootstrapless driving scheme for the proposed KY-S hybrid buck-boost converter

TABLE I.   SWITCH V_{GS} / V_{DS} STRESS

Switch	V_{GS}	V_{DS}
S_1	$2V_O$ - V_{IN}	V_{IN}
S_2, S_3	V_{IN}	V_{IN}
S_4	$2V_{IN}$ - V_O	V_O
S_5	V_O	V_O
S_6	V_{IN}	V_O

IV. INPUT VOLTAGE FEEDFORWARD CONTROL SCHEME

Traditional voltage mode control provides advantages such as design simplicity and high noise tolerance. However, it suffers from slow line transient response due to the absence of input voltage (V_{IN}) information. Specifically, any line transient is detected as a change in output voltage (ΔV_O), which subsequently causes a variation in the output of the compensator (V_C). This V_C is then compared to a sawtooth or triangular ramp signal (V_{ramp}) to derive the desired duty cycle (D). Consequently, when a change in V_{IN} occurs, the entire

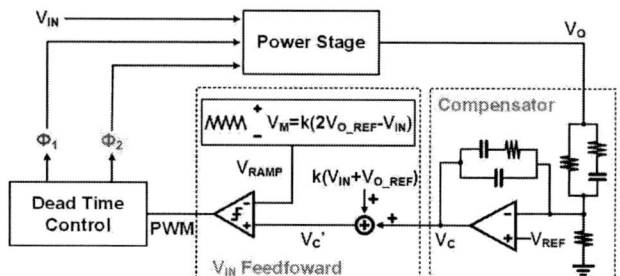

Fig. 6. The proposed line-transient feedforward control scheme.

control loop must complete its response before the system can adjust accordingly [10] – [16].

To address this issue, a line transient feedforward framework for hybrid converters utilizing voltage mode control has been developed, as illustrated in Fig. 6. V_{IN} and V_O are utilized in a feedforward manner to enhance the line transient response. The objective of this control scheme is to minimize ΔV_C and utilize V_{IN} and V_O in a feedforward manner to enhance the response of the system following a line transient, thereby reducing ΔV_O.

For any voltage mode hybrid converters, D can be expressed as (3), where $f(V_{IN},V_O)$ and $g(V_{IN},V_O)$ are derived from the relationship between conversion ratio and D. For instance, in the proposed KY-S converter, $f(V_{IN},V_O)$ and $g(V_{IN},V_O)$ are defined as $2V_O-V_{IN}$ and $V_{IN}+V_O$, respectively.

$$D = \frac{V_C}{\Delta V_{ramp}} = \frac{f(V_{IN},V_O)}{g(V_{IN},V_O)} \left(= \frac{2V_O - V_{IN}}{V_{IN} + V_O}\bigg|_{KY-S} \right) \quad (3)$$

The concept of the proposed generalized V_{IN}-Feedfoward scheme is shown in Fig. 7. We can simply design $V_C' = V_C + k \times f(V_{IN},\ V_O)$ and $V_{ramp}= k \times g(V_{IN},\ V_O)$. Consequently, equation (3) is modified to equation (4).

$$D = \frac{V_C'}{V_{ramp}} = \frac{V_C + k \times f(V_{IN},V_O)}{k \times g(V_{IN},V_O)} \left(= \frac{V_C + k \times (2V_O - V_{IN})}{k \times (V_{IN} + V_O)}\bigg|_{KY-S} \right) \quad (4)$$

As a result, the line-perturbation-induced control voltage perturbation could be expressed as (5) and (6). Based on observations from (5) and (6), it can be concluded that V_C is independent of V_{IN}. In other words, V_C exhibits minimal variation during line transients, resulting in only slight fluctuations in V_O.

$$\frac{\partial V_C'}{\partial V_{IN}} = \frac{\partial (V_{ramp} \times D)}{\partial V_{IN}} = \frac{\partial \left(k \times g(V_{IN},V_O) \times \frac{f(V_{IN},V_O)}{g(V_{IN},V_O)}\right)}{\partial V_{IN}} = \frac{\partial (k \times f(V_{IN},V_O))}{\partial V_{IN}} \quad (5)$$

$$\frac{\partial V_C}{\partial V_{IN}} = \frac{\partial (V_C' - k \times f(V_{IN},V_O))}{\partial V_{IN}} = \frac{\partial V_C'}{\partial V_{IN}} - \frac{\partial (k \times f(V_{IN},V_O))}{\partial V_{IN}} = 0 \quad (6)$$

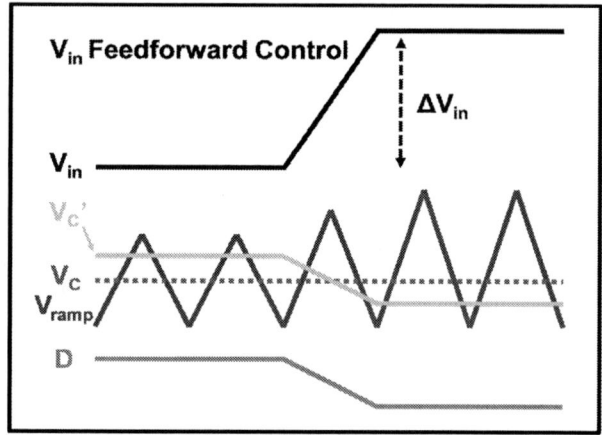

Fig. 7. Operating principle of the proposed control scheme.

Fig. 8 presents a comparison of the simulated gains for $G_{vg\_closed\_ff}(s)$ and $G_{vg\_closed\_vm}(s)$, where $G_{vg\_closed\_ff}(s)$ and $G_{vg\_closed\_vm}(s)$ represent the closed loop line-to-output transfer function of proposed feedforward control and voltage mode control, respectively. It is obvious that $G_{vg\_closed\_ff}(s)$ demonstrates a peak value that is 20 dB lower, which will result in significantly improved line transient response.

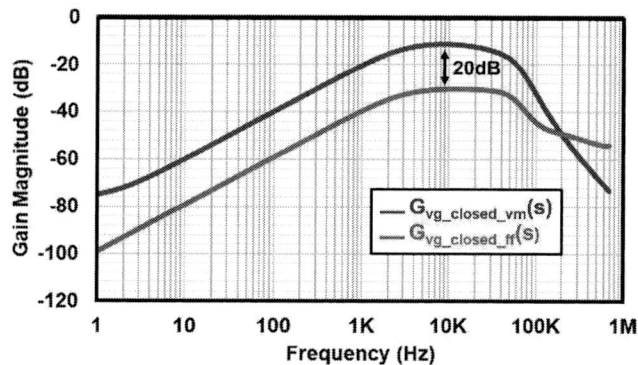

Fig. 8. Comparison of $G_{vg\_closed\_ff}(s)$ (blue line) and $G_{vg\_closed\_vm}(s)$ (red line).

TABLE II.   POWER STAGE SPECIFICATIONS

Component	Value
V_{IN}	4.2 V ~ 2.5 V
V_O	3.3 V
I_{OUT}	1 A
F_{SW}	1 MHz
L	6.8 µH
C_O	10.8 µF
C_{F1}	10.8 µF
C_{F2}	10.8 µF

979-8-3315-1612-3/25 $31.00 © 2025 IEEE

V. MEASUREMENT RESULT

The performance analysis of the KY-S hybrid buck-boost converter has been verified by the simulation with the fabrication using 180nm BCD process and measurements are currently in progress. Table II illustrates the specification of the power stage of the converter. With the advantage of low inductor current with output current ratio, the inductor could be chosen down to 6.8 µH with two flying capacitors C_{F1} and C_{F2}, which are both 10.8µF. The input voltage is designed to be in the range of 2.5 V to 4.2 V. Fig. 9 shows the efficiency of the proposed converter while loading current sweeps from 0.1 A to 1A, V_{IN} equals 4.2 V and V_O equals 3.3 V. The peak efficiency achieves 95.5% while the loading current is about 0.7 A. Fig. 10 illustrates the line transient response for input voltage step-up/down transitions from 4.2 V to 2.7 V, both with and without the proposed V_{IN}-feedforward scheme at the transistor level. The output ripple is reduced by up to 400% compared to conventional voltage-mode control. Furthermore, observed from the detailed line transient response shown in Fig. 11(a) and Fig. 11(b), the proposed control scheme minimizes the deviation in the error amplifier output V_c. Consequently, all line perturbations are immediately forwarded to V_c' and V_{ramp}, thereby minimizing output voltage perturbations, as analyzed in the previous section.

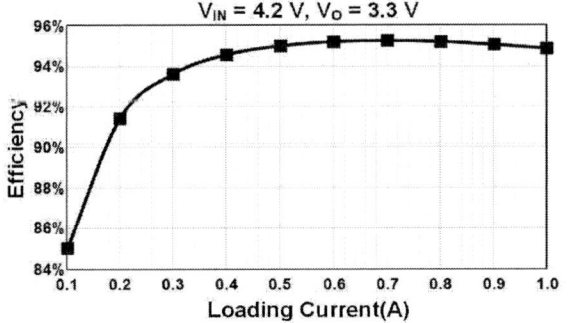

Fig. 9. Efficiency performance (V_{IN} = 4.2 V, V_O = 3.3 V, loading current = 0.1 A ~ 1 A)

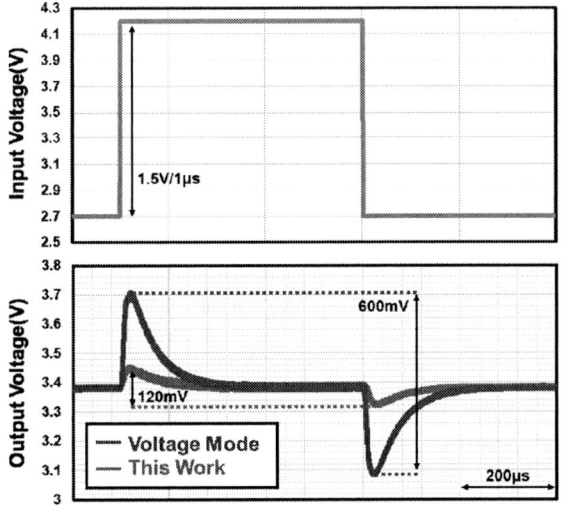

Fig. 10. Line transient response comparison between inserting V_{IN}-feedforward scheme and conventional voltage-mode control.

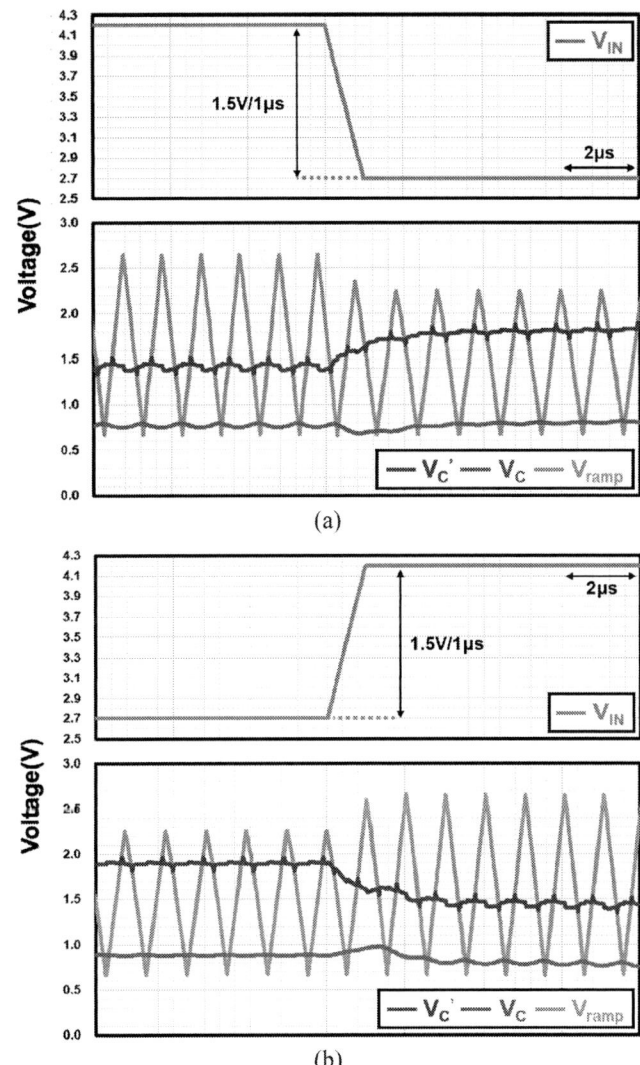

Fig. 11. Detailed line transient response of proposed control scheme: (a) line step-down from 4.2V to 2.7V (b) line step-up from 2.7V to 4.2V.

The load transient responses exhibit undershoot and overshoot within 20 mV for a 0.9 A load step-up/down event over a 5 µs interval, as shown in Fig. 12(a) and Fig. 12(b).

VI. CONCLUSION

In this work, we introduced a KY-S hybrid buck-boost converter with a bootstrapless driving scheme, eliminating the need for a bootstrap capacitor. Additionally, we proposed a generalized V_{IN}-feedforward scheme that enhances the line transient response, achieving a 400% reduction in output voltage (V_O) deviation compared to conventional voltage-mode control feedback in the proposed KY-S converter. The converter addresses major challenges related to high inductor current by incorporating flying capacitors, C_{F1} and C_{F2}, which create dual current paths to supply the load. This topology supports V_{IN} variations within the range of common lithium battery voltages, which is 4.2 V to 2.5 V, and maintains continuous operation without mode-switching. The simulated

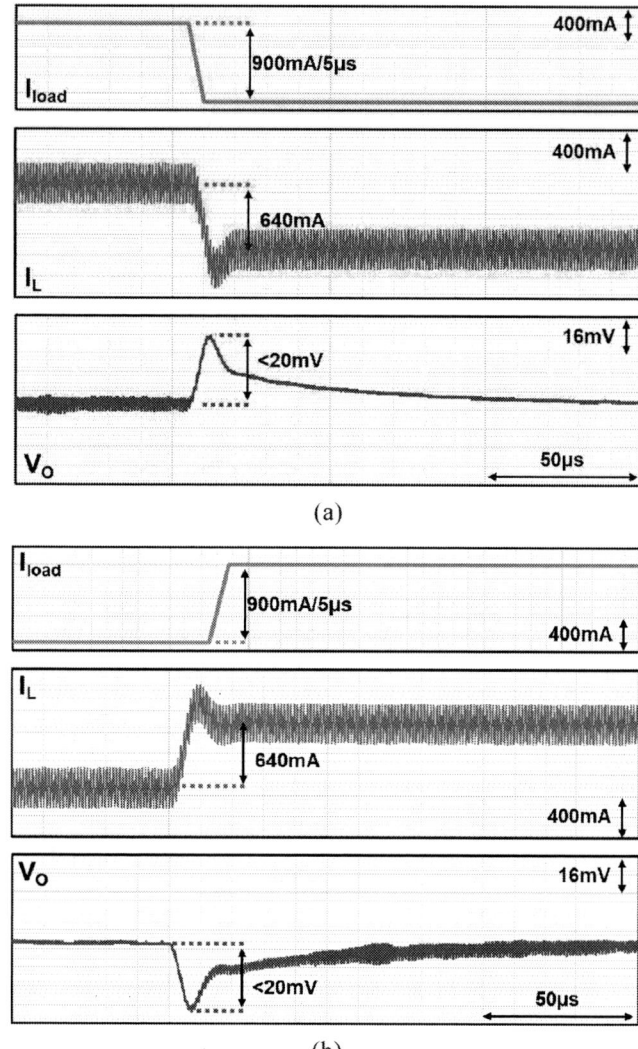

(a)

(b)

Fig. 12. Load transient response of proposed control scheme. (a) load step-down from 1A to 0.1A (b) load step-up from 0.1A to 1A.

peak efficiency reached 95.5%, with the chip fabrication using 180nm BCD process and measurement are currently in progress.

ACKNOWLEDGMENT

The authors gratefully acknowledge the support of the National Science and Technology Council (NSTC), Taiwan, as well as the educational funding for chip fabrication provided by TSRI, Taiwan. Additionally, the authors extend their appreciation to SIMPLIS Technologies for the provision of the SIMPLIS simulation tool.

REFERENCES

[1] K. I. Hwu and Y. T. Yau, "KY Converter and Its Derivatives," *IEEE Trans. on Power Electronics*, vol. 24, no. 1, pp. 128-137, Jan. 2009.

[2] G. -S. Seo and H. -P. Le, "S-Hybrid Step-Down DC–DC Converter—Analysis of Operation and Design Considerations," *IEEE Trans. Ind. Electronics*, vol. 67, no. 1, pp. 265-275, Jan. 2020.

[3] X. Li, Y. Liu and Y. Xue, "Four-Switch Buck–Boost Converter Based on Model Predictive Control With Smooth Mode Transition Capability," *IEEE Trans. Ind. Electronics*, vol. 68, no. 10, pp. 9058-9069, Oct. 2021.

[4] J. Jin, Y. Zhou, C. Chen, X. Han, W. Xu and L. Cheng, "30.6 A 98.6%-Peak-Efficiency 1.47A/mm2-Current-Density Buck-Boost Converter with Always Reduced Conduction Loss," in *Proc. 2023 IEEE ISSCC, San Francisco*, CA, USA, pp. 448-450.

[5] H. Shin et al., "A 96.6%-Efficiency Continuous-Input-Current Hybrid Dual-Path Buck-Boost Converter with Single-Mode Operation and Non-Stopping Output Current Delivery," in *Proc. IEEE 2021 Symp. on VLSI*, Kyoto, Japan, 2021, pp. 1-2.

[6] A. Mishra, W. Zhu, B. Wicht and V. D. Smedt, "An All-1.8-V-Switch Hybrid Buck–Boost Converter for Li-Battery-Operated PMICs Achieving 95.63% Peak Efficiency Using a 288-m DCR Inductor," *IEEE Trans. Power Electronics*, vol. 38, no. 3, pp. 3444-3454, March 2023.

[7] C.-J. Tsai, C.-H. Hsu, C.-J. Chen, Y.-T. Hung, C.-Y. Hsieh, "A Monolithic All-1.8V-Thin-Gate-NMOS KY-Boost Converter With Reused Flying-Capacitor Bootstrap Gate Driver Achieving 94.42% Peak Efficiency," *IEEE Trans. Power Electronics*, vol. 39, no. 6, pp. 7238-7251, Jun. 2024.

[8] Dae-Hyeon Kim and Hyun-Sik Kim, "A 96.9%-Peak-Efficiency Bilaterally-Symmetrical Hybrid Buck-Boost Converter Featuring Seamless Single-Mode Operation, Always-Reduced Inductor Current, and the Use of All CMOS Switches," in *Proc. 2024 IEEE ISSCC*, pp. 146-147, Feb. 2024.

[9] J. Jin, Y. Zhou, W. Xu and L. Cheng, "A 97.3%-Peak-Efficiency Always-Dual-Path Buck-Boost Converter with Single-Mode Operation and Fast Transient Responses," in *Proc. 2024 IEEE CICC*, Denver, CO, USA pp. 1-2.

[10] Y. Zhou, J. Li, X. Liu, Y. Zheng, and K. N. Leung, "Bidirectional buck–boost converter with reduced power loss and no right-half-plane zero," *IEEE Trans. Power Electron.*, vol. 38, no. 2, pp. 2127–2142, Feb. 2023.

[11] C. Pan, C. Zhan, R. P. Martins, and C.-S. Lam, "A continuous-output current buck–boost converter without right-half-plane-zero (RHPZ)," *IEEE Trans. Circuits Syst. I, Reg. Papers*, vol. 70, no. 12, pp. 4719–4728, Dec. 2023.

[12] X. Ren, X. Ruan, H. Qian, M. Li, and Q. Chen, "Dual-edge modulated four-switch buck-boost converter," in *Proc. IEEE Power Electron. Spec. Conf.*, Jun. 2008, pp. 3635–3641.

[13] C. Zheng and D. Ma, "A 10-MHz green-mode automatic reconfigurable switching converter for DVS-enabled VLSI systems," *IEEE J. Solid State Circuits*, vol. 46, no. 6, pp. 1464–1477, Jun. 2011.

[14] X.-E. Hong, J.-F. Wu, and C.-L. Wei, "98.1%-efficiency hysteretic current-mode noninverting buck–boost DC–DC converter with smooth mode transition," *IEEE Trans. Power Electron.*, vol. 32, no. 3, pp. 2008–2017, Mar. 2017.

[15] L. Callegaro, M. Ciobotaru, D. J. Pagano, E. Turano, and J. E. Fletcher, "A simple smooth transition technique for the noninverting buck–boost converter," *IEEE Trans. Power Electron.*, vol. 33, no. 6, pp. 4906–4915, Jun. 2018.

[16] K.-D. Kim, H.-M. Lee, S.-W. Hong, and G.-H. Cho, "A noninverting buck–boost converter with state-based current control for Li-ion battery management in mobile applications," *IEEE Trans. Ind. Electron.*, vol. 66, no. 12, pp. 9623–9627, Dec. 2019.

979-8-3315-1612-3/25 $31.00 © 2025 IEEE

Digital Control of a 600-V to 28-V 20-kW Two-Stage DC-DC Converter

Shreyas B Shah*, Rachit Pradhan*, Jiaqi Yuan*, Mohamed Ibrahim[†], Ahmed Elezab[†], Samuel Hemming*,
Giorgio Pietrini*, Piranavan Suntharalingam[‡], Mario F Cruz[†], and Ali Emadi*

*Department of Electrical and Computer Engineering, McMaster University, Hamilton, ON, Canada
[†]Aerospace Group, Fluid, and Electrical Distribution Division, Eaton Corporation, Torrance, CA, USA
[‡]Advanced Technology and Innovation Group, Eaton Corporation, Raleigh, NC, USA
Email: shahs135@mcmaster.ca

Abstract—A trend of increasing powertrain voltages is seen in electric vehicles and more electric aircraft. However, the systems close to humans have stayed at 12 to 48 V for safety reasons. A combination of control and converter architecture is needed to efficiently meet the required gain over all line and load conditions. This work briefly introduces a converter architecture and, in detail, demonstrates a digital control technique for a 600 V to 28 V 20 kW two-stage DC-DC converter. A modular input-parallel output-parallel (IPOP) configuration with two identical 10 kW modules is used. Each module consists of a buck stage for regulation followed by a series-resonant DC transformer (SR-DCX) for isolation and a fixed gain. A current sharing compensator is designed to ensure equal current sharing between IPOP modules. The control technique considers constraints due to signal-chain processing and minimizes digitization delays. The scheme experimentally demonstrates robust control over the required line and load variations for the 10 kW module. The devised control technique demonstrates equal current-sharing amongst the paralleled modules using a PLECS simulation.

Index Terms—Aerospace, average current mode control, buck converter, current sharing, dc-dc converter, digital control, series-resonant dc transformer (SR-DCX), timing diagrams.

I. INTRODUCTION

The powertrain bus levels of aircraft and electric vehicles are rising to reduce cabling weight and losses. However, the systems near humans have stayed around 12-48 V for safety reasons. Recent trends show that for an aircraft's Electrical Power Distribution System (EPDS), the main DC bus voltages may range from ±270 V to 3 kV [1]. The current generation of the electrical power system (EPS) EPS-A4 in aircraft uses a ±270 V DC bus on the high-voltage (HV) side, and 28 V DC bus on the low-voltage (LV) end [1]. Similar EPS have also found their way to military ground vehicular applications [2]. The typical bus voltages for EPS-A4 are governed by MIL-STD-704F [3] for both HV and LV buses and by MIL-PRF-GCS600A [4] on the HV bus for military ground vehicles. Given the gain, typical power level, and reliability requirements, this poses an exciting challenge for an interlinking converter between the two buses. The present state-of-the-art presents a topological analysis and improvements for similar converters in [5]–[7]. In [8], the authors describe efficiency improvement through improvements in modulation techniques. There have been contributions for controlling 270 V to 28 V converters in [9], [10]. However, the literature is limited

to topological improvements, focusing only on module-level improvements. Furthermore, average efficiencies over the total input voltage span are unknown, given design optimizations and demonstration at nominal operating voltages. Similarly, a gap in the demonstration of control techniques exists for such converters. The practical challenges arising in sensing signal chains due to the converter size are also not discussed.

Thus, this work addresses the gaps in the literature using a two-stage converter topology and demonstrates a digital control approach for a high-gain converter tailored to such applications. Given the power rating of the converter, an input-parallel output-parallel (IPOP) two-stage DC-DC converter architecture is used in this work. The buck converter is regulated by regulating the average inductor current, which is set by an outer voltage regulation loop. However, the paralleled modules bring another layer of control challenge of ensuring equal current sharing (CS) amongst both modules. The review in [11] highlights various approaches for controlling paralleled converters. The work in [12] is used as a starting point and expanded on to design a compensator for ensuring equal power sharing by both modules.

The article is split into five Sections: Section II details the converter structure, key components, operational requirements, and basic operating principles. The control problem is defined in Section III. Herein, the overall control law is derived, the analysis for regulating the first stage, and solutions for ensuring equal CS between paralleled modules are explored. Furthermore, signal chain constraints and discretization penalties are also added to the control problem, and a compensator design is derived. In Section IV, based on the established control law, the timing sequencing arranged for easy and improved microcontroller implementation is discussed. Finally, simulation results for the 20 kW converter and experimental results for a 10 kW module over the entire line and load conditions are demonstrated in Section V.

II. CONVERTER TOPOLOGY, DESIGN, AND OPERATION

The converter schematic, topology, and the values of critical components can be seen in Fig. 1. The converter is constructed using two identical modules of 10 kW each, connected in an IPOP configuration. Each module comprises a single 10 kW HV power processing end. The LV end, however, is split into

979-8-3315-1612-3/25 $31.00 © 2025 IEEE

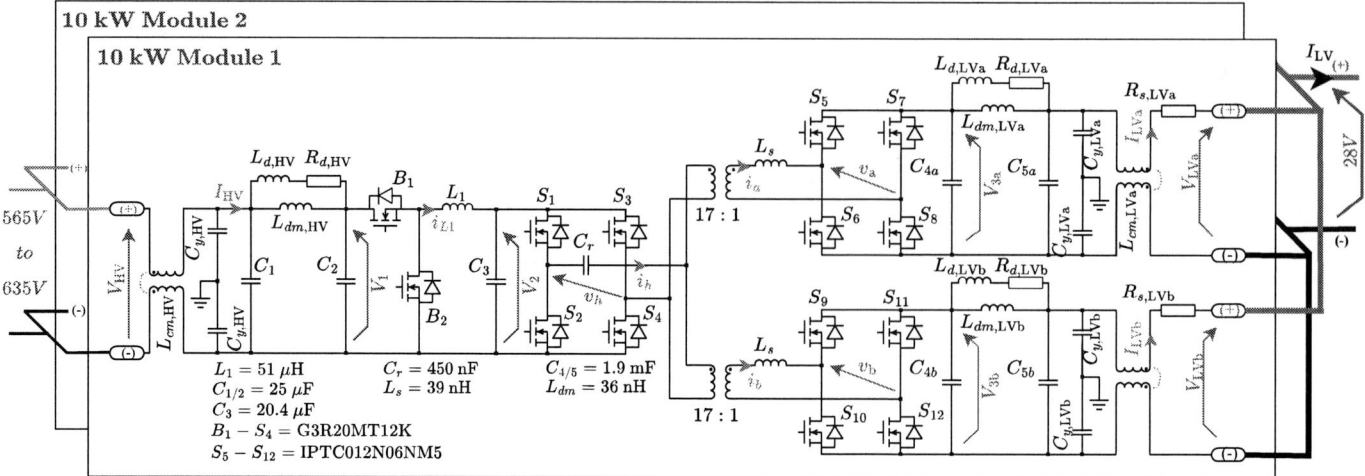

Fig. 1: Schematic of the 600-V to 28-V 20-kW two-stage DC-DC converter.

two 5 kW sub-modules. The HV end uses Silicon-Carbide (SiC) switches without paralleling on all nodes. The LV end is constructed using silicon switches and uses three devices in parallel to meet the necessary current ratings. Each of the 10 kW modules consists of a two-stage converter. The first stage is a synchronous-buck converter, and the second is a series-resonant DC transformer (SR-DCX). The buck stage is responsible for meeting the converter's full line and load regulation requirements. The buck converter's design and operating range ensure a continuous-current mode operation at all operating conditions except at very light loads. In contrast, the SR-DCX operates near its resonant frequency in an unregulated mode. The operation near resonance results in the SR-DCX stage exhibiting an almost fixed gain, irrespective of load variations. However, a small but linear reduction in the gain occurs with increasing power delivered by the converter [13], which is compensated by the buck converter.

Given the current levels on the LV end, the key reason for selecting an SR-DCX topology is to operate close to zero-current switching at the LV side of the converter. This approach relaxes thermal constraints on the LV side and has the added benefit of inherently lower EMI impact [14]. The SR-DCX on the LV side enables two key design improvements: Firstly, the sub-module's rated current decreases, reducing the number of devices needed in parallel and the challenges associated with device paralleling. Secondly, the

split allows for magnetics design with standard cores for handling the required currents and PCB design to meet package constraints. The partial paralleling structure also allows the using two independent 5 kW low-voltage modules to be fed from a single 10 kW HV end [15]. The partial paralleling approach also allows optimal utilization of space and rating of high-voltage components. The SR-DCX also enables near-symmetric current-sharing between the two LV modules [16]. The partial paralleling is achieved by having a common feed capacitor while using the inherently high system leakage inductance. The leakage inductance is high given the high turns ratio of the transformer at $N = 17$, thus amplifying the effect of transformer and PCB leakage inductance by a factor of N^2. This leakage inductance is also relatively unaffected by piece-to-piece variations due to PCBs' tight geometric manufacturing capabilities, further aiding symmetric CS. It should be noted that symmetric current-sharing is achieved in the SR-DCX stage without having any control degree of freedom between the two LV modules. The buck converter and SR-DCX are synchronously operated at the same switching frequency f_s of around 100 kHz. It should be noted that the switching frequency is similar but different in the two paralleled 10 kW modules due to the variations in the SR-DCX's tank components.

III. CONTROL STRATEGY, MODELING AND ANALYSIS

The converter structure selected for its various benefits imposes a layered control challenge. Firstly, the IPOP connection introduces a static and dynamic CS challenge between the two modules. Secondly, the two-stage nature of the module introduces modeling peculiarities. In order to address these, the following control strategy and model simplifications are done to analyze the control performance of the converter.

A. 10-kW Module Control

The module-level control strategy can be seen in Fig. 2. The buck stage is regulated using a fixed frequency variable duty cycle control approach. The employed control technique

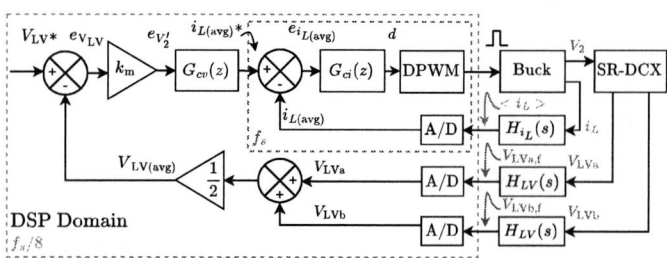

Fig. 2: Control scheme for a two-stage 10-kW module.

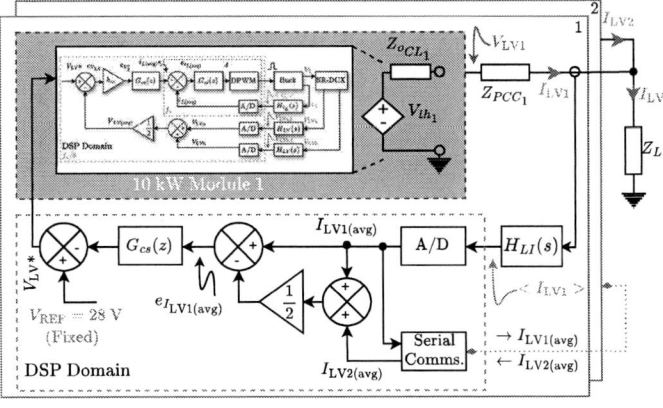

Fig. 3: Control scheme for the 600-V to 28-V 20-kW DC-DC converter.

is the dual-loop average current mode control approach [17]. The buck stage is controlled to directly regulate the average voltage seen on $V_{LVa/b}$ nodes. The SR-DCX's unregulated operation at a constant switching frequency near resonance is tuned offline for optimal full-load efficiency. Given the SR-DCX's operation near resonance, this stage exhibits an almost constant gain and only minor phase penalties. This constant gain effectively acts as a proportional controller k_m for the SR-DCX stage embedded within the first stage's control mechanism.

$$G_{id}(s) = \left(\frac{V_2}{dR_{eq}}\right)\frac{1 + sR_{eq}C_3}{s^2 L_1 C_3 + s\frac{L_1}{R_{eq}} + 1} \qquad (1)$$

$$G_{vd}(s) = \left(\frac{V_2}{d}\right)\frac{1}{s^2 L_1 C_3 + s\frac{L_1}{R_{eq}} + 1} \qquad (2)$$

where, d is the operating point duty ratio, R_{eq} is reflected load resistance by the SR-DCX on the buck converter, C_3, L_1, and V_2 are as referenced in Fig. 1.

$$G_{cv}(z) = \frac{0.1073z - 0.1046}{z - 1} \qquad (3)$$

In order to use the small-signal approximation for modeling the buck-converter, the modeling involves making two key approximations. Firstly, the EMI filter feeding the buck stage of the converter is designed to satisfy the Middlebrook stability criteria [18]. Thus, it is assumed that this circuital element does not impact the closed-loop analysis. Secondly, given that the second stage operates close to resonance, the first-harmonic approximation accurately refers the load resistance to the primary side [16]. Thus, as seen by the buck stage, the load is purely resistive. The stability analysis of the buck stage of the module is now carried out over the entire line and load regulation range.

$$G_{vc}(s) = \left(\frac{1}{H_{i_L}}\right)\left(\frac{G_{vd}}{G_{id}}\right)\frac{T_i}{1 + T_i} \qquad (4)$$

where, H_{i_L} is the averaging filter with a bandwidth of 12 kHz, T_i is the current loop gain, a product of G_{ci}, G_{id},

H_{i_L}, and discretization delay e^{-st_d} assuming the gain of the DPWM block is unity.

The average inductor current of the buck converter is tightly regulated with the help of a Type-3 compensator. The control-to-inductor current transfer function $G_{id}(s)$ of the plant is described by (1). The current controller is designed to have a bandwidth of 12.5 kHz around $\frac{1}{8}^{th}$ of the switching frequency. The current measurement is done with the help of a high-bandwidth hall-effect current transducer, and a third-order analog filter averages the signal to a bandwidth of 12 kHz. The Type-3 compensator $G_{ci}(z)$ accounts for the averaging filter (5). The compensator is designed for operation at the 650 V V_{HV} 10 kW operating point and discretized at the converter switching frequency using the Tustin approximation.

$$G_{ci}(z) = \frac{0.03055z^3 - 0.05192z^2 + 0.03131z - 0.006769}{z^3 - 0.4193z^2 - 0.4999z - 0.08083}$$
$$(5)$$

After the inner current loop is closed, the plant $G_{vc}(s)$, as observed by the outer voltage loop, is described by (4). Herein, the control-to-output voltage behavior $G_{vd}(s)$ of the plant is represented by (2). The outer voltage regulation loop uses a PI compensator $G_{cv}(z)$ with anti-windup capability and is band-limited at 300 Hz (3). The voltage loop runs at $\frac{1}{8}^{th}$ the switching frequency of 12.5 kHz and is discretized at this rate again using Tustin approximation. Both the compensators are also evaluated to ensure that the input voltage fluctuations are not dominant in governing the converter output. Given the digital control approach, discretization delays due to loop execution and analog-to-digital conversion (ADC) are considered in these loop gain plots using second-order Pade approximation of e^{-st_d}. The resulting continuous time margin plots for a 600 V V_{HV} operating point with different power levels can be seen in Fig. 4(a) and Fig. 4(b).

B. Paralleled Module Control

The 20 kW control block diagram is shown in Fig. 3. The paralleling of power modules using bus-bars to a point-of-common-coupling (PCC) introduces a connecting impedance Z_{PCCx} where $x \in \{1, 2\}$. In order to ensure equal CS between the paralleled modules, the Thevenin voltage V_{thx} must be regulated such that it overcomes both Z_{oCL} and Z_{PCCx} instantaneously to deliver the necessary current. Thus, the voltage reference at the input of Fig. 3 V_{LV}^* is used to overcome various system impedances using a CS compensator $G_{cs}(z)$. The compensator $G_{cs}(z)$ generates an error voltage added to the fixed reference to regulate the converter output voltage. The compensator is fed by the difference of: Firstly, the sum of the sensed output current from both the 5 kW sub-modules and, secondly, the average current flow out of both the 10 kW modules. This difference effectively corrects any current-sharing imbalances between the two modules. The current measured by each module I_{LV1} and I_{LV2} are communicated over a full-duplex RS-485 serial bus dedicated for control purposes.

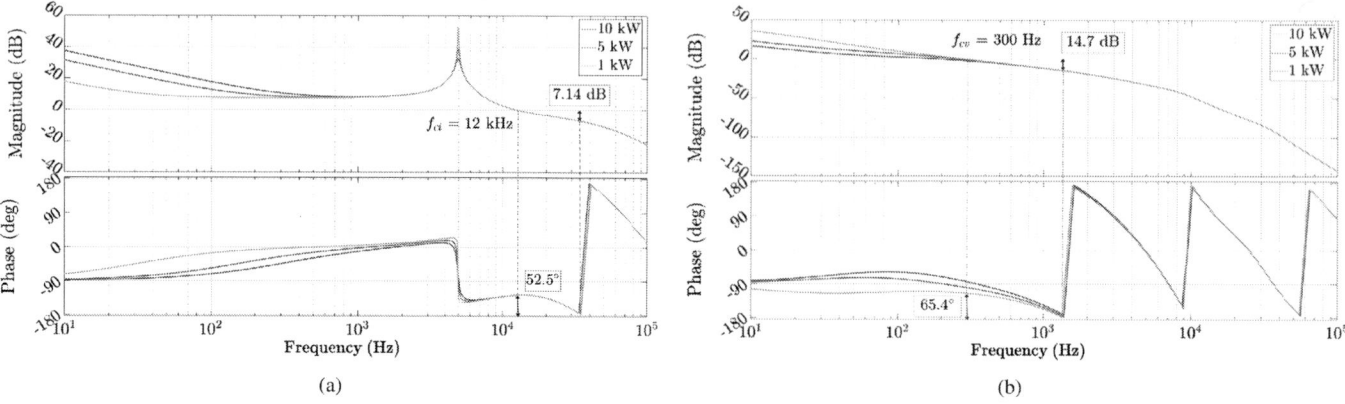

Fig. 4: Bode plot showing (a) current loop gain (T_i) and (b) voltage loop gain with inner loop closed (T_v).

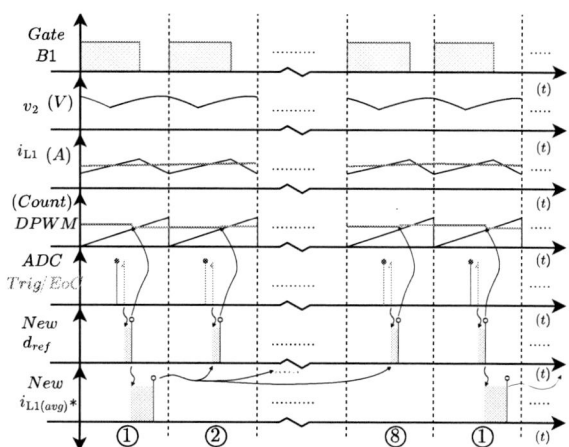

Fig. 5: Control logic timing diagram for a 10-kW module.

In order to ensure equal current-sharing between the two 10 kW modules, firstly, the impact on the converter's output impedance with a closed current loop is analyzed and is given by $Z_{o_{CL}}$ [17]. Then, the module level transfer function is achieved by block reduction of both the current and the voltage loops, representing the module's Thevenin voltage V_{thx}. This information is used to construct a Thevenin equivalent circuit. In order to overcome the impedances, the compensator $G_{cs}(z)$ must be analyzed to ensure that the changes to the voltage reference it makes do not make either of the modules unstable. The transfer functions from the reference voltage to the module output current for the same module is given by (6) and for the other module is given by (7) [12]. The transfer functions depend on the Thevenin parameters F and, crucially, the gain of the CS compensator $G_{cs}(z)$. Thus, designing $G_{cs}(z)$ such that G is stable while tuning it to a desired bandwidth and margins [12] enables a stable CS operation. In this work, pole-zero placement is done to achieve a desired loop response from G. The compensator $G_{cs}(z)$ is also discretized using Tustin approximation and is executed at the same rate as the voltage compensator loop T_v.

$$\frac{I_{lv_j}}{V_{\mathrm{REF}j}} = \frac{F\left(1 + H + \frac{G}{2}\right)}{(1+G)(1+2H)} \tag{6}$$

$$\frac{I_{lv_k}}{V_{\mathrm{REF}j}} = \frac{F\left(\frac{G}{2} - 1\right)}{(1+G)(1+2H)} \tag{7}$$

where, $F = \dfrac{1}{H_{lv}}\dfrac{T_v}{(1+T_v)Z}$, $Z = Z_{PCC_x} + \dfrac{Z_{\mathrm{OL}}}{1+T_i}$

$$G = F G_{cs}, \text{ and } H = \frac{Z_L}{Z}$$

IV. DISCRETE IMPLEMENTATION

As designed in Section III-A, the compensator difference equations are executed on a custom control board based on TMS320F28377D. Figure 5 shows the control sequence and how various digital signal processor (DSP) resources emulate a close-to-analog behavior. Firstly, all the control elements are linked to a base PWM module, which generates the gating signals for the buck converter stage. The analog-to-digital converter (ADC) conversion for sampling the average inductor current is triggered at a fixed 50% duty based on a compare-capture event on the base PWM module. The sampling point is selected to represent a noise-immune approximation of the average inductor current for the nominal operating points. Two samples with an acquisition window of 0.6 μs are picked and averaged to minimize operating point sensitivity. It should be noted that the position of sampling is selected based on the nominal operating duty range. Given that the duty range under regular operation ranges between 72 to 90%, there is approximately a 1 μs window for computing new duty value. The end-of-conversion notification from the ADC triggers the current compensator difference equation computation. Finally, the new duty value is immediately loaded for comparison, resulting in a worst-case discretization delay of approximately 0.4 cycles or e^{-st_d}, where $t_d = \frac{0.4}{f_s}$. Since most of the above actions are enabled by hardware configurations, very little CPU overhead is added while maintaining a symmetric cycle-by-cycle operation. Thus, the implementation enables a minimal phase margin penalty despite the discrete operation.

979-8-3315-1612-3/25 $31.00 © 2025 IEEE

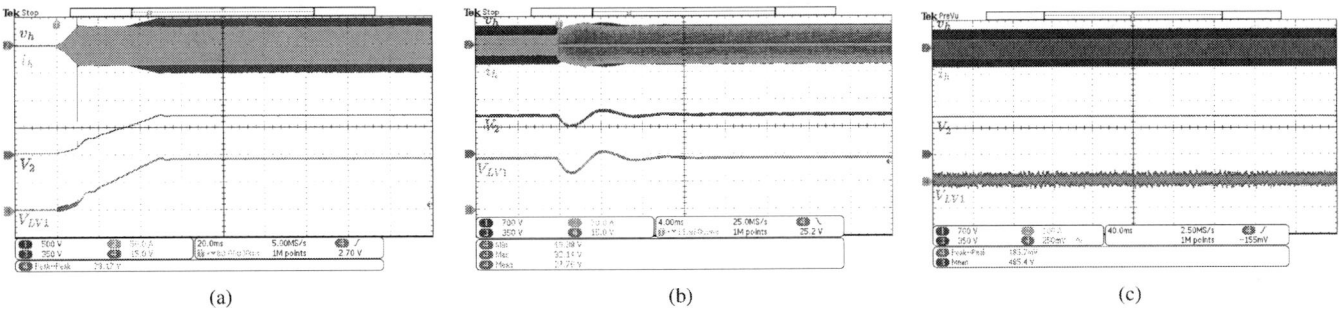

(a)   (b)   (c)

Fig. 6: Experimental results of a 10-kW module at V_{HV} = 650-V : (a) Soft-start at 10-kW, (b) Step load addition 5-kW → 10-kW, (c) Steady-state AC coupled response at 10-kW.

Similarly, the voltage compensator loop is executed by a separate PWM module linked to the base PWM module for synchronous operation. Here, the clock feed is prescaled by a clock divider of $\frac{1}{8}^{th}$, thereby resulting in an execution rate that is 8 times slower than the current compensator loop. Given all the PWM modules are linked, the prescaled clock feed to the voltage loop ensures a synchronized operation between current and voltage loops. The ADC conversion is triggered at a fixed 10% duty of the voltage loop timer to provide a new $i^*_{L(avg)}$ reference between two consecutive current loop executions. Furthermore, the same voltage loop PWM module samples the other analog signals used for over-current and over-voltage protection mechanisms. The SR-DCX stage's gating commands are also generated in alignment with the buck converter's base PWM module. This enables a synchronous operation between the two converters. It should be noted that the resonant frequency is tuned offline at full power, and start-up calibration of the resonant frequency is not considered in this work.

V. RESULTS

The experimental setup for the 600 V to 28 V 20 kW two-stage DC-DC converter is seen in Fig. 7. The discussed digital control technique is demonstrated experimentally on a 10 kW module. The CS demonstration for the full 20 kW converter control scheme is done using PLECS simulations. The experimental results using the above-designed compensator and control scheme can be seen in Fig. 6 for an input voltage V_{HV} of 650V. The soft-start operation for the converter is done by linearly increasing the reference voltage command from 0 to

28 V, as shown in Fig. 6(a) with the load set to 10 kW when performing the soft-start sequence. This result empirically confirms the control approach's large-signal stability. In order to evaluate the dynamic performance of the control scheme, a step load addition test is carried out. In this test, the load is increased from 5 kW to 10 kW with a current slew rate of $1\frac{A}{\mu s}$. Results can be seen in Fig. 6(b) shows a voltage sag of about 30%, within MIL-STD-704F limit [3]. The results in Fig. 6(c) show the steady-state AC coupled results with a peak-to-peak ripple of about 0.6%. The steady-state ripple results are shown with the load configured to 10 kW. The scheme is experimentally verified similarly for the entire input range of 550 to 650 V at the V_{HV} node as highlighted in Table I.

The control approach for paralleling two modules, described in Section III-B, is evaluated using a discrete step solver in PLECS with a fixed step execution rate of 100 ns based on Tustin approximation. The CS compensator, voltage- and current-control loops are simulated to match the DSP behavior as in Fig. 5. Also, a noise component is injected on all sense lines to simulate a more realistic system. The noise content is modeled by a line frequency sinusoidal component, a triangular wave-based impulse content for switching instances, and a random noise content. The noise component is kept at 5% of the signal swing range, as seen by the ADC. The operating point of the converter is set at 650 V at the V_{HV} node and a load of 14 kW at start-up. The results are shown in Fig. 8. The left half of the results show converter behavior

Fig. 7: Experimental setup of the two-stage 20-kW DC-DC converter.

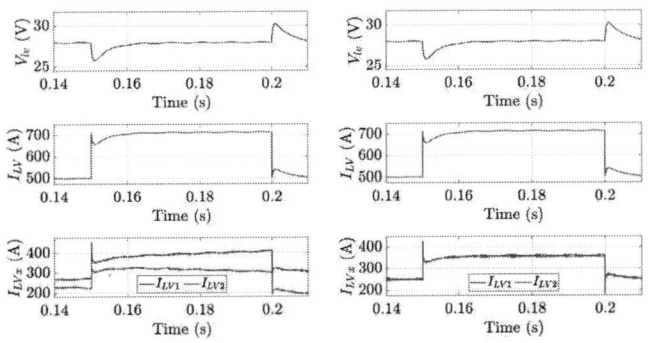

Fig. 8: PLECS simulation results of 20-kW closed-loop operation. Left: Without current sharing scheme and Right: With current sharing scheme.

979-8-3315-1612-3/25 $31.00 © 2025 IEEE

TABLE I: Experimental results for a 10-kW module under digital closed-loop control.

Input (V)	Steady State AC ripple (mV$_{pp}$)			$5 \rightarrow 10$ kW	$10 \rightarrow 5$ kW
V_{HV}	1 kW	5 kW	10 kW	Sag (V)	Swell (V)
550	172	288	198	8.6	6.8
600	165	197	1200	8.6	9.1
650	180	176	185	8.3	10

under no CS control, and the right half shows the behavior with the CS control enabled. The test sequence involves stepping the load from 14 kW to 20 kW at 150 ms and removing the load at 200 ms. The static CS is significantly off by a factor of 25% when the CS controller is not enabled. The CS controller ensures the current is balanced within $\pm$ 10A of the current demanded by the load.

VI. CONCLUSION

The paper discussed a digital control approach to regulate a 600 V to 28 V 20 kW two-stage converter. A brief introduction to the converter structure was provided. The control approach and analysis were discussed. A simple modeling approach for regulating the two-stage module is taken and experimentally verified for line and load operating conditions. Efficient DSP implementation for minimizing digitization delays is highlighted. The control analysis is experimentally verified on a 600 V to 28 V 10 kW module with a steady state AC ripple of 0.6%. A current sharing compensator is designed by modeling the closed-loop behavior of the modules and using the interconnecting impedances. A PLECS study is carried out to check and verify the effect of the current sharing loop, which ensures static current sharing stays within $\pm$ 10A of either module. The next revision to this work will experimentally demonstrate symmetric current sharing under dynamic and static loading scenarios. It will also discuss the mathematical framework for dynamic current sharing challenges and delays resulting from the digital communication link. Furthermore, improvements in the dynamic performance of the 10 kW module will be explored by phase compensation for the SR-DCX stage.

ACKNOWLEDGEMENTS

This research was undertaken, in part, thanks to funding from Mitacs and Eaton Corporation. The authors gratefully acknowledge Plexim GmBH's support with PLECS software and Mathworks' support with MATLAB/Simulink software used in this research.

REFERENCES

[1] V. Madonna, P. Giangrande, and M. Galea, "Electrical power generation in aircraft: Review, challenges, and opportunities," *IEEE Transactions on Transportation Electrification*, vol. 4, no. 3, pp. 646–659, 2018.

[2] M. Miller, G. Brinton, G. Roden, G. Hamilton, E. Jochum, R. Eddins, C. Milford, and R. Shiver, "U.s. army, bae systems p&s, and ge aviation jointly execute an m109a7 vehicle level demonstration of a ge silicon carbide converter," in *Vehicle Electronics and Architecture (VEA) & Ground Systems Cyber Engineering (GSCE) Technical Session*, 2018.

[3] U. S. Department of Defense, "Mil-std-704f w/change1: Aircraft electrical power characteristics," 2016. [Online]. Available: http://everyspec.com/MIL-STD/MIL-STD-0700-0799/MIL-STD-704F_CHG-1_55461

[4] U. S. Department of Defense, "Mil-prf-gcs600a: Characteristics of 600 volt dc electrical systems for military ground vehicles," 2010. [Online]. Available: https://ntrl.ntis.gov/NTRL/dashboard/searchResults/titleDetail/ADA538869.xhtml

[5] R. Pradhan, M. I. Hassan, Z. Wang, J. Yuan, G. Pietrini, P. Sunthar-alingam, M. F. Cruz, and A. Emadi, "Design of a 20 kw bidirectional dual active bridge converter for aerospace applications," in *2023 IEEE Applied Power Electronics Conference and Exposition (APEC)*, 2023, pp. 1024–1030.

[6] Y. Zhu, T. Yang, Z. Wang, X. Yan, S. Bozhko, and P. Wheeler, "Parasitic inductance impact of a high-turn-ratio half bridge active clamped converter for more-electric aircraft applications," in *2024 IEEE Applied Power Electronics Conference and Exposition (APEC)*, 2024, pp. 2226–2231.

[7] F. Giuliani, G. Buticchi, M. Liserre, N. Delmonte, P. Cova, and N. Pignoloni, "Gan-based triple active bridge for avionic application," in *2017 IEEE 26th International Symposium on Industrial Electronics (ISIE)*, 2017, pp. 1856–1860.

[8] S. Rahman, I. Khan, S. Dey, and A. Mallik, "Triple-active bridge-based dynamic power balancing solution for minimizing overdesigning in military aircraft power system," *IEEE Transactions on Vehicular Technology*, vol. 73, no. 3, pp. 3329–3339, 2024.

[9] Y. Zhu, Z. Wang, T. Yang, T. Dragicevic, S. Bozhko, and P. Wheeler, "Model predictive control for dc offset suppression of dual active bridge converter for more-electric aircraft applications," in *2021 IEEE 30th International Symposium on Industrial Electronics (ISIE)*, 2021, pp. 1–6.

[10] X. Yan, Y. Zhu, Z. Wang, T. Yang, S. Bozhko, and P. Wheeler, "Simplified modelling and control of dual active bridge converter for future electrified aerospace application," in *2023 IEEE International Conference on Electrical Systems for Aircraft, Railway, Ship Propulsion and Road Vehicles & International Transportation Electrification Conference (ESARS-ITEC)*, 2023, pp. 1–6.

[11] T. Dragičević, X. Lu, J. C. Vasquez, and J. M. Guerrero, "Dc microgrids—part i: A review of control strategies and stabilization techniques," *IEEE Transactions on Power Electronics*, vol. 31, no. 7, pp. 4876–4891, 2016.

[12] V. Thottuvelil and G. Verghese, "Analysis and control design of paralleled dc/dc converters with current sharing," *IEEE Transactions on Power Electronics*, vol. 13, no. 4, pp. 635–644, 1998.

[13] F. Flores-Bahamonde, H. Renaudineau, A. M. Llor, A. Chub, and S. Kouro, "The dc transformer power electronic building block: Powering next-generation converter design," *IEEE Industrial Electronics Magazine*, vol. 17, no. 1, pp. 21–35, 2023.

[14] J. Huber, G. Ortiz, F. Krismer, N. Widmer, and J. W. Kolar, "η-ρ pareto optimization of bidirectional half-cycle discontinuous-conduction-mode series-resonant dc/dc converter with fixed voltage transfer ratio," in *2013 Twenty-Eighth Annual IEEE Applied Power Electronics Conference and Exposition (APEC)*, 2013, pp. 1413–1420.

[15] H. Wang, Y. Chen, Y. Qiu, P. Fang, Y. Zhang, L. Wang, Y.-F. Liu, J. Afsharian, and Z. Yang, "Common capacitor multiphase llc converter with passive current sharing ability," *IEEE Transactions on Power Electronics*, vol. 33, no. 1, pp. 370–387, 2018.

[16] A. Elezab, O. Zayed, A. Abuelnaga, and M. Narimani, "High efficiency llc resonant converter with wide output range of 200–1000 v for dc-connected evs ultra-fast charging stations," *IEEE Access*, vol. 11, pp. 33 037–33 048, 2023.

[17] D. Maksimovic and R. W. Erickson, *Fundamentals of Power Electronics*. Cham, Switzerland: Springer International Publishing, 2020. [Online]. Available: https://doi.org/10.1007/978-3-030-43881-4

[18] A. Riccobono and E. Santi, "Comprehensive review of stability criteria for dc power distribution systems," *IEEE Transactions on Industry Applications*, vol. 50, no. 5, pp. 3525–3535, 2014.

Self-Calibrated Digital Current Emulation for High-Frequency Hysteretic Current-Mode Control in GaN PFC Converters

Mohammad Shawkat Zaman and Olivier Trescases

The Edward S. Rogers Sr. Department of Electrical & Computer Engineering, University of Toronto, Canada

E-mail: Shawkat.Zaman@utoronto.ca

Abstract—**This work presents a digital emulation technique that reproduces the switching-frequency behaviour of the inductor current in a power converter, enabling high-frequency (HF) cycle-by-cycle hysteretic current-mode control for applications with limited current-sensing opportunities. The proposed technique leverages lower-bandwidth current sensors to opportunistically self-calibrate the emulator, alleviating its dependence on preset system parameters. Removing the need for high-bandwidth sensors enables low-inductance power loops, critical for maximizing performance in HF converters. The design is demonstrated in a 240-V$_{\text{rms}}$-to-450-V all-GaN totem-pole PFC converter operating at 4 kW.**

Index Terms—**Digital Current Emulation, Current Observer, Self Calibration, Current Mode Control, Power Factor Correction, Totem Pole, Bridgeless Rectifier, GaN.**

I. INTRODUCTION

The market for gallium nitride (GaN) power transistors is expected to grow from \$46M in 2020 to \$1B in 2026, thanks to their superior electrical and thermal performance [1], with promising outlooks in electric-vehicle (EV) applications in the 400-V range [2]. Grid-facing EV power systems, like the onboard chargers (OBCs) with power-factor correction (PFC), benefit from current-mode control (CMC) for precise current-shaping, enhancing efficiency and minimizing input-filter size [3], [4]. While GaN devices operating in the medium- to high-voltage (HV) range (≥ 200 V) are capable of faster switching speeds compared to their Silicon (Si) counterparts [2], current-sensing becomes increasingly challenging at higher switching frequencies (≥ 300 kHz) and power levels (≥ 500 W). High-bandwidth sensing requires the insertion of current sensors in the power loop, which worsens parasitic inductances and limits the achievable switching frequency. Furthermore, while resistive current sensors can have bandwidths in the GHz range (compared to a few MHz for magnetic ones), they suffer from significant conduction losses in high-power applications [5]. SenseFET-like indirect sensing schemes can avoid these issues [6], [7], but must contend with isolation and noise constraints at HV.

To circumvent these current-sensing challenges, various aspects of the desired current waveform have been emulated using more easily sensed quantities, such as input and output voltages [5], [8]–[15]. However, many such designs only emulate the average inductor current [8]–[12], without capturing the switching-frequency ripple, preventing their use in high-frequency control techniques. Moreover, emulators that do reproduce the ripple [5], [13]–[15] have only been demonstrated for lower-voltage and -power (≤ 200 V, 200 W) systems, unsuitable for the 400-V, kW-level converters in EV onboard chargers (OBCs). In particular, [5] is implemented in an ac-dc PFC converter, but operating at only 100 kHz, while [13] is demonstrated at a fixed frequency of 500 kHz, but only for dc-dc conversion. They also rely on the precise inductance value (which may vary by up to $\pm 20\%$), requiring manual adjustment of the emulator parameters for every sample.

This work presents a digital emulator that reproduces the switching-frequency behaviour of the inductor current. To alleviate the need for precise knowledge of system parameters, a novel calibration technique leveraging current sensors with relatively low bandwidth is proposed. The emulator design is validated in a 240-V$_{\text{rms}}$-to-450-V all-GaN totem-pole PFC converter operating at 4 kW by enabling cycle-by-cycle hysteretic current-mode control (HCMC).

II. PROPOSED DESIGN

Fig. 1 shows the proposed system with the emulator-based control architecture. It consists of an all-GaN totem-pole PFC converter and an FPGA-based digital controller employing HCMC via inductor emulation.

A. Totem-Pole PFC converter with HCMC

The proposed PFC converter is implemented using an all-GaN totem-pole topology, as shown in Fig. 1, to leverage the low ON-resistance and switching loss of GaN devices and eliminate the lossy diode bridge required for conventional boost PFC converters [16]. $M_{x\text{S-LF}}$ form the low-frequency (LF) leg of the totem pole that commutates based on the polarity of the ac-line voltage, v_{ac}, while $M_{x\text{S-HF}}$ form the high-frequency (HF) leg that switches at a frequency, $f_{\text{sw}} \gg$ the line frequency, f_{ac}, to shape the inductor current, i_L. v_{ac} and the dc-link voltage, V_{link}, are digitally sampled using independent analog-to-digital converters (ADCs). A Hall-effect-based sensor placed at $V_{X\text{-LF}}$, along with a comparator and a digital reference, is used to sample the inductor current and correct emulator mismatches.

The PFC design employs variable-frequency HCMC, which controls both peak and valley of the inductor current and avoids the need for slope compensation inherent with fixed-frequency peak- or valley-current-mode control. The controller automatically switches between boundary and continuous conduction modes (BCM and CCM) to optimize efficiency, similar to the Silicon-Carbide-based design in [4], which operated

979-8-3315-1612-3/25 \$31.00 © 2025 IEEE

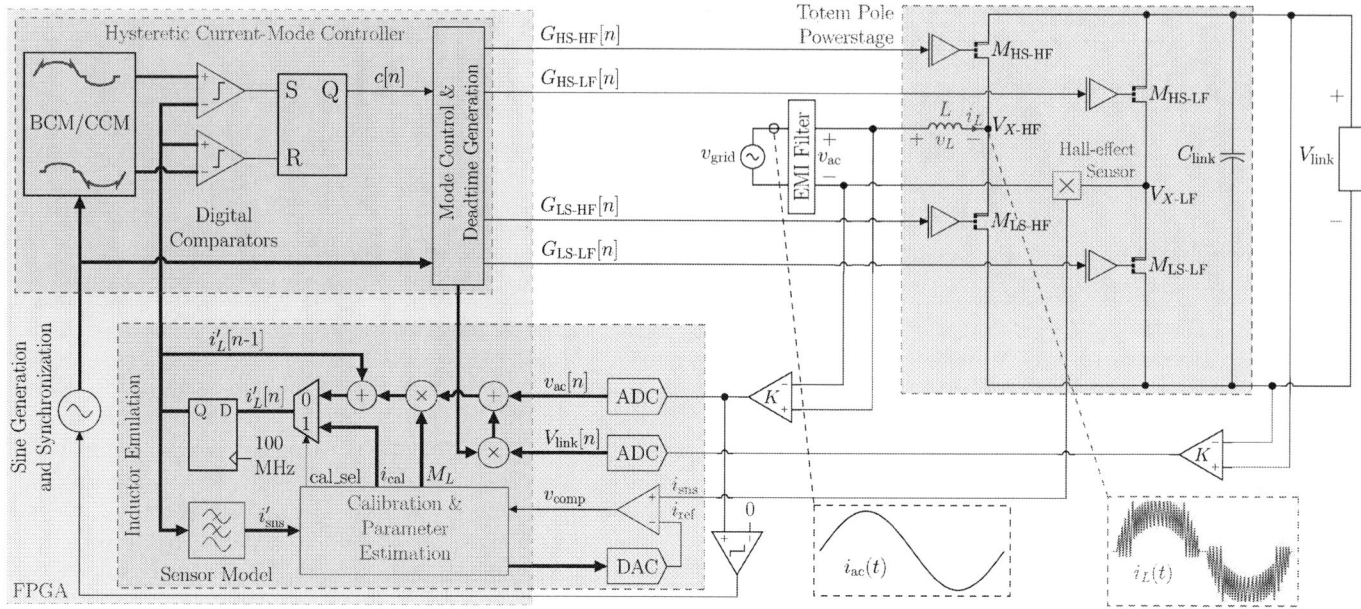

Fig. 1. PFC converter system diagram showing the power-stage, sensing circuitry, and the FPGA-based inductor-current emulator and controller. A thick line indicates a multi-bit digital signal. The outer voltage loop that regulates the PFC input-current amplitude is not shown for simplicity.

at up to 300 kHz. BCM operation at low input currents reduces switching losses by enabling zero-voltage switching (ZVS) for both HF switches, while CCM at high currents optimizes conduction losses. This design utilizes the high-resolution emulated current to perform HCMC, which enables higher-frequency operation, as the emulated waveform is free of the ubiquitous switching noise that otherwise limits the performance of converters with CMC, especially in ac-dc conversion with wide duty-cycle ranges.

B. Inductor Emulator

Fundamentally, the inductor current, $i_L(t)$, is related to the integral of its voltage, $v_L(t)$:

$$i_L(t) = i_L(t_0) + \frac{1}{L} \int_{t_0}^{t} v_L(\tau) d\tau \quad (1)$$

For the totem-pole topology, $v_L(t)$ is given by

$$v_L(t) = v_{ac}(t) + \left[S_{LF}(t) - S_{HF}(t) \right] V_{link}(t) \pm V_{loss}(t) \quad (2)$$

where $S_{LF}(t)$ and $S_{HF}(t)$ represent the states of the LF and HF legs of the totem pole, respectively:

$$S_x(t) = \begin{cases} 1 & \text{if } M_{HS\text{-}x} \text{ is conducting} \\ 0 & \text{if } M_{LS\text{-}x} \text{ is conducting} \end{cases}, x \in \{LF, HF\}$$

given that the $\left(M_{HS\text{-}x}, M_{LS\text{-}x} \right)$ pairs always switch in a complementary fashion. During deadtime (i.e., when both switches in a leg are OFF), conduction states can be inferred from the direction of i_L and the reverse-conduction capability of the switches.

$V_{loss}(t)$ represents the current-dependent conduction losses due to component resistances. In this system, i_L scales with v_{ac} (due to PFC operation), and the total resistance contributing to V_{loss} is $< 100\,\text{m}\Omega$. Consequently, this term is ignored in the proposed design for computational simplicity.

For a sufficiently small computation interval, T_{comp}, $v_L(t)$ can be assumed to be constant, and estimated using the sampled v_{ac} and V_{link} voltages:

$$v'_L[n] = v_{ac}[n] + \left(S_{LF}[n] - S_{HF}[n] \right) V_{link}[n] \quad (3)$$

The integral from (1) can then be approximated by a product, and $i_L(t)$ can be emulated by

$$\begin{aligned} i'_L[n] &= i'_L[n-1] + \frac{1}{L} \left(v'_L[n] \right) \left(T_{comp} \right) \\ &= i'_L[n-1] + M_L v'_L[n] \end{aligned} \quad (4)$$

where $M_L = \frac{T_{comp}}{L}$ is used as the adjustable slope parameter, instead of L, for ease of computation.

In this design, $T_{comp} = 10\,\text{ns}$ is chosen to ensure 0.5% or better temporal resolution at the highest targeted f_{sw} of 500 kHz. $S_x[n]$ can be determined from the digital gating signals, $\left(G_{HS\text{-}x}[n], G_{LS\text{-}x}[n] \right)$, generated by the current-mode controller, after accounting for gate-driver delays. The proposed design samples $v_{ac}(t)$ and $V_{link}(t)$ at $f_{samp} \approx 1\,\text{MHz}$ using 10-bit serial ADCs in order to balance voltage and timing resolutions, leading to a worst-case change in the measured voltages $\ll V_{LSB} \approx 0.7\,\text{V}$ between samples under typical operating conditions. Since $T_{comp} \ll 1/f_{samp}$, a sample-and-hold approach is utilized for generating the intermediate voltage samples between consecutive measurements. While extrapolation using past samples can improve emulation accuracy, the added computational complexity was deemed unjustified given the slow voltage dynamics in this application.

C. Emulator Calibration

In order to limit the deviation of (4) from reality over time due to measurement noise and component variations, occasional sensing is required to calibrate i'_L and M_L. Unlike direct

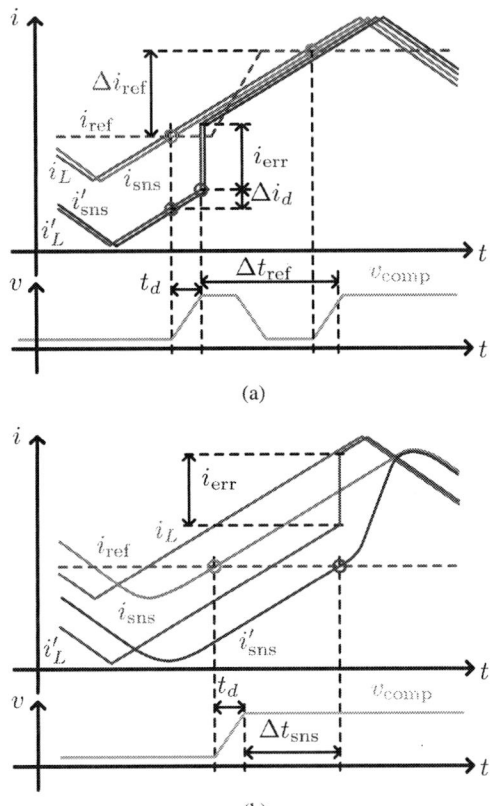

(a)

(b)

Fig. 2. Idealized waveforms of (a) direct and (b) indirect calibration of the emulator with high- and low-bandwidth sensors, respectively. While a positive correction is shown in each case, both positive and negative errors can be corrected in either approach.

current-sensing for conventional CMC, these calibrations need not be performed every cycle, though there is a direct relationship between the frequency of calibration and the maximum accumulated error in the emulated waveform. For example, a 1-LSB error in the voltage measurement ($\approx 0.7\,\mathrm{V}$ in this design) leads to a current error of approximately $177\,\mathrm{mA}$ over one period at the lowest targeted f_{sw} of $200\,\mathrm{kHz}$, which is less than 0.5% of the rated operating current. In a variable-frequency control scheme such as PFC with HCMC, calibration can be performed during lower-frequency periods in the operating cycle, enabling higher-frequency operation without direct sensing elsewhere.

If the current sensor has a bandwidth, $f_{\mathrm{BW}} > 5f_{\mathrm{sw}}$, it preserves the most significant harmonics of the triangular inductor current waveform, i_L, and the sensed waveform, i_{sns}, closely matches i_L, as shown in Fig. 2(a). This allows *direct calibration* by comparing i_{sns} with a reference level, i_{ref}. When i_{sns} crosses i_{ref}, as detected by the comparator output, v_{comp}, the emulated current, i_L', is set to

$$i_{\mathrm{cal}} = i_{\mathrm{ref}} + \Delta i_d \qquad (5)$$

where Δi_d accounts for the comparator delay, t_d, as given by

$$\Delta i_d = M_L \sum^{t_d} v_L'[n] \qquad (6)$$

where the summation term represents the sum of all v_L' samples over the t_d duration preceding the detection instant. This notation is used in the following equations as well.

By changing i_{ref} during the same switching state and performing a second comparison, M_L can also be estimated, as shown in Fig. 2(a). If i_{ref} is changed by Δi_{ref}, and the duration between the two detection instants is Δt_{ref}, M_L is approximated by

$$M_L \approx \frac{\Delta i_{\mathrm{ref}}}{\frac{\Delta t_{\mathrm{ref}}}{\sum} v_L'[n]} \qquad (7)$$

While this measurement requires the switching state to be sufficiently long to allow the changing and settling of i_{ref}, it is only needed infrequently, and can be performed opportunistically in operating conditions with the largest ripple.

In contrast, if $f_{\mathrm{BW}} \leq 5f_{\mathrm{sw}}$, as in this design, loss of the higher harmonics introduces peak distortion and phase delay in i_{sns}, as shown in Fig. 2(b). This makes direct calibration infeasible, and requires an *indirect calibration* approach to be used for correcting i_L'. The current error, i_{err}, is defined as

$$i_{\mathrm{err}} = i_L - i_L' \qquad (8)$$

To estimate i_{err}, the low-pass current-sensor behaviour must also be emulated, as shown in Fig. 2(b). This is modelled as a digital transfer function that replicates the frequency response of the sensor, which can be obtained from the manufacturer datasheet, or experimentally measured. Both i_{sns} and the emulated sensor waveform, i_{sns}', is compared against the same i_{ref} during one switching state, and the duration between the two crossings, Δt_{sns}, is measured. Note that detection of the i_{sns}' crossing is effectively instantaneous, since it is an internally generated signal and free from analog delays. Then i_{err} is given by

$$i_{\mathrm{err}} \approx M_L \left(\sum^{\Delta t_{\mathrm{sns}}+t_d} v_L'[n] \right) \qquad (9)$$

i_L' is then set to

$$i_{\mathrm{cal}} = i_L' + i_{\mathrm{err}} \qquad (10)$$

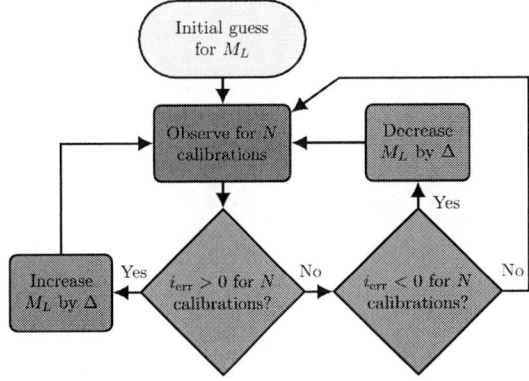

Fig. 3. Perturb-and-observe approach for estimating the slope parameter, M_L, with a low-bandwidth sensor. N and Δ are design parameters dependent on converter operating modes.

To estimate M_L with a low-bandwidth sensor, the converter can be occasionally operated in an LF mode to employ the direct-calibration method, or a perturb-and-observe approach may be used, as shown in Fig. 3. For example, in PFC operation with HCMC, the lowest operating frequencies occur in the BCM region near the BCM-CCM crossover point, as shown in Fig. 1, which presents a convenient opportunity for M_L estimation.

III. EXPERIMENTAL RESULTS

A prototype converter with the specifications shown in Table I was built to demonstrate the proposed current-emulation technique, as shown in Fig. 4. The power-stage uses the 650-V-rated `GS66516T` devices to leverage their near-chip-scale packaging with convenient top-side cooling surface, which enables heat extraction without interfering with the electrical layout. The `ACS733` isolated hall-effect sensor, having a bandwidth of 1 MHz, is used for sensing the inductor current.

TABLE I
PFC CONVERTER SPECIFICATIONS

Parameter	Value	Unit
AC-Line Voltage, v_{ac}	120-240	V_{rms}
AC-Line Frequency, f_{ac}	60	Hz
AC-Line Inductance, L	6×3.3	μH
DC-Link Voltage, V_{link}	300-450*	V_{dc}
DC-Link Capacitance, C_{link}	5×120	μF
Rated Output Power, P_{out}	4	kW
Switching Frequency, f_{sw}	200-500	kHz
Power Device, M_x	650 V, 25 mΩ	

* Dependent on ac-line voltage.

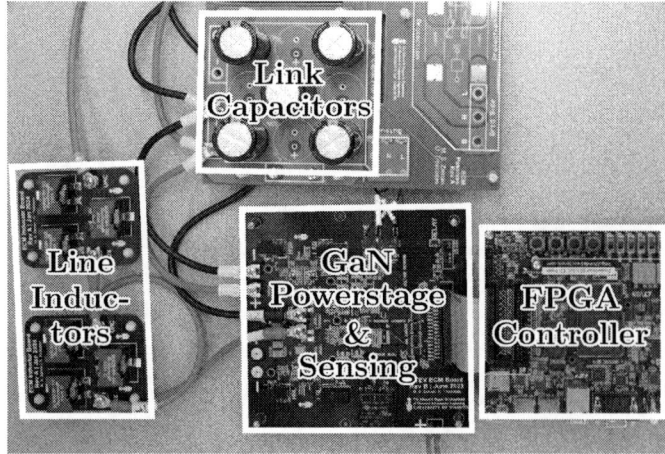

Fig. 4. Prototype PFC converter with emulator-based HCMC. The ac-line inductance is implemented with series-connected 3.3-μH SER2918H units in order to achieve the required saturation current. The power-stage is located on the underside of the PCB, along with the off-the-shelf liquid-cooled copper heatsink.

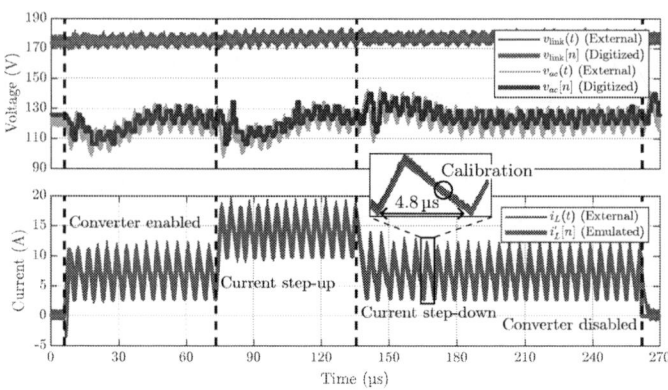

Fig. 5. Comparison between external (as measured by an oscilloscope) and digitized internal waveforms of the converter from an approximately 260-μs test run in dc-dc boost mode with $v_{ac} = 120 \, V_{dc}$, $V_{link} = 175 \, V_{dc}$, and current steps between 7 A and 14 A.

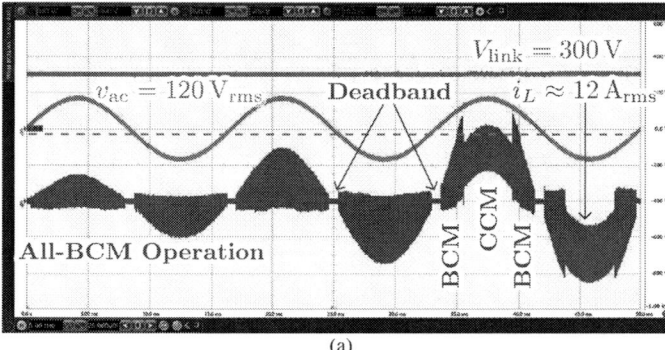

(a)

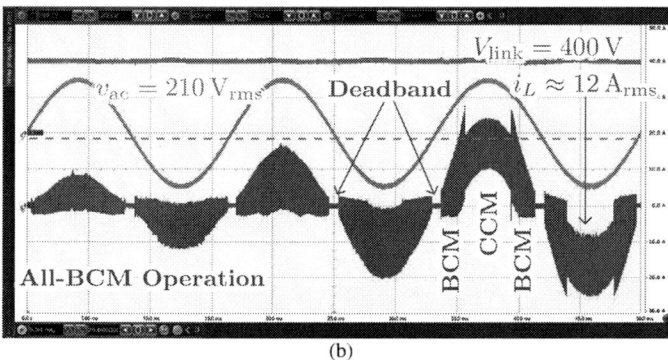

(b)

Fig. 6. Measured converter waveforms in ac-dc mode with an input current of up to 12 A_{rms}, and (a) $v_{ac} = 120 \, V_{rms}$, $V_{link} = 300 \, V$, and (b) $v_{ac} = 210 \, V_{rms}$, $V_{link} = 400 \, V$, showing sinusoidal current shaping, and automatic BCM-CCM transition based on operating current.

To verify the effectiveness of the proposed emulator and control technique, the converter was first operated in dc-dc boost mode, as shown in Fig. 5. The controller is able to reach steady-state operation in a single switching cycle, both at startup and subsequent current up- and down-steps, demonstrating the advantages of HCMC. The emulator makes this possible without peak- and valley-current sensing by providing a faithful representation of the inductor current, despite transient variations in v_{ac}.

The measured waveforms of the converter when operating in ac-dc PFC mode are shown in Figs. 6 and 7. In all cases, the inductor current is sinusoidally shaped and synchronized to

979-8-3315-1612-3/25 $31.00 © 2025 IEEE

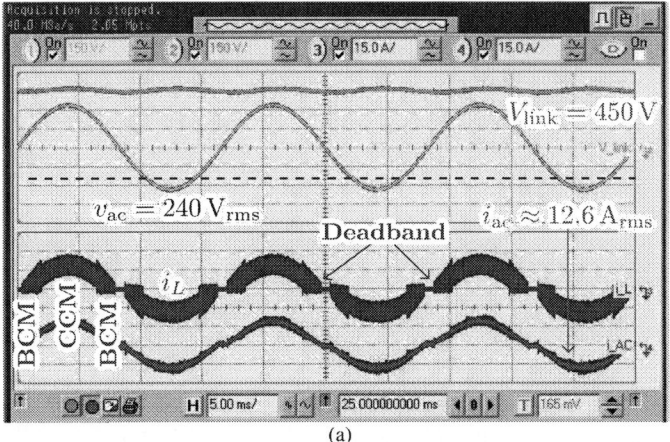

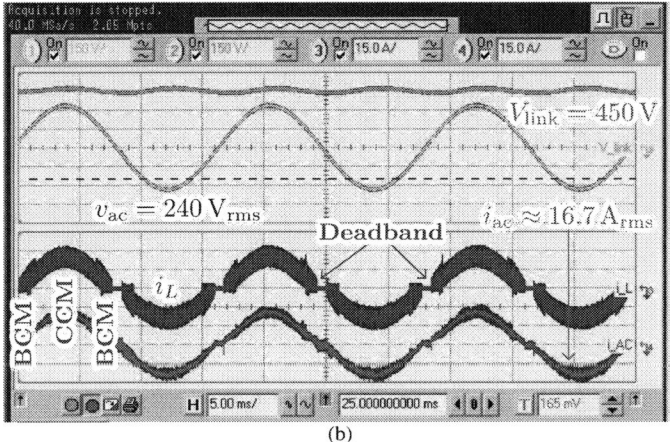

Fig. 7. Measured converter waveforms in steady-state ac-dc operation with $v_{ac} = 240\,V_{rms}$, $V_{link} = 450\,V$, and an input power of approximately (a) 3 kW, and (b) 4 kW. BCM-CCM transitions occur at different points in the cycle for different power levels due to the fixed current threshold. The ac-line current, i_{ac}, is measured with a 5-μF filter capacitance.

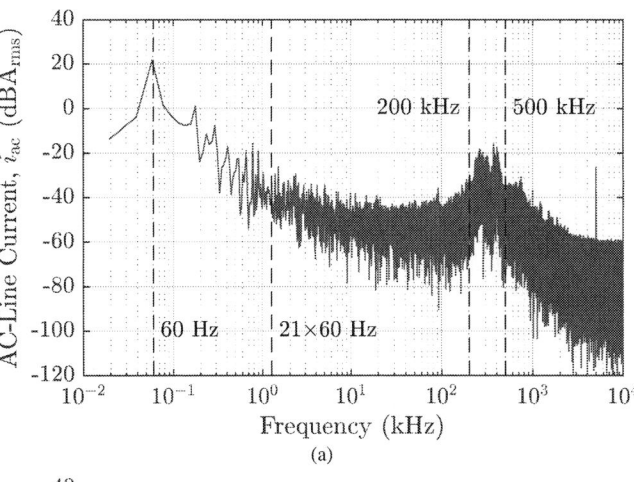

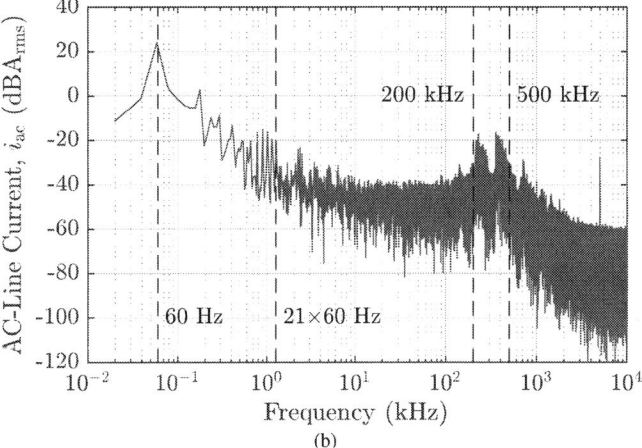

Fig. 8. Magnitude spectrum of the ac-line current, i_{ac}, when the converter is operated with $v_{ac} = 240\,V_{rms}$, $V_{link} = 450\,V$, and an input power of approximately (a) 3 kW, and (b) 4 kW. i_{ac} is measured with a 5-μF filter capacitance.

the input voltage, thanks to cycle-by-cycle CMC. Automatic mode transitions between BCM and CCM operation occurs based on a preset average-current threshold, and a slightly negative valley current is used in BCM to enable ZVS on all switches. The converter is disabled near the zero-crossings of the line cycle (the *deadband*) to avoid inefficient operation at very high frequencies and input-to-output ratios.

Thanks to the sinusoidal current-shaping, the unwanted frequency content of the current drawn from the ac line, i_{ac}, is relatively low, even with minimal filtering (only 5 μF in this case), as shown in Fig. 8. In particular, the total harmonic distortion (THD) at 4 kW is approximately 10.3%, using the first 20 harmonics. For comparison, the all-GaN totem-pole topology in [17] operating exclusively in BCM achieved a THD between 7.5%–18% over 0.3-3.3 kW, but with additional filtering. The variable-frequency operation results in a pair of distributed peaks between the designed minimum and maximum f_{sw}, which are approximately 40 dB below the fundamental component at 4 kW. This implies the need for reduced input filtering to meet electromagnetic interference (EMI) standards compared to fixed-frequency modulation schemes with a high-amplitude peak at a single

frequency. The power-stage achieved a peak efficiency of approximately 95.5% at 3 kW.

IV. CONCLUSION

This work presents a digital emulator that reproduces the switching-frequency ripple of the power inductor, and leverages a 1-MHz-bandwidth current sensor to opportunistically self-calibrate the emulated waveform and avoid the impact of noise from converter switching transitions. The HF emulation enables cycle-by-cycle CMC, and allows the power loop to be better optimized since no high-bandwidth sensors are needed. A 240-V_{rms}-to-450-V all-GaN totem-pole PFC converter, operating at 4 kW with HCMC and automatic BCM-CCM transitions for loss optimization over the full operating range, demonstrates the effectiveness of the proposed emulation technique. To the best of the authors' knowledge, this work provides the first demonstration of HF emulation in an ac-dc converter at 450 V and above 2 kW power levels.

ACKNOWLEDGMENT

This work was supported by the Natural Sciences and Engineering Research Council (NSERC) of Canada and the Taiwan Semiconductor Manufacturing Company (TSMC).

References

[1] A. B. Slimane and P. Chiu. (2021, May) GaN Power 2021: Epitaxy, Devices, Applications and Technology Trends. [Online]. Available: https://www.yolegroup.com/press-release/the-gan-power-market-will-surpass-1-billion-in-2026/

[2] M. Buffolo et al., "Review and Outlook on GaN and SiC Power Devices: Industrial State-of-the-Art, Applications, and Perspectives," IEEE Transactions on Electron Devices, vol. 71, no. 3, pp. 1344–1355, Mar. 2024.

[3] R. Fernandes and O. Trescases, "A Multimode 1-MHz PFC Front End With Digital Peak Current Modulation," IEEE Transactions on Power Electronics, vol. 31, no. 8, pp. 5694–5708, Aug. 2016.

[4] M. Nasr, K. Gupta, C. da Silva, C. H. Amon, and O. Trescases, "SiC Based On-Board EV Power-Hub with High-Efficiency DC Transfer Mode through AC Port for Vehicle-to-Vehicle Charging," in Applied Power Electronics Conference and Exposition (APEC), San Antonio, TX, USA, Mar. 2018, pp. 3398–3404.

[5] M. E. Ahmad, F. Schafmeister, and J. Böcker, "Digital Implementation of a Current Observer with On-Line Current Sample Correction for PFC Rectifiers," in Applied Power Electronics Conference and Exposition (APEC), Mar. 2023, pp. 2867–2873.

[6] M. Biglarbegian, N. Kim, T. Zhao, and B. Parkhideh, "Development of Isolated SenseGaN Current Monitoring for Boundary Conduction Mode Control of Power Converters," in Applied Power Electronics Conference and Exposition (APEC), San Antonio, TX, USA, Apr. 2018, pp. 2725–2729.

[7] M. S. Zaman et al., "Integrated SenseHEMT and Gate-Driver on a 650-V GaN-on-Si Platform Demonstrated in a Bridgeless Totem-pole PFC Converter," in International Symposium on Power Semiconductor Devices and ICs (ISPSD), Vienna, Austria, Sep. 2020.

[8] Q. Zhang, Q. Tong, and H. Zhang, "An Inductor Current Observer Based on Improved EKF for DC/DC Converter," in International Symposium on Computer, Consumer and Control, Jun. 2014, pp. 892–895.

[9] J.-M. Choe, B.-J. Byen, S. Moon, and J.-S. Lai, "A capacitor current control for stand-alone inverters using an inductor current observer," in International Conference on Power Electronics and ECCE Asia (ICPE-ECCE Asia), Jun. 2015, pp. 1143–1148.

[10] L. Liu, Y. Zhao, Y. Yin, and J. You, "Current Sensor-less Control for Boost DC-DC Converter Based on Switched Observer," in Annual Conference of the IEEE Industrial Electronics Society (IECON), Oct. 2018, pp. 1122–1127.

[11] J. Lin and G. Weiss, "Current sensorless control of bidirectional converters under mixed conduction mode," in Conference on Decision and Control (CDC), Dec. 2019, pp. 8118–8123.

[12] A. Aillane et al., "An Observer-Based Inductor Current Control for a Bifunctional Three-Phase DG-Inverter," in International Ccnference on Sciences and Techniques of Automatic Control and Computer Engineering (STA), Dec. 2022, pp. 566–571.

[13] F. Mezger and D. Killat, "Digital observer based current loop control for buck converters - Prototype implementation on an FPGA," in International Symposium on Industrial Electronics (ISIE), Jun. 2014, pp. 1336–1341.

[14] Y. Lee et al., "A High-Efficiency High-Voltage-Tolerant Buck Converter With Inductor Current Emulator for Battery-Powered IoT Devices," IEEE Transactions on Power Electronics, vol. 38, no. 9, pp. 10917–10932, Sep. 2023.

[15] E. Pazouki, J. A. De Abreu-Garcia, and Y. Sozer, "Fault Diagnosis Method for DC-DC Converters Based on the Inductor Current Emulator," in Energy Conversion Congress and Exposition (ECCE), Milwaukee, WI, USA, Sep. 2016, pp. 1–6.

[16] H. Ademane, R. Attanasio, and G. Vitale, "A GaN based Totem Pole Bridgeless Power Factor Correction Circuit," in Applied Power Electronics Conference and Exposition (APEC), Long Beach, CA, USA, Feb. 2024.

[17] J. K. Han, "Efficiency and PF Improving Techniques with a Digital Control for Totem-Pole Bridgeless CRM Boost PFC Converters," Energies, vol. 17, no. 2, p. 369, Jan. 2024.

High-Frequency Flying Capacitor Four-Level Drain Supply Modulator

Audrey Cheshire, Paul Flaten, Zoya Popović, and Dragan Maksimović

Department of Electrical, Computer, and Energy Engineering
University of Colorado Boulder
Boulder, CO 80309, USA
Email: {audrey.cheshire, paul.flaten, zoya.popovic, maksimov}@colorado.edu

Abstract—This paper presents a high-frequency, four-level drain supply modulator (DSM) based on a flying capacitor multilevel (FCML) architecture, designed for envelope tracking in radio-frequency power amplifier applications. Operating from a single dc supply voltage, the proposed FCML-DSM employs a simple state-machine-based active balancing of the flying capacitors and control of the output voltage levels at switching rates up to 25 MHz. A hardware prototype of the FCML-DSM is presented, utilizing an FPGA controller and GaN transistors as switching devices in the power stage. The experimental prototype has an output voltage range from 0 V to 20 V, discretized into equally spaced levels, and achieves an average efficiency of 95.4% for output powers exceeding 10 W while tracking a 35 MHz 16-QAM signal envelope.

Index Terms—Envelope tracking, flying capacitor, multilevel converter, drain supply modulator

I. INTRODUCTION

Modulation techniques, such as quadrature amplitude modulation (QAM), are an integral part of modern communication systems in order to achieve high data transmission rates. Increasingly complex modulation techniques are being employed, which result in non-uniform signal envelopes with high peak-to-average-power ratio (PAPR) [1]–[4]. To amplify these signals, the power amplifier (PA) frequently operates well below its peak output power, where its efficiency is typically highest.

To address the issue of low efficiency at lower output power levels, envelope tracking (ET), also referred to as drain supply modulation, is utilized. This technique varies the supply voltage to the PA according to the signal envelope such that the PA is driven close to its highest efficiency point [5]. In ET systems, the drain supply modulator (DSM) then becomes a crucial component to ensure high overall transmitter efficiency, which is defined as the product of the DSM efficiency and the PA efficiency. Traditional approaches to drain supply modulation include step-down switch-mode power converters that continuously track the signal envelope [4], [6]–[10]. Continuous tracking (CT) approaches operate at high switching frequencies to accurately track the envelope signal, typically several times the bandwidth of the envelope signal [10]–[12].

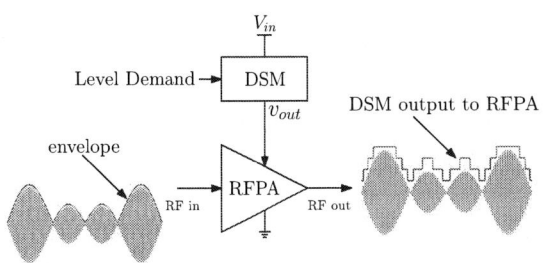

Fig. 1: Radio-frequency power amplifier (RFPA) with a multi-level drain supply modulator (DSM).

With increasing switching frequencies demanded by complex modulation techniques, it becomes difficult to design for both high tracking bandwidth and high efficiency due to switching loss in the system. To increase bandwidth without suffering from low efficiency, hybrid structures have been introduced such as linear-assisted switch-mode converters [3], [13], [14], and switched-capacitor based topologies [1], [15], [16], all of which still attempt to track the envelope continuously and accurately.

An alternative to continuous tracking is multi-level drain supply modulation (ML-DSM), illustrated in Fig. 1. This approach provides discrete levels to the PA instead of strictly following the envelope signal. Although the overall efficiency of the transmitter may be lower than continuous approaches, ML-DSM allows for the switching frequency requirements to be lower than the signal envelope bandwidth [17], [18], resulting in higher DSM efficiencies while still improving PA efficiency. ML-DSM can offer favorable trade-offs compared to a continuous tracking approach: a low-complexity and highly efficient DSM design, decreasing the number of components, physical system size, and cost.

A drawback of ML-DSM is that typical multi-level architectures require multiple dc input supplies to achieve a multi-level output [11], [12], [17]–[20]. This work is an extension of [21], which presented a *single-supply* four-level DSM based on the flying capacitor multilevel (FCML) architecture, though validated only for low-frequency operation. Also of note is that the FCML-DSM, in contrast to previous designs [1], [4],

979-8-3315-1612-3/25 $31.00 © 2025 IEEE

[8], [9], excludes the output LC filter, thus eliminating the bandwidth limitations associated with low-pass filtering. This paper focuses on the simple state-machine-based controller that enables active balancing of the flying capacitors, high-frequency operation, and wide-bandwidth tracking.

This paper is organized as follows: control of the FCML-DSM is discussed in Section II, hardware and experimental results are presented in Section III, followed by a discussion on composite efficiency in Section IV. Finally, conclusions are summarized in Section V.

II. OPERATION AND HIGH-FREQUENCY CONTROL OF THE FCML-DSM

The four-level FCML-DSM, shown in Fig. 2, consists of three pairs of complemenetary switches: (Q_1, Q_4), (Q_2, Q_5), and (Q_3, Q_6) and flying capacitors, C_1 and C_2. Only a single dc supply, V_{in}, is necessary to discretize the output into four levels: 0, $(1/3)V_{in}$, $(2/3)V_{in}$, and V_{in}. A resistive bias network composed of $R_1 - R_6$ is added to aid in the start-up process of the converter and to act as a passive balancing technique during long intervals where no switching occurs.

Direct sensing of the flying capacitor voltages v_{fly1} and v_{fly2} is employed using two comparators that continuously monitor and compare each flying capacitor to their ideal values. In this four-level case, the ideal capacitor values are: $v_{fly1} = (2/3)V_{in}$ and $v_{fly2} = (1/3)V_{in}$. The comparators output a logic zero when the capacitor voltage is below its ideal value, and a logic one when above. Two additional bits are used to demand one of the four available output levels. With only digital signals as inputs, the controller behaves as a finite state machine (FSM) with eight states, corresponding to the eight possible switch configurations dictated by the three complementary pairs of switches. A block diagram of the FPGA controller is shown in Fig. 3. The controller is driven by a main clock source and separated into three clocked processes using a phase-locked loop (PLL). The state machine and its four inputs ($REF_1, REF_2, CMP_1, CMP_2$) exist in separate clock domains and are therefore classified as fully asynchronous. CMP_1 and CMP_2 are comparator outputs sent from the power stage to the FPGA. REF_1 and REF_2 together define the desired output level, and in practical implementation with a PA, will also come from an external source. For hardware results in Section III, these signals are generated by the FPGA to verify controller operation.

It is well understood that asynchronous signals may cause failures due to metastability, which refers to logic signals that waver between 0 or 1 during a clock edge, violating clock setup and hold requirements. A common solution to mitigate metastability is the use of flip-flops (FFs) to re-synchronize the inputs into the new clock domain and allow time for the signal to settle before propagating to the rest of the circuit [22]–[24]. This technique is used at both the input and output of the controller to ensure stable and reliable state-machine behavior.

During converter start-up, the FSM remains inactive with $Q_4 - Q_6$ turned on such that $v_{out} = 0\,\text{V}$ until both flying

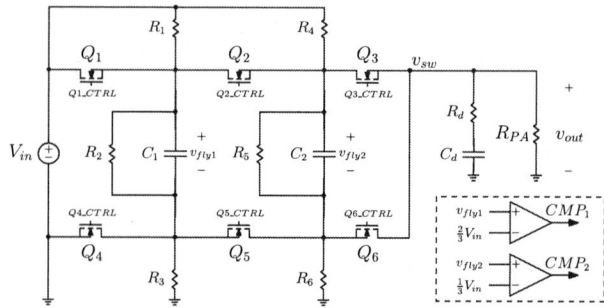

Fig. 2: Four-level FCML-based drain supply modulator.

TABLE I: FSM controller decision table. The control signals represent $Q1\_CTRL$, $Q2\_CTRL$, and $Q3\_CTRL$. The remaining signals are the complements.

Demand	B_3	B_2	B_1	B_0	State	Control signals
0	0	0	0	0	0	0 0 0
	0	0	0	1	0	0 0 0
	0	0	1	0	0	0 0 0
	0	0	1	1	0	0 0 0
$\frac{1}{3}V_{in}$	0	1	0	0	4	1 0 0
	0	1	0	1	1	0 0 1
	0	1	1	0	2	0 1 0
	0	1	1	1	1	0 0 1
$\frac{2}{3}V_{in}$	1	0	0	0	6	1 1 0
	1	0	0	1	5	1 0 1
	1	0	1	0	3	0 1 1
	1	0	1	1	3	0 1 1
V_{in}	1	1	0	0	7	1 1 1
	1	1	0	1	7	1 1 1
	1	1	1	0	7	1 1 1
	1	1	1	1	7	1 1 1

capacitors have reached their respective ideal values through the bias network. Once this condition is met, the state machine is enabled and operates to actively balance the flying capacitors while outputting the requested level demand. The controller timing diagrams are shown in Fig. 4, and a complete decision table is shown in Table I.

The FSM takes a 4-bit bus input, $[B_3B_2B_1B_0]$, which represents the synchronized input signals and outputs six control signals according to Table I. The main clock is a fixed

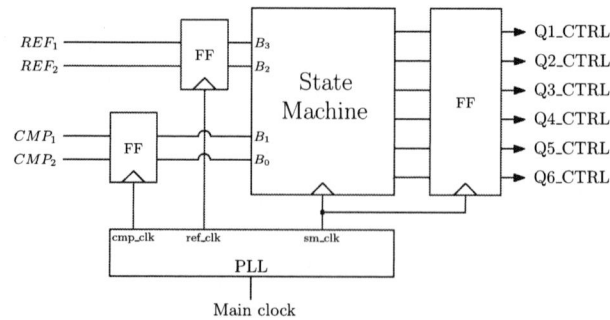

Fig. 3: Block diagram of the FPGA controller.

979-8-3315-1612-3/25 $31.00 © 2025 IEEE

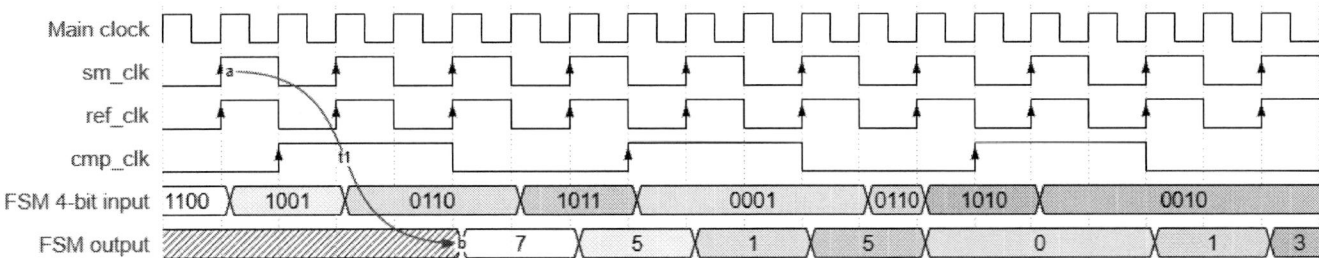

Fig. 4: Timing diagrams for the FSM controller.

system clock source, while sm_clk, ref_clk, and cmp_clk are derived clock sources. The FSM switches at a sampling rate set by sm_clk, which is also used for output synchronization. For simplicity, the timing diagram represents the FSM output as the corresponding state instead of the control signals. In the simulation, ref_clk emulates the rapid changes in an RF envelope and switches the level demand at that rate. As seen in Fig. 4, sm_clk and ref_clk are set to have the same frequency, as it is a priority to be able to track the envelope as accurately as possible. In contrast, cmp_clk, which samples the capacitor voltages, can be set lower than the FSM switching frequency. Compared to the reference inputs, the flying capacitor voltages are not as critical, as they change more slowly. Consequently, bits B_1 and B_0 are updated on the rising edge of cmp_clk, while B_3 and B_2 are updated at the rising edge of sm_clk. With this control scheme, the added synchronizers at both the input and the output introduce a two-cycle delay in the converter output, shown in Fig. 4 from node a to node b. Given that the envelope signal is typically known in advance, this delay may be re-tuned in the actual implementation of the system.

III. EXPERIMENTAL RESULTS

The hardware prototype is designed as detailed in [21] and is shown in Fig. 5 with a close-up of the power stage in Fig. 6. The prototype uses EPC 2014C GaN FETs as the switching devices in the power stage. Included in this prototype are additional connections for the FPGA, RFPA, an external clock source, as well as analog deadtime circuitry and the resistive bias network.

A. Output damping network

The switching behavior of the FCML-DSM is affected by the presence of parasitic elements L_p and C_p that originate from interconnect inductances and switch output and trace capacitances. For any given switch configuration, an approximate equivalent circuit comprising L_p, C_p, and R_p is shown in Fig. 7, where R_p represents the effects of switch on-resistances and capacitor ESR. R_p is estimated from datasheet values, while the parasitic inductance and capacitance values are found experimentally by first measuring the oscillation frequency f_{osc} of the ringing in the undamped waveforms. A small capacitance C_s is added at the output node until the oscillation frequency is half of its original value. Equations (1) and (2)

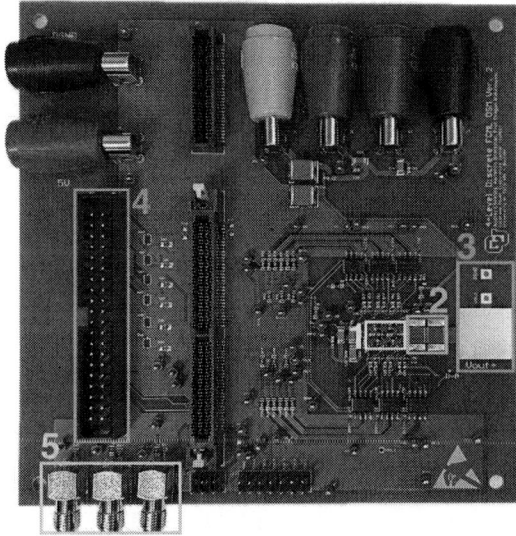

Fig. 5: 4-level FCML-DSM prototype: (1) power stage, (2) output damping network, (3) load connector, (4) FPGA connector, (5) connectors for external clock and reference signals.

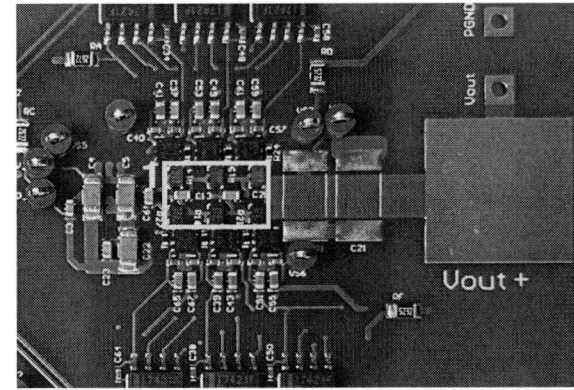

Fig. 6: Close-up of the power stage (1) of the FCML-DSM. Flying capacitors are on the reverse side of the board, not shown in this image.

describe the relationship between the added capacitance and parasitic capacitance, under the assumption that the oscillation frequency has been halved. L_p is solved for accordingly using (3) where $f_{osc} = 150\,\mathrm{MHz}$.

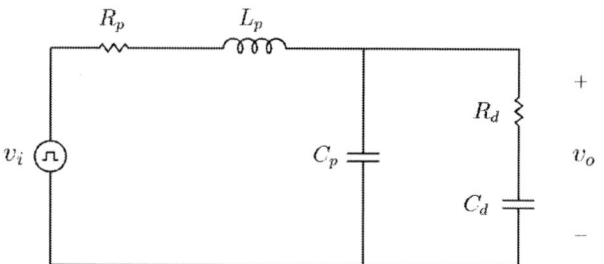

Fig. 7: Equivalent circuit with parasitic elements.

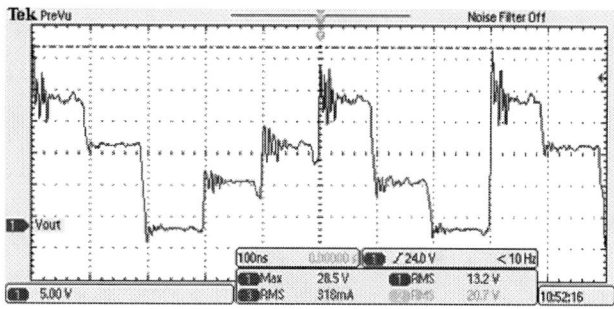

(a) Undamped response

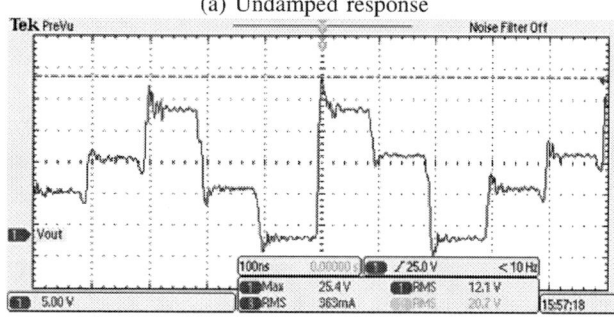

(b) Damped response with $R_d = 3.9\,\Omega$ and $C_d = 1.5\,\mathrm{nF}$

Fig. 8: Converter output voltage waveforms for (a) undamped and (b) damped cases at $f_{sm} = 10\,\mathrm{MHz}$. Time axis: 100 ns/div.

$$f_{osc,0.5} = \frac{1}{2\pi\sqrt{L_p(C_p + C_s)}} \tag{1}$$

$$C_p = \frac{C_s}{3} \tag{2}$$

$$L_p = \frac{1}{C_p(2\pi f_{osc})^2} \tag{3}$$

The parasitic element values are estimated to be $L_p = 7.2\,\mathrm{nH}$, $C_p = 152\,\mathrm{pF}$, and $R_p = 0.2\,\Omega$. The interaction between the parasitic elements results in undesired ringing during switching transitions and overshoot, as shown in Fig. 8(a). To minimize this undesired response, a damping network composed of $R_d - C_d$ in Fig. 7 is designed to suppress voltage overshoot and reduce ringing.

The dominant loss mechanism of the FCML-DSM stems from the output damping network, specifically the charge/discharge of the damping capacitor C_d. The power loss associated with C_d is proportional to the sampling rate and

capacitance value. As such, to minimize power loss at higher switching rates, it is important to design for a minimum C_d and optimum R_d such that efficiency is not significantly affected, while achieving desired output waveforms.

The design of the damping network follows the root-locus method to find values for the damping resistance, capacitance, and corresponding Q-factor. The characteristic equation $D(s)$ of the equivalent circuit in Fig. 7 is shown in (4), where R_d is considered to be the root-locus parameter. We observe $D(s)$ as R_d varies for various values of C_d. Each $R_d - C_d$ pair is evaluated to meet a specific Q-factor design constraint. For any given Q-factor constraint, the $R_d - C_d$ pair with the lowest capacitance value is chosen. For an allowed $Q = 0.7$, the following values are chosen: $R_d = 3.9\,\Omega$ and $C_d = 1.5\,\mathrm{nF}$. An experimental comparison between the undamped and damped responses is shown in Fig. 8 for $f_{sm} = 10\,\mathrm{MHz}$, given an arbitrary output waveform pattern. One may observe a near-30 % reduction in peak voltage overshoot and improved waveform quality at the expense of slightly reduced efficiency.

$$D(s) = 1 + R_d \frac{sC_d + s^2 R_p C_d C_p + s^3 L_p C_d C_p}{1 + sR_p(C_p + C_d) + s^2 L_p(C_p + C_d)} \tag{4}$$

B. FCML-DSM tracking

To emulate envelope tracking with the FCML-DSM, the FGPA generates the reference demands associated with a 35 MHz 16-QAM signal envelope. The nominal input voltage is $V_{in} = 20\,\mathrm{V}$, nominal load $R_{PA} = 25\,\Omega$, and deadtime $t_d = 10\,\mathrm{ns}$. The controller is set such that $f_{cmp} = 1\,\mathrm{MHz}$ remains constant for all experiments, while the state-machine sampling rate f_{sm} is varied from 1 MHz to 25 MHz with $f_{ref} = f_{sm}$. At $f_{sm} = 1\,\mathrm{MHz}$, the FCML-DSM acts as a constant supply, outputting only the maximum level, as the signal envelope changes too rapidly for the FCML-DSM to follow. At $f_{sm} = 10\,\mathrm{MHz}$, only two levels are used. By increasing f_{sm} to 25 MHz, this allows the FCML-DSM to use three available levels to further enhance the PA efficiency. Simulated and experimental results are shown in Fig. 9 for $f_{sm} = 25\,\mathrm{MHz}$. In Fig. 9(b), the DSM output coincides with the signal envelope during some transitions, which is undesired. In this work, the minimum output level is set at 0 V, but in practical implementation, the minimum voltage can be set above 0 V by adding a dc bias supply in series with the output. This ensures the FCML-DSM output is always above the signal envelope, and the FCML-DSM is able to use all four output levels.

C. Efficiency measurements

Efficiency measurements using the same 35 MHz envelope signal are recorded for nominal load and $R_{PA} = 50\,\Omega$ in Table II. As previously discussed, the efficiency drop as the sampling rate increases can be mainly attributed to the RC damping network with some loss associated with the on-resistance of the GaN switches and capacitor ESR. Efficiencies remain relatively constant for the different loads, and small differences may be due to physical measurement errors.

979-8-3315-1612-3/25 $31.00 © 2025 IEEE   685

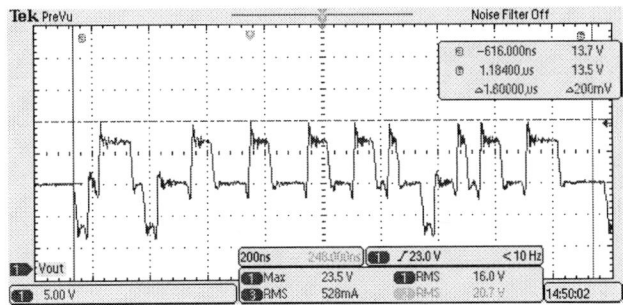

(a) Measured v_{out} (5 V/div). Time-axis: 200 ns/div.

Tracking waveforms for $f_{sm} = 25$ MHz

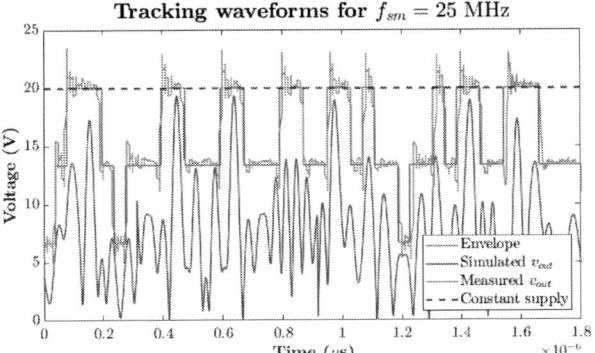

(b) Measured v_{out} post-processed and aligned with signal.

Fig. 9: Output waveforms for $f_{sm} = 25$ MHz.

TABLE II: Measured efficiencies of the FCML-DSM for the 35 MHz 16-QAM signal envelope.

f_{sm}	η_{DSM} for $R_{PA} = 25\,\Omega$	η_{DSM} for $R_{PA} = 50\,\Omega$
1	97.3	98.5
10	96.7	97.7
15	95.0	95.8
20	94.1	92.8
25	93.7	89.3

IV. COMPOSITE EFFICIENCY ESTIMATION

Composite efficiency, also known as transmitter efficiency, is given as the product of the DSM and supply-modulated PA efficiency, shown in (5). With respect to the FCML-DSM design, optimizing factors such as the voltage output levels, the number of output levels, and the switching rate, can lead to further improvement in PA and composite efficiency.

$$\eta_T = \eta_{DSM} \times \eta_{PA,mod} \tag{5}$$

In this analysis, composite and modulated PA efficiencies are estimated based on static PA efficiency and power gain curves reported in [25], which are digitized and shown in Fig. 10. The same 16-QAM 35 MHz signal used in hardware experiments is used for the following simulations. From Fig. 10, it can be seen that by supply modulating the PA, the peak efficiency point is moved to lower output powers which is desired for high PAPR signals. For the same output power level, the power-added-efficiency (PAE) is increased

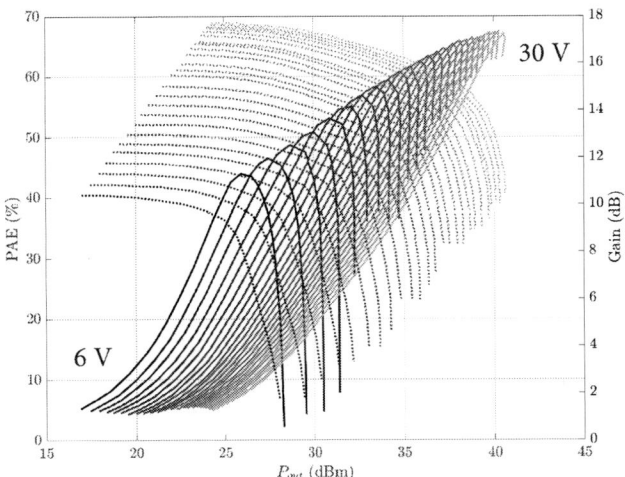

Fig. 10: Static PA characterization curves for PAE and power gain as drain voltage is varied from 6 V to 30 V in 1 V steps [25]. Gain curves are represented with dotted lines (right y-axis), and PAE curves are the solid lines (left y-axis).

as the drain voltage is decreased. For comparison, composite efficiency is first estimated under ideal tracking conditions, defined as the case when the signal envelope is tracked continuously and $\eta_{DSM} = 1$. This is calculated under the assumption that the minimum drain voltage is set at $V_{min} = 6$ V, and DSM output levels are in steps of 1 V. At each sample point k of the envelope signal, the ideal DSM output is chosen such that it is just above the envelope. This is illustrated in Fig. 11 for a small interval of the full signal. The instantaneous output power P_{out} of the envelope signal is calculated using

$$P_{out} \, [\text{dBm}] = 10 \log_{10} \left(\frac{v_{env}^2}{2R} \right) + 30 \tag{6}$$

assuming a standard $50\,\Omega$ environment. The PAE is interpolated from Fig. 10 for the corresponding DSM output voltage. Similarly, the input RF power can be found by $P_{in} = P_{out} - G$, where G is the power gain also found from Fig. 10. Finally, the dc input power can be found from the definition of PAE,

$$P_{dc} = \frac{P_{out} - P_{in}}{PAE} \tag{7}$$

The overall modulated PA efficiency is then estimated by the total output energy and total input energy: $\eta_{PA,mod} = E_{out}/E_{in}$, where E_{in} includes both RF and dc components. Energy values are found by trapezoidal integration of all sample points K of the full signal using (8) and (9) where $\Delta t = 0.6$ ns. With this procedure, the ideal tracking efficiency is found to be $\eta_{ideal} \approx 47.1\,\%$. With a constant drain supply of 20 V, the predicted PA efficiency is 23.74 %. One may note that the predicted improvement under ideal DSM conditions is 23.4 %.

$$E_{out} = \frac{\Delta t}{2} \sum_{k=1}^{K} \left(P_{out_k} + P_{out_{k+1}} \right) \tag{8}$$

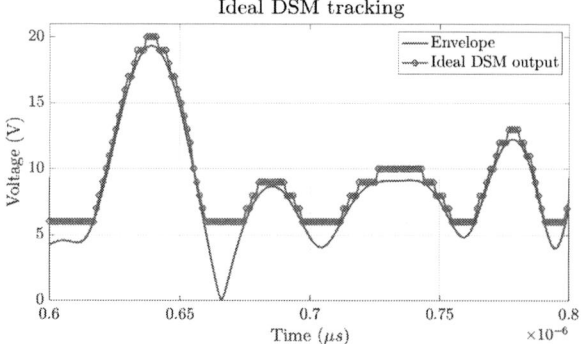

Fig. 11: Envelope signal and an idealized DSM output.

$$E_{in} = \frac{\Delta t}{2} \sum_{k=1}^{K} \left[\left(P_{dc_k} + P_{dc_{k+1}} \right) + \left(P_{in_k} + P_{in_{k+1}} \right) \right] \quad (9)$$

To predict how the FCML-DSM can improve overall system performance, estimation of the modulated PA efficiency and composite efficiency can be repeated in non-ideal conditions ($\eta_{DSM} \neq 1$) for each of the sampling rates tested in Section III-B. Assuming $R_{PA} = 25\,\Omega$ and a four-level output, estimated efficiencies are shown in Table III. The FCML-DSM is predicted to improve PA efficiency by 6 % and composite efficiency by 4.8 %. The maximum predicted composite efficiency is still below η_{ideal} which implies that further improvements can be made to the DSM design.

It is also of interest to see how much improvement ML-DSM can achieve compared to ideal continuous tracking. This is investigated by varying the sampling rate f_{sm} and the number of output levels n. Varying the number of output levels consequently changes the output voltage levels. A fixed difference between each of the output levels is assumed in this analysis. Considering an ideal FCML-DSM, we can observe how much ML-DSM can improve PA performance. Fig. 12 shows the resulting efficiency curves as n is varied from 2 to 10, and f_{sm} is varied up to 150 MHz. At the current design point, $n = 4$ and $f_{sm,max} = 25$ MHz, there is an improvement of 9.7 % compared to the static, unmodulated PA efficiency. Even at sampling rates well below the signal bandwidth, a significant improvement is observed. By further increasing n and f_{sm}, composite efficiency will ultimately approach η_{ideal}, though the trade-offs become less desirable

TABLE III: Measured and predicted efficiencies at various sampling rates.

f_{sm} [MHz]	Measured	Predicted	
	η_{DSM} [%]	$\eta_{PA,mod}$ [%]	η_T [%]
1	97.3	23.74	23.11
10	96.7	26.73	25.84
15	95.0	28.84	27.39
20	94.1	29.47	27.73
25	93.7	29.77	27.89

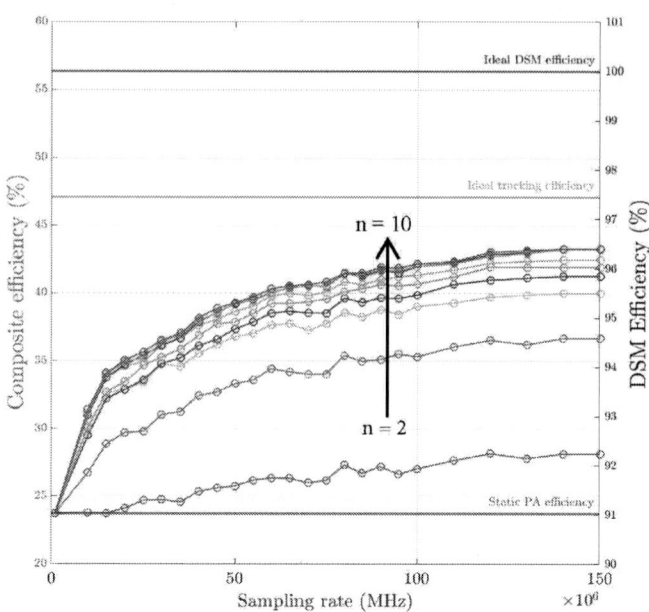

Fig. 12: Estimated composite efficiency curves for various values of n output levels and sampling rate f_{sm}, assuming $\eta_{DSM} = 1$.

for small improvements in efficiency as converter design and control complexity also increase. This analysis considers the output levels of the DSM to have a fixed difference between the minimum and maximum output levels. While not investigated in this work, additional efficiency improvements may be achievable by removing the fixed-step constraint.

V. CONCLUSIONS

This work presents a high-frequency four-level drain supply modulator (DSM) for radio-frequency power amplifiers based on the flying capacitor multilevel (FCML) architecture. A simple state-machine-based controller is implemented to maintain charge balance across the flying capacitors while supplying the correct output voltage level. The FCML-DSM achieves a maximum sampling rate of 25 MHz and demonstrates a high tracking bandwidth, using a 35 MHz 16-QAM test signal for experimental verification. For output powers greater than 10 W, the FCML-DSM achieved an average efficiency of 95.4%. Care is taken to minimize power loss by optimizing the output $R_d - C_d$ damping network. A general approach to estimating composite efficiency is also introduced to evaluate trade-offs between DSM complexity and achievable efficiency improvements. The single-supply four-level FCML-DSM, both in design and control, is relatively simple while achieving high efficiency and tracking bandwidth.

REFERENCES

[1] Q. Jin and M. Vasić, "Optimized design of gan switched-capacitor-converter-based envelope tracker for satellite application," *IEEE Journal of Emerging and Selected Topics in Power Electronics*, vol. 5, no. 3, pp. 1346–1355, 2017.

[2] C. Florian, T. Cappello, R. P. Paganelli, D. Niessen, and F. Filicori, "Envelope tracking of an rf high power amplifier with an 8-level digitally controlled gan-on-si supply modulator," *IEEE Transactions on Microwave Theory and Techniques*, vol. 63, no. 8, pp. 2589–2602, 2015.

[3] J. H. Kim, H. S. Son, W. Y. Kim, and C. S. Park, "Envelope amplifier with multiple-linear regulator for envelope tracking power amplifier," *IEEE Transactions on Microwave Theory and Techniques*, vol. 61, no. 11, pp. 3951–3960, 2013.

[4] M. Rodriguez, P. Miaja, A. Rodriguez, and J. Sebastian, "Multilevel converter for envelope tracking in rf power amplifiers," in *2009 IEEE Energy Conversion Congress and Exposition*, 2009, pp. 503–510.

[5] P. Asbeck and Z. Popovic, "Et comes of age: Envelope tracking for higher-efficiency power amplifiers," *IEEE Microwave Magazine*, vol. 17, no. 3, pp. 16–25, 2016.

[6] V. Mehrotra, A. Arias, C. Neft, J. Bergman, M. Urteaga, and B. Brar, "Gan hemt-based 1-ghz speed low-side gate driver and switch monolithic process for 865-mhz power conversion applications," *IEEE Journal of Emerging and Selected Topics in Power Electronics*, vol. 4, no. 3, pp. 918–925, 2016.

[7] Y. Zhang, J. Strydom, M. de Rooij, and D. Maksimović, "Envelope tracking gan power supply for 4g cell phone base stations," in *2016 IEEE Applied Power Electronics Conference and Exposition (APEC)*, 2016, pp. 2292–2297.

[8] S. Yerra and H. Krishnamoorthy, "Multi-phase three-level buck converter with current self-balancing for high bandwidth envelope tracking power supply," in *2020 IEEE Applied Power Electronics Conference and Exposition (APEC)*, 2020, pp. 1872–1877.

[9] J. Rodríguez, J. R. García-Meré, D. G. Aller, and J. Sebastián, "Pulsewidth modulated three-level buck converter based on stacking switch-cells for high power envelope tracking applications," *IEEE Transactions on Power Electronics*, vol. 37, no. 5, pp. 5786–5800, 2022.

[10] M. Krellmann, O. Bengtsson, and W. Heinrich, "Gan-hemts as switches for high-power wideband supply modulators," in *2013 European Microwave Conference*, 2013, pp. 553–556.

[11] A. Disserand, P. Bouysse, A. Martin, R. Quéré, O. Jardel, and L. Lapierre, "A new high speed and high efficiency gan hemt switching cell for envelope tracking modulators," in *2016 46th European Microwave Conference (EuMC)*, 2016, pp. 281–284.

[12] A. Bräckle, L. Rathgeber, F. Siegert, S. Heck, and M. Berroth, "Power supply modulation for rf applications," in *2012 15th International Power Electronics and Motion Control Conference (EPE/PEMC)*, 2012, LS8d.3-1-LS8d.3-5.

[13] Y. Li, X. Ruan, Y. Wang, and C. Zhang, "Digitally-assisted hysteresis voltage prediction control for series-form switch-linear hybrid envelope-tracking power sup-

ply," in *2019 IEEE Energy Conversion Congress and Exposition (ECCE)*, 2019, pp. 1348–1352.

[14] D. Kimball, J. Jeong, C. Hsia, *et al.*, "High-efficiency envelope-tracking w-cdma base-station amplifier using gan hfets," *IEEE Transactions on Microwave Theory and Techniques*, vol. 54, no. 11, pp. 3848–3856, 2006.

[15] H. Xi, Y. Xu, Y. Zhu, N. Liu, and J. Cao, "High bandwidth and compact envelope tracking power supply utilizing switched capacitor topology," *IEEE Access*, vol. 7, pp. 105 462–105 469, 2019.

[16] A. Cervera and M. M. Peretz, "Envelope tracking power supply for volume-sensitive low-power applications based on a resonant switched-capacitor converter," in *2016 IEEE Applied Power Electronics Conference and Exposition (APEC)*, 2016, pp. 2298–2303.

[17] C. Nogales, Z. Popović, and G. Lasser, "An 800-w four-level supply modulator for efficient envelope tracking of rf transmitters," *IEEE Journal of Emerging and Selected Topics in Power Electronics*, vol. 11, no. 3, pp. 3251–3260, 2023.

[18] A. Sepahvand, P. Momenroodaki, Y. Zhang, Z. Popović, and D. Maksimović, "Monolithic multilevel gan converter for envelope tracking in rf power amplifiers," in *2016 IEEE Energy Conversion Congress and Exposition (ECCE)*, 2016, pp. 1–7.

[19] Z. Wang, "Demystifying envelope tracking: Use for high-efficiency power amplifiers for 4g and beyond," *IEEE Microwave Magazine*, vol. 16, no. 3, pp. 106–129, 2015.

[20] N. Wolff, W. Heinrich, and O. Bengtsson, "A novel model for digital predistortion of discrete level supply-modulated rf power amplifiers," *IEEE Microwave and Wireless Components Letters*, vol. 26, no. 2, pp. 146–148, 2016.

[21] A. Cheshire, A. Bharathan, and D. Maksimović, "Flying capacitor four-level supply modulator with active balancing for rf power amplifier applications," in *2023 IEEE 24th Workshop on Control and Modeling for Power Electronics (COMPEL)*, 2023, pp. 1–7.

[22] S. Friedrichs, M. Függer, and C. Lenzen, "Metastability-containing circuits," *IEEE Transactions on Computers*, vol. 67, no. 8, pp. 1167–1183, 2018.

[23] H. Veendrick, "The behaviour of flip-flops used as synchronizers and prediction of their failure rate," *IEEE Journal of Solid-State Circuits*, vol. 15, no. 2, pp. 169–176, 1980.

[24] J. Stephenson, D. Chen, R. Fung, and J. Chromczak, "Understanding metastability in fpgas," Altera Corporation, Tech. Rep., 2009.

[25] M. R. Duffy, "Efficient supply modulated MMIC PAs for broadband linear amplification," Ph.D. dissertation, University of Colorado Boulder, 2020.

Discontinuous Modulation Strategy for Voltage and Temperature Balancing of MMCs

Davide D'Amato
ISIT
Fraunhofer Institute for Silicon Technology
Itzehoe, Germany
davide.damato@isit.fraunhofe.de

Stayner Nóbrega Barros
Chair of Power Electronics
Christian-Albrechts- Universität zu Kiel
Kiel, Germany
stu248322@mail.uni-kiel.de

Jun-Hyung Jung
ISIT
Fraunhofer Institute for Silicon Technology
Itzehoe, Germany
junhyung.jung@isit.fraunhofer.de

Marco Liserre
ISIT
Fraunhofer Institute for Silicon Technology
Itzehoe, Germany
marco.liserre@isit.fraunhofer.de

Abstract—**Modular multilevel converters are becoming attractive for many high- and medium-voltage (MV) applications due to the serial connection of submodules (SMs). Since these converters contain many power devices, however, it is important to ensure a highly reliable operation of the converter for uninterrupted operation with short maintenance time. Particularly, technical challenges such as the unbalance of SM capacitor voltage and thermal stress on power semiconductors must be addressed because they have a significant impact on system reliability. Despite these reasons, few studies have been conducted to balance SM voltages and thermal stress together for MV MMCs. Therefore, this paper proposes a novel discontinuous modulation strategy that balances both the thermal stress and capacitor voltage in SMs. Indeed, the uniqueness of the proposed approach is that two independent clamping angles are controlled to achieve multi-objective balancing. The effectiveness of the proposed strategy is validated with simulation and experimental results on a five-level prototype.**

Keywords—*Modular Multilevel Converter, voltage balancing, discontinuous modulation, junction temperature.*

I. INTRODUCTION

Modular multilevel converters (MMC) attract considerable interest for high- and medium-voltage (MV) DC applications due to their modular structure, and scalability for different voltage and power levels [1-3]. However, the converter structure with series connected SMs faces several technical challenges such as balancing SM voltages and mitigating circulating current [4-6]. Indeed, to ensure the operation of this converter, the voltages across the SM capacitors need to be balanced to provide high-quality output with reduced total harmonic distortion.

In the literature, several studies have proposed methods for balancing SM voltages and thermal conditions of the MMC. These techniques are based on hardware approaches, advanced control solutions, and various modulation strategies. An active thermal control (ATC) based on a hardware solution is proposed in [7] for a hybrid MMC consisting of half-bridge (HB) and full-bridge (FB) submodules. The authors proposed a solution based on the addition of a thyristor in parallel for each SMs to actively bypass it and a symmetrical modulation for the

full-bridge SMs to equally distribute the thermal stress among the power devices. However, the addition of an extra hardware circuit leads to an increase in the overall cost of the converter as the number of sub-modules increases. An approach to equalize the thermal distribution between the top and bottom devices that compose the HB SM of the MMC is proposed in [8]. Through an analytical analysis, the authors succeed in estimating the junction temperature of the devices and reducing the power losses of the single power device. However, these control methods do not aim to uniformly reduce the junction temperature of all SMs within the MMC. Rather, it seeks to manage and balance the thermal stress across the entire converter. This approach is specifically addressed in the modulation technique presented in this work.

Additionally, ATC techniques based on modulation strategies can be distinguished into nearest-level modulation (NLM) based techniques or phase-shifted pulse width modulation (PS-PWM) based techniques. In [9], a thermal balancing control method based on a sorting algorithm is applied to NLM in MMCs for HVDC applications. Subsequently, a reinforcement learning-based NLM modulation optimization algorithm has been proposed with the aim of balancing the temperatures and voltages of the SMs [10]. Even though these methods effectively achieve thermal balancing, NLM is not suitable for MMCs with a lower number of SMs due to severe harmonic distortion.

To balance the thermal stress, a PS-PWM strategy with variable frequency has been proposed in the work [11]. However, the authors do not analyze the effect of switching frequency variation on the voltage balance of SMs, as addressed in [12]. An active thermal control based on discontinuous pulse width modulation (DPWM) has been proposed in [13] to reduce the thermal stress of cascade-half-bridge (CHB) multilevel converters. However, additional consideration of the internal dynamic of the MMC is required when DPWM is applied.

Therefore, in this paper, a modified DPWM strategy is proposed that offers the unique feature of independently controlling two clamping angles, one for the positive half-period and the other for the negative half-period of the SM reference signal. In this way, it is possible to address the two

979-8-3315-1612-3/25 $31.00 © 2025 IEEE

control objectives: balancing capacitor voltages and junction temperatures T_j of power devices to ensure the overcurrent capability of the MMC. In addition, the submodules not affected by the clamping signal continue to operate based on the original reference signal, ensuring that the quality of the voltage and current waveforms is maintained. The objective of the proposed modulation is to pursue the precise balancing of SM voltages and junction temperature of power semiconductors through a simple and effective modulation strategy with minimal impact on the performance of the MMC.

The rest of the paper is organized as follows. Section II presents the various modulation techniques used for MMCs, delving into discontinuous modulation. Section III introduces the proposed modulation method and strategies to estimate the junction temperature. In Section VI, simulations and experimental results are presented to validate the proposed technique, and the device power losses of the proposed method are compared with the conventional phase-shift modulation strategy. Finally, in the last section, conclusions are drawn.

II. DISCONTINUOUS MODULATION

The phases of the MMC consist of an upper and lower arm connected by two inductors. Each arm consists of N HB and/or FB submodules connected in series, as shown in Fig. 1. There are many PWM techniques for the modular multilevel converter which can be distinguished into continuous or discontinuous modulation techniques. Continuous modulation is divided into phase-shifting pulse-width modulation (PS-PWM) and level-shifting pulse-width modulation (LS-PWM). PS-PWM employs several triangular carriers one for each submodule, which share the same frequency and amplitude but are differentiated by varying phase shift angle between adjacent carriers. The phase shift angle is closely related to the number of submodules in the arm [14]. Depending on the selected phase shift angle between the carriers of the submodules of the lower arm and those of the upper arm of a phase, it is possible to generate an output waveform of N+1 levels or 2N + 1 voltage levels [15]. Conversely, LS-PWM consists of triangular carriers that maintain identical amplitude and frequency, but they are arranged vertically from each other based on the number of submodules, resulting in an output voltage with N+1 levels [16]. In both PS-PWM and LS-PWM techniques, the modulating reference signal is a sinusoidal signal at the fundamental frequency, supplemented by the control actions to handle the internal dynamics of the MMC.

The discontinuous PWM (DPWM) techniques employ the generation of a clamping reference signal. This signal is created by modifying the sinusoidal reference signal of the MMC. The modification is achieved by adding a positive zero-sequence component signal. As a result, the reference signal will contain stretches with a unity modulation index, called the clamping region. However, to maintain the same power factor of the converter to the other SMs, the non-clamped reference signal obtained from the sum of the reference signal and the negative zero sequence is applied, as shown in Figure 2.

The clamped m_c and non-clamped references m_{nc} are derived from the expression of the fundamental reference signal as follows:

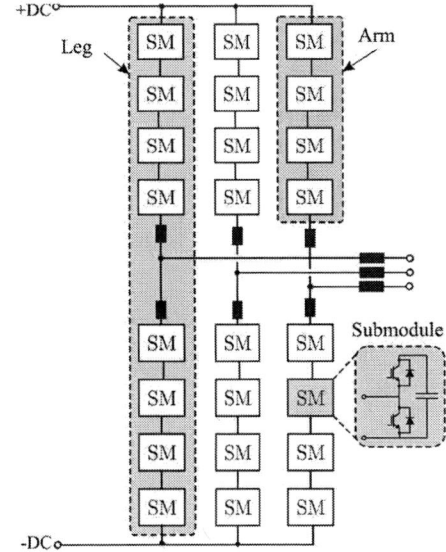

Figure 1: Structure of a three-phase MMC.

$$m_c = m \cos(\omega t) + m_{off} \qquad (1)$$

$$m_{nc} = m \cos(\omega t) - m_{off} \qquad (2)$$

where m is the modulation index and ω is the fundamental angular frequency. The reference signals m_c and m_{nc} are obtained by adding and subtracting from the fundamental signal, respectively, a zero-sequence signal m_{off}. The clamping angles for clamping and non-clamping modulation are identical, and the sum of their amplitudes must be zero. In this way, all submodules in an arm maintain the same power factor overall.

A. Power Losses

When discontinuous modulation is used to drive a converter, it occurs that during the clamping region, the power devices are not switching. Indeed, in this region, the only power losses that occur are conduction losses while switching losses are null. Consequently, adjusting the clamping angle can change the

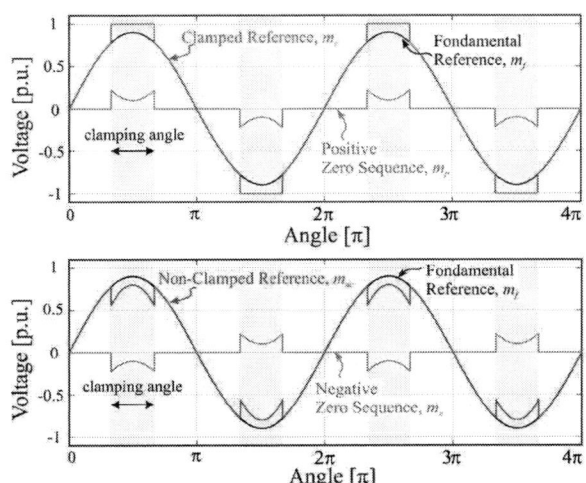

Figure 2: Generation of clammed and non-clammed references for discontinuous modulation.

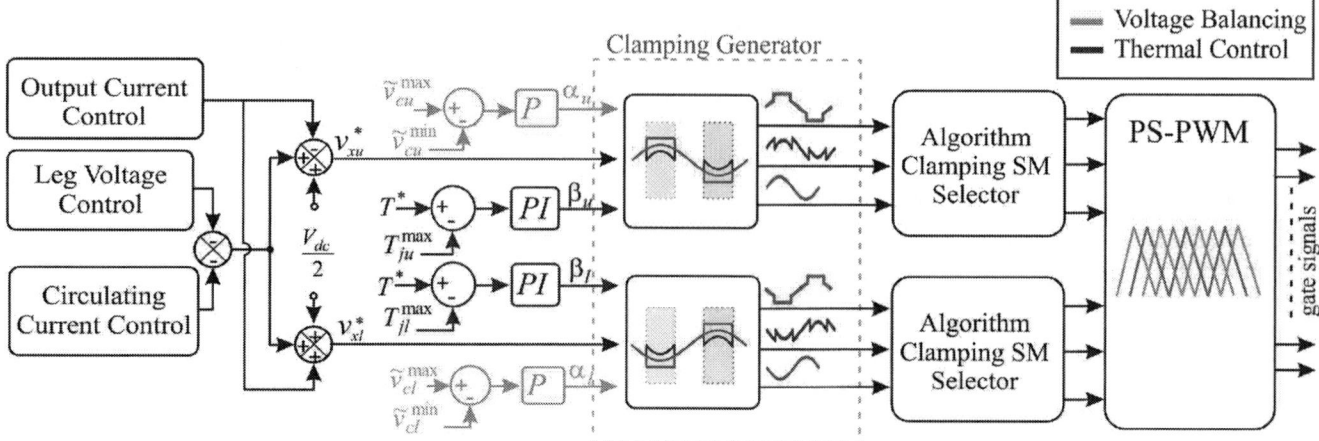

Figure 4: Simplified block diagram of the proposed DPWM.

duration of the clamping region, further contributing to varying switching power losses.

It is possible to define the total losses, L_{tot}, as expressed in eq. (3), where the switching loss L_{sw} contributions are strongly dependent on both load power P_{load} and clamping angle α. Accordingly, discontinuous modulation is mainly used to reduce switching losses in the converter [17]. In the work proposed in [18], the switching losses function (SLF) is proposed to describe the relationship between the variation of the alpha clamping angle and the switching losses as expressed in eq. (4).

$$L_{tot}\ (P_{load}\ ,\ \alpha) = L_{cd}\ (P_{load}) + L_{sw}\ (P_{load}\ ,\ \alpha) \qquad (3)$$

$$SFL = \frac{L_{sw}\ \left[0;\frac{\pi-\alpha}{2}\right] + L_{sw}\ \left[\frac{\pi+\alpha}{2};\pi\right]}{L_{sw}\ [0;\pi]} = \frac{L_{sw,DPWM}}{L_{sw,PWM}} \qquad (4)$$

$L_{sw,DPWM}$ and $L_{sw,PWM}$ are the switching losses for discontinuous and continuous modulation respectively.

The DPWM technique presented in this work applies clamping, non-clamping, and original reference signals alternately among the submodules of the MMC arms. This method efficiently controls the charging and discharging of the capacitances of the SMs and balances the junction temperature of the power devices by reducing their switching losses without worsening the total harmonic distortion of the output waveforms.

III. PROPOSED DPWM STRATEGY

The proposed DPWM strategy adjusts the clamping angle between the modulation references of the submodules to achieve two objectives: balancing SM voltages and reducing thermal stress on power semiconductors. The principle of the proposed modulation is shown in Figure 3. In particular, the SM voltage can be balanced by controlling the clamping angle α during the positive half-period of the submodule reference signal. In the second half, when the modulation signal is negative, the clamping angle β is controlled to achieve thermal control. Indeed, in the negative half-period, the switching losses are reduced by keeping the capacitor voltage constant, while in the positive cycle the clamped SMs charge or discharge according to the arm current. The proposed DPWM strategy provides excellent dynamic performance of balancing SM voltages, which reduces submodule thermal stress without negatively affecting the overall performance of the converter and achieves total harmonic distortion comparable to PS-PWM. The simplified block diagram of the proposed method is shown in Figure 4.

A. Voltage Balancing Control

The proposed voltage balancing strategy receives the reference voltage for each arm and generates the different references for each SM, which can be the original, clamped, or non-clamped references. These references affect the voltage of each SM differently, allowing the charge or discharge of the capacitors to be managed. The selection of the SMs to be balanced is determined by the sorting algorithm based on the voltage measurements. Once the submodules with the highest and lowest arm voltage are identified. The degree of unbalance (DOU) is calculated through the following equation:

$$DOU = max(V_{cxj}) - min(V_{cxj}) \qquad (5)$$

The DOU indicates how much the SMs of an arm are unbalanced with each other. To balance the submodule voltages, a clamping reference is generated by applying a clamping angle

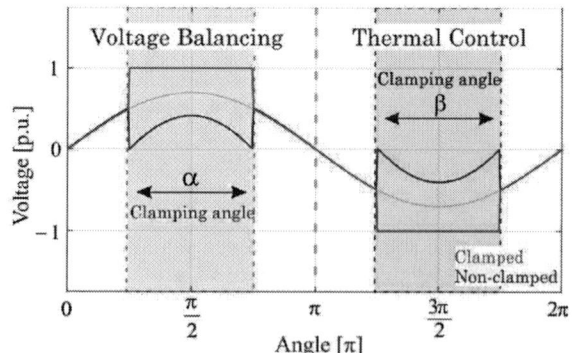

Figure 3: Principle of the proposed DPWM strategy.

during the positive half-cycle of the reference signal. The submodule receiving this clamping reference signal is always inserted during the clamping period. As a result, depending on the direction of the arm current, it can be charged or discharged more than the other SMs. Meanwhile, the submodule driven by the nonclamped signal tends to be charged less and its charge variation occurs more slowly. A proportional regulator controls the clamping angle α, increasing voltage convergence at high unbalances and reducing it in the steady state to achieve low harmonic distortion. The voltage balancing strategy is performed at each switching period to improve the dynamics of voltage balancing; in each arm, one SM receives the clamped reference, another receives the non-clamped reference, and the remaining SMs the original control reference. Therefore, individual references are generated at the controller output for each submodule.

B. Temperature Balancing Control

The temperature balancing control is applied in the second half of the reference signal period, bypassing the chosen SM and reducing its power losses, and its thermal stress. The structure of the thermal controller is similar to that of voltage balancing, with a controller managing the value of the β clamping angle and a generator of clamped and non-clamped signals. The clamping angle β is applied in the negative half-period of the reference signal. In this way, the clamped reference bypasses the SM, in the clamping region, thus reducing its power losses. As a result, this leads to a decrease in the junction temperature of the devices. By varying β, therefore, the junction temperature of the devices can be appropriately controlled. The sorting algorithm always selects the SM with the highest temperature at which to apply the clamped reference signal. The information of junction temperatures is provided through the estimate obtained from Foster's thermal chain and the device losses obtained by its characterization. The proposed discontinuous modulation strategy has the particularity to control the two clamping angles independently to achieve two distinct control objectives. However, it is relevant to highlight that the two controls are concomitant and it is necessary to coordinate them to ensure the balancing of voltages and junction temperatures.

In this work, coordination between the two controls is carried out by applying the voltage-balancing strategy in the positive half-period of the reference signal, while in the negative

half-period, the thermal control is applied, the flowchart of the proposed algorithm is shown in Figure 5. The activation of the thermal control can be managed according to different operating conditions for example to increase the overcurrent capability of the MMC in fault conditions or to perform power routing operations in the event of internal converter faults.

IV. SIMULATIONS AND EXPERIMENTAL VALIDATION

Simulations are carried out in the MATLAB/Simulink environments, the three-phase model of the MMC consists of eight SMs for each phase, a dc-link voltage of 4kV, and a nominal power of 250kW. Table I shows the simulation parameters and the experimental MMC used in the laboratory. The dynamic of the thermal distribution is simulated in PLECS Blockset using the Foster thermal Chain and the characterization of the power losses of the IGBT and diode.

Waveforms of the electrical quantities of the MMC derived from the simulation results are shown in Figure 6(a)-(d). As shown in Figure 6(a), the voltages of the submodules are rapidly balanced after the voltage balancing technique is applied at 1s, achieving equilibrium within approximately 500 ms. Upon the proposed DPWM is applied, all the SM voltages become perfectly aligned. At 4s, the strategy for balancing the submodule voltages is coordinated with the balancing of the junction temperature of the IGBTs within the MMC. In particular, when both strategies are activated, there is a slight deterioration in the capacitor voltage balance, indeed, the two controls are simultaneous and it is necessary to coordinate them to ensure the balance of voltages and junction temperatures. However, the voltages remain relatively balanced with each other, as illustrated in the enlarged section of the figures. All the junction temperatures of the power devices are balanced and reduced when the proposed thermal control is applied at 4s, as shown in Figure 6(b). The reduction in average temperature is 5°C. The maximum temperature is reduced by around 8°C. The imbalance in junction temperatures is reduced by 75 % when the proposed technique is applied. The waveforms of the voltages and currents in the arms and the output of the MMC when the proposed technique is applied are shown in Figure 6(c) and (d), respectively.

Figure 7 shows a comparison between the proposed method and the conventional PS-PWM, which employs a sorting algorithm for the voltage balance as described in [19,20], in the conduction and switching losses of the IGBT and the diode. The loss distribution of the MMC is calculated through simulations

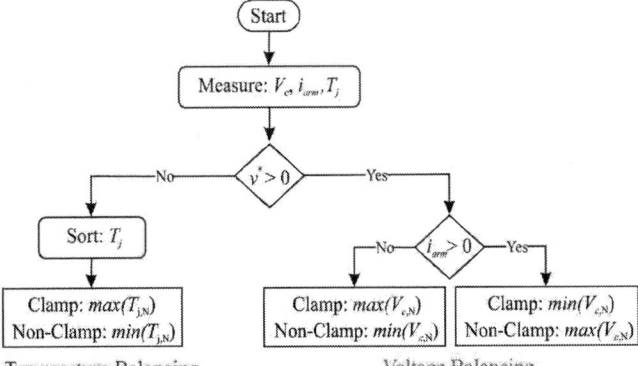

Figure 5: Flowchart of the proposed algorithm of clamping selection.

TABLE I – SPECIFICATION OF THE MMC

Parameter	Value
Number of SMs per Arm	4
SM capacitance	0.94 mF
Arm inductance	5 mH
dc-link voltage, Vdc	4kV – 600 V
Peak Output Current	100 A – 8 A
Switching Frequency	2 kHz
Modulation Index	0.9

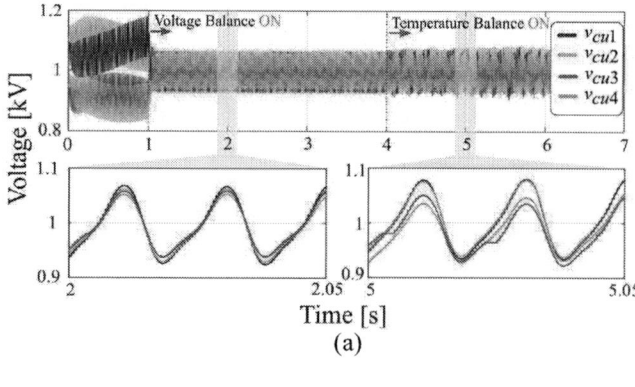

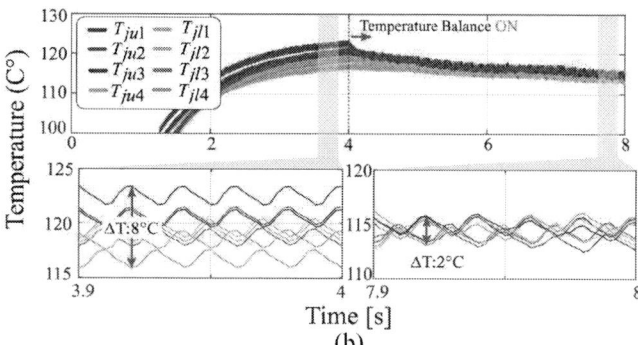

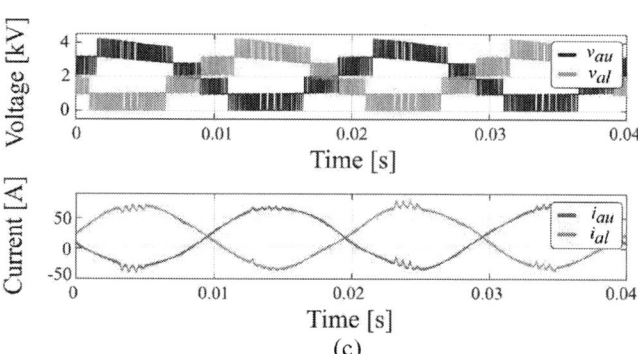

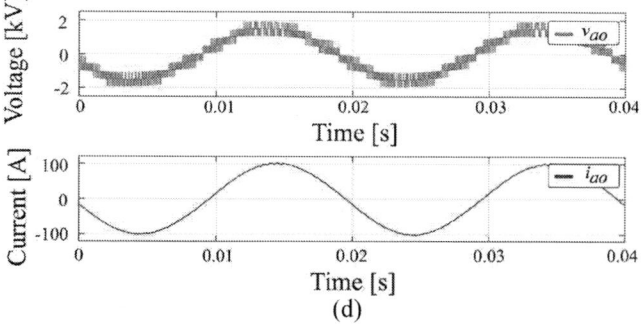

Figure 6: Simulation waveforms of the MMC: (a) SM capacitor voltages of the upper arm before and after activating the proposed DPWM, (b) behaviour of junction temperatures, (c) arm electrical quantities, and (d) output waveforms when the proposed method is activated.

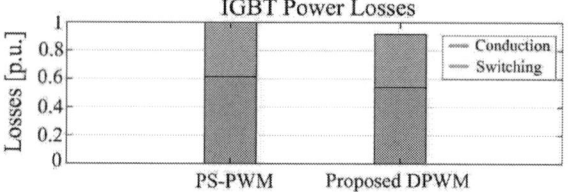

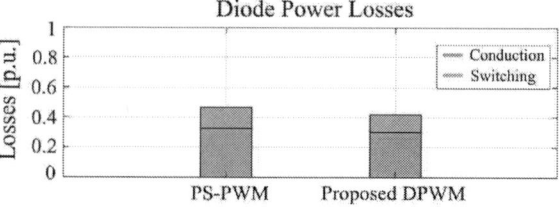

Figure 7: Power Losses for different modulation methods.

with PLECS software. The loss model is derived from a multidimensional table provided by the manufacturer for each power device and its corresponding antiparallel diode. Figure 7 shows losses for IGBT and diode of the upper arm. Notably, the proposed DPWM method achieves an approximate total loss reduction of 315W in all MMC. It is worth noticing that the proposed technique using discontinuous modulation reduces switching losses more, while conduction losses remain almost the same in both methods.

The experimental tests are conducted on a three-phase MMC laboratory prototype with four SMs per arm. Figure 8 shows a picture of the experimental prototype, and its specifications are provided in Table I. The control technique is implemented using a control board based on a Zynq system-on-chip (SoC) from Xilinx. Experimental tests are conducted to validate the voltage balancing of the SMs. The dc-link voltage is set to 600 V and the output current is controlled at 8 A peak value through the implementation of output current control based on proportional-integral controllers and reducing the circulating current through proportional-resonant controllers. The switching frequency is set to 2kHz, and the modulating signal is configured with a modulation index of 0.9.

Figure 9 depicts the clamped, unclamped, and original reference signals produced by the proposed DPWM for the four upper arm submodules. The clamping angle varies with each

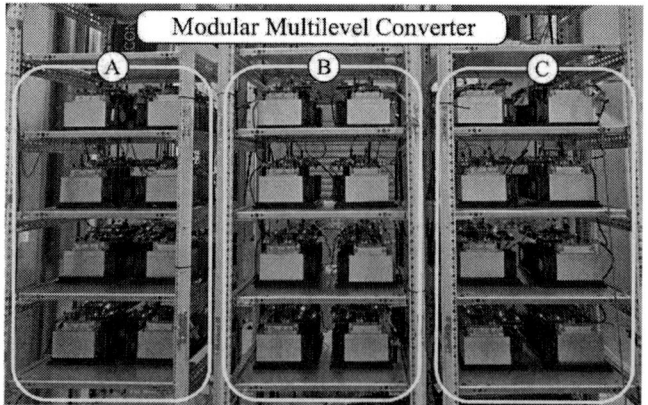

Figure 8: Picture of the experimental setup of the MMC.

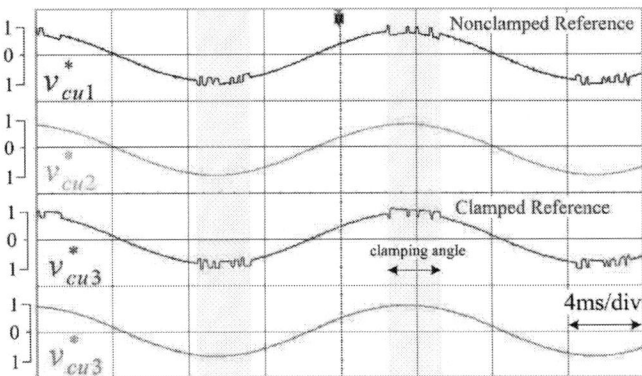

Figure 9: Experimental reference signals for upper arm submodules derived from the proposed technique.

switching period, increasing at high unbalance and decreasing at steady state to achieve low total harmonic distortion of the output waveforms. At this moment, SM₁ and SM₃ alternately receive clamping and non-clamping signals, while the remaining submodules are driven by signals from the output current

The upper arm submodule voltages before and after the proposed DPWM is applied, are shown in Figure 10. The experimental results are consistent with those obtained from simulations. The voltages of the SMs, once the proposed method is applied, are balanced to their reference voltage of 150 V with rapid dynamics, approximately in 500 ms. The proposed voltage balancing strategy exhibits fast and accurate dynamic performance. The submodule voltages start from an unbalance of 80V and, after the application of the proposed DPWM, the voltages balance each other maintaining a stable steady-state balancing condition. It is important to emphasize that the DPWM strategy based on different references for each sub-module with varying clamping angles has no negative impact on the control dynamics of the circulating current of the MMC, as demonstrated by the electrical quantities of phase voltage and current as shown in Figure 11. Indeed, the ac part of the circulating current is perfectly reduced to its dc value. Furthermore, the THD of the output current and voltage when the proposed method is applied is maintained at approximately 2.6% and 19.8%, respectively, comparable values when the PS-PWM strategy is utilized.

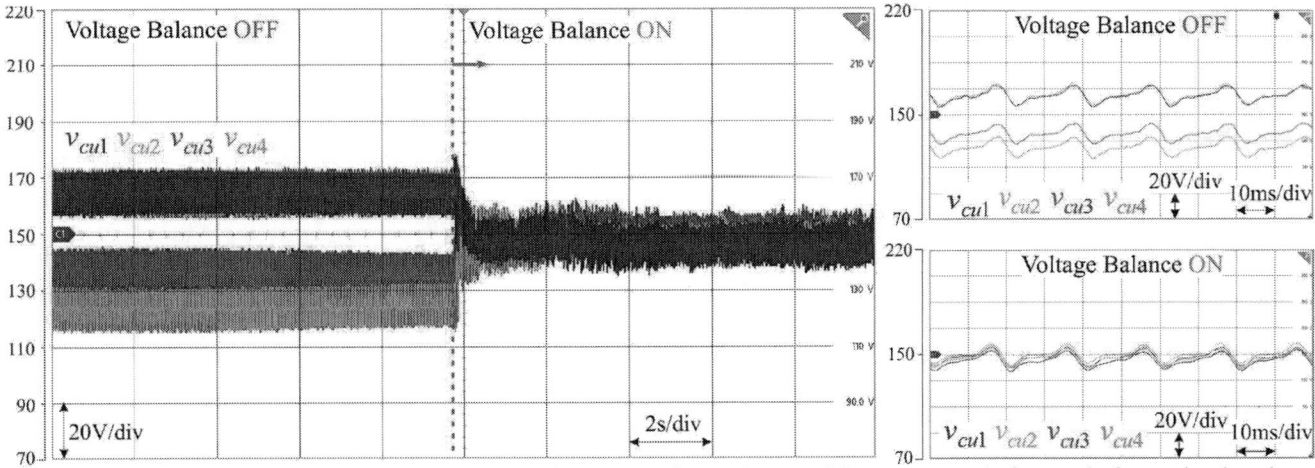

Figure 10: Dynamic performance of the experimental SM capacitor voltages of the upper arm before and after activating the proposed DPWM, on the right side the two details of the SM voltage in both unbalanced and balanced conditions.

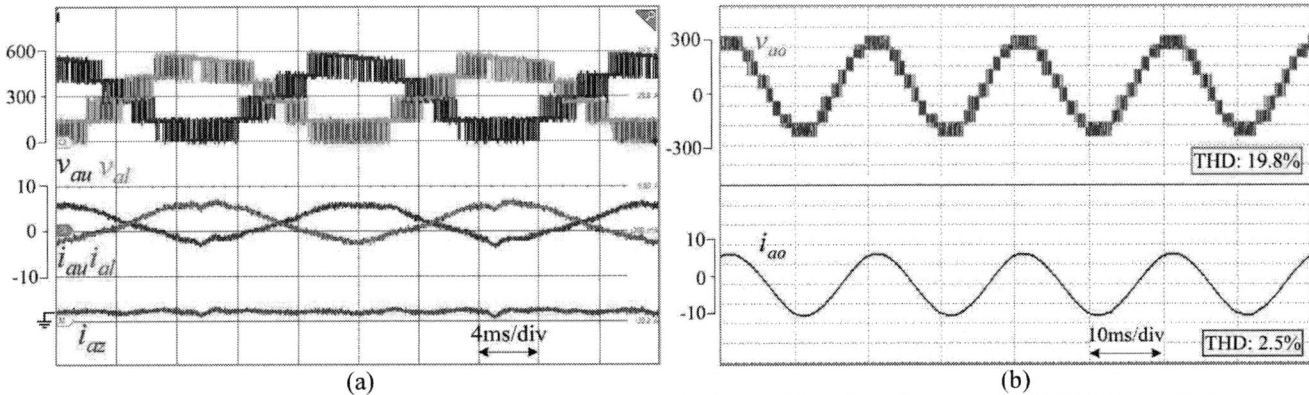

Figure 11: Experimental waveforms of the arm and output electrical quantities when the proposed DPWM is applied.

controller.

V. CONCLUSION

In this paper, a novel DPWM strategy is proposed to balance the capacitor voltages and thermal stress in the SMs of the MMC by adjusting actively the clamping angles of the modulating signal. The proposed technique based on DPWM reduces the switching losses of the SMs in the MMC while maintaining the THD of the output voltage and current comparable to that of PS-PWM by approximately 20% and 2.5% respectively. The uniqueness of the proposed DPWM strategy is that SMs voltages and thermal stresses are controlled by two independent clamping angles to achieve multi-objective balancing. The capacitor voltages across the SMs are balanced at about 500 ms, the junction temperature unbalanced is reduced by 75%, and the maximum temperature is reduced by 8°C. The total power losses are reduced by 15% compared to the conventional PS-PWM, demonstrating the effectiveness of the proposed DPWM technique in improving the performance of the converter.

ACKNOWLEDGMENT

This research work was kindly funded by the Federal Ministry of the Economics and Climate Protection (BMWK) of Germany, grant no. 03EI6105D (Project AC2DC – Phase II).

REFERENCES

[1] S. Debnath, J. Qin, B. Bahrani, M. Saeedifard and P. Barbosa, "Operation, Control, and Applications of the Modular Multilevel Converter: A Review," in IEEE Transactions on Power Electronics, vol. 30, no. 1, pp. 37-53, Jan. 2015.

[2] P. Bakas et al., "Review of Hybrid Multilevel Converter Topologies Utilizing Thyristors for HVDC Applications," in IEEE Transactions on Power Electronics, vol. 36, no. 1, pp. 174-190, Jan. 2021.

[3] Y. Chen, Z. Li, S. Zhao, X. Wei and Y. Kang, "Design and Implementation of a Modular Multilevel Converter With Hierarchical Redundancy Ability for Electric Ship MVDC System," in IEEE Journal of Emerging and Selected Topics in Power Electronics, vol. 5, no. 1, pp. 189-202, March 2017.

[4] D. D'Amato, R. Leuzzi and V. G. Monopoli, "An Innovative Single-Cell-Based Injection Method to Improve Efficiency and Reliability of MMC With Low Implementation Burden," in IEEE Transactions on Power Electronics, vol. 39, no. 10, pp. 12815-12825, Oct. 2024.

[5] Q. Song, W. Liu, X. Li, H. Rao, S. Xu and L. Li, "A Steady-State Analysis Method for a Modular Multilevel Converter," in IEEE Transactions on Power Electronics, vol. 28, no. 8, pp. 3702-3713, Aug. 2013.

[6] M. Hagiwara and H. Akagi, "Control and Experiment of Pulsewidth-Modulated Modular Multilevel Converters," in IEEE Transactions on Power Electronics, vol. 24, no. 7, pp. 1737-1746, July 2009

[7] J. Sheng et al., "Active Thermal Control for Hybrid Modular Multilevel Converter Under Overmodulation Operation," in IEEE Transactions on Power Electronics, vol. 35, no. 4, pp. 4242-4255, April 2020.

[8] H. Ding, F. Ma, R. Han, L. Wang and M. -C. Wong, "An Internal Thermal Distribution Balance Control Between the Top and Bottom Devices in Modular Multilevel Converter," in IEEE Transactions on Industrial Electronics, vol. 71, no. 9, pp. 10716-10726, Sept. 2024.

[9] F. Hahn, M. Andresen, G. Buticchi and M. Liserre, "Thermal Analysis and Balancing for Modular Multilevel Converters in HVDC Applications," in IEEE Transactions on Power Electronics, vol. 33, no. 3, pp. 1985-1996, March 2018.

[10] J. -H. Jung, E. Hosseini, M. Liserre and L. M. Fernández-Ramírez, "Reinforcement Learning Based Modulation for Balancing Capacitor Voltage and Thermal Stress to Enhance Current Capability of MMCs," 2022 IEEE 13th International Symposium on Power Electronics for Distributed Generation Systems (PEDG), Kiel, Germany, 2022, pp. 1-6.

[11] F. Deng, J. Zhao, C. Liu, Z. Wang, X. Cai, and F. Blaabjerg, "Temperature-Balancing Control for Modular Multilevel Converters Under Unbalanced Grid Voltages," in IEEE Transactions on Power Electronics, vol. 37, no. 4, pp. 4614-4625, April 2022.

[12] Y. Li, E. A. Jones and F. Wang, "The impact of voltage-balancing control on switching frequency of the modular multilevel converter," in IEEE Transactions on Power Electronics, vol. 31, no. 4, pp. 2829-2839, April 2016.

[13] Y. Ko, M. Andresen, G. Buticchi and M. Liserre, "Thermally Compensated Discontinuous Modulation Strategy for Cascaded H-Bridge Converters," in IEEE Transactions on Power Electronics, vol. 33, no. 3, pp. 2704-2713, March 2018.

[14] D. Grahame Holmes; Thomas A. Lipo, "CarrierBased PWM of Multilevel Inverters," in Pulse Width Modulation for Power Converters: Principles and Practice, IEEE, 2003, pp.453-530.

[15] Y. Li, Y. Wang and B. Q. Li, "Generalized Theory of Phase-Shifted Carrier PWM for Cascaded H-Bridge Converters and Modular Multilevel Converters," in IEEE Journal of Emerging and Selected Topics in Power Electronics, vol. 4, no. 2, pp. 589-605, June 2016.

[16] S. Sarkar and A. Das, "An isolated single input-multiple output dc–dc modular multilevel converter for fast electric vehicle charging", IEEE J. Emerg. Sel. Topics Ind. Electron., vol. 4, no. 1, pp. 178-187, Jan. 2023.

[17] R. Picas, S. Ceballos, J. Pou, J. Zaragoza, G. Konstantinou and V. G. Agelidis, "Closed-Loop Discontinuous Modulation Technique for Capacitor Voltage Ripples and Switching Losses Reduction in Modular Multilevel Converters," in IEEE Transactions on Power Electronics, vol. 30, no. 9, pp. 4714-4725, Sept. 2015.

[18] N. -V. Nguyen, B. -X. Nguyen and H. -H. Lee, "An Optimized Discontinuous PWM Method to Minimize Switching Loss for Multilevel Inverters," in IEEE Transactions on Industrial Electronics, vol. 58, no. 9, pp. 3958-3966, Sept. 2011.

[19] A. Dekka, B. Wu, N. R. Zargari and R. L. Fuentes, "Dynamic Voltage Balancing Algorithm for Modular Multilevel Converter: A Unique Solution," in IEEE Transactions on Power Electronics, vol. 31, no. 2, pp. 952-963, Feb. 2016.

[20] Q. Tu, Z. Xu and L. Xu, "Reduced Switching-Frequency Modulation and Circulating Current Suppression for Modular Multilevel Converters," in IEEE Transactions on Power Delivery, vol. 26, no. 3, pp. 2009-2017, July 2011.

Damping Control and Improvement of Grid-Forming Inverter from a Wideband Stability Perspective

Rui Kong, Subham Sahoo, Yubo Song, and Frede Blaabjerg
Department of Energy
Aalborg University
Aalborg, Denmark
ruko@energy.aau.dk, sssa@energy.aau.dk, yuboso@energy.aau.dk, fbl@energy.aau.dk

Abstract—Small-signal stability assessment and damping control of grid-forming (GFM) converter are limited to specific frequency bands, lacking comprehensive verification of wideband stability improvement. This paper aims to reveal the impact of different damping control methods from a wideband stability perspective and provide an integrated solution. An impedance model considering coupled control loops is developed in the dq frame to extract dominant oscillation modes in both low and high-frequency bands. Then, candidate damping control methods such as virtual resistor and feedforward compensation are compared to reveal their different damping impacts on wideband oscillation modes and possible side effects. Based on the assessment results, an appropriate solution is selected for wideband stability improvement. The findings of this work are verified under experimental conditions.

Index Terms—Grid-tied converter, grid-forming control, small-signal stability, damping control

I. INTRODUCTION

With the massive integration of renewable energy sources into the grid, the GFM control evolves as a promising solution due to its grid supportive services [1]. The grid-tied GFM inverter has a cascaded control structure with different bandwidths and timescales, causing instability issues over wide frequency ranges in stiff grids [2].

Existing studies commonly build a small-signal linearized model of the system for stability analysis [3] and propose corresponding damping control measures to mitigate oscillations [4], where the outer power loop and inner voltage/current loop in the GFM controller are usually considered decoupled due to differences in their bandwidth and dynamic response [5]. For instance, the low-frequency stability dominated by the outer power loop is focused in [6], and inner loops are considered to achieve ideal voltage tracking, thereby being neglected in the modeling and damping analysis. In [7], the outer power loop with slower dynamics is neglected since only high-frequency stability is concerned. However, it is stated in [8] that there is a non-negligible coupling between the inner and outer loops. On the other hand, [9] points out that control-loop interaction might introduce new oscillation modes. An impedance model of the GFM inverter considering both the outer and inner control loop is built in [10], but as only low-frequency oscillations are focused on, control delay is neglected, which is a key factor that affects the high-frequency stability. Thus, it is necessary to develop a detailed model of the GFM inverter to reveal all possible oscillation modes over wide frequency ranges.

On the other hand, various damping control methods by modifying the controller have been applied to improve the small signal stability of grid-tied inverters [11], where the damping of low-frequency oscillation is enhanced in [9] by tuning control parameters. Moreover, for mitigating low-frequency oscillations, [12] designs an additional feedforward damping branch, while the virtual impedance method is leveraged in [13], [14]. As for stability enhancement in the high-frequency range, the virtual admittance method is applied in [15], while the virtual impedance method is used in [7] to eliminate the negative output resistance up to half of the sampling frequency. However, existing studies commonly account only for instability issues in specific frequency ranges, being insufficient to justify their damping effectiveness for wideband oscillation modes. It also needs to be verified whether damping control methods for specific frequency bands have side effects on the stability of other frequency bands.

To fill these gaps, a small-signal closed-loop transfer function model considering detailed control loops of the GFM inverter is developed in this paper, thereby identifying the wideband dominant modes covering both the low and high-frequency ranges. Then, the impacts of candidate damping control methods on different oscillation modes are unveiled to adequately assess the damping performance. Based on the assessment results, a preferable wideband stability enhancement solution is provided. All analysis results are validated by experiments.

II. SYSTEM MODELING AND PROBLEM FORMULATION

Fig. 1(a) shows system diagram of a three-phase grid-tied GFM inverter, where the line resistance is neglected and the DC-side voltage of the inverter is assumed to be constant. The main downscaled system parameters are given in Table I,

979-8-3315-1612-3/25 $31.00 © 2025 IEEE

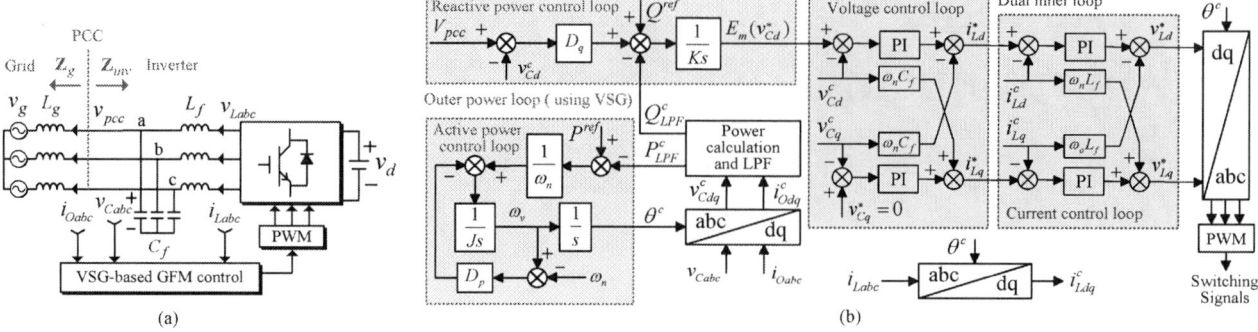

Fig. 1. System diagram of the three-phase grid-tied inverter — (a) Main circuit and (b) Control diagram of virtual synchronous generator (VSG)-based grid-forming (GFM) control.

TABLE I
MAIN PARAMETERS OF EXPERIMENTAL SYSTEM

Parameter	Description	Value
V_g	Grid phase voltage (RMS)	110 V
S_N	Rated power	1.5 kVA
L_g	Grid-side inductance	8 mH
L_n	Filter inductance	4 mH
C_n	Filter capacitance	10 uF
V_d	DC link voltage	500 V
f_s	Switching frequency	20 kHz
D_p	Active power damping coefficient	1.5
J	Virtual inertia	$0.001\ Kg \cdot m^2$
D_q	Reactive power droop coefficient	48
K	Reactive power inertia coefficient	1
K_{pvc}	Proportional gain of the voltage control loop	0.14
K_{ivc}	Integral gain of the voltage control loop	4
K_{pcc}	Proportional gain of the current control loop	12
K_{icc}	Integral gain of the current control loop	60

which are also used for the experiment verification. To reveal the oscillation modes over a wide frequency range, in addition to the grid line inductance and LC filter of converters, detailed control structure as shown in Fig. 1(b) are considered to build the small-signal model of the system in the dq frame, involving outer-loop control using virtual synchronous generator (VSG), dual inner loop control using voltage and current control loops, PWM delay, and first-order low-pass filter (LPF) for power calculation.

The steps for the small signal dq-impedance modeling can be summarized into deriving the differential equations based on the physical circuit and control block diagram, performing the Park transform (from abc to dq frame) and the Laplace transform, and then performing the small-signal linearization, where Δ represents a small-signal perturbation and capital letters with superscript '0' denotes a steady-state operating point. Thus, the small-signal linearized model of line inductance and LC filter in the dq frame and frequency domain can be given in a matrix form as:

$$\begin{bmatrix} \Delta v_{Cd} \\ \Delta v_{Cq} \end{bmatrix} - \begin{bmatrix} \Delta v_{gd} \\ \Delta v_{gq} \end{bmatrix} = \underbrace{\begin{bmatrix} sL_g & -\omega_n L_g \\ \omega_n L_g & sL_g \end{bmatrix}}_{\mathbf{M}_{Lg}} \begin{bmatrix} \Delta i_{Od} \\ \Delta i_{Oq} \end{bmatrix} \quad (1)$$

$$\begin{bmatrix} \Delta v_{Ld} \\ \Delta v_{Lq} \end{bmatrix} - \begin{bmatrix} \Delta v_{Cd} \\ \Delta v_{Cq} \end{bmatrix} = \underbrace{\begin{bmatrix} sL_f & -\omega_n L_f \\ \omega_n L_f & sL_f \end{bmatrix}}_{\mathbf{M}_{Lf}} \begin{bmatrix} \Delta i_{Ld} \\ \Delta i_{Lq} \end{bmatrix} \quad (2)$$

$$\begin{bmatrix} \Delta i_{Ld} \\ \Delta i_{Lq} \end{bmatrix} - \begin{bmatrix} \Delta i_{Od} \\ \Delta i_{Oq} \end{bmatrix} = \underbrace{\begin{bmatrix} sC_f & -\omega_n C_f \\ \omega_n C_f & sC_f \end{bmatrix}}_{\mathbf{M}_{Cf}} \begin{bmatrix} \Delta v_{Cd} \\ \Delta v_{Cq} \end{bmatrix} \quad (3)$$

where, ω_n is the fundamental angular frequency, and other variables are the voltage/current of the main circuit after transformation into the ideal dq frame. However, when the three-phase AC signals are sampled from the physical circuit and transformed into the dq frame for control, the control phase angle θ^c generated by the active power loop is actually used, where θ^c has a small deviation $\Delta\theta$ from the true phase angle θ at the point of common coupling (PCC) in the dynamic state, i.e., $\theta^c = \theta + \Delta\theta$. Thus, the transformation relationship between variables in the system dq frame (without superscript) and the control dq frame (with superscript 'c') can be obtained as:

$$\begin{bmatrix} x_d^c \\ x_q^c \end{bmatrix} = \begin{bmatrix} \cos\Delta\theta & \sin\Delta\theta \\ -\sin\Delta\theta & \cos\Delta\theta \end{bmatrix} \begin{bmatrix} x_d \\ x_q \end{bmatrix} \approx \begin{bmatrix} 1 & \Delta\theta \\ -\Delta\theta & 1 \end{bmatrix} \begin{bmatrix} x_d \\ x_q \end{bmatrix} \quad (4)$$

where, x can represent arbitrary control variables, i.e., v_C, v_L, i_L and i_O. After small-signal linearization, (4) can be expressed in a matrix form as:

$$\begin{bmatrix} \Delta v_{Cd}^c \\ \Delta v_{Cq}^c \end{bmatrix} = \begin{bmatrix} \Delta v_{Cd} \\ \Delta v_{Cq} \end{bmatrix} + \underbrace{\begin{bmatrix} V_{Cq}^0 & 0 \\ -V_{Cd}^0 & 0 \end{bmatrix}}_{\mathbf{M}_{VC}} \begin{bmatrix} \Delta\theta \\ \Delta E_m \end{bmatrix} \quad (5)$$

$$\begin{bmatrix} \Delta v_{Ld}^c \\ \Delta v_{Lq}^c \end{bmatrix} = \begin{bmatrix} \Delta v_{Ld} \\ \Delta v_{Lq} \end{bmatrix} + \underbrace{\begin{bmatrix} V_{Lq}^0 & 0 \\ -V_{Ld}^0 & 0 \end{bmatrix}}_{\mathbf{M}_{VL}} \begin{bmatrix} \Delta\theta \\ \Delta E_m \end{bmatrix} \quad (6)$$

$$\begin{bmatrix} \Delta i_{Ld}^c \\ \Delta i_{Lq}^c \end{bmatrix} = \begin{bmatrix} \Delta i_{Ld} \\ \Delta i_{Lq} \end{bmatrix} + \underbrace{\begin{bmatrix} I_{Lq}^0 & 0 \\ -I_{Ld}^0 & 0 \end{bmatrix}}_{\mathbf{M}_{IL}} \begin{bmatrix} \Delta\theta \\ \Delta E_m \end{bmatrix} \quad (7)$$

$$\begin{bmatrix} \Delta i_{Od}^c \\ \Delta i_{Oq}^c \end{bmatrix} = \begin{bmatrix} \Delta i_{Od} \\ \Delta i_{Oq} \end{bmatrix} + \underbrace{\begin{bmatrix} I_{Oq}^0 & 0 \\ -I_{Od}^0 & 0 \end{bmatrix}}_{\mathbf{M}_{IO}} \begin{bmatrix} \Delta\theta \\ \Delta E_m \end{bmatrix} \quad (8)$$

where, E_m is the generated voltage amplitude of the reactive power control loop. It is worth mentioning that, E_m is written here to unify the form with the power and voltage control loop mentioned in the subsequent modeling.

When using the equal amplitude Park transformation, the instantaneous active power P^c and reactive power Q^c of the inverter can be expressed as [6]:

$$P^c = \frac{3}{2}(v_{Cd}^c i_{Od}^c + v_{Cq}^c i_{Oq}^c)$$

$$Q^c = \frac{3}{2}(v_{Cq}^c i_{Od}^c - v_{Cd}^c i_{Oq}^c) \quad (9)$$

The small-signal linearized expression of (9) can be obtained as:

$$\begin{bmatrix} \Delta P^c \\ \Delta Q^c \end{bmatrix} = \frac{3}{2}\underbrace{\begin{bmatrix} I_{Od}^0 & I_{Oq}^0 \\ -I_{Oq}^0 & I_{Od}^0 \end{bmatrix}}_{\mathbf{M}_{PQV}}\begin{bmatrix} \Delta v_{Cd}^c \\ \Delta v_{Cq}^c \end{bmatrix} + \frac{3}{2}\underbrace{\begin{bmatrix} V_{Cd}^0 & V_{Cq}^0 \\ V_{Cq}^0 & -V_{Cd}^0 \end{bmatrix}}_{\mathbf{M}_{PQI}}\begin{bmatrix} \Delta i_{Od}^c \\ \Delta i_{Oq}^c \end{bmatrix}$$

$$(10)$$

The calculated active and reactive power are commonly filtered by a first-order LPF with cut-off frequency ω_{LPF}, so there is:

$$\begin{bmatrix} \Delta P_{LPF}^c \\ \Delta Q_{LPF}^c \end{bmatrix} = \underbrace{\begin{bmatrix} \dfrac{\omega_{LPF}}{s+\omega_{LPF}} & 0 \\ 0 & \dfrac{\omega_{LPF}}{s+\omega_{LPF}} \end{bmatrix}}_{\mathbf{M}_{LPF}}\begin{bmatrix} \Delta P^c \\ \Delta Q^c \end{bmatrix} \quad (11)$$

Then, according to the control diagram as shown in Fig. 1(b), the small-signal linearized model of active/reactive power controller, voltage control loop and current control loop can be derived as:

$$\begin{bmatrix} \Delta\theta \\ \Delta E_m \end{bmatrix} = -\underbrace{\begin{bmatrix} \dfrac{1}{(J\omega_n s^2 + D_p\omega_n s)} & 0 \\ 0 & \dfrac{1}{Ks} \end{bmatrix}}_{\mathbf{M}_{OPQ}}\begin{bmatrix} \Delta P_{LPF}^c \\ \Delta Q_{LPF}^c \end{bmatrix}$$

$$(12)$$

$$-\underbrace{\begin{bmatrix} 0 & 0 \\ \dfrac{D_q}{Ks} & 0 \end{bmatrix}}_{\mathbf{M}_{OVC}}\begin{bmatrix} \Delta v_{Cd}^c \\ \Delta v_{Cq}^c \end{bmatrix}$$

$$\begin{bmatrix} \Delta i_{Ld}^* \\ \Delta i_{Lq}^* \end{bmatrix} = \underbrace{\begin{bmatrix} 0 & K_{pvc} + \dfrac{K_{ivc}}{s} \\ 0 & 0 \end{bmatrix}}_{\mathbf{M}_{VE}}\begin{bmatrix} \Delta\theta \\ \Delta E_m \end{bmatrix}$$

$$(13)$$

$$-\underbrace{\begin{bmatrix} K_{pvc} + \dfrac{K_{ivc}}{s} & \omega_n C_f \\ -\omega_n C_f & K_{pvc} + \dfrac{K_{ivc}}{s} \end{bmatrix}}_{\mathbf{M}_{VPI}}\begin{bmatrix} \Delta v_{Cd}^c \\ \Delta v_{Cq}^c \end{bmatrix}$$

$$\begin{bmatrix} \Delta v_{Ld}^* \\ \Delta v_{Lq}^* \end{bmatrix} = \underbrace{\begin{bmatrix} 0 & -\omega_n L_f \\ \omega_n L_f & 0 \end{bmatrix}}_{\mathbf{M}_{CL}}\begin{bmatrix} \Delta i_{Ld}^c \\ \Delta i_{Lq}^c \end{bmatrix}$$

$$+ \underbrace{\begin{bmatrix} K_{pcc} + \dfrac{K_{icc}}{s} & 0 \\ 0 & K_{pcc} + \dfrac{K_{icc}}{s} \end{bmatrix}}_{\mathbf{M}_{CPI}}\left(\begin{bmatrix} \Delta i_{Ld}^* \\ \Delta i_{Lq}^* \end{bmatrix} - \begin{bmatrix} \Delta i_{Ld}^c \\ \Delta i_{Lq}^c \end{bmatrix}\right)$$

$$(14)$$

where, superscript '*' denotes the reference value output by controllers. The transfer function of control delay is $G_d = e^{-sT_d}$ [7], where T_d is the equivalent delay time, $T_d = 1.5T_s = 1.5/f_s$, and f_s is the switching frequency. Further, G_d can be approximated by a first-order inertial link, so the impact of the control delay on the output of the whole controller can be expressed as:

$$\begin{bmatrix} \Delta v_{Ld}^c \\ \Delta v_{Lq}^c \end{bmatrix} = \underbrace{\begin{bmatrix} 1/(1+sT_d) & 0 \\ 0 & 1/(1+sT_d) \end{bmatrix}}_{\mathbf{M}_{DEL}}\begin{bmatrix} \Delta v_{Ld}^* \\ \Delta v_{Lq}^* \end{bmatrix} \quad (15)$$

By combining (2)-(15), the equivalent inverter-side impedance $\mathbf{Z}_{inv}$ as shown in Fig. 1(a) can be derived with $-\Delta i_{Odq}$ as input and Δv_{Cdq} as output, which is given as:

$$\begin{bmatrix} \Delta v_{Cd} \\ \Delta v_{Cq} \end{bmatrix} = \underbrace{(\mathbf{M}_G - \mathbf{M}_{Cf})^{-1}(\mathbf{M}_H - \mathbf{I})}_{\mathbf{Z}_{inv}}\begin{bmatrix} -\Delta i_{Od} \\ -\Delta i_{Oq} \end{bmatrix} \quad (16)$$

where $\mathbf{I}$ is a second-order identity matrix, and there are

$$\mathbf{M}_H = \mathbf{M}_A^{-1}[\mathbf{M}_F\mathbf{M}_{OPQ}\mathbf{M}_{LPF}\mathbf{M}_{PQI} - \mathbf{M}_C]$$
$$\mathbf{M}_G = \mathbf{M}_A^{-1}[(\mathbf{M}_F(\mathbf{M}_{OPQ}\mathbf{M}_{LPF}\mathbf{M}_{PQV} + \mathbf{M}_{OVC})$$
$$\qquad - \mathbf{M}_B - \mathbf{I}]$$
$$\mathbf{M}_F = (\mathbf{M}_B\mathbf{M}_{VC} + \mathbf{M}_C\mathbf{M}_{IO} - \mathbf{M}_D)\mathbf{M}_E^{-1}$$
$$\mathbf{M}_E = \mathbf{I} + \mathbf{M}_{OPQ}\mathbf{M}_{LPF}(\mathbf{M}_{PQV}\mathbf{M}_{VC} + \mathbf{M}_{PQI}\mathbf{M}_{IO})$$
$$\qquad + \mathbf{M}_{OVC}\mathbf{M}_{VC}$$
$$\mathbf{M}_D = \mathbf{M}_{DEL}(\mathbf{M}_{CL} - \mathbf{M}_{CPI})\mathbf{M}_{IL} - \mathbf{M}_{VL}$$
$$\mathbf{M}_C = \mathbf{M}_{DEL}\mathbf{M}_{CPI}\mathbf{M}_{VE}\mathbf{M}_{OPQ}\mathbf{M}_{LPF}\mathbf{M}_{PQI}$$
$$\mathbf{M}_B = \mathbf{M}_{DEL}\mathbf{M}_{CPI}(\mathbf{M}_{VE}\mathbf{M}_{OPQ}\mathbf{M}_{LPF}\mathbf{M}_{PQV}$$
$$\qquad + \mathbf{M}_{VE}\mathbf{M}_{OVC} + \mathbf{M}_{VPI})$$
$$\mathbf{M}_A = \mathbf{M}_{Lf} - \mathbf{M}_{DEL}(\mathbf{M}_{CL} - \mathbf{M}_{CPI})$$

$$(17)$$

The matrix $\mathbf{M}_{Lg}$ in (1) can be regarded as the equivalent grid-side impedance $\mathbf{Z}_g$ as shown in Fig. 1(a). By combining (1) and (16), the closed-loop transfer function $\mathbf{H}_{CL}$ of the whole grid-tied inverter system with Δv_{gdq} as input and $-\Delta i_{Odq}$ as output can be derived as:

$$\begin{bmatrix} -\Delta i_{Od} \\ -\Delta i_{Oq} \end{bmatrix} = \underbrace{(\mathbf{Z}_{inv} + \mathbf{Z}_g)^{-1}}_{\mathbf{H}_{CL}}\begin{bmatrix} \Delta v_{gd} \\ \Delta v_{gq} \end{bmatrix} \quad (18)$$

After solving the poles of the closed-loop transfer function $\mathbf{H}_{CL}$, the system stability can be assessed by observing the presence of poles in the right half-plane (RHP). Furthermore, the impacts of the additional damping control methods on the system stability can be judged by observing the movement of the closed-loop poles.

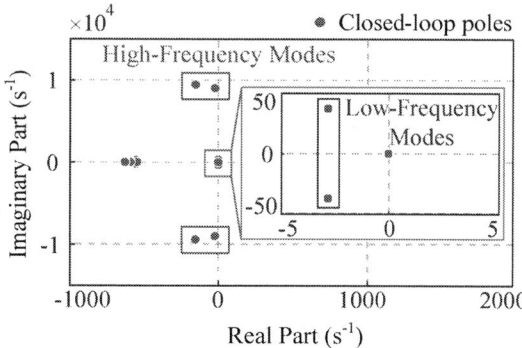

Fig. 2. Wideband oscillation modes of the grid-tied GFM inverter system — Dominant low/high-frequency modes exist simultaneously.

However, based on the built small-signal model considering the detailed control structure, the closed-loop pole plot of $\mathbf{H}_{CL}$ as shown in Fig. 2 reveals the simultaneous existence of dominant low-frequency oscillation (LFO) modes and high-frequency oscillation (HFO) modes in the stiff grids. Existing stability assessment and damping control for specific frequency bands [5]–[7], [12], [13], [15] cannot be demonstrated to be effective for both low and high-frequency oscillation modes, which even might have conflicting impacts on the damping of modes in different frequency ranges. Therefore, a comparative analysis of different damping control methods is implemented in this paper to evaluate their effectiveness within a wide frequency band, and the analysis results are verified using the downscaled experimental platform of the three-phase grid-tied inverter as shown in Fig. 3, where system parameters are consistent with Table I.

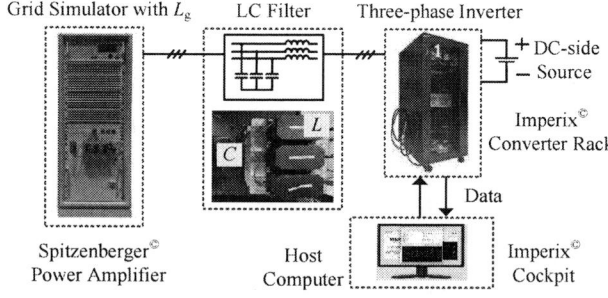

Fig. 3. Experimental setup of down-scaled grid-tied three-phase inverter.

III. COMPARATIVE ASSESSMENT OF DAMPING CONTROL METHODS FOR WIDEBAND STABILITY IMPROVEMENT

Three damping control strategies are analyzed in detail and compared from a wideband stability improvement perspective as follows:

1) Control parameter tuning [9], [16].
2) Output current feedback-based virtual resistor [5], [17].
3) Port current feedforward-based compensation [12], [18].

These damping methods are applied in the existing damping enhancement studies but they focus on oscillation issues in limited frequency bands.

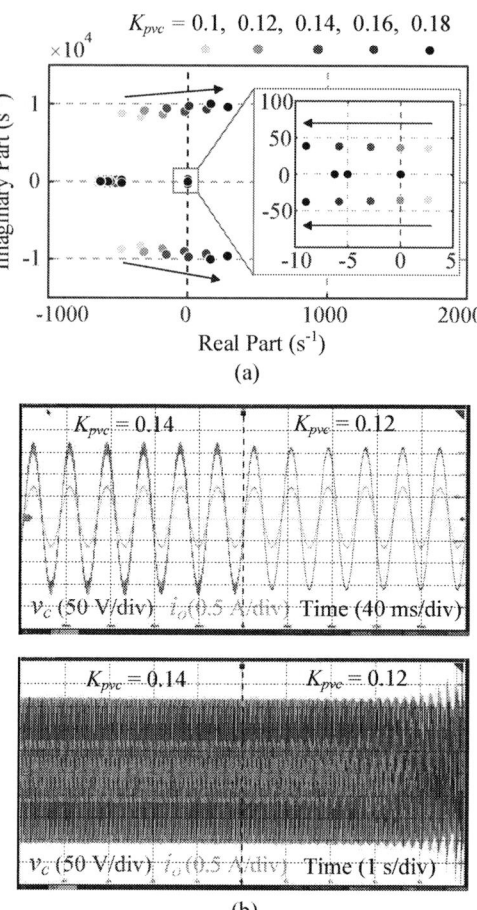

Fig. 4. Impact of tuning the proportional gain of voltage loop K_{pvc} on system stability — (a) Model-based analysis and (b) Experimental verification.

A. Control Parameter Tuning

Tuning the control parameters is a low-cost method for damping enhancement. For instance, it is pointed out in [9] that increasing the proportional gain K_{pvc} of the voltage control loop is beneficial for mitigating the LFO in GFM control inverter systems, but only LFO is concerned in [9]. Considering wideband oscillations, as shown in Fig. 4(a), when K_{pvc} increases from 0.1 to 0.18 and other parameters are the same as given in Table I, low-frequency modes (around 6 Hz) move to the left half plane with increased damping but the damping of high-frequency modes (around 1.5 kHz) decreases significantly. Thus, increasing K_{pvc} can mitigate LFO, which is consistent with the results in [9], but high-frequency mode damping is simultaneously compromised, thereby posing a risk of HFO. When decreasing the K_{pvc}, the experimental waveforms of the PCC voltage v_C and port current i_O are shown in Fig. 4(b), which match the theoretical analysis. This characteristic also exists for the other control parameters such as the proportional gain of current loop K_{pcc}, reactive power droop coefficient D_p, etc. Therefore, control parameter tuning is not a reliable wideband stability enhancement solution.

B. Output Current Feedback-Based virtual resistor

The virtual resistor (VR) method as shown in Fig. 5 is commonly used to provide additional system damping [5], [17], which is imposed on the inverter output current i_L and feedbacks to the PCC voltage reference v^*_{Cdq} as the input of voltage control loop. The VR contains a high-pass filter (HPF) (the cutoff frequency ω_{HPF} is typically selected as $0.2\omega_n$ [5], where ω_n is the fundamental angular frequency) and a proportional gain K_{vr}, which determines the damping level of VR. Although VR is usually leveraged for high-frequency oscillation mitigation, its impact on wideband stability is unclear.

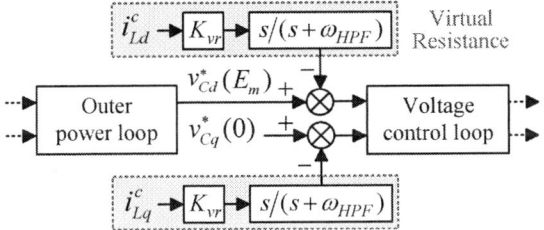

Fig. 5. Control diagram of output current feedback-based virtual resistor.

According to Fig. 5, when VR is enabled, the small-signal model of the voltage control loop in (13) will be updated as:

$$\begin{bmatrix} \Delta i^*_{Ld} \\ \Delta i^*_{Lq} \end{bmatrix} = \mathbf{M}_{VE} \begin{bmatrix} \Delta\theta \\ \Delta E_m \end{bmatrix} - \mathbf{M}_{VPI} \begin{bmatrix} \Delta v^c_{Cd} \\ \Delta v^c_{Cq} \end{bmatrix}$$

$$- \underbrace{\begin{bmatrix} (K_{pvc} + \dfrac{K_{ivc}}{s})H_{vr}(s) & 0 \\ 0 & (K_{pvc} + \dfrac{K_{ivc}}{s})H_{vr}(s) \end{bmatrix}}_{\mathbf{M}_{VR}} \begin{bmatrix} \Delta v^c_{Cd} \\ \Delta v^c_{Cq} \end{bmatrix}$$

(19)

where, $H_{vr}(s)$ is the transfer function of VR, and there is $H_{vr}(s) = k_{vr}s/(s+\omega_{HPF})$. Correspondingly, the $\mathbf{M}_A$ in (17) of the inverter-side impedance $\mathbf{Z}_{inv}$ is updated as:

$$\mathbf{M}_A = \mathbf{M}_{Lf} - \mathbf{M}_{DEL}(\mathbf{M}_{CL} - \mathbf{M}_{CPI} - \mathbf{M}_{CPI}\mathbf{M}_{VR}) \quad (20)$$

The closed-loop poles of the updated model are shown in Fig. 6(a), indicating that VR can mitigate HFO effectively and high-frequency mode damping increases as K_{vr} increases. However, low-frequency mode damping is compromised simultaneously. The experimental waveforms as shown in Fig. 6(b) verify that enabling the VR is effective for suppressing the HFO, but the LFO is triggered at the same time when having a large gain K_{vr}.

C. Port Current Feedforward-Based Compensation

Adding a compensation branch to the controller is another effective damping strategy [12], [18], as shown in Fig. 7, port current i_O is fed forward through the proportional gain K_{co} to the reference output current i^*_{Ldq} as the input of the current control loop.

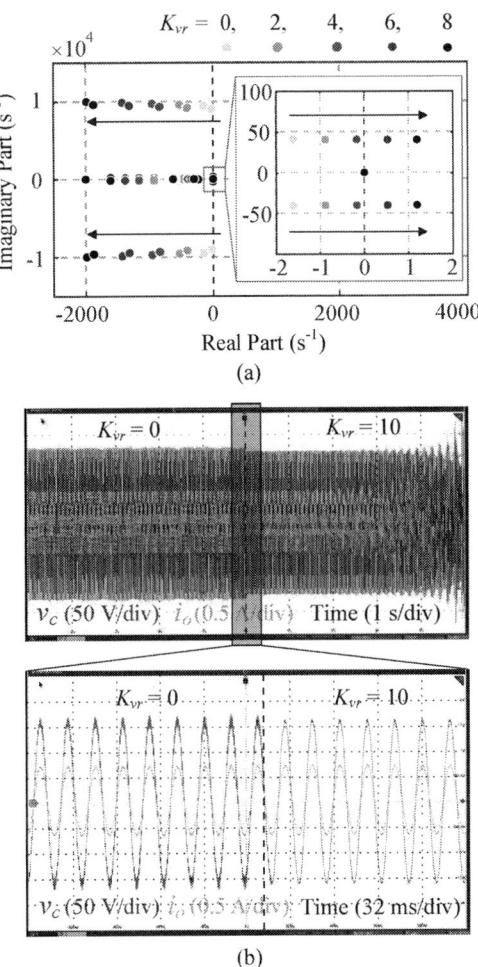

Fig. 6. Impact of increasing proportional gain of virtual resistor K_{vr} on system stability — (a) Model-based analysis and (b) Experimental verification.

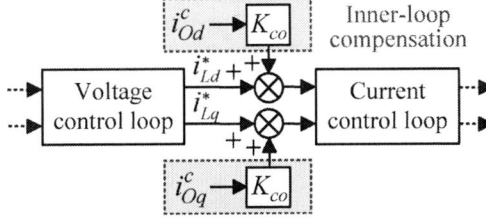

Fig. 7. Control diagram of port current feedforward-based compensation.

To verify the impact of compensation on wideband stability, the small-signal model of the current control loop in (14) needs to be updated as:

$$\begin{bmatrix} \Delta v^*_{Ld} \\ \Delta v^*_{Lq} \end{bmatrix} = M_{CL} \begin{bmatrix} \Delta i^c_{Ld} \\ \Delta i^c_{Lq} \end{bmatrix}$$

$$+ M_{CPI} \left(\begin{bmatrix} \Delta i^*_{Ld} \\ \Delta i^*_{Lq} \end{bmatrix} - \begin{bmatrix} \Delta i^c_{Ld} \\ \Delta i^c_{Lq} \end{bmatrix} + \underbrace{\begin{bmatrix} k_{co} & 0 \\ 0 & k_{co} \end{bmatrix}}_{M_{CO}} \begin{bmatrix} \Delta i^c_{Od} \\ \Delta i^c_{Oq} \end{bmatrix} \right)$$

(21)

Correspondingly, the $\mathbf{M}_C$ in (17) of the inverter-side impedance $\mathbf{Z}_{inv}$ is updated as:

$$\mathbf{M}_C = \mathbf{M}_{DEL}\mathbf{M}_{CPI}\mathbf{M}_{VE}(\mathbf{M}_{OPQ}\mathbf{M}_{LPF}\mathbf{M}_{PQI} - \mathbf{M}_{CO}) \tag{22}$$

Fig. 8(a) shows the closed-loop poles of the updated model as the compensation gain K_{co} increases. It can be seen that port current feedforward-based compensation can enhance the damping of oscillation modes in both low and high-frequency ranges, and the damping level improves as K_{vr} increases. However, there is an undesirable side effect that the damping of another oscillation mode at around 40 Hz will decline rapidly as K_{vr} increases. This oscillation mode is defined as synchronous resonance (SR) mode in [11], which can be effectively damped by the P-controller of the current control loop, and thus cannot be found as a dominant mode in Fig. 2 with the original control structure. However, the damping of the SR mode is affected by the compensation as shown in Fig. 7 and it decreases as K_{vr} increases. The experimental waveforms as shown in Fig. 8(b) verify the aforementioned analysis results. When K_{vr} is set as 0.8, the enabling compensation can mitigate both LFO and HFO. However, if K_{vr} increases to 1.1, a new instability issue near the fundamental frequency will be provoked.

D. Comparison and Discussion

In addition to the three aforementioned damping control strategies, there are some other reported methods such as port current feedback-based virtual impedance [14], virtual admittance replacing voltage control loop [15], [19], etc. Similarly, based on the small-signal benchmark model built in this paper, the model can be updated correspondingly according to the adopted damping control methods, and then their impacts on the wideband oscillation modes can be assessed. Due to space constraints, the details are not elaborated. The aforementioned damping control methods are summarized and compared in terms of impacts on wideband modes (whether there are conflicting effects on low/high-frequency oscillation modes), modification costs (required additional control branches), and undesired side effects, as shown in Table II.

In summary, for wideband small-signal stability enhancement, port current feedforward-based compensation is a better solution due to its low modification costs and capability of mitigating both low- and high-frequency oscillations, but the damping coefficient K_{vr} cannot be too large for the sake of a high damping level to avoid exciting synchronous resonance modes. Furthermore, by utilizing the mode identification technique based on sampled signals [20], [21], the oscillations in the system can be detected in real-time. The oscillation frequency and damping information can also be provided for adaptive damping branch triggering and damping gain tuning, thereby avoiding the side effects caused by the fixed control parameters, which is one of the future scopes of this work.

IV. CONCLUSIONS

In this paper, a small-signal closed-loop transfer function model of grid-tied inverters is developed considering the

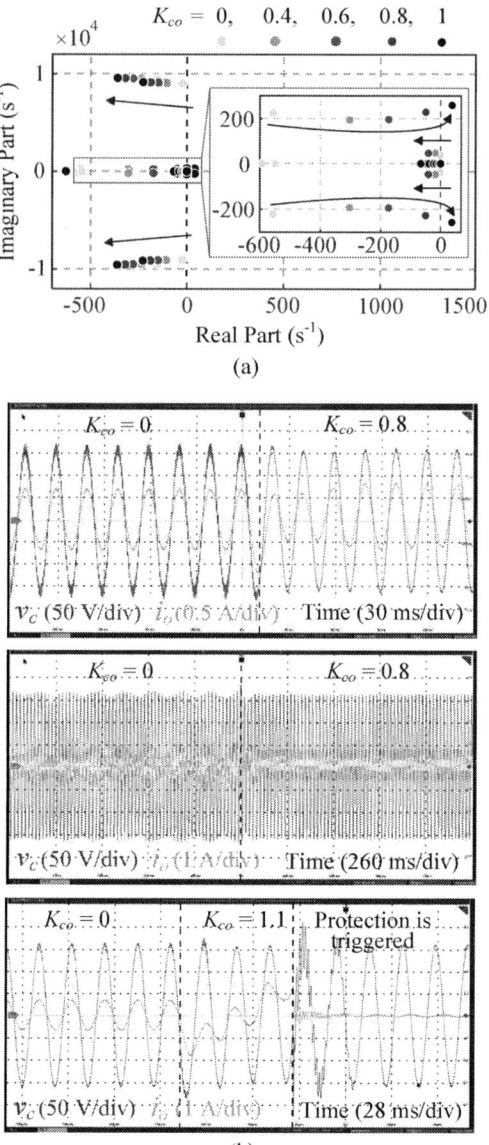

Fig. 8. Impact of increasing compensation gain K_{co} on system stability — (a) Model-based analysis and (b) Experimental verification.

detailed VSG-GFM control structure. By extracting the closed-loop poles, the simultaneous presence of dominant low and high-frequency oscillation modes is revealed. From a wideband stability perspective, candidate damping control methods are tested to explicate their possible conflict impacts on oscillation mitigation regarding low and high-frequency modes. Among them, port current feedforward-based compensation with a reasonable gain coefficient is assessed as a preferable solution for wideband stability improvement. To extend the scope of this article in the future, more damping control methods can be considered for a comprehensive comparison, while data-driven oscillation mode identification algorithms can be applied to achieve optimal design of damping strategies for wideband stability improvement.

TABLE II
COMPARISON OF DAMPING CONTROL METHODS USED IN STABILITY IMPROVEMENT OF GRID-TIED CONVERTER SYSTEMS.

Damping Control Methods	Consistency of Impacts on Wideband Modes	Modification Costs	Potential Side Effects
Control Parameter Tuning [9], [16]	×	No structural modifications	Impact on control loop bandwidth
Output Current Feedback-Based virtual resistor [5], [17]	×	One additional feedback branch	Cutoff frequency setting for HPF
Port Current Feedforward-Based Compensation [12], [18]	✓	One additional feedforward branch	Damping decrease of SR mode
Port Current Feedback-Based Virtual Impedance [13], [14]	×	One additional feedback branch	Amplifying high-frequency noise
Virtual Admittance Replacing Voltage Control Loop [15], [19]	✓	Entire voltage control loop	Steady state error in PCC voltage

✓: Yes, ×: No, HPF: High-Pass Filter, SR: Synchronous Resonance, PCC: Point of Common Coupling

REFERENCES

[1] R. H. Lasseter, Z. Chen, and D. Pattabiraman, "Grid-forming inverters: A critical asset for the power grid," *IEEE J. Emerg. Sel. Top. Power Electron.*, vol. 8, no. 2, pp. 925–935, 2020.

[2] L. Kong, Y. Xue, L. Qiao, and F. Wang, "Review of small-signal converter-driven stability issues in power systems," *IEEE Open Access J. Power Energy*, vol. 9, pp. 29–41, 2022.

[3] Q. Peng, G. Buticchi, N. M. L. Tan, S. Guenter, J. Yang, and P. Wheeler, "Modeling techniques and stability analysis tools for grid-connected converters," *IEEE Open J. Power Electron.*, vol. 3, pp. 450–467, 2022.

[4] D. B. Rathnayake, M. Akrami, C. Phurailatpam, S. P. Me, S. Hadavi, G. Jayasinghe, S. Zabihi, and B. Bahrani, "Grid forming inverter modeling, control, and applications," *IEEE Access*, vol. 9, pp. 114781–114807, 2021.

[5] L. Harnefors, M. Hinkkanen, U. Riaz, F. M. M. Rahman, and L. Zhang, "Robust analytic design of power-synchronization control," *IEEE Trans. Ind Electron.*, vol. 66, no. 8, pp. 5810–5819, 2019.

[6] X. Xiong, X. Li, B. Luo, M. Huang, C. Zhao, and F. Blaabjerg, "An additional damping torque method for low-frequency stability enhancement of virtual synchronous generators," *IEEE Trans. Power Electron.*, vol. 39, no. 12, pp. 15858–15869, 2024.

[7] Y. Liao, X. Wang, and F. Blaabjerg, "Passivity-based analysis and design of linear voltage controllers for voltage-source converters," *IEEE Open J. Ind. Electron. Soc.*, vol. 1, pp. 114–126, 2020.

[8] J. Liu, Y. Xia, W. Wei, Q. Feng, and P. Yang, "Effect of control damping on small-signal stability of grid-forming vscs considering interaction between inner and outer loops," *IEEE Trans. Power Electron.*, vol. 39, no. 6, pp. 7685–7695, 2024.

[9] F. Zhao, T. Zhu, L. Harnefors, B. Fan, H. Wu, Z. Zhou, Y. Sun, and X. Wang, "Closed-form solutions for grid-forming converters: A design-oriented study," *IEEE Open J. Power Electron.*, vol. 5, pp. 186–200, 2024.

[10] Y. Xiao, Z. Zhang, H. Luo, Y. Zhu, Y. Yang, M. Molinas, and D. Xu, "Eur. conf. power electron. appl., epe ecce europe," in *2023 25th European Conference on Power Electronics and Applications (EPE'23 ECCE Europe)*, 2023, pp. 1–9.

[11] F. Zhao, T. Zhu, Z. Li, and X. Wang, "Low-frequency resonances in grid-forming converters: Causes and damping control," *IEEE Trans. Power Electron.*, vol. 39, no. 11, pp. 14430–14447, 2024.

[12] Z. Liu, J. Liu, and Y. Zhao, "A unified control strategy for three-phase inverter in distributed generation," *IEEE Trans. Power Electron.*, vol. 29, no. 3, pp. 1176–1191, 2014.

[13] Y. Tao, Q. Liu, Y. Deng, X. Liu, and X. He, "Analysis and mitigation of inverter output impedance impacts for distributed energy resource interface," *IEEE Trans. Power Electron.*, vol. 30, no. 7, pp. 3563–3576, 2015.

[14] J. He and Y. W. Li, "Analysis, design, and implementation of virtual impedance for power electronics interfaced distributed generation," *IEEE Trans. Ind. Appl.*, vol. 47, no. 6, pp. 2525–2538, 2011.

[15] W. Zhang, A. M. Cantarellas, J. Rocabert, A. Luna, and P. Rodriguez, "Synchronous power controller with flexible droop characteristics for renewable power generation systems," *IEEE Trans. Sustain. Energy*, vol. 7, no. 4, pp. 1572–1582, 2016.

[16] Z. Qu, J. C.-H. Peng, H. Yang, and D. Srinivasan, "Modeling and analysis of inner controls effects on damping and synchronizing torque components in vsg-controlled converter," *IEEE Trans. Energy Convers.*, vol. 36, no. 1, pp. 488–499, 2021.

[17] S. I. Nanou and S. A. Papathanassiou, "Grid code compatibility of vsc-hvdc connected offshore wind turbines employing power synchronization control," *IEEE Trans. Power Syst.*, vol. 31, no. 6, pp. 5042–5050, 2016.

[18] Y. Tao, Y. Deng, G. Li, G. Chen, and X. He, "Evaluation and comparison of the low-frequency oscillation damping methods for the droop-controlled inverters in distributed generation systems," *J. Power Electron.*, vol. 16, no. 2, pp. 731–747, 2016.

[19] J. D. V. Leon, A. Tarraso, J. I. Candela, J. Rocabert, and P. Rodriguez, "Grid-forming controller based on virtual admittance for power converters working in weak grids," *IEEE Journal of Emerging and Selected Topics in Industrial Electronics*, vol. 4, no. 3, pp. 791–801, 2023.

[20] E. Barocio, B. C. Pal, N. F. Thornhill, and A. R. Messina, "A dynamic mode decomposition framework for global power system oscillation analysis," *IEEE Trans. Power Syst.*, vol. 30, no. 6, pp. 2902–2912, 2015.

[21] M. Zhao, H. Yin, Y. Xue, X.-P. Zhang, and Y. Lan, "Coordinated damping control design for power system with multiple virtual synchronous generators based on Prony method," *IEEE Open Access J. Power Energy*, vol. 8, pp. 316–328, 2021.

A Grid-Forming Split-Phase Three-Leg Inverter with Unbalanced Loading and Active Power Decoupling

Namwon Kim, Renata Kimpara, Michael Starke
Grid Systems Architecture
Oak Ridge National Laboratory
Oak Ridge, TN, USA
{kimn1, rezendedacor, starkemr}@ornl.gov

Abstract—**This paper presents a split-phase three-leg inverter that regulates 180-degree phase-shifted AC voltages to support residential house loads under unbalanced loading conditions and decouples double line frequency power pulsation by utilizing the third half-bridge switching leg. The active power decoupling function is added without extra active switching components. The control feasibility and effectiveness are verified through MATLAB Simulink simulation and controller hardware-in-the-loop test.**

Keywords—active power decoupling, split phase inverter, unbalanced loading condition.

I. INTRODUCTION

Integration of distributed energy resources (DERs) such as solar photovoltaic (PV) and battery energy storage to residential applications continues to rise. With the increase in electric vehicles (EV) purchases and inclement weather due to climate change, interest in using an EV battery as an energy backup solution has arisen as well. In order to meet all these demands, research and development of more efficient and advanced residential power distribution systems has become essential. A DC microgrid shown in Fig. 1 is one DER integration solution that has design features that include: a) DC link - many of renewable energy sources produce DC energy, b) fewer converter stages resulting in higher energy conversion efficiency and less system cost, and c) simpler power controls within the local DC bus (no reactive power control required) [1], [2].

In North America and Japan, a residential DC microgrid needs a split-phase DC-AC inverter to support house loads. This split-phase inverter handles both balanced and unbalanced loading conditions while regulating two 180-degree phase-shifted AC line voltages tied to the neutral. In addition, as a split-phase inverter is still a single-phase inverter, the converter inherently adds double line frequency (DLF) power pulsation to the DC power. This DLF power pulsation needs to be decoupled, or the performance of other devices connected to the DC bus will be affected [3]. These functional requirements of a split-phase inverter and their operating waveform are illustrated in Fig. 2.

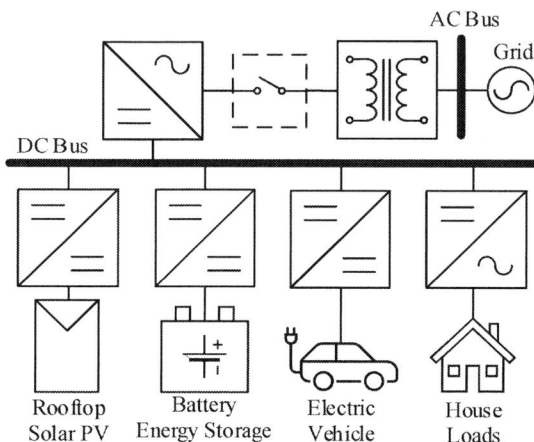

Fig. 1. A DC microgrid integrating DERs and supporting residential electric loads.

A DC-AC inverter supplying unbalanced loads has $N_{phase}+1$ wires to provide a path for the neutral current, where N_{phase} is the number of AC phases. This additional wire can be configured in three ways, a) split DC link capacitors [4], b) the neutral leg [5], and c) split DC link capacitors + the neutral leg [6]. Split phase inverters adopting these three methods are shown in Fig. 3. Among these three methods, the third one is the most promising solution because a) the neutral leg—the third half-bridge (HB) switching leg—can be controlled independently, b) the required capacitance of the split DC link capacitors can be reduced significantly, and c) the inverter input and output are at a fixed voltage potential with respect to the neutral [6].

The principle of the DLF power decoupling is bypassing the DLF component of the instantaneous inverter AC power through an energy storage element. Various DLF power decoupling circuits have been introduced to add an energy storage unit into a single-phase inverter circuit and control the DLF pulsation power/current effectively [7], [8]. Fig. 4 shows three different power decoupling circuits that can be connected in parallel to the DC input of a split-phase inverter: a) a passive decoupling

---

This manuscript has been authored by UT-Battelle, LLC, under contract DE-AC05-00OR22725 with the US Department of Energy (DOE). The US government retains and the publisher, by accepting the work for publication, acknowledges that the US government retains a non-exclusive, paid-up, irrevocable, world-wide license to publish or reproduce the submitted manuscript version of this work, or allow others to do so, for US government purposes. DOE will provide public access to these results of federally sponsored research in accordance with the DOE Public Access Plan (https://energy.gov/doe-public-access-plan).

capacitor bank, b) an active power decoupling (APD) buck converter [9], and c) an APD series-stacked buffer architecture [10]. Among these, the buck-type APD circuit can be a preferrable method because it can provide a higher DLF voltage swing that leads to a smaller capacitance of the decoupling capacitor while offering a simpler control and a more intuitive power processing than other power decoupling methods.

In literature, a split-phase three-leg inverter compensating for the neutral current caused by unbalanced loading conditions has been presented, but no APD function is addressed [11], [12]. On the other hand, a split-phase three-leg PV inverter with a bidirectional buck-boost-type APD circuit has been proposed, but no unbalanced loading condition is tested, and the DC bus voltage is not fixed, which is not suitable for DC microgrid applications [13].

This paper presents a split-phase three-leg inverter that can handle unbalanced loading conditions and provide APD function for the DLF power pulsation without additional active switching components. The neutral leg added to provide a path for the fundamental line frequency (FLF) neutral current can also be used to provide a path for the DLF pulsation current to a decoupling capacitor. The decoupling capacitor is located in between the neutral and the DC negative by adopting the buck-type APD circuit. This decoupling capacitor not only absorbs and discharges the DLF pulsation current but also builds a fixed or low-frequency voltage potential between the neutral and the DC negative, so the inverter input and output are not floating each other. Instead of a top split DC link capacitor, a DC link capacitor is added to decouple switching frequency current ripple. Thus, minor design modification is required, and this capacitor sizing and comparison with other APD implementation methods are discussed in this paper. Although the required capacitance of the decoupling capacitor of the proposed inverter is four times higher than that of a separate buck-type APD converter due to the DLF voltage swing restriction, the proposed inverter is more cost-effective and power-dense because neither additional HB leg nor its operating circuits are required. Besides, the capacitance value is still 20 times lower than the one for the passive method. This paper also presents the closed-loop control algorithm of the proposed split-phase inverter. Proportional-integral-resonant (PIR) controllers are employed to compensate for sinusoidal voltage and current control errors at FLF and DLF effectively and independently. The operation principle and control responses of the proposed inverter under step load changing conditions are tested and validated through MATLAB Simulink simulation and controller hardware-in-the-loop (CHIL) results.

II. PROPOSED INVERTER TOPOLOGY AND CONTROL SCHEME

A. Split-Phase Three-Leg Inverter

The proposed split-phase three-leg inverter topology is illustrated in Fig. 5. This inverter is a combination of a split-phase three-leg inverter shown in Fig. 3(c) and an APD buck converter shown in Fig. 4(b). Due to the structural similarity between the neutral leg of the split-phase inverter and the buck converter, only minor modifications to the DC link capacitors are required. Instead of the top-side split DC link capacitor, C_{s1}

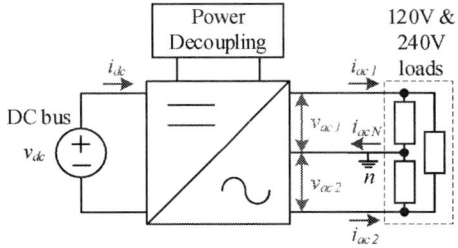

(a)   Split-phase inverter

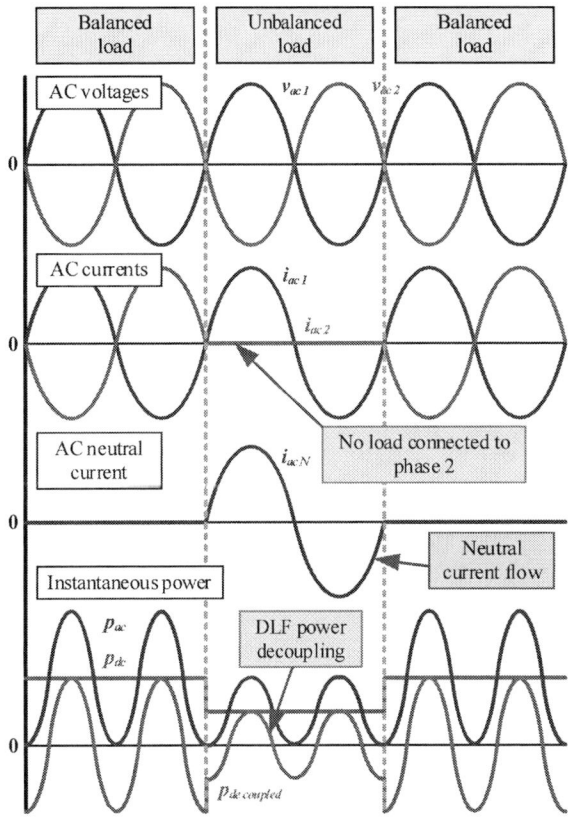

(b)   Instantaneous voltage/current/power plot

Fig. 2. Split-phase inverter function requirements.

in Fig. 3(c), a DC link capacitor, C_{dc} in Fig. 4(b), is connected to the DC bus, as shown in Fig. 5. Remaining components and connections are not changed from the split-phase inverter. Therefore, almost seamless integration of an APD circuit into a split-phase three-leg inverter is achieved. Neither additional semiconductor switching devices, gate driving circuits, nor voltage or current sensors are required. The inverter design, especially the decoupling capacitor, C_d, and the DC link capacitor, is different from the conventional split-phase three-leg inverter, and it is discussed in the next section. In this section, the topological functions of each component are explained.

The inverter consists of three HB legs. Two HB switching legs, HB1 and HB2, are connected to individual split phases. The third leg, HBn, is connected to the neutral to provide the neutral current path. Two LCL filters are employed to minimize the switching current ripple passing through the AC loads, $R_{load,1}$

979-8-3315-1612-3/25 $31.00 © 2025 IEEE   704

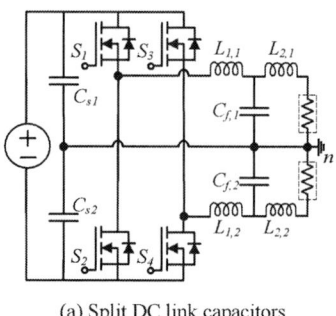

(a) Split DC link capacitors

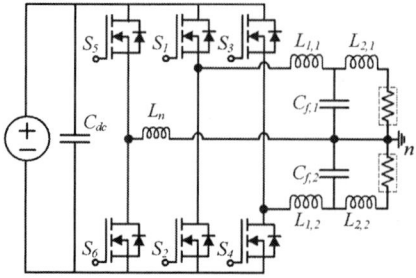

(b) The neutral leg

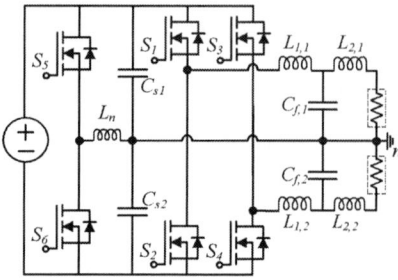

(c) Split DC link capacitors + the neutral leg

Fig. 3. Split-phase three-wire inverters.

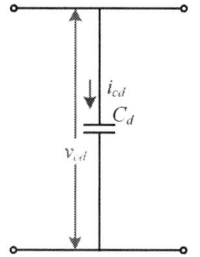

(a) Passive power decoupling – capacitor bank

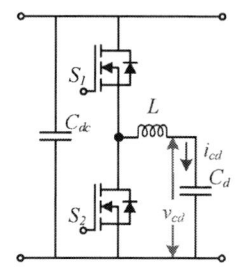

(b) Active power decoupling – buck

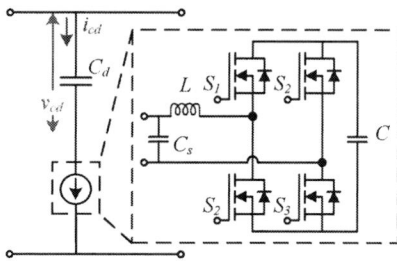

(c) Active power decoupling – series-stacked

Fig. 4. Power decoupling circuits.

and $R_{load,2}$, and regulate line-to-neutral phase voltages, $v_{cf,1}$ and $v_{cf,2}$, across the AC filter capacitors, $C_{f,1}$ and $C_{f,2}$. A filter inductor, L_n, is connected to the midpoint of the neutral leg to reduce the switching current ripple and control the neutral inductor current, i_{Ln}. The decoupling capacitor is used to build $\pm 0.5 V_{dc,bus}$ voltage potential between the neutral and the DC positive/negative. The capacitor also works as a main energy storage element for the DLF pulsation current. The DC link capacitor, C_{dc}, holds the DC bus voltage and also works as a switching current ripple decoupling capacitor.

B. Phase Voltage Control

The control algorithm is developed to provide three primary functions: a) AC voltage regulation, b) unbalanced load supply, and c) APD. The complete control diagram is shown in Fig. 6.

The two phase legs and their modulation indexes, m_1 and m_2, are configured to control sinusoidal split-phase 180-degree phase-shifted AC voltages under both balanced and unbalanced loading conditions. In order to control each phase leg's switching independently under unbalanced loading conditions, the phase voltage control is implemented per phase with separate AC voltage control references, $v_{cf,1,ref}$ and $v_{cf,2,ref}$, based on

$$y = V_{mag} \sin(2\pi f t + \theta \frac{\pi}{180°}) \qquad (1)$$

where y is the instantaneous sinusoidal voltage reference output, V_{mag} is the voltage magnitude, f is the voltage frequency, and θ is the degree of the phase-shift angle. Two PIR controllers are employed as the outer-loop AC capacitor voltage ($v_{cf,x}$) controller to compensate for the sinusoidal FLF control errors effectively. The proportional-resonant (PR) term is designed to respond FLF, ω_0, (PRω_0) as

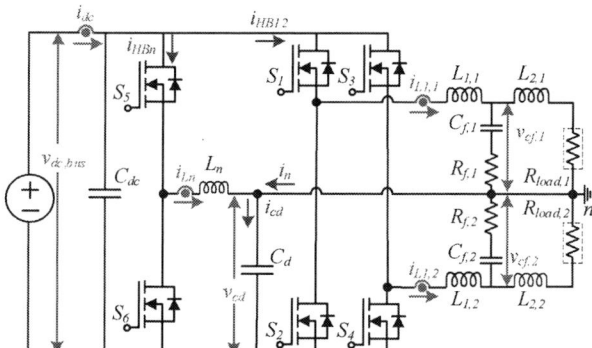

Fig. 5. The proposed split-phase three-leg inverter topology with the decoupling capacitor, C_d.

$$G_{PR\omega 0}(s) = K_p + \frac{2K_r \omega_c s}{s^2 + 2\omega_c s + \omega_0^2} \qquad (2)$$

where $G_{PR\omega 0}(s)$ is the transfer function of the PRω_0 term, K_p is the proportional gain, K_r is the resonant gain, and ω_c is the angular frequency representation of the resonant control bandwidth [14]. We selected $\omega_c = 2\pi \cdot (10 \text{ Hz})$ rad/s for our controller design. In addition, the proportional-integral (PI) term is designed to compensate for DC offsets caused by system's parasitic components or transients, as

$$G_{PI}(s) = K_p + \frac{K_i}{s} \qquad (3)$$

where $G_{PI}(s)$ is the transfer function of the PI term, and K_i is the integral gain. The same K_p is shared with the PRω_0 and PI terms, which makes the controller transfer function as

979-8-3315-1612-3/25 $31.00 © 2025 IEEE

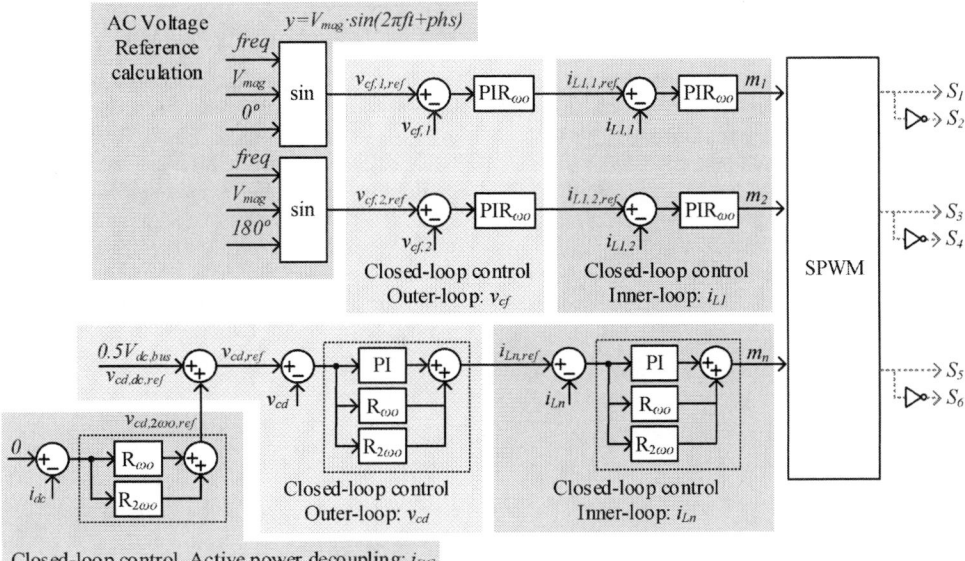

Fig. 6. The proposed inverter control with split-phase AC voltage regulation, DC voltage balancing for unbalanced loading conditions, and active power decoupling of DLF power pulsation.

$$G_{PIR\omega0}(s) = K_p + \frac{K_i}{s} + \frac{2K_r\omega_c s}{s^2 + 2\omega_c s + \omega_0^2}. \quad (4)$$

The voltage controllers derive the inner-loop AC inductor current control references, $i_{L1,1,ref}$ and $i_{L1,2,ref}$, automatically according to the corresponding loading condition. The same PIR controllers are used for the AC inductor current ($i_{L1,x}$) control. Two PIR controllers are used to control each phase current independently under unbalanced loading conditions. The modulation indexes are updated by the inner-loop current controllers with DC and FLF changes.

C. Neutral Leg Control

The neutral leg and corresponding modulation index, m_n, are configured to control the decoupling capacitor voltage, v_{cd}. Controlling the decoupling capacitor voltage in appropriate form and magnitude allows the neutral leg to pass the neutral current, i_n, caused by unbalanced loading conditions and the DLF pulsation current for the APD function.

Keeping the decoupling capacitor voltage to be constant makes the decoupling capacitor current zero. Under this condition, we can derive the neutral inductor current control reference, $i_{Ln,ref}$, as

$$i_{Ln,ref} = -i_n \quad (5)$$

according to the Kirchhoff's current law at the neutral node. To derive the proper magnitude and phase angle of the current control reference in FLF automatically, a PIR controller centered at FLF, ω_0, (PIRω_0) is used as the outer-loop decoupling capacitor voltage controller. The DC voltage control reference, $v_{cd,dc,ref}$, is set to the half of the DC bus voltage, $v_{dc,bus}$, to not only keep the decoupling capacitor voltage to be constant

but also guarantee equal distribution of the DC voltages for the AC voltage regulation of both phases. The integral term compensates for DC offsets and transients. To control the neutral inductor current, i_{Ln}, as same as a sinusoidal FLF neutral current, the same PIRω_0 controller is implemented as the inner-loop controller.

Swinging the decoupling capacitor voltage in the form of a sinusoidal DLF waveform with a constant DC offset makes the pulsation current or power flow into an energy storage unit or the decoupling capacitor. To achieve this, a DLF voltage control reference, $v_{cd,2\omega0,ref}$, is added to the DC voltage control reference, $v_{cd,dc,ref}$. Derivation of the proper magnitude and phase angle of the DLF voltage control reference is done based on the resonant controller. The control objective is making the DLF component in the DC link current zero. By comparing the zero current reference with the DC link current measurement, i_{dc}, the resonant controller centered at DLF, $2\omega_0$, (R$_{2\omega0}$) can derive the DLF voltage control reference automatically. The feedback input, i_{dc}, is effectively filtered at DLF because of its R$_{2\omega0}$ term. In this R$_{2\omega0}$ controller, an additional resonant term centered at FLF (R$_{\omega0}$) is added to improve the APD performance. Besides, the R$_{2\omega0}$ terms are added to both the outer-loop voltage controller and the inner-loop current controller as well to compensate for the corresponding DLF components effectively. The modulation index is updated by the inner-loop current controller with all DC, FLF, and DLF changes.

III. INVERTER DESIGN AND COMPARISON

A. Inverter Design

Sizing the decoupling capacitor, C_d, is critical to the control performance of the proposed inverter. The capacitance has to be big enough to absorb and discharge the DLF neutral inductor current with the DLF voltage swing. The equation for the required capacitance is expressed as

$$C_d = \frac{2 \cdot I_{pk}}{\omega \cdot dV} \qquad (6)$$

where I_{pk} is the peak of the DLF neutral inductor current, ω is the angular frequency of the DLF, and dV is the peak-to-peak value of the DLF voltage swing [9]. The magnitude of the DLF voltage swing is restricted so that the instantaneous DC voltages between the DC positive and the neutral, $v_{dc,bus} - v_{cd}$, and the DC negative and the neutral, v_{cd}, are always bigger than the minimum requirement, which equals to the absolute values of the AC voltages, $|v_{cf,1}|$ and $|v_{cf,2}|$. This restriction is expressed as

$$v_{cd} = 0.5 v_{dc,bus} + \frac{dV}{2} \sin \left(2\omega_0 t + \theta \frac{\pi}{180°} \right) \qquad (7)$$

$$\left(v_{dc,bus} - v_{cd} \right) \geq \left| v_{cf,1} \right| \qquad (8)$$

$$v_{cd} \geq \left| v_{cf,2} \right|. \qquad (9)$$

These conditions are corresponding to the operating condition where the modulation indexes of the phase legs do not reach to 1.0 or -1.0.

Another design change is the DC link capacitor, C_{dc}. The DC link capacitance is determined to decouple all switching current ripples generated by the three HB legs. The capacitance is calculated by considering 0.5 duty cycle ratio, the peak of the DC link current (the pulsation current) as the average capacitor current per cycle, and 1% of the DC bus voltage as the change of the capacitor voltage during the duty cycle ratio period. The equation is

$$C_{dc} = \frac{S_{1\phi}}{v_{dc,bus}^2 A_v f_{sw}} \qquad (10)$$

TABLE I. INVERTER CIRCUIT PARAMTERS USED IN DESIGN, SIMULATION, AND CHIL

Description	Symbol	Value
Rated power	P_{rated}	30 kW
Switching frequency	f_{sw}	20 kHz
Line frequency	f_0	60 Hz
AC voltage (LN/LL)	V_{ac}	120/240 V_{RMS}
DC bus voltage	$V_{dc,bus}$	400 V
AC load resistor	R_{load}	0.96 Ω
Inverter-side AC filter inductor	L_1	141 uH
AC filter capacitor	C_f	138 uF
AC damping resistor	R_f	57 mΩ
Grid-side AC filter inductor	L_2	4 uH
DC link capacitor	C_{dc}	469 uF
Neutral leg filter inductor	L_N	141 uH
DLF power decoupling capacitor	C_d	4.4 mF

where $S_{1\phi}$ is the apparent power of a single phase, A_v is the capacitor voltage ripple ratio, and f_{sw} is the switching frequency.

Using the parameters listed in Table I, the passive components are designed. The values of the passive components are also listed in Table I. The LCL filter is designed based on the basic rules of thumb, which are a) the peak-to-peak value of the switching-frequency current ripple of the inverter-side inductor equals to 20% of the AC current peak, b) 0.05 p.u. reactive power circulation through the filter capacitor, and c) the peak-to-peak value of the switching-frequency current ripple of the grid-side inductor equals to 2% of the AC current peak. The neutral leg filter inductance is the same as the inverter-side filter inductance. The calculated capacitance of the decoupling capacitor is 4.4 mF, and the DC link capacitance is 469 uF.

B. Comparison

The decoupling capacitors required for different decoupling circuits are calculated for comparison. Table II lists different values of the design parameters and the results. The decoupling circuits considered for the comparison are a capacitor bank shown in Fig. 4(a) and a separate buck converter shown in Fig. 4(b).

Both the separate buck and the proposed method use the same peak current, 150 A. However, they consider different voltage swing, 360 V and 90 V, respectively, because the proposed method has the voltage swing restriction as mentioned in the previous subsection, whereas the separate buck does not. Therefore, the capacitance of the decoupling capacitor for the separate buck is four-time smaller than the one for the proposed method. However, the separate buck method requires an additional HB leg with an LC filter, which is rated at 150 Apk, along with its operating circuits including gate drivers, voltage and current sensors, analog signal conditioning circuits, digital signal ports, etc. The passive method, in contrast, requires a huge capacitance, 88 mF. This is because the voltage swing at the DC bus is strictly limited. To achieve the same performance that the proposed method provides, 1.5 V voltage swing with 50 A peak current is considered.

According to the comparison, it is found that the proposed method can reduce the system cost and improve power density while providing the APD function in a split-phase inverter.

TABLE II. COMPARISON OF THE DECOUPLING CAPACITORS USED IN DIFFERENT DECOUPLING CIRCUITS

Parameter	Passive (Fig. 4(a))	Separate buck (Fig. 4(b))	Proposed (Fig. 5)
I_{pk}	50 A	150 A	150 A
ω	120 Hz	120 Hz	120 Hz
dV	1.5 V	360 V	90 V
C_d	88 mF	1.1 mF	4.4 mF
Other requirements	-	A HB leg (150 Apk rated) and its operating circuits	-

979-8-3315-1612-3/25 $31.00 © 2025 IEEE

IV. SIMULATION AND CHIL RESULTS

A. Simulation Results

The proposed inverter and control algorithm were modeled and simulated in MATLAB Simulink.

Fig. 7 shows the inverter control response under different operation modes; a) at 0.05 seconds, the AC voltage regulation stared without loads and with the APD control disabled, b) at 0.1 seconds, $R_{load,1}$ and $R_{load,2}$ were connected (balanced loading condition), c) at 0.2 seconds, $R_{load,2}$ was disconnected (unbalanced loading condition), d) at 0.3 seconds, the APD control was enabled, and e) at 0.4 seconds, $R_{load,2}$ was reconnected (balanced loading condition).

As shown in Mode b, $\pm I_{dc}$ of DLF pulsating current was observed in i_{dc}. In Mode c, the neutral current, i_n, caused by the unbalanced loads was compensated by i_{Ln}. The neutral leg maintained v_{cd} as $0.5V_{dc,bus}$ to achieve this compensation. In Mode d and e, the APD control was activated. Therefore, the neutral leg controlled the DLF component of i_{Ln} as well. The DLF current magnitude was determined by v_{cd} with corresponding DLF voltage oscillations, which was determined by the capacitance of C_d and the resonant controller trying to make the DLF component in i_{dc} zero.

Fig. 8 shows the simulation results that can represent the operating margin of the inverter. The operation mode was Mode e. The DC split voltages, $v_{dc,bus}$ - v_{cd} and v_{cd}, and the absolute values of the AC phase voltages, $|v_{cf,1}|$ and $|v_{cf,2}|$, are plotted. By comparing the DC split voltages and the AC phase voltages (the blue line vs. the red line and the yellow line vs. the purple line), it is observed that the designed decoupling capacitor made the decoupling capacitor voltage and the AC phase 2 voltage touch each other around 0.42 seconds. This corresponds to the minimum modulation index, -1.0, for the phase leg 2, m_2. Therefore, the DLF voltage swing is restricted to this level, $45 \cdot \sin(2\omega_0 t)$ V. This also means that there is no inverter operating

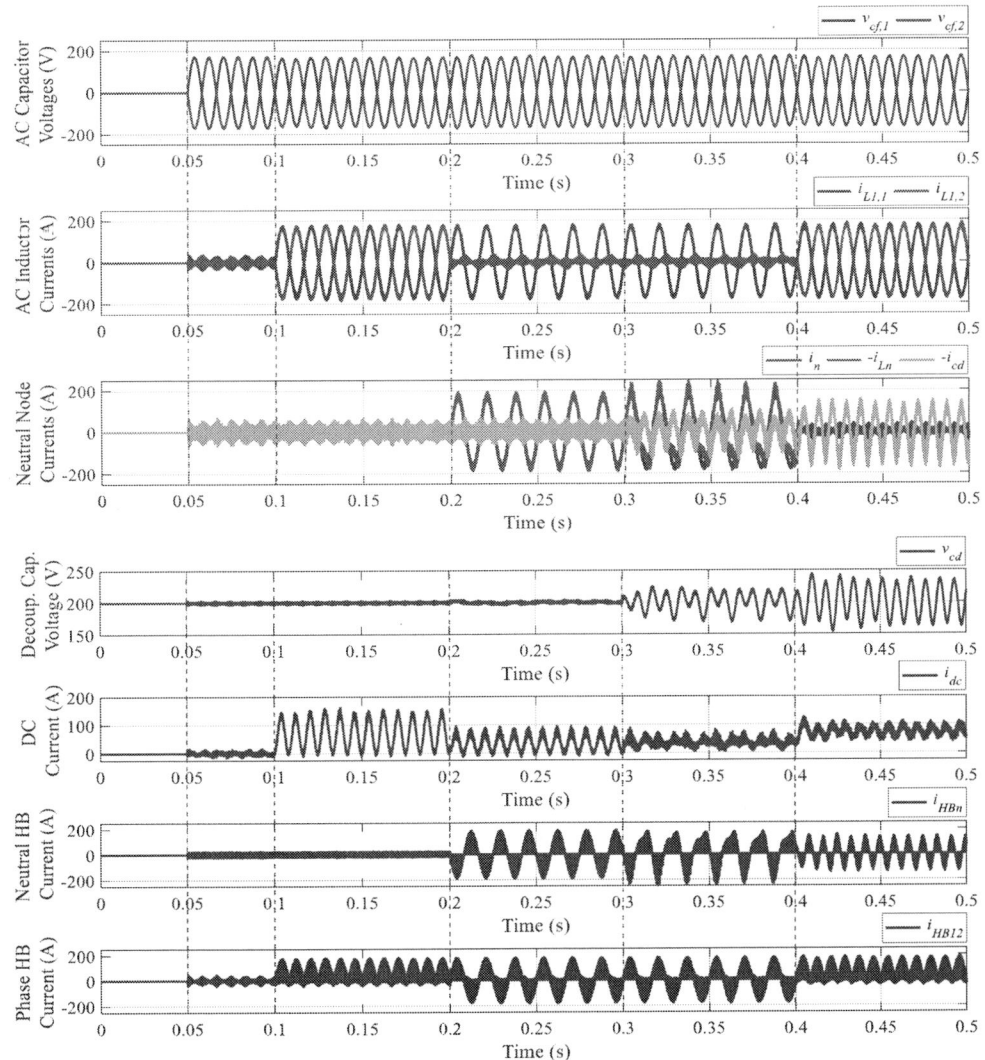

Fig. 7. MATLAB Simulink simulation results, inverter control response under different operation modes: at 0.05 seconds, start AC voltage regulation with APD control disabled, at 0.1 seconds, connect $R_{load 1}$ and $R_{load 2}$, at 0.2 seconds, disconnect $R_{load 2}$, at 0.3 seconds, enable APD control, at 0.4 seconds, reconnect $R_{load 2}$.

margin at this 400 V DC bus voltage operating condition. To guarantee operating margin for compensation for transients, 450 V DC bus voltage can be considered as the nominal operating condition.

B. CHIL Results

The proposed inverter and the control algorithm were tested in a Typhoon HIL CHIL testbed shown in Fig. 9. The electrical circuit models were developed in a HIL604 real-time simulator. The control functions were implemented in a TI DSP F28937D LaunchPad. Fig. 10 shows the CHIL results under balanced/unbalanced step load conditions and with/without the

APD control enabled. The results were directly screen-captured from the Capture widget in the Typhoon HIL Control Center software.

Unlike the simulation results, the DC average voltage control reference of the decoupling capacitor, $v_{cd,dc,ref}$, had to be increased from 200 V to 215 V in order to regulate sinusoidal AC voltages and currents without unwanted oscillations. At 200 V, high-frequency oscillations were observed on the AC voltage and current waveforms because the modulation indexes reached to their maximum and minimum values, 1.0 and -1.0, respectively. A more practical control implementation in the CHIL environment—including not only digital control

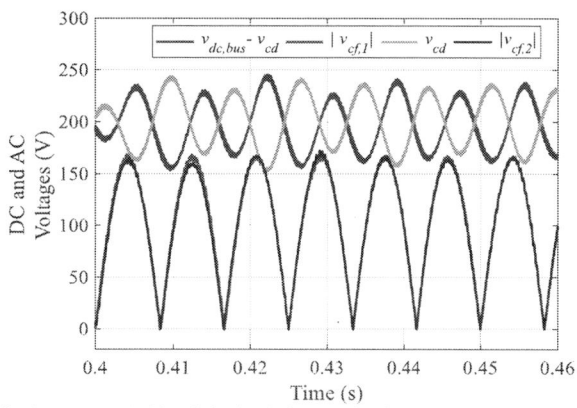

Fig. 8. MATLAB Simulink simulation results, inverter operating margin: the decoupling capacitor voltage vs. the inverter output voltages.

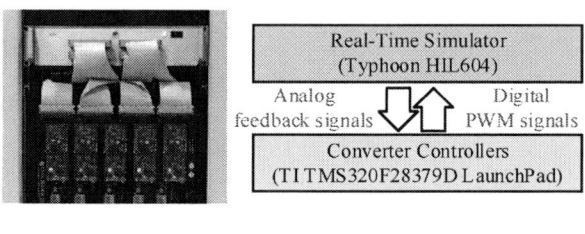

Fig. 9. CHIL Typhoon HIL604 Setup

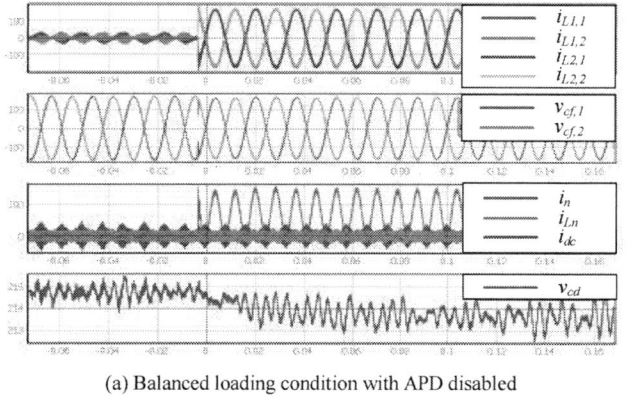

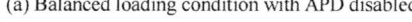

(a) Balanced loading condition with APD disabled

(b) Balanced loading condition with APD enabled

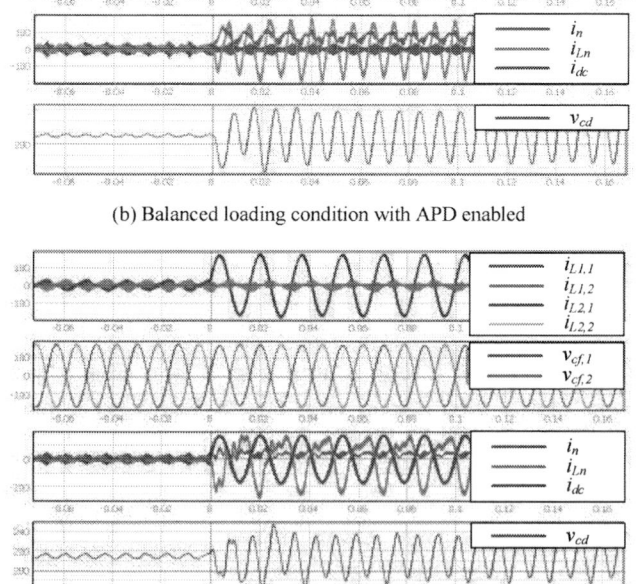

(c) Unbalanced loading condition with APD disabled

(d) Unbalanced loading condition with APD enabled

Fig. 10. CHIL results, inverter control response with a step load change under different operation modes.

processing with analog-to-digital conversion, interrupt service routine, and sequential computation order in the DSP, but also external device and signal delays such as real-time simulator delays and analog and digital signal delays—made the controller in the DSP use more modulation indexes than the controller in the simulation under the same operating condition. As shown in Fig. 8, the phase leg 1, m_1, has more operating margin than the phase leg 2, m_2; $(v_{dc,bus} - v_{cd})$ was always higher than $|v_{cf,1}|$. Therefore, pushing the decoupling voltage up helped to solve this issue. We observed that the available DC offset voltage range is 0 V $< V_{cd,offset} <$ 30 V, and we selected the middle point, 15 V.

After troubleshooting this, the CHIL results showed very similar results with the simulation results.

- The split-phase voltages were well regulated at 120 Vrms and 180-degree phase-shifted.

- The neutral leg compensated for the FLF neutral current under the unbalanced loading condition (Fig. 10(c),(d)).

- The neutral leg processed the DLF pulsation current effectively by regulating v_{cd} as *215 + (45 or 25) · sin(2ω$_0$t)* V (Fig. 10(b),(d)).

V. CONCLUSION

This paper presents a split-phase three-leg inverter that is capable of handling unbalanced loading conditions and providing APD feature without adding extra active switching components. The existing neutral leg controls both the FLF and DLF components to pass the neutral current to the DC bus and the pulsation current to the decoupling capacitor. The size of the decoupling capacitor is designed by considering available DLF voltage swing, which does not interfere with the AC voltage regulation function. The circuit design comparison showed that the proposed method is more cost-effective and power-dense because no significant circuit modification is required. The developed control scheme can automatically derive the required control references in both the FLF and DLF. The simulation and CHIL results validated its effectiveness and feasibility.

REFERENCES

[1] T. Dragičević, X. Lu, J. C. Vasquez and J. M. Guerrero, "DC Microgrids—Part I: A Review of Control Strategies and Stabilization Techniques," in *IEEE Transactions on Power Electronics*, vol. 31, no. 7, pp. 4876-4891, July 2016, doi: 10.1109/TPEL.2015.2478859.

[2] T. Dragičević, X. Lu, J. C. Vasquez and J. M. Guerrero, "DC Microgrids—Part II: A Review of Power Architectures, Applications, and Standardization Issues," in *IEEE Transactions on Power Electronics*, vol. 31, no. 5, pp. 3528-3549, May 2016, doi: 10.1109/TPEL.2015.2464277.

[3] Y. Tang, F. Blaabjerg, P. C. Loh, C. Jin and P. Wang, "Decoupling of Fluctuating Power in Single-Phase Systems Through a Symmetrical Half-Bridge Circuit," in *IEEE Transactions on Power Electronics*, vol. 30, no. 4, pp. 1855-1865, April 2015, doi: 10.1109/TPEL.2014.2327134.

[4] P. Verdelho and G. D. Marques, "Four- wire current-regulated PWM voltage converter," in *IEEE Transactions on Industrial Electronics*, vol. 45, no. 5, pp. 761-770, Oct. 1998, doi: 10.1109/41.720333.

[5] R. Zhang, V. H. Prasad, D. Boroyevich and F. C. Lee, "Three-dimensional space vector modulation for four-leg voltage-source converters," in *IEEE Transactions on Power Electronics*, vol. 17, no. 3, pp. 314-326, May 2002, doi: 10.1109/TPEL.2002.1004239.

[6] Q. . -C. Zhong, J. Liang, G. Weiss, C. M. Feng and T. C. Green, "HinftyControl of the Neutral Point in Four-Wire Three-Phase DC–AC Converters," in *IEEE Transactions on Industrial Electronics*, vol. 53, no. 5, pp. 1594-1602, Oct. 2006, doi: 10.1109/TIE.2006.882014.

[7] H. Hu, S. Harb, N. Kutkut, I. Batarseh and Z. J. Shen, "A Review of Power Decoupling Techniques for Microinverters With Three Different Decoupling Capacitor Locations in PV Systems," in *IEEE Transactions on Power Electronics*, vol. 28, no. 6, pp. 2711-2726, June 2013, doi: 10.1109/TPEL.2012.2221482.

[8] Y. Tang and F. Blaabjerg, "Power decoupling techniques for single-phase power electronics systems — An overview," *2015 IEEE Energy Conversion Congress and Exposition (ECCE)*, Montreal, QC, Canada, 2015, pp. 2541-2548, doi: 10.1109/ECCE.2015.7310017.

[9] R. Wang *et al.*, "A High Power Density Single-Phase PWM Rectifier With Active Ripple Energy Storage," in *IEEE Transactions on Power Electronics*, vol. 26, no. 5, pp. 1430-1443, May 2011, doi: 10.1109/TPEL.2010.2090670.

[10] S. Qin, Y. Lei, C. Barth, W. -C. Liu and R. C. N. Pilawa-Podgurski, "A High Power Density Series-Stacked Energy Buffer for Power Pulsation Decoupling in Single-Phase Converters," in *IEEE Transactions on Power Electronics*, vol. 32, no. 6, pp. 4905-4924, June 2017, doi: 10.1109/TPEL.2016.2601309.

[11] T. Tanaka, T. Sekiya, Y. Baba, M. Okamoto and E. Hiraki, "A new half-bridge based inverter with the reduced-capacity DC capacitors for DC micro-grid," *2010 IEEE Energy Conversion Congress and Exposition*, Atlanta, GA, USA, 2010, pp. 2564-2569, doi: 10.1109/ECCE.2010.5617983.

[12] Y. Baba, M. Okamoto, E. Hiraki and T. Tanaka, "A half-bridge inverter based current balancer with the reduced dc capacitors in single-phase three-wire distribution feeders," *2011 IEEE Energy Conversion Congress and Exposition*, Phoenix, AZ, USA, 2011, pp. 4233-4239, doi: 10.1109/ECCE.2011.6064347.

[13] L. C. Breazeale and R. Ayyanar, "A Photovoltaic Array Transformer-Less Inverter With Film Capacitors and Silicon Carbide Transistors," in *IEEE Transactions on Power Electronics*, vol. 30, no. 3, pp. 1297-1305, March 2015, doi: 10.1109/TPEL.2014.2321760.

[14] R. Teodorescu, F. Blaabjerg, M. Liserre, and P. C. Loh, "Proportional-resonant controllers and filters for grid-connected voltage-source converters," *IEE Proc.-Electr. Power Appl.*, vol. 153, no. 5, pp. 750–762, Sep. 2006.

Completely Decentralized Active and Reactive Power Control of Grid-connected Cascaded H-Bridge Inverters with Integrated Battery Storage

Soham Dutta and Brian Johnson

Department of Electrical and Computer Engineering, University of Texas at Austin, TX 78751, USA
Corresponding author email: *sdutta@utexas.edu*

Abstract—**AC-DC converters with cascaded units are gaining popularity for medium voltage level grid connection however, as the number of cascaded stages increase, their widespread application are limited by existing centralized and distributed control schemes. In this article, a completely decentralized control scheme has been proposed for cascaded-type ac-dc converters with integrated energy storage. The proposed control method is capable of locally controlling both the active and reactive power processed by an individual unit and thereby achieve State-of-Charge (SOC) balancing of the energy storage units integrated with each converter. In this framework, the reactive power control loops achieve voltage synchronization among the converters and the grid, and the active power control loops modulate the terminal voltage amplitudes that influence power delivery. The proposed control is inspired from conventional droop control used for parallel connected converters, however, it has been modified by incorporating state feedback of voltage angle in order to facilitate stable bidirectional power flow in series connected systems. The stability analysis and simulation results have been presented along with relevant experimental results.**

I. Introduction

Medium Voltage (MV)-level grid connected DC-AC converter systems are recently gaining popularity for various applications, example- Ultra-Fast Charging Stations for Electric Vehicles. Among existing power conversion systems used for medium voltage grid interfacing, series stacked DC-AC power converters are an excellent solution as they offer the advantage of using low-voltage semiconductor switches for the converters and provide high voltage gain by stacking the dc-ac converters on the MVAC-side. Another advantage is that these units can also have integrated local energy storage to store energy and supplement power from the grid during peak loads. Such a system is shown in Fig.1 where each module in the stack comprises of a local Battery Energy Storage System (BESS) which is connected to the ac side grid through an isolated dc-dc active bridge converter followed by a H-bridge inverter.

Funding provided by the U.S. Department of Energy Office of Energy Efficiency and Renewable Energy Solar Energy Technologies Office grant number DE-EE0008346. This work was authored in part by the National Renewable Energy Laboratory, operated by Alliance for Sustainable Energy, LLC, for the U.S. DOE under Contract No. DE-AC36-08GO28308. The views expressed in the article do not necessarily represent the views of the DOE or the U.S. Government. The U.S. Government retains and the publisher, by accepting the article for publication, acknowledges that the U.S. Government retains a nonexclusive, paid-up, irrevocable, worldwide license to publish or reproduce the published form of this work, or allow others to do so, for U.S. Government purposes.

However, as the number of units in the stack goes up, control becomes challenging as a central controller in incapable of managing so many modules due to computational, PWM and sensing resource limitation and reliability issues [1]–[3]. To overcome this, distributed controllers with a common communication bus have been used [4]–[11]. In most cases, information such as, grid voltage amplitude and phase, PWM phase shifts are broadcasted to the units by a master controller over the communication bus or else synchronization signals are routed over the entire stack. These not only require high speed communication channels, but also increase wiring complexity, cost, chances of communication failure and make high module counts impractical. These issues are collectively overcome by decentralized solutions where each module in the stack has its own autonomous controller.

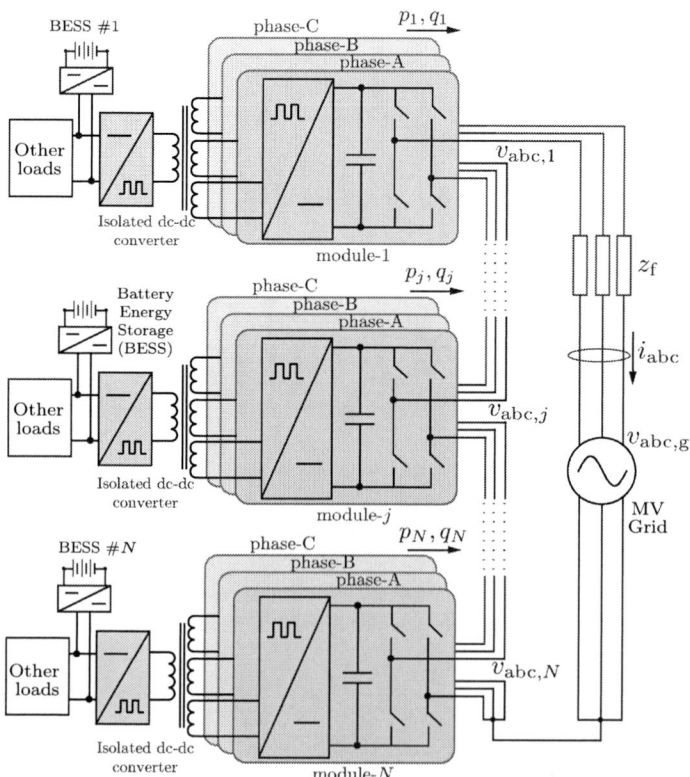

Fig. 1. System of cascaded dc-ac modules with local energy storage connected across a MV grid.

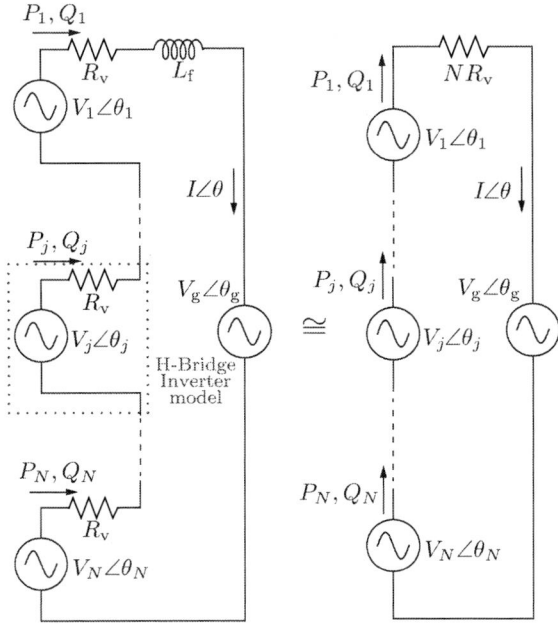

Fig. 2. Simplified system model for modeling and analysis.

To date, state-of-the-art decentralized control methods have following shortcomings. Decentralized methods in [12]–[15] are applicable to islanded systems only. Grid connected systems are considered in [16]–[18] where each inverter uses an active power versus frequency droop law, but reactive power control is unaddressed and stability only holds for unidirectional power flow. The methods proposed in [19], [20] requires one inverter in the stack to be controlled in current control mode and the others are voltage controlled. In this setup, the current-controlled inverter needs to be of higher transient power rating than the other inverters and [19] still require grid voltage zero-crossing information to be broadcast. In [21], resistance emulation was introduced along with integral controllers for both active and reactive power loops. However, the active power controller exhibits instability under reverse power transfer from grid.

To overcome these shortcomings, we propose a controller that gives decentralized control of both active and reactive power processed by each module as well as communication-free voltage synchronization among the inverters and grid. The proposed method is inspired from conventional droop control for parallel connected converters however, a) here active power versus voltage ($P - v$) and reactive power versus frequency ($Q - \omega$) droop has been used opposed to conventional parallel connected inverters. This is supported by altering the network type from an originally inductive network to a resistive network through virtual resistance emulation, and b) the $Q - \omega$ loop is modified by introducing a state feedback of the frequency deviation which changes the system closed loop pole locations to ensure stability during bi-directional active power flow. With the proposed control strategy, each unit in the cascaded system is controlled independently only by using its local information, and the active power output of each unit

is regulated according to, 1) SOC level of the local BESS and 2) required power set-points for other loads. In this proposed way, the SOC levels of the BESS units balance automatically without disturbing the net power delivered to/absorbed from the grid. The system is independent of the number of modules in the stack, which enables flexible plug-and-play operation.

II. POWER FLOW ANALYSIS & PROPOSED DECENTRALIZED CONTROL

For the purpose of modelling, the Multi-Active Bridge based isolated DC-DC stage is not considered as it is controlled to work as a DC transformer (DCX) [22]. From the output terminals, each H-bridge inverter is modeled as a controllable voltage source behind an emulated resistance R_v. As a result, the system under consideration as shown in Fig. 2 is composed of N series-connected H-bridge inverters which collectively deliver current, $I\angle\theta$, into the grid voltage, $V_g\angle\theta_g$, via a network impedance, $Z_f\angle\theta_f = NR_v + j\omega_oL_f$. The output voltage phasor, active power, and reactive power output of the jth inverter are denoted as $V_j\angle\theta_j$, P_j, and Q_j, respectively, for $j = 1, \cdots, N$, where, $\angle\theta_j$ is the angle of the jth inverter voltage. Accordingly, the active and reactive power output of the jth inverter can be expressed as

$$P_j = \sum_{k=1}^{N} \frac{V_jV_k}{Z_f} \cos(\theta_{jk} + \theta_f) - \frac{V_jV_g}{Z_f} \cos(\theta_{jg} + \theta_f), \quad (1)$$

$$Q_j = \sum_{k=1}^{N} \frac{V_jV_k}{Z_f} \sin(\theta_{jk} + \theta_f) - \frac{V_jV_g}{Z_f} \sin(\theta_{jg} + \theta_f), \quad (2)$$

where, θ_{jk} denotes angle between the jth and k^{th} inverter and θ_{jg} denotes the angle between the jth inverter and the grid. Note that in (1)–(2), the properties of $Z_f\angle\theta_f$ dictate the nature of cross-coupling of active and reactive power with the voltage amplitudes and angles of each inverter. To mitigate this cross-coupling and facilitate a decentralized control strategy, the virtual resistor R_v is emulated at each set of terminals via digital control such that $NR_v \gg \omega_oL_f$, where ω_o is the nominal line frequency (see Fig. 2(b)). This ensures that network impedance is predominantly resistive such that $Z_f = NR_v + j\omega_oL_f \approx NR_v$ and $\theta_f = \arctan(\omega_oL_f/(NR_v)) \approx 0$. Note that, in (1)–(2), the first summation term captures the power exchanged via cross-coupling between inverters and the second term equals the power exchanged with the grid which implies that if we can synchronize the inverter voltages ($\theta_{jk} \approx 0$), we can get rid of the cross coupling term.

1) Proposed Reactive Power Control and Output Voltage Synchronization:: As shown in block diagram Fig. 3, the output of the active power controller provides the required inverter terminal voltage amplitude and the reactive power control loop provides the required voltage frequency and angle. The proposed reactive power control law in time domain is,

$$\omega_j = \omega_o - K_Q(Q_j^\star - Q_j) - K_QK_F \int_0^{t^-} (\omega_j - \omega_o)dt \quad (3)$$

where, $Q_j^\star$ is the reactive power set point, Q_j is the measured reactive power at the inverter, K_Q is the droop co-efficient

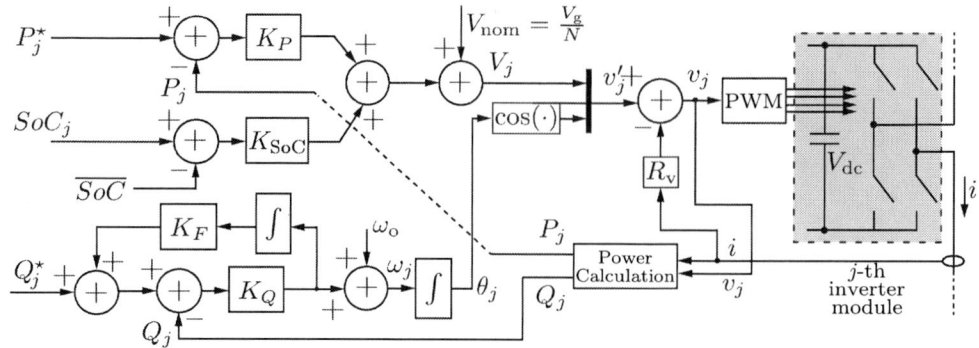

Fig. 3. Control block Diagram for the jth inverter module.

of the $Q - \omega$ droop law and K_F is the feedback gain of the frequency deviation from the nominal, the necessity of which will be explained in the next section on stability analysis. Initially, the $Q_j^\star$ is set to 0, to achieve voltage phase synchronization among the inverters and the grid.

2) Proposed SoC-based Active Power Control: For the active power control we propose the following control law that modulates the output voltage amplitude,

$$V_j = V_{\text{nom}} + K_P(P_j^\star - P_j) + K_{\text{SoC}}(SoC_j - \overline{SoC}), \quad (4)$$

where, $\overline{SoC}$ is the mean SOC of all the units which can be computed as,

$$\overline{SoC} = \frac{1}{N} \sum_{j=1}^{N} SoC_j \quad (5)$$

$$= \frac{1}{N} \sum_{j=1}^{N} SoC_{jo} - \frac{1}{NE_o} \int \left(\sum_{j=1}^{N} P_j \right) dt \quad (6)$$

$$= \overline{SoC}_{jo} - \int (V_{\text{stack}} I) dt \quad (7)$$

where, $\overline{SoC}_{jo}$ is the mean initial SOC of all the units which can be pre-measured and stored in every micro-controller. Moreover, since the stack voltage V_{stack} is almost equal to the nominal grid voltage, or, $V_{\text{stack}} \approx V_{\text{g}}$, (7) can be approximated as, $\overline{SoC} \approx \overline{SoC}_{jo} - \int (V_{\text{g}} I) dt$. The nominal grid voltage can be measured at start of operation and updated inside the system. The current is measured locally which eliminates the need of global communication inside the system. Hence, the mean SOC of the system can be computed in a local and decentralized manner.

III. Small Signal Stability Analysis

We consider, the system is perturbed around a steady state where all the units share equal active and reactive power as, $P_1 = \cdots = P_N = P_o$ and $Q_1 = \cdots = Q_N = Q_o$. Hence, their voltage amplitudes and angles are also equal, $V_1 = \cdots = V_N = V_o$ and $\theta_1 = \cdots = \theta_N$.

Injecting small angle perturbations around the steady-state point in (2), the perturbation in reactive power of the jth inverter is approximated as

$$\tilde{Q}_j \approx \frac{V_o}{Z_f} \left(\sum_{k=1,\neq j}^{N} V_o - V_g \cos \theta_{j\text{go}} \right) \tilde{\theta}_j - \frac{V_o}{Z_f} \sum_{k=1,\neq j}^{N} V_o \tilde{\theta}_k$$
$$- \frac{V_g \sin \theta_{j\text{go}}}{Z_f} \tilde{v}_j. \quad (8)$$

Small values of Q_o imply $\sin \theta_{j\text{go}} \approx 0$; hence, (8) becomes

$$\tilde{Q}_j \approx \frac{V_o}{Z_f} \left((N-1)V_o - V_g \cos \theta_{j\text{go}} \right) \tilde{\theta}_j - \frac{V_o}{Z_f}(N-1)V_o \tilde{\theta}_k \quad (9)$$

Next, we consider the system state as the vector of voltage angles $\tilde{\theta} = [\tilde{\theta}_1, \ldots, \tilde{\theta}_N]^\top$ and output as $\tilde{Q} = [\tilde{Q}_1, \ldots, \tilde{Q}_N]^\top$. The steady state voltages are represented as a fraction of grid voltage, $V_o = V_g/M$ and angles $\theta_{j\text{go}} = \theta_o$, for all $j \in \{1, \ldots, N\}$. From (9), we get small signal reactive power perturbation as $\tilde{Q} = C\tilde{\theta}$ where,

$$C = \frac{V_g^2}{Z_f M^2} \begin{bmatrix} N - 1 - M \cos \theta_o & \cdots & -1 \\ \vdots & \ddots & \vdots \\ -1 & \cdots & N - 1 - M \cos \theta_o \end{bmatrix}. \quad (10)$$

Next, considering small signal perturbations in the control law $\dot{\tilde{\theta}} = \tilde{\omega} = K_Q(\tilde{Q} - \widetilde{Q^\star} - K_F \tilde{\theta})$, and using $\tilde{Q} = C\tilde{\theta}$ gives the following state dynamics

$$\dot{\tilde{\theta}} = A\tilde{\theta} + B\widetilde{Q^\star}, \quad (11)$$

where $A = K_Q(C - K_F I_N)$ and $B = -K_Q I_N$. I_N denotes the $N \times N$ identity matrix. The eigenvalues of A are

$$\lambda_1(A) = -KM \cos \theta_o - K_Q K_F, \quad \text{and,} \quad (12)$$
$$\lambda_2(A) = \cdots = \lambda_N(A) = K(N - M \cos \theta_o) - K_Q K_F, \quad (13)$$

where $K = \frac{K_Q V_g^2}{Z_f M^2} > 0$. To achieve voltage synchronization with the grid, we need $\theta_o = -\sin^{-1}\left(\frac{Z_f Q_o}{V_o V_g}\right) \approx 0$ in (13) as the desired equilibrium point. In the absence of proposed state feedback ($K_F = 0$), this will be a stable equilibrium

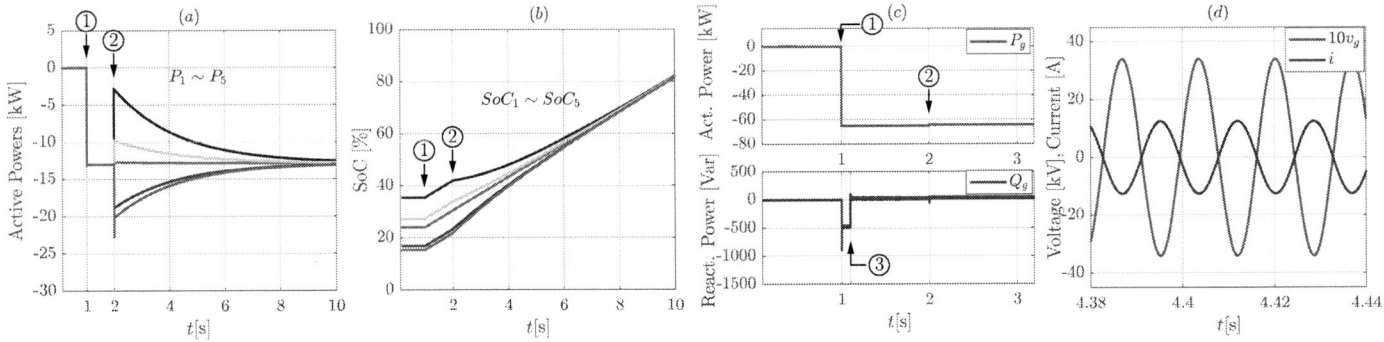

Fig. 4. Simulation results during charging. (a) Active powers P_1 to P_5, (b) Battey SoC levels SoC_1 to SoC_5, (c) Active and reactive power delivered to grid, P_g & Q_g, and (d) Grid voltage v_g and current i. Events ①, ②, and ③ correspond to: ① active power set point increased to -14 kW, ② SoC balancing turned ON, and ③ reactive power set point updated to correct delivered Q_g to ≈ 0 kVA.

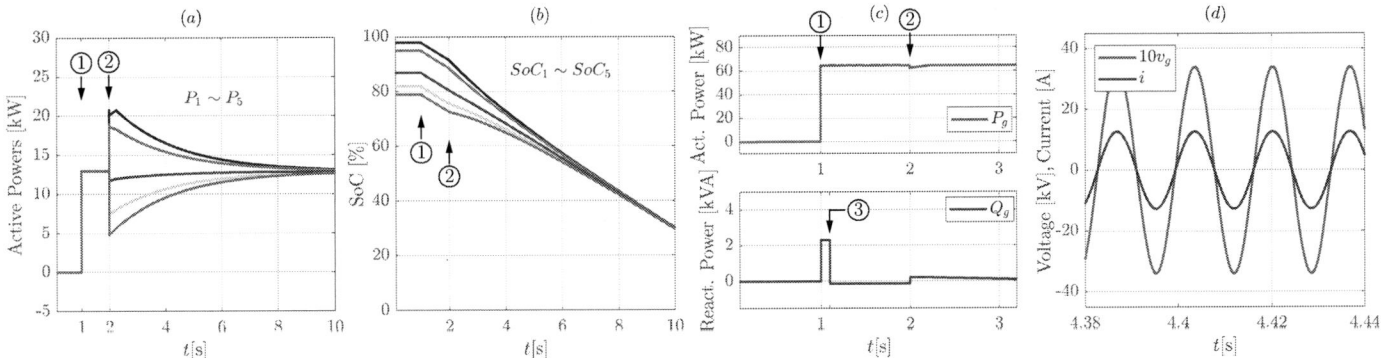

Fig. 5. Simulation results during discharging. Signals in each plot and events are identical to those in Fig. 4. As can be seen, in this case, the unit with higher SoC process higher active power to discharge faster as opposed to the charging case.

point if and only if $M \cos \theta_o > N$ or $M > N$ (near unity p.f.). However, from (1) it follows that in steady state $P_j = \frac{V_g^2}{M^2 Z_f}(N - M) < 0$ or, the inverters must absorb active power for stable reactive power delivery and voltage synchronization.

To bypass this restriction and enable bidirectional power transfer, the proposed state feedback is necessary and it is chosen as, $K_F = Km/K_Q = (V_g^2 m/Z_f M^2)$ where, m should be selected such that, the criteria $M \cos \theta_o + m > N$ is satisfied under any condition. From (13), the eigen values are now

$$\lambda_2(A) = \cdots = \lambda_N(A) = K(N - m - M \cos \theta_o) \quad (14)$$

M is minimum when V_j is maximum or the active power delivered by all inverters is maximum. Therefore, we must choose m such that, $M_{\min} \cos \theta_o + m > N$ is satisfied. This allow for control of active power delivery over the desired range.

IV. SIMULATION AND EXPERIMENTAL RESULTS

The proposed control method was first simulated in PLECS for a system of 5 series connected H-Bridge Inverters (HBI) connected to a 4.16 kV_{rms} lower medium voltage grid. For higher voltage grids the number of stacked modules in series can be increased. All the inverters are interfaced to battery modules at the DC side. The system and controller parameters

TABLE I
SYSTEM AND CONTROLLER PARAMETERS FOR SIMULATIONS

Symbol	Description	Value	Units
V_{batt}	Nominal Battery voltage	450	V
V_{dc}	H-Bridge input (DAB sec. side)	900	V
E	Nominal battery capacity	10	kWh
	Battery cap. (faster simulation)	55	Wh
V_{go}	L-L Grid voltage	4.16	kV$_{rms}$
ω_o	Nominal grid frequency	60	Hz
P_o	Rated power/module	20	kW
L_f	Filter inductance	5	mH
R_f	Filter resistance	0.25	Ω
Controller Parameters			
R_v	Virtual resistance/module	4	Ω
K_Q	Reactive power controller gain	-1	rad/sec.Var
K_P	Active power controller gain	0.05	V/W
K_F	State feedback gain	231	kVar/rad
K_{SoC}	SoC balancing controller gain	500	V

used in the simulation are given in Table-I. The results for active power control, SoC balancing and reactive power control during charging and discharging are shown in Fig. 4 and Fig. 5, respectively. In both these simulations, the reactive power controller is turned on first with reference $Q_{ref} = 0$, which controls the generated voltage phases and synchronizes

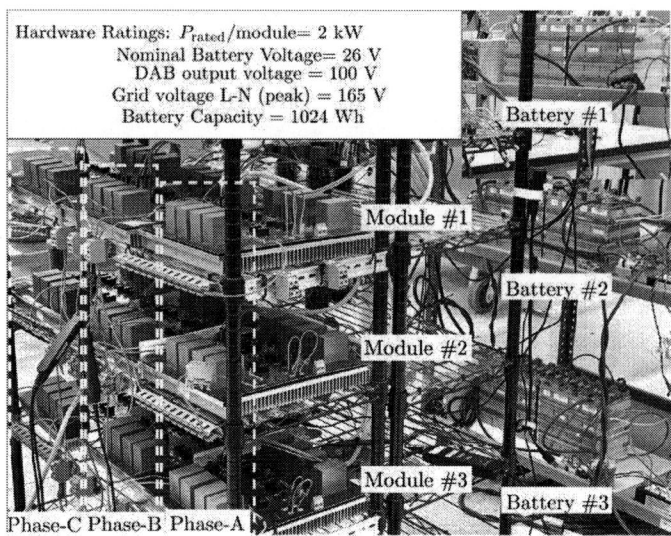

Fig. 6. Hardware setup of 3 series connected dc to 3-phase ac converter modules with Li-ion batteries. Each module has a QAB-based dc to dc stage followed by a H-Bridge inverter. Topology is shown in Fig. 1.

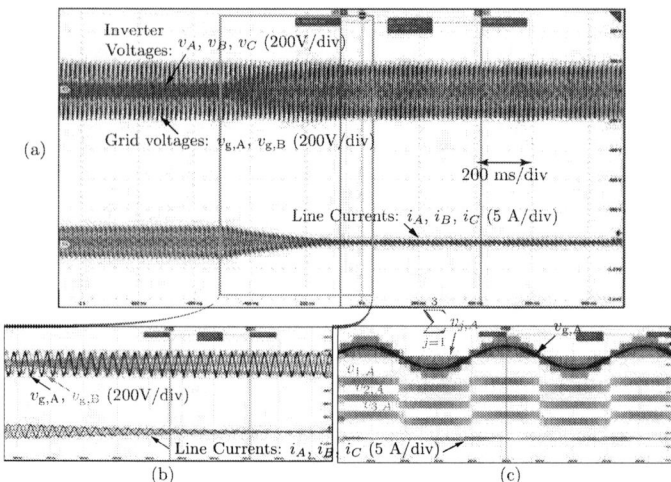

Fig. 7. Transient response of the reactive power control loop as it achieves voltage synchronization among the modules and grid after start-up.

the modules with the grid. Then the active power reference is changed from 0 to 14 kW per module and each module shares equal power. At this point the SOC balancing of the storage units is turned on. As a result of mismatched SoCs, the modules start to process power according to ones SoC level. During charging (discharging), the module with higher SoC process lower (higher) power than the module with lower SoC according to control law in equation (4) and as a result the SoCs are balanced. This can be seen in Figures 4 and Fig. 5, respectively, for charging and discharging scenarios.

The experimental results have been obtained on a system of 3 series connected dc to 3-phase ac converters with integrated battery storage as shown in Fig. 6. Each dc-ac converter has 2 stages in cascade- 1) A dc-dc Quadruple Active Bridge (QAB) converter which provide isolation, active power decoupling and 3 independent dc-links, followed by, 2) 3 H-Bridge Inverters (HBI) which connect to the 3-phase grid. The experimental setup corresponds to the topology shown in Fig. 1. For decentralized control each converter is equipped with a TMS320F28379D digital signal processor. On the dc side, each converter is connected to a 25.6 V LiFeMnPO4 battery with 40 Ah capacity from Batteryspace. The experimentally used physical and control parameters are given in Table II.

1) Communication-free Grid Synchronization: In order to synchronize the inverters with the grid, the active and reactive power references for each converter are set to 0 and a resistance is introduced into the circuit to limit the grid inrush current. In this condition, the reactive power controller synchronizes the phases of each individual inverter voltage with the grid voltage, the overall inverter stack voltage build up and the current decays to zero. This transient is shown in Fig. 7(a) and zoomed in Fig. 7(b). Fig. 7(c) shows the synchronized output voltages (switch level) of the 3 inverters and the grid after the controller is turned on. Note that a PLL

is not required in this case to achieve grid synchronization. After the currents come to zero, the startup resistance is cut off from the circuit and the inverters can begin exchanging power with the grid.

2) Reverse power transfer: Charging of batteries & SoC Balancing: Fig. 8(a) shows the transient response of the three phases of inverter stack voltages and the grid currents after a $P_1^\star = P_2^\star = P_3^\star = -200$ W power step command is given to the active power controllers. The grid currents are 180° out of phase with the grid voltage which implies that power is flowing from the grid to the batteries and the reactive power is almost zero. The stack voltage is the sum of the individual inverter output voltages, $v_{\text{stack}} = v_1 + v_2 + v_3$, and hence it has multilevel waveform. Fig. 8(b) shows the corresponding battery currents $i_{\text{batt},1-3}$ during the step change, which are almost equal as they process equal power. In this case, $P_1 \approx P_2 \approx P_3 \approx -180$ W.

At this point, the SOC balancing algorithm is turned on and the inverters start to process power according to the need of

TABLE II
SYSTEM AND CONTROLLER PARAMETERS FOR EXPERIMENTS

Symbol	Description	Value	Unit
V_{batt}	Nominal Battery voltage	25.6	V
E	Nominal battery capacity	40	Ah
V_{dc}	QAB input (primary side)	50	V
V_{dc}	H-Bridge input (QAB sec. side)	100	V
V_{go}	L-N Grid voltage	120	V_{rms}
ω_o	Nominal grid frequency	60	Hz
P_o	Rated power/module	500	W
L_{f}	Filter inductance	5	mH
R_{f}	Filter resistance	0.25	Ω
Controller Parameters			
R_v	Virtual resistance/module	6	Ω
K_Q	Reactive power controller gain	−5	rad/sec.Var
K_P	Active power controller gain	0.1	V/W
K_F	State feedback gain	500	Var/rad
K_{SoC}	SoC balancing controller gain	100	V

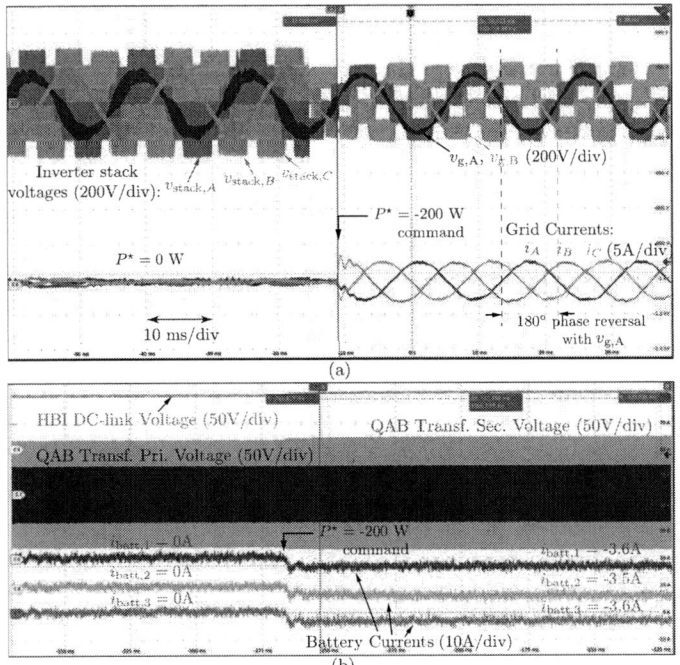

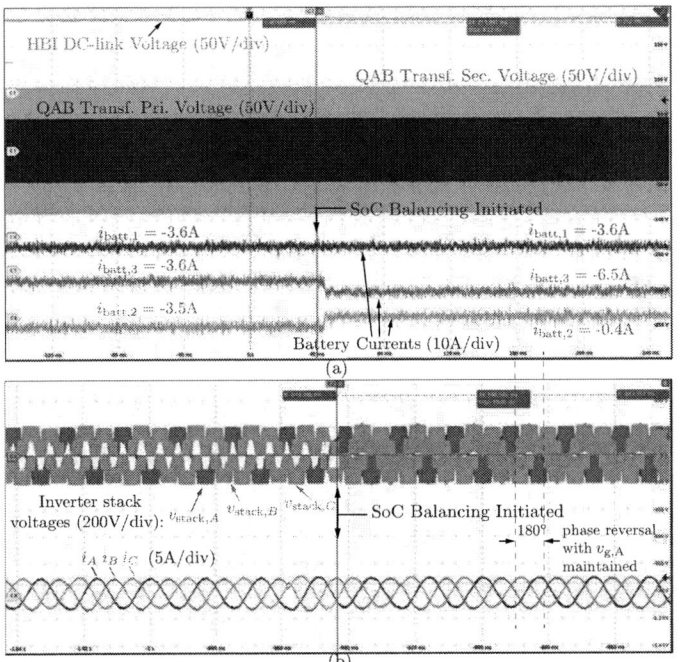

Fig. 8. Experimental results during battery charging: (a) Inverter stack voltages, grid voltages and currents when a $P^* = -200$ W negative power command is given. (b) Battery currents, QAB transformer primary and secondary voltages and HBI dc-link voltages during the same event. Converters track the power command and batteries charge with equal power ≈ -180 W.

Fig. 9. Experimental results during battery charging: (a) Battery currents before and after SoC balancing control is initiated. The 3rd battery with lowest SoC starts charging faster as seen by the difference in the currents. (b) Overall active power taken from the grid and power factor remains almost same before and after the control is initiated.

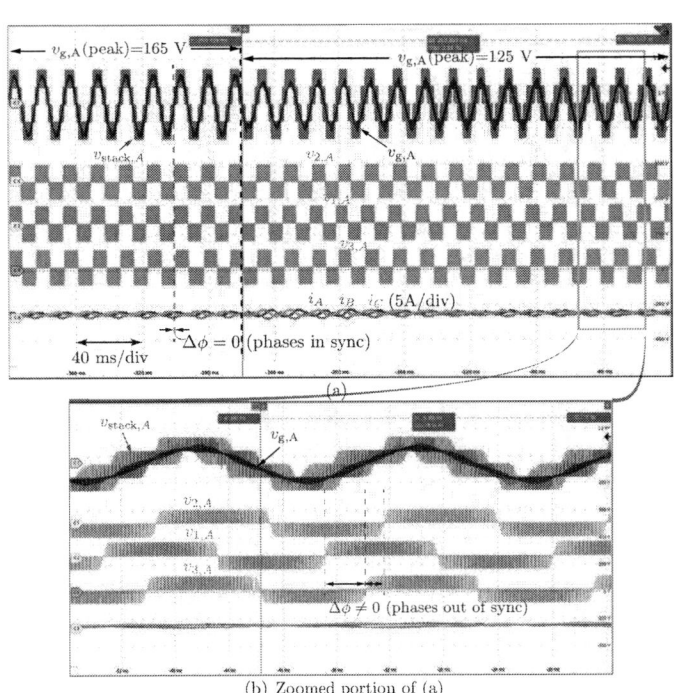

Fig. 10. (a) Insatbility of the reactive power controller when stack voltage becomes higher than grid voltage or $M < N$. Reactive power control no longer able to synchronize modules and help follow positive active power command after grid voltage is reduced. (b) Zoomed view of (a).

their corresponding storage batteries. The battery SoC levels are initially at, $SoC_{1o} = 0.34, SoC_{2o} = 0.51, SoC_{3o} = 0.2$. As can be seen in Fig. 9(a), the 3rd inverter starts drawing more current and the 2nd inverter almost stops drawing power due to it's higher SoC level. The power level of the 1st inverter does not change much as it's SoC level is close to the mean SoC of the system. This verifies the unequal power sharing capability of the units during SoC balancing. At the initial stage, $P_1 \approx -180$ W, $P_2 \approx -20$ W and $P_3 \approx -325$ W. As the batteries charge and the SoCs increase, the mismatch in the SOC levels and absorbed power gradually decrease. Fig. 9(b), verifies that the overall power delivered by the grid before and after the SoC balancing control is initiated remains same. However, the stack voltage waveform changes, as the units share unequal voltages.

3) Instability of Reactive power controller during Forward Power Transfer: Fig. 10(a) & (b) demonstrates the instability of the reactive power controller when $M < N$ or the stack voltage voltage is higher than the grid voltage, $V_{\text{stack}} > V_g$ to enable forward power transfer. This is theoretically pointed out in Section-III and experimentally verified in Fig. 10 by intentionally lowering the grid simulator voltage from 165 V to 125 V which is below the inverter stack voltage (150 V) corresponding to a reduction in M from 3.25 to 2.5 compared to $N = 3$. Fig. 10(a) & (b) shows the initially synchronized inverter voltages falling out of sync and unable to maintain synchronism with grid once the inverter stack voltage is higher than the grid voltage ($M < N$).

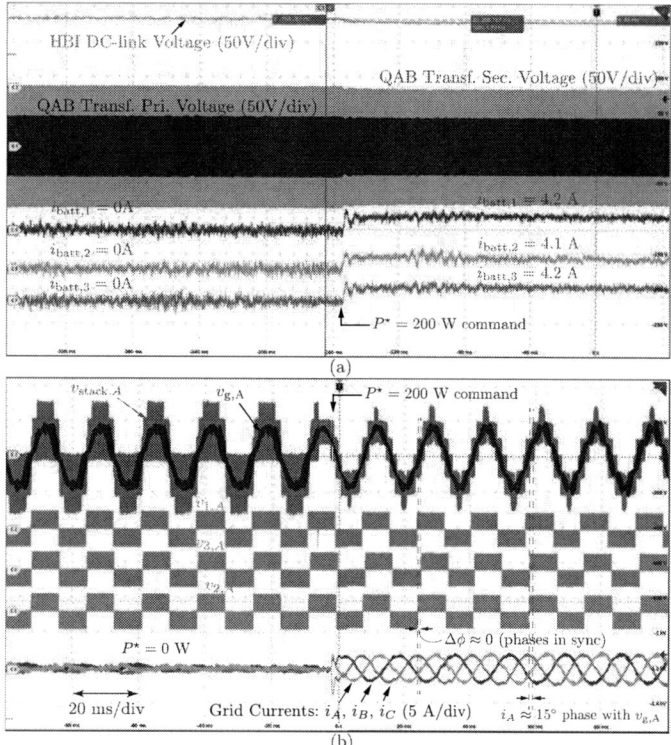

(a)

(b)

Fig. 11. Experimental results during battery discharging: (a) Battery currents and QAB voltages when a $P^\star = 200$ W positive power command is given. (b) A-phase inverter stack voltage, module switch node voltages, grid voltage and currents during the same event. Converters track the power command and batteries discharge with equal power ≈ 210 W.

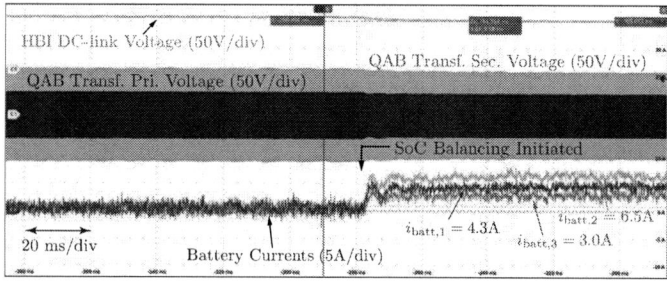

Fig. 12. Battery currents before and after SoC balancing control is initiated during discharging. The 2nd battery with highest SoC starts discharging faster compared to others.

4) Forward power transfer: Discharging of batteries & SoC Balancing: After implementing the proposed state feedback in the reactive power controller, the inverters are now able to support forward power flow into the grid. Fig. 11(a) shows the battery currents when a $P_1^\star = P_2^\star = P_3^\star = 200$ W power step command is given to the active power controllers. In this case, the batteries discharge with equal power ($P_1 \approx P_2 \approx P_3 \approx 210$ W) and the inverter stack voltages rise above the grid voltage to enable power flow from the stack to the grid. As the stack voltage rise, the voltage waveform becomes 7-level ($2N + 1$) which is seen in Fig. 11(b). The state feedback

controller assures that the reactive power loop is stable and individual inverter voltages (v_{1-3}) remain synchronized with the grid voltage. Fig. 11(b) also shows that the grid currents are almost in phase with a $15°$ phase lag with the grid voltage, which gives a power factor of 0.966.

Fig. 12 shows the battery currents once the SoC balancing is initiated during discharging. As mentioned previously, in this case, the unit with highest SoC level (unit-2) discharges fastest and unit with the lowest SoC level (unit-3) discharges slowest to help balance the SoCs eventually.

V. CONCLUSION

In this paper, we proposed and validated a control method to achieve decentralized control of active and reactive power processed by each inverter in a system of series-connected inverters tied to the grid. Each inverter has a storage battery at the dc port followed by an isolated dc-dc stage which is again followed by a dc-ac cascaded H-bridge which connects to the grid. The proposed method is superior compared to existing centralized and distributed control methods in a way that it does not require any real-time high speed communication among the series connected modules. Each module controller processes only the locally sensed current signal and synthesizes the output voltage of each inverter using two decoupled control loops for active and reactive power. The reactive power control loop generates the output voltage angle and maintains synchronism among the series-connected inverters and grid, whereas the active power control loop generates the voltage amplitude that governs the active power processed by each inverter and achieve SoC balancing of the batteries. The design method and small-signal stability analysis of both control loops are provided.

The proposed approach was validated in a simulation of a system comprising 5 inverter units connected in series and delivering power to a medium-voltage grid. Experimental results performed on a hardware platform of 3 series connected dc-ac converter modules are also provided which demonstrate the effectiveness of proposed method for decentralized power sharing. Future work will be directed towards in-detail analysis of non-uniform power sharing across the stack and considering the effect of grid voltage harmonics and clock drifts of digital signal processors into the design of the control loops and it's performance.

REFERENCES

[1] A. Dell'Aquila, M. Liserre, V. G. Monopoli, and P. Rotondo, "Overview of PI-based solutions for the control of DC buses of a single-phase H-bridge multilevel active rectifier," *IEEE Trans. Ind Appl.*, vol. 44, no. 3, pp. 857–866, 2008.

[2] E. Villanueva, P. Correa, J. Rodríguez, and M. Pacas, "Control of a single-phase cascaded h-bridge multilevel inverter for grid-connected photovoltaic systems," *IEEE Trans. Ind. Electron.*, vol. 56, no. 11, pp. 4399–4406, 2009.

[3] T. Zhao, G. Wang, S. Bhattacharya, and A. Q. Huang, "Voltage and power balance control for a cascaded H-bridge converter-based solid-state transformer," *IEEE Trans. Power Electron.*, vol. 28, no. 4, pp. 1523–1532, 2012.

[4] B. P. McGrath, D. G. Holmes, and W. Y. Kong, "A decentralized controller architecture for a cascaded H-bridge multilevel converter," *IEEE Trans. Ind. Electron.*, vol. 61, no. 3, pp. 1169–1178, 2013.

[5] P. K. Achanta, B. B. Johnson, G.-S. Seo, and D. Maksimovic, "A multilevel DC to three-phase AC architecture for photovoltaic power plants," *IEEE Trans. Energy Convers.*, vol. 34, no. 1, pp. 181–190, 2018.

[6] L. Zhang, K. Sun, Y. W. Li, X. Lu, and J. Zhao, "A distributed power control of series-connected module-integrated inverters for pv grid-tied applications," *IEEE Trans. Power Electron.*, vol. 33, no. 9, pp. 7698–7707, 2017.

[7] P. K. Achanta, D. Maksimovic, and M. Ilic, "Decentralized control of series stacked bidirectional DC-AC modules," in *2018 IEEE Applied Power Electronics Conference and Exposition*, pp. 1008–1013, 2018.

[8] J. He, Y. Li, C. Wang, Y. Pan, C. Zhang, and X. Xing, "Hybrid microgrid with parallel-and series-connected microconverters," *IEEE Trans. Power Electron.*, vol. 33, no. 6, pp. 4817–4831, 2017.

[9] H. Jafarian, S. Bhowmik, and B. Parkhideh, "Hybrid current-/voltage-mode control scheme for distributed ac-stacked PV inverter with low-bandwidth communication requirements," *IEEE Trans. Ind. Electron.*, vol. 65, no. 1, pp. 321–330, 2017.

[10] B. Xu, H. Tu, Y. Du, H. Yu, H. Liang, and S. Lukic, "A distributed control architecture for cascaded H-bridge converter with integrated battery energy storage," *IEEE Trans. Ind Appl.*, vol. 57, no. 1, pp. 845–856, 2020.

[11] X. Hou, K. Sun, X. G. Zhang, Y. Sun, and J. Lu, "A hybrid voltage/current control scheme with low-communication burden for grid-connected series-type inverters in decentralised manner," *IEEE Trans. Power Electron.*, 2021.

[12] J. He, Y. Li, B. Liang, and C. Wang, "Inverse power factor droop control for decentralized power sharing in series-connected-microconverters-based islanding microgrids," *IEEE Trans. Ind. Electron.*, vol. 64, no. 9, pp. 7444–7454, 2017.

[13] P. Achanta, M. Sinha, B. Johnson, S. Dhople, and D. Maksimovic, "Self-synchronizing series-connected inverters," in *Workshop on Control and Modeling for Power Electronics*, pp. 1–6, 2018.

[14] M. Lu, S. Dutta, and B. Johnson, "Self-synchronizing cascaded inverters with virtual oscillator control," *IEEE Transactions on Power Electronics*, vol. 37, no. 6, pp. 6424–6436, 2021.

[15] D. F. Frost and D. A. Howey, "Completely decentralized active balancing battery management system," *IEEE Transactions on Power Electronics*, vol. 33, no. 1, pp. 729–738, 2017.

[16] X. Hou, Y. Sun, X. Zhang, G. Zhang, J. Lu, and F. Blaabjerg, "A self-synchronized decentralized control for series-connected H-bridge rectifiers," *IEEE Trans. Power Electron.*, vol. 34, no. 8, pp. 7136–7142, 2019.

[17] X. Hou, Y. Sun, H. Han, Z. Liu, W. Yuan, and M. Su, "A fully decentralized control of grid-connected cascaded inverters," *IEEE Trans. Sustain. Energy*, vol. 10, no. 1, pp. 315–317, 2018.

[18] S. Dutta, M. Lu, B. Majmunovic, R. Mallik, G.-S. Seo, D. Maksimovic, and B. Johnson, "Grid-connected self-synchronizing cascaded h-bridge inverters with autonomous power sharing," in *2021 IEEE Energy Conversion Congress and Exposition (ECCE)*, pp. 2806–2813, IEEE, 2021.

[19] H. Jafarian, R. Cox, J. H. Enslin, S. Bhowmik, and B. Parkhideh, "Decentralized active and reactive power control for an AC-stacked PV inverter with single member phase compensation," *IEEE Trans. Ind Appl.*, vol. 54, no. 1, pp. 345–355, 2017.

[20] S. Farzamkia, H. S. K. Kedlaya, and A. Q. Huang, "A decentralized control system for series-connected grid-integrated photovoltaic inverters," in *2024 IEEE 15th International Symposium on Power Electronics for Distributed Generation Systems (PEDG)*, pp. 1–6, IEEE, 2024.

[21] S. Dutta, M. Lu, R. Mallik, B. Majmunovic, S. Mukherjee, G.-S. Seo, D. Maksimovic, and B. Johnson, "Decentralized control of cascaded H-bridge inverters for medium-voltage grid integration," in *2020 IEEE 21st Workshop on Control and Modeling for Power Electronics*, pp. 1–6, 2020.

[22] H. Qin and J. W. Kimball, "Closed-loop control of dc–dc dual-active-bridge converters driving single-phase inverters," *ieee transactions on power electronics*, vol. 29, no. 2, pp. 1006–1017, 2013.

Small-Signal Modeling and Damping Design of Unfolding-Based Single Stage AC-DC Converter Using the Extra Element Theorem

Dakota Goodrich[*], Aditya Zade, Shubhangi Gurudiwan, Mahmoud Mansour, Regan Zane, and Hongjie Wang

Electrical and Computer Engineering Department, Utah State University

Logan, Utah – USA 84341

Email: [*]dakota.goodrich@usu.edu

Abstract—**Unfolder-based quasi-single-stage ac-dc power converter has been widely used for high-power electric vehicle (EV) charging systems for its high efficiency and power density. However, the resonance between the grid inductance (impedance) and the capacitors on the soft-dc-link of the converter impacts the system stability and significantly limits the system control bandwidth and dynamic response performance. A quasi-single-stage ac-dc converter with unfolder plus T-bridge series resonant converter (T-SRC) is studied in this work. The small-signal modeling and plant transfer function derivation of the T-SRC is presented in this paper. A damping filter design using the extra element theorem (EET) is then proposed to achieve high-bandwidth and stable operation of the quasi-single-stage ac-dc converter. Simulation and hardware results from an 18 kW module for high-power EV charging are provided to validate the proposed modeling and damping filter design.**

Index Terms—**Ac-dc converter, battery charger, extra element theorem (EET), grid-tied system, high-bandwidth control, T-type converter, unfolder, input filter, damping circuit.**

I. INTRODUCTION

Grid-tied ac-dc power converter is a key equipment for electric vehicle (EV) charging system and renewable energy integration. A two-stage ac-dc converter design with an active front-end stage for power factor correction (PFC) followed by a second dc-dc stage for output power control and isolation has been commonly used [1], [2]. To achieve higher efficiency and power density, a quasi-single-stage ac-dc converter with unfolder plus T-bridge series resonant converter (T-SRC) is studied in this work, similar to the unfolding-based topologies in [3]–[6]. The circuit topology of the converter studied in this paper is shown in Fig. 1 and consists of a three-phase grid connection with grid inductance (impedance), an unfolder, a T-type bridge, a series resonant tank, and active secondary bridge with battery load.

The back-to-back switches of the unfolder (Q_a, Q_b, Q_c) are operated at $2 \times f_{grid} = 120$ Hz and the unfolder diodes (D_{a1}, D_{b1}, D_{c1}, D_{a2}, D_{b2}, D_{c2}), as shown in Fig. 1, operate at grid frequency such that the most positive grid phase voltage is seen at the p-port, the most negative voltage is connected to the n-port, and the remaining phase is connected to the o-port. The T-SRC, which is introduced in [7], connects to

the p-, o-, and n-ports with the soft-dc-link capacitors C_{po}, C_{on}, and C_{pn}. As detailed in [7], the T-SRC controls the PFC and output power with intermediate control variables $D_{p/n}$ and V_{1q}, respectively, where $D_{p/n}$ is calculated to control the ratio of input currents $I_{p/n} = I_p / I_n$, and V_{1q} is calculated for a given output current reference. These intermediate control variables are then used to calculate the modulation variables D_p, D_n, and D_ϕ, which are the duty cycle of the p-port and n-port bridges and the shift between the primary and secondary switching waveforms, respectively. Fig. 2 shows the large-signal waveforms governing the operation of the unfolder and T-SRC as a function of grid angle. Due to the low-frequency operation of the unfolder, the system exhibits a resonance between the grid inductance and the soft-dc-link capacitance of the T-SRC. This resonance has undesirable impacts on the system transfer functions and consequently on the closed-loop stability of the converter.

In [3], a resistor added in parallel with the grid inductance is used to dampen the resonance. Similarly, a network composed of two resistors and one capacitor is connected in parallel with the input inductance in [8]. However these approaches only work for low power applications where access to the physical inductor of the input filter is available. In high power applications that utilize the inherent grid inductance, the damping filter must be added in parallel to the entire input impedance, similar to the RC network proposed in [9]. This parallel passive damping approach is the chosen damping method for this application.

In [5], to address the input filter resonance and implement high-bandwidth control, an EET-based approach has been used to analyze the grid inductance to soft-dc capacitance resonance of an unfolder-based 21 kW ac-dc converter with a T-bridge, LCC resonant tank, and passive secondary diode bridge. However, it focuses on active damping control and the analysis cannot be easily applied to the T-SRC studied in this work due to different power converter topologies with different resonant tanks and secondary sides. In [7], the steady-state operation, modulation, and low-bandwidth control strategy of the T-SRC has been presented. However, the detailed small-signal modeling and high-bandwidth control are completely missing.

The key contributions of this paper include: (1) a detailed

This work is based in part upon work supported by the National Science Foundation (NSF) through the ASPIRE Engineering Research Center under Grant EEC-1941524 and CAREER Award under Grant 2239169.

979-8-3315-1612-3/25 $31.00 © 2025 IEEE

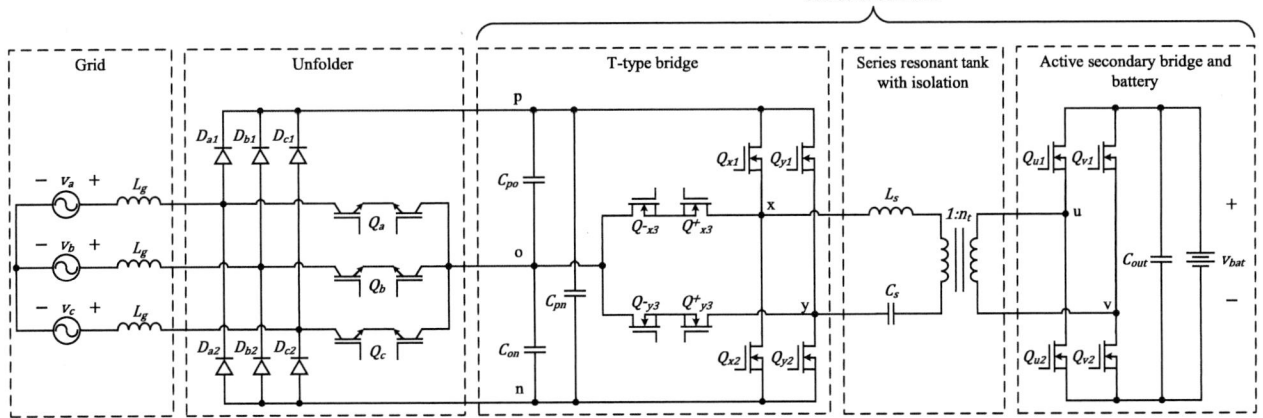

Fig. 1: Circuit diagram of the quasi-single-stage ac-dc converter with unfolder plus T-bridge series resonant converter (T-SRC).

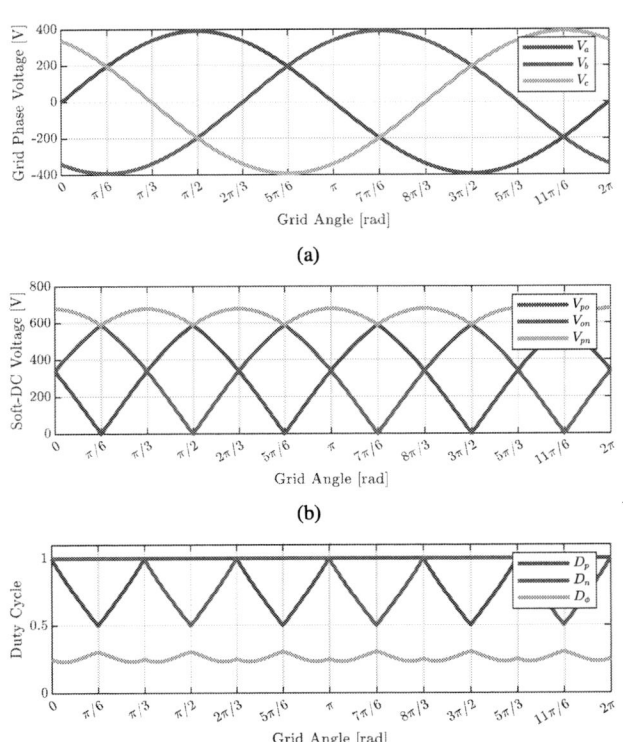

Fig. 2: Large-signal waveforms of the unfolder and T-SRC showing (a) grid voltages, (b) soft-dc-link voltages, and (c) duty cycles of the T-SRC as a function of grid angle.

small-signal modeling and analysis of a T-SRC with three critical open-loop transfer functions derived, (2) an investigation of stability issues caused by the grid inductance and soft-dc capacitors, and (3) a passive damping filter design using EET for achieving high-bandwidth control.

In this paper, Section II details the derivation of the small-signal model of the T-SRC and its use in finding three system transfer functions. Section III reviews the extra element theorem and utilizes the small-signal model to derive the EET

impedances and analyze their impacts on the system transfer functions. Section IV describes the design of the chosen damping method to mitigate the effects of the resonance. Section V then validates the damping method with simulation and hardware results, and Section VI concludes the paper.

II. SMALL-SIGNAL MODELING OF T-BRIDGE SERIES RESONANT CONVERTER

A. Notation

Throughout this paper, the following conventions are used:

- Large-signal, steady-state quantities are denoted with capital letters, such as D_p.
- Small-signal quantities representing perturbations are denoted with lower-case letters and a "hat", such as $\hat{d}_p$.
- Phasor or complex quantities are represented with an arrow symbol, such as $\vec{s}_p$.
- If a perturbed quantity has a variable superscript, it indicates the variable that was perturbed to obtain that quantity, such as $\hat{i}_{t,env}^{d_p}$, which indicates the perturbation in the envelope of the tank current I_t as a result of a perturbation in the duty cycle D_p.

B. Transfer Function Derivation

In this paper, the small-signal equivalent circuit of the T-SRC, which is the soft dc-dc converter as shown in Fig. 1, is first derived and presented in Fig. 3. Small signal phasor transformation and modeling approach for series resonant converters presented in [10], [11] has been utilized in this work to derive the small-signal equivalent circuit model.

From the derived small-signal equivalent circuit model, this paper derives three transfer functions of interest for which the EET was utilized. These are $\hat{i}_p/\hat{d}_p$, $\hat{i}_n/\hat{d}_n$, and $\hat{i}_{out}/\hat{d}_\phi$, as illustrated in Fig. 3. These functions were chosen because $\hat{i}_p/\hat{d}_p$ and $\hat{i}_n/\hat{d}_n$ most directly influence the PFC control loop, while $\hat{i}_{out}/\hat{d}_\phi$ most influences the output power control loop.

To derive these transfer functions, the fundamental harmonic approximation is used to derive switching functions $\vec{s}_p$, $\vec{s}_n$ and $\vec{s}_s$ as the complex transformation between the p-port, n-port,

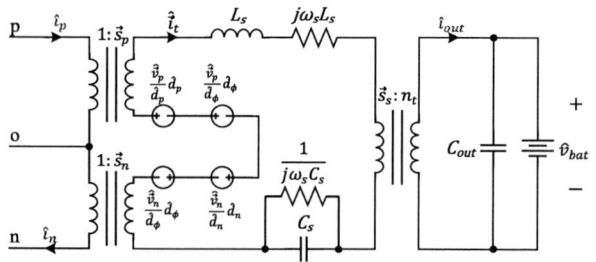

Fig. 3: Small-signal circuit diagram of T-SRC utilizing complex switching function transformers and phasor modeling of tank impedance.

and output port, respectively, with the resonant tank voltages and current. With the secondary active bridge duty cycle held at 50% and the secondary switching function taken as the reference angle, these are calculated as

$$\vec{s_p} = \frac{4}{\pi} \sin\left(\frac{\pi}{2} D_p\right) \angle \left(\pi D_\phi + \frac{\pi}{2}(1 - D_p)\right), \quad (1)$$

$$\vec{s_n} = \frac{4}{\pi} \sin\left(\frac{\pi}{2} D_n\right) \angle \left(\pi D_\phi + \frac{\pi}{2}(1 - D_n)\right), \quad (2)$$

$$\vec{s_s} = \frac{4}{\pi} \angle 0. \quad (3)$$

The tank voltages can be expressed as

$$\vec{V_{1p}} = V_p \vec{s_p}, \quad (4)$$

$$\vec{V_{1n}} = V_n \vec{s_n}, \quad (5)$$

$$\vec{V_{1s}} = V_{out} \frac{\vec{s_s}}{n_t}, \quad (6)$$

where n_t is the transformer turns ratio. By perturbing each respective duty cycle, the small-signal relationship between the p-port, n-port, and output port voltages and the three duty cycles can be found (e.g., $\hat{v}_p/\hat{d}_p$). These control-to-voltage responses are modeled as voltage sources in the small-signal circuit model. The resonant tank is also represented using the phasor transormation, which add complex-valued resistors in series and parallel with the tank inductor and capacitor, respectively. This small-signal ac equivalent circuit model shown in Fig. 3 allows the duty cycle-to-tank current transfer functions to be calculated using traditional circuit analysis, which can be separated into their envelope and phase responses (e.g., $\hat{i}_{t,env}^{d_p}/\hat{d}_p$ and $\hat{\theta}_t^{d_p}/\hat{d}_p$).

Next, using the complex switching functions, the time-domain response of the p-port, n-port, and output port currents are found as

$$\begin{aligned} I_p(t) &= I_t(t) \cdot s_p(t) \\ &= I_{t,env} \cos(w_s t + \theta_t) \cdot \Re\left(\vec{s_p} e^{j w_s t}\right), \end{aligned} \quad (7)$$

$$\begin{aligned} I_n(t) &= I_t(t) \cdot s_n(t) \\ &= I_{t,env} \cos(w_s t + \theta_t) \cdot \Re\left(\vec{s_n} e^{j w_s t}\right), \end{aligned} \quad (8)$$

$$\begin{aligned} I_{out}(t) &= I_t(t) \cdot \frac{s_s(t)}{n_t} \\ &= I_{t,env} \cos(w_s t + \theta_t) \cdot \frac{\Re\left(\vec{s_s} e^{j w_s t}\right)}{n_t}. \end{aligned} \quad (9)$$

By perturbing the duty cycles in the complex switching functions, the tank current envelope and phase are also perturbed. By linearizing these responses, the response due to changes in tank current envelope, tank current phase, and duty cycle (e.g., $\hat{i}_p^{d_p}/\hat{i}_{t,env}$, $\hat{i}_p^{d_p}/\hat{\theta}_t$, and $\hat{i}_p^{d_p}/\hat{d}_p$) are found.

Finally, the small-signal responses are combined to derive the transfer functions. The derived transfer functions are summarized as

$$\frac{\hat{i}_p}{\hat{d}_p} = \frac{\hat{i}_p^{d_p}}{\hat{i}_{t,env}} \frac{\hat{i}_{t,env}^{d_p}}{\hat{d}_p} + \frac{\hat{i}_p^{d_p}}{\hat{\theta}_t} \frac{\hat{\theta}_t^{d_p}}{\hat{d}_p} + \frac{\hat{i}_p^{d_p}}{\hat{d}_p}, \quad (10)$$

$$\frac{\hat{i}_n}{\hat{d}_n} = \frac{\hat{i}_n^{d_n}}{\hat{i}_{t,env}} \frac{\hat{i}_{t,env}^{d_n}}{\hat{d}_n} + \frac{\hat{i}_n^{d_n}}{\hat{\theta}_t} \frac{\hat{\theta}_t^{d_n}}{\hat{d}_n} + \frac{\hat{i}_n^{d_n}}{\hat{d}_n}, \quad (11)$$

$$\frac{\hat{i}_{out}}{\hat{d}_\phi} = \frac{\hat{i}_{out}^{d_\phi}}{\hat{i}_{t,env}} \frac{\hat{i}_{t,env}^{d_\phi}}{\hat{d}_\phi} + \frac{\hat{i}_{out}^{d_\phi}}{\hat{\theta}_t} \frac{\hat{\theta}_t^{d_\phi}}{\hat{d}_\phi} + \frac{\hat{i}_{out}^{d_\phi}}{\hat{d}_\phi}. \quad (12)$$

III. EXTRA ELEMENT THEOREM ANALYSIS AND SIMULATION

A. Extra Element Theorem Introduction

The final result of the EET for a series-added element is

$$G_{new} = G_{old} \left(\frac{1 + \dfrac{Z_{LC}}{Z_n}}{1 + \dfrac{Z_{LC}}{Z_d}} \right) = (G|_{Z_{LC}=0}) G_{cf}, \quad (13)$$

where G_{new} and G_{old} are the transfer function with and without the extra element included, respectively, Z_{LC} is the impedance of the extra element, which for the present application is the parallel combined impedance of the grid inductance L_g and wye-reflected capacitance of the dc-link capacitors $C_\curlywedge$ ($= 3C_{po/on/pn}$), and Z_n and Z_d are the double null injection and single injection driving point impedances, respectively [5], [12]–[15]. The three impedances Z_{LC}, Z_n, and Z_d combine to define the correction factor G_{cf}. For an extra element to not significantly alter a desired transfer function, the correction factor must be near unity for all frequencies, which is accomplished when

$$\|Z_{LC}\| \ll \|Z_n\| \quad \text{and} \quad \|Z_{LC}\| \ll \|Z_d\| \quad (14)$$

hold true at all frequencies. Using small-signal phasor modeling of the converter described below, Z_n and Z_d are calculated as

$$Z_n = \left. \frac{\hat{v}_{test}}{\hat{i}_{test}} \right|_{output \xrightarrow[null]{} 0}, \quad (15)$$

and

$$Z_d = \left. \frac{\hat{v}_{test}}{\hat{i}_{test}} \right|_{input=0}, \quad (16)$$

where $\hat{v}_{test}$ and $\hat{i}_{test}$ are measured at the p-port or n-port, "output" is chosen from $\{\hat{i}_p, \hat{i}_n, \hat{i}_{out}\}$, and "input" is chosen from $\{\hat{d}_p, \hat{d}_n, \hat{d}_\phi\}$, depending on the transfer function of interest. It is important to note that Z_n and Z_d can be analyzed at either the p-port or n-port, due to the symmetric operation of the two ports. This is demonstrated in the symmetry of the plots in Fig. 5.

979-8-3315-1612-3/25 $31.00 © 2025 IEEE

B. Calculation of Z_n

It is noteworthy to first explain that Z_n is infinite for all frequencies when the transfer function of interest is $\hat{i}_p/\hat{d}_p$ or $\hat{i}_n/\hat{d}_n$. This is because the output $\hat{i}_p$ or $\hat{i}_n$ corresponds with $\hat{i}_{test}$, which is nulled to zero when Z_n is measured at the p-port or n-port as shown in (15), respectively. On the other hand, Z_n is not necessarily infinite when the transfer function of interest is $\hat{i}_{out}/\hat{d}_\phi$. Therefore, analysis using small-signal phasor modeling was used to find the magnitude of Z_n at resonant frequency throughout the grid cycle, as measured at the p-port.

Firstly, the steady-state equation relating the input voltages (V_{po} and V_{on}) and complex tank voltages ($\overrightarrow{V_{1p}}$ and $\overrightarrow{V_{1n}}$) is perturbed with

$$V_{po} \rightarrow V_{po} + \hat{v}_{test} \text{ and } D_\phi \rightarrow D_\phi + \hat{d}_\phi, \tag{17}$$

and the result is linearized to

$$\frac{\widehat{\overrightarrow{v}}_{1p}}{\hat{v}_{test}} = \frac{4}{\pi} \sin\left(\frac{\pi}{2}D_p\right) e^{j\left(\pi D_\phi + \frac{\pi}{2}(1-D_p)\right)} \\ + \frac{\hat{d}_\phi}{\hat{v}_{test}} 4V_{po} \sin\left(\frac{\pi}{2}D_p\right) e^{j\left(\pi D_\phi + \frac{\pi}{2}(1-D_p) + \frac{\pi}{2}\right)}, \tag{18}$$

$$\frac{\widehat{\overrightarrow{v}}_{1n}}{\hat{v}_{test}} = \frac{\hat{d}_\phi}{\hat{v}_{test}} 4V_{on} \sin\left(\frac{\pi}{2}D_n\right) e^{j\left(\pi D_\phi + \frac{\pi}{2}(1-D_n) + \frac{\pi}{2}\right)}. \tag{19}$$

Using this result and the phasor transformed resonant circuit, the relationship between the perturbed test voltage $\hat{v}_{test}$ and the tank current is found, which can be separated into its envelope and phase response ($\hat{i}_{t,env}^{d_\phi}/\hat{v}_{test}$ and $\hat{\theta}_t^{d_\phi}/\hat{v}_{test}$). The envelope and phase response are both functions of the unknown term $\hat{d}_\phi/\hat{v}_{test}$. This term is solved for using the following expression for $\hat{i}_{out}/\hat{v}_{test}$, which is nulled to zero by definition of Z_n,

$$\frac{\hat{i}_{out}}{\hat{v}_{test}} = \frac{\hat{i}_{out}^{d_\phi}}{\hat{i}_{t,env}^{d_\phi}} \frac{\hat{i}_{t,env}^{d_\phi}}{\hat{v}_{test}} + \frac{\hat{i}_{out}^{d_\phi}}{\hat{\theta}_t^{d_\phi}} \frac{\hat{\theta}_t^{d_\phi}}{\hat{v}_{test}} = 0. \tag{20}$$

Upon solving for the unknown term, it can be back-substituted into the expressions of $\hat{i}_{t,env}^{d_\phi}/\hat{v}_{test}$ and $\hat{\theta}_t^{d_\phi}/\hat{v}_{test}$, which can then be used in the desired expression of Z_n as calculated at the p-port:

$$Z_{n-p-port} = \frac{1}{\dfrac{\hat{i}_{test}^{d_\phi}}{\hat{i}_{t,env}^{d_\phi}} \dfrac{\hat{i}_{t,env}^{d_\phi}}{\hat{v}_{test}} + \dfrac{\hat{i}_{test}^{d_\phi}}{\hat{\theta}_t^{d_\phi}} \dfrac{\hat{\theta}_t^{d_\phi}}{\hat{v}_{test}}}. \tag{21}$$

The analysis was performed in MATLAB and evaluated for steady-state operating points throughout the grid cycle. These results are shown in Fig. 4.

C. Calculation of Z_d

To calculate Z_d, a similar analysis to that of Z_n is utilized with the difference that the input variable is set to zero, as in (16). Since all other inputs are zero, the only perturbation is that of $\hat{v}_{test}$ added to V_{po} or V_{on} when Z_d is calculated at the p-port or n-port, respectively. Thus for Z_d measured at the p-port, the relevant tank voltage relationship is

$$\frac{\widehat{\overrightarrow{v}}_{1p}}{\hat{v}_{test}} = \frac{4}{\pi} \sin\left(\frac{\pi}{2}D_p\right) e^{j\left(\pi D_\phi + \frac{\pi}{2}(1-D_p)\right)}, \tag{22}$$

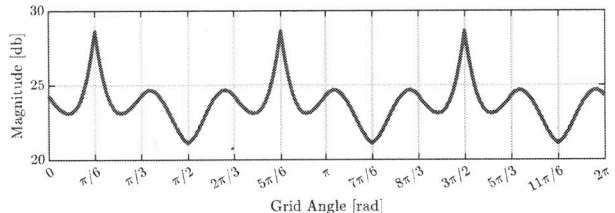

Fig. 4: Magnitude of $Z_{n-p-port}$ at input filter resonant frequency f_{LC} for $G_{old} = \hat{i}_{out}/\hat{d}_\phi$ using small-signal phasor modeling.

and for Z_d measured at the n-port, the tank voltage relationship is

$$\frac{\widehat{\overrightarrow{v}}_{1n}}{\hat{v}_{test}} = \frac{4}{\pi} \sin\left(\frac{\pi}{2}D_n\right) e^{j\left(\pi D_\phi + \frac{\pi}{2}(1-D_n)\right)}. \tag{23}$$

Once again, this tank voltage is used with the phasor transformed circuit elements to find the tank current response from the input voltage perturbation, which is separated into its envelope and phase response.

Finally, using the relationship between the p-port current and the perturbed tank envelope and phase, the final expression for Z_d calculated at the p-port is

$$Z_{d-p-port} = \frac{1}{\dfrac{\hat{i}_{t,env}}{\hat{i}_p} \dfrac{\hat{v}_{test}}{\hat{i}_{t,env}} + \dfrac{\hat{\theta}_t}{\hat{i}_p} \dfrac{\hat{v}_{test}}{\hat{\theta}_t}}, \tag{24}$$

and when calculated at the n-port is

$$Z_{d-n-port} = \frac{1}{\dfrac{\hat{i}_{t,env}}{\hat{i}_n} \dfrac{\hat{v}_{test}}{\hat{i}_{t,env}} + \dfrac{\hat{\theta}_t}{\hat{i}_n} \dfrac{\hat{v}_{test}}{\hat{\theta}_t}}. \tag{25}$$

The calculation of Z_d was performed using MATLAB for all operating points in a grid cycle. The analysis was also compared to a PLECS Blockset multitone analysis, and the results are shown in Fig. 5. Additionally, Fig. 6 shows the bode plot of $Z_{d-p-port}$ at the specific grid angle of $\pi/3$ radians.

D. Impact of LC Resonance Using EET

Without damping, the infinite magnitude of Z_{LC} at f_{LC} interacts with the magnitude of Z_n and/or Z_d, as shown in Fig. 7(a) for the transfer function $\hat{i}_p/\hat{d}_p$ (recall Z_n is infinite for this transfer function). As indicated in (13), the interaction of these magnitudes results in G_{cf} deviating from unity, affecting the plant transfer function G_{old}. This is shown in Fig. 7(b) as an additional 360° phase shift. This phase shift reduces the open-loop phase angle at the desired crossover frequency, which causes instability with closed-loop control bandwidths above the LC resonant frequency.

IV. DAMPING FILTER DESIGN

Whereas an undamped LC circuit has theoretically infinite impedance at resonance, it is desirable to determine the magnitude of Z_n and Z_d at the resonant frequency to know to what extent the resonance must be damped, thus satisfying (14). As seen in Figs. 4 and 5, these magnitudes change depending on the large-signal operating point, which varies throughout

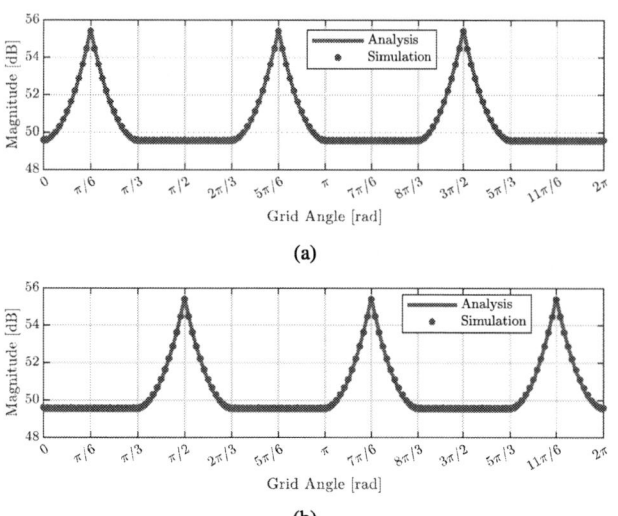

(a)

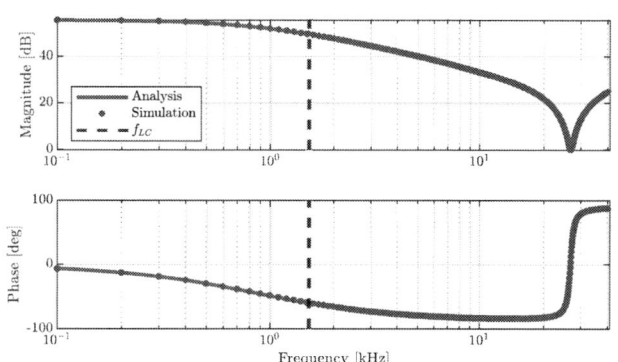

(b)

Fig. 5: Comparison of small-signal phasor modeling to PLECS Blockset/MATLAB simulation showing magnitude of (a) $Z_{d-p-port}$ and (b) $Z_{d-n-port}$ at input filter resonant frequency f_{LC}.

Fig. 6: Comparison of small-signal phasor modeling to PLECS Blockset/MATLAB simulation showing bode plot of $Z_{d-p-port}$ at steady-state operating point of grid angle equal to $\pi/3$ radians.

the grid cycle. The minimum magnitude of $Z_{n-p-port}$ (for transfer function $\hat{\imath}_{out}/\hat{d}_\phi$) is 21.1 dB at grid angle of $\pi/2$ radians (see Fig. 4). The minimum magnitude of $Z_{d-p-port}$ and $Z_{d-n-port}$ (for all transfer functions) is 49.5 dB at grid angle of $\pi/3$ radians (see Fig. 5). These worst case magnitudes are the maximum value to which Z_{LC} must be damped to sufficiently meet the requirements in (14).

For the present application, the LC impedance was damped to >20 dB (i.e., 1/10) below the 49.5 dB of Z_d. The impedance Z_d was prioritized as the damping limit even though it has a greater minimum than Z_n because it impacts the correction factor G_{cf} for all transfer functions, while Z_n is only finite for the transfer function $\hat{\imath}_{out}/\hat{d}_\phi$, which is the transfer function most directly impacting the output power control loop. The output power control loop has a bandwidth below the LC resonant frequency, while the PFC control loop utilizes high-bandwidth control, so the design allows for Z_{LC} to approach Z_n near the resonant frequency.

The chosen resistor value to meet the damping requirements is $R_{damp} = 10\ \Omega$, which corresponds to 20 dB. This value of

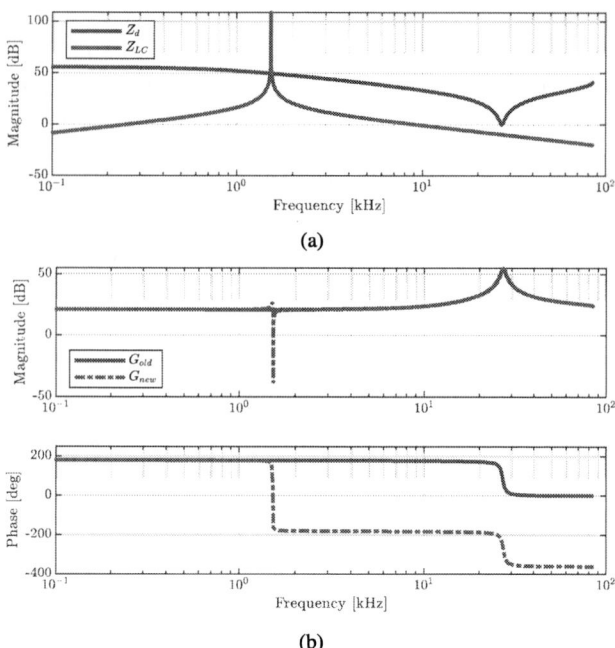

(a)

(b)

Fig. 7: (a) Comparison of Z_{LC} and Z_d as measured at the p-port for the $\hat{\imath}_p/\hat{d}_p$ transfer function and (b) the change in the transfer function due to the interaction of Z_{LC} and Z_d seen in (a), which includes a magnitude spike and $360°$ phase shift at the LC resonant frequency.

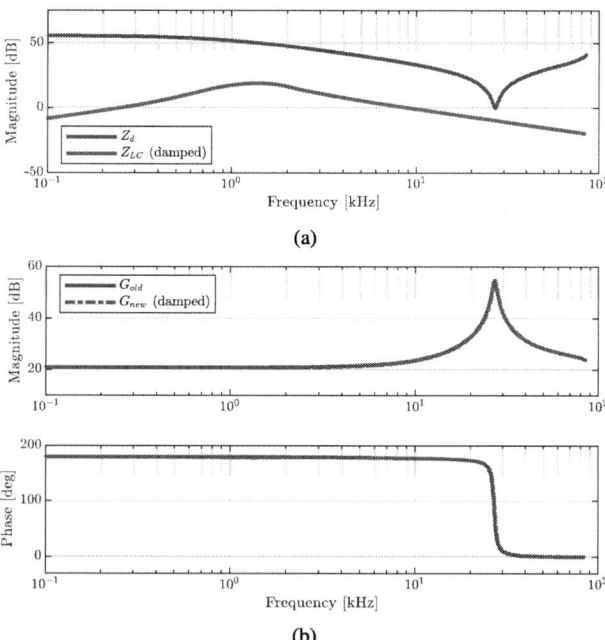

(a)

(b)

Fig. 8: (a) Comparison of damped Z_{LC} and Z_d as measured at the p-port for the $\hat{\imath}_p/\hat{d}_p$ transfer function and (b) the change in the transfer function due to G_{cf}. Because Z_{LC} does not interact with Z_d, G_{new} closely matches G_{old}.

R_{damp} is deliberately kept 20 dB lower than the lowest value of Z_d to avoid any interaction between the two impedances Z_d and Z_{LC}. To minimize the power loss at the 60 Hz grid frequency, an additional inductor and capacitor are added in parallel and series, respectively, with the resistor, as shown in the single-phase equivalent diagram in Fig. 9. The series

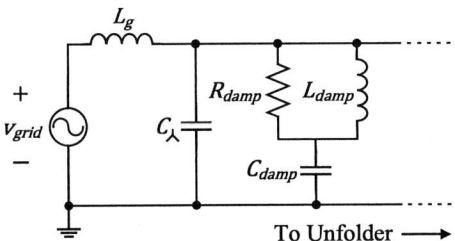

Fig. 9: Single-phase equivalent circuit of input to unfolder, including damping circuit.

capacitor C_{damp} minimizes the total current flow at low frequencies in the damping circuit branch, and the inductor L_{damp} allows low frequency current to bypass the R_{damp} branch specifically. Also, the inductor L_{damp} is chosen such that it has a higher impedance than R_{damp} at f_{LC} to ensure that Z_{LC} is damped to the chosen value of R_{damp}. For the same reason, C_{damp} is chosen to have a lower impedance than R_{damp} at f_{LC}.

With the selected damping filter component values, Z_{LC} now has a peak magnitude of 19.14 dB, thereby not interacting with Z_d. Thus, the requirements in (14) are met, and G_{cf} remains near unity at all frequencies and dc operating points. This is shown in Fig. 8, which shows the same Z_{LC} and Z_d plots as Fig. 7, but with the designed input damping filter.

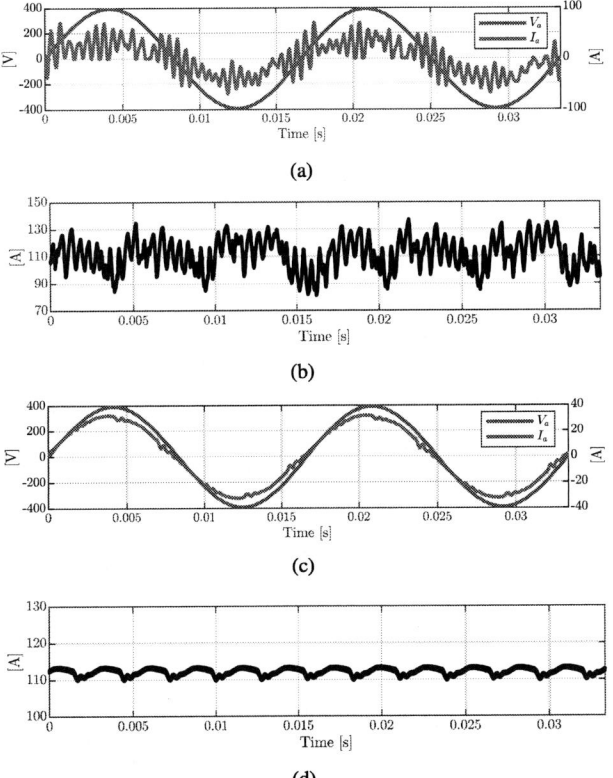

Fig. 10: (a) Phase A voltage and current and (b) output current when the damping filter is disconnected. Waveforms show high THD and distortion. (c) Phase A voltage and current and (d) output current with the damping filter connected, resulting in lower THD and closed-loop stability.

TABLE I: Summary of operating point and component values for simulation and hardware validation.

Parameter	Description
Nominal input voltage (V_{ll})	480 V, 60 Hz
Nominal battery (output) voltage (V_{bat})	160 V
Output power	18 kW
Grid inductance (L_g)	600 μH
Soft-dc-link capacitance ($C_{po/on/pn}$)	6.024 μF
Reflected grid-side capacitance ($C_{\perp}$)	18.072 μF
L_g to $C_{\perp}$ resonant frequency (f_{LC})	1.53 kHz
Unfolder diode ($D_{a1/b1/c1/a2/b2/c2}$)	WT1263Y200, 2000 V, 1263 A
Unfolder back-to-back IGBT ($Q_{a/b/c}$)	FZ1200R17KF6C-B2, 1700 V, 1200 A
T-SRC primary H-bridge MOSFET ($Q_{x1/y1/x2/y2}$)	GeneSiC G3R30MT12K 1200 V, 50 A, 2 in parallel
T-SRC back-to-back MOSFET ($Q_{x3/y3}^-$, $Q_{x3/y3}^+$)	GeneSiC G3R30MT12K 1200 V, 50 A
T-SRC secondary MOSFET ($Q_{u1/v1/u2/v2}$)	Infineon IRFP4768PBF 250 V, 93 A, 2 in parallel
T SRC switching frequency (f_s)	93.3 kHz
Transformer turns ratio (n_t)	0.25
Resonant inductance (L_s)	48 μH
Resonant capacitance (C_s)	120 nF
Output capacitance (C_{out})	195.2 μF
Damping resistance (R_{damp})	10 Ω
Damping inductance (L_{damp})	2.5 mH
Damping capacitance (C_{damp})	25 μF

V. SIMULATION AND EXPERIMENTAL VALIDATION

To validate the proposed damping technique, simulations using PLECS were ran with and without the damping filter. Table I shows the operating point and component values used in the simulation. The results, shown in Fig. 10, show high total harmonic distortion (THD) and poor PFC and output power regulation when the input impedance is not damped (Figs. 10(a) and (b)) and significantly improved waveforms with the damping filter included (Figs. 10(c) and (d)).

Additionally, the simulation was used to investigate the impact on the converter's efficiency when the damping filter is included. In steady-state operation, the simulation showed the shunt damper sinking an average of 4.24 W of power. With a rated output power of 18 kW, and a peak simulated efficiency of 97.9%, the shunt damper contributes a maximum of 1.09% of the system losses, or a 0.023% reduction in efficiency.

The system was tested in hardware using the setup shown in Fig. 11 and the same values as in Table I. As shown in Fig. 11,

979-8-3315-1612-3/25 $31.00 © 2025 IEEE

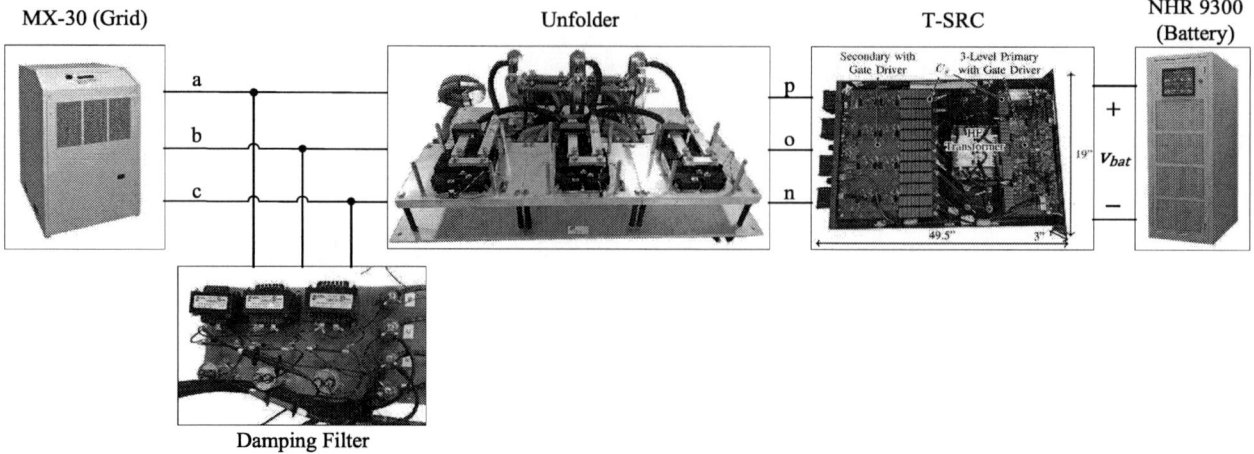

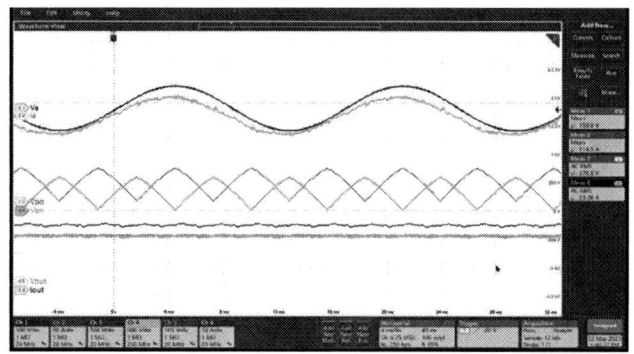

Damping Filter

Fig. 11: Block diagram of hardware setup for T-SRC with the damping filter at input.

Fig. 12: Oscilloscope capture of input and output waveforms. Waveform colors correspond with the following signals: black is V_a voltage, cyan is I_a current, pink is V_{po} voltage, green is V_{on} voltage, blue is battery current, and orange is battery voltage.

the testing assembly comprises a California Instruments MX-30 as a three-phase grid emulator, an unfolder, an 18 kW T-SRC module, and an NH Research 9300 in battery mode as the battery emulator. Connected in parallel to the three-phase input is the damping filter described in Section IV. With the damping filter connected, the converter can utilize a PFC control loop bandwidth greater than the LC resonant frequency without becoming unstable, as shown in Fig. 12.

VI. CONCLUSION

To conclude, this paper shows the effect of the resonance due to the equivalent input filter made from grid inductance and the dc-link capacitance of a three-phase ac-dc converter. First, the small-signal phasor modeling of the converter is explained, and three key transfer functions impacting the PFC and output power control loop are presented. Using similar modeling techniques, the EET impedances Z_n and Z_d of each transfer function are derived. Knowing the minimum magnitude of the EET impedances, a damping filter is designed to lower the peak magnitude of Z_{LC}. With the proposed damping filter, the three plant transfer functions are not affected, thereby

allowing PFC closed-loop operation above the LC resonant frequency. The proposed damping filter is validated in both simulation and hardware experimentation. The experimentation results indicate significant distortion without the damping filter and reduced grid current THD with the damping filter. This is accomplished with only a very minor (0.023%) reduction in converter efficiency.

REFERENCES

[1] O. A. A. Amer, I. Abdelsalam, and M. I. Marei, "An ev on-board charger based on dual-boost ac-dc converter," in *2024 International Telecommunications Conference (ITC-Egypt)*, 2024, pp. 263–268.

[2] R. Dwivedi, S. Singh, B. Singh, A. Chandra, and M. Rezkallah, "Three-phase ac/dc converter fed two parallel interleaved dc-dc converters for fast charging applications with improved power quality," in *2023 IEEE 14th International Conference on Power Electronics and Drive Systems (PEDS)*, 2023, pp. 1–6.

[3] L. Schrittwieser, J. W. Kolar, and T. B. Soeiro, "Novel swiss rectifier modulation scheme preventing input current distortions at sector boundaries," *IEEE Transactions on Power Electronics*, vol. 32, no. 7, pp. 5771–5785, 2017, doi: 10.1109/TPEL.2016.2609935.

[4] M. A. Mansour, "Control development for a medium-voltage three-phase unfolding input-series-output-parallel ac-dc converter," M.S. thesis, Utah State University, Logan, UT, May 2022, available at https://digitalcommons.usu.edu/etd/8475.

[5] A. Zade, S. Gurudiwan, D. Maksimović, and R. Zane, "High-bandwidth control of a 20-kw single-stage unfolding-based ac-dc converter using the extra element theorem and current emulation technique," *IEEE Transactions on Power Electronics*, pp. 1–19, 2024, doi: 10.1109/TPEL.2024.3423705.

[6] X. Li, J. Sun, L. Guo, M. Gao, H. Hu, and M. Xu, "A three-phase single-stage ac/dc converter based on swiss rectifier and three-level llc topology," *IEEE Transactions on Power Electronics*, vol. 38, no. 2, pp. 1958–1972, 2023.

[7] S. Gurudiwan, R. Hatch, M. Mansour, H. Wang, and R. Zane, "An 18 kw battery charger module for extreme fast charging applications using an unfolding-based ac-dc topology," in *2023 IEEE Applied Power Electronics Conference and Exposition (APEC)*, 2023, pp. 1715–1722, doi: 10.1109/APEC43580.2023.10131542.

[8] K. Kroics and J. Zarembo, "Lc input filter passive damping improvement for traction drive application," in *2021 IEEE 2nd International Conference on Smart Technologies for Power, Energy and Control (STPEC)*, 2021, pp. 1–4.

[9] P. N. A. Megat Yunus, A. Jusoh, and M. K. Hamzah, "Passive damping network for a single phase matrix converter (spmc) operating as a rectifier," in *2011 IEEE Symposium on Industrial Electronics and Applications*, 2011, pp. 173–177.

[10] C. Rim and G. Cho, "Phasor transformation and its application to the dc/ac analyses of frequency phase-controlled series resonant converters (src)," *IEEE Transactions on Power Electronics*, vol. 5, no. 2, pp. 201–211, 1990, doi: 10.1109/63.53157.

[11] D. Seltzer, L. Corradini, D. Bloomquist, R. Zane, and D. Maksimović, "Small signal phasor modeling of dual active bridge series resonant dc/dc converters with multi-angle phase shift modulation," in *2011 IEEE Energy Conversion Congress and Exposition*, 2011, pp. 2757–2764, doi: 10.1109/ECCE.2011.6064139.

[12] R. W. Erickson and D. Maksimovic, "Techniques of Design-Oriented analysis: Extra element theorems," in *Fundamental of Power Electronics*, 3rd ed. Cham, Switzerland: Springer, 2020, ch. 16, pp. 625–632.

[13] R. Middlebrook, "Null double injection and the extra element theorem," *IEEE Transactions on Education*, vol. 32, no. 3, pp. 167–180, 1989, doi: 10.1109/13.34149.

[14] ——, "The two extra element theorem," in *Proceedings Frontiers in Education Twenty-First Annual Conference. Engineering Education in a New World Order*, 1991, pp. 702–708, doi: 10.1109/FIE.1991.187583.

[15] R. Middlebrook, V. Vorperian, and J. Lindal, "The n extra element theorem," *IEEE Transactions on Circuits and Systems I: Fundamental Theory and Applications*, vol. 45, no. 9, pp. 919–935, 1998, doi: 10.1109/81.721258.

Methods to Enhance Cybersecurity of Multiple Inverters in Large Grid connected PV / Battery Energy Storage Systems

Hasan Ibrahim *Member, IEEE*, Jaewon Kim, *Member, IEEE*, Peng-Hao Huang, *Member, IEEE*,
Vishwam Raval, *Member, IEEE*, Prasad Enjeti, *Fellow, IEEE*
Power Electronics & Power Quality Laboratory, Dept of ECE, Texas A&M University,
College Station, TX 77845

Abstract—In this paper, methods for enhancing the cyber-security of multi-inverter systems in large-scale grid following photovoltaic (PV) and battery energy storage systems (ESS) are investigated. The study uses data from a 2.5 MW solar PV system and a 1.1 MW battery ESS, both interfaced with the grid through CenterPoint Energy Inc. at the Volkman location. The paper examines several attack scenarios, focusing particularly on False Data Injection (FDI) attacks targeting sensors within Cyber-Physical Systems (CPS). These inverters are especially vulnerable to cyber-attacks that compromise the integrity of the Programmable Logic Controller (PLC) and the transmitted sensor data. Attackers exploit vulnerabilities, including side-channel attacks on PLCs and sensor data spoofing, which threaten system security. To mitigate these risks, a dual-layer defense strategy is proposed, incorporating side-channel monitoring of the PLC and the embedding of a unique "watermark" (a low-magnitude, randomly fluctuating voltage signal) into the inverter's DC link voltage. This watermark serves as a diagnostic tool to detect unauthorized changes in sensor data, enhancing the system's ability to identify and respond to potential intrusions. Preliminary results from a hardware-in-the-loop (HIL) emulation, utilizing real-world data from the CenterPoint Energy solar farm, demonstrate the effectiveness of combining side-channel monitoring with watermarking in detecting previously undetectable attacks in multi-inverter systems.

Index Terms—Cyber Physical Security, PV, PLC, Watermarking, Side-Channel, hardware-in-the-loop (HIL)

I. INTRODUCTION

Cybersecurity concerns have evolved to encompass both Information Technology (IT) and Operational Technology (OT) systems within critical infrastructure sectors. This development is due to the increasing interconnectivity and reliance on Internet Protocol-based networks, which, while enhancing productivity, also introduce notable cyber vulnerabilities [2]. Inadequate cybersecurity measures have been estimated to cost the global economy $945 billion in 2020 alone [3]. The increasing use of information and communication technologies in smart grids amplifies the cybersecurity vulnerabilities of smart inverters, underscoring the urgency for enhanced cybersecurity measures to safeguard smart inverter operations and maintain grid integrity [4]. Large solar farms commonly employ multi-inverter systems and utilize network communication. While these farms commonly employ multi-inverter systems for enhanced grid flexibility, they also face increased vulnerability

to cyber attacks. Intrusions on such installations can severely impact grid reliability and stability [5]. Additionally, plant controllers utilizing SCADA and PLCs are venerable to cyber attacks such as false data injection attacks (FDIA) on their communication system [6], [7] and payload attack on PLCs control logic.

In recent years, various techniques have been proposed to enhance the security of Cyber-Physical Systems (CPS) against cyber-attacks. Traditional methods, highlighted in [8], often rely on network-based intrusion detection systems (IDS) and firewalls, which focus on monitoring traffic and identifying anomalous behaviors. In [9] a promising approach utilizes electromagnetic (EM) side-channel monitoring, where changes in EM emissions are analyzed to detect unauthorized activities in industrial control systems, such as those involving Programmable Logic Controllers (PLCs). This method enables the profiling of PLC programs and helps identify potentially malicious actions based on side-channel leakage. In [10] researchers have explored frequency-based signatures to defend against replay attacks, where a sinusoidal signal with a time-varying frequency is injected into the system to detect unauthorized data injections. In [11] an implementation of stochastic, time-varying parameters to enhance CPS security by limiting an attacker's ability to model the system accurately is explored. These methods, while effective in theory, face significant challenges in real-world applications due to system noise, increased complexity, computational overhead, and the potential for attackers to adapt to dynamic behaviors, making integration and performance difficult. Our proposed method combines the injection of a watermarking signal with side-channel monitoring for PLCs, aiming to enhance security by embedding unique identifiers within the system's operations and concurrently monitoring side-channel to detect anomalies. This integrated approach offers a more resilient defense against sophisticated cyber-attacks targeting CPS.

Fig.1 shows CenterPoint Energy *Volkman* solar PV/battery grid interface system diagram. The plant controller, such as a PLC, collects data from multiple sensors and system operators, then adjusts the set points for each PV/Battery grid interface. In [3], [4], [6], [12], numerous complex cyber attacks are documented targeting similar systems such as the notorious

979-8-3315-1612-3/25 $31.00 © 2025 IEEE

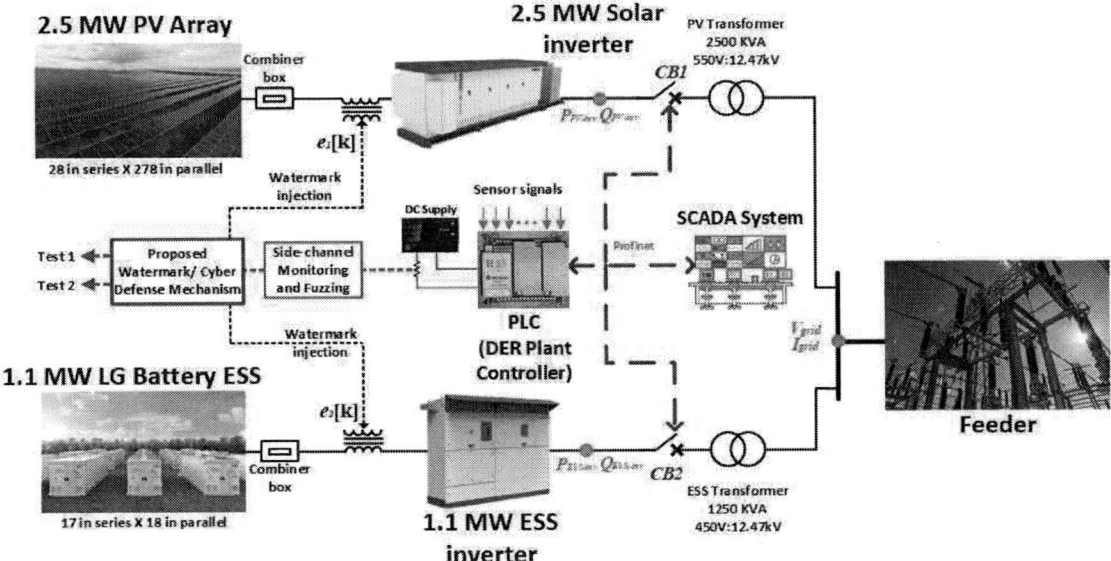

Fig. 1. Block diagram of CenterPoint Energy *Volkman* solar PV / Battery energy storage system interfaced to the grid. Such systems are monitored/controlled via a "DER plant controller". The Proposed Watermark/ Cyber Defense Mechanism [1] is shown as an add-on option to observe the system's signals and side channel monitoring to aid rapid detection of possible cyber intrusions (false data injection) in measurements employing commercial inverter systems interfaced to the gid and also connected to plant OT networks.

replay attack and several methods have been proposed to detect such attacks. In [13], to detect replay attacks in cyber-physical systems, an approach is employed that designs an authentication signal optimized to ensure minimal disturbance while maintaining effective attack detection. In [14], a method for detecting replay attacks in cyber-physical systems uses a frequency-based signature by injecting a sinusoidal signal with a time-varying frequency (authentication signal) into the system is introduced. However, such methods are solely validated through simulations and in real-world scenarios factors such as system noise may disrupt the functionality of such authentication signals.

As depicted in Fig. 1, our proposed Watermark/ Cyber Defense Mechanism utilizes watermarking to fortify ICS cybersecurity against sensor data manipulation through embedding a "watermark" - a randomly fluctuating voltage signal with a zero mean average and small magnitude - into the DC voltage connected to the PWM inverter. Simultaneously, it integrates PLC side-channel analysis to monitor power consumption through input current signature examination.

II. PROPOSED SYSTEM DEFENSE MECHANISM

In Fig. 1 , we present the block diagram illustrating Center point energy's multi-inverter PV/Battery system interface with the grid. The system's effective operation relies heavily on accurately reported sensor measurements, providing essential state information to the plant controller. However, the security of these systems faces a significant risk of compromise through false data injection into sensor measurements or corrupted controller logic, leading to improper set points for PV/battery subsystems. To address these security challenges, our approach

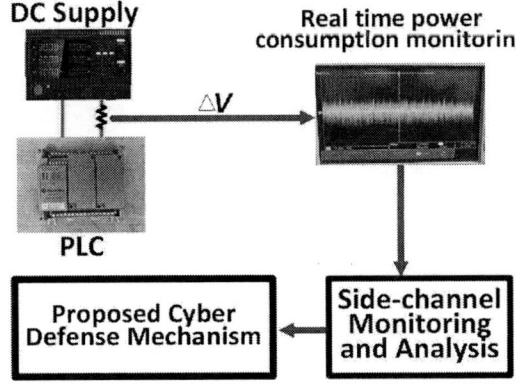

Fig. 2. Block diagram of CenterPoint Energy Power side-channel system illustration for a PLC

involves two key detection methods: side-channel monitoring of PLCs and sensor validation through system identification and watermark injection.

A. PLC operation and power side channel monitor

The proposed approach is evaluated by implementing cyber-attacks and defensive measures on a laboratory-scale prototype of an industrial control system (ICS) centered around an Allen-Bradley Micro820 PLC [15]. This PLC manages circuit breakers (CBs) for both the solar inverter and ESS inverter (see Fig. 1 and Fig. 4), integrating data from photovoltaic (PV) arrays and Energy Storage Systems (ESS) to regulate power flow and meet grid demand. The PLC compares PV array output with power demand, activating or deactivating the ESS circuit breaker as needed. To ensure the accurate delivery of power from both sources to the grid, the proposed defense mechanism employs a two-step verification process to secure the PLC control system. First, sensor data undergoes legitimacy verification through a cyber-shield. If sensor corruption is detected, a protection mode is triggered, ensuring the PLC operates based on legitimate sensor readings. However, benign sensor data input to corrupted PLC control logic can still lead to incorrect CB switching, destabilizing power output. To address this, the integrity of the PLC control logic is reinforced using a power side-channel monitor. This involves observing power consumption by measuring the voltage drop across a shunt resistor placed between the power supply and the PLC, as illustrated in Fig. 2. The proposed detector calculates the root-mean-square (RMS) of power usage using equations (1) and (2) with a fixed finite moving window size N.

Root-mean-square power consumption detection

$$P_{\text{inst}}[k] = V_{\text{supply}} \times \frac{\Delta V[k]}{R_{\text{shunt}}} \quad (1)$$

$$P_{\text{rms}}[N] = \sqrt{\frac{1}{N} \sum_{k=1}^{N} P_{\text{inst}}[k]^2} \quad (2)$$

$$\text{Decision} = \begin{cases} Benign & \text{when } \mu - 3\sigma \leq P_{\text{rms}}[N] \leq \mu + 3\sigma, \\ Malicious & \text{otherwise.} \end{cases}$$

B. Dynamic Watermark Architecture and tests

Based on the dynamic watermarking theory outlined in [16], and validated in the author's previous work [17]–[20], a small random varying signal, referred to as a watermark $e[k]$, is introduced into the DC input shown in Fig.1. Two variance tests, Test 1 and Test 2, are executed to calculate variance and identify potential cyber intrusions (attacks) aiming to manipulate sensor data through false data injection.

Test 1:

$$\lim_{T \to \infty} \frac{1}{T} \sum_{k=0}^{T-1} e_i[k] \big(z[k+1] - Az[k] - Bu[k] \big) = B_i \sigma_e^2 \quad (3)$$

Test 2:

$$\lim_{T \to \infty} \frac{1}{T} \sum_{k=0}^{T-1} \big(z[k+1] - Az[k] - Bu[k] \big) \quad (4)$$

$$\big(z[k+1] - Az[k] - Bu[k] \big)^T = \sigma_e^2 BB^T + \sigma_w^2 I_n$$

C. System Identification

To effectively conduct the two variance tests, a robust system identification methodology is crucial. This approach not only controls multi-scale systems but also analyzes the impact of external disturbances and potential faults. The system identification utilizes a *Least Squares Method (LSM)* for Multiple Input Single Output (MISO) system, where we determine the form of the prediction model as follows:

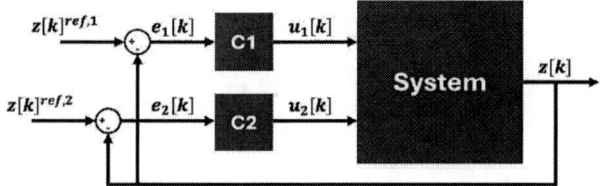

Fig. 3. Multiple Input Single Output (MISO) system.

$$\begin{aligned}
\hat{z}[k+1] = {} & \alpha_0 z[k] + \alpha_1 z[k-1] + \cdots + \alpha_N z[k-N] \\
& + \beta_0 u_1[k] + \beta_1 u_1[k-1] + \cdots + \beta_{M_1} u_1[k-M_1] \\
& + \gamma_0 u_2[k] + \gamma_1 u_2[k-1] + \cdots + \gamma_{M_2} u_2[k-M_2]
\end{aligned}$$
$$(5)$$

where $z[k]$ is the system output from sensors, $u_i[k]$ is the inputs for the system, $A_N = [\alpha_0 \; \alpha_1 \cdots \alpha_N]^T$, and $B_M = [\beta_0 \; \beta_1 \cdots \beta_{M_1} \gamma_0 \; \gamma_1 \cdots \gamma_{M_2}]^T$ are the parameters associated with input and output of the prediction model. The dimensions of the input and output vectors M and N are also unknown for the MISO system [21].

1) System Identification for Grid-Tied PV System: To model the grid-tied PV system shown in Fig. 1, we apply the system identification approach using the Least Squares Method (LSM). The prediction model for the system can be written as follows:

$$\begin{aligned}
\hat{z}[k+1] = f\big(& z[k-N:k], P_{\text{PV-inv}}[k-M_1:k], Q_{\text{PV-inv}}[k-M_2:k], \\
& P_{\text{ESS-inv}}[k-M_3:k], Q_{\text{ESS-inv}}[k-M_4:k], \\
& V_{\text{PV}}[k-M_5:k], V_{\text{ESS}}[k-M_6:k] \big)
\end{aligned}$$
$$(6)$$

Where:

- $z[k]$ is the system output (e.g., grid voltage or current) at time step k,
- $P_{\text{PV-inv}}[k]$ is the real power from the PV inverter at time k,
- $Q_{\text{PV-inv}}[k]$ is the reactive power from the PV inverter at time k,
- $P_{\text{ESS-inv}}[k]$ is the real power from the ESS inverter at time k,
- $Q_{\text{ESS-inv}}[k]$ is the reactive power from the ESS inverter at time k,
- $V_{\text{PV}}[k]$ is the voltage of the PV system at time k,
- $V_{\text{ESS}}[k]$ is the voltage of the ESS at time k,
- $N, M_1, M_2, M_3, M_4, M_5, M_6$ are the number of past time steps considered for each respective input or output.

Expanding the equation for the system identification model,

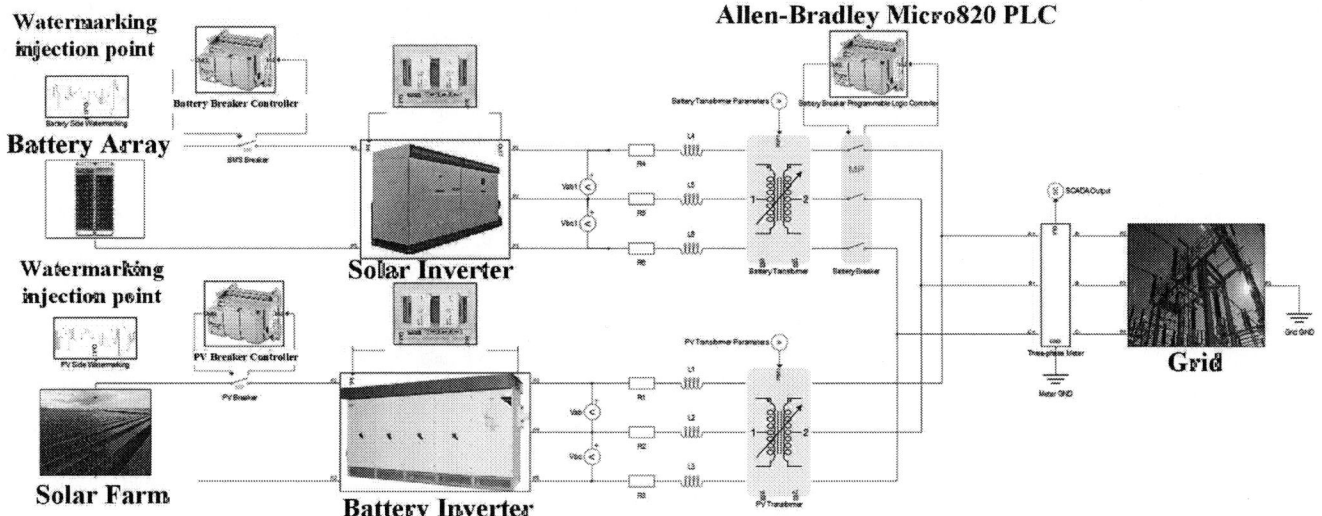

Fig. 4. Scaled-down Hardware-in-the-Loop (HIL) model of CenterPoint's solar farm and battery array, including their inverters connected to the grid. The figure illustrates the controllers for the breakers associated with both systems and the Programmable Logic Controller (PLC) used for monitoring side-channel attacks. Additionally, the points of dynamic watermarking injection are highlighted.

we have:

$$\hat{z}[k+1] = \alpha_0 z[k] + \alpha_1 z[k-1] + \cdots + \alpha_N z[k-N]$$
$$+\beta_0 P_{\text{PV-inv}}[k] + \beta_1 P_{\text{PV-inv}}[k-1] + \cdots + \beta_{M_1} P_{\text{PV-inv}}[k-M_1]$$
$$+\gamma_0 Q_{\text{PV-inv}}[k] + \gamma_1 Q_{\text{PV-inv}}[k-1] + \cdots + \gamma_{M_2} Q_{\text{PV-inv}}[k-M_2]$$
$$+\delta_0 P_{\text{ESS-inv}}[k] + \delta_1 P_{\text{ESS-inv}}[k-1] + \cdots + \delta_{M_3} P_{\text{ESS-inv}}[k-M_3]$$
$$+\lambda_0 Q_{\text{ESS-inv}}[k] + \lambda_1 Q_{\text{ESS-inv}}[k-1] + \cdots + \lambda_{M_4} Q_{\text{ESS-inv}}[k-M_4]$$
$$+\mu_0 V_{\text{PV}}[k] + \mu_1 V_{\text{PV}}[k-1] + \cdots + \mu_{M_5} V_{\text{PV}}[k-M_5]$$
$$+\nu_0 V_{\text{ESS}}[k] + \nu_1 V_{\text{ESS}}[k-1] + \cdots + \nu_{M_6} V_{\text{ESS}}[k-M_6]$$
$$(7)$$

Where:

- $A_N = [\alpha_0, \alpha_1, \ldots, \alpha_N]^T$ are the coefficients for the system output terms $z[k-N:k]$,
- $B_{M_1} = [\beta_0, \beta_1, \ldots, \beta_{M_1}]^T$ are the coefficients for the real power from the PV inverter,
- $B_{M_2} = [\gamma_0, \gamma_1, \ldots, \gamma_{M_2}]^T$ are the coefficients for the reactive power from the PV inverter,
- $B_{M_3} = [\delta_0, \delta_1, \ldots, \delta_{M_3}]^T$ are the coefficients for the real power from the ESS inverter,
- $B_{M_4} = [\lambda_0, \lambda_1, \ldots, \lambda_{M_4}]^T$ are the coefficients for the reactive power from the ESS inverter,
- $B_{M_5} = [\mu_0, \mu_1, \ldots, \mu_{M_5}]^T$ are the coefficients for the voltage of the PV system,
- $B_{M_6} = [\nu_0, \nu_1, \ldots, \nu_{M_6}]^T$ are the coefficients for the voltage of the ESS.

This model, based on the Least Squares Method (LSM), provides a framework for predicting the behavior of a grid-tied PV system by incorporating critical inputs such as real and reactive powers from the PV and ESS inverters and system voltages. As shown in Fig. 1, the system is monitored and controlled by a plant controller, which ensures grid stability. The performance of this model will be further validated in the **Experimental Results** section, where real data from the

CenterPoint Energy solar farm will be used to assess its effectiveness under actual operational conditions.

III. Experimental Results

In this section, a hardware-in-the-loop (HIL) emulation of *Volkman's* CenterPoint Energy solar farm is developed using the Typhoon HIL platform. The emulation utilizes real data collected over a period of 7 days from the facility, including multiple sensor readings such as total real power, ambient temperature, and DC voltage, as shown in Fig. 5. This real-world data captures all the system nonlinearities, noise, and operational characteristics, ensuring an accurate representation of photovoltaic (PV) panel operation. Fig. 6 illustrates the hardware-in-the-loop (HIL) test setup for the grid-tied inverter system. The system is modeled using the Typhoon HIL hardware emulator, while a physical Programmable Logic Controller (PLC) manages operations. A Texas Instruments Digital Signal Processor (TI DSP) is responsible for injecting watermark signals, monitoring side-channel data, and performing system identification and variance tests. To track system responses and detect anomalies or cyber-attacks in real-time, an oscilloscope is used. The setup includes a scaled-down 2.5 MW PV array and a 1.1 MW Battery Energy Storage System (ESS) to accommodate the Typhoon HIL system. The collected data forms the basis for modeling the dynamics of both the PV panels and the ESS inverter.

A. Cyber Attack detection

In this experiment, two attacks targeting the ICS are conducted: a replay attack is executed on the sensor data, where previously recorded data is replayed to manipulate the system's sensor readings, and a malicious payload attack is performed on the PLC control logic. For sensor attack detection, Variance

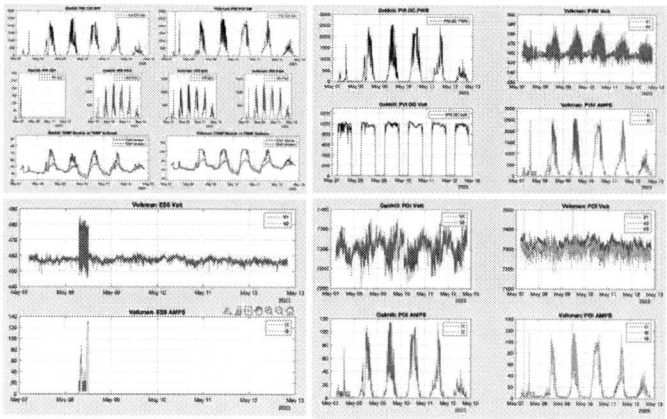

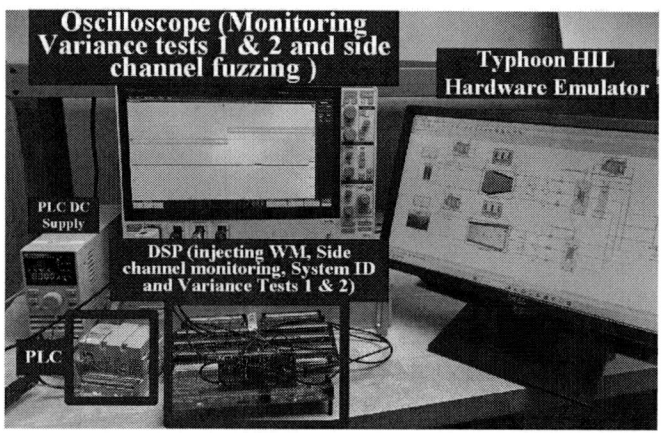

Fig. 5. Sensor data collected from the Volkman CenterPoint Energy solar farm over a 7-day period. The figure illustrates key performance metrics, including total real power output, ambient temperature, and DC voltage levels, capturing the system's real-world operational characteristics, nonlinearities, and environmental fluctuations.

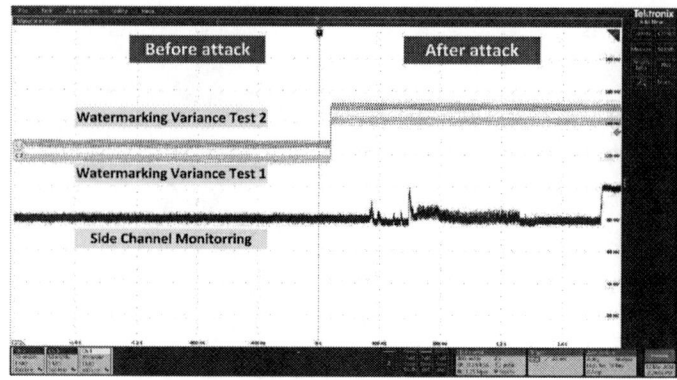

Fig. 7. Proposed watermarking/cyber defense mechanism Variance Tests 1 and 2 and side-channel monitoring signals. The figure shows the system's response before and after an attack, highlighting the detection of anomalies via watermark injection and side-channel signal monitoring.

IV. CONCLUSION

In this paper, a robust method for detecting Denial of Service (DoS) and Sensor False Data Injection (SFDI) attacks targeting sensors within Cyber-Physical Systems (CPS) is presented, focusing on grid-following PV inverters and battery system inverters in solar farms. These systems are particularly vulnerable to attacks on the Programmable Logic Controller (PLC), highlighting the critical need for effective countermeasures. By integrating side-channel monitoring of the PLC and embedding a unique "watermark" signal into the inverter's DC link voltage, the proposed approach offers a dual mechanism for detecting and identifying intrusions, thereby enhancing the security and integrity of the system. Preliminary results from a hardware-in-the-loop (HIL) system, based on real data collected from the Center Point Energy solar farm, validate the efficacy of this combined approach in detecting previously undetectable SFDI attacks in multi-inverter systems. Further evaluations will expand on these results, addressing multiple false data injection scenarios in a grid-connected PV/battery management system, focusing on both sensor data and PLC performance to ensure comprehensive defense against cyber threats.

REFERENCES

[1] P. Enjeti, P. R. Kumar, and L. Xie, "Methods and system for detecting compromised sensors using dynamic watermarking," Patent, (Pending) Texas A&M University System.

[2] "Ieee approved draft guide for cybersecurity of distributed energy resources interconnected with electric power systems," *IEEE P1547.3/D3.12, March 2023*, pp. 1–158, 2023.

[3] F. Cremer and Others, "Cyber risk and cybersecurity: a systematic review of data availability," *Journal of Cybersecurity Research*, vol. 10, no. 3, pp. 123–145, 2022.

[4] Y. Li and J. Yan, "Cybersecurity of smart inverters in the smart grid: A survey," *IEEE Transactions on Power Electronics*, vol. 38, no. 2, pp. 2364–2383, 2023.

[5] R. H. Lasseter, Z. Chen, and D. Pattabiraman, "Grid-forming inverters: A critical asset for the power grid," *IEEE Journal of Emerging and Selected Topics in Power Electronics*, vol. 8, no. 2, pp. 925–935, 2020.

[6] W. Alsabbagh, S. Amogbonjaye, D. Urrego, and P. Langendörfer, "A stealthy false command injection attack on modbus based scada systems," in *2023 IEEE 20th Consumer Communications & Networking Conference (CCNC)*, 2023, pp. 1–9.

Fig. 6. Hardware-in-the-loop (HIL) test setup for the grid-tied inverter system shown in Fig. 1. The Typhoon HIL hardware emulator models the system, while a physical PLC manages operations. The DSP injects watermark signals, monitors side-channel data, and performs system identification and variance tests. An oscilloscope tracks system responses, enabling real-time detection of anomalies and cyber-attacks during PV and Battery Energy Storage System (ESS) emulation.

Tests 1 and 2, as shown in Fig. 7, identify the attack by analyzing statistical deviations between the expected and actual data patterns. Both tests successfully detect the injected anomaly, confirming the presence of the attack. Concurrently, side-channel monitoring of RMS power consumption, using a fixed window size, is employed to verify the legitimacy of the PLC control logic. The distinctive pattern of power consumption during the attack, as illustrated in the figure, enables the system to detect unauthorized alterations in the control logic. This combined approach, utilizing both variance-based detection and side-channel analysis, effectively identifies the replay attack and underscores the robustness of the proposed defense mechanism.

979-8-3315-1612-3/25 $31.00 © 2025 IEEE

[7] Z. Wang, Y. Zhang, Y. Chen, H. Liu, B. Wang, and C. Wang, "A survey on programmable logic controller vulnerabilities, attacks, detections, and forensics," *Processes*, vol. 11, no. 3, 2023. [Online]. Available: https://www.mdpi.com/2227-9717/11/3/918

[8] K. Scarfone and P. Mell, "Guide to intrusion detection and prevention systems (idps)," 2007, accessed: 2024-11-15. [Online]. Available: https://csrc.nist.gov/pubs/sp/800/94/final

[9] C. M. Ahmed, M. Calder, S. Gunawan, J. Prakash, S. Nagaraja, and J. Zhou, "Time constant: Actuator fingerprinting using transient response of device and process in ics," 2024. [Online]. Available: https://arxiv.org/abs/2409.16536

[10] A. Humayed, J. Lin, F. Li, and B. Luo, "Cyber-physical systems security—a survey," *IEEE Internet of Things Journal*, vol. 4, no. 6, pp. 1802–1831, 2017.

[11] P. Griffioen, S. Weerakkody, and B. Sinopoli, "A moving target defense for securing cyber-physical systems," *IEEE Transactions on Automatic Control*, vol. 66, no. 5, pp. 2016–2031, 2021.

[12] A. Ginter, *Engineering-Grade OT Security: A manager's guide.* Abterra Technologies Inc., 2023. [Online]. Available: https://books. google.com/books?id=yLrrEAAAQBAJ

[13] T. Chen, L. Wang, X. Ren, Z. Liu, and H. Su, "Replay attack detection for cyber-physical systems with sensitive states," in *2023 62nd IEEE Conference on Decision and Control (CDC)*, 2023, pp. 2821–2826.

[14] H. S. Sánchez, D. Rotondo, T. Escobet, V. Puig, J. Saludes, and J. Quevedo, "Detection of replay attacks in cyber-physical systems using a frequency-based signature," *Journal of the Franklin Institute*, vol. 356, no. 5, pp. 2798–2824, 2019. [Online]. Available: https://www.sciencedirect.com/science/article/pii/S0016003219300134

[15] R. Automation. (2019) Micro820 programmable controllers. March, 2019. [Online]. Available: https://literature.rockwellautomation.com/idc/ groups/literature/documents/um/2080-um005_-en-e.pdf

[16] B. Satchidanandan and P. R. Kumar, "Dynamic watermarking: Active defense of networked cyber–physical systems," *Proceedings of the IEEE*, vol. 105, no. 2, pp. 219–240, 2017.

[17] W.-H. Ko, J. A. Ramos-Ruiz, T. Huang, J. Kim, H. Ibrahim, P. N. Enjeti, P. R. Kumar, and L. Xie, "Robust dynamic watermarking for cyber-physical security of inverter-based resources in power distribution systems," *IEEE Transactions on Industrial Electronics*, vol. 71, no. 7, pp. 7106–7116, 2024.

[18] H. A. J. Ibrahim, J. Kim, J. A. Ramos-Ruiz, W. H. Ko, T. Huang, P. N. Enjeti, P. R. Kumar, and L. Xie, "Detection of cyber attacks in grid-tied pv systems using dynamic watermarking," *IEEE Transactions on Industry Applications*, vol. 60, no. 1, pp. 819–827, 2024.

[19] H. Ibrahim, J. Ramos-Ruiz, J. Kim, W. H. Ko, T. Huang, P. Enjeti, P. R. Kumar, and L. Xie, "An active detection scheme for sensor spoofing in grid-tied pv systems," in *2021 IEEE Energy Conversion Congress and Exposition (ECCE)*, 2021, pp. 1433–1439.

[20] P.-H. Huang, J. Kim, P. R. Kumar, J. Rajendran, and P. Enjeti, "Enhancing cybersecurity for industrial control systems: Innovations in protecting plc-dependent industrial infrastructures," *IEEE Internet of Things Journal*, vol. 11, no. 22, pp. 36 486–36 493, 2024.

[21] J. Kim, H. Ibrahim, S. Wang, A. Mete, L. Xie, P. Enjeti, and P. Kumar, "Cyber-secure and safe operation of solar photovoltaic power distribution systems," in *2024 IEEE Applied Power Electronics Conference and Exposition (APEC)*. IEEE, 2024, pp. 1280–1287.

Optimal DC-DC Converter Topology and Control Algorithm for Fuel Cell Electric Vehicle with Series-Connected Supercapacitor

Hyeon Soo Kim
Department of Electrical and Computer Engineering
Sungkyunkwan University
Suwon, Korea
gustn6543@g.skku.edu

Yun Seong Hwang
Department of Electrical and Computer Engineering
Sungkyunkwan University
Suwon, Korea
hyshms@g.skku.edu

Seung Hyun Kang
Department of Electrical and Computer Engineering
Sungkyunkwan University
Suwon, Korea
dhsfkdls6258@g.skku.edu

Man Jae Kwon
Department of Electrical and Computer Engineering
Sungkyunkwan University
Suwon, Korea
akswo1234@skku.edu

Byoung Kuk Lee[†]
Department of Electrical and Computer Engineering
Sungkyunkwan University
Suwon, Korea
bkleeskku@skku.edu

Abstract— **This paper proposes a novel and optimized dc-dc converter employing a series-connected configuration of power sources and its control algorithm for fuel cell electric vehicles (FCEVs) with a supercapacitor (SC). Applying the series-connected configuration of power sources can decrease the voltage range of the FC and SC, leading to reduced volume of each power source and enhanced system efficiency. The proposed system intended for 120kW-rated FCEVs is designed and evaluated under various load conditions, and the feasibility is verified through experimental results on a DSP-based prototype downscaled to a 12kW-rated system.**

Keywords— *Fuel Cell Electric Vehicle (FCEV), Supercapacitor (SC), Series configuration of power sources, DC-DC converter*

I. INTRODUCTION

The conventional powertrain of fuel cell electric vehicles (FCEVs) widely uses a parallel-connected configuration between a fuel cell (FC) as the main power source and a supercapacitor (SC) as an auxiliary power source to support the dynamic characteristics of the FC, as shown in Fig. 1 [1-5]. Recently, research on the feasibility of a series-connected configuration of power sources, as shown in Fig. 2, is being studied and reviewed by several automotive manufacturing companies, and the key advantages of the series-connected configuration are as follows [6]:

1. When the power sources are connected in series, the voltage ratings of each power source satisfying the desired DC-link voltage (V_{DC}) can be reduced compared to the parallel-connected configuration. As a result, a reduction in the volume of the FC and SC can be effectively achieved since the volume of each power source is proportional to their rated voltage.

2. A series-connected power source can operate when high output power (P_{out}) is required. The series connection

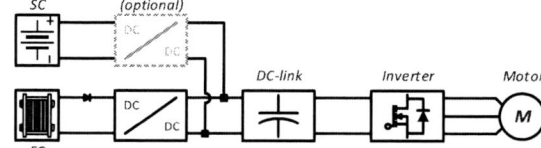

Fig. 1. Conventional structure [1-5].

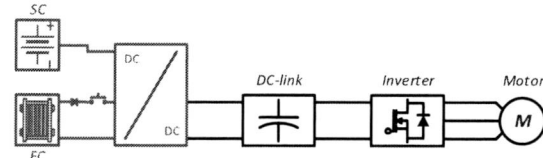

Fig. 2. Series-connected structure.

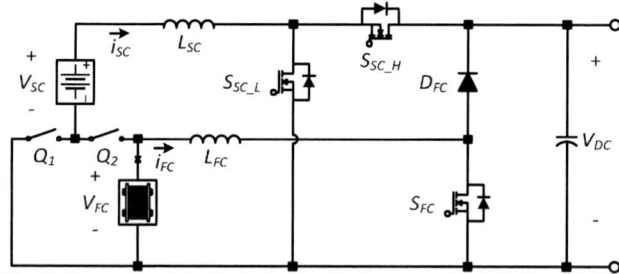

Fig. 3. The proposed dc-dc converter.

between the FC and SC results in increased input voltage compared to a single power source dealing with the load. In this case, the current stress of the power converter can be suppressed under the same P_{out} conditions, thereby enhancing system efficiency and being advantageous in the high P_{out} range [7].

To obtain the aforementioned advantages of the series-connected configuration and benefits of dynamic response in the

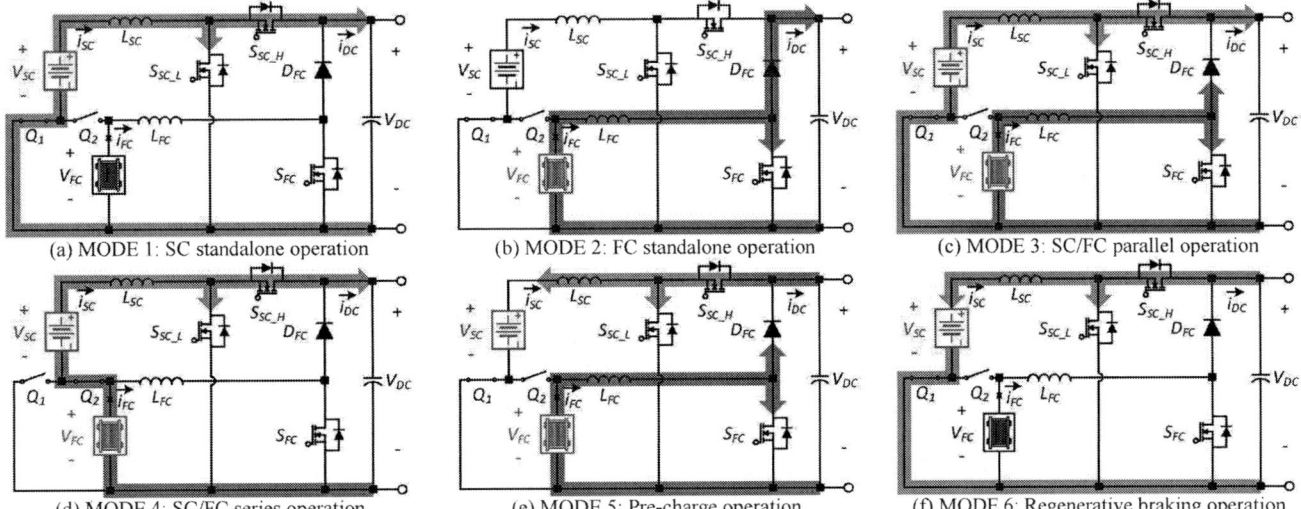

(a) MODE 1: SC standalone operation   (b) MODE 2: FC standalone operation   (c) MODE 3: SC/FC parallel operation

(d) MODE 4: SC/FC series operation   (e) MODE 5: Pre-charge operation   (f) MODE 6: Regenerative braking operation

Fig. 4. The key operation modes of the proposed dc-dc converter.

parallel-connected SC structure, it is necessary to consider an optimal dc-dc converter utilizing a series-connected power source, which can perform not only the series operation but also parallel and standalone operations of each power source [8]. Therefore, in this paper, the new optimized dc-dc converter based on the series-connected power source for FCEVs and its control algorithm which can effectively cover the entire operation modes are proposed and designed. The proposed dc-dc converter intended for 120kW-rated FCEV systems is designed, and an analysis of each operation mode and the control algorithm is conducted. To validate the proposed system, a downscaled 12kW-rated prototype based on the TMS320F28x DSP is implemented. The feasibility of the proposed dc-dc converter and control algorithm is verified through experimental results.

II. MODE ANALYSIS OF PROPOSED DC-DC CONVERTER

The proposed dc-dc converter as shown in Fig. 3 is composed of a pair of relays (Q_1, Q_2) and two boost converters. Q_1 and Q_2 provide a path of power flow and determine an operation mode according to frequently changeable load conditions. Since the V_{DC} should be regulated to supply a constant voltage to an inverter, one of the boost converters should conduct a voltage mode control (VMC) in any operation including standalone, parallel, series operations, and even a transition between the modes. The operation modes consist of a six types, with four modes delivering power to the load and two modes that charge the SC, as depicted in Fig. 4. The detailed analysis of each operation is as follows:

Fig. 4(a) and (b) represent the standalone operation of the SC and FC respectively. The standalone operations are performed such as during the initial startup of the converter and under constant-speed driving conditions. MODE 1 and MODE 2 can be carried out by activating Q_1 and applying VMC to the corresponding boost converters for each power source. In MODE 3, the SC and FC operate in parallel, simultaneously delivering power from each source, as shown in Fig. 4(c). During MODE 3, VMC is performed through the SC side while a current mode control (CMC) is carried out on the FC side, and

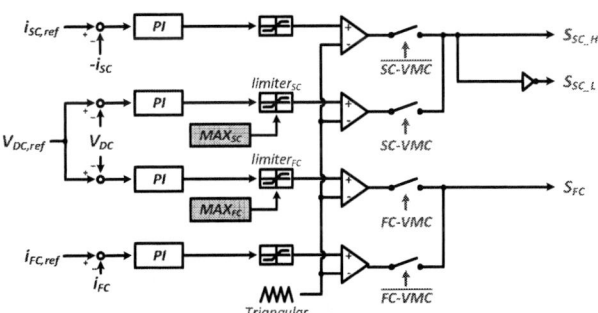

Fig. 5. Block diagram of control algorithm.

the distribution ratio of the load can be determined by controlling FC current (i_{FC}). MODE 4 represents series-connected operation mode of the SC and FC as shown in Fig. 4(d). It can be performed by activating the Q_2 and operating VMC to manage a constant V_{DC} by the boost converter on the SC side. When the high-power demand exceeding the rated power of each power source is required, the target P_{out} can be supplied through the combined input voltage of the FC and SC (V_{FC+SC}) without increasing input current. Fig. 4(e) illustrates the pre-charge operation mode. When the vehicle starts, the FC regulates the V_{DC} by performing VMC under no-load conditions. Subsequently, the SC is charged through CMC. The SC charging operation mode is additionally executed during regenerative braking, where the SC is charged by harnessing the energy generated from vehicle deceleration, as shown in Fig. 4(f). In the pre-charge and regenerative braking operation, the system operates in buck converter mode transferring power from the load to the SC side, with S_{SC_H} and S_{SC_L} conducting complementary switching.

III. PRINCIPLE OF CONTROL ALGORITHM

In case of switching the standalone or parallel operation into series operation, and in the reverse, a substantial overshoot of V_{DC} may occur during the transient state of the mode transitions, since the input voltage changes considerably. Thus, when switching from one mode to another, the flexible transition

979-8-3315-1612-3/25 $31.00 © 2025 IEEE

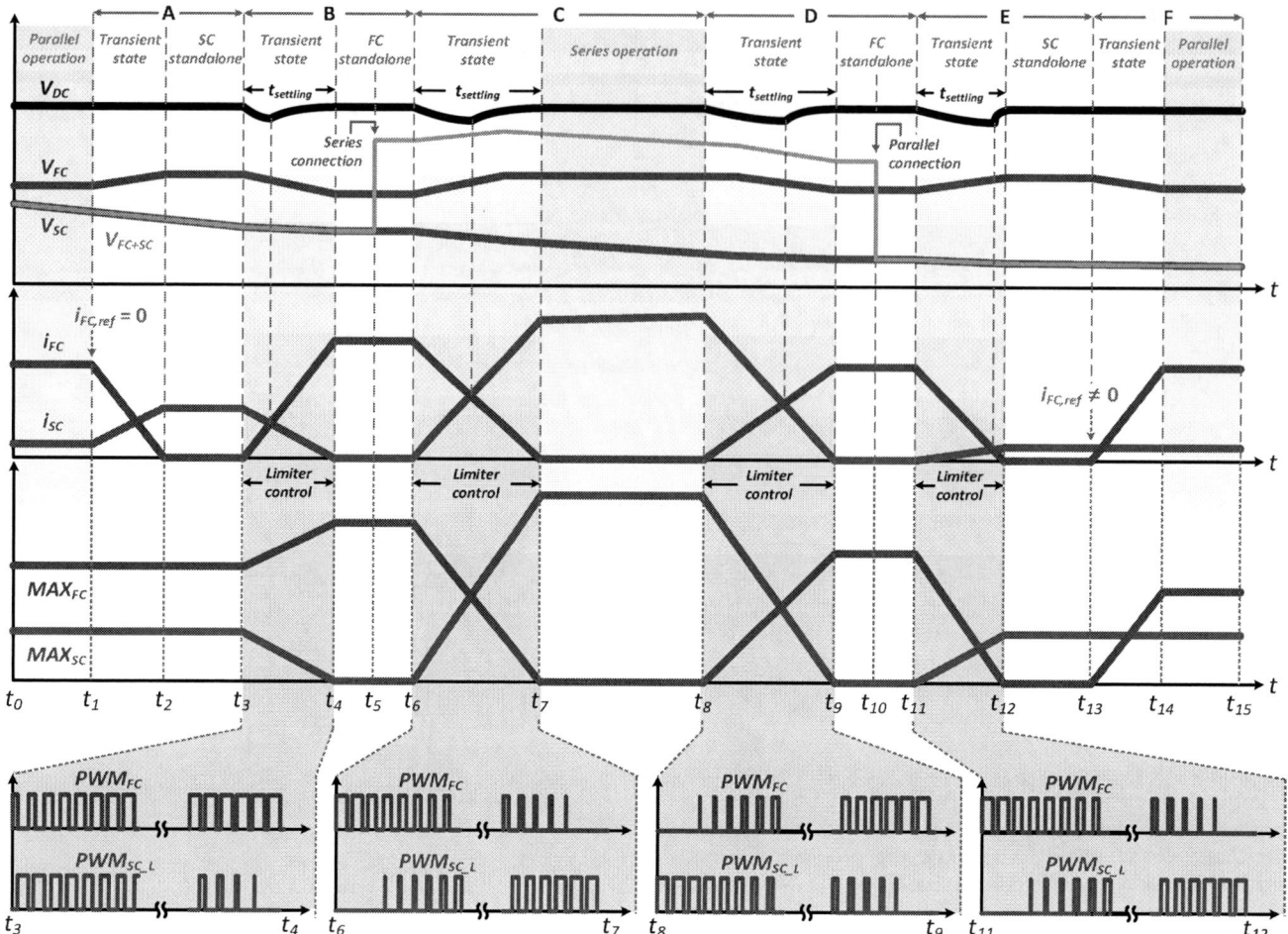

Fig. 6. The key waveforms for mode transition.

$$\Delta limiter = \frac{f_{SYSCLK} \cdot (V_{DC} - V_{in})}{n \cdot V_{DC} \cdot t_{settling} \cdot f_{sw}^2} \quad n = \begin{cases} 1, & (Sawtooth\ waveform) \\ 2, & (Triangular\ waveform) \end{cases} \quad (1)$$

process between VMC and CMC applied to boost converters is required to enhance the transient state. Therefore, the transition of VMC between the boost converters should be gradually performed to suppress drastic V_{DC} ripple. To achieve this, a control block diagram of the proposed dc-dc converter can be configured with MAX_{FC} and MAX_{SC} to enforce the controller output, as shown in Fig. 5. Fig. 6 demonstrates the proposed seamless transition process between configuration types of the FC and SC, allowing for both parallel-to-series and series-to-parallel transitions. The detailed analysis according to the transition sequence (t_1–t_{15}) is as follows:

A. Transition from parallel to SC standalone: t_1-t_3

Since the SC side converter performs VMC in both the SC standalone operation and the series connection, VMC of the SC side needs to be deactivated prior to establishing the interface for the power sources in series. Therefore, VMC on the FC side should be activated to maintain a stable V_{DC} during the deactivation of VMC on the SC side for seamless mode transitions. In sequence A, CMC on the FC side converter is

disabled by setting the reference current of the FC ($i_{FC,ref}$) to zero, allowing the SC side converter to take over VCM. When the i_{FC} reaches zero, the duty ratio of PWM for the FC side switch becomes nearly zero, resulting in converting parallel operation into the SC standalone operation. Maintaining a constant V_{DC} can be achieved by conducting VMC of the SC during this sequence.

B. Transition from SC standalone to FC standalone: t_3-t_6

In sequence B, VMC applied to the SC side converter should be gradually deactivated and VMC on the FC side converter should be gradually and simultaneously activated to ensure a seamless transition from SC-to-FC standalone operation. Adjusting the output value of each converter's controller (limiter$_{SC}$ and limter$_{FC}$) facilitates gradual VMC transition between converters. By applying limiter handling at t_3, the upper limit (MAX_{SC}) of the limiter$_{SC}$ decreases to zero, while the upper limit (MAX_{FC}) of the limiter$_{FC}$ increases to a target value to meet the V_{DC} in Fig. 6. As shown in Fig. 5 and 7, the settling time ($t_{settling}$) of the V_{DC} is determined by the linear incremental and decremental value ($\Delta limiter$) of the limiter$_{SC}$ and limiter$_{FC}$. The $\Delta limiter$ for each sampling period can be calculated through (1) to satisfy the user-defined $t_{settling}$ in the TMS320F28x DSP-based system [9]. When the MAX_{FC} and MAX_{SC} reach the target

Fig. 7. The 12kW-rated prototype of the proposed dc-dc converter.

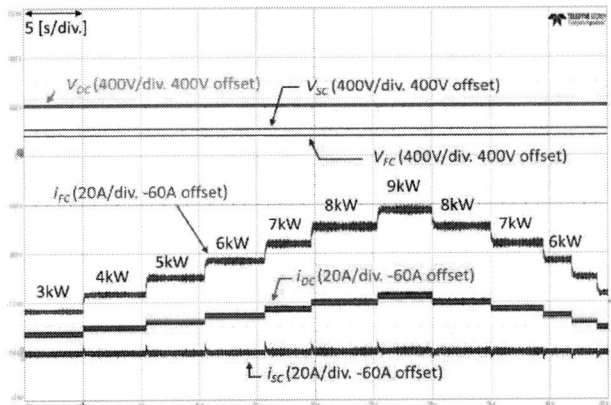

Fig. 8. The experimental results of parallel-connected configuration.

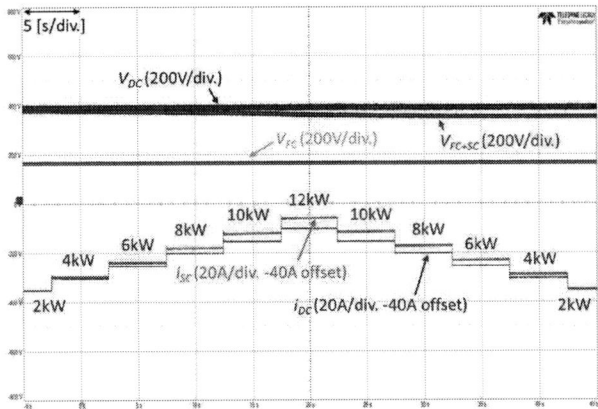

Fig. 9. The experimental results of series-connected configuration.

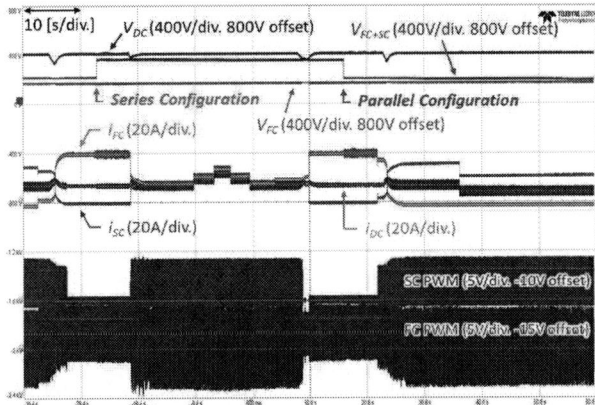

Fig. 10. The experimental results of parallel-series-parallel transition.

value and zero respectively at t_4, there is no current flow on the SC side as the operation mode is converted from the SC standalone to FC standalone mode. By activating Q_2 at t_5, the input voltage can be increased up to V_{FC+SC}.

C. Transition from FC standalone to series operation: t_6-t_8

In this sequence, since the switches of the SC side should finally be activated when the power sources are connected in series, VMC of the FC should be deactivated. Thus, the MAX$_{FC}$ decreases linearly, while the MAX$_{SC}$ increases linearly using (1) as shown in Fig. 7. Through this process, the seamless transition from the parallel to series operation between the FC and SC can be achieved.

D. Transition from series operation to FC standalone: t_8-t_{11}

The transition from the series to parallel operation can be achieved through the inverse process of sequence A to C. In order to achieve the inverse transition, the SC side VMC should be deactivated with Q_2 remaining active, and the FC side VMC should be activated. Thus, limiter handling is required through the linear decrease of MAX$_{SC}$ and the linear increase of MAX$_{FC}$. When MAX$_{SC}$ and MAXFC reach their respective target values at t_9, the SC side converter terminates operation, while the FC side converter operates VMC to maintain a constant V_{DC}. By deactivating Q_2 and activating Q_1 at t_{10}, the series connection between the SC and FC is disabled, meeting the conditions for parallel operation mode.

E. Transition from FC standalone to SC standalone: t_{11}-t_{13}

In parallel operation, as the FC operates under CMC while the SC performs VMC, the controller should terminate VMC on the FC between t_{11} and t_{13} by linearly adjusting MAX$_{FC}$ and MAX$_{SC}$ to satisfy the user-defined $t_{settling}$ using (1). When limiter$_{SC}$ and limiter$_{FC}$ reach their target values at t_{12}, the converter on the FC side ceases operation, and the converter on the SC side exclusively undertakes VMC.

F. Transition from SC standalone to parallel operation: t_{13}-t_{15}

During the standalone execution of VMC on the SC side, CMC of the FC side converter should be implemented to transition to parallel operation. By clearing the FC-VMC signal in Fig. 5 and feeding $i_{FC,ref}$ at t_{13}, CMC can be triggered, and once the steady state is reached at t_{14}, the system can fully return to its original parallel operation state. Since the proposed transition algorithm has a structure in which various mode transitions are included, all possible transitions between the modes can be achieved.

IV. EXPERIMENTAL VALIDATION OF PROPOSED CONVERTER

To validate the proposed dc-dc converter and its control algorithm for the parallel-series and series-parallel transition of power sources, a 12kW-rated prototype, as a scaled-down model of 120kW FCEV system, is configured as shown in Fig. 7. 3-phase interleaved boost converters are utilized on the SC and FC

979-8-3315-1612-3/25 $31.00 © 2025 IEEE

sides to meet the design specifications of current ripple and system volume, using an FC simulator and SC [10]. In parallel operation with load regulation, the SC current hardly changes as it performs VMC, while the FC current varies according to the load due to CMC, as shown in Fig. 8. Fig. 9 shows the series operation of the SC and FC under varying load conditions from 2kW to 12kW, and the V_{FC+SC} decreases as the V_{SC} falls due to the discharging. In addition, the transition from the parallel to series connection and the reverse transition can be flexibly achieved as shown in Fig. 10. Experimental results confirm the feasibility and validity of the proposed converter and control algorithm for seamless mode transition.

V. CONCLUSION

This paper proposed the novel and optimized dc-dc converter in FCEV with SC applications. Considering various driving scenarios, six operating mode are specified, and its control algorithm that enables reconfiguration of the series-parallel connections of power sources is developed. To experimentally validate the proposed dc-dc converter and control algorithm, a 12kW scaled-down system based on the TMS320F28x DSP was designed. As shown in Fig. 8 and 9, the feasibility of the proposed dc-dc converter was verified through experiments on the series and parallel configurations of the power sources. Furthermore, as illustrated in Fig. 10, the effectiveness of the control algorithm was validated through experiments on transitions from parallel to series and back to parallel configurations.

ACKNOWLEDGMENT

This work was supported by Korea Institute of Energy Technology Evaluation and Planning (KETEP) grant funded by the Korea government (MOTIE) (RS-2024-00422103, EV Smart Charging Platform Innovation Research Center). This work was also supported by the Announcement of Materials/Parts Technology Development Program (20024898, Development of 600kW Battery Emulator for dynamometer system) funded By the Ministry of Trade, Industry & Energy(MOTIE, Korea).

REFERENCES

[1] Q. Xun, Y. Liu and E. Holmberg, "A Comparative Study of Fuel Cell Electric Vehicles Hybridization with Battery or Supercapacitor," *2018 International Symposium on Power Electronics, Electrical Drives, Automation and Motion (SPEEDAM)*, Amalfi, Italy, 2018, pp. 389-394.

[2] A. Emadi, S. S. Williamson and A. Khaligh, "Power electronics intensive solutions for advanced electric, hybrid electric, and fuel cell vehicular power systems," *in IEEE Transactions on Power Electronics*, vol. 21, no. 3, pp. 567-577, May 2006.

[3] M. Li, P. Yu, Y. Wang, Z. Sun and Z. Chen, "Topology Comparison and Sensitivity Analysis of Fuel Cell Hybrid Systems for Electric Vehicles," *in IEEE Transactions on Transportation Electrification*, vol. 9, no. 4, pp. 5111-5121, Dec. 2023.

[4] H. El Fadil, F. Giri, J. M. Guerrero and A. Tahri, "Modeling and Nonlinear Control of a Fuel Cell/Supercapacitor Hybrid Energy Storage System for Electric Vehicles," *in IEEE Transactions on Vehicular Technology*, vol. 63, no. 7, pp. 3011-3018, Sept. 2014.

[5] T. Azib, O. Bethoux, G. Remy, C. Marchand and E. Berthelot, "An Innovative Control Strategy of a Single Converter for Hybrid Fuel Cell/Supercapacitor Power Source," *in IEEE Transactions on Industrial Electronics*, vol. 57, no. 12, pp. 4024-4031, Dec. 2010.

[6] 2010J. Prasad and R. K. Tripathi, "Series and hybrid connection of sediment microbial fuel cell for powering Led," *2017 14th IEEE India Council International Conference (INDICON)*, Roorkee, India, 2017, pp. 1-5.

[7] W. -S. Liu, J. -F. Chen, T. -J. Liang and R. -L. Lin, "Multicascoded Sources for a High-Efficiency Fuel-Cell Hybrid Power System in High-Voltage Application," *in IEEE Transactions on Power Electronics*, vol. 26, no. 3, pp. 931-942, March 2011.

[8] Y. Wu and H. Gao, "Optimization of Fuel Cell and Supercapacitor for Fuel-Cell Electric Vehicles," *in IEEE Transactions on Vehicular Technology*, vol. 55, no. 6, pp. 1748-1755, Nov. 2006.

[9] Texas Instruments, "TMS320F2837xS Real-Time Microcontrollers Technical Reference Manual," SPRUHX5I, October 2023. [Online]. Available: https://www.ti.com/lit/ug/spruhx5i/spruhx5i.pdf. [Accessed: April. 12, 2024.

[10] A. Laddha and S. Neeli, "Intelligent Control of a Two-phase Interleaved Boost Converter-interfaced Fuel Cell Electric Vehicle," *2022 International Conference on Intelligent Controller and Computing for Smart Power (ICICCSP)*, Hyderabad, India, 2022, pp. 1-6.

Reliability-Constrained Design of a High-Gain Power Optimizer based on a Real Mission Profile

Stefano Cerutti*, Francesco Iannuzzo§, Ariya Sangwongwanich§, Tamas Kerekes§, Mario Giuseppe Pavone†,
Francesco Gennaro†, Natale Aiello‡, Francesco Musolino*, Paolo Stefano Crovetti*

*Department of Electronics and Telecommunications, Politecnico di Torino, Torino, 10129, Italy
{stefano.cerutti, francesco.musolino, paolo.crovetti}@polito.it

§Department of Energy, Aalborg University, Aalborg, 9220, Denmark
Email: {fia, ars, tak}@energy.aau.dk

†ST Microelectronics srl, Catania, 95121, Italy
Email: {mario.pavone, francesco.gennaro}@st.com

‡ST Microelectronics srl, Agrate Brianza, 20864, Italy
Email: natale.aiello@st.com

Abstract—This article proposes a methodology for the design of a step-up solar optimizer based on the minimization of the cost-to-output energy ratio and constrained by reliability criteria. Real-world annual mission profiles are used in the optimization, allowing to a more accurate lifetime analysis and output energy estimation. The proposed approach is applied to the case study of an Input-Parallel-Output-Series power optimizer and exploits analytical and empirical loss and cost models for the converter components. The algorithm is implemented in Matlab and is based on the Particle Swarm Optimization (PSO) method, which discards the solutions not meeting the reliability requirements. The optimal solution found by the PSO is then compared with the exhaustive search in the variables space. A converter is designed based on the optimal solution, and its efficiency and accumulated damage are validated by LTSpice and Matlab simulations.

Keywords—Design optimization, Reliability, Power optimizer, Particle Swarm Optimization.

I. INTRODUCTION

Module-level power converters (MLPC), such as microinverters and power optimizers, are becoming widespread solutions to maximize the energy harvesting from small-scale distributed photovoltaic (PV) systems [1]–[3]. The shift from centralized or string configurations to MLPC, however, requires careful design considerations in terms of multiple conflicting objectives such as cost, power density, efficiency and reliability [4]. To assist the off-line design of power converters, deterministic or Artificial Intelligence (AI)-based optimization algorithms have been proposed [5].

Many works focus on the multi-objective optimization of MLPC in terms of weighted efficiency and power density [3], [6], volume [7] or cost [8], [9]. These works do not consider at all the converter reliability in the optimization: this may lead to a late re-design or costly replacements of faulty parts in the field.

This publication is part of the project PNRR-NGEU which has received funding from the MUR – DM 352/2022.

Other works take into account reliability requirements, expressed in terms of the Mean Time Between Failures (MTBF), within the objective function [10]–[12]. This approach, however, has two fundamental limitations: 1) it may result in noncompetitive design solutions from the cost point of view, and 2) it does not model the physical degradation of the converter components due to the repeated thermal stresses. The mission profile-based lifetime consumption (LC) of the power modules was considered in the objective function only in the large-scale PV systems design in [13], however it did not consider the cost implications of a reliability-oriented optimal solution. The inclusion of the mission profile in the optimization approach allows for a more accurate estimation of the converter working points, thermal stresses and expected harvested energy [14].

To overcome the above-mentioned limitations, this article proposes a new design optimization methodology based on the minimization of a modified definition of Levelized Cost of Energy (LCOE) including a mission-profile based lifetime estimation of the converter switches as a constraint. The methodology is applied to the case study of an asymmetric Input-Parallel-Output-Series Power Optimizer (IPOS-PO) [15]. The proposed methodology can be generalized to different converter topologies upon characterization of the electrical stresses of the target converter.

The rest of the article is divided as follows: Section II introduces the converter topology considered as case study of the proposed optimization algorithm, defines the objective function, and describes the modelling and constraints of the methodology; Section III illustrates the Particle Swarm Optimization (PSO) method adopted as search algorithm, presents the impact of different mission profiles on the optimal solutions and the simulation results of a converter designed according to the optimal variable set; finally, conclusions and future work are drawn in Section IV.

979-8-3315-1612-3/25 $31.00 © 2025 IEEE

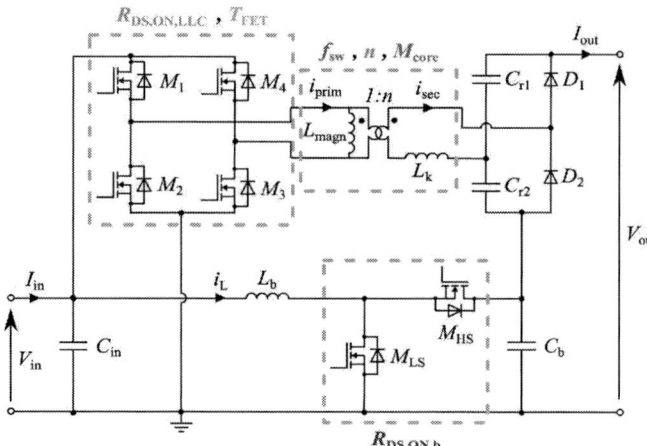

Fig. 1: Schematic of the IPOS-PO converter considered as case study for the optimization.

II. PROPOSED OPTIMIZATION METHODOLOGY

A. Case study

Before describing the details of the proposed optimization algorithm, the converter topology adopted as case study is introduced. IPOS converters are recently becoming attractive solutions for applications requiring high voltage gains with reduced electrical stresses [16]. In this work, the target converter is the asymmetric IPOS–PO shown in Fig. 1, consisting of a fixed-gain resonant LLC with Voltage Doubler Rectifier and a synchronous boost operating in Boundary Conduction Mode responsible for the Maximum Power Point Tracking (MPPT). In general, for improved performance, the two stages are designed so that the largest fraction of power is processed by the LLC. The detailed description of the converter operation and the derivation of the electrical stresses are out of the scope of this work and can be found in [15]. It is worth noticing that the proposed algorithm optimizes the design of both the conversion stages simultaneously.

In this work, the reduced-order modelling approach [17] has been adopted to identify the minimum size of the solution space, in order to optimize the computational complexity and time. With this approach, the minimum set of variables significantly affecting the converter cost and output energy were identified and are highlighted in Fig. 1:

- for the low-voltage Field-Effect Transistors (FETs) M_{1-4} of the LLC, the conduction resistance $R_{\mathrm{DS,ON,LLC}}$ and technology T_{FET};
- for the LLC transformer, the operating frequency f_{sw}, turns ratio n and core material M_{core};
- for the high-voltage boost MOSFETs, the conduction resistance $R_{\mathrm{DS,ON,b}}$.

It is worth observing that the selected design variables exhibit multiple effects on the converter design: for instance, n determines, for a specific input voltage operation, the required voltage gains and the fraction of input power processed by each conversion stage, with significant impact on all the power

devices. $R_{\mathrm{DS,ON,LLC}}$ not only has an impact on the cost, conduction and switching losses of the LLC FETs, but also on the magnetizing inductance requirement of the transformer, in turns affecting its size and cost. More detailed considerations on the modelling are provided in Section II-C.

B. Definition of the objective function

The LCOE is a widely adopted figure of merit to compare different technologies used in energy production in terms of the cost-to-benefit ratio over the system lifetime [18]. The complete definition of LCOE involves the computation of the Net Present Value (NPV) of the lifetime costs (including initial investment, operation and maintainance costs, and potential fuel costs) and the NPV of the total output energy produced during the lifetime [19]. In this work, it is of interest to optimize the design of the Parallel Power Optimizer (PPO) in Fig. 1, assuming that the topology, panel specifications and mission profile are design constraints.

The proposed design optimization is based on the search inside the N-dimensional solution space of the optimal variable set, i.e. the combination of design variables optimizing the objective function. As a consequence, all the cost contributions that are fixed independently of the converter design, such as the PV module or the installation costs, are not of interest in this discussion and are neglected from the objective function. Since one of the objectives of the proposed optimization is to ensure, through a reliability constraint, that the converter lifetime matches the PV module lifetime, no replacement costs are considered.

Thus, the system-level definition of LCOE is here reduced and simplified to take into account only the contributions of cost and power conversion losses related to the converter design. For this reason, the objective function is here explicitly called Converter Cost-Energy Ratio (CCER). Denoting by S the trial design solution and by M the mission profile, the definition of CCER becomes:

$$\mathrm{CCER} = \frac{C_{\mathrm{tot}}(S)}{E_{\mathrm{tot}}(S,M)} = \frac{C_{\mathrm{tot}}(S)}{\sum\limits_{k=0}^{25\,\mathrm{yrs}} (P_{\mathrm{MPP},k} - P_{\mathrm{loss,tot},k})\,\Delta t_k} , \quad (1)$$

where $C_{\mathrm{tot}}(S)$ is the cost of components of the design S, $E_{\mathrm{tot}}(S,M)$ is the total harvested energy at the output of the converter in a 25-year operation (for this case study), P_{MPP} is the PV module power and $P_{\mathrm{loss,tot}}$ is the total loss, dependent on both the specific design S and on the mission profile M. The minimization of the scalar objective function expressed in (1) is the goal of the proposed methodology.

The block diagram of the proposed optimization scheme is shown in Fig. 2. For any trial solution S^* inside the solution space, the corresponding CCER* is computed through cost models and loss models applied to the 25-year converter operation. For the converter transistors, the instantaneous losses and junction temperature profiles rely on electro-thermal models, considering the package-dependent thermal resistance $R_{\mathrm{th,j-amb}}$ and the temperature-dependent $R_{\mathrm{DS,ON}}(T_{\mathrm{j}})$. The trial solution S^* is discarded if the reliability criteria are not

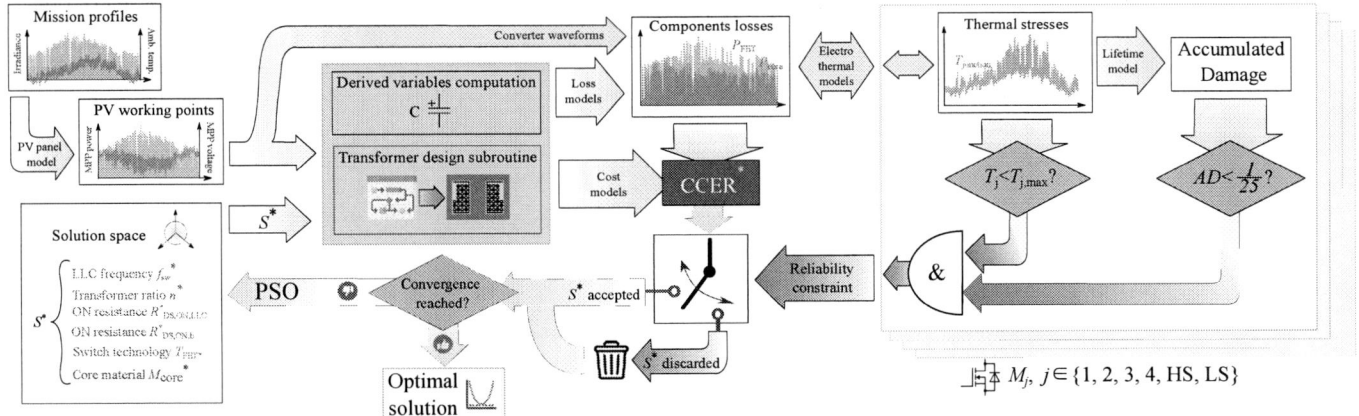

Fig. 2: Block diagram of the proposed optimization methodology. The Converter Cost-Energy Ratio (CCER*) of the trial solution S^* is highlighted in green.

met. The PSO algorithm is exploited to search within the solution space for the design that minimizes the objective function. Further details on the employed models and the reliability constraint are given in the next sections.

C. Modelling

As shown in Fig. 2, the proposed algorithm relies on multiple models to compute the objective function CCER* associated to a certain mission profile and trial solution S^*. A linear PV module model is used to derive the evolution of the working points from irradiance and panel temperature profiles. For simplicity, the MPPT efficiency is not taken into account. The analytical expressions of the converter waveforms in [15] are combined with the loss models for each component of the converter. It is worth noticing that the boost inductor L_b is sized according to the desired frequency window for the boundary conduction mode [15], independently of the rest of the design, and is thus assumed to be fixed. The same applies to the rectifier diodes D_1 and D_2, since their conduction losses only depend on the total power and not on the specific selection of n and f_{sw}.

For all the other components, specific cost models were derived from the analysis of commercial parts [20]. The computation of the total losses, instead, relied on well-established loss models. Table I lists the adopted loss- and cost models, whose empirical parameters are reported in Table II.

For the LLC switches M_{1-4}, commercial 60 V-rated Silicon MOSFETs and Gallium Nitride Field-Effect Transistors (GaNFETs) were both considered for the modelling. For the boost switches M_{LS-HS}, instead, the 250 V voltage rating, set according to the maximum expected voltage stress, only Si MOSFETs were considered, since the adoption of 650 V-rated GaNFETs would not be justified in terms of increased cost for comparable conduction and switching performances. A double exponential function of the conduction resistance was found to well-fit the unit-price for both the LLC and boost switches,

as shown in Fig. 3a. Since all the switches of the converter are assumed to turn ON at zero voltage, only the OFF switching losses are modelled through the OFF time t_{OFF}:

$$t_{OFF} = R_{gate} \cdot \frac{Q_{sw}(R_{DS,ON})}{V_{Miller}}, \qquad (2)$$

where R_{gate}, $Q_{sw}(R_{DS,ON})$ and V_{Miller} represent the gate resistance, the switching charge involved in the commutation, and the Miller plateau voltage, respectively [21]. The fitting of commercial components turn OFF time for the LLC switches is shown in Fig. 3b, highlighting the superior switching performances of GaNFETs for the same $R_{DS,ON}$.

As shown in Fig. 2, for each trial solution S^*, the junction temperature profiles of the corresponding six converter transistors are computed. To do this, the temperature increase due to the switch losses is taken into account in the algorithm through the simplified electro-thermal circuit shown in Fig. 4. The instantaneous junction temperature T_j is fed back through the non-linear function $\alpha_R(T_j)$, modelling the increase of conduction resistance with increasing temperature. No heat sink is adopted in the converter, thus the junction-to-ambient thermal resistance $R_{th,j-amb}$ was selected from the worst-case values of the switches datasheets. Notice that, since the time-step of the mission profiles is significantly longer (1 min) than the typical thermal time constants of the discrete components (without heat sink), any thermal capacitance is neglected from Fig. 4.

Film capacitors were selected for both C_{in} and C_b, with different voltage ratings, i.e. 63 V and 300 V, respectively. Their minimum capacitance value is, at each iteration of the algorithm, computed from the design equations in [15]. Both the cost and conduction loss models were derived by fitting the price per unit and Equivalent Series Resistance (ESR) as functions of the capacitance value.

As shown in Fig. 2, a specific subroutine of the algorithm was developed to extract the main geometry- and

TABLE I: List of the main cost and loss models adopted in the optimization algorithm.

Component	Primary variables	Derived variables	Cost model	Loss model
LLC FETs M_{1-4}	$R_{\mathrm{DS,ON,LLC}}$, T_{FET}, f_{sw}	t_{OFF}, α_R	$b_1 e^{d_1 R_{\mathrm{DS,ON,LLC}}} + b_2 e^{d_2 R_{\mathrm{DS,ON,LLC}}}$	$R_{\mathrm{DS,ON,LLC}} \alpha_R(T_j) I_{\mathrm{RMS}}^2 + \frac{1}{2} t_{\mathrm{OFF}} f_{\mathrm{sw}} V_{\mathrm{OFF}} I_{\mathrm{ON}}$
Boost FETs $M_{\mathrm{LS-HS}}$	$R_{\mathrm{DS,ON,b}}$	t_{OFF}, α_R	$b_3 e^{d_3 R_{\mathrm{DS,ON,b}}} + b_4 e^{d_4 R_{\mathrm{DS,ON,b}}}$	$R_{\mathrm{DS,ON,b}} \alpha_R(T_j) I_{\mathrm{RMS}}^2 + \frac{1}{2} t_{\mathrm{OFF}} f_{\mathrm{sw}} V_{\mathrm{OFF}} I_{\mathrm{ON}}$
Input capacitor C_{in}	f_{sw}, n	C_{in}, $ESR(C_{\mathrm{in}})$	$c_1 + c_2 \cdot C_{\mathrm{in}}$	$ESR(C_{\mathrm{in}}) I_{\mathrm{RMS}}^2$
Boost capacitor C_{b}	n	C_{b}, $ESR(C_{\mathrm{b}})$	$c_3 + c_4 \cdot C_{\mathrm{b}}$	$ESR(C_{\mathrm{b}}) I_{\mathrm{RMS}}^2$
Transformer core and bobbin	f_{sw}, n, M_{core}	L_{magn}, AP, N_1, V_{eff}, α_{SE}, β_{SE}, ρ_{SE}	$g_1 + g_2 AP^{\frac{1}{2}}$	$\frac{\pi}{4} K_{\mathrm{core}} V_{\mathrm{eff}}(AP) \rho_{\mathrm{SE}} f_{\mathrm{sw}}^{\alpha_{\mathrm{SE}}} \Delta B^{\beta_{\mathrm{SE}}}$
Transformer windings	f_{sw}, n	d_{s}, ω_{Cu}, N_1, N_2, R_{p}, F_{p}, R_{s}, F_{s}	$g_3 + \omega_{\mathrm{Cu}} \cdot \left[g_4 + g_5 e^{g_6 d_{\mathrm{s}}} \right]$	$R_{\mathrm{p}} F_{\mathrm{p}} I_{\mathrm{p,RMS}}^2 + R_{\mathrm{s}} F_{\mathrm{s}} I_{\mathrm{s,RMS}}^2$

TABLE II: Empirical parameters of the cost models in Table I.

Parameter	Value	Parameter	Value	Parameter	Value
b_1	23.84 €(Si) / 39.36 €(GaN)	d_3	$-0.0393\,\mathrm{m\Omega}^{-1}$ (Si)	g_1	0.704 €
b_2	3.236 €(Si) / 2.346 €(GaN)	d_4	$-0.003\,\mathrm{m\Omega}^{-1}$ (Si)	g_2	2.83 €/cm²
b_3	7.62 €(Si)	c_1	2.128 €	g_3	0.5 €
b_4	2.791 €(Si)	c_2	0.328 €/μF	g_4	19.04 €/kg
d_1	$-2.467\,\mathrm{m\Omega}^{-1}$ (Si) / $-0.821\,\mathrm{m\Omega}^{-1}$ (GaN)	c_3	2.017 €	g_5	1323 €/kg
d_2	$-0.091\,\mathrm{m\Omega}^{-1}$ (Si) / $-0.013\,\mathrm{m\Omega}^{-1}$ (GaN)	c_4	0.194 €/μF	g_6	$-886\,\mathrm{cm}^{-1}$

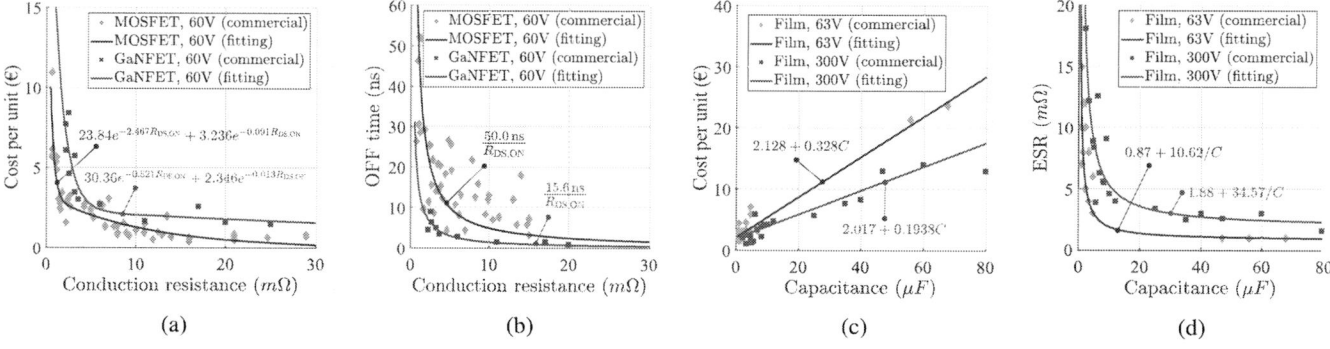

Fig. 3: Cost and key parameters of the loss models for some of the converter components, extracted from the analysis of commercial parts. (a) Cost per unit of 60 V Silicon MOSFETs and GaNFETs as function of $R_{\mathrm{DS,ON}}$. (b) Turn-OFF time of 60 V Silicon MOSFETs and GaNFETs as a function of $R_{\mathrm{DS,ON}}$. (c) Cost per unit of 63 V film capacitors (C_{in}) and 300 V capacitors (C_{boost}) as a function of the capacitance. (d) Equivalent Series Resistance (ESR) of 63 V and 300 V film capacitors as a function of the capacitance.

material-related parameters of the transformer core, such as the area product AP, the number of primary turns N_1, the effective volume V_{eff} or the Steinmetz Equation coefficients α_{SE}, β_{SE}, ρ_{SE} [22]. The cost model was derived from commercial EE and ETD cores, for their widespread adoption in power electronics applications and their availability in different sizes and materials. For the transformer windings, the cost model for Litz wire coils is based on the overall windings weight ω_{Cu} and diameter d_{s} of the copper strands [23]. The

loss model takes into account the increase of the winding resistance due to the proximity effect through the coefficients F_{p} and F_{s} [24]. To limit the impact of the AC losses in the windings, the strand diameter is selected to be $1/3$ of the skin depth. For a specific variable set S^*, the most impactful degrees of freedom on the transformer losses and cost are AP and N_1, from which many derived variables follow. At each iteration of the PSO algorithm, thus, the transformer design subroutine is responsible for identifying the domain of all

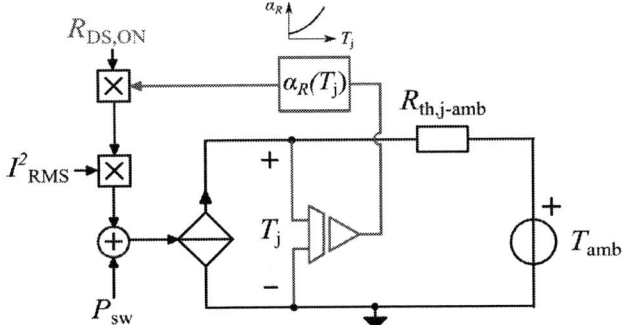

Fig. 4: Steady-state electro-thermal circuit considered in the optimization algorithm for the computation of the temperature-dependent losses of the converter switches.

the potential solutions $\{AP^*, N_1^*\}$, intersecting the following constraints:

- for a specific AP^*, N_1 must be sufficiently large to keep the flux density ΔB across the magnetic cross-section A_e lower than the maximum desired value (in this case, the $\Delta B_{\max}$ associated to $100\,\mathrm{mWcm}^{-1}$ loss density [25]):

$$N_1 > \frac{V_{\mathrm{in,max}}}{2 f_{\mathrm{sw}} \Delta B_{\max} A_e(AP^*)} \tag{3}$$

- for a specific AP^*, the primary and secondary coils, with conductor cross section A_{cond} must fit in the available window area A_w [26]:

$$N_1 < \frac{K_w A_w(AP^*)}{2 A_{\mathrm{cond}}} \tag{4}$$

- the operation of the LLC switches in Zero-Voltage Switching (ZVS) requires a certain magnetizing inductance L_{magn}, resulting in a triangular magnetizing current superimposed to the resonant sinusoidal current [15], [27]. For a specific AP^* and inductance factor A_L, N_1 should be large enough to obtain the required L_{magn}:

$$N_1 > \sqrt{\frac{L_{\mathrm{magn}}}{A_L(AP^*)}}. \tag{5}$$

The intersection of (3), (4) and (5) defines the domain of all the potential solutions $\{AP^*, N_1^*\}$. The subroutine is responsible for searching the optimal solution $\{AP_{\mathrm{opt}}, N_{1,\mathrm{opt}}\}$ that minimizes the *ad hoc* cost-output power ratio CPR_T objective function:

$$CPR_T(AP, N_1) = \frac{C_{\mathrm{core}} + C_{\mathrm{bobbin}} + C_{\mathrm{coils}}}{P_{\mathrm{rated}} - P_{\mathrm{core}} - P_{\mathrm{windings}}}, \tag{6}$$

where the numerator consists of the sum of the core, bobbin and coils costs, whereas the denominator is obtained by subtracting the total transformer losses from the rated input power.

Additional constraints are added in the transformer design subroutine: the transformer temperature should never exceed $T_{\max} = 100\,°C$, and the maximum flux density should be kept below the saturation B_{sat}.

D. Constraints

In order to limit the 6-dimensional solution space, the range of the design variables are constrained: $f_{\mathrm{sw}} \in [50\,\mathrm{kHz}, 500\,\mathrm{kHz}]$, $n \in [3.2, 4.0]$, $R_{\mathrm{DS,ON,LLC}} \in [1\,\mathrm{m}\Omega, 20\,\mathrm{m}\Omega]$, $R_{\mathrm{DS,ON,b}} \in [10\,\mathrm{m}\Omega, 150\,\mathrm{m}\Omega]$, $T_{\mathrm{FET}} \in \{\mathrm{Si}, \mathrm{GaN}\}$, $M_{\mathrm{core}} \in \{\mathrm{N27, N87, N97}\}$.

Contrarily to other works [9], in which the converter topology is a degree of freedom, here the IPOS topology in Fig. 1 is given as constraint, as well as the PV module [28].

As mentioned in Section I, one of the novelties of this article is the inclusion of a mission profile-based reliability constraint on the converter switches. Electrolytic capacitors, which are among the most critical components in terms of reliability [29], are not adopted in this converter. The reliability constraint consists of a set of two conditions to be satisfied by each converter switch: the junction temperature must never exceed the maximum datasheet value $T_{\mathrm{j,max}}$, and the accumulated damage (AD) due to the repeated thermal stresses in one year must be low enough to ensure a 25-years lifetime. The lifetime computation follows the approach described in [30]. Firstly, the yearly junction temperature profiles are decomposed into elementary thermal cycles applying the rainflow counting method; then, the lifetime model expressed in (7) [31] is applied to derive the number of cycles to failure $N_{\mathrm{f},i}$ associated to each thermal stress; finally, all the damage contributions are summed up to compute the Accumulated Damage (AD), as expressed in (8):

$$N_{\mathrm{f},i} = A \cdot \Delta T_{\mathrm{j},i}^B \cdot \exp\left(\frac{C}{\overline{T}_{\mathrm{j},i} + 273}\right) \tag{7}$$

$$AD = \sum_i \frac{N_i}{N_{\mathrm{f},i}}, \tag{8}$$

where $A = 4.9283 \cdot 10^{13}$, $B = -5.2776$, $C = 812$ and N_i is the number of cycles counted for the specific thermal stress i [31]. The trial solution S^* is discarded in case at least one of the converter switches exceeds $T_{\mathrm{j,max}}$ or its $AD > \frac{1}{25}$, meaning that the converter is likely to fail before 25 years.

III. RESULTS

A. Impact of mission profile on optimal solution

A PSO search algorithm was chosen among other meta-heuristic approaches because of its low complexity and the limited number of user-defined parameters [5]. The algorithm was implemented in Matlab considering a population of 30 solutions moving in the 6-dimensional solution space: the algorithm stops when the objective function CCER converges to a stable value within $\pm 0.5\%$.

Table III reports the results of the optimization algorithm for three different mission profiles, i.e. Aalborg (Denmark), Arizona (USA), and Turin (Italy). Independently of the mission profile, the PSO always converges to $n = 4$: according to [15], this transformer ratio ensures that, at the rated PV panel voltage $V_{\mathrm{in}} = 36.5\,V$, around 83% of the total input power is processed by the LLC stage, while only around 17% by the boost. A strongly unbalanced power splitting,

TABLE III: Results of the optimization algorithm for three different mission profiles: Aalborg (Denmark), Turin (Italy), Arizona (USA).

Mission profile	f_{sw}	n	M_{core}	T_{FET}	$R_{DS,ON,LLC}$	$R_{DS,ON,b}$	CCER
Aalborg	290 kHz	4	N97	Si	10 mΩ	130 mΩ	9.69 €/MWh
Arizona	240 kHz	4	N87	Si	10 mΩ	50 mΩ	3.43 €/MWh
Turin	250 kHz	4	N87	Si	13 mΩ	120 mΩ	6.19 €/MWh

TABLE IV: Selected components of the converter design for the Arizona case, used for the simulation results.

Component	Part number
LLC switches M_{1-4}	IPB090N06N3
Transformer core	ETD 44/22/15, N97
Transf. primary coil	5 turns, 2400x44 μm
Transf. secondary coil	20 turns, 600x44 μm
Rectifier diodes D_{1-2}	STTH30R04
Boost switches M_{HS-LS}	IPB600N25N3
Boost inductor L_b	74437529203330
Boost capacitor C_b	R75MW51004030J
Input capacitor C_{in}	4x CB182D0475JBC

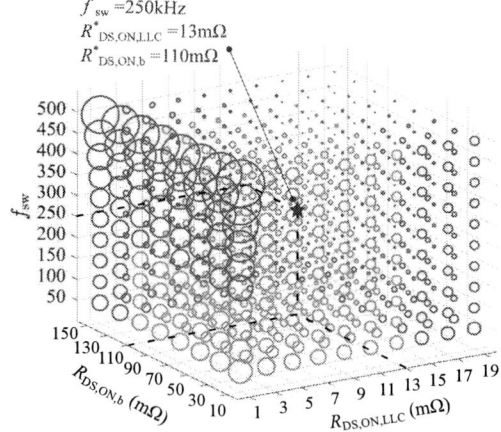

Fig. 5: Results of CCER with a constrained exhaustive search ($n = 4$, T_{FET} =Si, M_{core} =N97), for Turin mission profile. The markers size is proportional to the corresponding CCER. The optimal solution is highlighted with a blue star, while the red markers identify discarded solutions.

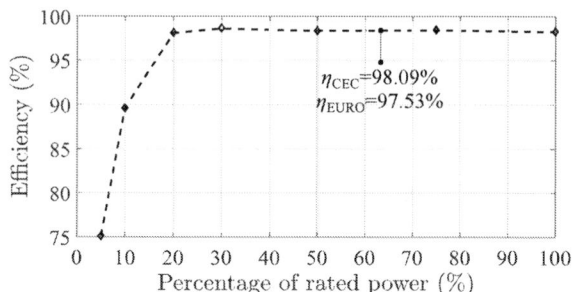

Fig. 6: Simulation results of converter efficiency as a function of operating power, at the rated voltage $V_{in} = 36.5$ V.

indeed, significantly reduces the current and voltage stresses on the boost switches, in terms improving their reliability, and helps reducing the size and cost of the filtering capacitors. In addition, the optimal variable sets always prefer Silicon MOSFETs over Gallium Nitride FETs: for the same $R_{DS,ON}$, the improved switching performances of the latter are not sufficient to justify the larger cost.

For the latter case, a constrained exhaustive search was performed by fixing transformer ratio $n = 4$, core material $N87$ and switch technology T_{FET} =Si, and sweeping the remaining design inputs. The results are graphically shown in Fig. 5, in which the size of the circular markers at each triplet of coordinates is directly proportional to the corresponding CCER. The green and red markers represent acceptable and non-acceptable solutions, respectively. It is relevant to observe that three different regions of the variable space do not satisfy the reliability constraints: 1 mΩ MOSFETs fail for $f_{sw} \geq 250$ kHz due to the high switching losses, whereas $R_{DS,ON,LLC} \geq 15$ mΩ and $R_{DS,ON,b} \geq 130$ mΩ are both unacceptable regions because of the high conduction losses. The minimum CCER solution is highlighted in blue and is consistent with the optimal solution found by the PSO (refer to Table III). Despite the larger conduction losses, the reduced cost of higher $R_{DS,ON}$ transistors is decisive to decrease the objective function. As a result, the minimum CCER is found in correspondence of the highest $R_{DS,ON,b}$ and $R_{DS,ON,LLC}$ still

meeting the reliability constraints. In addition, independently of the mission profile, the PSO always converges to $n = 4$: according to [15], this transformer ratio ensures that, at the rated PV panel voltage $V_{in} = 36.5$ V, around 83% of the total input power is processed by the LLC stage, while only around 17% by the boost. A strongly unbalanced power splitting, indeed, significantly reduces the current and voltage stresses on the boost switches, in terms improving their reliability, and helps reducing the size and cost of the filtering capacitors.

B. Simulation results of optimal solution

From the worst-case solution of Table III, which corresponds to the harshest environment conditions (Arizona), a converter was designed with the components in Table IV and simulated in Matlab and LTspice. The transformer was designed according to the procedure illustrated in Section II.

Fig. 6 shows the simulated converter efficiency as a function of the output power at the rated input voltage condition, $V_{in} = 36.5$ V, and assuming $25\,°C$ ambient temperature. The simulations were performed in LTspice with Spice models for both the active and passive devices. The calculated California Energy Commission (CEC) and European (EURO) efficiency are 98.09% and 97.53%, respectively.

Fig. 7 shows the annual junction temperature profiles for the selected converter transistors, referred to Aalborg (Fig. 7a), Arizona (Fig. 7b) and Turin (7c) mission profiles. The resulting annual AD is also reported for each temperature profile. The results were extracted in Matlab considering the electro-thermal circuit in Fig. 4. It is worth noticing that the selected MOSFET for the boost low-side switch does not meet

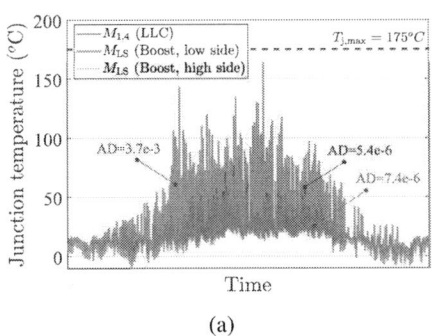 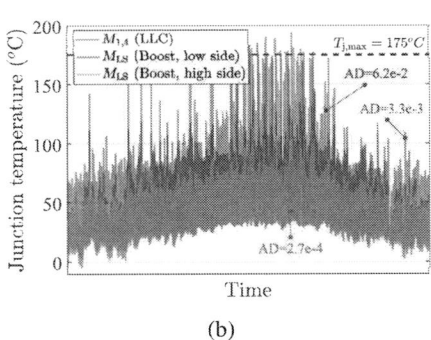 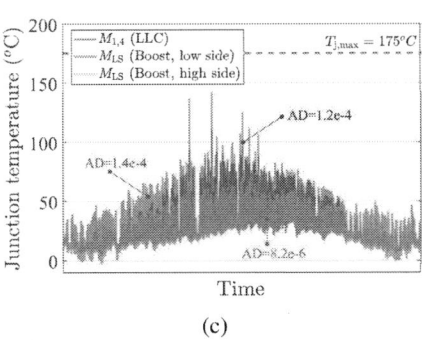

| (a) | (b) | (c) |

Fig. 7: Annual profile of the switches junction temperatures corresponding to three different mission profiles. (a) Aalborg (Denmark); (b) Arizona (USA); (c) Turin (Italy).

the reliability constraint in the Arizona mission profile: its worst-case junction-to-ambient thermal resistance is indeed higher than the one considered in the optimization ($62\,\mathrm{K/W}$ against $40\,\mathrm{K/W}$), and its $R_{\mathrm{DS,ON,b}}$ is slightly larger than the recommended $50\,\mathrm{m\Omega}$. A more conservative components selection and a proper thermal design at a layout level would be essential to meet the reliability constraint with a sufficient safety margin. Notice that, in general, a higher ambient temperature in the mission profile (Turin or Arizona) shifts the MPP of a PV panel to lower voltages, thus increasing the gain requirement of the boost and the electrical stresses of its devices.

IV. CONCLUSIONS

This article proposes a new optimization methodology for the design of power converters that is based on the minimization of the cost-energy ratio and that includes a mission profile-based reliability constraint. The proposed methodology is suited for power converters designed for photovoltaic applications, where cost, efficiency and reliability are crucial tradeoffs. The approach is applied to the specific case study of an asymmetric IPOS power optimizer consisting of a LLC and a synchronous boost stage. At each step, the proposed algorithm exploits multiple analytical and empirical models to compute the cost and losses of the trial converter solution, and discards it if at least one of the converter switches does not meet the reliability constraints. The PSO method is used to search for the optimal solution inside a 6-dimensional solution space. The algorithm is run for three different mission profiles, highlighting that the optimal solutions privilege a strongly unbalanced power splitting between the two stages and Silicon devices over Gallium Nitride ones. From the worst case optimal solution, a converter is designed and simulated, showing a California Energy Commission efficiency above 98%. The future steps of this work include the design of a physical prototype and its experimental validation.

REFERENCES

[1] S. M. MacAlpine, R. W. Erickson and M. J. Brandemuehl, "Characterization of Power Optimizer Potential to Increase Energy Capture in Photovoltaic Systems Operating Under Nonuniform Conditions," in IEEE Transactions on Power Electronics, vol. 28, no. 6, pp. 2936-2945, June 2013.

[2] K. Alluhaybi, I. Batarseh and H. Hu, "Comprehensive Review and Comparison of Single-Phase Grid-Tied Photovoltaic Microinverters," in IEEE Journal of Emerging and Selected Topics in Power Electronics, vol. 8, no. 2, pp. 1310-1329, June 2020.

[3] M. Kasper, D. Bortis and J. W. Kolar, "Classification and Comparative Evaluation of PV Panel-Integrated DC–DC Converter Concepts," in IEEE Transactions on Power Electronics, vol. 29, no. 5, pp. 2511-2526, May 2014.

[4] G. Spagnuolo, S. Kouro, and D. Vinnikov, "Photovoltaic Module and Submodule Level Power Electronics and Control," in IEEE Transactions on Industrial Electronics, vol. 66, no. 5, pp. 3856-3859, May 2019.

[5] S. Zhao, F. Blaabjerg and H. Wang, "An Overview of Artificial Intelligence Applications for Power Electronics," in IEEE Transactions on Power Electronics, vol. 36, no. 4, pp. 4633-4658, April 2021.

[6] M. D'Antonio, C. Shi, B. Wu and A. Khaligh, "Design and Optimization of a Solar Power Conversion System for Space Applications," in IEEE Transactions on Industry Applications, vol. 55, no. 3, pp. 2310-2319, May-June 2019.

[7] S. Vighetti, J. -P. Ferrieux and Y. Lembeye, "Optimization and Design of a Cascaded DC/DC Converter Devoted to Grid-Connected Photovoltaic Systems," in IEEE Transactions on Power Electronics, vol. 27, no. 4, pp. 2018-2027, April 2012.

[8] L. Schmitz et al., "Design optimization of a high step-Up DC-DC converter for photovoltaic microinverters," 2017 IEEE International Telecommunications Energy Conference (INTELEC), Broadbeach, QLD, Australia, 2017, pp. 432-437

[9] E. O. Prado, P. C. Bolsi, L. Aleixo, H. C. Sartori and J. R. Pinheiro, "Optimized Design of Non-Isolated DC-DC Converters for PV System Applications," 2023 15th Seminar on Power Electronics and Control (SEPOC), Santa Maria, Brazil, 2023, pp. 1-6.

[10] G. Marsala and A. Ragusa, "Reliability and Efficiency Optimization Assisted by Genetic Algorithm to Design a Quadratic Boost DC/DC Converter," 2018 IEEE International Conference on Industrial Engineering and Engineering Management (IEEM), Bangkok, Thailand, 2018, pp. 1687-1692.

[11] G. Adinolfi, G. Graditi, P. Siano and A. Piccolo, "Multiobjective Optimal Design of Photovoltaic Synchronous Boost Converters Assessing Efficiency, Reliability, and Cost Savings," in IEEE Transactions on Industrial Informatics, vol. 11, no. 5, pp. 1038-1048, Oct. 2015.

[12] M. Mirjafari, S. Harb and R. S. Balog, "Multiobjective Optimization and Topology Selection for a Module-Integrated Inverter," in IEEE Transactions on Power Electronics, vol. 30, no. 8, pp. 4219-4231, Aug. 2015.

[13] T. Dragičević, P. Wheeler and F. Blaabjerg, "Artificial Intelligence Aided Automated Design for Reliability of Power Electronic Systems," in IEEE Transactions on Power Electronics, vol. 34, no. 8, pp. 7161-7171, Aug. 2019.

[14] M. Sandelic, A. Sangwongwanich, S. Peyghami and F. Blaabjerg, "Reliability Modelling of Power Electronics with Mission Profile Forecasting for Long-Term Planning," 2022 IEEE 13th International Symposium on Power Electronics for Distributed Generation Systems (PEDG), Kiel, Germany, 2022, pp. 1-6.

979-8-3315-1612-3/25 $31.00 © 2025 IEEE

[15] S. Cerutti, M. Pavone, F. Gennaro, N. Aiello, F. Musolino, and P. Crovetti, "Design of a Multi-Mode Input-Parallel-Output-Series Power Optimizer for Wide-Voltage Range Photovoltaic Applications," *2024 IEEE Energy Conversion Congress & Exposition (ECCE)*, pp. 444-451, 2024.

[16] J. Zhang, Z. He, R. Han and Y. Liu, "Hybrid Structure for an Input-Parallel Output-Serial DC-DC Combined Converter with High Efficiency and High Power Density," *2021 IEEE 12th Energy Conversion Congress & Exposition - Asia (ECCE-Asia)*, pp. 432-437, 2021.

[17] M. Mirjafari and R. S. Balog, "Multi-objective optimization of the energy capture and boost inductor mass in a module-integrated converter (MIC) photovoltaic energy system," *2012 27th Annual IEEE Applied Power Electronics Conference and Exposition (APEC)*, Orlando, FL, USA, 2012, pp. 2002-2007.

[18] Y. Son, S. Mukherjee, R. Mallik, B. Majmunovi'c, S. Dutta, B. Johnson, D. Maksimović, and G.-Su Seo, "Levelized Cost of Energy-Oriented Modular String Inverter Design Optimization for PV Generation System Using Geometric Programming," in IEEE Access, vol. 10, pp. 27561-27578, 2022.

[19] J. S. Saranyaa and P. F. A, "A Comprehensive Survey on the Current Trends in Improvising the Renewable Energy Incorporated Global Power System Market," in IEEE Access, vol. 11, pp. 24016-24038, 2023.

[20] Digikey electronic parts distributor. [Online.] Available: https://www.digikey.it/en.

[21] F. Iannuzzo, *Modern Power Electronic Devices: Physics, applications, and reliability*. Institution of Engineering and Technology Engineering Series Vol. 152, 2020.

[22] H. E. Tacca, "Core Loss Prediction in Power Electronic Converters Based on Steinmetz Parameters," *2020 IEEE Congreso Bienal de Argentina (ARGENCON)*, Resistencia, Argentina, 2020, pp. 1-8

[23] R. M. Burkart, (2016). Advanced Modeling and Multi-Objective Optimization of Power Electronic Converter Systems [Doctoral dissertation, ETH Zurich].

[24] W. . -J. Gu and R. Liu, "A study of volume and weight vs. frequency for high-frequency transformers," *1993 IEEE Power Electronics Specialist Conference (PESC)*, Seattle, WA, USA, 1993, pp. 1123-1129.

[25] "Section 4 – Power Transformer Design". Texas Instruments. https://www.ti.com/lit/ml/slup126/slup126.pdf (accessed October 15, 2024).

[26] C. W. T. McLyman, *Transformer and Inductor Design Handbook – Third Edition*, CRC Press, Boca Raton, 2004.

[27] U. Kundu, K. Yenduri and P. Sensarma, "Accurate ZVS Analysis for Magnetic Design and Efficiency Improvement of Full-Bridge LLC Resonant Converter," in IEEE Transactions on Power Electronics, vol. 32, no. 3, pp. 1703-1706, March 2017.

[28] 3SUN-M40 module datasheet. [Online.] Available: https://www.3sun.com/content/dam/threesun/documents/technical/3SUN_M40_file.pdf.

[29] S. Peyghami, Z. Wang and F. Blaabjerg, "A Guideline for Reliability Prediction in Power Electronic Converters," in IEEE Transactions on Power Electronics, vol. 35, no. 10, pp. 10958-10968, Oct. 2020

[30] H. Huang and P. A. Mawby, "A Lifetime Estimation Technique for Voltage Source Inverters," in IEEE Transactions on Power Electronics, vol. 28, no. 8, pp. 4113-4119, Aug. 2013

[31] S. Dusmez, H. Duran and B. Akin, "Remaining Useful Lifetime Estimation for Thermally Stressed Power MOSFETs Based on on-State Resistance Variation," in IEEE Transactions on Industry Applications, vol. 52, no. 3, pp. 2554-2563, May-June 2016.

Submodule Voltage Balancing Technique of Solar MMC for Firing The Switches Using Integrated PWM Modules

Ahmed Elsanabary[1], Saad Mekhilef[2], Mokhtar Aly[3], Jose Rodriguez[3]

[1]*Department of Electrical Engineering, Faculty of Engineering, Port Said University,* Port Fouad 42526, Egypt
[2]*School of Science, Computing and Engineering Technologies, Swinburne University of Technology,* Hawthorn, 3122, VIC, Australia
[3]*Facultad de Ingeniería, Arquitectura y Diseño, Universidad San Sebastián,* Bellavista 7, Santiago 8420524, Chile
eng_san88@eng.psu.edu.eg , smekhilef@swin.edu.au , mokhtar.aly@uss.cl , jose.rodriguezp@uss.cl

Abstract— **Submodule (SM) voltage balancing of a modular multilevel converter (MMC) is crucial for stable operation of the converter. The conventional sorting algorithm is usually used to achieve this balance through instantaneous measuring of SM voltages and arm currents. However, in practice, this process utilizes digital out pins from the controller to operate the half-bridge SM switches, which are slower and inaccurate, compared to the embedded PWM interface modules provided in most commercial controllers. This work proposes a voltage balancing technique using a modified sorting algorithm to be compatible with the PWM interface module's output ports for any type of controllers. It performs the required balancing of the capacitors by identifying the required reference voltage signal. The technique is validated experimentally through a three-phase grid connected solar MMC system and the obtained results are presented to show the performance of the proposed technique to balance the SM voltages.**

Keywords—MMC, Voltage Balancing, Solar systems, Grid.

I. INTRODUCTION

The Modular multilevel converter (MMC) is currently gaining attention in the application of medium voltage (MV) photovoltaic (PV) plants due to its modularity, scalability, power balancing capability, and energy yield efficiency [1-5]. To efficiently utilize its benefits, the conventional MMC's structure is modified to accommodate multiple PV strings in each of its submodules (SM)s [6, 7]. Certain features are gained from this connection which include achieving independent MPPT control for each string yielding better energy harvesting, scalability through direct connection of the PV system to the MV network, modularity of the system allowing flexibility in connection and maintenance; and better efficiency and power density by abandoning the bulky line frequency transformer in favor of multiple high frequency transformers placed in each SM circuit [6, 8, 9].

Several studies have discussed the capability of MMC in this application resulting in key research points: (1) PV power imbalance, which occurs between the arms, legs or SMs of the MMC [10-15]. This research point is primarily targeted due to the complexity of controlling the MMC power flow during different solar irradiances of the PV strings connected to each SM circuit. Balancing control strategies and circuit configurations are proposed to address the power imbalance occurring between the arms and legs of the MMC during such conditions. (2) SM circuit optimization: enhancing power

density, component count, control simplicity, power loss, and cost [6]. The selection of the SM topology has been discussed on many occasions to achieve higher efficiency yield and control independence, especially for MPPT. At this level, converters such as Flybacks, dual active bridges (DABs) and Single active bridges (SABs) have been considered [8, 16]. Additionally, multiport converters contribute to new balancing features [14, 17]. (3) Fault detection and mitigation, requiring advanced control strategies for both internal and external faults [18]. The PV-MMC system should maintain operation under faulty conditions as anticipated in such large-scale systems. Advanced control techniques for mitigating internal SM faults and grid imbalances are of interest; (4) SM voltage balancing, ensuring balance between the SM voltages under any operating condition [19, 20]. The balancing process may also be required during SM power imbalances which can be problematic if high PV power disparity is observed.

Apart from that, the latter is crucial in preserving stable operation of the MMC in case of different SMs having unequal PV power generation. Various balancing methods have been used to address this issue. Among these, the traditional sorting algorithm shows satisfactory results in balancing the SMs' power when their generated PV powers are not equal. However, its practical operation is hindered because the sorting algorithm is typically a low switching frequency process due to the way signals are generated. In other words, using the common digital out ports of any controller (such as *Texas Instruments LaunchPads, dSPACE,* or *RT Box* platforms) to generate the signals reduces the accuracy and speed of the generated signals. Instead, these controllers offer Pulse width modulation (PWM) interface modules with superior accuracy and speed.

This paper proposes a technique for SM voltage balancing based on a modified sorting algorithm to be applicable to the PWM interface module of various types of controllers. The idea is to generate a modulated sinusoidal reference which works as the duty cycle to be fed to the PWM interface module. In order to realize this technique, the sorting process is required to select a suitable voltage reference to balance the SM capacitor voltages in each arm. The resulting modulated signal is then fed to the PWM interface module in the designated controller. In this work, a PV-MMC system is considered as the system under study and the SM voltage balancing efficacy is practically validated using *MicroLabBox*'s PWM interface control board.

979-8-3315-1612-3/25 $31.00 © 2025 IEEE

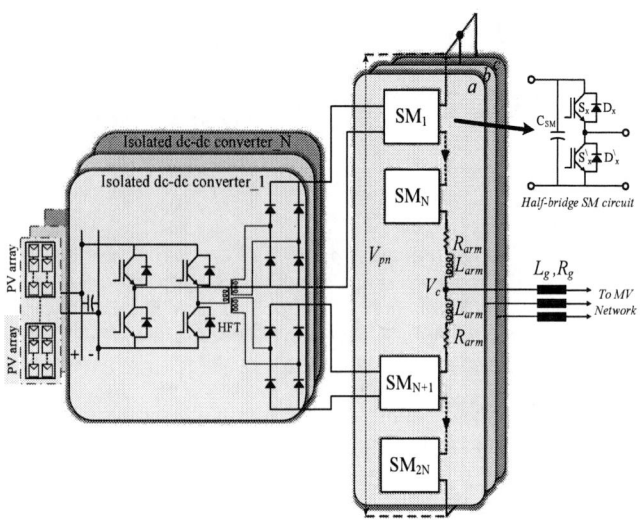

Fig. 1   The solar MMC system for grid integration.

II. System Description

The conventional MMC has a common DC-link, allowing power to transfer from the DC terminals to each converter's leg equally. Contrarily, the PV-MMC configuration consists of a three-phase MMC connected to the PV arrays without requiring a common DC-link, as depicted in Fig. 1. Power generation instead originates from the SMs themselves, allowing power transfer from the PV arrays to the utility grid through the MMC. The PV strings are connected to the SM capacitors through DC-DC converters to ensure the MPPT control, galvanic isolation and efficient power transfer. The SMs are connected in series to form the arms, while the upper and lower SMs of each arm are connected in parallel through isolated DC-DC converters. So that, the power in the arms of the same phase is equal and follows this equation.

$$P_{u,x} = P_{l,x} = v_{u,x}\, i_{u,x} = v_{l,x}\, i_{l,x} = \frac{P_{pv,x}}{2} \qquad (1)$$

where $P_{u,x}$, $v_{u,x}$, $i_{u,x}$ are the upper arm power, voltage and current, and $P_{l,x}$, $v_{l,x}$, $i_{l,x}$ are the lower arm power, voltage and current. $P_{pv,x}$ is the PV power generated at the input side; x denotes the phase number (a, b, c).

One of the major concerns of this technology is the unequal generation of power from the PV arrays due to intermittent irradiance. This issue results in power imbalances between the legs, arms and SMs of the MMC circuit. As proposed in [17], this circuit configuration has inherent arm balancing features that can eliminate the arm imbalances by connecting the same PV arrays set to both the upper and lower SMs of each arm as depicted in Fig. 1. Circulating current control can be achieved by regulating its DC component through each phase to balance the three phases of the MMC during varying PV power generation conditions.

III. Proposed Voltage Balancing Technique

The SM voltage balancing is essential in the operation of the MMC as it controls the process of charging and discharging to equalize all the SM voltages in each arm. Logical function-based algorithms are commonly used to achieve voltage balancing at the modulation stage [4]. The conventional sorting algorithm is widely used to achieve this balancing by utilizing the measured SM capacitor voltages and arm currents to take binary control actions in such a way to keep the SM voltages equalized and fluctuating around their nominal value. However, this algorithm suffers a low switching sequence due to binary commands which can only operate the switch signals through digital out ports. Digital out ports are known by their low switching frequency operation and unsatisfying signal accuracy.

To avoid this issue, this work is proposing a voltage balancing technique based on a modified sorting algorithm. Fig. 2 explains the processing of the proposed voltage balancing technique for its implementation to the PWM interface board. The technique requires measuring the arm currents and sorting the SM voltages in ascending order. The capacitor voltages for each phase ($V_{ci,x}$) should keep tracking to the desired reference value of the virtual DC-link voltage (V_{pn}).

$$V_{ci,x} = \frac{V_{pn}}{N}; \qquad (2)$$

where $\{V_{c1,x}, V_{c2,x}, \dots V_{c(N-1),x}\ V_{cN,x}\}$ are the SM voltages and to be sorted in according to their instantaneous values, and N is the number of SMs per arm. Afterwards, the selection criteria is based on the direction of the upper and lower arm currents to ensure the SMs with the lower voltages are charged when inserted and vice versa, which is similar to the conventional MMC. The equations of these currents in this system are assumed to be free of fundamental and double frequency circulating current components and they can be expressed as follows.

$$i_{u,x} = I_{circ,x} + \frac{I_{s,x}}{2} cos(\omega t - \varphi_i)\,,$$
$$\qquad (3)$$
$$i_{l,x} = I_{circ,x} - \frac{I_{s,x}}{2} cos(\omega t - \varphi_i)$$

Where $I_{circ,x}$, $I_{s,x}$ are the circulating and grid current magnitudes and φ_i: is the grid current phase angle. Level-shifted references of the arm voltages in each of the upper and lower arms with a number of ($N-1$) with an offset equal to $\left(\frac{1}{N-1}\right)$ are required to be fed to the algorithm. The upper and lower arm voltage references in each phase are expressed as follows.

$$v_{u,x} = \frac{V_{pn}}{2}\big(1 - m_a\, cos(\omega t)\big)\,,$$
$$\qquad (4)$$
$$v_{l,x} = \frac{V_{pn}}{2}\big(1 + m_a\, cos(\omega t)\big)$$

where m_a: is the modulation index of the converter voltage.

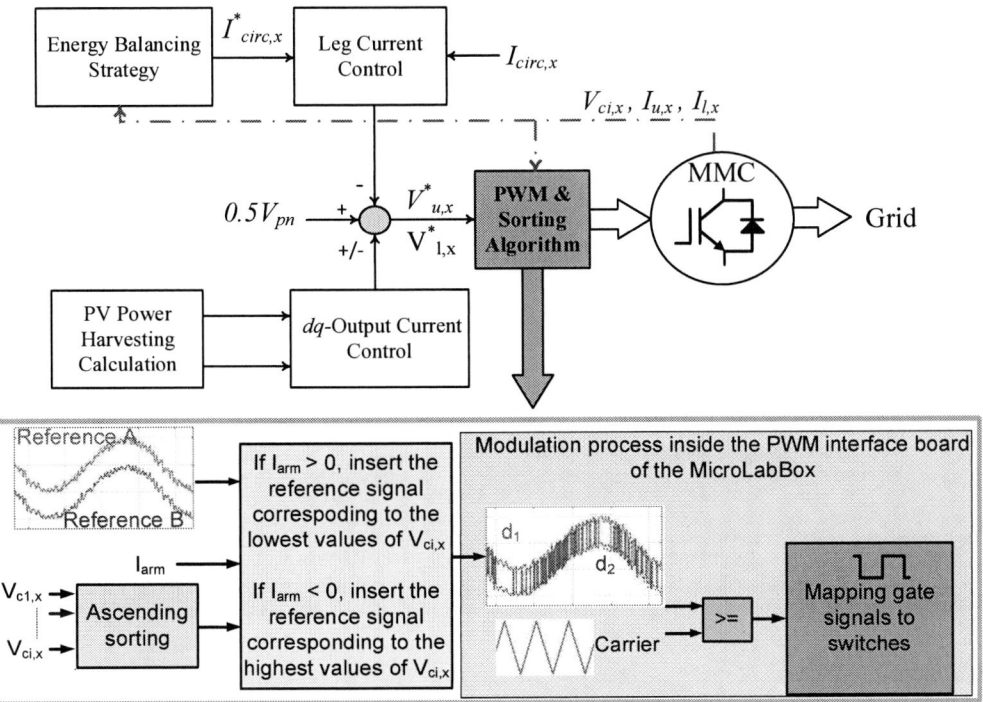

Fig. 2 Proposed voltage balancing technique to be applied to the PWM interface of the MicroLabBox and its function in the MMC control diagram

Afterwards, the algorithm selects the corresponding reference signal to the SM voltages in each arm based on the instantaneous arm currents and SM voltages. The output of this algorithm is a $(N-1)$ dimensional signal with duty cycles identified as $(d_1, \ldots d_{N-1})$ which are then given to the PWM interface for producing the corresponding gate signals for each SM switches. Thus, for example, the duty cycle (d_l) is a combination of all the listed references and can be expressed as.

$$d_1 \in \begin{cases} \left\{V_{u,x}, \ldots \ldots . V_{u,x}\left(1 + \dfrac{1}{N-1}\right)\right\}; \; for \; the \; upper \; arm \\ \left\{V_{l,x}, \ldots \ldots . V_{l,x}\left(1 + \dfrac{1}{N-1}\right)\right\}; \; for \; the \; lower \; arm \end{cases} \quad (5)$$

IV. EXPERIMENTAL RESULTS

A downscaled version of the PV-MMC circuit configuration is constructed in the laboratory. The built system consists of a three-phase grid connected MMC connected to multiple DC-DC converters through its SM. These converters are fed from controlled switched power supplies to emulate the characteristics of the PV strings. The designed configuration can be determined as depicted in Fig. 3 and the circuit parameters are given in Table I.

A *MicroLabBox* controlling board is utilized to execute the system control and processing actions through its analog inputs (for measuring) and PWM outputs (for driving the gate signals). To test the proposed balancing technique, the control system

runs at a sampling time 50 μs with switching frequency of 5 kHz. The downscaled system can effectively validate the operation of the MMC circuit, and the balancing control technique as presented in the obtained results.

The PV-MMC system is tested under unbalanced phase power generation condition. In this test, the PV power generation in *phase-A* is changed from 1 to 0.2 p.u while the other phases remained at full power capacity. As seen in Fig. 4 (a), the inverter voltage and grid currents show the reduction in the power generated due to the reduction occurred in *phase-A*, achieving a three-phase balanced set of grid currents as a primary control objective. Fig. 4 (b) shows the injected DC circulating current components into each of the MMC's phases in order to eliminate the power imbalances between the phases. It is noticed that the current value in *phase-A* is positive while the other phases have negative currents to compensate for the power disparity between the phases.

Fig. 5 includes zooming of the resulting voltage and current waveforms which shows a good balancing performance of the grid currents. In addition, it shows the effectiveness of the proposed balancing technique in balancing the SM voltages in both balanced and unbalanced phase power generation conditions. During the balanced power condition, the capacitor voltages are all equalized to each other and tracking the reference value which in this case is equal to 60 V. During the power imbalance, the SM voltages are regulated to their new references in each phase to allow the power transfer between the phases as per the control objectives.

The obtained results shows that the proposed SM voltage balancing technique is working effectively at 5 kHz enabling the capability of the PWM interface module in *MicroLabBox*. This application may also be considered for higher switching frequencies if required, which will contribute to better tracking accuracy and less harmonic contents of the voltage and current waveforms.

TABLE I.   Circuit Parameters of the Practical Test

Flybacks		MMC	
Input voltage	48 V	DC link voltage (V_{dc})	120 V
Output voltage	60 V	Line-Line AC Voltage	54 V
Input Capacitor	1 mF	SM capacitor (C_{sm})	3.3 mF
Transformer core	PQ50/50	Arm inductance (L_{arm})	4.5 mH
Flyback Frequency	30 kHz	MMC frequency	5 kHz

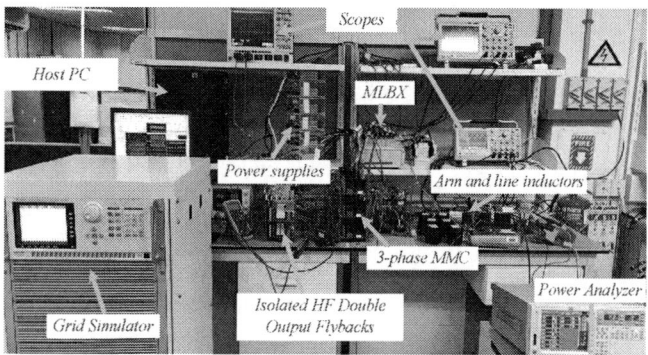

Fig. 3   A photograph of the experimental test bench.

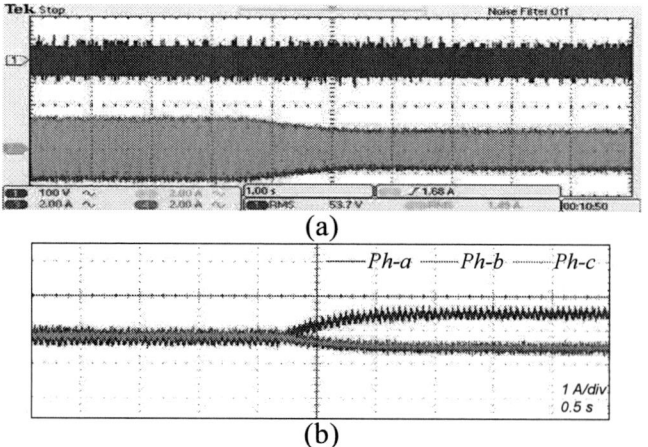

Fig. 4 Unbalanced Phase power generation (a) Voltage and current waveforms, (b) Circulating Current injection.

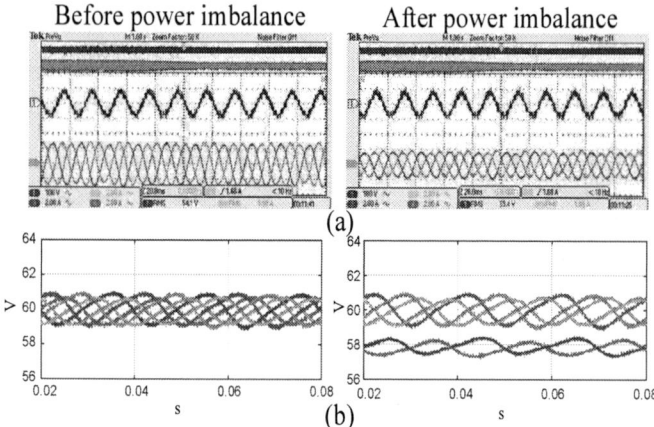

Fig. 5   Unbalanced Phase power generation (a) Zooming of the voltage and current waveforms; and (b) SM voltage balancing; before and after the power imbalance.

V.   Conclusion

In this work, a voltage balancing technique is proposed to give a solution for the implementation of the capacitor voltage balancing using the PWM interface board of any controller. This technique avoids the use of the slow and inaccurate digital out ports. The technique utilizes multiple arm voltage references with offsets to be compared inside the algorithm according to the SM voltages and arm current readings. The output of this controller is then compatible with the PWM interface and can be implemented with higher switching frequencies without hindering the accuracy of the balancing. The technique is practically validated using the PV-MMC system and the obtained SM voltage waveforms show a stable regulation at different PV power conditions.

Acknowledgment

J. Rodriguez acknowledges the support of ANID through projects FB0008, 1210208, and 1221293. This work was also supported by ANID, Chile FONDECYT 1230250, FONDECYT Iniciacion 11230430, and SERC-Chile ANID/FONDAP/1523A0006.

References

[1]   A. Dekka, B. Wu, R. L. Fuentes, M. Perez, and N. R. Zargari, "Evolution of Topologies, Modeling, Control Schemes, and Applications of Modular Multilevel Converters," *IEEE Journal of Emerging and Selected Topics in Power Electronics*, vol. 5, no. 4, pp. 1631-1656, 2017, doi: 10.1109/JESTPE.2017.2742938.

[2]   M. R. Islam, A. M. Mahfuz-Ur-Rahman, K. M. Muttaqi, and D. Sutanto, "State-of-the-Art of the Medium-Voltage Power Converter Technologies for Grid Integration of Solar Photovoltaic Power Plants," *IEEE Transactions on Energy Conversion*, vol. 34, no. 1, pp. 372-384, 2019, doi: 10.1109/TEC.2018.2878885.

[3]   A. B. Acharya, M. Ricco, D. Sera, R. Teoderscu, and L. E. Norum, "Performance Analysis of Medium-Voltage Grid Integration of PV Plant Using Modular Multilevel Converter," *IEEE Transactions on Energy Conversion*, vol. 34, no. 4, pp. 1731-1740, 2019, doi: 10.1109/TEC.2019.2930819.

[4]   M. A. Perez, S. Bernet, J. Rodriguez, S. Kouro, and R. Lizana, "Circuit Topologies, Modeling, Control Schemes, and Applications of Modular Multilevel Converters," *IEEE Transactions on Power Electronics*, vol. 30, no. 1, pp. 4-17, Jan 2015, doi: 10.1109/TPEL.2014.2310127.

[5] H. Nademi, A. Das, R. Burgos, and L. E. Norum, "A New Circuit Performance of Modular Multilevel Inverter Suitable for Photovoltaic Conversion Plants," *IEEE Journal of Emerging and Selected Topics in Power Electronics*, vol. 4, no. 2, pp. 393-404, Jun 2016, doi: 10.1109/Jestpe.2015.2509599.

[6] A. I. Elsanabary, G. Konstantinou, S. Mekhilef, C. D. Townsend, M. Seyedmahmoudian, and A. Stojcevski, "Medium Voltage Large-Scale Grid-Connected Photovoltaic Systems Using Cascaded H-Bridge and Modular Multilevel Converters: A Review," *IEEE Access*, vol. 8, pp. 223686-223699, 2020, doi: 10.1109/ACCESS.2020.3044882.

[7] T. Soong and P. W. Lehn, "Internal Power Flow of a Modular Multilevel Converter With Distributed Energy Resources," *IEEE Journal of Emerging and Selected Topics in Power Electronics*, vol. 2, no. 4, pp. 1127-1138, Dec 2014, doi: 10.1109/Jestpe.2014.2342656.

[8] S. Rivera, B. Wu, R. Lizana, S. Kouro, M. Perez, and J. Rodriguez, "Modular Multilevel Converter for Large-scale Multistring Photovoltaic Energy Conversion System," in *2013* IEEE *Energy Conversion Congress and Exposition (ECCE)*, pp. 1947-1952, 2013.

[9] S. Kouro, J. I. Leon, D. Vinnikov, and L. G. Franquelo, "Grid-Connected Photovoltaic Systems: An Overview of Recent Research and Emerging PV Converter Technology," *IEEE Industrial Electronics Magazine*, vol. 9, no. 1, pp. 47-61, Mar 2015, doi: 10.1109/MIE.2014.2376976.

[10] F. Rong, G. Xichang, and S. Huang, "A Novel Grid-Connected PV System Based on MMC to Get the Maximum Power Under Partial Shading Conditions," *IEEE Transactions on Power Electronics*, vol. 32, no. 6, pp. 4320-4333, Jun 2017, doi: 10.1109/TPEL1.2016.2594078.

[11] A. Elsanabary, S. Mekhilef, M. Seyedmahmoudian, and A. Stojcevski, "An energy balancing strategy for modular multilevel converter based grid-connected photovoltaic systems," *IET Power Electronics*, vol. 14, no. 12, pp. 2115-2126, April 2021, doi: 10.1049/pel2.12113.

[12] H. Bayat and A. Yazdani, "A Power Mismatch Elimination Strategy for an MMC-Based Photovoltaic System," *IEEE Transactions on Energy Conversion*, vol. 33, no. 3, pp. 1519-1528, Sep 2018, doi: 10.1109/TEC.2018.2819982.

[13] B. E. d. O. B. Luna, C. B. Jacobina, and A. C. Oliveira, "Internal Energy Balance of a Modular Multilevel Cascade Converter Based on Chopper-Cells With Distributed Energy Resources for Grid-Connected Photovoltaic Systems," *IEEE Transactions on Industry Applications*, vol. 59, no. 2, pp. 1935-1943, 2023, doi: 10.1109/TIA.2022.3225122.

[14] X. Pan et al., "Decoupling Capacitor Minimization of an MMC-Based Photovoltaic System With Three-Winding Power Channel," *IEEE Transactions on Power Electronics*, vol. 37, no. 1, pp. 1012-1026, 2022, doi: 10.1109/TPEL.2021.3100341.

[15] S. Nayak and A. Das, "A Power Balancing Strategy in a PV-Based Modular Multilevel Converter During Mismatched Power Generation," *IEEE Transactions on Industrial Electronics*, pp. 1-9, 2023, doi: 10.1109/TIE.2023.3303654.

[16] T. M. Parreiras, A. P. Machado, F. V. Amaral, G. C. Lobato, J. A. S. Brito, and B. C. Filho, "Forward Dual-Active-Bridge Solid-State Transformer for a SiC-Based Cascaded Multilevel Converter Cell in Solar Applications," *IEEE Transactions on Industry Applications*, vol. 54, no. 6, pp. 6353-6363, 2018, doi: 10.1109/TIA.2018.2854674.

[17] A. Elsanabary, S. Mekhilef, M. Seyedmahmoudian, and A. Stojcevski, "A Novel Circuit Configuration for The Integration of Modular Multilevel Converter with Large-Scale Grid-Connected PV Systems," *IEEE Transactions on Energy Conversion*, pp. 1-14, 2023, doi: 10.1109/TEC.2023.3318565.

[18] A. Elsanabary, S. Mekhilef, and N. F. A. Aziz, "Internal Power Balancing of an MMC-Based Large-Scale PV System under Unbalanced Voltage Sags," *IEEE Journal of Emerging and Selected Topics in Power Electronics*, pp. 1-1, 2024, doi: 10.1109/JESTPE.2024.3399395.

[19] M. López, F. Briz, A. Zapico, A. Rodríguez, and D. Diaz-Reigosa, "Control strategies for MMC using cells with power transfer capability," in *2015 IEEE Energy Conversion Congress and Exposition (ECCE)*, 20-24 Sept. 2015 2015, pp. 3570-3577, doi: 10.1109/ECCE.2015.7310165.

[20] Z. Wang, H. Lin, and Y. Ma, "Improved capacitor voltage balancing control for multimode operation of modular multilevel converter with integrated battery energy storage system," *IET Power Electronics*, vol. 12, no. 11, pp. 2751-2760, 2019, doi: 10.1049/iet-pel.2019.0033.

Single-Stage High-Frequency-Link Split-Phase Microinverter With High Voltage Gain Based on Buck-Boost AC Chopper

Xuewen Li
School of Electrical Engineering
Xi'an Jiaotong University
Xi'an, China
xuewen.li@stu.xjtu.edu.cn

Jia Liu
School of Electrical Engineering
Xi'an Jiaotong University
Xi'an, China
jia.liu@xjtu.edu.cn

Jinjun Liu
School of Electrical Engineering
Xi'an Jiaotong University
Xi'an, China
jjliu@mail.xjtu.edu.cn

Abstract—This paper proposes a novel single-stage high-frequency link (HFL) split-phase microinverter. It incorporates two buck-boost ac choppers to form an innovative secondary-side cycloconverter, enabling split-phase output with double the voltage gain compared to conventional topologies. Therefore, the required turns ratio for the high-frequency transformer (HFT) can be halved, resulting in high efficiency and low voltage stress. The microinverter supports split-phase output using a two-winding HFT, while retaining VSI characteristics for both grid-tied and islanded operations. A corresponding soft-switching modulation and dual-mode control strategy are also introduced. Experimental validation using a 600-W prototype demonstrates the efficacy of the proposed topology, achieving a peak efficiency of 96.5%.

Keywords—Microinverter, photovoltaic, high-frequency link, high voltage gain

I. INTRODUCTION

In distributed residential photovoltaic (PV) systems, microinverters have gained significant attention due to their component-level maximum power point tracking (MPPT), plug-and-play flexibility, and enhanced safety [1]. With advancements in PV technologies, the development of higher-power-rating microinverters to meet the demands of next-generation 600-W PV modules has become a critical focus for distributed household PV systems. Meanwhile, the capability to operate in both grid-tied and islanded modes is essential to expand the range of applications for microinverters [2-4].

In microinverter applications, flyback-based topologies are widely studied for their simple structure and low cost. However, a single flyback converter is typically suitable only for low-power applications below 150W because of the low utilization factor of the magnetic core [5]. To achieve higher power ratings, multiple flyback converters are usually paralleled, significantly increasing both the cost and volume of the microinverter. Consequently, flyback-based microinverters are not well-suited for next-generation 600-W class PV modules. In contrast, single-stage high-frequency-link (HFL)-based microinverters offer a high transformer utilization factor and facilitate straightforward implementation of soft switching, making them a promising solution for achieving both high power ratings and high efficiency. Among HFL-based topologies, matrix-type topologies are particularly suited for both grid-tied and islanded operations, due to their voltage-source-inverter (VSI) characteristics [6-9].

In conventional HFL-based microinverter, both the primary- and secondary-side converters are typically of the buck type. Therefore, the high-frequency transformer (HFT) must provide step-up capability with a high turns ratio to accommodate the low and variable voltage of PV modules. However, this high turns ratio introduces two main issues: first, it causes significant voltage stresses on the secondary-side devices at high input voltages; second, it requires more turns in the HFT, which exacerbates parasitic effects, increases the transformer's volume and copper losses, and further compromises conversion efficiency.

In grid fault or no-gird scenarios, microinverters are expected to support local loads by switching to islanded operation. Moreover, single-phase three-wire systems are widely employed in countries such as the USA and Japan, which represent key markets for residential PV systems [10, 11]. Microinverters deployed in these regions are expected to be directly compatible with the single-phase three-wire systems. To address this need, a single-stage split-phase matrix-type HFL microinverter was introduced in [12], which supports both grid-tied and islanded operations while maintaining compatibility with both single-phase two-wire and three-wire configurations. However, this topology still requires an high-turns-ratio HFT, limiting efficiency improvements.

To address these challenges, this paper proposes a novel single-stage split-phase HFL microinverter. In the proposed microinverter, the secondary-side cycloconverter consists of two buck-boost ac choppers, which effectively double the voltage gain compared to the cycloconverters in [12]. This allows the HFT turns ratio to be reduced by half, thereby decreasing voltage stresses on secondary-side devices and copper losses. Additionally, the microinverter supports split-phase output while maintaining VSI characteristics, making it suitable for both grid-tied and islanded modes, and directly compatible with single-phase two-wire and three-wire power systems. A corresponding soft-switching modulation strategy and closed-loop control strategy are also proposed to ensure high conversion efficiency and robust performance. Finally, experimental results from a 600-W prototype validate the effectiveness of the proposed topology and strategies.

II. OPERATION PRINCIPLES OF THE PROPOSED HFL MICROINVERTER

Fig. 1 illustrates the proposed HFL microinverter topology. The primary-side circuit consists of a full-bridge dc-ac inverter using switches $T_1\sim T_4$. On the secondary side, two input-parallel-output-series (IPOS) buck-boost ac choppers form a novel cycloconverter to realize the split-

---

This work was supported in part by the National Natural Science Foundation of China under Grant 52107204

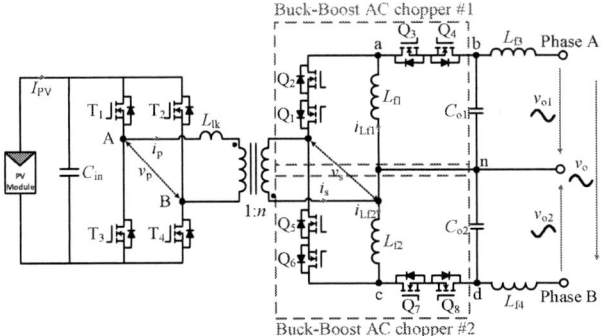

Fig. 1. Proposed single-stage HFL split-phase microinverter based on buck-boost ac choppers.

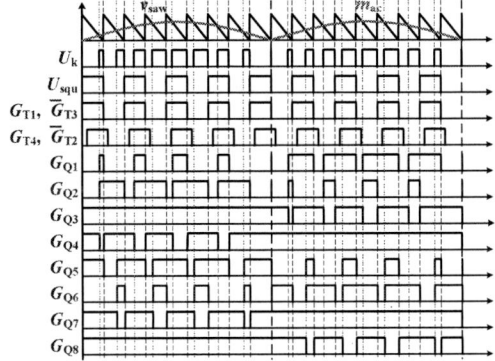

Fig. 2. Proposed soft-switching SSM strategy.

phase output, comprising switches $Q_1 \sim Q_8$ and inductors L_{f1}, L_{f2}. Inductors $L_{f1} \sim L_{f4}$ and capacitors C_{o1}, C_{o2} form a symmetrical LCL output filter. A two-winding HFT links the primary and secondary circuits. This configuration allows the two phase-to-neutral output voltages of the proposed HFL microinverter to naturally balance.

A. Modulation Strategy and Operating Modes

A soft-switching secondary-side modulation (SSM) strategy is proposed for the microinverter, with the corresponding driving signals depicted in Fig. 2. In this strategy, the primary-side full-bridge converter generates a square wave with a 50% duty cycle, while sinusoidal pulse-width modulation (SPWM) is implemented on the secondary-side cycloconverter. The reference m_{ac} is compared with a sawtooth carrier to produce the SPWM signal U_k, and the rising edge of sawtooth carrier is processed through a double-frequency divider to generate the signal U_{squ}. These signals U_k, U_{squ} are used to derive the driving signals for switches $T_1 \sim T_4$ and $Q_1 \sim Q_8$. During each half-line frequency cycle, two of the switches in Q_3, Q_4, Q_7, and Q_8 remain on to reduce switching losses. A dead time is introduced in the primary-side arm to prevent shoot-through, while an overlapping time between the secondary-side arms ensures a proper freewheeling path for the inductor current. Due to the use of buck-boost ac choppers, the output voltage is not directly proportional to the modulated waveform. Therefore, to achieve a desired sinusoidal output, the reference is defined as follow:

$$m_{ac} = \frac{\left| V_{o\text{-ref}} \sin\left(\omega_0 t\right) \right|}{n V_{in} + 0.5 \left| V_{o\text{-ref}} \sin\left(\omega_0 t\right) \right|}, \tag{1}$$

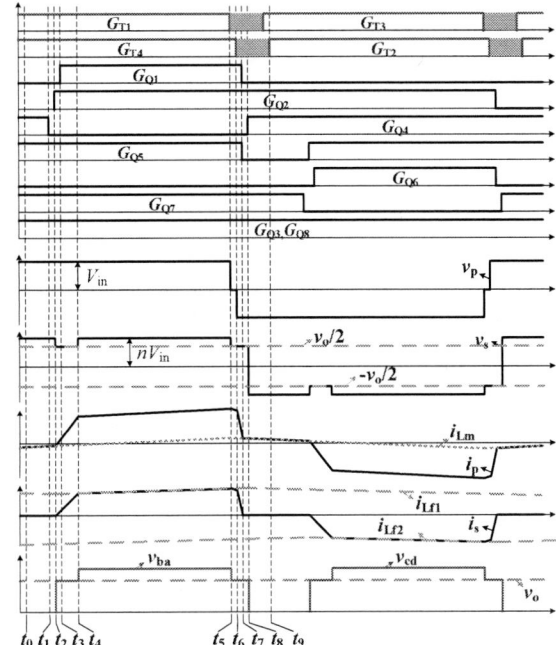

Fig. 3. Key waveforms for one switching period during the positive half-line-frequency cycle.

where $V_{o\text{-ref}}$ is the amplitude of the desired output voltage and ω_0 is the output angular frequency.

Fig. 3 illustrates the key waveforms for one switching period during the positive half-line-frequency cycle. Fig. 4 shows the equivalent circuits for each operating mode.

Mode 1a $[t_0 \sim t_1]$: As shown in Fig. 4(a), before t_0, the microinverter reaches a steady state, with the output capacitances of switches T_2, T_3, Q_2, and Q_6 fully charged. During this mode, the primary-side switches T_1, T_4, and secondary-side switches Q_3, Q_4, Q_7, Q_8 and Q_5 are conducting. i_p and i_{Lm} freewheel through T_1, T_4, and the primary-side winding, while i_{Lf1} and i_{Lf2} freewheel through Q_3, Q_4, and Q_7, Q_8, respectively. Both terminal voltages v_{ba} and v_{cd} are clamped to zero. This mode ends when Q_4 turns off.

Mode 1b $[t_1 \sim t_2]$: At t_1, Q_4 turns off, causing i_{Lf1} to freewheel through the body diode of Q_4, while the other switches maintain the same operation as in Mode 1a.

Mode 2a $[t_2 \sim t_3]$: At t_2, Q_2 turns on, forward-biasing the body diode of Q_1, which then turns on. The body diode of Q_4 turns off due to the reverse voltage, and i_{Lf1} continues to freewheel through Q_2, the body diode of Q_1, Q_5 and the body diode of Q_6, causing a further decrease in i_{Lf1}.

Mode 2b $[t_3 \sim t_4]$: At t_3, Q_1 turns on under ZVS, as its body diode is already conducting. Other switches continue to operate as in Mode 2a.

In Mode 2, energy begins to transfer from the primary side, causing both i_p and i_s increase linearly. Despite the path for L_{f1} charging is established, magnetization of L_{f1} does not occur due to the presence of leakage inductor L_{lk}, resulting in duty cycle loss. This mode ends when i_s equals i_{Lf1}.

979-8-3315-1612-3/25 $31.00 © 2025 IEEE

Mode 3 [$t_4 \sim t_5$]: As shown in Fig. 4(e), the circuit enters an active state, and energy from the PV module is stored in inductor L_{f1}. This mode ends when T_1 is turned off.

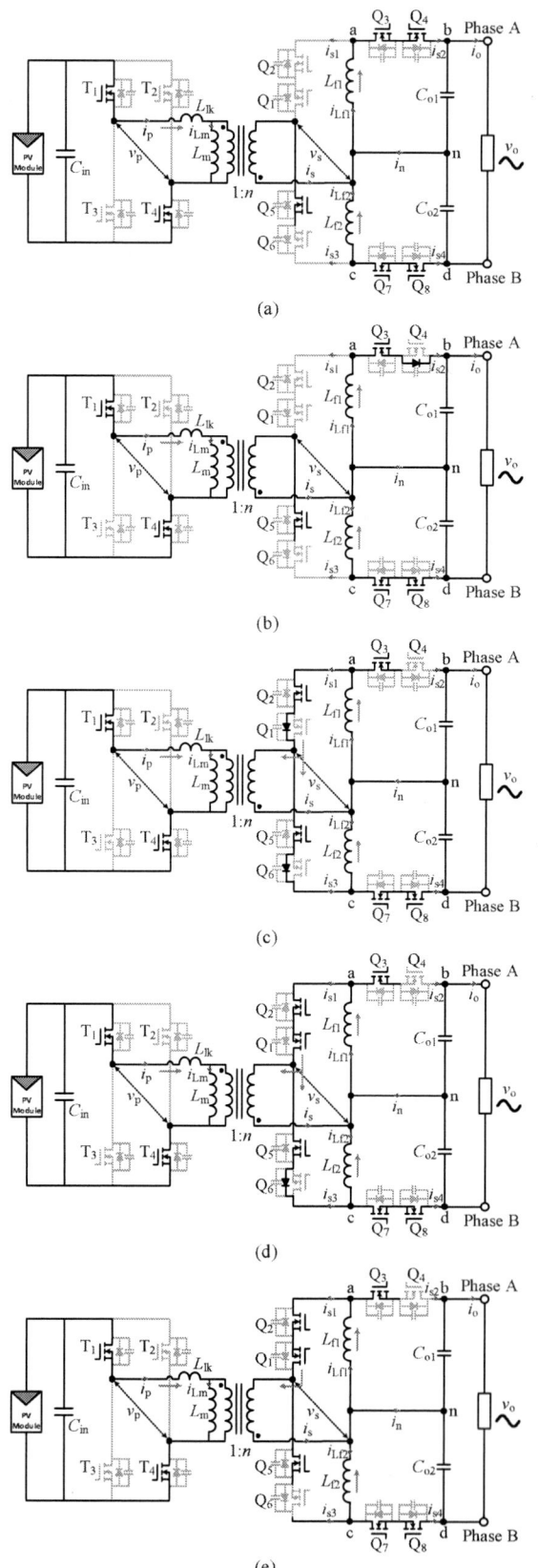

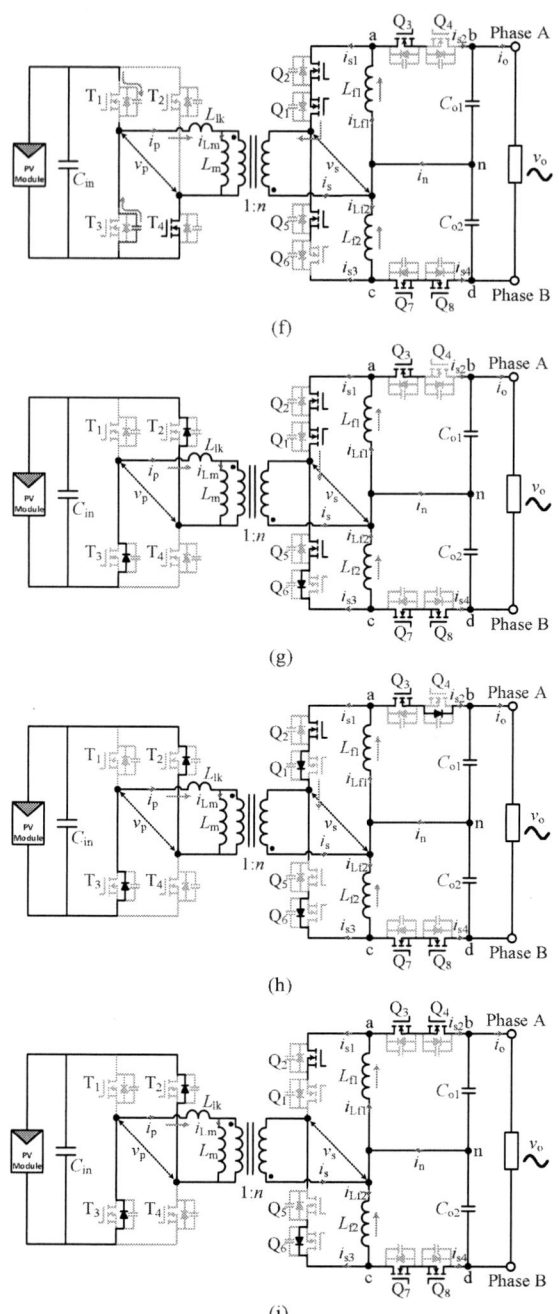

Fig. 4. Equivalent circuits of each mode in one switching cycle. (a) Mode 1a [$t_0 - t_1$]. (b) Mode 1b [$t_1 - t_2$]. (c) Mode 2a [$t_2 - t_3$]. (d) Mode 2b [$t_3 - t_4$]. (e) Mode 3 [$t_4 - t_5$]. (f) Mode 4 [$t_5 - t_6$]. (g) Mode 5 [$t_6 - t_7$]. (h) Mode 6 [$t_7 - t_8$]. (i) Mode 7 [$t_8 - t_9$].

Mode 4 [$t_5 \sim t_6$]: As shown in Fig. 4(f), at t_5, T_1 turns off. The output capacitor of T_1 is charged, causing a voltage rise, while the output capacitor of T_3 is discharged. Since i_p is large, the charging/discharging of capacitors C_{oss1_T1}/C_{oss1_T3} occurs quickly. Once C_{oss1_T3} is fully discharged, the body diode of T_3 turns on, forming a freewheeling path for i_p. Meanwhile, the output capacitors of Q_4 and Q_6 are discharged in this mode. After this process, the voltage across Q_4 is clamped to $v_{o1} - v_{o2}$, the voltage across Q_6 drops to zero, and v_s is clamped to $-v_{o2}$. At t_6, T_4 turns off, transitioning the circuit to the next mode.

Mode 5 [t_6~t_7]: At t_6, T$_4$ turns off. The output capacitor of T$_4$ is charged, causing a voltage rise, while the output capacitor of T$_2$ is discharged. Once C_{oss1_T2} is fully discharged, the body diode of T$_2$ turns on. The current i_p decreases under the influence of V_{in} and $-v_{o2}$.

Mode 6 [t_7~t_8]: At t_7, both Q$_1$ and Q$_5$ turn off. Their body diodes turn on due to forward voltage bias. During this mode, i_{s1} decreases, while i_{s2} increases. At the end of this mode, Q$_4$ is turned on under ZVS.

Mode 7 [t_8~t_9]: As shown in Fig. 4(i), in this mode, ZVS-on is achieved for both T$_2$ and T$_3$. After these modes, the circuit enters the negative half-switching period, which operates similarly to the positive half-switching period; thus, further discussions are omitted.

B. Voltage Gain Analysis

In the proposed HFL microinverter, the input dc voltage is converted to high-frequency square wave with a duty cycle of 50%, which is then transmitted through the HFT to the secondary side. Consequently, the maximum duty cycle for each buck-boost ac chopper is also 50%. During the positive half-line-frequency cycle, the voltage gains of the two buck-boost ac choppers can be derived as:

$$\begin{cases} M_1 = \dfrac{D}{1-D} \\ M_2 = \dfrac{-D}{1-D} \end{cases} \quad (2)$$

$$M = M_1 - M_2 = \frac{2D}{1-D} \quad (3)$$

where, M_1, M_2 and M represent the voltage gains of buck-boost ac choppers #1, #2 and the overall cycloconverter, and D is the equivalent duty cycle of arms Q$_{12}$ and Q$_{56}$ during one switching cycle.

From (2) and (3), the relationship between the voltage gains and duty cycle D is plotted in Fig. 5. Since the duty cycle is always less than 0.5, both buck-boost ac choppers operate in buck mode, with their maximum voltage gain reaching 1 at $D = 0.5$. Consequently, the maximum voltage gain of the novel cycloconverter reaches 2 at $D = 0.5$.

Compared with the topology in [12], which consists of two half-wave ac choppers, the maximum voltage gain of the proposed topology is double that of the previous design. Thus, under the same input and output conditions, the HFT turns ratio in the proposed microinverter can be halved, improving efficiency and reducing size.

C. Closed-loop Control Strategy

The proposed microinverter exhibits VSI characteristics, making it suitable for both grid-tied and islanded applications. The closed-loop control block diagrams for grid-tied and islanded modes are illustrated in Fig. 6.

In buck-boost converters, the inherent right-half-plane zero (RHPZ) poses challenges for direct output current control. To address this, the proposed control strategy regulates the buck-boost inductor current, thereby eliminating the RHPZ effect. The equivalent controlled current $i_{\text{L-eq}}$ is derived from the sampled values of i_{Lf1} and i_{Lf2}, as shown below:

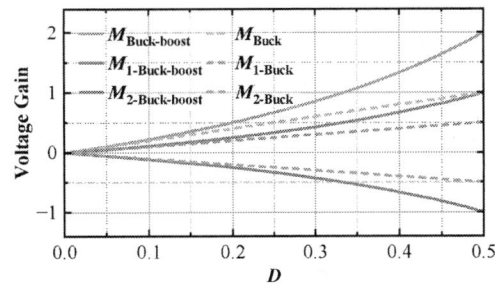

Fig. 5. Relationship between D and voltage gain compared with the topology in [12].

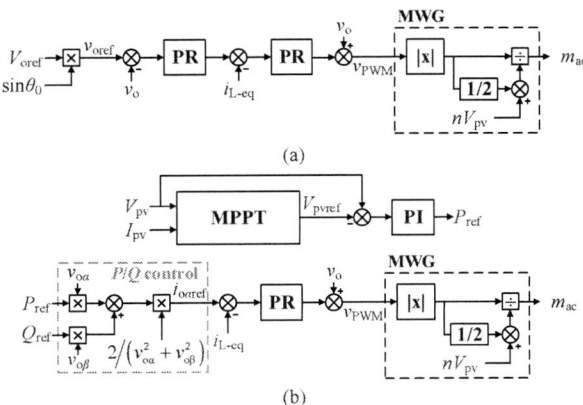

Fig. 6. Closed-loop control block diagrams (a) of islanded mode and (b) of grid-tied mode.

$$i_{\text{L-eq}} = \frac{nV_{pv}}{nV_{pv} + |0.5v_o|} \frac{(i_{Lf1} - i_{Lf2})}{2}. \quad (4)$$

Additionally, the input voltage V_{pv}, input current i_{pv}, and output voltage v_o are sampled to implement the closed-loop control. A second-order generalized integrator (SOGI) is used to extract the orthogonal components of the output voltage v_o. For grid-tied mode, a perturb and observe (P&O) algorithm is applied for MPPT, generating the reference voltage V_{pvref}. The error between V_{pvref} and the actual V_{pv} is processed by a voltage controller, which produces the active power command P_{ref}. To decouple the control of active and reactive power, the current reference is computed as:

$$i_{o\alpha ref} = \frac{2\left(P_{ref}v_{o\alpha} + Q_{ref}v_{o\beta}\right)}{v_{o\alpha}^2 + v_{o\beta}^2}. \quad (5)$$

A proportional-resonant (PR) controller is then employed to track the sinusoidal reference accurately. Subsequently, the modulated wave m_{ac} is generated using a modulation waveform generator (MWG).

In islanded mode, the outer voltage control loop shifts from PV voltage regulation to output voltage regulation, allowing the PV operating point to adjust automatically to balance the load power. The PR controller is retained in the inner current loop to maintain precise control.

III. EXPERIMENTAL RESULTS

To validate the proposed HFL microinverter, a 600-W prototype is constructed, as shown in Fig. 7. The detailed specifications of the prototype are listed in Table I.

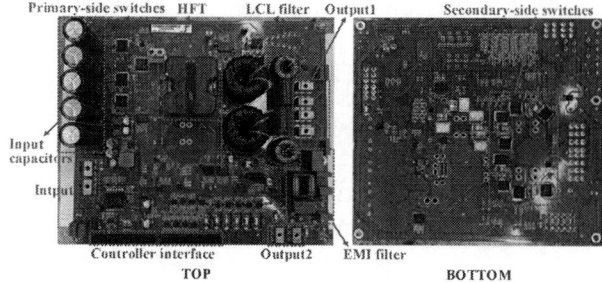

Fig. 7. Photograph of the 600-W prototype.

TABLE I. SPECIFICATIONS OF THE PROTOTYPE

Parameter	Symbol	Value
Input voltage	V_{pv}	35~55V
Input capacitor	C_{in}	16mF
Input voltage	V_{in}	35~55V
Output voltage	v_o	220V
Output voltage	v_{o1}, v_{o2}	110V
Rated power	P_0	600W
Primary-side switches	$T_1\sim T_4$	AGM12T02LL
Secondary-side switches	$Q_1\sim Q_8$	C3M0060065J
Input capacitor	C_{in}	16mF
Switching frequency	f_{sw}	50kHz
Transformer turns ratio	n	5
Leakage inductance	L_{lk}	0.25μH
Magnetizing inductance	L_m	225μH
Filtering inductor	L_{f1}, L_{f2}	1800μH
Filtering capacitor	C_{o1}, C_{o2}	3.3μF
Filtering inductor	L_{f3}, L_{f4}	120μH

Fig. 8 depicts the steady-state waveforms in islanded operation at V_{in} = 45V and R_o=100Ω. The output voltage remains sinusoidal and stable, validating the proposed topology and modulation strategy. Key waveforms at the switching-frequency timescale during the positive half-line-frequency cycle are shown in Fig. 9, illustrating the proposed soft-switching SSM strategy. In this strategy, the primary-side full-bridge converter only generates high-frequency waveforms with a fixed 50% duty cycle. The secondary-side switches modulate the high-frequency square wave v_s to produce SPWM terminal voltages v_{ab} and v_{cd}.

The dynamic response waveforms in islanded operation are shown in Fig. 10, demonstrating the response when the load on Phase B is alternately added and removed. During this transient process, Phase A current remains stable, and both phase voltages maintain balance, reflecting the robust dynamic response and auto-balancing capabilities of the proposed microinverter.

Fig. 11 presents the steady-state waveforms for grid-tied operation. The proposed control strategy indirectly regulates the output current by controlling the inductor current, simplifying sampling and addressing the RHPZ issue. Consequently, the inductor current i_{Lf1} exhibits a non-sinusoidal waveform, while the grid current i_g retains a sinusoidal shape.

To evaluate the performance of prototype with a PV module, the input was replaced by a PV simulator with a constant ambient temperature of 25°C. The dynamic waveforms for grid-tied operation are shown in Fig. 12. At 11 seconds, a decrease in solar irradiance from 1000 W/m² to 700 W/m² caused the maximum PV power (P_{mp}) to drop

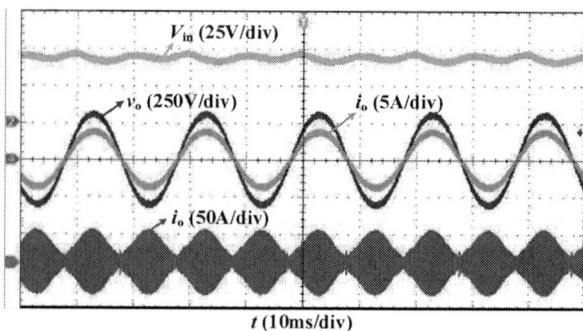

Fig. 8. Closed-loop steady-state waveforms in islanded operation with resistive load at V_{in} = 45V.

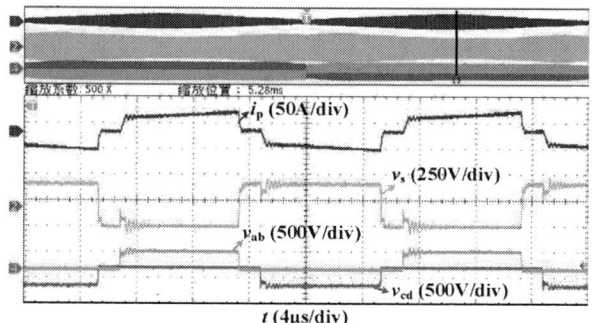

Fig. 9. Key waveforms illustrating the proposed SSM strategy at the positive half-line-frequency cycle.

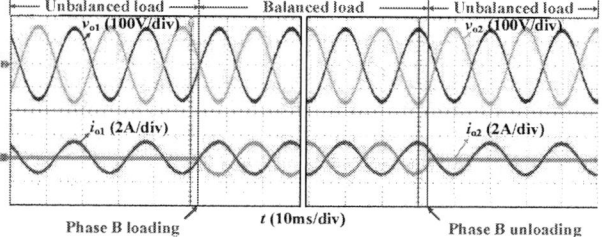

Fig. 10. Dynamic response waveforms for islanded operation.

from 550 W to 400 W. The actual PV power (P_{pv}) swiftly tracks the new maximum power point (MPP) after a brief transition. The MPPT efficiency exceeds 99% at steady state, demonstrating the effectiveness of the proposed microinverter and control strategy.

Fig. 13 compares the efficiency of the proposed topology with that of topology in [12]. Due to the low HFT turns ratio and two-winding design in the proposed microinverter, the copper loss is significantly reduced. Additionally, the soft-switching strategy and reduced voltage stress further decrease switching losses. As a result, the proposed microinverter demonstrates a noticeable efficiency improvement across the entire output power range, with a peak efficiency of 96.5%.

IV. CONCLUSIONS

This paper introduces a novel single-stage split-phase HFL microinverter that employs a secondary-side cycloconverter based on buck-boost ac choppers and a soft-switching modulation strategy to significantly reduce the HFT turns ratio and improve system efficiency. The microinverter retains VSI characteristics, enabling operation

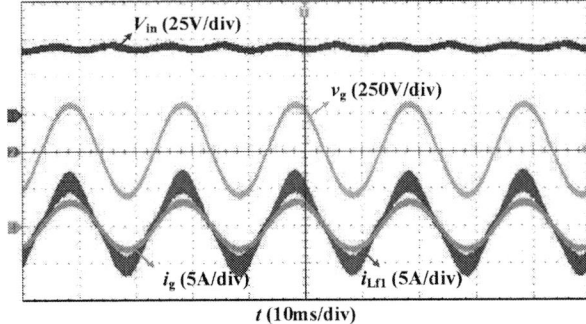

Fig. 11. Closed-loop steady-state waveforms for grid-tied operation.

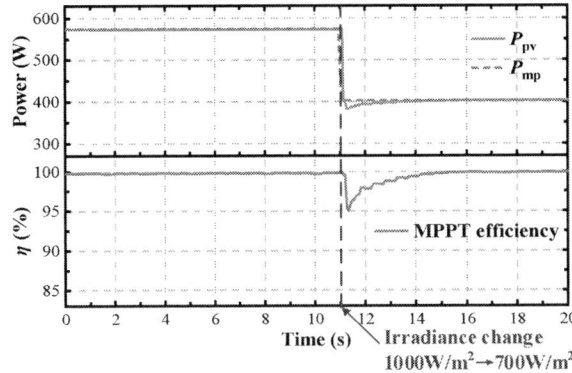

Fig. 12. Dynamic response waveforms for grid-tied application connected to PV arrays simulator under solar irradiance change condition.

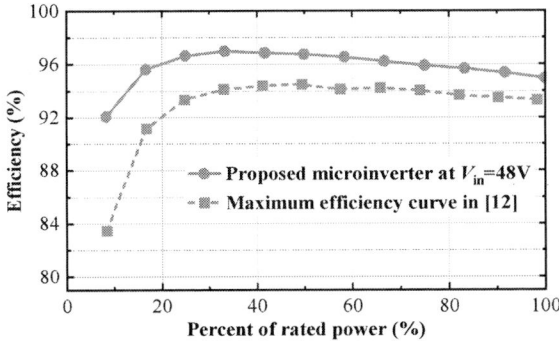

Fig. 13. Efficiency comparison between the proposed topology and the topology in [12] (P_0=600W).

in both grid-tied and islanded applications. A dual-mode control strategy is developed to ensure robust performance across operating modes. Experimental validation using a 600-W, 50-Hz prototype demonstrates a peak efficiency of 96.5%. Steady-state and dynamic test results confirm the efficacy of the proposed topology and control strategies.

REFERENCES

[1] Y. Chen and D. G. E. M. Xu, "Review of Soft-Switching Topologies for Single-Phase Photovoltaic Inverters," *IEEE Trans. Power Electron.*, pp. 1-1, 2021.

[2] J. Svarc, "Most powerful solar panels 2023," 2024. Accessed: Oct. 16, 2023. [Online]. Available: https://www.cleanenergy-reviews.info/blog/most-powerful-solar-panels

[3] K. Alluhaybi, I. Batarseh, and H. Hu, "Comprehensive review and comparison of single-phase grid-tied photovoltaic microinverters,"

IEEE J. Emerg. Sel. Topics Power Electron., vol. 8, no. 2, pp. 1310–1329, Jun. 2020.

[4] D. Dong *et al.*, "A PV residential microinverter with grid-support function: design, implementation, and field testing," *IEEE Trans. Ind. Appl.*, vol. 54, no. 1, pp. 469-481, Jan./Feb. 2018.

[5] Z. Zhang, M. Chen, W. Chen, C. Jiang, and Z. Qian, "Analysis and implementation of phase synchronization control strategies for BCM interleaved flyback microinverters," *IEEE Trans. Power Electron.*, vol. 29, no. 11, pp. 5921–5932, Nov. 2014.

[6] K.-S. Kim, S.-G. Jeong, O. Kwon, and B.-H. Kwon, "Weighted efficiency enhancement for single-power-conversion microinverters using hybrid mode modulation strategy," *IEEE Trans. Ind. Electron.*, vol. 67, no. 12, pp. 10243–10252, Dec. 2020.

[7] A. K. Bhattacharjee and I. Batarseh, "Sinusoidally modulated AC-link microinverter based on dual-active-bridge topology," *IEEE Trans. Ind. Appl.*, vol. 56, no. 1, pp. 422–435, Jan./Feb. 2020.

[8] Y. Wang, H. Liu, and P. Wheeler, "Research on the voltage spike suppression strategy for three-phase high frequency link matrix-type inverter," *IEEE J. Emerg. Sel. Topics Power Electron.*, vol. 10, no. 5, pp. 6070–6083, Oct. 2022.

[9] P. Nayak, K. Rajashekara, and S. K. Pramanick, "Soft-switched modulation technique for a single-stage matrix-type isolated DC–AC converter," *IEEE Trans. Ind. Appl.*, vol. 55, no. 6, pp. 7642–7656, Nov./Dec. 2019.

[10] H. L. Jou, G. R. Chen, J. C. Wu, K. D. Wu, and J. M. Jhang, "Three-port single-phase three-wire power converter interface for micro grid," *Renewable Energy*, vol. 85, pp. 524-533, Jan. 2016.

[11] N. S. de Moraes Lima Marinus, E. C. dos Santos, C. B. Jacobina, N. Rocha, and N. B. de Freitas, "A bridgeless controlled rectifier for single split-phase systems," *IEEE Trans. Ind. Appl.*, vol. 53, no. 5, pp. 4708–4717, Sep./Oct. 2017.

[12] X. Li, J. Liu, F. Ji, X. Cao, Y. Wang and J. Liu, "A Single-Stage High-Frequency-Link Split-Phase Microinverter for Both Grid-Tied and Islanded Operation," *IEEE Trans. Power Electron.*, vol. 39, no. 8, pp. 10409-10423, Aug. 2024.

Fault Diagnosis and Tolerant Strategy for Triple-Port Hydrogen Converter Using SSA-Optimized Random Forest Algorithm

Shiqi Zhang
Department of Electrical Engineering
Yanshan University
Qinhuangdao, 066004, China
s.zhang@stumail.ysu.edu.cn

Yiyina Teng
Department of Electrical Engineering
Yanshan University
Qinhuangdao, 066004, China
tengyiyina@stumail.ysu.edu.cn

Naizhe Diao
Department of Electrical Engineering
Yanshan University
Qinhuangdao, 066004, China
diaonaizhe@ysu.edu.cn

Xiaoqiang Guo
Department of Electrical Engineering
Yanshan University
Qinhuangdao, 066004, China
gxq@ysu.edu.cn

Vladimir Terzija
School of Engineering
Newcastle University
Newcastle upon Tyne, United Kingdom
Vladimir.Terzija@newcastle.ac.uk

Lichong Wang
Shenke Electronic Co., Ltd.
Shijiazhuang, China
wanglichong@snkgroup.cn

Abstract—Multi-port hydrogen converters are crucial for hydrogen production power sources, requiring high reliability. However, current fault diagnosis and fault-tolerant solutions for such converters are inadequate, relying on numerous sensors, offering limited diagnostic accuracy, and struggling with multiple switch fault detection. Additionally, existing fault-tolerant strategies cannot ensure stable power for electrolyzers or enable power-sharing between ports. To address these challenges, this paper presents a method for diagnosing open-circuit faults and implementing fault tolerance in triple-active bridge (TAB) hydrogen converters. The method uses a sparrow search algorithm (SSA) optimized random forest (RF) algorithm to achieve high accuracy, diagnosing faults in multiple switches using fewer sensors. For fault-tolerant operation, auxiliary switches maintain stable system output and enable power-sharing between ports after a fault. This ensures the hydrogen electrolyzer operates efficiently and reliably. Simulations validate the solution's effectiveness, demonstrating its economic viability and reliability, providing a foundation for safe and efficient green hydrogen production.

Keywords—*Hydrogen converter, Fault diagnosis, Fault-tolerant, Sparrow search algorithm, Random forest*

I. INTRODUCTION

The application of multi-port hydrogen converters in hydrogen production has attracted significant attention worldwide and is a critical direction for the power industry's future. As the core component of a hydrogen production system, the electrolyzer demands high reliability [1]. Faults in multi-port hydrogen conversion systems can cause fluctuations in the electrical energy supplied to the electrolyzer, disrupting its stable operation. This significantly impacts the hydrogen production system's operational efficiency and service life and, in extreme cases, may lead to system shutdown, causing irreversible damage [2]. Dependable fault diagnosis and fault-tolerant functionality are essential for maintaining the stability

of renewable energy electrolysis and optimizing electrolyzer energy output, performance, and expenses [3]. Yang utilizes five sensors to diagnose faults through logical analysis based on the average midpoint voltage of the bridge arm and the inductor current [4]. Another study identifies faults by leveraging the average midpoint voltage and the duty cycle [5]. Additionally, Rastogi applies the vector method for single-switch fault diagnosis in dual-active-bridge converters [6]. Overall, existing methods require many sensors, limited accuracy, and struggle with multi-switch fault diagnosis [7]. It cannot meet the diagnostic requirements of the hydrogen conversion system [8]. Moreover, existing fault-tolerant studies achieve fault diagnosis of dual-active-bridge converters through dual-loop fault-tolerant control [9]. Another researcher has also implemented fault-tolerant operation by blocking capacitors on both transformer sides [10]. These current fault-tolerant strategies fail to ensure continuous power supply and cannot share power between ports, it fails to maintain system stability under varying operating conditions [11]. This paper presents a high-precision, high-speed diagnostic method for multi-tube faults; fault diagnosis is performed by optimizing the RF model with the SSA. A fault-tolerant control approach is presented, which utilizes auxiliary switches and enables power sharing between ports. The proposed solution ensures that stable power can still be supplied to the electrolyzer in the event of a converter fault, maintaining the safe and stable operation of the electrolyzer. It also allows power-sharing adjustments between ports based on power supply requirements, thereby improving energy utilization efficiency.

The proposed topology of the TAB hydrogen converter is illustrated in Figure 1. Port 1 connects to the photovoltaic (PV) unit, Port 2 to the battery module, and Port 3 to the electrolyzer module. The electrolyzer module and battery module contain DC-DC converters. In case of a switch failure, the faulty switch is locked, and power is redirected from working ports to the affected load via S_1, maintaining stability. Furthermore, the

---

This work was supported by the National Natural Science Foundation of China (52377199 and 52477199), Science Research Project of Hebei Education Department (JZX2024027).

power generated by Ports 2 and 3 can be reroted through the auxiliary switch S_1 when necessary, enabling power sharing across the ports.

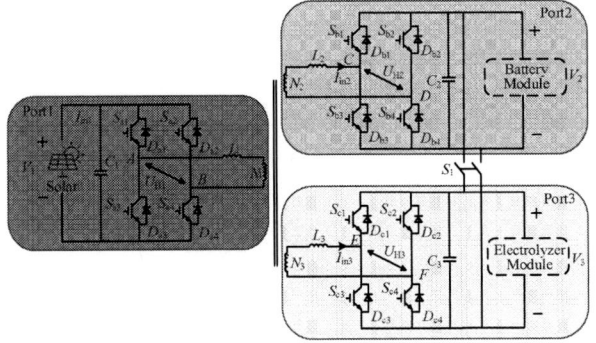

Fig. 1. Topology of fault-tolerant TAB converter.

The proposed topology of the TAB hydrogen converter is illustrated in Figure 1. Port 1 connects to the photovoltaic (PV) unit, Port 2 to the battery module, and Port 3 to the electrolyzer module. The electrolyzer module and battery module contain DC-DC converters. In case of a switch failure, the faulty switch is locked, and power is redirected from working ports to the affected load via S_1, maintaining stability. Furthermore, the power generated by Ports 2 and 3 can be reroted through the auxiliary switch S_1 when necessary, enabling power sharing across the ports.

II. FAULT DIAGNOSIS AND TOLERANCE CONTROL SCHEME

The overall steps using the SSA-RF algorithm are shown in Figure 2. The SSA-RF algorithm is used to analyze and filter each port's voltage and current data of the three-port hydrogen production converter. Figure 3 shows the filtering results, and the most critical features I_{i1}, I_{i2}, and I_{i3} are selected for training. The fault diagnosis through SSA-RF collects input current I_{i1}, I_{i2}, and I_{i3} from different ports. The specific steps are described as follows:

1) Collect the current I_{i1}, I_{i2}, and I_{i3} as training data, preprocess the data, and train the RF model. 2) Optimize the model using the SSA, setting the sparrow population size to 30 and initializing each sparrow's position, where the position represents the relevant parameters of the RF. 3) Calculate the fitness value of each sparrow individual, select the best and worst individuals in the current population, and randomly choose some individuals in the population as foragers, with the proportion of foragers set to 0.8 and the safety threshold set to 0.8. Modify the locations of sparrow individuals based on the fitness values and the safety threshold. 4) Iteratively update the sparrow positions and output the individual's position with the highest fitness value, representing the RF model parameters optimized by the SSA. In this paper, the upper limit of iterations is set to 500.

The improved RF model is used to perform fault diagnosis testing on the three-port hydrogen production converter, and Figure 4 displays the results; the diagnostic accuracy is as high as 99.7451%.

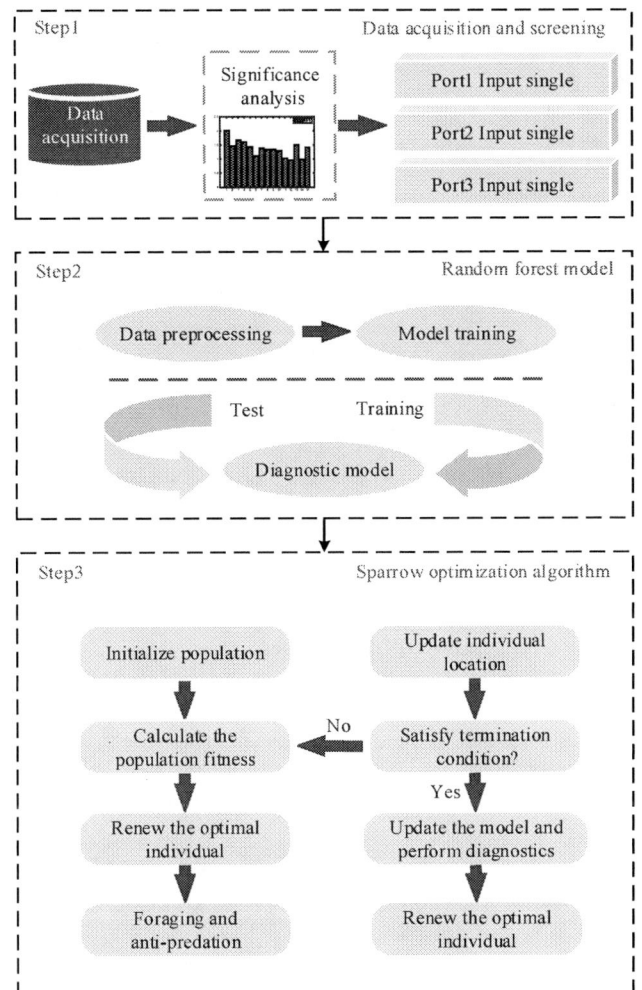

Fig. 2. Algorithm flow chart.

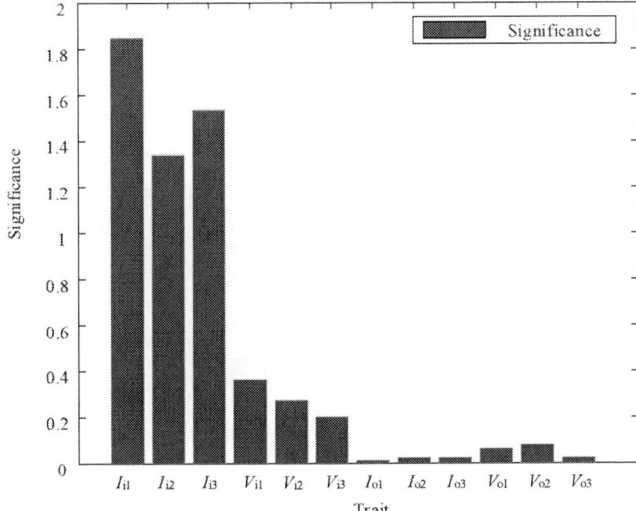

Fig. 3. Feature data selection analysis.

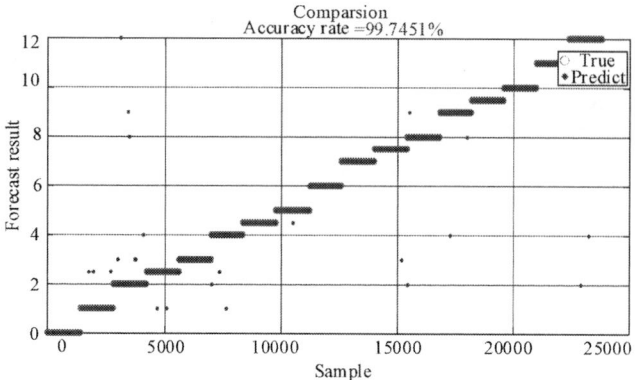

Fig. 4. Diagnostic results.

The proposed solution enables fault-tolerant operation after a fault in any part of the TAB converter and capacity sharing between the Ports 2 battery module and 3 electrolyzer module. Taking the electrolyzer port fault and capacity sharing scenario as an example, as shown in Figure 5, the fault-tolerant control process is as follows: in case of a fault on the electrolyzer side, switch S_1 is turned off by the system, redirecting power from the PV side to both the battery and electrolyzer via Port 2, ensuring the system continues to function normally. As depicted in Figure 6, S_1 can be closed depending on the need when the battery does not require charging, allowing Ports 2 and 3 to jointly supply power to the electrolyzer, thus enhancing its power supply.

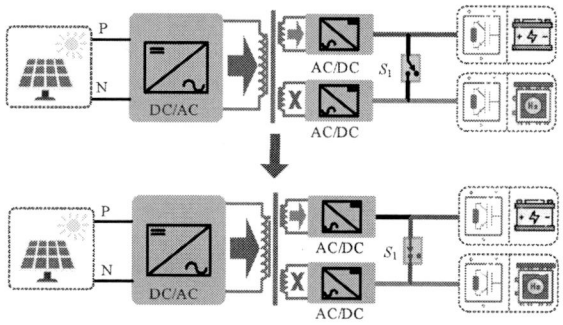

Fig. 5. Fault tolerance diagram.

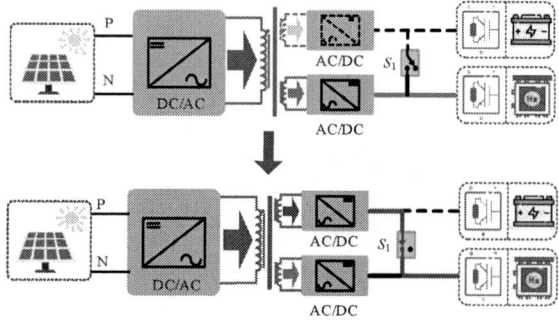

Fig. 6. Power sharing process.

III. SIMULATION VERIFICATION

To verify the proposed operation strategy, simulations were conducted under specific conditions: an input voltage of 144V, a battery voltage of 30V, a transformer turns ratio of 6:1:1, transformer-rated power of 10 kW, filter inductance of 100μH, and filter capacitance of 470 μF. The accuracy and effectiveness of the proposed fault diagnosis and tolerance approach are confirmed by the results.

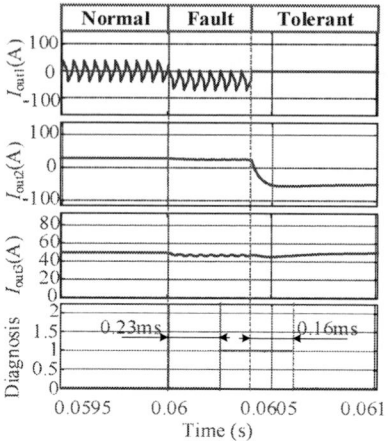

Fig. 7. Port 1 failure experiment.

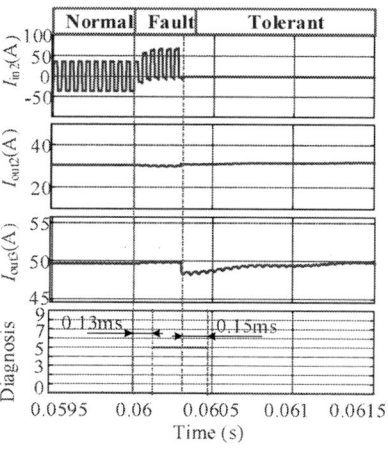

Fig. 8. Port 2 failure experiment.

Figure 7 shows the simulated waveform of fault diagnosis and fault-tolerant control when the switch S_{a1} in the Port 1 unit of the TAB hydrogen converter experiences an open-circuit fault. Figure 8 displays the simulated waveform during an open-circuit fault of the switch S_{b1} in the Port 2 unit. Figure 9 illustrates the simulated waveform when S_{c1} and S_{c2} in the Port 3 unit simultaneously experience open-circuit faults. Figure 10 demonstrates the simulated waveform under power-sharing conditions. The simulation outcomes verify the efficacy of the proposed fault diagnosis and fault-tolerant control strategy and the auxiliary switch's ability to enable power sharing. After a fault occurs, the fault signal is detected within 0.13 to 0.28 milliseconds, and a fault signal is output; fault-tolerant control is implemented 0.05 milliseconds later. The fault signal is eliminated within 0.15 to 0.2 milliseconds.

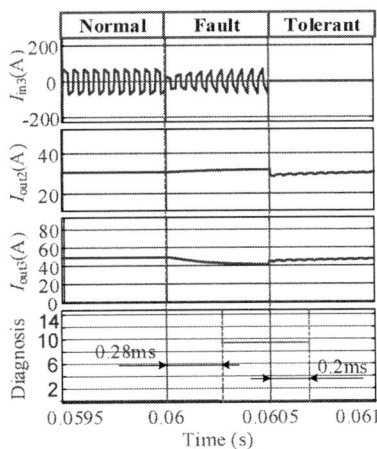

Fig. 9. Port 3 failure experiment.

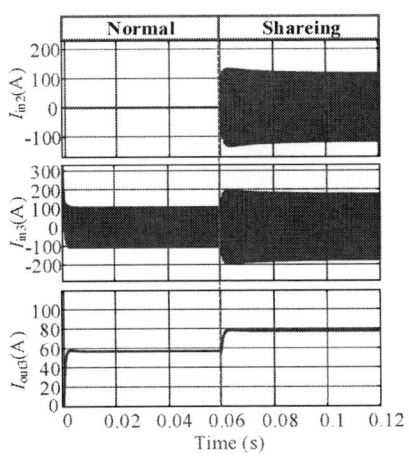

Fig. 10. Power sharing experiment.

Simulation results demonstrate that the method can accurately diagnose faults and quickly implement effective fault-tolerant control, enabling power sharing and maintaining stable power in the electrolyzer unit. This ensures the reliable operation of the TAB hydrogen converter.

IV. CONCLUSIONS AND FUTURE WORK

This paper presents an open-circuit fault diagnosis and fault-tolerant strategy for a TAB hydrogen converter, including a detailed analysis and validation. 1) The fault diagnosis approach employs SSA to optimize the RF model, enhancing both fault diagnosis speed and accuracy. 2) The fault diagnosis requires only three current sensors, which utilize the input current data from the converter's three ports, I_{i1}, I_{i2}, and I_{i3}, as diagnostic parameters. 3) The method can diagnose both single switch faults and multiple switch faults. 4) The fault-tolerant strategy uses auxiliary switches to enable fault-tolerant operation of the TAB hydrogen converter. 5) If a switch fails, the strategy effectively maintains stable output power while ensuring fault-tolerant operation in the event of a switch failure. 6) Through power sharing, flexible capacity scheduling between ports is enabled, boosting the electrolyzer side's power. Enhancing the hydrogen production rate makes the process more suitable and efficient for green hydrogen production. Given the difficulties faced by green hydrogen production, this research presents a design for a TAB hydrogen converter with improved economy and dependability, establishing a foundation for the dependable and efficient production of green hydrogen.

REFERENCES

[1] Y. Xia, H. Cheng, H. He, Z. Hu, and W. Wei, "Efficiency Enhancement for Alkaline Water Electrolyzers Directly Driven by Fluctuating PV Power," *IEEE Trans. Ind. Electron.*, vol. 71, no. 6, pp. 5755-5765, Jun. 2024.

[2] H. P. C. Buitendach, R. Gouws, C. A. Martinson, C. Minnaar, and D. Bessarabov, "Effect of a ripple current on the efficiency of a PEM electrolyser," *Results in Engineering*, vol. 10, pp. 100216, Jun. 2021.

[3] Q. Zhang, S. Lu, L. Xie, W. Xu, and H. Su, "Dynamic fault detection and diagnosis of industrial alkaline water electrolyzer process with variational Bayesian dictionary learning," *Int. J. Hydrogen Energy*, vol. 71, pp. 1492-1506, Jun. 2024.

[4] W. Yang, J. Ma, M. Zhu, and C. Hu, "Open-Circuit Fault Diagnosis and Tolerant Method of Multiport Triple Active-Bridge DC-DC Converter," IEEE Trans. Ind. Appl., vol. 59, no. 5, pp. 5473-5487, Sept.-Oct. 2023.

[5] C. Song, A. Sangwongwanich, Y. Yang, and F. Blaabjerg, "Open-Circuit Fault Diagnosis and Tolerant Control for 2/3-Level DAB Converters," IEEE Trans. Power Electron., vol. 38, no. 4, pp. 5392-5410, Apr. 2023.

[6] S. K. Rastogi, S. S. Shah, B. N. Singh, and S. Bhattacharya, "Vector-Based Open-Circuit Fault Diagnosis Technique for a Three-Phase DAB Converter," IEEE Trans. Ind. Electron., vol. 71, no. 7, pp. 8207-8211, Jul. 2024.

[7] D. Xie, and X. Ge, "A State Estimator-Based Approach for Open-Circuit Fault Diagnosis in Single-Phase Cascaded H-Bridge Rectifiers," IEEE Trans. Ind. Appl., vol. 55, no. 2, pp. 1608-1618, Mar.-Apr. 2019.

[8] Z. Li, B. Wang, Y. Ren, J. Wang, Z. Bai, and H. Ma, "L- and LCL-Filtered Grid-Tied Single-Phase Inverter Transistor Open-Circuit Fault Diagnosis Based on Post-Fault Reconfiguration Algorithms," IEEE Trans. Power Electron., vol. 34, no. 10, pp. 10180-10192, Oct. 2019.

[9] Y. Pan, Y. Yang, J. He, A. Sangwongwanich, C. Zhang, Y. Liu, and F. Blaabjerg, "A Dual-Loop Control to Ensure Fast and Stable Fault-Tolerant Operation of Series Resonant DAB Converters," IEEE Trans. Power Electron., vol. 35, no. 10, pp. 10994-11012, Oct. 2020.

[10] T. T. Le, M. K. Nguyen, T. D. Duong, C. Wang, and S. Choi, "Open-Circuit Fault-Tolerant Control for a Three-Phase Current-Fed Dual Active Bridge DC–DC Converter," IEEE Trans. Ind. Electron., vol. 70, no. 2, pp. 1586-1596, Feb. 2023.

[11] Y. Guan, Y. Xiao, L. Qin, X. Liu, H. Deng, and W. Wu, "A Path-Based Switch Open Circuit Fault-Tolerant Method for Three-Phase DAB Converter," IEEE Trans. Power Electron., vol. 39, no. 1, pp. 1577-1595, Jan. 2024.

Resilient Operation for Grid-connected Cascaded H-bridge Multilevel Inverter with Improving PV Source Stress

Jinli Zhu
Electrical & Computer Engineering
University of Pittsburgh
Pittsburgh, USA
jinli.zhu@pitt.edu

Yuan Li
Electrical & Computer Engineering
University of Pittsburgh
Pittsburgh, USA
yuan.li@pitt.edu

Hector Akuta
Electrical & Computer Engineering
University of Pittsburgh
Pittsburgh, USA
hoa43@pitt.edu

Jeonghun Kim
Electrical & Computer Engineering
University of Pittsburgh
Pittsburgh, USA
jeonghun.kim@pitt.edu

Uthandi Selvarasu
Electrical & Computer Engineering
Northeastern University
Boston, USA
selvarasu.u@northeastern.edu

Shumeng Wang
Electrical & Computer Engineering
Northeastern University
Boston, USA
wang.shum@northeastern.edu

Vikram Roy Chowdhury
National Renewable Energy Laboratory
Golden, CO, USA
vikram.roychowdhury@nrel.gov

Brad Lehman
Electrical & Computer Engineering
Northeastern University
Boston, USA
lehman@ece.neu.edu

Fang Z. Peng
Electrical & Computer Engineering
University of Pittsburgh
Pittsburgh, USA
fangzpeng@pitt.edu

Abstract—**In distributed power systems increasingly dominated by PV generation and grid-connected converters, cascaded H-bridge (CHB) multilevel inverters are particularly well-suited for high-power, high-voltage applications. However, CHB topologies are prone to unbalanced conditions due to submodule faults, which can be triggered by factors such as extreme weather, equipment aging, or operational stresses. While redundant submodules are commonly employed to mitigate these issues, they lower efficiency and increase the costs and complexity of circuits. This paper addresses unbalanced grid currents caused by faulty submodules in CHB multilevel inverters through a source stress-improved control strategy based on a neutral offset approach. When a submodule is bypassed due to faults or hot-swapping, all remaining submodules, across both normal and faulty phases, share the voltage lost from the bypassed submodule. By distributing the voltage contribution across all submodules in both faulty and non-faulty phases, this method reduces system costs and downtime while alleviating voltage stress and enhancing the operation efficiency of the inverter system.**

Index Terms—**CHB, grid-connected inverter, unbalanced, symmetric, resilient, control, optimization, neutral point, cascaded, multilevel, HIL**

I. Introduction

Future energy grids will increasingly rely on renewable energy sources, energy storage systems, and advanced power electronics to deliver sustainable, high-quality, and resilient power [1]. The cascaded H-bridge (CHB) multilevel inverter system has become a critical technology in photovoltaic (PV)

applications and other grid-connected systems, offering distinct advantages in modularity and scalability [2]. By cascading multiple low-voltage H-bridge submodules, CHB inverters can generate high-quality output voltages, making them ideal for large-scale PV and battery applications that require reliable power integration and grid support [3]–[5].

Despite the benefits of modular multilevel converter topologies, significant reliability challenges persist, particularly in demanding operational environments [6]–[8]. One of the primary challenges for CHB systems arises in scenarios involving natural disasters, extreme weather conditions, or human-induced faults, which can lead to failures in individual submodules [9]. In such cases, faulty submodules must often be bypassed or hot-swapped to maintain operation, resulting in an imbalance between the affected and unaffected phases [10], [11], and the submodules in the bypassed faulty phase are different from that of the normal phase, which will result in an abnormal operation for the inverter system, potentially causing abnormal power output and serious issues for the power grid, including voltage instability and harmonic distortion [12].

To address these reliability concerns, redundancy is an essential design element for ensuring system reliability, facilitating seamless submodule replacement during faults without interrupting operation. Redundancy schemes are generally classified into hot and cold categories [13]–[17]. In cold redundancy, it involves activating and charging the redundant submodule only after a fault has occurred. In contrast, hot re-

979-8-3315-1612-3/25 $31.00 © 2025 IEEE

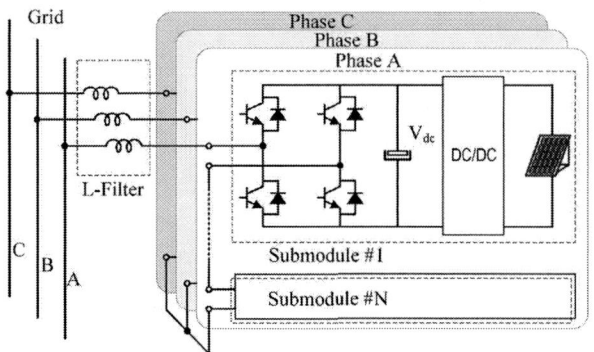

Fig. 1. Structure of the grid-connected cascaded H-bridge multilevel inverter.

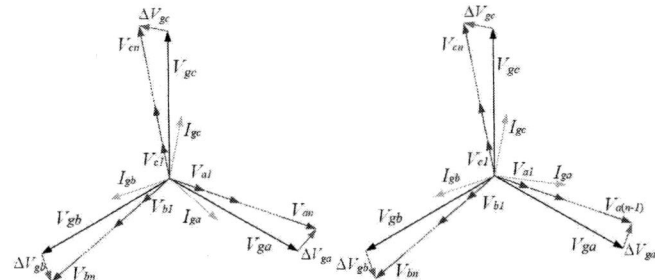

Fig. 2. Vector diagram of system voltage and current. (a) Normal operation. (b) Submodule fault in phase A.

dundant submodules run concurrently with active submodules and adjust to new operating points following a fault, resulting in reduced transition times. An advanced redundancy scheme is introduced for CHB electronic power transformers, enabling submodule replacement within microseconds [18], however, the redundancy configuration scheme significantly increase the costs and complexity of circuits and lower efficiency because the system must maintain additional components in standby mode.

In addition to redundancy strategies, there is considerable research focused on leveraging CHB topologies to mitigate grid voltage imbalances and harmonics [19]–[21]. Existing approaches for CHB topologies in power systems include reactive power control techniques aimed at ensuring stable grid interaction. For example, a method utilizing barycentric coordinates has been proposed to address power imbalances in grid-connected PV systems by modulating reactive power [6]. Another technique, offset-free discontinuous PWM, has shown promise in static compensators, effectively reducing power losses, even in unbalanced grid conditions [22]. Additionally, a power balance strategy for three-phase CHB inverters is explored in [23] to maintain continuous power output and stable grid support during grid imbalances. While these methods address power imbalances at a high level, they generally fall short of addressing inter-bridge imbalances due to specific submodule faults. In such cases, H-bridge voltage stress increases, leading to potential reliability issues that these conventional strategies do not adequately mitigate. While zero-sequence injection strategies can be used to alleviate unbalanced grid conditions/operations [24], it is not easy to effectively quantify its improvement for submodule source stresses.

In this paper, a novel operation approach for H-bridge source voltage stress improvement based on neutral offset is presented. Through accurate mathematical modeling combined with submodule states, this approach enables source voltage stress to be shared equally for all H-bridges, including both normal and faulty phases, ensuring consistent voltage levels across the system without excessive cost. As a result, the synthetic line voltage vector at the output of the CHB inverter is always in the shape of an equilateral triangle, which ensures a high-quality three-phase balanced grid current.

II. System Configuration

Fig. 1 presents the topology of a CHB multilevel inverter system in a PV application. Each phase of the inverter system consists of n H-bridge converters connected in series, and their neutral is connected in a star configuration. The DC link of the H-bridge is powered by one PV panel or a short string of PV panels and then fed to the grid. The DC capacitor voltage V_{dc} can be regulated using DC/DC circuits with voltage-oriented control so that each H-bridge operates at a suitable voltage level. The inverter system is connected to the 480 V grid via the L-filter that mitigates switching frequency harmonics. Here carrier phase shifted (CPS) modulation [25], [26] strategy is used to generate PWM signals. For the CHB inverter system with n sources, its output voltage has $(2n + 1)$ levels to synthesize the AC output [27]. The CHB topology not only facilitates the size reduction of the required output filter, but also lowers the voltage stress on the semiconductor switches.

III. Vector Analysis for Submodule Faults and Bridge Bypassing

During the normal operation of the CHB multilevel inverter system, as shown in the vector Fig. 2(a), the number of submodules is identical, and the output voltages of each phase are symmetrical. V_{ga}, V_{gb}, and V_{gc} represent the grid phase voltages, and I_{ga}, I_{gb}, and I_{gc} represent the grid currents. The H-bridge output voltages in three phases are V_{ai}, V_{bi}, and V_{ci}, respectively, $i \in 1, 2, ..., n$. In this work, the classical d-q decoupling control strategy [28]–[31] is used to realize the regulation of the output grid current. Based on Fig. 2(a), by adjusting the d-axis and q-axis current references in the inverter's controller, the CHB inverter can be easily controlled and outputs a balanced current. Under such an operation, when a common IGBT failure occurs in the CHB, it will result in an abnormal operation for the inverter system. During the unbalanced operation after the faulty submodules are bypassed, the number of submodules in the faulty phase is obviously different from that of the normal phase. As shown in Fig. 2(b), the output voltage (line-to-line) leads to an unbalanced output current and further a series of issues.

IV. Source Stress-improved Strategy

To maintain a three-phase balanced grid current, an intuitive solution is to use the residual submodule of the faulty phase

979-8-3315-1612-3/25 $31.00 © 2025 IEEE

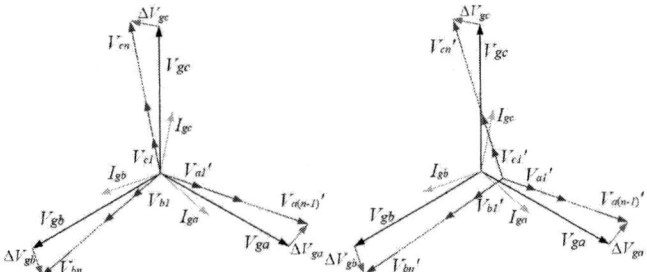

Fig. 3. Diagram of voltage and current for unbalanced CHB. (a) Phase A share voltage lost. (b) All H-bridges share voltage lost.

to share the voltage lost due to bypassing faulty submodules, as shown in Fig. 3(a). The residual H-bridge output voltages of the faulty A-phase are simultaneously increased to balance the line voltage. Another method is to control one of the H-bridges in the non-faulty phase to be bypassed, thus equalizing the number of submodules in the three phases. Both methods increase the submodules' source stress which can be reflected as the output voltage of each H-bridge or its modulation ratio. Of course, this can be achieved through three-phase submodules in which all components can be designed with redundancy to withstand higher voltage, which increases the cost. Therefore, we propose a source stress-improved control strategy based on the neutral offset. When a submodule is bypassed due to faults or hot-swapping, all residual submodules, including normal and faulty phases, share the voltage lost due to bypassing. All submodules maintain identical voltage stress, as shown in Fig. 3(b). The output voltages of each H-bridge are $V_{ai} = v$, $V_{bi} = v$, and $V_{ci} = v$. The inverter output line voltage magnitude is V_l, which is the synthetic vector voltage of H-bridges. There exists a mathematical relationship

$$2\pi = \cos^{-1} \frac{(N_b^2 + N_c^2)v^2 - V_l^2}{2N_b N_c v^2} +$$
$$\cos^{-1} \frac{(N_a^2 + N_c^2)v^2 - V_l^2}{2N_a N_c v^2} + \quad (1)$$
$$\cos^{-1} \frac{(N_a^2 + N_b^2)v^2 - V_l^2}{2N_a N_b v^2}$$

where N_a, N_b, and N_c denote the number of submodules in the three phases, and the output voltages of each H-bridge should satisfy the relationship below

$$\sum_{i=1}^{N_a} |V_{a,i}| = V_l/\sqrt{3} \quad (2)$$

$$\sum_{i=1}^{N_b} |V_{b,i}| = V_l/\sqrt{3} \quad (3)$$

$$\sum_{i=1}^{N_c} |V_{c,i}| = V_l/\sqrt{3} \quad (4)$$

During normal operation for the CHB inverter system ($N_a = N_b = N_c = n$), the output voltage of each H-bridge is solved from Eq. (1) as $v = V_l/\sqrt{3}n$.

1) Assume that one or more submodules of phase A are

bypassed due to faults or hot swapping ($N_a < n$). If the neutral point of the CHB inverter system output voltage is not moved, the remaining individual H-bridge output voltage in phase A should be given by

$$\hat{v} = V_l/\sqrt{3}N_a \quad (5)$$

Whereas, if the remaining H-bridges in phases B and C are all involved in this voltage regulation, then the output voltage of each H-bridge is rewritten as

$$\hat{v}' = \frac{1}{N_a} \frac{V_l}{\sqrt{k^2 + \frac{1}{2} + \sqrt{3k^2 - \frac{3}{4}}}} \quad (6)$$

where $k = n/N_a$. It can be seen that Eq. (6) is a monotonically decreasing function that takes its maximum value when k = 1. Through the neutral offset implemented with sharing the same voltage stress by all remaining submodules, the voltage stress ratio (VSR) in the faulty phase is defined as

$$\frac{\hat{v}}{\hat{v}'} = \frac{1}{\sqrt{3}} \sqrt{k^2 + \frac{1}{2} + \sqrt{3k^2 - \frac{3}{4}}} \quad (7)$$

2) Consider faults that happen in two phases, so the three phases are completely asymmetrical. Assume that one or more submodules in phases B and C are bypassed ($N_a < n$ and $N_c < n$). Eq. (1) can be rewritten as

$$\left(a_1 - \frac{b_1}{\hat{v}'}\right)^2 + \left(a_2 - \frac{b_2}{\hat{v}'}\right)^2 + \left(a_3 - \frac{b_3}{\hat{v}'}\right)^2$$
$$- 2\left(a_1 - \frac{b_1}{\hat{v}'}\right)\left(a_2 - \frac{b_2}{\hat{v}'}\right)\left(a_3 - \frac{b_3}{\hat{v}'}\right) = 1$$

where
(8)

$$a_1 = \frac{N_b^2 + N_c^2}{2N_b N_c}, b_1 = \frac{V_l^2}{2N_b N_c} \quad (9)$$

$$a_2 = \frac{n^2 + N_c^2}{2n N_c}, b_2 = \frac{V_l^2}{2n N_c} \quad (10)$$

$$a_3 = \frac{n^2 + N_c^2}{2n N_b}, b_3 = \frac{V_l^2}{2n N_b} \quad (11)$$

The output voltage $\hat{v}'$ of each H-bridge can be derived according to Eq. (8). Figs. 4 and 5 exhibit the variation trend of VSR $\hat{v}/\hat{v}'$ versus the number of submodules being bypassed ($n = 20$). The multiple trend curves indicate the quantity of submodules in phase C (increasing from 1 to 20 in the direction indicated by the arrows). The variation trend of the fault phase is shown in Fig. 4, where the y-axis represents the VSR of submodules in phase B. It can be observed that VSR reaches the maximum when the quantity of submodules in phase B is 1 while in phase C is 20, indicating that the voltage stress can be reduced by 12 times for all the modules involved in sharing the lost voltage compared to that of the single-phase approach. The variation trend of the normal phase is shown in Fig. 5, where the vertical axis represents the voltage stress factor of submodules in phase A. It can be seen that although

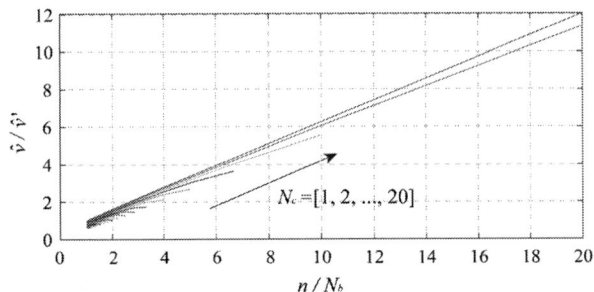

Fig. 4. $\hat{v}/\hat{v}'$ variation trend of the H-bridges in phase B.

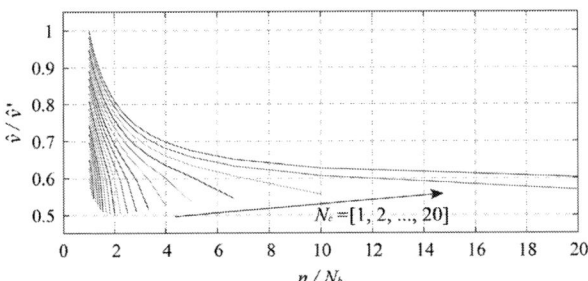

Fig. 5. $\hat{v}/\hat{v}'$ variation trend of the H-bridges in phase A.

the voltage stress in the normal phase increases by 1.6 times, this strategy enhances the reliability of the entire CHB system.

V. SIMULATION VERIFICATION RESULTS

To verify the effectiveness of the proposed strategy, a CHB inverter with 10 H-bridges for each phase is constructed. The inverter system is connected to a 480 V grid through a 4 mH filter. The dc link voltage is 75 V and the H-bridges switching frequency is 1 kHz.

Fig. 6 shows the simulation results of the unbalanced operation for the CHB inverter, in which 2 submodules in phase A are bypassed. The top plot shows the CHB output voltage for each phase. The output voltage for each H-bridge in three phases is 29.8 V. The VSR for phase A is 1.16, and the VSR for non-faulty phases is 0.92. The middle plot represents the PCC voltage. The bottom plot displays the CHB output current for each phase. Even with the unbalanced configuration, the output current waveforms are balanced and maintain a consistent amplitude, indicating the resilience of the control strategy in achieving current regulation and ensuring stable power output under unbalanced module conditions.

Fig. 7 illustrates the performance of the grid-connected CHB inverter under unbalanced conditions, where 2 submodules in phase A are bypassed and 1 submodule in phase B is bypassed. It can be seen that the output voltage of the CHB multilevel inverter system is noticeably unbalanced, however, the balanced output current is strictly maintained. The output voltage for each H-bridge in three phases is 30.9 V. The VSRs for phase A and phase B are 1.12 and 0.996, respectively, while the VSR for phase C is 0.897. Synthesizing the VSRs of Figs. 6 and 7, the voltage stress of the phase that has the least submodule can be greatly reduced relatively although

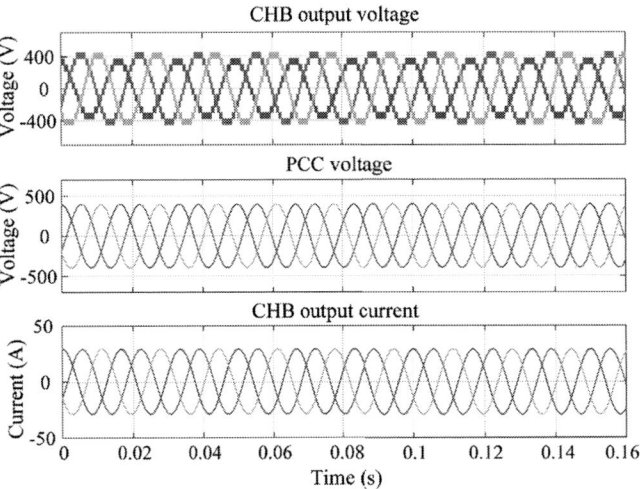

Fig. 6. Unbalanced operation (8 modules in phase A, 10 modules in phases B and C).

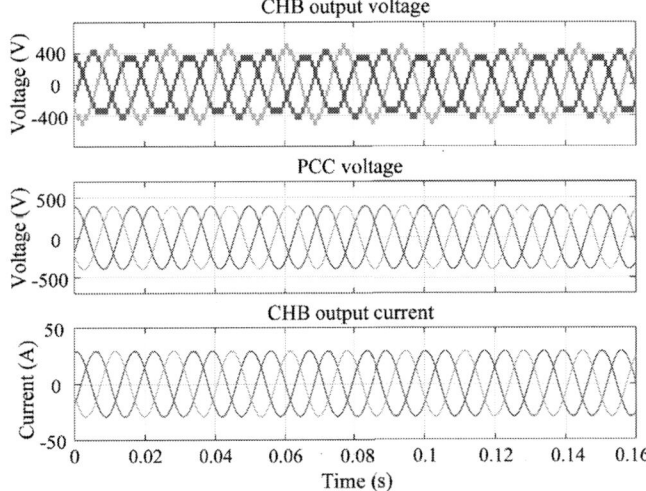

Fig. 7. Unbalanced operation (8 modules in phase A, 9 modules in phase B, and 10 modules in phase C).

the VSR of the other phase might be less than 1, which is corresponding to Figs. 4 and 5.

VI. EXPERIMENTAL RESULTS

To verify the performance of the proposed source stress improvement method, the controller hardware-in-the-loop (HIL) tests were conducted. The control algorithm was implemented on a TMS320F28379 microcontroller, which provided real-time control for the grid-connected CHB multilevel Inverter system. The main circuit, comprising the CHB inverter, filter, and grid source, was modeled and executed on a Typhoon HIL404 hardware system. Because of a limitation of single core's computation for the CHB topology that consists of multiple H-bridges/switching devices, the circuit model was partitioned into four subcircuits thanks to its four processing cores. Each phase of the CHB multilevel Inverter was allocated one processing core, and the fourth core was utilized for the

TABLE I
SYSTEM PARAMETERS

Parameter	Value
Grid line voltage	480 V rms /60 Hz
DC link voltage	350 V
Submodules of each phase	2 or 1
Switching frequency	5 kHz
Sampling period	100 μs
Filter inductance	4 mH
Filter resistance	0.1 Ω
Grid inductance	5 μH

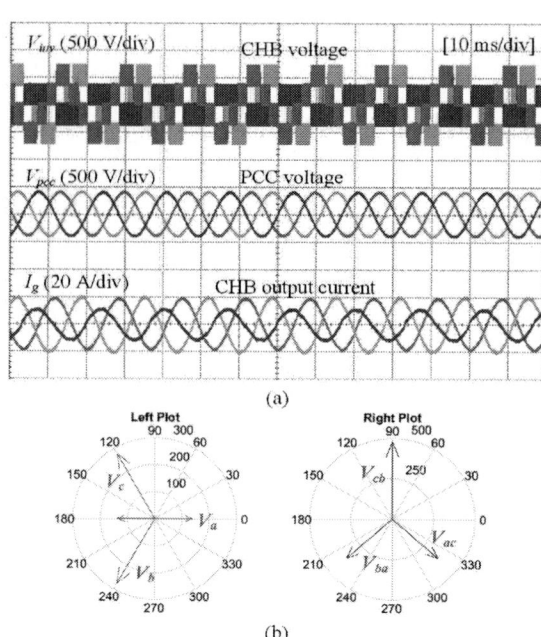

(a)

(b)

Fig. 9. Experimental results of the CHB system using traditional neutral point unshifted strategy under unbalanced operation (1 module in phase A, 2 modules in phases B and C). (a) Waveforms. (b) Voltage vector analysis.

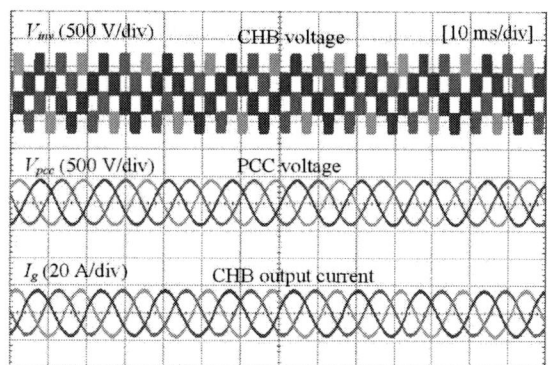

Fig. 8. Waveforms of the CHB inverter system under balanced operation (2 modules for each phase).

filter circuit and grid source. The system test parameters are shown in Table I.

Fig. 8 shows the C-HIL experimental waveforms of the CHB inverter system operating under balanced conditions, with two modules per phase. Each phase of the multilevel inverter consists of 2 H-bridges. The sample frequency is 10 kHz. It can be seen that the CHB inverter system produces a multilevel voltage waveform, with distinct voltage steps achieved by H-bridge modules' switching. The current waveforms are smooth and sinusoidal, indicating proper current regulation in the output current. The current amplitudes are balanced across the phases, verifying the effectiveness of the proposed source stress improvement method in achieving stable current output during balanced operation.

Fig. 9 presents the experimental waveforms of the CHB inverter system operating based on neutral point unshifted strategy under unbalanced conditions, with 1 module in phase A and 2 modules in phases B and C. In Fig. 9(a), the CHB output voltage steps are lost in phase A due to the reduced module count. The unbalanced number of modules across phases resulted in that the current amplitudes of three phase are clearly imbalanced, indicating the limitations of the traditional control strategy in maintaining current symmetry across phases during unbalanced operation. Fig. 9(b) provides phasor diagrams for both line-to-neutral (left plot) and line-to-line (right plot) voltages. In the line-to-neutral phasor diagram, the voltage vectors V_a, V_b, and V_c, show noticeable phase and magnitude imbalances, particularly in phase A, and also the line-to-line phasor diagram highlights the asymmetry between V_{ab}, V_{bc}, and V_{ca}.

Fig. 10 presents the experimental waveforms of the CHB inverter system operating based on proposed strategy under unbalanced conditions, with 1 module in phase A and 2 modules in phases B and C. From the CHB output voltage waveforms, the A-phase voltage still exhibits three levels, while the B-phase and C-phase have five levels, respectively. The output voltage for each H-bridge in three phases is 170.9 V, and VSR is 1.62. By analyzing their voltage vector diagram, the amplitude of phases B and C is twice that of phase A, which indicates that all the submodules share the same voltage stress. Meanwhile, the grid current is always balanced since the triangle formed by the inverter output line voltage is still presented as an equilateral triangle.

VII. CONCLUSION

A resilient operation method for the CHB multilevel inverter is proposed for fault- or inter-bridge imbalance-tolerant operation. The efficiency is improved by sharing the voltage lost among all the submodules, which significantly optimizes H-bridge source stresses compared to the single-phase internal submodule sharing and fault-free phase-bypass balanced bridge method. This approach also saves costs compared to redundancy schemes and guarantees excellent current-quality output. The proposed approach provides the CHB multilevel inverter system with boosted tolerance to inter-bridge imbalances or faults, with enhanced uninterrupted operation capability, leading to an enhancement of the system's resilience.

ACKNOWLEDGMENT

This project has been made possible through funding provided by the U.S. Department of Energy's Office of Energy Efficiency and Renewable Energy (EERE), specifically

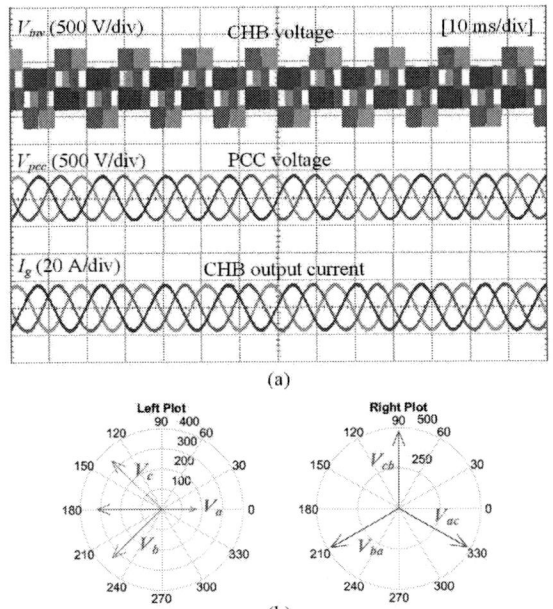

(a)

(b)

Fig. 10. Experimental results of the CHB system using proposed neutral offset strategy under unbalanced operation (1 module in phase A, 2 modules in phases B and C). (a) Waveforms. (b) Voltage vector analysis.

under the Solar Energy Technologies Office Award Number DEEE0010427 . It is important to note that the perspectives and conclusions presented by the authors in this document are their own and do not necessarily reflect the official policy or position of the United States Government or any of its affiliated agencies.

REFERENCES

[1] J. Zhu, Y. Li, F. Z. Peng, B. Lehman, and H. Huang, "From active resistor to lossless and virtual resistors: A review, insights, and broader applications to energy grids," *IEEE Journal of Emerging and Selected Topics in Power Electronics*, DOI 10.1109/JESTPE.2024.3481440, pp. 1–1, 2024.

[2] S. Nyamathulla and D. Chittathuru, "A review of multilevel inverter topologies for grid-connected sustainable solar photovoltaic systems," *Sustainability*, vol. 15, no. 18, p. 13376, 2023.

[3] E. Rodriguez *et al.*, "Closed-loop analytic filtering scheme of capacitor voltage ripple in multilevel cascaded h-bridge converters," *IEEE Transactions on Power Electronics*, vol. 35, DOI 10.1109/TPEL.2020.2966305, no. 8, pp. 8819–8832, 2020.

[4] Y. He, J. Zhu, B. Zhou, and X. Zhao, "Sensorless vector control for induction motor startup using multilevel converters," in *2023 IEEE 6th Student Conference on Electric Machines and Systems (SCEMS)*, pp. 1–7, 2023.

[5] B. Liu, D. Wang, J. Zhu, B. Gu, and C. Mao, "A rule-based bionic energy management system for grid-connected community microgrid using peer-to-peer trading with rapid settlement," *IEEE Transactions on Sustainable Energy*, vol. 15, no. 1, pp. 215–235, 2024.

[6] R. Sharma and A. Das, "Enhanced active power balancing capability of grid-connected solar pv fed cascaded h-bridge converter," *IEEE Journal of Emerging and Selected Topics in Power Electronics*, vol. 7, DOI 10.1109/JESTPE.2019.2890984, no. 4, pp. 2281–2291, 2019.

[7] P. Denholm *et al.*, "The challenges of achieving a 100% renewable electricity system in the united states," *Joule*, vol. 5, no. 6, pp. 1331–1352, 2021.

[8] Z. Zhang *et al.*, "High-efficiency silicon carbide-based buck-boost converter in an energy storage system: Minimizing complexity and maximizing efficiency," *IEEE Industry Applications Magazine*, vol. 27, DOI 10.1109/MIAS.2020.3024495, no. 3, pp. 51–62, 2021.

[9] A. Abuelnaga and M. Narimani, "Open circuit igbt fault classification using phase current in a chb converter," in *45th Annual Conference of the IEEE Industrial Electronics Society*, vol. 1, pp. 4636–4641, 2019.

[10] N. Bisht and A. Das, "A multiple fault-tolerant topology of cascaded h-bridge converter for motor drives using existing precharge windings," *IEEE Journal of Emerging and Selected Topics in Power Electronics*, vol. 9, no. 2, pp. 2079–2087, 2021.

[11] B. F. Brown, Z. Zhang, J. H. Enslin, and C. Wan, "Universal short-circuit and open-circuit fault protection and detection for a power converter's power device," *IEEE Transactions on Industry Applications*, DOI 10.1109/TIA.2024.3462910, pp. 1–9, 2024.

[12] J. Zhu *et al.*, "Design and field application of dc and harmonic suppression system for neutral current in 220 kv substation," *IEEE Transactions on Industrial Electronics*, vol. 69, no. 8, pp. 7560–7570, 2022.

[13] P. Hu *et al.*, "Energy-balancing control strategy for modular multilevel converters under submodule fault conditions," *IEEE Transactions on Power Electronics*, vol. 29, DOI 10.1109/TPEL.2013.2284919, no. 9, pp. 5021–5030, 2014.

[14] G. Liu, Z. Xu, Y. Xue, and G. Tang, "Optimized control strategy based on dynamic redundancy for the modular multilevel converter," *IEEE Transactions on Power Electronics*, vol. 30, no. 1, pp. 339–348, 2015.

[15] G. T. Son *et al.*, "Design and control of a modular multilevel hvdc converter with redundant power modules for noninterruptible energy transfer," *IEEE Transactions on Power Delivery*, vol. 27, no. 3, pp. 1611–1619, 2012.

[16] B. Li *et al.*, "Seamless transition control for modular multilevel converters when inserting a cold-reserve redundant submodule," *IEEE Transactions on Power Electronics*, vol. 30, no. 8, pp. 4052–4057, 2015.

[17] S. Sahoo and I. Ahmed, "Common mode voltage reduction in npc multilevel inverter by svpwm using gh-coordinate system," in *2020 International Conference on Computational Intelligence for Smart Power System and Sustainable Energy (CISPSSE)*, pp. 1–6. IEEE, 2020.

[18] J. Tian, C. Mao, D. Wang, S. Nie, and Y. Yang, "A short-time transition and cost saving redundancy scheme for medium-voltage three-phase cascaded h-bridge electronic power transformer," *IEEE Transactions on Power Electronics*, vol. 33, no. 11, pp. 9242–9252, 2018.

[19] R. Sen Goopta and A. Bhattacharya, "Asymmetrically switched chb multilevel inverters with harmonic mitigation techniques applied to photovoltaic power generation," *IETE Journal of Research*, vol. 68, no. 6, pp. 4360–4382, 2022.

[20] J. Tanguturi and S. Keerthipati, "Panel arrangement technique to mitigate power imbalance arising due to partial shading in a pv single-phase chb inverter system," in *2024 4th International Conference on Smart Grid and Renewable Energy (SGRE)*, pp. 1–6. IEEE, 2024.

[21] S. Badawi *et al.*, "Utilization of a reduced switch-count topology in regenerative cascaded h-bridge (chb) medium-voltage drives," *IEEE Journal of Emerging and Selected Topics in Power Electronics*, vol. 10, no. 6, pp. 6938–6949, 2022.

[22] Q. Liu *et al.*, "Minimum voltage operation of cascaded h-bridge statcoms using discontinuous modulation," *IEEE Transactions on Power Electronics*, vol. 39, no. 5, pp. 5014–5019, 2024.

[23] M. Wang *et al.*, "Module power balance control strategy for three-phase cascaded h-bridge pv inverter under unbalanced grid voltage condition," *IEEE Journal of Emerging and Selected Topics in Power Electronics*, vol. 9, no. 5, pp. 5657–5671, 2021.

[24] J. Zhu *et al.*, "A novel suppression method for grounding transformer against earth current from urban rail transit," *IEEE Transactions on Industrial Electronics*, vol. 68, no. 12, pp. 11 651–11 662, 2021.

[25] B. Pattanaik and S. Murugan, "Cascaded h-bridge seven level inverter using carrier phase shifted pwm with reduced dc sources," *International Journal of MC Square Scientific Research*, vol. 9, no. 3, pp. 30–39, 2017.

[26] V. G. Monopoli, Y. Ko, G. Buticchi, and M. Liserre, "Performance comparison of variable-angle phase-shifting carrier pwm techniques," *IEEE Transactions on Industrial Electronics*, vol. 65, no. 7, pp. 5272–5281, 2017.

[27] C. Zang *et al.*, "Research on the application of cps-spwm technology in cascaded multilevel inverter," in *2009 International Conference on Electrical Machines and Systems*, pp. 1–4. IEEE, 2009.

[28] Z. Guan *et al.*, "Control strategy and implementation of seamless closed-loop load transfer mobile prototype for 400 v distribution network," *IEEE Access*, vol. 12, pp. 12 279–12 294, 2024.

[29] B. Bahrani, S. Kenzelmann, and A. Rufer, "Multivariable-pi-based dq current control of voltage source converters with superior axis decou-

pling capability," *IEEE Transactions on Industrial Electronics*, vol. 58, no. 7, pp. 3016–3026, 2011.

[30] M. Lu, M. Qin, W. Mu, J. Fang, and S. M. Goetz, "A hybrid gallium-nitride–silicon direct-injection universal power flow and quality control circuit with reduced magnetics," *IEEE Transactions on Industrial Electronics*, vol. 71, no. 11, pp. 14161–14174, 2024.

[31] M. Lu *et al.*, "A novel direct-injection universal power flow and quality control circuit," *IEEE Journal of Emerging and Selected Topics in Power Electronics*, vol. 11, no. 6, pp. 6028–6041, 2023.

A Medium Voltage Grid-connected PV Inverter with a New Modular High Voltage Gain Converter Featuring Internal Modified Voltage Doubling Balancers

Kajanan Kanathipan, IEEE Member
Dept. of Electrical Engineering and Computer Science
York University
Toronto, Canada
kajanank@yorku.ca

Muhammad Ali Masood Cheema, *IEEE Member*
Northern Transformer
Toronto, Canada
alimasood_rcet1@hotmail.com

John Lam, IEEE Senior Member
Dept. of Electrical Engineering and Computer Science
York University
Toronto, Canada
john.lam@lassonde.yorku.ca

Abstract— This work proposes a medium voltage grid-connected inverter with modular high voltage gain converters for PV energy applications. The proposed topology utilizes (1) PV arrays interfaced with modular DC-DC converters in an input independent output series (IIOS) configuration, (2) internal modified voltage doubling balancers to achieve balanced output voltage under power mismatch scenarios, and (3) a high voltage neutral point clamped (NPC) 5-level grid-connected inverter system. The presented DC-DC converters are able to perform maximum power point tracking operation through phase-shift control on their full-bridge switches and ensure balanced output voltage under all scenarios with duty ratio control on the internal voltage doublers. In addition, a high step-up gain and soft-switching operation on all module switches are achieved through the use of an LLC resonant stage. A five-level three-phase inverter interfaces the system with the MVAC grid. The steady-state and dynamic performance of the proposed modular system are validated through simulation results on a 27kVrms (L-L) 60Hz output, four-module 60kW system and preliminary experimental results on a two-module proof-of-concept 1kV DC-link laboratory prototype.

Keywords— Medium Voltage, Grid Connected Inverter, PV energy, Power Balancer, Soft-switching

I. INTRODUCTION

There has been increasing interest in grid connected PV energy systems thanks to the advancement of high-power high-voltage converters using wide band gap semiconductors [1]. Solid state transformer (SST) based topologies have also seen increased interest due to their benefits over line-frequency transformers including better control flexibilities [2]-[3]. One benefit of an SST is its capability of providing medium voltage DC (MVDC) output which can then be interfaced with high voltage inverters to provide power to medium voltage ac (MVAC) grids thus providing increased reliability while decrease system size and cost. The operating voltage of a typical PV array module is on the order of 280-690V which implies the need for modular based step-up converters to help reach the grid voltage level. Various types of step-up power converters for renewable and PV energy applications have been presented in the literature [5]–[9] and one of the topologies that stands out is the LLC resonant converter due to its high efficiency, high

power density, and high operating frequency [6],[8]. However, modular configurations can lead to power mismatch and overvoltage scenarios in cases of varying atmospheric conditions. This can impact the system operation and lead to both damaged components and module failure [9-10]. To circumvent this issue, topologies make use to additional conversion stages which can have a detrimental impact on the system size, cost, and efficiency. In this paper, a medium voltage grid-connected inverter with modular high voltage gain converters is proposed for PV energy applications. The modular converters are connected in an input independent output series (IIOS) configuration to allow the total module output voltage to reach the MVDC level while internal modified voltage doubling balancers allow for balanced module output voltage under power mismatch scenarios without the need of external conversion stages. The five-level grid connected inverter functions with a phase locked loop (PLL) controller in a grid following fashion. The steady-state and dynamic performance of the proposed modular system are validated through simulation results on a 27kVrms (L-L) 60Hz output, four-module 60kW system and preliminary experimental results on a two-module proof-of-concept 600W, 1kV DC-link laboratory prototype.

II. DESCRIPTION OF THE PROPOSED MV INVERTER BASED PV SYSTEM

The proposed topology is shown in Fig. 1(a) and Fig. 1(b) for the DC/DC stage and the DC/AC stage respectively. It consists of input-independent modular DC/DC converters with their outputs connected together in series to form an IIOS system. Each module consists of PV arrays connected to a full bridge inverter which feeds into an LLC resonant circuit. The phase-shift between the full-bridge inverter legs are regulated to perform individual maximum power extraction for each module. This is implemented through the use of a simpler perturb and observe (P&O) maximum power point tracking (MPPT) controller. The use of an LLC resonant circuit allows for soft-switching operation for all of the module switches while simultaneously achieving a large step-up gain. The output voltage of the resonant stage is rectified through the use of a voltage doubler (VD) which provides an additional step-up

979-8-3315-1612-3/25 $31.00 © 2025 IEEE

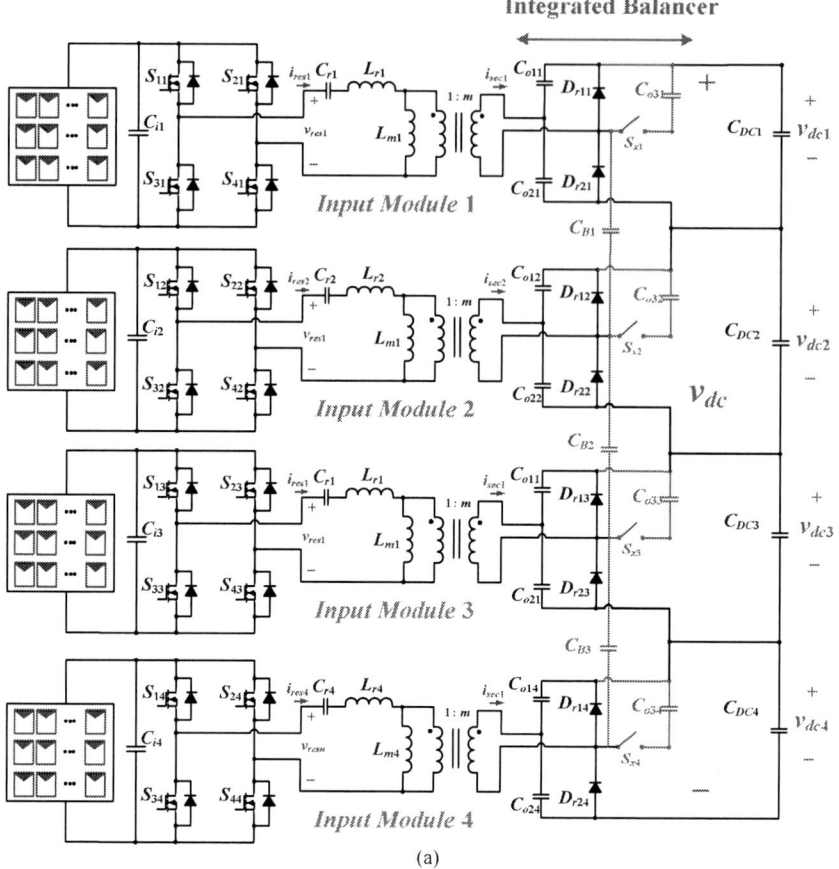

(a)

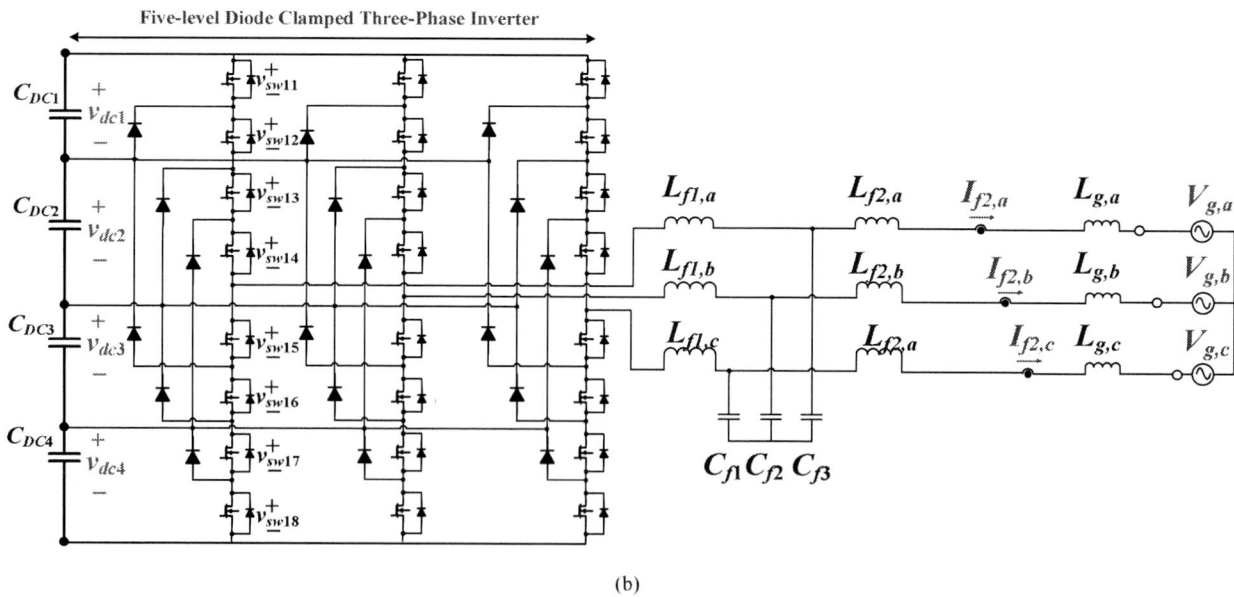

(b)

Fig. 1: (a) Proposed modular LLC resonant converters with internal modified voltage doubler based voltage balancer (b) Grid tied

gain. Due to the IIOS topology, varying modular PV power will result in unbalanced output voltages. To address this issue, a power balance unit consisting of a capacitor and switch is integrated within the secondary side of each module as highlighted in red in Fig 1(a). Controlling the duty-ratio of the switch relative to the neighboring modules charges and

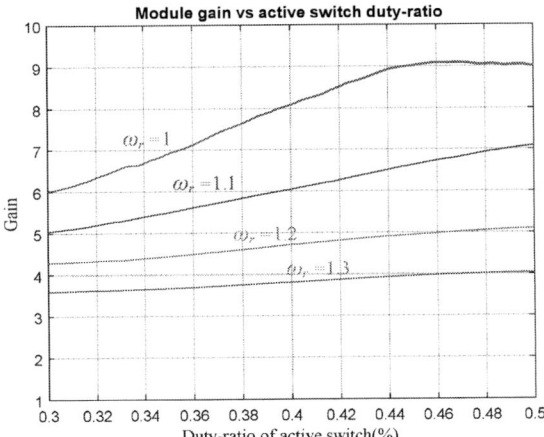

Fig. 2: Gain of a single module as a function of the power balancing active switch

discharges the interconnected capacitor C_{Bx}. This allow for the power flow between modules to be regulated which in turn results in balanced output voltages while maintaining an output series connection. The gain of one module of the proposed converter is provided in (1) where ω_r is the relative angular operating frequency between the operating frequency and the resonant frequency (2), Q is the quality factor as defined in (3), and k is the inductance ratio between the modular magnetizing inductance and the resonant inductance (4).

$$\frac{V_{pn}}{V_{resn}} = \frac{1}{\sqrt{\left(1 + \frac{1}{k}\left(1 - \frac{1}{\omega_r^2}\right)\right)^2 + \left(Q\left(\omega_r - \frac{1}{\omega_r}\right)\right)^2}} \tag{1}$$

$$\omega_0 = \frac{1}{\sqrt{L_{rn}C_{rn}}} \tag{2}$$

$$Q = \frac{R_{eq}}{L_m \omega_0} \tag{3}$$

$$k = \frac{L_{mn}}{L_m} \tag{4}$$

As shown in (2) the voltage gain varies based on the quality factor. By controlling the duty-ratio (D_x) of the balancing switch, the equivalent resistance in (3) changes and in turn the converter gain varies. Fig. 2 provides a plot of this gain at different D_x and ω_r. The control of D_x allows higher operating power modules to transfer excess power to neighboring modules. An example of the operating states of a two-module system is provided in Fig. 3 which displays only the secondary side of each module under the assumption that module 2 operates at a higher power level compared to module 1.

[$t_0 < t < t_1$]: At time t_0, the gate signal is applied to switch S_{x2}. As module 1 operates at a lower power the switch S_{x1} is inactive. The negative resonant current that flows through the anti-parallel diode which will facilitate ZVS operation in the next stage. As shown in by the green KVL loop Fig. 3(a), the

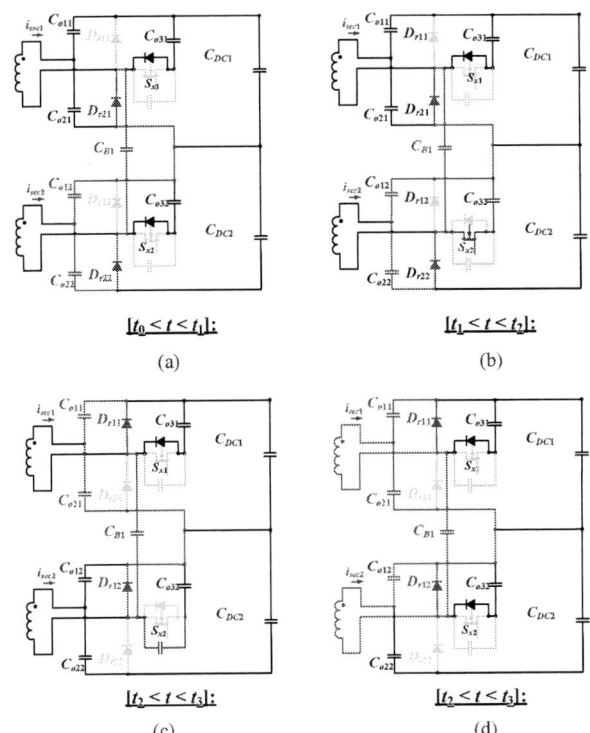

Fig. 3: Operating stages of the proposed system's secondary side illustrated for two modules

balancer capacitor, C_{B1}, is connected across the VD capacitors of module 2 through the diodes D_{r21} and D_{r22} which results in (5).

$$v_{CB1} = v_{Co12} + v_{Co22} \tag{5}$$

[$t_1 < t < t_2$]: At time t₁, S_{x2} switches on under ZVS condition. The secondary side current i_{sec2} splits between C_{B1} and S_{x2} which allows energy to be stored (6). This continues until the gate signal is removed from S_{x2}.

$$i_{CB1} = i_{sec2} - i_{sx2} \tag{6}$$

[$t_2 < t < t_3$]: At time t_2, the gate signal is removed from S_{x2} and the voltage polarity of the secondary side reverses. As a result, C_{B1} is connected across the VD capacitors of module 1 through the diodes D_{r11} and D_{r22} as shown by the green KVL loop in Fig. 3(c) and the voltage across C_{B1} is (7). The secondary side resonant current flows through S_{x2}'s snubber capacitor which allows it to charge to its maximum value (8). Once the capacitor has fully charged the system transitions to the next stage.

$$v_{CB1} = v_{Co11} + v_{Co12} \tag{7}$$

$$v_{sx2} = v_{cdc2} \tag{8}$$

[$t_3 < t < t_4$]: At time t_3, the secondary side resonant currents of each module begin to flow through the active diodes respectively. During this time the resonant currents are negative and increasing This stage ends once the gate signal is applied to S_{x2} and the system transitions back to the original stage.

Due to DC/DC modules being interfaced to the neutral point clamped (NPC) 5-level inverter, the peak voltage across each active switch of the internal modified voltage doubler is a fourth of the total DC-link voltage (9). The diodes of each voltage doubler are still related to the capacitors as provided in (10) and (11) respectively where "n" is the module number.

$$v_{sxn\_max} = \frac{v_{dc}}{4} \tag{9}$$

$$v_{Dr1n\_max} = v_{co1n} \tag{10}$$

$$v_{Dr2n\_max} = v_{co2n} \tag{11}$$

The output series connection feeds into a NPC 5-level grid connected inverter system. The inverter switches are controlled through the use of sinusoidal pulse width modulation (SPWM) along with a closed-loop phase locked loop (PLL) based controller which regulates the real and reactive power fed into the grid. The closed loop transfer function for the PLL's PI controller is provided in (12) and from here the proportional and integral gain is provided in (13) and (14) respectively. An example of this controller is provided in Fig. 2.

$$G_{CL} = \frac{v_g \left(k_p s + k_i \right)}{s^2 + v_g k_p s + v_g k_i} \tag{12}$$

$$k_i = \frac{\omega n^2}{v_g} \tag{13}$$

$$k_p = 2\zeta \sqrt{\frac{k_i}{v_g}} \tag{14}$$

As a PLL controller obtains the phase angle, the dq frame parameters can be used to calculate system's reference active and reactive current and are shown (15) and (16) respectively where v_d, v_q, i_d, and i_q are the system's voltage and current in the

dq frame, P_{ref} and Q_{ref} are the required active and reactive power respectively. When operating under the MPP, P_{ref} would be the sum of all modules maximum operating point. As the maximum value is already obtained from the P&O MPPT controllers,

for this implementation. Once the controller has determined the required active and reactive current, they are used to calculate the voltage reference signals as seen in (6) and (7) where Δi_q and Δi_d represent the correction from the PI controller and x_{Ls} is the filter inductance represented as its reactance.

$$i_{d\_ref} = \frac{\frac{2}{3} P_{ref} - v_q i_q}{v_d} \tag{15}$$

$$i_{q\_ref} = \frac{v_q i_q - \frac{2}{3} Q_{ref}}{v_d} \tag{16}$$

$$E_d = v_d + x_{Ls} \times i_q + \Delta i_q \tag{17}$$

$$E_q = v_q + x_{Ls} \times i_d + \Delta i_d \tag{18}$$

The generated voltage reference signals in the dq frame are referred back to the abc frame using the DQ to ABC

TABLE I: Simulation and Hardware Prototype Parameters

	Simulation	Scaled-down Prototype
Total Operating Power (P_o)	60kW	600W
Total Output Voltage (V_o)	36kV	1000V
PV Output Voltage (V_i)	600V	100V
Resonant Inductance (L_{mN})	211µH	211µH
Transformer (L_{MN})	1mH	1.5mH
Input Capacitance (C_{mN})	10µF	5µF
Resonant Capacitance (C_{rN})	100nF	100nF
Output Capacitance (C_{oN})	5µF x2	25µF x2
MOSFET Switch (S_N, S_{qN})	-	CAS120M12BM2
Switching Frequency (f_s)	60kHz	40kHz
DC/DC Module DSP Controller	-	TMSF28335
5-level Inverter DSP Controller	-	TMSF28379D

DQ to ABC Transform Grid Current Control and Five-level Inverter Signal Generation

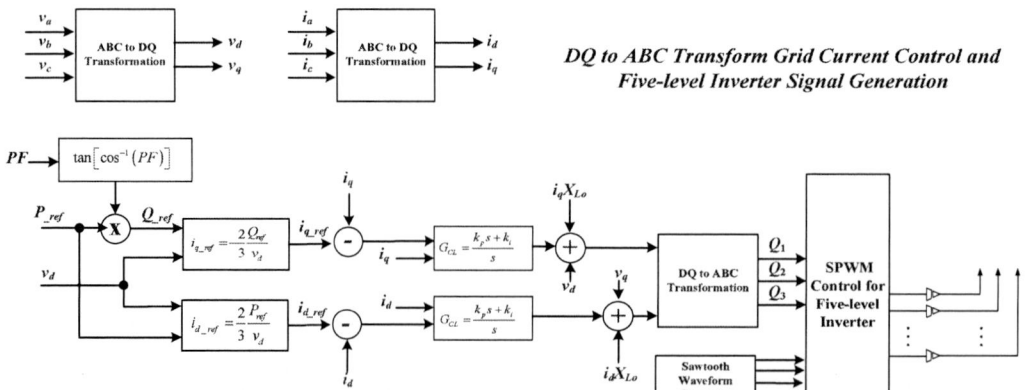

Fig. 4: NPC control system for the desired active and reactive power level

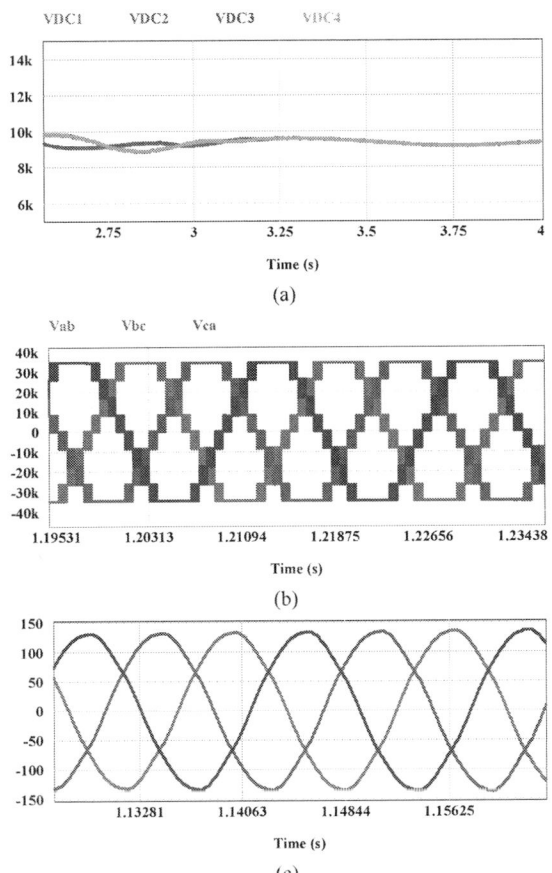

Fig. 5: (a) per-module output voltage (b) five-level inverter output voltage before filter, (c) grid current

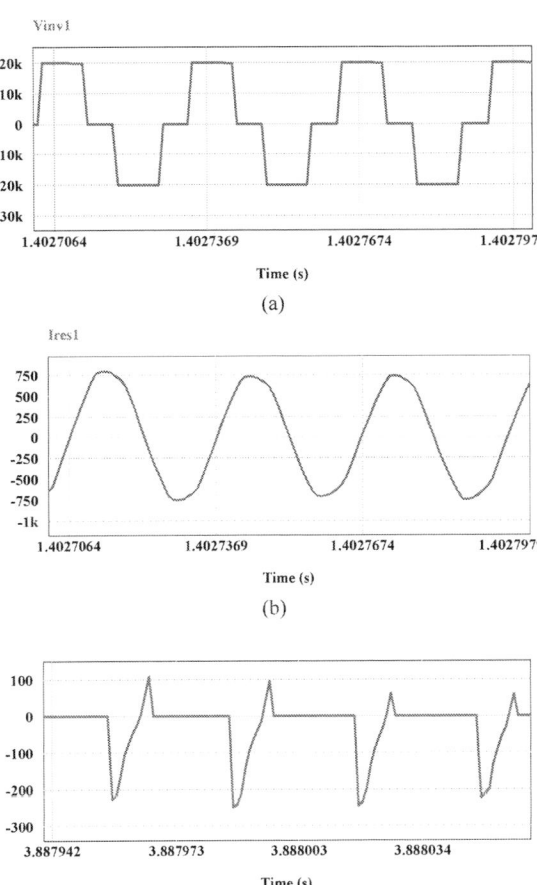

Fig. 6: (a) inverter voltage, (b) inverter current, (c) balancer switch current

transformation as shown in Fig. 4. From here, the output is provided to a sinusoidal pulse-width modular which along with sawtooth reference signals are used to generate the gate signals for each switch of the 5-level inverter.

III. SIMULATION AND EXPERIMENTAL RESULTS

To confirm the accuracy of the proposed system, simulation on a 27kVrms (L-L) / 60kW system was performed using the circuit simulation software PowerSim (PSIM). Four DC/DC full-bridge LLC resonant converters were connected in an IIOS configuration with different PV arrays connected to the input of each converter such that the system was subjugated to power mismatch situations. From here the module outputs were interfaced to the input of a 5-level inverter that feeds to a grid. The parameters used are provided in table I. The per-module output voltage is shown in Fig. 5(a) which can be seen to be balanced at approximately 10kV each. Fig. 5(b) shows the five-level inverter voltage while the current is provided in Fig. 5(c). The resonant voltage and current of module 1 are shown in Fig. 6(a) and (b) respectively, which can be seen to operate under soft-switching condition as current begins to flow through each switch before it transitions from negative to positive. Soft-switching operating of the internal modified voltage doubler's active switch is confirmed in Fig. 6(c).

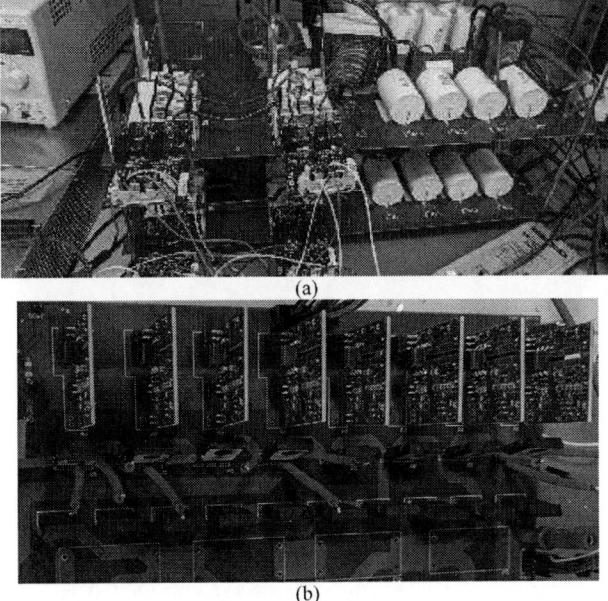

Fig. 7: A picture of (a) the modular DC/DC experimental prototype (b) one leg of the 5-level inverter

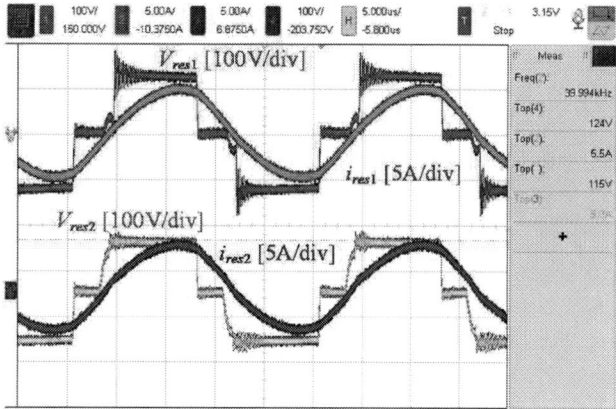

Fig. 8: Resonant input voltage and current per module

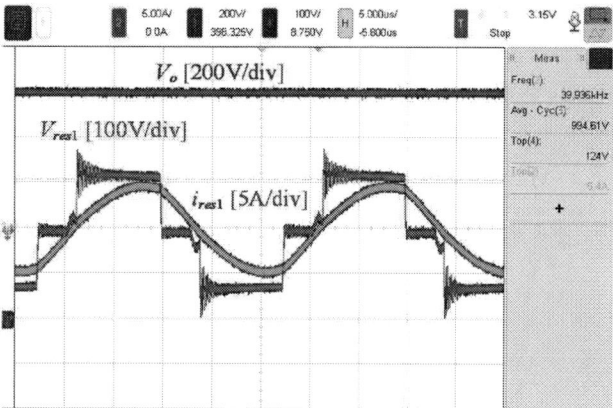

Fig. 9: Measured switching waveforms and 1kVDC total system output voltage

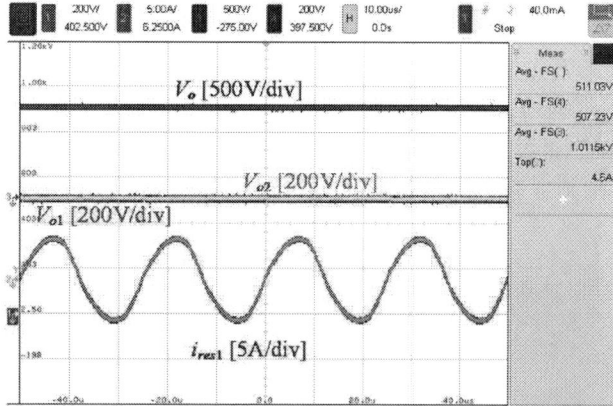

Fig. 10: Measured module resonant current and per-module output voltage along with the total system output voltage.

A 600W, 100V / 1kV proof-of-concept prototype system consisting of two modules has been designed and tested to validate the functionalities of the proposed DC/DC converter. The parameters used in the prototypes are provided in table I. A picture of the prototype system is shown in Fig. 7(a) while a picture of one leg of the developed five-level inverter is provided in Fig. 7(b). Two channels of a Keysight Solar Array Simulator E4360A were used to emulate PV array operation at different

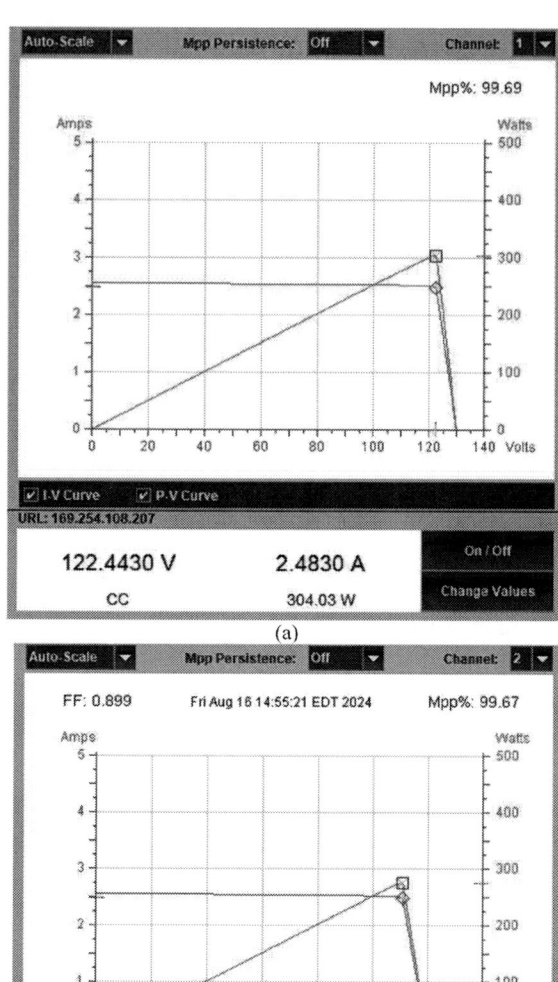

Fig. 11: MPP operation per module

power levels and were connected to the input of each DC/DC module.

The individual module resonant input voltage and current are shown in Fig. 8. It can be seen that soft-switching is achieved on the module switches as the resonant current lags the voltage as well as that phase-shift modulation is applied. The resonant waveforms of the first module as well as the total system output voltage are shown in Fig. 9. The system is confirmed to operate at 1kV under the previous condition which corresponds to a gain of approximately 5 per module. The total output voltage and the individual module voltage of the hardware prototype are shown in Fig. 7 which are seen to be balanced at 500V.

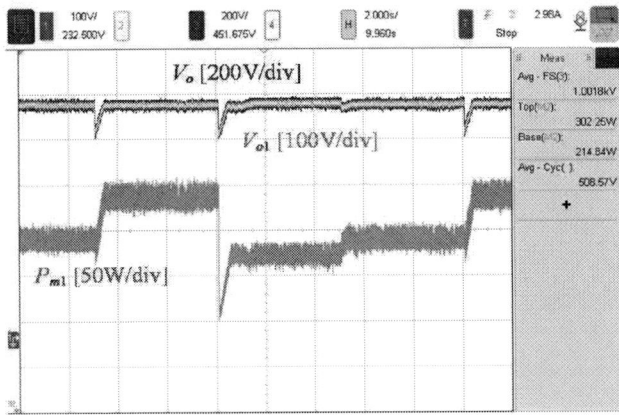

Fig. 12: Operating power and output voltage of module 1 and total system output voltage

The operating power of each module is shown in Fig. 11(a) and (b) and it can be seen that each module operated at approximately 99% of their maximum power. The module operating power was 304W and 274W respectively. The operating power of both modules were then varied between 300W and 200W respectively every five seconds. Fig. 12 shows the operating power of module 1 as well its corresponding output voltage. It can be seen that every five seconds the system moved towards the new maximum power point and the balancer regulated the output voltage at 500V. The modular output voltage is seen to be half of the total output voltage which confirm that both modules achieve balanced output voltage regardless of the power mismatch at their input. This confirms the proposed balancer was able to regulate the power flow such that all modules operated with balanced output voltage regardless of their input power level.

IV. CONCLUSION

This paper presented a medium voltage grid-connected inverter with modular high voltage gain converters for PV energy applications. The proposed topology is capable of balancing each module of an IIOS system's output voltage while simultaneously providing individual modular MPPT operation. Analysis of the proposed topologies DC/DC and DC/AC stages and converters have been provided. Experimental results on the modular DC/DC experimental prototype as well as the designed 5-level NPC inverter have been provided to confirm the designed system operation.

REFERENCES

[1] R. L. Da Silva, V. L. F. Borges, C. E. Possamai, and I. Barbi, "Solid-State Transformer for Power Distribution Grid Based on a Hybrid Switched-Capacitor LLC-SRC Converter: Analysis, Design, and Experimentation," IEEE Access, vol. 8, pp. 141182–141207, 2020.

[2] J. E. Huber and J. W. Kolar, "Solid-State Transformers: On the Origins and Evolution of Key Concepts," IEEE Ind. Electron. Mag., vol. 10, no. 3, pp. 19–28, 2016.

[3] Q. Ye, R. Mo, and H. Li, "Impedance Modeling and DC Bus Voltage Stability Assessment of a Solid-State-Transformer-Enabled Hybrid AC–DC Grid Considering Bidirectional Power Flow," IEEE Trans. Ind. Electron., vol. 67, no. 8, pp. 6531–6540, 2020.

[4] A. Q. Huang, "Medium-Voltage Solid-State Transformer: Technology for a Smarter and Resilient Grid," IEEE Ind. Electron. Mag., vol. 10, no. 3, pp. 29–42, 2016.

[5] A. Giannakis and D. Peftitsis, "MVDC Distribution Grids and Potential Applications: Future Trends and Protection Challenges," in 2018 20th European Conference on Power Electronics and Applications (EPE'18 ECCE Europe), 2018, p. P.1-P.9.

[6] F. Wang, Z. Zhang, T. Ericsen, R. Raju, R. Burgos, and D. Boroyevich, "Advances in Power Conversion and Drives for Shipboard Systems," Proc. IEEE, vol. 103, no. 12, pp. 2285–2311, 2015.

[7] Y. Yang, K. A. Kim, F. Blaabjerg, and A. Sangwongwanich, "Power electronic technologies for PV systems," Woodhead Publishing, 2019, pp. 15–43.

[8] M. Forouzesh, Y. P. Siwakoti, S. A. Gorji, F. Blaabjerg, and B. Lehman, "Step-Up DC–DC Converters: A Comprehensive Review of Voltage-Boosting Techniques, Topologies, and Applications," IEEE Trans. Power Electron., vol. 32, no. 12, pp. 9143–9178, 2017.

[9] M. R. Islam, Y. Guo, and J. Zhu, Power Converters for Medium Voltage Networks. New York: Springer, 2014.

[10] Y. Huang et al., "Bidirectional Buck-Boost and Series LC-Based Power Balancing Units for Photovoltaic DC Collection System," IEEE J Emerg Sel Top Power Electron, vol. 9, no. 6, pp. 6726–6738, Dec. 2021, doi: 10.1109/JESTPE.2021.3074575.

[11] X. Li, M. Zhu, M. Su, J. Ma, Y. Li, and X. Cai, "Input-Independent and Output-Series Connected Modular DC-DC Converter with Intermodule Power Balancing Units for MVdc Integration of Distributed PV," IEEE Trans Power Electron, vol. 35, no. 2, pp. 1622–1636, Feb. 2020, doi: 10.1109/TPEL.2019.2924043

Split-Source Common-Ground Inverter for Photovoltaic Applications

Mahmoud A. Gaafar
Aswan Power Electronics Applications Research Center (APEARC)
Aswan University
Aswan, Egypt
mgaafar@apearc.aswu.edu.eg

Mohamed Orabi
Aswan Power Electronics Applications Research Center (APEARC)
Aswan University
Aswan, Egypt
morabi@apearc.aswu.edu.eg

Samir Kouro
Department of Electronic Engineering
Universidad Tecnica Federico Santa Maria
Valparaiso, Chile
samir.kouro@ieee.org

Ahmed Ibrahim
Electrical Department, Faculty of Technology and Education, Sohag University
Sohag, Egypt;
ahmed-ibrahim22@techedu.sohag.edu.eg

Eltaib Abdeen D. Ibrahim
Electrical Engineering Department, High Institute for Engineering and Technology
Sohag, Egypt
eabdeen@ieee.org

Abstract- A common-ground (CG) inverter based on the split-source topology is proposed (CG-SSI). The proposed inverter offers the following features: 1) complete leakage current mitigation, 2) voltage boosting capability, 3) continuity of input current, 4) capability of reactive power support, 5) unipolar modulation mechanism; thus, low total harmonic distortion can be achieved. The operating states of the proposed inverter along with the stresses of semiconductors are analyzed. A closed loop control system is developed to achieve maximum power point tracking (MPPT) and load current regulation. A comparative study with other CG inverters is introduced. Experimental and real-time simulation results are provided to verify the performance of the proposed inverter.

Keywords: common-ground, photovoltaic, split-source.

I. INTRODUCTION

The common-mode voltage (CMV) produced over the parasitic capacitors of PV modules and the resultant leakage current are issues of great concern for both PV systems' owners and electricity companies. With the increased penetration of PV systems into the grid, leakage currents can result in [1]-[3]: 1) Wrong tripping of residual current protection devices, 2) degradation of electromagnetic compatibility (EMC), 3) increasing the temperature of PV modules, and in turn, performance degradation and decreasing the module's lifetime, 4) increasing the total harmonic distortion of the grid current, and 5) electrical hazard for people touching the PV modules.

Common-ground (CG) configurations have been recently gaining a lot of attention as an outstanding solution for leakage current mitigation in PV systems. CG topologies offer direct connection between the negative terminal of PV modules and the neutral of the grid side. Thus, the parasitic capacitors of PV modules are bypassed. In addition, to complete mitigation for the leakage current, grounding of PV modules can significantly minimize the effects of lighting and other surges [4], [5]. Compared to the other solutions for leakage current mitigation, CG configurations bring together the following merits:

- Transformer-less configuration. Thus, it offers decreased cost, better efficiency, and high-power density [6], [7].

- Unipolar modulation strategy is usually adopted. Thus, small filter size can be used. Therefore, compared with using full-bridge inverter with bipolar modulation, higher power density and reduced cost can be realized. Although modified full-bridge configurations, such as the HERIC inverter, and the H5 inverter [8], [9], have been introduced to facilitate the using of unipolar modulation schemes without producing leakage current, but these configurations utilize extra switches which increases the cost. In addition, all full-bridge configurations do not offer voltage boosting capability. Therefore, front-end boost DC-DC converter or series connection of PV modules should be adopted to fulfill the required load voltage; especially for grid connected applications. The first solution increases the overall cost and reduces the power density. While the second solution causes inefficient operation of MPPT; especially, during partial shading [10], [11].

Generally, the following factors should be considered for efficient, cost-effective, and reliable operation of CG topologies:

- Utilization of low number of semiconductors, and passive components.
- Continuity of input current. Therefore, efficient MPPT can be achieved [11], [12].
- Voltage boosting capability. Therefore, no need for using extra DC-DC converter or series connection of PV modules.
- Offering high-voltage DC-link. Thus, efficient power decoupling between the PV module and the grid side can be easily achieved using low capacitance value. Thus, high reliable operation can be guaranteed [13].
- Bidirectional power flow; thus, reactive power support can be achieved [14].

To realize these factors, a number of CG inverters have been introduced in the literature. Voltage-bucking CG inverters are presented in [15]-[19]. In these inverters, front-end boosting DC-DC converter or series connection of PV modules should be adopted. Furthermore, these inverters do not offer continuous input current. Thus, inefficient MPPT is highly expected. Voltage-boosting CG inverters are presented in [20]-

[28]. One of the following techniques is used to achieve the boosting capability:

- Coupled inductors are utilized in [20], [21]. However, this results in increased cost, weight, and size. In addition, these inverters use bipolar modulation scheme. Accordingly, large output filter should be used.
- The switched-capacitors (SC) configuration has been employed in [22]-[25]. However, SC configuration usually utilizes high number of semiconductors. In addition, the switches of this configuration are usually over-rated to withstand the inrush current during switching between the capacitors [26].

The Split-source inverter (SSI) has been recently introduced [27], [28] as a single-stage DC-AC inverter. The single-phase realization of an SSI is shown in Fig. 1. The SSI combines the following features: 1) SSI offers the same boosting gain of the conventional two-stage configuration with reduced number of active switches, therefore, no need for using extra DC-DC converter nor series connection of PV modules. Also, low capacitance can be employed for power decoupling; thus, higher reliability can be realized; 2) SSI offers continuous input current, therefore, MPPT can be easily implemented; 3) SSI offers reactive power support; 4) it employs unipolar modulation scheme; thus, low total harmonic distortion can be achieved. Although these are very desirable features of SSI, high common-mode voltage, and hence, large amount of leakage current is expected when used with PV systems. This is because its operation is based on unipolar modulation scheme.

This work proposes a common-ground configuration of SSI (CG-SSI). While maintaining the advantages of the conventional SSI, the CG-SSI offers common-ground connection between the negative terminal of the PV module and the neutral of the load side. Fig. 2 shows the proposed CG-SSI topology. It utilizes four switches (S_{1-4}) and two diodes ($D_{1,2}$). The DC-link capacitor (C) is used for power decoupling; its voltage is denoted as V_{dc}. An input inductance (L_{dc}) is connected in series with the input source.

II. OPERATION DESCRIPTION OF THE PROPOSED CG-SSI

In the proposed CG-SSI, the boost circuit inductor (L_{dc}) can be charged through one of these paths: a path formed by S_1 and D_2, or another path formed by S_2 and D_1. Changing between these paths is determined according to the required inverter voltage state. Discharging of the boost circuit inductor into the DC-link capacitor is achieved through the path formed by the diodes D_1 and D_2.

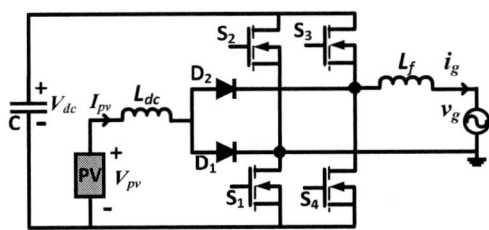

Fig. 1: Basic single-phase SSI topology.

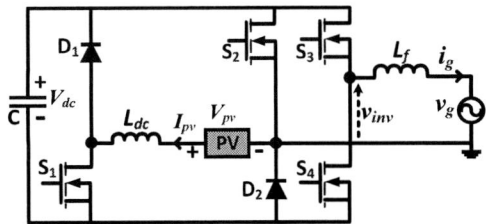

Fig. 2: Proposed CG-SSI topology

A. Modulation Strategy

Simultaneous DC-AC inversion and charging and discharging of the boost inductor is realized by comparing three signals: a carrier signal (v_c with frequency of F_{sw}), an inverter modulation signal (v_m with frequency of F_0), and a boost circuit modulation signal (d), which are shown in Fig. 3. To achieve simultaneous operation of the DC-DC boosting (from the input source to the DC-link) and the DC-AC inversion (between the DC-link and the output inverter ports), the condition expressed in (1) should be maintained. This condition is inherited from the conventional SSI topology [29]:

$$M_{max} < d \qquad (1)$$

where M_{max} is the peak value of v_m ($v_m = M_{max} \sin (2\pi F_0 t)$). Fig. 4 shows a flowchart to illustrate the switching mechanism. The switching mechanism can be explained as follow:

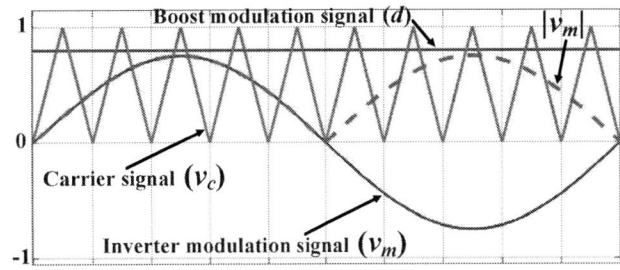

Fig. 3: Carrier and modulation signals of proposed CG-SSI.

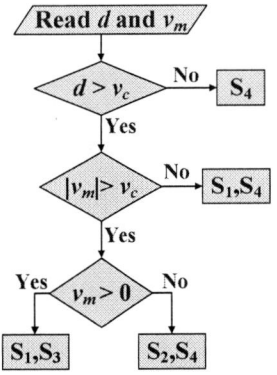

Fig 4: A flowchart of the switching pulses generation of DSSI.

979-8-3315-1612-3/25 $31.00 ©2025 IEEE

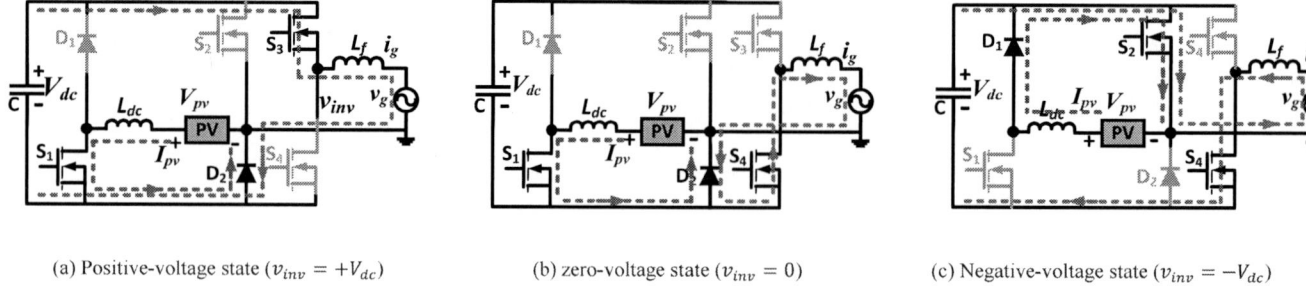

(a) Positive-voltage state ($v_{inv} = +V_{dc}$)   (b) zero-voltage state ($v_{inv} = 0$)   (c) Negative-voltage state ($v_{inv} = -V_{dc}$)

Fig. 5: Inverter voltage states when $d > v_c$ (boost inductor charging mode).

- When $d > v_c$: during this period, a closed path should be offered to charge the boost circuit inductor (L_{dc}) from the input source. Simultaneously, unipolar modulation for DC-AC inversion should be realized. Thus, during positive half-cycle of v_m, the output inverter voltage (v_{inv}) should be either (V_{dc}) or (0). On the other hand, during negative half-cycle of v_m, the inverter voltage (v_{inv}) should be either ($-V_{dc}$) or (0). This simultaneous operation of inductor charging, and DC-AC inversion is achieved as follow:

  o During the positive half-cycle of v_m, the switches S_1 and S_3 are turned ON when $v_m > v_c$. Thus, positive inverter voltage ($v_{inv} = +V_{dc}$) is realized through the switch S_3 and diode D_2. Concurrently, the inductor (L_{dc}) is charged through the switch S_1 and diode D_2. Fig. 5 (a) shows the corresponding circuit. On the other hand, when $v_m \leq v_c$, the switch S_3 is turned OFF and the switch S_4 is turned ON. Thus, zero-voltage ($v_{inv} = 0$) is realized through S_4 and D_2. Concurrently, as the switch S_1 is still turned ON, the inductor is still charged through the switch S_1 and diode D_2. Fig. 5 (b) shows the corresponding circuit connection of this state. It is worth stating here that the current flowing through the diode D_2 is practically the difference between two currents: the boost circuit current (I_{pv}) and the load current (i_g); this is according to the assumed currents' directions. Because of the voltage boosting operation (corresponding to current bucking operation) from the input to output ports, this difference is always positive at unity power factor (PF). For other PFs, this difference is positive within specified range of PFs.

  o During negative half-cycle of v_m, the switches S_2 and S_4 are turned ON when $|v_m| > v_c$. Thus, negative inverter voltage ($v_{inv} = -V_{dc}$) is realized. Concurrently, the inductor (L_{dc}) is charged through S_2 and D_1. Fig. 5 (c) shows the corresponding circuit. On the other hand, when $|v_m| \leq v_c$, the switch S_2 is turned OFF and the switch S_1 is turned ON. Fig. 4 (b) shows the circuit connection of this state. Thus, as the switch S_4 is still turned ON, zero-voltage ($v_{inv} = 0$) is realized through S_4 and D_2. Concurrently, the inductor is charged through S_1 and D_2. Thus, zero-voltage state is realized using the same circuit connection during positive and negative half-cycles.

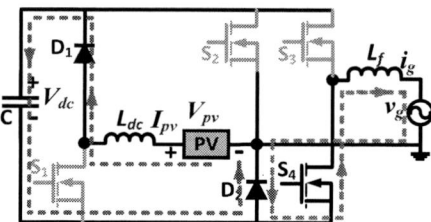

Fig. 6: Zero-voltage state ($v_{inv} = 0$) during inductor discharging mode.

- When $d \leq v_c$: according to the expression in (1), this period is always corresponding to zero-voltage state of the inverter; this can be implied from the signals shown in Fig. 2. To enable capacitor charging during this period while realizing the zero-voltage state of the inverter, the switch S_4 is only turned ON. Thus, zero-voltage state is realized through S_4 and D_2. Concurrently, the inductor discharges into the DC-link capacitor through the diodes D_1 and D_2. Fig. 6 shows the circuit connection during this state.

Based on the above modulation strategy, Figs. 7 (a) and (b) show the switching pulses produced during positive and negative half-cycles of v_m, respectively. The magnitude of v_c is plotted in dashed line for clarification. According to these switching pulses, Table I illustrates the switches' and diodes' states over two switching periods of carrier signal (v_c); one switching period during positive-half-cycle of v_m, and another switching period during negative half-cycle of v_m. The switching period is denoted as T_s. It is shown that 4 different sectors can always be identified over one T_s. These sectors can be used to derive expressions for power losses over the semiconductor devices.

Table I:
Switches states during positive and negative half-cycles of v_m

Semiconductor device	Positive half-cycle of v_m				Negative half-cycle of v_m			
	t_{p1}	t_{p2}	t_{p3}	t_{p4}	t_{n1}	t_{n2}	t_{n3}	t_{n4}
S_1	1	1	0	1	0	1	0	1
S_2	0	0	0	0	1	0	0	0
S_3	1	0	0	0	0	0	0	0
S_4	0	1	1	1	1	1	1	1
D_1	0	0	1	0	1	0	1	0
D_2	1	1	1	1	0	1	1	1

979-8-3315-1612-3/25 $31.00 © 2025 IEEE   777

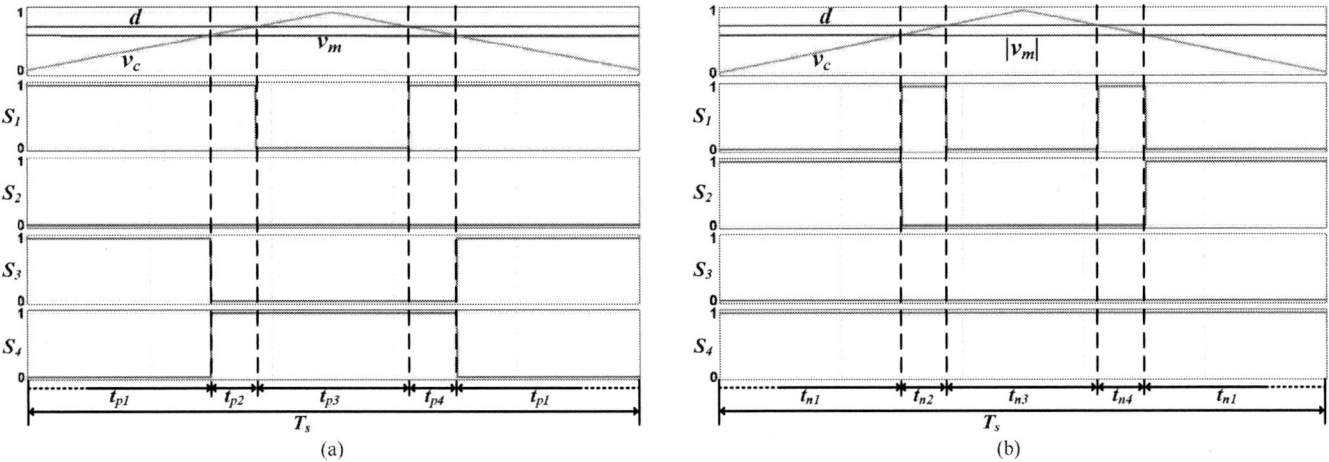

(a)   (b)

Fig. 7: Switching pulses generated, (a) during positive half-cycle of v_m. (a) during negative half-cycle of v_m.

B. Gain analysis and devices stresses

According to the above modulation strategy, the boosting gain (G_i) from the input to the DC-link voltage is determined as in (2). According to the constraint in (1) along with the expression in (2), the maximum gain ($B_{i\text{-}max}$) from the input source to the output inverter voltage is expressed in (3) where V_{max} is the amplitude of the fundamental component of output inverter voltage (v_{inv}).

$$G_i = \frac{V_{dc}}{V_{pv}} = \frac{1}{1-d} \qquad (2)$$

$$B_{i-max} = \frac{V_{max}}{V_{pv}} = \frac{d}{1-d} \qquad (3)$$

Practically, when the PV module is operated at MPP, the variation of the PV voltage is usually small. Accordingly, by proper choice of the DC-link voltage (considering the grid voltage), the condition in (1) can always be realized. Both the proposed CG-SSI and the conventional SSI need the same number of semiconductors. To perform a fair cost comparison between the two configurations, the total VA stress is considered. The current and voltage stresses of the semiconductors in the proposed CG_SSI and the conventional SSI are listed in Table II and Table III, respectively. I_{mpp} is the current at MPP of PV source, and I_{max} is the corresponding maximum load current.

Table II. Semiconductors stresses of the proposed CG-SSI

Semiconductors stresses	S_1	S_2	S_3	S_4	D_1	D_2
Current rating	I_{mpp}	$I_{mpp}+I_{max}$	I_{max}	I_{max}	I_{mpp}	$I_{mpp}+I_{max}$
Blocking voltage	V_{dc}					
Total VA stress [switches (VA_{sw})&diodes (VA_d)]	$V_{dc}(2I_{mpp} + 3I_{max})$				$V_{dc}(2I_{mpp} + I_{max})$	

Table III. Semiconductors stresses of the conventional SSI

Semiconductors stresses	S_1	S_2	S_3	S_4	D_1	D_2
Current rating	$I_{mpp}+I_{max}$				I_{mpp}	
Blocking voltage	V_{dc}					
Total VA stress [switches (VA_{sw})&diodes (VA_d)]	$4V_{dc}(I_{mpp} + I_{max})$				$2V_{dc}I_{mpp}$	

According to Tables II and III, the ratio of switches' VA stresses of CG-SSI to SSI is expressed in (4); this ratio is less than 1. Accordingly, CG-SSI needs lower VA stresses for the switches.

$$\frac{(VA_{sw})_{CG-SSI}}{(VA_{sw})_{SSI}} = \frac{(2I_{mpp}+3I_{max})}{4(I_{mpp}+I_{max})} < 1 \qquad (4)$$

III. CONTROL SYSTEM

Three main objectives should be fulfilled by the control system;
1. Input power control.
2. DC link voltage regulation for proper operation of DC/AC inversion.
3. Load current control for power quality issues.

A generalized control system for the proposed CG-SSI to be used for any input source is illustrated Fig. 8 (a).

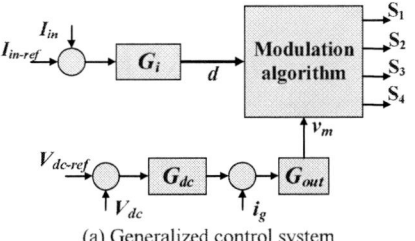

(a) Generalized control system

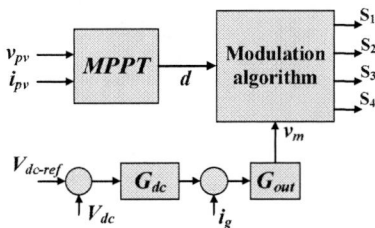

(b) Control system for PV systems

Fig. 8: Control system of the proposed CG-SSI.

979-8-3315-1612-3/25 $31.00 © 2025 IEEE   778

G_i is the input current control. The output of this block determines the modulation signals of the boost circuits (d). In addition, outer and inner control loops are used to regulate the DC-link voltage and the load current, respectively; their transfer functions are denoted as G_{dc} and G_{out}, respectively. Proportional Integral (PI) controller is employed for G_i and G_{dc} as expressed in (5) and (6), respectively. On the other hand, Proportional + Resonant (PR) controller is employed for G_{out} as expressed in (7).

$$G_i(s) = K_{pi} + \frac{K_i}{s} \tag{5}$$

$$G_{dc}(s) = K_{p-dc} + \frac{K_{i-dc}}{s} \tag{6}$$

$$G_{out}(s) = K_{p-out} + \frac{K_r}{s^2+\omega^2} \tag{7}$$

where K_{pi} and K_i are respectively the proportional and integral gains of the input current controller. K_{p-dc} and K_{i-dc} are respectively the proportional and integral gains of the DC-link voltage controller. K_{p-out} and K_r are respectively the proportional and resonant gains of the output current controller. In (7), ω is the fundamental frequency of the output inverter voltage in rad/sec. For PV systems, the input icontroller is replaced with MPPT block as illustrated in Fig. 8 (b). According to the MPPT method, voltage and/or current measurements of the PV sources will be required.

IV. RESULTS

An experimental prototype, with the parameters listed in Table IV, is built to verify the performance of the proposed CG-SSI. A DC supply is utilized as input source and resistive load is connected to the output. For open-loop operation, Figs. 9 (a) and (b) show the waveforms of the inverter voltage (v_{inv}), the DC link voltage (V_{dc}), the load current (i_l), the load voltage (v_l), and the input current (I_{in}). The waveforms show the proper operation for both DC-DC boosting along with DC-AC inversion; this ensures the well operation of the switching mechanism of the proposed CG-SSI.

Real-time simulation using Opal-RT (OP4510) is used to verify the performance of the proposed topology with grid-connected conditions. Table IV lists the system parameters used. The corresponding waveforms are illustrated in Fig. 10.

TABLE III
SYSTEM PARAMETERS FOR OPEN-LOOP CONDITIONS

Symbol	Quantity	Value
V_{in}	Input voltage	25 V
d	Duty cycle	0.78
M	Modulation signal amplitude	0.75
F_o	Modulation signal frequency	50 Hz
C	DC-link capacitors	220 μF
L_{dc}	Inductance of boosting circuit	1.5 mH
F_{sw}	Switching frequency	20 kHz
L_f	Inductance of output filter	6 mH

TABLE V
SYSTEM PARAMETERS FOR CLOSED-LOOP OPERATION

Symbol	Quantity	Value
V_{in}	Input voltage	55 V
V_g	Grid Voltage (RMS)	100 V
f_g	Grid frequency	50 Hz
C	DC link capacitors	470 μF
L_{dc}	Inductance of boosting circuits (i=1,2)	2.5 mH
L_f	Inductance of output filter	6 mH
F_{sw}	Switching frequency	10 kHz
V_{dc-ref}	Reference of DC-link voltage	230 V
I_{in-ref}	Reference of input current	11 A

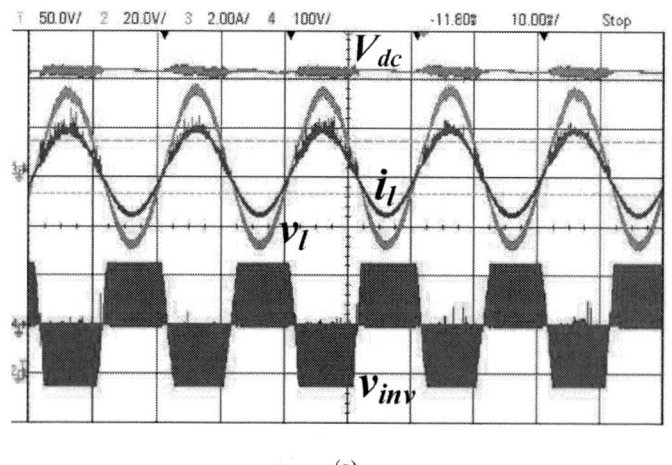

(a)

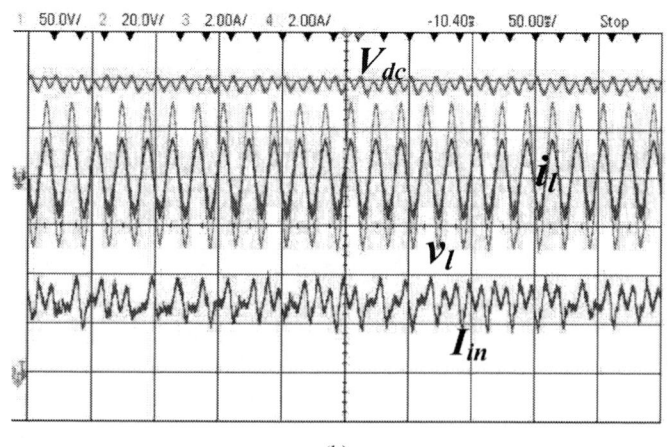

(b)

Fig. 9. Waveforms for the proposed CG-SSI with open-loop operation.

979-8-3315-1612-3/25 $31.00 © 2025 IEEE

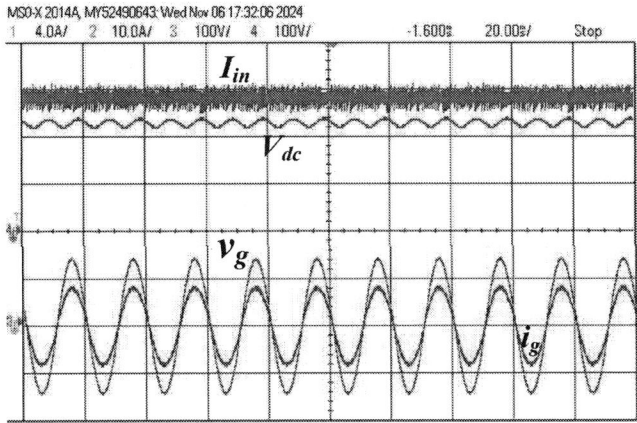

Fig. 10. Waveforms for the proposed CG-SSI with closed-loop operation.

V. COMPARATIVE STUDY

Table VI introduces a comparison between the proposed CG-SSI with other reported CG inverters. This comparison considers 1) the number of semiconductors; this reveals about the cost, size, and operation complexity, 2) continuoty of input current, 3) reactive power support, and 4) unipolar SPWM. According to Table V, the following remarks can be reported:

- Compared to CG inverters introduced in [15]-[17], [35], [37], and [40], the proposed CG-SSI can offer continous input current. Thus efficient MPPT operation can be easly realized for PV systems. In addition, compared to the CG inverter in [37], the proposed CG-SSI can offer reactive power support.

- Compared to the CG inverter in [16], [30]-[34], [37], [38], and [40], the proposed CG-SSI needs lower number of switches. Thus, lower cost and less-complexty can be realized.

Compared to the CG inverters in [20], [21], [38], and [39], the proposed CG-SSI can offer 3 levels of output inverter voltage. Thus low output filter size can be used for the same THD level.

Table VI.
Comparison of proposed CG-SSI with other CG inverters

Topology	No. of Components (S: switches, D: diodes)		Reactive power support capability	Continuity of input current	Unipolar SPWM
	S	D			
[15]	4	2	Yes	No	3
[16]	5	0	Yes	No	3
[17]	4	0	Yes	No	3
[20]	3	5	Yes	Yes	2
[21]	3	0	Yes	Yes	2
[30]	7	1	Yes	Yes	3
[31]	7	3	Yes	Yes	3
[32], [33]	5	0	Yes	Yes	3
[34]	5	0	Yes	Yes	3
[35]	3	6	Yes	No	3
[36]	4	3	Yes	Yes	3
[37]	6	3	No	No	3
[38]	6	0	Yes	Yes	2
[39]	4	0	Yes	Yes	2
[40]	5	1	Yes	No	3
Proposed CG-SSI	4	2	Yes	Yes	3

VI. CONCLUSION

This paper presents a novel common-ground (CG) inverter; it is abbreviated CG-SSI. The proposed inverter inherits the brilliant features of the split-source inverter (SSI). In addition, it offers complete leakage current mitigation. The modulation strategy and the different operating states of CG-SSI are explained. It is shown that the CG-SSI needs lower total VA stresses compared to that needed for the conventional SSI. A closed loop control system is presented to achieve maximum power point tracking (MPPT) and load current regulation. The results of the proposed CG-SSI show its capability to achieve both DC-DC boosting and DC-AC inversion. Compared to other CG inverters, the proposed CG-SSI brings together continuity of input current, using reduced number of switches, unipolar modulation, and capability of reactive power support.

ACKNOWLEDGMENT

This work is funded in part by the Egyptian Science and Technology Development Funds (STDF) ID: 41945, and in part by AC3E (ANID/BASAL/AFB240002) And SERC (ANID/FONDAP/1523A0006). Any opinions, findings, and conclusions or recommendations expressed in this material are those of the authors and do not necessarily reflect the views of the funding agencies.

REFERENCES

[1]. S. Iturriaga-Medina *et al.*, "Leakage-Ground Currents Compensation in a Transformerless HB-NPC Topology Using a DC-Link-Tied *LC* Filter for Photovoltaic Applications," *IEEE J. Emerg. Sel. Top. Power Electron.*, vol. 10, no. 4, pp. 4725–4737, Aug. 2022, doi: 10.1109/JESTPE.2021.3132239.

[2]. U. A. Khan, A. A. Khan, F. Akbar, and J.-W. Park, "Single-Stage Single-Phase H6 and H8 Non-Isolated Buck-Boost Photovoltaic Inverters," *IEEE J. Emerg. Sel. Top. Power Electron.*, vol. 10, no. 4, pp. 4865–4878, Aug. 2022, doi: 10.1109/JESTPE.2022.3153840.

[3]. E. Abdeen, Mahmoud A. Gaafar, M. Orabi, and M. Youssef, "A Novel High Gain Single-phase Transformer-less Multi-level Micro-inverter," in 2019 IEEE Applied Power Electronics Conference and Exposition (APEC), Anaheim, CA, USA, Mar. 2019, pp. 3263–3268. doi: 10.1109/APEC.2019.8722315.

[4]. Mahmoud A. Gaafar, M. Orabi, A. Ibrahim, R. Kennel, and M. Abdelrahem, "Common-Ground Photovoltaic Inverters for Leakage Current Mitigation: Comparative Review," *Appl. Sci.*, vol. 11, no. 23, p. 11266, Nov. 2021, doi: 10.3390/app112311266.

[5]. W. I. Bower and J. C. Wiles, "Analysis of grounded and ungrounded photovoltaic systems," in Proceedings of 1994 IEEE 1st World Conference on Photovoltaic Energy Conversion - WCPEC (A Joint Conference of PVSC, PVSEC and PSEC), Waikoloa, HI, USA: IEEE, 1994, pp. 809–812. doi: 10.1109/WCPEC.1994.520083.

[6]. M. N. H. Khan, M. Forouzesh, Y. P. Siwakoti, L. Li, T. Kerekes, and F. Blaabjerg, "Transformerless Inverter Topologies for Single-Phase Photovoltaic Systems: A Comparative Review," *IEEE J. Emerg. Sel. Top. Power Electron.*, vol. 8, no. 1, pp. 805–835, Mar. 2020, doi: 10.1109/JESTPE.2019.2908672.

[7]. E. A. D. Ibrahim, M. A. Gaafar, M. Orabi, A. Sheir, and M. Z. Youssef, "A Novel Dual-Input High-Gain Transformerless Multilevel Single-Phase Microinverter for PV Systems," *IEEE Trans. Power Electron.*, vol. 35, no. 5, pp. 4703–4714, May 2020, doi: 10.1109/TPEL.2019.2941387.

[8]. H. Schmidt, C. Siedle, and J. Ketterer, "DC/AC converter to convert direct electric voltage into alternating voltage or into alternating current," U.S. Patent 7046534 B2, May 16, 2006

[9]. M. Victor, F. Greizer, S. Bremicker, and U. Hubler, "Method of converting a direct current voltage from a source of direct current voltage, more specifically from a photovoltaic source of direct current voltage, into a alternating current voltage," U.S. Patent 7411802 B2, Aug. 12, 2008

[10]. Mustafa Abu-Zaher, Fang Zhuo, Mohamed Orabi, Alaaeldien Hassan, Mahmoud A. Gaafar, "Dual-input configuration of three-phase split-source inverter for photovoltaic systems with independent maximum power point tracking", Electric Power Systems Research, Volume 232, 2024, 110375, ISSN 0378-7796, doi:10.1016/j.epsr.2024.110375

[11]. Mahmoud A. Gaafar, Mustafa Abu-Zaher, Mohamed Orabi, Fang Zhuo, "Single-phase dual-input split-source inverter for photovoltaic systems", Electric Power Systems Research, Volume 221, 2023, ISSN 0378-7796, doi: 10.1016/j.epsr.2023.109397.

[12]. A. Sarikhani, B. Allahverdinejad, and M. Hamzeh, "A Nonisolated Buck–Boost DC–DC Converter With Continuous Input Current for Photovoltaic Applications," IEEE J. Emerg. Sel. Top. Power Electron., vol. 9, no. 1, pp. 804–811, Feb. 2021, doi: 10.1109/JESTPE.2020.2985844.

[13]. A. Jamatia, V. Gautam, and P. Sensarma, "Power Decoupling for Single-Phase PV System Using C´uk Derived Microinverter," IEEE Trans. Ind. Appl., vol. 54, no. 4, pp. 3586–3595, Jul. 2018, doi: 10.1109/TIA.2018.2812140.

[14]. T. K. S. Freddy, J.-H. Lee, H.-C. Moon, K.-B. Lee, and N. A. Rahim, "Modulation Technique for Single-Phase Transformerless Photovoltaic Inverters With Reactive Power Capability," IEEE Trans. Ind. Electron., vol. 64, no. 9, pp. 6989–6999, Sep. 2017, doi: 10.1109/TIE.2017.2686366.

[15]. J. F. Ardashir, M. Sabahi, S. H. Hosseini, F. Blaabjerg, E. Babaei, and G. B. Gharehpetian, "A Single-Phase Transformerless Inverter With Charge Pump Circuit Concept for Grid-Tied PV Applications," IEEE Trans. Ind. Electron., vol. 64, no. 7, pp. 5403–5415, Jul. 2017, doi: 10.1109/TIE.2016.2645162.

[16]. Y. Gu, W. Li, Y. Zhao, B. Yang, C. Li, and X. He, "Transformerless Inverter With Virtual DC Bus Concept for Cost-Effective Grid-Connected PV Power Systems," IEEE Trans. Power Electron., vol. 28, no. 2, pp. 793–805, Feb. 2013, doi: 10.1109/TPEL.2012.2203612.

[17]. A. Kadam and A. Shukla, "A Multilevel Transformerless Inverter Employing Ground Connection Between PV Negative Terminal and Grid Neutral Point," IEEE Trans. Ind. Electron., vol. 64, no. 11, pp. 8897–8907, Nov. 2017, doi: 10.1109/TIE.2017.2696460.

[18]. F. B. Grigoletto, "Five-Level Transformerless Inverter for Single-Phase Solar Photovoltaic Applications," IEEE J. Emerg. Sel. Top. Power Electron., vol. 8, no. 4, pp. 3411–3422, Dec. 2020, doi: 10.1109/JESTPE.2019.2891937.

[19]. F. B. Grigoletto, "Multilevel Common-Ground Transformerless Inverter for Photovoltaic Applications," IEEE J. Emerg. Sel. Top. Power Electron., vol. 9, no. 1, pp. 831–842, Feb. 2021, doi: 10.1109/JESTPE.2020.2979158.

[20]. A. Sarikhani, M. H. Ghaderi, and M. Hamzeh, "A Common-Ground Quazi-Z-Source Single-Phase Inverter Suitable for Photovoltaic Applications," IEEE Trans. Energy Convers., vol. 36, no. 2, pp. 594–601, Jun. 2021, doi: 10.1109/TEC.2020.3013115.

[21]. K. Kim, H. Cha, and H.-G. Kim, "A New Single-Phase Switched-Coupled-Inductor DC-AC Inverter for Photovoltaic Systems," IEEE Trans. Power Electron., vol. 32, no. 7, pp. 5016–5022, Jul. 2017, doi: 10.1109/TPEL.2016.2606489.

[22]. R. Barzegarkhoo, Y. P. Siwakoti, N. Vosoughi, and F. Blaabjerg, "Six-Switch Step-Up Common-Grounded Five-Level Inverter With Switched-Capacitor Cell for Transformerless Grid-Tied PV Applications," IEEE Trans. Ind. Electron., vol. 68, no. 2, pp. 1374–1387, Feb. 2021, doi: 10.1109/TIE.2020.2967674.

[23]. R. Barzegarkhoo, Y. P. Siwakoti, R. P. Aguilera, Md. N. H. Khan, S. S. Lee, and F. Blaabjerg, "A Novel Dual-Mode Switched-Capacitor Five-Level Inverter With Common-Ground Transformerless Concept," IEEE Trans. Power Electron., vol. 36, no. 12, pp. 13740–13753, Dec. 2021, doi: 10.1109/TPEL.2021.3074517.

[24]. S. S. Lee, Y. Yang, and Y. P. Siwakoti, "A Novel Single-Stage Five-Level Common-Ground-Boost-Type Active Neutral-Point-Clamped (5L-CGBT-ANPC) Inverter," IEEE Trans. Power Electron., vol. 36, no. 6, pp. 6192–6196, Jun. 2021, doi: 10.1109/TPEL.2020.3037720.

[25]. J. S. Mohamed Ali, A. Hota, N. Sandeep, and D. J. Almakhles, "A Single-Stage Common Ground-Type Transformerless Five-Level Inverter Topology," IEEE J. Emerg. Sel. Top. Power Electron., vol. 10, no. 1, pp. 837–845, Feb. 2022, doi: 10.1109/JESTPE.2021.3095125.

[26]. H. K. Jahan, M. Sarhangzadeh, J. F. Ardashir, and F. Blaabjerg, "A Symmetric Switched-Capacitor-Based Basic Inverter Unit for Grid-Connected PV Systems," IEEE Trans. Power Electron., vol. 37, no. 12, pp. 15594–15604, Dec. 2022, doi: 10.1109/TPEL.2022.3184757.

[27]. A. Abdelhakim, P. Mattavelli, and G. Spiazzi, "Three-Phase Split-Source Inverter (SSI): Analysis and Modulation," IEEE Trans. Power Electron., vol. 31, no. 11, pp. 7451–7461, Nov. 2016, doi: 10.1109/TPEL.2015.2513204.

[28]. A. Abdelhakim, P. Mattavelli, P. Davari, and F. Blaabjerg, "Performance Evaluation of the Single-Phase Split-Source Inverter Using an Alternative DC–AC Configuration," IEEE Trans. Ind. Electron., vol. 65, no. 1, pp. 363–373, Jan. 2018, doi: 10.1109/TIE.2017.2714122.

[29]. A. Abdelhakim, P. Mattavelli, V. Boscaino, and G. Lullo, "Decoupled Control Scheme of Grid-Connected Split-Source Inverters," IEEE Trans. Ind. Electron., vol. 64, no. 8, pp. 6202–6211, Aug. 2017, doi: 10.1109/TIE.2017.2677343.

[30]. N. Vosoughi, S. H. Hosseini, and M. Sabahi, "A New Single-Phase Transformerless Grid-Connected Inverter With Boosting Ability and Common Ground Feature," IEEE Trans. Ind. Electron., vol. 67, no. 11, pp. 9313–9325, Nov. 2020, doi: 10.1109/TIE.2019.2952781.

[31]. J. Roy, Y. Xia, and R. Ayyanar, "High Step-Up Transformerless Inverter for AC Module Applications With Active Power Decoupling," IEEE Trans. Ind. Electron., vol. 66, no. 5, pp. 3891–3901, May 2019, doi: 10.1109/TIE.2018.2860538.

[32]. X. Hu, P. Ma, B. Gao, and M. Zhang, "An Integrated Step-Up Inverter Without Transformer and Leakage Current for Grid-Connected Photovoltaic System," IEEE Trans. Power Electron., vol. 34, no. 10, pp. 9814–9827, Oct. 2019, doi: 10.1109/TPEL.2019.2895324.

[33]. S. S. Lee, Y. P. Siwakoti, C. S. Lim, and K.-B. Lee, "An Improved PWM Technique to Achieve Continuous Input Current in Common-Ground Transformerless Boost Inverter," IEEE Trans. Circuits Syst. II Express Briefs, vol. 67, no. 12, pp. 3133–3136, Dec. 2020, doi: 10.1109/TCSII.2020.2967899.

[34]. V. Gautam and P. Sensarma, "Design of Ćuk-Derived Transformerless Common-Grounded PV Microinverter in CCM," IEEE Trans. Ind. Electron., vol. 64, no. 8, pp. 6245–6254, Aug. 2017, doi: 10.1109/TIE.2017.2677352.

[35]. H. Heydari-doostabad, M. Pourmahdi, M. Jafarian, A. Keane, and T. O'Donnell, "Three-Switch Common Ground Step-Down and Step-Up Single-Stage Grid-Connected PV Inverter," IEEE Trans. Power Electron., vol. 37, no. 7, pp. 7577–7589, Jul. 2022, doi: 10.1109/TPEL.2022.3145193.

[36]. F. Peng, G. Zhou, N. Xu, and S. Gao, "Zero Leakage Current Single-Phase Quasi-Single-Stage Transformerless PV Inverter With Unipolar SPWM," IEEE Trans. Power Electron., vol. 37, no. 11, pp. 13755–13766, Nov. 2022, doi: 10.1109/TPEL.2022.3180287.

[37]. L. Wang and M. Shan, "A Novel Single-Stage Common-Ground Zeta-Based Inverter With Nonelectrolytic Capacitor," IEEE Trans. Power Electron., vol. 37, no. 9, pp. 11319–11331, Sep. 2022, doi: 10.1109/TPEL.2022.3167450.

[38]. F. Akbar, A. Elkhateb, H. F. Ahmed, A. A. Khan, H. Cha, and J.-W. Park, "Single-Phase Virtual-Ground Transformerless Buck–Boost Inverters," IEEE Trans. Power Electron., vol. 38, no. 9, pp. 11585–11600, Sep. 2023, doi: 10.1109/TPEL.2023.3283927.

[39]. S. Gangavarapu, M. Verma, and A. K. Rathore, "A Novel Transformerless Single-Stage Grid-Connected Solar Inverter," IEEE J. Emerg. Sel. Top. Power Electron., vol. 11, no. 1, pp. 970–980, Feb. 2023, doi: 10.1109/JESTPE.2020.3007556.

[40]. H. Tian, M. Chen, G. Liang, and X. Xiao, "A Single-Phase Transformerless Common-Ground Type PV Inverter With Active Power Decoupling," IEEE Trans. Ind. Electron., vol. 70, no. 4, pp. 3762–3772, Apr. 2023, doi: 10.1109/TIE.2022.3181361.

979-8-3315-1612-3/25 $31.00 © 2025 IEEE

Comprehensive Investigation and Proposal of a New Wireless Charging Road Structure Using Low-Environmental-Impact Magnetic Concrete

Shuntaro Inoue, Yuko Kano, Shin Tajima

Toyota Central R&D Labs., Inc., Nagakute, Aichi, Japan

Email: s-inoue@mosk.tytlabs.co.jp

Abstract—Previous research on dynamic wireless power transfer (DWPT) systems for electric vehicles has used traditional magnetic concretes like Mn-Zn ferrite to improve power transfer efficiency. However, these materials pose environmental risks, such as the potential release of manganese into soil and water. This study proposes a new road structure using Sendust-based concrete and aluminum plates as a safer and more eco-friendly alternative. While Sendust has lower magnetic permeability than Mn-Zn ferrite, simulations and experiments show it can reduce rebar-related losses by 88% and total system losses by 28% during 20 kW power transmission when combined with aluminum plates. The structure also maintains key self-inductance and mutual inductance levels, ensuring efficient power transfer. This research supports the development of sustainable DWPT systems, with future work focusing on heat management, improving efficiency, and creating materials to further reduce Sendust concrete losses.

I. Introduction

Dynamic Wireless Power Transfer (DWPT) systems for electric vehicles (EVs) provide a way to charge vehicles while they are moving [1]–[4]. This technology reduces the need for stationary chargers and extends driving ranges, making it a promising solution to promote EV adoption. However, achieving high efficiency in DWPT systems is challenging, especially when designing road structures that reduce energy losses and have minimal environmental impact.

Magnetic concretes are one approach to improving DWPT efficiency by focusing and strengthening the magnetic fields from transmitter coils [5], [6]. Materials like Mn-Zn ferrite are widely used for their excellent magnetic properties, but they pose environmental risks. Manganese from these materials can leach into the soil and groundwater, raising concerns about long-term sustainability [7]–[9]. These environmental issues need to be addressed for DWPT systems to be widely adopted.

Another major challenge is the rebar inside road structures. Rebar, commonly used to reinforce concrete, interacts with the magnetic fields from DWPT systems, causing energy losses through eddy currents and disrupting the magnetic field. This reduces power transfer efficiency and increases energy losses, which can be difficult to manage. While research on magnetic concretes has improved DWPT performance, few studies have focused on the specific problems caused by rebar. These losses lower system efficiency and raise operational costs, making DWPT systems less practical and economically feasible.

This study proposes a new road structure that combines Sendust-based concrete, an alloy of iron (Fe), silicon (Si), and aluminum (Al), with aluminum plates to reduce rebar-related losses. Sendust, with its Fe-10Si-5Al composition [10], offers good magnetic properties and is more environmentally friendly than Mn-Zn ferrites. The combination of Sendust concrete, which reduces energy losses, and aluminum plates, which shield the rebar, helps address these challenges.

To test this solution, we conducted material fabrication and characterization, simulations, and system-level experiments. These methods evaluate how well the proposed road structure reduces rebar losses, improves DWPT system performance, and minimizes environmental risks. By offering a sustainable and efficient alternative, this study aims to advance DWPT technology for future EV infrastructure.

II. Fabrication and Evaluation of Environmentally Friendly Magnetic Concrete

In this study, a geopolymer binder was used instead of traditional cement to create magnetic concrete. Geopolymers were chosen for their ability to support carbon neutrality, as they avoid the CO_2 emissions typically associated with cement production [11], [12]. This makes them a sustainable choice for infrastructure.

Two types of magnetic concrete were developed: one with Mn-Zn ferrite and another with Sendust. Sendust, an alloy of iron (Fe), silicon (Si), and aluminum (Al), is valued for its strong magnetic properties, low coercivity, and high resistance to corrosion and wear, making it suitable for demanding applications. However, since Sendust is a metallic material, it exhibits higher eddy current loss under high-frequency conditions compared to ferrite, resulting in increased core loss. This characteristic makes Sendust less efficient than Mn-Zn ferrite at higher frequencies, but it offers the advantage of reduced environmental impact and satisfactory structural performance.

Figure 1 shows examples of Mn-Zn ferrite-based and Sendust-based concrete. The magnetic and mechanical properties were tested by varying the volume of magnetic material. Ring-shaped samples were used to measure magnetic characteristics, while bar-shaped samples were used for mechanical testing. As shown in Fig. 1, increasing the magnetic material content makes the surface rougher, reducing the bending

(a) Mn-Zn ferrite concrete sample in a ring shape

(b) Sendust concrete sample in a ring shape

Fig. 1: Comparison of fabricated magnetic concrete samples with Mn-Zn ferrite and Sendust

(a) Sample with 74 vol% Mn-Zn ferrite content

(b) Sample with 78 vol% Mn-Zn ferrite content

Fig. 2: Surface condition of Mn-Zn ferrite concrete samples with different volume fractions

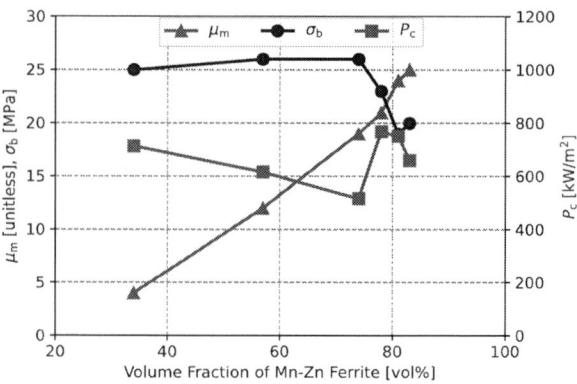

(a) Mn-Zn Ferrite Concrete

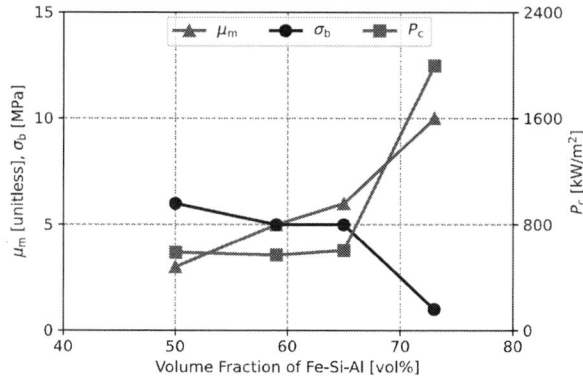

(b) Sendust (Fe-Si-Al) Concrete

Fig. 3: Measurement results of magnetic and mechanical properties

The results in Fig. 3 show that Mn-Zn Ferrite Concrete, with a 74% volume fraction, achieves a relative permeability μ_{m} of 19, a bending strength σ_{b} of 26 MPa, and a core loss P_{c} of 516 kW/m³. In comparison, Sendust Concrete, at a 65% volume fraction, has a lower relative permeability μ_{m} of 6 but offers reduced environmental impact. Although its magnetic performance is lower, Sendust maintains a bending strength σ_{b} of 5 MPa, meets structural requirements, and has a manageable core loss P_{c} of 605 kW/m³. Choosing a 65% volume fraction for Sendust balances structural integrity and loss management while minimizing environmental risks.

III. SIMULATION ANALYSIS OF ELECTRIFIED ROAD STRUCTURES WITH MAGNETIC CONCRETE

Simulations were conducted to study the effects of rebar on power transfer efficiency in DWPT systems and to evaluate the performance of proposed magnetic concrete materials. The 3D simulation model, shown in Fig. 4, represents a realistic scenario where a transmitter coil is placed on a concrete surface with embedded rebar, covered by an 80 mm asphalt layer. This setup assumes the DWPT system is installed during routine asphalt replacement, avoiding significant changes to the underlying concrete. A receiver coil is positioned 180 mm

strength (σ_{b}), which highlights the trade-off between magnetic properties and mechanical strength.

Figure 1 compares the visual appearance of Mn-Zn ferrite concrete and Sendust concrete, both fabricated as ring-shaped samples. This figure highlights the differences in texture and finish between the two types of magnetic concrete materials, which reflect their respective compositions and fabrication processes.

Figure 2, on the other hand, illustrates the effect of varying the volume fraction of magnetic material on the surface condition of Mn-Zn ferrite concrete. Bar-shaped samples with 74% and 78% volume fractions are shown for comparison. As the volume fraction increases, the surface becomes noticeably rougher, indicating a reduction in bending strength and increased brittleness. This highlights the trade-off between enhancing magnetic properties and maintaining mechanical integrity.

979-8-3315-1612-3/25 $31.00 © 2025 IEEE

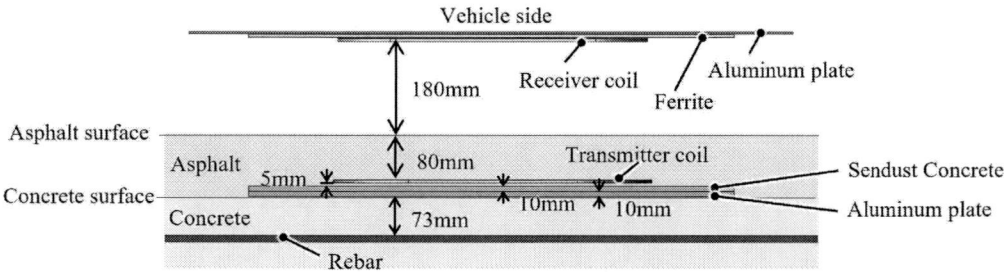

Fig. 4: Dimensional relationship of the 3D FEM model

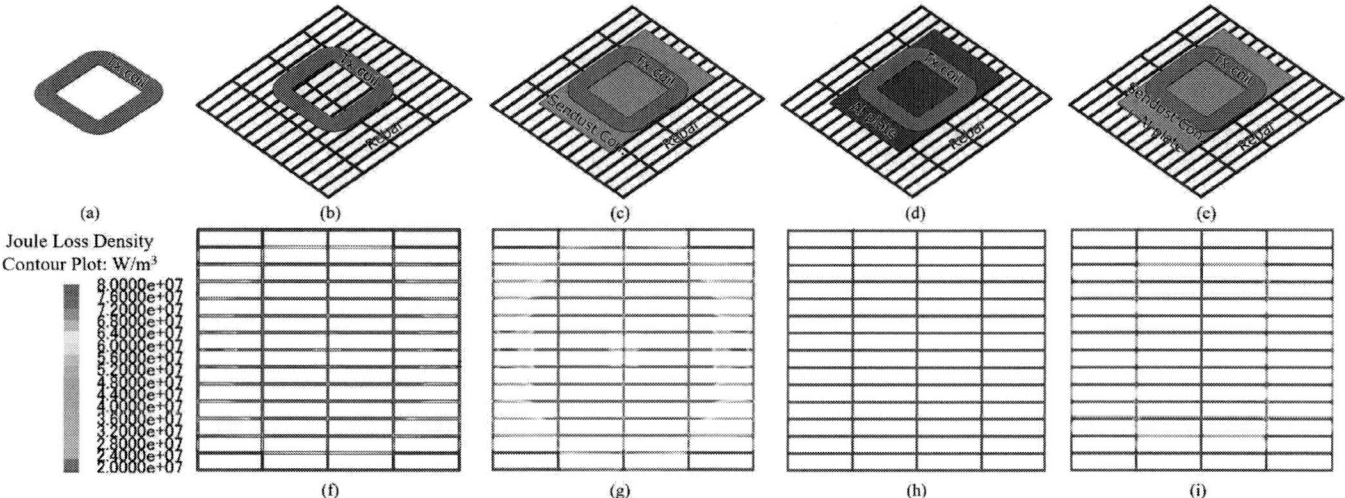

Fig. 5: The five simulation models and their results

above the asphalt, attached to the underside of a vehicle, simulating real-world operation.

The goal of the simulations was to compare five road structures, shown in Fig. 5, with three main objectives: (1) reduce energy losses in the rebar using environmentally friendly materials, (2) minimize the negative effects of rebar on power transfer efficiency, and (3) maintain overall system reliability and performance under real-world conditions. These objectives are critical for developing efficient and sustainable DWPT systems.

Figure 5 presents the five road structures studied in the simulations:

- **Model (a):** This model includes only the transmitter coil, without any rebar, representing an ideal scenario with no rebar-related losses. It serves as the benchmark for maximum efficiency.
- **Model (b):** This model introduces rebar below the transmitter coil without any additional materials. Significant energy losses occur in the rebar due to eddy currents, making this the baseline for evaluating the impact of rebar on DWPT systems.
- **Model (c):** A Sendust plate is placed between the transmitter coil and the rebar. However, because Sendust has relatively low magnetic permeability, it cannot effectively

shield the rebar from the magnetic field. As a result, the energy losses in this configuration are similar to those in Model (b).
- **Model (d):** An aluminum plate is placed between the transmitter coil and the rebar. The high conductivity of aluminum effectively reduces energy losses in the rebar by shielding it from the magnetic field. However, this shielding weakens the magnetic coupling between the transmitter and receiver coils, reducing power transfer efficiency.
- **Model (e):** This model combines a Sendust layer with an aluminum plate between the transmitter coil and the rebar. The aluminum plate reduces rebar-related losses, while the Sendust layer partially restores the magnetic coupling. This configuration achieves a balance between minimizing losses and maintaining efficiency, making it the most effective of the five models.

The results for the energy losses in the rebar under these conditions are shown in Fig. 6. Model (d), which uses an aluminum plate alone, significantly reduces rebar losses but compromises magnetic coupling. In contrast, Model (e), with both Sendust and aluminum layers, effectively reduces rebar-related losses while maintaining better magnetic coupling than

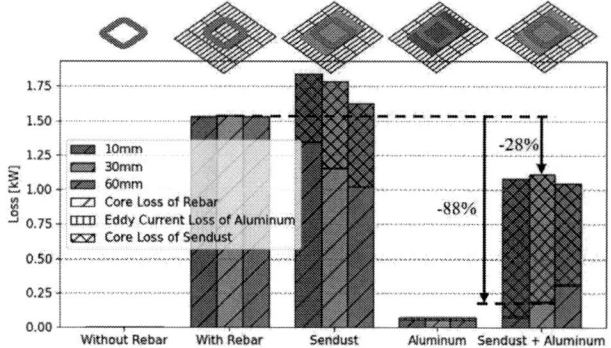

(a) Loss results of the 3D simulation model

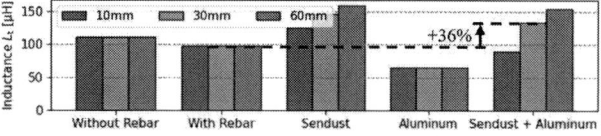

(b) Transmitter coil inductances L_t

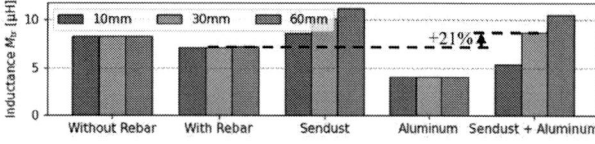

(c) Mutual inductances $M_{t,r}$

Fig. 6: Simulation results of the five wireless charging road structures

Model (d). This makes Model (e) the optimal configuration for balancing efficiency and performance.

Figure 6(a) shows the energy losses in the roadside structure. During 20 kW power transmission, rebar adds 1.5 kW of extra losses, reducing the system's efficiency by about 5%. However, the proposed structure in Model (e), which uses a 30 mm Sendust plate and a 1 mm aluminum plate, reduces these rebar losses by 88% and total system losses by 28%. This improvement is critical for maintaining system efficiency in practical applications.

Figures 6(b) and (c) highlight the recovery of magnetic properties with the proposed structure. Figure 6(b) shows a 36% recovery in the transmitter coil's self-inductance (L_t), while Fig. 6(c) shows a 21% recovery in the mutual inductance ($M_{t,r}$). These recoveries are essential to ensure effective power transfer even in the presence of rebar.

IV. EXPERIMENTAL RESULTS

The experimental setup used in this study is shown in Fig. 7. The rebar structure was built to mimic real-world conditions by connecting junctions with aluminum wires, simulating how rebar behaves in concrete. To ensure accurate measurements, the spacing between the transmitter coil and the rebar was carefully maintained using polystyrene foam and wood plates. These materials were chosen for their non-conductive and supportive properties. This setup was essential for isolating the impact of rebar on the system's inductance and Q-factor.

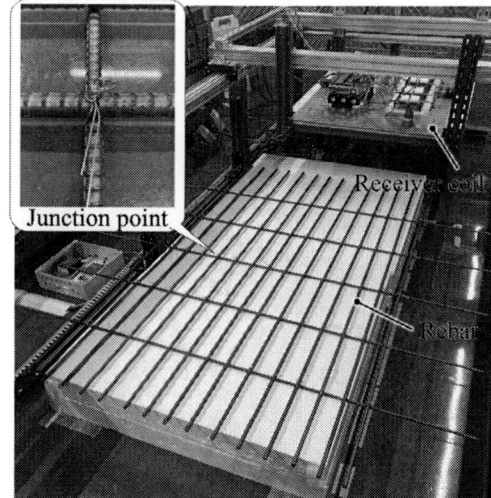

(a) Overall view of the experimental setup

(b) Side view of the DWPT system with rebar

Fig. 7: Constructed experimental setup for inductance and Q-factor measurement.

The self-inductance of the transmitter coil was measured in two conditions: with and without rebar. As shown in Table I, the presence of rebar reduced the self-inductance by 6% compared to when rebar was absent. While this reduction was smaller than the 12% predicted by simulation, the results follow the expected trend, showing the clear impact of rebar on the system's magnetic properties. This highlights the need to address rebar effects to ensure stable performance in real-world DWPT systems.

In addition to self-inductance, the Q-factor, an important measure of coil efficiency, showed a strong sensitivity to the presence of rebar. As shown in Table I, the Q-factor indicates how well the system minimizes energy losses and reflects its performance under different conditions. The results highlight that rebar has a major impact on DWPT efficiency.

The Q-factor is defined as $Q = \omega L_t / R_{Lt}$, where ω is the inverter switching angular frequency, L_t is the self-inductance, and R_{Lt} is the series parasitic resistance of the coil. A high

979-8-3315-1612-3/25 $31.00 © 2025 IEEE

TABLE I: Experimental and simulated results of inductance and Q-factor.

| | Inductance (L_t) [μH] | | Q-factor (Q_t) | |
	Without Rebar	With Rebar	Without Rebar	With Rebar
Measured	105.8	99.3 (-6%)	439.3	25.8 (-94%)
Simulated	111.0	97.8 (-12%)	-	-

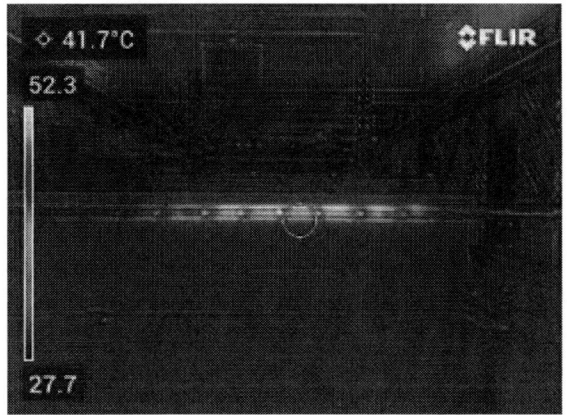

(a) Thermal distribution captured using an infrared camera

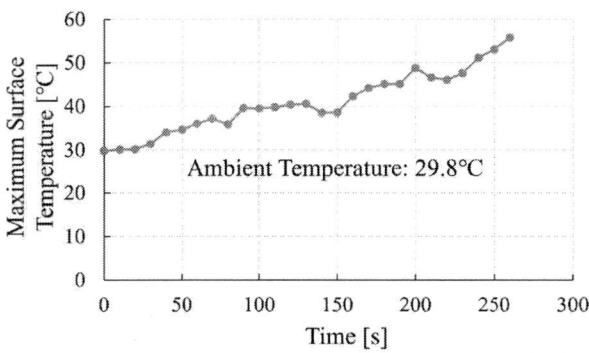

(b) Time variation of the maximum surface temperature of the experimental setup

Fig. 8: Measurement results of thermal distribution and temperature variation.

Q-factor, typically above 400, is crucial for efficient wireless power transfer with minimal losses. In this experiment, the presence of rebar caused the Q-factor to drop sharply from 439 to 26, a dramatic 94% reduction. This drop shows how severely rebar affects system efficiency by increasing energy losses, requiring higher input power and better thermal management to compensate. These findings emphasize the need for new design solutions to reduce the negative effects of rebar.

The thermal performance of the experimental setup was evaluated using an infrared camera, as shown in Fig.8(a), to measure the temperature distribution. A transmitter current of 10A, equal to one-tenth of the current needed to transmit 20 kW, was applied to the system. This lower current was chosen to safely observe thermal trends without exceeding the system's limits.

The results of the thermal distribution, depicted in Fig.8(a),

clearly show areas of significant heat generation, particularly around the rebar. To complement this analysis, the maximum surface temperature of the setup was tracked over time, as illustrated in Fig.8(b). During the 260-second experiment, the surface temperature steadily increased and exceeded 52°C.

These observations highlight the thermal challenges associated with DWPT systems, particularly in configurations involving rebar. Identifying and addressing these heat generation patterns is critical to improving the system's reliability and efficiency in practical applications.

During the 260-second experiment, the surface temperature of the setup steadily increased, exceeding 52°C. This rise in temperature highlights the thermal challenges in DWPT systems, especially when rebar is present. The heat concentrated near the rebar shows the effects of eddy current losses. Addressing these thermal issues is crucial for ensuring the system's reliability and efficiency in real-world applications.

One of the key findings from the thermal measurements was the significant heat generated around the rebar. This heat was primarily caused by eddy currents in the rebar, induced by the alternating magnetic fields from the transmitter coil. This localized heating reduces system efficiency and could lead to material degradation and increased maintenance needs over time, highlighting a key challenge for DWPT systems.

The presence of rebar creates two major challenges for DWPT systems. First, it reduces the system's inductance and Q-factor, as shown by the experimental and simulated results, which lowers the efficiency of power transfer. Second, the rebar generates significant heat, which affects system performance and speeds up the wear and tear of key components. This heat requires advanced cooling systems, making the design more complex and raising costs. These results highlight the need for new solutions, such as using Sendust-based materials, to reduce rebar-related losses and make DWPT systems more practical.

V. Conclusion

This study analyzed the challenges caused by rebar in DWPT systems and proposed a new road structure using environmentally friendly materials. While both Mn-Zn ferrite and Sendust-based concretes were evaluated, Sendust proved to be a safer and more sustainable option. The proposed structure combines Sendust concrete with aluminum plates, offering strong magnetic performance, durability for road use, and reduced environmental risks.

Our experiments and simulations showed that rebar significantly reduces system efficiency, mainly through a large drop in the Q-factor. Rebar also caused noticeable heat generation, leading to energy losses and potential material damage. The proposed structure reduced rebar-related losses by 88% and overall system losses by 28% during 20 kW power transmission, demonstrating its ability to improve efficiency and reliability.

Thermal analysis revealed the importance of managing heat generation near rebar to avoid material degradation and maintenance costs. The proposed design addresses these issues, improving system performance and durability.

In future work, we will evaluate the proposed structure's heat generation and power transmission efficiency under 20 kW, comparing results with and without rebar. We also plan to develop advanced materials to further reduce losses in Sendust concrete, improving the efficiency and sustainability of DWPT systems.

This design is a step toward more efficient and sustainable electrified roads, solving key technical and environmental challenges. It advances DWPT technology and supports the transition to greener and more electrified transportation infrastructure.

REFERENCES

[1] B. J. Limb, Z. D. Asher, T. H. Bradley, E. Sproul, D. A. Trinko, B. Crabb, R. Zane, and J. C. Quinn, "Economic viability and environmental impact of in-motion wireless power transfer," *IEEE Transactions on Transportation Electrification*, vol. 5, no. 1, pp. 135–146, 2019.

[2] C. C. Mi, G. Buja, S. Y. Choi, and C. T. Rim, "Modern advances in wireless power transfer systems for roadway powered electric vehicles," *IEEE Transactions on Industrial Electronics*, vol. 63, no. 10, pp. 6533–6545, 2016.

[3] T. Gardner, "Wireless power transfer roadway integration," Master's thesis, Utah State University, 2017.

[4] R. Tavakoli, U. Pratik, E. M. Dede, C. Chou, and Z. Pantic, "Minimizing the rebar impact on power dissipation in dynamic wireless power transfer systems," in *2021 IEEE Applied Power Electronics Conference and Exposition (APEC)*. IEEE, 2021, pp. 1599–1603.

[5] R. Tavakoli, A. Echols, U. Pratik, Z. Pantic, F. Pozo, A. Malakooti, and M. Maguire, "Magnetizable concrete composite materials for road-embedded wireless power transfer pads," in *2017 IEEE Energy Conversion Congress and Exposition (ECCE)*. IEEE, 2017, pp. 4041–4048.

[6] A. Agostino, "Multi-physics models to support the design of dynamic wireless power transfer systems," Master's thesis, Purdue University, 2022.

[7] W. C. Y. Do, "Secondary drinking water standards: Guidance for nuisance chemicals," 2017.

[8] H. B. Bradl, "Adsorption of heavy metal ions on soils and soils constituents," *Journal of colloid and interface science*, vol. 277, no. 1, pp. 1–18, 2004.

[9] D. C. Adriano, W. Wenzel, J. Vangronsveld, and N. Bolan, "Role of assisted natural remediation in environmental cleanup," *Geoderma*, vol. 122, no. 2-4, pp. 121–142, 2004.

[10] H. Masumoto, "On a new alloy "sendust" and its magnetic and electric properties," *J. Japan Inst. Metals*, vol. 1, pp. 127–135, 1937.

[11] J. Davidovits, "Geopolymers: inorganic polymeric new materials," *Journal of Thermal Analysis and calorimetry*, vol. 37, no. 8, pp. 1633–1656, 1991.

[12] P. Duxson, A. Fernández-Jiménez, J. L. Provis, G. C. Lukey, A. Palomo, and J. S. van Deventer, "Geopolymer technology: the current state of the art," *Journal of materials science*, vol. 42, pp. 2917–2933, 2007.

979-8-3315-1612-3/25 $31.00 © 2025 IEEE

Design of a Bidirectional High Power Inductive Power Transfer System with Auxiliary Winding for Automotive Applications

Luis Ruiz Chamorro
Centro de Electónica Industrial
Universidad Politécnica de Madrid
Madrid, Spain
luis.ruiz.chamorro@upm.es

Nikola Mirković
Centro de Electónica Industrial
Universidad Politécnica de Madrid
Madrid, Spain
n.mirkovic@upm.es

Alberto Delgado Expósito
Centro de Electónica Industrial
Universidad Politécnica de Madrid
Madrid, Spain
a.delgado@upm.es

Pedro Alou Cervera
Centro de Electónica Industrial
Universidad Politécnica de Madrid
Madrid, Spain
pedro.alou@upm.es

Miroslav Vasić
Centro de Electónica Industrial
Universidad Politécnica de Madrid
Madrid, Spain
miroslav.vasic@upm.es

Abstract—This article presents a high-power, highly efficient, bidirectional inductive power transfer (IPT) system for automotive applications using an intermediate coil to boost the system efficiency. It outlines the benefits of a three-coil system, namely, high power transfer capability and high robustness to clearance and misalignment. An equivalent circuit model is derived to enhance system understanding and tuning methodology for the compensation circuit is proposed. The coils and compensation circuitry were optimized for a resonant dual active bridge (DAB) topology. A 30 kW prototype was developed and tested, transferring power between 3.7 and 37.4 kW across an air gap of up to 150 mm, achieving efficiencies higher than 90% in all conditions. System performance was evaluated in both grid-to-vehicle and vehicle-to-grid modes.

Index Terms—Inductive Power Transfer, Bidirectional, Three Windings

This work was supported by the Ministerio de Ciencia, Ennovación y Universidades de España under Grants PID2023-151803OA-I00.

I. INTRODUCTION

Wireless chargers present as an alternative to conventional cable chargers. In these systems, energy is transferred using coupled coils with low coupling factor due to the absence of a magnetic core [1], so compensation is needed to overcome this problem [2]. In addition to the simplest two coil systems, three [3], [4], [5], [6] and four coils [7], [8] IPT systems are also present throughout the literature. While four coil systems offer greater efficiency, they are more suitable for low power applications [9], while three coil systems are capable of high-power transfer, even under misalignment conditions [10], [11]. Most of the mentioned works are unidirectional and highly complex. In this context, a tuning method and a design procedure for the S–S–S compensated three-coil bidirectional IPT system intended for the EV charging applications that is given in Fig. 1 is presented. Number of added compensation

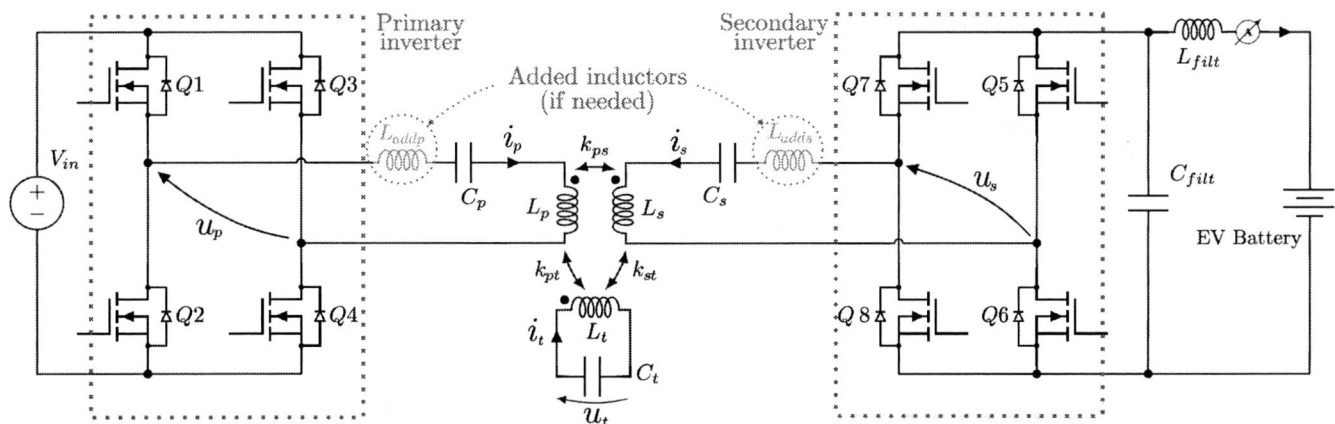

Figure 1: Considered three coil IPT system

979-8-3315-1612-3/25 $31.00 © 2025 IEEE

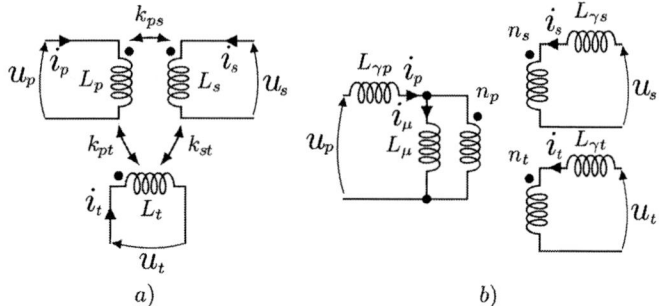

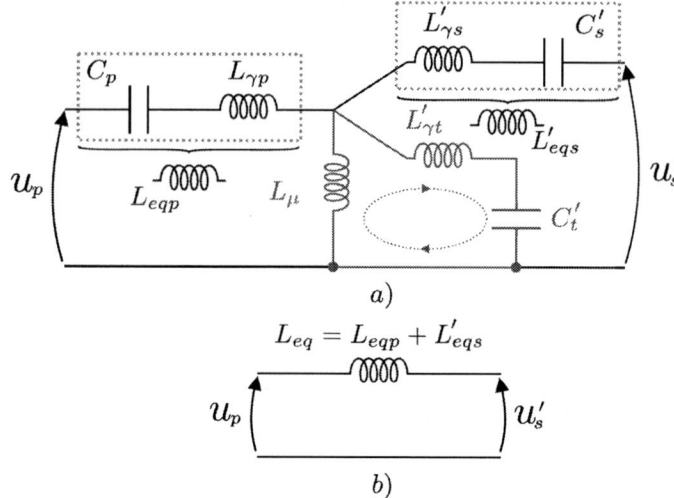

Figure 2: a) Three coupled inductors b) Equivalent circuit of three coupled inductors

elements in the proposed system is lower comparing to most of the previously mentioned articles, thus achieving high simplicity and low volume of the system.

II. MODEL AND WORKING PRINCIPLE OF THE THREE COIL IPT SYSTEM

Working directly with coupled inductors can be tedious and complex, so deriving an electrically equivalent circuit of the three coupled inductors given in Fig. 2(a) using only individual inductors instead of coupled ones as in Fig. 2(b) will facilitate an easier understanding. By comparing both systems and equating corresponding parts, the relationship between the real and equivalent circuit can be obtained, as given in equations (1)–(6) [12]:

$$\frac{n_p}{n_s} = \frac{L_{pt}}{L_{st}} \tag{1}$$

$$\frac{n_p}{n_t} = \frac{L_{ps}}{L_{st}} \tag{2}$$

$$L_\mu = \frac{L_{ps} \cdot L_{pt}}{L_{st}} \tag{3}$$

$$L_{\gamma p} = L_p - L_\mu \tag{4}$$

$$L_{\gamma s} = L_s - \left(\frac{L_{st}}{L_{pt}}\right)^2 L_\mu \tag{5}$$

$$L_{\gamma t} = L_t - \left(\frac{L_{st}}{L_{ps}}\right)^2 L_\mu \tag{6}$$

S-S-S compensation will be used in this work due to its simplicity and good results in the current state-of-the-art [12], [13]. In this topology, as shown in Fig. 3(a), tertiary capacitor is tuned to cancel magnetizing inductance (blue circle), whereas primary and secondary capacitors are tuned to partially cancel primary and secondary leakage inductances (red rectangles) as in equations (7)-(9), resulting in the equivalent circuit of Fig. 3(b). In this way, the reactive energy of the equivalent magnetizing inductance circulates within the magnetic structure rather than through the primary or secondary bridge, improving system efficiency.

Moreover, the equivalent circuit from Fig. 3(b) allows for a fully controllable system that operates as a classical Dual

Figure 3: a) Compensated circuit. b) Equivalent circuit after compensation

Active Bridge (DAB) DC-DC converter. If needed, additional series inductance may be added to further reduce harmonic content and resonant capacitors must account for that. Due to the resonant nature of these systems, first harmonic approximation can be used [14], and the power transferred between primary and secondary is given by equation (10), where U_p and U_s' are RMS values of first harmonic of primary and secondary voltages referred to primary, φ the phase shift between primary and secondary voltages and L_{eq} the inductance that is not compensated by the primary and secondary capacitors.

$$C_p = \frac{1}{\omega^2(L_{\gamma p} + L_{addp} - L_{eqp})} \tag{7}$$

$$C_s = \frac{1}{\omega^2(L_{\gamma s} + L_{adds} - L_{eqs})} \tag{8}$$

$$C_t = \frac{1}{\omega^2 L_t} \tag{9}$$

$$P = \frac{U_p U_s'}{\omega L_{eq}} \sin \varphi \tag{10}$$

III. DESIGN OF THE SYSTEM

The system is intended to provide a nominal power of 30kW with nominal primary and battery (secondary) voltages being 800V. Geometry of the coils will be based on SAE standard for a Z2 class with an outer diameter of 350mm and air gap

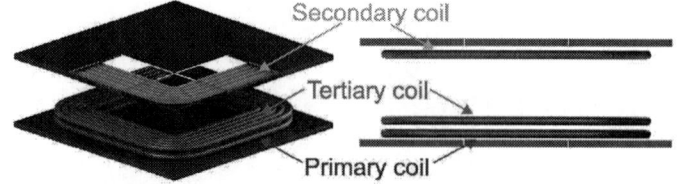

Figure 4: 3D model of the three IPT coils

979-8-3315-1612-3/25 $31.00 © 2025 IEEE

Table I: Parameters and their values

Parameter	Value	Parameter	Value	Parameter	Value
N_p	4	$L_t[\mu H]$	13.9	n_p/n_t	0.77
N_s	13	k_{ps}	0.21	$L_\mu[\mu H]$	8.2
N_t	5	k_{pt}	0.88	$L_{\gamma 1}[\mu H]$	2.4
$L_p[\mu H]$	10.6	k_{st}	0.24	$L_{\gamma 2}[\mu H]$	99.9
$L_s[\mu H]$	106	n_p/n_s	1.15	$L_{\gamma 3}[\mu H]$	0.06

between ground assembly and clearance assembly between 110 and 150mm [1]. Tertiary winding will be placed on top of the primary winding to have as little elements onboard as possible, as it is shown in Fig. 4.

The detailed design workflow is given in the flowchart of Fig. 5 and is briefly outlined in the following. Coil geometry is optimized to obtain maximum coupling coefficients between the coils of the ground and vehicle assemblies while minimizing the added vehicle weight by keeping the number of the secondary turns low.

$$\frac{n_p}{n_s} = \frac{L_{pt}}{L_{st}} = \frac{k_{pt}\sqrt{l_p}N_p}{k_{st}\sqrt{l_s}N_s} \qquad (11)$$

$$I_t = \frac{U_p}{\omega k_{pt}N_p N_s \sqrt{l_p l_t}} \qquad (12)$$

To gain a better understanding of this optimization process, equation (11), a reformulated version of equation (1), is shown, where l_p and l_s are respectively the per-turn inductances of the primary and secondary coils, determined by windings geometry. This equation also indicates that to reduce the secondary turns, while achieving the desired transform ratio it is necessary to keep the primary turns as low as possible, however, this will increase tertiary current as given by equation (12). The maximum allowable current depends mainly on the characteristics of the available Litz wire. With these considerations addressed, a compromise solution was found and final coil design parameters are listed in Table I.

Once the IPT coils are defined, it is necessary to determine the primary, secondary, and tertiary capacitor values. To do this, first, the value of the equivalent inductance L_{eq} has to be defined, which can be calculated using (13) once nominal phase shift is defined. If the simplified equivalent circuit is used, it can be shown that the reactive power in the inductor is given by equation (13).

$$Q_L = \frac{U_p^2 - 2U_p U_s' \cos\varphi_n + U_s'^2}{2\pi f L_{eq}} \qquad (13)$$

Which, for the considered ideal case of $U_p = U_s'$ becomes

$$Q_L = P_n \tan\frac{\varphi_n}{2} \qquad (14)$$

Here a trade off must be made, because ideally phase shift should be zero to eliminate reactive power, but that also would make the system very reactive to small changes in phase shift,

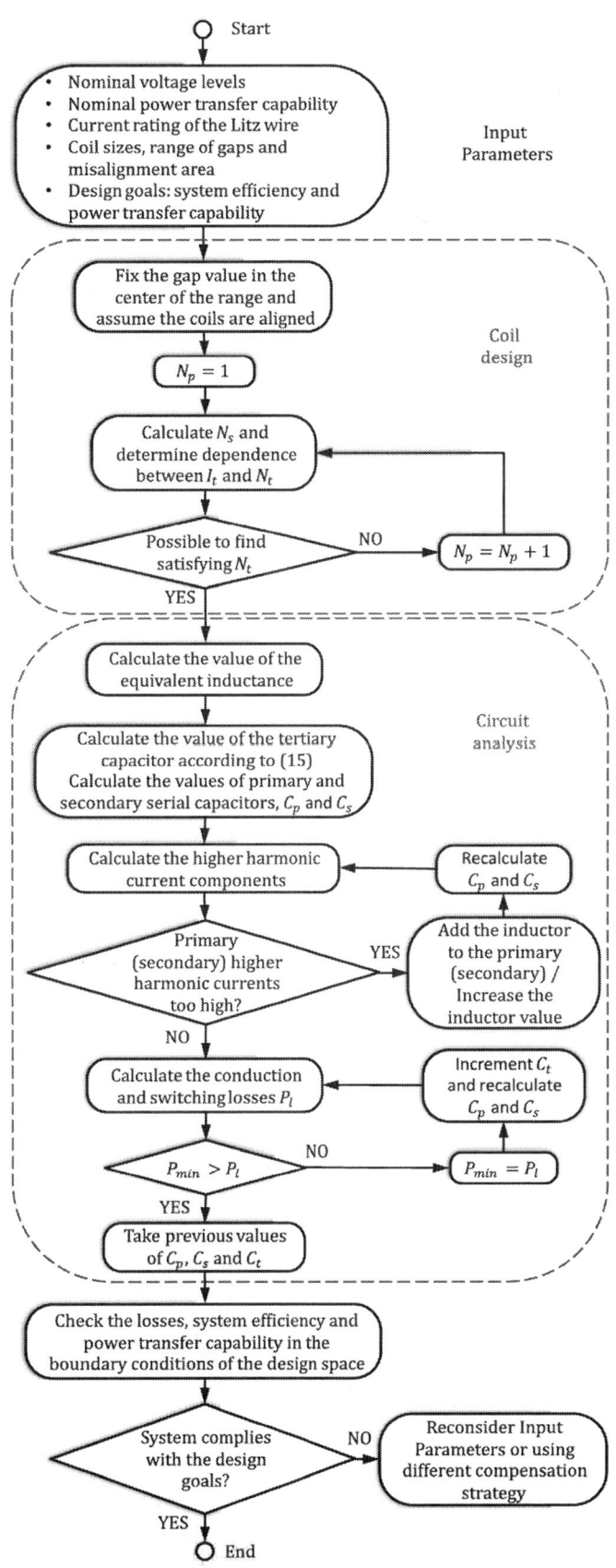

Figure 5: Design flowchart

making the system hard to operate. The change in power near the nominal operating point is given by

$$\frac{dP}{d\varphi} = P_n \cot \varphi_n \qquad (15)$$

Given that frequency is around 85 kHz for EV IPT chargers, to keep the potential power deviation at around 3% a potential phase mismatch of 50 ns between primary and secondary, nominal phase shift should be at least 30°. Once nominal phase shift is defined, equation (10) can be used to calculate equivalent inductance, which, for the considered case, the equivalent inductance is calculated to be 16 μH.

Since the system is bidirectional, the inductance is equally distributed between the primary and secondary to maintain symmetry. After defining the coil geometry, nominal phase-shift and the equivalent inductance, capacitors and the additional inductor are designed. The low self-inductance of the coils requires checking for higher harmonics in the primary and secondary currents. Fig. 6(a) shows primary and secondary currents and voltages obtained via PLECS simulations, where instantaneous switching currents reach high values, highlighting the significant impact of higher-order harmonics on the switching current.

To reduce these harmonic currents, a 30 μH inductor was added to the primary, with capacitance adjusted to maintain equivalent inductance. The chosen inductance value depends on permissible switching current and available space, as higher inductance values can suppress more harmonics but are limited by inductor size. Fig. 6(b) shows that with this 30 μH added inductor switching currents are reduced. The final component values are summarized in Table II.

Table II: Component values

Component	Value
C_p	144.4 nF
C_s	37.4 nF
C_t	274 nF
L_{add}	30 μH

IV. EXPERIMENTAL RESULTS

To verify the proposed ideas, a purpose-built experimental setup was constructed, as shown in Fig. 7. Detailed views of the primary, secondary, and tertiary coils, and also the inverters used, are provided in Fig. 8 (a), (b), (c), (d). The setup was tested under various conditions. For simplicity, coefficient ξ is introduced as a measurement of the coupling. The position and sizes of the coils suggest that the mutual inductances L_{ps} and L_{st} will change in the same proportion respective to their nominal values with the change of the secondary coil position. This allows L_{ps} and L_{st} to be expressed as their nominal values multiplied by ξ, where ξ equal to 1 indicates nominal coupling, values greater than 1 indicate higher coupling than nominal, and values less than 1 indicate lower coupling than nominal. The system was tested in three different positions/cases detailed in table III in order to validate

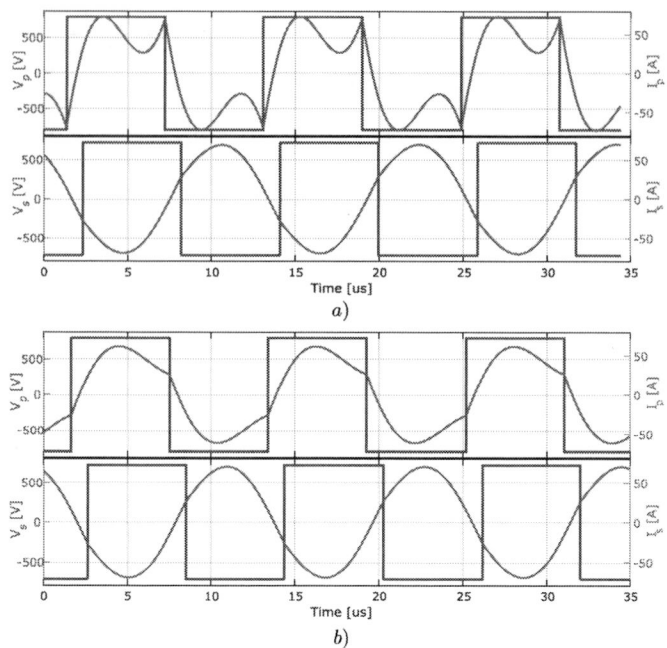

Figure 6: Simulated primary and secondary voltage and current waveforms a) $L_{addp} = 5.7 \mu H$ b) $L_{addp} = 30 \mu H$

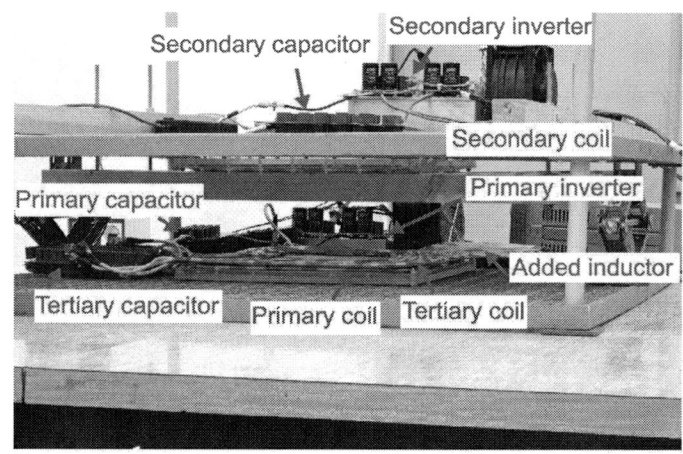

Figure 7: Experimental setup

its functionality with perfect coil alignment and the cases with an important misalignment.

The test scenarios included a relative coupling coefficient of 0.6, achieved with a 150 mm air gap and 50 mm of misalignment in both the x and y axes; a coefficient of 1, with a 130 mm air gap and no misalignment; and a coefficient of 1.3, with a 110 mm air gap and no misalignment. Each position was tested in constant power (CP), constant current (CC) and constant voltage modes (CV). The system was assessed with three secondary voltages that represent different stages during the charging process: 630, 710 and 830V. Table IV summarizes results in CP and CC modes. In constant power mode (CP), tests were carried out at the rated power of 30 kW. In case that this nominal power is not achievable, power is chosen so

Parameter	$\xi_{sim} = 1.3$	$\xi_{sim} = 1$	$\xi_{sim} = 0.6$
Air gap	110 mm	130 mm	150 mm
Clearance	145 mm	165 mm	185 mm
Misalignment (x,y)	(0 mm,0 mm)	(0 mm,0 mm)	(50 mm,50 mm)
ξ_{meas}	1.27	0.99	0.62
$L_p\ [\mu H]$	11.49	11.49	11.49
$L_s\ [\mu H]$	105.43	105.67	105.81
$L_t\ [\mu H]$	14.89	14.89	14.89
k_{ps}	0.258	0.207	0.126
k_{pt}	0.84	0.84	0.84
k_{st}	0.286	0.226	0.140

Table III: Measured parameters of the system in the three tested conditions

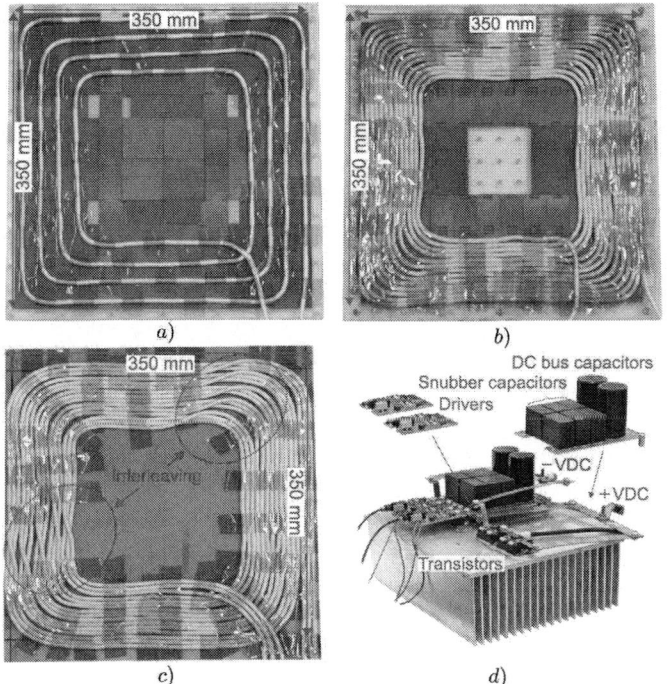

Figure 8: a) Primary coil b) Secondary coil c) Tertiary coil d) Inverter

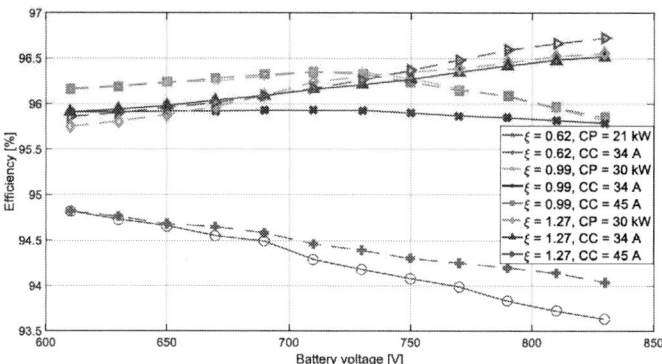

Figure 9: a) Primary coil b) Secondary coil c) Tertiary coil d) Inverter

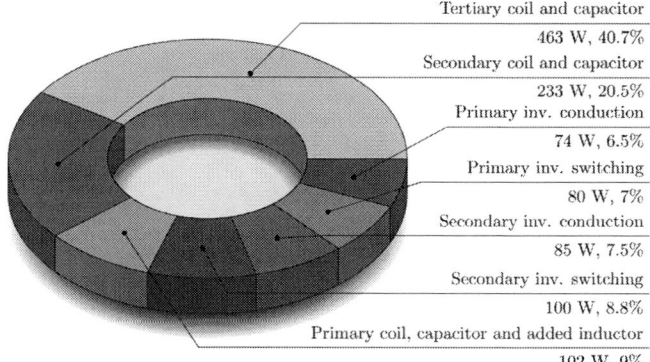

Figure 10: a) Primary coil b) Secondary coil c) Tertiary coil d) Inverter

that it is possible to be achieved in all battery voltage ranges as it is the case with maximum clearance and misalignment (ξ = 0.6). The system was tested in constant current (CC) mode at two current levels, 34 A and 45 A.

As it can be seen in the results of Table IV, CC mode allows for higher power transfer at higher battery voltage levels. In contrast, CP mode (constant power) limits power output to the level achieved at the lowest battery voltage, with the maximum current that can be achieved under these conditions being the limiting factor for the transferred power.

Figure 9 shows the system efficiency for all tested conditions in CP and CC modes and a losses breakdown of the system working in nominal conditions can be seen in Fig. 10. Losses in the tertiary winding are dominant because of the high currents that circulate through it, making it the most critical part of the system thermally, although temperatures reached are still far below the limit of the isolation of the

used Litz wire. Nevertheless, these losses are still significantly lower than in the case when the magnetizing current circulates through the primary bridge. Besides, tertiary winding is a convenient place to concentrate losses because of its large size and dissipation area and its good placement, being the element that is placed on top of the ground assembly.

Waveforms for the system in nominal coupling conditions in CP and CC modes can be seen in Fig.11. According to equation 11, transform ratio changes depending on coupling conditions. To keep reactive power, given by equation 13, as low as possible, first harmonic of U_p should be equal to first harmonic of U'_s. To achieve that, duty cycle of the highest between U_p and U'_s is reduced to make it equal to the lowest of them.

Following the completion of the CC and CP charging tests, the efficiency of the system was evaluated under constant voltage (CV) charging conditions. These conditions are relevant in because CV mode is present in the later part of both the CC/CV and CP/CV charging methods.

In these tests, the system was assessed using the same relative coupling coefficients as in the previous tests to ensure consistency. Throughout the tests, the battery voltage was maintained at 830 V, while the battery current varied between the maximum current applicable for each specific case and

ξ	Constant Power Charging (CP)			Constant Current Charging (CC)		
	Battery Current	System Efficiency Range	Charging Power	Battery Current	System Efficiency Range	Charging Power
1.27	36.14 A - 49.12 A	95.75% - 96.54%	30 kW	34 A	95.91% - 96.52%	20.7 kW - 28.2 kW
				45 A	95.86% - 96.72%	7.5 kW - 37.4 kW
0.99	36.14 A - 49.12 A	95.82% - 96.35%	30 kW	34 A	95.72% - 95.93%	20.7 kW - 28.2 kW
				45 A	95.85% - 96.35%	27.5 kW - 37.4 kW
0.62	25.30 A - 34.43 A	93.63% - 94.82%	21 kW	34 A	94.04% - 94.82%	20.7 kW - 28.2 kW

Table IV: Overview of results for the system in CP and CC modes

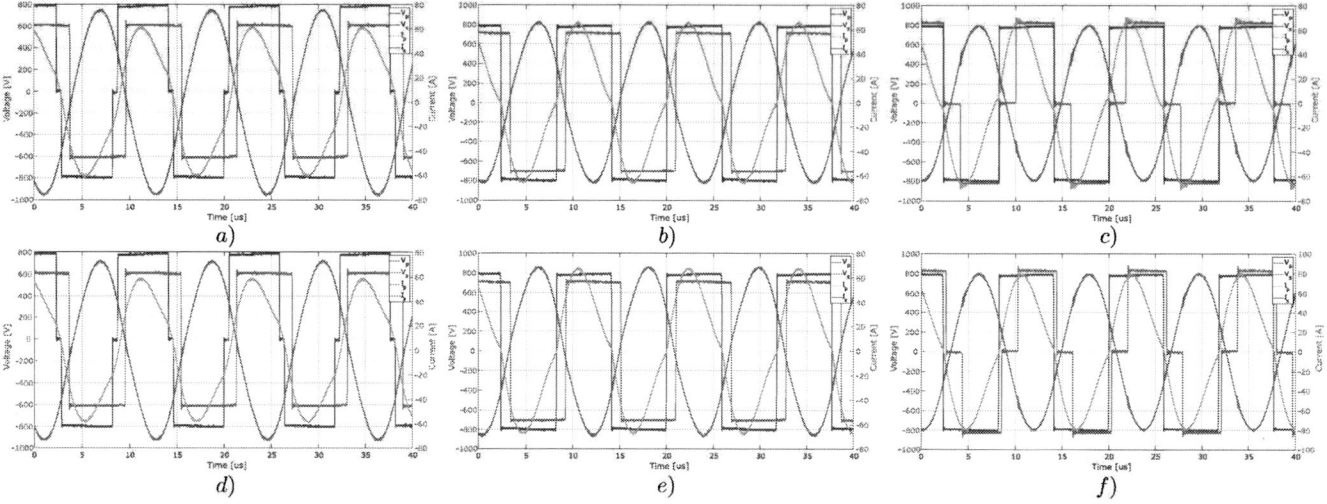

Figure 11: Currents and voltages of the primary and secondary inverters for $\xi = 0.99$, a) $V_{batt} = 610$ V , $P = 30$ kW b) $V_{batt} = 710$ V, $P = 30$ kW c) $V_{batt} = 830$ V, $P = 30$ kW d) $V_{batt} = 610$ V, $I = 45$ A e) $V_{batt} = 710$ V, $I = 45$ A f) $V_{batt} = 830$ V, $I = 45$ A

ξ	Constant Voltage Charging (CV)		
	Battery Current	System Efficiency Range	Charging Power
1.27	4.5 A - 45 A	92.22% - 96.72%	3.7 kW - 37.5 kW
0.99	4.5 A - 45 A	91.95% - 95.85%	3.7 kW - 37.5 kW
0.62	4.5 A - 34 A	91.29% - 94.04%	3.7 kW - 28.2 kW

Table V: Overview of the CV charging results

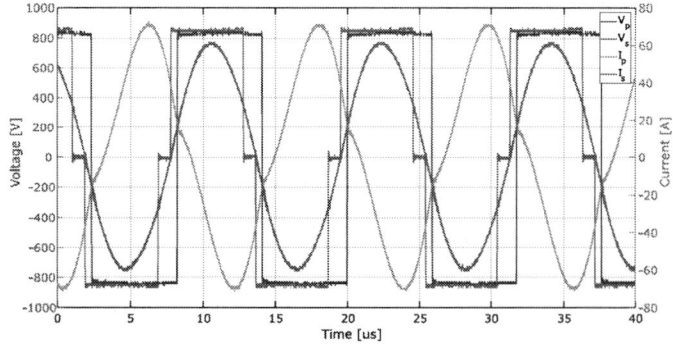

Figure 12: System operating in vehicle to grid mode

10% of the maximum current of 45 A. This 10% threshold is defined as the point below which the battery is considered fully charged, indicating the completion of the charging process. Table V summarizes the results of the tests in CV mode

Finally, vehicle to grid mode was checked for $\xi = 1$ and for battery voltage equal to 830 V while the power of 30 kW was being transferred to the primary side. Waveforms obtained during this test are given in Fig.12. Measured efficiency of the system was 95.76%.

V. CONCLUSIONS

This work presents a comprehensive equivalent circuit and tuning method for a three-coil IPT system designed for high power, high-efficiency bidirectional EV charging. The proposed S–S–S compensated three-coil IPT charger meets current SAE standards for coil sizes and system efficiency [1], with a relative coupling coefficient swing due to clearance and misalignment lower than predicted by the SAE standard. The tuning method enables seamless analysis and control of the system, similar to a bidirectional DAB converter, by removing magnetizing inductance, which boosts efficiency by preventing reactive current from passing through semiconductor elements. Experiments demonstrated that the system can be easily controlled in CC, CV, and CP modes. Testing showed efficiencies of over 90% across all battery voltages, clearances, misalignments, and power levels from 3.7 kW to 37.5 kW.

REFERENCES

[1] Hybrid - EV Committee, "Wireless Power Transfer for Light-Duty Plug-in/Electric Vehicles and Alignment Methodology."

[2] W. Zhang and C. C. Mi, "Compensation Topologies of High-Power Wireless Power Transfer Systems," *IEEE Transactions on Vehicular Technology*, vol. 65, pp. 4768–4778, June 2016.

[3] P. Darvish, S. Mekhilef, and H. A. B. Illias, "A Novel S–S– *LCLCC* Compensation for Three-Coil WPT to Improve Misalignment and Energy Efficiency Stiffness of Wireless Charging System," *IEEE Transactions on Power Electronics*, vol. 36, pp. 1341–1355, Feb. 2021.

[4] Y. Li, Q. Xu, T. Lin, J. Hu, Z. He, and R. Mai, "Analysis and Design of Load-Independent Output Current or Output Voltage of a Three-Coil Wireless Power Transfer System," *IEEE Transactions on Transportation Electrification*, vol. 4, pp. 364–375, June 2018.

[5] E. S. Lee, D. Kim, and S. Y. Jeong, "Triangular DQ Tx Coils of Wireless EV Chargers for Large Misalignment Tolerances," *IEEE Transactions on Vehicular Technology*, vol. 72, pp. 14179–14188, Nov. 2023.

[6] A. Bilal, S. Kim, F. Lin, and G. Covic, "Analysis of IPT Intermediate Coupler System for Vehicle Charging Over Large Air Gaps," *IEEE Journal of Emerging and Selected Topics in Industrial Electronics*, vol. 3, pp. 1149–1158, Oct. 2022.

[7] S. Seshadri, M. Kavitha, and P. B. Bobba, "Effect of coil structures on performance of a four-coil WPT powered medical implantable devices," in *2018 International Conference on Power, Instrumentation, Control and Computing (PICC)*, (Thrissur), pp. 1–6, IEEE, Jan. 2018.

[8] Runhong Huang and Bo Zhang, "Frequency, Impedance Characteristics and HF Converters of Two-Coil and Four-Coil Wireless Power Transfer," *IEEE Journal of Emerging and Selected Topics in Power Electronics*, vol. 3, pp. 177–183, Mar. 2015.

[9] M. Kiani, Uei-Ming Jow, and M. Ghovanloo, "Design and Optimization of a 3-Coil Inductive Link for Efficient Wireless Power Transmission," *IEEE Transactions on Biomedical Circuits and Systems*, vol. 5, pp. 579–591, Dec. 2011.

[10] F. Yang, Y. Liu, Y. Han, J. Chen, S. Cheng, Z. Tian, and H. Zhao, "Misalignment Tolerance Improvement for Loosely Coupled Transformer of IPT Systems via an Intermediate Coil With Detuned Compensation," *IEEE Access*, vol. 11, pp. 90181–90189, 2023.

[11] N. Mirkovic, A. Delgado, and M. Vasic, "Increasing Power Transfer Capability of Wireless Battery Charger Under Misalignment Conditions," (DE), VDE VERLAG GMBH, June 2023.

[12] N. R. Mirković, D. M. Stojić, A. Delgado, P. Alou, and M. Vasić, "Novel Three-Phase to Single-Phase Matrix Converter Modulation Strategy for Bidirectional Inductive Power Transfer," *IEEE Transactions on Power Electronics*, vol. 38, pp. 14830–14846, Dec. 2023.

[13] H. Wang, K. W. E. Cheng, X. Li, and J. Hu, "A Special Magnetic Coupler Structure for Three-Coil Wireless Power Transfer: Analysis, Design, and Experimental Verification," *IEEE Transactions on Magnetics*, vol. 57, pp. 1–8, Nov. 2021.

[14] R. W. Erickson and D. Maksimović, "Resonant Conversion," in *Fundamentals of Power Electronics*, pp. 933–993, Cham: Springer International Publishing, 2020.

Mutual Inductance and Load Identification Method based on the Voltage Transients of WPT Systems

Xiaosheng Wang
Department of Electrical Engineering
City University of Hong Kong
Hong Kong, China
xiaoswang9-c@my.cityu.edu.hk

Yibo Wang
Department of Electrical Engineering
City University of Hong Kong
Hong Kong, China
yibo.wang@my.cityu.edu.hk

C. Q. Jiang
Department of Electrical Engineering
City University of Hong Kong
Hong Kong, China
chjiang@cityu.edu.hk

Liping Mo
Department of Electrical Engineering
City University of Hong Kong
Hong Kong, China
lipingmo@cityu.edu.hk

Abstract—**The identification of the mutual inductance and load voltage is important to the control and optimization of the wireless power transfer (WPT) systems. The paper proposes a novel identification method based on the voltage transients. Connecting an equivalent sensor inductor in the primary resonant tank, the voltage transient values caused by the rising and falling edges of the inverter output voltage and the rectifier input voltage are relevant to the mutual inductance and load voltage. By detecting the voltage transient amplitudes, the mutual inductance and load voltage can be identified directly.**

Keywords—*sensor inductor, wireless power transfer, identification, mutual inductance, voltage transient*

I. INTRODUCTION

The mutual inductance and load voltage are important for optimizing system efficiency, controlling output power, designing system control parameters, and collaborative control without communication, and they will vary with coil misalignment and the state of charge. Consequently, the identification of mutual inductance and load voltage is very important.

One common approach is to short the active receiver and perform a frequency sweep [1], [2], [3]. By tracking the system's series resonant frequencies where the input impedance angle is zero, the coupling coefficient can be calculated through its relationship with the series resonant frequencies. Nonetheless, this method requires frequency scanning, lacks online operation capability, and is only suitable for WPT systems with active rectifiers.

Building and solving system impedance equations to identify mutual inductance based on the voltage and current is also the mainstream method to identify mutual inductance and load. In [4], the mutual inductance and load are identified by utilizing the system's Kirchhoff's Voltage Law (KVL) equations and the characteristic that the imaginary part of the resistive load is zero. However, this approach is prone to large errors when operating at the resonant frequency, making it unsuitable for online operation. In [5], the identification is based on the input impedance equation with a known load resistance, but this method is challenging to apply since the load cannot be predetermined without communication. In [6], the parameters are estimated by solving nine equations involving the fundamental and third harmonics, but this approach requires highly accurate initial

values of the resonant capacitor. Similarly, in [7], the multi-harmonics caused by pulse density modulation are utilized, but the computation is complex. Furthermore, in [8] and [9], a time-domain analysis is used to match the input current waveform using the least squares approach, allowing for the identification of mutual inductance, albeit with complex computations. In [10], a parameter method based on the second harmonic is proposed, however, it cannot be used in the symmetric modulation. Adding the additional circuit is also the main method to estimate the parameters [11], [12], [13].

Although there are many identification methods, they either require frequency sweeps and cannot achieve online identification or require a lot of calculations or additional complex circuits. To solve these problems, a very competitive method is presented in the paper. Connecting an equivalent sensor inductor in the primary resonant tank, the voltage transient value caused by the rising and falling edges of the inverter output voltage and the rectifier input voltage is used to identify mutual inductance and load voltage.

II. THE PARAMETER IDENTIFICATION PRINCIPLE AND THE VOLTAGE TRANSIENT ENHANCEMENT

A. The Basic Identification Principle based on the Voltage Transients

Fig. 1 illustrates the typical configuration of an S-S compensated WPT system. The power switches in the primary inverter and secondary rectifier are designated as S_x and D_x, respectively. The DC voltage supplies for the primary and secondary circuits are denoted as Vin and Vo, respectively. The output voltage of the primary inverter and the input voltage of the secondary rectifier are represented by u_P and u_S, respectively. The components L_1-L_2, C_1-C_2, and R_1-R_2 correspond to the resonant inductance, capacitance, and AC equivalent resistance. The mutual inductance is represented by M, and the coupling coefficient k is defined as:

$$k = \frac{M}{\sqrt{L_1 L_2}} \qquad (1)$$

The configuration depicted in Fig. 1 can be converted into a T-shaped equivalent circuit, as shown in Fig. 2(a), where the inverter and rectifier are represented as simplified square wave voltage sources. The equivalent inductances L_{1t} and L_{2t}

979-8-3315-1612-3/25 $31.00 © 2025 IEEE

are expressed as follows:

$$\begin{cases} L_{1t} = L_1 - M \\ L_{2t} = L_2 - M \end{cases} \quad (2)$$

In the T-shaped equivalent circuit, considering the extremely short rise and fall times of u_P and u_S, which are on the order of tens of nanoseconds, it is valid to approximate u_P and u_S as step functions when examining the transient voltages across inductors L_{1t} and L_{2t}, as shown in Fig. 2(b). Given the brief duration of these events, the voltage fluctuations in capacitors C_1 and C_2 can be safely neglected.

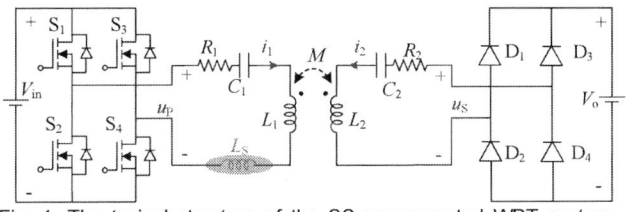

Fig. 1. The typical structure of the SS-compensated WPT system with uncontrolled rectifier and battery load.

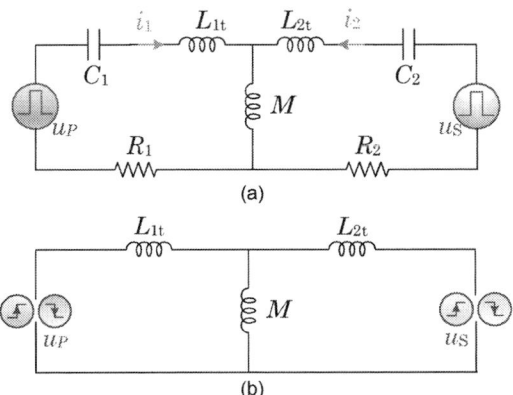

Fig. 2. T-shaped equivalent circuit of the WPT system. (a) T-shaped equivalent circuit. (b) Only consider voltage transient time.

In Fig. 2, when u_P rises from $-V_{in}$ to V_{in}, the voltage transient value on the inductor L_{1t} can be expressed as:

$$U_{L_{1t}}^P = 2V_{in} \frac{L_{1t}(M + L_{2t})}{L_{1t}(M + L_{2t}) + ML_{2t}} \quad (3)$$

When u_S rises from $-V_o$ to V_o, the voltage transient value on the inductor L_{1t} is:

$$U_{L_{1t}}^S = 2(V_o + 2V_d) \frac{ML_{1t}}{L_{2t}(M + L_{1t}) + ML_{1t}} \quad (4)$$

Building on the previous analysis, if the transient voltage across L_{1t} can be determined, the mutual inductance M can be identified using equation (3), assuming other parameters are known. Once M is obtained, the battery voltage can be further identified using equation (4). However, in practical applications, only the voltage across L_1 can be directly measured, as L_{1t} is coupled with M. To leverage the voltage transient phenomenon for parameter identification, a small sensor inductor L_S is introduced, connected in series with coil L_1, as depicted in Fig. 1. The transient voltage across the sensor inductor can be represented as:

$$U_{LS}^P = 2V_{in} \frac{L_{1t}'(M + L_{2t})}{L_{1t}'(M + L_{2t}) + ML_{2t}} \frac{L_S}{L_{1t}'} \quad (5)$$

$$U_{LS}^S = 2(V_o + 2V_d) \frac{ML_{1t}'}{L_{2t}(M + L_{1t}') + ML_{1t}'} \frac{L_S}{L_{1t}'} \quad (6)$$

where $L_{1t}' = L_{1t} + L_S$, V_d is the forward voltage drop of the diode at the receiver side.

Fig. 3 illustrates the voltage and current waveforms under various coupling coefficients when the sensor inductor L_s is employed. The voltage transients U_{LS}^P, U_{LS}^S are marked at time instants t_1 and t_2, respectively. When the coupling coefficient k is 0.2, the voltage transient U_{LS}^S is relatively small compared to the peak-to-peak value of u_{Ls}, making it challenging to accurately detect U_{LS}^S directly. In contrast, as shown in Fig. 3(b), the voltage transient value becomes more pronounced when the coupling coefficient k increases to 0.4, making it easier to detect.

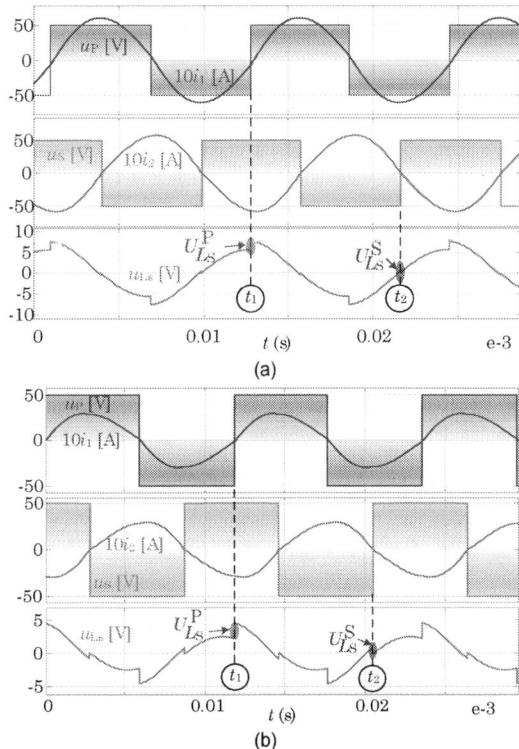

Fig. 3. The voltage and current waveforms of the WPT system ($V_{in}=V_o=50$ V, $L_1=L_2=100$ μH, $L_S=2$ μH, $f_s=f_r=85$ kHz). (a) $k=0.2$. (b) $k=0.4$.

As illustrated in Fig. 4, which is derived from equations (5) and (6), the relationship between U_{LS}^P, U_{LS}^S, and k reveals that the voltage transient value becomes more pronounced as the coupling coefficient k increases. Additionally, it is observed that the voltage transient value induced by changes in u_P is larger than that caused by changes in u_S. Notably, when k is smaller than 0.4, the variation in k has a minimal impact on U_{LS}^S, indicating that the identification results of mutual inductance are highly sensitive to the detection accuracy of U_{LS}^P.

979-8-3315-1612-3/25 $31.00 © 2025 IEEE

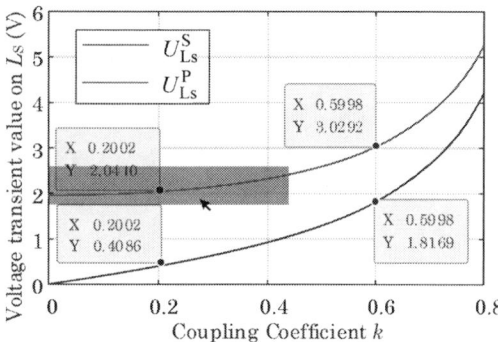

Fig. 4. The relationship between the voltage transient on the sensor inductor and the coupling coefficient k ($V_{in}=V_o=50$ V, $L_1=L_2=100$ μH, $L_S=2$ μH).

However, in strong coupling scenarios, such as those found in AGV and AUV applications, the voltage transient values on L_s exhibit satisfactory sensitivity to k. As a result, the value of U_{Ls}^P can be directly utilized to identify M using equation (5), and the value of U_{Ls}^P can be used to identify V_o further using equation (6). A significant advantage of this method is that the identification process can be completed regardless of the WPT system's working state, whether in a dynamic process or steady state, since the voltage transient values U_{Ls}^P and U_{Ls}^P are unaffected by the system's operating conditions. Furthermore, because capacitors C_1 and C_2 are not present in the simplified model shown in Fig. 2(b), the impact of system detuning on the identification results can be ignored. These two advantages are the most significant benefits of the proposed identification method.

B. Voltage Transient Enhancement Method

As illustrated in Fig. 3, the voltage transient values caused by the rising and falling of u_S are challenging to detect, particularly in weak coupling cases. The primary reason for this difficulty is that the large peak-to-peak value of u_{Ls} overshadows the voltage transient, resulting in U_{Ls}^P and U_{Ls}^S comprising a small proportion of the u_{Ls} waveform. To improve the detection of these transients, a notch filter can be employed to eliminate the fundamental component of u_{Ls}, thereby enhancing the visibility of the voltage transients. By removing the dominant frequency component, the notch filter can help to accentuate the transient features, making it easier to detect and analyze the voltage transients.

It is recommended to set the notch frequency to approximately the operating frequency of the WPT system. When the fundamental component of u_{Ls} is attenuated to below 5% ($g_{min}=0.05$) of its original value through the notch filter, the transient enhancement effect is deemed acceptable. As shown in Fig. 5, the voltage waveforms of u_{Ls} and u_{Lse} (u_{Lse} being the voltage after u_{Ls} passes through the notch filter) demonstrate a significant improvement in the voltage transient feature after crossing the notch filter. Since the notch filter does not affect the voltage transient values, the proportion of U_{Ls}^S in the peak-to-peak value of uLs increases from 2.6% to 11.8% of the peak-to-peak value of u_{Lse}, making it easier to detect U_{Ls}^P, U_{Ls}^S. This enhancement in transient detection facilitates the identification of mutual inductance and the battery voltage.

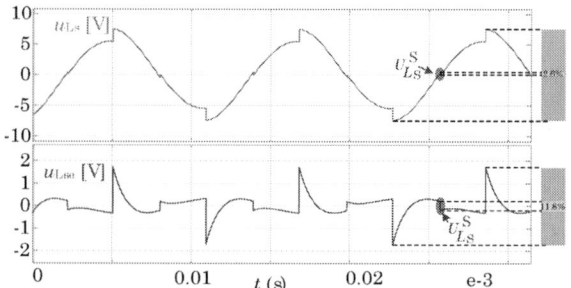

Fig. 5. The voltage transient enhancement results crossing a notch filter ($\omega=85*2\pi$ kHz, $f_s=85$ kHz, $k=0.2$, $g_{min}=0.01$).

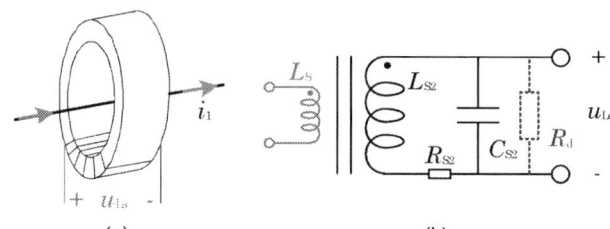

Fig. 6. The structure of the equivalent sensor inductor. (a) Physical structure. (b) Equivalent circuit of the sensor inductor.

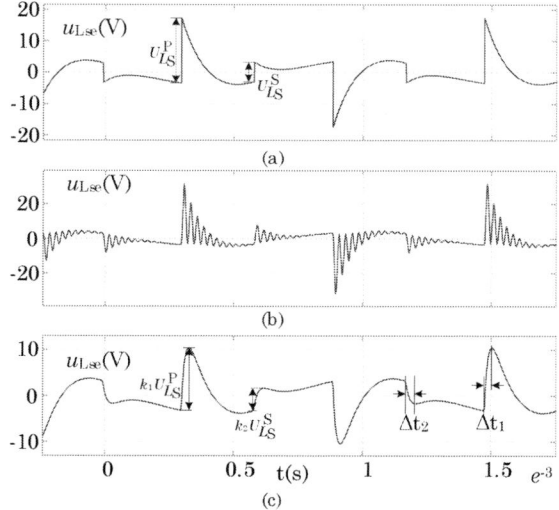

Fig. 7. The waveforms of u_{Lse} before and after damping ($L_S=1$ μH, $L_{S2}=20$ μH, $R_{S2}=50$ Ω, $C_{S2}=100$ pF, $k=0.3$) (a) Idea waveform without parasitic parameters. (b) Before damping with parasitic parameters. (c) After damping with $R_d=100$ Ω.

III. THE PARAMETER IDENTIFICATION SCHEME

By connecting the sensor inductor L_s in series with the resonant tank on the primary side and detecting the voltage transients on L_s, parameter identification can be achieved using equations (5), (6). However, to ensure isolation from the main circuit, the detection circuit should employ isolation chips. As shown in Fig. 6, a small magnetic ring with a multi-turn coil wound around it can be used as a voltage transformer, as depicted in Fig. 6(b). The primary side of the transformer has a single turn of the coil, which is connected in series between the transmitting coil and the inverter. By utilizing a nickel-zinc magnetic ring with low magnetic permeability, the equivalent series inductance L_s can be kept small, and the additional power loss introduced by the magnetic ring can be considered negligible. Assuming the coil on the magnetic ring has N turns, the voltage across the coil is Nu_{Ls}. For simplicity, the notation will omit N, and it will be uniformly referred to as u_{Ls}. However, the multi-turn winding on the magnetic ring introduces parasitic capacitance C_{S2}, and the parasitic resistance R_{S2} is very small, forming a resonant loop with the self-inductance of the winding. As a result, when u_{P} and u_{S} rise or fall due to switching operations, u_{Ls} will exhibit voltage oscillations, causing deviations between the actual voltage transient detection and theoretical values. Furthermore, during the switching process, especially with hard switching, u_{P} will also fluctuate, interfering with the amplitude detection of the voltage transients $U_{\mathrm{Ls}}^{\mathrm{P}}$ and $U_{\mathrm{Ls}}^{\mathrm{S}}$. To mitigate this issue, a damping resistor R_{d} is necessary to limit the oscillation of u_{Ls}.

Fig. 7 shows the u_{Ls} waveform before and after damping. After applying damping, the voltage oscillation in Fig. 7(b) is effectively suppressed, but the voltage transient phenomenon becomes smoother. Therefore, it is necessary to analyze the relationship between the actual $U_{\mathrm{Ls}}^{\mathrm{P}}$, $U_{\mathrm{Ls}}^{\mathrm{S}}$, and the calculated values in equations (5) and (6). Since the time-domain expression of u_{Ls} after passing through the notch filter is difficult to determine, only a qualitative explanation is provided. The waveforms of u_{Lse} under different mutual inductances exhibit similar characteristics at the moment of voltage transient. Under the same damping conditions, it can be approximately considered that the voltage transient value after damping satisfies $k_1 U_{\mathrm{Ls}}^{\mathrm{P}}$ and $k_2 U_{\mathrm{Ls}}^{\mathrm{S}}$, where k_1 and k_2 are approximately constant. These two coefficients can be calibrated during the product design process by comparing the theoretical value of the voltage transient at a certain M with the actual measured value to obtain k_1 and k_2. Then, the corresponding theoretical values $U_{\mathrm{Ls}}^{\mathrm{P}}$ and $U_{\mathrm{Ls}}^{\mathrm{S}}$ can be calculated based on the measured value at other coupling conditions. To implement this method in hardware and software, a sliding time window should be set to detect $U_{\mathrm{Ls}}^{\mathrm{P}}$ and $U_{\mathrm{Ls}}^{\mathrm{S}}$. As shown in Fig. 7(c), the maximum value detected within the sliding window will be $k_1 U_{\mathrm{Ls}}^{\mathrm{P}}$ around the moment of u_{P} rises and falls. The lengths Δt_1 and Δt_2 of this time window can also be calibrated during the product design stage.

The flow chart in Fig. 8 outlines the process of this identification method. Initially, calibration of the coefficients k_1 and k_2 is necessary. Once calibrated, the measured values of $k_1 U_{\mathrm{Ls}}^{\mathrm{P}}$ and $k_2 U_{\mathrm{Ls}}^{\mathrm{S}}$ can be used to calculate the theoretical values of $U_{\mathrm{Ls}}^{\mathrm{P}}$, $U_{\mathrm{Ls}}^{\mathrm{S}}$. By applying equations (5) and (6), the mutual inductance M and load voltage V_{o} can be identified.

IV. EXPERIMENTAL VERIFICATION

To validate the proposed parameter identification method, several experiments were conducted. The experimental parameters are listed in Table I, where the values of L_1, L_2, C_1, and C_2 were measured using an LCR meter. The experimental prototype is shown in Fig. 9. A nickel-zinc ferrite magnetic ring with a magnetic permeability of 100 was used, and the wire wrapped around the ring had 22 turns. After inserting the magnetic core, the inductance of L_{s} was measured to be approximately 70 nH. It is worth noting that if the sensor inductor is directly connected in series in

the circuit without the magnetic ring, a value ranging from nH to μH is also feasible, as long as it does not significantly increase the loss and affect the system operation. This alternative approach does not affect the analysis and conclusions drawn from the experiment.

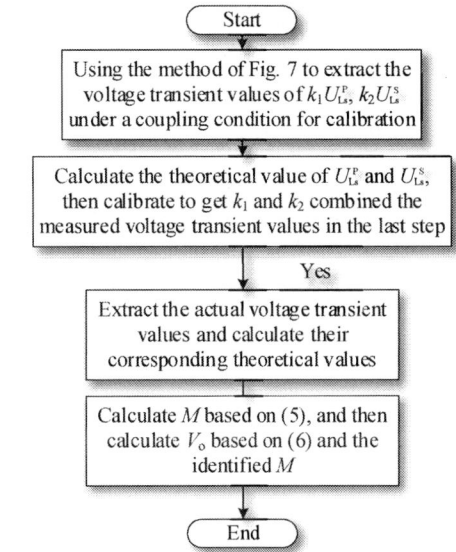

Fig. 8. The flowchart of the identification method.

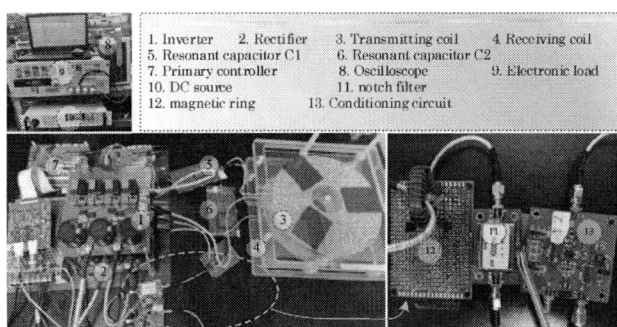

Fig. 9. Experiment prototype for the identification method.

TABLE I
Simulation and Experimental Parameters of the WPT System

Parameter	Symbol	Value
Input voltage of the inverter	V_{in}	30 V
Output voltage of the rectifier	V_{o}	30 V
Resonant capacitor of the primary-side coil	C_1	49.5 nF
Resonant capacitor of the secondary-side coil	C_2	49.5 nF
Coupling coefficient	k	0.15-0.68
Equivalent sensor inductor	L_{s}	70 nH

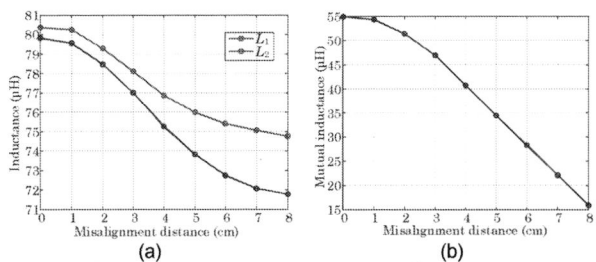

Fig. 10. The relationship between coil self-inductance, mutual inductance, and misalignment distance when the gap distance is 2.5 cm. (a) L_1 and L_2. (b) M.

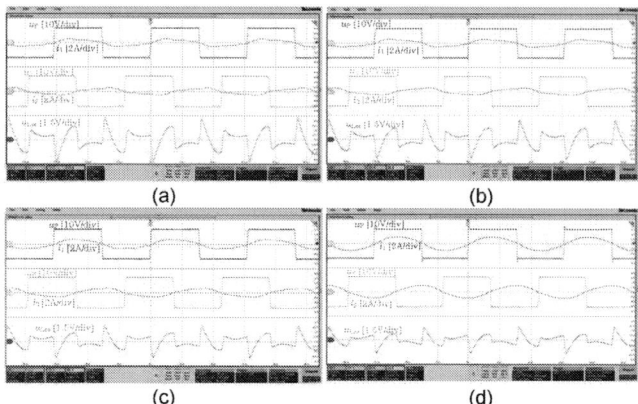

(a)　　　　　　　　(b)

(c)　　　　　　　　(d)

Fig. 11. Voltage and current waveforms for different misalignment distances with a gap distance of 2.5 cm. (a) Without misalignment. (b) Misalignment distance 2 cm. (c) Misalignment distance 4 cm. (d) Misalignment distance 6 cm.

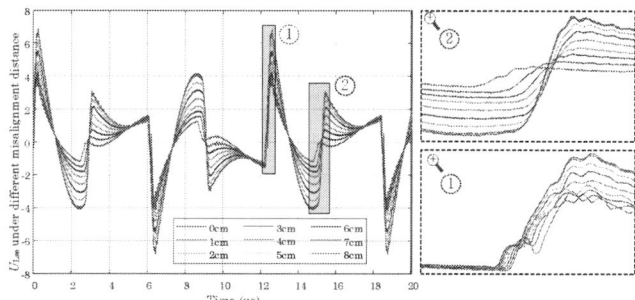

Fig. 12. The waveforms of u_{Lse} for different misalignment distances (gap distance of 2.5 cm).

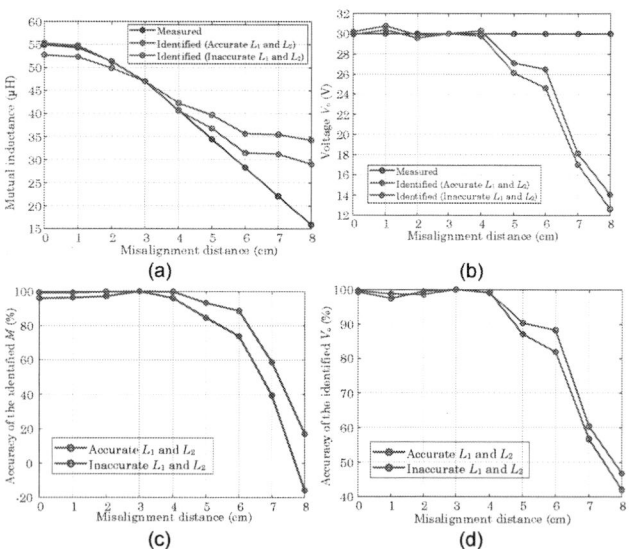

(a)　　　　　　　　(b)

(c)　　　　　　　　(d)

Fig. 13. Mutual inductance and load voltage identification results. (a) Measured M and identified M. (b) Measured V_o and identified V_o. (c) The accuracy of the identified M. (d) The accuracy of the identified V_o.

As discussed in Section II and Fig. 4, strong coupling cases exhibit sufficient sensitivity to enable direct identification of M and V_o using $U_{\mathrm{LS}}^{\mathrm{P}}$ and $U_{\mathrm{LS}}^{\mathrm{S}}$. In this instance, the calibration of $U_{\mathrm{LS}}^{\mathrm{P}}$ and $U_{\mathrm{LS}}^{\mathrm{S}}$ is based on a misalignment distance of 3 cm. Fig. 10 reveals that L_1 and L_2 undergo significant changes due to coil misalignment.

To accommodate the decrease in L_1 and L_2, the switching

frequency is adjusted to 81 kHz to meet the requirements of the actual application. Fig. 11 shows the waveforms when the gap distance is 2.5 cm. As the misalignment distance increases, the values of $k_1 U_{\mathrm{LS}}^{\mathrm{P}}$ and $k_2 U_{\mathrm{LS}}^{\mathrm{S}}$ decrease. To visualize this trend more clearly, the waveforms of u_{Lse} for different misalignment distances are collected and presented in Fig. 12. As M varies from 54.9 μH to 15.9 μH, the amplitudes of $k_1 U_{\mathrm{LS}}^{\mathrm{P}}$ and $k_2 U_{\mathrm{LS}}^{\mathrm{S}}$ undergo significant changes. By choosing the case with a misalignment distance of 3 cm as the calibration reference, k_1 and k_2 can be calculated, allowing for the determination of $U_{\mathrm{LS}}^{\mathrm{P}}$ and $U_{\mathrm{LS}}^{\mathrm{S}}$ for all misalignment distances. Once $U_{\mathrm{LS}}^{\mathrm{P}}$ is determined, M can be directly identified using equation (5). Substituting the identified M and $U_{\mathrm{LS}}^{\mathrm{S}}$ into equation (6) enables the further identification of V_o. The experimental results are presented in Fig. 13. As shown in Fig. 13(a) and Fig. 13(c), when accurate values of L_1 and L_2 are considered, the accuracy exceeds 99% for misalignment distances less than 4 cm. Even when using inaccurate values of L_1 and L_2, the accuracy remains above 95% for misalignment distances less than 4 cm. However, as the coupling coefficient decreases, particularly for misalignment distances over 6 cm, the identification accuracy is affected, consistent with the analysis in Fig. 4. The identification performance of V_o is also satisfactory for strong coupling cases.

V. CONCLUSION

This paper presents a novel mutual inductance and load voltage identification method based on voltage transient detection. The parameters can be identified by directly using the voltage transient values. Since the instantaneous voltage transient will not be affected by the capacitor, the method will not be affected by the system detuning. Compared with the traditional method, the method can be widely used in WPT systems due to its simplicity and practicability. In the future, this method will be improved to solve the problem of inaccurate identification under weak coupling.

ACKNOWLEDGEMENT

This work was supported in part by the Research Grants Council, Hong Kong SAR, China, under CRF-YCRG C1002-23Y, and in part by a grant from the Science Technology and Innovation Committee of Shenzhen Municipality, China, under project SGDX2021082310400 3034. Also, the authors are with Joint Laboratory of Energy Saving and Intelligent Maintenance for Modern Transportations.

REFERENCES

[1] J. Zeng, S. Chen, Y. Yang and S. Y. R. Hui, "A Primary-Side Method for Ultrafast Determination of Mutual Coupling Coefficient in Milliseconds for Wireless Power Transfer Systems," IEEE Trans. Power Electron, vol. 37, no. 12, pp. 15706-15716, Dec. 2022.

[2] Y. Yang, S. C. Tan and S. Y. R. Hui, "Fast Hardware Approach to Determining Mutual Coupling of Series–Series-Compensated Wireless Power Transfer Systems with Active Rectifiers," IEEE Trans. Power Electron, vol. 35, no. 10, pp. 11026-11038, Oct. 2020.

[3] Y. Yang, S. C. Tan and S. Y. R. Hui, "Front-end Parameter Monitoring Method based on Two-Layer Adaptive Differential Evolution for SS-Compensated Wireless Power Transfer Systems," IEEE Trans. Ind. Informat., vol. 15, no. 11, pp. 6101–6113, Nov. 2019.

[4] J. Yin, D. Lin, T. Parisini and S. Y. Hui, "Front-End Monitoring of the Mutual Inductance and Load Resistance in a Series–Series Compensated Wireless Power Transfer System," *IEEE Trans. Power Electron*, vol. 31, no. 10, pp. 7339-7352, Oct. 2016.

[5] V. Jiwariyavej, T. Imura and Y. Hori, "Coupling Coefficients Estimation of Wireless Power Transfer System via Magnetic Resonance Coupling Using Information From Either Side of the System," *IEEE J. Emerg. Sel. Topics Power Electron*, vol. 3, no. 1, pp. 191-200, March 2015.

[6] J. Liu, G. Wang, G. Xu, J. Peng and H. Jiang, "A Parameter Identification Approach With Primary-Side Measurement for DC–DC Wireless-Power-Transfer Converters With Different Resonant Tank Topologies," *IEEE Trans. Transp. Electrification*, vol. 7, no. 3, pp. 1219-1235, Sept. 2021.

[7] R. Dai, R. Mai and W. Zhou, "A Pulse Density Modulation Based Receiver Reactance Identification Method for Wireless Power Transfer System," *IEEE Trans. Power Electron*, vol. 37, no. 9, pp. 11394-11405, Sept. 2022.

[8] J. P. -W. Chow, H. S. -H. Chung and C. -S. Cheng, "Use of Transmitter-Side Electrical Information to Estimate Mutual Inductance and Regulate Receiver-Side Power in Wireless Inductive Link," *IEEE Trans. Power Electron*, vol. 31, no. 9, pp. 6079-6091, Sept. 2016.

[9] L. Mo, X. Wang, Y. Wang, B. Zhang, C. Jiang. Mutual Inductance Estimation of SS-IPT System through Time-Domain Modeling and Nonlinear Least Squares. *Energies* 2024, 17, 3307.

[10] X. Wang, L. Mo, C. Q. Jiang, J. Zhou, Y. Wang, and H. Zhao, "Utilization of Virtual Short-Circuit in Wireless Power Transfer System: Mutual Inductance Identification via Second Harmonic," IEEE Transactions on Industrial Electronics, early access, DOI:10.1109/TIE.2024. 3468619, 2024.

[11] S. Liu, Y. Feng, W. Weng, J. Chen, J. Wu and X. He, "Contactless Measurement of Current and Mutual Inductance in Wireless Power Transfer System Based on Sandwich Structure," *IEEE J. Emerg. Sel. Topics Power Electron*, vol. 10, no. 5, pp. 6345-6357, Oct. 2022.

[12] Y. -G. Su, H. -Y. Zhang, Z. -H. Wang, A. Patrick Hu, L. Chen and Y. Sun, "Steady-State Load Identification Method of Inductive Power Transfer System Based on Switching Capacitors," IEEE Trans. Power Electron, vol. 30, no. 11, pp. 6349-6355, Nov. 2015.

[13] X. Sheng and L. Shi, "Mutual Inductance and Load Identification Method for Inductively Coupled Power Transfer System based on Auxiliary Inverter,"IEEE Trans. Veh. Technol., vol. 69, no. 2, pp. 1533–1541, Feb. 2020.

Digitally Controlled Misalignment-Tolerant Inductive Power Transfer System with Adaptive Hybrid Compensation for CC/CV Charging of E-Scooter

Niranjan Shrestha
Department of Electrical, Computer, and Software Engineering
Ontario Tech University
Oshawa, Canada
niranjan.shrestha@ontariotechu.net

V.S.R.Varaprasad Oruganti
Department of Electrical, Computer, and Software Engineering
Ontario Tech University
Oshawa, Canada
varaprasad.oruganti@ieee.org

Sheldon Williamson
Department of Electrical, Computer, and Software Engineering
Ontario Tech University
Oshawa, Canada
sheldon.williamson@ontariotechu.ca

Abstract—This paper presents a digitally controlled Inductive power transfer (IPT) system for charging Electric scooters (E-scooters), utilising an adaptive hybrid compensation network design. This system ensures constant current (CC) and constant voltage (CV) charging, with misalignment. It leverages the advantages of double-sided inductor-capacitor-capacitor (DS-LCC) and LCC-series (LCC-S) topologies, which respectively provide CC and CV output. A Type-II digital anti-windup PI control method with a selectable compensation network is introduced, allowing both CC and CV modes to operate with zero voltage switching (ZVS) under misalignment conditions. This approach reduces system losses and enhances the stability and efficiency of the IPT system. A 270W/85-kHz IPT-based E-scooter charger was designed and simulated in the MATLAB/Simulink environment, and an experimental prototype was developed to evaluate the performance of the proposed charger as per SAE J2954 standards. Testing under three different coupling conditions perfect alignment, 5 cm misalignment, and 10 cm misalignment demonstrated that the output parameters remained constant across all conditions, confirming the effectiveness of the proposed control technique.

Keywords—Constant Current, constant Voltage, e-scooter, inductive power transfer, type-II controller, zero voltage switching.

I. INTRODUCTION

In recent years, electric scooters (E-scooters) have emerged as a favoured and fast-growing mode of urban transportation. Their popularity stems from their lightweight construction paired with effective propulsion systems, making them an efficient option for short commutes. These features, combined with their convenience and affordability, have positioned E-scooters as a transformative alternative to traditional transportation methods in densely populated cities. The rise in their adoption has been fuelled by the growing demand for eco-friendly, cost-effective, and flexible urban mobility solutions [1-2].

However, despite their advantages, the charging process for E-scooters presents significant operational and logistical challenges. To maintain functionality, operators are often required to collect scattered scooters from various locations across the city and transport them to centralised charging stations. At these facilities, scooters are either connected to multi-plug chargers or have their batteries replaced manually. While this approach ensures functionality, it comes with substantial drawbacks. These include high operational costs, potential incompatibilities between different chargers, and safety concerns such as the risk of electric shock during battery handling. These issues highlight the need for a more efficient and safer charging method [3].

One promising solution lies in inductive power transfer (IPT) technology, which enables wireless charging. By eliminating the need for physical connectors, IPT offers a contactless method of charging E-scooter batteries, reducing reliance on manual intervention and minimizing safety risks [4-6]. This innovative approach simplifies the charging process and can enhance the convenience and scalability of E-scooter operations [7-8]. However, the practical implementation of IPT is not without challenges. One significant issue is the precise alignment required between the transmitter (embedded in the charging pad) and the receiver (mounted on the E-scooter). Misalignment can cause variations in mutual inductance, leading to inconsistencies in power transfer and efficiency.

Furthermore, the effectiveness of IPT for E-scooters relies heavily on ensuring that the battery charging adheres to specific charging profiles—namely, the constant current (CC) and constant voltage (CV) profiles. These profiles are critical for maintaining battery health, optimizing energy delivery, and ensuring safety during the charging process. Addressing these technical and operational challenges is vital to unlocking the full potential of IPT technology in revolutionizing E-scooter charging systems. With continued advancements, IPT could pave the way for a seamless and sustainable charging infrastructure, further solidifying E-scooters as a key component of future urban transportation systems [9-11].

To address these issues, improving the misalignment tolerance of the IPT system while maintaining CC and CV charging is crucial. This paper adopts a Type-II Proportional-Integral (PI) controlled IPT system with anti-windup for E-scooter charging, featuring an adaptive hybrid compensation

979-8-3315-1612-3/25 $31.00 © 2025 IEEE

network design. This system ensures consistent CC and CV charging even with misalignment, utilizing the benefits of double-sided inductor-capacitor-capacitor (DS-LCC) and LCC-series (LCC-S) topologies, which provide reliable CC and CV outputs respectively. The paper is organized as follows: the IPT system configuration with proposed adaptive switching logic is presented in Section II. The digital control for CC/CV mode is reported in Section III. The simulation studies of the 270-watt IPT system for E-scooter charging are provided in Section IV. The hardware validation of 270-Watt IPT system based E-scooter charging is described in Section V. Finally, conclusions are drawn in Section VI.

II. INDUCTIVE POWER TRANSFER SYSTEM CONFIGURATION

The proposed IPT system for the E-scooter, illustrated in Fig. 1, comprises two distinct sections: the Grid Assembly (GA) and the Vehicle Assembly (VA), which are electrically separated by inductive coupling. The GA section includes a DC voltage source, a high-frequency inverter built with Silicon Carbide (SiC) technology, an LCC compensation network, and a primary coil (L_1). On the VA side, the system features a secondary coil (L_2), an LCC or series compensation network, a full-bridge diode rectifier, a filter capacitor, and a battery load represented by a variable resistor to simulate the E-scooter's battery [12-15].

To regulate the charging process, the VA incorporates two AC switches, S_1 and S_2, which operate in a complementary fashion under digital control using adaptive switching logic. This configuration allows the system to switch seamlessly between constant current (CC) and constant voltage (CV) charging modes. By designing the compensation networks in both the GA and VA sections to achieve Zero Voltage Switching (ZVS), the DS-LCC and LCC-S compensation networks ensure stable and efficient CC and CV charging, even under dynamic operating conditions.

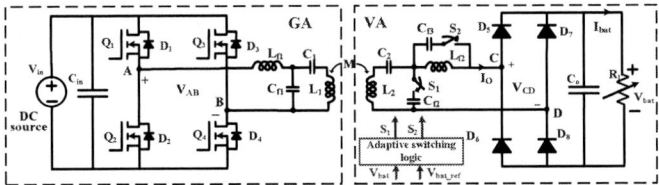

Fig. 1. IPT charging system for E-Scooter with adaptive hybrid compensation.

III. DIGITAL CONTROL FOR CC/CV OUTPUT MODE

In the proposed system, the mutual inductance between the GA and VA coils significantly impacts the load current and voltage, assuming a constant input voltage, resonant frequency, and compensation parameters. Reduced coupling due to misalignment causes a decline in both load current and voltage, falling below their rated levels. To address this, a closed-loop control mechanism is required to maintain the rated output in constant current (CC) and constant voltage (CV) modes. A Type-II PI-based digital control is employed on the primary-side inverter to stabilize the output and ensure consistent performance in CC/CV modes. In an IPT system for E-scooter charging, achieving stable and efficient constant current (CC) and constant voltage (CV) modes is critical for battery health and operational safety. A Type-II PI controller is particularly

suited for such systems due to its ability to enhance system stability, eliminate steady-state errors, and respond effectively to dynamic conditions such as misalignment between transmitter and receiver coils.

Key Functions of the Type-II PI Controller in IPT Systems:

- *Load Regulation under Misalignment*: Misalignment in IPT systems leads to variations in mutual inductance, affecting the output voltage and current. The Type-II PI controller compensates for these variations by dynamically adjusting the inverter's switching characteristics.

- *Support for ZVS Operation*: ZVS is critical for reducing switching losses and improving efficiency. The Type-II PI controller ensures precise regulation of the inverter's operation to maintain ZVS during both CC (5A) and CV (54V) modes.

- *Improved Dynamic Response*: The additional pole in the Type-II PI controller's transfer function enhances the system's phase margin, improving its ability to handle transient conditions and maintain stability at the 85 kHz switching frequency.

- *Seamless Transition Between CC and CV Modes*: The controller facilitates smooth transitions between CC and CV modes by using a switching logic that governs compensation network adjustments, ensuring stability in power delivery to the battery.

A. Type-II-based PI controller

The Type-II-based PI controller consists of a PI with a pole. The transfer function of a modified PI controller is defined as:

$$G(s) = k \frac{1+sT}{sT} \cdot \frac{1}{1+sT_c} \qquad (1)$$

Where $T_c = \frac{1}{\omega_c}$ and $\omega_c = 2\pi f_c$, k is the gain of the PI controller, T is the Time constant of the PI portion of the controller, and f_c Frequency of the pole, in Hz [16-18].

Given the E-scooter IPT system inverter operating frequency 85 kHz, the pole frequency is 10 times of the operating frequency. Therefore, the pole frequency f_c is 8.5kHz. By considering this frequency bandwidth the controller effectively attenuates high-frequency noise while maintaining responsiveness in the frequency range relevant for current regulation. A phase margin of 45°–60° ensures stability and robustness. The time constant (T) value is determined based on the bandwidth i.e. 8.5kHz. Therefore, the time constant is 18.7μs. Further, the optimum gain values for the maximum power transfer are tuned by using the MATLAB/Simulink tuning tool. Moreover, an anti-windup Type-II PI controller prevents integrator saturation while ensuring precise control and stability in dynamic systems. The detailed schematic diagram of the closed-loop control is illustrated in Fig. 2. To select the mode of operation and its corresponding compensation under misalignment conditions is driven by an adaptive selection logic. Here the E-scooter battery voltage and current values are considered to drive the inverter.

979-8-3315-1612-3/25 $31.00 © 2025 IEEE

The battery voltage and current relations with the compensation network parameters are described in the following equations [14]:

$$I_{bat} = \frac{2\sqrt{2}MV_{AB}}{\pi\omega L_{f1}L_{f2}} \qquad (2)$$

$$V_{bat} = \frac{\pi M V_{AB}}{2\sqrt{2}L_{f1}} \qquad (3)$$

From the above equations, V_{AB} is the RMS value of the fundamental component of inverter output voltage.

$$V_{AB} = \frac{2\sqrt{2}V_{DC}cos\left(\frac{\theta}{2}\right)}{\pi} \qquad (4)$$

Here The Input DC voltage of the inverter is denoted as V_{DC} and the inverter phase shift angle is denoted as θ.

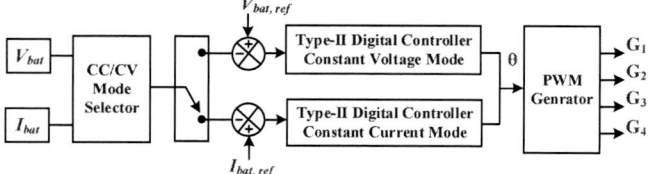

Fig. 2. Schematic of digital control of IPT system for E-Scooter charging.

B. Adaptive Mode Selection Switching Logic

The selection of the mode is governed by the switching logic shown in Fig. 3.

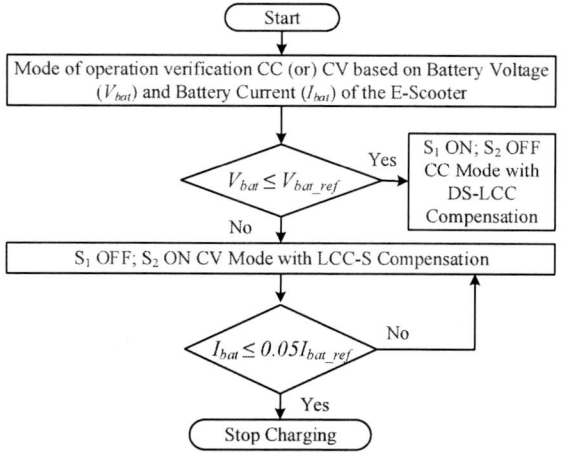

Fig. 3. Adaptive CC/CV Mode compensation switching logic.

The selection of the compensation and also the CC or CV mode operation is governed by measuring the E-scooter battery voltage and current for perfectly aligned conditions as depicted in Fig. 3. During misalignment, the inverter is operated by the proposed controller illustrated in Fig. 2. The simulation study of the proposed digital control for CC/CV mode of operation of E-scooter presented in the next section.

IV. IPT SYSTEM FOR E-SCOOTER CHARGING SIMULATION STUDY

A 270-watt IPT system suitable for E-scooter charging was simulated in a MATLAB/Simulink environment. The CC/CV output operations for hybrid compensation networks were simulated and validated. The parameters used in the simulations are detailed in Table 1.

A. CC/CV Output under a completely aligned position

The DS-LCC and LCC-S networks were combined into a hybrid compensation circuit and simulated in MATLAB Simulink to evaluate its CC/CV performance under a variable resistance load, emulating dynamic battery behaviour. The equivalent resistance, representing the E-scooter battery load during CC/CV charging, was varied from 7.2 Ω to 216 Ω, as depicted in Fig. 4. Results confirmed that the proposed network maintains CC/CV output regardless of load variations, validating its CV operation. Relevant input and output characteristics are illustrated in Figs. 5 through Fig. 8.

TABLE I. PARAMETERS OF 270-WATT E-SCOOTER IPT SYSTEM

Symbol	Description	Value
V_{DC}	Input DC voltage	60 V
f_r	Resonant frequency	85 kHz
L_1	Primary coil self-inductance	417.4 µH
L_2	Secondary coil self-inductance	119.4 µH
M	Mutual inductance in perfect alignment (16 cm air gap)	39.64 µH
L_{f1}	Primary compensation inductance	44 µH
C_{f1}	Primary parallel compensation capacitance	79.6 nF
C_1	Primary series compensation capacitance	9.39 nF
L_{f2}	Secondary compensation inductance	16.4 µH
C_{f2}	Secondary parallel compensation capacitance	214 nF
C_2	Secondary series compensation capacitance	34.1 nF
C_{f3}	Secondary parallel compensation capacitance	0.427 µF
I_{bat}	Current in CC mode	5 A
V_{bat}	Voltage in CV mode	54 V

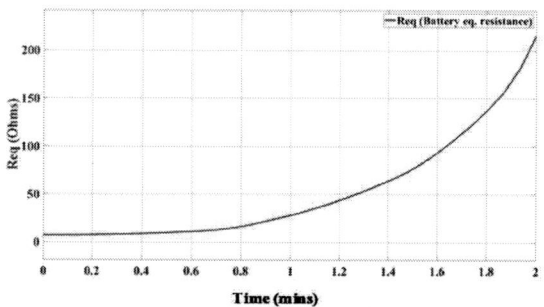

Fig. 4. Battery R_{eq} load (Emulating CC/CV charging behaviour).

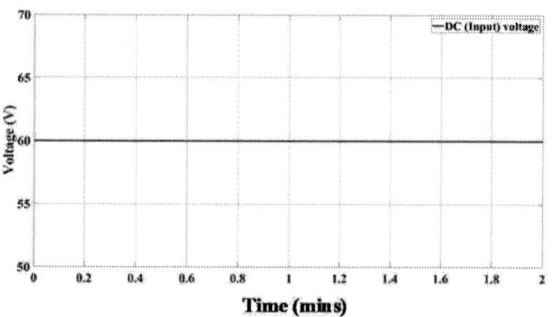

Fig. 5. Input DC voltage to the inverter.

Fig. 5 shows the input-side DC voltage, while Fig. 6 presents the inverter output voltage and current waveforms along with the rectifier input voltage and current waveforms during CC mode. Similarly, Fig. 7 illustrates the inverter and rectifier waveforms under CV mode. As shown in Fig. 8, despite variations in load resistance, the primary coil current remains constant, and the load current and voltage are maintained at the designed values of 5 A (CC mode) and 54 V (CV mode). This confirms that the proposed hybrid compensation circuit reliably delivers CC/CV output under dynamic loads when the transmitter and receiver coils are fully aligned.

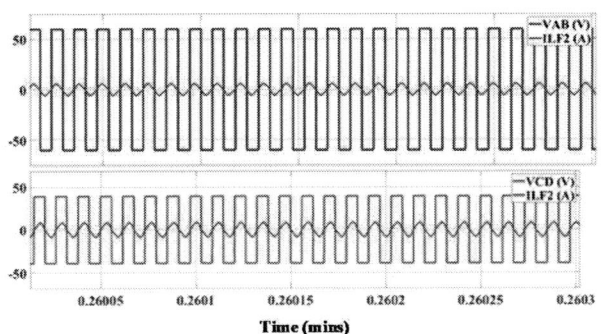

Fig. 6. CC Mode - Inverter output voltage (V_{AB}) and current (I_{Lf1}), Rectifier input voltage (V_{CD}) and current (I_{LF2}).

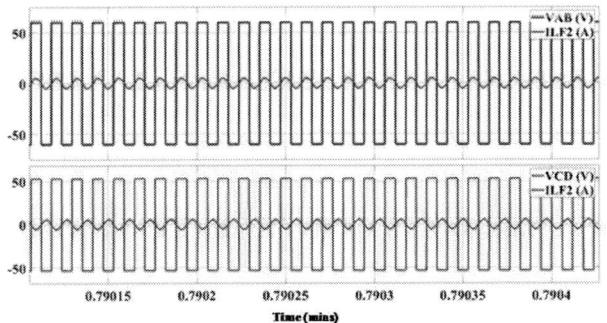

Fig. 7. CV Mode - Inverter output voltage (V_{AB}) and current (I_{Lf1}), Rectifier input voltage (V_{CD}) and current (I_{LF2}).

B. CC/CV Output under Misalignment and Dynamic Load – 270 W

A phase shift control is applied to the primary-side inverter to regulate its output voltage and maintain the rated current and voltage in CC/CV modes under misalignment conditions. As misalignment increases, the controller reduces the phase shift angle to achieve the required average inverter output voltage, necessitating a higher input voltage. For this purpose, a 100 V DC input was used, ensuring sufficient phase-shifted voltage even in extreme misalignment (k = 0.1 in this study). Three misalignment scenarios were analyzed: perfect alignment (k = 0.177), 5 cm misalignment (k = 0.14), and 10 cm misalignment (k = 0.1). Coupling coefficients for these scenarios were measured in the lab using a Keysight E4980AL LCR meter.

C. Coupling coefficient (k – 0.1)

The simulation was conducted with a coupling coefficient of 0.1, representing an extreme misalignment of 10 cm between

the transmitter and receiver coils. Thanks to the phase shift control logic, the output current and voltage remain stable at 5A and 54 V, as shown in Fig. 9(c). Figures 9(a) and 9(b) indicate that the phase shift angle is nearly 0°, resulting in the inverter producing an almost ideal rectangular waveform in CC mode under this severe misalignment condition.

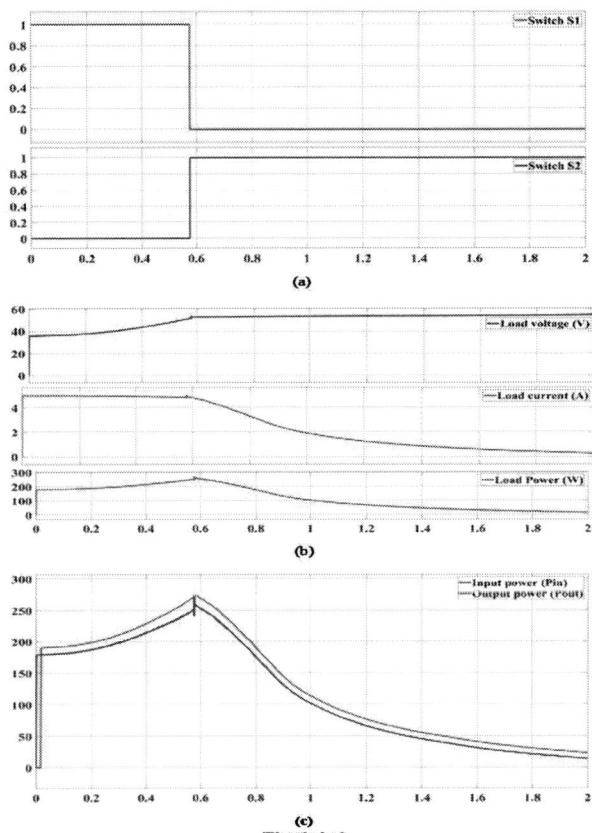

Fig. 8. (a) Switching sequence of switches S_1 and S_2, (b) Output (load) voltage, current, and power, (c) Input power vs output power.

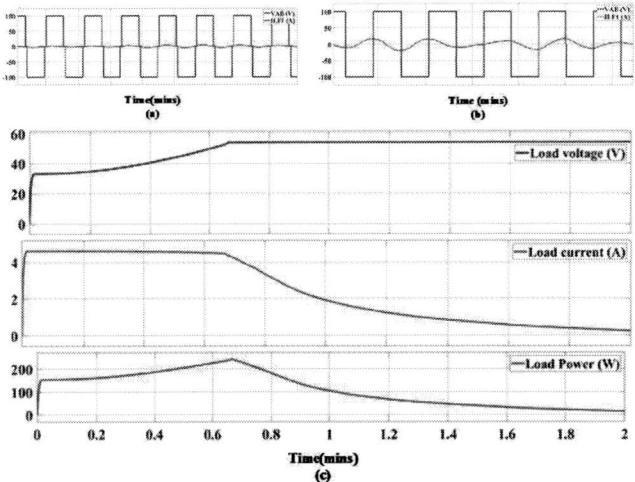

Fig. 9. Inverter output and load output parameters (a) CC Mode for k = 0.1 (b) CV modes for k = 0.1 (c) Load voltage, current, and power (CC/CV during misalignment) for k =0.1.

D. Coupling coefficient (k-0.14)

The simulation was performed with a coupling coefficient of 0.14, representing a 5 cm misalignment between the transmitter and receiver coils. Due to the phase shift control logic, the output current and voltage remain stable at 5 A and 54 V, as shown in Fig. 10(c). As seen in Fig. 10(a) and 10(b), the phase shift angle has increased compared to the k=0.1 case, causing the inverter to adjust its output voltage accordingly in CC mode under this condition.

E. Coupling coefficient (k-0.177)

The simulation was conducted with a coupling coefficient of 0.177, representing a perfectly aligned condition between the transmitter and receiver coils. Thanks to the phase shift control logic, the output current and voltage are maintained at 5 A and 54 V, as shown in Fig. 11(c). Figures 11(a) and 11(b) show that the phase shift angle has increased compared to the k=0.14 case, resulting in the inverter adjusting its output voltage accordingly in CC mode.

Simulation results for three different coupling coefficients in both CC and CV modes are presented in Figs. 9 through Fig. 11. In the perfectly aligned condition, the coupling coefficient between the GA and VA coils is 0.177. For the misaligned cases, the coupling coefficients for 5 cm and 10 cm misalignments, as measured by the LCR meter, are 0.14 and 0.1, respectively. Simulations were performed for all these coupling conditions, including perfect alignment.

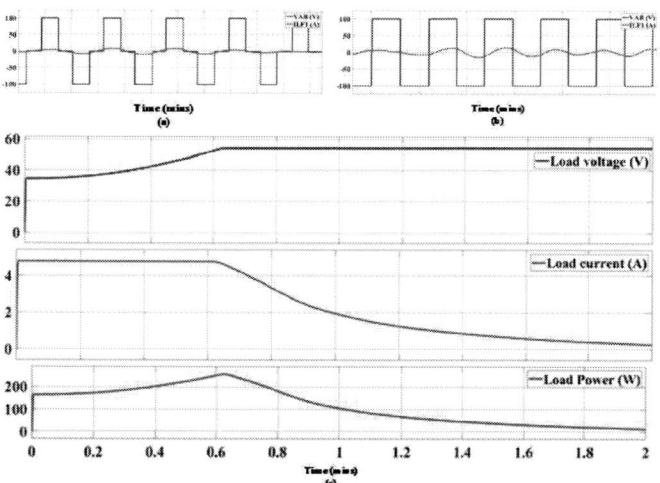

Fig. 10. Inverter output and load output parameters (a) CC Mode for k = 0.14 (b) CV modes for k = 0.14 (c) Load voltage, current, and power (CC/CV during misalignment) for k =0.14.

Figs. 9 to 11 show that the proposed closed-loop controller changes the phase shift angle accordingly for different coupling conditions. This produces the required corresponding output voltage from the inverter, maintaining CC (5 A) and CV (54 V) output on the load side.

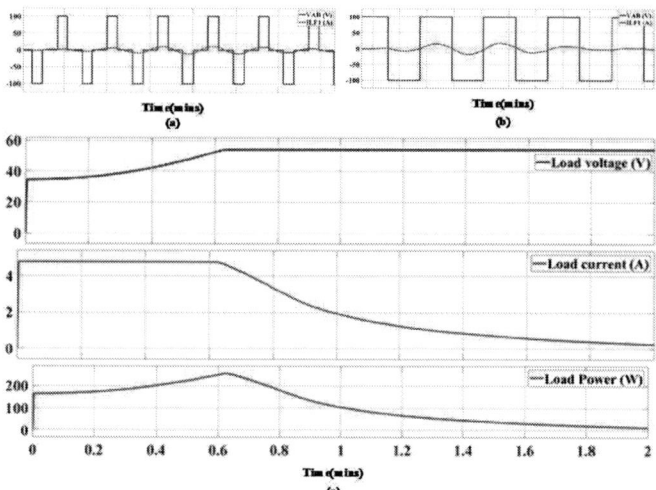

Fig. 11. Inverter output and load output parameters (k-0.177) (a) CC Mode for k = 0.177 (b) CV modes for k = 0.177 (c) Load voltage, current, and power (CC/CV during perfect alignment) for k =0.177).

V. HARDWARE VALIDATION OF IPT SYSTEM FOR E-SCOOTER CHARGING

The hardware setup for validating the 270 W hybrid compensation prototype includes a Keysight 10 kW regenerative DC power supply, a high-frequency SiC MOSFET inverter (CREE KIT8020CRD8FF1217P-1 [19]), primary compensation components, a transmitter coil, and a TI TMS320F28379D microcontroller [20] on the primary side. On the secondary side, it comprises a receiver coil, secondary compensation elements, an APT2X61D100J full-bridge rectifier [21], a Kemet C4DEHPQ6220A8TK filter capacitor [22], and a Chroma 63804 AC/DC electronic load. A 4-channel relay (SRD-05VDC-SL-C [23]), controlled by the same microcontroller, switches between DS-LCC and LCC-S compensation networks.

The DC supply powers the inverter, converting DC to high-frequency AC. Compensation networks on both sides, with components specified in Table 1, ensure CC or CV output. The rectifier converts AC to DC, and the Chroma load simulates the battery's CC/CV charging profile using programmable resistance. Real-time voltage and current data are acquired into MATLAB Simulink via a VISA API during testing.

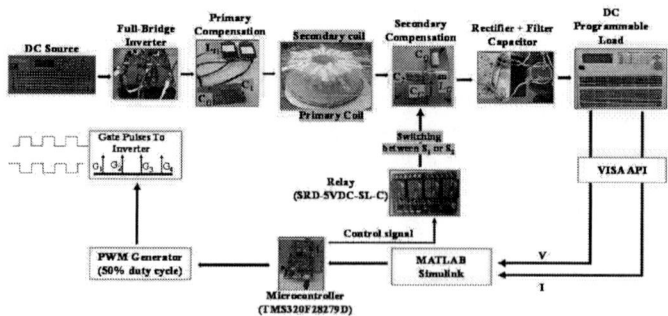

Fig. 12. Experimental setup for the hybrid compensated IPT system for an E-Scooter.

979-8-3315-1612-3/25 $31.00 © 2025 IEEE

The Keysight DC source is powered on, supplying the inverter, which receives 50% duty cycle, 85 kHz pulses from the TI microcontroller through specific GPIO pins. The inverter output is connected to the primary compensation elements (L_{f1}, C_{f1}, and C_1), which are linked to the transmitter (primary) coil that wirelessly transmits power to the secondary coil. The secondary coil output is connected to secondary compensation elements (C_2, C_{f2}, C_{f3}, and L_{f2}), which are attached to the full-bridge rectifier, followed by a filter capacitor and the DC programmable load. The complete setup is shown in Fig. 1. A 4-channel relay is used as an AC switch to toggle between DS-LCC and LCC-S compensation, with two channels activated for the required AC switches.

Initially, the goal is to enable LCC-LCC compensation for CC charging. In this configuration, relay R1 is activated (S_1 on, S_2 off), connecting C_{f2} in parallel with the secondary coil, as shown in Fig. 4, forming the LCC-LCC compensation topology. The microcontroller provides the control signal for relay activation and deactivation. Capacitor C_{f3} is not connected in this case. With the WPT system powered, the DS-LCC compensation network supplies a constant current to the load. During charging, the microcontroller continuously monitors the voltage and current, acquiring real-time data from the MATLAB Simulink interface via the DC programmable load. In CC mode, the load voltage and power increase while the current remains constant. When the voltage reaches the predetermined maximum, the system switches to CV charging mode.

To transition to CV mode, relay R_1 is deactivated, and relay R_2 is activated by the microcontroller (S_1 off, S_2 on). A 2-second delay is introduced between deactivating R_1 and activating R_2 to prevent sudden voltage and current surges. Once R_1 is turned off and R_2 is turned on, C_{f2} is isolated, and C_{f3} is connected in parallel with L_{f2}, as shown in Fig. 7, forming the LCC-S compensation. In this mode, the system provides CV charging, where the load current and power decrease while the voltage remains constant. Once the current drops to around 5% of the charging current, the CC/CV charging process is complete. Thus, the proposed hybrid compensation circuit successfully provides CC/CV charging to the load.

A. Experimental results and analysis of DS-LCC Compensation for Dynamic Load

Initially, The DS-LCC compensated WPT system for a 270 W peak load with variable load resistance was initially tested in the experimental setup to evaluate its CC output performance. While the compensation elements were designed for a 60 V input, it was found that 60 V was insufficient to deliver 5 A in CC mode and 54 V in CV mode. This discrepancy arose because the actual values of the compensation inductors and capacitors did not precisely match the design specifications, due to the unavailability of the required component values and variations in those specified in the datasheets. As a result, an 80 V DC input was needed to achieve the desired current (5 A) and voltage (54 V) in both CC and CV modes. Consequently, all experimental tests were conducted with an 80 V DC input.

The equivalent battery resistance, which simulates the CC charging operation, ranged from 7.2 Ω to 10.8 Ω, as shown in

Fig. 13(c). The DS-LCC compensation effectively acted as a constant current source on the load side, maintaining constant current regardless of the dynamic load, thus confirming its CC operation as demonstrated in the simulations. Relevant input and output characteristic graphs are presented in Fig. 13. As shown, despite changes in load resistance, the rectifier input current and load current on the secondary side remained constant at 5 A, while the load voltage increased. This validates that the DS-LCC compensation network provides a stable CC output under dynamic load conditions, assuming constant mutual inductance.

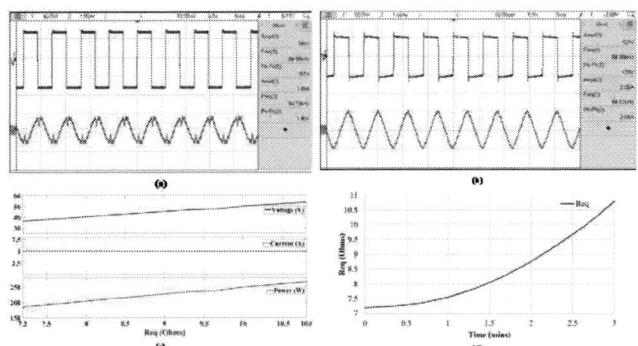

Fig. 13. (a) Inverter output voltage (V_{AB}) and current (I_{Lf1}) (b) Rectifier input voltage (V_{CD}) and current (I_{LF2}) (c) Output voltage, output current, and output power (d) Battery equivalent load (CC mode).

B. LCC-S Compensation for Dynamic Load

The LCC-S compensated WPT system was subsequently tested in the experimental setup to evaluate its CV output behaviour under variable load conditions. The equivalent battery resistance, simulating CV charging, varied from 11.6 Ω to 216 Ω, as shown in Fig. 14(c). The results demonstrated that the LCC-S compensation maintained a constant voltage on the load side, irrespective of the dynamic battery load, confirming its CV output performance. Relevant input and output characteristic graphs are presented in Fig. 14. As shown, despite variations in load resistance, the load voltage remained constant, while the load current decreased over time, reflecting the reduction in load power.

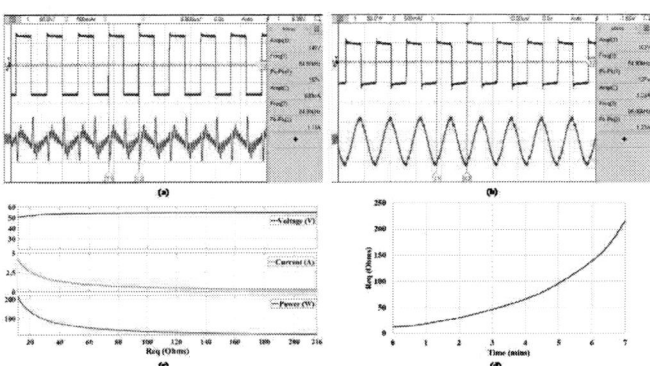

Fig. 14. (a). Inverter output voltage (V_{AB}) and current (I_{Lf1}) (b) Rectifier input voltage (V_{CD}) and current (I_{LF2}) (c) Output voltage, output current, and output power (d) Battery equivalent load (CV mode).

This validates that the LCC-S compensation circuit provides a stable CV output (54 V) under dynamic load conditions in a perfectly aligned setup.

C. Hybrid Compensation for Dynamic Load

The DS-LCC and LCC-S compensation networks were combined into a hybrid compensation circuit and tested in the hardware setup depicted in Fig. 12 with a variable resistance load to observe its CC/CV behaviour. The equivalent resistance, simulating the battery load in CC/CV charging, varied from 7.2 Ω to 216 Ω as shown in Fig. 15(f). The results show that the proposed network provides stable CC/CV output regardless of dynamic load variations, validating its performance. Relevant input and output characteristic graphs are shown in Fig. 15.

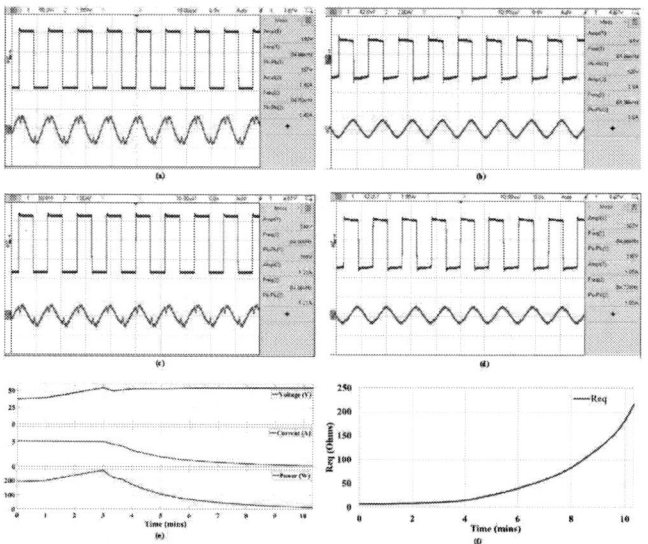

Fig. 15. (a). Inverter output voltage (V_{AB}) and current (I_{Lf1}) [CC mode] (b) Rectifier input voltage (V_{CD}) and current (I_{LF2}) [CC mode] (c) Inverter output voltage (V_{AB}) and current (I_{Lf1}) [CV mode] (d) Rectifier input voltage (V_{CD}) and current (I_{LF2}) [CV mode] (e) Output (load) voltage, current, and power(f) Battery R_{eq} load (Emulating CC/CV charging behaviour).

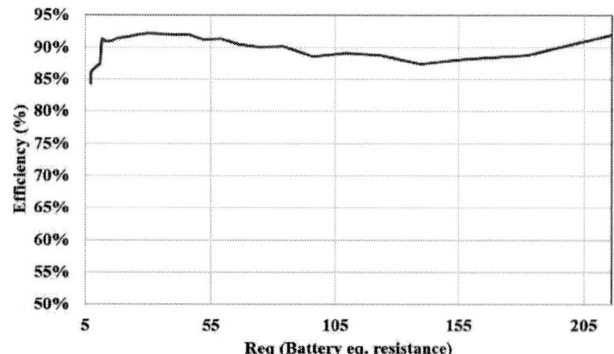

Fig. 16. Efficiency vs Battery equivalent resistance.

As shown in Fig. 15, despite variations in load resistance, the primary coil current remains constant throughout both CC and CV modes, with load current and voltage maintained at the desired design values of 5 A (CC) and 54 V (CV). This demonstrates that the hybrid compensation circuit effectively delivers CC/CV output under dynamic load conditions, provided the mutual inductance between the primary and secondary coils remains constant. Additionally, the DC-DC efficiency stays above 85%, peaking at nearly 92% as depicted in Fig. 16, with efficiency variations due to changes in current or voltage corresponding to load resistance fluctuations and associated losses.

VI. CONCLUSIONS

This article demonstrates the superior performance of the proposed Type-II digital anti-windup PI control method in an IPT system designed for charging electric scooters. By leveraging a selectable hybrid compensation network, the system successfully maintains stable constant current (CC) and constant voltage (CV) outputs under various misalignment conditions, including 5 cm and 10 cm misalignment, alongside perfect alignment. The Type-II controller ensures zero voltage switching operation, significantly reducing system losses and enhancing overall efficiency. The designed system achieved a conversion efficiency of 92% under dynamic load changes and misalignment, highlighting the robustness and effectiveness of the Type-II PI controller in optimizing system stability and performance. The experimental results, validated according to SAE J2954 standards, confirm the reliability and efficiency of the proposed IPT charger across all tested coupling conditions.

REFERENCES

[1] G. Li, H. Zhang, Y. Chen, J. Xie, C. H. Jo, and D. H. Kim, "3-D Misalignment Tolerant E-scooter IPT System with Hybrid Control Based on Three-coil Design for Load-independent CC/CV Outputs," *IEEE Trans. Transp. Electrification*, (Early Access), 2024.

[2] O. Altintasi and S. Yalcinkaya, "Siting charging stations and identifying safe and convenient routes for environmentally sustainable e-scooter systems," *Sustainable Cities and Society*, vol. 84, Article no. 104020, 2022.

[3] J. S. Hu, F. Lu, C. Zhu, C. Y. Cheng, S. L. Chen, T. J. Ren, and C. C. Mi, "Hybrid energy storage system of an electric scooter based on wireless power transfer," IEEE Transactions on Industrial Informatics, vol. 14, no. 9, pp. 4169–4178, Sep. 2018.

[4] N. Shrestha, A. Samanta, F. C. Fietosa, and S. Williamson, "State-of-the-Art Wireless Charging Systems for E-Bikes: Technologies and Applications," in *Proc. IEEE Int. Conf. Power Electron. Drive Syst. (PEDS)*, Aug. 2023, pp. 1–6.

[5] P. S. Huynh, D. Ronanki, D. Vincent, and S. Williamson, "Overview and comparative assessment of single-phase power converter topologies of inductive wireless charging systems," *Energies*, vol. 13, no. 9, p. 2150, 2020.

[6] E. Gomaa, A. Shawky, and M. Orabi, "Wireless charging techniques and converter topologies for light EVs, E-bikes, E-chairs and E-scooters: A review," in *Proc. IEEE Conf. Power Electron. Renew. Energy (CPERE)*, Feb. 2023, pp. 1–8.

[7] G. P. R. Vaddemani, D. Ronanki, A. Dekka, and A. R. Beig, "Design and Control of a New Single-Stage Wireless Charger with Interoperable Power Level Capability," in *Proc. IEEE Appl. Power Electron. Conf. Expo. (APEC)*, 2024, pp. 1294–1299.

[8] F. Pellitteri, N. Campagna, V. Castiglia, A. Damiano, and R. Miceli, "Design, implementation and experimental results of an inductive power transfer system for electric bicycle wireless charging," *IET Renew. Power Gener.*, vol. 14, no. 15, pp. 2908–2915, 2020.

[9] C. H. Kwan, J. M. Arteaga, D. C. Yates, and P. D. Mitcheson, "Design and construction of a 100 W wireless charger for an E-scooter at 6.78

MHz," in *Proc. IEEE PELS Workshop Emerging Technol. Wireless Power Transfer (WoW)*, Jun. 2019, pp. 186–190.

[10] P. K. Abraham, D. Mary, M. V. Jayan, and N. Paulson, "Design, Implementation, and Experimental Verification of a Solar PV Charge Controller for a Low-Speed E-Scooter Home Charging Station," in *Proc. Int. Conf. Smart Comput. Commun. (ICSCC)*, Aug. 2023, pp. 754–759.

[11] A. Triviño-Cabrera, J. C. Quiró, J. M. González-González, and J. A. Aguado, "Optimized design of a wireless charger prototype for an E-scooter," *IEEE Access*, vol. 11, pp. 33014–33026, 2023.

[12] V. Shevchenko, O. Husev, R. Strzelecki, B. Pakhaliuk, N. Poliakov, and N. Strzelecka, "Compensation topologies in IPT systems: Standards, requirements, classification, analysis, comparison and application," *IEEE Access*, vol. 7, pp. 120559–120580, 2019.

[13] S. S. Andure, A. G. Thosar, and D. S. Yeole, "Performance Analysis of Double-Sided LCC Compensation Topology," in *Proc. IEEE Int. Conf. Intell. Technol. (CONIT)*, Jun. 2024, pp. 1–5.

[14] N. Shrestha and S. Williamson, "Hybrid Compensation Network for Misalignment-Tolerant Constant Current/Constant Voltage Charging for Wireless Power Transfer System," in *Proc. IEEE Transp. Electrification Conf. Expo. (ITEC)*, Jun. 2024, pp. 1–6.

[15] M. Feliziani, T. Campi, S. Cruciani, and F. Maradei, Wireless Power Transfer for E-Mobility: Fundamentals and Design Guidelines for Wireless Charging of Electric Vehicles. Amsterdam, Netherlands: Elsevier, Nov. 2023.

[16] A. Dash and S. Chatterjee, "Design and Development of a Battery Charger using P–channel MOSFET and Type–III Compensator," in *Proc. IEEE Appl. Signal Process. Conf. (ASPCON)*, Nov. 2023, pp. 315–320.

[17] O. V. S. R. Varaprasad, A. S. Bubshait, D. V. S. S. S. Sarma, and M. G. Simões, "Real-time control of hybrid active power filter using conservative power theory in industrial power system," *IET Power Electron.*, vol. 10, no. 2, pp. 196–207, 2017.

[18] T. S. Sasmal, K. Yenduri, and P. Das, "Single-stage saturable inductive-link half-bridge point of load converter," in *Proc. IEEE Energy Convers. Congr. Expo. (ECCE)*, Oct. 2021, pp. 2042–2047.

[19] Cree Inc., "KIT8020CRD8FF1217P evaluation kit datasheet," accessed Aug. 11, 2024. [Online]. Available: https://www.mouser.ca/datasheet/2/90/CreeInc_KIT8020CRD8FF1217P_1_SS-1160766.pdf

[20] Texas Instruments, "LAUNCHXL-F28379D: C2000 Delfino LaunchPad development kit," accessed Aug. 11, 2024. [Online]. Available: https://www.ti.com/tool/LAUNCHXL-F28379D#tech-docs

[21] Microchip Technology Inc., "APT2X61D100J product datasheet," accessed Aug. 11, 2024. [Online]. Available: https://www.microchipdirect.com/product/APT2X61D100J

[22] KEMET Electronics Corporation, "C4DEHPQ6220A8TK capacitor datasheet," accessed Aug. 11, 2024. [Online]. Available: https://search.kemet.com/component-documentation/download/specsheet/C4DEHPQ6220A8TK

[23] Circuit Basics, "SRD-05VDC-SL-C relay datasheet," accessed Aug. 11, 2024. [Online]. Available: https://www.circuitbasics.com/wp-content/uploads/2015/11/SRD-05VDC-SL-C-Datasheet.pdf

On/off Control of Modular Inductive Power Transfer System

Kunxiao Zhou
ShanghaiTech University
Shanghai, China
zhoukx2022@shanghaitech.edu.cn

Guangdong Ning
ShanghaiTech University
Shanghai, China
ninggd@shanghaitech.edu.cn

Heyuan Li
ShanghaiTech University
Shanghai, China
lihy4@shanghaitech.edu.cn

Xinlin Wang
ShanghaiTech University
Shanghai, China
wangxl3@shanghaitech.edu.cn

Minfan Fu
ShanghaiTech University
Shanghai, China
fumf@shanghaitech.edu.cn

Abstract—This paper explores a planar and modular wireless charger utilizing distributed on/off control. By reconfiguring multiple non-overlapping transmitting (TX) coils to form different charging regions, it requires the examination of the real and reactive power of each module to clarify the cross-coupling effect. The voltage of the compensation capacitor is used as a reference to deactivate modules when the cross-coupling effect becomes significant. Utilizing zero-current detection, the phase information of each module is employed to achieve distributed on/off control, thereby avoid significant cross-coupling effect. This method ensures that each TX operates independently, facilitating the module's design. In the experiment, a four-TX system is set up to charge a mobile receiver. The effective detection of the capacitor voltage phase enables successful distributed on/off control, resulting in a peak system efficiency of 86%.

Index Terms—Wireless power transfer, multiple transmitters, cross coupling,on/off control

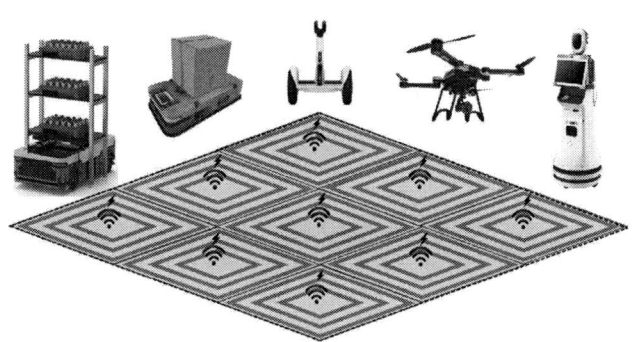

Fig. 1: Target proposed IPT system configuration.

I. INTRODUCTION

Inductive power transfer (IPT) holds promise for mid-range high-efficiency power transmission. On the transmitter (TX) side, power is converted into a magnetic field and subsequently absorbed by the receiver (RX) [1]. Its convenience has led to widespread adoption in the charging of mobile phones, medical implant devices, and automatic robots [2]–[7]. However, diverse applications exhibit varying terminal characteristics and mobility requirements. While a customized design might aid in defining a well-performing system, its scalability is somewhat limited. There is potential in utilizing basic standard and modular charging cells to delineate a TX area that can be user-reconfigured, as depicted in Fig. 1. Such a modular design approach necessitates consideration of cross-coupling. In previous years, techniques like coil overlapping or orthogonal systems have been successful in decoupling two or three coils, but their scalability is still restricted [8], [9]. Additionally, the requisite overlapping is not favorable for modular designs. Hence, there is a need to develop decoupling techniques for planar non-overlapping coils.

Another approach to decoupling involves addressing the adverse effects of cross coupling from a circuit perspective rather than focusing solely on the field perspective. This method acknowledges the continued existence of coil coupling. By leveraging circuit theory, the analysis of cross coupling effects is detailed in [10], and compensation is achieved through the addition of extra components or circuit branches [11]–[13]. However, these methods typically necessitate physical connections between adjacent cells, which is not particularly conducive to a modular design.The concept of multiple frequencies has found broad application in decoupling strategies across various scenarios. In single-load applications, diverse frequency components are utilized to establish three separate decoupled power transfer channels, thereby augmenting the output power capacity [14]. In multi-load scenarios, previous studies have employed the multi-frequency multi-amplitude (MFMA) superimposition technique for inverter design and control, tuning each RX to resonate at specific frequencies to receive the multi-band energy in a decoupled manner [15]–[17]. A reconfigurable and modular design, employing a dual-band design approach, has been proposed in [10], showcasing considerable scalability. Nonetheless, this method demands a substantial number of compensation components to generate several resonance frequencies, indicating a need for a simpler

979-8-3315-1612-3/25 $31.00 © 2025 IEEE

solution in modular IPT systems.

This paper focuses on the coupling sensing and on/off control aspects of a planar and modular IPT system. Rather than entirely resolving cross coupling issues, the primary objective is to confine their adverse effects to an acceptable level. To illustrate this concept, a four-TX and one-RX system is employed as an example. Utilizing circuit theory, the system's state equations are formulated and utilized to assess the impact of cross coupling [18]. In this approach, the circulating energy within each TX acts as an effective indicator to gauge cross coupling effects. Specifically, the real power delivered to the RX represents efficient power transfer, while the reactive power is indicative of cross coupling effects. By sensing and analyzing this information, each TX can be dynamically and independently controlled in an on/off manner once cross coupling reaches a significant level. This sensing and control logic relies solely on the information gathered from each TX and can be readily extended to a modular IPT.

II. ON/OFF CONTROL BASED ON VOLTAGE PHASE

The depicted modular charger is illustrated in Fig. 1. On the transmitter (TX) side, the planar TX charging area is comprised of square and modular cells, with each cell being powered by its dedicated inverter. Crucially, these cells require physical isolation, forbidding any coil overlapping. To align with the modular concept, individual TX units should operate autonomously based on their internal information, without relying on commands from a central controller. Notably, managing the cross coupling among multiple TXs poses the most significant challenge in this context.

This paper would use a simple four-TX case to explain the proposed intelligent on/off control. In Fig. 2, multiple TX coils are compensated by the same LCC networks, i.e., L_{tx}, C_t, C_{tx} are in series resonance, L_t, C_t are in series resonance. Series compensation is employed for the receiver. The state variables of all the components are denoted in the figure.

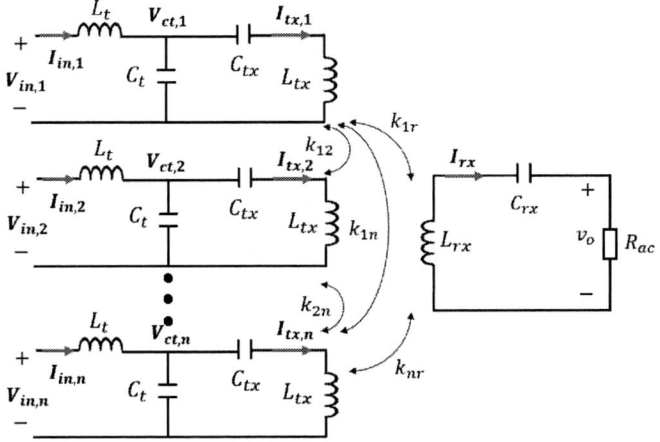

Fig. 2: Target modular transmitter circuit.

Under resonance, the state equations of the whole system is

$$
\begin{cases}
\mathbf{V_{in,1}} = \mathbf{I_{in,1}}(\frac{1}{j\omega C_t} + j\omega L_t) - \frac{1}{j\omega C_t}\mathbf{I_{tx,1}}, \\
\mathbf{V_{in,1}} = j\omega L_t\mathbf{I_{in,1}} + (j\omega C_{tx} + j\omega L_{tx})\mathbf{I_{tx,1}} \\
\qquad + j\omega M_1\mathbf{I_r} + \sum_{i=2}^{n} j\omega M_{1i}\mathbf{I_{tx,i}}, \\
\vdots \\
\mathbf{V_{in,n}} = \mathbf{I_{in,n}}(\frac{1}{j\omega C_t} + j\omega L_t) - \frac{1}{j\omega C_t}\mathbf{I_{tx,n}}, \\
\mathbf{V_{in,n}} = j\omega L_t\mathbf{I_{in,n}} + (j\omega C_{tx} + j\omega L_{tx})\mathbf{I_{tx,n}} \\
\qquad + j\omega M_1\mathbf{I_r} + \sum_{i=1}^{n-1} j\omega M_{1i}\mathbf{I_{tx,i}},
\end{cases} \quad (1)
$$

The input currents are solved as

$$
\begin{cases}
\mathbf{I_{in,1}} = \mathbf{V_{in}}(\frac{M_1 \sum_{i=1}^{n} M_i}{L_t^2 R} + j\frac{\sum_{i=2}^{n} M_{1i}}{w L_t^2}), \\
\vdots \\
\mathbf{I_{in,n}} = \mathbf{V_{in}}(\frac{M_n \sum_{i=1}^{n} M_i}{L_t^2 R} + j\frac{\sum_{i=1}^{n-1} M_{ni}}{w L_t^2}),
\end{cases} \quad (2)
$$

Since the coil currents are clamped by the inverter output voltage

$$
\begin{cases}
\mathbf{I_{tx,1}} = \frac{\mathbf{V_{in}}}{jw L_t}, \\
\vdots \\
\mathbf{I_{tx,n}} = \frac{\mathbf{V_{in}}}{jw L_t},
\end{cases} \quad (3)
$$

and the currents of C_{t1} and C_{t2} are

$$
\begin{cases}
\mathbf{I_{ct,1}} = \mathbf{I_{in,1}} - \mathbf{I_{tx,1}} \\
\vdots \\
\mathbf{I_{ct,n}} = \mathbf{I_{in,n}} - \mathbf{I_{tx,n}}
\end{cases} \quad (4)
$$

The capacitor voltage $\mathbf{V_{ct}}$s are further derived as

$$
\begin{cases}
\mathbf{V_{ct,1}} = \mathbf{V_{in}}\frac{(\sum_{i=1}^{n} M_{1i}+L_t)}{L_t} - j\frac{w M_1(\sum_{i=1}^{n} M_i)}{L_t R}, \\
\vdots \\
\mathbf{V_{ct,n}} = \mathbf{V_{in}}\frac{(\sum_{i=1}^{n} M_{ni}+L_t)}{L_t} - j\frac{w M_n(\sum_{i=1}^{n} M_i)}{L_t R},
\end{cases} \quad (5)
$$

With $\mathbf{V_{in,1}}$, $\mathbf{V_{in,2}}$...$\mathbf{V_{in,n}}$ equal to $\mathbf{V_{in}}$. When the phase of inverter output voltage is used as the reference, the phase of $\mathbf{V_{ct,1}}$, $\mathbf{V_{ct,2}}$, $\mathbf{V_{ct,3}}$, $\mathbf{V_{ct,4}}$ are derived as

$$
\begin{cases}
\phi_{ct,1} = -tan^{-1}\frac{\frac{w M_1(\sum_{i=1}^{n} M_i)}{L_t R}}{\mathbf{V_{in}}(\frac{\sum_{i=1}^{n} M_{1i}+L_t}{L_t})}, \\
\vdots \\
\phi_{ct,n} = -tan^{-1}\frac{\frac{w M_n(\sum_{i=1}^{n} M_i)}{L_t R}}{\mathbf{V_{in}}(\frac{\sum_{i=1}^{n} M_{ni}+L_t}{L_t})},
\end{cases} \quad (6)
$$

It is clearly the phase above is coupling dependent, which help evaluate the coupling effects.

The fundamental control logic involves the closure of the relevant TX once its induced circulating energy from M_{12}, M_{13}, M_{23}, M_{24} becomes notable. Fig. 3 compares two example coupling conditions. In Fig. 3(a), when the RX is positioned between four TXs, all TXs should remain activated for power transmission. In this scenario, embracing and effectively utilizing the cross coupling effect is necessary for the smooth switching of the inverters. It's important to note that this cross coupling introduces a capacitive element to the coil, subsequently resulting in an inductive load from the inverter's perspective. If cross coupling equals zero, the inverter would encounter a purely resistive load during perfect resonance. This position means the undesirable cross coupling effect is limited, and the system would utilize it.

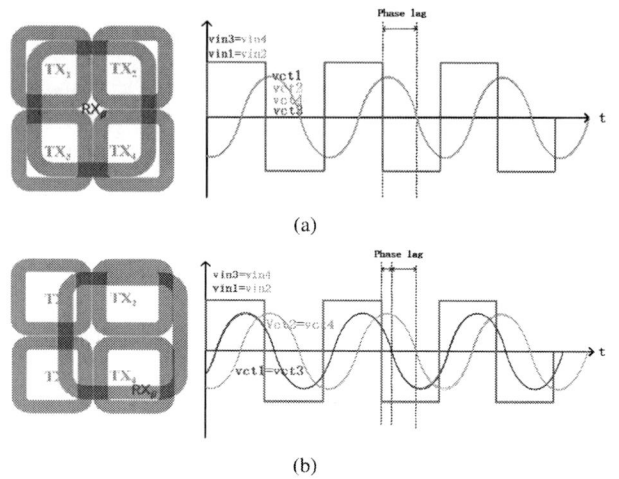

(a)

(b)

Fig. 3: Typical waveform when RX moves. (a) RX is placed at the center. (b) RX is placed at right.

Once the RX relocates to the top of TX1, TX4, it results in increasing M_1, M_4 and decreasing M_2, M_3, and relatively unchanged cross coupling M_{12}, M_{13}, M_{14}, M_{24}. When all TXs operate simultaneously, the inverter of TX1, TX3 confronts a substantial reactive load, indicated by the phase difference between $\mathbf{V_{in,1}}$, $\mathbf{V_{in,3}}$ and $\mathbf{I_{in,1}}$, $\mathbf{I_{in3}}$ in Fig. 3(b). From a physical coupling perspective, TX1 and TX3 is distanced from the RX and should not be activated for power transmission. Forcing its activation would amplify conduction losses due to the high circulating energy. Instead, maintaining TX2 and TX4 active for power transmission is advisable owing to its strong M_2, M_4. Although cross coupling remains relatively constant, its role or significance alters with RX movement, evident through the phase of inverter output currents, illustrating this cross coupling effect. Considering the expense associated with current sensing, the capacitor voltage phase $\phi_{\mathbf{ct,1}}$, $\phi_{\mathbf{ct,2}}$, $\phi_{\mathbf{ct,4}}$ and $\phi_{\mathbf{ct,3}}$ can serve as indicators to determine when to switch on/off the TX. The control of each TX merely depends on its own information, i.e, the phase of the capacitor voltage, and this does benefit the modular design.

III. IMPLEMENTATION OF ON/OFF CONTROL

A four-TX system is depicted in Fig. 4 to implement the proposed on/off control. Four identical TX coils are positioned adjacently to form a square area. Each TX coil is equipped with identical compensation and a full-bridge inverter. All these TX modules share a common central microcontroller (MCU). The voltage across the compensation capacitors is sensed and sent to a zero-crossing detection (ZCD) circuit. Consequently, four ZCD signals are transmitted to the central MCU. By utilizing the MCU's clock, the phase difference of the capacitor voltages can be determined.

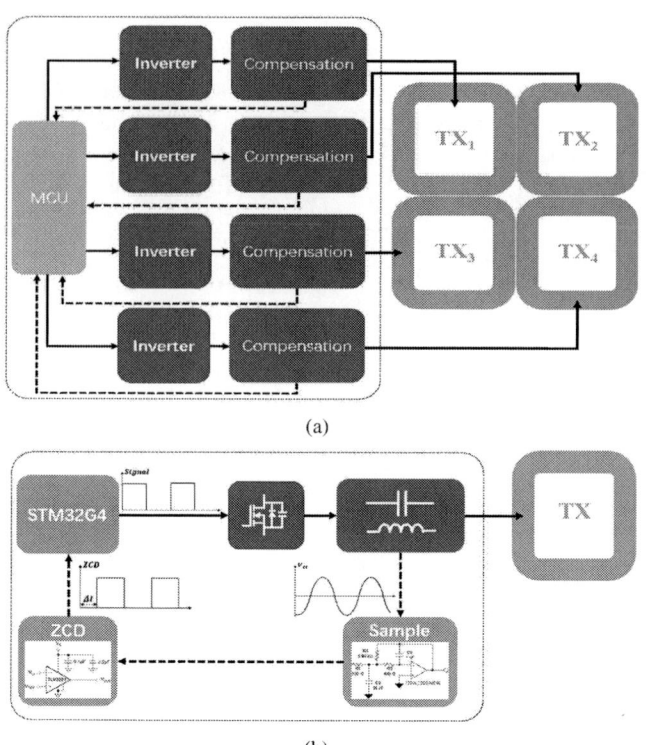

(a)

(b)

Fig. 4: (a)System configuration.(b)Sample circuit

Based on the phase information, the system control diagram is summarized in Fig. 5. The objective of the proposed control is to ensure that the TXs with relatively strong coupling to the RX are enabled. Therefore, in Fig. 5, once the four phases are obtained by the MCU, the ratio between each pair of phases is first calculated. This ratio is sufficient to determine the relative coupling condition.

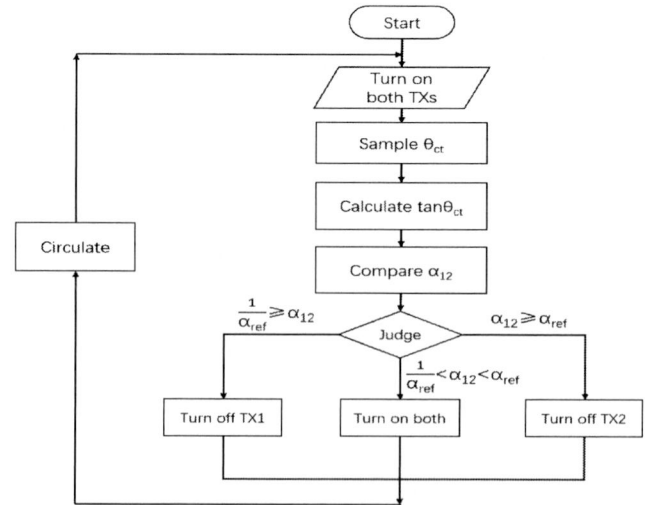

Fig. 5: Proposed control flow chart.

For example, in order to judge the on/off status of two adjacent coils, TX1 and TX2, the controller senses θ_{ct1}

979-8-3315-1612-3/25 $31.00 © 2025 IEEE   811

and θ_{ct2}. The controller then calculates their ratio $\alpha_{12} = |\tan(\theta_{ct1})| / |\tan(\theta_{ct2})|$. This ratio is compared to a reference value, α_{ref}. When $\alpha_{12} > \alpha_{ref}$ (where $\alpha_{ref} = 2$), TX1 should be turned on and TX2 turned off. When $\alpha_{ref} > \alpha_{12} > \frac{1}{\alpha_{ref}}$, both TX1 and TX2 should remain on. When $\frac{1}{\alpha_{ref}} > \alpha_{12}$, TX1 should be turned off and TX2 kept on. Using this logic, when α_{ref} equals 1, the system turns off the one with stronger coupling. The proposed system sets α_{ref} as 2. The system is able to enable both coils when the RX is placed in the adjacent areas of the target two TXs. The same comparison logic is applied to any other pair of adjacent coils. When the RX moves during the charging process, the whole system should reinitialize the judging logic after a certain period, such as 1 second.

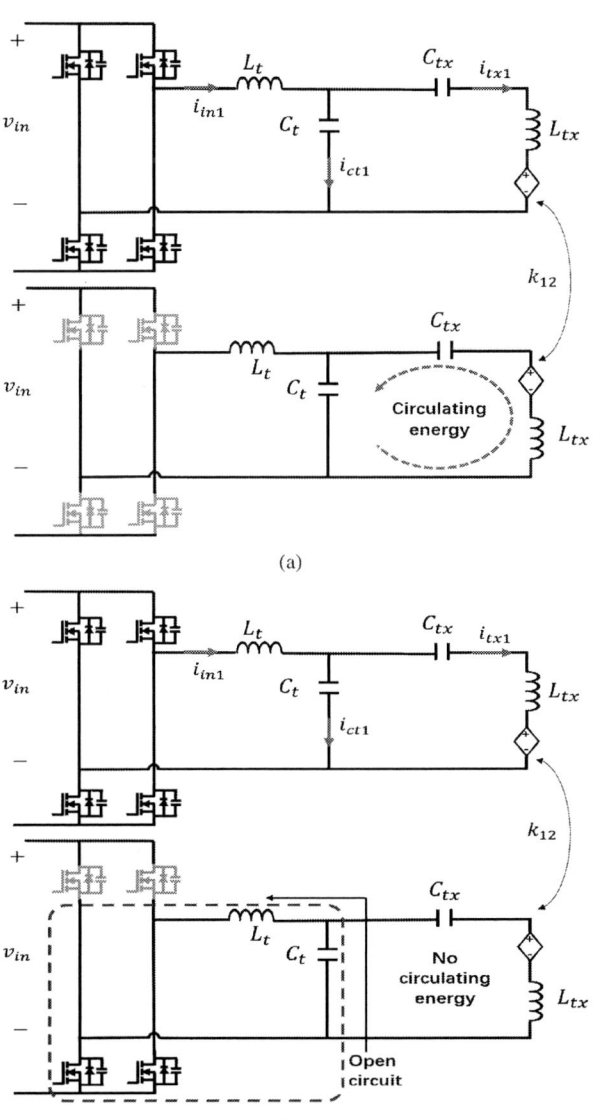

(a)

(b)

Fig. 6: Different turn-off mechanism. a) Turn off the inverter. b) Shoring the input AC terminal of TX2.

The off control of each TX cannot simply be achieved by turning off the inverters due to the coupling between TX coils. To illustrate the turn-off mechanism, a two-TX system is shown in Fig. 6 to compare two different turn-off methods. The example system is designed to enable TX1 and disable TX2. In order to achieve the desired on/off status, both Fig. 6(a) and Fig. 6(b) will enable the inverter for TX1 as usual. However, disabling the PWM signal of TX2's inverter—meaning all the switches are off—will result in the coil induced voltage of TX2 (due to TX1) causing a significant circulating current in the resonant tank. Note that both the MOSFET body diode and junction capacitance provide the conduction path. Therefore, TX2 cannot be turned off in the manner shown in Fig. 6(a).

When a full-bridge inverter is used, this paper proposes to disable TX2 by keeping the two lower devices on. In this configuration, the AC input terminals of TX2 are effectively shorted by the inverter. From the perspective of the coil, the parallel resonance of L_t and C_t behaves as an open circuit. This implies that no power is absorbed by TX2 from TX1.

IV. VERIFICATION OF PROPOSED CONTROL

A verification platform is built as shown in Fig. 7, which includes four TXs and one RX. A MCU is used to sense the zero cross point of capacitor voltage (i.e., telling the voltage phase) and control the TX in a on/off manner (i.e., controlling the inverters). Note that the inverter driving signal would right serve as the reference to determine the phase of the capacitor voltage.

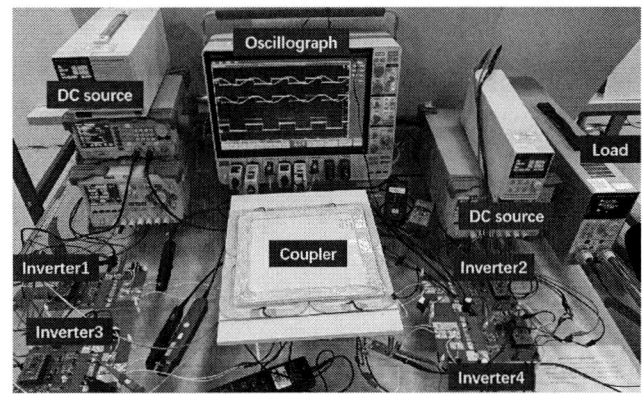

Fig. 7: Experiment setup.

When the RX coil is positioned at the central location, it implies a limited influence of cross coupling, and $M_1 = M_2 = M_3 = M_4$. The system waveforms are depicted in positionA of Fig. 8. The measured $\mathbf{V_{ct}}$ (via voltage probe) is almost in phase with $\mathbf{V_{in}}$. The proposed on/off control solely relies on phase information, utilizing a zero cross detection (ZCD) circuit for each capacitor voltage and obtaining corresponding signals. Notably, there exists no phase difference between four $\mathbf{V_{ct}}$s. However, when the RX shifts to the top of TX2 and TX4 (i.e., the edge position), there emerges a distinct phase delay (approximately $40°$) for $\mathbf{V_{ct,2}}$ and $\mathbf{V_{ct,4}}$ concerning $\mathbf{V_{ct,1}}$ and $\mathbf{V_{ct,3}}$.The waveform is illustrated in Fig. 3(b). Utilizing this information, the MCU would deactivate TX1 and TX3

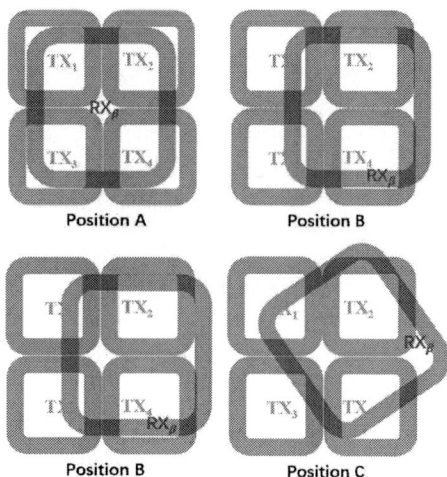

Fig. 8: Typical positions of the system.

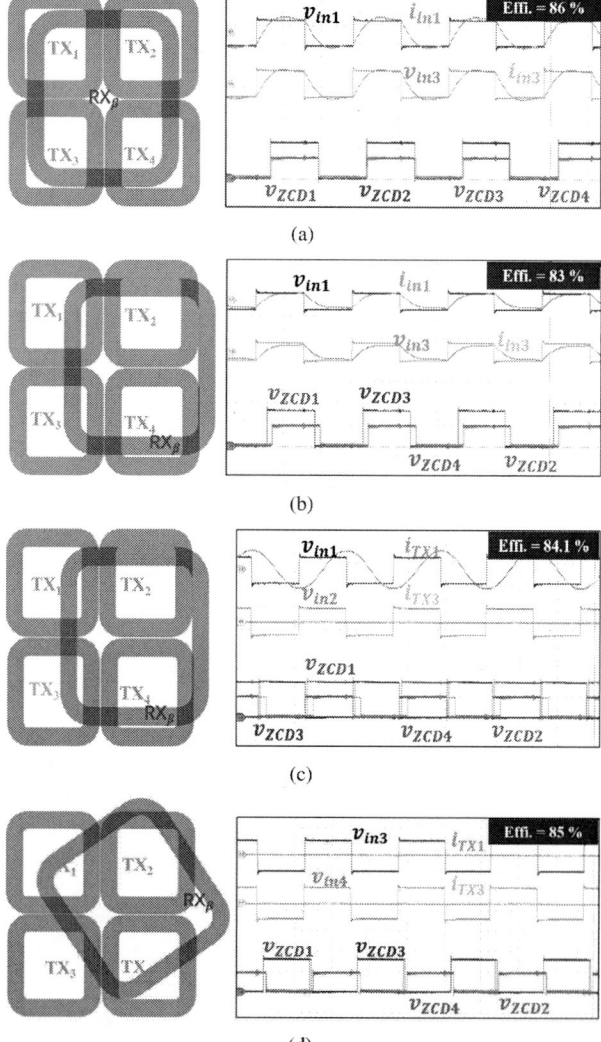

Fig. 9: Experiment results for the network.(a)PositionA (b)PositionB (c)PositionB (d)PositionC

while maintaining TX2 and TX4 activated. The outcome of the on/off control reveals the status as illustrated in positionB of Fig. 8, demonstrating the successful deactivation of TX1 and TX3 by the MCU. While the coupler is placed at position C in Fig. 8, with three strong coupling TX1,TX2,TX4 and one weak coupling TX3, the system would choose to deactivate the TX3.

V. CONCLUSION

This study delves into the on/off control mechanisms within the context of a modular and planar Inductive Power Transfer (IPT) system. Utilizing a four-TX system as an example, the paper elucidates the control logic and underscores the necessity of a modular design. The coupling effect is meticulously examined through a circuit model, delineating discussions on coupling-induced real and reactive power. Derivation of the phase relationship between the inverter current and the compensation capacitor voltage aids in the analysis of the coupling effect, serving as a benchmark for the on/off control of individual modules. In the experimental phase, the implementation of Zero Cross Detection (ZCD) circuits is introduced to realize the proposed sensing and control logic.

VI. ACKNOWLEDGEMENT

This work was supported by National Natural Science Foundation of China under Grant 52477013 and Lingang Laboratory under Grant NO. LG-GG-202402-06-10. (Corresponding Author: Minfan Fu).

REFERENCES

[1] J. Feng, Q. Li, F. C. Lee, and M. Fu, "Transmitter coils design for free-positioning omnidirectional wireless power transfer system," *IEEE Transactions on Industrial Informatics*, vol. 15, no. 8, pp. 4656–4664, 2019.

[2] A. Kamineni, G. A. Covic, and J. T. Boys, "Analysis of coplanar intermediate coil structures in inductive power transfer systems," *IEEE Transactions on Power Electronics*, vol. 30, no. 11, pp. 6141–6154, 2015.

[3] Y. Jang and M. M. Jovanovic, "A contactless electrical energy transmission system for portable-telephone battery chargers," *IEEE Transactions on Industrial Electronics*, vol. 50, no. 3, pp. 520–527, 2003.

[4] D. Ahn and S. Hong, "Wireless power transmission with self-regulated output voltage for biomedical implant," *IEEE Transactions on Industrial Electronics*, vol. 61, no. 5, pp. 2225–2235, 2014.

[5] H. Hu, T. Cai, S. Duan, X. Zhang, J. Niu, and H. Feng, "An optimal variable frequency phase shift control strategy for zvs operation within wide power range in ipt systems," *IEEE Transactions on Power Electronics*, vol. 35, no. 5, pp. 5517–5530, 2020.

[6] H. Matsumoto, Y. Shibako, and Y. Neba, "Contactless power transfer system for agvs," *IEEE Transactions on Industrial Electronics*, vol. 65, no. 1, pp. 251–260, 2018.

[7] V. B. Vu, D. H. Tran, and W. Choi, "Implementation of the constant current and constant voltage charge of inductive power transfer systems with the double-sided lcc compensation topology for electric vehicle battery charge applications," *IEEE Transactions on Power Electronics*, vol. 33, no. 9, pp. 7398–7410, 2018.

[8] A. Zaheer, G. A. Covic, and D. Kacprzak, "A bipolar pad in a 10-khz 300-w distributed ipt system for agv applications," *IEEE Transactions on Industrial Electronics*, vol. 61, no. 7, pp. 3288–3301, 2014.

[9] S. Kim, G. A. Covic, and J. T. Boys, "Tripolar pad for inductive power transfer systems for ev charging," *IEEE Transactions on Power Electronics*, vol. 32, no. 7, pp. 5045–5057, 2017.

[10] G. Ning, K. Zhou, J. Liang, H. Wang, and M. Fu, "Reconfigurable and modular wireless charger based on dual-band design," *IEEE Transactions on Circuits and Systems II*, vol. 70, no. 9, pp. 3524–3528, 2023.

[11] M. Fu, T. Zhang, X. Zhu, P. C. K. Luk, and C. Ma, "Compensation of cross coupling in multiple-receiver wireless power transfer systems," *IEEE Transactions on Industrial Informatics*, vol. 12, no. 2, pp. 474482, 2016.

[12] M. L. G. Kissin, J. T. Boys, and G. A. Covic, "Interphase mutual inductance in polyphase inductive power transfer systems," *IEEE Transactions on Industrial Electronics*, vol. 56, no. 7, pp. 2393–2400, 2009.

[13] R. Mai, Y. Luo, B. Yang, Y. Song, S. Liu, and Z. He, "Decoupling circuit for automated guided vehicles ipt charging systems with dual receivers," *IEEE Transactions on Power Electronics*, vol. 35, no. 7, pp. 6652–6657, 2020.

[14] I. Ghotbi, M. Najjarzadegan, H. Sarfaraz, S. J. Ashtiani, and O. Shoaei, "Enhanced power-delivered-to-load through planar multiple-harmonic wireless power transmission," *IEEE Transactions on Circuits and Systems II*, vol. 65, no. 9, pp. 1219–1223, 2018.

[15] C. Qi, S. Huang, X. Chen, and P. Wang, "Multifrequency modulation to achieve an individual and continuous power distribution for simultaneous mr-wpt system with an inverter," *IEEE Transactions on Power Electronics*, vol. 36, no. 11, pp. 12440–12455, 2021.

[16] F. Liu, Y. Yang, Z. Ding, X. Chen, and R. M. Kennel, "A multifrequency superposition methodology to achieve high efficiency and targeted power distribution for a multiload mcr wpt system," *IEEE Transactions on Power Electronics*, vol. 33, no. 10, pp. 9005–9016, 2018.

[17] J. Wu, L. Bie, W. Kong, P. Gao, and Y. Wang, "Multi-frequency multi-amplitude superposition modulation method with phase shift optimization for single inverter of wireless power transfer system," *IEEE Transactions on Circuits and Systems I*, vol. 68, no. 5, pp. 2271–2279, 2021.

[18] K. Yue, Y. Liu, X. Zhang, M. Fu, J. Liang and H. Wang, "Transmitter Side Voltage Based Mutual Inductances and Load Tracking for Two-Transmitter LCC-S Compensated Wireless Power Transfer Systems," *IEEE Journal of Emerging and Selected Topics in Power Electronics*, early access.

979-8-3315-1612-3/25 $31.00 © 2025 IEEE

Receiver Side Regulation of LCC Wireless Power Transfer System with Variable Notch Filter

Hsin-Che Hsieh
The Bradley Department of Electrical and Computer Engineering
Virginia Polytechnic Institute and State University
Blacksburg, VA, USA
hchsieh@vt.edu

Jih-Sheng Lai
The Bradley Department of Electrical and Computer Engineering
Virginia Polytechnic Institute and State University
Blacksburg, VA, USA
laijs@vt.edu

Abstract—The robustness and relative insensitivity to misalignments and mismatches make the LCC topology popular in wireless power transfer (WPT) applications, including more challenging dynamic or multi-receiver WPT systems. The output of LCC topology based WPT systems is typically regulated by an additional dc-dc converter stage, either before or after the WPT stage. The additional dc-dc stage increases the cost and complexity of the system and would require wireless communication for output regulation if placed on the transmitter side, making the system susceptible to electromagnetic interference. In this paper, a receiver side modulation scheme based on a variable inductor controlled notch filter is proposed. By modulating the variable inductor, the output can be independently adjustable in a wide-range. The variable inductor is controlled with simple pulse width modulation (PWM). The proposed control scheme is especially suitable for multi-receiver WPT systems; where transmitter based regulation schemes cannot simultaneously accommodate multiple receivers. The proposed system is verified experimentally with a 48-$V_{o(max)}$, 72-W prototype WPT system with multi-receiver capability.

Keywords—Wireless Power Transfer, Variable Inductor, Notch Filter, LCC Topology

I. INTRODUCTION

The LCC topology, attributting to its tolerance and robustness against misalignments, mistuning or circuit parameter variations [1-4] , is a popular choice when desiging wireless power transfer (WPT) systems, especially WPT systems whose receiver side is constantly moving, or systems that have to support multiple receivers simutaneously, as in [5-9]. Previously presented LCC based WPT systems typically control output with an additional dc-dc regulator stage, either before the transmitter or after the receiver stage [6-7, 10-13]. However, these popular methods have the disadvantage of increasing the system size, cost and complexity. For portable or moving receivers, incuding an additional regulator stage may not be favorable as this increases the size and weight, affecting portability or movability. However, if the pre-regulator stage is placed on the transmitter side, the receiver side would have to send the output information back to the transmitter side for closed-loop regulation, mandating the use of wireless communication [10-13]. Employing wireless communication for closed-loop control makes the system susectible to electromagnetic noise and interference, which could disrupt or interrupt the wireless communication link, impacting reliability of the WPT system in high-noise environment.

Alternative methods for controlling LCC based WPT systems without an additional dc-dc stage have been proposed. In [14-15], extra semiconductor switch(s) are added for periodically shorting the current source output of double-side LCC WPT stage to regulate the power. In [16-17], the duty cycle or phase of receiver side active rectifier or synchronous rectifier switches are adjusted for output regulation. These approaches are less bulky and costy than having an additional dc-dc regulator stage, since no extra passive components are required. However, adding multiple semiconductor switches that must handle full otuput voltage/current, as well as their associated driving circuits still considerably increase the cost and complexity of the system. Especially in systems similar to [14], the two back-to-back-connected switches at the receiver side would both need independently isolated gate drivers.

Variable inductors and similar devices (e.g. magnetic amplifiers or saturable cores) have also been employed to control WPT systems as well [18-23]. In [18-19], the transmitter side resonant inductor in LCC-S WPT system is replaced with a variable inductor for output voltage control, while in [20], both transmitter side resonant capacitor and inductor are replaced with controllable ones. Receiver side based or receiver-transmitter side cooperative control utilizing variable inductor–like devices have been proposed as well. In [21-22], magnetic amplifiers or saturable cores are used to control the tuning of the reveicer, while in [23], a cooperative control involving transmitter side variable capacitor and receiver side variable inductor is reported.

It should also be noted that, apart from the burden of communication requirements, any output regulation methods based only on the transmitter side modulation would be inherently limited to regulating the output of one single reviever. Since it would not be possible for a single transmitter to simutaniously accomodate multiple receivers that have vastly differet coupling coefficients, output requirements, and load conditions.

This paper proposes a receiver side regulation scheme based on variable inductor controlled notch filter for LCC WPT systems. By modulating the inductance of the variable inductor, the equivalent impedance of the receiver side notch filter can be adjusted from infinity to close to zero, allowing wide-range output voltage regulation under varying load conditions. The modulator that controls the variable inductor is isolated from the high voltage or current of WPT system output or resonant tank. Comparing with control schemes that relies on transmitter side modulation, the proposed method requires neither wireless communcation nor coordination of the transmitter side for output regulation, making the system immune to electromagnetic interference and simplifying the design of the transmitter side, allowing the transmitter side to operate under open-loop condition. Comparing with previously reported variable inductor based reciever side control methods requiring sophisticated tuning algorithms implemeted with specialized digital controllers [21-22], the control scheme proposed in this paper only requires a simple pulse width modulation (PWM) control performed by a generic and inexpensive, commercial off-the-shelf PWM dc-dc controller. The proposed control scheme is especially suitable for multi-receiver WPT systems, where it would not be feasable for one transmitter to regulate the output of multiple receivers simutaniously. A 48-$V_{o(max)}$, 72-W prototpye WPT system capable of supporting multiple receivers is implemented to experimentally verify the method presented in this paper.

II. II. PROPOSED VARIABLE NOTCH FILTER BASED RECEIVER SIDE REGULATION

An LCC WPT system incorporating the proposed variable inductor controlled notch filter based receiver side regulation is shown in Fig.1. An unique characterisitc of the LCC transmitter topology is maintainging a constant transmitter coil current, unaffected by changes in load conditions or coupling coefficients [1, 3-5, 7-8, 10-12, 14-15]. This constant transmitter coil current would induce a voltage source, in porportion with the mutual inductance, on the receiver coil in series with the self inductance of the coil. The receiver side equivalent circuit, highlighting the adjustable notch filter is given in Fig. 2 (a). The equivalent circuit can be further simplified to Fig. 2 (b), where all receiver side reactance is lumped into and represented by one total variable impedance, Z_v. By modulating the variable inductor (L_v), the effective reactance of the notch filter is modulated, and in turn, Z_v between the voltage induced on the WPT receiver coil and the load is adjusted. The use of notch filter allows wide-range adjustment of Z_v from infinity to zero through relative small inductance variations in L_v to accomodate different load conditions and output voltage requirements.

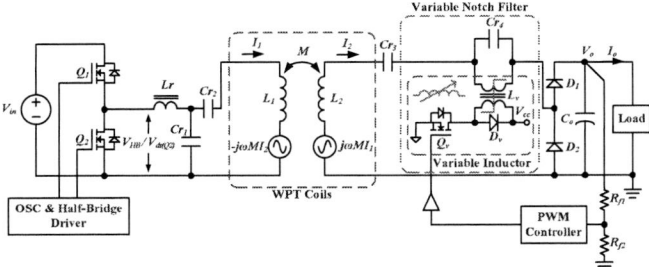

Fig. 1 LCC WPT system with proposed variable inductor controlled notch filter based receiver side regulation

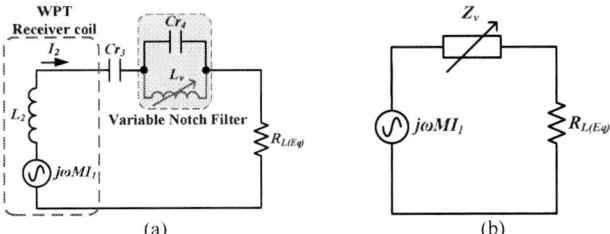

Fig. 2 (a) Receiver side equivalent circuit, highlighting the notch filter (b) All receiver side reactance lumped into one adjustable impedance

The inductance of the variable inductor can be modulated by adjusting the dc current into the control winding. In the proposed system, the control winding current of the variable inductor is adjusted by modulating the duty cycle of Q_v while D_v acts as a freewheeling path. The duty cycle is controlled by the PWM controller. The two receiver side resonant capacitors, Cr_3 and Cr_4 are selected so that the Z_v is infinity when the variable inductance is at maximum ($L_{v(max)}$), while minimum variable inductance ($L_{v(min)}$) would result in Z_v equals to zero. By modulating the duty cycle of Q_v to control impedance, the output voltage can be stabilized at the targeted value. Under light load condition, lower Q_v duty cycle from the PWM controller would drive L_v closer to $L_{v(max)}$, resulting in higher Z_v between voltage induced on the receiver coil and the load to prevent overvoltage. On the other hand, the PWM controller would impose a higher Q_v duty cycle under heavier load to drive L_v closer to $L_{v(min)}$, resulting in lower Z_v between induced voltage on the coil and the load to compensate for the output voltage drop.

Assuming $L_{v(max)}$ and $L_{v(min)}$ of the variable inductor are known, C_{r3} and C_{r4} should be selected with (1) and (2), where L_2 is the self inductance of receiver coil and f_{sw} is the switching frequency. When C_{r4} is selected according to (1), the notch filter impedance (and Z_v) would be infinity when L_v is at $L_{v(max)}$, resulting in zero gain. This feature guarantees safety as an uncontrolled variable inductor is always at maximum inductance, inherently forcing the gain to be zero if the PWM controller is inoperative. On the other hand, when L_v is at an inductance lower than $L_{v(max)}$, the reactance of the notch filter would be inductive. Selecting C_{r3} according to (2) would allow it to cancel out all inductive reactance in the receiver side circuit, which includes the reactance of the notch filter as well as L_2, under the condition that L_v is driven to $L_{v(min)}$. By driving L_v to $L_{v(min)}$, Z_v would be zero, resulting in the maximum gain.

$$C_{r4} = \frac{1}{4\pi^2 f_{sw}^2 L_{v(Max)}} \qquad (1)$$

$$C_{r3} = \frac{1}{\dfrac{4\pi^2 f_{sw}^2 L_{v(Min)}}{1 - 4\pi^2 f_{sw}^2 L_{v(Min)} C_{r4}} + 4\pi^2 f_{sw}^2 L_2} \qquad (2)$$

A distinct characteristic of the LCC topology is keeping the transmitter coil current (I_1) constant regardless of load or coupling conditions [1, 3-5, 7-8, 10-12, 14-15]. The selection of I_1 needs to satisfy the maximum output voltage ($V_{o(Max)}$) requirement, as in (3) where M is the mutual inductance:

$$I_1 > \frac{\sqrt{2}V_{o(Max)}}{2\pi^2 f_{sw}M} \tag{3}$$

Other parameters of the transmitter side LCC network can then be determined following the determination of I_1, as in (4)-(6) :

$$L_r = \frac{\sqrt{2}V_{in}}{2\pi^2 f_{sw}I_1} \tag{4}$$

$$C_{r1} = \frac{1}{4\pi^2 f_{sw}^2 L_r} \tag{5}$$

$$C_{r2} = \frac{1}{4\pi^2 f_{sw}^2 (L_1 - L_r)} \tag{6}$$

Theoretically, the proposed modulation scheme allows the output voltage of the receiver side to be adjusted all the way down to zero volts from nominal output voltage. In realistic WPT systems, the available range of output voltage would be limited by component voltage stress. The ac voltage stress of C_{r3} is only a function of dc output current, as in (7), while the voltage stress of notch filter components (C_{r4}, L_v) is also a function of dc output voltage, as in (8). It can be seen from (8) that under the same output current, a lower output voltage would result in higher voltage stress on the notch filter components. Due to the half-bridge rectifier utilized in the proposed circuit, there would also be a dc offset imposed on C_{r3}, as in (9).

$$V_{Cr3,rms} = \frac{I_o\sqrt{2}}{4f_{SW}C_{r3}} \tag{7}$$

$$V_{Notch,rms} = I_o\left(\frac{\sqrt{2}}{4f_{SW}C_{r3}} - \sqrt{2}\pi^2 f_{SW}L_2\right) + \sqrt{4\pi^2 f_{SW}^2 M^2 I_1^2 - \frac{2V_o^2}{\pi^2}} \tag{8}$$

$$V_{Cr3,dc} = \frac{V_o}{2} \tag{9}$$

Fig. 3 illustrates the structure of a variable inductor, which consists of two control windings and one main winding on a pair of E cores. The inductance of the main winding can be modulated by adjusting the dc current into the two series connected control windings, which partially saturate the E cores. By separating the control winding into two series connected sections on the outer legs of the E cores, dc flux from the control windings and ac flux from the main winding are decoupled. This prevents the main winding ac flux from inducing an undesirable ac voltage on the control winding [24-25]. The design and operation of variable inductor are detailed in [18-20, 22-25].

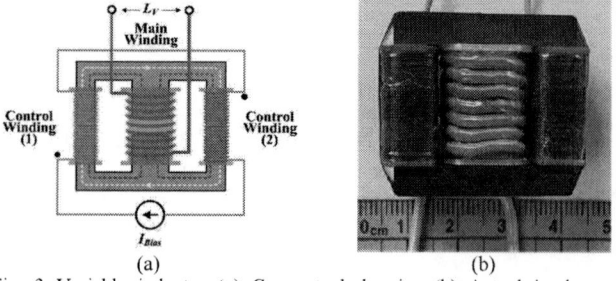

Fig. 3 Variable inductor (a) Conceptual drawing (b) Actual implemented variable inductor

In a variable inductor, by partially saturating the core via control winding current, the slope of the BH curve is altered to modulate the effective permeability of the core, as in (10). The inductance can then be represented in (11) [18-19]. The partially saturated region, where the variable inductor operates in, is illustrated in Fig. 4.

$$\mu_e = \frac{1}{\mu_0}\frac{dB}{dH} \tag{10}$$

$$L_v = N^2 \frac{A_e}{l}\mu_0\mu_e \tag{11}$$

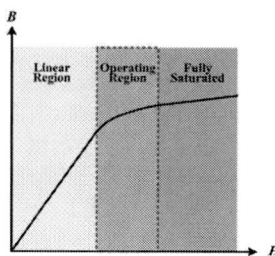

Fig. 4. Partially saturated region of the BH curve where the variable inductor operates

III. EXPERIMENTAL VERIFICATION

A 48-$V_{o(Max)}$, 72-W LCC WPT system is designed and built to experimentally verify the proposed variable notch filter based receiver side regulation. Key parameters of the prototype are given in Table I while the images of the prototype WPT transmitter and receiver are given in Fig. 5. The inductance vs. control current relationship of the variable inductor used in the experiments is given in Fig. 6. To demonstrate the wide output range capability, the prototype is tested at different output voltages and load conditions. When setting different output voltages, the only change made to the prototype receiver is the feedback resistor. The efficiencies of prototype at different output voltage settings and load conditions are plotted in Fig.7. The efficiency calculation includes all auxiliary power consumptions, such as gate drivers, PWM controllers and circuit for modulating the variable inductor.

Table I. Parameter of Prototype

V_{in}	48 V
L_r	4.56 μH
C_{r1}	88.5 nF
C_{r2}	13.9 nF
C_{r3}	19.9 nF
C_{r4}	21.9 nF
L_v	Max: 15.2 μH Min: 3.4 μH
L_1	33 μH
L_2	15.6 μH
M	5 μH
f_{SW}	250 kHz
Q_1, Q_2	FDB035AN06A0
D_1, D_2, D_v	STPS3150UF
Q_v	IRFZ34NS
Clearance	10 mm
Gate Drivers	UCC20520DWR (Q_1, Q_2) UCC27511ADBVR (Q_v)
PWM Controller	TL5001ID

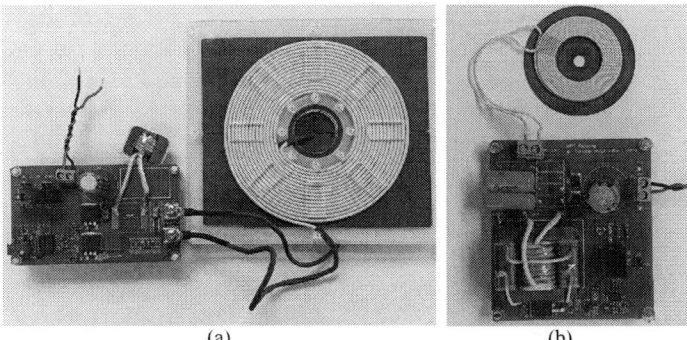

Fig. 5 Images of prototype WPT system (a) Transmitter (b) receiver

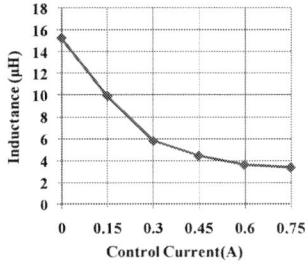

Fig. 6 Inductance vs. control current of the variable inductor

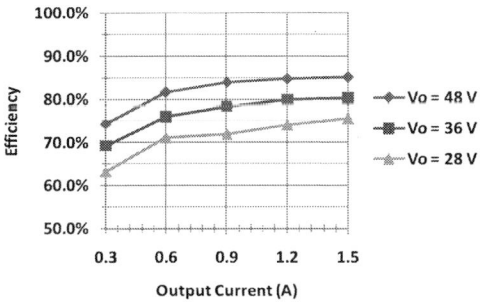

Fig. 7 Efficiency of prototype under different output voltage and current conditions

The transmitter side switching waveforms under various output voltages and load conditions are given in Fig 8-10. From the switching waveforms, it can be verified that the proposed modulation scheme does not impair the soft-switching of transmitter side semiconductor devices. Soft-switching is maintained under all output voltage settings and load current conditions.

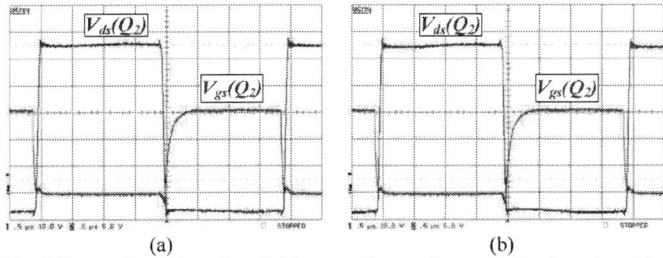

Fig. 8 Transmitter side soft switching waveforms when output voltage is set to 48 V (a) 0.3 A load current (b) 1.5 A load current

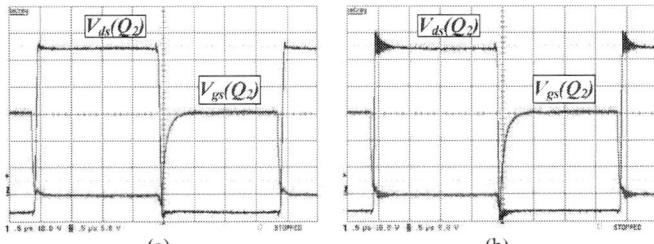

Fig. 9 Transmitter side soft switching waveforms when output voltage is set to 36 V (a) 0.3 A load current (b) 1.5 A load current

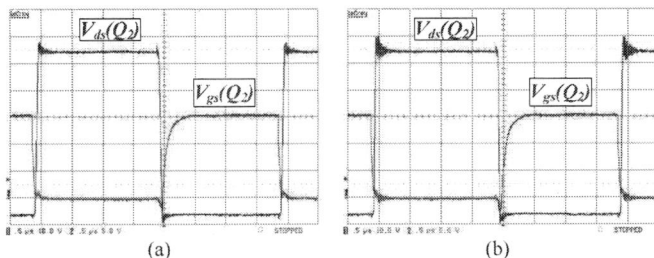

Fig. 10 Transmitter side soft switching waveforms when output voltage is set to 28 V (a) 0.3 A load current (b) 1.5 A load current

To verify output voltage regulation capability of the proposed regulation scheme, dynamic load testings under different output voltage conditions are performed using an electronic load. The output transients of one receiver under dynamic load test conditions are shown in Fig. 11-13. It can be seen from the test results that under different output voltage settings, the proposed regulation scheme steadily maintains output voltage at set values against changes in load current.

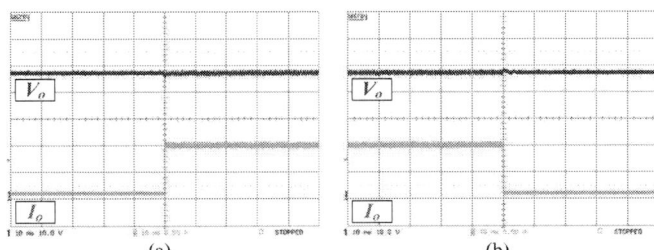

Fig. 11 Output transients under dynamic load condition when output voltage is set to 48 V (a) load current step-up from 0.1 A to 1 A (b) load current step-down from 1 A to 0.1 A

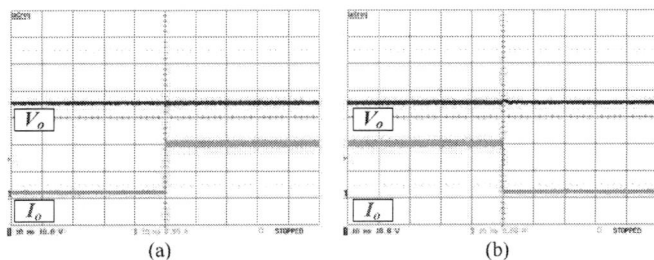

Fig. 12 Output transients under dynamic load condition when output voltage is set to 36 V (a) load current step-up from 0.1 A to 1 A (b) load current step-down from 1 A to 0.1 A

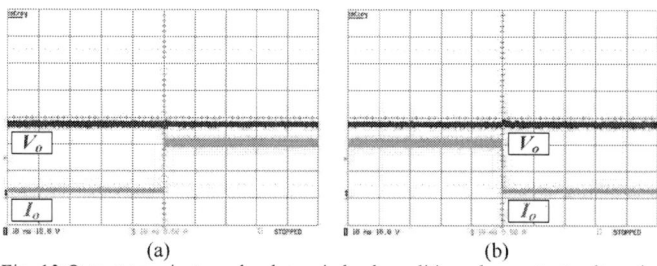

Fig. 13 Output transients under dynamic load condition when output voltage is set to 28 V (a) load current step-up from 0.1 A to 1 A (b) load current step-down from 1 A to 0.1 A

To demonstrate multi-receiver operation capability of the prototype WPT system, dynamic load testings are also performed when two receivers are receiving power from the transmitter coil at the same time. Fig. 14 shows the experimental setup for the multi-receiver testing. The two receivers are set to output different voltages to demonstrate independent wide-range output capability. The output voltage and current of two receivers under load transients are shown in Fig. 15. It can be seen from the two receiver dynamic load test results that increasing or decreasing the load current of one receiver does not impact the output voltage of the receiver itself, or the other receiver, verifying the independent controllability.

Fig. 14 Prototype WPT system under two-receiver test condition

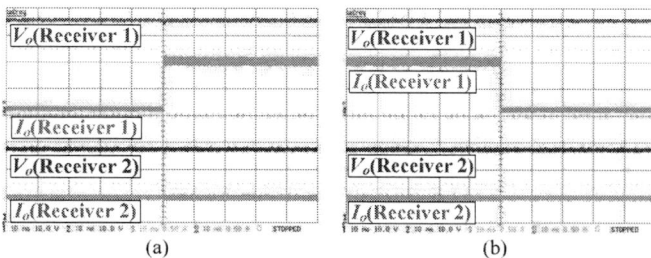

Fig. 15 Dynamic load testing with two receivers (a) Load step-up (b) Load step-down

IV. CONCLUSION

In this paper, a variable notch filter based receiver side regulation scheme for WPT systems with LCC transmitter side topology is proposed. By modulating the inductance of the receiver side varaible inductor, the impedance of the notch filter at receiver side can be adjusted in a wide-range to regulate output under different output voltage settings and load conditions. The modulator for controlling the variable inductor

is inherently isolated from the output or resonant circuit of the WPT system and does not handle high voltage or current. This approach eliminates the need of additional dc-dc converter stages or wireless communication requirements for output regulation, simplifies the WPT system, allowing the transmitter side to operate under constant frequency/duty cycle and improves reliability in high-noise environments by eliminating the possibility of system failure resulting from communication interruptions.

Furthermore, the proposed receiver side regulation scheme is particularly useful in multi-receiver WPT systems, where it would not be feasable for one transmitter to simutaniously regulate the outputs of multiple receivers that have different output/load conditions. A 48-$V_{O(max)}$, 72-W prototype WPT system is designed and built to verify the proposed regulation scheme, which reaches a peak efficiency of 85 % under full-load condition. The efficiency calculation includes the loss of all the auxiliary power supplies and gate drives. The dynamic load testings are performed under both single and dual receiver conditions, verifying the independent receiver side output voltage regulation as well as the multi-receiver operation capability of the prototype WPT system.

REFERENCES

[1] W. Li, H. Zhao, J. Deng, S. Li and C. C. Mi, "Comparison Study on SS and Double-Sided LCC Compensation Topologies for EV/PHEV Wireless Chargers," in *IEEE Transactions on Vehicular Technology*, vol. 65, no. 6, pp. 4429-4439, June 2016

[2] G. Zhu, D. Gao, S. Wang and S. Chen, "Misalignment tolerance improvement in wireless power transfer using LCC compensation topology," 2017 IEEE PELS Workshop on Emerging Technologies: Wireless Power Transfer (WoW), Chongqing, China, 2017, pp. 1-7, doi: 10.1109/WoW.2017.7959411.

[3] Q. Bo, Y. Zhang, Y. Guo, L. Wang, Z. Liu and S. Li, "Sensitivity Analysis to Parameter Variations in LCC-S Compensated Inductive Power Transfer Systems," *2020 IEEE PELS Workshop on Emerging Technologies: Wireless Power Transfer (WoW)*, Seoul, Korea (South), 2020, pp. 233-237

[4] Z. She, S. Chen, Y. Chen, Y. Zhang, H. Li and Y. Tang, "Efficiency Analysis of LCC-S and S-S Inductive Power Transfer Considering Switching Device and Component Losses," *2020 IEEE 9th International Power Electronics and Motion Control Conference (IPEMC2020-ECCE Asia)*, Nanjing, China, 2020, pp. 2956-2960

[5] J. Feng, Q. Li and F. C. Lee, "Coil and Circuit Design of Omnidirectional Wireless Power Transfer System for Portable Device Application," *2018 IEEE Energy Conversion Congress and Exposition (ECCE)*, Portland, OR, USA, 2018, pp. 914-920

[6] S. Wang, Z. Wang, J. Deng, Y. Yang and D. G. Dorrell, "Analysis of Multi-Pickup Inductive Power Transfer System with LCC Compensation for Maglev Train," 2019 IEEE Energy Conversion Congress and Exposition (ECCE), Baltimore, MD, USA, 2019, pp. 6762-6767

[7] T. Tan, K. Chen, Y. Jiang, Z. Zhao and L. Yuan, "Dynamic Modeling and Analysis of Multi-receiver Wireless Power Transfer System," *2019 IEEE PELS Workshop on Emerging Technologies: Wireless Power Transfer (WoW)*, London, UK, 2019, pp. 391-395

[8] M. Fan, J. Yang, L. Shi, W. Tang and Z. Yin, "A Novel Inductive Power Transfer System for Medium-low Speed Maglev Train Based on Double-ended Inverter," *2022 25th International Conference on Electrical Machines and Systems (ICEMS)*, Chiang Mai, Thailand, 2022, pp. 1-5, doi: 10.1109/ICEMS56177.2022.9983421

[9] M. Zavrel, V. Kindl and M. Tyrpekl, "Dynamic Wireless Charging Using LCC-S Compensation Topology in Low and Medium Power Applications," *2023 IEEE 32nd International Symposium on Industrial Electronics (ISIE)*, Helsinki, Finland, 2023, pp. 1-6

[10] M. Kim, D. -M. Joo and B. K. Lee, "Design and Control of Inductive Power Transfer System for Electric Vehicles Considering Wide Variation of Output Voltage and Coupling Coefficient," in *IEEE Transactions on Power Electronics*, vol. 34, no. 2, pp. 1197-1208, Feb. 2019

[11] J. Deng, Fei Lu, S. Li, T. -D. Nguyen and C. Mi, "Development of a high efficiency primary side controlled 7kW wireless power charger," 2014 IEEE International Electric Vehicle Conference (IEVC), Florence, Italy, 2014, pp. 1-6

[12] A. Ramezani and M. Narimani, "A New Wireless EV Charging System With Integrated DC–DC Magnetic Element," in IEEE Transactions on Transportation Electrification, vol. 5, no. 4, pp. 1112-1123, Dec. 2019

[13] F. Wu, Y. Wei, J. Su, K. Zhao and H. Liu, "Dual-Side Closed-Loop Control, Stability Analysis, and Parameter Design of Two-Stage LCC-LCC WPT," in *IEEE Journal of Emerging and Selected Topics in Power Electronics*, vol. 12, no. 1, pp. 305-315, Feb. 2024

[14] B. Pang, J. Deng, P. Liu and Z. Wang, "Secondary-side power control method for double-side LCC compensation topology in wireless EV charger application," *IECON 2017 - 43rd Annual Conference of the IEEE Industrial Electronics Society*, Beijing, China, 2017, pp. 7860-7865

[15] B. Chen, G. Yao, L. Zhou and M. Zhao, "Study on Secondary Side Control of Wireless Power Transfer Based on LCC Compensation," *2021 IEEE 3rd International Conference on Circuits and Systems (ICCS)*, Chengdu, China, 2021, pp. 196-202

[16] M. Wu *et al.*, "A New Control Method of Semi-bridgeless Active Rectifier for Wireless Power Transfer System Based on the Double-sided LCC Compensation," *2020 IEEE Energy Conversion Congress and Exposition (ECCE)*, Detroit, MI, USA, 2020, pp. 5514-5518

[17] S. Ann and B. K. Lee, "Analysis of Impedance Tuning Control and Synchronous Switching Technique for a Semibridgeless Active Rectifier in Inductive Power Transfer Systems for Electric Vehicles," in *IEEE Transactions on Power Electronics*, vol. 36, no. 8, pp. 8786-8798, Aug. 2021

[18] L. Solimene, F. Corti, S. Musumeci, A. Reatti and C. S. Ragusa, "A controlled variable inductor for an LCC-S compensated Wireless Power Transfer system," *IECON 2022 – 48th Annual Conference of the IEEE Industrial Electronics Society*, Brussels, Belgium, 2022, pp. 1-6

[19] V. S. Meshram, F. Corti, L. Solimene, S. Musumeci, C. S. Ragusa and A. Reatti, "Variable Inductor Control Strategy in LCC-S Compensated Wireless Power Transfer Application," 2023 AEIT International Annual Conference (AEIT), Rome, Italy, 2023, pp. 1-6

[20] C. -M. Lai, S. -J. Tsai, H. -E. Liu and D. -T. Lin, "Study of a Load-Independent LCC-S compensated WPT System with Variable-Inductor Variable-Capacitor (VIVC) Techniques," 2023 IEEE Transportation Electrification Conference and Expo, Asia-Pacific (ITEC Asia-Pacific), Chiang Mai, Thailand, 2023, pp. 1-8

[21] J. -U. W. Hsu, A. P. Hu and A. Swain, "A Wireless Power Pickup Based on Directional Tuning Control of Magnetic Amplifier," in IEEE Transactions on Industrial Electronics, vol. 56, no. 7, pp. 2771-2781, July 2009

[22] M. Zaheer, N. Patel and A. P. Hu, "Parallel tuned contactless power pickup using saturable core reactor," *2010 IEEE International Conference on Sustainable Energy Technologies (ICSET)*, Kandy, Sri Lanka, 2010, pp. 1-6

[23] X. Wang, M. Leng, L. He and S. Lu, "An Improved LCC-S Compensated Inductive Power Transfer System With Wide Output Voltage Range and Unity Power Factor," in *IEEE Transactions on Transportation Electrification*, vol. 10, no. 2, pp. 2342-2354, June 2024

[24] Z. H. Meiksin, "Comparison of Orthogonal-and Parallel-Flux Variable Inductors," in *IEEE Transactions on Industry Applications*, vol. IA-10, no. 3, pp. 417-423, May 1974

[25] D. Medini and S. Ben-Yaakov, "A current-controlled variable-inductor for high frequency resonant power circuits," *Proceedings of 1994 IEEE Applied Power Electronics Conference and Exposition - ASPEC'94*, Orlando, FL, USA, 1994, pp. 219-225 vol.1

84.7 percent Peak Efficiency Stress Tolerant DC DC Buck Converter For Li Ion Battery Driven Standby Circuits In 18nm FDSOI

Gautam Dey Kanungo
STMicroelectronics Pvt Ltd
DIG FEM TR&D
G Noida, India
gautam.kanungo@st.com

Pijush Kanti Panja
SiTime B.V.
(ex-STMicroelectronics Pvt Ltd)
Hague, Netherlands
pijushp@gmail.com

Vikas Bugade
STMicroelectronics Pvt Ltd
DIG FEM TR&D
Bengaluru, India
vikas.bugade@st.com

Kallol Chatterjee
STMicroelectronics Pvt Ltd
DIG FEM TR&D
G Noida, India
kallol.chatterjee@st.com

Abstract— A peak-current (I_{PK}) limited Pulse-Frequency Modulation (PFM)-based SoC-compatible buck converter is designed in scaled-CMOS technology. In an Li-ion compatible input supply range of 2.7V-4.2V, it delivers a 0.55V output. It employs a low-power Bandgap reference, in-built push-pull voltage-regulators with adaptive-biasing, duty-cycled sensors and a failsafe asynchronous block. This architecture alleviates the need for a clock and limits inrush current at startup. Made using 3.3V bulk-MOSFETs, device voltage stress is managed for each sub-block. It only needs external L & C of 4.7µH & 4.7µF. Occupying an area of 660µm x 1465µm, it targets *Standby* mode of operation for low-power applications like Internet-of-Things (IoT) and Bluetooth Low Energy (BLE), supporting a load-current of 0.5mA to 10mA with >80% efficiency and has >70% efficiency till 0.1mA. It achieves a peak efficiency of 84.7% at 5mA load.

Keywords— *Li-ion, SoC integrable, dc-dc, buck, PFM, IoT, Standby-mode, 18nm-FDSOI, stacked power stage, device stress-management, PMU,*

I. INTRODUCTION

IoT applications have penetrated almost all domains viz., from medical implants, wearables, smart home, to industrial sensors. Battery-driven and operating wirelessly, they require operating capacity from 10µWs to ~100mWs while being able to self-sustain for years – so high efficiency (>80%) is a must. They usually have four primary sub-blocks – *Sensors* to acquire the data, a *Processor* that can be a microcontroller, microprocessor, Tiny Machine-Learning Neural Processing Units (TinyML NPU), etc. to digest the data, a *Wireless Module* (e.g., WUR, WiFi, etc.) to communicate with the external world, and a power management unit (*PMU*) to power the preceding blocks as per their input supply requirement.

These SoCs have three distinct modes of operation. In *Retention* or *Deep-Sleep*, only the sensors are active. Current consumption is <100µA. When triggered, the processor turns up. Consuming 0.5mA to 10mA, this is the *Standby* mode. Finally, if the Wireless Module needs to communicate, the SoC is in *Active* mode. Current consumed can reach 100mA.

Today's NPUs and WURs are built in scaled CMOS technologies (≤10x nm) whose input supply gets as low as 0.5V (especially in Standby-mode) to get speed, power & area benefits from the scaling. The external power source remains the Li-Ion battery that shows a range of 2.7V to 4.2V over its lifetime. This necessitates a step-down PMU interface between the battery and the processor.

The PMU usually comes as a separate Power Management IC (PMIC) that is built in older technologies (≥180nm) where high voltage (e.g., 5V) devices are available [1]-[7]. This however increases the product cost, incurs multi-process dependence and causes board-integration losses for the IoT device. Thus the alternate trend to build SoC-integrable PMUs [8]-[13], in the same technology.

The battery's voltage decreases in a continuous manner, showing a Conversion Ratio (CR) – the ratio of the output voltage V_O (0.55V) by input voltage V_{IN} (4.2V to 2.7V) – of 0.13 to 0.2 over its lifetime. An ideal regulator can only have an efficiency as high as the CR, and thus is unacceptable. So, only switching-mode converters can work – Switched-cap (SC), inductor-based buck, or a hybrid of the two. An SC converter can be expected to have sawtooth efficiencies over the CR-range, peaking at certain CRs [14],[15]. Among the latter two, two architectures are good candidates while using low-voltage MOSFETs – stacked [8], [11]-[13] in case of buck, and Flying-Capacitor Multilevel (FCML) [9],[10] for hybrid. The stacked architecture is simpler to realize, as FCML brings in SC issues like charge-balancing [16].

A stacked dc-dc buck converter is made to cater to Standby-mode (0.5m to 10mA load) IoT blocks that operate at an input supply of 0.5-0.55V. It employs simple, well-established solutions and pushes them to their limits to get competitive (>80%) efficiency figures in its league, handling such low CRs.

The paper is organized as follows. Section II describes stacked architectures followed by the PFM scheme used at a top level. Section III goes into block-wise detail of the circuit.

979-8-3315-1612-3/25 $31.00 © 2025 IEEE

Section IV discusses silicon results followed by conclusions in Section V.

II. STACKED BUCK & PFM ARCHITECTURE

A. Stacked Buck Architecture

To avoid oxide damage and device degradation, low-voltage MOSFETs must not see terminal voltages higher than their rating. For the DC-DC's power stage, series devices are inserted, and their gates are driven by level-shifters with safe, limited voltage swing. Fig. 1 illustrates the modification on the right, when the technology provides 3.3V maximum rated devices for a high input supply say, V_{IN} = 5V.

The switching node V_X goes from V_{IN} to GND over one cycle of operation. So, the insertion of a second MOS on either side with their gates held at constant, suitable voltages protects the stacked MOSFETs from seeing the full supply-range. These gate voltages (VL & VH) are also conveniently used as power or ground supplies for the level-shifted driver-circuitry and other sub-blocks, ensuring them a safe voltage level. Although popular in literature [8]-[13], only few [8], [10] realize these supplies internally.

B. PFM Architecture

PFM operation [17] ensures high efficiency at Standby-mode power levels by linearly varying the buck-converter's switching-frequency (f_{sw}) with the load and making the conduction loss (P_{cond}) proportional to f_{sw} too (1), along with the switching losses (P_{sw}) – the two primary losses at such power levels. In (1), $R_{esr,L}$ is the DC resistance of the external inductor, R_{HI} & R_{LO} are the high-side & low-side on-resistances (Fig. 1), I_{PK} is the maximum inductor current level and f_{sw} (=1/T_{sw}) is the frequency of operation (Fig. 3).

$$P_{cond,Total} = \left(I_{Hi,rms}^2 R_{HI} + I_{Lo,rms}^2 R_{LO} + I_{L,rms}^2 R_{esr,L}\right),$$

$$I_{L,rms}^2 = \left(\frac{I_{PK}^3 L f_{sw}}{3}\right)\frac{V_{IN}}{V_O(V_{IN} - V_O)},$$

$$I_{Hi,rms}^2 = \left(\frac{I_{PK}^3 L f_{sw}}{3}\right)\frac{1}{(V_{IN} - V_O)},$$

$$I_{Lo,rms}^2 = \left(\frac{I_{PK}^3 L f_{sw}}{3}\right)\frac{1}{V_O} \tag{1}$$

PFM scheme can be implemented in various ways – I_{PK} limited [5], constant on-time (T_{HI}, Fig. 3) based, or adaptive on-time based [6]. Some schemes employ a low-frequency clock [1], [4] where the output voltage V_O is checked at the clock rising edge. The conversion is turned on only if required.

Fig. 2 and Fig. 3 show conceptually the PFM scheme implemented here. Three sensors work in a round-robin way at steady-state of operation. No clock is required this way.

a) Voltage-Sensor (VSENSE) that triggers when the output falls below a reference voltage (V_{Ref}), turning on the high-side switch (HI-SW),

b) a Current-Sensor (ISENSE) that trips when the inductor-current reaches a certain peak value (I_{PK}). It turns off the HI-

SW and turns on the low-side switch, LO-SW, causing the inductor current to start falling, and

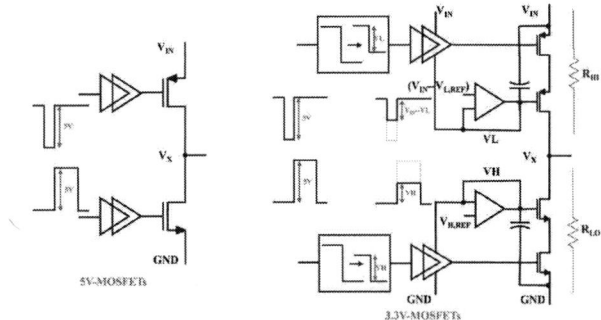

Fig. 1. Normal vs Stacked DC-DC Power Stage

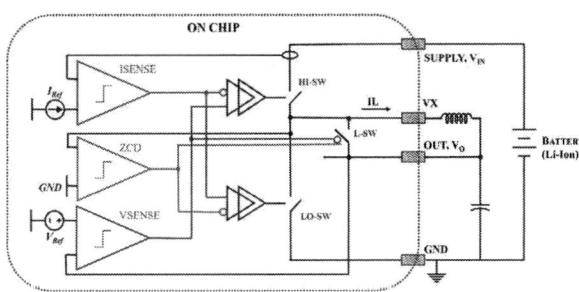

Fig. 2. PFM Top Level Architecture

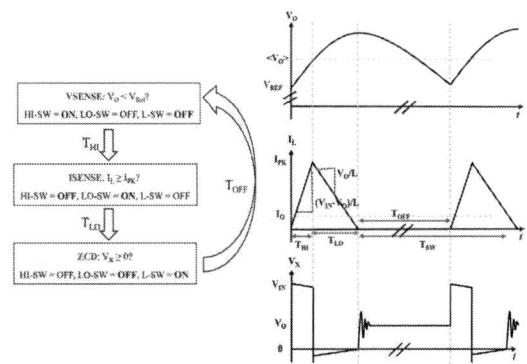

Fig. 3. PFM Top Level Functionality

c) a zero-current detector (ZCD) that trips when the inductor-current touches zero and is about to change direction and draw stored energy from the external capacitor. It turns off the LO-SW and shorts the inductor through the switch L-SW, de-magnetizing it with minimal oscillations at V_X node. This makes the converter enter discontinuous conduction mode (DCM) – zero current through the inductor, waiting for the output voltage to fall over time as per the load and restart the cycle from a). The asynchronous block ensures that.

An advantage of this scheme is that the peak current limiter ISENSE automatically limits the inrush current at startup too, alleviating the need for soft-start circuitry.

979-8-3315-1612-3/25 $31.00 © 2025 IEEE   822

III. LOW CR, HIGH VOLTAGE TOLERANT PFM BUCK

The circuit is tuned around a load of 1mA, the most likely consumption level of the target Standby block. The output power at V_O of 0.55V is 550µW. This amounts to an average current of about 130µA from a VIN of 4.2V. This gives us an idea of 1% efficiency loss figures – 5.5µW of power or 1.3µA of current from V_{IN} of 4.2V.

A. Power Stage & Regulators VH, VL

A stack of two 3.3V bulk MOSFETs is able to handle a maximum input voltage of V_{IN} = 4.2V. Assuming their maximum threshold voltage being less than 1V, $1V \leq VH \leq 3.3V$ ensures both LO-SW MOSFETs are safe. Similar offsets with respect to V_{IN} suffice for VL i.e., $3.2V \geq VL \geq 0.9V$ at 4.2V supply, protecting the HI-SW MOSFETs.

Considering choice for VH, there is a tradeoff between on-resistance of the LO-SW versus its CV^2f_{sw} switching losses. VH = 2.5V instead of 3.3V increases its on-resistance by 20%, while reducing the switching losses by 43%. It also ensures voltage safety during VH's transients, which are to be discussed shortly. Similarly, VL = (V_{IN} − 2.5V) is chosen for HI-SW.

Next, the width of the driver transistors was varied so that their on-resistances incurred P_{cond} in the range of 1% to 5% of efficiency. Optimal P_{cond} versus P_{sw} balance i.e., highest efficiency was achieved for Pcond = 3% for each side. P_{sw} is around 3%.

As we expect our circuit to work under low CR (0.13 to 0.2), an appropriate ratio of high- to low-side was kept so that both incur the same amount of P_{cond}. This means a seven times lower R_{LO} versus R_{HI} (2). Prioritizing R_{LO} helps optimize efficiency at such CRs.

$$I_{Hi,rms}^2 R_{HI} = I_{Lo,rms}^2 R_{LO}$$

$$\Rightarrow \frac{R_{HI}}{R_{LO}} = \frac{(V_{IN} - V_O)}{V_O} = \frac{(4.2V - 0.55V)}{0.55V} \simeq 7 \quad (2)$$

Coming to the VH & VL regulators, as shown in Fig. 4, they need to handle transient load currents in either direction. Thus, they need to have a push-pull output stage. When the HI-SW turns on, VX touches VIN. All nodes of the power-stage see a voltage jump (in green). This injects charges through C_{GS} & C_{DS} of the power devices to VH & VL. As the high-side gate node UP_BUF turns low, charges are dumped by the driver circuit to VL as well. Thus, VL experiences a higher jump.

When LO-SW turns on and HI-SW turns off, all power-stage nodes experience a voltage dip (in blue), with extra charges drawn from VH by the DN_BUF driver. Thus, VH sees a larger dip. Both VH & VL recover during the long period when the ZCD trips.

Thus, both VH & VL need 100pFs of decoupling capacitors to limit the voltage transients to safe level. So, both regulators are output-pole dominant. Further, as the static load consumption is very low (µAs), adaptive biasing is employed

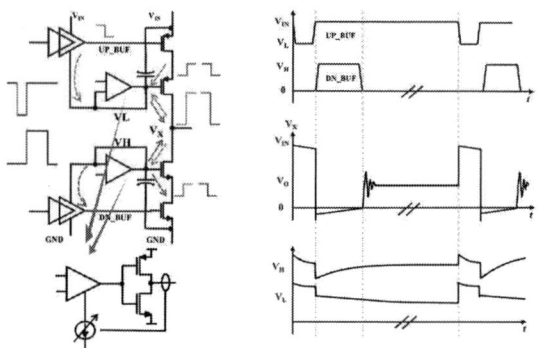

Fig. 4. VH & VL Regulators for Power Stage and their Operation

to increase their bandwidth when they experience increased output loads. These two regulators consume < 1% of the efficiency.

B. Failsafe Asynchronous Block, Level Shifters, and Drivers

Fig. 5 shows the complete block-wise implementation of the PFM converter. The circuit saves on using a clock by using an asynchronous block. It needs to handle all possible states of the circuit. Further, it ensures duty-cycled turn-on of current-heavy sensor blocks to drastically cut down their average power consumption (Fig. 6). Details follow in respective sensor sections.

Apart from the three states (triggered by the sensors VSENSE, ISENSE, and ZCD) of the circuit working in round-robin while at steady state or during startup, a fourth state is possible – when the output voltage is much higher (+20%) than V_{Ref} – an OUT_HI state.

Four states can be managed by *two* latches (Q1 & Q2). The digital asynchronous block is clearly demarcated into *three* sections that only have signals flowing in one direction between any two of them (Fig. 5): *Pulse-Generation* (VS, IS, ZCD, and OUT_HI), *State-Generation* (Q1, Q2), and *Signal-Conditioning* (RST_VS, RST_IS, RST_ZCD, UP, and DN).

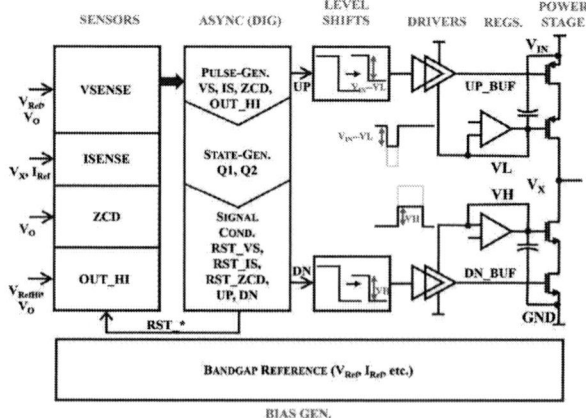

Fig. 5. PFM Buck, Block-Level

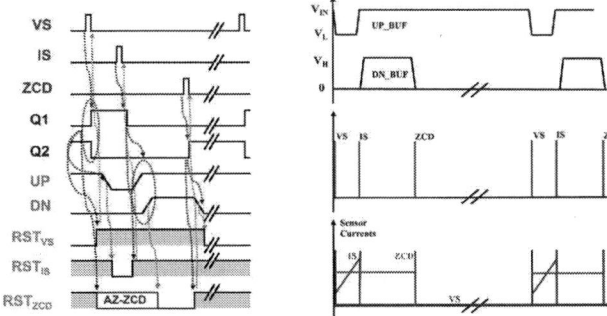

Fig. 6. Async. Block Timing Diagram & Sensors' Duty-Cycled Current Consumption

The signal-conditioning block ensures dead-zone between signals UP and DN. The reset signals (RST_*) are important in that they convert the sensor outputs from step to pulses, as required for correct operation of Q1 & Q2.

The UP & DN signals are suitably level-shifted and buffered to act as power-stage's gate signals UP_BUF & DN_BUF. The Level Shifters are switched-cap based to ensure low quiescent consumption like [18].

The Drivers consist of a series of sharply tapered buffers to reduce their switching power consumption. A few ns rise-fall slopes of UP_BUF & DN_BUF signals is acceptable.

The asynchronous block, the level shifters and the drivers consume about 1% of efficiency.

C. Output Sensor VSENSE

In the range of load currents under consideration, of the three sensors, the VSENSE must remain active for the longest time. Thus, its consumption must be minimized.

Further, it should resolve a V_O to V_{Ref} difference of about a fraction of the output ripple voltage. A 2% (= 11mV) ripple over the output supply is allowable. Thus, the VSENSE needs to resolve 1-2mV of voltage difference, consuming minimal current. This is ensured by implementing a low-power, high-gain comparator followed by an inverter in positive loop (Fig. 7). Efficiency hit is < 0.5%.

Fig. 7 shows the different states of VSENSE. De-assertion of RST_VS resets the comparator output, after which the comparator-to-output path is enabled. A small difference at its input results in sharp transition at the inverter's output. VS assertion triggers RST_VS assertion by the asynchronous block, which resets VSENSE's internal & output node.

D. Peak Current Limiter, ISENSE

High-side (HI_SW) current measurement technique (Fig. 8) is well established [19]. A small mirror copy of the HI_SW is placed in close proximity with the HI_SW. When active, the node VX_COPY tracks the VX node. The differential branches being biased by current sources, the extra ramping current, a scaled-down version of the HI_SW current, flows through the N_{Track1} & N_{Track2} MOSFETs. A current comparison occurs with

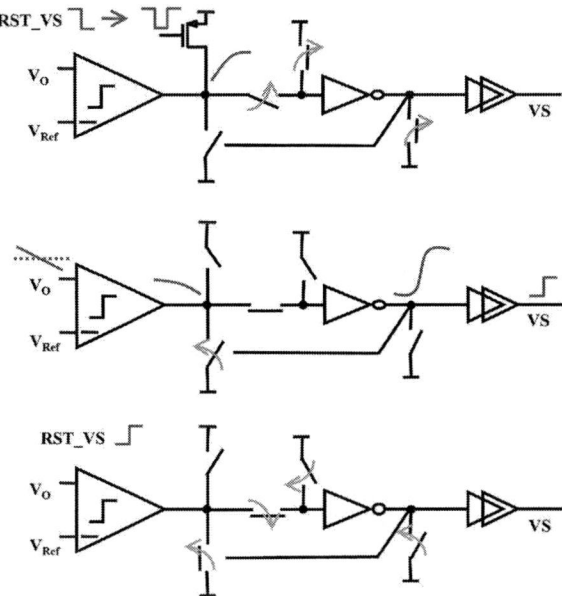

Fig. 7. Different States of VSENSE

the I_{Ref} trimmable reference current that trips the output IS

$$i_L \text{ rising slope,} \frac{\Delta i_L}{\Delta t} = \frac{(V_{IN} - V_O)}{L} = \frac{(4.2V - 0.55V)}{4.7uH}$$
$$= 77.6mA/100ns \tag{3}$$

when I_{PK} is reached.

To take care of voltage rating and save an extra op-amp, the main comparison differential branches themselves are modified to get a high gain using a cascode. Suitable protective devices (VL-biased) are inserted in the branches.

The ISENSE is the fastest of all three sensors (3), and burns high current. Thus, it is turned on only when HI_SW is on (Fig. 6). Fig. 8 shows it in off-state. This way, although reaching 100μA peak current consumption, it takes slightly more than 0.5% from efficiency.

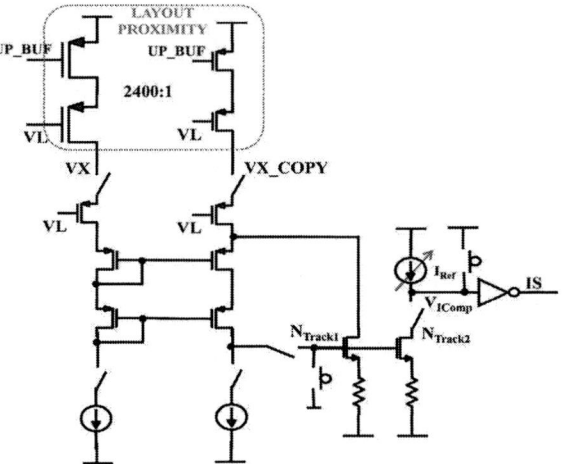

Fig. 8. ISENSE in Off-State

E. Analog ZCD

The ZCD sensor is the most challenging one of the three sensors, and is an active topic of research [20]. Complex and area & power-hungry auto-correcting digital & analog versions are pursued. Here, a trimmable analog ZCD is implemented (Fig. 9).

Any remnant current in the inductor when LO_SW turns off gets dissipated and is a loss in efficiency. As (4) indicates, the ZCD must respond in 10ns or less – fast decision needing high current. This is the first challenge.

$$i_L \text{ falling slope}, \frac{\Delta i_L}{\Delta t} = \frac{V_O}{L} = \frac{0.55V}{4.7uH} \approx 1.2mA/10ns \quad (4)$$

Second, as the ZCD monitors the voltage across the low-side rising above 0V by observing the V_X node, the lower the low-side's resistance, the tougher is its job. For example, a 1mA remnant current through a 0.2Ω R_{LO} means detecting a $200\mu V$ voltage difference. This necessitates the use of auto-zeroing (AZ).

Fig. 9 illustrates the different modes of the ZCD. The ZCD turns on and enters AZ mode when high-side is turned on, and AZ_ZCD is asserted. Offset is stored across the AZ-cap. When low-side turns on, AZ_ZCD is de-asserted. The input of the ZCD starts tracking V_X-node. A high-gain comparator is implemented to trip the ZCD output for low-level input differences.

Again, although its active mode consumption approaches $100\mu A$, duty-cycling (Fig. 6) manages to reduce its efficiency cost to slightly over 1%. In total, the three sensors consume slightly over 2% of efficiency.

F. Bias Generator

A conventional Bandgap reference consuming < 100nA was implemented, with requisite voltage-protections. It also provides current-references and all the different voltage biases required by various comparators and sensors. Bias references for ISENSE and ZCD although duty cycled still increase the block's consumption significantly so that it costs about 1% of the efficiency.

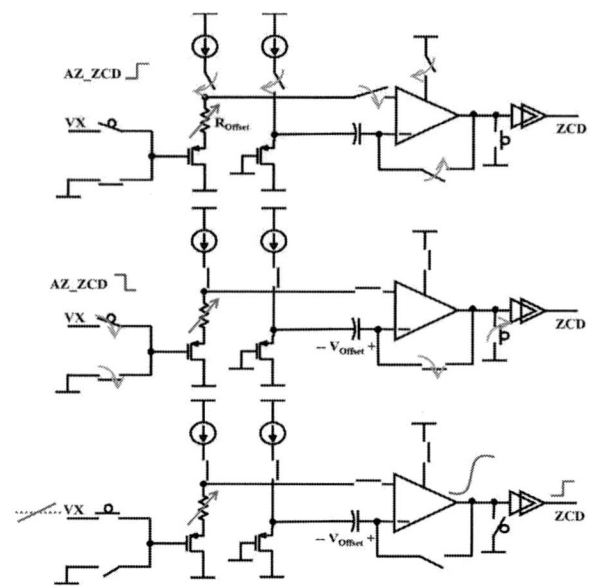

Fig. 9. Different States of ZCD

IV. SILICON MEASUREMENTS

The PFM Buck IP has a 660µm X 1465µm dimension. A testchip was built for it and fabricated. An inductor of 4.7µH and a capacitor of 4.7µF were used. Fig. 10 shows the IP layout and the chip micrograph.

The circuit was tested for VIN from 2.7V to 4.2V. Efficiencies for load current over two orders of magnitude – from 0.1mA to 10mA are plotted in Fig. 11. The test setup is shown in Fig. 12. Current drawn from the input supply source was noted. An external current sink drew different current loads, and the output voltage was noted. For this range, the buck converter shows an efficiency > 70%, achieving a peak efficiency of 84.7% for 5mA load at 2.7V.

Functionality of the PFM operation was tested by checking the behavior of the output in tandem with the inductor current and the V_X node (Fig. 13). Load steps between 0.5mA to 5mA

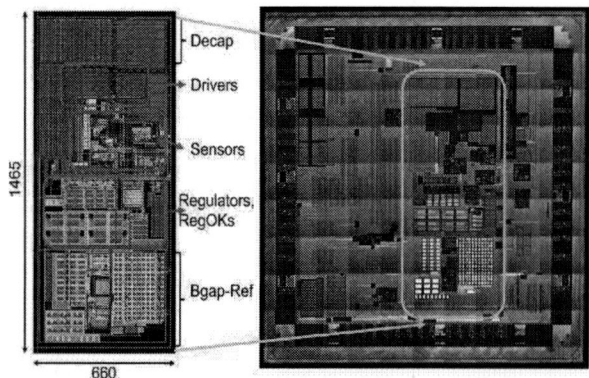

Fig. 10. Chip Layout & Micrograph

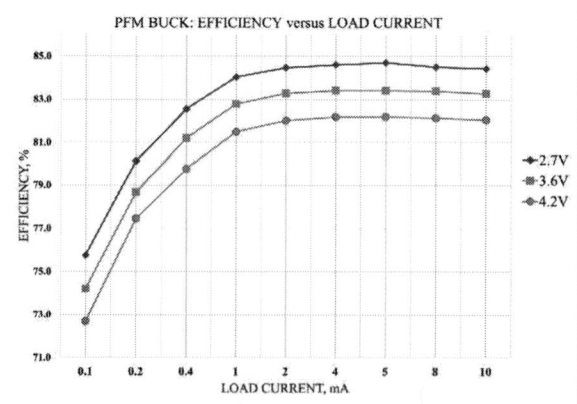

Fig. 11. PFM Buck Efficiency

979-8-3315-1612-3/25 $31.00 © 2025 IEEE

Table I Comparison with State-of-the-Art Buck Converters

	TPE17, [1]	TCAS20,[2]	TPE12, [8]	JSSC19,[9]	TPE23,[11]	*This Work*
Technology	130nm	180nm	350nm	28FDSOI	28FDSOI	*18FDSOI*
Stacked	N	N	Y	Y	Y	*Y*
Input, V	2.2-3.3	2.0-5.0	2.5-5.0	2.8-4.2	2.8-4.2	*2.7-4.2*
Output, V	1.7	0.8-3.0	1.0-4.0	0.6-1.2	1	*0.54*
Fsw, Hz	150k-1M		1.3M	200M	200k	*30k-200k*
Area, mm2	0.656	1.1	3.1	1.44	0.84	*0.97*
Load Range	0.5m-10mA	1m -50mA	10m-400mA	10m -40mA	500n-1mA	*0.1m-10mA*
L, H	3μ	2.2μ	4.7μ	3n	47μ	*4.7μ*
C, F	3μ	4.7μ	4.7μ	2x5n	4μ	*4.7μ*
Peak-Eff. % @(1/CR)	90.4@1.8	92.7@1.2	94@1.4	78@3.6	75@2.8	*84.7@5*
I_{LOAD}	10mA	1mA	60mA	10mA	0.1mA	*5mA*
Eff. % @max(1/CR)	82.7@1.8	76@4.5	73@1.4	48@7	67.8@4.2	*81.5@7.8*
I_{LOAD}	1mA	1mA	1mA	1mA	0.1mA	*1mA*

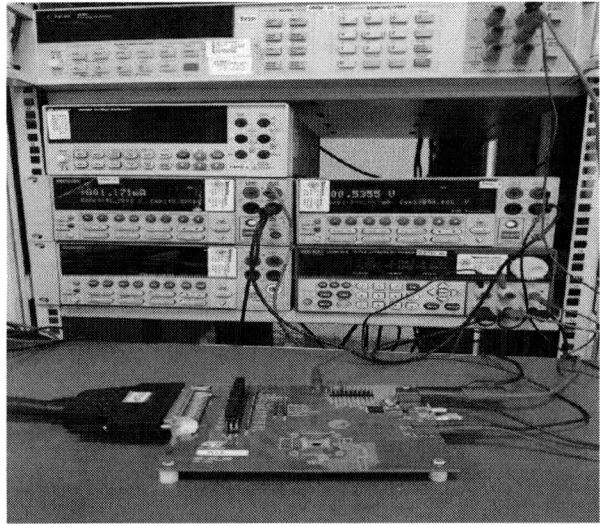

Fig. 12. Test Setup for measuring Efficiency

were applied at a 20μs time-slope (Fig. 14). The PFM circuit shows a load-regulation of 3mV for this step-size, without showing any over- or under-shoot.

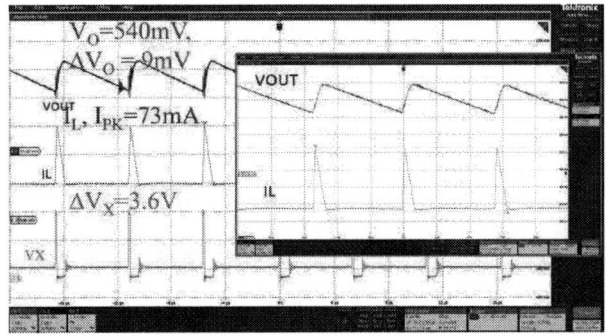

Fig. 13. PFM in Steady State

Lastly, the ZCD functionality was validated. Post-trim, the residual current in the inductor is 1mA. The VX node settles

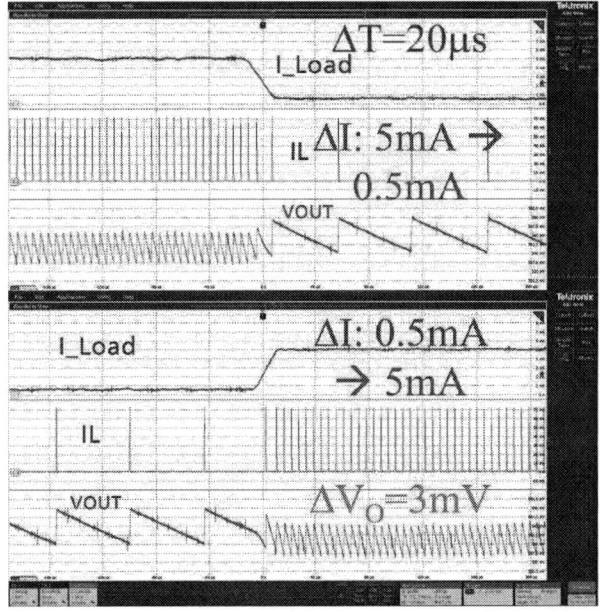

Fig. 14. Response to Load Transient Steps

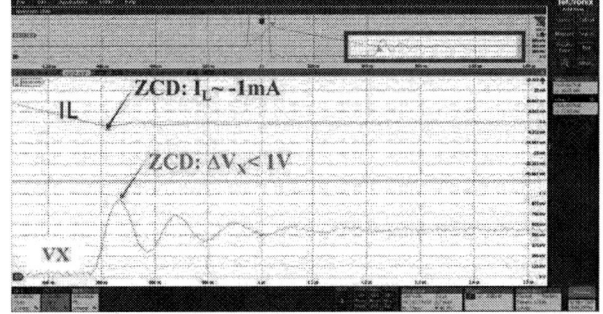

Fig. 15. ZCD Functionality Check

within 3 oscillations, with its peak remaining below 1V (Fig. 15).

Table I shows the comparison of this PFM Buck converter with other state of the art IoT-targeting converters. These are converters working in 0.1 to 10mA load currents, and/or have stacked architectures. The last row lists performance of all the converters at similar currents at their maximum 1/CR. Leaving [9] aside as it uses on-chip L & C, we can see that this converter has the highest 1/CR and rivals the efficiency of non-stacked one [1].

V. CONCLUSION AND FUTURE WORK

A SoC-integrable PFM dc-dc buck converter is built in scaled CMOS technology. Using judicious, uncomplicated design, strict power-budgeting, and duty-cycled sensors, it demonstrates better efficiency than many existing solutions at a higher (1/CR) demanded by upcoming Li-Ion operated Standby circuits. It does so while building the required intermediate supplies inside itself to avoid voltage stress of low-rated devices.

For future improvements, one is to replace the reliable but area-heavy conventional Bandgap reference by compact solutions. Another is to further optimize the power of different blocks and work towards 90% efficiency.

ACKNOWLEDGMENT (Heading 5)

We would like to thank our layout team including Varun Kumar Dwivedi, Madhvi Sharma, Neha Gupta, and Keyur R Bhatia for careful execution of the converter. Thanks go to our Testing team including Kapil Sharma and Alok Varshney for exhaustive testing of the IP. Finally thanks to Paras Garg for enabling us to explore this exciting and challenging venture.

REFERENCES

[1] Young-Jun Park, et al., "A design of a 92.4% efficiency triple mode control DC–DC buck converter with low power retention mode and adaptive zero current detector for IoT/wearable applications," IEEE Transactions on Power Electronics, vol. 32, no. 9, pp. 6946-6960, Sep. 2017.

[2] Wen-Liang Zeng, et al., "A 470-nA quiescent current and 92.7%/94.7% efficiency DCT/PWM control buck converter with seamless mode selection for IoT application," IEEE Trans. Circuits Syst. I, Reg. Papers, vol. 67, no. 11, pp. 4085-4098, Nov. 2020.

[3] W. Huang, L. Liu, X. Liao, C. Xu, and Y. Li, "A 240-nA quiescent current, 95.8% efficiency AOT-controlled buck converter with A^2-comparator and Sleep-Time Detector for IoT application," IEEE Transactions on Power Electronics, vol. 36, no. 11, pp. 12898-12909, Nov. 2021.

[4] Z. Gao, Y. Hao, H. Wei, Y. Li, and M. Chen, "A 96% peak efficiency adaptively controlled PSM buck converter with low-quiescent current and wide dynamic range for IoT applications," IEEE Solid-State Letters, vol. 5, no. 9, pp. 276-279, Nov. 2022.

[5] Jong-Seok Kim, J. Yoon, and B. Choi, "A high-light-load-efficiency low-ripple-voltage PFM buck converter for IoT applications," IEEE Transactions on Power Electronics, vol. 37, no. 5, pp. 5763-5772, May 2022.

[6] Menglian Zhao, et al., "An ultra-low quiescent current tri-mode dc-dc buck converter with 92.1% peak efficiency for IoT applications," IEEE Trans. Circuits Syst. I, Reg. Papers, vol. 69, no. 1, pp. 428-439, Jan. 2022.

[7] Tian Guo, Q. Duan, M. Lee, and J. Roh, "A 464-nA quiescent current buck converter with 93.7% peak efficiency for IoT devices," IEEE Transactions on Industrial Electronics, vol. 71, no. 7, pp. 8197-8201, Jul. 2024.

[8] Hyunseok Nam, Y. Ahn, and J. Roh, "5-V buck converter using 3.3-V standard CMOS process with adaptive power transistor driver increasing efficiency and maximum load capacity," IEEE Transactions on Power Electronics, vol. 27, no. 1, pp. 463-471, Jan. 2012.

[9] Sally Safwat Amin, and Patrick P. Mercier, "A fully integrated Li-Ion-compatible hybrid four-level dc-dc converter in 28-nm FDSOI," IEEE Journal of Solid-State Circuits, vol. 54, no. 3, pp. 720-732, Mar. 2019.

[10] Jeong-Il Seo, B. Lim, W. Choi, Y. Noh, and S. Lee, "A 95.1% efficiency hybrid hysteretic reconfigurable 3-level buck converter with improved load transient response," IEEE Transactions on Power Electronics, vol. 37, no. 12, pp. 14916-14925, Dec. 2022.

[11] Yunho Lee, et al., "A high-efficiency high-voltage-tolerant buck converter with inductor current emulator for battery-powered IoT devices," IEEE Transactions on Power Electronics, vol. 38, no. 9, pp. 10917-10932, Sep. 2023.

[12] Kim B. Ostman, J. K. Jarvenhaara, S. S. Broussev, and I. Viitaniemi, "A 3.6-to-1.8-V cascode buck converter with a stacked LC filter in 65-nm CMOS," IEEE Trans. Circuits Syst. II, Express Briefs, vol. 61, no. 4, pp. 234-238, Apr 2014

[13] Florian Neveu, B. Allard, C. Martin, Bp. Bevilacqua, and F. Voiron, "A 100 MHz 91.5% peak efficiency buck converter with a three-MOSFET cascode bridge," IEEE Transactions on Power Electronics, vol. 31, no. 6, pp. 3985-3988, Jun. 2016.

[14] Soyoung Bang, D. Blaauw, D. Sylvester, "A successive-approximation switched-capacitor dc-dc converter with resolution of vin/2n for a wide range of input and output voltages," IEEE Journal of Solid-State Circuits, vol. 51, no. 2, pp. 543-556, Feb. 2016

[15] Loai G. Salem, and P. P. Mercier, "A recursive switched-capacitor dc-dc converter achieving 2n-1 ratios with high efficiency over a wide output voltage range," IEEE Journal of Solid-State Circuits, vol. 49, no. 12, pp. 2773-2787, Dec.. 2014

[16] Xun Liu, Philip K. T. Mok, Junmin Jiang, and Wing-Hung Ki, "Analysis and design considerations of integrated 3-level buck converters," IEEE Trans. Circuits Syst. I, Reg. Papers, vol. 63, no. 5, pp. 671-682, May 2016.

[17] B. Arbetter, R. Erickson, and D. Maksimovic, "DC-DC converter design for battery-operated systems," IEEE PESC'95, vol. 1, pp. 103-109, Jun. 1995

[18] Zhidong Liu, L. Cong, and H. Lee, "Design of on-chip gate drivers with power-efficient high-speed level shifting and dynamic timing control for high-voltage synchronous switching power converters," IEEE Journal of Solid-State Circuits, vol. 50, no. 6, pp. 1463-1477, Jun. 2015

[19] C. F. Lee and Philip K. T. Mok, "A monolithic current-mode CMOS dc-dc converter with on-chip current-sensing technique," IEEE Journal of Solid-State Circuits, vol. 39, no. 1, pp. 3-14, Jan. 2004

[20] Ki-Chan Woo, J-M Oh, and B-D Yang, "DC-DC buck converter using analog coarse-fine self-tracking zero-current detection scheme," IEEE Trans. Circuits Syst. II, Express Briefs, vol. 66, no. 11, pp. 1850-1854, Nov. 2019

Leveraging Ultrasound and Neural Networks for Non-Invasive Power Converter Efficiency Estimation

Youssof Fassi
CEA-Leti, Université Grenoble
Alpes, F-38000 Grenoble, France
youssof.fassi@cea.fr

Vincent Heiries
CEA-Leti, Université Grenoble
Alpes, F-38000 Grenoble, France
vincent.heiries@cea.fr

Jérôme Boutet
CEA-Leti, Université Grenoble
Alpes, F-38000 Grenoble, France
jerome.boutet@cea.fr

Julien Marianne
SERMA INGENIERIE F-38000
Grenoble, France
julien.marianne@gmail.com

Sébastien Martin
CEA-Leti, Université Grenoble
Alpes, F-38000 Grenoble, France
sebastien-p.martin@cea.fr

Mathilde Chareyron
CEA-Leti, Université Grenoble
Alpes, F-38000 Grenoble, France
mathilde.chareyron@grenoble-inp.org

Clément Chambon
CEA-Leti, Université Grenoble
Alpes, F-38000 Grenoble, France
Mines Saint-Etienne, France
clement.chambon@cea.fr

Sébastien Boisseau
CEA-Leti, Université Grenoble
Alpes, F-38000 Grenoble, France
sebastien.boisseau@cea.fr

Abstract—**Measuring the efficiency of power converters traditionally involves invasive sensors or computationally heavy indirect methods. We propose as an emerging topic, a novel, contactless approach using ultrasound measurements to estimate power converter efficiency. Our method employs a multitask CNN-based neural network that simultaneously compresses and analyzes high-frequency ultrasound data. The network utilizes an AutoEncoder for efficient data compression, preserving essential features while reducing storage requirements. A custom loss function based on the characteristics of the signals optimizes both signal reconstruction and efficiency prediction, ensuring precise and reliable results. Experimental validation on a custom-built test bench demonstrates the effectiveness of this technique with a root mean squared error of 0.78%, being the first known exploitation of ultrasound measurements for efficiency estimation in power converters.**

Keywords— *Ultrasound-based monitoring, non-invasive testing, power converter, transformer, efficiency measurement, 1D-CNN, attention mechanism*

I. INTRODUCTION

The massive deployment of power converters across various domains, driven by the ongoing electrification of uses, has made them integral to modern infrastructure. However, efficiency losses in such widespread systems result in significant energy waste, and highlights the importance of accurately measuring power converter efficiency.

Nevertheless, measuring the efficiency of power converters presents several challenges. Traditional methods often involve invasive sensors to measure input and output currents and voltages, adding bulk to the system and complicating implementation on existing, on-the-field power converters [1-3]. Indirect methods are computationally heavy and rely on simulations, which can lack accuracy due to system complexity and operational variability. Calorimetric measurements, while accurate, are limited to offline estimations [4-6].

An emerging approach involves the use of non-invasive sensors for monitoring power converters [7]. Unlike electrical sensors, these probes can be positioned at a distance, reducing the risk of interference and the need for isolation. Ultrasound-based condition monitoring has been successfully applied in various industries, such as non-destructive evaluation (NDE) and monitoring of rotating machinery [8-12]. Studies used bandwidths ranging from a few hundred ok kHz to the MHz range, employing either stationary probes for point-based A-scans or moving probes for area-based B-scans. Despite its potential, its application to power converters remains relatively unexplored mainly due to the large data volumes involved, which poses significant challenges for storage and analysis. Data compression techniques, particularly through deep learning models like Convolutional Neural Networks (CNNs), offer promising solutions to data management challenges. CNN-based AutoEncoders can efficiently compress high-dimensional data while preserving local features [13-15]. This paper investigates a multitask neural network that compresses data and infers efficiency. Our analysis is validated through real-world measurements acquired using a custom-built test bench, demonstrating the practical applicability of our approach. To the best of the authors' knowledge, this is the first time that ultrasound measurements have been proposed for estimating power converter efficiency.

II. EXPERIMENTAL SETUP

As the power converter operates at varying power levels, changes in the magnetic field within its transformer induce electromagnetic forces that cause mechanical vibrations in the core components. These vibrations generate ultrasound signals, which vary depending on the level of mechanical stress and electromagnetic activity within the converter.

In this study, we explore the use of ultrasound signals generated intrinsically by a power converter's magnetic core during operation as a non-invasive tool for condition monitoring. Specifically, we employed a flyback topology rectifier as the power converter (UCC24630EVM-636), where the ultrasound signals arise naturally from the vibrations within magnetic components. The used test bench is illustrated in Figure 1.

979-8-3315-1612-3/25 $31.00 © 2025 IEEE

These signals were captured using a Brüel & Kjær 4138-A-015 probe, with a sensitivity of -65 dB relative to 1 V/Pa, operating within a frequency range of 6.5 Hz to 140 kHz and a dynamic range of 52.2 to 168 dB, positioned at 9 cm from the transformer of the power converter to maintain non-invasiveness. Unlike traditional approaches in Prognostics and Health Management (PHM), this method does not require active ultrasound injection; instead, it passively records acoustic emissions, providing a non-invasive and low-power diagnostic solution, compared with active signal injection methods.

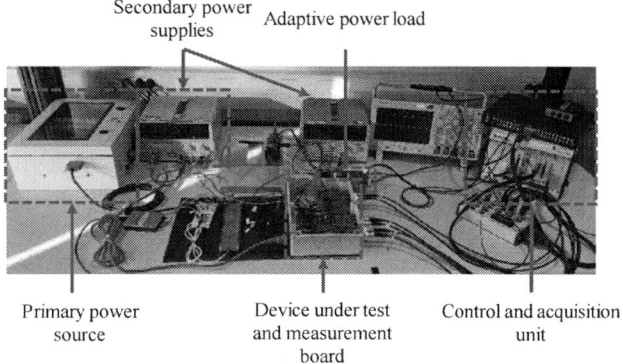

Figure 1: Experimental setup, the ultrasound probe goes on top of the device under test that is covered during operation

The generated ultrasound signals (Figure 2) span a continuous frequency range of 50-130 kHz, and acquisition was carried out at a sampling rate of 500 kHz. These signals are primarily influenced by mechanical vibrations within the power converter, such as the magnetic core, and are centered around 120 kHz—significantly higher than the switching frequency of most power converters, which typically operate in the lower kHz range, thereby lowering the impact of interference from switching operations.

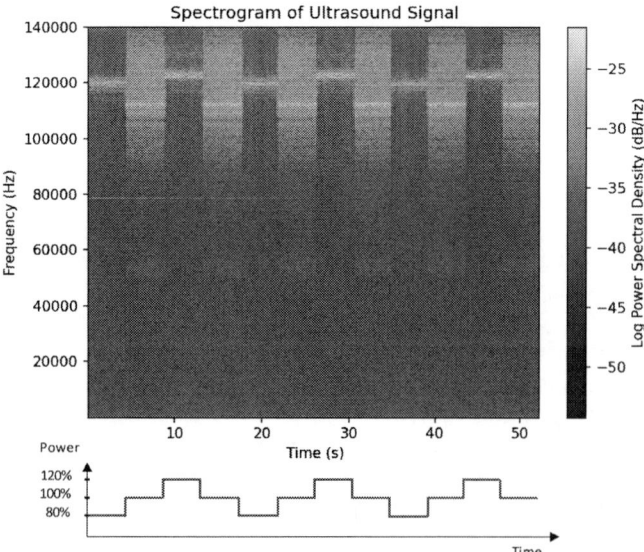

Figure 2: Spectrogram of ultrasound signal in three different power levels (the power is expressed as a percentage of the rated nominal power)

To examine the effects of operating power on the ultrasound signal characteristics, we used a custom adaptive power load circuit, emulating an RLC-type load. This configuration emulates more realistic loads compared to a purely resistive load, capturing the dynamic responses typical of power converters under varying conditions. This adaptive load setup allowed us to control and measure the ultrasound signals over a finely grained range of operating points, capturing the dynamic response as the converter transitions between different power levels. We specifically recorded signals at 80%, 100%, and 120% of nominal output power, each over a duration of 5 seconds, with stepwise changes to see power level fluctuations influence on ultrasound characteristics.

The data acquisition is conducted using a Spectrum MX4963 card controlled by a LabVIEW script. This script manages both the command and control operations: it sets the power converter to the desired power levels and simultaneously acquires the ultrasound signals.

III. EFFICIENCY INFERENCE MODEL

We introduce a CNN-based multitask network to enhance the generalization capability of the neural network, allowing it to extract features from the input streams that represent the efficiency of the power converter. The first task aims to reduce the volume of data, helping in data storage and processing, and the second infers the efficiency from that compressed representation.

A. Neural Network Architecture

The first task, compression, uses an AutoEncoder (AE) that consists of three main components: an encoder, which compresses the input into a lower-dimensional latent space; and a decoder (Figure 3-a) , which reconstructs the input from this latent representation using transposed convolution (deconvolution) layers [13-16]. In our approach, we use a 1D CNN architecture for both the encoder and the decoder due to its ability to preserve local features in time-series. We stack convolutional layers with padding and strides to achieve a compression ratio of 64. By comparison, the most commonly known audio signal compression ratios range from 6 to 10.

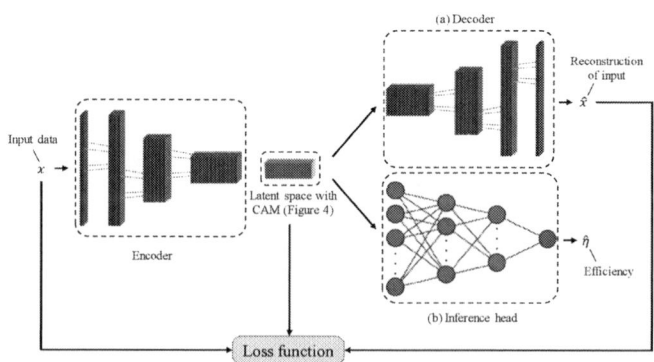

Figure 3: Architecture of the multitask neural network for efficiency estimation

979-8-3315-1612-3/25 $31.00 © 2025 IEEE

The second task, inference (Figure 3-b), takes the compressed set of features from the latent space and guides them through a fully connected network to infer the efficiency of the power converter.

B. Neural Network Training

To train the NN, we designed a custom loss function to optimize both tasks comprehensively. The objective function consists of two loss functions, one for each task of the neural network.

For the AutoEncoder component of the neural network, responsible for signal reconstruction, the loss function is composed of three primary terms that together promote signal reconstruction.

The first term in the loss function is the Mean Squared Error (MSE), which quantifies the squared differences between the reconstructed signal, denoted as $\hat{x}$, and the original signal x and can be expressed as in (1):

$$MSE(x, \hat{x}) = \frac{1}{N} \sum_{i=1}^{N} (x_i - \hat{x}_i)^2 \qquad (1)$$

To enhance the network's ability to reconstruct high-frequency components, a correlation coefficient term is included as the second component of the loss function (2). This term ensures alignment between the reconstructed and original signals by maximizing the linear correlation between them.

$$corr(x, \hat{x}) = \left(1 - \frac{\sum_{i=1}^{N}(x_i - \mu_x)(\hat{x}_i - \mu_{\hat{x}})}{\sqrt{\sum_{i=1}^{N}(x_i - \mu_x)^2 \sum_{i=1}^{N}(\hat{x}_i - \mu_{\hat{x}})^2}}\right)^2 \qquad (2)$$

Lastly, a frequency-domain loss term is incorporated to address errors specific to the spectral characteristics of the ultrasound signals. This term penalizes discrepancies between spectral decompositions of the original and reconstructed signals, with particular focus on meaningful frequency bands. Calculated using the Fast Fourier Transform (FFT) of both the reconstructed and original signals, the frequency-domain loss specifically targets errors within designated frequency bands, particularly within the 100-130 kHz range, as shown in (Figure 2). The FFT-based loss function, denoted as $\mathcal{L}_{FFT}$ in (3) combines two aspects: magnitude-based loss and complex-valued (real and imaginary) loss and is structured as follows:

$$\mathcal{L}_{FFT} = \sum_{(f_{start}, f_{end}) \in \mathcal{B}} \mathcal{L}_{mag}(f_{start}, f_{end}) \\ + \mathcal{L}_{comp}(f_{start}, f_{end}) \qquad (3)$$

where $\mathcal{B}$ is the set of frequency bands of interest. In our study, the primary band of interest that has been the analyzed is $\mathcal{B} = \{(100, 130)\}\ kHz$, and f_{start} and f_{end} denote the start and end frequencies of each band.

1) Magnitude-based Loss:
This component calculates the mean squared error between the magnitudes of the reconstructed and original signals within each frequency band as expressed in (4):

$$\mathcal{L}_{mag} = \sum_{band} \frac{1}{N_b} \sum_{f \in \mathcal{B}} \left\| |X(f)| - |\hat{X}(f)| \right\|^2 \qquad (4)$$

where $|X(f)|$ is the magnitude of the FFT, and N_b is the number of frequency components within each band.

2) Complex-valued Loss:
This term captures both the real and imaginary parts of the FFT within each frequency band, and evaluates the mean squared error between the neural network outputs and the ground truth target signal as in (5):

$$\mathcal{L}_{comp} = \sum_{band} \frac{1}{N_b} \sum_{f \in \mathcal{B}} \left(Re(X(f)) - Re(\hat{X}(f)) \right)^2 \\ + \left(Im(X(f)) - Im(\hat{X}(f)) \right)^2 \qquad (5)$$

where Re and Im denote the real and imaginary components, respectively.

The second one is the efficiency loss component, where the MSE minimization reduces the error between the predicted efficiencies $\hat{\eta}$ and the target ones η, which is expressed by (6):

$$MSE(\eta, \hat{\eta}) = \frac{1}{N} \sum_{i=1}^{N} (\eta_i - \hat{\eta}_i)^2 \qquad (6)$$

To compute the target efficiency η, we first determine the average power for both the input and the output of the system. The input power P_{in} is derived from the product of the input voltage $V_{in}(t)$ and input current $I_{in}(t)$, while the output power P_{out} is similarly derived from $V_{out}(t)$ and $I_{out}(t)$. To smooth the instantaneous power values over time, we apply a moving average with a window size w_s as represented in (7):

$$\overline{P_{in}}(t) = \frac{1}{w_s} \sum_{i=t-w_s+1}^{t} V_{in}(t).I_{in}(t) \\ \overline{P_{out}}(t) = \frac{1}{w_s} \sum_{i=t-w_s+1}^{t} V_{out}(i).I_{out}(t) \qquad (7)$$

The efficiency η is then calculated as in (8); the ratio of the average output power to the average input power:

$$\eta = \frac{\overline{P_{out}}}{\overline{P_{in}}} \qquad (8)$$

To focus on the most relevant features and improve the network's performance, we employ a Channel Attention Module (CAM) [20], which enhances learning by prioritizing specific channels in the latent space based on their importance. In convolutional neural networks, each channel of a feature map acts as a unique feature detector, and in this sense, the CAM exploits the inter-channel relationships by generating a channel attention map that emphasizes channels most relevant to the network's objectives and improving task-specific performance.

979-8-3315-1612-3/25 $31.00 © 2025 IEEE

The CAM (Figure 4) begins by aggregating spatial information from each feature map channel to generate descriptors that reflect the global context. This aggregation is achieved through both average pooling and max pooling, which create two distinct descriptors: the average-pooled feature and the max-pooled feature. These descriptors capture complementary information about the input data's spatial distribution, with average pooling emphasizing general intensity patterns and max pooling highlighting the most salient features.

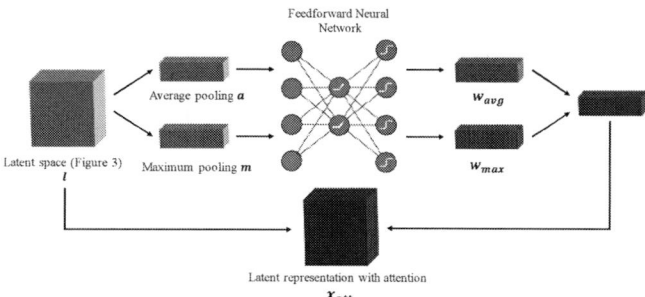

Figure 4: Diagram of the Channel Attention Mechanism applied on the latent space of Figure 3

Both pooled descriptors are then processed by a shared feedforward neural network with a hidden layer to produce the channel attention map. The network applies dimensionality reduction to limit parameter count and maintain computational efficiency. The operation can be expressed as in (9):

$$w_{avg} = \sigma\big(W_2 \cdot ReLU(W_1 \cdot a)\big)$$
$$w_{max} = \sigma\big(W_2 \cdot ReLU(W_1 \cdot m)\big) \tag{9}$$

where W_1 and W_2 are tuned weight matrices, $a = AvgPool(l)$ is the average pooling output for input signal l representing the latent features, $m = MaxPool(l)$ is the max pooling output, and σ is the sigmoid activation function.

The two processed feature vectors are then merged to form the final attention map, which is applied back to the input channels through element-wise multiplication, to scale each channel in the latent representation by its learned attention weight (10), thereby improving its task-specific contribution.

$$x_{att} = l \odot (w_{avg} + w_{max}) \tag{10}$$

where $\odot$ denotes the Hadamart product for element-wise multiplication.

IV. RESULTS AND DISCUSSIONS

To provide a clearer understanding of the neural network configuration used in this study, the main parameters of the neural network model and the input configuration are summarized in TABLE I.

TABLE I. NEURAL NETWORK AND INPUT PARAMETERS

Parameter	Value
Input signal duration	100 ms
Overlap of the sliding window	0.8
Number of conv layers	5
Kernel size (per layer)	[309, 229, 159, 69, 9]
Stride	4
Number of connected layers	4
Activation Function for the AutoEncoder	Sigmoid
Activation function for the Inference	ReLU

The proposed architecture was trained and tested on a single device with a 80%-20% split, then validated on a distinct unit of the same power converter model, separate from the one used during training, to assess the generalization capability of the neural network.

The architecture demonstrates strong predictive capability for efficiency estimation (**Figure 5**) across various high-load conditions (80% to 120% of nominal load), achieving a mean root mean square error (RMSE) of 0.78% ± 0.03% on the validation unit.

Achieving an RMSE of less than 1% on this unseen unit demonstrates the model's robustness with respect to the device-to-device variability.

An important contributor to this precision is the channel attention mechanism (CAM), which re-weights feature channels to focus the network's attention on relevant information while reducing noise from less informative features. This targeted attention has enhanced the model's efficiency prediction accuracy by 30%, further boosting generalization across varying operational conditions.

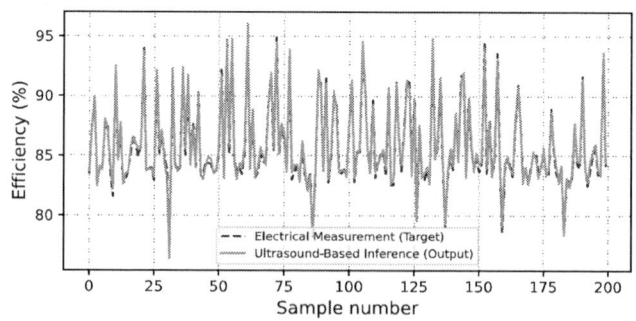

Figure 5: Generalization results comparing the electrical-based efficiency measure with the neural network estimates, the sample were drawn randomly from the validation dataset

Beyond prediction accuracy, the model's inference time is notably faster than indirect traditional efficiency measurement methods, which often rely on repeated simulations and intensive parameter sweeps. Once trained, the neural network can make predictions in as little as 3 ms on a desktop-based run (Intel i7-13800H CPU). In practical applications, we can tolerate longer inference times—up to three orders of magnitude more—allowing flexibility for future efforts in embeddability. Additionally, during inference, the encoding process used in the architecture enables compact storage of high-resolution efficiency data, reducing the storage and transmission demands typical in field conditions with bandwidth constraints.

Moreover, while this ultrasound-based monitoring approach may initially seem more intricate than traditional methods, it offers distinct advantages for on-the-field testing. Rather than requiring dedicated measurement equipment for each power converter, the ultrasound probe and processing unit can be shared across multiple converters, optimizing asset utilization. This mutualization of hardware makes the approach cost-effective in large-scale deployments, as a single sensor setup can be reused for monitoring several units in the field. Additionally, the ultrasound-based approach facilitates a non-invasive, passive monitoring process that is ideal for on-site conditions, eliminating the need for extensive load setups and reducing the logistics associated with traditional efficiency measurement equipment.

Although the current dataset focuses on high-load conditions, future work aims to extend the range to include smaller load increments from low to maximum load levels, providing a comprehensive efficiency profile across the entire operational range. This expanded dataset will allow a more detailed validation of the model's robustness, particularly at lower load levels where efficiency variations may be more sensitive. In summary, these results highlight the feasibility of the proposed ultrasound-based monitoring method as a scalable, low-cost solution for accurate efficiency estimation in field applications.

V. Conclusion

The results highlight the potential of this passive ultrasound-based monitoring technique for non-invasive, on-field diagnostics of power converters. By learning fundamental efficiency characteristics tied to intrinsic operational features rather than unit-specific details, the proposed architecture demonstrates promising generalization within the same converter model.

Currently, the model demonstrates effective efficiency estimation in high-load setting and sets the stage for a more comprehensive characterization of power converters across a full load spectrum. Planned data collection at lower load points will enable a more comprehensive evaluation of power converter performance across the full load spectrum, especially at low load currents, where achieving low error rates can be challenging.

In terms of scalability, exploiting this approach is promising to prove the suitability for diverse converter models and operational conditions. As datasets expand to include multiple power converters across different topologies, and sensor technology advances, we anticipate further accuracy gains at all load levels. Innovations in data processing ICs may also reduce costs, increasing competitiveness with traditional methods for field operations.

One potential limitation is the influence of probe placement on measurement accuracy. Future studies could investigate a wider range of placements, particularly in varied environments, to address any potential impact on diagnostics.

Acknowledgment

This work is part of the IPCEI Microelectronics and Connectivity and was supported by the French Public Authorities within the frame of France 2030.

References

[1] L. Aarniovuori, H. Kärkkäinen, A. Anuchin, J. J. Pyrhönen, P. Lindh, and W. Cao, "Voltage-Source Converter Energy Efficiency Classification in Accordance With IEC 61800-9-2," IEEE Trans. Ind. Electron., vol. 67, no. 10, pp. 8242–8251, Oct. 2020, doi: 10.1109/TIE.2019.2949526.

[2] Z. Zhao, P. Davari, W. Lu, H. Wang, and F. Blaabjerg, "An Overview of Condition Monitoring Techniques for Capacitors in DC-Link Applications," IEEE Trans. Power Electron., vol. 36, no. 4, pp. 3692–3716, Apr. 2021, doi: 10.1109/TPEL.2020.3023469.

[3] M. Crescentini, S. F. Syeda, and G. P. Gibiino, "Hall-Effect Current Sensors: Principles of Operation and Implementation Techniques," IEEE Sens. J., vol. 22, no. 11, pp. 10137–10151, Jun. 2022, doi: 10.1109/JSEN.2021.3119766.

[4] C. Xiao, G. Chen, and W. G. H. Odendaal, "Overview of Power Loss Measurement Techniques in Power Electronics Systems," IEEE Trans. Ind. Appl., vol. 43, no. 3, pp. 657–664, May 2007, doi: 10.1109/TIA.2007.895730.

[5] A. Kadavelugu, H. Suryanarayana, L. Liu, Z. Pan, C. Belcastro, and E.-K. Paatero, "A simple and accurate efficiency measurement method for power converters," in 2017 IEEE Applied Power Electronics Conference and Exposition (APEC), Mar. 2017, pp. 3265–3270, doi: 10.1109/APEC.2017.7931165.

[6] S. Schönewolf, A. Nagel, and M.-M. Bakran, "Power Dissipation Measurements at Power Pulsation Compensators for Single-Phase Grid-Connected Converters," IEEE Trans. Power Electron., vol. 39, no. 7, pp. 8202–8214, Jul. 2024, doi: 10.1109/TPEL.2024.3375652.

[7] Y. Fassi, V. Heiries, J. Boutet, and S. Boisseau, "Toward Physics-Informed Machine-Learning-Based Predictive Maintenance for Power Converters—A Review," IEEE Trans. Power Electron., vol. 39, no. 2, pp. 2692–2720, Feb. 2024, doi: 10.1109/TPEL.2023.3328438.

[8] Y. Yao, Y. Pan, and S. Liu, "Power ultrasound and its applications: A state-of-the-art review," Ultrason. Sonochem., vol. 62, p. 104722, Apr. 2020, doi: 10.1016/j.ultsonch.2019.104722.

[9] F. König, C. Sous, A. Ouald Chaib, and G. Jacobs, "Machine learning based anomaly detection and classification of acoustic emission events for wear monitoring in sliding bearing systems," Tribol. Int., vol. 155, p. 106811, Mar. 2021, doi: 10.1016/j.triboint.2020.106811.

[10] M. T. Pham, J.-M. Kim, and C. H. Kim, "Rolling Bearing Fault Diagnosis Based on Improved GAN and 2-D Representation of Acoustic Emission Signals," IEEE Access, vol. 10, pp. 78056–78069, 2022, doi: 10.1109/ACCESS.2022.3193244.

[11] S. Cantero-Chinchilla, P. D. Wilcox, and A. J. Croxford, "Deep learning in automated ultrasonic NDE – Developments, axioms and opportunities," NDT E Int., vol. 131, p. 102703, Oct. 2022, doi: 10.1016/j.ndteint.2022.102703.

[12] H. Sun, P. Ramuhalli, and R. E. Jacob, "Machine learning for ultrasonic nondestructive examination of welding defects: A systematic review," Ultrasonics, vol. 127, p. 106854, Jan. 2023, doi: 10.1016/j.ultras.2022.106854.

[13] L. Cavigelli, P. Hager, and L. Benini, "CAS-CNN: A deep convolutional neural network for image compression artifact suppression," in 2017 International Joint Conference on Neural Networks (IJCNN), May 2017, pp. 752–759. doi: 10.1109/IJCNN.2017.7965927.

[14] S. Bai, J. Z. Kolter, and V. Koltun, "An Empirical Evaluation of Generic Convolutional and Recurrent Networks for Sequence Modeling," Apr. 19, 2018, arXiv: arXiv:1803.01271. doi: 10.48550/arXiv.1803.01271.

[15] L. Ren, T. Wang, Z. Jia, F. Li, and H. Han, "A Lightweight and Adaptive Knowledge Distillation Framework for Remaining Useful Life Prediction," IEEE Trans. Ind. Inform., vol. 19, no. 8, pp. 9060–9070, Aug. 2023, doi: 10.1109/TII.2022.3224969.

[16] H. Shao, H. Jiang, H. Zhao, and F. Wang, "A novel deep autoencoder feature learning method for rotating machinery fault diagnosis," Mech. Syst. Signal Process., vol. 95, pp. 187–204, Oct. 2017, doi: 10.1016/j.ymssp.2017.03.034.

[17] H. Liu, J. Zhou, Y. Zheng, W. Jiang, and Y. Zhang, "Fault diagnosis of rolling bearings with recurrent neural network-based autoencoders," ISA Trans., vol. 77, pp. 167–178, Jun. 2018, doi: 10.1016/j.isatra.2018.04.005.

[18] X. Wu, Y. Zhang, C. Cheng, and Z. Peng, "A hybrid classification autoencoder for semi-supervised fault diagnosis in rotating machinery," Mech. Syst. Signal Process., vol. 149, p. 107327, Feb. 2021, doi: 10.1016/j.ymssp.2020.107327.

[19] M. Russell and P. Wang, "Physics-informed deep learning for signal compression and reconstruction of big data in industrial condition monitoring," Mech. Syst. Signal Process., vol. 168, p. 108709, Apr. 2022, doi: 10.1016/j.ymssp.2021.108709.

[20] S. Woo, J. Park, J.-Y. Lee, and I. S. Kweon, "CBAM: Convolutional Block Attention Module," Jul. 18, 2018, arXiv: arXiv:1807.06521. doi: 10.48550/arXiv.1807.06521.

A Load-Independent Multi-Relays Wireless Power Transfer with Self-Regulation and Single Compensation Network

Jong-Hun Kim
Graduate School of Semiconductor Technology
Pohang University of Science and Technology (POSTECH)
Pohang, Republic of Korea
kimjonghoon8@postech.ac.kr

Najam Ul Hassan
Department of Electrical Engineering
University of Michigan
Dearborn, USA
najam141@gmail.com

Seogyong Jeong
Samsung Electronics
Seoul, Republic of Korea
seogy.jeong@samsung.com

Myeong-Ho Kim
Graduate School of Semiconductor Technology
Pohang University of Science and Technology (POSTECH)
Pohang, Republic of Korea
mhkim98@postech.ac.kr

Min-Sik Kim
Department of Electrical Engineering
Pohang University of Science and Technology (POSTECH)
Pohang, Republic of Korea
minsik99@postech.ac.kr

Jee-Hoon Jung
Department of Electrical Engineering
Ulsan Institute of Science and Technology (UNIST)
Ulsan, Republic of Korea
jhjung@unist.ac.kr

Byunghun Lee
Department of Electronic/Biomedical Engineering
Hanyang University
Seoul, Republic of Korea
blee22@hanyang.ac.kr

Se-Un Shin
Department of Electrical Engineering
Pohang University of Science and Technology (POSTECH)
Pohang, Republic of Korea
seuns@postech.ac.kr

Abstract— **Wireless power transfer (WPT) technology is gaining prominence as a research focus in both industry and academia, particularly for applications where delivering power via metallic conductors is impractical. This paper introduces a WPT system designed for easy implementation near real-world demands, achieved through reduced circuit components and manufacturing costs. The system efficiently supplies power to various electrical devices that require constant-voltage (CV), constant-current (CC), or both CV/CC modes simultaneously like lithium-ion (Li-Ion) batteries. A single compensation network enables straightforward attainment of either CV or CC output, facilitated by a load-independent power-stage and control methodology. Experimental results demonstrate a PWM-based resonant regulating rectifier (3R), achieving a maximum power transfer efficiency (PTE) of 74.5% under two loads positioned 300 mm apart from a half-bridge class-D power transmitter (Tx).**

Keywords—Wireless power transfer, multi-relays, load-independent, constant-voltage, constant-current, compensation network

I. INTRODUCTION

Wireless power transfer (WPT) is widely used from low-power applications such as implantable medical devices (IMDs), including artificial heart, retinal implant, and deep brain stimulation, to high-power applications including online monitoring systems, electric vehicles, and underwater devices [1-4]. Among these, the most representative load is the lithium-ion battery, as shown in Fig. 1, which illustrates the internal resistance and charging characteristics of the lithium-ion battery over time [2]. Depending on its charge or discharge state, a lithium-ion battery requires constant-current (CC) mode charging in the early stages to prevent excessive current inrush, while in the later stages, it should be charged using constant-voltage (CV) mode. Therefore, the charger must have load-independence to deliver regulated power regardless of the load resistance and must be able to select either CC or CV output characteristics.

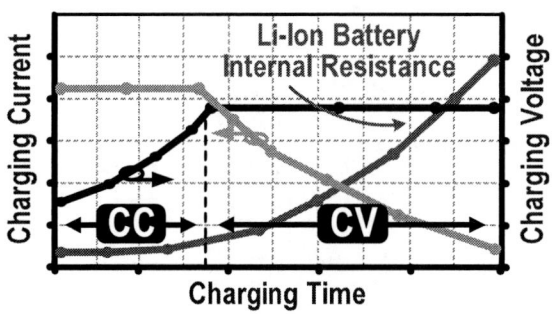

Fig.1. Charging characteristics of a lithium-ion battery.

Fig.2 illustrates conventional, previous, and the proposed wireless power receivers (Rxs). A conventional Rx consists of two stages: an AC-DC rectifier and a DC-DC converter for output regulation. It offers the advantage of flexibility in selecting the compensation network. However, additional circuit components are required for load-independence, leading to bulkiness, increased cost, and cascaded power losses [1-2]. Recent studies have proposed a hybrid-compensation network to achieve load-independent CV/CC outputs [2-3]. In [2], a hybrid topology that can switch between series-series-series (S-S-S) and series-series-LCC (S-S-T) compensation was proposed. However, this work employed back-to-back (B2B) power MOSFETs in the high-current compensation network path to prevent the conduction of body diode, which degrades power conversion efficiency (PCE). Although load-independent CV/CC output can be achieved, an unnecessary intermediate LC-tank is used, and an additional DC-DC converter is required to regulate the output. In [3], a wireless receiver was proposed to deliver power to multiple loads through the switchable compensation networks between relays, while still achieving load-independent CV/CC characteristics. However, this structure not only includes auxiliary LC-tank and anti-series MOSFETs in the power paths like [2], but requires a DC regulation circuits.

Fig. 2(c) shows the proposed wireless power receiver. This structure offers the following advantages:

1) Load-independent CV/CC characteristics are achieved using a series compensation network composed of only a single capacitor. Unlike previous studies, no additional L, C components or power switches are required in the compensation network, which improves power conversion efficiency and reduces system production costs.

2) The elimination of the intermediate LC-tank removes unnecessary circuit components and simplifies the analytical process, avoiding the need for complex calculations.

3) A resonant-regulating rectifier (3R) is selected as the power conversion topology, allowing both AC-DC rectification and DC-DC conversion to be performed in a single power conversion stage [5-7]. This reduces the power loss, which occurred twice in conventional or previous studie's power conversion circuits, to a single occurrence.

4) By repeating this relay structure, power can be delivered to multiple independent loads from a single Tx without any mutual interference in the outputs.

The rest of this paper is organized as follows. Section II presents the proposed load-independent CV/CC wireless power transfer system and explains the individual blocks. Section III provides experimental verification of the prototype and analyzes the results. Finally, Section IV concludes the paper.

II. THE PROPOSED MULTI-RELAYS WIRELESS RECEIVER

Fig.3 shows the schematic of the proposed load-independent CV/CC WPT system with single capacitor-only compensation network. The Rx features a simple compensation network using only one capacitor C_1 without any power-consuming switch for hybrid topology, and it does not require an intermediate coil. In addition, the 3R and a PWM controller are applied to regulate the output, achieving load-independent CV/CC output as in previous works. Ultimately, the same relays are repeated to extend the transfer distance and deliver power to multiple loads.

A. Resonant-Regulating Rectifier (3R)

The 3R, consisting of two Schottky diodes D_1, D_2, and two SiC MOSFETs S_3, S_4, enables AC-DC rectification and DC regulation in a single stage. On the low-side, MOSFETs S_3 and S_4 are controlled by a single digital signal, *DUTY*. When these MOSFETs are off, D_1, D_2, and the body diodes of S_3, S_4 form a full-wave rectifier (FWR) to deliver power to the load R_{OUT}. Conversely, when the MOSFETs are on, the LC-tank freewheels to prevent excessive power delivery to the load. Through this operation of the 3R and PWM control, both AC-DC rectification and output voltage or current regulation can be achieved in a single stage.

B. PWM Controller

Feedback resistors R_{FB}, a Hall-effect current sensor, a mode selection switch, an OP-AMP, a type-2 compensator, and a comparator constitute the PWM controller, and it generates the

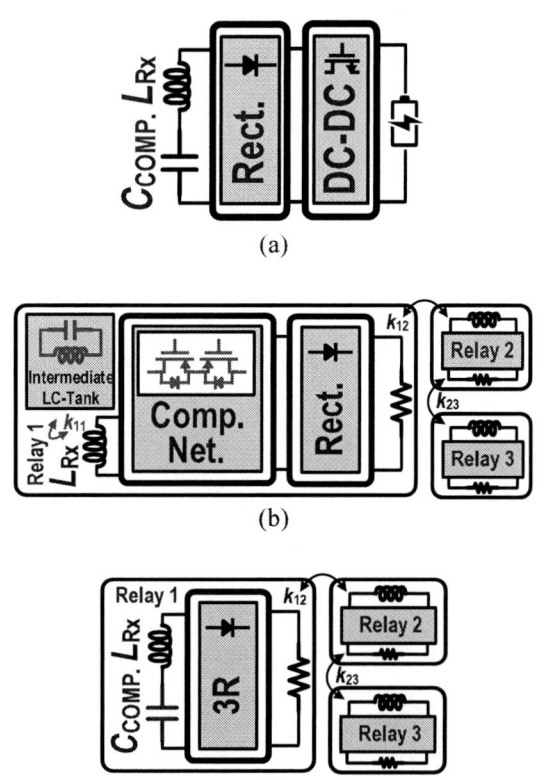

Fig.2. Wireless power receivers: (a) conventional, (b) previous work, (c) proposed work.

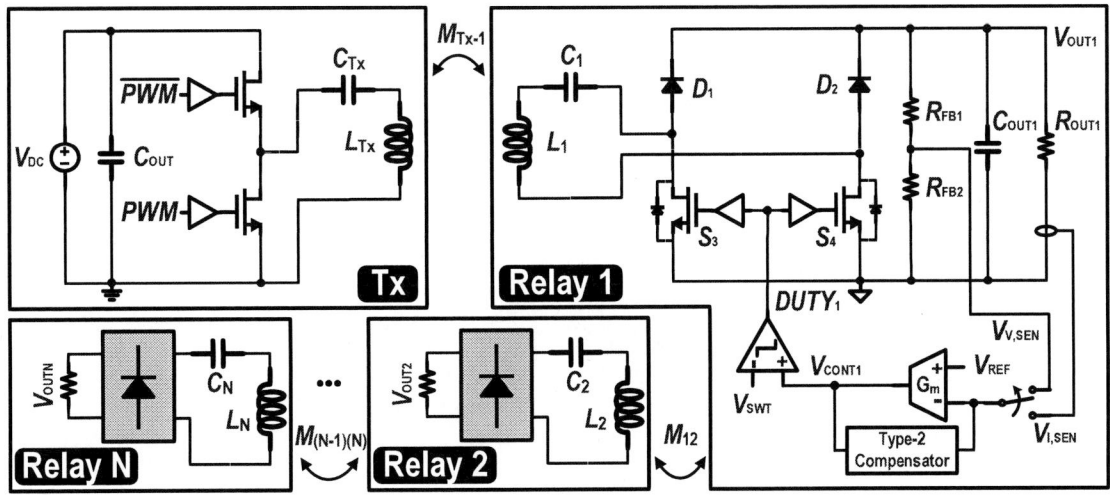

Fig.3. Schematic of the proposed wireless power transfer system.

duty ratio of the wireless receiver. The duty ratio, *DUTY*, controls the gate drivers of MOSFETs for CV/CC output regulation. By selecting the voltage $V_{V,SEN}$ from the feedback resistors or the output voltage from the current sensor $V_{I,SEN}$, either CV or CC mode can be chosen. Unlike previous studies, which placed anti-series MOSFETs in the compensation network along the high-current path to select CV or CC output, the controller in this structure uses a switch to select the feedback signal, significantly reducing power loss.

C. Compensation Network and Transmitter

In a three-coil wireless power transfer system with an intermediate LC-tank, a S-S-S compensation configuration can achieve CV characteristics, while an S-S-T compensation configuration results in CC characteristics [2]. Generalizing this, by configuring S-S-S, S-S-T, S-T-S, and S-T-T compensations, CC-CV, CV-CV, CC-CC, and CV-CC output characteristics can be achieved for two separate loads [3]. As mentioned, previous studies proposed hybrid topologies using power semiconductors to switch between S or T compensation networks in order to achieve constant-voltage or constant-current characteristics. However, this method required a large number of L and C components at the power receiver, making the system bulky and reducing practicality. Furthermore, the efficiency was lowered by the back-to-back power semiconductors connected in series. In contrast, this study applied a series compensation network using only a single capacitor, eliminating the need for complex compensation networks and mathematical analysis, thereby improving design simplicity and reducing costs. Additionally, since no power semiconductors are placed in series within the high-current power path, power conversion efficiency was improved.

A half-bridge class-D power amplifier roles the wireless transmitter (Tx). For high power transfer efficiency (PTE), a gallium-nitride (GaN) half-bridge power module and soft-switching technique are employed, operating at 200 kHz.

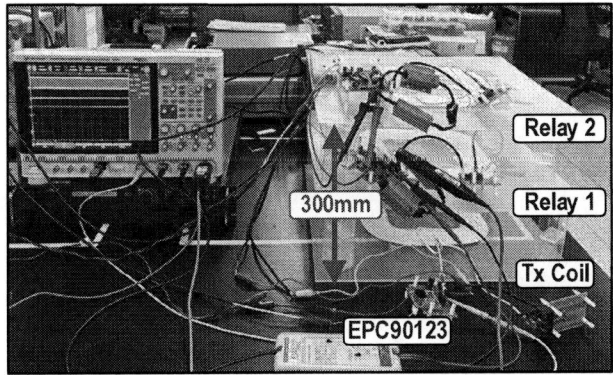

Fig.4. Photograph of the experimental setup.

TABLE I.   CIRCUIT IMPLEMENTATION

Component	Value		
Power module	EPC90123, GaN half-bridge		
MOSFET	C3M0160120D, SiC MOSFET		
Diode	MBR40250TG, Schottky		
L_{Tx}, L_{Rx1}, L_{Rx2}	Litz wire 0.07 mm/1560, 15 turns		
	64.8 [µH]	66.1 [µH]	66.2 [µH]
C_{Tx}, C_{Rx1}, C_{Rx2}	Metalized polypropylene film		
	9.770 [nF]	9.570 [nF]	9.570 [nF]
Gate driver IC	ADuM4121, Isolated, 2 A		
Current sensor IC	ACS730, 1 MHz bandwidth, Isolated		

III. MEASUREMENT RESULTS

Fig. 4 shows the experimental setup and specifications of the components used. The Tx/Rx coils are wound with 0.07 mm/1560 litz wire in 15 turns, providing a total power transfer distance of 300 mm when the distance of the adjacent coil is 150 mm. The coupling coefficient between the coils was from 0.38 to 0.43. Additionally, Table I presents the components used to build the circuit and their corresponding values.

Fig. 5 shows how the duty ratio ($DUTY$) of the 3R is generated by the feedback controller and how the output is regulated. When the DC link voltage (V_{DC}) of the Tx is 25 V, one relay is set to CV mode, and the $DUTY$ of 47% duty ratio is generated by the output of the Type-2 compensator (V_{CONT}) and the sawtooth signal (V_{SAW}). When the target output is set to 20 V, approximately 19.7 V is achieved, confirming that the output is regulated.

each relay was 44 Ω, set to 20 V for CV mode and 0.5 A for CC mode. The total power delivered to the load (PDL) was 18.2 W and 20.0 W, respectively. Fig.8(a) illustrates the output voltage or current with respect to V_{DC}. When V_{DC} is low, insufficient power is transmitted from the Tx, and neither relay is regulated, with the 3R operating as a full-wave rectifier. As V_{DC} increases, Relay 1 is regulated first. As V_{DC} increases further and sufficient power is supplied to both Relay 1 and Relay 2, the output of Relay 1 no longer increases thanks to $DUTY_1$, and Relay 2 also becomes regulated. In Fig.8(b), the output power and PTE are shown against the input power P_{IN}. Both experiments confirm that output is regulated when there is sufficient input power, with a maximum PTE of 74.5%. Table II compares the prior load-independent CV/CC wireless power transfer systems with the proposed system.

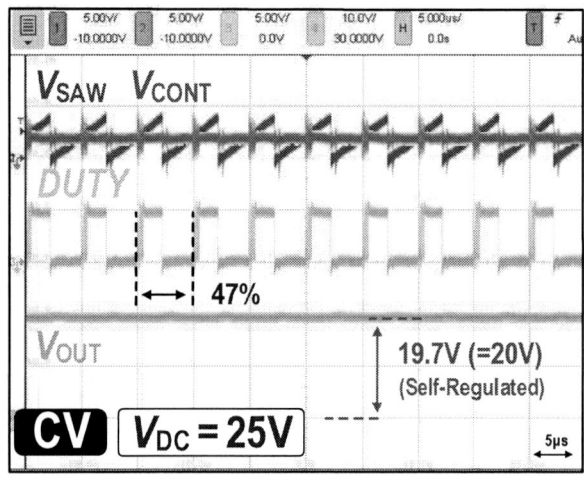

Fig.5. Output regulation by $DUTY$ signal in CV mode.

Fig. 6 shows the Tx inverter voltage V_{Tx} and the output of two relays set to CC and CV modes when the DC link voltage of the transmitter is varied. The load resistances, R_{OUT1} and R_{OUT2}, are 44 Ω and 22 Ω, respectively. When the output current of the first relay, I_{OUT1}, is set to 0.68 A, the measured values are 0.668 A and 0.659 A, respectively. Similarly, when the output voltage of the second relay, V_{OUT2}, is set to 10 V, the average output voltage was observed to be 9.65 V in both cases.

The load transient response of a single CV relay is depicted in Fig.7. When switching from 44 Ω to 22 Ω, an 8.1 V undershoot occurred, taking 58 ms for the voltage recover. Conversely, an overshoot of 10.9 V occurred from heavy to light load transition, requiring 64 ms for recovery. $DUTY$ generated by the PWM controller varies to regulate the output to the desired level and the stability of the controller is verified.

Two experimental results where Relay 1 operates in CV mode and Relay 2 operates in either CV or CC mode are presented in Fig.8. In both experiments, the load resistance for

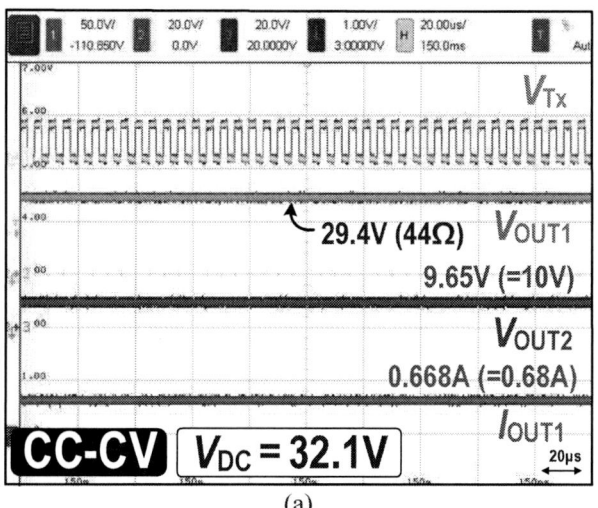

(a)

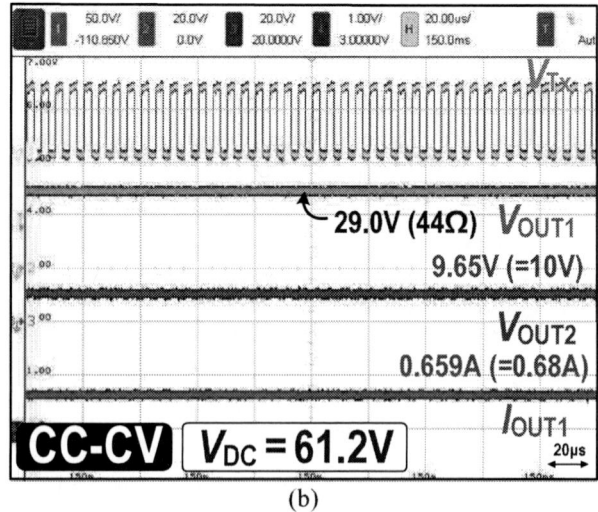

(b)

Fig.6. Line regulation measurements under CC-CV modes at (a) V_{DC}=32.1 V, (b) V_{DC}=62.1 V.

979-8-3315-1612-3/25 $31.00 © 2025 IEEE

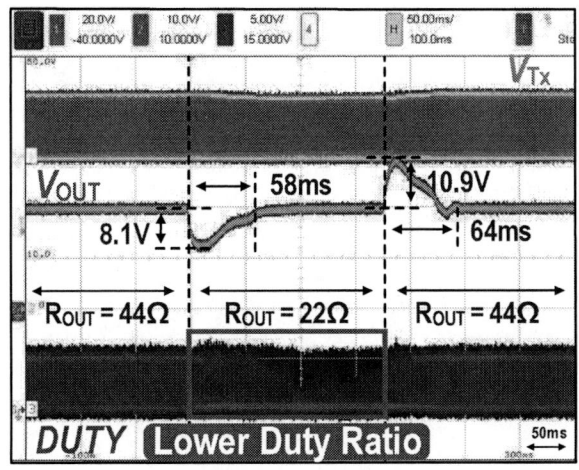

Fig.7. Measured waveforms of load transients with the changes in *DUTY*.

IV. CONCLUSION

This paper proposes a load-independent CV/CC WPT system using a single compensation network. Furthermore, multiple loads can receive regulated power from a long distance by multiple relays. The absence of series switches for selecting compensation networks can enhance PCE. Moreover, integrating AC-DC rectification and DC regulation into a single power stage mitigates cascaded power losses. A prototype was presented to verify the results, which were consistent with simulations. Achieving a PTE of 74.5%, 20.0 W was delivered to two loads at a maximum distance of 300 mm.

ACKNOWLEDGMENT

This work was supported by the National Research Foundation of Korea(NRF) grant funded by the Korea government(MSIT) (RS-2023-00219443). This work was

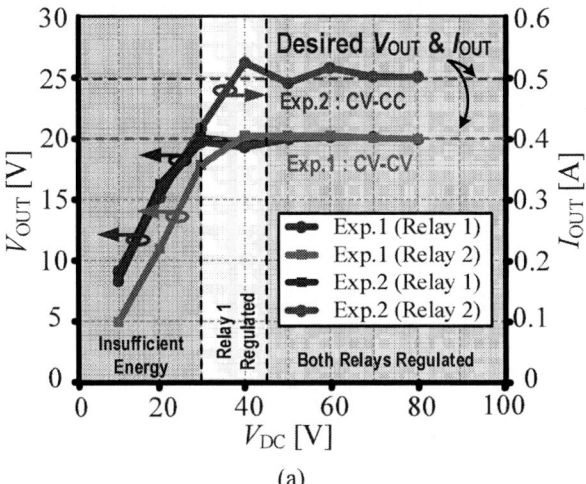

(a)

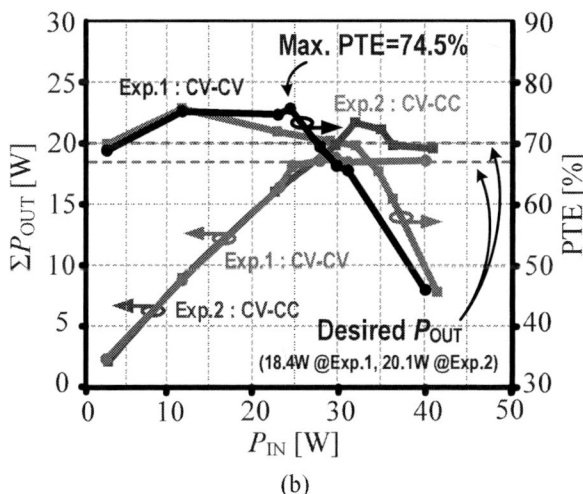

(b)

Fig.8. (a) Output of two relays for CV-CV and CV-CC modes, (b) power analysis and PTE.

TABLE II. COMPARISON WITH PREVIOUS WORKS

	TTE'18 [2]	TTE'20 [3]	TIE'24 [4]	This work
Load-independent CV/CC	Yes	Yes	Yes	Yes
Maximum PTE (PDL)	90.8% (384 W)	83% (N/A)	51.4% (28.8 mW)	74.5% (20.0 W)
# of SW/L/C in compensation network and auxilliary LC-tank	4/3/4	4/3/3	4/2/3	0/1/1
Carrier frequency [kHz]	200	400	4,000	200
Additional DC regulator	Yes	Yes	Yes	No
Intermediate coil	Yes	Yes	Yes	No
Tx/Rx coil distance [cm]	20	40	N/A	30

supported by Institute of Information & communications Technology Planning & Evaluation (IITP) grant funded by the Korea government(MSIT) (No.2022-0-00720, Development of W-band compact, high-efficiency, novel RF/power components for next-generation high-speed low-orbit satellite communications). Glocal University 30 Projects.

REFERENCES

[1] M. Choi, T. Jang, J. Jeong, S. Jeong, D. Blaauw and D. Sylvester, "A Resonant Current-Mode Wireless Power Receiver and Battery Charger With −32 dBm Sensitivity for Implantable Systems," *IEEE Journal of Solid-State Circuits*, vol. 51, no. 12, pp. 2880-2892, Dec. 2016.

[2] Y. Li, Q. Xu, T. Lin, J. Hu, Z. He and R. Mai, "Analysis and Design of Load-Independent Output Current or Output Voltage of a Three-Coil Wireless Power Transfer System," *IEEE Transactions on Transportation Electrification*, vol. 4, no. 2, pp. 364-375, June 2018.

[3] X. Xie, C. Xie and L. Li, "Wireless Power Transfer to Multiple Loads Over a Long Distance With Load-Independent Constant-Current or Constant-Voltage Output," *IEEE Transactions on Transportation Electrification*, vol. 6, no. 3, pp. 935-947, Sept. 2020.

[4] J. Lee, B. Bae, B. Kim, J. Lim and B. Lee, "A Load-Independent Battery Charging System for Multiple Wearable Devices Using Conductive Textile," *IEEE Transactions on Industrial Electronics*, Feb. 2024.

[5] J. -H. Choi, S. -K. Yeo, S. Park, J. -S. Lee and G. -H. Cho, "Resonant Regulating Rectifiers (3R) Operating for 6.78 MHz Resonant Wireless Power Transfer (RWPT)," *IEEE Journal of Solid-State Circuits*, vol. 48, no. 12, pp. 2989-3001, Dec. 2013.

[6] L. Cheng, W. -H. Ki, T. -T. Wong, T. -S. Yim and C. -Y. Tsui, "21.7 A 6.78MHz 6W wireless power receiver with a 3-level $1\times$ / $\frac{1}{2} \times$ / $0\times$ reconfigurable resonant regulating rectifier," *2016 IEEE International Solid-State Circuits Conference (ISSCC)*, San Francisco, CA, USA, 2016, pp. 376-377.

[7] J. Fuh, F. -B. Yang and P. -H. Chen, "A 69.3% Efficiency, 6.78-MHz Wireless Power Delivery System with 0X/1X Regulating Rectifier and Reconfigurable Power Amplifier," *2019 IEEE Asian Solid-State Circuits Conference (A-SSCC)*, Macau, Macao, 2019, pp. 37-38.

A GaN-based Single-Stage Solid-State Transformer Replacement for 40 VA Class 2 Line-Frequency Transformers

Allen T. Nguyen, Charles R. Sullivan
Thayer School of Engineering at Dartmouth College
Hanover, NH 03755 USA
Allen.T.Nguyen.th@dartmouth.edu, Charles.R.Sullivan@dartmouth.edu

Abstract—Class 2 transformers are common among several low-power applications, including HVAC control systems, doorbells, automatic doors, and much more. These transformers have low peak efficiencies and very high standby losses. We propose a power-electronic alternative with an emphasis on reducing the yearly average power loss based on a use-case load cycle. Using GaN power devices, a single-stage solid-state transformer is designed, focused on reduced overall solution control and complexity, while maximizing standby efficiency through reduced switching and core losses. An optimization model was developed, providing both passive component recommendations and an optimum operating condition for the selected topology. The 40 VA, 120 VAC to 24 VAC, prototype achieves a peak efficiency of 98.8%, a standby loss of 106 mW, and a use-case average yearly power loss of 207 mW; a 17x reduction in the yearly average power loss compared to Class 2 line-frequency transformers.

Index Terms—line-frequency transformer, solid-state transformer, standby loss, Class 2, light-load efficiency, low-power, ac-ac

I. INTRODUCTION

Low-power (less than 100 VA) line-frequency (50-60 Hz) transformers (LFTs), are found in several applications requiring low-voltage (typically 24 VAC or less); including HVAC controls, doorbells, automatic door controls, security systems, and more [1]. Under UL standards [2] and the U.S. National Electric Code [3], these are known as Class 2 transformers.

These low-power LFTs have low peak efficiencies and large standby losses. Measurements of 40 VA Class 2 transformers reported an average peak efficiency of 84.4% with an average standby loss of 2.8 W [4]. The latter is particularly problematic, as typical applications require that these transformers are connected to line voltage continuously, only intermittently providing full-load power. Thus, a superior replacement must reduce this standby loss [5], while also achieving both low cost and low volume to compete with these already simple LFTs.

An isolated ac-ac converter, commonly known as a solid-state transformer (SST), can potentially provide a competitively priced switch-mode replacement for these low-power LFTs while also achieving higher power densities and increased efficiencies across all load ranges, but particularly at no-load.

This research was supported by the Power Management Integration Center, an NSF I/UCRC, under grant number 1822140.

Typically categorized based on the number of conversion stages or dc-links [6]–[11], high-power (greater than 100 kVA) SSTs have seen great success.

In previous work, we compared SST categories in the context of standby-dominated low-power LFTs [12]. In summary, with the goal of keeping costs and complexity low for a practical Class 2 LFT replacement, two- and single-stage SST designs have more advantages than three-stage (ac-dc-dc-ac) SSTs.

Two-stage (ac-dc-ac) SST topologies are quite viable, as the dc link can buffer any bidirectional loading and provide a wider control range, while minor features, such as power factor correction, can be easily implemented [13]. Another potential benefit of a two-stage SST is to use large-volume-production power supplies as one of the stages, as implemented in [12]. This strategy can allow for reduced design time and can take advantages of features already included in the power supply, including a stable regulated dc-link, other protection features (surge and EMI filters), and very low standby loss, at low cost.

However, since the dc-link needs to buffer single-phase line-frequency current, it occupies substantial volume while the two-stage conversion results in reduced loaded efficiency. Also, low-cost high-efficiency commercial-off-the-shelf (COTS) power supplies in the needed power range and voltage range needed to take advantage of this benefit are limited. A two-stage SST can be designed without taking advantage of cheap COTS power supplies, but the increased complexity and reduced efficiency are still trade-offs needed to be explored in the context of Class 2 LFTs.

Due to the simplicity of their design and applications, Class 2 LFTs do not regulate their output voltage, leading low-power LFT replacement topologies more towards the single-stage direct-conversion (ac-ac) SST variations. SSTs in this category have reduced component count, lower control complexity, high power density, and high conversion efficiency compared to their two- and three-stage counterparts [14]–[18]. This results in overall reduced solution cost and reduced volume/weight. Allowing additional features, such as EMI filters, over-circuit protection, surge-protection, etc. to be made more complex while still being cost competitive with Class 2 LFTs.

A successful implementation of a low-power single-stage SST has been reported in [12]. Although it exceeded both the efficiency and standby loss metrics set by the benchmarking done in [4], its performance was not optimized for the application. In this work, we optimize the combination of both standby and full-load loss, while introducing robustness features in both design and operation; with the goal of informing future decisions that impact cost, weight, and volume, and thus can further make the case for a single-stage SST replacement to Class 2 line-frequency transformers.

The paper is organized as follows: Section II provides an overview of the selected single-stage topology and its changes on [12], Section III explains a yearly-average power loss optimization model that also provides an optimum operating condition and passive selection based on time-domain analysis, and Section IV provides experimental results of the optimized design.

II. TOPOLOGY OVERVIEW

The single-stage topology selected for optimization is shown in Fig. 1. It consists of two fixed-frequency fixed-duty cycle synchronous half-bridges isolated by the high-frequency transformer. To reduce active component count and to ensure bidirectional voltage blocking (four-quadrant operation), back-to-back source connected devices are used.

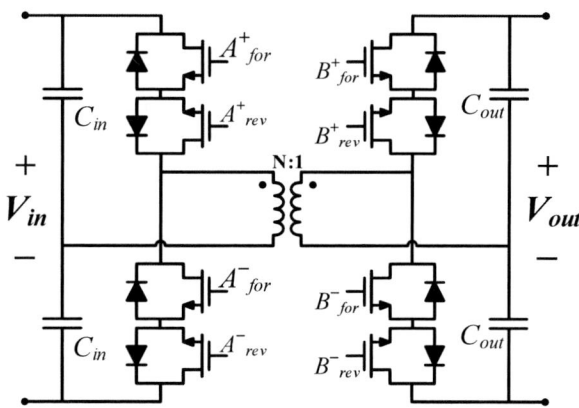

Fig. 1: Simplified schematic where each active device follows the switching scheme shown in Fig. 2.

The converter employs both zero-voltage switching (ZVS) and zero-current switching (ZCS), ensuring that standby loss of the SST can be driven very low. ZVS occurs in the primary side half bridge realized through a fixed-finite deadtime and magnetizing current, designed only for the peak of the line. Elsewhere, particularly towards the line-voltage zero-crossing, only partial ZVS occurs. Variable deadtime over the line cycle could be implemented to effectively reduce the overall primary side switching loss to near zero, but at this low voltage (120 VAC) and given the already low efficiency of the Class 2 LFTs, the benefit is nearly negligible. ZCS in the secondary half-bridge, through resonant operation driven by the leakage inductance (L_{lk}) of the transformer and an output capacitor (C_{out}), occurs throughout the varying line, thus there should be no switching/overlap loss that occurs there. When used for dc-dc conversion, the circuit topology is similar in operation to resonant or *LLC*-based dc transformer topologies [19]–[25].

Due to the nature of the switching scheme, there is no need for any large passives in the resonant tank to buffer any large voltage differences between the two bridges. Thus, the leakage inductance of the transformer is enough to shape the current waveform for ZCS purposes. However, as this leakage inductance is typically small, the Q of the resonant tank is lower, and thus basic resonant frequency estimations are not applicable, but rather an iterative approach is needed to ensure either the accurate resonant frequency or resonant capacitor for operation at a given frequency. Section III will explain in further detail.

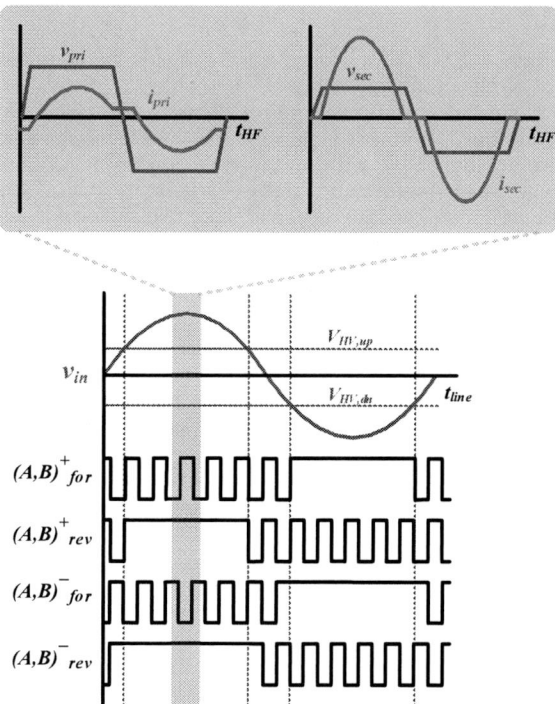

Fig. 2: Simplified switching scheme over the full line input; highlighted are single-switching-cycle transformer voltages and currents.

A key difference with respect to the single-stage SST in [12] is the control of the switches in a back-to-back pair. Unlike SSTs that have both switches in a pair switching at high frequency throughout the line cycle [12], [14], [15], the switches in a back-to-back pair will be synced to the line-voltage polarity. As shown in Fig. 2, when the input voltage is above a certain threshold, the *reverse* switches (*rev*) are on, while the *forward* switches (*for*) are high-frequency switching in a complementary fashion. The switching scheme is reversed when the input is below a certain threshold. This control scheme increases solution cost through the addition of the high-voltage sensing circuit. However, this control approach offers three significant advantages. First, gate-drive power is reduced up to 50%, depending on where the threshold limits are placed. Second, self-driven third-quadrant behavior (that could be characterized as body diode behavior) is reintroduced, increasing circuit robustness by ensuring that the switching node's voltage is limited at the peak of the line. Lastly, the output switch capacitance is well defined to allow for full ZVS of both switches in a pair (see Appendix A).

This single-active line-synced switching in a back-to-back pair is common in switching topologies with four-quadrant switching [16], [17], [26]. However, we introduce a margin around the zero-crossing where we allow the pair to synchronously switch, allowing for any variation in phase or accuracy between the sensed voltage and the input voltage while reducing the need for more complex control. The main impact on the converter operation during this synchronous switching period is on ZVS. As the switch node capacitance abruptly lowers and the body diode is removed, the switching node could be over or under charged, compared to the line voltage, increasing switching losses. However, depending on where the margin is placed, this only occurs in the low-voltage domain. Implementation details are in Appendix B.

III. Optimization and Design Workflow

An optimization process was developed to design the SST such that it minimizes its yearly average power loss (P_Y), a combination of the full-load (P_{FL}) and no-load (P_{NL}) power loss:

$$P_Y = P_{FL} D_{on} + P_{NL}(1 - D_{on}) \quad (1)$$

where the power loss is weighed based on an estimated use-case full-load to standby loss time-on ratio D_{on}. This allows for unique designs for different applications (doorbells to HVAC to automatic door, etc.), appropriately prioritizing no-load vs. full-load efficiency.

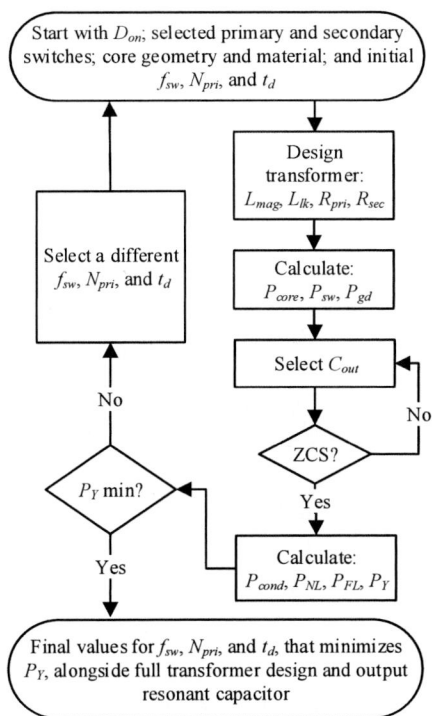

Fig. 3: Flow diagram of iterative design process.

A. Inputs and Outputs

The inputs to the optimization, driven by particle swarm minimization, include the fixed parameters: D_{on}, core geometry, core material, and the power switches' parameters (R_{on}, C_{iss}, and C_{oss}); and the optimized parameters: switching frequency (f_{sw}), operating deadtime (t_{dead}), and transformer primary turns (N). One can optimize additional variables, for example, a switch size factor that can span the space of power switches in a product family. However, fixing power switch selection early in the process can be beneficial as a proxy for cost; similarly, fixing the core size and material constrains the magnetics volume and cost.

B. Workflow/Design Procedure

1) Transformer Design: The magnetizing inductance (L_{mag}) of the transformer is driven by the f_{sw}, t_{dead}, and primary side C_{oss}. This is only calculated for the peak input line voltage. Once the magnetizing inductance is set, the primary bridge switching loss (P_{sw}) is calculated by characterizing the nonlinear C_{oss} and the magnetizing current (i_{mag}) over the full line period (T_{line}), and calculating the leftover energy in the C_{oss} at the end of each high-frequency switching period (T_{sw}).

$$v_{left}[\Delta t] = v_{in}[\Delta t] - \frac{v_{in}[\Delta t] t_{dead}}{8 L_{mag} f_{sw} C_{oss}[v_{in}[\Delta t]]} \quad (2)$$

$$P_{sw} = f_{line} \sum_{\Delta t = T_{sw}, 2T_{sw}, ..., T_{line}} C_{oss}[v_{in}[\Delta t]] \left(v_{left}[\Delta t]\right)^2 \quad (3)$$

Next, extract other transformer parameters based on assumptions of the transformer construction. Since the transformer mainly processes the resonant current of the ZCS resonant operation, a transformer design prioritizing ac resistance is preferred, for example, a single-interleaved litz-wire barrel-wound construction method. The ac resistance and leakage can be estimated as outlined in [27].

The core loss (P_{core}) of the transformer due to the magnetizing current is also calculated here. Similar to the switching losses, the core loss is estimated in every high-frequency switching period then averaged over the line-cycle. Using the iGSE formulation,

$$P_{core,HF}[\Delta t] = V_c k_i \left(\frac{v_{in}[\Delta t]}{4 f_{sw} A_c N} \right)^{\beta - \alpha} \left(\frac{v_{in}[\Delta t]}{2 A_c N} \right)^{\alpha} \quad (4)$$

$$P_{core} = \frac{f_{line}}{f_{sw}} \sum_{\Delta t = T_{sw}, 2T_{sw}, ..., T_{line}} P_{core,HF}[\Delta t] \quad (5)$$

where α and β are Steinmetz parameters, V_c and A_c are core parameters, and k_i is a constant calculated in [28].

2) Resonant Capacitance Selection: Although the transformer winding is designed for low ac resistance, the single-interleaved method also reduces its leakage, resulting in a potentially low-Q resonant tank. This affects the calculation of the resonant capacitance needed, as does deadtime, which is an optimized variable in the particle swarm optimization. The lower Q requires that the capacitance be lower to swing the resonant current more aggressively to counteract the damping of the higher series resistance [29], and higher deadtime reduces the resonant period compared to the switching period. Thus, assuming the resonant frequency is equal to the switching frequency is incorrect and would result in operating off of the optimum operating point or inaccurate capacitance prediction.

Thus, within the design routine, the low-Q time domain equations [29] are used to accurately predict the value of C_{out} such that ZCS is achieved. These time domain equations then allow for accurate predictions of the current in the converter, at each switching period over the full line-frequency period. Here we are then able to estimate the RMS current (for conduction loss purposes) across the entire line cycle despite the resonant operation in every high-frequency switching cycle. For example, the RMS current in the secondary terminals of the transformer ($I_{sec,RMS}$) over the line cycle is estimated as,

$$I^2_{sec,RMS,HF}[\Delta t] = \overline{i_{sec}(\Delta t \to \Delta t + T_{sw})^2} \quad (6)$$

$$I_{sec,RMS} = \sqrt{\frac{f_{line}}{f_{sw}} \sum_{\Delta t = ...} I^2_{sec,RMS,HF}[\Delta t]} \quad (7)$$

where the sum is from $\Delta t = 0, T_{sw}, 2T_{sw}, ..., T_{line}$, $I_{sec,RMS,HF}[\Delta t]$ is the RMS value of i_{sec} in an individual switching cycle, and i_{sec} can be calculated following [29].

3) Calculate Remaining Losses and P_Y: Switching, core, and conduction losses are calculated in the previous steps. With the remaining loss being gate drive loss (P_{gd}) which is based on the gate capacitance of the GaNFETs and the gate drive voltage. Any other losses are assumed to be negligible or at least independent of the optimized parameters.

Lastly, P_Y can now be calculated based on the calculation of the full load (P_{FL}) and the standby loss (P_{NL})

$$P_{FL} = P_{cond,fl} + P_{core} + P_{gd} \quad (8)$$

$$P_{NL} = P_{cond,nl} + P_{core} + P_{gd} \quad (9)$$

where $P_{cond,fl}$ is the full-load conduction loss and $P_{cond,nl}$ is the no-load conduction loss over the line cycle.

C. Simulation Verification

Fig. 4 shows the model P_Y results as compared to circuit simulations of the model output in SIMPLIS, where the fixed variables are held constant (switch selection, transformer volume, and material) and the model is run to find the optimum P_Y at a specified frequency.

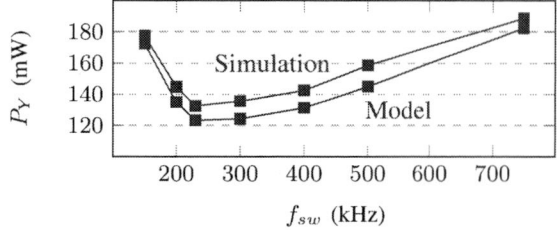

Fig. 4: Model compared to simulation across frequency, maximum measured error is 9.1%. Note the difference between 230 kHz and 300 kHz is 1 mW.

IV. EXPERIMENTAL VERIFICATION

Given the application of 40 VA Class 2 LFTs, 120 to 24 VAC, and an D_{on} of 0.28, a selected transformer core geometry of PQ20/16, a MnZn ferrite core material of ML91S from Proterial, and the selected primary and secondary side switches of EPC 2050 and EPC 2619, the model produced an optimal frequency of 227.6 kHz. The theoretical optimum transformer design specified 29.5 primary turns, rounded up to 30 in practice, a target magnetizing inductance of 316 μH and a target leakage inductance of 143 nH. Note that we are specifying 30 VA as the full load power loss for P_Y optimization purposes.

A. Hardware Implementation

The transformer primary was wound with 66/46AWG and the secondary with 360/46AWG. Table I shows the final prototype values and Fig. 5 shows a prototype image. GaN devices were selected for this optimization and implementation due to their higher figure-of-merit and lower size compared to their silicon counterparts [30], ensuring that the efficiency and size of the prototype are minimized.

Symbol	Description	Value
f_{sw}	Switching Freq.	227.2 kHz
V_{in}	Input Voltage	120 VAC
V_{out}	Output Voltage	24 VAC
P	Rated Power	40 VA
	Pri. Switches	EPC2050
	Sec. Switches	EPC2619
	Core Material	ML91S
	Core Geometry	PQ20/16
	Pri. Winding	30 turns of 66/46AWG
	Sec. Winding	6 turns of 360/46AWG
L_{mag}	Mag. Ind.	306 μH
L_{lkg}	Sec. Leak. Ind.	152 nH
C_{in}	Input Cap.	5.7 μF
C_{out}	Output Cap.	1.57 μF

TABLE I: System parameters and final measured values.

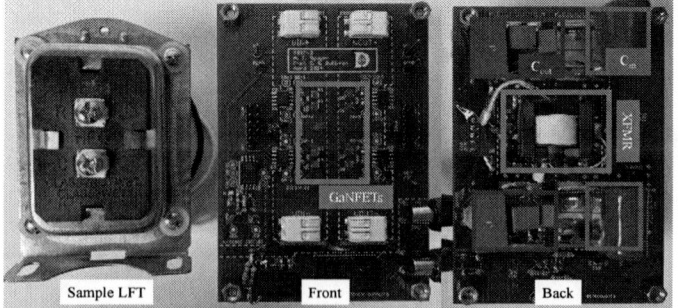

Fig. 5: Image of prototype; the board dimensions are 7 cm by 10 cm. The volume of the combined passives and devices in SST$_{2025}$ is estimated to be 20 cm^3. The sample LFT pictured has an estimated volume of 140 cm^3. Note that the FPGA used to supply the gate signals is not included in the volume estimate.

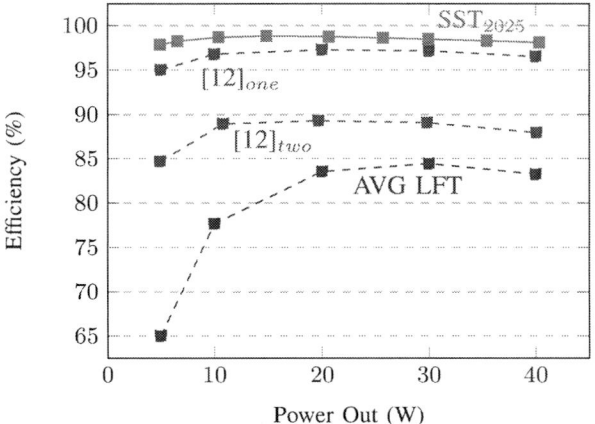

Fig. 6: Efficiency comparison. The peak efficiency of SST_{2025} was found to be 98.8% ($\pm 0.25\%$) at 15 VA out. This was measured using the Voltech PM6000 power analyzer.

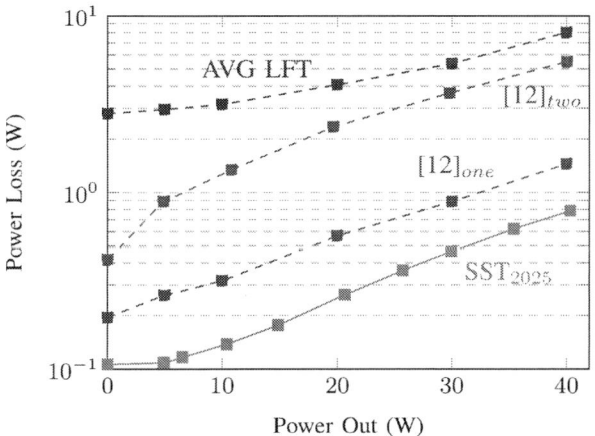

Fig. 7: Power loss measurements shown in a log scale. SST_{2025} achieves a 45% reduction in the power loss at full load compared to $[12]_{one}$.

V. RESULTS

The performance of the prototype (SST_{2025}) will be compared with the average benchmarking data in [4] (AVG LFT), along side both the two-stage and single-stage prototypes presented in [12] designated $[12]_{two}$ and $[12]_{one}$ in the figures.

A. Efficiency

Efficiency comparisons are shown in Fig. 6. SST_{2025} achieves higher efficiency across all load ranges compared to the leading design in [12].

B. Power Loss and P_Y Comparisons

Power loss comparisons are shown in Fig. 7, and Fig. 8 shows standby power loss across a $\pm 10\%$ line voltage variation. Compared to the average LFT, SST_{2025} achieves a 26.4x reduction in standby loss and an 8x reduction in full load loss, with a large reduction in passive volume and weight. Compared to $[12]_{one}$, this improvement is 1.85x in both standby and full-load loss with comparable volume.

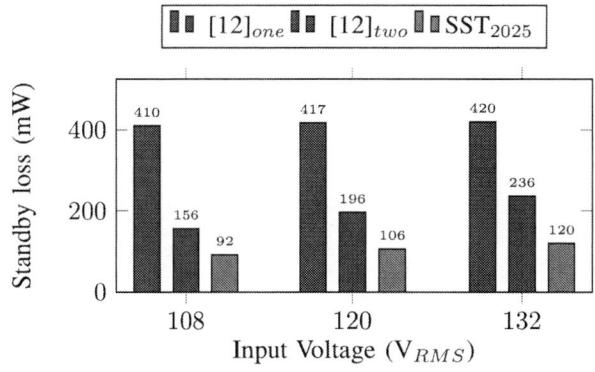

Fig. 8: Standby loss measured across input voltage variation. Note that the average LFT has a nominal standby loss of 2.8 W.

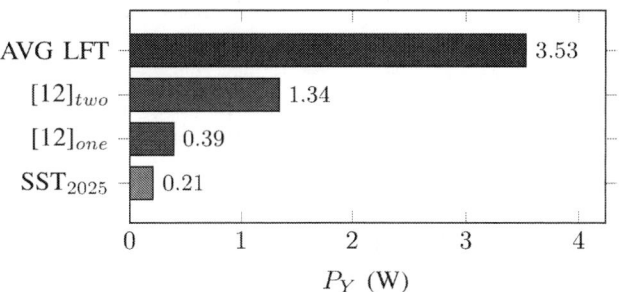

Fig. 9: P_Y comparisons.

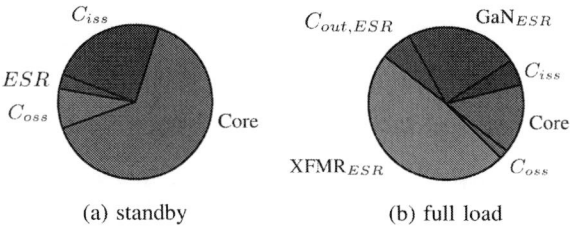

Fig. 10: Estimated power loss breakdowns for the standby and full-load conditions.

Fig. 9 shows the annual average power loss, P_Y. SST_{2025} achieves a 17x improvement compared to the average LFT, and a 1.85x improvement compared to $[12]_{one}$.

C. Power Loss Breakdown

The estimated power losses during the standby and full-load power point are outlined in Fig. 10. During the standby condition, the only current in the converter is the transformer magnetizing current, resulting in the core loss being the main contributor to power losses. The second main contributors are the switching losses associated with the gate drive and the GaN output capacitance.

Then, during the full-load condition, the transformer experiences the full resonant current of the ZCS operation as well as the load current, resulting in its ESR losses dominating the full-load power losses. Although small, the ESR losses of C_{out} are non-negligible; including an estimation of this loss can result in a circuit operation that could potentially reduce loss further.

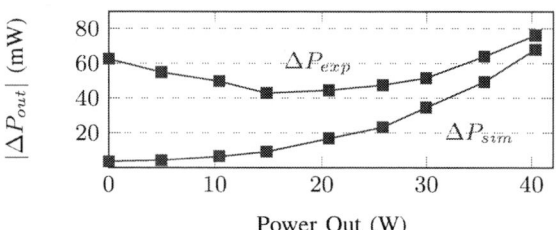

Fig. 11: Difference in power loss as estimated by the model compared to simulation and experimental data.

From these estimations, there are three main areas that present opportunities for potential power loss improvements; all of which are a function of the fixed variables. A different selection of the switches can further reduce switching losses, and the transformer can be made bigger to reduce ESR losses as well as core losses. But as previously stated, all these options impact cost and size; both of which might have priority over power loss in a successful SST replacement.

D. Experimental Error to Simulation and Model

To evaluate the accuracy of the model used in optimization, we compared its predictions to the measured performance, as well as to simulations. For this purpose, we used measured transformer parameters and included additional parasitics such as the ESR of C_{out}, in the model and the simulation.

Fig. 11 shows the difference between the loss predicted by the model and either a) the loss predicted by the simulation (ΔP_{sim}) or b) the measured loss (ΔP_{exp}). The error between the model and simulation is less than 10%.

ΔP_{exp} is relatively constant across the load range, suggesting a static source of experimental loss (e.g., core and/or C_{oss}) is higher than predicted. The manufacturer's core loss data doesn't include our actual operating frequency, so this is a plausible reason for this small discrepancy. In terms of the model, more accurate core loss data can further increase the accuracy of the optimization. Experimentally, SST_{2025}'s high efficiency challenges the accuracy limits of the measuring equipment, so this also may also explain ΔP_{exp}.

E. Experimental Waveforms

Fig. 12 shows the input and output voltage and currents during a resistive loaded condition. Note the capacitive loading on the input, a result of the input and output capacitors on the ac lines. No noticeable crossover distortion is found on the input/output waveforms due to the margin-based asynchronous switching scheme of the back-to-back GaNFETs.

Fig. 13 looks at a single switching cycle at the top of the line. This shows ZVS in the primary switching node and ZCS in the secondary transformer current. Also due to this ZCS, there is minimal ringing in the transitionary period of the secondary side transformer voltage.

F. Testing Under an Inductive Load

Lastly, SST_{2025} is tested under an example of more complex load that is typical of the applications of Class 2 LFTs: a

Description	Manufacturer	Part #
Inductive Load	Atlantic Valves	RSSM-3-24VAC
Sample Class 2 LFT	Resideo	AT72D-1683

TABLE II: Products used for inductive load comparison.

Description	AT72D	SST_{2025}
Input Power (W)	10.90	6.74
Input VA (VA)	20.33	8.58
Input PF	0.536	0.782
Output Power (W)	7.47	6.56
Output VA (VA)	14.73	12.67
Output PF	0.507	0.518
Output Voltage (V_{RMS})	25.17	24.10
Efficiency	68.53%	97.33%

TABLE III: Key performance results driving a solenoid valve.

solenoid valve rated for an input of 24 VAC. For comparison, this load will also be powered by a 40 VA sample Class 2 line-frequency transformer outlined in Table II. This inductive load is expected to draw 10 VA once energized.

Fig. 15 and 16 shows the input and output waveforms of both the sample Class 2 LFT and SST_{2025} when the inductive load is energized. Table III outline key performance metrics as measured from a power analyzer.

Due to the capacitive nature of SST_{2025}, the power factor (PF) is greatly improved compared to the sample Class 2 LFT, as the capacitive storage provides some PF correction with an inductive load. Similarly, the magnetizing inductance of the Class 2 LFTs degrades the PF.

SST_{2025} also achieves high efficiency with an inductive load, but also note that this load is also quite non-linear. Probably due to the design of the solenoid valve, it shows some saturation behavior, highlighting the converter's performance with even non-sinusoidal loads. These high-harmonic current excitations also may further degrade the conduction performance of the sample Class 2 LFT, as it doesn't use litz wire or other high-frequency winding techniques to mitigate eddy-current losses.

Lastly, we acknowledge the high-frequency ripple on the input current of the SST, a byproduct of the high-frequency switching and resonant operation. Work to reduce this ripple in either converter design/operation or in the EMI filter design will be discussed in future work.

VI. CONCLUSION

This paper introduces a 40 VA Class 2 LFT replacement in the form of a single-stage solid state transformer. This SST requires minimal control overhead and is designed using an optimization model based on the yearly average power loss. The presented SST's standby loss is 3.8% that of the Class 2 LFTs, and its yearly average standby loss is 5.9% of that of the LFTs. The presented SST also improves on low-power SST replacements as designed in previous work even with similar passive volume.

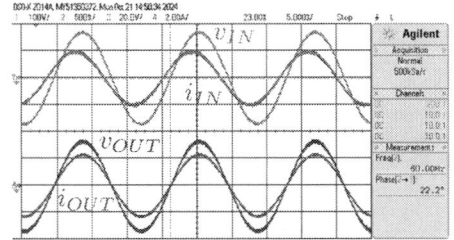

Fig. 12: 60 Hz input and output waveforms during loaded condition.

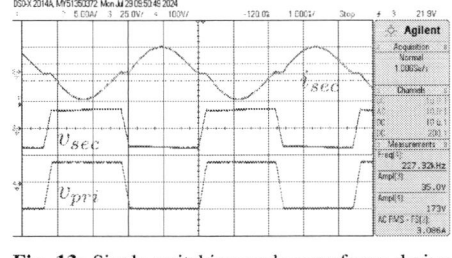

Fig. 13: Single-switching cycle waveforms during loaded condition, including the secondary side transformer current (i_{sec}) and the transformer primary (v_{pri}) and secondary (v_{sec}) voltages.

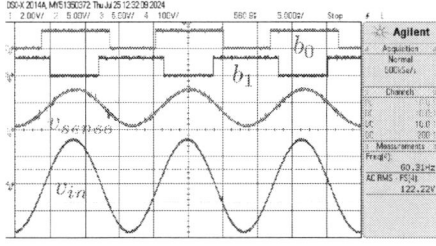

Fig. 14: Control waveforms; respective signals (b_0 & b_1) are high when the input voltage (v_{in}) is above or below a certain threshold. v_{sense} is a low-voltage control signal representing v_{in}.

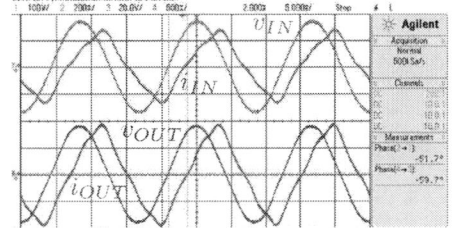

Fig. 15: 60 Hz input and output waveforms of the sample Class 2 LFT supplying an solenoid valve.

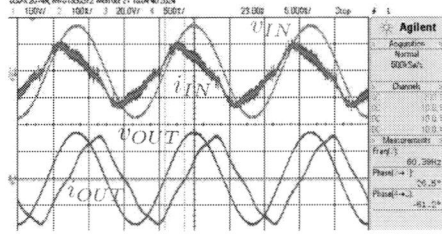

Fig. 16: 60 Hz input and output waveforms of SST_{2025} supplying a solenoid valve.

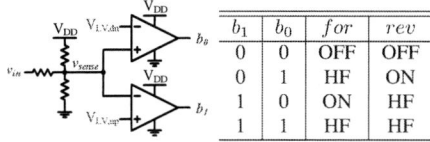

Fig. 17: State table and simplified diagram of control signals; indicating the states of the forward and reverse switches.

APPENDIX A
OUTPUT CAPACITOR BEHAVIOR OF SYNCHRONOUSLY-SWITCHED SOURCE-CONNECTED POWER DEVICES

Driving both switches in a pair synchronously rather than line-synced has the main benefit of control simplicity and reduced circuit complexity, at the cost of losing the body-diode functionality. However, the overall capacitance of the switches also isn't well defined, and unless sophisticated transient switching schemes are implemented [14], [15], true ZVS only occurs once a cycle (during the charging transient).

Fig. 18 shows this phenomenon in example, from the transition from the conduction of the − pair to the the + pair, with a positive input voltage. Ideally, initial conditions is as follows: the peak voltage (V_{pk}) sits on the A^+_{for}, with all other capacitors at a voltage of zero. The end goal is to discharge A^+_{for}'s capacitor to zero, while charging the A^-_{for} capacitor to V_{pk} using the peak magnetizing current (I_{mag}).

However, since the switches are switched synchronously, I_{mag} must also flow through the A^+_{rev} capacitor; resulting in charging/discharging of three distinct capacitors rather than the typical case of two. If designed to raise V_A to V_{pk}, I_{mag} would successfully charge A^-_{for} up. However, both A^+_{for} and A^+_{rev} will still have remaining voltage (V_{remain}) across their output capacitors, resulting in CV^2_{remain} energy loss on both of these switches.

If the output charge characteristics of the selected power switch is well known, the total charge (ΔQ_{ZVS}) needed during the deadtime to achieve this *pseudo* ZVS can be calculated as 1.5 times the charge needed to charge one output capacitor to V_{pk} ($Q_{0 \to Vpk}$). As only $\frac{1}{2}Q_{0 \to Vpk}$ is needed to satisfy that

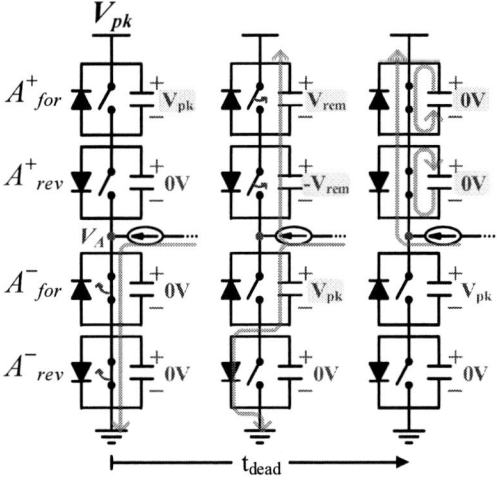

Fig. 18: Switch transition state and voltages of the primary half-bridge before and after the deadtime.

the change in charge is equal in A^+_{for} and A^+_{rev}, and that their remaining voltages sum to zero.

V_{remain} is simply the voltage obtained after charging to $\frac{1}{2}Q_{0 \to Vpk}$. If the capacitors were linear, V_{remain} should just be $\frac{1}{2}V_{pk}$. But with the typical non-linear characteristics of these output capacitance's, V_{remain} is lower than $\frac{1}{2}V_{pk}$.

For high-voltage systems, even this reduced hard switching is unacceptable; thus, they either opt for the line-synced pair switching, an unfolding/rectification stage to never have a back-to-back pair, or complex switching schemes during the switching transient. But low-voltage systems could tolerate these minor hard-switching transients as implemented in [12]. Lastly, compared to the line-synced switching (traditional two capacitor ZVS), achieving this pseudo ZVS requires only 75%

of the current needed to raise V_A to V_{pk}. So there lies a potential in choosing between higher switching losses or higher magnetizing current (or longer deadtime).

APPENDIX B
HARDWARE IMPLEMENTATION DETAILS OF THE MARGIN-BASED SYNCHRONOUS-SWITCHING SCHEME

The gate drive signals of each switching device is generated by an FPGA that was fed a two-bit signal that represented what section the input voltage is currently in; as shown in Fig. 2, either above $V_{HV,up}$, below $V_{HV,dn}$, or in between.

To implement this two-bit ($b_1 b_0$) signal, the input voltage was sensed through a resistive divider that shifted the high-voltage (HV) domain (-170 to +170 V) to the FPGA low-voltage (LV) domain (0 to 3.3 V). This LV signal is then compared to two threshold voltages, implemented by resistive dividers to set the desired margin, and the two-bit signal in passed to a digital isolator then to the FPGA for processing.

Within the programming, $b_1 b_0$ are used to mux in between a HF switching signal or a constant on state. Note, that there is an bit-code configuration that doesn't correspond to any section of the voltage waveform. This is an error state, that once sensed by the FPGA will shut down the converter. Fig. 14 shows these signals in real time, and Fig. 17 summarizes implementation details.

Lastly, one practical consideration is to consider the loading on the gate driving circuit. As the switching configuration changes several times a cycle, the gate drive will see either the full C_{iss} load or half the C_{iss} load. Ensuring that the gate drive voltage is robust to transients is essential.

REFERENCES

[1] Triad Magnetics. Understanding UL Class 2 Power Transformers. [Online]. Available: https://info.triadmagnetics.com/lp-download-understanding-ul-class-2-power-transformers

[2] Underwriters Laboratories, "UL 5085-3 Low Voltage Transformers - Part 3: Class 2 and Class 3 Transformers," Jan. 2022.

[3] *NFPA 70: National Electrical Code 2020.* National Fire Protection Association, 2019.

[4] A. T. Nguyen and C. R. Sullivan, "Class 2 transformers: ubiquitous, hidden, and inefficient," in *IEEE Conference on Technologies for Sustainability (SusTech)*, 2023, pp. 157–160.

[5] W. Wang, J. Su, Z. Hicks, and B. Campbell, "The Standby Energy of Smart Devices: Problems, Progress, & Potential," in *IEEE/ACM Fifth International Conference on Internet-of-Things Design and Implementation (IoTDI)*, 2020, pp. 164–175.

[6] S. Falcones, X. Mao, and R. Ayyanar, "Topology Comparison for Solid State Transformer Implementation," in *IEEE PES General Meeting*, Jul. 2010, pp. 1–8.

[7] M. A. Hannan, P. J. Ker, M. S. H. Lipu, Z. H. Choi, M. S. A. Rahman, K. M. Muttaqi, and F. Blaabjerg, "State of the Art of Solid-State Transformers: Advanced Topologies, Implementation Issues, Recent Progress and Improvements," *IEEE Access*, vol. 8, pp. 19 113–19 132, 2020.

[8] A. Q. Huang, "Medium-Voltage Solid-State Transformer: Technology for a Smarter and Resilient Grid," *IEEE Industrial Electronics Magazine*, vol. 10, no. 3, pp. 29–42, Sep. 2016.

[9] J. E. Huber and J. W. Kolar, "Solid-State Transformers: On the Origins and Evolution of Key Concepts," *IEEE Industrial Electronics Magazine*, vol. 10, no. 3, pp. 19–28, Sep. 2016.

[10] F. Ruiz, M. A. Perez, J. R. Espinosa, T. Gajowik, S. Stynski, and M. Malinowski, "Surveying Solid-State Transformer Structures and Controls: Providing Highly Efficient and Controllable Power Flow in Distribution Grids," *IEEE Industrial Electronics Magazine*, vol. 14, no. 1, pp. 56–70, 2020.

[11] L. Heinemann and G. Mauthe, "The Universal Power Electronics Based Distribution Transformer, an Unified Approach," in *IEEE 32nd Annual Power Electronics Specialists Conference*, vol. 2, 2001, pp. 504–509 vol.2.

[12] A. T. Nguyen and C. R. Sullivan, "Low-Power Solid-State Transformers to Replace Line-Frequency Class 2 Transformers," in *IEEE Applied Power Electronics Conference and Exposition (APEC)*, 2024, pp. 3163–3170.

[13] J.-S. Lai, W.-H. Lai, S.-R. Moon, L. Zhang, and A. Maitra, "A 15-kV Class Intelligent Universal Transformer for Utility Applications," in *IEEE Applied Power Electronics Conference and Exposition (APEC)*, 2016, pp. 1974–1981.

[14] H. Qin and J. W. Kimball, "Solid-State Transformer Architecture Using AC–AC Dual-Active-Bridge Converter," *IEEE Transactions on Industrial Electronics*, vol. 60, no. 9, pp. 3720–3730, 2013.

[15] M. Kang, P. Enjeti, and I. Pitel, "Analysis and Design of Electronic Transformers for Electric Power Distribution System," *IEEE Transactions on Power Electronics*, vol. 14, no. 6, pp. 1133–1141, 1999.

[16] W. Xu, S. Rajendran, Z. Guo, A. Vctrivelan, and A. Q. Huang, "7.2kV/100kVA Solid State Transformer Based on Half Bridge LLC Resonant Converter and 15kV SiC AC Switch," in *IEEE Applied Power Electronics Conference and Exposition (APEC)*, 2023, pp. 1516–1522.

[17] W. Khan, J. A. Dar, K. S. Parihar, and M. K. Pathak, "Analysis of a New Single-Stage AC/AC Converter for Solid-State Transformer," *IEEE Transactions on Industry Applications*, vol. 60, no. 2, pp. 3359–3372, 2024.

[18] K. Basu, A. Shahani, A. K. Sahoo, and N. Mohan, "A Single-Stage Solid-State Transformer for PWM AC Drive With Source-Based Commutation of Leakage Energy," *IEEE Transactions on Power Electronics*, vol. 30, no. 3, pp. 1734–1746, 2015.

[19] Y. Ren, M. Xu, J. Sun, and F. Lee, "A Family of High Power Density Unregulated Bus Converters," *IEEE Transactions on Power Electronics*, vol. 20, no. 5, pp. 1045–1054, 2005.

[20] S. Mukherjee, A. Kumar, and S. Chakraborty, "Comparison of DAB and LLC DC–DC Converters in High-Step-Down Fixed-Conversion-Ratio (DCX) Applications," *IEEE Transactions on Power Electronics*, vol. 36, no. 4, pp. 4383–4398, 2021.

[21] W. Qin, X. Wu, and J. Zhang, "A Family of DC Transformer (DCX) Topologies Based on New ZVZCS Cells With DC Resonant Capacitance," *IEEE Transactions on Power Electronics*, vol. 32, no. 4, pp. 2822–2834, 2017.

[22] G. Deng, Y. Sun, G. Xu, X. Chen, S. Xie, S. Yan, M. Su, and Y. Liao, "ZVS Analysis of Half Bridge LLC-DCX Converter Considering the Influence of Resonant Parameters and Loads," in *IEEE Energy Conversion Congress and Exposition (ECCE)*, 2020, pp. 1186–1190.

[23] Y. Wei, Q. Luo, and A. Mantooth, "A Hybrid Half-bridge LLC Resonant Converter and Phase Shifted Full-bridge Converter for High Step-up Application," in *IEEE Workshop on Wide Bandgap Power Devices and Applications in Asia (WiPDA Asia)*, 2020, pp. 1–6.

[24] K. Tomas-Manez, Z. Zhang, and Z. Ouyang, "Unregulated series resonant converter for interlinking DC nanogrids," in *IEEE 12th International Conference on Power Electronics and Drive Systems (PEDS)*, 2017, pp. 647–654.

[25] X. Wu and H. Shi, "High Efficiency High Density 1 MHz 380–12 V DCX With Low FoM Devices," *IEEE Transactions on Industrial Electronics*, vol. 67, no. 2, pp. 1648–1656, 2020.

[26] J. E. Huber, J. Böhler, D. Rothmund, and J. W. Kolar, "Analysis and Cell-Level Experimental Verification of a 25 kW all-SiC Isolated Front End 6.6 kV/400 V AC-DC Solid-State Transformer," *CPSS Transactions on Power Electronics and Applications*, vol. 2, no. 2, pp. 140–148, 2017.

[27] J. D. Pollock and C. R. Sullivan, "Design Considerations for High-Efficiency Leakage Transformers," in *IEEE Applied Power Electronics Conference and Exposition (APEC)*, 2015, pp. 162–169.

[28] K. Venkatachalam, C. Sullivan, T. Abdallah, and H. Tacca, "Accurate Prediction of Ferrite Fore Loss with Nonsinusoidal Waveforms Using Only steinmetz Parameters," in *IEEE Workshop on Computers in Power Electronics, 2002. Proceedings.*, 2002, pp. 36–41.

[29] A. T. Nguyen and C. R. Sullivan, "Optimization of an Unregulated Low-Q Resonant Isolated DC-DC Stage for Low-Power AC-AC Converters," in *IEEE 25th Workshop on Control and Modeling for Power Electronics (COMPEL)*, 2024.

[30] D. Reusch, A. Lidow, J. Strydom, and M. de Rooij, "Emerging Applications for GaN Transistors," Jan. 2022, Efficient Power Conversion Corporation.

Survey of Components and Topologies for High-Efficiency and High-Power Density 48V DC-DC Converters

Joseph Winkler, Niklas Deneke, Bernhard Wicht

Institute of Microelectronic Systems
Leibniz University Hannover
Hannover, Germany
{joseph.winkler, niklas.deneke, bernhard.wicht}@ims.uni-hannover.de

Abstract—**Vast design spaces with numerous topologies and parameter requirements make optimal DC-DC converter design choices and comparisons challenging. Despite their variety, DC-DC converters share fundamental components. To this end, trends and relationships between passive components are investigated in the survey presented in this paper. The potential for 200 times higher energy density in multilayer ceramic capacitors (MLCCs) compared to discrete power inductors is confirmed. The survey also shows higher volumetric conductivity in discrete GaN switches compared to Si switches. Recent publications on 48 V DC-DC converters are investigated in order to identify topologies capable of achieving high efficiency and high power density. Hybrid switched-capacitor converters emerge as promising candidates, demonstrating the potential to achieve both goals simultaneously, while also meeting the demands of high-power applications.**

Index Terms—**Survey, 48 V, Energy Density, Passive Components, MLCC, DC-DC Converter, GaN, Discrete Components**

I. INTRODUCTION

The rise of electronic systems in cars has significantly increased power consumption, with 12 V systems reaching currents over 250 A, resulting in substantial conduction losses [1]. To reduce currents at given output power, the board net voltage is elevated to 48 V (e.g., Tesla [1]). Similarly, AI and data centers drive the adoption of 48 V rack voltage [2]. Both fields require efficient, high-power density DC-DC converters. The range of design options complicates the selection of optimal components and configurations. Achieving high efficiency and power density requires understanding the realistic device parameters. This survey examines trends in passive components as well as in Silicon (Si) and Gallium Nitride (GaN) transistors, and analyzes recent 48 V converter publications. Previous publications have focused on isolated aspects (e.g., passive components [3], [4], topologies [5], topologies and magnetics [6]), whereas this survey puts the focus on a comprehensive analysis. The dataset is publicly available on *GitHub* [7] to ensure transparency and support further research.

Section II provides a comparison of the energy densities of discrete capacitors and inductors, with capacitance and inductance densities, as well as related parasitic elements, being analyzed in respect to their effect on the converter design. In Section III, commercially available Si and GaN switches for 48 V DC-DC converters are examined, and a volume-related figure of merit (FOM) is introduced as a basis for optimization. In Section IV, 48 V converters from literature are analyzed, and key parameters are extracted to identify topologies that achieve the highest efficiency and power density.

II. PASSIVE COMPONENTS

In physical devices, capacitance and inductance values are limited to the discrete series of components offered by manufacturers, forming a finite space of design options. However, solutions require multi-dimensional trade-offs concerning parameters such as voltage and current rating, volume, and resistance, affecting power density and efficiency. This section compares the energy densities of capacitors and inductors, as well as other relevant parameters.

A. Maximum Energy Density

In Fig. 1, the maximum energy density of multilayer ceramic capacitors (MLCCs) from *Murata* [8], *TDK* [9] and *Wuerth Elektronik* [10], as well as inductors from *Vishay* [11] and *Coilcraft* [12] are compared.

$$\frac{E_{C,\max}}{\text{Vol.}} = \frac{1}{2} \frac{1}{\text{Vol.}} C(V) V^2 \qquad (1)$$

Each data point in Fig. 1 (a) is calculated according to (1). The investigated capacitors cover a range of voltage ratings from 2.5 V up to 100 V. For comparison, their energy densities are calculated for each voltage up to their own rating; hence, capacitors with higher ratings generate more points in the graph. Accounting for capacitance derating, each energy density is calculated using the derated capacitance value $C(V)$ (see (1)) at the respective bias voltage, based on information provided by the manufacturer.

$$\frac{E_{L,\max}}{\text{Vol.}} = \frac{1}{2} \frac{1}{\text{Vol.}} L I_{\max}^2 \qquad (2)$$

979-8-3315-1612-3/25 $31.00 © 2025 IEEE

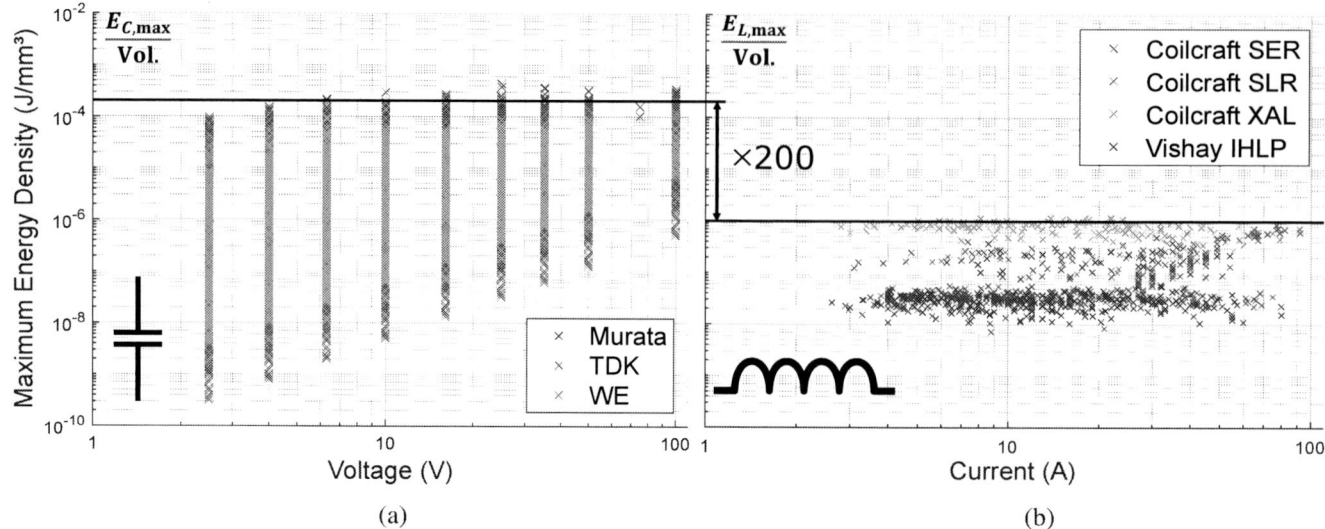

Fig. 1: Maximum energy density for (a) consumer grade MLCCs from *Murata* [8], *TDK* [9] and *Wuerth Elektronik* [10] and (b) power inductor families from recent literature with saturation currents of 2 A-200 A from *Vishay* [11] and *Coilcraft* [12].

The maximum inductor energy densities, shown in Fig. 1 (b), are calculated using (2). This figure illustrates the correlation between power inductors and current rating. The maximum current I_{max} is defined as the lesser of the saturation current or current rating, considering inductance degradation from bias current. All data used in the analysis were sourced directly from manufacturers that provide detailed component specifications, ensuring a comprehensive and accurate survey.

The data points in Fig. 1 (a) form vertical lines at the calculated voltage rating values offered by manufacturers. Despite the significant bias voltage capacitance derating in Class II MLCCs [8]–[10], the observed energy density tends to increase with voltage. The inductor data presented in Fig. 1 (b) form horizontal point clouds, illustrating the constant scaling of the energy density of inductors to higher currents. It is confirmed that the energy density of inductors is significantly lower than that of MLCCs [4], with capacitors achieving up to 200 times higher energy density in this survey. This information guides the partitioning of system volume into capacitive and inductive components when designing a hybrid DC-DC converter with MLCCs and power inductors.

B. Capacitors

To provide further insight into the evaluated characteristics, Fig. 2 shows three scatter plots of MLCC SMD capacitors from *Murata*, *TDK* and *Wuerth Elektronik*. The plots depict the volumetric capacitance density ρ_C in F/mm^3, over (a) voltage rating, (b) volume in mm^3 and (c) nominal capacitance. The volumetric capacitance density values exhibit a broad range from 0.3 nF/mm^3 to 0.4 mF/mm^3. Unlike the energy density analysis in Section II-A, the plots in Fig. 2 do not account for capacitance derating under bias voltage, ensuring consistent comparability across all subfigures. Instead, the capacitance density is based on the nominal capacitance value, providing a basis for comparison that is independent of specific operating conditions.

In Fig. 2 (a), the majority of data points is confined around vertical lines at the discrete voltage rating values. For voltage ratings below 7 V, the maximum volumetric capacitance density ρ_C remains stable at 90 µF/mm^3, but it gradually decreases for higher voltage ratings.

In Fig. 2 (b), the data points also accumulate around vertical lines, reflecting discrete volume steps of SMD package sizes noted on the x-axis. The highest volumetric capacitance densities are found in the smallest evaluated package sizes (i.e. 0201 and 0402). Additionally, parasitics are closely linked to package size: Smaller MLCC packages tend to have reduced series inductance (ESL) due to shorter current paths. However, this benefit comes at the cost of increased series resistance (ESR), whereas larger packages typically exhibit lower ESR [7], [10]. This trade-off highlights the balance between minimizing parasitic inductance and resistance, based on design requirements.

Fig. 2 (c) illustrates the relationship between capacitance density and nominal capacitance. The data formation corresponds to Fig. 2 (b), but tilted to the right, showing that the highest volumetric capacitance density ρ_C is achieved at the largest capacitance value for a given package size. In circuit design, selecting the smallest possible package for the required capacitance and voltage rating maximizes power density. Multiple MLCCs with high energy density can be connected in parallel to achieve the desired capacitance value. They can be soldered together both vertically and horizontally on the PCB to form bricks for optimal utilization of the available height (e.g., in [13]).

MLCCs with capacitance values in the microfarad range typically exhibit high ESL and resonant frequencies in the (several) megahertz range [8]. 48 V DC-DC converters with

979-8-3315-1612-3/25 $31.00 © 2025 IEEE

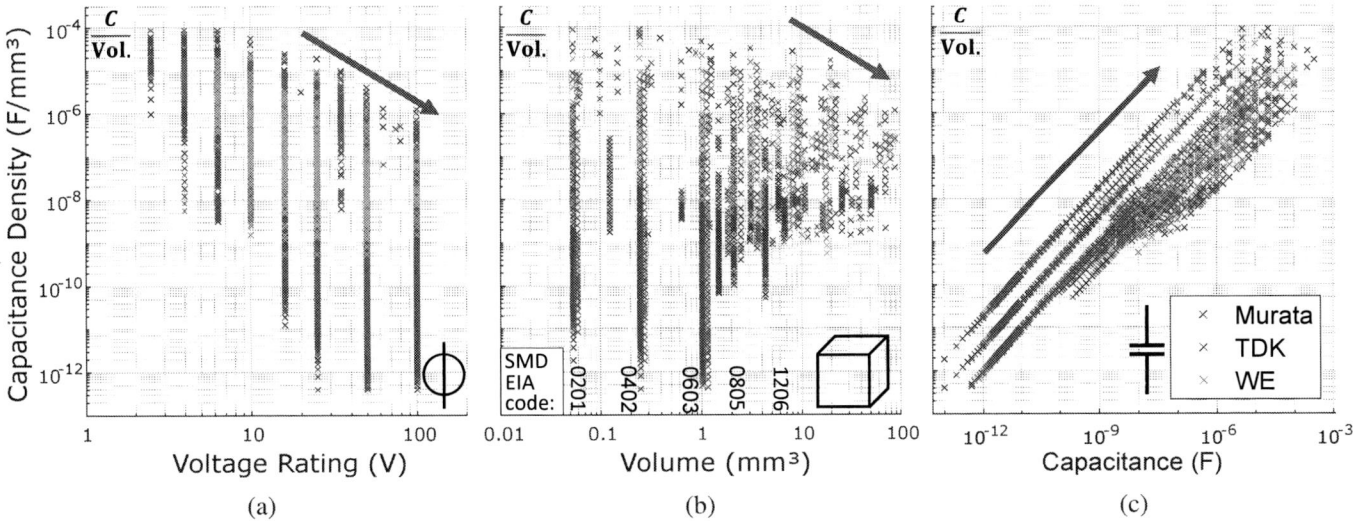

Fig. 2: Volumetric capacitance density per (a) maximum blocking voltage rating, (b) volume and (c) capacitance. Comparison of consumer grade MLCC rated up to $100\,\mathrm{V}$ from *Murata* [8] and *TDK* [9].

discrete power stages often have significant parasitic inductance due to PCB layout, leading to typical operating frequencies below $1\,\mathrm{MHz}$ [13]–[15]. Therefore, carefully chosen MLCCs remain the preferred capacitor technology for discrete $48\,\mathrm{V}$ converters, owing to their superior energy density.

The survey confirms that capacitance density tends to decrease with rated voltage and that the biggest capacitance values correlate with the highest capacitance density. By selecting components with the lowest rated voltage and smallest package size that meet the specific requirements of the application, the highest possible capacitance density and therefore high converter power density can be achieved.

C. Inductors

To illustrate the challenge in inductor design, an inductor with a thin wire and fixed number of turns is considered. The inductor has a small volume and therefore a high inductance density. Due to finite conductance and small wire width, the permissible current is low because the inductor's DC resistance R_L is high. This leads to conduction losses and thus to a degradation of overall efficiency. Using a thicker wire lowers DC resistance and increases current capacity, but also increases volume. This trade-off is explored in the following analysis.

Four common SMD power inductor families from *Coilcraft* [12] and *Vishay* [11] are examined to determine strategies for component selection, simultaneously targeting high power density and efficiency.

Fig. 3 (a) shows volume-related inductance density ($\mathrm{H/mm^3}$) over saturation current, ranging from $2\,\mathrm{A}$ to $200\,\mathrm{A}$. The data indicate a clear decrease in inductance density as saturation current increases.

Fig. 3 (b) displays the volumetric inductance density over the maximum DC resistance R_L, with the same y-axis scaling as in Fig. 3 (a). The uppermost data point in each data series in Fig. 3 (a) corresponds to the same device as the top most

data point in Fig. 3 (b). This indicates that achieving high inductance density typically comes at the cost of higher DC resistance R_L and lower saturation current, confirming the previously mentioned trade-off inherent in the design process.

Combining the previous parameters, Fig. 3 (c) shows the inductance per DC resistance in $\mathrm{H/\Omega}$ over volume in $\mathrm{mm^3}$. The data indicate that larger inductor volumes improve the inductance-to-DC-resistance ratio, confirming that package volume can be traded for lower DC resistance R_L. Thus, there is a compromise between power density and efficiency when choosing an inductor. Selecting an inductor that closely adheres to the application's specifications ensures high inductance density and subsequently high power density.

III. POWER SWITCH SELECTION

Different transistor types can be used for power switches, with Si-MOSFETs and GaN-FETs being common choices. The size of a power semiconductor relates to its blocking voltage and on-state resistance. High efficiency requires low on-state resistance and minimal parasitics (C_{iss}, C_{oss}, ...). The ratio of on-state conductivity ($S_{\mathrm{DS,on}} = 1/R_{\mathrm{DS,on}}$) per volume serves as a FOM which is shown in Fig. 4 for GaN-FETs (EPC [16]) and Si-MOSFETs (Infineon Optimos [17] and others from recent publications). Si transistors achieve a high FOM at voltages up to $30\,\mathrm{V}$, but GaN devices surpass them as voltage rating increases. In recent $48\,\mathrm{V}$ converter literature (e.g., [15], [18], [19]), GaN transistors are frequently chosen. Currently, the market offers a broader range and greater availability of Si devices compared to GaN devices. Both factors should be taken into account during converter design, with the evaluation in this work clearly supporting the selection of GaN devices for achieving high efficiency and power density in the $48\,\mathrm{V}$ design space. The switch choice is related to the various converter topology studies in the following section.

979-8-3315-1612-3/25 $31.00 © 2025 IEEE

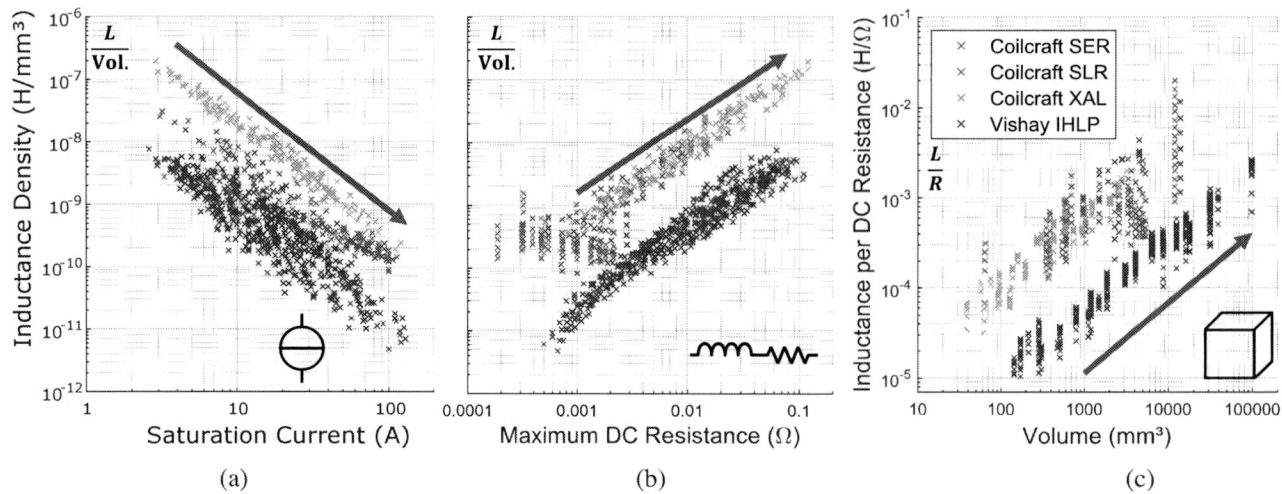

Fig. 3: Volumetric inductance density vs. (a) saturation current and (b) maximum DC resistance; (c) Inductance per DC resistance vs. volume. Comparison of power inductors (2 A-200 A) from recent literature by *Vishay* [11] and *Coilcraft* [12].

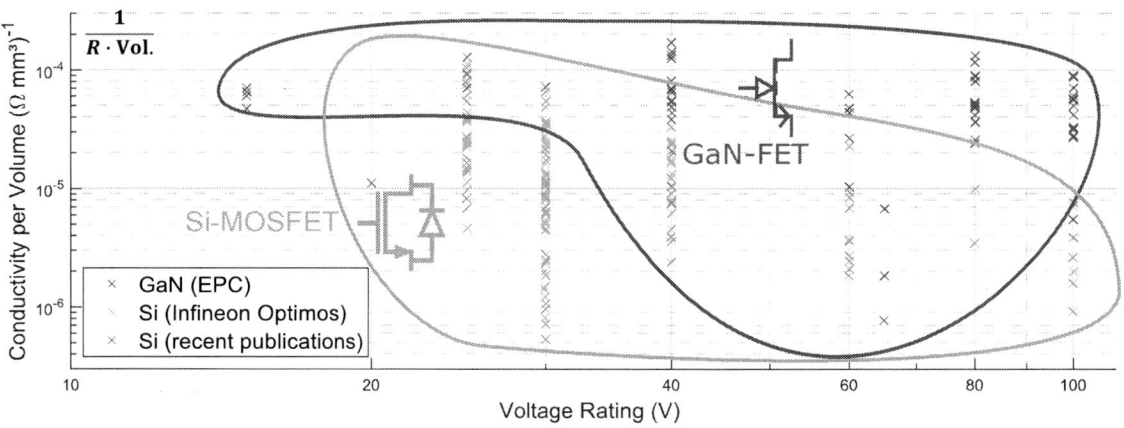

Fig. 4: Conductivity per volume of discrete N-type GaN and Si-FETs (drain-source on-resistance specified at $V_{GS} = 5\,V$)

IV. 48 V CONVERTER TOPOLOGIES

In this section, 48 V converters from various publications are examined. A fair comparison is challenging due to differences in performance characteristics and the self-reported nature of the data. Variations in efficiency and power density measurement methods include peripheral loss definitions, cooling application for maximum current ratings, and volume calculation. Furthermore, the transient response capabilities are not considered in this section due to the inhomogeneous nature of the available data.

Despite inherent discrepancies, this simplified comparison highlights general trends and technological progress, allowing a relative performance assessment of different topologies. It serves as a practical benchmark for evaluating their potential, even with variations in data collection methods. The focus of this survey lies in assessing the potential of the topologies for achieving the highest power density and efficiency.

Table I lists abbreviations utilized in Fig. 5 (b) and Fig. 6 of this survey.

A. Peak Efficiency

Fig. 5 shows scatter plots of recent 48 V DC-DC converters. Both subfigures use the same data points, showing the reported peak efficiency over the maximum output current, while encoding different additional information. Note that peak efficiency typically does not occur at the maximum output current. All converters included in this comparison achieve peak efficiencies above 80%.

Fig. 5 (a) displays the author and year of publication next to each data point. Colors encode conversion ratios, with a heat bar linking colors to ratios. Red indicates a high conversion ratio (up to 48), while blue represents a low ratio (as small as 4). Ratios of 4 to 12 yield the highest peak efficiencies.

Fig. 5 (b) includes the topology name and a color-coded legend indicating integrated or discrete power stages. Converters exceeding 5 A output currents tend to use discrete switches, suitable for higher currents. Both GaN and Si switches are widely adopted. Efficient topologies include hybrid switched-capacitor (hySC) and transformer-based topologies such as

LLC-tank converters. The topologies that have been considered, which exhibit maximum output currents ranging from 200 A to 2000 A and feature high conversion ratios of over 24, all utilize hySC topologies. Furthermore, the topologies in this current range employ more than one inductor and predominantly operate in more than two phases in order to distribute the current load.

B. Peak Power Density

Power density is typically defined as the ratio of output power to volume. To ensure meaningful comparisons, consistency in calculation methods is essential. A common approach uses the box volume, which encloses the power stage within a cuboid, as this method closely mirrors real-world applications. In contrast, cumulative volume yields higher density values but only allows for relative comparisons. Accurate assessments also require clear specifications of the components included within the evaluated volume. Despite inhomogeneous determination methods, the power density values given in recent publications are compared below to identify trends, technological progress, and the comparative performance of different topologies.

Fig. 6 shows the power density in $\mathrm{W/in}^3$ over the maximum output current. The power density values in Fig. 6 are self-reported in the publications. The data point colors indicate switch material (GaN or Si). Interestingly, no clear superiority between Si and GaN switches is observed, which may change in the future. The topology as well as the first author's last name and the publishing year are shown next to each data point. High power densities are achieved by hySC topologies (e.g. [20], [21]). Transformer-based topologies fall behind in this regard. This fits the observed difference in energy density between inductors and MLCCs (see Fig. 1 in Section II-A).

V. CONCLUSION

This survey examines fundamental components of 48 V DC-DC converters and recent publications. It compares properties of common discrete MLCCs and power inductors, confirming capacitors' potential for up to 200 times higher energy density. Discrete GaN power transistors offer better volumetric conductivity than Si switches. Recent publications show highest efficiencies at low conversion ratios (4:1 or 6:1). Converters with output currents above 5 A typically use discrete power stages. Hybrid switched-capacitor (hySC) topologies dominate at high conversion ratios (>24) and high output currents (200–2000 A), utilizing multiple inductors and multi-phase operation to effectively distribute the current load. Efficient topologies include transformer-based and hySC converters. HySC converter topologies using discrete components demonstrate the highest efficiency and power density simultaneously.

The data set for all components and compared publications is made accessible on *GitHub* [7].

ACKNOWLEDGMENT

This work is funded as part of the research project KI4BoardNet in the funding program MANNHEIM (BMBF)

(Grant number 16ME0784). The responsibility for this publication is held by the authors only.

REFERENCES

[1] Tesla. Tesla Investor Day 2023. [Online]. Available: https://digitalassets.tesla.com/tesla-contents/image/upload/IR/Investor-Day-2023-Keynote

[2] R. Hintemann and S. Hinterholzer, "Energy Consumption of Data Centers Worldwide," in *2019 Conference for Information and Communication Technologies for Sustainability (ICT4S)*, pp. 1–5.

[3] J. Wittmann, *Integrated High-Vin Multi-MHz Converters*. Springer International Publishing.

[4] J. Zou, N. C. Brooks, S. Coday, N. M. Ellis, and R. C. Pilawa-Podgurski, "On the Size and Weight of Passive Components: Scaling Trends for High-Density Power Converter Designs," in *2022 IEEE 23rd Workshop on Control and Modeling for Power Electronics (COMPEL)*, pp. 1–7.

[5] M. Gong, X. Zhang, and A. Raychowdhury, "Non-isolated 48V-to-1V Heterogeneous Integrated Voltage Converters for High Performance Computing in Data Centers," in *2020 IEEE 63rd International Midwest Symposium on Circuits and Systems (MWSCAS)*, pp. 411–414.

[6] M. Chen, S. Jiang, J. A. Cobos, and B. Lehman, "Design Considerations for 48-V VRM: Architecture, Magnetics, and Performance Tradeoffs," in *2023 Fourth International Symposium on 3D Power Electronics Integration and Manufacturing (3D-PEIM)*. IEEE, pp. 1–9.

[7] J. Winkler. Data Set Survey of Components and Topologies for High-Efficiency and High-Power Density 48V DC-DC Converters. [Online]. Available: https://github.com/josephwinkler/48V_DCDC_Converter_Survey

[8] Murata. Murata MLCC Catalog. [Online]. Available: https://www.murata.com/en-global/search/productsearch?cate=luCeramicCapacitorsSMD&stype=2& realtime=1

[9] TDK. TDK MLCC Catalog. [Online]. Available: https://product.tdk.com/en/search/capacitor/ceramic/mlcc/catalog

[10] W. Elektronik. RedExpert MLCC Catalog. [Online]. Available: https://redexpert.we-online.com/we-redexpert/en/#/home

[11] Vishay. Vishay Inductor Catalog. [Online]. Available: https://www.vishay.com/en/inductors/

[12] Coilcraft. Coilcraft Inductor Catalog. [Online]. Available: https://www.coilcraft.com/en-us/products/power/#/

[13] Y. Zhu, T. Ge, N. M. Ellis, L. Horowitz, and R. C. N. Pilawa-Podgurski, "The Switching Bus Converter: A High-Performance 48-V-to-1-V Architecture With Increased Switched-Capacitor Conversion Ratio," vol. 39, no. 7, pp. 8384–8403.

[14] Y. Chen, H. Cheng, D. M. Giuliano, and M. Chen, "A 93.7% Efficient 400A 48V-1V Merged-Two-Stage Hybrid Switched-Capacitor Converter with 24V Virtual Intermediate Bus and Coupled Inductors," in *2021 IEEE Applied Power Electronics Conference and Exposition (APEC)*, pp. 1308–1315.

[15] M. E. Blackwell, S. Krishnan, N. M. Ellis, and R. C. Pilawa-Podgurski, "Direct 48 V to 6 V Automotive Hybrid Switched-Capacitor Converter with Reduced Conducted EMI," in *2022 IEEE 23rd Workshop on Control and Modeling for Power Electronics (COMPEL)*. IEEE, pp. 1–8.

[16] EPC. EPC GaN Product Catalog. [Online]. Available: https://epc-co.com/epc/products/gan-fets-and-ics

[17] Infineon. OptiMOS™ and StrongIRFET™ MOSFETS Selection guide 2020. [Online]. Available: https://www.infineon.com/dgdl/Infineon-Selection_guide_MOSFET_OptiMOS_and_StrongIRFET_2019-SG-v01_00-EN.pdf?fileId=5546d46269e1c019016a11d00c523f76

[18] R. Das and H.-P. Le, "A Two-Stage 110VAC-to-1VDC Power Delivery Architecture Using Hybrid Converters for Data Centers and Telecommunication Systems."

[19] M. H. Ahmed, F. C. Lee, and Q. Li, "Two-Stage 48-V VRM With Intermediate Bus Voltage Optimization for Data Centers," vol. 9, no. 1, pp. 702–715.

[20] T. Ge, Z. Ye, and R. C. Pilawa-Podgurski, "A 48-to-12 V Cascaded Multi-Resonant Switched Capacitor Converter with 4700 W/in^3 Power Density and 98.9% Efficiency," in *2021 IEEE Energy Conversion Congress and Exposition (ECCE)*. IEEE, pp. 1959–1965.

[21] Z. Ye, Y. Lei, and R. C. N. Pilawa-Podgurski, "The Cascaded Resonant Converter: A Hybrid Switched-Capacitor Topology With High Power Density and Efficiency," vol. 35, no. 5, pp. 4946–4958.

TABLE I: Abbreviations used in Fig. 5 and 6.

Abbr.	Explanation	Abbr.	Explanation	Abbr.	Explanation
C	Capacitor	CP	Charge-Pump	stg.	stage
(2/6)L	(two/six) Inductor(s)	SP	Series-Parallel	interl.	interleaved
hySC	Hybrid switched-capacitor	DPh/MPh	Dual/Multi-Phase	casc.	cascaded
ML	Multi-level	LEGO	Linear Extendable Group Oper-	mod.	modified
HB	Half-Bridge		ated Point-of-Load architecture		

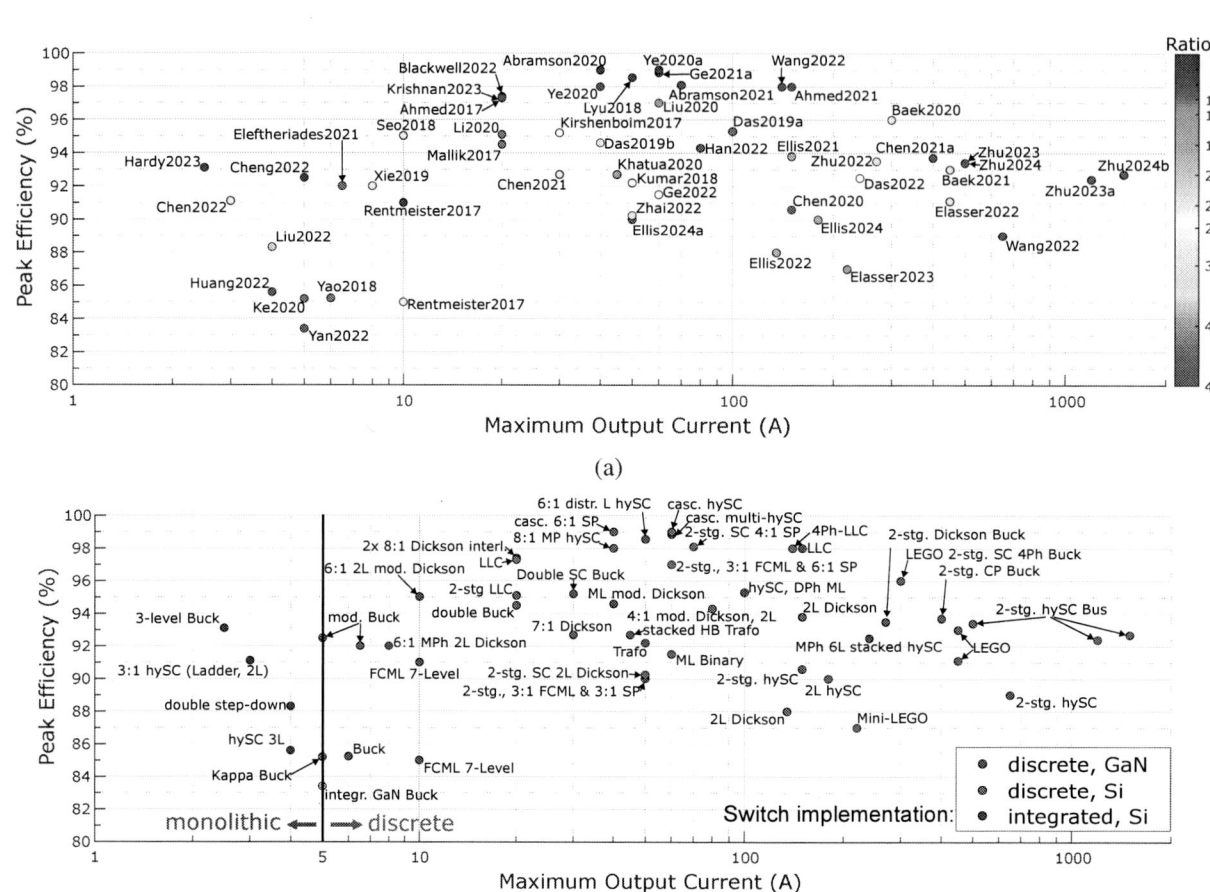

(a)

(b)

Fig. 5: Peak efficiencies for recent converter topologies over maximum output current with (a) corresponding conversion ratio, first author and publication year; (b) implementation details (topology, discrete or integrated, GaN or Si switches). Axes and points are the same between the figures but the encoded information differs.

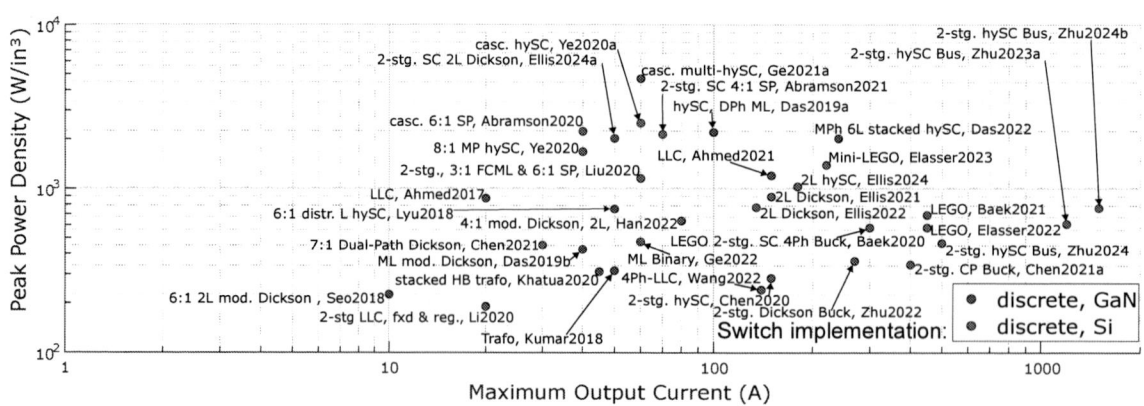

Fig. 6: Peak power density for recent converter topologies over maximum output current with implementation details.

A Novel Solid-State Circuit Breaker Using B-TRAN™

Mudit Khanna
Ideal Power, Inc.
Austin, Texas, USA
mudit.khanna@idealpower.com

Ruiyang Yu
Ideal Power, Inc.
Austin, Texas, USA
ryan.yu@idealpower.com

Milad Tayebi
Ideal Power, Inc.
Austin, Texas, USA
milad.tayebi@idealpower.com

Jiankang Bu
Ideal Power, Inc.
Austin, Texas, USA
jiankang.bu@idealpower.com

Jeffery Knapp
Ideal Power, Inc.
Austin, Texas, USA
jeff.knapp@idealpower.com

Abstract— Solid-State Circuit Breakers (SSCBs) use a power semiconductor device to break the short circuit current and hence offer ultra-fast response time compared to mechanical breakers. However, a reliable and low loss bidirectional power semiconductor switch is essential for commercial viability of SSCB. This article provides an overview of a novel solid-state circuit breaker design using B-TRAN™ discrete device. The paper describes design considerations like B-TRAN™ cascode switching circuit setup, driver design, peak current capability of the device and layout optimization. The impact of these parameters on switch performance is analyzed. Then system level design considerations such as voltage clamping methods and overcurrent detection techniques are discussed. Test results for 400V, 200A breaking current are also presented. To conclude, the paper provides a useful framework to develop breaker designs using B-TRAN™ and how it can be compared with other power semiconductor switch technologies.

Keywords—Solid-State Circuit Breaker (SSCB), Circuit Breaker, Bidirectional Switch, EV Charging, DC microgrids, EV Contactor, Circuit Protection, B-TRAN, Power Semiconductor, SiC, GaN.

I. INTRODUCTION

With the rise in renewable energy source and battery energy storage systems (BESS), DC distribution systems have gained rapid adoption due to their higher efficiency and power density. Not only in power generation and distribution, DC systems can be found in multiple applications such as microgrids, data centers, residential/commercial applications, electric vehicles, electric aircrafts etc. [1]. One of the main challenges of DC systems is the lack of fast-acting protection devices. The fault current in DC systems can rise rapidly and due to the lack of zero crossing in current waveforms, the breaking of high fault current can become extremely challenging. SSCBs offer ultra-fast response as it can turn-off in microseconds compared to tens of milliseconds for mechanical breakers and they are also more reliable due to fewer moving components [2].

The downside of using a SSCB is large losses in power semiconductor devices. In most applications, the power semiconductor device must also be a bidirectional switch which further increases the losses. Recently, wide bandgap devices (WBG) such as SiC and GaN MOSFETs have helped reduce the losses in high voltage applications but the cost of these technologies and reliability under overcurrent/short circuit conditions poses a restriction in wide scale commercial adoption [10]. B-TRAN™ is a novel bidirectional switch architecture which offers ultra-low conduction losses at relatively lower cost and high reliability with readily available silicon process. A comparison of commercially available bidirectional switches is presented in Section II. The section also introduces the design considerations for B-TRAN™ based SSCB such as driver design, peak current capability etc.

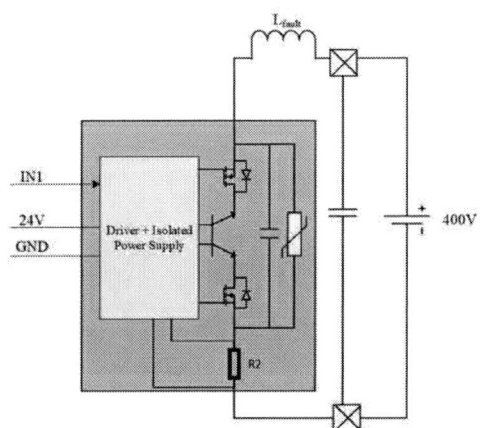

Fig 1. SSCB block diagram using B-TRAN™.

400V is a widely used bus voltage in DC systems such as in data centers, electric vehicles and microgrids. SSCB design for such applications have been presented before using various devices such as SiC MOSFETs, SiC JFETs, GaN MOSFETs, Superjunction Si MOSFETs etc. [6][8][9][11]. This paper

979-8-3315-1612-3/25 $31.00 © 2025 IEEE   854

presents a B-TRAN™ based 400V/50A SSCB reference design highlighting advantages of using B-TRAN™ for this application. For the first time, design considerations of implementing a B-TRAN™ discrete device using cascode MOSFETs are presented. The test results, shown in section IV, highlight ultra-low conduction losses in B-TRAN™, rugged device operation and fast response time. The block diagram of a solid-state circuit breaker design using B-TRAN™ is shown in Fig 1.

II. POWER DEVICE SELECTION AND DESIGN CONSIDERATIONS

A low loss bidirectional power semiconductor switch is essential for SSCB implementation. For 400V applications, >900V blocking voltage is desirable since the fault inductance causes an overshoot across the power semiconductor device. There a multiple bidirectional switches available in the market today and some monolithic bidirectional switches have also been proposed but a high power bidirectional switch has not been commercialized yet [4]. B-TRAN™ is one of the industry's first 1200V, high power monolithic bidirectional switch commercially available. Fig 2 shows some of the other bidirectional switches. A common-emitter IGBT configuration is readily available from multiple vendors, however, the losses become very high in this configuration. Common-source or common-drain SiC MOSFETs are also gaining popularity as a SiC MOSFETs offer low on-state resistance and high current capability compared to GaN.

A comparison of various power semiconductors with their conduction performance has been studied in [3]. The SiC JFET has the lowest losses among all technologies but also has the 5x the $/A compared to IGBT [3]. Fig 3 shows a comparison of losses across multiple commercially available bidirectional switches. The losses in B-TRAN™ (IPAD01205A04) are much lower compared to other devices even after considering driving and cascode MOSFET losses. A 12mΩ SiC die is used in back-to-back configuration for bidirectional switch comparison resulting in a 24mΩ bidirectional switch. Although, this can be further lowered by paralleling multiple dies but that can increase the cost significantly.

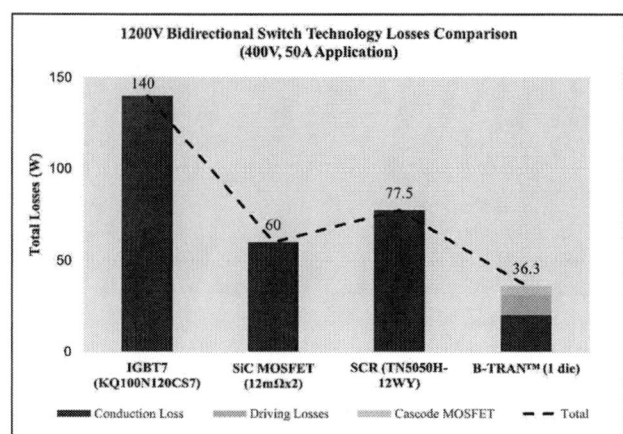

Fig 3. Comparison of conduction losses in commercially available bidirectional switches.

A. B-TRAN™ Device and Switching Operation

The TO-264 B-TRAN™ discrete device is normally on and hence a dual cascode implementation is used for nomrally-off operation as shown in Fig 5. The operation and characterization of the device was explained earlier [5]. Referring to Fig 5, Q3A and Q3B are low voltage (60V) Si MOSFETs which allow blocking voltage up to 1200V on each side. Q2A and Q2B are used to ground/float B1-B2 terminals during the switching operation. Q1A/Q1B are turned ON when the device is ON to inject base current into B1/B2. 12V-20V MOSFETs can be used in both cases.

To understand the switching operation, the switching waveform during turn off with the cascode FET is shown in Fig 4. When the cascode FETs Q3A/B turn OFF, Q2A/B are turned ON shorting the B1/B2 terminal to S1/S2 respectively (*time t_0*). The voltage across the switch S1-S2 starts to rise (*time t_1*). At the same time, the E2-S2 voltage also rises to create the depletion region across E2-B2 region while the load current flow transfers to B2. The E1-B2 diode, which is now reverse biased, starts blocking high voltage (*time t_2*) and a reverse recovery current flows out of B2 causing the tail current.

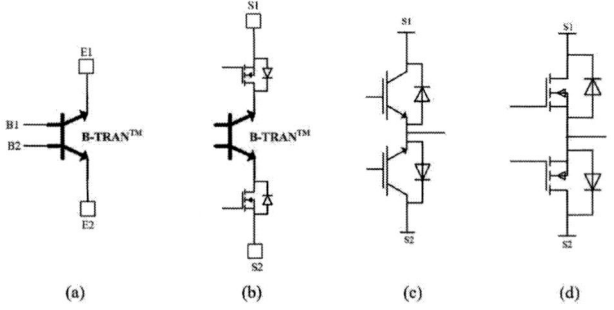

Fig 2. Different configurations of bidirectional switches. (a) Normally on B-TRAN™ device (b) Normally-Off B-TRAN™ with cascode FETs, (c) common-emitter IGBT (d) common-source MOSFETs

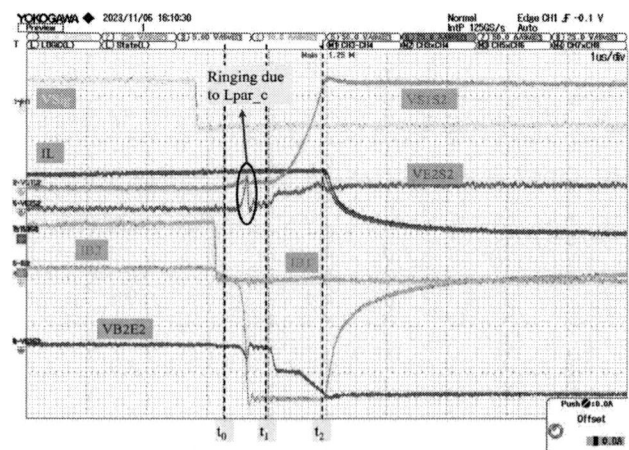

Fig 4. Turn Off waveforms for cascode B-TRAN™ switching.

979-8-3315-1612-3/25 $31.00 © 2025 IEEE   855

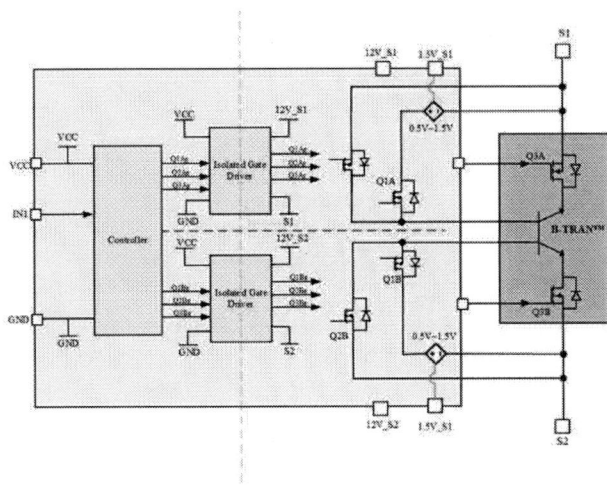

Fig 5. B-TRAN™ driver block diagram.

B. B-TRAN™ Driver Design for Breaker Applications

The B-TRAN™ driver circuit needs to drive the 6 low voltage FETs as shown in Fig 5. Since it is a bidirectional switch, two floating supplies (w.r.t S1 and S2) are needed to drive the high side FETs (Q1A-Q3A) and low side FETs (Q1B-Q3B). Two low voltage and high current floating supplies (1V-1.5V) are also needed to inject current into the base B1 and B2. The block diagram for the driver implementation is shown in Fig 5. The IPAD01205A04 does not come with integrated cascode MOSFETs and hence those are also added externally in the driver board. The base current information of the device can be monitored for diagnostics and overcurrent protection as well.

C. Driver Bias Power Supply Design

The bias supply can be generated from an external low voltage supply or can be directly tapped from the high voltage DC bus. In the current design, a low voltage 24V supply is used to generate the floating rails. An LLC or a flyback converter can be used to generate 12V-S1 and 12V-S2 supplies (w.r.t S1 and S2 respectively) followed by a buck converter to step down the 12V rails to 1.5V needed to drive current into the base as shown in Fig 6. It is important that the current rating of the buck converter should be chosen as per the maximum load current. For Gen1 B-TRAN™ the current gain is 5 which means base current should be at least 1/5th of the load current.

D. Peak Current Capability and Layout Considerations

The peak current capability of B-TRAN™ is highly dependent on the parasitic inductance between the E2-S2 (L_{par_c}) and B2-S2 (L_{par_p}). For the device used in this testing (IPAD01205A04), the peak current capability is typically rated to 3x of the rated current. However, with careful optimization of the Q3B/Q3A and Q2B/Q2A MOSFETs layout, this can be increased up to 7x. Considering S1 is the positive high voltage terminal, during turn-off when Q3B is turned off and Q2B is turned on, the load current flowing through Q3B and L_{par_c} is interrupted and current transfers to B2 and L_{par_p}. Because there is no path for L_{par_c} current to flow, it causes a voltage spike

across E2-S2 and E2-B2. This can cause the breakdown of E2-B2 diode at very high voltage and currents. Fig 4 shows this ringing but because of low voltage and current, it does not cause the breakdown of device. The impact of this inductance was studied experimentally by increasing the distance between the E2 terminal of the device and Q3B/Q2B MOSFETs in a clamped inductive switching test (Fig 7). The peak current capability increased from 294A to 361A with smaller distance i.e. lower L_{par_c} and L_{par_p} as shown in Table I. In future work, characterization of the parasitic inductance and subsequent development of a co-pack module is planned to improve the performance even further.

TABLE I.   COMPARISON OF PEAK CURRENT VS PARASITIC INDUCTANCE

Distance ('d')	Peak turn off current	Rated Current	Turn Off Voltage
5 mm	294 A	50 A	800 V
3 mm	334 A	50 A	800 V
1 mm	361 A	50 A	800 V

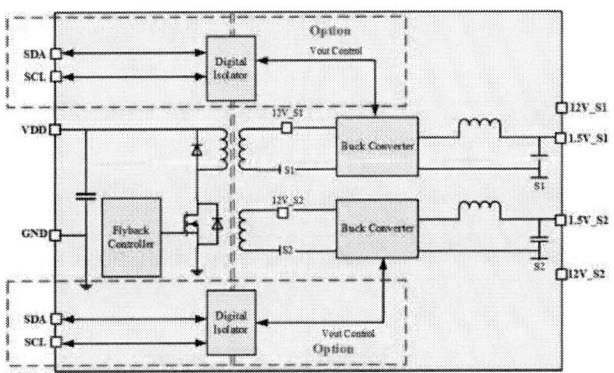

Fig 6. Isolated bias power supply design for B-TRAN™ driver. A low voltage high current buck converter is used to generate base current driving supply.

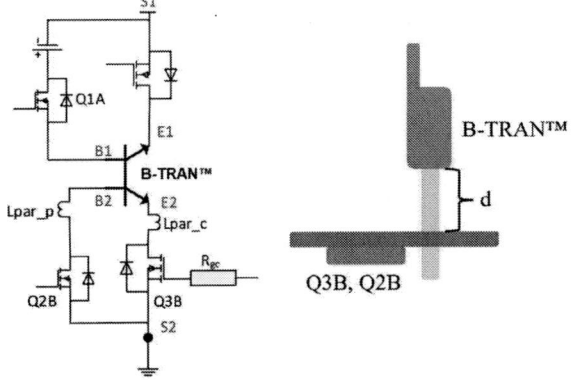

Fig 7. Partial schematic block diagram showing parasitic inductance. The board layout used for testing is also shown. The disatnce 'd' is varied to change the parasitic inductance 'Lpar_c' and the impact of It on peak current capability of the device is reported.

979-8-3315-1612-3/25 $31.00 © 2025 IEEE   856

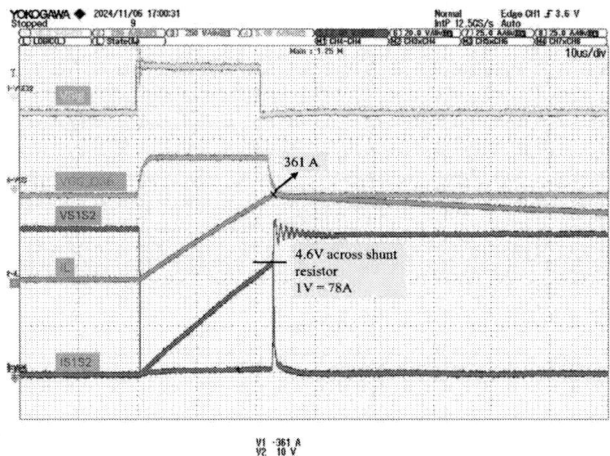

Fig 8. 800V, 361A turn off waveform for a 50A B-TRAN™ die.

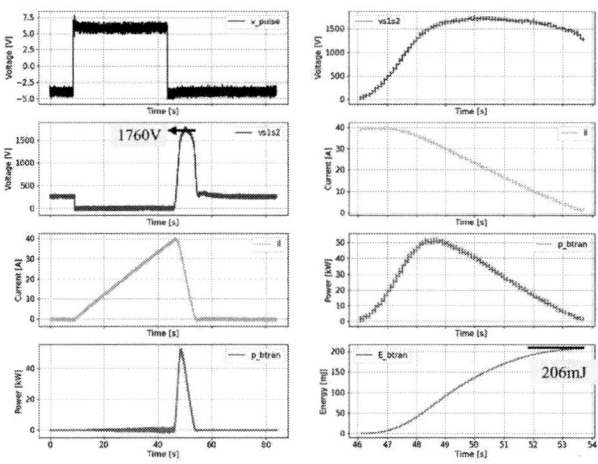

Fig 9. Unclamped Inductive Switching (UIS) test setup and results.

E. Avalanche Capability of B-TRAN™

The avalanche rating of power semiconductor device is an important parameter for SSCB applications. To investigate a typical avalanche energy, unclamped inductive switching (UIS) test was conducted on the device. The input voltage was set to 260V and a 250μH inductor was used with a single pulse of 35μs. Preliminary results, shown in Fig 9, show single pulse avalanche rating 206mJ with maximum voltage reaching 1760V with 40A without failure. Since there is no co-packed diode, like IGBT, the rating of the device is solely dependent on the B-TRAN™ die and the cascode MOSFETs. Further work on the avalanche ruggedness of the device and different factors it can be impacted by is ongoing.

III. System Design Considerations

In addition to designing the B-TRAN™ switch and driver, the other components in the SSCB design are selected based on the specifications listed in Table II. Voltage clamping device selection and design and overcurrent protection technique used are described.

TABLE II. SPECIFICATIONS OF SSCB REFERENCE DESIGN

Parameters	Value
Nominal Blocking Voltage	400 V
Continous current	50 A
Maximum Breaking Current	200 A
Cooling	Forced Air
Low Voltage Power Supply, Nominal Voltage	24 V
Low Voltage Power Supply Current Rating	2 A
B-TRAN™ Device Rating	1200 V, 50A

A. Voltage Clamping

The voltage clamping is needed to protect the semiconductor device from overvoltage and to absorb the residual energy in the fault inductance. MOVs are widely used as voltage clamping devices and can be selected depending the maximum clamping voltage, maximum continuous operating voltage and the expected energy absorption needed [6][7]. In addition to using MOVs, an RC snubber can be added in parallel to limit dv/dt and voltage spikes during device turn-off. A resistor is typically included in series with the capacitor to prevent a large current surge into the semiconductor device during turn-on. However, a larger resistor slows down the turn-on and turn-off processes due to longer capacitor charge/discharge times. In B-TRAN™ designs, a capacitor-based snubber can be utilized without a series resistor, leveraging the device's adjustable on-state resistance (Fig 10). When no base current is injected, B-TRAN™ exhibits a high on-state resistance (typically 2-3Ω), which can be used to limit turn-on surge current. The B-TRAN™ is initially turned on in this high-resistance mode, limiting peak current, and then base current is injected to reduce the on-state resistance for efficient conduction. This technique limits peak turn-on current to 150A, as shown in Fig 13, reducing the need for traditional RC snubber resistors and improving overall system efficiency.

B. Overcurrent Protection

One of the main functions of a SSCB is sensing and trip electronics which constantly measures the breaker current. When fault or overload current is detected, the overcurrent protection turns off the power semiconductor device through the gate driver. Various current sensing mechanisms can be used as discussed in [2] providing different levels of accuracy and speed.

In this SSCB reference design, the breaker current is measured through a shunt resistor. The overload and fault current limits are set at 50A and 200A, respectively. In both overload and fault conditions, the microcontroller turns off the gate driver. The gate driver stays off until the overcurrent protection fault is cleared in the microcontroller. Fig 11 shows the tripping capability of the SSCB when a fault occurs. In this figure, a short circuit event was tested with a 10μs pulse, 400V, and a 8μH inductor in series with B-TRAN™ and cascode FETs. It takes 680ns from when the breaker's current reaches 200A till B-TRAN™'s current reaches zero current, and the breaker current reaches zero after 6μs.

979-8-3315-1612-3/25 $31.00 © 2025 IEEE   857

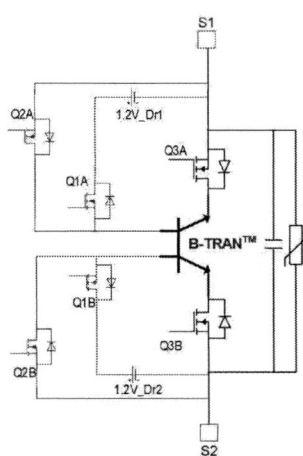

Fig 10. B-TRAN™ switch assembly with MOV and snubber capacitor for overvoltage clamping. The snubber capacitor can be used without a series resistance which can speed-up the turn off process the switch.

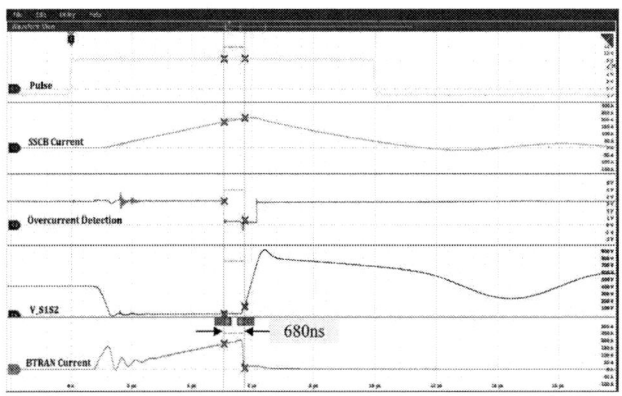

Fig 11. Overcurrent detection and reaction time. The gate resistance of cascode FET (Rg_Q3B) is 100Ω for this test.

IV. TEST RESULTS

The SSCB reference design board with discrete device (IPAD01250A04) is shown in Fig 13. The controller board includes gate driver, overcurrent protection circuit and the microcontroller to control the switch and gate signals. Power Supply board contains the isolated power supply and the buck converters for base current driving.

To test the short circuit detection and protection, the switch is turned ON under short circuit conditions at 400VDC voltage. A fault inductance of 8µH is added which is results in a 50A/µs di/dt. Once the device current reaches ~200A, the overcurrent protection circuit is tripped and the microcontroller turns of the B-TRAN™ switch. The response time for the control circuit to initiate the turn off process <100ns. The total fault clearing time is 6µs with 900V peak voltage across the switch.

TABLE III. RESULTS SUMMARY

Parameters	Value
Overcurrent detection threshold	200A
Turn off response time	< 100ns
Total fault clearing time	6µs
Peak voltage across the device	900V
Fault inductance	8 µH
Di/dt tested	50 A/µs

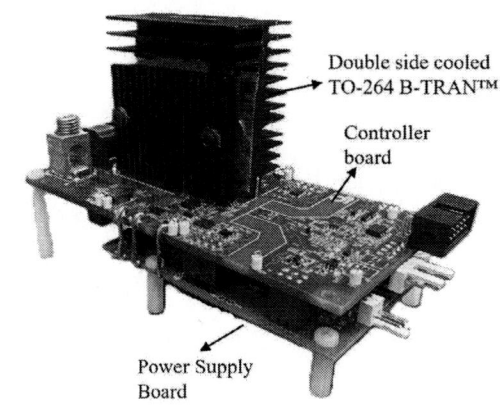

Fig 12. SSCB reference design board used for testing.

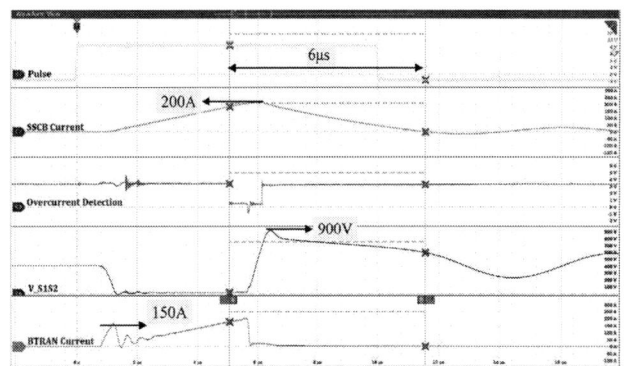

Fig 13. 400V, 200A breaking test results. The peak voltage is 900V across the device with 150A during turn-on due to 'soft-turn' feature of B-TRAN™.

V. CONCLUSION AND FUTURE WORK

B-TRAN™ was introduced in the past as one of the lowest on-state resistance monolithic bidirectional switch in using silicon. This paper successfully demonstrated the application of B-TRAN™ in a SSCB application and highlighted its merits. The paper discussed in detail the design considerations needed when designing an SSCB using B-TRAN™. A discrete normally-on device was used to demonstrate the capabilities of the switch. Some unique advantages of B-TRAN™ such as soft-turn on to eliminate snubber resistance were also presented.

In the future, a higher current SSCB design using multi-chip module SymCool® is planned which can increase the current rating, switching performance and demonstrate parallel operation of B-TRAN™.

REFERENCES

[1] DC Distribution Network Market Size & Share Analysis - Growth Trends & Forecasts (2024 - 2029) Source: https://www.mordorintelligence.com/industry-reports/global-dc-distribution-networks-market-industry

[2] R. Rodrigues, Y. Du, A. Antoniazzi and P. Cairoli, "A Review of Solid-State Circuit Breakers," in *IEEE Transactions on Power Electronics*, vol. 36, no. 1, pp. 364-377, Jan. 2021, doi: 10.1109/TPEL.2020.3003358.

[3] A. Giannakis and D. Peftitsis, "Conduction performance evaluation of power semiconductor device technologies for solid-state DC breakers," *2023 IEEE Applied Power Electronics Conference and Exposition (APEC)*, Orlando, FL, USA, 2023, pp. 2266-2271, doi: 10.1109/APEC43580.2023.10131135.

[4] J. Huber and J. W. Kolar, "Monolithic Bidirectional Power Transistors," in *IEEE Power Electronics Magazine*, vol. 10, no. 1, pp. 28-38, March 2023, doi: 10.1109/MPEL.2023.3234747.

[5] M. Dong, R. Yu, Y. Jiang, J. Bu, J. Knapp and D. Brdar, "B-TRAN™ Optimization and Performance Characterization," *2023 IEEE Applied Power Electronics Conference and Exposition (APEC)*, Orlando, FL, USA, 2023, pp. 1835-1839, doi: 10.1109/APEC43580.2023.10131453.

[6] L. Ravi, D. Dong, R. Burgos, X. Song and P. Cairoli, "Evaluation of SiC MOSFETs for Solid State Circuit Breakers in DC Distribution Applications," *2021 IEEE Applied Power Electronics Conference and Exposition (APEC)*, Phoenix, AZ, USA, 2021, pp. 2237-2242, doi: 10.1109/APEC42165.2021.9487070.

[7] X. Song, Y. Du and P. Cairoli, "Survey and Experimental Evaluation of Voltage Clamping Components for Solid State Circuit Breakers," *2021 IEEE Applied Power Electronics Conference and Exposition (APEC)*, Phoenix, AZ, USA, 2021, pp. 401-406, doi: 10.1109/APEC42165.2021.9487424.

[8] K. Askan, M. Bartonek and F. Stueckler, "Bidirectional Switch Based on Silicon High Voltage Superjunction MOSFETs and TVS Diode Used in Low Voltage DC SSCB," PCIM Europe 2019; International Exhibition and Conference for Power Electronics, Intelligent Motion, Renewable Energy and Energy Management, Nuremberg, Germany, 2019, pp. 1-8.

[9] Z. J. Shen *et al.*, "First experimental demonstration of solid state circuit breaker (SSCB) using 650V GaN-based monolithic bidirectional switch," *2016 28th International Symposium on Power Semiconductor Devices and ICs (ISPSD)*, Prague, Czech Republic, 2016, pp. 79-82, doi: 10.1109/ISPSD.2016.7520782.

[10] J. Sun, J. Wei, Z. Zheng, Y. Wang and K. J. Chen, "Short Circuit Capability and Short Circuit Induced VTH Instability of a 1.2-kV SiC Power MOSFET," in *IEEE Journal of Emerging and Selected Topics in Power Electronics*, vol. 7, no. 3, pp. 1539-1546, Sept. 2019, doi: 10.1109/JESTPE.2019.2912623.

[11] W. Wang *et al.*, "A 400V/300A Ultra-Fast Intelligent DC Solid State Circuit Breaker Using Parallel Connected SiC JFETs," *2019 IEEE Energy Conversion Congress and Exposition (ECCE)*, Baltimore, MD, USA, 2019, pp. 1899-1904, doi: 10.1109/ECCE.2019.8912780.

Development of A Supercritical Fluid-Insulated Fast Mechanical Switch for MVDC Hybrid Circuit Breakers

Zhiyang Jin*, Qichen Yang†, Alfonso Cruz*, and Lukas Graber*

*School of Electrical and Computer Engineering, Georgia Institute of Technology, Atlanta, Georgia 30332
Email: {zjin44, acruz, lgraber3}@gatech.edu
†School of Electrical and Computer Engineering, University of Central Florida, Orlando, Florida 32816
Email: qichen.yang@ucf.edu

Abstract—**The Medium Voltage Direct Current (MVdc) circuit breaker is one of the enabling technologies for the increasingly popular MVdc grids. The MVdc hybrid circuit breaker EDISON ("Efficient DC Interrupter with Fault Protection") has been developed, exhibiting superior properties compared to other dc hybrid circuit breakers, including higher efficiency, faster switching speed, and increased power density. The key component of EDISON is its fast mechanical switch (FMS), which features a compact design utilizing a piezoelectric actuator and supercritical CO_2 as the dielectric medium. This paper focuses on the design of the FMS, presenting results from finite element analysis (FEA) and experimental studies across mechanical, dielectric, thermal, and magnetic domains.**

I. INTRODUCTION

Medium Voltage Direct Current (MVdc) grids have gained popularity because of the rapid development of Distributed Energy Resources (DERs) [1], all-electric ships [2], and the fast growth of dc loads, such as for electric transportation [3] and data centers [4]. The technology of MVdc grids is expected to modernize the power distribution system towards enhanced reliability, flexibility, and efficiency [5]. It requires reliable, fast, and efficient MVdc circuit breakers [6]. One solution of the MVdc circuit breaker technology is the hybrid circuit breaker, where the fault current is commutated to a solid-state path connected in parallel to a main mechanical switch [7], [8], [9]. A new variation of the MVdc hybrid circuit breaker, the EDISON ("Efficient DC Interrupter with Fault Protection") has been developed [10], as shown in Figure 1. The major components of EDISON include the fast mechanical switch (FMS) as the main current path, the fault current commutation circuit (FC3) as a controllable voltage source to initialize the fault current commutation process, and the power stack for absorbing the residual energy after the fault current is commutated. The power stack is composed of N series-connected submodules. Each submodule consists of parallel connected metal-oxide varistors (MOVs) and semiconductor switches [11]. One of the innovative and enabling

This work was supported by the Advanced Research Projects Agency-Energy (ARPA-E) of U.S. Department of Energy under Award DE-AR0001113 in the BREAKERS program.

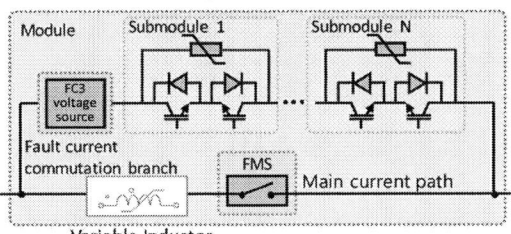

(a)

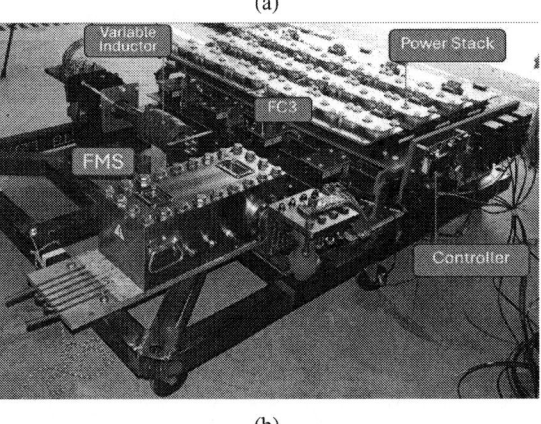

(b)

Fig. 1: MVdc hybrid circuit breaker EDISON. (a) Circuit diagram; (b) Prototype.

technologies for EDISON is the supercritical fluid-insulated FMS that features the combination of a piezoelectric actuator with supercritical CO_2 as the dielectric medium.

Many concepts for mechanical switches rely on electrodynamic actuation such as the so-called Thomson coil [13], [14], [15]. These devices often suffer from limited controllability and a non-optimal contact travel curve. Piezoelectric actuators are based on a stack of crystalline material with a significant piezoelectric effect such as lead zirconate titanate $Pb[Zr_xTi_{1-x}]O_3$ (PZT) [16], [17]. Electrodes connected to the piezoelectric ceramic control the electrostatic field, which

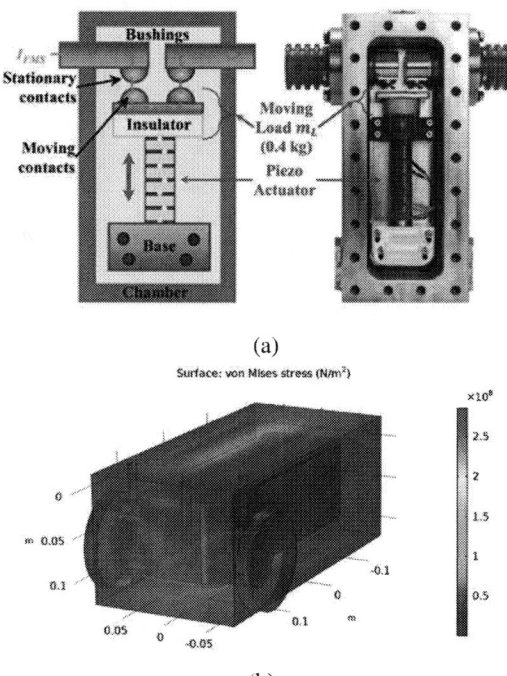

(a)

Surface: von Mises stress (N/m²)

(b)

Fig. 2: Piezoelectric actuating mechanism of 12 kV, 2 kA FMS. (a) FMS assembly without the lid[12]; (b) COMSOL simulation of the mechanical stress of FMS at a rated pressure of 12 MPa.

controls the expansion and contraction of the material along the electric field lines. The displacement of the actuator can be controlled precisely down to nanometer resolution. This high accuracy of piezoelectric actuators can be used to shape the travel curve, to minimize the travel time, avoid contact bouncing, and minimize contact wear during switching cycles. However, piezoelectric actuators have sub-millimeter contact travel and require a medium with high dielectric strength to withstand the transient and system voltage levels when the FMS is open. Literature [18] suggests that certain supercritical fluids have low viscosity, they are partially compressible, their density can be adjusted, and they provide excellent dielectric and heat transfer characteristics. The breakdown experiment with supercritical CO_2 (scCO_2) in [19] shows the average breakdown strength is above 200 kV/mm, under a uniform electric field at a 0.1 mm gap. Such high dielectric strength of scCO_2 enables the use of actuator technologies with limited displacement like the piezoelectric actuator. The combination of supercritical fluid insulation and a piezoelectric actuator makes the FMS more compact and facilitates faster switching, both of which enhance the performance of the EDISON circuit breaker, resulting in greater power density, higher efficiency, and faster fault current interruption.

This paper starts with the design considerations of FMS, followed by experimental results including the contact actuation and arc quenching tests, static breakdown tests with different contact gap distances, and high current tests. In addition, test results during the fault current commutation process of the EDISON circuit breaker are also presented.

II. FMS DESIGN

At rated conditions of EDISON, the scCO_2 has a pressure of approximately 8 MPa, which is more than ten times the pressure of a SF_6 insulated disconnect switch. Figure 2 shows the high-pressure chamber of FMS, rated for 12 MPa. The chamber (with the lid) has the dimensions of 373×166×148 mm³. It is made of stainless steel 17-4 PH 900, which has a yield strength of 1187 MPa, and an ultimate strength of 1397 MPa. The high-pressure chamber has passed the hydrostatic testing with a pressure of 24 MPa. The electric power is introduced through a pair of bushings that are designed to contain not only the high-pressure scCO_2 inside the chamber, but also to withstand a high voltage of 30 kV dc and a high current of 2 kA dc (continuous). Between the two bushings, there are two identical contact pairs electrically connected in series, as shown in Figure 2a. The moving contacts are dielectrically insulated from the actuator (electrically grounded) with a dielectric disk. At the other end of the actuator, a stainless steel block is pressed against the inner wall of the chamber by four bolts. The position adjustment of the stainless steel block can adjust the contact gap distance. The operating temperature of FMS should stay above the critical temperature of the insulation medium to maintain its supercritical state. For pure CO_2, its critical temperature is around 31 °C. The material of the insulation disk ("Insulator" in Figure 2a) is polyetheretherketone (PEEK), which is selected to compensate the thermal expansion of the other components due to the temperature increase during the assembling process and the operation. The open contact gap distance is kept at 100 μm. A water cooling and heat plate is placed under the FMS. The temperature of the FMS is regulated by controlling the temperature of a water bath connected to the plate.

TABLE I: Key parameters of FMS of the EDISON circuit breaker

Parameter	Value	Units
Rated voltage	12	kV
Rated current (continuous)	2	kA
Rated operating temperature	35	°C
Maximum allowable working pressure	12	MPa
Opening time	<500	μs
Dimensions	373×166×148	mm³

III. EXPERIMENTAL RESULTS

A. Contact actuation and arc quenching tests

After the fault current being successfully commutated from the FMS branch to the solid-state path, the FMS starts to open the contacts and needs to quench a current with the amplitude around 3 A which is determined by a variable inductor in series [20]. Figure 3 shows the experimental setup to investigate the speed of the contact movement, and the arc quenching

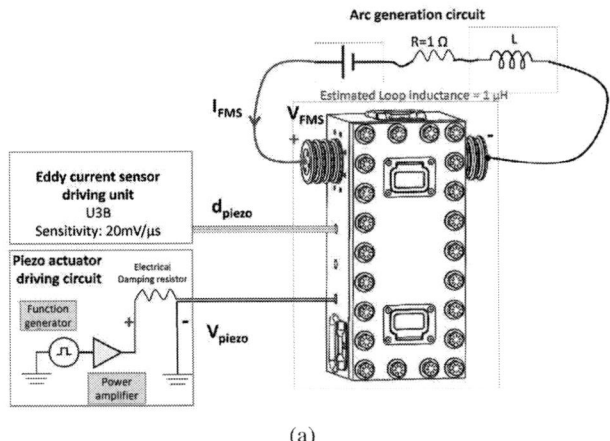

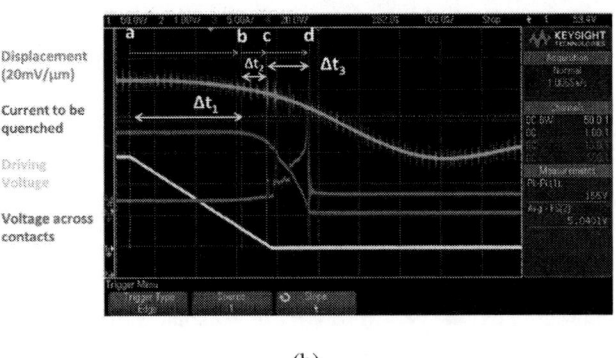

(a)

(b)

Fig. 3: (a) Schematic diagram of experimental setup for arc quenching tests in scCO$_2$, L=1.31 mH; (b) Typical waveforms captured by the oscilloscope.

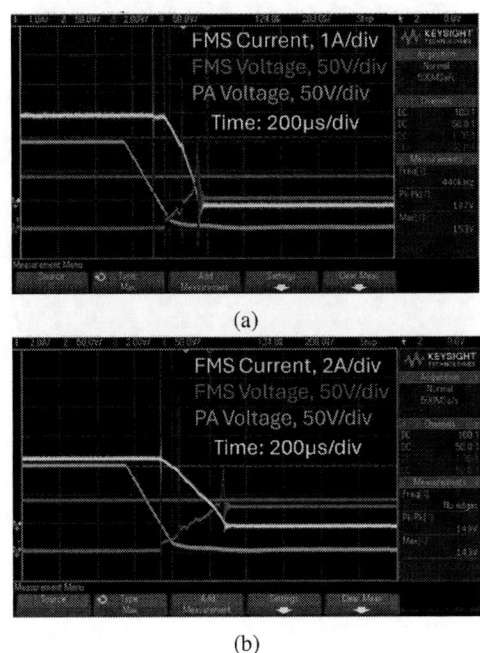

(a)

(b)

Fig. 4: Test results of arc quenching in scCO$_2$. (a) Voltage between FMS after arc quenching (amplitude of current is 3.0 A) is 50 V; (b) Voltage between FMS after arc quenching (amplitude of current is 4.5 A) is 75 V.

capability of FMS. It features a dc arc generation circuit, and a piezoelectric actuator driving circuit. The time duration Δt_3 in the typical captured waveform indicates the time from the onset of arcing to the quench of the arc. The sum of Δt_1, Δt_2, and Δt_3 gives the time needed to extinguish the arc from the start of the movement of the piezoelectric actuator. Two different experiment results are shown in Figure 4. With an inductance of 1.31 mH (L) in the arc generation circuit, it took approximately 400 μs to quench an arc with an amplitude of 3.0 A, and 525 μs to quench an arc with an amplitude of 4.5 A There are several options to further speed it up, including optimization of the actuation circuit and the piezoelectric actuator itself [21], and reduction of the moving mass [22].

Moreover, literature [23] indicates that the stainless steel 17-4 PH is magnetic. An experiment was carried out to study the magnetic properties of the FMS chamber, as shown in Figure 5a. A pulsed current, with a peak valve of 3.6 kA, was generated by a capacitor bank. A cable carrying this current was fed through the bushing holes of the chamber. Based on the measurement results in Figure 5b, the equivalent inductance in the circuit is estimated to be 1.5 μH. The test without the FMS chamber in the circuit was also conducted and the measured inductance in the circuit is about 1.4 μH.

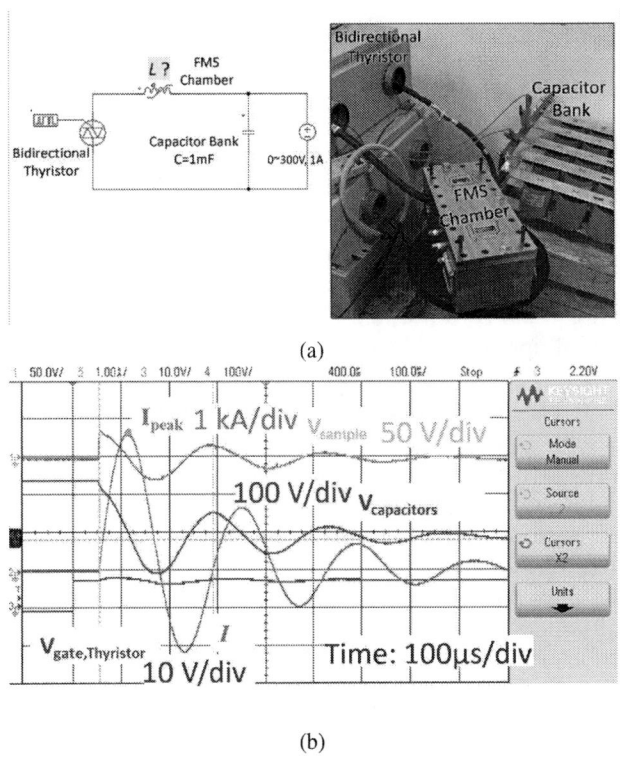

(a)

(b)

Fig. 5: (a) Experimental setup for investigating the magnetic properties of the FMS chamber; (b) Measurement results with the FMS chamber in the circuit.

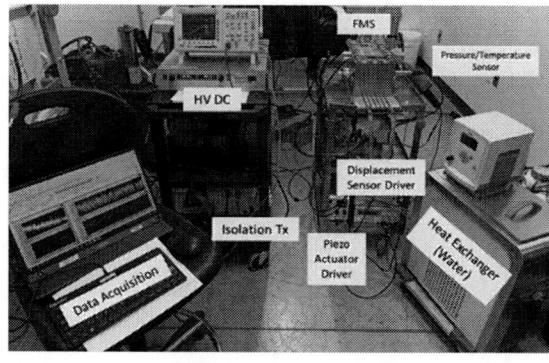

(a)

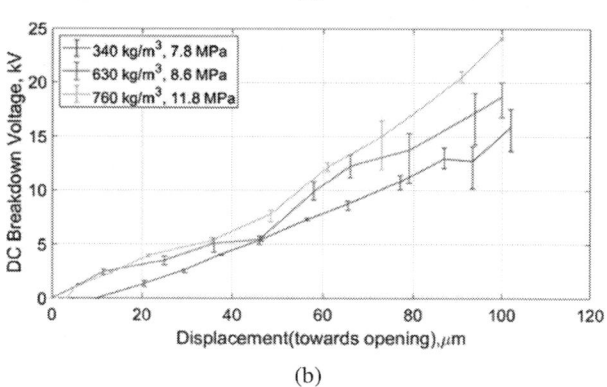

(b)

Fig. 6: (a) Experimental setup of dc breakdown test in FMS with different gap distance (constant temperature of 35 °C); (b) dc breakdown voltage of different contact gap distance in scCO$_2$ with different densities.

Therefore, the impact of the magnetic FMS chamber to the circuit breaker performance can be neglected.

B. Static breakdown results with different gap distances

Once the current flowing through the FMS is quenched, the process of submodule insertion can be initiated to dampen the fault current. However, the voltage across the FMS also starts to build up due to the voltage established across the inserted MOVs. The peak voltage the FMS needs to withstand is about 24 kV based on the unique technique of sequential power stack insertion scheme [11], [24]. DC breakdown tests of FMS have been conducted with scCO$_2$ and different gap distances. The experimental setup and results are shown in Figure 6. Five tests were performed for each data point. The voltage is applied at a rate of 400 V/s until the breakdown occurs. It shows that a dc breakdown voltage of 24 kV can be achieved in scCO$_2$ with a density of 760 kg/m^3. The breakdown strength is less than the value in [19] due to the different electrode geometries and hence different electric field distribution. The contacts are made of copper and have the geometry of Rogowski profile. The previous study [25] compared the contacts with different profiles and materials, and found this combination works the best for the FMS application with good electric field distribution as well as improved mechanical performance. It should be also noted that during the sequential insertion

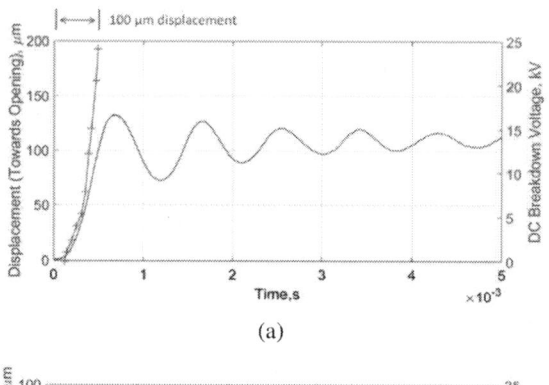

(a)

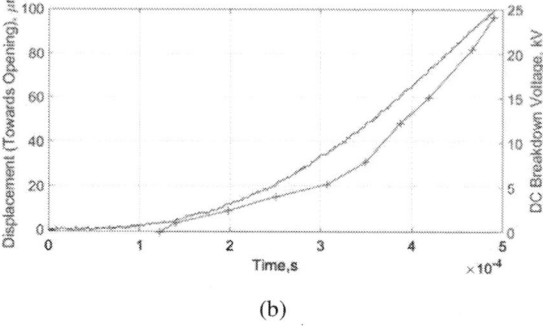

(b)

Fig. 7: (a) Plot of static breakdown voltage of FMS (with the scCO$_2$ density of 760 kg/m^3) together with a travel curve of the moving contacts; (b) Zoom in (displacement from 0 to 100 μm).

of MOVs, the duration of the peak voltage of 24 kV is approximately several hundreds of microseconds. Whereas, this experiment studies the static dc breakdown voltage that was applied with a longer duration.

Figure 7 combines the dc breakdown voltage with different gap distances with the travel curve of piezoelectric actuator. The breakdown voltage profile of the first 100 μm displacement can be a valuable guidance to the MOV insertion strategy – the voltage built-up from the inserted MOVs should stay below to prevent the arc reignition between the contacts of FMS, which leads to the failure of fault current interruption. It could be also observed from Figure 7a that the undershoot of the contact movement can decrease the breakdown voltage across FMS, and hence excessive undershoot should be avoid. It can be achieved either by closed-loop control schemes [12] or advanced open-loop control techniques such as controlling individual stacks of the piezoelectric actuator instead of controlling them simultaneously [26], [27].

C. 2 kA High current test

Compared to solid-state MVdc circuit breakers during the normal operations, the hybrid circuit breaker such as EDISON only has the joule losses due to the contact resistance in FMS. Four identical dc power supplies (Agilent/Keysight 6681A) were connected in parallel to generate a current of 2 kA, which was injected into the closed FMS via the two bus bar terminations as shown in Figure 8a. Figure 8b marks the

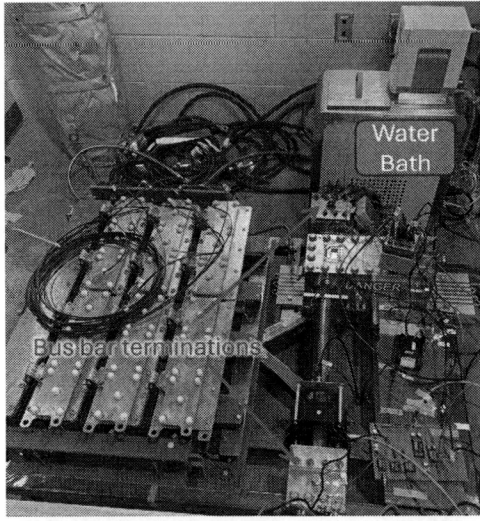

(a)

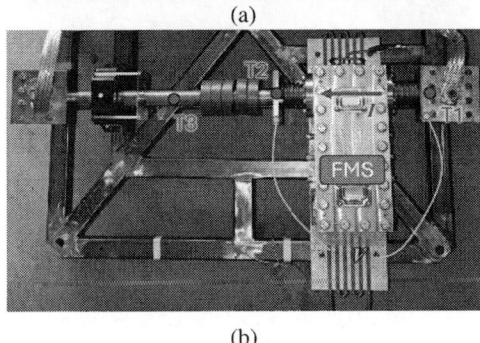

(b)

Fig. 8: (a) Experimental setup of the 2 kA test; (b) Locations of the thermocouple along the bus bar.

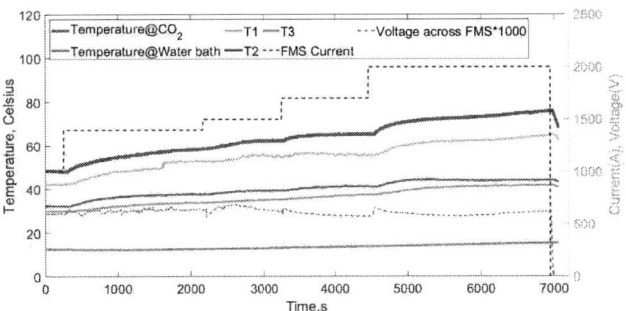

Fig. 9: Experimental results of the 2 kA test.

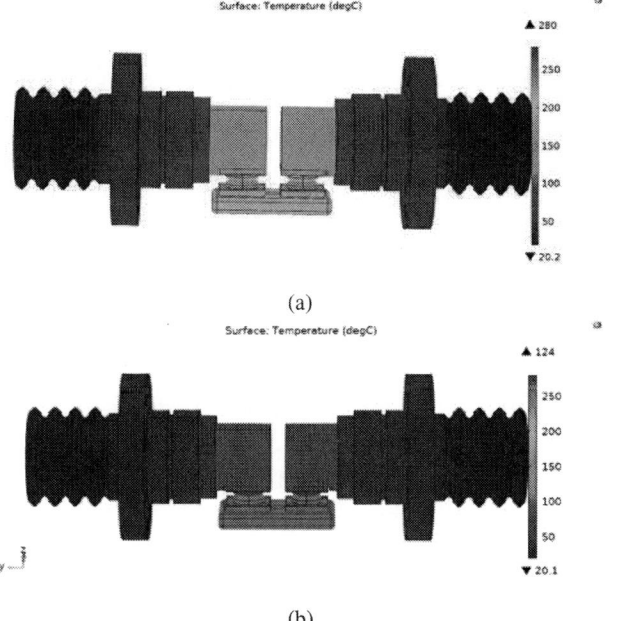

(a)

(b)

Fig. 10: (a) Finite element analysis (FEA) results of steady-state temperature distribution with boundary conditions from experimental measurement results; (b) FEA results with modified boundary conditions.

locations of three thermocouples along the current path. A temperature/pressure transducer mounted on the high-pressure chamber of FMS monitored the conditions of scCO$_2$ inside. The voltage across FMS was also recorded during the test. Figure 9 presents the measured results for the process of increasing the current from 1 kA to 2 kA in steps. At t=0 s, the conditions of scCO$_2$ are 9.1 MPa, 49 °C (hence with a density of about 310 kg/m^3). The test was stopped at t=6947 s because the scCO$_2$ has reached 12 MPa, which is the maximum allowable working pressure of FMS. During the test, the water bath (Thermo Scientific^{TM} ARCTIC A25) was running at its maximum cooling power of 500 W and the water bath temperature was kept around 13 °C.

Before the end of this high-current test (before t=6947 s), the joule losses across the FMS is about 1.2 kW. The total power loss, including the 1.2 kW from the FMS and the cooling power of 500 W from the water bath, is 1.7 kW that is 0.007 % of the rated power of EDISON of 24 MW. In addition, a COMSOL model has been built to calculate the temperature increase inside the FMS with the boundary conditions from the experimental results. The boundary conditions include,

- Temperature of scCO$_2$ inside FMS is 76 °C.
- Temperature at end of bushings, T1=65 °C and T2=44 °C.

The difference is due to that there is a long bus bar between one of the bushings and the bus bar termination, and hence larger surface area for convective heat transfer in the ambient air.
- Constant convective heat transfer rate is 5 W/(m^2·K) for the ambient air, and 250 W/(m^2·K) for the scCO$_2$ [28].
- Contact resistance of 154 μΩ for each contact pair.

Figure 10a shows the steady-state temperature distribution with the above boundary conditions. With a current of 2 kA flowing through the FMS, the maximum temperature rise locates between contact pairs. The moving contact plate has a temperature of about 230 °C. However, the contact resistance between contact pairs can be further reduced by techniques such as silver-plating. In Figure 10b, the temperature rise inside FMS is lower with the contact resistance set to 60 μΩ for each pair, and reduced temperature of 40 °C at the end of

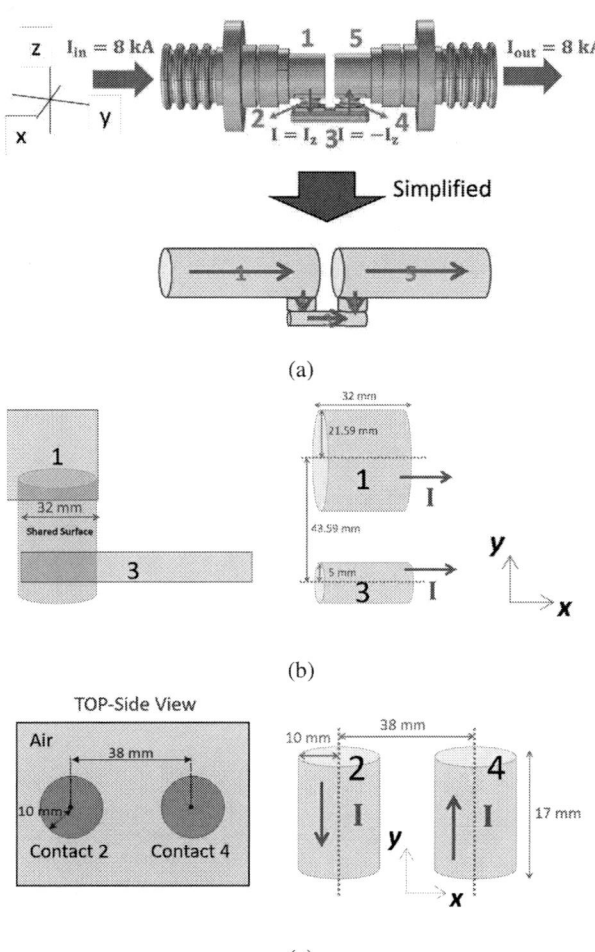

(a)

(b)

(c)

Fig. 11: (a) Simplified geometries for studying the magnetic force when the FMS is carrying 8 kA; (b) Geometries studied in FEA for analyzing the magnetic force between conductor 1 and 3;(c) Geometries studied in FEA for analyzing the magnetic force between conductor 2 and 4.

bushings. Moreover, the scCO$_2$ temperature is set to 40 °C, which means that the FMS can be filled with a higher density of scCO$_2$, advantageous for the arc quenching, voltage withstand capability, and heat dissipation.

In addition, when the FMS is carrying 2 kA, there is the magnetic force being induced on the conductors. This force needs to be analyzed due to the compact design of FMS. Finite element analysis has been conducted to investigate the induced magnetic force on the conductors inside FMS when they are carrying 8 kA, which is the rated peak current during the fault current interruption. Figure 11 lists the simplified geometries for the FEA studies. The magnetic force between the conductor of one bushing and the moving contact plate, as shown in Figure 11b, is about 9.21 N in the y direction. The force between the two contact pairs, as shown in Figure 11c, is estimated of 5.7 N in the x direction. Compared to the contact

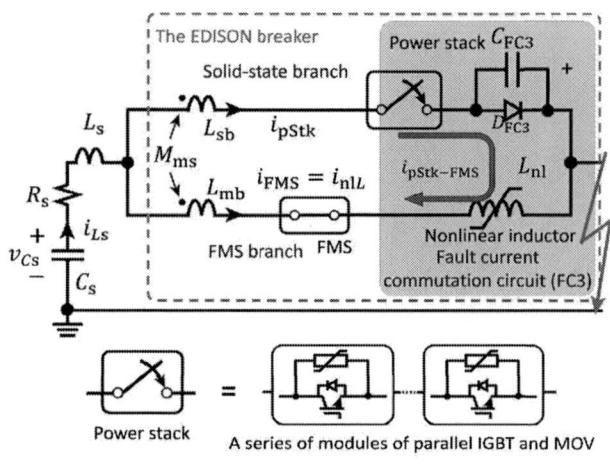

(a)

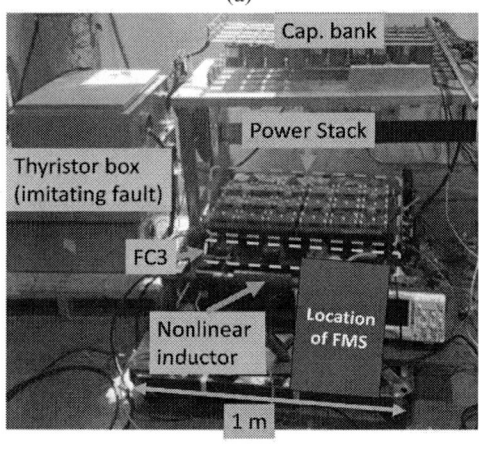

(b)

Fig. 12: Commutation test setup: (a) Schematic diagram; (b) Capacitor bank, thyristor to imitate fault, and overall setup layout (without FMS).

force of approximately 500 N for each contact pair, none of the above induced magnetic force can affect the performance of FMS when closed.

D. Commutation test

To verify the functionality of the FMS in conditions closely approximating actual operation, a comprehensive commutation test was conducted, presenting the zero-current opening process of the FMS after the fault current commutation period. The setup schematic is illustrated in Figures 12a and the parameters of the test setup are in Table II. Figure 12b shows the capacitor bank, the thyristor used for fault imitation, and the overall layout of the setup without the FMS. Figure 13 illustrates the voltage and current of the FMS during the opening process, following the fault current commutation. At t_1, the fault was initiated and the current through the FMS, i.e., i_{FMS} increases rapidly after t_1. The IGBTs in the power stack were turned on and EDISON breaker starts to break the fault current at t_2, at which i_{FMS} reaches its peak value of

979-8-3315-1612-3/25 $31.00 © 2025 IEEE   865

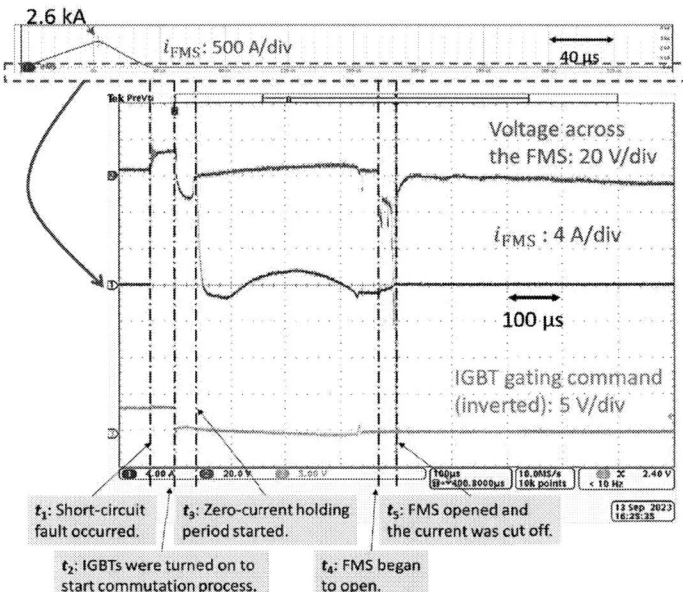

Fig. 13: FMS opening waveforms subsequent to fault current commutation.

approximately 2.6 kA. Since the changing rate of i_{FMS} is more than 40 A/µs, i_{FMS}, i.e., the blue line in Figure 13, appears nearly vertical around t_1 and t_3. When i_{FMS} decreased to around 12 A at t_3, the nonlinear inductor in series with the FMS transitioned to its linear region. As a result, the large inductance in this region reduced the changing rate of i_{FMS}, effectively holding the current close to 0 A [20]. During the zero-holding period, the contacts in the FMS start to separate at t_4 and completely broke the current at t_5. At t_4, the FMS current is about 1 A. Between t_4 and t_5, a voltage pulse with an 80 V peak value was observed in v_{FMS}, i.e., the red line in Figure 13. The voltage rise is caused by the inductance in the loop (approximately 800 µH) with the FMS current change, i.e. di_{FMS}/dt, during the opening of FMS. The loop is consisted of the solid-state branch and the paralleled FMS branch. The experimental results verifies the functionality of the FMS, together with the commutation system.

TABLE II: Parameters of setup for the commutation test

Parameter	Value	Units
Capacitance of the capacitor bank	880	µF
Source inductance	16	µH
Initial voltage of the capacitor bank	1	kV
FC3 capacitance	3.2	mF
Initial voltage of the FC3 capacitor	177	V

IV. CONCLUSION

This paper presents the world's first supercritical CO_2 insulated fast mechanical switch (FMS) for a MVdc hybrid circuit breaker, with the following experimental and FEA simulation results.

- The maximum allowable working pressure of FMS is 12 MPa. The chamber of FMS passed the 24 MPa hydrostatic test.
- The FMS has achieved an opening time of 400 µs while it quenched an arc with an amplitude of 3.0 A and a 1.3 mH inductor in series. It can be sped up with a more powerful piezoelectric actuator.
- The dc breakdown tests shows a maximum voltage of 24 kV when FMS is fully open and the contact gap distance is 100 µm. The FMS is designed to withstand a maximum voltage of 24 kV when fully opened, but with a very short duration (on the order of 10 µs). In addition, the profile of dc breakdown voltage with different contact gap distances benefits control design of the MOV insertion after the fault current commutation.
- The FMS is tested with a dc current of 2 kA, and its thermal performance could be further improved with reduced contact resistance between two contact pairs.
- The commutation test with the whole EDISON circuit breaker demonstrated the successful fault current interruption, with the peak current of 2.6 kA. The experimental results verifies the functionality of the FMS, together with the commutation system.

ACKNOWLEDGMENT

The authors wish to acknowledge the late Dr. Michael (Mischa) Steurer for his invaluable contributions to the design and testing of the FMS of the EDISON circuit breaker, which greatly influenced this work. His legacy continues to inspire our research.

The authors would also like to thank the former members of the Plasma and Dielectrics Lab at Georgia Tech—Dr. Maryam Tousi, Dr. Chunmeng Xu, Dr. Jia Wei, and Kevin Whitmore—as well as the Center for Advanced Power Systems at Florida State University for their insightful discussions, technical assistance, and collaborative efforts throughout this project.

REFERENCES

[1] R. Sarrias-Mena, L. M. Fernández-Ramírez, C. A. García-Vázquez, C. E. Ugalde-Loo, N. Jenkins, and F. Jurado, "Modelling and control of a medium-voltage dc distribution system with energy storage," in *2016 IEEE International Energy Conference (ENERGYCON)*. IEEE, 2016, pp. 1–6.

[2] Y. Chen, S. Zhao, Z. Li, X. Wei, and Y. Kang, "Modeling and control of the isolated dc–dc modular multilevel converter for electric ship medium voltage direct current power system," *IEEE Journal of Emerging and Selected Topics in Power Electronics*, vol. 5, no. 1, pp. 124–139, 2016.

[3] A. Verdicchio, P. Ladoux, H. Caron, and C. Courtois, "New medium-voltage dc railway electrification system," *IEEE Transactions on Transportation Electrification*, vol. 4, no. 2, pp. 591–604, 2018.

[4] J. Cui, J. Yan, X. Han, Y. Zhang, and L. Qu, "Design, evaluation and simulation of low-carbon integrated energy supply system for data center based on medium voltage dc technology," in *IET Conference Proceedings CP880*, vol. 2024, no. 6. IET, 2024, pp. 282–288.

[5] Z. Ma, R. Li, P. Lürkens, M. Han, S. Kim *et al.*, "Medium voltage direct current (MVDC) grid feasibility study," *Technical brochure TB793, CIGRE*, 2020.

979-8-3315-1612-3/25 $31.00 © 2025 IEEE

[6] M. Heidemann, G. Nikolic, A. Schnettler, A. Qawasmi, N. Soltau, and R. W. De Donker, "Circuit-breakers for medium-voltage dc grids," in *2016 IEEE PES Transmission & Distribution Conference and Exposition-Latin America (PES T&D-LA)*. IEEE, 2016, pp. 1–6.

[7] J. Hafner, "Proactive hybrid HVDC breakers-a key innovation for reliable HVDC grids," in *Proc. CIGRE Bologna Symposium*, 2011, pp. 1–8.

[8] C. Peng, I. Husain, A. Q. Huang, B. Lequesne, and R. Briggs, "A fast mechanical switch for medium-voltage hybrid dc and ac circuit breakers," *IEEE Transactions on Industry Applications*, vol. 52, no. 4, pp. 2911–2918, 2016.

[9] C. Peng, X. Song, A. Q. Huang, and I. Husain, "A medium-voltage hybrid dc circuit breaker—part ii: Ultrafast mechanical switch," *IEEE Journal of Emerging and Selected Topics in Power Electronics*, vol. 5, no. 1, pp. 289–296, 2016.

[10] L. Graber, T. Damle, C. Xu, J. Wei, J. Sun, M. Mehraban, Z. Zhang, M. Saeedifard, S. Grijalva, J. Goldman *et al.*, "Edison: A new generation dc circuit breaker," *Cigre Paris Session*, 2020.

[11] Z. J. Zhang and M. Saeedifard, "Overvoltage suppression and energy balancing for sequential tripping of hybrid dc circuit breakers," *IEEE Transactions on Industrial Electronics*, vol. 70, no. 7, pp. 6506–6517, 2022.

[12] C. Xu, Z. Jin, M. Tousi, and L. Graber, "Critical damping in travel curves of piezoelectrically actuated fast mechanical switches for hybrid circuit breakers," *IEEE Transactions on Power Delivery*, vol. 37, no. 5, pp. 3873–3884, 2022.

[13] A. Bissal, J. Magnusson, and G. Engdahl, "Comparison of two ultra-fast actuator concepts," *IEEE Transactions on Magnetics*, vol. 48, no. 11, pp. 3315–3318, 2012.

[14] D. Vilchis-Rodriguez, R. Shuttleworth, and M. Barnes, "Double-sided thomson coil based actuator: Finite element design and performance analysis," 2016.

[15] V. Puumala and L. Kettunen, "Electromagnetic design of ultrafast electromechanical switches," *IEEE Transactions on Power Delivery*, vol. 30, no. 3, pp. 1104–1109, 2014.

[16] H. Jin, X. Gao, K. Ren, J. Liu, L. Qiao, M. Liu, W. Chen, Y. He, S. Dong, Z. Xu *et al.*, "Review on piezoelectric actuators based on high-performance piezoelectric materials," *IEEE Transactions on ultrasonics, ferroelectrics, and frequency control*, vol. 69, no. 11, pp. 3057–3069, 2022.

[17] X. Zhou, S. Wu, X. Wang, Z. Wang, Q. Zhu, J. Sun, P. Huang, X. Wang, W. Huang, and Q. Lu, "Review on piezoelectric actuators: materials, classifications, applications, and recent trends," *Frontiers of Mechanical Engineering*, vol. 19, no. 1, p. 6, 2024.

[18] J. Zhang, E. van Heesch, F. Beckers, A. Pemen, R. Smeets, T. Namihira, and A. Markosyan, "Breakdown strength and dielectric recovery in a high pressure supercritical nitrogen switch," *IEEE Transactions on Dielectrics and Electrical Insulation*, vol. 22, no. 4, pp. 1823–1832, 2015.

[19] J. Wei, C. Park, and L. Graber, "Breakdown characteristics of carbon dioxide–ethane azeotropic mixtures near the critical point," *Physics of Fluids*, vol. 32, no. 5, 2020.

[20] Q. Yang, M. Steurer, S. Song, M. Pickles, Y. He, J. Hauer, M. Bosworth, N. Bonaventura, M. Coleman, Y. Shi *et al.*, "Analysis, modeling, and experiments of a nonlinear inductor-based fault current commutation strategy for a hybrid dc circuit breaker," *IEEE Transactions on Transportation Electrification*, 2024.

[21] M. Tousi, G. Mansuy, M. Thomachot, A. Pages, E. Karimi, Z. Jin, K. Whitmore, and L. Graber, "A piezoelectric actuator optimized for fast mechanical switch applications," in *2022 IEEE 67th Holm Conference on Electrical Contacts (HOLM)*. IEEE, 2022, pp. 1–6.

[22] A. Cruz-Feliciano, Z. Jin, T. Damle, M. Tousi, N. Lee, and L. Graber, "Comprehensive study on the contact materials and geometry of an MVDC piezoelectrically actuated switch," in *2022 IEEE 67th Holm Conference on Electrical Contacts (HOLM)*. IEEE, 2022, pp. 1–7.

[23] A. Dash, S. Bose, and A. Bandyopadhyay, "Additively manufactured 17–4 ph stainless steels for fracture management devices," *Virtual and Physical Prototyping*, vol. 19, no. 1, p. e2397698, 2024.

[24] Z. J. Zhang and M. Saeedifard, "Analysis and experimental verification of overvoltage suppression in a hybrid dc circuit breaker," in *2022 IEEE Applied Power Electronics Conference and Exposition (APEC)*. IEEE, 2022, pp. 792–797.

[25] T. Damle, "Design of electrical contacts for fast mechanical disconnect switches," *Diss. Georgia Institute of Technology*, 2020.

[26] A. Ghosh, Z. Jin, K. Whitmore, M. Tousi, and L. Graber, "Response and control of individual stacks of a multi-stack piezoelectric actuator for dc fast disconnect switches," in *2023 IEEE 68th Holm Conference on Electrical Contacts (HOLM)*. IEEE, 2023, pp. 1–5.

[27] X. Ba, Q. Pan, B. Ju, and Z. Feng, "Ultrafast displacement actuation of piezoelectric stacks with time-sequence," *IEEE Transactions on Industrial Electronics*, vol. 64, no. 4, pp. 2955–2961, 2016.

[28] H. Tokanai, Y. Ohtomo, H. Horiguchi, E. Harada, and M. Kuriyama, "Heat transfer of supercritical CO_2 flow in natural convection circulation system," *Heat transfer engineering*, vol. 31, no. 9, pp. 750–756, 2010.

Dynamic Impedance Matching for a Variable Reluctance Energy Harvesting Application with Constrained Space

Fernando Pérez, Alejandro Redondo, Airán Francés and Gabriel Mujica

Centro de Electrónica Industrial

Universidad Politécnica de Madrid

Madrid, Spain

Email: fernando.ptorrero@upm.es, alejandro.redondo.ayala@upm.es, airan.frances@upm.es, gabriel.mujica@upm.es

Abstract—**Variable Reluctance Energy Harvesters (VREH) are scavenging devices that provide a sinusoidal voltage of variable frequency and have an inductive output impedance. Thus, to maximize power extraction, a variable capacitor is required to perform impedance matching at each frequency. For space-constrained applications, the technology of the capacitor plays an important role, since it determines the physical size and possible capacitance deviations in the presence of high voltages. In this paper, an adaptive capacitance circuit based on a custom capacitor array is designed for VREH applications with space constraints. A modeling methodology that considers the capacitance changes of class II ceramic capacitors under high voltages is proposed. With this technology, a 91% reduction in capacitor volume is achieved when compared to solutions that use film capacitors. The system is tested in a real setup obtaining output power levels that are in accordance with the simulated model.**

Index Terms—**Capacitor Voltage Characteristic, Class II Ceramic capacitors, Modeling, Impedance matching, Variable Reluctance Energy Harvesting**

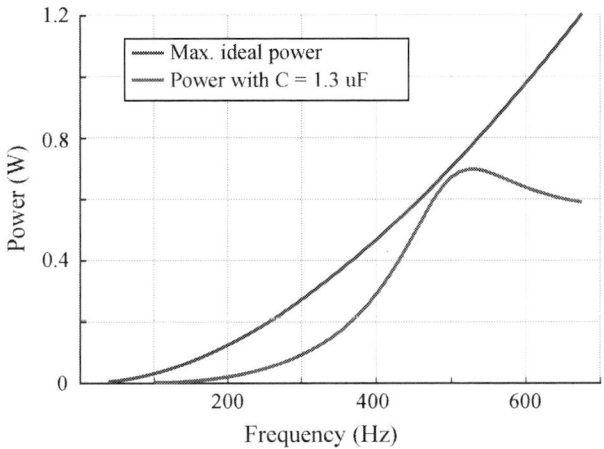

Fig. 1. Extracted power from a VREH adding a series capacitor.

I. INTRODUCTION

The use of wireless Internet of Things (IoT) devices has experienced a massive growth in recent years, being found in multitude of applications, such as asset tracking and condition monitoring. Due to the nature of these applications, it is often impractical to replace the batteries that power the nodes after they are deployed [1]. Energy harvesting offers an efficient solution to power them and extend battery lifetime by extracting energy from the environment [2], [3]. Energy can be harvested from many sources, such as photovoltaic [4], piezoelectric [5], thermoelectric, and electromagnetic [6], [7], among others.

Variable Reluctance Energy Harvesters (VREH) are an interesting solution to obtain energy in applications where rotating elements are present, such as in vehicles and industrial machines [8]. Their main advantage over other types of electromagnetic transducers is their robustness and low complexity, since both the magnet and pickup coil are stationary [9]. Applications of VREH include powering sensors, used for example in train passage detection, and structural health monitoring [10]. VREH can be modeled as a sinusoidal voltage source, whose frequency and amplitude grow linearly with the rotational speed, in series with a resistance and an inductance. They can be composed of one or multiple phases, which can help reduce the cogging torque at low speeds [11]. Regardless of the number of phases, the inductive behavior creates a phase mismatch between voltage and current that decreases the effective power. To maximize power extraction, impedance matching must be performed by adding a series capacitor, whose value depends on the inductance and the frequency [12], as shown in:

$$C = \frac{1}{(2\pi f)^2 L}.$$ (1)

The power vs. frequency curves of a VREH are shown in Fig. 1. The maximum power curve corresponds to the power obtained if impedance matching is ideally performed at every frequency. However, if a single capacitor is used, the system is only efficient at the resonant frequency. Thus, to obtain the maximum power at every frequency, a system that provides a variable capacitance would be required.

In previous VREH works implementing experimental setups, two scenarios can be distinguished. In the first one, the inductance is neglected, as in [9]–[11], [13]–[16], because the rotational speed of the harvester is low, and the behavior

979-8-3315-1612-3/25 $31.00 © 2025 IEEE

is mainly resistive. In the second one, a single capacitor is added [8], [12], [17]–[19], selected for a fixed operating point, whereas the case in which the speed of the VREH changes dynamically is not addressed.

Additionally, some VREH applications, such as smart bearing systems [17]–[19], are space-constrained because both the harvesting and the power conversion electronics must fit within the mechanical component. This needs to be considered for the choice of technology of the matching capacitors. Class II ceramic capacitors are usually available in small package sizes; however, their capacitance decreases as the voltage between their terminals increases [20], a phenomenon commonly known as DC bias. If this is not considered during the design process, the efficiency of the power extraction might be diminished. Although film capacitors do not suffer from this effect, they are usually much larger in size, which might make them unsuitable for this type of applications. Class I ceramic capacitors are not affected by DC bias either, but they are usually available only in small values and voltage ratings, which rules them out if large capacitances are needed. Polarized technologies are also discarded, since the capacitors must withstand AC voltages.

This paper proposes an adaptive impedance matching circuit for VREH in variable speed conditions, specifically targeted for space-constrained applications. It is based on an array of N class II ceramic capacitors that are combined with switches to obtain $2^N - 1$ capacitances. The appropriate capacitor combination is configured in real time by sensing the harvester frequency. A detailed model that includes the DC bias effect is proposed for the design of the system. The model provides accurate simulations that can be used to select the most suitable capacitors for a given range of operating frequencies.

The rest of the article is organized as follows. In Section II, the VREH system and the simulation model are described in detail. In Section III, an experimental setup is presented, and results of the experiments are shown and compared with simulations. Section IV contains some conclusions and future work.

II. ADAPTIVE IMPEDANCE MATCHING CIRCUIT AND SYSTEM MODELING

A. VREH system description

The methodology presented in this paper is applied to a VREH with the specifications collected in Table I.

TABLE I
SUMMARY OF VREH SPECIFICATIONS

Parameter	Value	Description
L_H	77.5 mH	Equivalent inductance
R_H	60 Ω	Equivalent resistance
f_H	20 − 675 Hz	Voltage frequency range
V_H^{RMS}	0.5 − 16.95 V	Generated RMS voltage range
C_H	816 μF − 717 nF	Matching capacitor range
$V_{C_H}^{pk}$	17 mV − 70 V	C_H peak voltage range

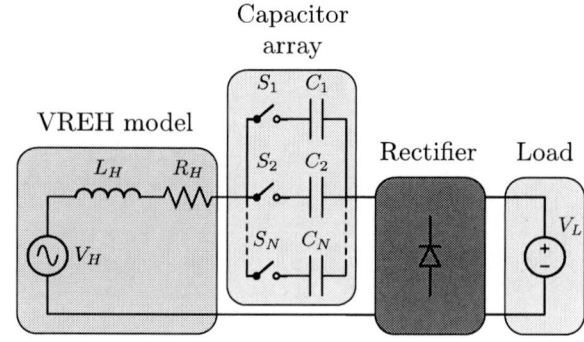

Fig. 2. Schematic of the system.

The amplitude of the VREH voltage grows linearly with the frequency, as shown in:

$$V_H = 0.0355 * f_H \qquad (2)$$

Apart from the impedance matching circuit, the complete system under study contains a rectification stage, and a constant voltage load that absorbs the generated power (Fig. 2). In the final implementation for a real application, a power management unit would be placed between rectifier and load, performing tasks such as delivering a regulated voltage, charging a battery, and implementing some type of Maximum Power Point Tracking (MPPT) to optimize power extraction. However, for the purpose of this research, focused on the impedance matching, a controlled voltage load is directly connected to the output of the rectifier. A common technique in energy harvesting systems to maximize power extraction is to regulate the output voltage of the harvester, usually to a fixed ratio of its Open Circuit Voltage (OCV). For photovoltaic panels, the optimal ratio is between 70% and 80%, while it is around 50% for thermoelectric harvesters. In order to find the optimal ratio for the VREH, simulations have been

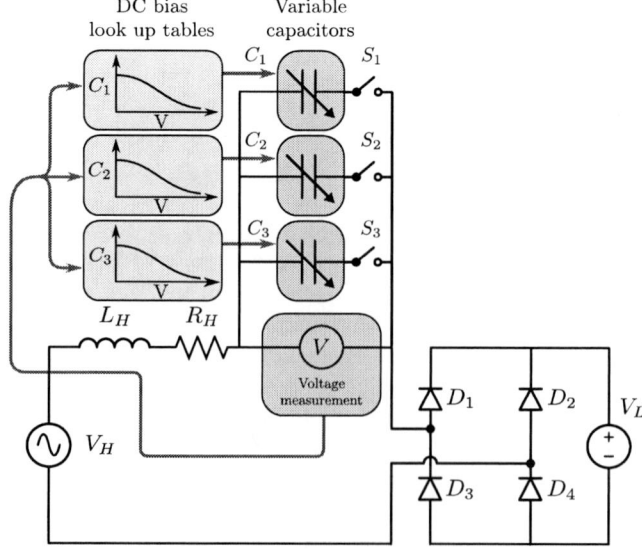

Fig. 3. Proposed model of VREH and impedance matching circuit.

performed with the model shown in Fig. 2. For a given harvester frequency, a sweep of load voltages V_L from zero to the peak OCV voltage of the harvester V_H has been performed. The results show that the maximum power is obtained when V_L is set to around 40% of the peak OCV. This ratio will be used for the experiments.

B. Modeling of VREH with ceramic capacitors

The drop in capacitance caused by DC bias depends on many factors, the most important ones being the dielectric material, package size, and voltage rating. Most capacitor manufacturers provide the change in capacitance with DC voltage in the shape of curves. Modeling this effect is critical for choosing which capacitors to use and how to combine them at each frequency because the changes in capacitance shift the resonant frequencies, thereby modifying the power vs. frequency curves. Since the capacitance varies instantaneously with the voltage, it is not straightforward to predict the real behavior analytically, but simulations can provide accurate power profiles. The proposed model includes the harvester model, a full wave rectifier, a constant voltage load, and an array of variable capacitors (Fig. 3). The DC bias curves are implemented as look up tables, providing the capacitance values fed to the variable capacitors from voltage measurements across their terminals.

For the current application, three commercial capacitors have been selected (Table II) with the goal of approximately covering the entire range of frequencies considered. Their DC bias curves (Fig. 4) show that the capacitance decreases significantly with the applied voltage, dropping to 20% of the nominal value for voltages close to the rated one. The model has been run in MATLAB/Simulink for all seven combinations of capacitors, obtaining their power vs. frequency curves (Fig. 5b). The same curves without modeling the DC bias effect, i.e., using ideal capacitors, are shown in Fig. 5a.

The maximum theoretical power is obtained with (3), by applying the maximum power transfer theorem. It is computed as the power absorbed by a resistor equal to the series resistance of the harvester R_H subjected to a sinusoidal voltage with an amplitude half of that of the harvester.

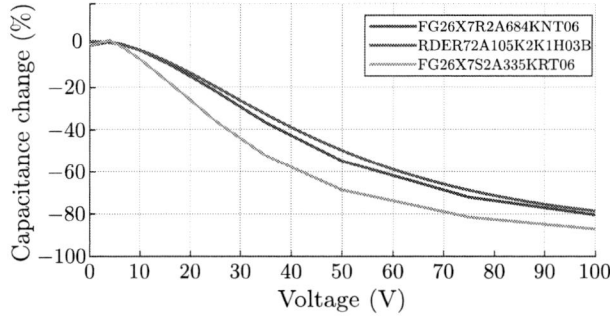

Fig. 4. DC bias curves of the three selected capacitors.

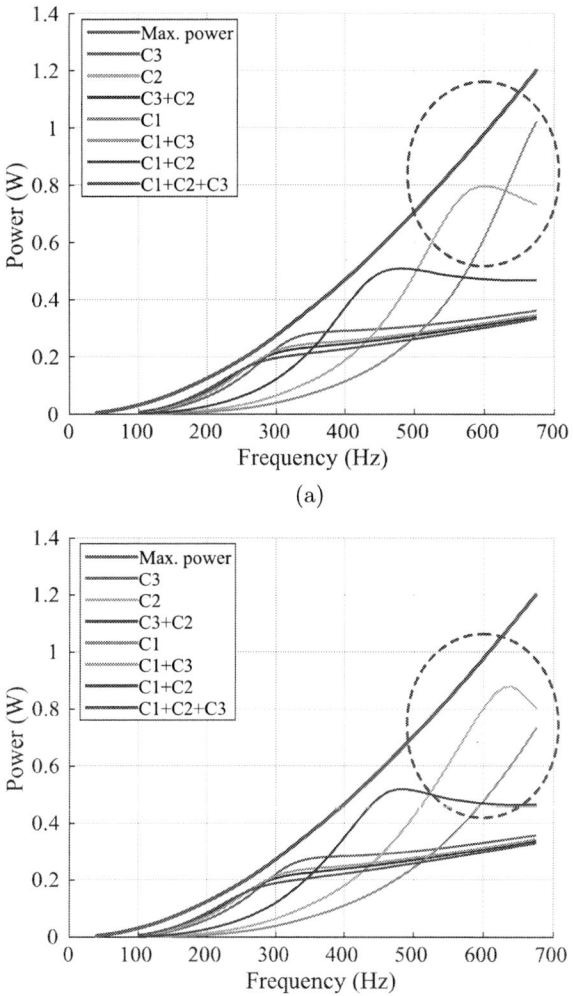

Fig. 5. Results of the model simulations. (a) with ideal capacitors. (b) with DC bias model.

$$P_{max} = \left(\frac{V_H}{2\sqrt{2}}\right)^2 \frac{1}{R_H} \qquad (3)$$

It can be observed that, for the smaller capacitors, especially the 0.68 μF one, the resonant frequencies are higher than expected due to their capacitance being reduced. Indeed, the simulations show that the voltages on the capacitors reach up to 60 V at high frequencies. The time domain waveforms of a simulation at 675 Hz with the 0.68 μF capacitor enabled are shown in Fig. 6. The capacitor voltage waveform is visibly distorted due to the DC bias, resembling a triangular shape instead of a sinusoidal one. The instantaneous value of capacitance is also plotted, which drops to 0.26 μF, or 32% of the nominal value when the voltage reaches 60 V. The proposed circuit switches on and off capacitors depending on the operating frequency, therefore, if the DC bias is not considered for the design of the switching algorithm, the efficiency would considerably decrease. These results allow

TABLE II
CHARACTERISTICS OF SELECTED CERAMIC CAPACITORS.

Capacitor	Value	Commercial part	Rated voltage	Dielectric	Theoretical resonant frequency
1	0.68 μF	FG26X7R2A684KNT06	100 V	X7R	693 Hz
2	1 μF	RDER72A105K2K1H03B	100 V	X7R	571 Hz
3	3.3 μF	FG26X7S2A335KRT06	100 V	X7S	314 Hz

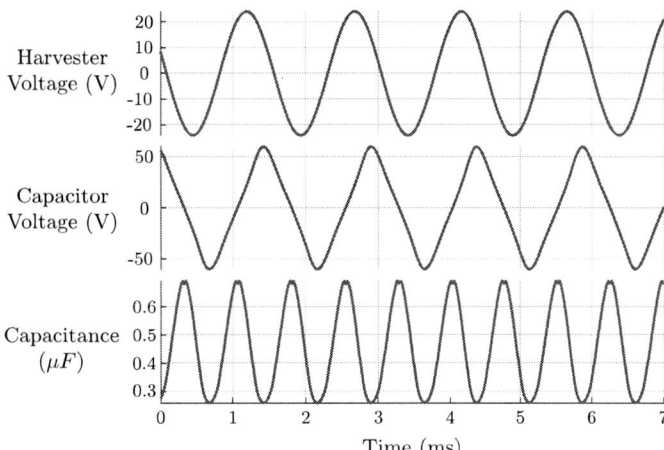

Fig. 6. Time domain waveforms of a simulation at 675 Hz.

to craft an optimized look up table of capacitor configurations vs. operating frequency, as well as optimizing the capacitor selection in advance.

III. EXPERIMENTAL RESULTS

The simulated system was tested with the setup shown in Fig. 8. The VREH was emulated with a custom PCB that implements its electrical model, a full bridge diode rectifier, and additional measurement circuits, including a frequency sensor. Schottky diodes with low forward voltage are used for the rectifier with the goal of minimizing power losses. An AC source provides the sinusoidal voltage of the harvester, and an electronic load is connected to the output of the rectifier imposing a voltage that maximizes the extracted power, as detailed in Section II. To facilitate prototyping, two capacitor arrays are implemented on additional PCBs, one using film capacitors and the other ceramic ones, whose details are shown in Table III. Both PCBs can be connected to the main board and have been tested in order to show a comparison between the technologies. The film capacitors have been selected to

TABLE III
SUMMARY OF USED CAPACITOR SIZES

Capacitance	Commercial part	Volume
0.68 μF (film)	B32529C1684J000	472.5 mm³
1 μF (film)	B32529D1105J000	790.9 mm³
3.3 μF (film)	B32522C1335J289	2218.5 mm³
0.68 μF	FG26X7R2A684KNT06	115.5 mm³
1 μF	RDER72A105K2K1H03B	69.3 mm³
3.3 μF	FG26X7S2A335KRT06	115.5 mm³

match the characteristics of the ceramic ones in terms of value, voltage, and temperature rating.

Each bidirectional switch consists of two MOSFETs (Infineon's BSC120N12LS) with their sources connected together to block AC currents, and they are controlled with a single isolated driver (Skyworks's SI8751AB-ISR). The enable signals for the switches are generated by an external microcontroller (STMicroelectronics's STM32L031K6) that can read frequency measurements from a frequency sensor implemented in the main PCB. The frequency sensor uses a comparator to generate a square wave from the sinusoidal voltage of the harvester, which is sent to the microcontroller to compute the frequency by measuring the time between rising edges.

The power obtained with both PCBs for the entire range of harvester frequencies has been measured. The results with the ceramic capacitors are shown in Fig. 7, which are close to the obtained ones with simulations. The dashed lines are the envelopes of the power curves of all the capacitor configurations, for both the experimental and simulation results. They are obtained by picking the maximum power out of all the possible configurations at each frequency.

Efficiency results are shown in Fig. 9 for both PCBs. It can be seen that the efficiency of the ceramic capacitors solution is around 10% less than the film one for frequencies higher than 200 Hz. However, a big reduction in volume is achieved. While the volume of the film capacitors is 3481.9 mm³, the ceramic capacitors ones is 300.3 mm³, therefore, a 91.4% size reduction is achieved. Considering also the volumes of the

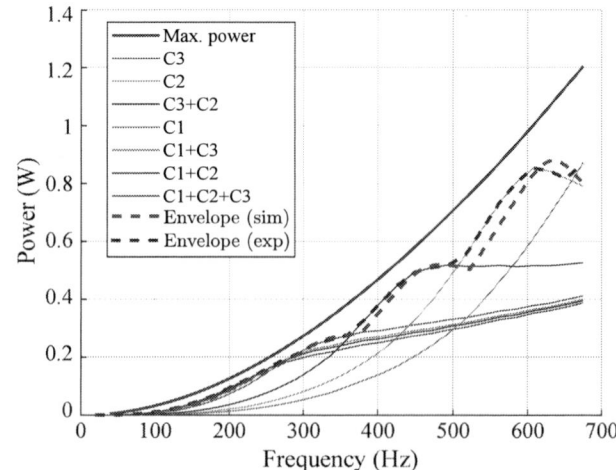

Fig. 7. Measured output power using ceramic capacitors and comparison with simulations.

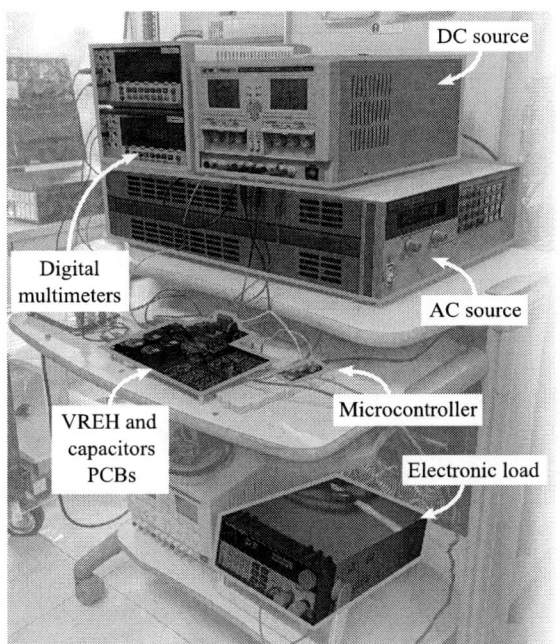 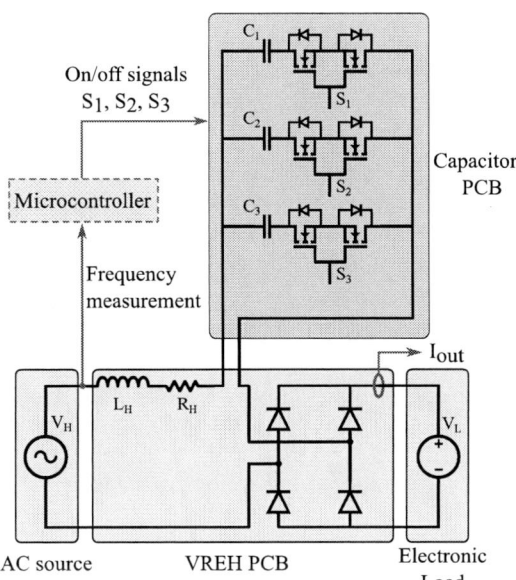

Fig. 8. Experimental setup.

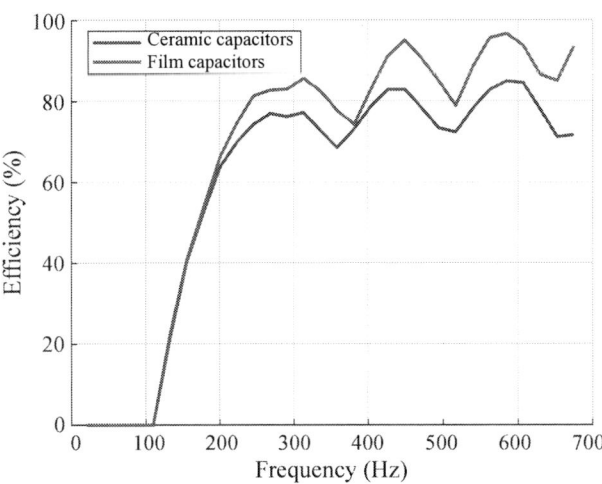

Fig. 9. Measured efficiency with ceramic and film capacitors.

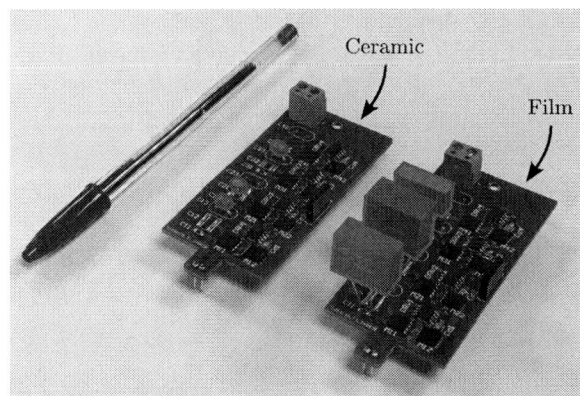

Fig. 10. Size comparison between ceramic and film capacitors PCBs, with the capacitors highlighted.

PCBs and the rest of the components, the total volumes are 8942.7 mm^3 and 5761.1 mm^3, which translates to a 35.6% size reduction relative to the entire PCB. The board height is also an interesting metric, since it might be the limiting factor in applications in which the board must fit in thin spaces. In this case, the height would be reduced from 15 mm to 5 mm. In this research, the same board has been used for both options, changing just the capacitors, but it is worth noting that the ceramic capacitors PCB could be further reduced and adapted specifically to the size of the capacitors. A picture of both PCBs side by side is shown in Fig. 10 to highlight the size differences. Moreover, when using ceramic capacitors at frequencies below 200 Hz, the size reduction in relation to film capacitors is maintained but the efficiency does not decrease, which makes this technology the best option in low speed applications.

Regarding the power required to operate the circuitry that performs the switching, it can be reduced to the MOSFET conduction losses and the static consumption of the drivers, which require 1.5 mA of current each. In this setup they are powered with 3.3 V provided by a bench power supply, but they would be powered by a battery of a similar voltage in a final application. Therefore, the total power consumption is 14.85 mW, which represents 1.2% of the total maximum obtainable power. All these losses are included in the efficiency curves shown in Fig. 9. The switching losses of the MOSFETs have been neglected since they are not switching continuously at a high frequency, but they are occasionally turned on or off in response to slow changes in the harvester frequency.

IV. CONCLUSIONS AND FUTURE WORK

This paper presents the design of a capacitor array using class II ceramic capacitors for dynamic impedance matching in VREH applications. A model that considers the voltage bias effect on the effective capacitance is proposed. This allows for the design of the system with this capacitor technology instead of using others that feature larger physical sizes, resulting in a suitable solution for applications with space restrictions. A 90% reduction in capacitor volume is achieved with only a 10% drop in efficiency at high frequencies. For low speed applications, the efficiencies are the same as the ones obtained with other capacitor technologies, however, the size reduction is maintained. Future work includes the integration of optimization algorithms to select the optimal capacitors for a specific VREH and a range of operating frequencies.

ACKNOWLEDGMENT

This work is supported by the LoLiPoP IoT project, Grant PCI2023-143407, funded by MICIU/AEI /10.13039/501100011033 and cofunded by European Comission. LoLiPoP IoT has received funding from the Chips Joint Undertaking (Chips JU) under grant agreement No 101112286. The JU receives support from the European Union's Horizon Europe research and innovation programme and accordingly from the participating countries.

REFERENCES

[1] J. Portilla, G. Mujica, J.-S. Lee, and T. Riesgo, "The extreme edge at the bottom of the internet of things: A review," *IEEE Sensors Journal*, vol. 19, no. 9, pp. 3179–3190, 2019.

[2] F. Akhtar and M. H. Rehmani, "Energy replenishment using renewable and traditional energy resources for sustainable wireless sensor networks: A review," *Renewable & Sustainable Energy Reviews*, vol. 45, pp. 769–784, 2015.

[3] T. Sanislav, G. D. Mois, S. Zeadally, and S. C. Folea, "Energy harvesting techniques for internet of things (iot)," *IEEE Access*, vol. 9, pp. 39 530–39 549, 2021.

[4] H. Zhang, X. Liu, M. Kedia, and R. S. Balog, "Photovoltaic hybrid power harvesting system for emergency applications," in *2013 IEEE 39th Photovoltaic Specialists Conference (PVSC)*, 2013, pp. 2902–2907.

[5] M. Shirvanimoghaddam *et al.*, "Towards a green and self-powered internet of things using piezoelectric energy harvesting," *IEEE Access*, vol. 7, pp. 94 533–94 556, 2019.

[6] C.-L. Chang and T.-C. Lee, "An thermoelectric and rf multi-source energy harvesting system," in *2016 2nd International Conference on Intelligent Green Building and Smart Grid (IGBSG)*, 2016, pp. 1–5.

[7] I. D. Bougas, M. S. Papadopoulou, A. D. Boursianis, K. Kokkinidis, and S. K. Goudos, "State-of-the-art techniques in RF energy harvesting circuits," *Telecom*, vol. 2, no. 4, pp. 369–389, 2021.

[8] Y. Xu, S. Bader, M. Magno, P. Mayer, and B. Oelmann, "System implementation trade-offs for low-speed rotational variable reluctance energy harvesters," *Sensors*, vol. 21, no. 18, 2021, Art. no. 6317.

[9] Y. Xu, S. Bader, and B. Oelmann, "A survey on variable reluctance energy harvesters in low-speed rotating applications," *IEEE Sensors Journal*, vol. 18, no. 8, pp. 3426–3435, Apr. 2018.

[10] M. Kroener, N. Moll, S. K. T. Ravindran, P. Mehne, and P. Woias, "Characterization of a variable reluctance harvester," *Journal of Physics: Conference Series*, vol. 557, no. 1, nov 2014, Art. no. 012035.

[11] Y. Xu, Y. Zhang, S. Bader, B. Oelmann, and J. Cao, "Three-phase variable reluctance energy harvesting," *Energy Conversion and Management: X*, vol. 14, 2022, Art. no. 100211.

[12] Y. Xu, S. Bader, and B. Oelmann, "Design, modeling and optimization of an m-shaped variable reluctance energy harvester for rotating applications," *Energy Conversion and Management*, vol. 195, pp. 1280–1294, Sep. 2019.

[13] Y. Zhang *et al.*, "Enhanced variable reluctance energy harvesting for self-powered monitoring," *Applied Energy*, vol. 321, 2022, Art. no. 119402.

[14] Y. Zhang, X. Wu, Y. Lei, J. Cao, and W.-H. Liao, "Self-powered wireless condition monitoring for rotating machinery," *IEEE Internet of Things Journal*, vol. 11, no. 2, pp. 3095–3107, Jan. 2024.

[15] Y. Zhang, H. Zhu, Y. Xu, J. Cao, S. Bader, and B. Oelmann, "Theoretical modeling and experimental verification of rotational variable reluctance energy harvesters," *Energy Conversion and Management*, vol. 233, 2021, Art. no. 113906.

[16] M. Kroener, S. K. T. Ravindran, and P. Woias, "Variable reluctance harvester for applications in railroad monitoring," *Journal of Physics: Conference Series*, vol. 476, no. 1, Dec. 2013, Art. no. 012091.

[17] Y. Gong *et al.*, "A variable reluctance based rotational electromagnetic harvester for the high-speed smart bearing," *Smart Materials and Structures*, vol. 31, no. 4, Mar. 2022, Art. no. 045023.

[18] Y. Gong, S. Wang, Z. Xie, and W. Huang, "Parameter study of the variable reluctance energy harvester for smart railway axle box bearing," in *2020 International Conference on Sensing, Measurement & Data Analytics in the era of Artificial Intelligence (ICSMD)*, Xi'an, China, Oct. 2020, pp. 464–469.

[19] Y. Gong *et al.*, "Self-powered wireless sensor node for smart railway axle box bearing via a variable reluctance energy harvesting system," *IEEE Transactions on Instrumentation and Measurement*, vol. 70, pp. 1–11, 2021, Art. no. 9003111.

[20] D. Zakzewski, Y. Shen, A. Hasnain, R. Resalayyan, and A. Khaligh, "Class ii ceramic capacitor voltage characteristic modeling and compensation for ac-connected applications," *IEEE Journal of Emerging and Selected Topics in Industrial Electronics*, vol. 5, no. 4, pp. 1582–1592, 2024.

Renewable energy-powered DC-converted refrigerator based on a supercapacitor-assisted technique

Nirashi Polwaththa Gallage[1], Nihal Kularatna[1], Alistair Steyn-Ross[1], Dulsha Kularatna-Abeywardana[2],

[1]School of Engineering
University of Waikato, Hamilton 3240, New Zealand
Email:np138@students.waikato.ac.nz, nihal.kularatna@waikato.ac.nz, asr@waikato.ac.nz
[2]Department of Electrical, Computer and Software
University of Auckland,Auckland 1010, New Zealand
Email: d.abeywardana@auckland.ac.nz

Abstract—**In six ICDCM conferences, the value of future DC systems, and DC homes for reducing losses in our electrical systems was discussed indicating the need for a DC-based future. DC outputs from renewable sources contribute to reducing CO2 emissions and lowering electrical energy consumption. In modern white goods, internal circuits are based on DC power rails derived from AC mains followed by a rectifier. Powering AC households from renewable requires maximum power point tracking based on battery chargers and inverters. End-to-end efficiency is related to the number of individual converters inside an equipment. Here we describe a low-frequency supercapacitor-assisted converter system coupled with an "inverter-driven" refrigerator converted to DC operation by eliminating the rectifiers. With this modification, whiteware becomes more tolerant of variable DC sources, allowing continuous operation without battery buffers in solar or wind energy systems. A modified 230 V, 50 Hz refrigerator for DC operation was connected with a supercapacitor module rated at 110 V DC, allowing it to run directly from a renewable source and eliminating the need for battery buffering. This is another supercapacitor-assisted technique based on the relatively new supercapacitor-assisted loss management concept and it eliminates the requirement for electromagnetic compatibility filtering.**

Index Terms—**DC refrigerator; supercapacitors; power converters; low frequency switching circuit; renewable energy**

I. INTRODUCTION

Electrical motor-driven equipment accounts for the largest share of electricity consumption in both the residential and industrial sectors: approximately 43–46 % in the domestic sector and 60–70 % in the industrial sector [1–4]. Additionally, it contributes around 13 % of global CO_2 emissions [1]. One of the key strategies for reducing electricity consumption in electric motors is improving efficiency through speed regulation and load-based voltage control methods [1, 5]. For constant torque loads, the power consumption is directly proportional to the speed of the motor. Therefore, energy savings can be achieved by using motors that adjust their speed according to the load. Brushless DC (BLDC) motors, paired with advanced motor controllers, have emerged as a solution to meet these requirements, driven by advancements in power

electronics [6, 7]. BLDC motors have gained popularity in modern household appliances due to their compact size, high efficiency, ease of control, low starting current (with no inrush current), and reduction of noise levels [8]. Today most whiteware such as refrigerators, washing machines, dryers and heat pumps use BLDC motors rather than single-phase induction motors as illustrated in Fig. 1. They are built with the BLDC motor driven by three-phase power electronic inverter modules powered by DC rail derived from the AC main supply via the rectifier stage.

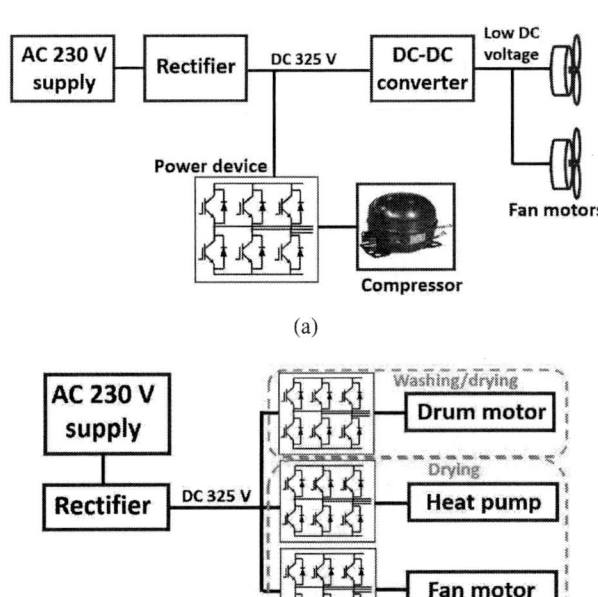

(a)

(b)

Fig. 1: block diagram of the whiteware having BLDC motors (a) refrigerator with variable speed compressor; (b) speed controlled motor-driven washing machine (washer and dryer combination) [9]

Refrigerators, heat pumps, etc. carry variable capacity compressors (VCCs) which consume less than 65 % of power than old induction-motor driven compressors [10–12]. Also modern efficient LED lighting systems, infotainment systems, and computers are all powered by the internal DC rail derived from AC mains. When directly powering appliances from the DC supply by removing the rectifier stage, almost 5 % of energy can be saved in the AC-DC converter stage [13]. Once we convert a household appliance to DC sources directly with minimum or no power converters, which allows us to acheive higher end-to-end efficiency (ETEE). With the decremental cost of photovoltaic panels and advancement in power electronic components, solar energy usage from rooftop-mounted PV panels has significantly increased. The Renewable 2023 Global status report states that 243 GW of electricity was used from solar PV from 348 GW in renewables in 2022 [14].

When powering the available AC-powered household appliances from the solar DC rail, usage of a DC-AC converter (inverter) is mandatory. Typically, to buffer the fluctuating nature of renewable energy, a maximum power point tracking (MPPT) controller with a battery bank is essential as shown in Fig. 2. In this example, if each of the four converter modules has 95 % efficiency, ETEE is given by the product of the individual module efficiencies $= 0.95^4 = 0.81$. If the same equipment without a front-end AC-DC converter, is supplied by a suitable DC source, removing three converter stages marked in the purple box in Fig. 2, the effective ETEE becomes, achieving an efficiency improvement of approximately 14 %.

When directly powering modified DC-operable whiteware with a solar PV system, an energy storage system is essential to buffer the fluctuations of the energy supply. In this research, we develop a supercapacitor (SC) based low-frequency converter system to power the inverter-driven refrigerator with a solar PV system that provides buffering and eliminates the need for environmentally unfriendly batteries.

II. BATTERY ELIMINATION BY THE USE OF SC ENERGY STORAGE

Supercapacitors (SCs) are a newly emerged family of energy storage devices (ESD) in the watt-hour range, offering high power density and moderate energy density. Commercially available SCs are divided into four distinct families, and their comparison is illustrated in Table I. All these families exhibit longer cycle life and capacitance values approximately one million times greater than those of electrolytic and film capacitors of the same canister size. Given that SCs are have DC voltage ratings in the range of 2.5 to 4.2 V to build as energy storage module we need to stack them in series as in the case of rechargeable batteries. Recent advancements in electrode and electrolyte materials for supercapacitors aim to enhance energy density and enable them to replace batteries. Fig. 3 shows the Ragone plot for various types of energy storage devices (ESDs) and demonstrates that hybrid capacitors are approaching the energy density of Li-ion batteries. With their longer cycle life, supercapacitors (SCs) can be considered a 'fit-and-forget' solution for the system having long-term reliability.

One important attribute of SCs is that they come with relatively low equivalent series resistance (ESR) compared to similar size Li-ion cells and this ESR is independent of depth of discharge. This property translates to a case of high-power density which is useful in delivering the inrush current of motors and electronic circuits. Fig. 4 depicts a quantitative comparison of features of SCs and rechargeable batteries.

III. MODIFICATION OF AC POWERED REFRIGERATOR TO DC OPERATION

Figure 5a depicts the typical block diagram of a AC operated BLDC-based refrigerator such as RF402B from Fisher & Paykel which was used in the research. In this case we identified two separate rectifier stages powering BLDC and the associated inverter stage and the auxiliary circuits controlled by a microprocessor. Figure 5b shows the case of modified refrigerator by the removal of two rectifier stages for direct DC operation. Moment we power the modified system we

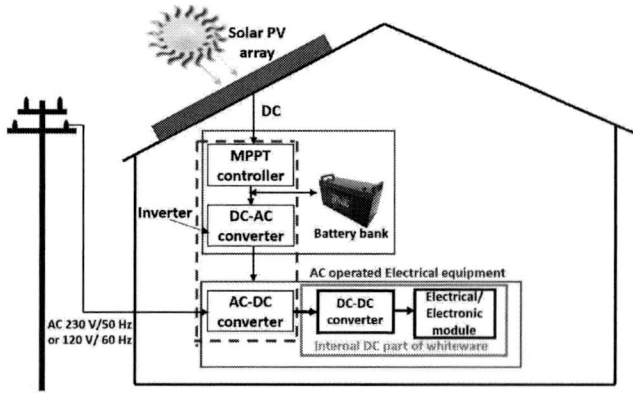

Fig. 2: AC operated whiteware fed from AC and DC simultaneously

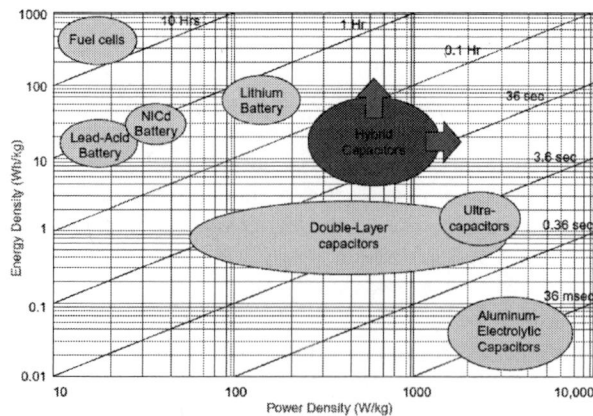

Fig. 3: Ragone plot of various ESDs [17]

979-8-3315-1612-3/25 $31.00 © 2025 IEEE

TABLE I: Comparison of commercial SC types and batteries [15–18]

Type	Electrode material	Energy density	Power density	Cycle life
Electrical Double Layer Capacitor (EDLC)	Activated carbon, graphene	Low to Moderate	Very High	Very High
Pseudocapacitor	ransition metal oxides, conducting polymers	Moderate to High	Moderate	Moderate
Hybrid capacitor	Activated carbon + pseudocapacitive material	Moderate to High	High	High
Li-ion capacitor	Graphitic carbon + activated carbon	Moderate	Very High	High
Li-ion battery	Graphite carbon, ransition metal oxides	High	Medium	Medium
Lead-Acid batteries	Lead, Lead Dioxide	Low to Medium	Medium	Short

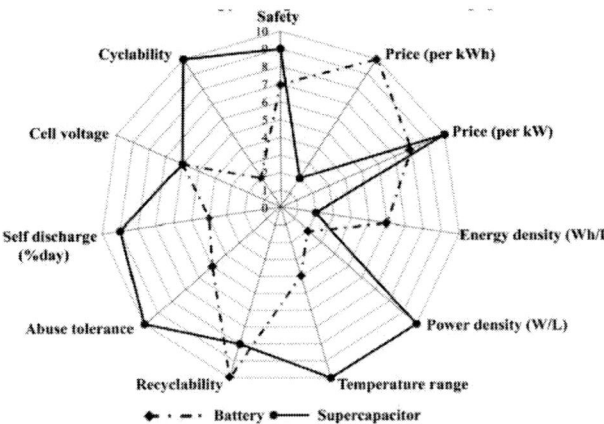

Fig. 4: Comparison of SCs and batteries[19]

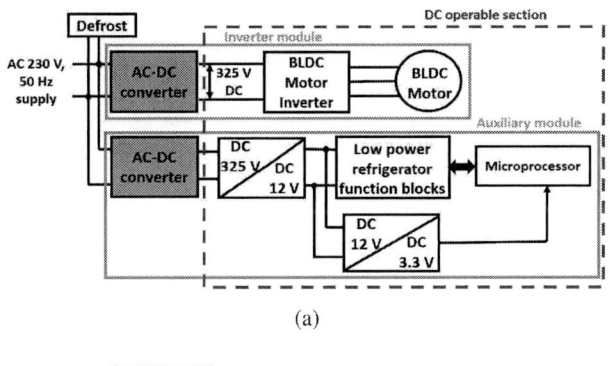

(a)

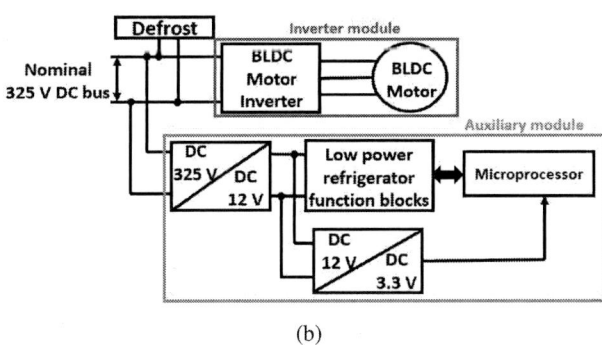

(b)

Fig. 5: Block diagram of an inverter-driven refrigerator powered from (a) AC main; (b) DC source [21]

learnt that the system could operate from a wide range of DC voltage varying from 255 to over 325 V DC. As in [13, 20], we learn that starting cycle of the compressor consumes about 100 W and after temperature settles it consumes around 40 W per compressor cycle. Another observation was that by the removal of rectifiers we save 4 - 5 % of energy making the overall efficiency higher. Table II provides a summary of power consumption under DC supply conditions.

IV. INTEGRATING THE SC MODULES WITH A LOW FREQUENCY CONVERTER FOR THE REFRIGERATOR

As stated in [22–24], our research team have developed several supercapacitor-assisted techniques (SCA) based on a new theory published as supercapacitor-assisted loss management (SCALoM) [22]. We have extended the same theory here for a

TABLE II: Average power consumption of the refrigerator under AC and DC input voltage with efficiency improvement

Input voltage (V)	Average power consumption (W)					
	AC 230	DC 325	Efficiency gain (%)	AC 180	DC 255	Efficiency gain (%)
Initial cycle	102.3	96.4	5.8	104.1	98.6	5.3
Steady-state cycle	42.7	40.7	4.7	41.1	39.2	4.6

SCA converter suitable for powering the DC refrigerator with following attributes.

1) Replacing the MPPT based battery charger and associated inverter with SC modules (SCMs).
2) Sizing the SCMs to run one complete compressor cycle under steady state conditions.
3) Using the SCMs to buffer fluctuations of renewable energy source
4) Making use of the wide operational voltage range of the DC refrigerator to optimise the size of SCMs.

Figure 6a is a simplified representation of refrigerator combined with the SCM adopting the SCALoM concept where we used a refrigerator load as a useful resistor (PEBB) in a modified resistor-capacitor loop for improved energy efficiency. For more details on SCALoM theory and its applications available in [17]. In this overall configuration with an extra low-frequency switching cycle SCM is sized to discharge an energy content of approximately Pt where P is the average power consumption of the refrigerator and t is the time of the compressor ON time under steady state condition.The SCM is constructed by connecting n number of individual cells in series to achieve the required operational DC voltage for the modified refrigerator. Each supercapacitor cell has a voltage rating of 3 V. A total of 135 cells, each with a capacitance of 500 F, were selected to operate within a voltage range of 330 V to 270 V during the discharge phase, while maintaining an energy capability of 20 Wh. To ensure safety, each cell operates at a maximum voltage of 2.7 V, providing an adequate safety margin for the SCM.

Fig. 6a refers to a case of a solar PV system of 600 V feeding the refrigerator (with a nominal DC supply of 300 V) working with a SC module of 300 V nominal operational voltage. When S_1 and S_3 are closed SCM works in series with the refrigerator under charging phase based on a long-time constant process as shown in Fig. 6b . While SCM charges from 300 to 320 V, the refrigerator keeps working from a DC rail of 300 V to 280 V, where SCM keeps accumulating energy. When SCM reaches 320 V, S_2 and S_4 are switched on and other switches off, moving SCM to discharge mode, powering the refrigerator, and isolating the solar supply as depicts in Fig. 6c. This system can power the refrigerator using an input DC voltage range of 580 V to 620 V, with added buffering.

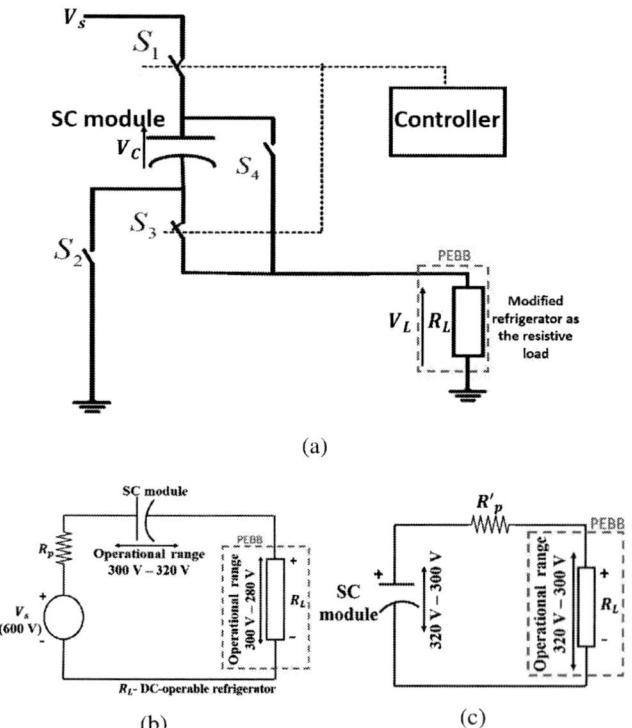

Fig. 6: Concept of SCA-refrigerator (a) Switching circuit; (b) refrigerator working in series with SCM; (c) discharging the SCM into refrigerator isolating the DC source.

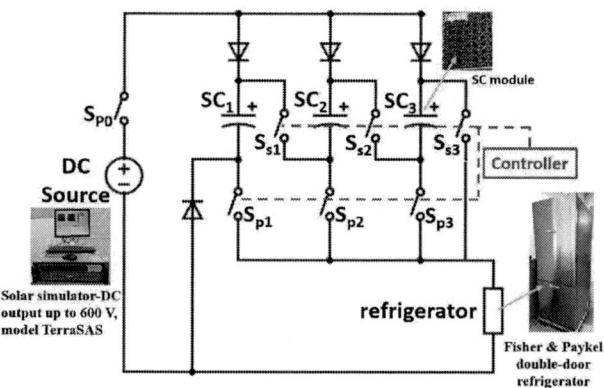

Fig. 7: Concept of scaref

However, this requires a solar PV system with a nominal output of 600 V and a larger SCM, which adds significant cost to the system. In this research, we aim to utilize a medium DC voltage rail as the input source by splitting the SCM into three identical series strings of SCs and arranging them in parallel and series strings during the charging and discharging phases.

V. IMPLEMENTATION OF SCA-REFRIGERATOR

By configuring three SCMs, each consisting of 45 individual cells and charging from 90 V to 110 V, the input DC voltage source can be reduced to 400 V. This approach significantly reduces the number of solar panels required in the new SCA-refrigerator (SCARef) concept, as illustrated in Fig. 7.

When switches S_{p0}, S_{p1}, S_{p2}, and S_{p3} are closed, the SCM charge in parallel from the DC power supply from 90 to 110 V, while the DC refrigerator operates within a voltage range of 310 to 290 V. Once the SCM voltage reaches 110 V, the DC source is disconnected by switching off the aforementioned switches, and switches S_{s1}, S_{s2}, and S_{s3} are turned on to power the refrigerator in SCM discharging mode, with all SCM arranged in series. Fig. 7 shows the overall system with SCMs under charge and discharge phases, for a solar PV system of 400 V.

In this SCARef configuration based on modified RC circuit loaded refrigerator (PEBB) becomes the dominating resistor element combined with the SC as the capacitance in the loop. As per highlights of SCALoM theory typical resistive losses in an RC circuit are mostly consumed by the refrigerator. When loop parasitic resistances (combination of switch ON resistances , SC ESR, and connection resistances) are much smaller compared to the effective input resistance of the refrigerator, it increases the efficiency of charging.

Fig 8 shows the laboratory setup of the refrigerator and the SCMs in the overall system powered by a solar simulator.. yellow square in Fig 8(a) indicates the PCB modules where two rectifier stages are bypassed. Fig 9 shows the voltages across the SCM and the refrigerator in one complete switching cycle associated with the initial cycle of the compressor. While this initial compressor cycle completes within 45 minutes approximately, under steady state compressor cycle switching

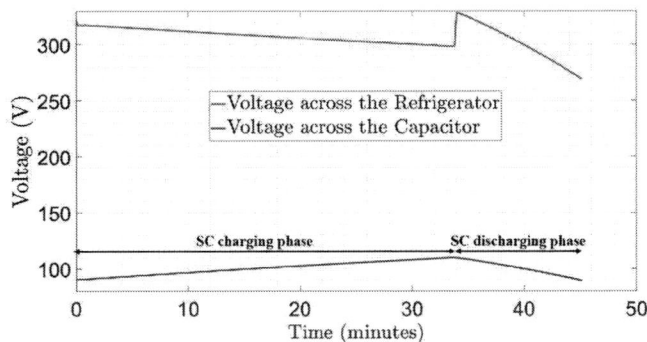

Fig. 9: Voltage variation across the SC bank and the refrigerator with time

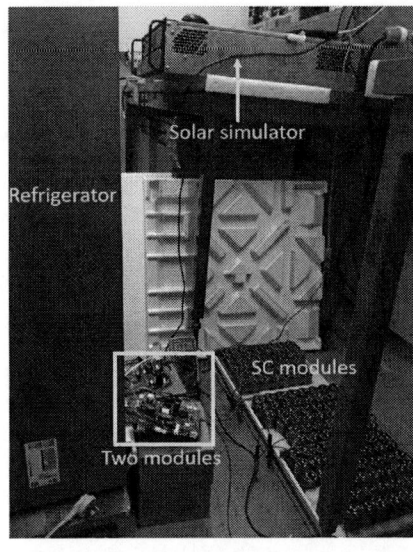

(a)

(b)

Fig. 8: Laboratory setup (a) SCM charging phase by paralleling; (b) SCM Discharging mode in series

may take over 100 minutes depending on the load and the door opening. Additionally, the SCA converter operates at a very low frequency which helps eliminate the dynamic losses associated with the diodes and switching transistors.

VI. CONCLUSIONS

The supercapacitor-assisted DC refrigerator system described here eliminates the need for a battery bank, MPPT controller, and the associated inverter, etc. in a solar-powered home system with whiteware. Rectifier-stage free, DC refrigerator becomes more efficient and SCM becomes an energy buffer working under a relatively new SCALoM concept where EMC issues are also eliminated due to an extra low-frequency switching cycle based on the long-time constant circuit. Future work will show how a DC home and DC whiteware can be fully configured using SCA low-frequency converters, with localized energy buffers on a distributed resource basis with minimal or no batteries.

ACKNOWLEDGEMENT

This research receives support from the Ministry of Business, Innovation, and Employment New Zealand through the Future Architecture Network (FAN) project of the Advanced Energy Technology Program (AETP). We also extend our gratitude to Fisher & Paykel Appliances Holdings Ltd for generously donating an inverter-driven refrigerator for the research.

REFERENCES

[1] V. S. Santos, J. J. C. Eras, and M. J. C. Ulloa, "Evaluation of the energy saving potential in electric motors applying a load-based voltage control method.," *Energy*, p. 132012, 2024.

[2] E. Abdelaziz, R. Saidur, and S. Mekhilef, "A review on energy saving strategies in industrial sector," *Renewable and sustainable energy reviews*, vol. 15, no. 1, pp. 150–168, 2011.

[3] R. Saidur, S. Mekhilef, M. B. Ali, A. Safari, and H. A. Mohammed, "Applications of variable speed drive (VSD) in electrical motors energy savings," *Renewable and sustainable energy reviews*, vol. 16, no. 1, pp. 543–550, 2012.

[4] A. T. De Almeida, F. J. Ferreira, and J. A. Fong, "Standards for efficiency of electric motors," *IEEE Industry Applications Magazine*, vol. 17, no. 1, pp. 12–19, 2010.

[5] D. M. Ionel, "High-efficiency variable-speed electric motor drive technologies for energy savings in the us residential sector," in *2010 12th International Conference on Optimization of Electrical and Electronic Equipment*, pp. 1403–1414, IEEE, 2010.

[6] P. Yedamale, "Brushless DC (BLDC) motor fundamentals." Available online at microchip, 2003. Accessed: 2024-11-01.

[7] H. Omori and A. Kanouda, "A trend of home and consumer appliances in japan: The past 50 years and the future," *IEEJ Transactions on Electrical and Electronic Engineering*, vol. 16, no. 3, pp. 340–348, 2021.

979-8-3315-1612-3/25 $31.00 © 2025 IEEE

[8] R. Gambhir and A. K. Jha, "Brushless dc motor: Construction and applications," *Int. J. Eng. Sci*, vol. 2, no. 5, pp. 72–77, 2013.

[9] A. H. Sabry, A. H. Shallal, H. S. Hameed, and P. J. Ker, "Compatibility of household appliances with dc microgrid for pv systems," *Heliyon*, vol. 6, no. 12, 2020.

[10] T. Qureshi and S. Tassou, "Variable-speed capacity control in refrigeration systems," *Applied thermal engineering*, vol. 16, no. 2, pp. 103–113, 1996.

[11] T. R. DuMoulin and D. A. Collings, "Higher efficiencies by means of variable-speed technology in a domestic refrigeration application," *ASHRAE Transactions*, vol. 104, p. 652, 1998.

[12] A. H. Sabry and P. J. Ker, "Improvement on energy consumption of a refrigerator within pv system including battery storage," *Energy Reports*, vol. 7, pp. 430–438, 2021.

[13] N. P. Gallage, N. Kularatna, A. Steyn-Ross, and D. Kularatna-Abeywardana, "Comparison of energy consumption of an inverter-driven refrigerator under dc operation vs ac operation–preliminary investigation on dc operability of whiteware," in *2023 IEEE Fifth International Conference on DC Microgrids (ICDCM)*, pp. 1–6, IEEE, 2023.

[14] Renewables in Energy Supply, Secretariat: Paris, France, *REN21. Renewables 2023 Global Status Report collection*, 2023. ISBN 978-3-948393-08-3.

[15] W. Raza, F. Ali, N. Raza, Y. Luo, K.-H. Kim, J. Yang, S. Kumar, A. Mehmood, and E. E. Kwon, "Recent advancements in supercapacitor technology," *Nano Energy*, vol. 52, pp. 441–473, 2018.

[16] M. E. Şahin, F. Blaabjerg, and A. Sangwongwanich, "A comprehensive review on supercapacitor applications and developments," *Energies*, vol. 15, no. 3, p. 674, 2022.

[17] N. Kularatna and K. Gunawardane, *Energy Storage Devices for Renewable Energy-Based Systems: Rechargeable Batteries and Supercapacitors*. Academic Press, 2021.

[18] B. E. Conway, *Electrochemical Supercapacitors: Scientific Fundamentals and Technological Applications*. Springer Science & Business Media, 1999.

[19] V. Muralidharan, R. Narayanan, J. Nithish, *et al.*, "A novel supercapacitor assisted regenerative braking scheme for electric vehicle," in *2022 International Virtual Conference on Power Engineering Computing and Control: Developments in Electric Vehicles and Energy Sector for Sustainable Future (PECCON)*, pp. 1–7, IEEE, 2022.

[20] N. P. Gallage, D. C. T. Sirimanne, N. Kularatna, A. Steyn-Ross, and D. Kularatna-Abeywardana, "Supercapacitor-buffered dc-operable refrigerators for dc homes," in *2024 IEEE Applied Power Electronics Conference and Exposition (APEC)*, pp. 2965–2971, IEEE, 2024.

[21] Fisher and Paykel, *ActiveSmart Refrigerator*, version 4, 5 and 6 models ed., 2022.

[22] N. Kularatna, K. Subasinghage, K. Gunawardane, D. Jayananda, and T. Ariyarathna, "Supercapacitor-assisted techniques and supercapacitor-assisted loss management concept: New design approaches to change the roadmap of power conversion systems," *Electronics*, vol. 10, no. 14, p. 1697, 2021.

[23] N. Kularatna and D. Jayananda, "Supercapacitor-based long time-constant circuits: A unique design opportunity for new power electronic circuit topologies," *IEEE Industrial Electronics Magazine*, vol. 14, no. 2, pp. 40–56, 2020.

[24] N. Kularatna, "Supercapacitors improve the performance of linear power-management circuits: Unique new design options when capacitance jump from micro-farads to farads with a low equivalent series resistance," *IEEE Power Electronics Magazine*, vol. 3, no. 1, pp. 45–59, 2016.

Design and Evaluation of Flexible Inductors for Wearable Power Electronics

Sean Logi, F. Selin Bagci and Katherine A. Kim
Department of Electrical Engineering
National Taiwan University
Taipei, Taiwan
email: r12921104@ntu.edu.tw, d09921018@ntu.edu.tw, kakim@ntu.edu.tw

Abstract—**Wearable electronics and soft robotics require adaptable circuit components that maintain electrical performance under mechanical stress. For these applications, rigidity components would impair functionality, so it is essential to explore physically flexible power converters, especially the inductor. This paper presents the design, fabrication, and performance evaluation of flexible inductors using commercially available ferrite sheets as magnetic cores and Litz wire for windings. Four grades of ferrite cores were used, each tested under bending and folding conditions to simulate realistic scenarios. Electrical characteristics, including inductance retention and DC resistance (DCR), were evaluated at frequencies ranging from 10 MHz, demonstrating stability in the target operating range of 200–300 kHz. The prototypes showed consistent performance despite mechanical stress, showing a worst-case inductance decrease of 88% from the original value when folded twice. The impedance sweeps further confirmed minimal variance in the DCR among the different core grades, indicating strong mechanical resilience. This study highlights the feasibility of flexible inductors for low-power wearable applications, emphasizing their potential for integration into physically pliable applications.**

Index Terms—**Flexible inductors, inductor fabrication, flexible magnetic cores, wearable technology, e-textiles, soft robotics**

I. INTRODUCTION

Wearable technology has emerged as a rapidly growing field, with applications ranging from medical devices and activity tracking to gesture recognition and advancements in the fashion industry [1]–[3]. Wearable electronics and soft robotics require designs that can conform to curved or moving surfaces without compromising performance. These technologies require solutions that are both flexible and conformable, capable of withstanding constant mechanical stress—such as bending, stretching, and twisting—while preserving electrical integrity. Consequently, there is a growing interest in developing highly flexible and mechanically compliant circuit boards and components to meet the unique demands of these applications [4]–[6]. Among these components, inductors are essential in power electronics circuits, as they enable smooth and controlled current flow while storing energy in a magnetic field [7]. Unlike conventional rigid inductors, flexible inductors have the ability to bend, stretch, and conform to different shapes [8], [9]. This property makes them effective for use in advanced sensor systems, flexible displays, and wearable electronics [10].

In the literature, fully flexible inductors have previously been demonstrated, which were fabricated on plastic-based substrates such as polyimide and polyester [9], [11]. Furthermore, recent advances in electronic textile (e-textile) technology have enabled the manufacturing of fully deformable, pliable inductors, allowing the use of fiber-based materials and eliminating the stiffness associated with plastic [12], [13].

Another innovative approach presented in the literature involves the design of stretchable inductors utilizing a liquid magnetic core. Replacement of traditional solid magnetic cores with a liquid alternative enhances flexibility and adaptability, allowing these inductors to maintain inductance while undergoing significant physical deformations. Such liquid-core inductors perform reliably even under stretching and bending, expanding the possibilities for flexible power components in wearable electronics, soft robotics, and other applications [14]. Another approach involves magnetic elastomers, where magnetic particles are embedded within an elastomer matrix, allowing inductors to endure mechanical stress while preserving inductive properties. These magnetic elastomer-based inductors have been shown to exhibit minimal degradation in performance under mechanical stress, supporting their functionality in flexible and dynamic applications [15]. However, the development of flexible inductors for power electronics applications still presents significant challenges, particularly in achieving high inductance and low losses while ensuring mechanical durability under stress.

II. PROBLEM STATEMENT: CHALLENGES IN DESIGNING INDUCTORS FOR WEARABLE CONVERTERS

The development of inductors suitable for wearable power converters presents several challenges in wearability and mechanical resilience. One approach to overcome these challenges is to manufacture inductors that are very small on plastic substrates [11]. As researchers strive to achieve smaller inductor volumes, primarily, to meet the demands of modern miniaturized electronics [11], a common trade-off arises: reducing the size of the inductor usually decreases its inductance. Although this approach can achieve relatively high inductances when connected in series and is suitable for high-volume production, it results in decreased mechanical strength and flexibility, limiting the overall wearability of the device.

To address this challenge, we focus on designing inductors that, while larger, retain flexibility and mechanical resilience, making them suitable for applications in wearable technology where both performance and durability are critical. Larger but fully flexible inductors can be made using unconventional components such as e-textiles and flexible magneto-dielectric materials. E-textiles hold promise for creating soft and conformable inductors. However, they currently face significant limitations for power electronics applications, including high resistivity and difficulties with soldering, which complicate their

979-8-3315-1612-3/25 $31.00 © 2025 IEEE

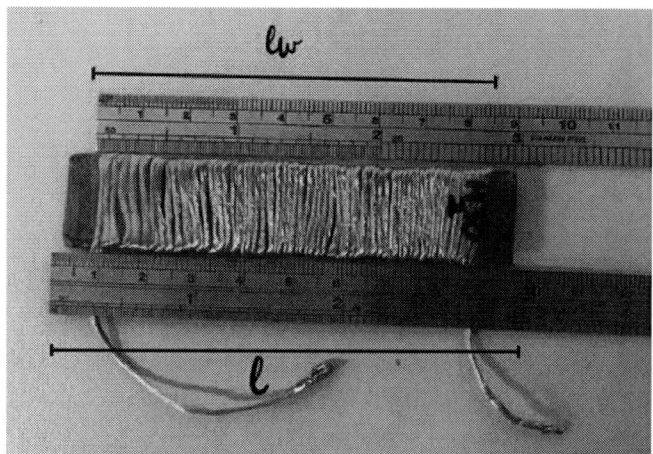

Fig. 1: Flexible PCB used to compare connection methods.

TABLE I: Flexible Inductor Specifications.

Core Number	Length [cm]	Width [cm]	Core Thickness [mm]	Turns Number	μ_r [H/m]	d [mm]
M10	9.5	2	0.13	60	20	1
M30	9.5	2	0.13	60	60	1
M40	9.5	2	0.13	60	120	1
M50	9.5	2	0.13	60	140	1

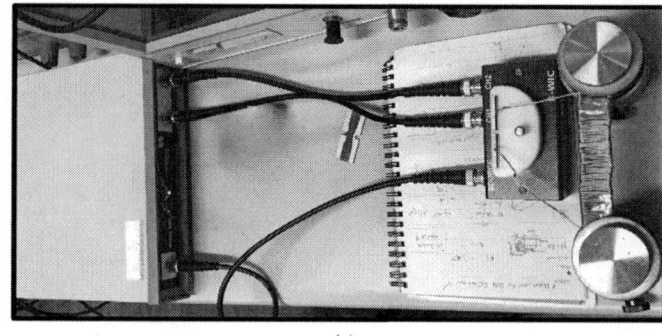

(a)

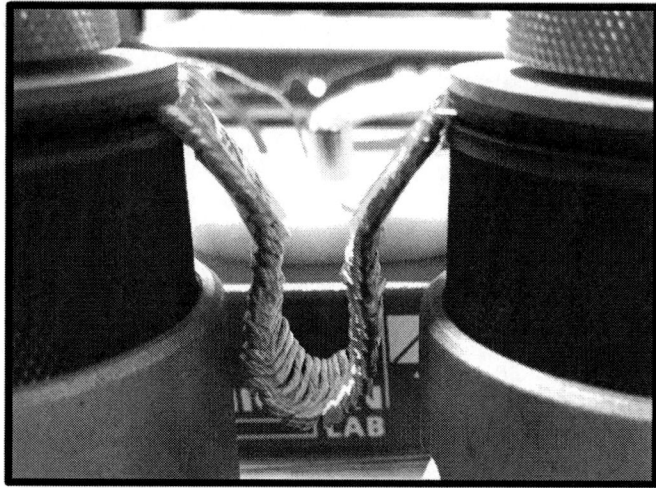

(b)

(c)

Fig. 2: (a) Experiment setup during the unfolded state, flexible inductor when (b) folded once and (c) folded twice.

integration into circuits. On the other hand, nanostructured flexible magneto-dielectric materials in combination with traditional copper coils have been successfully used in making flexible magnetic cores in [8]. This approach was validated with a 10-W boost converter, but the trade-offs between flexibility, inductance, and durability have not been fully addressed.

This paper aims to expand on this approach by using commercially-available flexible ferrite sheets as magnetic cores and provide a characterization of their performance in terms of mechanical durability and flexibility. The final goal is to design a flexible inductor in the 5 to 20 μH range that is suitable for a low-power boost converter. The size constraints of the inductor can be overcome by integrating it into the garment itself (e.g., into the strap of a watch or cuffs of a shirt).

III. Proposed Flexible Inductor Design Method

The flexible inductors used in this study were designed and constructed specifically for low-power applications, utilizing commercially available flexible ferrite sheets (38M10 Series, Fair Rite) as the magnetic core with different permeabilities and Litz wire for the windings, as shown in Fig. 1. For this study, flexible magnetic cores available in four different permeability grades (M10, M30, M40, and M50) were used. These cores have relative permeability values (μ_r) of 20, 60, 120, and 140 H/m, respectively, as shown in Table I.

To evaluate the performance of each permeability value and ensure the consistency of the results, at least two prototypes were constructed for each core permeability. Since the datasheet of the ferrite sheet does not specify the saturation flux density

(B_{sat}), a uniform value was assumed for all cores. Based on [16], a B_{sat} of 0.3 T was selected. This value represents the typical B_{sat} value for most ferrites, which provides an acceptable magnetic saturation margin under typical operating conditions. The core dimensions are specified as 9.5 cm in length, 2 cm in width, and 0.13 mm in thickness. These dimensions are chosen to ensure that the magnetic path length (l_c) of the inductor

TABLE II: Electrical Characteristics of the Prototypes.

Core Number	μ_r [H/m]	Inductance [μH] (at 200 kHz)			DCR [mΩ] (at 100 Hz)			Winding Length [cm]
		Unfolded	Folded 1x	Folded 2x	Unfolded	Folded 1x	Folded 2x	
M10 ♯1	20	3.79	3.68	3.65	110.14	109.61	109.68	8.8
M10 ♯2	20	3.87	3.82	3.82	107.51	105.66	105.38	8.9
M30 ♯1	60	7.14	6.85	6.69	117.58	117.24	118.79	8.0
M30 ♯2	60	7.38	7.08	6.89	102.08	101.67	101.35	7.8
M40 ♯1	120	10.88	10.10	9.80	109.46	109.17	109.46	8.3
M40 ♯2	120	12.05	11.32	10.82	108.58	108.52	108.09	7.5
M50 ♯1	140	17.66	16.57	15.89	102.06	101.55	98.65	8.2
M50 ♯2	140	18.04	16.83	16.02	116.13	116.69	115.33	7.8

is sufficiently long to facilitate easier manufacturability while preserving the flexibility of the core.

The Litz wire, composed of 50 strands, was hand-wound around the ferrite core. A total of 60 turns was selected for the winding, based on inductance requirements and physical space available on the 9.5-cm core strip. Insulating tape was applied at the ends of each sample to maintain winding stability and positioning throughout the testing. The insulation on the ends of the Litz wire was removed through soldering to create electrical connections for testing.

Before fabrication, key design parameters were calculated to ensure that the inductors did not saturate during operation. Using the magnetic field relationship in (1) [17], the number of turns, N, required to prevent saturation was determined by calculating the magnetomotive force (MMF) with respect to operating current, I_{op}, relative permeability, μ_r, and permeability of free space μ_0. The calculations indicated that the maximum current of 1 A would reach the saturation of the inductor if the winding number exceeded 480 for the M10 prototype, 160 for M30, 80 for M40, and 68 for M50, which confirms that the 60-turn winding will not saturate the core [17], [18]. The inductance value L was then calculated with (2) [8], which incorporates parameters such as the number of coil layers (m), cross-sectional area (A), and average wire-to-wire distance (d). For practical purposes, the distance between the wires was assumed to be approximately the thickness of the Litz wire, as the wires were wound tightly. The specifications are detailed in Table I.

$$N = \frac{MMF}{I_{op}} = \frac{B_{sat}l_c}{\mu_0\mu_r I_{op}} \tag{1}$$

$$L = \frac{mN\mu_r\mu_0 A}{d} \tag{2}$$

IV. EXPERIMENTAL RESULTS

The flexible inductors that have been made were then tested using an impedance analyzer (Bode100, Omicron Lab) to find the inductor's electrical characteristics, such as inductance and DC resistance (DCR) with the setup shown in Fig. 2a. The tests were repeated for each prototype under different folding conditions, shown in Fig. 2b and Fig. 2c, to assess the variation in inductance under different relative permeabilities and mechanical stresses. The goal of these experiments is to evaluate the performance of the inductors under conditions relevant to wearable applications.

The impedance sweeps were performed from 100 Hz to 10 MHz, and the results are summarized in Table II. However, the results should be evaluated in two parts: (a) at low frequencies to find the DCR and (b) at mid-to-high frequencies to find the inductance. This is based on the total impedance of an inductor, which is equal to $j\omega L + R$; where ω is the angular frequency, L is the inductance, and R corresponds to the DCR of the inductor. It is also important to consider the self-resonant frequency, where capacitive characteristics begin to dominate the impedance. However, in this case, no self-resonance was observed within the 10 MHz range of the sweep. Consequently, the DCR was measured at 100 Hz and the inductance was evaluated at 200 kHz, the target converter operating frequency.

The DCR of the M10, M30, M40, and M50 prototype inductors was measured at 100 Hz under various folding conditions. Theoretically, the DCR is determined by the length of the Litz wire and is expected to remain around 110 mΩ per 10 cm. Because DCR is independent of core permeability and mechanical stress, all prototype inductors are expected to exhibit similar values. The experimental results confirm that most prototypes maintain consistent DCR values, suggesting that neither the relative permeability of the core nor mechanical stress significantly influences the DCR. Among the prototypes, the M10 first prototype demonstrated the most stable performance, with DCR values of 110.14 mΩ in the unfolded state, 109.61 mΩ when folded once and 109.68 mΩ when folded twice. In contrast, the first M50 prototype exhibited the largest deviation from the theoretical value. Its DCR was 102.06 mΩ in the unfolded state, decreasing slightly to 101.55 mΩ when folded once and then dropping to 98.65 mΩ when folded twice. The M30 first prototype showed minor fluctuations, with DCR values of 117.58 mΩ in the unfolded state, 117.24 mΩ when folded once, and 118.79 mΩ when folded twice.

These results suggest that mechanical stress does not directly affect the DCR value. However, slight variations observed during testing may result from contact inconsistencies at the points where the inductor connects to the Bode 100 impedance analyzer. Overall, differences in DCR values across the prototypes are primarily attributed to hand-winding variations and minor measurement errors caused by these contact inconsistencies. The general findings of the DCR are summarized in Table II.

Next, the inductance over the frequency was calculated from

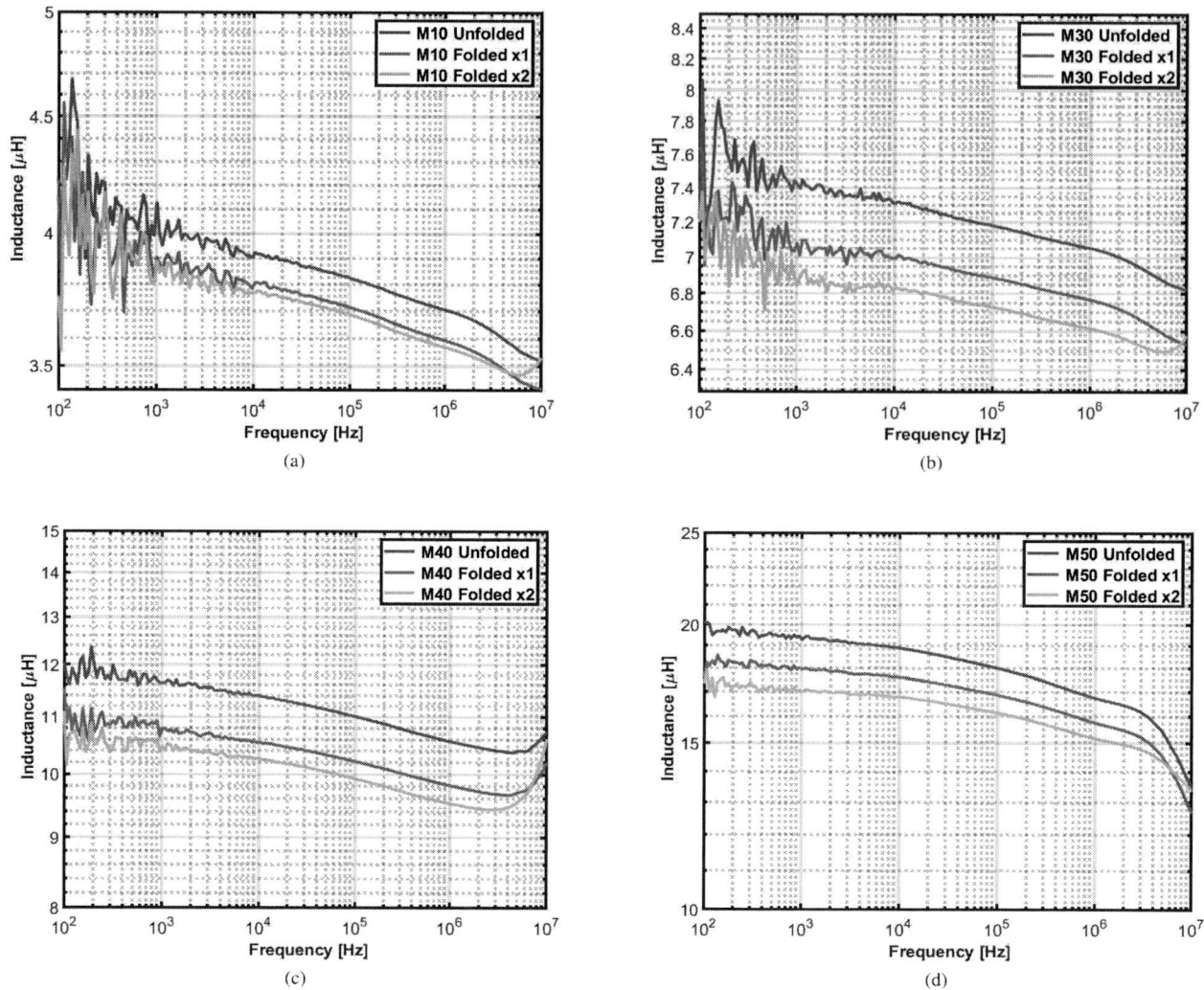

Fig. 3: Inductance vs Frequency from 100Hz to 10MHz (a) M10, (b) M30, (c) M40, and (d) M50.

an impedance sweep. The results for all M10, M30, M40 and M50 prototypes, under both folding conditions, are shown in Fig. 3a, 3b, 3c, and 3d, respectively. As mentioned, inductance measurements were taken at 200 kHz. Fig. 3 shows that inductors with higher core permeability exhibit higher inductance and lower noise at lower frequency measurements. This behavior arises because higher-permeability cores reduce the influence of frequency on inductance, leading to earlier stabilization. As shown in Fig. 3a, the M10 prototype, with the lowest core permeability, experiences significant noise in its inductance values at low frequencies, stabilizing only around 10 kHz. In comparison, Fig. 3b illustrates that the M30 prototype achieves stability at approximately 1 kHz. For cores with higher permeability, stability is observed at even lower frequencies. As depicted in Fig. 3c, the M40 prototype stabilizes around 400 Hz, while Fig. 3d shows that the M50 prototype, with the highest permeability, maintains stable inductance starting from 100 Hz. These results highlight that increasing core permeability enhances inductance stability, reducing the impact of frequency variations, and enabling consistent performance at lower frequencies.

In theory, the two prototypes with the same relative permeability should have consistent inductance values. However, slight variations are observed in the M40 and M50 prototypes between the first and second samples. For the M40 prototype in the unfolded state, the first prototype has an inductance of 10.88 μH, while the second prototype measures 12.05 μH. Similarly, for the M50 prototype, the first prototype shows an inductance of 17.66 μH, compared to 18.04 μH for the second. These variations are likely due to minor differences from hand winding the inductors, which affects the winding length and, consequently, the average wire-to-wire distance, as shown in Fig. 4. A shorter winding length reduces the wire-to-wire distance, increasing inductance (L), according to (2). These findings indicate that, while slight inductance variations occur, they are primarily a result of physical winding inconsistencies under the same core material properties. A summary of these inductance measurements is provided in Table II.

To further analyze the performance of the inductors, Fig. 5 shows the comparison between the experimentally measured

979-8-3315-1612-3/25 $31.00 © 2025 IEEE

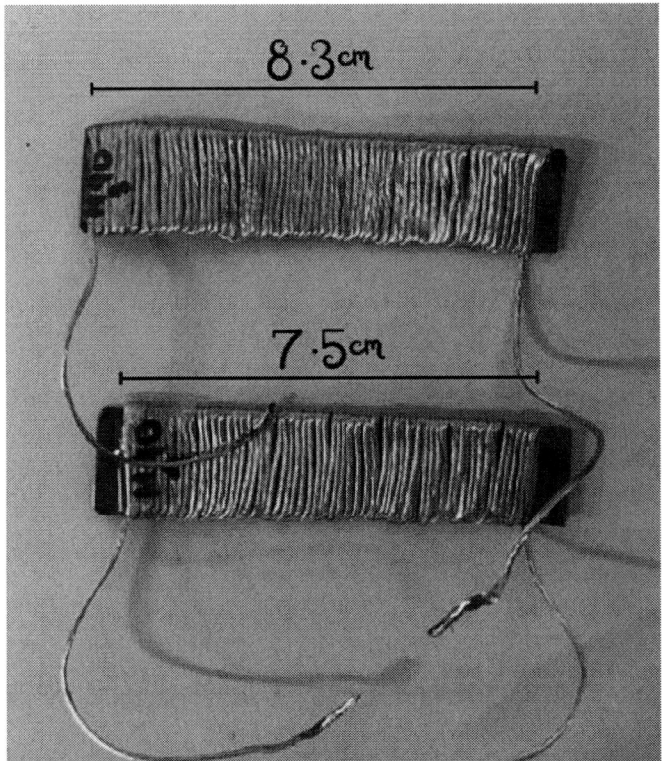

Fig. 4: Difference Length Of the Winding Under Same Permeability.

inductance with the theoretically calculated values, derived using (2). As shown in Fig. 5, there is a noticeable discrepancy between the theoretical and experimental inductance values in most prototypes. The M10 inductor, with a relative permeability of 20 H/m, exhibited values closest to the theoretical predictions, demonstrating a minimal deviation. However, for the M30 and M50 inductors, the experimental values showed approximately a 30% difference from the theoretical calculations. In particular, the M40 inductor showed the largest deviation, with an experimental inductance nearly 50% different from the calculated theoretical value. This discrepancy across prototypes is attributed to the inherent non-idealities in the hand-wound inductor fabrication process, slight inconsistencies in winding geometry and the theoretical calculation being based on solenoid inductors, which differ from the inductors made in this study. However, the specific reasons for the greater variance in the M40 inductor remain unclear and may require further investigation.

Table III summarizes the percentages of inductance retention for M10, M30, M40, and M50 inductors under varying mechanical stress conditions compared to the unfolded state condition. All prototypes demonstrate a similar trend of decreasing inductance under mechanical stress. The second M10 prototype showed the best retention, with only a 1.29% reduction in the folded-once state and 1.42% under the folded-twice condition. In contrast, the second M50 prototype experienced the most significant reduction, at 11.22%, in the folded-twice state. Notably, the first M40 prototype exhibited the largest drop in the folded-once condition, with a decrease of 6.26%, which

deviated from the trend observed in other prototypes.

These results suggest that inductors with higher core permeability tend to experience greater inductance losses under mechanical stress. Also, while slight inductance reductions occur under folding, the flexible inductors maintain relatively consistent performance even under extreme flexion. Even under twice-folded conditions, the inductors retain a high percentage of their original inductance, confirming their suitability for applications requiring mechanical flexibility.

TABLE III: Percentage of the Inductance Compare to Straight Inductor.

Core Number	μ_r [H/m]	Folded 1x [%]	Folded 2x [%]
M10 ♮1	20	97.01	96.33
M10 ♮2	20	98.71	98.58
M30 ♮1	60	95.93	93.71
M30 ♮2	60	95.84	93.31
M40 ♮1	120	92.84	90.13
M40 ♮2	120	93.97	89.78
M50 ♮1	140	93.86	89.98
M50 ♮2	140	93.26	88.78

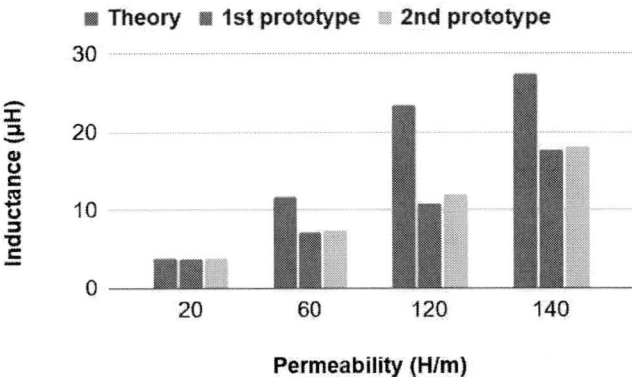

Fig. 5: Comparison of theoretical inductance with experimental results.

For further experimental evaluation, the first prototype of the flexible M40 inductor was selected, as its inductance value of approximately 10 μH closely matched the desired target for the boost converter. The inductor was connected to a power supply and a load, forming a series R-L circuit. To monitor the inductor's performance during the charging and discharging phases, an oscilloscope was employed to record the current and voltage across the inductor. The power supply was configured to supply 2 V and the load resistance was set at 2 Ω. Under these conditions, the theoretical current for the inductor was expected to peak at 1 A. However, the maximum current observed was approximately 842 mA. This discrepancy was attributed to the additional resistance present in the flexible inductor, measured at around 110 mΩ, along with a wire resistance of 0.25 Ω. Together with the 2-Ω load, this brought the total circuit resistance to approximately 2.35 Ω, aligning well with the observed 842 mA peak current. The experiments were repeated under different folding conditions, those depicted in Fig. 2b and 2c, respectively. These mechanical deformations were intended to simulate real-world conditions in which wearable and flexible electronics might encounter bending and folding. Fig. 6 presents the results of the discharging current under these stress conditions. Despite additional mechanical strain, the inductor's performance remained consistent, with negligible changes in current levels observed during discharge. This finding indicates that the inductance value and overall electrical properties of the inductor remained stable, even under multiple folding states. Specifically, the stable current profiles suggest that the magnetic and electrical properties of the inductor are effectively maintained, likely due to the inherent flexibility of the material.

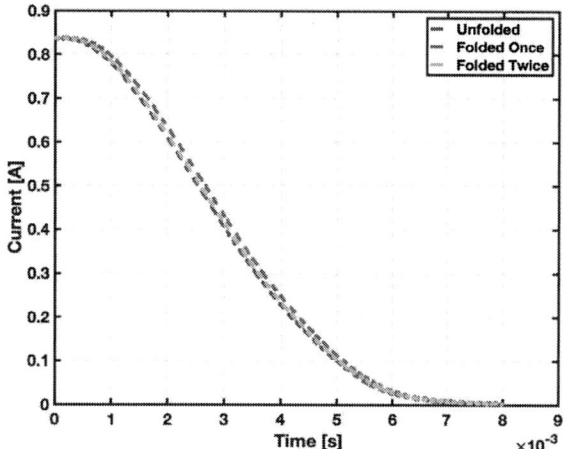

Fig. 6: Discharging inductor current under different folding conditions.

V. CONCLUSION

Fabrication of physically pliable circuit components is important for the integration of power electronics in wearables and soft robotics applications. This study demonstrated the design and evaluation of flexible inductors constructed with commercially available flexible ferrite sheets and Litz wire. The flexible ferrite cores demonstrated good tolerance to bending, and even with substantial mechanical stress (e.g., multiple folding), the inductors maintained high levels of inductance retention, decreasing down to 92.84% when folded once and to 88.78% when folded twice. In addition, the inductors exhibited stable DCR values decreasing down to 1.73% when folded once and to 3.34% when folded twice across different permeability grades (M10, M30, M40, and M50), indicating consistent behavior up to 10 MHz, well within the target operating range of 100–300 kHz. This consistency across prototypes highlights the robustness of the design despite potential hand-winding variations. The comparison between theoretical and experimental inductance revealed deviations of approximately 30% for M30 and M50 and nearly 50% for M40, indicating fabrication non-idealities. M10 showed the closest match with 0.03% error, highlighting better alignment with theoretical predictions.

The study's focus on using flexible ferrite cores with distinct permeability grades further allowed us to observe how relative permeability affects inductance value retention under mechanical stress. Moreover, practical charging and discharging tests reinforced the viability of prototypes for low-power converters. The experimental results indicate that, during various folding conditions, the current difference in the M40 flexible inductor was minimal. This research demonstrates the potential of flexible inductor designs using ferrite cores for wearable electronics and other applications that require conformable components. The prototypes' low variation in inductance value under mechanical stress, combined with their stable performance metrics, shows their promise for integration into flexible electronic systems.

ACKNOWLEDGMENT

This work was supported in part by the National Science and Technology Council (NSTC) of Taiwan under Grant 113-2628-E-002-008. The author acknowledges the use of ChatGPT as a tool to assist in refining the formality of certain sentences within this paper. All content generated or restructured using this tool was fully reviewed and revised by the authors to ensure technical accuracy and alignment with the intended meaning.

REFERENCES

[1] V. Kartsch, S. Benatti, M. Mancini, M. Magno, and L. Benini, "Smart wearable wristband for emg based gesture recognition powered by solar energy harvester," in *2018 IEEE International Symposium on Circuits and Systems (ISCAS)*, 2018, pp. 1–5.

[2] K. M. Diaz, D. J. Krupka, M. J. Chang, J. Peacock, Y. Ma, J. Goldsmith, J. E. Schwartz, and K. W. Davidson, "Fitbit®: An accurate and reliable device for wireless physical activity tracking," *International journal of cardiology*, vol. 185, pp. 138–140, 2015.

[3] J. S. Heo, J. Eom, Y.-H. Kim, and S. K. Park, "Recent progress of textile-based wearable electronics: A comprehensive review of materials, devices, and applications," *Small*, vol. 14, no. 3, p. 1703034, 2018.

[4] M. A. Butt, N. L. Kazanskiy, and S. N. Khonina, "Revolution in flexible wearable electronics for temperature and pressure monitoring—a review," *Electronics*, vol. 11, no. 5, 2022.

[5] F. S. Bagci, R. Angsetya, S. Logi, and K. A. Kim, "Flexible pcb connection methods for wearable energy harvesting applications," in *2023 IEEE Applied Power Electronics Conference and Exposition (APEC)*, 2023, pp. 1881–1887.

[6] M. Stoppa and A. Chiolerio, "Wearable electronics and smart textiles: A critical review," *Sensors*, vol. 14, no. 7, pp. 11 957–11 992, 2014.

[7] W. G. Hurley and W. H. Wölfle, *Transformers and inductors for power electronics: theory, design and applications*. John Wiley & Sons, 2013.

[8] Y. Tang, B. Lee, M. Vural, P. Kofinas, and A. Khaligh, "Toward flexible ferromagnetic-core inductors for wearable electronic converters," in *2015 IEEE Applied Power Electronics Conference and Exposition (APEC)*, 2015, pp. 2540–2545.

[9] Y. Yao, X. Qiu, Y. Cheng, J. C. C. Lo, S. W. R. Lee, W.-H. Ki, and C.-Y. Tsui, "A flexible thin-film inductor for high-efficiency wireless power transfer," *IEEE Electron Device Letters*, vol. 44, no. 3, pp. 504–507, 2023.

[10] A. R. Plamootil Mathai, T. Stalin, and P. Valvivia y Alvarado, "Flexible fiber inductive coils for soft robots and wearable devices," *IEEE Robotics and Automation Letters*, vol. 7, no. 2, pp. 5711–5718, 2022.

[11] E. Waffenschmidt, B. Ackermann, and M. Wille, "Integrated ultra thin flexible inductors for low power converters," in *2005 IEEE 36th Power Electronics Specialists Conference*, 2005, pp. 1528–1534.

[12] Y. Kim, H. Kim, and H.-J. Yoo, "Electrical characterization of screen-printed circuits on the fabric," *IEEE Transactions on Advanced Packaging*, vol. 33, no. 1, pp. 196–205, 2010.

[13] Y. Liu, M. Wang, M. Yu, B. Xia, and T. T. Ye, "Embroidered inductive strain sensor for wearable applications," in *2020 IEEE International Conference on Pervasive Computing and Communications Workshops (PerCom Workshops)*, 2020, pp. 1–6.

[14] N. Lazarus and C. D. Meyer, "Stretchable inductor with liquid magnetic core," *Materials Research Express*, vol. 3, no. 3, p. 036103, mar 2016. [Online]. Available: https://dx.doi.org/10.1088/2053-1591/3/3/036103

[15] N. Lazarus, C. D. Meyer, S. S. Bedair, G. A. Slipher, and I. M. Kierzewski, "Magnetic elastomers for stretchable inductors," *ACS Applied Materials & Interfaces*, vol. 7, no. 19, pp. 10 080–10 084, 2015, pMID: 25945395.

[16] S. Maniktala, "Chapter 2 - dc–dc converter design and magnetics," in *Switching Power Supplies A - Z (Second Edition)*, second edition ed., S. Maniktala, Ed.   Oxford: Newnes, 2012, pp. 61–121.

[17] P. T. Krein, *Elements of Power Electronics.*   New York: Oxford Univ. Press, 1998.

[18] J. Kaiser and T. Dürbaum, "An overview of saturable inductors: Applications to power supplies," *IEEE Transactions on Power Electronics*, vol. 36, no. 9, pp. 10 766–10 775, 2021.

Design of Boost Power Factor Corrector and Asymmetrical Half-Bridge Flyback Converter for USB-PD Applications

Yun-Keng Cheng[*], Tsorng-Juu Liang[*], Kai-Hui Chen[*], Ming-Chang Tsou[†]

[*]Green Energy Electronics Research Center, National Cheng Kung University, Tainan, Taiwan

[†]Leadtrend Technology Corporation, Taiwan

mailto: tjliang@mail.ncku.edu.tw

Abstract –USB power delivery (USB-PD) chargers have a wide output voltage range for various charging applications. In this paper, a two-stage converter for USB-PD is implemented to meet the requirement of harmonic standard IEC61000-3-2. The front stage is a boost power factor corrector (PFC) to improve the power factor, and the second stage is an asymmetrical half-bridge flyback (AHBF) converter which can achieve soft-switching operation for output voltage regulation. The operating principles and control methods of AHBF for achieving primary-side zero-voltage switching and secondary-side zero-current switching are presented. In addition, the key parameters for AHBF converter are optimized, and the PFC stage is disable in light load condition to improve the system efficiency. Finally, a 240 W USB-PD 3.1 experimental prototype is implemented with ac input voltage of 90 ~ 264 Vrms with the output voltage/current of USB-PD 3.1. The experimental results show that the highest efficiency of the boost PFC converter is 97.2%. The highest power factor is 0.99 at full load condition. The maximum efficiency of AHBF is 91.3%, and the overall system efficiency can reach 88.6%.

Keywords: power factor correction, asymmetrical half-bridge flyback converter, wide range, PD charger

I. INTRODUCTION

With the rapid growth of demand for consumer electronics products over the past few decades, large number of adapters with different specifications and interfaces are in demand. This trend not only causes waste of material resources and environmental pollution, but also troubles users in their daily lives. The Universal Serial Bus Implementer Forum (USB-IF) published USB Power Delivery (USB-PD) standards to define voltage and current specifications for consumer electronics products. Over the past years, with the increasing demand for fast charging and the need to accommodate more charging devices, the past USB-PD specifications have become no longer sufficient to meet these requirements. In the latest release of USB-PD 3.1 in 2021, the power range was extended and the maximum power was increased to 240 W (48 V/5 A), as shown in Table 1 [1]. For the wide output voltage/current, from 5 V/3 A to 48 V/5 A, USB-PD is more flexible and friendly for multiple application usages. As a portable device, the input terminal of the PD charger is usually an ac utility power. Thus, the rectifiers and filters are adopted for ac to dc voltage converting. The use of rectification and filter circuits causes high current harmonic components and distortion, resulting in low power factor. Low power factor not only leads to energy loss on the transmission line but also increases the burden on the power system. For the harmonic current level standard created by the European Power Supply Manufacturers Association (EPSMA), the EN 61000-3-2 standard specifies the harmonic current limit for non-lighting equipment with power consumption above 75 W [2]. Under the expansion of power specifications in USB-PD, a power factor correction (PFC) circuit is required in order to comply with current harmonic limitations [3]-[5].

Table 1 USB-PD 3.1 specifications.

Power Range	Available Voltage	Available Current	Maximum Power
Standard Power Range	5 V	3 A	15 W
Standard Power Range	9 V	3 A	27 W
Standard Power Range	15 V	3 A	45 W
Standard Power Range	20 V	5 A	100 W
Extended Power Range	28 V	5 A	140 W
Extended Power Range	36 V	5 A	180 W
Extended Power Range	48 V	5 A	240 W

Discussions regarding the two-stage converters that incorporate the PFC stage for PD 3.1 expanded power specifications are essential. Firstly, some background research and control methods regarding PFC will be addressed [6]-[13], followed by an introduction to the topology of isolated dc-dc converters [14]-[17]. Various topologies based on flyback converters have been introduced for increasing output power and efficiency. The quasi-resonant flyback converter, active clamped flyback converter and resonant-forward-flyback converter have all been revealed. In recent years, the AHBF converter has been invented and studied [18], [19]. Compared to traditional flyback converters, the AHBF converter is characterized by lower voltage stress across the switches, which equals the input voltage. It retains the merits of the flyback converter in using a magnetizing inductor to store energy and using the leakage inductor of the transformer as a resonant inductor. This allows it to have the characteristic of a resonant converter and increase power density with fewer circuit components. Additionally, the AHBF can achieve ZVS on the primary-side switches and ZCS on the secondary-side output diode. It is controlled by PWM, making it suitable for wide range applications. These advantages have made the AHBF converter highly regarded in recent research on low-power applications. Nevertheless, there is still a lack of comprehensive discussion regarding the control method for the AHBF converter. Before discussing the control methods for AHBF, steady-state analyses and characteristics of AHBF will be explained.

II. STEADY STATE ANALYSES OF ASYMMETRICAL HALF-BRIDGE FLYBACK CONVERTER

The circuit topology and key waveforms of the AHBF converter with ZCS control are shown in Fig. 1 (a) and (b).

979-8-3315-1612-3/25 $31.00 © 2025 IEEE

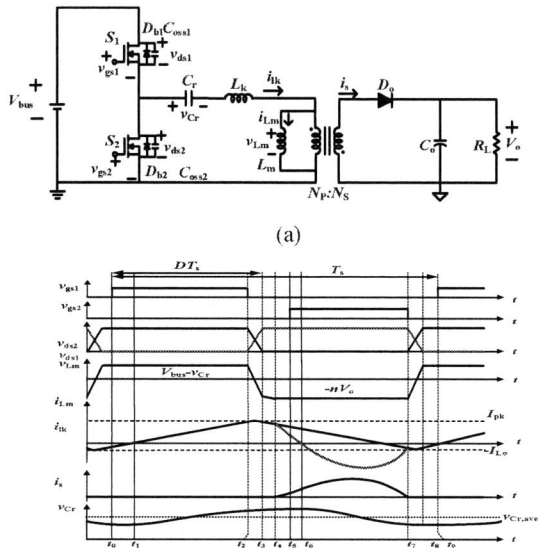

(a)

(b)

Fig. 1 AHBF converter with ZCS control. (a) Circuit diagram. (b) Key waveforms.

Fig. 2 illustrates the simplified waveforms for steady-state analyses of the AHBF converter under BCM operation with ZCS control. The transient periods are ignored and the BCM operation of AHBF converter is defined by a continuous voltage across L_m. Normally, the effect of L_k during S_1 on period can be ignored because it is much smaller than L_m, and the resonance frequency of C_r and L_m is significantly lower than the S_1 on period. Therefore, the voltage across C_r is analyzed as a constant value V_{Cr} as expressed in (1).

$$V_{Cr} = \frac{1}{T_s}\int_0^{T_s} v_{Cr}(t)dt = \frac{1}{T_s}\left[\int_0^{DT_s}(V_{bus} - v_{Lm})dt + \int_{DT_s}^{T_s} v_{Lm}dt\right]$$

$$= \frac{1}{T_s}\int_0^{DT_s} V_{bus}dt = DV_{bus} \qquad (1)$$

By applying volt-second balance of magnetizing inductor L_m, (2) can be obtained:

$$\int_0^{DT_s}(V_{bus} - V_{Cr})dt + \int_{DT_s}^{T_s}(-nV_o)dt = 0 \qquad (2)$$

From (2), (3) is obtained.

$$(V_{bus} - V_{Cr})DT_s + (-nV_o)(1 - D)T_s = 0 \qquad (3)$$

From (3), the voltage gain is derived as (4).

$$\frac{V_o}{V_{bus}} = \frac{D}{n} \qquad (4)$$

In the steady-state condition, the secondary-side current i_s is the difference between the magnetizing current and the leakage inductor current on the primary side, multiplied by the turns ratio, as expressed in (5):

$$\frac{1}{T_s}\int_0^{T_s} i_s\, dt = \frac{n}{T_s}\int_{DT_s}^{T_s}(i_{Lm} - i_{lk})dt = \frac{1}{T_s}\int_0^{T_s}(i_{Co} + i_R)dt \qquad (5)$$

According to the charge balance of C_o, the average current flowing through C_o is zero, therefore (5) can be modified as (6):

$$\frac{n}{T_s}\int_{DT_s}^{T_s}(i_{Lm} - i_{lk})dt = \frac{1}{T_s}\int_0^{T_s} i_R\, dt = I_o \qquad (6)$$

As i_{lk} equals to i_{Lm} during DT_s, (6) can be modified as (7).

$$\frac{n}{T_s}\int_{DT_s}^{T_s}(i_{Lm} - i_{lk})dt = \frac{n}{T_s}\int_0^{T_s}(i_{Lm} - i_{lk})dt = \frac{n}{T_s}\int_0^{T_s} i_{Lm}\, dt -$$

$$\frac{n}{T_s}\int_0^{T_s} i_{lk}\, dt = I_o \qquad (7)$$

Because ilk also flows through resonant capacitor C_r, it therefore produces zero average current during steady-state operation. As such, (7) can be reformulated into (8):

$$\frac{n}{T_s}\int_0^{T_s} i_{Lm}\, dt = nI_{Lm,ave} = I_o \qquad (8)$$

This reveals that output current I_o is proportional to the average value of magnetizing inductor current i_{Lm}. According to Fig. 2, the expressions for the peak and valley of magnetizing inductor currents ($I_{Lm,Pk}$, $I_{Lm,Lo}$) can be expressed as (9) and (10), respectively:

$$I_{Lm,Pk} = I_{Lm,avg} + \frac{\Delta I_{Lm}}{2} = \frac{I_o}{n} + \frac{nV_o}{2L_m}\cdot(1 - D)T_s \qquad (9)$$

$$-I_{Lm,Lo} = I_{Lm,avg} - \frac{\Delta I_{Lm}}{2} = \frac{I_o}{n} - \frac{nV_o}{2L_m}\cdot(1 - D)T_s \qquad (10)$$

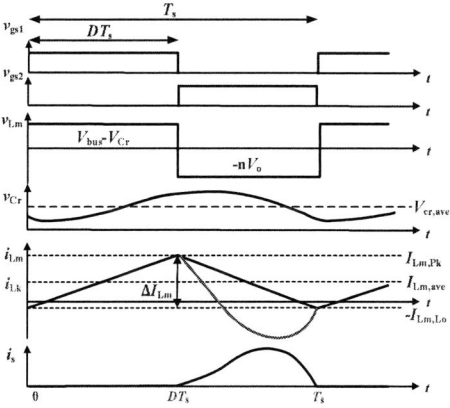

Fig. 2 Simplified waveforms of AHBF converter.

III. ZCS CONTROL FOR AHBF CONVERTER

The ZCS control of the AHBF converter is achieved by turning S_2 off when the secondary-side output diode current becomes zero, it removes the unnecessary circulating current interval. As shown in Fig. 3, when the load gradually decreases, the time point when the current reaches zero advances, and the switching frequency of the converter gradually increases.

Fig. 4 shows the risk issue of shoot-through under specific load. As the output current decreases, $I_{Lm,Lo}$ will also decrease vertically; this avoids the risk of shoot-through and helps achieve ZVS for S_1. However, when the load current decreases, the energy releasing time of C_r and L_m are mismatched due to the decrease in the charging current of capacitor C_r; this slows down the rising of v_{Cr} and causes v_{Cr} to be lower than nV_o when v_{gs2} becomes high. When v_{gs2} becomes high, the transformer remains decoupled, and C_r continues to charge until v_{cr} exceeds nV_o. This delays the coupling time of the transformer and results in the occurrence of a circulating energy interval. The energy circulates on the primary side and affect the efficiency performance.

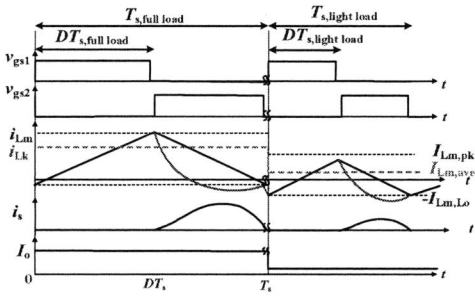

Fig. 3 Waveforms of AHBF with ZCS control under various loads.

979-8-3315-1612-3/25 $31.00 © 2025 IEEE

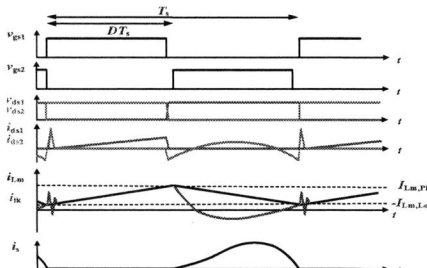

Fig. 4 Half-bridge shoot-through issue of AHBF with ZCS control.

As shown in Fig. 2, the peaks and valleys of the magnetizing inductor current are not symmetrical. $|I_{Lm,pk}|$ is usually much larger than $|-I_{Lm,Lo}|$, this demonstrates that when v_{gs1} becomes low, there is sufficient energy to discharge the C_{oss2} of S_2 within a short dead time. On the contrary, a longer dead time is required to discharge C_{oss1} of S_1 when v_{gs2} becomes low. The ZVS conditions of S_1 and S_2 are described in (11) and (12):

$$|I_{Lm,pk}| \geq \frac{2C_{oss} \cdot V_{bus}}{t_{d1}} \tag{11}$$

$$|-I_{Lm,Lo}| \geq \frac{2C_{oss} \cdot V_{bus}}{t_{d2}} \tag{12}$$

By using equations (10) and (12), the value of magnetizing inductance can be determined to achieve ZVS while operating the converter in BCM, as derived in (13):

$$L_{m,BCM} = \frac{n^2 \cdot V_o(1-D)}{2 \cdot (nI_{Lm,Lo}+I_o) \cdot f_s} \tag{13}$$

Additionally, considering the energy stored in the magnetizing inductor, as shown in equation (14).

$$\frac{1}{2}L_m \cdot I_{Lm,Lo}^2 \geq \frac{1}{2} \cdot 2 \cdot C_{oss} \cdot V_{bus}^2 \tag{14}$$

To ensure that the magnetizing current rises linearly and steadily during the on time of S_1, the resonant period T_{r1} of C_r, L_k and L_m should at least eight times larger than $D_{max}T_s$ to ensure a linear rising of i_{Lm} during $D_{max}T_s$, as expressed in (15) and (16) respectively.

$$T_{r1} = 8 \cdot D_{max}T_s = 2\pi\sqrt{(L_m+L_k) \cdot C_r} \tag{15}$$

$$f_{r1} = \frac{1}{T_1} = \frac{1}{2\pi\sqrt{(L_m+L_k) \cdot C_r}} \tag{16}$$

From (15), the minimum capacitance of C_r can be calculated as (17):

$$C_{r,ZCS} = \frac{((1-D_{max})T_s)^2}{4 \cdot \pi^2 \cdot L_k} \tag{17}$$

IV. ZVS CONTROL FOR AHBF CONVERTER

The key waveforms of AHBF with ZVS control are shown in Fig. 5. Compared to ZCS control, the operating mode is almost the same. After $i_{lk} = i_{Lm}$ ($t_7 \sim t_8$), there is an additional interval of transformer decoupling. During this interval, the valley of i_{Lm} is detected to ensure the achievement of ZVS operation for S_1. To simplify the analyses, the repeated operating mode analysis will not be discussed again.

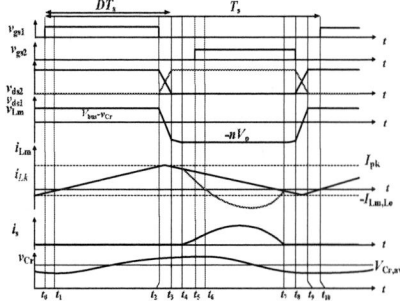

Fig. 5 Key waveforms of AHBF with ZVS control.

As shown in the Fig. 6, when the output current decreases, the switching frequency increases to maintain the same negative current $-I_{Lm,Lo}$ to ensure the ZVS operation, compared with fixed switching frequency control, the ZVS control reduces the proportion of circulating current on the primary side while achieving zero voltage switching. From (10) and (12), if $|-I_{Lm,Lo}|$ is fixed, the function of frequency and output current can be rewritten as (18).

$$f_{s,ZVS}(I_o) = \frac{nV_o(1-D)}{2L_m(\frac{I_o}{n} - \frac{2C_{oss} \cdot V_{bus}}{t_{d2}})} \tag{18}$$

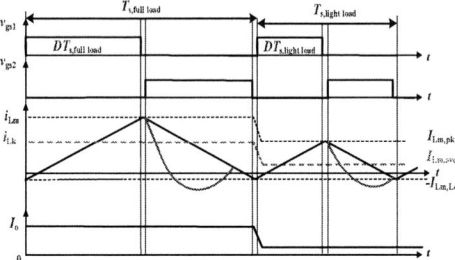

Fig. 6 Waveforms of AHBF with ZVS control under various loads.

From (18), the switching frequency corresponding to output current under different output voltages is illustrated in Fig. 7. To maintain the same negative current valley $-I_{Lm,Lo}$, the switching frequency increases as the load current decreases. It can also be observed that significant frequency variations and high switching frequency under light load conditions make it unsuitable for light load control.

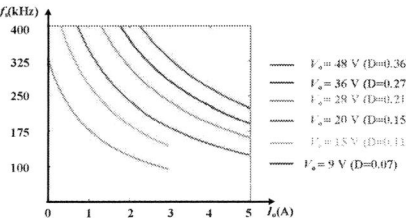

Fig. 7 Relationship between frequency and output current of the AHBF at different output voltages.

With the ZVS control method, it is crucial to ensure that the current valley $I_{Lm,Lo}$ can be accurately detected even with the load changes. Otherwise, incorrect current values will be detected, leading to incorrect switching frequency. Therefore, the resonant interval of the leakage inductance and resonant capacitor should be designed to be shorter than the conduction time of S_2 during full load condition. Additionally, as shown in Fig. 8, when the varying duty cycle causes the moment where i_{lk} and i_{Lm} are equal to no longer be fixed, the moment when i_{Lm} and i_{lk} have equal currents will occur earlier. However, to meet the ZVS condition for the next cycle, S_2 remains conducting, causing the circulating current on the primary side and it may cause multiple resonant cycles of L_k and C_r, this multiple resonant cycle not only causes the secondary side diode to lose ZCS but also distorts the current detection on the primary side.

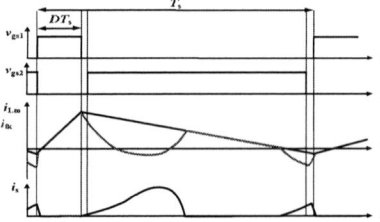

Fig. 8 Multiple resonance cycle waveforms of AHBF converter.

V. SPECIFICATION AND KEY PARAMETERS DESIGN

The system specifications of the built experimental prototype for USB-PD application are shown in Table 2. The input voltage of the boost converter is universal ac voltage 90~264 V_{rms}. The output modes of the USB-PD charger are 5 V/ 3 A, 9 V/ 3 A, 15 V/ 3 A, 20 V/ 5 A, 28 V/ 5 A, 36 V/ 5 A and 48 V/ 5 A. The rated output power $P_{o,max}$ is 240 W. The rated output power of the first stage boost PFC converter is set to 280 W with consideration of conversion efficiency. To further enhance light-load efficiency, the front-end PFC stage is disable during light-load operation. According to IEC 61000-3-2 [2], the output power boundary is considered as 65 W for shutting down the PFC stage, as illustrated in Fig. 9.

The shutdown of the PFC stage will affect the duty cycle of the downstream converter. The Table 3 illustrates the variation in duty cycle of the downstream converter when the PFC stage is disabled. In the case of converting low input line voltage to high output voltage, excessively high duty cycle in the downstream converter will result in a failure to achieve voltage conversion. Under this condition, the PFC stage cannot be disable. Furthermore, when low output voltages such as 5 V is required, high input line voltage corresponds to extremely short duty cycles. In such cases, a burst mode with a fixed minimum on-time is employed to achieve low voltage conversion.

Table 2 Specifications of USB PD charger system

Boost PFC	
Circuit Parameters	**Value**
Input rms voltage(v_{ac})	90 ~ 264 V
Bus voltage (V_{bus})	400 V
Output power (P_{bus})	280 W
Input line frequency (f_{line})	60. Hz
PFC switching frequency ($f_{PFC,min}$)	100 kHz
AHBF converter	
Output mode (V_o/ I_o)	5 V/3 A, 9 V/3 A, 15 V/3 A, 20 V/5 A, 28 V/5 A 36 V/5 A, 48V/5A
Rated output power ($P_{o,max}$)	240 W
Full load switching frequency	200 kHz

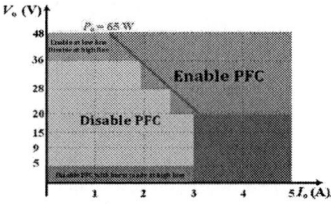

Fig. 9 PFC ON/OFF condition for PD charger.

Table 3 Duty variation when PFC stage is shut down.

V_{in} (v_{ac}) / V_o (V)	110 V_{rms}	220 V_{rms}
48	>1	0.51
36	0.77	0.39
28	0.6	0.3
20	0.43	0.21
15	0.32	0.16
9	0.19	0.1
5	0.11	0.05

Experimental Results and Discussions of Boost PFC Converter

The boost PFC circuit controlled by CCM PFC controller IC ICE2PCS01 is shown in Fig. 10. The controller is operated with fixed frequency, and variable duty control. With internal current averaging circuit and nonlinear gain block, ICE2PCS01 achieves the power factor correction in CCM without input sinusoidal reference sensing. To optimize efficiency and duty cycle of the downstream converter, the PFC stage is turned off when the output power is below 65 W, except for the scenario where the output voltage is required 48 V and the input voltage is at the low line voltage. The key parameters for PFC stage are shown in Table 4.

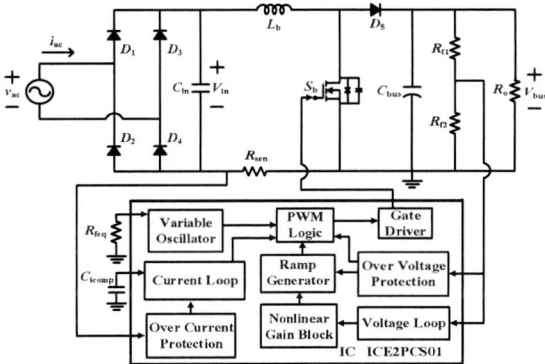

Fig. 10 Boost power factor correction circuit with ICE2PCS01.

Table 4 Key parameters for PFC stage.

Parameters	**Value**
Input capacitance (C_{in})	680 nF
Inductance (L_b)	470 µH
Power MOSFET (S_b)	IPA65R225C7XKSA1 (V_{DSS}= 650 V, I_D= 7 A)
Diode (D_b)	MUR260 (V_R= 600 V, I_F= 2 A)
Bus capacitance (C_{bus})	100 µF

Experimental Results and Discussions of Boost PFC Converter

The measured waveforms of v_{ac}, i_{ac}, i_{Lb} and v_{gsb} with full load and V_{bus}= 400 V under v_{ac}= 90 V_{rms}, v_{ac}= 110 V_{rms}, v_{ac}= 220 V_{rms}, v_{ac}= 264 V_{rms} are showns in Fig. 11 (a), (b), (c), and (d), respectively. From the measured waveforms, the input currents are in phase sinusoidally-with input voltage, achieving high power factor.

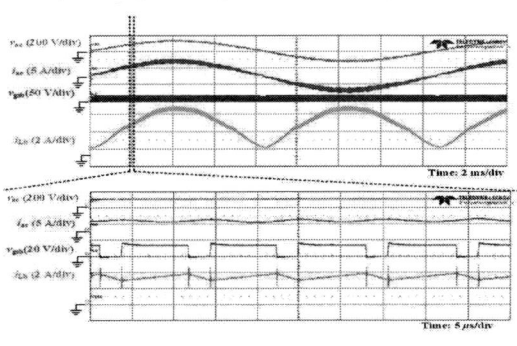

(a)

979-8-3315-1612-3/25 $31.00 © 2025 IEEE

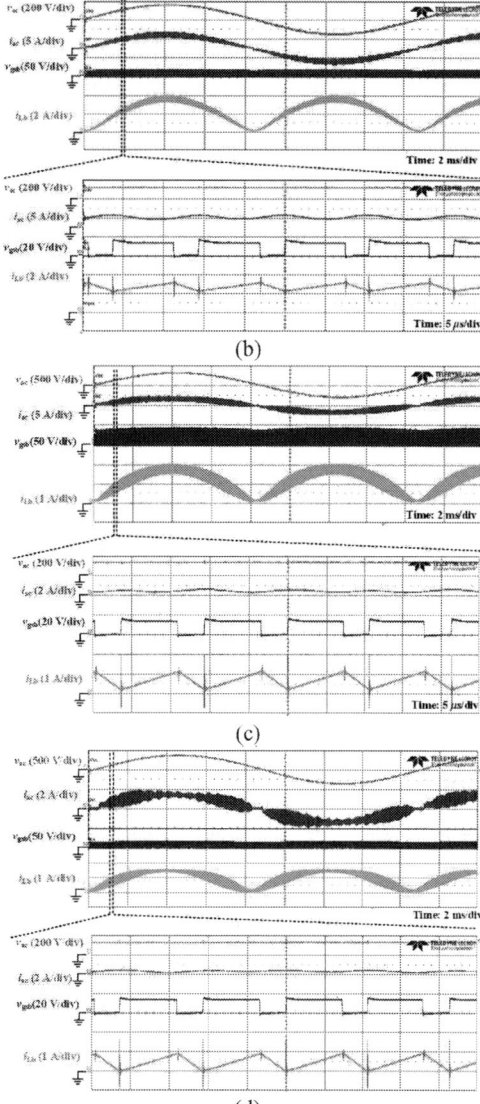

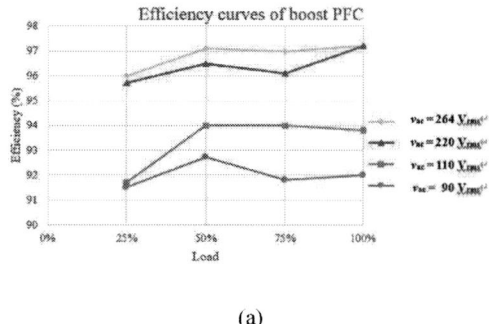

(d)

Fig. 11 Measured waveforms of v_{ac}, i_{ac}, v_{gsb} and i_{Lb} in PFC under (a) v_{ac}= 90 V_{rms} (b) v_{ac}= 110 V_{rms} (c) v_{ac}= 220 V_{rms} (d) v_{ac}= 264 V_{rms}, V_{bus}= 400 V, I_{bus}= 0.675 A.

Fig. 12 (a) and Fig. 22 (b) show the efficiency and the power factor curves of the boost PFC under different input ac voltage. The maximum power factor is 0.997 and the maximum efficiency is 97.2% under v_{ac} = 264 V_{rms}.

(a)

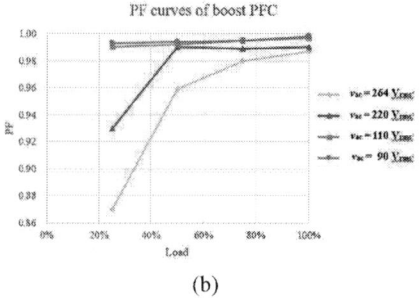

(b)

Fig. 12 (a) Measured efficiency and (b) Measured power factor of boost PFC under different vac.

Experimental Results and Discussions of AHBF Converter

The specifications of AHBF are shown in Table 2, and the control scheme of the AHBF converter, which utilizes the digital signal controller dsPIC33CK256MP506, is presented in Fig. 13. By using voltage mode control, feedback from the output voltage is utilized to modulate the duty cycle. To regulate the output voltage and minimize the circulating energy, S_2 is turned off precisely when the secondary side current reaches zero. Additionally, the PFC stage is turned off when the output power is lower than 65 W.

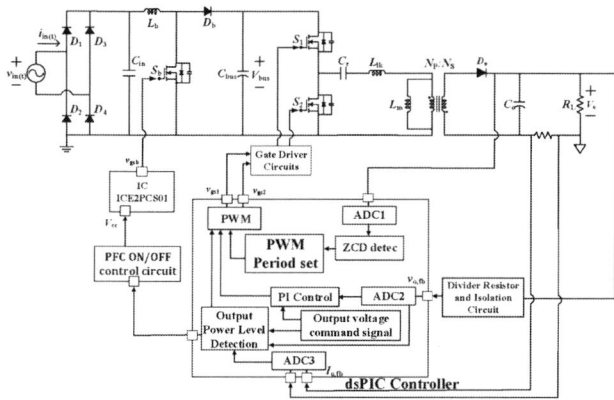

Fig. 13 Control scheme of AHBF converter.

Transformer turns ratio selection n

The design of the turns ratio will not only affect the maximum and minimum duty, but also affect the rms value of the primary-side current and the voltage stress of the output diode. From the voltage conversion ratio (4), by defining the maximum duty cycle, D_{max}, as 0.4, the turns ratio n is equal to 3.33.

Magnetizing inductance design L_m

To ensure the ZVS operation for switch S_1 and S_2, during the dead time, the absolute value of the inductor current must be large enough. Considering the operating frequency, rising time, and falling time of the component, dead times t_{d1} and t_{d2} are set to 150 ns and 300 ns, respectively. Therefore, the minimum $I_{Lm,pk}$ and $I_{Lm,Lo}$ for ZVS can be calculated from (11) and (12),as (19) and (20).

$$|I_{Lm,pk}| \geq \frac{2 \cdot C_{oss} \cdot V_{bus}}{t_{d1}} = \frac{2 \cdot 350 \cdot 10^{-12} \cdot 400}{150 \cdot 10^{-9}} = 1.86 \text{ (A)} \quad (19)$$

$$|-I_{Lm,Lo}| \geq \frac{2 \cdot C_{oss} \cdot V_{bus}}{t_{d2}} = \frac{2 \cdot 350 \cdot 10^{-12} \cdot 400}{300 \cdot 10^{-9}} = 0.93 \text{ (A)} \quad (20)$$

From (13), the magnetizing inductance L_m can be calculated as follows (21):

$$L_{m,BCM} = \frac{n^2 \cdot V_o(1-D_{max})}{2 \cdot (nI_{Lm,Lo}+I_o) \cdot f_{s,FL}} = \frac{3.33^2 \cdot 48 \cdot (1-0.4)}{2.(3.33 \cdot 0.93+5)200 \cdot 10^3} = 98.7 \ \mu H \quad (21)$$

Where the switching frequency under full load $f_{s,FL}$ is set at 200 kHz when $V_o = 48$ V, and L_m is selected as 95 μH to ensure the ZVS condition for the power switches S_1 and S_2. And the L_k is 1.23 μH, which is 1.3% of L_m.

Resonant capacitance design C_r:

When the maximum duty and magnetizing inductance are determined, the resonant capacitance can be calculated from (15) and (17) to achieve ZCS of output diode as (22) and (23):

$$C_{r,min} = \frac{(8 \cdot D_{max}T_s)^2}{4 \cdot \pi^2 \cdot (L_m + L_k)} = \frac{(16 \cdot 10^{-6})^2}{4 \cdot \pi^2 \cdot (85 + 1.23) \cdot 10^{-6}} = 67.3 \text{ nF} \quad (22)$$

$$C_{r,ZCS} = \frac{((1 - D_{max})T_s)^2}{4 \cdot \pi^2 \cdot L_k} = \frac{(3 \cdot 10^{-6})^2}{4 \cdot \pi^2 \cdot 1.23 \cdot 10^{-6}} = 185 \text{ nF} \quad (23)$$

Finally, based on simulation and experimental results, C_r is chosen as 200 nF, and the key parameters are listed in Table 5.

The following figures, Fig. 14 to Fig. 31, show the waveforms of AHBF converter. The ZCS control is adopted to optimize the efficiency when the output voltage exceeds 20 V. According to PFC on/off conditions shown in Fig. 9, starting from the highest output voltage of 48 V to the lowest output voltage 5 V. The waveforms for PFC enabled and disabled will be presented separately, and under PFC disabled, the waveforms will be further categorized into high input line voltage and low input line voltage for presentation.

Table 5 Key parameters of AHBF converter.

Parameters	Value
Turns ratio ($N_p : N_s$)	10:3
Magnetizing inductance (L_m)	95 μH
Leakage inductance (L_k)	1.2 μH
Resonant capacitance (C_r)	200 nF
Power MOSFET (S_1, S_2)	IPA65R225C7 ($V_{DSS} = 600$ V, $I_D = 9$ A, $R_{ds,on} = 0.225$ Ω)
Output Diode (D_o)	PFR20L200CTF ($V_R = 200$ V, $I_F = 20$ A)
Output capacitance (C_o)	220 μF

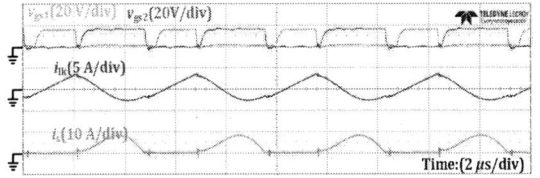

Fig. 14 Measured AHBF converter waveforms of v_{gs1}, v_{gs2}, i_{lk} and i_s under $v_{ac} = 110$ V and $V_o = 48$ V, $I_o = 1.25$ A, $f_s = 215$kHz.

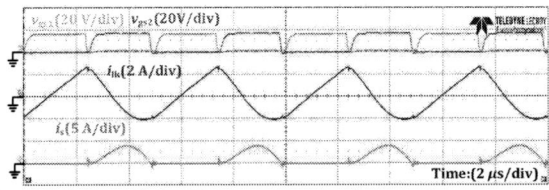

Fig. 15 Measured AHBF converter waveforms of v_{gs1}, v_{gs2}, i_{lk} and i_s under $v_{ac} = 220$ V and $V_o = 48$ V, $I_o = 1.25$ A, $f_s = 200$ kHz.

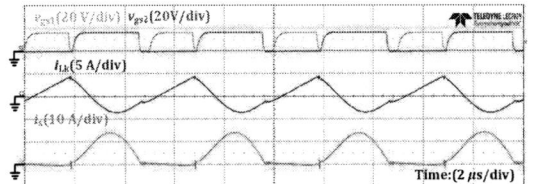

Fig. 16 Measured AHBF converter waveforms of v_{gs1}, v_{gs2}, i_{lk} and i_s under $V_{bus} = 400$ V and $V_o = 48$ V, $I_o = 5$ A, $f_s = 200$ kHz.

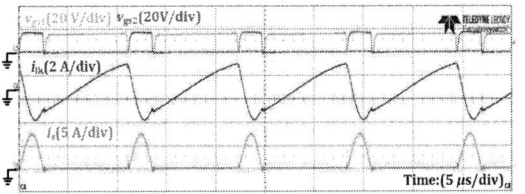

Fig. 17 Measured AHBF converter waveforms of v_{gs1}, v_{gs2}, i_{lk} and i_s under $v_{ac} = 110$ V$_{rms}$ and $V_o = 36$ V, $I_o = 1.25$ A, $f_s = 90$ kHz.

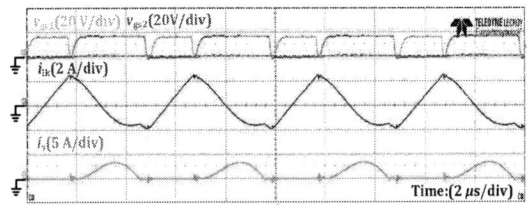

Fig. 18 Measured AHBF converter waveforms of v_{gs1}, v_{gs2}, i_{lk} and i_s under $v_{ac} = 220$ V$_{rms}$ and $V_o = 36$ V, $I_o = 1.25$ A, $f_s = 200$ kHz.

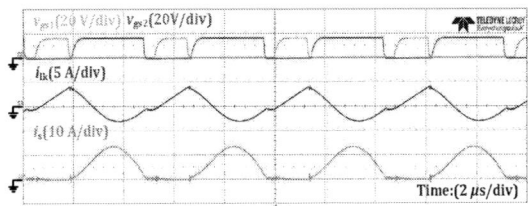

Fig. 19 Measured AHBF converter waveforms of v_{gs1}, v_{gs2}, i_{lk} and i_s under $V_{bus} = 400$ V and $V_o = 36$ V, $I_o = 5$ A, $f_s = 208$ kHz.

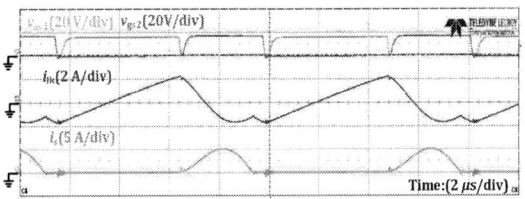

Fig. 20 Measured AHBF converter waveforms of v_{gs1}, v_{gs2}, i_{lk} and i_s under $v_{ac} = 110$ V$_{rms}$, $V_o = 28$ V, $I_o = 1.25$ A, and $f_s = 120$ kHz.

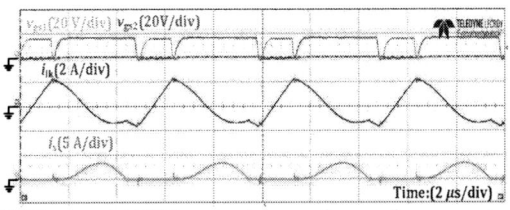

Fig. 21 Measured AHBF converter waveforms of v_{gs1}, v_{gs2}, i_{lk} and i_s under $v_{ac} = 220$ V$_{rms}$, $V_o = 28$ V, $I_o = 1.25$ A, and $f_s = 200$ kHz.

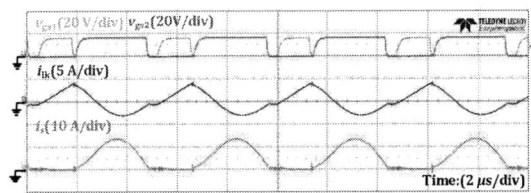

Fig. 22 Measured AHBF converter waveforms of v_{gs1}, v_{gs2}, i_{lk} and i_s under $V_{bus} = 400$ V$_{rms}$ and $V_o = 28$ V, $I_o = 5$ A, $f_s = 215$ kHz.

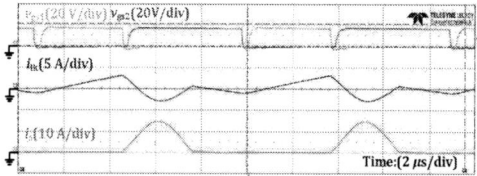

Fig. 23 Measured AHBF converter waveforms of v_{gs1}, v_{gs2}, i_{lk} and i_s under v_{ac} = 110 V$_{rms}$, V_o = 20 V, I_o = 2.5 A, f_s = 110 kHz.

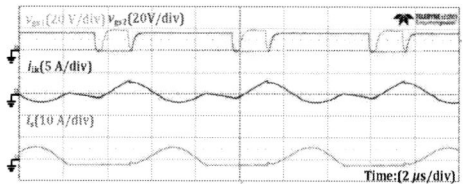

Fig. 24 Measured AHBF converter waveforms of v_{gs1}, v_{gs2}, i_{lk} and i_s under v_{ac} = 220 V$_{rms}$, V_o = 20 V, I_o = 2.5 A, f_s = 155 kHz.

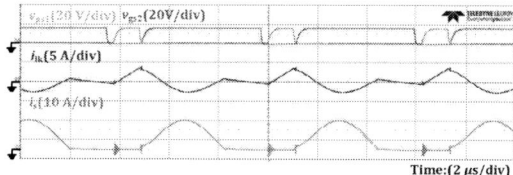

Fig. 25 Measured AHBF converter waveforms of v_{gs1}, v_{gs2}, i_{lk} and i_s under V_{bus} = 400 V and V_o = 20 V, I_o = 5 A, f_s = 160 kHz.

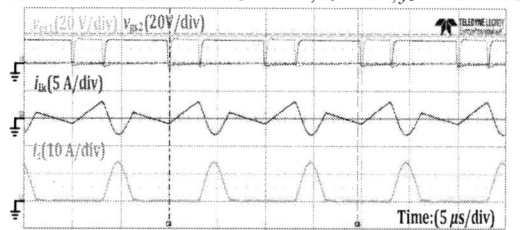

Fig. 26 Measured AHBF converter waveforms of v_{gs1}, v_{gs2}, i_{lk} and i_s under v_{ac} = 110 V$_{rms}$ and V_o = 15 V. I_o = 3 A, f_s = 100 kHz.

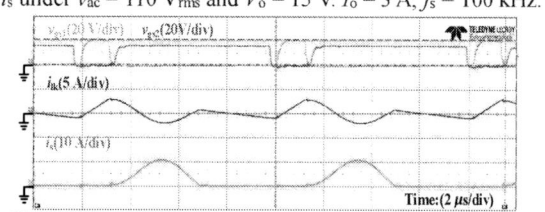

Fig. 27 Measured AHBF converter waveforms of v_{gs1}, v_{gs2}, i_{lk} and i_s under v_{ac} = 220 V$_{rms}$ and V_o = 15 V, I_o = 3 A, f_s = 123 kHz.

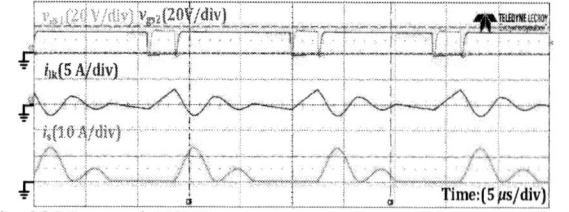

Fig. 28 Measured AHBF converter waveforms of v_{gs1}, v_{gs2}, i_{lk} and i_s under v_{ac} = 110 V$_{rms}$ and V_o = 9 V. I_o = 3 A, f_s = 72 kHz.

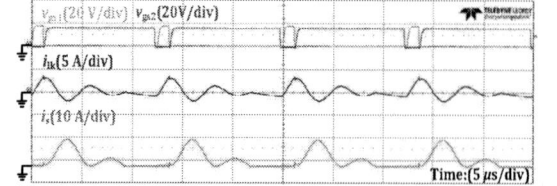

Fig. 29 Measured AHBF converter waveforms of v_{gs1}, v_{gs2}, i_{lk} and i_s under v_{ac} = 220 V$_{rms}$ and V_o = 9 V, I_o = 3 A, f_s = 80 kHz.

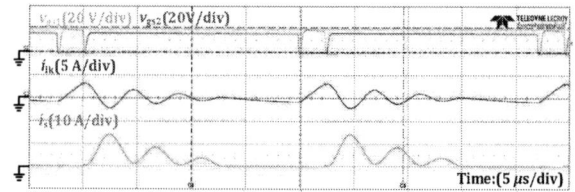

Fig. 30 Measured AHBF converter waveforms of v_{gs1}, v_{gs2}, i_{lk} and i_s under v_{ac} = 110 V$_{rms}$ and V_o = 5 V, I_o = 3 A, f_s = 45 kHz.

When the V_o is above 20 V, ZCS control is used to reduce circulating energy losses. As the duty cycle decreases, $|-I_{Lm,Lo}|$ reduces gradually. To prevent shoot-through issue at low output voltages, maximum frequency clamping operation is employed, moreover, as the duty cycle decreases, as previously analyzed, multiple resonant cycles will appear. When the 5 V output voltage is required, the burst mode operation is used under v_{ac} = 220 V$_{rms}$. With consideration for the audible frequency range of the human ear, the minimum switching frequency is set at 25 kHz, as shown in the Fig. 31.

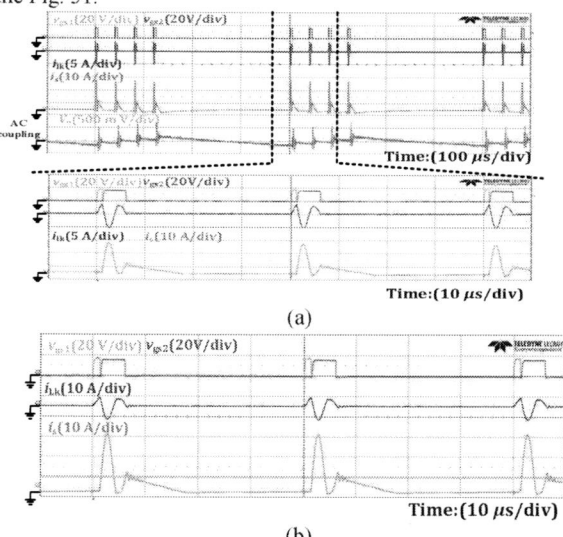

Fig. 31 Measured AHBF converter waveforms of v_{gs1}, v_{gs2}, i_{lk} and i_s under v_{ac} = 220 V$_{rms}$ and V_o = 5 V. (a) I_o = 0.75 A, f_s = 25 kHz with burst mode control. (b) I_o = 3 A, f_s = 25 kHz.

Fig. 32 (a) and Fig. 32 (b) show the measured efficiency of AHBF converter with different output voltage under low line and high line voltage, respectively. The shutdown of PFC stage are indicated by dashed lines. The maximum efficiency is 91.3 % when V_o = 48 V under full load condition. As the output current decreases, the efficiency of the AHBF stage exhibits a decreasing trend due to the high switching losses and the increased proportion of circulating energy losses caused by frequency clamping during light load conditions. Additionally, when the front stage is turned off during light load, the different V_{bus} cause changes in the duty cycle and operating switching frequency of the second stage, resulting in a sudden efficiency improvement. This trend is particularly evident at low line voltages because the lower input voltage reduces switching losses.

Fig. 33 (a) and (b) show the system efficiency of two-stage converter. The efficiency curves of the two-stage converter are primarily influenced by the secondary AHBF converter. A maximum efficiency of 88.6% is achieved at full loas condition. Due to the efficiency characteristics of the PFC stage, the two-stage system exhibits better heavy load efficiency under high-line voltage. Conversely, when the PFC stage is shut down under light load conditions, the efficiency of the secondary stage is affected by the decrease in V_{bus} voltage.

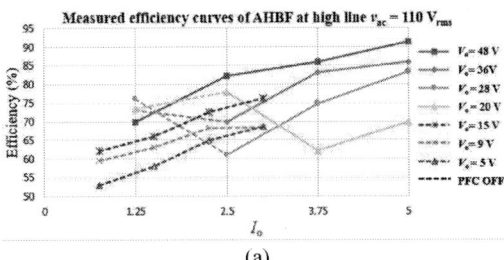

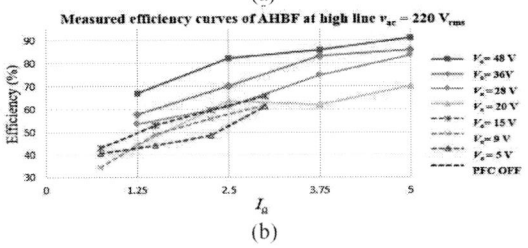

Fig. 32 Measured efficiency of AHBF converter with different output voltage (a) Under low line. (b) Under high line.

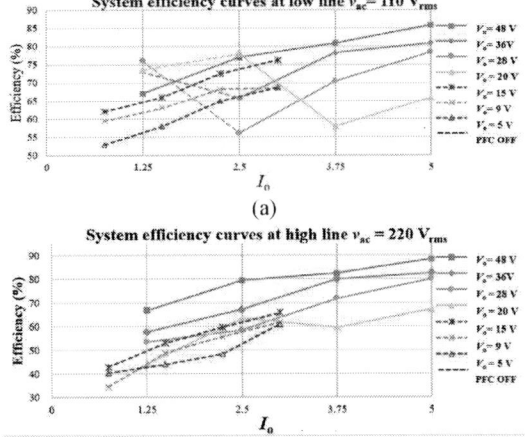

Fig. 33 Measured efficiency of AHBF converter with different V_o. (a) Under $v_{ac} = 110$ V. (b) Under $v_{ac} = 220$ V_{rms}.

VI. CONCLUSIONS

In this paper, an experimental prototype for a USB-PD with an input voltage of universal ac voltage 90~264 V_{ac}, output modes of 5 V/ 3 A, 9 V/ 3 A, 15 V/ 3 A, 20 V/ 5 A, 28 V/ 5 A, 36 V/5 A and 48 V/ 5 A, and rated output power 240 W is implemented. The boost PFC circuit stage achieves high power factor and is shut down when the output power is lower than 65 W. In the second stage, an asymmetrical half-bridge flyback converter is employed. This configuration allows the power devices on the primary side to achieve zero voltage switching, while on the secondary side, zero current switching is achieved through the sensing of the current of output diode. Ultimately, the experimental findings demonstrate that the boost PFC converter achieves peak efficiency at 97.2% with input voltage of 220 V_{rms}, and power factor reaches 0.99. The highest efficiency of AHB flyback converter is 91.3% at full load condition, and the overall system efficiency can reach 88.6%.

ACKNOWLEDGEMENT

This paper was financially supported by the National Science and Technology Council, Taiwan under project number 113-2640-E-006-003- and 113-2221-E-006-160-MY2. Additionally, the authors express their deepest gratitude to Leadtrend Technology Corp. for their support.

REFERENCE

[1] USB.org, "Universal serial bus power delivery specification," May 2021.

[2] EPSMA.org, "Harmonic current emissions – guidelines to the standard EN61000-3-2," Nov. 2010.

[3] J. C. Crebier, B. Revol, and J. P. Ferrieux, "Boost-chopper-derived PFC rectifiers: interest and reality," IEEE Trans. Ind. Appl., vol. 52, no. 1, pp. 36-45, Feb. 2005.

[4] B. Singh et al. "A review of single-phase improved power quality AC-DC converters," IEEE Trans. Ind. Electron., vol. 50, no. 5, pp. 962-981, Oct. 2003.

[5] Power Factor Correction (PFC) Handbook, 5th ed., ON Semiconductor Co., 2014.

[6] J. Itoh and F. Hayashi, "Ripple current reduction of a fuel cell for a single-phase isolated converter using a dc active filter with a center tap," IEEE Trans. Power Electron., vol. 25, no. 3, pp. 550-556, March 2010.

[7] X. Zhang and J. W. Spencer, "Analysis of boost PFC converters operating in the discontinuous conduction mode," IEEE Trans. Power Electron., vol. 26, no. 12, pp. 3621-3628, Dec. 2011.

[8] H. S. Athab, "A duty cycle control technique for elimination of line current harmonics in single-stage DCM boost PFC circuit," in Proc. IEEE Region Conf., 2008, pp. 1-6.

[9] "L6561," STMicroelectronics, Jun. 2004.

[10] Y. J. Ke, Y. F. Zhou, and J. N. Chen, "Control Bifurcation in PFC Boost Converter under Peak Current-Mode Control," in Proc. IEEE Int. Conf. Power Electron. Motion Control, 2006, pp. 1-5.

[11] "ICE2PCS01 standalone power factor correction (PFC) controller in continuous conduction mode (CCM)," Infineon Technologies, Nov. 2019.

[12] K. M. Smedley and S. Cuk, "One-cycle control of switching converters," IEEE Trans. Power Electrons., vol. 10, no. 6, pp. 625-633, Nov. 1995.

[13] "PFC converter design with IR1150 one cycle control IC," International Rectifier, Jun. 2005.

[14] R. Brown and M. Soldano, "One cycle control IC simplifies PFC designs," in Proc. 20th Annu. IEEE Appl. Power Electron. Conf. Expo., 2005, pp. 825-829.

[15] G. C. Huang, T. J. Liang, and K. H. Chen, "Losses analysis and low standby losses quasi-resonant flyback converter design," in Proc. IEEE Int. Symp. Circuits Syst., 2012, pp. 217-220.

[16] Y. Panov and M. M. Jovanovic, "Adaptive off-time control for variable-frequency, soft-switched flyback converter at light loads," IEEE Trans. Power Electron., vol. 17, no. 4, pp. 596-603, July 2002.

[17] B. R. Lin, H. K. Chiang, K. C. Chen, and D. Wang, "Analysis, design and implementation of an active clamp flyback converter," in Proc. Int. Conf. Power Electron. Drives Syst., 2005, pp. 424-429.

[18] M. Li, Z. Ouyang and M. A. E. Andersen, "Analysis and optimal design of high-frequency and high-efficiency asymmetrical half-bridge flyback converters," IEEE Trans. Ind. Electron., vol. 67, no. 10, pp. 8312-8321, Oct. 2020.

[19] L. Huber and M. M. Jovanović, "Analysis, design, and performance evaluation of asymmetrical half-bridge flyback converter for universal-line-voltage-range applications," in Proc. IEEE Appl. Power Electron. Conf. Expo., 2017, pp. 2481-2487.

[20] A. Medina-Garcia, F. J. Romero, D. P. Morales, and N. Rodriguez, "Advanced control methods for asymmetrical half-bridge flyback," IEEE Trans. Power Electron., vol. 36, no. 11, pp. 13139-13148, Nov. 2021.

979-8-3315-1612-3/25 $31.00 © 2025 IEEE

Computationally Efficient Current Sensorless Predictive Control for PMSM Drive fed by a Matrix Converter with CMV-Free Operation

1st Ali Sarajian
chair of high-power converter systems
Technical university of Munich
Munich, Germany
ali.sarajian@tum.de

2nd Ibrahim Harbi
chair of high-power converter systems
Technical university of Munich
Munich, Germany
Ibrahim.harbi@tum.de

3rd Quanxue Guan
School of Intelligent Systems Engineering
Sun Yat-Sen University
Shenzhen, China
guanqx3@mail.sysu.edu.cn

4th Davood Arab Khaburi
Department of Electrical Engineering
Iran University of Science and Technology
Tehran, Iran
khaburi@iust.ac.ir

5th Ralph Kennel
chair of high-power converter systems
Technical university of Munich
Munich, Germany
ralph.kennel@tum.de

6th Jose Rodriguez
Facultad de Ingeniería,
Arquitectura y Diseño
Universidad San Sebastián
Santiago, Chile
jose.rodriguezp@uss.cl

7th Patrick Wheeler
School of Electrical and Electronic Engineering
University of Nottingham
Nottingham, UK
pat.wheeler@nottingham.ac.uk

8th Mokhtar Aly
Facultad de Ingeniería,
Arquitectura y Diseño
Universidad San Sebastián
Santiago, Chile
mokhtar.aly@uss.cl

Abstract—Model predictive control (MPC) for matrix converters (MCs) is typically computationally intensive due to the need to evaluate all possible switching states, limiting its industrial application. Recent advances have reduced this burden by cutting the required predictions to just two, improving computation speed and lowering costs. However, the need for multiple sensors to measure various currents and voltages still drives up overall expenses. This paper introduces two look-up tables that reduce candidate switching states from 27 to 5 to address these issues. In addition, It incorporates a Luenberger observer to eliminate current sensors while maintaining system robustness. To further enhance the overall performance of the MC, a common mode voltage (CMV) elimination technique utilizing only 6 rotating vectors is integrated into the current sensorless control scheme. Experimental results on a PMSM drive system fed by an MC demonstrate significant reductions in computational load and hardware costs, enabling high-frequency operation with enhanced performance.

Index Terms—matrix converter, predictive control, computational burden

I. INTRODUCTION

Matrix converters (MCs) are AC-AC converters that directly connect a three-phase power supply to a three-phase load without needing a bulky DC link or large reactive components. This configuration offers high power density, a compact design, sinusoidal input/output currents, controllable input power factor, and bidirectional power flow capabilities [1].

Model predictive control (MPC) has emerged as a promising MC control strategy, optimizing switching states by minimizing prediction errors. However, traditional Finite Control Set MPC (FCS-MPC) requires evaluating 27 possible switching states, leading to significant computational complexity, which limits its industrial application [2]. Recent approaches have sought to reduce this burden by decreasing the number of predictions and evaluations required [3, 4]. Despite these advancements, challenges remain, particularly with the need for multiple sensors to measure currents and voltages, which is increasing system costs.

Reducing costs remains a significant challenge for the industrial adoption of matrix converters (MCs). Although encoderless methods represent progress, the current sensors, such as those in Yaskawa products with rated currents ranging from 27A to 520A, still contribute substantially to overall costs [5]. Moreover, achieving high system reliability is critical in the design and operation of MCs. While various safety measures, including protections against overvoltage, overspeed, overheating, and overcurrent, are typically in place, most electrical drives respond to faults by cutting output voltage or decelerating the motor below a predefined threshold. However, overcurrent protection does not address potential faults within the current sensors themselves, which can still result in system failures. Sensorless control methods, particularly for current estimation, offer a way to reduce both costs and system com-

979-8-3315-1612-3/25 $31.00 © 2025 IEEE

plexity by eliminating the need for expensive and failure-prone current sensors [6]. Various strategies have been proposed for fault detection and tolerance in PMSM drives [7, 8]. These methods still rely on three-phase current and rotor position data. However, current sensorless control has received less attention. The sliding mode observer, the Luenberger observer, and the Kalman filter are some examples of observers in this context [9]. While position/speed estimation techniques are well-studied, sensorless current control in MCs remains relatively unexplored.

This paper builds on the concepts in [2, 10] and proposes a computationally efficient sensorless MPC for MC-fed PMSMs, utilizing a Luenberger observer to eliminate the need for physical current sensors. The number of candidate switching states is reduced through the use of active and rotating vectors, significantly lowering computational effort without compromising control performance. Current sensorless Zero CMV MPC is also achieved using only 6 rotating vectors of MC. Experimental results validate that the observer accurately estimates both source and load currents across various operating conditions, demonstrating the effectiveness of the proposed method.

II. SYSTEM DESCRIPTION AND MODELING

The system under investigation comprises a three-phase matrix converter, a PMSM, and an input filter (Fig. 1). The matrix converter must prevent input-side short circuits and output-side open circuits to ensure safe operation. Given the converter's connection to a voltage source and the inductive load, 27 feasible switching states arise (Table I). These states can be categorized as active, rotating, or zero vectors based on their switching patterns. Active vectors exhibit variable magnitude and constant direction while rotating vectors have constant magnitude and variable direction. Zero vectors produce zero output magnitude.

According to the conventional FCS-MPC, a matrix converter feeding a PMSM load is controlled by predicting load currents and source currents for all possible switching states, comparing the predicted variables with their respective reference values, and determining the best switching state [9]. The representation of a PMSM drive fed by a matrix converter within a synchronous dq-frame, aligned with the rotor position angle, is outlined as follows:

$$f(\cdot) =$$
$$\begin{pmatrix} -(R_s/L_s)i_d + p\omega_r i_q + (1/L_s)v_d \\ -(R_s/L_s)i_q - p\omega_r i_d - (\psi_{pm}/L_s)p\omega_r + (1/L_s)v_q \\ (3/2J_m)\psi_{pm}pi_q - (B_m/J_m)\omega_r - (1/J_m)T_L \\ \omega_r \end{pmatrix}$$
$$x = [i_d \; i_q \; \omega_r \; \theta_r]^T,$$
$$u = [v_d \; v_q]^T. \tag{1}$$

The above equations can be used to predict the load currents at instant $k+1$ as, A discrete-time model of the input filter

TABLE I
POSSIBLE SWITCHING CONFIGURATIONS OF MCS

Switching State			Output Voltage		Input Current	
State	A	B	C	v_o	α_o	i_i	β_i
+1	a	b	b	$\frac{2}{3}v_{ab}$	0	$\frac{2}{\sqrt{3}}i_{oa}$	$-\frac{pi}{6}$
-1	b	a	a	$-\frac{2}{3}v_{ab}$	0	$-\frac{2}{\sqrt{3}}i_{oa}$	$-\frac{pi}{6}$
+2	b	c	c	$\frac{2}{3}v_{bc}$	0	$\frac{2}{\sqrt{3}}i_{oa}$	$\frac{pi}{2}$
-2	c	b	b	$-\frac{2}{3}v_{bc}$	0	$-\frac{2}{\sqrt{3}}i_{oa}$	$\frac{pi}{2}$
+3	c	a	a	$\frac{2}{3}v_{ca}$	0	$\frac{2}{\sqrt{3}}i_{oa}$	$\frac{7pi}{6}$
-3	a	c	c	$-\frac{2}{3}v_{ca}$	0	$-\frac{2}{\sqrt{3}}i_{oa}$	$\frac{7pi}{6}$
+4	b	a	b	$\frac{2}{3}v_{ab}$	$\frac{2\pi}{3}$	$\frac{2}{\sqrt{3}}i_{ob}$	$-\frac{pi}{6}$
-4	a	b	a	$-\frac{2}{3}v_{ab}$	$\frac{2\pi}{3}$	$-\frac{2}{\sqrt{3}}i_{ob}$	$-\frac{pi}{6}$
+5	c	b	c	$\frac{2}{3}v_{bc}$	$\frac{2\pi}{3}$	$\frac{2}{\sqrt{3}}i_{ob}$	$\frac{pi}{2}$
-5	b	c	b	$-\frac{2}{3}v_{bc}$	$\frac{2\pi}{3}$	$-\frac{2}{\sqrt{3}}i_{ob}$	$\frac{pi}{2}$
+6	a	c	a	$\frac{2}{3}v_{ca}$	$\frac{2\pi}{3}$	$\frac{2}{\sqrt{3}}i_{ob}$	$\frac{7pi}{6}$
-6	c	a	c	$-\frac{2}{3}v_{ca}$	$\frac{2\pi}{3}$	$-\frac{2}{\sqrt{3}}i_{ob}$	$\frac{7pi}{6}$
+7	b	b	a	$\frac{2}{3}v_{ab}$	$\frac{4\pi}{3}$	$\frac{2}{\sqrt{3}}i_{oc}$	$-\frac{pi}{6}$
-7	a	a	b	$-\frac{2}{3}v_{ab}$	$\frac{4\pi}{3}$	$-\frac{2}{\sqrt{3}}i_{oc}$	$-\frac{pi}{6}$
+8	c	c	b	$\frac{2}{3}v_{bc}$	$\frac{4\pi}{3}$	$\frac{2}{\sqrt{3}}i_{oc}$	$\frac{pi}{2}$
-8	b	b	c	$-\frac{2}{3}v_{bc}$	$\frac{4\pi}{3}$	$-\frac{2}{\sqrt{3}}i_{oc}$	$\frac{pi}{2}$
+9	a	a	c	$\frac{2}{3}v_{ca}$	$\frac{4\pi}{3}$	$\frac{2}{\sqrt{3}}i_{oc}$	$\frac{7pi}{6}$
-9	c	c	a	$-\frac{2}{3}v_{ca}$	$\frac{4\pi}{3}$	$-\frac{2}{\sqrt{3}}i_{oc}$	$\frac{7pi}{6}$
0_a	a	a	a	0	-	0	-
0_b	b	b	b	0	-	0	-
0_c	c	c	c	0	-	0	-
+10	a	b	c	v_{imax}	α_i	i_{omax}	β_o
-10	a	c	b	v_{imax}	$-\alpha_i$	i_{omax}	$-\beta_o$
+11	c	a	b	v_{imax}	$\alpha_i + \frac{2\pi}{3}$	i_{omax}	$\beta_o + \frac{2pi}{3}$
-11	b	a	c	v_{imax}	$-\alpha_i + \frac{2\pi}{3}$	i_{omax}	$-\beta_o + \frac{2pi}{3}$
+12	b	c	a	v_{imax}	$\alpha_i + \frac{4\pi}{3}$	i_{omax}	$\beta_o + \frac{4pi}{3}$
-12	c	b	a	v_{imax}	$-\alpha_i + \frac{4\pi}{3}$	i_{omax}	$-\beta_o + \frac{4pi}{3}$

can be used to predict the source current. The input filter can be represented as a continuous-time space-state model:

$$\begin{bmatrix} \dot{v}_i \\ \dot{i}_s \end{bmatrix} = A \begin{bmatrix} v_i \\ i_s \end{bmatrix} + B \begin{bmatrix} v_s \\ i_i \end{bmatrix} \tag{2}$$

where

$$A = \begin{bmatrix} 0 & \frac{1}{C_f} \\ -\frac{1}{L_f} & -\frac{R_f}{L_f} \end{bmatrix}, \; B = \begin{bmatrix} 0 & -\frac{1}{C_f} \\ \frac{1}{L_f} & 0 \end{bmatrix} \tag{3}$$

where C_f and L_f are the input filter capacitance and inductance, respectively, and R_f is the leakage resistance of L_f. The discrete-time state-space form of (2) is as follows

$$\begin{bmatrix} v_i^{k+1} \\ i_s^{k+1} \end{bmatrix} = A \begin{bmatrix} v_i^k \\ i_s^k \end{bmatrix} + B \begin{bmatrix} v_s^k \\ i_i^k \end{bmatrix} \tag{4}$$

$$A_d = e^{AT_s}, \; B_d = \int_0^{T_s} e^{A(T_s-\tau)} B d\tau \tag{5}$$

The prediction of the source current is obtained by solving i_s^{k+1} from (2) as

$$i_{s|sw}^{k+1} = A_d(2,1)v_i^k + A_d(2,2)i_s^k + B_d(2,1)v_s^k + B_d(2,2)i_{i|sw}^k \tag{6}$$

III. Proposed FCS-MPC Method without current sensor

The proposed method intends to simplify FCS-MPC. Instead of using 27 load current predictions and 27 source current predictions for instant $k+1$, it uses the desired output voltage vector v_o^* and the input current vector i_i^* for prediction as,

$$v_o^{*k} = \frac{L_s}{T_s}[i_o^{*k+1} - i_o^k] + R_s i_o^k + e^k, \tag{7}$$

$$i_i^{*k} = \frac{i_s^{*k+1} - A_d(2,1)v_i^k - A_d(2,2)i_s^k - B_d(2,1)v_s^k}{B_d(2,2)}, \tag{8}$$

where $e(k)$ represents the back electromotive force (EMF) vector, which is incorporated into (7) when the load is a PMSM.

To compute the source current reference in Equation (8), the negligible voltage drop across the input filter allows for the approximation of MC input voltage (v_i) as equal to source voltage (v_s). Given the absence of energy storage components in the MC and negligible iron loss of PMSM, the power balance ($P_{in} = P_{out}$) holds under lossless conditions. Assuming unity power factor operation, the source current reference generation can be calculated as:

$$\frac{3}{2}V_{sa} I_{sa}^* = \omega_r T_e^* \longrightarrow I_{sa}^* = \omega_r T_e^* \times \left(\frac{2}{3V_{sa}}\right) \times \left(\frac{v_{sa}}{V_{sa}}\right)$$

$$\longrightarrow i_{sa}^* = \omega_r T_e^* \times \left(\frac{v_{sa}}{v_{sa}^2 + v_{sb}^2 + v_{sc}^2}\right). \tag{9}$$

where $V_{sa}^2 = (2/3)(v_{sa}^2 + v_{sb}^2 + v_{sc}^2)$.

It should be mentioned that sensors need to measure i_s, v_s, and v_i. A Leunberger observer will be employed to remove current sensors on the source side and load side of MC. Two sets of the system equations of the estimator for the MC system feeding a PMSM load can be expressed as,

$$\frac{d}{dt}\begin{bmatrix} \hat{i}_s \\ \hat{v}_i \end{bmatrix} = \begin{bmatrix} \dfrac{1}{L_f}(v_s - v_i - R_f i_s) \\ \dfrac{1}{C_f}(i_s - i_i) \end{bmatrix} + \begin{bmatrix} L_1 \\ L_2 \end{bmatrix}(v_i - \hat{v}_i), \tag{10}$$

$$\frac{d}{dt}\begin{bmatrix} \hat{i}_d \\ \hat{i}_q \\ \hat{\omega}_r \end{bmatrix} = \begin{bmatrix} -\dfrac{R_s}{L_s}i_d + p\omega_r i_q + \dfrac{v_d}{L_s} \\ -\dfrac{R_s}{L_s}i_q - p\omega_r i_d + \dfrac{v_q}{L_s} - \dfrac{p\omega_r \psi_m}{L_s} \\ \dfrac{1}{J_m}\left(\dfrac{3}{2}p\psi_m i_q - B_m\omega_r - T_L\right) \end{bmatrix} + \begin{bmatrix} L_3 \\ L_4 \\ L_5 \end{bmatrix}(\omega_r - \hat{\omega}_r), \tag{11}$$

where $\hat{i}_s$, $\hat{v}_i$, $\hat{i}_d$, $\hat{i}_q$ and $\hat{\omega}_r$ represent the estimated source currents, input voltage, load currents and rotor speed of MC, and L_1, L_2, L_3, and L_4 are constant coefficients for the Luenberger estimator. Similarly, this system is linearized

TABLE II
Active Switching State Selection Based on Input Voltage Sector and Desired Voltage Vector Sector

Sector sU \ sV	I	II	III	IV	V	VI
I	+1,+2,-3	-1,+2,-3	-1,+2,+3	-1,-2,+3	+1,-2,+3	+1,-2,-3
II	-7,-8,+9	+7,-8,+9	+7,+8,-9	+7,+8,-9	-7,+8,-9	-7,+8,+9
III	+4,+5,-6	-4,+5,-6	-4,+5,+6	-4,-5,+6	+4,-5,+6	+4,-5,-6
IV	-1,-2,+3	+1,-2,+3	+1,-2,-3	+1,+2,-3	-1,+2,-3	-1,+2,+3
V	+7,+8,-9	-7,+8,-9	-7,+8,+9	-7,-8,+9	+7,-8,+9	+7,-8,-9
VI	-4,-5,+6	+4,-5,+6	+4,-5,-6	+4,+5,-6	-4,+5,-6	-4,+5,+6

TABLE III
Rotating Switching State Selection Based on Input Voltage Sector and Desired Voltage Vector Sector.

Sector sU \ sV	I	II	III	IV	V	VI
I	+10	-11	+11	-12	+12	-10
II	-11	+10	-12	+11	-10	+12
III	+12	-12	+10	-10	+11	-11
IV	-12	+12	-10	+10	-11	+11
V	+11	-10	+12	-11	+10	-12
VI	-10	+11	-11	+12	-12	+10

around an operating point by placing the poles on the left of the imaginary axis, which determines the overall range of the constants. The following values, when adjusted in the simulation, lead to the highest precision: $L_1 = 0.001$, $L_2 = 1$, $L_3 = 0.01$ and $L_4 = 10$.

Finally, In order to control the load current and the source current simultaneously, a cost function is required to decide the best switching state, which can be defined as,

$$CF_{sw} = |v_o^*(k) - v_{o|sw}| + k_I |i_i^*(k) - i_{i|sw}| \tag{12}$$

A further simplification of the proposed method is to reduce the number of candidate switching states for prediction and evaluation in the cost function from 27 to 5. After calculating the output voltage reference and input current reference, it becomes possible to determine the sector within the corresponding hexagons where these references lie. Therefore, three active vectors, one rotating vector, and one zero vector, totaling 5 switching states, can be used as candidate switching states (Tables II and III). CMV can be eliminated by utilizing only six rotating vectors instead of all 27 voltage vectors, which trace a unique input phase at each output phase. In this case, not only is full controllability of MC attained by two predictions of the output voltage vector and input current vector calculation, but zero CMV is also ensured.

TABLE IV
SYSTEM SPECIFICATIONS

Parameter	Explanation	Value
L_s(mH)	Stator inductance	6.5
$R_s(\Omega)$	Stator resistance	0.36
ψ_{pm} (Wb)	Rotor flux	0.15
$J_m(Kg.m^2)$	moment of inertia	0.025
p	Pole pairs	4
v_{rms} (v)	Source line to line voltage	230
ω_r (rpm)	Rated speed	1200
T_r (Nm)	Rated torque	8
L_f (mH)	Input filter inductance	1
C_f (μF)	Input filter capacitance	27
f_{sw} (kHz)	Switching frequency	16

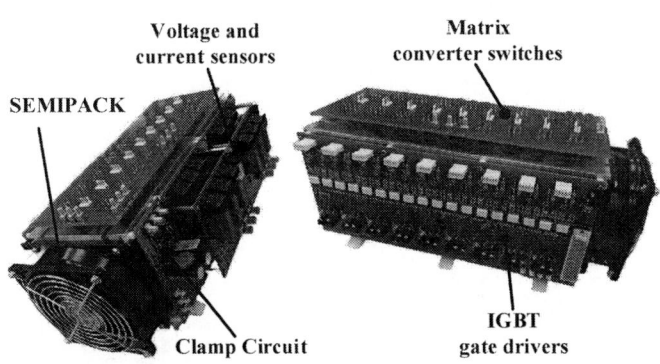

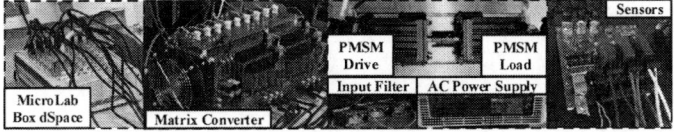

Fig. 1. Implementation of PMSM drive fed by a matrix converter

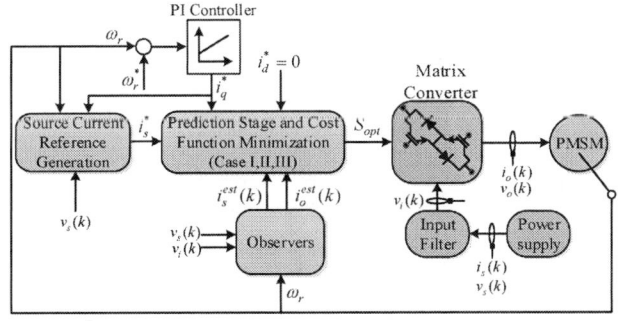

Fig. 2. The proposed FCS-MPC control scheme for PMSM Drive Using a Luenberger Observer.

IV. EXPERIMENTAL RESULTS

A 2 kW matrix converter-fed PMSM drive system was experimentally implemented (Fig. 1) to evaluate the proposed sensorless current control methods (Case I is current sensorless MPC with 5 vectors, and Case II is CMV-free current sensorless MPC with 6 vectors). Fig. 2 shows the proposed method in conjunction with the computationally efficient FCS-MPC. The matrix converter utilized IGBT modules driven by isolated gate drivers and a three-phase LC filter connected the converter to the 230V (RMS), 50 Hz grid. A protection clamp circuit was integrated. A dSPACE system equipped with RTI and Control Desk software controlled the system, with PWM pulses generated from Simulink-based C code. The PMSM and input filter parameters are detailed in Table IV. All algorithms operated at a sampling frequency of 16 kHz. The PMSM was controlled in speed mode at a reference of 1200 rpm, while the load PMSM, driven by a commercial inverter, operated in torque mode for variable load torque generation.

To assess the performance of the sensorless proposed control algorithms (Cases I and II) in mitigating common mode voltage, their effectiveness is evaluated under a source voltage of $230V$(RMS) at 50. The results are presented in Fig. 3. The results demonstrate satisfactory performance, with both the load and source currents achieving a sinusoidal waveform and the input power factor approaching unity. However, the zero CMV method (Case II) introduces some degradation in current performance compared to the method without zero CMV (Case I). This degradation is manifested by increasing THD of source/load currents. Therefore, a trade-off between eliminating common mode voltage and maintaining current quality should be considered when selecting the appropriate control algorithm for a particular application.

However, the proposed methods achieved significant advancements, as listed below,

- Reduced Computational Burden: The control scheme substantially decreased computational complexity by requiring only two predictions and utilizing two lookup tables to reduce candidate switching states.
- Full MC Controllability and Zero CMV: The reduced candidate switching states enabled full MC controllability. Additionally, employing six rotating vectors as candidate-switching states facilitated the achievement of zero CMV.
- Improved THD with Trade-off: The simplification resulted in a slightly increased THD in current waveforms, which a shorter sampling frequency can compensate for. However, a shorter time interval typically translates to a higher switching frequency. While a higher switching frequency can reduce THD, it leads to increased switching losses within the converter. Therefore, a practical application should find a reasonable trade-off between achieving lower THD and maintaining converter efficiency based on specific requirements.
- Decreased system cost by removing the current sensor: The proposed method has utilized a Luenberger observer for current estimation. The control system then used the estimated currents to track the desired reference currents with high precision. This functionality has been made possible because the studied system is fundamentally observable. This means that all system states can be determined from the measurable outputs and the estimated

979-8-3315-1612-3/25 $31.00 © 2025 IEEE

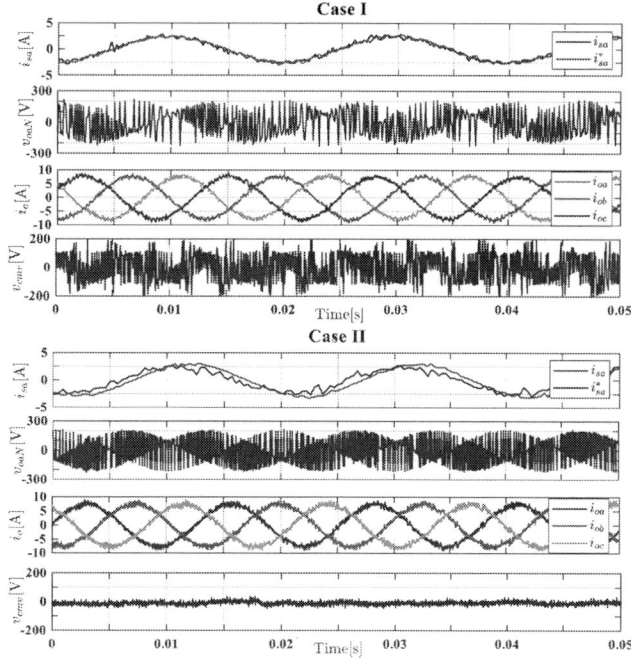

Fig. 3. Steady state performance of the system for Case I and II

currents. Consequently, the need for physical current sensors has been eliminated, leading to a simpler system design.

V. CONCLUSION

This paper presented a computationally efficient current sensorless MPC for a PMSM drive fed by a matrix converter. The proposed method addresses the computational challenges of traditional FCS-MPC by reducing the number of candidate switching states from 27 to 5, significantly lowering the computational burden. The integration of a Luenberger observer effectively eliminates the need for current sensors, reducing hardware costs and enhancing system reliability. Experimental results demonstrate that the sensorless MPC maintains robust control performance while accurately estimating both source and load currents under various operating conditions. Additionally, the method achieves CMV-free operation, making it highly suitable for industrial applications where reliability is paramount.

ACKNOWLEDGMENT

J. Rodriguez acknowledges the support of ANID through projects FB0008, 1210208, and 1221293. This work was also supported by ANID, Chile FONDECYT 1230250, FONDECYT Iniciacion 11230430, and SERC-Chile ANID/FONDAP/1523A0006. Additionally, this work was partly supported by the Key-Area Research and Development Program of Guangdong Province under Grant 2023B0909050006, and partly supported by the

Shenzhen Science and Technology Program under Grant JCYJ20220530150005011.

REFERENCES

[1] P. W. Wheeler, J. Rodriguez, J. C. Clare, L. Empringham and A. Weinstein, "Matrix converters: a technology review," in IEEE Transactions on Industrial Electronics, vol. 49, no. 2, pp. 276-288, April 2002, doi: 10.1109/41.993260.

[2] M. Siami, D. Arab Khaburi and J. Rodriguez, "Simplified Finite Control Set-Model Predictive Control for Matrix Converter-Fed PMSM Drives," in IEEE Transactions on Power Electronics, vol. 33, no. 3, pp. 2438-2446, March 2018, doi: 10.1109/TPEL.2017.2696902.

[3] T. Geyer and D. E. Quevedo, "Multistep Finite Control Set Model Predictive Control for Power Electronics," in IEEE Transactions on Power Electronics, vol. 29, no. 12, pp. 6836-6846, Dec. 2014, doi: 10.1109/TPEL.2014.2306939.

[4] M. Habibullah, D. D. -C. Lu, D. Xiao and M. F. Rahman, "A Simplified Finite-State Predictive Direct Torque Control for Induction Motor Drive," in IEEE Transactions on Industrial Electronics, vol. 63, no. 6, pp. 3964-3975, June 2016, doi: 10.1109/TIE.2016.2519327.

[5] Varispeed AC INSTRUCTION MANUAL, MANUAL NO. TOEP C710636 00D, Matrix Converter for Environmentally Friendly Motor Drives Model: CIMR-ACA, Yaskawa, available at: https://www.yaskawa.com, viewed at 01 Oct 2024.

[6] D. Xu, B. Wang, G. Zhang, G. Wang and Y. Yu, "A review of sensorless control methods for AC motor drives," in CES Transactions on Electrical Machines and Systems, vol. 2, no. 1, pp. 104-115, March 2018, doi: 10.23919/TEMS.2018.8326456.

[7] C. Wu, C. Guo, Z. Xie, F. Ni and H. Liu, "A Signal-Based Fault Detection and Tolerance Control Method of Current Sensor for PMSM Drive," in IEEE Transactions on Industrial Electronics, vol. 65, no. 12, pp. 9646-9657, Dec. 2018, doi: 10.1109/TIE.2018.2813991.

[8] Jankowska, Kamila, and Mateusz Dybkowski. "Design and analysis of current sensor fault detection mechanisms for PMSM drives based on neural networks." Designs 6.1 (2022): doi.org/10.3390/designs6010018

[9] A. Sarajian, D. A. Khaburi, and M. Rivera, "Using Extended Kalman Filter and Adaptive Filter for sensorless predictive torque control of PM-Assissted Synchronous Reluctance Motor," 2016 7th Power Electronics and Drive Systems Technologies Conference (PEDSTC), Tehran, Iran, 2016, pp. 64-69, doi: 10.1109/PEDSTC.2016.7556839.

[10] M. Siami, M. Amiri, H. K. Savadkoohi, R. Rezavandi and S. Valipour, "Simplified Predictive Torque Control for a PMSM Drive Fed by a Matrix Converter With Imposed Input Current," in IEEE Journal of Emerging and Selected Topics in Power Electronics, vol. 6, no. 4, pp. 1641-1649, Dec. 2018, doi: 10.1109/JESTPE.2018.2837109.

PMSM Motor Drive with Current Direct Digital Control and Near 1st-Order Speed Control

Po-Chang Lee
Photovoltaic Inverter Business
Unit in Energy Infrastructure &
Industrial Solutions Business
Group
Delta Electronics Inc. (Tainan
Factory I)
line 4: Tainan, Taiwan
Email: john40770@gmail.com

Tsai-Fu Wu
Elegant Power Electronics
Applied Research Laboratory in
Department of Electrical
Engineering
National Tsing Hua University
Hsinchu, Taiwan
Email: tfwu@ee.nthu.edu.tw

Han Ku
Electric Mobility Power
Business Unit in EV Powertrain
System Business Group
Delta Electronics Inc. (Chungli
Factory I)
line 4: Taoyuan, Taiwan
Email: eric7317564@gmail.com

Chien-Chih Hung
Elegant Power Electronics
Applied Research Laboratory in
Department of Electrical
Engineering
National Tsing Hua University
Hsinchu, Taiwan
Email: jeff80121@gmail.com

Jui-Yang Chiu
Elegant Power Electronics
Applied Research Laboratory in
Department of Electrical
Engineering
National Tsing Hua University
Hsinchu, Taiwan
Email: s110061802@m110.nthu.edu.tw

Abstract— This paper presents a permanent-magnet-synchronous-motor (PMSM) drive with the three-phase three-wire inverter current direct digital control (DDC). The DDC achieves high accuracy and fast response, and expresses duty ratios as analytical functions without any parameter tuning. Thanks to the proposed current DDC, the fast response can simplify the conventional double-loop speed-current control system to a single-loop one, allowing a simple speed control design. Thus, this paper also presents how to design the simple speed-control law based on deadbeat and disturbance-observer-based control, achieving near 1st-order response. Besides, the low-speed cogging torque issue is addressed with the virtual cogging torque (VCT) control. Experimental results confirm the accurate and fast current/torque tracking, near 1st-order speed response, and stable low-speed VCT control.

Keywords— PMSM drive, direct digital control (DDC), D-Σ process, and disturbance-observer-based (DOB) control

I. INTRODUCTION

Effective current control in permanent-magnet synchronous motors (PMSMs) is crucial for motor torque and speed controls [1]. Conventional three-phase three-wire (3Φ3W) inverter controls, such as PID control [2]-[4] and hysteresis control [5], [6] are simple, but require experienced parameter tuning. They are also difficult to manage non-linear or variable loads. Alternatives like fuzzy control [7] and model predictive control [8], [9] can handle complex systems but require tedious implementation. Instead of a complicated control, this paper presents a simple current direct digital control (DDC) derived through the division-summation (D-Σ) process [10]-[13], featuring high accuracy and fast response for current tracking. Unlike PID controller, the DDC formulates the PWM duty ratios as analytical functions without parameter tuning. The DDC has

been relied on 3Φ grid-connected inverters [10]-[13], but has not been applied to PMSM. Thus, the DDC is used to control the motor current, thereby fulfilling precise and fast field-oriented torque control (FOC) [1], [14]-[16]. Additionally, the fast current control can simplify the double-loop speed-current control system to a single-loop one, allowing a simple design.

This paper first derives the current DDC laws through the D-Σ process. A simple speed-control law is also derived, like the deadbeat control [17], [18] and the disturbance-observer-based (DOB) control [19]-[22], achieving a near 1st-order speed response. Then, the cogging torque issue in low-speed operation is resolved by the VCT control [23]. Finally, the experiments show that the accurate and fast current/torque tracking and near 1st-order speed response can be achieved.

II. CONTROL-LAW DERIVATIONS

Fig. 1 shows the PMSM drive configuration, consisting of the 3Φ3W inverter, the controller and the mechanical load.

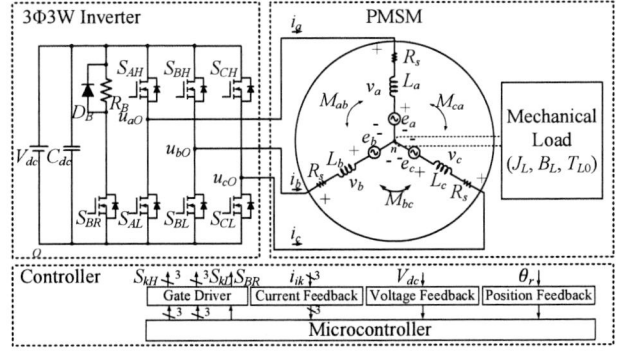

Fig. 1. System configuration.

To simplify the analysis and fulfill FOC, it is necessary to transform the inverter-PMSM circuit into that in rotor-reference frame [24], as shown in Fig. 2. The current control laws and the speed control law are derived as follows:

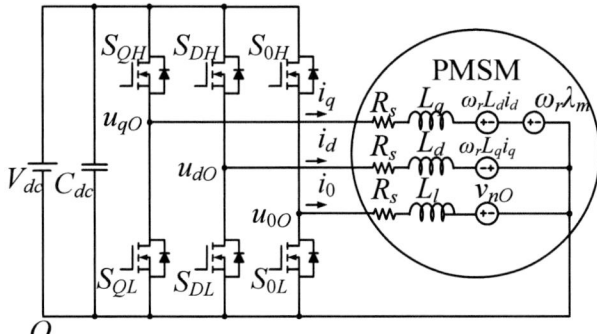

Fig. 2. Inverter-PMSM circuit in rotor-reference frame.

A. Current DDC

Fig. 2 shows all phases becoming independent, allowing the D-Σ process to be relied on each phase individually. For q-phase, based on Kirchhoff's voltage law (KVL), u_{qO} can be expressed as

$$u_{qO} = R_s i_q + L_q \frac{d}{dt} i_q + \omega_r L_d i_d + \omega_r \lambda_m, \tag{1}$$

where R_s is the stator resistance, L_q and L_d are the PMSM constant inductances, ω_r is the motor speed with respect to electrical angle, and λ_m is the maximum flux of the permanent magnet. We next apply the D-Σ process as follows:

Division (D): One switching cycle T_s is divided into magnetizing interval $d_q T_s$ and demagnetizing interval $(1 - d_q) T_s$. By moving the terms of (1), their corresponding inductor current variations are expressed as follows:

$$\Delta i_{q,mag} = \frac{1}{L_q} \left(V_{dc} - R_s i_q - \omega_r L_d i_d - \omega_r \lambda_m \right) d_q T_s \tag{2}$$

and

$$\Delta i_{q,demag} = \frac{1}{L_q} \left(0 - R_s i_q - \omega_r L_d i_d - \omega_r \lambda_m \right) \cdot (1 - d_q) T_s. \tag{3}$$

Summation (Σ): Both individual variations (2) and (3) are summarized to obtain the total inductor current variation:

$$\Delta i_q = \frac{1}{L_q} \left(V_{dc} d_q - R_s i_q - \omega_r L_d i_d - \omega_r \lambda_m \right) T_s, \tag{4}$$

from which the duty ratio d_q can be finally found as follows:

$$d_q = L_q \frac{\Delta i_q}{V_{dc} T_s} + \frac{R_s i_q}{V_{dc}} + \frac{\omega_r L_d i_d}{V_{dc}} + \frac{\omega_r \lambda_m}{V_{dc}}, \tag{5}$$

where $\Delta i_q = i_{q,ref}[n+1] - i_q[n]$. With the current reference $i_{q,ref}$ and the feedback currents i_q and i_d, i_q can track the reference in every switching cycle. Similarly, the control laws of other phases can also be derived as

$$d_d = L_d \frac{\Delta i_d}{V_{dc} T_s} + \frac{R_s i_d}{V_{dc}} - \frac{\omega_r L_q i_q}{V_{dc}} \tag{6}$$

and

$$d_0 = L_l \frac{\Delta i_0}{V_{dc} T_s} + \frac{R_s i_0}{V_{dc}} + \frac{\overline{v_{nO}}}{V_{dc}}, \tag{7}$$

where $\overline{v_{nO}}/V_{dc}$ is an arbitrary value as long as the duty ratio is not over modulation. For SPWM, it is set to 1/2 [13]. Through inverse Park transformation of (d_q, d_d, d_0), the duty ratios (d_a, d_b, d_c) can be found. These control laws in (5)-(7) allow an accurate and fast FOC torque control by setting $i_{d,ref} = i_{0,ref} = 0$ and using the following equation:

$$i_{q,ref} = \frac{2}{N_p} \frac{2}{3\lambda_m} T_{ind,ref}, \tag{8}$$

where N_p is the number of poles.

B. Speed Control

Most speed controls utilize a double-loop system [1], [15] with the inner current/torque control and the outer speed control, respectively. Since the current DDC provides a faster response for the inner loop, it can be simplified to a single-loop speed control, as illustrated in Fig. 3.

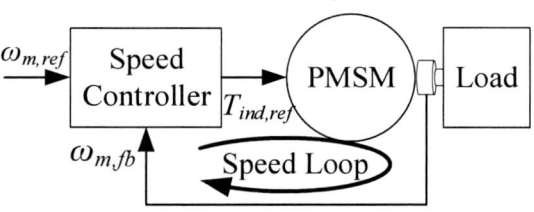

Fig. 3. Simplified speed control block diagram.

We first formulate the following mechanical differential equation:

$$T_{ind} = J_L \frac{d\omega_m}{dt} + B_L \omega_m + T_{L0}, \tag{9}$$

where ω_m is the motor speed. J_L, B_L and T_{L0} are the load inertia, viscosity and the exerted torque, respectively. By discretizing (9) using the forward-Euler method with the sampling period $T_{s\omega}$, and assuming $T_{ind}[n] \approx T_{ind}[n+1]$, we have

$$T_{ind}[n+1] = J_L \frac{\omega_m[n+1] - \omega_m[n]}{T_{s\omega}} + B_L \omega_m[n] + T_{L0}[n], \tag{10}$$

where $B_L \omega_m[n] + T_{L0}[n]$ can be replaced by $T_{ind}[n] - J_L \alpha_m[n]$, in which $\alpha_m[n]$ is the angular acceleration. Furthermore, the first term in (10) should be multiplied by a scalar K_ω ($0 < K_\omega < 1$) since the bandwidth is limited within half of its sampling frequency based on Nyquist–Shannon sampling theorem. Thus, the control law becomes

$$T_{ind}[n+1] = K_\omega J_L \frac{\omega_m[n+1] - \omega_m[n]}{T_{s\omega}} + T_{ind}[n] - J_L \alpha_m[n]. \tag{11}$$

By treating [n+1] terms as the references, while [n] terms as the feedback signals, equation (11) becomes the basic control law. However, as depicted in Fig. 4, three low-pass filters (LPF), $H_1(z)$, $H_2(z)$ and $H_3(z)$, should be introduced to filter out the variations in ω_m, α_m and T_{ind}, resembling the DOB control [19]-[22], [25]. They can be designed by analyzing the following characteristic equation (assuming $H_2(z) = H_3(z)$ makes $J_L\alpha_m(z)$ insignificant):

$$1 + K_\omega \frac{J_L}{T_{s\omega}} \frac{z^{-1}(1-a_m)}{B_L(1-a_m z^{-1})} H_1(z) \frac{(1-a_3 z^{-1})}{a_3(1-z^{-1})} = 0, \quad (12)$$

where the mechanical pole $a_m \approx (1 - (B_L/J_L)T_{s\omega})$. $H_3(z)$ and $H_2(z)$ can be designed by setting $a_3 = a_2 = a_m$ to cancel out pole a_m. After simplification, it yields

$$1 + K_\omega \frac{z^{-1}}{a_m(1 - z^{-1})} H_1(z) = 0. \quad (13)$$

$H_1(z)$ is designed to filter out the noise $n(z)$ in the speed sensor, as the following cut-off frequency condition:

$$a_1 = e^{-2\pi(0.1 f_{n,min})T_{s\omega}}. \quad (14)$$

where $f_{n,min}$ is the minimum noise frequency. Finally, the root locus of (13) shown in Fig. 5 illustrates that the system is stable and can fulfill 1st order response if $0 < K_\omega < K_{\omega,max}$, where $K_{\omega,max}$ corresponds to the breakaway point $\sqrt{a_1}$ and $K_{\omega,max} = (1 - \sqrt{a_1})/(1 + \sqrt{a_1})$. Furthermore, it is easy to obtain the transfer function, as $\frac{K_\omega z^{-1}}{1+(1-K_\omega)z^{-1}}$ from which pole $(1 - K_\omega)$ determines the respond time by setting K_ω with the desired time constant τ_ω, as follows:

$$K_\omega = 1 - e^{-T_{s\omega}/\tau_\omega}, \ni 0 < K_\omega < K_{\omega,max}. \quad (15)$$

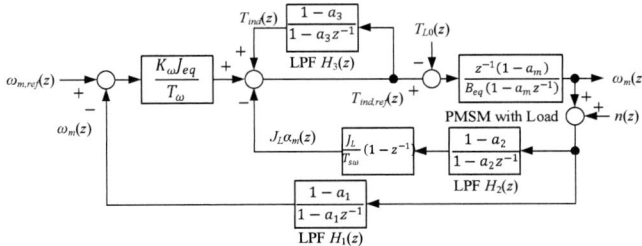

Fig. 4. Block diagram of proposed control law.

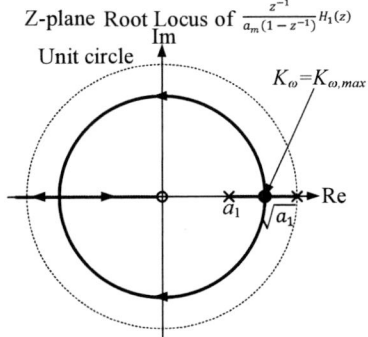

Fig. 5. Root locus of the system.

III. COGGING TORQUE ISSUE IN LOW-SPEED OPERATION

In fact, PMSM cannot operate smoothly in low speed owing to the cogging torque. As shown in Fig. 6, there are several stable equilibrium points (SEP) where the rotor tends to settle down. Reference [23] came up with a large virtual cogging torque (VCT) method which reduces to a position-programmable virtual SEP, as shown in Fig. 7.

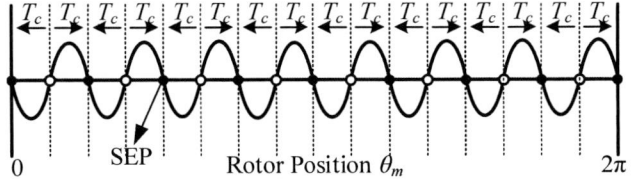

Fig. 6. Cogging torque waveform.

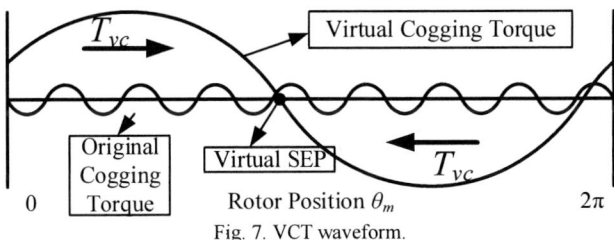

Fig. 7. VCT waveform.

VCT covers the original cogging torque, allowing the rotor attracted by the virtual SEP. It can be formulated as

$$T_{vc} = T_{vc,peak} \sin(\theta_{m,ref} - \theta_m), \quad (16)$$

where $T_{vc,peak}$ is the peak value of VCT and $\theta_{m,ref}$ is the desired position of the virtual SEP. This method needs to be realized with a good current control. Thanks to the proposed DDC, and by setting $T_{ind,ref} = T_{vc}$, VCT can be realized. Thus, in the low-speed operation, we can integrate $\omega_{m,ref}$ to obtain $\theta_{m,ref}$, as

$$\theta_{m,ref}[n + 1] = \theta_{m,ref}[n] - \omega_{m,ref}T_{s\omega}. \quad (17)$$

IV. EXPERIMENTAL RESULTS

This paper implements a dc100 V/5 kW inverter to drive a PMSM, model MVWF7K6-102, up to the output torque of 13 N-m and rated speed of 262 rad/s. Fig. 8 shows the 3Φ current steady-state waveforms of the current control under rated speed with $(i_{q,ref}, i_{d,ref}, i_{0,ref}) = (34, 0, 0)$ A and $(67, 0, 0)$ A. Apparently, the three-phase currents tracks the desired current with the accuracy up to 99.56% but without large ripple, verifying the accurate and low-ripple performances. Fig. 9 shows the 3Φ current transient responses with $i_{q,ref} = 0 \rightarrow 34$ A and $67 \rightarrow 34$ A. Both results show the fast rising and falling responses of 2 to 3 times switching period of 50 μs. Fig. 10 demonstrates the experiment of torque control by setting $T_{out,ref}$ to 13 N-m and -4 N-m. The motor accelerations are measured and can verify that the accurate torques ($T_{out} = J_L\alpha_m$) are 12.7 N-m and -3.85 N-m, close to the references with the accuracy up to 96.25 %. Fig. 11 presents the experiments of the speed control under $J_L = 0.2$ kg-m^2 and $B_L = 0.01$ N-m-s with the references of $250 \rightarrow 260$ rad/s and $260 \rightarrow 250$ rad/s. Ignoring the sensor noise, the results confirm the near 1st-order speed responses. In addition, under 10 rad/s, the motor enters the low-speed operation. Fig. 12 verifies that the low-speed control is stable with VCT, while diverges without VCT. These results show that the proposed current direct digital control is able to handle motor torque, speed, and VCT controls.

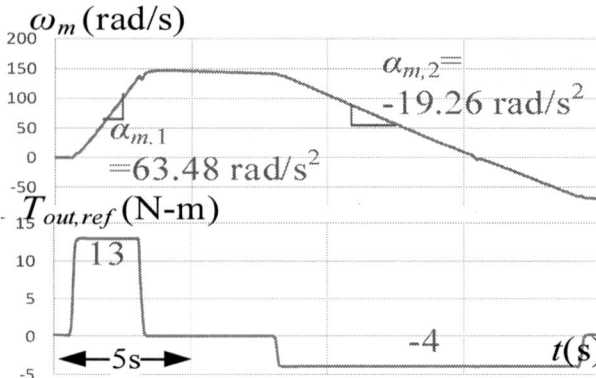

Fig. 10. Torque control experiment under $J_L = 0.2$ kg-m^2.

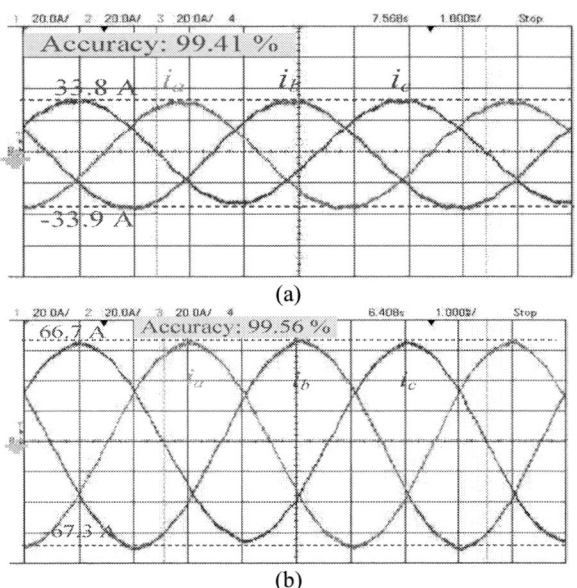

(a)

(b)

Fig. 8. Steady state 3Φ currents experiments under (a) $i_{q,ref} = 34$ A and (b) $i_{q,ref} = 67$ A.

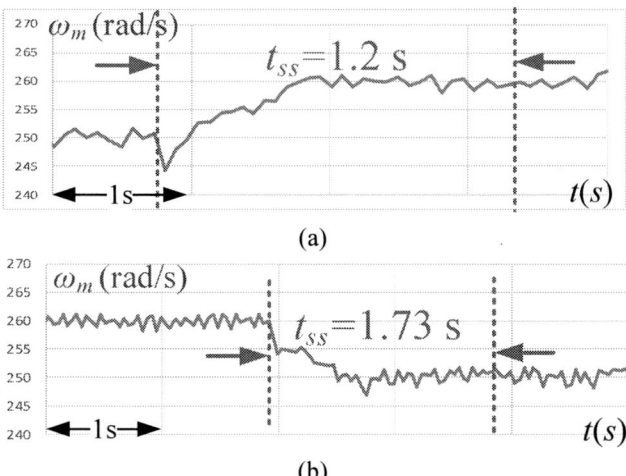

(a)

(b)

Fig. 11. Speed control experiment: (a) $\omega_{ref} = 250 \rightarrow 260$ rad/s and (b) $\omega_{ref} = 260 \rightarrow 250$ rad/s.

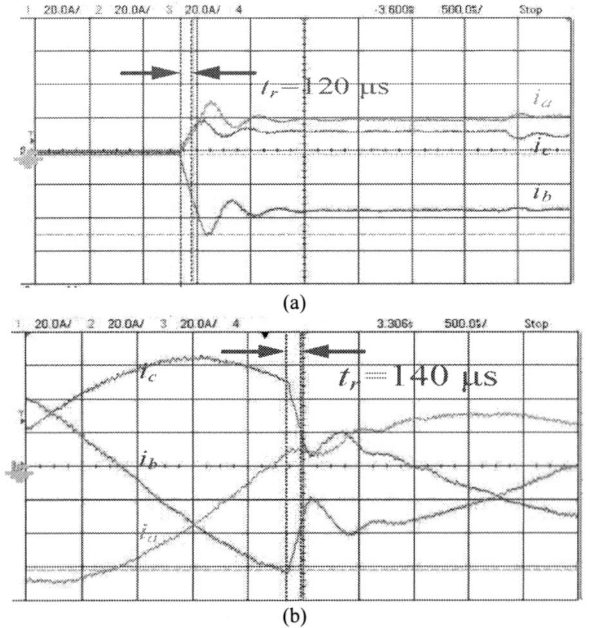

(a)

(b)

Fig. 9. Transient 3Φ currents experiments under (a) $i_{q,ref} = 0 \rightarrow 34$ A and (b) $i_{q,ref} = 67 \rightarrow 34$ A.

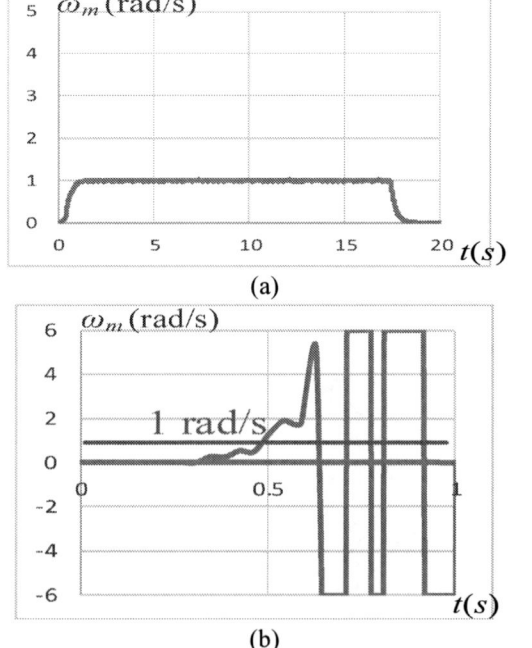

(a)

(b)

Fig. 12. Low-speed control (a) with VCT and (b) without VCT.

V. CONCLUSIONS

This paper has applied the proposed current DDC for a PMSM drive, suitable for the simple speed control and the VCT control. The simple speed-control laws have also been derived based on the concept of deadbeat control and DOB control. Furthermore, the cogging torque issue in the low-speed operation has been resolved by VCT. The experiments have verified that the proposed current DDC can achieve high accuracy of 99.56%, fast response of 2 to 3 times switching period and low torque ripple, resulting in the accurate torque control and stable speed control. The experiments of speed control have verified that the proposed control law can reach near 1st-order response. In addition, the experiment of low-speed VCT control with the DDC has resolved the cogging torque issue. According to the experiments, we have shown that the current DDC can handle motor torque, speed and VCT controls.

REFERENCES

[1] S. S. R. Kiranmayi and P. N. Mandadi, "A Review on Speed Control of Permanent Magnet Synchronous Motor Drive Using Different Control Techniques," presented at the *2018 International Conference on Power, Energy, Control and Transmission Systems (ICPECTS)*, Chennai, India, 22-23 February, 2018.

[2] C. He, H. Xiong and Z. Shen, "Current Loop Control Strategy of PMSM Based on Fractional Order PID Control Technology," presented at the *2021 6th Asia Conference on Power and Electrical Engineering (ACPEE)*, Chongqing, China, 08-11 April, 2021.

[3] R. Kumar, R. A. Gupta and B. Singh, "Intelligent Tuned PID Controllers for PMSM Drive - A Critical Analysis," presented at the *2006 IEEE International Conference on Industrial Technology*, Mumbai, India, 15-17 December, 2006.

[4] S. Fukuda and R. Imamura, "Application of a sinusoidal internal model to current control of three-phase utility-interface converters," *IEEE Transactions on Industrial Electronics*, vol. 52, no. 2, pp. 420 - 426, 04 April 2005.

[5] B. K. Bose, "An Adaptive Hysteresis-Band Current Control Technique of a Voltage-Fed PWM Inverter for Machine Drive System," *IEEE Transactions on Industrial Electronics*, vol. 37, no. 5, pp. 402 - 408, October 1990.

[6] L. Malesani and P. Tenti, "A Novel Hysteresis Control Method for Current-Controlled Voltage-Source PWM Inverters with Constant Modulation Frequency," *IEEE Transactions on Industry Applications*, vol. 26, no. 1, pp. 88 - 92, Jan.-Feb. 1990.

[7] L. A. Zadeh, "Fuzzy Sets," *Information and Control*, vol. 8, no. 3, pp. 338-353, 1965.

[8] Y. Zhang and H. Lin, "Simplified Model Predictive Current Control Method of Voltage-Source Inverter," presented at the *8th International Conference on Power Electronics - ECCE Asia*, Jeju, Korea (South), 30 May - 03 June, 2011.

[9] S. J. Qin and T. A. Badgwell, "A Survey of Industrial Model Predictive Control Technology," *Control Engineering Practice*, vol. 11, no. 8, pp. 733-764, July 2003.

[10] T.-F. Wu, C.-H. Chang, L.-C. Lin, Y.-C. Chang, and Y.-R. Chang, "Two-Phase Modulated Digital Control for Three-Phase Bidirectional Inverter With Wide Inductance Variation," *IEEE Transactions on Power Electronics*, vol. 28, no. 4, pp. 2598-1607, 10 May 2013.

[11] T.-F. Wu, C.-H. Chang, L.-C. Lin, G.-R. Yu, and Y.-R. Chang, "A D-Σ Digital Control for Three-Phase Inverter to Achieve Active and Reactive Power Injection," *IEEE Transactions on Industrial Electronics*, vol. 61, no. 8, pp. 3879 - 3890, 22 October 2013.

[12] T.-F. Wu, H.-C. Hsieh, C.-H. Chang, L.-C. Lin, and Y.-R. Chang, "Improvement of Control Law Derivation and Region Selection for D−Σ Digital Control," *IEEE Transactions on Industrial Electronics*, vol. 62, no. 10, 24 April 2015.

[13] T.-F. Wu, Y.-H. Huang, Y.-T. Liu, and M. Misra, "Decoupled Direct Digital Control with D-Σ Process and Average Common-Mode Voltage Model for 3Φ3W LCL Converters," presented at the *2019 IEEE Applied Power Electronics Conference and Exposition (APEC)*, Anaheim, CA, USA, 27 May, 2019.

[14] V. M. Bida, D. V. Samokhvalov and F. S. Al-Mahturi, "PMSM Vector Control Techniques — A Survey," presented at the *2018 IEEE Conference of Russian Young Researchers in Electrical and Electronic Engineering (EIConRus)*, Moscow and St. Petersburg, Russia, 29 January - 01 February, 2018.

[15] D. Mohanraj, J. Gopalakrishnan, B. Chokkalingam, and L. Mihet-Popa, "Critical Aspects of Electric Motor Drive Controllers and Mitigation of Torque Ripple—Review," *IEEE Access*, vol. 10, pp. 73635 - 73674, 30 June 2022.

[16] S. Bouchiker, G.-A. Capolino and M. Poloujadoff, "Vector Control of a Permanent-Magnet Synchronous Motor Using AC-AC Matrix Converter," *IEEE Transactions on Power Electronics*, vol. 13, no. 6, pp. 1089 - 1099, November 1998.

[17] L. C. Westphal, *Handbook of Control Systems Engineering*. NY: Springer New York, 2001.

[18] S. Wang and S. Wan, "Full Digital Deadbeat Speed Control for Permanent Magnet Synchronous Motor with Load Compensation," *IET Power Electronics*, vol. 6, no. 4, pp. 634 - 641, 2013.

[19] M. Iwasaki and N. Matusi, "Robust Speed Control of IM with Torque Feedforward Control," *IEEE Transactions on Industrial Electronics*, vol. 40, no. 6, pp. 553-560, 1993.

[20] K. Kyeong-Hwa and Y. Myung-Joong, "A Nonlinear Speed Control for a PM Synchronous Motor Using a Simple Disturbance Estimation Technique," *Transactions on Industrial Electronics*, vol. 49, no. 3, pp. 524-535, 2002.

[21] J. Han, H. Kim, Y. Joo, N. H. Jo, and J. H. Seo, "A Simple Noise Reduction Disturbance Observer and Q-Filter Design for Internal Stability," presented at the *2013 13th International Conference on*

979-8-3315-1612-3/25 $31.00 © 2025 IEEE

Control, Automation and Systems, Gwangju, Korea (South), 20-23 October, 2013.

[22] W.-H. Chen, J. Yang and L. G. S. Li, "Disturbance-Observer-Based Control and Related Methods—An Overview," *IEEE Transactions on Industrial Electronics*, vol. 63, no. 2, pp. 1083 - 1095, 14 September 2016.

[23] F. Bu *et al.*, "Speed Ripple Reduction of Direct-Drive PMSM Servo System at Low-Speed Operation Using Virtual Cogging Torque Control Method," *IEEE Transactions on Industrial Electronics*, vol. 68, no. 1, pp. 160 - 174, 03 January 2021.

[24] R. H. Park, "Two-Reaction Theory of Synchronous Machines Generalized Method of Analysis-Part I," *Transactions of the American Institute of Electrical Engineers*, vol. 48, no. 3, pp. 716-727, 1929.

[25] S. Li, J. Yang, W.-H. Chen, and X. Chen, *Disturbance Observer-Based Control Methods and Applications*. CRC Press, 2016.

Fault-Tolerant Multilevel Converter for Multiphase Switched Reluctance Motor Drives Based on q+2 Converter

Mahmoud A. Gaafar
Aswan Power Electronics Applications Research Center (APEARC)
Aswan University
Aswan, Egypt
mgaafar@apearc.aswu.edu.eg

Mohamed Orabi
Aswan Power Electronics Applications Research Center (APEARC)
Aswan University
Aswan, Egypt
morabi@apearc.aswu.edu.eg

Hao Chen
School of Electrical and Power Engineering
China University of Mining and Technology
Xuzhou, China
hchen@cumt.edu.cn

Mostafa Dardeer
Aswan Power Electronics Applications Research Center (APEARC)
Aswan University
Aswan, Egypt
mdardeer@aswu.edu.eg

Abstract— This paper proposes a new multilevel switched reluctance motor (SRM) converter based on *q+2* converter. The proposed converter can offer the three operating modes of SRM (magnetization, regenerative demagnetization, and freewheeling modes). In addition, it can offer simultaneous operation of SRM phases at any of these modes. Thus, it can be used for multiphase SRM. Furthermore, it offers redundant switching states for the SRM operating modes. Thus, it has good fault-tolerance capability. Comparative study between the proposed converter and other SRM converters is introduced. Compared to other multilevel converters (T-type and NPC), the proposed converter uses the lowest number of semiconductors. Real-time simulation results using OPAL-RT are presented to verify the proper operation of the proposed converter. The operation of the proposed converter is investigated also under open-circuit fault condition to verify its fault-tolerance capability.

Keywords— *Switched reluctance motor, converter, multilevel*

I. INTRODUCTION

Switched reluctance motors (SRMs) do not need permanent magnets nor conductors on the rotor, and its stator windings are simple. This structure brings many attractive features for SRMs compared to other types of machines. These features include low manufacturing cost, easy cooling, robust operation at high speeds and temperatures, and inherent fault tolerance. Thus, SRMs are recently gaining increased interest in many applications [1]-[5]. However, due to high levels of torque ripple and acoustic noise, SRM drive system needs further development for wide adoption especially in high-performance applications such as EV [6]-[10]. Increasing the number of phases can be an effective way to reduce the torque ripple and to increase the torque per phase current ratio [11]-[15]. For instance, the torque ripple in the six-phase SRM is one-third of that in three-phase SRM. In addition, increasing the number of SRM phases can significantly enhance the motor fault-tolerance capability. For instance, in case of one phase fault, the developed torque in six-phase SRMs is higher than that in three-phase SRMs [11], [14], [16]. However, increasing the number

of SRM phases increases the number of semiconductor devices required by SRM converter. This increases the cost and volume [6], [13].

Ideally, the SRM converter should offer the following features [6], [17]:1) It should offer three modes of operation for the SRM phases; magnetization, regenerative demagnetization, and freewheeling modes. 2) It should offer independent control for each phase. Thus, many phases can be simultaneously exited; this can significantly facilitate proper application of torque ripple minimization algorithms, such as torque sharing function (TSF). In multiphase SRM, simultaneous excitation of more than one phase is usually required. 3) It should have fault-tolerance capability through offering redundant switching states. 4) It should use low number of semiconductor devices. 5) It should offer multilevel operation; this can give better flexibility for current control, especially at high speeds.

Asymmetric Half-bridge (AHB) converter – shown in Fig. 1(a) – is one of the earliest used SRM topologies. It needs $2q$ active switches and $2q$ diodes; where q is the number of SRM phases [6], [17]. This high number of semiconductors along with the absence of any redundant switching states for phases magnetization makes AHB not the optimum choice for multiphase SRMs. The split dc converter, introduced in [18], utilizes one active switch and one diode per phase. However, this converter does not offer freewheeling mode. In addition, only half of the input voltage can be used to excite SRM phases. Also, it does not offer fault-tolerance capability. Number of SRM converters utilize common semiconductor devices to be shared by SRM phases. Among these converters, the $(q + 1)$ switch converter (known as Miler converter) uses a common switch and only one power switch per phase [19]. On the other hand, the $(q+2)$ converter utilizes two common switches and one switch per phase [13], [20], [21]. A q-switch converter is used with six-phase SRM in [15], [22]; this converter requires only one switch per phase, each phase shares one half bridge arm with the outgoing phase and another half bridge arm with the incoming phase.

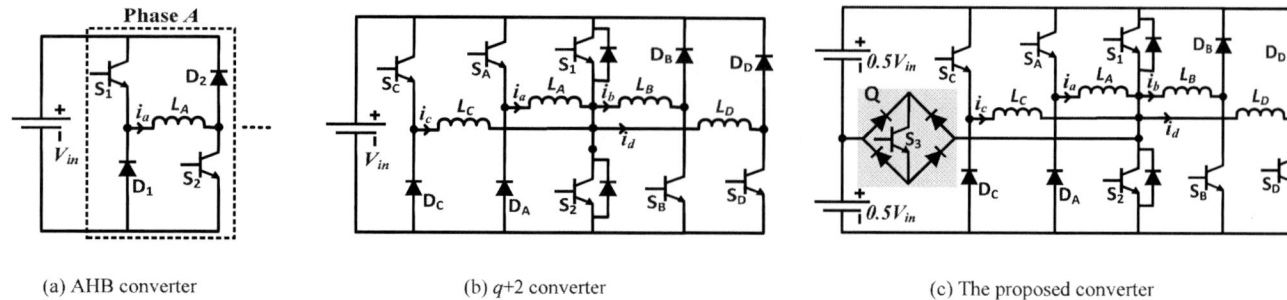

(a) AHB converter (b) *q+2* converter (c) The proposed converter

Fig. 1 Topologies for AHB converter, *q+2* converter, and the proposed converter

However, in (*q+1*) and (*q+2*) converters [13], [19]-[21], failure of any of the common switches will cause complete disable for the SRM. On the other hand, in q-switch converter [15], [22], failure of any of the switches will cause disable for 2 phases. Accordingly, the reliability of these converters is significantly degraded compared to AHB converter in which failure of any switch disables only the corresponding SRM phase. In addition, independent control of each phase cannot be realized in these converters.

This paper proposes a new multilevel SRM converter based on q+2 topology shown in Fig. 1(b). The proposed converter is shown in Fig. 1(c) for 4-phase SRM. Compared to q+2 converter, a split source is used in the input of the proposed converter. In addition, a four-quadrant switch (denoted as Q) is used. This switch is realized by one active switch (S3) and four diodes bridge. The midpoint of the split source is connected into the common point of the SRM phases through Q.

II. OPERATION DESCRIPTION OF THE PROPOSED CONVERTER

For 4-phase SRM, Figs. 2 and 3 show the different operational circuits of the proposed converter.

- Figs. 2 (a-d) shows the circuits for the three operating modes (magnetization, regenerative demagnetization, and freewheeling) during single phase excitation with the complete bus voltage (V_{in}). On the other hand, Figs. 3 (a, b) shows the circuits for simultaneous magnetization, and regenerative demagnetization modes of two successive phases (A and B) with the complete bus voltage (V_{in}). It is worth to state that the circuits of Figs. 2 (a-d) and Figs. 3 (a, b) are inherited from q+2 converter.

- For single phase excitation, the proposed converter offers additional circuits for magnetization and regenerative demagnetization modes with the half bus voltage ($0.5V_{in}$); these circuits are shown in Figs. 2 (e, f). Also, the proposed converter offers additional circuits for simultaneous magnetization, and regenerative demagnetization modes of two successive phases with the half bus voltage ($0.5V_{in}$); these circuits are shown in Figs. 3 (c, d). In Figs 3 (c, d), the current direction in the four-quadrant switch connected to the midpoint of the input source is determined according to the values of the phase currents.

- Figs. 3 (e, f) show the circuits for simultaneous magnetization of two successive phases and simultaneous regenerative demagnetization of two successive phases, respectively. It is worth to state that q+2 converter cannot offer these simultaneous modes in two successive phases.

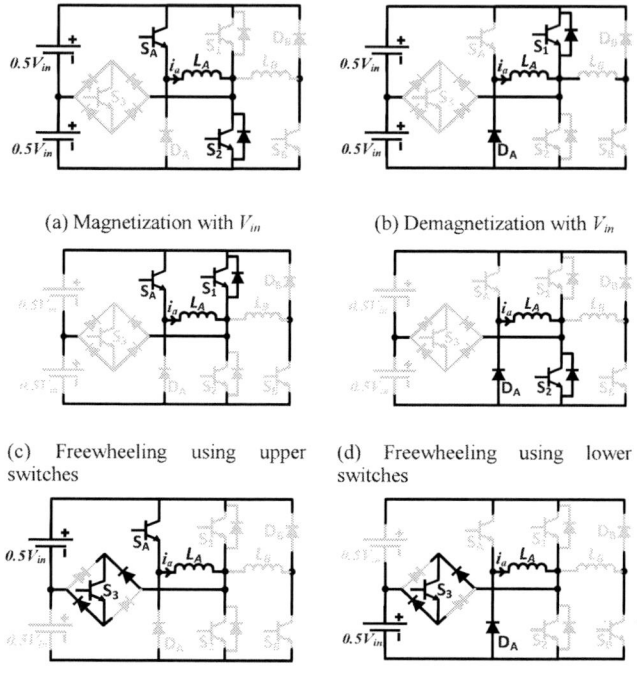

(a) Magnetization with V_{in} (b) Demagnetization with V_{in}

(c) Freewheeling using upper switches (d) Freewheeling using lower switches

(e) Magnetization with $0.5V_{in}$ (f) Demagnetization with $0.5V_{in}$

Fig. 2 Operational circuits during single phase excitation (phase A only)

The modes in Figs. 3 (e, f) are usually needed in multiphase motors. In Figs. 3 (e, f), the current direction in the four-quadrant switch Q connected to the midpoint of the input source is determined according to the value of the phase currents.

III. CONTROL SYSTEM OF THE PROPOSED CONVERTER

Fig. 4 shows the control system adopted for the proposed converter. It consists of PI controller (expressed in (1) as *G*) to regulate the SRM speed. The output of the speed controller along with the values of turn-on and turn-off angles (θ_{on} and θ_{off}) of the SRM phases, the required current waveforms of the different SRM phases are estimated. According to the estimated phase reference currents and the measurements of the rotor position and the phases currents, Hysteresis controller is employed to determine the required phases voltages (V_{in}, $0.5V_{in}$, 0, $-0.5V_{in}$, or $-V_{in}$), and accordingly generate the appropriate switching pulses.

979-8-3315-1612-3/25 $31.00 © 2025 IEEE   907

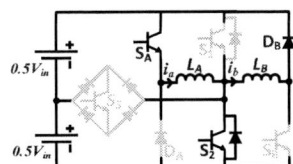

(a) Simultaneous magnetization of phase A and demagnetization of phase B (with V_{in})

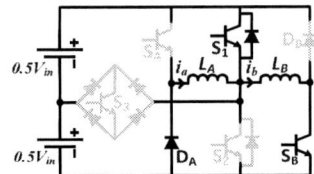

(b) Simultaneous magnetization of phase B and demagnetization of phase A (with V_{in})

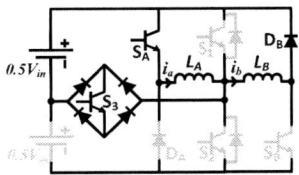

(c) Simultaneous magnetization of phase A and demagnetization of phase B (with $0.5V_{in}$)

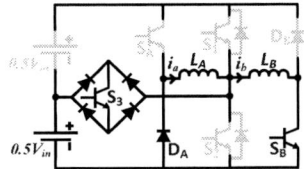

(d) Simultaneous demagnetization of phase A and magnetization of phase B (with $0.5V_{in}$)

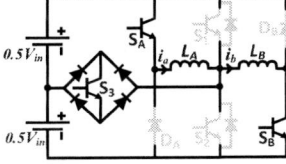

(e) Simultaneous magnetization of phase A and phase B (with $0.5V_{in}$)

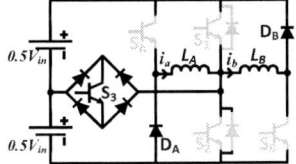

(f) Simultaneous demagnetization of phase A and phase B (with $0.5V_{in}$)

Fig. 3 Operational circuits during simultaneous magnetization and demagnetization of two phases.

$$G = K_p + \frac{K_I}{s} \qquad (1)$$

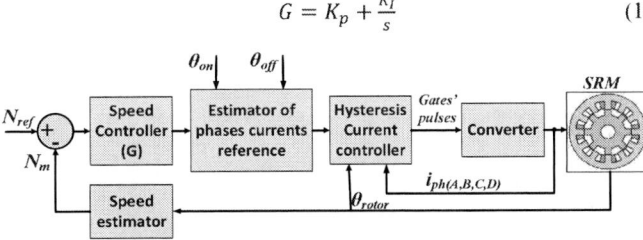

Fig. 4 Control system of the SRM drive system; where θ_{on} and θ_{off} are the turn of and turn off angles, respectively.

IV. RESULTS

The simulation parametrs are listed in Table I. An 4-phase 8/6 SRM is consider. The simulation is carried out under steady state conditions at two reference speeds of 1000 rpm and 600 rpm. For these conditions, the simulation is carried out under healthy mode (where there is no faulty switches) and switch fault mode (where one of the common switches is faulty). For each case, the phase conduction angles (θ_{on} and θ_{off}) are selected to reduce the torque ripple as possible. The torque ripple is determined for each case as in (2) and the results are listed in Table II where T_l is the load torque, and T_{max} and T_{min} are the maximum and minimum values of the motor torque, respectively.

Table I. Specifications of the simulation system

Parameter	Value
SRM configuration	8/6 (4 phases)
Input voltage (V_{in})	400 V
Load torque (T_L)	15 Nm

$$T_{ripple} = \frac{T_{max}-T_{min}}{T_l} x100 \qquad (2)$$

A. Healthy mode Results

The waveforms are shown in Figs 5 and 6 for reference speeds of 1000 rpm and 600 rpm, respectively. The waveforms show that the proposed converter can operate the SRM properly at the two speeds. Toque ripples of 30% and 73% are determined for the speeds 1000 rpm and 600 rpm, respectively. Although the torque ripple value at speed of 600 rpm is relatively high, sophisticated algorithms can be applied for torque ripple reduction; e.g. torque sharing function (TSF). Real-time simulation using Opal-RT (OP4510) is carried out also to verify the performance of the proposed converter. The results for 1000 rpm are shown in Figs. 7 and 8.

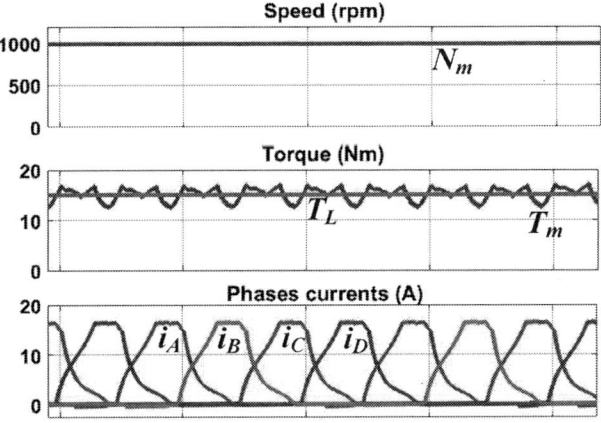

Fig. 5 Waveforms in healthy mode at reference speed of 1000 rpm

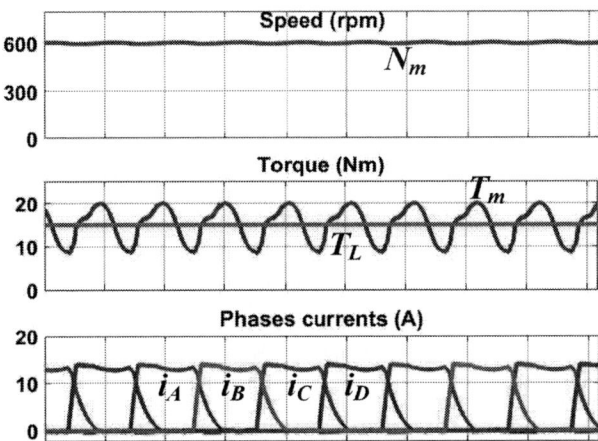

Fig. 6 Waveforms in healthy mode at reference speed of 600 rpm.

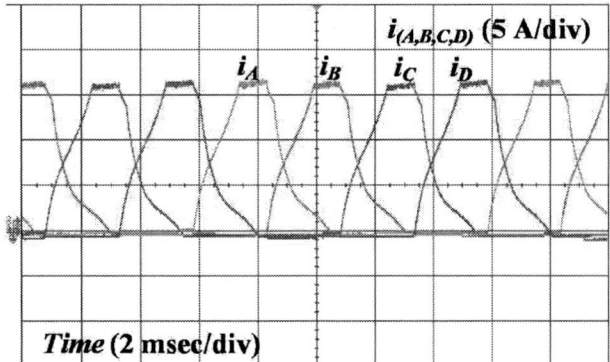

Fig. 7 waveforms for the motor phases' currents (i_A, i_B, i_C, i_D).

Fig. 8 waveforms for the motor torque (T_m), the load torque (T_L), and the motor speed (N_m)

B. Fault-tolerance study

For $q+2$ converter, failure of one of the common switches (S_1 and S_2) will disable the magnetization mode for half of the SRM phases, and disable the demagnetization mode for the other half of SRM phases. This can be recognized from the operational circuits shown in Fig. 2 (a, b) and Fig. 3 (a, b). Thus, complete disable for the SRM is inevitable in this case.

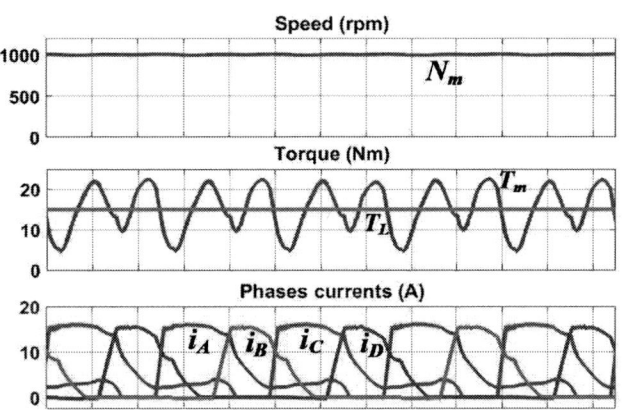

Fig. 9 Waveforms under open-circuit fault of switch S_1 at reference speed of 1000 rpm.

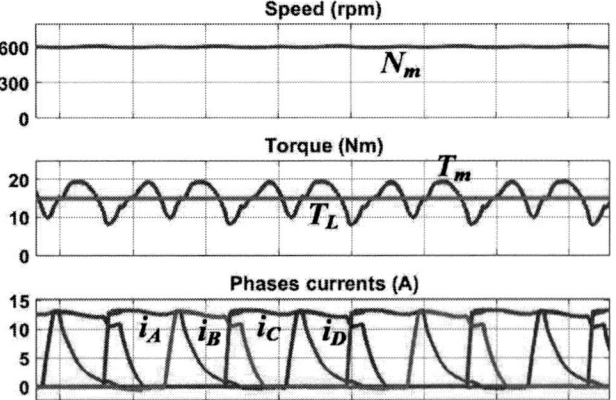

Fig. 10 Waveforms under open-circuit fault of switch S_1 at reference speed of 600 rpm.

On the other hand, the proposed converter can still provide magnetization and demagnetization modes for all the SRM phases in case of any of the common switches is faulty. In case of fault of the four-quadrant switch Q, the converter can be operated as $q+2$ converter where three voltage levels of (V_{in}, 0, $-V_{in}$) can be offered to achieve the magnetization, regenerative demagnetization, and freewheeling modes. On the other hand, in case of fault of any of the switches S_1 or S_2, the proposed converter can still provide 4 voltage states for each phase. Table III summarizes the available voltage states for each phase in case of fault of the switches S_1 or S_2. Thus, the proposed converter can still achieve the magnetization, regenerative demagnetization, and freewheeling modes.

To verify the fault tolerance ability of the proposed converter, the above simulations are repeated when open-circuit fault is considered for the switch S_1. The waveforms are shown in Figs 9 and 10 for reference speeds of 1000 rpm and 600 rpm, respectively. Under this fault mode, toque ripples of 113% and 73% are determined for the speeds 1000 rpm and 600 rpm, respectively. Although the torque ripple is increased at 1000 rpm speed, the waveforms show that the motor is still able to operate with the load torque at the required speeds. On the other hand, it is shown that the torque ripple value is not changed at speed of 600 rpm.

Table II. Torque ripples for the simulation waveforms

Operating condition	1000 rpm			600 rpm		
	T_{min}	T_{max}	T_{ripple}	T_{min}	T_{max}	T_{ripple}
Healthy condition	12.5	17	30%	9	20	73%
Fault condition	5	22	113%	8	19	73%

Table III. Available phase voltage states with open-circuit fault in switch S_1 or S_2

SRM Phases	Open-circuit fault in S_1					Open-circuit fault in S_2				
	V_{in}	$0.5V_{in}$	0	$-0.5V_{in}$	$-V_{in}$	V_{in}	$0.5V_{in}$	0	$-0.5V_{in}$	$-V_{in}$
A & C	√	√	√	√	×	×	√	√	√	√
B & D	×	√	√	√	√	√	√	√	√	×

979-8-3315-1612-3/25 $31.00 © 2025 IEEE

V. Comparative Study

Table IV summarizes the features of the proposed converter compared to other SRM converters. The number of switches is expressed in terms of the number of phases. Accordingly, the number of switches is calculated for 4-phase SRM. The features of the proposed SRM can be summarized as follow:

- Compared to $q+2$ converter, the proposed converter offers alternative circuits for magnetization and regenerative demagnetization. Thus, failure of one of the common switches will not cause complete disable of SRM. In addition, it offers simultaneous magnetization and simultaneous regenerative demagnetization of two successive SRM phases. This simultaneous operation is not offered by $q+2$ converter. Finally, it offers multilevel operation.

- Compared to AHB converter, the proposed converter uses a lower number of active switches for SRMs with 4 phases or more. In addition, it offers multilevel operation.

- Compared to other multilevel converters (T-type and NPC), the proposed one uses the lowest number of semiconductors (active switches and diodes).

Table IV. Comparison between the proposed SRM converter and other converters

Factor of comparison		AHB	T-Type	NPC	q+2	Proposed converter
No. of switches (4-phase SRM)		2q (8)	4q (16)	4q (16)	q+2 (6)	q+3 (7)
No of diodes (4-phase SRM)		2q (8)	2q (8)	4q (16)	q+2 (6)	q+6 (10)
No. of voltage levels		3	5	5	3	5
Availability of alternative switching states (in case of failure of one of the common switches)		No common switches			x	√
Availability of simultaneous operation of two successive phases	SMR	√	√	√	√	√
	SM	√	√	√	x	√
	SR	√	√	√	x	√

SMR: simultaneous magnetization and regenerative demagnetization of two phases.
SM: simultaneous magnetization of two phases.
SR: simultaneous regenerative demagnetization of two phases.

VI. Conclusion

This paper proposes a new multilevel converter to be used with SRM. The operational circuits of the proposed converter are illustrated. It is shown that it can offer the following merits: 1) It offers three modes of operation; magnetization, regenerative demagnetization, and freewheeling modes. 2) It offers simultaneous magnetization and/or demagnetization of SRM phases. 3) It uses the lowest number of semiconductors compared to other multilevel SRM converters. 4) It offers good fault-tolerant characteristics compared to other converters which employ common switches to be shared by SRM phases. These characteristics makes it good candidate to be used with multiphase SRMs. The performance of the proposed converter is verified under healthy and fault conditions. The results show that the converter can operate the SRM properly at different speeds under both healthy and fault modes.

Acknowledgment

This work was supported by Academy of Scientific Research and Technology (ASRT), Egypt under grant program for Egypt-China collaboration project entitled "Basic Research on Switched Reluctance Motor Drive System for Electric Vehicle".

References

[1] P. Azer, B. Bilgin, and A. Emadi, "Comprehensive Analysis and Optimized Control of Torque Ripple and Power Factor in a Three-Phase Mutually Coupled Switched Reluctance Motor With Sinusoidal Current Excitation," IEEE Trans. Power Electron., vol. 36, no. 6, pp. 7150–7164, Jun. 2021, doi: 10.1109/TPEL.2020.3038741.

[2] L. Ge, H. Xu, Z. Guo, S. Song, and R. W. De Doncker, "An Optimization-Based Initial Position Estimation Method for Switched Reluctance Machines," IEEE Trans. Power Electron., vol. 36, no. 11, pp. 13285–13292, Nov. 2021, doi: 10.1109/TPEL.2021.3081618.

[3] Z. Yu, C. Gan, K. Ni, Y. Chen, and R. Qu, "Dual-Electric-Port Bidirectional Flux-Modulated Switched Reluctance Machine Drive With Multiple Charging Functions for Electric Vehicle Applications," IEEE Trans. Power Electron., vol. 36, no. 5, pp. 5818–5831, May 2021, doi: 10.1109/TPEL.2020.3029822.

[4] Z. Yu, C. Gan, Y. Chen, and R. Qu, "DC-Biased Sinusoidal Current Excited Switched Reluctance Motor Drives Based on Flux Modulation Principle," IEEE Trans. Power Electron., vol. 35, no. 10, pp. 10614–10628, Oct. 2020, doi: 10.1109/TPEL.2020.2975121.

[5] B. Bilgin, J. W. Jiang, and A. Emadi, Eds., Switched Reluctance Motor Drives: Fundamentals to Applications, 1st ed. First edition. | Boca Raton, FL : CRC Press/Taylor & Francis Group, 2018.: CRC Press, 2019. doi: 10.1201/9780203729991.

[6] M. A. Gaafar, A. Abdelmaksoud, M. Orabi, H. Chen, and M. Dardeer, "Switched Reluctance Motor Converters for Electric Vehicles Applications: Comparative Review," IEEE Trans. Transp. Electrific., vol. 9, no. 3, pp. 3526–3544, Sep. 2023, doi: 10.1109/TTE.2022.3192429.

[7] D. Ronanki, K. R. Pittam, A. Dekka, P. Perumal, and A. R. Beig, "Phase Current Reconstruction Method With an Improved Direct Torque Control of SRM Drive for Electric Transportation Applications," IEEE Trans. on Ind. Applicat., vol. 58, no. 6, pp. 7648–7657, Nov. 2022, doi: 10.1109/TIA.2022.3196329.

[8] V. Shah and S. Payami, "Integrated Converter With G2V, V2G, and DC/V2V Charging Capabilities for Switched Reluctance Motor Drive-Train Based EV Application," IEEE Trans. on Ind. Applicat., vol. 59, no. 3, pp. 3837–3850, May 2023, doi: 10.1109/TIA.2023.3242636.

[9] M. A. Gaafar, A. Abdelmaksoud, M. Orabi, H. Chen and M. Dardeer, "Performance Investigation of Switched Reluctance Motor Driven by Quasi-Z-Source Integrated Multiport Converter with Different Switching Algorithms". Sustainability. Vol. 13, no. 17, doi:10.3390/su13179517.

[10] A. Abdelmaksoud, M. A. Gaafar, M. Orabi, H. Chen and M. Dardeer, "Performance Investigation of Switched Reluctance Motor Drive System Under Firing Angles Variation," *2023 IEEE Conference on Power Electronics and Renewable Energy (CPERE)*, Luxor, Egypt, 2023, pp. 1-8, doi: 10.1109/CPERE56564.2023.10119624.

[11] S. Han, K. Diao, and X. Sun, "Overview of multi-phase switched reluctance motor drives for electric vehicles," Advances in Mechanical Engineering, vol. 13, no. 9, p. 168781402110451, Sep. 2021, doi: 10.1177/16878140211045195.

[12] X. Deng, B. Mecrow, R. Martin, and S. Gadoue, "Effects of Winding Connection on Performance of a Six-Phase Switched Reluctance Machine," IEEE Trans. Energy Convers., vol. 33, no. 1, pp. 166–178, Mar. 2018, doi: 10.1109/TEC.2017.2750804.

[13] M. Guan, C. Liu, P. Liu, and X. Sun, "Fault Diagnosis Scheme for Open-Circuit Fault in N+2 Converter of Six-Phase SRM Using Midpoint Current Spectrum Analysis," IEEE Access, vol. 10, pp. 105983–105992, 2022, doi: 10.1109/ACCESS.2022.3211977.

[14] X. Sun, Z. Xue, S. Han, X. Xu, Z. Yang, and L. Chen, "Design and Analysis of a Novel 16/10 Segmented Rotor SRM for 60V Belt-Driven Starter Generator," Journal of Magnetics, vol. 21, no. 3, pp. 393–398, Sep. 2016, doi: 10.4283/JMAG.2016.21.3.393.

[15] X. Deng, B. Mecrow, H. Wu, and R. Martin, "Design and Development of Low Torque Ripple Variable-Speed Drive System With Six-Phase Switched Reluctance Motors," IEEE Trans. Energy Convers., vol. 33, no. 1, pp. 420–429, Mar. 2018, doi: 10.1109/TEC.2017.2753286.

[16] A. Dorneles Callegaro, J. Liang, J. W. Jiang, B. Bilgin, and A. Emadi, "Radial Force Density Analysis of Switched Reluctance Machines: The Source of Acoustic Noise," IEEE Trans. Transp. Electrific., vol. 5, no. 1, pp. 93–106, Mar. 2019, doi: 10.1109/TTE.2018.2887338.

[17] B. Bilgin, J. W. Jiang, and A. Emadi, Eds., Switched Reluctance Motor Drives: Fundamentals to Applications, 1st ed. First edition. | Boca Raton, FL : CRC Press/Taylor & Francis Group, 2018.: CRC Press, 2019. doi: 10.1201/9780203729991.

[18] K. Ha, C. Lee, J. Kim, R. Krishnan, and S.-G. Oh, "Design and Development of Low-Cost and High-Efficiency Variable-Speed Drive System With Switched Reluctance Motor," IEEE Trans. on Ind. Applicat., vol. 43, no. 3, pp. 703–713, 2007, doi: 10.1109/TIA.2007.895744.

[19] C. Pollock and B. W. Williams, "Power convertor circuits for switched reluctance motors with the minimum number of switches," IEE Proc. B Electr. Power Appl. UK, vol. 137, no. 6, p. 373, 1990, doi: 10.1049/ip-b.1990.0046.

[20] S. Dai, C. Zhou, and C. Liu, "DC-biased SPWM voltage control for six-phase switched reluctance motor drive based on N+2 power electronic converter," in IECON 2015 - 41st Annual Conference of the IEEE Industrial Electronics Society, Yokohama: IEEE, Nov. 2015, pp. 000025–000030. doi: 10.1109/IECON.2015.7392959.

[21] M. Guan, C. Liu, S. Han, and X. Sun, "Analysis of Midpoint Current Characteristics for Novel Six-Phase N+2 Power Converter in Different Working Condition," IEEE Access, vol. 8, pp. 105104–105117, 2020, doi: 10.1109/ACCESS.2020.2999933.

[22] Y. Hu, T. Wang, and W. Ding, "Performance Evaluation on a Novel Power Converter With Minimum Number of Switches for a Six-Phase Switched Reluctance Motor," IEEE Trans. Ind. Electron., vol. 66, no. 3, pp. 1693–1702, Mar. 2019, doi: 10.1109/TIE.2018.2840480

Uncertainty-Aware Artificial Intelligence for Gear Fault Diagnosis in Motor Drives

Subham Sahoo*, Huai Wang and Frede Blaabjerg

Department of Energy
Aalborg University
Aalborg, Denmark
e-mail: {`sssa, hwa, fbl`}@energy.aau.dk

Abstract—This paper introduces a novel approach to quantify the uncertainties in fault diagnosis of motor drives using Bayesian neural networks (BNN). Conventional data-driven approaches used for fault diagnosis often rely on point-estimate neural networks, which merely provide deterministic outputs and fail to capture the uncertainty associated with the inference process. In contrast, BNNs offer a principled framework to model uncertainty by treating network weights as probability distributions rather than fixed values. It offers several advantages: (a) improved robustness to noisy data, (b) enhanced interpretability of model predictions, and (c) the ability to quantify uncertainty in the decision-making processes. To test the robustness of the proposed BNN, it has been tested under a conservative dataset of gear fault data from an experimental prototype of three fault types at first, and is then incrementally trained on new fault classes and datasets to explore its uncertainty quantification features and model interpretability under noisy data and unseen fault scenarios.

Index Terms—Power electronics, Artificial intelligence, Fault diagnosis, Uncertainty-aware AI, Uncertainty quantification

I. INTRODUCTION

Models developed using deep learning are widely used in all types of inference and decision making in the field of power electronics [1]. In other words, it is becoming increasingly important to assess the reliability and effectiveness of artificial intelligence (AI) models before putting them into practice. This is because the predictions of AI are usually affected by noise and model output errors, leading to unexplainable results [2]. These uncertainties arise when a mismatch between the testing and training data is encountered. Although these uncertainties can have a significant impact on the trained AI model's estimation capabilities, it is difficult to compensate for the uncertainties arising out of model knowledge uncertainty. As a result, such uncertainties in data and models can be segmented into two categories, namely *aleatoric* and *epistemic* uncertainty.

A. Classification of uncertainties in AI

By definition, statistical inconsistencies in data leading to prediction uncertainty by an AI model is called as *aleatoric* uncertainty (commonly referred as data uncertainty). This type

This work was supported by Innovation Fund Denmark through the project AI-Power [19]: Artificial Intelligence for Next-Generation Power Electronics. *(Corresponding author: Subham Sahoo)*

of uncertainty is an inherent property of inconsistent data distribution, which ultimately becomes a barrier in distinguishing between overlapping data groups.

In contrast, *epistemic* uncertainty (commonly referred as model uncertainty) occurs due to inadequate knowledge of the model. Even in scenarios when the data is sufficient, their valuation can still be deemed as information-poor from a contextual data collection perspective. In such cases, AI-based methods are usually referred to characterize the emergent features of the data. However, since the data required for developing AI-based methods can be rather incomplete, noisy, discordant, the predictions are not always accurate. This aspect has been accounted in Fig. 2(a) performed on a sinusoidal signal, and corresponding data is extracted to map the *true signal*. Based on our definitions above, it can be seen that some data do not really characterize themselves close to the sinusoidal variations and naturally increase the aleatoric uncertainty (around t = 3.5 sec), whereas the epistemic uncertainty is seen around t = [5, 6] sec, where missing data doesn't transcend to the actual model information.

B. Key reasons behind uncertainties in AI predictions

Limited data: Going beyond limited training to have an highly accurate ensemble model, there are many practical scenarios which does not invoke more data because of the risks. One such example for power electronic applications is fault based scenarios. Since fault data is limited and risky to be emulated in the lab, NNs can easily provide overconfident decisions when trained over limited set of data. This has been clearly illustrated in Fig. 1(b), where overfitting over less data can cause large deviations from the actual polynomial model trajectory (in red).

Unseen data/scenarios: The state-of-the art methods rely on the training data with a strong assumption that it is well connected with out-of-distribution (OOD) data. However, it can be a very strong assumption specifically for research in fault diagnosis, where the nature, type and properties of a new fault can significantly vary from the past training data. Basically, the state-of-the-art methods rely on a *frequentist* deep learning approach, that represents emergent behavior of data merely as deterministic values to deliver a point-estimate prediction. These predictions, in turn, can often be over-

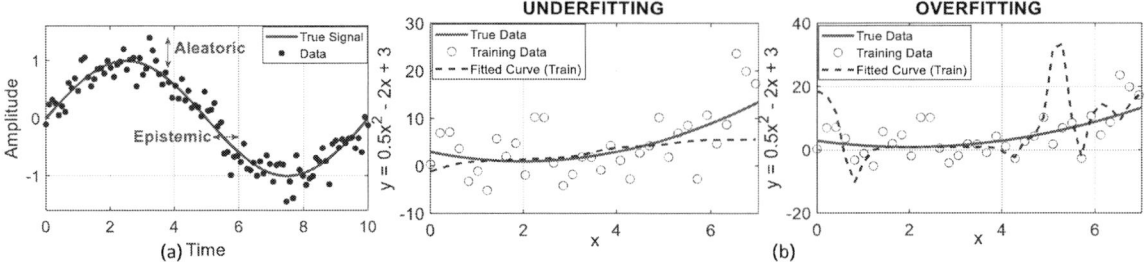

Fig. 1. (a) Schematic view of the categorical differences between aleatoric and epistemic uncertainty – the former is aimed at noisy diverging data and the latter focused on missing information, (b) Overfitting issue caused by NNs due to training over limited data – as the training data corresponding to the polynomial $y = 0.5x^2 - 2x + 3$ is collected aimed at regressing over the true data, overfitting over minimal points can cause a large deviation from the actual model.

confident and project a false representation about the system due to unseen data.

This gap has been illustrated in Fig. 2, which highlights the training data as the independent identically distributed variables being the foundation behind the learning policy of neural networks (NN). One of the primary pre-requisite behind training a NN is that the training and testing data must follow independently identically distribution (IID) pattern, as shown in Fig. 2. However, this is not intuitive in real-world applications where the data on which a NN is trained and finally tested might differ significantly. This leads to an out-of-distribution (OOD) problem that often leads to unreliable decisions for unforeseen data. As evident, the OOD or testing samples in Fig. 2 accounting for an unseen condition/scenario will be seen as an adversarial condition by the NN, and can easily lead to under/overconfident decisions merely based on limited statistical insights.

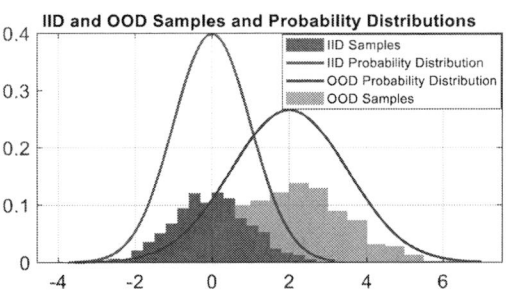

Fig. 2. Out of distribution (OOD) samples correspond to the unseen data/conditions, that ultimately aggravates the uncertainty in deep learning predictions.

C. Literature survey

As unpredictable faults in general can be, mechanical faults in machines pose a significant risk which necessitates intelligent fault diagnostic methods and health monitoring tools [3], [4]. The automation of such diagnostic procedures has been made more intelligent by using AI to set an alarm for dynamic contingent conditions. Although model-based signal processing approaches, such as wavelet analysis, empirical mode decomposition (EMD) have been vital in extracting faulty signatures [5], [6], they are not directly compatible with

advanced and intelligent data-driven methods. Furthermore, they only provide a low automation degree.

Since fault diagnosis is particularly a feature engineering task, data-driven approaches in the form of AI tools are well-equipped and pronounced to determine any fault signatures and can seamlessly update the library of faults simply by introduction of new faulty data [7]. Many deep learning models, such as stacked encoders (SAE), convolutional neural networks (CNN) have been utilized for fault diagnosis before [8]- [10]. To compensate for the lack of interpretability in these models, many grey-box models have been developed to bridge the gap between domain knowledge and data-driven predictions [11]- [12]. Since these are point-estimate algorithms that only infer with deterministic values, new probabilistic data-driven approaches in the form of Bayesian neural networks [13] have been exploited for fault diagnosis of mechanical devices. However, they still lack a principled way of testing and model structuring – without providing any direct guideline on an optimal selection of the number of layers, neurons or variational inferences.

D. Main contributions

Based on the issues discussed above, this paper exploits uncertainty-aware Bayesian neural networks (BNN) and its effectiveness for fault diagnosis of gear box faults emulated on a fault simulator. A preliminary case study has been thoroughly covered in [14] using point-estimate neural networks, which only generalizes the decisions based on accuracy without any explainability measures. Some of the key features of BNN that favor in minimizing the uncertainty of predictions for power electronics are:

- **Probabilistic outputs**: BNNs provide probabilistic outputs rather than point estimates. This means that instead of predicting a single value, they provide a distribution over possible outcomes, capturing uncertainty in predictions.
- **Uncertainty estimation**: They offer a principled way to estimate uncertainty associated with predictions. This uncertainty can be categorized into aleatoric uncertainty (inherent randomness or noise in the data) and epistemic uncertainty (uncertainty due to limited data or model uncertainty).

- **Bayesian inference**: Bayesian neural networks use Bayesian inference techniques to learn model parameters. Instead of finding a single set of parameters that maximize a likelihood function (as in traditional neural networks), they learn a distribution over parameters given the data, incorporating prior knowledge and updating beliefs based on evidence.

II. SYSTEM PRELIMINARIES & PROBLEM STATEMENT

To emulate such faults, SpectraQuest's Gearbox Dynamics Simulator (GDS), shown in Fig. 3, is used to simulate industrial gearboxes in both educational and experimental applications. It is highly precise and comprehensive, featuring a two-stage parallel shaft gearbox with rolling bearings and a magnetic brake system. This intricate design provides an ideal platform for advanced insights into the multifaceted dynamics and acoustic behavior of gearboxes. More details on this setup can be found in [14].

Fig. 3. Gearbox Dynamics Simulator used for collecting fault data. Detailed setup specifications can be found in [14].

Acquisition: The GDS has five potential fully developed faults on the spur and helical gears. Gear faults typically manifest as cracks on the gear or wear and tear of the gear teeth.

One of the methods employed in various reports is gear fault detection using vibration analysis. For vibration analysis, gearboxes are typically mounted with an acceleration sensor on the gearbox housing. A healthy gearbox theoretically has the dominant vibration mode in the axial direction; the vibration frequency is also called the gear-mesh frequency. More details on the gear mesh spectrum can be found in [14].

A. Data & Setup Specifications

The setup in Fig. 3 consists of the following actuators and sensors:

1) Actuators:

- Motor (Reconfigurable) – This unit is multiple copies of a 3-phase motor which is configurable to the following states:
  1) No fault (3HP)
  2) rotor unbalance fault (1HP)
  3) rotor misalignment fault (1HP)
  4) bowed rotor fault(1HP)
  5) broken rotor fault (1HP)
  6) stator winding fault (1HP)
  7) voltage unbalance and single phasing (1HP)

- Gearbox (Reconfigurable) – The experimental set-up houses a gearbox, which can be easily swapped. They can be reconfigured to the following faults:
  1) Missing tooth gear
  2) Chipped tooth gear
  3) Root crack gear
  4) Surface wear gear
  5) Eccentricity
- Brake (Reconfigurable load) – A programmable magnetic brake (24.8567 N-m) that emulates gearbox loading.

2) Sensors: Two types of sensors were used in this project, intrinsic and extrinsic:

- *Intrinsic Sensors*: The setup in Fig. 3 was modified with a Danfoss VLT Drive FC-103 to provide the following intrinsic parameters as outputs:
  1) Speed
  2) Motor Torque
  3) DC-link voltage
  4) Reactive stator current
  5) Active stator current
  6) Motor power

  sampled at a frequency of 5 kHz.
- *Extrinsic Sensors*: The gearbox was outfitted with a pair of orthogonally aligned analogue accelerometers ADXL1001.

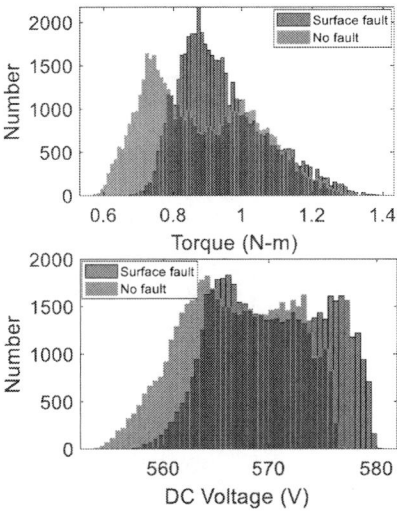

Fig. 4. Fault signatures for the same loading profile – surface fault vs. no fault. The overlapping region for both torque as well as DC voltages can lead to over-confident decisions from AI models due to conventional point-estimate deterministic learning approaches.

B. Challenges with current diagnostic approaches

Considering a simple comparative example of extrinsic sensor data on surface faults against the datasets representing no fault (see Fig. 5), it can be seen that there is no visible statistical difference for torque or DC voltage profile, since the distribution shifts on the application of load. This could potentially be due to the lack of further application of filters or

979-8-3315-1612-3/25 $31.00 © 2025 IEEE

little understanding of how the parameters interact with each other. As explained before, this affects the decisions taken by AI, which is primarily driven by these statistical attributes.

III. UNCERTAINTY-AWARE AI

Despite of promising applications offered by deep learning (DL) methods for power electronics and motor drives, the lack of interpretability and uncertainty quantification in their decisions is a significant barrier with their implementation. Hence, we propose a customized uncertainty-aware Bayesian neural network (BNN) for quantifying the uncertainty in fault diagnosis of the gear boxes in Fig. 3. Since the diagnosis is performed on extrinsic signals, the variational inference is tweaked with model information to achieve high generaltion.

A. Theory

Before discussing Bayesian modeling principles, let us start with the preliminaries of a simple feed forward neural network (NN) to understand uncertainty modeling in detail.

Consider a preliminary structure of a NN [16] having multiple layers with $\mathbf{x}$ be a D-dimensional input vector, bias b and a linear mapping function W_1 for its transformation into a vector of Q elements, given by $W_1\mathbf{x} + b$. On top, activation function $\sigma(.)$ to smoothen the output of hidden layers. As a result, a multi-layer inference with another cascaded linear function W_2 can be given by:

$$\hat{y} = \sigma(W_1\mathbf{x} + b)W_2. \tag{1}$$

Since fault diagnosis is primarily a classification task relying on feature engineering to determine intrinsic faulty signatures, we exploit a probabilistic approach to determine the possibility of $\mathbf{x}$ exclusively belonging to a certain class $\{1,...,C\}$. Finally, the score is obtained by computing the model output $\hat{y}$ with a softmax function $\hat{p}_d = \exp(\hat{y}_d)/(\sum_{d'} \exp(\hat{y}_d))$. Hence, the softmax loss is calculated using:

$$E^{W_1, W_2, b}(X, Y) = -\frac{1}{N} \sum_{i=1}^{N} \log(\hat{p}_{i,c_i}). \tag{2}$$

where, $X = \{\mathbf{x}_1,...,\mathbf{x}_N\}$ and $Y = \{y_1,...,y_N\}$ are the model's input and output vectors, respectively.

B. Uncertainty Modeling

Predictive uncertainty (PU) consists of two parts: (i) epistemic uncertainty (EU), and (ii) aleatoric uncertainty (AU), and can be represented as their sum:

$$PU = EU + AU. \tag{3}$$

Let $D_{tr} = \{X, Y\} = \{(\mathbf{x}_i, y_i)\}_{i=1}^{N}$ denote a training dataset with inputs $\mathbf{x}_i \in \mathbb{R}^D$ and outputs $y_i \in \{1, ..., C\}$, where C represents the number of classes. Given both types of uncertainties in (3), the objective is to optimize the parameters ω in $y = f^\omega(\mathbf{x})$ and obtain the desired output. To achieve this using a probabilistic approach, Bayesian methodologies

define a model likelihood, $p(y|\mathbf{x}, \omega)$. For classification, *softmax* likelihood will be obtained using:

$$p(y = c|\mathbf{x}, \omega) = \frac{\exp(f_c^\omega(\mathbf{x}))}{\sum_{c'} \exp(f_{c'}^\omega(\mathbf{x}))}. \tag{4}$$

For a given test sample $\mathbf{x}^*$, the probability of identifying a class label with regard to $p(\omega|X, Y)$ can be predicted using:

$$p(y^*|\mathbf{x}^*, X, Y) = \int p(y^*|\mathbf{x}^*, \omega)p(\omega|X, Y)d\omega \tag{5}$$

However, $p(\omega|X, Y)$ in (5) cannot be computed analytically. Hence, we exploit variational inference algorithms [17] to approximate the variational parameters, i.e., $q_\theta(\omega)$. Hence, the rationale behind using variational inference algorithms is to approximate a distribution for each neuron, such that it is close to the posterior distribution obtained by the model.

It is worthy notifying that the Bayesian inferencing in statistics is strongly correlated with the frequentist paradigm, primarily used for hypothesis testing of training data. In simple terms, it can be summarized by the following:

- probability is a measure of trust in the occurrence of events, rather than a limit in the frequency of occurrence.
- prior inferences influence posterior output, as stipulated in the Bayes' theorem, which can be mathematically stated as:

$$P(H|D) = \frac{P(D|H)P(H)}{P(D)} = \frac{\text{Likelihood} * \text{Prior}}{\text{Evidence}} \tag{6}$$

It is worth notifying that H and D in (6) are considered as the sets of outcomes. The Bayesian inferencing rules considers H to be a hypothesis about which one holds some prior belief, and D to be some data that will update one's belief about H. Catering back to the fault diagnosis example in this paper, H would imply on faulty signatures collected from iteratively processed feature engineering on different conditions. Whereas, D would be the dataset that confirms this hypothesis. Following this logic over multiple layers in BNN, the probability distribution $P(D|H)$ in (6) is commonly termed as *likelihood*, since it is basically an evaluation stage for the hypothesis over the given dataset. It encodes the aleatoric uncertainty in the model. Finally, $P(H|D)$ is termed as the *posterior*, which ultimately encodes the epistemic uncertainty.

C. Inferential Algorithm

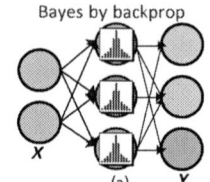

Bayesian Convolutional Layers	5 (2-D network)
	Max-pooling: 2x1
Network Size (Neurons)	2 hidden (128 & 64)
	1 output layer (3)
Optimizer	Stochastic gradient descent
	Learning rate: 0.001

Fig. 5. (a) Probabilistic assignment of neuronal weights to formalize a variational inference approach, (b) Model specifications of the designed bayesian neural network (BNN).

979-8-3315-1612-3/25 $31.00 © 2025 IEEE

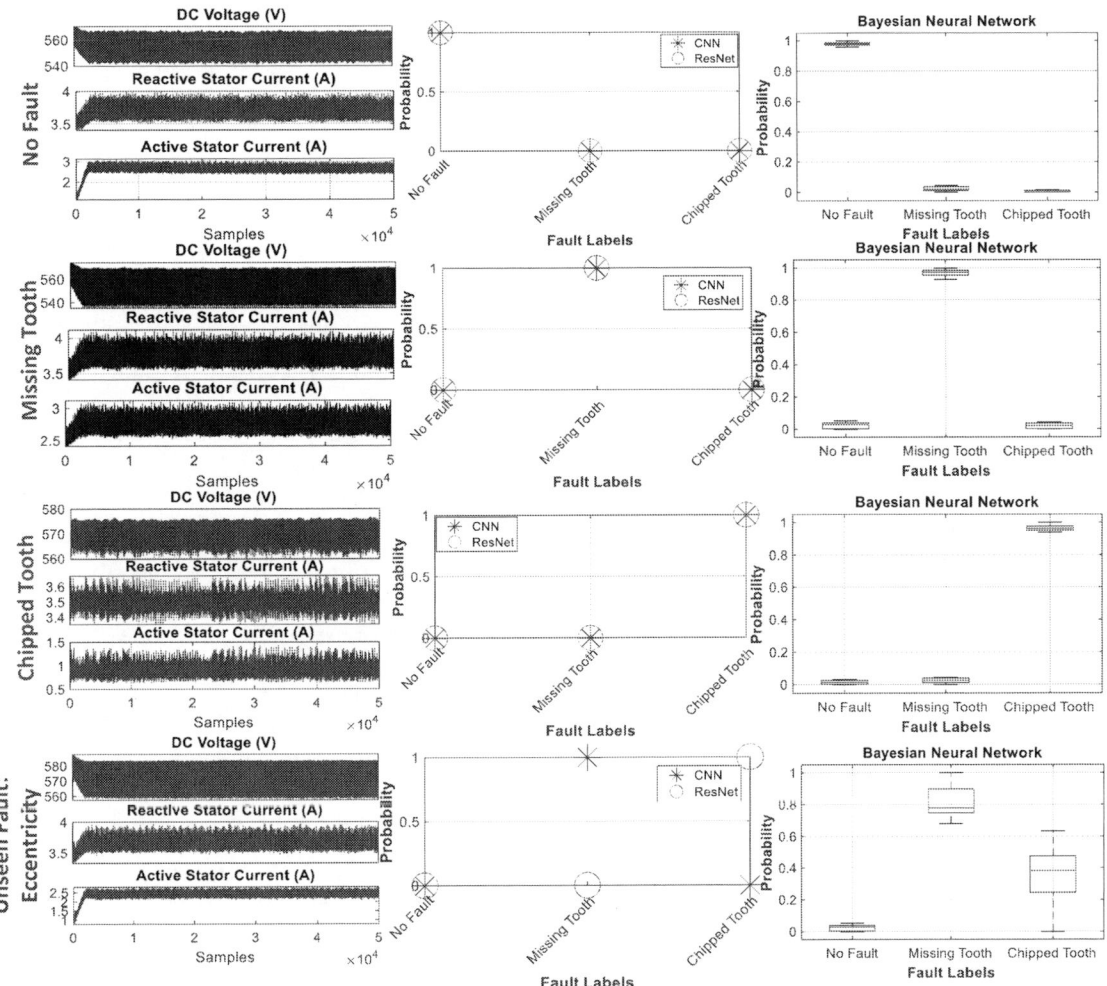

Fig. 6. Diagnosis of the testing samples of seen domain and unseen fault with benchmarking models and probabilistic Bayesian NN: For the extrinsic measurements considered for the three fault labels {No Fault, Missing Tooth, Chipped Tooth}, the diagnostic results of benchmarking models trained in the seen domain are fairly accurate for all the three candidate classification algorithms. However, a big discrepancy arises for an unseen fault where the uncertainty in BNN predictions rises to an alarming value.

In this paper, we will exploit the variational posterior using the well-known Bayes-by-backprop (BBB) method [18]. Basically, this algorithm introduces a random variable ϵ having a given probability density and a deterministic transform $t(\theta, \epsilon)$, such that the weights of the BNN w can be equivalent to $t(\theta, \epsilon)$. This formalizes a variational inference approach to carry out a probabilistic evaluation of uncertainty introduced by AI models, as shown in Fig. 5(a).

The key principled mechanism behind BBB method is that the random variable ϵ can define the variational distribution by assigning different values to each set-point in the distribution function, and consequently shape the weights w as an indirect deterministic transformation of ϵ. As a result, instead of a point-estimate inference for each condition in a conventional NN, BNN offers a range of weights that can be used to check the likelihood from input to the output. Indeed, by writing w

as $w = t(\theta, \epsilon)$, in place of evaluating:

$$\frac{\partial}{\partial \theta} \mathcal{E}_{q(w|\theta)}[f(w, \theta)] = \mathcal{E}_{q(\epsilon)}\left[\frac{\partial}{\partial \theta} f(t(\theta, \epsilon), \theta)\right] \qquad (7)$$

$$= \mathcal{E}_{q(\epsilon)}\left[\frac{\partial f(w, \theta)}{\partial w}\frac{\partial w}{\partial \theta} + \frac{\partial f(w, \theta)}{\partial \theta}\right] \qquad (8)$$

we guarantee that the convergence over probabilistic variational posteriors can be achieved. The reason behind using backpropagation using Bayesian paradigm over commonly used Markov Chain Monte Carlo (MCMC) approach is due to the faster convergence of predictions and high interpretability.

IV. PERFORMANCE EVALUATION

In this section, we will evaluate the performance of the modeled BNN from two different perspectives: (a) its capability with handling seen data and its accuracy estimates, (b) its capability with handling unseen conditions and data. This will provide a multi-faceted leap not only by quantifying uncertainties in predictions by AI models, but also quantifying its confidence levels alongside its predictions as another dimension of decision making. The model specifications of the BNN used in the following studies can be found in Fig. 5(b). The BNN was trained with the datasets of fault labels: `No Fault`, `Missing Tooth`, `Chipped Tooth`. As a result, the rest of the fault labels will be considered as an *unseen* condition by the BNN.

To provide a comparative evaluation, we have considered three candidates for data-driven classification: BNN, Convolutional neural network (CNN), and ResNet, trained over the seen datasets. Since CNN and ResNet are deterministic tools, their output only indicates a single point-estimate of the probability of each fault, whereas Bayesian neural network provides a range of predictive uncertainty for the seen datasets, based on the hypothesis H and *likelihood* calculated over multiple BNN layers. However, when all the three algorithms are tested under the unseen `Eccentric Fault` in Fig. 6, their predictions differ by a very high degree, as CNN and ResNet provide conflicting probabilistic outputs. However, the same can also be implied for BNNs, where the confidence level behind the probabilistic estimates is low for all the three seen fault conditions. In any case, this visualization offered by BNN could be seen as a powerful prospect in differentiating that the new dataset doesn't fall into any of the fault category and require further investigations.

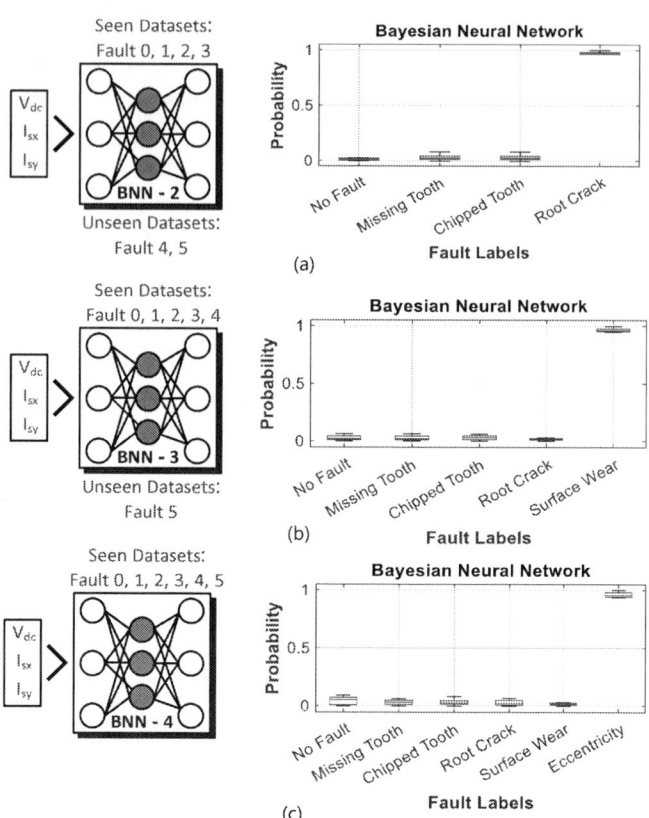

Fig. 8. Performance evaluation of updated seen datasets by transforming BNN into: (a) BNN–2 by adding fault label 3 into the training database onto BNN, (b) BNN–3 by adding fault label 4 into the training database onto BNN–2, (c) BNN–4 by adding fault label 5 into the training database onto BNN–3.

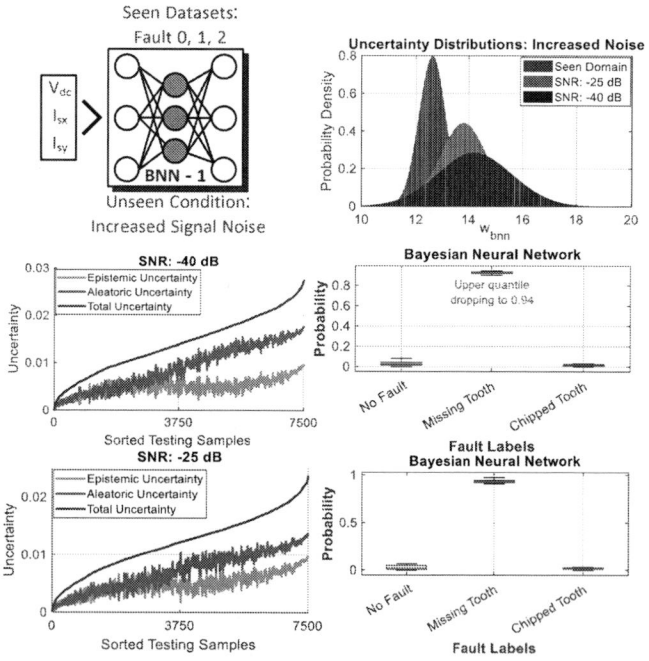

Fig. 7. Uncertainty estimation and decomposition using BNN for the testing samples of seen domain and different noisy environments: the upper quantile of the predictions (refer to the box plots) drop to 0.94. With increase in noise, the aleatoric uncertainty dominate over the epistemic uncertainty.

In Fig. 7, the modeled BNN is tested against different noise levels, which clearly indicates that the total uncertainty increases for higher noise level. When the SNR is less than -

25 dB which is indicative of a high noise profile, the proposed BNN outputs a significant rise in the aleatoric uncertainty. This is well aligned with the fact that noise, in particular, increases the level of data based uncertainties. As a result, the change in aleatoric uncertainty becomes highly prominent over epistemic uncertainty with a decrease in SNR.

BNN trained only with the seen datasets on fault label `No Fault`, `Missing Tooth`, `Chipped Tooth`} is reconfigured to include the new fault cases into the seen dataset. Hence, we incorporate the new fault datasets in a step-wise manner so that three new BNN designs can be made. As shown in Fig. 8(a), BNN–2 is updated with fault 3 in its seen environment, which automatically minimizes the diagnostic uncertainty and accurately identifies fault 3 (root crack fault) into the correct category, as opposed to the wrong predictions made initially in Fig. 6. On the other hand, BNN–2 trained only with the seen datasets on fault label `No Fault`, `Missing Tooth`, `Chipped Tooth`, `Root Crack`} is reconfigured to include the new fault cases into the seen dataset. We then incorporate the new fault datasets in a step-wise manner so that three new BNN designs can be made. As shown in Fig. 8(b), BNN–3 is updated with fault 4 in its seen environment, which automatically minimizes the diagnostic uncertainty

979-8-3315-1612-3/25 $31.00 © 2025 IEEE

and accurately identifies `Surface Wear` into the correct fault category. Finally, BNN–3 trained only with the seen datasets on fault label `No Fault`, `Missing Tooth`, `Chipped Tooth`, `Root Crack`, `Surface Wear`} is reconfigured to include the new fault cases into the seen dataset. We then incorporate the new fault datasets in a step-wise manner so that three new BNN designs can be made. As shown in Fig. 8(c), BNN–4 is updated with fault 5 in its seen environment, which automatically minimizes the diagnostic uncertainty and accurately identifies `Eccentricity fault` into the correct fault category.

In this way, the unseen conditions can be gradually augmented to improve an accurate diagnosis model that not only provide reliable predictions, but also highlight the confidence interval behind each prediction.

V. CONCLUSION

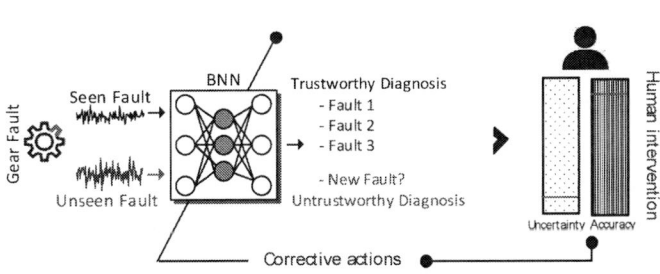

Fig. 9. Bayesian neural networks is a promising solution, that offers a multi-dimensional decision and allows a qualitative assessment of its predictions.

In conclusion, this paper delves into the application of uncertainty-aware AI algorithms for fault diagnosis of gear faults, with a particular focus on Bayesian neural networks (BNNs). Through the utilization of BNNs, we have effectively quantified uncertainties in predictions, providing probabilistic outputs that offer valuable insights into the reliability of diagnostic assessments. We have carried out rigorous test cases by considering different subsets of faults as the training data and identifying reasonable answers from BNN for the unseen faults in a structural way. The effect of noise variance, model parameters and unseen data has been covered in detail with key results based on theoretical foundations. This provides us with a formidable framework for a multi-dimensional decision making process (see Fig. 9) that requires human intervention before finalizing the data-driven algorithm for fault diagnosis of power electronics. This enhances a qualitative assessment of predictions from AI that often goes overlooked due to high accuracy as the sole decision metric. As a follow-up, the decision making process will ascertain corrective actions across the input stage such that further data analysis and investigations can be carried out.

As a result, this work provides a significant leap towards trustworthy machine learning in power electronics by quantifying the uncertainty in the predictions of AI, as a direct measure of either data-driven or model-driven uncertainties.

REFERENCES

[1] S. Zhao, F. Blaabjerg and H. Wang, "An Overview of Artificial Intelligence Applications for Power Electronics," *IEEE Trans. Power Electron.*, vol. 36, no. 4, pp. 4633-4658, April 2021.

[2] S. Sahoo, H. Wang, and F. Blaabjerg, "On the Explainability of Black Box Data-Driven Controllers for Power Electronic Converters," *2021 IEEE Energy Conversion Congress and Exposition (ECCE)*, Vancouver BC, Canada, 2021.

[3] E. Zio, "Prognostics and health management (PHM): Where are we and where do we (need to) go in theory and practice" *Reliab Eng Syst Saf*, vol. 218, pp. 108119, 2022.

[4] Y. Hu, X. Miao, Y. Si, E. Pan, E. Zio, "Prognostics and health management: A review from the perspectives of design, development and decision", *Reliab Eng Syst Saf*, vol. 217, pp. 108063, 2022.

[5] J. Jiao, M. Zhao, J. Lin, K. Liang, "Hierarchical discriminating sparse coding for weak fault feature extraction of rolling bearings" *Reliab Eng Syst Saf*, vol. 184, pp. 41-54, 2019.

[6] Y. Qin, Y. Mao, B. Tang, Y. Wang, H. Chen, "M-band flexible wavelet transform and its application to the fault diagnosis of planetary gear transmission systems", *Mech Syst Signal Process*, vol. 134, pp. 106298, 2019.

[7] Y. Lei, B. Yang, X. Jiang, F. Jia, N. Li, A. Nandi, " Applications of machine learning to machine fault diagnosis: A review and roadmap", *Mech Syst Signal Process*, vol. 138, pp. 106587, 2020.

[8] W. Mao, W. Feng, Y. Liu, D. Zhang, X. Liang, "A new deep auto-encoder method with fusing discriminant information for bearing fault diagnosis", *Mech Syst Signal Process*, vol. 150, pp. 107233, 2021.

[9] B. Zhao, X. Zhang, H. Li, Y. Yang, " Intelligent fault diagnosis of rolling bearings based on normalized CNN considering data imbalance and variable working conditions", *Knowl-Based Syst*, vol. 199, pp. 105971, 2020.

[10] B. Han, S. Ji, J. Wang, H. Bao, X. Jiang, "An intelligent diagnosis framework for roller bearing fault under speed fluctuation condition", *Neurocomputing*, vol. 420, pp. 171-180, 2021.

[11] D. Wang, Y. Chen, C. Shen, J. Zhong, Z. Peng, C. Li, "Fully interpretable neural network for locating resonance frequency bands for machine condition monitoring", *Mech Syst Signal Process*, vol. 168, pp. 108673, 2022.

[12] T. Li, Z. Zhao, C. Sun, L. Cheng, X. Chen, R. Yan, R. Gao, "WaveletKernelNet: An interpretable deep neural network for industrial intelligent diagnosis", *IEEE Trans Syst Man Cybern*, pp. 1-11, 2021.

[13] T. Zhou, T. Han, and E. L. Droguett, "Towards trustworthy machine fault diagnosis: A probabilistic Bayesian deep learning framework" *Reliability Engineering System Safety*, vol. 224, no. 108525, 2022.

[14] A. Biswas, "Intelligent motor fault detection", Master's thesis, University of South Denmark, 2023.

[15] J. Mukhoti, Y. Gal, "Evaluating Bayesian deep learning methods for semantic segmentation", *arXiv preprint arXiv:1811.12709*, 2018.

[16] D.E. Rumelhart, G.E. Hinton, R.J. Williams, "Learning Internal Representations by Error Propagation", *Tech. Rep., California Univ San Diego La Jolla Inst for Cognitive Science*, 1985.

[17] L. V. Jospin, H. Laga, F. Boussaid, W. Buntine and M. Bennamoun, "Hands-On Bayesian Neural Networks—A Tutorial for Deep Learning Users," *IEEE Computational Intelligence Magazine*, vol. 17, no. 2, pp. 29-48, May 2022.

[18] C. Blundell, J. Cornebise, K. Kavukcuoglu, D. Wierstra, "Weight uncertainty in neural networks", *arXiv preprint arXiv:1505.05424*, 2015.

[19] AI-Power, hhttps://www.ipower.ai/. Last accessed 24 Nov, 2024.

Neural Network Based Digital Twin Health Monitoring of BLDC Motor Drives for Robots

Mohamed Y. Metwly
*Department of Electrical Engineering
and Computer Science
University of Tennessee
Knoxville, TN, USA*
mohamed.metwly@utk.edu

Benjamin Luckett
*Department of Electrical and
Computer Engineering
University of Kentucky
Lexington, KY, USA*
ben.luckett@uky.edu

Landon Clark
*Department of Electrical and
Computer Engineering
University of Kentucky
Lexington, KY, USA*
landon.clark@uky.edu

JiangBiao He
*Department of Electrical Engineering
and Computer Science
University of Tennessee
Knoxville, TN, USA*
jiangbiao.he@utk.edu

Biyun Xie
*Department of Electrical and
Computer Engineering
University of Kentucky
Lexington, KY, USA*
biyun.xie@uky.edu

Abstract—**Robots have shown promising prospect in numerous applications, such as space exploration and disaster rescue. Due to the harsh environmental conditions (e.g., high temperature) in many applications, motor drive systems in the robotic arms are vulnerable to hardware failures such as inverter switching aging or faults. To address this challenge and avoid significant downtime cost, a digital twin based online health monitoring model, is developed for diagnosing potential switching faults that could occur to the robotic brushless DC (BLDC) motor drives. Specifically, the online digital twin health monitoring model is based on a dynamic neural network (DNN). Various DNN architectures have been tested to determine the best trade-off between the model accuracy and computational efficiency, which is to ensure that the proposed model can be embedded into a microprocessor and used in real-time applications. Finally, the efficacy of the proposed DNN-based digital twin approach is validated with testing data in a BLDC motor-drive prototype.**

Index Terms—**Digital twin, health monitoring, neural network, BLDC motor drives, robotic arm joints.**

I. INTRODUCTION

As industry embraces the increased benefits afforded by automation and electrification, robots have become an indispensable equipment for automated manufacturing and processing, especially for the applications where it is hazardous or dangerous for human being to complete the tasks. The robotic arms are typically constructed from multiple joint motor-drives which allow them to maneuver in a variety of directions. In numerous automation applications, these joints are facilitated by brushless DC (BLDC) motors which allow elevated levels of speed and position control [1]. For example, 7 BLDC motors have been utilized as the joint motors of the Kinova Gen3 robotic arm, as shown in Fig. 1. Low

This material is based upon work partially supported by the U.S. National Science Foundation under Grant No. 2205292.

maintenance costs and high torque density are among the main merits of the BLDC motors. Hardware failures of the power electronic drives and joint motors may occur since more and more robots are being utilized in harsh environmental conditions, such as high temperatures, high humidity, or high radiation [2], [3]. These switching failures may cause the robotic arms to malfunction. To avoid excessive downtime and maintenance cost, real-time monitoring of the motor-drive system's health condition allows potential problems to be identified and mitigated before any damage occurs [4], [5]. This can be accomplished through an online digital twin (DT) health monitoring model which transforms the physical BLDC motor-drive system's health information into its interactive computerized clone (i.e., digital replica).

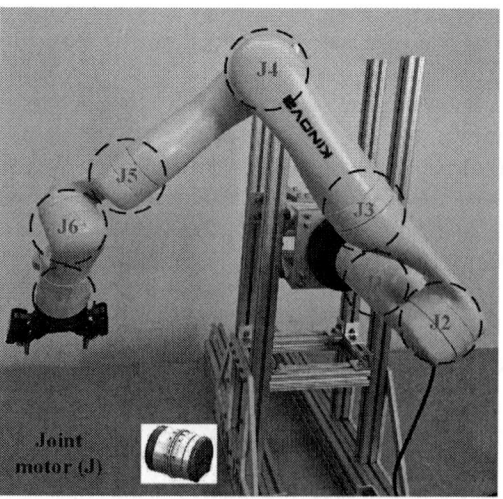

Fig. 1. Kinova Gen3 robotic arm with 7 BLDC motors.

TABLE I
COMPARISON OF DIGITAL TWIN MODELS FOR POWER ELECTRONIC CONVERTERS

Ref.	Converter Topology	DT Model	Implementation Complexity	Computational Burden	Model Accuracy	Real-time Operation	Applications
[6]	Buck converter	Probablistic	+++	+++	+++	✓	—
[7]	Boost converter	NARX	+	++	++	✓	—
[8]	2-L 3-P inverter	Feedforward NN	+	++	+++	×	Electrical drives
[9]	VSI	Feedforward NN	++	+	++	×	Grid-connected VSI
Prop.	2-L 3-P inverter	NARX	+	+++	+++	✓	Robotic arms

2-L 3-P: two-level three-phase, VSI: voltage source inverter, +: low, ++: medium, +++: high

DT modeling plays an essential role, as it is expected to represent a physical system on a digital platform to predict and replicate the health condition of the physical system [10]. In the literature, the DT concept has been investigated through various approaches, such as continuously updating the DT model by providing real-time sensory data from the physical systems. DT models have been developed for power system microgrids, aircraft propulsion systems, and robotic arms at the component, subsystem, and system levels [11], [12]. The utilization of DT technology facilitates condition monitoring of power converters and estimation of their remaining useful lifetime. Moreover, it becomes possible to conduct various testing scenarios including extreme conditions in a cost-effective and safe way. In [13], a DT model of a three-level active-neutral-point-clamped (ANPC) inverter has been introduced with parameter monitoring scheme. Firstly, the key parameters are initialized and transmitted to the DT model. Afterwards, the output phase current and capacitor voltage that represent the external characteristics are calculated, and the objective evaluation function is formulated. Finally, if the global optimal value is found and the ending condition is met, the optimal parameters are determined. This model can achieve noninvasive multi-parameter monitoring.

Furthermore, the DT concept has been investigated through machine learning (ML) approaches which is the main focus of this paper. The DT models based on neural networks (NN) are time-domain, real-time, and embeddable models, which can offer noninvasive and efficient health monitoring of power electronic converters [4], [14]. In [6], [15], the parameters of DC-DC converters are firstly characterized before using embedded state equations to estimate circuit outputs. They rely on Runge-Kutta methods which are prone to high iteration counts in order to avoid cascaded error accumulation. The method presented in [7] employs a nonlinear autoregressive exogenous (NARX) neural network (NN) to estimate the values of state variables in a boost converter. The proposed model is calculated in real-time and can be embedded into the converter's controller. The accuracy of the model under different operating conditions has been verified based on both time-domain and frequency-domain responses. On the contrary, the method using a switch-averaged model may negatively impact the accuracy of instantaneous voltage and current signals, which would be critical if the semiconductor temperature estimates are also included.

The work of [8] also employs a NN to monitor the nonlinear behavior of a 2-level 3-phase inverter at both low and high duty cycles. For the sake of comparison, a gray-box inverter model has been presented based on particle swarm optimization (PSO) and recorded datasets. Both models can effectively estimate the phase voltages per switching period. The RMS error is lower when the NN-based model is used with an error value of 0.65 V compared to an effort value of 1.1 V when a gray-box model is used. Another DT model of a 2-level 3-phase inverter has been presented based on the convolutional NN (CNN). It is worth mentioning that CNN is one of the most important deep learning (DL) models, as it has very powerful extraction capabilities. The proposed technique can achieve high accuracy at various noise conditions.

To address the aforementioned challenges, this paper proposes a NARX NN to monitor the voltage waveforms of a BLDC motor-drive system for robots. It utilizes instantaneous switch pattern information to calculate the progression of signals through time, circumventing the averaged models present in the literature. The proposed NN-based DT approach can be embedded in the controller of the power inverter to improve the robotic arm motion planning and enhance its reliability and robustness. Moreover, several DT models of power converters are compared, as listed in Table I. These DT models of power converters are compared considering the converter topology, implementation complexity, computational burden, model accuracy, real-time operation, and the targeted applications. It is clear that the proposed NN-based DT model exhibits superior performance when compared to the ones reported in the literature, since it supports real-time operation and low implementation complexity. The potential limitation of this model is the computational burden when embedded on a low-cost microprocessor platform. Thus, experimental validations by using the trained NN-based DT model on an embedded environment will be investigated in the future work. Finally, the accuracy of the proposed DT model has been verified using experimental datasets.

The rest of this paper is organized as follows. Section II describes the BLDC motor drive system and a detailed explanation of the artificial NN implementing the DT model.

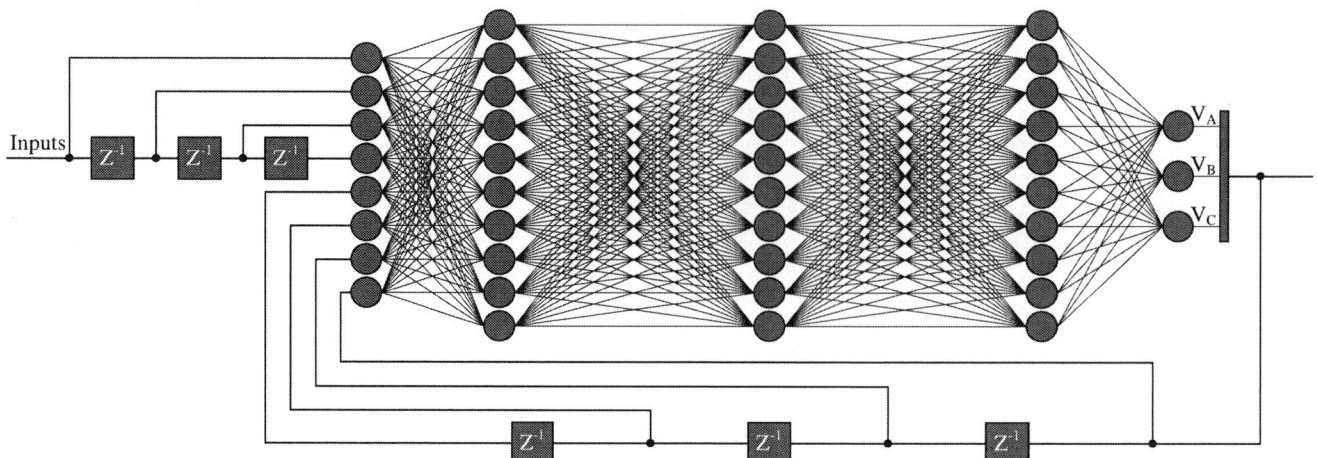

Fig. 2. The chosen NARX NN architecture (only 10 of the 20 hidden layer neurons are shown due to spacing).

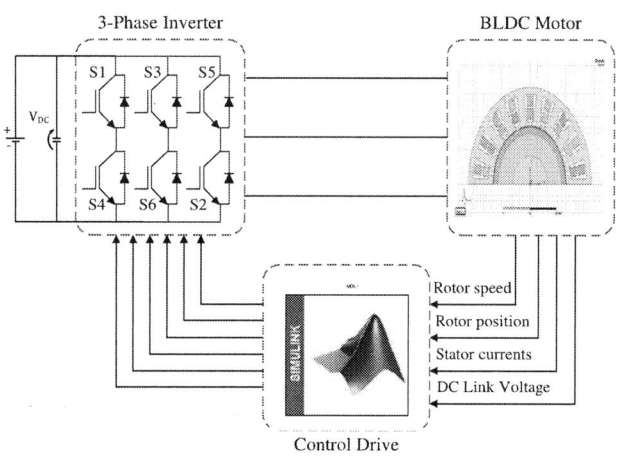

Fig. 3. Block diagram of a BLDC motor-drive system.

Section III presents results from a lab-scale BLDC motor drive setup, and final conclusions are drawn in Section IV.

II. BLDC MOTOR-DRIVE SYSTEM AND NEURAL NETWORK DIGITAL TWIN APPROACH

This section presents the system overview of the BLDC motor, the associated power inverter, and the utilized NN approach. Fig. 3 depicts the main topology of the BLDC motor drive system under investigation for robotic applications. As can be seen, the three-phase BLDC motor is supplied by a two-level three-phase inverter via phase resistance and leakage inductance, and it is controlled by a BLDC current controller. In essence, a trapezoidal back EMF is produced from the PWM drive that feeds the BLDC motor by exciting two phases simultaneously.

One area of artificial intelligence that shows promise for time series forecasting is the NARX model [16]. The NARX NN is a network that is recurrent in both output and input

values. Delay lines are added so that information from a set number of previous time steps assists in the current calculation process. This allows for improved modeling of systems which inherently have a memory component, such as the state variables of a circuit. Unlike the feedforward NN, the dynamic NARX NN is of paramount importance for modeling dynamic systems that are described by a set of differential equations. Overall, the NARX NN can be considered as a feedforward NN with tapped delay lines from the input and output.

A diagram depicting an example of the NARX NN architecture utilized in this paper is shown in Fig. 2. The NARX NN typically comprises input, hidden, and output layers. Each hidden layer consists of a number of neurons, i.e., the key element of the artificial NN alongside weights, biases, and activation function. Several training algorithms can be used, such as Levenberg-Marquardt and Bayesian regularization. Levenberg-Marquardt is used in this work since it is the fastest algorithm, although it requires more computational memory.

The first step of the selected NARX NN-based DT model is to determine the inputs and the outputs of the model. Avoiding any additional unnecessary measurements is desired for simplicity and ease of implementation on the NN, so the inputs to the proposed NN consist of three-phase inverter currents which are used in the control process, gate pulses of all 6 switching devices, and the value of DC voltage from the speed controller. Each of such data is necessary in the prediction process. Switching pulses determine the instantaneous circuit configuration and the system dynamics, while the output current and input DC voltage relate to the magnitude and phase of the output voltages. These 10 necessary variables are then fed into the NN which outputs the motor terminal voltages for all three phases. To improve the prediction accuracy, a delay depth of 3 samples was imposed upon both input and output values. In this paper, an experimental setup is used to generate the training datasets, as explained in the following section.

The proposed network needs to simultaneously possess high

979-8-3315-1612-3/25 $31.00 © 2025 IEEE

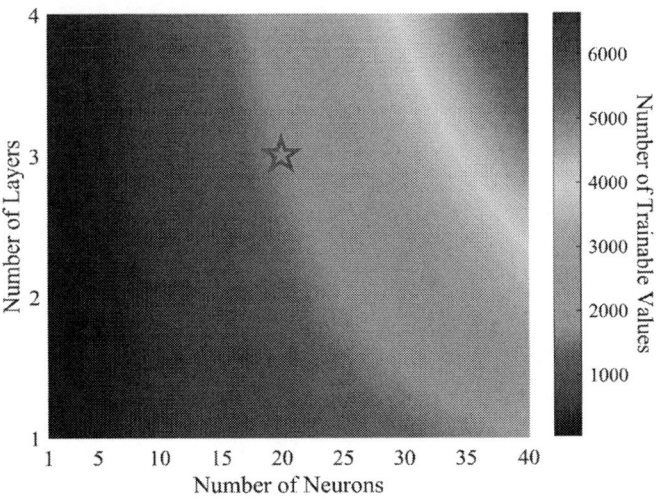

Fig. 4. Influence of neuron and layer count on trainable variables and computational complexity.

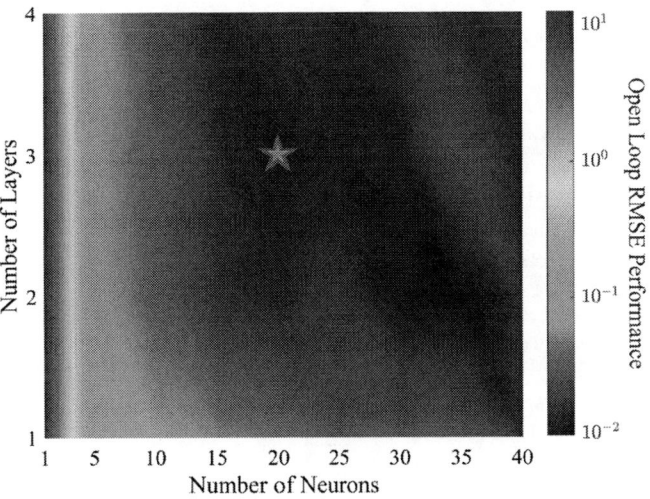

Fig. 5. Effect of neuron and layer count on output estimation performance.

speed and accuracy to achieve an acceptable DT model, so the NARX NN's hyperparameters (i.e., layer count and size) were swept in order to locate an acceptable trade-off between these performance indicators. The parametric study results are displayed in Figs. 4 and 5. Since one major goal of a DT is implementability onto a microcontroller, the proposed NN needs to limit the quantity of trainable parameters. These mainly consist of the entries in the weight matrices whose sizes are determined by the number of neurons in the hidden layers. As shown in Fig. 4, when more neurons are added, the trainable values increase nonlinearly. Additionally, NNs rely on matrix multiplication which has a computational complexity on the order of $\mathcal{O}(n^3)$ (n is the matrix size which corresponds to the number of neurons in a hidden layer), so even small changes in layer size and count can have large timing effects. The performance of the open loop NN using the same set of architectures is depicted in Fig. 5. As long as the NARX NN utilizes greater than approximately 10 neurons and 1 layer, the performance is adequate. For these reasons, an architecture of 20 neurons and 3 layers was chosen which possesses comparatively low trainable variables (i.e. lower order matrices) and acceptable open loop performance. It is marked with a star in Figs. 4 and 5. In certain situations, adding more layers and/or neurons to each layer above this size resulted in slight performance gains; however, this improvement must be balanced against higher computational burden.

III. EXPERIMENTAL VERIFICATION

To verify the efficacy of the proposed DT model, experiments were carried out on the motor-drive test bench shown in Fig. 6. Experiments were conducted at the rated speed of 2000 r/min and DC-link voltage of 24 V for robotic applications. The PWM-based BLDC speed control is implemented in a driving mode [17]. The BLDC motor is tested under various speed scenarios. The fundamental frequency is determined based on the speed of the utilized 4-pole motor, e.g., 50 Hz

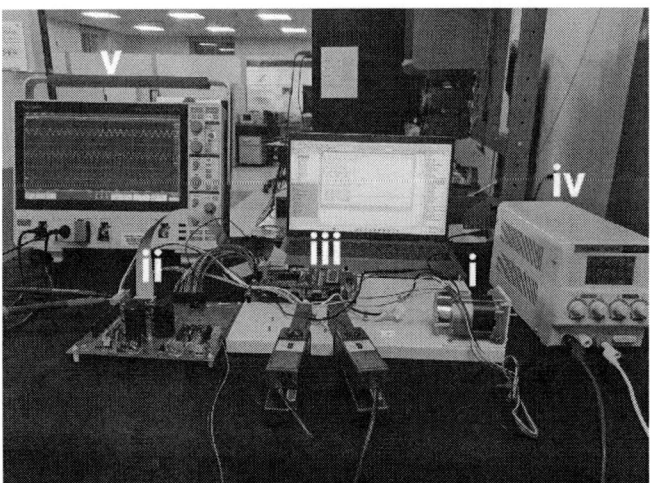

Fig. 6. Test setup for experimental validations: (i) BLDC motor, (ii) PWM inverter, (iii) control board, (iv) DC supply, and (v) oscilloscope.

for the speed of 1500 r/min. Datasets have been collected for training and testing of the proposed NN-based DT model using a sampling rate of 12.5 kS/s. Specifically, a high-frequency oscilloscope, high-frequency differential voltage probes, and hall-effect current sensors have been utilized to record and measure the three-phase currents, values of DC-link voltage, and motor terminal voltages concurrently.

In this paper, a NARX NN with 3 hidden layers and 20 neurons has shown excellent performance to predict the motor terminal voltages using the above-mentioned inputs. The NN is firstly trained in an open-loop condition based on large datasets, and then tested under both open-loop and closed-loop configurations. It is worth mentioning that the NN predicts the output based on the past values of both inputs and outputs in an open-loop configuration. However, the closed-loop form can predict the output based on the past values of the inputs and

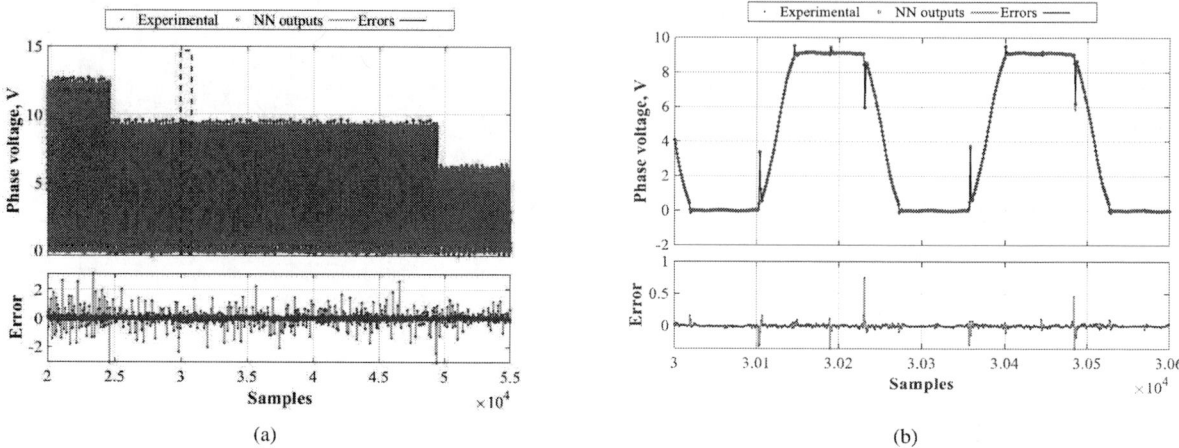

Fig. 7. Training results based on large dataset under various speeds. (a) waveforms of the motor phase voltage and error. (b) zoomed-in view.

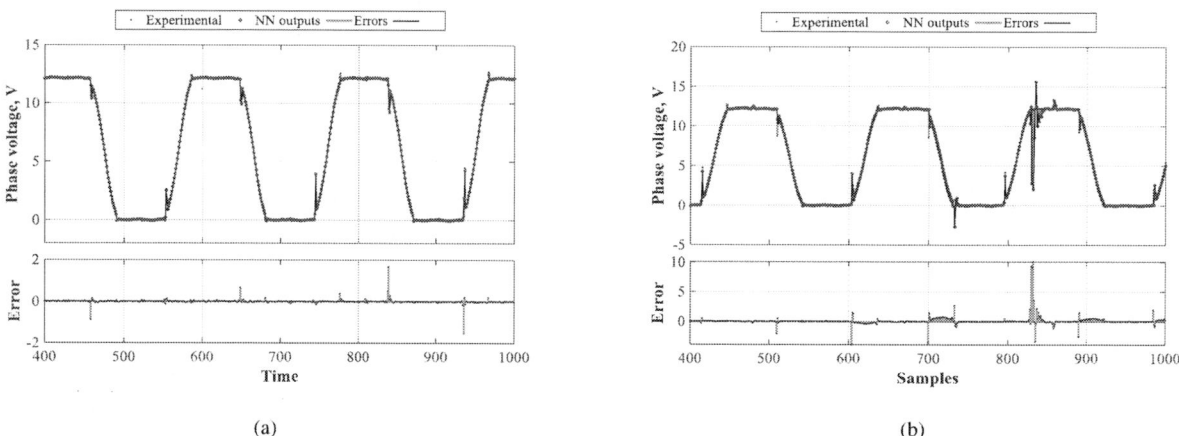

Fig. 8. Testing results (phase voltage and error) with different NN forms. (a) open loop form. (b) closed loop form.

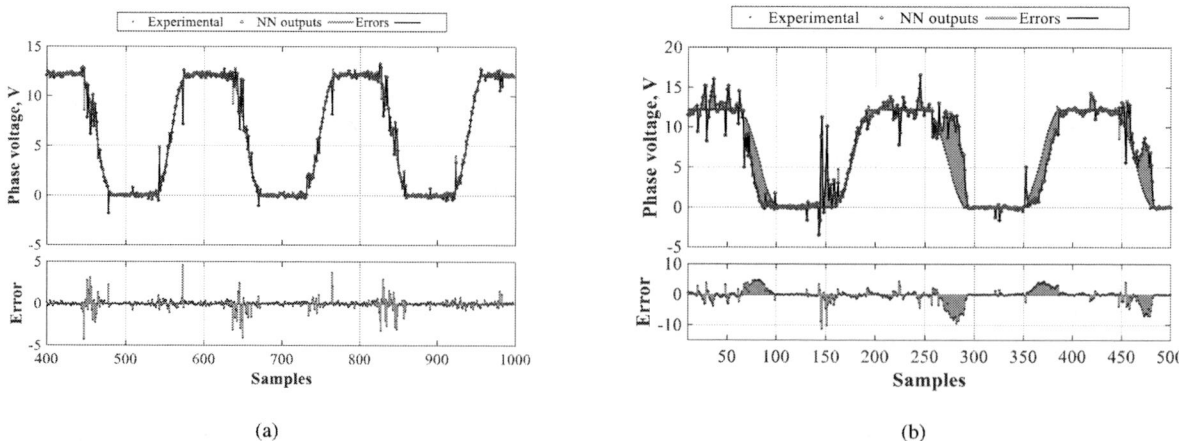

Fig. 9. Testing results (phase voltage and error) in the noisy case with different NN forms. (a) open loop. (b) closed loop.

the predicted values by the NN itself, as depicted in Fig. 2.

Fig. 7 shows the training results of the NN-based DT model with a zoomed-in view. It is clear that the NN outputs well match the measured phase voltages from the experimental setup. In that case, the mean squared error (MSE) between the physical model outputs and its DT model outputs at each timestep is minimized. After that, it is crucial to test the DT model based on the weights of the NN using a new dataset excluded in the training dataset. An absolute agreement between both the DT model and the experimental results with respect to the phase voltages has been highlighted in Fig. 8 at 2000 r/min using open- and closed-loop NN configurations. The open-loop NN is more accurate; however, the closed loop configuration is faster and facilitates the real-time application of the DT model. The same conclusion can be drawn for other motor speeds, e.g., 1500 and 1000 r/min. This validates the capability of the proposed DT model to learn the dynamic behavior of the BLDC motor drive system.

To further validate the proposed DT model, it is tested by adding noise to the signal, i.e., 30 signal-to-noise (SNR) ratio, since robotic arms might be operated in harsh environments. As a result, the proposed DT model can closely track its physical twin's outputs, as shown in Fig. 9.

IV. CONCLUSION

In this paper, a digital twin based health monitoring model of a BLDC motor drive is introduced for robotic arm joints based on a machine learning approach. The proposed model can be embedded in the inverter's microprocessor and integrated with the robotic arm models. The digital twin model aims to be used for fault prognosis and diagnosis and real-time co-simulation at the system level. A lab-scale prototype was utilized to generate the training datasets under various operating conditions. The proposed DT model is further tested on a new dataset excluded in the training datasets. Results verify the effectiveness of the NN-based DT model. Finally, more intensive experimental validations will be presented in the future work to further verify that the proposed NN-based digital twein model is accurate, robust, embeddable, and real-time.

ACKNOWLEDGMENT

This material is based upon work partially supported by the U.S. National Science Foundation under Grant No. 2205292.

REFERENCES

[1] B. Xie and A. A. Maciejewski, "Maximizing the probability of task completion for redundant robots experiencing locked joint failures," *IEEE Transactions on Robotics*, vol. 38, no. 1, pp. 616–625, 2022.

[2] W. Wang, W. Gao, S. Zhao, W. Cao, and Z. Du, "Robot protection in the hazardous environments," in *Robots Operating in Hazardous Environments*. IntechOpen, 2017, pp. 87–107.

[3] M. Y. Metwly, C. L. Clark, J. He, and B. Xie, "A review of robotic arm joint motors and online health monitoring techniques," *IEEE Access*, vol. 12, pp. 128 791–128 809, 2024.

[4] Y. Fassi, V. Heiries, J. Boutet, and S. Boisseau, "Toward physics-informed machine-learning-based predictive maintenance for power converters—a review," *IEEE Transactions on Power Electronics*, vol. 39, no. 2, pp. 2692–2720, 2024.

[5] F. Yüce and M. Hiller, "Condition monitoring of power electronic systems through data analysis of measurement signals and control output variables," *IEEE Journal of Emerging and Selected Topics in Power Electronics*, vol. 10, no. 5, pp. 5118–5131, 2022.

[6] M. Milton, C. D. L. O, H. L. Ginn, and A. Benigni, "Controller-embeddable probabilistic real-time digital twins for power electronic converter diagnostics," *IEEE Transactions on Power Electronics*, vol. 35, no. 9, pp. 9850–9864, 2020.

[7] A. Wunderlich and E. Santi, "Digital twin models of power electronic converters using dynamic neural networks," in *2021 IEEE Applied Power Electronics Conference and Exposition (APEC)*, 2021, pp. 2369–2376.

[8] M. Stender, O. Wallscheid, and J. Böcker, "Comparison of gray-box and black-box two-level three-phase inverter models for electrical drives," *IEEE Transactions on Industrial Electronics*, vol. 68, no. 9, pp. 8646–8656, 2021.

[9] M. Zhang, X. Wang, D. Yang, and M. G. Christensen, "Artificial neural network based identification of multi-operating-point impedance model," *IEEE Transactions on Power Electronics*, vol. 36, no. 2, pp. 1231–1235, 2020.

[10] M. T. Fard, B. J. Luckett, and J. He, "Digital twin enabled open-circuit fault diagnosis for five-level anpc multilevel converters," *IEEE Journal of Emerging and Selected Topics in Power Electronics*, pp. 1–1, 2024.

[11] Z. Wu, Y. Yao, J. Liang, F. Jiang, S. Chen, S. Zhang, and X. Yan, "Digital twin-driven 3-d position information mutuality and positioning error compensation for robotic arm," *IEEE Sensors Journal*, vol. 23, no. 22, pp. 27 508–27 516, 2023.

[12] A. S. Wunderlich and E. Santi, "Closed-form implicit models for efficient simulation of power electronics," *IEEE Journal of Emerging and Selected Topics in Power Electronics*, vol. 11, no. 2, pp. 1568–1577, 2023.

[13] W. Song, Z. Zhang, S. Zhang, C. Ma, and J. Li, "Digital twin modeling and multiparameter monitoring schemes of three-level anpc inverters," *IEEE Transactions on Power Electronics*, vol. 39, no. 12, pp. 16 596–16 608, 2024.

[14] V. S. Bharath Kurukuru, A. Haque, R. Kumar, M. A. Khan, and A. K. Tripathy, "Machine learning based fault classification approach for power electronic converters," in *2020 IEEE International Conference on Power Electronics, Drives and Energy Systems (PEDES)*, 2020, pp. 1–6.

[15] J. Nwoke, M. Milanesi, J. Viola, Y. Chen, and A. Visioli, "A reduced-order digital twin fpga-based implementation with self-awareness capabilities for power electronics applications," *IEEE Journal of Radio Frequency Identification*, vol. 8, pp. 493–505, 2024.

[16] R. W. Chan, J. K. Yuen, E. W. Lee, and M. Arashpour, "Application of nonlinear-autoregressive-exogenous model to predict the hysteretic behaviour of passive control systems," *Engineering Structures*, vol. 85, pp. 1–10, 2015.

[17] P. Suganthi, S. Nagapavithra, and S. Umamaheswari, "Modeling and simulation of closed loop speed control for bldc motor," in *2017 Conference on Emerging Devices and Smart Systems (ICEDSS)*. IEEE, 2017, pp. 229–233.

MTPA Control Using Predictive P&O Method for Dual Parallel Surface-Mounted Permanent Magnet Synchronous Motor Drives Fed by a Single Inverter

Jae-Seong Kim
Department of Electrical and Computer Engineering
Ajou University
Suwon, South Korea
idasgas@ajou.ac.kr

Kyo-Beum Lee
Department of Electrical and Computer Engineering
Ajou University
Suwon, South Korea
kyl@ajou.ac.kr

Abstract—This paper presents maximum torque per ampere (MTPA) control using predictive perturbation and observation (P&O) method for dual parallel surface-mounted permanent magnet synchronous motor (SPMSM) drives fed by a single inverter. This control method adaptively adjusts the *d*-axis current that minimizes copper losses based on the predicted model of the electrical characteristics of the single inverter dual parallel (SIDP) SPMSM drive system. This optimized *d*-axis current is varied in real-time to enhance the efficiency of the SIDP SPMSM drive system. The proposed method ensures that the two motors operate at synchronized speeds while the SIDP SPMSM drive system satisfies the MTPA conditions. The validity of the proposed strategy for MTPA control in the SIDP SPMSM drive system is verified by simulations using PSIM software.

Keywords—Maximum torque per ampere, optimal efficiency drive, PMSM, single inverter multi-motor

I. INTRODUCTION

Modern industrial applications require a compact and efficient motor drive system, and a multi-motor drive system fed by a single inverter has been adopted as an effective solution. Traditional multi-motor drive systems typically require multiple inverters, increasing the number of switching devices such as insulated gate bipolar transistors (IGBTs). A single inverter approach significantly reduces the number of these devices, resulting in a more efficient and cost-effective design by comparison [1]–[4]. In particular, permanent magnet synchronous motors (PMSMs) have higher efficiency and greater durability compared to induction motors (IMs), leading to an increase in research focused on applying the PMSM in multi-parallel systems fed by a single inverter [5]. However, existing research has primarily focused on ensuring system stability, with relatively little attention directed toward minimizing energy losses, such as copper losses, which are crucial for maximizing overall efficiency and performance in multi-parallel PMSM systems. The Lagrange multiplier method is a mathematical technique primarily used for the MTPA control in the SIDP SPMSM drive system. It enables the calculation of optimal control variables, such as the reference currents and the difference between the rotor positions of the two motors. However, this mathematical process of finding the optimal values is complex and challenging, often requiring intricate calculations and significant computational resources to achieve accurate results [6].

The proposed MTPA control method adjusts the *d*-axis current and performs predictions for the next control state by using the predictive P&O method to reduce copper losses without utilizing intricate mathematical methodologies. The conventional P&O method adjusts the control variable by periodically perturbing it and observing the change in output to determine whether to continue or reverse the adjustment direction, whereas the predictive P&O method leverages a system model of the electrical characteristics to predict the next state, optimizing efficiency by dynamically adjusting control inputs [7]. The SIDP SPMSM drive system achieves speed synchronization of the two motors while tracking the MTPA conditions when applying the proposed MTPA control method. The validity of the proposed method is verified through simulations.

II. ELECTRICAL CHARACTERISTICS OF THE SIDP SPMSM

A. Single inverter dual parallel SPMSM drive system

Fig. 1 shows the topology of the SIDP SPMSM drive system. In this configuration, the master motor is controlled using a closed-loop manner, where feedback mechanisms adjust parameters such as torque and speed. On the other hand, the slave motor is not controlled in terms of torque or speed, as it operates in an open-loop manner [8]. Furthermore, the two motors are connected in parallel to a single inverter, so the voltage applied to both motors is identical, satisfying the voltage equation of (1) in the synchronous reference frame of the two motors. The slave voltage can be expressed as the master voltage through this equation. θ_d is defined as the electrical angle difference between the two motors and can be expressed as (2). The rotor positions can be changed depending on the load torques applied to the individual motors, where the subscripts *m* and *s* represent the master and slave motors, respectively. v_{dm} and v_{qm} are the *d*-and *q*-axis master voltages, v_{ds} and v_{qs} are the *d*-and *q*-axis slave voltages, θ_m and θ_s are the electrical rotor positions. Using this voltage relationship, the currents of the slave motor can be obtained.

$$\begin{bmatrix} v_{ds} \\ v_{qs} \end{bmatrix} = \begin{bmatrix} \cos\theta_d & \sin\theta_d \\ -\sin\theta_d & \cos\theta_d \end{bmatrix} \begin{bmatrix} v_{dm} \\ v_{qm} \end{bmatrix}. \tag{1}$$

$$\theta_d = \theta_s - \theta_m. \tag{2}$$

B. Current relationships between the two motors

The currents of the slave motor is to be expressed as the currents of the master motor, as shown in (3).

$$\begin{bmatrix} i_{ds} \\ i_{qs} \end{bmatrix} = \begin{bmatrix} A & B \\ C & D \end{bmatrix} \begin{bmatrix} i_{dm} \\ i_{qm} \end{bmatrix} + \begin{bmatrix} E \\ F \end{bmatrix}, \qquad (3)$$

where A, B, C, D, E, and F are the coefficients that describe the relationship between the currents of the master motor and the slave motor. i_{dm} and i_{qm} are the master d-and q-axis currents, i_{ds} and i_{qs} are the slave d-and q-axis currents. They incorporate the electrical properties and dynamic factors of both motors. These coefficients enable the expression of the slave motor's currents as a function of the master motor's currents, allowing for indirect control of the slave motor. The coefficients are calculated as follows:

$$Z_s = R_s^2 + \omega_{es}^2 L_s^2, \, Z_a = \omega_{em} L_m, \, Z_b = \omega_{es} L_s, \qquad (4)$$

$$A = \frac{R_m}{Z_s}(R_s \cos\theta_d - Z_b \sin\theta_d) + \frac{Z_a}{Z_s}(R_s \sin\theta_d + Z_b \cos\theta_d), \qquad (5)$$

$$B = \frac{R_m}{Z_s}(R_s \sin\theta_d + Z_b \cos\theta_d) - \frac{Z_a}{Z_s}(R_s \cos\theta_d - Z_b \sin\theta_d), \qquad (6)$$

$$C = -\frac{R_m}{Z_s}(R_s \sin\theta_d + Z_b \cos\theta_d) + \frac{Z_a}{Z_s}(R_s \cos\theta_d - Z_b \sin\theta_d), \qquad (7)$$

$$D = \frac{R_m}{Z_s}(R_s \cos\theta_d - Z_b \sin\theta_d) + \frac{Z_a}{Z_s}(R_s \sin\theta_d + Z_b \cos\theta_d), \qquad (8)$$

$$E = \frac{\omega_{em}}{Z_s}\lambda_m(R_s \sin\theta_d + Z_b \cos\theta_d) - \frac{\omega_{es}}{Z_s}Z_b\lambda_s, \qquad (9)$$

$$F = \frac{\omega_{em}}{Z_s}\lambda_m(R_s \cos\theta_d - Z_b \sin\theta_d) - \frac{\omega_{es}}{Z_s}R_s\lambda_s, \qquad (10)$$

where R_m and R_s are the stator phase resistances, L_m and L_s are the stator phase inductances, λ_m and λ_s are the permanent magnet fluxes, ω_{em} and ω_{es} are the electrical rotating speeds. The calculations can be simplified by assuming identical speeds and parameters between the two motors.

III. PROPOSED MTPA CONTROL METHOD

The predictive P&O method is used in this paper as a strategy to reduce the copper losses of the SIDP SPMSM drive system. Iron losses and other losses are ignored, and only copper losses are considered when calculating efficiency. The proposed method assumes that the system is in steady-state condition.

A. Predictive P&O method for MTPA control

The predictive P&O in the SIDP SPMSM drive system, continuously adjusts the master d-axis MTPA current to minimize copper losses and find the optimal operation point. The term 'the master d-axis MTPA current' will be referred to simply as 'the d-axis MTPA current' for brevity in this paper. The initial d-axis MTPA current is set to zero to start generally.

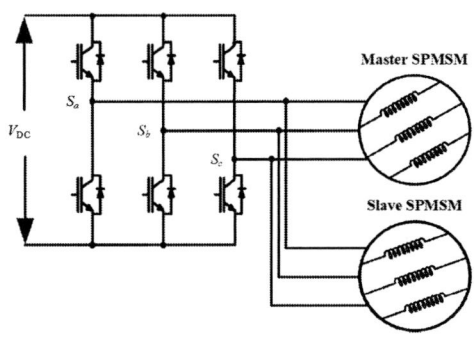

Fig. 1. Configuration of the SIDP SPMSM drive system.

The d-axis MTPA current increases or decreases by current increment (i_v) for each control state. The d-axis MTPA current for the next control state in each case is as in (11) to (13), where k represents the current state and $k+1$ represents the next control state.

Situation 1: The d-axis MTPA current remains fixed,

$$i_{dm}[k+1] = i_{dm}[k]. \qquad (11)$$

Situation 2: The d-axis MTPA current is increased by i_v,

$$i_{dm}[k+1] = i_{dm}[k] + i_v. \qquad (12)$$

Situation 3: The d-axis MTPA current is decreased by i_v,

$$i_{dm}[k+1] = i_{dm}[k] - i_v. \qquad (13)$$

The master q-axis current of the next control state is shown in (14).

$$i_{qm}[k+1] = \frac{T_{em}^*}{K_t}, K_t = \frac{3}{4}P\lambda_m, \qquad (14)$$

where T_{em}^* is the reference torque of the master motor, K_t is the torque constant of the master motor, and P is the number of poles. The slave currents of the next control state can be expressed as (15) using (3), where the coefficients can be obtained by substituting variables such as the present speed and the electrical angle difference between the two motors into (4) to (10) as the system is in the steady-state condition. This allows for calculating the slave currents by converting them to the master currents,

$$\begin{bmatrix} i_{ds}[k+1] \\ i_{qs}[k+1] \end{bmatrix} = \begin{bmatrix} A & B \\ C & D \end{bmatrix} \begin{bmatrix} i_{dm}[k+1] \\ i_{qm}[k+1] \end{bmatrix} + \begin{bmatrix} E \\ F \end{bmatrix}. \qquad (15)$$

The copper losses for the next control state can be predicted using the relationships between the master and slave currents. The q-axis currents are excluded from the copper loss calculation because they are determined by the load torque [9]. The copper losses (P_c) are calculated using only the d-axis currents as shown in equation (16), with the coefficient of 1.5 omitted for simplicity in the analysis.

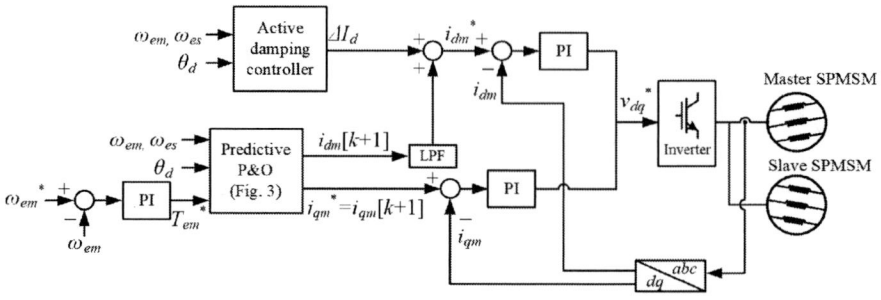

Fig. 2. A Block diagram of controllers for SIDP SPMSM drives.

$$P_c = R_m i_{dm}^2 + R_s i_{ds}^2. \tag{16}$$

For the three cases of equations in (11) to (13), the predicted copper losses for the next control state can be calculated as in (17) to (19).

Situation 1: The d-axis MTPA current remains fixed,

$$P_c[k+1] = R_m (i_{dm}[k])^2 \\ + R_s (Ai_{dm}[k] + Bi_{qm}[k+1] + E)^2. \tag{17}$$

Situation 2: The d-axis MTPA current is increased by i_v,

$$P_c[k+1] = R_m (i_{dm}[k] + i_v)^2 \\ + R_s (A(i_{dm}[k] + i_v) + Bi_{qm}[k] + E)^2. \tag{18}$$

Situation 3: The d-axis MTPA current is decreased by i_v,

$$P_c[k+1] = R_m (i_{dm}[k] - i_v)^2 \\ + R_s (A(i_{dm}[k] - i_v) + Bi_{qm}[k] + E)^2. \tag{19}$$

This algorithm assesses the predicted copper losses in each situation and selects the d-axis MTPA current that yields the lowest loss among the three. For example, if equation (19) yields the lowest loss, The d-axis MTPA current for the next control state is selected as $i_{dm}[k] - i_v$.

B. Design of controllers for SIDP SPMSM drives

Fig. 2 shows a block diagram of controllers for SIDP SPMSM drives. The controller architecture follows the structure proposed in [10]. The system's control scheme mainly consists of the active damping controller and the predictive P&O method. The reference torque of the master motor (T_{em}^*) is generated to control the master motor's speed to the reference speed (ω_{em}^*) by using a PI controller. The active damping controller ensures speed synchronization between the master and slave motors by generating the d-axis active damping current (ΔI_d) required to synchronize the speeds of the two motors. The predictive P&O method generates the d-axis MTPA current for the next control state ($i_{dm}[k+1]$) and the q-axis MTPA current for the next control state ($i_{qm}[k+1]$) required to achieve minimum copper losses of the SIDP system. The q-axis MTPA current for the next control state is directly applied as the master q-axis reference current (i_{qm}^*). The d-axis active damping

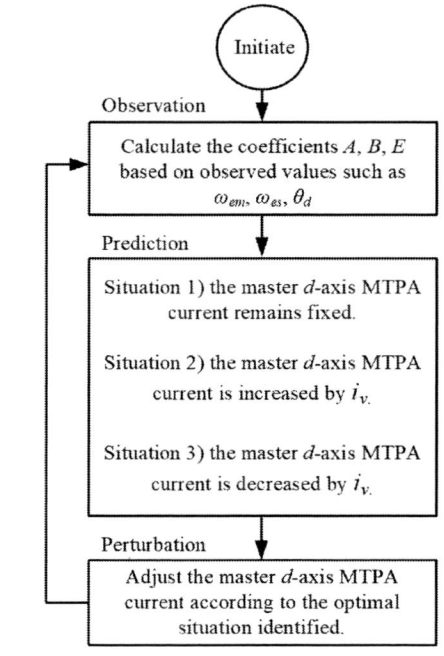

Fig. 3. Flowchart of predictive P&O method for MTPA control.

current and the d-axis MTPA current for the next control state are added to generate the master d-axis reference current (i_{dm}^*). The active damping current is a high-frequency current with a fast response characteristic, so a low-pass filter is applied to distinguish the d-axis active damping current and the d-axis MTPA current for the next control state. The master d-q axis reference currents are applied to the PI current controller to generate the d-q axis reference voltage (v_{dq}^*), which is then supplied to the inverter connected in parallel to two motors.

Fig. 3 shows a flowchart of the predictive P&O method for the MTPA control. Initially, the master d-axis MTPA current is set to zero. The process starts with the observation phase, where the key parameters like the motor speeds and the angle difference are measured to calculate the coefficients A, B, and E. In the prediction phase, the control strategy assesses three scenarios: fixing, increasing, or decreasing the d-axis MTPA current, predicting copper losses for each. In the perturbation phase, the situation with the lowest predicted copper loss is

979-8-3315-1612-3/25 $31.00 © 2025 IEEE

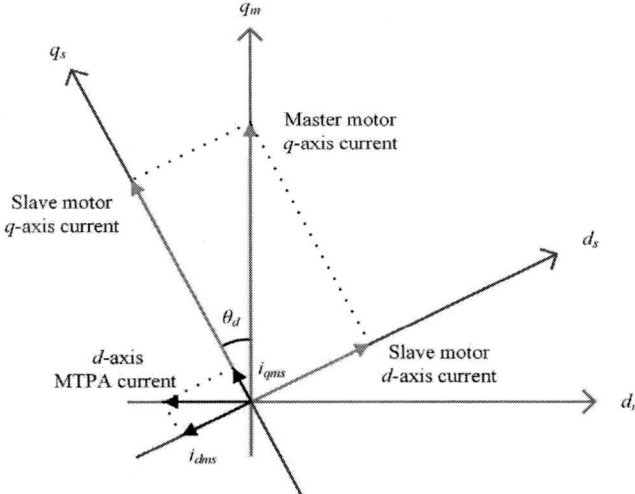

Fig. 4. A small signal diagram of the SIDP SPMSM drive system.

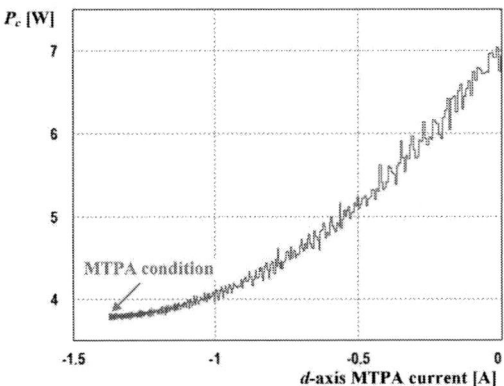

Fig. 5. PSIM simulation result for copper losses according to the d-axis MTPA current, when the master motor is under a load of 8.1 Nm, and the slave motor is under a load of 4.05 Nm.

chosen, and the master d-axis MTPA current is adjusted to optimize efficiency and minimize losses. This ensures real-time adaptation for efficient, low-loss operation.

C. Validation of the proposed method using the small-signal modeling

Fig. 4 shows a small-signal diagram of the SIDP SPMSM drive system. The synchronous frames of the master motor are d_m and q_m, while those of the slave motor are d_s and q_s [11]. In small-signal analysis, the d-axis MTPA current, whose magnitude is the current increment (i_v), can be interpreted as i_{dms} and i_{qms} in the synchronous frames of the slave motor. Referring to the green line in the figure, when the electrical angle difference (θ_d) is not zero, the master q-axis current induces the slave d-axis current. The slave d-axis current, as a loss component, is compensated by the i_{dms}. Equation (20) represents the relationship between the small-signal variation of the slave torque (ΔT_s) and i_{qms} that affects the slave torque. The sine function in equation (20) is bounded by 1, enabling the derivation of inequality (21), which imposes a constraint on the small-signal variation of the slave torque. If an appropriate i_v is selected by considering (21), the variation of the slave torque due to the d-axis MTPA current is negligible.

$$\Delta T_s = \frac{3}{4} P \lambda_s i_{qms} = \frac{3}{4} P \lambda_s i_v \sin \theta_d. \tag{20}$$

$$\left| \Delta T_s \right| \leq \frac{3}{4} P \lambda_s i_v. \tag{21}$$

IV. SIMULATION RESULTS

The proposed MTPA control method has been validated using simulation results. The motor parameters listed in Table I were used, assuming identical parameters for both the master and slave motors. The simulations were performed using PSIM, with all simulations conducted at the rated speed of 2100 rpm. Assuming that the small-signal variation of the slave torque is 0.01 % of the rated torque, i_v was selected as 0.001 A using (21).

TABLE I. SPMSM PARAMETERS

Parameter	Value
Rated power	1.78 [kW]
Rated current	10 [A]
Rated speed	2100 [r/min]
Rated torque	8.1 [N·m]
Stator phase resistance	0.588 [Ω]
Stator phase inductance	2.5 [mH]
Permanent magnet flux	0.15 [Wb]
Number of poles	8

Fig. 5 shows PSIM simulation result for copper losses according to the d-axis MTPA current, when the master motor is under a load of 8.1 Nm, and the slave motor is under a load of 4.05 Nm. When the MTPA condition is reached, the d-axis MTPA current does not change any further, maintaining the copper losses at their minimum value, with the system efficiency reaching 95.8 %, where the efficiency η can be obtained as follows [9]:

$$\eta = \frac{\frac{3}{2} \omega_{em} \lambda_m (i_{qm} + i_{qs})}{\frac{3}{2} \omega_{em} \lambda_m (i_{qm} + i_{qs}) + \frac{3}{2} R_m (i_{dm}^2 + i_{qm}^2 + i_{ds}^2 + i_{qs}^2)}. \tag{22}$$

Fig. 6 shows PSIM simulation results according to load torque conditions of the proposed MTPA control. It represents the speeds, output torques, copper losses, and d-axis currents under various load conditions, with only the d-axis currents considered when calculating the copper losses. The proposed MTPA control method is applied at $t=1$ second. The two motors generated output torques approximately equal to the load torques and maintained synchronized speeds before and after starting the MTPA control in all cases.

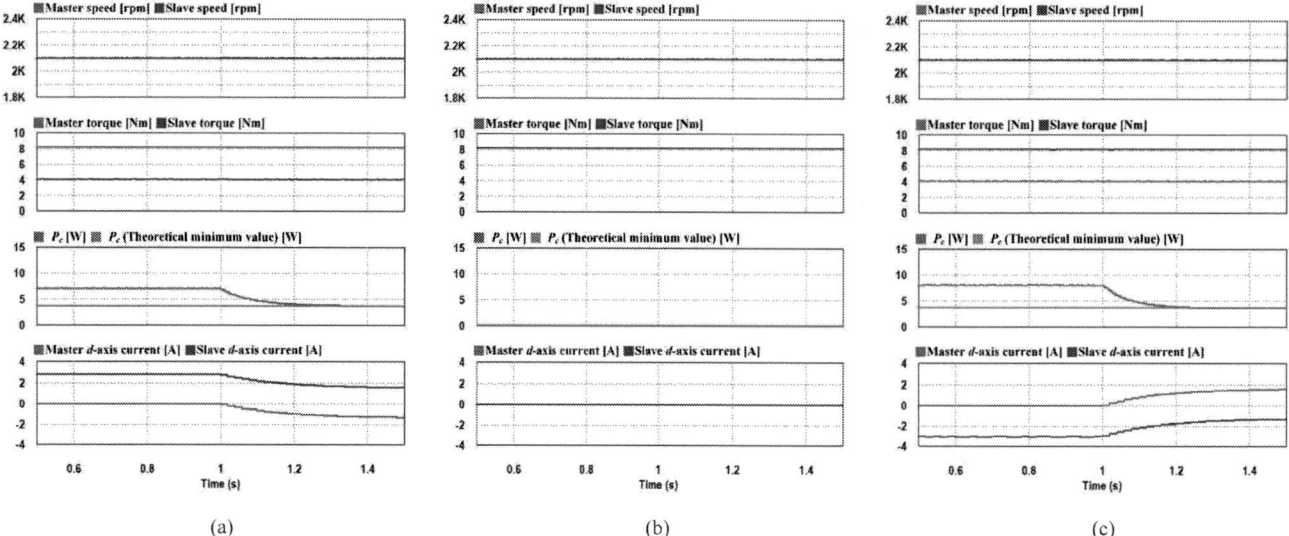

Fig. 6. PSIM simulation results according to load torque conditions of the proposed MTPA control. (a) the master motor is under a load of 8.1 Nm, and the slave motor is under a load of 4.05 Nm. (b) the master motor is under a load of 8.1 Nm, and the slave motor is under a load of 8.1 Nm. (c) the master motor is under a load of 4.05 Nm, and the slave motor is under a load of 8.1 Nm.

In Fig. 6(a), the master motor is under a load of 8.1 Nm, and the slave motor is under a load of 4.05 Nm. The master d-axis current decreased from 0 to -1.38 A, and the slave d-axis current decreased from 2.81 to 1.53 A, resulting in a decrease in copper losses from 6.96 to the theoretical minimum value of 3.74 W. Applying a negative d-axis current to the master motor reduces the magnitude of the slave motor's d-axis current, thereby further decreasing copper losses.

In Fig. 6(b), the master motor is under a load of 8.1 Nm, and the slave motor is under a load of 8.1 Nm. The MTPA trajectory is identical to that of a single motor in this case. Thus, the master and slave d-axis currents are theoretical zero, resulting in zero losses. The simulation results also showed nearly zero losses.

In Fig. 6(c), the master motor is under a load of 4.05 Nm, and the slave motor is under a load of 8.1 Nm. The master d-axis current increased from 0 to 1.58 A, and the slave d-axis current increased from -3 to -1.31 A, decreasing copper losses from 7.94 to the theoretical minimum value of 3.72 W. In this case, the magnitude of the slave d-axis current was reduced by adjusting the master d-axis current in the positive direction.

V. CONCLUSION

This paper proposed the MTPA control using the predictive P&O method for the SIDP SPMSM drives. The d-axis current is continuously adjusted by the predictive copper losses to maximize the efficiency of the SIDP SPMSM drive system. The proposed method maintains the required torque output while achieving the synchronized speeds for both motors. The validity of the proposed strategy was demonstrated through PSIM simulations.

REFERENCES

[1] K. Matsuse, H. Kawai, Y. Kouno, and J. Oikawa, "Characteristics of speed sensorless vector controlled dual induction motor drive connected in parallel fed by a single inverter," *IEEE Trans. Ind. Appl.*, vol. 40, no. 1, pp. 153–161, Jan./Feb. 2004.

[2] V. Khanna, *Insulated Gate Bipolar Transistor IGBT Theory and Design.* Hoboken, NJ, USA: Wiley, 2003.

[3] A. Bouscayrol, M. Pietrzak-David, P. Delarue, R. Pena-Eguiluz, P. Vidal, and X. Kestelyn, "Weighted control of traction drives with parallel connected AC machines," *IEEE Trans. Ind. Electron.*, vol. 53, no. 6, pp. 1799–1806, Dec. 2006.

[4] R. Omata, K. Oka, A. Furuya, S. Matsumoto, Y. Nozawa, and K. Matsuse, "An improved performance of five-leg inverter in two induction motor drives," in *Proc. CES/IEEE 5th Int. Power Electron. Motion Control Conf.*, 2006, pp. 1–5.

[5] D. Bidart, M. Pietrzak-David, P. Maussion, and M. Fadel, "Mono inverter dual parallel PMSM - structure and control strategy," in *Proc. 34th Annu. Conf. IEEE Ind. Electron.*, 2008, pp. 268–273.

[6] T. Liu, M. Fadel, J. Li, and X. Ma, "A MTPA control strategy for mono-inverter multi-PMSM system," *IEEE Trans. Power Electron.*, vol. 36, no. 6, pp. 7165–7177, Jun. 2021.

[7] N. Femia, D. Granozio, G. Petrone, G. Spagnuolo, and M. Vitelli, "Predictive & adaptive MPPT perturb and observe method," *IEEE Trans. Aerosp. Electron. Syst.*, vol. 43, no. 3, pp. 934–950, Jul. 2007.

[8] Y. Lee and J. Ha, "Analysis of parameter variations on mono inverter dual parallel SPMSM drive system," in *Proc. 9th Int. Conf. Power Electron./IEEE Energy Convers. Congr. Expo. Asia*, 2015, pp. 1875–1880.

[9] J. Lee and J. -W. Choi, "MTPA control method for MIDP SPMSM drive system using angle difference controller and P&O algorithm," *IEEE Trans. Power Electron.*, vol. 37, no. 12, pp. 15382–15396, Dec. 2022.

[10] Y. Lee and J. Ha, "Control method for mono inverter dual parallel surface mounted permanent-magnet synchronous machine drive system," *IEEE Trans. Ind. Electron.*, vol. 62, no. 10, pp. 6096–6107, Oct. 2015.

[11] Y. Lee and J. Ha, "Minimization of stator currents for mono inverter dual parallel PMSM drive system," in *Proc. Int. Power Electron. Conf.*, 2014, pp. 3140–3144.

A Novel I-f Startup Strategy with Smooth Transition to Sensorless Control for CSI-Fed PMSM Drives Used in Submersible Pumps

Milad Bahrami-Fard, *Student Member, IEEE*, Majid Ghasemi Korrani, *Student Member, IEEE*, and Babak Fahimi, *Fellow, IEEE*
Renewable Energy and Vehicular Technology (REVT) lab
Department of Electrical Engineering, University of Texas at Dallas
Richardson, TX, USA
Email: milad.bahramifard@utdallas.edu

Abstract—This paper proposes a practical startup strategy for current source inverter (CSI)-fed Permanent Magnet Synchronous Motor (PMSM) drives in submersible pump applications, focusing on ensuring a seamless shift to sensorless field-oriented control (FOC). The method effectively manages the transition to sensorless operation without requiring precise current or alignment error calculations, thereby simplifying implementation. By addressing speed and current oscillations directly during the startup and transition stages, the approach significantly enhances overall system stability and responsiveness. Validation through simulation and experimental testing demonstrates the strategy's success in maintaining low oscillation levels across various operating conditions, confirming its reliability for high-performance industrial applications.

Keywords—*Current source inverter (CSI), field-oriented control (FOC), PMSM drives, sensorless control, submersible pumps.*

I. INTRODUCTION

In modern industrial applications, there is a continuous push for innovations that improve efficiency, precision, and operational reliability. Recent advancements in control strategies and power electronics have increasingly focused on enhancing the reliability and adaptability of Permanent Magnet Synchronous Motor (PMSM) drives in demanding industrial environments. PMSM drives are considered in industrial applications for their high efficiency, dynamic performance, and power density, making them ideal for tasks demanding precise speed and torque control [1]-[7]. To achieve these attributes, Field-Oriented Control (FOC) has become prevalent, as it decouples motor flux and torque components, enabling accurate control. Traditionally, high-performance FOC strategies rely on position sensors that provide accurate rotor position and speed measurements, optimally positioning the stator magnetic field with the rotor flux for optimal performance [8]. However, position sensors introduce higher costs and reduced reliability, particularly under extreme environmental conditions, such as submersible pumps that can operate several kilometers below the surface, where installing and maintaining sensors becomes extremely challenging [9].

Position sensorless control algorithms have therefore been developed to eliminate the need for physical sensors. These sensorless methods are categorized broadly into high-frequency signal injection (HFI) and model-based observer techniques. While HFI techniques excel at detecting rotor position at low or zero speeds by analyzing high-frequency current responses due to rotor magnetic saliency, they are highly sensitive to electrical noise and require complex filtering, complicating real-world implementation [10]-[13]. In contrast, model-based methods—better suited for medium to high speeds—estimate back electromotive force (EMF) and utilize a phase-locked loop (PLL) to track rotor position and velocity [14]-[17]. For comprehensive sensorless FOC across the full speed range, these methods are often combined. However, this dual-method approach increases controller complexity and can reduce reliability, particularly in demanding applications like submersible pumps, where simplicity and robustness are essential [18]. Furthermore, in submersible pump application, access to voltage and current at motor terminal is not trivial and injection of high frequency field components will face similar challenges across a few kilometers of cables that connect the inverter to the motor.

Current Source Inverters (CSIs) bring additional benefits to PMSM drives, particularly in industrial applications requiring durability and robustness. CSIs support four-quadrant operation, short-circuit protection, and voltage-boosting capability while providing motor-friendly waveforms with low dv/dt, all of which enhance the connected PMSM's reliability and longevity [19]-[21]. This makes CSI-fed drives well-suited for high-power, medium-voltage applications, especially in industries like petrochemicals, mining, and metal processing, as well as in pumping applications requiring long cable runs, where they outperform voltage source inverters (VSIs) that struggle with capacitive cable effects [22]. Although sensorless control research has largely focused on VSI-fed drives, studies on sensorless control for CSI-fed PMSMs are still relatively limited and call for more attention, keeping in mind that access to accurate motor currents in CSI is a key element for successful implementation of sensorless control.

To perform sensorless control at medium to high speeds, methods based on back EMF estimation are particularly effective, provided that there is a reliable startup strategy to reach a speed where the back EMF can be reliably sensed.

Conventional startup methods like V/f control, common in induction motors, are simple and open-loop but lack closed-loop current feedback, leading to instability under heavy loads [23]. The I-f method, first implemented in PMSMs in [24], offers an alternative with closed-loop current regulation and does not require position feedback. It approximates rotor position using a ramped speed command while maintaining constant reference-frame current, delivering a smoother startup and reduced torque ripples compared to V/f [25]. However, the I-f method can result in a misalignment between the synchronous and actual rotor frames, especially under high startup currents, as seen in submersible pump applications [26]. Accurate frame alignment is crucial for a successful transition from I-f startup to sensorless FOC, as misalignment introduces rotor position estimation errors. This, in turn, challenges the transition and potentially destabilizes the dynamics of speed [27]-[31].

Several methods have been proposed to manage this transition. For example, [24] used a first-order compensator to align the synchronous reference frame with the rotor position, facilitating a smooth shift from I-f control to EMF-based control, though the effectiveness depended heavily on compensator tuning. In [32], a gradual decrease in angle gradient deviation was achieved by adjusting it based on the reference current as the motor reached the target speed, allowing a smoother transition but extending the startup duration. Alternatively, [33] applied feedforward compensation within the current loop, enhancing stability but primarily prolonging the I-f starting period. Lastly, [34] proposed a speed-dependent weighting factor to align the synchronous reference frame to the rotor position, yet finding the correct parameters for convergence remained challenging, and no detailed stability analysis for I-f methods was provided.

In this paper, a new startup strategy is introduced using which a smooth transition from I-f startup to sensorless FOC for CSI-fed PMSM drives is obtained. Unlike previous methods that depend on precise frame alignment or tuning the q-axis current, our approach uses an error compensation strategy to eliminate frame alignment errors. By first aligning the estimated frame to a virtual reference and then applying gradual error correction, stable speed and current profiles have been achieved without requiring complex parameter calculations. This strategy allows for accurate rotor position estimation without calculating the minimum q-axis current or other parameters.

The rest of this article is structured as follows: Section II reviews sensorless control fundamentals for CSI-fed PMSM drives, including EMF-based observer derivation, and provides a detailed analysis of the I-f startup process, examining primary

non-ideal factors impacting rotor position estimation along with the proposed transition strategy. Section III presents simulation results validating the method, followed by experimental confirmation in Section IV. Section V concludes with a summary of contributions and potential future directions.

II. SENSORLESS CONTROL OF THE CSI-FED PMSM WITH AN IMPROVED STARTUP

Fig. 1 depicts the CSI-fed PMSM drive system with a long cable connection, comprising a passive rectifier, an asymmetric H-bridge, a current source inverter, a DC inductor, and output capacitor filters. The asymmetric H-bridge regulates the DC current to maintain a constant DC-link current. The motor speed and torque are controlled by a CSI using a FOC.

A. Vector Control of Sensorless CSI Fed PMSM

The block diagram of the sensorless FOC for the CSI-fed PMSM based on the back-EMF observer is illustrated in Figs. 2 and 3. The control sets the d-axis current to zero while regulating the q-axis current according to the set value of the speed controller.

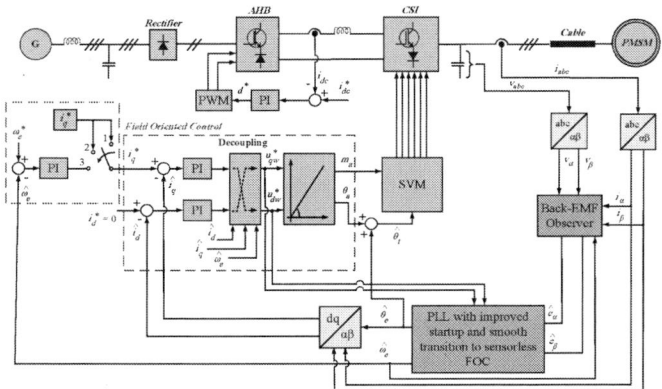

Fig. 2. FOC control of the CSI-fed PMSM drive system with proposed smooth transition strategy

B. Mathematical Model of the PMSM

To start the process of the position estimation for PMSMs, the mathematical model of the surface-mounted PMSM, expressed in a stationary reference frame (αβ), is given as follows.

$$L_s\left(\frac{di_\alpha}{dt}\right) = -R_s i_\alpha - e_\alpha + u_\alpha \tag{1}$$

$$L_s\left(\frac{di_\beta}{dt}\right) = -R_s i_\beta - e_\beta + u_\beta \tag{2}$$

where i_α, i_β and u_α, u_β are the stator currents and voltages, respectively, e_α and e_β are the back EMFs, R_s is the stator resistance, and L_s is the synchronous inductance. The back EMF for each phase can be described as

$$e_\alpha = -\frac{\sqrt{3}}{2}\psi_f P\omega_e \sin(\theta_e) \tag{3}$$

$$e_\beta = \frac{\sqrt{3}}{2}\psi_f P\omega_e \cos(\theta_e) \tag{4}$$

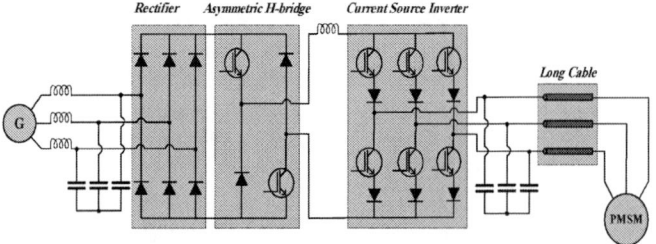

Fig. 1. CSI-fed PMSM drive system with a long cable connection for submersible pump applications.

979-8-3315-1612-3/25 $31.00 © 2025 IEEE

where ψ_f is the flux linkage due to rotor permanent magnet, P is the number of pole pairs, ω_e is the rotor angular speed, and θ_e is the actual rotor position. The α- and β-axis back EMF inherently contain the actual rotor position θ_e, but direct measurement of these values is not feasible. Therefore, an observer along with a PLL will be utilized to derive the position information.

To estimate the position of PMSMs, mathematical model of a surface-mounted PMSM, expressed in dq-frame, is given as follows.

$$L_s\left(\frac{di_d}{dt}\right) = -R_s i_d + \omega_e L_s i_q + u_d \tag{5}$$

$$L_s\left(\frac{di_q}{dt}\right) = -R_s i_q - \omega_e L_s i_d - \omega_e \psi_f + u_q \tag{6}$$

where, i_d, i_q and u_d, u_q are the stator currents and voltages, respectively, ω_e is the rotor electrical speed, ψ_f is the PM-originated flux linkage, R_s is the stator resistance, and L_s is the synchronous inductance.

Reference Generator

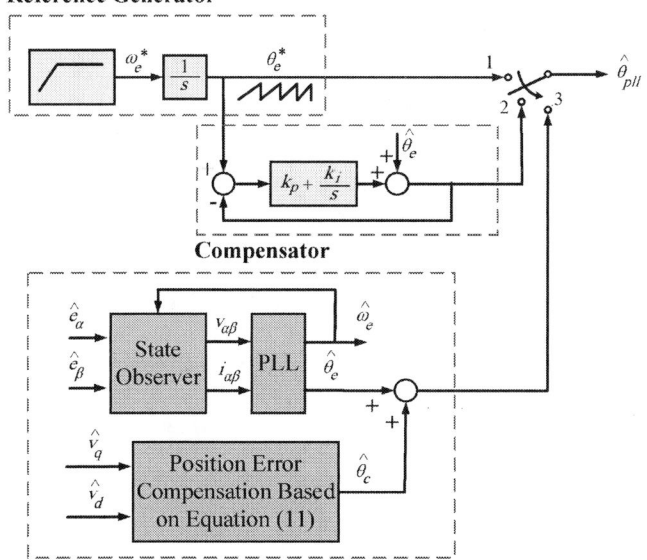

Fig. 3. Diagram of the PLL with improved startup and smooth transition to sensorless FOC

C. I-f Startup Analysis

During the speed ramp-up, the startup procedure uses a closed current loop and an open speed loop in the synchronous reference d^*q^*-frame, as shown in Figs. 4 and 5. In this process, the reference q^*-axis current (i_q^*) is maintained constant, and the i_d^* is set to zero. The reference θ_e^* is obtained by integrating the ramp speed command (ω_e^*) as follows:

$$\theta_e^* = \int \omega_e^* dt \tag{7}$$

$$\omega_e^* = K_\omega t \tag{8}$$

where K_ω is a constant. At the start of the procedure, the d^*q^*-frame is set to a $\delta_e^* = 90°$ phase lag relative to the real rotor

dq-frame so that the dq-frame can start moving smoothly, as shown in Fig. 4. The d^*q^*-frame rotates as the ramp speed command is integrated. As angle δ_e^* decreases, the q-axis current (i_q) increases proportional to $i_q^* \cos \delta_e^*$. Once i_q generates sufficient torque to exceed the load torque, the rotor begins to rotate, and δ_e^* converges to a fixed value depending on the load and acceleration. The relationship between the d^*q^*-axis currents and the actual currents is defined as follows:

$$i_q = i_q^* \cos \delta_e^* - i_d^* \sin \delta_e^* \tag{9}$$

$$i_d = i_q^* \sin \delta_e^* + i_d^* \cos \delta_e^* \tag{10}$$

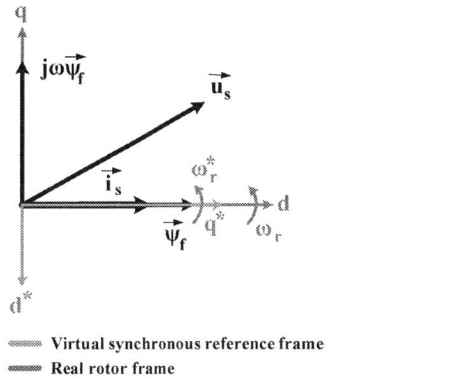

Fig 4. Diagram of the virtual synchronous d^*q^*-reference frame and real rotor dq frame at initial startup

If angle δ_e^* becomes negative, the motor's self-stabilization is lost, leading to a rapid decrease in speed and potentially stall of the rotor. To prevent this, it is crucial to keep the d^*q^*-axis lagging behind the rotor dq-axis with sufficient margin in angle δ_e^*. This is typically achieved by maintaining a high enough q^*-axis current. However, when i_q^* is greater than i_q, it creates an angle error (δ_e^*) between the synchronous reference frame and the real rotor frame. Therefore, an immediate switching of the speed controller from open-loop startup to closed-loop FOC control mode can cause large torque and current pulses.

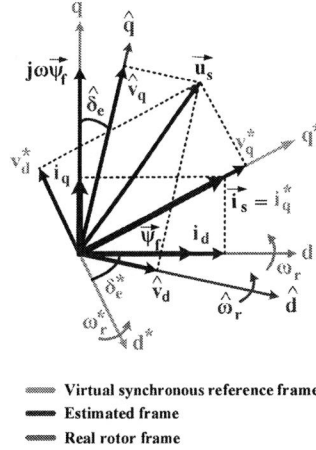

Fig. 5. Diagram of the virtual synchronous d^*q^*-reference frame, real rotor dq frame, and estimated $\hat{d}\hat{q}$ frame during acceleration

The estimated rotor axis is denoted by the $\hat{d}\hat{q}$-axis, and Fig. 5 illustrates its relationship to the dq-axis. The estimated position is shown by $\hat{\theta}_e$, and the actual position is denoted by θ_e. As previously discussed, performance of FOC is directly influenced by the precision of rotor position estimation. There are multiple non-ideal elements that contribute to position estimation error. These include, saturation-related deviations in L_s, computation time, feedback-related control delays in the motor drive system, and Pulse Width Modulation (PWM) of the inverter, all of which lead to errors proportionate to rotor speed. It is difficult to quantify these inaccuracies, especially those caused by control delay (T_d). The impact of iron loss is also highly significant at high speeds, however its quantification is challenging since it depends on multiple non-ideal criterions. Furthermore, because of the substantial deadtime ratio in a single switching period at high speeds, the inverter's insertion of deadtime needs a greater command voltage than the actual voltage, leading to severe position estimate inaccuracies.

Defining the difference between the estimated and the real positions as position error ($\hat{\delta}_e = \theta_e - \hat{\theta}_e$), the equations (4) and (5) can be transformed into the following expressions in $\hat{d}\hat{q}$-axis under steady state condition.

$$\hat{u}_q = \omega_e L_s \hat{i}_d + \omega_e \psi_f \cos(\hat{\delta}_e) \qquad (11)$$

$$\hat{u}_d = -\omega_e L_s \hat{i}_q + \omega_e \psi_f \sin(\hat{\delta}_e) \qquad (12)$$

The difficulty in accurately measuring the entire estimation error caused by many non-ideal elements makes it difficult to directly compensate for $\hat{\delta}_e$. Therefore, it is necessary to engineer a method that can compensate position errors δ_e^* and $\hat{\delta}_e$ before transitioning from open-loop startup to closed-loop FOC control mode.

TABLE I.   System Parameters

Parameters	Value
equivalent resistance R_s	2.16 Ω
equivalent inductance L_s	4.56 mH
cable equivalent resistance R_c	11.76 Ω
cable equivalent inductance L_c	9.7 mH
cable equivalent capacitance C_c	111 nF
number of pole pairs P	6
grid voltage V_g	480 V
grid frequency f_g	60 Hz
DC-link inductance L_{dc}	10 mH
output filter C_o	50 uF

D. Proposed Smooth Transition to Senesorless Control Strategy

The block diagram in Figs. 2 and 3 illustrates a sensorless FOC system for a CSI-fed PMSM, designed with a new approach for smooth transition to sensorless control. This system is implemented through three consecutive steps, each represented by a distinct reference frame configuration across the startup and transition phases.

In the initial step, the I-f method activates the motor from a standstill with controlled acceleration to reach a stable target speed. This provides sufficient back EMF for position estimation using a PLL and observer, with terminal 1 selected. During this period, the q^*-axis current remains constant while the d^*-axis current is zero, setting up the synchronous d^*q^*-reference frame to lag with respect to the actual rotor frame by 90 degrees. As the synchronous frame begins its rotation, the real q-axis current (i_q) increases, and an angle offset (δ_e^*) emerges between the frames. Once torque from i_q overcomes the load, rotor rotation aligns the rotor and synchronous frames.

Next, to minimize the position offset between the estimated and the actual reference frames, a compensator brings the error to zero, preparing the system for terminal 2. Here, only the position information will update while i_q^* remains fixed. The alignment of the estimated $\hat{d}\hat{q}$-frame with the virtual synchronous d^*q^*-reference frame enables a seamless terminal switch, avoiding any fluctuations in motor speed or current.

In the final step, transitioning to terminal 3 introduces an initial position error from the compensator. This error is systematically reduced using an error correction strategy outlined in equations (11) and (12). This shift maintains continuity in speed and current due to stable outputs from the speed PI controller and position compensation. As a result, the position error diminishes gradually, achieving smooth operation after switching terminals.

Based on (4), to completely correct the position error after shifting to terminal 3, u_q' must be at its maximum value (i.e., $\cos(\hat{\delta}_e) = 1$). The estimated position is adjusted using the proposed method to maintain u_q' at its maximum and to make sure that the position error is compensated for. 'θ_c' is designated as a compensation factor utilized to rectify the position error estimation unit. The algorithm starts by initializing θ_c and assuming a variable step size '$d\theta$'. By setting ' h ' as the number of control cycles amplitudes of u_q' and u_d' are calculated. If u_q' increases, $d\theta[h]$ remains constant. Otherwise, the sign of $d\theta[h]$ is altered (i.e., $-d\theta$) and the magnitude of the voltages are recalculated. This method will guarantee the position error will be compensated in a relatively short time without requiring complex models.

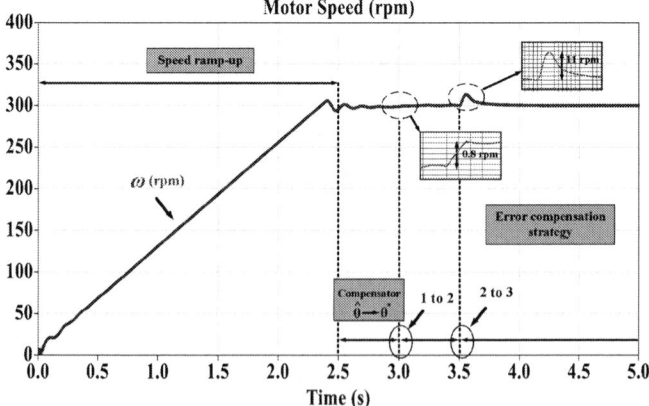

Fig. 6. Simulation waveform of the motor speed using the proposed startup strategy with a smooth transition to sensorless FOC control

III. SIMULATION RESULTS

The proposed startup strategy for the CSI-fed PMSM drive has been validated through simulation studies. Table I outlines the key parameters of the CSI-fed PMSM drive, while Fig. 6 presents the simulation results, illustrating a seamless transition to sensorless FOC. Using the I-f method, the motor successfully starts from standstill and accelerates to a stable speed of 300 rpm, where the back EMF becomes sufficiently large for position estimation via the PLL and observer. At t=2.5s, the compensator aligns the estimated frame with the virtual reference frame. At t=3s, the terminal transitions from 1 to 2 without inducing any tangible speed or current oscillations. Finally, at t=3.5s, the terminal shifts from 2 to 3, utilizing the proposed compensation strategy from Equation (11) to align the estimated frame with the real frame. The results indicate minimal speed oscillation of about 11 rpm, demonstrating the effectiveness of the smooth transition strategy.

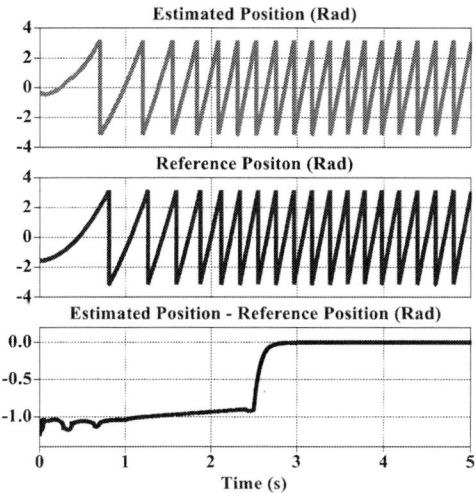

Fig. 7. Simulation results of the proposed startup strategy; Estimated position, Virtual synchronous reference position, Angle error between estimated $\hat{d}\hat{q}$ frame and virtual synchronous d^*q^*-reference frame.

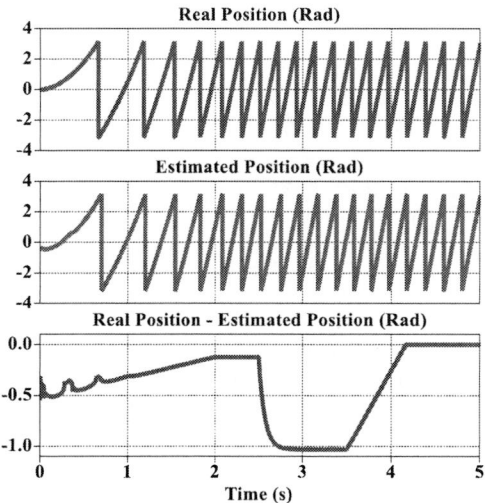

Fig. 8. Simulation results of the proposed startup strategy; Real position, Estimated position, Angle error between actual rotor dq frame and estimated $\hat{d}\hat{q}$ frame

In Fig. 7, the estimated position, virtual synchronous position, and the error between them are plotted. At t=2.5s, the compensator aligns the estimated frame with the virtual reference frame. Once alignment is achieved, the terminal switches from 1 to 2 at t=3s without introducing transients in position, current, or speed. Fig. 8 illustrates the real and estimated positions along with the error between them. As shown, the error between the estimated and real positions is eliminated using the proposed compensation strategy.

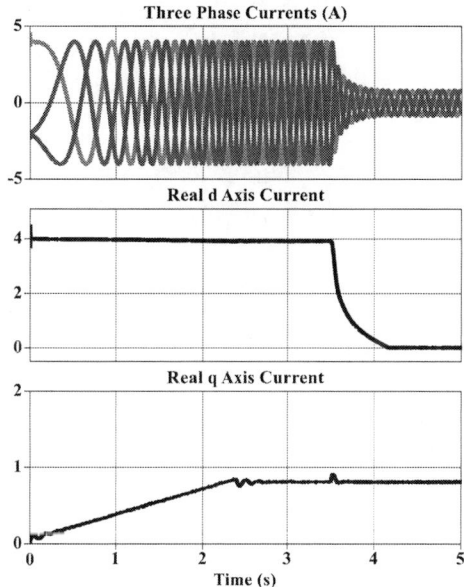

Fig. 9. Simulation results of the proposed startup strategy (a) Three phase currents. (b) Real dq-axis currents

Figs. 9 and 10 present the simulation waveforms of the three-phase, dq-axis, and $\hat{d}\hat{q}$-axis currents during various stages of the proposed startup strategy. As shown in Fig. 9, the transition from terminal 1 to 2 occurs without any oscillation in the dq-axis currents. Furthermore, during the transition from terminal 2 to 3, when the proposed compensation strategy is applied, the dq-axis and $\hat{d}\hat{q}$-axis currents converge and become identical after 0.7 s, with minimal oscillation in current (0.09 A).

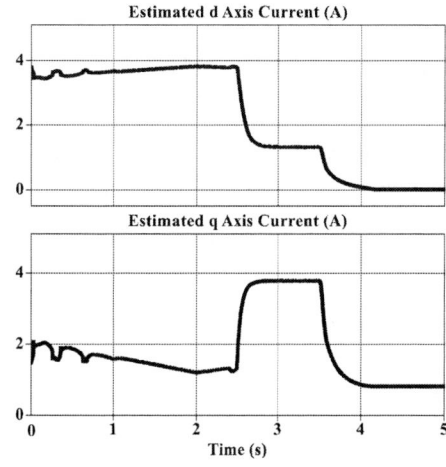

Fig. 10. Simulation results of the proposed startup strategy; Estimated $\hat{d}\hat{q}$-axis currents.

IV. EXPERIMENTAL RESULTS

To experimentally validate the proposed startup strategy, tests were performed on a CSI-fed PMSM drive system with a 1.8 km long cable connection, using a TMS320F28388D processor, as shown in Fig. 11. An optical encoder was employed exclusively to obtain the actual position for comparison purposes.

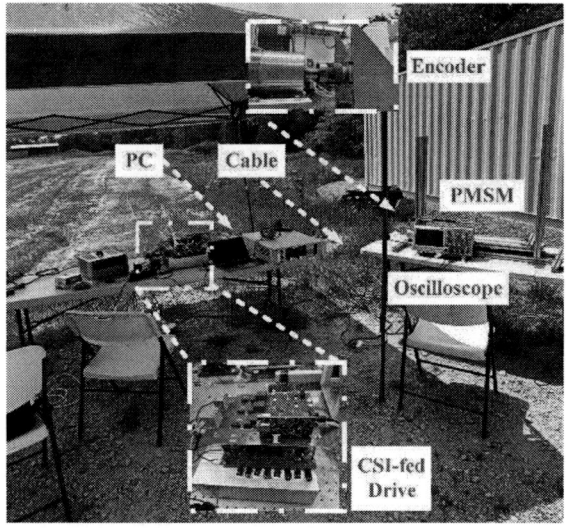

Fig. 11. Experimental setup of the CSI-fed PMSM drive system with a long cable connection for submersible pump.

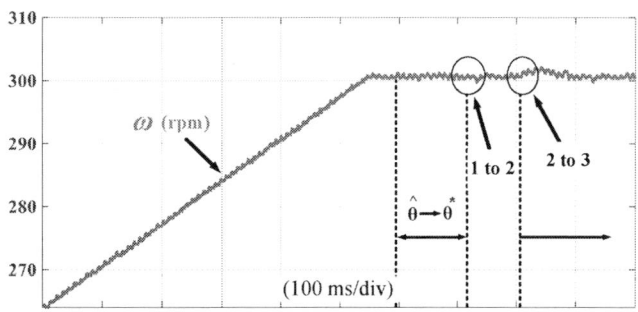

Fig. 12. Experimental results of the motor speed during different stages of the proposed startup strategy.

Fig. 12 shows the experimental results of motor speed during the entire transition from the I-f method to sensorless FOC, with the application of the proposed smooth transition strategy. Similar to the simulation tests, the motor accelerates to 300 rpm, where the estimated speed is accurately captured. The compensator then aligns the estimated position with the virtual reference frame. Following this, the terminal transitions from 1 to 2 without causing any speed or current oscillations. Finally, the transition from terminal 2 to 3 occurs, and the position error between the estimated and actual position is corrected using the proposed strategy. The experimental results validate the effectiveness of the proposed startup method, showing only minimal speed oscillations.

Figs. 13 and 14 present the experimental results of the proposed startup strategy throughout all stages of the transition from the I-f method to sensorless FOC. Initially, the I-f method is used to start the motor from standstill, gradually accelerating to a stable speed where the back EMF becomes high enough to estimate the position using the PLL and observer. At this point, both switches are connected to terminal 1. The compensator reduces the error between the estimated and virtual synchronous positions to zero, preparing the system for the transition to terminal 2. As shown in Fig. 13(a) and (b), the virtual synchronous position and the estimated position are aligned, and the error angle between them is minimized, as demonstrated in Fig. 13(c). After switching to terminal 3, the position error between the estimated and virtual synchronous positions is corrected using the proposed compensation strategy, as described in the equations. The angle error between the estimated and virtual synchronous frames is effectively reduced, demonstrating the success of the error compensation strategy. Additionally, as the system transitions from terminal 2 to 3, the angle error between the actual rotor frame and the estimated frame is eliminated, as shown in Fig. 14(c). The alignment between the estimated and real positions is achieved, as shown in Fig. 14(a) and (b). This confirms the effectiveness of the compensation strategy, with the estimated frame aligning with the real rotor position and achieving minimal error. This alignment is maintained throughout the transition, showcasing the robustness and accuracy of the proposed method.

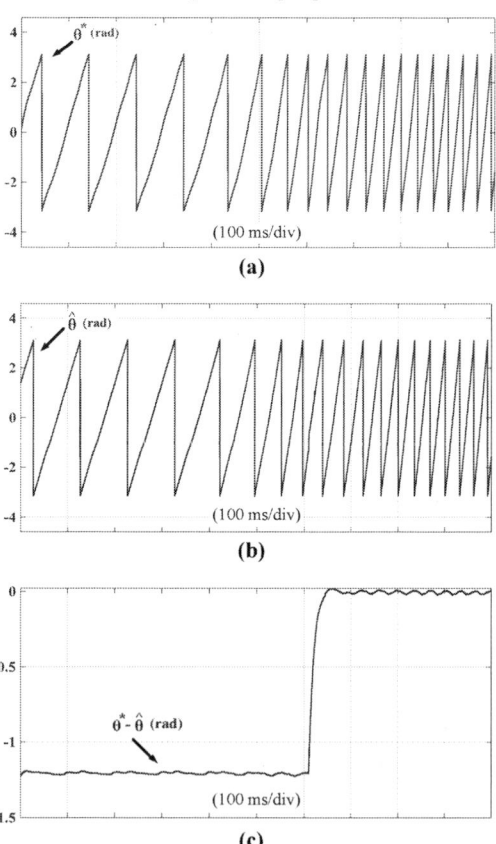

Fig. 13. Experimental results of the proposed startup strategy throughout all stages of the transition from the I-f method to sensorless FOC (a) Virtual synchronous position. (b) Estimated position. (c) Angle error between estimated $\hat{d}\hat{q}$ frame and virtual synchronous d^*q^*-reference frame.

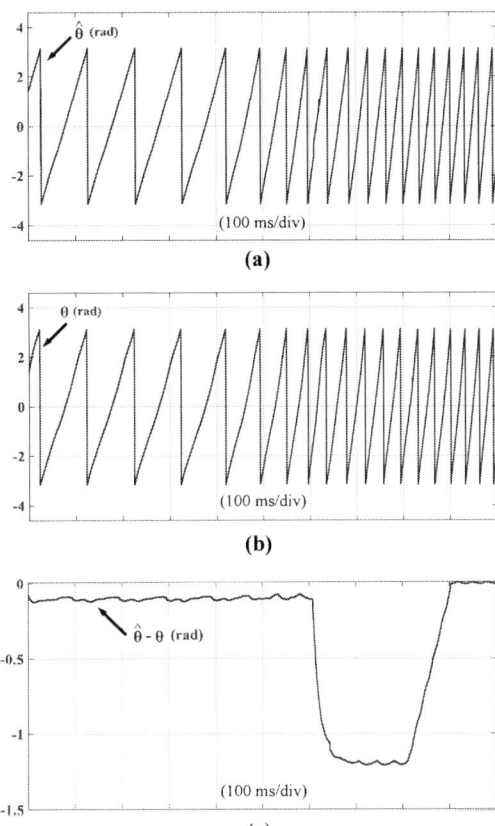

Fig. 14. Experimental results of the proposed startup strategy throughout all stages of the transition from the I-f method to sensorless FOC (a) Estimated position. (b) Real position. (c) Angle error between actual rotor dq frame and estimated $\hat{d}\hat{q}$ frame.

V. CONCLUSION

This paper introduces a novel startup strategy for CSI-fed PMSM drives, ensuring a smooth transition to sensorless FOC in submersible pump applications. The proposed method addresses the challenges of accurately calculating the minimum q-axis current and correcting real-estimated frame alignment errors, offering a practical and reliable solution. Using an error compensation strategy, the transition from I-f startup to sensorless FOC is achieved seamlessly, progressively correcting the error between the estimated and real rotor positions. This results in minimal speed and current oscillations, improving system performance and stability. Both simulation and experimental results validate the strategy's effectiveness in reducing torque fluctuations and enhancing efficiency during critical transition stages. This work provides a robust solution for sensorless control in CSI-fed PMSM drives, supporting high reliability and smooth operation in industrial applications.

REFERENCES

[1] G. Xu, F. Xiao and C. Lian, "A Position Sensorless Control Strategy for PMSM Drives With Single-Phase Current Sensor," in IEEE Transactions on Transportation Electrification, vol. 10, no. 3, pp. 4678-4688, Sept. 2024.

[2] H. Sun, X. Zhang, X. Liu and H. Su, "Adaptive Robust Sensorless Control for PMSM Based on Improved Back EMF Observer and Extended State Observer," in IEEE Transactions on Industrial Electronics, vol. 71, no. 12, pp. 16635-16643, Dec. 2024.

[3] W. Xu, S. Qu, L. Zhao, and H. Zhang, "An improved adaptive sliding mode observer for middle- and high-speed rotor tracking," IEEE Trans. Power Electron., vol. 36, no. 1, pp. 1043–1053, Jan. 2021.

[4] D. D. Patel, I. Boldea, B. Fahimi and M. B. Fard, "High Torque Density Double Stator PM Synchronous Machine," 2024 IEEE Transportation Electrification Conference and Expo (ITEC), Chicago, IL, USA, 2024, pp. 1-6.

[5] Y. Zuo, C. Lai and K. L. V. Iyer, "A Review of Sliding Mode Observer Based Sensorless Control Methods for PMSM Drive," in IEEE Transactions on Power Electronics, vol. 38, no. 9, pp. 11352-11367, Sept. 2023.

[6] M. Bahrami-Fard, B. M. Mosammam, M. Hassan Ghaderi, D. D. Patel, P. Balsara and B. Fahimi, "An Effective Cooling System for High Torque Electric Motors Using Microchannels and Two-phase Coolants," 2024 IEEE Transportation Electrification Conference and Expo (ITEC), Chicago, IL, USA, 2024, pp. 1-6.

[7] M. Ghasemi, A. Honarbakhsh, M. Saradarzadeh and M. Hamzeh, "Ultra-Wide Voltage Range Control of DC-DC Full-Bridge Converter with Hysteresis Controller," 2022 13th Power Electronics, Drive Systems, and Technologies Conference (PEDSTC), Tehran, Iran, Islamic Republic of, 2022, pp. 624-629.

[8] G. Wang, M. Valla and J. Solsona, "Position Sensorless Permanent Magnet Synchronous Machine Drives—A Review," in IEEE Transactions on Industrial Electronics, vol. 67, no. 7, pp. 5830-5842, July 2020.

[9] G. Bi et al., "High-Frequency Injection Angle Self-Adjustment Based Online Position Error Suppression Method for Sensorless PMSM Drives," in IEEE Transactions on Power Electronics, vol. 38, no. 2, pp. 1412-1417, Feb. 2023.

[10] Y. -R. Lee and S. -K. Sul, "Switching Frequency Signal-Injection Sensorless Control in Dual Three-Phase PMSM Robust to Nonideal Characteristics of Inverter System," in IEEE Transactions on Power Electronics, vol. 39, no. 7, pp. 8540-8552, July 2024.

[11] G. Wang, D. Xiao, G. Zhang, C. Li, X. Zhang, and D. Xu, "Sensorless control scheme of IPMSMs using HF orthogonal square-wave voltage injection into a stationary reference frame," IEEE Trans. Power Electron., vol. 34, no. 3, pp. 2573–2584, Mar. 2019.

[12] G. Wang, L. Yang, G. Zhang, X. Zhang, and D. Xu, "Comparative investigation of pseudorandom high-frequency signal injection schemes for sensorless IPMSM drives," IEEE Trans. Power Electron., vol. 32, no. 3, pp. 2123–2132, Mar. 2017.

[13] A. Benevieri, A. Formentini, M. Marchesoni, M. Passalacqua and L. Vaccaro, "Sensorless Control With Switching Frequency Square Wave Voltage Injection for SPMSM With Low Rotor Magnetic Anisotropy," in IEEE Transactions on Power Electronics, vol. 38, no. 8, pp. 10060-10072, Aug. 2023.

[14] M. Bahrami-Fard, M. Ghasemi Korrani, M. Rastegar, P. Balsara and B. Fahimi, "Error Compensation Strategy in Encoderless Surface Mounted PMSM Drives," 2024 IEEE Energy Conversion Congress and Exposition (ECCE), Phoenix, AZ, USA, 2024.

[15] A. Andersson and T. Thiringer, "Motion sensorless IPMSM control using linear moving horizon estimation with Luenberger observer state feedback," IEEE Trans. Transp. Electrific., vol. 4, no. 2, pp. 464–473, Jun. 2018.

[16] Y. Zhong, H. Lin, J. Wang, and H. Yang, "Improved adaptive sliding mode observer based position sensorless control for variable flux memory machines," IEEE Trans. Power Electron., vol. 38, no. 5, pp. 6395–6406, May 2023.

[17] M. Bahrami-Fard, M. Ghasemi Korrani, M. Rastegar, P. Balsara and B. Fahimi, "Adaptive PLL-based Sensorless Control for CSI-Fed PMSM Drives Used in Submersible Pumps," IECON 2024- 50th Annual Conference of the IEEE Industrial Electronics Society, Chicago, IL, USA, 2024.

[18] M. Seilmeier and B. Piepenbreier, "Sensorless control of PMSM for the whole speed range using two-degree-of-freedom current control and HF test current injection for low-speed range," IEEE Trans. Power Electron., vol. 30, no. 8, pp. 4394–4403, Aug. 2015.

[19] L. Ding, Y. W. Li, N. R. Zargari and R. Paes, "Sensorless Control of CSC-Fed PMSM Drives With Low Switching Frequency for Electrical Submersible Pump Application," in IEEE Transactions on Industry Applications, vol. 56, no. 4, pp. 3799-3808, July-Aug. 2020.

[20] S. Yang, Z. Yin, C. Tong, Y. Sui and P. Zheng, "Active Damping Current Control for Current-Source Inverter-Based PMSM Drives," in IEEE Transactions on Industrial Electronics, vol. 70, no. 4, pp. 3549-3560, April 2023.

[21] Z. Wang, Y. Xu, P. Liu, Y. Zhang and J. He, "Zero-Voltage-Switching Current Source Inverter Fed PMSM Drives With Reduced EMI," in IEEE Transactions on Power Electronics, vol. 36, no. 1, pp. 761-771, Jan. 2021.

[22] L. Ding, Y. W. Li and N. R. Zargari, "Discrete-Time SMO Sensorless Control of Current Source Converter-Fed PMSM Drives With Low Switching Frequency," in IEEE Transactions on Industrial Electronics, vol. 68, no. 3, pp. 2120-2129, March 2021.

[23] Y. Xu, C. Lin, J. Xing, Q. Zeng and J. Sun, "I-f Starting Rapid and Smooth Transition Method of Full-Speed Sensorless Control for Low Current Harmonic Ultra-high-speed PMSM," 2022 IEEE Applied Power Electronics Conference and Exposition (APEC), Houston, TX, USA, 2022, pp. 1820-1826.

[24] M. Fatu, R. Teodorescu, I. Boldea, and G. D. Andreescu, "I-F starting method with smooth transition to EMFbased motion-sensorless vector control of PM synchronous motor/generator," in in Conf. Rec. Power Electron.Spec. Conf., 2008, pp. 1481–1487.

[25] S. V. Nair, K. Hatua, N. V. P. R. D. Prasad and D. K. Reddy, "A Quick I-f Starting of PMSM Drive With Pole Slipping Prevention and Reduced Speed Oscillations," in IEEE Transactions on Industrial Electronics, vol. 68, no. 8, pp. 6650-6661, Aug. 2021.

[26] X. Song, J. Fang, B. Han, and S. Zheng, "Adaptive compensation method for high-speed surface PMSM sensorless drives of EMF-based position estimation error," IEEE Trans. Power Electron., vol. 31, no. 2, pp. 1438–1449, Feb. 2016.

[27] G. Wang, T. Li, G. Zhang, X. Gui, and D. Xu, "Position estimation error reduction using recursive-least-square adaptive filter formodel-based sensorless interior permanent-magnet synchronous motor drives," IEEE Trans. Ind. Electron., vol. 61, no. 9, pp. 5115–5125, Sep. 2014.

[28] G. Zhang, G. Wang, D. Xu, and N. Zhao, "ADALINE-network-based PLL for position sensorless interior permanent magnet synchronous motor drives," IEEE Trans. Power Electron., vol. 31, no. 2, pp. 1450–1460, Feb. 2016.

[29] T. Mannen and H. Fujita, "Dead-time compensation method based on current ripple estimation," IEEE Trans. Power Electron., vol. 30, no. 7, pp. 4016–4024, Jul. 2015.

[30] S. Jung, H. Kobayashi, S. Doki, and S. Okuma, "An improvement of sensorless control performance by a mathematical modelling method of spatial harmonics for a SynRM," in Proc. Power Electron. Conf., 2010, pp. 2010–2015.

[31] Y. Park and S. Sul, "Implementation schemes to compensate for inverter nonlinearity based on trapezoidal voltage," IEEE Trans. Ind. Appl., vol. 50, no. 2, pp. 1066–1073, Mar./Apr. 2014.

[32] Z. Wang, K. Lu and F. Blaabjerg, "A Simple Startup Strategy Based on Current Regulation for Back-EMF-Based Sensorless Control of PMSM," in IEEE Transactions on Power Electronics, vol. 27, no. 8, pp. 3817-3825, Aug. 2012.

[33] M. Rho and S. Kim, "Development of robust starting system using sensorless vector drive for a microturbine," IEEE Trans. Ind. Electron., vol. 57, no. 3, pp. 1063-1073, 2010.

[34] A. Stirban, I. Boldea, G.-D. Andreescu, D. Iles, and F. Blaabjerg, "Motion sensorless control of BLDC PM motor with offline FEM info assisted state observer," in 12th Int. Conf. OPTIM, 2010, pp. 321–328.

979-8-3315-1612-3/25 $31.00 © 2025 IEEE

Simulation-assisted Design and Implementation of an Electrically Excited Synchronous Motor Drive System

Shih-Gang Chen
Department of Electrical Engineering
National Taipei University of Technology
Taipei, Taiwan
chensg@mail.ntut.edu.tw

Jun-Ming Hsu
Department of Electrical Engineering
National Taipei University of Technology
Taipei, Taiwan
t113319006@ntut.edu.tw

Chun-Yen Chen
Department of Electrical Engineering
National Taipei University of Technology
Taipei, Taiwan
t113318014@ntut.edu.tw

Ming-Shi Huang
Department of Electrical Engineering
National Taipei University of Technology
Taipei, Taiwan
mingshi.huang@gmail.com

Abstract—The electrically excited synchronous motor (EESM), also called the wound field synchronous motor, does not use permanent magnets and provides advantages such as a wide constant power range, high efficiency, and easy control at high speeds. Moreover, the rotor coil is used to instead of rotor magnet and generate the required rotor magnetic field through current control. This article presents the preliminary design of an EESM rated at 3.4 kW/3000 rpm using dual finite element analysis tools, ANSYS Motor-CAD and Maxwell 2D. Wherein the rotor current is generated by a full-bridge circuit and the current is through carbon brushes and slip rings to rotor coil. Finally, a digital signal processor-based motor drive is constructed to verify design results of the EESM. The efficiency is 90.9% at rated point (@3.4kW/3000 rpm). Moreover, the errors in electrical parameters and generated torque between simulated and experimental results are less than 10%.

Keywords—*Electrically excited synchronous motor (EESM), Finite element analysis (FEA), Motor design, Wound field synchronous motor*

I. INTRODUCTION

The interior permanent magnet synchronous motor (IPMSM) with rare earth magnets is widely used in electric carriers due to its high efficiency and high power density. However, the price of rare earth magnets fluctuates and exists the demagnetization issue of the magnet. In recent years, electrically excited synchronous motor (EESM) has become one of the alternative selections to IPMSM due to magnetic field is generated by rotor coil current. Therefore, EESM has the advantages of no rare-earth magnets, a wide range of constant power, and higher efficiency in the field-weakening speed compared to the IPMSM [1-3].

In order to improve the EESM's performance, [1], [5-6] propose flux barrier placed in the rotor to increase the q-axis reluctance. This enhances the saliency ratio, thereby increasing the reluctance torque and improving generated torque. In [2], [7], additional permanent magnets are used in rotor to enhance power and torque density. In [8-9], multi-objective optimization and the metamodel of optimal prognosis are used for optimal design of motor. The control strategy of EESM has been presented in [10-12], achieving methods such as maximum torque per ampere (MTPA) and copper loss minimization by distributing dq-axis currents and field currents.

In this paper, an ANSYS Motor-CAD simulation software is used to initially design an EESM rated at 3.4 kW/3000 rpm. The initial design of the EESM will focus on the rotor, and the stator will use the existing 18-slot, 8-pole IPMSM's stator. Moreover, ANSYS Maxwell is used to perform detailed optimization to improve the output performance with fewer parameters in section III. In section IV, the characteristics of the designed EESM are verified by experiment.

II. FORMULATION OF THE EESM

Figure 1 shows the definition of the coordinate axis and the phasor diagram of the EESM in synchronous frame respectively, where I_s is the peak value of stator current, and θ_i is current angle. Therefore, the governing equations of the EESM under the synchronous frame can be expressed as [13].

$$\begin{bmatrix} v_{qs}^r \\ v_{ds}^r \\ v_{fr} \end{bmatrix} = \begin{bmatrix} r_s + pL_{qs} & \omega_r L_{ds} & \omega_r L_{fd} \\ -\omega_r L_{qs} & r_s + pL_{ds} & pL_{fd} \\ 0 & 1.5pL_{fd} & r_f + pL_{fr} \end{bmatrix} \begin{bmatrix} i_{qs}^r \\ i_{ds}^r \\ i_{fr} \end{bmatrix} \quad (1)$$

$$T_e = \frac{3}{2}\frac{P}{2}[L_{fd}i_{fr}i_{qs}^r + (L_{ds} - L_{qs})i_{qs}^r i_{ds}^r] \quad (2)$$

where p is differential operator. v_{qs}^r, v_{ds}^r and v_{fr} are the voltages in q-axis, d-axis and field winding respectively; i_{qs}^r, i_{ds}^r and i_{fr} are the currents in q-axis, d-axis and field winding respectively; L_{qs}, L_{ds} and L_{fr} are the inductances in q-axis, d-axis and field winding respectively; r_s and r_f are the phase resistance and field winding resistance respectively; L_{fd} is the mutual inductance between field winding and d-axis; ω_r is the rotor electrical angular speed; T_e, P is generated torque and number of poles respectively.

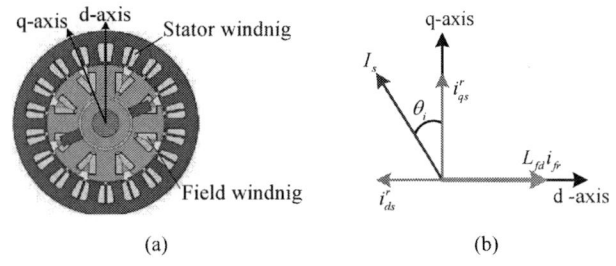

(a)   (b)

Fig. 1. (a) d-q axis of EESM (b) phasor diagram of EESM.

III. SIMULATION-ASSISTED DESIGN EESM

A. Design procedure

The EESM design procedure is shown in Fig. 2, and described as follows.

1) Preliminary Design of EESM

In order to reduce costs, the existing 18-slot, 8-pole IPMSM stator is used for the EESM stator, and the slot pole combination of the EESM remains the same as the original IPMSM, which helps to provide low spatial harmonics [14]. Based on the specifications shown in Table I, the rotor of the EESM is preliminarily designed using Motor-CAD. After confirming that it meets the required specifications, a detailed analysis of the rotor is subsequently performed.

2) Detailed design of the rotor shape

The detailed design of the rotor shape is performed using ANSYS Maxwell 2D. The key mechanical parameters of the rotor, which are d_1, d_2, θ_1, θ_2, are defined in Fig. 3. Figure 4 shows that the pole arc offset (d_2) makes the air gap flux density nearly sinusoidal, reducing its torque ripple and harmonic distortion of the no load back-electromotive force (EMF) [15].

Expect for d_2, the other parameters (d_1, θ_1, θ_2) will be used to find optimal characteristic point of EESM through a parameter sweep. However, it is necessary to consider the rotor winding path and maintain a sufficient winding area to avoid the problem of increased rotor copper loss caused by an excessively small winding area. Additionally, the sweep range of the parameters should be limited. Table II shows the limitation of parameter sweep region.

The results of the rotor parameter sweep are shown in Fig. 5, which illustrate the influence of torque and torque ripple under different d1 values with changes in θ_1 and θ_2. Based on Fig. 5, the maximum torque point ($d_1 = 0.12$, $\theta_1 = 75°$, $\theta_2 = 140°$) is selected for analyzing the effect of d_2 on T_e, and the analyzed results are shown in Fig. 6. Wherein $d_2 = 0.175$ is chosen for the final design due to minimal decreased torque and lower torque ripple. In addition, the air gap is reduced from 0.7 mm to 0.6 mm to prevent insufficient rotor field in the EESM rotor, which is caused by the limitations of the current resistance in the existing slip rings and carbon brushes.

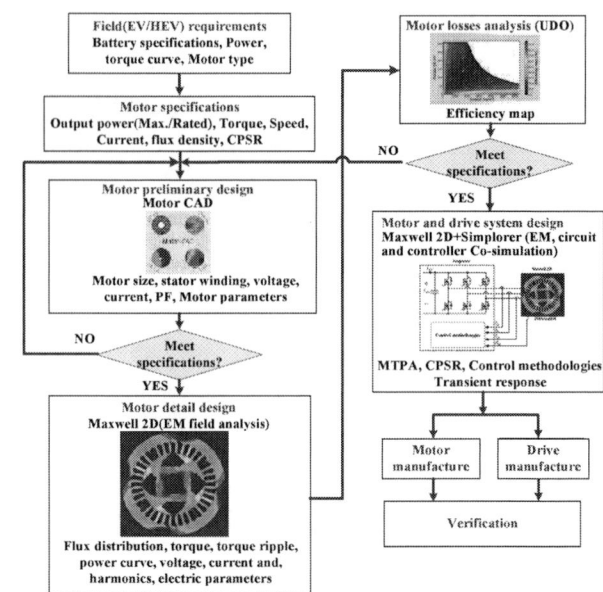

Fig. 2. Design procedure of the EESM.

TABLE I. EESM SPECIFICATIONS

Parameter	Value	Parameter	Value
Rated power	3.4 kW	Stator outer / inner	25.0 /109.4 mm
Rated torque	10.8Nm	Rotor outer / inner	110.5 /173 mm
Rated speed	3000 rpm	Air gap length	100.0 mm
Peak phase current	29.7 A	Stack length	0.7 mm
Peak field current	8 A	Stator/Rotor Turns	22/21turns
Line voltage	220V$_{rms}$	Poles/Slots	18/8

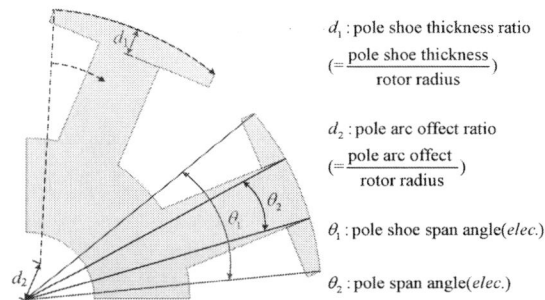

Fig. 3. Key mechanical parameter definition of rotor structure.

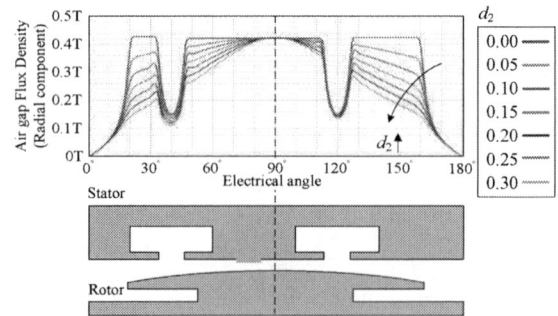

Fig. 4. Radial air gap flux density vs. d_2.

979-8-3315-1612-3/25 $31.00 © 2025 IEEE

TABLE II. THE LIMITATIOM OF PARAMETER SWEEP REGION

Parameter	Parameter sweep region
d_1	0.1~0.12
θ_1	50°~75°
θ_2	120°~140°

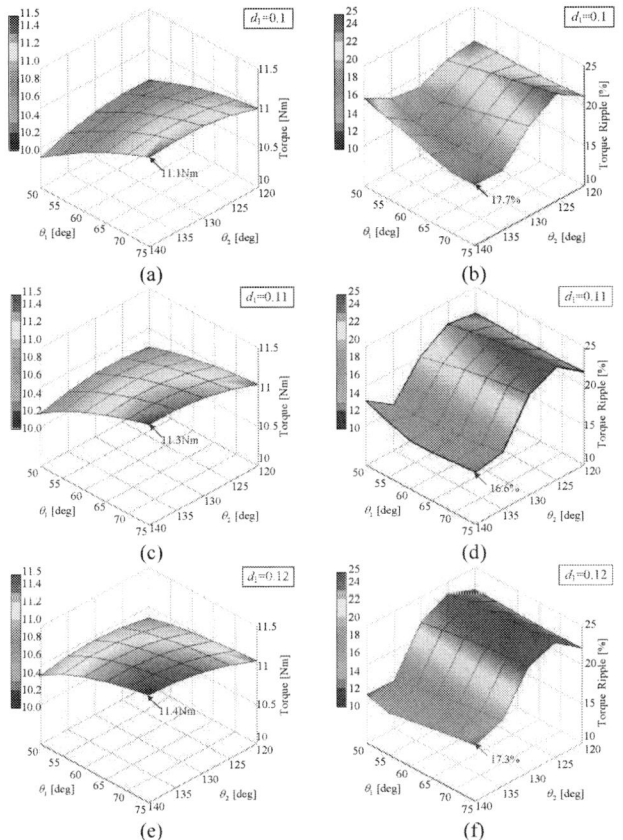

(a) (b)

(c) (d)

(e) (f)

Fig. 5. The influence of θ_1 and θ_2 on generated torque under different d_1: (a) Average torque (d_1=0.10) (b) Torque ripple (d_1=0.10) (c) Average torque (d_1=0.11) (d) Torque ripple (d_1=0.11) (e) Average torque (d_1=0.12) (f) Torque ripple (d_1=0.12).

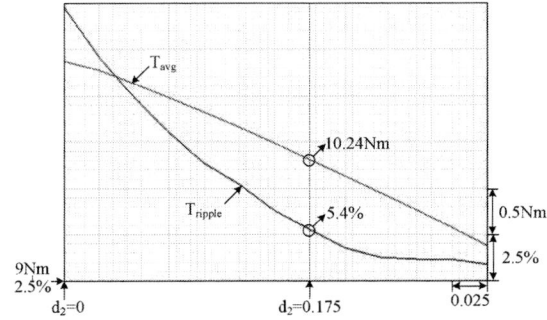

Fig. 6. Average toque and torque ripple vs. d_2 (d_1=0.12, θ_1=75°, θ_2=140°)

B. Design result

Figure 7(a)-(d) shows the comparison of rotor's structure, T_e verus θ_i under maximal i_{fr} and I_s, maximal T_e waveforms, and torque-speed (T-N) and power-speed (P-N) curve of the EESM before and after improvement, respectively. The maximum torque increased by 2.7Nm after improvement. Fig. 7(e), (f) shows the efficiency map before and after the improvement, and its overall efficiency is improved by about 1%. Table III is the comparison of output characteristics before and after improvement of the EESM at rated point which shows that the improved version has significant performance. Fig. 8(a), (b) is the flux density distribution map of EESM before and after improvement. After the improvement, the magnetic saturation of the rotor teeth decreases, which is related to the decrease of torque ripple [16].

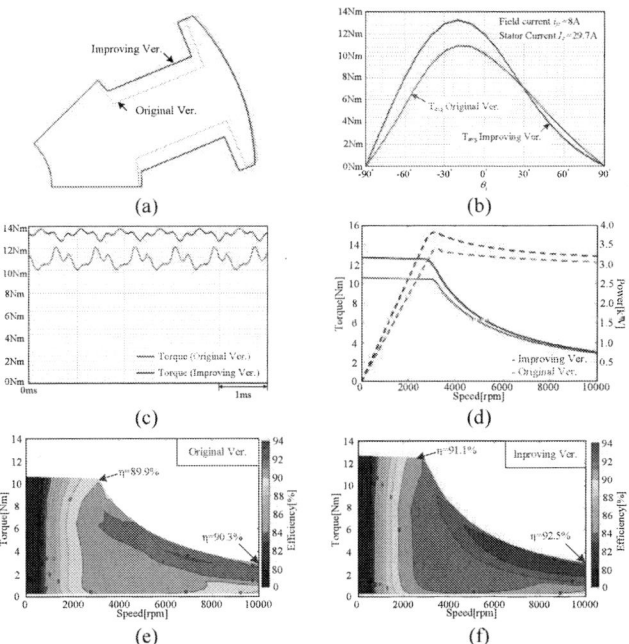

(a) (b)

(c) (d)

(e) (f)

Fig. 7. Comparison of EESM's characteristics before and after improvement : (a) Rotor structure (b) T_e vs. θ_i (@i_{fr} = 8A, I_s = 29.7A, 3000rpm) (c) Maximum torque waveform (@i_{fr} = 8A, I_s = 29.7A, θ_i =-20 °, 3000rpm) (d) T-N and P-N curve (e) Efficiency map before improvement : (f) Efficiency map after improvement.

TABLE III. EESM SIMULATED RESULTS AT RATED POINT

	Original	Improved
Rated power	3.2kW	3.4kW
Rated torque (T_{rated})	10.2Nm@θ_i = -20°	10.8Nm@θ_i = -20°
Efficiency	89.9%	90.9%
Torque ripple ($T_{pk\text{-}pk}$ / T_{rated})	18.4%	6.4%

979-8-3315-1612-3/25 $31.00 © 2025 IEEE

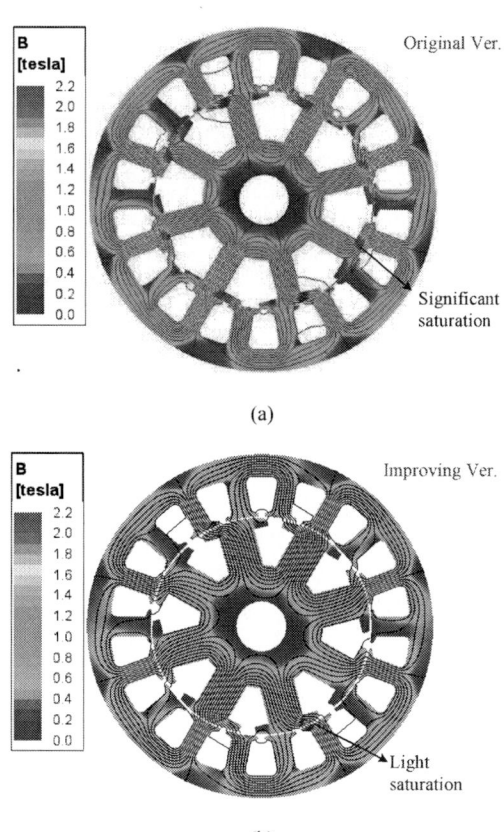

Fig. 8. Flux density distribution map of EESM before and after improvement (@i_{fr} = 8A, I_s = 29.7A, 3000rpm): (a) before improvement (b) after improvement.

IV. EXPERIMENTAL RESULT

In order to verify the characteristics of the designed EESM, this paper constructs a digital signal processor (DSP)-based (TMS320F28075) controller and uses silicon carbide (SiC) to build the power stage. Figure 9 shows the test setup and function block of the EESM. Table III shows the measured and simulated results of the EESM's electrical parameters. Fig.10(a) is the simulated and measured waveforms of the no load back-EMF and the maximal error is less than 10% at i_{fr} = 8A and 1000rpm. The simulated and experimental results of the T_e vs. θ_i (@1000 rpm) of the proposed EESM with different i_{fr} and I_s shown in Fig. 10(b).

TABLE IV. MEASURED AND SIMULATED RESULTS OF THE EESM'S ELECTRICAL PARAMETERS

	Measurement	Simulation	Error
phase resistance	0.09Ω	0.10Ω	11.1%
field winding resistance	0.92Ω	1.03Ω	11.9%
d-axis inductance	5.4mH	5.0mH	8%
q-axis inductance	3.8mH	3.6mH	4.8%
field winding inductance	31.6mH	31.7mH	0.3%

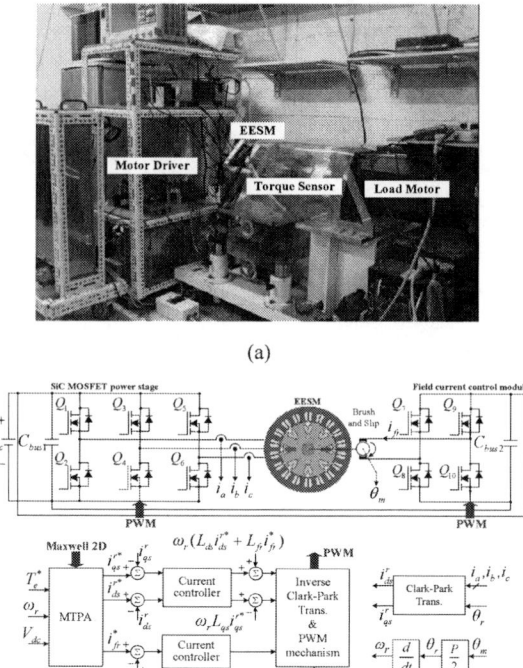

Fig. 9. (a)Test setup (b)Function block of the EESM.

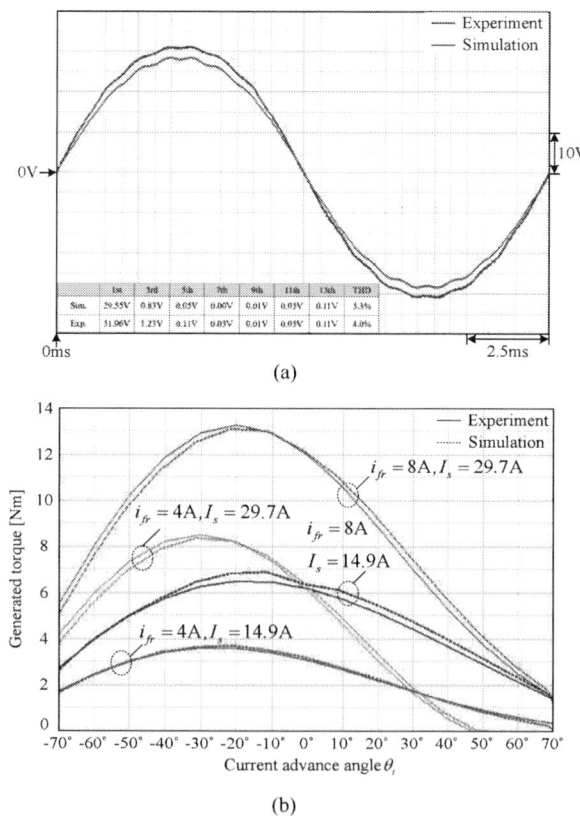

Fig. 10. EESM simulated and measured results: (a) no load back-EMF (i_{fr} = 8A@1000rpm) (b) Torque characteristic under various I_s, i_{fr}, and θ_i (@1000rpm).

V. CONSCLUSION

This paper uses ANSYS Motor-CAD to design an EESM that meets specifications, and ANSYS Maxwell to optimize the rotor's mechanical shape to reach maximum T_e and lower torque ripple. After the design improvements using Maxwell 2D, the maximum T_e and high efficiency zone of EESM are enhanced and the torque ripple is reduced compared with the original design. Among them, the torque in the constant torque region is increased by 2.7Nm, the efficiency is increased by 1%, the torque ripple is reduced from 18.4% to 6.4% at peak torque. The efficiency is 90.9% at rated point (@3.4kW/3000 rpm). Subsequently, a DSP-based motor drive with 300V DC link voltage is constructed to verify the characteristics of the EESM. The maximal errors between simulated and measured no load back-EMF and d-q axis inductance are all less than 10%. Furthermore, the simulated and measured are consistent under different I_s, i_{fr}, and θ_i.

ACKNOWLEDGMENT

This work is support by National Science and Technology Council, R.O.C. (MOST 110-2221-E207-050)

REFERENCES

[1] Illiano, E. "Design of a highly efficient brushless current excited synchronous motor for automotive purposes," ETH dissertation, 2014.

[2] L. R. Huang, Z. Q. Zhu and W. Q. Chu, "Optimization of electrically excited synchronous machine for electrical vehicle applications", *in Proc. 8th IET Int. Conf. Power Electron. Mach. Drives*(PEMD), pp. 1-6, Apr. 2016

[3] T. Grelle, C. Schmülling, and P. Zimmerschied, "Magnet-free HV traction drives with contactless power transmission," *MTZ Worldwide*, vol. 82, no. 4, pp. 28–33, Apr. 2021.

[4] W. Chai, W. Zhao and B.-I. Kwon, "Optimal design of wound field synchronous reluctance machines to improve torque by increasing the saliency ratio", *IEEE Trans. Magn.*, vol. 53, no. 11, pp. 1-4, Nov. 2017.

[5] W. Liu ,and T. A. Lipo, "Saliency enhancement of salient pole wound field synchronous machines for variable speed applications", *in Proc. IEEE Int. Electr. Mach. Drives Conf.* (IEMDC), pp. 1-7, May. 2017.

[6] M. Märgner and W. Hackmann, "Control challenges of an externally excited synchronous machine in an automotive traction drive application", *in Proc. Emobility—Elect. Power Train*, pp. 1-6, Nov. 2010.

[7] S. W. Hwang, J. H. Sim, J. P. Hong and J. Y. Lee, "Torque improvement of wound field synchronous motor for electric vehicle by PM-assist", *IEEE Trans. Ind. Appl.*, vol. 54, no. 4, pp. 3252-3259, Jul./Aug. 2018

[8] A. Di Gioia et al., "Design and demonstration of a wound field synchronous machine for electric vehicle traction with brushless capacitive field excitation," *IEEE Trans. Ind. Appl.*, vol. 54, no. 2, pp. 1390-1403, Mar./Apr. 2018

[9] N. Tang, D. Sossong, N. Krause, X. Hou, M. J. T. Liben, D. C. Ludois, et al., "Implementation of a metamodel-based optimization for the design of a high power density wound field traction motor", *IEEE Trans. Ind. Appl.*, vol. 59, no. 6, pp. 6726-6735, Nov. 2023

[10] Y. Kim and K. Nam, "Copper-loss-minimizing field current control scheme for wound synchronous machines," *IEEE Trans. Power Electron.*, vol. 32, no. 2, pp. 1335-1345, Feb. 2017

[11] J. Tang, B. Jiang, H. Chen, Y. Liu and S. Lundberg, "Dynamic current reference determination of electrically excited synchronous machines based on torque gradients of copper losses," *IEEE Trans. Power Electron*, vol. 39, no. 6, pp. 7423-7433, June 2024

[12] Y. Han, C. Gong, G. Chen, Z. Ma and S. Chen, "Robust MTPA control for novel EV-WFSMs based on pure SM observer based multistep inductance identification strategy", *IEEE Trans. Ind. Electron.*, vol. 69, no. 12, pp. 12390-12401, Dec. 2022.

[13] S. K. Sul, Control of electric machine drive systems IEEE Press, 2012.

[14] J. Wang, V. I. Patel and W. Wang, "Fractional-Slot Permanent Magnet Brushless Machines with Low Space Harmonic Contents", *IEEE Trans Magn*, vol. 50, no. 1, pp. 1-9, Jan. 2014.

[15] Evans, "Salient pole shoe shapes of interior permanent magnet synchronous machines", *Proc. Int. Conf. Elect. Mach.*, pp. 1-6, 2010.

[16] S. H. Do, B. H. Lee, H. Y. Lee ,and J. P. Hong, "Torque ripple reduction of wound rotor synchronous motor using rotor slits," i*n Proc 2012.15th Int. Conf. Electr. Mach. Syst.* (IEMDCS), pp. 1-4. 2012

Implementation and Analysis of Direct Torque Control on High-Speed PMSMs: A Comparative Study of Commercial and Laboratory-Developed Motors

Md Moniruzzaman
*Department of Electrical and
Computer Engineering
Mississippi State University*
Starkville, USA
mm5111@msstate.edu

Kishor Joshi
*Department of Electrical and
Computer Engineering
Mississippi State University*
Starkville, USA
kj1377@msstate.edu

Md Rashedul Rahman
*Department of Electrical and
Computer Engineering
Mississippi State University*
Starkville, USA
mr2575@msstate.edu

Md Khurshedul Islam
Sr. Mechatronics Design Engineer,
ASML, Wilton, CT 06897, USA
mdkhurshedul.islam@asml.com

Seungdeog Choi
*Department of Electrical and
Computer Engineering
Mississippi State University*
Starkville, USA
seungdeog@ece.msstate.edu

Masoud Karimi Ghartemani
*Department of Electrical and
Computer Engineering
Mississippi State University*
Starkville, USA
karimi@ece.msstate.edu

Abstract— This paper analyzes the Direct Torque Control (DTC) algorithm applied to ultra-high-speed (UHS) Permanent Magnet Synchronous Motors (PMSMs) for applications up to 500,000 RPM. A sensor-less DTC algorithm is formulated and implemented using MATLAB-based automatic code generation, incorporating optimized speed, torque, and flux control loops for enhanced dynamic performance and computational efficiency. To ensure stable and precise speed transitions, an adaptive transition control mechanism is incorporated, which considers thermal and electrical constraints such as motor winding temperature, inverter limitations, and phase current regulation. This mechanism assists in managing acceleration profiles while maintaining system integrity at high speeds. Simulations have been conducted for two motors up to 500,000 RPM. Experimentally, the commercial motor operated at 110,000 RPM, and the lab-built---A Mechanically Based Antenna (AMEBA)--- PMSM motor operated at 70,000 RPM with closed-loop operation. With modulation index regulation, AMEBA motor was also operated up to 150,000 RPM in open-loop operation. AMEBA has been developed for communications with submarines, underground mines, and underwater facilities. The study also compared DTC with Field-Oriented Control (FOC), under ultra-high-speed conditions. Furthermore, the effects of current sensor delays are analyzed, and a phase-advancing method is employed to mitigate their impact. Future work will involve operating both motors across their full speed range and addressing additional delays through hardware testing.

Keywords—Permanent Magnet Synchronous Motor (PMSM), Direct Torque Control (DTC), Ultra-High-Speed Motors, Adaptive Rate Controller, Sensor Delay Compensation

I. INTRODUCTION

Ultra-high-speed Permanent Magnet Synchronous Motors (UHS-PMSMs) have garnered significant attention due to their applications in compact, high-efficiency systems such as turbo compressors, gas turbines, and aerospace systems [1]. These motors offer high power density and improved efficiency, making them suitable for applications requiring high rotational speeds. However, controlling PMSMs at ultra-high speeds introduces considerable challenges, including control delays, torque ripple, and harmonic distortion [2].

Advancements in power electronics and high-performance processors have facilitated the operation of PMSMs at ultra-

high speeds. Nevertheless, these technological improvements have not entirely mitigated the control challenges associated with high-speed operations. At these speeds, the fundamental frequency of the motor increases dramatically, necessitating extremely short control periods and fast response times for both estimators and controllers. Conventional vector control methods, such as Field-Oriented Control (FOC), require accurate rotor position information for coordinate transformations and struggle to maintain stability due to the significant phase lag introduced by control delays.

Direct Torque Control (DTC) is a promising solution to these challenges [3]. Unlike vector control, DTC operates in a stationary α-β reference frame, eliminating the need for rotor position sensors and enabling sensor-less operation with reduced computational overhead. This attribute makes DTC especially suitable for ultra-high-speed applications, where rapid torque and flux control are essential.

However, implementing DTC in ultra-high-speed (UHS) PMSMs presents its own set of challenges. Control delays become more pronounced at higher speeds, leading to phase lag and instability. Additionally, torque ripples and harmonic distortions can adversely affect motor performance and durability. Therefore, optimizing the DTC algorithm for ultra-high-speed applications is crucial.

This paper analyzes the DTC algorithm applied to UHS-PMSMs, specifically targeting applications up to 500,000 RPM. Using MATLAB code generation methods, a sensor-less DTC algorithm optimized for high-speed PMSM control is developed with the precise tuning of speed, torque, and flux Proportional-Integral (PI) controllers. A adaptive rate (reference) controller is used to enhance DTC performance to ensure smooth speed transitions, and lower overshoot and undershoot while achieving high ramp rates from low to maximum speeds. The adaptive rate controller offers superior performance compared to traditional step speed commands.

Simulations are conducted for a commercial PMSM and a custom UHS PMSM up to 500,000 RPM. Experimentally, the commercial motor reached 110,000 RPM, while the lab-built motor achieved 70,000 RPM in closed-loop and 150,000 RPM in open-loop with modulation index regulation. DTC and FOC are compared under UHS conditions to evaluate performance. Furthermore, the effects of current sensor delays are analyzed, and a phase-advancing method is employed to mitigate their impact. The phase-advancing method effectively compensated for the delays, improving the accuracy of torque and flux control at high speeds [4-5].

Future work will involve operating both motors across their full speed range and addressing additional delays through enhanced hardware testing. The implementation of advanced compensation methods and optimization of the control algorithm will further enhance DTC's stability and precision in ultra-high-speed applications.

The remainder of this paper is structured as follows. Section II describes the DTC algorithm and its application to UHS-PMSMs. Section III outlines the modeling and parameter derivation of the motors. Section IV discusses the methodology, including modulation index regulation, adaptive rate controller, and delay compensation techniques. Section V presents the experimental setup. Section VI details simulation and hardware results, and Section VII concludes the paper with key findings and future work.

II. DIRECT TORQUE CONTROL (DTC) ALGORITHM

DTC provides fast, precise torque and flux control, making it suitable for high-speed PMSM applications.

A. Overview of Sensor-less Direct Torque Control

Direct Torque Control (DTC) is a control strategy that directly regulates motor torque and flux without requiring complex coordinate transformations [6]. DTC offers fast dynamic response and robustness to parameter variations, making it suitable for high-speed Permanent Magnet Synchronous Motor (PMSM) applications, particularly in sensor-less configurations.

In DTC, torque and flux are directly controlled by selecting appropriate voltage vectors based on the errors between the reference and estimated values of torque and flux. The control algorithm operates within a stationary α-β reference frame, utilizing flux and torque estimators to select the optimal voltage vector. This eliminates the need for precise rotor position information. The sensor-less DTC approach further enhances the system by estimating rotor position and speed without relying on physical sensors, thus improving reliability and reducing system cost [7-8].

B. DTC Control Algorithm Development for UHS-PMSM

The DTC algorithm operates within a stationary α-β reference frame, utilizing flux and torque estimators to select the optimal voltage vector, and uses a dq-reference frame for torque control reference duty cycle generation. Fig. 1 presents the detailed control block diagram of the DTC algorithm. It operates within the stationary α-β reference frame, where the flux and torque of the PMSM are estimated using transient machine model equations. The current feedback from the PMSM is used to estimate the rotor flux in the α-β reference frame. The flux component along the α-β axes and the magnitude of the rotor flux linkage are computed as follows:

$$\lambda_\alpha = L_s\, i_\alpha + \lambda_{PM} \cos\theta \qquad (1)$$
$$\lambda_\beta = L_s\, i_\beta + \lambda_{PM} \sin\theta \qquad (2)$$

$$\lambda = \sqrt{\left(\lambda_\alpha\right)^2 + \left(\lambda_\beta\right)^2} \qquad (3)$$

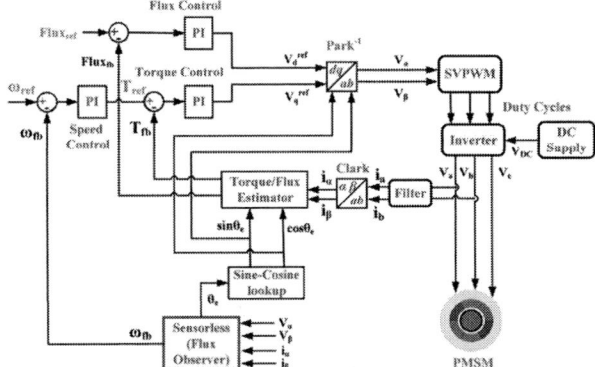

Fig. 1. Control block diagram of the sensor-less DTC Algorithm.

where λ_α and λ_β represent the rotor flux along α- and β-axes, λ denotes the rotor flux, λ_{PM} is its per-unit equivalent, L_s is stator inductance, i_α and i_β are the motor currents along α-β axes reference frame, and θ is the rotor position. Subsequently, the torque is calculated as

$$T = \frac{3}{2}P\left(\lambda_\alpha\, i_\beta - \lambda_\beta\, i_\alpha\right) \qquad (4)$$

where P is the number of pole pairs of the motor, and T represents the rotor torque. UHS PMSMs typically require rotor position sensors such as encoders or Hall sensors for closed-loop control, but these sensors pose cost and reliability challenges. Sensor-less control methods, such as rotor flux observers, address these challenges while maintaining accuracy by estimating stator flux linkage from current and voltage in the α-β reference frame. The rotor flux linkage components along the α-β axes and rotor position angle are

$$\lambda_\alpha = \int \left(v_\alpha - i_\alpha R\right)dt - L_s i_\alpha \qquad (5)$$
$$\lambda_\beta = \int \left(v_\beta - i_\beta R\right)dt - L_s i_\beta \qquad (6)$$
$$\theta_e = \tan^{-1}\left(\frac{\lambda_\beta}{\lambda_\alpha}\right) \qquad (7)$$

where R is the stator resistance and θ_e is the estimated rotor position angle. MATLAB code generation methods are employed to develop the sensor-less DTC algorithm for high-speed PMSM control. This approach allows for rapid prototyping and deployment on embedded systems suitable for high-speed motor control. Tuning of the PI controllers for speed, torque, and flux is critical for stable operation at ultra-high speeds. MATLAB's simulation environment allows for systematic tuning and validation of controller parameters.

III. MODELING OF ULTRA-HIGH-SPEED PMSM

A. UHS-PMSM Motor Specifications

Two ultra-high-speed PMSMs rated up to 500,000 RPM are considered: a commercially available motor and the laboratory-developed AMEBA motor. Although sharing similar operational objectives, they differ in design and parameters, detailed in Table I [9-10]. The AMEBA motor,

979-8-3315-1612-3/25 $31.00 © 2025 IEEE

TABLE I.	MOTOR PATAMETERS	
Parameter	AMEBA PMSM	COMMERCIAL PMSM
Rated Speed	500,000 RPM	500,000 RPM
Stator Resistance (R$_s$)	0.1382 Ω	0.37 Ω
Stator Inductance (L)	65×10^{-6} H	6.2×10^{-6} H
Inertial Coefficient (J)	3.8×10^{-6} kg-m^2	1.95×10^{-8} kg-m^2
Friction Coefficient(B)	2×10^{-7} kg-m^2/s	2×10^{-9} kg-m^2/s
Rated Current (I$_{rms}$)	6.3 A (rms)	4.24 A (rms)
Rated Torque (T)	0.040 Nm	0.002 Nm
Pole Pairs (P)	1	1
Rated Power	2000 W	100 W

illustrated in Fig. 2(a) and 2(b), is designed with a slot-less stator and surface-mounted permanent magnets optimized for high power density and minimal losses, making it suitable for compact applications like portable communication systems. In this motor, the d-axis and q-axis inductances are equal, effectively simplifying the model to $L_d = L_q = L_s$. This motor is integrated into a custom-designed dynamometer setup, operating alongside a matching generator unit. The commercial UHS-PMSM, shown in Fig. 2(c) and 2(d), incorporates a specialized mechanical rotor construction to withstand extreme stresses, high-speed ball bearings with permanent lubrication, and an efficient stator winding and core configuration.

B. Dynamic Model of PMSMs

The motor's dynamic behavior is described by the dq-axis voltage equations expressed as

$$v_d = L_s \frac{di_d}{dt} + R_s i_d - L_s P \omega i_q \tag{8}$$

$$v_q = L_s \frac{di_q}{dt} + R_s i_q + L_s P \omega i_d + P \omega \lambda_{pm} \tag{9}$$

where v_d is the d-axis voltage, v_q is the q-axis voltage, i_d is the d-axis current, i_q is the q-axis current, P is the number of magnetic pole pairs, λ_{pm} is the permanent magnet flux linkage, and ω is the rotor speed.

Essential parameters such as the back electromotive force (EMF) constant (k_e), torque constant (k_t), and flux linkage (λ_{pm}) are crucial for accurate modeling and control. These parameters are defined using

$$k_e = \frac{V_{pk\_LL}}{RPM \times 10^{-3}} \tag{10}$$

$$k_t = \frac{\sqrt{3}}{2} . k_e \tag{11}$$

$$\lambda_{pm} = \frac{k_e}{\sqrt{3} . 2\pi . 1000 . \frac{P}{60}} \tag{12}$$

where, V_{pk_LL} is the peak line-to-line voltage.

Table II presents these parameters for both motors, including the torque constant, flux linkage, and back EMF constant.

TABLE II.	CALCULATED MOTOR PARAMETERS	
Parameter	AMEBA PMSM	COMMERCIAL PMSM
Back EMF Constant (k$_e$)	0.5804 V$_L$/kRPM	0.136 V$_L$/kRPM
Torque Constant (k$_t$)	4×10^{-3} Nm/A	1.125×10^{-3} Nm/A

Flux Linkage (λ_{pm})	0.0032 Wb	0.00075 Wb

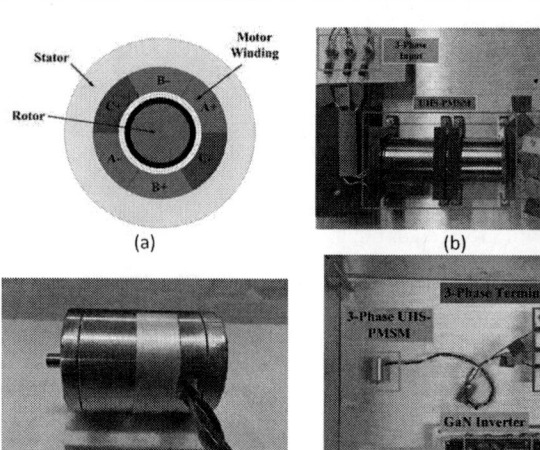

Fig.2. (a) AMEBA motor design (Ansys); (b) AMEBA physical design; (c) Commercial UHS-PMSM; and (d) Commercial PMSM testbed setup.

IV. METHODOLOGY

This section outlines the methodology for testing and enhancing the DTC algorithm for high-speed PMSMs, focusing on modulation index regulation, smooth speed transition using a adaptive rate controller, and delay compensation techniques.

A. Modulation Index Regulation for Open-loop Operation

Open-loop control, or scalar or Volts/Hz control, is used during the motor startup phase. This method operates without feedback and determines the frequency and amplitude of the stator voltages only based on the reference speed. By maintaining a consistent ratio of stator voltage (V_s) to stator frequency (f_s), the stator magnetic flux (λ_{pm}) is kept relatively constant, ensuring effective initial motor operation. To initiate motor rotation in open-loop control, the modulation index (m) is regulated as

$$m = \frac{2}{V_{dc}} \left(\frac{k_e}{\sqrt{3}} \times \frac{\omega_{mech}}{1000} \right) + V_{offset} \tag{13}$$

In (13), V_{dc} represents the DC link voltage, ω_{mech} is the mechanical angular velocity and V_{offset} compensates for the back EMF effects. At high speeds in an open loop, adjusting V_{offset} is necessary to overcome the effects of back EMF and ensure stable operation. However, a key challenge is the higher current draw compared to closed-loop operation.

During the open-loop operation, the motor is driven with a pre-defined reference speed. Once the motor achieves the minimum required speed, the control system shifts to closed-loop. To facilitate a seamless transition, the Proportional-Integral (PI) controllers are reset and initialized with the outputs from the open-loop phase, ensuring continuity and stability in the control system. Fig. 3 illustrates the modulation index regulation during open-loop operation and its integration with closed-loop control.

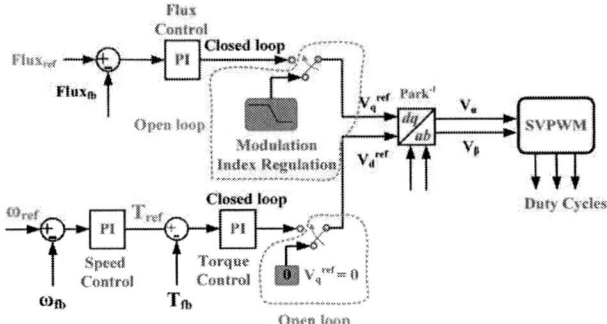

Fig. 3. Transition from open-loop to closed-loop control.

B. Adaptive Rate Controller

Ultra-high-speed permanent magnet synchronous motor (PMSM) drives can experience significant thermal and electrical stress if the motor is accelerated too quickly. To mitigate this issue, an Adaptive Speed Transition Control strategy is proposed. Unlike purely time-based or fixed ramp approaches, this approach continuously monitors three key variables, namely, the inverter temperature $T_{inv}(t)$, the motor-winding temperature $T_{mw}(t)$, and the phase current $I_{ph}(t)$ and adjusts the speed-ramp rate whenever any of these signals approaches its corresponding limit. Let the corresponding limits be $T_{inv,lim}$, $T_{wm,lim}$ and $I_{ph,lim}$. Each measured value is continuously monitored, and when it crosses its respective threshold, the system reduces the acceleration rate to mitigate thermal stress and electrical overload as shown in Fig. 5. Rate adjustment strategy based on system constraints are discussed below:

(a) Normal Operating Condition

When all monitored parameters remain within their safe limits, i.e., up to time t_1, the system follows the initial acceleration rate (rpm/s) and acceleration follows the profile of $\psi(t)$.

$$R_1 = R_0$$

(b) Inverter Temperature Exceeds Threshold

If at time t_1, the inverter temperature surpasses its limit, the system reduces the acceleration rate as follows:

$$R_2 = R_0 - k_1R_0 = (1 - k_1)R_0$$

Here, k_1 is a weighting factor based on the inverter temperature threshold ($T_{inv,lim}$). The transition to the new acceleration rate follows:

$$\psi(t) = \psi(f_2t_1 - \phi_1) \quad (14)$$

where the phase shift (ϕ_1) and transition time (T_a) are:

$$\phi_1 = (f_2 - f_1)t_1 \qquad T_a = \frac{f_2 - f_1}{f_1}t_1$$

The terms f_1 and f_2 represent frequencies associated with acceleration changes. The updated acceleration function becomes:

$$R_2\psi(f_2t - \frac{R_2 - R_1}{R_1}t)$$

At this stage, the system adapts to the increased inverter temperature by reducing the acceleration rate, thereby limiting thermal stress.

(c) Motor Winding Temperature Exceeds Threshold

If, at time t_2 the motor-winding temperature surpasses its threshold, the acceleration rate undergoes a further reduction:

$$R_3 = R_2 - k_2R_2 = (1 - k_2)R_2$$

where k_2 is a weighting factor controlling the rate adjustment based on the motor winding temperature limit. The transition follows:

$$\psi(f_2t_2) = \psi(f_3t_2 - \phi_2) \quad (15)$$

where the phase shift (ϕ_2) and transition time (T_b) are:

$$\phi_2 = (f_3 - f_2)t_2 - \phi_1 \qquad T_b = \frac{R_3 - R_2}{R_3}t_2 - \frac{R_2 - R_1}{R_3}t_1$$

The revised acceleration profile is updated as: $R_3\psi(f_3t - T_b)$

At t_2, the motor winding temperature surpasses its threshold, prompting a further reduction in the ramp rate to protect the winding insulation and prevent overheating.

For phase current, the adaptive control follows the same principle as for inverter temperature and motor-winding temperature.

(d) Final Rate Transition Computation

After multiple rate modifications, the cumulative time shift T_x is computed as:

$$T_x = \frac{R_n - R_{n-1}}{R_n}t_{n-1} - \frac{R_{n-1} - R_{n-2}}{R_{n-2}}t_{n-3}$$
$$+ \cdots \ldots \ldots \ldots + \frac{R_3 - R_2}{R_3}t_2 + \frac{R_2 - R_1}{R_2}t_1 \quad (16)$$

$$T_x = \sum_{i=2}^{n} \frac{R_i - R_{i-1}}{R_i}t_{i-1} \quad (17)$$

For each rate adjustment, the system updates the acceleration rate using:

$$Rate = R_i\psi(f_it - T_x) \quad (18)$$

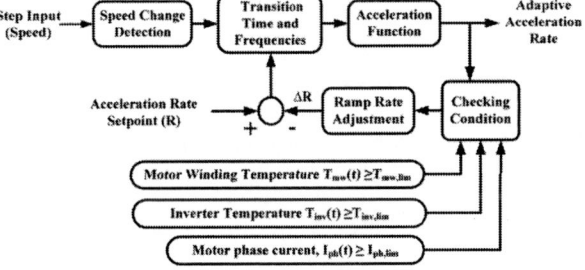

Fig. 5. Block diagram of the adaptive rate controller

C. Delay Compensation

In sensor-less DTC of PMSMs, accurate rotor position and speed estimation are essential for precise torque and flux control. However, delays introduced by current sensing can result in a phase lag in the estimated rotor position θ_e. This misalignment between the estimated and actual rotor positions adversely impacts control accuracy, especially at high speeds where the effects of delays are magnified due to faster system dynamics. To address this challenge, delay compensation techniques, specifically the phase advancing method, are implemented to counteract the lag caused by these delays. This method compensates for the phase lag by predicting and advancing the estimated rotor position using the known current sensing delay time (T_d) and rotor angular velocity (ω_e). The current sensing delay is modeled using a transfer function:

$$H(s) = \frac{K_s}{1 + T_d s} \qquad (19)$$

where K_s is sensor gain and T_d is the sensor delay time. The compensated rotor position ($\theta_{e,\,adv}$) is expressed as:

$$\theta_{e,adv} = \theta_e + \omega_e T_d \qquad (20)$$

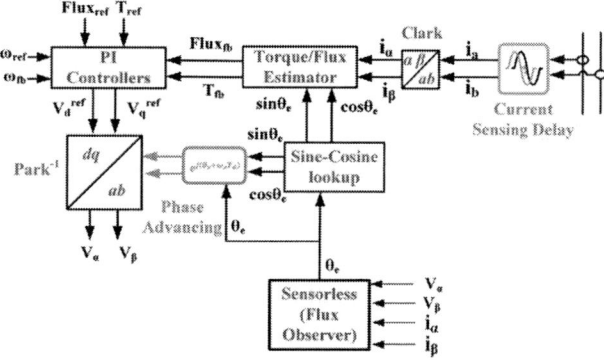

Fig. 6. Phase-advancing method for compensating current sensing delay in sensor-less DTC of PMSMs.

By incorporating $\theta_{e,\,adv}$ into the control algorithm, the stator flux orientation is better aligned with the actual rotor position, minimizing the adverse effects of delay-induced misalignments due to current sensing delays. The block diagram in Fig. 6 illustrates the implementation of the phase advancing method in a sensor-less DTC system, focusing specifically on current sensing delays.

D. PI Controller Tuning

Precise tuning of the PI controllers for speed, torque, and flux is critical for stable operation at ultra-high speeds. PI controller tuning is essential for stable Direct Torque Control (DTC) in ultra-high-speed PMSMs, requiring iterative refinement beyond simulations. Initial parameters, derived from motor characteristics, are optimized in simulations to minimize overshoot, steady-state error, and ripple. However, real-world testing highlighted additional system dynamics, such as sensor noise and delays, necessitating further adjustments. K_p and K_i are fine-tuned during hardware testing

to achieve stable operation. By properly tuning the PI controllers for simulation and hardware implementations, stable and precise control of the PMSMs at ultra-high speeds is achieved.

V. EXPERIMENTAL SETUP

For the lab-built AMEBA motor, the experimental setup is built with the CREE CGD15FB45P1 inverter, F28379D controller, and current sensors, as shown in Fig. 7. The CREE CGD15FB45P1 inverter, capable of higher switching frequencies (over 200 kHz), is chosen to drive the AMEBA at full rated speeds. For the commercial UHS-PMSM motor, the BOOSTXL-3PHGANINV inverter, equipped with integrated current sensing, is used for high-speed operation. This UHS-PMSM is tested up to 110,000 RPM in closed loop using the setup shown in Fig 2. (d). The Motor Control Blockset and Embedded Coder in MATLAB are utilized for code generation and deployment in this implementation. The developed model is deployed to the F28379D controller for real-time testing. Through the host model, speed commands and various debug signals (speed, current, torque, flux, etc.) can be monitored. Before the experiment started, calibrations are carried out to match the actual current with the host model's displayed current. The motor is operated at 500/1000 RPM in an open-loop mode first and then switched to the closed-loop mode for high-speed operation. A tachometer is used to verify the speed of the sensor-less control.

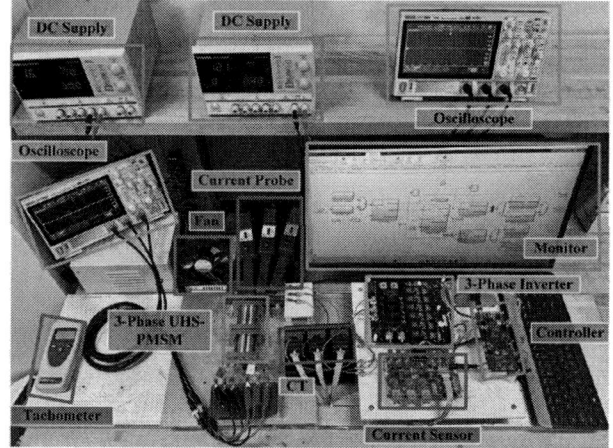

Fig. 7. Experimental setup for AMEBA motor

VI. RESULTS AND DISCUSSION

A. DTC Simulation up to 500,000 RPM

The DTC is developed and simulated for two ultra-high-speed PMSMs. The PI controllers' parameters are individually tuned for each motor. Fig. 8(a) shows the motor accelerating to maximum speed with excellent response and no significant overshoot.

A 5-sample delay (25 µs) has been introduced into the model to assess the impact of increased current sensor delays. As shown in Fig. 8(b), the motor failed to reach the reference speed, deviating from the speed command at approximately 6.7 seconds and failing to follow the reference speed beyond 412,000 RPM. Applying the phase-advancing delay compensation restored accurate tracking, enabling the motor to reach 500,000 RPM successfully, shown in Fig. 8(c).

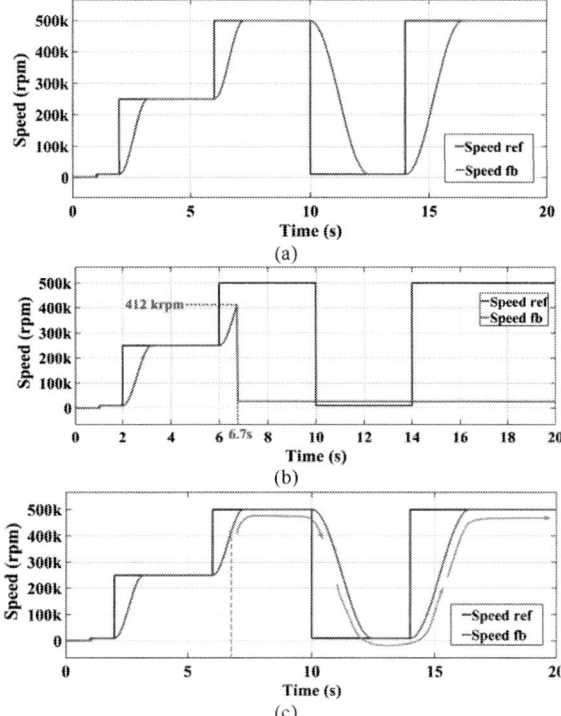

Fig.8. Speed Response at 500,000 RPM (a) 4-sample delay (20 μs) and (b) 5-sample delay (25 μs). (c) After delay compensation

B. Simulation Results of Adaptive Rate Controller

The simulation results validate the effectiveness of the proposed Adaptive Rate Limiter (ARL) in regulating acceleration while mitigating thermal stress in high-speed drives. As shown in Fig. 9, the reference speed follows an abrupt transition to 20k RPM, which may induce mechanical and electrical stress. The traditional rate limiter enforces a fixed acceleration rate but lacks adaptability to thermal constraints.

In contrast, the adaptive rate controller dynamically adjusts the acceleration profile based on real-time feedback from the inverter temperature $T_{inv}(t)$, motor-winding temperature $T_{mw}(t)$, and phase current $I_{ph}(t)$. At the first transition point $\omega_1(1.8s)$, the motor-winding temperature $T_{mw}(t)$ surpassed its threshold $T_{mw,lim}$, prompting a reduction in the acceleration rate to prevent excessive heating.

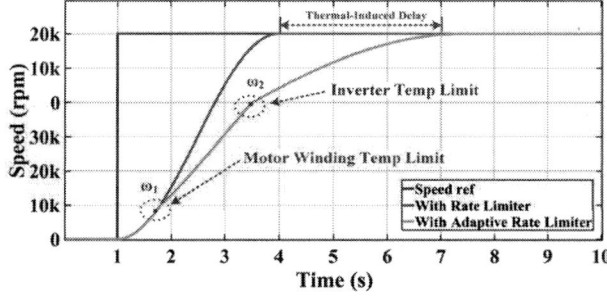

Fig. 9. Comparison of speed transition: reference speed, with general rate limiter, and with adaptive rate controller

Subsequently, at the second transition point $\omega_2(3.5s)$, the inverter temperature $T_{inv}(t)$ exceeds its limit $T_{inv,lim}$, further reducing acceleration.

This thermal-induced delay prevents excessive heating and electrical overload, ensuring a thermally constrained acceleration profile. The ARL enhances system longevity and operational reliability while maintaining smooth speed transitions in high-performance PMSM applications.

C. Open-loop Testing at High Speed

The AMEBA motor achieved up to 150,000 rpm in open-loop with modulation index regulation. Fig. 10 and 11 show the speed and current responses, respectively. The motor performed a one-step speed transition from 1,000 RPM to 150,000 RPM.

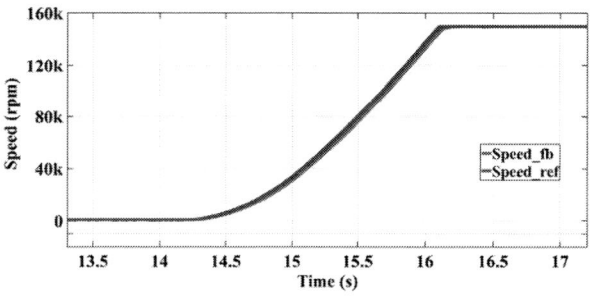

Fig.10. Speed response of the AMEBA motor in open-loop testing up to 150,000 RPM in hardware.

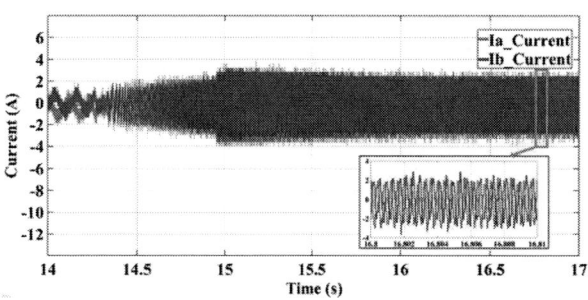

Fig.11. The current response of the AMEBA motor in open-loop testing up to 150,000 RPM in hardware.

D. Closed-loop Testing at High Speed

Closed-loop testing is conducted for both the AMEBA and commercial PMSMs, with PI controllers tuned for high-speed operation. A single set of PI controller gains was used across the speed range. However, for higher speeds, the gains were carefully tuned during the development process to minimize overshoot and steady-state error. Table III lists the PI parameter values for both PMSMs.

TABLE III.   PI CONTROLLER PARAMETERS FOR HIGH-SPEED OPERATION

Parameter	AMEBA PMSM	COMMERCIAL PMSM
Speed K_p	1.1097	3.0894
Speed K_i	0.0058	0.0041
Torque K_p	0.9569	0.0537
Torque K_i	0.0702	0.0847
Flux K_p	0.9569	0.0537
Flux K_i	0.0702	0.0847

Fig. 12 show the AMEBA motor's speed response during closed-loop testing up to 70,000 RPM with minimal overshoot and a steady-state ripple.

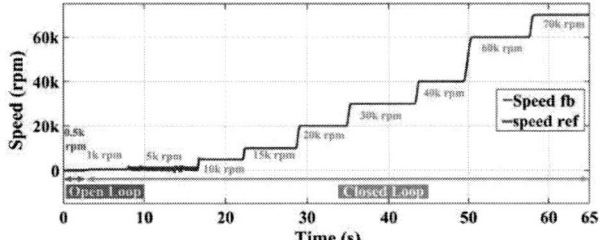

Fig. 12. Speed response of AMEBA up to 70,000 RPM with incremental speed jump in closed loop operation

Fig. 13 presents the current response ramp-up to 70,000 RPM and at the maximum speed, highlighting stable phase currents.

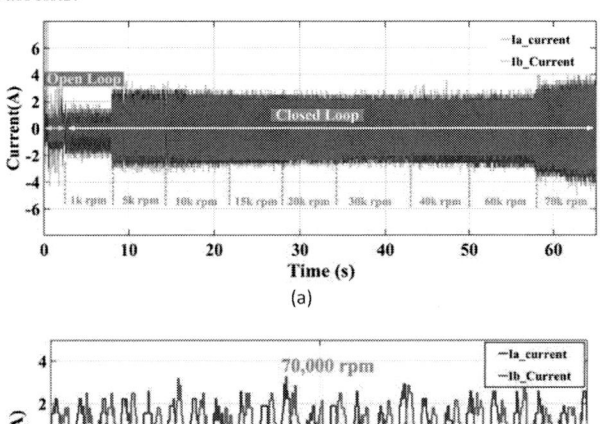

(a)

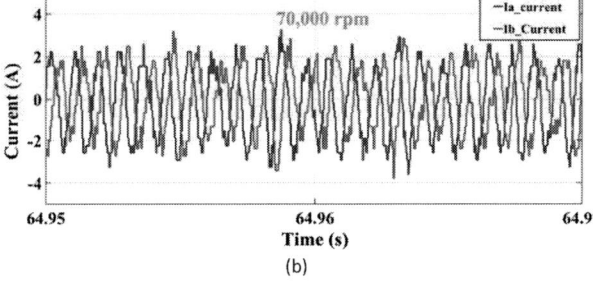

(b)

Fig. 13. Current response of the AMEBA motor: (a) Current response up to 70,000 RPM; and (b) Phase current at 70,000 RPM.

The commercial PMSM motor was operated up to 110,000 RPM in a closed-loop mode with a single-step speed increase using linear rate limiter, as shown in Fig.14. Compared to the adaptive rate controller, the linear rate limiter results in higher speed overshoot. The current responses (Phase A and Phase B) at 5,000 RPM and 110,000 RPM are presented in Figs. 15(a) and 15(b), respectively.

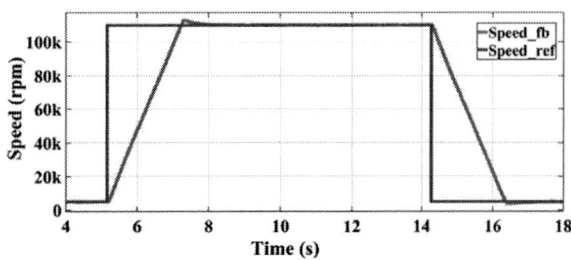

Fig.14. Speed Response at 110,000 RPM (Commercial Motor)

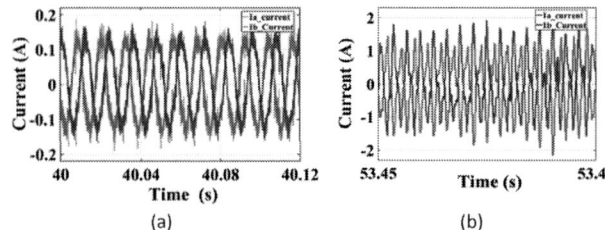

(a)   (b)

Fig.15. Current of commercial PMSM (a) at 5000 and (b) 110,000 RPM

To assess the accuracy of the flux observer during operation, position estimations were plotted across the testing range. The flux observer performed well, with the estimated position closely matching the actual position at both 5,000 RPM and 110,000 RPM. This indicates the effective performance of the sensor-less system.

E. Performance Analysis of the Two UHS-PMSM Motors

The performance of both the commercial PMSM and the AMEBA PMSM has been evaluated in terms of overshoot and steady-state speed ripple at their respective maximum tested speeds. Figures 17 and 18 illustrate the speed response for the commercial PMSM and the AMEBA PMSM at 70,000 RPM, respectively. The performance metrics are summarized in Table IV. The AMEBA PMSM demonstrated an overshoot of 0.5% and a steady-state speed ripple of ±50 RPM indicating precise control and stability.

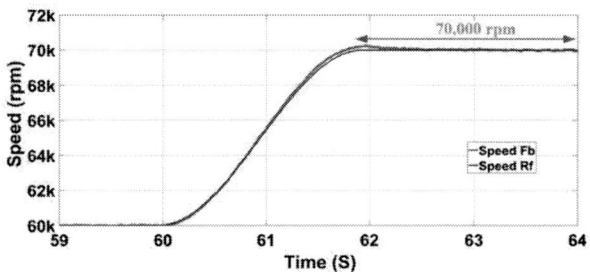

Fig. 17. Speed response and overshoot of the commercial PMSM at 70,000 RPM.

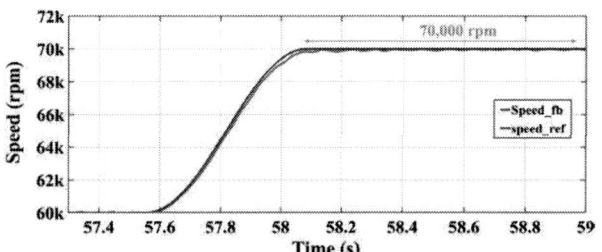

Fig. 18. Speed response of the AMEBA PMSM at 70,000 RPM.

Parameter	AMEBA PMSM (70k RPM)	COMMERCIAL PMSM (70k RPM)
Overshoot	0.5%	3%
Steady State Ripple	+/-50 RPM	+/-100 RPM

TABLE IV. OVERSHOOT AND STEADY RIPPLE

In comparison, the commercial PMSM exhibited an overshoot of 3% and a steady-state speed ripple of ±100 RPM.

F. Performance Analysis of DTC and FOC Control

A comparative simulation analysis of DTC and FOC control is conducted for the two UHS-PMSMs during a one-step speed transition from 1,000 RPM to 500,000 RPM, as shown in Fig. 19 and 20.

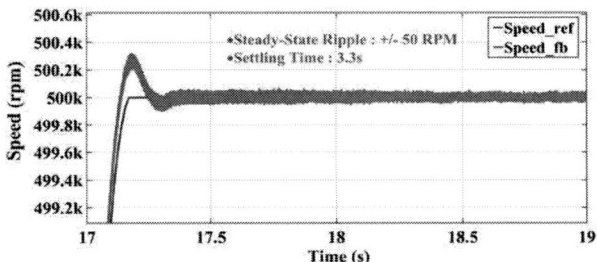

Fig.19. Speed response at 500,000 RPM for DTC control in simulation

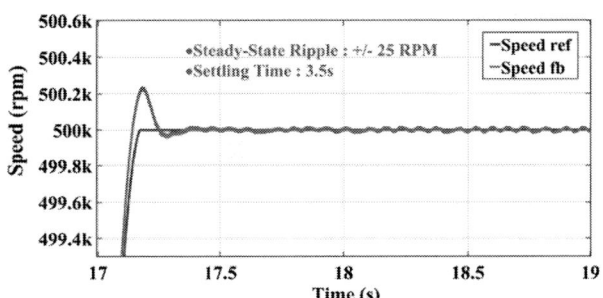

Fig.20. Speed response at 500,000 RPM for FOC control in simulation

The performance metrics in Table V indicate negligible differences between DTC and FOC in terms of overshoot, steady-state ripple, and settling time. FOC shows a lower steady-state ripple ($\pm$ 25 RPM vs. $\pm$ 50 RPM). However, DTC demonstrated a slightly faster settling time (3.3s vs. 3.5s). Performance metrics are summarized in Table V. FOC control was also tested on the AMEBA motor up to 30,000 RPM, with the speed response shown in Fig. 21. In hardware implementation, FOC exhibited higher oscillations compared to DTC control shown in Fig.12(a), indicating lower stability at high speeds.

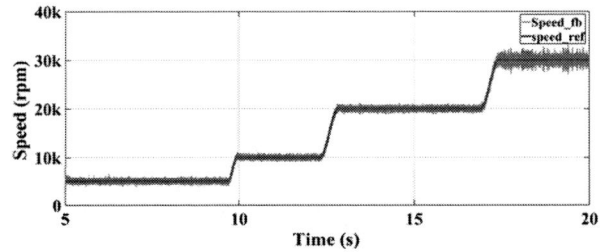

Fig.21. Speed response at 30,000 RPM for FOC control in hardware

TABLE V. COMPARISON OF DTC AND FOC CONTROL

Parameter	DTC (500k RPM)	FOC (500k RPM)
Overshoot	0.065%	0.045%
Steady State Ripple	+/-50 RPM	+/-25 RPM
Settling Time	3.3 s	3.5 s

VII. CONCLUSION AND FUTURE WORK

In this study, a sensor-less Direct Torque Control (DTC) algorithm for ultra-high-speed PMSMs is developed and analyzed. Implementing a adaptive rate controller and employing delay compensation techniques enhanced the DTC algorithm's performance and stability under ultra-high-speed conditions. Experimental results demonstrated the effectiveness of the proposed methods, with the commercial motor operating at 110,000 rpm and the lab-built AMEBA motor at 70,000 rpm in closed-loop control. Future work will involve operating both motors across their full speed range up to 500,000 rpm and addressing additional delays through advanced hardware testing. Implementing advanced compensation methods and further optimizing the control algorithm will enhance DTC's stability and precision in ultra-high-speed applications.

ACKNOWLEDGEMENT

This work is partially supported by the NSF Partnerships for Innovation - Research Partnerships (PFI-RP) program under Award 2234271.

REFERENCES

[1] M. K. Islam, S. Choi, Y. K. Hong, and S. Kwak, "Design of high-power ultra-high-speed rotor for portable mechanical antenna drives," IEEE Transactions on Industrial Electronics, vol. 69, no. 12, pp. 12610-12620, Dec. 2021.

[2] C. J. Volpato Filho, D. Xiao, R. P. Vieira, and A. Emadi, "Observers for high-speed sensorless PMSM drives: Design methods, tuning challenges and future trends," IEEE Access, vol. 9, pp. 56397-56415, 2021

[3] D. Mohanraj, J. Gopalakrishnan, B. Chokkalingam, and L. Mihet-Popa, "Critical aspects of electric motor drive controllers and mitigation of torque ripple," IEEE Access, vol. 10, pp. 73635-73674, 2022.

[4] B. H. Bae and S. K. Sul, "A compensation method for time delay of full-digital synchronous frame current regulator of PWM AC drives," IEEE Transactions on Industry Applications, vol. 39, no. 3, pp. 802-810, May/Jun. 2003

[5] S. Kim, K. Park, D. Kang, and G. H. Lee, "High-performance permanent magnet synchronous motor control with electrical angle delayed component compensation," IEEE Access, vol. 11, pp. 129467-129478, 2023.

[6] Yukinori Inoue, Shigeo Morimoto, Masayuki Sanada "Control Method Suitable for Direct-Torque-Control-Based Motor Drive System Satisfying Voltage and Current Limitations" in 2012 IEEE Transactions on Industry Applications, pp 03-15-2012

[7] A. Podder and D. Pandit, "Study of Sensorless Field-Oriented Control of SPMSM Using Rotor Flux Observer & Disturbance Observer Based Discrete Sliding Mode Observer," in 2021 IEEE 22nd Workshop on Control and Modelling of Power Electronics (COMPEL), pp. 1-6, Dec. 2021.

[8] C. J. V. Filho, D. Xiao, R. P. Vieira, and A. Emadi, "Observers for High Speed Sensorless PMSM Drives: Design Methods, Tuning Challenges and Future Trends," IEEE Access, vol. 9, pp. 56397-56415,2021.

[9] K. N. Tasnim, M. K. Islam, M. S. Haque, and S. Choi, "Networked control of multiple ultra-high-speed PMSMs for AMEBA," in 2022 IEEE Energy Conversion Congress and Exposition (ECCE), pp. 01-08, 2022.

[10] V. E. Vavilov, "Super high-speed electric motors," Russian Engineering Research, vol. 37, no. 11, pp. 991-994, 2017.

A Ferrite Based Carbon Reinforced Composite Wrapped IPM Rotor Design for High-Speed Traction Applications

MD Rashedur Rahman
Dept. of Electrical and Computer Engineering
Mississippi State University
Starkville, MS-39762, USA
mr2575@msstate.edu

Md Khurshedul Islam
Sr. Mechatronics Design Engineer
ASML
Wilton, CT 06897, USA
mdkhurshedul.islam@asml.com

MD Moniruzzaman
Dept. of Electrical and Computer Engineering
Mississippi State University
Starkville, MS-39762, USA
mm5111@msstate.edu

Seungdeog Choi
Dept. of Electrical and Computer Engineering
Mississippi State University
Starkville, MS-39762, USA
seungdeog@ece.msstate.edu

Han-Gyu Kim
Dept. of Aerospace Engineering
Mississippi State University
Starkville, MS-39762, USA
hkim@ae.msstate.edu

Andrew Walters
Dept. of Aerospace Engineering
Mississippi State University
Starkville, MS-39762, USA
aw3071@msstate.edu

Abstract— **This paper presents the design and analysis of a novel high-speed, multi-layer ferrite-based interior permanent magnet motor (HSFR) incorporating a reinforced carbon composite (RCC) rotor. While ferrite-based IPMs offer a cost-effective alternative to rare-earth (RE) magnet motors for traction applications, their mechanical limitations at high speeds necessitate innovative design approaches. Traditional central ribs, often employed to enhance rotor stability, compromise electromagnetic performance. Consequently, removing the ribs and introducing a retention sleeve around the rotor is a feasible choice until it offers a good-fit between electromagnetic and mechanical performances. However, this technique has been limitedly studied for ferrite magnets, which is mechanically weaker than RE ones. To address this, the RCC is strategically integrated in circumferential (sleeve) and radial (rib) directions to optimize rotor strength. Additionally, a six-phase winding configuration is implemented to augment torque output, compensating for the inherent lower magnetic energy density of ferrite magnets. A 4kW rated, maximum 30,000rpm, six-phase ferrite-based IPM with an RCC-wrapped internal permanent magnet motor (6PF-CCW-IPM) is developed through analytical optimization and finite element analysis (FEA). Performance comparisons with equivalent reference design validate the effectiveness of the proposed design in achieving high-speed operation while maintaining competitive power density**

Keywords—rare-earth free magnets, high-speed motor, carbon composites, retention sleeve, multi-phase winding

I. INTRODUCTION

The increasing geopolitical tensions and diminishing global supply of rare-earth (RE) elements have significantly disrupted the supply chain of high-performance permanent magnets, a critical component in modern traction applications [1]. While alternative machine topologies such as wound field synchronous machines (WSMs), induction machines (IMs), and synchronous/switched reluctance machines (Syn/SRMs) have been investigated, their performance characteristics often fall short of the superior power density and efficiency offered by conventional permanent magnet synchronous machines (PMSMs) [2]. In response to these challenges, researchers have actively pursued the development of rare-earth-free (REF) or reduced RE PMSM designs, aiming to mitigate the reliance on critical materials while maintaining competitive performance levels [3-6]. By employing advanced materials, innovative design techniques, and optimized control strategies, REF-PMSMs have demonstrated the potential to bridge the performance gap with conventional PMSMs. Although REF-PMSMs may exhibit a slight reduction in specific performance metrics compared to their RE-based counterparts, they offer a compelling balance of cost-effectiveness, environmental sustainability, and manufacturability [7]. These factors make REF-PMSMs a promising solution for the transportation industry, particularly in applications where high performance is not the sole driving factor.

Achieving high-speed operation in Permanent Magnet Synchronous Machines (PMSMs) is a complex engineering challenge requiring careful consideration of electromagnetic, thermal, and mechanical factors [8]. Increasing rotor rib thickness has traditionally been employed to enhance high-speed capability, but this approach often results in increased magnetic leakage and reduced machine performance. To mitigate this issue, retaining sleeves have been introduced in Interior Permanent Magnet (IPM) designs to improve mechanical strength and thermal management [9-11]. However, these studies primarily focus on rare-earth (RE) based magnets, neglecting the unique challenges associated with ferrite magnets. Due to their inherent mechanical brittleness, ferrite-

979-8-3315-1612-3/25 $31.00 © 2025 IEEE

based IPM rotors require specialized designs to withstand high-speed operation. Building upon previous research [12], a delta-shaped magnet rotor design incorporating carbon sleeves and bars has been proposed to optimize the rotor's structural integrity. Rigorous safety factor (SF) analysis has been conducted to ensure the mechanical reliability of the ferrite magnets under high-speed operating conditions. However, a comprehensive understanding of the intricate interplay between electromagnetic forces and mechanical stresses remains elusive, particularly in the presence of radially inserted carbon ribs and retaining sleeves. Further research is needed to quantify these effects and develop optimized design strategies for high-speed ferrite-based IPM rotors.

This paper presents a novel high-speed six-phase multi-layer ferrite-based IPM with a reinforced carbon composite (RCC)-wrapped internal permanent magnet motor (6PF-CCW-IPM), targeting a rated power of 4kW and a maximum speed of 30,000 rpm. The proposed rotor structure, incorporating both ferrites and 3D carbon composites, aims to deliver superior performance at high-speeds while maintaining cost-effectiveness, making it a promising candidate for electrified transportation applications.

II. PROPOSED DESIGN AND OPTIMIZATION

The design process for this study is bifurcated into electromagnetic and structural domains. While the former encompasses analytical and FEA calculations, the latter exclusively relies on FEA.

A. Electromagnetic Design

The electromagnetic design phase commences with a parameterization process encompassing material selection, stator and rotor geometry, magnet dimensions, sleeve thickness (considering air gap), and conductor count per phase. The optimization goal is maximum speed, minimum active material cost, and maximum power factor. To realize the desired performance, a stator configuration with two parallel three-phase windings, displaced by a standard 60° electrical angle, was adopted. The rotor magnets were selected to be NMF 12G+, a high-energy product ferrite renowned for its robust mechanical properties. Once satisfactory optimization results are obtained, the design will undergo rigorous electromagnetic finite element analysis (FEA) to facilitate detailed field calculations. Table I presents the materials employed throughout this work's electromagnetic, mechanical, and thermal properties.

B. Structural Design

After a comprehensive electromagnetic field analysis, if the design objectives are met, a structural finite element analysis (FEA) is performed to assess the mechanical integrity, ensuring a minimum safety factor (SF) to maximize the operational speed range. To minimize core losses and improve tensile strength, the stator and rotor cores are constructed from high-silicon steel. Fig. 1 visually represents the design optimization algorithm, while Table II outlines the initial optimization parameters.

Three rotor configurations were examined to investigate the impact of carbon components: Model-I, Model-II, and Model-III (Fig. 2). Model-I comprised a steel rotor with ribs for structural integrity but without a carbon sleeve. In contrast, model-II featured a carbon sleeve in lieu of ribs, while model-

III integrated both a carbon sleeve and carbon ribs, replacing the steel ribs. Moreover, the airgap bridge is also removed from the first layer flux barriers, while the second layer remains the same.

TABLE I. MATERIAL PROPERTIES

Properties	Steel core M250-35A	Magnetic material (NMF 12G+)	Carbon composites
Electrical Resistivity (Ω-m)	5.90E-07	100	3.03E-04
Thermal conductivity (W/mK)	22	~5	0.87
Density (Kg/m^3)	7650	5000	1780
Tensile Strength (MPa)	~610	Compressive~700 Flexural~60	Tangential~2723 Radial ~64.1
Yield Strength (MPa)	430	~35	−
Poisson's ratio	0.3	0.28	0.315
Elastic Modulus (GPa)	185	150	Tangential~162 Radial ~9.72
Thermal expansion coefficient (*10E-6/K)	12	15 (∥), 10(⊥)	Tangential~34 Radial ~0.14

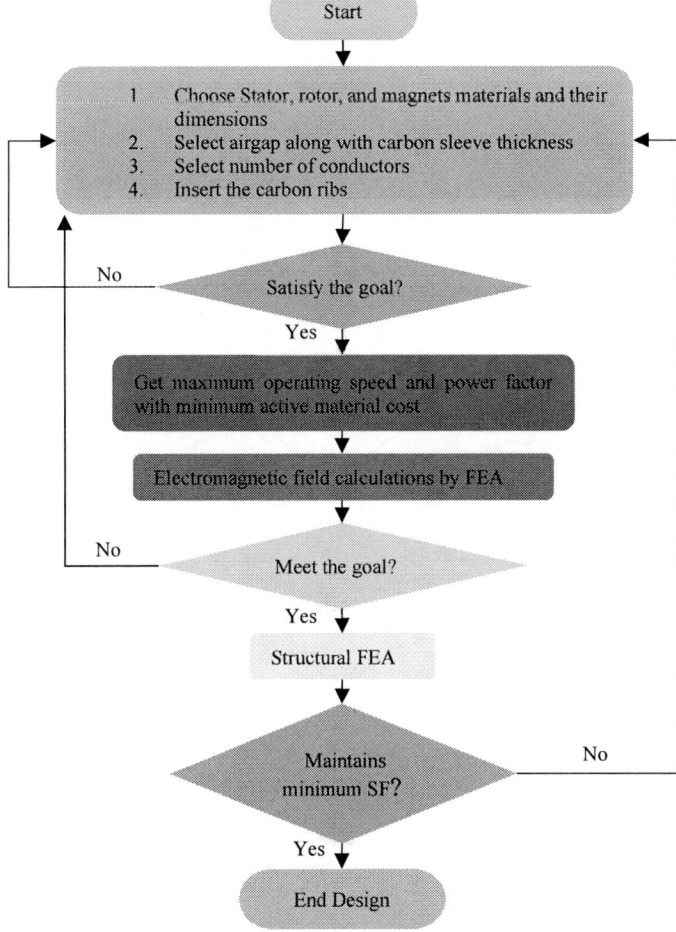

Fig. 1. Design and optimization algorithm.

It is essential to note that the magnet volume remained constant across all rotor configurations.

TABLE II. DESIGN VARIABLES

Parameters	Model-I	Model-II	Model-III
Pole/slot	4/ 24		
Magnetic material	Hitachi NMF 12G+		
Split Ratio	165/90		
Max current (Apeak)	20	25	25
DC bus voltage(V)	300		
Rated power (kW)	4		
Max Speed (rpm)	30000		
Airgap	0.7mm	1.7mm	1.7mm
Carbon fiber sleeve thickness	-	1.2mm	1.2mm
Axial length	50mm		

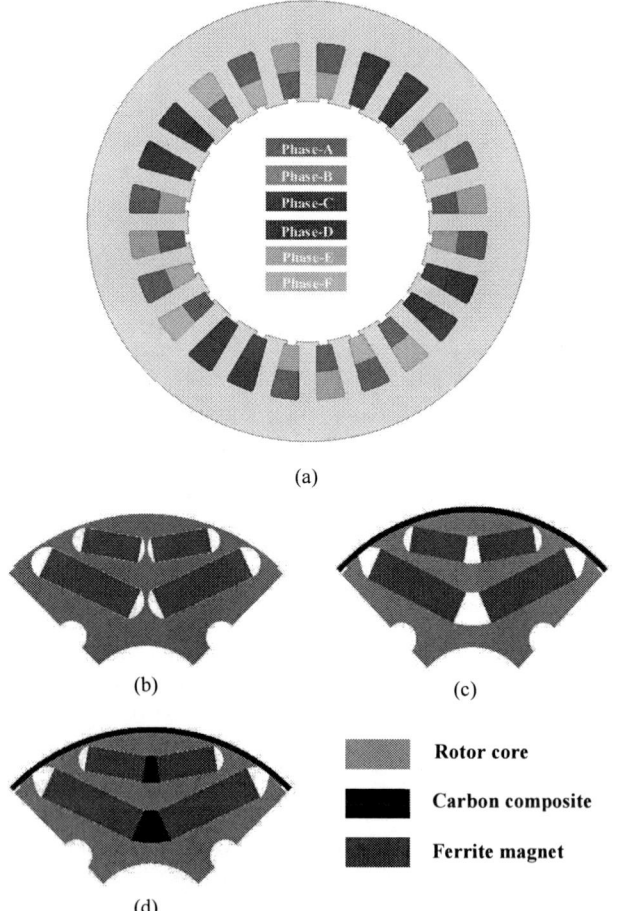

(a)

(b)   (c)

Rotor core

Carbon composite

Ferrite magnet

(d)

Fig. 2. Stator & rotor geomtries: (a) 6-φ Winding; (b) Model-I; (c) Model-II, and (d) Model-III

III. FEA RESULTS AND ANALYSIS

This section delves into the results obtained from finite element analyses (FEA) conducted for both electromagnetic and structural domains. Electromagnetic simulations were carried out using 2D Ansys Maxwell to investigate specific electromagnetic phenomena, e.g., electromagnetic field distributions and losses. Structural simulations were validated using Ansys Mechanical to assess specific structural behaviors, e.g., stress distribution, deformation.

A. Electromagnetic Analysis

Based on the data presented in Table II, three models were investigated under both no-load and load conditions. The load condition was assumed to be a nearly pure sinusoidal current excitation. Under no-load conditions, the flux linkages of all models were calculated. Subsequently, the back-EMF was determined using the following equation:

$$E = p\omega\phi_{pm} \qquad (1)$$

where p is the no. of pole pairs, ω is the rotational speed of the rotor in rad/s, and ϕ_{pm} is the flux linkage of permanent magnets in Wb. As depicted in Figs. 3 and 4, a notable enhancement of 34% and 31% is observed in the peak no-load flux linkage and back-emf, respectively, when transitioning from rotor model-I to rotor model-II. This improvement is attributed to the elimination of central and airgap bridges in rotor model-II. The reduction in flux leakage between these bridges and the optimized flux paths through the airgap contributes significantly to the elevated no-load flux linkage and back-emf values in rotor model-II.

In the d-q reference frame, the q-axis and d-axis inductances were determined via Finite Element Analysis (FEA) at a rotational speed of 8000 rpm. The d-axis inductances are 2.20 mH and 1.60 mH for models I and II, respectively. Consequently, the characteristic currents is calculated using the following equation:

$$I_{ch} = \frac{\phi_{pm}}{L_d} \qquad (2)$$

From (2), it is evident that model-II exhibits a higher peak characteristic current of 26.87A compared to model-I, which reaches 14.09 A. Although model-II experiences a reduction in peak constant torque, it maintains a higher torque level at elevated speeds (exceeding 20,000 rpm) than model-I, as illustrated in Fig. 5. Furthermore, the increased characteristic current in model-II results in enhanced flux-weakening capability at higher speeds, leading to a more favorable constant power-to-speed ratio (CPSR) as depicted in Fig. 6.

To enhance the maximum constant torque at low speeds, two primary strategies can be considered: (i) augmenting magnet volume and (ii) increasing the drive current supplied by the inverter. While the first approach, employed in [10], maintains an identical constant torque despite a larger air gap introduced by a carbon sleeve, it necessitates additional magnet material, incurring both cost and space overhead within the rotor. Additionally, the increased magnetic field strength may introduce challenges in terms of magnetic saturation and eddy current losses. Conversely, increasing the drive current can lead to elevated current densities and temperature rise within the motor, especially when the winding configuration remains unchanged. Moreover, this approach imposes additional demands on the inverter, potentially incurring additional costs.

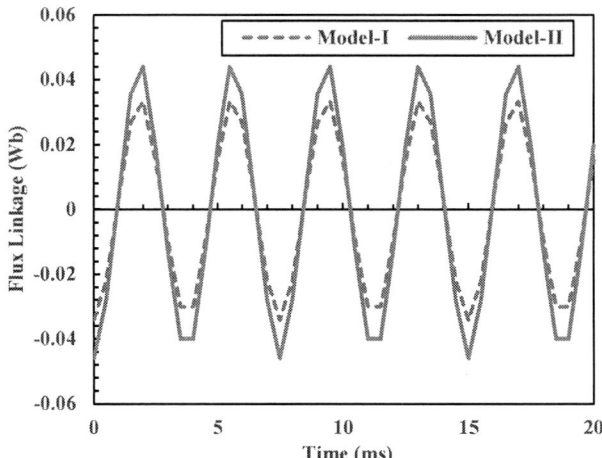

Fig. 3. No-load flux linkage at 8000rpm.

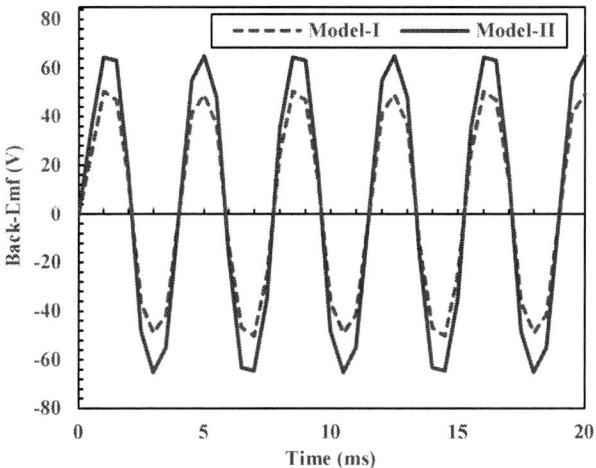

Fig. 4. No-load back-emf at 8000rpm.

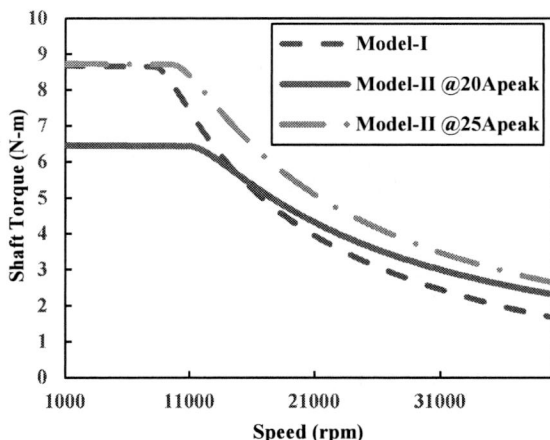

Fig. 5. Output torque characteristics over speed.

Given these limitations, this research focuses on the second strategy, leveraging the high-temperature capability of the double-layer ferrite magnet-based rotor design. By analyzing the maximum torque matching criterion, it was determined that a 25% increase in drive current, as illustrated by the green dotted lines in Fig. 5, is required to achieve the peak constant torque.

The power calculations presented in Fig. 6 show that model-II demonstrates superior performance beyond the base speed without additional drive currents. Nevertheless, it encounters difficulties in achieving the same power levels as model-I at lower speeds. By increasing the drive currents by 25%, model-II sustains high-speed performance and outperforms its base-speed counterparts.

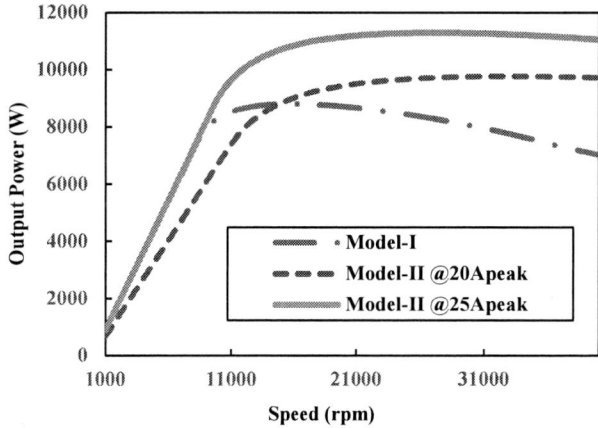

Fig. 6. Output power characteristics over speed.

Table II provides a detailed comparison of electromagnetic performances for various rotor designs. While achieving optimal torque constant and saliency ratio, model-I experiences slightly elevated torque ripple and reduced power factor. In contrast, despite a 25% decrease in torque constant, models II and III exhibit a 29% increase in back-emf constant and a 10% enhancement in peak power at 20A peak current. Including lightweight carbon ribs in model-III may slightly minimize overall electromagnetic performance relative to model-II.

TABLE III. PERFORMANCE COMPARISON

Parameters	Model-I	Model-II	Model-III
Peak Efficiency (%)	94.3	94.5	94.4
Total losses (W) @8000rpm	152	202	204
Power factor	0.74	0.85	0.85
Back-EMF constant (Vs/rad)	0.0596	0.077	0.077
Torque constant (Nm/A)	0.43	0.325	0.325
Torque ripple (%)	12.34	11.02	11.02
Peak airgap flux density (p.u.)	1	0.75	0.75
Saliency ratio (p.u.)	1	0.82	0.82
Active weight (p.u.)	1	0.988	0.992
Peak Power (p.u.) @20Apeak	1	1.1	1.09

B. Structural Analysis

A structural finite element analysis (FEA) was conducted on the rotor models, utilizing the material properties outlined in Table I. The investigation was divided into two primary phases. In the initial phase, only centrifugal loading was considered to determine the maximum equivalent stress distribution across the rotor models. Subsequently, the analysis was extended to incorporate thermal expansion coefficients, in conjunction with centrifugal loading, to provide a more accurate representation of the maximum equivalent stress under varying ambient temperature conditions.

Fig. 7 presents the equivalent stress distribution within the rotor models under the sole influence of centrifugal load induced by rotational speed. Rotor model-I exhibits a maximum stress concentration (Fig. 7a) in the central bridge of the first layer. This stress surpasses the yield strength of the rotor core steel at 30,000 rpm, resulting in a significantly low safety factor (SF), given that the SF, defined as the ratio of maximum permissible stress to equivalent stress, underscores the critical nature of this stress concentration. To mitigate this issue, the central and air-gap bridges in the first layer of magnets were removed, and a carbon sleeve was introduced to encapsulate the rotor. The subsequent stress distribution, depicted in Fig. 7b, reveals a shift in the maximum stress to the carbon sleeve. The higher yield strength of carbon compared to steel enables the structure to accommodate the increased stress. In a final modification, carbon bars were incorporated in place of the iron central bridges. As shown in Fig. 7c, this configuration substantially reduces stress compared to models I and II. The combined effect of the carbon sleeve and carbon bars significantly enhances the rotor's structural integrity, providing a substantial improvement in the safety factor.

A comprehensive structural finite element analysis (FEA) was conducted at a rotational speed of 30,000 rpm to evaluate the safety factor (SF) of three distinct rotor configurations. The primary objective was to maximize the SF while preserving optimal multi-physics performance, including electromagnetic characteristics. As detailed in Table IV, rotor model III demonstrated the highest SF among the three configurations, surpassing model II without compromising electromagnetic performance. The substitution of steel rotor ribs with a carbon sleeve and ribs proved instrumental in significantly enhancing mechanical stability at elevated rotational speeds. Furthermore, the inclusion of a 78K delta temperature resulted in a reduction of the SF across all configurations, as illustrated in Table IV. This finding underscores the critical influence of thermal effects on the structural integrity of high-speed rotors.

TABLE IV.   SAFETY FACTOR AT 30,000RPM

Parameters	Model-I	Model-II	Model-III
SF (Centrifugal load)	0.62	1.14	4.49
SF (Centrifugal load+ Delta Temp.)	0.52	1.11	3.41

Fig. 8 presents a comparative analysis of the performance of all rotor models across a range of speeds, considering only the influence of centrifugal loads. Models I and II exhibit a similar trend over the entire speed range, albeit model-II undergoes abrupt changes in stress distribution from low to high speeds. While models I and III share analogous performance characteristics, the locations of maximum equivalent stress differ, with each model employing materials with varying maximum permissible stress values. For example, the carbon composite material, oriented in a specific direction, demonstrates superior stress management capabilities compared to electrical steel. Notably, model-II, devoid of central bridges, experiences significantly higher stresses from the outset. However, these stresses are concentrated on the carbon sleeve, which maintains a higher safety factor than model-I

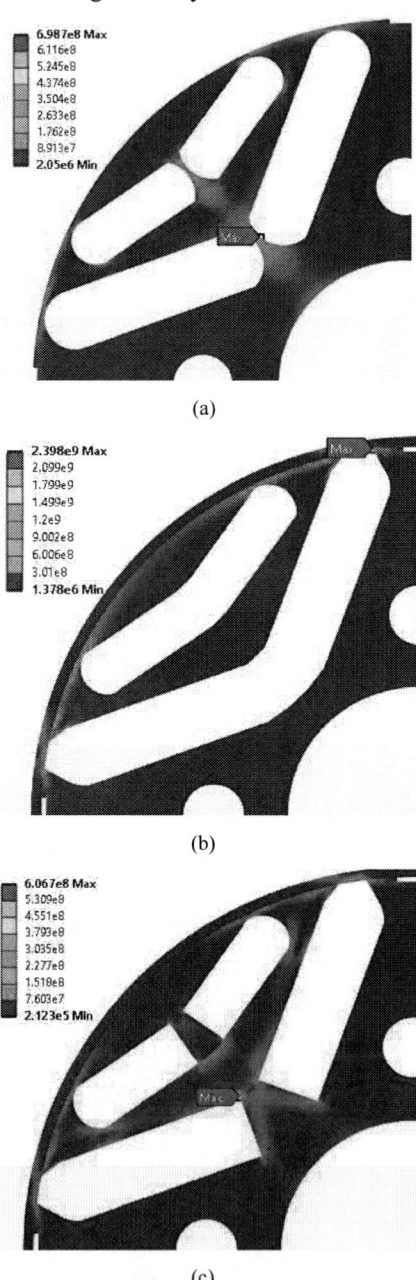

(a)

(b)

(c)

Fig. 7. Von-Mises stress (Pa) distribution at 30,000rpm (only centrifugal load): (a) Model-I; (b) Model-II, and (c) Model-III

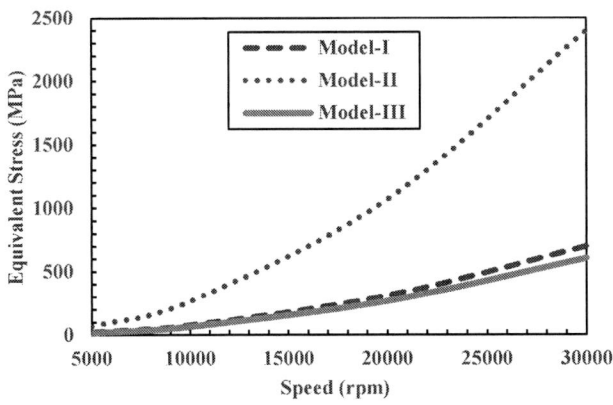

Fig. 8. Von-Mises stress (MPa) distribution over speed ranges.

IV. INTEGRATION OF CARBON COMPOSITES AND ITS MANUFACTURING

Rotor sleeves are structural components required to maintain permanent magnets in their position during high-speed rotation. Non-magnetic metallic sleeves have widely been adopted in conventional rotor designs; however, the high electrical conductivity of these sleeves induces eddy current losses, significantly reducing the efficiency of an electromagnetic field. To address this issue, carbon-fiber/epoxy composites are employed as an alternative to non-magnetic metals for a rotor sleeve in the proposed design, as shown in Fig. 9a.

Carbon composites have low electrical conductivity and thus minimize eddy current losses while contributing to light structural design with low density and high strength. However, the difference between the thermal conductivity of carbon sleeves and metallic rotor bodies would require additional structural design in the form of pre-tension in the sleeves. In the proposed method, carbon-fiber/epoxy composite rotor sleeves are manufactured by applying pre-tension during the fiber layup process, as shown in Fig. 9b. Commonly used pre-impregnated fibers (or prepreg) would not be a suitable choice for the manufacturing process as matrix (commonly epoxy resin) structures could easily be damaged during the pre-tension process. The adhesive character of prepreg could be a hindrance to the manufacturing process. To address this issue, dry carbon fibers will be adopted for the wrapping process, and the wrapped fibers will be infused with resin using a resin transfer molding process. For pre-tensioning, carbon fibers are winded on a mold at liquid nitrogen temperature. The severe cold temperature shrinks the rotor and provides pre-tension when the rotor returns to room temperature.

The orientation of carbon-fiber wrapping is determined to strengthen the critical stress directions based on stress analysis data from high-speed rotating scenarios. In addition to the sleeve structure, structural partitions (ribs) are made of carbon-fiber/epoxy composites and placed between magnets, as shown in Fig. 9a. These partitions are intended to release high-magnitude stress on the magnets under high-speed rotation and thereby enhance the strength of the rotor structure. The accurate dimensions of the partitions are strictly required to place the partitions between the magnets and the metallic rotor body

structure. To address this issue, the partitions are manufactured using an additive manufacturing technique with carbon-fiber-reinforced thermoplastics such as a stereolithography (SLA) 3D printing technique. Alternatively, the partitioners can be laid up with carbon-fiber/resin prepregs and the cured laminates can be post-processed to have the required shapes using a high-precision milling machine.

Lastly, thermal management is required for the carbon-composite sleeves and partitioners to control heat induced by high-speed rotation (i.e., heat at 150-200°C). For thermal management, the matrix material needs to be selected to have at least 200 °C of glass transition temperature (Tg) or melting temperature (Tm). Alternatively, thermal barrier coatings (TBCs) on the surfaces of the carbon-composite sleeves and partitioners can be considered.

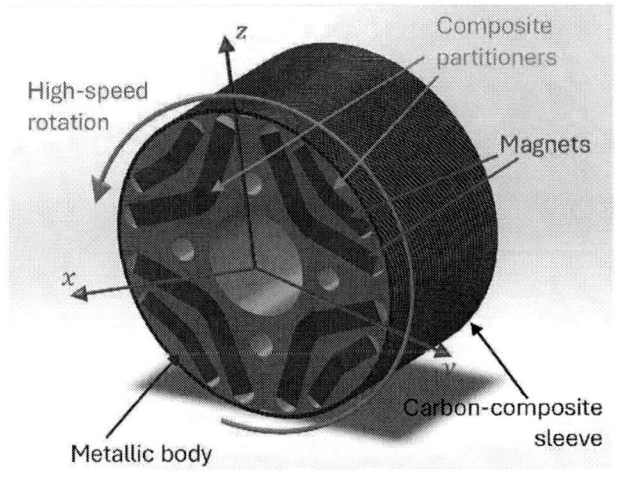

(a)

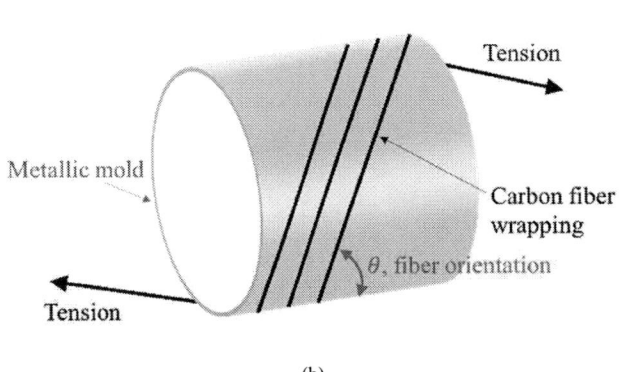

(b)

Fig. 9. (a) Design of a carbon fiber/epoxy composite rotor sleeve; (b) Schematic drawing for carbon fiber wrapping and pre-tensioning process.

Carbon fiber sleeves were fabricated using an X-Winder 4X-23 four-axis filament winding machine. AS4-GP 12k (0.9%) carbon fiber was utilized in conjunction with Fiberglast 4500 infusion resin and a two-hour hardener. The composite material was wound onto an 8-inch (203.2 mm) long, 6061 aluminum tube with an outer diameter of 4.25 inches (107.95 mm) and an inner diameter of 3.5 inches (88.9 mm). This aluminum tube served as the mold for all carbon fiber components produced on the filament winder.

Fig. 10a illustrates the manufacturing process, showcasing the X-Winder 4X-23 equipped with a roll of AS4-GP 12k (0.9%) carbon fiber-fed through a tow rig. The fibers were subsequently immersed in a resin bath containing Fiberglast 4500 before being deposited onto the rotating aluminum mold tool (Fig. 10a). The winder's main body moved axially along the mold length, controlled by the machine's program, to achieve the desired fiber orientation.

Upon completion of the winding process, the part was enclosed in heat shrink tape and cured in an oven at the resin's specified temperature of 150 °F (65 °C) for three hours. The finished carbon fiber sleeve was then removed from the aluminum mold (Fig. 10b).

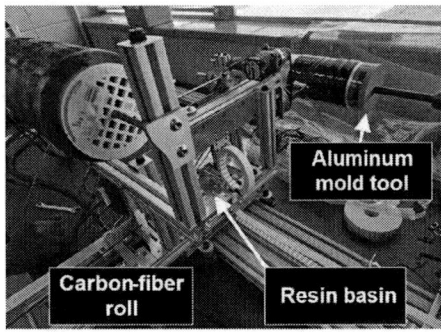

(a)

(b)

Fig. 10. Manufacturing of carbon fiber sleeves: (a) Manufacturing setup; (b) A manufactured carbon fiber sleeve.

V. Conclusion

This research presents a novel approach to enhance electric motors' mechanically constrained speed range utilizing ferrite magnets. The core innovation involves strategically integrating carbon composite components, specifically carbon sleeves and ribs, within the motor design. By modifying the electromagnetic circuit through the air gap and incorporating a carbon sleeve, a significant 10% increase in peak power at top speeds is achieved.

While carbon fiber's manufacturing complexity is acknowledged, its demonstrated potential in expanding speed range, as evidenced by previous research, justifies its incorporation into motor designs. This study proposes a further refinement by introducing carbon ribs to the design, aiming to augment the safety factor and, consequently, the speed range. The strategic placement of these ribs within the flux barrier path is anticipated to contribute to a substantial increase in safety factor, as illustrated by the improvement from 0.52 to 3.41, while considering thermal constraints.

Although the mechanical properties of carbon fiber are susceptible to manufacturing processes, its inherent strength advantage over traditional electrical steel warrants further investigation. Given the considerable costs associated with motor prototyping, a comprehensive multi-physics analysis, encompassing structural, thermal, and harmonic analyses, will precede extensive physical testing in future research. Additionally, future work will focus on the experimental characterization of various carbon fiber configurations, including variations in thickness and manufacturing techniques, while acknowledging the inherent limitations of the material.

Acknowledgment

This work was supported in part by the NSF Partnerships for Innovation - Research Partnerships (PFI-RP) program under Award 2234271.

References

[1] Smith, Braeton J., et al. "Rare-earth permanent magnets-Supply chain deep dive assessment, " USDOE Office of Policy (PO), 2022.

[2] James D. Widmer, Richard Martin, Mohammed Kimiabeigi, "Electric vehicle traction motors without rare earth magnets, " Sustainable Materials and Technologies, Volume 3, 2015, Pages 7-13, ISSN 2214-9937, https://doi.org/10.1016/j.susmat.2015.02.001.

[3] Vagati, Alfredo, et al. "Design of ferrite-assisted synchronous reluctance machines robust toward demagnetization." IEEE Transactions on Industry Applications 50.3 (2013): 1768-1779.

[4] Du, Zhentao S., and Thomas A. Lipo. "Cost-effective high torque density bi-magnet machines utilizing rare earth and ferrite permanent magnets." IEEE Transactions on energy conversion 35.3 (2020): 1577-1584.

[5] Al-ani, M., et al. "Modifications to PM-assisted Synchronous Reluctance Machine to Achieve Rare-Earth Free Heavy-duty Traction." IEEE Journal of Emerging and Selected Topics in Power Electronics 11.2 (2022): 2029-2038.

[6] Hoang, Khoa D., et al. "Design, analysis and experimental evaluation of a novel high-speed high-power Ferrite IPM Machine for Traction Applications." IEEE Transactions on Industry Applications (2024). (early access)

[7] Liu, Xiangdong, et al. "Research on the performances and parameters of interior PMSM used for electric vehicles." IEEE Transactions on Industrial Electronics 63.6 (2016): 3533-3545.

[8] Jannot, Xavier, et al. "Multiphysic modeling of a high-speed interior permanent-magnet synchronous machine for a multiobjective optimal design." IEEE Transactions on Energy Conversion 26.2 (2010): 457-467.

[9] Binder, Josef, et al. "High-speed IPM motors with rotor sleeve: structural design and performance evaluation." 2023 IEEE Workshop on Electrical Machines Design, Control and Diagnosis (WEMDCD). IEEE, 2023.

[10] Olsen, L. E., et al. "Permanent magnet motor with wrapping." WO2021225902 (A1) (2021).

[11] Clauer, M., Binder, A., "Investigation of permanent magnet synchronous machines with buried magnets and carbon fiber sleeve for automotive application. " Elektrotech. Inftech. 140, 302–313 (2023). https://doi.org/10.1007/s00502-023-01131-7.

[12] Rahman, MD Rashedur, Choi, Seungdeog, Kim, Han-Gyu, "A high-speed rare-earth free IPM with carbon composite wrapped rotor." 2024 IEEE Energy Conversion Congress and Expo (ECCE). IEEE, 2024, in press.

A Novel Phase-Mode Controller for Resonant Converters

Claudio Adragna
Application Specific Products (ASP) Division
STMicroelectronics
Agrate Brianza, Italy
ORCID 0000-0003-1215-2143

Daniele Cazzaniga
Application Specific Products (ASP) Division
STMicroelectronics
Agrate Brianza, Italy
daniele.cazzaniga@st.com

Stefano Manzoni
Application Specific Products (ASP) Division
STMicroelectronics
Agrate Brianza, Italy
stefano.manzoni@st.com

Abstract— **This work proposes an analog implementation of phase-mode control (PMC) for resonant converters, suitable for integration into a low-cost silicon chip. The PMC method, which is based on controlling the displacement angle (phase shift) between the voltage applied to the resonant tank and the resonant current, has been long overlooked in the industry despite its significant benefits. Control robustness (insensitivity to component tolerance), intrinsic ZVS operation (the control ensures that the resonant current lags the applied square-wave voltage), fast transient response, potential for high-efficiency designs, and system simplification are very well documented in the literature.**

This work, after describing the proposed phase-mode controller and determining the related control law, discusses the challenges of its integration into a control IC. Although PMC is applicable to any voltage-fed resonant converter, in this work the LLC converter is considered for the large and small-signal analysis as well as for the experiments. The good performance of the proposed controller is demonstrated, and the expected benefits are confirmed.

Keywords—Resonant converters, LLC converters, phase-mode control, Power Factor control, self-sustained oscillation.

I. INTRODUCTION

Phase-mode control (PMC) is a type of current mode control for resonant converters with a fast transient response. It has two nested control loops: an inner loop that controls the phase shift Φ between the square wave voltage applied to the tank circuit and the resonant current so that it equals a reference value Φ_{ref}, and an outer loop that regulates the dc output voltage or current by providing the reference phase shift Φ_{ref} to the inner loop, as shown in the block diagram in Fig. 1.

Controlling the voltage-current phase shift offers several benefits: intrinsic zero-voltage switching (ZVS) operation, fast transient response, control robustness, and opportunity for high efficiency designs, as widely reported in the literature [1]-[8].

There are some additional practical benefits not previously highlighted. They stem from the fact that the phase shift Φ lies in the interval (0, $\pi/2$) rad, i.e., $(0, 90°)$, irrespective of the resonant tank configuration (LLC, LCC, series, parallel, etc.), resonant tank parameters and input-to-output voltage gain.

A first practical consequence of directly controlling Φ is that the control variable of the outer loop (Xc, then $\Phi_{ref} = f(Xc)$) will change within a predetermined range of values set by the controller, thus simplifying the setup of the outer control loop.

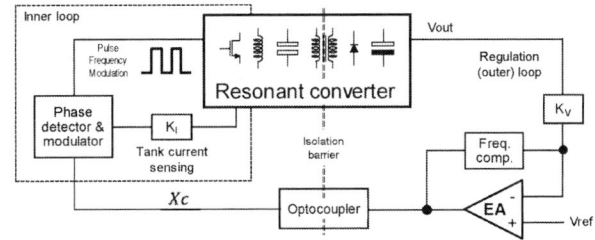

Fig. 1. Block diagram of a phase-mode controlled resonant converter.

A second, more significant consequence is the simple implementation of the burst-mode function to improve light load efficiency and meet standby requirements in energy-conscious applications such as PC power or ac-dc adapters and chargers.

In fact, as the load approaches zero, the resonant current will lag the applied voltage by $\pi/2$ rad. This makes the control variable Xc, near $\Phi = \pi/2$, closely related to the load and less to other quantities. As a result, an accurate load level for burst-mode onset can be set with minimal variation in mass production by comparing Xc to a reference, similar to PWM-controlled flyback converters. Compared to other control techniques, this simplifies the external circuitry needed to achieve the same goal.

The concept of controlling the voltage-current phase shift is almost as old as that of resonant conversion [1] and has been reconsidered over the years with different names: "Self-sustained Oscillation" [2]-[5], or "Power Factor Control" [6]-[7]. A control IC implementing a PLL-based PMC is disclosed in [8].

Self-sustained Oscillation has been applied to various resonant topologies: series, parallel, series-parallel and, more recently, LLC. An analog [2] controller has been proposed but has not achieved broad industrial success. Recently, Power Factor Control has been applied to the LLC converter. Reference [6] proposes a complex analog controller unsuitable for integration into a low-cost silicon chip, whereas [7] proposes a simpler digital implementation that still requires an MCU to be programmed.

The implementation of phase-mode control proposed in this work derives from the Time-shift Control (TSC) method [9]. It directly controls the phase shift Φ, unlike other controllers that control either the supplementary angle [2] or the complementary angle [7], and provides a linear control law ($\Phi_{ref} = K_c \, Xc$). Additionally, it can be easily integrated into a control chip.

979-8-3315-1612-3/25 $31.00 © 2025 IEEE

This paper is arranged as follows: Section II presents the proposed implementation of the phase-mode controller, derives the control law, explains how to set up the controller, and finally discusses challenges in implementing such a circuit in a low-cost analog integrated circuit; Section III addresses the steady-state behavior of LLC resonant converters with PMC while Section IV deals with their dynamic behavior; Section V shows and discusses the results of the bench evaluation of a prototype of a PMC-controlled LLC converter; finally, Section VI draws conclusions and touches on future work.

II. THE NOVEL PHASE-MODE CONTROLLER

A. Implementation and control law derivation

The proposed implementation of the phase-mode controller, whose basic schematic is shown in Fig. 2, is based on generating a two-slope ramp voltage across a capacitor C_T with a pair of current sources, Io and $k \cdot Ic$ (k is the transfer ratio of a current mirror). This ramp, synchronized to the toggling of the half/full bridge, is reset at the beginning of every switching half cycle by SW1, generating the sawtooth $V(C_T)$ shown in Fig. 3.

The outer control loop that regulates the dc output voltage (or current) of the converter sets the current Ic sourced by the reference voltage $Vref$ through pin FB. It is therefore $Xc = Ic$.

The instantaneous resonant current, represented by the voltage Vs (generated, e.g., across a sense resistor), is read by the comparator CO2 through pin CS to determine whether it is positive or negative. The two-slope ramp has a first slope in the initial part (Tz) of each switching half-period $Ts/2$, when the resonant current and the applied voltage $V(HB)$ have opposite signs, and a second steeper slope in the remaining part of the switching half-period ($Ts/2 - Tz$), when they have equal signs.

The first lower slope is associated to the current $k \cdot Ic$, which charges the capacitor C_T throughout the entire switching half-period; to obtain the second steeper slope, a fixed current Io, active only during the interval $Ts/2 - Tz$ when the output of the XNOR port is high, is added to $k \cdot Ic$.

With reference to the key waveforms shown in Fig. 3, the ramp peak voltage $V(C_T)_{pk}$ on the capacitor C_T is:

$$V(C_T)_{pk} = \frac{1}{C_T}\left[k \cdot Ic\frac{Ts}{2} + Io\left(\frac{Ts}{2} - Tz\right)\right]. \tag{1}$$

When $V(C_T)_{pk}$ equals the reference voltage Vp, the output of CO1 going high will toggle the outputs of the flip-flop FF, causing the half-bridge to toggle too, and $V(C_T)$ to be reset at zero by SW1. The logic gate AND1 enables this only during the interval $Ts/2 - Tz$, when the resonant current and $V(HB)$ have equal signs. In this way, the necessary condition for the converter to operate with ZVS is satisfied.

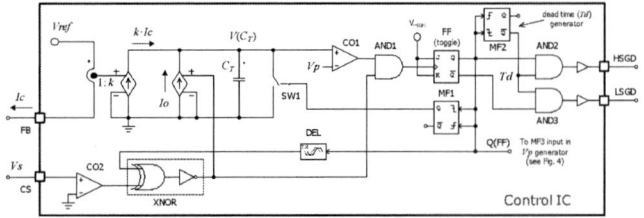

Fig. 2. Basic schematic of the proposed phase-mode controller.

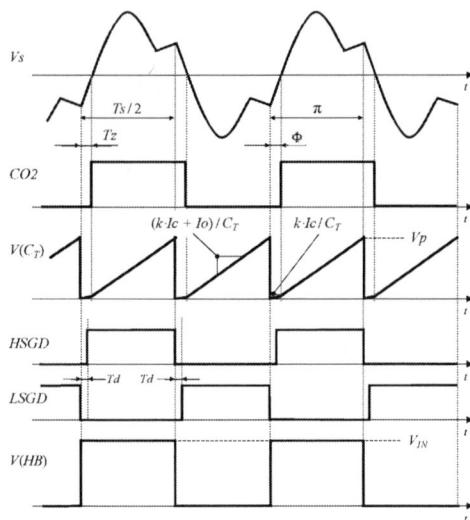

Fig. 3. Key waveforms of the circuit in Fig. 2.

Solving (1) for Ic we find the control law that links the control current to the controlled quantity. Since $V(C_T)_{pk} = Vp$:

$$Ic = \frac{1}{k}\left[\frac{Vp\, C_T}{Ts/2} - Io\left(1 - \frac{Tz}{Ts/2}\right)\right]. \tag{2}$$

The term $Tz/(Ts/2)$ is proportional to the voltage-current phase shift Φ. Expressing Φ in radians:

$$\Phi = \pi\frac{Tz}{Ts/2}. \tag{3}$$

The reference voltage Vp for CO1 in Fig. 2 is generated by the circuit shown in Fig. 4 (left) along with its key waveforms (right). The capacitor Cx is charged with a constant current Ix, generating a voltage ramp. Its peak value $Vx = Ix \cdot (Ts/2)/Cx$ is sampled and held on the edges of the signal Q of the toggle flip-flop FF in Fig. 2, $Q(FF)$, via the output pulses of the monostable MF3; right after sampling, the ramp is reset at zero by SW2.

The output voltage of the S/H circuit $Vp = Vx$ and reference voltage for CO1 is proportional to the switching half-period:

$$Vp = \frac{Ix}{Cx}\frac{Ts}{2}. \tag{4}$$

Substituting (4) in (2), and taking (3) into consideration:

$$Ic = \frac{1}{k}\left[Ix \cdot \frac{C_T}{Cx} - Io\left(1 - \frac{\Phi}{\pi}\right)\right]. \tag{5}$$

Finally, the circuit of Fig. 4 will be designed in such a way that $Ix\, C_T/Cx = Io$. Therefore (5) becomes:

$$Ic = \frac{Io}{k}\frac{\Phi}{\pi} \rightarrow \Phi_{ref} = \frac{k\pi}{Io}Ic. \tag{6}$$

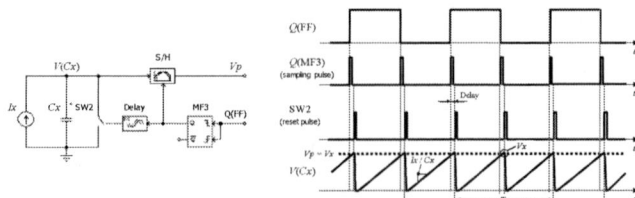

Fig. 4. Vp reference generator (left-hand side) and its key waveforms (right-hand side).

979-8-3315-1612-3/25 $31.00 © 2025 IEEE

Notice that (6) is consistent with the operation of a typical optocoupler-based feedback loop, where the feedback current is at a minimum when the converter outputs the maximum power with the minimum input voltage and at a maximum when the converter outputs the minimum power with the maximum input voltage. Also notice that the analysis that resulted in the control law (6) does not refer to a specific resonant topology.

Since $\Phi \in (0, \pi/2)$, it follows that $Ic \in (0, Io/2k)$, and this holds regardless of the resonant tank configuration (LLC, LCC, series, parallel, etc.), resonant tank parameters and input-to-output voltage gain. Therefore, it is reasonably expected that Φ and Ic are much less sensitive to the tolerance of the resonant tank parameters than the switching frequency f_{sw} when directly controlled (direct frequency control, DFC). This is confirmed, indirectly, in [3], which shows that a ±20% tolerance in L_S e C_R changes the input-to-output gain by up to 62.5% when f_{sw} is controlled and by only 4.7% when Φ is controlled. Therefore, unlike the traditional DFC, PMC exhibits robustness.

B. Phase-mode controller setup

When $Ic = 0$, the commanded Φ_{ref} is zero too. Therefore, if no current is sunk from $Vref$, the converter will work at the boundary between inductive and capacitive modes, an operating region not recommended because ZVS operation is not possible there. Defining a minimum Ic, Ic_{min}, will ensure a minimum phase shift Φ_{min} even when the optocoupler sinks no current. Φ_{min} will be chosen to ensure ZVS operation based on steady-state analysis. Ic_{min}, and Φ_{min}, can be set by connecting a resistor R_0 between the pin FB and ground, as shown in Fig. 5.

At start up, no current is sunk from the optocoupler, but the initial reference phase shift must be close to $\pi/2$ rad to prevent excessive currents, and then progressively reach the steady state under control by the outer feedback loop (soft start).

As depicted in Fig. 5, this can be done by connecting an RC series circuit in parallel to R_0, with the resistor value such that it commands $\Phi_{ref} = \pi/2$ rad when the capacitor is initially discharged. The capacitor is selected so that it is completely charged in Tss (in the ms), allowing the resonant current to increase smoothly until the feedback loop takes over, after which the resistor no longer affects the operation of the loop.

C. Integration challenges and phase errors

The accuracy of (6) must be carefully considered. It depends on the matching of C_T/Cx and of Io/Ix, as well as on the accuracy of Io and k. The target accuracy can be achieved, at IC design level, with ordinary matching and trimming techniques.

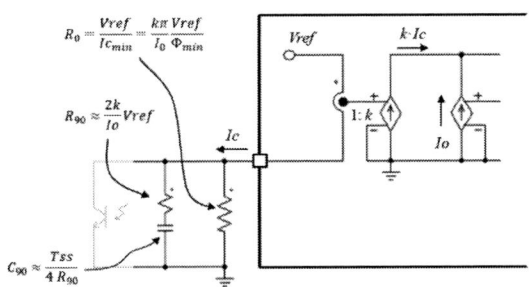

Fig. 5. Setup of the proposed phase-mode controller.

The sample-and-hold circuit that provides Vp is supposed to employ an open-loop architecture and does not pose special challenges other than the typical ones (charge injection and hold step minimization, droop rate and feedthrough minimization, etc.). The acquisition and settling times are not critical at all.

The biggest source of phase error is the delay between the internal signals used to compute the phase shift and the external waveforms applied to the resonant tank. Notice that the internal Tz starts when $Q(FF)$ changes state, whereas the real Tz starts when $V(HB)$ begins its swing. $V(HB)$ lags $Q(FF)$ due to the propagation delays of the logic chain and the gate driver, as well as the turn-off delay of the power switches and the delay time of $V(HB)$. Notice that the propagation delays of CO1 and CO2 tend to cancel each other.

The internal delay contributions can be estimated quite accurately, while the external contributions cannot, but an educated guess can be made. The purpose of the delay block DEL in Fig. 2 is to compensate for all these contributions. There will be a residual error depending on the difference between the actual overall delay and the internally added one, which will reasonably be significantly smaller than the uncompensated one.

III. STEADY-STATE BEHAVIOR OF A PHASE-MODE CONTROLLED LLC RESONANT CONVERTER

The First-Harmonic Approximation (FHA) method [10] can provide insight into the relationship between the operating conditions of the converter (input voltage V_{IN} and output current I_{OUT}) and the associated phase shift Φ.

In the FHA framework, Φ is the argument of the input impedance $Z_{IN}(j\omega)$ of the tank circuit. Due to the approximate nature of FHA, the numerical results will be rather inaccurate, especially when working below resonance (@ $V_{IN} < V_{INnom}$), but the trends are expected to be well represented.

Fig. 6 shows how Φ changes with the quality factor Q of the tank circuit (Q is proportional to I_{OUT}) for different V_{IN} values in the exemplary LLC converter specified in Table I. Q_{max} is the Q value that makes $\Phi = \angle Z_{IN}(j\omega) = 0$ @ $V_{IN} = V_{INmin}$.

Notice that all the curves converge to 90° ($\pi/2$ rad) as Q, i.e., I_{OUT}, tends to zero. Except when operating well above resonance at $V_{IN} = 450$ V, all curves are very close to one another for $Q < Q_{max}/2$ and nearly overlap for $Q < Q_{max}/10$. In that region, Φ and Ic are also little sensitive to the input voltage.

TABLE I. ELECTRICAL SPEC AND CHARACTERISTICS OF THE EXEMPLARY LLC HALF BRIDGE CONVERTER USED IN SIMULATIONS.

Parameter	Symbol	Value	Unit
Input voltage range	$V_{INmin} - V_{INmax}$	300 - 450	V
Nominal input voltage	V_{INnom}	400	V
Regulated output voltage	V_{OUT}	12	V
Maximum dc output current	I_{OUTmax}	20	A
Primary-to-secondary turns ratio	n	16:1	---
Parallel resonant inductance	L_P	560	μH
Series resonant inductance	L_S	105	μH
Resonant capacitance	C_R	39	nF
Maximum quality factor	Q_{max}	0.3146	---
Output capacitance	C_{OUT}	4000	μF
ESR of output capacitance	R_c	10	mΩ

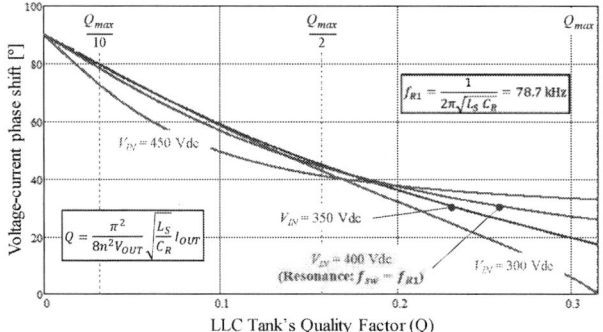

Fig. 6. Φ vs. Q ($\propto I_{OUT}$) for different V_{IN} values in the exemplary LLC converter specified in Table I (Mathcad® computations).

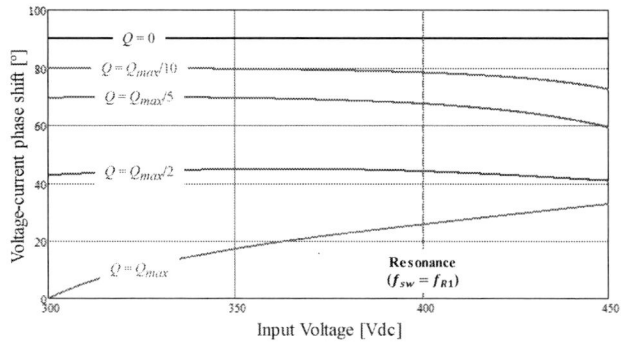

Fig. 7. Φ vs. V_{IN} for different Q ($\propto I_{OUT}$) values in the exemplary LLC converter specified in Table I (Mathcad® computations).

Fig. 7 shows how Φ changes with the input voltage V_{IN} for different values of Q (i.e., of I_{OUT}) in the same LLC converter.

The phase shift is maximally sensitive to the input voltage when $Q = Q_{max}$. This dependence becomes smaller in the region around $Q_{max}/2$, slightly increases in the above resonance region at lighter loads ($Q_{max}/5$, $Q_{max}/10$) and tends to vanish as the load approaches zero.

These features can be exploited to set an accurate burst-mode threshold Φ_{BM} by comparing Ic to a current reference value $Ic_{BM} = \Phi_{BM}Io/k\pi$, stopping the converter if $Ic > Ic_{BM}$, and letting it operate if $Ic < Ic_{BM}$, in the same simple way and with the same accuracy as in PWM-controlled flyback converters.

This is shown in the plots of Fig. 8, which illustrate the Φ vs. I_{OUT} relationship found through PSIM® simulation of the converter specified in Table I. Setting $\Phi_{BM} = 80°$, the typical burst-mode entry threshold is located at 8.3% of the rated load (i.e., 1.66 A) at the nominal voltage $V_{IN} = 400$ Vdc; this threshold changes from 6.8% (= 1.36 A) @ $V_{IN} = 300$ Vdc to 10% of the rated load (= 2 A) @ $V_{IN} = 450$ Vdc. Normally, however, the LLC converter is powered by a power factor corrector (PFC) pre-regulator, so the input voltage is essentially fixed at the nominal value and its tolerance (a few %) has negligible impact.

Fig. 9 shows the effect of the tolerance of the components of the resonant tank on the Φ vs. I_{OUT} relationship in the light load region where the burst-mode threshold is typically located: the burst-mode threshold is expected to be contained in the interval (7.7% – 9.3%) of the rated load due to those tolerances.

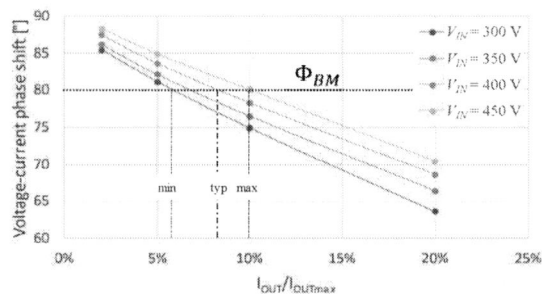

Fig. 8. Φ vs. normalized I_{OUT} for different V_{IN} values in the light load region in the LLC converter specified in Table I (PSIM® simulations results).

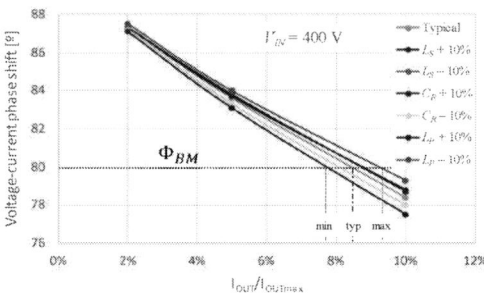

Fig. 9. Φ vs. normalized I_{OUT} @ $V_{IN} = 400$ Vdc in the light load region in the LLC converter specified in Table I considering a ±10% tolerance in the parameters of the LLC tank (PSIM® simulations results).

IV. DYNAMIC BEHAVIOR OF A PHASE-MODE CONTROLLED LLC RESONANT CONVERTER

Regarding the dynamic behavior of an LLC converter with PMC, the "Power Factor Control" method proposed in [6] shows an interesting analogy with a voltage-mode controlled buck converter operated in the Continuous Conduction Mode (CCM), with the power factor of the input current (i.e., cos Φ) playing the role of the duty cycle D.

This analogy and its consequences are valid under some assumptions that, unfortunately, are not generally satisfied. To obtain a more comprehensive picture, the dynamic properties of a phase-mode controlled LLC converter have been explored by means of simulations.

Specifically, the Bode plots of the control-to-output transfer function of the exemplary LLC converter specified in Table I have been derived using SIMPLIS®. Fig. 10 shows the controller setup used for the simulations, and Table II lists the parameters of the controller.

The Bode plots of the control-to-output transfer function $G(j\omega) = \hat{v}_{out}/\hat{v}_{FB}$ have been determined under the following operating conditions:

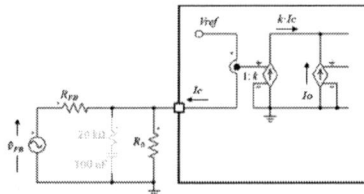

Fig. 10. Setup used for SIMPLIS® simulations of the LLC converter specified in Table I with the controller defined in Table II.

979-8-3315-1612-3/25 $31.00 © 2025 IEEE

- V_{IN} = 300 Vdc, 400 Vdc, 450 Vdc;
- I_{OUT} = 100%, 50%, and 10% of I_{OUTmax}.

The simulations results are shown in Figs. 11, 12 and 13.

TABLE II.   PARAMETERS OF THE PHASE-MODE CONTROLLER IN FIG. 10.

Parameter	Symbol	Value	Unit
Reference voltage	V_{ref}	2	V
Current mirror transfer ratio	k	1/30	A/A
Internal current generator	Io	6	μA
Minimum phase shift resistor	R_0	180	kΩ
Feedback resistor	R_{FB}	18	kΩ

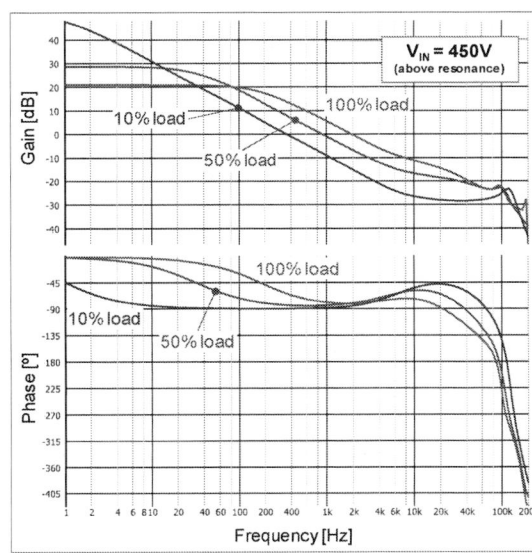

Fig. 13. Bode plots of the control-to-output transfer function $\hat{v}_{out}/\hat{v}_{FB}$ of the LLC converter specified in Table I with the phase-mode controller defined in Table II, @ V_{IN} = 450 Vdc and with different load conditions.

When operated at resonance (the typical operating condition by design [10]) or above resonance at full load, the system has a single-pole characteristic in a quite broad frequency range, above $f_{R1}/10$. Below resonance, this single-pole characteristic is maintained in a narrower range, about one fourth, after that both gain and phase start rolling off more steeply, presumably due to a low-Q pole pair. This pole pair, whose signs are still barely visible in Figs. 12 and 13, at lower loads shifts to higher frequency, their Q further lessens, and their effect disappears.

With a fixed load, the crossover frequency f_c does not change much with V_{IN} and is essentially the same at V_{IN} = 300 Vdc and 450 Vdc. However, at V_{IN} = 300 Vdc and 100% load the phase is larger and steeper, representing the worst-case condition for the design of the output-to-control transfer function that closes the outer loop. As the load is reduced, f_c decreases as well.

The next investigation concerns the response of the inner phase shift control loop. This analysis, carried out with PSIM® simulations, provides the intrinsic response speed of the control method, unaffected by how the outer loop is compensated, and can be used to compare different methods.

Fig. 14 shows the open-loop dynamic behavior of the converter specified in Table I. The output voltage V_{OUT} is fixed at the regulated value and the input voltage V_{IN} is set at the nominal value. A step change is applied to the control current I_C such that the output current changes from 50% to 100% of I_{OUTmax} and vice versa. In fact, with this setup the converter can be regarded as a programmed current source.

Notice that the currents through the secondary rectifiers (red and blue traces) reach the new steady-state in a few cycles and that the response is faster when going from 100% to 50% load. This is expected, as the bandwidth larger at 100% load. Also, notice that the dc output current I_{OUT} (dark green trace) is obtained by filtering the secondary currents with a 2nd order low-pass filter (BW = 25 kHz, ζ = 1), so its evolution in time in the 100% to 50% faster transition is affected by the filter's response.

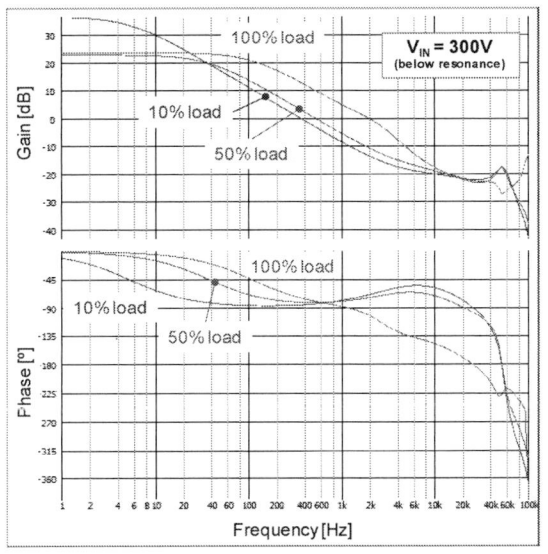

Fig. 11. Bode plots of the control-to-output transfer function $\hat{v}_{out}/\hat{v}_{FB}$ of the LLC converter specified in Table I with the phase-mode controller defined in Table II, @ V_{IN} = 300 Vdc and with different load conditions.

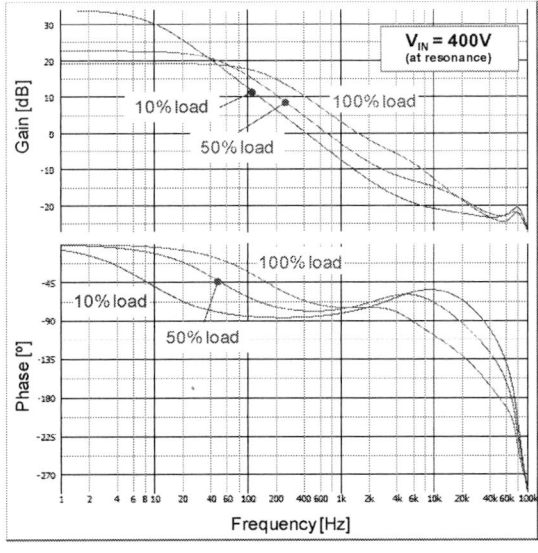

Fig. 12. Bode plots of the control-to-output transfer function $\hat{v}_{out}/\hat{v}_{FB}$ of the LLC converter specified in Table I with the phase-mode controller defined in Table II, @ V_{IN} = 400 Vdc and with different load conditions.

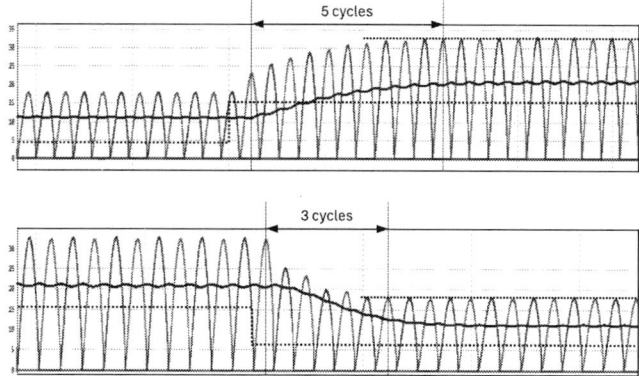

Fig. 14. Open-loop step response for the converter specified in Table I with PMC. Control step-change resulting in a 50% ↔ 100% load change. Conditions: V_{OUT} = 12 V, V_{IN} = 400 V. Red and blue traces: secondary currents; dark green trace: dc output current (I_{OUT}); dotted black line: control signal (I_c).

To appreciate the improvement of PMC over the traditional DFC, the open-loop step response in Fig. 14 can be compared to that of the same converter controlled with DFC.

The result of this simulation is depicted in Fig. 15. The DFC response is significantly slower than that of PMC: it takes 25 cycles for the secondary currents to reach the steady state, 5 times more in the 50% → 100% transition, 8+ times more in the 100% → 50% transition.

It is also interesting to compare the open-loop step response in Fig. 14 to that obtained in the same converter with other control methods that have been developed in recent years to enhance the dynamic properties [11].

Fig. 16 shows the open-loop step response of the LLC converter specified in Table I with Time-shift Control (TSC) [9], while Fig. 17 shows the same with Bang-bang Charge Control (BBCC) [12]. The response of PMC and TSC are almost the same; PMC response is very similar to that of BCC as well.

The conclusion of this comparison is that with PMC the dynamic performance of an LLC converter is essentially on par with other fast-dynamics control methodologies.

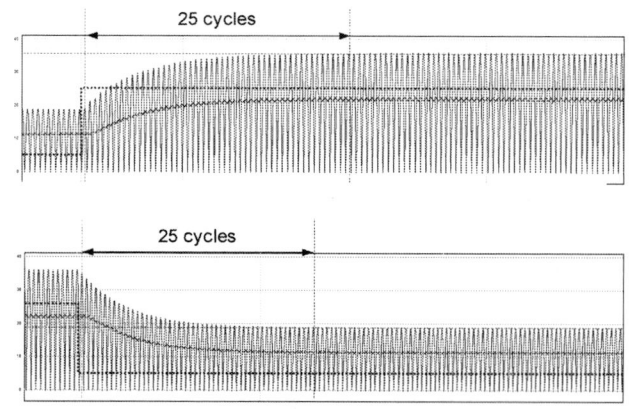

Fig. 15. Open-loop step response for the converter specified in Table I with DFC. Control step-change resulting in a 50% ↔ 100% load change. Conditions: V_{OUT} = 12 V, V_{IN} = 400 V. Red and blue traces: secondary currents; green trace: dc output current (I_{OUT}); dotted black line: control signal (I_c).

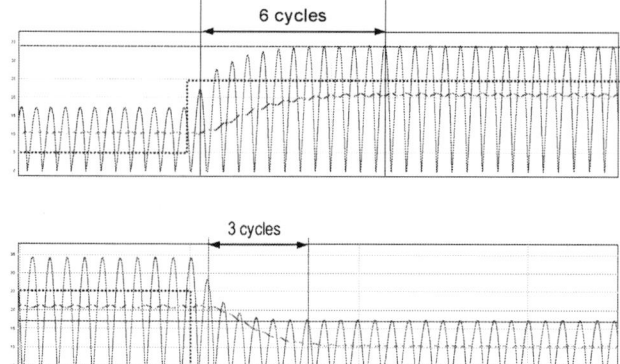

Fig. 16. Open-loop step response for the converter specified in Table I with TSC. Control step-change resulting in a 50% ↔ 100% load change. Conditions: V_{OUT} = 12 V, V_{IN} = 400 V. Red and blue traces: secondary currents; green trace: dc output current (I_{OUT}); dotted black line: control signal (I_c).

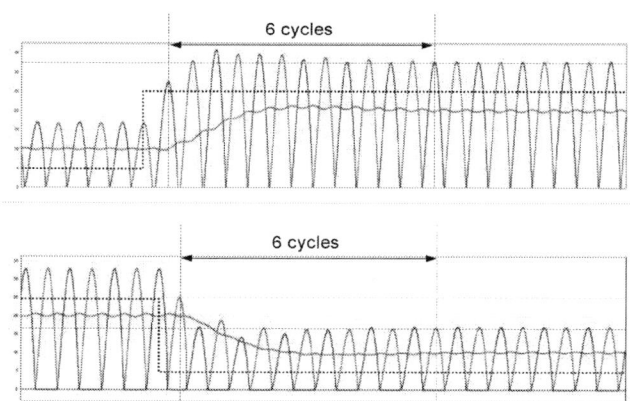

Fig. 17. Open-loop step response for the converter specified in Table I with BBCC. Control step-change resulting in a 50% ↔ 100% load change. Conditions: V_{OUT} = 12 V, V_{IN} = 400 V. Red and blue traces: secondary currents; green trace: dc output current (I_{OUT}); dotted black line: control signal (I_c).

V. EXPERIMENTAL RESULTS

The experiments described in this section have been carried out on the LLC resonant converter shown in Fig. 18 and whose electrical specification is given in Table III.

The objective of the experiments was to check the static and dynamic behavior of the converter with the proposed controller.

TABLE III. ELECTRICAL SPECIFICATION AND MAIN CHARACTERISTICS OF THE LLC HALF BRIDGE CONVERTER USED FOR THE EXPERIMENTS.

Parameter	Symbol	Value	Unit
Input voltage range	$V_{INmin} - V_{INmax}$	300 − 450	V
Nominal input voltage	V_{INnom}	400	V
Regulated output voltage	V_{OUT}	24	V
Maximum dc output current	I_{OUTmax}	10.4	A
Primary-to-secondary turns ratio	n	10	---
Parallel resonant inductance	L_P	170	μH
Series resonant inductance	L_S	40	μH
Resonant capacitance	C_R	2 × 4.7	nF
Output capacitance	C_{OUT}	3 × 470	μF
ESR of output capacitance	R_c	28 / 3	mΩ

979-8-3315-1612-3/25 $31.00 © 2025 IEEE

Fig. 18. Prototype of the LLC converter specified in Table III.

The diagrams depicted in Fig. 19 show the static relationship between the commanded phase shift and the control current Ic, compared to the theoretical control law (6).

The experimental data are in good agreement with (6), especially @ V_{IN} = 300 and 350 V and low Ic. At higher V_{IN} and larger Ic values, where the switching frequency is higher, they slightly deviate from (6). Reasonably, this is due to the greater impact of the errors described at point C in section II as the switching frequency gets higher.

Notice that Φ exceeds 90° at light load and V_{IN} = 450 V. This is supposedly due to nonidealities in the power circuit that would need further investigations to be fully understood.

Fig. 20 shows the Φ vs. I_{OUT} relationship for different V_{IN} values. The general trend resembles the theoretical one derived with the FHA approach and reported in Fig. 6. In the light load region, despite some irregularities presumably due to the high frequency noise that was observed on the current signal, the experimental data are not much different from those found by simulations and shown in Fig. 8.

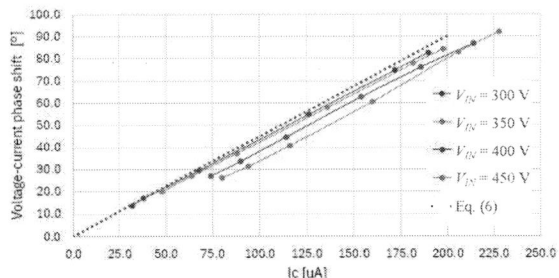

Fig. 19. Comparison of theoretical control law (6) and experimental values of Φ vs. I_C relationship for the converter specified in Table III.

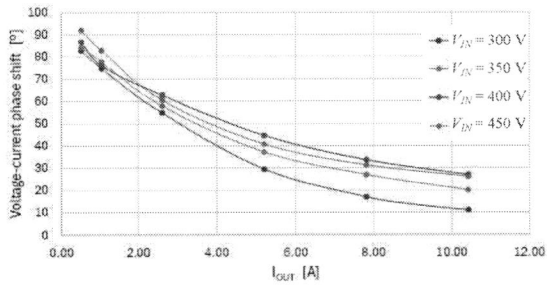

Fig. 20. Φ vs. I_{OUT} for different V_{IN} values in the LLC converter specified in Table III (experimental values).

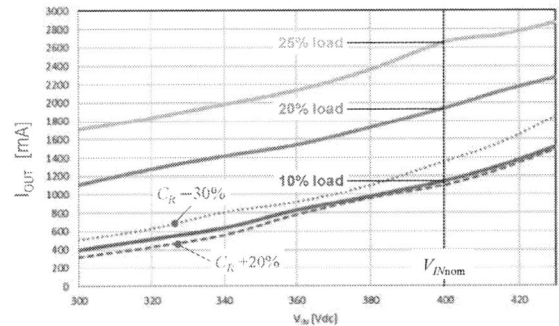

Fig. 21. Dependance of the load level for onset of burst-mode operation on the input voltage for different settings in the LLC converter specified in Table III.

Fig. 21 shows the threshold of burst-mode operation as a function of the input voltage for different settings. The data indicate quite a marked dependance on V_{IN}, which is expected based on the results shown in Fig. 20. However, this is not generally an issue: in most practical cases the LLC converter is powered by a PFC pre-regulator with a regulated voltage. Also, notice in Fig. 21 that the threshold does not change much even with a very large variation in the value of the resonant capacitor C_R (−30%, +20%). This confirms the robustness of PMC.

The dynamic behavior of the converter has been firstly analyzed in open loop to assess the intrinsic response speed of PMC. Again, a step change is applied to the control current I_C such that the output current changes from 50% to 100% of I_{OUTmax} and vice versa, with the output voltage V_{OUT} fixed at the regulated value and the input voltage V_{IN} at the nominal value.

The results are shown in the oscilloscope pictures of Fig. 22. PMC exhibits a very fast response in this case too, not much different from that found by simulations in the system defined in Table I and that is shown in Fig. 14.

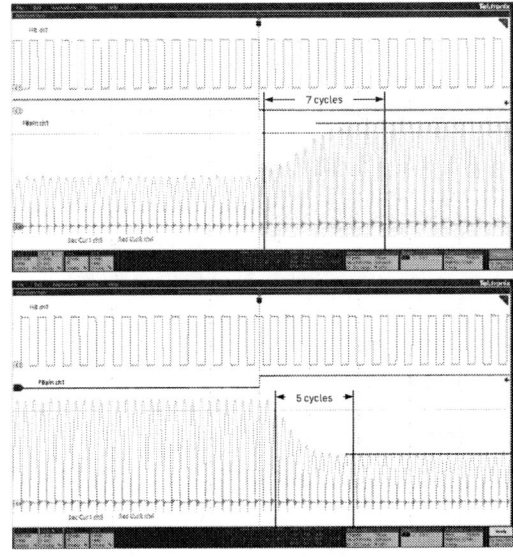

Fig. 22. Open-loop step response for the converter specified in Table III with PMC. Control step-change resulting in a 50% ↔ 100% load change. Conditions: V_{OUT} = 24 V, V_{IN} = 400 V. Orange and green traces: secondary currents; carmine trace: half-bridge midpoint; black trace: control signal (Ic).

979-8-3315-1612-3/25 $31.00 © 2025 IEEE

The closed-loop transient response has been checked under the same conditions (step-load change from 50% to 100% of I_{OUTmax} and vice versa). The results are shown in Fig. 23. The response is well damped; the overshoot and the undershoot in V_{OUT} do not exceed 1.2% of the regulated value.

Finally, the closed-loop PSRR (Power supply rejection ratio) of the converter has been checked to assess the residual voltage ripple on V_{OUT} caused by a 100 Hz sinusoidal voltage ripple superimposed on the dc input voltage. This is the case when the LLC converter is powered by a PFC pre-regulator.

The test has been performed with the converter fully loaded and a 100 Hz, 15 V rms (21 V peak) sinusoidal voltage on top of 400 V dc input voltage. The resulting low-frequency output ripple was completely obscured by the ripple at the switching frequency, so its amplitude was measured by FFT analysis.

Fig. 24 shows the result: 12 mV rms, which means a PSRR close to 62 dB, comparable to that in a discontinuous conduction mode flyback converter with current-mode control. This PSRR is approximately 20 dB more than the values typically attainable with DFC, essentially because of the larger 100 Hz gain of the output-to-control transfer function achievable with PMC.

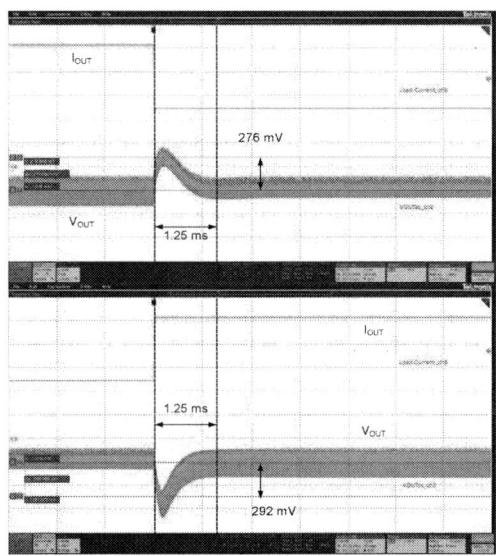

Fig. 23. Closed-loop step response for the converter specified in Table III with PMC. Load transient 50% ↔ 100% at V_{IN} = 400 V.

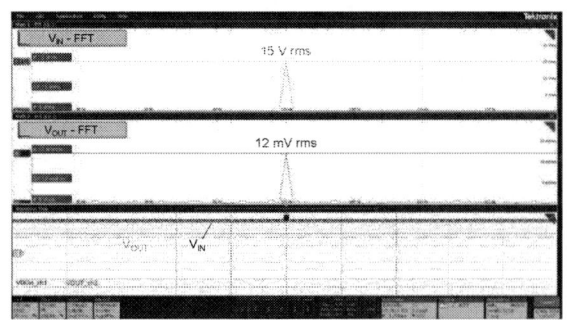

Fig. 24. FFT of V_{IN} and V_{OUT} to evaluate the PSRR at 100 Hz of the converter specified in Table III. V_{IN} = 400 Vdc + 15 Vrms, 100% load.

VI. CONCLUSIONS AND FUTURE WORK

A novel analog phase-mode controller for resonant converters has been proposed, its operating principle illustrated, and the challenges of its implementation in an integrated control circuit discussed. Its large and small signal behavior has been analyzed.

The main benefits – intrinsic ZVS operation, fast dynamics, control robustness, opportunity for high efficiency designs, light load monitoring through the control variable of the outer feedback loop, excellent closed-loop PSRR, and overall system simplification – have been theoretically analyzed and validated by simulations and experiments with an LLC resonant converter.

The experiments have shown some irregularities in the Φ vs. I_{OUT} relationship, and Φ exceeding 90° at light load. Although not bringing significant practical consequences, these anomalies will be the topic of further investigations to identify their origin.

The proposed phase-mode controller has been realized in a test chip that has been used for the experiments, and is planned to be soon integrated into a commercial product.

REFERENCES

[1] F. C. Schwarz, "An improved method of resonant current pulse modulation for power converters," 1975 IEEE Power Electronics Specialists Conference, Culver City, CA, USA, 1975, pp. 194-204.

[2] H. Pinheiro, P. Jain and G. Joos, "Self-sustained oscillating resonant converters operating above the resonant frequency," Proceedings of APEC 97 - Applied Power Electronics Conference, Atlanta, GA, USA, 1997, pp. 993-999 vol.2.

[3] M. Z. Youssef, P. K. Jain, "A Front End Self-Sustained LLC Resonant Converter," 2004 IEEE Power Electronics Specialists Conference, Aachen, Germany, 2004, pp. 2651-2656.

[4] M. B. Vandishi, A. Lahooti Eshkevari and A. Salemnia, "A New Neutral-Point Clamped Based LLC Converter with Self-Sustained Resonant Controller," 2019 International Power System Conference (PSC), Tehran, Iran, 2019, pp. 434-440.

[5] R. Bonache-Samaniego, C. Olalla, H. Valderrama-Blavi, and L. Martinez-Salamero. "Analysis and Design of Self-Oscillating Resonant Converters with Loss-Free Resistor Characteristics". Energies 2020, 13(14): 3743. [Online]. Available: https://www.mdpi.com. Accessed: Dec. 9, 2024.

[6] K. Yamada, K. Umetani, E. Hiraki and M. Ishihara, "Phase-Shift Based on Power Factor Control for LLC Converter with High Output Stability Against Load Fluctuation," 2022 IEEE 7th Southern Power Electronics Conference (SPEC), Nadi, Fiji, 2022, pp. 1-6.

[7] T. Zaitsu, Y. Yoshimura, K. Umetani, M. Ishihara, E. Hiraki and K. Horii, "Compact Hardware Implementation of Power Factor Control for LLC converter with Event-Driven-Timer Based Digital Controller," 2024 IEEE Applied Power Electronics Conference and Exposition (APEC), Long Beach, CA, USA, 2024, pp. 2015-2020.

[8] T. J. Ribarich, "Resonant Converter with Phase Delay Control," U.S. Patent 6,903,949 B2, June 7, 2005. [Online]. Available: https://worldwide.espacenet.com. [Accessed: Dec. 9, 2024].

[9] C. Adragna, "Time-shift control of LLC resonant converters," PCIM Europe 2010 Proceedings, Paper #113, pp. 661- 666, May 2010.

[10] S. De Simone, "LLC resonant half-bridge converter design guideline," AN2450, Rev 6, STMicroelectronics Application Note, 2014. [Online]. Available: https://www.st.com. [Accessed: Dec. 9, 2024].

[11] C. Basso, "Modern Control Methods for LLC Converters Simplify Compensator Design," How To Power Today, Exclusive Technology Feature, Dec. 2021. [Online]. Available: https://www.How2Power.com. [Accessed: Dec. 9, 2024].

[12] Z. Y. Hu, Y. F. Liu, and P. C. Sen, "Bang-bang charge control for LLC resonant converters," IEEE Transactions on Power Electronics, vol. 30, no. 2, pp. 1093-1108, Feb. 2015

A Regulated 36V-60V-Input V_{IN}-Insensitive Resonant Switched-Capacitor Converter with Large Voltage Conversion Ratio

Yichao Ji, Jingyi Yuan, Lin Cheng
School of Microelectronics
University of Science and Technology of China, Hefei, China
Email: jyc123@mail.ustc.edu.cn, yjy913@mail.ustc.edu.cn, eecheng@ustc.edu.cn

Abstract—This paper presents an input-voltage-insensitive regulated resonant switched-capacitor converter for converting 36 V-60 V to 3.3 V or 5 V. By introducing a regulating stage to decouple input voltage from resonant tanks, the resonant capacitor voltages are determined solely by the fixed output voltage. This design avoids resonant frequency variation caused by input voltage changes when applying Class II capacitors that offer higher capacitance but suffer significant capacitance derating due to voltage bias. Meanwhile, the voltage difference between two resonant capacitors is greatly reduced, resulting in only 4% capacitance derating difference, which ensures a better match between the resonant tanks. Unlike using Class I ceramic capacitors that have stable capacitance but lower energy density, this approach achieves higher power density. Compared with the conventional two-stage architecture (a switched tank converter and a regulated buck converter), this converter realizes a larger voltage conversion ratio without adding more passive components. A prototype using only Class II capacitors was designed with a constant resonant frequency of 340 kHz, immune to input voltage or capacitance derating. Measurement results show that the converter achieves a peak efficiency of 92.6% for 36 V-to-5 V conversion and a power density of 567 W/in³.

Keywords—*Capacitance derating, large voltage conversion ratio, resonant switched-capacitor converter, V_{IN}-insensitive, wide input voltage.*

I. INTRODUCTION

In order to reduce copper loss and accommodate higher power demands of modern CPUs/GPUs, data centers have increased their DC distribution bus voltage from 12 V to 48 V. By introducing a 12-V intermediate DC bus voltage with an unregulated 48 V-to-12 V fixed voltage-conversion-ratio (VCR) converter, a smooth transition from the original 12-V DC bus to the modern 48-V DC bus can be facilitated [1]. Recently, there has been a shift toward lowering the intermediate DC bus voltage to 3.3 V or 5 V to ease the requirements for the final conversion stage, allowing the implementation of high-frequency fully integrated voltage regulators (FIVRs) located on the processor substrate for enhanced performance [2], [3], [4].

Resonant switched-capacitor (ReSC) converters (e.g., switched tank converter, STC) have demonstrated great potential in achieving high efficiency for 48 V to 12 V due to their soft charging of capacitors and soft switching of power switches, which offer advantages over pure switched-inductor or switched-capacitor converters [5], [6]. However, to achieve a

This work was supported by the National Natural Science Foundation of China under Grant 92373203.

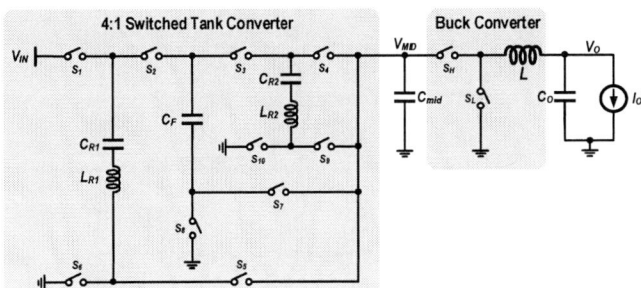

Fig. 1. Conventional two-stage architecture adopting switched tank converter and buck converter.

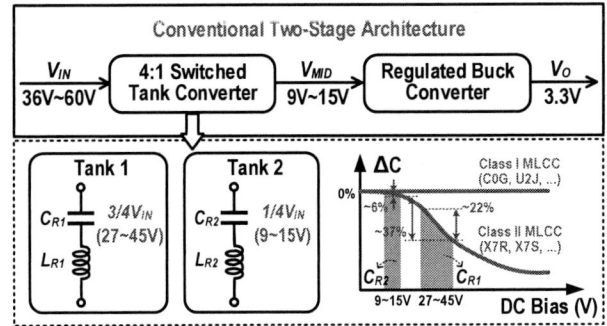

Fig. 2. Issues in a conventional two-stage architecture when applying Class II MLCCs and considering wide V_{IN} and mismatch between resonant tanks.

further step-down conversion from a 48-V rail, additional power switches and passive components are required, compromising both efficiency and power density. Hence, an inductive buck converter is typically employed as a second stage to regulate the output voltage (V_O) [7], as shown in Fig. 1.

Class I ceramic capacitors (C0G, U2J, etc.) are commonly adopted in ReSC converters due to their stable capacitance, low tolerance, and immunity to voltage bias [6]-[9]. These properties support tight capacitor matching and stable resonant operation. However, their low dielectric constant results in limited capacitance [6], necessitating the use of more capacitors and constraining power density. In contrast, cost-effective Class II ceramic capacitors (X7R, X7S, etc.) can offer higher capacitance, making them particularly favored in applications requiring high volumetric efficiency [10], [11]. Nevertheless, their performance is challenged by voltage bias effects throughout the entire input voltage (V_{IN}) range (typically

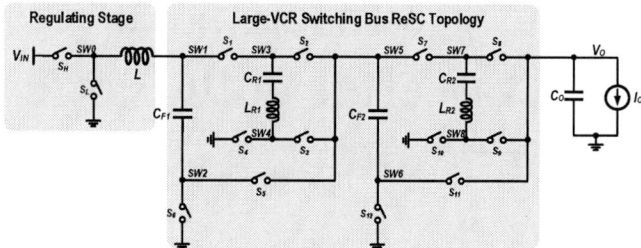

Fig. 3. The proposed regulated V_{IN}-insensitive ReSC converter.

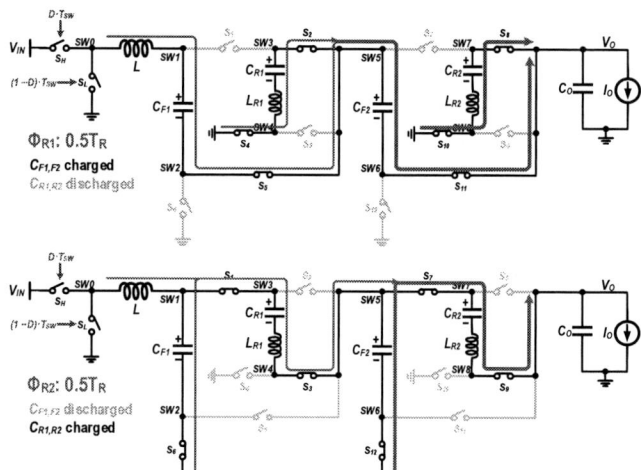

Fig. 4. Working principle of the proposed converter.

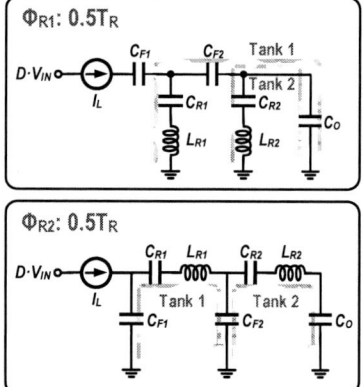

Fig. 5. Equivalent circuit of the proposed converter.

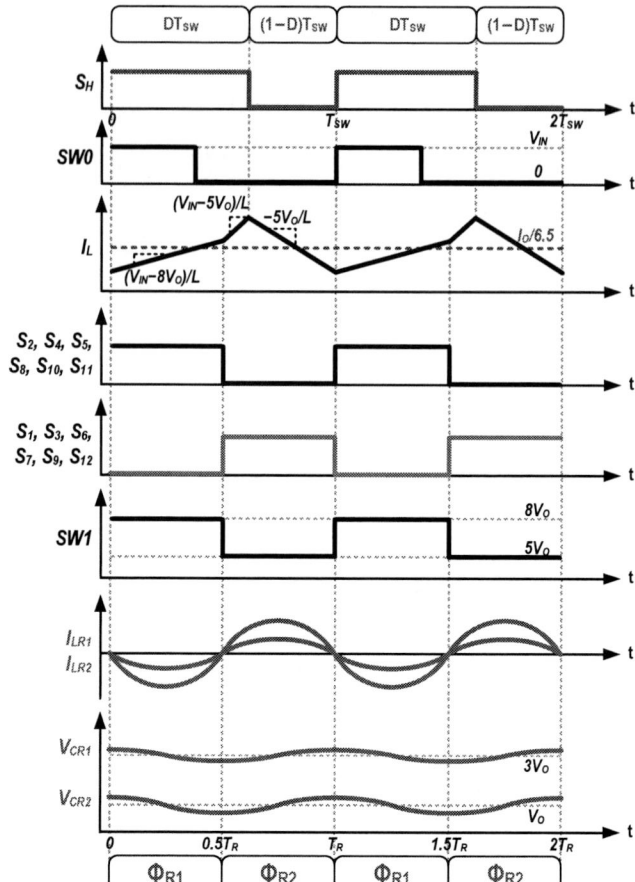

Fig. 6. Key waveforms of the proposed converter.

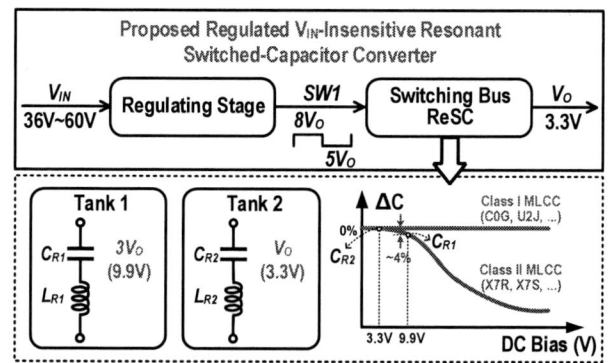

Fig. 7. Small capacitance difference in the proposed regulated V_{IN}-insensitive ReSC converter.

ranging from 36 V to 60 V). As depicted in Fig. 2, these capacitors can experience up to 22% derating in a single capacitor and a 37% derating difference between capacitors in different resonant tanks. This results in a significant change in resonant frequency (f_r) and mismatched resonant tanks, which lead to hard turn-off conditions for power switches and low power efficiency, and thus a complex controller achieving accurate zero-current switching (ZCS) is required, which will significantly increase costs.

In this work, we propose a regulated V_{IN}-insensitive ReSC converter with a large VCR. This converter alleviates resonant frequency variation and resonant tank mismatch caused by wide V_{IN} range and capacitance derating when only using Class-II ceramic capacitors. As a result, both high efficiency and high power density are achieved.

II. PROPOSED REGULATED ReSC CONVERTER

Fig. 3 shows the proposed regulated V_{IN}-insensitive ReSC converter, which comprises a regulating stage and a large-VCR switching bus ReSC topology. The regulating stage is responsible for final V_O regulation, utilizing two high-voltage (HV) devices to handle a wide V_{IN} range. Instead of employing a fixed 4:1 switched tank converter as a separate stage with an intermediate DC bus capacitor providing its input voltage, the proposed design introduces a switching bus ReSC topology. In this topology, SW1 alternates between two voltage levels rather than maintaining a fixed DC voltage. Compared with the

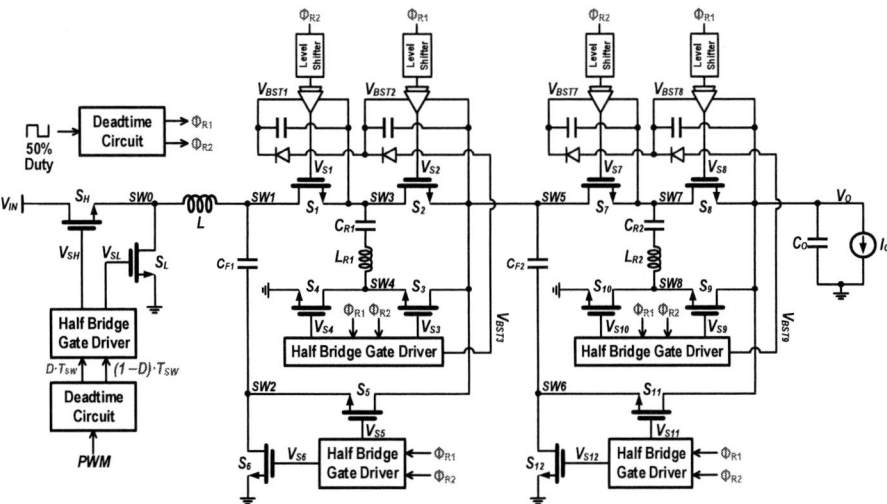

Fig. 8. System architecture of the proposed converter.

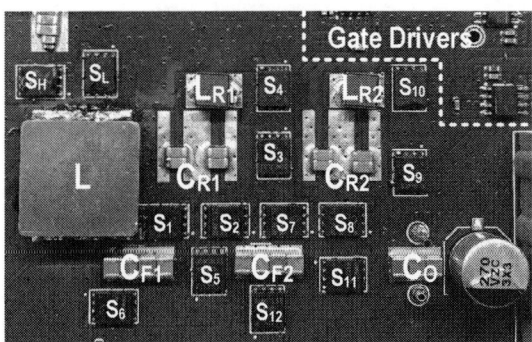

Fig. 9. Photograph of the hardware prototype.

TABLE I. CIRCUIT COMPONENTS AND SPECIFICATIONS

Components	Part Information	Parameters
S_H, S_L	Infineon BSC117N08NS5	80V, 11.7mΩ
S_1	Infineon BSC032N04LS	40V, 3.2mΩ
S_{2-12}	Infineon BSC009NE2LS5I	25V, 0.95mΩ
L	Vishay IHLP6767GZER100M01	10μH, 12mΩ, 25A(20% Isat)
L_{R1}, L_{R2}	Pulse PA4059.550HLT	55nH, 0.2mΩ, 61A(20% Isat)
C_{F1}, C_{F2}	Murata GRM32EC72A106KE05L	40μF (10μF×4, 100V, X7S, 1210)
C_{R1}, C_{R2}	Murata GRM31CR72A105KA01	4μF (1μF×4, 100V, X7R, 1206)
C_O	Murata GRM32ER71E226KE15L Panasonic EEHZC1V271P	66μF (22μF×3, 1210) 270μF
Gate Drivers	Texas Instruments UCC27282	120V, Half Bridge

switched tank converter, the right node of S_3 is connected to SW5 instead of V_O. Besides, the eliminated DC bus capacitor is repurposed as a flying capacitor C_{F1}, with two additional power switches connected to ground and SW5. This configuration allows the charge from C_{F1} and resonant tank 1 (C_{R1} and L_{R1}) to be collected and reused by the subsequent C_{F2} and resonant tank 2 (C_{R2} and L_{R2}), thus further increasing both the charge transferred to the output and the VCR.

Fig. 4, Fig. 5, and Fig.6 illustrate its working principle, equivalent circuit, and key waveforms, respectively. S_H and S_L operate similarly to a conventional buck converter with a duty cycle (D) controlling the energizing duration of the power inductor L. The switching bus ReSC topology operates at its resonant frequency with a fixed 50% duty cycle. In one half cycle of the period, L charges C_{F1}, while C_{F2} gets charged by L and tank 1 simultaneously. The output is charged through C_{F2} and tank 2 in parallel. SW1 rises to V_{CF1} + V_{CF2} + V_O. In the other half cycle, both tanks get charged, while C_{F1} and C_{F2} are discharged and SW1 switches to V_{CF1}. By applying Kirchhoff's Voltage Law (KVL), DC voltages across capacitors can be calculated:

$$V_{CF1} = 5V_O, \ V_{CR1} = 3V_O, \ V_{CF2} = 2V_O, \ V_{CR2} = V_O \quad (1)$$

As a consequence, SW1 switches between $8V_O$ and $5V_O$, giving the ReSC topology an equivalent voltage gain of 6.5, which is much higher than that of the switched tank converter

without adding more passive components. Thus the VCR of the proposed converter is D/6.5, and the ratio of inductor DC current (I_L) to output current (I_O) is only 1/6.5. Although HV devices in the regulating stage may incur significant power loss due to their large parasitic parameters, the reduced inductor current substantially lowers the power loss dissipated at them.

Additionally, the two flying capacitors are always soft charged or discharged by the resonant tanks or L, and thus charge redistribution loss is minimized. Since C_{F1}, C_{F2} and C_O are much larger than C_{R1} and C_{R2}, while L has a large inductance and can be assumed as a DC current source, the resonant frequencies in both tanks are approximately equal:

$$f_r \approx \frac{1}{2\pi\sqrt{L_{R1}C_{R1}}} \approx \frac{1}{2\pi\sqrt{L_{R2}C_{R2}}} \quad (2)$$

As shown in Fig. 7, the proposed regulated ReSC converter utilizes L to decouple V_{IN} from the resonant tanks, ensuring that the voltages across the resonant capacitors depend solely on V_O and are irrespective of V_{IN}. For a wide V_{IN} and a fixed V_O, this approach allows the use of Class II ceramic capacitors instead of Class I ones, achieving cost savings and higher power density while avoiding capacitance derating. Moreover, the voltage difference between two resonant tanks is $2V_O$, only inducing a capacitance derating mismatch of 4%. Therefore, this design establishes a specific resonant frequency to guarantee complete soft charging and renders the system insensitive to V_{IN} variations

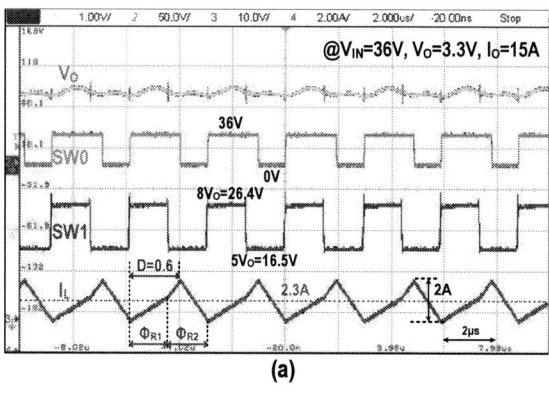

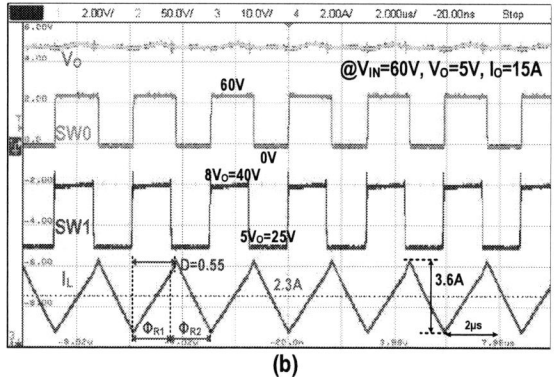

Fig. 10. Measured steady-state waveforms at (a) 36 V to 3.3 V and (b) 60 V to 5 V.

and capacitance derating, effectively avoiding the need for a complex ZCS control circuit.

III. IMPLEMENTATION AND EXPERIMENTAL RESULTS

Fig. 8 shows the system architecture of the proposed regulated V_{IN}-insensitive ReSC converter. The switches S_H/S_L, S_3/S_4, S_5/S_6, S_9/S_{10}, and S_{11}/S_{12} are referenced to ground and operate complementarily, which are similar to the high- and low-side transistors in a conventional buck converter, and hence they are directly driven by half-bridge gate drivers. S_H and S_L are controlled by a pulse-width modulation (PWM) signal to regulate the final V_O, while the power switches in the ReSC topology operate with a fixed 50% duty cycle. Due to their floating source voltages, stacked bootstrap circuits composed of bootstrap capacitors and Schottky diodes are adopted to drive S_1, S_2, S_7, and S_8, which are powered by the bootstrap voltages V_{BST2}, V_{BST3}, V_{BST8}, and V_{BST9}, respectively.

A 36 V-60 V-input and 3.3 V/5 V-output hardware prototype was designed and built to verify the functionality and performance of the proposed converter. Fig. 9 shows the photograph of the prototype, featuring a 6-layer power board with 4-oz copper. The key components are listed in Table I. All MOSFETs operate at 340 kHz, determined by the resonant frequency. A large inductance of 10 μH was chosen for L to accommodate higher V_{IN} and reduce inductor current ripple, so that it can be regarded as a DC current source for the resonant tanks. To minimize volume and cost, each resonant tank uses only four 1-μF X7R capacitors in a 1206 package.

Fig. 10 shows the measured steady-state waveforms of V_O,

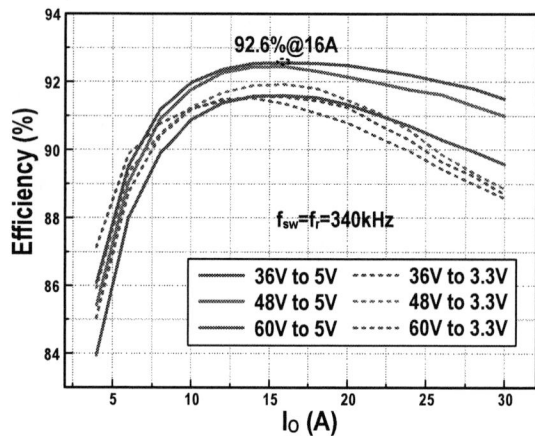

Fig. 11. Measured power efficiency.

I_L, and switching nodes on both sides of L at 36 V-to-3.3 V and 60 V-to-5 V conversions when I_O is 15 A. SW0 switches between V_{IN} and ground, while SW1 alternates between $8V_O$ and $5V_O$. Due to the difference in duty cycles between the regulating stage and the switching bus ReSC topology, I_L exhibits three distinct (de-)energized slopes. Unlike the conventional two-stage solution shown in Fig. 1, the power inductor in the proposed design handles a further reduced DC current in the regulating stage, rather than the whole output current, and hence power loss in HV devices is minimized. Fig. 11 displays the measured power efficiency over I_O for various input and output voltage scenarios. The converter realizes a peak efficiency of 92.6% at 36 V to 5 V. At the maximum supported V_{IN} of 60 V, it achieves a peak efficiency of 91.6%. Its high efficiency over a wide V_{IN} range is attributed to reduced inductor loss, resonant operation, and insensitivity to V_{IN}.

Table II summarizes and compares the performance. By using low-cost and high-capacitance-density Class II ceramic capacitors, the total volume of the several-μF resonant capacitors is reduced to less than one-third of that in other designs. A large inductance is selected for the regulating inductor to minimize ripple, which slightly limits improvement in power density. As a result, the proposed converter achieves a power density of 567 W/in³ when considering the total volume of all the power-stage components. This work is the only design that achieves V_{IN}-insensitive resonant frequency by only using X7R resonant capacitors, while still maintaining comparable peak efficiency and power density with prior works.

IV. CONCLUSIONS AND FUTURE WORK

This paper presents a regulated V_{IN}-insensitive resonant switched-capacitor converter with a large voltage conversion ratio from 36-60 V to 3.3/5 V. It mitigates resonant frequency variation and resonant tank mismatch caused by a wide V_{IN} range and capacitance derating, enabling the use of Class II ceramic capacitors (X7R, X7S, etc.) for higher density and lower cost. Measurement results show that the proposed converter achieves high efficiency and high power density by only using X7R resonant capacitors.

Note that this paper primarily focuses on preventing voltage-induced capacitance derating, thereby eliminating the need for a complex control circuit to maintain resonant ZCS operation.

979-8-3315-1612-3/25 $31.00 © 2025 IEEE

TABLE II. PERFORMANCE SUMMARY AND COMPARISON

Architecture	Two Stage[a]		Single Stage		
Publication	APEC 2018 [6]	MPM3695-25 [12]	ECCE 2021 [13]	APEC 2024 [7]	This work
Topology	STC	Buck	DPHD	DSDSTC	ReSC
V_{IN}	40 – 60V	4 – 16V	36 – 65V	48V	36 – 60V
V_O	10 – 15V	0.5 – 5.5V	1 – 2V	1 – 3.3V	3.3V/5V
Max. Output Power	750W	110W	60W	165W	150W
Switching Frequency	320kHz	600kHz	250kHz	111 – 367kHz	340kHz
Inductor	2 × 58nH	0.36µH	260nH	2µH + 2 × 30nH	10µH + 2 × 55nH
Resonant Capacitor / Total Volume	2 × 3.8µF (U2J) / 391.68mm³	– –	– –	2 × 1.9µF (U2J) / 195.84mm³	2 × 4µF (X7R) / 65.536mm³
Flying Capacitor	60µF	– –	6 × 6.8µF	50µF + 100µF	2 × 40µF
Output Capacitor	610µF	420µF	238µF	370µF	336µF
Peak Efficiency	93.2% (98.61% × 94.5%)		92.7%	94.7%	92.6%
Power Density[b]	533W/in³[c]		451W/in³	595W/in³	567W/in³
V_{IN}-Insensitive Resonant Frequency if only using Class II Capacitors?	No		– –	No	Yes

(a) The two-stage architecture is evaluated by adopting a STC in [4] as the first stage and a product of buck converter from MPS [9] as the second stage.
(b) Power Density = Max. output power / Total power stage component volume.
(c) To match the power level, multiple buck converters are applied in parallel for multi-phase operation in the second stage.

However, due to the wide tolerance of resonant capacitor over temperature variations and part-to-part differences in mass production, resonant frequency variation is inevitable. This issue warrants further attention in future work.

ACKNOWLEDGMENT

The authors would like to thank the Information Science Laboratory Center, University of Science and Technology of China (USTC), Hefei, China, for the hardware/software services.

REFERENCES

[1] X. Li and S. Jiang, "Google 48V Rack Adaptation and Onboard Power Technology Update," in *Open Compute Project (OCP) 2019 Summit*, 2019.

[2] Monolithic Power Systems, "48V Data Center Solutions," [Online]. Available: https://www.monolithicpower.com/en/products/powermanagement/48v-data-center.html.

[3] A. Dago, M. Balutto, S. Saggini, M. Leoncini, S. Levantino, and M. Ghioni, "A 260-A/48-V Bus Hybrid Resonant Converter with Large Conversion Ratio for Future Data Centers," in *2024 IEEE Applied Power Electronics Conference and Exposition (APEC)*, Long Beach, CA, USA, Feb. 2024, pp. 83–87.

[4] K. Tang, Y. Ji, and L. Cheng, "A 6-Phase Integrated Voltage Regulator With Multi-Phase Transient Optimization Technique in 28-nm CMOS Process," *IEEE Transactions on Circuits and Systems II: Express Briefs*, vol. 71, no. 10, pp. 4596–4600, Oct. 2024.

[5] T. Ge, Z. Ye, R. A. Abramson, and R. C. N. Pilawa-Podgurski, "A 48-to-12 V Cascaded Resonant Switched-Capacitor Converter Achieving 4068 W/in³ Power Density and 99.0% Peak Efficiency," in *2021 IEEE Applied Power Electronics Conference and Exposition (APEC)*, Phoenix, AZ, USA, Jun. 2021, pp. 1335–1342.

[6] S. Jiang, C. Nan, X. Li, C. Chung, and M. Yazdani, "Switched tank converters," in *2018 IEEE Applied Power Electronics Conference and Exposition (APEC)*, Mar. 2018, pp. 81–90.

[7] S. Y. Sim, X. Zhang, J. Jiang, K. Wei, and C. Huang, "A 94.7% Efficiency Direct-Step-Down Switched-Tank-Based 48V to 1V-3.3V Hybrid Converter with Constant-Resonant-Time Closed-Loop Control," in *2024 IEEE Applied Power Electronics Conference and Exposition (APEC)*, Long Beach, CA, USA, Feb. 2024, pp. 1344–1350.

[8] S. Zaffin, A. Dago, M. Leoncini, A. Gasparini, O. Zambetti, S. Levantino, and M. Ghioni, "A 60 A Switched Tank Converter with Buck-Boost Sigma Regulation for 48 V Bus Down-Conversion," in *2024 IEEE Applied Power Electronics Conference and Exposition (APEC)*, Long Beach, CA, USA, Feb. 2024, pp. 63–69.

[9] Y. He, S. Jiang, and C. Nan, "Switched tank converter based partial power architecture for voltage regulation applications," in *2018 IEEE Applied Power Electronics Conference and Exposition (APEC)*, Mar. 2018, pp. 91–97.

[10] D. Menzi, S. Ben-Yaakov, G. Zulauf, and J. W. Kolar, "ESR Modeling of Class II MLCC Large-Signal-Excitation Losses," *IEEE Transactions on Power Electronics*, vol. 38, no. 5, pp. 5711–5715, May 2023.

[11] Y. Jiang, B. Hu, B. Wen, Y. Shen, and T. Long, "Methodology for Large-signal Loss Characterization of Ferroelectric Class II MLCC in High-frequency Range," in *2022 IEEE Energy Conversion Congress and Exposition (ECCE)*, Oct. 2022, pp. 1–6.

[12] Monolithic Power Systems, *MPM3695-25 Datasheet*, "16V, 20A, Scalable DC/DC Power Module with PMBus," Sep, 2019. [Online]. Available: https://www.monolithicpower.com/en/documentview/productdocument/ index/version/2/document_type/Datasheet/lang/en/sku/MPM3695GRF-25/document_id/4474/.

[13] C. Chen, J. Liu, and H. Lee, "A 92.7%-Efficiency 30A 48V-to-1V Dual-Path Hybrid Dickson Converter for PoL Applications," in *2021 IEEE Energy Conversion Congress and Exposition (ECCE)*, Oct. 2021, pp. 1989–1994.

A Hybrid Switched Capacitor Converter Enabling Capacitive-Based Wireless Power Transfer for Battery Charging Applications

Jade Sund, *Student Member, IEEE,* and Samantha Coday, *Member, IEEE*

Department of Electrical Engineering and Computer Science, Massachusetts Institute of Technology

Email: jcsund@mit.edu

Abstract—This work investigates the feasibility of employing hybrid switched capacitor converters in wireless battery charging applications. A capacitively-isolated hybrid Dickson converter designed for use in low-power volume-constrained wireless power transfer applications is presented. The converter analysis, operation and design are detailed. Finally, a hardware prototype is presented and experimental results validate wireless power transfer, across varying load and gap distances.

Index Terms—hybrid switched capacitor converter, wireless power transfer, Dickson converter, resonant converter

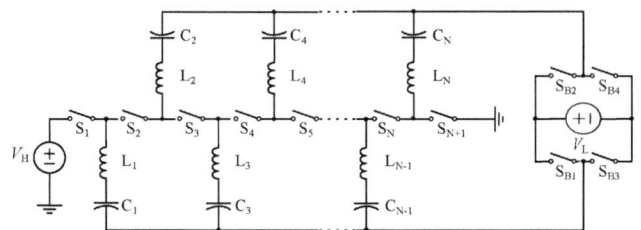

Fig. 1. Capacitively-isolated *N*:1 hybrid Dickson converter schematic.

I. INTRODUCTION

Wireless power transfer (WPT) is emerging for applications such as autonomous drones, electric vehicles, biomedical implants, and phone charging, where wired charging is inconvenient or impractical. Moreover, wired connections endure additional stresses on chargers, requiring user intervention and increasing failure rates. Inductive WPT (I-WPT), which utilizes magnetic coupling for power transfer, is dominant in on-market WPT solutions, with demonstration in industry [1] and academia [2] reaching high power density and efficiency. However, I-WPT systems suffer from high cost and rely on ferrite cores that are both fragile and heavy [3]. Demonstrations of I-WPT also face challenges in volume-constrained applications, as magnetics' sizing is limited by losses [4]. As such, capacitive WPT (C-WPT), which utilizes electric fields for power transfer, has been investigated as an alternative to I-WPT, in biomedical [2], electric vehicle [5], and autonomous underwater vehicles (AUV) [6] [7] charging applications. Previous research has focused on converter topologies composed of a dc source, inverter stage, resonant tank with capacitive coupling, rectifier stage, and load. These converter topologies are operated at high resonant frequencies (>1 MHz) to minimize the required capacitance and inductance. In high power (>1 kW) large air-gap applications, such as electric vehicle charging, this approach has achieved efficiencies of over 90% and power densities of over 49 kW/m² [5], [8]. In applications such as biomedical devices and underwater charging, the gap between plates is comprised of human tissue or water, both of which increase the dielectric constant by a factor of ten. However, these applications are extremely volume-constrained and previous demonstrations are limited in efficiency (< 40%) [2], [9], [10].

As a result, this work investigates alternative topologies which increase effective capacitance and decrease inductor requirements. Hybrid switched capacitor converters have been shown to have high power density and efficiency due to their use of energy dense capacitors and low switch stress [11], [12]. The Dickson converter, a type of switched capacitor converter, was developed for on-chip energy processing [13], and has become a popular candidate for power dense conversion due to its low switch stress compared to other hybrid switched capacitor converters [14]. Previous work [15] [16] originally presented a modified hybrid Dickson converter which utilizes the flying capacitors to process energy and provide dielectric isolation. Here utilizing the capacitively-isolated hybrid Dickson topology for use in WPT systems, to decrease passive component sizing and switch stress is explored. The remainder of this work presents the analysis of the capacitively-isolated Dickson converter and results for both the 12:1 and 4:1 converters demonstrating its viability as a WPT solution.

II. THEORETICAL ANALYSIS

A. Converter Operation

The capacitively-isolated Dickson converter, shown as an *N*:1 converter in Fig. 1, where *N* is the converter level count and conversion ratio, can be operated bidirectionally; therefore, the voltage sources will be referred to as the high-side (V_H) and low-side (V_L) voltages. The capacitively-isolated Dickson converter can be operated as a two phase 50% switching scheme while maintaining soft-charging of the capacitors. In phase one, the odd numbered string switches, S_{B2}, and S_{B3} are closed and the remaining switches are open. In phase two, the even numbered switches, S_{B1}, and S_{B4} are closed and the remaining switches are open. The drain-to-source blocking voltage of the switches are expressed in (1).

$$v_{ds,n} = \begin{cases} V_L & \text{for } n = 1, N+1 \\ 2V_L & \text{for } n = [2...N] \\ V_L & \text{for } B1 - B4 \end{cases} \quad (1)$$

When the capacitor and inductor values are matched, e.g $C_1 = C_2 = \cdots = C_N \triangleq C$ and $L_1 = L_2 = \cdots = L_N \triangleq L$, the resonant frequency of the converter is as expressed in (2).

$$f_{res} = \frac{1}{2\pi\sqrt{LC}} \quad (2)$$

The capacitively-isolated Dickson converter is operated at resonance; therefore, the inductor sizing depends on the chosen switching frequency and the size of capacitive coupling shown in (3).

$$L = \frac{1}{(2\pi f_{res})^2 C} \quad (3)$$

Moreover, like other resonant tank converters the capacitor ripple does not increase switch stress, allowing for sizing of switches independent of load and capacitance. The decreased switch stress also allows for the use of lower voltage switches, which have improved figures-of-merit [17]–[19].

The mid-range voltage analysis of the coupling capacitors results in (4) where V_{ISO} is the voltage difference between the ground of V_H and that of V_L and k is the index of the flying capacitor capacitor as seen in Fig. 1. In the capacitively-isolated Dickson converter, the energy processing capacitors C_1 - C_N, also provide isolation [16].

$$V_{C,k} = V_{ISO} + (N - k) V_L \quad (4)$$

There are multiple methods to achieve soft charging in the capacitively-isolated Dickson converter [20]; the method of distributed inductance, shown in Fig. 1, pairs each capacitor with a series inductor. The lumped inductance approach places an inductor at each of the rectifier switch nodes. While lumped inductance takes advantage of the fact that larger volume inductors have higher power density and efficiency [4], a distributed inductance approach was chosen to minimize the component volume on the receiving side of the converter. The component volume of the receive (RX) side is prioritized as in the applications of focus, AUV and biomedical charging, the RX volume is more critically constrained than the transmit (TX) volume.

B. Power Transfer Analysis

Power transfer analysis, following the method presented in [21], can be used to compare the maximum power transfer capabilities per gap distance of WPT topologies. This method requires the elimination of switching devices in the converter topology, which motivates the development of an ac equivalent circuit model for the N:1 capacitively-isolated Dickson converter. The model maintains the voltage characteristics of the LC tanks. The resultant model consists of replacing the string switches (S_2-S_{N+1}) with voltage sources whose values

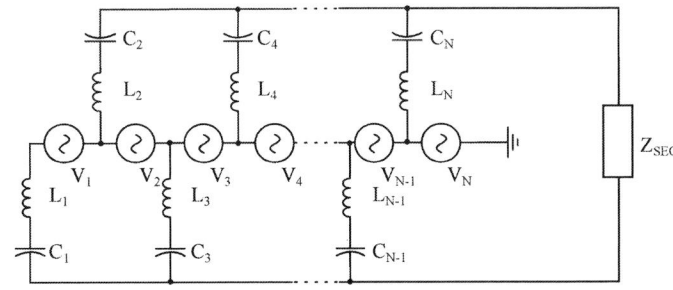

Fig. 2. Derived ac equivalent model for N:1 capacitively-isolated Dickson converter.

are the sum of the mid-range voltage experienced by capacitor (k) and the ac voltage ripple across capacitor, as seen in (5). The two phase operation of the converter results in the even and odd capacitors having a phase shift of 180 degrees.

$$\begin{aligned} V_k &= V_{dc_k} + v_{ac_k} \\ &= V_{C_k} + 2(-1)^k cos(\omega t + \phi) \end{aligned} \quad (5)$$

A load impedance, Z_{SEC}, replaces rectifier and secondary side passives [21]. The full model is presented in Fig. 2. The current delivered to the secondary by each voltage source is seen in (6), where C_T is the total coupling capacitance and $v_T(s)$ is the voltage across C_T where s is the complex frequency variable.

$$i_k(s) = \begin{cases} sC_T v_T(s) & k \text{ is even} \\ 0 & k \text{ is odd} \end{cases} \quad (6)$$

The total current delivered to the load is seen in (7), indicating that the total power delivered is as expressed in (8).

$$i_T = \frac{sN}{2}C_T v_T(s) \quad (7)$$

$$P_L(s) = [\frac{sN}{2}C_T v_T]^2 Z_{SEC} \quad (8)$$

When the coupling capacitor values are matched, e.g $C_1 = C_2 = \cdots = C_N \triangleq C$, (8) can be expressed in terms of the total capacitive area (A), the relative permeability, (ε), and the electric field strength of C_T where $C_T = \frac{N}{4}C$, ($E(s)$) as seen in (9).

$$P_L(s) = [\frac{sN}{8}A\varepsilon E(s)]^2 Z_{SEC} \quad (9)$$

In terms of the plate separation distance (d) and the ac voltage across the plates ($v_{ac}(s)$), (9) can be expressed as (10).

$$P_L(s) = [\frac{sN}{8}A\varepsilon \frac{v_{ac}(s)}{d}]^2 Z_{SEC} \quad (10)$$

Assuming a purely resistive secondary impedance (9) can be expressed in the time domain as (11).

$$P_L(t) = [\frac{\omega N}{8}A\varepsilon E(t)]^2 R_{SEC} \quad (11)$$

$$= [\frac{\pi f_{sw}N}{4}A\varepsilon E(t)]^2 R_{SEC} \quad (12)$$

The 4:1 capacitively-isolated Dickson converter has the same

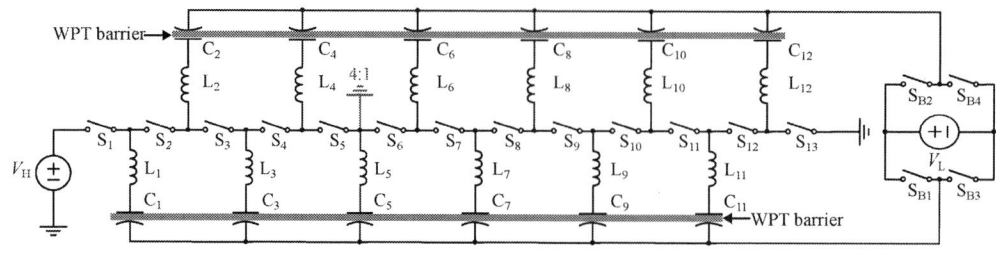

Fig. 3. Capacitively-isolated 12:1 hybrid Dickson converter schematic with WPT barrier shown. The ground connection (red) at the node between S_5 and S_6 converts the schematic to the 4:1 capacitively-isolated hybrid Dickson converter when $S_6 - S_{13}$, $L_5 - L_{12}$, and $C_5 - C_{12}$ are omitted.

power transfer capabilities as conventional C-WPT topologies, such as the resonant capacitor converter [21]. Moreover, unlike the conventional approaches, the proposed converter provides a N:1 voltage conversion. This conversion provides an interesting system benefit in many applications, such as biomedical devices charging, where a step down stage is necessary and usually employed with an additional converter. Furthermore, (11) suggests that the power transfer capabilities of the capacitively-isolated Dickson converter with $N > 4$ exceed those of conventional C-WPT approaches. The N dependence of (11) indicates that there is a nine-fold improvement in power transfer capabilities of the 12:1 capacitively-isolated Dickson converter over the 4:1 capacitive-isolated and 1:1 conventional C-WPT topologies with the same operating conditions. However, this conversion may not be preferable in many other applications which require 1:1 voltage; therefore, requiring further system level optimization depending on application space.

C. Passive Component Selection

The capacitors, C_1 - C_N, are realized through aligning exposed copper platting, seen coated in Kapton tape in Fig. 4b, on the TX and RX printed circuit boards (PCBs). Kapton tape has a similar dielectric to coatings commonly used for bio-compatibility and corrosion prevention in biomedical and AUV applications. The capacitance, seen in (13), is determined by the plate area (A), distance between plates (d_1), dielectric constant of WPT medium (ε_1), thickness of coating (d_2), and dielectric constant of coating (ε_2).

$$C = \frac{A\varepsilon_1\varepsilon_2}{d_1\varepsilon_2 + d_2\varepsilon_1} \quad (13)$$

The plate area, distance between plates, and dielectric constant of the medium are fixed by the application; the choice and thickness of the coating can be used to control the final capacitance. In conditions where $d_2\varepsilon_1 >> d_1\varepsilon_2$, the capacitance equation simplifies to (14).

$$C = \frac{A\varepsilon_2}{d_2} \quad (14)$$

III. HARDWARE

To validate the analysis presented in Section II, a hardware prototype of the capacitively-isolated Dickson converter which can be operated at either a 4:1 or 12:1 conversion ratio was developed.

For the initial hardware prototype the capacitance was estimated using (13) to be 1.01 nF when coated with 1 mil thick Kapton tape. As a trade-off in inductor sizing and switching losses, and initial switching frequency of 100 kHz was selected, resulting in a designed inductance of 2.5 mH. With the increased number of coupling capacitors, the parasitic capacitance between the plates becomes a larger challenge for design and analysis. As described in [22], parasitic capacitance is important to characterize for optimal performance in C-WPT systems. Parasitic capacitance can lead to decreased output power and is a focus of future work.

Components	Part Number	Description
S_1 - S_{13}, S_{B1} - S_{B4}	GSF3404B	30 V, 43 mΩ
L_1 - L_{12}	SDS680R	2.2 μH, 0.1 A
Gate Driver	NCP81074	5 V, 10 A
Signal Isolator	ADUM5241	Power and Signal

TABLE I
COMPONENTS SELECTED FOR CAPACITIVELY-ISOLATED DICKSON CONVERTER HARDWARE PROTOTYPE.

The components for the prototype are listed in Table I. This initial prototype uses an ADUM for power and signal isolation but bootstrapping methods can be implemented to improve end-to-end efficiency [23], [24].

Two PCBs were designed to realize the converter of Fig. 3; each PCB represents one side of the WPT barrier seen in Fig. 3. The TX PCB is designed such that one side of the board contains the string switches (S_1 - S_{13}), the gate drive for the string switches, the inductors (L_1 - L_{12}), and all control circuitry as seen in Fig. 4a and Fig. 5a. The other side of the board contains only the TX plate for the capacitors C_1 - C_{12}, whose layout is identical to the that shown in Fig. 4b. When operated as a 4:1 converter, a connection to ground is made, shown in red in Fig. 3, and the rest of the string switches ($S_6 - S_{13}$) and inductors ($L_5 - L_{12}$) are

979-8-3315-1612-3/25 $31.00 © 2025 IEEE

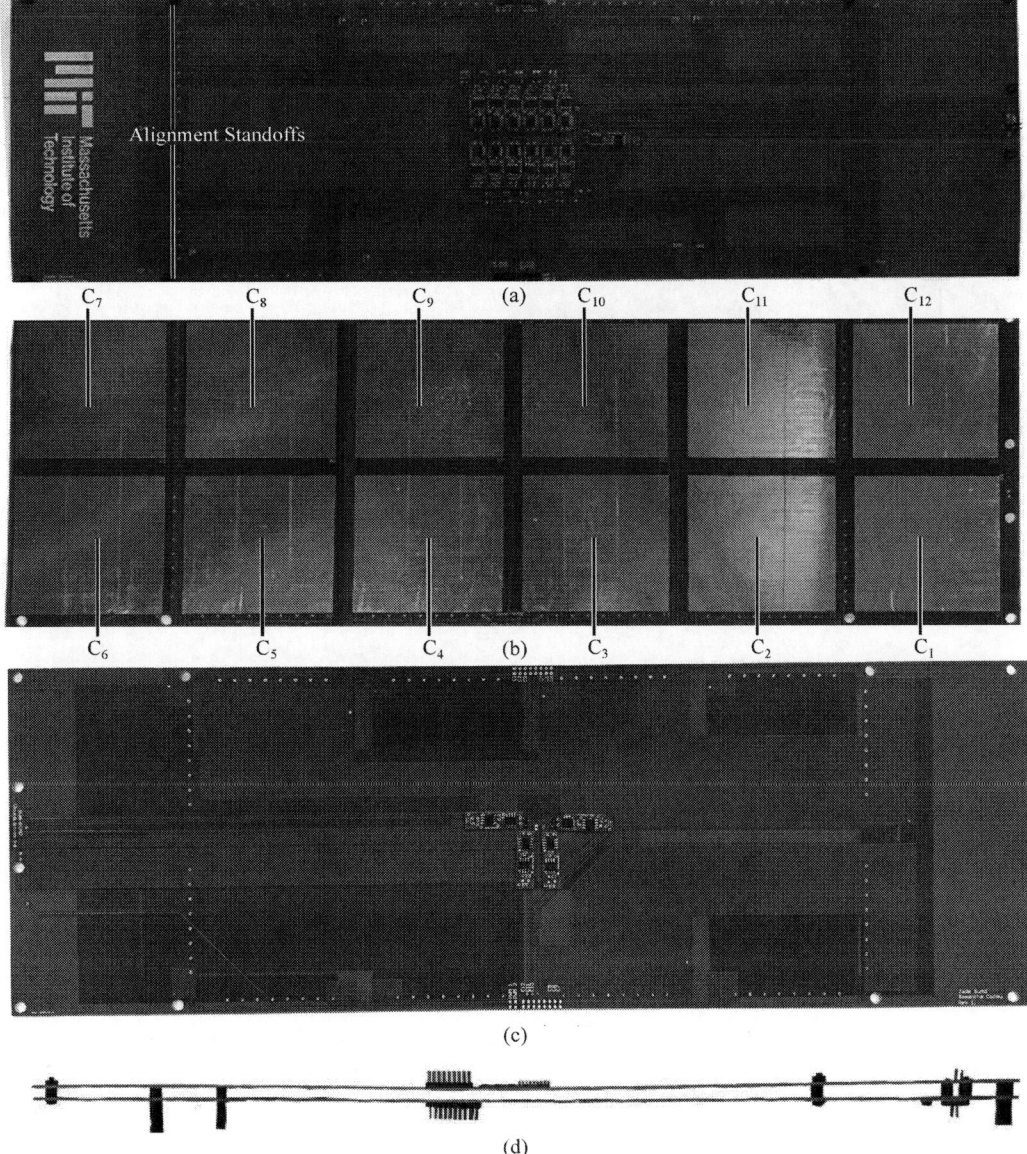

Fig. 4. Development boards. (a) Transmit boards. (b) Bottom of receive boards. (c) Receive Board. (d) Example of board coupling technique.

not populated. The RX PCB represents the low-side of the WPT barrier; the RX board is designed such that one side of the board contains the rectifier shown in Fig. 5b and the other contains the receive plate for capacitors C_1 - C_{12}. The two PCBs are coupled though the capacitive plates of the TX and RX boards. De-ionized water was selected for the dielectric, as the dielectric constant is similar to that of undersea AUV and biomedical charging, where the anticipated dielectric constants are ≈ 69 and $30-80$, respectively. For this preliminary prototype, proper alignment and spacing of the capacitive plates is achieved through standoffs placed along the edges of the board as seen in Fig. 4d.

A. Board Design

Minimizing commutation loop area and thereby parasitic inductance is paramount for ensuring low voltage overshoot, noise, and EMI, allowing for higher switching frequency operating and efficiency. In particular, due to the high number of switches and the comparatively large capacitors, the design of the Dickson converters' commutation loop is challenging. Here, seen in Fig. 5a, switches S_1 - S_{13} are placed in a rectangular shape on the board to reduce the distance between the drain and source of adjacent switches. Further reduction in the drain-to-source distance is limited by the associated loss of symmetry, which is utilized to minimize parasitic mismatch.

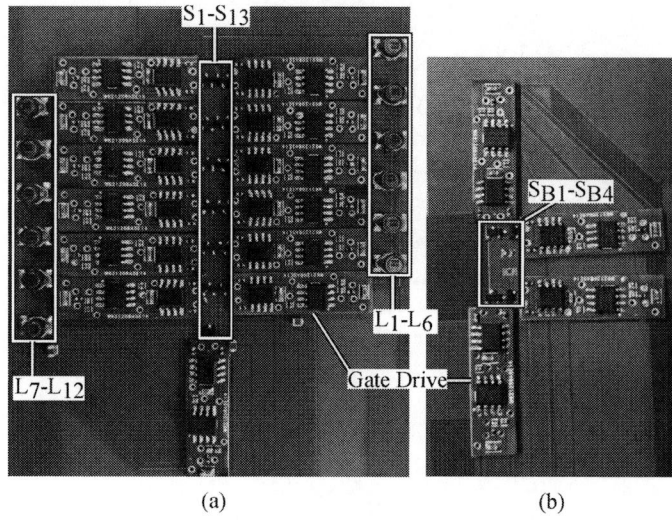

(a)   (b)

Fig. 5. Development boards. (a) Transmit board switching cells. (b) Receive board switching cell.

IV. RESULTS

The 12:1 isolated Dickson converter was tested at $V_H = 48$ V, switching frequency of 42.5 kHz and a dielectric of de-ionized water. The 4:1 isolated Dickson converter was tested at $V_H = 16$ V, switching frequency of 75 kHz, and a dielectric of de-ionized water. The output voltage of both converters with no loss is expected to be $V_L = 4$ V.

A. Resonant Frequency Tuning

For full soft charging of capacitors and thus maximum output power the converter needs to be operated at resonance. In the ideal case where the capacitors and inductors are perfectly matched, e.g $C_1 = C_2 = \cdots = C_N \triangleq C$ and $L_1 = L_2 = \cdots = L_N \triangleq L$, the capacitor voltage waveforms should be sinusoidal. The sinusoidal nature of a waveform was determined by comparing the RMS measured by the oscilloscope and the RMS calculated using the measured amplitude of the waveform. When the capacitor voltages are sinusoidal there should be little mismatch between these two values. The frequency that results in the minimum mismatch was determined to be the resonant frequency. Furthermore, the resonant frequency, calculated in this manner, should be the same for all capacitor voltages.

For the 4:1 capacitively-isolated Dickson converter, the frequency that minimizes the difference in measured RMS and calculated RMS for V_{C1} results in a resonant frequency of 88.9 kHz, shown in Fig. 6a. The output voltage, at this frequency is 0.29 V with $V_H = 12$ V and 6 kΩ load.

The desired operation frequency can also be determined by finding the frequency that maximizes the output voltage. The resultant capacitor voltage waveforms for operation of the 4:1 isolated Dickson converter at 75 kHz, the frequency that maximizes output voltage, are shown in Fig. 6b At this operation frequency, with a 6 kΩ load, the output voltage is

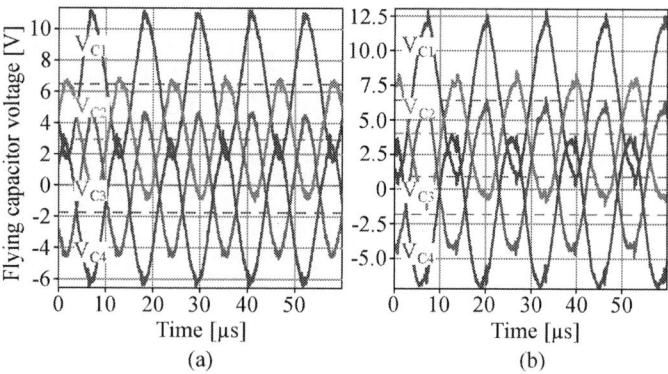

Fig. 6. 4:1 capacitively-isolated Dickson converter capacitor voltages with $V_H = 12$ V. (a) $f_{sw} = 88.9$ kHz. (b) $f_{sw} = 75$ kHz.

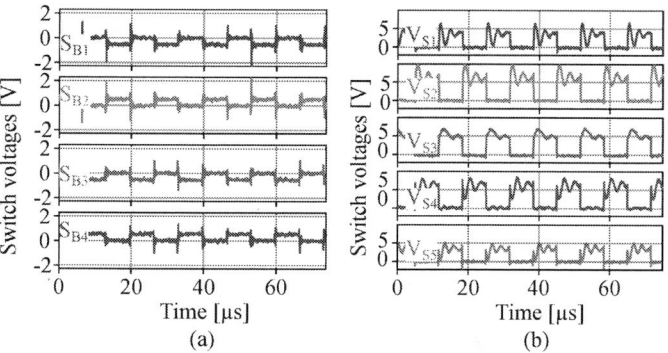

Fig. 7. 4:1 isolated Dickson converter switch waveforms for $V_H = 12$ V, $V_L = 0.31$ V. (a) RX bridge switches reference to input ground. (b) TX string switches.

0.58 V with $V_H = 12$ V. Note that the voltage ripple on C_1 is larger in Fig. 6b than Fig. 6a. In contrast, there is a larger discontinuity at the peaks and troughs of V_{C1} in Fig. 6b than Fig. 6a. In the case where operation frequency is determined by output voltage, the smoothness of the voltage waveform is traded for and increased waveform amplitude. The resonant frequency, 88.9 kHz in the case of the 4:1 isolated Dickson converter, is used to calculate the coupling capacitance of $C_1 = 3.2$ nF. This is significantly larger than the theoretical value calculated from (13), which implies the presence of large parasitic capacitance.

B. Switch Voltages

As discussed in Section III, minimizing the commutation loop is important in reducing voltage overshoot at the switch nodes. Figure 7b shows the voltages across S_1 - S_5 for the 4:1 converter operated at $V_H = 16$ V. The ripple seen on these switch voltages is a function of the capacitor mismatch as well as the commutation loop inductance. Further design to minimize the commutation loop may decrease the ripple seen here. The voltage across the bridge switches, shown in Fig. 7a, indicates that the rectifier switches are functioning as expected.

979-8-3315-1612-3/25 $31.00 © 2025 IEEE

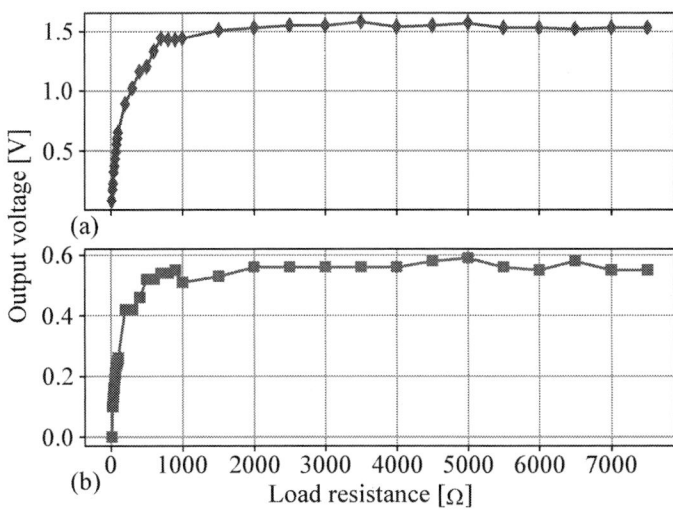

Fig. 8. Load resistance versus output voltage for (a) 12:1 capacitively-isolated Dickson converter operated at $V_H = 48$ V and $f_{sw} = 46.5$ kHz and (b) 4:1 capacitively-isolated Dickson converter operated at $V_H = 16$ V and $f_{sw} = 75$ kHz.

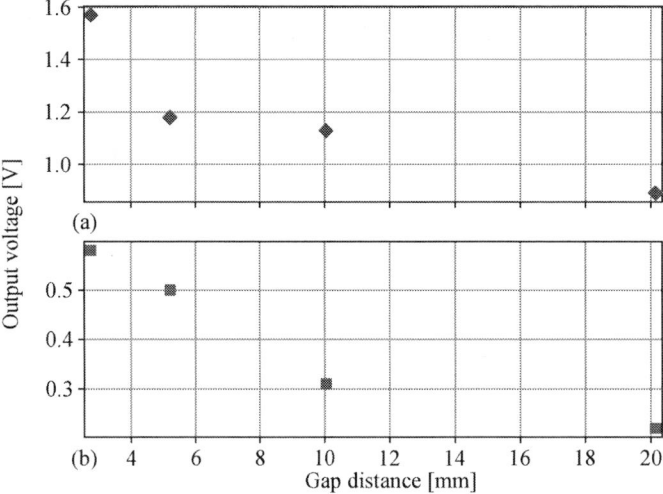

Fig. 9. Output voltage as a function of gap distance (d_2) for (a) 12:1 capacitively-isolated Dickson converter operated at $V_H = 48$ V, $f_{sw} = 46.5$ kHz and 6 kΩ load and (b) 4:1 capacitively-isolated Dickson converter operated at $V_H = 16$ V, $f_{sw} = 75$ kHz and 6 kΩ load.

C. Capacitor Voltages

The sinusoidal nature of the capacitor voltages, in Fig. 6a, indicates coupling between the TX and RX board capacitor plates was achieved and the converter is operating at resonance with minimal hard charging of the capacitors. Compared to the analysis in Section II-A, the experimental capacitor waveforms have variable amplitudes and phase shifts. Furthermore, the 12:1 converter voltage waveforms shows increased capacitor mismatch when compared to the 4:1 converter. The variations in amplitude and phase are likely due to parasitic capacitance throughout the converter. Techniques to minimize and mitigate the effects of parasitic capacitance need to be investigated. The value of the series inductance can be independently tuned to allow for equivalent resonant frequencies across LC tanks with variable capacitance to mitigate the effect of capacitor mismatch.

D. Output

The output voltages of the 4:1 and 12:1 capacitively-isolated Dickson converters are shown in Fig. 8 as a function of load and in Fig. 9 as a function of gap distance. The output voltage of the 4:1 remains constant over a wide range of load resistances. The output voltage decays as a function of gap distance.

CONCLUSION

This work presents a hybrid switched capacitor approach to capacitive wireless power transfer in battery charging applications. The analysis and design of an capacitively-isolated Dickson converter is considered. A hardware prototype is presented which validates the capacitively-isolated Dickson converter's ability to transfer power wirelessly.

REFERENCES

[1] A. Kurs and G. Lestoquoy, "A framework for evaluating multi-kilowatt highly-resonant wireless power transfer systems," *2015 IEEE 82nd Vehicular Technology Conference (VTC2015-Fall)*, pp. 1–6, 2015.

[2] R. Erfani, F. Marefat, A. M. Sodagar, and P. Mohseni, "Transcutaneous capacitive wireless power transfer (c-wpt) for biomedical implants," *IEEE International Symposium on Circuits and Systems (ISCAS)*, pp. 1–4, May 2017.

[3] G. A. Covic and T. Boys, J, "Modern trends in inductive power transfer for transportation applications," *IEEE Journal of Emerging and Selected Topics in Power Electronics (JESTPE)*, vol. 1, pp. 28–41, Mar. 2013.

[4] C. Sullivan, B. A. Reese, A. L. F. Stein, and K. P. A., "On size and magnetics: Why small efficient power inductors are rare," *2016 International Symposium on 3D Power Electronics Integration and Manufacturing (3D-PEIM)*, pp. 1–23, 2016.

[5] F. Lu, H. Zhang, H. Hofmann, and C. Mi, "A double-sided lclc-compensated capacitive power transfer system for electric vehicle charging," *IEEE Transactions on Power Electronics*, vol. 30, pp. 6011 – 14, Nov. 2015.

[6] L. Yang, L. Ma, J. Huang, and Y. Fu, "Characteristics of undersea capacitive wireless power transfer system," *2020 IEEE 9th International Power Electronics and Motion Control Conference (IPEMC2020-ECCE Asia)*, p. 2952–2955, Nov. 2020.

[7] H. Li, G. Li, X. Jin, J. Li, and G. Xu, "A lc-cll compensated capacitive wireless power transfer system in fresh water," *5th International Conference on Power and Energy Applications (ICPEA)*, pp. 130–137, Nov. 2022.

[8] S. Regensburger, B. Sinha, A. Kumar, and K. K. Afridi, "A 3.75-kw high-power-transfer-density capacitive wireless charging system for evs utilizing toroidal-interleaved-foil coupled inductors," *IEEE Transportation Electrification Conference and Expo (ITEC)*, p. 839–843, June 2020.

[9] R. Erfani, F. Marefat, A. M. Sodagar, and P. Mohseni, "Modeling and experimental validation of a capacitive link for wireless power transfer to biomedical implants," *EEE Trans. Circuits Syst. II*, vol. 65, pp. 923–927, July 2018.

[10] A. Koruprolu, S. Nag, R. Erfani, and P. Mohseni, "Capacitive wireless power and data transfer for implantable medical devices," *IEEE Biomedical Circuits and Systems Conference (BioCAS)*, pp. 1–4, Oct. 2018.

[11] Z. Ye, S. R. Sanders, and R. C. N. Pilawa-Podgurski, "Modeling and comparison of passive component volume of hybrid resonant switched-capacitor converters," *IEEE Trans. Power Electron.*, vol. 37, pp. 10903–10919, Sept. 2022.

[12] N. M. Ellis, N. C. Brooks, M. E. Blackwell, R. A. Abramson, S. Coday, and R. C. N. Pilawa-Podgurski, "A general analysis of resonant switched

capacitor converters using peak energy storage and switch stress including ripple," *IEEE Trans. Power Electron.*, pp. 1–21, 2023.

[13] J. F. Dickson, "On-chip high-voltage generation in mnos integrated circuits using an improved voltage multiplier technique," *IEEE J. Solid-State Circuits*, vol. 11, pp. 374–378, June 1976.

[14] R. C. N. Lei, Y. Pilawa-Podgurski, "A general method for analyzing resonant and soft-charging operation of switched-capacitor converters," *IEEE Transactions on Power Electronics*, vol. 30, no. 10, pp. 5650–5664, 2015.

[15] A. Jackson, N. M. Ellis, and R. C. N. Pilawa-Podgurski, "A capacitively-isolated dual extended lc-tank hybrid switched-capacitor converter," *IEEE Applied Power Electronics Conference and Exposition (APEC)*, vol. 15, pp. 1279 –1283, Mar. 2022.

[16] S. Coday, E. Krause, M. E. Blackwell, N. M. Ellis, A. Barchowsky, and R. C. N. Pilawa-Podgurski, "Design and implementation of a gan-based capacitively-isolated hybrid dickson switched-capacitor dc-dc converter for space applications," *2023 IEEE Applied Power Electronics Conference and Exposition (APEC)*, pp. 3154–3159, 2023.

[17] J. Stauth, "Pathways to mm-scale dc-dc converters: Trends opportunities and limitations," *Proc. IEEE Custom Integr. Circuits Conf. (CICC)*, pp. 1–8, 2018.

[18] M. Guacci, D. Bortis, and J. W. Kolar, "High-efficiency weight-optimized fault-tolerant modular multi-cell three-phase gan inverter for next generation aerospace applications," *2018 IEEE Energy Conversion Congress and Exposition (ECCE)*, pp. 1334–1341, Sept. 2018.

[19] R. A. Abramson, S. J. Gunter, D. M. Otten, K. K. Afridi, and D. J. Perreault, "Design and evaluation of a reconfigurable stacked active bridge dc/dc converter for efficient wide load-range operation," *2017 IEEE Applied Power Electronics Conference and Exposition (APEC)*, pp. 3391–3401, Mar. 2017.

[20] P. H. McLaughlin, J. S. Rentmeister, M. H. Kiani, and J. T. Stauth, "Analysis and comparison of hybrid-resonant switched-capacitor dc–dc converters with passive component size constraints," *IEEE Trans. Power Electron.*, vol. 36, pp. 3111–3125, Mar. 2021.

[21] S. Sinha, S. Maji, and K. K. Afridi, "Comparison of large air-gap inductive and capacitive wireless power transfer systems," *IEEE Applied Power Electronics Conference and Exposition (APEC)*, 2021.

[22] S. Sinha, A. Kumar, B. Regensburger, and K. K. Afridi, "A new design approach to mitigating the effect of parasitics in capacitive wireless power transfer systems for electric vehicle charging," *IEEE Trans. Transp. Electrific.*, vol. 5, p. 1040–1059, Dec. 2019.

[23] Z. Ye, Y. Lei, W. C. Liu, P. S. Shenoy, and R. C. N. Pilawa-Podgurski, "Improved bootstrap methods for powering floating gate drivers of flying capacitor multilevel converters and hybrid switched-capacitor converters," *IEEE Transactions on Power Electronics*, vol. 35, pp. 5965–5977, June 2020.

[24] R. K. Iyer, N. M. Ellis, Z. Ye, and R. C. N. Pilawa-Podgurski, "A high-efficiency charge-pump gate drive power delivery technique for flying capacitor multi-level converters with wide operating range," *2021 IEEE Energy Conversion Congress and Exposition (ECCE)*, pp. 5360–5365, 2021.

A 48V to 50-110V Resonant Power-Bus Charger with Reduced Conduction Loss for MHz-Frequency Long-Range LiDAR Driver

Hangxiao Ma[1,2], Xuchu Mu[1], Yang Jiang[1], Weihang Zhang[3], Jincheng Zhang[3], Rui P. Martins[1,4], and Pui-In Mak[1]

[1]State Key Laboratory of Analog and Mixed-Signal VLSI, IME, University of Macau, Macau, China
[2]UM Hetao IC Research Institute, Shenzhen, China
[3]State Key Laboratory of Wide Band-Gap Semiconductor Materials and Devices, Xidian University, Xi'an, China
[4]Instituto Superior Técnico, Universidade de Lisboa, Lisbon, Portugal
Email: {mc36107, yc37448, timjiang}@um.edu.mo, {whzhang, jchzhang}@xidian.edu.cn, {rmartins, pimak}@um.edu.mo

Abstract—The paper presents a novel power-supply bus charger for Light Detection and Ranging (LiDAR) driver systems in automotive applications. The charger utilizes a proposed synchronous resonant boost topology that responds swiftly to rapid drops in the power bus voltage during periodic laser pulse emission. In contrast to conventional hard-charging-based bus chargers, this approach eliminates hard-charging losses while maintaining robust output level regulation. This technique improves LiDAR driving efficiency across a wide supply bus range, ensuring the generation of high-amplitude, narrow-width pulses. Experimental results with a 48-V input confirm that the proposed design can regulate an output from 50V to 110V. When supplying a 100V bus, the charger attains a 250-ns bus recovery time while supporting a peak driving current of 49A with 2-ns pulse-width for the laser diode, enabling LiDAR emission frequencies to exceed the MHz-level. In addition, measured results demonstrate a significant reduction in overall system power consumption, from 10.2W to 5.09W, compared to the hard-charging-based approach.

Keywords—bus charger, boost conversion, laser diode, light detection and ranging (LiDAR), resonant

I. INTRODUCTION

With the rapid development of advanced driver assistance systems (ADAS), LiDAR technology based on time-of-flight principle has become essential for future automotive sensor systems [1]. To ensure adequate detection range and accuracy, automotive LiDAR transmitters require nanosecond-pulse driving currents with amplitudes in tens of amperes and operation frequencies in the MHz range. These requirements pose significant challenges for the power module responsible for generating the supply bus for the transmitter. There are primarily two types of LiDAR drivers: FET-controlled [2][3] and capacitive-discharge [4-8]. FET-controlled drivers are typically employed in scenarios with lower voltages (e.g., below 20V) due to the risk of severe voltage stress on the laser diode, limiting their application mainly to short-range detection. In contrast, capacitive-discharge drivers operate in a resonant state, making them more suitable for long-range detection as they can significantly reduce the risk of overstressing the laser diode.

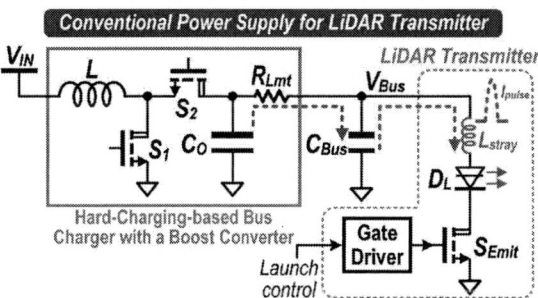

Fig. 1. Conventional power supply for LiDAR transmitter.

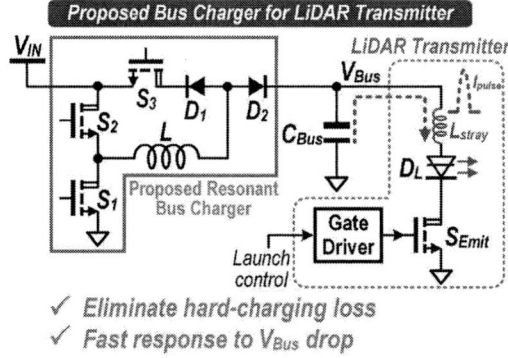

Fig. 2. Proposed bus charger for LiDAR transmitter.

Fig. 1 illustrates a conventional capacitive-discharge LiDAR power supply scheme utilizing a boost converter with a large output capacitor C_O to generate the required bus supply voltage V_{Bus} [8]. During laser pulse emission, V_{Bus} drops from tens of volts to negative tens of volts, followed by charge redistribution between C_O and C_{Bus}, restoring V_{Bus} to its original level. To mitigate excessive hard-charging current from C_O to C_{Bus}, R_{Lmt} is introduced. However, this scheme suffers from substantial hard-charging loss, several times higher than the power

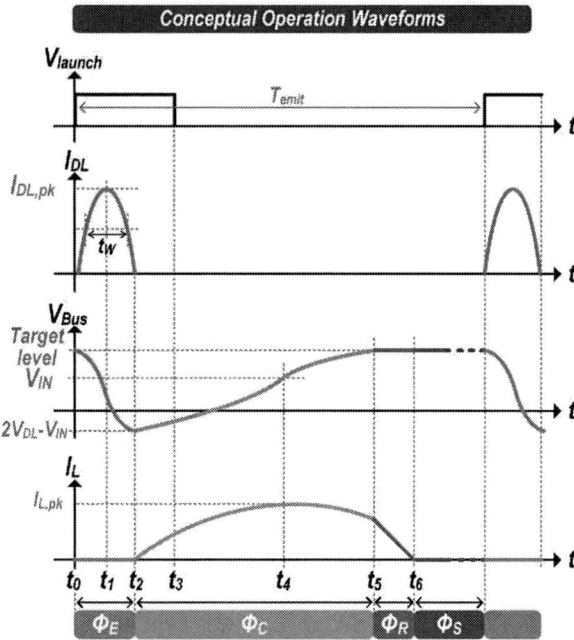

Fig. 3. Conceptual operation waveforms.

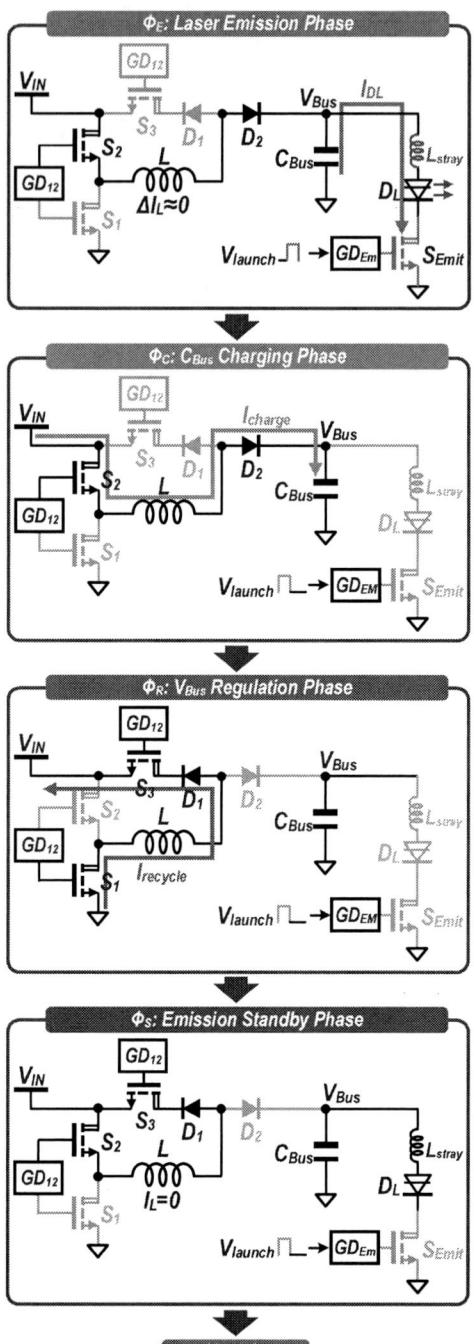

Fig. 3. Power stage operation phases.

consumed by the laser diode, leading to low energy efficiency, especially under high pulse repetition frequencies and elevated V_{Bus} conditions.

II. PROPOSED REGULATABLE RESONANT BUS CHARGER

A. Topology and Operations

Addressing the above issue, we propose a regulatable resonant supply-bus charger shown in Fig. 2. This charger eliminates the intermediate DC bus maintained by C_O and directly charges C_{Bus} by inductor current based on resonant operations. In this way, not only can the boosting function and the rapid recovery of V_{Bus} be achieved, but also it is a soft charging operation that avoids the hard charging between C_o and C_{Bus} in Fig. 1. The charger operates in four distinct phases. The conceptual waveforms of the proposed converter are shown in Fig. 3.

In the laser emission phase, a control signal V_{launch} triggers the switching of S_{Emit}, forward biasing D_L, and energizing the stray inductor L_{stray}, as illustrated in Fig. 4(a). Consequently, a resonant current pulse discharges C_{Bus} to $2V_{\text{DL}}-V_{\text{IN}}$, where V_{DL} denotes the forward conduction voltage of D_L. Throughout this phase, whose duration is relatively short, inductor current I_L changes very little, even if S_2 is on at the beginning of the phase. The theoretical resonant period of C_{Bus}-L_{stray} tank circuit is given by $T_{\text{disch}} = 2\pi\sqrt{L_{\text{stray}}C_{\text{bus}}}$. The half-maximum pulse-width t_W and the peak current $I_{\text{DL,pk}}$ of the current pulse through D_L can be derived as follows:

$$t_w = T_{\text{disch}} \frac{\pi - 2\sin^{-1}\frac{1}{2}}{2\pi} = \frac{2\pi}{3}\sqrt{L_{\text{stray}}C_{\text{Bus}}} \qquad (1)$$

$$I_{\text{DL,pk}} = \left(V_{\text{Bus,target}} - V_{\text{DL}}\right)\sqrt{\frac{C_{\text{Bus}}}{L_{\text{stray}}}} \qquad (2)$$

A rapid decline in V_{Bus} initiates the power stage transition into C_{Bus} charging phase Φ_C. During Φ_C, power switch S_2 turns on, establishing an L-C_{Bus} resonant loop with a resonant period of $T_{\text{ch}} = 2\pi\sqrt{LC_{\text{Bus}}}$, causing L to magnetize sharply between t_2

to t_4, as illustrated in Fig. 4(b). The resonant current then charges C_{Bus}. At $t = t_4$, V_{Bus} exceeds V_{IN}, leading I_L to commence its decrease while continuing to charge V_{Bus}.

To effectively regulate V_{Bus} to a targeted level, an active regulation phase Φ_R is introduced. During Φ_R, S_2 is turned off while S_1 and S_3 are activated, allowing I_L to recycle back to the input rather than charging C_{Bus}. This mechanism enables adjustment of the supply level for the LiDAR driver system by controlling the duration of Φ_C. During Φ_R, I_L decreases linearly at a rate of $-V_{IN}/L$, as depicted in Fig. 4(c). Once I_L reaches 0, D_1 blocks the reverse current path automatically, marking one cycle of I_L and V_{Bus} completed. Subsequently, the system transitions into a standby phase Φ_S in preparation for the next laser pulse emission, as shown in Fig. 4(d).

From the above, compared to conventional hard-charging-based structure, the proposed charger theoretically eliminates hard-charging losses during C_{Bus} charging while retaining V_{Bus} level regulation capability. In addition, the introduction of Φ_R enables the recovery of residual energy from the inductor, thereby enhancing overall efficiency.

B. Loss Analysis

In a conventional hard-charging-based scheme shown in Fig. 1, the power loss due to the hard charging between C_O and C_{Bus} (where $C_O \gg C_{Bus}$) during a laser emission period T_{emit} can be estimated by [9]:

$$P_{loss,\,hc} = \frac{2C_{Bus}\left(V_{Bus,target} - V_{DL}\right)^2}{T_{emit}} \quad (3)$$

Referring to Fig. 3, from t_0 to t_2, the effective power supplied to the laser diode is given by:

$$P_{DL} = \frac{2C_{Bus}V_{DL}\left(V_{Bus,target} - V_{DL}\right)}{T_{emit}} \quad (4)$$

Neglecting other losses, the driving efficiency of the hard-charging-based approach can be expressed as:

$$\eta_{hc} = \frac{P_{DL}}{P_{loss,\,hc} + P_{DL}} = \frac{V_{DL}}{V_{Bus,target}} \quad (5)$$

In a practical scenario, such as $C_{Bus} = 1\text{nF}$, $T_{emit} = 1\mu s$, $V_{DL} = 15V$, and $V_{target} = 70V$, the hard-charging loss is $P_{loss,hd} = 6.05W$ and the laser diode power is $P_{DL} = 1.65W$, resulting in a system power efficiency of only 21.4%. In contrast, the proposed charger inherently avoids hard charging. Although other common conduction losses from power devices and metal paths still impact efficiency, the proposed technique achieves an evident efficiency improvement.

III. EXPERIMENTAL RESULTS

Fig. 5 displays the PCB implementation of the proposed power-bus charger, designed for a 48-V input source and capable of generating a V_{Bus} level exceeding 100V for long-distance laser transmission in automotive applications. Component selections are summarized in the table shown in Table I. To validate the performance advantages of the proposed scheme, we also implemented and tested a scenario involving hard-charging losses with a voltage source and R_{Lmt} to charge C_{Bus} directly.

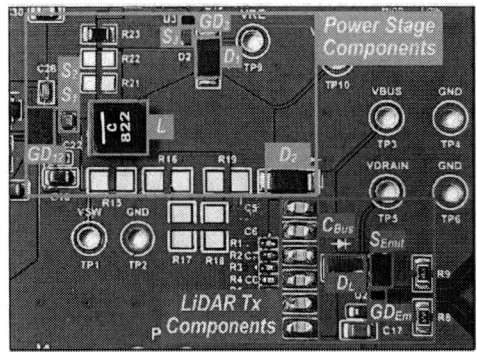

Fig. 5. Prototype PCB implementation.

TABLE I.   COMPONENT INFORMATION SUMMARY

Components	Part number	Parameters
S_1, S_2	EPC2106 (Half-Bridge)	100 V, 70 mΩ, 1.7 A
S_3	EPC2012C	200 V, 100 mΩ, 5 A
S_{Emit}	EPC2010C	200 V, 25 mΩ, 22 A
L	XGL5050-822MEC	8.2 uH
C_{Bus}	GRM21A5C2E101FW01	100 pF, 250 V $\times$ 6
D_1, D_2	RSX058LAP2S	200 V, 3 A
D_L	TPGAD1S09H	Laser Diode
GD_{12}	LMG1210 (Half-Bridge)	Driver
GD_3, GD_{Em}	LMG1020	Driver

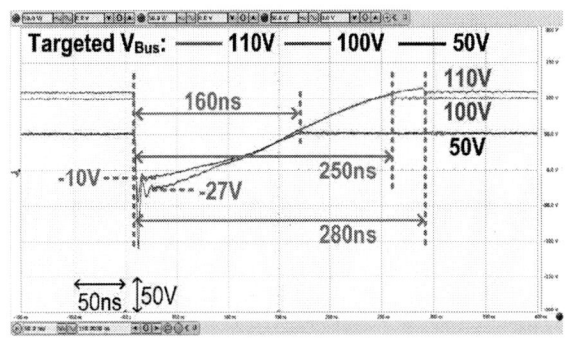

Fig. 6. Measured waveforms of the bus voltage.

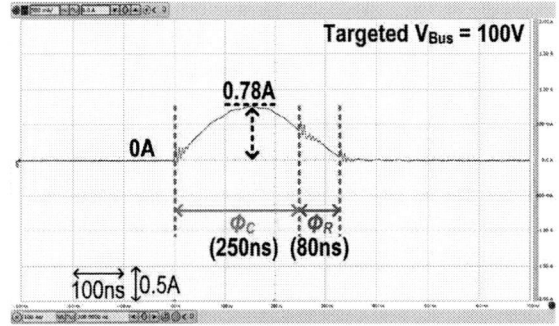

Fig. 7. Measured waveforms of the inductor current.

Fig. 6 shows the measured waveform of V_{Bus} voltage variation following laser diode emission at different targeted

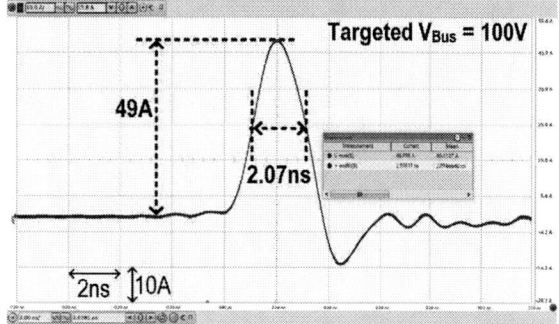

Fig. 8. Measured waveforms of the LiDAR emission driving current.

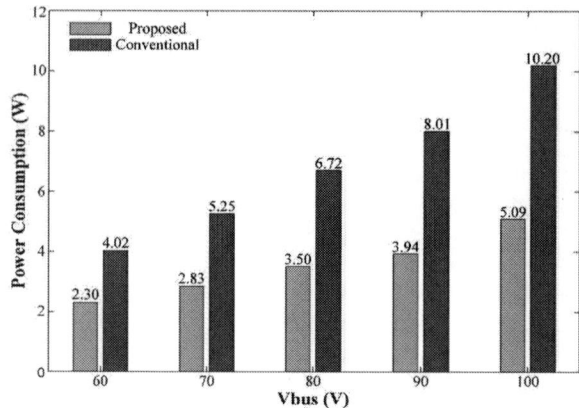

Fig. 9. Measured power loss comparison between the proposed and conventional chargers.

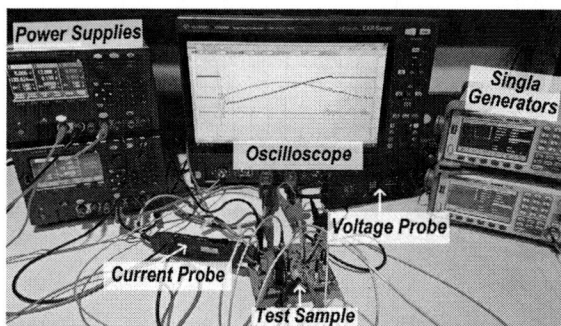

Fig. 10. Test environment.

V_{Bus} levels. The results indicate that as the targeted V_{Bus} level increases, the post-emission V_{Bus} drop becomes more pronounced. Consequently, the recovery time of V_{Bus} also lengthens with higher $V_{Bus,target}$. When $V_{Bus,target} = 110V$, the measured recovery time is 280ns, enabling support for a laser emission repetition frequency of over 3MHz. Fig. 7 shows the measured I_L waveform, highlighting the two switching phases of the inductor. At $V_{Bus} = 100V$, the duration of Φ_C aligns with the V_{Bus} recovery time observed in Fig. 6. These operating states are consistent with theoretical analysis. Fig. 8 shows the measured pulse current of the laser diode during a single emission. As observed, with a V_{Bus} of 100 V, the pulse peak current reaches 49A, and the half-maximum pulse width is 2.07 ns, fulfilling the requirements for long-distance object detection.

The power consumption of the emulated LiDAR transmitter system was tested under the conditions of $T_{emit} = 1\mu s$ and $C_{Bus} = 0.3nF$. As shown in Fig. 9, the system power consumption becomes more obvious with the increase of V_{Bus}. At $V_{Bus} = 100V$, this work reduces the system power consumption by approximately 50% compared to the conventional structure, significantly enhancing efficiency. Table II shows a performance summary and comparison with prior-art works, highlighting that this design offers a wider bus voltage range, faster bus recovery time, and lower power-stage losses. Fig. 10 shows the testing environment.

IV. CONCLUSION

This paper proposes a resonant bus charger for LiDAR driver systems. Compared to conventional hard-charging-based structures, this work demonstrates significant advantages. Experimental results clearly confirm the excellent performance of the work. With a 48V input, the output voltage can be regulated between 50V and 110V. When $V_{Bus} = 100V$, the charger recharges C_{Bus} in just 250ns, while providing a peak driving current of 49A for the laser diode with a pulse width of up to 2ns. Additionally, the overall system power consumption is significantly reduced.

TABLE II.   PERFORMANCE SUMMARY AND COMPARISION

	[5] PCIM Europe'18	[6] PRIME'22	[7] EPC9126HC	[8] JSSCC'23	Direct Bus Supply (Test)	This Work
Input Voltage	75V[#]	100V[#]	75V[#]	4V-40V	100V[#]	**48V**
Output Voltage	75V	100V	75V	60V-80V	100V	**50V-110V**
C_{Bus} Charging Type	Hard-Charging	Hard-Charging	Hard-Charging	Hard-Charging	Hard-Charging	**Soft-Charging**
Generated V_{Bus}	75V	100V	75V	60V-80V	100V	**50V-110V**
Peak of Laser Current Pulse	26A	44A	63A	40A	48A	**49A**
Half-Maximum Pulse-Width	1.8ns	2.1ns	5ns	N/A	2ns	**2ns**
Total Input Power	N/A	N/A	N/A	N/A	10.2W*	**5.09W**
V_{Bus} Recovery Time	N/A	N/A	N/A	500ns	600ns*	**250ns**

[#]: No converter used

ACKNOWLEDGMENT

This work was funded by the Science and Technology Development Fund, Macau SAR (FDCT) under Grant 0075/2024/AGJ, 0140/2023/RIA2 and 004/2023/SKL, Hetao SZ-HK S&T Innovation Cooperation Zone Project under Grant HTHZQSWS-KCCYB-2023030, and University of Macau under Grant MYRG-GRG2023-00147-IME, and MYRG-GRG2024-00196-IME

REFERENCES

[1] J. Glaser, "How GaN Power Transistors Drive High-Performance Lidar: Generating ultrafast pulsed power with GaN FETs," in IEEE Power Electronics Magazine, vol. 4, no. 1, pp. 25-35, March 2017.

[2] E. Abramov, M. Evzelman and M. M. Peretz, "Low-Voltage Sub-Nanosecond Pulsed Current Driver IC for High-Speed LIDAR Applications," in IEEE Journal of Emerging and Selected Topics in Power Electronics, vol. 8, no. 3, pp. 3001-3013, Sept. 2020.

[3] Y. -S. Ma et al., "29.6 A Digital-Type GaN Driver with Current-Pulse-Balancer Technique Achieving Sub-Nanosecond Current Pulse Width for High-Resolution and Dynamic Effective Range LiDAR System," 2019 IEEE International Solid-State Circuits Conference (ISSCC), San Francisco, CA, USA, 2019, pp. 466-468.

[4] X. Cui, C. Keller and A. -T. Avestruz, "Cycle-by-Cycle Digital Control of a Multi-Megahertz Variable-Frequency Boost Converter for Automatic Power Control of LiDAR," 2019 IEEE Energy Conversion Congress and Exposition (ECCE), Baltimore, MD, USA, 2019, pp. 702-711.

[5] J. Glaser, "High Power Nanosecond Pulse Laser Driver Using an GaN FET," PCIM Europe 2018; International Exhibition and Conference for Power Electronics, Intelligent Motion, Renewable Energy and Energy Management, Nuremberg, Germany, 2018, pp. 1-8.

[6] A. Bettini et al., "Analysis and Design of a Fully-Integrated Pulsed LiDAR Driver in 100V-GaN IC Technology," 2022 17th Conference on Ph.D Research in Microelectronics and Electronics (PRIME), Villasimius, SU, Italy, 2022, pp. 273-276.

[7] Efficient Power Conversion Corp, 2022. "Getting the Most Out of eGaN FETs and Your EPC9126 Laser Driver," [Online]. Available: https://epc-co.com/epc/Portals/0/epc/documents/application-notes/AN027%20Getting-the-Most-out-of-eGaN-FETs.pdf

[8] S. -Y. Li et al., "A 4–40 V Wide Input Range Boost Converter With the Protection Re-Cycling Technique for 200 W High Power LiDAR System in a Long-Distance Object Detection," in IEEE Journal of Solid-State Circuits, vol. 58, no. 7, pp. 1850-1859.

[9] Y. Lei and R. C. N. Pilawa-Podgurski, "A General Method for Analyzing Resonant and Soft-Charging Operation of Switched-Capacitor Converters," in IEEE Transactions on Power Electronics, vol. 30, no. 10, pp. 5650-5664, Oct. 2015

A Trajectory Controlled 48-to-24 V Resonant Switched Capacitor Converter with 98.7% Efficiency and Ultrafast Dynamic Response

Hélène T.W. Ma Yang
Edward S. Rogers Sr.
Department of Electrical &
Computer Engineering
University of Toronto
Toronto, Canada
helenetw.mayang@mail.utoronto.ca

Liang Wang
School of Information Science
and Technology

ShanghaiTech University
Shanghai, China
wangliang1@shanghaitech.edu.cn

Haoyu Wang
School of Information Science
and Technology

ShanghaiTech University
Shanghai, China
wanghy@shanghaitech.edu.cn

Wai Tung Ng
Edward S. Rogers Sr.
Department of Electrical &
Computer Engineering
University of Toronto
Toronto, Canada
ngwt@ece.utoronto.ca

Abstract—**Resonant converters have high power conversion efficiency and can achieve soft switching over a wide range of load conditions. However, during load transients, mismatches between the current provided by the resonant tank and the load current can cause abrupt voltage changes, affecting stability. Maintaining stable output voltage requires a large output capacitance, increasing cost and size. By controlling the trajectory of the resonant tank, rapid recovery after each load transient can be achieved. This work proposes a trajectory control method for a 48-to-24 V resonant switching capacitor converter to improve the transient response performance with a maximum load current of 20 A and output power of 480 W, a 90% reduction in output capacitance for the same output voltage ripple can be achieved with a peak efficiency of 98.7% and a very fast transient recovery of under 300 ns.**

Keywords—dynamic response, load transient, resonant series capacitor, trajectory control

I. INTRODUCTION

The Information Age is ushering in an unprecedented increase in data generation, processing, and storage requirements. This trend has been intensified by the recent rise of cloud computing, big data processing, and artificial intelligence (AI). The surge in data volume has placed immense pressure on data centers, which form the core of the digital infrastructure [1]. In 2022, data centers consumed no less than 500 TWh of energy globally. The International Energy Agency (IEA) conservatively projects that this consumption will increase to more than 800 TWh by 2026 [2]. With current power grids nearing their capacity limits, upgrading the existing infrastructure is a costly venture. Therefore, data centers must leverage technological advances to meet the growing demands for ever higher processing speeds while maintaining energy efficiency and reliability. Moreover, the conventional 12 V bus architecture in data centers is gradually being replaced by the 48 V bus architecture [3]. This upgrade is part of the shift to next-generation data centers. The 48 V architecture reduces losses but also introduces additional challenges, particularly for onboard power supplies (OPS).

Fixed-ratio buck converters are crucial in two-stage data center OPS. These power architectures feature a primary stage DC converter. This primary stage reduces the step-down requirement for secondary stage converters. The primary stage DC converter uses simple, efficient topologies with high efficiency, high power density, and simple control schemes [4]. Improving the transient response speed is crucial for improving the power density of the entire converter system. Faster response reduces the number of bus capacitors between the two stages, hence increasing the power density [5]. Common resonant converters in data center power supplies include LLC converters and resonant switching capacitor (RSC) converters. For isolated LLC converters, control methods have been proposed by Virginia Tech researchers in [6]-[9] to enhance its transient response. However, the transformer limits its efficiency. Non-isolated topologies such as the RSC converters are simpler and feature fewer components, allowing higher efficiency and power density. Optimized control can also enable zero current switching (ZCS) for soft switching. A single resonant unit in RSC converters achieves a 2:1 step-down ratio. Multiple units in series or switched tank converters can achieve higher ratios. However, additional components limit power density and efficiency. Optimized control for resonant RSC converters can modulate the output voltage and enhance performance [10], [11]. A few studies have addressed trajectory control for non-

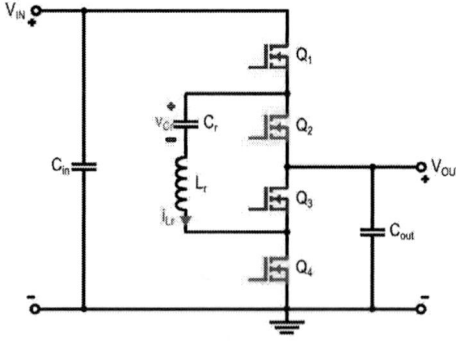

Fig. 1. The proposed resonant switched capacitor converter topology.

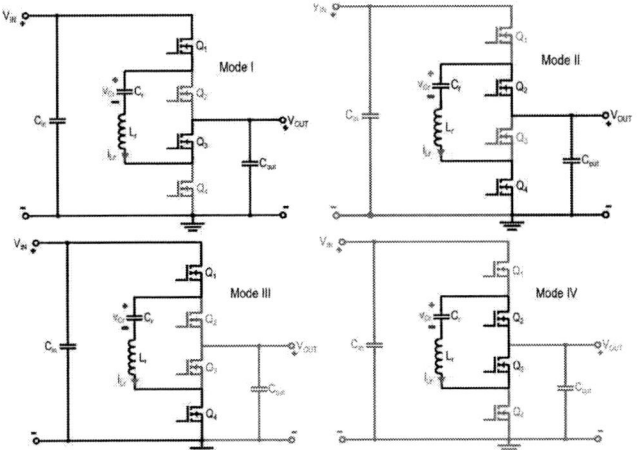

Fig. 2. Operation modes of the resonant switched capacitor converter.

isolated resonant RSC converters [12], [13]. Work by Sim *et al.* has reported V_{OUT} regulation using such control [14].

This paper proposes a trajectory control method for RSC converters to improve transient response performance. The organization is as follows: Section II introduces the circuit topology, describes the RSC converter's operation principles, and explains the trajectory control method. Section III presents the simulation results. Section IV provides the experimental results. Section V concludes the paper.

II. CIRCUIT TOPOLOGY, OPERATION PRINCIPLES, AND TRAJECTORY CONTROL

A. Circuit Topology

The topology of an RSC converter is shown in Fig. 1. It features four series-connected MOSFETs $Q_1 - Q_4$. A resonant tank is formed by the resonant capacitor C_r and resonant inductor L_r. The input and output capacitors C_{in} and C_{out} are sufficiently large such that the input and output voltages can be assumed to be constant during steady-state and transient operations.

B. Steady State Analysis

The RCS converter has two steady-state modes of operation, shown in Fig. 2. In Mode I, switches Q_1 and Q_3 conduct. The resonant tank is connected to the input power source and output load. The voltage across the resonant tank is $V_{IN} - V_{OUT}$. In Mode II, switches Q_2 and Q_4 conduct. The resonant tank is connected across output load. The voltage across the resonant tank is V_{OUT}. Fig. 3 shows the steady-state waveforms of the RCS converter. The duty cycle D for transistors $Q_1 - Q_4$ is 50%. The adjacent switches conduct in a complementary manner with a certain deadtime to prevent shoot through current.

The resonant tank, comprising of resonant capacitor C_r and resonant inductor L_r, has a resonant frequency f_r which can be expressed as:

$$f_r = \frac{1}{2\pi\sqrt{L_r \cdot C_r}}. \quad (1)$$

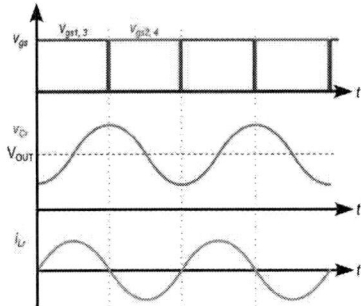

Fig. 3. Steady-state waveforms of the resonant switched capacitor converter.

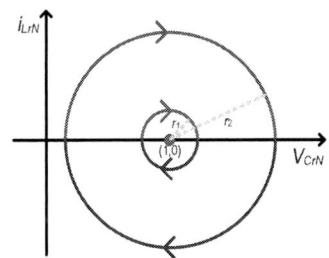

Fig. 4. Steady-state trajectory waveform of the resonant switched capacitor converter.

During mode transitions, the resonant inductor current $i_{Lr}(t)$ decreases to zero. This behavior achieves ZCS, reduces switching losses, and improves efficiency. The starting times of Mode I and II are defined as t_0 and t_1, respectively. The resonant inductor current $i_{Lr}(t)$ and capacitor voltage $v_{Cr}(t)$ over a full cycle can be expressed as:

$$i_{Lr}(t) = i_{Lr}(t_0) \cos[\omega(t - t_0)] \\ - \frac{v_{Cr}(t_0) - V_{OUT}}{Z_0} \sin[\omega(t - t_0)], \quad (2)$$

$$v_{Cr}(t) = i_{Lr}(t_0)Z_0 \sin[\omega(t - t_0)] \\ + (v_{Cr}(t_0) - V_{OUT}) \cos[\omega(t - t_0)] + V_{OUT}. \quad (3)$$

The state-plane analysis can be performed by normalizing the resonant inductor current and capacitor voltage. (2) and (3) are obtained after normalizing all voltages with respect to V_{OUT} and all currents with respect to V_{OUT}/Z_0. The normalized resonant trajectory is plotted in Fig. 4.

$$I_{LN} = I_{LON} \cos \theta - (V_{CON} - 1) \sin \theta, \quad (4)$$

$$V_{CN} = I_{LON} \sin \theta + (V_{CON} - 1) \cos \theta + 1. \quad (5)$$

$$I_{LN}{}^2 + (V_{CN} - 1)^2 = I_{LON}{}^2 + (V_{CON} - 1)^2 = (V_{CON} - 1)^2. \quad (6)$$

In the normalized resonant trajectory plotted in Fig. 4, the horizontal axis represents the normalized capacitor voltage V_{CrN}. The vertical axis represents the normalized inductor current I_{LrN}. The trajectories are circles centered at $(1, 0)$. The radii are determined by different load conditions. The radius r of the circle can be expressed as:

$$r = 1 - V_{CON} = \frac{Z_0 I_{OUT} \pi}{2 V_{OUT}}. \quad (7)$$

979-8-3315-1612-3/25 $31.00 © 2025 IEEE

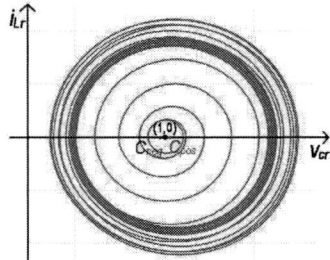

Fig. 5. Trajectory waveforms of the resonant switched capacitor converter with open-loop fixed-frequency control during load step-up.

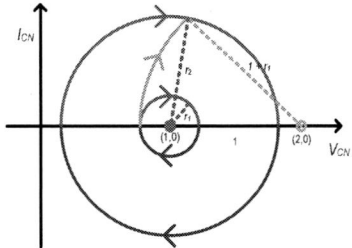

Fig. 6. Optimized trajectory waveform of the resonant switched capacitor converter during load step-up.

C. Transient State Analysis

During a load transient, the output voltage response is influenced by the converter's output impedance. Although the output voltage can eventually stabilize, a sudden load increase or decrease can cause mismatches between the inductor current and the load demand. This mismatch leads to temporary drops or overshoots in the output voltage. These output voltage changes affect stability and require additional output capacitors to maintain a steady voltage. However, adding capacitors increases cost and system size, which is undesirable for data center OPS.

The output voltage V_{OUT} deviates from the steady-state output voltage $V_{OUT,st}$ during load transients. The trajectory of the RSC converter during a load step-up transient is shown in Fig. 5. As the load current suddenly increases, V_{OUT} decreases. The trajectory on the positive half-cycle is thus $(V_{IN} - V_{OUT})/V_{OUT,st} > 1$, shifting to the right. Similarly, the trajectory on the negative half-cycle is $V_{OUT}/V_{OUT,st} < 1$, shifting to the left. In both cases, the radius will increase as per (7). After multiple cycles, the trajectory expands outward from smaller inner circles to larger outer circles. The resonant tank can balance the energy mismatch between the input and load demands. This mismatch leads to temporary ringing and overshoot at the output voltage. Therefore, traditional open-loop control with fixed frequency typically exhibits slow transient recovery and large voltage fluctuations during load changes.

To improve the transient performance of the RSC converter, a trajectory control scheme is proposed. During steady-state operation, an open-loop fixed-frequency control method is used to maintain the switching frequency to be the same as the resonant frequency. During transient operation, an optimized trajectory control method predicts the trajectory before and after the load variation. It adjusts the control signal immediately after switching and accurately generates the corresponding control signal. This allows the resonant tank state to follow the optimal trajectory in the shortest time possible. Doing so ensures fast switching between LO- and HI-load states, reduces the transient response time, minimizes output voltage ripple, and smooths the trajectory. Fig. 6 shows the optimized transient trajectory.

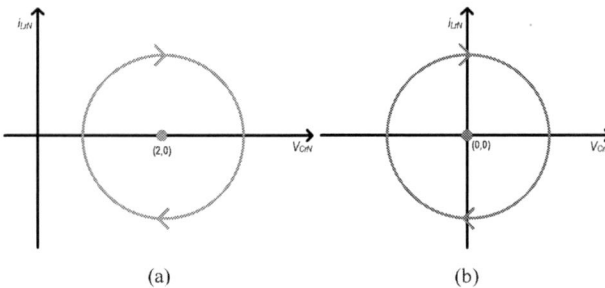

Fig. 7. Trajectory waveforms for (a) Mode III and (b) Mode IV.

To implement this trajectory control, Modes III and Mode IV are introduced, as shown in Fig. 7. In Mode III, switches Q_1 and Q_4 conduct. The resonant tank is connected to the input voltage. The resonant tank's voltage of $V_{IN} = 2\,V_{OUT}$ allows its energy to change rapidly. In Mode IV, switches Q_2 and Q_3 conduct. There is zero voltage across the resonant tank. The starting times of Mode III and IV are defined as t_0 and t_1, respectively. The resonant inductor current $i_{Lr}(t)$ and capacitor voltage $v_{Cr}(t)$ over a full cycle can be expressed as:

$$i_{Lr}(t) = i_{Lr}(t_0) \cos[\omega(t - t_0)] \\ - \frac{V_{IN} - v_{Cr}(t_0)}{Z_0} \sin[\omega(t - t_0)], \tag{7}$$

$$v_{Cr}(t) = i_{Lr}(t_0) Z_0 \sin[\omega(t - t_0)] \\ + (v_{Cr}(t_0) - V_{IN}) \cos[\omega(t - t_0)] + V_{IN}. \tag{8}$$

Normalizing all voltages with respect to V_{OUT} and all currents with respect to V_{OUT}/Z_0, (7) and (8) can be rewritten and combined in (9). The trajectory with its center at $(2, 0)$ is plotted in Fig. 7 (a). Equation (10) shows the trajectory radius.

$$I_{LN}{}^2 + (V_{CN} - 2)^2 = I_{L0N}{}^2 + (V_{C0N} - 2)^2. \tag{9}$$

$$r = \sqrt{I_{L0N}{}^2 + (V_{C0N} - 2)^2}. \tag{10}$$

Similarly for Mode IV, the resonant inductor current $i_{Lr}(t)$ and capacitor voltage $v_{Cr}(t)$ over a full cycle and the trajectory radius can be expressed as (11)-(13). Fig. 7 (b) shows the Mode IV's trajectory.

$$i_{Lr}(t) = i_{Lr}(t_0) \cos[\omega(t - t_0)] \\ - \frac{v_{Cr}(t_0)}{Z_0} \sin[\omega(t - t_0)], \tag{11}$$

$$v_{Cr}(t) = i_{Lr}(t_0) Z_0 \sin[\omega(t - t_0)] \\ + v_{Cr}(t_0) \cos[\omega(t - t_0)]. \tag{12}$$

$$I_{LN}{}^2 + V_{CN}{}^2 = I_{L0N}{}^2 + V_{C0N}{}^2. \tag{13}$$

$$r = \sqrt{I_{L0N}{}^2 + V_{C0N}{}^2}. \tag{14}$$

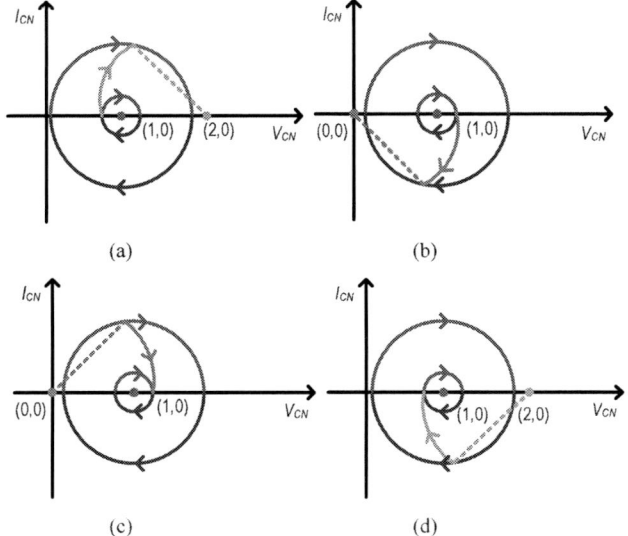

(a) (b)

(c) (d)

Fig. 8. Proposed transient trajectory methods for (a) positive half-cycle load step-up, (b) negative half-cycle load step-up, (c) positive half-cycle load step-down, and (d) negative half-cycle load step-down.

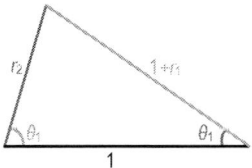

Fig. 9. Geometric relations for the transient trajector control strategy

D. Trajectory Control

Fig. 9 illustrates the proposed trajectory control principle. The control strategy can be divided into four cases depending on the load variation and the switching within the positive or negative half-cycle.

a) *Load step-up during positive half-cycle:* When the inductor current reaches zero, the RSC converter switches from Mode II to Mode III. The resonant tank energy increases rapidly until it matches the required energy. Then, the trajectory returns to the steady-state trajectory Mode I and switches between Modes I and II.

b) *Load step-up during negative half-cycle:* When the inductor current reaches zero, the RSC converter switches from Mode I to Mode IV. The trajectory continues in Mode IV until it reaches the steady-state trajectory Mode II and switches between Modes I and II.

c) *Load step-down during positive half-cycle:* After some time in Mode I, the system transitions to Mode IV. The resonant tank trajectory then follows Mode IV until it aligns with the steady-state trajectory of Mode II. When the inductor current reaches zero, the trajectory returns to Mode II and completes the transient response.

d) *Load step-down during negative half-cycle:* After some time in Mode II, the system transitions to Mode III. The resonant tank trajectory then follows Mode III until it aligns with the

steady-state trajectory of Mode I. When the inductor current reaches zero, the trajectory returns to Mode I and completes the transient response.

Precise timing during transient switching is crucial to trajectory control, as shown in Fig. 8. Accurately calculating the durations of different modes during the transient process is critical for achieving rapid trajectory transitions. Taking the load step-up scenario as an example, Modes III and I's duration must be calculated accurately. By analyzing the geometrical relationships in Fig. 8, the geometric relationships of Fig. 9 can be obtained. r_1 and r_2 represent the radii of the trajectories in the steady-state Modes II and I, respectively. I_{OUT1} represents the load current during LO-load. I_{OUT2} represents the load current during HI-load. Note that LO- and HI-loads refer to the relative load currents before and after the change.

$$r_1 = \frac{Z_0 I_{OUT1} \pi}{2 V_{OUT}}, \qquad r_2 = \frac{Z_0 I_{OUT2} \pi}{2 V_{OUT}} \qquad (15)$$

The triangle's sides are $(1 + r_1)$, r_2, and 1. Using the cosine rule, the angles θ_1 and θ_2 can be derived with θ_1 corresponding to the duration of the transition mode.

$$\theta_1 = arccos \frac{1 + (1 + r_1)^2 - r_2^2}{2(1 + r_1)},$$
$$\theta_2 = arccos \frac{1 + r_2^2 - (1 + r_2)^2}{2 r_2} \qquad (16)$$

The time corresponding to Mode III (t_1) and to Mode I (t_2) during the load step-up transient process can be derived as:

$$t_1 = \frac{\theta_1}{2\pi} T_s, \qquad t_2 = \frac{\theta_2}{2\pi} T_s \qquad (17)$$

Therefore, with knowledge of the load current before and after transient switching, the control duration of each mode can be determined. This allows for rapid transient response and helps reduce the output capacitor size. Fig. 10 shows the control strategy. Fig. 11 depicts the trajectory control flowchart. When the output voltage is stable and the load current ripple is not severe, steady-state control is used. This maintains the converter's operation at steady-state, ensuring stability at the output voltage without being affected by load variations.

When significant load current changes are detected, transient trajectory control is activated. First, the load current is sampled to obtain the values for I_{OUT1} and I_{OUT2}. Based on the corresponding equations, the duration of each transient mode at both ends of the process is calculated. The system determines whether the converter is undergoing a load step-up or step-down process by comparing I_{OUT1} and I_{OUT2}. Next, the system checks whether it is in a positive half-cycle or a negative half-cycle based on the current operating state. These actions require real-time current monitoring. After completing transient trajectory control, the system exits transient mode and returns to steady-state fixed-frequency operation until the next transient load disturbance is detected.

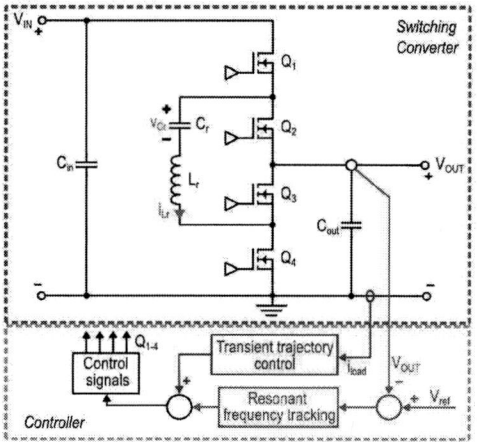

Fig. 10. Control strategy of the resonant switched capacitor converter.

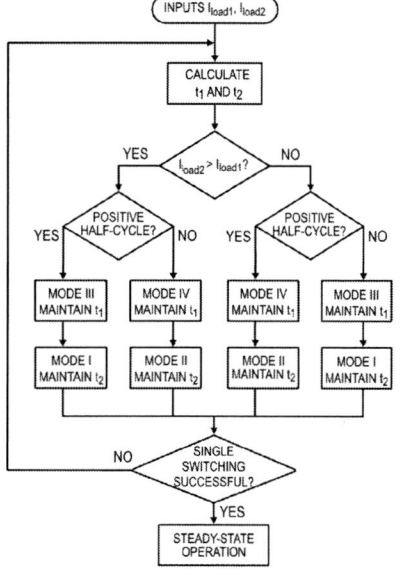

Fig. 11. Flowchart of the transient trajectory control.

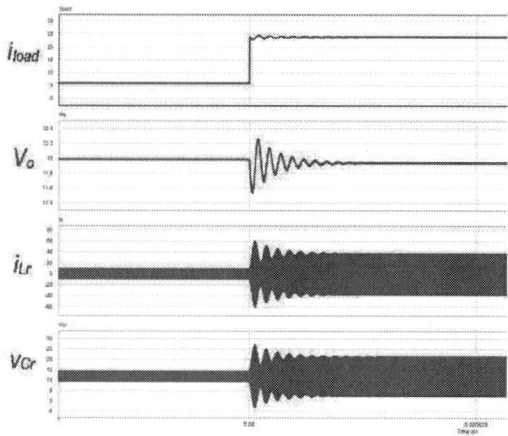

Fig. 12. Transient time-domain waveforms with conventional fixed-frequency during load step-up.

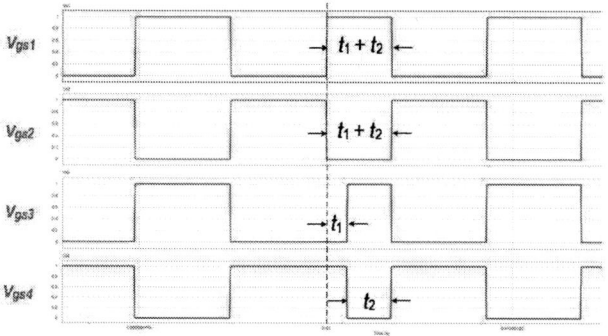

Fig. 13. Control signals during transient for trajectory control.

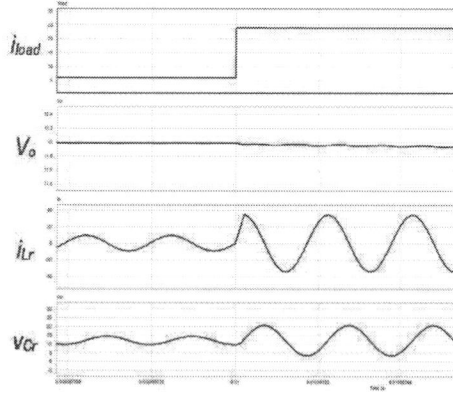

Fig. 14. Simulated time-domain transient waveforms with trajectory control.

III. SIMULATION RESULTS

The proposed transient trajectory control is verified through PSIM simulations. The verification uses the example of load step-up from 6 A to 24 A during the positive half cycle. Fig. 12 shows the simulated waveforms using the traditional fixed-frequency control method. The simulation results show that when the load changes significantly under open-loop fixed-frequency control, the output voltage fluctuates considerably. The peak-to-peak amplitude is about 0.5 V. It takes about 700 µs for the voltage to return to steady state. The inductor current and capacitor voltage exhibit large oscillations. These oscillations could affect system lifespan and stability.

Fig. 13 shows the transient switching control signals. The resulting circuit waveforms are illustrated in Fig. 14. Using trajectory control, the inductor current quickly charges and transitions from LO- to HI-load in under 300 ns. There is little fluctuation in the voltage across the output capacitor. The

proposed trajectory control method greatly reduces the output voltage fluctuation amplitude during sudden load changes.

Comparing the proposed trajectory control with traditional fixed-frequency open-loop control through simulation, the proposed method significantly reduces the voltage ripple during load step changes. It allows a 90% reduction in the output capacitance compared to the conventional control method. These reductions further lower the costs and the converter sizes.

979-8-3315-1612-3/25 $31.00 © 2025 IEEE

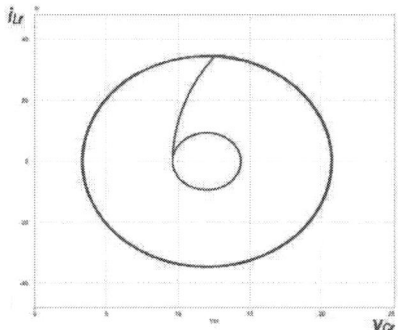

Fig. 15. Simulated transient trajectory waveform during load step-up.

Using time-domain waveforms, the trajectory waveform is shown in Fig. 15. Trajectory control rapidly transitions the resonant tank from the inner LO-load circle to the outer HI-load circle in a single-step operation. With precise control and matching, there is ideally no overshoot nor oscillation. This improves component stability.

IV. EXPERIMENTAL RESULTS

To verify the effectiveness of the proposed control scheme, an RSC converter with a rated input of 48 V and a rated output of 24 V was designed. The circuit parameters are shown in Table I. The prototype converter is shown in Fig. 16. Two half-bridge drivers are used. Bootstrapping supplies the gate drive circuits for the switching devices.

Fig. 17 shows the steady-state experimental waveforms of the resonant capacitor voltage and resonant inductor current. Since the resonant frequency matches the switching frequency, the resonant inductor current and resonant capacitor voltage exhibit ideal sinusoidal waveforms. This ensures zero current during switching transitions, achieving ZCS soft-switching. In low-voltage and high-current applications, ZCS soft-switching helps to reduce switching losses and improve efficiency.

The RSC converter's steady-state efficiency achieves a peak efficiency of 98.7%, shown in Fig. 18. Under full load conditions, the converter reaches an efficiency of 97.5%, despite increased conduction losses. This is due to its simple topology, non-isolated structure, and soft switching characteristics.

TABLE I.   RSC CONVERTER PARAMETERS

Components	Values
Input voltage V_{IN}	48 V
Output voltage V_{OUT}	24 V
Resonant capacitor C_r	5.2 µF
Resonant inductor L_r	100 nH
Switching frequency f_s	220 kHz
Output capacitance C_{OUT}	100 µF
Load current I_{OUT}	0-20 A

The output voltage waveform during transient load changes with trajectory control was verified experimentally. Fig. 19 shows the transient time-domain waveforms and trajectory for a load step-up from 5 A to 15 A during the positive half-cycle. The settling time after load transient is 260 ns. The experimental results show that the resonant inductor current rapidly charges

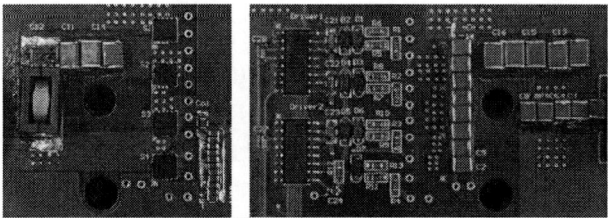

Fig. 16. The proposed resonant switched capacitor converter.

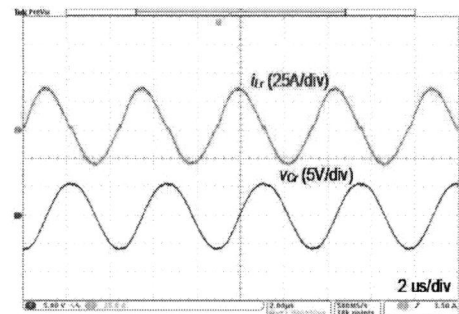

Fig. 17. Experimental steady-state waveforms.

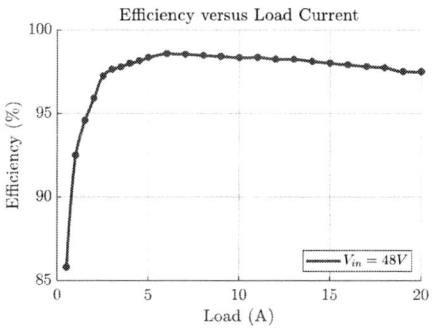

Fig. 18. A plot of the power conversion efficiency versus load current.

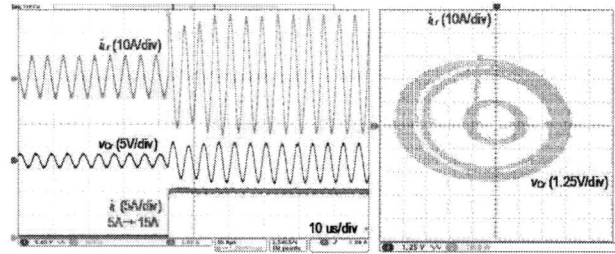

Fig. 19. Experimental waveforms of the trajectory control during negative half-cycle at load step-up.

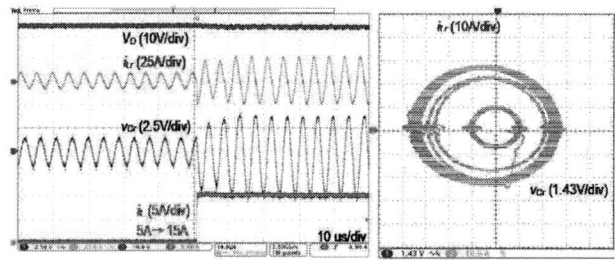

Fig. 20. Experimental waveforms of the trajectory control during negative half-cycle at load step-up.

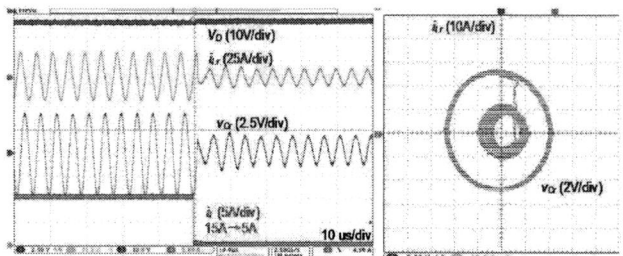

Fig. 21. Experimental waveforms of the trajectory control during positive half-cycle at load step-down.

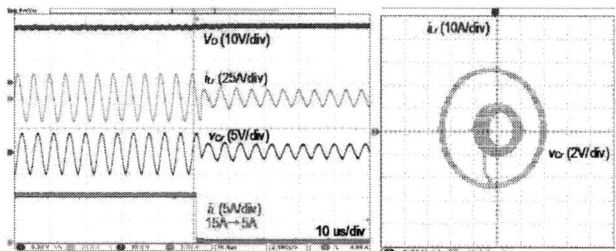

Fig. 22. Experimental waveforms of the trajectory control during negative half-cycle at load step-down.

during the transient mode to match the load current. The trajectory diagram shows a rapid trajectory transition during the LO-load to HI-load jump. Minor oscillations were observed during the experiment, but the output voltage returned to the set value after a short period.

Similarly, Fig. 20 shows the transient time-domain waveforms and trajectory for a load step-up from 5 A to 15 A during the negative half-cycle. Fig. 21 illustrates the transient time-domain waveforms and trajectory for a load step-down from 15 A to 5 A during the positive half-cycle. Finally, Fig. 22 presents the transient time-domain waveforms and trajectory for a load step-down from 15 A to 5 A during the negative half-cycle. All four experiments demonstrate excellent transient response performance. The trajectory control maintains a stable output voltage with no significant fluctuations during large load transients.

V. CONCLUSIONS

A trajectory control strategy for the primary-stage RSC converter is proposed to improve transient response performance in data center applications. Two transitional modes are introduced to improve transient response speed. Control strategies are proposed for the four scenarios of transitions between positive and negative half-cycles and LO- and HI-load conditions. The mathematical principles underlying trajectory control provide a theoretical foundation for precise control. Experimental results verified the feasibility of the proposed control concept and compared it with the traditional open-loop fixed-frequency control strategy. Under transient load conditions, the proposed control method demonstrated a 90% reduction in output capacitance for the same output voltage ripple, highlighting the validity of the proposed control strategy.

ACKNOWLEDGMENT

This project was made possible by financial support from the Natural Sciences and Engineering Research Council of Canada.

REFERENCES

[1] A. Shehabi, S. J. Smith, E. Masanet, and J. Koomey, 'Data center growth in the United States: decoupling the demand for services from electricity use', *Environ. Res. Lett.*, vol. 13, no. 12, p. 124030, Dec. 2018, doi: 10.1088/1748-9326/aaec9c.

[2] 'Electricity 2024 – Analysis and forecast to 2026', IEA. Accessed: Sep. 09, 2024. [Online]. Available: https://www.iea.org/reports/electricity-2024

[3] P. Sandri, 'Increasing Hyperscale Data Center Efficiency: A Better Way to Manage 54-V/48-V-to-Point-of-Load Direct Conversion', *IEEE Power Electron. Mag.*, vol. 4, no. 4, pp. 58–64, Dec. 2017, doi: 10.1109/MPEL.2017.2760113.

[4] K. Kesarwani and J. T. Stauth, 'The direct-conversion resonant switched capacitor architecture with merged multiphase interleaving: Cost and performance comparison', in *2015 IEEE Applied Power Electronics Conference and Exposition (APEC)*, Mar. 2015, pp. 952–959. doi: 10.1109/APEC.2015.7104464.

[5] A. Borzooy, S. A. Khajehoddin, M. Karimi-Ghartemani, and M. Ebrahimi, 'Alternative Control Approach to Achieve Fast Load-Transient Responses in DC–DC Converters', *IEEE Trans. Ind. Electron.*, vol. 68, no. 12, pp. 12668–12678, Dec. 2021, doi: 10.1109/TIE.2020.3040675.

[6] R. Oruganti, J. J. Yang, and F. C. Lee, 'Implementation of optimal trajectory control of series resonant converter', in *1987 IEEE Power Electronics Specialists Conference*, Jun. 1987, pp. 451–459. doi: 10.1109/PESC.1987.7077214.

[7] W. Feng, F. C. Lee, and P. Mattavelli, 'Optimal Trajectory Control of Burst Mode for LLC Resonant Converter', *IEEE Trans. Power Electron.*, vol. 28, no. 1, pp. 457–466, Jan. 2013, doi: 10.1109/TPEL.2012.2200110.

[8] C. Fei, Q. Li, and F. C. Lee, 'Digital Implementation of Light-Load Efficiency Improvement for High-Frequency LLC Converters With Simplified Optimal Trajectory Control', *IEEE J. Emerg. Sel. Top. Power Electron.*, vol. 6, no. 4, pp. 1850–1859, Dec. 2018, doi: 10.1109/JESTPE.2018.2832135.

[9] A. Nabih, M. H. Ahmed, Q. Li, and F. C. Lee, 'Transient Control and Soft Start-Up for 1-MHz LLC Converter With Wide Input Voltage Range Using Simplified Optimal Trajectory Control', *IEEE J. Emerg. Sel. Top. Power Electron.*, vol. 9, no. 1, pp. 24–37, Feb. 2021, doi: 10.1109/JESTPE.2020.2973660.

[10] Y. P. B. Yeung, K. W. E. Cheng, S. L. Ho, K. K. Law, and D. Sutanto, 'Unified analysis of switched-capacitor resonant converters', *IEEE Trans. Ind. Electron.*, vol. 51, no. 4, pp. 864–873, Aug. 2004, doi: 10.1109/TIE.2004.831743.

[11] K. K. Law, K. W. E. Cheng, and Y. P. B. Yeung, 'Design and analysis of switched-capacitor-based step-up resonant converters', *IEEE Trans. Circuits Syst. Regul. Pap.*, vol. 52, no. 5, pp. 943–948, May 2005, doi: 10.1109/TCSI.2004.840482.

[12] J. Caro, J. Mayo-Maldonado, J. Valdez-Resendiz, A. Alejo-Reyes, F. Beltran-Carbajal, and O. Lopez-Santos, 'An Overview of Non-Isolated Hybrid Switched-Capacitor Step-Up DC–DC Converters', *Appl. Sci.*, vol. 12, p. 8554, Aug. 2022, doi: 10.3390/app12178554.

[13] Y. Dong-Ying, K. Shih-Hao, S. Yong-Long, C. Huang-Jen, L. Kuo-Chi, and Y. Ta-Yung, 'An Improved 48-to-12V Series-Parallel Resonant Switched-Capacitor Converter for Data Center Applications', in *2021 IEEE International Future Energy Electronics Conference (IFEEC)*, Nov. 2021, pp. 1–6. doi: 10.1109/IFEEC53238.2021.9661888.

[14] S. Y. Sim, X. Zhang, J. Jiang, K. Wei, and C. Huang, 'A 94.7% Efficiency Direct-Step-Down Switched-Tank-Based 48V to 1V-3.3V Hybrid Converter with Constant-Resonant-Time Closed-Loop Control', in *2024 IEEE Applied Power Electronics Conference and Exposition (APEC)*, Feb. 2024, pp. 1344–1350. doi: 10.1109/APEC48139.2024.10509251.

Low Power, Non-Isolated, Extremely-High Step-Up, Quasi-Resonant Hybrid dc–dc Converter

Kumar Joy Nag ©, Aleksandar Prodić ©

The Edward S. Rogers Sr. Department of Electrical & Computer Engineering
University of Toronto
10 King's College Road, Toronto, ON, M5S 3G4, Canada
kumarjoy.nag@mail.utoronto.ca, prodic@ece.utoronto.ca

Abstract—**This paper introduces a low-power, non-isolated, step-up quasi-resonant converter and a complementary mixed-signal controller. The introduced converter is capable of generating conversion ratios exceeding 50x, while maintaining high power processing efficiency. At low power levels, ranging from a fraction of a watt to a few watts, conventional ultra-high step-up solutions operate at efficiencies that are not exceeding 45%, mostly limited by switching losses. The introduced solution exhibits efficiencies exceeding 70%. The improvements are achieved by strategically incorporating a N-stage charge pump with a previously introduced fully soft-switching quasi-resonant boost converter. The solution exhibits load-independent soft-switching for all the active switches, at all voltage conversion ratios. The operation and performance of the introduced converter are experimentally verified with a discrete prototype with input voltage 3.6 V, boosting it to high output voltages ranging from 55 V to 250 V, while supplying load currents ranging from 1 mA to 20 mA and achieves peak efficiency of around 85%.**

Index Terms—**Charge Pump, Flying Capacitor, High Voltage, Piezoelectric, Quasi Resonant, Soft-Switching, Switching Loss**

I. INTRODUCTION

Low-power, high-voltage converters are becoming increasingly prevalent in applications such as piezoelectric actuation, bio-medical stimulators, energy harvesting, etc. Piezoelectric actuators are widely used in micro air vehicles (MAVs), micro-scale robots, propulsion systems for small-sized satellites, etc., owing to their ability to deliver the highest power density of competitor actuators in milligram-level [1], [2]. However, the most commonly used piezo actuator material, lead zirconate titanate (PZT), require actuation voltages of 100 V-300 V. Work done in [3], utilizes piezoelectric resonators to miniaturize plasma generation systems. Wearable Electric stimulators (ESs) are employed to inject charge balanced low-energy pulses to stimulate the highly resistive muscle tissues [4]. For this, high voltages up-to 200 V need to be generated. Here, the input source is typically a lithium-ion battery providing 3 V-3.6 V. A highly-efficient and high power-density dc–dc step-up converter is required for extended battery operation and light-weight design. Purely-magnetic-based boost converter operating in discontinuous conduction mode (DCM) can be considered to be the simplest solution. Since the main power switch needs to conduct the inductor charging current and block the full output voltage, it will have a large V-A product resulting in huge voltage-current overlap losses during switch turn-OFF. Purely switched-capacitor-based topologies [5], [6], [7] (e.g., Dickson,

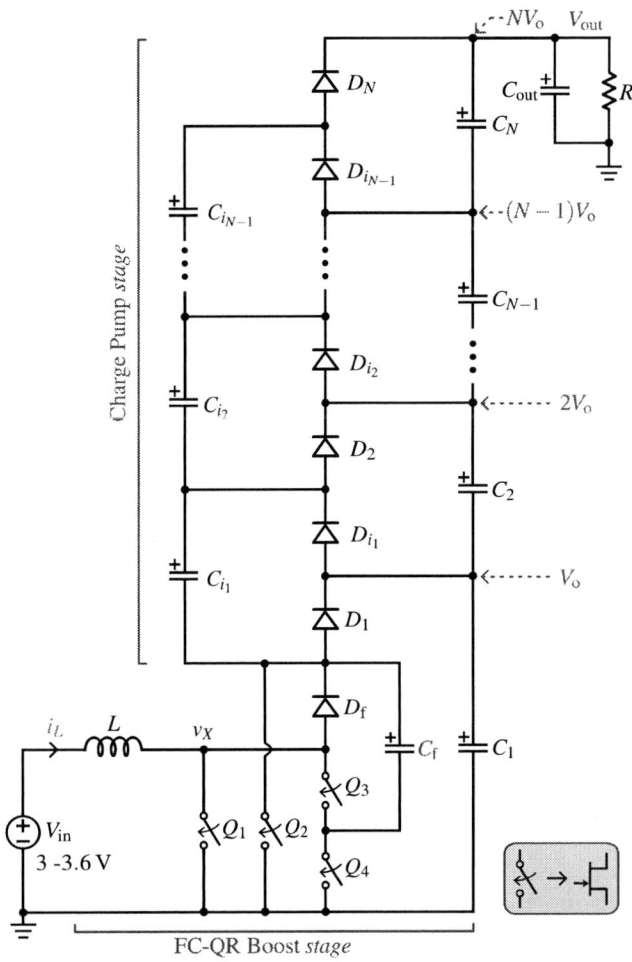

Fig. 1. Schematic of the introduced hybrid flying-capacitor based quasi-resonant (FC-QR) boost converter.

Fibonacci, series-parallel, etc.) are a potential alternative for low-power applications, as they offer reduced switch-voltage-stresses and higher power densities. However, they suffer from non-adiabatic capacitive energy transfer, exhibit regulation within a discrete set of voltage-conversion-ratios (e.g., 1:2, 1:4, etc.) [8], [9] and, would require prohibitively large number of components for a large step-up conversion ratio. Hybrid solutions [3], [7], [10], [11], etc., strategically incorporate both

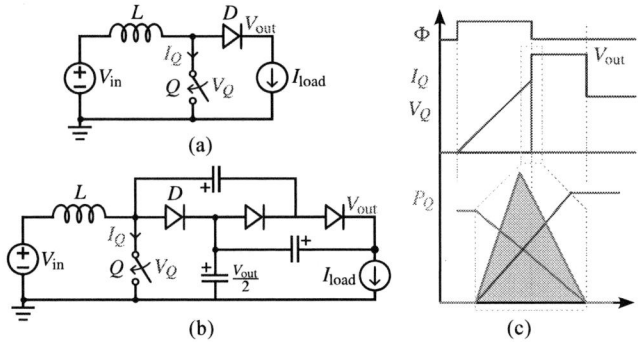

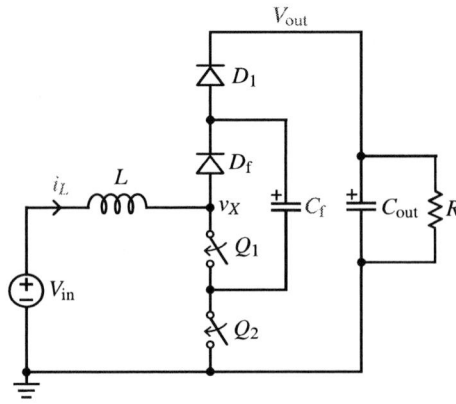

Fig. 2. (a) Schematic of a purely-magnetic boost converter, (b) schematic of a hybrid switched-capacitor boost converter and, (c) associated main switch voltage-current waveforms.

Fig. 3. Schematic of fully soft-switching quasi-resonant boost converter [12].

capacitive and magnetic elements into the converter topology. Work in [10], uses a class-E inverter to step up the input voltage to an intermediate bus voltage, which is used to drive a capacitor based voltage multiplier for a further boost. [7] incorporates a hybrid switched capacitor boost converter to step up coin cell battery voltage to output voltages as high as 200 V. Even though the switch stress can be reduced, it can't be completely eliminated. In low-power applications, even a single hard-switched edge can degrade the power efficiency. This paper introduces a new topology, shown in Fig. 1, and presents the design and implementation of the low-power, non-isolated, extremely high step-up dc–dc converter.

Fig. 2 shows a magnetic boost converter and a 2-stage hybrid switched-capacitor boost converter [7], both operating in DCM. The primary switch (Q) of the boost converter, undergoes hard turn-OFF which results in large voltage-current overlap losses for high output voltages, V_{out}. To reduce switch stress, the hybrid switched capacitor boost converter [7] utilizes a boost converter to generate an intermediate voltage and to drive a passive diode-capacitor-based charge pump to generate higher output voltage. The rectangular waveform across the rectifying diode, D, allows the charge pump to operate without any active components. This further allows the usage of lower voltage-rated switches, capacitors and diodes – all rated for the boost converter output voltage. Even though the voltage stress across the main switch is reduced, turn-OFF of Q is still hard causing significant voltage-current-overlap losses.

II. INTRODUCED SOLUTION

A fully soft-switching flying-capacitor-based quasi-resonant boost converter was introduced in [12] and shown in Fig. 3 (upper two switches are replaced by diodes D_f and D_1). There, a small flying capacitor, C_f, is resonantly charged (to V_{out}) and discharged (to zero), to produce zero voltage across the switches during turn-OFF. Due to this operation, all the components are rated for the full output voltage (V_{out}). In high-output-voltage scenarios, this presents challenges, as the switch's ON-resistance (R_{ON}) increases with the square of the blocking voltage (V_{BV}), for a given switch area. Moreover, for on-chip implementation, there may not be widespread availability of

high-voltage fabrication process node. The maximum inductor current and V_{BV} of the topology in [12] can be expressed as

$$i_{L,\max} = \frac{V_{out} - V_{in}}{\sqrt{L/C_f}}, \quad V_{BV} = V_{out}. \tag{1}$$

It is evident that, with increase in the input-to-output conversion ratio, $i_{L,\max}$ increases causing large core and conduction losses. L can be increased to reduce $i_{L,\max}$. This is not admissible as it would degrade the power density of the solution. Therefore, it is desired to maintain the soft-switching characteristics with reduced V_{BV} and $i_{L,\max}$.

The introduced hybrid flying-capacitor based quasi-resonant boost converter topology is shown in Fig. 1. It can be viewed as a 3-level quasi-resonant boost converter (L, C_f, Q_3, Q_4, D_f and D_1), strategically merged with a N-stage charge pump (rest of the components in Fig. 1). Capacitors $C_1 - C_N$ are the output bus capacitors (OBC), $C_{i_1} - C_{i,N-1}$ are the intermediate bus capacitors (IBC). Diodes $D_1 - D_N$ are utilized to charge the OBCs and $D_{i1} - D_{i,N-1}$ are utilized to charge the IBCs. The introduced topology integrates the advantages of both magnetic and capacitive topologies: (a) the fully soft-switching QR-boost converter steps-up the low input voltage to any high intermediate bus voltage, V_o and, (b) the N-stage charge pump further multiples V_o to a higher voltage (NV_o) depending on the number of stages in the charge pump, here $N = 4$. The underlying principle is to charge and discharge a small C_f quasi-resonantly, such that zero voltage appears across the switches across multiple switch transitions, therefore achieving zero voltage switching (ZVS). Moreover, L operates in DCM, therefore providing zero current switching (ZCS) at multiple switch transitions. Combining the discontinuous voltage mode (DVM) of operation of the flying capacitor and DCM of the inductor, full soft-switching is achieved for all the active switch transitions. All the components are rated for the minimum intermediate bus voltage (V_o). Therefore, lower V_{BV} switches with better figure of merit (FOM) and lower energy storage capacitors in the charge pump can be utilized.

979-8-3315-1612-3/25 $31.00 © 2025 IEEE

III. Principle of Operation

The operation of this converter can be described with the help of Figs. 4 and 5. Switch configurations during different switching states are shown in Fig. 4 and key waveforms are shown in Fig. 5. The operation of the N-stage charge pump can be better explained with a smaller number of stages. As there are basically two operating states of the charge pump: (a) lower output bus capacitors charging the intermediate bus capacitors and, (b) intermediate bus capacitors charging the upper output bus capacitors. For simplicity, a FC-QR boost with a 2-stage charge pump, $N = 2$ in Fig. 1, has been considered for explaining the operating principle.

A. Switching Sequence

The switching sequence happens over six intervals, labeled with numbered purple circles in Fig. 5, where the converter is going through four circuit configurations (states) **a-d**, Fig. 4, as explained earlier. At the beginning of a switching period, inductor current, i_L, is zero and voltage across the flying capacitor, v_{C_f} is also zero. The converter sits idle and no switching action takes place as long as voltage at the output of the charge pump stage, V_{out}, is less than the reference output voltage, V_{ref}.

- **Interval 1**: During this interval, converter is in state **a**, Fig. 4(a). Switches Q_1, Q_2 and Q_4 are turned-ON with ZCS. The inductor, L, is linearly charged by the input source (through Q_1) and the intermediate bus capacitor, C_{i_1}, is charged by the output bus capacitor, C_1, through the diode D_{i_1} and switch Q_2. No current flows through Q_4; the reason for turning-ON this switch during this interval will be explained later. During this interval, the inductor current, $i_L(t)$, and voltage across the flying capacitor, $v_{C_f}(t)$, can be expressed as

$$i_L(t) = \frac{V_{in}}{L} t \qquad (2)$$

$$v_{C_f}(t) = 0. \qquad (3)$$

The duration of this interval, termed as ON-time: T_{ON}, should be sufficiently long such that, L stores enough energy, E_f, to fully charge C_f in the next switching interval. T_{ON}, which will be quantified later, can be expressed as

$$T_{ON} = \frac{1}{\omega_0} \sqrt{\left(\frac{V_{out}}{NV_{in}} - 1 \right)^2 - 1}, \qquad (4)$$

where $\omega_0 = 1/\sqrt{LC_f}$ is the resonant frequency of the $L - C$ tank.

- **Interval 2**: During this interval, converter is in state **b**, Fig. 4(b). Q_1 is turned-OFF with ZVS. ZVS is achieved due to the state of voltage across the C_f. At the instance of switch transition, $v_{C_f} = 0$, resulting in $V_{DS_1} = 0$ before and after the switch transition. Q_2 is turned-OFF with both ZCS and ZVS because the capacitive charge transfer takes a much shorter time when compared to T_{ON} and thus current flowing through Q_2 is already zero. Moreover, C_f imposes zero voltage across Q_2 during the transition. During this

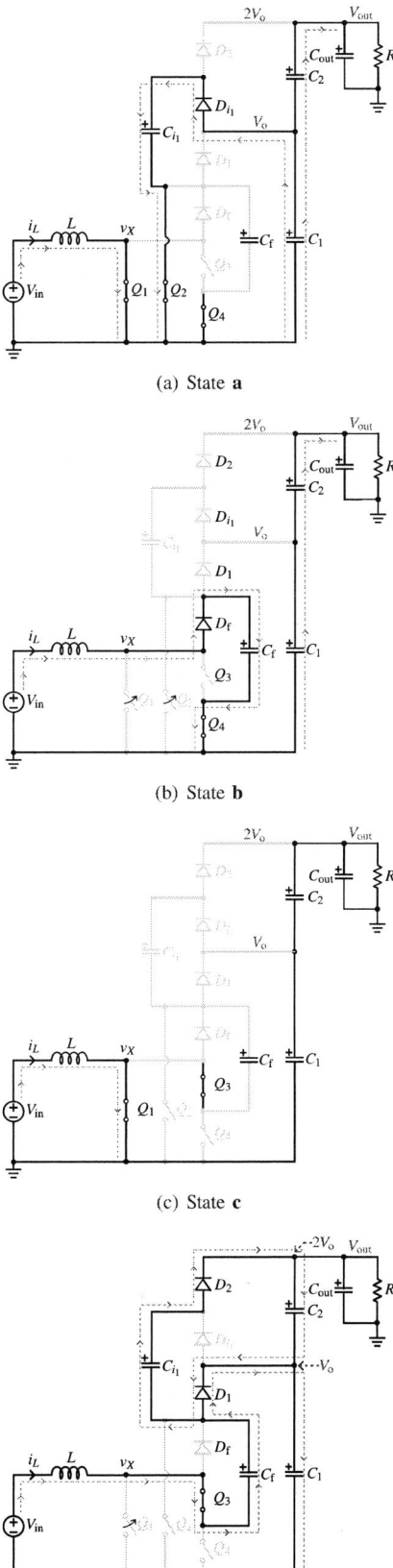

(a) State **a**

(b) State **b**

(c) State **c**

(d) State **d**

Fig. 4. Circuit configurations of the introduced hybrid FC-QR boost converter.

interval, energy stored in L is resonantly transferred to C_f and, $i_L(t)$ can be expressed as

$$i_L(t) = \frac{V_{\text{in}}}{L}T_{\text{ON}}\cos(\omega_0 t) + \frac{V_{\text{in}}}{L\omega_0}\sin(\omega_0 t) \quad (5)$$

The duration of this interval is termed as OFF-time, T_{OFF}: $i_L(T_{\text{OFF}}) = 0$ and $v_{C_f}(T_{\text{OFF}}) = V_o$. Solving (5) for $i_L(t) = 0$ and $i_{L,\max}$ (peak inductor current) results in

$$T_{\text{OFF}} = \frac{1}{\omega_0}[\pi - \tan^{-1}(\omega_0 T_{\text{ON}})], \quad (6)$$

$$i_{L,\max} = \frac{V_{\text{out}}/N - V_{\text{in}}}{\sqrt{L/C_f}}. \quad (7)$$

Comparing $i_{L,\max}$ of the single stage solution in [12] (expressed in (1)) to this work, the maximum inductor current is reduced by approximately the number of stages, N, in the charge pump (if $V_{\text{out}} \gg V_{\text{in}}$). Moreover, since C_f is charged to $V_o(= V_{\text{out}}/N)$ and all the bus capacitors and intermediate bus capacitors maintain a voltage of V_o across them, voltage rating of all the switches and diodes is reduced N-times. Therefore, better FOM switches, lower voltage rated diodes (with lower V_F and Q_{rr}) and, capacitors can be utilized to build the converter.

- **Interval 3**: At the beginning of this interval, Q_4 is turned-OFF with both ZVS and ZCS. This is a resting interval, active for a fixed time span termed as T_{DF}, and no energy transfer takes place. During interval 2, the parasitic capacitance at the switch node, C_X, gets charged to $v_{C_f}(= V_o)$. Therefore, i_L and v_X have small and damped oscillations when all the switches are turned-OFF, Fig. 8. The purpose of this interval is to allow the inductor current to settle back to zero before the next switching interval.

- **Interval 4**: During this interval, converter is in state **c**, Fig. 4(c). Q_1 and Q_3 are both turned-ON with ZCS, since $i_L = 0$ before the start of this interval. L is again linearly charged by V_{in} through Q_1. No current flows through Q_3. This switch will conduct current in the next switching interval and turning-ON Q_3 now provides ZCS to it. This will be further explained later. During this interval:

$$i_L(t) = \frac{V_{\text{in}}}{L}t \quad (8)$$

$$v_{C_f}(t) = V_o. \quad (9)$$

The duration of this interval, termed as $T_{\text{ON}}^\diamond$, should be sufficiently long such that, L stores enough energy: {a} E_f to fully discharge C_f to 0 and, {b} E_{CP} to provide the ripple energy to the output bus capacitors which is transferred through the intermediate bus capacitors. Since, E_{CP} is much smaller than E_f, but not negligible, L has to store some extra energy during this interval when compared to interval 1, therefore $T_{\text{ON}}^\diamond \gtrsim T_{\text{ON}}$.

- **Interval 5**: During this interval, converter is in state **d**, Fig. 4(d). Q_1 is turned-OFF with ZVS because $V_{DS_1} = v_X = V_o - v_{C_f} = 0$ during turn-OFF. Energy stored in L and C_f is resonantly transferred to C_1 and, C_{i_1} charges up C_2. During this interval, $i_L(t)$ has the same variation as

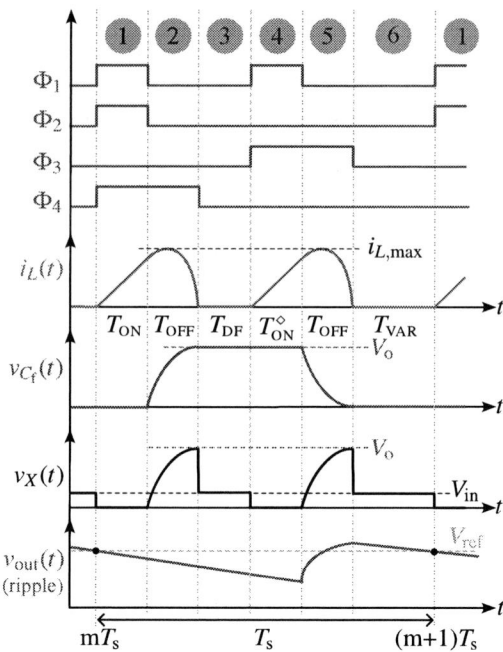

Fig. 5. Ideal switching waveforms of the hybrid FC-QR boost converter.

in interval 2 and v_{C_f} resonantly settles down to 0, Fig. 5. The duration of this interval is T_{OFF}, as was derived for interval 2.

- **Interval 6**: At the beginning of this interval, Q_3 is turned-OFF, with ZCS. Similar to interval 3, this is a resting period to allow i_L to settle back to zero. The main difference being that this is a variable duration interval, T_{VAR}, used for output voltage regulation, which will be further explained in §IV.

B. Improvements

The voltage conversion ratio, M, peak inductor current, $i_{L,\max}$, and rated voltage of the switches, diodes and capacitors of the introduced N-stage FC-QR boost converter can be expressed as

$$M = \frac{V_{\text{out}}}{V_{\text{in}}} = N\left(1 + \sqrt{1 + \omega_0^2 T_{\text{ON}}^2}\right) \quad (10)$$

$$i_{L,\max} = \frac{V_{\text{out}}/N - V_{\text{in}}}{\sqrt{L/C_f}}, \quad (11)$$

$$V_{\text{BV}} = \frac{V_{\text{out}}}{N}. \quad (12)$$

By strategically merging a N-stage charge pump to the previously introduced fully soft-switching boost converter in [12]: {a} the conversion ratio is increased N-times, {b} inductor peak current is reduced approximately N-times ($V_{\text{out}} \gg V_{\text{in}}$) for the same voltage conversion ratio, which in turn reduces the core losses in the inductor, and conduction losses in the inductor and switches, {c} voltage rating of the switches is reduced N-times, allowing the use of better FOM switches and therefore further reducing the conduction losses.

TABLE I
RELEVANT COMPONENTS USED IN THE EXPERIMENTAL DISCRETE PROTOTYPE

Item	Manufacturer	Part Number	Qty.	Notable Specifications †
Switches, Q	Efficient Power Conversion	EPC2110	1	120 V, R_{ON} = 110 mR, Q_g = .8 nC @ 5 V_{gs}, 1.35 × 1.35 mm
	Efficient Power Conversion	EPC2054	2	200 V, R_{ON} = 32 mR, Q_g = 2.9 nC @ 5 V_{gs}, 1.3 × 1.3 mm
Inductor, L	Murata Electronics	DFE32CAH4R7MR0L	1	4.7 μH ± 20%, 79 mΩ, 1210, 1.9 A_{DC}, 4.0 A_{sat}
Flying capacitor, C_f	KEMET	C0603C682KARACAUTO	2	6800 pF ± 10%, 250 V, X7R, 0603
Input capacitors, C_{in}	TDK Corporation	C1005X5R1V225K050BC	8	2.2 μF ± 10%, 35 V, X5R, 0402
Bus capacitors	Holy Stone Enterprise	C0805X104K251T	3	0.1 μF ± 10%, 250 V, X7R, 0805
	KEMET	C1210X474KARACAUTO	14	0.47 μF ± 10%, 250 V, X7R, 1210
Diodes, D	Micro Commercial Co	SMD1150PL-TP	9	150 V, 1 A, V_F = 0.85 V @ 1 A, SOD-123FL
Gate Drivers, GD	Texas instruments	LMG1210RVRT	1	200 V, 1.5 A_{src}, 3 A_{snk}, 5 V_{gs}, 3 × 4 mm
	Texas instruments	UCC27524DSDT	1	5 V, 5 A_{src}, 5 A_{snk}, 5 V_{gs}, 3 × 3 mm

† Listed capacitances of MLCCs are their nominal values at zero dc-bias.

IV. CONTROLLER

To ensure soft-switching and output voltage regulation, a simple dual-function controller is introduced here. The controller has two main functional blocks: the "output voltage controller" block regulates V_{out} through hysteretic control and, the timing of all the switching intervals are set by the "ON/OFF-time generator" block. To achieve soft-switching, it is required that L stores just enough energy during the charging intervals (1 and 4 in Fig. 5) to fully charge C_f to V_o and discharge to zero in intervals 2 and 5 respectively. Meaning that, the switch node voltage, v_X, reaches V_o at the same instant the inductor current, i_L, reaches zero at the end of intervals 2 and 5, Fig. 5. A soft-switching scheme that ensures this, was presented in [12]. There, samples of i_L and v_X are taken at the end of interval 2 and compared with zero and V_o respectively. T_{ON} and T_{OFF} are iteratively adjusted according to the comparison results, to meet the above mentioned criteria. This control scheme can be simplified if the conversion ratio of the boost stage is very high $V_o \gg V_{in}$. Eq. (4) and (6) can be simplified to

$$T_{ON} = \frac{1}{\omega_0} \frac{V_o}{V_{in}}, \tag{13}$$

$$T_{OFF} = \frac{1}{\omega_0} \frac{\pi}{2}. \tag{14}$$

In [12], all the switches in the QR-boost are active, so during the inductor discharge intervals (2 and 5), i_L will go negative if Q_3 and Q_4 are not turned-OFF at the right instants. In this introduced converter, the upper two active switches have been replaced by diodes D_f and D_1. This would prevent any negative current flow and thus timing of Q_3 and Q_4 turn-OFF can be relaxed. Durations of intervals 2 and 4, T_{OFF}, just needs to be more than the T_{OFF} calculated in (14).

V. PRACTICAL IMPLEMENTATION AND EXPERIMENTAL RESULTS

A discrete prototype, FC-QR hybrid boost converter was fabricated (Fig. 6) to validate the operation of the introduced converter and controller. Key component details and part numbers are listed in Table I. The controller is implemented using an FPGA board, terasIC DE10 standard, using simple counters to generate proper ON/OFF-time gating signals based on

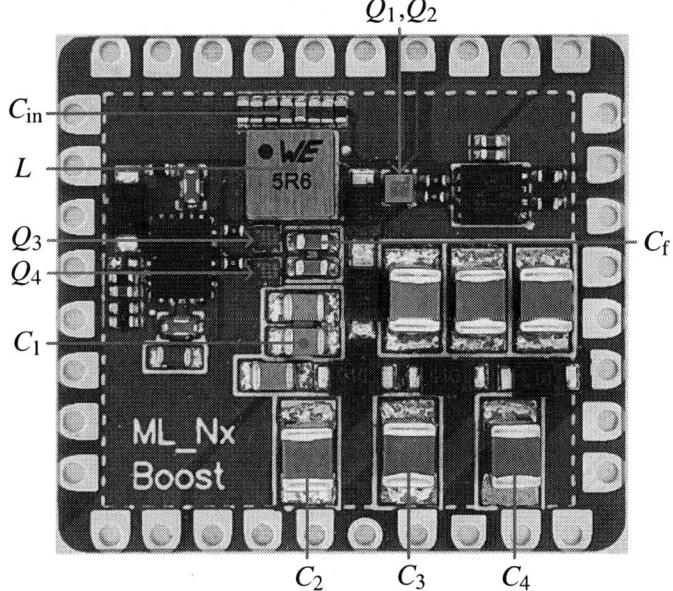

Fig. 6. Annotated photograph of the discrete experimental prototype of QR-FC boost with 4-stage charge pump used in this work. Entire power converter topology and associated gate driving circutry are enclosed within the dotted rectangle (23 mm × 20 mm).

the feedback from the PCB mounted output voltage comparator (not shown in Fig. 6). One of the other goals, apart of achieving high power processing efficiency, is to have high power density too. To achieve high power density, smaller form factor inductor, with worse conduction and core loss performance when compared to bigger wire-wound counterpart, was used. A chip inductor from Murata (listed in Table I) with a typical DCR of 79 mΩ was used. For the switches, 120 V and 200 V GaN devices from Efficient Power Conversion (EPC) were used because GaN offer lower ON-resistance (R_{ON}) when compared to MOSFETs, for the same switch size and blocking voltage (V_{BV}).

Fig. 8 shows the basic operating waveforms of the introduced converter, stepping up 3.6 V to 120 V while supplying a load current of 10 mA. The waveforms closely follow the theoretical

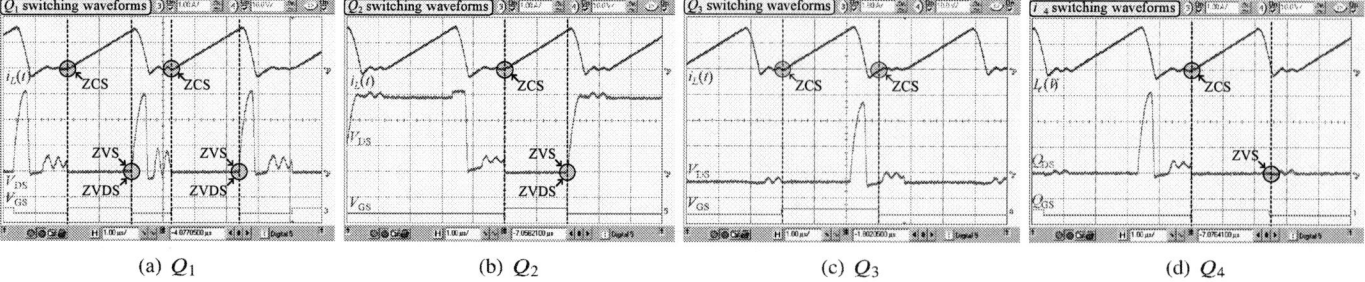

(a) Q_1 (b) Q_2 (c) Q_3 (d) Q_4

Fig. 7. Switch waveforms demonstrating ZCS, ZVS and ZVDS. Ch. 3 (purple) is the inductor current, Ch. 4 (pink) is the drain-source voltage and, digital signals are the gating signals.

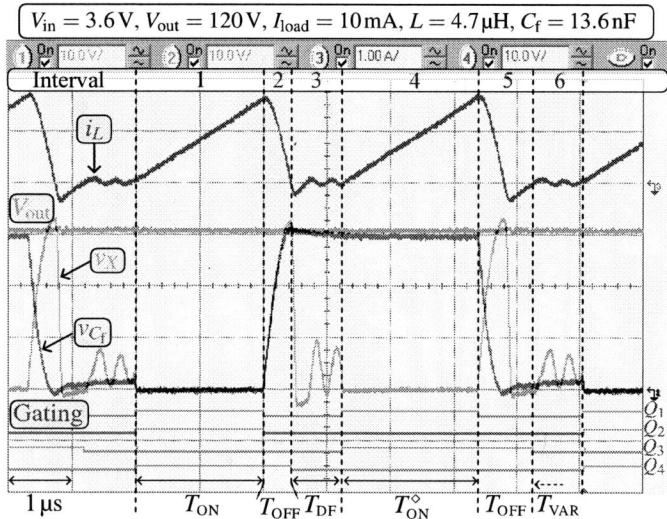

Fig. 8. Annotated oscilloscope screenshot of the converter stepping up 3.6 V input to 120 V output while supplying a load current of 10 mA. Ch. 1 (yellow) is the switch node voltage, Ch. 2 (green) is the output voltage, Ch. 3 (purple) is the inductor current and, Ch. 4 (pink) is the flying capacitor voltage.

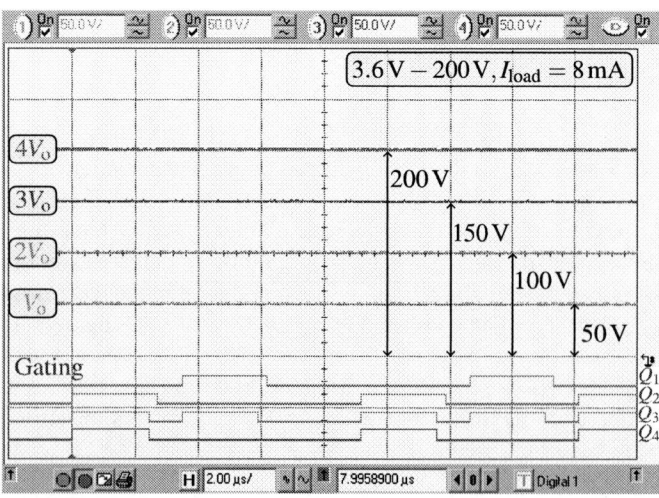

Fig. 9. Annotated oscilloscope screenshot of the bus voltages. Ch. 1 (yellow) is the output voltage of the FC-QR stage, Ch. 2 (green) is the second stage, Ch. 3 (purple) is the third stage and, Ch. 4 (pink) is the fourth stage bus capacitor voltage.

waveforms as explained in §III-A, and as shown in Fig. 5. Inductor current is initially zero and L first linearly charged and then C_f is resonantly charged to V_o. Again L is linearly charged and then energy stored in L and C_f is resonantly transferred to the output bus capacitor C_1. Fig. 9 shows the output bus voltages of the charge pump for the converter operating at 3.6 V input and 200 V at the output while supplying 8 mA of load current. Voltage measurements are tapped at V_o, $2V_o$, $3V_o$ and $4V_o (= V_{out})$.

Fig. 7 shows the drain-source voltage (V_{DS}), gating signal (V_{GS}) and, drain-source current ($I_{DS}(= i_L(t))$) of all the active switches in steady state operation for 3.6 V - 120 V supplying 10 mA of load current. All the switches undergo either ZCS or ZVS or both during switch transitions. An interesting feature is the zero voltage derivative switching (ZVDS) characteristics of Q_1 and Q_2 turn-OFF. The ZVDS is a significant feature as it allows variations in T_{ON} without loosing soft-switching. The experimental waveforms thus confirm the soft-switching operation as explained in §III-A.

To demonstrate the power processing effectiveness of the introduced topology, for extremely-high step-up conversion ratios and low loads, converter efficiency results with an input voltage of 3.6 V (nominal lithium-ion battery voltage), stepping it up to output voltages ranging from 55 volt to 250 V, and load currents ranging from 1 mA to 18 mA are presented. Fig. 10 shows the power efficiency results for V_{out} up-to 150 V and $C_f = 13.6$ nF. By virtue of soft-switching, the converter demonstrates high and practically flat efficiencies that surpasses best-in-class industry piezoelectric drivers like [13]. Since, C_f is fully charged (to V_o) and discharged (to zero) by L in every switching cycle, as V_{out} increases, L needs to store increasingly more energy in intervals 1 and 4 and $i_{L,max}$ also increases (12). Therefore incurring more conduction and core losses due to longer T_{ON} time intervals. For higher conversion ratios (above 40x), a smaller $C_f = 3.9$ nF was used. Power efficiencies for V_{out} up-to 250 V (conversion ratio up-to 70x) are shown in Fig. 11. Fig. 12 shows the power efficiency comparison for two choices of C_f: 3.9 nF and 13.6 nF. Due to higher T_{ON} requirements for a bigger C_f (4), conduction losses are higher in case of bigger C_f for the same conversion ratio and L.

979-8-3315-1612-3/25 $31.00 © 2025 IEEE   995

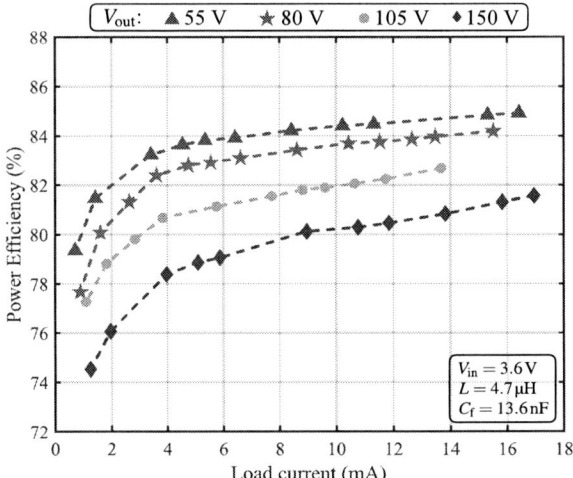

Fig. 10. Power efficiency of the discrete prototype at various output voltages for $C_f = 13.6\,$nF.

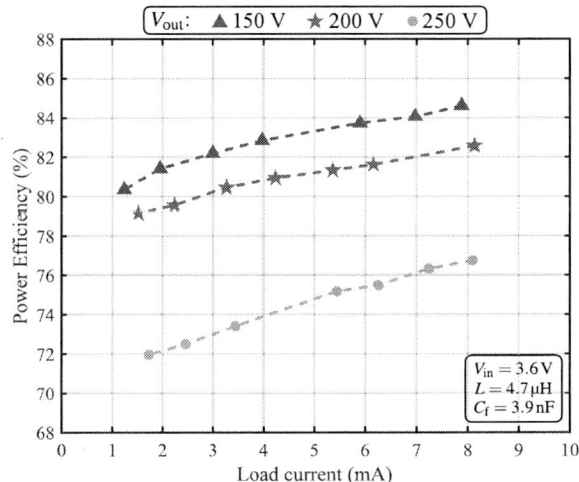

Fig. 11. Power efficiency of the discrete prototype at various output voltages for $C_f = 3.9\,$nF.

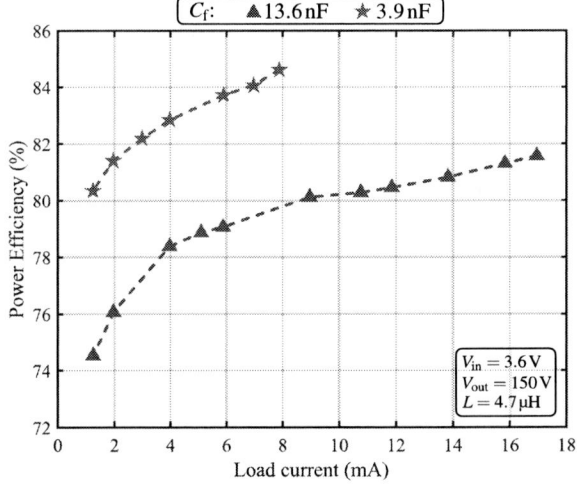

Fig. 12. Influence of C_f on power efficiency for the same conversion ratio.

VI. CONCLUSIONS

A hybrid flying-capacitor based quasi-resonant converter topology suitable for low power, extremely-high step-up conversion ratios is introduced. It combines a fully soft-switching quasi-resonant boost converter with a charge pump to achieve small peak inductor currents and reduced switch blocking voltages; both crucial for targeted applications. The presented experimental results correlate well with the theoretical analysis, confirming: (1) soft-switching for all the switches, (2) switched-capacitor charge pump operation, and (3) efficiency curves which are kept high even at the lowest load currents (~1 mA) and highest conversion ratios (~ 70x).

REFERENCES

[1] R. Wood, E. Steltz, and R. Fearing, "Optimal energy density piezoelectric bending actuators," *Sensors and Actuators A: Physical*, vol. 119, no. 2, pp. 476–488, 2005, doi: 10.1016/j.sna.2004.10.024. [Online]. Available: https://www.sciencedirect.com/science/article/pii/S0924424704007757

[2] H. Jin, X. Gao, K. Ren, J. Liu, L. Qiao, M. Liu, W. Chen, Y. He, S. Dong, Z. Xu, and F. Li, "Review on piezoelectric actuators based on high-performance piezoelectric materials," *IEEE Transactions on Ultrasonics, Ferroelectrics, and Frequency Control*, vol. 69, no. 11, pp. 3057–3069, 2022, doi: 10.1109/TUFFC.2022.3175853.

[3] M. Yang, E. A. Stolt, Z. Ye, and J. M. Rivas-Davila, "Piezoelectric based class-e resonant inverter for driving surface dielectric barrier discharge plasma," in *2024 IEEE 10th International Power Electronics and Motion Control Conference (IPEMC2024-ECCE Asia)*, 2024, pp. 367–372, doi: 10.1109/IPEMC-ECCEAsia60879.2024.10567170.

[4] M. A. Trout, A. T. Harrison, M. R. Brinton, and J. A. George, "A portable, programmable, multichannel stimulator with high compliance voltage for noninvasive neural stimulation of motor and sensory nerves in humans," National Center for Biotechnology Information, Tech. Rep., 2023, doi: 10.1038/s41598-023-30545-8, PMID: 36859464, PMCID: PMC9977866.

[5] G. Zhu and A. Ioinovici, "Switched-capacitor power supplies: Dc voltage ratio, efficiency, ripple, regulation," in *1996 IEEE International Symposium on Circuits and Systems (ISCAS)*, vol. 1, 1996, pp. 553–556 vol.1, doi: 10.1109/ISCAS.1996.540007.

[6] M. Makowski and D. Maksimovic, "Performance limits of switched-capacitor dc-dc converters," in *Proceedings of PESC '95 - Power Electronics Specialist Conference*, vol. 2, 1995, pp. 1215–1221 vol.2, doi: 10.1109/PESC.1995.474969.

[7] M. D. Seeman, "A design methodology for switched-capacitor dc–dc converters," Ph.D. Dissertation, Dept. Electrical Engineering & Computer Sciences, *Univ. Califonia, Berkeley*, May 2009, [Online.] Available: https://www2.eecs.berkeley.edu/Pubs/TechRpts/2009/EECS-2009-78.pdf.

[8] P. Lin and L. Chua, "Topological generation and analysis of voltage multiplier circuits," *IEEE Transactions on Circuits and Systems*, vol. 24, no. 10, pp. 517–530, 1977, doi: 10.1109/TCS.1977.1084273.

[9] G. Palumbo and D. Pappalardo, "Charge pump circuits: An overview on design strategies and topologies," *IEEE Circuits and Systems Magazine*, vol. 10, no. 1, pp. 31–45, 2010, doi: 10.1109/MCAS.2009.935695.

[10] S. Park, A. Goldin, and J. Rivas-Davila, "Miniature high-voltage dc-dc power converters for space and micro-robotic applications," in *2019 IEEE Energy Conversion Congress and Exposition (ECCE)*, 2019, pp. 2007–2014, doi: 10.1109/ECCE.2019.8913249.

[11] M. Forouzesh, Y. P. Siwakoti, S. A. Gorji, F. Blaabjerg, and B. Lehman, "Step-up dc–dc converters: A comprehensive review of voltage-boosting techniques, topologies, and applications," *IEEE Transactions on Power Electronics*, vol. 32, no. 12, pp. 9143–9178, 2017, doi: 10.1109/TPEL.2017.2652318.

[12] K. J. Nag and A. Prodic, "Fully soft-switching flying capacitor based quasi-resonant boost converter," in *2023 IEEE Applied Power Electronics Conference and Exposition (APEC)*, 2023, pp. 919–925, doi: 10.1109/APEC43580.2023.10131658.

[13] *DRV2700 Industrial Piezo Driver with Integrated Boost Converter*, Texas Instruments, 2023, [Online]. Available: https://www.ti.com/product/DRV2700.

Isolated Soft-Switching Flying-Capacitor based Quasi-Resonant Step-Up Converter

Kumar Joy Nag ⊚, Aleksandar Prodić ⊚

The Edward S. Rogers Sr. Department of Electrical & Computer Engineering
University of Toronto
10 King's College Road, Toronto, ON, M5S 3G4, Canada
kumarjoy.nag@mail.utoronto.ca, prodic@ece.utoronto.ca

Abstract—**An isolated quasi-resonant step-up converter, and a complementary mixed-signal controller suitable for isolated low-power ultra-high step-up applications, with voltage conversion ratios exceeding 60x and sub-one-watt to a few watts load, are introduced. The converter has the same structure as the previously reported isolated flying capacitor multilevel flyback converter (FCMFC), but operates in a completely different manner. With modification to this typology's component parameters, and a newly introduced control scheme, load-independent soft-switching is achieved on all the switches on the secondary side and voltage stress across the primary switch has been reduced to the input voltage, for any input-to-output voltage conversion ratio. These improvements are achieved by fully charging and discharging a small flying capacitor on the secondary side, as a resonant element, through a specific technique that ensures proper gating timing for achieving soft-switching. The operation of the converter is experimentally verified using a discrete prototype with input voltage of 3 V and common output voltages of 90 V, 120 V and 180 V, and load currents ranging from 1 mA to 15 mA, demonstrating efficiencies up to 81%.**

Index Terms—**Flying Capacitor, High Step-Up, Isolated dc–dc Converter, Quasi Resonant, Soft-Switching, Switching Loss**

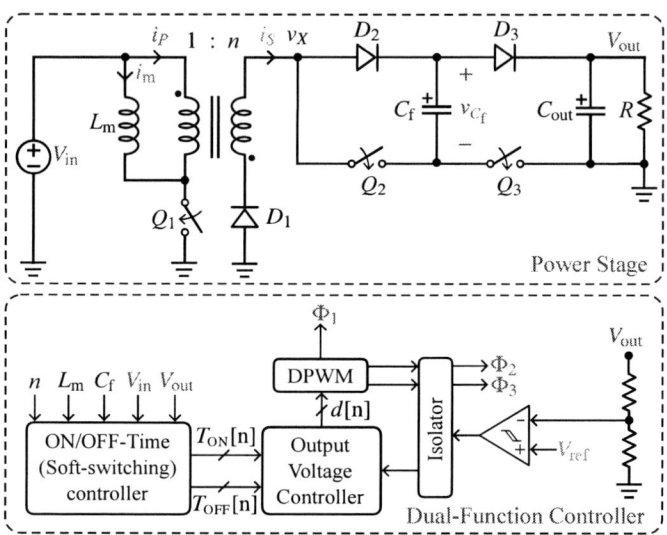

Fig. 1. Schematic of the introduced flying-capacitor based isolated quasi-resonant step-up converter and its complementary controller.

I. INTRODUCTION

In many emerging niche applications, low-power, isolated, high-voltage dc–dc power converters are increasingly becoming prevalent. Functional electrical stimulation (FES) is a technique by which electric stimulators (ESs) inject charge balanced low-energy pulses for generation of muscle contractions [1], [2]. Most notable applications for FES include therapies for spinal cord injury, muscle relaxation, restoration of lost motor control functions etc. To cover the functional needs of a FES and address the safety of the consumer, the dc–dc converters for portable ES need to have the following properties: (a) depending on the tissue/muscle health state, resistance can be really high and therefore, a high-voltage is required for stimulation, (b) for safety purposes and to keep the stimulation less painful, output current levels of the electric pulses (on an average) must be kept low, (c) galvanic isolation is required by safety standards [3], (d) Additional design challenge comes from the need for the ESs to be portable, which implies that the power source should be a small lithium-ion cell, small outline and light-weight. Therefore, from the perspective of the dc-dc converter, it needs to be an isolated high-voltage step-up converter operating off a small input voltage and delivering very low loads (sub 1 W). In low-power high-voltage step-up applications, voltage-current

overlap losses take up a major percentage in the net loss of the system [4], [5]. Low power, isolated step-up converters are also used for stand-by power supplies in industrial sensors and automotive applications.

Flyback converter, with high turns ratio, operating in quasi-resonant mode [6], at first look, may seem to be a viable choice of the dc–dc step-up stage. Quasi-resonant mode of operation reduces the turn-ON switching losses of the primary switch but not the turn-OFF switching losses. High turns ratio in the transformer is preferred to reduce the primary switch stress during turn-OFF, this is a major contributor to the overall converter losses in low power regime. [7] introduces a class of isolated flying capacitor multilevel converters (FCMFC) which reduces the primary switch voltage stress and increases the conversion ratio without increasing the turns ratio of the transformer. For a small converter solution, a custom planar transformer with high turns ratio and PCB windings laid around a small core would be required. Power magnetic components are especially challenging to miniaturize as magnetic scaling laws fundamentally work against achieving high efficiency for smaller form factors [8]. An efficient, high turns-ratio flyback in a small form factor operating in high-voltage conditions is

979-8-3315-1612-3/25 $31.00 © 2025 IEEE

challenging to design due to the following reasons.

- Transformer design: In high step-up transformers, a high secondary turns count is required. [9] discusses the challenges in designing high step-up ratio PCB planar transformers. High number of turns is difficult to accommodate in a miniaturized form factor owing to turn-to-turn spacing requirements, since the windings are implemented on a PCB. The poor window fill factor in planar PCB transformers further aggravates the winding layout, leading to design of thinner PCB trace windings eventually leading to higher conduction losses. Windings can be spread across multiple PCB layers but will eventually give rise to inter- and intra-winding capacitance. Moreover, the height of the small transformer core is low and only a limited number of layers can be accommodated, usually up to 6. Therefore, designing a high turn-ratio, small form factor transformer is complex and expensive to manufacture.

- Component design and selection: High turns ratio lowers the primary switch voltage rating as the reflected voltage across the switch during OFF-state is $V_{in} + V_{out}/n$. As a consequence, contrarily increases the voltage rating of the output diode during OFF-state: $nV_{in} + V_{out}$. High blocking voltage diodes have worse forward voltage drops and junction capacitance, which affect the efficiency and introduce ringing during power transfer.

- Other losses: Higher turns ratio of the transformer leads to higher leakage inductance, which adversely affects system efficiency. Losses in snubber increase with leakage inductance.

One attractive solution to address these challenges is to decompose the transformer into multiple smaller parallel-primary and series-secondary transformers, as done in matrix transformers (MT). MT may be a viable choice in high power scenarios [10], but may not be a suitable candidate for miniaturized converters where size is one of the major concerns. Multiphase and fractional turns transformers [11], [12] can reduce the number of windings on the high voltage side but are exacting to design. Therefore, a different approach needs to be developed to reduce the primary switch voltage stress, in high voltage and low power scenarios, during turn-OFF, and not requiring high turns ratio.

The topology introduced in this paper, Fig. 1, reduces the turn-OFF switching loss of the primary switch and eliminates the need of a snubber by strategically generating zero voltage across the transformer windings during primary switch turn-OFF. The switch voltage during turn-OFF transition is reduced from $V_{in} + V_{out}/n$, in case of conventional flyback, to just V_{in}.

II. INTRODUCED SOLUTION

Consider the primary side switch, Q, turn-OFF event: right before, Fig. 3(a), and right after, Fig. 3(b). Let's assume that the secondary side winding has a voltage V_X across it. In the event of turn-OFF: voltage across Q, V_Q, jumps from 0 to $V_{in} + V_X/n$. In case of a conventional flyback converter, $V_X = V_{out}$. This causes significant voltage-current-overlap losses during switch turn-OFF, especially when V_{out} is high and turns

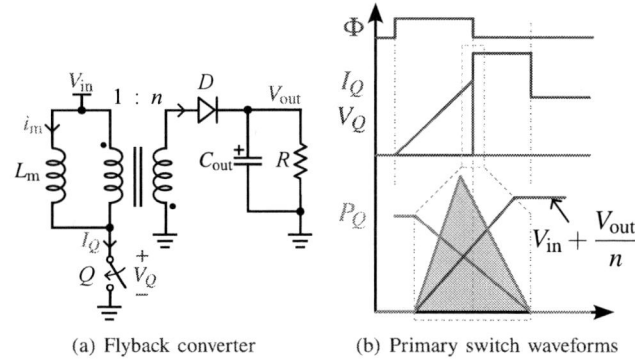

(a) Flyback converter   (b) Primary switch waveforms

Fig. 2. Flyback converter and primary switch waveforms.

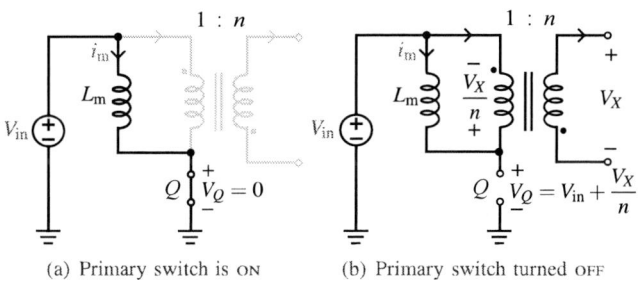

(a) Primary switch is ON   (b) Primary switch turned OFF

Fig. 3. Basis of switch voltage reduction during switch turn-OFF.

ratio (n) is small. If somehow, V_X could be forced to be 0 during switch turn-OFF transitions, reducing V_Q to just V_{in} rather than $V_{in} + V_X/n$, the overlap switching losses can be greatly reduced. This can be achieved (explained in detail in later sections) by introducing two extra switches, two diodes and a small flying capacitor (C_f) on the secondary side. In low power, high step-up applications: {a} conduction losses are already low on the high-voltage (secondary) side, and {b} switching losses dominate due to increased voltage-current overlap losses. Therefore, it needs to be ensured that the extra circuitry introduced on the secondary side does not incur any switching losses. This can be achieved by adhering to the fundamentals principles for achieving load independent soft-switching in a 3-level boost converter introduced in [13], and slightly modifying them here:

- fully charge/discharge C_f quasi-resonantly to generate zero-voltage-switching (ZVS) on the switches in the secondary side and for the highest possible energy/volume utilization: effectively operating C_f in discontinuous voltage mode (DVM).
- operate the magnetizing inductance (L_m) in discontinuous current mode (DCM) to provide zero current switching (ZCS) for primary switch soft turn-ON.
- turning-ON primary switch when $V_Q = 0$, to further minimize C_{oss} losses.

Since voltage stress of Q is reduced to V_{in} ($V_{out} > V_{in}$), there is enough room for overshoot due to ringing in v_Q. Therefore, no snubber is required for ringing suppression. The DVM operation of C_f and DCM of L_m, results in an input-to-output

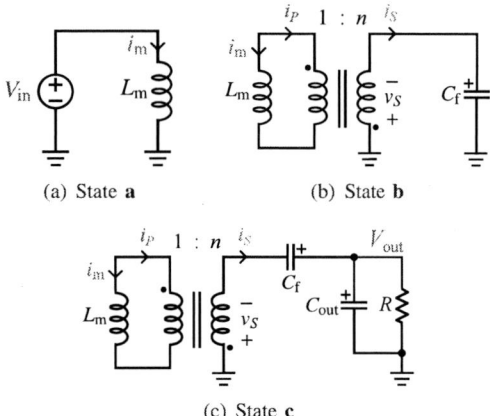

(a) State **a**　　(b) State **b**

(c) State **c**

Fig. 4. Converter switching states.

voltage conversion ratio to be independent of duty ratio and number of turns of the transformer. In the following sections, operation and control of the introduced converter are discussed in detail and finally concluded with experimental results.

III. PRINCIPLE OF OPERATION

The operation of this converter can be described with the help of Fig. 4, 5 and 6. Fig. 4 shows the three basic states through which the converter goes through during a switching interval.

- **State a**: The magnetizing inductance of the transformer, L_m, is linearly charged by the input voltage source, V_{in}.
- **State b**: Energy stored in L_m is resonantly transferred to the flying capacitor, C_f, on the secondary side.
- **State c**: Energy stored in L_m and C_f are resonantly transferred to the output capacitor, C_{out}, and load.

A. Switching Sequence

Switch configurations during different intervals are shown in Fig. 5 and ideal switching waveforms are shown in Fig. 6. The switching sequence happens over six sub-intervals (1-6), labeled with numbered purple circles in Fig. 5, where the converter is going through three switching states (**a-c**), Fig. 4. At the beginning of the switching period, $i_{L_m}(t) = 0$ and $v_{C_f}(t) = 0$, Fig. 6.

- **Interval 1**: During this interval, Fig. 5(a), converter is in state **a**, Fig. 4(a). Primary switch Q_1 and secondary side switch Q_3 are both turned-ON with ZCS, as initially no energy was stored in the reactive elements, L_m and C_f. L_m is connected in series with V_{in}, thus charging it linearly. During this interval, the voltage across Q_1, V_{Q_1}, voltage across the magnetizing inductance of the transformer, v_{L_m} and secondary winding current, $i_S(t)$, can be expressed as

$$V_{Q_1}(t) = 0 \tag{1}$$

$$v_{L_m}(t) = L_m \frac{di_m}{dt} = V_{in} \tag{2}$$

$$i_S(t) = 0. \tag{3}$$

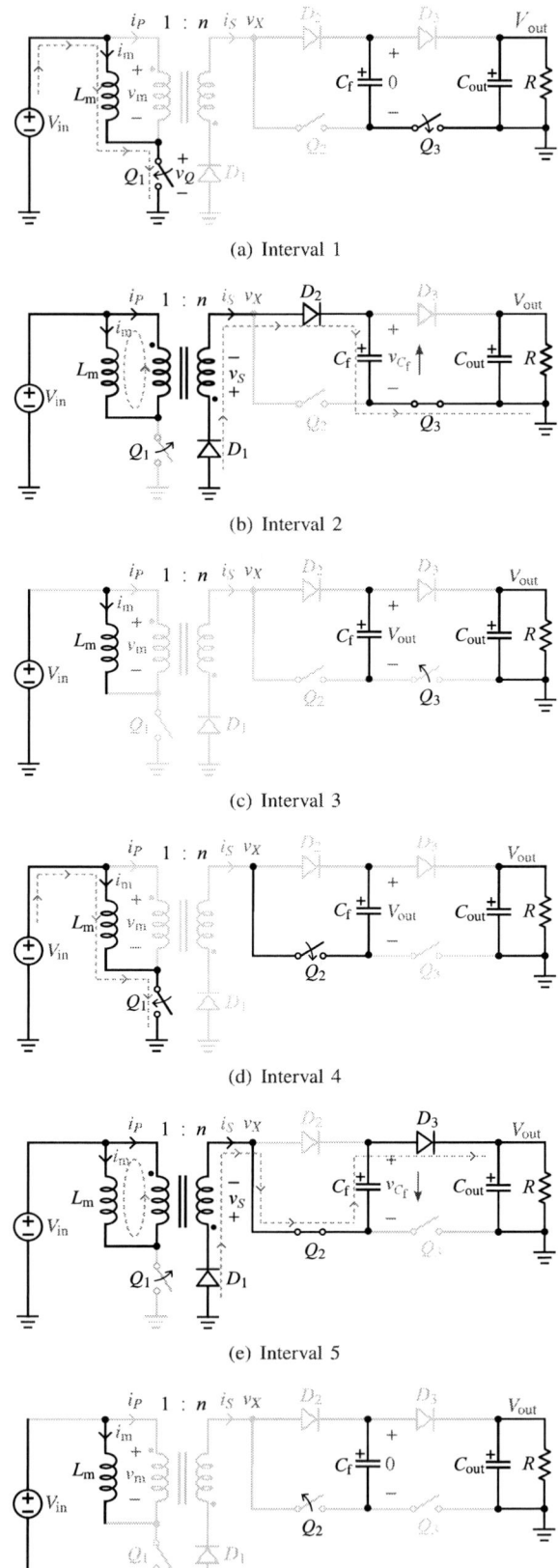

(a) Interval 1

(b) Interval 2

(c) Interval 3

(d) Interval 4

(e) Interval 5

(f) Interval 6

Fig. 5. Switching intervals of the converter switching action.

Therefore, $i_m(t)$ and $v_{C_f}(t)$ during this interval can be expressed as

$$i_m(t) = \frac{V_{in}}{L_m}t, \tag{4}$$

$$v_{C_f}(t) = 0. \tag{5}$$

The duration of this interval is termed as ON-time, T_{ON}. The ON-time is adjusted such that L_m stores sufficient energy during this interval to fully charge C_f to V_{out} in the subsequent interval. T_{ON}, which will be quantified later in section , can be expressed as

$$T_{ON} = \frac{V_{out}}{\omega_0 V_{in}}. \tag{6}$$

- **Interval 2**: During this interval, Fig. 5(b), converter is in state **b**, Fig. 4(b). During this interval, $v_{L_m}(t)$, $i_S(t)$ and, current through the flying capacitor, $i_{C_f}(t)$ can be expressed as

$$v_{L_m}(t) = L_m \frac{di_m(t)}{dt} = -\frac{v_{C_f}(t)}{n} \tag{7}$$

$$i_{C_f}(t) = C_f \frac{dv_{C_f}(t)}{dt} = \frac{i_m(t)}{n} \tag{8}$$

Solving these differential equations, give us the equations based on which $i_m(t)$ and $v_{C_f}(t)$ vary with time during this interval:

$$i_m(t) = \frac{V_{in}}{L_m}T_{ON}\cos\left(\frac{\omega_0}{n}t\right) \tag{9}$$

$$v_{C_f}(t) = V_{in}\omega_0 T_{ON}\sin\left(\frac{\omega_0}{n}t\right). \tag{10}$$

Where $\omega_0 = 1/\sqrt{L_m C_f}$ is defined as the resonant angular frequency of the circuit. The duration of this interval is defined as OFF-time, T_{OFF}. The OFF-time is adjusted such that, at the end of this interval, C_f is fully charged to V_{out} and L_m is fully discharged to zero. Solving (9) for $i_m(T_{OFF}) = 0$, gives us

$$T_{OFF} = \frac{n}{\omega_0}\frac{\pi}{2}. \tag{11}$$

- **Interval 3**: Q_3 is turned-OFF with ZCS, Fig. 5(c). This interval can be considered as a resting interval when no switching and energy transfer takes place. The circuit sits idle until the first valley in V_{Q_1} is detected.
- **Interval 4**: During this interval, Fig. 5(d), converter is again in state **a**, Fig. 4(a). Q_1 and Q_2 are turned-ON with ZCS, as L_m was previously fully discharged in interval 2 and i_m slowly ramps up after Q_1 turn-ON. L_m is again linearly charged by V_{in} for time T_{ON}. During this interval,

$$i_m(t) = \frac{V_{in}}{L_m}t, \tag{12}$$

$$v_{C_f}(t) = V_{out}. \tag{13}$$

- **Interval 5**: During this interval, Fig. 5(e), converter is in state **c**, Fig. 4(c). During this interval,

$$v_{L_m}(t) = L_m \frac{di_m(t)}{dt} = \frac{v_{C_f}(t) - V_{out}}{n} \tag{14}$$

$$i_{C_f}(t) = C_f \frac{dv_{C_f}(t)}{dt} = -\frac{i_m(t)}{n} \tag{15}$$

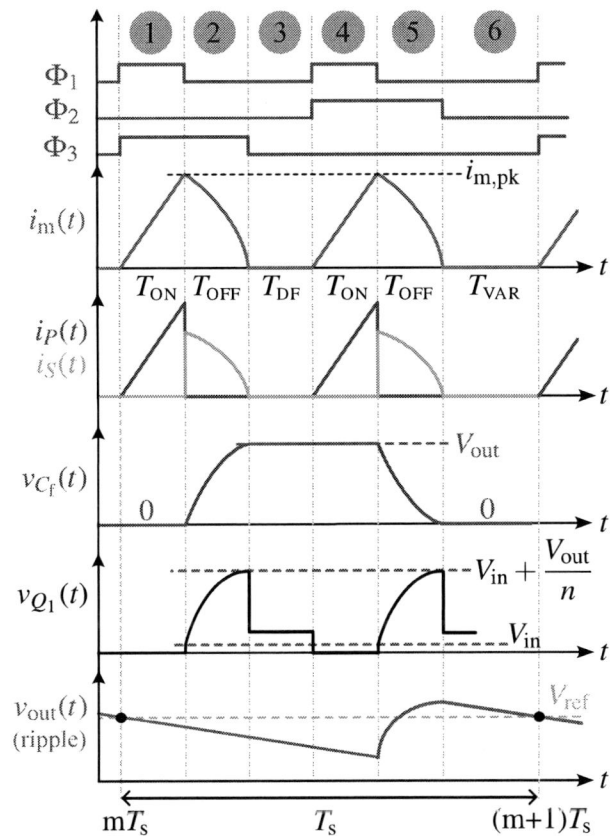

Fig. 6. Schematic of the introduced flying-capacitor based isolated quasi-resonant step-up converter and its complementary controller.

Solving these differential equations, result in exactly the same time domain equations for $i_m(t)$ as in interval 2 (9) and, $v_{C_f}(t)$ can be expressed as

$$v_{C_f}(t) = V_{out} - V_{in}\omega_0 T_{ON}\sin\left(\frac{\omega_0}{n}t\right). \tag{16}$$

At the end of this interval, C_f needs to be fully discharged, therefore solving $v_{C_f}(t) = 0$ results in exactly the same T_{OFF} as calculated for interval 2 in (11). Energy stored in L_m and C_f is resonantly transferred to the output capacitor, C_{out}, which was supplying the load current throughout all the previous intervals.

- **Interval 6**: Q_2 is turned-OFF with ZCS, Fig. 5(f). The converter sits idle and no switching action take place similar to interval 3, with the main difference that a fresh cycle begins only when V_{out} is less than the reference voltage, V_{REF}, rather than triggering the next switching interval at the first valley point of V_{Q_1}. This technique is used for output voltage regulation.

B. Voltage Conversion Ratio

Voltage conversion ratio, $M = V_{out}/V_{in}$, can be derived using the principle of conservation of energy. Energy is transferred by the input source only during intervals 1 and 4: V_{in} linearly charges L_m for T_{ON} time. Therefore, energy delivered by the input source in one switching cycle is: $E_{in} = 2(E_1 + E_4)$, where

$E_1 = E_4 = (V_{in}^2 T_{ON}^2)/(2 L_m)$. Assuming 100 % efficiency, from energy conservation

$$P_{in} = P_{out} \tag{17}$$

$$\frac{V_{in}^2 T_{ON}^2}{L_m} f_{sw} = \frac{V_{out}^2}{R}. \tag{18}$$

Furthermore, since C_f is fully charged to V_{out} and fully discharged to zero every switching cycle,

$$C_f V_{out}^2 f_{sw} = \frac{V_{out}^2}{R} \tag{19}$$

Therefore, comparing (18) and (19), we get

$$M = \frac{V_{out}}{V_{in}} = \omega_0 T_{ON} \tag{20}$$

Eq. (20) indicates that, for a fixed resonant frequency ω_0, input-to-output conversion ratio is independent of the load and turns ratio, n, of the transformer and, can be varied solely by adjusting T_{ON}.

IV. CONTROLLER

To achieve the aforementioned benefits of the introduced converter: {a} to reduce the overlap losses in the primary switch and, {b} to ensure soft-switching for all the added circuitry in the secondary side, it is required that L_m stores just enough energy during the linear charging intervals 1 and 4, in Fig. 5, to fully charge C_f to V_{out} in interval 2 and fully discharge C_f to zero in interval 5. L_m charging time is T_{ON} and C_f charging and discharging time is T_{OFF}. Eq. (6) indicates that T_{ON} is dependent on L_m, C_f, V_{in} and V_{out}. Once the converter has been designed, meaning that values of L_m and C_f are chosen, depending on the values of V_{in} and V_{out}, T_{ON} can be adjusted. Eq. (11) indicates that T_{OFF} is independent of any voltage/current variables, e.g. V_{in}, V_{out} etc., and solely depends on the choice of L_m, C_f and n. Again, once the converter has been designed, T_{OFF} can be calculated. A possible implementation of the controller is shown in Fig. 1. For simple implementation and control, the controller is implemented on the primary side. Output voltage is regulated using hysteretic control: converter switching is activated when $V_{out} < V_{ref}$. A more accurate and soft-switching scheme, as introduced previously in [13], can be employed if none of the system parameters are known or if they are changing with time. Sensing of the switch node voltage, v_X, and zero current crossing detection of the secondary current will be required to iteratively adjust T_{ON} and T_{OFF}.

V. PRACTICAL IMPLEMENTATION AND EXPERIMENTAL RESULTS

A discrete prototype, shown in Fig. 7, was fabricated to validate the operation of the introduced converter and controller operation. Key component details and part numbers are listed in Table I. The controller is implemented using a FPGA board using simple counters to generate proper ON/OFF-time gating signals based on the PCB mounted comparator output (not shown in Fig. 7). For shorter design turn-over time, an off-the-shelf high voltage (300 V), medium turns ratio (1:10) transformer was used in the current prototype. Work on a

Fig. 7. Annotated photograph of the discrete experimental prototype of the isolated step-up converter used in this work.

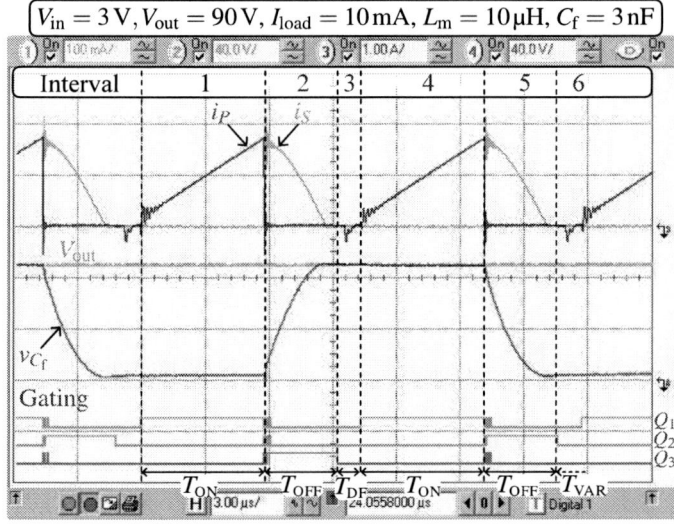

Fig. 8. Annotated oscilloscope screenshot of the converter stepping up 3 V input to 90 V output while supplying a load current of 10 mA. Ch. 1 (yellow) is the secondary side current, Ch. 2 (green) is the output voltage, Ch. 3 (purple) is the primary side current and, Ch. 4 (pink) is the flying capacitor voltage. Vertical scaling of the primary and secondary side current waveforms are adjusted according to the turns ratio of the transformer (1:10).

miniaturized prototype which incorporates a high-voltage planar transformer utilizing a smaller core and even smaller turns ratio is under progress.

Fig. 8 shows the basic operating waveforms of the introduced isolated converter, stepping up 3 V to 90 V while supplying a load current of 10 mA. The experimental waveforms closely follow the theoretical waveforms as explained in §III, and as shown in Fig. 6. L_m is first linearly charged by the input source, and voltage across C_f stays at zero. After T_{ON}-time, Q_1 is turned-OFF and circulating current in the core charges up C_f resonantly to the output voltage, V_{out}. L_m is again linearly charged and finally energy stored in C_f and core of the transformer is resonantly transferred to the output capacitor. Both L_m and C_f

TABLE I
Relevant Components Used in the Experimental Discrete Prototype

Item	Manufacturer	Part Number	Qty.	Notable Specifications †
Switches, Q	Efficient Power Conversion	EPC2054	3	200 V, R_{ON} = 32 mR, Q_g = 2.9 nC @ 5 V_{gs}, 1.3 × 1.3 mm
Transformer	Wurth Elektronik	750032051	1	10 µH @ 10 kHz, 1:10
Flying capacitor, C_f	KEMET	C0603C152JBRACAUTO	2	1500 pF ± 5%, 630 V, X7R, 0603
Input capacitors, C_{in}	TDK Corporation	C1005X5R1V225K050BC	8	2.2 µF ± 10%, 35 V, X5R, 0402
	Murata Electronics	GRM188R61E106KA73D	2	10 µF ± 10%, 25 V, X5R, 0603
Output capacitors	KEMET	C1210X474KARACAUTO	9	0.47 µF ± 10%, 250 V, X7R, 1210
Diodes, D	Micro Commercial Co	SMD1150PL-TP	4	150 V, 1 A, V_F = 0.85 V @ 1 A, SOD-123FL
Gate Drivers, GD	Texas instruments	LMG1210RVRT	1	200 V, 1.5 A_{src}, 3 A_{snk}, 5 V_{gs}, 3 × 4 mm
	Texas instruments	LMG1025QDEERQ1	1	5 V, 7.5 A_{src}, 5 A_{snk}, 5 V_{gs}, 2 × 2 mm

† Listed capacitances of MLCCs are their nominal values at zero dc-bias.

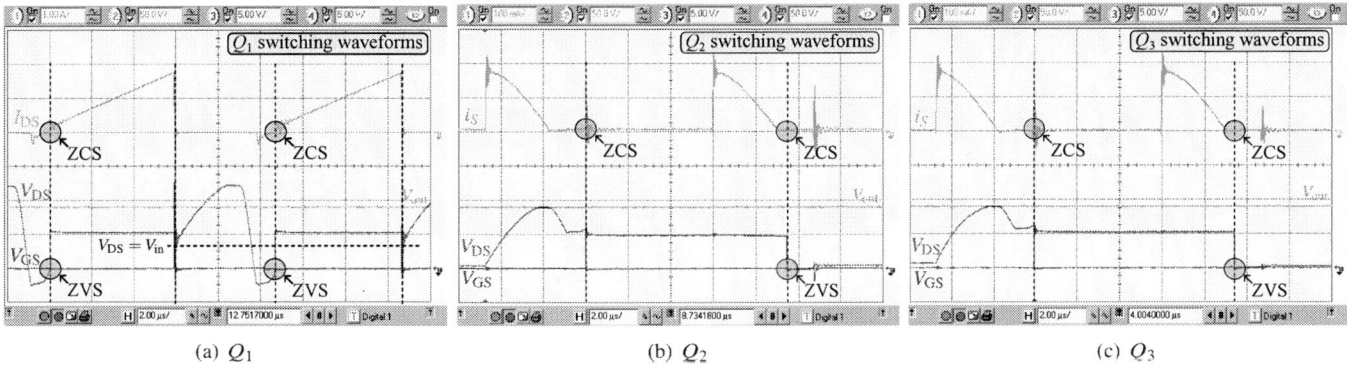

(a) Q_1 (b) Q_2 (c) Q_3

Fig. 9. Switch waveforms demonstrating ZCS, ZVS operation for all the active switches. Ch. 1 (yellow) is the current through the switch (when gating signal is high), Ch. 2 (green) is the output voltage, Ch. 3 (purple) is the gate-to-source signal and, Ch. 4 (pink) is the drain-to-source voltage across the switch.

are fully discharged at the end of the switching period.

Fig. 9 shows the drain-source voltage (V_{DS}), gating signal (V_{GS}) and, drain-source current (I_{DS}, either equal to the primary/secondary currents when turned-ON) of all the active switches in steady state operation for 3 V-80 V and supplying 8 mA of load current. Q_1 turns-ON with ZCS and ZVS and turns-OFF with non-zero current and a voltage of only V_{in} across it, instead of a much higher voltage of $V_{\text{in}} + V_{\text{out}}/n$ in case of a conventional flyback. This drastically reduced switching losses of the primary switch. Q_2 and Q_3 undergo either ZCS or ZVS or both during every switching transition. These experimental results thus confirm the soft-switching operation as explained in § III-A.

To demonstrate the power processing effectiveness of the introduced converter for low output power levels and high step-up conversion ratios, converter power efficiency results are shown in Figs. 10 and 11. Fig. 10 shows the efficiency results for the converter operating with a low input voltage of 3 V (lower range of lithium-ion battery voltage), stepping it up-to voltages of 90 V and 120 V. Since the flying capacitor, C_f, is fully charged and discharged every switching cycle, with increase in V_{out}, energy stored and transferred through C_f per cycle increases (= $C_f V_{\text{out}}^2$). Therefore the maximum power supply capability of the converter increases with increase in V_{out}, Fig. 10. Efficiencies of around 80 % are achieved for sub-1 W power levels. Fig. 11 shows the efficiency results

for converter operating with V_{in} as low as 2 V and stepping it up-to V_{out} =180 V at output power levels less than 1.5 W. Since, voltage across the primary switch, Q_1, is equal to V_{in} during turn-OFF and all the other switches undergo full soft-switching, the only switching losses (which are lower than a conventional flyback) in the converter are due to Q_1 turn-OFF. Therefore, efficiencies for 90x (2 V-180 V) conversion ratio is higher than 60x (3 V-180 V) because V_{in} is lower for 90x conversion senario.

VI. Conclusions

An isolated quasi-resonant converter, and a complementary mixed-signal controller, for low power ultra-high gain applications are introduced. Switching losses, which are dominant in low power regime, are reduced drastically by reducing the primary switch voltage-current overlap losses during turn-OFF, and ensuring full soft-switching for the added circuitry for achieving this effect. A small flying capacitor works as a resonant element which gets fully charged to the output voltage and discharged to zero. This discontinuous voltage mode operation of the flying capacitor ensures zero voltage across the primary winding during switch turn-OFF, therefore reducing the switching losses in the primary switch. This operation further decouples the input-to-output conversion ratio independent of the load and turns ratio of the transformer. The reduced voltage stress in the primary switch allows the operation of the

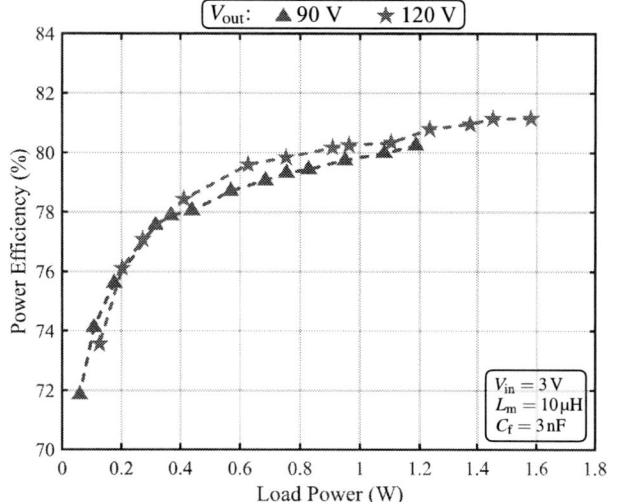

Fig. 10. Power efficiency of the discrete prototype at various output voltages for $C_f = 3\,\text{nF}$.

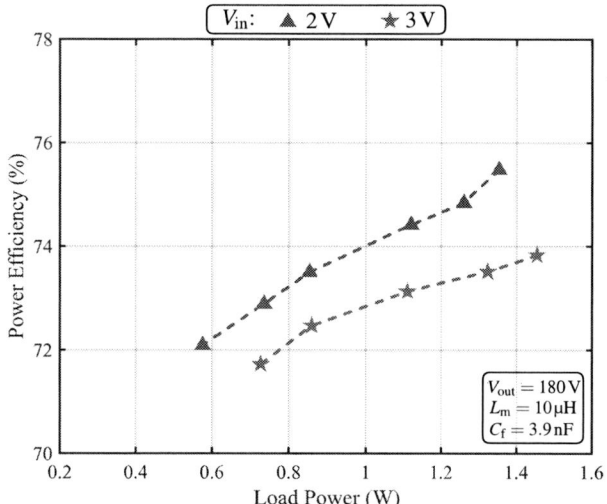

Fig. 11. Power efficiency of the discrete prototype at various input voltages for $C_f = 3.9\,\text{nF}$.

converter without a snubber. The power processing effectiveness of the introduced topology is experimentally verified for voltage conversion ratios exceeding 60x, sub-2 W output power levels and achieving peak efficiencies around 80 %.

REFERENCES

[1] C. Marquez-Chin and M. R. Popovic, "Functional electrical stimulation therapy for restoration of motor function after spinal cord injury and stroke: a review," *BioMedical Engineering OnLine*, vol. 19, no. 34, 2020, doi: 10.1186/s12938-020-00773-4.

[2] M. Tarulli, S. C. Huerta, A. Prodić, M. R. Popović, and P. W. Lehn, "Merged flyback–sc-based output stage for versatile portable transcutaneous electrical stimulators," *IEEE Journal of Emerging and Selected Topics in Power Electronics*, vol. 4, no. 1, pp. 318–331, 2016, doi:10.1109/JESTPE.2015.2509604.

[3] "Medical eletrical equipment- part 1-8: General requirements for basic and essential performance- collateral standard: General requirements, tests and guidance for alarm systens in medical electrical equipment and medical electrical systems," in *International Standard IEC 60601-1-8*, 1st ed., Aug 2003.

[4] M. Forouzesh, Y. P. Siwakoti, S. A. Gorji, F. Blaabjerg, and B. Lehman, "Step-up dc–dc converters: A comprehensive review of voltage-boosting techniques, topologies, and applications," *IEEE Transactions on Power Electronics*, vol. 32, no. 12, pp. 9143–9178, 2017, doi:10.1109/TPEL.2017.2652318.

[5] T. Halder, "Comprehensive power loss model of the main switch of the flyback converter," in *2013 International Conference on Power, Energy and Control (ICPEC)*, 2013, pp. 792–797, doi:10.1109/ICPEC.2013.6527763.

[6] H.-P. Park and J.-H. Jung, "Design methodology of quasi-resonant flyback converter with a divided resonant capacitor," *IEEE Transactions on Industrial Electronics*, vol. 68, no. 11, pp. 10796–10805, 2021, doi:10.1109/TIE.2020.3029481.

[7] S. F. Graziani, T. V. Cook, and B. M. Grainger, "Isolated flying capacitor multilevel converters," *IEEE Open Journal of Power Electronics*, vol. 3, pp. 197–208, 2022, doi: 10.1109/OJPEL.2022.3160049.

[8] C. R. Sullivan, B. A. Reese, A. L. F. Stein, and P. A. Kyaw, "On size and magnetics: Why small efficient power inductors are rare," in *2016 International Symposium on 3D Power Electronics Integration and Manufacturing (3D-PEIM)*, 2016, pp. 1–23, doi:10.1109/3DPEIM.2016.7570571.

[9] A. Muneeb, J. Kaplun, D. Singh, M. Ul Hassan, and F. Luo, "Design challenges of high frequency high step ratio pcb-based planar transformer for gan based dual active bridge converter," in *2023 IEEE Energy Conversion Congress and Exposition (ECCE)*, 2023, pp. 5564–5571, doi:10.1109/ECCE53617.2023.10362737.

[10] M. Mu and F. C. Lee, "Design and optimization of a 380–12 v high-frequency, high-current llc converter with gan devices and planar matrix transformers," *IEEE Journal of Emerging and Selected Topics in Power Electronics*, vol. 4, no. 3, pp. 854–862, 2016, doi:10.1109/JESTPE.2016.2586964.

[11] Y.-C. Liu, C. Chen, K.-D. Chen, Y.-L. Syu, D.-J. Lu, K. A. Kim, and H.-J. Chiu, "Design and implementation of a planar transformer with fractional turns for high power density llc resonant converters," *IEEE Transactions on Power Electronics*, vol. 36, no. 5, pp. 5191–5203, 2021, doi:10.1109/TPEL.2020.3029001.

[12] M. K. Ranjram and D. J. Perreault, "A 380-12 v, 1-kw, 1-mhz converter using a miniaturized split-phase, fractional-turn planar transformer," *IEEE Transactions on Power Electronics*, vol. 37, no. 2, pp. 1666–1681, 2022, doi:10.1109/TPEL.2021.3103434.

[13] K. J. Nag and A. Prodic, "Fully soft-switching flying capacitor based quasi-resonant boost converter," in *2023 IEEE Applied Power Electronics Conference and Exposition (APEC)*, 2023, pp. 919–925, doi: 10.1109/APEC43580.2023.10131658.

Accurate Small-Signal Phasor Transformation-Based Modeling of Secondary-Side Diode-Bridge Rectifiers for Battery Charging Applications

Aditya Zade and Regan Zane

Department of Electrical and Computer Engineering, Utah State University

Logan, Utah – USA 84341

Email: aditya.p.zade@usu.edu

Abstract—This paper analyzes a dc-dc resonant converter in the small-signal domain with a secondary-side diode-bridge rectifier for battery charging applications. The conventional approach that models the diode bridge circuit as an equivalent resistor results in excessive damping of high-frequency resonance in the case of a battery load, leading to inaccuracies in small-signal modeling. This paper proposes an improved method for small-signal modeling, representing the diode bridge with a battery load as a dependent voltage sink of constant magnitude and variable phase, to achieve accurate results. The effectiveness of this method is verified through phasor transformation-based small-signal modeling of a T-type bridge-based dc-dc converter with an LCC tank and a diode bridge connected to a battery. The proposed method is validated through hardware testing on a 4 kW, 85 kHz battery charger prototype, demonstrating the accuracy of the modeling technique for perturbation frequencies up to 55 kHz.

Index Terms—Battery chargers, dc-dc converters, diode-bridge rectifiers, electric vehicles (EVs), LCC, phasor transformation, resonant converters, small-signal modeling, T-type.

I. INTRODUCTION

A single active bridge converter, comprising an active controllable bridge on the primary side and a passive diode-bridge rectifier on the secondary side of isolation, is widely used in applications requiring unidirectional power flow, such as electric vehicle (EV) battery charging [1], [2]. Replacing the secondary active bridge in a dual active bridge converter with a diode bridge significantly simplifies the circuit structure, reducing the complexity of both hardware and control strategies. This design choice not only enhances power density and improves system reliability, but also lowers manufacturing costs [3], [4]. Furthermore, zero current switching is inherently achieved as the diodes naturally turn on when the current reaches zero crossing points, leading to a substantial reduction in overall switching losses of the dc-dc converter [5].

In addition, applications such as wireless power transfer for EV charging, where electronics on the secondary side of isolation are located within the vehicle, face challenges in receiving control signals from the primary side due to isolation requirements. This isolation complicates control signal

---

This material is based in part upon work supported by the United States Department of Energy (DOE) under Award Number DE-EE0008803 Falcon and by the National Science Foundation (NSF) through the ASPIRE Engineering Research Center under Grant 1941524.

communication, making a passive diode-bridge rectifier on the secondary side a practical choice for reliable power delivery without requiring complex communication systems [6]–[10].

Considering the advantages of the circuit configuration with a diode-bridge rectifier, this study utilizes a dc-dc resonant converter with a secondary-side diode bridge for battery charging application and conducts a comprehensive small-signal analysis of this converter. It has been observed that the conventional approach of modeling the diode bridge circuit as an equivalent resistor using the fundamental harmonic approximation, as described in [11], [12], results in excessive and incorrect damping of the higher-frequency resonance for a battery load, leading to inaccuracies in small-signal modeling. This paper introduces an alternative method for modeling the diode bridge with a battery load as a dependent voltage sink, offering a more accurate representation. This voltage sink undergoes phase deviations that align with the phase variations in the diode bridge input current, which occur in response to perturbations introduced in the control input. However, the magnitude of the voltage sink remains constant. The proposed method yields precise small-signal modeling results and effectively captures higher-frequency resonance.

The proposed small-signal modeling approach for the secondary-side diode-bridge rectifier is validated using a T-type bridge-based dc-dc resonant converter, featuring an LCC tank, a secondary-side diode bridge, and a battery load. The higher-order LCC resonant tank, designed with an appropriate quality factor, ensures sinusoidal current or voltage waveforms on both the primary and secondary sides of the dc-dc converter. In this context, the use of a phasor transformation-based small-signal modeling approach, as discussed in [13]–[16], is particularly well-suited. The overall modeling is verified through hardware testing on a 4 kW, 85 kHz battery charger prototype, demonstrating the accuracy of the modeling process for perturbation frequencies up to 55 kHz.

II. NOTATION

In this paper, lowercase letters with a hat (e.g., $\hat{i}_{\text{load}}$) denote small-signal quantities, while capital letters (such as I_{load}) indicate steady-state quantities. Lowercase letters without a hat (e.g., $i_{\text{load}} = I_{\text{load}} + \hat{i}_{\text{load}}$) represent full-signal quantities, encompassing both steady-state and small-signal components.

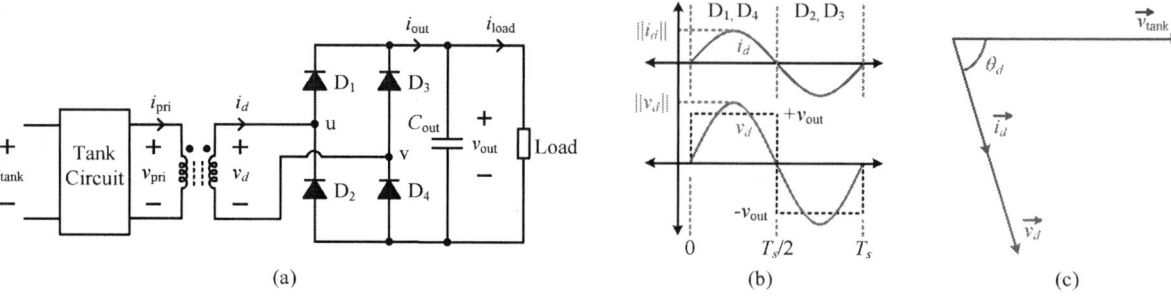

Fig. 1: (a) Typical circuit configuration of a resonant dc-dc converter with a diode bridge on the secondary side connected to a generic load; (b) current and voltage waveforms at the input of the diode bridge; and (c) a phasor diagram of current and voltage quantities in the dc-dc converter.

Phasors are marked with an arrow (e.g., $\vec{i_d}$), and magnitudes (peak values) are denoted with double bars (as in $\|\vec{i_d}\|$), or specified with 'mag' in the subscript. Finally, angled brackets (e.g., $\langle i_{\text{out}} \rangle$) are used for average values.

III. SECONDARY-SIDE DIODE-BRIDGE RECTIFIER IN THE SMALL-SIGNAL DOMAIN

In resonant dc-dc converters, as shown in Fig. 1(a), with an appropriate quality factor of the tank circuit, the tank current entering the diode-bridge rectifier, denoted as i_d in this work, exhibits a sinusoidal shape [12]. During the positive half-cycle of this current, diodes D_1 and D_4 conduct, resulting in a positive voltage ("+v_{out}") across the input of the diode bridge. On the other hand, during the negative half-cycle of the current, diodes D_2 and D_3 conduct, causing a negative voltage ("-v_{out}") to appear across the input of the diode bridge. As a result, this switching pattern of diodes generates a square-wave voltage at the input of the diode bridge that is in phase with the current i_d, as depicted in Fig. 1(b). Furthermore, the sinusoidal fundamental component of this square-wave voltage, denoted as v_d, also remains in phase with the current i_d. Both i_d and v_d can be represented as phasor quantities:

$$\vec{i_d} = \|\vec{i_d}\| e^{-j\theta_d}, \tag{1}$$

$$\vec{v_d} = \|\vec{v_d}\| e^{-j\theta_d} = \frac{4}{\pi} v_{\text{out}} e^{-j\theta_d}, \tag{2}$$

where v_{out} represents the output dc voltage across the load and θ_d denotes the amount of phase lag of the diode-bridge rectifier input current relative to the voltage applied across the tank input, as illustrated in Fig. 1(c). It can be observed from (2) that the magnitude of $\vec{v_d}$ depends upon the output dc voltage, while its phase always aligns exactly with the phase of $\vec{i_d}$, as depicted in the phasor diagram given in Fig. 1(c).

A. Resistive Load

In a typical configuration with a resistive load connected to the output of the diode-bridge rectifier, as shown in Fig. 2(a), small-signal perturbations introduced in the tank input voltage (by perturbing the duty ratios of the primary bridge) lead to small-signal deviations in both the magnitude and phase of $\vec{i_d}$. These small-signal deviations are also proportionally reflected in the output dc voltage, v_{out}, due to its dependency on $\vec{i_d}$,

as given by

$$v_{\text{out}} = i_{\text{load}} R_{\text{load}} = \frac{2}{\pi} \|\vec{i_d}\| R_{\text{load}}. \tag{3}$$

Now, using (2) and (3), $\vec{v_d}$ can be calculated as

$$\vec{v_d} = \frac{8}{\pi^2} R_{\text{load}} \|\vec{i_d}\| e^{-j\theta_d}$$

$$= R_{\text{eq}} \|\vec{i_d}\| e^{-j\theta_d}$$

$$= R_{\text{eq}} \vec{i_d}. \tag{4}$$

Therefore, the diode bridge input voltage in the small-signal domain, in the case of a resistive load, is given as

$$\hat{\vec{v_d}} = R_{\text{eq}} \hat{\vec{i_d}}, \tag{5}$$

where

$$R_{\text{eq}} = \frac{8}{\pi^2} R_{\text{load}}, \tag{6}$$

represents the equivalent resistor that can be used to model the diode-bridge rectifier with a resistive load in the small-signal domain, as depicted in Fig. 2(a).

B. Battery Load

In the case where a battery is connected to the output of the diode-bridge rectifier, as depicted in Fig. 2(b), small-signal perturbations in the tank input voltage (by perturbing the duty ratios of the primary bridge) induce corresponding small-signal deviations in both the magnitude and phase of $\vec{i_d}$. However, the output dc voltage, v_{out}, exhibits negligible deviation in the small-signal domain due to the significantly higher time constant of the battery State of Charge [17]. Therefore, during small-signal analysis, the battery voltage, which is the output dc voltage, can be considered constant [16]. As a result, the input voltage phasor of the diode bridge, $\vec{v_d}$, given by (2), has a constant magnitude, but its phase varies during small-signal analysis,

$$\vec{v_d} = \frac{4}{\pi} V_{\text{batt}} e^{-j\theta_d}. \tag{7}$$

Therefore, in the case of a battery load, the diode bridge input voltage phasor in the small-signal domain is given as

$$\hat{\vec{v_d}} = -j \frac{4}{\pi} V_{\text{batt}} e^{-j\Theta_d} \hat{\theta}_d, \tag{8}$$

979-8-3315-1612-3/25 $31.00 © 2025 IEEE

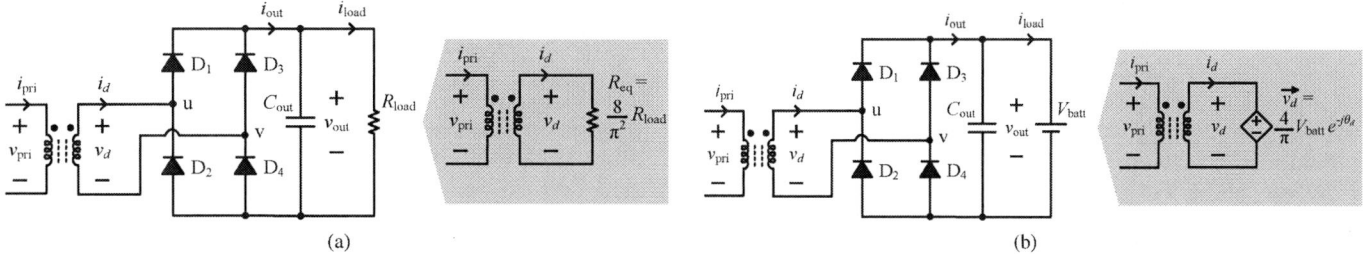

Fig. 2: (a) The diode-bridge rectifier with a resistive load can be modeled as an equivalent resistor in the small-signal domain; (b) the diode-bridge rectifier with a battery load needs to be modeled as a dependent voltage sink in the small-signal domain.

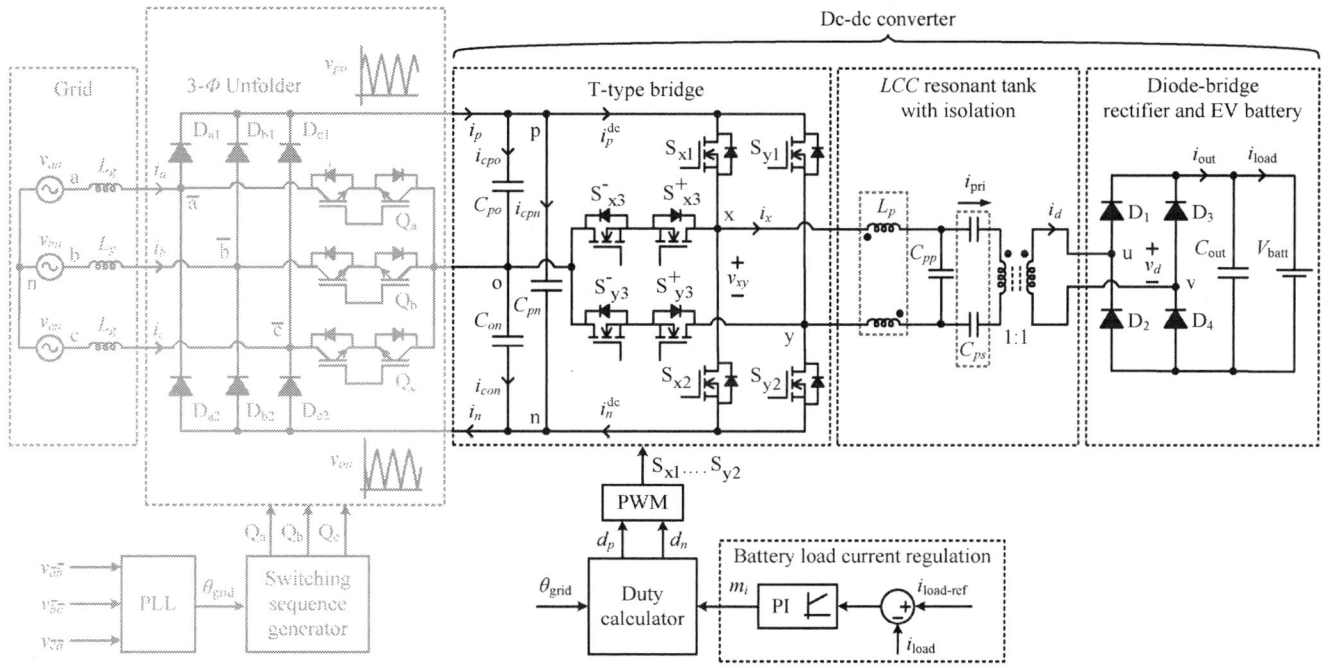

Fig. 3: Circuit diagram of a T-type bridge-based dc-dc converter consisting of an LCC tank with isolation and a diode bridge in an unfolding-based battery charger topology. The closed-loop control regulates the battery load current.

where V_{batt} is the battery voltage, and Θ_d is the steady-state phase lag of the diode bridge input current phasor, $\vec{I_d}$, relative to the tank input voltage. It can be observed from (8) that the diode bridge circuit cannot be simply modeled as a resistor in the case of a battery-connected load. The circuit needs to be modeled as a dependent voltage sink with a constant magnitude but a variable phase that depends upon the phase deviations of the diode bridge input current. The circuit diagram for this scenario is shown in Fig. 2(b). Incorrectly modeling this circuit as a resistor excessively damps higher-frequency resonance, as will be discussed in the subsequent experimental results section, leading to inaccuracies in small-signal modeling.

On the other hand, since the phase lag angle, θ_d, of the diode bridge input voltage in a dc-dc converter with a secondary-side diode bridge is not directly regulated by a controller, one might consider treating this angle as a steady-state constant, Θ_d, by neglecting its small-signal variations. This would render the diode bridge input voltage an independent quantity with con-

stant magnitude and phase. However, this approach introduces excessive resonances and leads to inaccurate modeling results, as discussed in the experimental results section.

IV. SMALL-SIGNAL MODELING OF A DC-DC CONVERTER WITH A DIODE-BRIDGE RECTIFIER AND A BATTERY LOAD

To validate the hypothesis concerning the small-signal modeling of a secondary-side diode bridge with a battery load as a dependent voltage sink, as defined by (8), this study examines a dc-dc resonant converter comprising a primary T-type bridge, an LCC tank, and a secondary diode bridge connected to a battery load, as shown in Fig. 3. The dc-dc converter in this study is connected to a grid-tied Unfolder operating in open-loop mode, switching at a frequency up to twice the grid frequency [18]–[24]. The Unfolder rectifies each negative segment of the ac input voltages into a positive polarity, generating two time-varying dc link voltages, v_{po} and v_{on}, which pulsate at three times the grid frequency at the input of the subsequent T-type bridge, as depicted in Fig. 3.

The closed-loop control architecture is specifically designed for the high-frequency (HF) T-type bridge-based dc-dc converter to regulate the battery load current. The duty calculator, as illustrated in Fig. 3, precisely calculates the duty ratios (d_p and d_n) based on the modulation index (m_i) obtained from the closed-loop control and the grid angle (θ_{grid}), which is derived from the phase-locked loop (PLL). The duty calculator, in combination with the closed-loop control, ensures the desired load current while maintaining power factor correction (PFC) on the grid side. A comprehensive description of the circuit configuration, modulation strategy, and closed-loop control is provided in [18]. Since the Unfolder operates in open-loop mode, the small-signal analysis presented in this paper focuses on the closed-loop controlled T-type bridge-based dc-dc conversion system, which incorporates an LCC resonant tank with isolation, and a diode-bridge rectifier connected to a battery load. More details on the small-signal modeling can be found in [15].

The implementation of a higher-order LCC resonant tank with an appropriate quality factor enables the achievement of sinusoidal current or voltage waveforms on both the primary and secondary sides of the dc–dc conversion system. As a result, the subsequent analysis considers only the fundamental components of the tank voltage and current quantities. The fundamental component of the voltage at the output of the T-type bridge, denoted as v_{xy} and depicted in Fig. 4, is determined from the two dc input voltages, v_{po} and v_{on}, along with their corresponding duty ratios, d_p and d_n:

$$v_{xy}(t) = v_{xy\text{-}po}(t) + v_{xy\text{-}on}(t)$$

$$= \frac{4}{\pi}\left[v_{po}\sin\left(\frac{\pi d_p}{2}\right) + v_{on}\sin\left(\frac{\pi d_n}{2}\right)\right]\cos(\omega_s t), \tag{9}$$

where

$$v_{xy\text{-}po}(t) = \frac{4}{\pi}v_{po}\sin\left(\frac{\pi d_p}{2}\right)\cos(\omega_s t), \tag{10}$$

$$v_{xy\text{-}on}(t) = \frac{4}{\pi}v_{on}\sin\left(\frac{\pi d_n}{2}\right)\cos(\omega_s t), \tag{11}$$

and ω_s denotes the angular switching frequency of the T-type bridge, which is kept constant (f_s = 85 kHz) for this application. The output voltage of the T-type bridge, as defined in (9), can be represented in phasor form as

$$\overrightarrow{v_{xy}} = \frac{4}{\pi}\left[v_{po}\sin\left(\frac{\pi d_p}{2}\right) + v_{on}\sin\left(\frac{\pi d_n}{2}\right)\right]\angle 0, \tag{12}$$

which can be further solved by substituting the following expressions for v_{po}, v_{on}, d_p, and d_n during $0 < \theta_{\text{grid}} \leq \pi/3$:

$$v_{po} = V_{g\text{-mag}}\sin\left(\theta_{\text{grid}} + 2\pi/3\right), \tag{13}$$

$$v_{on} = V_{g\text{-mag}}\sin\left(\theta_{\text{grid}}\right), \tag{14}$$

$$d_p = (2/\pi)\sin^{-1}\left(m_i\sin\left(\theta_{\text{grid}} + \pi/2\right)\right), \tag{15}$$

$$d_n = (2/\pi)\sin^{-1}\left(m_i\sin\left(\theta_{\text{grid}} + \pi/6\right)\right), \tag{16}$$

where $V_{g\text{-mag}}$ represents the peak value of the line-to-line ac input voltage of the Unfolder, and m_i denotes the modulation

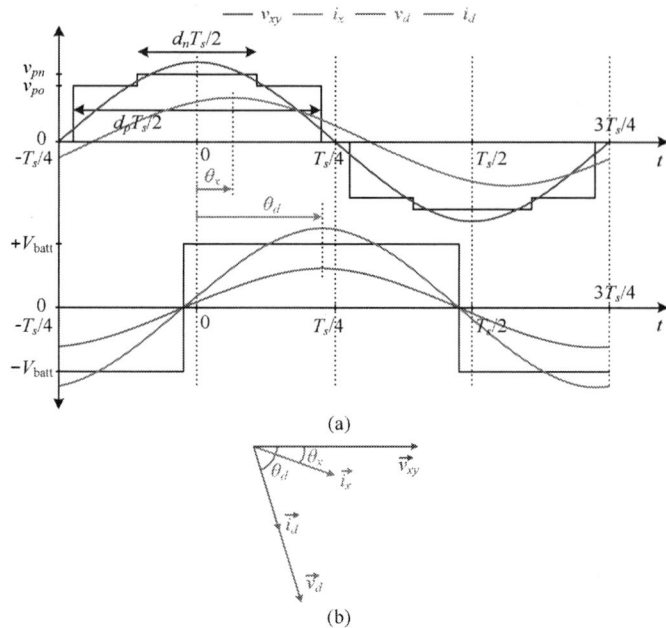

Fig. 4: (a) Fundamental components of the T-type bridge output voltage (which is the tank input voltage), v_{xy}, and current, i_x, as well as the diode bridge input voltage, v_d, and current, i_d. The operating region where $d_p > d_n$ is shown, with the two quasi-square voltages generated at the T-type bridge output using these duty ratios being center-aligned. (b) Phasor diagram of the voltage and current quantities in the T-type bridge-based dc-dc converter.

index that governs the magnitude of the duty ratios. A detailed discussion of the above equations can be found in [18]. By combining (12)–(16), the T-type bridge output voltage phasor, $\overrightarrow{v_{xy}}$, can be written as

$$\overrightarrow{v_{xy}} = \frac{2\sqrt{3}}{\pi}m_i V_{g\text{-mag}} \angle 0 = \frac{2\sqrt{3}}{\pi}m_i V_{g\text{-mag}}. \tag{17}$$

Since the modulation index m_i governs the magnitude of the T-type bridge output voltage, which defines the load current i_{load} charging the battery, a small-signal analysis is performed on the plant transfer function derived between the small-signal load current $\hat{i}_{\text{load}}$ and the small-signal modulation index $\hat{m}_i$:

$$G_{\text{plant}} = \frac{\hat{i}_{\text{load}}}{\hat{m}_i}. \tag{18}$$

A. Small-Signal Phasor-Transformed Circuit

To derive the small-signal phasor-transformed circuit, the tank inductors and capacitors are phasor-transformed as outlined in [13]. The inductor is modeled as a series combination of the dynamic impedance (sL) and the steady-state impedance ($j\omega_s L$) while the capacitor is represented as a parallel combination of the dynamic impedance ($\frac{1}{sC}$) and the steady-state impedance ($\frac{1}{j\omega_s C}$). The output voltage of the T-type bridge is linearized around the steady-state value of the control input m_i, allowing the steady-state and small-signal components of $\overrightarrow{v_{xy}}$ to be expressed as

$$\overrightarrow{v_{xy}} = \overrightarrow{V_{xy}} + \overrightarrow{\hat{v}_{xy}} = \overrightarrow{V_{xy}} + \left(\frac{\partial\overrightarrow{v_{xy}}}{\partial m_i}\right)\hat{m}_i$$

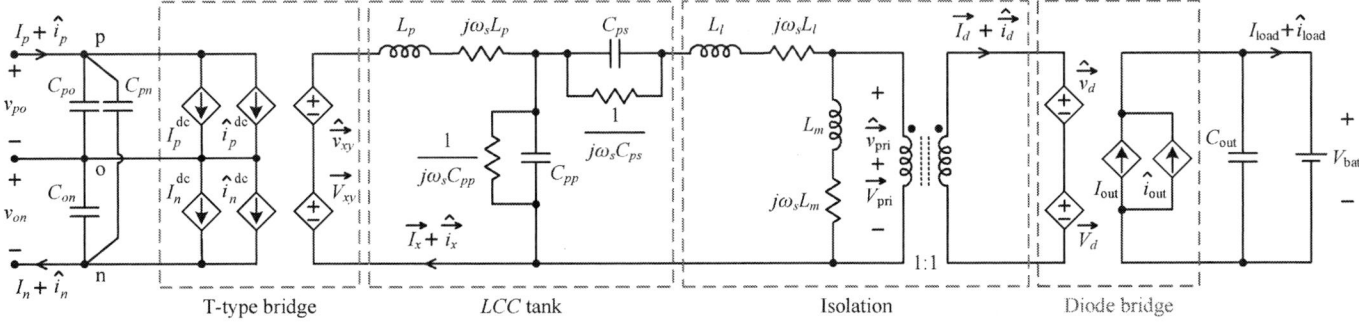

Fig. 5: Phasor-transformed circuit for a T-type bridge-based dc-dc converter with an LCC resonant tank, cantilever-modeled isolation, secondary diode bridge, and output capacitive filter connected to the battery. This circuit is a combination of both the small-signal and steady-state models.

$$= \frac{2\sqrt{3}}{\pi} M_i V_{g\text{-mag}} + \frac{2\sqrt{3}}{\pi} \hat{m}_i V_{g\text{-mag}} . \tag{19}$$

Therefore,

$$\frac{\vec{\hat{v}_{xy}}}{\hat{m}_i} = \frac{2\sqrt{3}}{\pi} V_{g\text{-mag}} . \tag{20}$$

Now, the diode bridge input voltage, written as

$$\vec{v_d} = \frac{4}{\pi} V_{\text{batt}} e^{-j\theta_d} , \tag{21}$$

can be expressed in terms of the small-signal component by taking the partial derivative with respect to the control input, m_i. The hypothesis presented in Section III regarding the small-signal representation of the diode bridge with a battery-connected load, modeled as a dependent voltage sink with a constant magnitude but a phase that varies based on the phase deviations of the diode bridge input current, can now be applied to (21). Consequently, the small-signal variations in the diode bridge input voltage phasor, in relation to the perturbations introduced in the control input, can be derived using (8) as

$$\frac{\vec{\hat{v}_d}}{\hat{m}_i} = -j\frac{4}{\pi} V_{\text{batt}} e^{-j\Theta_d} \frac{\hat{\theta}_d}{\hat{m}_i} . \tag{22}$$

The complete linearized steady-state and small-signal model of the T-type bridge-based dc-dc converter with an LCC tank is depicted in Fig. 5. The isolation is modeled using the cantilever model, which includes leakage and magnetizing inductances. To enhance the accuracy of the model, the parasitic resistances of inductors and capacitors are represented as series and parallel resistances, respectively, as employed in the simulation and experimental results.

B. Small-Signal Diode Bridge Input Current

As the load current i_{load} depends on the magnitude of the diode bridge input current i_d, the small-signal deviations in the diode bridge input current phasor, $\vec{\hat{i}_d}$, are calculated in relation to perturbations introduced in the control input, $\hat{m}_i$, to derive the plant transfer function. The small-signal diode bridge input current phasor is calculated using Fig. 5 as

$$\vec{\hat{i}_d} = \frac{\left(\vec{\hat{v}_{xy}} - \vec{\hat{v}_d}(f_1 f_3 - f_4) \right)}{f_2 f_3}$$

$$= \frac{\left(\frac{2\sqrt{3}}{\pi} \hat{m}_i V_{g\text{-mag}} + j\frac{4}{\pi} V_{\text{batt}} e^{-j\Theta_d} \hat{\theta}_d (f_1 f_3 - f_4) \right)}{f_2 f_3} , \tag{23}$$

where f_1, f_2, f_3, and f_4 are functions of the tank parameters:

$$f_1 = \frac{Z_{ps} + Z_l + Z_m}{Z_m} , \tag{24}$$

$$f_2 = Z_{ps} + Z_l , \tag{25}$$

$$f_3 = 1 + \frac{Z_p}{Z_{pp}} + \frac{Z_p}{Z_{ps} + Z_l} , \tag{26}$$

$$f_4 = \frac{Z_p}{Z_{ps} + Z_l} , \tag{27}$$

$$Z_p = sL_p + j\omega_s L_p + r_{L_p} , \tag{28}$$

$$Z_{pp} = \frac{r_{C_{pp}}}{1 + r_{C_{pp}}(sC_{pp} + j\omega_s C_{pp})} , \tag{29}$$

$$Z_{ps} = \frac{r_{C_{ps}}}{1 + r_{C_{ps}}(sC_{ps} + j\omega_s C_{ps})} , \tag{30}$$

$$Z_l = sL_l + j\omega_s L_l + r_{L_l} , \tag{31}$$

$$Z_m = sL_m + j\omega_s L_m , \tag{32}$$

where r_{L_p}, $r_{C_{pp}}$, $r_{C_{ps}}$, and r_{L_l} represent the parasitic resistances of the tank inductors and capacitors. Equation (23) can be modified into a transfer function using (20) and (22) as

$$\frac{\vec{\hat{i}_d}}{\hat{m}_i} = \frac{\left(\frac{\vec{\hat{v}_{xy}}}{\hat{m}_i} - \frac{\vec{\hat{v}_d}}{\hat{m}_i}(f_1 f_3 - f_4) \right)}{f_2 f_3}$$

$$= \frac{\left(\frac{2\sqrt{3}}{\pi} V_{g\text{-mag}} + j\frac{4}{\pi} V_{\text{batt}} e^{-j\Theta_d} \frac{\hat{\theta}_d}{\hat{m}_i}(f_1 f_3 - f_4) \right)}{f_2 f_3} . \tag{33}$$

An alternative, more useful form of the result (33) is derived by splitting this equation into its linearized magnitude and phase components using a first-order Taylor series approximation. This approach enables the complex phasor transfer function to be represented by two transfer functions with real coefficients,

979-8-3315-1612-3/25 $31.00 © 2025 IEEE   1008

which correspond to the magnitude and phase variations of the diode bridge input current. These transfer functions are calculated as

$$\frac{\hat{i}_{d\text{-mag}}}{\hat{m}_i} = \frac{\Re\left[\frac{\vec{\hat{i}_d}}{\hat{m}_i}\right]\Re[\vec{I_d}] + \Im\left[\frac{\vec{\hat{i}_d}}{\hat{m}_i}\right]\Im[\vec{I_d}]}{\|\vec{I_d}\|}, \quad (34)$$

$$\frac{\hat{\theta}_d}{\hat{m}_i} = \frac{\Re\left[\frac{\vec{\hat{i}_d}}{\hat{m}_i}\right]\Im[\vec{I_d}] - \Im\left[\frac{\vec{\hat{i}_d}}{\hat{m}_i}\right]\Re[\vec{I_d}]}{\|\vec{I_d}\|^2}. \quad (35)$$

It is important to note that (33) and (35) are interdependent and must be solved as simultaneous equations.

C. Small-Signal Load Current

The output current of the diode bridge is a dc quantity and, thus, is not representable in phasor form. To determine the full output current, comprising both dc and HF components, a time-domain analysis is performed. The time-domain output current can be obtained using the circuit model illustrated in Fig. 5 as

$$i_{\text{out}}(t) = \frac{4}{\pi}\cos(\omega_s t - \theta_d)\, i_{d\text{-mag}}\cos(\omega_s t - \theta_d)$$
$$= \frac{2}{\pi}(I_{d\text{-mag}} + \hat{i}_{d\text{-mag}})(1 + \cos(2\omega_s t - 2\Theta_d - 2\hat{\theta}_d)). \quad (36)$$

The battery load current, which is the averaged component of the output current of the diode bridge, can now be written as

$$i_{\text{load}} = \langle i_{\text{out}} \rangle = I_{\text{load}} + \hat{i}_{\text{load}} = \frac{2}{\pi}I_{d\text{-mag}} + \frac{2}{\pi}\hat{i}_{d\text{-mag}}. \quad (37)$$

Therefore, the small-signal deviations in i_{load} that directly depend on $\hat{i}_{d\text{-mag}}$ and $\hat{\theta}_d$ are

$$\frac{\hat{i}_{\text{load}}}{\hat{i}_{d\text{-mag}}} = \frac{2}{\pi}, \quad (38)$$

$$\frac{\hat{i}_{\text{load}}}{\hat{\theta}_d} = 0, \quad (39)$$

respectively.

Now, by combining (34), (35), (38), and (39), the plant transfer function can be calculated as

$$G_{\text{plant}} = \frac{\hat{i}_{\text{load}}}{\hat{m}_i} = \frac{\hat{i}_{\text{load}}}{\hat{i}_{d\text{-mag}}}\frac{\hat{i}_{d\text{-mag}}}{\hat{m}_i} + \frac{\hat{i}_{\text{load}}}{\hat{\theta}_d}\frac{\hat{\theta}_d}{\hat{m}_i}. \quad (40)$$

V. EXPERIMENTAL RESULTS

To validate the phasor transformation-based small-signal modeling in hardware, tests are conducted on the T-type bridge-based dc-dc conversion system illustrated in Fig. 6. This system comprises a single-sided LCC tank with isolation and a secondary-side diode-bridge rectifier connected to a battery load. Two dc-voltage sources, manufactured by REGATRON, are connected to the inputs of the T-type bridge, while a dc-voltage sink (NHR 9300) emulates a battery at the output. Experimental parameters are provided in TABLE I. Parasitic resistances of inductors and capacitors are expressed in terms of series and parallel resistances, respectively. The modeling verification is carried out for the plant transfer function, $G_{\text{plant}} = \frac{\hat{i}_{\text{load}}}{\hat{m}_i}$, at an output power of 4 kW. In this verification, the modulation index, m_i, is sinusoidally perturbed by $\pm$ 0.05 around a steady-state value of 0.95. The perturbation is performed across a range of frequencies: 50 Hz, 100 Hz, 500 Hz, 1 kHz, 3.7 kHz, 5 kHz, 10 kHz, 20 kHz, 30 kHz, 32.5 kHz, 36.6 kHz, 40 kHz, 42.5 kHz, and 55 kHz. The control input, m_i, is updated at twice the switching frequency (170 kHz) using an up-down carrier, implemented with the TMS320F28379D microcontroller. The resulting magnitude and phase of the perturbation-frequency component present in the load current, i_{load}, are measured to derive the Bode plots.

Fig. 7 illustrates the phase deviations in the diode bridge input square-wave voltage when m_i is perturbed, while its magnitude remains unchanged, as previously proven mathematically. Fig. 8, Fig. 9, and Fig. 10 provide a comparison of analytical Bode plots of G_{plant} with hardware results. In Fig. 8, modeling the diode bridge with a battery load as a resistor results in excessive damping of the higher-frequency resonance, leading to inaccurate analytical plots. In Fig. 9, treating the diode bridge with a battery load as an independent voltage sink with constant magnitude and phase while ignoring the phase deviations results in excessive resonances

TABLE I: Experimental parameters.

Parameter	Value
P_{out}, f_s	4 kW, 85 kHz
$V_{g\text{-mag}}, V_{\text{batt}}$	$208\sqrt{2}$ V, 316 V
v_{po}, v_{on}	208 V, 76 V
L_p, r_{L_p}	29.3 μH, 116.4 mΩ
$C_{pp}, r_{C_{pp}}$	119.9 nF, 14.9 kΩ
$C_{ps}, r_{C_{ps}}$	112.4 nF, 15.6 kΩ
L_l, r_{L_l}	37.1 μH, 31.5 mΩ
L_m	802.2 μH

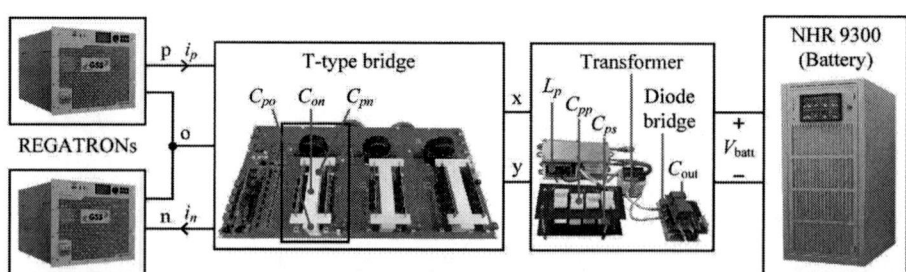

Fig. 6: A 4 kW hardware setup of a dc-dc conversion system featuring a T-type bridge, a single-sided LCC tank, and a diode bridge on the secondary side of the isolation. Two dc power supplies from REGATRON serve as dc input sources, while the NHR 9300 functions as a battery load. Small-signal perturbations are applied to the modulation index m_i using the TMS320F28379D microcontroller.

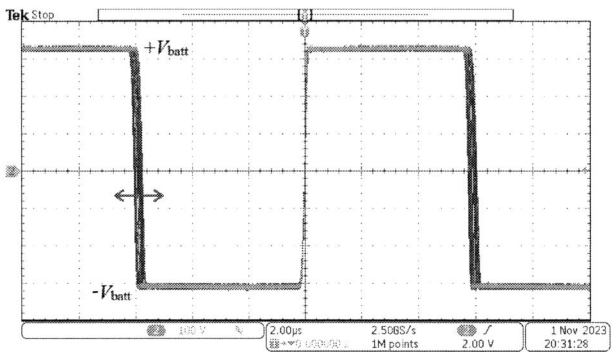

Fig. 7: Phase deviations ($\hat{\theta}_d$), as shown by the arrow, in the diode bridge input square-wave voltage due to perturbations introduced in the control input, $\hat{m}_i$, while the magnitude remains constant.

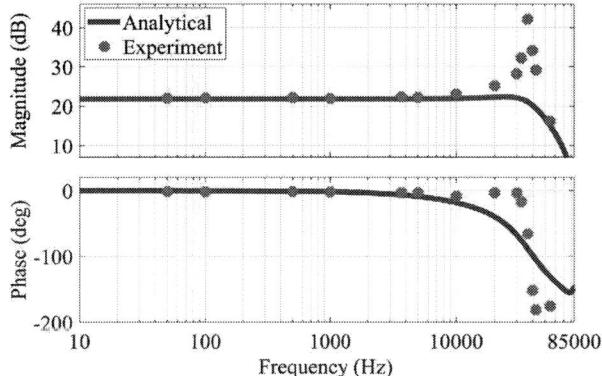

Fig. 8: Analytical Bode plots of G_{plant} compared with hardware results, where the diode bridge with a battery load is modeled as a resistor. The analysis results in excessive damping and fails to capture the high-frequency resonance.

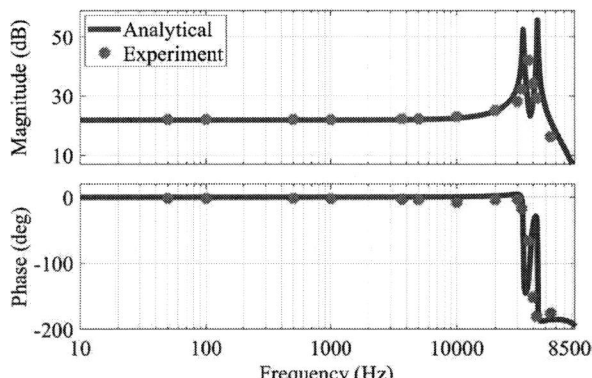

Fig. 9: Analytical Bode plots of G_{plant} compared with hardware results, where the diode bridge with a battery load is modeled as a constant independent voltage sink. The analysis results in excessive resonances.

and inaccuracies. Modifications are made to Fig. 5 and (33) to model the diode bridge with a battery load as a resistor and as an independent voltage sink.

Finally, Fig. 10 models the diode bridge with a battery load as a dependent voltage sink, incorporating small-signal phase deviations and constant magnitude, which produces accurate analytical Bode plots consistent with hardware results across all perturbation frequencies. Moreover, a step response of the diode bridge output current, i_{out}, has been measured during

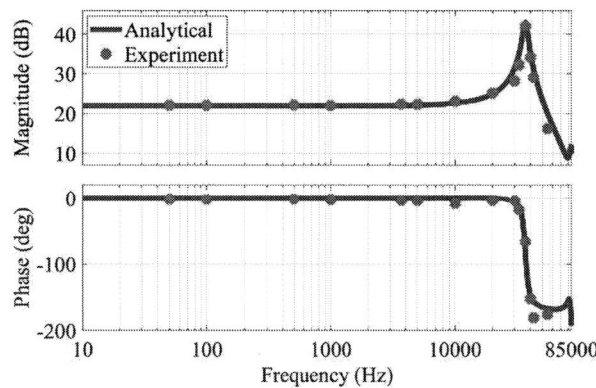

Fig. 10: Analytical Bode plots of G_{plant} compared with hardware results, where the diode bridge with a battery load is modeled as a dependent voltage sink with a constant magnitude and variable phase, as proposed in this paper. The analysis accurately matches the hardware results.

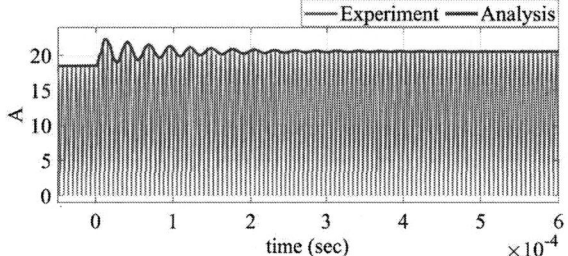

Fig. 11: Experimental waveform of the diode-bridge rectifier output current, i_{out}, for the T-type bridge-based dc-dc converter, superimposed on the envelope predicted by the small-signal model during a step change in the modulation index, m_i, from 0.9 to 1. The diode bridge with a battery load is modeled here as a dependent voltage sink with a constant magnitude and variable phase, as proposed in this paper.

the hardware testing by introducing a step change in the modulation index from 0.9 to 1. The envelope observed during this step change in the hardware has been compared with the step response obtained from the small-signal modeling with a dependent voltage sink, as presented in Fig. 11. The results demonstrate that the modeling precisely predicts the envelope of the diode bridge output current during this step change, thereby confirming the accuracy of the modeling. Accurate small-signal modeling that captures the high-frequency dynamics of the dc-dc resonant converter with a diode bridge connected to a battery load enhances the design and stability of closed-loop control, leading to more reliable and efficient power delivery.

VI. CONCLUSION

This paper presents a small-signal analysis of a dc-dc resonant converter with a secondary-side diode-bridge rectifier for battery charging applications. It reveals that the conventional way of modeling the diode bridge as an equivalent resistor introduces inaccuracies, particularly excessive damping that fails to capture high-frequency resonance when a battery load is connected. To address this, the paper proposes modeling the diode bridge with a battery load as a dependent voltage sink, with constant magnitude and variable phase, yielding

979-8-3315-1612-3/25 $31.00 © 2025 IEEE

more accurate results. This approach provides a more realistic depiction of the system dynamics, capturing high-frequency behavior and resonance with improved precision.

The proposed methodology is validated through hardware testing on a 4 kW battery charger circuit, featuring a T-type bridge-based dc-dc converter with a secondary-side diode bridge connected to a battery load. Experiments show that the dependent voltage sink model aligns closely with hardware results, making it a reliable approach for modeling and control of dc-dc resonant converters with a diode bridge in battery charging applications.

REFERENCES

[1] A. Rodríguez, A. A. Gomez, M. M. Hernando, D. G. Lamar and J. Sebastian, "A Dynamic study of the Single Active Bridge Converter," in IEEE Transactions on Power Electronics, doi: 10.1109/TPEL.2024.3426590.

[2] A. Averberg and A. Mertens, "Characteristics of the single active bridge converter with voltage doubler," 2008 13th International Power Electronics and Motion Control Conference, Poznan, Poland, 2008, pp. 213-220, doi: 10.1109/EPEPEMC.2008.4635269.

[3] Rodriguez, Alberto, Javier Sebastian, Diego G. Lamar, Marta M. Hernando, Iban Ayarzaguena, Igor Larrazabal, David Ortega, Jose M. Bermejo, and Francisco Vazquez. 2022. "An Overall Analysis of the Static Characteristics of the Single Active Bridge Converter" Electronics 11, no. 4: 601. https://doi.org/10.3390/electronics11040601.

[4] C. Fontana, M. Forato, M. Bertoluzzo and G. Buja, "Design characteristics of SAB and DAB converters," 2015 Intl Aegean Conference on Electrical Machines & Power Electronics (ACEMP), 2015 Intl Conference on Optimization of Electrical & Electronic Equipment (OPTIM) & 2015 Intl Symposium on Advanced Electromechanical Motion Systems (ELECTROMOTION), Side, Turkey, 2015, pp. 661-668, doi: 10.1109/OPTIM.2015.7427025.

[5] J. Deng et al., "Frequency and Parameter Combined Tuning Method of LCC–LCC Compensated Resonant Converter With Wide Coupling Variation for EV Wireless Charger," in IEEE Journal of Emerging and Selected Topics in Power Electronics, vol. 10, no. 1, pp. 956-968, Feb. 2022, doi: 10.1109/JESTPE.2021.3077459.

[6] S. Li and C. C. Mi, "Wireless Power Transfer for Electric Vehicle Applications," in IEEE Journal of Emerging and Selected Topics in Power Electronics, vol. 3, no. 1, pp. 4-17, March 2015, doi: 10.1109/JESTPE.2014.2319453.

[7] A. Mahesh, B. Chokkalingam and L. Mihet-Popa, "Inductive Wireless Power Transfer Charging for Electric Vehicles–A Review," in IEEE Access, vol. 9, pp. 137667-137713, 2021, doi: 10.1109/ACCESS.2021.3116678.

[8] G. A. Covic and J. T. Boys, "Inductive Power Transfer," in Proceedings of the IEEE, vol. 101, no. 6, pp. 1276-1289, June 2013, doi: 10.1109/JPROC.2013.2244536.

[9] W. Zhang, S. -C. Wong, C. K. Tse and Q. Chen, "Load-Independent Duality of Current and Voltage Outputs of a Series- or Parallel-Compensated Inductive Power Transfer Converter With Optimized Efficiency," in IEEE Journal of Emerging and Selected Topics in Power Electronics, vol. 3, no. 1, pp. 137-146, March 2015, doi: 10.1109/JESTPE.2014.2348558.

[10] S. -Y. R. Hui, Y. Yang and C. Zhang, "Wireless Power Transfer: A Paradigm Shift for the Next Generation," in IEEE Journal of Emerging and Selected Topics in Power Electronics, vol. 11, no. 3, pp. 2412-2427, June 2023, doi: 10.1109/JESTPE.2023.3237792.

[11] A. C. Bagchi, H. Wang, T. Saha and R. Zane, "Small-Signal Phasor Modeling of an Underwater IPT System in Constant Current Distribution," 2019 IEEE Applied Power Electronics Conference and Exposition (APEC), Anaheim, CA, USA, 2019, pp. 876-883, doi: 10.1109/APEC.2019.8721899.

[12] R. W. Erickson and D. Maksimovic, "Resonant Conversion", Fundamental of Power Electronics, 3rd Ed., pp. 933-993, 2020.

[13] C. T. Rim and G. H. Cho, "Phasor transformation and its application to the DC/AC analyses of frequency phase-controlled series resonant converters (SRC)," in IEEE Transactions on Power Electronics, vol. 5, no. 2, pp. 201-211, April 1990, doi: 10.1109/63.53157.

[14] C. T. Rim, "Unified General Phasor Transformation for AC Converters," in IEEE Transactions on Power Electronics, vol. 26, no. 9, pp. 2465-2475, Sept. 2011, doi: 10.1109/TPEL.2011.2107920.

[15] A. Zade, S. Gurudiwan and R. Zane, "Small-Signal Phasor Modeling of T-Type Bridge-Based Single-Sided and Double-Sided *LCC* Resonant Converters for WPT Applications," 2025 IEEE Applied Power Electronics Conference and Exposition (APEC), Atlanta, GA, USA, 2025.

[16] W. Han and L. Corradini, "Analytical Small-Signal Transfer Functions for Phase Shift Modulated Dual Active Bridge Converters Using Phasor Transformation," 2018 IEEE Energy Conversion Congress and Exposition (ECCE), Portland, OR, USA, 2018, pp. 1442-1448, doi: 10.1109/ECCE.2018.8558437.

[17] M. Cacciato, G. Nobile, G. Scarcella and G. Scelba, "Real-Time Model-Based Estimation of SOC and SOH for Energy Storage Systems," in IEEE Transactions on Power Electronics, vol. 32, no. 1, pp. 794-803, Jan. 2017, doi: 10.1109/TPEL.2016.2535321.

[18] A. Zade et al., "A 21-kW Unfolding-Based Single-Stage AC–DC Converter for Wireless Charging Applications," in IEEE Journal of Emerging and Selected Topics in Power Electronics, vol. 12, no. 1, pp. 8-27, Feb. 2024, doi: 10.1109/JESTPE.2023.3309588.

[19] S. GURUDIWAN, A. Zade, H. Wang and R. Zane, "Accurate ZVS Analysis of a Full-Bridge T-Type Resonant Converter for a 20-kW Unfolding-Based AC-DC Topology," in IEEE Open Journal of Power Electronics, vol. 5, pp. 692-708, 2024, doi: 10.1109/OJPEL.2024.3400256.

[20] A. Zade, S. Gurudiwan, D. Maksimović and R. Zane, "High-Bandwidth Control of a 20-kW Single-Stage Unfolding-Based AC–DC Converter Using the Extra Element Theorem and Current Emulation Technique," in IEEE Transactions on Power Electronics, vol. 39, no. 11, pp. 14411-14429, Nov. 2024, doi: 10.1109/TPEL.2024.3423705.

[21] B. Zhang, S. Xie, Z. Li, P. Zhao and J. Xu, "An Optimized Single-Stage Isolated Swiss-Type AC/DC Converter Based on Single Full-Bridge With Midpoint-Clamper," in IEEE Transactions on Power Electronics, vol. 36, no. 10, pp. 11288-11297, Oct. 2021, doi: 10.1109/TPEL.2021.3073742.

[22] A. Blinov, D. Zinchenko, J. Rabkowski, G. Wrona and D. Vinnikov, "Quasi Single-Stage Three-Phase Filterless Converter for EV Charging Applications," in IEEE Open Journal of Power Electronics, vol. 3, pp. 51-60, 2022, doi: 10.1109/OJPEL.2021.3134460.

[23] N. H. Pham, T. Mannen and K. Wada, "Power Factor Operation of a Boost Integrated Three-Phase Solar Inverter using Current Unfolding and Active Damping Methods," 2018 IEEE Energy Conversion Congress and Exposition (ECCE), Portland, OR, USA, 2018, pp. 2896-2903, doi: 10.1109/ECCE.2018.8558360.

[24] L. Schrittwieser, J. W. Kolar and T. B. Soeiro, "Novel SWISS Rectifier Modulation Scheme Preventing Input Current Distortions at Sector Boundaries," in IEEE Transactions on Power Electronics, vol. 32, no. 7, pp. 5771-5785, July 2017, doi: 10.1109/TPEL.2016.2609935.

High-Efficiency Isolated Piezoelectric Transformers for Magnetic-less DC-DC Power Conversion

Sourav Naval, Wentao Xu, Mustapha Touhami, and Jessica D. Boles

Berkeley Power and Energy Center
University of California, Berkeley
Berkeley, CA, USA
sourav_naval@berkeley.edu

Abstract—**Piezoelectric components are promising alternative passive components for power conversion, offering high efficiency and high energy densities at small scales. In this paper, we model, design, and demonstrate isolated piezoelectric transformers (PTs) capable of high efficiencies and galvanic isolation as primary passive components in magnetic-less dc-dc converters. We present a design framework for radial mode PTs that enables them to simultaneously achieve peak efficiency and zero voltage switching in a power converter at a specified operating point. The proposed PT design strategy is validated in a dc-dc converter prototype that demonstrates efficiencies of >97% at multiple voltage levels. The converter achieves a peak efficiency of 97.6%, which represents a 17x reduction in loss ratio compared to previous isolated magnetic-less PT-based dc-dc converter designs.**

Index Terms—**piezoelectric transformers (PTs), isolated DC-DC converters, radial mode**

I. INTRODUCTION

The miniaturization of power electronics has long been constrained by passive components, especially magnetic elements like inductors and transformers. Magnetics scale poorly to small sizes with fundamentally decreasing power densities and efficiencies at low volume [1], [2]. The demand for high-performance, miniaturized power electronics motivates exploration of alternative passive components for future power conversion. One promising alternative passive component technology is piezoelectric components, which exhibit inductive impedance above their resonant frequencies and fundamentally increasing power handling densities at small scales [3].

Piezoelectric components store energy in the mechanical compliance and inertia of an acoustic wave, with quality factors and energy densities significantly greater than those of magnetics at small scales [4], [5]. Further, piezoelectrics offer practical benefits including planar form factors, batch fabrication, potential for integration, and no generation of stray magnetic fields. Recent magnetic-less dc-dc converter designs based on single-port piezoelectric resonators (PRs) have demonstrated power stage efficiencies of 99% and PR power handling densities of up to 5.7 kW/cm³ [6]–[10]. While these demonstrations mark tremendous milestones, such performance has only been achieved in non-isolated dc-dc converters with mild voltage conversion ratios. Capacitive

This material is based upon work supported in part by Enphase Energy and the Berkeley Power and Energy Center.

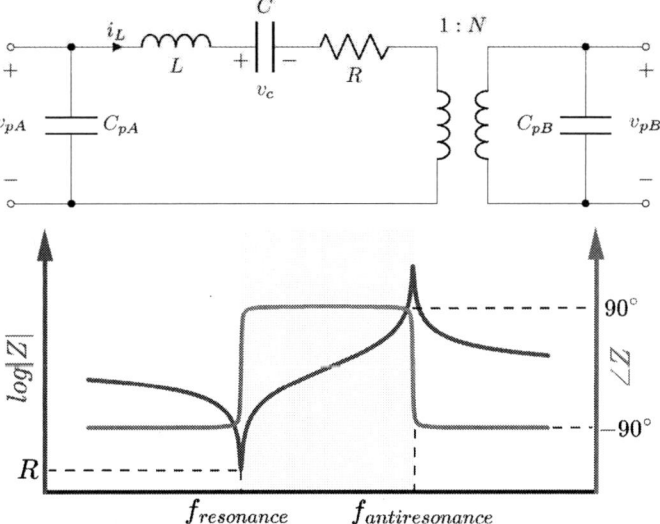

Fig. 1. (Top) Mason lumped circuit model of a PT, which describes the the PT's behavior in the proximity of its resonant frequency. (Bottom) Magnitude and phase of the impedance of a PT in the vicinity of a resonant mode with output port short-circuited.

isolation may be realized using multiple PRs [11], but this approach entails significant capacitive coupling across the isolation barrier and is sensitive to mismatch between the PRs. Thus, the utility of piezoelectric-based power conversion is confined to a narrow subset of power electronics applications.

Piezoelectrics may be expanded to a broader set of applications through use of multi-port piezoelectric transformers (PTs), which offer the same advantages as PRs but with the added potential for galvanic isolation and inherent voltage transformation [12]–[14]. When AC voltage is applied to one port of a PT, it induces mechanical deformation in the piezoelectric material (i.e., the converse piezoelectric effect) that cascades throughout the component and causes charge displacement at the PT's other port(s) (i.e., the direct piezoelectric effect). Thus, all energy traveling between ports is first converted from the electrical domain to the mechanical domain, stored mechanically, and then extracted from the PT electrically at the other port. Similar to magnetic coupling, this mechanical coupling provides galvanic isolation with minimal

979-8-3315-1612-3/25 $31.00 © 2025 IEEE

TABLE I
COMPARISON WITH PREVIOUS MAGNETIC-LESS ISOLATED PT-BASED DC-DC CONVERTERS

Reference	[27]	[28]	This Work	
Efficiency (%)	68	62	97.6	97.2
Power (W)	6.5	1	0.09	4.93
Input Voltage (V)	230	20-75	10	100
Resonant Frequency (kHz)	114	400	90	90

capacitance, and therefore minimal direct charge flow, between ports. PTs are typically operated near the resonant frequency of a specific vibration mode, where their behavior is accurately represented by the Mason lumped circuit model as illustrated in Fig. 1. This model includes physical port capacitances (C_{pA}, C_{pB}), an ideal voltage transformation (1 : N), and an LCR branch modeling the PT's mechanical resonance and loss properties [15]–[17], .

While PTs have seen decades of development for power conversion, widespread commercial use of PTs has been generally limited to low-power, high voltage step-up applications such as CCFL backlight drivers [13]. Early PT-based dc-dc converter designs often included auxiliary magnetic components, which limited their potential to achieve high power densities [18]–[21]. PTs have also been utilized to design magnetic-less dc-dc converters, but their efficiencies and power densities have been limited to ~90% and <5 W/cm^3, respectively, in non-isolated designs [14], [22]–[26]. Performance has been drastically lower for isolated magnetic-less PT-based dc-dc converters, in which efficiency has been limited to 68% as summarized in Table I. This highlights a significant gap in achievable efficiency between presently available PRs and PTs.

In this paper, we investigate how isolated PTs may be designed to achieve high efficiencies as primary passive components for isolated dc-dc converters. We conduct this study in the context of full-bridge/full-bridge (FB-FB) converter topology as visualized in Fig. 2 using a switching sequence that most efficiently utilizes the PT's resonant cycle as developed in [14]. We note that the proposed design strategies are applicable to a wide variety of isolated converter topologies.

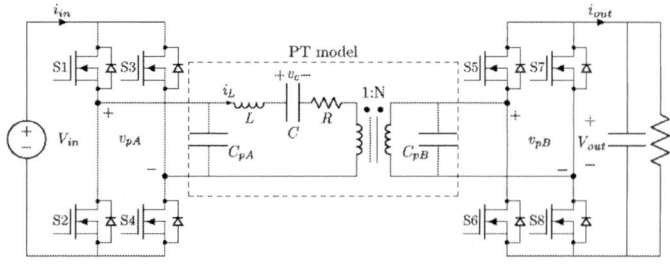

Fig. 2. Full-bridge/Full-bridge (FB-FB) topology for isolated PT-based dc-dc converters.

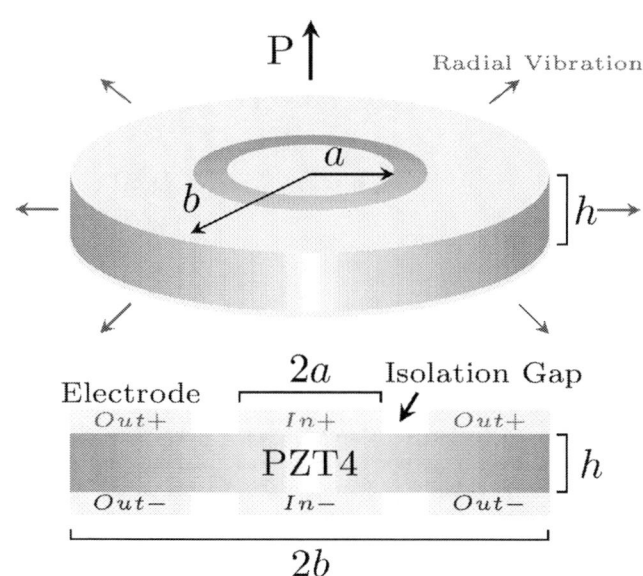

Fig. 3. (Top) 3D model of a ring-dot PT structure. (Bottom) Cross-section view of the PT with dimensions and ports labeled. P denotes the polarization direction of the piezoelectric material.

II. RADIAL MODE PT STRUCTURE AND MODEL

A. Structure of Radial Mode PT

Piezoelectric component design begins with selection of a piezoelectric material and vibration mode. The radial vibration mode of lead zirconate titanate (PZT) has demonstrated the highest efficiencies to date in PRs [29], [30], so we investigate how the PZT radial mode may be similarly leveraged to design high-efficiency isolated PTs. The radial (k_p) vibration mode is commonly realized with a disc poled homogeneously in its axial direction. In k_p mode, charge displacement occurs in the axial direction while the acoustic wave propagates in the radial direction. The structure of a "ring-dot" radial-mode PT ($k_p - k_p$ mode) is illustrated in Fig. 3 [3]. It consists of a disc-shaped monolithic layer of PZT, and the top and bottom surfaces of the disc are metalized with identical patterns of conductive electrodes. This electrode pattern contains an inner disc corresponding the PT's input port and an outer ring corresponding to the PT's output port. The electrode gap between these regions electrically isolates the two ports.

B. Equivalent Circuit Model of Radial Mode PT

To derive the equivalent circuit model of the radial mode PT, we conceptualize it as a disc and ring resonator mechanically coupled at their interface. We solve for the radial displacement and the radial velocity for each port as a function of the radial forces, applied voltages, currents and material properties. The relevant material properties are defined in Table II. To establish the Mason model for the complete PT, we equate the radial forces ($[F_{disc}]_{r=a} = [F_{ring}]_{r=a}$) and the radial velocities ($[v_{disc}]_{r=a} = [v_{ring}]_{r=a}$) of the two ports at their interface [17], assuming the gap between them to be negligible. The

TABLE II
MATERIAL AND PROPERTY DEFINITIONS

Parameter	Definition
ϵ_{33}	Dielectric Constant
s_{11}	Elastic Compliance Constant
d_{31}	Piezoelectric Charge Constant
k_p	Planar Coupling Factor
c_a	Acoustic Velocity
κ_r	Resonant Wavenumber
ν	Poisson Ratio
Q	Quality Factor
ρ	Mass Density
α	Radius Ratio (a/b)

TABLE III
RING-DOT PT MASON MODEL PARAMETERS

Parameter	Expression
C_{pA}	$\dfrac{\epsilon_{33}\pi a^2 (1-k_p^2)}{h}$
C_{pB}	$\dfrac{\epsilon_{33}\pi (b^2-a^2)(1-k_p^2)}{h}$
L	$\dfrac{h\psi^2(\kappa_r^2+\nu^2-1)}{2\pi\alpha^2 k_p^2\epsilon_{33}(1+\nu)\kappa_r^2 c_a^2\phi^2}$
C	$\dfrac{2\pi a^2 k_p^2\epsilon_{33}(1+\nu)\phi^2}{h\psi^2(\kappa_r^2+\nu^2-1)}$
R	$\dfrac{h\psi^2(\kappa_r^2+\nu^2-1)}{2\pi a\alpha k_p^2\epsilon_{33}Q\kappa_r c_a(1+\nu)\phi^2}$
N	$\dfrac{\alpha\phi}{\alpha\phi-\psi}$

$\psi = J_1(\kappa_r)$ and $\overline{\phi = J_1(\kappa_r\alpha)}$, where $J_1(x)$ represents the first-order Bessel function of the first kind.

combined circuit model is illustrated in Fig. 4. This model is then simplified into a lumped parameter model that represents the PT when it is vibrating near its resonant frequency. The resonant frequency is expressed as a function of the resonant wavenumber, acoustic velocity and the radius of PT as follows:

$$\omega_r = 2\pi f_r = \frac{\kappa_r c_a}{b} \tag{1}$$

The resonant wavenumber is a function of the material properties, and is determined by solving the following transcendental equation:

$$\frac{\kappa_r J_0(\kappa_r)}{J_1(\kappa_r)} + \nu - 1 = 0 \tag{2}$$

The solution of this equations yields multiple values corresponding to various harmonics. We derive the lumped model by performing Taylor series expansion around the first positive solution, which corresponds to the fundamental resonance mode. The resulting reduced Mason model is shown in Fig. 1, and the parameters are specified in Table III. A detailed derivation of this circuit model for the ring-dot structure is provided in the Appendix.

III. PT DESIGN OPTIMIZATION

Using the Mason model developed in Section II, we now derive design conditions for optimizing the PT dimensions to simultaneously achieve maximum efficiency and zero voltage switching (ZVS) in an isolated dc-dc converter.

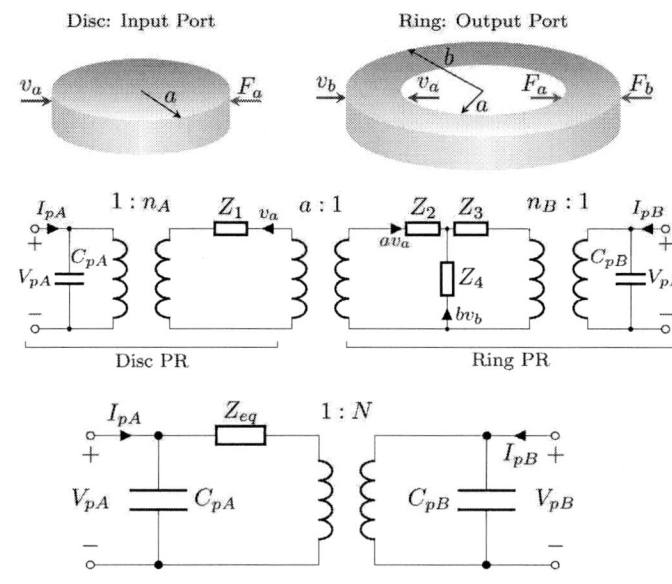

Fig. 4. (Top) Illustration of radial-mode velocity and forces on input disc PR and output ring PR. (Middle) Combination of input and output PR models to develop PT model. (Bottom) Simplified PT circuit model at resonant frequency.

A. Assumed Converter Switching Sequence

For this investigation, we assume the FB-FB topology visualized in Fig. 2 operating with the high-efficiency, isolated switching sequence $[V_{in}, -V_{in}|V_{out}, -V_{out}, Zero]$ detailed in [14]. Within this switching sequence, both the input and output bridges nominally operate with 50% duty cycle and appropriate dead time for ZVS, with a phase shift between the two bridges' switching. To regulate the output voltage, the output bridge is controlled to short-circuit the PT's output port (rather than send energy to the load) for a brief period of time during the switching cycle. This period is referred to as a "Zero" stage as detailed in [14].

Throughout this operating regime, the PT has the following amplitude of resonance (i.e., amplitude of i_L), which is derived from the required charge transfer during each stage of the converter's switching sequence [14]:

$$I_L = \pi\left(\frac{P_{out}}{2V_{in}} + 2f C_{pA}V_{in}\right) \tag{3}$$

This amplitude of resonance serves as a model for the PT's behavior throughout the assumed switching sequence. With it, we now systematically derive conditions for a PT to achieve maximum efficiency and ZVS at a nominal converter operating point [31].

B. Maximum Efficiency Geometry Condition

PT efficiency can be calculated based on the loss ratio expression $\left(\frac{P_{loss}}{P_{out}}\right)$ as follows:

$$\eta = \frac{1}{1 + \frac{P_{loss}}{P_{out}}} \tag{4}$$

979-8-3315-1612-3/25 $31.00 © 2025 IEEE

To maximize PT efficiency, the power loss ($P_{loss} = \frac{1}{2}I_L^2 R$) in the PT must be minimized for a given output power P_{out}. The P_{loss}/P_{out} ratio can be expanded by substituting the value of I_L from (3):

$$\frac{P_{loss}}{P_{out}} = \frac{\pi^2 R}{2P_{out}}\left(\frac{P_{out}^2}{4V_{in}^2} + 4V_{in}^2 f^2 C_{pA}^2 + 2P_{out} f C_{pA}\right) \quad (5)$$

To conduct this minimization with respect to the PT's geometry, we separate the geometry terms from the expressions of resistance R and frequency-capacitance product $B = fC_{pA}$ of the PT.

$$R = \left(\frac{J_1^2(\kappa_r)(\kappa_r^2 + \nu^2 - 1)}{2\pi k_p^2 \epsilon_{33} Q \kappa_r c_a (1+\nu) J_1^2(\kappa_r \alpha)}\right)\left(\frac{bh}{a^2}\right) = \frac{R_0}{J_1^2(\kappa_r \alpha)G}$$

$$B = fC_{pA} = \left(\frac{\kappa_r c_a \epsilon_{33}(1 - k_p^2)}{2}\right)\left(\frac{a^2}{bh}\right) = B_0 G \quad (6)$$

For the radial-mode PT, R_0 and B_0 are geometry-normalized terms with respect to a term defined as $G = \frac{a^2}{bh}$. Then, the loss ratio expression ($\frac{P_{loss}}{P_{out}}$) can be expressed in terms of the material constants R_0 and B_0, geometry term G, and operating point information as follows:

$$\frac{P_{loss}}{P_{out}} = \frac{\pi^2 R_0}{2J_1^2(\kappa_r \alpha)}\left(\frac{P_{out}}{4V_{in}^2 G} + \frac{4V_{in}^2 B_0^2 G}{P_{out}} + 2B_0\right) \quad (7)$$

Minimizing the loss ratio with respect to a geometry term defined as G, by setting $\partial \frac{P_{loss}}{P_{out}}/\partial G = 0$, yields the following PT geometry design condition for achieving maximum efficiency, and corresponding peak efficiency, at a specified converter operating point:

$$G = \frac{a^2}{bh} = \frac{P_{out}}{4B_0 V_{in}^2} \quad (8a)$$

$$\eta_{peak} = \frac{1}{1 + \frac{2\pi^2 R_0 B_0}{J_1^2(\kappa_r \alpha)}} \quad (8b)$$

Thus, if a PT design adheres to the condition in (8a) for its nominal operating point, it can be expected to achieve the peak efficiency in (8b) at that operating point.

C. ZVS Geometry Condition

This maximum efficiency condition assumes that the converter is capable of ZVS for all switches, as well as all-positive instantaneous power transfer. To ensure that the converter is capable of both, we derive an additional geometry condition for ZVS that translates to sizing the PT's input and output capacitances such that these high-efficiency behaviors are possible at a nominal converter operating point. To derive this condition, we equate the total charge that must be displaced by i_L throughout the resonant cycle from the perspective of each port. These expressions consider both energy-transfer stages

and dead time, assuming ZVS and all-positive instantaneous power transfer are achieved:

$$\frac{P_{out}}{fV_{in}} + 4C_{pA}V_{in} = N\left(\frac{P_{out}}{fK_B V_{out}} + 4C_{pB}V_{out}\right) \quad (9)$$

The charge transfer utilization factor K_B is defined as the proportion of charge displacement during energy transfer stages at the output port that actually delivers energy to the load (i.e., charge displaced while the PT is connected to the load rather than when its terminals are shorted in a "Zero" stage). For this switching sequence, $\frac{1}{2} \leq K_B \leq 1$, where $K_B = 1$ corresponds to the unregulated sequence with no "Zero" stage, and $K_B < 1$ corresponds to the output port spending part of its cycle in a "Zero" stage for voltage regulation purposes.

To achieve ZVS simultaneously with the maximum efficiency derived in (8b), the output power operating point derived from (8a) is substituted into (9). This results in the following constraint on α for which these high-efficiency behaviors are achieved:

$$\alpha = \frac{1}{\sqrt{\frac{V_{in}}{V_{out}}\left(\frac{2}{N} - \frac{V_{in}}{K_B V_{out}}\right) + 1}} \quad (10)$$

Thus, a PT can be expected to achieve ZVS over the range operating points that can satisfy (10) with $\frac{1}{2} \leq K_B \leq 1$.

IV. EXAMPLE PT DESIGN FOR MAXIMUM EFFICIENCY

Within the design framework proposed in Section III, we design a high-efficiency isolated PT for power conversion by carefully choosing the PT's geometric dimensions. First, we select a value of α using (10). We can obtain a range of α for which this equation is satisfied based on the range of K_B for the assumed switching sequence. The peak efficiency value also depends on α, as indicated in (8). To demonstrate maximum efficiency, we choose to maximize the value of α, which occurs as K_B approaches the extremum of its range. Thus, we select the value of $\alpha = 0.42$.

Having determined α, there are two degrees of freedom remaining in the PT's geometric dimensions that may be selected to ensure the PT achieves maximum efficiency at a nominal operating point by satisfying (8a). The value of disc radius b determines the PT's resonant frequency, and therefore the switching frequency range, and a higher resonant frequency tends to yield a higher power handling density for a given vibration mode [5]. The value of disc thickness h should be constrained for radial mode operation by assuming a value such that $h \ll b$.

In this work, we choose PT dimensions that correspond to a maximum-efficiency converter operating point of $P_{out}/V_{in}^2 = 0.5\ mW/V^2$, not considering the effects of switch parasitics. We choose $b = 12.7$ mm, which corresponds to a resonant frequency of 90 kHz from (1), and $h = 0.52$ mm to satisfy (8a). We calculate input port radius a from the value of α to be $a = 5.35$ mm. We also include a 0.6 mm wide isolation ring between the input and output ports to establish isolation between ports and ensure that the peak electric field

979-8-3315-1612-3/25 $31.00 © 2025 IEEE

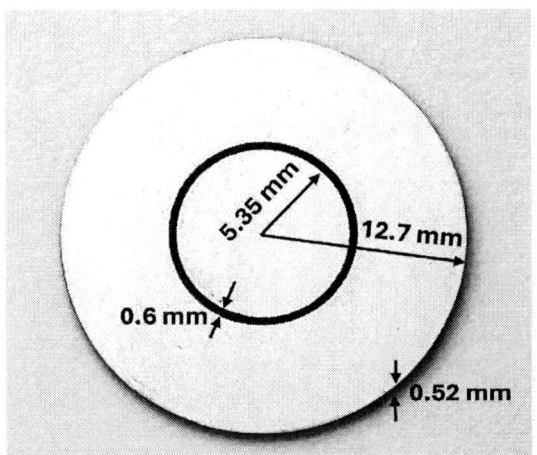

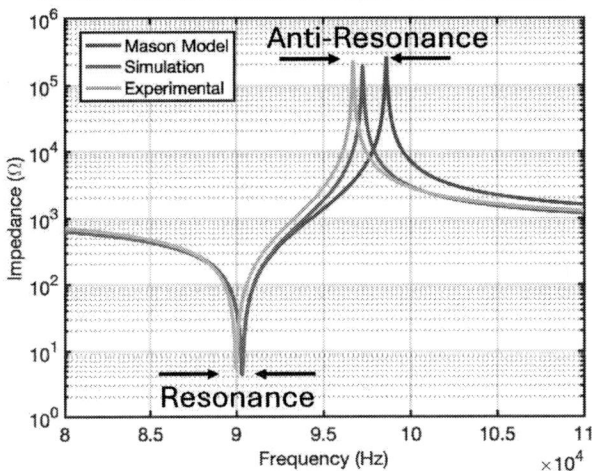

Fig. 5. (Top) Custom-patterned PT design fabricated by APC International. (Bottom) Comparison between modeled, simulated, and experimental PT impedance characteristics.

in the isolation gap is within the coercive field limit of the piezoelectric material. In this manner, all dimensions of the high-efficiency PT are determined.

V. SIMULATION AND EXPERIMENTAL VALIDATION

A. PT Mounting and Characterization

We acquire custom-manufactured PTs from APC International (APC 841 PZT Material) based on the dimensions selected in Section IV. The PT component with labeled dimensions is shown in Fig. 5. The Mason model parameters for this PT calculated using the expressions derived in Table III are provided in Table IV. The frequency response of the

TABLE IV
CALCULATED PT PARAMETERS

Parameter	C_{pA} (nF)	C_{pB} (nF)	R (Ω)	L (mH)	C (nF)	N	f_r (kHz)
Value	1.37	6.33	5.36	13.2	0.23	-0.41	90.317

Fig. 6. DC-DC converter prototype based on mounted radial-mode PT.

derived Mason Model is validated via Finite Element Method (FEM) simulation performed in COMSOL Multiphysics 6.1.

To characterize the PT's experimental performance, we mount the PT between two PCBs secured together with three spring-loaded M2.5 screws. Between the PT and each PCB are spring contact SMT micro clips (Alps Alpine SCTA2A0100) to serve as electrical connections for the PT and hold it in place with minimal pressure-induced damping of the PT. The same mounting structure is used to later mount the PT on the converter board. The impedance response of the mounted PT is measured using a Keysight E4990A impedance analyzer and aligns closely with analytical and simulated responses near the resonant frequency as shown in Fig. 5. This validates the proposed model in Section II as an accurate representation of the PT's impedance characteristic.

B. DC-DC Converter Implementation and Performance

After characterizing the PT, we implement it in a dc-dc converter topology of Fig. 2 on a four-layer 1-oz copper board as shown in Fig. 6, with the parts listed in Table V. We operate this prototype with a modified version of the $[V_{in}, -V_{in}|V_{out}, -V_{out}, Zero]$ high-efficiency switching sequence detailed in [14], with the output-port "Zero" stage split between both halves of the cycle to simplify control. All switching times are tuned manually for each operating point, and the converter is tested with a constant-voltage electronic load. Since the PT parameters are temperature dependent, we use a 650RPM 12V fan to ensure that the PT remains at a relatively constant temperature during testing.

To evaluate the prototype's performance, we conduct data sweeps over various voltage and power levels, keeping a constant $\frac{V_{out}}{V_{in}}$ conversion ratio of 0.5. For a given input and output voltage, we vary the output power by modulating the switching frequency, taking advantage of the PT's frequency-dependent gain visualized in Fig. 1, and then tune other parameters such

TABLE V
PROTOTYPE PARTS LIST

Component	Part
PT	APC Custom - 841 Double Bullseye Dia-25.4 mm, Thk-0.52 mm
Switch	EPC 2215 GaN FET
Gate Driver	Texas Instruments LMG1210
Microcontroller	Texas Instruments TMDSCNCD28379D

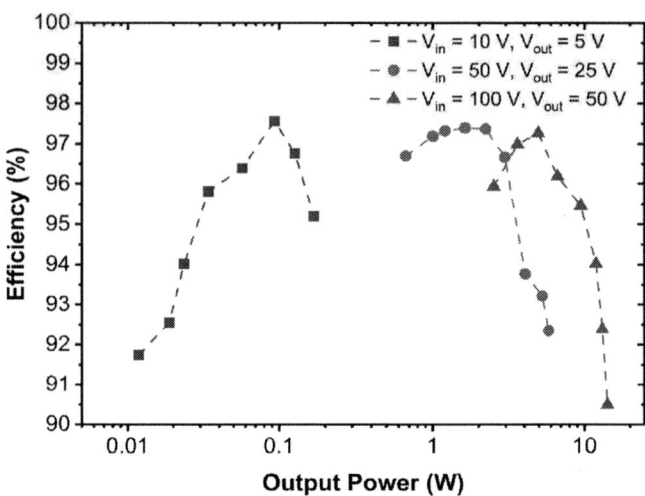

Fig. 8. Converter power stage efficiency with respect to output power for different input voltages at constant $\frac{V_{out}}{V_{in}} = 0.5$.

as phase shift and dead time for high-efficiency behaviors. To measure efficiency, the input and output current values are recorded using Keysight 34465A digital multimeter ($2\mu A$ accuracy), and the input and output voltages are recorded with Keysight PA2201A power analyzer. Thus, the resulting efficiency calculation ($\eta = V_{out}I_{out}/V_{in}I_{in}$) can be assumed to have an accuracy of $\pm 0.1\%$.

Fig. 8 displays the converter's experimental power stage efficiency vs. output power. The PT attains its peak efficiency of 97.6% at 93 mW for $V_{in} = 10$ V, and achieves efficiencies as high as 97.2% at $V_{in} = 100$ V. The waveforms of the peak efficiency operating point at $V_{in} = 10$ V, $V_{out} = 5$ V, and $P_{out} = 93$ mW are shown in Fig. 7 and demonstrate zero-voltage switching and all-positive instantaneous power transfer. The modeled peak efficiency for this PT design is calculated using the design framework established in (8). To consider the effect of switch capacitances on the PT performance, we modify C_{pA} and C_{pB} to include parasitic switch capacitance for each relevant segment when computing efficiency capability. The modeled peak efficiency for this prototype is 97.6%, which is expected to occur at a power level of 92.7 mW for $V_{in} = 10$ V. This aligns closely with the measured peak efficiency. A power-stage loss breakdown

estimate at the peak efficiency operating point reveals that the dominant loss ($\sim$ 2.3 mW) occurs in the PT, and a small fraction of losses occur in the switches ($\sim$ 0.05 mW). At this peak efficiency point, the converter demonstrates a 17x reduction in loss ratio compared to previous magnetic-less PT-based isolated dc-dc converter designs, validating the proposed models and framework for designing high-efficiency isolated PTs.

VI. CONCLUSION

This paper reveals isolated PTs to be capable of significantly greater efficiencies than previously demonstrated in magnetic-less dc-dc converters. Such high efficiency is achieved through selection of a high-performance PT vibration mode and specific PT geometry design conditions for simultaneously achieving maximum efficiency and ZVS at a nominal converter operating point. A dc-dc converter based on a PT designed using the proposed framework demonstrates a peak PT efficiency of 97.6%, which represents a 17x reduction in loss compared to previous work and validates the proposed design framework. To the authors' knowledge, these results are the first to experimentally demonstrate the potential of PTs to enable high-efficiency isolated dc-dc power conversion without magnetics. Thus, the proposed design framework extends the utility of piezoelectrics to a wide variety of power conversion applications requiring galvanic isolation.

APPENDIX

Equivalent Circuit Model Derivation

To derive the equivalent circuit model for the radial-mode PT, we first develop the Mason model for the individual input (disc) and output (ring) PRs. Then, we obtain the Mason model for the PT by connecting the Mason models of individual PRs using appropriate boundary conditions. We begin with the following piezoelectric material constitutive equations in strain-charge form for cylindrical coordinates:

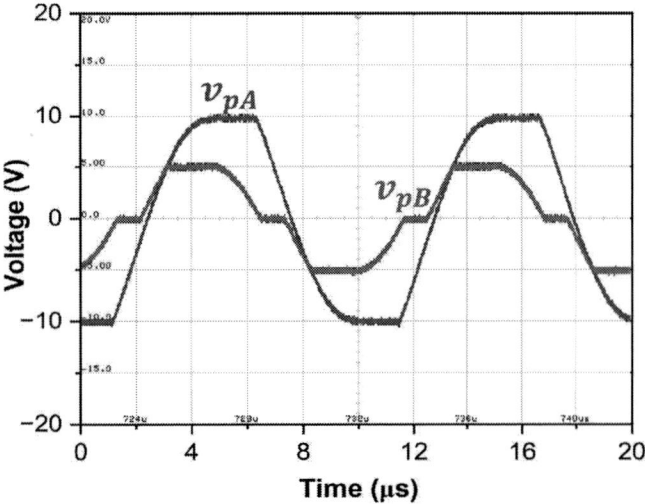

Fig. 7. Oscilloscope waveforms of differential input and output switch nodes of PT at $V_{in} = 10V, V_{out} = 5V$, and $P_{out} = 93$ mW. Labels correspond to the topology in Fig. 2.

$$S_r = s_{11}T_r - \nu s_{11}T_\theta + d_{31}E_z; \tag{11a}$$

$$S_\theta = s_{11}T_\theta - \nu s_{11}T_r + d_{31}E_z; \tag{11b}$$

$$D_z = d_{31}(T_r + T_\theta) + \epsilon_{33}E_z; \tag{11c}$$

where S and T denote the mechanical strain and stress, respectively, and D and E are the electric displacement and electric field, respectively. The subscripts r, θ, and z, correspond to the radial, azimuthal and axial directions of the disc. Relevant material properties are defined in Table II.

The acoustic wave propagating along the radial dimension can be expressed using the following wave equation:

$$\rho\frac{\partial^2 u_r}{\partial t^2} = \frac{\partial T_r}{\partial r} + \frac{T_r - T_\theta}{r} \tag{12}$$

For sinusoidal excitation, the solution to the differential equation (12) is expressed in terms of the first order Bessel functions of first ($J_1(x)$) and second kind ($Y_1(x)$) as follows:

$$u_r = AJ_1(\kappa r) + BY_1(\kappa r) \tag{13}$$

u_r denotes the radial displacement, A and B are constants which can be solved by applying the boundary conditions for the radial velocity ($v_r = \partial u_r/\partial t$). κ is the radial wavenumber, which is a function of the excitation frequency ω and the acoustic wave velocity in the piezoelectric material c_a:

$$\kappa = \omega\sqrt{\rho s_{11}(1-\nu^2)} = \frac{\omega}{c_a} \tag{14}$$

Assuming a sinusoidal voltage of magnitude V_{pA} is applied to input port (disc of radius a), parallel to the polarization direction, we calculate the current flowing through the input port I_{pA} and the radial force F_a at the disc circumference.

The current flowing through the disc is calculated as $I_{pA} = j\omega \int_0^a D_z 2\pi r dr$. Substituting the expression for D_z from (11), we get

$$I_{pA} = j\omega C_{pA}V_{pA} - n_A v_a \tag{15}$$

where, $C_{pA} = \frac{\epsilon_{33}\pi a^2}{h}(1-k_p^2)$ is the input capacitance, $n_A = \frac{2\pi a d_{31}}{s_{11}(1-\nu)}$ is the electrical to mechanical transformation ratio at the input port, and $k_p = \frac{\sqrt{2}d_{31}}{\sqrt{\epsilon_{33}s_{11}(1-\nu)}}$ is defined as the planar electromechanical coupling factor.

The radial force is expressed as the product of the stress and the curved surface area: $F_a = -T_{r=a} \times 2\pi a h$. Combining (11) and (13), we can solve for the radial force F_a is expressed as follows:

$$F_a = Z_1 v_a + n_A V_{pA} \tag{16}$$

The mechanical acoustic impedance Z_1 is defined as follows:

$$Z_1 = \frac{2\pi\rho c_a^2 h}{j\omega}\left(\frac{\kappa a J_0(\kappa a)}{J_1(\kappa a)} + \nu - 1\right) \tag{17}$$

Likewise, to develop a circuit model for the output port (ring), we write the equations for the current flowing through the ring, radial force at the inner ($r = a$) and outer ($r = b$) boundaries of the ring.

$$I_{pB} = j\omega C_{pB}V_{pB} - n_B(av_a + bv_b) \tag{18}$$

$$F_a = (Z_2 + Z_3)a^2 v_a + Z_3 ab v_b + an_B V_{pB} \tag{19}$$

$$F_b = Z_3 ab v_a + (Z_3 + Z_4)b^2 v_b + bn_B V_{pB} \tag{20}$$

where, $C_{pB} = \frac{\epsilon_{33}\pi(b^2-a^2)}{h}(1-k_p^2)$ is the output capacitance, $n_B = \frac{2\pi d_{31}}{s_{11}(1-\nu)}$ is the electrical to mechanical transformation ratio at the output port. The mechanical acoustic impedance Z_2, Z_3 and Z_4 are defined as follows:

$$Z_2 = \frac{2\rho c_a^2 h[\pi\kappa ab J_1(\kappa b)Y_0(\kappa a) - \pi\kappa ab Y_1(\kappa b)J_0(\kappa a) + \pi b(\nu-1)(Y_1(\kappa a)J_1(\kappa b) - Y_1(\kappa b)J_1(\kappa a)) - 2a]}{j\omega a^2 b(Y_1(\kappa b)J_1(\kappa a) - J_1(\kappa b)Y_1(\kappa a))} \tag{21}$$

$$Z_3 = \frac{4\rho c_a^2 h}{j\omega ab(Y_1(\kappa b)J_1(\kappa a) - J_1(\kappa b)Y_1(\kappa a))} \tag{22}$$

$$Z_4 = \frac{2\rho c_a^2 h[\pi\kappa ab J_1(\kappa a)Y_0(\kappa b) - \pi\kappa ab Y_1(\kappa a)J_0(\kappa b) + \pi a(\nu-1)(Y_1(\kappa b)J_1(\kappa a) - Y_1(\kappa a)J_1(\kappa b)) - 2b]}{j\omega ab^2(Y_1(\kappa b)J_1(\kappa a) - J_1(\kappa b)Y_1(\kappa a))} \tag{23}$$

To derive the the Mason model for the complete PT, we assume the continuity of radial forces, radial velocities at the interface of disc and ring, assuming the gap between them to be negligible [17]. We also assume the outer boundary of the ring to be stress free ($F_b = 0$). Combining the circuit model of the input disc and output ring based on (15)-(23) and applying these boundary conditions lead to the equivalent circuit shown in Fig. 4.

This circuit can be further simplified into reduced Mason model for a PT operating near the resonance. This is done by performing a Taylor expansion of the impedance around the fundamental radial resonance (ω_r). On performing the Taylor expansion and simplifying the impedance network, we get the following equivalent impedance:

$$Z_{eq} \approx \frac{jhJ_1^2(\kappa_r)(\kappa_r^2 + \nu^2 - 1)}{\pi\alpha^2 k_p^2 \epsilon_{33}(1+\nu)\kappa_r^2 c_a^2 J_1^2(\kappa_r\alpha)}(\omega - \omega_r) \tag{24}$$

While simplifying the impedance network, the transformers in the circuit model are lumped to obtain a single transformer with net transformation ratio $N = \frac{\alpha J_1(\kappa_r\alpha)}{\alpha J_1(\kappa_r\alpha) - J_1(\kappa_r)}$. The simplified circuit model is shown in Fig 4. This simplified circuit can be further converted to a reduced Mason Model illustrated in Fig 1 by equating the equivalent impedance to that of a $L-C$ series lumped circuit network. The Taylor series expansion of a $L-C$ network around its resonance frequency $\omega_r = 1/\sqrt{LC}$ is expressed as $Z_{LC} \approx 2jL(\omega - \omega_r)$. On equating this with (24), we can calculate the lumped inductance and capacitance values. The mechanical damping losses constitute the dominant loss mechanism in the PT can be modeled by adding a resistance in series with the $L - C$ network [32]. Internal mechanical losses are described by the mechanical quality factor Q and the series resistance is inversely proportional to the quality factor as $R = \frac{1}{Q}\sqrt{\frac{L}{C}}$. The expressions for the Mason Model parameters of this PT are mentioned in Table III.

REFERENCES

[1] D. J. Perreault, J. Hu, J. M. Rivas, Y. Han, O. Leitermann, R. C. Pilawa-Podgurski, A. Sagneri, and C. R. Sullivan, "Opportunities and challenges in very high frequency power conversion," in *Proc. IEEE Applied Power Electronics Conference and Exposition*, Washington, DC, USA, Feb. 2009, pp. 1–14.

[2] C. R. Sullivan, B. A. Reese, A. L. Stein, and P. A. Kyaw, "On size and magnetics: Why small efficient power inductors are rare," in *Proc. IEEE Intl. Symposium on 3D Power Electronics Integration and Manufacturing*, Raleigh, NC, USA, Jun. 2016, pp. 1–23.

[3] J. Erhart, P. Půlpán, and M. Pustka, *Piezoelectric Ceramic Resonators*. Springer, 2017.

[4] P. A. Kyaw and C. R. Sullivan, "Fundamental examination of multiple potential passive component technologies for future power electronics," in *2015 IEEE 16th Workshop on Control and Modeling for Power Electronics (COMPEL)*, 2015, pp. 1–9.

[5] J. D. Boles, J. J. Piel, E. Ng, , J. E. Bonavia, J. H. Lang, and D. J. Perreault, "Piezoelectric-based power conversion: Recent progress, opportunities, and challenges," in *Proc. IEEE Custom Integrated Circuits Conference*, Newport Beach, CA, Apr. 2022, pp. 1–8.

[6] J. D. Boles, J. E. Bonavia, J. H. Lang, and D. J. Perreault, "A piezoelectric-resonator-based dc–dc converter demonstrating 1 kw/cm³ resonator power density," *IEEE Transactions on Power Electronics*, vol. 38, no. 3, pp. 2811–2815, 2023.

[7] J. D. Boles, J. J. Piel, and D. J. Perreault, "Enumeration and analysis of dc-dc converter implementations based on piezoelectric resonators," *IEEE Transactions on Power Electronics*, vol. 36, no. 1, pp. 129–145, 2021.

[8] M. Touhami, G. Despesse, and F. Costa, "A new topology of dc-dc converter based on piezoelectric resonator," *IEEE Transactions on Power Electronics*, 2022.

[9] E. Stolt, W. D. Braun, L. Gu, J. Segovia-Fernandez, S. Chakraborty, R. Lu, and J. Rivas-Davila, "Fixed-frequency control of piezoelectric resonator dc-dc converters for spurious mode avoidance," *IEEE Open Journal of Power Electronics*, vol. 2, pp. 582–590, 2021.

[10] E. Stolt, W. Braun, K. Nguyen, V. Chulukhadze, R. Lu, and J. Rivas-Davila, "A spurious-free piezoelectric resonator based 3.2 kw dc–dc converter for ev on-board chargers," *IEEE Transactions on Power Electronics*, vol. 39, no. 2, pp. 2478–2488, 2024.

[11] V. Breton, E. Bigot, G. Despesse, and F. Costa, "A new isolated topology of dc–dc converter based on piezoelectric resonators," *IEEE Transactions on Power Electronics*, vol. 38, no. 8, pp. 10012–10025, 2023.

[12] Y. Hou, M. Garud, Q. Ji, A. Persaud, P. Seidl, T. Schenkel, A. Lal, and K. Afridi, "Vertically stacked piezoelectric transformer for high-frequency power amplifier," in *2023 IEEE Applied Power Electronics Conference and Exposition (APEC)*, 2023, pp. 392–396.

[13] A. Vazquez Carazo, "Piezoelectric transformers: An historical review," in *Actuators*, vol. 5, no. 2. MDPI, 2016, p. 12.

[14] J. D. Boles, E. Ng, J. H. Lang, and D. J. Perreault, "Dc-dc converter implementations based on piezoelectric transformers," *IEEE Journal of Emerging and Selected Topics in Power Electronics*, 2022.

[15] W. P. Mason, *Electromechanical Transducers and Wave Filters*. D. Van Nostrand Co., 1948.

[16] L. Wang and R. P. Burgos, "Comprehensive analysis of models and operational characteristics of piezoelectric transformers," in *Proc. IEEE Applied Power Electronics Conference and Exposition*, New Orleans, LA, USA, Mar. 2020, pp. 1422–1429.

[17] C.-Y. Lin, "Design and analysis of piezoelectric transformer converters," Ph.D. dissertation, Virginia Tech, 1997.

[18] L. Wang, Q. Wang, M. Khanna, R. P. Burgos, K. D. Ngo, and A. Vazquez Carazo, "Design and control of tunable piezoelectric transformer based dc/dc converter," in *2018 IEEE Energy Conversion Congress and Exposition (ECCE)*, Portland, OR, USA, Sep. 2018, pp. 5987–5993.

[19] Y.-P. Liu, D. Vasic, F. Costa, W.-J. Wu, and D. Schwander, "Fixed frequency controlled piezoelectric 10w dc/dc converter," in *2010 IEEE Energy Conversion Congress and Exposition*. IEEE, 2010, pp. 3030–3037.

[20] L. Wang, K. Sun, and R. Burgos, "Planar piezoelectric transformer-based high step-down voltage-ratio dc-dc converter," *IEEE Transactions on Power Electronics*, 2022.

[21] J.-h. Park, S. Choi, S. Lee, and B. H. Cho, "Gain-adjustment technique for resonant power converters with piezoelectric transformer," in *2007 IEEE Power Electronics Specialists Conference*, 2007, pp. 2549–2553.

[22] M. Ekhtiari, Z. Zhang, and M. A. Andersen, "State-of-the-art piezoelectric transformer-based switch mode power supplies," in *IECON 2014-40th Annual Conference of the IEEE Industrial Electronics Society*. IEEE, 2014, pp. 5072–5078.

[23] M. Rodgaard, T. Andersen, and M. A. Andersen, "Empiric analysis of zero voltage switching in piezoelectric transformer based resonant converters," in *Proc. IET International Conference on Power Electronics, Machines and Drives*, Bristol, UK, Mar. 2012.

[24] G. Ivensky, I. Zafrany, and S. Ben-Yaakov, "Generic operational characteristics of piezoelectric transformers," *IEEE Transactions on Power Electronics*, vol. 17, no. 6, pp. 1049–1057, 2002.

[25] E. Ng, "Design of high-performance piezoelectric transformer-based dc-dc converters," 2022.

[26] E. A. Stolt, C. Daniel, and J. M. Rivas-Davila, "A stacked radial mode lithium niobate transformer for dc-dc conversion," in *2024 IEEE Workshop on Control and Modeling for Power Electronics (COMPEL)*, 2024, pp. 1–7.

[27] M. S. Rødgaard, M. Weirich, and M. A. Andersen, "Forward conduction mode controlled piezoelectric transformer-based pfc led drive," *IEEE Transactions on Power Electronics*, vol. 28, no. 10, pp. 4841–4849, 2012.

[28] M. Sanz, P. Alou, A. Soto, R. Prieto, J. Cobos, and J. Uceda, "Magnetic-less converter based on piezoelectric transformers for step-down dc/dc and low power application," in *Eighteenth Annual IEEE Applied Power Electronics Conference and Exposition, 2003. APEC '03.*, vol. 2, 2003, pp. 615–621 vol.2.

[29] J. D. Boles, J. E. Bonavia, P. L. Acosta, Y. K. Ramadass, J. H. Lang, and D. J. Perreault, "Evaluating piezoelectric materials and vibration modes for power conversion," *IEEE Transactions on Power Electronics*, 2022.

[30] A. K. Jackson, J. W. Perreault, J. H. Lang, and D. J. Perreault, "Large-signal characterization of piezoelectric resonators for power conversion," in *2024 IEEE Applied Power Electronics Conference and Exposition (APEC)*, 2024, pp. 2637–2643.

[31] E. Ng, J. D. Boles, J. H. Lang, and D. J. Perreault, "Piezoelectric transformer component design for dc-dc power conversion," in *2023 IEEE 24th Workshop on Control and Modeling for Power Electronics (COMPEL)*, 2023, pp. 1–8.

[32] K. Uchino and S. Hirose, "Loss mechanisms in piezoelectrics: how to measure different losses separately," *IEEE Transactions on Ultrasonics, ferroelectrics, and frequency control*, vol. 48, no. 1, pp. 307–321, 2001.

First Characterization of GaN Power Device and IC at Deep Cryogenic Temperatures down to 100 mK

Xin Yang
Center for Power Electronics Systems (CPES)
Virginia Tech
Blacksburg, USA
Email: xxxyang@vt.edu

Matthew Porter
Center for Power Electronics Systems (CPES)
Virginia Tech
Blacksburg, USA
Email: maporter@vt.edu

Zineng Yang
Center for Power Electronics Systems (CPES)
Virginia Tech
Blacksburg, USA
Email: zinengy@vt.edu

Zichen Xi
Bradley Department of Electrical and Computer Engineering
Virginia Tech
Blacksburg, USA
Email: xizichen@vt.edu

Liyang Jin
Bradley Department of Electrical and Computer Engineering
Virginia Tech
Blacksburg, USA
Email: liyangjin@vt.edu

Liyan Zhu
Center for Power Electronics Systems (CPES)
Virginia Tech
Blacksburg, USA
Email: liyanz@vt.edu

Linbo Shao
Bradley Department of Electrical and Computer Engineering
Virginia Tech
Blacksburg, USA
Email: shaolb@vt.edu

Yuhao Zhang
Department of Electrical and Electronic Engineering
The University of Hong Kong
Hong Kong, China
Email: yuhzhang@hku.hk

Abstract— Electrical energy conversion at deep cryogenic temperatures (T < 4.2 K) is desirable for space, quantum, and astronomical applications. However, the functionality of power semiconductors within this temperature range remains uncertain. Particularly, high-voltage or dynamic switching tests have not been reported for GaN devices below 77 K. For the first time, we evaluate the static and dynamic performances of discrete GaN HEMT and GaN power IC at temperatures as low as 100 mK. We employ a dilution refrigerator connected to a custom circuit setup that integrates double-pulse testing and dynamic on-resistance (R_{ON}) measurements. Key findings at temperatures below 1 K include: 1) GaN HEMTs maintain normally-off operation and high breakdown voltage; 2) discrete GaN HEMT and GaN power IC are both capable of hard switching under gate control; and 3) the dynamic R_{ON} are approximately ~4.5 times lower than at 295 K. These results pave the way for developing power converters operational at temperatures below the current boundaries. [1]

Keywords—power semiconductor device, cryogenic measurement, dilution refrigerator, GaN HEMT, monolithic power IC, switching performances, dynamic on-resistance.

I. INTRODUCTION

Many applications in space [1], [2], [3], healthcare[4], [5], metrology [6], and quantum computing [7], [8] require electronics that can function at extremely low cryogenic temperatures, ranging from ~4 K down to a few millikelvins. Power electronics are especially important in these scenarios for efficient electrical energy conversion [9], [10], [11], [12], [13], [14], [15], [16], [17] and are essential for future grids involving superconducting power delivery and energy storage. However, significant knowledge gaps exist regarding the functionality of power semiconductor devices at this temperature range. For power devices that include a lightly-doped drift region, such as metal-oxide-semiconductor field-effect transistors (MOSFETs) and insulated-gate bipolar transistors (IGBTs), there is a concern that carrier freeze-out could occur at deep cryogenic

---

[1] This work is supported in part by the Office Naval Research monitored by Lynn Petersen (Award N000142412227) and in part by the CPES Industry Consortium.

temperatures [18], [19], [20]. In contrast, gallium nitride (GaN) power devices, which rely on a two-dimensional electron gas (2DEG) channel for current conduction, are known to be resistant to carrier freeze-out down to cryogenic temperatures [18].

Fig. 1 provides a summary of the lowest temperatures reported in cryogenic studies of GaN high-electron-mobility transistors (HEMTs) [21], [22], [23], [24], [25] in the literature, covering static low-voltage tests (e.g., output I-V), high-voltage breakdown tests, and inductive switching tests. To date, GaN device characterization below 4.2 K has not been reported, likely due to limitations in conventional cryogenic probe stations that cannot reach these lower temperatures. Moreover, the lowest temperature reported for the high-voltage test and switching test is only 77 K, leaving significant gaps on their low-T characteristics.

To achieve cryogenic test conditions, various types of cryogenic refrigerators are employed. The fridge that uses liquid nitrogen is commonly used in cryogenic power electronics testing due to its convenience and wide access [24]. However, the base temperature of this type of fridge is limited to around 77 K, as determined by the boiling point of liquid nitrogen. The cryogen-free dilution fridge, which utilizes a pulse tube

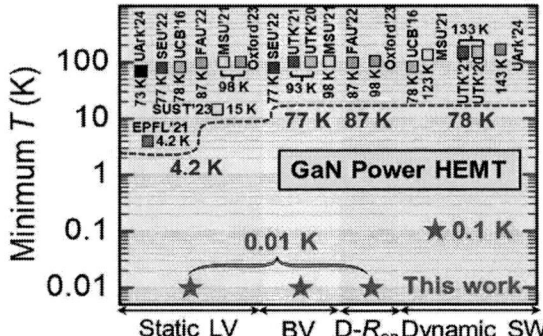

Fig. 1. Reported cryogenic studies of GaN power HEMT highlighting the unexplored range in the prior literature. Studies are grouped into the static low-voltage (LV) test, breakdown voltage (BV) test, dynamic R_{ON} (D-R_{ON}) test, and dynamic switching (SW) test.

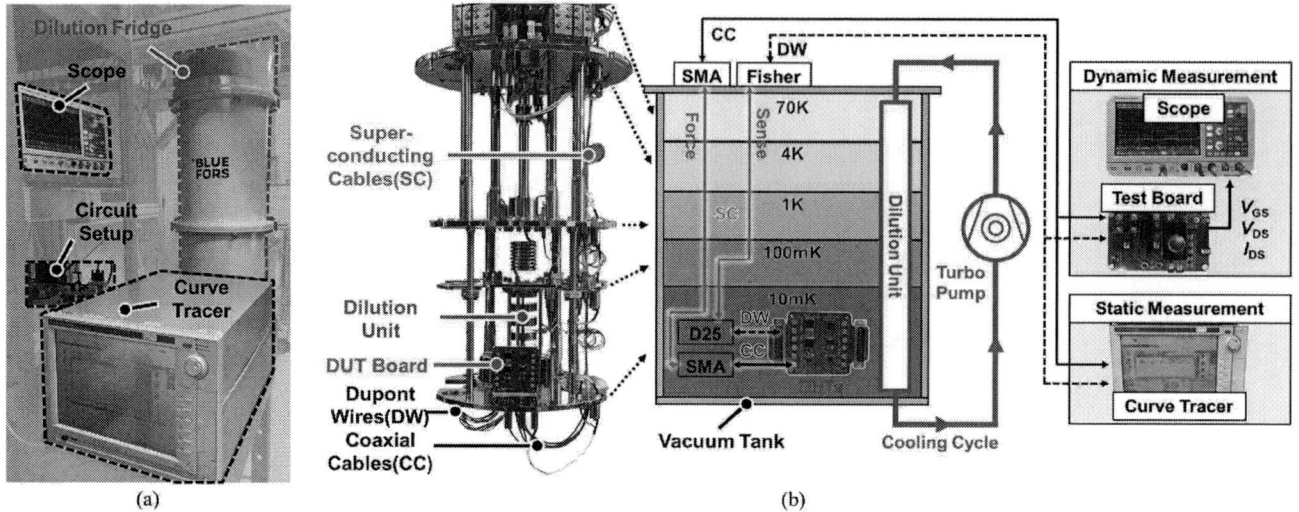

Fig. 2. (a) Photograph and (b) schematic of cryogenic test setup, which consists of a BlueFors dilution fridge, a Keysight B1505 curve tracer, and a customized circuit test setup. The daughter board with DUTs is mounted on the 10 mK plate, and other components and equipment are placed outside the fridge.

cryocooler and a helium-3/helium-4 (He3/He4) dilution unit, represents the most advanced cryogenic measurement system [26], offering large experimental space, high cooling power, and enabling a base temperature below 10 mK. While this equipment is widely used in various fields, it has not been applied to power device testing. Recently, we for the first time characterized the characteristics of Si, SiC and GaN devices using the dilution fridge [27].

This study extends our previous work [27] by characterizing both the static and dynamic characteristics of GaN power devices and GaN power ICs down to 100 mK using a dilution refrigerator. The circuit setup includes double-pulse testing (DPT) and dynamic R_{ON} (D-R_{ON}) testing. These circuit-based tests enable the assessment of power device performance at record-low temperatures, and the newly revealed device physics is thoroughly discussed.

II. DEEP-CRYOGENIC TEST SYSTEM

Fig. 2(a) shows a photo of the cryogenic test system, which is divided into two main parts: the cooling system based on a cryogen-free dilution refrigerator and the measurement system, which includes the circuit setups and a curve tracer. The schematic of the test system is shown in Fig. 2(b).

The Bluefors LD250 dilution refrigerator has a multi-plate structure that provides a range of temperatures from 70 K to 10 mK for measurement. The temperatures are measured by a well-calibrated sensor (CX-1010-SD-HT-P). The entire system is enclosed in four cans to ensure temperature isolation and maintain a vacuum condition. A dilution unit with a turbo pump is used to maintain the cooling cycle during operation. The base temperature in the mixing chamber (located at the bottom of the fridge) can be stabilized to temperatures below 10 mK. A customized daughter board, mounted in the mixing chamber with thermal paste, carries the devices under test (DUTs) and connectors.

The connection of measurement signal connections follows the Kelvin connection (4-wire connection). From the daughter board to connectors inside the fridge, the coaxial cables (CCs) are used for high-current signals, while dupont wires (DWs) handle control and sense signals. The fridge provides several superconducting cables (SCs), which become superconducting at cryogenic temperatures (below 4 K). These SCs transmit signals from each plate to the top of the fridge via SMA and Micro-D25 (M-D25) connectors. Several SMA and Fisher connectors are located at the top exterior of the fridge for user connections. The CCs are used from the SMA connectors to the motherboard or curve tracer for power and high-frequency signals, and DWs with Fisher-D25 connectors are applied for low-current and sense signals.

The measurement system, located outside the fridge, is divided into dynamic and static sections. To measure dynamic switching performance and dynamic on-resistance (R_{ON}), a double-pulse test (DPT) circuit board is designed, comprising driver, power, and measurement circuits. The static characteristics of the DUTs are measured using a Keysight B1505A curve tracer.

To analyze the DUT's switching performance and on-resistance at various temperatures, a DPT mother board and a DUT daughter board were carefully designed. Fig 3(a) and (b) show the circuit schematic and photo of the DPT board. The mother board is located outside the fridge, while only the DUTs are inside the fridge. The power loop consists of a DC power supply (V_{DC}), DC input capacitors (C_{DC}), an inductor (L), and a free-wheeling diode (FWD). The driver circuit, based on an external driver IC (Si8271), is also located on the mother board. The DUTs used in this study include a GaN HEMT (GS065011L)

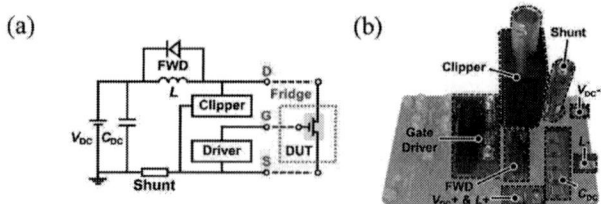

Fig. 3. (a) Circuit schematic and (b) photograph of DPT setup.

979-8-3315-1612-3/25 $31.00 © 2025 IEEE   1021

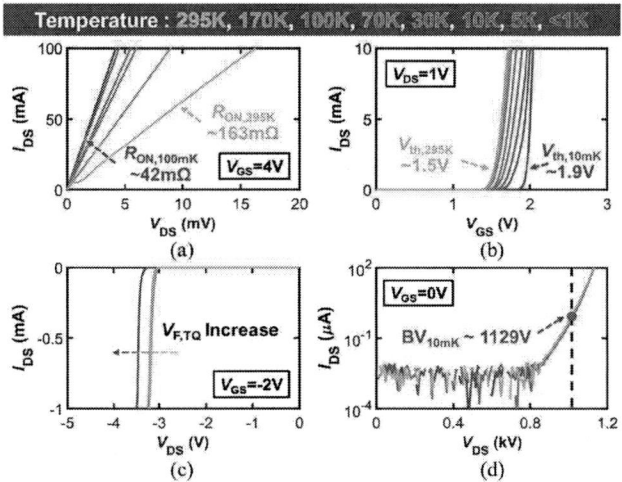

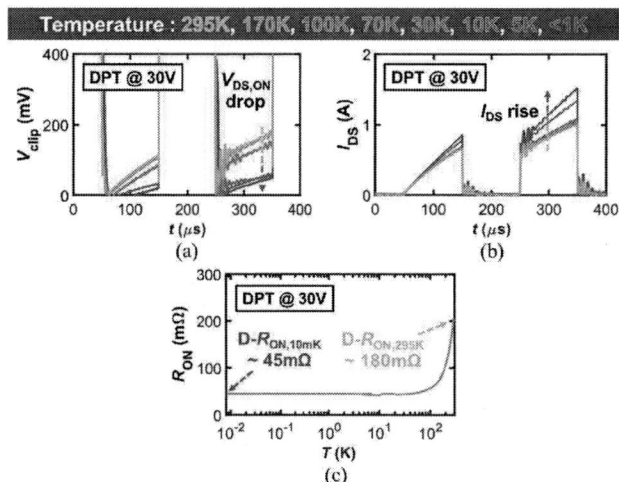

Fig. 4. (a) Output, (b) transfer, (c) third-quadrant, and (d) off-state I_{DS}-V_{DS} characteristics of GaN HEMT at various temperatures.

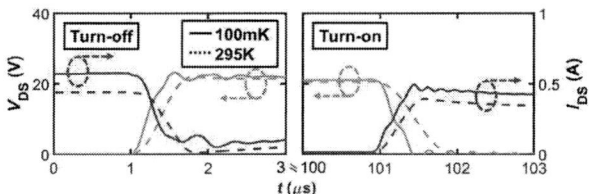

Fig. 5. Zoomed turn-off and turn-on waveforms of GaN HEMT under 20V/0.5A DPT at 100 mK (solid lines) and 295 K (dashed lines).

and a GaN Power IC (NV6125). For switching tests, V_{DC} is fixed at 20 V under all test conditions. For dynamic R_{ON} tests, V_{DC} varies from 0 V to 50 V to assess the dynamic R_{ON} performance of the GaN HEMT. The V_{GS} and V_{DS} are measured using a passive probe (RT-ZP10), and an oscilloscope (RTM3004) captures all signals. A 0.1 Ω current shunt and a voltage clipper (CLP1500V15A1) are used to measure the device's current (I_{DS}) and on-state voltage drop ($V_{DS,ON}$). The clipper blocks high voltage when the DUT is off and outputs $V_{DS,ON}$ when the device is on, which is widely employed in dynamic R_{ON} measurement [11].

III. CRYOGENIC PERFOKRMANCES OF GAN DEVICES

A. Test Results of GaN HEMT

The static performance of the GaN HEMT was first characterized at various temperatures, ranging from 295 K to below 1 K. All temperatures were recorded after testing, accounting for thermal influences caused by the measurements.

Fig. 4(a) shows the output characteristics at a 4 V V_{GS}. The R_{ON} decreases from 163 mΩ at 295 K to 42 mΩ at 100 mK, demonstrating a 4x reduction in R_{ON} at cryogenic temperatures. Fig. 4(b) shows the transfer characteristics at 1 V V_{DS}, revealing a ~0.4 V increase in threshold voltage (V_{TH}) as the temperature drops to 10 mK. For third-quadrant performance with V_{GS} at -2 V, as shown in Fig. 4(c), the third-quadrant forward voltage ($V_{F,TQ}$) increases as the temperature decreases. Fig. 4(d) shows the off-state I_{DS}-V_{DS} characteristics at 0 V V_{GS}, indicating a temperature-independent breakdown voltage (BV) of approximately 1129 V, extracted at I_{DS}=1 μA.

Fig. 6. (a) Clipper and (b) current shunt output waveforms of GaN HEMT under 30V/0.7A DPT at selected temperature. (c) Dynamic R_{ON} as a function of temperature extracted from the 30V/0.7A DPT

Fig. 5 shows a zoomed switching waveforms of the GaN HEMT during a 20 V/0.5 A DPT. The test results at 100 mK and 295 K are depicted as solid and dashed lines, respectively. The fall time (t_f) is 283.2 ns at 100 mK and 403.2 ns at 295 K, while the rise time (t_r) is 303.0 ns and 576.0 ns at those temperatures, respectively. It should be clarified that the reduction in switching times does not directly indicate improved switching performance for the GaN HEMT. The loop resistances, particularly R_{SC}, also decrease with temperature, which directly affects the DPT results. However, the turn-on and turn-off waveforms at 100 mK demonstrate good gate control and fast switching speeds for the GaN HEMT, highlighting its potential for future applications at cryogenic temperatures.

Fig. 6(a) and Fig. 6(b) show the on-state voltage and current waveforms during a 30 V/0.7 A DPT at selected temperatures. As the temperature decreases, the results indicate a lower on-state voltage drop and higher current, suggesting a reduction in R_{ON} at low temperatures. To confirm this, the extracted R_{ON} values from the DPT results are plotted as a function of temperature in Fig. 6(c). The relationship between R_{ON} and temperature shows a decreasing trend, reaching a saturation value of approximately 45 mΩ when the temperature drops below ~10 K, which aligns with the static measurement results.

B. Test Results of GaN IC

Fig. 7(a) presents the internal schematic of the GaN power IC, showing its logic circuits, integrated gate driver, and GaN power switch. The typical application circuit is shown in Fig. 7(b), with additional components positioned on the motherboard

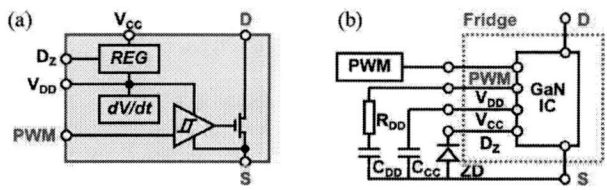

Fig. 7. (a) Internal schematic of GaN Power IC and (b) typical application circuit schematic.

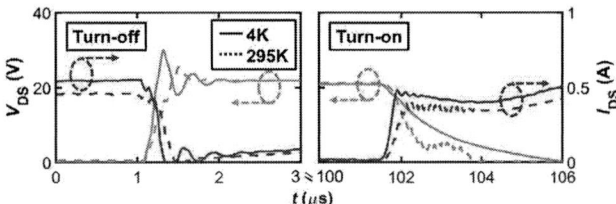

Fig. 8. Zoomed turn-off and turn-on waveforms of GaN Power IC under 20V/0.5A DPT at 4K (solid lines) and 295 K (dashed lines).

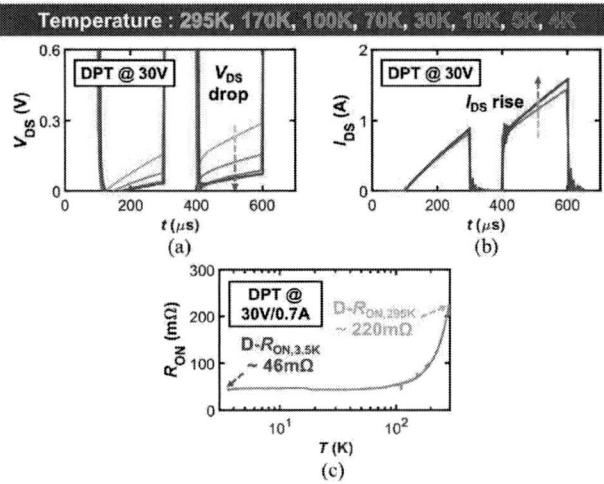

Fig. 9. (a) Clipper and (b) current shunt output waveforms of GaN Power IC under 30V/0.7A DPT at selected temperature. (c) Dynamic R_{ON} as a function of temperature extracted from the 30V/0.7A DPT

outside the fridge. A V_{CC} range of 10 V to 24 V is recommended for driving the device, with a quiescent current typically in the milli-amp range, leading to a quiescent power loss of approximately 10 to 30 mW. This quiescent power consumption restricts the minimum achievable temperature of the GaN power IC, which is set to 4 K in this test.

Fig. 8 shows zoomed-in switching waveforms of the GaN power IC under a 20 V/0.5 A DPT. Solid and dashed lines represent the results at 4 K and 295 K, respectively. The t_f is 115.2 ns at 4 K compared to 254.4 ns at 295 K, while the t_r is 3.2 μs at 4 K and 1.03 μs at 295 K. These switching waveforms demonstrate effective gate control and fast switching at 4 K. The longer rise time at 4 K may result from the performance degradation in the logic and gate driver circuits at low temperatures, which requires scrutinization in the future work.

Fig. 9(a) and (b) show the on-state voltage and current waveforms of the GaN power IC during a 30 V/0.7 A DPT at selected temperatures, and Fig. 9(c) plots the extracted R_{ON} values from the DPT as a function of temperature. Similar to the discrete GaN HEMT, the R_{ON} of the low-side device in the power IC decreases at lower temperatures, reaching ~46 mΩ at cryogenic temperatures, indicating reduced conduction losses for the GaN power IC at low temperatures.

PHYSICAL MECHANISMS

As 2DEG originates from the net positive polarization charge at AlGaN/GaN heterostructures rather than intentional doping, free electrons in the 2DEG channel are known to be immune to carrier freeze-out. The static I-V and dynamic switching characterization results of GaN HEMTs and power

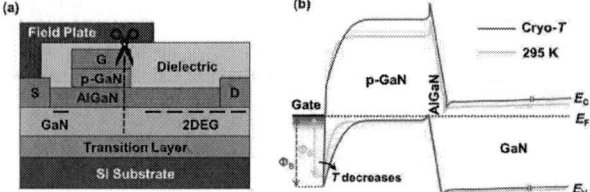

Fig. 10. (a) Schematic of a Schottky p-GaN-gated power HEMT. (b) Comparison of band diagram along the gate/p-GaN/AlGaN/GaN heterostructure at 295 K and cryo-T, showing the increased Φ_B at cryo-T.

ICs in this work confirm that this immunity holds down to 0.1 K, extending previous findings reported at 4.2 K [21] and 2 K [25], as well as previous works only on discrete GaN devices [15]. The GaN HEMT shows superior static and dynamic characteristics at deep cryo-T, including a decrease in R_{ON}, high BV at 10 mK, minimal D-R_{ON}, faster switching speed and a small V_{TH} shift.

The 4 to 5-fold reduction in R_{ON} at $T < 1$ K can be explained by increased 2DEG mobility due to the reduced electron-phonon interactions compared to room temperature [28]. The high BV at $V_{GS} = 0$ V suggests the functionality of the p-n junction under the gate; otherwise, punch-through would occur. The D-R_{ON} typically originates from charge trapping effects at different locations in GaN HEMTs [29]. The elimination of D-R_{ON} is attributed to the trap freeze-out [23], suggesting the stability issue of GaN HEMTs could be largely resolved at cryo-T. Additionally, the faster switching speed is due to the high transconductance as a result of higher electron mobility at cryo-T.

Cryo-T-induced V_{TH} instability presents a stability concern in GaN HEMTs, as shown by a positive V_{TH} shift of ~0.4 V (extracted by linear extrapolation) when T decreases to 10 mK (see Fig. 9(b)). Prior reports indicate bidirectional V_{TH} shift in GaN HEMTs from various vendors at cryo-T, depending on the specific p-GaN gate technology employed [21]. The electrostatic V_{TH} model for Schottky-type p-GaN-gated GaN HEMTs [see Fig. 10(a)] is given by [30]

$$V_{TH} \approx \phi_B - E_G/q + \psi_{bi} + \psi_s - \Delta V_b \qquad (1)$$

where ϕ_B, ψ_{bi}, ψ_s, E_G, and ΔV_b denote the Schottky barrier height in relative to the valance band edge, the built-in potential of the p-GaN Schottky contact, surface potential at the p-GaN/AlGaN barrier interface, bandgap of the p-GaN layer and the voltage drop across the AlGaN barrier.

In practical device operation, the effective p-GaN Schottky barrier height is lower than ϕ_B shown in the band diagram in Fig. 10(b) due to the interfacial trap-assisted carrier tunneling, which is sensitive to temperature and can potentially lead to V_{th} instability [31], [32]. At cryo-T, however, trap freeze-out suppresses the trapping dynamics [33], resulting in an increased Schottky barrier height. The elevated Schottky barrier height causes a more positive V_{TH} in GaN HEMTs due to the increased depletion width across the metal/p-GaN junction [34], [35].

The rise in V_{TH} at cryo-T also impacts the third-quadrant I-V characteristics of the GaN HEMT (see Fig. 9(c)). At $V_{GS} < V_{TH}$, the device operates in third quadrant when the gate-to-drain bias (V_{GD}) exceeds V_{TH}. Due to the steep I-V characteristics, R_{ON} can

979-8-3315-1612-3/25 $31.00 © 2025 IEEE

be neglected, and the source-to-drain $V_{F,TQ}$ can be approximated by the following relation

$$V_{F,TQ} \approx V_{TH} - V_{GS} \qquad (2)$$

This equation suggests that the T-dependence of $V_{F,TQ}$ is directly tied to that of V_{TH}, which increases with lower T.

IV. CONCLUSION

In this work, we present the first static and dynamic characterizations of a discrete GaN HEMT down to 0.1 K, as well as the dynamic switching performance of an all-GaN Power IC down to 3 K, using a dilution refrigerator integrated with a custom circuit setup. At T < 1 K, the GaN HEMT demonstrated the ability to block high breakdown voltages and perform hard switching under gate control. At T down to 3 K, the GaN power IC retains the full functionality. Both the discrete GaN HEMT and GaN power IC exhibited reduced conduction losses, highlighting their potential for cryogenic temperature applications.

REFERENCES

[1] B. Collaudin and N. Rando, "Cryogenics in space: a review of the missions and of the technologies," Cryogenics, vol. 40, no. 12, pp. 797–819, Jan. 2000, doi: 10.1016/S0011-2275(01)00035-2.

[2] R. R. Ward et al., "Power diodes for cryogenic operation," in IEEE 34th Annual Conference on Power Electronics Specialist, 2003. PESC '03., Jun. 2003, pp. 1891–1896 vol.4. doi: 10.1109/PESC.2003.1217741.

[3] R. L. Patterson, A. Hammoud, J. E. Dickman, S. Gerber, M. Elbuluk, and E. Overton, "Electronics for deep space cryogenic applications," in Proceedings of the 5th European Workshop on Low Temperature Electronics, Jun. 2002, pp. 207–210. doi: 10.1109/WOLTE.2002.1022482.

[4] D. H. Johansen, J. D. Sanchez-Heredia, J. R. Petersen, T. K. Johansen, V. Zhurbenko, and J. H. Ardenkjær-Larsen, "Cryogenic Preamplifiers for Magnetic Resonance Imaging," IEEE Trans. Biomed. Circuits Syst., vol. 12, no. 1, pp. 202–210, Feb. 2018, doi: 10.1109/TBCAS.2017.2776256.

[5] J. D. Sánchez-Heredia, R. Baron, E. S. S. Hansen, C. Laustsen, V. Zhurbenko, and J. H. Ardenkjær-Larsen, "Autonomous cryogenic RF receive coil for 13C imaging of rodents at 3 T," Magn. Reson. Med., vol. 84, no. 1, pp. 497–508, 2020, doi: 10.1002/mrm.28113.

[6] J. M. Williams, "Cryogenic current comparators and their application to electrical metrology," IET Sci. Meas. Technol., vol. 5, no. 6, pp. 211–224, Nov. 2011, doi: 10.1049/iet-smt.2010.0170.

[7] E. Charbon et al., "Cryo-CMOS for quantum computing," in 2016 IEEE International Electron Devices Meeting (IEDM), Dec. 2016, p. 13.5.1-13.5.4. doi: 10.1109/IEDM.2016.7838410.

[8] J. M. Hornibrook et al., "Cryogenic Control Architecture for Large-Scale Quantum Computing," Phys. Rev. Appl., vol. 3, no. 2, p. 024010, Feb. 2015, doi: 10.1103/PhysRevApplied.3.024010.

[9] Y. Zhang, F. Udrea, and H. Wang, "Multidimensional device architectures for efficient power electronics," Nat. Electron., vol. 5, no. 11, pp. 723–734, Nov. 2022, doi: 10.1038/s41928-022-00860-5.

[10] E. A. Jones, F. F. Wang, and D. Costinett, "Review of Commercial GaN Power Devices and GaN-Based Converter Design Challenges," IEEE J. Emerg. Sel. Top. Power Electron., vol. 4, no. 3, pp. 707–719, Sep. 2016, doi: 10.1109/JESTPE.2016.2582685.

[11] X. Yang et al., "Evaluation and MHz Converter Application of 1.2-kV Vertical GaN JFET," IEEE Trans. Power Electron., pp. 1–11, 2024, doi: 10.1109/TPEL.2024.3445667.

[12] M. Porter, X. Yang, H. Gong, B. Wang, Z. Yang, and Y. Zhang, "Switching figure-of-merit, optimal design, and power loss limit of (ultra-) wide bandgap power devices: A perspective," Appl. Phys. Lett., vol. 125, no. 11, p. 110501, Sep. 2024, doi: 10.1063/5.0222105.

[13] T.-S. Li, M. Ngo, R. Burgos, and D. Dong, "Modeling and Analysis of Voltage Overshoot in Bidirectional Phase-Shift Full Bridge Converters," in 2024 IEEE Sixth International Conference on DC Microgrids (ICDCM), Aug. 2024, pp. 1–7. doi: 10.1109/ICDCM60322.2024.10665239.

[14] M. Wu et al., "Modeling of Litz-Wire DD Coil With Ferrite Core for Wireless Power Transfer System," IEEE Trans. Power Electron., vol. 38, no. 5, pp. 6653–6669, May 2023, doi: 10.1109/TPEL.2022.3222228.

[15] Y. Jiang et al., "A Dynamic Efficiency Optimization Method under ZVS Conditions in the Series-Series Type Wireless Power Transfer System," in 2020 IEEE Energy Conversion Congress and Exposition (ECCE), Oct. 2020, pp. 995–1001. doi: 10.1109/ECCE44975.2020.9235452.

[16] Z. Xiao, W. Lei, Z. Xiang, Y. Yin, and W. Wang, "Full-state discrete-time model-based stability analysis and parameter design of dual active bridge converter in energy storage system," IET Power Electron., vol. 15, no. 13, pp. 1229–1248, 2022, doi: 10.1049/pel2.12319.

[17] C. Zhao et al., "A Load Detection Method in Multi-transmitter Dynamic Wireless Power Transfer Systems without Extra Sensors and Communication," in 2023 11th International Conference on Power Electronics and ECCE Asia (ICPE 2023 - ECCE Asia), May 2023, pp. 2573–2578. doi: 10.23919/ICPE2023-ECCEAsia54778.2023.10213968.

[18] H. Gui et al., "Review of Power Electronics Components at Cryogenic Temperatures," IEEE Trans. Power Electron., vol. 35, no. 5, pp. 5144–5156, May 2020, doi: 10.1109/TPEL.2019.2944781.

[19] O. Olanrewaju, Z. Yang, N. Evans, A. Fayyaz, T. Lagier, and A. Castellazzi, "Investigation of Temperature Distribution in SIC Power Module Prototype in Transient Conditions," in 2019 20th International Symposium on Power Electronics (Ee), Oct. 2019, pp. 1–5. doi: 10.1109/PEE.2019.8923270.

[20] F. Stella, O. Olanrewaju, Z. Yang, A. Castellazzi, and G. Pellegrino, "Experimentally validated methodology for real-time temperature cycle tracking in SiC power modules," Microelectron. Reliab., vol. 88–90, pp. 615 619, Sep. 2018, doi: 10.1016/j.microrel.2018.07.072.

[21] L. Nela, N. Perera, C. Erine, and E. Matioli, "Performance of GaN Power Devices for Cryogenic Applications Down to 4.2 K," IEEE Trans. Power Electron., vol. 36, no. 7, pp. 7412–7416, Jul. 2021, doi: 10.1109/TPEL.2020.3047466.

[22] C. Zhang et al., "Unclamped-Inductive-Switching Behaviors of p-GaN HEMTs at Cryogenic Temperature," IEEE Trans. Power Electron., vol. 37, no. 10, pp. 11507–11510, Oct. 2022, doi: 10.1109/TPEL.2022.3173725.

[23] Z. Jiang et al., "Roles of Hole Trap on Gate Leakage of p-GaN HEMTs at Cryogenic Temperatures," IEEE Electron Device Lett., vol. 44, no. 10, pp. 1612–1615, Oct. 2023, doi: 10.1109/LED.2023.3311395.

[24] Y. Wei, M. M. Hossain, and H. A. Mantooth, "Cryogenic Overcurrent Characteristic of GaN HEMT and Converter Evaluation," IEEE Trans. Ind. Appl., vol. 60, no. 4, pp. 6479–6487, Jul. 2024, doi: 10.1109/TIA.2024.3379490.

[25] Y. H. Ng et al., "p-GaN gate power HEMT heterostructure as a versatile platform for extremely wide-temperature-range (X-WTR) applications," Appl. Phys. Lett., vol. 124, no. 4, p. 043504, Jan. 2024, doi: 10.1063/5.0184784.

[26] L. Shao et al., "Electrical control of surface acoustic waves," Nat. Electron., vol. 5, no. 6, pp. 348–355, Jun. 2022, doi: 10.1038/s41928-022-00773-3.

[27] X. Yang, M. Porter, Z. Yang, Z. Xi, Q. Li, L. Shao, Y. Zhang, "First Characterization of Si, SiC and GaN Power Deivces at Deep Cryogenic Temperatures down to 0.1 K," in 2024 IEEE International Electron Devices Meeting (IEDM), Dec. 2024.

[28] D. Zanato, S. Gokden, N. Balkan, B. K. Ridley, and W. J. Schaff, "The effect of interface-roughness and dislocation scattering on low temperature mobility of 2D electron gas in GaN/AlGaN," Semicond. Sci. Technol., vol. 19, no. 3, p. 427, Jan. 2004, doi: 10.1088/0268-1242/19/3/024.

[29] J. P. Kozak et al., "Stability, Reliability, and Robustness of GaN Power Devices: A Review," IEEE Trans. Power Electron., vol. 38, no. 7, pp. 8442–8471, Jul. 2023, doi: 10.1109/TPEL.2023.3266365.

[30] B. Bakeroot, A. Stockman, N. Posthuma, S. Stoffels, and S. Decoutere, "Analytical Model for the Threshold Voltage of p -(Al)GaN High-

Electron-Mobility Transistors," IEEE Trans. Electron Devices, vol. 65, no. 1, pp. 79–86, Jan. 2018, doi: 10.1109/TED.2017.2773269.

[31] S. Li et al., "High-temperature electrical performances and physics-based analysis of p-GaN HEMT device," IET Power Electron., vol. 13, no. 3, pp. 420–425, 2020, doi: 10.1049/iet-pel.2019.0510.

[32] Y. Shi et al., "Bidirectional threshold voltage shift and gate leakage in 650 V p-GaN AlGaN/GaN HEMTs: The role of electron-trapping and hole-injection," in 2018 IEEE 30th International Symposium on Power Semiconductor Devices and ICs (ISPSD), May 2018, pp. 96–99. doi: 10.1109/ISPSD.2018.8393611.

[33] X. Wang et al., "Threshold Voltage Instability of Schottky-type p-GaN Gate HEMT down to Cryogenic Temperatures," in 2023 35th International Symposium on Power Semiconductor Devices and ICs (ISPSD), May 2023, pp. 115–118. doi: 10.1109/ISPSD57135.2023.10147433.

[34] I. Hwang et al., "p-GaN Gate HEMTs With Tungsten Gate Metal for High Threshold Voltage and Low Gate Current," IEEE Electron Device Lett., vol. 34, no. 2, pp. 202–204, Feb. 2013, doi: 10.1109/LED.2012.2230312.

[35] F. Lee, L.-Y. Su, C.-H. Wang, Y.-R. Wu, and J. Huang, "Impact of Gate Metal on the Performance of p-GaN/AlGaN/GaN High Electron Mobility Transistors," IEEE Electron Device Lett., vol. 36, no. 3, pp. 232–234, Mar. 2015, doi: 10.1109/LED.2015.2395454.

Dynamic Environment-Aware Lifetime Prediction of SiC MOSFET Modules Through LSTM

Md Zakir Hasan
*Department of Electrical and
Computer Engineering
Mississippi State University
Starkville, USA
mh3482@msstate.edu*

Seungdeog Choi
*Department of Electrical and
Computer Engineering
Mississippi State University
Starkville, USA
seungdeog@ece.msstate.edu*

Youssef Aider
*Department of Mechanical,
Aerospace & Biomedical
Engineering
University of Tennessee
Knoxville, USA
yaider@vols.utk.edu*

Prashant Singh
*Department of Mechanical,
Aerospace & Biomedical
Engineering
University of Tennessee
Knoxville, USA
psingh15@utk.edu*

Chun-Hung Liu
*Department of Electrical and
Computer Engineering
Mississippi State University
Starkville, USA
chliu@ece.msstate.edu*

Abstract— The state-of-the-art determines the remaining useful lifetime (RUL) through a steady-state, fixed power cycling tests (PCT) without considering the impact of dynamically changing environmental conditions. It has resulted in considerable RUL prediction errors in the real world. However, the dynamic changing conditions (e.g., large temperature swings) may affect the degradation evolution of SiC MOSFET, which could eventually result in RUL changes. Thus, it must be integrated to make accurate predictions. To precisely understand the RUL variation complexity, the junction temperature (T_j) has been measured with a Negative Thermal Coefficient (NTC) thermistor, Temperature Sensitive Electrical Parameter (TSEP), and these profiles have been modeled through the thermal model RC foster network using Extended Kalman Filter (EKF). Then, the on-state resistance (Rds,on) variations and Degradation Acceleration Factor (DAF) under the dynamic environment conditions are integrated into a lifetime prediction model to accurately predict the RUL through the Long Short-Term Memory (LSTM) machine learning algorithm.

Keywords— *SiC Module, Environment Conditions, PCT, DAF, Junction Temperature Estimation, Remaining Useful Lifetime (RUL), EKF, LSTM*

I. INTRODUCTION

Wide Band Gap (WBG) power switches offer several advantages compared to other semiconductor switches [1-2] (e.g., enhanced power density, improved thermal management, and higher operating temperature). However, power switches (SiC MOSFET, GAN FET, IGBT) are one of the most vulnerable components in modern power converters [3]. Many industrial applications expose the SiC MOSFETs to harsh environments, such as high environment temperature and high humidity [4]. Such continuous exposure could be subjected to

unpredictable thermomechanical stress, which may cause SiC MOSFETs to prematurely fail or significantly reduce their lifetime. Thus, without considering dynamic environmental conditions, state-of-the-art could result in considerable RUL prediction errors [5]. Common practices assuming fixed, steady-state power/thermal cycling testing [6] could result in a significant deviation in the lifetime of SiC MOSFETs in the real-world applications.

The critical challenge is to precisely model the degradation precursors (e.g., junction temperature, on-state resistance) of SiC MOSFET under dynamic conditions. Because the characteristics will change if the junction temperature varies [7]. The dynamic environment condition will impact the SiC MOSFET for the junction temperature deviation, eventually changing the characteristics and impacting the remaining useful lifetime (RUL). Many methods accurately estimate junction temperature [8-10]. Foster/ Cauer RC thermal network is used to estimate junction temperature. However, it shows very low accuracy in measurement due to some uncertainties in the thermal interfacing and heatsinks. Case temperature is also used to estimate junction temperature. However, this approach is inaccurate because the case and junction temperature difference significantly affect high power [11]. Temperature Sensitive Electrical Parameter (TSEP) is another approach to estimating junction temperature [12]. This method uses TSEPs (e.g., on-state resistance, threshold voltage, gate resistance) to estimate the junction temperature. This approach heavily depends on the load conditions, but the sensitivity of the TSEPs is low. Another approach is NTC sensor measurement to estimate T_j [13]. This approach is not sufficiently accurate due to the dependence on the response time of NTC.

This paper measures the junction temperature precisely by combining the NTC sensor measurement and TSEP method, considering dynamic environmental conditions. Combining these two approaches with the proposed EKF method [14-15] provides accurate junction temperature estimation in the presence of degradation effect and dynamic environment conditions effect which changes the TSEP parameters value, effectively eliminating the measurement noise sensitivity. EKF can handle non-linear data, which will be effective for T_j estimation.

There are various approaches to estimating the lifetime of SiC MOSFETs. One such approach is the power cycling test, which provides a rapid assessment of the device's lifespan [16-17]. The Rds,on is widely used as the aging and degradation indicators of SiC MOSFETs [18]. However, threshold voltage (V_{th}), drain current (I_d), rise time (t_r), and fall time (t_f) are also used as the precursor for determining the lifetime of SiC MOSFETs [19]. As dynamic environmental conditions impact junction temperature, it can lead to faster or slower degradation. Thus, in real-world applications, dynamic environment conditions must be considered when determining the lifetime of SiC MOSFET. The degradation acceleration factor (DAF) must be incorporated for accurate RUL prediction based on dynamic environment conditions and the on-state resistance.

There are both physical and data-driven models for predicting the Remaining Useful Life (RUL) [20]. Physical models, like the Coffin-Manson and Bayer models, are commonly used for estimating the lifetime of SiC MOSFETs, though they tend to have low accuracy and limited efficiency. In contrast, data-driven approaches predict lifetime using available data, offering greater accuracy and efficiency. This paper employs the proposed LSTM model to predict the remaining useful life (RUL). Within the LSTM framework, Rds,on is used as the key RUL parameter, while the Degradation Acceleration Factor (DAF) is incorporated as a feature to enhance prediction accuracy. This approach enables the LSTM model to effectively predict the lifetime of SiC MOSFETs under dynamic environmental conditions.

The structure of this paper is organized as follows. Section II introduces the analysis of junction temperature estimation of SiC MOSFET. Section III provides RUL predictions of SiC MOSFETs. In section IV, simulation results are provided. Section V provides an experimental study, and section VI gives results and a discussion. Finally, Section VII concludes this paper.

II. JUNCTION TEMPERATURE ESTIMATION OF SiC MOSFET

Junction temperature estimation must be precise to predict the RUL of SiC MOSFET. Thus, the RUL mapping and junction temperature estimation will integrate dynamic environment conditions. Fig. 1 shows the impact of environmental conditions on junction temperature through the RC Foster network. The RC parameters of the Foster network represent the transient thermal behavior of the SiC MOSFET. Junction temperature is the function of switch loss, which can be predicted through the RC Foster model as follows:

$$T_j(t) = T_c(t) + P_{loss}(t)Z_{th}(t) \qquad (1)$$

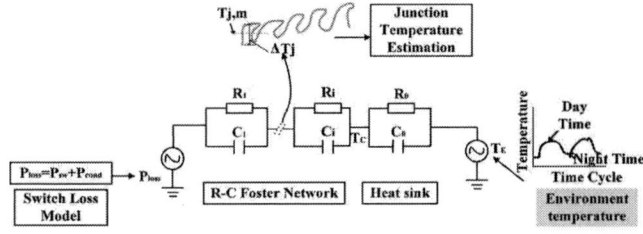

Fig. 1. Impact of environmental temperature on junction temperature through R-C Foster Network.

where Z_{th} will be changed by the varying environmental conditions. If the dynamic environment temperature is taken into account, then the thermal impedance, Z_{th} as follows:

$$Z_{th}(s) = \sum_{i=1}^{n} \frac{R_i}{(1+R_iC_is)} + \frac{R_\theta}{(1+R_\theta C_\theta s)} + \frac{R_{ambient}}{(1+R_{ambient}C_{ambient}s)} \qquad (2)$$

where R_i and C_i are the RC Foster network model parameters, R_θ, C_θ are the heatsink parameters, $R_{ambient}, C_{ambient}$ are the environment temperature parameters. Fig. 2 shows the magnitude gain deviation with different environment temperature conditions, which indicates the impact of junction temperature changes due to environmental conditions.

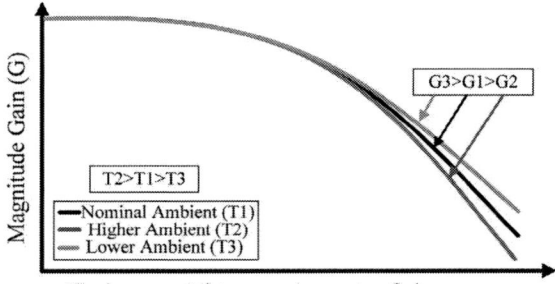

Fig. 2. Junction temperature deviation in different environment temperature.

Fig. 3 shows the junction temperature estimation for SiC MOSFETs. TSEP and NTC sensor measurement methods is used for the junction temperature estimation.

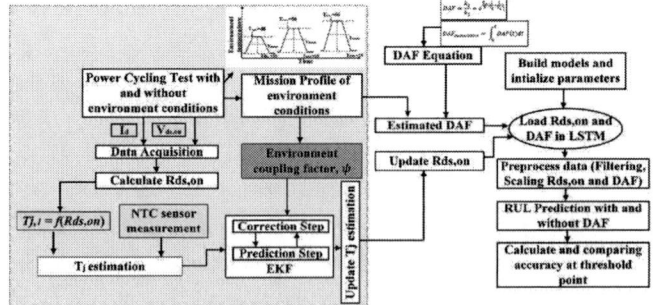

Fig. 3. Junction temperature estimation for SiC MOSFET by integrating the environmental coupling factor.

The relationship between $R_{ds,on}$ and T_j can be described by

$$T_j = f(R_{ds,on}) \qquad (3)$$

where $R_{ds,on}$ is taken as TSEP to measure T_j. Three different load conditions (5A, 10A, and 15A) are taken to measure the junction temperature in Fig.4 where $R_{ds,on}$ taken for seven different temperature points. By using these measured $R_{ds,on}$ values, a

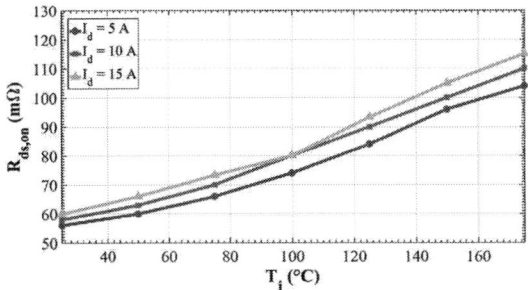

Fig. 4. $R_{ds,on}$ vs T_j with different load conditions.

generalized equation is made by using curve fitting in MATLAB.

$$R_{ds,on} = 0.32T_j + 50.75 \tag{4}$$

Then, T_j can be measured on $R_{ds,on}$ function which expressed by:

$$T_j = 3.125R_{ds,on} - 158.6 \tag{5}$$

This equation is used in the proposed EKF method along with the NTC sensor measurement to estimate junction temperature accurately while considering environmental conditions. The environment coupling factor (ψ) quantifies the sensitivity of the junction temperature variations due to environment temperature. The environment coupling factor (ψ) should be integrated to make accurate prediction.

$$T_j(t) = T_{jc}(t) + \psi (T_E) \tag{6}$$

where T_{jc} is the previous temperature estimation and T_j is the updated junction temperature estimation after integrating the coupling factor in the EKF model. The EKF is the non-linear version of the KF that linearizes the current mean and covariance [21]. The EKF estimates a non-linear system into a linear system using Taylor series expansion. The state equation and measurement equation of KF is shown as follows:

$$\Delta T_{k+1} = A_k \Delta T_k + B_k u_k + w_k \tag{7}$$

$$T_{j_k} = C_k \Delta T_k + D_k u_k + v_k \tag{8}$$

where u_k is the input vector, ΔT is the temperature difference between $R_{th,jc}$ to $R_{th,heatsink}$, and A_k, B_k, C_k, and D_k are the coefficient matrices. w_k and v_k are the process and measurement noises. The KF is a two-step method. One is the prediction step, and another is the correction step. The prediction step:

$$\Delta T_k^- = A_{k-1} \Delta T_{k-1}^+ + B_{k-1} u_{k-1} \tag{9}$$

$$P_k^- = A_{k-1} P_{k-1}^+ A_{k-1}^T + Q \tag{10}$$

where supercript "-" and "+" denote a estimated and corrected covariance. The correction step:

$$G_k = P_k^- C_k^T [C_k P_k^- C_k^T + R]^{-1} \tag{11}$$

$$\Delta T_k^+ = \Delta T_k^- + G_k[T_{j_k} - (C_k \Delta T_k^- + D_k u_k)] \tag{12}$$

$$P_k^+ = (I - G_k C_k) P_k^- \tag{13}$$

where G_k is the Kalman gain. KF improves the performance by optimizing Q and R. The EKF algorithm principle is generally the same as the KF algorithm. In this case, A, B, and C are replaced with partial time derivatives. The EKF prediction step:

$$\Delta T_k^- = f(\Delta T_{k-1}^+, u_{k-1}) \tag{14}$$

$$P_k^- = A_{k-1} P_{k-1}^+ A_{k-1}^T + Q \tag{15}$$

Correction Step:

$$G_k = P_k^- C_k^T [C_k P_k^- C_k^T + R]^{-1} \tag{16}$$

$$\Delta T_k^+ = \Delta T_k^- + G_k[T_{j_k} - g(\Delta T_k^-, u_k)] \tag{17}$$

$$P_k^+ = (I - G_k C_k) P_k^- \tag{18}$$

III. RUL PREDICTION OF SiC MOSFET

The degradation acceleration factor (DAF) is a crucial aspect of devices being subject to varying environmental stresses. So, DAF needs to consider the degradation tests and the effect of dynamic environmental conditions. DAF will be changed with dynamic environmental conditions, and the load will be unchanged. To formulate DAF, considering the basic Arrhenius equation is follows:

$$k = A * e^{\frac{-E_a}{RT}} \tag{19}$$

where k is the rate constant, E_a is the activation energy, and R is the Boltzmann constant. This equation is temperature-dependent reaction rates in chemical kinetics and physical changes in materials. This equation is relevant to electronics devices also, where higher temperature can accelerate wear-out mechanisms like diffusion, electromigration, and oxide breakdown. By modifying the Arrhenius equation to compare the degradation rate under two different temperatures:

$$DAF = \frac{k_1}{k_2} = e^{\frac{E_a}{R}(\frac{1}{T_2} - \frac{1}{T_1})} \tag{20}$$

If the environment temperature is dynamic, then DAF must be calculated for each interval and integrated into the RUL prediction model to get the device's operational lifetime.

$$DAF_{cumulative} = \int_0^T DAF(t)dt \tag{21}$$

where T is the total operational lifetime. So, the effective lifetime will be:

$$RUL_{effective} = \frac{RUL_{nominal}}{DAF_{cumulative}} \tag{22}$$

where $RUL_{nominal}$ is the expected lifetime of the SiC Module under normal operating conditions and $RUL_{effective}$ is the adjusted lifetime, considering the DAF due to environmental temperature stress.

In this paper, the LSTM model incorporates On-state resistance and DAF to get accurate RUL. Fig.5 shows the RUL prediction algorithm of dynamic environment conditions. LSTM retains long-term dependencies and information over time through their cell states [22]. The LSTM network's main features are the gates that allow selective memory updates and ignore irrelevant data, including input, forget, and output gates. Fig.6 shows the LSTM architecture. The forward calculation process of LSTM network can be specifically expressed as:

$$f_t = \sigma(W_f \cdot [y_{t-1}, x_t] + b_f) \quad (23)$$

$$i_t = \sigma(W_i \cdot [y_{t-1}, x_t] + b_i) \quad (24)$$

$$\tilde{C}_t = tanh(W_C \cdot [y_{t-1}, x_t] + b_c) \quad (25)$$

$$C_t = f_t \cdot C_{t-1} + i_t \cdot \tilde{C}_t \quad (26)$$

$$o_t = \sigma(W_o \cdot [y_{t-1}, x_t] + b_o) \quad (27)$$

$$y_t = o_t \cdot tanh(C_t) \quad (28)$$

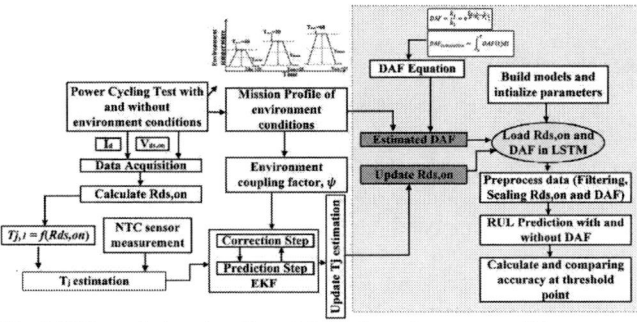

Fig. 5. RUL prediction algorithm of dynamic environment conditions.

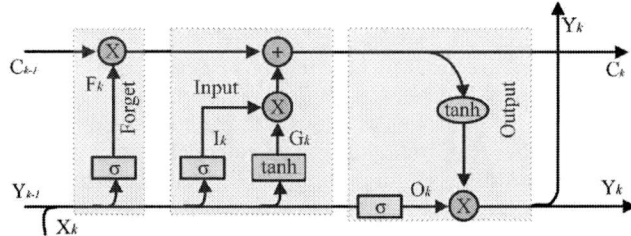

Fig. 6. Schematic architecture of LSTM network

where x_t and y_t are the input and output information at time t, σ and $tanh$ are activation functions. W_f, W_i, W_C, and W_C are weight matrixes, and b_f, b_i, b_c, and b_o are bias matrixes.

To be more convenient in training and predicting the LSTM model, the Savitzky-Golay (Savgol) filter is used for the smoothing technique that fits a low-degree polynomial to a moving window of data points using the least squares method. For each data point y_i, a polynomial of degree p is fitted overall window of $2k+1$ neighboring points. The smoothed value y_i' is calculated by colvolving the data with precomputed filter coefficients C_j as follows:

$$y_i' = \sum_{j=-k}^{k} C_j * y_{i+j} \quad (29)$$

where C_j are derived from the polynomial fit. The parameters control the smoothness of the result, allowing noise reduction while preserving important signal features.

IV. SIMULATION RESULTS

A three-dimensional transient heat transfer study is conducted with COMSOL software on a representative domain

modeled as SiC MOSFET. The governing equation for the heat diffusion is given as,

$$\frac{\partial^2 T}{\partial x^2} + \frac{\partial^2 T}{\partial y^2} + \frac{\partial^2 T}{\partial z^2} + \dot{q}_\forall = \frac{1}{\alpha}\frac{\partial T}{\partial t} \quad (30)$$

where T represents the environment temperature, $\dot{q}_\forall$ is the volumetric heat generation term, and α is the thermal diffusivity. The computational domain is shown in Fig. 7. The top plane is subjected to zero heat flux. The four side planes are subjected to radiative heat loss from the SiC to the ambient. The bottom plane is subjected to convective cooling, where the net thermal resistance from the junction to the heat sink ($R_{j-HS} = 1.5$ K/W) is considered along with the convective heat transfer resistance. Total thermal resistance and effective heat transfer coefficient applied on the bottom wall is given as,

$$R_{eff} = R_{j-HS}\{a*b\}[\frac{m^2K}{W}] + \frac{1}{h_s}[\frac{m^2K}{W}] = \frac{1}{h_{eff}}[\frac{m^2K}{W}] \quad (31)$$

TABLE I. SIMULATION PARAMTERS

Parameters	Expression	Value
Length	a	0.0625 m
Width	b	0.0338 m
Thickness	c	0.0164 m
Volume	$a*b*c$	3.4645e-05 m³
Supplied power (heat)	Q	120 W
Supplied power/volume	Q/vol	3.51277e5 W/ m³
Heat transfer coefficient	h_s	200 W/ (m².K) (sample value)
Top and bottom area	$a*b$	0.0021125 m²
Effective heat transfer coefficient	$1/R\_eff$	122.42 W/(m²·K)
Environment temperature	T_ref	25, 35, 45, 55 °C

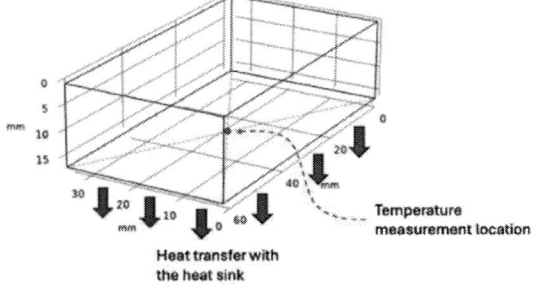

Fig. 7. The computational domain and the temperature sampling location for the transient heating-cooling phase

where the h_s was parametrically varied to study the influence of convective heat transfer levels on the transient heating and cooling phases of the solid block modeled as SiC in Fig. 7. The T_j shown in Fig. 7 has been presented in a non-dimensional form given as

$$\theta = \frac{T_j - T_i}{T_{j,max} - T_i} \quad (32)$$

where T_i is the minimum, and $T_{j,max}$ is the maximum junction temperature during the heating phase. The simulations are conducted for environment temperature variation between 25-55°C. The temporal variation of the non-dimensional junction temperature (θ) is shown in Fig. 8, and the simulation parameters are provided in Table I. In the heating phase, the temporal evolution of the junction temperature is not sensitive towards the convective cooling rates. However, during the cooling phase, the convective heat transfer coefficient distinctly affects temperatures. The cooling curves became less sensitive to convective heat transfer coefficient results for $h_s >$ 800 $W/m^2 K$. So, there is a significant effect on SiC lifetime when it always operates in high environmental conditions. Fig.

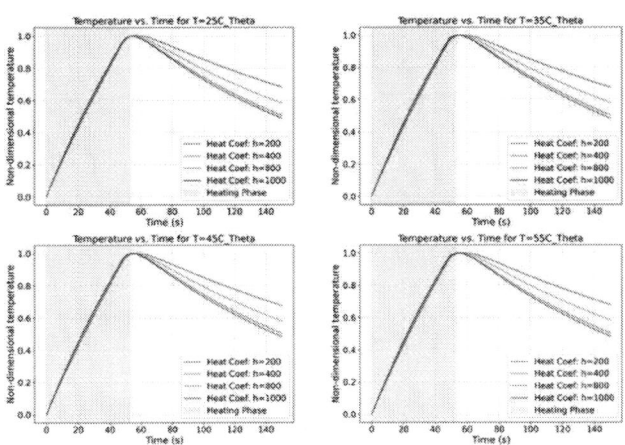

Fig. 8. Temporal variation of the non-dimensional junction temperature (θ) for different convective heat transfer coefficients, as well as different ambient temperatures.

9 shows the DAF changes with dynamic environment conditions, keeping the same temperature swing. It shows that with the temperature increase of the environmental conditions, the DAF will also increase, leading to faster degradation, and the device may fail earlier.

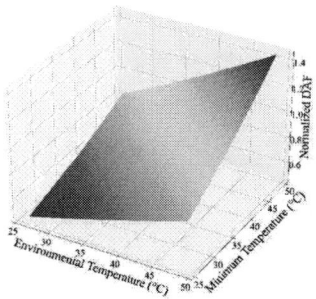

Fig. 9. DAF changes with the dynamic environment conditions.

V. Experimental Study

Integrating dynamic environment conditions in the lifetime testing of SiC MOSFETs is challenging. A climate chamber has been used to create dynamic conditions. The degradation testbed diagram is shown in Fig. 10, and the testing environment is provided in Table II. DUTs (Device Under Test)

are used in the experiment is a commercial CoolSiC™ MOSFET half-bridge module with rated voltage 1200V, rated current 15A, typical on-state resistance 55 mΩ, and the maximum junction temperature under overload conditions with up to 175 °C. The degradation circuit comprises DC-link capacitors, DC supply, and resistive loads. The TMS320F28335 microcontroller from Texas Instruments generates all the control signals. The temperature swing (125°C) is kept same for all the tests by properly selecting load current (I_{dc}), on/off time (t_{on}/t_{off}), and the temperature of climate chamber. With the degradation of SiC MOSFET Module, changes of temperature swing can be possible.

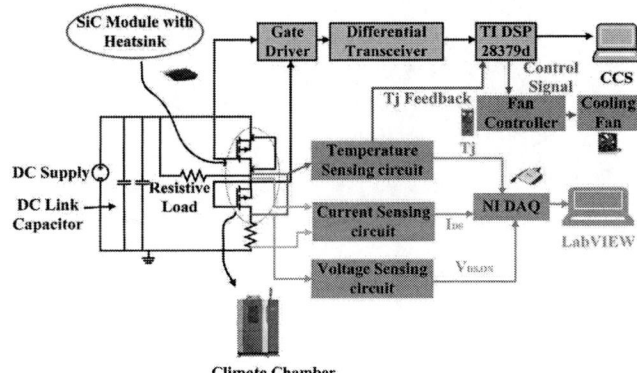

Fig. 10. Degradation testbed with measurement circuits

TABLE II. Degradation Testing Environment

Parameters	Symbol	Value
Load Current	I_{load}	15 A
Switching Frequency	f_{sw}	10 kHz
Gate-source Voltage	V_{GS}	-5/+20 V
Junction Temperature Swing	ΔT_j	125 °C
Heating time/Cooling time	T_{on}/T_{off}	45s/45s.

The $R_{ds,on}$ is measured at the minimum targeted junction temperature to reduce the impact of measurement errors during

degradation studies. The recorded $R_{ds,on}$ is used as a TSEP for estimating the junction temperature of the SiC MOSFET. Data is collected automatically using an NI DAQ 9252 and LabVIEW software, and saved in CSV format.

TABLE III. Degradation Testing Environment Groups

Conditions	Test 1	Test 2	Test 3
Environment Temperature	25 °C	25~45 °C	25~55 °C
Temperature swing	125 °C	125 °C	125 °C
Minimum Junction Temperature	25 °C	25 °C	25 °C
Load Conditions	15A	15A	15A

The test is divided into three categories, as shown in Table III, and the experimental degradation test setup is illustrated in Fig. 11. The dynamic environmental conditions considered include 25°C (fixed/reference), 25°C ~ 45°C and 25°C ~ 55°C. Each experiment is conducted under identical load conditions (15A) and a temperature swing of 125°C. These dynamic conditions are chosen to demonstrate can be aged under different temperature ranges and temperature control schemes. This paper used constant on/off time control strategy. It is open loop control strategy which needs temperature sensing circuit only. In the experiment, temperature sensing circuit is used to extract temperature from NTC pin and TSEP method is used to

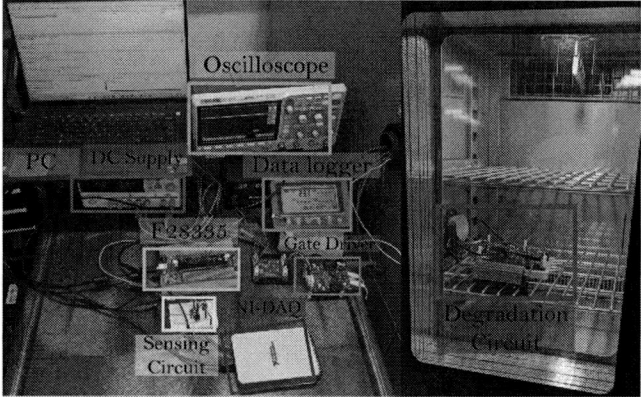

Fig. 11. Testbed of power cycling test at the controlled environment chamber.

estimate junction temperature. These temperatures are calibrated for the constant time control strategy which is recommended by IEC-60749-34 [23]. As degradation can alter aging parameters such as $R_{ds,on,}$ and the T_j swing, the on and off times need to be periodically adjusted.

VI. RESULT AND DISCUSSION

A. Junction Temperature Estimation by EKF

The performance of the junction temperature estimation method is experimentally validated by the experimental setup shown in Fig. 11, where the EKF algorithm is used to estimate junction temperature. The NTC sensor measurement, the TSEP method, is used for junction temperature estimation. Fig.4 shows the profile of junction temperature in different load conditions, while Fig. 12 shows the performance of EKF for junction temperature estimation. Fig. 12(b) shows the zoomed portion of temperature from 30 to 30.175s, and the T_j error between the EKF method and measured data is 0.12°C.

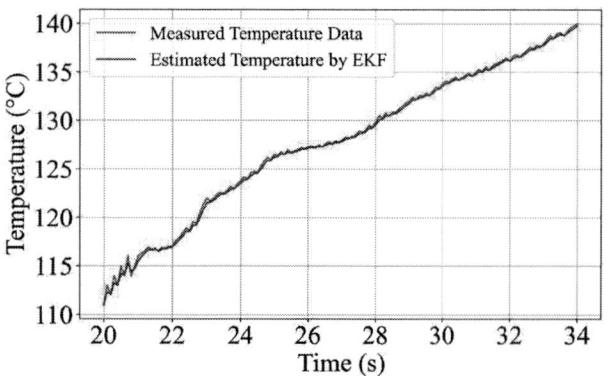

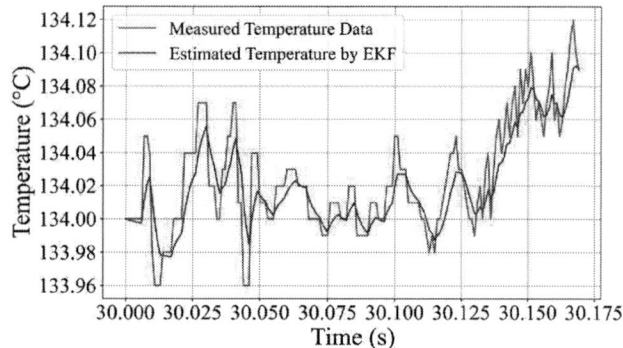

Fig. 12. (a) Measured transient temperature estimation by EKF and (b) zoomed portion

B. RUL Prediction

Fig.13 shows the $R_{ds,on}$ experimental data profile for three dynamic environment conditions. The experimental results indicate that the RUL of the SiC MOSFET decreases as the environmental temperature increases, highlighting the impact of high temperatures on the device's lifespan. Two switch datasets are taken for each of the conditions, and the RUL is defined as a 20% increase in the $R_{ds,on}$ value from its initial value.

The LSTM model aims to predict the degradation in on-state resistance as the device is subject to power and thermal cycling, accounting for environmental conditions through the DAF. In this model, on-state resistance is taken as input parameter, DAF, and temperature conditions are taken as input features. The Adam algorithm with a learning rate 0.1 has been used to train the neural network. Mean Squared Error (MSE) loss function is used in the model, and the optimal model configuration is selected based on the validation loss.

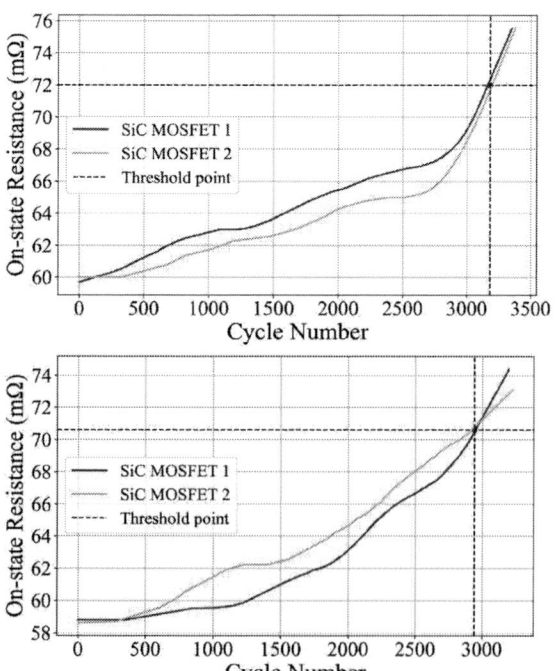

979-8-3315-1612-3/25 $31.00 © 2025 IEEE

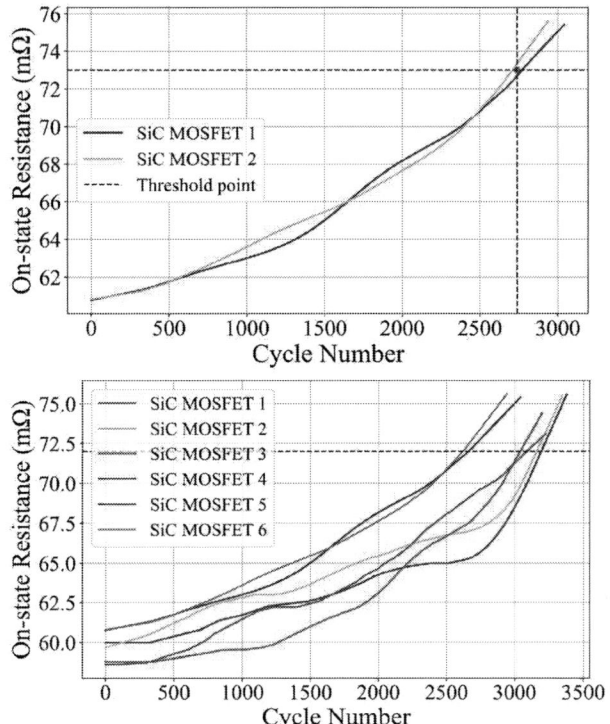

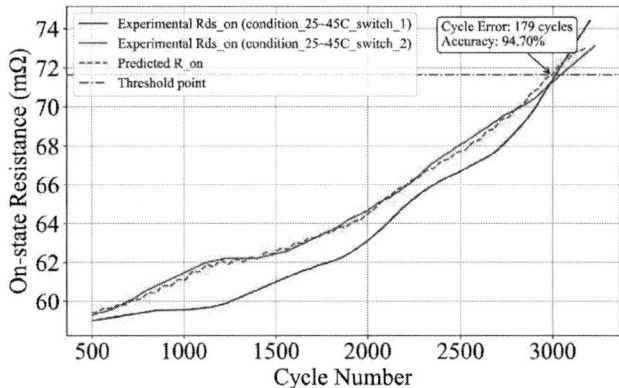

Fig. 14. (a) RUL prediction without DAF features and (b) with DAF features at 25~45 °C environment condition

VII. CONCLUSION

The influence of real-world thermal environmental conditions on T_j has been incorporated into an accurate RUL prediction model based on LSTM, using $R_{ds,on}$ as a precursor and DAF as a feature for SiC MOSFETs. DAF has been thoroughly measured under fixed and dynamic temperatures to determine RUL. Simulations and experimental results indicate that the degradation rate increased by 10-15% under 25~55°C conditions compared to fixed 25°C conditions, highlighting the effect of environmental conditions on SiC MOSFET lifetime prediction. The correlation between DAF and RUL has been thoroughly explored under various dynamic temperature conditions within a climate chamber, and a detailed mathematical model has been derived. The LSTM model, incorporating DAF, has achieved a 30% improvement in lifetime prediction accuracy. This in-depth research and the developed RUL estimation approach that considers real-world environmental conditions could serve as a dependable solution for various system applications.

ACKNOWLEDGMENT

This work was supported in part by the NSF Electrical, Communications and Cyber Systems (ECCS) Program under Award 2210106.

REFERENCES

[1] Zhang, Qinghao, and Pinjia Zhang, "A novel model of the aging effect on the on-state resistance of sic power mosfets for high-accuracy package-related aging evaluation," IEEE Transactions on Industrial Electronics, 70.9 (2022), pp. 9495-9504.

[2] L. R. GopiReddy, L. M. Tolbert and B. Ozpineci, "Power Cycle Testing of Power Switches: A Literature Survey," in IEEE Transactions on Power Electronics, vol. 30, no. 5, pp. 2465-2473, May 2015, doi: 10.1109/TPEL.2014.2359015.

[3] Ni, Z., Lyu, X., Yadav, O. P., Singh, B. N., Zheng, S., and Cao, D., "Overview of real-time lifetime prediction and extension for SiC power converters," IEEE Transactions on Power Electronics, 35(8), 2019, pp.7765-7794.

[4] Y. Wang, E. Deng, L. Wu, Y. Yan, Y. Zhao, and Y. Huang, "Influence of Humidity on the Power Cycling Lifetime of SiC MOSFETs," in IEEE

Fig. 13. Experimental $R_{ds,on}$ data profile at (a) 25°C, (b) 25°C~45°C, (c) 25°C~55°C, and (d) three dynamic environment temperatures.

The dynamic DAF values for each condition are incorporated as a feature in the LSTM model. Integrating this dynamic DAF feature allows the model to learn how varying environmental conditions influence the rate of degradation, either accelerating or slowing it. For RUL prediction, two scenarios are considered: one incorporating DAF values as a feature and the other without DAF. Predictions are made for all conditions, and the results, shown in Fig. 14, a specific condition, demonstrate that incorporating DAF values yields better prediction accuracy compared to when they are not included.

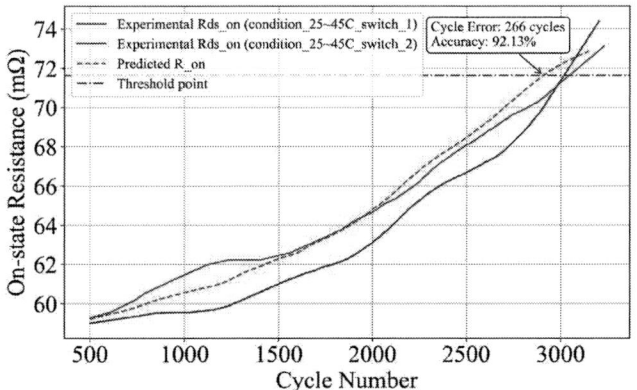

Transactions on Components, Packaging and Manufacturing Technology, vol. 12, no. 11, pp. 1781-1790, Nov. 2022, doi: 10.1109/TCPMT.2022.3223957.

[5] Md Z. Hasan, A. Amin, Md. Moniruzzaman, S. Choi, P. Singh, and C.H. Liu, "Impact of Environmental Conditions on the Remaining Useful Lifetime of SiC MOSFET," in press.

[6] Pu, S., Yang, F., Vankayalapati, B. T., and Akin, B., "Aging mechanisms and accelerated lifetime tests for SiC MOSFETs: An overview," IEEE Journal of Emerging and Selected Topics in Power Electronics, 10(1), 2021, pp.1232-1254.

[7] Zhang, Q., Lu, G., Yang, Y., and Zhang, P., "A high-frequency online junction temperature monitoring method for SiC MOSFETs based on on-state resistance with aging compensation," IEEE Transactions on Industrial Electronics, 70(7), 2022, pp. 7393-7405.

[8] D. Herwig and A. Mertens, "Junction Temperature Estimation of SiC MOSFETs During Inverter Operation Using Switching Times and On-State Voltages," 2021 IEEE Energy Conversion Congress and Exposition (ECCE), Vancouver, BC, Canada, 2021, pp. 2747-2754, doi: 10.1109/ECCE47101.2021.9595535.

[9] F. Stella, G. Pellegrino, E. Armando and D. Daprà, "Online Junction Temperature Estimation of SiC Power MOSFETs Through On-State Voltage Mapping," in IEEE Transactions on Industry Applications, vol. 54, no. 4, pp. 3453-3462, July-Aug. 2018, doi: 10.1109/TIA.2018.2812710.

[10] S. Mocevic et al., "Gate-Driver Integrated Junction Temperature Estimation of SiC MOSFET Modules," 2020 IEEE Energy Conversion Congress and Exposition (ECCE), Detroit, MI, USA, 2020, pp. 3761-3768, doi: 10.1109/ECCE44975.2020.9235607.

[11] Han, X., and Saeedifard, M., "Junction temperature estimation of SiC MOSFETs based on extended Kalman filtering," in 2018 IEEE Applied Power Electronics Conference and Exposition (APEC), 2018, pp. 1687-1694.

[12] M. Luo et al., "Load-Independent Junction Temperature Estimation Via Combined TSEPs Modeling for SiC MOSFETs," in IEEE Transactions on Power Electronics, doi: 10.1109/TPEL.2024.3473529.

[13] Aplication note , AN 2009-10: "Using the NTC inside a power electronic module," Warstein, Germany : Infineon Technologies AG, 2010-01-13 [2018-12-17].

[14] D. P. Nayak and S. K. Pramanick, "Implementation of an Electro-Thermal Model for Junction Temperature Estimation in a SiC MOSFET Based DC/DC Converter," in CPSS Transactions on Power Electronics and Applications, vol. 8, no. 1, pp. 42-53, March 2023, doi: 10.24295/CPSSTPEA.2023.00005.

[15] K. Sharma, S. Kamm, K. M. Barón and I. Kallfass, "Characterization of Online Junction Temperature of the SiC power MOSFET by Combination of Four TSEPs using Neural Network," 2022 24th European Conference on Power Electronics and Applications (EPE'22 ECCE Europe), Hanover, Germany, 2022, pp. 1-8.

[16] X. Ding, B. Wang and Y. Yang, "DC Power Cycling Test and Lifetime Prediction for SiC MOSFETs," 2023 26th International Conference on Electrical Machines and Systems (ICEMS), Zhuhai, China, 2023, pp. 4638-4643, doi: 10.1109/ICEMS59686.2023.10344327.

[17] Y. Wang, E. Deng, T. Wu, Y. Zhang, L. Xie and Y. Huang, "Thermo-Hygroscopic-Mechanical Coupling Simulation Method for Power Electronics Under Power Cycling Test," in IEEE Transactions on Power Electronics, vol. 38, no. 9, pp. 11521-11530, Sept. 2023, doi: 10.1109/TPEL.2023.3290749.

[18] T. I. Mannan, A. Amin and S. Choi, "Investigation of SiC MOSFET Aging Effects on Common-Mode EMI Emissions," 2023 IEEE Electric Ship Technologies Symposium (ESTS), Alexandria, VA, USA, 2023, pp. 477-483, doi: 10.1109/ESTS56571.2023.10220554.

[19] X. Lu, L. Wang, Q. Yang, F. Yang, Y. Gan and H. Zhang, "Investigation and Comparison of Temperature-Sensitive Electrical Parameters of SiC mosfet at Extremely High Temperatures," in IEEE Transactions on Power Electronics, vol. 38, no. 8, pp. 9660-9672, Aug. 2023, doi: 10.1109/TPEL.2023.3267472.

[20] M. Moniruzzaman, A. H. Okilly, S. Choi, J. Baek, T. I. Mannan and Z. Islam, "A Comprehensive Study of Machine Learning Algorithms for GPU based Real-time Monitoring and Lifetime Prediction of IGBTs," 2024 IEEE Applied Power Electronics Conference and Exposition (APEC), Long Beach, CA, USA, 2024, pp. 2678-2684, doi: 10.1109/APEC48139.2024.10509167.

[21] M. S. Haque, S. Choi and J. Baek, "Auxiliary Particle Filtering-Based Estimation of Remaining Useful Life of IGBT," in IEEE Transactions on Industrial Electronics, vol. 65, no. 3, pp. 2693-2703, March 2018, doi: 10.1109/TIE.2017.2740856.

[22] Vaccaro, A., Biadene, D., and Magnone, P., "Remaining Useful Lifetime Prediction of Discrete Power Devices by Means of Artificial Neural Networks," IEEE Open Journal of Power Electronics, 4, 2023, pp. 978-986.

[23] IEC-International Electrotechnical Commission, document IEC 60749-34, 2010

Guarding-Based C-V Characterization of 10 kV SiC MOSFET in Half-bridge Module Configuration

Nianzun Qi
AAU Energy
Aalborg University
Aalborg, Denmark
nqi@energy.aau.dk

Gao Liu
AAU Energy
Aalborg University
Aalborg, Denmark
gaol@energy.aau.dk

Zhixing Yan
AAU Energy
Aalborg University
Aalborg, Denmark
zhya@energy.aau.dk

Shaokang Luan
AAU Energy
Aalborg University
Aalborg, Denmark
slu@energy.aau.dk

Pawel Piotr Kubulus
AAU Energy
Aalborg University
Aalborg, Denmark
ppk@energy.aau.dk

Yuan Gao
AAU Energy
Aalborg University
Aalborg, Denmark
yuga@energy.aau.dk

Stefan Meyer
AAU Energy
Aalborg University
Aalborg, Denmark
stefanm@energy.aau.dk

Hongbo Zhao
AAU Energy
Aalborg University
Aalborg, Denmark
hzh@energy.aau.dk

Asger Bjørn Jørgensen
AAU Energy
Aalborg University
Aalborg, Denmark
abj@energy.aau.dk

Stig Munk-Nielsen
AAU Energy
Aalborg University
Aalborg, Denmark
smn@energy.aau.dk

Abstract—Capacitance-Voltage (C-V) characteristics of SiC MOSFETs are critical for understanding their dynamic performance. Direct C-V measurement on a 10 kV SiC MOSFET die is hindered by insulation challenges, necessitating the use of a specialized fixture. As an alternative, measurements through a half-bridge power module offer a feasible solution. However, parasitic capacitances within the power module, comparable in magnitude to the intrinsic capacitances of 10 kV SiC MOSFETs, can affect measurement accuracy. This paper proposes a guarding-based method and an analytical approach to characterize MOSFET capacitances in a power module configuration, enabling precise separation of intrinsic MOSFET capacitances from parasitic effects. A comprehensive comparison of the two methods is presented using C_{gd} - V_{ds} measurements for a 10 kV SiC MOSFET as a case study. Based on the experimental verifications, the proposed guarding-based method is accurate, straightforward and feasible.

Index Terms—10 kV SiC MOSFET, Capacitance-Voltage (C-V) characteristics, Guarding method.

I. INTRODUCTION

Medium-voltage (MV) Silicon Carbide (SiC) power devices have become promising technologies to further improve the power converters' performance [1]–[9]. Identifying the power MOSFET characteristics is the key to unlocking the potential of the SiC power devices, among which the C-V characteristic dominates the device's dynamic performance [10]. The C-V characteristics refer to the dependence of gate-drain capacitance C_{gd}, gate-source capacitance C_{gs}, and drain-source capacitance C_{ds} on the drain-source voltage (V_{ds}). It is normally shown in the datasheets of commercial MOSFETs, which determines the transient charging currents during the dv/dt switching events.

This work is co-supported by the MV-BASIC project and the MVolt project. The MV-BASIC project is co-funded by AAU Energy of Aalborg University, together with industrial project partners KK Wind Solutions and Siemens Gamesa. MVolt project is co-funded by AAU Energy of Aalborg University, Innovation Fund Denmark, Siemens Gamesa Renewable Energy, Vestas Wind Systems, and KK Wind Solutions.

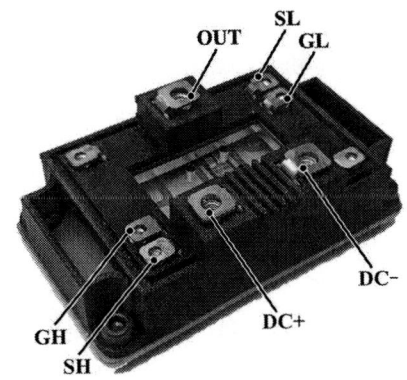

Fig. 1: Custom packaged medium-voltage half-bridge power module [14].

The commercial parametric curve tracer is adopted to measure the C-V characteristic [11]. The measurement circuit is based on the auto-balancing bridge theory [12]. With an AC voltage source at a fixed frequency and a superimposed DC bias, the capacitance can be calculated by analyzing the phase difference and amplitude ratio between the measured AC voltage and current [13]. When measuring the 10 kV SiC MOSFET C-V characteristics, the DC bias needs to reach several kVs. It proposes a high insulation requirement for the test fixture for bare die measurements.

Alternatively, extracting die C-V characterisitcs through a packaged power module can be feasible. Fig. 1 demonstrates a 10 kV SiC MOSFET power module using a conventional power module packaging structure [14]. The power module is encapsulated with silicone gel, which provides a higher electric breakdown strength than air [15]. Therefore, the measured vds range of die C-V characteristics based on the power module can be extended to device-rated voltage.

However, the parasitic capacitive coupling between the baseplate and the top copper islands can interfere with the die capacitance measurement, introducing extra components to capacitance under test (CUT). The guarding technique is introduced to bypass the components that are not intended for measurement and has now been integrated into Keysight commercial products [16]. When measuring the impedance of a specific component in a multi-port network, the other ports not involved in the measurement can be connected to the guard terminal to eliminate their influence on the measurement, thereby improving the accuracy during in-circuit tests [17].

In this paper, the guarding technique will be utilized to exclude the packaging impact during the C-V measurement on the 10 kV SiC MOSFET in half-bridge module configuration. Initially, the equivalent circuit of the power module and measurement loop is analyzed, providing insight into how deviation can be introduced. Then, the two methods to test the die C-V characteristic in the power module are proposed and compared with each other. One is through post-processing to compensate for the measurement without a guarding technique while the other is to measure with a guarding technique. The limitations and advantages of the two methods are analyzed based on the measurement results.

II. PROBLEM FORMULATION

The theoretical circuit of power device analyzer (Keysight B1505A) to measure C_{gd} - v_{ds} characteristics of power MOSFET discretes is shown in Fig. 2 [11]. The C_{gd} is taken as the capacitance under test (CUT) since C_{gd} is sensitive to high dv/dt and is in the pF level compared to the C_{gs} and C_{ds} [17].

As shown in Fig. 2, the auto-balancing bridge with the integrated Guard port is adopted. The measurement circuit consists of three parts: the excitation circuit, the measurement circuit, and the auto-balancing part. An AC voltage source (v_{ac}) operated at a fixed frequency (f), which is connected in series to a DC voltage source (V_{dc}), formulating the excitation. V_{dc} set the static bias voltage v_{ds} of the MOSFET, whereas the v_{ac} drives the small-signal current at the output port (H). The voltage meter before the H port, and the voltage meter cross the R_m can provide the v_m and i_m, enabling further derivation of the impedance. The voltage potential of the Low port (L) is nearly zero (virtual ground due to the auto-balancing bridge), which is the same as the Guard port [3].

By connecting source terminal (S) to the Guard port, the i_2 flow to the Guard directly and i_3 is nearly zero. The recorded signal current (i_m) is the current flowing through the C_{gd} (i_1). Therefore, combined with the vector of signal voltage (v_1), the C_{gd} can be calculated according to following equation:

$$C_{gd} = \frac{I_m}{V_m \cdot 2\pi f \cdot \cos(\theta)} \quad (1)$$

However, the equivalent circuit for the 10 kV SiC MOSFET half-bridge power module is more complicated than a MOSFET with three intrinsic capacitors. Inside the power module, the bare 10 kV SiC MOSFET dies are soldered onto the top copper layer of a DBC substrate, which features a 1 mm AlN ceramic layer, supported by a 5 mm AlSiC baseplate. Terminals are mounted on the copper planes, which are used

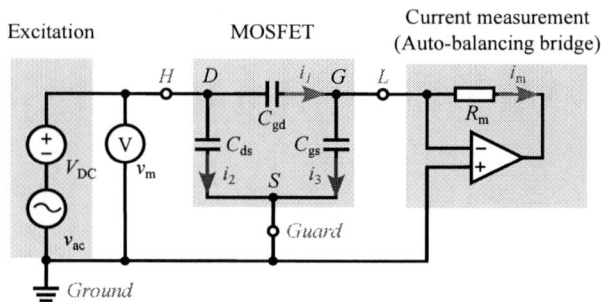

Fig. 2: Equivalent measurement circuit of curve tracer to measure C_{gd} - v_{ds} characteristics of MOSFET.

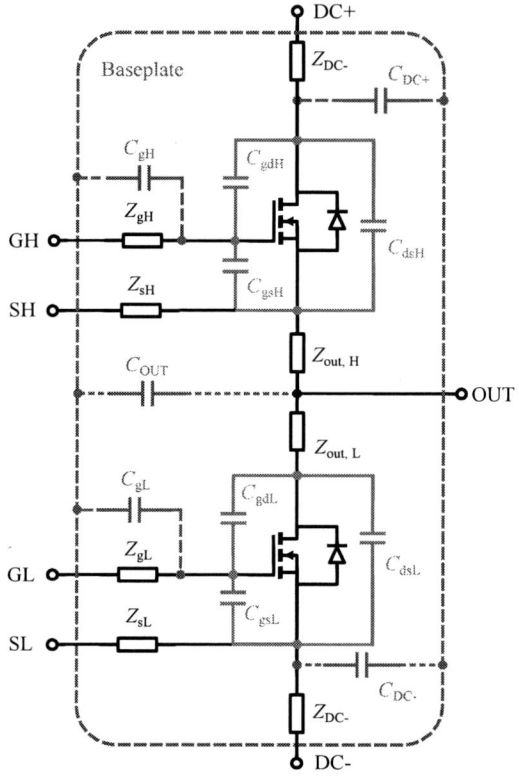

Fig. 3: Photo of the C-V characterization of the 10 kV SiC MOSFET half-bridge power module.

to connect the bare dies with the measurement circuits. Therefore, the capacitors in the power module consist of intrinsic capacitors and parasitic capacitances, as shown in Fig. 3. The top copper islands, DC+, DC-, OUT, gate-high (GH), and gate-low (GL), can have parasitic capacitive coupling with the baseplate. These parasitic capacitances (C_{DC+}, C_{gH}, C_{out}, C_{gL}, C_{DC-}) allow high-frequency signals or noise to couple between the copper islands and the baseplate, forming additional unintended current paths.

When characterizing the C_{gd} of 10 kV SiC MOSFET die inside the power module, the power module itself can introduce parasitic capacitance of a similar magnitude, which superimposes the measurement results. The gate-drain capacitance (C_{gd}) of 10 kV SiC MOSFET shifts significantly, dropping

979-8-3315-1612-3/25 $31.00 © 2025 IEEE   1035

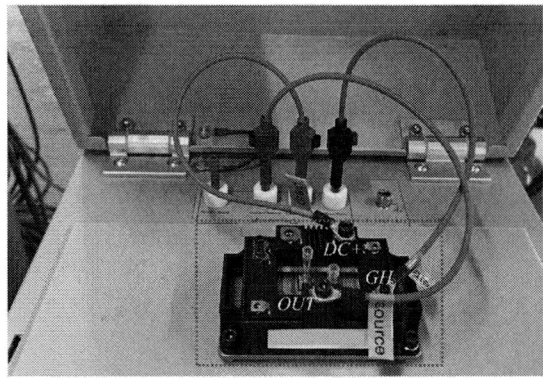

Fig. 4: Equivalent measurement circuit of curve tracer to measure C_{gd} - V_{ds} characteristics of 10 kV half-bridge SiC MOSFET module.

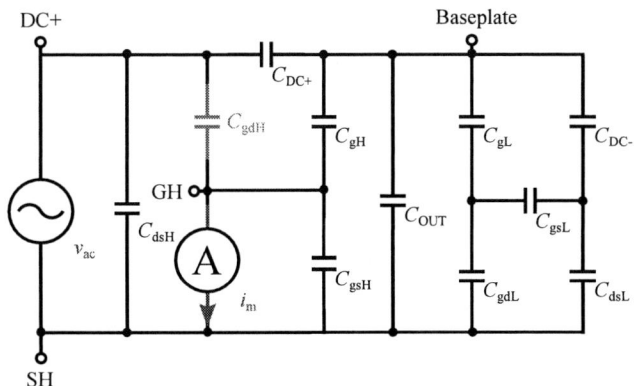

Fig. 5: Simplified circuit of C_{gdH} - V_{ds} measurement.

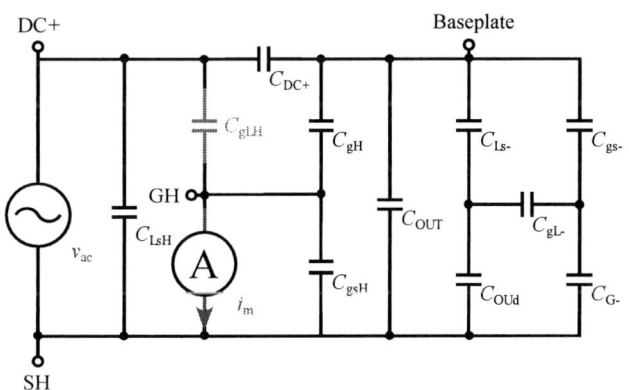

Fig. 6: Simplified circuit of C_{gdL} - V_{ds} measurement.

from the nF range to several pF as the drain-source voltage (v_{ds}) increases [18]–[20]. Meanwhile, the parasitic capacitance between top copper traces of direct bonded copper (DBC) and baseplate are in the pF range.

Fig. 4 shows the equivalent circuit and photo of the C_{gdH} measurement on the high side. The module parasitic capacitance provide an another current loop (i_4), which further contributes to the recorded current (i_m) through the C_{gH} branch (i_5), together with measurement loop current (i_1). Therefore, the measured capacitance value includes two parts: the C_{gdH} of the 10 kV SiC MOSFET, which is the purpose of the measurement, and the equivalent capacitance due to the additional loops. While assuming linear characteristic of the AlN ceramic, the introduced additional capacitance for high-side C_{gdH} measurement ($C_{add,HS}$) is constant and can be calculated if the power module structure is known. It needs to be compensated in the C_{gdH} - V_{ds} measurement.

Therefore, it is essential to develop a method for characterizing the C-V characteristics of medium-voltage MOSFET dies through the power module. This method should ensure high measurement accuracy, support high measurement voltages, and be easily implementable across most power modules.

III. PROPOSED METHODS TO CONDUCT C-V CHARACTERIZATION THROUGH POWER MODULE

This section presents two approaches for characterizing the C-V characteristics of medium-voltage SiC MOSFETs in power modules. The first is an analytical method, which utilizes detailed analysis and simulations to calculate the excess capacitance and compensate for the test results. The second is a guarding-based method, which minimizes parasitic influences by connecting the baseplate to the guard port.

A. Analytical method

The analytical method involves directly measuring the C-V characteristics using the setup shown in Fig. 4, while also calculating the additional capacitance contributing to the measured value of the CUT and subsequently deducting it from the results. It requires the pre-analysis of the module

capacitance and post-processing of the measurement data. These two steps needs the knowledge of the internal structure of power module.

The 10 kV SiC MOSFET module used in this paper is designed and packaged in-house. Therefore, the internal structure is available. For C_{gdH} and C_{gdL} of MOSFET characterization, the test circuits are further simplified to Fig. 5 and Fig. 6. Therefore, the measured capacitance is calculated based on the voltage (v_{ac}) and current (i_m), according to the Eq.(1). It should be noted that the measured capacitance includes not only the capacitance under test (CUT) but also the series and parallel effects of other capacitances. If the value of the capacitance at one condition is known, by using the star-delta transformation, the $C_{add,HS}$ and $C_{add,LS}$ can be calculated.

With Ansys Q3D simulations, the parasitic capacitance from the power module packaging can be extracted, shown in Tab. I [21]. Meanwhile, a power module without dies is manufactured. Without the interconnection, the parasitic capacitance from each DBC island to the baseplate can be measured through the impedance analyzer. The simulated and measured results are summarized into the Tab. I. Based on the data in Tab. I, the $C_{add,HS}$ and the $C_{add,LS}$ have been calculated and are presented in Tab I.

979-8-3315-1612-3/25 $31.00 © 2025 IEEE

TABLE I: The pre-analyses and post-processing of the 10 kV SiC MOSFET power module capacitances.

Parasitic capacitance	Simulated value	Measured value
C_{DC+}	41.30 pF	44.6 pF
C_{gH}	7.41 pF	8.65 pF
C_{out}	49.54 pF	53.90 pF
C_{gL}	21.86 pF	23.20 pF
C_{DC-}	19.80 pF	21.50 pF
C_{gd} ($V_{ds} = 0$ V)	/	3.5 nF
C_{gs} ($V_{ds} = 0$ V)	/	6.4 nF
C_{ds} ($V_{ds} = 0$ V)	/	7.7 nF
$C_{add,HS}$	2.19 pF	2.54 pF
$C_{add,LS}$	15.33 pF	16.34 pF

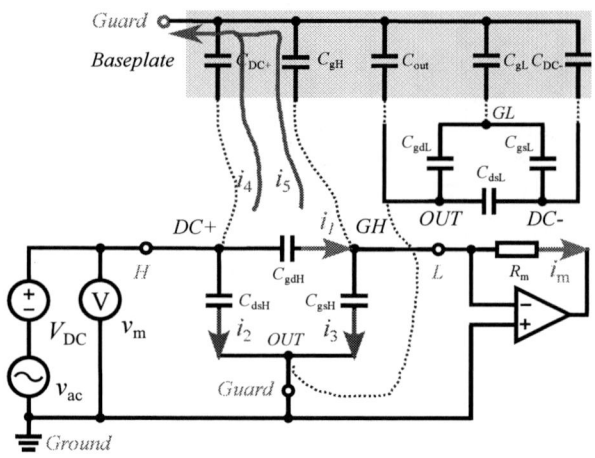

Fig. 7: The equivalent measurement circuit to measure C_{gdH} - V_{ds} characteristics of 10 kV half-bridge SiC MOSFET module with baseplate guarded.

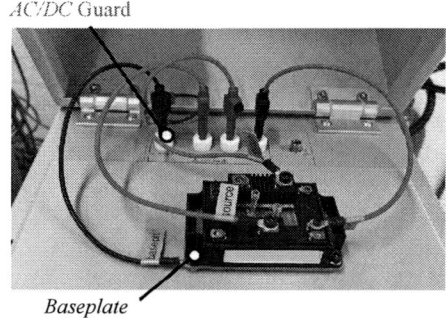

Fig. 8: Measurement circuit of curve tracer to measure C_{gdH} - v_{ds} characteristics of 10 kV SiC MOSFET with baseplate grounded.

B. Guarding-based method

The guarding technique eliminates the influences of C_{ds} and C_{gs}, when measuring C_{gd}, as shown in Fig. 2. This is achieved by connecting the source to the guard terminal, which isolates the measurement path from undesired interference. This functionality is automatically realized by the device capacitance selector, such as the N1272 [22].

With this principle, the proposed circuit measurement circuit is designed to further exclude the impact of parasitic capacitances from packaging, as shown in Fig. 7. In the proposed circuit, the baseplate terminal of the power module is connected to the Guard potential. The guard potential is referenced to the virtual ground at the L port of the circuit. This configuration ensures that the current associated with the parasitic capacitance C_{DC+} (i_4) flows directly to the virtual ground, bypassing the L port. By preventing i_4 from contributing to the measured current (i_m), the influence of C_{DC+} on the measurement is effectively eliminated. As a result, the measurement accuracy for C_{gdH} is improved, ensuring that the measured current reflects the behavior of the desired capacitance.

Furthermore, the GH port is maintained at a voltage potential near zero, effectively acting as a virtual ground. This condition causes the parasitic capacitance C_{gH} to be shorted, thereby allowing the associated current i_5 to be neglected. With these parasitic effects mitigated, the current i_1 becomes approximately equal to i_m. Thus, the measurement result captures the frequency response of the targeted capacitance, C_{gdH}.

The physical implementation of this setup is shown in Fig. 8. In this configuration, the baseplate of the power module is directly connected to the AC/DC guard port on the measurement panel using a cable. It is important to note that the cable used for this connection can introduce additional impedance, especially at higher measurement frequencies. Such impedance can lead to stray AC currents, which in turn may cause measurement errors. To minimize these effects, it is critical to use a cable that has low inductance. This careful selection ensures that the stray impedance is negligible, allowing for

accurate and reliable capacitance measurements.

IV. EXPERIMENTAL VERIFICATION

In this section, the guarding-based method and the analytical method are implemented simultaneously on the 10 kV SiC MOSFET C_{gd} - V_{ds} characterization. All these measurements are conducted by power device analyzer B1505A which are calibrated (open/short capacitance compensation). The ac source is set as 30 mV amplitude and frequency is 100 kHZ.

The C_{gdH} - V_{ds} and C_{gdL} - V_{ds} data measured by different methods is shown in Fig. 9 and Fig.10. The blue dashed line is directly from Fig. 4 measurement. After deducting the $C_{add,HS}$ and $C_{add,LS}$, the results after post-processing is shown in blue solid line in Fig. 9 and Fig.10. Meanwhile, the C_{gd} - V_{ds} curves measured by the guarding-based method is presented in red solid line. It shows a good match between two methods.

Both the guarding-based method and the analytical method are feasible to obtain the C-V characteristic with accuracy. If the internal structure of power module is known, for the analytical method, all the parasitic capacitances in the power module need to be tested before and be considered in the data

979-8-3315-1612-3/25 $31.00 © 2025 IEEE

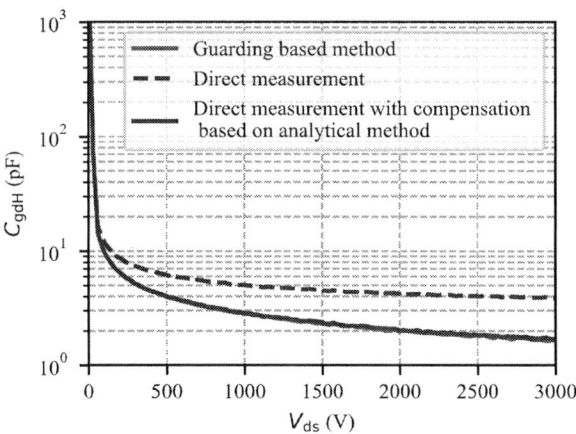

Fig. 9: Measured C_{gdH} - V_{ds} characteristics of 10 kV SiC MOSFET through power module.

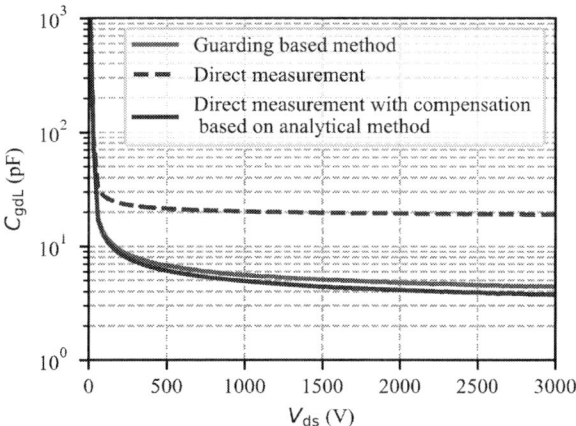

Fig. 10: Measured C_{gdL} - V_{ds} characteristics of 10 kV SiC MOSFET through power module.

post-processing. Circuit analysis can be time-consuming and prone to errors. In contrast, the guarding-based method simplifies the process with direct measurements and a consistent setup.

However in most cases, the power module functions as a black box, making its internal interconnections unknown and difficult to analyze. Consequently, the guarding-based method proves to be a valuable approach for directly extracting the die's C-V characteristics through the module.

V. CONCLUSION

This paper proposes two methods, analytical method and guarding-based method, which identify the internal capacitances of a 10 kV SiC MOSFET, without the effect from the parasitic capacitances from the power module. These two approaches leverage the module's encapsulation, which inherently provides better insulation and protection against high-voltage stress. The guarding-based method can directly measure the MOSFET capacitances and can be generalized

for device characterization with power module packaging. Without module pre-analyses and data post-processing, it maintains high accuracy with the analytical method.

REFERENCES

[1] B. F. Kjærsgaard, G. Liu, M. R. Nielsen, R. Wang, D. N. Dalal, *et al.*, "Parasitic Capacitive Couplings in Medium Voltage Power Electronic Systems: An Overview," *IEEE Transactions on Power Electronics*, vol. 38, no. 8, pp. 9793–9817, 2023.

[2] C. DiMarino, B. Mouawad, C. M. Johnson, M. Wang, Y.-S. Tan, *et al.*, "Design and Experimental Validation of a Wire-Bond-Less 10-kV SiC MOSFET Power Module," *IEEE Journal of Emerging and Selected Topics in Power Electronics*, vol. 8, no. 1, pp. 381–394, 2020.

[3] B. J. Baliga, "Silicon Carbide Power Devices: Progress and Future Outlook," *IEEE Journal of Emerging and Selected Topics in Power Electronics*, vol. 11, no. 3, pp. 2400–2411, 2023.

[4] J. Millán, P. Godignon, X. Perpiñà, A. Pérez-Tomás, and J. Rebollo, "A Survey of Wide Bandgap Power Semiconductor Devices," *IEEE Transactions on Power Electronics*, vol. 29, no. 5, pp. 2155–2163, 2014.

[5] S. Ji, Z. Zhang, and F. Wang, "Overview of high voltage sic power semiconductor devices: development and application," *CES Transactions on Electrical Machines and Systems*, vol. 1, no. 3, pp. 254–264, 2017.

[6] M. R. Nielsen, S. Deng, A. B. Mirza, B. F. Kjærsgaard, A. B. Jørgensen, *et al.*, "High-Power Electronic Applications Enabled by Medium Voltage Silicon-Carbide Technology: An Overview," *IEEE Transactions on Power Electronics*, vol. 40, no. 1, pp. 987–1011, 2025.

[7] S. K. Mazumder, L. F. Voss, K. M. Dowling, A. Conway, D. Hall, *et al.*, "Overview of Wide/Ultrawide Bandgap Power Semiconductor Devices for Distributed Energy Resources," *IEEE Journal of Emerging and Selected Topics in Power Electronics*, vol. 11, no. 4, pp. 3957–3982, 2023.

[8] D. Rothmund, T. Guillod, D. Bortis, and J. W. Kolar, "99.1% Efficient 10 kV SiC-Based Medium-Voltage ZVS Bidirectional Single-Phase PFC AC/DC Stage," *IEEE Journal of Emerging and Selected Topics in Power Electronics*, vol. 7, no. 2, pp. 779–797, 2019.

[9] Z. Lu, C. Li, A. Zhu, H. Luo, C. Li, *et al.*, "Medium Voltage Soft-Switching DC/DC Converter With Series-Connected SiC MOSFETs," *IEEE Transactions on Power Electronics*, vol. 36, no. 2, pp. 1451–1462, 2021.

[10] R. Stark, A. Tsibizov, N. Nain, U. Grossner, and I. Kovacevic-Badstuebner, "Accuracy of Three Interterminal Capacitance Models for SiC Power MOSFETs Under Fast Switching," *IEEE Transactions on Power Electronics*, vol. 36, no. 8, pp. 9398–9410, 2021.

[11] A. Wadsworth, *The Parametric Measurement Handbook*, 4th Edition. 2018.

[12] *Impedance Measurement Handbook*, Accessed: 2024-11-27, 2016. [Online]. Available: https://www.keysight.com/us/en/assets/7018-06840/application-notes/5950-3000.pdf.

[13] F. Wu and Y. Liu, "Extraction of Parasitic Inductances in Three-Phase SiC MOSFET Modules Through Enhanced Two-Port-S-Parameter Measurement," in *2024 IEEE Energy Conversion Congress and Exposition (ECCE)*, 2024.

[14] A. B. Jørgensen, N. Christensen, D. N. Dalal, S. D. Sønderskov, S. Bęczkowski, *et al.*, "Reduction of parasitic capacitance in 10 kV SiC MOSFET power modules using 3D FEM," in *2017 19th European Conference on Power Electronics and Applications (EPE'17 ECCE Europe)*, 2017, pp. 1–8.

[15] B. Zhang, X. Jiang, K. Li, Z. Yang, X. Li, *et al.*, "Dielectric properties characterization and evaluation of commercial silicone gels for high-voltage high-power power electronics module packaging," *IEEE Transactions on Dielectrics and Electrical Insulation*, vol. 30, no. 1, pp. 210–219, 2022.

[16] H. Zhao, D. N. Dalal, A. Bjørn Jørgensen, J. K. Jørgensen, X. Wang, *et al.*, "Physics-Based Modeling of Parasitic Capacitance in Medium-Voltage Filter Inductors," *IEEE Transactions on Power Electronics*, vol. 36, no. 1, pp. 829–843, 2021.

[17] S. Luan, S. Munk-Nielsen, B. Wakelin, M. Hortans, J. Schupp, and H. Zhao, "A General Method to Measure Parasitic Capacitance of Transformer Using Guarding Technique," in *European Conference on Power Electronics and Applications*, 2022, pp. 1–9.

[18] K. Chen, Z. Zhao, L. Yuan, T. Lu, and F. He, "The impact of nonlinear junction capacitance on switching transient and its modeling for SiC MOSFET," *IEEE Transactions on Electron Devices*, vol. 62, no. 2, pp. 333–338, 2014.

[19] N. Wang and J. Zhang, "Nonlinear capacitance model of SiC MOSFET considering envelope of switching trajectory," *IEEE Transactions on Power Electronics*, vol. 37, no. 7, pp. 7977–7988, 2022.

[20] Z. Duan, T. Fan, X. Wen, and D. Zhang, "Improved SiC power MOSFET model considering nonlinear junction capacitances," *IEEE Transactions on Power Electronics*, vol. 33, no. 3, pp. 2509–2517, 2017.

[21] A. B. Jørgensen, S. Munk-Nielsen, and C. Uhrenfeldt, "Overview of Digital Design and Finite-Element Analysis in Modern Power Electronic Packaging," *IEEE Transactions on Power Electronics*, vol. 35, no. 10, pp. 10 892–10 905, 2020.

[22] A. Aman, A. Chanekar, S. Anand, and A. Agarwal, "Impact of Operational Parameters on dVDS/dt of SiC MOSFET and a Scheme for Gate Driver Resistance Selection to Limit dVDS/dt," in *IEEE Energy Conversion Congress and Exposition (ECCE)*, 2024, pp. 1159–1165.

Automated Characterization Platform for Comprehensive Dynamic R_{dson} Assessment of GaN HEMTs From 50 K to 400 K

Tian Qiu*, Zheyu Zhang, Purushottam Khadka, Ahmed Siraj and Dilip Rana

Rensselaer Polytechnic Institute
Troy, USA
qiut2@rpi.edu

Abstract—**Gallium Nitride (GaN) HEMTs demonstrate superior performance at cryogenic temperatures, making them promising candidates for lunar missions where temperatures vary from 50 to 400 K. However, the current collapse effect increases dynamic on-state resistance (R_{dson}) during switching, leading to additional energy loss, especially in high-frequency applications. To conduct a comprehensive study with high efficiency, we developed an automated characterization platform for dynamic R_{dson} extraction under multi-stressor conditions. The design considerations include 1) programmable stressors such as temperatures, voltages, currents, etc., 2) electrothermal-mechanical compatibility with the cryocooler across a wide temperature range, and 3) software framework for full automation. Results for a commercial GaN HEMT show that dynamic R_{dson} varies non-monotonically with temperature increment, as it increases at $50 - 180$ K and then decreases a $180 - 400$ K. Dynamic R_{dson} follows this trend at both hard- and soft-switching, but is more sensitive to stressors at hard switching, while remaining nearly unchanged at soft switching, indicating that, under this testing plan that applies low stressing voltages, hot electron effect makes the major contribution to dynamic R_{dson}.**

Keywords—*GaN HEMT, dynamic characterization, cryogenic test, automated platform*

I. INTRODUCTION

Power electronics for space applications have attracted significant attention in recent years, such as the solar-powered systems for lunar mission. These systems, however, face extreme environmental challenges on the lunar surface, where temperatures range from 400 K during the day to 50 K at night [1]. Gallium Nitride (GaN) high electron mobility transistors (HEMTs) offer significant advantages over silicon devices, including fast switching speed and low conduction loss [2]. As a promising candidate for space applications, GaN HEMTs possess great potential especially at cryogenic temperatures, in terms of, for example, generally lower ON-state resistance (R_{dson}) at low temperatures [3].

Nevertheless, GaN HEMTs suffer from dynamic ON-state resistance (D-R_{dson}) effect: At the onset that the device is turned on, its actual R_{dson} is notably higher than the static R_{dson} [4]. As shown in Fig. 1, the extra R_{dson} introduces more energy losses and things will be even worse when the switching frequency becomes higher [5]. Therefore, it is essential to characterize the behavior of D-R_{dson} at a wide temperature range, in such way

NASA, 22-22-ECF-0025, under award No 80NSSC24K0179

that a comprehensive database can be established to support the modeling and design of GaN-based converters for space applications.

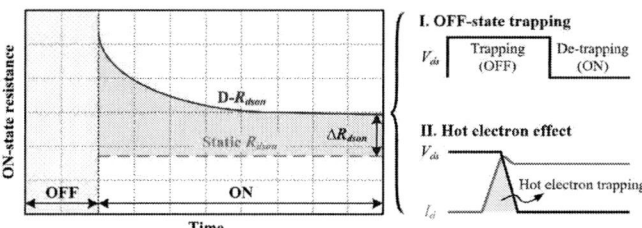

Fig. 1. Dynamic R_{dson} and its physical origins.

The major physical origins of D-R_{dson} includes two aspects: I. Electron trapping (capture) during the OFF state when the device is blocking a high DC bus voltage [6], and II. Hot electron effect that occurs during the switching transient when a large electric field and large current exist simultaneously in the drain region [7]. Thus, on top of temperature, D-R_{dson} can be impacted by multiple factors or stressors, including the amplitude of blocking voltage and conducted current, the stressing time (i.e., the duration of being biased by high voltage), as well as switching mode since the hot electron effect is more severe under hard-switching conditions. In addition, the switching speed is also impactful since it is related to switching loss and hot electron effect. As presented in TABLE I, existing research has performed comprehensive studies on D-R_{dson} [8]-[17]. Generally, it includes 1) designing circuit topology to enable flexible stressor configuration and reduce self-heating of the device under test (DUT) [8], 2) designing clamping circuit for fast and accurate switching transient measurement [10], 3) studying the physical mechanism and analyzing stressors' impact [11] and 4) modeling the dynamic behavior via equivalent circuit or analytical approaches [16].

However, the study of D-R_{dson} behavior under multi-stressor conditions, especially in a wide temperature range like $50 - 400$ K, has not been fully explored yet. In addition, as reported in Ref. [5], discrepancies exist among the results from different research studies due to variations in test setups. Unlike the static characterization that can be performed using a curve tracer, the dynamic characterization largely depends on specifically developed testing platforms. Despite the JEDEC standard that

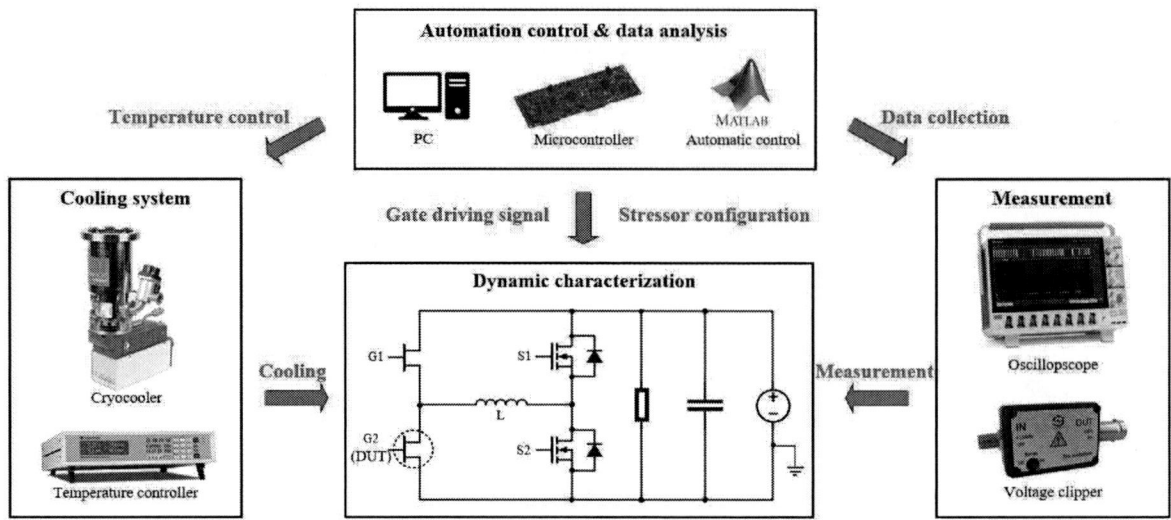

Fig. 2. Platform overview, including function modules of control, cooling, testing and measurement.

recommends double pulse test (DPT) for D-R_{dson} measurement [18], the difference in circuit parameters and PCB layout design still challenges the fair comparison among devices. On top of that, performing a comprehensive test accounting for as many stressors will significantly increase the testing workload, making it practically challenging for manual operation.

To address these issues, we propose to develop a fully automated platform for comprehensive D-R_{dson} measurement from 50 to 400 K. The platform automatically programs various stressors—temperatures, switching modes, voltages, currents, etc.—allowing comprehensive characterization of GaN devices with minimal human effort. In Section II, the platform overview is introduced, involving the manipulation of each stressor and assembly of functional modules at system level. In Section III, the design considerations of the testing circuit is discussed, in terms of both hardware and software. In Section IV, the platform is implemented to collect D-R_{dson} data, by which the impact of each stressor is analyzed. Section V concludes the work.

II. PLATFORM OVERVIEW

Fig. 2 shows the platform overview with functional module assembly. In this work, the investigated impacting factors / stressors are as follows:

- Temperature
- Switching mode
- Blocking voltage
- Drain current
- Stressing time

The configuration of each stressor is achieved by automatic control using the computer and MATLAB, which can be accessed with remote desktop. The details are given as follows.

(1) Temperature control

The cooling system comprises mainly a cryocooler and a temperature controller. The cryocooler cools down the DUT with a LHe$_2$-compressor performing closed-loop helium cycling. A silicon diode sensor is implemented to measure the temperature, and once it approaches the target, the temperature controller will activate the heater inside the cryocooler to balance the thermal process, such that the final temperature is stabilized around the target. The temperature controller is managed by a developed MATLAB software that sends out commands via serial interface, adjusting temperature target, slew rate, heater power, etc. at each test round.

(2) Switching mode and stressing time control

The control of switching mode and stressing time is achieved by configuring the sequence and duration of turn-on and turn-off of S1, S2, G1, and G2 (DUT) in Fig. 2. MATLAB generates corresponding gate control timing, and then sends commands

TABLE I. SUMMARY OF STATE-OF-THE-ART STUDIES OF D-R_{DSON}.

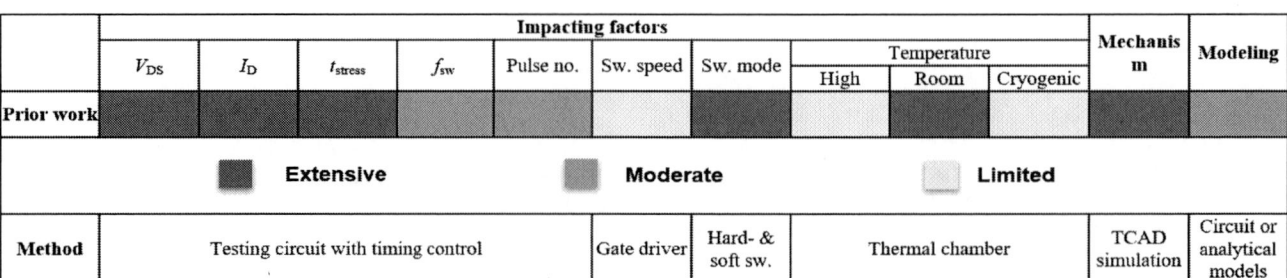

979-8-3315-1612-3/25 $31.00 © 2025 IEEE

through serial interface to the microcontroller (MCU) to conduct turn-on and -off of each transistor. The detailed timing design is presented in the next section.

(3) Blocking voltage control

A 600-V DC power supply is used to apply the bus voltage, which is controlled by the MATLAB command to accordingly adjust the output voltage. When performing the test, MATLAB leverages the pre-defined voltage output percentage and the DUT's rating to determine the actually required bus voltage.

(4) Drain current control

The current through the device is controlled by the charging voltage and duration of the load inductor in the same way as DPT [20]. During the test, MATLAB uses the pre-defined current output percentage, the bus voltage and inductance to determine the duration of turn-on and -off of each transistor to generate desired current. Details are given in the next section.

III. DESIGN CONSIDERATIONS FOR COMPREHENSIVE TEST

A. Circuit Design and timing control

According to JEDEC standard, the double pulse test (DPT) is used for D-R_{dson} extraction [18]. The circuit diagram is shown in Fig. 3. G1 and G2 are GaN HEMTs, with G2 acting as the DUT for D-R_{dson} extraction. Two additional SiC MOSFETs (S1 and S2) are used to adjust the switching modes. Each stressor (voltage, current, stressing time, etc.) is independently regulated by the computer controlling V_{DC} and switching timing of transistors.

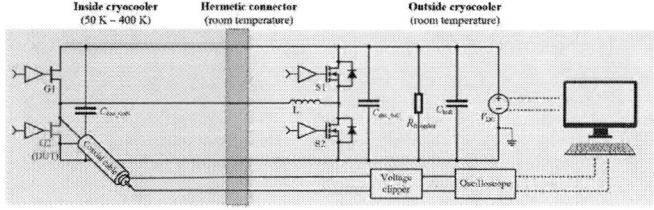

Fig. 3. Functional diagram of circuitry.

The stressors are adjusted by controlling the turn-on / -off of G1, G2, S1 and S2, which is implemented via ePWM function of the MCU. A detailed illustration is shown in Fig. 4. The scheme comprises three sections: de-trapping, trapping and stressing time. During the de-trapping section, only the DUT is turned on to avoid and unexpected bias affecting the results and to ensure the device is free of the trapping effect from the last test. Considering the time constants of de-trapping [16], [19], the duration of this section is set to be 10 seconds. As shown in Fig. 5, in hard-switching mode, after de-trapping, the operation is the same as the conventional DPT setup, where t_{charge} is determined by Eq. (1). In the trapping section, the DUT is biased by V_{DC} and I_L freewheels through G1 and S1, by changing the duration of which, different stressing time is applied in the test. On top of the extraction of D-R_{dson}, the hard-switching operation is also capable of measuring the static R_{dson} during the 1st pulse of G2, right after I_L starting increasing once S1 is turned on. The measured static R_{dson} is ensured to be free of trapping effect and hot electron effect because of the sufficiently long de-trapping time, which guarantees a fair result.

$$t_{\text{charge}} = \frac{LI_L}{V_{DC}} \quad (1)$$

In terms of soft-switching mode, however, additional pre-stressing time is required before charging the inductor, since the charging and trapping are coupled. As shown in Fig. 5, in the current charging stage, G1 and S2 are turned on to increase I_L, and simultaneously the DUT is biased by V_{DC}. Suppose the target stressing (trapping) time is t_{stress}, therefore, a pre-stressing time t_{pre}, which is determined in Eq. (2), is desired to have the DUT in the trapping section ahead of charging of I_L.

$$t_{\text{pre}} = t_{\text{stress}} - t_{\text{charge}} \quad (2)$$

By introducing t_{pre}, the stressing time and charging time can be decoupled, making it flexible to control the trapping process.

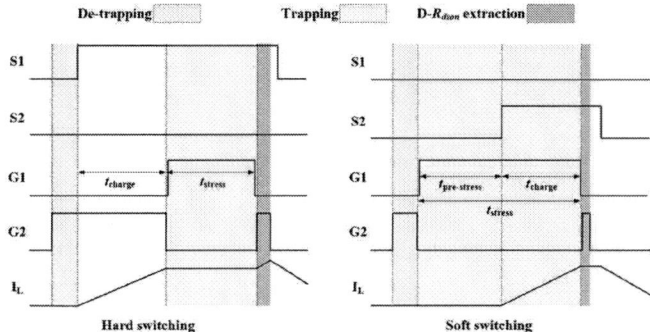

Fig. 4. Timing control scheme at hard- and soft-switching conditions.

The design methodology of DPT has been proposed in Ref. [20], which depicts the requirements for passive components such as energy storage capacitor (C_{bulk}), inductor (L), etc., as shown in Eqs. (3) and (4).

$$L \geq \frac{V_{DC(\max)} t_{sw}}{k_{\Delta i} I_{L(\min)}} \quad (3)$$

$$C_{\text{bulk}} \geq \frac{L I_{L(\max)}^2}{2 k_{\Delta v} V_{DC(\min)}^2} \quad (4)$$

Here, $k_{\Delta i}$ and $k_{\Delta v}$ are respectively the current and voltage variation percentage, and t_{sw} is switching time. For a single device, L and C_{bulk} can be determined in a straightforward way once the ranges of V_{DC} and I_L are decided, and t_{sw} is obtained from the datasheet. For a generic design purpose, however, V_{DC} and I_L may fall out of the design, and t_{sw} is also very different among devices, which may require a modification or update on the designed DPT.

For the sake of compatibility, the design is to have the capacitor bank soldered on the PCB to minimize parasitics, while the inductor is connected to the PCB via screws and ring connectors, making it more flexible for reassembly. Our approach is fixing the value of C_{bulk} and only designing L. Once the inductance in use cannot meet the requirements of $k_{\Delta i}$ and $k_{\Delta v}$, a new inductor will be redesigned and adopted. In addition, considering the pre-stressing time in Eq. (2), which ought to be a positive value. Hence, based on Eqs. (1) – (4), a new design consideration for L is proposed as follows:

979-8-3315-1612-3/25 $31.00 © 2025 IEEE   1042

$$\min\left[\frac{2k_{\Delta v}C_{\text{bulk}}V_{DC(\min)}^2}{I_{L(\max)}^2}, \frac{t_{\text{stress}}V_{DC(\min)}}{I_{L(\max)}}\right] \geq L \geq \frac{V_{DC(\max)}t_{sw}}{k_{\Delta i}I_{L(\min)}} \quad (5)$$

To perform the characterization in a wide temperature range, a design trade-off between higher reliability and lower parasitics should be considered. For instance, if only the GaN devices are placed in the cryogenic environment, long gate wires and DC bus wires have to be applied since the rest of the DPT components are at room temperature. This will ensure that the power stage functions well, but will inevitably introduce severe parasitic inductance in the gate and power loop. On the other hand, if the overall platform is moved in the cryocooler, the performance of the power stage cannot be guaranteed due to the temperature impact on the value of L and C_{bulk}. In addition, the cooling capacity of the cryocooler should be capable enough for such a large thermal load.

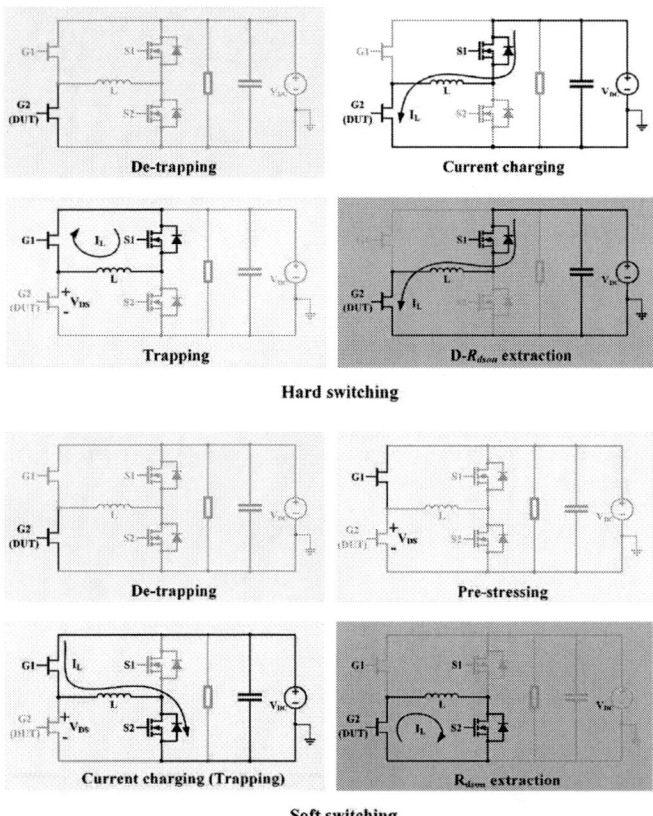

Fig. 5. Circuit operation in hard- and soft-testing modes.

In this paper, we separate the phase legs consisting of SiC and GaN into different PCBs. As shown in Fig. 3, the PCB that carries GaN devices, gate drivers and decoupling capacitors, is held inside the cryocooler, as the Gan board under test, while energy storage capacitors, inductor, SiC MOSFETs, etc. are at room temperature as the power stage, which will be detailed in the next subsection. Considering the wide temperature applied on the GaN board, the passive components should be carefully selected. It is suggested in Ref. [3] that NP0 capacitors and thin film resistors have good temperature consistency. However, the NP0 capacitors have relatively small capacitance compared to other types such as X7R. Since the energy storage capacitors are

placed at room temperature, therefore, NP0 capacitors will be only used for decoupling function which requires merely hundreds of nano-farad [20].

B. Electrothermal-mechanical Integration

To perform the cryogenic test down to 50 K, a LHe_2 cryocooler is utilized rather than LN_2. As shown in Fig. 6, dedicated design for the cryogenic test is conducted, including the vacuum enclosure, radiation shield and sample holder. The sample holder, which is made of copper, is mounted on the cryocooler's cooling station with indium as thermal interface. The radiation shield of polished aluminum is implemented on the cryocooler and surrounds the cooling station to block thermal radiation from the inner surface of vacuum enclosure. When performing the cryogenic test, a dry-type vacuum pump first operates to reduce the pressure to be ~ 10 mTorr, by which the heat convection is alleviated. Then, the compressor and cryocooler work to cool down the sample holder via the closed-loop cycling of LHe_2.

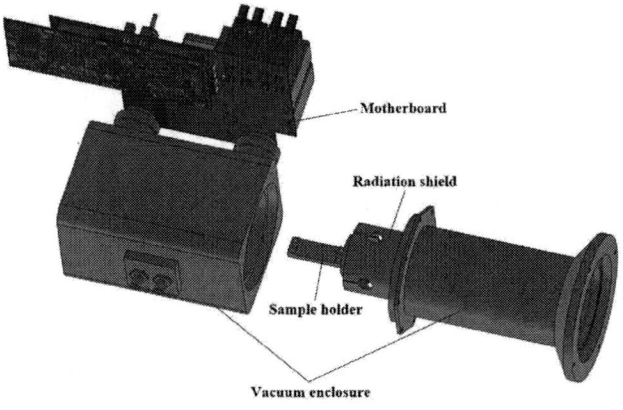

Fig. 6. Accessory design for the cryocooler.

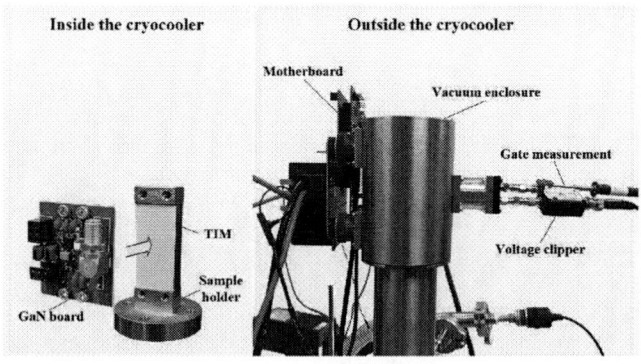

Fig. 7. Hardware implementation: DUT board, motherboard and assembly

Fig. 7 demonstrates the integration method of DPT hardware with the cryocooler setup. Inside the cryocooler, the GaN board is mounted on the sample holder made of copper, which actually acts as a cold head to cool down the device. Between the devices and sample holder, there is a thermal interface material to improve the cooling efficiency. Outside the cryocooler, to reduce the parasitics in the power loop, the motherboard carrying SiC MOSFETs, energy storage capacitors, etc. is directly fixed on the vacuum enclosure through two hermetic feedthroughs, one of which is for the gate signals and the 12-V auxiliary DC voltage, and the other is for bus voltage. A hermetic dual SMA feedthrough is adopted for the measurement of V_{GS} and V_{DS} of G2, and a voltage clipper (CLP1500V15A1) is implemented to capture the small value of V_{DS} during the ON-state to thereby extract D-R_{dson}.

C. Software Framework

Since multiple impacting factors are studied in this paper, the total number of tests N_{test} to be performed is determined as follows.

$$N_{\text{test}} = \prod n_{\sigma i} \qquad (6)$$

where n_{σ_i} is the number of testing points of the stressor σ_i. For instance, considering a testing plan that $V_{DC} = 40$ V, 60 V, 80 V, $T = 50$ K, 90 K, 120 K, ..., 400 K, $I_L = 19$ A, 28 A, 38 A, 48 A, $t_{\text{stress}} = 25$ µs, 50 µs, 100 µs, 200 µs, 400 µs, in both hard- and soft-switching modes, $N_{\text{test}} = 3 \times 14 \times 4 \times 5 \times 2 = 1680$, which is practically challenging to make it manually.

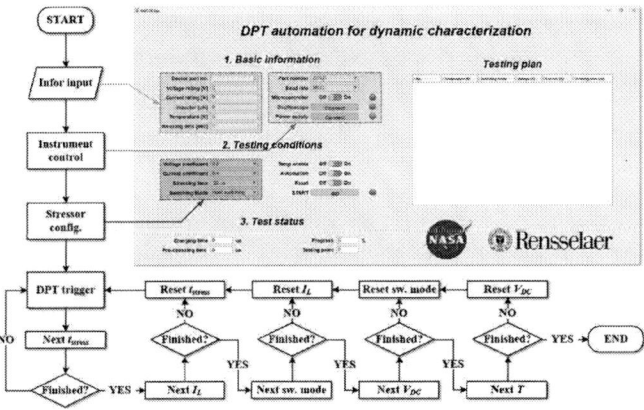

Fig. 8. Framework of the automated control.

To cope with this fact, a full automation function is essential to help save human effort and improve testing efficiency. As shown in Fig. 8, a graphical user interface (GUI) has been developed using MATLAB to realize automated configuration of stressors and measurement. The basic operation logic is: Firstly, the GUI establishes communication with the MCU, oscilloscope, DC power supply and temperature controller. Then, a table containing all required testing conditions is generated. The GUI will go through the table and convert stressors' value into commands for the MCU. For instance, if the current condition has $V_{DC} = 60$ V and $I_L = 28$ A, the GUI will use Eq. (1) to calculate t_{charge} and convert it into the comparator level for ePWM, which can be directly executed by MCU. After that, the GUI will send out the commands to all controlled instruments, setting up voltage in the DC power supply, and making the MCU generate desired gate signals for S1, S2, G1 and G2. Simultaneously, the vertical and horizontal scale and position of the oscilloscope are adjusted corresponding to V_{DC} and I_L. Once the DPT begins, the oscilloscope will be triggered to capture waveforms, which will be named with a specific format and saved. By repeating such a procedure, the platform can operate 24/7 and thousands of datasets will be collected

automatically. Roughly, performing the full test will take 18 – 20 hrs, 30% of which is for temperature setup.

IV. IMPLEMENTATION AND DATA COLLECTION

The implementation of the platform is illustrated in Fig. 9. To demonstrate the functionality of the platform, two commercial GaN e-mode HEMTs are used to build the phase leg for the GaN board. The DUT information and testing plan are given in TABLE II. The platform records waveforms of V_{ds} (through the voltage clipper) and I_L to obtain D-R_{dson} with $\frac{V_{ds}}{I_L}$. Simultaneously, V_{gs} is observed as well to check if the gate driver survives at cryogenic temperatures or not.

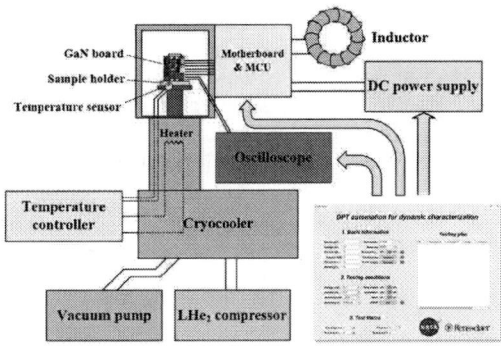

(a) Functional diagram

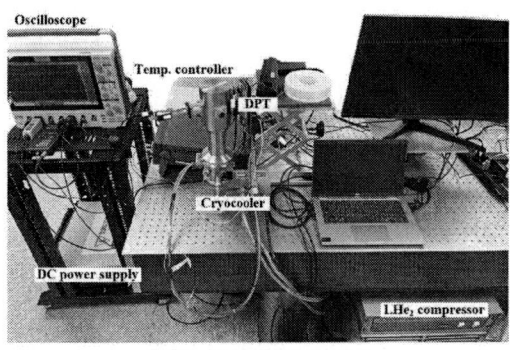

(b) Hardware implementation

Fig. 9. Implementation of the platform.

The rated gate driver output is 0 and +5 V, and the measured V_{gs} is presented in Fig. 10, showing that V_{gs} waveforms under different temperatures are almost overlapped. As temperatures vary from 50 K to 400 K, V_{gs} slightly changes between 4.97 V and 5.04 V, with a standard deviation of 0.023 V, indicating temperature impact on DUT's gate voltage is negligible.

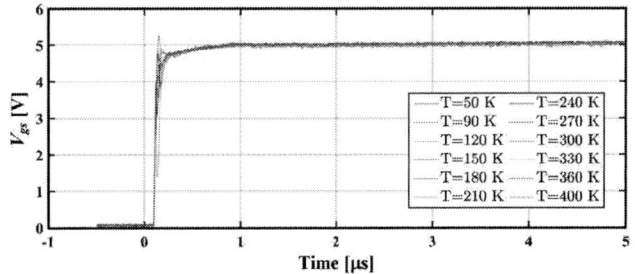

Fig. 10. V_{gs} at different temperatures (50 – 400 K).

TABLE II. DEVICE INFORMATION AND TESTING PLAN.

Package	Rated voltage [V]	Rated current [A]	Testing plan				
			T [K]	Sw. mode	V_{DC} [V]	I_L [A]	t_{stress} [µs]
QFN	200	48	50 – 400	Hard & soft	40 – 80	19 – 48	25 – 400

The raw data of D-R_{dson} are presented in Fig. 11. Simply observing from the trend of the results, D-R_{dson} under hard switching is overall higher than that at soft switching throughout the test, and is more sensitive to the applied stressors' value. This is revealed by the data dispersion, as D-R_{dson} at hard switching varies from $8 - 22$ mΩ, and at soft switching, D-R_{dson} varies from $6 - 14$ mΩ. The explanation is that due to the hot electron effect and higher switching losses at hard switching, which is related to $V_{ds} \times I_d$ at switching transient [13], [21], increasing testing voltages and currents enhances D-R_{dson} more effectively than it does at soft switching, where the hot electron effect is largely eliminated.

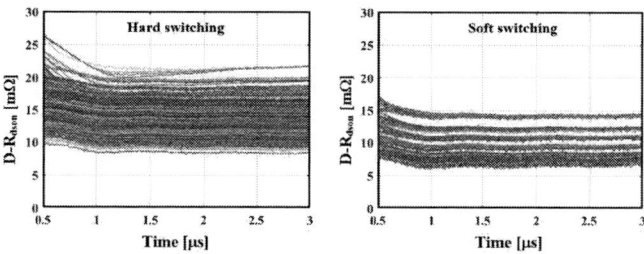

Fig. 11. Raw data of D-R$_{dson}$ at hard- and soft-switching modes, with various temperatures (50 – 400 K), voltages (40 – 80 V), currents (19 – 48 A) and stressing time (25 – 400 µs).

More quantitative comparison is made in Fig. 12, where D-R_{dson} data are averaged and normalized by the measured static R_{dson} at each temperature point, with Eq. (7).

$$\mathrm{DR}_{dson(\mathrm{norm})}(T) = \frac{1}{\Delta t} \int_{t_0}^{t_0+\Delta t} \frac{\mathrm{DR}_{dson}(T, t)}{\mathrm{SR}_{dson}(T)} dt \qquad (7)$$

where t_0 is the time after turn-on to record D-R_{dson} data ($t_0 = 0.5$ µs), Δt is the measurement window ($\Delta t = 2.5$ µs), $SR_{dson}(T)$ is the static R_{dson} as a function of temperature.

As shown in Fig. 12 (a), the static R_{dson} presents a non-monotonic variation with temperature increment: With temperature increasing, R_{dson} descends until around 180 K, and then keeps ascending. This is not common for GaN devices as they typically have the positive temperature coefficient of R_{dson} [3]. In Fig. 12 (b) – (d), normalized D-R_{dson} versus temperature is visualized at hard- / soft-switching modes and compared in terms of varying V_{DC}, I_L and t_{stress}, respectively. Overall, the impact of D-R_{dson} peaks at around 180 K, and at either higher or lower temperatures, D-R_{dson} becomes less impactful. In terms of the influence of each stressor, it is found that temperatures and switching modes play the most significant role. The hard-switching D-R_{dson} is around $2 - 4$ times of the static value, and soft-switching one is 1- 2 times of the static value, indicating the hard switching amplifies D-R_{dson} effect almost two times as soft switching does, reflecting the contribution of hot electron effect as well as switching losses. Stressors including V_{DC}, I_L and

t_{stress} basically can enhance D-R_{dson}, but are way more pronounced at hard switching, while the results under soft switching are almost the same.

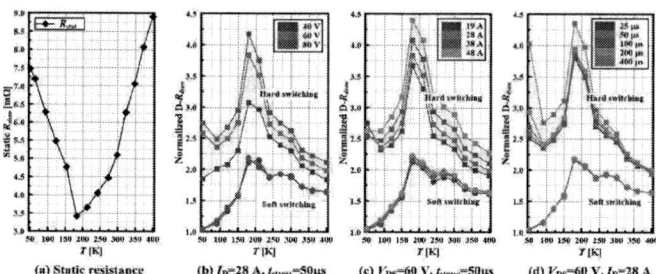

Fig. 12. D-R$_{dson}$ under different voltage, current and stressing time conditions.

The impacts of V_{DC}, I_L and t_{stress} are further quantified at cryogenic, room and high temperature conditions (respectively 50, 300 and 400 K), as shown in Fig. 13. In terms of hard-switching scenario, at room and high temperatures, separately increasing V_{DC} from 40 to 80 V and I_L from 19 to 48 A enhances D-R_{dson} by around 30% and 38% of static R_{dson}, respectively, while increasing t_{stress} has no influence on D-R_{dson}. However, at 50 K, increasing I_L has less impact on D-R_{dson} but increasing t_{stress} significantly contributes to D-R_{dson}.

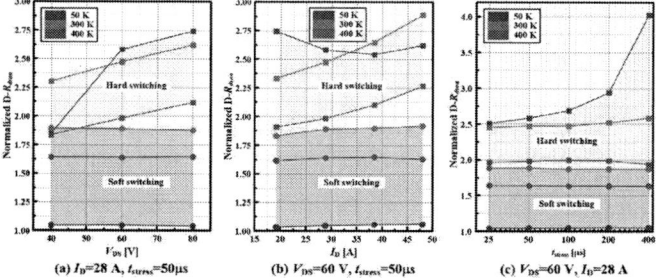

Fig. 13. Stressors' impacts on D-R$_{dson}$ at different temperatures (50 K, 300 K and 400 K).

Under soft switching, V_{DC}, I_L and t_{stress} almost imposes no impact on D-R_{dson} at either cryogenic or high temperatures, indicating the electron trapping effect during the OFF state is negligible. This is possibly due to the applied blocking voltage is not sufficient to form a strong electric field between the drain and substrate and/or the drain and gate, since the maximum V_{DC} during this test is only 40% of DUT's voltage rating. With a hypothesis that the OFF state trapping is not sufficiently activated at low voltage level, the observation that t_{stress} does not influence D-R_{dson} can be also explained. In this case, only hot electron effect contributes to D-R_{dson}.

979-8-3315-1612-3/25 $31.00 © 2025 IEEE

V. Conclusions

In this paper, an automated platform for characterizing D-R_{dson} of GaN HEMTs from 50 K to 400 K was developed. The platform features programmable stressors, including switching mode, voltage, current, and stress time, enabling comprehensive studies with high efficiency at which it automatically operates 24/7 and collects thousands of datasets within 20 hrs. Test results from a commercially available GaN HEMT reveal that D-R_{dson} effect can vary non-monotonically with temperature, peaking at 180 K. Among the influencing factors, temperature and switching modes are found to be the most pronounced to impact D-R_{dson}. At hard switching, D-R_{dson} is overall two times of that at soft switching, and it is more sensitively increased by the increment of voltages and current, especially at room and high temperatures. At soft switching, D-R_{dson} is way less impacted by the stressors and almost remains unchanged, revealing that OFF-state electron trapping is not significant with the applied stressing conditions that cannot form a strong electric field within the device. In this case, hot electron effect is the dominant origin for D-R_{dson} at hard switching. Therefore, soft switching is more preferrable for low stressing applications for the sake of mitigating D-R_{dson}.

Acknowledgment

This material is based on work supported by NASA under Award No. 80NSSC24K0179. The authors would also like to thank Mr. Richard C. Oeftering, Dr. Nicholas R. Uguccini, Dr. Brian A. Holler, and Mr. Frederick W. Van Keuls from NASA Glenn Research Center for their contributions to the design of the cryocooler platform. Additionally, we extend our gratitude to Prof. Ishwara Bhat from Rensselaer Polytechnic Institute for his assistance with the operation and maintenance of the cryogenic apparatus.

References

[1] Oeftering, Richard C. "Power Hibernation for Low-Cost Solar Powered Lunar Missions," In Future In-Space Operations (FISO) Telecon Seminar Presentations. 2023.

[2] T. P. Chow et al., "Smart Power Devices and ICs Using GaAs and Wide and Extreme Bandgap Semiconductors," in IEEE Transactions on Electron Devices, vol. 64, no. 3, pp. 856-873.

[3] H. Gui et al., "Review of Power Electronics Components at Cryogenic Temperatures," in IEEE Transactions on Power Electronics, vol. 35, no. 5, pp. 5144-5156, May 2020.

[4] R. Vetury, N. Q. Zhang, S. Keller and U. K. Mishra, "The impact of surface states on the DC and RF characteristics of AlGaN/GaN HFETs," in IEEE Transactions on Electron Devices, vol. 48, no. 3, pp. 560-566, March 2001.

[5] G. Zulauf, M. Guacci and J. W. Kolar, "Dynamic on-Resistance in GaN-on-Si HEMTs: Origins, Dependencies, and Future Characterization Frameworks," in IEEE Transactions on Power Electronics, vol. 35, no. 6, pp. 5581-5588, June 2020.

[6] K. Tanaka, M. Ishida, T. Ueda, and T. Tanaka, "Effects of deep trapping states at high temperatures on transient performance of AlGaN/GaN heterostructure field-effect transistors," Jpn. J. Appl. Phys., vol. 52, no. 4S, Apr. 2013.

[7] I. Rossetto et al., "Evidence of Hot-Electron Effects During Hard Switching of AlGaN/GaN HEMTs," in IEEE Transactions on Electron Devices, vol. 64, no. 9, pp. 3734-3739, Sept. 2017.

[8] R. Kumar et al., "H-Bridge Derived Topology for Dynamic On-Resistance Evaluation in Power GaN HEMTs," in IEEE Transactions on Industrial Electronics, vol. 70, no. 2, pp. 1532-1541, Feb. 2023.

[9] P. J. Martínez et al., "A Test Circuit for GaN HEMTs Dynamic RON Characterization in Power Electronics Applications," in IEEE Journal of Emerging and Selected Topics in Power Electronics, vol. 7, no. 3, pp. 1456-1464, Sept. 2019.

[10] G. Zulauf et al., "The Impact of Multi-MHz Switching Frequencies on Dynamic On-Resistance in GaN-on-Si HEMTs," in IEEE Open Journal of Power Electronics, vol. 1, pp. 210-215, 2020.

[11] R. Li et al., "Dynamic on-State Resistance Test and Evaluation of GaN Power Devices Under Hard- and Soft-Switching Conditions by Double and Multiple Pulses," in IEEE Transactions on Power Electronics, vol. 34, no. 2, pp. 1044-1053, Feb. 2019.

[12] T. Zhao et al., "Dynamic ON-Resistance Characterization of GaN HEMT Under Soft-Switching Condition," 2021 IEEE 8th Workshop on Wide Bandgap Power Devices and Applications (WiPDA), Redondo Beach, CA, USA, 2021, pp. 246-249.

[13] X. Geng et al., "Experimental Evaluation and Analysis of Dynamic On-Resistance in Hard- and Soft-switching Operation of a GaN GIT," PCIM Europe digital days 2020; International Exhibition and Conference for Power Electronics, Intelligent Motion, Renewable Energy and Energy Management, Germany, 2020, pp. 1-8.

[14] H. Zhu et al., "Accurate Measurement of Dynamic ON-Resistance in GaN Transistors at Steady-State," in IEEE Transactions on Power Electronics, vol. 38, no. 7, pp. 8045-8050, July 2023.

[15] Y. Wei et al., "Comprehensive Cryogenic Characterizations of a Commercial 650 V GaN HEMT," 2021 IEEE International Future Energy Electronics Conference (IFEEC), Taipei, Taiwan, 2021, pp. 1-6.

[16] K. Li et al., "Modelling GaN-HEMT Dynamic ON-state Resistance in High Frequency Power Converter," 2020 IEEE Applied Power Electronics Conference and Exposition (APEC), New Orleans, LA, USA, 2020, pp. 1949-1955.

[17] S. Li et al., "Physics-Based SPICE Modeling of Dynamic on-State Resistance of p-GaN HEMTs," in IEEE Transactions on Power Electronics, vol. 38, no. 7, pp. 7988-7992, July 2023.

[18] Dynamic On-Resistance Test Method Guidelines for GaN HEMT Based Power Conversion Devices, version 1.0, JEDEC Standard JEP173, Jan. 2019.

[19] M. C. J. Weiser, J. Hückelheim and I. Kallfass, "A Novel Approach for the Modeling of the Dynamic ON-State Resistance of GaN-HEMTs," in IEEE Transactions on Electron Devices, vol. 68, no. 9, pp. 4302-4309, Sept. 2021.

[20] Z. Zhang, B. Guo, F. F. Wang, E. A. Jones, L. M. Tolbert and B. J. Blalock, "Methodology for Wide Band-Gap Device Dynamic Characterization," in IEEE Transactions on Power Electronics, vol. 32, no. 12, pp. 9307-9318, Dec. 2017.

[21] Rossetto et al., "Evidence of Hot-Electron Effects During Hard Switching of AlGaN/GaN HEMTs," in IEEE Transactions on Electron Devices, vol. 64, no. 9, pp. 3734-3739, Sept. 2017.

A Gate Driving Scheme for GaN GIT with Enhanced Short Circuit Capability for Motor Drive Application

Zongjie Zhou
Department of Electronic and Computer Engineering
The Hong Kong University of Science and Technology
Hong Kong SAR, China
zzhoudb@connect.ust.hk

Yan Cheng
Department of Electronic and Computer Engineering
The Hong Kong University of Science and Technology
Hong Kong SAR, China
ychengbs@connect.ust.hk

Kevin J. Chen
Department of Electronic and Computer Engineering
The Hong Kong University of Science and Technology
Hong Kong SAR, China
eekjchen@ust.hk

Abstract—**Short circuit capability in power switching devices is highly desired for motor drive application, and it is a challenging issue for GaN power high-electron-mobility transistor (HEMT) devices because of concentrated heat generation and inefficient thermal dissipation that partially stems from much smaller die size. This paper proposes a current mode gate driving scheme with a shunt resistor in the gate loop for a GaN gate injection transistor (GIT) to prolong short circuit duration time and thus improve the system robustness and reliability. The shunt resistor not only can limit the short circuit peak current as a result of the source negative feedback but also can be used as a current sensing element that is indispensable for cost-sensitive motor control. The experiments demonstrate a greatly increased short circuit capability and no obvious compromise on other performances in motor drive application with the proposed gate driving scheme.**

Keywords—*short circuit capability, GaN GIT, gate driving, shunt resistor, motor drive*

I. INTRODUCTION

Recently, GaN power devices have gained much attention for its superior performance [1], particularly in motor drive applications [2], [3]. Normally-off operation is desirable in inverter systems for meeting the fail-safe requirement. Among several technology options, two kinds of approaches are commercially available in GaN HEMTs: one is to utilize a depletion mode GaN HEMT in series with a low voltage silicon MOSFET to form a cascode structure; the other is to fabricate an enhancement mode (E-mode) HEMT to provide a single-chip solution. The single-chip configuration is more suitable for hard-switching motor drive inverters, especially the low-power intelligent power module (IPM), due to its reduced packaging die counts and easier dv/dt control. Commercial E-mode GaN HEMTs deploy a p-GaN layer in the gate stack to modulate the threshold voltage (V_{TH}) to positive values, in two types of gate contact schemes. The type I features a conductive ohmic contact between the gate electrode and the p-GaN layer, referred to as current-driven gate injection transistor (GIT) [4], while type II features a highly resistive Schottky gate contact with an enlarged gate voltage swing, and is voltage driven.

Since both forward and reverse conductions are dominated by majority carriers (i.e., electrons), the GaN HEMTs are free of reverse recovery process and the related loss, leading to substantially lower switching losses and reduced electromagnetic interference (EMI). The absence of a turn-off current tail further contributes to reducing switching losses [5]. Besides, their much smaller specific on-resistance enables inverters to be packaged into a smaller-size IPM without the

need for a heatsink [6]. In addition, the high frequency switching capability and much reduced dead time would help to suppress the harmonic elements, leading to improvement in the motor efficiency and the performance of the noise and vibration.

Conversely, short circuit (SC) capability is highly desired for motor drive applications. This poses a severe challenge for GaN devices due to their heat confinement effect and much smaller die size [7]. Therefore, many efforts have been devoted to the development of ultra-fast SC detection and protection for GaN devices. A high-speed SC protection based on PCB-integrated inductive current sensor was presented in [8], but this method requires large PCB areas and extra high speed amplifiers, making it too complex to be implemented in low-cost applications. Another method involves fast detection of the voltage dip on the drain terminal, active current clamping of the gate driver, and de-saturation based fault confirmation for SC protection [9]. An overcurrent protection circuit on the basis of

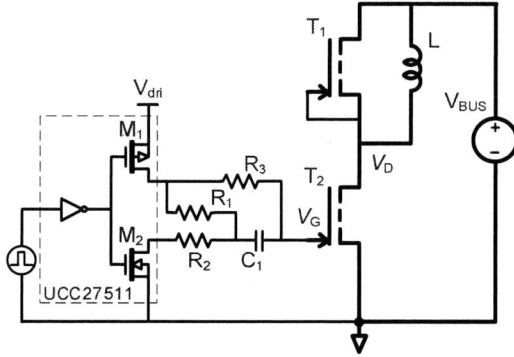

Fig. 1. Conventional gate driver for GaN GITs.

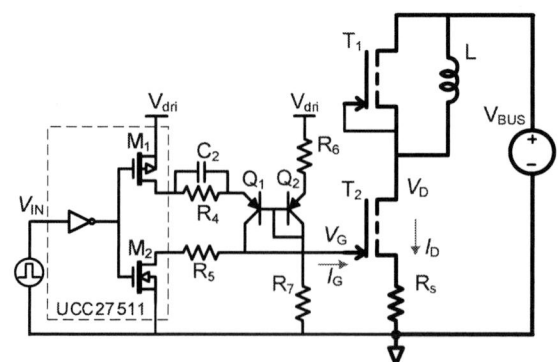

Fig. 2. Proposed GIT gate driver for motor drive applications.

This work is supported by Hong Kong RGC Strategic Topics Grant (STG) under STG3/E-602/23N.

979-8-3315-1612-3/25 $31.00 © 2025 IEEE

1047

detecting the gate voltage that is affected by the drain current is also developed for GaN GITs [10]. Nevertheless, these fast current detection and protection methods would be compromised by the blanking time (typically several hundred nanoseconds) required for noise immunity in real applications. It has been reported that a 600 V commercialized GaN device can only endure a few SC cycles before failure under specific conditions (e.g., V_D =400 V and 100 ns duration time) [7]. Hence, it is essential to improve the SC survival time of GaN devices to enhance their reliability and performance in motor drive applications.

In this paper, a novel gate driving scheme for GIT is proposed with enhanced SC capability. The gate loop incorporates a shunt resistor between the source and ground to limit the SC current by using source negative feedback. The current-mode driving is adopted to guarantee no degradation on R_{ON} during normal operation. The shunt resistor can also serve as a current sensor in motor drive applications [11], [12]. Experiments are conducted to validate the concept, and the results show a dramatic improvement in SC capability, from single-digit to over 10^4 in terms of SC events without degradation.

II. A GATE DRIVING SCHEME WITH ENHANCED SHORT CIRCUIT CAPABILITY

The metal/p-GaN/AlGaN/GaN gate-stack of GIT could be modelled as a resistor for the leaky metal/p-GaN ohmic contact in series with a p-GaN/AlGaN/GaN p-i-n heterojunction. The basic GIT gate driving requires a sufficiently large gate current during the turn-on process and only a relatively small current to hold the on-state and limit the power consumption in gate drive. This is quite different from the Schottky-type p-GaN gate HEMT that is more like a silicon MOSFET from the gate driving point of view. The conventional GaN GIT gate driver is shown in Fig. 1, where R_1 and R_2 are the instantaneous pull-up and pull-down resistors, together with the charge pump capacitor C_1, to provide relatively large source and sink current during the turn-on and turn-off transients. R_3 sets the on-state gate current.

Obviously, in the conventional gate driver as shown in Fig. 1, the short circuit capability of the GIT relies on the device itself. In order to prolong the SC duration time, a current shunt R_S is placed between the source and the ground in the gate driving loop. When the SC event occurs, the source potential will be elevated by the excessive SC current, pushing the gate voltage V_G toward V_{dri}. This trend results in a reduced gate current and thus less SC current (referred to as negative source feedback effect). The proposed gate driver scheme itself does not sacrifice inverter's efficiency, because the shunt resistors are often (in silicon device inverter) used for current sensing in motor control and over current protection, especially in low power applications (less than 500 W). In addition, the efficiency of low power motors is typically less than 90% [13], so the impact of the power consumption (typically less than 1 W) occurred in the shunt resistors is immaterial for the overall system.

In practical operation, the reduction on gate current due to the increased source voltage induced by shunt resistor Rs will lead to a compromised on-resistance or output current capability, especially under heavy loads at high temperature. To suppress

the potential risk, R_3 in Fig. 1 is replaced by a current source implemented by PNP bipolar transistor current mirror as illustrated in Fig. 2. In the proposed driving circuit, R_4 and R_6 with the same resistance, typically 50 - 100 Ω, are to reduce the mismatch between Q_1 and Q_2. The amplitude of the current source is set by R_7. The static bias current can be adjusted by tuning the ratio of the resistances of R_4 and R_6, or the ratio areas of Q_1 and Q_2 if implemented in integrated circuits. C_2 is optional to offer the current path to reduce the delay during the turn-on process. The GIT is turned off by R_5. By tuning R_7 and R_5, the turn on and turn-off speed or dv/dt can be readily programmed, so that better EMI performances can be achieved.

Fig. 3 illustrates the static turn-on IV curves of the proposed gate driver under different R_7 through LTspice DC sweep simulation. The other parameters are V_{dri} = 5 V, R_4 = R_6 = 100 Ω and BJT 2N3906 [14] is selected as the current mirror transistors. As shown in Fig.3, below the turning-point voltage, Q_1 works as a relatively perfect current source, of which the value I_{dri_on} is determined by R_7 as shown in equation (1). V_{BE_Q2} is typically 0.7 V.

$$I_{dri\_on}\left(R_7 + R_6\right) + V_{BE\_Q2} = V_{dri} \tag{1}$$

When the gate voltage V_G is higher than the turning-point, the gate driving current drops steeply as Q_1 enters the triode region. The turning-point decreases with a higher driving current because of a greater voltage drop on R_3. Adding the voltage of R_3 to the saturation voltage of Q_1 yields the headroom voltage (V_{head}). Therefore, it is not difficult to derive the relation between R_S and the short circuit current, which is given as

$$I_{SC}R_S + V_{GS\_ON} + V_{head} = V_{dri} \tag{2}$$

Where I_{SC} is the short circuit current, V_{GS_ON} is the on-state gate voltage of the GIT (typically 3.5 V). From equation (2), the SC or saturation current of GIT can be predetermined by R_S.

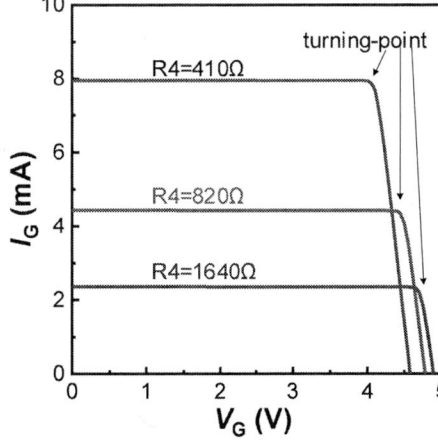

Fig. 3. Static gate I - V curves of the proposed gate driver under different R_7 for a GaN

979-8-3315-1612-3/25 $31.00 © 2025 IEEE

It is worth noting that the SC enhancement applicability of the proposed GIT gate driving is very similar to the conventional silicon device counterpart, as the proposed gate driver shares the current sensing resistor. Generally, only the low side shunt resistors need to satisfy the SC or over current protection requirements, which is valid for the most common cases, such as load SC or the output phase-to-phase SC, and other cases where shunt resistor is in the SC current path. However, if the output switching node is shorted to the ground (negative bus), by passing the shunt resistor, the high side turned-on switching would burn out. Therefore, it is recommended to incorporate source resistances to all switches for applications that demand extremely high reliability.

III. EXPERIMENTAL VERIFICATION

Experiments are carried out to verify the feasibility of the proposed gate driving scheme for GITs. The photographs of the prototype testing circuit are shown in Fig. 4. The layout mainly follows the flux cancelation methods in [15] to minimize the loop inductance that may cause unwanted ring oscillations. A 600-V/270-mΩ GIT IGLR60R340D1 [16] is selected as the device under test (DUT). A general purpose voltage-type gate driver UCC27511 [17] is adopted in the prototype and it features split pull-up and pull-down outputs. The turn-off resistance R_5 is 100 Ω. The driving loop shunt resistor R_S is 250 mΩ with the package size of 1210, and another 100 mΩ shunt resistor with the same package size and the accuracy of 1% is placed between ground and the negative bus to probe the instantaneous drain current. The other component values are given in the previous section.

Fig. 5 shows the V_G and I_G transient waveforms of GIT under different driving current with a drain voltage V_D of 0 V. V_{IN} is the gate signal with a 1-μs pulse width. The driving strength is programmed by different resistance values of R_7. The latency between input signal V_{IN} and V_G is negligible (as shown in Fig. 5), mainly introduced by the driver IC UCC27511. I_G begins with a current spike intentionally induced by the speed-up capacitor C_2 and then falls to the preset current I_{dri_on} which is slightly lower than that in Fig. 3 due to the discrepancy between actual devices and the device model. The V_G waveforms exhibit different rising slew rates and steady state levels under different driving current levels set by R_7.

Fig. 6 presents the SC waveforms of different driving methods under $V_D = 311$ V and a SC pulse of 1 μs. It is observed that the SC peak current is reduced by half when the gate driving scheme based on source negative feedback is deployed, suggesting a substantial mitigation of thermal stress. The SC

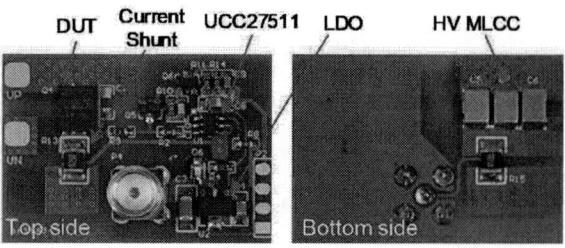

Fig. 4.   Photographs of the prototype testing circuit.

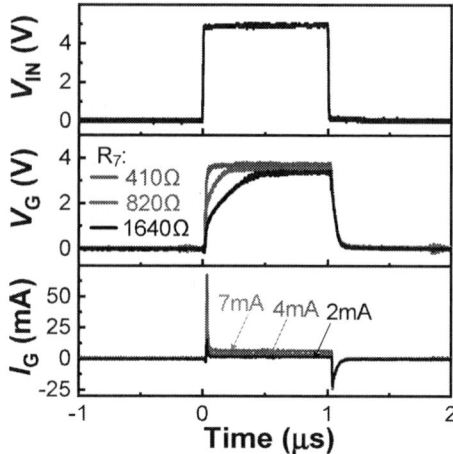

Fig. 5. V_G and I_G waveforms under different driving current ($V_D = 0$ V).

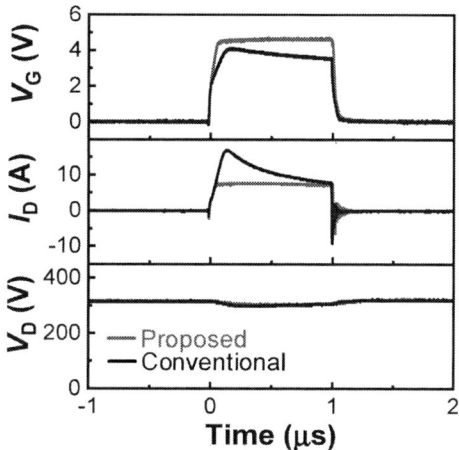

Fig. 6.   Short circuit waveforms of different driving schemes ($V_D = 311$ V, $T_P = 1$ μs).

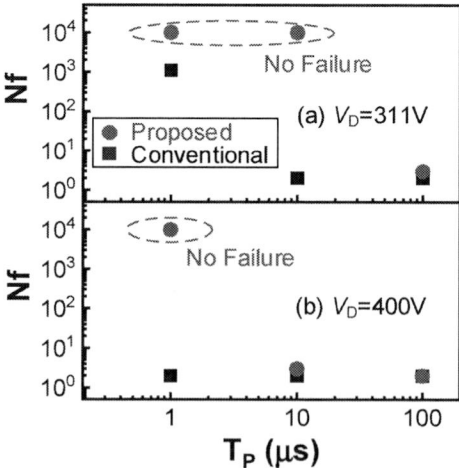

Fig. 7.   Number of short circuit cycles to failure (a) $V_D = 311$ V (b) $V_D = 400$V.

current remains unchanged with V_G approaching V_{dri} owing to the source negative feedback effect. The drain voltages show a slightly drop during the turn-on period, which is attributed to the resistance between the input bus of prototype and the external bulk capacitors.

Fig. 7 plots the number of short-circuit cycles the DUT can sustain under different drain biases, with a 5 s time interval between the SC events. By the proposed gate driving scheme, the DUT can survive 10^4 SC cycles without failure under 1μs and 10 μs SC duration time (T_P) and $V_D = 311$ V, while the DUT driven without source negative feedback only experienced single-digit SC cycles before failure under the same condition. The proposed gate driver also shows dramatic SC capability improvements under $T_P = 1$ μs and $V_D = 400$ V.

IV. CONCLUSIONS

In this work, a gate driving scheme based on source negative feedback is demonstrated for GaN GIT in motor drive applications. Experiments show reduced SC peak current and greatly increased SC cycles from single digit to over 10^4. With the proposed gate driver, the prolonged duration time under SC event allows more blanking or deglitching time for the system before taking protection action. Switching induced EMI noise or gate ringing would be allowed to settle down within a longer time window, leading to better immunity to the noisy transients.

REFERENCES

[1] K. J. Chen *et al.*, "GaN-on-Si Power Technology: Devices and Applications," *IEEE Trans. Electron Devices*, vol. 64, no. 3, pp. 779–795, Mar. 2017, doi: 10.1109/TED.2017.2657579.

[2] E. Persson and D. Wilhelm, "Gate Drive Concept for dv/dt Control of GaN GIT-Based Motor Drive Inverters," in *2020 IEEE International Electron Devices Meeting (IEDM)*, San Francisco, CA, USA: IEEE, Dec. 2020, p. 27.6.1-27.6.4. doi: 10.1109/IEDM13553.2020.9372095.

[3] T. Morita *et al.*, "99.3% Efficiency of three-phase inverter for motor drive using GaN-based Gate Injection Transistors," in *2011 Twenty-Sixth Annual IEEE Applied Power Electronics Conference and Exposition (APEC)*, Mar. 2011, pp. 481–484. doi: 10.1109/APEC.2011.5744640.

[4] Y. Uemoto *et al.*, "Gate Injection Transistor (GIT)—A Normally-Off AlGaN/GaN Power Transistor Using Conductivity Modulation," *IEEE Trans. Electron Devices*, vol. 54, no. 12, pp. 3393–3399, 2007, doi: 10.1109/TED.2007.908601.

[5] Y. Kobayashi, A. Nakagawa, M. Takei, Y. Onishi, and N. Fujishima, "Analysis for rapid tail current decay in IGBTs with low dose p-emitter," in *2012 24th International Symposium on Power Semiconductor Devices and ICs*, Jun. 2012, pp. 141–144. doi: 10.1109/ISPSD.2012.6229043.

[6] Texas Instruments Incorporated, "DRV7308 Three Phase 650V, 5A, GaN Intelligent Power Module," 2024. [Online]. Available: www.ti.com

[7] J. Sun, J. Wei, Z. Zheng, and K. J. Chen, "Short Circuit Capability Characterization and Analysis of p-GaN Gate High-Electron-Mobility Transistors Under Single and Repetitive Tests," *IEEE Trans. Ind. Electron.*, vol. 68, no. 9, pp. 8798–8807, Sep. 2021, doi: 10.1109/TIE.2020.3009603.

[8] J. Acuna, J. Walter, and I. Kallfass, "Very Fast Short Circuit Protection for Gallium-Nitride Power Transistors Based on Printed Circuit Board Integrated Current Sensor," in *2018 20th European Conference on Power Electronics and Applications (EPE'18 ECCE Europe)*, Sep. 2018, p. P.1-P.10. Accessed: Jul. 16, 2024. [Online]. Available: https://ieeexplore.ieee.org/document/8515547/?arnumber=8515547

[9] K. Wang *et al.*, "A Reliable Short-Circuit Protection Method with Ultra-Fast Detection for GaN based Gate Injection Transistors," in *2019 IEEE 7th Workshop on Wide Bandgap Power Devices and Applications (WiPDA)*, Oct. 2019, pp. 43–46. doi: 10.1109/WiPDA46397.2019.8998869.

[10] E. A. Jones, P. Williford, and F. Wang, "A fast overcurrent protection scheme for GaN GITs," in *2017 IEEE 5th Workshop on Wide Bandgap Power Devices and Applications (WiPDA)*, Oct. 2017, pp. 277–284. doi: 10.1109/WiPDA.2017.8170560.

[11] S. M. Billè, "Two or three shunt resistor based current sensing circuit design in 3-phase inverters," *STMicroelectronics application note AN4096*, 2012.

[12] D. Torres and J. Zambada, "Single-Shunt Three-Phase Current Reconstruction Algorithm for Sensorless FOC of a PMSM," *Microchip Technology application note AN1299*, 2009.

[13] C. Candelo-Zuluaga, J.-R. Riba, A. G. Espinosa, and P. T. Blanch, "Customized PMSM Design and Optimization Methodology For Water Pumping Applications," *IEEE Transactions on Energy Conversion*, vol. 37, no. 1, pp. 454–465, Mar. 2022, doi: 10.1109/TEC.2021.3088674.

[14] Semiconductor Components Industries, LLC, "2N3906 - General Purpose Transistors PNP Silicon," 2010, [Online]. Available: https://www.onsemi.com/pdf/datasheet/2n3906-d.pdf

[15] D. Reusch and J. Strydom, "Understanding the Effect of PCB Layout on Circuit Performance in a High-Frequency Gallium-Nitride-Based Point of Load Converter," *IEEE Trans. Power Electron.*, vol. 29, no. 4, pp. 2008–2015, Apr. 2014, doi: 10.1109/TPEL.2013.2266103.

[16] Infineon Technologies AG, "IGLR60R340D1 600V CoolGaN^{TM} enhancement-mode Power Transistor," 2022. [Online]. Available: www.infineon.com

[17] Texas Instruments Incorporated, "UCC2751x Single-Channel, High-Speed, Low-Side GateDriver (With4-APeakSource and 8-APeakSink)," 2014. [Online]. Available: www.ti.com

Online Detection and Reduction of the Influence of Parameter Tolerance of Paralleled SiC MOSFETs in an EV Inverter Environment

Hadiuzzaman Syed
Robert Bosch GmbH
Reutlingen, Germany
hadiuzzaman.syed@de.bosch.com

Jochen Streit
Robert Bosch GmbH
Reutlingen, Germany
jochen.streit@de.bosch.com

Robert Kragl
Robert Bosch GmbH
Reutlingen, Germany
robert.kragl@de.bosch.com

Muhammad Muneeb Alam
Robert Bosch GmbH
Reutlingen, Germany
muhammadmuneeb.alam@de.bosch.com

Alberto Martinez-Limia
Robert Bosch GmbH
Reutlingen, Germany
alberto.martinez-limia@de.bosch.com

Karl Oberdieck
Robert Bosch GmbH
Reutlingen, Germany
karl.oberdieck@de.bosch.com

Ertuğrul Sönmez
Reutlingen University
Reutlingen, Germany
ertugrul.soenmez@reutlingen-university.de

Abstract—The tolerance of SiC MOSFET parameters like threshold voltage, parasitic capacitances, and transconductance can cause large switching loss tolerances and asymmetrical current sharing in a power module. Online detection of dispersion of these parameters facilitates driving strategies that mitigate the effect of these large tolerances. A technique based on a simple slow rising gate-source voltage waveform at inverter no-load condition is used to detect the tolerances of the threshold voltage and input capacitance of the paralleled SiC MOSFETs. The detected parameters show an excellent correlation with measurements performed with a curve tracer under various operating conditions. The tolerances of the parameters detected online are used to derive a novel pre-defined driving strategy that reduces switching loss tolerance in paralleled SiC MOSFETs during an electric vehicle (EV) inverter's operation. The strategy is realized with a novel active voltage-source gate driver (AVSGD) having configurable gate resistances and delays. Double pulse test (DPT) results with the driving strategy show an improved current sharing and turn-on switching loss symmetry between the paralleled SiC MOSFETs.

Index Terms—silicon carbide MOSFET, online detection, current sharing, threshold voltage tolerance, active gate driver, driving strategy.

I. INTRODUCTION

Paralleling SiC MOSFETs is a common technique to increase the current carrying capacity of power modules in an electric vehicle (EV) inverter. However, the tolerances of the MOSFET parameters create an imbalance in current sharing and switching losses among the MOSFETs, which affects the lifetime and system reliability [1]. The detection methods as described in the literature senses voltage and current transients (di/dt or dv/dt) in the power path, therefore requiring high bandwidth and high precision electronics [2]–[5]. The detec-

tion is also combined with an active gate driver to correct the asymmetry during switching operation [3], [6]. Selectable external gate resistors employing push-pull stages are one of the techniques that have been employed previously to correct the voltage transitions [7]. Few other intelligent drivers without any active detection circuit based on turn-on and turn-off delay optimization are presented in [8]. An interesting approach is discussed in [9], where drain-source voltage or drain current transients are not sensed with high-speed circuitry in the power path, rather using a reference gate-source voltage to correct the asymmetry during normal operation. Yet, this approach still needs to deal with a high-speed transient gate-source waveform and the effect of gate loop inductance.

This work presents a detection method that directly measures the SiC MOSFET's threshold voltage and input capacitance exploiting the inverter condition when the electric vehicle is at rest and the load-current of the inverter is zero. Therefore, no broadband sensors are needed in the power path. The first part of this paper discusses the detection technique in detail and compares the resulting values with the same, extracted from the standard measurement technique with a curve tracer. A physics-based technology computer-aided design (TCAD) simulation is also shown for a deeper understanding of the device behaviour during the detection phase. The second part discusses three different strategies to correct the turn-on current asymmetry between two paralleled SiC MOSFETs based on the detected parameters. An active voltage-source gate driver (AVSGD) is developed to implement the detection and the corrective strategies. The active gate driver concept is also discussed in detail in this section. Finally, paralleled asymmetrical switches in a half-bridge configuration

979-8-3315-1612-3/25 $31.00 © 2025 IEEE

are tested under inverter operating condition in a double pulse test (DPT) setup. The results before and after implementation of the correction strategies are benchmarked and discussed.

II. PARAMETER DETECTION

A. Theory and dynamic TCAD simulation

The drain current I_D of a SiC MOSFET can be expressed as a function of gate-source voltage V_{GS} in the saturation region by the following simplified equation

$$I_D(V_{GS})\Big|_{V_{DS} > V_{GS} - V_{th}} \approx \frac{1}{2} \cdot \beta \cdot (V_{GS} - V_{th})^2 \,, \quad (1)$$

with the transconductance parameter β [10] and technological threshold voltage V_{th}. The equation can be rewritten as

$$V_{GS}(I_D)\Big|_{V_{DS} > V_{GS} - V_{th}} \approx \sqrt{\frac{2I_D}{\beta}} + V_{th} \,. \quad (2)$$

If the MOSFET is switched on very slowly with the output capacitances of the MOSFETs as capacitive load under a certain DC-link voltage $V_{DC\text{-link}}$, the plateau-start voltage $V_{GS(pl),start}$ can be considered as the threshold voltage of the SiC MOSFET as shown in (3), as $I_D \approx 0 \, A$.

$$V_{GS}(I_D \approx 0 \, A)\Big|_{V_{DS} > V_{GS} - V_{th}} \approx V_{GS(pl),start} \approx V_{th} \quad (3)$$

The following explanations relate to the half-bridge circuit depicted in Fig. 1, which employs two paralleled SiC MOS-FETs to increase the nominal load current. The drain current measurement is not used for the proposed current balancing method.

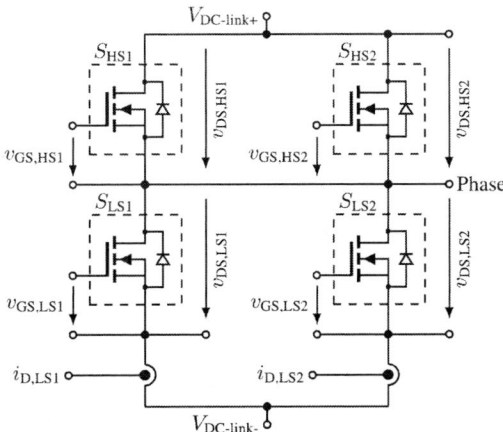

Fig. 1: Half-bridge circuit.

In Fig. 2 the switching waveforms of the low-side (S_{LS1}) and high-side (S_{HS1}) MOSFET of a half-bridge configuration (Fig. 1) are shown under zero-load condition. S_{HS1} is turned on at time $t = t_0$ so that S_{LS1} is stressed with the full $V_{DC\text{-link}}$. S_{LS1} is then turned on slowly at $t = t_2$. At $t = t_3$ the gate-source voltage $v_{GS,LS1}$ changes its slope (enters the

plateau phase) while the low-side drain-source voltage $v_{DS,LS1}$ starts to commutate simultaneously. In compliance with (3), the plateau-start voltage is defined as the threshold voltage indicator $V_{th,in}$. Between $t_3 < t < t_4$, the gate-source voltage $v_{GS,LS1}$ remains in the plateau phase (with a finite slope) and the drain-source voltages of both the switches commutate.

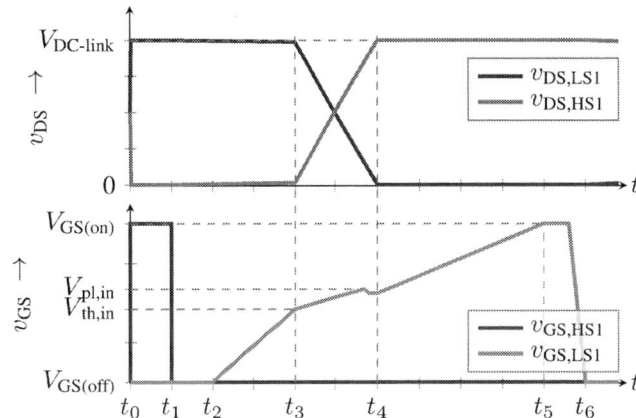

Fig. 2: Schematic waveforms of the parameter identification (PI) measurement of S_{LS1}.

A dynamic TCAD simulation is performed to understand the behaviour of the MOSFET S_{LS1} under zero-load condition. The cross section of the SiC trench MOSFET at three specific time instants defined in Fig. 2 are shown in Fig. 3.

The left most cross-section in Fig. 3 corresponds to the time $t_2 < t < t_3$ in Fig. 2, when the MOSFET is blocking $V_{DC\text{-link}}$ and the channel is completely depleted of electrons. At time instant $t = t_3$, when the gate-source waveform $v_{GS,LS1}$ enters the plateau phase, the cross-section in the middle shows that the MOSFET channel is full of electrons and

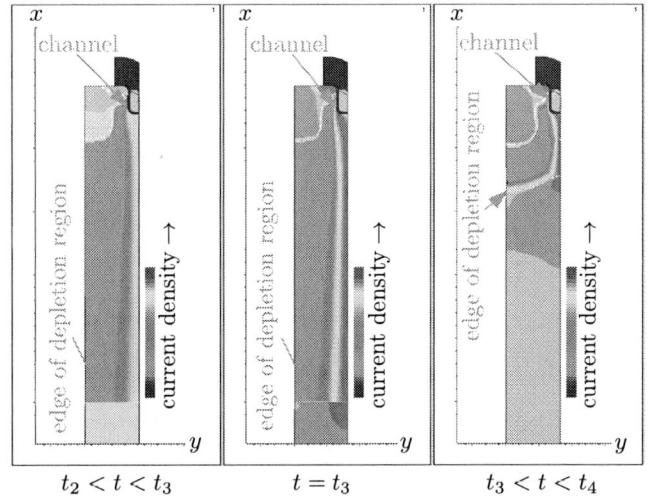

Fig. 3: Resulting cross-sections of SiC trench MOSFET from a TCAD simulation of S_{LS1} at time intervals corresponding to Fig. 2.

still blocks the full DC-link voltage. The DC-link voltage starts to commutate beyond time instant $t = t_3$, causing the displacement current from the output capacitances to flow as the MOSFET channel is now fully open. The right most cross section captures the time interval between $t_3 < t < t_4$. It shows that the depletion width is significantly reduced as $v_{\text{DS,LS1}}$ is decreasing across the MOSFET. The channel and the drift region have high conduction current density resulting from the displacement current of the output capacitances. The TCAD simulation clearly shows that the gate-source waveform $v_{\text{GS,LS1}}$ at $t = t_3$ corresponds to the threshold voltage of the MOSFET as the channel at that point is fully inverted and full of electrons. The opening of the channel at this time instant enables the drain-source voltage to commutate through the capacitive displacement current now flowing through the channel.

B. Implementation of the detection method

The detection method described in this paper is implemented with an AVSGD as discussed in section IV. The low-side switches of the paralleled SiC MOSFETs from the half-bridge circuit (S_{LS1} and S_{LS2} as shown in Fig. 1) are used as the devices under test (DUTs). During detection phase, the paralleled low-side DUTs of the half-bridge topology are sequentially turned on at no-load condition and full DC-link voltage applied by the AVSGD using a high gate resistor $R_{\text{G,PI}}$ in kΩ range. The slow rising v_{GS} waveform is sampled and analysed to detect the semiconductor characteristics parameters.

Fig. 2 describes various phases of the measured v_{GS} waveform and the corresponding drain-source voltage v_{DS}. Generally speaking, at threshold voltage indicator $V_{\text{th,in}}$ at $t = t_3$, the v_{GS} slope changes and v_{DS} starts to commutate. The $V_{\text{th,in}}$ value from the v_{GS} waveform and the corresponding time are extracted via the AVSGD. The measured threshold voltage $V_{\text{GS(th)}}$ of the same device is further determined with the Keysight B1505 curve tracer. The measurement results of $V_{\text{th,in}}$ and $V_{\text{GS(th)}}$ of eight devices are plotted in Fig. 4 for two DC-link voltages $V_{\text{DC-link}} = 400\,\text{V}$ and $V_{\text{DC-link}} = 850\,\text{V}$ as blue dots. The regression line is added in red. Consequently, the parameters $V_{\text{th,in}}$ and $V_{\text{GS(th)}}$ show a high Pearson correlation coefficient $r > 0.89$.

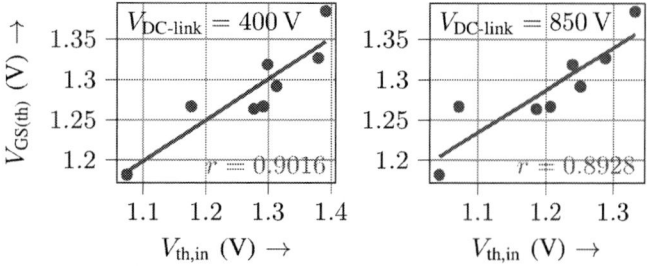

Fig. 4: Correlation of $V_{\text{th,in}}$ of parameter identification measurement and $V_{\text{GS(th)}}$ measurement with curve tracer.

The input capacitance indicator $C_{\text{iss,in}}$ is calculated by

$$C_{\text{iss,in}} \approx \frac{t_3 - t_2}{-R_{\text{G,PI}} \cdot \ln\left(1 - \dfrac{V_{\text{th,in}} - V_{\text{GS(off)}}}{V_{\text{GS(on)}} - V_{\text{GS(off)}}}\right)} \quad (4)$$

from the slope of the v_{GS} waveform between time t_2 and t_3 as shown in Fig. 2, with the turn-on ($V_{\text{GS(on)}}$) and turn-off ($V_{\text{GS(off)}}$) voltage levels of the gate driver. The measured input capacitance C_{iss} of the same device determined by a curve tracer also shows very high degree of Pearson correlation ($r > 0.9$) with $C_{\text{iss,in}}$, as depicted in Fig. 5.

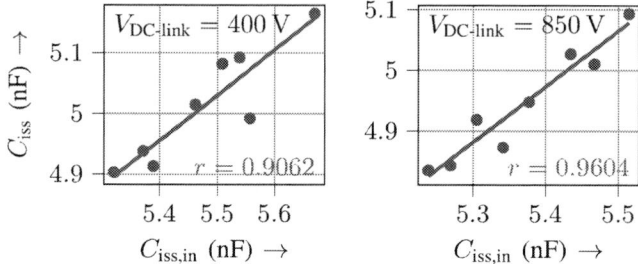

Fig. 5: Correlation of $C_{\text{iss,in}}$ of parameter identification measurement and $C_{\text{iss}}(V_{\text{DS}} = V_{\text{DC-Link}})$ measurement with curve tracer.

This approach for detecting parameters can be used for SiC MOSFETs in a power module with separate gate drivers or logical switches (containing parallel chips in common-gate configuration) of multiple paralleled power modules.

III. REDUCTION OF TURN-ON SWITCHING LOSS AND CURRENT SHARING ASYMMETRY

A simple algorithm is developed to correct the current-sharing asymmetry between the two paralleled SiC MOSFETs S_{LS1} and S_{LS2} based on the detected parameters as mentioned in the previous section. Three scenarios or correction procedures are described in detail in this section.

The current sharing asymmetry during the turn-on of the paralleled SiC MOSFETs is firstly affected by the asymmetric turn-on delay between the devices. It is defined as the time interval between the gate driver starting the charging of the gate and the moment the threshold voltage is reached. The turn-on delay Δt_{on} of a SiC MOSFET can be simply written as

$$\left.\Delta t_{\text{on}}\right|_{v_{\text{DS}} = \text{const.}} = -\ln\left(\frac{V_{\text{GS(th)}} - V_{\text{GS(on)}}}{V_{\text{GS(off)}} - V_{\text{GS(on)}}}\right) \cdot R_{\text{G}} C_{\text{iss}}(V_{\text{DC-link}}). \quad (5)$$

The input capacitance C_{iss} and threshold voltage $V_{\text{GS(th)}}$ asymmetry cause an asymmetric turn-on delay between the paralleled SiC MOSFETs with identical gate resistance and gate drive voltages.

A. First method: Compensation of turn-on delays

In the first method, the correction algorithm detects threshold voltage and input capacitance asymmetry through asymmetric turn-on delays between the switches as described in the previous section. The AVSGD in Fig. 8 in the detection mode with a pre-selected high-ohmic gate resistance $R_{G,PI}$ from the push-pull stage, records the time $\Delta t_{on,PI,i}$ of the DUT S_i when it reaches $V_{th,in,i}$ with the help of the AVSGD's integrated slope detector. Once all n parallel MOSFETs are characterized, the MOSFET with the highest turn on delay $\Delta t_{on,PI,ref}$ is chosen as reference MOSFET S_{ref}. The required turn-on delay for the DUT S_i with respect to the reference device S_{ref} is then given by

$$t_{d,on,i} = \frac{R_G}{R_{G,PI}} \cdot (\Delta t_{on,PI,ref} - \Delta t_{on,PI,i}) , \qquad (6)$$

where R_G is the gate resistance of the MOSFET S_i during normal operation. Equation (6) is valid, if all n parallel MOSFETs are switched with the same gate resistor value. The calculated delay forces the simultaneous start of the current commutation in all parallel switches.

B. Second method: Compensation of turn-on delays & gate charging time constants

The drain to source current commutation is mainly defined by the C_{iss} and β. The rise time of the current t_{rise} can be written as

$$t_{rise} = -R_G C_{iss} \cdot \ln \left(\frac{\sqrt{\frac{2I_D}{\beta}} - v_{drv} + V_{GS(th)}}{V_{GS(off)} - V_{GS(on)}} \right) . \qquad (7)$$

The second method reduces the current sharing asymmetry based on the detected difference in $C_{iss,in}$ between the paralleled SiC MOSFETs not taking differences in β into account. A suitable value of the external gate resistance R_G is calculated by

$$R_{G,i} = \frac{\tau_{ref}}{C_{iss,in,i}} . \qquad (8)$$

Accordingly, the calculated R_G aims to equalize the gate charging constant τ_i of S_i to τ_{ref} of S_{ref}, reducing the effect of C_{iss} mismatch during current commutation and thus improving the alignment of slopes of the rising drain currents between the two switches.

C. Third method: Compensation of turn-on delays, gate charging time constants & transconductances

The third correction method considers both C_{iss} and β asymmetry between the switches and selects a suitable R_G for normal operation. A predefined function between threshold voltage and β at $I_D = 80\,A$ as shown in Fig. 6 is used to estimate the variation of β between the paralleled switches. The detected $V_{th,in}$ as described in the previous section, is used

to estimate the β of the switches at $I_D = 80\,A$. The $R_{G,i}$ of S_i is calculated by

$$R_{G,i} \approx \frac{-t_{rise,ref}}{C_{iss,in,i} \cdot \ln \left(\frac{\sqrt{\frac{2I_{D\,ref}}{\beta_i}} - v_{drv} + V_{th,in,i}}{V_{GS(off)} - V_{GS(on)}} \right)} . \qquad (9)$$

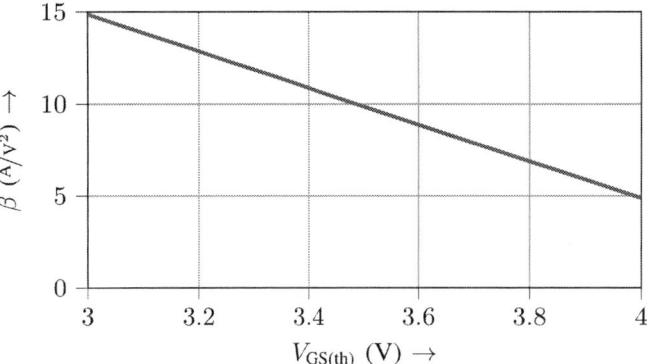

Fig. 6: β simplified as a function of $V_{GS(th)}$.

It must be noted that when a different R_G is selected to correct the di_D/dt of the DUT so that it aligns with the reference S_{ref}, the required turn-on delay to ensure the simultaneous start of the current commutation as shown in (6) must be rewritten as

$$t_{d,on,i} = \frac{R_{G,ref}}{R_{G,PI}} \cdot \Delta t_{on,PI,ref} - \frac{R_{G,i}}{R_{G,PI}} \cdot \Delta t_{on,PI,i} . \qquad (10)$$

The third method described above is captured in the flow chart in Fig. 7 and implemented with the AVSGD as described in detail in the following section. It is worthwhile to mention that this control concept can be implemented with any number of parallel SiC MOSFETs or power modules in parallel configuration, as long as each device can be controlled individually.

IV. CONCEPT AND CONSTRUCTION OF THE ACTIVE VOLTAGE-SOURCE GATE-DRIVER

The AVSGD consists of the following blocks:

- Five selectable gate resistors implemented with push-pull stages.
- An Analog-Digital-Converter (ADC) to measure the slow rising v_{GS}-waveform and a slope detector to detect its slope.

The printed circuit board (PCB) of the AVSGD is depicted in Fig. 8.

A. Push-pull stages

There are five push-pull stages which enables five separate values of turn-on and turn-off gate resistors. Among the five selectable resistance values, the high ohmic one is used for the detection of the semiconductor parameters as described in section II. The other four resistors are pre-defined through the analysis of the wafer level technology data.

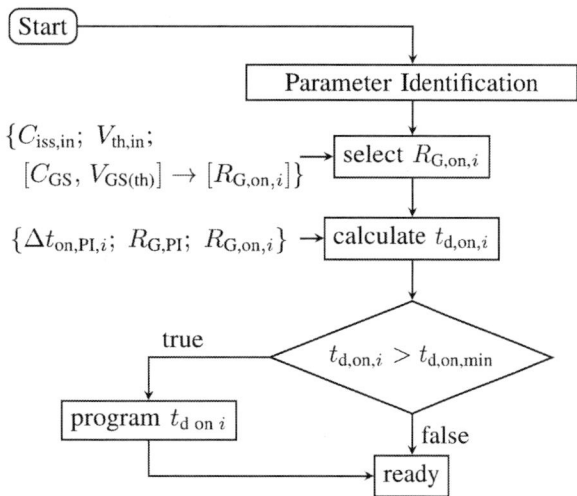

Fig. 7: Flow chart of the proposed driving concept.

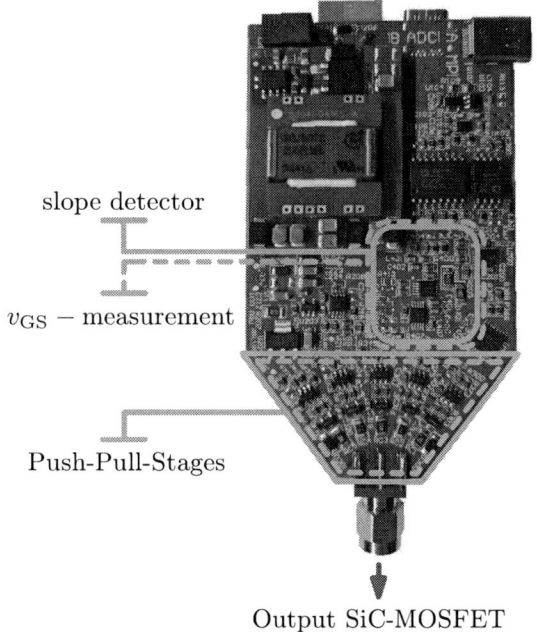

Fig. 8: Active voltage-source gate driver circuit board.

B. Sensing unit

The slope detector [7] detects the change of the slope of the gate-source waveform, dv_{GS}/dt, and records the turn on delay during the detection of the parameter routine. The time thus detected is used to calculate the required delay as discussed in section III as the first method, where a delay is used exclusively.

The ADC on the other hand is required if the second and third methods in section III need to be implemented. The required R_G is chosen from the pool of pre-defined resistors through the push-pull stage based on the current value of ADC and future calculation.

V. RESULTS AND DISCUSSION

The AVSGD, discussed in the previous section is used in conjunction with a specially designed PCB as shown in Fig. 9. The PCB is populated with two parallel 1200 V SiC MOSFETs in the high-side and the low-side of a half-bridge configuration. Similar to [11], the bare dies are bonded on a PCB substrate to enable the flexible parallel connection of devices with different MOSFET parameters. Drain current from each individual high-side and low-side MOSFET path can be measured. Each MOSFET is driven by the AVSGD separately. The paralleled MOSFETs are chosen deliberately with a threshold voltage difference $\Delta V_{GS(th)} = 760\,\text{mV}$. Added to the threshold voltage, the difference in input capacitance is $\Delta C_{iss}(V_{GS} = 0\,\text{V}, V_{DS} = 850\,\text{V}) = 210\,\text{nF}$. Tab. I shows the respective $V_{GS(th)}$ and C_{iss} of the paralleled MOSFETs. The device S_{LS1} has 26.03 % higher $V_{GS(th)}$ and 4.25 % higher C_{iss} respectively compared to S_{LS2}. This is a hypothetical scenario; in reality, chips to be parallelized are chosen with little parameter difference. Yet, in this case, the aim is the verification with the worst-case parallel connection.

TABLE I: Characteristics of S_{LS1} and S_{LS2} from curve tracer measurements.

Parameter	S_{LS1}	S_{LS2}	Difference	Δ/S_{LS1}
$V_{GS(th)}(I_D = 25\,\text{mA})$	3.68 V	2.92 V	0.76 V	20.65 %
$C_{iss}(V_{DS} = 850\,\text{V})$	5.15 nF	4.94 nF	0.21 nF	4.08 %

Tab. II shows the measurement results of the proposed in-situ PI-method with the AVSGD's integrated ADC. As explained in section II-B those parameters are different from those determined by a curve tracer measurement but show a high degree of correlation.

TABLE II: Characteristics of S_{LS1} and S_{LS2} from parameter identification measurements with AVSGD's ADC.

Parameter	S_{LS1}	S_{LS2}	Difference	Δ/S_{LS1}
$V_{th,in}$	1.79 V	1.18 V	0.61 V	34.08 %
$C_{iss,in}$	6.94 nF	6.69 nF	0.25 nF	3.60 %
$\Delta t_{on,PI}$	7.12 µs	6.41 µs	0.71 µs	9.97 %

Fig. 9: Half-bridge circuit board.

A typical operating point of the automotive inverter is chosen for the DPT with $V_{DC\text{-link}} = 850\,\text{V}$, $I_{load} = 175\,\text{A}$ and $T_j = 25\,°\text{C}$.

979-8-3315-1612-3/25 $31.00 © 2025 IEEE

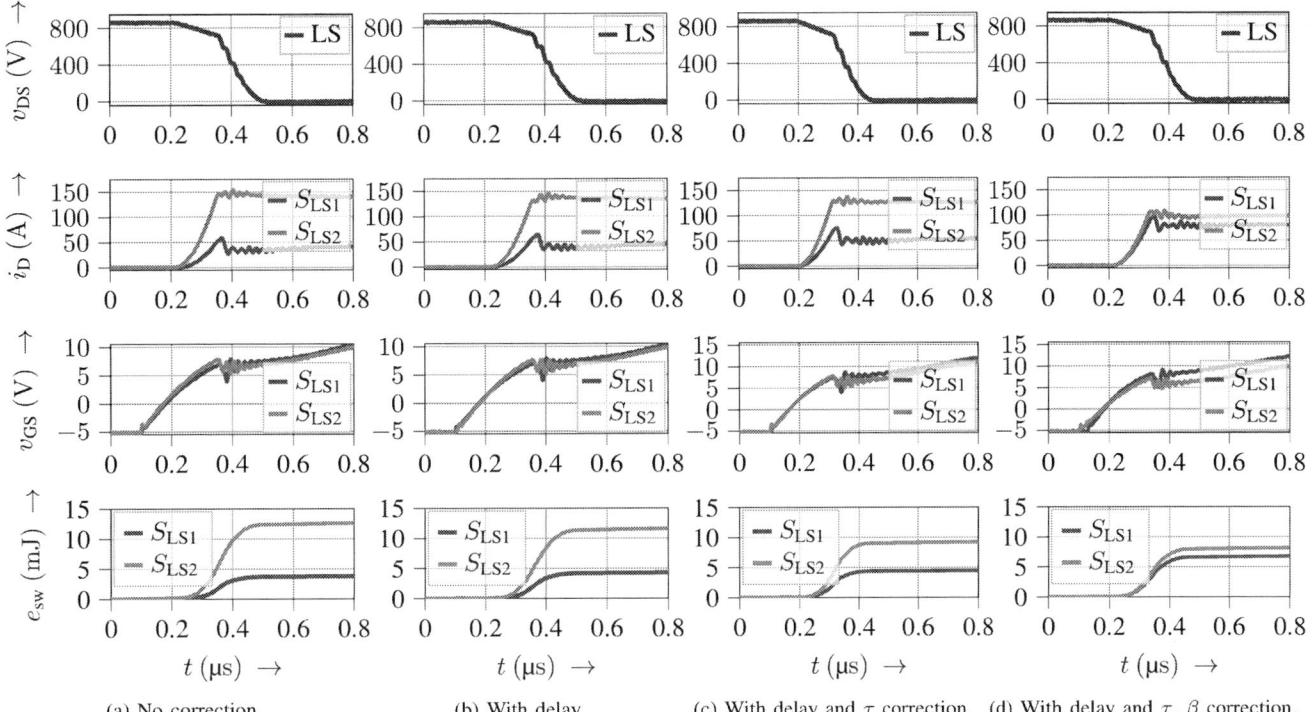

(a) No correction.   (b) With delay.   (c) With delay and τ correction.   (d) With delay and τ, β correction.

Fig. 10: Double pulse measurements with active turn-on at $V_{\text{DC-link}} = 850\,\text{V}$, $I_{\text{load}} = 175\,\text{A}$ and $T_{\text{j}} = 25\,°\text{C}$ with and without the proposed driving corrections.

A. Measurement: Reference (no correction)

As shown in Fig. 10(a) and Tab. III the reference measurement is performed without any correction measures and all the MOSFETs having the same $R_G = 39\,\Omega$. Due to the threshold voltage, input capacitance and transconductance differences, the switching loss asymmetry $\Delta E_{\text{sw}}/E_{\text{sw,tot}}$ between S_{LS1} and S_{LS2} is around $54\,\%$ (see Tab. VII).

TABLE III: AVSGD parameters without any correction.

Parameter	S_{LS1}	S_{LS2}
R_G	$39\,\Omega$	$39\,\Omega$
$t_{\text{d,on}}$	$0\,\text{ns}$	$0\,\text{ns}$

B. Measurement: First method (compensation of turn-on delays)

The corrective measures implemented according to the first method as described in section III calculate a relative delay of $9.8\,\text{ns}$ between the switches by usage of (6), while using the same R_G value, as is shown in Tab. IV. Fig. 10(b) shows that the delay forces simultaneous start of the current commutation in the two switches. The switching loss asymmetry between the switches reduced slightly to $45.86\,\%$ (see Tab. VII).

C. Measurement: Second method (compensation of turn-on delays & gate charging time constants)

The corrective measure that considers delay and different R_G values for both the switches according to the second

TABLE IV: AVSGD parameters with delay correction.

Parameter	S_{LS1}	S_{LS2}
R_G	$39\,\Omega$	$39\,\Omega$
$t_{\text{d,on}}$	$0\,\text{ns}$	$9.8\,\text{ns}$

method in section III is shown in Fig. 10(c). The parameters are calculated by using (8) and (10) and are depicted in Tab. V. The use of different R_G values reduces the asymmetry in the $\text{d}i_{\text{D}}/\text{d}t$ phase but as the transconductance asymmetry is not considered in this case the energy difference is reduced further to $35.14\,\%$ (see Tab. VII).

TABLE V: AVSGD parameters with delay and τ correction.

Parameter	S_{LS1}	S_{LS2}
R_G	$30\,\Omega$	$33\,\Omega$
$t_{\text{d,on}}$	$0.7\,\text{ns}$	$0\,\text{ns}$

D. Measurement: Third method (compensation of turn-on delays, gate charging time constants & transconductances)

The last correction measure recalculates the value of the R_G and delay based on the procedure described as the third scenario in section III and additionally considering the β asymmetry between the switches; this is depicted in Fig. 10(d), the AVSGD parameters are shown in Tab. VI. Those parameters were calculated with the help of (9) and (10). After the

979-8-3315-1612-3/25 $31.00 © 2025 IEEE

correction procedure is implemented, the asymmetry reduces to only 10% (see Tab. VII), almost equalizing the sharing of the current and the switching losses between the switches.

TABLE VI: AVSGD parameters with delay, τ and β correction.

Parameter	S_{LS1}	S_{LS2}
R_G	$27\,\Omega$	$39\,\Omega$
$t_{d,on}$	$26\,ns$	$0\,ns$

TABLE VII: Switching energy distribution between S_{LS1} and S_{LS2} with different correction methods.

Correction	$E_{sw,LS1}$	$E_{sw,LS2}$	ΔE_{sw}	$\Delta E_{sw}/E_{sw,tot}$
None	$3.81\,mJ$	$12.61\,mJ$	$8.80\,mJ$	0.54
Delay	$4.32\,mJ$	$11.64\,mJ$	$7.32\,mJ$	0.46
Delay & $R_G(\tau)$	$4.44\,mJ$	$9.25\,mJ$	$4.81\,mJ$	0.35
Delay & $R_G(\tau,\beta)$	$6.76\,mJ$	$8.19\,mJ$	$1.43\,mJ$	0.10

VI. CONCLUSION

The threshold voltage and input capacitance tolerances of the paralleled SiC MOSFETs are extracted using a novel technique compared to the state of the art, where broadband sensors in the power path are used. A dynamic TCAD simulation is also performed to understand the device behaviour in the detection phase. The detection technique spares high frequency, high resolution detection circuitry in the power path during normal inverter operation. The detected parameters show high degree of correlation with correlation coefficients $r > 0.9$ with a standard curve tracer measurement. Thus, the detected threshold voltage can also be used for online aging monitoring of the chip or module.

A novel AVSGD is developed to implement the detection of the tolerances of the parameters. The AVSGD implements three correction scenarios with added complexities using its programmable delay and selectable pre-defined gate resistors. The asymmetry in the switching energies $\Delta E_{sw}/E_{sw,tot}$ was reduced to only 10% with the third method for current balancing compared to 54% asymmetry without any correction measure.

ACKNOWLEDGMENT

This work is supported by the H2020 - KDT JU programme of the European Union under the grant of the TRANSFORM project 'Trusted European SiC Value Chain for a greener Economy' (KDT Grant No. 101007237).

REFERENCES

[1] J. Ewanchuk, J. Brandelero, and S. Mollov, "Lifetime extension of a multi-die sic power module using selective gate driving with temperature feedforward compensation," in *2017 IEEE Energy Conversion Congress and Exposition (ECCE)*, 2017, pp. 2520–2526. DOI: 10.1109/ECCE.2017.8096480.

[2] J. Henn, C. Fronczek, and R. W. De Doncker, "Design of sensors for real-time active electromagnetic-emission control in sic traction inverters," in *PCIM Europe digital days 2021; International Exhibition and Conference for Power Electronics, Intelligent Motion, Renewable Energy and Energy Management*, 2021, pp. 1–7.

[3] J. Henn, L. Heine, and R. W. De Doncker, "A high bandwidth active sic gate driver for dynamic adjustment of electromagnetic emissions in electric vehicles," in *PCIM Europe digital days 2020; International Exhibition and Conference for Power Electronics, Intelligent Motion, Renewable Energy and Energy Management*, 2020, pp. 1–7.

[4] K. Oberdieck, C. Lüdecke, G. Engelmann, and R. W. De Doncker, "Boost stage for gate current injection during igbt turn-on event," in *2017 IEEE Southern Power Electronics Conference (SPEC)*, 2017, pp. 1–6. DOI: 10.1109/SPEC.2017.8333577.

[5] C. Lüdecke, G. Engelmann, K. Oberdieck, D. Bundgen, and R. W. De Doncker, "Experimental comparison of voltage and current source gate drivers for igbts," in *2017 IEEE 12th International Conference on Power Electronics and Drive Systems (PEDS)*, 2017, pp. 460–466. DOI: 10.1109/PEDS.2017.8289280.

[6] S. Beushausen, F. Herzog, and R. W. De Doncker, "Gan-based active gate-drive unit with closed-loop du/dt-control for igbts in medium-voltage applications," in *PCIM Europe digital days 2020; International Exhibition and Conference for Power Electronics, Intelligent Motion, Renewable Energy and Energy Management*, 2020, pp. 1–8.

[7] M. Laumen, R. Kragl, C. Lüdecke, and R. W. De Doncker, "Closed-loop dv/dt control of sic mosfets yielding minimal losses and machine degradation," in *2020 IEEE Transportation Electrification Conference & Expo (ITEC)*, 2020, pp. 414–419. DOI: 10.1109/ITEC48692.2020.9161544.

[8] C. Lüdecke, M. Laumen, and R. W. De Doncker, "Balancing unequal temperature distributions of parallel-connected sic mosfets using an intelligent gate driver," in *2021 IEEE 8th Workshop on Wide Bandgap Power Devices and Applications (WiPDA)*, 2021, pp. 299–304. DOI: 10.1109/WiPDA49284.2021.9645080.

[9] J. Ao, Z. Wang, J. Chen, L. Peng, and Y. Chen, "The cost-efficient gating drivers with master-slave current sharing control for parallel sic mosfets," in *2018 IEEE Transportation Electrification Conference and Expo, Asia-Pacific (ITEC Asia-Pacific)*, 2018, pp. 1–5. DOI: 10.1109/ITEC-AP.2018.8433284.

[10] B. J. Baliga, *Fundamentals of Power Semiconductor Devices*. Springer International Publishing, 2019. DOI: 10.1007/978-3-319-93988-9.

[11] F. Jülich, R. Kragl, K. Oberdieck, K. Spanos, and R. W. De Doncker, "Fast eme characterization of bare die sic-mosfets," in *PCIM Europe 2024; International Exhibition and Conference for Power Electronics, Intelligent Motion, Renewable Energy and Energy Management*, 2024, pp. 1380–1387. DOI: 10.30420/566262190.

Dynamic Current Sharing Issues with Paralleling SiC Power MOSFETs

Ching-Yao Liu
Department of Mechanical Engineering
National Yang-Ming Chiao-Tung University
Hsinchu, Taiwan
liucy721.en10@nycu.edu.tw

Chen-Chan Lee
Department of Electrical Engineering
National Taiwan University
Taipei, Taiwan
f07921029@ntu.edu.tw

Jih-Sheng Lai
Future Energy Electronics Center
Virginia Tech
Blacksburg, VA, USA
laijs@vt.edu

Abstract— **This work comprehensively evaluates the key factors that impact the dynamic current sharing of paralleling silicon carbide (SiC) MOSFETs at the phase-leg circuit level. The power device matching is necessary and is well-known method to improve current balance. Stray inductance differences in the power loop, gate drive loop, and printed circuit board (PCB) layout are also well-known key factors. In addition to the above conventional sorting and passive matching methods, this paper proposes additional active matching approach by using the negative gate-off voltage, which can not only eliminate switching noise induced false turn on, but also help current sharing with gating signal delay matching similar to adjusting gate resistance. Impact of all key factors have been verified through experimental results.**

Keywords—SiC MOSFET, Paralleling, Current Sharing, Dynamic Current, Power Modules

I. INTRODUCTION

There are three levels of paralleling: (1) chip level, (2) device level, (3) phase-leg level, and (4) combination of any of above levels. In terms of device gate structure, there are three typical different device types including trench, trench-assisted planar, and planar [1-3]. Due to negative temperature coefficient (NTC) effect of the threshold voltage (V_{th}), paralleling SiC MOSFETs requires substantial efforts of sorting and matching at the chip level [4]. At the device level, additional effort will be at the PCB trace and path balancing [5]. At the phase-leg level, further challenges are gate driving delay matching [6].

The chip level paralleling is typically performed by the chip manufacturer, and the end result is a high-power module which makes it convenient for end users to put the circuit together. The main issue is the high cost associated with the packaging if the quantity is relatively small, and the nonrecurring engineering fee is high. Additional unseen issue is the package related parasitic components due to wire bonding and pin connections inside the module [7].

The device level paralleling can be performed by circuit designers using the commercially available discrete devices. With multiple devices in parallel, the gate driving current is also multiplied, and the output stage of the gate driver typically requires additional amplifying stages such as a pair of totem-pole transistors [8]. The phase-leg level paralleling is to have individual gate drive circuits for each phase leg to avoid excessive gate driving loop length that tends to cause different switching delays between devices.

The combination of device paralleling and phase-leg level paralleling are mainly for large current requirements [9]. In that case, one can increase number of cells per device and/or duplicate multiple sets of paralleled devices or phase legs that are driven by one gate drive circuit. An example case is the Tesla Model 3 traction motor drive inverter [10]. Its custom-made device or module consists of two SiC chips in parallel. A total of four modules are paralleled to form a switch. For such a multilevel paralleling to ensure equal current sharing is nontrivial due to chip size limit, gate driving capability limit, gate driving path and PCB layout balancing, and power loop parasitic delays [11].

This paper intends to quantify the impact of factors that cause dynamically unbalanced current distribution in SiC MOSFET paralleling. Three different types of gate structures were evaluated. A multiple phase-leg circuit board is designed for double-pulse testing (DPT). Experimental results will show the unbalanced current sharing waveforms due to different factors such as the threshold voltage (V_{th}), device type, negative turn-off voltage, gate driver delays, and PCB layout, etc. The bandwidth of voltage and current measurement will also be discussed to show how to achieve high bandwidth measurement while not injecting much power loop parasitic to avoid waveform corruption by the switching induced noises. Through extensive measurement results, the use of negative gating voltage is also proposed as a potential approach for dynamic current sharing.

II. ANALYSIS OF KEY FACTORS AFFECTING DYNAMIC CURRENT SHARING

Key factors of dynamic current sharing include mismatched V_{th}, gate driver delays, gate drive loop, PCB traces, and power loop parasitic. The negative temperature coefficient of the threshold voltage comes from device trapping and de-trapping that cause the shift of the transfer characteristic may be alleviated by sorting, but most other factors are all related to the circuit components and layout parasitic, which would also need sorting and matching.

Fig. 1(a) depicts the schematic circuit diagram of a typical paralleled test circuit with two sets of gating circuits to drive two devices under test, M_1 and M_2. The top two devices serve as freewheeling diodes to circulate the inductor current so their gate and source are shorted individually. Fig. 1(b) shows the switching timing under turn-on and -off conditions. Since the drain-source current i_{ds} starts rising after the gate-source voltage v_{gs} reaches V_{th} during turn on, any mismatch of V_{th} and delay of v_{gs} will result in different current rise rates. Similar mismatch will also cause different current fall rates during turn-off dynamic. In this example diagram, i_{ds1} shares more current during turn on because it may have a smaller V_{th}. During turn off, the small V_{th} delays i_{ds1} falling, so i_{ds2} drops first, and thus transferring the inductor current to i_{ds1}, which clearly shows a current bump. Overall, the smaller V_{th} device shares more current under both turn-on and turn-off transients.

979-8-3315-1612-3/25 $31.00 © 2025 IEEE

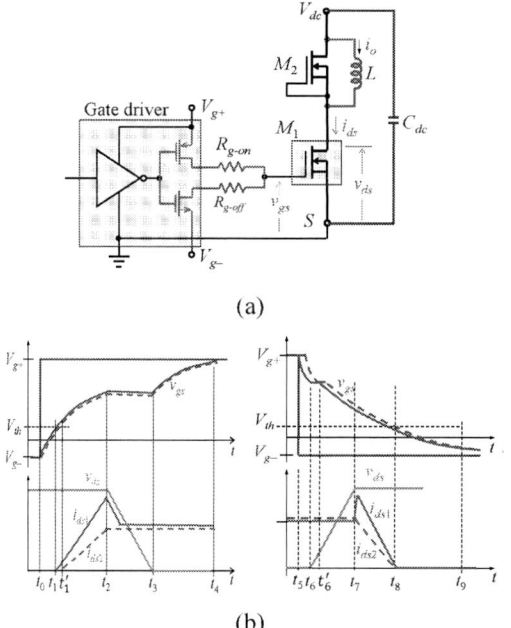

(a)

(b)

Fig. 1.   (a) A typical DPT circuit, (b) MOSFET device switching timing diagram during turn-on and -off dynamics.

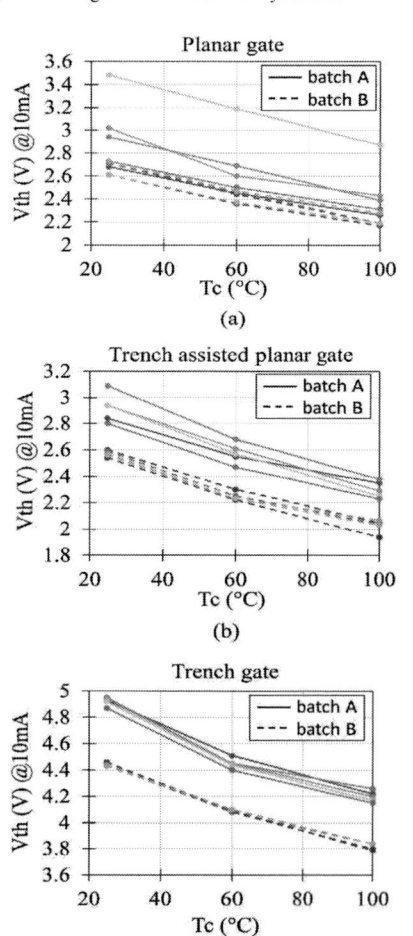

Fig. 2.   Threshold voltage test with NTC effect using (a) planar gate devices, (b) trench-assisted planar devices, and (c) trench devices.

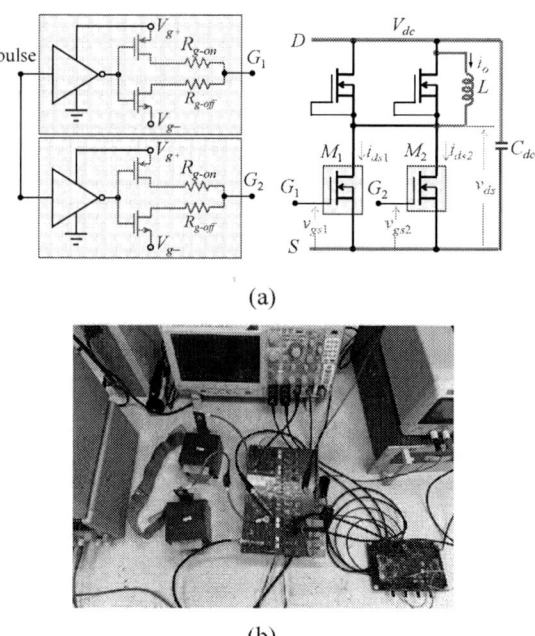

(a)

(b)

Fig. 3.   DPT with (a) two phase-legs gating circuit and (b) its experiment setups.

III. EXPERIMENTAL SETUP AND RESULTS

A. Static V_{th} Sorting Under Different Temperature Conditions

Before performing dynamic pulse testing, three different gate structure devices were sorted to see their V_{th} variation among 10 samples. Fig. 2 shows their test results with V_{th} measured at I_D =10 mA for all devices. The solid line and dashed line represent devices from different batches. In planar gate devices, shown in Fig. 2(a), one of the solid-line samples shows significantly deviated V_{th}, which can be a good sample for dynamic current sharing comparison under different V_{th} conditions. The V_{th} reduction rate as a function of temperature is –0.52 V/°C in average. In Fig. 2(b), the two batches of trench-assisted devices show two distinguishable V_{th} group among them with approximately 0.2 V between the two groups. This indicates that the two batches of devices might have come from different processes. The V_{th} reduction rate as a function of temperature is–0.60 V/°C in average. In Fig. 2(c), the two batches show even more distinctive gap on their V_{th} with about 0.5 V difference out of less than 5-V V_{th} under the room temperature and the difference maintains near constant throughout the entire temperature range. The V_{th} reduction rate as a function of temperature is –0.76 V/°C in average. It is also observed that devices from two different batches would have higher deviation which makes current balancing unignorable.

B. Dynamic Switching Experimental Setup

Fig. 3(a) depicts the double pulse test circuit diagram with two separate gate drive signals G_1 and G_2 controlling two gate source voltages v_{gs1} an v_{gs2}. Fig. 3(b) shows the picture of the experimental setup. The current is measured by a home-made toroid-core current transform (CT) with its output tied to a 2 GHz bandwidth current viewing resistor (CVR). The gate-source signal v_{gs} is measured with a 1-GHz probe, and the drain-source voltage v_{ds} is measured with an 800-MHz non-isolated high-voltage probe.

979-8-3315-1612-3/25 $31.00 © 2025 IEEE

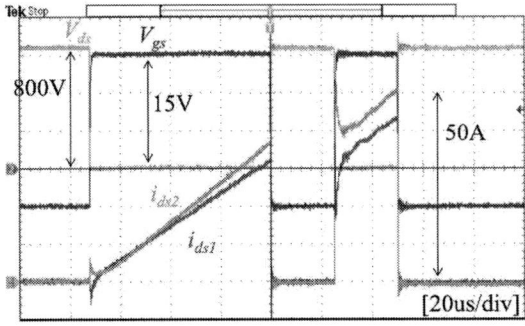

Fig. 4.   Current imbalance results with mismatched V_{th} devices.

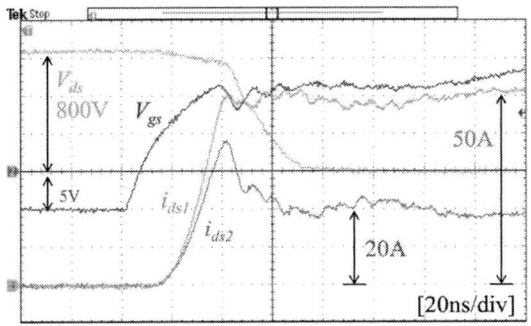

Fig. 5.   Exploded view for mismatched V_{th} under turn-on condition

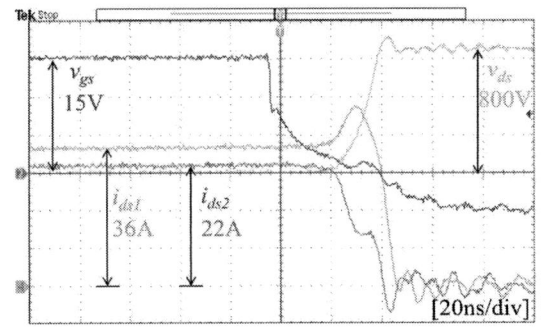

Fig. 6.   Exploded view for mismatched V_{th} under turn-off condition.

Fig. 4 illustrates the measured voltage and current waveforms of two paralleled planar-gate SiC MOSFETs under highly mismatched V_{th} condition. The two devices with highest and lowest V_{th} in Fig. 2(a) were selected for this test. This extreme case is to indicate that if the devices were not sorted before paralleling a severe current imbalance will occur. The result can be explained in below exploded views.

C. Current sharing imbalance due to V_{th} Mismatch

Figs. 5 and 6 show the exploded view of mismatched current under turn-on and -off transients. In this example diagram, i_{ds1} shares more current during turn on because it has a lower V_{th}, and thus turning on first to carry more current. In such a short turn-on period, the current was split into 50 A and 20 A between i_{ds1} and i_{ds2}, respectively. In other words, it is 70% versus 30% split, as indicated in Fig. 5.

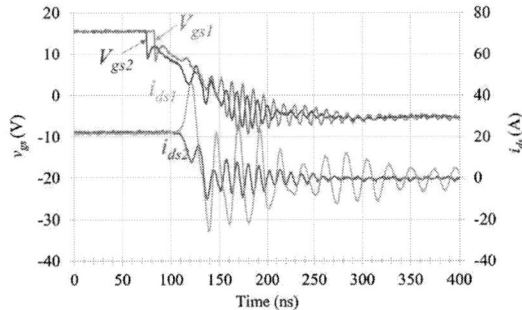

Fig. 7.   Current imbalance impact due to mismatched gate-off delay.

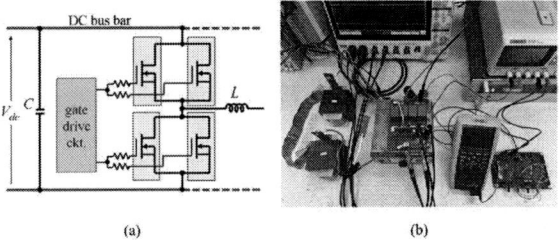

Fig. 8.   DPT with (a) two devices in parallel under one gating circuit driving both devices and (b)its experiment setups.

During turn off, with a higher V_{th}, to i_{ds2} drops first, and thus transferring more current to i_{ds1} and resulting a current bump on i_{ds1}. The detailed turn-off current sharing waveforms can be illustrated in Fig. 6. Before turning off, the total current was split into 36 A and 20 A between i_{ds1} and i_{ds2}, respectively. After turn off, i_{ds1} peaks to 48 A before dropping to zero because i_{ds2} drops early and transfers more current to i_{ds1}. Overall, the device with a lower V_{th} will incur significantly higher current during both turn-on and -off transients as explained in an early section and verified with hardware experiment.

Fig. 6 depicts a case with timing delay due to the gate driver IC mismatches under the turn-off conditions. The gate driver IC used in this test in Infineon 1ED3124MU12H. The datasheet indicated there is a 15 ns typical tolerance. margin of error for each component. In this specific set of ICs, there is 10-ns mismatch. Fig. 7 clearly indicates a significant difference between two device current. The gate drive waveforms indicate that v_{gs2} drops first, so i_{ds2} also drops before i_{ds1}, which transfers a big portion of load current to i_{ds1}. The test indicates that gate driver turn-on and -off delays also need to be matched before putting in the circuit.

D. Dynamic Current Sharing Varies with the Negative Gate Drive Voltage

Traditionally, the gate drive resistance was used to adjust the gating delay, and the negative gate drive voltage was used to prevent the noise during switching transition. There is a possibility that the negative gate voltage can help current balance especially during turn-off, because it can create an artificial delay and gate current slew rate which may coincidently match the timing delay for a pair of devices with the same or nearly the same V_{th}.

979-8-3315-1612-3/25 $31.00 © 2025 IEEE   1060

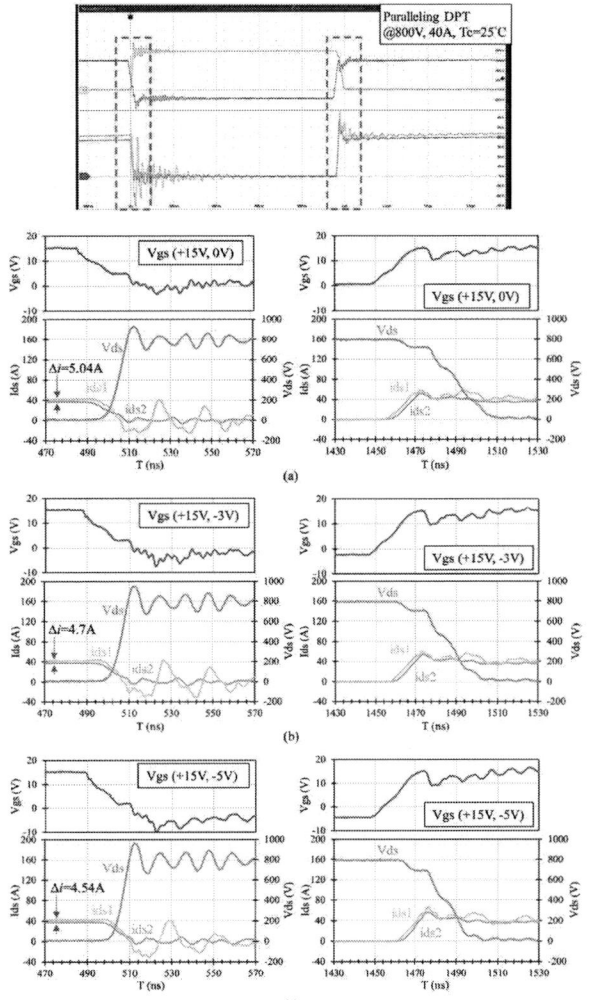

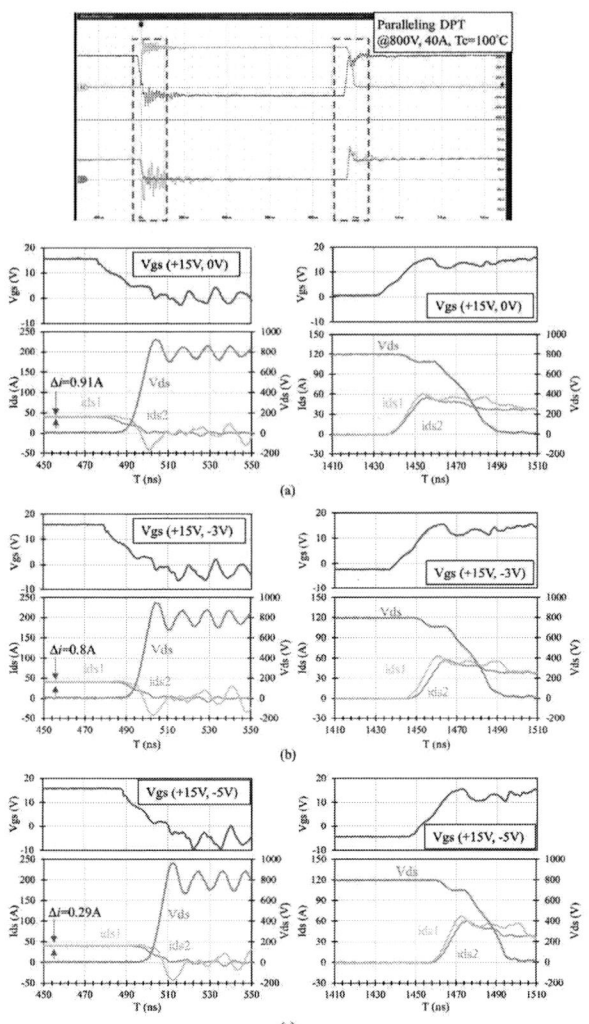

Fig. 9. Impact of turn-off negative voltage on current sharing under T_c=25°C and turn-off voltage condition: (a) v_{gs} = 0V, (b) v_{gs} = –3V, (c) v_{gs} = –5V.

Fig. 10. Impact of turn-off negative voltage on current sharing under T_c=100°C and turn-off voltage condition: (1) v_{gs} = 0V, (b) v_{gs} = –3V, (c) v_{gs} = –5V.

The experimental setup for the negative gate-off voltage is depicted in Fig. 8. Unlike the multiple gating circuit shown in Fig. 3, this setup intends to reduce the cost. Paralleling devices at the phase-leg level enables a higher switching frequency, thereby increasing power density. However, using individual gate drivers for each high-side and low-side legs increases costs and can introduce delay time mismatches. In contrast, Fig. 8 utilizes a single gate driver to each legs, with separate connections to the gate resistors for two paralleled devices.

Additionally, to observe the effects of temperature on current sharing, an external heater and thermometer are added in the setup. The trench device, IMZ120R060M1H, is used in this test. Fig. 9 and Fig. 10 compare the 800V, 40A DPT experimental results at the case temperature of T_c = 25°C and 100°C. The gate drive resistance for turn-on and turn-off are R_{g_on} = 10 Ω and R_{g_off} = 4.7 Ω, respectively. Under both temperature test conditions, v_{gs} varies from 0 to –5V.

Fig. 9 compares the turn-on and -off current waveforms under different negative gate-off voltage conditions when T_c=25°C. For this device, the manufacturer suggested the gate voltage is 18V for turn on and 0 V for turn off. The test condition here is to vary the gate-off voltage from 0 to –5V.

Under turn-on condition, the negative V_{gs} only causes the turn-on delay, which may have implications on turn-on loss but not on the dynamic current sharing. However, the scenario of turn-off conditions shows different results. As the negative V_{gs} varies from suggested 0V to –5V, not only the switching delay occurs, but also the current sharing changes. The current unbalance is improved as the gate-off voltage decreased. The turn-off current ringing appear noticeable which notes that the current oscillation is unavoidable due to power loop and inserted current sensing parasitic inductance. Nevertheless, the gate-off voltage magnitude is shown as a critical factor to the dynamic current sharing, which has not been extensively discussed in the literature.

It is also observed that the initial v_{ds} drop during the turn-on period, which is caused by the product of the power loop inductance and the device current rise rate, can be used to estimate the loop inductance. In this case, the loop inductances for the two devices are estimated at 28.7 nH and 25.4 nH, respectively. This is the main reason that causes different turn-on current slopes. It also indicates that symmetrical PCB layout is critical to equal current sharing.

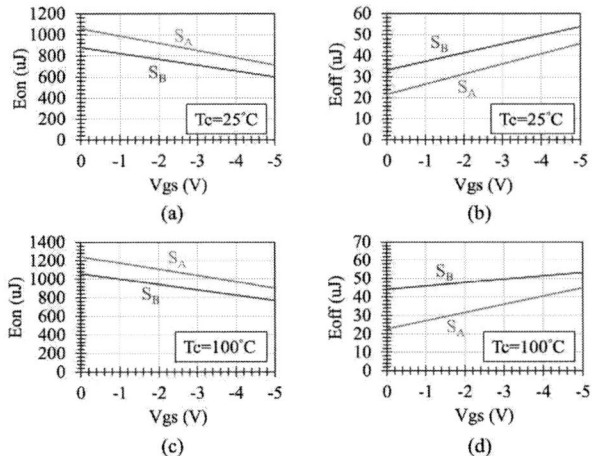

Fig. 11. Turn-on and Turn-off energy under (a), (b) T_c=25°C and (c), (d) T_c=100°C.

Comparing Fig. 9 and Fig. 10, the current sharing difference at V_{gs} = 0 decreases from 5.04A at T_c =25°C to 0.91A at T_c = 100°C. Such a better current sharing gets better due to the positive temperature coefficient (PTC) effect of $R_{ds(on)}$. However, the NTC effect of V_{th} still exists in switching transient period. As indicated in Fig. 9, the unbalanced current between two devices monotonically decreases from 5.04A to 4.54A as the V_{gs} varies from 0 to −5V. Similar phenomena is also found in Fig. 10, the unbalanced current decreases from 0.91A to 0.29A. Overall, both results demonstrate the current sharing is effectively improved via the applied negative V_{gs}.

Fig. 11(a) and (b) present the turn-on switching energy E_{on} and turn-off switching energy E_{off} under T_c=25°C conditions. The switching energy is usually used to quantize the dynamic current sharing improvement. The device that denotes as S_A is a device with a lower V_{th} and S_B represents a device with a higher V_{th}. As the V_{gs} decreases, E_{on} of both devices decreases, while E_{off} increases due to turn-off time delay. In Fig. 10(c) and (d), the results tested under T_c=100°C follow the same trend, though the difference in E_{off} becomes noticeably smaller. The E_{off} difference in Fig. 10(d) also reduces as the V_{gs} decreases. Overall, the experimental results show the negative turn-off voltage clearly helps share current in the turn-off period as well as a part of turn-on period.

IV. CONCLUSION

This paper comprehensively explored the dynamic current sharing issues with devices from different manufacturers and different key parameters that cause unequal current distribution at the circuit level. Using high bandwidth measurement setup, the paralleled SiC device currents were measured and characterized.

The major contribution is to quantify the current sharing with different key parameters that can impact the dynamic current sharing. A major finding is on the negative gate-off voltage impact to the current sharing. As the traditional approach of applying a negative gate voltage was to prevent the noise turn-on, this paper found that it can also help current sharing. Experimental results verified the findings under both room and a high case temperature conditions.

In addition to the dynamic current sharing, the impact to the switching energy between different V_{th} devices were also measured under different negative gate voltage and different temperature conditions. A low V_{th} device always draws more turn-on energy, but the negative gate voltage helps turn-on loss reduction. On the other hand, a high V_{th} device always draws more turn-off energy, but the negative voltage under high temperature condition helps balance out the share of turn-off loss due to the added PTC effect of $R_{ds(on)}$.

ACKNOWLEDGMENT

This research is supported in part by the Ministry of Education (MOE-Taiwan) under Yushan Fellow Program and by National Science and Technology Council (NSTC-Taiwan) under project number: NSTC 113-2640-E-A49-007. The authors would like to express their sincerely thanks to technical discussions and financial sponsorship of Chroma Inc., Taiwan.

REFERENCES

[1] H. Li, S. Zhao, X. Wang, L. Ding, and H. A. Mantooth, "Parallel connection of silicon carbide MOSFETs-challenges, mechanism, and solutions," *IEEE Trans. Power Electron.*, vol. 38, no. 8, pp. 9731–9749, Aug. 2023.

[2] R. Singh, "Advances in SiC Technologies Address High-Voltage Electrification Design Challenges," *IEEE Power Electronics Magazine*, vol. 11, no. 2, pp. 39-45, June 2024, doi: 10.1109/MPEL.2024.3393168.

[3] Salvatore La Mantia et al, "Design rules for paralleling of Silicon Carbide Power MOSFETs," STMicroelectronics, Catania, Italy.

[4] G. Wang, J. Mookken, J. Rice and M. Schupbach, "Dynamic and static behavior of packaged silicon carbide MOSFETs in paralleled applications," in *2014 IEEE Applied Power Electronics Conference and Exposition - APEC 2014*, 2014.

[5] Y. He, J. Zhang, and S. Shao, "Symmetric circuit layout with decoupled modular switching cells for multiparalleled SiC MOSFETs," *IEEE Trans. Power Electron.*, vol. 38, no. 6, pp. 7092–7106, Jun. 2023.

[6] J. Chen et al., "A novel power loop parasitic extraction approach for paralleled discrete SiC MOSFETs on multilayer PCB," *IEEE J. Emerg. Sel. Topics Power Electron.*, vol. 9, no. 5, pp. 6370–6384, Oct. 2021.

[7] B. Zhang and S. Wang, "Parasitic inductance modeling and reduction for wire-bonded half-bridge SiC multichip power modules," *IEEE Trans. Power Electron.*, vol. 36, no. 5, pp. 5892–5903, May 2021.

[8] Y. Wei, L. Du, X. Du, and A. Mantooth, "Multi-level active gate driver for SiC MOSFETs with paralleling operation," in *Proc. IEEE 22nd Workshop Control Modelling Power Electron.*, 2021, pp. 1–7..

[9] J. S. Knoll, G. Son, C. DiMarino, Q. Li, H. Stahr, and M. Morianz, "A PCB-embedded 1.2 kV SiC MOSFET half-bridge package for a 22 kW AC–DC converter," *IEEE Trans. Power Electron.*, vol. 37, no. 10, pp. 11927–11936, Oct. 2022, doi: 10.1109/TPEL.2022.3177369.

[10] A. Kumar et al., "Wide band gap devices and their application in power electronics," *Energies*, vol. 15, no. 23, p. 9172, Dec. 2022, doi: 10.3390/en15239172.

[11] G. Yadav and S. S. Nag, "Review of Factors Affecting Current Sharing and Techniques for Current Balancing in Paralleled Wide Bandgap Devices," in *2020 3rd International Conference on Energy*, 2021.

Integrated Short-Circuit Protection Design Based on Dual-Channel Gate Driver for Series Connected Medium-Voltage SiC MOSFETs

Rui Wang and Drazen Dujic
Power Electronics Laboratory - PEL
École Polytechnique Fédérale de Lausanne - EPFL
Lausanne, CH-1015, Switzerland
ru.wang@epfl.ch, drazen.dujic@epfl.ch

Abstract— **Series connection of Silicon carbide (SiC) MOSFETs is needed for increasing the blocking voltage of available power devices (limited to 3.3 kV in the market) and has been extensively studied, particularly in the context of voltage balancing design during normal switching operations. In addition, significant attention must be given to the short-circuit (SC) protection due to the lower SC withstanding capability of SiC MOSFETs and the asynchronous action of series-connected devices. Therefore, integrating SC protection by considering the series-connected devices inside the module as an equivalent single device is of great concern to avoid device SC failure. Given the limited coverage of this topic in the current literature, this paper proposes a novel SC protection strategy for the aggregated half-bridge SiC MOSFET power module. In this configuration, the slope detection-based SC protection of one device can complement and interact with the desaturation detection-based SC protection of the other, thus achieving fast and effective SC protection with minimal redundancy.**

Keywords—series connection, short-circuit protection, SiC MOSFET

I. INTRODUCTION

Silicon carbide (SiC) MOSFETs with superior electrical characteristic and switching performance hold significant potential to replace the current dominance of Silicon (Si) IGBTs in the industrial market. One major obstacle to this progress is the high cost of SiC devices, which is expected to be overcome as the supply chain and manufacturing processes mature. Another significant challenge is the relatively poor short-circuit (SC) withstand of SiC devices, which increases their usage risk [1]. Generally, the SC withstanding time of Si IGBT could reach 10 μs, while that of SiC MOSFET is much smaller due to the smaller size of the chip and the higher SC current. For example, 3.3 kV SiC MOSFETs module can only survive 3 μs SC time duration in the design case of [2]. Therefore, developing SC protection solutions for SiC devices is essential [3].

For a single SiC MOSFET, many efforts have been made regarding of the SC protection. Adding Rogowski current sensor is a reliable approach to detect the SC fault [4-6], but it is hard to be integrated into gate driver design and it increases the burden of integration and complexity. Using Kelvin to power source inductance to detect the current slope is another choice to indirectly monitor the SC fault [7-9], but the inductance is small and ultra-low current slope is hard to be detected. The *dv/dt* detection method can accelerate the detection [10], but less analysis is given as in low voltage case this method could lose the function since no obvious *dv/dt* could occur in both normal turn-on and SC cases. Desaturation method is still considered as the most cost-effective SC protection method [11-13], although it is sensitive to the gate driver parameters and converter parameters. Once the parameters are fixed as in the converter prototype, using desaturation method is a fair approach as long as the risk posed by the blanking time can be overcome. In summary, a single SC protection scheme has its own limitations, often requiring the combination of two methods to achieve optimal SC protection.

Compared with a single device, the research work about SC protection of series connected SiC MOSFETs is limited [14-17]. Series connection development is a vital technique to increase the blocking voltage of devices. While significant research has focused on voltage balancing during normal switching, SC protection also requires attention. Facing the challenges of protection detection and execution differences and device I-V characteristic tolerance causing severe SC voltage unbalancing issue, applying the individual SC protection to each device in series directly could cause the device breakdown if the design redundancy is not sufficient. Therefore, to achieve a better and more reliable SC protection while maximizing the capability of series connected devices, it is necessary to consider the series connected devices as a whole.

As a continuation of the previously proposed dual channel gate driver which aggregates the half-bridge SiC MOSFET power module into a single SiC MOSFET with double voltage rating [18], a novel SC protection strategy is proposed in this paper and the SC protection circuitry is further integrated into the dual channel gate driver. For two devices in series inside this half-bridge power module, the SC detection of one device is based on slope detection, and the other is based on desaturation detection. They can interact and compensate for each other to enable the synchronous protection action, thereby achieving the fast and effective SC protection of this aggregated half-bridge power module in a low-redundancy manner.

979-8-3315-1612-3/25 $31.00 © 2025 IEEE

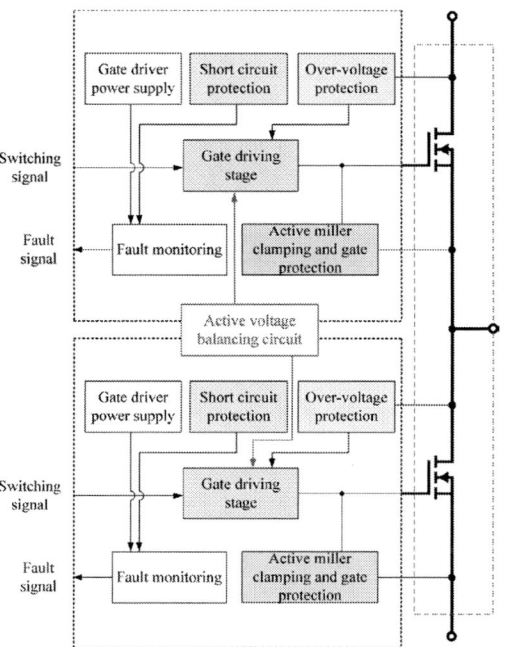

Fig. 1. Overall structure of the designed GD including voltage balancing, active miller clamping, short circuit protection, etc.

II. OVERALL GATE DRIVER DESIGN

The gate driver (GD) plays a crucial role in establishing the interconnection between the power semiconductor module and the upstream control system. For the illustration purpose, the conceptual structure of the designed GD is shown in Fig. 1, which serves as a dual-channel GD for two series-connected SiC MOSFETs in the half-bridge module, aggregating them into a single device with doubled voltage rating. It consists of driver power supply, driving stage, active miller clamping and protection, etc. The active voltage balancing circuit is indispensable to ensure the voltage balancing of series connected SiC MOSFETs, which is already presented and verified in [18]. Therefore, on the basis of realized well-balanced voltages of series connected SiC MOSFETs, the integrated short-circuit protection is presented.

III. PROPOSED INTEGRATED SHORT-CIRCUIT PROTECTION

Short-circuit protection is a critical aspect of gate driver design. To streamline the layout design, there is typically a preference for integrating it directly within the driver. In this case, the proposed SC protection takes the advantage of the three power terminals structure, combing the advantages of two types of SC detection methods to create a comprehensive SC protection solution for the series connected SiC MOSFETs. As shown in Fig. 2, the SC detection integrated in the lower channel of gate driver is based on desaturation detection. Once the drain-source voltage of the lower SiC MOSFET during on state exceeds the pre-defined threshold, the short-circuit protection will be triggered. It is a robust method as the design with the same principle has already been applied to the SC protection of IGBT in the commercial market for decades. It can provide effective SC protection in both fault under load (FUL) and hard switch fault (HSF) conditions. However, the blanking time requirement to avoid false-trigger for HSF increases the risk of SC failure, which becomes more significant when it is applied to SiC MOSFET. To overcome this drawback, the slope detection is integrated in the upper channel of gate driver. During the turn-on process of SiC MOSFET, the large magnitude of commutating current when overcurrent or HSF occurs will greatly influence the slope of the drain-source voltage. Taking it as the decision criterion, the SC protection can be triggered earlier and more reliably in HSF condition. In this manner, two SC detection parts respectively located at upper and lower channels of gate driver detect the SC fault individually. Through signal transfer, the error signal can be communicated between two devices when a SC fault is detected. Therefore, the coordination of two detection parts achieves a reliable and effective SC detection for the series connected SiC MOSFETs.

Once a SC fault is detected, the SC action parts for both devices get activated. As illustrated in Fig. 3, the synchronous fault signals ER1 and ER2 are turned into low-level from normal high-level, and the gate driving stages of both channels are disabled. In the meantime, "soft-turn-off" branches are activated to turn-off series connected SiC MOSFETs where the turn-off speed is controlled by the branch resistor. In order to avoid the overvoltage and the breakdown of device, the active clamping circuit is appended to each SiC MOSFET. Once the voltage of SiC MOSFET exceeds the clamping value set by the TVS diodes, the gate current will flow into the gate of SiC MOSFET to further slowdown the SC turn-off action. Therefore, series connected SiC MOSFETs can be reliably turned off without any damage in the SC fault occurrence.

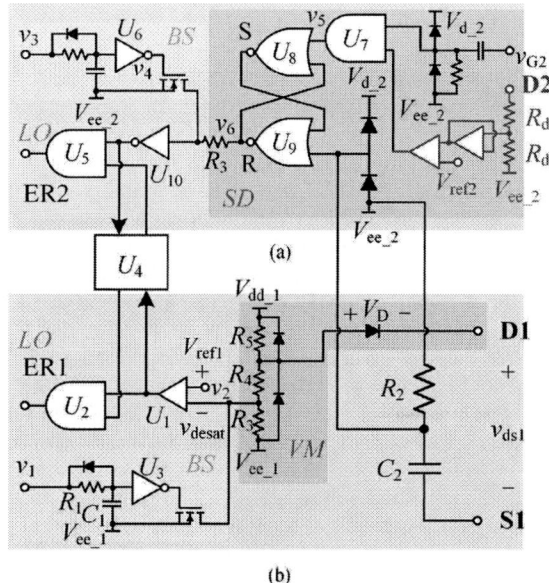

Fig. 2. Circuit implementation for short-circuit detection (a) slope detection for the upper channel (b) desaturation detection for the lower channel

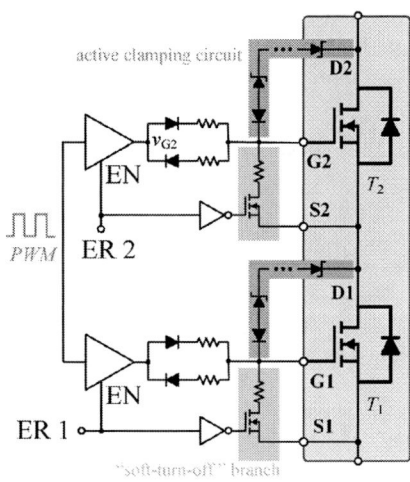

Fig. 3. Circuit implementation for short-circuit action

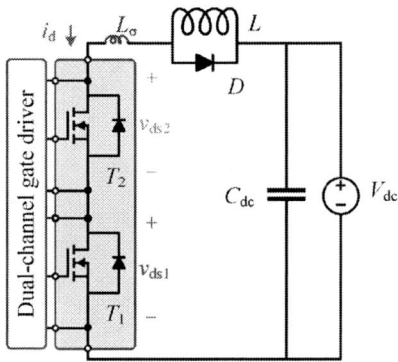

Fig. 4. Test setup principle with series connected SiC MOSFETs as device under test

A. Desaturation detection

The desaturation detection circuit consists of three parts: drain-source voltage monitoring part (VM), logic output part (LO) and blanking-time setting part (BS). To better illustrate its working principle, the balanced voltage sharing of two series connected SiC MOSFETs T_1 and T_2 is assumed, and a typical double-pulse test circuit with the inductive load is set up, which is depicted in Fig. 4. Before the turn-on signal arrives, BS part outputs logic "0" as it pulls down the inverting input of the comparator U_1 inside the LO part to T_1's negative driving voltage V_{ee_1}, and U_1 outputs "1" as a positive reference voltage V_{ref1} is set at the non-inverting input. If the status signal from the upper channel is "1" as well, the output signal ER1 of the AND gate U_2 is "1", which indicates no SC error. When the turn-on signal arrives, the driving voltage changes from V_{ee_1} to the positive voltage V_{dd_1}, and the voltage signal v_1 changed from "0" to "1". Determined by the following resistor-capacitor R_1C_1 branch and NOT gate U_3, BS part outputs Hi-Z after a pre-defined blanking time T_{b1}, which allows the working of VM part. When SiC MOSFET is on and the SC occurs, the voltage across the device becomes higher because of the large current, which in turn makes the output voltage v_2 of VM higher than V_{ref1} and

ER1 becomes "0". In addition, "0" is transferred to the upper channel through the isolation chip U_4, and ER1 becomes "0" as well. This type of SC fault is named as FUL, and desaturation detection can effectively detect the fault to make the gate driver turn-off the series connected SiC MOSFETs.

However, in the case of HSF, that is, the SC already occurs before the device is turned on. Since T_{b1} is set with margin for the reliable transition of SiC MOSFET from cut-off region to linear region, the current will go extremely high during this period once the turn-on signal arrives. As the bus voltage V_{dc} goes higher, the risk of device failure cause by heat accumulation is greatly increased.

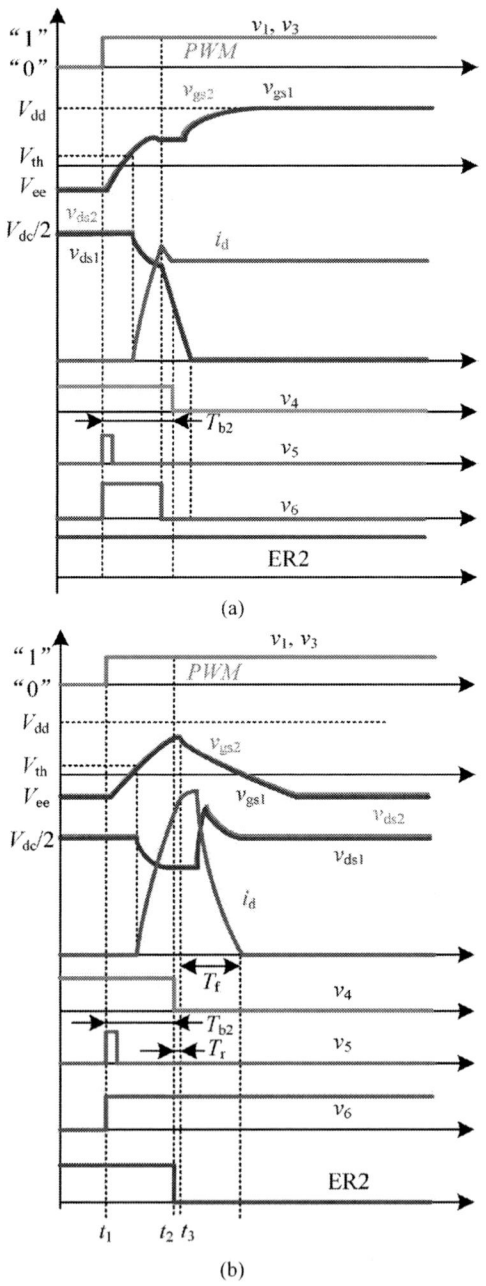

Fig. 5 Turn-on waveforms in (a) normal (b) SC conditions

B. Slope detection

As a compensation for SC detection, the proposed slope detection circuit is added in the upper channel of gate driver, which consists of three parts as well: slope detection part (SD), logic output part (LO) and blanking-time setting part (BS). In order to illustrate the working principle, the corresponding waveforms during the turn-on process of SiC MOSFETs are depicted in Fig. 5.

In the normal switching process as shown in Fig. 5(a), when the turn-on signal arrives, the voltage signal v_3 changes from "0" to "1", and the BS part plays the same role of setting blanking time T_{b2}. T_1's driving voltage v_{G2} changing from the negative voltage V_{ee_2} to the positive voltage V_{dd_2} enables a positive pulse by RC branch and AND gate U_7 to set the following SR latch consisting of two NOR gate U_8, U_9. Therefore, v_4 becomes "1". As the drain-source voltage v_{ds1} dramatically drops in the duration of T_{b2}, a positive voltage pulse is formed by the RC detection branch to reset the SR latch, and thus v_4 is returned back to "0". After T_{b2}, BS part outputs Hi-Z and v_6 as the input of NOT gate U_{10} through a resistor R_3 controls the logic output of U_{10}. In the case that the status signal from the lower channel is "1", the output logic ER 2 of AND gate U_5 becomes "0", which enables the normal working of gate driver.

By contrast, in the SC condition as shown in Fig. 5(b), since the drain-source voltage dropping slope of S_1 is not large enough to reset the SR latch, v_6 keeps as "1" after T_{b2}, and thus ER 2 becomes "0" to indicate the occurrence of SC. Afterwards, the drain current i_d flowing through T_1 and T_2 gets decreased to zero with SC protection action.

Since T_{b2} is theoretically smaller than T_{b1} as the falling stage period of v_{ds1} could not be considered, a faster response can be ensured. Moreover, the higher robustness of this branch enables a smaller margin to set T_{b2} for avoiding false turn-on, which further improves the SC detection speed.

However, one limitation should be considered in practice. The existence of leakage inductance L_σ results in a drain-source voltage pre-dropping in the initial stage during the turn-on process. If the dc bus voltage is too low, large voltage slope could not occur despite in the case of normal switching process, which cause the false detection of SC. Therefore, a voltage monitoring circuit indicated in brown line in Fig. 2(a) is designed to disable the slope detection when the dc bus voltage is too low. In the case of low dc voltage, desaturation detection works for the SC protection. Despite that T_{b1} is large than T_{b2}, the heat accumulation also becomes smaller to make the withstanding time of SiC MOSFET longer, which is beneficial for the effective SC protection with this gate driver design.

IV. PARAMETERS TUNNING PRINCIPLE

In this section, the parameter tunning criteria for SC protection circuit are given. Starting from the desaturation detection, when SiC MOSFETs are turned on, the voltage v_{desat} to U_1 is continuously monitored, and it has a following relationship with v_{ds1}:

$$v_{desat} = \begin{cases} (v_{ds1} + V_D - V_{ee\_1})\dfrac{R_3}{R_3 + R_4}, & v_{ds1} < V_D + (V_{dd\_1} - V_{ee\_1})\dfrac{R_3 + R_4}{R_3 + R_4 + R_5} \\[2mm] (V_{dd\_1} - V_{ee\_1})\dfrac{R_3}{R_3 + R_4 + R_5}, & v_{ds1} \geq V_D + (V_{dd\_1} - V_{ee\_1})\dfrac{R_3 + R_4}{R_3 + R_4 + R_5} \end{cases} \quad (1)$$

where R_3, R_4 and R_5 are the resistors indicated in Fig. 2 (b), and V_D is the forward voltage drop of the diode.

Once v_{desat} exceeds V_{ref1} as FUL occurs, the output of U_1 will flip from "1" to "0", and the output of U_2 becomes "0", indicating the occurrence of SC fault. Therefore, the threshold drain-source voltage $V_{ds}(th)$ to trigger the SC protection can be determined as:

$$V_{ds}(th) = V_{ref1}\frac{R_3 + R_4}{R_3} + V_{ee\_1} - V_D. \quad (2)$$

According to the desired SC current threshold for protection and I-V characteristic of device, the resistances and V_{ref1} in (2) can be adjusted accordingly to meet the requirement.

As already mentioned, since there is a transition time duration of device from cut-off region to linear region during the turn-on transient, in order to avoid the false trigger of protection, T_{b1} which is called blanking time is required, during which the v_{desat} monitoring function is disabled. T_{b1} is chosen according to the longest transition time in the case of the highest load current and lowest junction temperature, and the margin is normally given further to compensate the non-ideal factors and improve the noise immunity for fast switching SiC MOSFET.

However, in the case of HSF, the large T_{b1} will greatly increase the risk of device failure, and corresponding heat accumulation can be approximated as:

$$E = \int_0^{T_{b1}+T_r+T_f} i_d \cdot v_{ds} dt \approx \frac{T_{b1} + T_r + T_f}{2} \cdot (\frac{v_{dc}}{2} - L_\sigma \cdot \frac{di_d}{dt}) \cdot \frac{di_d}{dt} \cdot (T_{b1} + T_r) \quad (3)$$

where T_r is SC action latency caused by circuit delays, etc., and T_f is SC current falling period, which are also indicated in Fig. 5(b). The generated E can be used as the criterion of estimating whether the device is damaged or not during the SC event.

From (3), it is found that, in the case that V_{dc} is small, desaturation detection method can still effectively trigger the device protection. While as V_{dc} increases, E becomes larger which could result in the device failure, therefore the slope detection works to trigger the device protection in advance.

The slope detection only works when T_2's drain-source voltage v_{ds2} exceeds a certain value to avoid false trigger. Specifically, v_{dc} should be above the threshold voltage V_a, which meets the condition as:

$$V_a > L_\sigma \cdot \frac{di_d}{dt} \quad (4)$$

In the other hand, from the perspective of circuit design, V_a has the following relationship as:

$$\frac{V_a}{2} = V_{ref2}\frac{R_{d1} + R_{d2}}{R_{d1}} + V_{ee\_2} \quad (5)$$

where R_{d1} and R_{d2} are the voltage dividing resistors in Fig. 2(a).

Moreover, the threshold voltage slope $dv_{ds}/dt(th)$ to trigger SC protection can be expressed as:

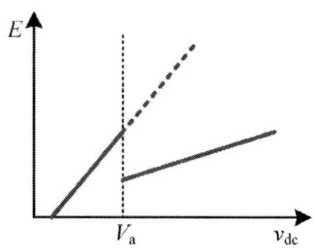

Fig. 6 The relationship between E and v_{dc}

$$\frac{dv_{ds}}{dt}(th) = \frac{V_{U\_th}}{R_2 C_2} \cdot \quad (6)$$

where V_{U_th} is the voltage to reset SD part, determined by the high-level input voltage of U_9.

Similar to the parameter identification of the desaturation detection, the parameter identification in the slope detection also relies on the specific condition of application. According to the relationship between the voltage slope and the turn-on current, which proves to be nearly irrelevant to the dc bus voltage and the load current in the considered cases by experiments, R_2 and C_2 in (6) can be chosen once the SC current protection threshold is set. Also, the parameters in (5) can be selected according to (4).

Compared with T_{b1}, since the choosing of T_{b2} does not need to consider the falling stage of v_{ds}, T_{b2} becomes much smaller. By drawing the relationship diagram of v_{dc} and E according to (3), a clear tendency variation can be found at V_a, as given in Fig. 6.

V. EXPERIMENTAL RESULTS

Following the previous presentation, the experimental platform for verifying the concept and design is developed, as shown in Fig. 7(a). It is based on the testing schematic depicted

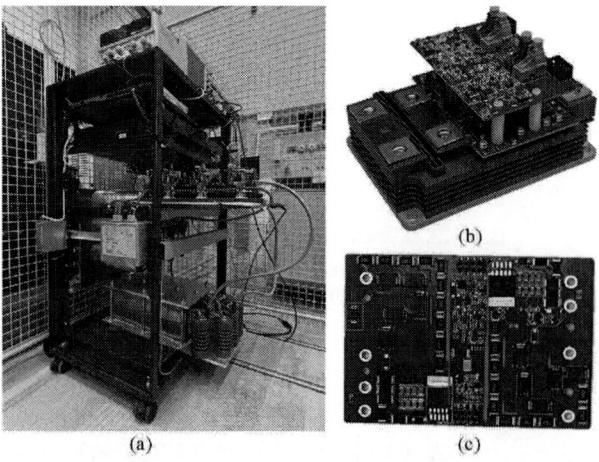

Fig. 7 (a)Testing platform (b) SC protection integrated dual channel gate driver (c) Proposed SC protection circuitry.

in Fig. 4, and the freewheeling diode D is replaced by the same series connected devices under test for symmetry. On the basis of 3.3kV/750A SiC MOSFET power module from Mitsubishi Electric, the short-circuit protection integrated GD is developed and assembled, as depicted in Fig. 7(b). The proposed short-circuit protection circuits are positioned on the bottom PCB of the GD, which is mounted to the surface of the module, as shown in Fig. 7(c). The applied oscilloscope to record the waveforms is DLM4058 (2.5GS/s / 500 MHz) from YOKAGAWA. For the accuracy, the gate source voltages are measured by optical-fiber isolated low-voltage probes with 800 MHz bandwidth, and the drain source voltages are measured by high voltage differential probes with 50 MHz bandwidth, and the current is measured by Rogowski current transducer with 10 MHz bandwidth.

A. Current rise relation with load current and dc bus voltage

Since the slope detection plays a significant role in limiting SC current in the occurrence of HSF, the basis of realizing its functionality is validated first. As shown in Fig. 8, in the case of V_{dc} is equal to 3.5 kV, the turn-on waveforms of series connected SiC MOSFETs with multiple load current I_L values are given. In all cases, T_1 and T_2 share the DC voltage equally. When turn-on signal arrives, v_{gsi} ($i=1, 2$) starts to increase from -5V, enabling the device transiting from cut-off region to linear region. Notably, as the commutating current increases, turn-on dv/dt becomes smaller, while turn-on di/dt can be considered as a constant value. In addition, the turn-on waveforms in the varied V_{DC} cases are also provided as shown in Fig. 9. It is found that turn-on di/dt is also nearly unchanged as V_{DC} increases from 1 kV to 3.5 kV. Therefore, the negligible influence of V_{DC} and I_L on turn-on di/dt is verified, which makes the current protection threshold designable once the converter is well-defined. Specifically, the turn-on di/dt can be estimated in advance. With the expected protection threshold, T_{b2} can be set accordingly.

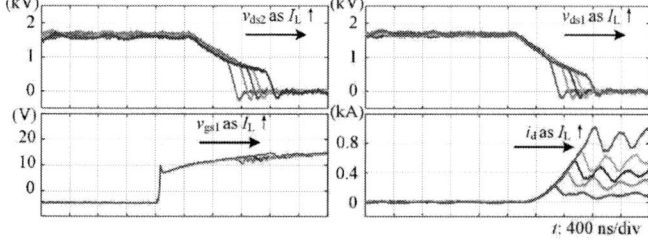

Fig. 8. Turn-on waveforms of series connected SiC MOSFETs as I_L varies

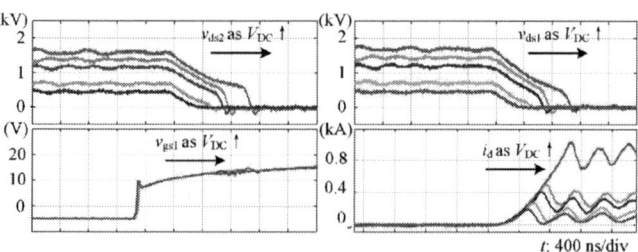

Fig. 9. Turn-on waveforms of series connected SiC MOSFETs as V_{DC} varies

From Fig. 9, it is observed that, due to the existence of parasitic inductance in the power loop, the positive turn-on di/dt will cause a voltage drop of v_{dsi} (i=1, 2) during the current rising stage. After this period, v_{dsi} starts to drop to zero abruptly, which helps to reset the S-R latch in slope detection and indicates the normal turn-on process. However, when V_{DC} is decreased to 1.5 kV, the voltage drop already makes v_{dsi} reach zero in this case, and no abrupt voltage slope occurs to reset the S-R latch. As a result, the SC protection by slope detection is falsely triggered although it is a normal turn-on process. Therefore, in the design, 1.8 kV is set as the threshold to enable the slope detection by properly choosing the parameters of voltage detector in SD part.

B. Fault Under Load

Firstly, the condition of FUL is emulated to verify the performance of the proposed design by setting the FUL protection threshold as 900 A. As shown in Fig. 10, when the turn-on signals arrives and v_{gsi} increases from -5V to 18V, i_d starts to increase from zero. After i_d reaches 400A, the used load inductor is saturated, and i_d continues to rise with a higher slope. The normal turn-off signal still does not arrive and i_d rises to a higher value which triggers the short-circuit protection. Hence the switching signal is locked and the gate driver enables the soft turn-off of two series connected SiC MOSFETs. It is worth noting that, in our case the soft turn-off resistor is the same value as the normal turn-off resistor to speed up the SC protection, and the active clamping circuits work to suppress the overvoltage

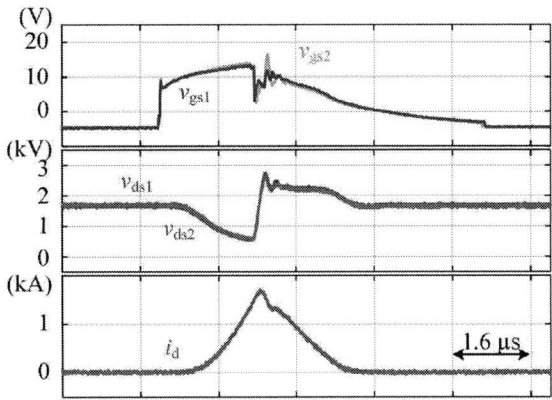

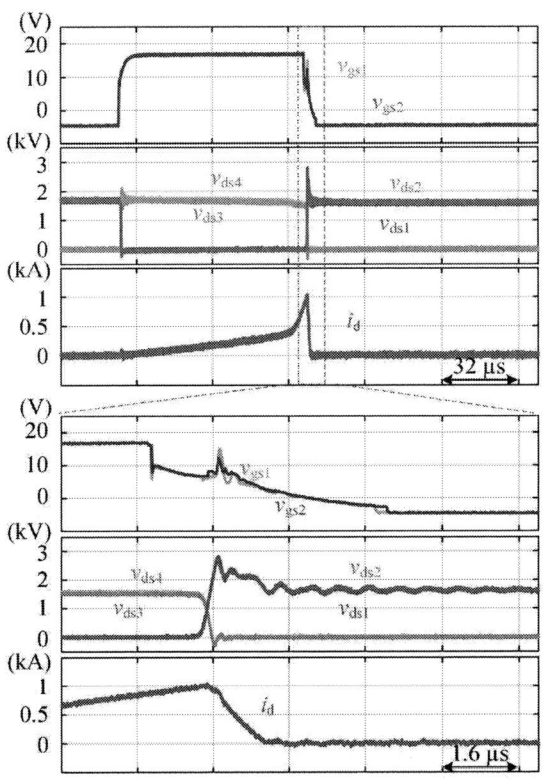

Fig. 10. FUL protection test of two series connected SiC MOSFETs

Fig. 11. HSF protection test of two series connected SiC MOSFETs

shoots of v_{dsi} (i=1, 2), which can be concluded as the voltage spikes of v_{gsi} are observed when the peak of v_{dsi} exceeds the clamping values.

It is found that in the case of protection triggered, the voltage balancing circuit and the SC protection circuit work well to balance v_{ds1} and v_{ds2}, while suppressing the voltage overshoots. As another half-bridge 3.3kV/750A SiC MOSFET power module is configured as the freewheeling diode, its dual-channel GD makes its inside SiC MOSFETs constantly off and its inside anti-parallel diodes works with series connection. v_{ds3} and v_{ds4} waveforms in Fig. 10 indicate the voltages across these two devices, which also shows good voltage balancing without any snubber circuit appended.

C. Hard Switch Fault

Next, the load inductor is short-circuited, and series connected SiC MOSFETs under test are forced to turn-on to emulate the condition of HSF. As shown in Fig. 11, the current increases abruptly as turn-on signal arrives, and the large di/dt causes the voltage drop of v_{dsi} (i=1, 2). Since this voltage slope is small, it can not work to reach the threshold to reset S-R latch. After the blanking time, the soft turn-off of gate driver gets activated and i_d is cut off to zero after reaching the peak value 1750A. In conclusion, the effectiveness of the proposed SC protection gets verified.

VI. CONCLUSION

In this paper, an integrated SC protection design is proposed on the basis of a dual-channel GD for series connected medium-voltage SiC MOSFETs. For two series connected devices inside the 3.3kV/750A SiC MOSFET half-bridge power module from Mitsubishi Electric, two different SC detection circuits are implemented on their GD, named as the slope detection and the desaturation detection respectively. The slope detection can accelerate the SC detection in HSF under higher voltages compared to the desaturation detection, in the meantime possessing the higher robustness. This hybrid SC detection design together with SC action parts contributes to a more effective SC protection for series connected SiC MOSFETs. The detailed working principles are parameter identification criteria are given. Finally, the experimental results show that the SC currents in both HSF and FUL conditions are cut off without any damage to power module, and the voltage balancing of series connected SiC MOSFETs are maintained during the SC protection process, which proves the effectiveness of the proposed design.

ACKNOWLEDGMENT

The results presented in this paper are a part of the FOR²ENSICS project that has received funding under the European Union's Horizon Europe research and innovation program (Grant Agreement No. 101075672). The work conducted at EPFL has received the funding directly from the Swiss State Secretariat for Education, Research, and Innovation (SERI).

REFERENCES

[1] J. Sun, H. Xu, X. Wu and K. Sheng, "Comparison and analysis of short circuit capability of 1200V single-chip SiC MOSFET and Si IGBT," 2016 13th China International Forum on Solid State Lighting: International Forum on Wide Bandgap Semiconductors China (SSLChina: IFWS), Beijing, China, 2016, pp. 42-45.

[2] Y. Cong et al., "Short-Circuit Ruggedness Characterization of State-of-the-Art 3.3 kV SiC MOSFETs," 2023 IEEE 10th Workshop on Wide Bandgap Power Devices & Applications (WiPDA), Charlotte, NC, USA, 2023.

[3] M. Zhang, H. Li, Z. Yang, S. Zhao, X. Wang and L. Ding, "Short Circuit Protection of Silicon Carbide MOSFETs: Challenges, Methods, and Prospects," in IEEE Transactions on Power Electronics, vol. 39, no. 10, pp. 13081-13095, Oct. 2024.

[4] J. Wang, S. Mocevic, R. Burgos and D. Boroyevich, "High-Scalability Enhanced Gate Drivers for SiC MOSFET Modules With Transient Immunity Beyond 100 V/ns," in IEEE Transactions on Power Electronics, vol. 35, no. 10, pp. 10180-10199, Oct. 2020.

[5] M. Stecca, P. Tiftikidis, T. B. Soeiro and P. Bauer, "Gate Driver Design for 1.2 kV SiC Module with PCB Integrated Rogowski Coil Protection Circuit," 2021 IEEE Energy Conversion Congress and Exposition (ECCE), Vancouver, BC, Canada, 2021, pp. 5723-5728.

[6] J. -A. Lee, D. H. Sim and B. K. Lee, "Short-Circuit Protection for SiC MOSFET Based on PCB-Type Rogowski Current Sensor: Design Guidelines, Practical Solutions, and Performance Validation," in IEEE Transactions on Power Electronics, vol. 39, no. 3, pp. 3580-3589, March 2024.

[7] C. Li, J. Sheng and D. Dujic, "Reliable Gate Driving of SiC MOSFETs With Crosstalk Voltage Elimination and Two-Step Short-Circuit Protection," in IEEE Transactions on Industrial Electronics, vol. 70, no. 10, pp. 10066-10075, Oct. 2023.

[8] K. Sun, J. Wang, R. Burgos and D. Boroyevich, "Design, Analysis, and Discussion of Short Circuit and Overload Gate-Driver Dual-Protection Scheme for 1.2-kV, 400-A SiC MOSFET Modules," in IEEE Transactions on Power Electronics, vol. 35, no. 3, pp. 3054-3068, March 2020.

[9] Y. Wu, C. Li, J. Liu and Z. Zheng, "An Ultrafast and Low-Invasive Short-Circuit Current Limiting Method by Gate Voltage Control for SiC MOSFETs With Kelvin-Source," in IEEE Transactions on Power Electronics, vol. 39, no. 12, pp. 15696-15708.

[10] Z. Guo and H. Li, "Dv/dt Sensing-Based Short-Circuit Protection for Medium-Voltage SiC mosfets," in IEEE Transactions on Power Electronics, vol. 38, no. 9, pp. 10554-10558, Sept. 2023.

[11] S. Ji et al., "Short-Circuit Characterization and Protection of 10-kV SiC mosfet," in IEEE Transactions on Power Electronics, vol. 34, no. 2, pp. 1755-1764, Feb. 2019.

[12] Z. Liu et al., "A Hybrid Desaturation Fast Detection Circuit for Bridge Leg Short-Circuit Faults," in IEEE Transactions on Power Electronics, vol. 39, no. 8, pp. 9221-9229, Aug. 2024.

[13] X. Huang, D. Li, M. Lin, L. M. Tolbert, F. Wang and W. Giewont, "Desat Protection With Ultrafast Response for High-Voltage SiC MOSFETs With High dv/dt," in IEEE Open Journal of Industry Applications, vol. 5, pp. 94-105, 2024.

[14] H. Du, I. Omura, S. Matsumoto and T. Arai, "Temperature-Dependent Mechanism of Short-Circuit Voltage Imbalance in Series-Connected SiC MOSFETs," 2024 36th International Symposium on Power Semiconductor Devices and ICs (ISPSD), Bremen, Germany, 2024, pp. 92-95.

[15] C. Liu, Z. Zhang, Y. Liu, Y. Si, M. Wang and Q. Lei, "A Comprehensive Short-Circuit Protection Scheme for Series-Connected SiC MOSFETs," in IEEE Open Journal of Power Electronics, vol. 3, pp. 115-130, 2022.

[16] A. Kumar, R. K. Kokkonda, S. Bhattacharya, J. Baliga and V. Veliadis, "Short Circuit Behavior of Series-Connected 10 kV SiC MOSFETs," 2021 23rd European Conference on Power Electronics and Applications (EPE'21 ECCE Europe), Ghent, Belgium, 2021, pp. P.1-P.10.

[17] R. Wang, A. B. Jørgensen, H. Zhao and S. Munk-Nielsen, "Short-Circuit Characteristic of Single Gate Driven SiC MOSFET Stack and Its Improvement With Strong Antishort Circuit Fault Capabilities," in IEEE Transactions on Power Electronics, vol. 37, no. 11, pp. 13577-13586, Nov. 2022.

[18] R. Wang and D. Dujic, "Active Voltage Balancing With Seamless Integration Into Dual Gate Driver for Series Connection of SiC Mosfets," in IEEE Transactions on Power Electronics, vol. 39, no. 6, pp. 6635-6639, June 2024.

Long-Term High-Temperature Dynamic Gate Stress Reliability of a Last-Generation, Automotive-Grade, Planar 1200 V SiC MOSFET

Giuseppe Mauromicale*, Alessandro Sitta*‡, Michele Fiore*, Michele Calabretta*, and Francesco Iannuzzo†

*Quality, Manufacturing and Technology R&D, STMicroelectronics, Catania, Italy
†Department of Energy Technology, University of Aalborg, Aalborg, Denmark
email: giuseppe.mauromicale@st.com, alessandro.sitta@st.com, michele.fiore01@st.com,
michele.calabretta@st.com, fia@energy.aau.dk
‡Corresponding author

Abstract—In this work, the dynamic gate stress (DGS) on a planar-gate silicon carbide MOSFET is performed based on an automotive mission profile. An extensive characterization was performed, even beyond automotive qualification guidelines, including the effect of the temperature. The results show that the temperature accelerates the threshold voltage drift, which ranges from 3 % at room temperature up to 6 % at 125 °C and 150 °C, with negligible impact on the on-state resistance. Moreover, we measured the gate-source voltage waveforms before and after the DGS stress, not finding significant variations in switching gradients.

Index Terms—Silicon Carbide, MOSFET, gate oxide, high-temperature dynamic gate stress, threshold voltage, on-state drain-source resistance, reliability

I. INTRODUCTION

Power electronic devices are nowadays employed in several applications automotive fields, where an electrical energy conversion occurs [1]. In the automotive environment, wide bandgap (WBG) materials, such as silicon carbide (SiC), are more and more used thanks to their better thermal and electrical performances compared to silicon (Si)-based devices [2], [3]. Therefore, SiC-based power electronic devices are nowadays employed in electric vehicles, mainly in onboard chargers and traction inverters [4]. In inverters, SiC MOSFETs can replace traditional Si-based insulated gate bipolar transistors (IGBT) [5]. However, differently from the more consolidated IGBT technology, WBG-based high-power devices, including SiC MOSFETs still need research, in particular to assess their long-term performance.

In this scenario, reliability assessment is needed to predict the expected life, bearing in mind that the possible failure mechanisms are not necessarily the same as in silicon counterparts [6], [7]. Among other relevant phenomena for SiC MOSFETs, static gate bias has been acquired more and more importance in comparison to previous Si power devices, because of the high interface trap density between SiC and SiO_2. Previous literature reports that a static positive gate bias leads to an increase of threshold voltage (V_{th}), as well as a negative gate-source voltage (V_{GS}) results in a decrease of V_{th} [8]. Concerning gate oxide reliability, along with the

static high-temperature gate bias, new dynamical tests are under development [9]. These are aimed at the defects in the gate structure. Although the root cause of this phenomenon remains still not fully understood, a recent paper has proposed an analytical model related to recombination-enhanced defect reactions phenomena [10].

A novel test concept proposes a dynamic stress of the V_{GS}, whose name is Dynamic Gate Stress (DGS) but can be also referred to as High-Temperature Gate Stress (HTGS), AC gate stress, or gate switching stress. This concept dates back to 2018 when the first application note from Infineon was published [9]. DGS is present in AQG 324 guideline from the european center for power electronics (ECPE), too, for the automotive qualification of power modules [11], as well as in the JEP 195 guideline from the joint electron device engineering council (JEDEC) [12]. The rationale of such a test is that electric parameters' degradation, mostly V_{th}, can happen because of DGS stress [13]–[17]. Moreover, the impact of possible on-state drain-source resistance $R_{DS(on)}$ drift must be ascertained [14]. It has been shown that this phenomenon is not recoverable, except by annealing at very high temperatures [15]. In addition, it depends on the number of cycles, independently from the switching frequency [9].

The impact of DGS has been shown to be dependent on the device structure. Both planar and trench gate structures are used in SiC MOSFETs [7], [18]. There exist various design versions of both trench and planar devices, from different vendors. More in detail, the gate oxide is protected in planar devices, at the JFET region, by the depletion layer, whereas in trench MOSFETs, even if a lower on-state resistance is achievable, the gate oxide suffers from a high electric field when in the off state with high drain-source voltage (V_{DS}) bias [19]. In some papers, the DGS test is more severe in trench-based devices, compared to planar ones [13], [17], [20]. Regarding the instability of the V_{th} induced by DGS, analytical models to evaluate the switching losses in SiC MOSFETs in both trench and planar technologies have been developed [21].

In recent years, few papers have investigated temperature

979-8-3315-1612-3/25 $31.00 © 2025 IEEE

effects in DGS, and, on top of that, their result appear rather inconsistent or, in some cases, even contradictory [9]. In some of them, no effects of the temperature on V_{th} drift are observed [16]. Instead, in some others, the temperature can either alleviate or worsen the parametric drift [15], [17].

In this paper, we evaluate the long-term reliability behavior of SiC MOSFETs under DGS. We tested automotive-grade planar-gate 1200 V SiC MOSFETs belonging to the latest generation technology available on the market, which permits a low $R_{DS(on)}$ per area unit and lower switching losses in comparison to the previous one. The results of this DGS test campaign aim at 1) shedding light on the dependence of this phenomenon on the temperature and 2) proposing a statistical characterization of the obtained data.

The paper is organized as follows. In Section II, the experimental setup is presented, while Section III deals with the results of the DGS test. Finally, Section IV presents the conclusions and possible future works.

II. EXPERIMENTAL SETUP

We tested discrete third generation, automotive-grade SiC MOSFETs, from STMicroelectronics, housed in TO-247 four-leaded package. Rising and falling gradients dV_{GS}/dt, as well as V_{GS} over- and under-shoots waveforms, were measured using the MSO54B oscilloscope from Tektronix. Rising gradients were obtained considering 10 % and 90 % of the maximum value of V_{GS}. A similar procedure was considered for the falling gradients. The switching frequency was 300 kHz, with a duty cycle of 50 %. V_{GS} range was set at - 5 V/18 V during the off and the on times, respectively.

Fig. 1: Photo of the equipment used in the test.

Fig. 2: Focus on experimental setup: drawer with DUTs and passive probe.

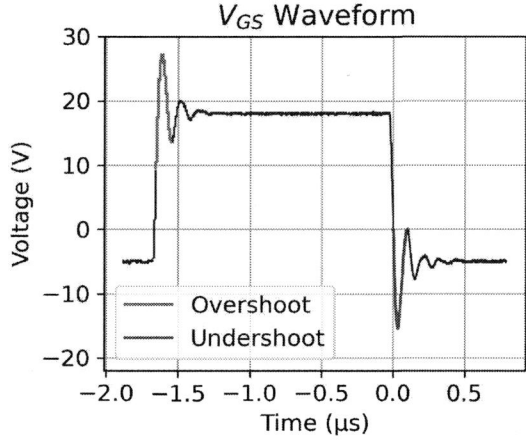

Fig. 3: Experimental V_{GS} waveform.

R_g was purposely set to 0 Ω to generate over- and under-shoots, which are not accounted for in AQG 324 but worsen the parametric drift [17]. This is needed to mimic a realistic scenario. V_{DS} was set to 0 V. Worth noting, in real-field scenarios V_{DS} variation might have a non negligible impact on V_{th}, like some papers argued [22], [20], [23]. For instance, [23] observed that the V_{th} drift in applicative-like case, a synchronous boost topology at light load and R_g equal to 4.7 Ω, is similar to V_{th} drift achieved by DGS with V_{DS} and R_g equal to zero.

The number of cycles was equal to $1.8e11$.

Sixty devices under test (DUTs) were tested at different temperatures, namely 20 DUTs at 25 °C, 20 at 125 °C, and

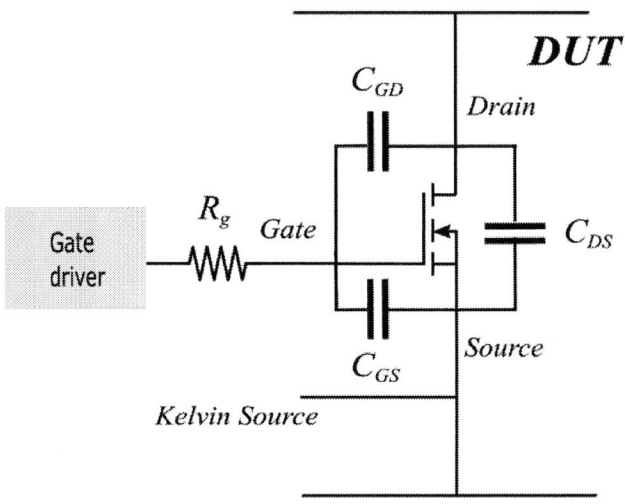

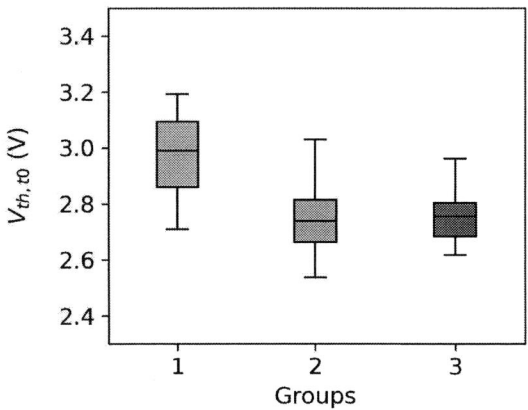

Fig. 5: V_{th} distribution among the three groups.

Fig. 4: Schematic of the DUT inside the DGS test setup, including the internal capacitances and the external gate resistor.

20 at 150 °C, hence, three groups of devices have been tested. We selected this number of samples as a trade-off among practical concerns, i.e. the equipment capacity and the statistical valence. It is worth noting that the V_{th} distributions were similar for the three test groups, and purposely spanning a large interval to consider a representative sample of the manufacturing process. Fig. 5 depicts the V_{th} distribution for the three groups. All the groups follow a Gaussian distribution. We observe that group 1, the one tested at 25 °C, has a slighty higher mean value in comparison to the other ones. Nevertheless, we see no correlation between V_{th} value at time zero and V_{th} drift after DGS. Therefore, the V_{th} at time zero does not influence the V_{th} drift.

The instrument used for the test is from SET-NI company. A photo of the experimental setup, including the testing machine, the oscilloscope, and the probes is depicted in Fig. 1. Fig. 2 shows a focus of the experimental setup. More specifically, a passive probe PR150B, placed between the gate and the Kelvin source of the MOSFET, was used for the tests. Fig. 4 depicts the electric schematic of the MOSFET inside the test setup, including the gate driver, the internal capacitors, and the gate resistor. The V_{GS} signal was filtered against electromagnetic interference by subtracting the background noise captured at a later stage when the probe was grounded. All the characterizations, like the sample one depicted in Fig. 3, were performed at 25 ° C, once we ensured that V_{GS} waveforms were not affected by temperature. V_{th} measurement, using preconditioning, was executed following JEP 183 guideline from JEDEC at the test temperature, i.e., V_{th} is measured at 150 °C for samples stressed at 150 °C.

III. EXPERIMENTAL RESULTS AND DISCUSSION

This section presents the results of the DGS test and the statistical data analysis. Fig. 6a shows the online V_{th} during the test, averaged among the 20 DUTs of each of the three groups. V_{th} decreases at high temperature as it should, as a negative correlation between V_{th} and temperature exists [24] and not for a different V_{th} behavior among groups. In Fig. 6b, a V_{th} drift is captured compared to time zero for the three tested temperatures. In the 25 °C case, we observe an increase of 102 mV at the end of the test, or 3 % in relative terms compared to the time zero value. Regarding the 125 °C and 150 °C cases, we observed 141 mV and 136 mV V_{th} drifts, respectively, or 6 % in both cases. Nevertheless, $R_{DS(on)}$ had no significant shift. In Fig. 6c, we report the behavior of this parameter during the test, showing the relative variation for the three groups at time zero (t_0).

As a first conclusion, it is worth highlighting that temperature speeds V_{th} drift up, which can be likely due to the interplay between electrons and holes capture phenomena [13]. A possible interpretation is that the electrons' capture time constant decreases with temperature more than the holes' one, which ends up polarizing the gate oxide with a net negative charge. This results in a higher V_{th} drift. It is worth noting that the temperature effect is similar between 125 °C and 150 °C.

We performed also a statistical analysis of the data obtained from DGS tests. The first step concerned the investigation of the normality of the data distribution. To implement this analysis, we chose the Shapiro-Wilk test, which is the most powerful according to the literature [25]. The outcome of the test does not validate the normality hypothesis. Accordingly, a non-parametric test was used to statistically assess the significant difference among the data (that is, statistical significance $\alpha = 0.05$) in the waveform parameters before and after the DGS test. Specifically, since paired observations are the ones under our analysis [26], we adopted the two-sample Wilcoxon signed rank test [27].

Fig. 7a and 7b show pre- and post-stress maximum and minimum V_{GS} values, respectively: we did not find any relevant variation, i.e., p-values are greater than $\alpha = 0.05$ for

979-8-3315-1612-3/25 $31.00 © 2025 IEEE

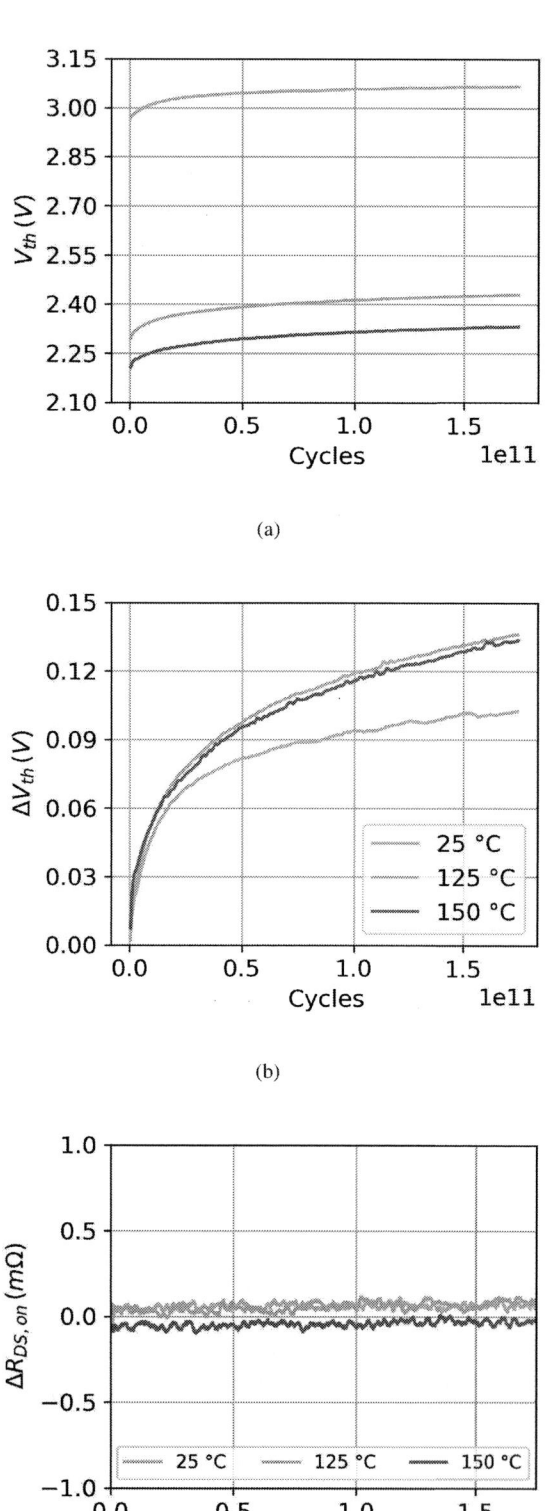

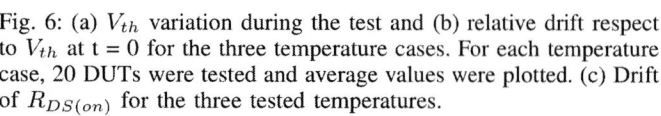

Fig. 6: (a) V_{th} variation during the test and (b) relative drift respect to V_{th} at t = 0 for the three temperature cases. For each temperature case, 20 DUTs were tested and average values were plotted. (c) Drift of $R_{DS(on)}$ for the three tested temperatures.

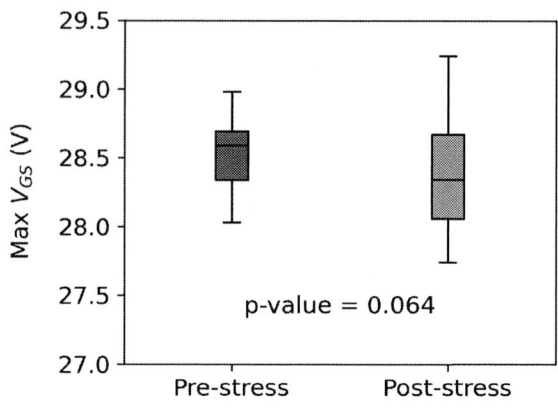

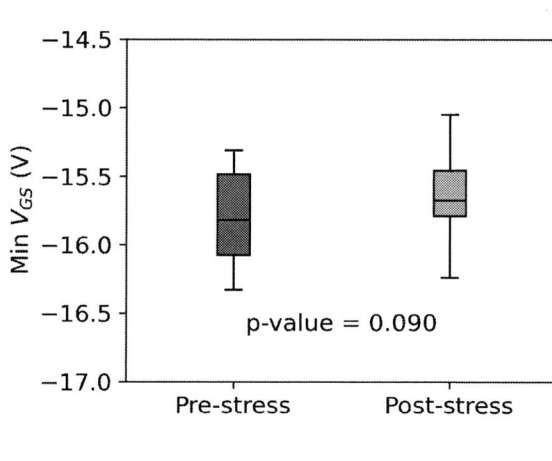

Fig. 7: Pre- and post-stress V_{GS} maximum (a) and minimum (b) values at 150 °C (20 samples per boxplot).

all parameters. More in detail, the p-value is equal to 0.064 for the maximum V_{GS} values, while is equal to 0.090 for the minimum V_{GS} values. The same conclusion can be drawn for rising and falling dV_{GS}/dt values, which are shown in Figs. 8a and 8b, respectively: in this case, we obtained p-values equal to 0.43 and 0.927, respectively.

Therefore, from an applicative point of view, the above results indicate that working at standard automotive conditions, i.e., $\geq 1e11$ cycles, DGS does not have a relevant impact.

IV. CONCLUSIONS AND FUTURE WORK

This paper has presented an experimental study of dynamic gate stress (DGS) on automotive-grade, planar-gate 1200 V SiC power MOSFETs. The temperature effect on threshold voltage (V_{th}) drift has been studied, showing an acceleration of V_{th} increase at high temperatures. On the other hand, no

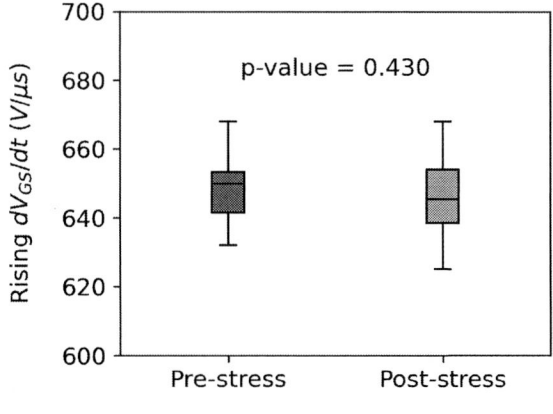

(a)

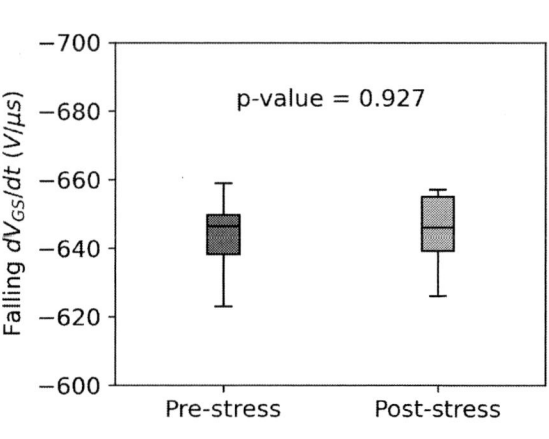

(b)

Fig. 8: Pre- and post-stress rising (a) and falling (b) dV_{GS}/dt values at 150 °C (20 samples per boxplot).

significant difference in gate-source (V_{GS}) voltage switching gradients has been observed, neither pre-, nor post-stress.

We conclude that long-term, high-temperature DGS at standard automotive conditions, i.e., $\geq 1e11$ cycles, has produced negligible alterations on the tested devices, i.e., 6 % for V_{th} at high temperatures, which do not have any relevant impact from an application point of view.

Possible extensions of this work would deal with the impact of V_{DS} being different from zero. We can achieve this goal by implementing a dynamic reverse bias (DRB) test with V_{GS} switching, as developed in [28]. In future applications, we will focus on a more accurate V_{th}, according to JEP 183, and employ different V_{DS} values, using the DRB setup.

REFERENCES

[1] W. Saito, "A future outlook of power devices from the viewpoint of power electronics trends," *IEEE Transactions on Electron Devices*, 2023.

[2] B. J. Baliga, "Silicon Carbide Power Devices: Progress and Future Outlook," *IEEE Journal of Emerging and Selected Topics in Power Electronics*, vol. 11, no. 3, pp. 2400–2411, Jun. 2023. [Online]. Available: https://ieeexplore.ieee.org/document/10072398/

[3] F. Di Giovanni, "Silicon carbide: Physics, manufacturing, and its role in large-scale vehicle electrification," *Chips*, vol. 2, no. 3, pp. 209–222, 2023.

[4] M. Calabretta, A. Sitta, S. M. Oliveri, and G. Sequenzia, "Silicon carbide multi-chip power module for traction inverter applications: thermal characterization and modeling," *IEEE Access*, vol. 9, pp. 76 307–76 314, 2021.

[5] X. Yuan, I. Laird, and S. Walder, "Opportunities, challenges, and potential solutions in the application of fast-switching sic power devices and converters," *IEEE Transactions on Power Electronics*, vol. 36, no. 4, pp. 3925–3945, 2020.

[6] S. Pu, F. Yang, B. T. Vankayalapati, and B. Akin, "Aging Mechanisms and Accelerated Lifetime Tests for SiC MOSFETs: An Overview," *IEEE Journal of Emerging and Selected Topics in Power Electronics*, vol. 10, no. 1, pp. 1232–1254, Feb. 2022. [Online]. Available: https://ieeexplore.ieee.org/document/9529341/

[7] J. Wei, Z. Wei, H. Fu, J. Cao, T. Wu, J. Sun, X. Zhu, S. Li, L. Zhang, S. Liu, and W. Sun, "Review on the Reliability Mechanisms of SiC Power MOSFETs: A Comparison Between Planar-Gate and Trench-Gate Structures," *IEEE Transactions on Power Electronics*, vol. 38, no. 7, pp. 8990–9005, Jul. 2023. [Online]. Available: https://ieeexplore.ieee.org/document/10098210/

[8] L. Yang and A. Castellazzi, "High temperature gate-bias and reverse-bias tests on sic mosfets," *Microelectronics Reliability*, vol. 53, no. 9-11, pp. 1771–1773, 2013.

[9] M. W. Feil, K. Waschneck, H. Reisinger, J. Berens, T. Aichinger, S. Prigann, G. Pobegen, P. Salmen, G. Rescher, D. Waldhoer, A. Vasilev, W. Gustin, M. Waltl, and T. Grasser, "Gate Switching Instability in Silicon Carbide MOSFETs—Part I: Experimental," *IEEE Transactions on Electron Devices*, vol. 71, no. 7, pp. 4210–4217, Jul. 2024. [Online]. Available: https://ieeexplore.ieee.org/document/10532137/

[10] T. Grasser, M. W. Feil, K. Waschneck, H. Reisinger, J. Berens, D. Waldhoer, A. Vasilev, M. Waltl, T. Aichinger, M. Bockstedte, W. Gustin, and G. Pobegen, "Gate Switching Instability in Silicon Carbide MOSFETs—Part II: Modeling," *IEEE Transactions on Electron Devices*, vol. 71, no. 7, pp. 4218–4226, Jul. 2024. [Online]. Available: https://ieeexplore.ieee.org/document/10538242/

[11] "AQG 324 ECPE Guideline - Qualification of Power Modules for Use in Power Electronics Converter Units in Motor Vehicles," ECPE European Center for Power Electronics, Nuremberg, DE, Tech. Rep., 05 2021. [Online]. Available: https://www.ecpe.org/research/working-groups/automotive-aqg-324/

[12] "Jep 195, "guideline for evaluating gate switching instability of silicon carbide metal-oxide-semiconductor devices for power electronic conversion"," JEDEC, Standard, 05 2021.

[13] H. Jiang, X. Qi, G. Qiu, X. Zhong, L. Tang, H. Mao, Z. Wu, H. Chen, and L. Ran, "A Physical Explanation of Threshold Voltage Drift of SiC MOSFET Induced by Gate Switching," *IEEE Transactions on Power Electronics*, vol. 37, no. 8, pp. 8830–8834, Aug. 2022. [Online]. Available: https://ieeexplore.ieee.org/document/9740407/

[14] J. R. Garcia-Mere, A. A. Gomez, J. Roig-Guitart, J. Rodriguez, and A. Rodriguez, "Aging Modeling and Simulation of the Gate Switching Instability Degradation in SiC MOSFETs," in *2024 IEEE Applied Power Electronics Conference and Exposition (APEC)*. Long Beach, CA, USA: IEEE, Feb. 2024, pp. 653–658. [Online]. Available: https://ieeexplore.ieee.org/document/10509430/

[15] D. B. Habersat and A. J. Lelis, "AC-Stress Degradation and Its Anneal in SiC MOSFETs," *IEEE Transactions on Electron Devices*, vol. 69, no. 9, pp. 5068–5073, Sep. 2022. [Online]. Available: https://ieeexplore.ieee.org/document/9837829/

[16] M. W. Feil, K. Waschneck, H. Reisinger, J. Berens, T. Aichinger, P. Salmen, G. Rescher, W. Gustin, and T. Grasser, "Towards Understanding the Physics of Gate Switching Instability in Silicon Carbide MOSFETs," in *2023 IEEE International Reliability Physics Symposium (IRPS)*. Monterey, CA, USA: IEEE, Mar. 2023, pp. 1–10. [Online]. Available: https://ieeexplore.ieee.org/document/10117740/

[17] X. Zhong, H. Jiang, G. Qiu, L. Tang, H. Mao, X. Chao, X. Jiang, J. Hu, X. Qi, and L. Ran, "Bias Temperature Instability of Silicon Carbide Power MOSFET under AC Gate Stresses," *IEEE*

979-8-3315-1612-3/25 $31.00 © 2025 IEEE

Transactions on Power Electronics, pp. 1–1, 2021. [Online]. Available: https://ieeexplore.ieee.org/document/9516960/

[18] S. Zhu, L. Shi, M. Jin, J. Qian, M. Bhattacharya, H. L. Rao Maddi, M. H. White, A. K. Agarwal, T. Liu, A. Shimbori, and C. Chen, "Reliability Comparison of Commercial Planar and Trench 4H-SiC Power MOSFETs," in *2023 IEEE International Reliability Physics Symposium (IRPS)*. Monterey, CA, USA: IEEE, Mar. 2023, pp. 1–5. [Online]. Available: https://ieeexplore.ieee.org/document/10117998/

[19] C.-C. Tu, C.-L. Hung, K.-B. Hong, S. Elangovan, W.-C. Yu, Y.-S. Hsiao, W.-C. Lin, R. Kumar, Z.-H. Huang, Y.-H. Hong *et al.*, "Industry perspective on power electronics for electric vehicles," *Nature Reviews Electrical Engineering*, pp. 1–18, 2024.

[20] Y. Cai, P. Sun, C. Chen, Y. Zhang, Z. Zhao, X. Li, L. Qi, Z. Chen, and H.-P. Nee, "Investigation on Gate Oxide Degradation of SiC MOSFET in Switching Operation," *IEEE Transactions on Power Electronics*, vol. 39, no. 8, pp. 9565–9578, Aug. 2024. [Online]. Available: https://ieeexplore.ieee.org/document/10508119/

[21] Y. Cai, P. Sun, Y. Zhang, C. Chen, Z. Zhao, X. Li, L. Qi, Z. Chen, and H.-P. Nee, "Dynamic analytical switching loss model of sic mosfet considering threshold voltage instability," *IEEE Transactions on Power Electronics*, 2024.

[22] S. Thiele, C. Herrmann, D. Yang, M. Neumeister, and T. Basler, "Gate switching instability of sic mosfets under simultaneously high drain-source voltage and high frequency acceleration," in *2024 36th International Symposium on Power Semiconductor Devices and ICs (ISPSD)*. IEEE, 2024, pp. 136–139.

[23] A. A. Gómez, J. R. García-Meré, A. Rodríguez, J. Rodríguez, C. Jimenez, and J. Roig-Guitart, "Deep investigation on sic mosfet degradation under gate switching stress and application switching stress," in *2024 IEEE Applied Power Electronics Conference and Exposition (APEC)*. IEEE, 2024, pp. 1067–1072.

[24] H. Li, X. Liao, Y. Hu, Z. Zeng, E. Song, and H. Xiao, "Analysis of SiC MOSFET dI/dt and its temperature dependence," *IET Power Electronics*, vol. 11, no. 3, pp. 491–500, Mar. 2018. [Online]. Available: https://onlinelibrary.wiley.com/doi/10.1049/iet-pel.2017.0203

[25] N. M. Razali and Y. B. Wah, "Power comparisons of shapiro-wilk, kolmogorov-smirnov, lilliefors and anderson-darling tests," *Journal of Statistical Modeling and Analytics*, vol. 2, no. 1, pp. 21–33, 2011. [Online]. Available: https://ieeexplore.ieee.org/document/10508119/

[26] T. J. Cleophas and A. H. Zwinderman, "Paired Continuous Data (Paired T-Test, Wilcoxon Signed Rank Test)," in *Clinical Data Analysis on a Pocket Calculator*. Cham: Springer International Publishing, 2016, pp. 31–36. [Online]. Available: http://link.springer.com/10.1007/978-3-319-27104-0_6

[27] F. S. Nahm, "Nonparametric statistical tests for the continuous data: the basic concept and the practical use," *Korean Journal of Anesthesiology*, vol. 69, no. 1, p. 8, 2016. [Online]. Available: http://ekja.org/journal/view.php?doi=10.4097/kjae.2016.69.1.8

[28] A. Sitta, G. Mauromicale, M. Fiore, and M. Calabretta, "Reliability assessment of sic power mosfets in dynamic reverse bias test," in *35th European Symposium on Reliability of Electron Devices, Failure Physics and Analysis (ESREF 2024)*, 2024, pp. 1–4.

Innovative gate driver structure achieving low time skew across isolation barrier for parallel connected SiC modules

1st Louison Gouy
IETR, Nantes Université
Nantes, France
louison.gouy@etu.univ-nantes.fr

2nd Anne-Sophie Descamps
IETR, Nantes Université
Nantes, France
anne-sophie.descamps@univ-nantes.fr

3rd Nicolas Ginot
IETR, Nantes Université
Nantes, France
nicolas.ginot@univ-nantes.fr

4th Christophe Batard
IETR, Nantes Université
Nantes, France
christophe.batard@univ-nantes.fr

Abstract—The increasing demand for high-power density applications has driven the need for parallel connection of power transistors. However, the development of wide bandgap semiconductors introduces significant challenges in parallel operation, primarily due to the need to synchronize sharp current edges during switching. One cause of current imbalance is the delay between gate signals arise from mismatches in gate drivers. This work introduces an innovative low-jitter gate driver architecture design for eight parallel 1.2kV 204A SiC modules (ref.BSM180D12P2E002). A novel approach leverages a planar transformer with dual windings, integrating level-shifting capabilities and minimizing components responsible for jitter. A dedicated synchronization channel ensures less than 100 ps timing precision between gate signals, while a modular and flexible chained structure supports scalability to various transistor configuration. Experimental validation demonstrates the proposed architecture's robustness across a wide temperature range (-20°C to 80°C) and its ability to achieve a maximum propagation delay skew of 5.15ns. This design represents a significant step forward in enabling efficient, high-speed switching in parallel transistor systems.

Index Terms—Gate driver, Jitter, Silicon carbide (SiC), MOSFET, paralleled power modules, gate driver topologies, current sharing

I. PARALLEL CONNECTION CHALLENGES

The advantages of silicon carbide (SiC) technology are now well established [1]. The maximum current rating of commercial SiC power modules is typically limited to 500A, parallel operation is therefore a common strategy for enhancing current capability [2]. In this configurations, it is crucial that the currents in parallel MOSFETs follow the same trajectory during switching phases to prevent unequal heating [3] and premature aging [4]. Current imbalances can arise from variations in device characteristics, such as threshold voltage (Vth) [5] [6], or arise from circuit differences including asymmetries in power loops [7] [8] [9], unequal thermal dissipation [10] [11], and gate signal delays [12] [13]. Another significant challenge in parallel operation is the existence of parasitic gate loop oscillations [14] [15] [16]. This phenomenon is caused by mismatched circuit inductances and current imbalances.

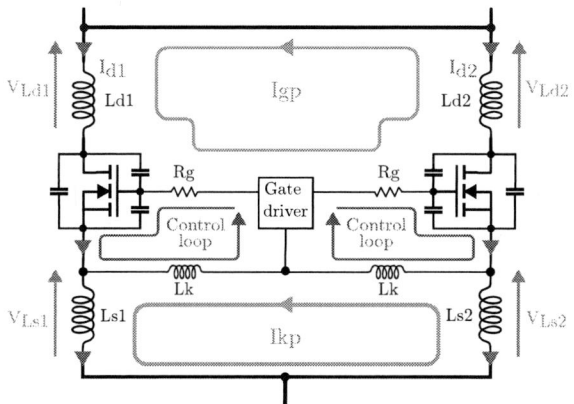

Fig. 1. Schematic of parasitic gate loop

During switching, the rising and falling edges of power current generate a voltage drop in stray inductances Lsx and Ldx. If an asymmetry exists in the current slope or in the inductance values, the resulting voltage difference will allow a current to flow through the parasitic loops (see figure 1). This phenomenon is primarily observable in the source loop due the low-impedance path. When the current flows through the Kelvin sources, it generates a voltage drop in the component that it crosses, such as the parasitic inductances Lk. In the control loop, this voltage can be added or subtracted from the driver voltage, thereby modifying the Vgs value. This will consequently impact the switching behaviors, increase power current imbalance, introduce oscillations, and potentially damage the module due to overvoltage on the gate or overcurrent in the Kelvin source [17]. This article proposes a novel low-jitter gate driver structure for paralleling eight 1.2kV 204A SiC modules (ref.BSM180D12P2E002). It addresses both the problem of current imbalance caused by propagation delays and the issue of gate loop oscillations.

979-8-3315-1612-3/25 $31.00 © 2025 IEEE

II. LITERATURE ON GATE DRIVER STRUCTURES

The literature documents several gate driver structures aimed at optimizing the parallel operation of transistors. At the gate driver level, the two principal challenges to address are synchronization and the mitigation of parasitic gate loop oscillations. Each gate driver structure exhibits specific benefits in different application contexts. The simplest approach to control transistors in parallel is to use a single gate driver (Figure 2-a). This configuration offers optimal synchronization, as all transistors receive the same gate signal simultaneously. However, this structure is primarily feasible for discrete transistors, where interconnection parasitic inductances are relatively low. In power modules, where gate connections are often separated by distances of several centimeters, the single gate driver structure becomes impractical due to the substantial parasitic inductances introduced by longer connections. Consequently, in such applications, it is essential to position a buffer as close as possible to each transistor gate (Figure 2-b). This arrangement preserves satisfactory synchronization since galvanic isolation—often a source of jitter—is located upstream in the circuit. In [18], these structures are classified into three distinct categories: non-isolated indirect push-pull gate drivers (a), non-isolated direct push-pull gate drivers (b) and isolated gate drivers (c). The second constraint associated

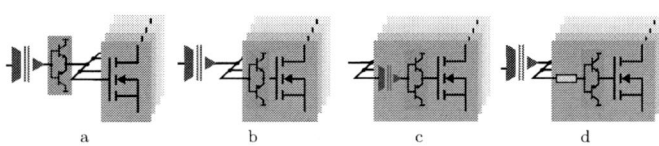

a   b   c   d

Fig. 2. Existing gate driver structure: non-isolated indirect push-pull (a), non-isolated direct push-pull (b), isolated (c) and non-isolated direct push-pull with damping impedance (d)

with parallel transistor operation is the presence of parasitic loops, which can induce oscillations. The two aforementioned configurations, classified as non-isolated gate driver structures, do not adequately address this issue, rendering them unsuitable for high-power applications. To mitigate this constraint, a third approach involves deploying one gate driver per transistor (Figure 2-c). This structure effectively breaks the parasitic loop, as the isolation is integrated within each gate driver. However, regarding synchronization, isolated gate driver structures present disadvantages due to variable propagation times across galvanic isolation methods—such as optical fibers, optocouplers, or magnetic isolators—which can fluctuate by tens of nanoseconds [19, 20]. Such variability can cause significant current imbalances under fast-switching conditions (typically <50 ns). Thus, a fundamental trade-off exists between synchronization and the control of gate loop oscillations. In recent studies, various methods have been explored to suppress oscillations without introducing additional sources of jitter. A common strategy is to place an impedance upstream of the buffer to attenuate oscillations while minimally impacting the gate command signal (Figure 2-d). For example, a resistor has been implemented with this intent in [18]. In other approaches,

a common-mode choke, connected between the gate and Kelvin source, is proposed [14, 21, 22]. This component is particularly effective, as it allows targeted filtering of common-mode currents, which are the primary drivers of oscillations. Despite these solutions demonstrate effectiveness in specific contexts, component sizing remains challenging, particularly for plug-and-play gate drivers. Common-mode chokes, for instance, face limitations in terms of bandwidth and saturation. Additionally, no studies to date have extensively addressed the reliability of these circuits under short-circuit conditions within power modules. Furthermore, few measurement-based studies exist to evaluate synchronization performance across these designs. The scalability of these configurations is another limitation, as circuits are typically designed for a fixed number of parallel transistors (usually two or three), making it difficult to expand or adapt them for larger configurations.

III. INNOVATIVE LOW-JITTER GATE DRIVER STRUCTURE

This article addresses the two main challenges of gate driver design for parallel transistors by proposing a novel structure. This design derived from the fourth configuration previously discussed (Figure 2-d) where isolation is applied upstream and a dedicated buffer is used for each module. However, instead of utilizing a damping impedance to mitigate oscillations, this solution employs a secondary isolation stage to entirely break the current loop. This isolation, specifically designed to maintain a constant propagation delay, is based on a planar transformer. The development of a galvanic isolation system that can withstand high dv/dt while maintaining low jitter represents a significant challenge. This is why the selected architecture (ref Figure 3) employs two different isolations. The first is a regular secure double isolation which follows rigorous specifications and norms. The second is a low-voltage (100V) and jitter-optimized isolation employed to suppress common-mode currents. The secure isolation is shared by all transistors, while the second separates each driver, enabling synchronous control across all transistors. The structure is illustrated in Figure 3, showing the overall prototype. The design comprises a predriver and individual driver stages (detailed in Figure 4). The predriver is divided into a primary section and two secondary sections, corresponding to the high-side and low-side switches of the half-bridge. The conventional secure and functional isolation stages are shown with double and single red lines, respectively, while the novel isolation is indicated by the dashed red line. The drivers are chained via a flat cable, which links them to the predriver, facilitating modular integration power stack. Due to the uncontrolled propagation delay in flat cables, the proposed solution integrates a dedicated synchronization channel. A synchronization pulse (Sync), which serves as a clock signal, is transmitted with each state change of the control signal through coaxial cables and the low-jitter isolation stage. The switching commands, along with auxiliary signals such as short-circuit detection and soft turn-off, are transmitted through the flat cable, passing through a low-cost digital isolator without strict timing constraints. Inspired by a processor data bus structure, the switching

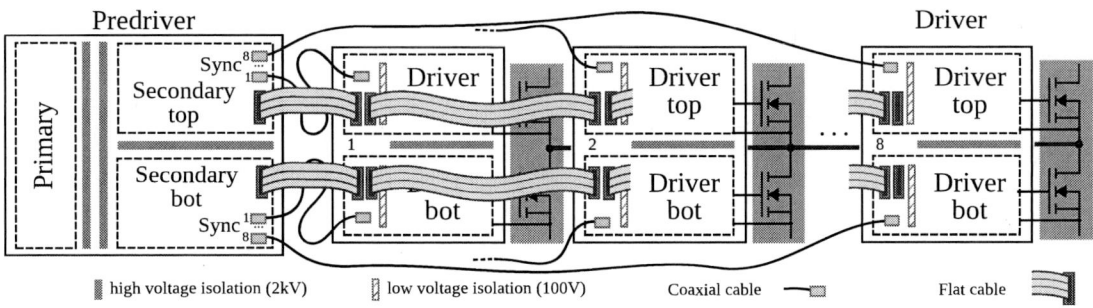

Fig. 3. Proposed structure of gate driver

commands are re-synchronized to the clock pulse's rising edge using a flip-flop. The synchronization stage across isolation barrier using a flipflop was initially proposed in [19] and [23] for other applications. It is extensively adapted in the context of parallelization. The novel low-jitter isolation design, based

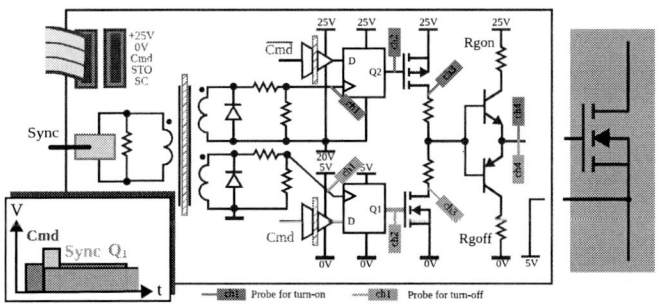

Fig. 4. Driver board electronic circuit

on a planar transformer with low part-to-part variation, enables a synchronous command signal across all drivers. To eliminate jitter sources, the driver avoids conventional level shifters, which typically convert the control voltage (5V) to the gate voltage (25V), as these can introduce up to 20ns of jitter [24]. The proposed solution (Figure 4) leverages the existing planar transformer by creating a double secondary winding. Thereby, it is enabling to address the two flip-flops, one referenced at the top and the other at the bottom of the driver power supply, achieving so the level conversion. Two diodes control the demagnetization voltage of the transformer. A push-pull complementary MOSFET preamplifier provides the base current for the bipolar transistors of the final buffer. The gate resistances R_{gon} and R_{goff} are placed on the collector, allowing different values for turn-on and turn-off without relying on diodes highly temperature-sensitive. It is interesting to note the low number of components employed after the resynchronization stage (after the flip-flop) in order to reduce the potential source of jitter. To validate the proposed architecture a rigorous measurement phase is conducted.

IV. EXPERIMENTATION

A. Method of measurement

The novel gate driver has been designed to eliminate the source of jitter. Subsequently, the performance of the circuit must be evaluated by comparing the operation of four gate drivers in parallel. Specifically, the time difference between identical signals is gauged at a predetermined level. This type of measurement is typically used in digital electronics, where it is referred to as jitter or time skew (figure 5). The jitter refers to time deviation of a controlled edge from its nominal position. It is employed in microprocessor clock generation systems. The time skew is defined as the magnitude of the time difference between two events that ideally would occur simultaneously. It can be expressed as the difference between two propagation times t_p.

$$t_{sk} = t_{p1} - t_{p2} \qquad (1)$$

This technique is applied to evaluate the functionality of clock trees in integrated circuits. The key differentiating factor between jitter and skew is the reference frame in which they are measured. Jitter is a measure of the discrepancy between a signal and its expected or historical value. In contrast, skew involves a comparison between two signals. Consequently, time skew represents an effective method for

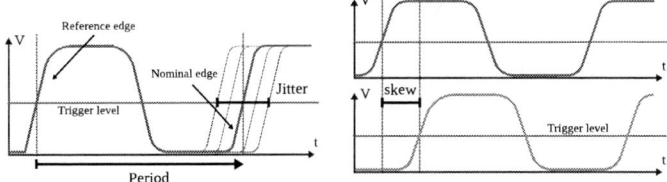

Fig. 5. Definition of jitter and time skew

evaluating the performance of gate driver synchronism. This definition is suitable for use in measuring two signals but need to be extended for more. In the literature and particularly in datasheets, time skew is typically measured between every channel, with the maximum value reported. This method would yield an overestimated result in our application. Indeed, the distribution of losses between transistors can occur on the scale of several switching cycles. The fastest gate driver can also be the slowest in the next switching, and the result is a good distribution. By retaining only the maximum time skew, we would fail to leverage this property. Therefore, it is preferable to calculate the mean time skew over Ns switching cycles and then select the highest. In the next experiment the

979-8-3315-1612-3/25 $31.00 © 2025 IEEE

time skew is calculated based on the propagation time. For this reason, it is interesting to note that mean time skew can also be expressed has difference of mean propagation times.

$$\langle t_{sk} \rangle = \frac{1}{N_s} \sum_{n=1}^{N_s} t_{p1_n} - t_{p2_n} = \frac{1}{N_s} \sum_{n=1}^{N_s} t_{p1_n} - \frac{1}{N_s} \sum_{n=1}^{N_s} t_{p2_n} \quad (2)$$

The second expression can be easily to visualize on propagation time histograms in Figure 10 to 13.

B. Synchronization channel skew

The synchronization (Sync) signal generated by the pre-driver and transmitted to the gate driver ensures proper synchronization between drivers. This signal is carried through a coaxial cable and passes through the transformer before being processed by a flip-flop. It consists of a 15V pulse with a duration of 50ns. It contains high-frequency harmonics with wavelengths comparable to the length of the transmission cable (1m). For this reason, a 50Ω resistor is placed at the cable output to ensure impedance matching and prevent signal reflections. Coaxial cables must be the same length. A tolerance is allowed, each centimeter of difference adds 53ps of delay. The innovation presented in this study lies in the use of low-voltage and jitter-optimized isolation techniques to suppress common-mode currents. The critical synchronization path, spanning from pulse generation on the pre-driver to the secondary side of the planar transformers, including the coaxial cable, is examined. To simulate conditions under maximum load for the synchronization generator, eight planar transformers are connected in parallel, with four of them monitored using a four-channel oscilloscope (figure 6). The primary objective is

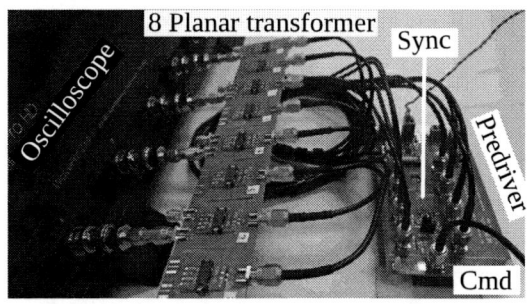

Fig. 6. Setup synchronization channel jitter measurement

to measure the time skew between signal edges, as described in method section IV-A. The trigger level is set at 3.5V, which corresponds to the typical input threshold of the flip-flop. Time skew measurements are conducted over 10,000 signal edges, with the maximum mean time skew observed being 68ps. The figure 7 illustrate the results on a sample of 500 edges. The results demonstrate the efficacy of the synchronization path in mitigating jitter. This is particularly interesting because the signal crosses an isolation barrier that also acts as a level transfer. The two aforementioned stages, identified as the primary source of jitter in the isolated gate driver, have been replaced by this structure. Additional testing can be conducted to assess the performance of the complete gate driver.

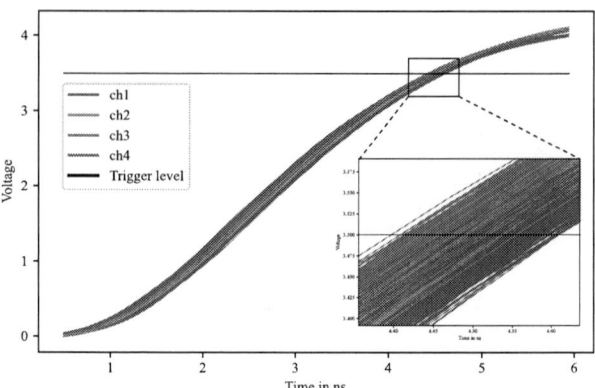

Fig. 7. Sync edge after 4 planar transformers for 500 samples

C. Propagation delay over temperature

This experiment examines the propagation delay of the driver board (Figure 8) at three temperatures -20°C, 25°C and 80°C at various points in the circuit. The aim is twofold: first,

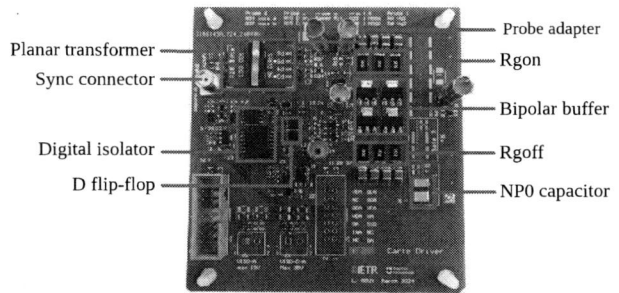

Fig. 8. Driver board with component caption

to assess the circuit's performance over varying temperatures, and second, to identify which components contribute the most to jitter generation. The board corresponds exactly to the driver section shown in the previous figures, where the placement of the probes is illustrated in Figure 4. The Sync connector visible on Figure 8 is followed by the transformer mounted in a castellated configuration with 3F36 ferrite. Below, the flat cable arrives on the blue connector, where it is connected to the non-jitter-optimized isolator before being routed to the flip-flops (74LVC1G74). Finally, the bipolar transistors (ref. 2SA2016/2SC5569) are located on the right, with gate resistances of 1.1Ω. In order to ensure an accurate analysis of the gate driver contribution, a fixed-value NP0 capacitor of 33nF is utilized instead of the power MOSFET. This capacitive value is selected based on the capacity Cgs of the targeted power module. The Figure 4 indicates the probe location. The time difference between channels 1 and 2 shows the flip-flop performance. Channels 2 and 3 observe the MOSFET. Channels 3 and 4 are for the bipolar transistors. Each signal is triggered at a different level depending on the following stage operation. Channel 1 is triggered at 3.5V, which corresponds to the Vt+ level of the flip-flop. Channel 2 is triggered at 2.5V,

979-8-3315-1612-3/25 $31.00 © 2025 IEEE

which is equal to the Vth of the signal MOSFET. Channel 3 is triggered at 50% of the output, as the bipolar transistor operates in linear mode without specific threshold. Channel 4 is triggered at 2.5V, which is equal to the Vth of the power MOSFET. The propagation delay for each of the three stages is calculated based on 10,000 observations. The board is placed in a thermal chamber with precise temperature control. The measurement is repeated for the three temperature, and it is observed that the propagation time increased with the temperature increase. The difference between the mean values μ at -20°C and 80°C represents the maximum excursion.

Stage\Temp	-20°C	25°C	80°C	$\mu_{80} - \mu_{-20}$
Flip flop	1.077ns	1.255ns	1.339ns	0.262ns
Mosfet N	6.652ns	6.709ns	6.835ns	0.183ns
Bipolar	21.438ns	21.643ns	22.051ns	0.613ns
All	29.167ns	29.607ns	30.226ns	1.059ns

TABLE I

EXPERIMENTAL RESULT: STAGES MEAN PROPAGATION DELAY OVER TEMPERATURE AT TURN-ON

Stage\Temp	-20°C	25°C	80°C	$\mu_{80} - \mu_{-20}$
Flip flop	1.039ns	1.072ns	1.283ns	0.244ns
Mosfet P	3.556ns	3.776ns	3.981ns	0.425ns
Bipolar	51.773ns	53.242ns	55.202ns	3.429ns
All	56.367ns	58.089ns	60.466ns	4.099ns

TABLE II

EXPERIMENTAL RESULT: STAGES MEAN PROPAGATION DELAY OVER TEMPERATURE AT TURN-OFF

The table II and I shows the numerical results. The stage most sensitive to temperature is the bipolar stage with the maximum excursion. The discrepancies become more evident at turn-off point, as the distance between the initial state and the threshold is greater. Indeed, as V_{th} is taken as trigger level, the gate voltage is observed to begin at -5V and end at 2.5V during the turn-on phase. Conversely, the turn-off phase is studied from 25V to 2.5V. With a same time constant, the time to travel across a wilder voltage difference is naturally higher. Nevertheless, the results show an excellent performance with a maximum difference in the mean propagation delay values of 4ns despite harsh temperature variations.

D. Propagation delay on 4 gate drivers

In the third experiment, the part-to-part propagation delay of the gate signals is evaluated. This analysis is conducted using four gate drivers connected in parallel, as shown in Figure 9. Firstly, the gate drivers are tested at 25°C, where their propagation times are measured and ranked accordingly. The propagation delay is defined as the time between the synchronization (Sync) signal generated by the predriver and the virtual gate voltage on the NPO capacitor. The trigger is chosen on the gate voltage threshold of the targeted power module. To accommodate the oscilloscope's (Teledyne Lecroy WavePro 254HD) four-channel limit, the external trigger is employed to start measurements on the rising edge of the Sync signal. This setup captures the total propagation delay across

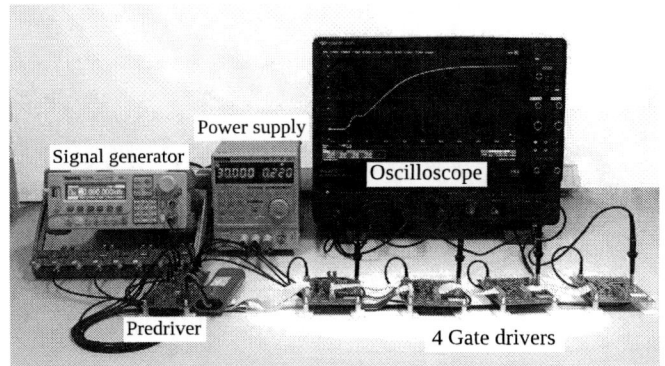

Fig. 9. Test setup for part-to-part measurement

the system. Each measurement is repeated 10,000 times to ensure statistical accuracy. The control signal is generated by an arbitrary signal generator operating at a fixed frequency of 20kHz. The predriver is loaded with eight planar transformers to emulate realistic conditions for eight driver boards. To minimize the influence of common-mode currents, the predriver is powered by a battery, although this precaution is not strictly necessary for this experiment. The recorded measurements are stored and subsequently processed on a computer. Figures 10 and 11 present the experimental results as histograms for turn-on and turn off respectively. The data are approximated by normal distributions, with the mean values and standard deviations reported. The mean propagation delay during turn-

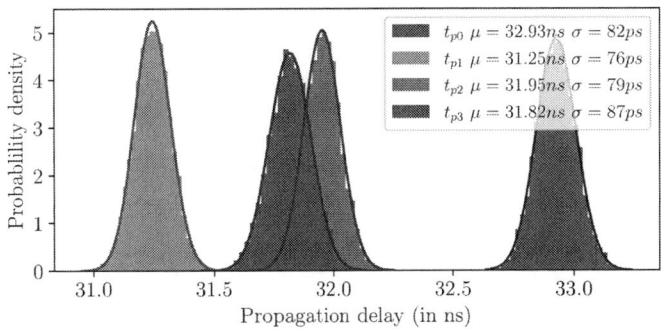

Fig. 10. Experimental result: Propagation delay of four gate drivers for turn-on at 25°C

on is observed to range from 31.25ns to 32.93ns, while during turn-off, it ranges from 61.14ns to 62.53ns. As expected, and consistent with the single-driver study, the turn-on delay is shorter compared to the turn-off delay. To highlight potential discrepancies, the drivers are exposed to extreme temperature conditions. The two drivers with the shortest propagation delays at 25°C are cooled to -20°C, while the other two are heated to 80°C. The ranking of the propagation delays differs between the turn-on and turn-off. The drivers are paired and placed in two thermal chambers configured to the specified temperatures. Temperature monitoring is performed using a digital thermometer, with the sensor attached directly to the driver boards for accurate readings. The experimental results

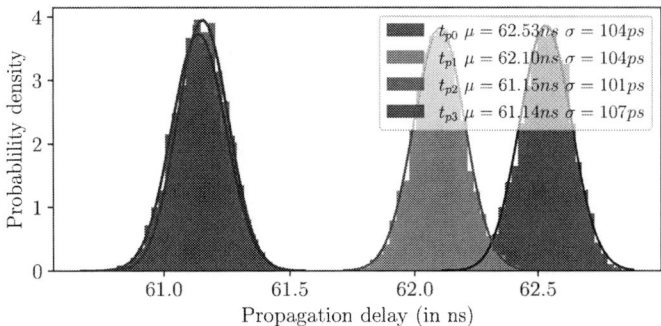

Fig. 11. Experimental result: Propagation delay of four gate drivers for turn-off at 25°C

are presented as histograms in Figures 12 and 13 for the second phase. The mean propagation time at turn-on is observed to be

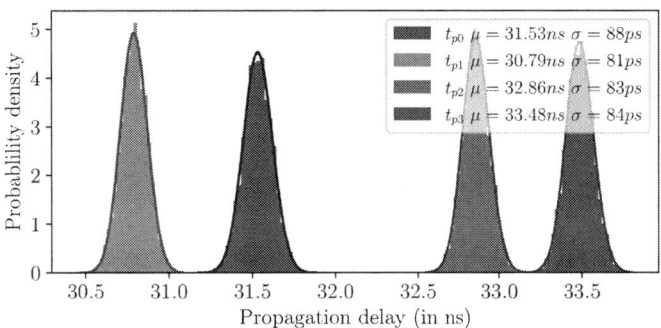

Fig. 12. Experimental result: Propagation delay of four gate driver at turn-on with ch1, ch2 at -20°C and ch3, ch4 at 80°C

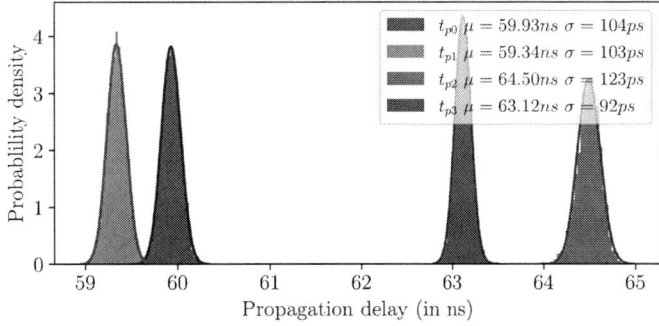

Fig. 13. Experimental result:Propagation delay of four gate driver at turn-ff with ch1, ch2 at -20°C and ch3, ch4 at 80°C

between 30.79ns and 33.48ns, while at turn-off, it is observed to be between 59.34ns and 64.50ns. While the absolute value is known, the most interesting indicator is the difference in propagation time between drivers. In order to achieve an equal repartition of current and switching losses between the power modules, the objective is to obtain identical mean values. The difference in maximum mean propagation times, also known as maximum mean time skew (see method section IV-A), is extracted from these results. At 25°C, the propagation delay difference between channel 0 and channel 1 is measured

as 1.68ns for the turn-on phase. Similarly, during turn-off at the same temperature, the difference between channel 3 and channel 0 is determined to be 1.39ns. Under varying temperature conditions, for the turn-on phase, subtracting the mean propagation delay of channel 3 from channel 0 yields a result of 2.69ns. For the turn-off phase, the delay difference is calculated by subtracting the mean value of channel 2 from channel 1, resulting in a value of 5.15ns. The results of this experiment demonstrate the influence of temperature and the impact of component disparities. According to the datasheet for the selected power module (BSM180D12P2E002), rise time t_r and fall time t_f are 45ns. To guarantee a dynamic current sharing, a small time difference is tolerate. Table IV-D presents the minimum, maximum, and variation in propagation delays t_p measured during turn-on and turn-off under uniform (25°C) and extreme (-20°C and 80°C) temperature conditions.

Conditions	$t_{p,\min}$	$t_{p,\max}$	Δt_p	% t_r, t_f
on, 25°C	31.25ns	32.93ns	1.68ns	3.73%
off, 25°C	61.14ns	62.53ns	1.39ns	3.09%
on, -20°C, 80°C	30.79ns	33.48ns	2.69ns	5.98%
off, -20°C, 80°C	59.34ns	64.50ns	5.15ns	13.28%

TABLE III
MEAN PROPAGATION DELAY AND DIFFERENCE COMPARISON OF
TEMPERATURE CONDITIONS

The results demonstrate that at 25°C, the time skew represents 3.73% of the power module switching time for turn-on and 3.09% for turn-off. In the context of extreme temperature conditions, the respective percentages are 5.98% and 13.28%. The low propagation delay variation, even under extreme temperature conditions, confirms that the proposed gate driver architecture introduces negligible timing discrepancies relative to the rise and fall times of the power module. This experiment, seldom documented in literature, demonstrates excellent gate driver synchronization.

V. CONCLUSION

The design of gate drivers for the parallel operation of power transistors must address both gate synchronization and the control of parasitic current loops. Existing structures highlight a trade-off between controlling propagation delays and mitigating parasitic loops. This article introduces an innovative low-jitter gate driver structure design to reduce current imbalances caused by gate signal delays in eight parallel SiC modules. Parasitic current loops are effectively eliminated using a dedicated isolation strategy. Level conversion, often a source of jitter, is implemented using a secondary winding on a planar transformer, minimizing the number of components. The proposed structure integrates a dedicated synchronization channel, enabling highly flexible configurations and easy scalability for varying numbers of transistors. The experimental validation follows a rigorous three-phase process. First, the synchronization channel alone is tested, demonstrating ultra-low jitter of 68ps. Second, measurements conducted under varying temperatures (-20°C to 80°C) on a single gate driver reveal the contribution of individual components to propagation delay. A difference of 4.099ns is observed between mean propagation delays, with the bipolar buffer stage responsible for 3.429ns. Lastly, four gate drivers are tested simultaneously, both at the same temperature and under harsh temperature differences. The results show a maximum mean time skew of 1.68ns at identical temperatures and 5.15ns with a significant temperature gradient. These findings confirm the excellent synchronization performance of the proposed gate driver. Notably, the measured propagation delay skews represent a negligible percentage of the rise and fall times of modern high-speed power modules, making this design suitable for next-generation applications where switching speeds continue to increase. A complete gate driver is currently being developed to confirm the solution in a power electronics context.

REFERENCES

[1] Xu She et al. "Review of Silicon Carbide Power Devices and Their Applications". In: *IEEE Transactions on Industrial Electronics* 64.10 (Oct. 2017), pp. 8193–8205. ISSN: 1557-9948. DOI: 10.1109/TIE.2017.2652401.

[2] Juan Colmenares et al. "High-Efficiency 312-kVA Three-Phase Inverter Using Parallel Connection of Silicon Carbide MOSFET Power Modules". In: *IEEE Transactions on Industry Applications* 51.6 (Nov. 2015), pp. 4664–4676. ISSN: 1939-9367. DOI: 10.1109/TIA.2015.2456422.

[3] Helong Li et al. "Parallel Connection of Silicon Carbide MOSFETs—Challenges, Mechanism, and Solutions". In: *IEEE Transactions on Power Electronics* 38.8 (Aug. 2023), pp. 9731–9749. ISSN: 1941-0107. DOI: 10.1109/TPEL.2023.3278270.

[4] Shi Pu et al. "Aging Mechanisms and Accelerated Lifetime Tests for SiC MOSFETs: An Overview". In: *IEEE Journal of Emerging and Selected Topics in Power Electronics* PP (Sept. 2021), pp. 1–1. DOI: 10.1109/JESTPE.2021.3110476.

[5] Roman Horff et al. "Current mismatch in paralleled phases of high power SiC modules due to threshold voltage unsymmetry and different gate-driver concepts". In: *2016 18th European Conference on Power Electronics and Applications (EPE'16 ECCE Europe)*. Sept. 2016, pp. 1–9. DOI: 10.1109/EPE.2016.7695409.

[6] Yang Xue et al. "Active Current Balancing for Parallel-Connected Silicon Carbide MOSFETs". In: *2013 IEEE Energy Conversion Congress and Exposition*. Denver, CO, USA: IEEE, Sept. 2013, pp. 1563–1569. ISBN: 978-1-4799-0336-8. DOI: 10.1109/ECCE.2013.6646891. (Visited on 03/21/2023).

[7] Jianing Wang et al. "Accurate Modeling of the Effective Parasitic Parameters for the Laminated Busbar Connected With Paralleled SiC MOSFETs". In: *IEEE Transactions on Circuits and Systems I: Regular Papers* 68.5 (May 2021), pp. 2107–2120. ISSN: 1558-0806. DOI: 10.1109/TCSI.2021.3064010.

[8] Angelo Raciti et al. "Effects of the Device Parameters and Circuit Mismatches on the Static and Dynamic Behavior of Parallel Connections of Silicon Carbide MOSFETs". In: *2018 IEEE Energy Conversion Congress and Exposition (ECCE)*. Sept. 2018, pp. 1846–1852. DOI: 10.1109/ECCE.2018.8557676.

[9] Gangyao Wang et al. "Dynamic and static behavior of packaged silicon carbide MOSFETs in paralleled applications". In: *2014 IEEE Applied Power Electronics Conference and Exposition - APEC 2014*. Mar. 2014, pp. 1478–1483. DOI: 10.1109/APEC.2014.6803502.

[10] Alessandro Borghese et al. "Statistical Analysis of the Electrothermal Imbalances of Mismatched Parallel SiC Power MOSFETs". In: *IEEE Journal of Emerging and Selected Topics in Power Electronics* 7.3 (Sept. 2019). parallelisation, pp. 1527–1538. ISSN: 2168-6777, 2168-6785. DOI: 10.1109/JESTPE.2019.2924735. (Visited on 03/21/2023).

[11] Yasushige Mukunoki et al. "Electro-thermal co-simulation of two parallel-connected SiC-MOSFETs under thermally-imbalanced conditions". In: *2018 IEEE Applied Power Electronics Conference and Exposition (APEC)*. Mar. 2018, pp. 2855–2860. DOI: 10.1109/APEC.2018.8341422.

[12] Chen Wang et al. "Analytical Model of the Parallel-Connected Silicon Carbide MOSFET Turn-ON Switching Behavior Under Asynchronous Gate Signals". In: (Oct. 2022), pp. 1–6. DOI: 10.1109/ITECAsia-Pacific56316.2022.9941859.

[13] Jianing Wang et al. "Comprehensive Analysis of Paralleled SiC MOSFETs Current Imbalance Under Asynchronous Gate Signals". In: *IEEE Journal of Emerging and Selected Topics in Power Electronics* 11.5 (Oct. 2023), pp. 4850–4866. ISSN: 2168-6785. DOI: 10.1109/JESTPE.2023.3290935.

[14] Ziqing Zheng. *A practical example of hard paralleling SiC MOSFET modules*. en. PCIM Asia, 2019. ISBN: 978-3-8007-4970-6.

[15] Toshiba Electronic. *MOSFET Parallening(Parasitic Oscillation between ParallelPower MOSFETs)*. Tech. rep. 20180726 AKX00067. Toshiba Electronic, July 2018.

[16] Dongxin Jin et al. "Analysis and Reduction of Turn-on Gate-source Voltage Oscillation on Paralleled SiC MOSFETs Application". In: Oct. 2022, pp. 1–7. DOI: 10.1109/ECCE50734.2022.9947878.

[17] Ye Zhu et al. "Influence of Paralleled SiC MOSFET on Turn-off Gate Voltage Oscillation". In: *2020 IEEE Energy Conversion Congress and Exposition (ECCE)*. Oct. 2020, pp. 683–689. DOI: 10.1109/ECCE44975.2020.9235763.

[18] Yan Li et al. "Comparative Analysis and Evaluation of Gate Driver Topologies for Paralleling Silicon Carbide (SiC) Power Modules". In: *2023 11th International Conference on Power Electronics and ECCE Asia (ICPE 2023 - ECCE Asia)*. May 2023, pp. 2010–2017.

DOI: 10.23919/ICPE2023-ECCEAsia54778.2023.10213664.

[19] M. Mauerer, A. Tüysüz, and J. W. Kolar. "Gate signal jitter elimination and noise shaping modulation for high-SNR Class-D power amplifiers". In: (Mar. 2016), pp. 1198–1205. DOI: 10.1109/APEC.2016.7468021.

[20] Le hanh Long. "Isolation galvanique intégrée pour nouveaux transitors de puissance". Theses. Université Grenoble Alpes, Nov. 2015. URL: https://theses.hal.science/tel-01265549.

[21] Eddy Aeloiza et al. "Evaluation and Characterization of Parallel Connected Ultra-Low-Inductance 400A SiC MOSFET Modules". In: (Sept. 2019), pp. 1934–1940. ISSN: 2329-3748. DOI: 10.1109/ECCE.2019.8912921.

[22] Jiye Liu and Zedong Zheng. "Switching Current Imbalance Mitigation for Paralleled SiC MOSFETs Using Common-mode Choke in Gate Loop". In: *2020 IEEE Energy Conversion Congress and Exposition (ECCE)*. Oct. 2020, pp. 705–710. DOI: 10.1109/ECCE44975.2020.9236412.

[23] Mario Mauerer, Arda Tüysüz, and Johann W. Kolar. "Low-Jitter GaN E-HEMT Gate Driver With High Common-Mode Voltage Transient Immunity". In: *IEEE Transactions on Industrial Electronics* 64.11 (Nov. 2017), pp. 9043–9051. ISSN: 1557-9948. DOI: 10.1109/TIE.2017.2677354.

[24] Integrated Circuit Division. *IXD609 9-Ampere Low-SideUltrafast MOSFET Drivers Datasheet*. Tech. rep. Littelfuse, 2017. URL: https://www.littelfuse.com/media?resourcetype=datasheets&itemid=63dc5a9e-6cea-4349-bf4e-4540dadb0b53&filename=ixd-609.

Fully Integrated Closed-Loop Active Gate Driver IC With Real-Time Control of Gate Current Change Timing by Gate Current Sensing

Yaogan Liang
The University of Tokyo
Tokyo, Japan
liangy@iis.u-tokyo.ac.jp

Katsuhiro Hata
Shibaura Institute of Technology
Tokyo, Japan
khata@shibaura-it.ac.jp

Makoto Takamiya
The University of Tokyo
Tokyo, Japan
mtaka@iis.u-tokyo.ac.jp

Abstract— In order to provide a low-cost closed-loop active gate driver (AGD) that is applicable to both 3-pin and 4-pin power devices, a fully integrated closed-loop gate current sensing (GCS) AGD IC for IGBTs is proposed that controls the timing of gate current changes in real time. In an active gate driving where the gate current is changed three times from large to small to large at turn-on, GCS AGD has the function of changing the time of the first and second slots in real time by detecting the change in gate current using two comparators. Double pulse test measurements of IGBTs are performed at 600 V and under nine combinations of conditions including load current of 20 A, 50 A and 80 A, and temperature of 25 °C, 75 °C and 125 °C, using GCS AGD IC fabricated with 180 nm BCD process. The results show that the proposed GCS AGD reduces the switching loss (E_{LOSS}) and the collector current overshoot ($I_{OVERSHOOT}$) in all nine conditions compared to the conventional single-step gate driving. Specifically, in the nine conditions, with the same $I_{OVERSHOOT}$, the E_{LOSS} is reduced by 16% to 30%, and without increasing E_{LOSS}, the reduction in $I_{OVERSHOOT}$ range from 14% to 21%.

Keywords—active gate driver, gate current sensing, switching loss, current overshoot, trade-off curve

I. INTRODUCTION

Active gate driver (AGD) [1-28], which changes the gate driving strength multiple times in fine time slots during the switching period of power devices, is attracting attention as a technology that can solve the trade-off problem between energy loss and noise during power device switching. AGD with closed-loop [9-28] instead of open-loop [1-8] is essential to cope with operating condition variations such as load current and temperature. The target of this paper is to develop a fully integrated closed-loop AGD IC with automatic control of gate current (I_G) change timing that senses only the gate terminal of power devices at any reasonable gate resistance (R_G) value. Table I shows a comparison table of the closed-loop AGD IC with automatic control of I_G change timing. Drain-to-source voltage (V_{DS}) sensing cannot be fully integrated on IC because of the high voltage of V_{DS} [26] or requires a high voltage IC process [23], resulting in high AGD costs. Sensing using a Kelvin emitter or source [24, 26] can be used with 4-pin power devices, but not with 3-pin power devices. While gate-to-source voltage (V_{GS}) sensing [22, 28] provides gate terminal-only sensing, [22] is not fully integrated because it requires an FPGA, and [28] has been demonstrated to operate at high R_G of 125 Ω,

and [28] is expected to malfunction when used under fast switching conditions at low R_G. To solve the problems, a fully integrated closed-loop gate current sensing (GCS) AGD IC with no restriction on R_G value for IGBTs is proposed controlling the timing of I_G changes in real time, which is applicable to both 3-pin and 4-pin power devices.

TABLE I. COMPARISON TABLE OF CLOSED-LOOP AGD IC WITH AUTOMATIC CONTROL OF I_G CHANGE TIMING

	APEC'23 [24]	TIE'23 [23]	ISPSD'24 [26]	ISPSD'22 [22]	TPEL'24 [28]	This work
Target power device	IGBT	GaN	SiC	SiC	SiC	IGBT
Measured conditions	600 V, 80 A	400 V, 10 A	Simulated	Simulated	400 V, 40 A	600 V, 80 A
Sensor input	V_{Ee} [1]	V_{DS}	V_{DS}, V_{Ss} [2]	V_{GS}	V_{GS}	I_G
Sensing gate terminal only	No	No	No	Yes	Yes	Yes
Arbitrary R_G value	Yes	Yes	Yes	Yes	No [3]	Yes
Real-time control	Yes	Yes	No	No	Yes	Yes
Number of states per switching	3	4	4	2	3	3
I_G levels	6 bit	3	NA	5	NA	6 bit
Fully integrated on IC	Yes	Yes	No [4]	No	Yes	Yes
IC Process	180 nm BCD	500 nm, 600 V SOI [5]	180 nm BCD	180 nm BCD	180 nm BCD	180 nm BCD

(1) Voltage between power emitter and Kelvin emitter
(2) Voltage between power source and Kelvin source
(3) R_G = 125 Ω
(4) Voltage divider for V_{DS} is not integrated.
(5) Requires IC process with breakdown voltage higher than the main circuit voltage.

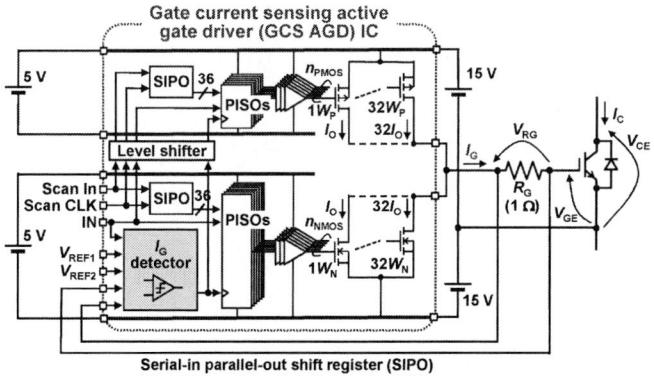

Fig. 1: Circuit schematic of proposed GCS AGD IC.

979-8-3315-1612-3/25 $31.00 © 2025 IEEE

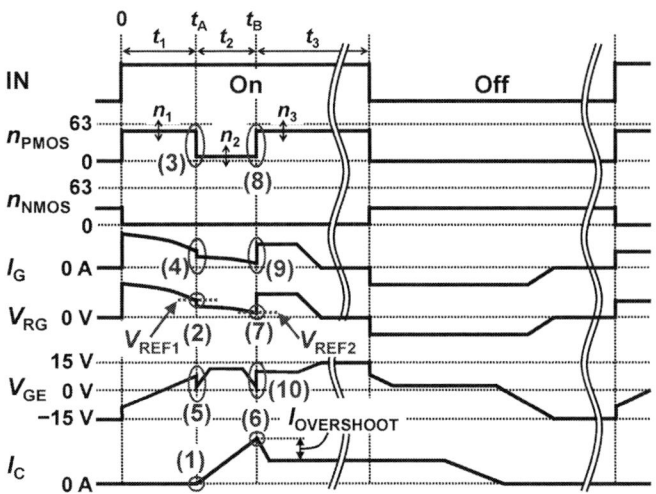

Fig. 2: Timing chart of proposed GCS AGD IC.

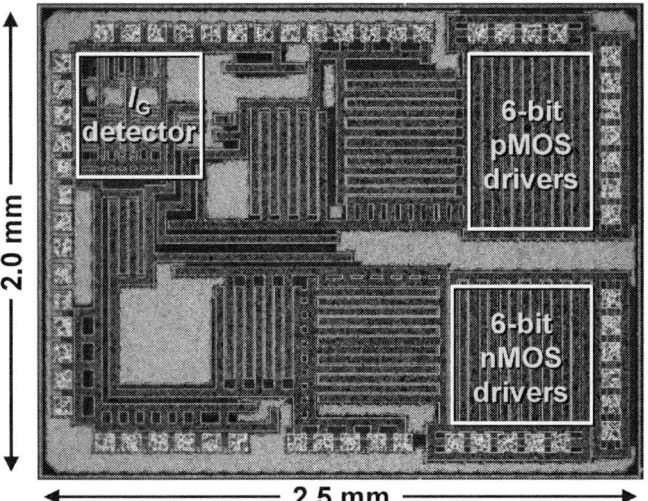

Fig. 3: Die photo of GCS AGD IC.

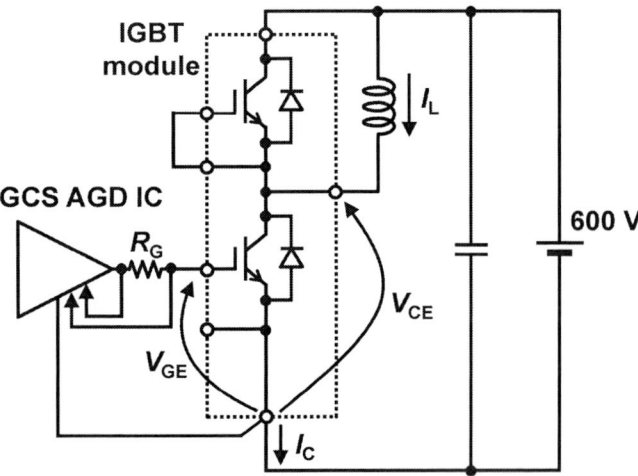

Fig. 4: Circuit schematic of double pulse test.

Fig. 5: Measurement setup.

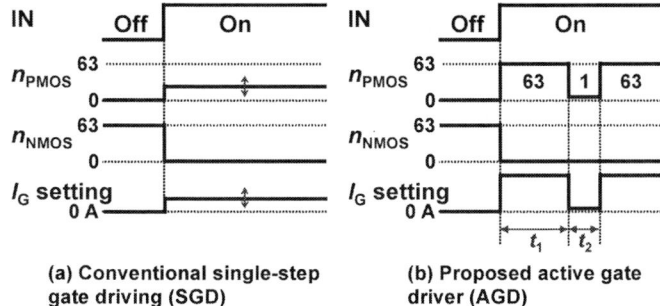

(a) Conventional single-step gate driving (SGD)

(b) Proposed active gate driver (AGD)

Fig. 6: Timing charts for turn-on.

II. PROPOSED GATE CURRENT SENSING ACTIVE GATE DRIVER IC

Figs. 1 and 2 show a circuit schematic and a timing chart of the proposed GCS AGD IC, respectively. The IC includes I_G detector to determine the timing of I_G changes, controller for the state change, and a 6-bit digital gate driver with variable I_G in 64 levels, where $I_G = n_{PMOS} \times 95$ mA and n_{PMOS} is an integer from 0 to 63. At turn-on, an active gate driving is performed in three slots from t_1 to t_3 with different I_G of strong (n_1) -weak (n_2) - strong (n_3). n_1 to n_3 are preset by a digital input (Scan In), while t_1 and t_2 are automatically determined by I_G detector. To sense I_G, the voltage drop (V_{RG}) due to I_G and R_G is given to I_G detector, and V_{RG} is compared with two reference voltages (V_{REF1} and V_{REF2}) by two comparators inside I_G detector, respectively. Fig. 2 shows the sequence for determining the end timing of t_1 (t_A) and the end timing of t_2 (t_B). t_A is the timing when the collector current (I_C) rises from 0 A and t_B is the timing when I_C peaks. The role of I_G detector is to detect t_A and t_B from V_{RG}. V_{REF1} and V_{REF2} should be adjusted in advance so that t_A and t_B can be detected, respectively. When gate-to-emitter voltage (V_{GE}) reaches the IGBT threshold voltage and I_C rises from 0 A (1), V_{RG} crosses V_{REF1} and I_G detector detects t_A (2). As the controller changes n_{PMOS} from n_1 to n_2 (3), I_G decreases (4) and V_{GE} also decreases (5). When I_C peaks (6), V_{RG} crosses V_{REF2} and I_G detector detects t_B (7). As the controller changes n_{PMOS} from n_2 to n_3 (8), I_G increases (9) and V_{GE} also increases (10). Fig. 3

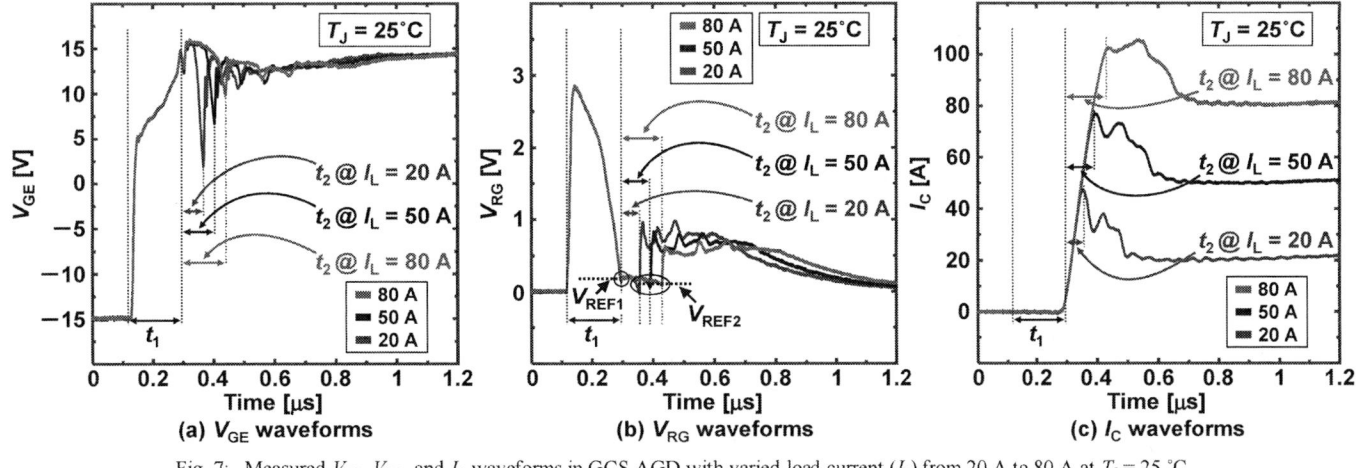

Fig. 7: Measured V_{GE}, V_{RG}, and I_C waveforms in GCS AGD with varied load current (I_L) from 20 A to 80 A at $T_J = 25$ ℃.

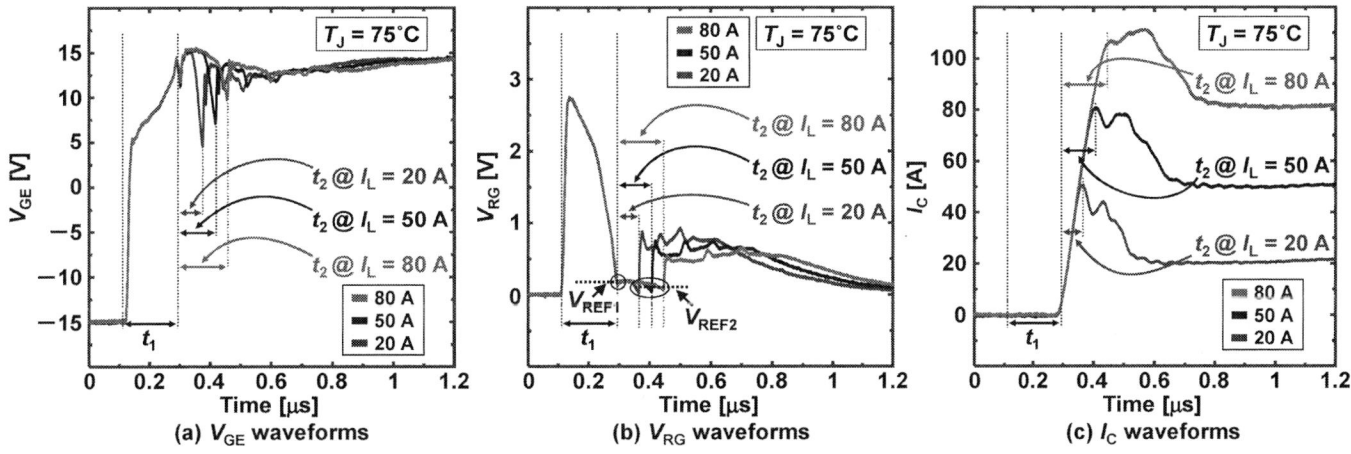

Fig. 8: Measured V_{GE}, V_{RG}, and I_C waveforms in GCS AGD with varied load current (I_L) from 20 A to 80 A at $T_J = 75$ ℃.

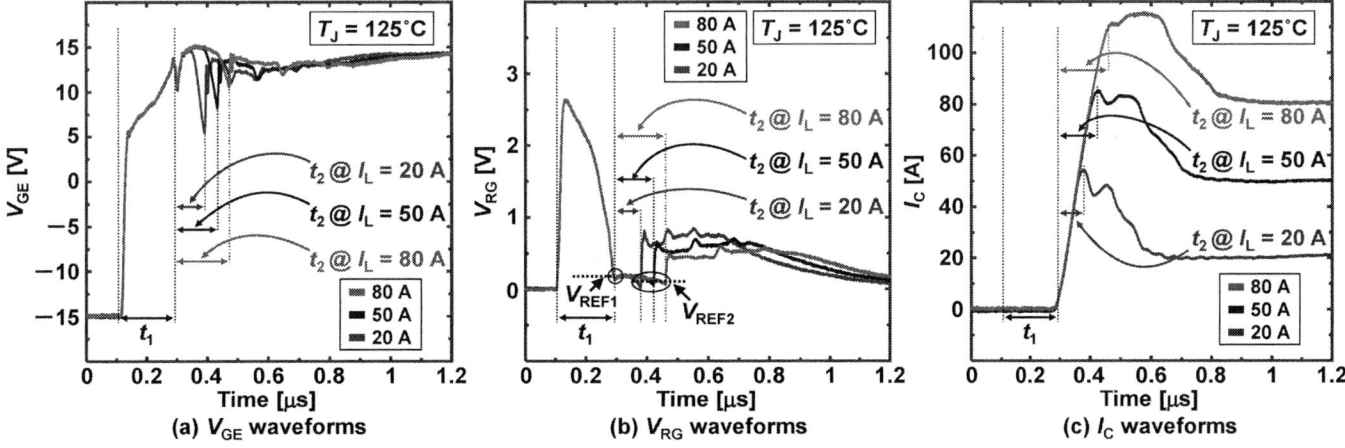

Fig. 9: Measured V_{GE}, V_{RG}, and I_C waveforms in GCS AGD with varied load current (I_L) from 20 A to 80 A at $T_J = 125$ ℃.

shows a die photo of GCS AGD IC fabricated with 180-nm BCD process.

III. MEASURED RESULTS

Figs. 4 and 5 show a circuit schematic and a measurement setup of the double pulse test at 600 V using the developed GCS AGD IC and an IGBT module (FS100R12N2T4, 1200 V, 100

A), respectively. R_G is 1 Ω. Note that the role of R_G in this paper is not to adjust the gate driving strength but to be a sense resistor for I_G measurement. In this paper, the gate driving strength adjustment is done by digitally controlling n_{PMOS} in Fig. 2. This IGBT module has a Kelvin emitter pin, but since the purpose of this paper is to demonstrate the effectiveness of the proposal in a 3-pin IGBT, the Kelvin emitter pin is left floating for the

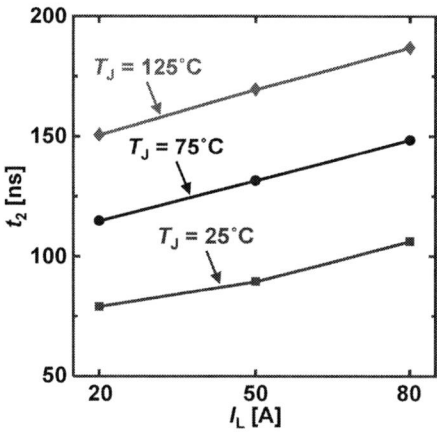

Fig. 10: Measured t_2 vs. I_L at T_J = 25 °C, 75 °C, and 125 °C.

measurements. Figs. 6 (a) and (b) show timing charts of the conventional single-step gate driving (SGD) and the proposed AGD for comparison, respectively. In SGD, n_{PMOS} is varied, which emulates a conventional gate driver with varied gate resistance. In AGD, (n_1, n_2, n_3) are preset to (63, 1, 63), and t_1 and t_2 are automatically determined by GCS.

Figs. 7, 8 and 9 show the measured V_{GE}, V_{RG}, and I_C waveforms in GCS AGD with varied load current (I_L) from 20 A to 80 A and with different T_J from 25 °C to 125 °C, respectively. In Fig. 7 (b), Fig. 8 (b) and Fig. 9 (b), it is clearly observed that t_1 is determined at the intersection of V_{RG} and V_{REF1}, and t_2 is determined at the intersection of V_{RG} and V_{REF2}. In Figs. 7, 8 and 9, as I_L increases, t_1 remains constant at the same T_J, and as T_J becomes higher, t_1 slightly increases because I_G decreases as gate driver IC temperature rises, increasing gate capacitance charging time. On the other hand, Fig. 10 shows the measured t_2 vs. I_L relationship of the measured results at different T_J. Because t_2 represented the time that I_C changes from zero to its peak value, t_2 is automatically changed by GCS AGD for best performance under different conditions. It can be observed that t_2 increases approximately linearly with I_L at the same T_J. And for the same I_L, t_2 duration increases as T_J increases, because according to the characteristics of the tested IGBT module, $I_{OVERSHOOT}$ increases as T_J increases. Therefore, the proposed

GCS AGD can automatically extend the time slots (t_2) with weak driving strength to suppress the surge current under different I_C and T_J conditions.

Figs. 11 (a) to (c) show the measured switching loss (E_{LOSS}) vs. the collector current overshoot ($I_{OVERSHOOT}$) of the conventional SGD and the proposed AGD under nine combinations of conditions including I_L = 20 A, 50 A and 80 A, and T_J = 25 °C, 75 °C and 125 °C. All the trade-off curves of the conventional SGD is measured with varied n_{PMOS} from 4 to 63. The result points of proposed AGD is marked as colored stars. Because the proposed AGD breaks the trade-off relationship between E_{LOSS} and $I_{OVERSHOOT}$ under all conditions by automatically changing t_1 and t_2, Fig. 11 (a) to (c) clearly show that the proposed AGD is on the lower left side compared to the trade-off curve of the conventional SGD. At I_L = 20 A in Fig, 11 (a), compared with the conventional SGD, the proposed AGD reduces E_{LOSS} by 16 %, 16 %, and 16 % under $I_{OVERSHOOT}$-aligned condition and reduces $I_{OVERSHOOT}$ by 14 %, 17 %, and 21 % under E_{LOSS}-aligned condition at T_J = 25 °C, 75 °C, and 125 °C, respectively. At I_L = 50 A in Fig, 11 (b), compared with the conventional SGD, the proposed AGD reduces E_{LOSS} by 30 %, 26 %, and 29 % under $I_{OVERSHOOT}$-aligned condition and reduces $I_{OVERSHOOT}$ by 18 %, 20 %, and 21 % under E_{LOSS}-aligned condition at T_J = 25 °C, 75 °C, and 125 °C, respectively. At I_L = 80 A in Fig, 11 (c), compared with the conventional SGD, the proposed AGD reduces E_{LOSS} by 25 %, 22 %, and 26 % under $I_{OVERSHOOT}$-aligned condition and reduces $I_{OVERSHOOT}$ by 18 %, 14 %, and 17 % under E_{LOSS}-aligned condition at T_J = 25 °C, 75 °C, and 125 °C, respectively.

In Figs. 11 (a) and (c), Points A1 to D1 and Point A2 to D2 are defined, where Point B1 and B2 are the proposed AGD, Point A1 and A2 are SGD with closest $I_{OVERSHOOT}$ comparing to Point B1 and B2, and Point C1 and C2 are SGD with the closest E_{LOSS} comparing to Point B1 and B2. Point D1 and D2 are SGD with n_{PMOS} = 63, which are equivalent to t_2 = 0 ns in AGD, and thus worth comparing with Point B1 and B2. Figs. 12 (a) to (d) and Figs. 13 (a) to (d) show the measured waveforms of Points A1 to D1 and A2 to D2 in Figs. 11 (a) and (c) at I_L = 20 A, T_J = 25 °C and I_L = 80 A, T_J = 125 °C, respectively. Point B1 in Fig. 12 (b) and Point B2 in Fig 13 (b) clearly demonstrate the time slot (t_1 and t_2) change operation of the proposed GCS AGD IC.

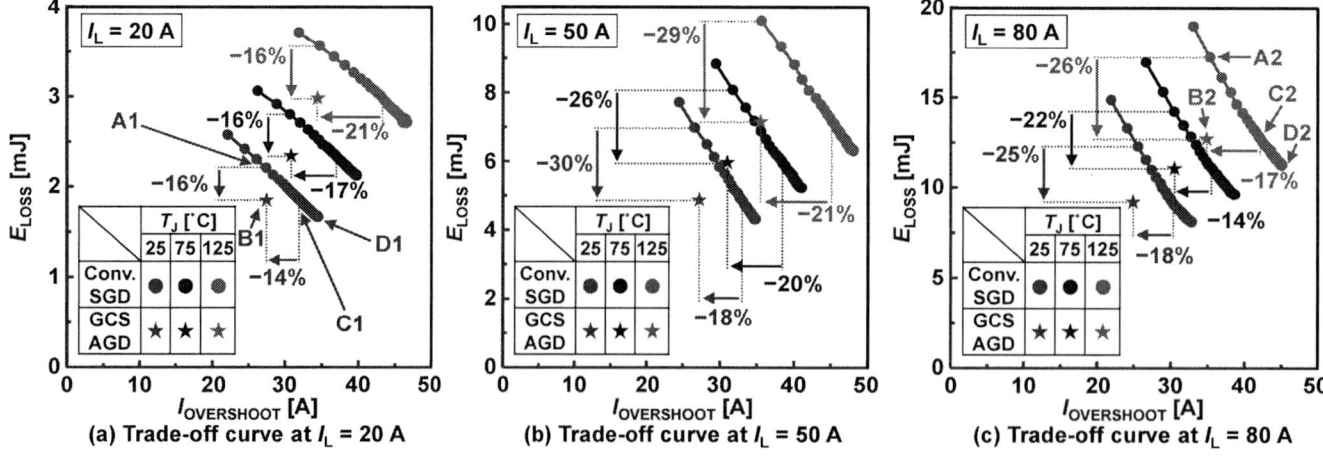

Fig. 11: Measured E_{LOSS} vs. $I_{OVERSHOOT}$ of conventional SGD and proposed AGD at I_L = 20 A, 50 A, and 80 A and T_J = 25 °C, 75 °C, 125 °C.

979-8-3315-1612-3/25 $31.00 © 2025 IEEE

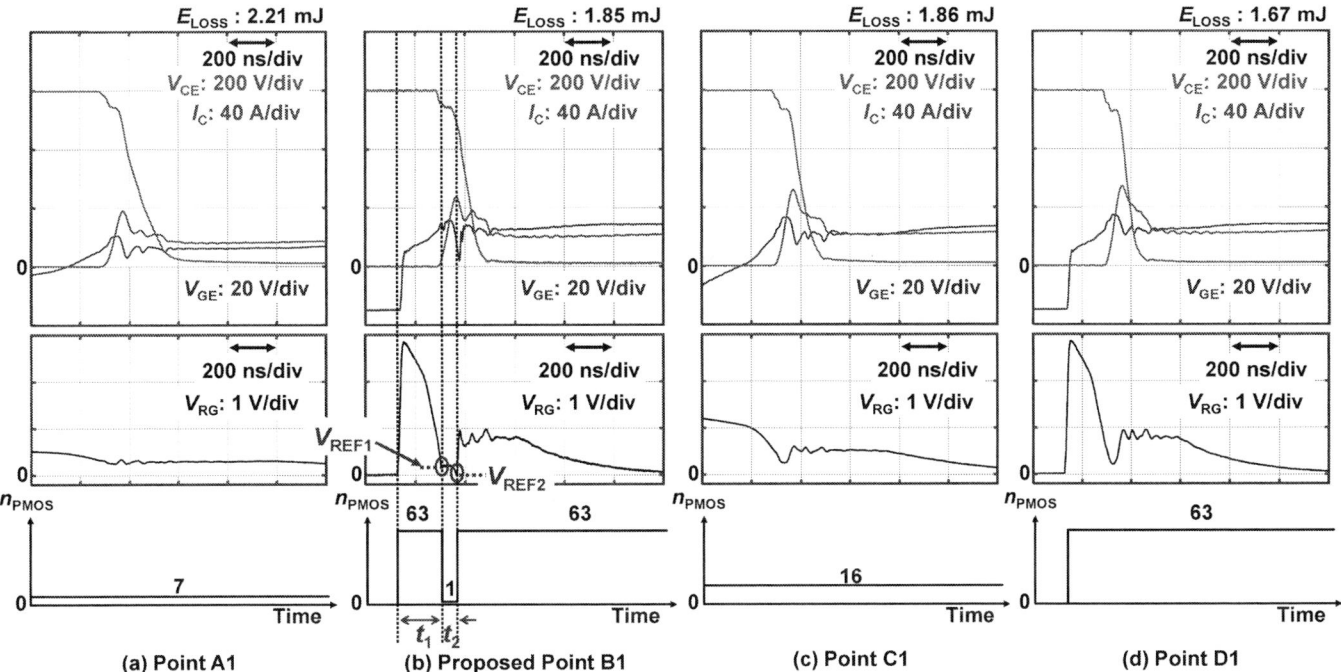

Fig. 12: Measured waveforms of Points A1 to D1 in Fig. 11 (a) at $I_L = 20$ A and $T_J = 25$ ˚C.

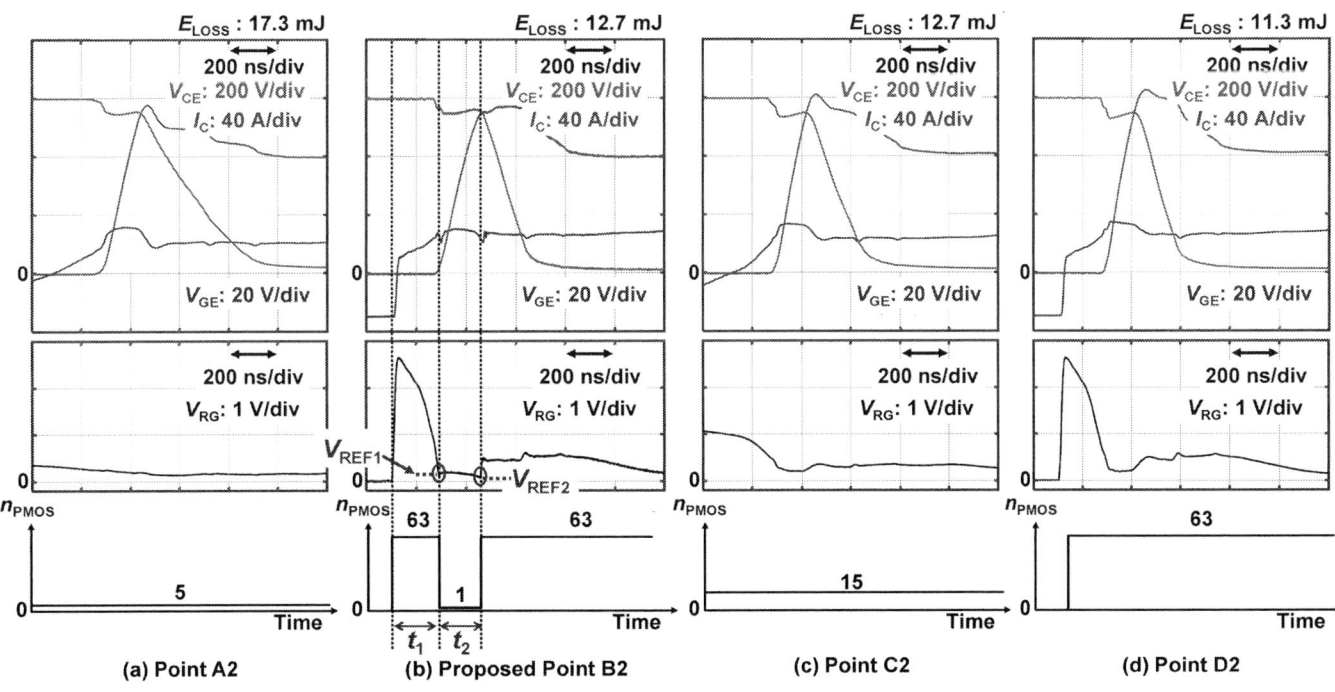

Fig. 13: Measured waveforms of Points A2 to D2 in Fig. 11 (c) at $I_L = 80$ A and $T_J = 125$ ˚C.

Point A1 and A2 have similar $I_{OVERSHOOT}$ as B1 and B2, but the I_C and V_{CE} overlaps are larger. The I_C and V_{CE} overlaps in C1 and C2 are similar as B1 and B2, but the $I_{OVERSHOOT}$ is higher. Compared with D1 and D2, the proposed point B1 and B2 greatly reduced the $I_{OVERSHOOT}$ and dI_C/dt for the strong-weak-strong I_G configuration. As shown in Table I, this paper is the first work achieving sensing only the gate terminal, no restriction on R_G value, real-time control, and fully integrated on

IC in the closed-loop AGD IC with automatic control of I_G change timing.

IV. CONCLUSIONS

In this work, the first fully integrated GCS AGD IC is proposed to break the trade-off relationship between E_{LOSS} and $I_{OVERSHOOT}$ in real time and without restriction on R_G value. The proposed IC automatically determines the timing of I_G changes and varies the driving strength in the order of strong-weak-strong to suppress surge current without greatly increasing E_{LOSS}. For verification, double pulse test experiments are conducted under nine condition combinations with $I_L = 20$ A, 50 A and 80 A and $T_J = 25$ °C, 75 °C and 125 °C. Compared with the conventional SGD, the results of AGD move to the lower-left corner of the trade-off curve at all conditions. The proposed AGD reduces E_{LOSS} by 16 %, up to 30 % and up to 26 % under $I_{\text{OVERSHOOT}}$-aligned condition at 20 A, 50 A and 80 A, respectively. With almost the same E_{LOSS}, the reduction in $I_{\text{OVERSHOOT}}$ achieves up to 21 %, up to 21 % and up to 18 % at 20 A, 50 A and 80 A, respectively.

ACKNOWLEDGMENT

This work was partly supported by NEDO (JPNP21009).

REFERENCES

[1] K. Miyazaki, S. Abe, M. Tsukuda, I. Omura, K. Wada, M. Takamiya, and T. Sakurai, "General-purpose clocked gate driver IC with programmable 63-level drivability to optimize overshoot and energy loss in switching by a simulated annealing algorithm," *IEEE Trans. Ind. Appl.*, vol.53, no.3, pp. 2350—23–57, May-June 2017.

[2] T. Sai, K. Miyazaki, H. Obara, T. Mannen, K. Wada, I. Omura, M. Takamiya, and T. Sakurai, "Load current and temperature dependent optimization of active gate driving vectors," in *Proc. IEEE Energy Conversion Congress & Exposition*, Sep. 2019, pp. 3292-3297.

[3] T. Sai, K. Miyazaki, H. Obara, T. Mannen, K. Wada, I. Omura, T. Sakurai, and M. Takamiya, "Stop-and-go gate drive minimizing test cost to find optimum gate driving vectors in digital gate drivers," in *Proc. IEEE Applied Power Electronics Conf. and Expo.*, March 2020, pp. 3096-3101.

[4] W. J. Zhang, J. Yu, Y. Leng, W. T. Cui, G. Q. Deng, and W. T. Ng, "A segmented gate driver for E-mode GaN HEMTs with simple driving strength pattern control," in *Proc. IEEE Int. Symp. Power Semiconductor. Devices ICs*, Sep. 2020, pp. 102-105.

[5] R. Katada, K. Hata, Y. Yamauchi, T. -W. Wang, R. Morikawa, C. -H. Wu, T. Sai, P. -H. Chen, and M. Takamiya, "5 V, 300 MSa/s, 6-bit digital gate driver IC for GaN achieving 69 % reduction of switching loss and 60 % reduction of current overshoot," in *Proc. IEEE Int. Symp. Power Semiconductor. Devices ICs*, May 2021, pp. 55-58.

[6] D. Liu, H. C. P. Dymond, S. J. Hollis, J. Wang, N. McNeill, D. Pamunuwa, and B. H. Stark, "Full custom design of an arbitrary waveform gate driver with 10-GHz waypoint rates for GaN FETs," *IEEE Trans. Power Electron.*, vol. 36, no. 7, pp. 8267-8279, July 2021.

[7] W. J. Zhang, J. Yu, W. T. Cui, Y. Leng, J. Liang, Y.-T. Hsieh, H.-H. Tsai, Y.-Z. Juang, W.-K. Yeh, and W. T. Ng, "A smart gate driver IC for GaN power HEMTs with dynamic ringing suppression," *IEEE Trans. on Power Electronics*, vol. 36, no. 12, pp. 14119-14132, Dec. 2021.

[8] K. Horii, K. Hata, R. Wang, W. Saito, and M. Takamiya, "Large current output digital gate driver using half-bridge digital-to-analog converter IC and two power MOSFETs," in *Proc. IEEE Int. Symp. Power Semiconductor. Devices ICs*, May 2022, pp. 293-296.

[9] V. John, B.-S. Suh, and T. A. Lipo, "High performance active gate drive for high power IGBTs," in *Proc. IEEE Industry Applications Conf.*, Oct. 1998, pp. 1519-1529.

[10] Y. Sun, L. Sun, A. Esmaeli, and K. Zhao, "A novel three stage drive circuit for IGBT," in *Proc. IEEE Conf. on Industrial Electronics and Applications*, May 2006, pp. 1-6.

[11] N. Idir, R. Bausiere, and J. J. Franchaud, "Active gate voltage control of turn-on di/dt and turn-off dv/dt in insulated gate transistors," *IEEE Tran. on Power Electron.*, vol. 21, no. 4, pp. 849-855, July 2006.

[12] Z. Wang, X. Shi, L. M. Tolbert, F. Wang, and B. J. Blalock, "A di/dt feedback-based active gate driver for smart switching and fast overcurrent

protection of IGBT modules," *IEEE Trans. on Power Electron.*, vol. 29, no. 7, pp. 3720-3732, July 2014.

[13] Y. Lobsiger and J. W. Kolar, "Closed-loop di/dt and dv/dt IGBT gate driver," *IEEE Trans. on Power Electron.*, vol. 30, no. 6, pp. 3402-3417, June 2015.

[14] F. Zhang, X. Yang, Y. Ren, L. Feng, W. Chen, and Y. Pei, "Advanced active gate drive for switching performance improvement and overvoltage protection of high-power IGBTs," *IEEE Trans. on Power Electron.*, vol. 33, no. 5, pp. 3802-3815, May 2018.

[15] S. Kawai, T. Ueno, and K. Onizuka, "A 4.5V/ns active slew-rate-controlling gate driver with robust discrete-time feedback technique for 600V super junction MOSFETs," in *Proc. IEEE International Solid-State Circuits Conf.*, Feb. 2019, pp. 252-254.

[16] Y. Wen, Y. Yang, and Y. Gao, "Active gate driver for improving current sharing performance of paralleled high-power SiC MOSFET modules," *IEEE Trans. on Power Electron.*, vol. 36, no. 2, pp. 1491-1505, Feb. 2021.

[17] Y. Ling, Z. Zhao, and Y. Zhu, "A self-regulating gate driver for high-power IGBTs," *IEEE Trans. on Power Electron.*, vol. 36, no. 3, pp. 3450-3461, March 2021.

[18] E. Raviola and F. Fiori, "Experimental investigations on the tuning of active gate drivers under load current variations," in *Proc. International Conf. on Applied Electronics*, Sep. 2021, pp. 1-4.

[19] M. Sayed, S. Araujo, F. Carraro, and R. Kennel, "Investigation of gate current shaping for SiC-based power modules on electrical drive system power losses," in *Proc. European Conf. on Power Electronics and Applications*, Sep. 2021, pp. 1-10.

[20] J. Zhu, D. Yan, S. Yu, W. Sun, G. Shi, S. Liu, S. Zhang, "A 600V GaN active gate driver with dynamic feedback delay compensation technique achieving 22.5% turn-on energy saving," in *Proc. IEEE International Solid-State Circuits Conf.*, Feb. 2021, pp. 462-464.

[21] D. Han, S. Kim, X. Dong, H. Li, J. Moon, Y. Li, and F. Peng, "An integrated multi-level active gate driver for SiC power modules," in *Proc. IEEE Transportation Electrification Conf. & Expo.*, June 2022, pp. 727-732.

[22] W. T. Cui, W. J. Zhang, J. Y. Liang, H. Nishio, H. Sumida, H. Nakajima, Y. Hsieh, H. Tsai, Y. Juang, W. Yeh, and W. T. Ng, "A dynamic gate driver IC with automated pattern optimization for SiC power MOSFETs," in *Proc. IEEE International Symposium on Power Semiconductor Devices and ICs*, May 2022, pp. 33-36.

[23] S. Yu, G. Shi, T. Wu, J. Zhu, L. Zhang, and W. Sun, "A 400-V half bridge gate driver for normallyoff GaN HEMTs with effective dv/dt control and high dv/dt immunity," *IEEE Trans. Ind. Electron.*, vol. 70, no. 1, pp. 741–751, 2023.

[24] D. Zhang, K. Horii, K. Hata, and M. Takamiya, "Digital gate driver IC with fully integrated automatic timing control function in stop-and-go gate drive for IGBTs," in *2023 IEEE Applied Power Electronics Conference and Exposition (APEC)*, 2023, pp. 1225–1231.

[25] D. Zhang, K. Horii, K. Hata, and M. Takamiya, "Digital gate driver IC with real-time gate current change by sensing drain current to cope with operating condition variations of SiC MOSFET," in *2023 11th International Conference on Power Electronics and ECCE Asia (ICPE 2023 - ECCE Asia)*, 2023, pp. 374–380.

[26] R. Lai, Y. Yang, Y. Dai, J. Wu, Y. Shi, Z. Zhou, B. Zhang, H. Li, and X. Peng, "A quad-slope smart gate driver with mixed-signal auto-timing technique for power devices segment control," in *2024 36th International Symposium on Power Semiconductor Devices and ICs (ISPSD)*, 2024, pp. 347–350.

[27] C.-W. Kuo, T.-W. Wang, H.-C. Kuo, C.-C. Tu, and P.-H. Chen, "Closed-loop gate-sensing active driver IC with adaptive delay compensation technique for silicon carbide power MOSFETs," in *2024 36th International Symposium on Power Semiconductor Devices and ICs (ISPSD)*, 2024, pp. 462–465.

[28] T.-W. Wang, L.-C. Chen, M. Takamiya, and P.-H. Chen, "Active gate driver IC integrating gate voltage sensing technique for SiC MOSFETs," *IEEE Trans. Power Electron.*, vol. 39, no. 7, pp. 8562–8571, 2024.

Analyze and Design of Digitally Load Current Modulated Active Gate Driver for GaN HEMTs based Buck DC-DC

Wentao Liu
School of Microelectronics
Fudan University
Shanghai,China
wtliu22@m.fudan.edu.cn

Zhina Lian
School of Microelectronics
Fudan University
Shanghai,China
znlian23@m.fudan.edu.cn

Taotao Wu
School of Microelectronics
Fudan University
Shanghai,China
ttwu22@m.fudan.edu.cn

Xiaochuan Peng
School of Microelectronics
Fudan University
Shanghai,China
xcpeng23@m.fudan.edu.cn

Hao Min
School of Microelectronics
Fudan University
Shanghai,China
hmin@fudan.edu.cn

Abstract—The active gate driver (AGD) reduces instability and electromagnetic interference (EMI) resulting from rapid switch-node slew rates in the half-bridge driver of gallium nitride (GaN) high electron mobility transistors (HEMTs) while maintaining the switching speed. This paper seeks to improve the effectiveness of the AGD method by integrating the influence of load current on the voltage slew rates. Consequently, we present a tri-current AGD prototype that employs a 4-bit current-based DAC to accurately regulate the drive current during the Miller Plateau (MP) phase. By measuring the load current, we implement the feedback control of the MP drive strength via a digital loop on the FPGA. Experimental measurements validate that the suggested AGD circuit sustains an overshoot voltage of 0.3V/A when the load current fluctuates between 0.8A and 4A and achieves a driving speed of 25.6% faster than the conventional methods.

Keywords—Active gate driver (AGD), Gallium Nitride (GaN), Current-based DAC, Digital loop control, Stability, Load current, Electromagnetic interference (EMI).

I. INTRODUCTION

Gallium nitride (GaN) high electron mobility transistors (HEMTs) hold great promise for high-speed and high-power DC-DC converters because of their fast switching speeds, high power density, and low conduction losses, all of which contribute to smaller circuit sizes and improved efficiency. [1]-[4] However, the rapid transitions in voltage and current associated with GaN HEMTs can lead to increased instability, especially during the Miller plateau. This instability results in issues such as self-oscillation, cross-modulation, and ringing, which limit their application in high-speed environments. [5,6] Additionally, electromagnetic interference (EMI) effects are more pronounced in these devices. Conventional methods for driving power switch face great challenge on the trade-off between high speed and high stability.

In the recent decade, by applying piecewise linear fitting to the nonlinear switching process, researchers have discovered that dynamically regulating the drive current at various stages of the switching process can both increase switching speed and inhibit ringing and oscillation, which is called active gate driver (AGD). [5]-[9] During different switching phases, current transients (dI/dt) and voltage transients (dV/dt) create significant noises, affecting the driver circuit's stability and EMI the most. Thus, in articles [7,8], the driving currents are separated into three levels for push-pull driving to minimize the gate drive current at dV/dt and dI/dt while increasing the gate drive current at other switching processes. However, these implementations are of limited use in practical DC-DC because they use constant drive levels during each phase, neglecting the impact of variable load current and failing to effectively suppress instability when the load current changes. To enhance generalizability, several works suggest the digital active gate driver (DAGD) [9,10] using N-bits programmable gate resistors controlled by digital inputs. However, the digital inputs that control the drive intensity are pre-stored in the LUT. Consequently, they also fail to realize on-time control of driving strength, rendering them incapable of adjusting to variations in operating conditions.

To address the previous issues, we present a load current-modulated tri-current active gate driver (AGD) for GaN buck DC-DC converters using digital loop control. By theoretically analyzing the relationship between dV/dt and load current during each switching phase, we discover a square proportional relationship during the Miller plateau period. Based on this, we design an AGD prototype, which includes load current sampling, a control algorithm, and the usage of a 4-bit current DAC to adjust the drive current. Finally, we test the system to prove its effectiveness.

979-8-3315-1612-3/25 $31.00 © 2025 IEEE

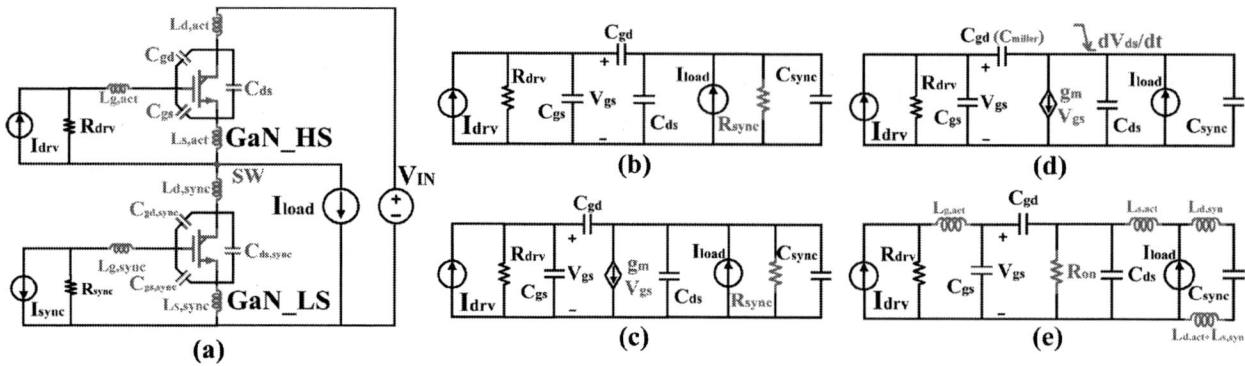

Fig. 1. Equivalent circuit of (a) synchronized GaN buck DC-DC converter and small signal circuits in (b) delay, (c) dI/dt, (d) dV/dt and (e) oscillation period.

II. Analysis of dV/dt

In GaN HEMTs, the on and off processes have four stages: delay, dI/dt, dV/dt (also called the Miller plateau period) and oscillation. [10] Despite the modeling of the switching process during the dV/dt interval [6], a thorough study of how stability affects each switching process or how variations in load affect each switching phase remains absent, which will be done in the next section.

A. Modeling of Four Phases of Switching

Figure 1(a) displays the equivalent circuit of a synchronized GaN buck DC-DC converter at active switch turn-on. A similar examination can be performed upon turning off. GaN_HS refers to an actively controlled switch, whereas GaN_LS refers to a synchronous switch. SW is the switching node in the half-bridge topology. Each power switch has three parasitic capacitors: C_{gs}, C_{gd} and C_{ds}. Parasitic inductances are taken into account from packaging and interconnections. Because the load current varies considerably more slowly than the switching speed, it can be substituted by an equivalent current source.

Figures 1(b) and (e) show the small signal circuit during the delay and oscillation period. R_{sync} indicates the on-resistance of LS in reverse conduction during the dead time, while R_{on} refers to the resistance of HS in the deep-triode region. Due to the low resistive path between the drain and source in (b) and (d), the gate drive current has little effect on the voltage on the switching node. A similar situation also exists in dI/dt, shown in Fig 1(c). As the HS turns on, the current flowing through the LS decreases. Therefore, R_{sync} increases, yet still able to keep the voltage between the drain and source nearly constant. However, in dV/dt phase, the low resistive path vanishes, and high dV_{ds}/dt is directly applied to C_{gd} which is shown in Figure 1(d). Thus, with large g_m and output impedance, the Miller effect is significant. This is the phase where stability issues are most probable to arise.

B. Theoretical Analysis of Relation between Optimal dV/dt and Load Current

According to the fig. 1(d), We can list the KCL equations of dV/dt period as below:

$$I_{drv} = C_{gd}\frac{d(V_{gs} - V_{ds})}{dt} + \frac{V_{gs}}{R_{drv}} + C_{gs}\frac{dV_{gs}}{dt} \tag{1}$$

$$C_{gd}\frac{d(V_{gs} - V_{ds})}{dt} + I_{load} = g_m V_{gs} + C_{ds}\frac{dV_{ds}}{dt} \tag{2}$$

By neglecting dV_{gs}/dt [11], we can conclude

$$\frac{dV_{sw}}{dt} = \frac{-I_{load} + g_m R_{drv} I_{drv}}{2C_{oss} + g_m R_{drv} C_{RSS}} \tag{3}$$

where $C_{OSS} = C_{ds} + C_{gd}$, $C_{RSS} = C_{gd}$, and g_m indicates the transconductance of the GaN HEMT, which is proportional to the square root of the current flowing through it. After dI/dt, the switching current is equal to I_{load}. Thus, by eliminating the g_m,

$$\frac{dV_{sw}}{dt} = \frac{-I_{load} + \sqrt{2I_{load}K_{GaN}}R_{drv}I_{drv}}{2C_{oss} + \sqrt{2I_{load}K_{GaN}}R_{drv}C_{RSS}} \tag{4}$$

where K_{GaN} is the process parameter of GaN HEMTs. Because under most load conditions, $I_{load} \ll 2K_{GaN}R_{drv}^2 I_{drv}^2$, the first term of the numerator $-I_{load}$ is negligible. For denominator, $C_{RSS} \ll C_{OSS}$. Thus, (4) can be simplified as

$$\frac{dV_{sw}}{dt} \approx \sqrt{\frac{K_{GaN}}{2}I_{load}}\frac{I_{drv}R_{drv}}{C_{OSS}} \tag{5}$$

It can be concluded that $\frac{dV_{sw}}{dt}$ is approximately square root proportional to I_{load} and proportional to I_{drv}. This square root relationship deviates under heavy load conditions, though a positive correlation typically remains for buck DC-DC converters. Thus, to remain a stable dV/dt during miller plateau period, the drive current I_{drv} should be inversely proportional to the square root of load current I_{load}.

979-8-3315-1612-3/25 $31.00 © 2025 IEEE

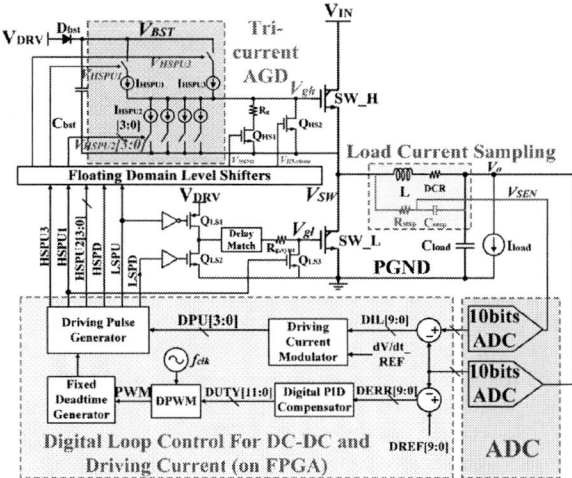

Fig. 2. System architecture of the load current modulated AGD driver.

C. Effectiveness of Controlling dV/dt

After the dV/dt interval, the power loop transitions into the RLC series resonant state, which operates independently of the gate drive loop. Due to the low R_{on}, the loop might readily become under-damped. The output voltage at SW node is

$$V_{sw} = V_{IN} - R_{on}C_{OSS}\left(\frac{dV_{sw}}{dt}\right)_0 e^{-\alpha t}\left(\cos\beta t + \frac{\alpha}{\beta}\sin\beta t\right) \quad (6)$$

Where $\alpha = R/2L$, $\beta = \sqrt{1/L_{loop}C_{OSS} - \alpha^2}$. Although the damp factor α and β cannot be changed, reduced dV_{sw}/dt suppresses the maximum overshoot or undershoot voltage, which reduces EMI and improves circuit operating stability. Depending on the requirements of various application scenarios, different dV/dt maximum values can be set.

III. PROPOSED LOAD CURRENT MODULATED AGD DRIVER

Fig. 2 displays the architecture of the proposed load current modulated tri-current active gate driver for a GaN buck DC-DC converter. It consists of three main components: a reactor-based inductor current sampling circuit, a digital control loop, and a tri-current active gate driver with a 4-bit current-based DAC. The whole system is based on an all-digital voltage-mode PWM modulated buck DC-DC. By load current sampling circuit, the varied load current is converted to voltage V_{sen} and sent to the

digital control circuit through a 10-bit ADC with the digitalized output voltage V_{out}. By trimming a reference dV/dt with load current, the driving current modulator sets an optimal drive current code DPU [3:0] and transmits it to the pulse generator to provide the required 4-bit pulse signal for AGD control. The digital PID compensator ensures loop stability.

The principle of active gate driver that controls the gate charging current of high-side switch is displayed in fig. 3(a). Similar to the work in [9], three distinct current sources regulate the driving strength in three phases: before the Miller plateau (MP), during MP, and in the oscillation phase. The I_{HSPU1} and I_{HSPU3} are positive and relatively large to speed up switch-on, whereas the I_{HSPU2} is negative and small to reduce the driving strength. The switches QHS2 and QLS3 in figure 2 are used to clamp the gate voltage during turn-off. As an improvement, I_{HSPU2} is controlled by a 4-bit digital input HSPU2[3:0], which is generated by digital control loop.

Fig. 3(b) shows the detailed circuit of the tri-current AGD. It utilizes a 4-bit current digital-to-analog converter (DAC) with a current source and four emitter-resistive feedback NPN current mirrors for temporary pulling down, and two discrete NMOS switches for fast pulling up. Before MP, only M1 is on to provide sufficient pull-up current . Following a delay, the MP is activated, at which point the 4-bit current-based DAC engages, generating a minor pull-down current. This approach can yield a rapid response of under 10 nanoseconds, even at the board level. After MP, the M3 is on, bringing the gate driving strength to the maximum.

Fig. 3(c) shows the timing diagram of pulse generator for controlling half-bridge circuit. Assuming the switch is fully opened during the pulse duration, the power consumption of the driver circuit can be substantially reduced using pulse control while ensuring fast operation. However, considering the duration of each switching step roughly 10ns and the difficulty of producing such short pulses at the board level, this may result in incorrect activation or deactivation of the switch. Thus, we achieve rapid gate control by modulating the delay of the rising edges between each pulses. In detailed, the delay t_{d1} between HSPU1 and HSPU2 represents the time between initial turn-on and MP, while the t_{d2} between HSPU2 and HSPU3 indicates the duration of MP.

Table I enumerates the model numbers of the discrete components utilized, along with their respective functions.

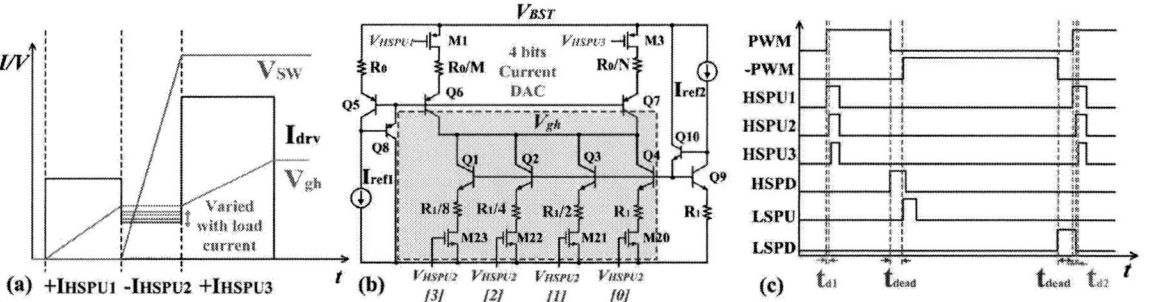

Fig. 3. Principle (a) and circuits (b) of proposed tri-current AGD, with timing diagram of pulse generator (c) for controlling AGD.

979-8-3315-1612-3/25 $31.00 © 2025 IEEE

TABLE I. MODEL NUMBERS AND FUNCTIONS OF DISCRETE CPMPONENTS

Device Type	Name	Device Number	Function
NPN	PMBT2222	Q1, Q2, Q3, Q4, Q9, Q10	Pull-down Current Mirror
PNP	PMBT3906	Q5, Q6, Q7, Q8	Pull-up Current Mirror
NMOS	PMZ450UNE	M20, M21, M22, M23, QHS1, QHS2	Pull-down Switch
PMOS	PMZ1200UPE	M1,M3	Pull-up switch and Clamp
Current Source	PSSI2021SAY	Iref1, Iref2	PVT Invariant Current Source

IV. EXPERIMENT RESULT

A. Test Set-up

The test circuit schematic for the proposed AGD-driven Buck DC-DC converter is shown in Fig. 4(a). The circuit uses two enhanced mode GaN transistors (INN100W032A) in a split-capacitor layout to reduce power loop inductance [12]. The Buck DC-DC converter steps down the voltage from 12V to 3.3V, and handles load currents between 0.8A and 4A. The switching frequency is 2MHz. The entire digital loop control module is realized by using a development board containing a Xilinx Artix-7 FPGA chip and an external dual-channel 10-bit ADC chip. The sampling rate of the ADC and the operating frequency of the digital loop is consistent with the switching frequency of the DC-DC. Figure 4(b) illustrates the top view of the printed circuit board (PCB) for the entire tested system.

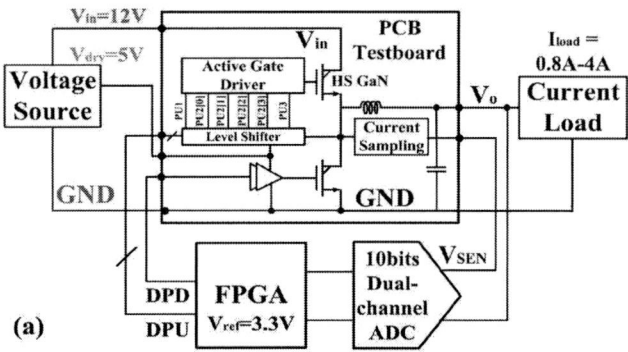

Fig. 4. Test set-up. (a) Test circuit schematic. (b) Top view of PCB test board.

In order to verify the effect of load-modulated AGD on the driving speed and stability, four driving modes are set: slow mode with drive current 210mA, fast mode with drive current 650mA, AGD with constant MP drive current 210mA and proposed load modulated MP drive current.

B. Experiment Result

To demonstrate the capacity of AGD to suppress overshoot on the switching node, the time-domain waveform of switching node Vsw and gate node Vgh in slow mode, fast mode, and AGD mode is compared and illustrated in Fig. 5. The load current is set as 1A. Compared to the quick mode in 5(c), the overshoot on the switching node is greatly suppressed by 5.6V, for an 11ns increase of turn-on speed. Meanwhile, compared to the slow mode in 5(b), the turn-on speed increases by 10ns with a 0.9V overshoot voltage increase. The result proves that the proposed AGD achieves high speed and low overshoot.

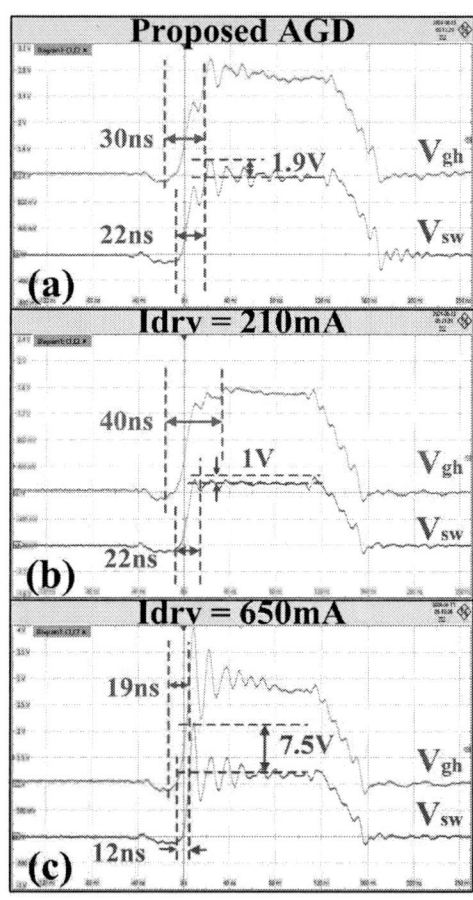

Fig. 5. Test waveform of switching node voltage and high-side gate voltage. (a) Under proposed AGD; (b) under Idrv = 210mA; (c) under Idrv = 650mA.

Figure 6 further analyzes the variation of switching speed versus ringing as the load current increases. To prove the feasibility of the proposed AGD, four drive modes are compared altogether. The driving speed of the proposed load-modulated AGD is 25.6% faster with less than 0.8V overshoot voltage larger than slow mode drive. Meanwhile, compared to the fast mode, the overshoot voltage is suppressed by more than 5.5V while the switching period increases by less than 14ns. Most

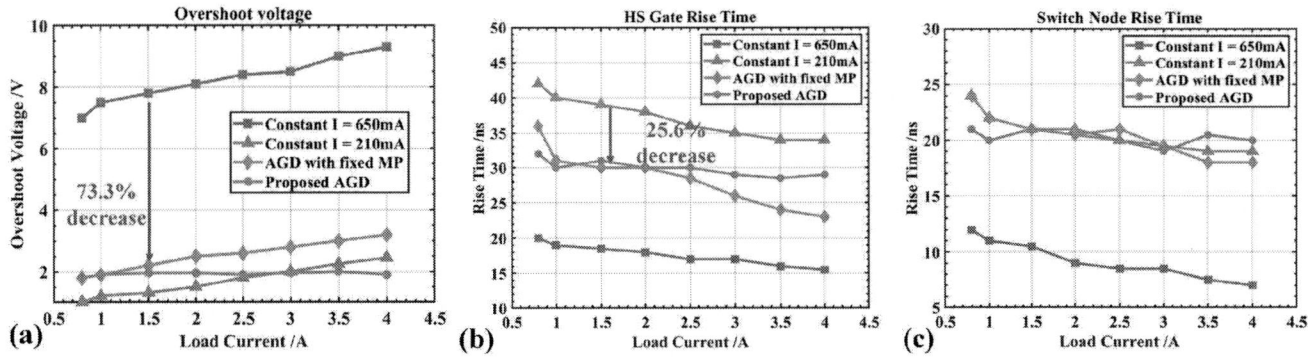

Fig. 6. Comparison of overshoot voltage (a) and rise time of HS gate (b) and switching node (c).

TABLE II. COMPARISON BETWEEN THIS WORK AND PRIOR ARTS

Design Metrics	JSSC 2018 [7]	JSSC 2022 [8]	TPE 2018 [10]	CICC 2023 [9]	This Wock
Power Switch	GaN HEMT	GaN HEMT	GaN HEMT	IGBT	GaN HEMT
Input Voltage	3 – 40 V	48 V	12 V	200 V	12 – 20V
Load Current	0.1 – 1 A	0.5 – 5 A	Nor Mentioned	10 A – 80 A	0.8 A – 4 A
Design of Driver	Tri - slope	Tri - slope	8 bits DAC	8 bits DAC	Tri – current with 4 bits (MP)
Control Method in MP	Presetting	Temperature compensation	Presetting	dIc/dt Modulated	Load Current Modulated
Real-time Control	No	No	No	Yes	Yes
Switching Speed	10MHz	1MHz – 10MHz	Nor Mentioned	< 1MHz	2MHz
Overshoot of Vsw	< 4.5V	Nor Mentioned	Nor Mentioned	Nor Mentioned	< 1.8V
Overshoot Voltage versus Load Current	3.3 V/A	Nor Mentioned	Nor Mentioned	Nor Mentioned	< 0.2V/A
Implementation	IC(0.35um BCD)	Monolithic GaN	PCB	IC(0.18um BCD)	PCB

importantly, by utilizing the proposed AGD, the variation of overshoot voltage remains under 0.2V/A and the switching time remains 30ns when load current varies, compared to 0.6V/A in fixed MP AGD. In summary, the proposed AGD maintains a stable voltage transition and overshoot, thus achieving both fast speed and high stability in different load conditions.

Table II compares our AGD work with prior arts.

V. CONCLUSIONS AND FUTURE WORK

In this work, we propose a tri-current active gate driver (AGD) with digital load-current modulated Miller Plateau strength control for GaN buck DC-DC converters. We initially analyze that the dV/dt at the switching node during the Miller Plateau is proportional to the square root of the load current for moderate load conditions. Based on this, we design a prototype that utilizes a tri-current active driver and a 4-bit current-source DAC to maintain constant dV/dt control during the Miller Plateau. The test results prove that the proposed AGD can maintain low overshoots and relatively high speed for various load conditions. This work further refines the effectiveness of the active gate driver and can be easily realized on a digital control loop for IC implementation, enabling higher switching speed and more precise switching control.

REFERENCES

[1] E. A. Jones, F. F. Wang and D. Costinett, "Review of Commercial GaN Power Devices and GaN-Based Converter Design Challenges," in IEEE Journal of Emerging and Selected Topics in Power Electronics, vol. 4, no. 3, pp. 707-719, Sept. 2016.

[2] Millán, P. Godignon, X. Perpiñà, A. Pérez-Tomás and J. Rebollo, "A Survey of Wide Bandgap Power Semiconductor Devices," in IEEE Transactions on Power Electronics, vol. 29, no. 5, pp. 2155-2163, May 2014.

[3] A. I. Emon and F. Luo, "A Review of High-Speed GaN Power Modules: State of the Art, Challenges, and Solutions," in IEEE Journal of Emerging and Selected Topics in Power Electronics, vol. 11, no. 3, pp. 2707-2729, June 2023.

[4] X. Mu et al., "Floating-Domain Integrated GaN Driver Techniques for DC–DC Converters: A Review," in IEEE Transactions on Circuits and Systems I: Regular Papers, vol. 70, no. 9, pp. 3790-3805, Sept. 2023.

[5] Y. Yorozu, M. Hirano, K. Oka, and Y. Tagawa, "Electron spectroscopy studies on magneto-optical media and plastic substrate interface," IEEE Transl. J. Magn. Japan, vol. 2, pp. 740–741, August 1987 [Digests 9th Annual Conf. Magnetics Japan, p. 301, 1982].

[6] N. Idir, R. Bausiere and J. J. Franchaud, "Active gate voltage control of turn-on di/dt and turn-off dv/dt in insulated gate transistors," in *IEEE Transactions on Power Electronics*, vol. 21, no. 4, pp. 849-855, July 2006, doi: 10.1109/TPEL.2007.876895.

[7] X. Ke, J. Sankman, Y. Chen, L. He and D. B. Ma, "A Tri-Slope Gate Driving GaN DC–DC Converter With Spurious Noise Compression and Ringing Suppression for Automotive Applications," in *IEEE Journal of Solid-State Circuits*, vol. 53, no. 1, pp. 247-260, Jan. 2018.

[8] T. -W. Wang *et al.*, "Monolithic GaN-Based Driver and GaN Switch With Diode-Emulated GaN Technique for 50-MHz Operation and Sub-0.2-ns Deadtime Control," in *IEEE Journal of Solid-State Circuits*, vol. 57, no. 12, pp. 3877-3888, Dec. 2022.

[9] D. Zhang, K. Horii, K. Hata and M. Takamiya, "Digital Gate ICs for Driving and Sensing Power Devices to Achieve Low-Loss, Low-Noise, and Highly Reliable Power Electronic Systems," 2023 IEEE Custom Integrated Circuits Conference (CICC), San Antonio, TX, USA, 2023.

[10] H. C. P. Dymond *et al.*, "A 6.7-GHz Active Gate Driver for GaN FETs to Combat Overshoot, Ringing, and EMI," in *IEEE Transactions on Power Electronics*, vol. 33, no. 1, pp. 581-594, Jan. 2018.

[11] N. Perera *et al.*, "Evaluation of GaN HEMT dv/dt Immunity and dv/dt Induced False Turn-On Energy Loss," *2024 IEEE Applied Power Electronics Conference and Exposition (APEC)*, Long Beach, CA, USA, 2024.

[12] K. Wang, L. Wang, X. Yang, X. Zeng, W. Chen and H. Li, "A Multiloop Method for Minimization of Parasitic Inductance in GaN-Based High-Frequency DC–DC Converter," in *IEEE Transactions on Power Electronics*, vol. 32, no. 6, pp. 4728-4740, June 2017.

Impact of Real-Time Variable Gate-Drive Strength on Drive Cycle Efficiency in SiC Inverter-Fed PMSM Traction Drives

Matteo Pizzuto
*Department of Electrical and
Computer Engineering
University of Windsor*
Windsor, Canada
pizzutom@uwindsor.ca

Aiswarya Balamurali
R&D Americas, Schaeffler
Chatham, Canada
aiswarya.balamurali
@vitesco.com

Aniket Anand
R&D Americas, Schaeffler
Chatham, Canada
aniket.anand@vitesco.com

Narayan C. Kar
*Department of Electrical and
Computer Engineering
University of Windsor*
Windsor, Canada
nkar@uwindsor.ca

Abstract— **This paper investigates an analysis of variable gate-drive strength (GDS) in Silicon Carbide (SiC) inverters with interior permanent magnet synchronous machines (IPMSMs) for electric vehicle (EV) traction drives. While SiC inverters offer superior efficiency and performance at high switching frequencies, their fast switching transients result in high dV_{ds}/dt and di_{ds}/dt, leading to overshoot, electromagnetic interference (EMI), and increased harmonic losses in the motor. This study evaluates the impact of strong and weak GDS configurations on both inverter switching losses and motor efficiency using a co-simulation process that integrates circuit simulation, control modeling, and finite element analysis (FEA). Experimental validation on a 300 kW SiC inverter and IPMSM using a dynamometer further demonstrates the trade-offs between inverter and motor performance under varying operating conditions. By addressing the system-level effects of GDS selection, an insight into achieving a balanced and efficient design for SiC-based EV powertrains is presented.**

Keywords—Gate-drive strength, gate resistance, harmonic losses, interior permanent magnet synchronous machine, SiC inverter, switching transients

I. INTRODUCTION

Electric vehicles (EVs) are favoured for their sustainability, high performance, and energy efficiency compared to conventional internal combustion engine vehicles. The efficiency of an electric powertrain is heavily dependent on the electric drive unit, primarily composed of an electric machine and an inverter. Even marginal inverter improvements can yield performance enhancements, leading to longer driving range, lower operating costs, and reduced heat generation [1]. Modern EVs have adopted silicon carbide (SiC) power switching devices in the inverter for their high switching speed and loss reduction, though increased switching speed can also produce high drain-source voltage V_{ds} overshoot, oscillation, and electromagnetic interference (EMI) [2], [3]. These effects can be detrimental and require design considerations to combat eventual component damage. The conventional gate drivers (CGDs) used to power the switching devices use a fixed gate resistance R_g that is robust and effective over its entire working range [1]. Selecting a higher R_g can reduce the V_{ds} overshoot, but also reduces switching speed and increases switching losses, thus, opposing the main advantages of SiC MOSFETs [4]. To mitigate these unwanted switching waveforms, active gate driver (AGD)

methods have been proposed to dynamically alter the gate voltage V_{gs}, gate current I_{gs}, and R_g in a 1 to 50 ns range [5]–[8]. This approach aims to optimize switching behavior and achieve a weight-based balance between performance and efficiency [9]. However, the numerous AGD topologies presented face complexities in implementation with respect to the timing resolution and the number of voltage/current levels during the switching transient [10]. Despite multiple studies on the effects of AGD on inverter performance, a thorough analysis of its impact at the motor and drive system level is missing.

Understanding the system-level implications of altering the gate-drive strength (GDS) is crucial, as performance improvements must extend beyond the inverter to the entire electric drive unit. Moreover, reducing R_g increases overshoot and oscillations, thus increasing the magnitude of high-frequency voltage harmonics. Generally, when analyzing the impact of inverter switching harmonics, high-frequency current harmonics above three times the switching frequency f_c are neglected due the inductive nature of electric machines [11]. Having said this, the voltage transients (very high-frequency, MHz range) generated from the fast-switching speeds in SiC devices require further analysis to examine how their high magnitudes can affect the harmonic spectrum and losses [4] [12]. In [3], the overshoot produced in SiC MOSFETs peaks at double the DC bus voltage, though this depends on capacitor configurations, inductance magnitude, and load. Nonetheless, it is crucial to optimize such transients as an increased voltage total harmonic distortion (THD) can lead to additional motor core losses [13]. Furthermore, added losses equate to lower efficiency, potentially impacting the thermal requirements of the electric drive system.

In this paper, a comprehensive analysis of various GDSs is conducted to evaluate the impact on traction drive efficiency and system-level performance using SiC inverters. The GDS is adjusted by varying R_g on a μs scale to balance CGD robustness with minimized inverter switching losses. Detailed simulations model the switching transients and harmonic content associated with different GDS levels, highlighting how these factors influence losses and high-frequency harmonics in an interior permanent magnet synchronous machine (IPMSM). Furthermore, the analysis is experimentally validated using a 300 kW SiC inverter and IPMSM system mounted on a dynamometer, allowing for real-world performance assessment across key operating points. In Section II, the impact of GDS on

979-8-3315-1612-3/25 $31.00 © 2025 IEEE

switching characteristics is discussed, along with an overview of SiC MOSFET switching phenomena. Section III presents the co-simulation methodology to capture switching-induced effects on IPMSM performance. Simulation results in Section IV illustrate the effects of different GDS levels on switching losses, EMI, and harmonic distortion within the inverter and motor. In Section V, experimental results validate the simulation process and findings across critical operating conditions. Lastly, conclusions are drawn in Section VI.

II. Gate Driver Impact on System Performance and Losses

Gate drivers play a crucial role in the operation of power switching devices within an inverter, impacting the performance and efficiency of the system. The choice of R_g directly alters the switching characteristics, including drain-source voltage slew rate dV_{ds}/dt, drain-source current slew rate di_{ds}/dt, overshoot, switching loss, and reliability. The turn-off transient of SiC MOSFETs is where the drain-source voltage overshoot $V_{ds,os}$ becomes maximum. The $V_{ds,os}$ is calculated from (1) [1]:

$$ V_{ds,os} = V_{dc} - L_{loop} \left(\frac{V_{gg} - V_{th} - \dfrac{i_L}{g_m}}{\dfrac{C_{iss} R_g}{g_m} + L_s} \right) + V_{FWD} \qquad (1) $$

where V_{dc} is the DC-link voltage, V_{gg} denotes the gate-drive voltage supply, V_{th} represents the threshold voltage, and V_{fwd} is the diode forward voltage. L_{loop} indicates the circuit's loop inductance, while L_s specifies the MOSFET source inductance. C_{iss} is the input capacitance, i_L is the load current, and g_m is the transconductance.

Unlike AGD methods where R_g is modified between turn-on and turn-off intervals in the ns range, dV_{ds}/dt and di_{ds}/dt can be approximated using (2) with a fixed R_g [1], [5]:

$$ \frac{dV_{ds}}{dt} = \frac{V_{gg} - V_{th} - \dfrac{i_L}{g_m}}{C_{gd} R_g} ; \frac{di_{ds}}{dt} = g_m \frac{V_{th} + \dfrac{i_L}{g_m} - V_{gg}}{\dfrac{C_{iss} R_g}{g_m} + L_s} \qquad (2) $$

where C_{gd} is the gate-drain capacitance.

The SiC MOSFET switching losses E_{sw} are a sum of the energy losses experienced at both the turn-on t_{on} and turn-off t_{off} switching times. The primary sources of losses during the turn-on transient are produced by current rise and voltage decay, whereas losses in the turn-off transient are predominantly dependent on voltage rise and current decay [5]. Nonetheless, increasing R_g will reduce the switching speed, thus, increasing the turn-on and turn-off switching time, resulting in larger E_{sw} in (3):

$$ E_{sw} = \frac{1}{2} V_{dc} \left[i_L (t_{off} + t_{on}) + C_{oss} V_{dc} \right] \qquad (3) $$

where C_{oss} is the output capacitance.

The total inverter losses can be defined as a sum of E_{sw} and conduction losses E_{cond} in the SiC MOSFETs and freewheeling diodes (4) [1]. The inverter energy losses can be expressed as power losses using (6):

$$ E_{cond} = \int I_s^2 R_{on} \, dt \qquad (4) $$

$$ E_{loss,inv} = \frac{1}{2} V_{dc} \left[i_L (t_{off} + t_{on}) + C_{oss} V_{dc} \right] + \int I_s^2 R_{on} \, dt \qquad (5) $$

$$ P_{loss,inv} = E_{loss,inv} \cdot f_{sw} \qquad (6) $$

The PWM harmonics from the inverter lead to sideband harmonic currents that cause harmonic losses in the stator and rotor of the IPMSM, as well as increased torque pulsations in the machine [14], [15]. Thus, the harmonic dq-axis flux linkage λ_{dqh} with harmonic order h is derived in (7) and used to verify the losses [15]:

$$ \lambda_{dh}(i_{dh}, i_{qh}) = \frac{V_{qh} - R_s i_{qh}}{\omega_e} ; \lambda_{qh}(i_{dh}, i_{qh}) = \frac{-V_{dh} + R_s i_{dh}}{\omega_e} \qquad (7) $$

where V_{dqh} are the dq-axis voltage, i_{dqh} are the dq-axis current, R_s is stator resistance, and ω_e is the electrical angular speed of the rotor. The h represents base band harmonics of order $m=5$, 7, 11, etc., switching harmonics with order $m_1 f_c$ and side-band harmonics of order $m_1 f_c \pm m_2 f_s$, where $m_1 = 1, 2, 3$, etc., and $m_2 = \pm 1, \pm 2, \pm 3$, etc., where f_s is the fundamental frequency [15]. The copper losses P_{Cu} of an IPMSM are expressed as a function of the stator current and R_s in (8). The core losses P_{core} consist of eddy current losses P_{eddy} and hysteresis losses P_{hys}, seen in (9) and (10), where f_h is the frequency at h. Together, the total controllable losses of the IPMSM $P_{loss,mot}$ are given as the sum of P_{Cu} and P_{core}, as shown in (11):

$$ P_{Cu} = \frac{3}{2} R_s (T_s) \left[\sum_{h=1}^{\infty} \left(i_{dh}^2 + i_{qh}^2 \right) \right] \qquad (8) $$

$$ P_{eddy} = k_{edq} \sum_{h=1}^{\infty} f_h^2 \left[\lambda_{dh}(i_{dh}, i_{qh}) + \lambda_{qh}(i_{dh}, i_{qh}) \right]^2 \qquad (9) $$

$$ P_{hys} = k_{hydq} \sum_{h=1}^{\infty} f_h \left[\lambda_{dh}(i_{dh}, i_{qh}) + \lambda_{qh}(i_{dh}, i_{qh}) \right]^2 \qquad (10) $$

$$ P_{loss,mot} = k_{edq} \sum_{h=1}^{\infty} f_h^2 \left[\lambda_{dh}(i_{dh}, i_{qh}) + \lambda_{qh}(i_{dh}, i_{qh}) \right]^2 $$
$$ + k_{hydq} \sum_{h=1}^{\infty} f_h \left[\lambda_{dh}(i_{dh}, i_{qh}) + \lambda_{qh}(i_{dh}, i_{qh}) \right]^2 \qquad (11) $$
$$ + \frac{3}{2} R_s (T_s) \left[\sum_{h=1}^{\infty} \left(i_{dh}^2 + i_{qh}^2 \right) \right] $$

where k_{edq} and k_{hydq} are the eddy current and hysteresis coefficients, respectively, and T_s is stator temperature. Note, losses in the magnet are not considered in this study. Finally, the system level power losses $P_{loss,sys}$ are computed by combining $P_{loss,inv}$ and $P_{loss,mot}$, in (12):

$$ P_{loss,sys} = P_{loss,inv} + P_{loss,motor} \qquad (12) $$

Lastly, the inverter, motor, and system level efficiency values η_{inv}, η_{mot}, and η_{sys} are calculated as follows in (13):

$$ \left(\eta_{inv} = \frac{P_{inv}}{P_{DC}} ; \eta_{mot} = \frac{P_{mot}}{P_{inv}} ; \eta_{sys} = \frac{P_{mot}}{P_{DC}} \right) \times 100\% \qquad (13) $$

where P_{DC}, P_{inv}, and P_{mot} is DC power, inverter power and motor power, respectively.

III. Inverter and Motor Modelling

Pre-built MOSFET models in Simulink are insufficient in modelling the true overshoot and ringing of SiC MOSFETs. Consequently, to model the impact of switching transients and

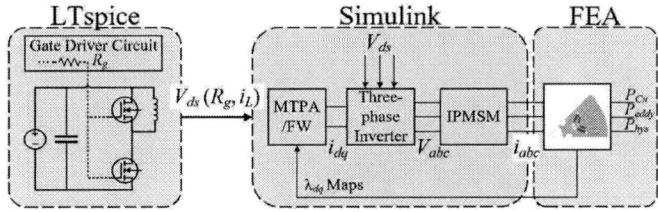

Fig. 1. Co-simulation workflow utilizing LTspice, Simulink, and FEA

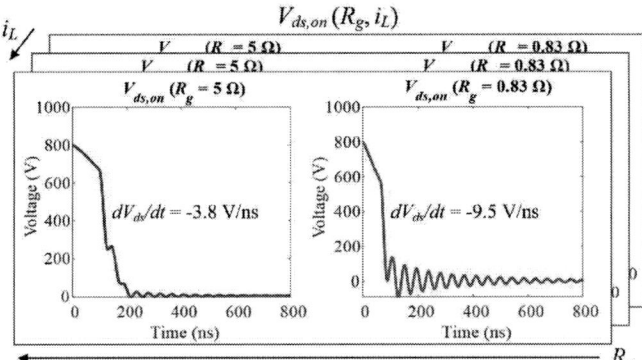

(a)

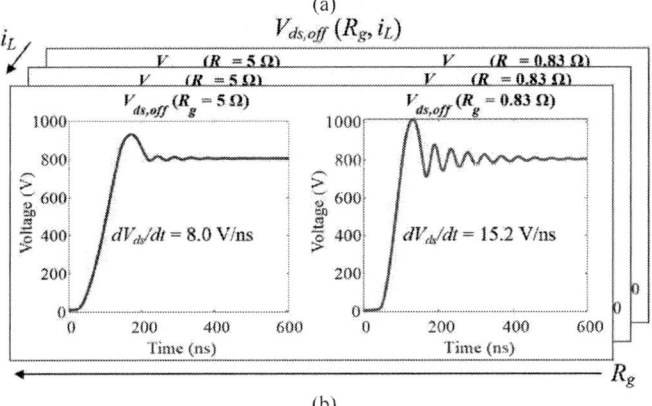

(b)

Fig. 2. Switching waveform LUTs that vary with i_L and R_g. (a) $V_{ds,on}$ sequence. (b) $V_{ds,off}$ sequence.

overcome the limitations at the motor and system level, a simulation process was developed. A half-bridge inverter LTspice model of the Wolfspeed CAB450M12XM3 power module was used to simulate double pulse tests with two different R_g: 0.83 Ω (strong GDS) and 5 Ω (weak GDS), and output the V_{ds} waveforms at various loads. During the switch on sequence, the voltage drops as the SiC MOSFET begins conducting, where the speed of the drop is over two times as fast in the strong GDS. This rapid drop is indicative of reduced switching time but also introduces higher EMI due to the sharp transition [16]. Similarly, the switch off sequence shows the voltage rise as the SiC MOSFET switches to its off state, and the rise is almost twice as fast for the strong GDS.

The combined co-simulation process is shown in Fig. 1. The V_{ds} waveforms were segregated and reformatted into switch on $V_{ds,on}$ and off $V_{ds,off}$ waveforms and were implemented into a Simulink inverter model as look-up tables (LUTs) as shown in Fig. 2, where custom logic enables the application of the waveforms at appropriate intervals within each switching cycle. Subsequently, the phase current waveforms were transferred to

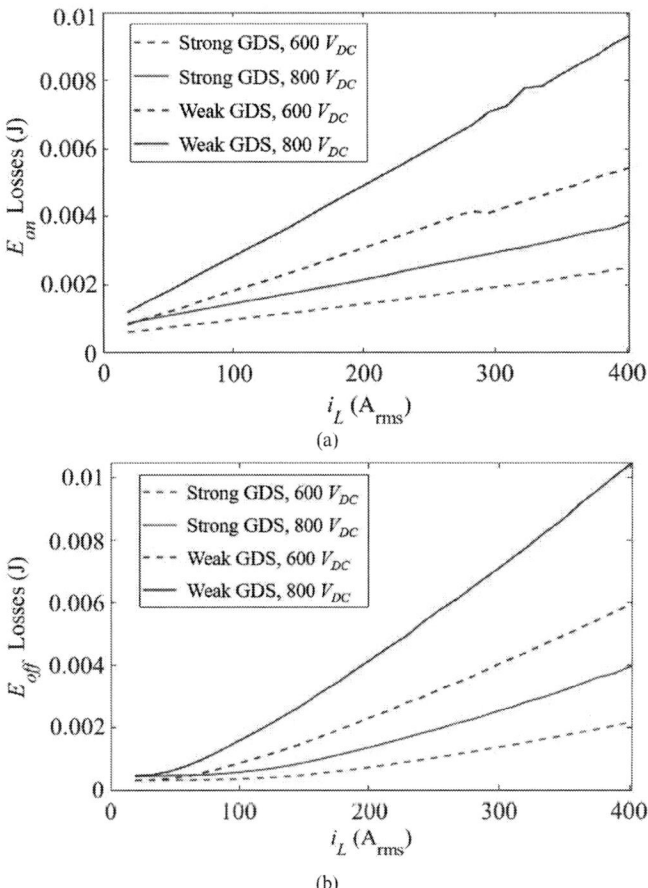

(a)

(b)

Fig. 3. Loss comparison at different i_L and V_{DC} with varying R_g. (a) E_{on} losses vs. i_L. (b) E_{off} losses vs. i_L.

Ansys Maxwell, where a finite element analysis (FEA)-based loss evaluation was performed. The FEA loss model integration considers the influence of harmonics generated during the rapid switching events. By this integrated process, realistic inverter behavior under both strong and weak GDS conditions can be simulated and fed into a high fidelity IPMSM model. Thus, the proposed method enables a detailed analysis of switching-induced phenomena like overshoot, ringing, and transient losses on the machine, and optimization of drive system-level performance with varying GDS.

IV. LOSS AND HARMONIC ANALYSIS – SIMULATION RESULTS

The co-simulation process discussed in section III was performed to characterize the impact of varying GDSs on both inverter and motor losses across a wide range of operating points. First, using LTspice, the turn-on losses E_{on} and turn-off losses E_{off} for strong and weak GDS were evaluated. These simulation results were obtained across a range of i_L between 20 and 400 A_{rms} and V_{DC} of 600 V and 800 V, as seen in Fig. 3. The data demonstrates a substantial reduction in $P_{loss,inv}$, where E_{on} decreases by up to 50% and E_{off} by up to 40% when utilizing the strong GDS.

Using the waveforms in Fig. 2, the Simulink inverter and IPMSM control model was simulated at equivalent operating points for both strong and weak GDSs. Fig. 4 reveals the line-to-line voltage harmonic spectrum across a 25 MHz range as the

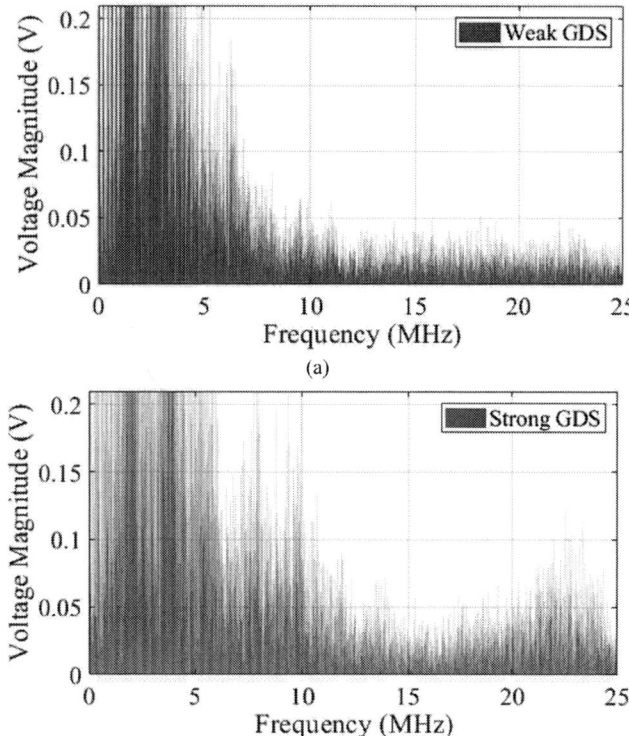

Fig. 4. FFT of line-to-line voltage at 2000 rpm and 300 Nm. (a) Weak GDS. (b) Strong GDS.

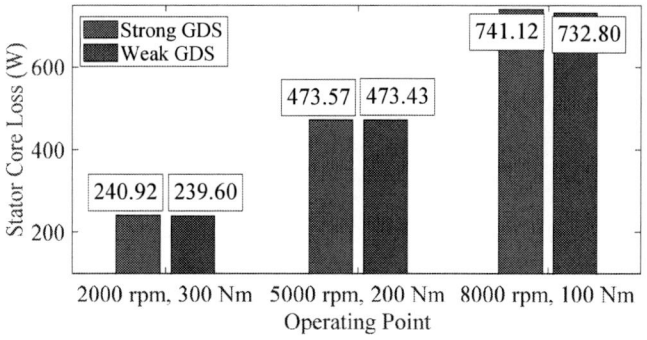

Fig. 5. Stator core loss of IPMSM using strong and weak GDS at different operating points.

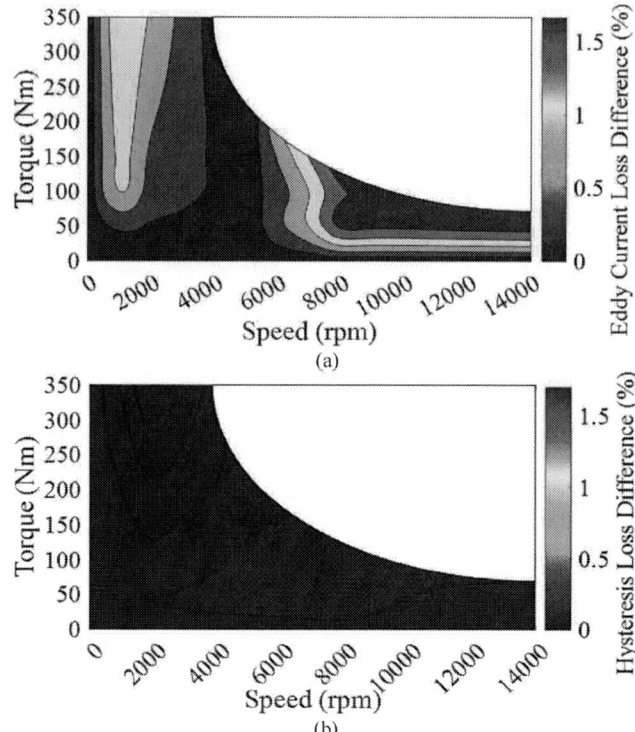

Fig. 6. Stator core loss differences between strong and weak GDS.(a) Eddy current loss difference. (b) Hysteresis loss difference.

TABLE I.	DRIVE SYSTEM PARAMETERS	
Parameter	**Symbol**	**Value**
Nominal Bus Voltage	V_{DC}	800 V
Number of Pole Pairs	P	4
Stator Winding Resistance	R_s	0.035 Ω
Base Speed	n_{base}	4,500 rpm
Maximum Speed	n_{max}	15,000 rpm
Peak Torque	T_{max}	350 Nm
Peak Phase Current	I_{max}	230 A$_{rms}$
Switching Frequency	f_c	10 kHz
Gate Resistance	R_g	0.83 Ω or 5 Ω
DC Link Capacitance	C_{DC}	300 µF
Parasitic Inductance	L_{PL}	5.3 nH

as the SiC switching oscillations occur between 20-25 MHz. It is evident that the strong GDS spectrum introduces these high frequency hamonics due to faster transitions and ringing, while the weak GDS presents a smoother signal and voltage spectrum. As a result of these differences, there is a significant difference in voltage THD of 1.32% at low speed and high torque, as calculated using (14):

$$THD = \frac{\sqrt{\sum_{n=2}^{\infty} V_n^{2}}}{V_1} \times 100\% \qquad (14)$$

Fig. 5 provides insight into motor-specific losses, comparing P_{core} for strong and weak GDSs across different operating points. The P_{core} for strong GDS reaches 741.12 W at 8,000 rpm and 100 Nm, while weak GDS results a lesser amount of 732.80 W, a loss difference of 1.14%. Similarly, below base speed nearing maximum load, a difference exists but nearly half as significant

at 0.55%. This loss difference can be attributed to the characteristics of P_{eddy}, where f_h is a squared factor of λ_{dqh}, thus, resulting in more significant variation at high frequencies. This can be better visualized in Fig. 6(a). As seen in both Fig. 5 and 6, however, the mid speed range of the machine presents a minimal difference in losses. Fig. 6(b) shows negligible differences in P_{hys}, where f_h contributes less on the total value. Furthermore, the P_{Cu} are dependent on current, where high-order harmonics are filtered and negligible, leaving almost no difference between weak and strong GDS. Overall, the discrepancies underscore slightly higher electromagnetic stress imposed by strong GDS on the machine's core [17].

V. EXPERIMENTAL VALIDATION

The experimental setup in Fig. 7(a) was designed to validate the simulation findings and evaluate the studied impact of variable GDS on a SiC inverter-fed traction drive system. The test bench consisted of a 300 kW SiC inverter equipped with Wolfspeed CAB450M12XM3 power modules, a Hyundai Ioniq 5 traction IPMSM, and a 15,000 rpm, 300 kW dynamometer with integrated energy storage at the CHARGE facility. The dynamometer was operated in speed control mode, while the IPMSM ran in torque/current control mode. The inverter utilized in this study incorporates a built-in real-time adjustable GDS via software, enabling seamless transition between weak and strong GDS. This feature allowed for direct side-by-side comparisons of performance metrics under different GDS configurations without requiring hardware modifications. Fig. 7(b) illustrates the control block diagram implemented for real-time GDS adjustment. The system integrates LUT-based maximum torque per ampere/flux weakening (MTPA/FW) control with feedback from torque, speed, and temperature sensors.

To ensure high accuracy in data collection, various data acquisition systems were connected to AVL PUMA, interfacing real-time data with the dynamometer and the inverter control system. Power measurements were obtained using a Yokogawa WT3000 Power Analyzer, paired with LEM IT-700s current transducers for precision in capturing high-frequency voltage and current harmonics. The Yokogawa WT3000 also computes losses and η_{inv}, η_{mot}, and η_{sys} in real time. Stator temperature was monitored via K-type thermocouples, where $T_s = 35°C$ was used as the operating condition to eliminate thermal variability during testing.

Fig. 8 presents the experimental motor loss differences where strong GDS is divided by weak GDS $P_{loss,mot}$ for equivalent operating conditions. Motor loss differences are

seen to follow the trends set in Fig. 6, where the most significant differences are seen at low speed and high torque, and high speeds. However, around the base speed of the machine, as specified in the system parameters in Table I, the differences are minor.

Fig. 9 depicts the values of η_{inv}, η_{mot}, and η_{sys} at points inside the main operating region of the IPMSM. Consistent with the presented results in simulation, there is an improvement in η_{inv} when using strong GDS for all cases, however, η_{mot} counteracts these differences due to increased high-frequency λ_{dqh}, thus, P_{eddy}.

Table II is the proposed decision matrix for optimized drive cycle efficiency and trade-offs, split in terms of inverter benefit and motor benefit. To support the decision-making process for optimized GDS usage with emphasis on the system-level impacts, the studies done in [18] are used. As demonstrated in [18], limiting strong GDS to lower load currents can effectively mitigate high voltage overshoot, enhancing inverter performance while maintaining reliability. Further, in [18], an efficiency improvement of approximately 4.6% is seen across the Worldwide Harmonized Light Vehicles Test Cycle (WLTC) procedure. It is to be noted that this improvement is attributed solely to inverter benefits and does not cover the potential motor-level implications. Thus, it is crucial to extend these insights by addressing system-level performance. While variable GDS enhances inverter efficiency, our findings reveal its potential to exacerbate motor harmonic losses and reduce system-level efficiency under certain operating conditions. In [19], a WLTC cycle is discussed for design optimizations of a Hyundai Ioniq 5 IPMSM for the respective all wheel drive EV. Experimentally performed WLTC results for the equivalent

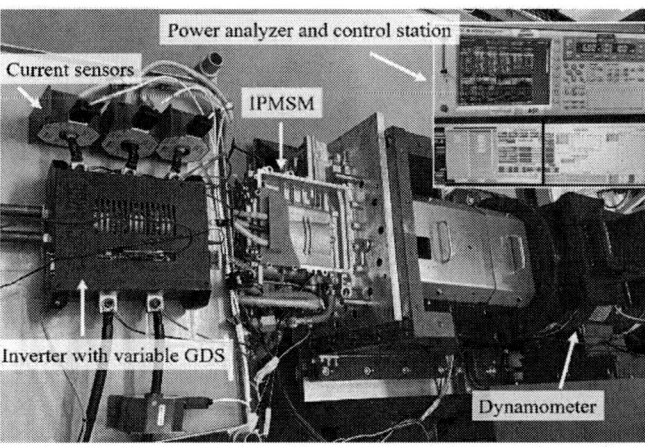

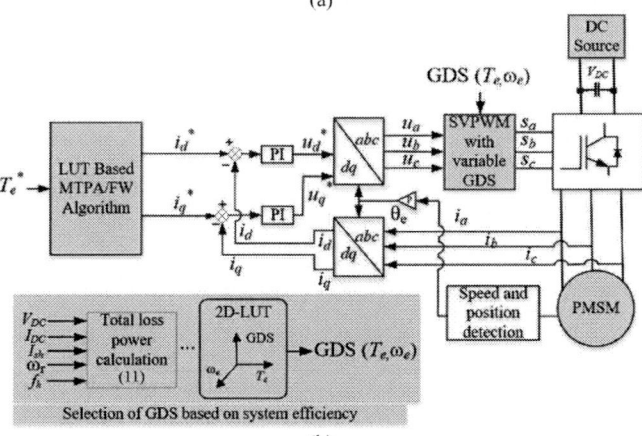

Fig. 7. Control algorithm and experimental setup. (a) Experimental setup used to determine system-level losses and parameters. (b) Control block diagram with variable GDS based on losses and tradeoffs.

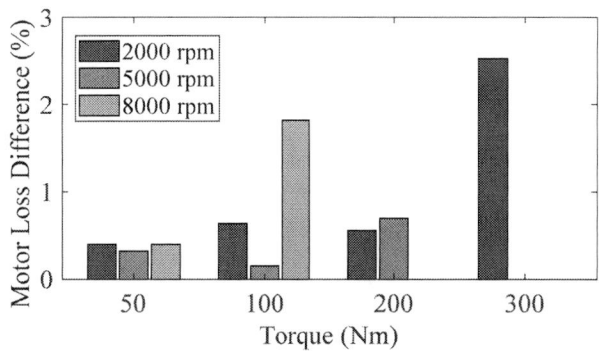

Fig. 8. Motor loss differences between strong and weak GDS.

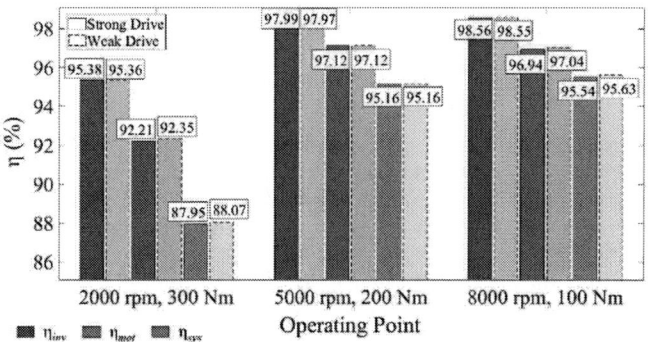

Fig. 9. Effect of GDS on η_{inv}, η_{mot}, and η_{sys} at different operating points.

TABLE II. DRIVE CYCLE DECISION MATRIX

IONIQ 5 Operating Region	Use case selection based on losses and trade-offs			
	Inverter [18]		Motor	
	Weak GDS	*Strong GDS*	*Weak GDS*	*Strong GDS*
Low speed Low torque		•		•
Low speed High torque	•		•	
Mid speed Low torque		•		•
Mid speed High torque	•			•
High speed Low torque		•	•	
High speed Mid torque		•	•	
Max. speed Low torque		•	•	
Speed: Low: < 4500 rpm, Mid: 4500–8000 rpm, High: > 8000 rpm **Torque**: Low: < 100 Nm, Mid: 100–250 Nm, High: > 250 Nm				

IPMSM showcase the motor operating within less than 50% of the motors peak operating envelope [19]. Therefore, the motor does not operate in the low-to-mid-speed and high-torque region, or high-speed and mid-torque region. In fact, it spends majority of the WLTC in the low-to-mid-speed and low-torque region, where the suggested inverter and motor GDS selections are equivalent. Additionally, for the high-speed region, the load is small enough where the differences are negligible, leaving the option of both weak and strong GDS selections viable. For these reasons, there are negligible energy savings at the motor level between weak and strong GDS for the WLTC. Nonetheless, the effects have proven to be impactful at other major operating points and should be considered when optimizing both high-speed and high-torque drive-cycle system level efficiency for other types of drive cycles, such as US06 or user-based highway and urban cycles.

VI. CONCLUSION

This paper investigates the role of real-time variable GDS in optimizing the performance of SiC inverters and IPMSMs for EV traction systems. By integrating both simulation and experimental validation, the study reveals the trade-offs between strong and weak GDS configurations at the inverter, motor, and system levels. Strong GDS effectively minimizes switching losses and enhances inverter efficiency, particularly at lower load currents. However, its application introduces high-frequency harmonics, leading to increased motor core losses and reduced system-level performance under specific operating conditions. Conversely, weak GDS mitigates these harmonic effects, preserving motor efficiency but at the expense of higher switching losses in the inverter. The findings underscore the necessity of dynamically balancing GDS selection to accommodate varying load conditions and operational demands. Furthermore, the framework of this study can be extended by incorporating a wider range of drive cycles and machines to validate the long-term efficiency gains and further refine real-time GDS control strategies, as even minimal energy savings can be crucial due to stringent targets. Through this, the study provides a basis for achieving higher overall efficiency and reliability in SiC-based EV powertrains, addressing both inverter and motor-level challenges.

REFERENCES

[1] L. Wen, W. Yu, J. Geiger, and I. Husain, "Selective gate driver in SiC inverter to improve fuel economy of electric vehicles," in *Proc. 2022 IEEE Energy Convers. Congr. Expo.*, Detroit, MI, USA, 2022, pp. 1-7.

[2] S. Zhao *et al.*, "Adaptive multi-level active gate drivers for SiC power devices," *IEEE Trans. Power Electron.*, vol. 35, no. 2, pp. 1882-1898, Feb. 2020.

[3] P. Yi, Y. Cui, A. Vang, and L. Wei, "Investigation and evaluation of high power SiC MOSFETs switching performance and overshoot voltage," in *Proc. 2018 IEEE Appl. Power Electron. Conf. Expo.*, San Antonio, TX, 2018, pp. 2589-2592.

[4] H. Li, Y. Jiang, C. Feng, and Z. Yang, "A voltage-injected active gate driver for improving the dynamic performance of SiC MOSFET," in *Proc. IEEE Energy Convers. Congr. Expo.*, 2019, pp. 6943-6948.

[5] Y. Yang, Y. Wen, and Y. Gao, "A novel active gate driver for improving switching performance of high-power SiC MOSFET modules," *IEEE Trans. Power Electron.*, vol. 34, no. 8, pp. 7775-7787, Aug. 2019.

[6] H. C. P. Dymond *et al.*, "A 6.7-GHz active gate driver for GaN FETs to combat overshoot, ringing, and EMI," *IEEE Trans. Power Electron.*, vol. 33, no. 1, pp. 581–594, Jan. 2018.

[7] R. Grezaud, F. Ayel, N. Rouger, and J. Chebier, "An adaptive output impedance gate drive for safer and more efficient control of wide bandgap devices," in *Proc. IEEE Workshop Wide Bandgap Power Devices Appl.*, Oct. 2013, pp. 68–71.

[8] M. Sasaki, H. Nishio, and W. T. Ng, "Dynamic gate resistance control for current balancing in parallel connected IGBTs," in *Proc. 2013 IEEE Appl. Power Electron. Conf. Expo.*, Mar. 2013, pp. 244–249.

[9] X. Chen, H. Peng, S. Song, Q. Tong, and Y. Kang, "A novel control strategy for optimal tradeoff between overshoot and switching loss based on double closed-loop self-regulating active gate driver," *IEEE Trans. Power Electron.*, vol. 39, no. 10, pp. 13033-13043, Oct. 2024.

[10] M. Parker, I. Sahin, R. Mathieson, S. Finney, and P. D. Judge, "Investigation into active gate-driving timing resolution and complexity requirements for a 1200 V 400 A silicon carbide half bridge module," *IEEE Open J. Power Electron.*, vol. 4, pp. 161-175, 2023.

[11] W. Liang, J. Wang, P. C. Luk, W. Fang, and W. Fei, "Analytical modeling of current harmonic components in PMSM drive with voltage-source inverter by SVPWM technique," *IEEE Trans. Energy Convers.*, vol. 29, no. 3, pp. 673-680, Sept. 2014.

[12] Z. Wang, Y. Wu, M. H. Mahmud, Z. Yuan, Y. Zhao, and H. A. Mantooth, "Busbar design and optimization for voltage overshoot mitigation of a silicon carbide high-power three-phase T-type inverter," *IEEE Trans. Power Electron.*, vol. 36, no. 1, pp. 204-214, Jan. 2021.

[13] J. H. Seo, T. K. Chung, C. G. Lee, S. Y. Jung, and H. K. Jung, "Harmonic iron loss analysis of electrical machines for high-speed operation considering driving condition," *IEEE Trans Magn.*, vol. 45, no. 10, pp. 4656-4659, Oct. 2009.

[14] W. Liang, W. Fei, and P. C. Luk, "An improved sideband current harmonic model of interior PMSM drive by considering magnetic saturation and cross–coupling effects," *IEEE Trans. Ind. Electron.*, vol. 63, no. 7, pp. 4097–4104, Jul. 2016.

[15] A. Balamurali, A. K. Anik, W. Clandfield, and N. C. Kar, "Non-invasive parameter and loss determination in PMSM considering the effects of saturation, cross-saturation, time harmonics, and temperature variations," *IEEE Trans. Magn.*, vol. 57, no. 2, pp. 1-6, Feb. 2021.

[16] K. Lee, M. Fereydooonian, M. Benson and W. Lee, "Partial discharge and electromagnetic interference under repetitive voltage pulses with high slew rate in AC machine drives," in *Proc. 2022 IEEE Transp. Electrific. Conf. Expo.*, Anaheim, CA, USA, 2022, pp. 917-922.

[17] N. G. Minh Thao, S. Zhong, K. Fujisaki, F. Iwamoto, T. Kimura and T. Yamada, "Experimental assessment of motor core loss, inverter loss and ringing phenomenon under SiC-MOSFET inverter excitation," in *Proc. 2019 IEEE Int. Elect. Mach. Drives Conf.*, San Diego, CA, USA, 2019, pp. 1634-1640

[18] G. Lakkas, "Improve traction inverter system efficiency at lower cost," 2023. Available: https://www.ti.com/lit/ml/slyp888/slyp888.pdf.

[19] H. Mohammadi, S. Saini, R. Nasirizarandi, and A. Balamurali, "Drive cycle-based design optimization of traction motor drives for battery electric vehicles using data-driven approaches," SAE Technical Paper, 2024.

Demonstration of Efficiency Increase of 350 V-to-13.3 V Isolated DC-DC Converters for Electric Vehicles by Active Gate Driving

Yohei Sukita
The University of Tokyo
Tokyo, Japan
ysukita@iis.u-tokyo.ac.jp

Katsuhiro Hata
Shibaura Institute of Technology
Tokyo, Japan
khata@shibaura-it.ac.jp

Hiroki Kondo
Toyota Industries Corporation
Aichi, Japan

Kenichi Watanabe
Toyota Industries Corporation
Aichi, Japan

Kenichi Nagayoshi
Toyota Industries Corporation
Aichi, Japan

Makoto Takamiya
The University of Tokyo
Tokyo, Japan
mtaka@iis.u-tokyo.ac.jp

Abstract—An active gate driving (AGD) is applied to a power converter product, and the efficiency increase of the power converter by AGD is demonstrated by measurements under switching noise aligned condition compared to a conventional gate driving. Specifically, the gate driver for a single Si power MOSFET comprising a 165 kHz, 350 V-to-13.3 V isolated DC-DC converter product for electric vehicles was replaced with a developed time-domain stop-and-go active gate driver, and the DC-DC converter efficiency and the spectrum amplitude at 27 MHz of the power MOSFET drain current of the conventional single-step gate driving and the proposed AGD are compared in measurements at 1.2 kW (= 13.3 V, 90 A) output. The results show that the proposed AGD reduces switching loss by 45% compared to the conventional single-step gate driving under the drain current spectrum amplitude alignment condition, resulting in a 10% reduction in total DC-DC converter loss and a 0.9% increase in DC-DC converter efficiency from 90.3% to 91.2%.

Keywords—active gate drive, switching loss, DC-DC converter, efficiency

I. INTRODUCTION

Active gate driving (AGD), which changes the gate driving strength multiple times in fine time slots during the switching period of power devices, is attracting attention as a technology that can solve the trade-off problem between loss and noise during power device switching. Most papers on AGD, however, are limited to measurements of double-pulse tests in half-bridge circuits [1-16], and very few papers have quantitatively demonstrated the advantages of AGD to the whole power converter with measurements [17-19]. Table I shows previous papers that have applied AGD to power converters. In conventional gate driving, which varies the gate resistance, loss and noise during power device switching are in a trade-off relationship. Therefore, to demonstrate the superiority of AGD over the conventional gate driving, it is necessary to either align noise and compare loss or align loss and compare noise. [17] and [18], however, do not show the measured efficiency increase under noise-aligned conditions. [19] shows the measured efficiency increase under noise-aligned conditions, however, it

TABLE I.   COMPARISON TABLE OF POWER CONVERTERS WITH AGDs APPLIED

Reference	[17]	[18]	[19]	This work
Power device	GaN	IGBT	SiC	Si MOSFET
Power converter	Boost converter	Buck converter	3-phase inverter	Isolated DC-DC
Types of Power Converters	Prototype	Prototype	Prototype	Product
V_{IN}	12 V (DC)	300 V (DC)	282 V (DC)	360 V (DC)
V_{OUT}	24 V (DC)	50 V (DC)	200 V (AC)	13.3 V (DC)
Output power	NA	500 to 5 kW	NA	1.2 kW
Switching frequency	NA	10 kHz	2 kHz, 20 kHz	165 kHz
Efficiency increase by AGD	Loss: 9.79 W ➔1.55 W (Simulated)	92.4% ➔93.1% (0.7%)	95.1%➔95.4% (0.3%) @ 2 kHz 83.9%➔85.2% (1.3%) @ 20 kHz	90.3% ➔91.2% (0.9%)
Types of noise	V_{DS} spectrum	NA	Surge voltage	I_D spectrum
Measured efficiency increase under noise-aligned conditions	No	No	Yes	Yes

lacks reality because the power converter is not a commercial product, but a prototype. To solve the problems, in this paper, AGD is applied to a power converter product, and the efficiency increase of the power converter by AGD is demonstrated by measurements under noise-aligned conditions.

II. TIME-DOMAIN STOP-AND-GO ACTIVE GATE DRIVER

Fig. 1 shows a circuit schematic [20] and specifications of a 350 V-to-13.3 V isolated DC-DC converter product for electric vehicles to which AGD is applied. The time-domain stop-and-go active gate driver (TD AGD) [1-16], which changes the gate driving strength three times from "strong to high-Z to strong", is applied only to the turn-on of Si power MOSFET (S_{P1}), which has the largest hard switching loss in the DC-DC converter. Figs. 2 and 3 show the circuit schematic of TD AGD and IN1 generator, which generates t_1 and t_2, the key parameters determining the performance of TD AGD, respectively. IN2 generator is the dummy delay circuit of IN1 generator. Figs. 4 and 5 show the timing chart and PCB photo of TD AGD, respectively. In the timing chart in Fig. 4, the delay of the

979-8-3315-1612-3/25 $31.00 © 2025 IEEE

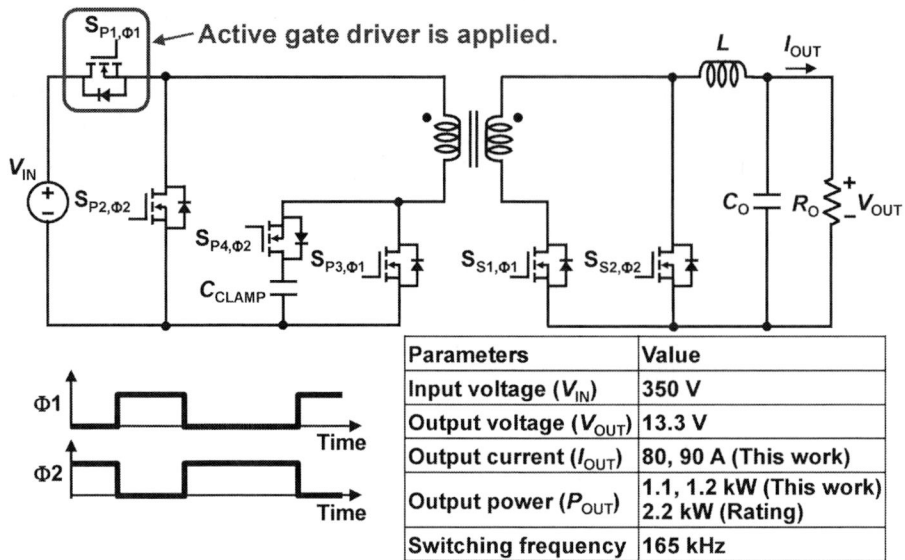

Fig. 1: Isolated DC-DC converter for EVs to which AGD is applied.

Parameters	Value
Input voltage (V_{IN})	350 V
Output voltage (V_{OUT})	13.3 V
Output current (I_{OUT})	80, 90 A (This work)
Output power (P_{OUT})	1.1, 1.2 kW (This work) 2.2 kW (Rating)
Switching frequency	165 kHz

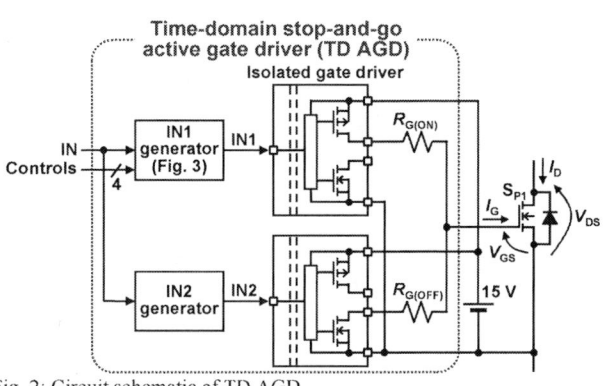

Fig. 2: Circuit schematic of TD AGD.

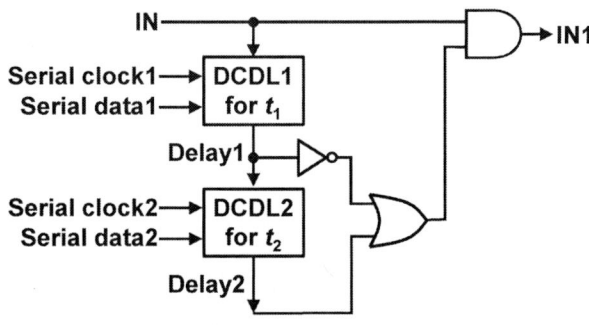

Fig. 3: Circuit schematic of IN1 generator.

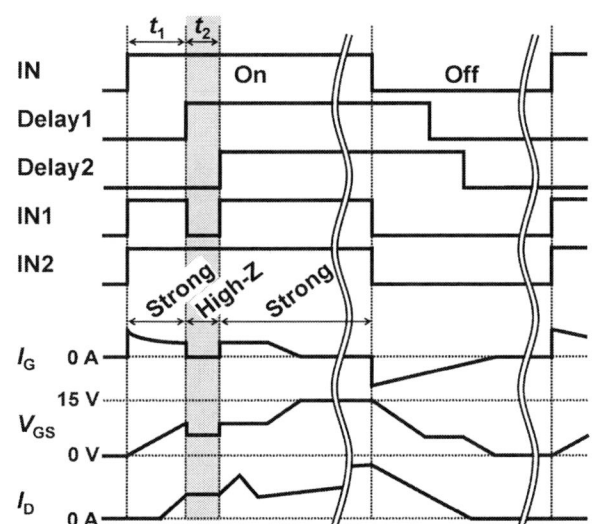

Fig. 4: Timing chart of TD AGD.

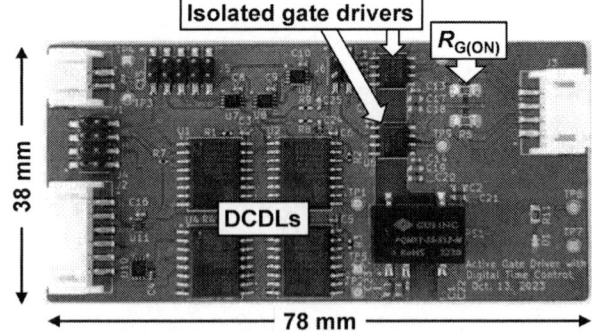

Fig. 5: PCB photo of TD AGD.

isolated gate drivers and the logic gates is assumed to be zero. As shown in Fig. 4, the first strong drive period after turn-on is defined as t_1, and the next high-Z drive period as t_2. TD AGD, implemented with two isolated gate drivers (UCC5320SCDR) and two 8-bit digitally controlled delay lines (DCDLs) (DS1023S-200+), allows t_1 and t_2 to be changed in 2 ns steps

through digital control signals of "Serial data1" and "Serial data2" in Fig. 3.

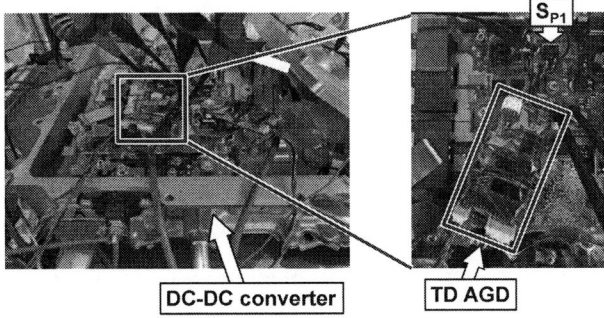

Fig. 6: Measurement setup of DC-DC converter.

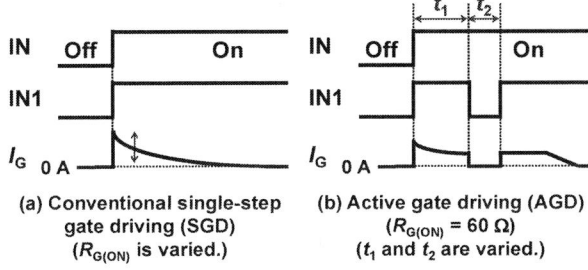

(a) Conventional single-step gate driving (SGD) ($R_{G(ON)}$ is varied.)

(b) Active gate driving (AGD) ($R_{G(ON)} = 60\ \Omega$) (t_1 and t_2 are varied.)

Fig. 7: Timing charts of the conventional SGD and proposed AGD at turn-on for comparison.

III. MEASURED RESULTS

A. Measurement Setup

Fig. 6 shows the measurement setup of the DC-DC converter with the gate driver of S_{P1} modified to TD AGD. The measurements were performed under the conditions shown in Fig. 1. Figs. 8 to 14 show the measured results at output current (I_{OUT}) = 90 A, and Figs. 15 to 17 show the measured results at I_{OUT} = 80 A. In this paper, the performance improvement effect of AGD is investigated in I_{OUT} range where the DC-DC converter is in hard-switching operation instead of soft-switching. Figs. 7 (a) and (b) show timing charts of the conventional single-step gate driving (SGD) and the proposed AGD at turn-on for comparison, respectively. In SGD, the operation is performed by setting $t_1 = t_2 = 0$ ns in TD AGD and the gate resistance $R_{G(ON)}$ shown in Fig. 2 is varied in 6 ways from 30 Ω to 200 Ω. In AGD, $R_{G(ON)} = 60\ \Omega$ is fixed and t_1 and t_2 are varied using DCDLs.

B. Measured Results at $I_{OUT} = 90\ A$

Figs. 8 and 9 show the measured gate-to-source voltage (V_{GS}), drain-to-source voltage (V_{DS}), and drain current (I_D) waveforms and I_D spectrums of S_{P1} in SGD, respectively. Increasing $R_{G(ON)}$ reduces switching speed of S_{P1} (Fig. 8) and I_D spectrum amplitude (Fig. 9). Since peaks of I_D spectrums are observed near 27 MHz in Fig. 9, the maximum value of I_D spectrum amplitude in the range of 26 MHz to 28 MHz is defined as I_D spectrum amplitude at 27 MHz in this paper, and it is used as an index of the switching noise.

Fig. 10 shows the measured DC-DC converter efficiency vs. I_D spectrum amplitude of S_{P1} at 27 MHz in SGD and AGD. The SGD curve is obtained from Figs. 8 and 9. To find the best AGD, t_1 and t_2 of TD AGD are adjusted in 2 ns steps to maximize the

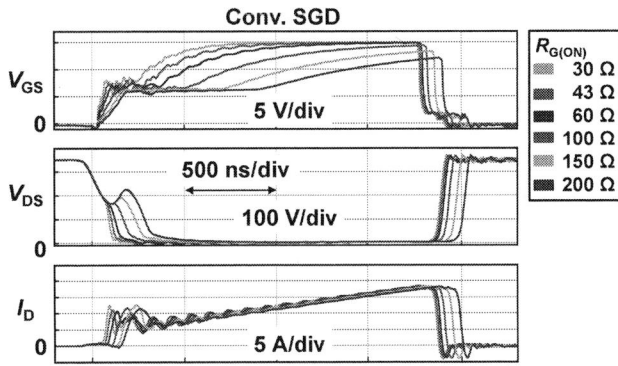

Fig. 8: Measured waveforms of S_{P1} in SGD at I_{OUT} = 90 A.

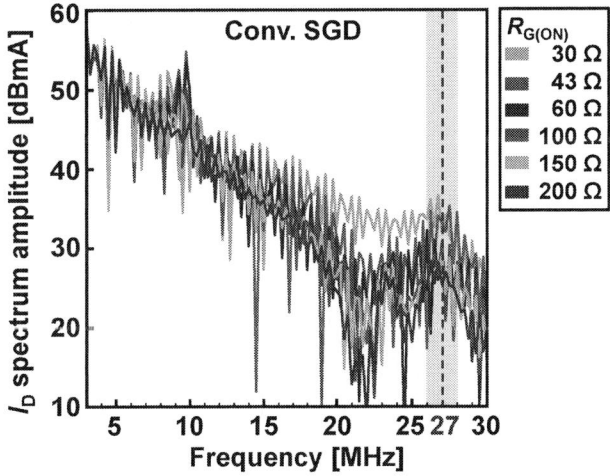

Fig. 9: Measured I_D spectrums of S_{P1} in SGD at I_{OUT} = 90 A.

efficiency increase of AGD relative to SGD under I_D spectrum amplitude alignment condition. As an example of t_1-dependent measured results of AGD, in Fig. 10, t_2 = 50 ns is fixed and t_1 is varied from 90 ns to 98 ns in 2 ns increments. The AGD with t_1 = 94 ns and t_2 = 50 ns is defined as the best AGD.

Fig. 11 shows the measured DC-DC converter efficiency vs. I_D spectrum amplitude of S_{P1} at 27 MHz in SGD and the best AGD. SGD shows a clear trade-off curve between efficiency and I_D spectrum amplitude. Points A and C are the conventional SGD points with efficiency and I_D spectrum amplitude approximately the same as the best AGD (Point B), respectively. Compared with Points A and C, the best AGD (Point B) increases efficiency by 0.9 % under I_D spectrum amplitude-aligned condition and reduces I_D spectrum amplitude by 2.9 dBmA under efficiency-aligned condition.

Figs. 12 and 13 show the measured waveforms and I_D spectrums of Points A to C in Fig. 11 for S_{P1}, respectively. The best AGD (Point B) with t_1 = 94 ns, t_2 = 50 ns reduces I_D spectrum amplitude at 27 MHz by splitting the I_D overshoot into two peaks by inserting t_2 compared to Point A under efficiency-aligned condition. On the other hand, the best AGD (Point B) ($R_{G(ON)} = 60\ \Omega$) achieves higher efficiency by driving the gate

979-8-3315-1612-3/25 $31.00 © 2025 IEEE   1104

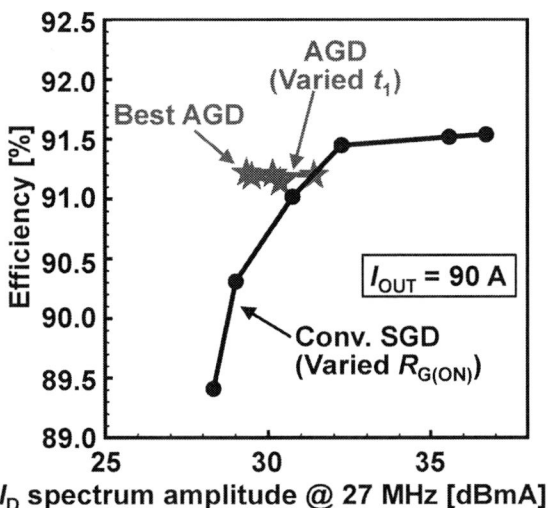

Fig. 10: Measured DC-DC converter efficiency vs. I_D spectrum amplitude of S_{P1} at 27 MHz at $I_{OUT} = 90$ A. In AGD, $t_2 = 50$ ns is fixed and t_1 is varied from 90 ns to 98 ns in 2 ns increments.

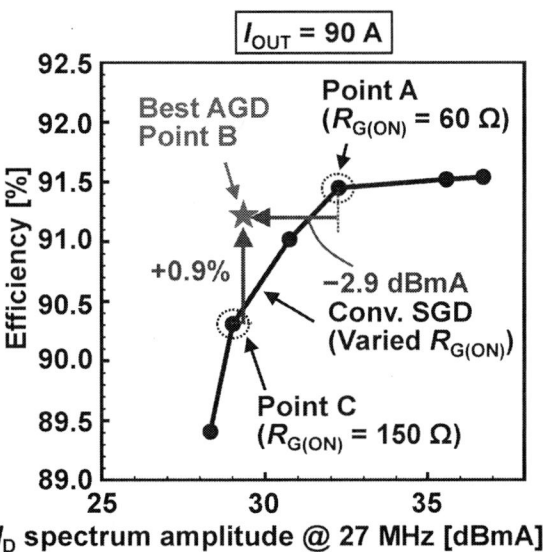

Fig. 11: Measured DC-DC converter efficiency vs. I_D spectrum amplitude of S_{P1} at 27 MHz at $I_{OUT} = 90$ A.

more strongly with lower $R_{G(ON)}$ than Point C ($R_{G(ON)} = 150$ Ω) under I_D spectrum amplitude-aligned condition.

Fig. 14 shows an analysis of the power loss breakdown for Points B and C. Compared to the conventional Point C, the proposed best AGD (Point B) reduces switching losses by 45% from 27.9 W to 15.4 W, resulting in 10% reduction in total DC-DC converter losses from 128.8 W to 115.5 W and a 0.9% increase in converter efficiency from 90.3% to 91.2% under I_D spectrum amplitude-aligned condition.

C. Measured Results at $I_{OUT} = 80$ A

To investigate the I_{OUT} dependence of the parameters (t_1, t_2) of the best AGD and the performance improvement effect of

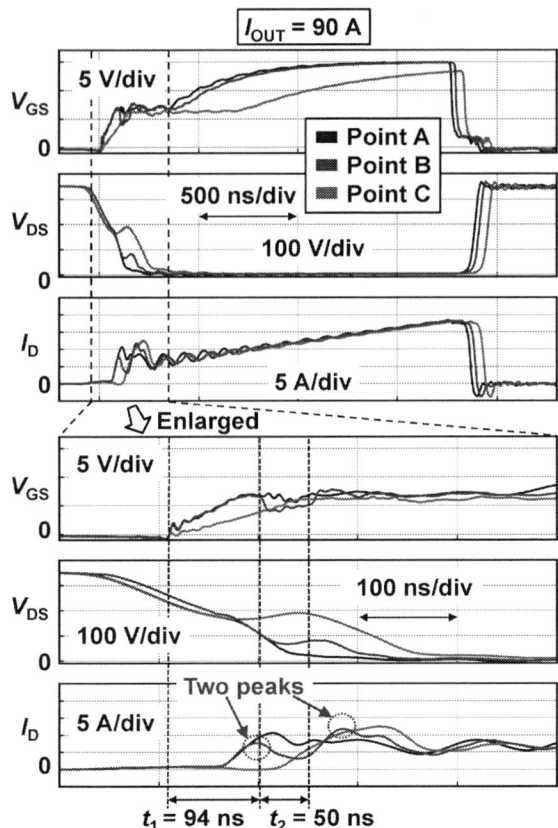

Fig. 12: Measured waveforms of Points A to C in Fig. 11 at $I_{OUT} = 90$ A.

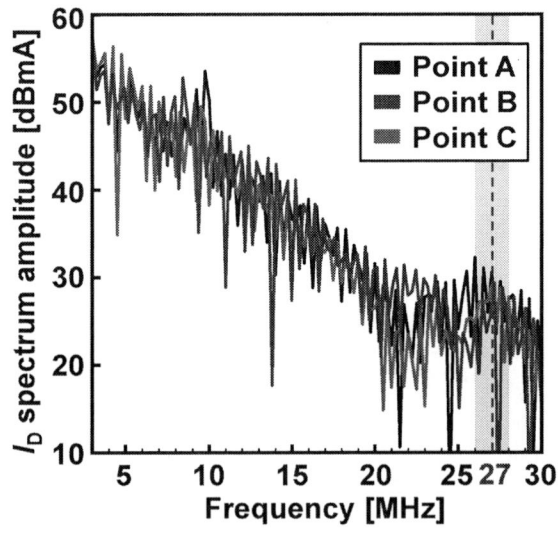

Fig. 13: Measured I_D spectrums of Points A to C in Fig. 11 at $I_{OUT} = 90$ A.

AGD, the measurement results at $I_{OUT} = 80$ A are presented in this section. Fig. 15 shows the measured DC-DC converter efficiency vs. I_D spectrum amplitude of S_{P1} at 27 MHz in SGD and the best AGD at $I_{OUT} = 80$ A. The similar AGD performance improvement effect is obtained for $I_{OUT} = 80$ A as for $I_{OUT} = 90$ A (Fig. 11). Points D and F are the conventional SGD points

979-8-3315-1612-3/25 $31.00 © 2025 IEEE   1105

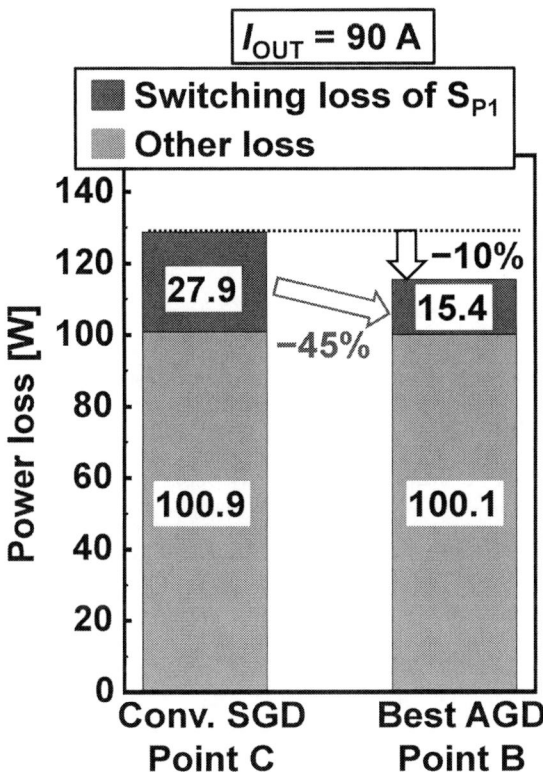

Fig. 14: Power loss breakdown for Points B and C at I_{OUT} = 90 A.

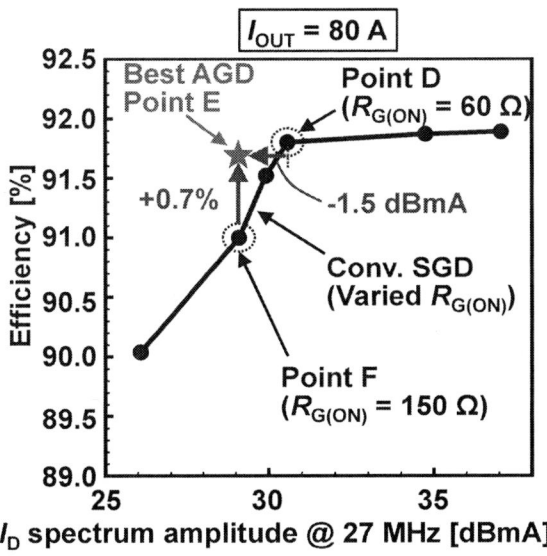

Fig. 15: Measured DC-DC converter efficiency vs. I_D spectrum amplitude of S_{P1} at 27 MHz at I_{OUT} = 80 A.

with efficiency and I_D spectrum amplitude approximately the same as the best AGD (Point E), respectively. Compared with Points D and F, the best AGD (Point E) increases efficiency by 0.7 % under I_D spectrum amplitude-aligned condition and reduces I_D spectrum amplitude by 1.5 dBmA under efficiency-aligned condition.

Fig. 16 shows the measured waveforms of Points D to F in Fig. 15 for S_{P1}. In the best AGD, (t_1, t_2) = (72 ns, 40 ns) at I_{OUT} = 80A in Fig. 16, while (t_1, t_2) = (94 ns, 50 ns) at I_{OUT} = 90A in Fig. 12. Note that (t_1, t_2) of the best AGD is different for I_{OUT} = 80 A and 90 A, which suggests that (t_1, t_2) of AGD needs to be changed depending on I_{OUT}. The theory of how to determine (t_1, t_2) is a future research challenge.

Fig. 17 shows an analysis of the power loss breakdown for Points E and F at I_{OUT} = 80 A. Compared to the conventional Point F, the proposed best AGD (Point E) reduces switching losses by 41% from 24.4 W to 14.4 W, resulting in 8% reduction in total DC-DC converter losses from 105.2 W to 96.4 W and a 0.9% increase in converter efficiency from 91.0% to 91.7% under I_D spectrum amplitude-aligned condition.

IV. CONCLUSIONS

TD AGD is applied to the 165 kHz, 350 V-to-13.3 V isolated DC-DC converter product for electric vehicles, and the efficiency increase of the converter by AGD is demonstrated by measurements at 1.2 kW (= 13.3 V, 90 A) output. Compared to the conventional SGD, the proposed TD AGD with (t_1, t_2) = (94 ns, 50 ns) reduces switching losses by 45% from 27.9 W to 15.4 W, resulting in 10% reduction in total DC-DC converter losses from 128.8 W to 115.5 W and a 0.9% increase in converter efficiency from 90.3% to 91.2% under I_D spectrum amplitude-aligned condition.

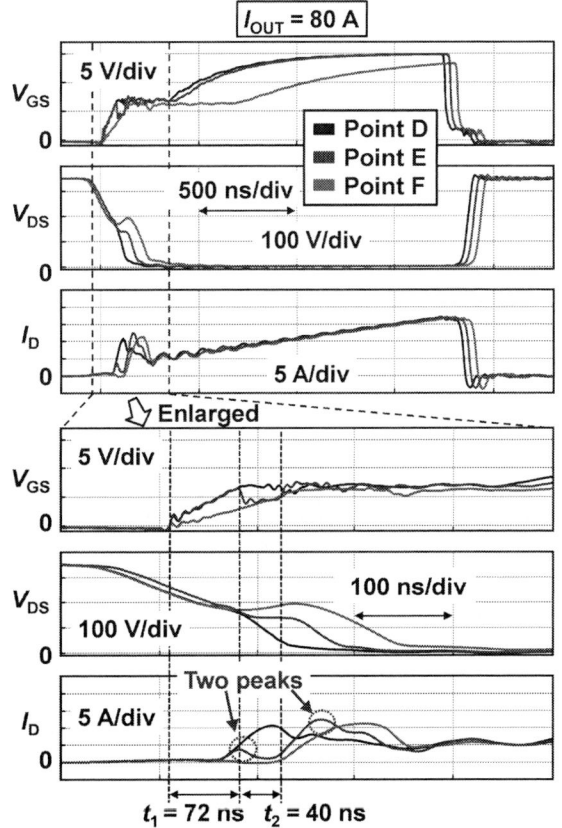

Fig. 16: Measured waveforms of Points D to F in Fig. 15 at I_{OUT} = 80 A.

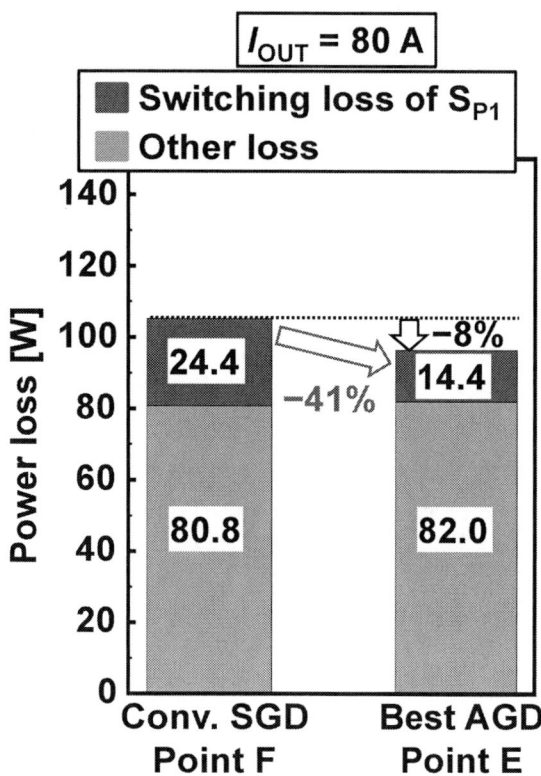

Fig. 17: Power loss breakdown for Points E and F at $I_{OUT} = 80$ A.

REFERENCES

[1] V. John, B.-S. Suh, and T. A. Lipo, "High-performance active gate drive for high-power IGBT's," *IEEE Trans. Ind. Appl.*, vol. 35, no. 5, pp. 1108–1117, Sep-Oct 1999.

[2] S. Takizawa, S. Igarashi, and K. Kuroki, "A new di/dt control gate drive circuit for IGBTs to reduce EMI noise and switching losses," in *PESC 98 Record. 29th Annual IEEE Power Electronics Specialists Conference*, 1998, vol. 2, pp. 1443–1449.

[3] F. Zhang, X. Yang, Y. Ren, L. Feng, W. Chen, and Y. Pei, "Advanced active gate drive for switching performance improvement and overvoltage protection of high-power IGBTs," *IEEE Trans. Power Electron.*, vol. 33, no. 5, pp. 3802–3815, 2018.

[4] Y. Ling, Z. Zhao, and Y. Zhu, "A self-regulating gate driver for high-power IGBTs," *IEEE Trans. Power Electron.*, vol. 36, no. 3, pp. 3450–3461, 2021.

[5] A. P. Camacho, V. Sala, H. Ghorbani, and J. L. R. Martinez, "A novel active gate driver for improving SiC MOSFET switching trajectory," *IEEE Trans. Ind. Electron.*, vol. 64, no. 11, pp. 9032–9042, 2017.

[6] J. Zhu *et al.*, "A 600V GaN active gate driver with dynamic feedback delay compensation technique achieving 22.5% turn-on energy saving," in 2021 IEEE International Solid- State Circuits Conference (ISSCC), 2021, vol. 64, pp. 462–464.

[7] S. Yu et al., "A 400-V half bridge gate driver for normally-off GaN HEMTs with effective dv/dt control and high dv/dt immunity," IEEE Trans. Ind. Electron., vol. 70, no. 1, pp. 741–751, 2023.

[8] Y. Sun, L. Sun, A. Esmaeli, and K. Zhao, "A Novel Three Stage Drive Circuit for IGBT," in 2006 1ST IEEE Conference on Industrial Electronics and Applications, 2006, pp. 1–6.

[9] M. Riefer, J. Winkler, S. Strache, and I. Kallfass, "Implementation of current-source gate driver with open-loop slope shaping for SiC-MOSFETs," in PCIM Europe digital days 2021; International Exhibition and Conference for Power Electronics, Intelligent Motion, Renewable Energy and Energy Management, 2021, pp. 1–8.

[10] S. Kawai, T. Ueno, and K. Onizuka, "A 4.5V/ns active slew-rate-controlling gate driver with robust discrete-time feedback technique for 600V superjunction MOSFETs," in 2019 IEEE International Solid- State Circuits Conference - (ISSCC), 2019, pp. 252–254.

[11] Y. Yang, Y. Wen, and Y. Gao, "A novel active gate driver for improving switching performance of high-power SiC MOSFET modules," IEEE Trans. Power Electron., vol. 34, no. 8, pp. 7775–7787, 2019.

[12] Y. Wen, Y. Yang, and Y. Gao, "Active gate driver for improving current sharing performance of paralleled high-power SiC MOSFET modules," IEEE Trans. Power Electron., vol. 36, no. 2, pp. 1491–1505, 2021.

[13] M. Sayed, S. Araujo, F. Carraro, and R. Kennel, "Investigation of gate current shaping for SiC-based power modules on electrical drive system power losses," in 2021 23rd European Conference on Power Electronics and Applications (EPE'21 ECCE Europe), 2021, pp. 1–10.

[14] K. Horii, R. Morikawa, K. Hata, K. Morokuma, Y. Wada, Y. Obiraki, Y. Mukunoki, and M. Takamiya, "Sub-0.5 ns step, 10-bit time domain digital gate driver IC for reducing radiated EMI and switching loss of SiC MOSFETs," in 2022 IEEE Energy Conversion Congress and Exposition (ECCE), 2022, pp. 1–8.

[15] D. Zhang, K. Horii, K. Hata, and M. Takamiya, "Digital gate driver IC with fully integrated automatic timing control function in stop-and-go gate drive for IGBTs," in 2023 IEEE Applied Power Electronics Conference and Exposition (APEC), 2023, pp. 1225–1231.

[16] D. Zhang, K. Horii, K. Hata, and M. Takamiya, "Digital gate driver IC with real-time gate current change by sensing drain current to cope with operating condition variations of SiC MOSFET," in 2023 11th International Conference on Power Electronics and ECCE Asia (ICPE 2023 - ECCE Asia), 2023, pp. 374–380.

[17] C. Krause and S. Frei, "Narrowband frequency domain optimized gate driving signals for power transistors of DC/DC converters," in 2023 International Symposium on Electromagnetic Compatibility – EMC Europe, 2023, pp. 1–6.

[18] G. Zhang, B. J. Zhang, S. Shao, and W. Qu, "A load adaptive intelligent IGBT gate drive," in 2023 11th International Conference on Power Electronics and ECCE Asia (ICPE 2023 - ECCE Asia), 2023, pp. 3294–3299.

[19] V.-L. Pham, H. Obara, and K. Hata, "A partial active gate control for improvement of a trade-off relation between surge voltage and efficiency in a three-phase inverter," IEEE Trans. Ind. Appl., vol. 60, no. 3, pp. 4239–4250, May-June 2024.

[20] K. Watanabe, "DC-DC converter," Japanese patent, P2020-202679A, Dec. 17, 2020.

A Multi-Level Active Gate Driver for Achieving Thermal Balance in Parallel Connected Power MOSFETs

Jingyuan Liang
The Edward S. Rogers Sr.
Department of Electrical &
Computer Engineering
University of Toronto
Toronto, Canada
jingyuan.liang@mail.utoronto.ca

Lingwei Sun
The Edward S. Rogers Sr.
Department of Electrical &
Computer Engineering
University of Toronto
Toronto, Canada
lingwei.sun@mail.utoronto.ca

Wen Tao Cui
The Edward S. Rogers Sr.
Department of Electrical &
Computer Engineering
University of Toronto
Toronto, Canada
wt.cui@mail.utoronto.ca

Wai Tung Ng
The Edward S. Rogers Sr.
Department of Electrical &
Computer Engineering
University of Toronto
Toronto, Canada
ngwt@ece.utoronto.ca

Motomitsu Iwamoto
Fuji Electric Co., Ltd.
Matasumoto, Japan
Iwamoto-motomitsu@fujielectric.com

Haruhiko Nishio
Fuji Electric Co., Ltd.
Matasumoto, Japan
nishio-haruhiko@fujielectric.com

Abstract—**In high power applications, the temperature differences between parallel connected power MOSFETs can be significant if the cooling condition is non-uniform. Although power MOSFETs have an inherent negative temperature coefficient to balance their drain-to-source currents in parallel connection, their temperatures may still be different due to non-uniform cooling air or liquid flow, compromising the system reliability. With the adoption of multi-level active gate driver, applied gate voltage of the MOSFETs can be dynamically changed to adjust their on resistances. As a result, thermal balance can be achieved with a minimal efficiency reduction. For parallel connected MOSFETs, the cooler MOSFET can be switched with a nominal gate drive voltage of 15 V. For the warmer MOSFET, a lower gate drive voltage, such as 11 V, can be applied. Under this condition, the on-resistance of the warmer MOSFET will be increased, leading to decreased current flow. The conduction loss of the warmer MOSFET is then reduced, lowering its temperature. In this multi-level active gate driver, the power MOSFET's gate-to-source capacitance and the flying capacitor form a voltage divider. The gate drive voltage can be switched between the nominal 15 V to a designed lower voltage, e.g., 11 V. By activating and deactivating this multi-level switching function, a thermal balance can be achieved between the parallel connected MOSFETs without any noticeable degradation in power conversion efficiency.**

Keywords—multi-level, active gate driver, dynamic gate driver, thermal balance, parallel connected MOSFETs, flying capacitor

I. INTRODUCTION

Thermal imbalance between parallel power devices in high power applications is a common issue [1]. In such applications, thermal imbalance often arises due to non-uniform cooling conditions. Several methods have been applied to mitigate this issue. Lucid Motors uses a heat exchanger with parallel cooling channels as shown in Fig. 1 (a) to deliver symmetrical coolant flow under the power devices [2]. In this configuration, each cooling channel has equal cooling capacity, therefore the amount of heat will be extracted for both power devices and the temperature difference is minimized. However, this type of heat exchanger may require a higher overall flow rate to achieve the same total cooling capacity compared to a traditional heat exchanger with serial cooling channel shown in as Fig. 1 (b).

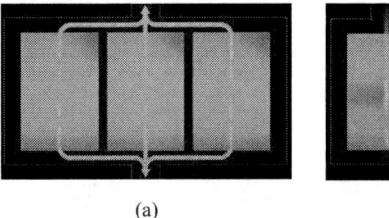 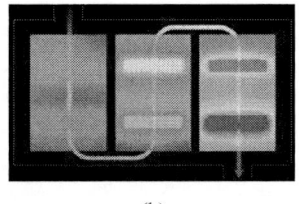

| (a) | (b) |

Fig. 1. Examples of heat exchanger coolant flow in a (a) parallel and (b) serial arrangement [2].

Another method [3], shown in Fig. 2 (a) and (b), is to cut out the substrate under the cooler device and force it to receive less cooling capacity. However, the physical modifications on the circuit substrate requires extensive electrical thermal simulation and this approach is more tailored towards temperature difference caused by the thermal coupling effect.

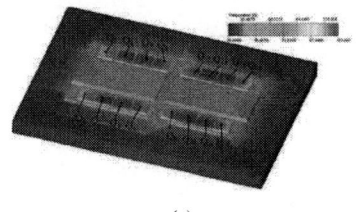

(a)

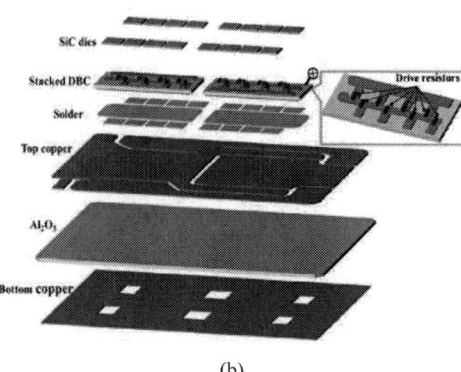

(b)

Fig. 2. Substrate cutout method: (a) simulation showing cooler dies, (b) substrate with cutouts [3].

The same goal can also be achieved by implementing a thermal model with a dedicated algorithm, shown in Fig. 3 (a), to achieve thermal balance between different phases [4]. Despite the effectiveness as shown in Fig. 3 (b) and (c), the thermal balance control loop is embedded in the converter's output voltage regulation loop. Therefore, the control structure cannot be universally applied to other circuit topologies.

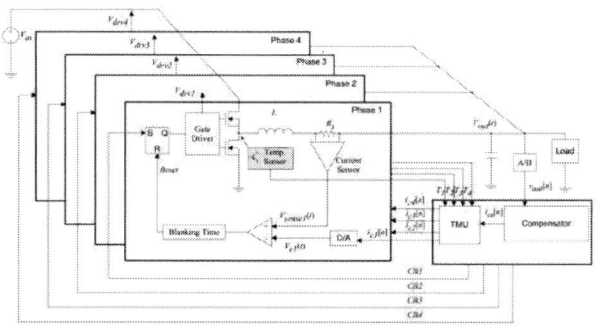

(a)

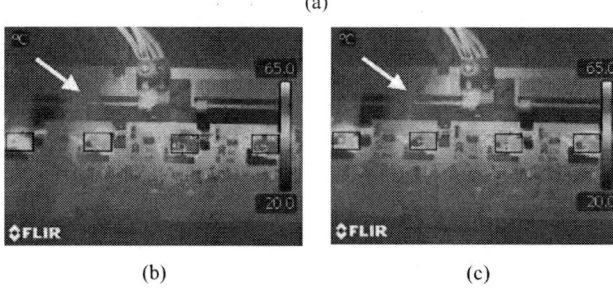

(b)   (c)

Fig. 3. A control method for thermal balance in a multi-phase converter: (a) system block diagram, (b) balance feature OFF, (c) balance feature ON [4].

In this paper, a multi-level active gate driver (AGD) with a flying capacitor (C_{FLY}) is proposed as a solution for thermal imbalance issues. For a device that is overheating relative to other paralleled devices, this gate driver will generate a lower V_{GS} by using C_{FLY} to form a voltage divider with the power MOSFET's gate-to-source capacitance C_{GS}. With a lower V_{GS} for the warmer device on a temporary basis, its drain-to-source on resistance ($R_{DS(on)}$) will be increased, decreasing the current flow. As a result, its conduction loss can be reduced, lowering its temperature. By toggling between the normal and higher $R_{DS(on)}$ state, the temperature difference between parallel connected MOSFETs is minimized.

Similar multi-level AGDs have been implemented to address other issues, such as dynamic gate switching for power devices [5], [6], [7]. However, these AGDs require multiple gate driver power supplies to realize different V_{GS}, making integration difficult. The proposed multi-level AGD with a flying capacitor can generate a lower gate drive voltage from only one power supply. The optimal value of the flying capacitor can be determined based on different system setups, addressing thermal imbalanced issues caused by non-uniform cooling capacity between MOSFETs connected in parallel.

II. THERMAL IMBALANCE

A. Root Cause

Circuits using more than one MOSFETs either in parallel or multi-phase configuration often suffer from thermal imbalance. For example, system with an enclosure requires a cooling fan to deliver cooler air from an inlet. The air flow absorbs dissipated heat from power devices on the way to the outlet. Due to the heat accumulation, the power device closer to the inlet is typically cooler than the power device closer to the outlet. With higher power dissipation and longer cooling path, this non-uniform thermal behavior can often lead to a large temperature difference between parallel or multi-phase connected MOSFETs. The warmer device not only limits the system's maximum output power, but compromises the system's long term reliability.

B. Effects on Reliability

In power electronics, the lifetime of the power device is proportional to its junction temperature swing ΔT_j. The Coffin-Manson model is a common method to estimate a device's lifetime. An example study using this model is discussed in [8].

$$N_f \propto \Delta T_j^{-m} \tag{1}$$

- N_f: the number of thermal cycle to failure

- ΔT_j: junction temperature swing

- m: Coffin-Manson exponent, typical value is around 2

By applying the Coffin-Manson model and assuming an initial ΔT_j of 100 °C, an estimation on lifetime coefficient can be calculated as follows:

$$N_{f\_initial} = 1/100^2 = 1 \times 10^{-4} \tag{2}$$

If thermal balance is achieved and temperature of the warmer device is reduced by 10 °C, the improved lifetime coefficient is now changed to:

$$N_{f\_improved} = 1/90^2 = 1.23 \times 10^{-4} \tag{3}$$

Comparing improved and initial estimated device lifetime, there is a 23 % of increase. This improvement is especially important for power modules with multiple devices connected in parallel, because even if only the warmest device fails, the entire power module can fail subsequently.

979-8-3315-1612-3/25 $31.00 © 2025 IEEE   1109

III. MULTI-LEVEL ACTIVE GATE DRIVER DESIGN

The multi-level topology with a flying capacitor circuit is well-known in power converters and inverters. Instead of driving an inductive load, this circuit is designed to drive a capacitive load for a gate driver design. In this case, the capacitive load is the C_{GS} of a power MOSFET.

In this multi-level active gate driver with a flying capacitor, there are four series-connected MOSFETs (M_1 to M_4) in the gate driver's output stage as shown in Fig. 4. Each of the MOSFETs is driven by a pre-driver (Infineon 1EDN7550U) featuring true differential inputs (TDI), eliminating the need of level shifting circuits for M_2, M_3 and M_4. Conventional asynchronous boot-strap circuits are used to generate the supply voltages for the three floating pre-drivers VDD_1, VDD_2 and VDD_3. The flying capacitor is connected between the switching nodes SW_1 and SW_3. During a thermal imbalance event, the gate voltage can be changed to a lower voltage using the configuration shown in Fig. 5. If the multi-level mode is deactivated, V_{GS} of the power MOSFET is switched normally between 0 and 15 V. When charging the gate, switches M_3 and M_4 are turned on as shown in Fig. 6 (a). When discharging the gate, switches M_1 and M_2 are turned on as shown in Fig. 6 (b). If the multi-level mode is activated, switches M_2 and M_4 are turned on to charge the serial connected C_{FLY} and C_{GS} as shown in Fig. 6 (c). Before the next switching cycle, both C_{FLY} and C_{GS} need to be discharged by turning on M_1, M_2, M_3 and turning off M_4 as shown in Fig. 6 (d).

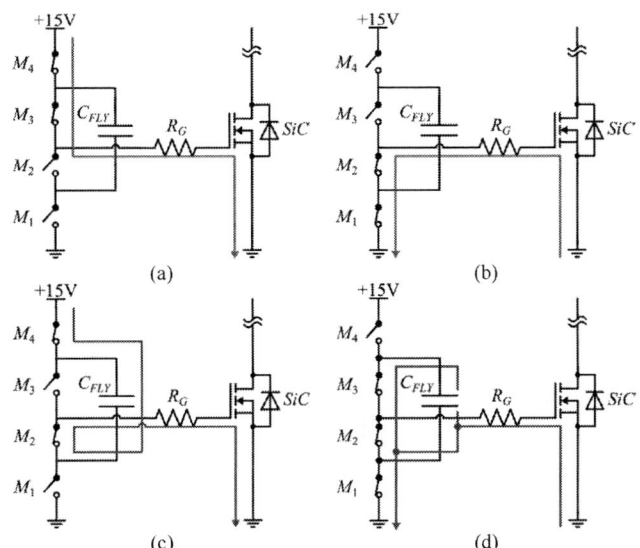

Fig. 6. Gate drive sequences, (a) normal turn-on, (b) normal turn-off, (c) multi-level turn-on, (d) multi-level turn-off.

During the multi-level turn-on, C_{GS} and any parasitic capacitances C_{SS} between the gate and source nodes are connected in series with C_{FLY}, as shown in Fig. 7 (a). Generally, C_{FLY} is chosen between a few nano Farads to a few hundred nano Farads. A smaller C_{FLY} value will increase the $R_{DS(on)}$ faster to quickly reach thermal balance, but it will cause more efficiency loss. The location of C_{FLY} is indicated in Fig. 7 (b). In systems requiring thermal balancing features, the value of C_{FLY} can be selected based on how large a temperature difference between the MOSFETs needs to be reduced.

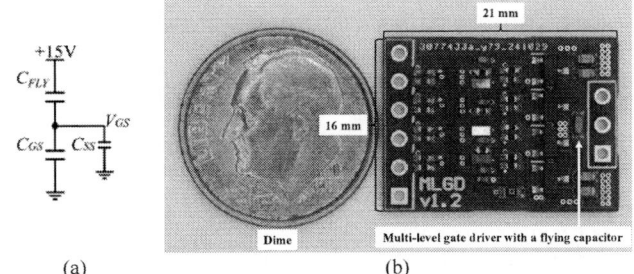

Fig. 7. (a) An equivalent capacitive voltage divider circuit, (b) multi-level gate driver with a flying capacitor.

IV. EXPERIMENT SETUP AND RESULTS

The thermal balancing feature is verified using a 300 W asynchronous boost converter with two parallel connected low-side SiC MOSFETs. The input voltage is set to 24 V and the output voltage is about 60 V. The circuit schematic is shown in Fig. 8 (a), and the board layout with forced air-cooling (48 CFM) direction is shown in Fig. 8 (b). The top parallel connected low-side MOSFET M_{LS1} is closer to the cooling fan. Therefore, it is cooler than M_{LS2}, which is further from the cooling fan.

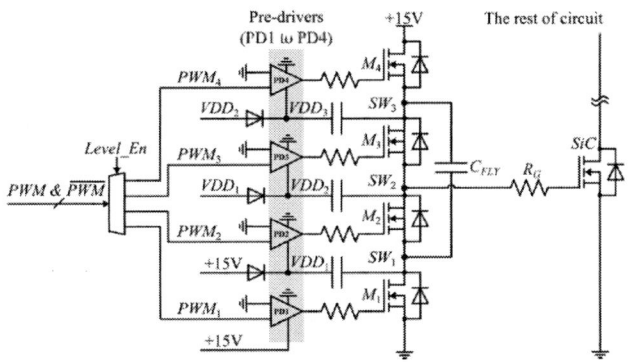

Fig. 4. Schematic of the multi-level active gate driver with a flying capacitor.

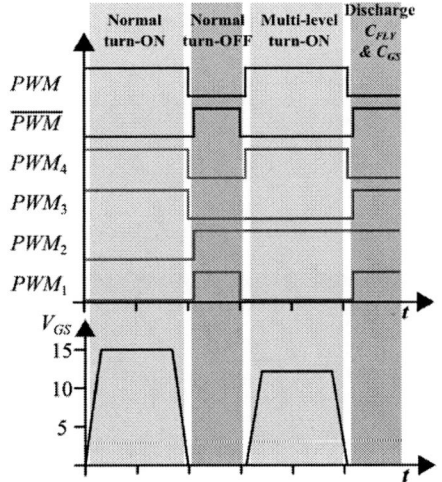

Fig. 5. The gate drive sequence used to generate two different gate voltages.

979-8-3315-1612-3/25 $31.00 © 2025 IEEE

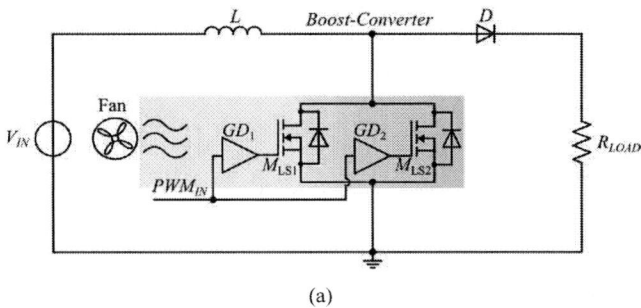

(a)

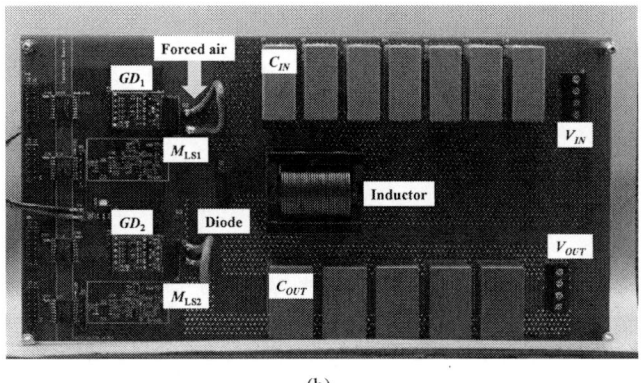

(b)

Fig. 8. An asynchronous boost converter (a) schematics showing two parallel connected low-side MOSFETs, (b) the boost converter board with forced air cooling direction indicated.

The experiment setup is shown in Fig. 9. PWM signals are provided to the gate driver from an FPGA board. During the test, the temperatures of the two MOSFETs are monitored using T-type thermal couples. A manual activation and deactivation of multi-level switching is used to achieve thermal balance. An infrared (IR) camera is used to observe the temperature distribution before and after the thermal balance is achieved.

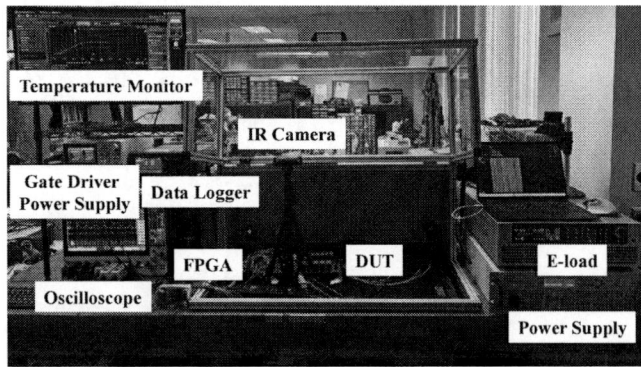

Fig. 9. Experiment setup.

A. Gate Voltage Reduction

During normal gate turn-on, both parallel connected MOSFETs are turned on using a nominal voltage of 15 V as shown in Fig. 10 (a) and (c) with C_{FLY} of 100 nF and 22 nF. During multi-level gate turn-on, the gate voltage is at 13.2 V and 11 V for C_{FLY} equals to 100 nF and 22 nF, respectively. Referring to capacitive voltage divider, a larger C_{FLY} would lead to a higher V_{GS}.

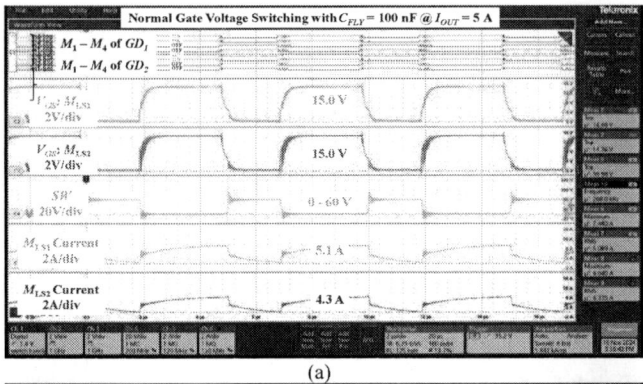

(a)

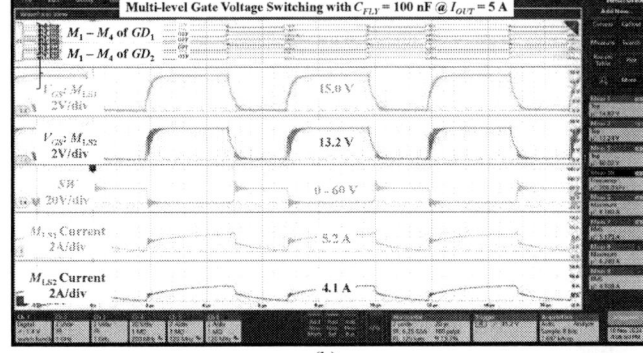

(b)

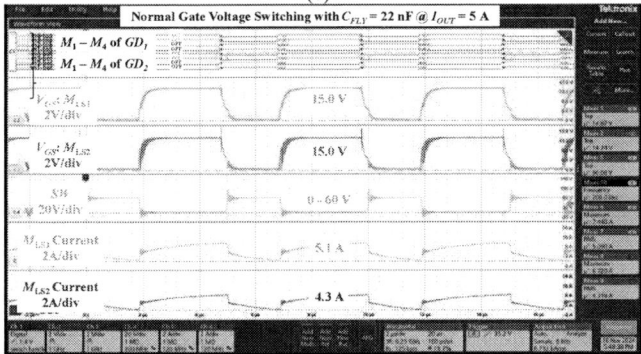

(c)

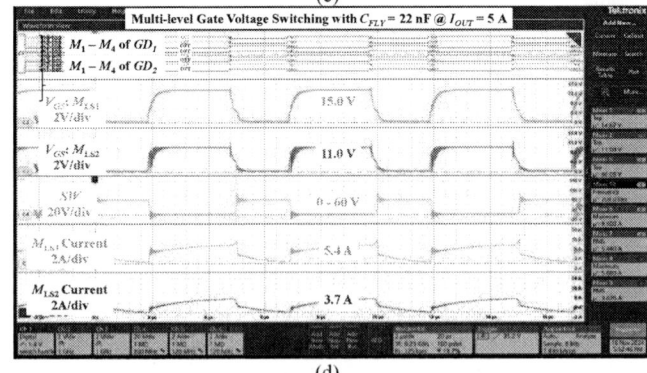

(d)

Fig. 10. Voltage waveforms for (a) normal gate turn-on with C_{FLY} = 100 nF, (b) multi-level gate turn-on with C_{FLY} = 100 nF, (c) normal gate turn-on with C_{FLY} = 22 nF, (d) multi-level gate turn-on with C_{FLY} = 22 nF.

B. Drain-to-Source Current

When running the boost converter, the drain-to-source current (I_{DS}) of each MOSFETs is measured. The I_{DS} changing between normal and multi-level turn-on is plotted in Fig. 11 below. Because the total current is determined by the converter's load, the current flowing through each parallel connected MOSFET is determined by its $R_{DS(on)}$. If the warmer MOSFET M_{LS2} is driven by a multi-level V_{GS}, its $R_{DS(on)}$ will increase and push more current to the cooler MOSFET M_{LS1}, as indicated by the two short arrows.

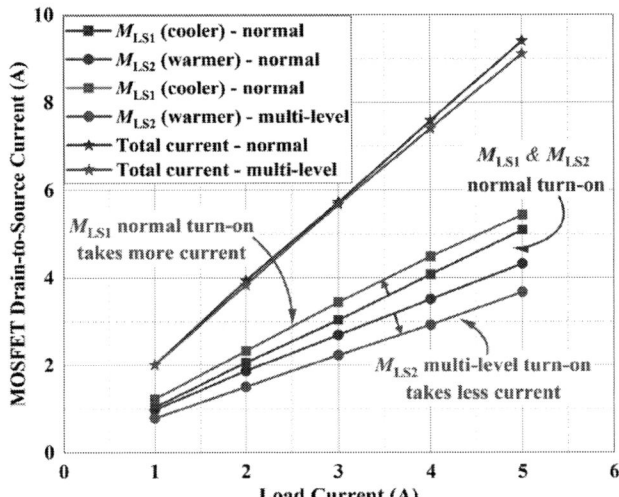

Fig. 11. Paralleled MOSFETs I_{DS} current relationship.

C. MOSFETs Temperature Distribution Profile

Based on the system setup above, two sets of transient tests are performed with C_{FLY} = 100 nF and 22 nF. The case temperature of the two parallel connected MOSFETs are monitored and shown in Fig. 12. The converter is tested up to 293 W. At each output power level, the multi-level AGD function is activated during the time window highlighted in green. With C_{FLY} = 100 nF as shown in Fig. 12 (a), the temperature difference is reduced but not eliminated. This behavior indicates that the lower gate voltage of 13.2 V is not sufficient to increase the $R_{DS(on)}$ to a level that can bring the two MOSFETs to thermal balance. Therefore, a lower C_{FLY} value of 22 nF is tried, and the results are shown in Fig. 12 (b). In this case, the gate voltage with the multi-level AGD function activated is further reduced to 11 V. Thermal balance is achieved with almost no temperature difference between the two MOSFETs. At the highest output power of 293 W, the temperature difference is 19.3 °C without activating the multi-level AGD. During the multi-level AGD activation window, the temperature difference is eventually reduced to 0.4 °C. Fig. 13 (a) and (b) shows the temperature distribution with and without the multi-level AGD function activated. Regardless of different output power levels, thermal balance can all be achieved with the multi-level AGD function. A statistics contains means (μ) and standard deviations (σ) on temperature difference is summarized in Table I.

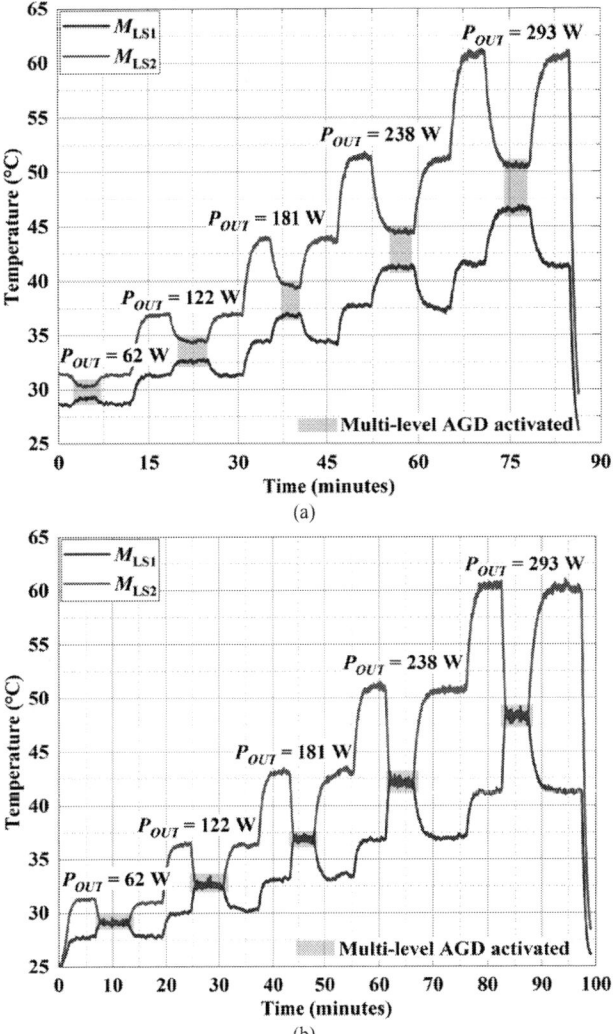

Fig. 12. Temperature transient tests with C_{FLY} of (a) 100 nF and (b) 22 nF.

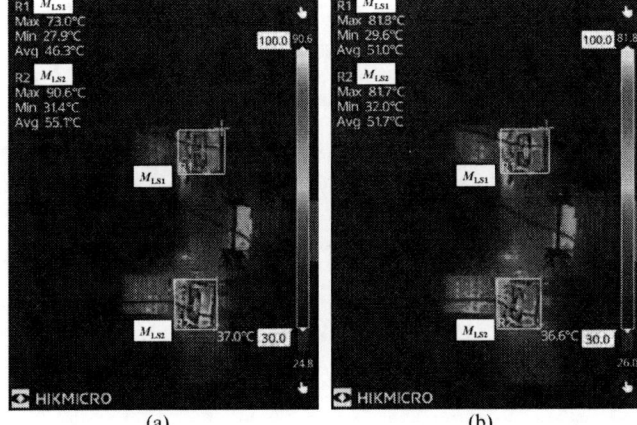

Fig. 13. Temperature distributions of two parallel connected SiC MOSFETs at output power of 293 W, (a) without and (b) with multi-level AGD activation.

TABLE I   TEMPERATURE DATA FROM THE MULTI-LEVEL AGD TESTS

P_{OUT} (W)	Without Activation		With Activation	
	μ (°C)	σ (°C)	μ (°C)	σ (°C)
62	3.5	0.1	0.2	0.2
122	6.3	0.1	0.3	0.2
181	10.0	0.1	0.3	0.2
238	14.2	0.2	0.3	0.2
393	19.3	0.2	0.4	0.3

D. Effects on System's Efficiency

During the multi-level AGD function activation, there are two factors that contribute to the overall system efficiency loss. First is the increase in $R_{DS(on)}$ for the warmer MOSFET M_{LS2} due to the decrease on turn-on gate voltage. Second is the increase in gate driver power consumption because the gate driver needs to charge and discharge the C_{FLY}. During the normal gate turn-on and turn-off process C_{FLY} is never charged or discharged. Despite these extra power consumptions, their combined negative effects on system's overall efficiency are only about 0.1 % across the entire output range, as shown in Fig. 14.

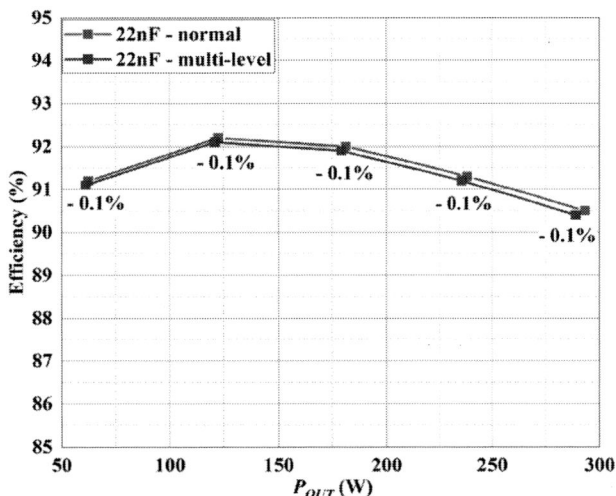

Fig. 14. Efficiency vs. P_{OUT} between normal and multi-level gate drive voltages.

V. CONCLUSIONS

This work presents a multi-level active gate driver with a flying capacitor that can generate a lower voltage on the gate of a power MOSFET. The proposed method achieves thermal balance between parallel connected power MOSFETs. The flexibility of choosing different flying capacitor allows this driver to be used for different operating and cooling conditions. In the boost converter setting at an output power of 293 W, the temperature difference between two parallel connected MOSFETs is reduced from 19.3 °C to 0.4 °C. With this feature activated, the proposed method only incurs a small system efficiency loss of 0.1%.

ACKNOWLEDGMENT

This project is made possible by the generous support from Fuji Electric Co., Ltd. and the Natural Sciences and Engineering Research Council of Canada.

REFERENCES

[1] X. Li *et al.*, "Thermal Imbalance Among Paralleling Chips in Power Modules and the Impact From Traction Inverter System View," *IEEE Transactions on Industrial Electronics*, vol. 70, no. 3, pp. 2231–2240, Mar. 2023, doi: 10.1109/TIE.2022.3159920.J. Clerk Maxwell, A Treatise on Electricity and Magnetism, 3rd ed., vol. 2. Oxford: Clarendon, 1892, pp.68–73.

[2] Eric Bach, "Tech Talks - Episode 5 Inverter," Lucid Motors. Accessed: Nov. 12, 2024. [Online]. Available: https://lucidmotors.com/en-ca/tech-talks.

[3] C. Zhao, L. Wang, Y. Xu, F. Yang, J. Wang, and Z. Qi, "A Method to Minimize Junction Temperature Difference of Dies in Multichip Power Modules," in *IEEE Energy Conversion Congress and Exposition (ECCE), Baltimore, MD, USA*, IEEE, 2019, pp. 3318–3324.Y. Yorozu, M. Hirano, K. Oka, and Y. Tagawa, "Electron spectroscopy studies on magneto-optical media and plastic substrate interface," IEEE Transl. J. Magn. Japan, vol. 2, pp. 740–741, August 1987 [Digests 9th Annual Conf. Magnetics Japan, p. 301, 1982].

[4] P. (Ke) Cao, W. T. Ng, and O. Trescases, "Thermal Management for Multi-Phase Current Mode Buck Converters," in *Applied Power Electronics Conference and Exposition (APEC)*, 2011.

[5] H. C. P. Dymond, D. Liu, J. Wang, J. J. O. Dalton, and B. H. Start, "Multi-level active gate driver for SiC MOSFETs," in *Energy Conversion Congress & Expo*, 2017.

[6] S. Zhao *et al.*, "Adaptive Multi-Level Active Gate Drivers for SiC Power Devices," *IEEE Trans Power Electron*, vol. 35, no. 2, pp. 1882–1898, Feb. 2020, doi: 10.1109/TPEL.2019.2922112.

[7] Y. Wen, Y. Yang, and Y. Gao, "Active Gate Driver for Improving Current Sharing Performance of Paralleled High-Power SiC MOSFET Modules," *IEEE Trans Power Electron*, vol. 36, no. 2, pp. 1491–1505, Feb. 2021, doi: 10.1109/TPEL.2020.3006071.

[8] H. Cui and R. Micro Devices, "Accelerated Temperature Cycle Test and Coffin-Manson Model for Electronic Packaging,"

A Fast Short-Circuit Protection Method for Ohmic gate P-GaN HEMT Based on Gate Charge

Yue Wu
Xidian University Key Laboratory of
Wide Bandgap Semiconductor
Materials Ministry of Education
Xidian University
Guangzhou, China
wyue6570@163.com

Xi Jiang
Xidian University Key Laboratory of
Wide Bandgap Semiconductor
Materials Ministry of Education
Xidian University
Guangzhou, China
xjiang@xidian.edu.cn

Song Yuan
Xidian University Key Laboratory of
Wide Bandgap Semiconductor
Materials Ministry of Education
Xidian University
Guangzhou, China

Xiaowu Gong
Xidian University Key Laboratory of
Wide Bandgap Semiconductor
Materials Ministry of Education
Xidian University
Guangzhou, China

Zhaoheng Yan
Xidian University Key Laboratory of
Wide Bandgap Semiconductor
Materials Ministry of Education
Xidian University
Guangzhou, China

Jiahong Chen
Xidian University Key Laboratory of
Wide Bandgap Semiconductor
Materials Ministry of Education
Xidian University
Guangzhou, China
24181214460@stu.xidian.edu.cn

Yun Xu
Xidian University Key Laboratory of
Wide Bandgap Semiconductor
Materials Ministry of Education
Xidian University
Guangzhou, China
24181214299@stu.xidian.edu.cn

Jinjie Liu
Xidian University Key Laboratory of
Wide Bandgap Semiconductor
Materials Ministry of Education
Xidian University
Guangzhou, China
24181214458@stu.xidian.edu.cn

Abstract—GaN HEMT devices exhibit lower short-circuit capability, posing significant risks in practical applications. This paper presents a fast short circuit (SC) protection circuit for ohmic gate P-GaN HEMT under hard-switch fault (HSF) condition and fault under load (FUL) condition, based on the gate charge characteristics. The proposed SC protection technique utilizes gate current integration to measure gate charge, enabling the detection of SC faults in the device. Experimental results demonstrate that the proposed method allows the SC protection circuit to detect HSF or FUL SC fault events within 50 ns and to shuts down the GaN HEMT device within 150 ns, offering rapid and reliable protection against short-circuit faults.

Keywords—*P-GaN HEMT, short-circuit protection circuit, hard-switch fault, fault under load*

I. INTRODUCTION

The advent of Gallium Nitride (GaN) device has revolutionized the field of power electronics by offering superior electrical and thermal performance compared to traditional silicon (Si)devices. As wide-bandgap (WBG) materials, GaN devices enable higher efficiency, greater power density, and operation at elevated temperatures, which are critical for applications in renewable energy systems, electric vehicles, and aerospace technologies. Despite their numerous advantages, reliability concerns remain a critical challenge for GaN high-electron-mobility transistors (HEMTs). Unlike conventional silicon devices, GaN HEMT devices generally exhibit a lower short-circuit (SC) withstand time, presenting high risks in practical applications. For instance, at a DC bus voltage of 350V, the GaN HEMT short circuit withstand time is as low as 400ns [1]. Moreover, the repetitive SC capability of p-GaN gate HEMTs is substantially inferior to that of silicon carbide (SiC) power MOSFETs [2].

The GaN device can endure approximately 20 cycles at 20 ns SC pulse under 400V bus voltage, whereas SiC MOSFETs can withstand hundreds of cycles at 4 μs SC pulse. Given the relatively weak repetitive SC capability of GaN devices, the short-circuit safe operating area (SCSOA) of GaN HEMTs must be careful considered in the design of SC detection and protection circuits.

The short-circuit detection methods commonly used for silicon-based devices are not suitable for GaN HEMTs, as their response times fail to meet the stringent sub-microsecond requirements under high bus voltage conditions. Due to the fast switching speed of GaN HEMT devices, most of the traditional protection methods with short SC detection times are highly susceptible to interference from the high dv/dt switching noise characteristic of GaN devices. This susceptibility often results in erroneous protection actions during normal switching transients.

To address these challenges, researchers have proposed various improved desaturation (Desat) SC protection methods. The SC protection method based on the voltage dip on the phase-leg dc voltage after an HSF fault occurs is prone to false triggering, although the device can be turned off within 280 ns[3]. Additionally, gate voltage-based detection methods, which monitor the gate voltage during the on-state of the device is proposed [4]. However, their protection time under hard switching fault (HSF) condition remains relatively long at 620 ns. A summary of SC protection methods for GaN HEMTs and their corresponding response times is provided in Table I. Despite their improvements, most existing SC protection methods primarily focus on addressing HSF conditions while paying less attention to fault under load (FUL) conditions. These SC protection methods often fail to handle both HSF and FUL conditions effectively.

979-8-3315-1612-3/25 $31.00 © 2025 IEEE

Table I SC protection methods for GaN HEMT

Method	$V_{DC}(V)$	$T_{HSF}(ns)$	$T_{FUL}(ns)$
Phase-leg voltage dip [3]	400	281	-
Detection of gate voltage [4]	300	620	-
Three-step solution [5]	400	2000	-
Desat [6]	400	110	-
Low pass filter [7]	400	224	-
Rogowski coil [8]	400	-	70
SSCL [9]	430	800	-
PCB layout parasitic[10]	400	370	-

This paper proposed a fast SC detection and protection method for ohmic gate P-GaN HEMT devices. The proposed method utilizes the principle of gate charge reduction caused by the decreased input capacitance of GaN HEMTs under SC fault condition. The proposed SC protection method utilizes gate current integration to calculate the gate charge and detects the occurrence of SC faults in the device. This approach significantly reduces the SC protection time and demonstrates strong immunity to the high dv/dt switching noise characteristic of GaN HEMTs, ensuring fast and reliable protection. Furthermore, the proposed method is capable of addressing both hard switching fault (HSF) and fault under load (FUL) conditions, making it a versatile solution for GaN HEMT applications.

II. SHORT-CIRCUIT DETECTION METHOD BASED ON GATE CHARGE CHARACTERISTIC

In this section, the gate charge characteristics of GaN HEMT under SC conditions are analyzed. In particular, the differences in gate charge behavior between normal turn-on, HSF short-circuit, and FUL short-circuit conditions are discussed. The purpose of this analysis is to provide a comprehensive understanding of SC protection principles and circuit design method.

The Ohmic gate GaN HEMT is a normally off type current drive device, as shown in Fig. 1. The non-isolated gate structure of p-ohmic-gate GaN HEMT exhibits a diode-like input characteristic, requiring continuous gate current injection to ensure low on-state resistance and stable operation [1]. This unique gate structure significantly influences the device's response under various operating conditions, including SC faults. Therefore, it is critical to understand the gate structure to develop effective SC protection strategy and circuit design.

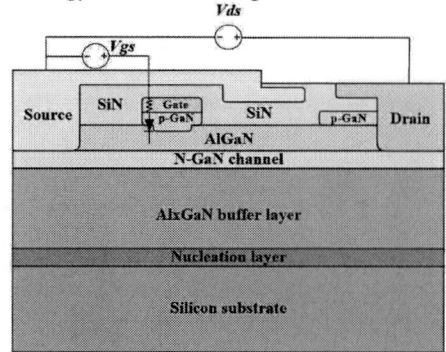

Fig. 1 The structure of ohmic gate P-GaN HEMT

A. HSF Condition

Fig. 2 shows the simplified waveform of an Ohmic-gate GaN HEMT in normal condition and HSF short circuit. The gate current has a significant effect on both the normal condition and the HSF short circuit of the device. When the GaN HEMT is turn on, the gate current simultaneously charges C_{gs} and C_{gd}. When the gate voltage reaches V_{th}, I_{ds} increases, V_{ds} decreases, and the decrease of V_{ds} causes the increase of C_{gd}. The gate current continues to charge the C_{gd} while the V_{gs} remains unchanged. This period of time when the V_{gs} is unchanged is called the Miller platform (V_{gp}). After the Miller platform is charged, V_{ds} stops falling and V_{gs} begins to rise again.

However, when an HSF short circuit occurs, the drain voltage does not decrease as it does under normal conditions. Instead, the V_{ds} of the device under HSF remains much higher than in normal operation condition. Therefore, the C_{gd} under HSF condition is lower than C_{gd} under normal condition. The reduction of the gate capacitance results in the absence of the Miller platform during the gate charging process, which in turn reduces the overall gate charge. As shown in Fig. 2, Q_g in HSF condition lacks the gate-drain charge (Q_{gd}) present in normal condition. Fig. 3 shows the difference between the gate charge of the Ohmic gate GaN HEMT in the case of normal condition and HSF condition. The peak value of the gate transient current (I_{g-peak}) under normal condition is lower and occurs later than in HSF condition. Therefore, the gate charge in HSF conditions is reduced due to the absence of the Miller plateau. Additionally, when HSF occurs, the drain current quickly rises. Unlike normal switching condition, the lack of a Miller plateau leading to a peak in V_{gs} under HSF conditions [11].

From the above analysis, two main criteria can be obtained for HSF protection circuit to detect the SC condition: (1) When HSF SC occurs, the gate turn-on peak voltage ($V_{gs-peak}$) is higher than that under normal turn-on condition. (2) The gate charge at the beginning of HSF condition is less than that during normal operation.

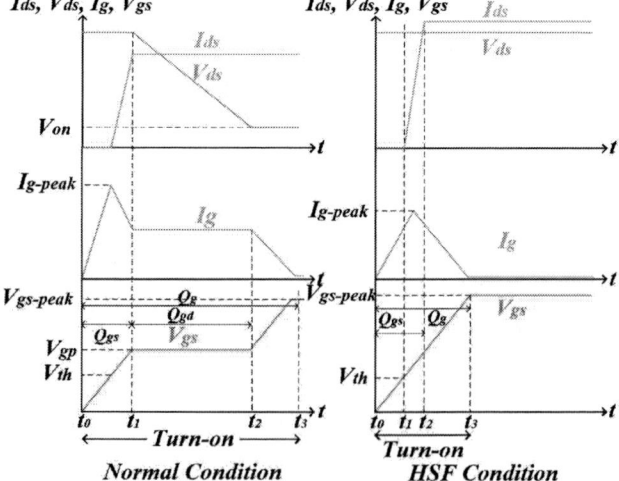

Fig. 2 Waveform of GaN HEMT in HSF condition.

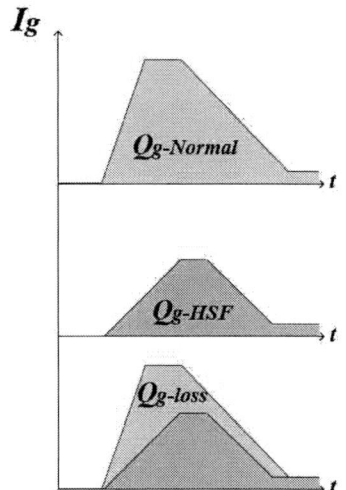

Fig. 3 The difference of the gate charge under normal operation condtions and HSF condition

B. FUL Condition

When FUL SC occurs, the device is short-circuited due to an external load fault during conduction. Under this condition, both V_{ds} and V_{gs} increase sharply when the SC fault occurs, which result in a decrease in C_{gd}, but with almost no significant change in C_{gs}. The drain current of the device rises rapidly along with the rapidly increasing V_{ds}. The rise in the device's drain-source voltage generates a large dV_{ds}/dt. An increase in drain voltage leads to a decrease in C_{gd}. Although the gate capacitance C_{gs} does not change significantly, dV_{gs}/dt increases rapidly. Therefore, according to the gate current equation (1), it can be concluded that the gate current of the device is reduced under FUL SC condition [12].

$$I_g = C_{gd}(dV_{gd}/dt) - C_{gs}(dV_{gs}/dt) \quad （1）$$

Fig. 4 shows the simplified waveform of an Ohmic-gate GaN HEMT in normal condition and FUL condition. Fig. 5 shows the absence of gate charge in FUL condition. As shown in Fig. 4, the V_{gs} spikes and the gate charge decreases under FUL conditions. The SC protection criteria for FUL condition are similar to those for HSF condition: (1) When FUL occurs,

the gate voltage V_{gs} spikes to a level (V_{peak}) much higher than that during normal conduction. (2) The amount of gate charge during an FUL condition is less than the amount of gate charge during normal operation.

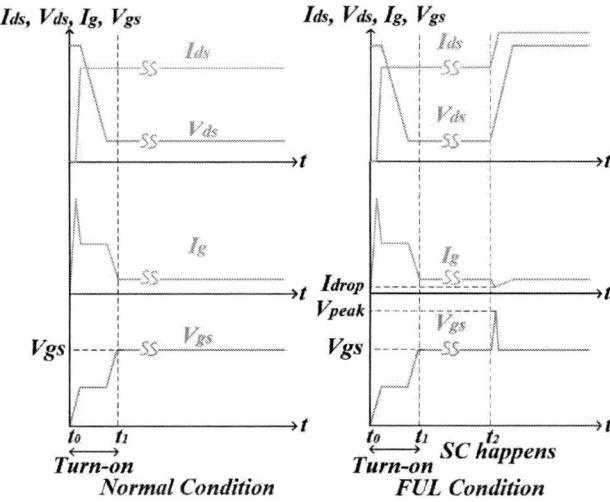

Fig. 4 Waveform of GaN HEMT in FUL condition

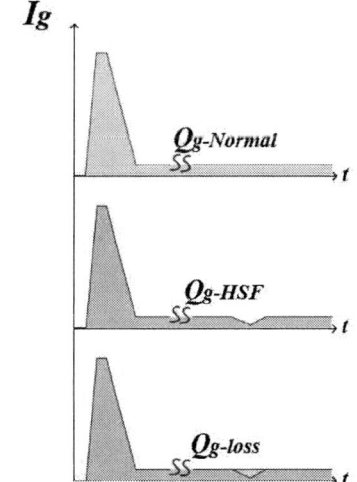

Fig. 5 The difference between the gate charge

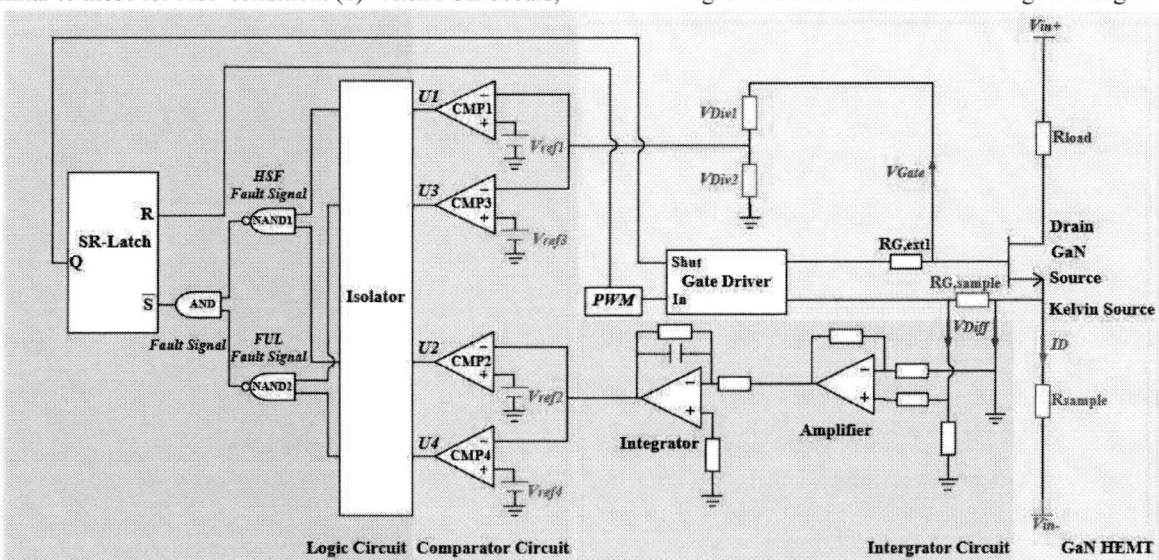

Fig. 6 The SC protection circuit based on gate charge

III. WORKING PRINCIPLE OF SHORT CIRCUIT PROTECTION CIRCUIT

As shown in Fig 6, the protection circuit mainly consists of four parts. The first part is the GaN HEMT circuit, including: GaN HEMT and gate driver. The second part is integrator circuit for determine the amount of gate charge. This part includes a sampling amplifier that samples and amplifies the gate current by measuring the differential voltage across the gate sample resistance ($R_{G,sample}$, located at the Kelvin source terminal of the device), and an integrator that processes this sampled voltage to determine the gate charge value. The third part is the comparator circuit. This part of the circuit is composed of four comparators, which are used to output SC fault signals. The fourth part of the circuit is logic circuit, which uses *PWM* signals and comparator signals to decide whether to turn off the GaN HEMT device.

A. The SC Protection Circuit Working Principle under HSF condition

When HSF fault occurs, the peak gate current of GaN HEMT device is lower and later than that under normal condition, and the gate current peak duration is shorter, resulting in lower gate charge than that in normal condition. In HSF condition, integrator outputs lower voltage than that in normal condition. The lower voltage is compared by comparator 2 with the reference voltage V_{ref2} set for the HSF. If HSF happens, the comparator 2 outputs signal *U2* which is high level voltage. When HSF causes a surge in V_{gs}, comparator 1 outputs signal *U1* which is high level voltage. After passing through the digital isolator, signals *U1* and *U2* are input into NAND 1. When the signals *U1* and *U2* are high level voltage at the same time and the GaN device is turn-on, the S-R latch outputs high level voltage to the Gate driver to turn off the GaN device. The signal timing logic diagram of normal condition and HSF short-circuit condition is shown in Fig. 7.

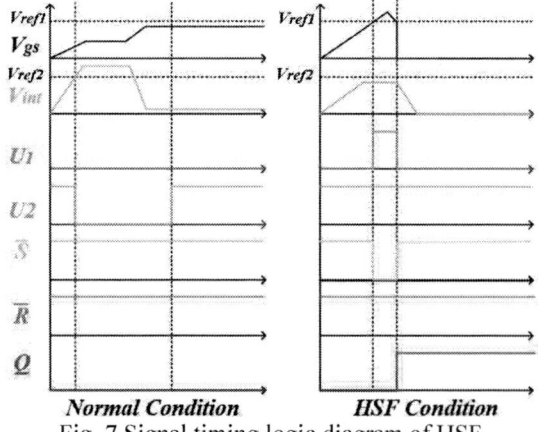

Normal Condition   **HSF Condition**
Fig. 7 Signal timing logic diagram of HSF

B. The SC Protection Circuit Working Principle under FUL condition

When the FUL SC occurs, there is a drop in the gate current, which can also be captured by the integrated circuit. The gate charge is lower than under normal conditions. After comparing it with a preset reference voltage V_{ref4} set by comparator 4 for FUL. A high level voltage signal *U4* is output. Simultaneously, the V_{gs} comparator 3 judge whether the gate voltage spikes occurs. If V_{gs} is higher than V_{ref3} set for FUL, comparator 3 outputs another high-level voltage signal

U3. Both signals, *U3* and *U4*, are then passed through a digital isolator and input into NAND gate 2. When the signals *U3* and *U4* both are high level and the *PWM* signal is high, the S-R latch outputs high level voltage to the gate driver to turn off the GaN HEMT device. The signal timing logic diagram of normal condition and FUL short-circuit condition is shown in Fig. 8.

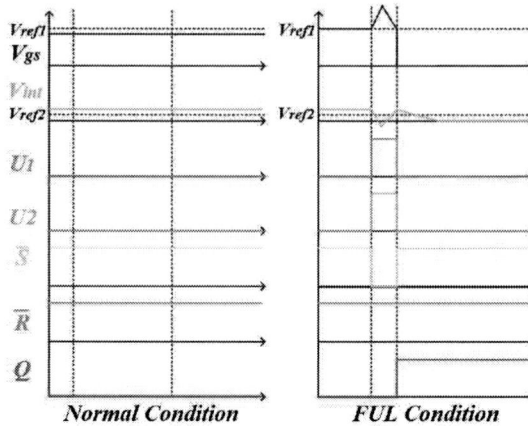

Normal Condition   **FUL Condition**
Fig. 8 Signal timing logic diagram of FUL

IV. EXPERIMENTAL RESULT

As shown in Figure 9, a SC testbench was customized to verify the functionality of the SC protection circuit. In order to verify that the SC protection circuit can still work at 150 °C condition, a heating patch was used to heat the GaN HEMT to 150°C. The key specifications of the SC test conditions are summarized in Table II, while the details of the experimental instruments used in the setup are provided in Table III.

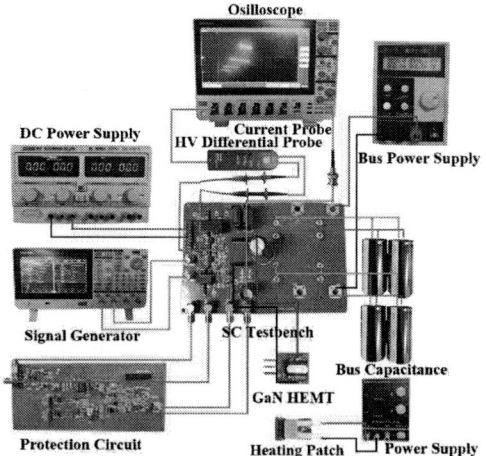

Fig. 9 SC test condition and SC protection circuit

Table II Key specification of SC test conditions

Parameter	Value
GaN HEMT	IGOT60R070D1AUMA3
DUT DC voltage	400V
DUT gate external resistance	560Ω
DUT gate resistance	15Ω
DUT gate capacitance	3.3nF
Gate driving voltage	9V
DC-link capacitance	2200μF
Tested case temperature T_c	25°C or 150°C
DUT short circuit time	1.5μs

Table III Information of experimental instrument

Parameter	Value
HV Differential Probe	Tektronix THDP0200
Osilloscope	Tektronix 4 series
DC power supply	Zhaoxin RXN-3010D-II
Bus power supply	Hanshengpuyuan HSPY-600
Power supply	Zhaoxin KXA-3205D

A. Protection Circuit under HSF Conditon

Fig. 10 and Fig. 11 shows the waveform of the protection circuit and DUT when HSF short circuit occurs under 400V bus voltage. Fig. 10 corresponds to a case temperature of 25℃, while Fig.11 shows a T_c of 150℃. In both scenarios, the protection circuit successfully shuts down the DUT, achieving fault isolation within 100 ns after the HSF short-circuit occurs.

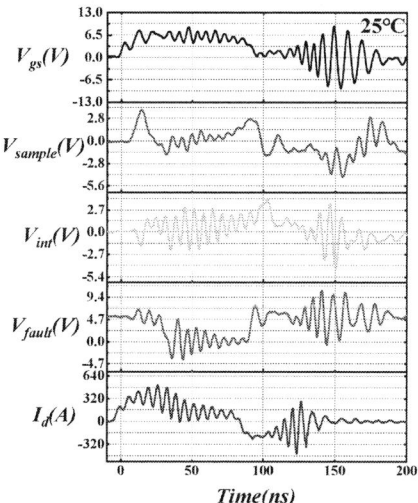

Fig. 10 The waveform of HSF short circuit under 400V at T_c of 25°C

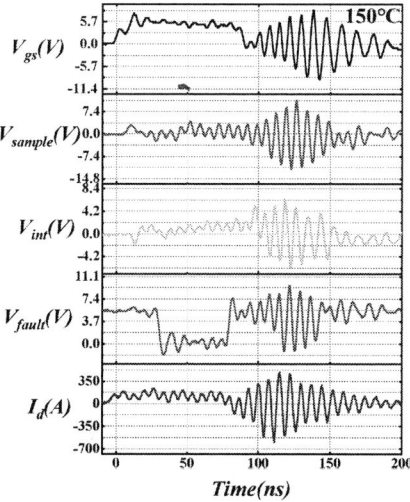

Fig. 11 The waveform of HSF short circuit under 400V at T_c of 150°C

Fig. 12 shows an oscilloscope screen capture of the HSF short circuit at 400V at T_c of 25°C. When the HSF short circuit occurs under 400V condition and the protection circuit is activated. Fault signals $U1$ and $U2$ are processed through NAND1. Then AND gate generates the fault signal (V_{fault}) approximately 35 ns after the HSF short circuit happens. The S-R latch receives low-level V_{fault}, identifies the HSF condition, and generates a high-level output signal. This signal instructs the gate driver to turn off the DUT, thereby isolating the fault and protecting the device.

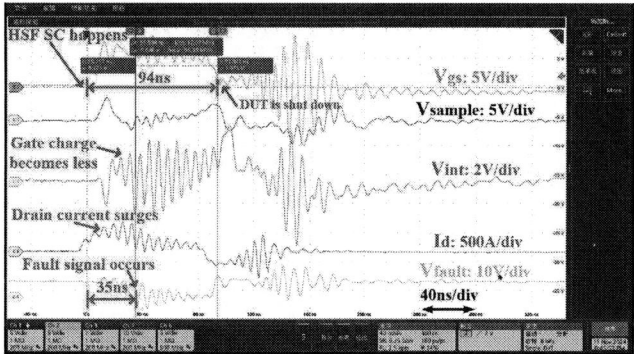

Fig. 12 The waveform of HSF SC at 400V at T_c of 25°C

B. Protection Circuit under FUL Condition

Fig. 13 shows how the SC testbench achieves a FUL short circuit by controlling the turn-on of the IGBT connected in parallel with the load resistance (R_{load}). During the DUT turn-on period, the IGBT is suddenly turned on to short-circuit the load resistance R_{load} and thereby achieving the FUL short-circuit condition.

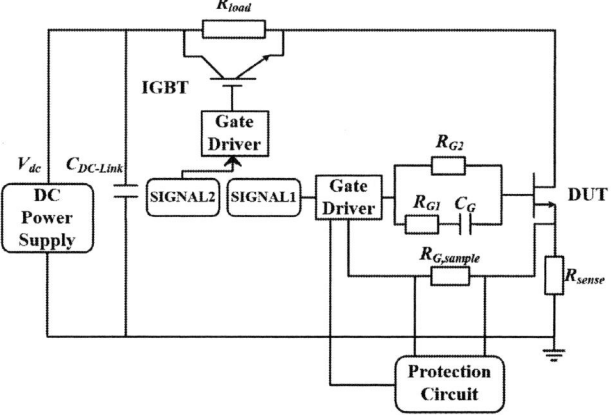

Fig. 13 FUL testbench

Fig. 14 and Fig. 15 shows the waveform of the protection circuit and DUT when FUL short circuit occurs under 400V bus voltage. Fig. 14 corresponds to a case temperature T_c of 25℃, Fig.15 represents a T_c of 150℃. Whether at T_c of 25°C or at T_c of 150 °C, the DUT is successfully protected within 80ns after the FUL short circuit occurs.

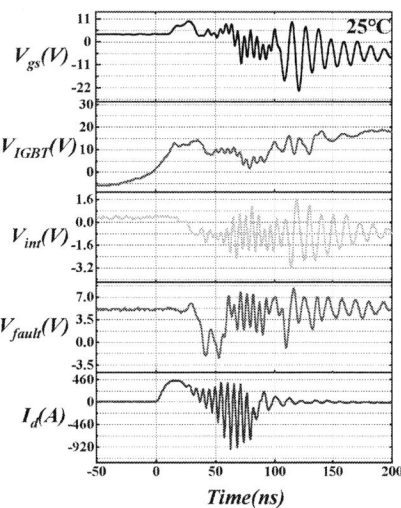

Fig. 14 The waveform of FUL short circuit
under 400V at T_c of 25°C

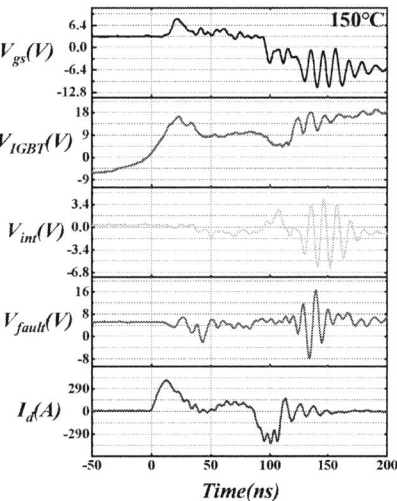

Fig. 15 The waveform of FUL short circuit
under 400V at T_c of 150°C

Fig. 16 shows the waveform of the FUL short circuit at 400V at T_c of 25°C. When the FUL short circuit occurs under 400V condition and the protection circuit is activated. Fault signals $U3$ and $U4$ are processed through NAND2. Then AND gate generates the fault signal (V_{fault}) approximately 40 ns after the HSF short circuit happens. The S-R latch receives the low-level V_{fault}, identifies the fault condition, and generates a high-level output signal. This signal controls the gate driver to turn off the DUT, effectively isolating the fault and protecting the device from further damage.

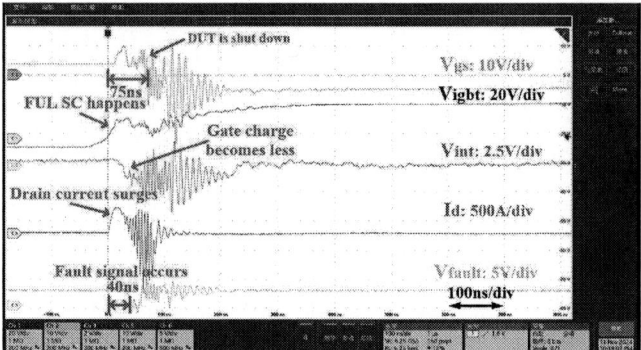

Fig. 16 The waveform of FUL SC at 400V at T_c of 25°C

V. CONCLUSION

In this paper, a fast short-circuit protection method for ohmic gate P-GaN HEMT was proposed and experimentally validated. The protection circuit utilizes gate charge and gate voltage to effectively address both HSF and FUL short-circuit conditions. Furthermore, the method demonstrates reliable operation at a case temperature of 150°C. The experimental results confirm that the protection circuit can detect the fault within 40ns after the short circuit occurs and turn off the GaN device within 100ns with low risk of false triggering.

ACKNOWLEDGMENT

This work was supported in part by the Natural Science Foundation of Guangzhou under Grant 2023A04J1064 and in part by the Young Scientists Fund of National Natural Science Foundation of China under Grant 62404166.

REFERENCES

[1] H. Li et al., "E-mode GaN HEMT short circuit robustness and degradation," 2017 IEEE Energy Conversion Congress and Exposition (ECCE), Cincinnati, OH, USA, 2017, pp. 1995-2002, doi: 10.1109/ECCE.2017.8096401.

[2] N. Badawi, A. E. Awwad and S. Dieckerhoff, "Robustness in short-circuit mode: Benchmarking of 600V GaN HEMTs with power Si and SiC MOSFETs," 2016 IEEE Energy Conversion Congress and Exposition (ECCE), Milwaukee, WI, USA, 2016, pp. 1-7, doi: 10.1109/ECCE.2016.7855410.

[3] H. Li et al., "An Ultra-Fast Short Circuit Protection Solution for E-mode GaN HEMTs," 2018 1st Workshop on Wide Bandgap Power Devices and Applications in Asia (WiPDA Asia), Xi'an, China, 2018, pp. 187-192, doi: 10.1109/WiPDAAsia.2018.8734686.

[4] X. Zhan, P. Sun, X. Huang and Y. Wang, "A New Simple and Low Cost Short Circuit Protection Method for p-GaN HEMT," 2021 IEEE 2nd China International Youth Conference on Electrical Engineering (CIYCEE), Chengdu, China, 2021, pp. 1-6, doi: 10.1109/CIYCEE53554.2021.9676798.

[5] X. Lyu et al., "A Reliable Ultrafast Short-Circuit Protection Method for E-Mode GaN HEMT," in IEEE Transactions on Power Electronics, vol. 35, no. 9, pp. 8926-8933, Sept. 2020, doi: 10.1109/TPEL.2020.2968865.

[6] J. Wu, W. Meng, F. Zhang, G. Dong and J. Shu, "A Short-Circuit Protection Circuit With Strong Noise Immunity for GaN HEMTs," in IEEE Transactions on Power Electronics, vol. 36, no. 2, pp. 2432-2442, Feb. 2021, doi: 10.1109/TPEL.2020.3013984.

[7] K. Wang et al., "A Reliable Short-Circuit Protection Method with Ultra-Fast Detection for GaN based Gate Injection Transistors," 2019 IEEE 7th Workshop on Wide Bandgap Power Devices and Applications (WiPDA), Raleigh, NC, USA, 2019, pp. 43-46, doi: 10.1109/WiPDA46397.2019.8998869.

[8] J. Le Leslé, J. Morand, R. Perrin and G. Lefevre, "A Fast Short-Circuit Detection and Protection Method for Wide Band-gap Devices based on Current Derivative Sensing," 2021 23rd European Conference on Power Electronics and Applications (EPE'21 ECCE Europe), Ghent,

Belgium, 2021, pp. 1-9, doi: 10.23919/EPE21ECCEEurope50061.2021.9570568.

[9] D. Bisi *et al.*, "Short-Circuit Protection for GaN Power Devices with Integrated Current Limiter and Commercial Gate Driver," *2022 IEEE Applied Power Electronics Conference and Exposition (APEC)*, Houston, TX, USA, 2022, pp. 181-185, doi: 10.1109/APEC43599.2022.9773446.

[10] O. S. Alemdar, F. Karakaya and O. Keysan, "PCB Layout Based Short-Circuit Protection Scheme for GaN HEMTs," *2019 IEEE Energy Conversion Congress and Exposition (ECCE)*, Baltimore, MD, USA, 2019, pp. 2212-2218, doi: 10.1109/ECCE.2019.8913081.

[11] X. Jiang *et al.*, "Short-Circuit Failure Modes and Mechanism Investigation of Ohmic-Gate GaN HEMT," in *IEEE Transactions on Electron Devices*, vol. 71, no. 3, pp. 1455-1463, March 2024, doi: 10.1109/TED.2023.3314401.

[12] S. Musumeci, R. Pagano, A. Raciti, F. Frisina and M. Melito, "Transient behavior of IGBTs submitted to fault under load conditions," *Conference Record of the 2002 IEEE Industry Applications Conference. 37th IAS Annual Meeting (Cat. No.02CH37344)*, Pittsburgh, PA, USA, 2002, pp. 2182-2189 vol.3, doi: 10.1109/IAS.2002.1043834.

Comparison of Ultrafast-Rise-Time Gate Drivers for Wide-Bandgap Devices in Sub-Microsecond Pulsed Power Applications

Soham Roy ⓘ, Duy T. Nguyen ⓘ, Neeraj Anantha and Alex J. Hanson ⓘ

The University of Texas at Austin

2501 Speedway, Austin, TX 78712, USA

{soham.roy, daniel.nguyen2304, neerajanantha, ajhanson} @utexas.edu

Abstract—**Pulsed power applications such as high-resolution medical imaging and transient plasma generation demand well-regulated pulses that are a few hundred nanoseconds in duration. To achieve pulse rise times on the order of a few ns, these systems favor the use of fast-switching semiconductor devices, such as GaN HEMTS and SiC MOSFETs. The switching performance of these transistors is partly dependent on ultrafast gate drivers that can generate very high v_{gs} slew rates. There are a variety of discrete ultrafast gate-driver topologies in the state of the art, which can generally be classified as voltage-source, current-source, and gate-boosting approaches. This paper presents a framework for a fair comparison of these different gate-driver architectures, alongside the best-in-market commercial gate-driver ICs as references. Through simulations and experiments using a simple pulse-generation circuit, the advantages and trade-offs of each topology are highlighted.**

Index Terms—**Pulsed power, GaN gate driver, SiC gate driver, ultrafast rise time, slew rate, modular pulse generator**

I. INTRODUCTION

Systems requiring power in precisely controlled sub-μs pulses, such as transient plasma generation and electroporation-based medical procedures, rely on very rapidly rising pulses. Fast-switching solid-state devices developed over the past few decades are becoming increasingly favorable in such pulse generators [1]–[3] – taking the place of traditional gas/spark-gap switches, which are harder to control and suffer from the high costs of low production volumes. The increasing use of semiconductor switches in high-voltage systems has been enabled by modular topologies such as the Marx generator [4] and the inductive voltage adder (IVA) or linear transformer driver (LTD) [1], in which individual modules and switches can be rated for much lower voltages. With sufficient modularity, high-energy pulses can be achieved with wide-bandgap (WBG) devices such as GaN and SiC, with excellent figures of merit. Taking full advantage of these devices for fast-rising pulses requires gate drivers capable of driving them at ultra-high slew rates under hard switching conditions.

SiC power devices have been in development for a longer time than GaN, and the availability of up to 3.3-kV SiC devices has recently led them to gain popularity in pulsed power

Research reported in this publication was supported by the Air Force Office of Scientific Research Award #: FA9550-22-1-0071

applications. Most prior studies on semiconductor-based pulse generators and their ultrafast gate-driver circuits have therefore used SiC MOSFETs as switching devices [5]–[9]. In contrast, GaN HEMTs have seen rapid development of high-voltage commercial devices (up to 1200 V currently available [10]) in the more recent past. Having an even higher saturation velocity and breakdown electric field than SiC [11], GaN transistors have inherently better electrical figures of merit than SiC. The superiority in their electrical characteristics, paired with the fast-paced innovation in their development, could make GaN devices a favorable choice for modular pulsed power applications in the near future [12].

Several ultrafast gate-driver circuits built using discrete components have been proposed in past works to potentially overcome the slew-rate limitations posed by commercial integrated gate drivers [8], [9], [12]–[17]. However, the relative performance of these methods has not been studied well, and a thorough comparison of their benefits and trade-offs is missing in the literature. Therefore, this work presents a setup for a direct comparison of the effectiveness of some of these different architectures (in some cases with slight modifications, for better functioning and for a fair comparison) under near-identical conditions. Their operation with both types of WBG devices is examined here: a commercial 650-V GaN device and a commercial 650-V SiC device (with comparable current ratings), using the same gate-driver circuits with appropriate driving voltages.

II. APPROACHES FOR ULTRAFAST SWITCHING

Gate drivers capable of generating an extremely high v_{gs} turn-on slew rate ($\frac{dv_{gs}}{dt}$) can help achieve ultrafast pulse rise times at the load; whereas the pulse fall time tends to depend largely on the load current, especially under low load conditions [8], [18]. Additionally, ultra-low rise times are a far more critical requirement than fall times in pulsed applications. For these reasons, the main emphasis of this work is on minimizing rise times using the topologies considered.

We can broadly classify gate-driver circuits into voltage-source, current-source, and gate-boosting topologies. Typical integrated drivers fit into the voltage source category, where a nominally low-impedance supply voltage is abruptly connected to the gate to turn on the device. It is critical that

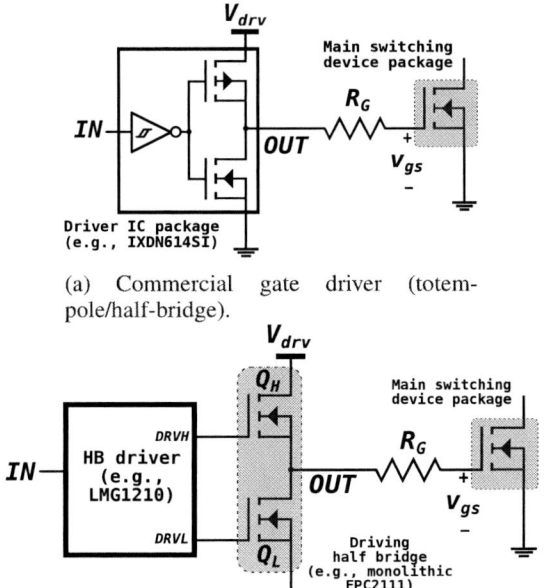

(a) Commercial gate driver (totem-pole/half-bridge).

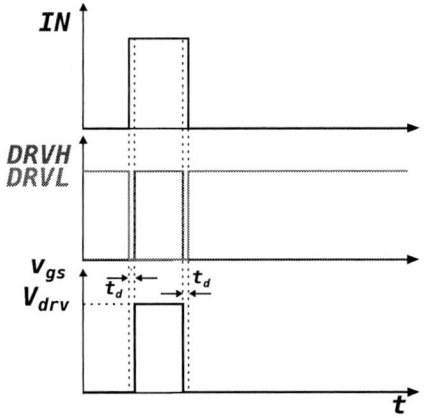

(b) External half-bridge (HB) driver of [12]–[14].

Fig. 1. Voltage-source gate-driving approaches considered in this work.

Fig. 2. Waveforms of relevant signals in the gate-pulse generation chain for the voltage-source external half-bridge (HB) driver of Fig. 1b.

the output node of the voltage-source drivers have minimal impedance during the on- and off-transitions. By contrast, current-source topologies first build current into an external magnetic-energy-storage element (e.g., inductor), which acts as a high-impedance current source for a short duration, and can be rapidly connected to the gate (during its turn-on) to provide a high charging current. In current-source circuits, it is critical that the output node present maximum impedance during the transition. These drivers require extra time (possibly on the order of the pulse length, for narrow pulses) to energize the inductors, which could limit the repetition rate of the pulse generator. Lastly, gate-boosting drivers use a high driving voltage and take advantage of the underdamped behavior of the gating loop to create a temporary voltage overshoot; additional circuit components help the gate voltage settle at a lower value

after the transition.

The switches used in driver circuits can generally be schematically rearranged (without affecting functionality) into half- or full-bridge structures. The topologies used for comparison here are presented after re-configuring them in this manner, which allows for a more direct comparison.

A. Voltage-Source Gate Drivers

The voltage-source gate drivers considered here are shown in Fig. 1. Most commercial gate-driver ICs fall into this category, including those used for baseline comparison in this work. As depicted in Fig. 1a, they typically comprise a tapered chain of smaller totem-pole/half-bridge stages of driving devices (having complementary switching with a defined dead time t_d) inside them. The gate driver output is the switching node of the final stage of the tapered chain, and the integrated devices used in this stage are designed to be able to generate large peak charging and discharging currents, e.g., up to $\sim$30 A in IXDN630YI [19].

Extending this notion, one methodology seen in recent works [12]–[14] is the external half-bridge (HB) driver approach, shown in Fig. 1b – it uses a discrete packaged part with a half bridge of (usually) GaN devices to generate a potentially higher drive current than integrated devices. The external half bridge may itself be driven by a half-bridge driver IC (as depicted in the figure), or two separate single-output driver ICs. The switching waveforms for this approach are illustrated in Fig. 2 – the two driving devices (Q_H and Q_L in Fig. 1b) are fed complementary driving signals, $DRVH$ and $DRVL$ respectively, with a finite dead time t_d. The v_{gs} voltage obtained at the switching device has the same parity as the input signal IN.

B. Current-Source Gate Drivers

The current-source driver considered here is illustrated in Fig. 3. This is a full-bridge (FB) inductive driver proposed in [20], which can be implemented as two half bridges with an external inductor L_{ext} connected between their switching nodes. The switching waveforms for this topology are demonstrated in Fig. 4. It should be noted that Q_{L1} and Q_{H2} are fed identical signals, so IN_L1 and IN_H2 are presented using the same color. Also, Q_{L2} is always kept on, except for the duration of the pulse – this allows the switching device's gate to be pulled down to ground whenever it is required to be off. Q_{H1} is turned on for a duration t_{chg} before the pulse, allowing the inductor L_{ext} to build a current $I_{chg} \approx (\frac{V_{drv}}{L_{ext}})t_{chg}$ prior to the switching device's turn-on. At the turn-on instant, a part of the initial pre-built energy in the inductor is dumped into the gate of the switching device to turn it on rapidly. As a natural consequence of the finite dead time at the beginning of the pulse, v_{gs} rises to a level higher than the driving voltage V_{drv} for the brief duration t_d. The magnitude of this increase corresponds to the source-to-drain voltage $V_{sd,neg}$ of the driving GaN devices in their third-quadrant operation. Since GaN devices do not have a body diode, $V_{sd,neg}$ depends

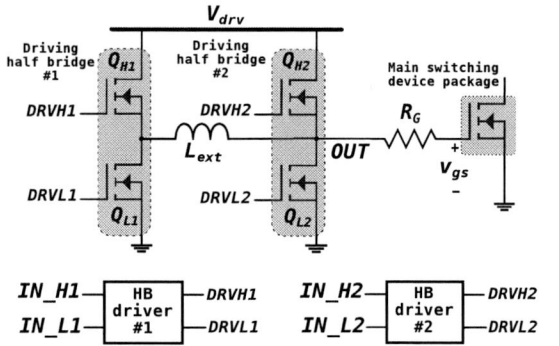

Fig. 3. Current-source gate-driving approach considered in this work – full-bridge (FB) inductive driver of [20].

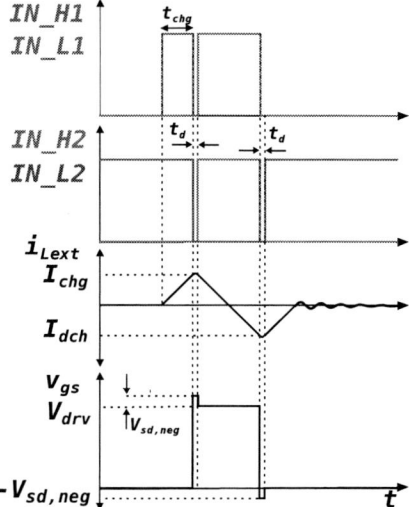

Fig. 4. Waveforms of relevant signals in the gate-pulse generation chain for the current-source full-bridge (FB) inductive driver of Fig. 3.

on the drain current and is typically in the 2-3 V range (higher than the body diode's $0.7\text{-}V_f$ drop for Si/SiC) [21].

By the end of the pulse, the inductor current is able to ramp down to a negative value I_{dch} – this discharging current can increase the turn-off slew rate of v_{gs}. To allow the inductor current to be ramped back to zero over the course of the pulse, t_{chg} should be chosen to be less than the pulse duration. A relatively longer t_{chg} enables a higher I_{chg}, but an I_{dch} of lower magnitude. Similar to the temporary voltage rise at the beginning of the pulse, a dip in v_{gs} below zero is seen at the end of the pulse for a duration t_d – due to third-quadrant operation of the driving devices. After the pulse, the surplus energy in the inductor is fed back to the source – this is an advantageous feature offered by this topology, unlike other current-source drivers in the literature [15], [16].

C. Gate-Boosting Drivers

A class of techniques called gate boosting [8], [9], [17] has recently gained popularity in pulsed applications. The gate-boosting circuit presented in [8], [9], as shown in Fig. 5, can be visualized as the external half-bridge driver of Fig. 1b

with some added components that are highlighted in red. The additions include a coupling capacitor C_C placed in series with the C_{gs} of the switching device to act as a high-frequency capacitive divider, along with a parallel dc-restoration diode which also prevents unintended gate charging during the steady-off state. This circuit uses a high driving voltage V'_{drv} (higher than the target voltage) in order to apply a high initial voltage at OUT, thereby forcing more current to flow through the gate impedance represented by R_G, L_g, and $R_{g,int}$. This current flows through C_{gs}, charging the gate more rapidly. Ignoring gating loop inductance, the voltage at OUT initially spikes to V'_{drv} and then settles to the final target value $V_{drv,fin}$ which is based on the capacitor divider formed by C_C, C_{gs} and C_{gd}. The settling time is based on the $(R_G + R_{g,int}) \times C_{gs}$ time constant.

Users of this technique have also leveraged an underdamped gating loop to generate a temporary overshoot that helps lower the v_{gs} rise time. This allows for taking maximum advantage of the fast initial rise in v_{gs} (assuming a first-order response) and avoids the latter (slower) part of a first-order response.

For a given switching device with known C_{gs}, its gate-boosting power supply voltage V'_{drv} should be selected to a value that is higher than the final target $v_{gs} = V_{drv,fin}$, while carefully ensuring that the initial overshoot at the device die is not too high to damage the gate. The value of C_C is then

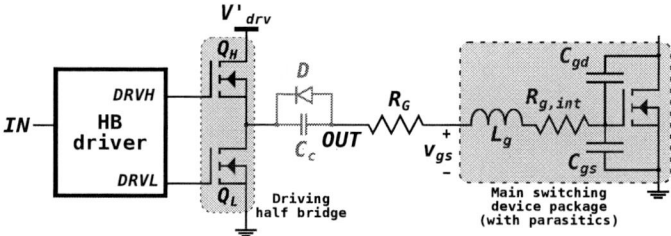

Fig. 5. Gate-boosting approach of [8], [9] considered in this work. Aside from the modified V'_{drv} driving voltage and additional red components, this is identical to the circuit of Fig. 1b.

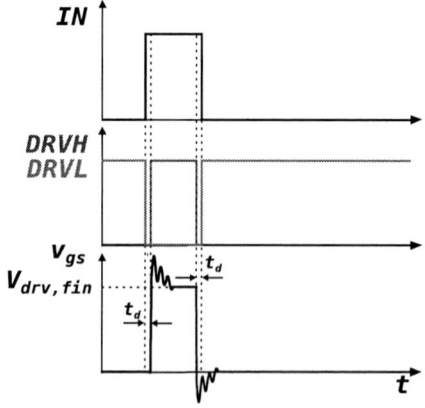

Fig. 6. Waveforms of relevant signals in the gate-pulse generation chain for the gate-boosting driver of Fig. 5.

979-8-3315-1612-3/25 $31.00 © 2025 IEEE   1123

selected such that v_{gs} settles at the desired value after the overshoot. Based on the capacitive voltage-divider ratio, the total gate charge Q_g (which must flow through C_C during device turn-on) can be written as: $Q_g = C_C(V'_{drv} - V_{drv,fin})$. Then, applying the conservation of gate charge over C_{gs} and C_{ds}, we get: $Q_g = Q_{gs} + q_{gd}(V_{dc})$, where $Q_{gs} = C_{gs}V_{drv,fin}$ and $q_{gd}(V_{dc})$ is a function of V_{dc}. Thus, the value of C_C can be calculated by equating the two expressions for Q_g:

$$C_C(V'_{drv} - V_{drv,fin}) = C_{gs}V_{drv,fin} + q_{gd}(V_{dc})$$
$$\implies C_C = \frac{C_{gs}V_{drv,fin} + q_{gd}(V_{dc})}{V'_{drv} - V_{drv,fin}} \qquad (1)$$

It is known that for any switching device, $q_{gd}(V_{dc})$ varies non-linearly with V_{dc} – it increases rapidly from $V_{dc} = 0$ up to a certain value of V_{dc}, and then becomes constant. For a chosen value of C_C, the deviation in $V_{drv,fin}$ (with V_{dc}) can be accounted for, by rearranging (1):

$$V_{drv,fin} = \frac{C_C V'_{drv} - q_{gd}(V_{dc})}{C_C + C_{gs}} \qquad (2)$$

III. DESIGN OF THE EVALUATION CIRCUIT

The circuit used for simulations and experiments is shown in Fig. 7 – it is electrically equivalent to a module of a single series stage of a modular pulse generator like the inductive voltage adder (IVA) [13]. A dc voltage V_{dc} of up to 450 V is applied to the main switching devices, through the pre-charged parallel dc-link capacitance $C_{dc} = 3$ μF for supplying high-frequency currents. A load resistance R_L of 25 Ω and higher (for different load currents) is used. 100-ns long gate pulses are applied with each gate driver to turn on the devices. An external gate resistor $R_G = 0$ Ω is used for all topologies.

Three commercial gate driver parts LMG1020, IXDN630YI and IXDN614SI are picked as references for comparison against the discrete topologies: LMG1020 ($\sim$7 A peak) and IXDN630YI ($\sim$30 A peak) are the highest-output-current, single-output commercial gate drivers for GaN and SiC de-

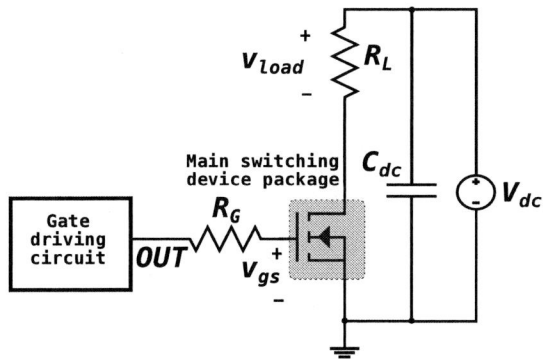

Fig. 7. Schematic of the circuit used for simulations and experiments, which is electrically equivalent to one module of a single series-stage of a modular pulse generator. The 'gate driving circuit' block corresponds to the different driver topologies considered.

vices respectively, while IXDN614SI ($\sim$14 A peak) supports the driving voltages of both GaN and SiC devices.

For each topology, the low-voltage monolithically integrated GaN part EPC2111 is chosen as the half bridge(s) of driving FETs – this allows minimizing unwanted parasitic effects. Each EPC2111 is driven by the fastest commercial half-bridge driver currently available, LMG1210.

The LMG1210 chip in the external HB and gate-boosting topologies is used in the 'PWM mode' and provided a single input signal; the dead time between its two output signals is programmed to be its minimum allowed $t_d = 0.8$ ns using 1.8-MΩ dead-time resistors at the DHL and DLH pins [22].

Both the LMG1210 chips in the FB inductive driver are used in the 'Independent Input Mode' to allow independent inputs to be provided to each driving device. The high-side floating voltage of each LMG1210 is provided using a separate 5-V isolated power supply – this bypasses its in-built bootstrap capability and avoids its associated duty-cycle limitation.

To study the effects on different kinds of WBG devices, a 650-V GaN device GS66508B ($R_{ds,on} \sim$50 mΩ, $Q_g \sim$6.1 nC) and a 650-V SiC device C3M0120065D ($R_{ds,on} \sim$120 mΩ, $Q_g \sim$28 nC) are tested as the main switching devices. Based on the static v_{gs} ratings in their datasheets, their V_{drv} values are selected as 7 V and 15 V respectively [23], [24]. Since the commercial GaN driver LMG1020 is designed only for a 5-V supply voltage [25], $V_{drv} = 5$ V is used with it.

For the FB inductive driver, a charging time $t_{chg} = 50$ ns, i.e., half of the pulse length of 100 ns, is selected – to generate I_{dch} of a similar magnitude as I_{chg}. The value of L_{ext} is picked as 47 nH – this is a low enough inductance such that it can build several amperes of current within 50 ns, but also not too low to compete against the parasitic inductances of the layout and the device. The dead time (t_d in Fig. 4) is chosen to be 5 ns, which is half of the length of the 10-ns clock cycle of the microcontroller used in the experiments.

For the gate-boosting driver, a low-V_f Schottky barrier diode is used as the dc-restoration diode. The V'_{drv} values for the GaN and the SiC device are chosen as 11 V and 22.5 V respectively (a little higher than their datasheet transient v_{gs} ratings, which are specified for up to $\sim$1 μs). Their C_C values are accordingly determined using (1) as 1.2 nF and 1.8 nF respectively, with $V_{drv,fin}$ (at $V_{dc} = 0$) set equal to their earlier chosen V_{drv} values. Although v_{gs} overshoots could be a common concern (especially for GaN), a recent work [26] shows that GaN HEMTs can handle transient voltage overshoots of up to $\sim$13 V under hard switching conditions without any impact on device lifetime. Moreover, the actual overshoot at the gate of the device die would always be lower than the measured v_{gs} at the device pin, which includes the effect of package inductance.

IV. SIMULATION RESULTS

The gate driver topologies are simulated in LTspice using complete Spice models, including parasitics, of the selected drivers and switching devices obtained from the manufacturers [27]–[29]. The simulation models for IXDN614SI and

979-8-3315-1612-3/25 $31.00 © 2025 IEEE

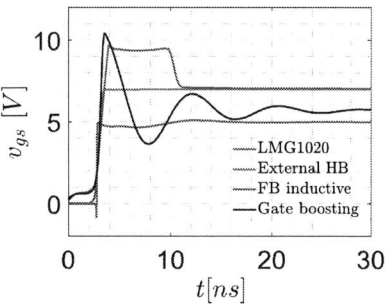

(a) GaN device GS66508B.

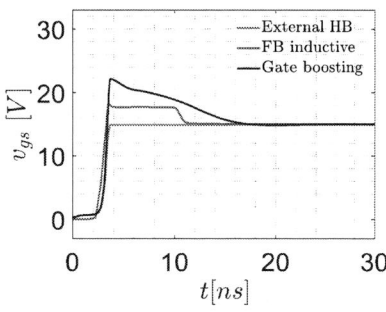

(b) SiC device C3M0120065D.

Fig. 8. Rising transient of the simulated gate-source voltage v_{gs} waveforms with V_{dc} = 450 V and R_L = 25 Ω waveforms using actual device/driver models in LTspice.

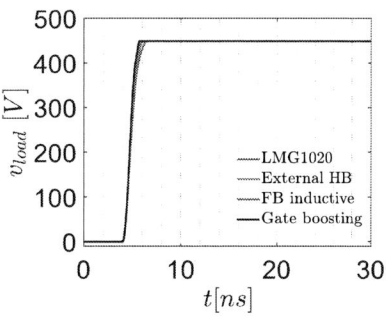

(a) GaN device GS66508B.

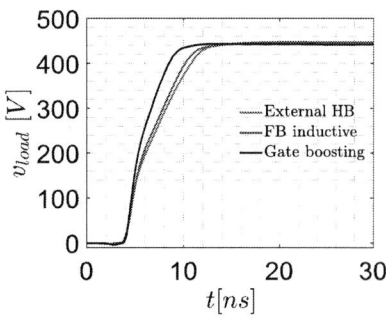

(b) SiC device C3M0120065D.

Fig. 9. Rising transient of the simulated load pulse voltage v_{load} waveforms with V_{dc} = 450 V and R_L = 25 Ω using actual device/driver models in LTspice.

IXDN630YI are not provided by IXYS, so they are not included in the simulations.

The rising transients of the simulated gate-source voltage v_{gs} and load pulse voltage v_{load} waveforms with V_{dc} = 450 V and R_L = 25 Ω are presented in Figs. 8 and 9 respectively. For the gate-boosting driver with the GaN device (shown in black), the settled v_{gs} is noticeably lower than 7 V in Fig. 8a. Using (2), this phenomenon can be explained – $V_{drv,fin}$ at V_{dc} = 450 V is expected to be lower than its chosen value of 7 V at V_{dc} = 0 due to the $-q_{gd}(V_{dc})$ term in the numerator, which increases in magnitude non-linearly with V_{dc}. The non-linear variation of $q_{gd}(V_{dc})$ is less prominent the SiC device, so the reduction in its $V_{drv,fin}$ value is harder to distinguish (more so, on the 0-30 V scale) in Fig. 8b.

From Fig. 9, it is evident that with the low-Q_g GaN device, the various gate drivers give essentially identical pulse rise times. With the relatively high-Q_g SiC device, the difference is more pronounced, with the gate-boosting circuit providing the fastest rise time, followed by the FB inductive driver and then the external HB driver.

V. EXPERIMENTAL SETUP & RESULTS

The PCB prototypes for the implementation of the different gate-driver circuits are shown in Fig. 10. The parasitics of the driving loop play a crucial role in impacting switching performance at the ns-level [30], [31] – therefore, careful consideration is taken to ensure that the gate loop for every layout has the tightest possible layout to minimize parasitics, and the layout differences across topologies are minimized. Due to the high degree of commonality between the external HB driver and the gate-boosting circuit, the same PCB prototype is used for both, as shown in Fig. 10c.

The dc voltage V_{dc} across the load and switching device is supplied by means of a PVS60085 dc power supply. 630-V SMD ceramic capacitors are used for the dc-link capacitance C_{dc}. Parallel 100-Ω thick film resistors TKH45P1R00FE-TR are used for R_L. The input signals for all driving devices and commercial drivers are generated using a LAUNCHXL-F280049C evaluation board (clock frequency f_{clk} = 100 MHz, i.e., clock cycle length 10 ns). For the FB inductive driver, an off-the-shelf 47-nH inductor FP2-S047-R is used. A 5-volt-output isolated dc-dc converter NME1S0505SC is used as the high-side supply for LMG1210.

The experimental waveforms are obtained using an MSO58 series 2-GHz oscilloscope. In each prototype, the gate-source voltage v_{gs} and load pulse voltage v_{load} waveforms are measured very close to the switching device and load resistors respectively – v_{gs} using a 1-GHz passive probe with a common mode (CM) choke to avoid CM currents, and v_{load} using an optically isolated 800-MHz probe. The rising transients of the v_{gs} waveforms obtained on the devices at V_{dc} = 450 V with R_L = 25 Ω are shown in Fig. 11. The discrete topologies achieve very high v_{gs} turn-on slew rates, following a trend similar to the simulations.

Fig. 12 shows the rising transients of the obtained load pulse voltage waveforms. The pulse rise times (calculated as the time

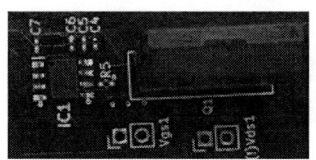
(a) Commercial driver IXDN614SI prototype (SiC device C3M0120065D).

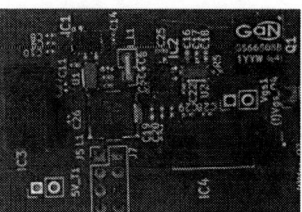
(b) Full-bridge (FB) inductive driver prototype (GaN device GS66508B).

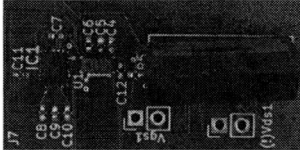
(c) Half-bridge/gate-boosting driver prototype (SiC device C3M0120065D).

Fig. 10. Photographs of the gate-driver circuit components of the PCB prototypes used for testing the various topologies. For each kind, only one version (either the one with GS66508B or with C3M0120065D) is shown.

taken to rise from 10% to 90% of the pulse amplitude) plotted as a function of the load current I_{load} are shown in Fig. 13. The rise times obtained with the GaN device are 5-10× faster than for the SiC device (for a given load and gate driver topology). The variation across load is almost imperceptible with the GaN device, whereas the high-Q_g SiC device shows a noticeable increase in t_{rise} as I_{load} increases. An interesting observation from Fig. 11b, Fig. 12b and Fig. 13b for the SiC device is that the external HB driver (shown in light blue) yields a lower v_{load} slew rate and higher rise time than IXDN630YI (shown in dark blue), despite the former showing a visually higher v_{gs} slew rate initially.

The pulse fall times are shown in Fig.14. Unlike rise times, the fall times for GaN and SiC devices are similar – indicating that the fall time depends most strongly on the load, especially at light load. Gate drivers that generate a brief negative v_{gs} during turn-off (e.g., the gate-boosting and FB inductive circuits) to accelerate the v_{gs} turn-off slew rate can help reduce fall times relative to other drivers at heavy load. This is most noticeable with the IXDN630YI driver. However, t_{fall} remains over 10 ns at maximum load, and further curtailment requires additional circuitry in the pulse generator, like tail-biter switches [32].

VI. DISCUSSION

Table I lists the v_{load} slew rates (at the highest dc voltage and load current) obtained experimentally and in simulation with each gate driver, and summarizes their advantages and disadvantages/limitations. Most of the discrete gate drivers are found to experimentally show an improvement in load-pulse-voltage slew rates compared to the best-in-market integrated drivers, with both the GaN and the SiC device. While the simulation results for the GaN device obtained with discrete

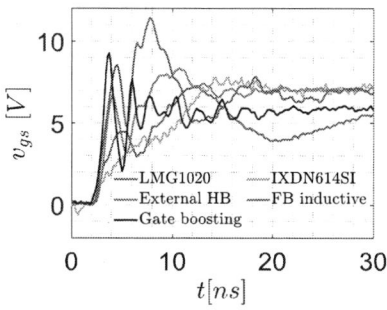
(a) GaN device GS66508B.

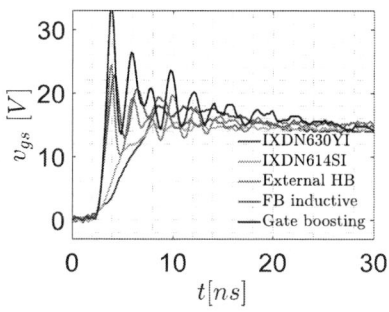
(b) SiC device C3M0120065D.

Fig. 11. Rising transient of experimental gate-source voltage v_{gs} waveforms with $V_{dc} = 450$ V and $R_L = 25\ \Omega$ obtained using a 1-GHz passive probe on a 2-GHz oscilloscope.

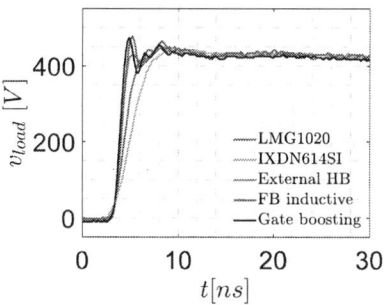
(a) GaN device GS66508B.

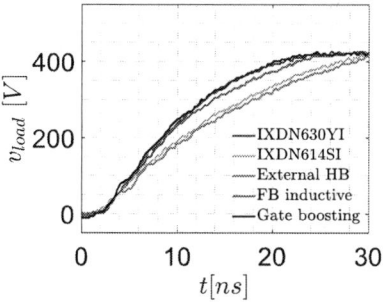
(b) SiC device C3M0120065D.

Fig. 12. Rising transient of experimental load pulse voltage v_{load} waveforms with $V_{dc} = 450$ V and $R_L = 25\ \Omega$ obtained using an 800-MHz optically isolated probe on a 2-GHz oscilloscope.

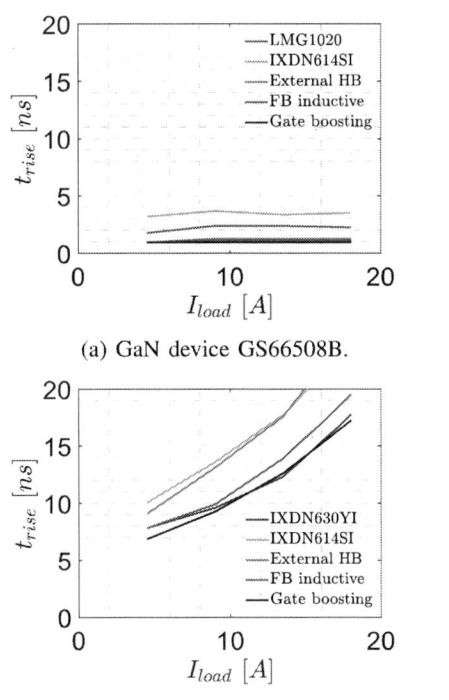

(a) GaN device GS66508B.

(b) SiC device C3M0120065D.

Fig. 13. Variation of rise time t_{rise} as a function of the load current I_{load} at V_{dc} = 450 V.

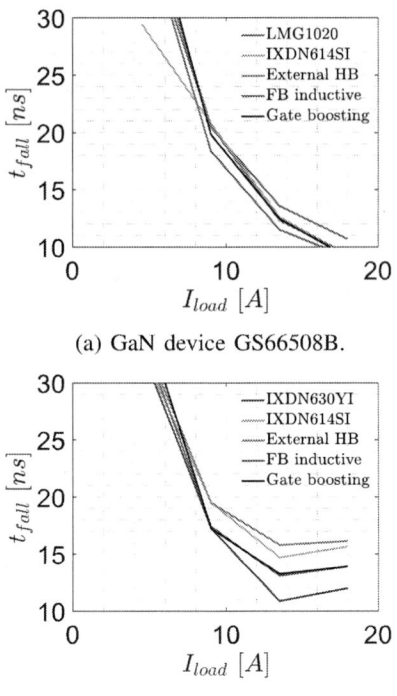

(a) GaN device GS66508B.

(b) SiC device C3M0120065D.

Fig. 14. Variation of fall time t_{fall} as a function of the load current I_{load} at V_{dc} = 450 V.

drivers line up extremely well with the experiments, those obtained with the SiC device also follow the same trend (in comparative performance) as the experiments.

The gate-boosting circuit helps achieve the fastest load-pulse-voltage slew rates for both devices, followed by the FB inductive driver, and then the external HB driver. Previous works on gate boosting [8], [9], [17] have used extremely

high V'_{drv} voltages (up to 80 V) on SiC devices to obtain exceptionally high v_{gs} slew rates. In this work, gate boosting outperformed other techniques even with a modest V'_{drv} voltage (just a little higher than the rated transient v_{gs}), which was chosen to ensure the reliability of the devices. The rise-time improvement with higher V'_{drv} voltages could be even greater (especially at high load currents), but at the potential cost of impacting the devices' lifespan.

Table I: Summary of the different gate drivers compared in this work.

| Type | Part / Topology | $(\frac{dv_{load}}{dt})_{rise}$ [V / ns] @ 450 V, 18 A | | | | Advantages | Disadvantages / Limitations |
| | | GS66508B (GaN) | | C3M0120065D (SiC) | | | |
		Exp	Sim	Exp	Sim		
Voltage-source	LMG1020 (IC)	161	99	–	–	Simple design	Cannot drive GaN devices at > 5 V
	IXDN614SI (IC)	102	–	15	–	Simple design	Very slow for GaN devices
	IXDN630YI (IC)	–	–	20	–	Simple design, fastest IC for SiC	Cannot be used for GaN devices
	External half-bridge	281	298	14	54	Simplest discrete topology	Slowest among discrete options
Current-source	Full-bridge inductive	321	342	18	60	Faster than external HB	Might limit pulse repetition rate
Gate-boosting		375	356	21	79	Fastest for GaN and SiC	v_{gs} overshoot may shorten device lifespan

Extra losses and EMI generated by gate drivers are most often insignificant in the context of physically large, relatively low-efficiency, sparsely pulsed modular pulse generators like the IVA (efficiency up to $\sim$70% reported in [33]). The v_{load} slew rate is the most important metric, which is used here.

VII. Conclusion

This paper describes a setup used for a like-for-like implementation and comparison among different discrete gate-driver architectures which are commonly used in sub-μs pulsed power generators. Their simulated and experimental performance is studied on two commercial 650-V, $\sim$20-A WBG devices (one each of GaN and SiC) in a simple pulse generator circuit. We found that any of the discrete drivers explored here can achieve similar load-pulse rise times when driving a GaN device, while the improvement obtained with gate boosting is significantly more with the SiC device.

The test circuit here is electrically equivalent to a single series stage of a modular pulse generator. Rise times can be further degraded in multi-stage systems as the size of the system adds parasitic inductance and as turn-on instants between modules are not perfectly synchronized.

References

[1] W. Jiang, H. Sugiyama, and A. Tokuchi, "Pulsed Power Generation by Solid-State LTD," *IEEE Transactions on Plasma Science*, vol. 42, no. 11, pp. 3603–3608, 2014.

[2] T. Huiskamp and J. J. Van Oorschot, "Fast Pulsed Power Generation With a Solid-State Impedance-Matched Marx Generator: Concept, Design, and First Implementation," *IEEE Transactions on Plasma Science*, vol. 47, no. 9, pp. 4350–4360, 2019.

[3] D. Malviya and M. Veerachary, "Voltage Gain Enhancement of Solid-State Pulse Modulator by Adapting Sequential cum Parallel Charging of Capacitors," *IEEE Transactions on Industrial Electronics*, vol. 69, no. 7, pp. 6838–6849, 2022.

[4] W. Jiang, W. Diao, and X. Wang, "Marx generator using power mosfets," in *2009 IEEE Pulsed Power Conference*, 2009, pp. 408–410.

[5] D. Woog, M. Barnes, J. Holma, and T. Kramer, "Prototype Inductive Adder for the Proton Synchrotron at CERN," in *2018 IEEE International Power Modulator and High Voltage Conference (IPMHVC)*, 2018, pp. 464–468.

[6] L. M. Redondo, M. Zahyka, and A. Kandratsyeu, "Solid-State Generation of High-Frequency Burst of Bipolar Pulses for Medical Applications," *IEEE Transactions on Plasma Science*, vol. 47, no. 8, pp. 4091–4095, 2019.

[7] L. Pang, T. Long, K. He, Y. Huang, and Q. Zhang, "A Compact Series-Connected SiC MOSFETs Module and Its Application in High Voltage Nanosecond Pulse Generator," *IEEE Transactions on Industrial Electronics*, vol. 66, no. 12, pp. 9238–9247, 2019.

[8] M. Azizi, J. J. Van Oorschot, and T. Huiskamp, "Ultrafast Switching of SiC MOSFETs for High-Voltage Pulsed-Power Circuits," *IEEE Transactions on Plasma Science*, vol. 48, no. 12, pp. 4262–4272, Dec. 2020.

[9] M. Feizi, T. Huiskamp, and B. Vermulst, "An Improved Gate-Boosting Gate Driver for Ultrafast Switching of GaN Transistors for Nanosecond Pulse Generation," in *2024 10th Euro-Asian Pulsed Power Conference, 25th International Conference on High-Power Particle Beams and 20th International Symposium on Electromagnetic Launch Technology (EAPPC/BEAMS/EML)*, 2024, pp. 1–2.

[10] GaNPower. (2021) GaNPower Demonstrates Industry's First 1200 V Single-Die E-Mode GaN Power Devices. [Online]. Available: https://iganpower.com/ganpower-demonstrates-industrys-first-1200-v-single-die-e-mode-gan-power-devices

[11] J. Millán, P. Godignon, X. Perpiñà, A. Pérez-Tomás, and J. Rebollo, "A Survey of Wide Bandgap Power Semiconductor Devices," *IEEE Transactions on Power Electronics*, vol. 29, no. 5, pp. 2155–2163, 2014.

[12] S. Roy, C. Deng, and A. J. Hanson, "A Simple, High-Speed Measurement Technique for Dynamic on-resistance of GaN Devices for Hard-Switched Pulsed Power Applications," in *2023 IEEE 24th Workshop on Control and Modeling for Power Electronics (COMPEL)*, Ann Arbor, MI, USA, Jun. 2023, pp. 1–8.

[13] R. Risch and J. Biela, "Low Voltage GaN-Based Gate Driver to Increase Switching Speed of Paralleled 650 V E-mode GaN HEMTs," in *2020 22nd European Conference on Power Electronics and Applications (EPE'20 ECCE Europe)*, Sep. 2020, pp. 1–11.

[14] J. Ma, L. Yu, L. Ren, W. Xu, J. Ma, S. Dong, and C. Yao, "Ultrafast Gate Driver With GaN HEMTs for ns-Pulse Generator Using SiC MOSFET," *IEEE Transactions on Plasma Science*, vol. 51, no. 10, pp. 2771–2780, Oct. 2023.

[15] L. Collier, T. Kajiwara, J. Dickens, J. Mankowski, and A. Neuber, "Fast SiC Switching Limits for Pulsed Power Applications," *IEEE Transactions on Plasma Science*, vol. 47, no. 12, pp. 5306–5313, Dec. 2019.

[16] P. Anthony, N. McNeill, and D. Holliday, "High-speed resonant gate driver with controlled peak gate voltage for silicon carbide mosfets," in *2012 IEEE Energy Conversion Congress and Exposition (ECCE)*, 2012, pp. 2961–2968.

[17] M. Hochberg, M. Sack, and G. Mueller, "Analyzing a gate-boosting circuit for fast switching," in *2016 IEEE International Power Modulator and High Voltage Conference (IPMHVC)*, San Francisco, CA, USA, Jul. 2016, pp. 171–175.

[18] B. Li, J. Bai, L. Fan, J. He, J. Tu, F. Yang, H. Wang, S. Liu, and Z. Zhang, "Tailing phenomenon in high voltage solid-state pulse modulators: Analyzing and modeling," *AIP Advances*, vol. 13, no. 8, pp. 1–10, Aug. 2023.

[19] IXYS, "IXD_630 30-Ampere Low-Side Ultrafast MOSFET Drivers," IXD_630 datasheet, Rev. R04.

[20] W. Eberle, Z. Zhang, Y.-F. Liu, and P. C. Sen, "A Current Source Gate Driver Achieving Switching Loss Savings and Gate Energy Recovery at 1-MHz," *IEEE Transactions on Power Electronics*, vol. 23, no. 2, pp. 678–691, 2008.

[21] J. Styles — GaN Systems, "Common misconceptions about the body diode," Power & Efficiency Handbook, Oct. 2019.

[22] Texas Instruments, "LMG1210 200-V, 1.5-A, 3-A half-bridge MOSFET and GaN FET driver with adjustable dead time for applications up to 50 MHz," LMG1210 datasheet, Nov. 2018 [Revised Jan. 2019].

[23] GaN Systems, "GS66508B Bottom-side cooled 650 V E-mode GaN transistor," GS66508B datasheet, Apr. 2020.

[24] Wolfspeed, "C3M0120065D Silicon Carbide Power MOSFET C3M^{TM} MOSFET Technology N-Channel Enhancement Mode," C3M0120065D datasheet, Dec. 2023 [Rev. 03].

[25] Texas Instruments, "LMG1020 7-A/5-A single channel gate driver with 5-V UVLO for nano second input pulses," LMG1020 datasheet, Feb. 2018 [Revised Oct. 2018].

[26] B. Wang, R. Zhang, Q. Song, H. Wang, Q. He, Q. Li, F. Udrea, and Y. Zhang, "Gate Robustness and Reliability of P-Gate GaN HEMT Evaluated by a Circuit Method," *IEEE Transactions on Power Electronics*, vol. 39, no. 5, pp. 5576–5589, 2024.

[27] Texas Instruments. LMG1020 PSpice Transient Model (Rev. C). [Online]. Available: https://www.ti.com/lit/zip/snom618

[28] ——. LMG1210 Unencrypted PSPICE Transient Model. [Online]. Available: https://www.ti.com/lit/zip/snom677

[29] Efficient Power Conversion. EPC2111 - Enhancement-Mode GaN Power Transistor Half Bridge LTspice Device Model. [Online]. Available: https://epc-co.com/epc/documents/spice-files/LTSPICE/EPCGaNLibrary.zip

[30] J. Wang, H. S.-h. Chung, and R. T.-h. Li, "Characterization and Experimental Assessment of the Effects of Parasitic Elements on the MOSFET Switching Performance," *IEEE Transactions on Power Electronics*, vol. 28, no. 1, pp. 573–590, 2013.

[31] R. Risch and J. Biela, "Solid-State Marx Generator vs. Linear Transformer Driver: Comparison of Parasitics and Pulse Waveforms for Nanosecond Pulsers," in *2021 IEEE Pulsed Power Conference (PPC)*, 2021, pp. 1–12.

[32] S.-M. Park and H.-J. Ryoo, "Pulsed Power Modulator With Active Pull-Down Using Diode Reverse Recovery Time," *IEEE Transactions on Power Electronics*, vol. 35, no. 3, pp. 2943–2949, Mar. 2020.

[33] Y. Feng, T. Sugai, A. Tokuchi, and W. Jiang, "Solid-State Linear Transformer Driver Using Inductive Energy Storage," *IEEE Transactions on Plasma Science*, vol. 48, no. 9, pp. 3188–3192, 2020.

979-8-3315-1612-3/25 $31.00 © 2025 IEEE

A Discrete Multilevel Active Gate Driver for GaN HEMTs to Optimize the Switching Behavior

1st Celine Lawniczak
Chair of Energy Conversion
TU Dortmund University
Dortmund, Germany
celine.lawniczak@tu-dortmund.de

2nd Martin Pfost
Chair of Energy Conversion
TU Dortmund University
Dortmund, Germany
martin.pfost@tu-dortmund.de

Abstract–**A simple implementation of a cost-effective multilevel gate driver employing solely commercially available components is introduced. Using the proposed driver, the system is able to adjust the applied gate voltage and gate resistance individually with five stages to successfully control the switching transient of GaN HEMTs. This is beneficial to experimentally determine optimized gate switching schemes in the laboratory. An investigation of seven different multilevel driving schemes with varying timing in the switching patterns and resistance combinations was conducted. The results demonstrate that the discrete multilevel gate driver exhibits superior performance in terms of faster turn-on transients with reduced turn-on energy losses as well as overshoots and undershoots, in comparison to conventional gate drivers with a constant gate resistance.**

Index Terms—**Active gate driver, switching slew rate control, multi-level gate driver, switching energy, threshold voltage, Miller plateau**

I. INTRODUCTION

Wide bandgap power semiconductors, in particular gallium nitride high electron mobility transistors (GaN HEMTs), have a favorable trade-off between blocking voltage capability and area-specific on-state resistance [1]. GaN HEMTs exhibit reduced parasitic capacitances compared to their Si- and SiC-based counterparts, resulting in superior switching characteristics and faster transitions. Nevertheless, elevated switching frequencies give rise to considerable dv/dt and di/dt transients, which in turn give rise to increased voltage and current overshoot as well as electromagnetic interference (EMI) issues [2].

In order to fully exploit the potential of GaN HEMTs, it is necessary to implement an active gate driver, which can be achieved by modifying the gate resistance [3], gate current [4], [5], or gate voltage [6] in order to control the switching waveform. Such gate drivers can be implemented as an integrated circuit (IC) as demonstrated in [3], [7], [8], or with discrete components [6], [9], [10].

The majority of multilevel gate drivers has been developed for SiC to date. They are not applicable to GaN due to the differing requirements concerning time resolution and driving

strength. Driver concepts employed in GaN applications typically aim to minimize reverse conduction losses, rather than actively influencing the switching curve during the transient. Moreover, fast switching transients of GaN HEMTs are in the range of 10 ns, making high speed a critical requirement for the gate driver [11].

Therefore, this paper proposes a discrete gate driver design that offers high flexibility in the gate signal generation. It offers a low cost, fast and easy-to-implement way to enhance the switching performance, solely through the use of commercially available components. The proposed gate driver is able to control the applied voltage level and the gate resistance with a minimal pulse width of 7 ns and a pulse width resolution of 150 ps. The multilevel gate driver represents a particularly promising approach for optimization in the laboratory to minimize switching losses while maintaining an acceptable EMI performance.

II. PROPOSED MULTILEVEL GATE DRIVER

The multilevel gate driver principle is shown in Fig. 1. It uses a TMS320F280049 microcontroller (µC) to generate multiple pulses with a high temporal resolution as small as 150 ps between the rising and falling edge. However, as the µC has a clock frequency of 100 MHz, the minimum PWM pulse width is 10 ns and thus limits the precise control of the switching waveform of the gate. Therefore, a CPLD (XCR3064XL) is used to combine different µC signals so that pulses with a total length of as small as 7 ns can be generated.

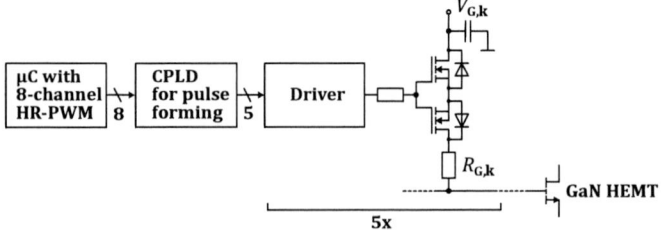

Fig. 1. Schematic of the proposed multilevel gate driver. Note that five separate driver stages are available, each with its own (positive or negative) voltage $V_{G,k}$ and gate resistance $R_{G,k}$ for highest flexibility.

The CPLD controls five separate driver stages, each comprising a driver and two NMOS transistors. The output voltage of the µC and CPLD is 3.3 V. Accordingly, a driver stage is incorporated to elevate the voltage, thus guaranteeing a dependable transition between the on and off states of the NMOS transistors. The selection of driver stages is based on the criterion of minimal device-to-device propagation delay. Nevertheless, the propagation delay between the individual devices is compensated in a calibration step.

As stated, the driver stage is followed by two NMOS transistors arranged in an anti-parallel configuration to prevent unwanted current from flowing through the body diodes. The NMOS stage supplies the individual $V_{G,k}$ and $R_{G,k}$ to ultimately trigger the switching of the GaN HEMT.

The µC is capable of generating a total of 8 PWM outputs, which can be combined in multiple configurations to generate different drive schemes, controlling the effective gate resistance R_G and gate voltage V_G during the transient of the gate-source voltage V_{GS}. One typical driving scheme and its idealized waveforms are presented in Fig. 2, which demonstrates how R_G can be increased or decreased during the switching transient.

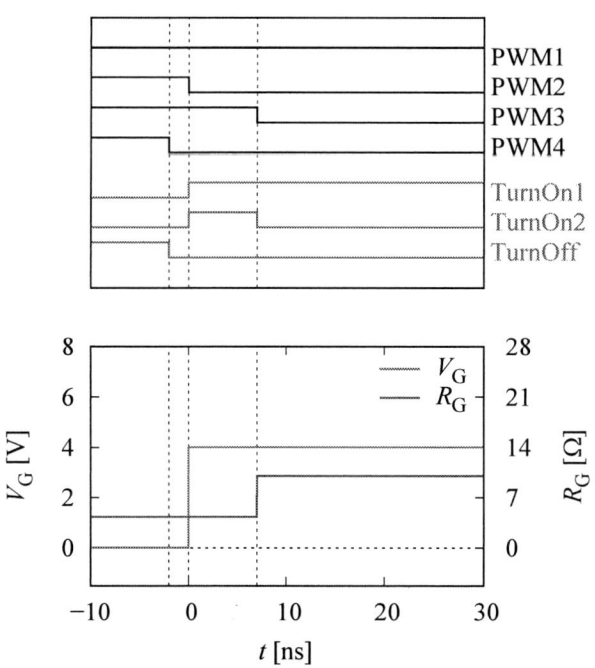

Fig. 2. Top: Generation of the turn-on and turn-off signals for driving scheme M1. Bottom: Resulting effective gate voltages V_G and resistances R_G for the driving scheme.

In order to generate the *TurnOn1* signal for driving scheme M1, illustrated in Fig. 2, the CPLD is programmed to combine the signals *PWM1* and *PWM2* using XOR logic. The *TurnOn2* signal is then generated by *PWM2* XOR *PWM3*. In order to generate the *TurnOff* signal, the XNOR function is applied to *PWM1* and *PWM4*. Due to the delay between *PWM2* and *PWM4*, a dead time between the turn-on and turn-off signals can be adjusted according to the switching circuit.

All components can be programmed easily. One of the five stages of the driver network acts as a continuous signal supply

to ensure that the gate of the GaN HEMT never floats between switching operations of the other driver stages, which are either activated or deactivated as necessary to accommodate a specific gate signal. The number of gate driving stages provides a high degree of flexibility during the testing and characterization of the gate driver.

III. SWITCHING PATTERNS FOR OPTIMIZED GATE DRIVING

The combination of the gate-source capacitance, the parasitic inductance of the gate loop, and the gate resistance form a resonant circuit. To obtain fast drain-source voltage V_{DS} and source current I_S transitions in hard-switching applications, three different gate driving schemes, denoted as M1, M2, and M3, were investigated. The effective gate voltages V_G and gate resistances R_G over timeare shown in Fig. 2 for M1 and in Fig. 3 for M2 and M3.

It is often suggested that the gate resistance must start low and then be increased to higher values to minimize current overshoot and oscillation. Driving scheme M1, presented in Fig. 2, is designed to accomplish this by choosing the point of tranistion of R_G at the time when V_{GS} is slightly above the Miller-plateau voltage V_{Miller}. Upon reaching V_{Miller}, the switching current I_S is at its nominal value. Fig. 4 illustrates the switching behavior of a GaN HEMT showing that the applied R_G during the start and along the Miller-plateau phase determines the overshoot of I_S and the transition time of V_{DS}.

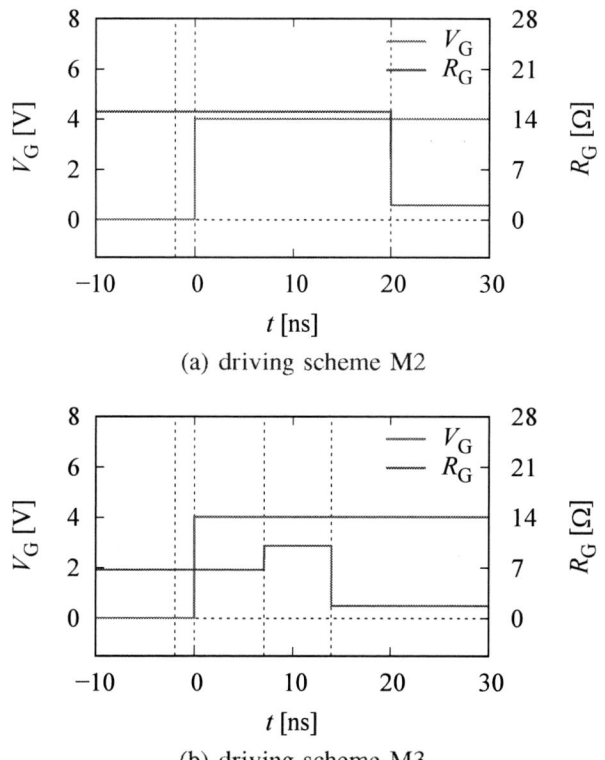

(a) driving scheme M2

(b) driving scheme M3

Fig. 3. Resulting effective gate voltages V_G and resistances R_G for the two driving schemes M2 (a) and M3 (b).

Driving scheme M2 in Fig. 3(a) follows the approach to reduce E_{on} by lowering R_G shortly after reaching V_{Miller} as shown in Fig. 4, whereby the overshoot of I_S is maintained at a relatively small level, in alignment with the initial higher gate resistance. The subsequent reduction of R_G allows to increase the rise of V_{GS} and thus the fall of V_{DS}. Consequently, the overall switching time of the GaN HEMT is reduced.

Driving scheme M3 represents a synthesis of the preceding approaches, now with two changes of R_G, see Fig. 4. As illustrated in Fig. 3(b), driving scheme M3 commences with a relatively low turn-on resistance in order to achieve a fast di/dt and is afterwards increased to dampen the current overshoot. However, the increase in resistance must be reversed in the subsequent phase in order to reduce the value of E_{on}, as was achieved in driving scheme M2. This reduction of R_G allows V_{GS} to rise more rapidly to the nominal value of 4 V, thereby facilitating an earlier stabilization of the transient, decreasing E_{on}.

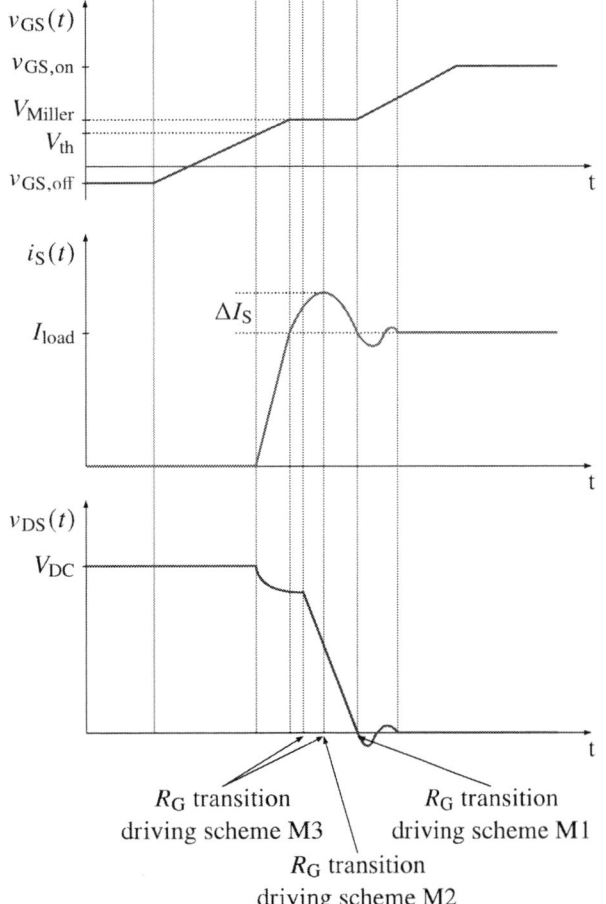

Fig. 4. Schematic of the switching curve of V_{GS}, I_S and V_{DS} during turn-on and the transition point of R_G for driving schemes M1, M2, and M3 in relation to the switching behavior of a GaN HEMT.

IV. MEASUREMENT RESULTS

The proposed discrete multilevel gate driver (see Fig. 5) is tested in a double pulse setup. The tests were performed up to 400 V and 20 A.

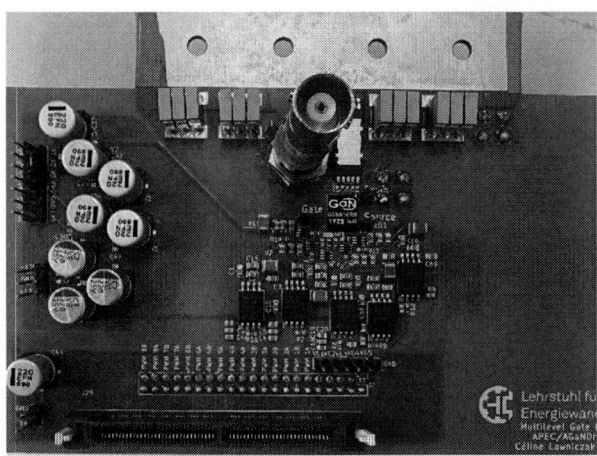

Fig. 5. Discrete gate driver

The discrete multilevel gate driver comprises one *TurnOff* and four *TurnOn* stages, each with a fixed resistor value (see Tab. I). When switching the stages separately or in parallel, a large range of effective resistance values for R_G can be obtained as shown in Tab. II.

TABLE I
RESISTANCE VALUES OF THE FOUR DIFFERENT TURNON STAGES $R_{G,1}$ TO $R_{G,4}$.

$R_{G,1}$	$R_{G,2}$	$R_{G,3}$	$R_{G,4}$
2 Ω	7.5 Ω	10 Ω	20 Ω

TABLE II
EFFECTIVE RESISTANCE VALUES R_G THAT CAN BE OBTAINED THROUGH THE PARALLELIZATION OF THE INDIVIDUAL STAGES.

$R_{G,1} \parallel R_{G,2}$	$R_{G,1} \parallel R_{G,3}$	$R_{G,1} \parallel R_{G,4}$
1.6 Ω	1.7 Ω	1.8 Ω

$R_{G,2} \parallel R_{G,3}$	$R_{G,2} \parallel R_{G,4}$	$R_{G,3} \parallel R_{G,4}$
4.3 Ω	5.5 Ω	6.7 Ω

Fig. 6 shows the result of a multilevel driving scheme based on M1, see Fig. 2, and a conventional driving with $R_G = 2\,\Omega$. For M1, the gate resistance transitions from 4.3 Ω to 10 Ω after 7 ns. The point at which V_{GS} reaches approximately 3.2 V in Fig. 6 and is thus above V_{Miller} which is at approximately 3 V.

In Fig. 6, the current overshoot of I_S is reduced from 38.3 A to 36.3 A with a negligible decrease in di/dt. The slope rate of V_{DS} is damped by approximately 9 V/ns. However, due to the combination of a damped switching current and voltage transition, the turn-on energy E_{on} remains at a constant level between the reference of 2 Ω and the active resistance of 4.3 Ω in driving scheme M1. Upon reaching the end of the Miller-plateau phase, with the potential of V_{DS} at 0 V for the first time, the *TurnOn2* signal is deactivated, and the applied R_G increases to 10 Ω. This demonstrates the most significant enhancement in driving scheme M1 with a reduction in the undershoot of V_{DS} to -1.1 V compared to -20.9 V for the $2\,\Omega$ gate resistor.

Finally, a reduction of the switching energy during turn-on from $56\,\mu J$ to $36\,\mu J$ is achieved while maintaining a similar peak value of the current.

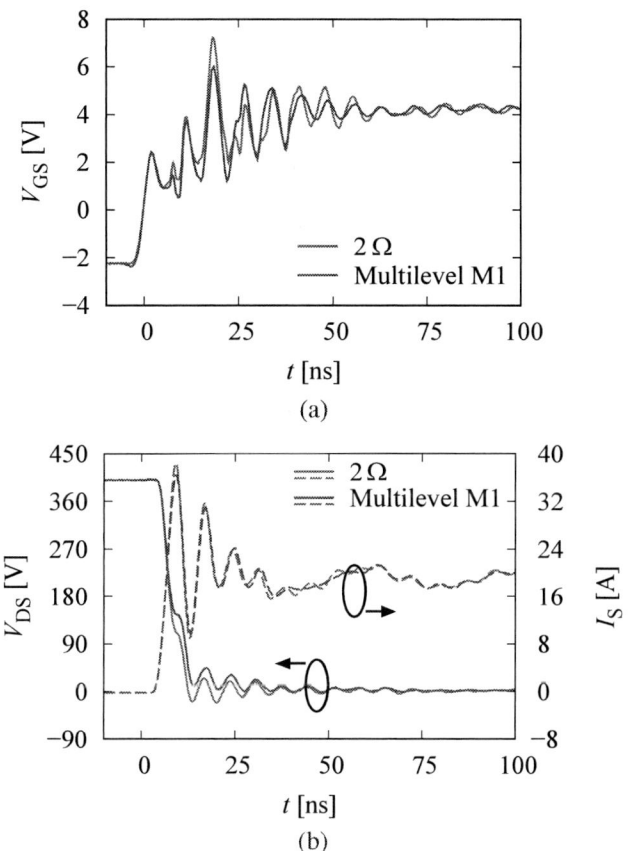

Fig. 6. (a) Gate-source voltage V_{GS} and (b) drain-source voltage V_{DS} (solid) and source current I_S (dashed) during turn-on for continuous gate driving with a $2\,\Omega$ on-state resistance and driving scheme M1 as shown in Fig. 2.

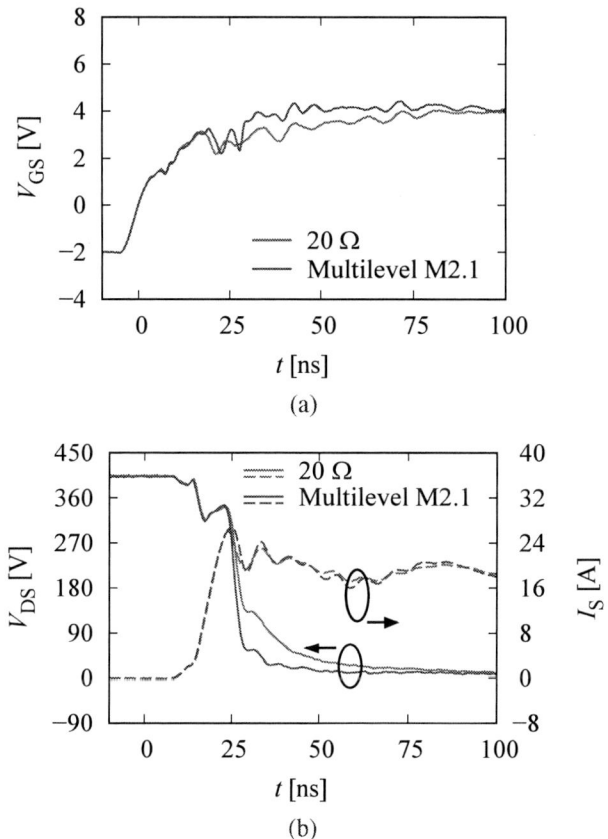

Fig. 7. As Fig. 6, but now with a $20\,\Omega$ on-state resistance and multilevel driving scheme M2.1 of Fig. 3(a)

A reduction in turn-on energy while maintaining a constant peak current can be achieved by decreasing the gate resistance once the Miller-plateau is reached, see Fig. 4, thereby accelerating dv/dt. Therefore, several approaches based on the driving scheme M2 are investigated as shown in Tab. III.

TABLE III
ACTIVATED RESISTANCE VALUES R_G FOR FIVE DIFFERENT MULTILEVEL DRIVING SCHEMES BASED ON FIG. 3(A)

	M2.1	M2.2	M2.3	M2.4	M2.5
Before transition	$20\,\Omega$	$20\,\Omega$	$20\,\Omega$	$10\,\Omega$	$10\,\Omega$
After transition	$6.7\,\Omega$	$5.5\,\Omega$	$1.8\,\Omega$	$4.3\,\Omega$	$1.7\,\Omega$

In driving scheme M2.1, R_G is reduced from $20\,\Omega$ to $6.7\,\Omega$ after 22 ns to maintain the overshoot of I_S at the peak of a driver with constant $R_G = 20\,\Omega$, see Fig. 7. Lowering R_G to $6.7\,\Omega$ afterwards accelerates the rise of V_{GS} and thus the fall of V_{DS} as can be seen in Fig. 7(a) and (b). The total rise time of V_{GS} is shortened by 30 ns, resulting in a switching time of 31 ns, thus accelerating the dv/dt transient of V_{DS} by 8 V/ns.

Fig. 8 illustrates the impact on the switching curve of V_{DS} and I_S of executing the decrease of R_G at an earlier or later point in time, as opposed to shortly after reaching the Miller plateau as shown in Fig. 7. The earlier the transition, the sooner R_G is reduced to $6.7\,\Omega$, the larger is the overshoot of I_S see Fig. 8(a). Similarly, Fig. 8(b) illustrates that if the reduction to $6.7\,\Omega$ is initiated too late, the advantage of minimizing switching energy may diminish as the switching behavior of $20\,\Omega$ assumes a more significant role and the turn-on energy continues to increase. The optimal transition point is precisely when I_S reaches its maximum with a $20\,\Omega$ driving. This can also be seen in Fig. 9 and Fig. 10.

E_{on} is being compared to the peak value of I_S at 200 V, 300 V, and 400 V for driving schemes M2.1 to M2.5. As has already been identified, applying just $20\,\Omega$ for an insufficient duration results in an increased current overshoot. Subsequent activation allows for a further reduction in current overshoot. This is accompanied by an increase in turn-on energy, yet the reduction remains significant in comparison to the corresponding reference value of $20\,\Omega$ in Fig. 9 and $10\,\Omega$ in Fig. 10.

It can be observed that, regardless of the load conditions, the multilevel driving approach M2 consistently exhibits an

979-8-3315-1612-3/25 \$31.00 © 2025 IEEE

inflection point at which the optimal reduction in E_{on} and a minimal overshoot can be achieved. When the multilevel approach is consistently applied at the optimal time, E_{on} can be minimized by at least 18.9% and up to 46.9% .

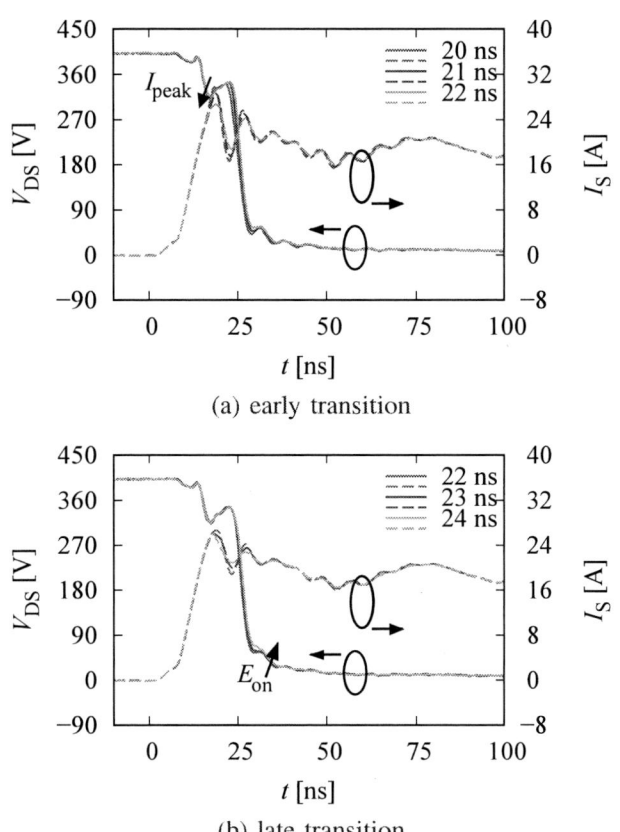

(a) early transition

(b) late transition

Fig. 8. Impact on V_{DS} and I_S when the transition is executed (a) earlier or (b) later compared to Fig. 7(b)

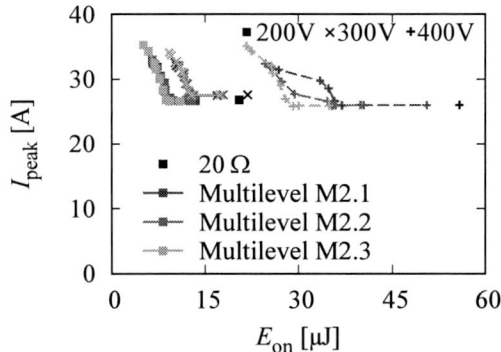

Fig. 9. E_{on} compared to the peak value of I_S for 200 V, 300 V, and 400 V at 20 A load current for driving schemes M2.1, M2.2 and M2.3 at different points of transition for R_G and 20 Ω

In the context of driving schemes M1 and M2, it was established that the overshoot and the turn-on energy can be reduced through the utilization of the multilevel driver. Driving scheme M3 starts with a turn-on resistance of 6.7 Ω through the parallelization of *TurnOn1* and *TurnOn2*. Following a

period of 7 ns, *TurnOn2* is deactivated, resulting in an increase in the active resistance at the gate to 10 Ω. This enables the current overshoot to be maintained at the peak of a 10 Ω reference value, while simultaneously increasing the di/dt by 39%, see Fig. 11.

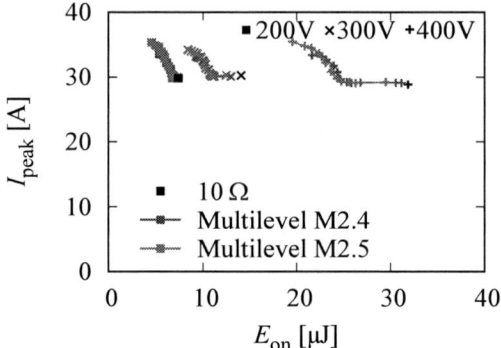

Fig. 10. E_{on} compared to the peak value of I_S for 200 V, 300 V, and 400 V at 20 A load current for driving schemes M2.4 and M2.5 at different points of transition for R_G and 10 Ω

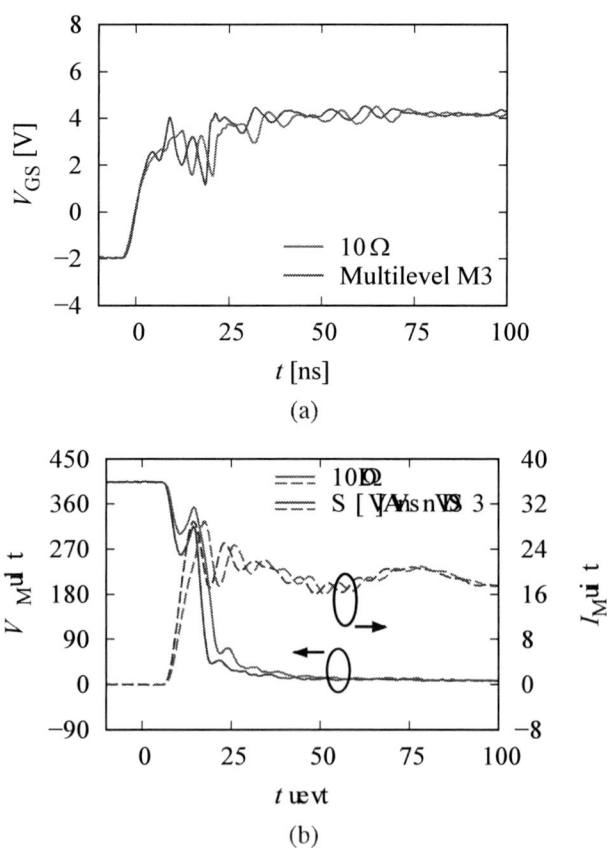

(a)

(b)

Fig. 11. As Fig. 6, but with a 10 Ω on-state resistance and the multilevel driving scheme M3 as shown in Fig. 3(b)

979-8-3315-1612-3/25 $31.00 © 2025 IEEE   1133

TABLE IV
Evaluation of multilevel driving approaches at 400 V and 20 A load current

Driving Approach / Parameter	I_{peak} [A]	E_{on} [µJ]	di/dt [A/ns]	dv/dt [V/ns]
2 Ω	38.3	15.4	7.49	53.1
M1	36.3	15.3	7.4	44.1
20 Ω	26.0	55.9	2.3	9.7
M2.1	25.9	35.5	2.2	17.2
M2.2	25.9	35.5	2.2	17.7
M2.3	25.8	29.2	2.4	23.8
10 Ω	28.8	31.9	3.3	17.0
M2.4	29.0	25.9	3.0	24.4
M2.5	29.2	24.8	3.1	26.0
M3	28.8	26.7	4.6	20.7

In order to reduce the value of E_{on}, the resistance is reduced shortly afterwards, as shown in Fig. 4, by activating *TurnOn3* in parallel to *TurnOn1*, which results in a reduction of the effective resistance at the gate to 1.67 Ω, thus reducing E_{on} by 5.2 µJ.

Tab. IV presents the evaluation of multiple switching parameters for the investigated multilevel gate driver schemes and the conventional gate driver references. In comparison to a conventional gate driver, all control methods result in an improved switching transition. Approach M1 dampens the undershoot of V_{DS} by 95% and the overshoot of I_S by 5% while maintaining a similar level of E_{on}. Furthermore, approach M2 is centered on the reduction of E_{on} with a distinct emphasis on regulating the di/dt and dv/dt rate. The losses of E_{on} can be decreased by up to 46.9% for M2.1 to M2.3 in comparison to a conventional gate driver, as the dv/dt transient is increased by a factor of 2.5. Method M3 introduces an additional switching event, thus enabling both to enhance the acceleration rate of di/dt and dv/dt while simultaneously reducing E_{on} without any adverse effect on the overshoot of I_S.

V. Conclusion

A simple and fast implementation of a cost-effective multilevel gate driver employing solely commercially available components is introduced. Using the proposed driver, the system is able to adjust V_G and the applied R_G for each of the five stages to successfully control the switching transient of GaN HEMTs. An investigation of seven different multilevel approaches with different switching patterns and resistance combinations was conducted. The results demonstrate that the discrete multilevel gate driver exhibits superior performance in terms of faster turn-on transients with reduced turn-on energy losses as well as overshoots and undershoots, in comparison to conventional gate drivers with a constant gate resistance.

References

[1] M. Ishida, T. Ueda, T. Tanaka, and D. Ueda, "GaN on Si technologies for power switching devices," *IEEE Transactions on Electron Devices*, vol. 60, no. 10, pp. 3053–3059, Oct. 2013.

[2] D. Reusch and J. Strydom, "Understanding the effect of PCB layout on circuit performance in a high-frequency gallium-nitride-based point of load converter," vol. 29, no. 4, pp. 2008–2015, Apr. 2014.

[3] H. Dymond, J. Wang, D. Liu, J. Dalton, N. McNeill, D. Pamunuwa, S. Hollis, and B. Stark, "A 6.7-GHz active gate driver for GaN FETs to combat overshoot, ringing, and EMI," vol. 33, no. 99, pp. 581 – 594, Jan. 2018.

[4] C. Geng, D. Zhang, X. Wu, W. Shen, and R. Dong, "A novel active gate driver with auxiliary gate current control circuit for improving switching performance of high-power SiC MOSFET modules," in *2020 IEEE 1st China International Youth Conference on Electrical Engineering (CIYCEE)*, Nov. 2020, pp. 1–7.

[5] P. Nayak and K. Hatua, "Active gate driving technique for a 1200 V SiC MOSFET to minimize detrimental effects of parasitic inductance in the converter layout," in *2016 IEEE Energy Conversion Congress and Exposition (ECCE)*, Sep. 2016, pp. 1–8.

[6] H. Takayama, S. Fukunaga, and T. Hikihara, "Binary-weighted modular multi-level digital active gate driver," in *2023 25th European Conference on Power Electronics and Applications (EPE'23 ECCE Europe)*, Sep. 2023, pp. 1–8.

[7] T. Liu, R. P. Martins, and Y. Lu, "A 600-V GaN active gate driver with level shifter common-mode noise sensing for built-in dv/dt self-adaptive control," in *2023 35th International Symposium on Power Semiconductor Devices and ICs (ISPSD)*, May 2023, pp. 195–198.

[8] W. J. Zhang, J. Yu, W. T. Cui, Y. Leng, J. Liang, Y.-T. Hsieh, H.-H. Tsai, Y.-Z. Juang, W.-K. Yeh, and W. T. Ng, "A smart gate driver IC for GaN power HEMTs with dynamic ringing suppression," *IEEE Transactions on Power Electronics*, vol. 36, no. 12, pp. 14 119–14 132, Dec. 2021.

[9] J. Nagao, J. Furuta, and K. Kobayashi, "Capacitor-based three-level gate driver for GaN HEMT only with a single voltage supply," in *2020 IEEE 21st Workshop on Control and Modeling for Power Electronics (COMPEL)*, 2020, pp. 1–7.

[10] R. Das and H.-P. Le, "Gate driver circuits with discrete components for GaN-based multilevel multi-inductor hybrid converter," *IEEE Transactions on Industrial Electronics*, vol. 70, no. 2, pp. 1105–1114, Feb. 2023.

[11] Y. Chen and D. B. Ma, "A 10-MHz closed-loop EMI-regulated GaN switching power converter using emulated Miller plateau tracking and adaptive strength gate driving," *IEEE Journal of Solid-State Circuits*, vol. 56, no. 2, pp. 531–540, Feb 2021.

Attenuation of Fundamental Component of Differential Mode Noise Using Active EMI Filter

Guru Abhilash Mulumudi, *Graduate Student Member IEEE,*
Naveed Ishraq, *Student Member IEEE,* Ayan Mallik, *Senior Member IEEE,*
Power Electronics and Control Engineering Laboratory (PEACE), Arizona State University, Mesa,
USA Email: gmulumud@asu.edu, nishraq@asu.edu, amallik3@asu.edu

Abstract—This paper introduces a boost converter in conjunction with a synchronized switch mode Active EMI Filter (AEF), which reduces energy storage requirements compared to a passive EMI filter. The proposed AEF operating at a frequency of 30 MHz effectively mitigates additional EMI into the system as its operational frequency lies beyond the typical range of conducted EMI. The AEF is realized using a synchronous buck converter with a series resonant tank and the current configuration is designed to counteract the fundamental ripple component of the boost converter. Duty-controlled switching is implemented in the proposed AEF using high-speed analog circuits to generate pulse width modulated (PWM) signals for the filter. The proposed methodology controls the magnitude of AEF current to a desired value using a modulating signal ($\hat{m}$) in an open loop to reduce the complexity of the circuit. The AEF is employed in a (6-12)V-to-24V boost converter switching at 150 kHz and has been found to attenuate the fundamental ripple component by 23dB based on the experimental results.

Keywords— *Active EMI filter (AEF), Electromagnetic Interference (EMI), Differential Mode (DM) noise, Fundamental Attenuation of the input ripple current*

I. INTRODUCTION

Switched-mode power converters operate at frequencies ranging from tens to hundreds of kilohertz and tend to generate significant conducted EMI within the regulated frequency band of 0.15–30 MHz.. Converters typically require an input filter to comply with electromagnetic compatibility standards to prevent high-frequency currents from traveling through the power source conductors. The traditional EMI filters, usually made of passive components, which are substantial in size, sometimes occupying as much as one-third to one-fourth of the total volume, limiting the power density of the power converters [1]. The considerable size of EMI filters stems from various factors. Firstly, the need for a high attenuation level arises from the substantial currents and stringent filtering standards [2]. Secondly, when parasitics are considered [3], the significant bandwidth required for filtering is influenced by the band-pass or band-stop characteristics exhibited by all physical reactive components. Moreover, the regulated frequency range spans two orders of magnitude, contributing to the complexity. Lastly, filter components often face challenges blocking high DC input voltages or handling large DC input currents. It poses difficulties for inductors (which already occupy significant space) and capacitors (which encounter voltage de-rating issues with high DC voltages). An alternative to bulky passive EMI filters is the utilization of active electronics, which inject voltages or currents to counteract the interference signal. These types of active filters usually consist of linear amplifiers in a feedback loop that measures input ripple voltage or current and injects an opposing voltage or current to compensate [4]–[12]. In addition, these conventional AEF with linear amplifier circuits have high power consumption as it depends on the maximum current to be filtered. The methodology described in [13] uses a synchronous switch mode AEF, which results in a smaller EMI filter than passive filters, higher efficiency than AEFs based on linear amplifiers, and very low circuit complexity. It is attributed to the AEF being connected to a much lower voltage than the primary converter voltage, effectively reducing its size.

The study conducted by [14] discusses operating the active EMI filter within a high-frequency range (30-300 MHz), where a switching amplifier would offer significantly enhanced efficiency without introducing its own conducted EMI. The emergence of fast and efficient gallium nitride (GaN) transistors, along with recently discovered magnetic materials exhibiting excellent performance at tens of MHz, enables the realization of such an amplifier. In [14], the methodology incorporates a fractional order filter (FOF) in the compensator design to achieve the desired attenuation for fundamental and other harmonic components. However, there is still a need to develop an AEF with less complexity in control and size.

The papers above discuss the attenuation of noise utilizing active components by injecting current from the filter circuit and highlight that injecting current in phase with the noise current leads to effective cancellation. This paper proposes an AEF utilizing a switch-mode amplifier circuit in place of traditional linear amplifiers, achieving higher efficiency. It replaces the closed-loop compensator with a duty-controlled filter model that aims to synchronize currents in an open loop. The key contributions of the research are:

- Proposed a differential mode noise cancellation mechanism using an active EMI filter operating at a high frequency of 30 MHz.
- Designed an analytical modeling of the filter resonant circuit to determine the filter parameters.
- Designed analog circuit to generate PWM signals using crystal oscillators, differential amplifiers, and high-speed comparators; the combined functionalities of which were verified using LTspice and PLECS simulation, followed by hardware implementation and integration with AEF network.

979-8-3315-1612-3/25 $31.00 © 2025 IEEE

Section II covers the design and operation of the proposed AEF, highlighting its key features and advantages over traditional linear amplifier designs. Section III details the fully analog implementation of the AEF control and modulation system, which enables efficient operation and control of the filter. In Section IV, we present steady-state simulation results alongside experimental findings to validate the performance of the AEF. Section V discusses the volumetric advantage of the proposed AEF, comparing its compactness and efficiency to conventional filter designs. Finally, Section VI provides conclusions drawn from this work and suggests directions for future research, focusing on further improvements in the attenuation of higher order harmonics.

II. PROPOSED SWITCHED MODE AEF

Fig. 1 illustrates the proposed AEF with the main converter which in our study is selected to be a boost converter operating in continuous conduction mode at a switching frequency of 150 kHz. The fundamental structure of AEF is realized by a synchronous buck converter switching at 30 MHz, utilizing wide-bandgap (WBG) GaN devices. The choice of AEF operating frequency is made in a way that no additional conducted EMI is introduced into the system due to AEF switching. Our proposed approach specifically targets the attenuation of the fundamental component of the main converter ripple current by injecting current from the filter. The key idea behind the proposed AEF design lies in ensuring that the current through the AEF (I_f) must be in phase with the main converter inductor ripple current (I_L) and with equal magnitudes. When synchronized, this approach tends to cancel out the superposition of AEF and the main converter fundamental component, resulting in the input supply current (I_S) with an attenuation in its switching frequency component.

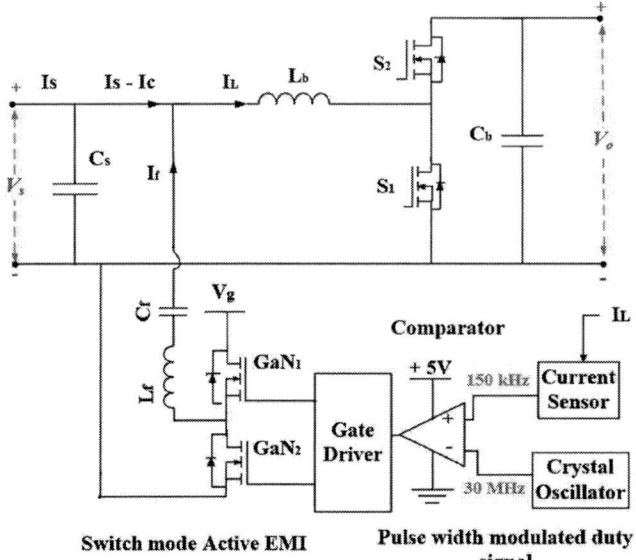

Fig. 1. Schematic diagram of proposed switch mode AEF

Two current components are generated during switching at the resonant tank: (a) one with the resonant frequency and (b) the other with the switching frequency. The magnitude of the switching frequency component is negligible, given its high

frequency compared to the resonant frequency of the series tank, and thus, it is not addressed in this study. The resonant frequency component eventually decays to zero due to internal parasitic resistance in the circuit.

Table I Boost Converter Parameters		Table II AEF Parameters	
Parameter	**Value**	**Parameter**	**Value**
V_{in}	(6-12) V	V_g	12 V
V_{out}	24 V	F_d	150 kHz
D_b	0.5	F_{sw}	30 MHz
Power	75 W	L_f	1 μH
ΔI_L	1.76A	C_f	1 μF
L_b	22.7 μH	$\overline{D}$	0.3
C_b	6.54 μF	$\widetilde{D}$	0.15

To generate a current at 150 kHz, the duty cycle of the input to the filter must be adjusted accordingly with a frequency of 150 kHz by appropriately tuning the values of inductance (L), capacitance (C), and input filter voltage. After synchronization, the generated AEF current matches the magnitude of the IL, effectively attenuates the fundamental component to a desired level. The parameters of the boost converter and the AEF at their rated conditions are listed in the table below.

A. Modeling of AEF

In Fig. 2, the filter circuit consists of a series LC resonant tank whose gain depends on switching frequency. The LC parameters are to be designed based on the desired filter current. The input voltage across the filter can be assumed as the superposition of voltage due to the 30 MHz switching component, average DC voltage from duty cycle-controlled input, and a sinusoidal component with frequency equal to that of duty cycle variation..

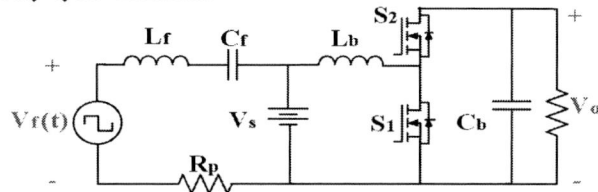

Fig. 2. Single line diagram of Test System

$$V_f(t) = \overline{D} + \widehat{D} \cdot \sin(W_d t) \tag{1}$$

Where D(t) is the duty cycle input to the filter consisting of the offset value, $\overline{D}$ and $\widetilde{D}$ is the amplitude of the sinusoidal component, W_d is the frequency of duty cycle variation ($2\pi \times 150 kHz$). The duty cycle equation in (1) is obtained by comparing a modulating signal, $\widetilde{m}$ with a 30 MHz triangular signal both having a DC offset. The voltage generated across the half-bridge of AEF with duty D(t) & input voltage V_g can be written as

$$V_f(t) = V_g * [\overline{D} + \widehat{D} \cdot \sin(W_d t)] \tag{2}$$

979-8-3315-1612-3/25 $31.00 © 2025 IEEE   1136

Assuming the filter circuit to be equivalent to a series RLC circuit, the filter current can be found by using phasor analysis,

$$I_f = \frac{V_g * \widehat{D}}{|Z|} \quad \text{where} \quad |Z| = \sqrt{R_p^2 + (W_d L_f - \frac{1}{W_d C_f})^2} \quad (3)$$

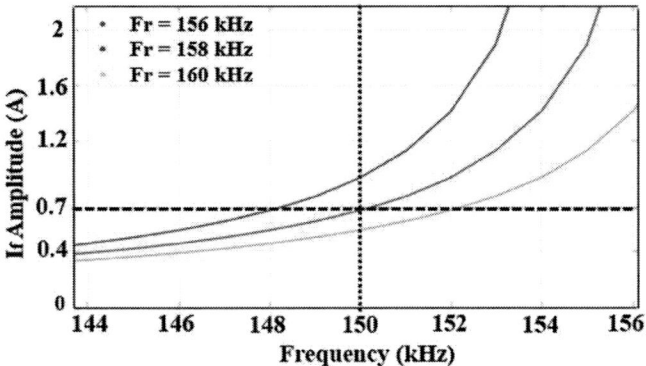

Fig. 3. Variation of AEF current to operating frequency

As the half-bridge is switched at a very high frequency compared to the resonant frequency, it can be assumed that the high-frequency components in the filter current can be neglected. The AEF current in equation (3) is generated due to input source $V_f(t)$. It can be observed that the circuit offers infinite impedance to the DC component and the resulting I_f consists of a continuous sinusoidal component whose amplitude depends on V_g, $\widehat{D}$, and $|Z|$. Assuming the parasitic resistance R_p to be 1mΩ, and L_f to be 1µH, the amplitude of I_f is plotted in Fig. 3 by varying operating frequencies at different resonant frequencies (F_r) of AEF series tank. It can be clearly stated from equation (3) that the I_f magnitude can be adjusted as needed by linear variation of V_g and $\widehat{D}$.

The parameters are tuned to achieve an AEF current of 0.77A, which corresponds to the fundamental component of the current I_L at rated conditions. To attain an AEF current of 0.77A at a frequency of 150kHz, F_r of (150kHz) is selected, resulting in a capacitance value of 1µF. It is found that the phase of I_f is leading $\widehat{m}$ by almost 90 degrees due to the nature of the series LC tank and operating frequency.

B. Mathematical Model of Input Inductor Current in a Boost Converter and its FFT Analysis

The input side current spectrum of the boost converter is required to be derived for designing an active EMI filter. This section presents the mathematical modeling of the input side current FFT analysis. The equation for the input current can be expressed as,

$$I_L = I_{L,fund} + I_{L,ripple} \quad (3)$$

Where the fundamental component is given by,

$$I_{L,fund} = I_{pk} * \sin(W_d t) \ \& \ W_d = 2\pi * 60 \ rad/s \quad (4)$$

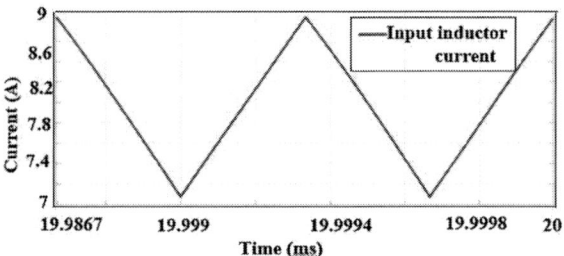

Fig. 4. Input Inductor Current Waveform for Two Switching Cycles

Based on Fig. 5, the input inductor current is a triangular waveform with an average value shown for two switching cycles. The spectrum of this waveform in Fig. 6 clearly shows the average component, fundamental switching frequency component, and other harmonics. The objective of the AEF in this work, as mentioned earlier, is only to attenuate this fundamental component.

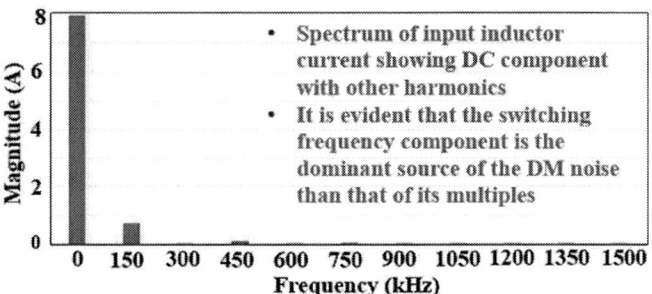

Fig. 5. Spectrum of Input Inductor Current

C. Phase Synchronization of Currents for Cancellation of Noise Component

The complex task involved in the implementation is the perfect cancellation of both the I_L and I_f which are to be synchronized. From the equations derived for the filter current, the phase of the filter current corresponding to the 150 kHz component from duty cycle variation must be determined. For deriving the phase angle of the inductor current to its voltage, it is obvious that the time when the increasing triangular current waveform meets its average value will be the required phase. The phase angle of the input inductor ripple current to inductor voltage can also be derived analytically which results in finding out the delay to be given to the PWM pulses of the boost converter for synchronization. This implies that both currents will be in phase and tend to cancel out effectively. As the currents in phase are triangular and sinusoidal, it can be stated that this only helps in eliminating fundamental components of the ripple current at 50% duty with a desired attenuation.

$$I_{L,v} = I_{L,avg} - \frac{\Delta I_L}{2}$$

$$= I_{L,avg} - \frac{\Delta I_L}{2} \quad (5)$$

During this time,

$$I_L(t) = I_{Lv} + \frac{\Delta I_L}{D * T} * t \qquad (6)$$

At $\alpha = W_d \times t$,

$$I_L(t) = \overline{I_L} = \frac{I_o}{1-D} \qquad (7)$$

By using (5) and substituting (6) in (7),

$$\alpha = \Pi \times D_b \qquad (8)$$

For deriving the phase difference between the inductor current and its voltage, it is obvious that the angle corresponding to the instant when the increasing triangular current waveform meets its average value will be the required phase. This states that the phase angle of I_L to V_L depends on the converter duty ratio. The key waveforms in the system depicted in Fig. 6 implies that the phase difference will be $D_b\Pi$ as derived in (8), which is 90° for a converter with D_b=0.5. In closed loop realization with input voltage transient, the phase can be adjusted based on required duty ratio, updated by the input voltage sensor data.

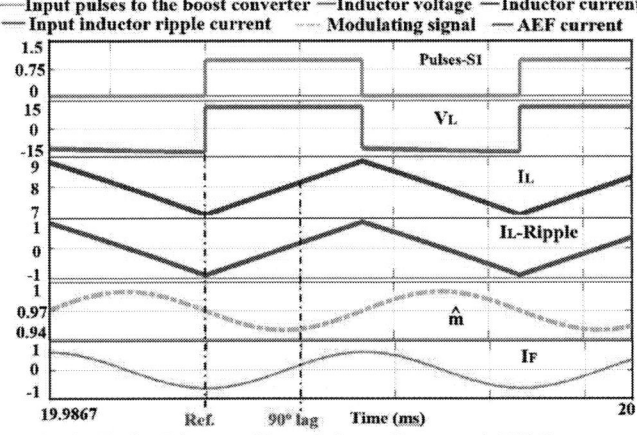

Fig. 6. Timing Diagram of Input Inductor Current and AEF Current

Observation reveals that aligning I_L and $\hat{m}$ in phase leads to the filter current consisting of 150 kHz component, to lead $\hat{m}$ by 90°, as demonstrated in the waveform above. It becomes apparent that adjusting the generated m^to lag by 90° relative to I_L results in the perfect synchronization of I_f and I_L. This is because when the AEF half-bridge is switched with PWM pulses, the phase angle of AEF current depends on operating frequency, and in this case the frequency is 150kHz. The resonant frequency of the AEF is selected in such a way that the operating frequency falls below the resonant frequency resulting in the current leading duty signal by 90°.

III. FULLY ANALOG IMPLEMENTATION OF THE AEF CONTROL AND MODULATION

The proposed AEF is designed to filter the DM current from a boost converter operating with an input voltage of (6-12) V and 50% duty ratio. The converter is operated in CCM mode and 150 kHz switching. Since the generated filter current is sinusoidal, it is evident that this concept only helps in the cancellation of the fundamental component in I_L despite having other harmonics in its spectrum. The main objective of this implementation is to drive the half bridge of AEF with PWM duty signals. The analog circuit is developed to generate $\hat{m}$ and 30MHz carrier signals.

The generation of the $\hat{D}$ involves comparing $\hat{m}$ with a 30 MHz triangular wave to achieve pulse control for switching the AEF half-bridge. Due to the high-frequency requirement of 30 MHz and the duty cycle variation of 150 kHz, utilizing a DSP controller for comparing these waveforms becomes challenging. The period value at such high frequencies becomes significantly short, leading to phase resolution issues when comparing both signals.

To overcome this issue, an analog circuit is designed and developed which generates a modulating signal and a 30 MHz signal to obtain duty-controlled pulses for the AEF as shown in Fig. 7. For the duty input, generating a DC offset value of 1V along with $\hat{m}$ with a peak-peak voltage of 495 mV is necessary to produce a filter current with a peak-to-peak ripple of 1.6A, matching that of the input inductor's fundamental ripple current. The DC offset value is decided based on the amplitude of the unipolar triangular signal generated by the crystal oscillator followed by the RC circuit. The amplitude of $\hat{m}$ is decided based on the current amplitude to be cancelled, which is directly calculated from the magnitude of $\hat{D}$, and the peak value of 30 MHz triangular signal. A current sensor CT431-HSWF20DR, with a resolution of 0.1V/A, is utilized to sense I_L, and its output is passed through a DC blocking unity gain amplifier circuit that makes the output signal with zero bias. The obtained output is fed into a differential amplifier to adjust the gains of output modulating accordingly. In this stage, an inverter is connected to one of the inputs of differential amplifier that helps in having DC bias in the modulating signal. The DC bias is introduced to compare with triangular 30 MHz signal from crystal oscillator which has a default DC bias. TL081H operational amplifier is used in above mentioned circuits.

For generating a 30 MHz signal, a crystal oscillator is used that produces the required signal with square wave output with amplitude equal to 3.3V (supply VDD of the oscillator). This is passed through an RC filter to get triangular output which is then compared with a $\hat{m}$ to obtain pulse width modulated duty cycle. The IC used for the comparator is LTC6752-3 as shown in Fig. 8 works at very high frequency with hysteresis control. The additional advantage of this comparator is that it produces complementary outputs for switching GaN devices in the half-bridge AEF circuit. To verify the hardware performance of the analog circuit and AEF, the board is powered and the signal level outputs of the circuit along with switch node voltages are captured.

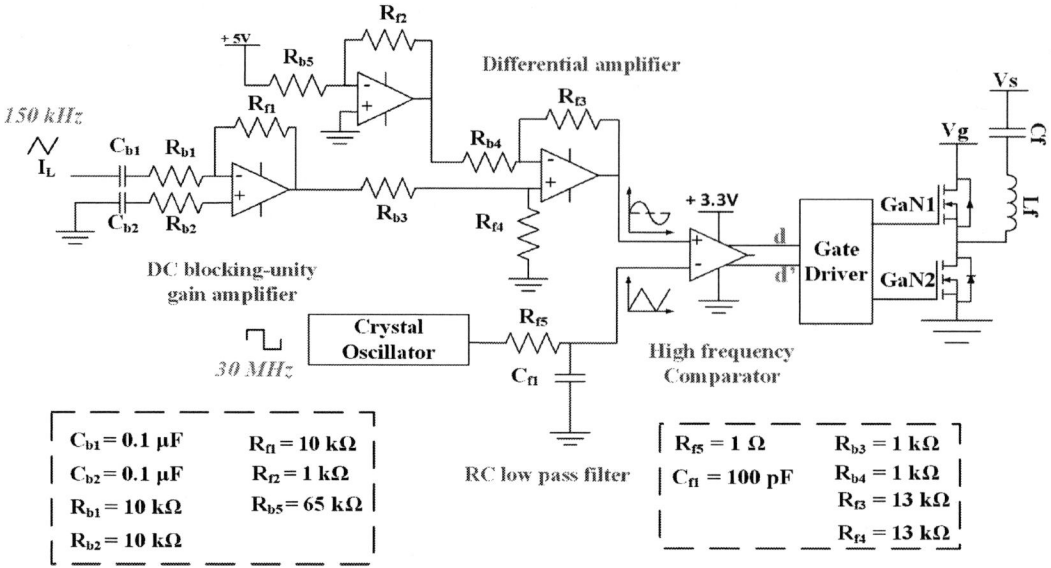

Fig. 7. Analog Circuit for Obtaining Pulses to the AEF

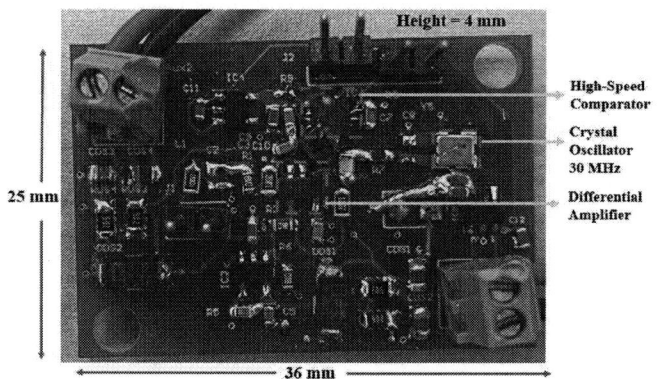

Fig. 8. Analog Circuit PCB for Realizing AEF Control

IV. STEADY STATE SIMULATION AND EXPERIMENTAL RESULTS

In this section, the simulation results are presented at two different operating conditions and are compared at rated conditions as mentioned.

A. Simulation Results

The simulation is carried out in the PLECS software to analyse the nature of the proposed system modeling. It can be seen in Fig. 9 that the input inductor current without the DC component is in phase with the filter current resulting in the cancellation of its fundamental switching component. From the simulation results shown, it can be observed that the I_L and I_f are synchronized. It should be noted that the currents are compared after removing the DC value to only observe AC ripple currents. It is observed that the ripple current amplitude of the resulting DC source current is reduced significantly. It should be noted that AEF current needs to have the magnitude

that is equal to fundamental component of I_L for perfect cancellation. To calculate the attenuation, spectrum information of both the currents is compared and the results are shown in Fig. 10.

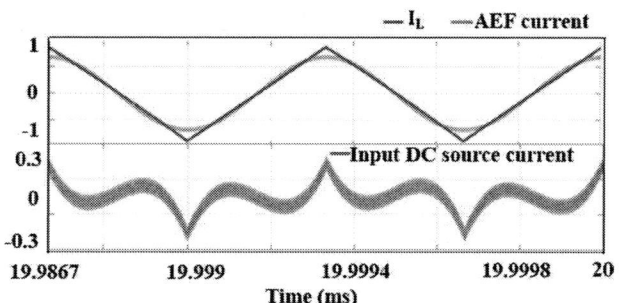

Fig. 9. Simulation Result of Effective Cancellation of Fundamental Switching Harmonic Component

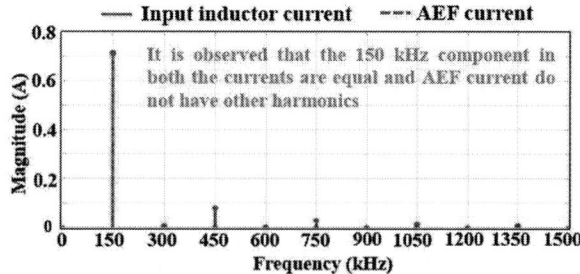

Fig. 10. FFT Spectrum of IL and AEF Current

The fundamental magnitudes are made equal and resulting in cancellation of DM noise current. The spectrum of input source mains current is shown in Fig. 11 consisting of attenuated fundamental noise current leaving other order harmonics.

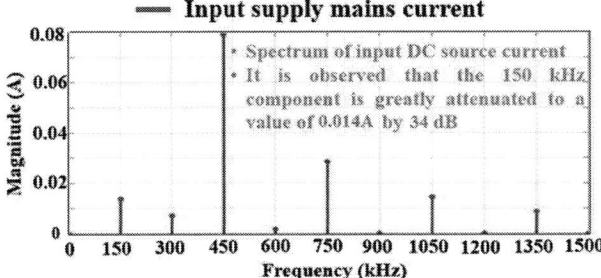

Fig. 11. FFT Spectrum of Input supply mains current

B. Hardware prototyping and Results

To verify the performance of the filter designed, a converter proof-of-concept rated at 75W is developed. Pulses for the boost converter are generated using a DSP microcontroller, Texas Instruments TMS320F28335. The converter is operating at a switching frequency of 150 kHz. For switching the AEF at 30 MHz, GaN devices EPC8004 are used that are driven by high-speed gate driver IC LMG1210 which can operate up to 50 MHz. The analog circuit designed is combined with AEF board, and integrates to the boost converter. Fig. 12 shows the hardware prototype implementation including different components used. The different components used in the hardware prototype are highlighted. In Fig. 13, AEF and boost converter developed in this work are presented along with some of the key components used.

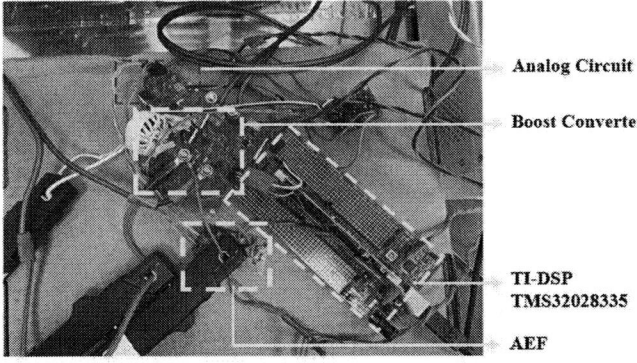

Fig. 12. Hardware Test Setup of AEF along with the Boost Converter

The experimental waveform for a 12V-24V conversion, at 75W load is illustrated in Fig. 14. It can be clearly seen that the AEF current synchronizes with I_L fundamental (of ~0.77A), hence resulting in ~100% attenuation of the input mains current (I_S) fundamental. The FFT spectrum of (I_S) with and without AEF is presented in Fig. 15.

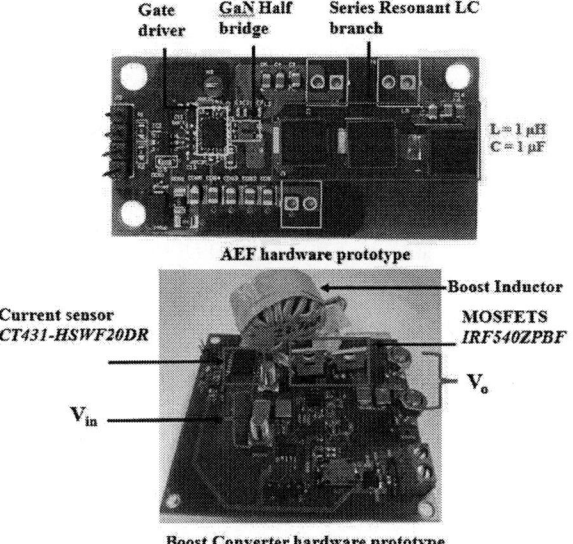

Fig. 13. AEF and Boost Converter in the Proposed System

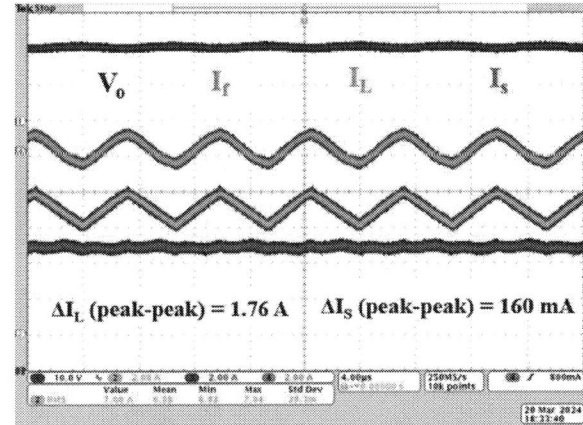

Fig. 14. Experimental Results of the system at 50% Duty, 75W Power (i) Output Voltage (Vo), (ii) AEF Current (If), (iii) Inductor Current (IL), (iv) Input Supply mains current (Is)

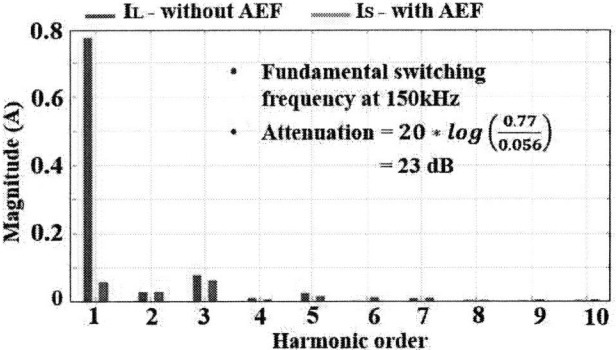

Fig. 15. FFT of Input Mains DC Current with and without AEF

979-8-3315-1612-3/25 $31.00 © 2025 IEEE

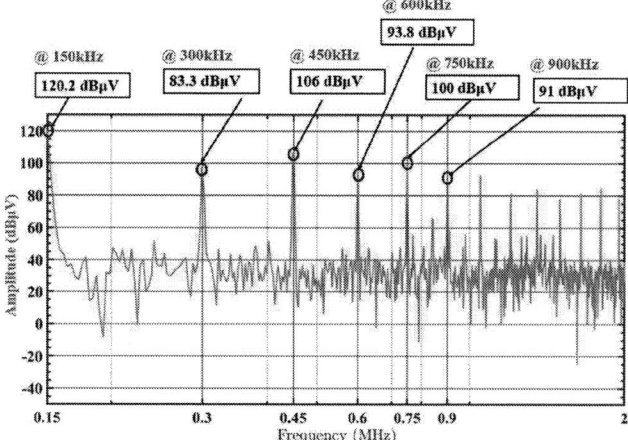

Fig. 16. EMI Spectrum of Input Mains DC Current without AEF

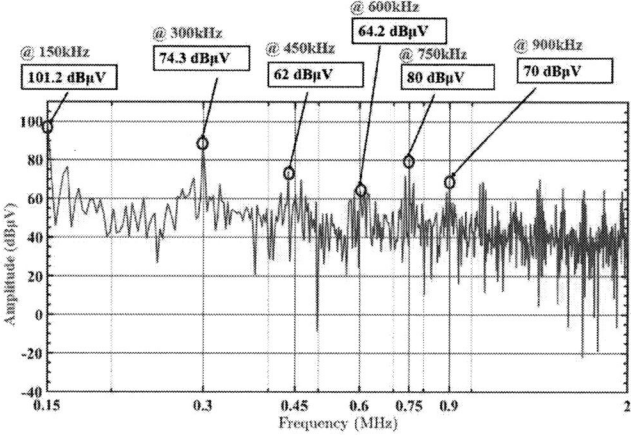

Fig. 17. EMI Spectrum of Input Mains DC Current with AEF

Fig. 16 and Fig. 17 shows the EMI spectrum of I_S without AEF and with AEF respectively. These are captured using Line Impedance Stabilization Network (LISN) and Spectrum analyzer at the positive and neutral lines of the input of main converter. The fundamental switching frequency component in I_S is clearly attenuated by nearly 20dB, aligning closely with the calculated current magnitudes. It should be highlighted that in the absence of AEF, no capacitor is integrated at the input of the boost converter. This configuration is intentional as the presence of a substantial input inductor, when paired with such a capacitor, could form a passive EMI filter. This filter significantly attenuates the switching frequency component, making it impossible to isolate the effect of the AEF.

Upon the inclusion of the AEF, a small capacitor (100nF) is added across the input, a requirement for the LISN to function optimally. Consequently, a passive EMI filter is established, with a corner frequency closely aligned with the fundamental switching frequency, thus providing a mild attenuation to this component. However, the primary attenuation of the fundamental component is accomplished by the AEF, while the attenuation of other harmonics is due to the presence of passive filter. It is crucial to note that the fundamental frequency component's attenuation is directly attributable to the AEF, whereas the additional harmonics are mitigated through the passive filtering mechanism.

V. VOLUMETRIC ANALYSIS OF THE PROPOSED AEF

The calculation shows that nearly 23dB of fundamental attenuation is achieved using AEF. To verify the impact of AEF design, we present a comparative analysis of AEF and a passive EMI filter in terms of component size. The designed AEF yielded inductance (L) and capacitance (C) values of 1μH and 1μF, respectively. In contrast, the passive EMI filter is utilized for the same attenuation as provided by AEF, L=10μH, and C=1.5μF. The obtained volume reduction does not account for higher-order harmonics in the designed AEF. To address this, we calculated volumes demonstrating the benefits of a Hybrid Active EMI Filter (HEF) configuration, which combines an Active EMI Filter (AEF) for fundamental differential mode (DM) noise attenuation with a passive filter to handle higher-order harmonics, in comparison to a fully passive EMI Filter (PEF). Using a 23dB attenuation benchmark set by the AEF, we evaluate the volume of HEF relative to that of PEF. Specifically for passive filter used in HEF, attention is given to attenuating the second-order component at 300kHz within the I_L spectrum as the fundamental is taken care by AEF. As there is no defined standard for required attenuation levels in this context, a benchmark of 23dB provided by the AEF is used, representing a worst-case scenario design for the passive filter within HEF. It is worth noting that in practice, the size of this filter obtained is typically smaller than the calculated value.

Table III
Volumetric comparison of AEF vs PEF

| Description | HEF | | PEF |
	AEF	Passive Filter	Fully Passive filter
Frequency of interest	150 kHz	300 kHz	150 kHz
Attenuation required	23 dB	23 dB	23 dB
LC product	1e-12	3.98e-12	1.59e-11
Capacitor volume	3.625 mm³	3.625 mm³	3.625 mm³
Inductor volume	0.2 cm³	1.18 cm³	6.174 cm³
Total volume	6.25 cm³	1.2 cm³	6.177 cm³

The comparative analysis presented in the table explores the volume considerations of Hybrid Electric Filters (HEF) and Passive Electric Filters (PEF) by factoring in how attenuation levels are managed during component design. For the passive element in the HEF, inductor and capacitor values are specified at 2.8 μH and 1.5 μF, respectively, aligned with the targeted corner frequency. These components are chosen based on their current and voltage capabilities from off-the-shelf suppliers. It should be noted that the volume of the HEF aligns closely with that of the PEF, thus offering no significant volumetric benefit at this power level when using an AEF. However, the desired attenuation of 23dB does not meet certain standards at the operating frequency of 150kHz. According to CISPR25, the frequency component should be below 70dBuV. The AEF in this configuration does not sufficiently attenuate the fundamental frequency, necessitating further attenuation by the HEF. This could potentially increase the overall volume of the combined HEF system.

The real potential of the AEF is recognized when adjustments are made to increase attenuation beyond 23dB. This can be achieved through careful tuning of the magnitude and phase synchronization of the interference current I_f. Such enhancements would allow for a reduction in HEF volume relative to the PEF, as the HEF would only need to attenuate non-fundamental harmonics. This adjustment shifts the attenuation burden away from large passive components, which is critical for the feasibility of the AEF, particularly at higher power levels where the volume of AEF circuitry does not scale as significantly as that of fully passive components.

Although the current study does not demonstrate volume effectiveness, the proposed approach of integrating an AEF to handle Differential Mode (DM) noise could potentially surpass traditional passive solutions by offering higher attenuation levels without a corresponding increase in component volume. This prospect underscores the importance of further research and development in AEF technology to enhance its application in noise attenuation scenarios.

VI. CONCLUSION

This paper presents the design and open-loop implementation of an Active EMI Filter (AEF) using a fully analog circuit, comparing it with conventional passive EMI filters in terms of passive component size for a given power level. To validate EMI noise attenuation through AEF, a 75W prototype of a (6-12)V-24V boost converter was developed, operating at a switching frequency of 150kHz with a fundamental inductor current magnitude of 0.77A. The AEF generates a 0.77A sinusoidal current, effectively attenuating ripple current by 23dB. This approach leverages open-loop digital control, eliminating the need for closed-loop bandwidth constraints, resulting in a simpler, cost-effective, and phase resolution-free system.

The current work only attenuates a portion of fundamental component of DM mode noise using active filtering. Further work can made on attenuating the fundamental to a maximum extent just by tuning the AEF parameters and this can be effectively performed at higher power levels for significant attenuation. Other harmonic components can be attenuated using multi-stage active switch mode filter. Multiple AEF's corresponding to each harmonic is employed while elevating volume constraints over passive EMI filters. Internal parasitic resistance of the AEF is assumed to be Multiple AEF stages are employed to attenuate harmonics in I_L. Each stage corresponds to respective harmonic in IL, and

with properly tuned magnitude and phase of AEF currents, attenuation of all harmonics can be possible.

ACKNOWLEDGMENT

This material is based upon work supported by the National Science Foundation under Grant Number 2236846.

REFERENCES

[1] Y. Yang, "EMI Noise Reduction Techniques for High Frequency Power Converters," Ph.D. dissertation, Virginia Tech, 2018. [Online]. Available: http://hdl.handle.net/10919/83372.

[2] A. Singh, A. Mallik and A. Khaligh, "A Comprehensive Design and Optimization of the DM EMI Filter in a Boost PFC Converter," in IEEE Transactions on Industry Applications, vol. 54, no. 3, pp. 2023-2031, May-June 2018, doi: 10.1109/TIA.2018.2789859.

[3] S. Wang, J. D. van Wyk and F. C. Lee, "Effects of Interactions Between Filter Parasitics and Power Interconnects on EMI Filter Performance," in IEEE Transactions on Industrial Electronics, vol. 54, no. 6, pp. 3344-3352, Dec. 2007, doi: 10.1109/TIE.2007.906126.

[4] P. Wang, C. Tao, and J. Zhang, "Research and design of a common mode hybrid EMI filter for switch-mode power supply," in 2009 3rd International Conference on Power Electronics Systems and Applications (PESA), 2009, pp. 1-4.

[5] W. Chen, X. Yang, and Z. Wang, "An active EMI filtering technique for improving passive filter low-frequency performance," IEEE Transactions on Electromagnetic Compatibility, vol. 48, no. 1, pp. 172-177, 2006. doi: 10.1109/TEMC.2006.870803.

[6] M. Ali, E. Labouré, and F. Costa, "Integrated hybrid EMI filter: Study and realization of the active part," in 2013 15th European Conference on Power Electronics and Applications (EPE), 2013, pp. 1-8. doi: 10.1109/EPE.2013.6634697.

[7] A. Kumar, Y. Hou, Y. Ramadass, T. Merkin, T. Hegarty and A. Obidat, "An Active EMI Filter for High-Power Off-Line Applications," 2023 IEEE Applied Power Electronics Conference and Exposition (APEC), Orlando, FL, USA, 2023, pp. 2063-2067, doi: 10.1109/APEC43580.2023.10131427.

[8] C.E. Kurien, T.K. Sindhu, and D.E. Koshy, "Modelling and Implementation of an Active EMI Filter for Conducted EMI Noise Reduction," in 2021 IEEE 5th International Conference on Condition Assessment Techniques in Electrical Systems (CATCON), 2021, pp. 205-210. doi: 10.1109/CATCON52335.2021.9670535.

[9] B. Narayanasamy, H. Peng, Z. Yuan, F. Luo, and Y. Chu, "Zero-Phase-Filtering based Digital Active EMI Filter," in 2020 IEEE 9th International Power Electronics and Motion Control Conference (IPEMC2020-ECCE Asia), 2020, pp. 1910-1917. doi: 10.1109/IPEMC-ECCEAsia48364.2020.9367786.

[10] D. Hamza and P. K. Jain, "Conducted EMI noise mitigation in DC-DC converters using active filtering method," 2008 IEEE Power Electronics Specialists Conference, Rhodes, Greece, 2008, pp. 188-194, doi: 10.1109/PESC.2008.4591923.

[11] A. Vedde, M. Neuburger, C. Cheshire and F. Gliese, "Optimization of a Passive Common Mode EMI Filter by Adding an Active Feedback Loop," 2021 IEEE Southern Power Electronics Conference (SPEC), Kigali, Rwanda, 2021, pp. 1-6, doi: 10.1109/SPEC52827.2021.9709490.

[12] R. Perraud et al., "Active EMI Filter for High Power Converters," 2022 ESA Workshop on Aerospace EMC (Aerospace EMC), Virtual, 2022, pp. 1-6, doi: 10.23919/AerospaceEMC54301.2022.9828852.

[13] D.T. Nguyen, C. Deng, E. Macias, and A.J. Hanson, "Synchronously Switched Active EMI Filter," in 2022 IEEE Energy Conversion Congress and Exposition (ECCE), 2022, pp. 1-8. doi: 10.1109/ECCE50734.2022.9948006.

[14] D.T. Nguyen, E. Macias, and A.J. Hanson, "Active EMI Filter with Switch-Mode Amplifier for High Efficiency," in 2022 IEEE Applied Power Electronics Conference and Exposition (APEC), 2022, pp. 443-450. doi:10.1109/APEC43599.2022.9773582

Graph Neural Network Based Performance Modeling for the Dual Active Bridge Converter with Operational Generalization

Weihao Lei
School of Electrical Engineering
Zhejiang University
Hangzhou, China
22310129@zju.edu.cn

Fanfan Lin
ZJU-UIUC Institute
Zhejiang University
Haining, China
fanfanlin@intl.zju.edu.cn

Xinze Li
Department of Electrical Engineering
University of Arkansas
Fayetteville, AR, USA
xinzel@uark.edu

Xiaokun Bao
School of Electrical Engineering
Zhejiang University
Hangzhou, China
22310151@zju.edu.cn

Xin Zhang
School of Electrical Engineering
Zhejiang University
Hangzhou, China
zhangxin_ieee@zju.edu.cn

Abstract—**Existing Artificial Intelligence (AI) assisted performance modeling approach for the power converter generally has limited generalization capability, restricted by the training data. To address this challenge, this paper proposes an innovative graph neural networks (GNN)-based modeling approach, leveraging on both GNN and Multi-Layer Perceptron. Taking the dual active bridge converter as a design case, the proposed performance modeling approach can be generalized to different operational conditions and modulation strategies. And the mean absolute percentage error for generalizing to out-of-domain cases is no higher than 1% compared to the models specifically trained for those domains. The feasibility of the proposed approach is also validated by 1kW hardware experiments.**

Keywords—*power converter design, artificial intelligence, graph neural network, power converter modeling, current stress.*

I. INTRODUCTION

The dual active bridge (DAB) is a dc-dc converter widely used in power electronics due to its high-power density, high efficiency, bidirectional power flow capability [1], [2]. The topology of DAB is shown in Fig. 1. The common modulation strategies for DAB include extended phase shift (EPS) modulation [3] and triple phase shift (TPS) modulation[4]. The waveforms of the gate driving signals, v_p and v_s of DAB under TPS modulation are shown in Fig. 2, where D_0 is the outer phase shift angle, D_1 and D_2 are the inner phase shift angles. EPS is a simplified version of TPS with D_1 or D_2 equal to 0.

With the development of AI technology, more and more researchers are applying AI algorithms like artificial neural networks (ANN), support vector machines (SVM), and random forests (RF) in data-driven modeling [5]. Despite improvements in accuracy and efficiency, these AI models still exhibit poor generalization capabilities and lack robustness to operational diversity in design outcomes. Specifically, when operating

conditions or the modulation strategies change, it is necessary to retrain a new model from scratch to adapt to the new conditions.

Graph neural networks (GNN) is an emerging neural network model specialized in graphical data [6], which has the potential to extract the information from the topology and node characteristics for complex information learning and representation [7]. The transfer path and node information embedded into GNN enhances its interpretability and generalization. Therefore, this paper is the first to propose a GNN-based performance modeling method that combines GNN and Multi-Layer Perceptron (MLP) [8] for current stress (i_{pk}) modeling of DAB converters. With this method, it is possible to model DAB converters under different operating scenarios by performing a few-shot learning on an existing model with only a small amount of data without training from scratch.

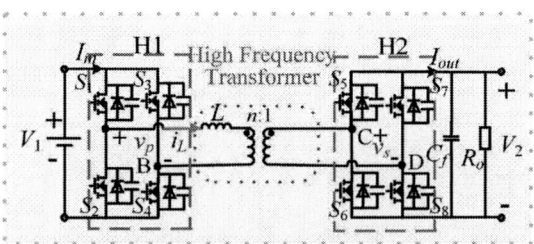

Fig. 1. The topology of DAB converter.

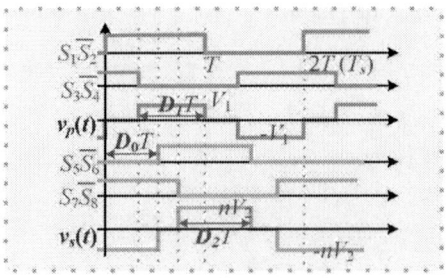

Fig. 2. Waveforms in TPS.

II. METHODOLOGY OF THE PROPOSED GNN-BASED MODELING METHOD

The modeling process of the peak inductor current (i_{pk}) can be divided into three steps. Firstly, the circuit is converted into bond graph to provide standardized graphical input for the GNN. Then, based on the type of circuit components and the circuit values obtained from the simulation, the node feature matrix of the graph is formed. Finally, the node feature matrix and the standardized graphical data are fed into GNN and MLP for modeling i_{pk}. To optimize the current stress to enhance reliability and efficiency, the output of the GNN-based model can serve as function evaluator for the particle swarm algorithm (PSO) till finding the optimal modulation parameters [9].

A. Graph Representation for GNN

Since GNN mainly process graphical data, it is necessary to convert the circuit into graph representation. The key to this process is the abstraction of the components of the circuit and their connections to the nodes and edges of the graph. With bond graph representation [10], [11], as shown in Fig. 3, circuit components (e.g., resistors, capacitors, inductors, etc.) and

series-parallel structures can be represented as nodes, and the connections between these components can be represented as edges. This method not only clearly defines the interactions between the various components but also provides a standardized graph structure for subsequent modeling, significantly enhancing interpretability of the model to a large extent.

B. Feature Assignment for Bond Graph

After converting circuit into graph, the next step is to assign features to each node, creating a node feature matrix that captures key information about the circuit components. As shown in Fig. 4, the node features are concatenated by a one-hot encoding indicating the type of node and an analog value indicating the magnitude of the element value. The analog value can include the parameter values of the circuit components, such as resistance, capacitance as well as the voltage and current values of the nodes, the modulation phase shift angle, and other circuit attributes. As for the nodes with values represented as *xx*, they need to be determined according to the physical significance of the circuit.

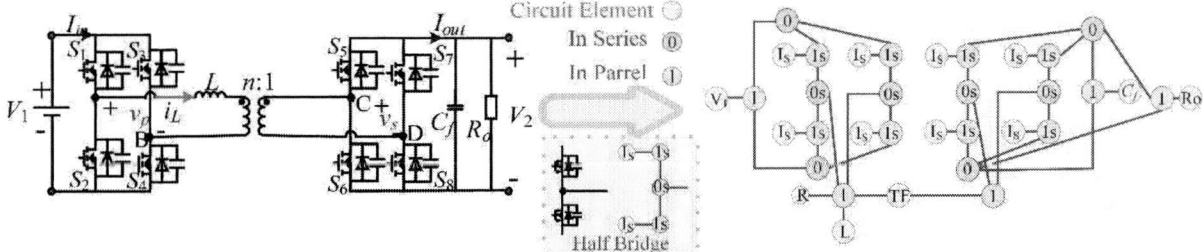

Fig. 3. Bond graph representation.

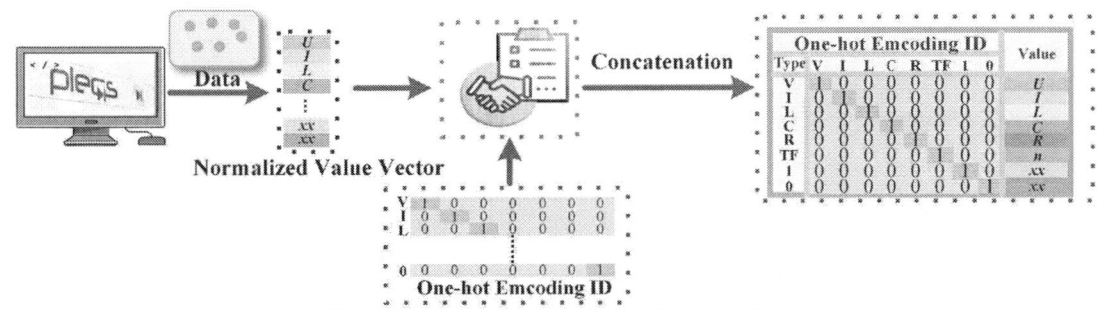

Fig. 4. The process of constructing the feature matrix.

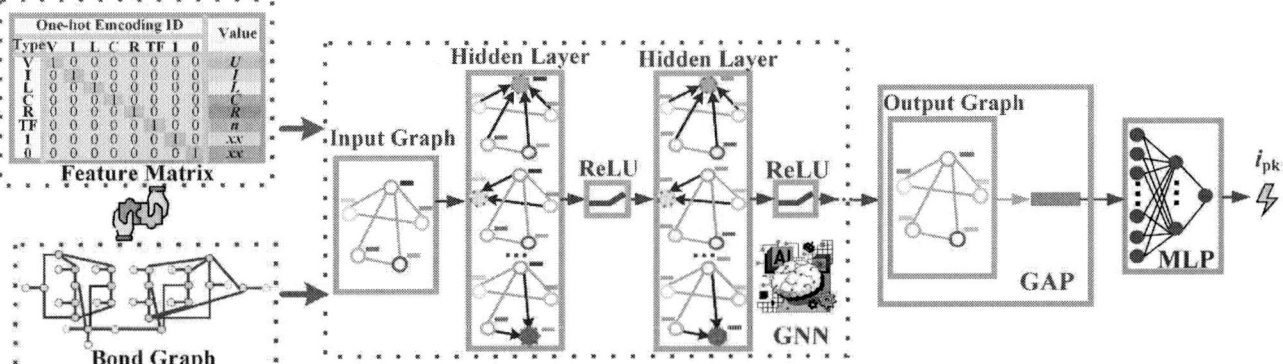

Fig. 5. The structure of GNN-based model.

979-8-3315-1612-3/25 $31.00 © 2025 IEEE
1144

C. Graph Neural Network Based Modeling

The structure of GNN-based model is shown in Fig. 5. GNN is an emerging neural network based on graph theory and consists of three main features: node, edge, and graph. By feeding graphical data with node feature matrix into the GNN, it can efficiently analyze the connectivity and interactions between components (such as inductors, capacitors, switching devices, etc.), extracting the effective feature representation of the circuit. The learned circuit node features are then processed by global average pooling (GAP) layer to obtain the graph feature. The MLP performs complex nonlinear transformations and feature extraction on graph feature data through its multilayer neuron structure.Through the synergistic combination of GNN and MLP, the i_{pk} of the DAB can be accurately predicted.

III. DESIGN CASE AND HARDWARE VALIDATION

In this section, a deign case will be presented to illustrate the modeling process using the GNN-based method. The operating conditions of the DAB converter is presented in Table I. The hardware platform is shown in Fig. 6.

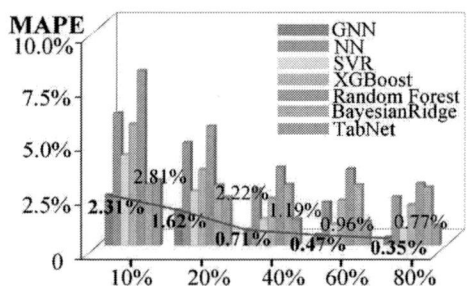

Fig. 7. MAPE of models trained with different size of TPS data.

B. Generalization Capability

The generalization ability is verified from two perspectives: operating conditions and modulation strategies, respectively. For the modulation generalization, the Fine-tuned Model1 is obtained from Model1 by using a small amount (10%) of EPS1, EPS2, TPS data to perform a few-shot learning. For the generalization of operating conditions, the Model6 is subjected to a small amount (10%) of 160V, 200V, and 240V data by

TABLE I. DESIGN SPECIFICATIONS AND MODEL SETTINGS.

Model	Training data	Purpose
Fine-tuned Model 1	EPS1(10%) +EPS2(10%) +TPS (10%)	
Model1	EPS1(80%)	➢ Test **modulation generalization**
Model2	EPS2(80%)	capability of GNN-based method
Model3	TPS (80%)	
Model4	EPS1(80%) + EPS2(80%) +TPS (80%)	
Fine-tuned Model 2	160V (10%) +200V (10%) +240V (10%)	
Model5	160V (80%)	➢ Test **operating generalization**
Model6	200V (80%)	capability of GNN-based method
Model7	240V (80%)	
Model8	160V (80%) +200V (80%) +240V (80%)	
Power range $P_L\in$[100W,1000W]		Output Voltage $V_2\in$[160V,240V]
Modulation Parameters		$D_1\in$[0.68,1] $D_2\in$[0.68,1]
HARDWARE PLATFORM CONFIGURATIONS		
Switching Device	Series C2M0080120D	Dead time 200ns
Transformer	Leakage inductor 61μH	Turn ratio n 1
Control Platform	dSPACE1202	

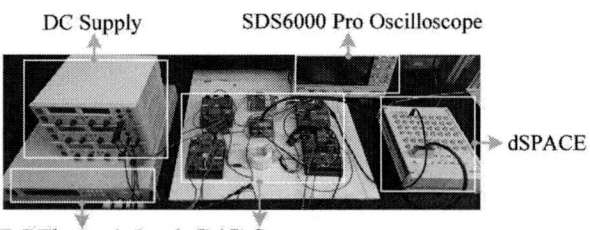

Fig. 6. Hardware Platform

A. Modelling in Single-Modulation Strategy

Fig. 7 shows the mean absolute percentage error (MAPE) of the prediction results from models trained with prevalent AI algorithms using TPS datasets of different sizes. As shown in the Fig. 6, with different amount of data for training, the MAPE of GNN-based method are always lower than other algorithms and the MAPE is lower than 0.5% when the amount of trained data is increased.

means of few-shot learning to obtain the Fine-tuned Model2. The results, shown in Fig. 8 and Fig. 9, demonstrate that the fine-tuned models can be applied to various operating scenarios with a MAPE of less than 1% compared to models trained on specific scenarios.

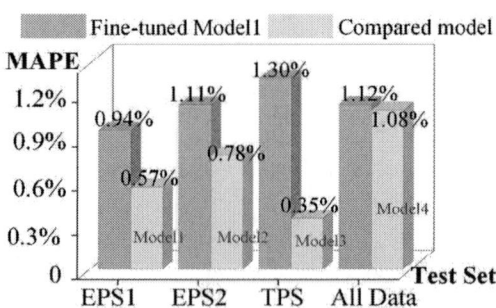

Fig. 8. MAPE for models trained with different modulation data.

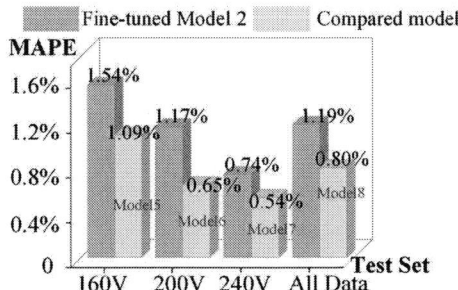

Fig. 9. MAPE for models trained with different operating conditions data.

To verify that the Fine-tuned Model1 can be applied to a variety of modulation strategies, PSO algorithm is employed to interact with the Fine-tuned Model1 in order to obtain the modulation parameters that achieve the lowest current stress for particular operating condition. Fig. 10 shows the steady state operation waveforms corresponding to the optimal modulation parameters of the three modulation modes at P=100W, V_2=160V.

Similarly, to demonstrate the applicability of Fine-tuned Model2 across various operating conditions, Fig. 11 presents the steady state operation waveforms under the optimal modulation parameters at V_2=160V, 200V, and 240V with different power levels (P=100W, 1000W) under TPS modulation. The average difference between the predicted value (i_{pk_gnn}) and measured i_{pk} is 5.78%, which is acceptable due to the inherent bias between the simulation model and the hardware experiment.

IV. CONCLUSION

This paper proposes a GNN-based performance modeling method, taking the DAB converter as a design case. The modeling method has three steps, starting with circuit representation with bond graph, followed by node feature construction and ending with predicting the performance metric i_{pk} with GNN and MLP. Compared with the prevailing AI modeling methods, it exhibits significant advantages in excellent modeling accuracy with MAPE less than 0.5%. And it also presents good generalization capability to new operating conditions and new modulation strategies with MAPE less than 1% compared to the models trained specifically for these scenarios.

ACKNOWLEDGMENT

This work was supported by the Young Scientists Fund of the National Natural Science Foundation of China (Grant No. 52407225).

REFERENCES

[1] S. Shao et al., "Modeling and Advanced Control of Dual-Active-Bridge DC–DC Converters: A Review," in *IEEE Transactions on Power Electronics*, vol. 37, no. 2, pp. 1524-1547, Feb. 2022

[2] C. Sun, X. Zhang and X. Cai, "A Step-Up Nonisolated Modular Multilevel DC–DC Converter With Self-Voltage Balancing and Soft Switching," in *IEEE Transactions on Power Electronics*, vol. 35, no. 12, pp. 13017-13030, Dec. 2020

[3] B. Liu, P. Davari and F. Blaabjerg, "An Optimized Hybrid Modulation Scheme for Reducing Conduction Losses in Dual Active Bridge Converters," in *IEEE Journal of Emerging and Selected Topics in Power Electronics*, vol. 9, no. 1, pp. 921-936, Feb. 2021.

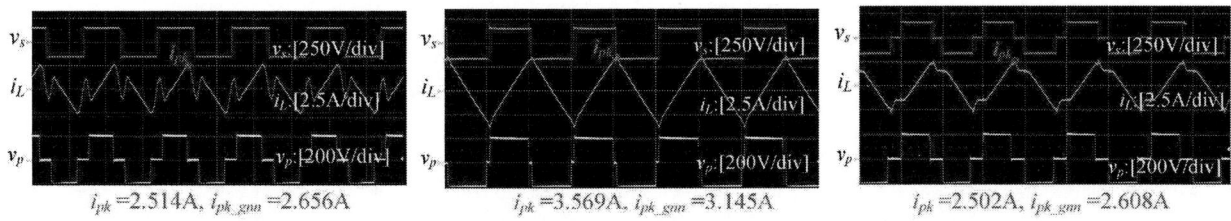

Fig. 10. Waveforms of DAB in different modulations at P=100W, V_2=160V. (a)EPS1 (b)EPS2 (c)TPS.

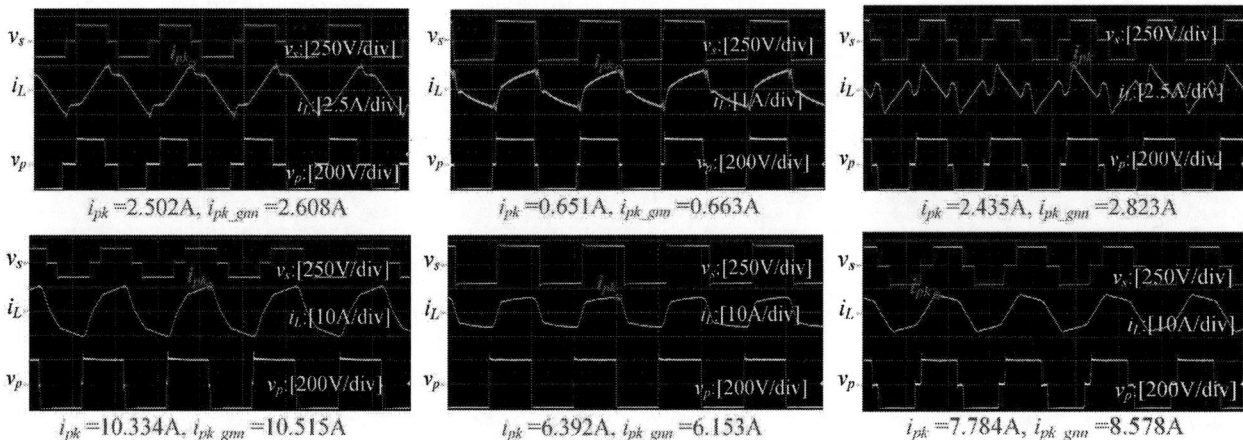

Fig. 11. Waveforms of DAB in different operating conditions. (a) P=100W, V_2=160V (b) P=100W, V_2=200V (c)P=100W, V_2=240V (d)P=1000W, V_2=160V (e) P=1000W, V_2=200V (f)P=1000W, V_2=240V.

[4] Q. Gu, L. Yuan, J. Nie, J. Sun, and Z. Zhao, "Current Stress Minimization of Dual-Active-Bridge DC–DC Converter Within the Whole Operating Range," *IEEE J. Emerg. Sel. Top. Power Electron.*, vol. 7, no. 1, pp. 129–142, Mar. 2019.

[5] H. S. Krishnamoorthy and T. Narayanan Aayer, "Machine Learning based Modeling of Power Electronic Converters," 2019 IEEE Energy Conversion Congress and Exposition (ECCE), Baltimore, MD, USA, 2019.

[6] F. Scarselli, M. Gori, A. C. Tsoi, M. Hagenbuchner and G. Monfardini, "The Graph Neural Network Model," in *IEEE Transactions on Neural Networks*, vol. 20, no. 1, pp. 61-80, Jan. 2009.

[7] A. K. Khamis and M. Agamy, "Circuit Dynamics Prediction via Graph Neural Network & Graph Framework Integration: Three Phase Inverter Case Study," *IEEE Open J. Power Electron.*, vol. 5, pp. 987–1001, 2024.

[8] Tolstikhin, Ilya O., et al. "Mlp-mixer: An all-mlp architecture for vision." Advances in neural information processing systems 34 (2021): 24261-24272.

[9] X. Li, X. Zhang, F. Lin, C. Sun, and K. Mao, "Artificial-Intelligence-Based Triple Phase Shift Modulation for Dual Active Bridge Converter With Minimized Current Stress," *IEEE J. Emerg. Sel. Top. Power Electron.*, vol. 11, no. 4, pp. 4430–4441, Aug. 2023.

[10] Gawthrop, P. J., & Bevan, G. P. (2007). Bond-graph modeling: a tutorial introduction for control engineers. *IEEE Control Systems*, 27(2), 24-45. https://doi.org/10.1109/MCS.2007.338279

[11] A. K. Khamis and M. Agamy, "Comprehensive Mapping of Continuous/Switching Circuits in CCM and DCM to Machine Learning Domain Using Homogeneous Graph Neural Networks," in *IEEE Open Journal of Circuits and Systems*, vol. 4, pp. 50-69, 2023.

An Augmented State Space Modelling Approach for DC-DC Converter Start-Up in Closed Loop

Waseah Anjum
Chair of Power Electronics
Kiel University
Kiel, Germany
waan@tf.uni-kiel.de

Arkadeb Sengupta
Chair of Power Electronics
Kiel University
Kiel, Germany
arsa@tf.uni-kiel.de

Marco Liserre
Chair of Power Electronics
Kiel University
Kiel, Germany
ml@tf.uni-kiel.de

Abstract—**The continuity of power supply for disturbance-sensitive loads is ensured by replacing faulty converters while the load remains operational. The start-up transient of the incoming converter is crucial for uninterrupted power delivery to the load as well as the stability of the overall system. Existing small-signal approaches for DC-DC converters are of insufficient for modelling this start-up accurately, due to its inherent large-signal nature. This paper addresses this issue by proposing a modelling approach for the closed-loop start-up of DC-DC converters, using state space augmentation. The proposed method appends the state space of a system operating in closed loop by considering the control input to be an additional state. This appended or augmented model facilitates accurate prediction of the transient behaviour during closed-loop start-up. An analytical modelling algorithm is presented for a generic system. Subsequently, the proposed approach is instantiated for a boost converter operating in closed loop. The accuracy of the proposed approach is demonstrated through simulation results as well experiments on a laboratory prototype.**

Index Terms—**DC-DC converter, closed-loop, modelling**

I. INTRODUCTION

In many electrical systems, ensuring a constant power supply is critical, especially in applications like uninterruptible power supplies (UPS) for disturbance-sensitive loads [1]. These systems need to operate smoothly, even during disturbances and faults, by quickly switching from faulty DC-DC converters to healthy ones, online. However, due to their nonlinear nature, the start-up transient behaviour of the DC-DC converters can be tricky to predict [2], especially when they need to start-up in closed loop.

Traditionally, linearized small signal models are used for modelling the dynamic behaviour of DC-DC converters [3]. This modeling approach assumes that variations in the system's states, inputs and disturbances, are small. This approach works well as long as this assumption of small variations holds true, but is rendered inaccurate when the changes become too large, as often happens during start-up or recovery from a big fault. Such cases, with large variations, require a different modelling approach which can accurately model the transient behaviour of the system even when there are large signal variations. Although, there are several published works that pertain to dealing with such large signal variations, that typically occur during start-up [4]–[7], none of them propose a

method for modelling the closed-loop start-up dynamics such as the one presented here.

The key contributions of this paper are:

(i) a novel modelling approach, which takes into account the dynamics of the control signal in closed-loop operation, facilitating simple and computationally cheap numerical simulations of closed-loop system without approximations like linearization,

(ii) a comparison with the traditionally used small signal linearization approach by means of simulations,

(iii) an exprimental validation of the proposed approach.

The organization of this paper is as follows: the proposed augmented state space modelling approach is demonstrated for a generic nonlinear system in section II. In section III, a boost converter operating in closed-loop is used to exemplify the approach. Then, the accuracy of the proposed approach is tested against PLECS simulations for a closed-loop start-up scenario. The experimental setup for hardware verification of the proposed approach and its results are discussed in section IV. Section V concludes the paper.

II. CLOSED-LOOP STARTUP MODELING BASED ON STATE SPACE AUGMENTATION

For demonstration, a system with two states, x_1 and x_2, is considered as described in (1). Here, u is the control input and d can be considered as a disturbance. All variables, here, are large signal quantities.

$$\begin{pmatrix} \dot{x}_1 \\ \dot{x}_2 \end{pmatrix} = \begin{pmatrix} 0 & -(1-u) \\ 1-u & -1 \end{pmatrix} \begin{pmatrix} x_1 \\ x_2 \end{pmatrix} + \begin{pmatrix} 1 \\ 0 \end{pmatrix} d \quad (1)$$

Considering u to be provided by a PI controller, $\dot{u}$ can be expressed as shown in (2), where k_P and k_I are the proportional gain and the integral gain, respectively.

$$\dot{u} = -k_P \dot{x}_2 - k_I x_2 \quad (2)$$

Since, the dynamics of the closed-loop control input can be expressed in terms of the states and the parameters of the system, it can be regarded as a state of the system, itself. The state space can, thus, be augmented with u as a new state. This yields a nonlinear augmented state space model of the system as shown in (3), below.

979-8-3315-1612-3/25 $31.00 © 2025 IEEE

$$\begin{pmatrix} \dot{x}_1 \\ \dot{x}_2 \\ \dot{u} \end{pmatrix} = \begin{pmatrix} -(1-u)x_2 \\ (1-u)x_1 - x_2 \\ -k_P((1-u)x_1 - x_2) - k_I x_2 \end{pmatrix} + \begin{pmatrix} d \\ 0 \\ 0 \end{pmatrix} \quad (3)$$

The traditionally used, small signal linearized model can also be derived in a similar fashion by linearizing the dynamics of u, as shown in (4), below. Here, $\hat{x}_1$, $\hat{x}_2$ and $\hat{u}$ are small signal values and X_1, X_2 and U are steady state average values.

$$\begin{pmatrix} \dot{\hat{x}}_1 \\ \dot{\hat{x}}_2 \\ \dot{\hat{u}} \end{pmatrix} = \begin{pmatrix} 0 & -(1-U) & X_2 \\ (1-U) & -1 & -X_1 \\ -k_P(1-U) & k_P - k_I & -k_P X_1 \end{pmatrix} \begin{pmatrix} \hat{x}_1 \\ \hat{x}_2 \\ \hat{u} \end{pmatrix}$$
$$+ \begin{pmatrix} 1 \\ 0 \\ 0 \end{pmatrix} \hat{d}$$
$$(4)$$

As mentioned before, the linearized model is derived by approximating the system's behaviour around a fixed operating point, in the vicinity of which it is assumed to behave linearly. This approach simplifies the model, facilitating easier analysis and control design.

However, since the actual system is inherently nonlinear, the linearized model's accuracy diminishes if the states diverge significantly from the linearization point. During a closed-loop start-up, it is unreasonable to assume only small signal changes in the. In fact, large transient changes are typically expected in such scenarios.

In contrast, the proposed augmented state-space approach retains the system's nonlinear characteristics within the model. This allows accurate prediction the system's behaviour even under substantial deviations in the states. The inclusion of the control input as an additional state facilitates the numerical simulation of the system's start-up transient behaviour in closed loop, where the control input also undergoes transient changes.

It is worth noting that the proposed approach can be implemented using free and open-source tools. This reduces dependency on proprietary software that are usually paid and not open-source. Moreover, these proprietary software tend to have several layers of abstraction, which can obscure the underlying code/components.

III. START-UP TRANSIENT MODELING: AN EXAMPLE

A. Modeling of a Boost Converter

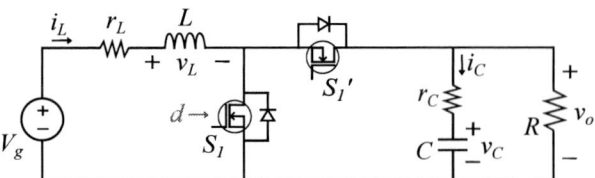

Fig. 1: Boost converter circuit schematic

A boost converter, whose circuit schematic is shown in figure 1, is considered as an example to showcase the proposed approach. An integral controller is used such that the duty cycle, $d = k_I \int (V_{o,ref} - v_o) \, dt$ where $V_{o,ref}$ is the reference output voltage. The augmented state space model for the resulting system is shown in (5) and (6).

$$\begin{pmatrix} \dot{i}_L \\ \dot{v}_C \\ \dot{d} \end{pmatrix} = \begin{pmatrix} -\frac{r_L + a(1-d)r_C}{L} i_L - \frac{a(1-d)}{L} v_C + \frac{V_g}{L} \\ \frac{a(1-d)}{C} i_L - \frac{a}{RC} v_C \\ k_I(V_{o,ref} - (a(1-d)r_C i_L + a v_C)) \end{pmatrix} \quad (5)$$

$$y = v_o = a(1-d)i_L r_C + a v_C \quad (6)$$

Where, $a = \frac{R}{R + r_C}$.

The linearized augmented state space model is given in (7) and (8). Small signal variations in the source voltage, $\hat{v}_g$ are neglected, assuming it is very small during the start-up transient period.

$$\begin{pmatrix} \dot{\hat{i}}_L \\ \dot{\hat{v}}_C \\ \dot{\hat{d}} \end{pmatrix} = \begin{pmatrix} -\frac{r_L + a(1-D)r_C}{L} & -\frac{a(1-D)}{L} & \frac{a(V_C + I_L r_C)}{L} \\ \frac{a(1-D)}{C} & -\frac{a}{RC} & -\frac{aI_L}{C} \\ -k_I a(1-D)r_C & -k_I a & k_I a I_L r_C \end{pmatrix} \begin{pmatrix} \hat{i}_L \\ \hat{v}_C \\ \hat{d} \end{pmatrix}$$
$$(7)$$

$$\hat{y} = \hat{v}_o = \begin{pmatrix} a(1-D)r_C & -k_I a & -a r_C I_L \end{pmatrix} \begin{pmatrix} \hat{i}_L \\ \hat{v}_C \\ \hat{d} \end{pmatrix} \quad (8)$$

B. Simulation

A simulation setup is prepared in MATLAB/Simulink with the help of the PLECS toolbox. The simulation scenario is as follows: the boost converter starts-up in closed loop, with the load voltage already at the required reference level. However, the inductor is not energized. While the inductor charges, the capacitor discharges across the load leading to a transient in both, the inductor current and the capacitor voltage. Additionally, since the converter starts-up in closed-loop, the duty cycle also undergoes transient changes.

This same scenario is modelled in MATLAB using the proposed augmented state space approach ((5) and (6)) as well as the linearized approach ((7) and (8)). The ode45 solver is used for both models as well as the PLECS simulations.

The PLECS model is used as a reference to evaluate the performance of the proposed modelling approach. Table I shows the simulation parameters, including the initial conditions.

The simulation results are shown in figures 2, 3 and 4. As expected, a dip in the capacitor voltage is seen intially, while the inductor charges-up. In response, the controller increases the duty cycle. Thus, leading to transients in all the states, including the duty cycle.

As was foreseen, the proposed nonlinear model, which preserves the nonlinearities of the system, aligns very well with the reference PLECS simulation. On the other hand, the linearized model shows significant deviations. Evidently, the linearized model becomes inaccurate when changes in the

TABLE I
CIRCUIT PARAMETERS AND INITIAL CONDITIONS

Source voltage, V_g	5 V
Load reference voltage, V_o	10 V
Load resistance, R	50 Ω
Inductance, ESR, L, r_L	62 μH, 57 mΩ
Capacitance, ESR, C, r_C	250 μF, 400 mΩ

(a) CIRCUIT PARAMETERS

Switching frequency, f_s	50 kHz
Controller integral gain, k_I	75
Initial output voltage, V_o	10 V
Initial inductor current, I_L	0
Initial duty cycle, D	0.5

(b) INITIAL CONDITIONS

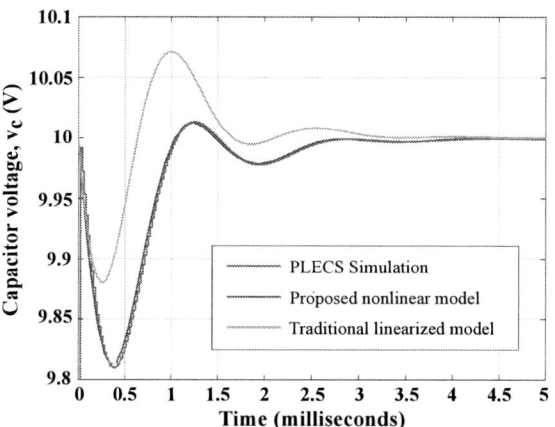

Fig. 2: Simulation result: Capacitor voltage, v_C

states are no longer small signal in nature. Thus, it is clear that the proposed model is superior for simulating closed-loop start-up behaviour of a system.

Tables II, III and IV summarize the performance of both, the proposed nonlinear augmented model and the traditional linearized model, against the reference PLECS simulations.

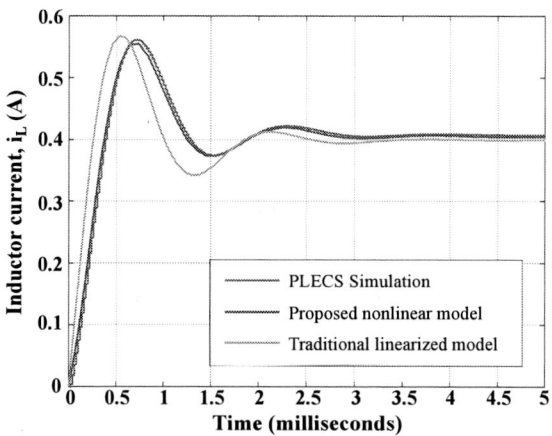

Fig. 3: Simulation result: Inductor current, i_L

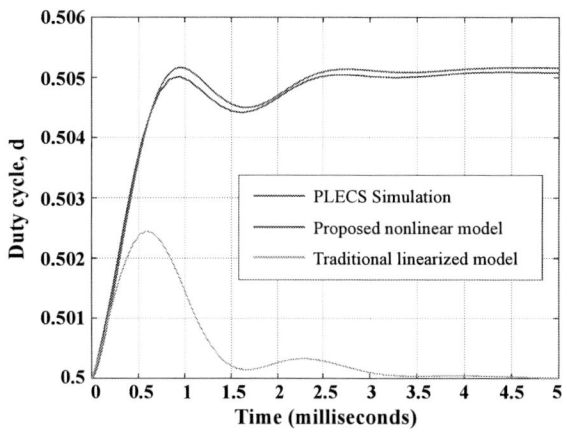

Fig. 4: Simulation result: Duty cycle, d

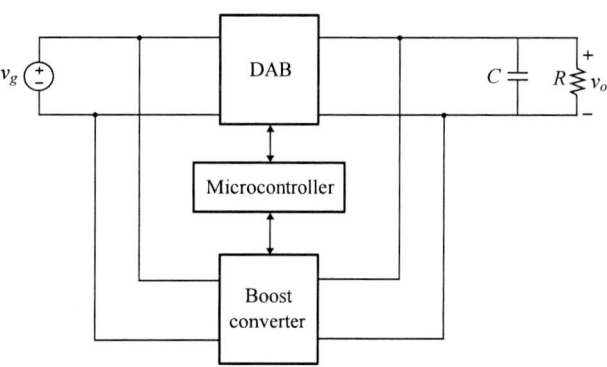

Fig. 5: Block diagram of the experimental setup

IV. EXPERIMENTAL RESULTS

For experimental verification of the proposed modelling approach, a setup is designed as can be seen in figure 6. A block diagram representation of the same is shown in figure 5.

The aim is to simulate the same conditions as done in section III and capture the closed-loop start-up transients of the boost converter. To achieve this, a Dual-Active Bridge (DAB) is used to maintain the load voltage at the reference value of 10V. The DAB is, then, switched off and at the same time the boost converter is started up in closed loop. The start-up transients are captured using an oscilloscope.

The results, shown in figures 7 and 8, demonstrate the accuracy of the proposed modeling approach. There seem to be slight errors in the prediction of the transient behaviour.

This can be attributed to non-idealities such as uncertainty in the ESR value of the capacitor, non-idealities in the inductor and the switches, which are unmodelled. Moreover, there are parasitic elements such as noise in the voltage and current probes and parasitic inductance of the long connecting wires in the experimental set-up.

Nonetheless, despite these elements of uncertainty, the proposed model provides a good prediction of the closed loop start-up transient behaviour of the boost converter. The

TABLE II
SUMMARY OF SIMULATION RESULTS: CAPACITOR VOLTAGE, v_C

Model	Steady state value	Minimum value		Peak value		Settling time (1% of final value)
		Value	Time	Value	Time	
PLECS Simulation	10 V	9.84 V	0.4 ms	10.023 V	1.34 ms	0.72 ms
Proposed Model	10 V	9.84 V	0.4 ms	10.023 V	1.3 ms	0.72 ms
Linearized Model	10 V	9.89 V	0.266 ms	10.07 V	1.1 ms	0.375 ms

TABLE III
SUMMARY OF SIMULATION RESULTS: INDUCTOR CURRENT, i_L

Model	Steady state value	Peak value		Settling time (5% of final value)
		Value	Time	
PLECS Simulation	0.406 A	0.572 A	0.78 ms	2 ms
Proposed Model	0.403 A	0.566 A	0.73 ms	2 ms
Linearized Model	0.4 A	0.575 A	0.59 ms	1.9 ms

TABLE IV
SUMMARY OF SIMULATION RESULTS: DUTY CYCLE, d

Model	Steady state value	Peak value	
		Value	Time
PLECS Simulation	0.5052	0.5052	0.96 ms
Proposed Model	0.505	0.505	0.94 ms
Linearized Model	0.5024	0.5	0.59 ms

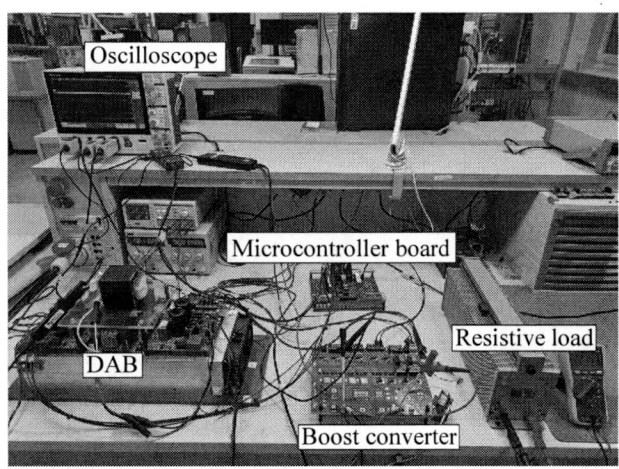

Fig. 6: Experimental setup

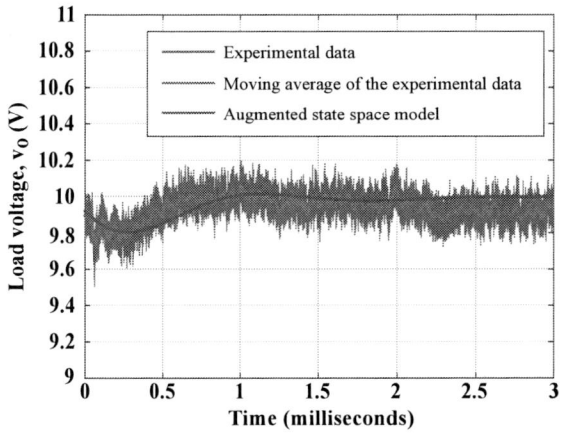

Fig. 7: Experimental result: Load voltage, v_o

performance of the proposed approach against experimental data for the inductor current is summarized in Table V.

TABLE V
SUMMARY OF EXPERIMENTAL RESULTS: INDUCTOR CURRENT, i_L

Model	Peak value	Peaking time
Experimental result	0.572 A	0.44 ms
Augmented state space model	0.566 A	0.72 ms

V. CONCLUSION

In this paper, a novel augmented state space modeling approach is proposed for modeling the closed-loop start-up dynamics for DC-DC converters. This model is simple and can be implemented on free-to-use platforms such as python, eliminating the need for proprietary software. Its simplicity

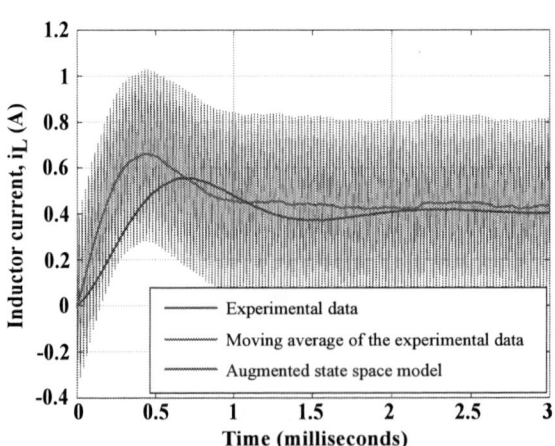

Fig. 8: Experimental result: Inductor current, i_L

also makes it computationally less expensive than its equivalent models in environments like PLECS and Simulink. This augmented state space approach is tested against an equivalent PLECS simulation as well as a traditional, linearized model. The proposed model is found to be as accurate as the PLECS model for predicting the startup dynamics of DC-DC converters. The proposed model is, also, experimentally verified.

ACKNOWLEDGMENT

This research has received funding from the European Innovation Council (EIC) under the European Union's Horizon 2020 research and innovation programme, vide grant agreement no. 101057679 within the framework of the project Super-HEART.

REFERENCES

[1] J. Chen, C. Wang, and J. Chen, "Investigation on the selection of electric power system architecture for future more electric aircraft," *IEEE Transactions on Transportation Electrification*, vol. 4, no. 2, pp. 563–576, 2018.

[2] W. Yuanbin, "Dynamic behavior of pwm dc-dc converters in starting process," in *2009 IEEE International Symposium on Industrial Electronics*, 2009, pp. 1608–1611.

[3] R. W. Erickson and D. Maksimović, *Fundamentals of Power Electronics*, 2nd ed. Kluwer Academic Publishers, 2001.

[4] X. Liu, Z. Zhang, M. Han, Z. Li, Y. Yin, Z. Dong, and J. Wang, "A simple and fast start-up strategy for dual-active-bridge converters with dc bias suppression," *IEEE Transactions on Power Electronics*, vol. 38, no. 9, pp. 10 629–10 639, 2023.

[5] L. Gong, Y. Peng, C. Cui, J. Chen, L. Jiang, X. Fang, J. Wang, J. Xu, and Y. Wang, "An ultra-fast and wide-safe-range start-up method of dab converters with straightforward frequency-phase closed-loop control," *IEEE Transactions on Power Electronics*, pp. 1–6, 2024.

[6] J. Hu, S. Cui, and R. W. De Doncker, "Closed-loop black start-up of dual-active-bridge converter with boosted dynamics and soft-switching operation," *IEEE Transactions on Power Electronics*, vol. 36, no. 10, pp. 11 009–11 013, 2021.

[7] M. Ishaq, M. Waqar, and M. H. Afzal, "Design of double closed-loop boost converter controller to reduce transient voltage dip for sudden load connection," in *2023 International Conference on Emerging Power Technologies (ICEPT)*, 2023, pp. 1–5.

The Utilization of a Parallel Computing Algorithm for Accelerating Switching-Level Modeling of Power Electronics Simulations in a T-Type PV Inverter

Buck F. Brown, III
Department of Electrical and Computer Engineering
Clemson University
North Charleston, SC, USA
buckb@clemson.edu

Liwei Wang
School of Computing
Clemson University
North Charleston, SC, USA
liweiw@clemson.edu

Zheyu Zhang
Department of Electrical, Computer, and Systems Engineering
Rensselaer Polytechnic Institute
Troy, NY, USA
zhangz49@rpi.edu

Johan Enslin
Department of Electrical and Computer Engineering
Clemson University
North Charleston, SC, USA
jenslin@clemson.edu

Yi Li
Department of Electrical and Computer Engineering
Clemson University
North Charleston, SC, USA
yli26@clemson.edu

Abstract—High-fidelity, switching-level modeling of power electronics simulations can be computationally intensive and time-consuming. This computational burden has recently escalated further due to the increased number of converters implemented in a system, e.g., renewable energy systems. Unlike switching modeling, average modeling can alleviate the computational burden and adequately represent a converter's behavior for controller design. However, average modeling does not encapsulate switching harmonics, which are crucial for power electronics design, operation, and reliability studies. This paper proposes a novel way to accelerate the switching-level modeling of simulations by leveraging parallel computing techniques and showcases this technique in a specific case study. This new technique is called the average-to-switching (A2S) model. First, the basic concept of the A2S model is outlined. Second, a case study is designed for a two-stage, T-type PV inverter. Third, electrical and thermal waveforms produced by the case study are given and discussed. The results of testing the A2S model showcase a speedup of greater than 22 times the simulation benchmark while maintaining an R^2 value of greater than 0.99.

Keywords—Parallel computing, power electronics, PV applications, simulation accuracy, simulation speedup, switching model.

I. INTRODUCTION

Power electronics have become more widespread than ever. Regardless of the application, simulation almost always occurs before hardware development in power electronics systems, and it is one of the fundamental steps to reduce design costs [1]-[4].

Two primary models are used for power electronics simulations: average and switching. Average modeling is popularly used for general system behavior modeling due to its reduced complexity through state-space averaging methods [5]-[9]. While average models can be simulated relatively quickly, the tradeoff to this faster simulation speed is reduced simulation accuracy. In contrast, switching modeling is popularly used for harmonic-level modeling. Since switching models consider the switching functions of individual power devices in the converter, they provide insight for accurate passive design, loss estimation, and thermal management, which are crucial for converter design, operation, and reliability studies [10], [11].

The two main power electronics simulation models have their respective downfalls. An average model lacks the high-fidelity information of a switching model, and a switching model lacks the fast computation speed of the average model. As simulations become more complex with the advancement of new technologies and ever-growing system sizes, producing a new type of power electronics model becomes essential. Based on a complete understanding of both modeling approaches and the utilization of advanced computing techniques, this paper proposes and utilizes a novel type of power electronics simulation model called the average-to-switching (A2S) model. The idea of the A2S model is initially given in [12] and [13]. In this paper, the idea is reintroduced and implemented in a concrete system.

This paper begins by examining the basic concept of the A2S model, illustrating how the usage of parallel computing is possible. Afterward, a two-stage, three-level inverter circuit is considered, and concrete parameters corresponding to the circuit are given. Then, results are presented and discussed, highlighting the accuracy and speed of the A2S model when used in such a circuit.

979-8-3315-1612-3/25 $31.00 © 2025 IEEE

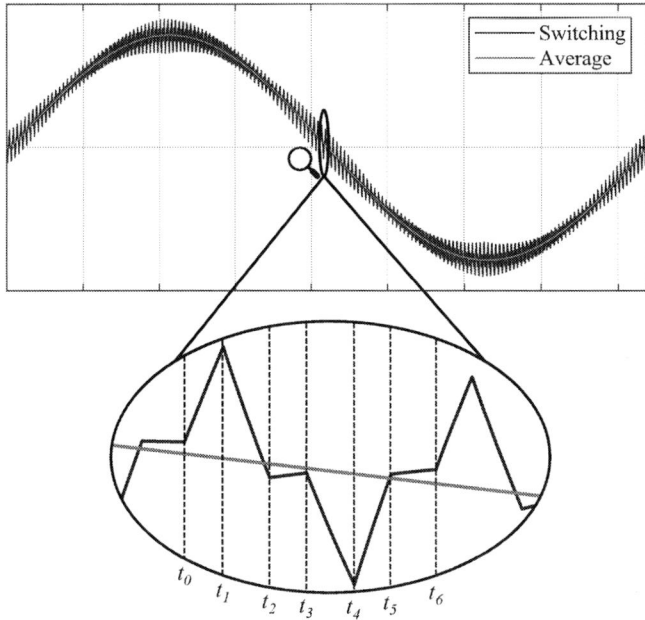

Fig. 1. Switching versus average model resultant waveforms.

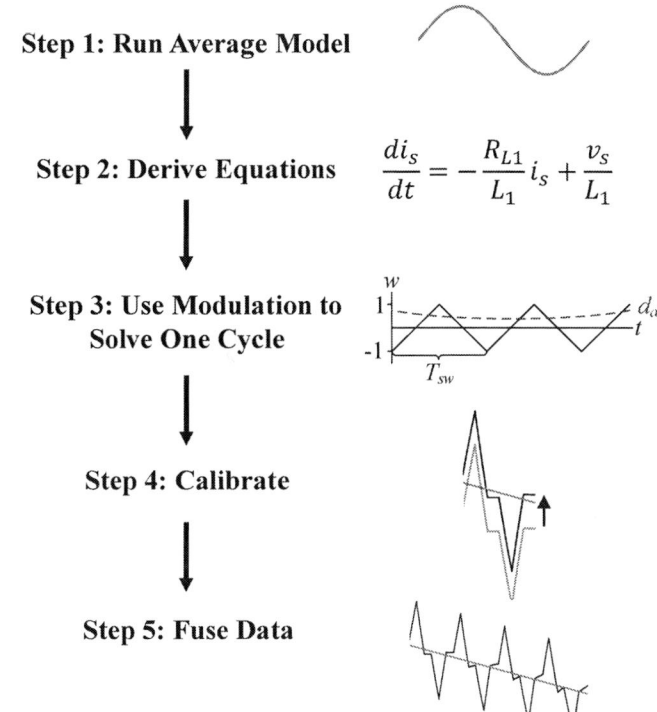

Fig. 2. Essential steps used in the A2S model.

II. BASIC CONCEPT

Fig. 1 shows the waveforms of a switching-based and an average-based simulation under the same operating conditions. Two takeaways are observed: (1) the switching waveform is simulated sequentially by taking the final value from the previous simulation time step as the initial value for the upcoming simulation time step; as a result, the dependency between simulation time steps prevents the usage of high-speed parallel computing techniques; (2) within one switching cycle, the average waveform value is equal to the mean value of the switching waveform. Therefore, the average model (faster and regardless of switching frequency) can provide the mean value for each switching cycle, allowing each switching cycle to be solved independently with parallel computing. In other words, each switching cycle can be solved separately from another by leveraging the waveforms produced by the average model. This understanding is the basic concept of the proposed A2S model.

The steps required for the A2S model are shown in Fig. 2. The first step in the A2S model is to run the average model. While this step requires sequential computing, since the average model is usually at least 50 times faster than the switching model, the amount of time it adds is negligible. The running of the average model is essential since it provides an initial condition that is used to solve the differential equations in Step 3 and calibrate the waveform in Step 4. The second step in the A2S model is to derive the differential equations of the circuit, i.e., deriving the state-space equations. After the derivation of the differential equations, the third step is to use the specific modulation scheme of the circuit (which gives the values of the switching functions) to solve one switching cycle. Following the solution of the differential equations in Step 3, the results are calibrated so that the average value of the newly constructed A2S waveform has the same average value as the average model

waveform. Steps 3 and 4 are repeated in parallel for all switching periods. In Step 5, the A2S waveforms for each switching period are fused to produce one resultant A2S waveform.

III. CASE STUDY

One of the most common types of inverters is a two-stage, three-level, T-type inverter. Such an inverter is a typical inverter for photovoltaic (PV) applications [14]. A depiction of a two-stage, three-level, T-type inverter is given in Fig. 3. Due to the inverter's popularity, if the A2S model can be used to speedup simulations involving two-stage, three-level, T-type inverters, then the A2S model can have a profound impact on PV system simulations.

A case study is proposed to showcase the usage of the A2S model in a two-stage, three-level, T-type inverter. The parameters for the case study are listed in Table I. The circuit in Fig. 3 is simulated in MATLAB/Simulink using these parameters in an open-loop setting. The waveforms produced by the simulation are given in Fig. 4–Fig. 6, whereas the speed of the simulation is given in Fig. 7. A note is given here that the waveforms shown in Fig. 4–Fig. 6 pertain specifically to the a-phase values of the circuit.

A. Accuracy Verification

The DC electrical waveforms produced by the A2S model, switching model, and average model are shown in Fig. 4. These waveforms include the current flowing through the boost inductor, along with the upper and lower DC-link voltages across the upper and lower DC-link capacitors. On the left-hand side of Fig. 4, it is qualitatively apparent that the overall trend of

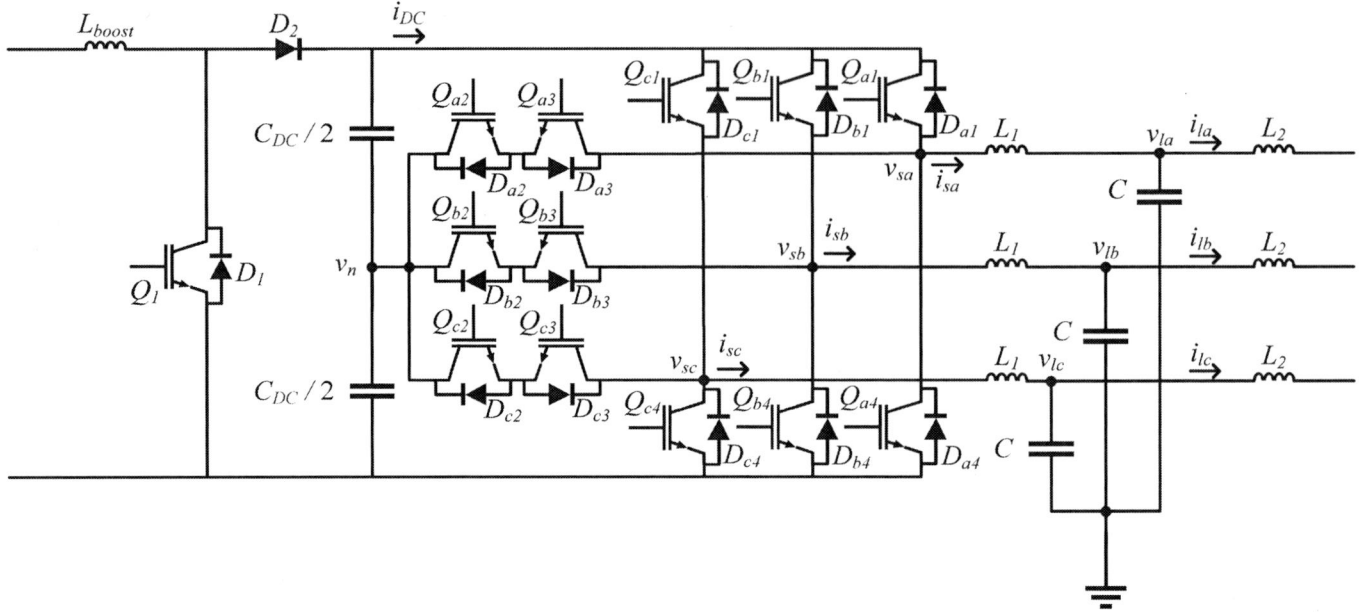

Fig. 3. Two-stage, three-level, T-type inverter used in the case study.

TABLE I
SIMULATION PROPERTIES FOR CASE STUDY

Simulation Properties	Values
Run/Simulation Time	0.3125 s
AC Power	100 kVA
DC Voltage	1000 V
AC Voltage	480 V
Time Step	312.5 nanoseconds
Switching Frequency	16 kHz
Modulation Scheme	SVM
L_{boost}	500 µH
C_{DC1}	10.351 mF
C_{DC2}	10.351 mF
L_1	300 µH
C	97.86 µF
L_2	931.83 µH
Power Module Reference	SEMiX305MLI12E4V2
T_a	25°C

the A2S and switching waveforms are very similar. The main difference between the waveforms is due to harmonics in the switching waveforms that are not encapsulated in the average model. If these harmonics are considered in the average model, the A2S waveform would overlap with the switching waveform almost perfectly. The switching ripple present in the A2S and switching waveforms are shown on the right-hand side of Fig. 4. Not only are the shapes of the A2S and switching waveforms almost identical, but the boost inductor currents are always within 2 A of each other, and the upper and lower DC-link voltages are always within 0.5 V of each other. Due to the high level of similarity between the A2S waveforms and the switching waveforms, the DC electrical waveforms of the two-stage, three-level, T-type inverter are accurately produced by the A2S model.

The AC electrical waveforms produced by the A2S model, switching model, and average model are shown in Fig. 5. These waveforms include the first-stage inductor current, output capacitor voltage, and second-stage inductor current. On the left-hand side of Fig. 5, it appears as though the steady-state values of the A2S and switching waveforms are identical. Unlike with the DC electrical waveforms, there are no additional harmonics present leading to a mismatch in steady-state values. The switching ripple present in the A2S and switching waveforms are shown on the right-hand side of Fig. 5. The switching ripple is most evident in Fig. 5(a), corresponding to the first-stage inductor current. Based on this illustration, the shapes of the A2S and switching waveforms are almost identical. Despite some discrepancies in the specific values of the waveforms, the first-stage inductor currents are within 2 A of each other, the output capacitor voltages are within 2 V of each other, and the second-stage inductor currents are within 2 A of each other. Due to the high level of similarity between the A2S waveforms and the switching waveforms, the AC electrical waveforms of the two-stage, three-level, T-type inverter are accurately produced by the A2S model.

In many applications, it is desired to encapsulate the switching ripple of a circuit to ensure accurate thermal modeling of the circuit. As a result, the junction temperature of each IGBT in Fig. 3 is simulated based on the A2S model and the switching model. The thermal waveforms produced by the A2S model and the switching model are shown in Fig. 6. These thermal waveforms are simulated through the reference of the SEMiX305MLI12E4V2 datasheet and the PLECS Blockset platform in MATLAB/Simulink [15]. As shown in Fig. 6, the similarities between the junction temperatures produced by the A2S model and the switching model are remarkable. In each case, the R^2 score is greater than 0.99. As a result, the A2S model provides the necessary electrical information that yields accurate thermal modeling of the two-stage, three-level, T-type inverter.

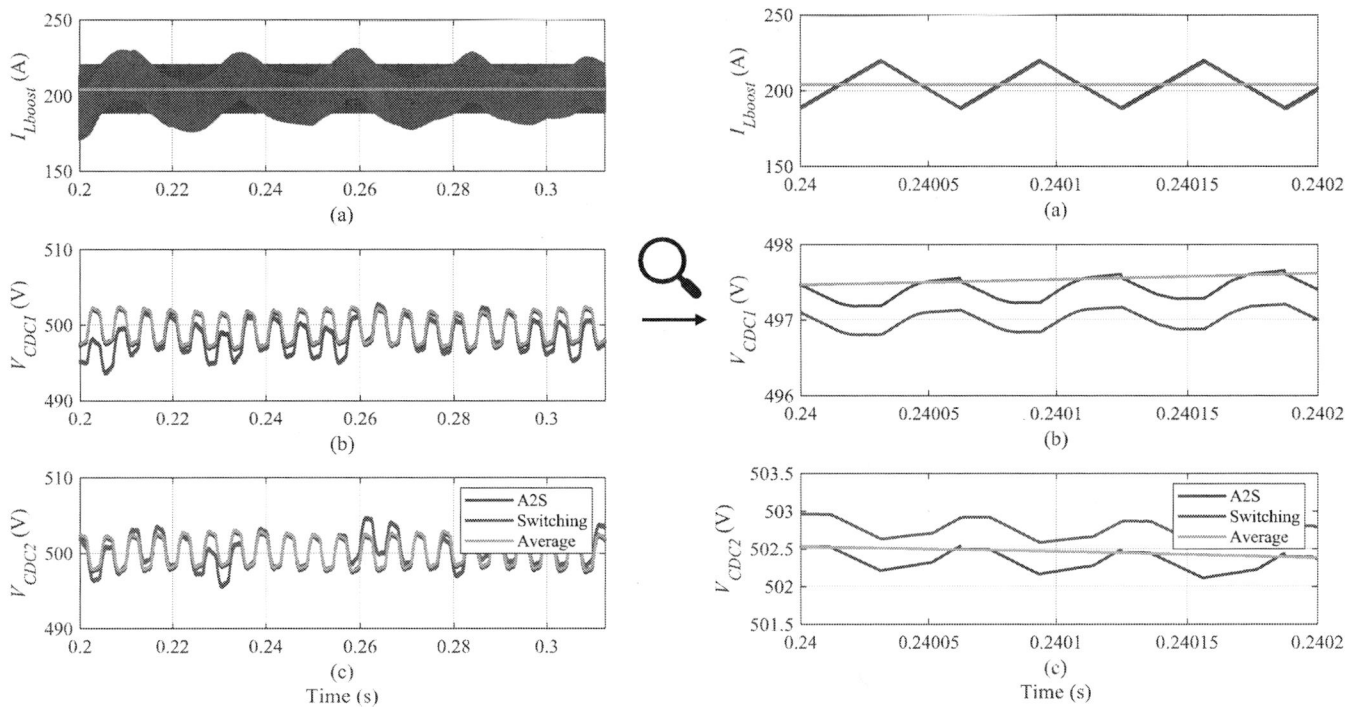

Fig. 4. DC electrical waveforms of the (a) boost inductor current, (b) upper DC-link capacitor voltage, and (c) lower DC-link capacitor voltage.

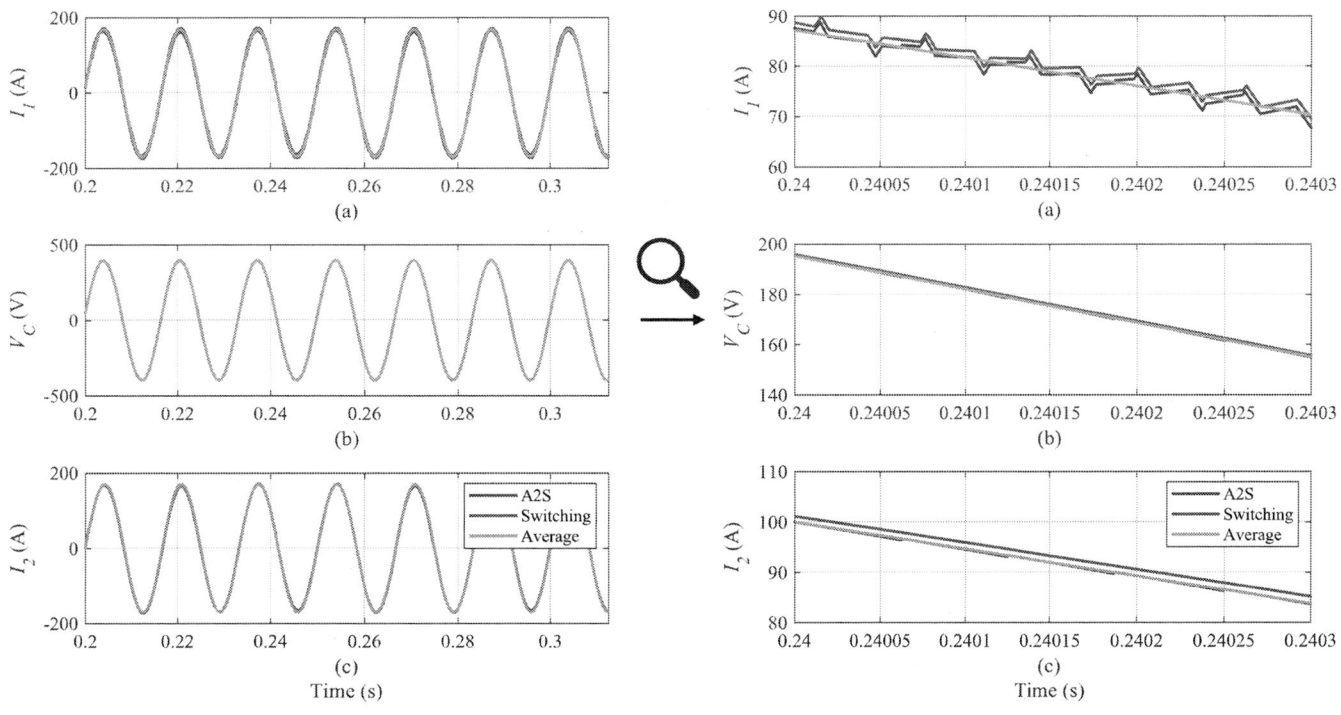

Fig. 5. AC electrical waveforms of the (a) first-stage inductor current, (b) output capacitor voltage, and (c) second-stage inductor current.

B. Speed Verification

Based on Table I, the amount of time being simulated in this case study is 0.3125 s. Fig. 7 shows the amount of run-time necessary to simulate the case study for 0.3125 s when using the switching model, average model, and A2S model. As shown in Fig. 7, the switching model takes 33.5 s to run the simulation, the average model takes 0.35 s to run the simulation, and the

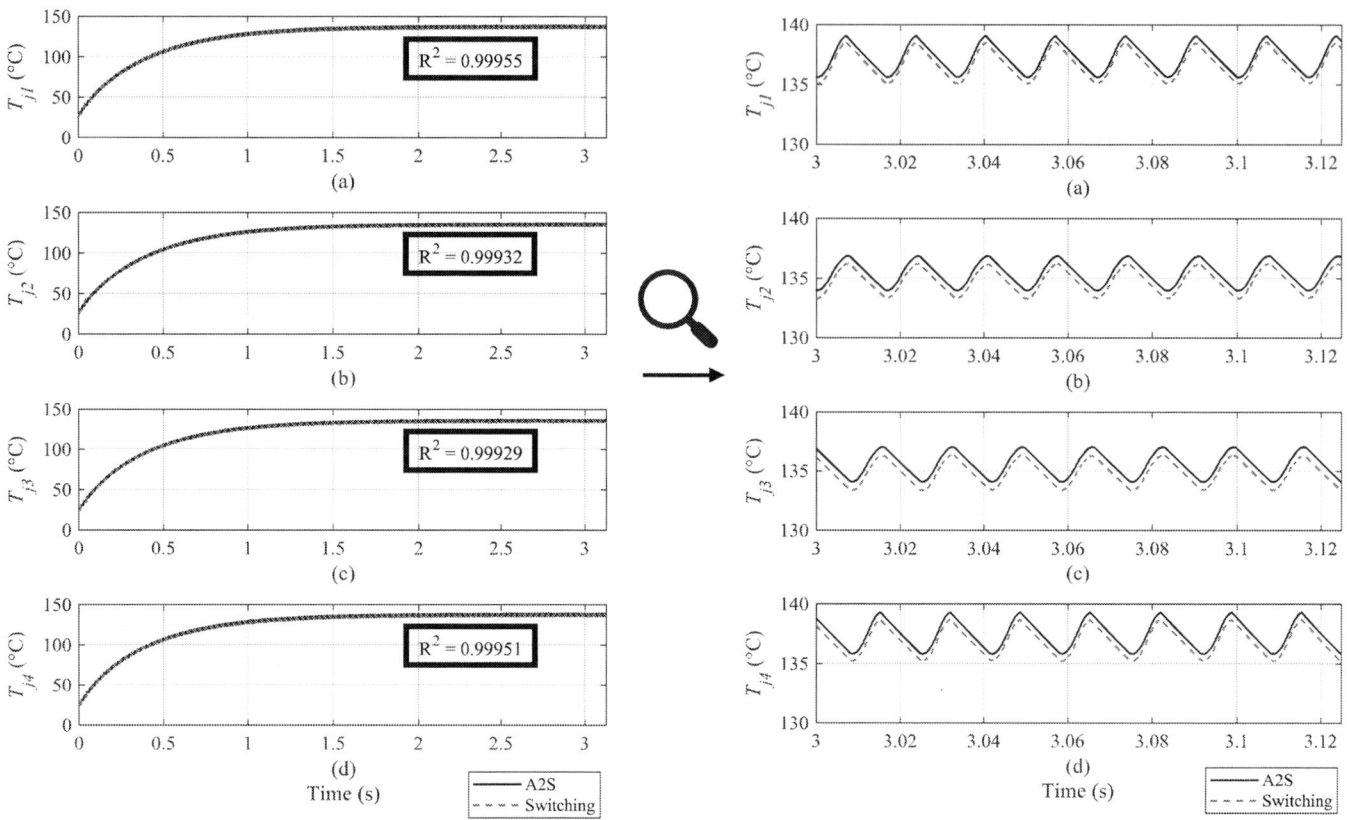

Fig. 6. Thermal waveforms showcasing the junction temperatures of (a) Q_{a1}, (b) Q_{a2}, (c) Q_{a3}, and (d) Q_{a4}.

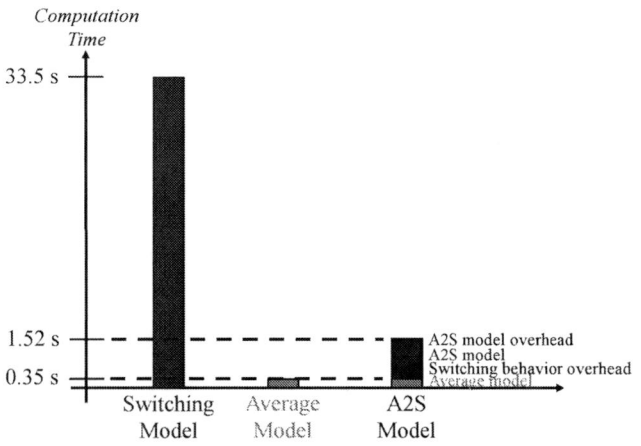

Fig. 7. Computation times of the switching, average, and A2S models.

A2S model takes 1.52 s to run the simulation (assuming 40 cores are available for parallel computing purposes). Therefore, the A2S model is over 22 times faster than the switching model in this case study. Considering the accuracy of the A2S model, this amount of speedup is extremely significant.

IV. CONCLUSION

This paper reinforces the potential of the newly proposed A2S model. While the A2S model has been introduced previously, this paper presents it for the first time being utilized to model a two-stage, three-level, T-type PV inverter. Such an application of the A2S model could be vital when designing future renewable energy systems, as most PV inverters are multilevel inverters. In the specific case study conducted, the A2S model is found to be highly accurate, with an R^2 score of above 0.99 when compared to the switching model, and extremely fast, with a speed 22 times faster than the switching model. Despite this significant performance improvement, the A2S model has room for further development. In the future, by deriving a more comprehensive average model and applying the A2S model in numerous case studies, more capabilities of the novel A2S model can be highlighted.

ACKNOWLEDGMENT

The authors would like to thank the Department of Energy (DOE) Solar Energy Technologies Office (SETO) for their continued support via Award Number DE-EE0009348.

REFERENCES

[1] D. Maksimovic, A. M. Stankovic, V. J. Thottuvelil and G. C. Verghese, "Modeling and simulation of power electronic converters," in Proceedings of the IEEE, vol. 89, no. 6, pp. 898-912, June 2001, doi: 10.1109/5.931486.

[2] S. Bacha, I Munteanu, and AI Bratcu. "Power electronic converters modeling and control." Advanced textbooks in control and signal processing 454 (2014): 454.

[3] U. -M. Choi, F. Blaabjerg and K. -B. Lee, "Study and Handling Methods of Power IGBT Module Failures in Power Electronic Converter

Systems," in IEEE Transactions on Power Electronics, vol. 30, no. 5, pp. 2517-2533, May 2015, doi: 10.1109/TPEL.2014.2373390.

[4] F. Blaabjerg and K. Ma, "Future on Power Electronics for Wind Turbine Systems," in IEEE Journal of Emerging and Selected Topics in Power Electronics, vol. 1, no. 3, pp. 139-152, Sept. 2013, doi: 10.1109/JESTPE.2013.2275978.

[5] A. Emadi, "Modeling and analysis of multiconverter DC power electronic systems using the generalized state-space averaging method," in IEEE Transactions on Industrial Electronics, vol. 51, no. 3, pp. 661-668, June 2004, doi: 10.1109/TIE.2004.825339.

[6] J. Mahdavi, A. Emaadi, M. D. Bellar and M. Ehsani, "Analysis of power electronic converters using the generalized state-space averaging approach," in IEEE Transactions on Circuits and Systems I: Fundamental Theory and Applications, vol. 44, no. 8, pp. 767-770, Aug. 1997, doi: 10.1109/81.611275.

[7] P. T. Krein, J. Bentsman, R. M. Bass and B. C. Lesieutre, "On the use of averaging for the analysis of power electronic systems," 20th Annual IEEE Power Electronics Specialists Conference, Milwaukee, WI, USA, 1989, pp. 463-467 vol.1, doi: 10.1109/PESC.1989.48523.

[8] N. Vukadinović, A. Prodić, B. A. Miwa, C. B. Arnold and M. W. Baker, "Extended wide-load range model for multi-level Dc-Dc converters and a practical dual-mode digital controller," 2016 IEEE Applied Power Electronics Conference and Exposition (APEC), Long Beach, CA, USA, 2016, pp. 1597-1602, doi: 10.1109/APEC.2016.7468080.

[9] S. (1995). Modeling and control of three-phase PWM converters (Order No. 9626120). Available from ProQuest Dissertations & Theses A&I. (304251813). Retrieved from https://www.proquest.com/dissertations-theses/modeling-control-three-phase-pwm-converters/docview/304251813/se-2

[10] B. Wen, D. Boroyevich, R. Burgos, P. Mattavelli and Z. Shen, "Small-Signal Stability Analysis of Three-Phase AC Systems in the Presence of Constant Power Loads Based on Measured d-q Frame Impedances," in IEEE Transactions on Power Electronics, vol. 30, no. 10, pp. 5952-5963, Oct. 2015, doi: 10.1109/TPEL.2014.2378731.

[11] Y. Wang, O. Lucia, Z. Zhang, S. Gao, Y. Guan and D. Xu, "A Review of High Frequency Power Converters and Related Technologies," in IEEE Open Journal of the Industrial Electronics Society, vol. 1, pp. 247-260, 2020, doi: 10.1109/OJIES.2020.3023691.

[12] Y. Li, C. Wagner, C. Edrington, S. Jin and Z. Zhang, "Quantitative Analysis of Accelerated Power Electronics Simulation Using Advanced Computing Technology," 2022 IEEE Applied Power Electronics Conference and Exposition (APEC), Houston, TX, USA, 2022, pp. 274-278, doi: 10.1109/APEC43599.2022.9773649.

[13] Y. Li, Z. Zhang, C. Edrington and S. Jin, "Accelerating Switching Model-Based Simulation Through Parallel Computing," 2023 IEEE Energy Conversion Congress and Exposition (ECCE), Nashville, TN, USA, 2023, pp. 2771-2775, doi: 10.1109/ECCE53617.2023.10362655.

[14] S. Chellappan and J. Rangaraju, "Power Topology Considerations for Solar String Inverters and Energy Storage Systems," Oct. 2020. Available: https://www.ti.com/lit/an/slla498/slla498.pdf?ts=1723809787204. [Accessed: Aug. 16, 2024]

[15] Semikron, "3-Level NPC IGBT-Module," SEMiX305MLI12E4V2 datasheet, Feb. 2021.

A New Reduced Order Analytical Switching Model for eGaN HEMTs

Ruqi Li   Douglas Arduini   Phen Lumod   Shobhana Punjabi   River Lin   Harold Gutierrez

Central Engineering Department
Cisco Systems, Inc.
San Jose, CA 95134, U.S.A.

Abstract—This paper presents a new reduced order, closed form analytical model for predicting the switching behavior of eGaN HEMTs. Design-oriented analyses of the equivalent circuit for each switching sub-interval during both top side device turn on and turn off are carried out to derive a set of simple and user-friendly equations for all electrical quantities. These equations are such that they are expressed explicitly with the eGaN HEMT parasitic and equivalent circuit parameters, revealing more physical insight into the switching characteristics of the eGaN HEMTs. Unlike the existing numerical models, the new model can be used to perform eGaN HEMT switching loss computations on the closed form analytical basis. The model is validated with both simulation and experimentation.

Keywords—eGaN HEMT, Modeling, Reduced Order Switching Transient Model, Simulation

I. Introduction

The new eGaN HEMTs have substantially less parasitic capacitances; therefore, they possess superior ultra-fast switching characteristics suitable for high-frequency and high-density power conversion. To better understand the new switching characteristics of the eGaN HEMTs, vast amount of research has been devoted to developing transient models, which are inherently high-order state equations. Solving these equations analytically is very difficult, if not at all impossible. So far, numerical solution techniques are commonly adopted in almost all the studies [1]-[7]. To improve numerical solution accuracy, non-linear device parasitic elements, such as drain-source voltage-dependent capacitances, drain-current dependent transconductance, temperature effects and other conceivable factors are included in the already complex models [1]-[7]. This further adds much more sophistication in the model and calculation process.

It should be emphasized that even though the numerical solution to the high-order state equations could be considered to be very accurate in theory by considering every aspect in the modeling process, practicality of these models in engineering designs is questionable. It is highly desirable to develop a switching model that is both accurate and simple to be used in practical applications. To achieve this goal, the focus of this paper is on the state equation solution process while inclusion of the non-linear properties of the eGaN HEMTs parasitic elements in the model is in the secondary consideration. Based on the well-recognized switching characteristics of the eGaN HEMTs, two insightful and simplified assumptions are made in the solution process such that the high-order state equations are reduced to the third order. Compared with the third order turn on switching model in [9], the new model presented in this paper is more complete and easily understandable to reveal more physical insight into the eGaN HEMT switching characteristics. The model is validated against both simulation and experimental data with very good correlation.

II. Reduced Order Model Development

The Top Device (TD) turn on and turn off switching process of the eGaN HEMTs in a hard switching, half bridge topology is well understood and reported in the literature [1]-[7]. The turn on and turn off equivalent circuit is depicted collectively in Fig. 1(a). During both TD turn on and turn off, because of the dead time between the two device gate drive signals, the Bottom Device (BD) is freewheeling the inductor current. I_o is equal to I_{Lmin} when TD is turned on, and I_o is equal to I_{Lmax} when TD is turned off. V_r represents the eGaN HEMT reverse conduction voltage drop. Both TD turn on and turn off process can be divided into four sub-intervals. Fig.1(b) illustrates the main switching waveforms during the TD turn on process. Fig. 2 shows the equivalent circuit for each of the four sub-switching intervals which are derived from Fig. 1(a). The TD turn off process is simply the reverse of the turn on process.

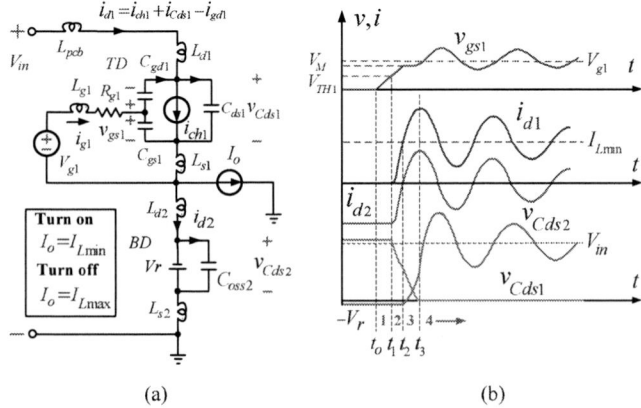

(a)   (b)

Fig. 1(a) Consolidated TD turn on and turn off equivalent circuit from which four sub-interval equivalent circuits for both TD turn on and turn off can be constructed. Fig. 1(b) TD turn on main switching waveforms. Four Sub-intervals labeled as 1, 2, 3 and 4 are shown along with switching instants from t_o to t_3.

979-8-3315-1612-3/25 $31.00 © 2025 IEEE   1159

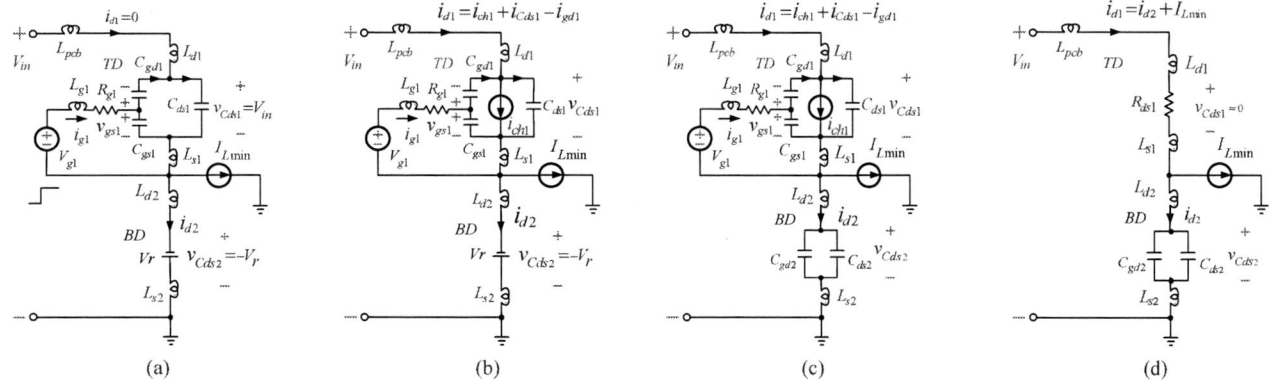

(a) (b) (c) (d)

Fig. 2 TD turn on equivalent circuits for the four Sub-intervals labeled from 1 to 4 in Fig. 1(b).

With the equivalent circuit for each Sub-interval, a set of first-order, simultaneous state equations, $\frac{dX}{dt} = AX + BU$, can be established. The state variable vector X includes a number of independent unknowns $X = [\; i_{g1} \; v_{gs1} \; v_{Cds1} \; i_{d1} \; v_{Cds2} \;]^{\mathrm{T}}$. Because the set of state equations are mutually coupled, deriving a single variable equation using any variable in the state vector X results in a high-order differential equation, which generally has no closed form solution. In this Section, a simplified third-order closed form analytical model is derived based on two key simplified assumptions for both TD turn on and turn off.

A. The New Reduced Order Turn On Switching Model

As stated previously, the TD turn on switching process is divided into four sub-intervals, each of which is discussed in detail.

1) Sub-interval 1: Turn on Delay for $t_o \leq t \leq t_1$

The equivalent circuit in this Sub-interval is shown in Fig. 2(a). This sub-interval starts when the gate drive voltage is applied to the TD gate-source terminals at $t = t_o$ with the drain current equal to zero. The applied voltage charges the gate capacitance C_{iss1}, and the gate-source voltage $v_{gs1}(t)$ rises exponentially. The equations governing the TD gate loop in the Sub-interval 1 are given by

$$(L_{g1} + L_{s1})\frac{di_{g1}}{dt} + R_{g1}i_{g1} + v_{gs1} = V_{g1}, \quad (1)$$

$$i_{g1}(t) = (C_{gs1} + C_{gd1})\frac{dv_{gs1}}{dt} = C_{iss1}\frac{dv_{gs1}}{dt}. \quad (2)$$

Combing the two expressions, we derive the following second-order equation

$$(L_{g1} + L_{s1})C_{iss1}\frac{d^2v_{gs1}}{dt^2} + R_{g1}C_{iss1}\frac{dv_{gs1}}{dt} + v_{gs1} = V_{g1}. \quad (3)$$

Solution to the gate-source voltage $v_{gs1}(t)$ and current $i_{g1}(t)$ is given by

$$v_{gs1}(t) = V_{g1}[1 + (\frac{1}{r_a - r_b})(r_b e^{r_a(t-t_o)} - r_a e^{r_b(t-t_o)})], \quad (4)$$

$$i_{g1}(t) = V_{g1}C_{iss1}(\frac{r_a r_b}{r_a - r_b})[e^{r_a(t-t_o)} - e^{r_b(t-t_o)}], \quad (5)$$

where r_a and r_b are the roots of the characteristic equation (3).

The turn on delay in the Sub-interval 1 ends when the TD gate-source voltage $v_{gs1}(t)$ reaches the threshold V_{TH1} at $t = t_1$.

2) Sub-interval 2: TD and BD Current Commutation for $t_1 \leq t \leq t_2$

The equivalent circuit for the Sub-interval 2 is depicted in Fig. 2(b). As illustrated in Fig. 1(b), once the TD gate-source voltage $v_{gs1}(t)$ rises beyond the device threshold voltage V_{TH1}, the TD channel current $i_{ch1}(t)$, under the control of $v_{gs1}(t)$, begins to conduct current, and current commutation between TD and BD takes place. The state equations in this Sub-interval are

$$\begin{cases} L_{g1}\frac{di_{g1}}{dt} + L_{s1}\frac{d(i_{d1}+i_{g1})}{dt} + R_{g1}i_{g1} + v_{gs1} = V_{g1}, & (6) \\[2mm] L_p\frac{di_{d1}}{dt} + v_{Cds1} = V_{in} + V_r, & (7) \\[2mm] C_{iss1}\frac{dv_{gs1}}{dt} - C_{gd1}\frac{dv_{Cds1}}{dt} = i_{g1}(t), & (8) \\[2mm] i_{ch1}(t) = g_{fs1}[v_{gs1} - V_{TH1}], & (9) \\[2mm] i_{d1}(t) = i_{ch1} + C_{oss1}\frac{dv_{Cds1}}{dt} - C_{gd1}\frac{dv_{gs1}}{dt}, & (10) \end{cases}$$

where L_p represents collectively the sum of all inductances in the power loop, i.e. $L_p = L_{d1} + L_{s1} + L_{d2} + L_{s2} + L_{d2} + L_{pcb}$.

Expressions from (6) to (10) represent a minimum fourth-order system if any of the independent variable is used to derive a differential equation characterizing this Sub-interval. To obtain a closed form, reduced order model, two simplified assumptions are made. The first assumption is to assume that once the TD drain current $i_{d1}(t)$ starts to flow, its slew rate $\frac{di_{d1}}{dt}$ is significantly larger than the gate drive current $i_{g1}(t)$ slew rate $\frac{di_{g1}}{dt}$, i.e. $\frac{di_{g1}}{dt} = \frac{di_{d1}}{dt}$. As a result, it is assumed that $\frac{di_{g1}}{dt} = 0$ in (6). Similarly, the gate-source voltage $v_{gs1}(t)$ has risen in the vicinity or already in the Miller Plateau region.

979-8-3315-1612-3/25 $31.00 © 2025 IEEE   1160

Consequently, its slew rate $\frac{dv_{gs1}}{dt}$ can be assumed to be zero, that is $\frac{dv_{gs1}}{dt} = 0$ in (10). This second assumption implies that the capacitive current contribution $C_{gd1}\frac{dv_{gs1}}{dt}$ is insignificant compared with the term $C_{oss1}\frac{dv_{Cds1}}{dt}$ in (10), therefore the former is neglected.

With these two simplified assumptions, the following third-order differential equation with respect to the TD drain current $i_{d1}(t)$ is derived using the initial conditions at the start of the Sub-interval $t = t_1$

$$a_1\frac{d^3 i_{d1}}{dt^3} + b_1\frac{d^2 i_{d1}}{dt^2} + c_1\frac{di_{d1}}{dt} + i_{d1} = I_a, \qquad (11)$$

$$i_{d1}(t_1)=0, \ \frac{di_{d1}}{dt}\Big|_{t=t_1}=0, \ \frac{d^2 i_{d1}}{dt^2}\Big|_{t=t_1}=0. \qquad (12)\text{-}(14)$$

Coefficients in (11) are given by

$$\begin{cases} a_1 = R_{g1} L_p C_{iss1} C_{oss1}, & (15) \\ b_1 = L_p(R_{g1} g_{fs1} C_{gd1} + C_{oss1}), & (16) \\ c_1 = R_{g1} C_{iss1} + g_{fs1} L_{s1}, & (17) \\ I_a = g_{fs1}(V_{g1} - V_{TH1}). & (18) \end{cases}$$

APPENDIX A outlines the solution process to derive a closed form formulae. The drain current $i_{d1}(t)$ is approximately equal to

$$i_{d1}(t) \approx I_a + \sqrt{C_1^2 + C_2^2}\, e^{-\sigma_1(t-t_1)} \cos[\omega_{d1}(t-t_1) - \alpha_1]$$
$$+ C_3 e^{-p_1(t-t_1)}. \qquad (19)$$

It is seen that the drain current dynamics in this brief Sub-interval can be modeled by a combination of a cosine function and an exponential function. Once $i_{d1}(t)$ is known, other electrical quantities such as TD drain-source voltage $v_{Cds1}(t)$, channel current $i_{ch1}(t)$, gate-source voltage $v_{gs1}(t)$ or any other unknowns can be readily obtained from (6) to (10).

3) Sub-interval 3: TD Current Rises Beyond Inductor I_{Lmin} Until Drain-Source Voltage $v_{Cds1}(t)$ Reaches Zero for $t_2 \le t \le t_3$

The equivalent circuit for the Sub-interval 3 is shown in Fig.2(c). At the beginning of this Sub-interval, the entire inductor current has been transferred to the TD. Unlike the Si MOSFETs, eGaN HEMTs do not possess reverse recovery characteristics. However, the TD drain current continues to flow through BD as a result of discharging the TD drain-source capacitance C_{oss1} and charging the BD C_{oss2}. The TD channel current is still controlled by its gate-source voltage. The state equations in this Sub-interval are similar to the ones from (6) to

(10) in the previous Sub-interval, and only (7) needs to be modified according to Fig. 2(c)

$$L_p\frac{di_{d1}}{dt} + v_{Cds1} + v_{Cds2} = V_{in}. \qquad (20)$$

In addition, since BD no longer carries load current, the TD drain-source current is therefore equal to

$$i_{d1}(t) = I_{Lmin} + C_{oss2}\frac{dv_{Cds2}}{dt}. \qquad (21)$$

In (21), I_{Lmin} is the minimum (valley) inductor current during the TD turn on transition period. For simplicity, during this Sub-interval, the potential dv_{Cds2}/dt-induced voltage on the BD gate-source capacitance C_{gs2} is not considered, i.e. $v_{gs2}(t) = 0$. This effectively connects C_{gd2} in parallel with C_{ds2} or $C_{oss2} = C_{ds2} + C_{gd2}$ as shown in Fig. 2(c). During this very short Sub-interval, the two simplified assumptions still hold, and consequently, a similar third-order equation as in (11) but with a different set of coefficients can be obtained

$$\frac{a_1}{K_1}\frac{d^3 i_{d1}}{dt^3} + \frac{b_1}{K_1}\frac{d^2 i_{d1}}{dt^2} + c_2\frac{di_{d1}}{dt} + i_{d1} = I_b, \qquad (22)$$

with the coefficients in (22) given by

$$K_1 = (1 + \frac{C_{oss1}}{C_{oss2}} + R_{g1} g_{fs1}\frac{C_{gd1}}{C_{oss2}}), \qquad (23)$$

$$c_2 = (\frac{1}{K_1})[R_{g1} C_{iss1}(1 + \frac{C_{oss1}}{C_{oss2}}) + g_{fs1} L_{s1}], \qquad (24)$$

$$I_b = (\frac{1}{K_1})[g_{fs1}(V_{g1} - V_{TH1}) + I_{Lmin}(\frac{R_{g1} g_{fs1} C_{gd1} + C_{oss1}}{C_{oss2}})]. \qquad (25)$$

Solving (22) follows the same procedure as shown in the Sub-interval 2 and APPENDIX A. $i_{d1}(t)$ is given by

$$i_{d1}(t) \approx I_b - \sqrt{C_4^2 + C_5^2}\, e^{-\sigma_2(t-t_2)} \cos[\omega_{d2}(t-t_2) + \alpha_2]$$
$$+ C_6 e^{-p_2(t-t_2)}. \qquad (26)$$

Coefficients C_4, C_5 and C_6 in (26) are computed based on the initial conditions of $i_{d1}(t)$ and its first-order derivative derived in the previous Sub-interval evaluated at $t = t_2$. The parameters in (26) are

$$\omega_{o2} \approx \sqrt{\frac{c_2 K_1}{a_1}}, \ \sigma_2 = \sigma_1, \ \omega_{d2} = \sqrt{\omega_{o2}^2 - \sigma_2^2}, \ p_2 \approx \frac{1}{c_2}. \qquad (27)\text{-}(30)$$

With the drain current $i_{d1}(t)$ given in (26), the BD and TD drain-source voltages $v_{Cds2}(t)$, $v_{Cds1}(t)$ can be found by using (21) and (20), respectively. In particular, the BD drain-source voltage $v_{Cds2}(t)$ is equal to

$$v_{Cds2}(t) = \frac{1}{C_{oss2}}\int_{t_2}^{t}(i_{d1} - I_{Lmin})dt. \qquad (31)$$

Closed form $i_{ch1}(t)$ expressions for both Sub-interval 2 and 3 are computed next. For the Sub-interval 2, the TD channel

979-8-3315-1612-3/25 $31.00 © 2025 IEEE

current $i_{ch1}(t)$ can be evaluated by using (7), (10). It follows that

$$i_{ch1}(t)=i_{d1}-C_{oss1}\frac{dv_{Cds1}}{dt}=i_{d1}+L_pC_{oss1}\frac{d^2i_{d1}}{dt^2}, t_1\le t\le t_2 \quad (32)$$

Likewise, for the Sub-interval 3, $i_{ch1}(t)$ is found by using (10), (20) and (21)

$$i_{ch1}(t)=i_{d1}-C_{oss1}\frac{dv_{Cds1}}{dt}=(1+\frac{C_{oss1}}{C_{oss2}})i_{d1}+L_pC_{oss1}\frac{d^2i_{d1}}{dt^2}$$
$$-\frac{C_{oss1}}{C_{oss2}}I_{L\min}. \quad t_2\le t\le t_3 \quad (33)$$

In obtaining both (32) and (33), Equations (19) and (26) are utilized to compute the second derivative of $i_{d1}(t)$. The $C_{gd1}\dfrac{dv_{gs1}}{dt}$ term in (10) is neglected because of the second simplified assumption. It should be noted that having an analytical expression for the TD channel current is of particular interest and importance to reveal the TD turn on and turn off switching behavior because experimental measurement cannot uncover its time varying properties in these two Sub-intervals.

4) Sub-interval 4: Ringing Between the BD Output Capacitance C_{oss2} and Power Loop Inductance L_p for $t \ge t_3$

The equivalent circuit describing this Sub-interval is illustrated in Fig. 2(d). Once the BD drain-source voltage $v_{Cds2}(t)$ rises to the input voltage, TD is nearly or fully enhanced, and its drain-source is "shorted" if the voltage drop cross the TD on resistance is neglected. Resonance begins between the BD output capacitance C_{oss2} and the power loop inductance L_p. The BD drain-source voltage $v_{Cds2}(t)$ satisfies the following equation in the Sub-interval 4

$$L_pC_{oss2}\frac{d^2v_{Cds2}}{dt^2}+r_pC_{oss2}\frac{dv_{Cds2}}{dt}+v_{Cds2}=V_{in}, \quad (34)$$

with the following two initial conditions,

$$v_{Cds2}(t_3)=V_{in}, \quad i_{d1}(t_3)=C_{oss2}\frac{dv_{Cds2}}{dt}\Big|t=t_3. \quad (35)\text{-}(36)$$

Equ. (34) has the following solution

$$v_{Cds2}(t)=V_{in}+i_{d1}(t_3)Z_oe^{-\sigma_3(t-t_3)}\sin[\omega_{d3}(t-t_3)], \quad (37)$$

where $\omega_{d3}=\sqrt{\omega_{o3}^2-\sigma_3^2}$ in (37) and $\omega_{o3}=\dfrac{1}{\sqrt{L_pC_{oss2}}}$ are the natural radian frequencies with or without circuit damping, respectively. $\sigma_3=\dfrac{r_p}{2L_p}$ is the damping factor due to various low-frequency and high-frequency dependent loop resistances. $Z_o=\sqrt{L_p/C_{oss2}}$ represents the characteristic impedance of the RLC power loop.

Power loop ringing current $i_{d1}(t)$ can be found by differentiating $v_{Cds2}(t)$ in (37).

B. The New Reduced Order Turn Off Switching Model

Similar to the TD turn on process, the entire turn off process can also be divided into four sub-intervals illustrated in Fig. 3. The TD turn off key switching waveforms are also sketched in Fig. 3. Since TD turn off replicates the turn on process in reverse order, only differences need to be highlighted in this Section. Fig. 4 depicts the four TD turn off equivalent circuits.

1) Sub-interval 1: Turn off Delay for $t_4 \le t \le t_5$

The equivalent circuit is shown in Fig. 4(a). The TD turn off process starts when its gate drive is removed at $t = t_4$, i.e. $V_{g1} = 0$ in (6). The gate-source voltage $v_{gs1}(t)$ is equal to

$$v_{gs1}(t)=V_{g1}(\frac{1}{r_a-r_b})[r_ae^{r_b(t-t_4)}-r_be^{r_a(t-t_4)}]. \quad (38)$$

This Sub-interval ends when $v_{gs1}(t)$ reaches

$$v_{gs1}(t_4)=V_{TH1}+\frac{I_{Lmax}}{g_{fs1}}. \quad (39)$$

2) Sub-interval 2: TD Voltage $v_{Cds1}(t)$ Rising and Both the Channel Current and the Drain Current Falling for $t_5 \le t \le t_6$

The equivalent circuit is the same as that during the TD turn on transition in the Sub-interval 3 with $V_{g1} = 0$ as shown in Fig. 4(b). This Sub-interval replicates the turn on Sub-interval 3 with the instantaneous inductor current at its peak value I_{Lmax} rather than at its minimum value I_{Lmin}. The TD gate voltage loop equation is derived simply by setting $V_{g1} = 0$ in (6). Because the gate capacitance of eGaN HEMTs is significantly smaller than that of the comparable Si MOSFETs, discharging the gate-source capacitance C_{gs1} is much faster. Once the gate-source voltage $v_{gs1}(t)$ is below the value given by (39), the TD channel current $i_{ch1}(t)$ gets cut off rapidly within a very short time. The TD drain current $i_{d1}(t)$ decreasing slew rate, the on the other hand, is limited by the power loop inductance and the initial slow charging of the much larger drain-source capacitance C_{oss1} when TD drain-source voltage v_{Cds1} is low.

The third-order differential equation of $i_{d1}(t)$ predicting the transient behavior in this Sub-interval is the same as that in (22) except for replacing I_b in (25) by I_{b1} given by

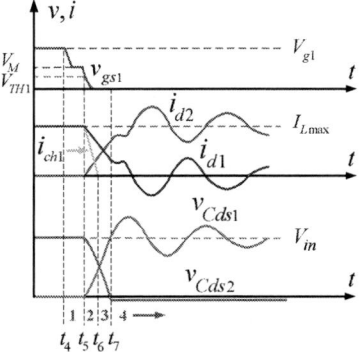

Fig. 3 TD turn off switching waveforms. Similar to turn on process, turn off process is also composed of four Sub-intervals labeled as 1 to 4. The switching instances are marked from t_4 to t_7.

979-8-3315-1612-3/25 $31.00 © 2025 IEEE   1162

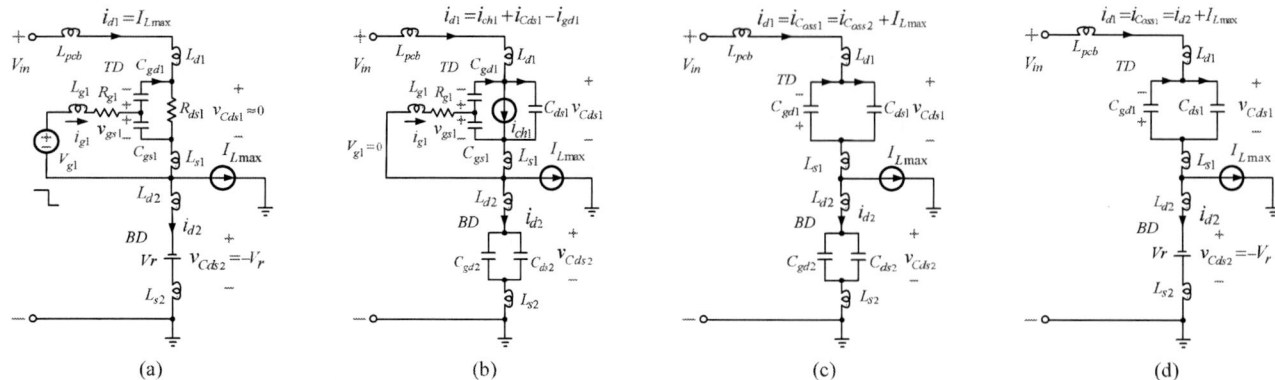

Fig. 4 eGaN HEMT turn off equivalent circuits for the four sub-intervals. Fig. 4(b) is similar to Fig. 2(c) with TD gate drive voltage V_{g1} and the inductor changed to zero and I_{Lmax}, respectively.

$$I_{b1} = (\frac{1}{K_1})[I_{Lmax}(\frac{R_g 1 g_{fs1} C_{gd1} + C_{oss1}}{C_{oss2}}) - g_{fs1} V_{TH1}]. \quad (40)$$

The drain current $i_{d1}(t)$ is found to be

$$i_{d1}(t) \approx I_{b1} - \sqrt{C_7^2 + C_8^2} e^{-\sigma_2(t-t_5)} \cos[\omega_{d2}(t-t_5) - \alpha_3]$$
$$+ C_9 e^{-p_2(t-t_5)}. \quad (41)$$

Calculation of the BD and TD drain-source voltages $v_{Cds2}(t)$, $v_{Cds1}(t)$, TD channel current $i_{ch1}(t)$ given $i_{d1}(t)$ in (41) utilizes the equations (31), (20) and (10). This Sub-interval ends when the TD channel current drops to zero while the drain current decreases.

It should be noted that there exists a second, much less frequent scenario where the BD drain-source voltage drops to zero before the TD channel current reaches zero. Which scenario takes place depends largely on the converter parasitics and eGaN HEMT parameters [1]. Since the analysis for the second case is very similar to the turn on Sub-interval 2 without the presence of the TD gate drive voltage, it is not repeated.

3) Sub-interval 3: TD Drain-Source Voltage $v_{ds1}(t)$ Rises to the Input Voltage for $t_6 \leq t \leq t_7$

The equivalent circuit in this Sub-interval is shown in Fig. 4(c). Because the TD channel no longer conducts current, it must flow through the TD output capacitance C_{oss1}. The current is composed of the inductor current I_{Lmax} and the power loop *RLC* resonant current. The equations modeling this Sub-interval are given by

$$r_p i_{d1} + L_p \frac{di_{d1}}{dt} + v_{Cds1} + v_{Cds2} = V_{in}, \quad (42)$$

$$i_{d1} = C_{oss1} \frac{dv_{Cds1}}{dt} = C_{oss2} \frac{dv_{Cds2}}{dt} + I_{Lmax}. \quad (43)$$

Combining the two expressions leads to

$$L_p C_{oss} \frac{d^2 i_{d1}}{dt^2} + r_p C_{oss} \frac{di_{d1}}{dt} + i_{d1} = \frac{C_{oss1}}{C_{oss1} + C_{oss2}} I_{Lmax}, \quad (44)$$

where $C_{oss} = (C_{oss1} // C_{oss2})$, and r_p is the equivalent power loop resistance.

It is noted that during this Sub-interval, Equation (44) predicts that the resonant frequency is higher owing to reduction of the equivalent capacitance of the LC resonant circuit of the power loop. If $C_{oss1} = C_{oss2}$, the resonant frequency is increased by a factor of $\sqrt{2}$. Solving this equation satisfying the initial conditions of $i_{d1}(t)$ and its first-order derivative at $t = t_6$ is straightforward.

Once $i_{d1}(t)$ is derived, the TD and BD drain-source voltages $v_{Cds1}(t)$ and $v_{Cds2}(t)$ can be found by integrating (43). This Sub-interval completes when the drain-source voltage $v_{ds1}(t)$ reaches V_{in} at $t = t_7$.

4) Sub-interval 4: Ringing Between TD Output Capacitance C_{oss1} and Power Loop Inductance L_p for $t \geq t_7$

The equivalent circuit for this Sub-interval is illustrated in Fig. 2(d). Unlike the previous Sub-interval where both drain-source capacitances are present, only the TD output capacitance C_{oss1} exists in this Sub-interval. The drain-source voltage $v_{Cds1}(t)$ equation is similar to (37) with $v_{Cds2}(t)$ replaced by $v_{Cds1}(t)$ and C_{oss2} replaced by C_{oss1}, respectively. The solution to $v_{Cds1}(t)$ is found to be

$$v_{Cds1}(t) = (V_{in} + V_r) + i_{d1}(t_7)\sqrt{\frac{L_p}{C_{oss1}}} e^{-\sigma_4(t-t_7)} \sin[\omega_{d4}(t-t_7)], \quad (45)$$

where $\omega_{d4} = \sqrt{\omega_{o4}^2 - \sigma_4^2}$ in (45) and $\omega_{o4} = \frac{1}{\sqrt{L_p C_{oss1}}}$ are the natural radian frequencies with or without circuit damping, respectively. $\sigma_4 = \sigma_3 = \frac{r_p}{2L_p}$ is the damping factor due to various low-frequency and high-frequency loop resistances, and $\sqrt{\frac{L_p}{C_{oss1}}}$ represents the characteristic impedance of the *RLC* power loop.

979-8-3315-1612-3/25 $31.00 © 2025 IEEE

III. PERFORMANCE EVALUATION OF THE NEW REDUCED ORDER ANALYTICAL MODEL

A 52V to 24V, 100W open-loop Buck converter was built and tested using an evaluation board from an IC manufacture as shown in Fig. 5 to validate the theoretical analysis. The eGaN HEMTs are at the bottom side of the PCB board. The converter specification, and the main parameters of the eGaN HEMT are listed in TABLE I. Two eGaN HEMTs used in the evaluation are the Infineon GS61008P. The device has a separate Kelvin source pin [9]. The function of this additional pin is to remove the negative feedback effect of the TD common-source inductance on the TD gate drive voltage to allow for even faster switching speed. To consider a more general eGaN HEMT structure, this pin function is not used. Because of the test board layout limitation, the TD drain current was not measured. Model validation is performed using measurable TD and BD voltage waveforms. Finally, in addition to experimentation, circuit simulation was performed using vendor-provided eGaN HEMT LTSpice model to provide additional evidence to support validation of the new analytical model.

In the following discussion, to demonstrate the capability of the new model to predict eGaN HEMT switching performance without adding additional complexity, fixed values provided in the component data sheet are used in the calculation. Fig. 6 shows the converter steady state inductor current and switch-node voltage waveforms from which detailed TD switching transition waveforms at both turn on and turn off are zoomed in to compare with the new analytical model prediction data.

A. eGaN HEMT Reduced Order Turn on Model Validation

To avoid having to many waveforms in one plot to cause confusion, the predicted and simulated TD switching waveform plots are split into two. Fig. 7(a) illustrates the predicted TD turn on switching waveforms. The upper plot shows the TD gate drive voltage $v_{gs1}(t)$ excluding the effect of the common-

Fig. 5 Photo of the evaluation board. The eGaN HEMTs are at the bottom side of the PCB board.

TABLE I.
POWER CONVERTER AND eGaN HEMT PARAMETERS

BUCK CONVERTER MAIN SPECIFICATIONS					
V_{in}(V)	V_{out} (V)	I_{out} (A)	I_{Lmax}(A)	I_{Lmin} (A)	f_{sw} (kHz)
52.0	24.0	4.0	5.0	3.0	250

GS61008P eGaN HEMT TYPICAL DATA SHEET PARAMETERS [9]								
$L_{d1,2}$ (nH)	$L_{s1,2}$ (nH)	R_{g1} (Ω)	C_{iss1} (pF)	C_{rss1} (pF)	C_{oss1} (pF)	L_{g1} (nH)	V_{TH1} (V)	g_{fs1} (S)
0.2	0.5	0.8	600	12	385	1.0	1.7	25

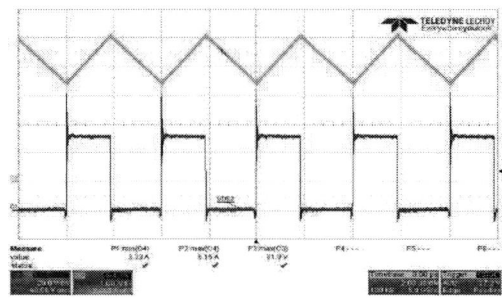

Fig. 6 Steady state inductor current (Green) and switch-node voltage (Blue) waveforms of the experimental 52V to 24V, 100W Buck DC-DC converter.

source inductance L_{s1}. This is the true drive voltage at the device internal gate-source terminals. The power loop TD drain current $i_{d1}(t)$ and the inductor current I_{Lmin} waveforms also plotted on the same figure. The lower Fig. 7(a) shows the predicted drain-source voltage $v_{Cds1}(t)$ and $v_{Cds2}(t)$ waveforms directly cross the TD and BD drain-source internal terminals. Plots of the drain-source voltage $v_{DS1}(t)$ and $v_{DS2}(t)$ waveforms including the inevitable TD and BD packaging inductances are superimposed on the same plot. The difference between $v_{Cds1}(t)$, $v_{Cds2}(t)$ and $v_{DS1}(t)$, $v_{DS2}(t)$ is the additional term given by

$$v_{DS1} = v_{Cds1} + (L_{d1} + L_{s1})\frac{di_{d1}}{dt}, \quad (46)$$

$$v_{DS2} = v_{Cds2} + (L_{d2} + L_{s2})\frac{di_{d1}}{dt}. \quad (47)$$

As shown in the predicted $v_{DS1}(t)$ switching waveforms in Fig. 7(a), when $v_{Cds1}(t)$ reaches zero or $v_{Cds2}(t)$ rises to the input voltage V_{in} at the beginning of the Sub-interval 4 at $t=t_3$, ringing begins owing to the resonance between the BD output capacitance C_{oss2} and power loop inductance L_p. The device external terminal drain-source voltage $v_{DS1}(t)$, which is measurable in the lab environment, is not zero because of the second term in (46). With regard to $v_{Cds2}(t)$ and $v_{DS2}(t)$, the second term of (47) and (37) predict that the induced voltage on the parasitic inductances L_{d2} and L_{S2} is out of phase with respect to $v_{Cds2}(t)$. As a result, the magnitude of $v_{Cds2}(t)$ is higher than what is usually measured in experiments. This clearly shows in Fig. 7(a) lower plot. This is an important design consideration when the eGaN HEMT voltage stress is determined, particularly when PCB parasitic inductances are large.

Fig. 7(b) shows LTSpice simulation waveforms under the same operating condition. With the assistance of simulation, additional insight into some of the unmeasurable electrical quantities in the equivalent circuits can be exposed to validate the proposed new model. Simulation accuracy depends largely on the Spice model accuracy. Compared with simulation and experimental data in Fig. 7(c), it is evident that the model can perform satisfactorily even fixed value eGaN HEMT parameters are utilized. It is observed that at the beginning of the BD drain-source voltage $v_{Cds2}(t)$ rise period, the voltage rise slew rate is quite low. This is because C_{oss} and C_{rss} have much larger values primarily at low drain-source voltages.

B. eGaN HEMT Reduced Order Turn off Model Validation

Fig. 8(a) shows the predicted TD turn off switching transient

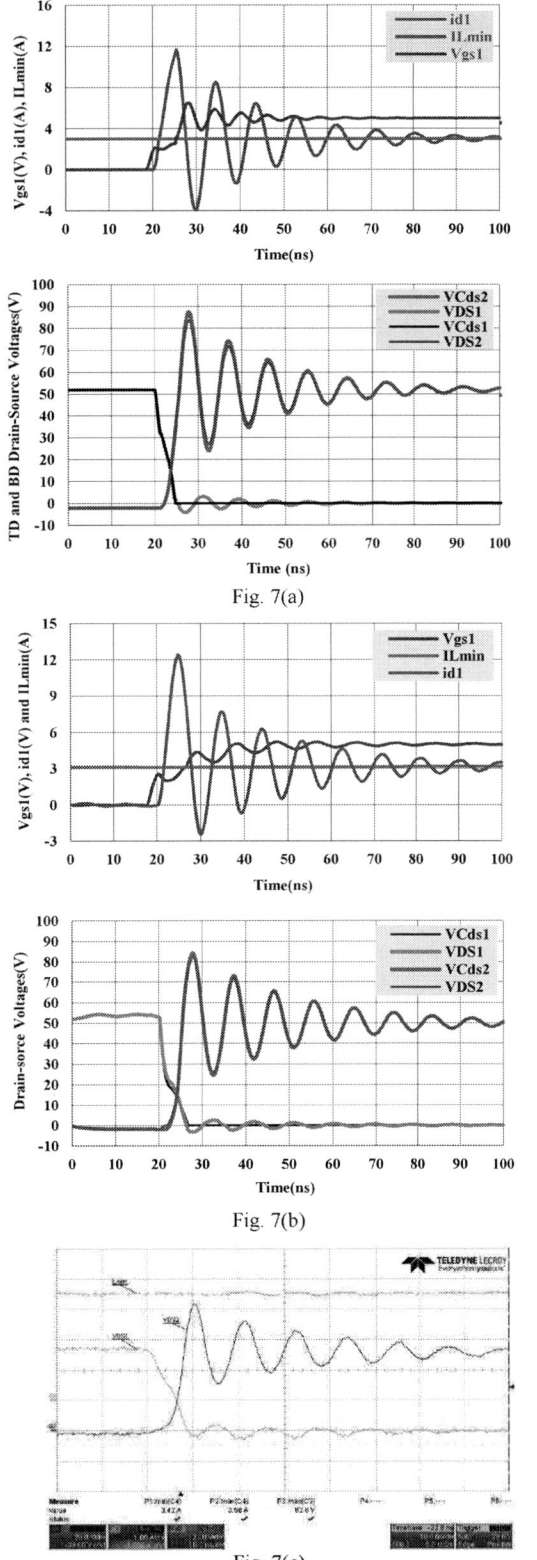

Fig. 7(a)

Fig. 7(b)

Fig. 7(c)

Fig. 7 TD turn on switching transition waveforms. Fig. 7(a) Upper and lower plots - model-predicted. Fig. 7(b) Upper and lower plots - LTSpice-simulated. Fig. 7(c) – Measured waveforms. Yellow trace – $v_{DS1}(t)$, Blue trace – $v_{DS2}(t)$, and Green trace – Minimum inductor current I_{Lmin}.

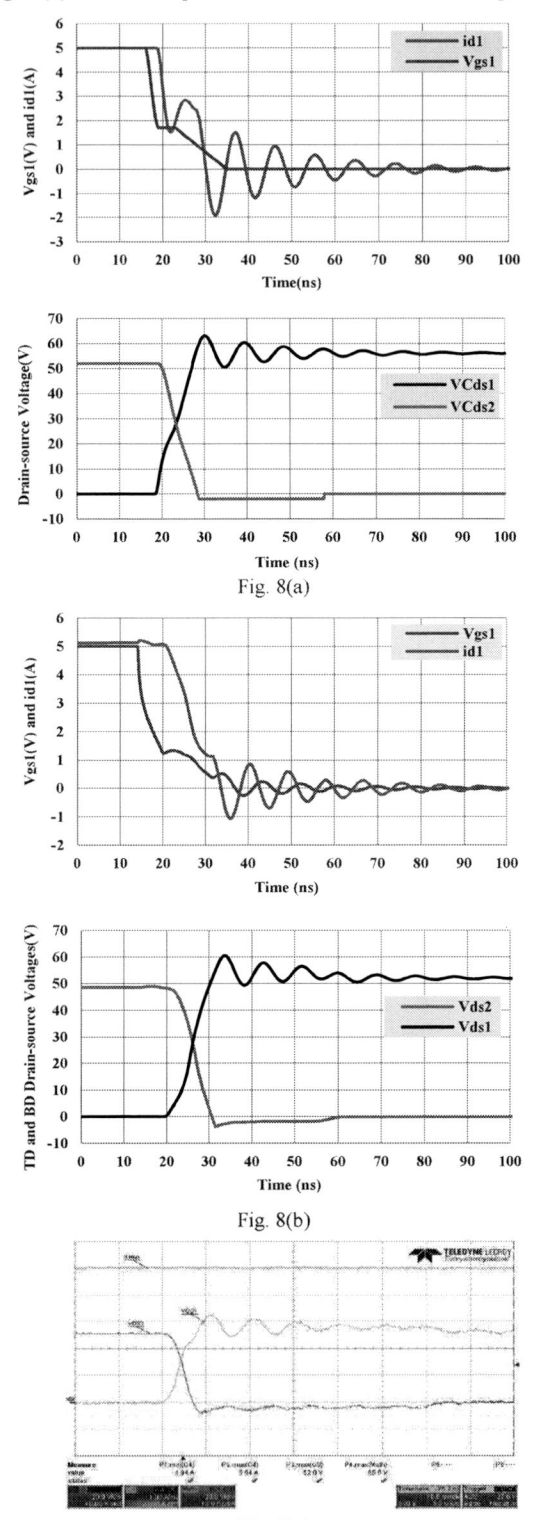

Fig. 8(a)

Fig. 8(b)

Fig. 8(c)

Fig. 8 TD turn off switching transition waveforms. Fig. 8(a) Predicted, Fig. 8(b) LTSpice simulated, and Fig. 8(c) – Measured. Yellow trace – $v_{DS1}(t)$, Blue trace – $v_{DS2}(t)$, and Green trace – Maximum inductor current I_{Lmax}

979-8-3315-1612-3/25 $31.00 © 2025 IEEE   1165

waveforms. For gate-source voltage $v_{gs1}(t)$, once the initial turn off delay time is reached, simple linear approximation is used. Except for some discrepancy between calculated and simulated waveforms at the start of the drain current decline period, good correlation between the two can be observed.

C. Prediction and Simulation of Other Voltage and Current Transient Waveforms

To demonstrate the advantage of having a closed form analytical model to better understand the eGaN HEMT transient switching behavior, Fig. 9 depicts the TD channel current $i_{ch1}(t)$, drain current $i_{d1}(t)$, output capacitance current $i_{Coss1}(t)$, and gate-source voltage $v_{gs1}(t)$ turn on switching waveforms. The plot pictures two ultra-fast channel current build up events first when the gate-source voltage rises above the threshold voltage, and second when the TD is nearly fully enhanced in the Miller Plateau region. The difference between $i_{ch1}(t)$ and $i_{Coss1}(t)$ is equal to the power loop drain current $i_{d1}(t)$. Once the TD gate gets nearly or fully enhanced, the drain-source voltage $v_{Cds1}(t)$ drops close to zero. The drain current flows through the TD on resistance, carrying both the resonant current and load current.

IV. CONCLUSIONS

A simple and design-oriented eGaN HEMT switching model is presented in this paper. Unlike the existing models which can only be solved by numerical means, the new model is expressed in the closed form, making it possible to predict eGaN HEMT switching loss on an analytical basis for practical designs. Model accuracy can be improved by considering the nonlinear properties of the device parasitics.

APPENDIX A

The characteristic equation of (11) is a third order polynomial, which can be approximately factored into a first-order term and a second-order term shown in (A1) provided that the certain criteria are met. The natural frequency and quality factor of the second order polynomial are related closely with the third order polynomial coefficients

$$a_1 r^3 + b_1 r^2 + c_1 r + 1 \approx (1 + \frac{r}{Q_1 \omega_{o1}} + \frac{r^2}{\omega_{o1}^2})(r + p_1). \text{(A1)}$$

where $\omega_{o1} \approx \sqrt{\dfrac{c_1}{a_1}}$ and $Q_1 \approx \dfrac{\sqrt{a_1 c_1}}{b_1}$. (A2)-(A3)

The three roots of the characteristic equation are equal to

$$r_1, r_2 \approx -\frac{\omega_{o1}}{2Q_1} \pm j\sqrt{\omega_{o1}^2 - \sigma_1^2} = -\sigma_1 \pm j\omega_{d1}, \ r_3 = -p_1 \approx \frac{1}{c_1}.$$
(A4)-(A5)

Once the general solution to (11) is found, its coefficients C_1 to C_3 are determined by the three initial conditions (12)-(14). The final solution to (11) is approximately given by

$$i_{d1}(t) \approx I_a + \sqrt{C_1^2 + C_2^2}\, e^{-\sigma_1(t - t_1)} \cos[\omega_{d1}(t - t_1) - \alpha_1]$$
$$+ C_3 e^{-p_1(t - t_1)}, \tag{A6}$$

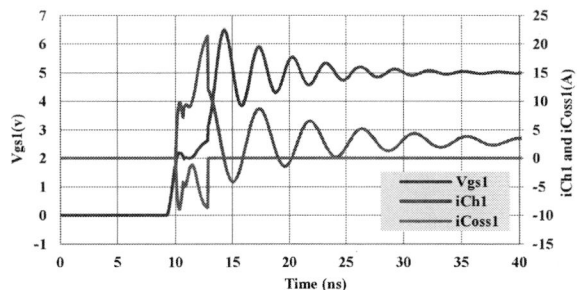

Fig. 9 Predicted TD channel current $i_{ch1}(t)$, output capacitive current $i_{Coss1}(t)$, and gate-source voltage $v_{gs1}(t)$.

with C_1 to C_3 and α_1 given by

$$C_1 = \frac{-I_a p_1(2\sigma_1 - p_1)}{p_1(2\sigma_1 - p_1) - \omega_{o1}^2}, \ C_2 = (\frac{-I_a p_1}{\omega_{d1}})[\frac{\sigma_1(2\sigma - p_1) - \omega_{o1}^2}{p_1(2\sigma_1 - p_1) - \omega_{o1}^2}],$$

$$C_3 = I_a [\frac{\omega_{o1}^2}{p_1(2\sigma_1 - p_1) - \omega_{o1}^2}], \ \alpha_1 = \tan^{-1}(\frac{C_2}{C_1}). \text{ (A7)-(A10)}$$

For the TD turn on Sub-interval 3, the same procedure can be applied to find the closed for solution to (22).

REFERENCES

[1] K. Wang, X. Yang, H. Li, H. Ma, X. Zeng, and W. Chen, "An Analytical Switching Process Model of Low-Voltage eGaN HEMTs for Loss Calculation," *IEEE Transactions on Power Electronics*, vol. 31, no. 1, pp. 635-647, Jan. 2016.

[2] K. Wang, M. Tian, H. Li, F. Zhang, X. Yang and L. Wang, "An improved switching loss model for a 650V enhancement-mode EGaN transistor," in *Proc. IEEE Annual Southern Power Electronics Conference (SPEC)*, 2016, pp. 1-6.

[3] M. R. Ahmed, R. Todd and A. J. Forsyth, "Predicting SiC MOSFET Behavior Under Hard Switching, Soft-Switching, and False Turn-On Conditions," in *IEEE Transactions on Industrial Electronics*, vol. 64, no. 11, pp. 9001-9011, Nov. 2017.

[4] [5] R. Xie, H. Wang, G. Tang, X. Yang and K. J. Chen, "An Analytical Model for False Turn-On Evaluation of High-Voltage Enhancement-Mode eGaN Transistor in Bridge-Leg Configuration," in *IEEE Transactions on Power Electronics*, vol. 32, no. 8, pp. 6416-6433, Aug. 2017, doi: 10.1109/TPEL. 2016.2618349.

[5] J. Chen, Q. Luo, J. Huang, Q. He and X. Du, "A Complete Switching Analytical Model of Low-Voltage eGaN HEMTs and Its Application in Loss Analysis," in *IEEE Transactions on Industrial Electronics*, vol. 67, no. 2, pp. 1615-1625, Feb. 2020, doi: 10.1109/TIE.2019.2891466.

[6] X. Li, Z. Xiong and Y. Liu, "A Charge-Based Analytical Model for Accurate Switching Transient Description of Half-Bridge eGaN HEMTs," *IECON 2023- 49th Annual Conference of the IEEE Industrial Electronics Society*, Singapore, 2023, pp. 1-6, doi: 10.1109/IECON5178. 2023.10312589.

[7] Y. Liu, X. Liu, X. Li and H. Yuan, "Analytical Model and Safe-Operation-Area Analysis of Bridge-Leg Crosstalk of GaN E-HEMT Considering Correlation Effect of Multi-Parameters," in *IEEE Transactions on Power Electronics*, vol. 39, no. 7, pp. 8146-8161, July 2024, doi: 10.1109/TPEL. 2024. 3381638.

[8] Y. Shen, H. Wang, Z. Shen, F. Blaabjerg and Z. Qin, "An analytical turn-on power loss model for 650-V GaN eHEMTs," *2018 IEEE Applied Power Electronics Conference and Exposition (APEC)*, San Antonio, TX, USA, 2018, pp. 913-918, doi: 10.1109/APEC.2018. 8341123.

[9] Infineon/GaN Systems GS61008P data sheet. https://www.infineon.com/cms/en/product/power/gan-hemt-gallium-nitride- transistor/gs61008p-tr/

Proposal of an Alternative Reverse Recovery Calculation Method

Brian DeBoi
Power Modules
Wolfspeed, Inc.
Fayetteville, USA
Brian.DeBoi@wolfspeed.com

Blake Nelson
Power Modules
Wolfspeed, Inc.
Fayetteville, USA
Blake.Nelson@wolfspeed.com

Austin Curbow
Power Modules
Wolfspeed, Inc.
Fayetteville, USA
Austin.Curbow@wolfspeed.com

Abstract— **Quantifying the reverse recovery characteristics of SiC MOSFETs is necessary for generating device datasheets and loss models. However, the existing methods for calculating reverse recovery are based on standards developed for silicon IGBTs do not provide consistent and accurate reverse recovery calculations for SiC MOSFETs. This presents a major challenge to semiconductor manufacturers applying automated processing to measured data. As such, new methods for calculating reverse recovery that are both consistent and provide similar estimations to existing methods are needed. This paper analyzes several calculation methods and applies them to a large empirical dataset of two SiC power modules from different manufacturers to determine their suitability for standardization. It is found that slope extrapolation techniques are the most effective method for calculating reverse recovery.**

Keywords—SiC, MOSFET, Module, Reverse Recovery, Measurement, Calculation, Automation

I. INTRODUCTION

The reverse recovery (RR) behavior of SiC MOSFET body diodes has a large impact on their overall turn-off performance during switching, particularly at high temperatures [1], [2]. For application designers, it is important to consider reverse recovery in the total loss calculations to accurately determine current carrying capability and cooling requirements [3]. For device manufacturers, this requires providing quantified reverse recovery charge and loss metrics in datasheets; in some cases, tabulated reverse recovery loss across operating conditions are also needed to generate accurate models for customers [4], [5]. In both cases, a method to accurately and consistently quantify the reverse recovery loss from the transient waveforms is needed.

There are two main challenges with calculating reverse recovery of SiC MOSFETs. First, the reverse recovery behavior between parts varies significantly based on the semiconductor design. The body diode behavior of a MOSFET has important implications not only in device losses, but also for voltage overshoots caused by the di/dt during the reverse recovery event [6], [7]. Manufacturers will often design their parts to have "soft-recovery" to minimize this overshoot. Overall, the design trade-offs between reverse recovery losses and di/dt yields devices that can have significantly different behaviors that present challenges when applying automation procedures.

The second challenge results from the currently accepted standard calculation method described JEDEC JESD24-10. The calculation method is straight-forward, but its strict definition makes it inconsistent when applied to a broad set of devices. As mentioned, device behavior can change based on the device itself (every semiconductor design will behave differently), and even based on the operating conditions or metrology used to evaluate the device. Employing a rigid definition to such a broad range of behaviors yields inconsistent results that require manual intervention.

The primary contribution of this work is the presentation of an alternative calculation method for reverse recovery that accomplishes two criteria. One, the method provides the same results as the currently accepted standard method in cases where the standard calculation definition is valid. This is important, as the goal of an alternative calculation method is not to create a new definition, but to calculate the original definition with improved robustness and automation capabilities. Two, the method is more flexible and provides a consistent estimation of the reverse recovery charge even in cases where the standard method fails.

To determine the efficacy of the proposed method, this paper applies multiple reverse recovery calculation techniques to two different SiC power modules across a wide range of bus voltages, load currents, gate resistances, and junction temperatures. Each technique is evaluated to determine 1) which provides results most similar to the standard method and 2) which is most consistent across various waveforms.

II. TRADITIONAL REVERSE RECOVERY CALCULATION METHOD

Reverse recovery can be measured using a clamped inductive load (CIL) test circuit, as shown in Fig. 1. Here, the load inductor is placed on the low-side switch so that a current viewing resistor (CVR) with a ground-referenced measurement can be used to measure the reverse recovery current, I_{RR}. The reverse recovery voltage, V_{RR}, is measured across Q2. The high-side device (Q1) is actively switched during the test, while Q2 is held OFF. During the transition from Q1 ON to Q1 OFF, the current will begin freewheeling through the body diode of Q2. Q1 is then turned back ON, and the body diode of Q2 will stop conducting, initiating the reverse recovery event that is measured through the CVR.

979-8-3315-1612-3/25 $31.00 © 2025 IEEE

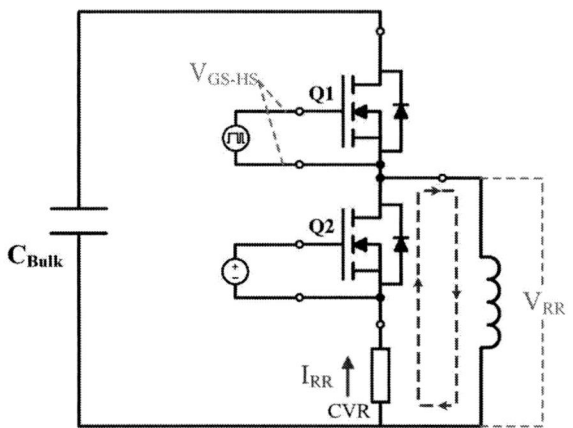

Fig. 2: General reverse recovery measurement circuit with a low-side current measurement

Fig. 2 shows a measured reverse recovery event and the traditional technique described by JEDEC JESD24-10 for quantifying the reverse recovery charge. The method integrates the reverse recovery current between the first two zero crossings, T_{RR1} and T_{RR2}. An example of a correctly determined reverse recovery charge (Q_{RR}) is shown in Fig. 2 (a). However, this method often has limitations for SiC MOSFETs and does not

consistently find the correct upper integration bound T_{RR2}. For example, consider the waveform for the same part (but at different conditions) in Fig. 2 (b). Due to measurement offsets and ringing, the current does not cross the 0 A threshold until well after the RR event. This causes the current to be integrated for a period well after the reverse recovery current is concluded. Note that the reverse recovery energy is determined by integrating the power between T_{RR1} and T_{RR2}, so any errors in these integration limits will translate to the energy calculation.

In the case of Fig. 2 (b), the incorrectly determined integration limit would result in an outlier of the calculated data and require manual adjustment to correct. Such measures are not only time consuming, but also create inconsistencies between the reported reverse recovery charge/energy and the actual definition employed. For example, in Fig. 2 (b), if T_{RR2} were manually moved to the "actual" end of the reverse recovery event, there would then be an inconsistency between the reported definition and the provided quantified value.

In this paper, multiple SiC power modules from different manufacturers are measured using similar laboratory setups (that vary depending on the package layout and the operating voltage/current). Fig. 3 shows an example CIL evaluation kit for 62 mm power modules. These CIL kits feature low-inductance layouts and high-bandwidth metrology. RR

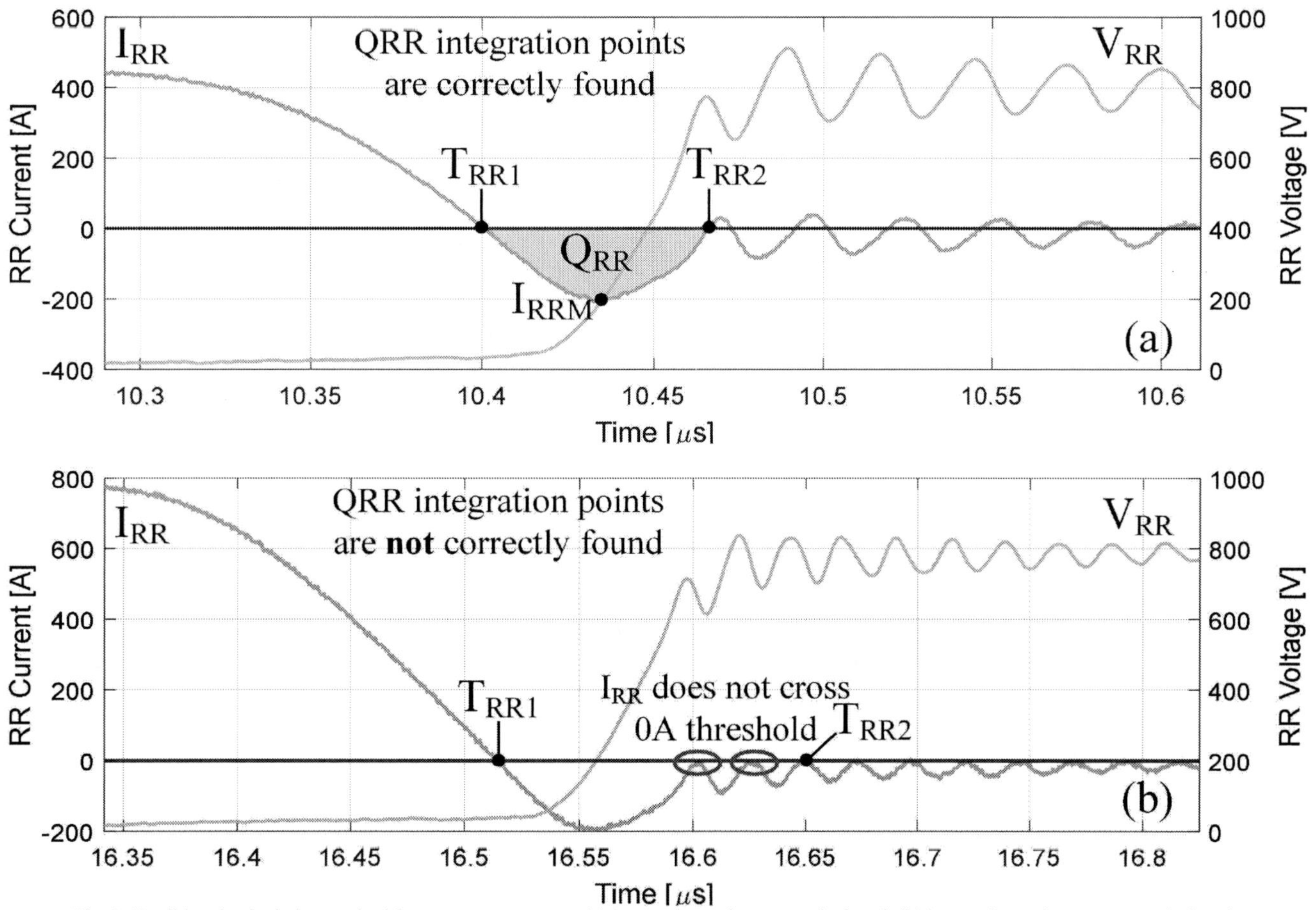

Fig. 1: Traditional calculation method for reverse recovery – (a) integration points correctly found, (b) integration points not correctly found

Fig. 3: Example CIL evaluation kit for reverse recovery measurements

current measurements are made with a T&M Research W-2-0025-4FC 2.5 mΩ current viewing resistor and the voltage measurements are made with a Tektronix THDP0200 1.5 kV differential probe. This configuration minimizes measurement errors and establishes the focus on the actual device behavior, and provides a similar measurement environment for each unique device under test (DUT) considered in this paper.

III. OVERVIEW OF ALTERNATIVE REVERSE RECOVERY CALCULATION METHODS

There are many proposed alternative methods for determining the reverse recovery integration limits in the literature. One of the most common methods is to decrease the threshold at which to find T_{RR2}. This threshold is usually base on the peak reverse recovery current, I_{RRM}. So, for example, if I_{RRM} is -100 A, then T_{RR2} would be found when the rising edge crosses -10 A to -20 A (10% to 20% are typical). This technique is simple and effective, but does come with some limitations. First, because the reverse recovery event is being truncated, these methods do underestimate the reverse recovery charge and energy. Second, these methods are not robust and can still have the same failure mode as the traditional method under some conditions.

The more complex reverse recovery calculation methods studied in this work are shown in Fig. 4. For all methods, the lower integration limit (T_{RR1}) is described by the first 0 A crossing of I_{RR}. Each method then differs in how it selects the upper integration limit (T_{RR2}). The traditional method, as discussed previously, uses the second 0 A crossing to determine T_{RR2}. The vd98 method, originally presented in [4], uses the point at which V_{RR} crosses 98 % of the nominal operating voltage. For example, for an 800 V bus, the point would be selected when V_{RR} is first greater than 784 V. The authors of [4] state that the method suppresses the influence of stray inductance in RR calculations.

The method proposed in this paper, herein described as the "Irrm7520extr" method, or 75:20 method, is performed as follows. First, the peak recovery current, I_{RRM}, is determined. Second, the first points at which the return current crosses 75% of this IRRM and 20% of I_{RRM} are found. In Fig. 4, I_{RRM} is -200 A, so the 75% and 20% points are -150 A and -40 A, respectively. Third, a line is drawn between the 75% and 20% points, and is extended until it crosses zero. The time at which this line crosses zero is selected as T_{RR2} on the waveform. In Fig. 4, this point occurs slightly after traditional method. It may be noted that this would result in integrating positive current, which would decrease the calculated Q_{RR}. To negate this, an added feature of this method is to use the traditional T_{RR2} integration point if a true zero crossing occurs before the 75:20 point.

Before discussing the efficacy of these methods, it is important to discuss what makes a "good" method. All dynamic calculations–switching losses, slew rates, reverse recovery, etc.– are methods to summarize device characteristics as a single quantity. To be useful, these calculations should:

- Represent a useful metric of device behavior
- Be applicable to a wide range of devices and behaviors
- Provide continuous behavior across conditions. For example, a small change in load current should not result in a large, unexplained change in the result
- Require the minimum amount of information to calculate
- Be unsensitive to measurement errors

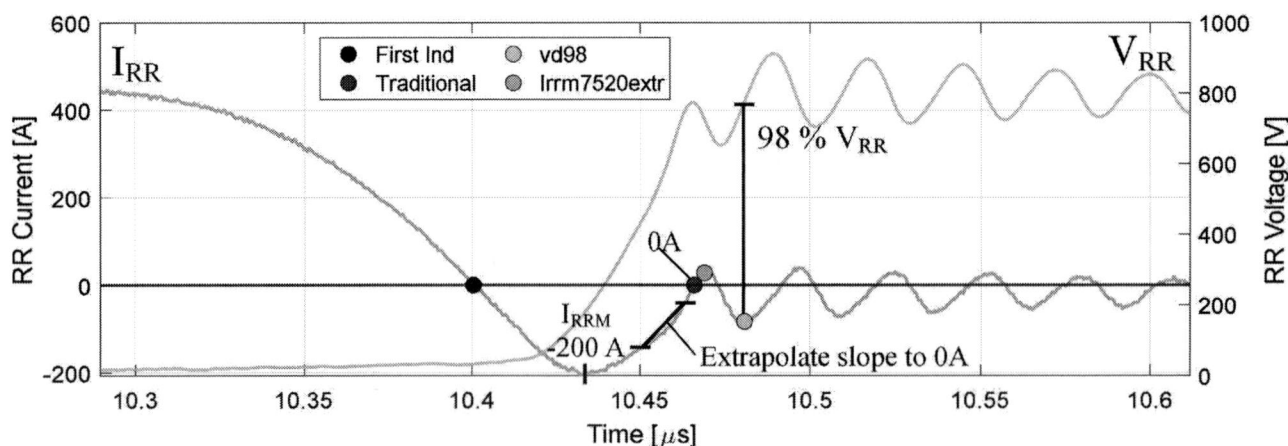

Fig. 4: Description of reverse recovery calculation methods

979-8-3315-1612-3/25 $31.00 © 2025 IEEE   1169

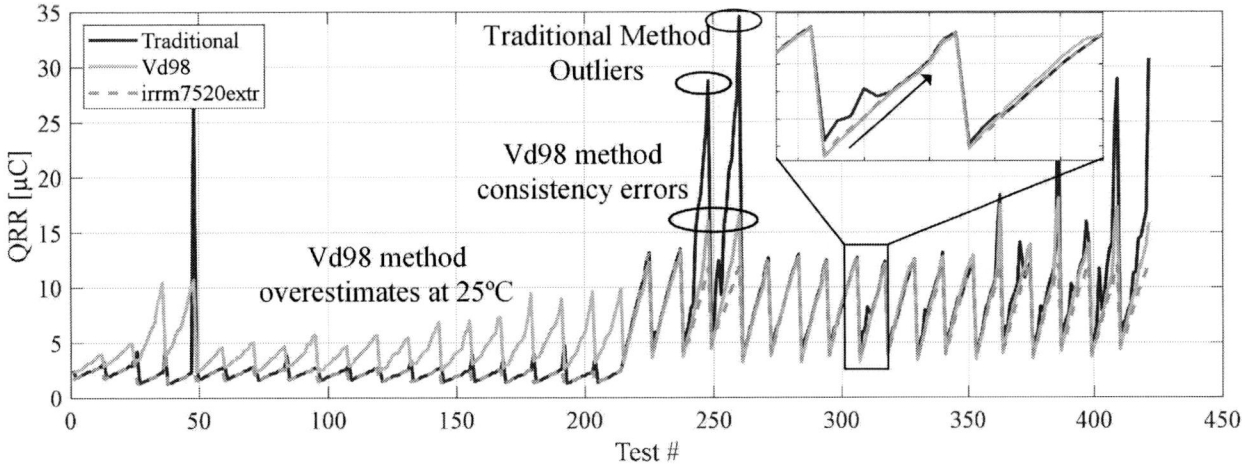

Fig. 5: CAB450M12XM3, overlay of each reverse recovery calculation for each evaluated test

It should be noted here that the goal of this paper is *not* to create a "new" metric, but create an alternative calculation method to derive the same quantity. To that end, the efficacy of the evaluated techniques is determined by how well they meet the listed criteria while providing roughly equivalent results to the traditional method. In this way, the new method can be used to quantify reverse recovery without updating the physical meaning or definition of the quantity.

IV. RESULTS

Two SiC power modules were selected for analyzing the reverse recovery calculation methods. The devices, tabulated in Table I, were selected to provide a variety of reverse recovery behavior. The CAB450M12XM3 power module uses Wolfspeed gen3 die technology, which has very different RR behavior from the Infineon FF3MR20M1H module. In addition, these modules use die developed for different nominal voltage operation.

All three modules were evaluated across a wide range of operating conditions. First, the CAB450M12XM3 power module was evaluated at approximately 400 operating conditions. These conditions vary from 25°C – 175°C, 0 Ω – 10 Ω RG, 50 A – 900 A, and 600 V – 800 V. Each RR calculation method in Fig. 4 was applied to the measured RR waveforms. A summary comparison of each method is shown in Fig. 5. Each datapoint corresponds to a unique set of test conditions; as the test number increases first current swept, then voltage, then gate resistance, and finally junction temperature. The sawtooth shape of the waveforms is given by each sweep in current. This overview provides a qualitative comparison of the techniques.

From the results, the traditional method shows clear outliers at multiple tests, indicative of a dataset where the upper integration limit was not found correctly. The vd98 method not only shows these similar outliers, but also shows a general overprediction in RR between tests 0 – 220 (which indicated 25 °C). On the other hand, the 75:20 technique shows a consistent sawtooth pattern and agrees well with the traditional method but with no outliers.

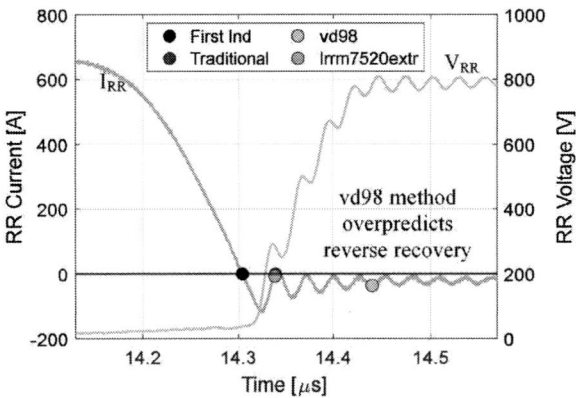

Fig. 6: CAB450M12XM3, comparison of reverse recovery calculation methods for a single condition

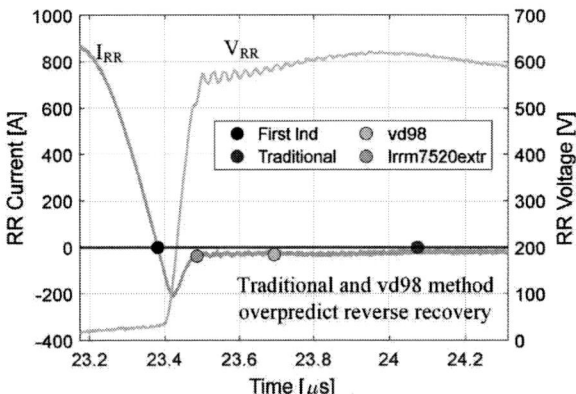

Fig. 7: CAB450M12XM3, comparison of reverse recovery calculation methods for a single condition

TABLE I
LIST OF MEASURED SiC POWER MODULES

Device	Manufacturer	Description
CAB450M12XM3	Wolfspeed	Gen3 1.2 kV, 450 A half-bridge SiC power module
FF3MR20M1H	Infineon	2 kV, 400 A half-bridge SiC power module

Fig. 6 and Fig. 7 show detailed comparisons of the three methods applied to an RR waveform at two different operating conditions. The lower integration bound is given by T_{RR1}, and the upper integration bound of each method is given by its respective marker. In Fig. 6, the 75:20 method provides the same result as the traditional method, but the vd98 method overestimates the reverse recovery. In Fig. 7, both the traditional and vd98 methods overestimate the reverse recovery. The traditional method fails due to a measurement offset, while the vd98 method fails due to ringing on the waveform that skewed finding the 98% nominal voltage point.

Similarly, the FF3MR20M1H module was evaluated at more than 500 conditions. A summary comparison of each method for this module is shown in Fig. 8. Here, the traditional method and the 75:20 method show identical behavior across the entire dataset (this device has a faster recovery event that makes the second zero crossing occur consistently). However, the vd98 method shows some differences but largely underestimates the actual reverse recovery. Inspecting an individual test condition in Fig. 9 shows that the module has a very fast di/dt during the reverse recovery event. This causes V_{RR} to have a large overshoot, which effects when the vd98 index is determined, and the method underestimates the reverse recovery charge. The vd98 index may also be effected by any deskew errors.

From these examples, the 75:20 method demonstrates good agreement with the traditional calculation method, while ignoring errors caused by measurement or parasitics, and is overall a much more consistent method to calculate the same physical quantity. It should be stated that the vd98 method is not necessarily wrong, but may not be a suitable replacement for the traditional reverse recovery calculation method. This leads to discussion on what the definition of "reverse recovery" is. Based strictly on the currently accepted standard for reverse recovery calculations, the 75:20 method provides a direct replacement with improved robustness. In most cases, this new proposed method will provide the same result as the traditional method, but many edge cases are resolved. The vd98 method may provide some utility for calculating a quantity related to reverse recovery, but caution should be taken when using the method as a direct replacement for calculating reverse recovery. The method also requires careful attention to the reverse recovery

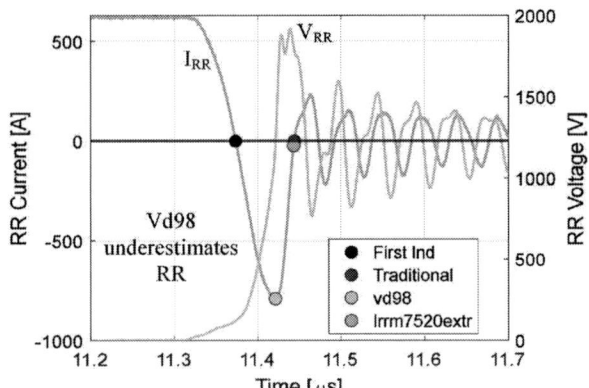

Fig. 9: FF3MR20M1H, comparison of reverse recovery calculation methods for a single condition

voltage measurement and its deskew relative to the reverse recovery current measurement.

V. CONCLUSIONS AND FUTURE WORK

This paper analyzes two existing and one proposed method for calculating reverse recovery and applies them across large datasets for three different SiC power modules. The results of the analysis show that the proposed method provides a more reliable and robust method for calculating reverse recovery, which will enable more accurate datasheet and model generation. Future endeavors should seek to continue improving these calculation methods to determine simpler and more standardized calculation methods that can accurately quantify the reverse recovery behavior of devices of increasing complexity.

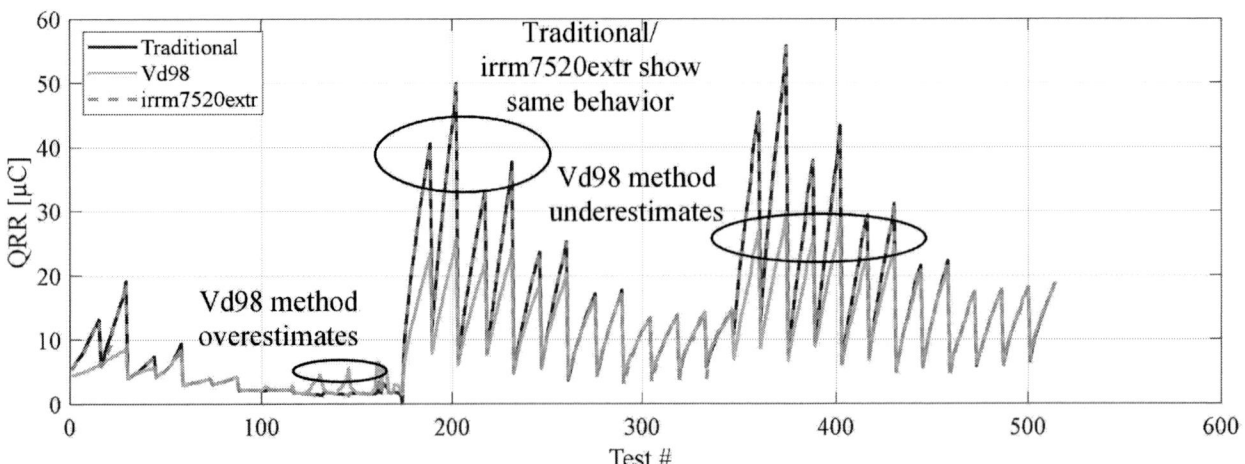

Fig. 8: FF3MR20M1H: overlay of each reverse recovery calculation for each evaluated test

REFERENCES

[1] Z. Wang, J. Ouyang, J. Zhang, X. Wu and K. Sheng, "Analysis on reverse recovery characteristic of SiC MOSFET intrinsic diode," 2014 IEEE Energy Conversion Congress and Exposition (ECCE), Pittsburgh, PA, USA, 2014, pp. 2832-2837, doi: 10.1

[2] C. Qian et al., "Investigation of Reverse Recovery Phenomenon for SiC MOSFETs in High-Temperature Applications," in IEEE Transactions on Power Electronics, vol. 38, no. 11, pp. 14375-14387, Nov. 2023, doi: 10.1109/TPEL.2023.3273351.

[3] A. März, S. Schönewolf, A. Nagel, M. Rauh and M. -M. Bakran, "Deadtime optimization eliminating snap-off of 3.3kV SiC MOSFET bodydiodes," 2023 25th European Conference on Power Electronics and Applications (EPE'23 ECCE Europe), Aalborg, Denmark, 2023, pp. 1-7, doi: 10.23919/EPE23ECCEEurope58414.2023.10264563.

[4] A. U. Rashid, B. Brooks, S. Manz, D. J. Lichtenwalner and S. -H. Ryu, "Modeling of the Snappy, and Soft Reverse Recovery of SiC MOSFET's Body Diode," 2023 IEEE BiCMOS and Compound Semiconductor Integrated Circuits and Technology Symposium (BCICTS), Monterey, CA, USA, 2023, pp. 310-313,

[5] M. Zhu, Y. Pei, F. Yang, Z. Cheng, D. Ma and L. Wang, "An Analytical Switching Loss Model for SiC MOSFET Considering Temperature-Dependent Reverse Recovery Over an Extremely Wide High-Temperature Range," in IEEE Transactions on Power Electronics, vol. 39, no. 6, pp. 7029-7044, June 2024, doi: 10.1109/TPEL.2024.3365467.

[6] P. Sochor, A. Huerner, M. Hell and R. Elpelt, "Understanding the Turn-off Behavior of SiC MOSFET Body Diodes in Fast Switching Applications," PCIM Europe digital days 2021; International Exhibition and Conference for Power Electronics, Intelligent Motion, Renewable Energy and Energy Management, Online, 2021, pp. 1-8.

[7] G. Di Luca Cardillo et al., "Investigation of the limit conditions of SiC MOSFET body diode reverse recovery," 2023 IEEE Energy Conversion Congress and Exposition (ECCE), Nashville, TN, USA, 2023, pp. 5763-5770, doi: 10.1109/ECCE53617.2023.10362298.

[8] JEDEC, "Test method to measure the reverse recovery characteristics of the drain source diode of a power MOSFET," August 1994

Improvement of CM EMI Attenuation Ability of Transformer with Negative Capacitor

Qinghui Huang, Yiming Li, Yirui Yang, Shuo Wang, Yanwen Lai, and Zhedong Ma
Power Electronics and Electrical Power Research Lab
University of Florida, Gainesville, FL, USA
Email: qinghuihuang@ufl.edu

Abstract- **The capacitance balance technique to reduce CM EMI noise for isolated converters would be less effective at high frequency. A third-order CM model of the transformer is derived and verified from the two coupled windings and the six-capacitor model, which has the advantage of a wider effective frequency range compared to the two-capacitor model. The failure of the capacitance balance technique can be explained by the proposed model. With the understanding of the failure mechanism, a negative capacitance is applied to extend the effective frequency range of the balancing technique, which is also verified by experiments.**

Index Terms—Common mode (CM), electromagnetic interference, isolated power converters, Negative capacitor

I. INTRODUCTION

Isolated converters are commonly used in applications that require galvanic isolation such as , charger stations for electric vehicles[1][2], and data center power supplies [3]-[5]. Wide-band-gap devices (GaN, SiC) play a very important role in improving the efficiency and power density of power electronics systems [6]-[8]. High dV/dt and dI/dt caused by fast switching would worsen the EMI performance [9]-[11] and require more EMI filters[12]. Parasitic parameters of magnetic components widely used in power converters[14]-[15] become important EMI noise paths, such as inductors[18] and transformers[19][23].

Transformers are essential parts in the isolated converters to provide galvanic isolation and high input/output voltage gain [16]. However, transformers are the main path for CM EMI noise propagation in isolated converters due to the intra-winding capacitors across primary and secondary windings. An accurate and simple CM EMI model of a transformer plays an important role in EMI prediction and mitigation.

The existing CM modeling methods for transformers can be categorized into low-order models and high-order models. For low-frequency range, the two-capacitor model is the most simplified CM model to represent the CM path of the transformer [16][17], which is only effective in the low-frequency range. With concerned frequency increasing, the two-impedance model[19] and multi-RLC ladder model[20] can be used to characterize the CM path of the transformer at a high-frequency range. These models can accurately simulate the characterization of the CM path. However, they show no explicit physical meaning and are hard to guide EMI noise mitigation.

Based on the two-capacitor CM model, the capacitance balance technique is widely used to reduce CM EMI noise in isolated converters due to its efficient cost and easy implementation [21][22]. However, the effectiveness of capacitance balance would gradually decrease with frequency increasing. The mechanism of failure of capacitance balancing and the improving approach are unknown, which causes high-frequency CM EMI noise through the transformer to be very hard to reduce. A bifilar structure is applied to primary winding to reduce the CM inductance of the transformer and then benefits the wider effective frequency range of the two-capacitor model and capacitance balance technique [23]. An RLC balance technique is proposed to extend the effective frequency range of the balance technique, which shows almost double the effective frequency range compared to the capacitance balance technique [24].

Negative capacitor techniques have been widely investigated to cancel the EPC (equivalent parallel capacitor) of inductors in EMI filters. In [26], several conductive layers are embedded into the inductors of EMI filters, which helps generate a negative capacitor parallel to the CM inductor and explains the negative capacitor caused by the transformation from star-connection to delta-connection. Similarly, a discrete capacitor can be added between the central tap of winding and grounding to generate a negative capacitor in parallel with EPC to improve CM EMI noise attenuation performance [27]. Negative capacitors are generated in parallel with DM (differential mode) inductors to improve the DM attenuation ability of EMI filters by imitating mutual capacitance.

In this paper, the third-order CM model of the transformer is derived from the two coupled windings and the six-capacitor model. The derived model can cover the two-capacitor model frequency range and a higher frequency range. The main contribution of this digest includes: 1. a third-order CM model of transformer is derived from the two coupled windings and six-capacitor model and has also been verified by experimental results. 2. The CM voltage gain is derived from the model, which helps to unveil the failure mechanism of the capacitance balance technique. 3. A negative capacitance generation approach is applied to the transformer to reduce CM voltage gain at high frequency. The advantages of the proposed method have been verified by experimental results.

II. CM EMI MODELING OF TRANSFORMER

A. Two-coupled-winding and Six-capacitor CM model

A typical transformer consists of two coupled windings, L_P and L_S with mutual inductance M_{ps} for power delivery, as shown in Fig. 1 (a). Six capacitors are commonly used to represent the capacitive coupling between two coupled windings, as shown in Fig. 1(a). As for CM EMI noise analysis, two impedances are enough to present the CM path of the transformer, as shown in Fig.

1(b) [19]. The CM impedance curves of the investigated transformer are shown in Fig. 2.

$$Gain_{CM} = \frac{a_1 s + s^3 a_2}{s b_1 + s^3 b_2} \quad a_1 = ((C_3 + C_4)) - (C_4 + (C_2 + C_\alpha))M_{PS}/L_P)$$

$$a_2 = L_{lks}(C_3 C_4 + C_3(C_2 + C_\alpha) + (C_3 + C_4)(C_S + C_\gamma))$$

$$b_1 = ((C_1 + C_\beta) + C_3 + C_4 + (C_2 + C_\alpha)) \tag{1}$$

$$b_2 = L_{lks}(((C_1 + C_\beta) + C_3)(C_4 + (C_S + C_\gamma) + C_2 + C_\alpha) + (C_S + C_\gamma)((C_2 + C_\alpha) + C_4)))$$

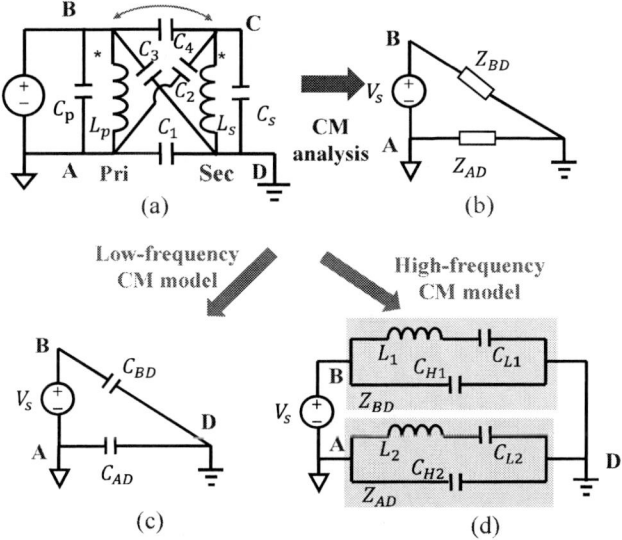

(a)　　　　　　　(b)

(c)　　　　　　　(d)

Fig. 1. CM EMI models of a transformer.

At low frequency, a two-capacitor model simplified from a six-capacitor model can be used to represent the CM path based on capacitive energy conservation and CM current conservation, as presented in Fig. 1 (c) [1]. Correspondingly, the CM impedances of the transformer are capacitively dominant at low frequencies.

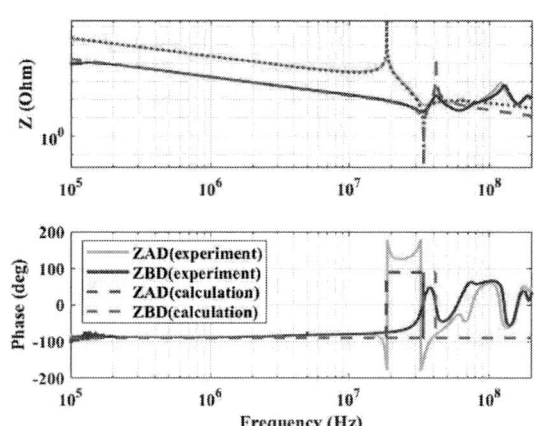

Fig. 2. Measured and calculated CM impedance curves.

However, resonances of CM impedance would occur with frequency increasing. To characterize the CM impedance Z_{BD} and Z_{AD} at a higher frequency, a third-order CM model can be developed from the two coupled windings with six capacitors in this paper. To calculate CM impedance Z_{BD}, points A and D can be shortened. Then the impedance seen from the V_s source is impedance Z_{BD}, as shown in (2). Similarly, CM impedance Z_{AD} can be calculated by shorting points B and D, as presented in (3).

$$Z_{BD} = \frac{1}{s\dfrac{(C_3 + C_4)L_P - (C_2 + C_4)M_{PS}}{L_P}} \frac{1 + (1 + (C_S + C_2 + C_4)L_{lks}s^2)}{1 + \dfrac{(C_3(C_2 + C_4 + C_S) + C_4 C_S)L_{lks}}{(C_3 + C_4) - (C_2 + C_4)M_{PS}/L_P}s^2} \quad L_{lks} = \frac{L_P L_S - M_{PS}^2}{L_P} \tag{2}$$

$$Z_{AD} = \frac{1}{s\dfrac{(C_1 + C_2)L_P + (C_2 + C_4)M_{PS}}{L_p}} \frac{1 + (C_S + C_2 + C_4)L_{lks}s^2}{1 + \dfrac{(C_1(C_2 + C_4 + C_S) + C_2 C_S)L_{lks}}{(C_1 + C_2) + (C_2 + C_4)M_{PS}/L_P}s^2} \quad L_{lks} = \frac{L_P L_S - M_{PS}^2}{L_P} \tag{3}$$

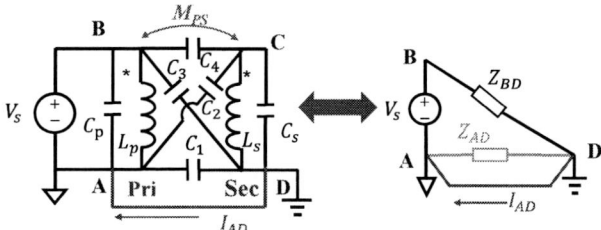

Fig. 3. Derivation of CM impedance Z_{BD}.

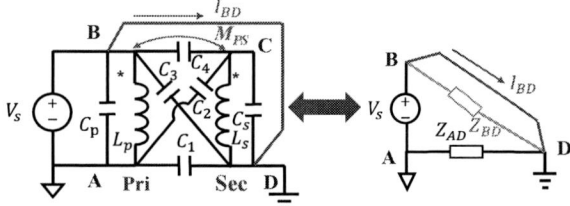

Fig. 4. Derivation of CM impedance Z_{AD}.

At the low-frequency range, the first-order term (blue part) is dominant. Thus, the third-order model can be equivalent to the two-capacitor model of the transformer. CM impedances Z_{AD} and Z_{BD} are capacitive at low frequency. With the concerned frequency increasing, the second-order term in the numerator (green part) or second-order term in the denominator (red part) would become dominant. The zeroes of Z_{AD} and Z_{BD} would cause impedance valleys as shown in impedance curves. The poles of Z_{AD} and Z_{BD} would cause impedance peaks.

The derived impedance expressions are verified by the CM impedances of the transformer, as shown in Fig. 2. The dotted lines are calculated based on (2) and (3). The calculated impedance curves match the measured impedance curves up to 50MHz. The leakage inductance L_{lks} can be calculated from the inductance L_p, L_s, and M_{ps}, which can be obtained by measuring the low-frequency impedance. L_{lks} of investigated transformer is 0.496uH. C_1, C_2, C_3, C_4, C_s can be derived from the capacitances of the two-capacitor models and three resonant frequencies. C_s, C_1, C_2, C_3, and C_4 are 18.59pF, 7.29pF, 19.5pF, 18.34pF and 55.07pF, respectively.

B. Analysis of Failure Mechanism of Capacitance Balance Technique

The converter circuit of the flyback converter is shown in Fig. 5 (a). Replacement theory is commonly used to build the EMI model of converters, as presented in Fig. 5 (b). The superposition theory is usually used to analyze the EMI noise contributed by each source. The current source of I_{D1} has little contribution to CM EMI noise at low frequency and then it can be neglected. Finally, the reduced CM EMI model is indicated in Fig. 5 (c). To reduce the EMI noise flowing through the LISN side, the voltage at the CM port should be decreased. The CM voltage gain is defined as the ratio between CM voltage across primary and secondary windings and the noise source, which is the main metric to evaluate the CM noise attenuation ability of the transformer. The expression of CM voltage gain is shown in (5). The capacitive balance technique has been widely used in the flyback converter to reduce CM EMI noise, as shown in Fig. 6 [1]. An extra discrete capacitor C_{bal} is applied parallel to the branch AC. Then the CM voltage gain can be updated as (5). The first term of numerator can be designed as zero by tuning the capacitance of C_{bal}. As a

result, the CM voltage gain at low frequency can be largely reduced. The CM voltage gain curves with and without the capacitance balance technique are shown in Fig. 7. The calculated CM voltage gain can match well with the measured CM voltage gain up to 50MHz. It can be clearly seen that the CM voltage gain at low frequency can be largely reduced up to 20MHz. However, the CM voltage gain would increase with the frequency further increasing. This can be explained by the second term of CM voltage gain becoming dominant with frequency increasing based on (5). The added balance capacitor is negative in the first term of CM voltage gain but positive in the second term. Thus, it cannot help reduce CM voltage when frequency increases.

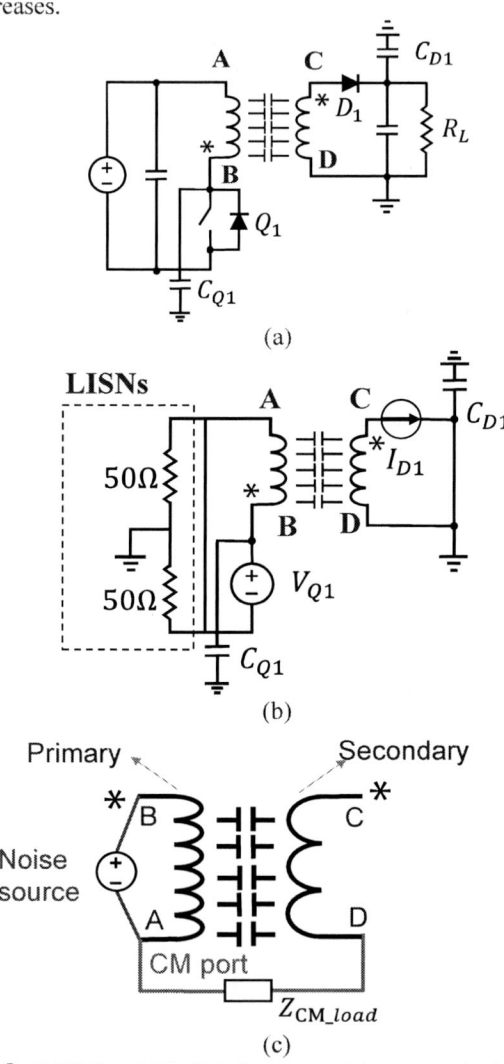

Fig. 5. CM EMI model for flyback converter: (a) converter circuit, (b) substitute nonlinear devices with voltage and current sources, (c) simplified CM model of the flyback converter.

$$Gain_{CM} = \frac{V_{CM}}{V_{noise}} = \frac{Z_{AD}}{Z_{AD}+Z_{BD}} = \frac{((C_3+C_4)-(C_4+C_2)M_{PS}/L_P)s + s^3 L_{lks}(C_3 C_4 + C_3 C_2 + (C_3+C_4)C_S)}{s(C_1+C_3+C_4+C_2) + s^3 L_{lks}((C_1+C_3)(C_4+C_S+C_2) + C_S(C_2+C_4))} \quad (4)$$

$$Gain_{CM} = \frac{((C_3+C_4)-(C_4+C_2+C_{bal})M_{PS}/L_P)s + s^3 L_{lks}(C_3 C_4 + C_3(C_2+C_{bal}) + (C_3+C_4)C_S)}{s(C_1+C_3+C_4+C_2+C_{bal}) + s^3 L_{lks}((C_1+C_3)(C_4+C_S+C_2+C_{bal}) + C_S(C_2+C_{bal}+C_4))} \quad (5)$$

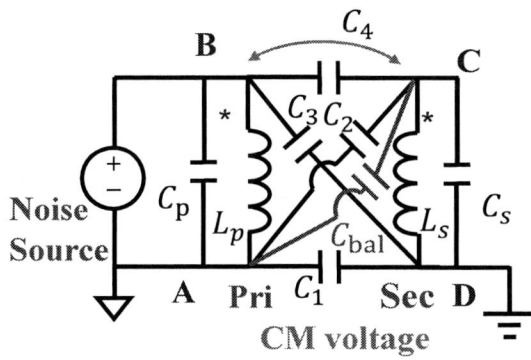

Fig. 6. Balance technique to reduce CM EMI noise.

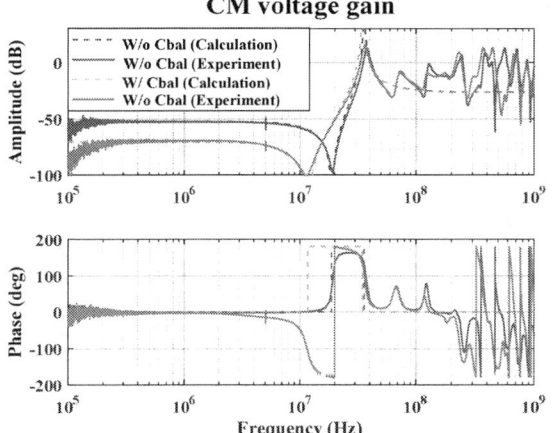

Fig. 7. CM voltage gain with balance capacitance.

III. APPLYING A NEGATIVE CAPACITOR TO IMPROVE CM EMI ATTENUATION

A. Principle of Generating Negative Capacitor

The effective way to reduce the CM voltage gain at high frequency is to reduce capacitance C2, C3, C4, Cp, Cs, and leakage inductance L_{lks} based on (5). However, reducing these values in a well-designed transformer with optimized efficiency and power density is hard.

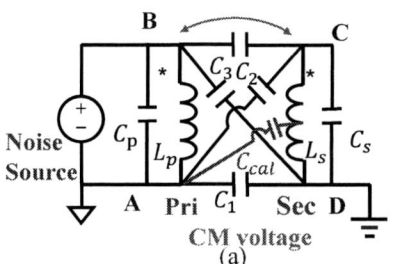

Fig. 8. Equivalent parallel capacitor cancellation in inductors.

A negative capacitor can be generated parallel to C_S. A cancellation capacitor is introduced between the central tap of the secondary winding and the primary ground, as shown in Fig. 9 (a). Then, with the transformation between star-connection and delta-connection, the cancellation capacitor can be equivalent to $C_\alpha = C_{cancel}$, $C_\beta = C_{cancel}$, and $C_\gamma = -\frac{1}{4} C_{cancel}$. C_α is added in parallel with C_2 functioning as a balance capacitor. C_β is added in parallel with C_1, functioning as a Y-cap, which helps to reduce the CM voltage across primary and secondary ground. C_γ is added in parallel with C_s. Furthermore, the previous voltage gain (5) can be changed to (1). The sign of C_γ is opposite to C_{cancel}, which means that C_γ is negative when the cancel

$$Gain_{CM} = \frac{a_1 s + a_2 s^3}{s b_1 + s^3 b_2} \quad a_1 = ((C_3 + C_4)) - ((C_4 + \Delta C_4) + (C_2 + C_\alpha)) M_{PS} / L_P)$$

$$a_2 = L_{lks}(C_3(C_4 + \Delta C_4) + C_3(C_2 + C_\alpha) + (C_3 + C_4 + \Delta C_4)(C_S + C_\gamma))$$

$$b_1 = ((C_1 + C_\beta) + C_3 + C_4 + \Delta C_4 + (C_2 + C_\alpha)) \tag{6}$$

$$b_2 = L_{lks}(((C_1 + C_\beta) + C_3)(C_4 + \Delta C_4 + (C_S + C_\gamma) + C_2 + C_\alpha) + (C_S + C_\gamma)((C_2 + C_\alpha) + C_4 + \Delta C_4))$$

Negative capacitance is proposed in this paper to reduce CM voltage gain at high frequency. The negative capacitance has been investigated to cancel the EPC (equivalent parallel capacitor) of inductors[27]. The basic idea is to utilize the transformation between star-connection and delta-connection to generate a negative capacitor, as shown in Fig. 8.

capacitor is positive. As a result, it can reduce the second term (red font) in the numerator of (1).

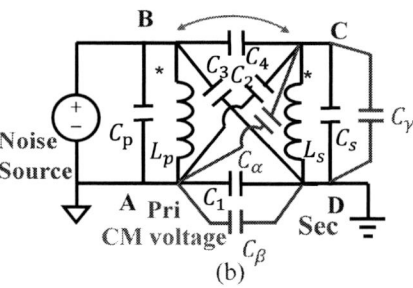

(b)

Fig. 9. Intra-winding capacitance cancellation in transformers.

B. Principle of improving CM EMI attenuation capability

As analyzed above, C_{cancel} can successfully introduce a negative capacitance in the second term of (1). To reduce CM voltage gain as much as possible, the first-term and second-term numerators of (1) can be designed as zero, as shown in (7) and (8).

$$(C_3 + C_4)) - (C_4 + (C_2 + C_\alpha))M_{PS}/L_P = 0 \quad (7)$$

$$C_3C_4 + C_3(C_2 + C_\alpha) + (C_3 + C_4)(C_S + C_\gamma) = 0 \ (8)$$

Since $C_\alpha = -4C_\gamma = C_{cancel}$, (8) can be revised to

$$C_3C_4 + C_3C_2 + 3/4C_3C_{cancel} + (C_3 + C_4)C_S - 1/4C_4C_{cancel} = 0 \quad (9)$$

Solving expression (9), the required cancellation capacitor can be obtained with

$$C_{cancel} = \frac{C_3C_4 + C_3C_2 + (C_3 + C_4)C_S}{(1/4C_4 - 3/4C_3)} \quad (10)$$

To make sure the designed C_{cancel} larger than zero, the condition presented in (11) should be met.

$$C_4 > 3C_3 \quad (11)$$

When (11) cannot be met in an initial transformer, another capacitor ΔC_4 can be added in parallel with C_4 to make (11) be satisfied.

V. EXPERIMENT

To validate the proposed idea of reducing CM voltage gain, a transformer with 2:1 is made as shown in Fig. 10. The turn number of the primary winding is 24. The turn number of the secondary winding is 12 and the secondary winding is bifilar wounded to reduce leakage inductance between the two windings of the secondary winding, which can help to extend the effective frequency range of the negative capacitor. The bobbin type is ATQ-2312. The core material is JPP-95. The core type is ATQ23.2/11.6C.

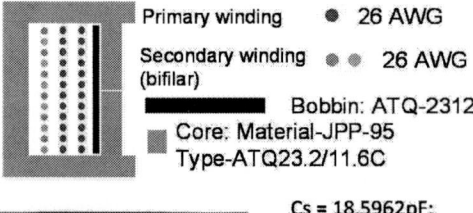

Cs = 18.5962pF;
Llks= 0.496uH;
C1 = 7.29pF;
C2 = 19.5pF;
C3 = 18.24pF;
C4 = 55.07pF;

Fig. 10. Transformer for experiments.

The extracted parasitic parameters have been included in Fig. 10. The parasitic parameters are extracted from the CM impedances of the transformer measured by (VNA)

vector network analyzer. The CM voltage gain with the variation of cancellation capacitors is shown in Fig. 11(a). It shows that applying cancellation capacitors helps to reduce the CM voltage gain at low frequency and high frequency. 37pF cancellation capacitor can achieve good reduction at low frequency (35dB) and good reduction at high frequency. Increasing the cancellation capacitor to 75pF, a higher frequency reduction (20dB@50MHz 60dB 100MHz) has been achieved but the over-compensation would occur at low frequency. Increasing C4 helps to achieve better balance at low frequencies, as shown in Fig. 11 (b). Almost 50dB reduction can be achieved at low frequency and about 30 dB reduction at high frequency can be achieved when 75pF cancellation capacitor is applied and 95pF capacitance is added to C_4. The results show that the proposed cancellation capacitor can effectively reduce the CM voltage gain at low and high frequencies

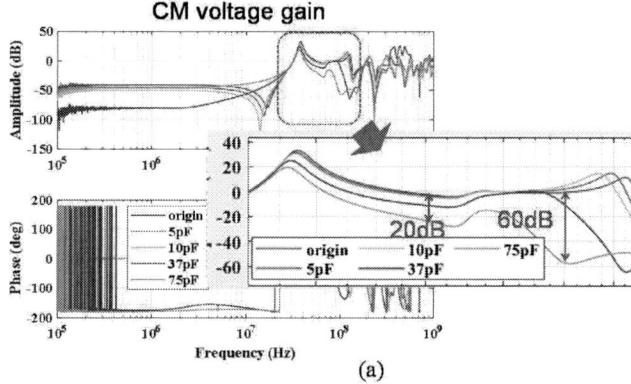

(a)

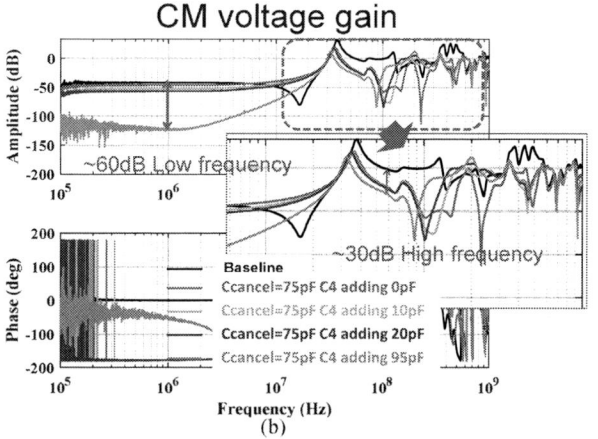

(b)

Fig. 11. CM voltage gain of transformer with cancellation capacitors.

VI. CONCLUSION

In this paper, a high-order CM model of the transformer is derived to characterize the CM path of a transformer. Based on the derived model, the failure mechanism of the capacitance balance technique is explained. Furthermore, a negative capacitor is introduced to decrease the CM voltage gain at a higher frequency. The advantages of the proposed idea are verified by experiments. The proposed technique will be applied to a flyback converter to show the EMI reduction on the converter level.

References

[1] S. Fan et al., "Optimal Circulant Modulation for Submodule Voltage Ripple Minimization With Inherent Balancing Capability in

979-8-3315-1612-3/25 $31.00 © 2025 IEEE   1177

Modular Multilevel DC–DC Converters," in IEEE Transactions on Power Electronics, vol. 39, no. 1, pp. 784-798, Jan. 2024, doi: 10.1109/TPEL.2023.3322976.

[2] X. Liu, M. Qiu, K. Hobbs, A. Dahneem, H. Meng and D. Cao, "A 500kW 1600V Zero-Voltage Switching Resonant Switched-Capacitor DC-DC Converter for Electric Trucks and Electric Aircraft Application," *2023 IEEE Energy Conversion Congress and Exposition (ECCE)*, Nashville, TN, USA, 2023, pp. 2084-2090.

[3] H. Meng, M. Qiu, Z. Sun, X. Liu and D. Cao, "A 9x Matrix Autotransformer Switched-Capacitor DC-DC Converter for Datacenter Application," 2023 IEEE Energy Conversion Congress and Exposition (ECCE), Nashville, TN, USA, 2023, pp. 3305-3312.

[4] H. Cao, L. Du, F. Guo, Z. Ma and Y. Zhao, "A Triple Active Bridge (TAB) Based Solid-State Transformer (SST) for DC Fast Charging Systems: Architecture and Control Strategy," 2023 IEEE Energy Conversion Congress and Exposition (ECCE), Nashville, TN, USA, 2023.

[5] X. Yang, Y. Lyu, K. Wang, U. Kim, Z. Zhang and K. -B. Park, "A Computationally Efficient FCS-MPC Imitator for Grid-Tied Three-Level NPC Power Converters Based on Sequential Artificial Neural Network," 2022 IEEE Energy Conversion Congress and Exposition (ECCE), Detroit, MI, USA, 2022, pp. 1-6, doi: 10.1109/ECCE50734.2022.9947831.

[6] Q. Yang, F. Jin and Q. Li, "An Accurate Temperature-Based Method for Fast Switching Loss Extraction of WBG Device," *2024 IEEE Applied Power Electronics Conference and Exposition (APEC)*, Long Beach, CA, USA, 2024, pp. 2183-2187.

[7] S. Fan; X. Xiang; J. Sheng; Y. Gu; H. Yang; W.L , , "Inherent SM Voltage Balance for Multilevel Circulant Modulation in Modular Multilevel DC–DC Converters," in IEEE Transactions on Power Electronics, vol. 37, no. 2, pp. 1352-1368, Feb. 2022.

[8] H. Gao *et al.*, "Improved Short-Circuit Protection Method for SiC MOSFET Split-Output Power Module Achieving Ultra-Low Fault Dissipated Energy," in *IEEE Transactions on Power Electronics*, vol. 39, no. 2, pp. 2270-2280, Feb. 2024.

[9] Z. Ma, S. Wang, Q. Huang and Y. Yang, "A Review of Radiated EMI Research in Power Electronics Systems," in IEEE Journal of Emerging and Selected Topics in Power Electronics, vol. 12, no. 1, pp. 675-694, Feb. 2024.

[10] Y. He, J. Zhu, B. Zhou and X. Zhao, "EMI Issues and Reduction in 800 V Electrical Vehicle Systems by Incorporating Cable Effects," *2023 IEEE 6th Student Conference on Electric Machines and Systems (SCEMS)*, HuZhou, China, 2023, pp. 1-5.

[11] H. Cao, H. Wang, P. Darvish, et al. "Double Line Frequency Voltage Ripple Reduction Control Strategy for Dual-Active-Bridge in DC/AC System," 2024 IEEE Energy Conversion Congress and Exposition (ECCE), Phoenix, AZ, USA, 2024.

[12] X. Guo, J. Zhou, R. He, X. Jia and C. A. Rojas, "Leakage Current Attenuation of a Three-Phase Cascaded Inverter for Transformerless Grid-Connected PV Systems," in IEEE Transactions on Industrial Electronics, vol. 65, no. 1, pp. 676-686, Jan. 2018, doi: 10.1109/TIE.2017.2733427.

[13] H. Jie et al., "VNA-Based Fixture Adapters for Wideband Accurate Impedance Extraction of Single-Phase EMI Filtering Chokes," in IEEE Transactions on Industrial Electronics, vol. 70, no. 8, pp. 7821-7831, Aug. 2023, doi: 10.1109/TIE.2023.3239906.

[14] M. Gao, L. Yi and J. Moon, "Mathematical Modeling and Validation of Saturating and Clampable Cascaded Magnetics for Magnetic Energy Harvesting," in IEEE Transactions on Power Electronics, vol. 38, no. 3, pp. 3455-3468, March 2023, doi: 10.1109/TPEL.2022.3218725.

[15] L. Yi and J. Moon, "Direct in-situ Measurement of Magnetic Loss in Power Electronic Circuits," in IEEE Transactions on Power Electronics, vol. 36, no. 3, pp. 3247-3257, March 2021, doi: 10.1109/TPEL.2020.3015182.

[16] H. Zhang, S. Wang, Y. Li, Q. Wang, and D. Fu, "Two-capacitor transformer winding capacitance models for common-mode EMI noise analysis in isolated DC-DC converters," IEEE Trans. Power Electron., vol. 32, no. 11, pp. 8458–8469, Nov. 2017.

[17] C. Xu, F. Zhang and G. Dong, "Wideband Modeling of Transformer Common-Mode Characteristics Using RLC Ladder Network," 2021 IEEE 1st International Power Electronics and Application Symposium (PEAS), 2021.

[18] H. Jie *et al.*, "High-Precision Broadband Impedance Measurements of Three-Phase Common-Mode Chokes Using Single-Port Circuit De-Embedding and Three-Port Network Calibration Methods," in *IEEE Transactions on Industrial Electronics*, vol. 71, no. 8, pp. 8248-8258, Aug. 2024.

[19] J. Yao, Y. Li, S. Wang, X. Huang and X. Lyu, "Modeling and Reduction of Radiated EMI in a GaN IC-Based Active Clamp Flyback Adapter," IEEE Trans. Power Electron., vol. 36, no. 5, pp. 5440-5449, May 2021.

[20] Ce Xu, Fanghua Zhang, "Reduced-Order Ladder Network Modeling for Common-Mode Characterization of Transformers", IEEE Journal of Emerging and Selected Topics in Power Electronics, vol.11, no.4, pp.3995-4009, 2023.

[21] P. J. Kong, S. Wang, F. C. Lee, and Z. J. Wang, "Reducing common-mode noise in the two-switch forward converter," IEEE Trans. Power Electron., vol. 26, no, 5, pp. 1522–1533, May 2011.

[22] D. B. Fu, S. Wang, P. J. Kong, F. C. Lee, and D. C. Huang, "Novel techniques to suppress the common-mode EMI noise caused by transformer parasitic capacitances in DC-DC converters," IEEE Trans. Ind. Electron., vol. 60, no. 11, pp. 4968–4977, Nov. 2013.

[23] Q. Huang, Y. Yang, Z. Ma, Y. Lai and S. Wang, "Transformer Structure of Bifilar Primary Winding with Advanced Common Mode Noise Attenuation Performance for Isolated DC-DC Converters," 2023 IEEE Applied Power Electronics Conference and Exposition (APEC), Orlando, FL, USA, 2023, pp. 441-448.

[24] Q. Huang, Y. Li, Z. Ma, Y. Yang, Y. Lai and S. Wang, "RLC Balance Technique of Transformer to Reduce CM EMI for Isolated DC-DC Converters," 2023 IEEE Energy Conversion Congress and Exposition (ECCE), Nashville, TN, USA, 2023, pp. 2945-2952.

[25] H. Jie, Z. Zhao, H. Li, C. Wang, Y. Chang and K. Y. See, "Characterization and Circuit Modeling of Electromagnetic Interference Filtering Chokes in Power Electronics: A Review," in *IEEE Transactions on Power Electronics*, doi: 10.1109/TPEL.2024.3454152.

[26] Rengang Chen, J. D. van Wyk, Shuo Wang and W. G. Odendaal, "Improving the Characteristics of integrated EMI filters by embedded conductive Layers," in IEEE Transactions on Power Electronics, vol. 20, no. 3, pp. 611-619, May 2005.

[27] S. Wang, F. C. Lee and J. D. van Wyk, "Design of Inductor Winding Capacitance Cancellation for EMI Suppression," in IEEE Transactions on Power Electronics, vol. 21, no. 6, pp. 1825-1832, Nov. 2006.

Damping Factor Based PCB Parasitic Inductance Value Optimization to Minimize Voltage Overshoot and Settling Time of Semiconductors

Reza Shahbazi
School of Electrical Engineering and
Computer Science
The Pennsylvania State University
University Park, PA, USA
rvs6198@psu.edu

Yunting Liu
School of Electrical Engineering and
Computer Science
The Pennsylvania State University
University Park, PA, USA
ypl5778@psu.edu

Abstract— **It is widely recognized that the parasitics of Printed Circuit Board (PCB) will influence the switching transients such as Drain-Source voltage V_{DS}/Collector-Emitter voltage V_{CE}, Gate-Source voltage V_{GS}/Gate-Emitter voltage V_{GE}, and Drain current I_D/Collector current I_C of MOSFETs/IGBTs. However, due to the packaging, the footprint of semiconductor modules, and PCB tracks, some stray inductance is unavoidable. Although designers' goal is to design a PCB with the minimized possible parasitics, an over-reduced parasitic inductance will lead to a longer settling time for switching transient and subsequently, the switching power loss will be increased. To avoid the potential suboptimized performance caused by a too small parasitic inductance, this paper proposes a method based on the power loop damping factor to find the optimal value for PCB parasitic inductance that improves switching transients such as minimizing turn-off voltage overshoot and settling time. Following the theory, the simulation and experimental results for multiple PCB designs for a SiC MOSFET switch and an IGBT module are presented. Overshoots of 7% and 6.5%, and settling times of 0.11 μs and 0.12 μs are achieved with the critically damped layout design compared to an overshoot of 29% and 22.5%, settling times of 0.78 μs and 0.47 μs using an underdamped PCB layout design for the SiC MOSFET switch and IGBT module, respectively.**

Keywords—**Parasitic inductance, PCB, damping factor, switching transient model, energy loss, wide bandgap device, semiconductor, power converter**

I. INTRODUCTION

In recent years, significant efforts have been made to reduce undesired switching oscillations in power switches [1]. Reducing the stray inductances in the power loop is crucial to fully leverage the properties of switches, as it constrains the rate of change of current (di/dt) and leads to voltage overshoots and oscillations [1]-[7]. Nevertheless, some internal stray inductance is expected due to the packaging. For instance, the common footprints of Semiconductors are TO-247 and Power modules SOT-227 where the minimal internal leads length is 7 mm and 36 mm, respectively, and thus, the minimum internal parasitic inductance range could be from 0.5 nH to 5 nH, respectively [4]. Hence, the summation of the semiconductor internal stray

inductance and the Printed Circuit Board (PCB) parasitic inductance is present in the converter circuit. Subsequently, if the PCB is not optimally designed and has a significant amount of parasitic inductance, the converter switching issues such as switching oscillations, overshoots, and shoot-through faults will appear [1], [3].

Many studies showed parasitics impacts on switching transients and introduced methods such as snubber circuits to mitigate these issues [1], [2]. Moreover, [8]-[10] have shown that by drastically reducing the stray inductance, the total switching power loss will increase. However, none has specified a specific PCB stray inductance value at which the voltage overshoot and settling time is minimized nor investigated the PCB optimization based on the circuit damping factor as a potential solution to improve switching transients. The aim of this paper is to introduce a method based on the circuit damping factor that finds the optimal PCB stray inductance value at which the voltage overshoot and settling time are as small as possible. It also challenges the common assumption that a smaller stray inductance is always preferable. To find the optimal value of the PCB parasitic inductance, the power loop damping factor theoretical analysis based on the PCB parasitic inductance as well as semiconductor package internal values is provided. Afterwards, simulation results for different types of semiconductors with different PCB parasitic inductances are given. Finally, experimental results for a SiC MOSFET switch and an IGBT module, which installed on multiple PCB designs with different parasitic inductance values, are presented. The analysis for suboptimal and optimal PCB designs confirms the presented theory and shows the improvement of the Drain-Source $V_{DS(off)}$ and Collector-Emitter turn-off voltage $V_{CE(off)}$ overshoot, settling time and switching energy loss.

II. TURN-OFF SWITCHING TRANSIENT MODFELING OF A SEMICONDUCTOR

A. Voltage overshoot and settling time

Fig. 1 shows the power loop switching schematic with parasitics for a phase consisting of two MOSFETs with an inductive load parallel with the upper switch. Fig. 2 presents the

979-8-3315-1612-3/25 $31.00 © 2025 IEEE

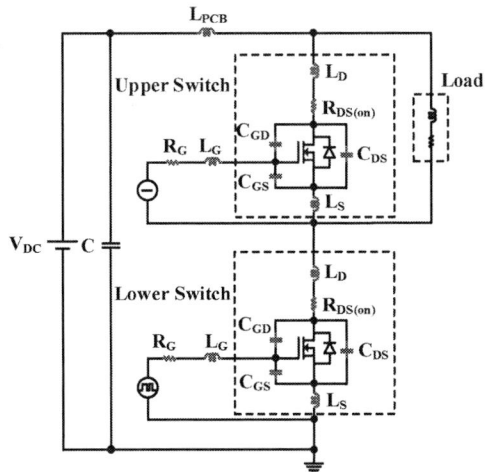

Fig. 1. power loop switching schematic.

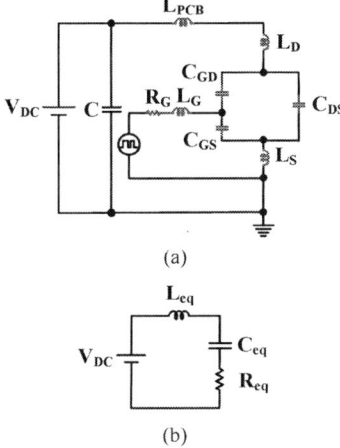

(a)

(b)

Fig. 2. inductive load switching transient modeling with parasitic elements. (a) turn-off switching model; (b) simplified switching model.

turn-off switching transient models. When the lower switch stops conducting and is fully off, the freewheeling diode of upper switch is forward biased, the output capacitance of upper switch is bypassed, and the Drain-Source Miller capacitance C_{DS} of the lower switch is being charged. Hence, the lower switch can be modeled as a resistor R_{sw} and C_{sw} where $C_{sw} = C_{GD} + C_{DS}$ Therefore, the simple RLC turn-off transient model can be shown as Fig. 2(b) where $L_{eq} = L_{PCB} + L_D + L_S$ and $C_{eq} = C_{GD} + C_{DS}$. Considering the second order RLC turn-on and turn-off transient circuits, we can write:

$$V_{R_{eq}} + V_{L_{eq}} + V_{C_{oss}} = V_{DC}(t) \tag{1}$$

where $V_{R_{eq}}$, $V_{L_{eq}}$, $V_{C_{oss}}$ and $V_{DC}(t)$ are the resistor, inductor, capacitor, and input voltages, respectively. By substituting the voltage equations ($V_{R_{eq}} = RI(t)$, $V_{L_{eq}} = L_{eq}\dfrac{dI(t)}{dt}$, and

$V_{C_{oss}} = V(0) + \dfrac{1}{C}\int_0^t I(\tau)d\tau$), taking the time derivative, and divide by L_{eq}, characteristic equation (2) will be:

$$s^2 + 2\alpha s + \omega_0^2 = 0 \tag{2}$$

where α is the neper frequency which determines how fast the switching transient response dies, and ω_0 is the angular resonance frequency. Hence, the damping factor is introduced as [11], [12]:

$$\varsigma \equiv \frac{\alpha}{\omega_0} = \frac{R_{eq}}{2}\sqrt{\frac{C_{eq}}{L_{eq}}} \tag{3}$$

The damping factor defines how fast the response reaches the final value. Among all potential outcomes for ς , the critically damped condition ($\varsigma = 1$) which means the response reaches the final value without going into oscillation is desire. Additionally, the the turn-off resonance frequency, overshoot percent and settling time based on the damping factor for a second order RLC circuit can be expressed as [11]:

$$\omega_{off} \approx \frac{1}{\sqrt{L_{eq} \cdot C_{eq}}} \tag{4}$$

$$overshoot\% = \exp\left(\frac{-\varsigma\pi}{\sqrt{1-\varsigma^2}}\right) \times 100 \tag{5}$$

$$t_s = \frac{4}{\varsigma\omega_{off}} \tag{6}$$

Fig. 3 shows the results of overshoot and settling time for different values of the damping factor using (3) to (6) and the datasheet given parameters of Wolfspeed SiC MOSFET C3M0040120K and Infineon IGBT F475R12KS4 as the tested devices in this paper.

B. PCB parasitics value optimization effect on switching energy loss

As it can be seen in Fig. 3, the damping factor of 1 has the lowest settling time. While the overshoot will reach zero when the damping factor is above 1 (overdamped condition), the settling time will increase and dv/dt will decrease. The longer settling time and smaller dv/dt when the switch works under overdamped condition influence turn-on and turn-off switching energy losses. During turn-on transients, the stray inductance reduction decreases the voltage sag while its fall duration remains unchanged. Also, the current rising rate increases and hence, the total turn-on energy loss increases.

On the other hand, during the turn-off transients, the stray inductance reduction decreases the voltage rising rate while the current fall rate remains almost unchanged. Therefore, the turn-off switching energy loss decreases. Nevertheless, the turn-on switching loss increase is higher than the turn-off switching loss reduction and consequently, the total switching power loss increases [9], [10]. This phenomenon is presented in the next section for Wolfspeed SiC MOSFET C3M0040120K and

979-8-3315-1612-3/25 $31.00 © 2025 IEEE

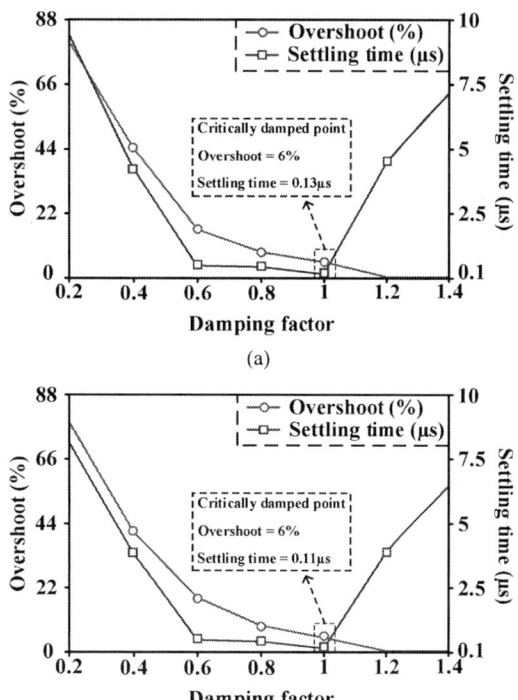

Fig. 3. Damping factor Vs Overshoot and Settling time for (a) Wolfspeed C3M0040120K and (b) Infineon F475R12KS4.

Infineon IGBT F475R12KS4. Therefore, the critically damped scenario is the ideal condition where the overshoot, settling time, and the total switching loss are minimized. Table I shows the optimized parasitic inductance value for the tested devices in this paper based on the given parameters in their datasheets.

III. SIMULATION AND EXPERIMENTAL RESULTS

Fig. 4 shows the simulation results for different semiconductor packages with different PCB parasitic inductances in the power loop working at 400 V and 20 A. Based on (3) and the semiconductor internal parameters, different values of the PCB parasitic inductance are designed to have the underdamped ($\varsigma < 1$), critically damped ($\varsigma = 1$), and overdamped ($\varsigma > 1$) conditions. As it can be seen in Fig. 4, the overshoot is improved from 114.4V and 80 V to 24 and 24 V for the SiC MOSFET and IGBT switches, respectively. Also, settling time is reduced from 0.75 µs and 0.43 µs to 0.11 µs and 0.10 µs, for the SiC MOSFET and IGBT switches, respectively. Therefore, the data in Fig. 3 is confirmed.

Fig. 5 shows the simulation results for the switching energy loss of Wolfspeed SiC MOSFET C3M0040120K and Infineon IGBT F475R12KS4 at different PCB parasitic values. As it can be seen in Fig. 5, turn-on and turn-off energy loss decrease when the circuit damping factor value increase from underdamped to critically damped condition. However, the turn-on energy loss drastically increases when the circuit damping factor value increases from critically damped to overdamped condition while the turn-off energy loss slightly decreases. Therefore, the total energy loss is increases when the circuit is in overdamped

TABLE I. OPTIMAL PCB PARASITIC INDUCTANCE BASED ON THE PACKAGING INTERNAL PARAMETERS

Device	Internal Resistance (mΩ)	C_{oss} (nF)	$L_{Package}$ (nH)	Optimal L_{PCB}
Wolfspeed C3M0040120K	40	0.10	5	14
Infineon F475R12KS4	140	0.71	11	24

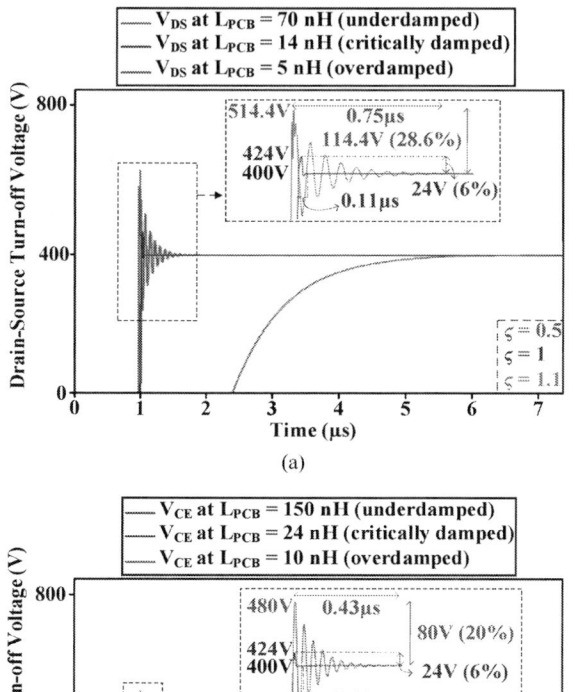

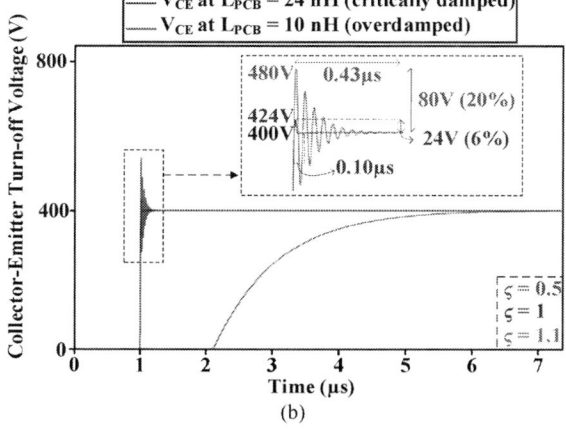

Fig. 4. Simulation results for voltage turn-off waveform of a suboptimal and optimal PCB designs for (a) Wolfspeed SiC MOSFET C3M0040120K V_{DS} and (b) Infineon IGBT F475R12KS4 V_{CE}.

condition compared to the critically damped point.

Fig. 6 presents the experimental results for Wolfspeed SiC MOSFET C3M0040120K and Infineon IGBT F475R12KS4 and proves the theory and simulation results. For experiment tests, two single layer PCB with the power loop length of 39 cm and 20 cm, and two double layer PCB with the power loop length of 12 cm and 8cm are designed. The parasitic inductances for designs are measured by Ansys Q3D to ensure that the parasitic inductance values are matched with the simulation

979-8-3315-1612-3/25 $31.00 © 2025 IEEE

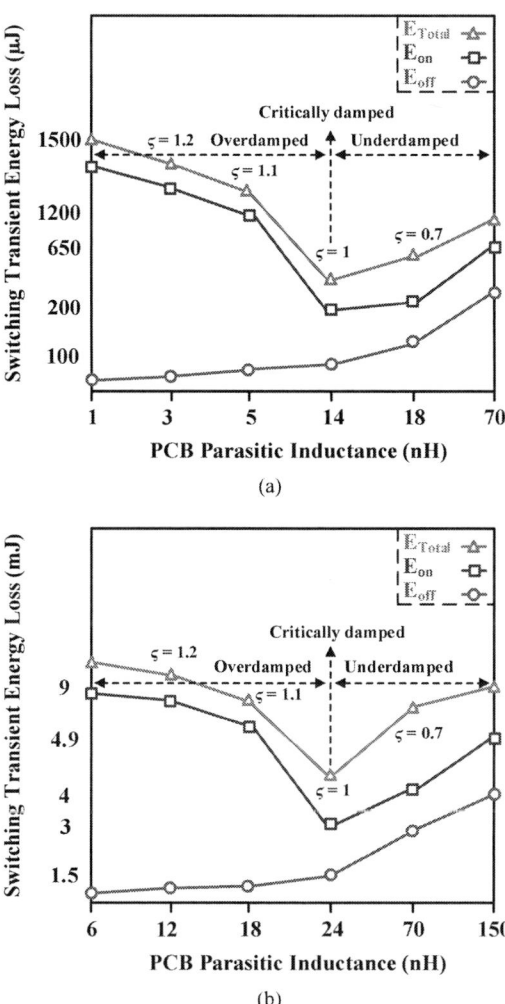

(a)

(b)

Fig. 5. Switching energy loss Simulation results for PCB parasitic values for (a) Wolfspeed SiC MOSFET C3M0040120K and (b) Infineon IGBT F475R12KS4.

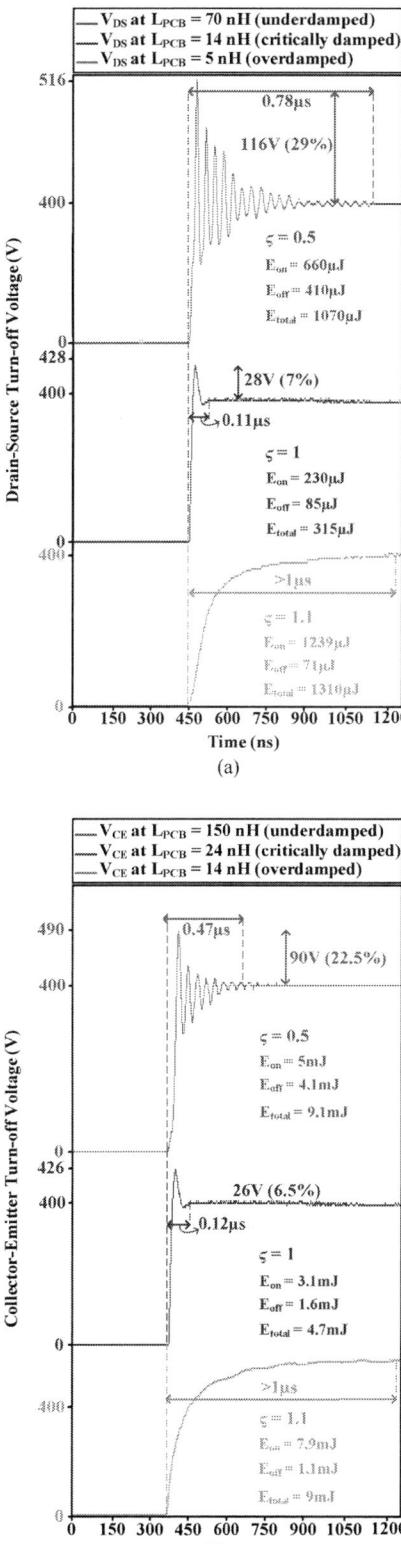

(a)

(b)

Fig. 6. Experimental results for voltage turn-off waveform of a suboptimal and optimal PCB designs for (a) Wolfspeed SiC MOSFET C3M0040120K V_{DS} and (b) Infineon IGBT F475R12KS4 V_{CE}.

results for both SiC MOSFET and IGBT switches. By reducing the cable length from the power supply to the PCB boards, the underdamped condition is achieved.

As it can be seen in Fig. 6, voltage overshoot is reduced from 116 V and 90 V to 28 V and 26 V for the SiC MOSFET and IGBT, respectively. Similarly, the turn-off voltage settling time is reduced from 0.78 μs and 0.47 μs to 0.11 μs and 0.12 μs for the SiC MOSFET and IGBT, respectively. Although the voltage overshoot is zero in overdamped results, the settling times are significantly increased. The switching energy loss calculations are also shown in Fig. 6. As expected, the critically damped condition has the lowest switching energy loss while operating in overdamped condition drastically increases the switching energy loss. This phenomenon is more intense in the SiC MOSFET compared to the IGBT switch, which confirms the simulation results.

979-8-3315-1612-3/25 $31.00 © 2025 IEEE   1182

IV. Conclusion

PCB parasitic inductance is one of the key factors that influences the switching waveforms such as the voltage overshoot and settling time as well as the switching energy loss. This paper presents a new method based on the switching circuit damping factor to find the optimal PCB parasitic value at which the switch has the lowest possible turn-off voltage overshoot, settling time, and energy loss. The circuit turn-off switching transient model is considered to find the relation between the PCB optimal value and circuit parameters. The simulation and experimental results for a SiC MOSFET switch and an IGBT module confirm the theory. Moreover, It has shown that the lowest possible PCB parasitic value is not always better since it increases the switch turn-off voltage settling time and consequently, the switching energy loss significantly.

References

[1] T. Liu, T. T. Y. Wong, and Z. J. Shen, "A Survey on Switching Oscillations in Power Converters," *IEEE J. Emerging Sel. Topics Power Electron.*, vol. 8, no. 1, pp. 893-908, Mar. 2020.

[2] T. Liu, R. Ning, T. T. Y. Wong, and Z. J. Shen, "Modeling and Analysis of SiC MOSFET Switching Oscillations," *IEEE J. Emerging Sel. Topics Power Electron.*, vol. 4, no. 3, pp. 747-756, Sept. 2016.

[3] Z. Zhang, B. Guo, F. F. Wang, E. A. Jones, L. M. Tolbert, and B. J. Blalock, "Methodology for Wide Band-Gap Device Dynamic Characterization," *IEEE Trans. Power Electron.*, vol. 32, no. 12, pp. 9307-9318, Dec. 2017.

[4] A. Kopta et al., "Next Generation IGBT and Package Technologies for High Voltage Applications," *IEEE Trans. Electron Devices*, vol. 64, no. 3, pp. 753-759, Mar. 2017.

[5] C. Abbate, G. Busatto, A. Sanseverino, F. Velardi, and C. Ronsisvalle, "Analysis of Low- and High-Frequency Oscillations in IGBTs During Turn-ON Short Circuit," *IEEE Trans. Electron Devices*, vol. 62, no. 9, pp. 2952-2958, Sept. 2015.

[6] Y. Yang, Y. Wen, and Y. Gao, "A Novel Active Gate Driver for Improving Switching Performance of High-Power SiC MOSFET Modules," *IEEE Trans. Power Electron*, vol. 34, no. 8, pp. 7775-7787, Aug. 2019.

[7] J. Chen, X. Du, Q. Luo, X. Zhang, P. Sun and L. Zhou, "A Review of Switching Oscillations of Wide Bandgap Semiconductor Devices," *IEEE Trans. Power Electron*, vol. 35, no. 12, pp. 13182-13199, Dec. 2020.

[8] M. Ando and K. Wada, "Design of Acceptable Stray Inductance Based on Scaling Method for Power Electronics Circuits," *IEEE J. Emerging Sel. Topics Power Electron.*, vol. 5, no. 1, pp. 568-575, Mar. 2017.

[9] G. Zou and Z, Zhao, "Research on impacts of different parameters on transient power loss of IGBT," in *in Proc. 2013 Int. Conf. Electrical Machines and Systems (ICEMS)*, Busan, Korea (South), Oct. 2013, pp. 490-495.

[10] S. Hain, M.M. Bakran, C. Jaeger, F. J. Niedernostheide, D. Domes, and D. Heer, "The effect of different stray inductances on the performance of various types of IGBTs — Is less always better?," in *Proc. 2015 17th Eur. Conf. Power Electron. Appl. (EPE'15 ECCE-Europe)*, Geneva, Switzerland, Sept. 2015, pp. 1-9.

[11] W. Bolton, "Instrumentation and Control Systems" 3rd edition, Chapter 10, pp. 227-256, 2021.

[12] M. Cupelli, A. Riccobono, A. Monti, "Modern Control of DC-Based Power Systems-A Problem-Based Approach", Chapter 7, pp. 249-259, 2018.

Hardware Implementation of Virtual Resistance Based FRT Logic in Programmable 3-Level ANPC Inverters

Mohammad Safayet Hossain, Shuvangkar Chandra Das, Paychuda Kritprajun, Amin Banaie, Tapas Barik, Deepak Ramasubramanian, Aboutaleb Haddadi, Evangelos Farantatos, and Ulrich Muenz

Abstract—This paper develops a hardware testbed to experimentally validate a virtual resistance-based fault ride-through (FRT) and current limiter logic for a grid-forming inverter (GFM). The testbed consists of three-level ANPC (active neutral point clamped) programmable inverters and a power circuit representing a transmission system equivalent. The paper presents the implementation details and sample experimental results. Firstly, the FRT logic with necessary GFM control are designed and simulated in MATLAB/Simulink software. Then, the inverter microcontrollers are programmed for the hardware test. Two GFM inverters are synchronized in parallel at the beginning of the experiment. Then, various balanced and unbalanced faults experiments are performed with the test system. A programmable relay is used to capture high resolution data during fault. The voltage and current sensor measurements of inverters are recorded and logged in real-time. Finally, the results from hardware and simulation are compared to justify the consistency of the hardware implementation with simulation.

Index Terms—Hardware testbed, 3-level ANPC inverter, programmable inverter, FRT algorithm

I. INTRODUCTION

Recent grid codes and inverter-based resources (IBRs) interconnection standards require IBRs to ride through a system fault and provide a specified performance [1]. These performance requirements are implemented through IBR controls commonly referred to as fault ride-through (FRT) control [2]–[10]. To protect power electronic switches against a sustained overcurrent during FRT operation, Grid forming (GFM) control system limits the current through a current limiting strategy. One of the current limiting techniques is the virtual impedance method. An improved virtual imepdance-based FRT is proposed in [3]. This algorithm integrates voltage information with threshold current magnitude to

---

M. S. Hossain is with Department of ECE at University of Central Florida, Orlando, FL, 32816, USA. mohammad.safayet.hossain@ucf.edu

S. C. Das is with Department of ECE at Clarkson University, Potsdam, NY 13699, USA. dassc@clarkson.edu

P. Kritprajun is with Department of ECE at University of Tennessee, Knoxville, TN 37996, USA. pkritpra@vols.utk.edu

A. Banaie, T. Barik, D. Ramasubramanian, A. Haddadi, and E. Farantatos are with Electric Power Research Institute, Knoxville, TN, 37932 USA. abanaie@epri.com, tbarik@epri.com, dramasubramanian@epri.com, ahaddadi@epri.com, efarantatos@epri.com

U. Muenz is with Siemens Technology, Princeton, NJ, 08540, USA. ulrich.muenz@siemens.com

The information, data, or work presented herein was funded in part by the Advanced Research Projects Agency-Energy (ARPA-E), U.S. Department of Energy, under Award Number DE-AR0001570. The views and opinions of authors expressed herein do not necessarily state or reflect those of the United States Government or any agency thereof.

increase the current limiting ability of GFM inverters during balanced fault. An adaptive FRT algorithm is designed by generating virtual resistance to protect GFM inverters from unbalanced faults in [4]. A GFM control system is designed by controlling both positive and negative sequence components to operate the inverter within a defined current limit during balanced and unbalanced faults in [5]. Another control scheme is designed for IBRs by leveraging negative sequence current to support the relay operation during unbalanced faults in [6]. In [7], the current control reference signals are updated based on priority among active power, reactive power, positive sequence current, and negative sequence current during unbalanced fault. In [8], the conventional FRT algorithm of a GFM inverter is improved by utilizing the power factor angle in current saturation technique during faults according to IEEE 2800-2022 requirements. Most of the available works in the literature focus only on the simulation results. However, the hardware experimental results are also important to validate the feasibility of the proposed design. Programmable inverters are useful devices for building a hardware testbed. Some of the research works of the state-of-the-art literature are reported by leveraging the programmable inverters for hardware experiment [11], [12]. However, the paralleling of multiple GFM inverters and fault experiments have not been investigated in those works.

In this paper, the control system of a GFM IBR that uses a virtual resistance-based FRT strategy is implemented in programmable 3-level ANPC inverters for hardware experiments. Two inverters are operated in parallel with the network based on droop control. The experimental set-up is tested with various fault scenarios including single-line-to-ground (SLG), line-to-line-to-ground (LLG), line-to-line (LL), and line-to-line-to-line-to-ground (LLLG) faults. The simulation and experimental results are compared to validate the successful hardware implementation of the design. The technical background of the work is explained in Section II. The simulation and hardware experimental results are explained in Section III. Finally, Section IV includes concluding remarks.

II. TECHNICAL BACKGROUND

The theoretical background of grid forming (GFM) control and fault ride-though (FRT) logic are discussed in this section. A reduced transmission network with four AC buses (North, South, East, and West) is considered for the test system. The single-line diagram of the testbed is shown in Fig. 1. The network parameters are given in Table I.

979-8-3315-1612-3/25 $31.00 © 2025 IEEE

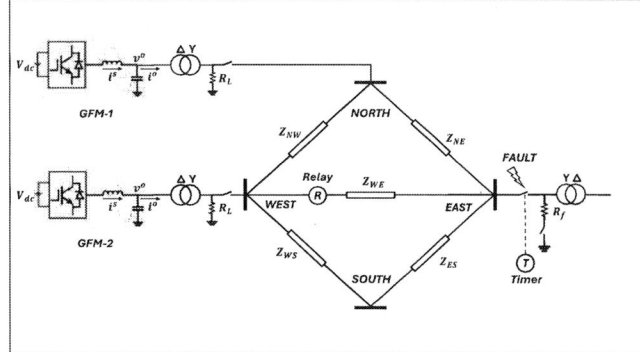

Fig. 1: The single-line diagram of a test setup.

TABLE I: Parameters of the Reduced Transmission Network

Parameter	Value
Nominal Voltage, V_0	$200\sqrt{2}/3V$
Nominal Current, I_0	$6.9982\sqrt{2}A$
Base Voltage (L-L) rms, V_{base}	$200V$
Base Power, S_{base}	$2424.2W$
Nominal/Base Frequency, f_0	$60Hz$
Load Resistance, R_L	33Ω
Line Impedance, Z_{NW}	$(0.1033 + j0.27633)\Omega$
Line Impedance, $Z_{NE}, Z_{WS}, Z_{ES}, Z_{WE}$	$(0.155 + j0.4145)\Omega$
Fault Resistance, R_f	10Ω
DC link voltage, V_{dc}	$400V$
Isolation Transformer	$480V - 480V, \Delta - Yg, 15KVA$

Two GFM inverters energize their local loads as well as the network from the North and West buses. Each GFM supplies power through an isolation transformer for the dynamic coupling. The fault is applied on East bus. A programmable relay is installed in the East-West corridor to record the current and voltage measurements with a high resolution.

A. GFM control and FRT logic

A GFM inverter is controlled as an AC voltage source with fixed voltage magnitude and frequency (f). Firstly, the three-phase voltage and current measurements are converted to $\alpha\beta0$ stationary reference frame from abc reference frame using Clarke transformation. Then, measurements of $\alpha\beta0$ reference frame are converted to $dq0$ rotating reference frame using Park transformation. The combination of Park-Clarke transformation is represented by matrix $M(\omega t)$ where the d-axis is $90°$ lag behind the α axis initially. In other words, the q-axis is aligned with the α axis when time, $t = 0$.

$$M(\omega t) = \frac{2}{3} \begin{bmatrix} \sin(\omega t) & \sin(\omega t - \delta) & \sin(\omega t + \delta) \\ \cos(\omega t) & \cos(\omega t - \delta) & \cos(\omega t + \delta) \\ -\cos(\delta) & -\cos(\delta) & -\cos(\delta) \end{bmatrix}$$

where,

$$\omega = 2\pi f, \quad \delta = \frac{2\pi}{3}$$

Grid forming control of an inverter is designed per unit (p.u.) by leveraging output voltage (v^o), output current (i^o), and inductor current (i^s) of output LC filter. The positive sequence components of v^o, i^o, and i^s measurements are

calculated using equations (1), (2), and (3). The negative sequence of i^s measurement is calculated using equation (4).

$$[V_d^{o+}, V_q^{o+}, V_0^{o+}]^\mathsf{T} = M(\omega t)[V_a^o, V_b^o, V_c^o]^\mathsf{T} \tag{1}$$

$$[I_d^{o+}, I_q^{o+}, I_0^{o+}]^\mathsf{T} = M(\omega t)[I_a^o, I_b^o, I_c^o]^\mathsf{T} \tag{2}$$

$$[I_d^{s+}, I_q^{s+}, I_0^{s+}]^\mathsf{T} = M(\omega t)[I_a^s, I_b^s, I_c^s]^\mathsf{T} \tag{3}$$

$$[I_d^{s-}, I_q^{s-}, I_0^{s-}]^\mathsf{T} = M(-\omega t)[I_a^s, I_b^s, I_c^s]^\mathsf{T} \tag{4}$$

The sensor measurements of an inverter are filtered before using in GFM control and FRT algorithm. The transfer function of the low pass filter is given below.

$$H(s) = \frac{1}{0.01s + 1}$$

where, $s = j\omega$.

The maximum magnitude of three-phase inductor current, I^p is computed from positive sequence and negative sequence components of i^s using a set of equations (5) - (12).

$$a_1 = \sqrt{3}(I_q^{s+} - I_q^{s-}) + I_d^{s-} \tag{5}$$

$$a_2 = \sqrt{3}(I_q^{s-} - I_q^{s+}) + I_d^{s-} \tag{6}$$

$$b_1 = \sqrt{3}I_d^{s-} + I_q^{s-} + I_q^{s+} \tag{7}$$

$$b_2 = \sqrt{3}I_d^{s-} - I_q^{s+} - I_q^{s-} \tag{8}$$

$$I_a^s = \sqrt{(I_q^{s+} + I_q^{s-})^2 + (I_d^{s+} - I_d^{s-})^2} \tag{9}$$

$$I_b^s = \frac{1}{2}\sqrt{(a_1 - I_d^{s+})^2 + (b_1 + \sqrt{3}I_d^{s+})^2} \tag{10}$$

$$I_c^s = \frac{1}{2}\sqrt{(a_2 - I_d^{s+})^2 + (b_2 + \sqrt{3}I_d^{s+})^2} \tag{11}$$

$$I^p = \max(I_a^s, I_b^s, I_c^s) \tag{12}$$

The I^p value is compared with a set value i.e. I_{max} and the tracking error is fed to a proportional integral (PI) controller which generates virtual resistance, R_v. Then the virtual voltage is calculated from inductor current, i^s and R_v.

$$R_v = K_p(I^p - I_{max}) + \frac{K_I}{s}(I^p - I_{max}) \tag{13}$$

$$[v_a^{vir}, v_b^{vir}, v_c^{vir}]^\mathsf{T} = R_v[i_a^s, i_b^s, i_c^s]^\mathsf{T} \tag{14}$$

The positive sequence components of v^o and i^o measurements are used to calculate active power (P) and reactive power (Q) as follows.

$$P = \frac{3}{2}(V_d^{o+}I_d^{o+} + V_q^{o+}I_q^{o+}) \tag{15}$$

$$Q = \frac{3}{2}(V_q^{o+}I_d^{o+} - V_d^{o+}I_q^{o+}) \tag{16}$$

The value of, P and Q are compared with reference active power, P_0 and reference reactive power, Q_0 to regulate the output voltage and frequency of a GFM inverter [13]. The reference frequency, f and the reference voltage, V_d^r of a GFM inverter are updated continuously based on following droop control loop.

$$f = f_0 - D_f(P - P_0) \tag{17}$$

$$V_d^r = V_0 - D_v(Q - Q_0) \tag{18}$$

where, D_f and D_v are droop gains of frequency and voltage respectively. For control input, the output voltage measurement, $[V_d^{o+}, V_q^{o+}]^T$ is compared with the voltage reference, $[V_d^r, V_q^r]^T$ and the tracking error is fed to a PI control loop.

$$u_d^r = K_{Pv}(V_d^r - V_d^{o+}) + \frac{K_{Iv}}{s}(V_d^r - V_d^{o+}) \tag{19}$$

$$u_q^r = K_{Pv}(V_q^r - V_q^{o+}) + \frac{K_{Iv}}{s}(V_q^r - V_q^{o+}) \tag{20}$$

The inverter control input reference vector in $dq0$ axes is converted back to abc axes using inverse transformation matrix, $M^{-1}(\omega t)$. Finally, the virtual voltage is subtracted from the control input before sending it to the PWM.

$$[m_a, m_b, m_c]^T = M^{-1}(\omega t)\frac{2\sqrt{2}V_{base}}{\sqrt{3}V_{dc}}[u_d^r, u_q^r, 0]^T - [v_a^{vir}, v_b^{vir}, v_c^{vir}]^T \tag{21}$$

The FRT algorithm consists of a set of conditions and some sequential steps for fault detection signal (FD) and virtual resistance (R_v) generation to protect an inverter. The detail of the procedure is discussed in Algorithm 1.

Algorithm 1 Virtual resistance based FRT logic

Input: $t_{set}, I_{max}, I^p, I_d^{s+}, I_q^{s+}$
Output: Rv, FD
1: **if** $I^p > (0.9I_{max}) \wedge (t > t_{set})$ **then**
2:     $FD = 1$ ▷set fault detect flag
3: **end if**
4: **if** $\sqrt{(I_d^{s+})^2 + (I_q^{s+})^2} > (0.9I_{max}) \wedge (t > t_{set})$ **then**
5:     $FD = 1$ ▷set fault detect flag
6: **end if**
7: **if** $FD = 1$ **then**
8:     send $FD = 1$ signal to voltage loop to freeze integrator
9:     send $FD = 1$ signal to droop loop to freeze frequency
10:     $R_v = K_p(I^p - I_{max}) + \frac{K_I}{s}(I^p - I_{max})$
11: **else**
12:     send $FD = 0$ signal to voltage loop
13:     send $FD = 0$ signal to droop loop
14:     $R_v = 0$
15: **end if**
16: return: R_v, FD

B. Hardware implementation of the test system

The hardware testbed is shown in Fig. 2. The parameters of GFM control and FRT algorithm are given in Table II. The 3-level ANPC inverter used in this research is provided by Siemens. The inverter has a TI microcontroller (TMS320F28379D) which can be programmed using a C2000 microcontroller blockset of MATLAB-Simulink software [14] and code composer studio software [15]. The details of the inverter parameters and the PWM switching topology are discussed in [11]. A unified inverter control model in a discrete-time domain is designed per unit (p.u.)

Fig. 2: Experimental setup of GFM inverters in EPRI Laboratory, Knoxville, TN, USA.

TABLE II: Parameters of Inverter Control and FRT Logic

Parameter	Value
Reference Power	$P_0 + jQ_0 = (0.5 + j0.0083)\,p.u.$
P-I Gain, Voltage Control Loop	$K_{Pv} = 0.35, K_{Iv} = 25$
P-I Gain, Virtual Resistance Loop	$K_P = 2, K_I = 100$
Voltage & Frequency Droop Gain	$D_v = 2\%, D_f = 2\%$
Maximum Inductor Current Limit	$I_{max} = 0.65\,p.u.$
Delay Time of Grid Stability	$t_{set} = 120\,s$
Fault Application Time	$t_{fault} = 0.2\,s$

based on multi-tasking scheduling in embedded system to integrate GFM control, FRT logic, PWM switching topology, ADC of sensor measurements, and real-time data acquisition. The code is optimized as much as possible to achieve a minimum run time that satisfies the microcontroller overrun constraint. The detail of frequency/angle generation is discussed in [16]. The carrier signal frequency is set as $24kHz$ for PWM switching. The triggering frequency of the ADC interrupt is $24kHz$ too. The sensor measurements of inverters are logged using the Modbus communication protocol. A Matlab script is developed to record and plot data in real-time. An RS485/TTL converter is used to interface a computer and an inverter for data acquisition.

A simulation model of the testbed is developed in which the unified model is integrated with average-value model of inverters. The simulation model is tested with various balanced and unbalanced fault scenarios as mentioned earlier. After running the successful simulation, the unified model is downloaded in the microcontroller of each 3-level ANPC inverter using a TI debug probe (XDS110) for the hardware experiment. A programmable relay (SIPROTEC 7SX82) from Siemens is installed in the East-West corridor to capture high-resolution data (voltage and current) during fault.

III. RESULTS AND DISCUSSION

The details of the hardware experiment procedure and results are explained in this section. Firstly, the GFM-1 inverter has been powered up to energize the transmission network with its local load. Then the GFM-2 has been started to supply power to its local load. After a while, GFM-2 has been connected to the network using the grid synchronization technique (same voltage, frequency, phase sequence, and near-zero phase difference). The system has been running for at least 2 minutes for the synchronization of GFM-2 and stable grid forming before the application of any fault. A timer with a magnetic contactor switch is operated to connect the fault module to the network. The fault has been applied for 0.2 seconds. Various balanced and unbalanced fault scenarios including single-line-to-ground (SLG), line-to-line-to-ground (LLG), line-to-line (LL), and line-to-line-to-line-to-ground (LLLG) faults have been created.

Firstly, the SLG fault is applied in phase-a on the East bus. The output voltage (rms), three-phase inductor current (peak), frequency, virtual resistance, active power, and reactive power measurements of each inverter are recorded and plotted in real time. The reaction data of inverters during a fault are shown in Fig. 3. The inductor currents of both inverters are restricted within the set limit. The frequency of both inverters is kept constant during fault. There is a little voltage sag, 10.19% in the SLG fault scenario. The voltage and current of the East-West corridor during the SLG fault are recorded by the protection relay and are shown in Fig. 4. The measured current is very low before the fault as there are no loads in the South and East buses. The rise of current and fall of voltage are visible in phase-a. After some time, the LLG fault scenario is created by applying fault in phase-a and phase-b. The inverter's responses are shown in Fig. 5. There is a significant sag, 32.08% in the output voltage of both inverters. The inductor currents and frequency are controlled as expected. The relay measurements in Fig. 6 show the voltage drop and current rise in faulty phases. After running the system briefly, phase-a and phase-b are connected with fault resistance without ground to create an LL fault scenario. The experimental results of the LL fault scenario are shown in Fig. 7. The voltage sag is 27.76% which is close to the LLG fault. The voltage of faulty phases shifted by 180° each other during LL fault as shown in Fig. 8. The inverters are connected in the primary (Δ) of isolation transformers and the relay position is in the secondary (Y) side of the transformers. Consequently, the inverters response in each phase and relay record are inconsistent during unbalanced fault scenarios because of the phase shift between line voltages in $\Delta - Yg$ transformer configuration.

Finally, the LLLG fault scenario is created by connecting three phases of the system with a grounded three-phase fault resistance module. The inverter sensor measurements are shown in Fig. 9. The voltage sag, 37.55% in the LLLG fault scenario is the largest among the four fault scenarios. The inductor current of all three phases is controlled within the limit. Each inverter has generated a significant amount of virtual resistance to limit the fault current. The relay record shows the three-phase balanced voltage dip and current rise during the LLLG fault in Fig. 10. The comparison between simulated and experimental results of SLG fault, LLG fault, LL fault, and LLLG fault scenarios are shown in Fig. 11, Fig. 12, Fig. 13, and Fig. 14 respectively. There is a good match in steady-state response between simulated and experimental results. The transient responses of simulated and experimental data are slightly different, which is expected.

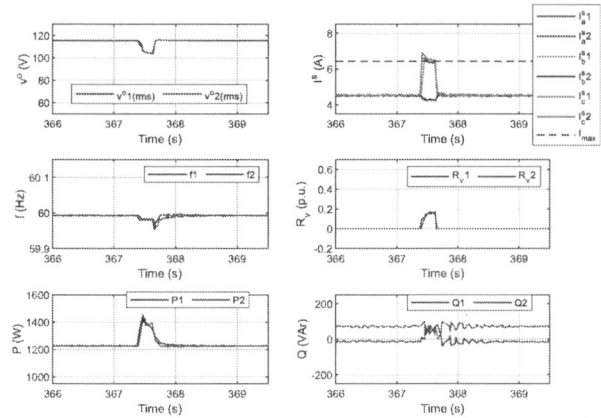

Fig. 3: The experimental result of two GFM inverters in FRT experiment with SLG fault scenario.

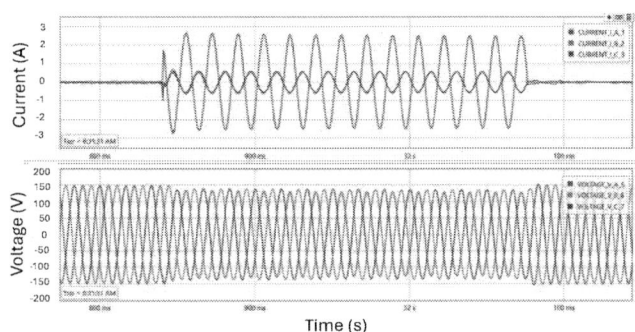

Fig. 4: Voltage and current measurements of East-West corridor by a programmable relay during the SLG fault.

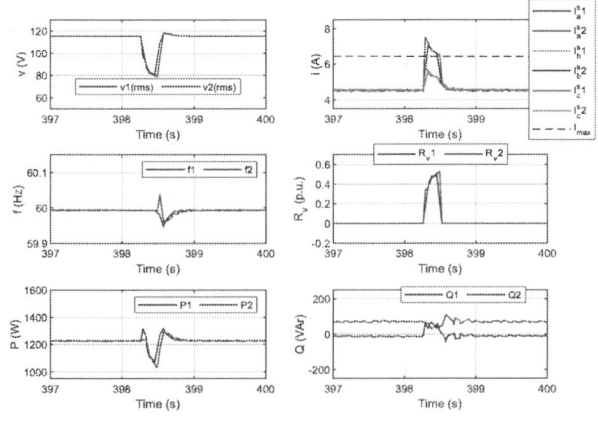

Fig. 5: The experimental result of two GFM inverters in FRT experiment with LLG fault scenario.

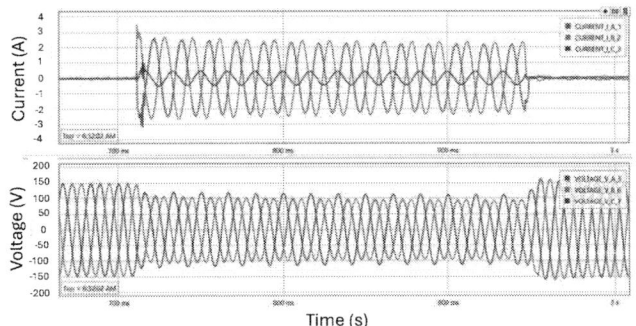

Fig. 6: Voltage and current measurements of East-West corridor by a programmable relay during the LLG fault.

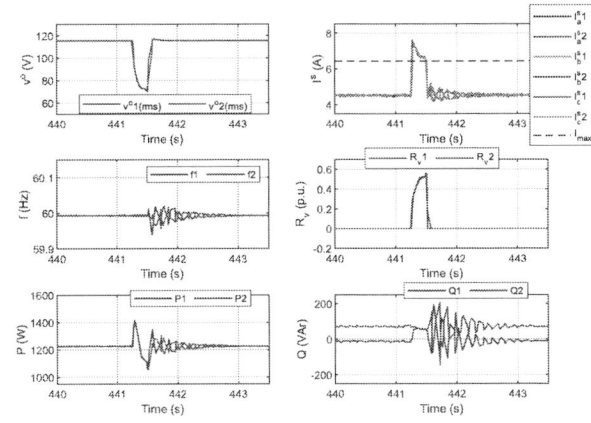

Fig. 9: The experimental result of two GFM inverters in FRT experiment with LLLG fault scenario.

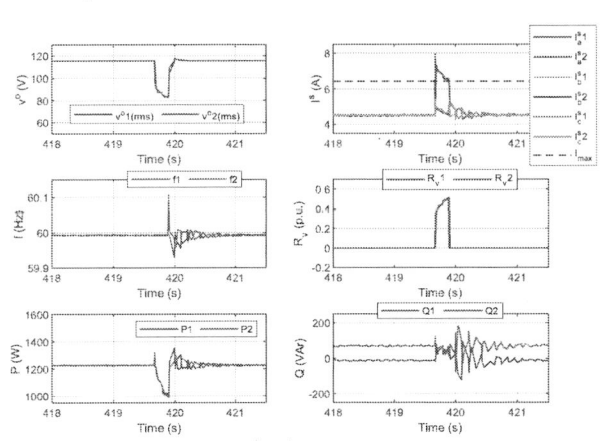

Fig. 7: The experimental result of two GFM inverters in FRT experiment with LL fault scenario.

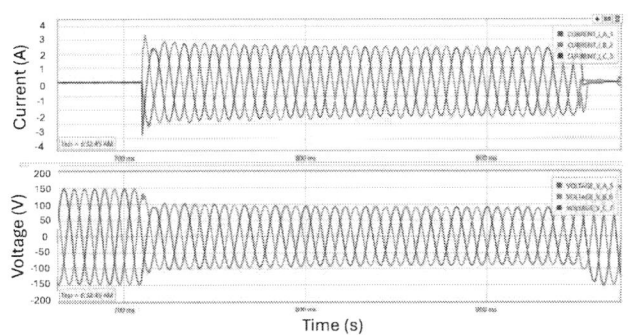

Fig. 10: Voltage and current measurements of East-West corridor by a programmable relay during the LLLG fault.

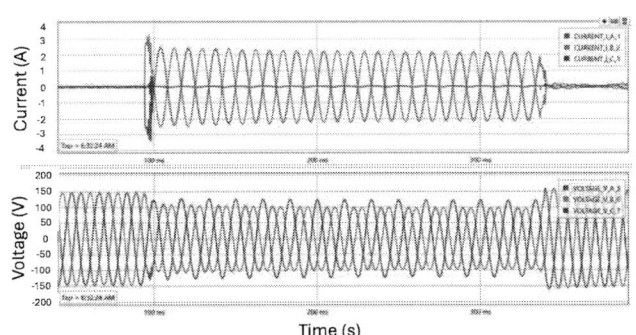

Fig. 8: Voltage and current measurements of East-West corridor by a programmable relay during the LL fault.

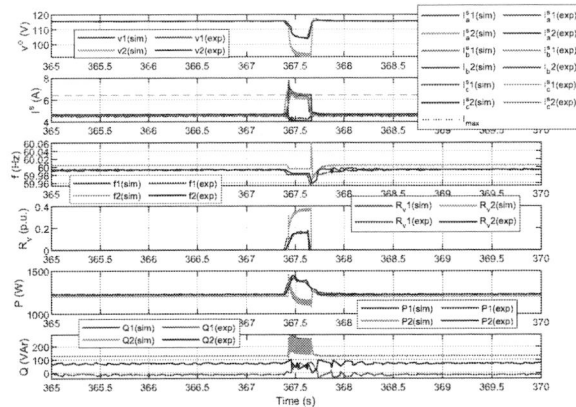

Fig. 11: The comparison between experimental and simulated results for SLG fault.

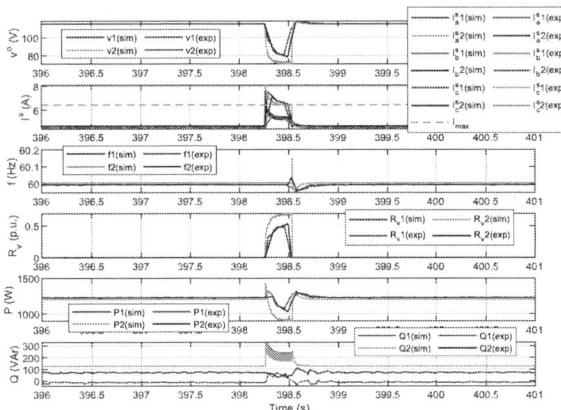

Fig. 12: The comparison between experimental and simulated results for LLG fault.

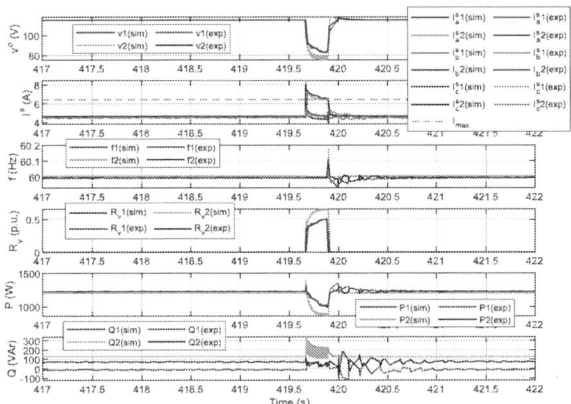

Fig. 13: The comparison between experimental and simulated results for LL fault.

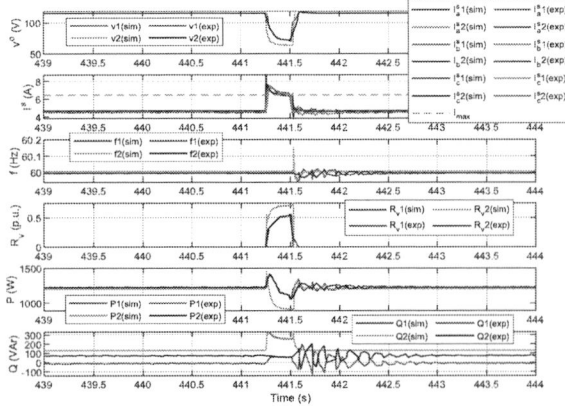

Fig. 14: The comparison between experimental and simulated results for LLLG fault.

IV. CONCLUSION

The FRT algorithm works successfully during fault. The inductor current of each GFM inverter is restricted by the set limit of maximum current based on preliminary results. The reference frequency of the inverter remains constant during fault. The largest output voltage sag is observed during the LLLG fault experiment. The hardware experimental and simulation results are close to each other, which justifies the consistency of the hardware implementation with the original design. The reduced transmission network will be populated with additional inverters, loads, and relays in the future.

REFERENCES

[1] "IEEE Standard for Interconnection and Interoperability of Inverter-Based Resources (IBRs) Interconnecting with Associated Transmission Electric Power Systems," in *IEEE Std 2800-2022*, vol., no., pp.1-180, 22 April 2022.

[2] I. Sadeghkhani, M. E. Hamedani Golshan, J. M. Guerrero and A. Mehrizi-Sani, "A Current Limiting Strategy to Improve Fault Ride-Through of Inverter Interfaced Autonomous Microgrids," in *IEEE Transactions on Smart Grid*, vol. 8, no. 5, pp. 2138-2148, Sept. 2017.

[3] Z. Zeng, P. Bhagwat, M. Saeedifard and D. Groß, "Hybrid Threshold Virtual Impedance for Fault Current Limiting in Grid-Forming Converters," *2023 IEEE Energy Conversion Congress and Exposition (ECCE)*, Nashville, TN, USA, 2023, pp. 913-9187.

[4] Z. Li, K. W. Chan, J. Hu and S. W. Or, "An Adaptive Fault Ride-Through Scheme for Grid-Forming Inverters Under Asymmetrical Grid Faults," in *IEEE Transactions on Industrial Electronics*, vol. 69, no. 12, pp. 12912-12923, Dec. 2022.

[5] B. Mahamedi, M. Eskandari, J. E. Fletcher and J. Zhu, "Sequence-Based Control Strategy With Current Limiting for the Fault Ride-Through of Inverter-Interfaced Distributed Generators," in *IEEE Transactions on Sustainable Energy*, vol. 11, no. 1, pp. 165-174, Jan. 2020.

[6] A. Banaiemoqadam, A. Hooshyar and M. A. Azzouz, "A Comprehensive Dual Current Control Scheme for Inverter-Based Resources to Enable Correct Operation of Protective Relays," in *IEEE Transactions on Power Delivery*, vol. 36, no. 5, pp. 2715-2729, Oct. 2021.

[7] W. Wes Baker et al., "Inverter Current Limit Logic based on the IEEE 2800-2022 Unbalanced Fault Response Requirements," *2023 IEEE Power & Energy Society General Meeting (PESGM)*, Orlando, FL, USA, 2023, pp. 1-5.

[8] A. Pal, D. Pal and B. K. Panigrahi, "A Current Saturation Strategy for Enhancing the Low Voltage Ride-Through Capability of Grid-Forming Inverters," in *IEEE Transactions on Circuits and Systems II: Express Briefs*, vol. 70, no. 3, pp. 1009-1013, March 2023.

[9] B. Wang, R. Burgos and B. Wen, "Grid-Forming Inverter Control Strategy with Improved Fault Ride Through Capability," *2022 IEEE Energy Conversion Congress and Exposition (ECCE)*, Detroit, MI, USA, 2022, pp. 1-8.

[10] Z. Zhang, R. Schürhuber, L. Fickert, G. Chen and Y. Zhang, "Low Voltage Ride Through Characteristics of Grid Forming inverters," *2020 21st International Scientific Conference on Electric Power Engineering (EPE)*, Prague, Czech Republic, 2020, pp. 1-6.

[11] M. S. Hossain et al., "Implementing a Hardware Testbed Using 3-Level ANPC Software Defined Inverters for Fault Analysis of a Transmission Network," *SoutheastCon 2024*, Atlanta, GA, USA, 2024, pp. 1456-1463.

[12] S. Nag et al., "A Unified Grid-Forming and Grid-Following Primary Control Design With Optimized Enforcement of Grid Operational Constraints," *IEEE Access*, vol. 11, pp. 57415-57427, 2023.

[13] K. De Brabandere, B. Bolsens, J. Van den Keybus, A. Woyte, J. Driesen and R. Belmans, "A Voltage and Frequency Droop Control Method for Parallel Inverters," *IEEE Transactions on Power Electronics*, vol. 22, no. 4, pp. 1107-1115, July 2007.

[14] C2000 Microcontroller Blockset. Online: *https://www.mathworks.com/products/ti-c2000-microcontroller.html*

[15] Code Composer Studio™ integrated development environment (IDE). Online: *https://www.ti.com/tool/download/CCSTUDIO/12.2.0*

[16] S. C. Das et al., "Ensuring Accurate Frequency in the Hardware Design of Grid-Forming (GFM) Inverter," *2024 56th North American Power Symposium (NAPS)*, El Paso, TX, USA, 2024, pp. 1-6.

Rad-Hard PSFB Controller for High-Voltage Space Applications

1st Reynaldo S. Gonzalez
Space Science Division
Southwest Research Institute
San Antonio, United States

2nd Robert E. Bolaños
Space Science Division
Southwest Research Institute
San Antonio, United States

Abstract—**The primary challenge in controlling a phase-shifted full-bridge (PSFB) converter for space applications is the absence of commercially available radiation-hardened (rad-hard) control ICs specifically designed for PSFB. To address this, this work implements a control scheme using discrete rad-hard components, including a PWM generator, a discrete dead-time circuit, and high-speed CMOS chips for phase-shifting logic. The design and implementation of a discrete PSFB controller are detailed, and its performance is validated on a 300W PSFB converter delivering a high-voltage output of 600V. Simulation and analysis demonstrate stable operation with minimal ripple, showcasing the feasibility and reliability of this approach for high-voltage space systems.**

Index Terms—**PSFB, radiation-hardened, rad-hard, space, high-voltage**

I. INTRODUCTION

With the rapid growth of private and commercial space exploration, the power demands of spacecraft and satellite systems are expected to increase significantly. To meet these demands while minimizing power system mass, spacecraft power distribution currents must be kept low. Higher currents result in increased cable sizing, switchgear requirements, and reduced overall efficiency—factors that contribute to greater system mass, which is a critical constraint in space applications. A practical solution to decreasing distribution currents while simultaneously increasing power is to raise the DC bus voltage [1], [2]. For systems such as the International Space Station (ISS), which operates at 80 kW, most of the mass-saving benefits are realized at voltages between 150 and 200 V [1]. However, higher power levels continue to benefit from increased bus voltages, enabling significant improvements in mass efficiency.

Furthermore, specific space applications, such as Hall effect thrusters, demand even higher voltages to achieve optimal thrust efficiency, often exceeding 600 V [3], [4]. These factors underscore the need for high-voltage DC-DC converters that can reliably deliver power at elevated voltages. Among available topologies, the phase-shifted full-bridge (PSFB) is well-suited for high-voltage applications due to its inherent advantages, including a second-order transfer function that can be effectively compensated using an analog Type II compensator.

However, the primary challenge in designing a PSFB converter for space lies in the lack of commercially available radiation-hardened controller ICs. This study addresses this challenge by presenting the design and analysis of a radiation-hardened PSFB controller constructed from discrete rad-hard components.

II. PHASE-SHIFTED FULL-BRIDGE FOR SPACE APPLICATIONS

The PSFB topology has been used in the design of Hall Effect Propulsion systems [3], [4]. Both works achieve relatively high output voltages of 400 V and 300 V, respectively. In [3], the authors achieve ZVS with efficiency of 92% for 1.5 kW. However, no focus is placed in making the design space qualifiable. In [4], the authors achieve efficiencies greater than 95% for a full load of 1 kW, and a laboratory prototype of the Hall thruster is demonstrated. However, the control of the PSFB is accomplished through the use of a UCC3895 controller which is not a space qualified part. Another PSFB for space is seen implemented in [5], using digital control methods, however, digital control requires increased cost and time in implementation compared to analog control. The use of GaNFETs and space-qualified analog control methods for PSFB converters is detailed in [6], [7], and [8]. Key components for rad-hard PSFB control include PWM, dead-time circuits, and PSFB logic. The converter in [6] utilizes synchronous rectification and achieves 95% efficiency at an output of 20 V, while [7] focuses on power density and efficiency comparisons for various planar transformers. Both works use secondary side switches in the synchronous current doubler PSFB variant and target low voltage outputs of 20 V. Previous work has predominantly focused on low-voltage, synchronous configurations, this work sets itself apart by addressing the standard diode bridge PSFB topology with higher output voltage. This study also deals with a higher input voltage of 200 V, which increases the voltage stress on the primary side switches, making component selection more challenging. None of the previous works have presented control analysis or measurements on the rad-hard PSFB controller. The contributions of this paper include the implementation of a rad-hard PSFB controller designed for high-input and high-output voltage applications, detailed component selection, and comprehensive control analysis and methodology. This work implements the controller on a 300 W converter, selected for its manageable scale and cost-effectiveness, to provide a practical proof of concept and validate the design principles.

979-8-3315-1612-3/25 $31.00 © 2025 IEEE

III. RAD-HARD PSFB CONTROLLER

A. Phase-Shifting Logic

The control scheme, shown in Fig. 1, utilizes two D flip-flops (DFF) to create a phase delay between the bridge legs of the PSFB converter. The PWM signal, is fed into the clock input of the DFFs, and directly determines the phase delay.

$$\Phi = D \cdot T_{PWM} \qquad (1)$$

A decrease in the PWM duty cycle leads to a reduction in the phase delay, which subsequently lowers the output voltage of the converter, and vice versa. The maximum achievable phase shift is determined by the period, T_{PWM}, of the PWM switching frequency. Since the controller is configured to change the state of each phase leg on every rising edge of the PWM signal (or its complement), the effective switching frequency of the phase legs is reduced to half of the PWM switching frequency.

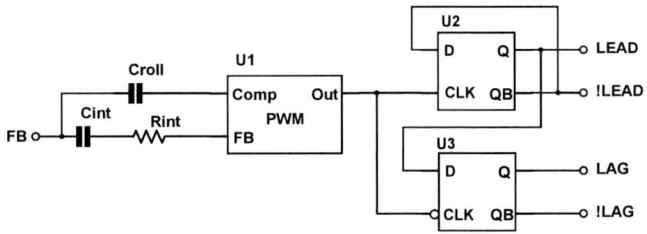

Fig. 1. Discrete PSFB Controller

The relationship between the PWM duty cycle and phase delay is illustrated further in Figs. 2, 3, and 4. Fig. 2 shows the phase delay for a 50% duty cycle. The red lines indicate the rising edge of the PWM signal, while the blue lines indicate the rising edge of the complementary PWM signal. For a 50% duty cycle, the delay between rising edges corresponds to half the period, T_{PWM}, which becomes the effective phase shift, ϕ, between each phase leg.

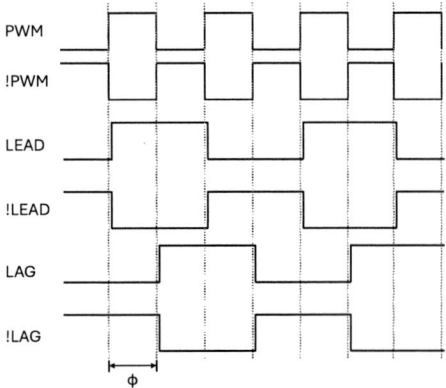

Fig. 2. Phase-Shift for 50% Duty Cycle

In the case of an increasing duty cycle, Fig. 3, the rising edge of the complementary PWM signal occurs later within each switching period. This increased delay corresponds to an increased phase shift, ϕ, which leads to a longer conduction overlap, increasing the energy transfer of the converter.

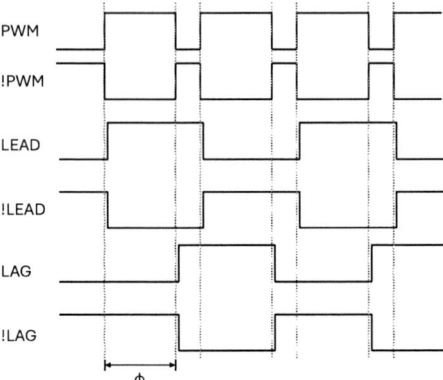

Fig. 3. Phase-Shift for Increasing Duty Cycle

Likewise, in the case of a decreasing duty cycle, Fig. 4, the rising edge of the complementary PWM signal occurs earlier within each switching period. This decreased delay corresponds to a decreased phase shift, ϕ, leading to shorter conduction overlap and energy transfer.

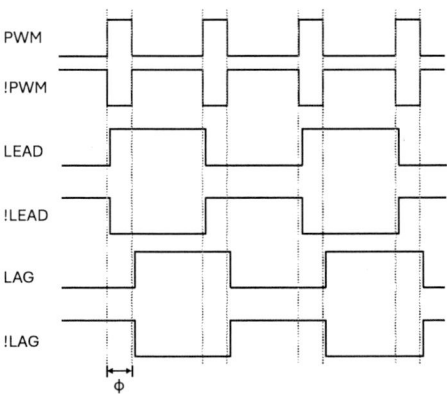

Fig. 4. Phase-Shift for Decreasing Duty Cycle

The four switching outputs of the DFFs are fed into gate transformers, which provide isolation and allow for high-side driving. The secondary sides of the gate transformers are connected to dead-time circuitry and gate driver stages.

B. Gate Transformer

Gate Transformer, Dead-time, and Driver Circuits for Phase-A Leg are shown in Fig. 5. The LEAD and !LEAD signals are connected to low-side drivers U4 and U5, respectively. These low-side drivers, which are readily available as rad-hard components, drive the gate transformer primaries. In a rad-hard PSFB converter, gate transformers are essential for controlling the switching of the power stage switches. They

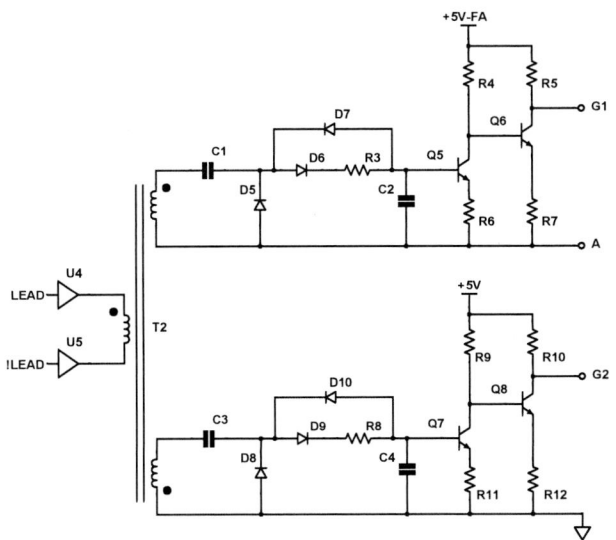

Fig. 5. High and Low Side Gate Transformer, Driver, and Dead-time Circuits for Phase-A Leg

provide galvanic isolation between the low-voltage control circuitry and the high-voltage power stage, ensuring safety and preventing high voltage from feeding back into sensitive control systems. Furthermore, gate transformers enable high-side driving, a necessity for each of the high-side switches, which must be referenced to their respective switch nodes. Moreover, gate transformers eliminate the need for high-side driver ICs, which are not available as radiation-hardened components. However, care must be taken in selecting a gate transformer to ensure that core saturation does not occur. The volt-time product, ET, can be calculated and compared to the datasheet value to check for core saturation.

$$ET = V \cdot t_{on} \qquad (2)$$

$$ET \leq ET_{datasheet} \qquad (3)$$

If $ET \leq ET_{datasheet}$, the core will not saturate, and the transformer is suitable for the application. On the secondary side of the gate transformer, each signal passes through a DC restore circuit that re-establishes the correct DC bias for proper gate drive operation.

C. Dead-Time Adjustment

Dead-time between high and low-side switches is also required. Dead-time is the intentional delay between turning off one switch and turning on the complementary switch to avoid cross-conduction or shoot-through. Following the DC restore stage is the dead-time adjustment circuit. This consists of feed-forward diodes (D6, D9) and RC networks (R3, C2, R8, C4), which create a delay to adjust the rising edge of the switching signal. The RC time constant determines the amount of delay introduced, allowing fine-tuning of the dead

time to prevent cross-conduction. The falling edge of the switching signal is designed to occur rapidly through to the use of feedback diodes (D7, D10), which bypass the RC delay for fast turn-off. This arrangement ensures precise control of switching transitions while maintaining the necessary dead-time for reliable operation. For a switching frequency of 100 kHz, a dead-time of 2% would be equal to 200 ns.

D. Gate Drive

The driver stage follows and consists of transistors Q5–Q8 and resistors R4–R12. The pull-up resistors on the output transistors (R5, R10) play a dual role: they ensure proper pull-up operation and can also be used to adjust the rise time of the output switching waveform. By selecting appropriate resistor values, the rise time can be tailored to balance switching speed and EMI considerations.

It is important to note that the high-side driver is referenced to switch node A, while the low-side driver is referenced to ground. This distinction is necessary to accommodate the floating nature of the high-side switch. The high-side driver is powered by a floating +5V supply (+5V-FA), which is achieved using an isolated charge pump converter. This floating supply ensures that the high-side driver operates correctly, even as the voltage at the switch node varies dynamically during switching transitions. Each high-side driver has its own independent floating supply, referenced to its respective switch node.

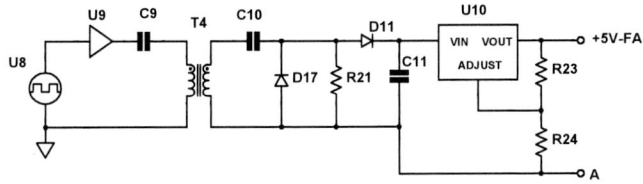

Fig. 6. Isolated Floating Charge-Pump Supply for Phase-A Leg

E. Floating Supplies

This circuit shown in Fig. 6 operates as a charge-pump with isolation, leveraging a transformer to transfer energy and generate floating voltages for high-side drivers. A square-wave signal, generated by an oscillator circuit (simplified as U8) and driven through U9, excites the transformer T4. This method of transferring energy in discrete pulses closely resembles the operation of a traditional charge-pump.

On the secondary side, C10 and D17 restore the dc bias to the signal generated by the transformer. Diode D11 and capacitor C11 store and smooth this energy, replicating the charge transfer and storage mechanism seen in charge-pump circuits. The rectified and smoothed voltage feeds into an LDO regulator (U10), which ensures a stable output voltage (+5V-FA) for powering the high-side driver circuitry.

This floating supply design provides isolated voltage sources for both phase legs (A and B) of the converter, ensuring

979-8-3315-1612-3/25 $31.00 © 2025 IEEE

proper operation of the high-side switches while maintaining isolation.

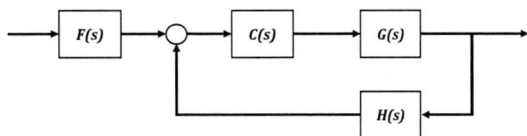

Fig. 7. Closed-Loop Control Architecture

F. Control Theory

The compensator utilized in this work is a Type II analog compensator, as depicted in Fig. 1. This type of compensator is widely used in power electronics control systems for its ability to provide a combination of proportional and integral control with phase boost, making it ideal for improving stability and dynamic response. The transfer function of the plant, which represents the power stage of the system, considers key components such as the output inductor L_{out}, output capacitor C_{out}, and the parasitic resistance of the capacitor R_{esr}. The plant transfer function is expressed as:

$$G(s) = \frac{1}{L_{\text{out}}C_{\text{out}}s^2 + (R_{\text{esr}}C_{\text{out}})s + 1} \tag{4}$$

This function describes the second-order dynamics of the plant, which include a resonance point determined by the inductor-capacitor combination and a damping factor influenced by R_{esr}. The resonance must be properly managed to prevent instability and achieve smooth output voltage regulation. The open-loop transfer function, which describes the system's behavior without feedback, is given by:

$$G_{OL}(s) = C(s) \cdot G(s) \cdot H(s) \tag{5}$$

Here:

- $C(s)$ represents the compensator's transfer function, which is responsible for shaping the system response, adding gain, and introducing phase boost.
- $G(s)$ is the plant transfer function that models the dynamics of the power stage.
- $H(s)$ is the sensor transfer function, which is defined as the gain of the resistive feedback network. It determines the scaling factor between the output voltage and the feedback voltage supplied to the controller.

The closed-loop transfer function, which incorporates the feedback mechanism to regulate the output, is given as:

$$T(s) = \frac{F(s) \cdot C(s) \cdot G(s)}{1 + H(s) \cdot C(s) \cdot G(s) \cdot F(s) \cdot C(s) \cdot G(s)} \tag{6}$$

In this equation, $F(s)$ is the pre-filter, which is equal to the reference voltage of the op-amp within the PWM. This ensures that the system output tracks the reference voltage. The closed-loop control architecture is shown in Fig. 7.

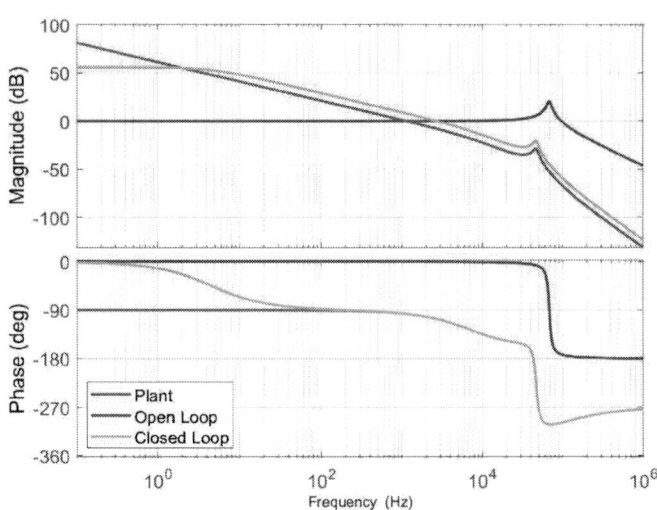

Fig. 8. Plant, Open-Loop, & Closed-Loop Transfer Functions

G. Compensating the Loop

The Type II compensator is used to stabilize the feedback loop, addressing the phase lag introduced by the plant's LC filter. It ensures that the system maintains sufficient phase margin, preventing oscillations or instability. By introducing a zero and a pole at appropriate frequencies, the compensator enhances the system's transient response. The general form of the Type II compensator is:

$$C(s) = k \cdot \frac{s + \omega_z}{s \cdot (s + \omega_p)} \tag{7}$$

where k is the compensator DC gain, ω_z is the frequency of the zero, and ω_p is the frequency of the pole. The power stage's LC filter introduces a resonant frequency, which is given by:

$$f_r = \frac{1}{\sqrt{L_{\text{out}}C_{\text{out}}}} \tag{8}$$

The compensator pole is placed below the resonant frequency to allow for some gain margin while the compensator zero is placed above the resonant frequency to retain phase margin. Once the pole/zero placement is deteremined, the values can be calculated. The compensation components, shown in Fig. 1, can be calculated using the following formulas:

$$\text{Choose } R_{\text{int}} \Rightarrow \quad C_{\text{int}} = \frac{1}{2\pi f_{z1} R_{\text{int}}} \tag{9}$$

$$C_{\text{roll}} = \frac{1}{2\pi f_{p1} R_{\text{int}}} \tag{10}$$

where f_{z1} and $fp2$ are the frequency locations of the zero and pole. These work as starting values and give a stable loop as shown in Fig. 8. However, once the converter is built, the actual loop should be measured to ensure the design achieves the desired stability and performance.

It is important to note that the PWM current sense pin should be supplied with a ramp signal to change the controller from current-mode to voltage-mode operation.

979-8-3315-1612-3/25 $31.00 © 2025 IEEE

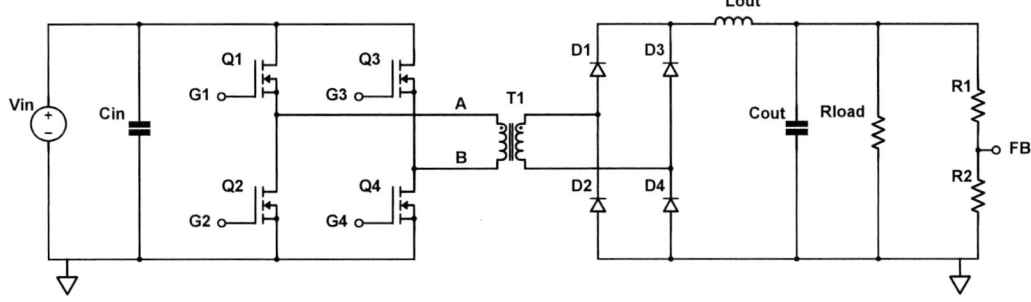

Fig. 9. Full-Bridge and Diode Rectifier Power Stage

IV. PSFB POWER STAGE

The design of the PSFB power stage emphasizes the selection of rad-hard and high-voltage components. Power stage specifications are detailed in Table I, while the corresponding calculated design values are outlined in Table II. A list of the key radiation-hardened components selected, along with their commercial counterparts used in the prototype, is provided in Table III. The average input current of the converter, critical for sizing the primary switches, is determined using the formula:

$$I_{\text{in,avg}} = \frac{P_{\text{out}}}{\eta \cdot V_{\text{in}}} \tag{11}$$

For the main power switches, GaNFETs were selected due to their inherent radiation tolerance and fast switching capabilities. Specifically, the EPC2050 was chosen for $V_{ds,max}$ of 350 V which satisfies $V_{ds,max} > V_{in}$. The continuous drain current is rated at 6.3 A, which exceeds the average input current of 1.6 A.

The output diode rectifiers were chosen based on their voltage and current ratings, which must exceed the output voltage and load current. For the rad-hard implementation, the JANTX1N5623US rectifier diode was selected, offering a 1 kV voltage rating and a 1 A current rating. Its commercial counterpart, the 1N5623, shares these specifications.

For supporting circuits, including the PWM controller, drivers, logic inverter, and logic buffer, appropriate rad-hard components and their commercial equivalents were chosen. The PWM controller is an ST1845, which has a direct commercial counterpart, the UC3845B. The gate driver, needed for the main gate drive circuit and floating supplies, is the RHRPM4422, with the MIC4422A serving as its commercial counterpart.

A. Transformer Design

The ETD34 core with N87 material is chosen for its compact design, efficient heat dissipation, and high-frequency performance. The ETD34 provides sufficient cross-sectional area for flux handling while minimizing transformer size and simplifying winding. Material N87 offers a high saturation flux density ($B_{sat} \approx 0.49$T at 25°), low core losses at high frequencies, and excellent thermal stability across a wide

TABLE I
DESIGN PARAMETERS

Input Voltage	200 V
Output Voltage	600 V
Output Current	0.5 A
Output Power	300 W
Switching Frequency	100 kHz
Efficiency Target	80 %

TABLE II
CHOSEN DESIGN VALUES

Turns ratio n	5
Primary Inductance	1.3 mH
Secondary Inductance	30 mH
Output Inductor	250 uH
Output Capacitor	5 uF

temperature range. A small gap is introduced to reduce the flux density. Primary turns are calculated using:

$$N_{\text{pri}} = \frac{V_{\text{in}} \cdot t_{\text{on}}}{B_{\text{max}} \cdot A_e} \tag{12}$$

A B_{max} value of 0.45 T is used, which gives 23 turns on the primary side and 115 turns on the secondary side. Inductance for primary and secondary sides can be calculated using:

$$L = \frac{N^2 \cdot \mu_0 \cdot \mu_r \cdot A_e}{l_m} \tag{13}$$

A small gap was placed on the transformer which increases the leakage inductance on the secondary side reducing the need for output inductance.

TABLE III
PARTS LIST

	Rad-hard Part	Commercial Part
GaNFET	EPC2050	EPC2050
Rectifier Diode	JANTX1N5623US	1N5623GP
D-Flip-Flop	HCC4013B	HCF4013Y
PWM Controller	ST1845	UC3845B
Driver	RHRPM4424	MIC4422A
Inverter	HCC4049UB	M74HC04
Buffer	HCC4050B	MM74HC126

979-8-3315-1612-3/25 $31.00 © 2025 IEEE   1194

TABLE IV
ESTIMATED LOSSES

Loss Value	(W)
Ppri	0.22
Psec	1.10
Pdiodes	0.80
Pfet,cond	0.89
Pfet,sw	3.33
Pcore	15.26
Ptotal	21.61

B. Efficiency Calculations

Various power losses within PSFB include transformer winding losses, diode conduction losses, MOSFET conduction and switching losses, and core losses. The resistive losses in the primary winding are calculated based on the RMS current through the primary winding:

$$P_{\text{pri}} = I_{\text{pri,rms}}^2 \cdot R_{\text{pri}} \tag{14}$$

Similarly, the resistive losses in the secondary winding depend on the secondary RMS current and the resistance of the secondary winding:

$$P_{\text{sec}} = I_{\text{sec,rms}}^2 \cdot R_{\text{sec}} \tag{15}$$

Each diode conducts half of the output current:

$$I_{\text{diode,avg}} = \frac{I_{\text{out}}}{2} \tag{16}$$

Diode conduction losses are based on the forward voltage drop and the average current:

$$P_{\text{diode}} = V_f \cdot I_{\text{diode,avg}} \tag{17}$$

For a full-wave rectifier with two conducting diodes per cycle, the total diode loss can be found using:

$$P_{\text{diodes}} = 2 \cdot P_{\text{diode}} \tag{18}$$

The conduction losses for each MOSFET are proportional to the primary RMS current and the on-state resistance:

$$P_{\text{MOSFET,cond}} = I_{\text{pri,rms}}^2 \cdot R_{\text{DS(on)}} \tag{19}$$

Switching losses depend on the drain-source voltage, the peak primary current, the combined rise and fall times, and the switching frequency:

$$P_{\text{MOSFET,sw}} = \frac{1}{2} \cdot V_{\text{ds}} \cdot I_{\text{pri,pk}} \cdot (t_{\text{on}} + t_{\text{off}}) \cdot f_{\text{sw}} \tag{20}$$

Core losses are estimated using the relative core loss chart from the manufacturer datasheet and then calculated using:

$$P_{\text{core}} = V_{\text{core}} \cdot P_{\text{density}} \tag{21}$$

A table of estimate losses is shown in Table IV. For an output power of 300 W, the estimated efficiency is 93%.

V. SPICE SIMULATION

A Type II compensator is designed and applied to obtain the open- and closed-loop transfer functions, as shown in Fig. 8. The compensator was implemented and simulated in TopSpice alongside a simplified behavioral model of the PSFB converter. An AC response simulation was performed to obtain the gain and phase characteristics of the open-loop system. The compensator's pole and zero were adjusted to achieve stable loop performance, as illustrated in Fig. 10. The resulting system demonstrates a gain margin of 15 dB and a phase margin of 80°, ensuring robust stability and reliable operation.

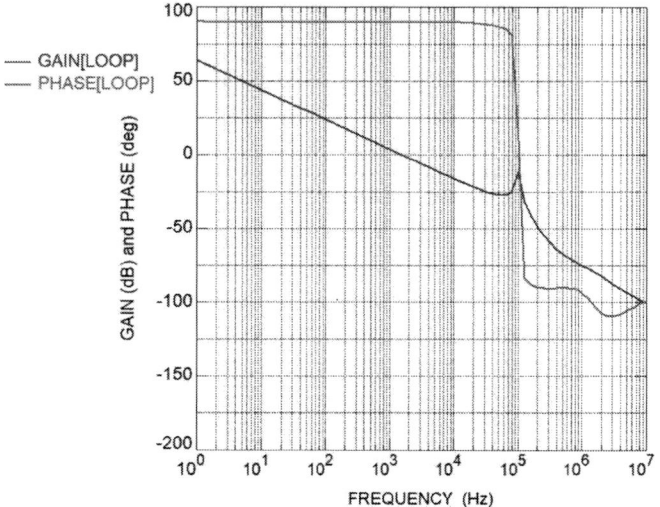

Fig. 10. AC Response Simulation

The PSFB and discrete controller, incorporating GaNFETs and rectifier models, were implemented and simulated in Top-Spice to validate performance using component models. The simulation results, illustrated in Fig. 11, show the feedback voltage and the comparator output. As the output voltage approaches regulation, the comparator output drops to zero, indicating that the feedback loop is actively maintaining the desired voltage.

Figure 12 demonstrates the output voltage and current ripple of the PSFB converter, measured at approximately 8 V and 7 mA, respectively, both remaining within 2% of their nominal values. The average input current is measured at 1.85 A, corresponding to an overall efficiency of 81%.

The simulation results demonstrate the controller's effectiveness in maintaining stable output regulation, even for a high-voltage application. The ability to achieve a tightly regulated 600 V output with minimal voltage and current ripple, both within 2% of the nominal values, show reliable operation for sensitive systems. Additionally, the robust startup behavior and steady-state stability highlight the controller's capacity to handle the challenges associated with high-voltage power systems, such as transient conditions and load variations. These features make the controller particularly suitable for critical high-voltage applications, such as spacecraft power

979-8-3315-1612-3/25 $31.00 © 2025 IEEE

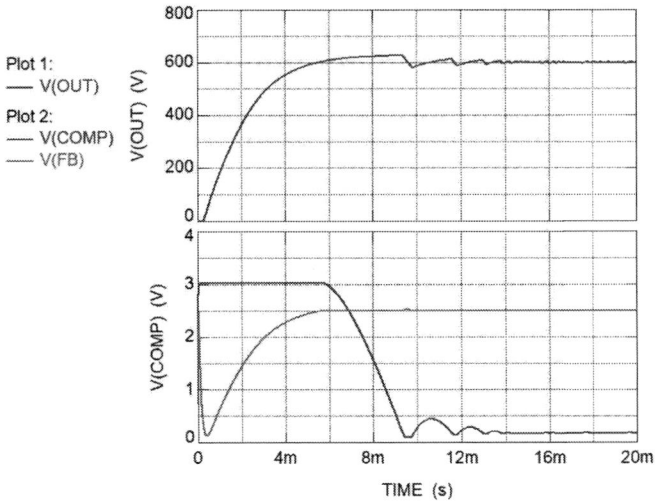

Fig. 11. Controller and Output Voltage

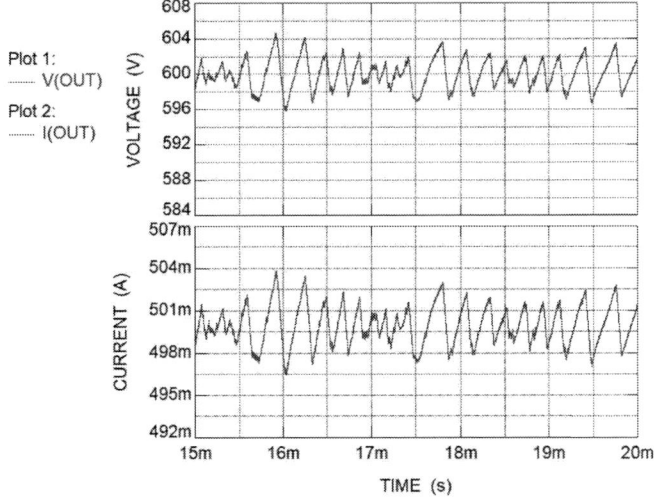

Fig. 12. PSFB Simulation Ripple Voltage/Current

systems and other mission-critical operations where precision and reliability are paramount.

VI. CONCLUSION AND FUTURE WORK

This work presents a control scheme for a phase-shifted full-bridge (PSFB) converter, designed for implementation using discrete radiation-hardened components. The theory of operation for the discrete PSFB controller is thoroughly discussed, along with the various circuits required to build its rad-hard implementation. The study includes an in-depth exploration of control theory, focusing on the modeling and compensation of the analog controller, as well as a detailed discussion on design considerations and component selection. SPICE simulation results of the converter are presented, with a comprehensive explanation of system losses. The significance of this work lies in demonstrating the practical feasibility of such a control scheme for high-risk space missions. The proposed converter is

applicable to both spacecraft bus power systems and spacecraft ion thrusters. Simulation results validate the control scheme's effectiveness for high-voltage converters.

Ongoing research is focused on the hardware implementation of the proposed control scheme and PSFB converter. A detailed analysis of the controller's performance is being conducted, with loop gain measurements planned for comparison against the simulation results to verify system accuracy.

Future work will emphasize scaling up power by paralleling multiple PSFB modules. Individual module output power could also be increased by raising the output current, with careful consideration to ensure that diode rectifiers do not experience excessive power dissipation. Similarly, the secondary winding must be capable of handling the increased output current without incurring significant losses. Paralleling PSFB converters for higher power output would necessitate a higher level of control, such as the use of an FPGA, to manage the coordinated turn-on and turn-off of multiple modules. Although higher-level control would require the use of an FPGA, the analog control implementation offers a reliable and robust method to manage individual modules, and it further retains simplicity while enabling the coordination of parallel operation. This approach would allow the development of significantly higher-power systems while maintaining scalability and efficiency.

REFERENCES

[1] R. Scheidegger and J. Soeder, "Spacecraft Bus Voltage Selection," NASA Glenn Research Center.

[2] I. M. Hackler, R. L. Robinson, and B. Hendrix, "A Power Management and Distribution Concept for Space Station," in INTELEC '84 - International Telecommunications Energy Conference, IEEE, Dec. 1984, pp. 124–129. doi: 10.1109/INTLEC.1984.4794107.

[3] X. He, Y. Dong, C. Lu, T. Liu, K. Gao, and S. Liu, "Research on phase-shifted full-bridge converter applied in the anode-supply of hall electric propulsion," Proc. Int. Conf. Power Electron. Drive Syst., vol. 2015-Augus, no. June, pp. 47–51, 2015, doi: 10.1109/PEDS.2015.7203535.

[4] L. R. Piñero, P. Y. Peterson, and G. E. Bowers, "High performance power module for hall effect thrusters," 38th AIAA/ASME/SAE/ASEE Jt. Propuls. Conf. Exhib., no. September, 2002, doi: 10.2514/6.2002-3947.

[5] P. K. Joseph, N. V. Bijeev, N. B. Sharma, and V. H. Jani, "Analysis of digital PWM design for Zero Voltage Transition Phase Shifted Full bridge converters for space applications," 2016 Int. Conf. Emerg. Trends Commun. Technol. ETCT 2016, pp. 1–5, 2017, doi: 10.1109/ETCT.2016.7882923.

[6] V. Turriate, B. Witcher, D. Borovevich, and R. Burgos, "Self-powered Gate Driver Design for a Gallium Nitride Based Phase Shifted Full Bridge DC-DC Converter for Space Applications," 2018 IEEE 6th Work. Wide Bandgap Power Devices Appl. WiPDA 2018, pp. 141–148, 2018, doi: 10.1109/WiPDA.2018.8569056.

[7] V. Turriate, B. Witcher, D. Boroyevich, and R. Burgos, "Practical implementation and efficiency evaluation of a phase shifted full bridge DC-DC converter using radiation hardened GaN FETs for space applications," Conf. Proc. - IEEE Appl. Power Electron. Conf. Expo. - APEC, vol. 2019-March, pp. 1587–1594, 2019, doi: 10.1109/APEC.2019.8722059.

[8] "Hi-Rel Phase-Shift Full Bridge Design," Frenetic.ai. Accessed: Nov. 29, 2023. [Online]. Available: https://frenetic.ai/magnetic-notes/hi-rel-phase-shift-full-bridge-design

979-8-3315-1612-3/25 $31.00 © 2025 IEEE

Modeling, Control and Digital Implementation of a Buck Converter Operating in Triangular Current Mode for a Wide Output Voltage Range Space application

Regina Ramos * [image], Sara Perez *, Guillermo Nuñez *, Pedro Alou* [image], and Javier Torres [†]

*Centro de Electrónica Industrial (CEI), Universidad Politécnica de Madrid (UPM) [†] Airbus Crisa

Email: regina.ramos@upm.es

Abstract—**This paper presents a simplified constant on time modeling and control of a buck converter operating in triangular current mode for space applications. The buck converter operates at a variable frequency to achieve ZVS in both transitions while minimizing the RMS current through the inductor. This control is implemented in a field-programmable gate array. Since the converter operates at a variable frequency, two different data acquisition alternatives are explored: sampling at a variable frequency (once per switching cycle) or sampling at a constant frequency (higher than the switching frequency). The proposed control is validated by simulation and experimental results with a 500 W prototype.**

Index Terms—**Digital control, space application, variable frequency, electric propulsion, radiofrequency generator, gridded ion technology thruster.**

I. INTRODUCTION

In recent years, the space industry has experienced a shift in trends. The commercial expansion of space missions has resulted in an annual growth rate of about 10% in this field. New space objectives, such as returning to the lunar surface [1], Mars exploration missions [2], and further exploration of the solar system [3], have driven the aerospace sector to develop innovative technological solutions to tackle these challenges [4].

Electric propulsion has emerged as a promising alternative to traditional chemical propulsion [5], offering distinct advantages [6]. It provides a higher specific impulse, greater efficiency, longer operational lifespans, increased payload capacity, and enhanced flexibility. For instance, electric thrusters can reach ion ejection velocities of up to 50 km/s, roughly ten times the speed of chemical thrusters.

This work is focused on electrostatic thrusters based on Gridded Ion Technology, where plasma is generated using radio frequency. In these systems, noble gases, typically xenon, are ionized to create the plasma used for propulsion. The plasma is then accelerated by an electric field formed between two grids with a voltage difference. The ionization process

DEEP PPU project ID 101082685 is co-funded by the European Union. Views and opinions expressed are however those of the author(s) only and do not necessarily reflect those of the European Union or HADEA. Neither the European Union nor the granting authority can be held responsible for them.

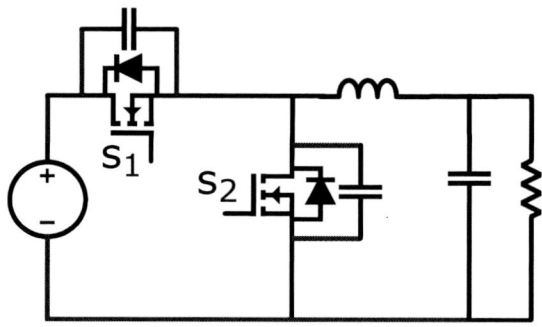

Figure 1: Synchronous buck converter topology

Table I: Specifications

Parameter	Value
Maximum Power	550 W
Input voltage	100 V
Output Voltage Range	$10 - 90$ V
Inductance	3.4 μH
Output Capacitance	30 μF

is driven by high-frequency magnetic fields, usually in the megahertz range.

To convert the spacecraft's electrical power into the high voltage needed to operate the ion thruster, a power processing unit (PPU) is utilized [7]. One of the components of the PPU is the radio frequency generator (RFG), which is used to ionize the propellant allowing to control the thrust. The RFG is composed of two stages: a DC-DC converter to regulate the power, and a resonant inverter that synchronizes with the load's frequency and phase, ensuring a high-efficiency power transfer. This paper is focused on the DC-DC stage, the buck converter aims to control the thrust by regulating its output voltage.

This document is organized as follows. The main specification and the topology selected are described in Section II. Section III derives small signal model of the presented topology. The control design, including the digital implementation, is provided in Section IV. The modulator implementation

Table II: Operating Points

Operating Point	V_{out}	P_{out}	T_{on}	f_{sw}
OP1	90 V	550 W	5.46 μs	164.66 kHz
OP2	85 V	552.5 W	3.8 μs	223.7 kHz
OP3	10 V	40 W	0.4675 μs	213.9 kHz
OP4	10 V	11.1 W	0.274 μs	364.93 kHz
OP5	50 V	12.33 W	0.856 μs	666.67 kHz

is discussed in sec.V. Simulation and experimental results presented in Section VI confirm the stability of the converter. Section VII concludes the digest.

II. STATEMENT OF WORK

Table I listed the main operating points for the DC-DC stage.

Due to the optimization analysis, it has been concluded that for this particular application, the most efficient solution is a synchronous buck converter operating (Fig. 1) in triangular current mode (TCM) achieving an efficiency over the 90% in the full output voltage range and 98.7% at full load and maximum output voltage.

This type of modulation presents a good balance between conduction and switching losses.

In a conventional synchronous buck converter, ZVS can be achieved in both transitions when having positive current to discharge the parasitic capacitance of the low side MOSFET during the second interval (switch S_1 ON and S_2 OFF Fig. 1) and enough negative current (I_{ZVS}) to discharge the parasitic capacitance of the high side MOSFET during the second interval (switch S_1 OFF and S_2 ON Fig. 1). The ZVS operation not only impacts the efficiency but also reduces the EMI noise. [8], [9].

To reduce the RMS current, and therefore, the conduction losses, TCM modulation adjusts the switching frequency dynamically to always switch with the same negative current. The soft switching transition under this operation is detailed in [10].

To assure the ZVS operation two main alternatives were analyzed, a shunt resistor and a comparator to detect the I_{ZVS} a cross-detection (ZCD) circuit. It has to be noticed that the peak current in this work can reach up to 10A, therefore when using a shunt resistor an important trade-off between the sensitivity and the losses exists. When having a low resistance to reduce the losses, even when operating in soft switching the noise level due to the transitions were comparable with the threshold configured in the comparator. Therefore, a zero cross-detection (ZCD) circuit with a saturable current transformer, based on the solution presented in [11], has been designed to assure the ZVS operation. This solution presents some drawbacks that are analyzed in the sec.V

In this work, a linear voltage control loop is proposed, the control parameter is the on-time on the high side MOSFET, while the switching frequency is adjusted based on the signal of the ZCD.

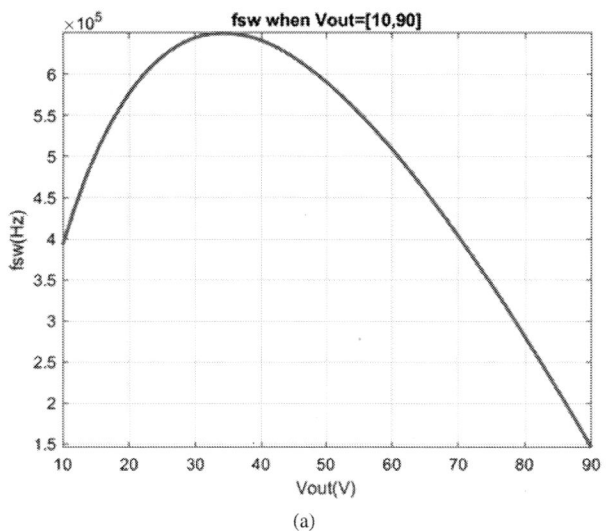

(a)

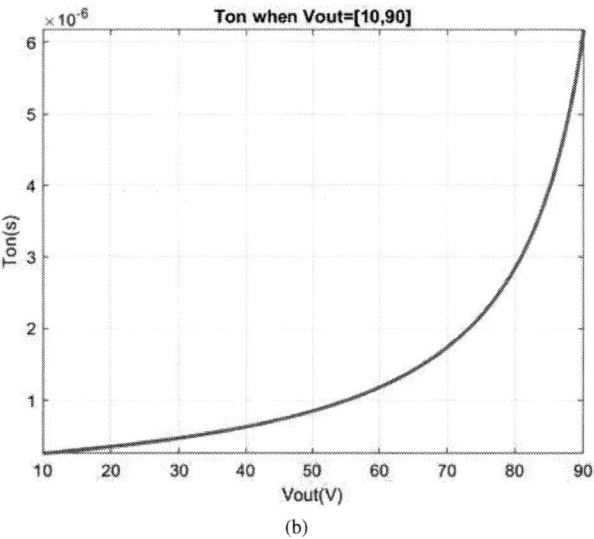

(b)

Figure 2: a) variation of the switching frequency as a function of the output voltage b) variation of the T_{on} as a function of the output voltage

The load after the RFG is plasma, which presents a complex behavior since its impedance depends on the operation mode of the thruster and other factors [12]. For the sake of simplicity, the load is modeled by a constant resistance which is accurate enough at this stage.

Eq.1 represent the average inductor, knowing that in steady state the average capacitor current is equal to zero and therefore $\langle i_L \rangle = I_{out}$, eq.2 and eq.3 can be obtained. Figure

2 represents the switching frequency and the on-time for the output voltage range with a constant resistive load. Both graphs are based on eq.2 and eq. 3, knowing that I_{ZVS} is the required current to achieve ZVS, in this case $-3\,A$, and T the switching period.

$$\langle i_L \rangle = -\frac{v_{out}T}{2L} + I_{ZVS} + \frac{t_{on}V_{in}}{L} - \frac{t_{on}^2 V_{in}}{2LT} \tag{1}$$

$$f_{sw} = 0.5\frac{\left(d - d^2\right)V_{in}}{L\left(I_{out} - I_{ZVS}\right)} \tag{2}$$

$$t_{on} = \frac{v_{out}}{f_{sw}V_{in}} \tag{3}$$

As can be concluded from the previous equation the load resistance also impacts the on-time and switching frequency.

III. SMALL SIGNAL MODEL

To obtain the average model of the converter under TCM operation the inductor average current has been obtained neglecting the dead time intervals, in steady state the average inductor current matches with the output current obtained in eq.1. That equation can be rewritten as:

$$\langle i_L \rangle = I_{ZVS} + \frac{T_{on}}{2L}\left(V_{in} - V_{out}\right) \tag{4}$$

The plant of the system is obtained by linearizing eq.4. As can be seen is a first order system. The main challenge to control this simple plant is that the gain and the pole depend on the load and the output voltage, the impact of the input voltage variation is not considered in this work as the input bus is a well regulated voltage.

$$\tilde{v}_{out} = \frac{R}{1 + RCs}\tilde{i}_L \tag{5}$$

$$\tilde{i}_L = \frac{V_{in} - V_{out}}{L}\tilde{t}_{on} - \frac{T_{on}}{L}\tilde{v}_{out} \tag{6}$$

$$\frac{\tilde{v}_{out}}{\tilde{t}_{on}} = \frac{\left(V_{in} - V_{out}\right)R}{T_{on}R + 2L + 2LCRs} \tag{7}$$

The analytical model has been validated by comparing the bode plot obtained in Matlab and Simplis. Only one operating point is shown as a reference (fig.3) the high-frequency resonance that appears in the Simplis simulation is due to the switching frequency, neglecting that effect is can be seen that the analytical model well represents the converter dynamic response.

IV. CONTROL DESIGN

The dynamic behavior of the thruster is not very demanding with output voltage steps up to 10V/ms, making it possible to simplify the control by having a unique regulator for the whole operation range.

In figure 4 the plant of the system for the listed operating point is shown. The maximum output voltage ripple does not restrict the design of the output capacitor but the position of the

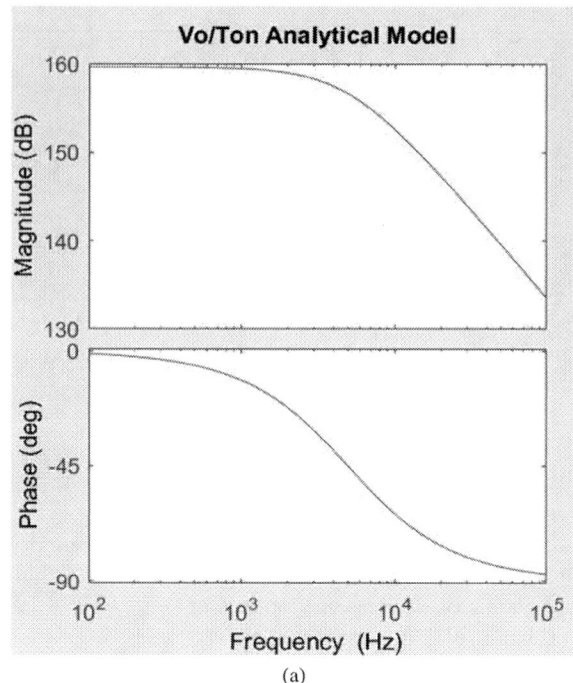

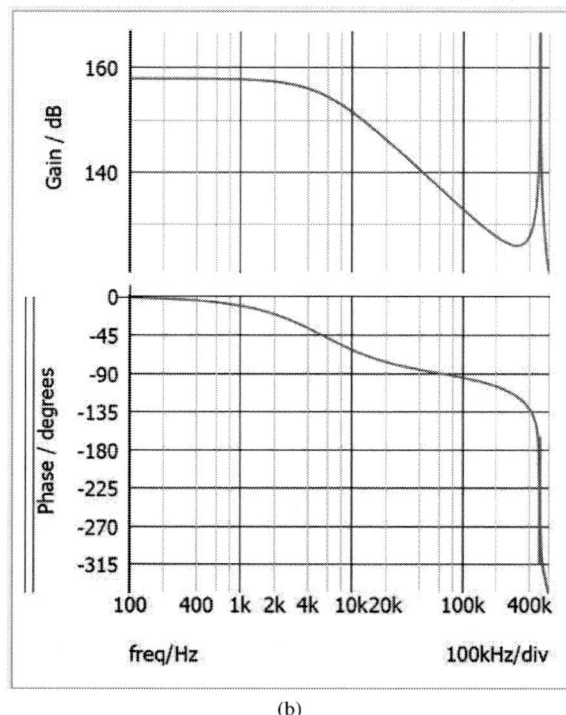

Figure 3: a) bode plot of the analytical model b) AC analysis response obtained in Simplis

pole of the system. Having a $C_{out} = 30\mu F$, and knowing that the switching frequency range is from $150\,kHz$ to $675\,kHz$, the design controller is a PI regulator

$$R(s) = \frac{10^{-4}s + 1}{330s} \tag{8}$$

Figure 5 represents the loop gain response of one of the

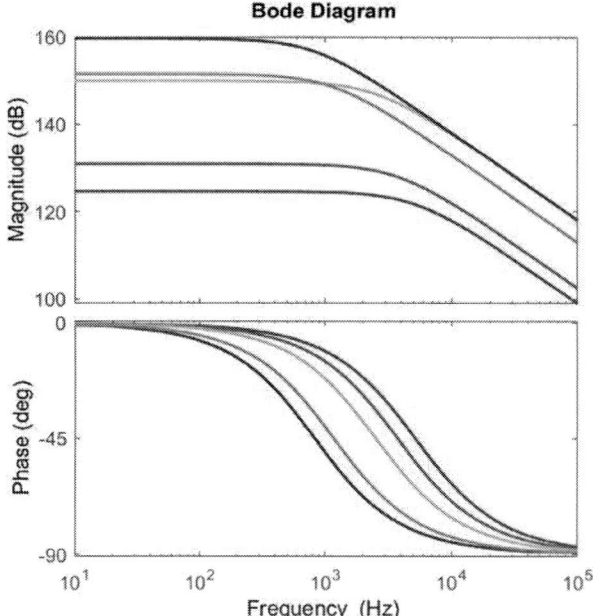

Figure 4: Frequency response for the different operating point listed in TableII

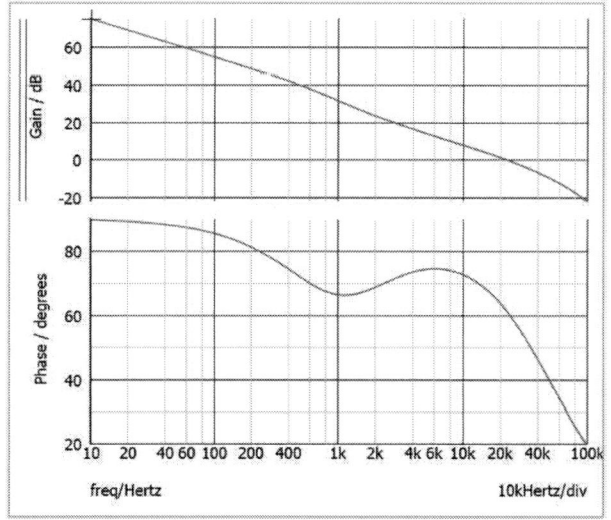

Figure 5: Loop gain frequency response for the OP3 listed in TableII

listed operating points.

Table III summarizes the main information of the plant for the different operating points and the bandwidth and phase margin obtained with the proposed controller. Due to the plant

variation, the bandwidth varies from 938 Hz up to 24.2 kHz obtaining a minimum phase margin of $83.1°$.

To sense the output voltage a voltage divider has been implemented with a low pass filter (RC filter with a cut off frequency set at 50 kHz), this additional filter compensates the non-idealities of the output capacitor ($C_{out} = 27.28 \ \mu F$, $ESR = 5.57 \ m\Omega$, $ESL = 29.79 \ nH$). The phase lag introduced by this filter was also computed in the regulator design.

The digital implementation of this control loop is not straightforward, the two key points are the the modulator and the variable switching frequency.

Two different alternatives were analyzed regarding the variable switching frequency: sample at constant frequency or sample at variable frequency, synchronized with the switching frequency. At the simulation level, there is no big impact. Still, due to the noise generated during the switching transition, the experimental performance is better when sampling at variable frequency setting the sampling event far away from the switching events.

For simplicity, the discretization method used was a Backward Euler as the phase drop introduced by this type of discretization method was not critical for this specific application. The discrete version of the proposed regulator takes into account the variable sampling time (T_s).

$$R(z) = \frac{(0.0001 + T_s)z - 0.0001}{330.033z - 330.033} \quad (9)$$

The digital control has been validated in Plecs including the delays and the quantization effects [13].

V. FPGA IMPLEMENTATION

This control has been implemented in a Basys 3 FPGA from Xilinx but instead of using the internal ADC, the ADC used has been ADC7380-4 from analog devices.

The VHDL structure is represented in Fig.7.

The ADC block records the data obtained by the external ADC and controls the sampling event. To be able to change the output voltage reference without using the ILA, the reference voltage can be programmed with the switches included in the Basic 3 board. When the DPWM block triggers the clk_sample signal the controller calculates the new T_{ON}

As mentioned before, to operate in TCM a ZCD circuit has been designed, but the period must end at $-3 \ A$ to assure ZVS, due to the variable output voltage the time between the zero current detection and the desired switching point is

Table III: Control parameter for the different operating points

	Pole frequency	DC Gain	fase margin	bandwidth
OP1	5.19 kHz	124.35 dB	109.6°	938 Hz
OP2	3.76kHz	130.545 dB	114.62°	2.73 kHz
OP3	2.53 kHz	149.39 dB	83.1°	24 kHz
OP4	831 Hz	157.76 dB	84.18°	24.2 kHz
OP5	1.18 kHz	150.17 dB	88.24°	13.4 kHz

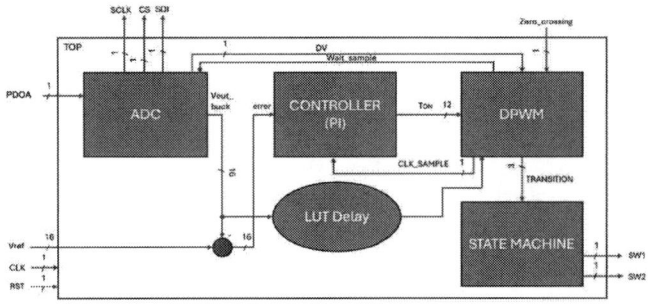

Figure 7: VHDL structure

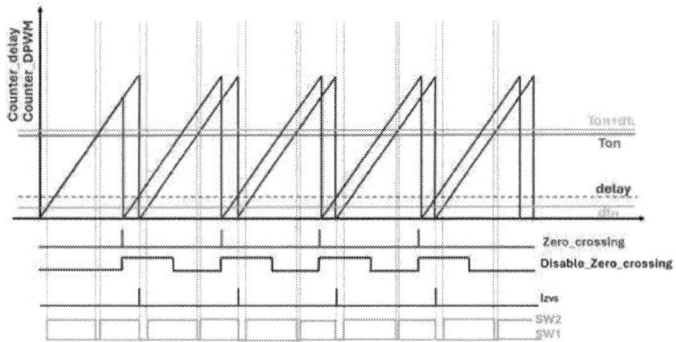

Figure 8: Modulator's workflow

variable. To solve this issue a look-up table is included to take that into account. Fig.8 represents how the modulator works. The main drawback of this solution is that it needs to know the exact value of the inductor, so either ensure it is manufactured with low tolerances and minimal aging or an additional average current sensor can be included to correct the drift from the ideal value, as this method controls the on time and the inductor current at the begging of the switching cycle, the look-up table could be dynamically adapted to adjust the inductor tolerances.

The DPWM block has several functions, as has been explained above, it controls the ADC acquisition to measure the output voltage at half of the T_{off} assuming the switching cycle does not change drastically. As the same time, once it receives a trigger from the *Zero_crossing* it launches a counter based on the measured output voltage and the LUT delay look-up table, once the counter finishes the switching cycle starts. The DPWM block also controls the state of the two switches taking into account the dead time for the high side and low side generating a signal (*transition*) that indicates the configuration of the switches ('001' dead time from low side to high side, '010' high side ON, '011' dead time from high side to low side and '100' low side ON, '000' is the reset state) Fig. 8 shows the workflow of the modulator.

Finally, the state machine block reads the *transition* signal and generates the two driving signals.

VI. SIMULATION RESULTS AND EXPERIMENTAL VALIDATION

Fig. 9 shows the experimental setup built to validate the control proposed. The prototype is implemented with commercial components as the aim of this work is to validate the control, anyhow the switches are COTS components and certified components for space are identified to reproduce the setup. The simulation analysis of the digital control has been run in PLECs to include the nonidealities of the digital converter, such as the delay introduced by the ADC and the quantization error. In fig.6 a close loop simulation proves the correct output voltage tracking for all the operating points listed above. It can be noticed that the converter keeps a stable behavior under load steps. The upper plot shows the reference output voltage and the actual output voltage that can accurately follow the

Figure 9: Experimental setup

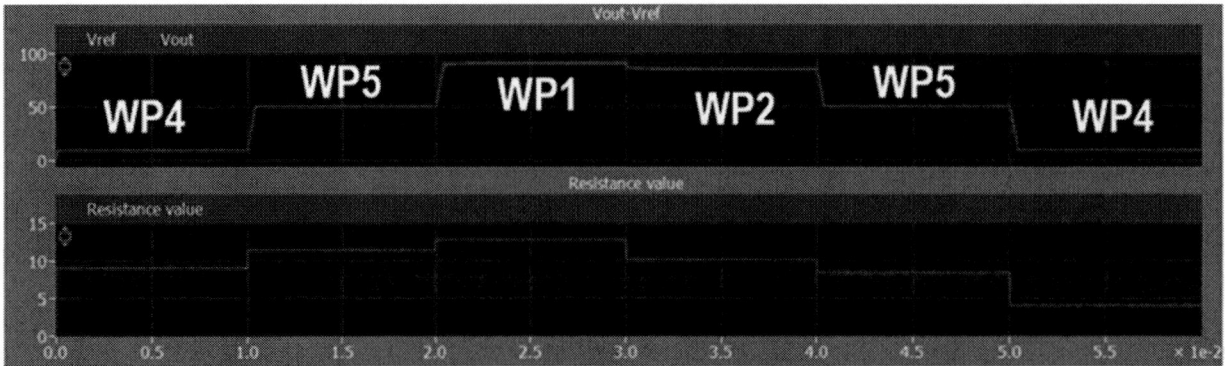

Figure 6: simulation results showing the full set of operating points

reference, in the bottom part the resistance value is represented to show the dynamic of the different load steps.

Finally, Fig.10 shows the experimental waveforms. Fig.10a shows a steady state operation in a close loop, the output voltage is represented in blue, the inductor current in yellow, the Ton signal in green, and in orange is represented the ZCD signal. Fig.10b and Fig.10c present output voltage reference steps of 10 V/ms, in these plots the output voltage is represented in blue.

VII. CONCLUSIONS

This article describes the control algorithm proposed for a buck converter operating in TCM. The converter works at a wide range of operating points (output voltage range from $10\ V$ to $90\ V$, and output power from $10\ W$ up to $550\ W$) with variable switching frequency (from $150\ kHz$ to $675\ kHz$). The dynamic of the converter depends on the load, the output, and the input voltage. The proposed control assures stability in the full range of operations.

To assure ZVS in all the transitions a ZCD circuit is designed and a look-up table is included in the FPGA to delay the start of the new switching cycle depending on the output voltage and therefore always switch with at $-3\ A$.

Simulation and experimental results have been presented to validate this control proposal.

REFERENCES

[1] NASA. Gateway overview. [Online]. Available: https://www.nasa.gov/gateway-space-station-news/

[2] NASA. Mars Sample Return. [Online]. Available: https://science.nasa.gov/mission/mars-sample-return/

[3] ESA.Juice. [Online]. Available: https://www.esa.int/Science_Exploration/Space_Science/Juice

[4] CRISA. Exploring the solar system. [Online]. Available: https://crisa.airbus.com/en/space-technology-from-spain/exploring-the-solar-system

[5] ESA. Electric Spacecraft Propulsion. [Online]. Available: https://sci.esa.int/web/smart-1/-/34201-electric-spacecraft-propulsion

[6] D. M. Goebel, I. Katz, and I. G. Mikellides, *Fundamentals of electric propulsion*. John Wiley & Sons, 2023.

[7] K. Holste, P. Dietz, S. Scharmann, K. Keil, T. Henning, D. Zschätzsch, M. Reitemeyer, B. Nauschütt, F. Kiefer, F. Kunze *et al.*, "Ion thrusters for electric propulsion: Scientific issues developing a niche technology into a game changer," *Review of Scientific Instruments*, vol. 91, no. 6, 2020.

[8] V. Vorpérian, "Quasi-square-wave converters: topologies and analysis," *IEEE Transactions on Power Electronics*, vol. 3, no. 2, pp. 183–191, 1988.

[9] D. Maksimovic, "Design of the zero-voltage-switching quasi-square-wave resonant switch," in *Proceedings of IEEE Power Electronics Specialist Conference-PESC'93*. IEEE, 1993, pp. 323–329.

[10] A. Vázquez, K. Martín, M. Arias, D. G. Lamar, M. R. Rogina, and J. Sebastián, "Small signal model of triangular current mode (tcm) operation for bidirectional source/sink buck and boost power converters," in *2020 IEEE Applied Power Electronics Conference and Exposition (APEC)*. IEEE, 2020, pp. 77–84.

[11] J. Biela, D. Hassler, J. Miniböck, and J. W. Kolar, "Optimal design of a 5kw/dm 3/98.3% efficient tcm resonant transition single-phase pfc rectifier," in *The 2010 International Power Electronics Conference-ECCE ASIA-*. IEEE, 2010, pp. 1709–1716.

[12] C. Volkmar and U. Ricklefs, "Implementation and verification of a hybrid performance and impedance model of gridded radio-frequency ion thrusters," *The European Physical Journal D*, vol. 69, pp. 1–17, 2015.

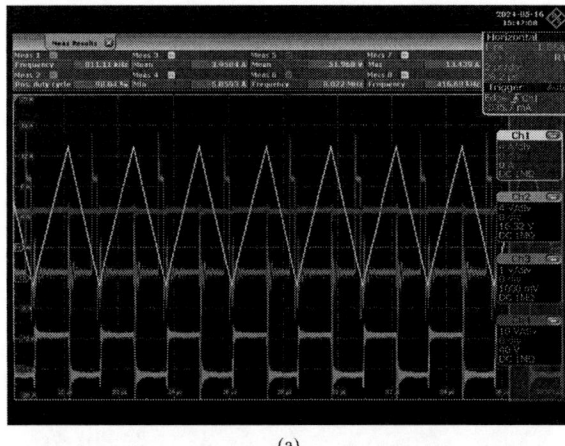

(a)

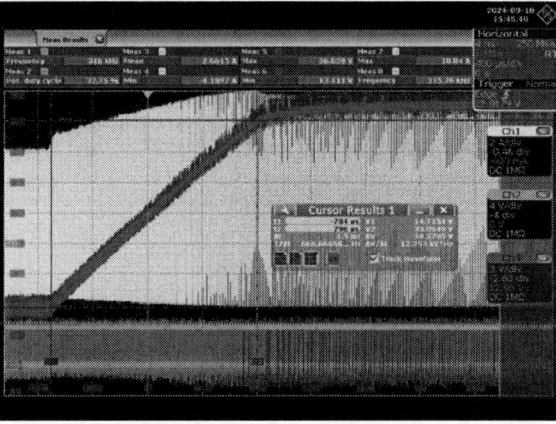

(b)

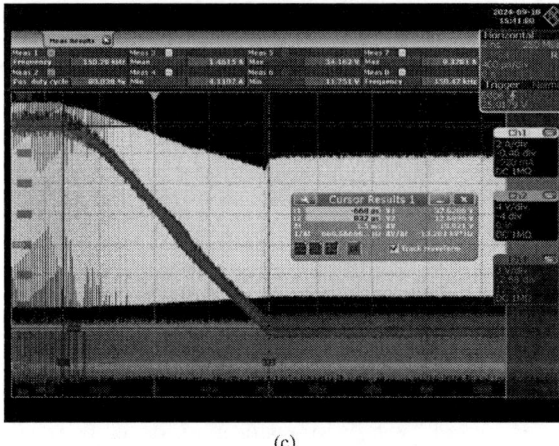

(c)

Figure 10: a) Experimental results at $V_{in} = 100\ V$ and $V_{out} = 40\ V$ b) Experimental reference voltage step at $V_{in} = 40\ V$ from $V_{out} = 15\ V$ to $V_{out} = 30\ V$ c) Experimental reference voltage step at $V_{in} = 40\ V$ from $V_{out} = 30\ V$ to $V_{out} = 15\ V$

[13] A. V. Peterchev and S. R. Sanders, "Quantization resolution and limit cycling in digitally controlled pwm converters," *IEEE Transactions on Power Electronics*, vol. 18, no. 1, pp. 301–308, 2003.

Thermal Model and Optimization of a Multi-winding Transformer for Lunar Surface Power Transmission

Zhining Zhang*, Yuzhou Yao*, Junchong Fan*, Juchen Yang*,
Robert Guenther†, Pengyu Fu*, Jin Wang*

*Center for High Performance Power Electronics (CHPPE),
The Ohio State University, Columbus, Ohio, USA
Emails: zhang.9895@osu.edu, yao.960@osu.edu, fan.867@osu.edu, yang.7091@osu.edu,
fu.901@osu.edu, wang.1248@osu.edu
†GPEM LLC., Marysville, Ohio, USA
Email: rnguenther@gmail.com

Abstract—In this paper, a steady-state thermal analysis is conducted for a multi-winding transformer designed for lunar surface power transmission applications. The study employs a lumped parameter network (LPN) method to derive a thermal equivalent circuit dominated by radiation and conduction for performance analysis and design optimization. A Finite Element Method (FEM) model is developed to accurately evaluate the thermal performance of two transformer design iterations under operating conditions of 77 K and vacuum. The FEM model is validated against experimental results conducted in a low-pressure chamber to ensure its reliability. To further demonstrate the design improvements, two transformer prototypes are evaluated using simulation results under cryogenic vacuum conditions and experimental results under low-pressure, room-temperature conditions. The experimental results confirm the improvements over the design iterations, while the FEM analysis highlights significant performance enhancements in the second design.

Index Terms—Thermal equivalent circuit, heat transfer, cryogenic and vacuum conditions, multi-winding transformer

I. INTRODUCTION

A lunar surface power transmission system is proposed in [1] for NASA's Watts on the Moon Challenge. The system consists of two modular multilevel isolated dc-dc converters for the lunar surface HVDC system. Each converter features a transformer with the same high voltage side winding voltage but different low voltage side winding voltages. To simplify development and standardize the design across both converters, a multi-winding transformer is developed to fit both with the configurable secondary windings, as illustrated in Fig. 1 [2]. The key specifications are listed in Table I. The proposed structure of the multi-winding transformer features four highly symmetrical windings. Since in the configuration of the load side converter, the four secondary windings will be connected in parallel to supply the full 1.5-kW power, the geometrically symmetric design and arrangement are required. Thus, a multilayer winding strucutre is proposed as shown in Fig. 2.

The transformer is required to operate at 77 K under vacuum conditions for lunar surface applications. The changes in the characteristics of magnetic components under cryogenic conditions have been extensively investigated in [2]–[4]. However, thermal dissipation is another major challenge in the design of power transformers for lunar applications. While the thermal performance of various power transformers at room temperature and under normal air pressure has been well studied for traditional structures [5]–[7], the absence of air in vacuum conditions changes the heat dissipation model. Despite the absence of convection, heat dissipation analysis remains complicated, as thermal radiation becomes the dominant mode of heat transfer. This shift necessitates detailed radiation modeling, significantly increasing the complexity of the thermal analysis.

For the proposed multilayer transformer, the multilayer structure introduces significant challenges for heat dissipation during vacuum operation. The structure obstructs thermal radiation between different parts and unfilled gaps hinder thermal conduction, both of which significantly affect the heat dissipation performance. To ensure reliable operation under cryogenic and vacuum conditions, it is essential to analyze the thermal model and implement design improvements to support the full-power operation as required.

This paper is organized as follows. Section II introduces the transformer thermal analysis, including loss calculation, heat transfer modeling, and the thermal equivalent circuit. Section III presents the proposed methods to improve the transformer's heat dissipation. Section IV discusses the transformer thermal performance of two designs, supported by simulation and experimental results. Finally, Section V concludes the paper, providing a summary and outlining directions for future work.

II. TRANSFORMER HEAT TRANSFER MODEL AT CRYOGENIC VACUUM CONDITIONS

To evaluate the transformer thermal performance under cryogenic and vacuum condition, it is necessary to develop the transformer heat transfer model, including the power loss modeling and thermal equivalent circuit analysis.

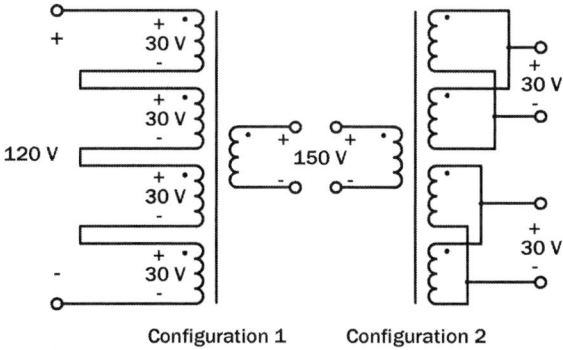

Fig. 1: Multi-winding transformer with two different winding configurations.

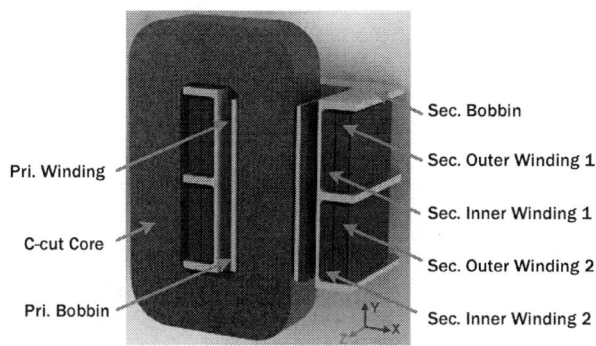

Fig. 2: Transformer section view in XY plane.

TABLE I: Transformers specifications.

Specifications	Transformer I		Transformer II	
	Primary	Secondary	Primary	Secondary
Current (A)	13.75	11	11	55
Voltage (V)	±120	±150	±150	±30
Turns ratio	4:5		5:1	
Power rating	1.5 kW			
Frequency	200 kHz			
Leakage inductance	3.7 μH			

A. Transformer Power Loss Calculation

1) Core loss: The core loss is calculated as 4.80 W using the simplified Steinmetz equation (2), based on a bipolar square wave.

$$P_{core} = V * k_i * 2^{\alpha+\beta} * f^\alpha * B_{max}^\beta \ (W) \quad (1)$$

where V is the volume of the core in m^3; k_i, α, β is the Steinmetz coefficients; f is the transformer frequency in Hz; B_{max} is the maximum operating flux density in Tesla.

2) Winding Loss: The primary and secondary winding losses (all 4 windings) are calculated as 2.41 W and 2.75 W respectively, using equation (1).

$$P_{copper} = I_{rms}^2 R = I_{rms}^2 \rho \frac{MLT * N}{A_w} \bullet \frac{S_w}{S_r} \ (W) \quad (2)$$

where I_{rms} is the winding current RMS value in Amps; ρ is the resistivity of copper; MLT is the mean length per turn of winding in m; N is the number of turns; A_w is the cross-section area of the conductor in m^2; S_w is the total surface of the conductor in m^2; S_r is the surface area of the conductor in m^2 through which current will flow due to the skin effect.

B. Thermal Equivalent Circuit

The thermal equivalent circuit can be derived based on the geometry of the transformer. In toroidal or planar transformers, the thermal equivalent circuit is typically simplified due to geometric symmetry [6]–[8]. However, the multi-winding structure under investigation, with primary and secondary windings located on the same limb of the transformer core, requires a detailed derivation for the entire structure. Thus, the thermal equivalent circuit using lumped parameter network (LPN) method is analyzed in both the XY plane and the XZ plane through the center of the transformer, as illustrated in Fig. 3.

For the thermal equivalent circuit in the XY plane, as shown in Fig. 3(a), thermal dissipation within the transformer core is considered along both the X-axis and Y-axis. In the primary windings, thermal radiation is modeled only between the transformer core and the secondary bobbin, as the outward-facing radiation area along the Y-axis is minimal due to the 0.4-mm thickness and 16-AWG litz wire. For the secondary bobbin, which is closely wrapped by the secondary winding, the temperature difference between the bobbin and the secondary winding is negligible compared to the difference between the winding and the ambient temperature. Therefore, the thermal radiation from the secondary bobbin is considered part of the secondary winding radiation. Consequently, the winding radiation area in R_{rad_W} includes the surfaces facing both positive and negative Y directions.

Regarding heat sources in the equivalent circuit, they are decomposed into distinct components based on the power loss analysis. Q_{Core} represents the transformer core loss, which is divided into two parts due to the symmetric shape of the C-cut core. The portion of Q_{Core} on Node T_{C_IN} corresponds to the core loss on the inner limb, dissipated primarily through the transformer windings. The remaining portion of Q_{Core} represents the core loss in the outer limb, which radiates more heat directly to the ambient. The total radiation thermal resistance of the core is denoted as R_{rad_C}. Similarly, Q_{Pri} and Q_{Sec} represent the power losses in the primary and secondary windings, respectively, and are distributed across the winding geometries reflected in the transformer's thermal equivalent circuit.

C. Thermal Resistance Breakdown

The thermal resistances in the equivalent circuit are analyzed individually. Under vacuum conditions, the transformer

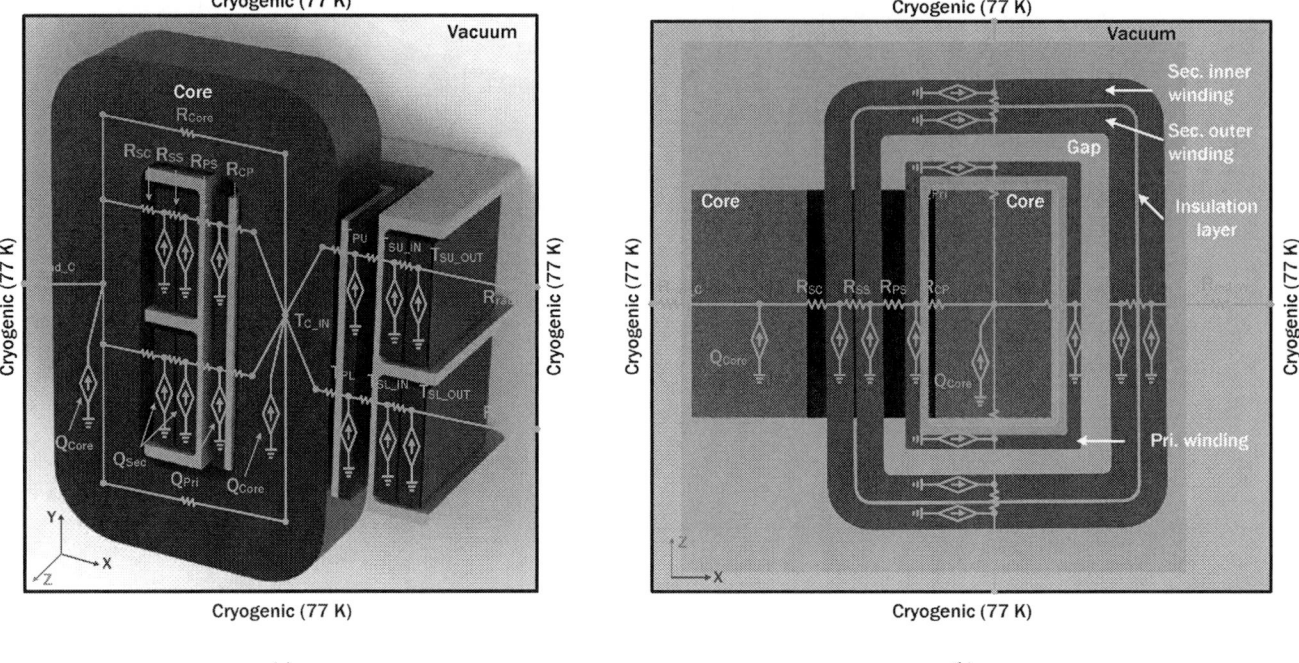

(a)   (b)

Fig. 3: Thermal equivalent circuit of the transoremr in different planes. (a) Thermal equivalent circuit in XY plane. (b) Thermal equivalent circuit in XZ plane.

dissipates heat through conduction and radiation. The equations for calculating conduction and radiation thermal resistances are provided below:

$$R_{cond} = \frac{L}{kA}; R_{rad} = \frac{1}{\alpha A}. \tag{3}$$

Where k is the thermal conductivity, α is the radiation coefficient, and L and A represent the conduction length and the heat transfer area, respectively.

R_{CP} represents the thermal resistance between the inner limb of the core and the primary winding. It comprises the thermal resistance of the primary bobbin, $R_{PB_{cond}}$, and the thermal resistance due to radiation between the inner core and the primary bobbin, $R_{CB_{rad}}$. This relationship can be expressed as:

$$R_{CP} = R_{PB_{cond}} + R_{CB_{rad}}. \tag{4}$$

R_{PS} represents the thermal resistance between the primary winding and the secondary winding. It consists of the thermal resistance of the secondary bobbin $R_{SB_{cond}}$ and the thermal resistance of the radiation between the primary winding and the secondary bobbin $R_{SB_{rad}}$. It can be written as:

$$R_{PS} = R_{SB_{cond}} + R_{SB_{rad}}. \tag{5}$$

R_{SS} represents the conduction thermal resistance of the insulation layer between the inner and outer secondary windings. R_{SC} represents the radiation thermal resistance between the secondary winding and the outer limb of the core. R_{Core} denotes the thermal resistance of the F3CC transformer core.

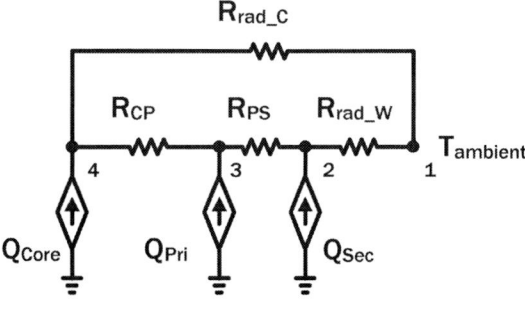

Fig. 4: The simplified thermal equivalent circuit for the multi-winding transformer under vacuum.

D. Simplified Thermal Equivalent Circuit

As shown in the LPN thermal model of the transformer, the asymmetric geometry of the multi-winding transformer and its multilayer structure make solving the equivalent circuit in its original form challenging. Given the current geometry, the thermal equivalent circuit must account for the entire structure.

For lunar surface applications, the environmental conditions are similar to those in space, featuring an isotropic vacuum and cryogenic ambient temperatures. Assuming the upper and lower windings are well-balanced in current, they can be considered thermally identical. Consequently, the thermal dissipation paths through the upper and lower windings form two parallel, identical branches. This symmetry enables the simplification of the thermal equivalent circuit, where the upper and lower sections of the transformer contribute equally to heat transfer.

In the equivalent circuit for the XZ plane, isotropic thermal radiation is assumed on all exposed surfaces without obstruction. The thermal resistance due to radiation in the windings along the X and Z axes is uniformly calculated and simplified into two primary paths: outward to the environment and inward to the core limb.

Since the insulation between the two secondary windings is a layer of tightly wrapped 10-mil Kapton tape, the thermal resistance R_{SS} is negligible. Therefore, the two layers of secondary windings can be approximated as a single layer for thermal analysis.

Based on the above analysis, the transformer's thermal equivalent circuit has been simplified to contain seven nodes. Considering the practical structure with the nanocrystalline core, the thermal conductivity of the core along the lamination layers is relatively high. By assuming the core thermal resistance, R_{Core}, is negligible compared to other components, the thermal equivalent circuit can be further simplified.

The final simplified thermal equivalent circuit for the multi-winding transformer is depicted in Fig. 4.

E. Thermal Radiation Modeling and FEM Method

With this simplified equivalent circuit, it is possible to derive the temperatures of the transformer core and winding temperature under cryogenic and vacuum conditions.

However, accurate calculation of thermal radiation resistance requires precise modeling of radiation. Specifically, modeling radiation is challenging as it varies with material properties, surface finishes, and other factors. Furthermore, radiation exchange occurs not only between the transformer components and the environment but also among the windings and the transformer core, further increasing the complexity.

In summary, the derived thermal model provides a practical approach for performance estimation and guiding the design process. To accurately evaluate the thermal performance of the transformer, the Finite Element Method (FEM) will be implemented to enhance the precision of thermal performance evaluation and optimization.

III. TRANSFORMER DESIGN IMPROVEMENT FOR CRYOGENIC AND VACUUM CONDITIONS

A. Thermal Hotspot Analysis

According to the derived thermal equivalent circuit, the primary winding poses a significant thermal challenge, as it lacks a direct radiation path to the ambient. Its heat is primarily dissipated through two paths: dissipating to the transformer core and dissipating to the secondary winding. However, the thermal radiation components in R_{CB} and R_{PS} are significantly higher than their conduction counterparts, limiting the heat dissipation efficiency and increasing the likelihood of the

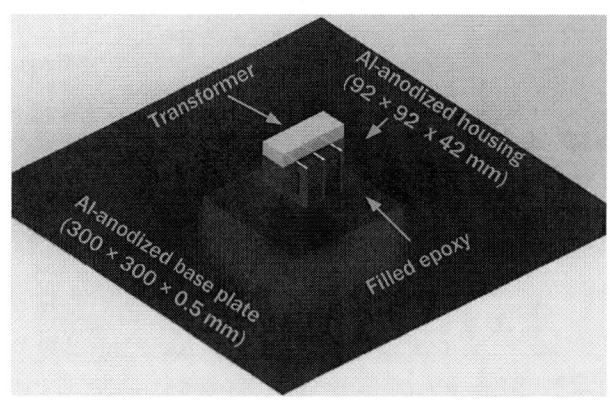

Fig. 5: Proposed version 2 transformer design with epoxy in Al-anodized housing and a Al-anodized base plate.

primary winding becoming a hotspot, potentially leading to overheating.

Beyond the primary winding, the overall heat dissipation capability of the transformer is constrained by R_{rad_C} and R_{rad_W}, as these are the only paths through which heat can be transferred to the ambient environment. Additionally, the limited radiation area of the windings and core further restricts the transformer's overall thermal performance under vacuum conditions.

To address these issues, two solutions are proposed. First, introducing gap fillers between transformer components can replace thermal radiation resistances with conduction resistances, significantly improving heat transfer efficiency between parts. Second, to enhance overall heat dissipation, an anodized aluminum radiator, commonly used in space applications, can be employed. This approach increases the effective radiation area and improves the thermal radiation efficiency of the system.

B. Improve Transformer Design

By incorporating these two methods, the Version 2 transformer design is proposed, as shown in Fig.5, while Fig.2 illustrates the Version 1 design.

The Version 2 transformer is encapsulated within a 92 mm × 92 mm × 42 mm aluminum-anodized enclosure filled with epoxy (EP770), which has a post-curing thermal conductivity of 2.5 W/m-K. The epoxy fills the gaps between transformer components, significantly improves the thermal distribution. However, due to the epoxy's relatively low viscosity, potential air gaps with high thermal resistance may form during encapsulation. To address this, additional copper strips are inserted into the insulation layers of the windings, creating low-resistance thermal paths that efficiently transfer heat from the windings to the epoxy. Furthermore, the transformer core's mechanical clamp is constructed with aluminum to further improve thermal conductivity.

979-8-3315-1612-3/25 $31.00 © 2025 IEEE

TABLE II: Inductance of Transformer Ver. 1 and turns ratio.

Windings	Inductance (µH) @ 200 kHz	Turns Ratio
Primary	1415.82	1
Sec. 1	56.86	4.99
Sec. 2	57.27	4.97
Sec. 3	57.27	4.97
Sec. 4	57.60	4.96
Leakage	4.1	-

TABLE III: Inductance of Transformer Ver. 2 and turns ratio.

Windings	Inductance (µH) @ 100 kHz	Turns Ratio
Primary	637.10	1
Sec. 1	25.80	4.97
Sec. 2	25.63	4.99
Sec. 3	25.68	4.98
Sec. 4	25.86	4.96
Leakage	4.93	-

The bottom of the transformer enclosure features a black anodized aluminum base plate, designed for secure mounting of the Version 2 transformer. This base plate, with dimensions of 300 mm × 300 mm and a thickness of 0.5 mm, is fastened to the enclosure using screws. The dimensions are chosen based on the footprint of the power converters designed in [1], ensuring structural stability while also enhancing thermal dissipation for both the transformer and the converters.

The introduction of epoxy encapsulation is expected to impact the transformer inductance. The inductance values for the Version 1 transformer are presented in Table II, while those for the Version 2 transformer are listed in Table III.

IV. TRANSFORMER THERMAL SIMULATION RESULTS AND EXPERIMENTAL VALIDATION

To accurately evaluate and compare the thermal performance of the two transformer designs, FEM simulations are conducted in ANSYS Icepak. Experimental results, obtained under controlled conditions emulating a vacuum environment, are used to validate the accuracy of the simulation models. Once validated, these models, along with the designated experiments, are utilized to analyze and quantify the performance improvements achieved in the Version 2 transformer design.

A. Thermal Simulation Model and Experimental Validation

Although the transformer is designed to operate at 77 K in a vacuum, the experimental validation is performed in a sealed chamber at room temperature and low pressure (30 Torr). This setup is chosen due to constraints in the available experimental equipment. The low-pressure condition effectively suppresses convection, closely replicating the intended operating environment, where heat transfer relies primarily on conduction and radiation.

In the experiment, DC currents are injected into both the primary and secondary windings to emulate winding losses, which are measured using an oscilloscope. Core losses are excluded from the experiment to avoid interference. Using AC excitation would significantly increase AC copper losses in the long interface cables, causing these losses to dominate

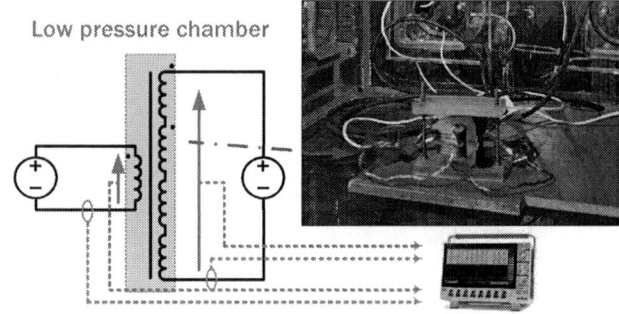

Fig. 6: Experimental setup for simulation model validation on version 1 transformer.

TABLE IV: Selected material properties for simulation model.

Parts	Material	Property	Value
Core surface	FINEMET	Emissivity	0.9
Core body	FINEMET	Ther. conductivity	20 W/m-K
Bobbin surface	SLA Resin	Emissivity	0.9
Bobbin body	SLA Resin	Ther. conductivity	0.15 W/m-K
Winding surface	Copper	Emissivity	0.05
Winding body	Copper	Ther. conductivity	387.6 W/m-K
Epoxy	EP770	Ther. conductivity	2.5 W/m-K

and making it challenging to quantify the copper losses in the windings. The experimental setup is illustrated in Fig. 6, where a thermocouple is used to measure the temperature of the primary winding.

The simulation model for the Version 1 transformer is modified to replicate the experimental conditions, including room temperature and a pressure of 30 Torr. The key material properties used in the simulation are listed in Table IV. Heat losses from the transformer windings are set according to experimental measurements, with 2.235 W on the primary winding and 3.96 W on the secondary winding. As radiation and conduction are the dominant heat dissipation mechanisms, these are modeled by appropriately defining surface radiation and solid conduction properties in the simulation. It should be noted that the low-pressure chamber is not an exact vacuum, and the remaining air still contributes to heat dissipation. Therefore, the simulation model accounts for this by enabling airflow for the validation study.

The simulation results are presented in Fig. 7. In the simulation model, the primary winding temperature reaches 98.1°C, while the measured primary winding temperature stabilizes at 98.9°C. This close agreement between the simulated and experimental results demonstrates that the transformer FEM model is reliable and can effectively evaluate the thermal performance of the transformer.

B. Simulation Results of Two Transformers

With the simulation model validated for the Version 1 transformer, additional simulations are conducted for both transformer designs under their intended operating conditions

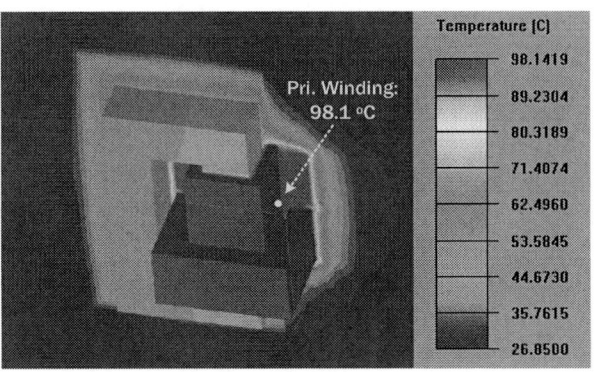

Fig. 7: Version 1 transformer simulation result at 30 Torr, 300 K.

TABLE V: Simulation results of the transformer temperatures under 77K and Vacuum.

Version #	Core	Pri. Winding	Sec. Winding
Ver. 1	150.3°C	236.6°C	149.1°C
Ver. 2	-21.1°C	-24.0°C	-27.0°C

of 77 K and vacuum. In this environment, heat transfer is limited to radiation and conduction, and all dissipation paths—through the core, windings, and enclosure—are considered and modeled. The power loss values used in the simulations are the calculation results from Section II.

The simulation results for the Version 1 transformer are illustrated in Fig.8, while the corresponding results for the Version 2 transformer are depicted in Fig.9. The temperatures of the transformer cores and windings for both simulation models are summarized in Table V.

When comparing the results, the Version 1 transformer

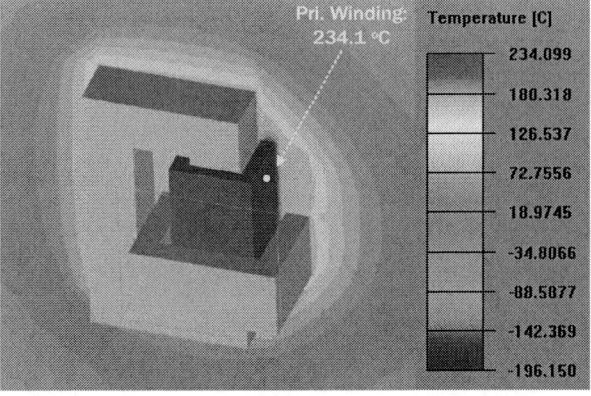

Fig. 8: Version 1 transformer simulation result at 77 K and vacuum.

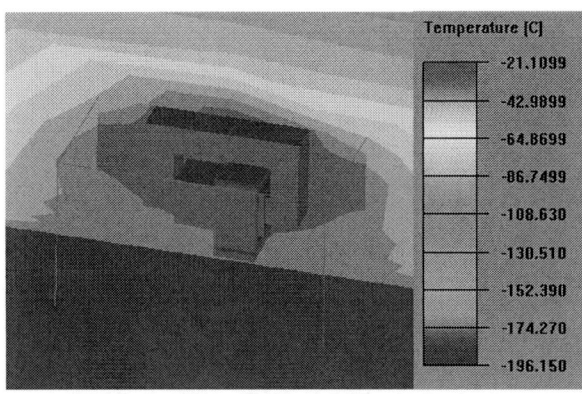

Fig. 9: Version 2 transformer simulation result at 77 K and vacuum.

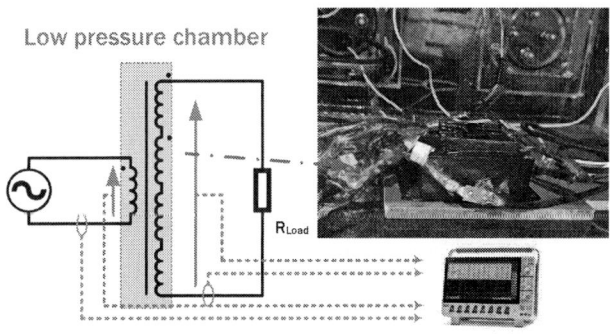

Fig. 10: Transformer thermal performance evaluation setup with Version 2 transformer.

exhibits a maximum primary winding temperature of 236.6°C, making it the hottest part of the transformer. This temperature exceeds the safe operating range of the litz wire, posing significant reliability concerns. Additionally, the temperatures of the secondary winding and core are nearly at the upper limits of their allowable ranges. In contrast, the improved Version 2 design demonstrates a substantial reduction in temperature, with the transformer core temperature dropping to -21°C, representing a remarkable 260°C difference. This comparison highlights that the Version 2 design is well-suited to operate under cryogenic and vacuum conditions, addressing the thermal challenges identified in the initial design.

C. Transformer Thermal Performance Experimental Results

To further evaluate the thermal performance differences between the Version 1 and Version 2 transformer designs, an experimental setup is developed to directly assess their thermal dissipation in a low-pressure chamber, as illustrated in Fig. 10.

In the experimental configuration, an AC source is connected to the primary winding of the transformer, while a small load resistor is connected to the secondary winding.

979-8-3315-1612-3/25 $31.00 © 2025 IEEE 1208

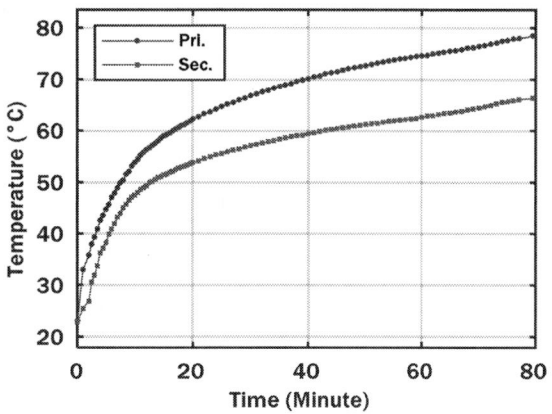

Fig. 11: Version 1 transformer thermal performance evaluation results.

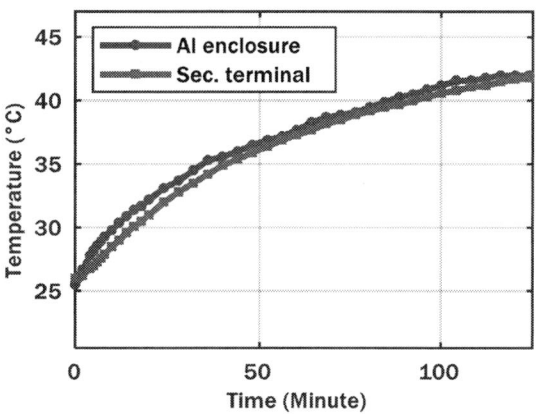

Fig. 12: Version 2 transformer thermal performance evaluation results.

The winding currents in both transformers are controlled to remain nearly identical, ensuring equivalent power losses in both transformers. Under these conditions, the thermal steady-state temperatures of the two transformers are compared to directly evaluate their thermal performance.

In the final evaluation, the Version 1 transformer operates with a primary-side RMS current of 8.711 A and a secondary-side RMS current of 7.425 A. Similarly, the Version 2 transformer operates with a primary-side RMS current of 8.732 A and a secondary-side RMS current of 8.074 A. Both tests are conducted under consistent environmental conditions of 30 Torr pressure and room temperature.

The temperature versus time curves are shown in Fig.11 and Fig.12. For the Version 1 transformer, the winding tempera-tures are directly measured. For the Version 2 transformer, the enclosure and secondary winding terminal temperatures are measured.

From the experimental results, it is observed that, despite the Version 2 transformer having a slightly higher secondary-side current—and consequently a marginally higher power loss—its terminal temperature remains significantly lower. After 120 minutes of operation, the terminal temperature of the Version 2 transformer stabilizes at only 42 °C, whereas the winding temperature of the Version 1 transformer reaches 67 °C. These results demonstrate that the improvements implemented in the Version 2 design have significantly enhanced its thermal performance, particularly under thermal radiation-dominant scenarios.

V. CONCLUSION

This paper presents the thermal analysis and design opti-mization of a multi-winding transformer designed to operate at 77 K in vacuum conditions, specifically for lunar surface power transmission applications. A lumped parameter network model was developed to analyze the heat dissipation mech-anisms of the transformer. The derived thermal equivalent circuit served as the foundation for performance evaluation and guided the proposed design improvements. Key improvement methods were identified through model analysis and evaluated using FEM simulations. The simulation models were validated against experimental results obtained under low-pressure con-ditions. Once verified, the models were used to compare the thermal performance of two transformer designs under the actual operating conditions of 77 K and vacuum. The results demonstrated significant thermal performance enhancements in the Version 2 design, attributed to epoxy encapsulation and anodized aluminum radiators. Future work will focus on extending the thermal model by incorporating a detailed radiation model with the radiation properties of materials.

REFERENCES

[1] Y. Yao, J. Fan, Z. Zhang, Y. Shi, J. Kwon, P. Fu, and J. Wang, "1.5-kV, 1.5-kW, High Voltage Ratio, Bidirectional Dc/dc Converters for Power Distribution on the Moon" 2024 IEEE Energy Conversion Congress and Exposition (ECCE), Phoenix, AZ, USA, 2024. (accepted)

[2] Z. Zhang et al., "A 200-kHz/1.5-kW Multi-winding Transformer for Lunar Surface Power Applications," 2023 IEEE Energy Conversion Congress and Exposition (ECCE), Nashville, TN, USA, 2023.

[3] M. Chen et al., "The magnetic properties of the ferromagnetic materials used for HTS transformers at 77 K," in IEEE Transactions on Applied Superconductivity, vol. 13, no. 2, pp. 2313-2316, June 2003.

[4] S. Yin et al., "Characterization of Inductor Magnetic Cores for Cryo-genic Applications," 2021 IEEE Energy Conversion Congress and Ex-position (ECCE), Vancouver, BC, Canada, 2021.

[5] S. Purushothaman and F. de Leon, "Heat-Transfer Model for Toroidal Transformers," in IEEE Transactions on Power Delivery, vol. 27, no. 2, pp. 813-820, April 2012.

[6] L. Nie, J. Yang and K. Tang, "Thermal Network Modeling of High Frequency Insulated Core Transformers," in IEEE Transactions on Applied Superconductivity, vol. 32, no. 6, pp. 1-5, Sept. 2022.

[7] L. A. Militão, B. Bertoldi, A. G. L. Furlan, M. L. Heldwein and J. Riso Barbosa, "Thermal Network Models for Toroidal and E-Core Inductors in Still Air," in IEEE Transactions on Power Electronics, vol. 38, no. 12, pp. 15879-15892, Dec. 2023.

979-8-3315-1612-3/25 $31.00 © 2025 IEEE

[8] R. Shafaei, M. Ordonez and M. A. Saket, "Three-Dimensional Frequency-Dependent Thermal Model for Planar Transformers in LLC Resonant Converters," in IEEE Transactions on Power Electronics, vol. 34, no. 5, pp. 4641-4655, May 2019.

[9] Y. Wang, G. Calderon-Lopez and A. Forsyth, "Thermal management of compact nanocrystalline inductors for power dense converters," 2018 IEEE Applied Power Electronics Conference and Exposition (APEC), San Antonio, TX, USA, 2018

[10] R. Raman and A. Thakur, "Thermal emissivity of materials," *Applied Energy*, vol. 12, no. 3, pp. 205–220, 1982.

[11] Ning, Jie & Zhang, Linjie & Yin, Xianqing & Zhang, Jianxun & Na, Suck-Joo. (2019). Mechanism study on the effects of power modulation on energy coupling efficiency in infrared laser welding of highly-reflective materials. Materials & design. 178. 1-16. 10.1016/j.matdes.2019.107871.

[12] J. Ji, M. Wang, M. Hu, L. Mao, Q. Wang, W. Zhou, M. Tian, J. Yuan, K. Hu, and Y. Wei, "3D printing AIE stereolithography resins with real time monitored printing process to fabricate fluorescent objects," *Composites Part B: Engineering*, vol. 206, p. 108526, 2021.

[13] S.P. Wong, C.F. Chow, J. Roller, Y.T. Chong, Q. Li, M.A. Lourenco, and K.P. Homewood, "Structures and light emission properties of nanocrystalline FeSi2/Si formed by ion beam synthesis with a metal vapor vacuum arc ion source," *Thin Solid Films*, vol. 515, no. 22, pp. 8122-8128, 2007.

Active Gate Driver Power Supply for High-Reliability Applications

Joseph P. Kozak, Juan Ramirez, Jesse Lin, Allison Orr, Alexander Martin, Hala Tomey
Johns Hopkins University - Applied Physics Lab
Laurel, MD, USA

Abstract—**High-reliability aerospace applications require strict derating of gate and drain biases on all power devices. With the integration of SiC MOSFETs, the derating protocols dramatically influence the electrical performance of power electronic circuits. This work introduces an active gate driver power supply to provide a varied gate driving voltage to meet both derating and high-performance operations. A push-pull power converter with a linear pre-regulator was designed to produce two discrete (derated and rated) output voltages. In this way, the power device will normally operate under the derated voltage, but under high-performance, pulsed applications will operate at the rated voltage. The overall reliability is therefore achieved, and electrical losses can be minimized when high-current outputs are necessary.**

Keywords— *aerospace, gate driver, power converter, SiC MOSFET, reliability.*

I. INTRODUCTION

One of the staple elements of power electronic converters is the gate driver network and gate driver power supply. For each converter, the power transistor requires an auxiliary power source and control signal to control the transition between the blocking and conducting modes. With the integration of new wide bandgap (WBG) power semiconductor devices to power converters, in particular silicon carbide (SiC) MOSFETs, many new methodologies have been proposed for more efficient and reliable switching [1-5]. To integrate SiC MOSFETs into high-reliability, aerospace applications, the gate and drain voltages must be derated to minimize the overall electrical stress on the device and excite various failure mechanisms [6-9]. The EEE-INST-002 standard which is used for many deep-space, high-reliability aerospace missions, has devices derated to 60% of the rated gate-source voltage (V_{GS}) [6]. With SiC MOSFETs, this can significantly limit performance as the on-resistance (R_{DSon}) can vary considerably between the derated and manufacturer-rated gate voltages.

Recently, adaptive gate driver networks have been introduced to provide different voltage levels onto a primary switching device while focusing on increasing the efficiency of the power converter. These systems have been especially integrated for SiC MOSFETs because of the opportunity for faster switching times in comparison to traditional silicon devices. Fig. 1 showcases where recent methodologies have implemented additional circuitry to develop the active gate driver network. The two primary instances to develop an active gate driver network include introducing changes to the supply

voltage to the gate driver IC (V_{CC}) [10-15], or changing the analog circuit network connecting to the primary device's gate terminal [16], [17]. With techniques that focus on changing the V_{CC} of the system, two particular methodologies have been demonstrated. The first is a multi-voltage level network. This utilizes multiple (external) power supply voltages and connects them in series, but includes a switch to bypass the network. In this way, the voltage magnitude can be stepped up or down at different time intervals to not overdrive the voltage of the system overall [10-13]. Additionally, different linear regulators have been introduced after the traditional gate driver power supply (GDPS) to tune the V_{CC} magnitude. This includes a feedback network that can be controlled from various factors including the input or output current, the device's operating temperature, or on-state resistance (R_{DSon}) [14], [15]. Both of these methodologies provide additional circuitry on top of the baseline required GDPS to operate the primary power circuit. Finally, other methodologies have been introduced to change the driving voltage or current directly in the analog circuitry connecting the gate driver IC and the device's gate. These methods primarily adjust the changes in the turn-on/off times based on the currents being sourced or sinked [16-17].

The work presented in this manuscript proposes to shift any circuits used for altering the operating gate voltage from the

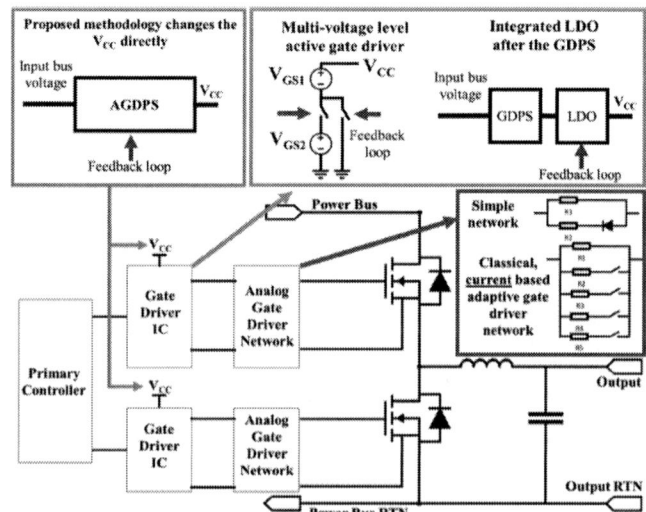

Fig. 1. Half bridge power circuit which showcases where traditional adaptations and where the proposed methodology is introduced to create an active gate driver network.

979-8-3315-1612-3/25 $31.00 © 2025 IEEE

driving network into the GDPS itself. As a result, this changes the need for additional analog circuitry in the gate network or after the GDPS, and a traditionally designed, multi-output power converter can be utilized. An active gate driver power supply (AGDPS) has been designed with multiple voltage outputs to supply the gate driver network on two primary transistors in a half-bridge configuration such that additional components to the driver circuitry are not added, and the operating V_{GS} can be adjusted based upon the operating conditions of the primary device.

The remainder of the paper is structured such that the derating values are analyzed, and the current transition point is determined in Section II. Section III discusses the selected AGDPS topology, and design considerations. Section IV details the experimental test setup, and results which showcase the minimized losses within the system.

II. MOTIVATION FOR ACTIVE GATE DRIVER NETWORK IN HIGH RELIABILITY SYSTEMS

High-reliability, aerospace applications require strict derating of both drain and gate biases to ensure overall lifetime and reliability of the power transistors. These standards have been primarily driven as a result of combined radiation and electrical stressors observed in traditional silicon transistors. One of the most common standards, EEE-INST-002, instructs that MOSFETs be derated to 75% the drain-source voltage (V_{DS}) rating, and 60% the rated V_{GS}. Additionally, the junction temperature of the device is another factor that requires derating; for MOSFETs the standard requires to adhere to 80°C or 75% the maximum junction temperature (whichever is lower) [6]. These ratings greatly impact the performance capabilities of newer devices like SiC MOSFETs which have a higher junction temperature rating (and higher thermal conductivity), as well as more variability in the R_{DSon} between on-state driving V_{GS} magnitudes.

Since these standards were derived from years of data acquired on silicon devices, the impact of the derating can lead to drastic changes in the operating conditions for new devices. For the aforementioned SiC MOSFETs, derated values can have a 50% R_{DSon} increase in comparison to the rated gate voltage. Additionally, SiC device manufacturers are now including both static and transient ratings. The decision on which V_{GS} magnitude to derate against (either static or dynamic) is causing disconnects between converter designers who are focusing on maximizing electrical performance, as well as reliability for future aerospace missions. Since newer aerospace applications, such as the NASA Dragonfly mission [7] apply a high-current through the semiconductor device, a realization to allow the device to operate closer to terrestrial applications has been made, and the derating has been applied to the transient V_{GS} rating. However, operating at this derated state still introduces losses that are higher than those seen in terrestrial applications.

Equations 1 and 2 show the overall losses within a transistor operating in a power converter (conduction and switching losses), and that the conduction losses are equal to the conduction current (I_D) squared, multiplied by the R_{DSon}, and duty cycle for a pulsating current waveform [18].

$$P_{loss} = P_{con} + P_{sw} \tag{1}$$

$$P_{con} = I_D{}^2 * R_{DSon} * D \tag{2}$$

To fully demonstrate the value of the AGDPS, and determine the feedback point where the output V_{GS} magnitude is increased, the losses for a primary SiC MOSFET with a 650 V, 16 A rating (C3M0120065K) were analyzed and used for the experimental design and verification of the proposed methodology [19].

As mentioned, many SiC manufacturers now include both a static and transient rating; this device consists of a transient rated gate-voltage of 19 V, and a static rated gate-voltage of 15 V [19]. Since the current derating standards were not created for SiC-based devices, a comparison of the derated values for both ratings was initially considered. With the standard gate-voltage rating (15 V) the resulting derated voltage would be 9 V, and for the transient gate-voltage rating (19 V) the derated value would be 11.4 V. The reported output characteristics on the datasheet [17] were examined such that the R_{DSon} values could be extracted based on load current and gate voltage. Table I shows the extracted values for V_{DS} across the device for different load currents from the datasheet. R_{DSon} was then calculated, and normalized in comparison to the lowest value (when V_{GS} = 15 V). The conduction losses, as expressed by (2) were then calculated. At lower conduction currents the loss difference for the chosen device is less than 2 W; however, when the conduction current increases, the difference in losses increase to multiple watts of power loss. The analysis was performed with a fixed duty cycle as differences in conduction loss due to variations in duty cycle would be negligible. Additionally, while the switching losses are impactful in the overall operation of the design, and do change with respect to operating voltage and current ratings, a full analysis for these losses was not considered as the operating point for aerospace applications will normally have P_{con} to be greater than P_{sw} [7].

Table I – Extracted electrical characteristics, normalized R_{DSon}, and calculated power losses for the selected device under test at 25°C.

V_{GS} (V)	V_{DS} (V)	I_D (A)	R_{DSon} (Ω)	R_{DSon} norm	P_{con} (W)
9	1.6	5	0.32	2.13	4
11	1	5	0.2	1.33	2.5
13	0.8	5	0.16	1.07	2
15	0.75	5	0.15	1	1.875
9	3.25	10	0.325	2.6	16.25
11	1.9	10	0.19	1.52	9.5
13	1.5	10	0.15	1.2	7.5
15	1.25	10	0.125	1	6.25
9	N/A	20	N/A	N/A	N/A
11	4.1	20	0.205	1.64	41
13	3	20	0.15	1.2	30
15	2.5	20	0.125	1	25

As a result of this analysis, the outputs of the designed active gate driver power supply were selected to output 15 V for the high value to match the manufacturer rated voltage, and 11 V as the low value to meet the derating guidelines. In this way, the 11 V will meet the derating values for the dynamic V_{GS} rating.

III. Active Gate Driver Power Supply Design

Traditional adaptive gate driver networks integrate additional components and methodologies into the gate driving network itself. This adds complexity and decreases the overall reliability to the complete circuit architecture. The proposed methodology shifts this adaptive nature to the gate driver power supply itself, which is a staple auxiliary circuit. To determine the best overall approach to achieve the AGDPS methodology, a trade was conducted which considered the circuit topology, control methodology, and feedback methodology and mechanisms. Throughout this approach, a multi-output design was considered such that the AGDPS would supply voltage to both high-side and low-side power devices in a half-bridge orientation.

Three primary topologies including a forward rectifier, flyback, and push-pull converter were considered. Additionally open-loop control, an open-loop with a pre-regulated input control, and full closed-loop control were all considered in association with the various circuit topologies. Finally, the feedback mechanisms included both the voltage sense for the GDPS itself, and the current feedback mechanism from the primary power circuit which is used to change the output voltage of the AGDPS from the low output (11 V) to the high output (15 V).

This trade and evaluation resulted in utilizing a cascaded methodology where a linear voltage regulator would feed an open-loop, push-pull converter with single winding secondary outputs and bridge rectifiers. The overall schematic of the

AGDPS is shown in Fig. 2, and the designed parameters are listed in Table II. It can also be seen that a difference amplifier was selected to sense the low-side output (to the half-bridge low-side device), as a feedback to the linear regulator.

Table II – Design parameters for the AGDPS

General Parameters	Value
Input Voltage	12 V
Switching Frequency	200 kHz
Output Voltage	15 V and 11 V
Max load current	20 mA
Transformer Design	
Turns Ratio	1:1:1.5:1.5
Max primary RMS current	50 mA
Min magnetizing inductance	100 µH
Input Filter	
Attenuation at 200 kHz	60 dB
Max output impedance	1.5 Ohm

To control the magnitude of the output of the gate driver power supply, a feedback loop was integrated with the primary power circuit (not depicted). While multiple types of current measurement techniques could be utilized in this methodology, a hall effect sensor was selected to measure the average input current of the power converter. This current is compared to a reference point, which changes the setpoint of a linear pre-regulator in-line with the push-pull converter. Once the current reaches a determined threshold, the gate driver power supply will alter the output voltage to increase the gate voltage. This results in decreasing the R_{DSon}, and minimizes the overall system losses.

Filtering on the current feedback signal is included such that the voltage change will only occur when the system's conducted current is maintained above the threshold for a significant period of time. Hysteresis is also included in the comparator circuit to prevent toggling as the current feedback gets close the threshold reference. A design consideration is to include preload to keep the outputs from drifting due to transformer leakage inductance. An additional feature of the circuit that is not shown in Fig. 2 is a resistive discharge circuit across the regulated supply into the push-pull converter that is momentarily enabled when the voltage setpoint is decreased to discharge the bulk capacitance on the primary side and improve output voltage response time. These parameters (Table II) include a 12 V input which can be taken from one of the primary buses in aerospace applications. The AGDPS transformer includes multiple secondary windings for both the high-side and low-side driving voltages. The transition from the 11 V to the 15 V output was selected to be conducted at around a 10 A output to best demonstrate the increase in performance of the driven power SiC MOSFET.

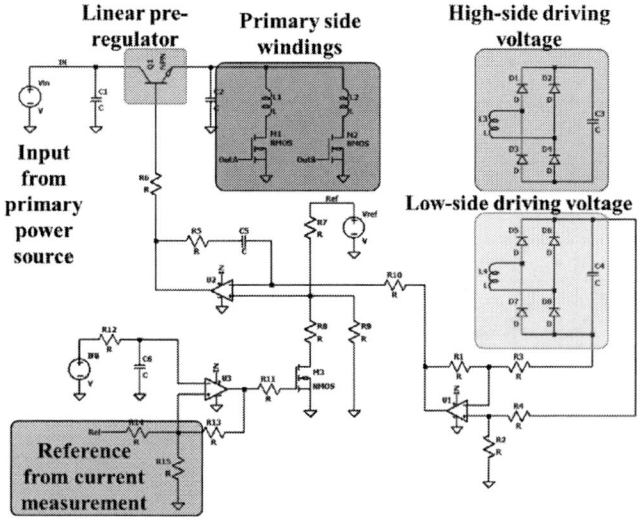

Fig. 2. Schematic of the push-pull converter with linear pre-regulator that incorporates a feedback reference which comes from the current measurement of the primary power circuit.

IV. EXPERIMENTAL INTEGRATION AND EVALUATION

To fully test the functionality of the AGDPS, a synchronous buck converter was designed and developed with the analyzed SiC MOSFET as the primary switching device. The converter was designed such that the input voltage could range up to 56 V to demonstrate a 28 V output which is a common power rail for aerospace applications. The general selected duty cycle was selected to be 50% such that the output voltage would be halved, and the input current would be doubled, ideally. The buck converter was also designed to operate with a switching frequency of 50 kHz up to 250 kHz. The fabricated and assembled buck converter and AGDPS are shown in Fig. 3. A custom frame was designed and manufactured such that the two boards would integrate seamlessly, and similarly to a final flight-grade board.

To demonstrate the overall changes in performance of the system with only changing the gate voltage, experimental tests were conducted to measure the overall system efficiency at different switching frequencies, load currents, and static/external V_{GS} amplitudes (11 V and 15 V). Experimental tests were conducted with a 24 V input, and input current of 2.5 A and 5 A (5 A and 10 A output). The switching frequency was swept between 50 kHz and 250 kHz. It can be seen in Fig. 4(a) and (b), that the synchronous converter operates as expected, with the overall decrease in efficiency as switching frequency increases. Additionally, a major shift can be seen between the 5 A and 10 A output current instances where the general efficiency when V_{GS} = 11 V was around 89% at the 5 A load, and down towards 81% at the 10 A load. Similarly, the efficiency made a dramatic decrease with respect to output load when V_{GS} = 15, however to a smaller degree. The efficiency dropped from roughly 92% towards 86%; only a 6% difference in comparison to an 8% difference. The changes can also be seen in Fig. 4, through the changes in the operating temperature of the high-side SiC MOSFET. In the 10 A load case alone, the temperature had a maximum of 44.1°C when V_{GS} = 11 V, where the maximum was only 40.2°C when the V_{GS} = 15 V. Considering

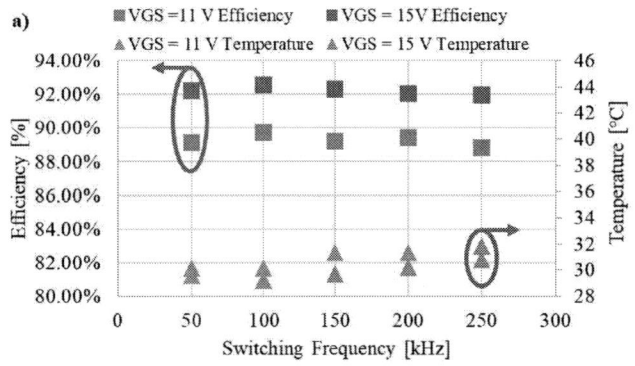

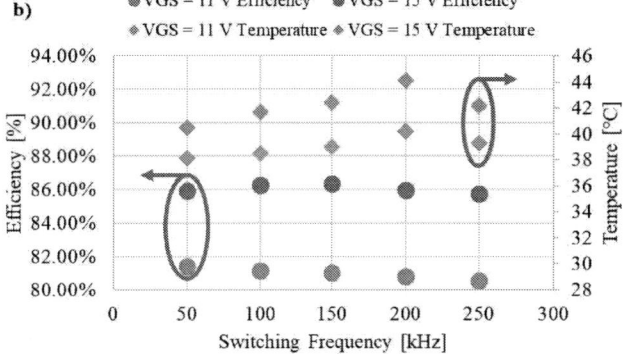

Fig. 4. Measured efficiency and high-side transistor case temperature of the synchronous buck converter operating at varying switching frequency, different V_{GS} values, and 5 A output current (a) and 10 A output current (b).

the overall load power of this demonstration is not too demanding, this 4°C delta shows the direct relationship and need for having the opportunity to increase the V_{GS} to the terrestrially rated value.

The AGDPS was then fully integrated with the buck converter and the startup routine was established and monitored before primary power was applied to the synchronous buck converter. The startup waveforms were recorded and are shown in Fig. 5. Once the primary input rail of the AGDPS has an applied voltage, the circuit transitioned to output the low voltage setpoint, and after a few milliseconds the outputs for the low and high sides were measured to be approximately 11.2 V.

The synchronous buck converter was then tested with the integrated AGDPS at 11 V, and the current sense loop was tuned and adjusted such that the transition point would be seen near the 10 A output current. Figure 6 shows the two switching waveforms of the AGDPS as a load step up and down was applied to the synchronous buck converter. The buck converter was operating with a 24 V input and at 100 kHz, and the electronic load was set to sink 5 A and step to 10 A and back down to 5 A. The transition time seen in Fig. 6(a) shows a relatively fast transition time only needing approximately 1 ms of total time between the load step and the transition from 11 V to 15 V. Figure 6(b) shows a slower transition time of approximately 6 ms to get from 15 V to 11 V. The difference in transition times between the load stepping up and down is due

Load inductor   **SiC FETs and diodes**

Fig. 3. Fabricated and assembled synchronous buck converter (parent-board) and AGDPS (child-board) integrated for full electrical testing.

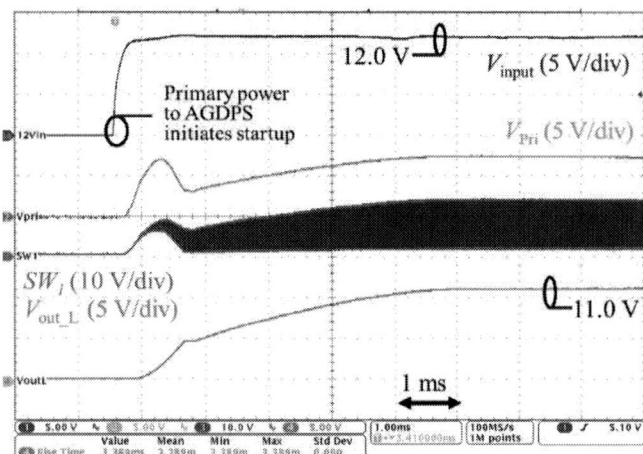

Fig. 5. Startup sequence of the AGDPS when primary power was applied to the input voltage rail, and output voltage was monitored to be 11.0 V.

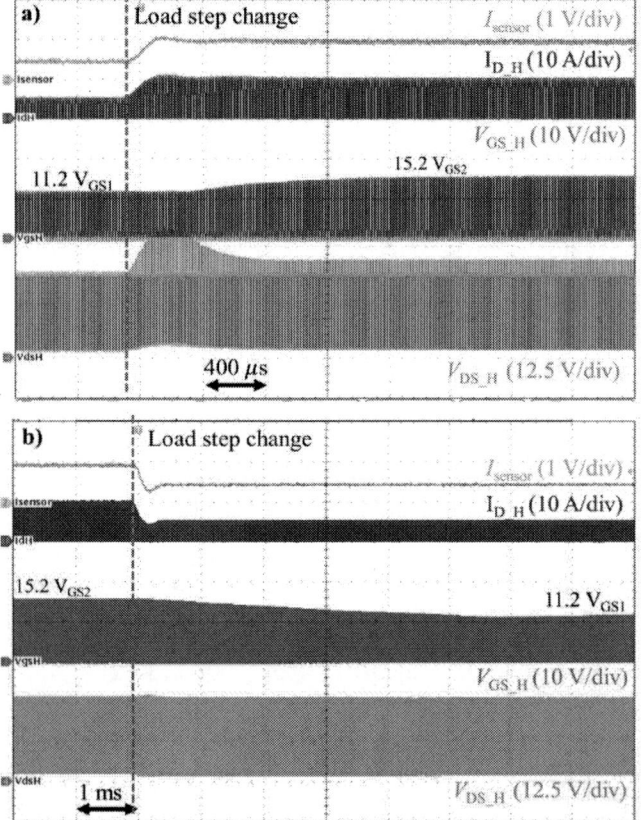

Fig. 6. Output waveforms of the AGDPS from the transition from the low-output to high-output states (a). Output waveforms of the AGDPS from the transition from the high-output state down to the low-output state (b).

to the light load currents seen by the AGDPS in relation to the capacitance across each output; the difference in transition time would be even larger without the resistive discharge circuit

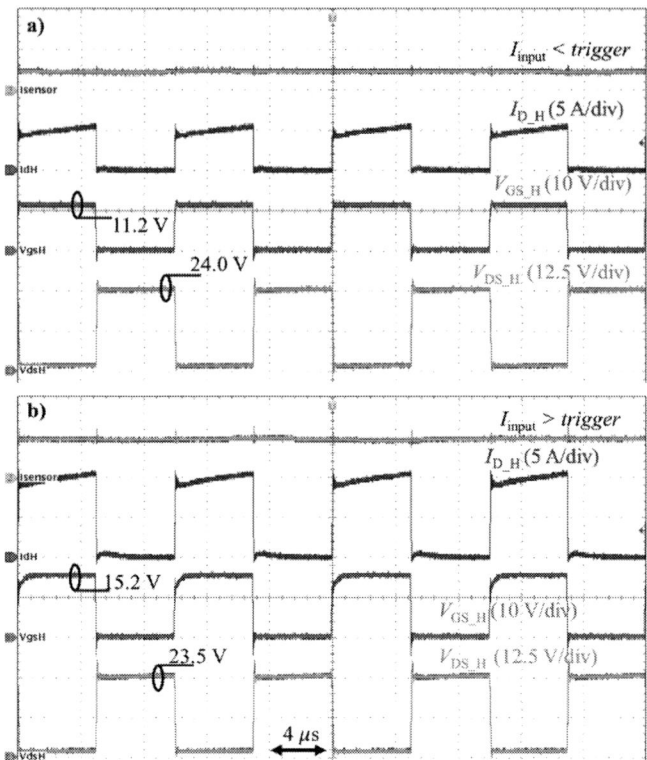

Fig. 7. Output waveforms for the synchronous buck converter operating at 100 kHz when the load current was below the current and $V_{GS} = 11$ V (a), and when the load current was above the current trigger and $V_{GS} = 15$ V (b).

across the regulated supply into the push-pull converter. The switching waveforms from the operating synchronous buck converter at both V_{GS} values just before and just after the transition can be seen in Fig. 7 (a) and (b).

CONCLUSION

High-reliability, aerospace applications require strict derating of both drain and gate biases to ensure overall lifetime of the power transistors. With the emergence of SiC MOSFETs into terrestrial applications, the integration into aerospace applications has been limited and many of the benefits are not realized because of the derating restrictions; however, an active gate driver network could be used to ensure the device operates at the derated value most of the time, and up to the terrestrially rated value under high-performance instances. This work has introduced a new methodology to utilize a multi-output gate driver power supply as the mechanism to shift the power device driving voltage. In this way, only limited extra circuitry is introduced in comparison to the traditional active gate driver solutions. The AGDPS was integrated into a synchronous buck converter which showcased the functionality of the proposed method by decreasing losses in the primary power SiC FETs. Results showed a fast increase in the V_{GS} once the load current exceeded the threshold, and a moderate decrease in V_{GS} when the load current fell below the threshold showing seamless operation within the half-bridge SiC MOSFET gate driver circuits and a promising solution for aerospace applications requiring higher performance for relatively short periods of time while meeting derating requirements the rest of time.

REFERENCES

[1] Chen J, Li Y, Liang M. A Gate Driver Based on Variable Voltage and Resistance for Suppressing Overcurrent and Overvoltage of SiC MOSFETs. Energies. 2019; https://doi.org/10.3390/en12091640

[2] Tan J, Zhou Z. An Optimized Switching Strategy Based on Gate Drivers with Variable Voltage to Improve the Switching Performance of SiC MOSFET Modules. Energies. 2023; 16(16):5984. https://doi.org/10.3390/en16165984

[3] J. Gottschlich and R. W. De Doncker, "A programmable gate driver for power semiconductor switching loss characterization," 2015 IEEE 11th International Conference on Power Electronics and Drive Systems, Sydney, NSW, Australia, 2015, pp. 456-461, doi: 10.1109/PEDS.2015.7203542.

[4] Z. Zhang, F. Wang, L. M. Tolbert and B. J. Blalock, "Active Gate Driver for Crosstalk Suppression of SiC Devices in a Phase-Leg Configuration," in IEEE Transactions on Power Electronics, vol. 29, no. 4, pp. 1986-1997, April 2014, doi: 10.1109/TPEL.2013.2268058.

[5] A. Paredes, H. Ghorbani, V. Sala, E. Fernandez and L. Romeral, "A new active gate driver for improving the switching performance of SiC MOSFET," 2017 IEEE Applied Power Electronics Conference and Exposition (APEC), Tampa, FL, USA, 2017, pp. 3557-3563, doi: 10.1109/APEC.2017.7931208.

[6] K. Sahu, "EEE-INST-002: Instructions for EEE Parts Selection, Screening, and Derating," National Aeronautics and Space Administration Goddard Space Flight Center, Greenbelt, MD, USA, 2008

[7] J. Neville, S. Pfeiffer and J. P. Kozak, "Design and Analysis of a SiC-MOSFET Based Three-Phase Motor Drive for an Off-World Application," 2024 IEEE Aerospace Conference, Big Sky, MT, USA, 2024, pp. 1-6, doi: 10.1109/AERO58975.2024.10521308.

[8] J. Wang and X. Jiang, "Review and analysis of SiC MOSFETS ruggedness and reliability," IET Power Electron., vol. 13, no. 3, pp. 445–455, 610 Aug. 2019. [Online]. Available: https://digital-library.theiet.org/content/611journals/10.1049/iet-pel.2019.0587

[9] J. P. Kozak, R. Zhang, J. Liu, K. D. T. Ngo and Y. Zhang, "Degradation of SiC MOSFETs Under High-Bias Switching Events," in IEEE Journal of Emerging and Selected Topics in Power Electronics, vol. 10, no. 5, pp. 5027-5038, Oct. 2022, doi: 10.1109/JESTPE.2021.3064288.

[10] H. C. P. Dymond, D. Liu, J. Wang, J. J. O. Dalton and B. H. Stark, "Multi-level active gate driver for SiC MOSFETs," 2017 IEEE Energy Conversion Congress and Exposition (ECCE), Cincinnati, OH, USA, 2017, pp. 5107-5112, doi: 10.1109/ECCE.2017.8096860.

[11] S. Zhao et al., "Adaptive Multi-Level Active Gate Drivers for SiC Power Devices," in IEEE Transactions on Power Electronics, vol. 35, no. 2, pp. 1882-1898, Feb. 2020, doi: 10.1109/TPEL.2019.2922112.

[12] Y. Yang, Y. Wen and Y. Gao, "A Novel Active Gate Driver for Improving Switching Performance of High-Power SiC MOSFET Modules," in IEEE Transactions on Power Electronics, vol. 34, no. 8, pp. 7775-7787, Aug. 2019, doi: 10.1109/TPEL.2018.2878779.

[13] J. Cao, Z. -K. Zhou, Y. Shi and B. Zhang, "An Integrated Gate Driver Based on SiC MOSFETs Adaptive Multi-Level Control Technique," in IEEE Transactions on Circuits and Systems I: Regular Papers, vol. 70, no. 4, pp. 1805-1816, April 2023, doi: 10.1109/TCSI.2023.3233956

[14] F. Ahmad, "Adaptive gate drive voltage circuit," WO2006023912A3, Jul. 13, 2006 Accessed: Dec. 01, 2024. [Online]. Available: https://patents.google.com/patent/WO2006023912A3/en?oq=WO+2006%2f023912+A3

[15] W. Qiu, B. Dowlat, R. Abou-Hamze, and S. Laur, "DC/DC converter with adaptive drive voltage supply," USRE44587E1, Nov. 12, 2013 Accessed: Dec. 01, 2024. [Online]. Available: https://patents.google.com/patent/USRE44587E1/en?oq=US+RE44%2c587+E

[16] P. Camacho, V. Sala, H. Ghorbani and J. L. R. Martinez, "A Novel Active Gate Driver for Improving SiC MOSFET Switching Trajectory," in IEEE Transactions on Industrial Electronics, vol. 64, no. 11, pp. 9032-9042, Nov. 2017, doi: 10.1109/TIE.2017.2719603.

[17] T. Ivaniš, M. Kovačić, M. Makar and Ž. Jakopović, "Cycle-by-Cycle Optimized Active Gate Drive for SiC MOSFETs," 2024 IEEE 21st International Power Electronics and Motion Control Conference (PEMC), Pilsen, Czech Republic, 2024, pp. 1-6, doi: 10.1109/PEMC61721.2024.10726349.

[18] R. W. Erickson and D. Maksimovic, Fundamentals of Power Electronics. Springer Science & Business Media, 2007.

[19] C3M0120065K Datasheet: https://assets.wolfspeed.com/uploads/2024/01/Wolfspeed_C3M0120065K_data_sheet.pdf

A Hybrid Energy Storage System for eVTOL Unmanned Aerial Vehicles Using Supercapacitors

Ali Alenezi
Electrical and Computer
Engineering Departement
Texas A&M University
College Station, Texas, USA
ali2@tamu.edu

PengHao Huang
Electrical and Computer
Engineering Departement
Texas A&M University
College Station, Texas, USA
alexamiiah79@tamu.edu

Prasad Enjeti
Electrical and Computer
Engineering Departement
Texas A&M UniversityCollege
Station, Texas, US
enjeti@tamu.edu

Abstract— **Electric vertical take-off and landing (eVTOL) aircraft have gained considerable interest for their potential to transform public services and meet environmental objectives. Designing an effective power supply for eVTOL is challenging due to the extreme power requirements during takeoff and landing. This work presents a power supply solution and energy management control for an all-electric hybrid energy storage system that integrates supercapacitors and batteries to enhance eVTOL endurance. The proposed system employs DC-DC converters to regulate power output from each source. This method has the potential to improve the efficiency of the energy storage system and enhances the overall lifetime of the battery system. The integration of supercapacitors reduced total energy losses by 25.24% per flight compared to the conventional battery-only system. The work is validated through simulations and hardware-in-the-loop real-time testing, demonstrating its effective performance.**

Keywords—Supercapacitors, eVTOL, UAV, Drones, Hybrid.

I. INTRODUCTION

Unmanned aerial vehicles (UAVs), or drones, are remotely piloted or autonomous aircraft that have transformed industries like agriculture, delivery, and search and rescue. They operate using electric vertical take-off and landing (eVTOL) technology, allowing them to take off and land in tight spaces. eVTOLs are capable of short missions with light payloads or longer missions carrying passengers and cargo. eVTOLs rely on electric motors, which are more efficient, eco-friendly, and cost-effective than combustion engines. They produce no emissions, generate less noise, and require less maintenance. However, 28% of an eVTOL's weight is due to its power sources [1], typically lithium-ion batteries. These batteries struggle to meet the high-power demands of takeoff and landing, requiring discharge rates as high as 60C, with peak power exceeding 1000 kW/kg. A study in [2] has shown that repeated high discharge rates can lead to severe battery cell failure which emphasizes the challenges in extending the eVTOL battery lifecycle. Efficiency can drop to 60% during hovering with high discharge rate [3] but can rise to 90% at lower rates.

Current attempts to address this challenge involve the use of oversized battery arrays, but they increase weight and cost, limiting flight range and payload. Hybrid-electric systems with batteries and engines emit pollutants, conflicting with zero-emission goals. All-electric hybrid storage with batteries and fuel cells is an alternative [3]-[4], but fuel cells have challenges such as low power density, limited refueling infrastructure, and a low tank-to-weight ratio.

A promising solution is the hybridization of batteries with supercapacitors [5]-[6]. Supercapacitors handle high power demands during takeoff and landing, while batteries power the cruise phase. Advances in supercapacitor technology have led to high power densities of several thousand W/Kg at a reasonable cost, along with better charge-discharge efficiency and a longer life cycle than batteries.

DC-DC converters are employed for each energy storage source to control the discharge rates of the battery and supercapacitors while regulating the output voltage to ensure stable operation of the eVTOL motors. Additionally, they manage the recharging of the supercapacitor after it delivers high power during demanding phases.

This work employs a hybrid energy storage system combining supercapacitors and batteries in eVTOL UAV to potentially extend battery lifespan [5], improve battery discharge efficiency [3], and minimize losses while increasing flight range [7].

II. BACKGROUND

Small eVTOL UAV are generally designed based on specific mission requirements and can vary by size, flight range, maximum take-off weight (MTOW), endurance, and fuselage type (rotary, fixed, or flapping wings). Their power needs depend on their classification, intended mission, and payload. During takeoff and landing, the aircraft operates in hovering state, during which it must generate enough lift to overcome gravity and MTOW. The hover power (W) can be calculated as [1]-[2]:

979-8-3315-1612-3/25 $31.00 © 2025 IEEE

Current Profile (A)

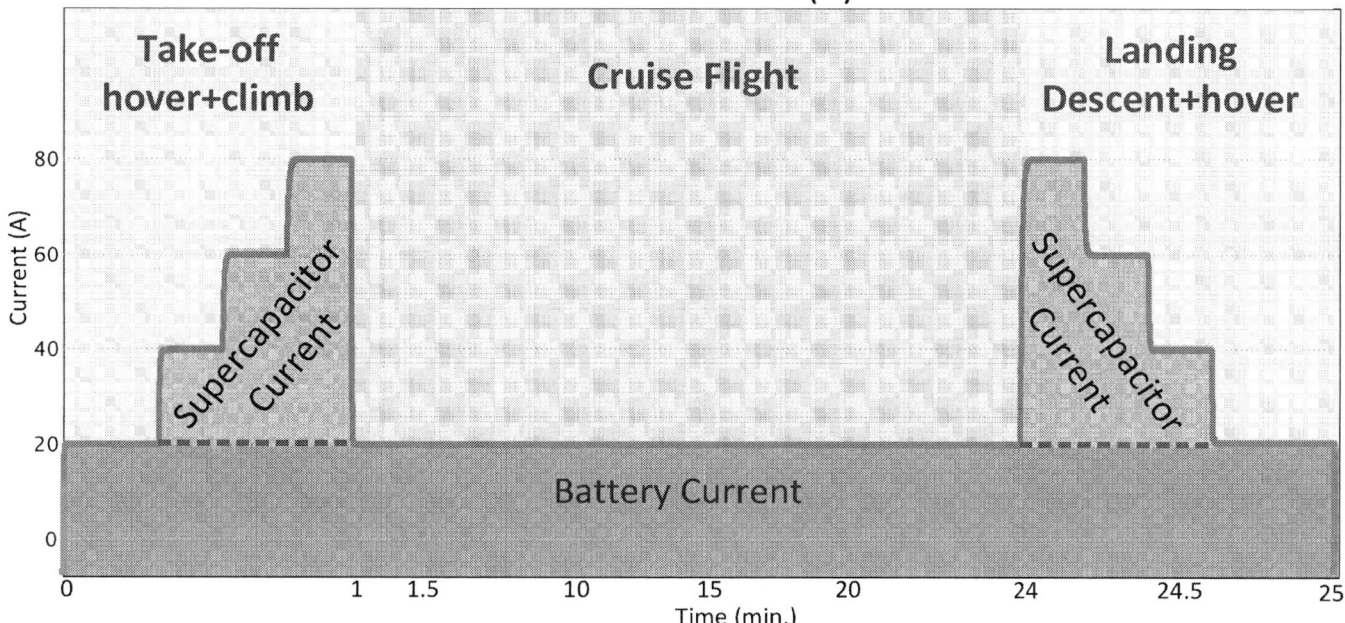

Fig. 1. Estimated current profile for the eVTOL load during a typical mission

$$P_{hover} = \frac{m(MTOW) \times g}{\eta(hover)} \sqrt{\frac{\delta}{2\rho}} \qquad (1)$$

While P_{hover} is hover power, m(MTOW) is the MTOW (kg), g is gravitational acceleration , η(hover) is hover efficiency, δ is disc loading (N/m²), and ρ is air density (kg/m³). Disc loading ranges from 200 N/m² to 1000 N/m². At cruising altitude, the eVTOL maintains a parallel orientation to ground, requiring less power than hovering. Cruise power (W) is calculated as:

$$P_{cruise} = \frac{m(MTOW)\, g}{\frac{L}{D}} \frac{V(cruise)}{\eta(cruise)} \qquad (2)$$

Where $\frac{L}{D}$ is the lift-to-drag ratio and $V(cruise)$ is cruise velocity (m/s). Due to the lack of real flight data, power profiles are often estimated [8]. In this work, a small eVTOL UAV with a maximum power of 4.4 kW is considered. With the output voltage is regulated, the estimated eVTOL current profile (in Amperes) is shown in Fig.1.

The electric motor propulsion system in an eVTOL is connected to a common DC bus through a DC-AC inverter. There are three types of battery-supercapacitor hybrids: passive, semiactive, and fully active. The passive configuration, where the battery and supercapacitors are directly connected to the load, is the simplest but offers lower performance because the output voltage is not regulated. The semiactive configuration improves performance by adding a DC-DC converter and control circuitry, while the fully active hybrid provides optimal performance with two DC-DC converters.

This work employs a fully active power management for supercapacitors and batteries as shown in Fig. 2. The DC-DC converters manage the charge discharge function while regulating the voltage and controlling the current supplied from each source.

A supercapacitor is used to meet high current demands during take-off and landing by taking advantage of its high-power density. The power required from the supercapacitor $P_{supercapacitor}(t)$ (W) can be calculated using:

$$P_{supercapacitor}(t) = P_{Load}(t) - P_{battery}(t) \qquad (3)$$

While $P_{Load}(t)$ is the eVTOL output power which varies across different flight phases, $P_{battery}(t)$ is the battery output power, all expressed in W. The total energy required from the supercapacitor $E_{supercapacitor}$ in joules is then can be calculated as:

$$E_{supercapacitor} = \int_{t1}^{t2} P_{supercapacitor}(t)\, dt \qquad (4)$$

Because the energy stored in a supercapacitor depends directly on the voltage, a 30% drop in voltage results in a 50% reduction in stored energy. The relationship can be expressed as:

$$E_{supercapacitor} = \frac{1}{2} C \left[V_f{}^2 - (0.7\, V_f)^2 \right] \qquad (5)$$

While C is the total required capacitance (F), V_f is the supercapacitor full voltage (V). When connecting a DC-DC converter (assumed ideal for simplicity) with the supercapacitor, it operates in constat power discharge mode. Here, the maximum current I_{max} that can be supplied from the supercapacitor to maintain a constant power output is given by:

$$I_{max} = \frac{P}{\sqrt{V_f{}^2 - \frac{2pT}{C}}} \qquad (6)$$

While P is the constant input/output power (W), C is the total capacitance, T is the discharge time in seconds [9].

979-8-3315-1612-3/25 $31.00 © 2025 IEEE   1218

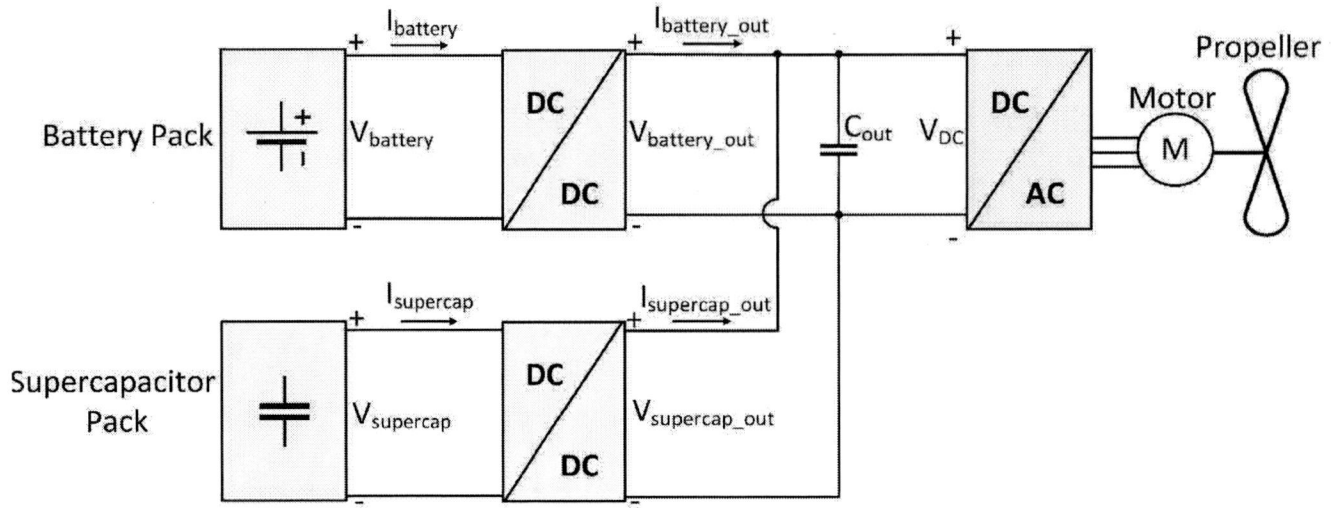

Fig. 2. Proposed hybrid battery and supercapacitor power supply with the propulsion system.

Additionally, it is important to account for the losses caused by the supercapacitor's equivalent series resistance (ESR) to ensure accurate performance predictions. The adjusted full voltage, accounting for the initial voltage drop due to the ESR, can be calculated using the following expression:

$$V_{f,n} = V_f - I_{initial}\ ESR \qquad (7)$$

While $I_{initial}$ refers to the current drawn from the supercapacitor at the beginning of the discharge cycle.

Table I compares multiple energy storage sources which shows the advantages of supercapacitors over batteries. Table II compares existing supercapacitor technologies, highlighting the 3,140 F from Maxwell for its highest specific power density and lowest ESR, which will be used in this work for designing the supercapacitor pack.

TABLE I
COMPARISON BETWEEN DIFFERENT ENERGY STORAGE DEVICES

Energy source	Specific Power density [w/kg]	Specific Energy density [Wh/kg]	Cycle Life	Efficiency [%]	Advantages	Disadvantages
Li-ion battery	600-2000	100-120	1000	Up to 90	-High voltage -High energy density -Long cycle life	-Lifetime decreases at high temperature -Non-overcharge -Non-over discharge -High security requirement
Li-Po battery	Up to 2800	180	500	Up to 90	-Low self-discharging rate -no memory effect -no pollution	-Lifetime decreases at high temperature -Non-overcharge -Non-over discharge -High security requirement - Low cycle life
Supercapacitor	1000-10,0000	4-15	100,000	Up to 98	-Fast charging and discharging speed - High power density -pollution-free -extremely long life	Low energy density

TABLE II
COMPARISON BETWEEN DIFFERENT SUPERCAPACITORS

Manufacturer	Maxwell Technologies	KYOCERA	Tecate Group	Eaton
Rated capacitance [F]	3,140	3,000	3,000	3000
Rated voltage [V]	3.0	2.7	3.0	3.0
Specific power density[kW/kg]	18.10	6.03	7.91	9.11
Gravimetric energy density[Wh/kg]	8.50	6.08	7.28	7.38
ESR [mΩ]	0.13	0.29	0.26	0.23
Mass [g]	460	500	525	525
Manufacture number	BCAP3000 P300 K04/05	SCCZ1EB3 08SWB	TPLH-3R0/3000SL60 X138	XL60-3R0308 T-R

PROPOSED SYSTEM

The proposed system includes a Li-ion battery and a supercapacitor, each represented by a simplified equivalent circuit. The battery's circuit consists of a DC source with an internal series resistor (ISR), while the supercapacitor's circuit includes a capacitor and an equivalent series resistor (ESR). Both the battery and supercapacitor are connected to separate DC-DC converters and linked in parallel to the DC bus through the output capacitor. Each DC-DC converter features a bi-directional buck-boost converter using a half-bridge leg and an inductor as shown in Fig. 3. SiC switches are used due to their strong potential to improve efficiency and handle higher currents [10]. Power management is handled by two dual-loop PI controllers, which offer high accuracy in tracking-controlled variables [11]. Since the supercapacitor's voltage discharges faster than the battery voltage, the supercapacitor controller dynamics is designed to respond more quickly than the battery controller. The battery's PI controller regulates the inductor L1 current ($I_{battery}$) and limits the output power to the cruise power using

979-8-3315-1612-3/25 $31.00 © 2025 IEEE

a variable limiter, as illustrated in Fig. 4 (a). Meanwhile, the supercapacitor's PI controller shown in Fig. 4 (b) manages the discharge process, ensuring that the supercapacitor provides power only during high demand in take-off and landing (Fig. 1). If the supercapacitor's voltage drops below 70% of its initial value, the supercapacitor's PI controller will limit the duty ratio (D_2) to stop discharging and switches the converter to buck mode to recharge the supercapacitor from the DC bus, as shown in Fig. 5, without needing an extra voltage sensor. This method features fewer computational processes as it uses simple control strategy [12].

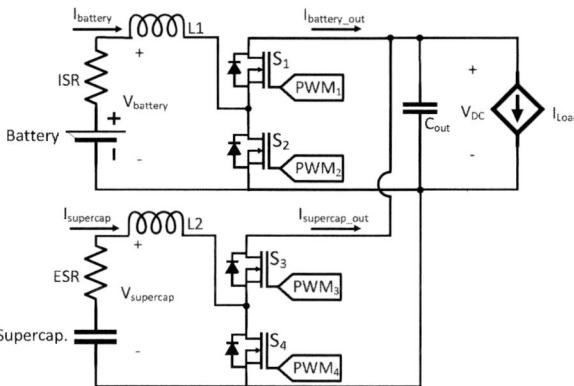

Fig. 3. Proposed DC-DC converters topology.

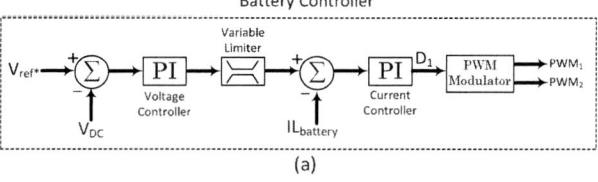

(a)

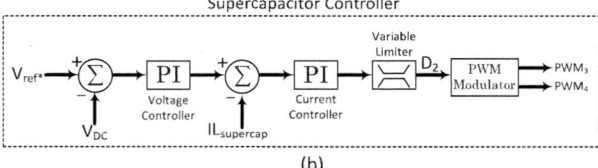

(b)

Fig. 4. Proposed power management control scheme.

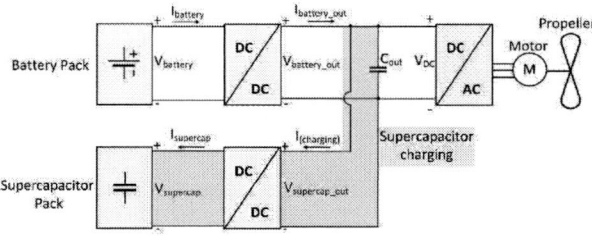

Fig. 5. Supercapacitor getting charged in buck-operation.

The parameters of the proposed eVTOL power supply system is outlined in Table III. The estimated power profile for the eVTOL, as depicted in Fig. 1, is detailed in Table IV. The current in the eVTOL gradually increases to 80 A (4.4 kW) during hovering before decreasing to 20 A (1.1kW) during cruising.

TABLE III
PROPPSED POWER SUPPLY PARAMETERS

Parameter	Value
Battery nominal voltage [V]	22.2
Battery ISR [mΩ]	30
Supercapacitor initial voltage [V]	22.2
Supercapacitor (C) [F]	785
ESR [mΩ]	0.52
DC bus [V]	55
Battery capacity [Ah]	14
Switching frequency [kHz]	100

TABLE IV
MISSION PROFILE SYSTEM PARAMETERS

Parameter	Value
Take-off [s]	45
Landing [s]	30
Cruise [s]	1245
MTOW [kg]	10
P_{hover} [W]	4400
P_{cruise} [W]	1100

RESULTS

A. Simulation Results

The simulation results for the DC-DC converters, using the estimated eVTOL load profile, were obtained using PSIM. The results focus on the first 55 seconds of the mission profile, covering the take-off and cruising power demands. Fig. 6 (a) illustrates the supercapacitor's voltage during discharge, where it drops from an initial 22.2V to 70% of its voltage at 15.54V, before being recharged by the DC bus during cruising. The proposed control scheme is designed to boost the input voltage to a regulated 55V DC bus throughout the mission, as depicted in Fig. 6 (b). The controller limits the battery inductor current, ensuring the converter output remains at a maximum of 20 A, even as the inductor input current increases due to the voltage boost ratio, as shown in Fig. 6 (c). The supercapacitor provides current only during high power demands when the load current exceeds the regulated battery current, as illustrated in Fig. 6 (c), with the overall load current detailed in Fig. 6 (d). The simulation results validate the effectiveness of the proposed control scheme in regulating the output power of both the battery and supercapacitor, ensuring a stable DC bus during the take-off and cruising phases, with similar performance observed during the landing phase.

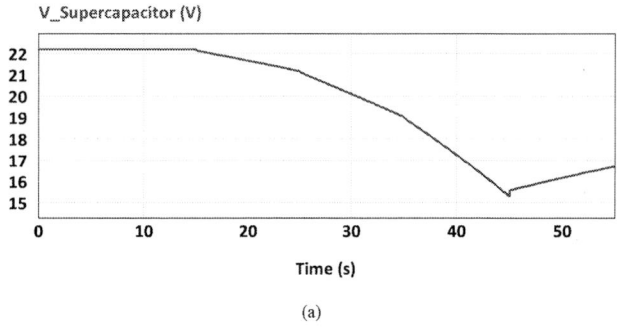

(a)

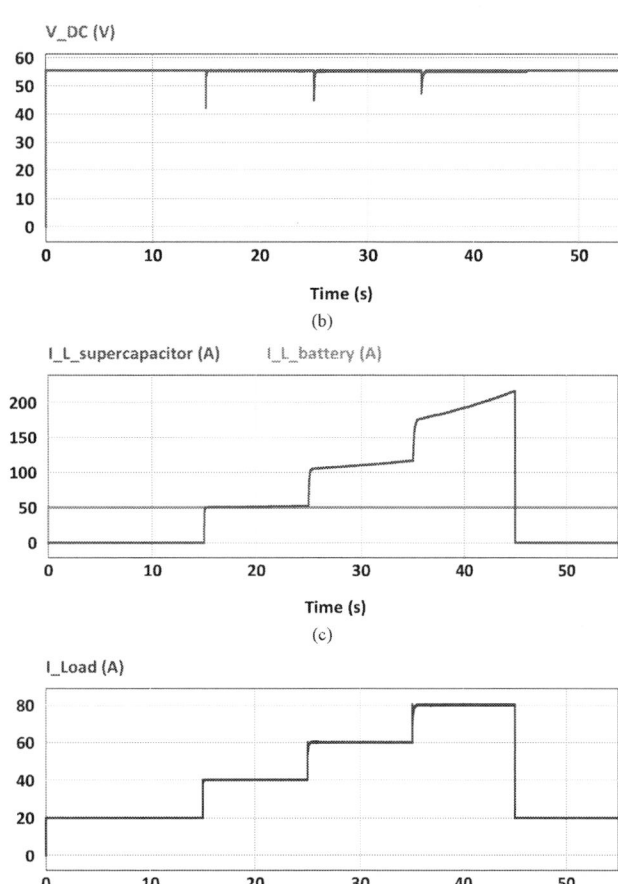

Fig. 6. Simulation results during take-off and cruising load demand (a) Supercapacitor voltage (b) DC bus voltage (c) Supercapacitor and battery inductor currents (d) Load current.

Fig. 7. Hardware in the loop experimental test set-up.

B. Hardware-in-the-Loop Experimental Results

The proposed topology was emulated using the Typhoon HL602+ hardware-in-the-loop (HIL) system, which was interfaced with Texas Instruments' TMS320F28379D digital signal processer (DSP). The HIL test setup is illustrated in Fig. 7. The HIL system enabled real-time responses to control signals, allowing for dynamic interaction with the emulated hardware. The control algorithm, implemented on the DSP, regulated the operation of the emulated system to ensure accurate modeling of the hybrid power supply. To accelerate the emulation process, the supercapacitor capacitance was scaled down, while a variable load was used to simulate the current demand during eVTOL takeoff. Each load variation within the step load was maintained for 1 second to comprehensively cover the eVTOL takeoff phase. The HIL emulation results closely match the simulation results, as shown in Fig. 8, demonstrating the effectiveness and accuracy of the proposed system.

C. Comparison of Energy Losses in Battery only and Hybrid Energy Storage systen with Supercapacitors

The total energy losses for the battery and supercapacitor packs are determined using the following equation:

$$E_{loss} = (I_{discharge})^2 \, R \, t \qquad (8)$$

Where E_{loss} represents the total energy losses in joules (J) caused by R; which is the internal resistance of the battery (ISR) or supercapacitor (ESR) both measured in Ω, $I_{discharge}$ is the discharging current in A, and t is the discharging time in seconds. The energy losses were calculated for a battery-only system and compared with the hybrid battery-supercapacitor system according to the parameters in Table III. Results show that the proposed eVTOL hybrid energy storage system reduces total energy losses by 25.24% over the entire flight mission. A summary of the energy loss comparison is provided in Table V.

TABLE V
COMPARISON OF TOTAL ENERGY LOSSES BETWEEN BATTERY
ONLY SYSTEM AND THE PROPOSESD HYBRID ENERGY
STORAGE SYSTEM

Energy Storage System	Total Energy Losses [J]
Battery only	24240.0
Battery with supercapacitor	18120.6
Reduction in total energy losses	25.24 %

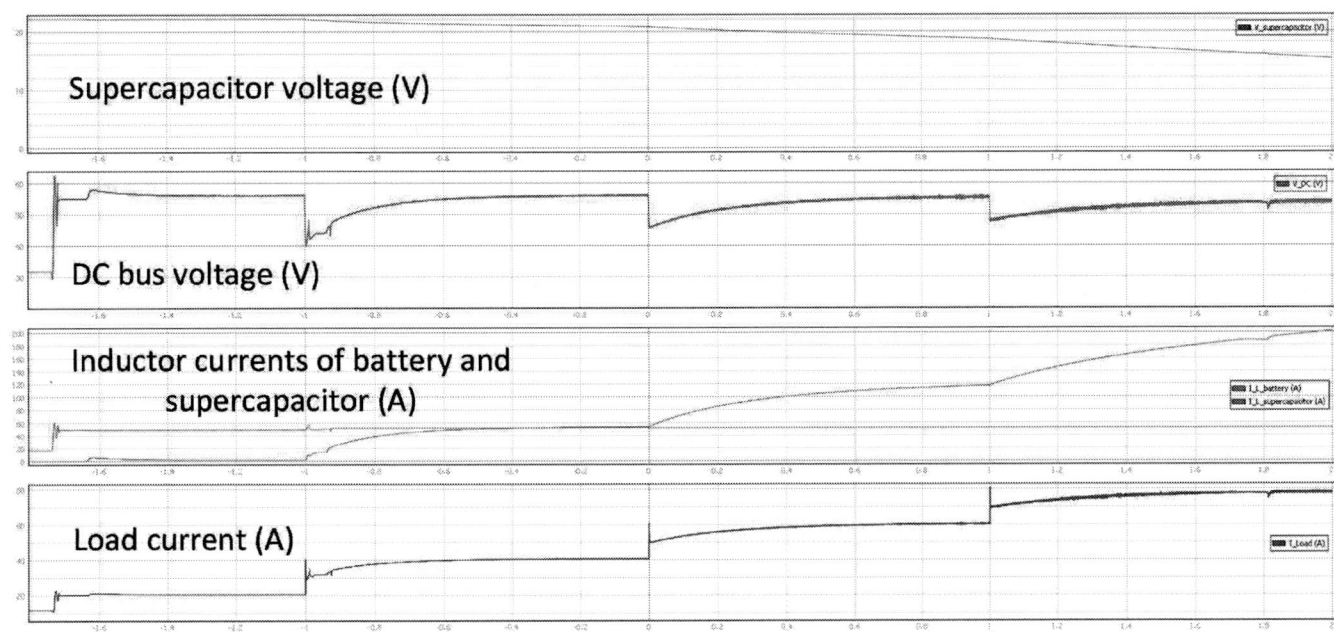

Fig. 8. Hardware in the loop results of the proposed system.

CONCLUSIONS AND FUTURE WORK

This work proposed an all-electric hybrid power supply system for eVTOL UAV, integrating batteries and supercapacitors to optimize performance. Each power source is managed by a dedicated DC-DC buck-boost converter: the battery current is regulated to supply only low discharge rates, minimizing stress on the battery, while the supercapacitor is controlled to handle high power demands during takeoff and landing. During cruising, the supercapacitor can be recharged from the DC bus. This approach improves the efficiency of the energy storage system and significantly enhances the overall lifetime of the battery by reducing its exposure to high discharge rates. The proposed approach can save approximately 25.24% losses per flight compared to the battery-only configuration. Simulation results and hardware-in-the-loop testing confirmed the system's robust performance and reliability. For future work, an experimental hardware implementation will be built, featuring the development of a high-power density DC-DC converter to support the proposed hybrid energy storage system for eVTOL UAVs.

REFERENCES

[1] N. Swaminathan, S. R. P. Reddy, K. RajaShekara and K. S. Haran, "Flying Cars and eVTOLs—Technology Advancements, Powertrain Architectures, and Design," in IEEE Transactions on Transportation Electrification, vol. 8, no. 4, pp. 4105-4117, Dec. 2022, doi: 10.1109/TTE.2022.3172960.

[2] M. Dixit, A. Bisht, R. Essehli, R. Amin, C.-B. M. Kweon, and I. Belharouak, "Lithium-Ion Battery Power Performance Assessment for the Climb Step of an Electric Vertical Takeoff and Landing (eVTOL) Application," *ACS Energy Letters*, vol. 9, no. 3, pp. 934-940, Mar. 2024, doi: 10.1021/acsenergylett.3c02385.

[3] D. Menzi, L. Imperiali, E. Bürgisser, M. Ulmer, J. Huber and J. W. Kolar, "Ultra-Lightweight High-Efficiency Buck-Boost DC-DC Converters for Future eVTOL Aircraft with Hybrid Power Supply," in IEEE Transactions on Transportation Electrification, doi: 10.1109/TTE.2024.3375026.

[4] J. J. Cooley et al., "Multiconverter System Design for Fuel Cell Buffering and Diagnostics Under UAV Load Profiles," in IEEE Transactions on Power Electronics, vol. 29, no. 6, pp. 3232-3244, June 2014, doi: 10.1109/TPEL.2013.2274600.

[5] R. Carter, A. Cruden and P. J. Hall, "Optimizing for Efficiency or Battery Life in a Battery/Supercapacitor Electric Vehicle," in IEEE Transactions on Vehicular Technology, vol. 61, no. 4, pp. 1526-1533, May 2012, doi: 10.1109/TVT.2012.2188551.

[6] M. -E. Choi, S. -W. Kim and S. -W. Seo, "Energy Management Optimization in a Battery/Supercapacitor Hybrid Energy Storage System," in IEEE Transactions on Smart Grid, vol. 3, no. 1, pp. 463-472, March 2012, doi: 10.1109/TSG.2011.2164816.

[7] L. Palma, P. Enjeti and J. W. Howze, "An approach to improve battery run-time in mobile applications with supercapacitors," IEEE 34th Annual Conference on Power Electronics Specialist, 2003. PESC '03., Acapulco, Mexico, 2003, pp. 918-923 vol.2, doi: 10.1109/PESC.2003.1218178.

[8] M. N. Boukoberine, T. Donateo and M. Benbouzid, "Optimized Energy Management Strategy for Hybrid Fuel Cell Powered Drones in Persistent Missions Using Real Flight Test Data," in IEEE Transactions on Energy Conversion, vol. 37, no. 3, pp. 2080-2091, Sept. 2022, doi: 10.1109/TEC.2022.3152351.

[9] H. Douglas and P. Pillay, "Sizing ultracapacitors for hybrid electric vehicles," 31st Annual Conference of IEEE Industrial Electronics Society, 2005. IECON 2005., Raleigh, NC, USA, 2005, pp. 6 pp.-, doi: 10.1109/IECON.2005.1569143.

[10] L. Dorn-Gomba, J. Ramoul, J. Reimers and A. Emadi, "Power Electronic Converters in Electric Aircraft: Current Status, Challenges, and Emerging Technologies," in IEEE Transactions on Transportation Electrification, vol. 6, no. 4, pp. 1648-1664, Dec. 2020, doi: 10.1109/TTE.2020.3006045.

[11] Y. Zhao, F. Qin and Z. Zhang, "Robust Tracking Control Design of Hybrid Battery-Supercapacitor Energy Storage System in Electric Vehicles," in IEEE Transactions on Transportation Electrification, doi: 10.1109/TTE.2024.3436650.

[12] S. K. Kollimalla, M. K. Mishra and N. L. Narasamma, "Design and Analysis of Novel Control Strategy for Battery and Supercapacitor Storage System," in IEEE Transactions on Sustainable Energy, vol. 5, no. 4, pp. 1137-1144, Oct. 2014, doi: 10.1109/TSTE.2014.2336896.

Evaluation of Retired Lithium-ion Batteries for Second-life Applications through Electrochemical Impedance Spectroscopy

Latha Anekal
Department of Electrical,
Computer and Software
Engineering
Ontario Tech University
Oshawa, Canada
latha.anekal@ontariotechu.net

Sheldon Williamson
Department of Electrical,
Computer and Software
Engineering
Ontario Tech University
Oshawa, Canada
sheldon.williamson@ontariotechu.ca

Abstract—**The rapid growth of electric transportation will generate a substantial number of retired batteries. Recycling the retired LIBs is being looked at worldwide to recover the expensive raw materials from used batteries. Alternatively, the retired electric vehicle (EV) batteries can be reused directly or refurbished for second-life applications. Nevertheless, Lithium-ion batteries (LIBs) degrade during their first-life usage, hence ensuring safety, efficiency, and reliability in second-life applications is crucial. This study investigates LIB degradation using electrochemical impedance spectroscopy (EIS) to assess the battery condition at the end of its primary use. A Samsung 21700 nickel cobalt aluminium (NCA) battery is cycled to 80% of its initial capacity and equivalent circuit model (ECM) parameters obtained through electrochemical impedance spectroscopy (EIS) are analyzed. Individual ECM parameters are evaluated to acquire the health indicators. Although the battery capacity decreased by 20%, resistive components such as bulk resistance, charge transfer resistance and Warburg resistance increased by 140%, 150% and 132% respectively. Given the relative capacity and resistance, the battery is well-suited for second-life applications such as grid frequency regulation, uninterruptible power supply, electric forklifts, and material handling equipment. Other health indicators such as Loss of lithium inventory (LLI) and loss of active materials (LAM) are also captured.**

Keywords— battery management system, lithium-ion battery recycling, reuse of battery, end-of-life

I. INTRODUCTION

Lithium-ion batteries (LIBs) have become ubiquitous in electric transportation due to their superior performance characteristics including high energy density, high power density, better efficiency and long cycle life. However, the LIBs are prone to degradation with aging. The end-of-life (EOL) for LIBs from EVs is defined as when the capacity is degraded by 20% to 30% from its original capacity. A typical retirement criterion, originally published by the United States Advanced Battery Consortium (USABC) in 1996, suggests replacing a battery pack when it has lost 20% of its original capacity [1]. While this limit is still largely regarded as a convincing benchmark, its applicability to present battery technology has become more uncertain. While batteries that have reached 80% of their original capacity are deemed unsuitable for high-speed traction applications due to their low power and energy

capabilities, they can still be reused for applications with lower power and energy requirements. This transition in LIB usage underscores the increased potential for second-life applications in lower-demand use cases. Based on the operating conditions during their first-life usage, the batteries are to be disposed of, recycled, reused or repurposed. Disposing of the LIBs in the landfill is hazardous therefore recycling, reusing and repurposing the LIBs are safer and have economic benefits [2]. Primary recycling methods that are currently under active research or employed in the industry are pyrometallurgy, hydrometallurgy, bio-metallurgy, and direct recycling [3]. Recycling of LIBs offers substantial economic and environmental advantages through resource conservation and supporting circular economy. Critical metals such as lithium, cobalt, and nickel are in short supply, necessitating sustainable exploration and effective recycling to mitigate resource constraints. Although recycling can save up to 51% of extracted materials, it has disadvantages, including high costs, the need for specialized equipment, and insufficient infrastructure. The life cycle of a LIB is depicted in Fig. 1.

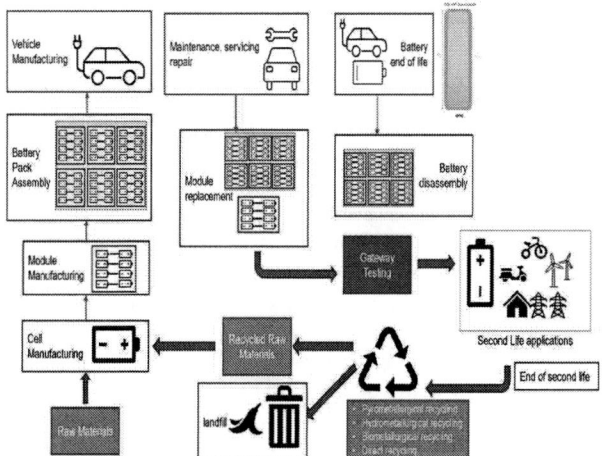

Fig. 1: Lithium-ion battery lifecycle

979-8-3315-1612-3/25 $31.00 © 2025 IEEE

Reusing and repurposing batteries by identifying retired functional battery packs, modules, and cells offers economic and sustainable advantages for second-life applications. The requirements for traction batteries and their second-life applications differ considerably based on their intended purposes and performance expectations. Automotive batteries demand high energy and power density for high performance. The first-life automotive applications require the battery to support fast charge-discharge capability, long cycle life, stringent safety criteria and optimal state of health. However, the performance criteria for second-life battery applications are typically more relaxed compared to their first-life application with reduced requirements for cycle life and performance rate.

The retired battery packs can be directly reused or refurbished for various applications including,

- Main grid utilization: supporting grid stabilization, energy arbitrage, frequency regulation, peak shaving, and power smoothing.

- Off-grid power solutions and renewable energy integration: supporting reliable energy storage for remote locations and enabling renewable energy integration.

- Residential and commercial building storage: serving as reserve power for homes, small commercial establishments and small industrial utilities.

- EV Charging Stations:

- Personal E-mobility: to be used in low-power vehicles such as golf carts and electric scooters

- Small electronic devices: repurposed for portable electronics and low-energy devices

Batteries with a state of health (SOH) of more than 80% are refurbished for use in electric vehicles. Cells with an SOH of 40% to 80% are recycled for energy storage systems, which support grid integration, renewable energy and residential and commercial building power backup applications. Batteries with SOH levels below 40% are recycled to recover important components [4].

To ensure grid stability and frequency regulation, the second-life batteries must be able to supply and absorb power and respond quickly to voltage and frequency fluctuations. Similarly for power smoothing, batteries must be able to absorb and release power quickly to level off the fluctuations from renewable energy sources such as wind and photovoltaic sources [5]. For energy arbitrage applications, long battery cycles are required to support several charge-discharge cycles. The battery should be designed to have high energy density to efficiently store energy during low-demand periods and release the energy when demand is high. Further, the capacity second life batteries should be adequate to handle large power demands for short duration to efficiently shave the peaks [6].

Furthermore, energy storage systems built using second-life batteries enable fast charging in rural and urban areas with limited grid capacity and reduce utility and vehicle charging costs through peak shaving. This also increases energy storage capacity to support vehicle charging during power outages. For such applications, the second-life battery needs to retain sufficient residual capacity to provide grid ancillary services and must be sized to meet the power and energy requirements accounting for factors such as depth of discharge (DOD), charge-discharge cycles and battery degradation.

Although, discharge capacity is the widely used indicator of EOL of first life, relative ohmic resistance is also employed as an additional indicator. The EOL with resistive indicator is set to 200% of the initial resistance. These EOL indicators are simple to implement, however, Estimating the lifespan of LIBs in EVs requires more complex battery EOL criteria based on the application and battery design characteristics, compared to simplified estimates. The internal operating conditions of a LIB is a complex phenomenon and the degradation is non-linear. Hence it is important to thoroughly assess the retired batteries to ensure safe, efficient and reliable operation in second-life applications. Typically, capacity fade is assessed to determine the state of health (SOH) of the batteries through the Coulomb counting method. Although measuring capacity is convenient, capacity-based SOH estimation techniques are inadequate in accounting for internal factors that contribute to the degradation mechanisms [6].

The capacity fade and increased internal resistance are considered crucial health indicators, however, other health indications evolve with battery aging. Reduced end-of-discharge voltage, open circuit voltage drift, increased reversible temperature, The loss of lithium-ion inventory (LLI), and the loss of active material (LAM) on the electrodes. This paper validates the above-listed health indicators of a Samsung 21700, 4000 mAh battery cycles for 680 charge-discharge cycles to determine the EOL of the battery and its suitability for second-life applications.

II. EXPERIMENTAL SETUP AND TEST PROCEDURE

The experimental setup and methodology are designed to investigate the battery degradation process as shown in Fig.3. A nickel cobalt aluminium (NCA) LIB with 4000 mAh is used to obtain the data through a D&V BCT-150 cell tester. The tester includes a battery cycler module, a high-frequency signal module for electrochemical impedance spectroscopy (EIS) tests, and an inbuilt temperature monitoring unit for real-time battery temperature measurement. The battery was subjected to 680 charge-discharge cycles with a stable thermal environment set at 25°C and EIS is performed after every 10 cycles. The cell is cycled with 1.3C constant-current constant-voltage (CC-CV) charging and 1C discharging currents. EIS is performed at 100% SOC and after a rest period of 60 minutes post-charging to ensure the equilibrium condition of the LIB. A frequency range of 10 kHz to 0.01 Hz is used to obtain the impedance spectrum from the EIS test. The capacity, internal resistance, open circuit voltage and discharge voltages are captured. LLI and LAM data are captured through EIS. The experimental data is collected to establish the ECM parameters that accurately model the electrochemical behaviour of the LIB through aging. This is crucial for understanding the long-term performance and degradation of LIB. The ECM parameters are estimated by fitting the simulated Nyquist plot to the actual Nyquist plot using commercially available ZView software.

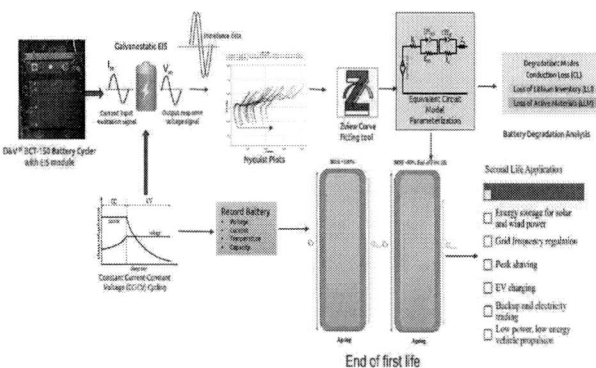

Fig. 2: Experimental set-up and procedure

III. RESULTS AND DISCUSSION

End of discharge (EOD) is the minimum cutoff voltage point at which a LIB should not be discharged. Discharging below this voltage limit will cause the battery to degrade. EOD is an effective health indicator that determines the EOL of LIB. Tracking the voltage trend for EOD over the life cycle enables early aging detection. The EOD for the battery under test was defined as 2.5V. From the experimental results, the voltage decreases gradually to EOD during preliminary cycles. The voltage trend during the initial and mid-cycles has a linear tendency with cycle time and with increased aging the EOD voltage dropped sharply to 2.5V. Fig. 3 shows the EOD for 680 discharge cycles.

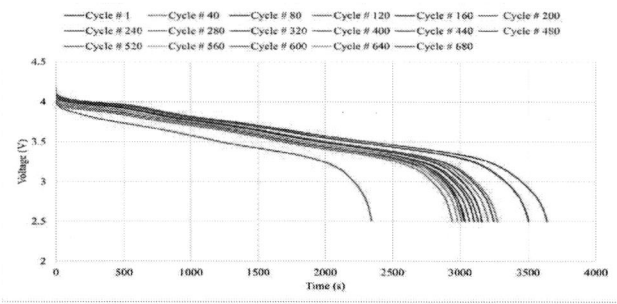

Fig. 3: End-of-Discharge voltage through cycling

Open circuit voltage for charge-discharge cycles has been captured for the first cycle, after the 250th cycle and for the last cycle as shown in Fig. 4 and Fig. 5 respectively. The cell was charged and discharged at a very slow rate of C/30. With the small current the cell will be at near equilibrium. The OCV profile has gradually changed over the life cycle. The degree of drift from the baseline voltage of the initial cycle indicates the cell is nearing EOL. This drift is due to the formation of a solid electrolyte interface (SEI) layer, loss of active materials, and structural changes in the electrodes [7]. Although, OCV drift is a better indicator of the health of a battery, voltage measurement conditions, requirement of relaxation time, long cycle time and temperature dependency are the challenges for the OCV to be considered to predict the EOL of the battery.

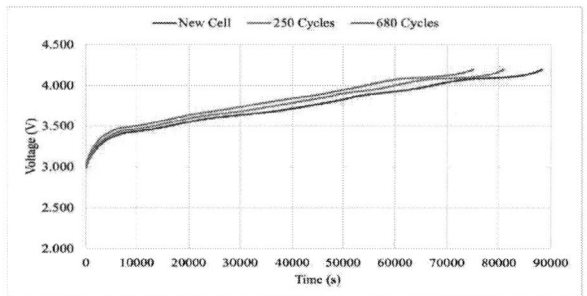

Fig. 4: Charging open circuit voltage

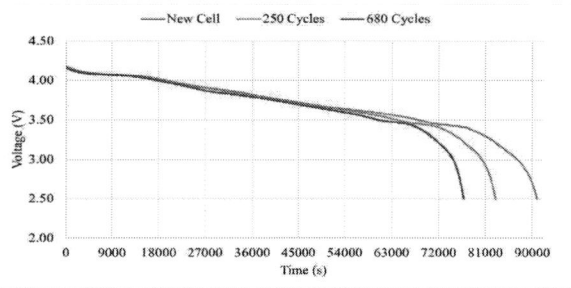

Fig. 5: Discharging open circuit voltage

Further, the temperature behaviour of the battery under test was analysed to study the effects of degradation. Temperature tracking for LIB is critical to ensure safety, reliability, and durability. The battery heats up from the core due to reversible and irreversible reactions during their normal operation. The heat generation due to entropy change is reversible, while the heat generation due to ohmic, kinetic and mass transport overpotentials is irreversible. The contact resistance, SEI resistance, and joule heating from resistance to electron and ionic flow within electrodes and electrolytes are the causes of the ohmic heat. Kinetic heat generation is due to the charge transfer reactions, availability of reaction species and reaction sites. Mass transport heat is attributed to mass transport limitations in the electrodes or electrolyte and temporary concentration gradients, which are linked to reaction sites. Due to the changes in the internal resistance of the LIB through aging, the internal temperature of the batteries will rise with the battery cycle life. From the experimental studies, the temperature rise as shown in Fig. 6 at the end of the charge through aging, increased from 29.4°C during the first charge cycle to 36.2°C during the 680th cycle. During discharge, the end temperature raised from 30.32°C to 35.02°C between the first and 680th discharge cycle.

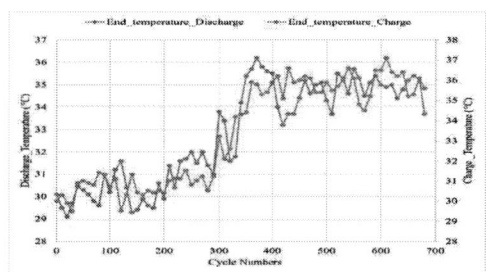

Fig. 6: Variation in charge-discharge temperature with cycle life

Various characterization tests including cyclic aging tests, voltage and capacity measurement tests, and current pulse-based tests, are used as diagnostic techniques to determine the battery degradation modes and capacity fade. However, these techniques do not capture all the degradation mechanisms in LIBs. Electrochemical impedance spectroscopy (EIS) is a potential diagnostic tool that is a non-invasive and non-destructive technique that can perform in-situ analysis to study LIB degradation behaviour [8]. EIS involves passing an excitation sinusoidal signal through an electrochemical system over a broad frequency spectrum to obtain a sinusoidal response from the system. The result from EIS characterization provides a comprehensive knowledge of battery behaviour through internal impedance data obtained at different frequencies. The Nyquist plot of the impedance spectrum for LIBs represents the key elements including ionic resistance, electron resistance, bulk electrolyte resistance, solid interface layer (SEI) resistive and capacitive characteristics, charge transfer resistance, and double-layer capacitance.

Fig. 7 shows the Nyquist plot of a used LIB. At high frequencies, the plot highlights the resistive behaviour of LIB. The semicircle in the mid-frequency region represents double-layer capacitance and charge transfer resistance. This region also considers the effects of charge carriers at the SEI layer. Initially, during the first few cycles, the impact of this layer is minimal. However, as the battery ages, the growth of the SEI layer increases, leading to the formation of a second semicircle in the mid-frequency region. As the frequency decreases towards zero, the Nyquist plot reflects the impedance associated with chemical reactions at the bulk electrode material, indicating the solid-state mass transport of lithium ions. This diffusion process is represented by a straight line with a 45° slope at the lower frequencies seen at the tail end of the plot. This study proposes the utilization of equivalent circuit (ECM) parameters obtained from the EIS technique to determine LIB degradation at EOL. The battery degradation mechanisms are interrelated and can be classified into three different modes namely, Conduction loss (CL): variations in the ohmic resistances within the battery components observed at high frequencies, Loss of lithium-ions (LLI): caused by side reactions, solid electrolyte (SEI) layer, and lithium plating, at mid-frequencies and Loss of active materials (LAM) results from structural electrode breakdown, and passivation at low frequencies.

This study uses a second-order equivalent circuit to represent the internal characteristics of LIB as shown in Fig. 7. R_s represents the ohmic resistance (electrode, electrolyte, separator, and current collectors). R_{ct} represents the charge transfer resistance for lithium-ions (Li$^+$) due to activation barriers at the electrode and electrolyte interface. C_{dl} represents the double-layer capacitance that influences the energy and power storage capacity. Z_w represents the Warburg impedance that reflects the mass transfer of Li$^+$ in the solid electrode. Further, the growth of the SEI layer as the battery ages is represented by one more RC parallel circuit. R_{SEI} is the resistance due to the growth of the SEI layer and C_{SEI} is the capacitance component of the SEI and electrolyte interface.

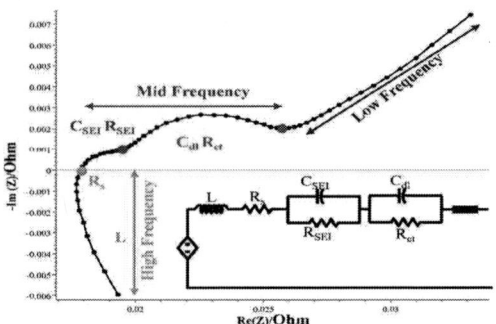

Fig. 7: Nyquist plot for used LIB

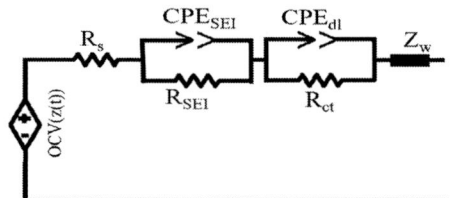

Fig. 8: 2-RC- Equivalent circuit model

IV. EQUIVALENT CIRCUIT PARAMETERS FOR SECOND-LIFE APPLICATIONS

The Nyquist plots for 680 cycles, displayed in Fig. 9, are used to track the degradation of the battery. At the end of the 680th cycle, the battery capacity reached nearly 80% of its original capacity as shown in Fig. 10 and was considered EOL for the battery. The EOL is also defined by relative resistance and is determined typically when the internal resistance of the battery is increased by 200% [9]. From the ECM parameter values, it is observed that the series resistance R_s has linearly increased from 0.0125 Ω to 0.0301 Ω from the first cycle to the 680th charge-discharge cycle, a 140.8% rise as shown in Fig. 10(a) resulting in conduction loss and capacity loss. The charge transfer resistance, R_{ct} reflecting the activation energy barrier for Li$^+$ at the electrode-electrolyte interface rose from 0.0032 Ω to 0.0091 Ω, a 150% rise as shown in Fig. 11 (b). A study [10] suggests that the stability of charge transport between the battery components is highly stable above 80% SOH but decreases below this threshold. This underscores the importance of monitoring the R_{ct} assess the battery health, diagnose safety issues and characterize LIBs for second-life applications and recycling.

R_{SEI} increased slightly from 0.0014 Ω to 0.0021 Ω. The Q_{dl} component represents the properties of the electrode surface, the electroactive species, and the non-ideal capacitance behaviour of LIB and increased with aging as shown in Fig. 11 (d). The Q_{SEI} capacitance also increased over the lifetime of the battery as shown in Fig. 11 (e). R_{ct} and Q_{dl} together represent the LAM degradation mechanism, this factor contributes to the estimation of the remaining useful life (RUL) of the battery. The resistive component R_w shown in Fig. 11 (c) of Warburg impedance increased by 150% at EOL. The SEI characteristics and R_w represents LLI degradation. This health indicator ensures sufficient lithium availability for energy storage applications.

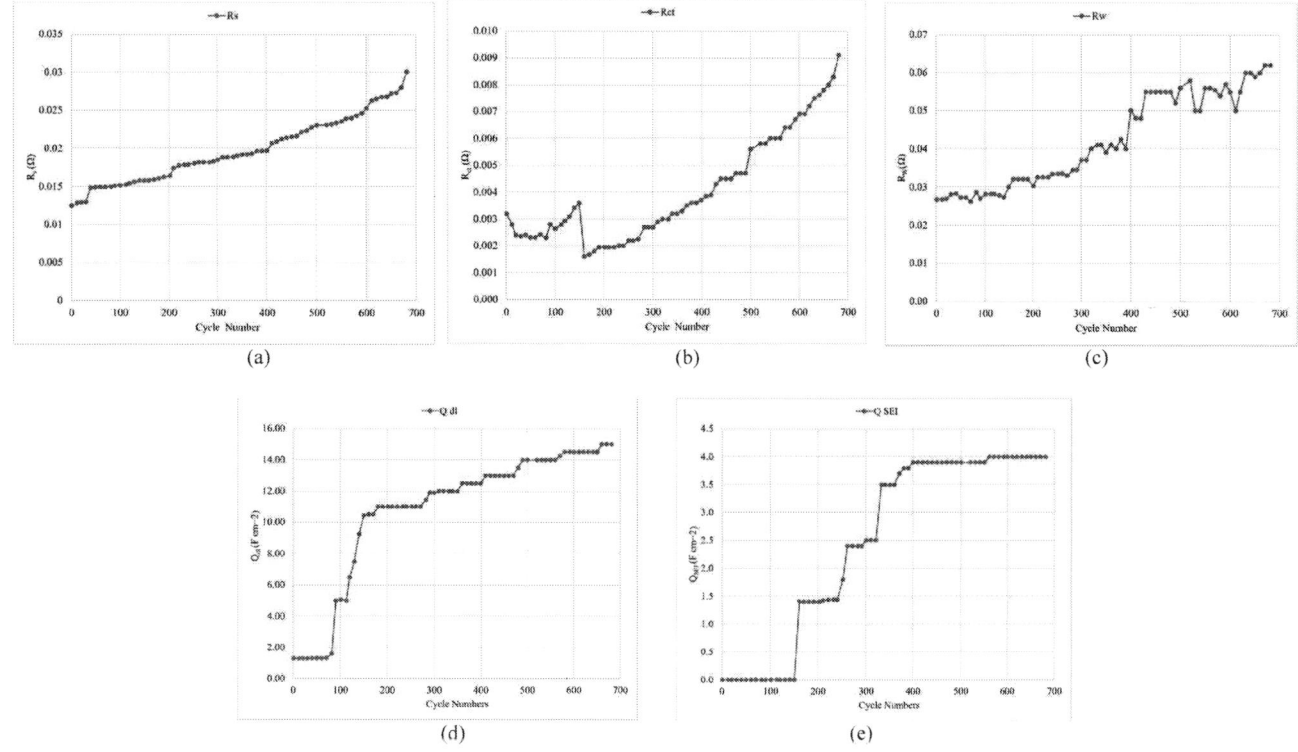

Fig. 10: ECM parameters (a) variation of Rs through aging, (b) variation of Rct through aging, (c)variation of Rw through (d) aging variation of Qdl through aging, (e) variation of QSEI through aging

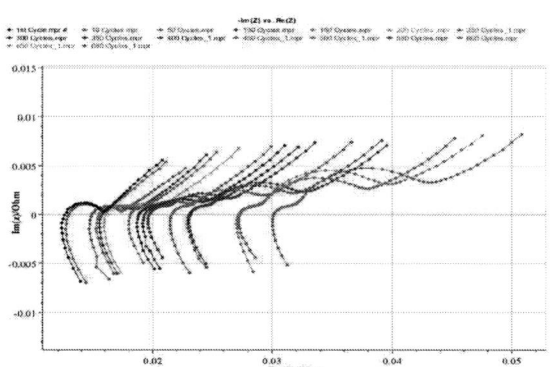

Fig. 9: Nyquist plots from experimental data

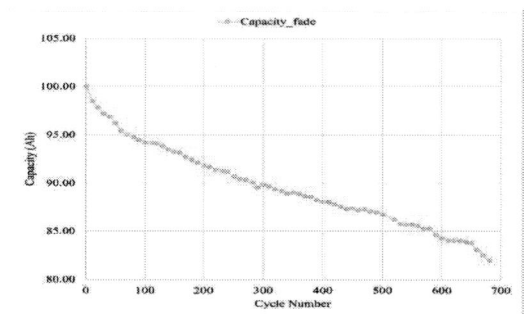

Fig. 10: Capacity fade

Conclusions

Understanding and accurately modelling ECM parameters are crucial for the effective deployment of retired EV batteries in second-life applications. Further, it is important to understand the degradation trend in retired batteries to maximize the lifespan of batteries as they transition to secondary applications. By analyzing the ECM parameter values at EOL, batteries can be effectively categorized for appropriate second-life applications. The OCV drift, end-of-discharge voltage and reversible temperature changes are studied. From the study, it can be seen that the resistive components including R_s, R_{ct} and, R_{SEI} rose maxim of 150%. Based on this inference, the battery is suitable for high-power applications such as grid stability and energy storage for backup systems. However, since the battery has a low LAM measured through R_{ct} and Q_{dl}, the battery is well suited for energy storage applications for renewable energy sources.

REFERENCES

[1] J. Zhu et al., "End-of-life or second-life options for retired electric vehicle batteries," Cell Reports Phys. Sci., vol. 2, no. 8, pp. 1–26, 2021, doi: 10.1016/j.xcrp.2021.100537.

[2] A. Rahman, R. Afroz, and M. Safrin, "Recycling and disposal of lithium batteries: An economical and environmental approach," IIUM Eng. J., vol. 18, no. 2, pp. 238–252, 2017, doi: 10.31436/iiumej.v18i2.773.

[3] Z. J. Baum, R. E. Bird, X. Yu, and J. Ma, "Lithium-Ion Battery Recycling—Overview of Techniques and Trends," ACS Energy Lett., vol. 7, no. 2, pp. 712–719, 2022, doi: 10.1021/acsenergylett.1c02602.

[4] M. T. Sarker, M. H. S. M. Haram, S. J. Shern, G. Ramasamy, and F. Al Farid, "Second-Life Electric Vehicle Batteries for Home Photovoltaic Systems: Transforming Energy Storage and Sustainability," Energies , vol. 17, no. 10, 2024, doi: 10.3390/en17102345.

[5] A. Saez-De-Ibarra, E. Martinez-Laserna, D. I. Stroe, M. Swierczynski, and P. Rodriguez, "Sizing Study of Second Life Li-ion Batteries for Enhancing Renewable Energy Grid Integration," IEEE Trans. Ind. Appl., vol. 52, no. 6, pp. 4999–5007, 2016, doi: 10.1109/TIA.2016.2593425.

[6] A. N. Patel et al., "Lithium-ion battery second life: pathways, challenges and outlook," Front. Chem., vol. 12, no. April, pp. 1–21, 2024, doi: 10.3389/fchem.2024.1358417.

[7] S. Barcellona, S. Colnago, E. Fedele, D. Iannuzzi, L. Piegari, and M. Ribera, "Cycle Aging Effect on the Open Circuit Voltage of a LiFePO4Battery," 2023 IEEE Veh. Power Propuls. Conf. VPPC 2023 - Proc., 2023, doi: 10.1109/VPPC60535.2023.10403323.

[8] L. A. Middlemiss, A. J. R. Rennie, R. Sayers, and A. R. West, "Characterisation of batteries by electrochemical impedance spectroscopy," Energy Reports, vol. 6, pp. 232–241, May 2020, doi: 10.1016/J.EGYR.2020.03.029.

[9] M. Arrinda et al., "Application dependent end-of-life threshold definition methodology for batteries in electric vehicles," Batteries, vol. 7, no. 1, pp. 1–20, 2021, doi: 10.3390/batteries7010012.

[10] Y. Bao and Y. Gong, "Li-ion battery charge transfer stability studies with direct current impedance spectroscopy," Energy Reports, vol. 9, pp. 34–41, 2023, doi: 10.1016/j.egyr.2023.03.002.

Uninterruptable Non-isolated Integrated Power Electronics Converter (UNIPEC) for Commercial Truck Auxiliary Power Unit

Pouya Zolfi
Department of Electrical and Computer Eng.
Marquette University
Milwaukee, USA
pouya.zolfi@marquette.edu

Ahmad Alzahrani
Department of Electrical and Computer Eng.
Marquette University
Milwaukee, USA
ahmad.alzahrani@marquette.edu

Ayman EL-Refaie
Department of Electrical and Computer Eng.
Marquette University
Milwaukee, USA
ayman.el-refaie@marquette.edu

Abstract— **A low component count high efficiency and high-power density three-port DC-DC converter named UNIPEC is introduced for commercial trucks auxiliary power unit (APU) application. The proposed UNIPEC is a single magnetic core topology which is achieved by integration and modification of switched capacitor technology and bidirectional multi-level DC-DC converter. With peak simulation efficiency of 97.3% and power density of 12.6 kW/L, proposed unit is an attractive candidate for conventional and refrigerated freight transportation applications. Comparison study and low power experimental validation results are presented. Reliability analysis of UNIPEC is also conducted and mean time to failure (MTTF) reported. A 1 kW PCB is designed for power density calculations.**

Keywords— *Auxiliary power unit (APU), DC-DC converter, Electrified transportation, Reliability*

I. INTRODUCTION

Auxiliary Power Units (APUs) are essential in the world of freight transportation and truck drivers, providing an important balance of comfort, efficiency, and environmental responsibility. These units are critical in modern logistics operations, providing power to a wide range of AC and DC loads inside trucks [1]. AC loads, such as heating, ventilation, and air conditioning (HVAC) units, are typically rated for 120 volts [2] while DC loads, such as lighting systems, onboard electronics, and other auxiliary equipment, typically operate at 12 or 24 volts, supporting critical functions for vehicle operation and driver convenience [3]. In recent years, integrated APU systems have increasingly used technological advancements to meet varying voltage requirements more efficiently [4-6]. Integrated APUs improve operational efficiency while also significantly lowering environmental impacts by optimizing power distribution and energy management. One of the most significant technological advancements in this context is the use of multiport converters (MPCs). MPCs provide numerous benefits for APU applications, most notably simplifying power distribution by integrating multiple input and output ports. This integration results in more compact and streamlined electrical systems, which reduces the complexity and physical footprint of the truck's power distribution network. In addition, MPCs improve system reliability and resilience by including redundancy features [7]. These features allow MPCs to dynamically adjust voltage levels in real time, ensuring that critical loads receive a continuous and stable power supply even

when components fail, or power demand fluctuates. MPCs' strategic management of power flow not only improves system dependability but also significantly contributes to increased fuel efficiency. The integration of advanced APUs and MPCs represents a game-changing step forward for the freight transportation industry.

The transition from conventional two-port converters to MPCs marks a significant advancement in power electronics technology, especially for the field of electrified transportation systems. Lower number of components, flexible topology, higher power density, shared packaging and thermal management system, and centralized control system are key advantages of the MPCs [8]. The non-isolated three-port converter (TPC) architecture enables efficient power distribution among critical subsystems such as propulsion, and auxiliary components. There are numerous MPCs introduced in literature for a wide range of applications such as dc microgrids and electrified transportation. A four-port non-isolated converter is proposed in [9] implementing boost and buck-boost building blocks for electric vehicle (EV) applications. In [10], a non-isolated MPC is proposed for integration of FC, battery, and PV. In [11], a topology of non-isolated TPC is presented which is derived by combining buck and super-lift Luo converter building blocks. A novel non-isolated MPC is presented in [12] for FC-based hybrid streetcar application, integrating grid connection and backup battery with the main energy source. A family of step-up MPCs is introduced in [13] for FC-based hybrid vehicular power systems. This multi-port converter family offers benefits such as reduced active and passive components, extendibility, simple power flow management, and low current and voltage ripples compared to the literature.

In this research, a three-port uninterruptable non-isolated integrated power electronics converter (UNIPEC) is proposed for commercial truck APU applications. UNIPEC has reduced number of passive components and higher power density while maintaining dynamic power flow among ports. A UNIPEC-based truck APU architecture is presented, as shown in Fig. 1. UNIPEC guarantees full utilization of the truck battery while enabling the back-up battery in the case of faulty/depleted main battery and idling periods. Smart DC circuit breakers (SCBs) controlled by load requirements are implemented the proposed architecture as a mechanism to manage power flow from batteries.

979-8-3315-1612-3/25 $31.00 © 2025 IEEE

II. PROPOSED UNIPEC TOPOLOGY

The proposed UNIPEC, as shown in Fig. 2, is a single magnetic core topology which is achieved by integration and modification of switched capacitor technology and bidirectional multi-level DC-DC converter [14 - 15]. UNIPEC consists of five semiconductor switching blocks (two fully directional and three unidirectional), two capacitors and one inductor and has continuous current for both input ports. Compact structure, high efficiency and simplicity of topology and control system makes UNIPEC an attractive solution for medium power APU systems for commercial trucks. The analytical analysis is conducted for the three main scenarios during continuous conduction mode (CCM). The following assumptions are made in order to simplify the steady-state analysis of the converter:

- The capacitors are large enough and their voltage can be considered constant during the switching period.
- ESR resistances of passive components are neglected in steady-state analysis but are considered for obtaining efficiency values.

A. Building Block #1: Three-Level Converter

This function consists of four operating modes as shown in figures 3(a)-(d). Related waveforms for the converter cell are also presented in Fig. 4. Switches S_1, S_4 and S_5 are always OFF during this mode while S_2, S_3 and diodes D_4 and D_5 participate in power conversion. Considering d_{eff} as duty cycle of switching for active switches, CCM voltage gain equation is obtained for the converter.

B. Building Block #2: Switched Capacitor Converter

Similar analysis can be done for this building block in CCM. The switched capacitor block has the bidirectional power flow capability which leads to step-up and step-down operations, as shown in Fig. 5.

C. Integration Scenarios and Power Flow Control Logic

The UNIPEC power converter is designed to enhance the efficiency and versatility of power management in vehicular systems by operating in three distinct modes (shown in Fig. 6(a)), each corresponding to a specific integration scenario.

Mode A, the converter channels power from the truck's main battery to the vehicle's loads, ensuring that all essential and auxiliary systems receive a steady and reliable power supply during operation. This mode is crucial for the normal functioning of the vehicle, allowing it to perform its primary duties without interruption. The power converter optimizes the flow of energy, ensuring that the main battery's power is utilized effectively, reducing losses, and extending the overall lifespan of the battery.

Mode B allows the secondary or backup battery to take over the supply of power to the loads. This mode is particularly useful in scenarios where the main battery is either depleted or needs to be preserved for critical functions. By tapping into the secondary battery, the UNIPEC converter provides an additional layer of security and operational continuity, ensuring that the vehicle's loads remain powered even if the main battery is compromised. This redundancy is vital for mission-critical applications where power interruptions can lead to significant issues.

Mode C, the focus of the UNIPEC converter shifts to recharging the secondary battery. This mode ensures that the backup battery is always maintained at optimal charge levels, ready to be deployed whenever needed. By efficiently managing the charging process, the converter minimizes energy losses and reduces the wear and tear on the battery, thereby extending its service life.

Together, these three operating modes enable the UNIPEC power converter to provide a robust, adaptable, and efficient solution for managing the complex power needs of modern vehicles, particularly those that require high reliability and continuous operation. Scenarios A and B present step-up operation of UNIPEC, while scenario C has step down functionality. CCM voltage gain equations for the operation scenarios are summarized as Table I.

The selection logic between the three power flow scenarios in the UNIPEC is governed by a carefully designed logical circuit, as depicted in Fig. 6(b). This logic circuit called power flow management integrated circuit (PFMIC) dynamically evaluates the availability of power sources and the current load requirements to determine the optimal mode of operation. By analyzing the status of the truck's main battery, the secondary battery, and the load demands, the circuit ensures that power is directed from the most appropriate source at any given time. For instance, if the main battery is fully charged and capable of meeting the load demand, the logic prioritizes Mode A, supplying power directly from the main battery. However, if the main battery's charge is low or needs to be conserved, the logic shifts to Mode B, allowing the secondary battery to take over. When the load demand is met, and the secondary battery requires recharging, the circuit seamlessly transitions to Mode C to replenish the backup battery. This logic-based approach ensures efficient power management, optimizing the performance and longevity of the vehicle's power systems by real-time switching between the three scenarios.

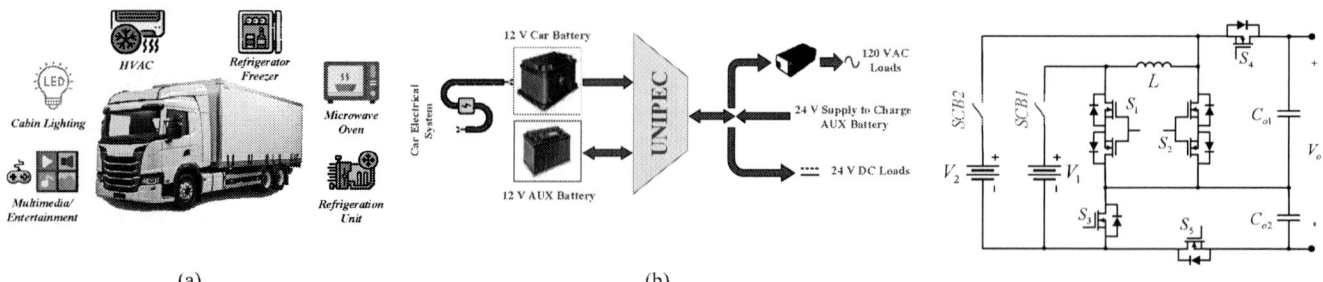

Fig. 1. (a) Loads in a commercial truck (b) UNIPEC-based APU

Fig. 2. Proposed UNIPEC

979-8-3315-1612-3/25 $31.00 © 2025 IEEE

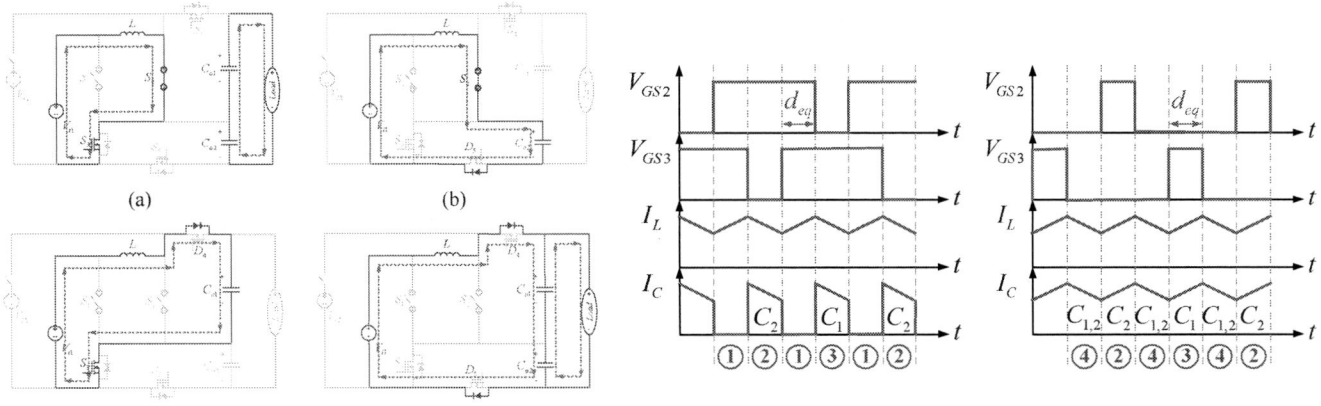

Fig. 3. UNIPEC operating modes for scenario A   Fig. 4. UNIPEC waveforms for scenario A

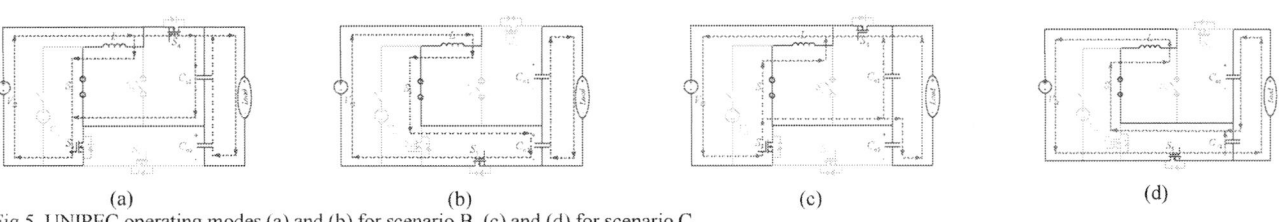

(a)   (b)   (c)   (d)

Fig.5. UNIPEC operating modes (a) and (b) for scenario B, (c) and (d) for scenario C

TABLE I - VOLTAGE GAIN EQUATIONS FOR VARIOUS OPERATION SCENARIOS

Scenario	A. Main Batt to Load	B. Back-up Batt to Load	C. Charging Back-up Batt
Voltage Gain	$M_a = \dfrac{V_o}{V_{i1}} = \dfrac{1}{1-d_a}$	$M_b = \dfrac{V_o}{V_{i2}} = \dfrac{2-d_b}{1-d_b}$	$M_c = \dfrac{V_{i2}}{V_o} = \dfrac{1-d_c}{2-d_c}$

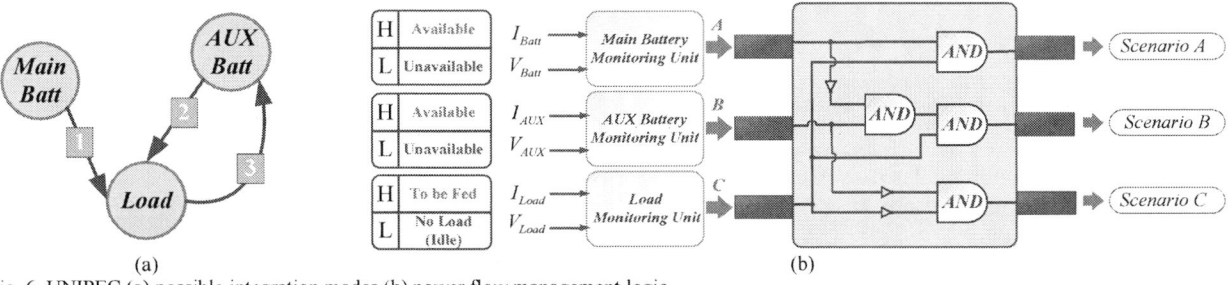

(a)   (b)

Fig. 6. UNIPEC (a) possible integration modes (b) power flow management logic

D. Efficiency Analysis

The efficiency of the proposed converter is determined by analyzing the losses in each component. Additionally, core losses and diode reverse recovery losses are included for comprehensive analysis. The analysis are provided for scenario A, but can be applied similarly to other modes as well.

- *Inductor Losses*

The inductor losses arise due to its DC resistance and the RMS current. The RMS current represents the effective value of the fluctuating inductor current, which consists of the average current and the ripple component. The inductor current ripple can be expressed as follows:

$$\Delta I_L = \frac{V_{i,1} d_a}{L f_s} \tag{1}$$

where $V_{i,1}$ is the input voltage, d_a is the duty cycle, L is the inductance, and f_s is the switching frequency. The average inductor current, which corresponds to the input current in a boost converter, is given by:

$$I_{L_{avg}} = I_{i,1} = \frac{P_{out}}{V_{i,1}} \tag{2}$$

The RMS current through the inductor, considering both average current and ripple, is:

$$I_{L_{RMS}} = \sqrt{I_{L_{RMS}}^2 + \frac{1}{12}\left(\frac{V_{i,1} d_a}{L f_s}\right)^2} \tag{3}$$

The DC resistance loss in the inductor is:

$$P_{inductor} = I_{L_{RMS}}^2 R_L \tag{4}$$

Additionally, the core losses in the inductor are modeled as the sum of hysteresis and eddy current losses [16]:

979-8-3315-1612-3/25 $31.00 © 2025 IEEE   1232

$$P_{core} = k_h B_{max}^2 f_s V_{core} + k_e (B_{max} f_s)^2 V_{core} \qquad (5)$$

where k_h and k_e are the hysteresis and eddy current coefficients, B_{max} is the peak magnetic flux density, and V_{core} is the core volume.

- *MOSFET Losses*

MOSFET losses are divided into conduction losses due to the on-state resistance $R_{DS(ON)}$ and switching losses during turn-on and turn-off events. The RMS current through the MOSFET is:

$$I_{S_{RMS}} = \sqrt{d_a(I_{L_{min}}^2 + I_{L_{min}} \frac{V_{i,1} d_a}{Lf_s} + \frac{1}{3}(\frac{V_{i,1} d_a}{Lf_s})^2)} \qquad (6)$$

The MOSFET conduction losses for two MOSFETs are given by:

$$P_{MOSFET_{cond}} = 2I_{S_{RMS}}^2 R_{DS(ON)} \qquad (7)$$

The switching losses are modeled as:

$$P_{MOSFET_{sw}} = 2\frac{1}{2}V_{out} I_{i,1}(t_{ON} + t_{OFF})f_s \qquad (8)$$

Thus, the total MOSFET losses are:

$$P_{MOSFET_{total}} = P_{MOSFET_{cond}} + P_{MOSFET_{sw}} \qquad (9)$$

- *Diode Losses*

Diode losses can be obtained similarly, and the equations are as follows:

$$I_{D_{avg}} = I_{i,1}(1 - d_a) \qquad (10)$$

$$I_{D_{RMS}} = \sqrt{(1-d_a)(I_{L_{max}}^2 - I_{L_{max}} \frac{V_{i,1} d_a}{Lf_s} + \frac{1}{3}(\frac{V_{i,1} d_a}{Lf_s})^2)} \qquad (11)$$

$$P_{diode} = 2V_f I_{D_{avg}} \qquad (12)$$

$$P_{rr} = 2V_{i,1} Q_{rr} f_s \qquad (13)$$

- *Capacitor Loss*

The capacitor losses are due to the equivalent series resistance (ESR) and are proportional to the square of the output current. The losses in the capacitors are:

$$P_{capacitor} = 2I_{out}^2 (ESR) \qquad (14)$$

The total power loss in the converter is the sum of all individual losses. The efficiency of UNIPEC is calculated as:

$$\%\eta = \frac{P_{out}}{P_{out} + P_{loss}} \times 100 \qquad (15)$$

III. COMPARISON STUDY

The UNIPEC converter proposed demonstrates superior technical improvements in comparison to current converters, as outlined in references [12-13] and [17-18]. An analysis was performed, specifically examining key design factors such as the quantity of active and passive components, with a particular emphasis on the magnetic core count, power density, and overall efficiency. Comparative data, as summarized in Table II, emphasizes the superior performance of UNIPEC. An important enhancement is the decrease in the quantity of components, which directly affects the power density of the converter. Through careful design, UNIPEC is able to minimize the number of magnetic cores and other passive components. This not only simplifies the circuit topology, but also decreases the overall system footprint. This decrease enables the creation of a smaller and lighter converter without sacrificing performance. The UNIPEC converter demonstrates superior efficiency compared to other MPCs discussed in recent literature. Figure 7 presents a comparison diagram to visually demonstrate these technical advantages. This diagram clearly compares the performance of UNIPEC with other converters, highlighting its superiority.

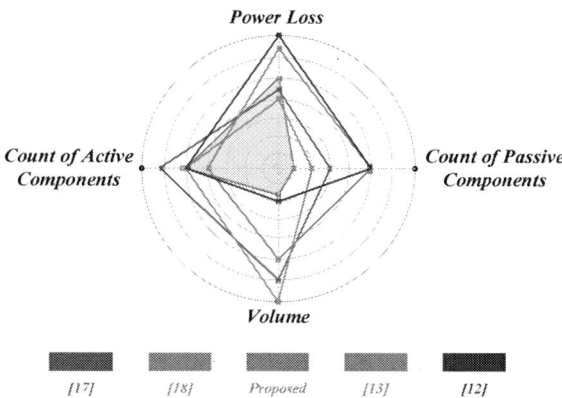

Fig. 7. Comparison diagram: Four main factors

IV. RELIABILITY EVALUATION OF UNIPEC

In this section an overview of the reliability of the proposed UNIPEC along with a reliability evaluation of the power electronic structure is presented. First, reliability equations are presented, and the reliability of the system is calculated using Markov chain. Finally, assuming a test system characteristic, a numerical perspective is illustrated.

TABLE II - COMPARISON AMONG PROPOSED AND TOPOLOGIES IN RECENT LITERATURE

Topology / Parameter	[17]	[18]	[13]	[12]	Proposed
Isolation	Semi	Semi	No	No	No
Application Power Level	≤ 5 kW	≤ 5 kW	≥ 20 kW	≥ 20 kW	≤ 5 kW
# of magnetic cores	4	3	2	3	1
# of capacitors	1	1	4	3	2
# of semiconductors	8	7	6	7	7
Peak Efficiency %	98.2	98.6	95.6	95.1	97.3
Power Density (kW/L)	6.7	5.7	7.9	12	12.6
Application	PV+Battery System	HESSs	FC+Batt Vehicle	Hybrid Streetcar	APU

Failure rate models for components of the proposed UNIPEC are shown where λ_b, π_T, π_A, π_E, π_Q and π_{ES} are basic failure rate, temperature parameter, environment of operation parameter, quality of material and electrical stress parameter, respectively. The most crucial parameter in the failure rate is π_T which depends on the power loss of a component and is calculated using the equations given in Table III. Other parameters can be considered as constant values. Thermal analysis of the components is formulated, using equations. $T_{junction}$ is the junction temperature of the MOSFET and diode. Also, T_{case} and $T_{ambient}$ are symbols for case and ambient temperatures, respectively.

TABLE III. FAILURE RATE MODELS FOR COMPONENTS

Component	Temperature factor
MOSFET	$\pi_T = \exp\left[-1925(\frac{1}{T_{junction}+273} - \frac{1}{298})\right]$
Diode	$\pi_T = \exp\left[-3091(\frac{1}{T_{junction}+273} - \frac{1}{298})\right]$
Inductor	$\pi_T = \exp\left[-1277(\frac{1}{T_{hotspot}+273} - \frac{1}{298})\right]$

Also, the thermal resistances of the case-to-ambient and junction-to-case are shown by ζ_{CA} and ζ_{JC}. For the inductor, $T_{hotspot}$ is the hotspot temperature can be obtained where A is the surface area of the case. The power loss of the components is a function of the current passing through them and some other characteristics of the component.

$$\lambda_{MOSFET} = \lambda_b.\pi_T.\pi_A.\pi_E.\pi_Q.\pi_{ES} \tag{16}$$

$$T_{junction} = T_{case} + \zeta_{JC}P_{Loss} \tag{17}$$

$$T_{case} = T_{ambient} + \zeta_{CA}P_{Loss} \tag{18}$$

$$T_{hotspot} = T_{ambient} + \frac{137.5 \times P_{Loss}}{A} \tag{19}$$

The Markov reliability model is used for reliability modeling of the converter. Fig. 8 shows Markov chain diagrams of the proposed UNIPEC where letters B, X and L stand for main battery, backup (auxiliary) battery and load stages,

respectively. In state BXL, all the interfaces are healthy, and, in each state, any eliminated letter shows the lack of proper functionality in that stage. State N demonstrates that there is no healthy interface. Also, λ is the transitioning rate from one state to another. Equation (20) shows the occupational probabilities of the states. Considering the BXL state as the initial state, the reliability of the proposed structure for supplying the load is the sum of P_{BXL}, P_{XL} and P_{BL} which presented in terms of (21). The value of the λ_s is dependent to the components of the respective stage in the converter. Any failure in the components results in failure of that stage.

In order to have a perspective of the reliability index for the proposed UNIPEC converter, some numerical calculations are presented. It is assumed that a 1kW load is connected to the batteries through UNIPEC. The batteries are in their maximum SOC. The failure rates of components of the system are calculated for scenario A and the reliability equation of the system (R_{sys}) is calculated. Using that, mean time to failure (MTTF) index for the UNIPEC at constant power of 1 kW, becomes as below:

$$MTTF = \int_0^\infty R_{sys}\,dt = 68329.7 \ hours \approx 7.8 \ yrs \tag{22}$$

V. UNIPEC VALIDATION

Simulations of the UNIPEC converter were carried out using PLECS software, specifically tailored for a power level of 1 kW, to validate its operational performance under various conditions. The input voltage levels for both the main and backup batteries are designed to fluctuate between 12 and 24 volts. This range ensures that the duty cycles of the power switches remain below 50%, optimizing the efficiency and reliability of the switching operations. The output voltage, corresponding to the DC-link, is regulated at 24 V, with the capability to increase up to 48 V to accommodate varying load requirements, thereby enhancing the converter's flexibility in different operational scenarios. Key components are selected to ensure high performance and reliability. The AGP4233-474ME inductor, with 470 µH inductance, 12.4 A current rating, and 11.5 mΩ DC resistance, minimizes conduction losses. The IXFQ72N30X3 MOSFET, with 300 V breakdown voltage, 72 A drain current, and 19 mΩ

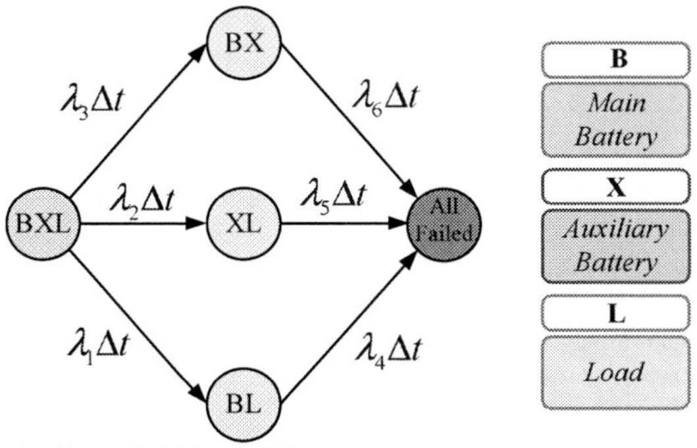

Fig. 8. Markov reliability model

$$\frac{d}{dt}\begin{bmatrix} P_{BXL} \\ P_{BX} \\ P_{XL} \\ P_{BL} \\ P_N \end{bmatrix} = \begin{bmatrix} -\lambda_1-\lambda_2-\lambda_3 & 0 & 0 & 0 & 0 \\ \lambda_1 & -\lambda_4 & 0 & 0 & 0 \\ \lambda_2 & 0 & -\lambda_5 & 0 & 0 \\ \lambda_3 & 0 & 0 & -\lambda_6 & 0 \\ 0 & \lambda_4 & \lambda_5 & \lambda_6 & 0 \end{bmatrix}\begin{bmatrix} P_{BXL} \\ P_{BX} \\ P_{XL} \\ P_{BL} \\ P_N \end{bmatrix} \tag{20}$$

$$R_{sys} = P_{BXL} + P_{XL} + P_{BL} = \left[1 - \frac{1}{\lambda_1+\lambda_2+\lambda_3-\lambda_5} - \frac{1}{\lambda_1+\lambda_2+\lambda_3-\lambda_6}\right]e^{-(\lambda_1+\lambda_2+\lambda_3)t} + \frac{1}{\lambda_1+\lambda_2+\lambda_3-\lambda_5}e^{-\lambda_5 t} + \frac{1}{\lambda_1+\lambda_2+\lambda_3-\lambda_6}e^{-\lambda_6 t} \tag{21}$$

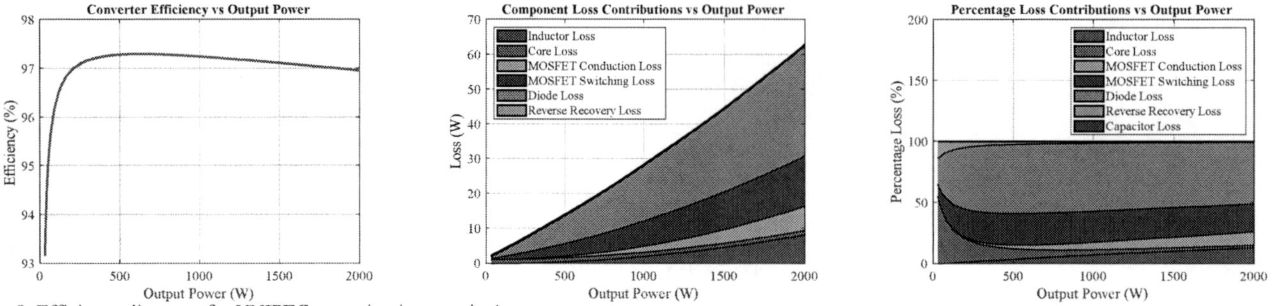

Fig.9. Efficiency diagrams for UNIPEC operating in scenario A

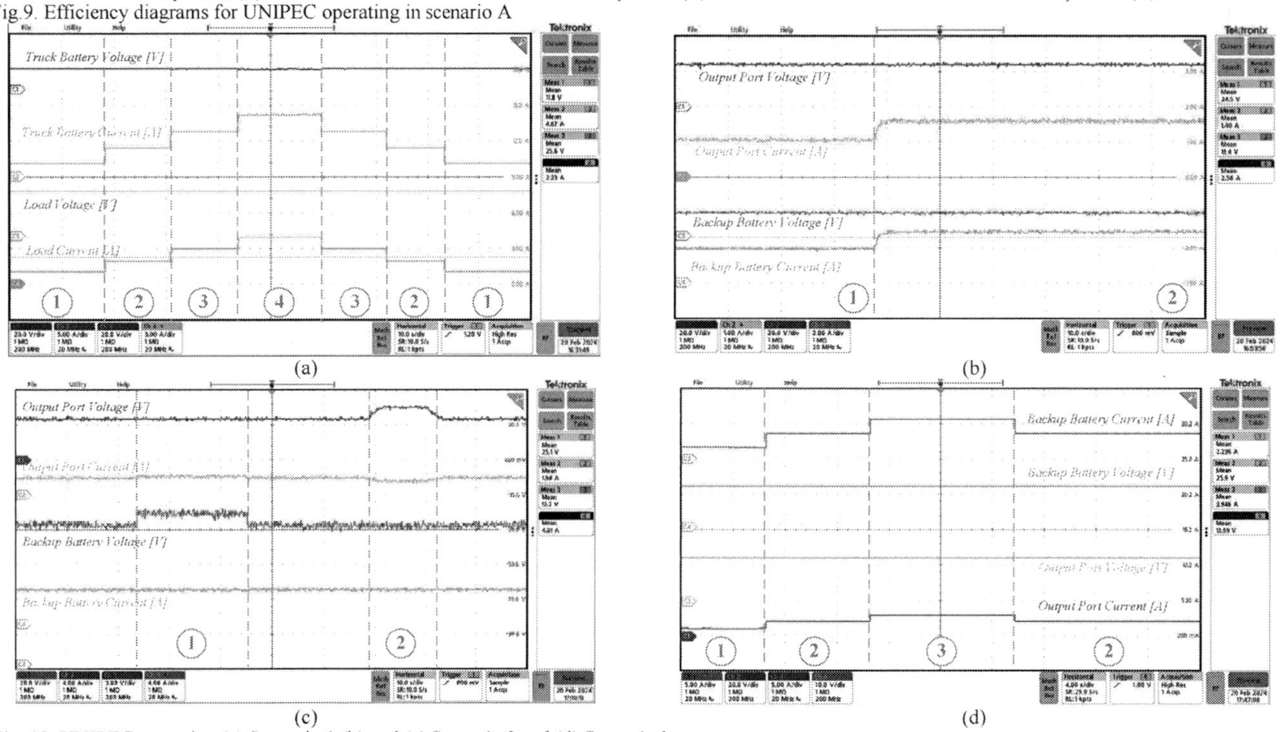

Fig. 10. UNIPEC operation (a) Scenario 1 (b) and (c) Scenario 3 and (d) Scenario 2

Fig.11. 1 kW UNIPEC designed PCB

TABLE IV - 1 KW PROTOTYPE SPECIFICATIONS

Spec/Comp	Value/Type	Spec/Comp	Value/Type
V_{i1}, V_{i2}	12 - 24 V	C_{o1}, C_{o2}	300 V – 470 µF
V_o	24 - 48 V	Switches	IXFQ72N30X3
L	AGP4233-474ME	Diodes	SBR60A300CT

$R_{DS(ON)}$, offers low conduction losses and fast switching (25 ns rise, 11 ns fall) at 100 kHz. The SBR60A300CT Schottky diode, rated for 60 A forward current and 300 V reverse voltage, reduces switching losses with a 0.94 V forward drop and 50 ns reverse recovery time. Output capacitors (470 µF, 2 mΩ ESR) smooth voltage ripple and further minimize losses. Effective thermal management through heatsinks ensures safe operation under high loads. This design achieves higher efficiency and reliability and meets the demands of applications requiring up to 1 kW of output power. The thermal behavior of MOSFETs, under various load conditions, was thoroughly considered to

obtain precise efficiency values, including peak efficiency, which is documented in Fig. 9 for scenario A. The simulations also consider the equivalent series resistance (ESR) values of the passive components, which significantly impact the overall efficiency and performance of the converter, especially under high-power conditions. In addition to simulations, low-power experimental tests were conducted using a 100 W prototype to validate the converter's dynamic behavior and operational scenarios. The results of these tests are presented in Fig.10. Fig.10(a) illustrates the waveforms and dynamic response of the UNIPEC converter during Scenario A, where the main battery supplies power to the load. In Scenario C, where the backup battery is charged, the tests were performed in both constant current (CC) and constant voltage (CV) modes during the step-down function of UNIPEC. The corresponding results are depicted in Figures 10(b) and 10(c). Notably, time period 2 in Fig. 8(c) demonstrates the stability of the output voltage, where fluctuations are effectively mitigated, ensuring that the battery is charged without adverse effects. Figure 10(d) showcases the dynamic waveforms for Scenario B, where the secondary battery powers the load, highlighting the converter's ability to manage transitions and maintain stable operation. To further extend the study, a 1 kW lab-scale prototype of the UNIPEC converter is currently under development and will be tested for the future extension of this research paper. The design and layout of the UNIPEC's PCB, which was also utilized in power density calculations, are presented in Figure 11. The detailed parameters and component specifications for this PCB are provided in Table IV.

VI. CONCLUSIONS

A three-port DC-DC converter named UNIPEC was presented in order to be utilized in commercial truck APUs. This converter is distinguished by its low component count, high efficiency, and high power density. The proposed UNIPEC is a singular magnetic core topology that is realized through the integration and modification of switched capacitor technology and a bidirectional multi-level DC-DC converter. The proposed converter demonstrated a peak simulation efficiency of 97.3% and a power density of 12.6 kW/L, which made it an appealing choice for applications involving conventional and refrigerated freight transportation. The presentation included both a comparative study as well as the findings of a low-power experimental validation. Following the completion of a reliability analysis of UNIPEC, which was documented, the expected mean time to failure was determined to be 7.8 years under scenario A operation conditions. In order to fully validate the system's performance, a printed circuit board (PCB) with a power rating of 1 kW has been designed and is currently under the development for future publications.

REFERENCES

[1] A. Hasnain, A. S. Gamwari, R. Resalayyan, K. F. Sadabadi and A. Khaligh, "Medium and Heavy Duty Vehicle Electrification: Trends and Technologies," in *IEEE Transactions on Transportation Electrification*, doi: 10.1109/TTE.2024.3444692.

[2] T. Gechev, "Progress in fuel cell usage as an auxiliary power unit in heavy-duty vehicles," *AIP Conf. Proc.* 2024; 3104 (1): 020006. doi: https://doi.org/10.1063/5.0198806

[3] Volodarets, M., I. Gritsuk, I. Taran, V. Volkov, M. Bulgakov, and M. Izteleuova. "Features of Modernization of A Truck With A Hybrid Power Transmission." *Natsional'nyi Hirnychyi Universytet. Naukovyi Visnyk* 1, 2023: 80-87.

[4] M. Dellermann, O. Gehring and O. Zirn, "Optimal Control of a Multi Voltage Powernet with Electrified Auxiliaries in Hybrid-Electric Trucks," *2020 IEEE International Conference on Environment and Electrical Engineering and 2020 IEEE Industrial and Commercial Power Systems Europe (EEEIC / I&CPS Europe)*, Madrid, Spain, 2020, pp. 1-6, doi: 10.1109/EEEIC/ICPSEurope49358.2020.9160600.

[5] B. Pregelj, A. Debenjak, G. Dolanc and J. Petrovčič, "A Diesel-Powered Fuel Cell APU—Reliability Issues and Mitigation Approaches," in *IEEE Transactions on Industrial Electronics*, vol. 64, no. 8, pp. 6660-6670, Aug. 2017, doi: 10.1109/TIE.2017.2674628.

[6] Y. Cho and J. -S. Lai, "High-Efficiency Multiphase DC–DC Converter for Fuel-Cell-Powered Truck Auxiliary Power Unit," in *IEEE Transactions on Vehicular Technology*, vol. 62, no. 6, pp. 2421-2429, July 2013, doi: 10.1109/TVT.2012.2227522.

[7] P. K. Pedapati, K. Murugan and A. H. Chander, "Reconfigurable Multiswitch Fault-Tolerant Multiport Converter," in *IEEE Transactions on Power Electronics*, vol. 39, no. 10, pp. 13274-13284, Oct. 2024, doi: 10.1109/TPEL.2024.3428852.

[8] R. Wang, H. Wang, W. Wang, S. Zhang and X. Wang, "Wide-Voltage Input Photovoltaic Microgrid System Based on Multi-Port Three-Level Converter," in *IEEE Access*, vol. 12, pp. 165678-165691, 2024, doi: 10.1109/ACCESS.2024.3491942.

[9] K. Suresh et al., "A Multifunctional Non-Isolated Dual Input-Dual Output Converter for Electric Vehicle Applications," in IEEE Access, vol. 9, pp. 64445-64460, 2021, doi: 10.1109/ACCESS.2021.3074581.

[10] F. Kardan, R. Alizadeh and M. R. Banaei, "A New Three Input DC/DC Converter for Hybrid PV/FC/Battery Applications," in IEEE Journal of Emerging and Selected Topics in Power Electronics, vol. 5, no. 4, pp. 1771-1778, Dec. 2017, doi: 10.1109/JESTPE.2017.2731816.

[11] B. Faridpak, M. Farrokhifar, M. Nasiri, A. Alahyari and N. Sadoogi, "Developing a super-lift luo-converter with integration of buck converters for electric vehicle applications," in CSEE Journal of Power and Energy Systems, vol. 7, no. 4, pp. 811-820, July 2021, doi: 10.17775/CSEEJPES.2020.01880.

[12] P. Zolfi, S. Vahid and A. EL-Refaie, "A Novel Non-Isolated Multi-Port DC-DC Converter for Hybrid Streetcar Application," IECON 2021 – 47th Annual Conference of the IEEE Industrial Electronics Society, Toronto, ON, Canada, 2021, pp. 1-7, doi: 10.1109/IECON48115.2021.9589573.

[13] P. Zolfi, S. Vahid and A. EL-Refaie, "Development of A Family of High Voltage Gain Step-Up Multi-Port DC-DC Converters for Fuel Cell-based Hybrid Vehicular Power Systems," 2022 24th European Conference on Power Electronics and Applications (EPE'22 ECCE Europe), Hanover, Germany, 2022, pp. 1-11.

[14] S. Li, K. M. Smedley, D. R. Caldas and Y. W. Martins, "Hybrid Bidirectional DC–DC Converter With Low Component Counts," in *IEEE Transactions on Industry Applications*, vol. 54, no. 2, pp. 1573-1582, March-April 2018, doi: 10.1109/TIA.2017.2785760.

[15] S. Dusmez, A. Hasanzadeh and A. Khaligh, "Comparative Analysis of Bidirectional Three-Level DC–DC Converter for Automotive Applications," in *IEEE Transactions on Industrial Electronics*, vol. 62, no. 5, pp. 3305-3315, May 2015, doi: 10.1109/TIE.2014.2336605.

[16] E.C. Snelling, Soft Ferrites: Properties and Applications, 2nd ed., London: Butterworths, 1988.

[17] S. Vahid and A. EL-Refaie, "A Novel Semi-Isolated Three-Port dc-dc Power Converter with Soft Switching Technique for Hybrid Energy Storage Applications," 2021 22nd IEEE International Conference on Industrial Technology (ICIT), 2021, pp. 278-284, doi: 10.1109/ICIT46573.2021.9453694.

[18] S. Vahid, M. Abarzadeh, N. Weise and A. EL-Refaie, "A Novel Three-Port dc-dc Power Converter with Adaptive Boundary Current Mode Controller for a Residential PV-Battery System," IECON 2020 The 46th Annual Conference of the IEEE Industrial Electronics Society, 2020, pp. 4263-4268, doi: 10.1109/IECON43393.2020.9255022.

Investigation of electrical safety for non-isolated single-phase on-board chargers used in BEV/PHEV

Soya Kataoka
Department of electrical and mechanical engineering
Nagoya Institute of Technology
Aichi, Japan
s.kataoka.859@stn.nitech.ac.jp

Shohei Funatsu
Department of electrical and mechanical engineering
Nagoya Institute of Technology
Aichi, Japan
s.funatsu.027@stn.nitech.ac.jp

Hiroaki Matsumori
Department of electrical and mechanical engineering
Nagoya Institute of Technology
Aichi, Japan
matsumori.hiroaki@nitech.ac.jp

Takashi Kosaka
Department of electrical and mechanical engineering
Nagoya Institute of Technology
Aichi, Japan
kosaka@nitech.ac.jp

Keisuke Nakamura
Aisin corporation
Aichi, Japan
keisuke.nakamura@aisin.co.jp

Subrata Saha
Aisin corporation
Aichi, Japan
subrata.saha@aisin.co.jp

Abstract— This paper proposes a non-isolated single-phase onboard charger (OBC) for use in EV/PHEVs to meet electrical safety which is mainly caused by common mode currents. The non-isolated OBC must not only pass the touch current test (human body impedance voltage drop less than 250 mVrms) but must also prevent the residual current (RCD) device trip (trip current is 30 mA) by large ground leakage current due to the low common mode impedance compared to the galvanic isolated OBC. Therefore, the authors use a T-type PFC and active power decoupling circuit. The effectiveness of human impedance voltage drop reduction and ground leakage current reduction is verified through suitable experiments. These values are confirmed as under the limit value defined by the relevant standard.

Keywords—component, on-board charger, human impedance voltage, ground leakage current, T-type PFC, active power decoupling

I. INTRODUCTION

The onboard charger (OBC) used in BEV/PHEV requires high-power density, high efficiency, and low battery charging current ripple [1]. Since a relevant standard for OBC such as UL 2202 [2] and IEC 61851 [3] do not require galvanic isolation and the non-isolated system has a potential to enhance power density and efficiency, various research has been carried out on non-isolated OBC over the recent years. However, the non-isolated OBC has tackled electrical safety issues such as human body touch current and residual current devices (RCDs) trip of ground leakage current which is caused by low common-mode (CM) impedance compared to isolated OBC [4,5]. There are several papers which have described it for three-phase systems [6,7], but there is still little discussion of single-phase systems. Therefore, the authors propose T-type based non-isolated OBC as shown in

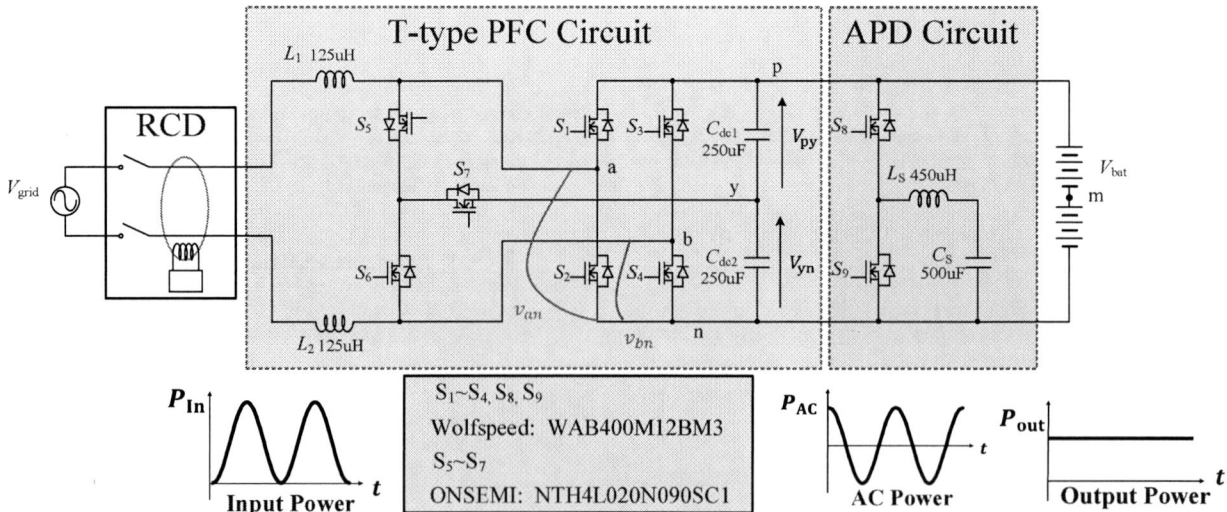

Fig. 1: T-type based non-isolated OBC.

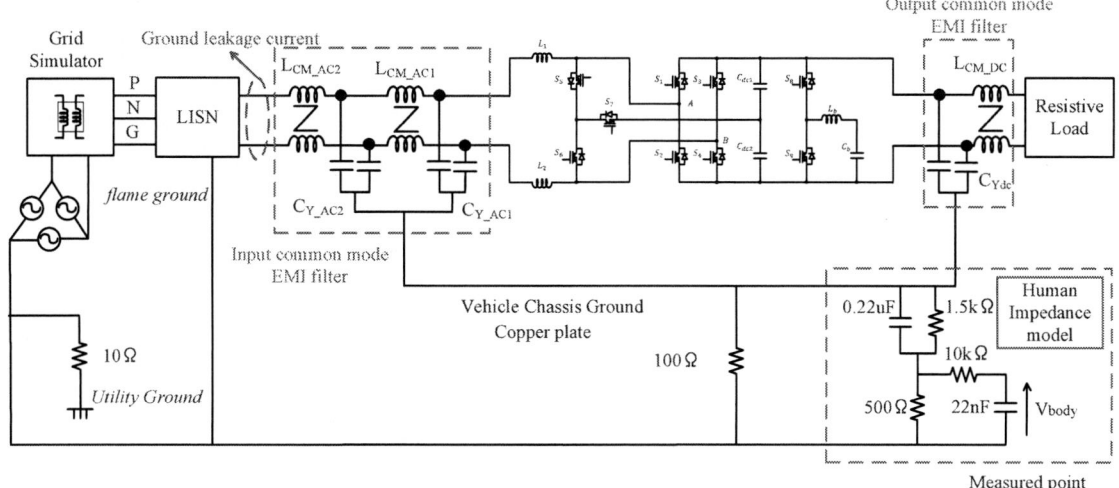

Fig.2 Electrical safety test circuit which simulating a vehicle.

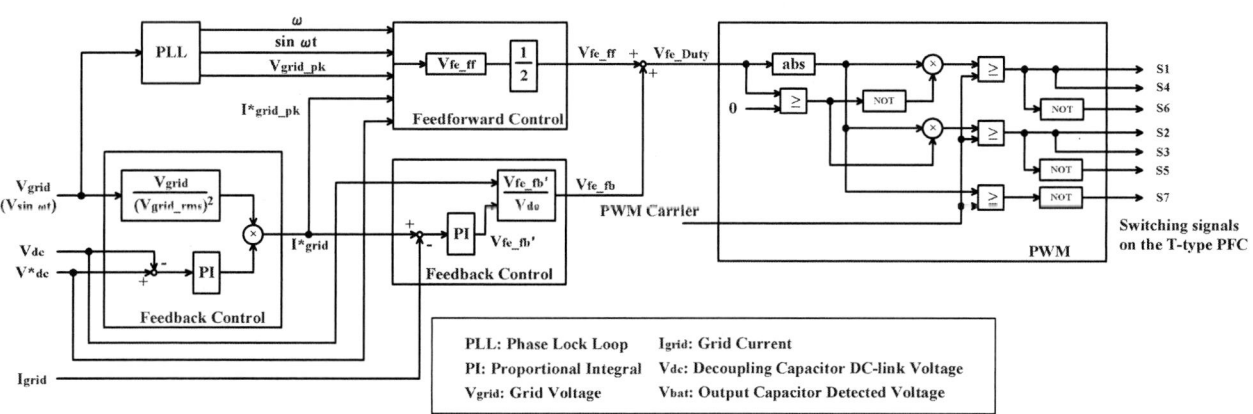

Fig. 3(a): Control block for the T-type PFC circuit

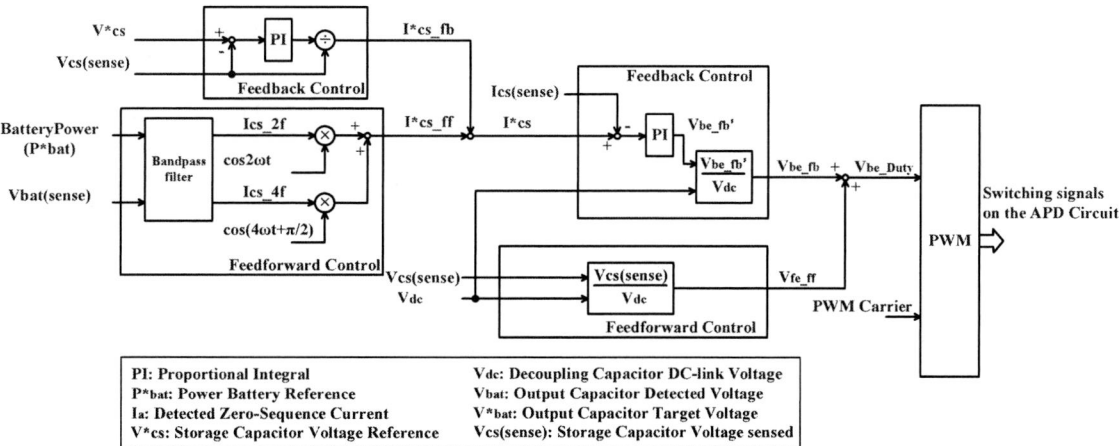

Fig. 3(b): Control block for the APD circuit

979-8-3315-1612-3/25 $31.00 © 2025 IEEE

Fig. 1 and investigate electrical safety under test conditions which simulating a vehicle as shown in Fig. 2. The contribution of this paper is that the proposed single-phase non-isolated OBC is able to meet the basic requirements of the utility grid while also meeting the electrical safety issues.

II. MECHANISM OF GROUND LEAKAGE CURRENT GENERATION

Electrical safety issues under non-isolated OBC are caused by ground leakage current due to low common mode impedance [8]. In the non-isolated single-phase OBC system, the common mode voltage, v_{CM}, in Fig. 1 at the input boost inductor $L_1 = L_2$ is expressed as follows:

$$v_{CM} = \frac{v_{an} + v_{bn}}{2} \tag{1}$$

Where v_{AN} and v_{BN} are the voltages of inverter terminal legs a and b to negative terminal n.

Therefore, if the fluctuation of the common mode voltage, v_{CM}, is suppressed, the ground leakage current will also be suppressed [8].

III. PROPOSED CIRCUIT

According to Ref. [9], the T-type-based PFC converter is a suitable solution for reducing the ground leakage current because the common-mode voltage is flattened without high-frequency pulses due to the active clamping method. Therefore, the midpoint voltage, v_{yn}, should be stabilized by a T-type switch control, external voltage balancer, connecting midpoint voltage, y, and battery midpoints, m, and so on.

However, the twice grid frequency battery current ripple still exists, and it will deteriorate the battery lifetime. Therefore, the authors combine the T-type PFC converter and an active power decoupling circuit as shown in Fig. 1. The control blocks of the proposed converter are shown in Fig. 3. The active power decoupling circuit is operated as DCM control to reduce inductance value and switching loss.

Table 1. EMI filter parameters.

Components	Descriptions	Values
$L_{CM\_AC1,2}$	AC side loop CM inductances	3 mH
$C_{Yac1,2}$	AC side loop Y capacitances	0.5 nF , 1.5 nF
L_{CM_DC}	DC side loop CM inductances	2 mH
C_{Ydc}	DC side loop Y capacitance	0.47 uF

Table 2. Experimental conditions for proposed OBC.

Experimental conditions	Values
Output power	1.8 kW
AC Grid supply voltage (V_{grid})	200 Vrms
Output voltage (V_{bat})	300 V
Storage capacitor (C_s) voltage reference (V_{cs}^*)	200 V
Switching frequency of the T-type PFC circuit	50 kHz
Switching frequency of the active power decoupling circuit	15 kHz
Control Frequency	100 kHz

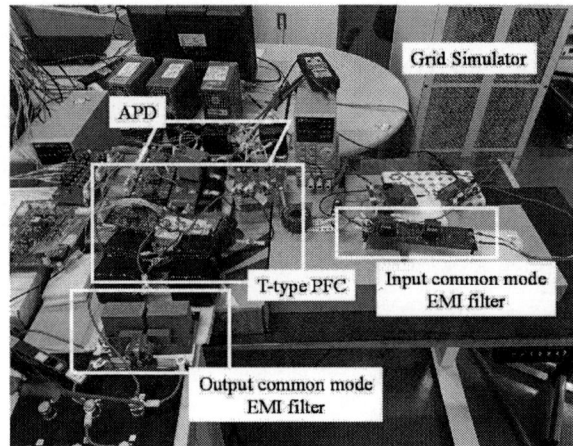

Fig. 4 The experimental setup.

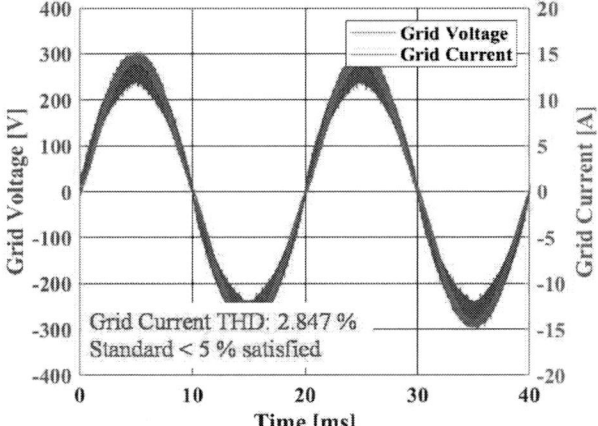

Fig. 5 Experimental result of the input grid current.

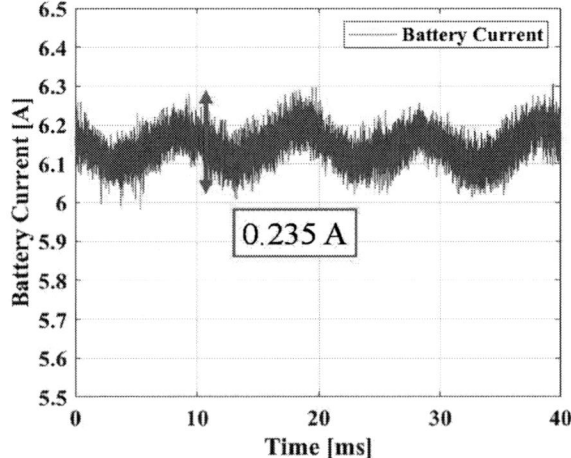

Fig. 6: Experimental result of the output load current.

IV. EXPERIMENTAL RESULT

In order to verify the ground leakage current and human body voltage reduction for electrical safety by the proposed T-type

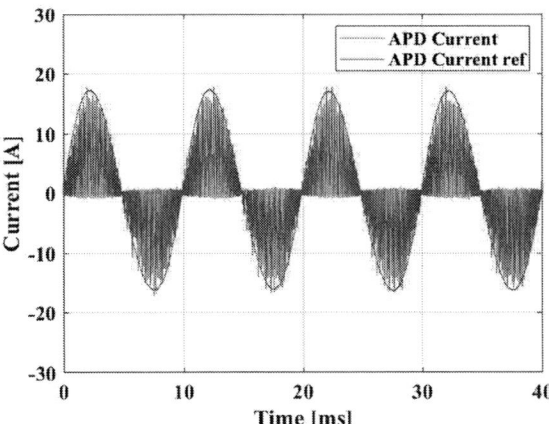

Fig. 7: Experimental result of the discontinuous storage capacitor current.

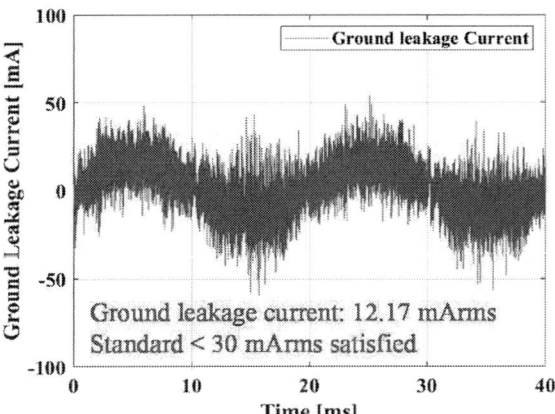

Fig. 8: Experimental result of the ground leakage current.

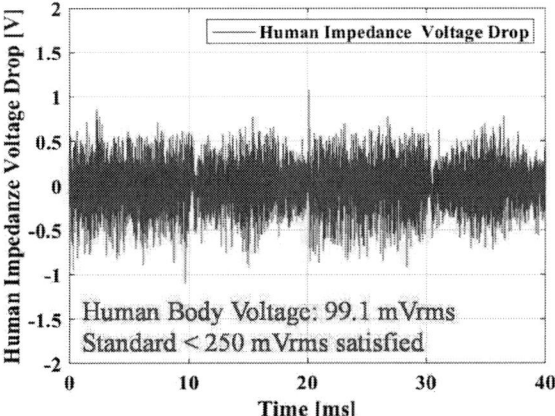

Fig. 9: Experimental result of the human body voltage.

based converter, the experiment is conducted under a 1.8 kW system. The experimental setup is shown in Fig. 4 and the parameters including the EMI filter are shown in Tables 1 and 2.

First, the basic performance required of OBCs is reviewed. The grid voltage and current waveforms are shown in Fig. 5 and the battery charging current waveform in Fig. 6. The grid current was well controlled and the THD value was only 2.85 %, which is below the Japanese power utility standard (the THD is less than 5 %). Furthermore, the peak-to-peak battery current ripple is only 0.235 A, which is thanks to the APD circuit absorbing the double-frequency power pulsation by DCM control as shown in Fig. 7.

Then, the ground leakage current and human body voltage for electrical safety are reviewed. The ground leakage current and human body voltage waveforms are shown in Figs 8 and 9. The experimental results show that the RMS value of the ground leakage current is 12.17 mArms and the human body voltage value is 99.1 mVrms. Both the ground leakage current value and the human body voltage value are within the acceptable range for relevant standards (ground leakage current value > (30 mArms) and human body voltage value < (250 mVrms) in UL 2202 and the touch current level is considered safe in UL2202).

V. CONCLUSION

This paper proposes the non-isolated on-board charger which consists of the T-type PFC and active power decoupling circuit to address safety issues. The actual device tests show that proper control of the T-type circuit stabilizes the common-mode voltage and addresses the harmonic ground leakage current and human body voltage issues.

VI. ACKNOWLEDGMENT

The authors would like to thank Dr. D. Menzi, Prof. J. Huber, and Prof. J. W. Kolar from ETH Zurich, Switzerland for their discussion and advice on non-isolated systems.

REFERENCES

[1] S. S. G. Acharige, M. E. Haque, M. T. Arif, N. Hosseinzadeh, K. N. Hasan and A. M. T. Oo, "Review of Electric Vehicle Charging Technologies, Standards, Architectures, and Converter Configurations," in IEEE Access, vol. 11, pp. 41218-41255, 2023.

[2] UL2202: Standard for safety: Electric vehicle (EV) charging system equipment, UL Std. 2202, 2012.

[3] IEC61851: Electric vehicle conductive charging system - Part 1: General requirements, IEC Std. 61 851-1, 2017.

[4] J. Wang et al., "Nonisolated Electric Vehicle Chargers: Their Current Status and Future Challenges," in IEEE Electrification Magazine, vol. 9, no. 2, pp. 23-33, June 2021.

[5] D. Zhang, D. Cao, J. Huber, J. Everts and J. W. Kolar, "Non-isolated Three-Phase Current DC-Link Buck–Boost EV Charger With Virtual Output Midpoint Grounding and Ground Current Control," in IEEE Transactions on Transportation Electrification, vol. 10, no. 1, pp. 1398-1413, March 2024.

[6] D. Zhang, C. Leontaris, J. Huber and J. W. Kolar, "Optimal Synergetic Control of Three-Phase/Level Boost–Buck Voltage DC-Link AC/DC Converter for Very-Wide Output Voltage Range High-Efficiency EV

Charger," in IEEE Journal of Emerging and Selected Topics in Power Electronics, vol. 12, no. 1, pp. 28-42, Feb. 2024.

[7] L. Zhou, M. Eull, W. Wang, G. Cen and M. Preindl, "Design of Transformerless Electric Vehicle Charger with Symmetric AC and DC Interfaces," 2021 IEEE Applied Power Electronics Conference and Exposition (APEC), Phoenix, AZ, USA, 2021, pp. 2769-2774.

[8] Wuhua Li, Yunjie Gu, Haoze Luo, Wenfeng Cui, Xiangning He, Changliang Xia, "Topology Review and Derivation Methodology of Single-Phase Transformerless Photovoltaic Inverters for Leakage Current Suppression" IEEE Transactions on Industrial Electronics, VOL. 62, NO. 7, July. 2015.

[9] Ye Mei, Senjun Hu, Lei Lin, Wuhua Li, Xiangning He, Fengwen Cao," Highly efficient and reliable inverter concept based transformerless photovoltaic inverters with tri-direction clamping cell for leakage current elimination," IET Power Electronics, 2016.

An 8-Level Flying Capacitor Multilevel Converter for Electric Aircraft Pulse Deicing

Nicole Stokowski, Andrew Freeman, Aidan Rodgers, Aria Delmar, Jonathan Sengstock, Alex Solecki, and Andrew Stillwell
Department of Electrical and Computer Engineering, University of Illinois Urbana Champaign
Email: nicole25@illinois.edu

Abstract—Commercial and small personal aircraft use a variety of methods to remove and/or prevent ice buildup on plane wings. Recent research into "pulse deicing" techniques on small passenger aircraft has gained traction. Pulse deicing calls for controlled power systems that can provide the necessary energy to remove accumulated ice. This work looks at the design of a 3 kW, 8-level flying capacitor multilevel (FCML) converter, with over 99% efficiency, that can be used as a controlled energy buffer between the aircraft battery and the deicing mechanism. The prototype developed in this work lends itself to an extremely lightweight design (15 kW/kg), with no need for a heat-sink, due to good thermal performance.

Index Terms—FCML converter, Pulse deicing, GaN

I. INTRODUCTION

Industry and academia find themselves in the midst of a push towards electric solutions to many of the world's needs, transportation being one of the most prevalent. The aviation industry is no exception [1]. Many onboard systems are transitioning to full electrification and these changes can have resounding effects on other subsystems, such as deicing infrastructure.

Ice formation on plane wings and other critical surfaces is a hazard in flight due to negative impacts on aerodynamic lift [2], [3]. Internal combustion engine (ICE) aircraft naturally have a means of removing this ice by utilizing hot, pressurized, air generated in the engine. This high-pressure, hot air, often referred to as "bleed air," can be piped to the wings and other critical parts of the aircraft for cabin heating, wing heating [4], or boot inflation [5]. Fully electric aircraft will not have access to this excess air, therefore alternate heating/deicing solutions must be explored for these applications. It should be noted that liquid-based solutions, such as ethylene glycol-water mixtures, do not rely on compressed hot air, and therefore are not reliant on ICE infrastructure. Unfortunately, these liquid-based systems are heavy and imprecise, in addition to being only a preventative measure.

Resistive heating presents an electric alternative. Research into resistive heating solutions has been around for a while and is suitable for many applications [6], including aircraft deicing [7]. For the aerospace sector in particular, resistive heating is desirable, as it relies on onboard electric systems and trades the weight of heavy air compressors or liquid tanks with the weight of converters, cabling, and control. Overall, there are fewer moving parts in an electrical system, making it more reliable than conventional mechanical methods (e.g. deicing boots). In addition, electrical-based solutions can be

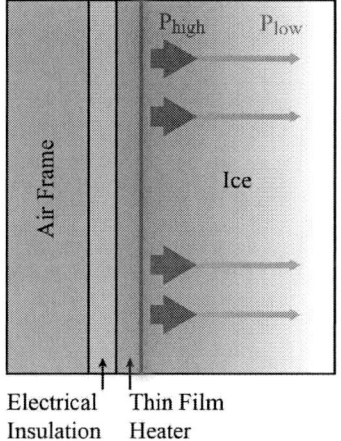

Fig. 1: Illustration of the cross-section of a wing with applied heater material and insulation layers, showcasing the difference between steady heating (P_{low}) and pulse deicing (P_{high})

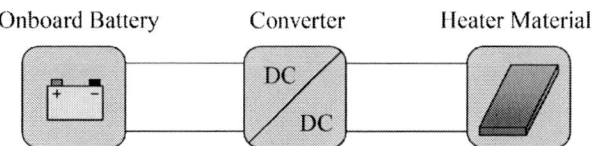

Fig. 2: Pulse deicing system architecture

finely controlled and modulated to target specific areas of the plane without wasting energy or having to carry energy in reserve, which tank-based systems, such as TKS, suffer from.

There are multiple approaches to resistive heating. Some heating solutions opt for low power over a longer period of activation time, while other solutions apply high power for a short duration in order to melt the inner layer of ice. The findings in [8] show that the latter pulsed high-power approach, paired with the use of super-hydrophobic materials, has the potential to remove a given volume of ice using less energy than steady heating solutions. This is the deicing method that will be considered in this paper—referred to as "pulse deicing". Pulse deicing utilizes high temperatures at the interface between the ice and heater layers to completely melt the innermost layer of ice, allowing wind shear and gravity to remove the remaining ice. In comparison, steady resistive heating does not generate enough heat to fully melt the ice initially. Rather, heat generated at the interface diffuses

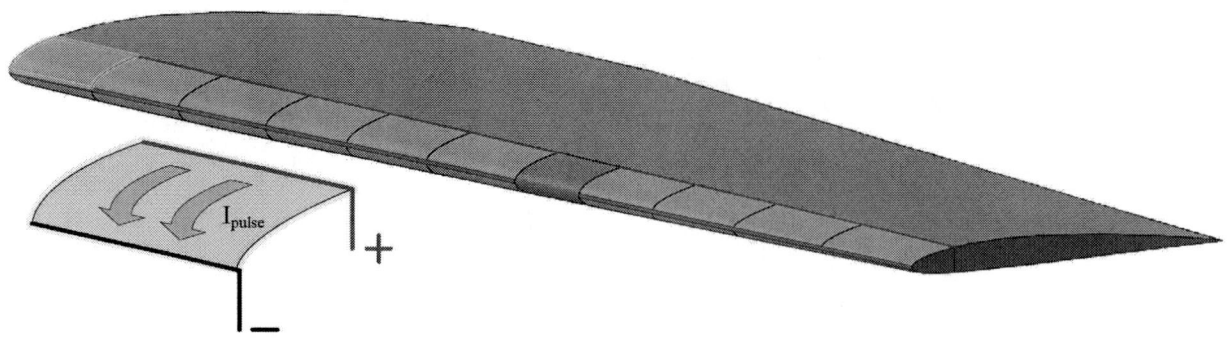

Fig. 3: Isometric view of Cessna 337 airfoil CAD model with proposed heater material layout

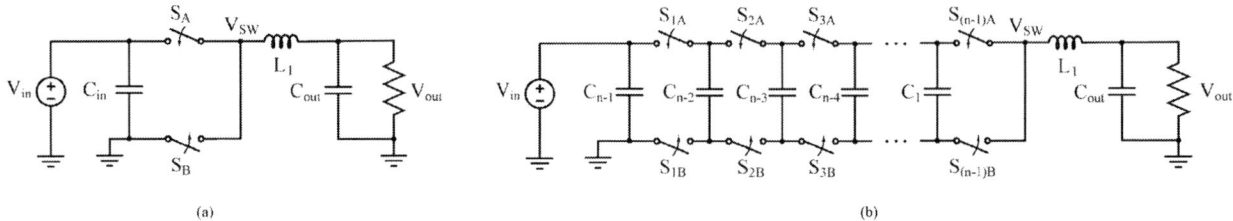

Fig. 4: (a) 2-level FCML converter (b) n-level FCML converter

through the layers of ice, slowly raising the temperature of the entire mass until the ice has melted. The difference between these heating methods is illustrated in Figure 1. As a result, the pulsed approach should require a constant amount of energy to deice regardless of thickness, whereas the energy required for steady heating will vary with thickness. This fixed energy usage leads to energy savings that are especially desirable in the aerospace sector. Electrified aircraft derive their energy from onboard batteries, which are currently less dense than aviation fuels [9], [10]: every kWh off added energy is added weight.

An example application of the heater material on a single Cessna 337 wing is shown in Figure 3. The converter presented in this work is designed to interface with this heater load and the plane's onboard Li-ion battery to facilitate pulse deicing (Figure 2). Considerations for the converter design as well as motivation for the inclusion of the converter stage are provided in the following section.

II. FCML THEORY AND MOTIVATION

The purpose of the dc dc converter stage is to provide an intermediate bus for voltage/current conversion, as well as regulation, between the onboard battery and the heater material placed along the wings. In addition to facilitating regulated discharge of the battery, which may help prolong battery life [11], the converter stage adds a level of control to the deicing architecture that can adapt to the power wants and needs of the system (e.g. changes in aircraft bus voltage, flight profile, energy requirements, heater material electrical properties, etc). By making the system adaptable, the converter is able to interface with different deicing materials and power sources without altering the architecture drastically.

One of the practical constraints in aviation is payload, or maximum weight of onboard systems. In electrical systems, a majority of the system weight consists of bulky devices, like inductors, and thermal management solutions (e.g. heat sinks). Therefore, for weight sensitive applications like aerospace, the size/use of these system components should be minimized. The FCML converter [12], [13] exhibits favorable magnetics sizing, as well as thermal performance, and has been shown to be a viable medium voltage, high power density, solution [14], [15]. With this in mind, the FCML converter was chosen as the topology for this application.

The FCML buck converter is an n-level buck converter where the "levels" are defined by $n-2$ intermediate voltage levels maintained by "flying" capacitors ($C_1 - C_{n-2}$), as shown in Figure 4b. For a given conversion ratio and input/output voltage, the n-level FCML converter, compared to the standard 2-level converter (Figure 4a), sees reduced blocking voltage (maximum voltage across drain to source). Switches S_A and S_B in the 2-level converter see a switch voltage stress of V_{in} meanwhile, the switches in an n-level FCML buck converter each see $\frac{V_{in}}{n-1}$. In high voltage applications, this voltage division among levels presents many benefits. The devices used for $S_{1A}, S_{1B} - S_{(n-2)A}, S_{(n-2)B}$ can have significantly lower voltage rating, opening up the design space to smaller, more efficient, switching technology such as GaN. Furthermore, the FCML converter experiences distributed losses among its levels; power that would normally be dissipated in a single device is distributed amongst the smaller, more efficient devices. Heat is dispersed across a wider area, mitigating the need for heatsinks or cooling loops, which add to system weight.

The FCML converter has a higher effective switching frequency (1) and reduced voltage ripple at the switch node than a conventional 2-level converter (for the same conversion ratio) due to it's use of phase-shifted PWM (PSPWM) and intermediate floating voltage levels. In (1), $f_{sw(eff)}$ is the switching frequency seen at the switch node, and f_{sw} is the switching frequency. This high effective switching frequency enables the use of a comparatively smaller filtering inductor, which is often the heaviest component. The equation for inductor current ripple is given in (2). Here, D_{eff} is the "effective duty ratio" seen by the inductor [16]. It is important to note that there is a factor of $(n-1)^2$ in the denominator of this expression, which leads to our beneficial magnetics sizing. A single factor of $(n-1)$ comes from the frequency multiplication we get from use of PSPWM and the other factor of $(n-1)$ comes from the reduced voltage ripple at the switch node.

$$f_{sw(eff)} = (n-1)f_{sw} \qquad (1)$$

$$\Delta i_L = \frac{V_{in}(D_{eff}(1-D_{eff}))}{Lf_{sw}(n-1)^2} \qquad (2)$$

There are drawbacks to using the FCML converter. Namely, the FCML converter uses $2*(n-1)$ switches, resulting in more points of failure than a lower level-count converter. For this reason, special care must be taken to ensure safe and reliable operation of the FCML converter.

Switch failure is often a result of applied voltage exceeding a switch's rating. Switches are especially susceptible to an over voltage condition during switching transitions, as well as during initial startup and shutdown of the converter. During switching transitions, excessive ringing or voltage ripple can occur across the V_{DS} of the switches. This high frequency ringing can be shunted away from the switches with the help of local decoupling flying capacitors, C_{local}, placed as close to the switches as physically possible (Figure 5). Meanwhile, sufficiently large bulk flying capacitors, C_{bulk}, ensure small voltage ripple at the effective switching frequency. Fundamentally, the high frequency ringing seen at a transition is a result of having parasitic inductance in the switching cell

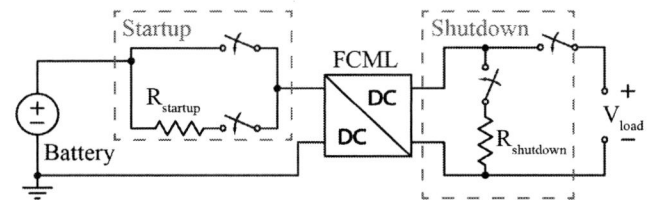

Fig. 6: Simplified schematic of startup and shutdown circuits connected to the 8-level FCML converter

commutation loop (Figure 5). Reduction in commutation loop area, A_{loop}, helps to minimize this parasitic inductance and subsequently the observed overshoot.

During startup, the converter's flying capacitors all start at 0 V. Unregulated, these capacitors individually will charge reasonably quickly when they are connected to high voltage, potentially leading to a case where the switches will see more than their blocking voltage (these switches are not rated to block full input voltage). A similar event can occur during shutdown if any of the capacitors are discharged faster than others. To prevent this from happening, startup and shutdown circuits are needed to control inrush and outrush current, regulating steady charging and discharging of all flying capacitors together. These regulating circuits are shown in Figure 6 [17]. The startup circuit utilizes temporary resistance in the power path to control inrush current at startup, which is switched out when the capacitors are fully charged. The shutdown circuit facilitates safe discharging of the flying capacitors, by either shunting current through the load or through a set of power resistors parallel to the load (if the load is not an option). The shutdown circuit is enabled once the converter is no longer in use and/or in an emergency shutdown situation [18].

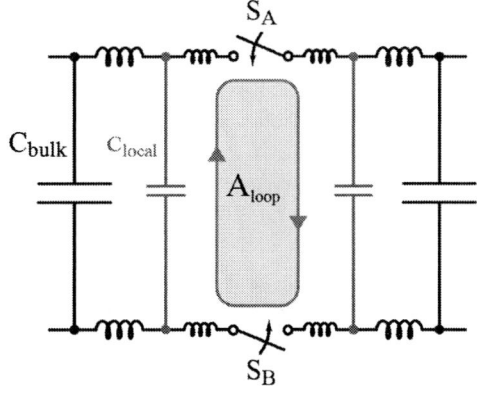

Fig. 5: Switching cell with local decoupling and bulk capacitors labeled

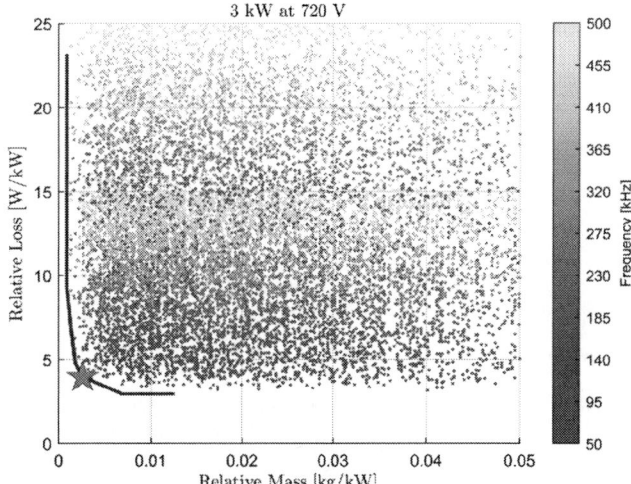

Fig. 7: Relative losses vs. relative mass of 25,000 randomly generated FCML converters. The green star labels an 8-level design at the knee of the Pareto front

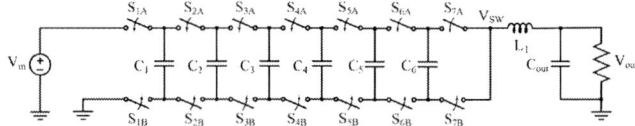

Fig. 8: 8-level FCML converter schematic

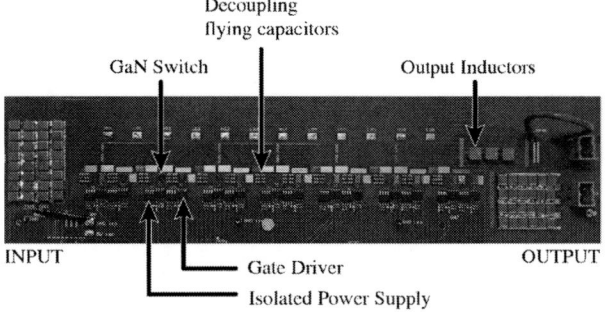

Fig. 9: Annotated power stage of prototype 8-level FCML converter

Component	PartNumber
Switch	EPC 2034C
Gate Driver	Si8271
Power Isolator	ADUM5240
Local Flying Capacitor	C2012X7T2W473K125AA
Bulk Flying Capacitor	C5750X6S2W225K250KA
Power Inductor	ETQ-P4M3R3KFN

TABLE I: Table of devices used in hardware prototype

Parameter	Value
Input Voltage	500–720 V
Output Voltage	400 V
Output Power	3 kW
Switching Frequency	100 kHz
Effective f_{sw}	700 kHz
Specific Power	15 kW/kg
Power Density	13.2 W/cm^3

TABLE II: Table of FCML hardware prototype operating parameters and physical characteristics

V). As shown in Figure 9, the output inductors implemented in the hardware prototype have a small form factor and are light-weight. Furthermore, in order to improve reliability of the switches at high power, switching cell commutation loop sizes are minimized using a "hybrid" commutation loop design [20]. This design has been shown to reduce losses and noise generated during switching events. A full photo of the board is shown in Figure 10, and a list of the noteworthy devices used for this prototype are given in Table I.

III. FCML CONVERTER DESIGN

With the benefits and the drawbacks of the FCML converter in mind, some clear trade-offs arise that need to be addressed before choosing the switching frequency, number of levels, etc. Namely, while increasing the number of switching cells leads to lower individual device ratings, it also means more devices, which will eventually add up weight-wise and loss-wise. Eventually, these additions will outweigh their benefits (literally and figuratively), so more thorough analysis is required. To inform a design decision, a Monte Carlo analysis was performed on 25,000 randomly generated FCML converter designs. Each generated design was allowed to randomly select device technologies, number of levels, and switching frequency, while keeping voltage and power operating conditions constant. From this generated dataset, total power loss and total board weight of each design were plotted, relative to converter power throughput, and an initial design was chosen at the knee point of the resulting Pareto front (Figure 7). This knee point is an 8 level FCML converter.

The chosen 8-level FCML converter (Figure 8) allows for the use of high figure-of-merit (FOM) EPC GaN, 200 V devices (EPC2034C) [19], while leaving ample voltage headroom, given the range of expected input voltages (500–720

IV. EXPERIMENTAL HARDWARE RESULTS

A summary of the power converter operating conditions is given in Table II. Table III includes the constraints placed on the operation of the power converter by the battery and pulsed load.

The hardware prototype was tested over the specified range of input voltage values (500–720 V) and at load powers exceeding 3 kW. A plot of the converter's efficiency at the minimum and maximum input voltage operating points is shown in Figures 11 and 12. The converter is over 99% efficient at 2.5 kW and above, with a maximum board power density of 13.2 W/cm^3 and specific power density of 15 kW/kg. This power density and specific power do not include the microcontroller used or the startup and shutdown circuits. Compared to converters in a similar voltage and power class, this performance and specific power density align closely with the state of the art FCML converter designs in literature [14].

The converter demonstrates good thermal performance at heavy load (3–4 kW) as shown in Figure 13. The GaN switches remain below 40 °C, which is well below rated temperature. The power isolators (ADUM5240) operate around 53 °C, but

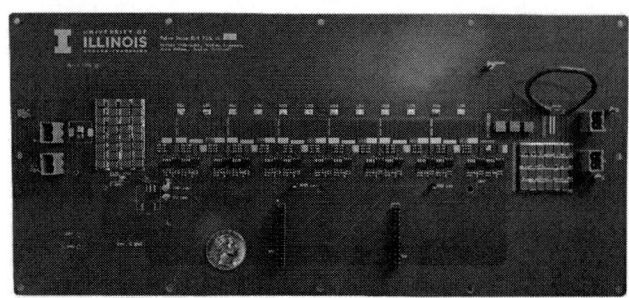

Fig. 10: Tested 8-level FCML converter hardware prototype

Parameter	Value
Battery Voltage	500–720 V
Battery Current Limit	168 A
Load Power req.	10 W/cm^2 [8]
Deicing Time	1 s/event

TABLE III: Onboard battery and heater material voltage, current, and power constraints that inform converter design

979-8-3315-1612-3/25 $31.00 © 2025 IEEE

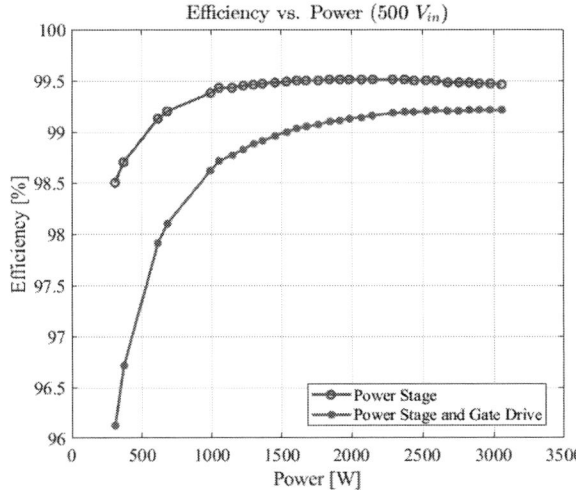

Fig. 11: Plotted efficiency for load sweep at 500 V_{in}

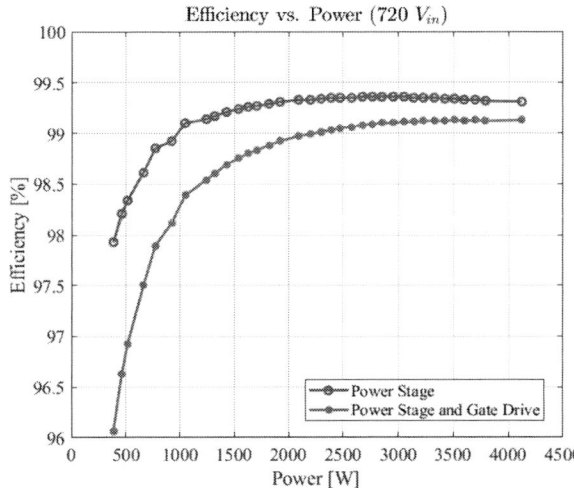

Fig. 12: Plotted efficiency for load sweep at 720 V_{in}

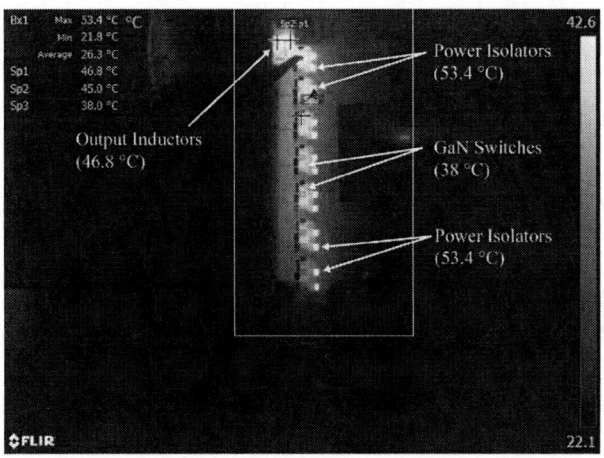

Fig. 13: Thermal image of power stage taken at 500 V_{in}, 3 kW

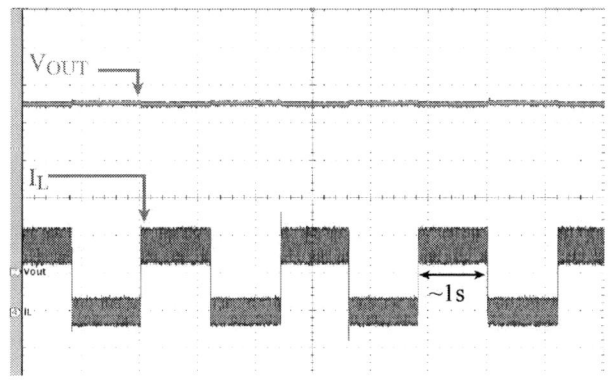

Fig. 14: Waveforms showcasing 1 s current pulses applied at full power, 720 V_{in}

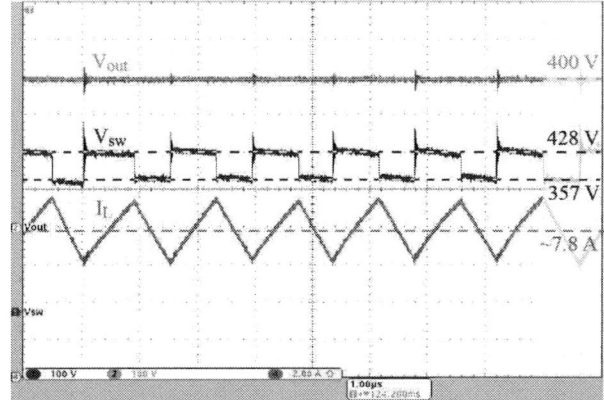

Fig. 15: Current and voltage waveforms of 8-level FCML converter operating in steady state, 3 kW output

are still well below their rated temperatures (150 °C), with some added margin for thermal impedances. Thermal measurements were taken with forced air cooling applied across the top layer of the board. With this thermal performance, bulky, heavy heatsinks are not necessary.

Waveforms acquired during pulsed operation at 3 kW, 720 V_{in}, are shown in Figure 14. As shown, during the load-step transitions, the converter experiences no output voltage droop, indicating sufficient output capacitance and good regulation. Waveforms recorded during steady state operation at 3 kW are shown in Figure 15. The switch node voltage remains well within the bounds of the expected intermediate voltage levels, signifying good balancing of flying capacitors [21].

V. CONCLUSION

Pulse deicing has been investigated as a low energy consumption, non ICE-dependent solution to plane deicing. This work outlines the design and testing of an 8-level flying capac-

979-8-3315-1612-3/25 $31.00 © 2025 IEEE

itor multilevel buck converter that can regulate pulsed power to the pulse deicing heater load. The converter exhibits high efficiencies at full load (>99%) as well as excellent thermal performance with minimal airflow and no added heatsinking, promoting a relatively lightweight (15 kW/kg) and power-dense (13.2 kW/cm^3) design fit for use in aerospace.

REFERENCES

[1] B. Sarlioglu and C. T. Morris, "More electric aircraft: Review, challenges, and opportunities for commercial transport aircraft," *IEEE Transactions on Transportation Electrification*, vol. 1, no. 1, pp. 54–64, 2015.

[2] J. S. Yodice, "Pilot counsel," Aug 2005. [Online]. Available: https://web.archive.org/web/20150101014911/http://www.aopa.org/News-and-Video/All-News/2005/August/1/Pilot-Counsel-(8)

[3] J. Steuernagle, K. Roy, D. Wright, and K. Hummel, "Aircraft icing," 2008. [Online]. Available: https://www.aopa.org/-/media/Files/AOPA/Home/Pilot-Resources/ASI/Safety-Advisors/sa11.pdf

[4] "Aircraft ice protection systems." [Online]. Available: https://skybrary.aero/articles/aircraft-ice-protection-systems

[5] N. Singer, "Air management 101," Jun 2022. [Online]. Available: https://www.aopa.org/news-and-media/all-news/2022/june/pilot/turbine-mentor-matters-bleed-air

[6] A. Rahmatmand, S. J. Harrison, and P. H. Oosthuizen, "An experimental investigation of snow removal from photovoltaic solar panels by electrical heating," *Solar Energy*, vol. 171, pp. 811–826, 2018. [Online]. Available: https://www.sciencedirect.com/science/article/pii/S0038092X18306789

[7] E. Brun, "Distribution of temperatures over an airplanewing with reference to the phenomena of ice formation," *Publica-tions Scientifiques et Techniques de l'Air, No. 119, 1938. Trans-lated in N.A.C.A. Technical Memorandum No. 883*, 1938.

[8] L. Li, S. Khodakarami, X. Yan, K. Fazle Rabbi, A. A. Gunay, A. Stillwell, and N. Miljkovic, "Enabling renewable energy technologies in harsh climates with ultra-efficient electro-thermal desnowing, defrosting, and deicing," *Advanced Functional Materials*, vol. 32, no. 31, p. 2201521, 2022. [Online]. Available: https://onlinelibrary.wiley.com/doi/abs/10.1002/adfm.202201521

[9] A. Bills, S. Sripad, W. L. Fredericks, M. Singh, and V. Viswanathan, "Performance metrics required of next-generation batteries to electrify commercial aircraft," *ACS Energy Letters*, vol. 5, no. 2, pp. 663–668, 2020. [Online]. Available: https://doi.org/10.1021/acsenergylett.9b02574

[10] A. Misra, "Energy storage for electrified aircraft: The need for better batteries, fuel cells, and supercapacitors," *IEEE Electrification Magazine*, vol. 6, no. 3, pp. 54–61, 2018.

[11] G. Yüksek and A. Alkaya, "Effect of the depth of discharge and c-rate on battery degradation and cycle life," in *2023 14th International Conference on Electrical and Electronics Engineering (ELECO)*, 2023, pp. 1–5.

[12] T. Meynard and H. Foch, "Multi-level conversion: high voltage choppers and voltage-source inverters," in *PESC '92 Record. 23rd Annual IEEE Power Electronics Specialists Conference*, 1992, pp. 397–403 vol.1.

[13] J.-S. Lai and F. Z. Peng, "Multilevel converters-a new breed of power converters," *IEEE Transactions on Industry Applications*, vol. 32, no. 3, pp. 509–517, 1996.

[14] S. Coday, N. Ellis, N. Stokowski, and R. C. Pilawa-Podgurski, "Design and implementation of a (flying) flying capacitor multilevel converter," in *2022 IEEE Applied Power Electronics Conference and Exposition (APEC)*, 2022, pp. 542–547.

[15] Y. Lei, C. Barth, S. Qin, W.-C. Liu, I. Moon, A. Stillwell, D. Chou, T. Foulkes, Z. Ye, Z. Liao, and R. C. N. Pilawa-Podgurski, "A 2-kw single-phase seven-level flying capacitor multilevel inverter with an active energy buffer," *IEEE Transactions on Power Electronics*, vol. 32, no. 11, pp. 8570–8581, 2017.

[16] A. Stillwell, E. Candan, and R. C. N. Pilawa-Podgurski, "Active voltage balancing in flying capacitor multi-level converters with valley current detection and constant effective duty cycle control," *IEEE Transactions on Power Electronics*, vol. 34, no. 11, pp. 11 429–11 441, 2019.

[17] A. Stillwell and R. C. N. Pilawa-Podgurski, "A five-level flying capacitor multilevel converter with integrated auxiliary power supply and start-up," *IEEE Transactions on Power Electronics*, vol. 34, no. 3, pp. 2900–2913, 2019.

[18] S. Coday, N. M. Ellis, and R. C. N. Pilawa-Podgurski, "Modeling and analysis of shutdown dynamics in flying capacitor multilevel converters," *IEEE Transactions on Power Electronics*, vol. 39, no. 8, pp. 9150–9159, 2024.

[19] J. Azurza Anderson, G. Zulauf, P. Papamanolis, S. Hobi, S. Mirić, and J. W. Kolar, "Three levels are not enough: Scaling laws for multilevel converters in ac/dc applications," *IEEE Transactions on Power Electronics*, vol. 36, no. 4, pp. 3967–3986, 2021.

[20] "Thermal management of egan fets," 2021. [Online]. Available: https://epc-co.com/epc/Portals/0/epc/documents/application-notes/How2AppNote012 - How to Get More Power Out of an eGaN Converter.pdf

[21] R. H. Wilkinson, T. A. Meynard, and H. du Toit Mouton, "Natural balance of multicell converters: The general case," *IEEE Transactions on Power Electronics*, vol. 21, no. 6, pp. 1658–1666, 2006.

Impact of Position Measurement Delay Angle on Performance of PMSM Drives for Electric Power Steering in A Wide Speed Range

Yingzhe Wu[1*], Hengbin Zhang[2], Yuxiang Xue[2], Lisheng Wang[1], Hui Li[2], Shan Yin[2]

[1] Beijing Research and Development Center, Shanghai Gatek Automotive Electronics Company Ltd, Beijing, China
[2] School of Aeronautics and Astronautics, University of Electronic Science and Technology of China, Chengdu, China
Email: microuestc@163.com*

Abstract—Permanent magnet synchronous motor (PMSM) provides the power source to electric power steering (EPS) system, which determines the steering feeling. Thus, the control performance of the PMSM drives is critical for EPS system. The position measurement delay angle plays a significant role in PMSM drives, especially when the motor rotating in the deep flux weakening (FW) region with higher speed. In this paper, impact of the position delay angle on performance of the PMSM drives for EPS in a wide speed range has been elaborated. The influence of such delay angle has been analyzed based on the voltage and current constraint boundaries of the motor in various rotation speed. Then, calibration and compensation of the position measurement delay angle have been carried out followed by. According to relative experiment results, it can be confirmed that the motor can operate stably in a wide speed range with an appropriate compensation of the delay angle. On the contrary, the motor will be out of control in deep FW region without position measurement delay angle compensation.

Index Terms—Position measurement delay angle, permanent magnet synchronous motor (PMSM), electric power steering (EPS), flux weakening (FW).

I. Introduction

Electric power steering is considered as an advanced technology in automobiles, and has been widely applied in both passenger cars and commercial vehicles [1]–[4]. Fig. 1 illustrates the structure of a typical column-type EPS (C-EPS) system. It is obvious that the motor provides the power source to the whole system, which significantly impacts the steering feeling. Therefore, the torque performance of the motor drives is critical for the EPS system [5]. Due to the merits of improved reliability, increased output torque, higher power density, and reduced vibration, the permanent magnet synchronous motor (PMSM) has become a potential alternative to the traditional dc motor in EPS applications [6], [7]. In order to ensure the torque performance of PMSM satisfying with the steering requirement, the field oriented control (FOC) strategy is usually recommended, which generates the electromagnetic torque of the motor by regulating the d and q axes current. It should be noted that the rotor position angle acquired by rotor position sensor (RPS) is one of the most important factors in realization of the FOC [8]–[10]. However, there is a measurement angle error existing between the actual position and the sampled position. It is obvious that the measurement angle error is augmented with the increased rotation speed, and

deteriorates the torque characteristics of the motor operating in high speed ranges. Therefore, impact of such angle error on performance of the PMSM drives should be taken into account.

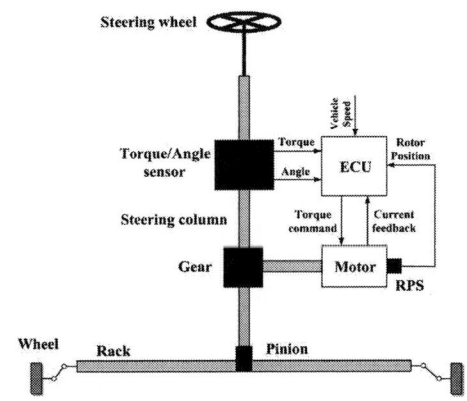

Fig. 1. Structure of a typical C-EPS system.

The measurement angle error can be classified as periodic error due to the signal distortion of the RPS [11]–[13] and the measurement delay angle caused by sampling and calculation process of the micro-controller unit (MCU) [14]–[17]. Since the high precision RPS (e.g. magneto-resistive sensor) has been adopted, the periodic error seems negligibly small in EPS system compared with other applications (see Fig. 2). Therefore, only the measurement delay angle should be considered in EPS. In [15], mechanism of the measurement delay angle has been demonstrated. It is noted that the PMSM can maintain a constant output torque in a wide speed range with an appropriate delay time compensation. However, the determination process of the delay time has not been elaborated. In [16], an unidirectional driving method was applied to calibrate the delay time, which shows an improved torque performance. In [17], the impact of the measurement delay angle on PMSM drives has been analyzed based on the root locus principle. But the method is suitable only for the motor operates in none flux weakening (FW) region. Compared with other applications, the PMSM applied in EPS is characterized by large torque (4~5 N.m), high current (100~150 A), low voltage (12~24 V), and low rated speed (around 100 rad/s) [6], [18]. Since the EPS motor usually enters deep FW region

979-8-3315-1612-3/25 $31.00 © 2025 IEEE

once the rotation speed exceeds 350 rad/s, the delay angle will cause the motor be out of control in such scenario. It is noted that most of existing researches mainly focus on analyzing influence of the delay angle in the aspects of torque ripples and torque capacity. But few of these works discuss the stability issue when the motor operates in FW region.

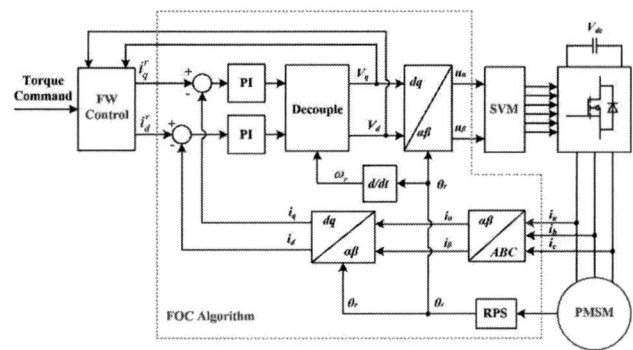

Fig. 3. Control strategy of the PMSM applied in EPS system.

where, i_d, i_q, V_d, V_q represent d and q axes current and voltage, respectively. I_{max} is the rated current of the motor, and V_{dc} is the dc-link voltage. By neglecting the influence of the stater resistance, the current and voltage boundaries for the SPMSM can be simplified as:

$$\begin{cases} \sqrt{i_d^2 + i_q^2} \leq I_{\max} \\ \sqrt{(i_d + \varphi_f/L_s)^2 + i_q^2} \leq \dfrac{V_{dc}}{\sqrt{3}L_s p \omega_r} \end{cases} \quad (2)$$

in which, L_s, φ_f, p and ω_r are inductance, magnetic flux leakage, pole pairs, and rotation speed of the motor. As a result, the current trajectory of a SPMSM operating in a wide speed range with the rated current (I_{max}) is illustrated below. According to Fig. 4, it can be observed that the motor operates in maximum torque per ampere (MTPA) region at the point '1' ($\omega_r \leq \omega_1$) and then toward the intersection point of the current and voltage constraint boundaries through curve '1-3'. Once the rotation speed exceeds ω_3, the motor enters maximum torque per voltage (MTPV) region (can also be considered as deep FW region) and operates toward the curve '3-4'. It should be noticed that the amplitude of the q-axis current is no less than I_{q_min} and the amplitude of the d-axis current is no more than I_{d_max} once the motor operates in deep FW region. In addition, ω_4 is considered as the maximum available speed of the motor applied in EPS (less than 500 rad/s).

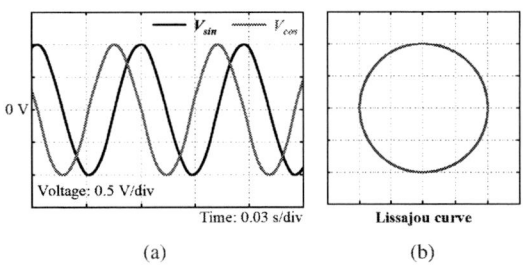

(a) (b)

Fig. 2. Output signal of the RPS when rotation speed of the motor is 100 rad/s, (a) Time domain waveform, (b) Lissajou curve.

In this paper, impact of the position measurement delay angle on performance of the PMSM drive operating in a wide speed range for EPS has been elaborated. In addition, calibration and compensation the position measurement delay angle have also been demonstrated. According to relative experiment results, it is noted that the motor can operate stably in a wide speed range with an appropriate compensation of the delay angle. On the contrary, the motor will lose control in the deep FW region without position delay angle compensation. The rest of this paper is organized as follows. Section II demonstrates the control strategy of the PMSM applied in EPS system. The mechanism and influence of the position measurement delay angle will be elaborated in Section III. Calibration and compensation of the delay angle and experiment verification will be given in Section IV and V, respectively. The conclusion is summarized in Section VI.

II. CONTROL STRATEGY OF THE PMSM FOR EPS

The control strategy of the PMSM applied in EPS system is illustrated in Fig. 3, which mainly consists of FW control strategy, FOC algorithm, and space vector modulation (SVM). The torque command is generated based on the torque control algorithm of the EPS, which can be viewed as the input variable of the motor control strategy. Since the output torque of the surface-mounted PMSM (SPMSM) is proportional to the q-axis current, the torque command can be equivalent to the current reference of the motor. It should also be noted that once the output voltage (V_d and V_q) of the FOC achieves the maximum available voltage (V_{max}) provided by inverter, FW control strategy should be implemented.

As stated in [19], [20], the current control trajectory of the PMSM is restricted by the current and voltage boundaries, which can be expressed as:

$$\begin{cases} \sqrt{i_d^2 + i_q^2} \leq I_{\max} \\ \sqrt{V_d^2 + V_q^2} \leq V_{\max} = V_{dc}/\sqrt{3} \end{cases} \quad (1)$$

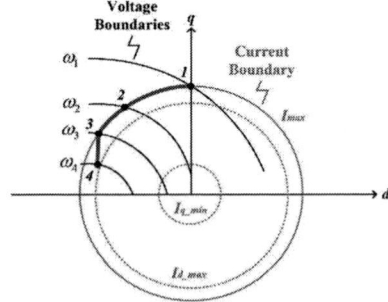

Fig. 4. Current trajectory of SPMSM operating in a wide speed range with the rated current (I_{max}).

In this paper, a FW control strategy based on phase lead angle (θ_{lead}) regulation is adopted (see Fig. 5). When $\omega_r \leq \omega_1$, θ_{lead} is considered as 0, and the motor operates in MTPA

979-8-3315-1612-3/25 $31.00 © 2025 IEEE   1249

region ($i_d^r = 0$ and $i_q^r = i_s$). With increase of ω_r, the output voltage generated by FOC exceeds the voltage boundary, θ_{lead} begins to increase through PI regulator, and the d and q axes current reference of the motor (i_d^r and i_q^r) is governed by:

$$i_d^r = -|i_s| \cdot \sin(\theta_{lead}) \tag{3}$$

$$i_q^r = i_s \cdot \cos(\theta_{lead}) \tag{4}$$

where, i_s is the current reference of the motor. When the motor enters deep FW region ($\omega_r \geq \omega_3$), i_d^r is maintained as $-I_{d\_max}$ and i_q^r can also be governed by (4). Finally, the amplitude of i_q^r is restricted by I_{q_min} when ω_r achieves ω_4.

$$I_{q\_min} = I_{max} \cdot \cos(\theta_{max}) \tag{5}$$

in which, θ_{max} means the maximum value of the phase lead angle.

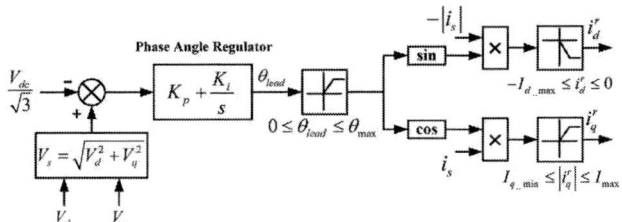

Fig. 5. FW control strategy based on phase lead angle regulation.

III. IMPACT OF POSITION MEASUREMENT DELAY ANGLE OF THE PMSM FOR EPS

A. Mechanism of the Delay Angle

Fig. 6 depicts mechanism of the delay angle in time sequence. It can be clearly observed that the delay time contains position delay time (t_{pd}), current delay time (t_{cd}), and voltage delay time (t_{vd}). Usually, t_{pd} is attributed to the delay caused

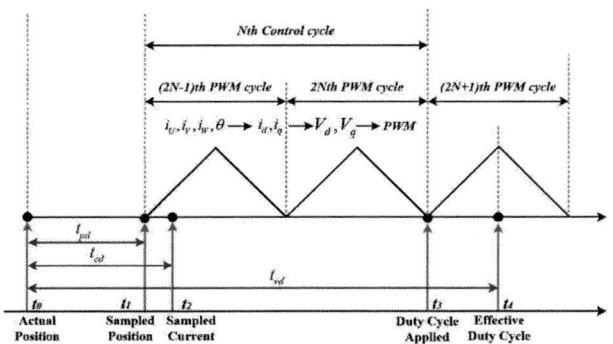

Fig. 6. Mechanism of the delay angle in time sequence.

by RPS, IC device, and circuit filter. t_{cd} and t_{vd} refer to the time delay between actual position and current/voltage vectors. The relationship between t_{pd}, t_{cd}, and t_{vd} is given as:

$$\begin{cases} t_{pd} = t_1 - t_0 \\ t_{cd} = t_{pd} + (t_2 - t_1) \\ t_{vd} = t_{pd} + T_c + 0.5 \cdot T_{PWM} \end{cases} \tag{6}$$

where, T_c and T_{PWM} represent period of the control cycle and PWM cycle, respectively. Then, the delay angle can be acquired accordingly.

$$\begin{cases} \theta_{pd} = p\omega_r \cdot t_{pd} \\ \theta_{cd} = p\omega_r \cdot t_{cd} \\ \theta_{vd} = p\omega_r \cdot t_{vd} \end{cases} \tag{7}$$

It is noted that the time interval between t_1 and t_2 varies in several micro-seconds (its influence is slight), which can be ignored for simplification. In addition, the time interval between t_1 and t_4 can be viewed as a constant value ($T_c + 0.5 \cdot T_{PWM}$). Thus, the main concern is the position delay angle (θ_{pd}), which will be analyzed in detail followed by.

B. Impact of the Position Delay Angle

Fig. 7 shows the current and voltage constraint boundaries (the motor operates in the motoring mode) when ω_r increases from ω_0 to ω_4 ($\omega_0 < \omega_1 < \omega_2 < \omega_3 < \omega_4$). The d^*-q^* reference frame is applied to represent the situation considering position measurement delay occurs in practical applications. It is noted that there is a position delay angle (θ_{pd}) existing between d^*-q^* and d-q reference frames. Since the PMSM applied in EPS always operates in the motoring condition, the analysis followed by will be carried out in such situation.

Case 0 ($\omega_r \leq \omega_0$): When $\omega_r \leq \omega_0$, the PMSM operates in point A^* due to the influence of the delay angle (see Fig. 7 a). In this situation, the current component of i_s^* on the actual d-q axis (i_d^r and i_q^r) can be expressed as (8) and (9). Since the rated speed of the motor applied in EPS ranges from 80~120 rad/s, the delay angle (θ_{pd}) is negligibly small in such speed region. Thus, it impacts the torque capacity of the motor slightly.

$$i_d^r = I_{max} \cdot \sin(\theta_{pd}) \tag{8}$$

$$i_q^r = I_{max} \cdot \cos(\theta_{pd}) \tag{9}$$

Case 1 ($\omega_0 < \omega_r \leq \omega_1$): When $\omega_0 < \omega_r \leq \omega_1$, the PMSM can operate in point A due to the phase lead angle (θ_{lead}) regulated by FW strategy illustrated in Fig. 5, which can effectively compensate the position delay angle. It is interesting to observe that the current component of i_s^* on the actual q-axis is the same with i_s in this condition (see Fig. 7 b), which manifests that the delay angle does not affect the output torque of the PMSM anymore.

Case 2 ($\omega_1 < \omega_r \leq \omega_2$): The PMSM operates around the intersection point of the current and voltage constraint boundaries when the rotation speed varies from ω_1 to ω_2. Similarly, the component of i_s^* on q-axis is the same with that of i_s because that θ_{lead} can completely compensate the delay angle (see Fig. 7 c). As a result, the output torque of the PMSM is also not affected in this case.

Case 3 ($\omega_2 < \omega_r \leq \omega_3$): As $\omega_r > \omega_2$, the current component of i_s^* on the d^* axis reaches $-I_{d\_max}$. It can be figured out that the phase angle (θ^*) between i_s and q^* axis exceeds θ_d [see Eq. (15)]. As a result, the delay angle (θ_{pd}) cannot be fully compensated in this case, and the operation point of PMSM at ω_3 moves from C to C^* (see Fig. 7 d). It is clearly noted

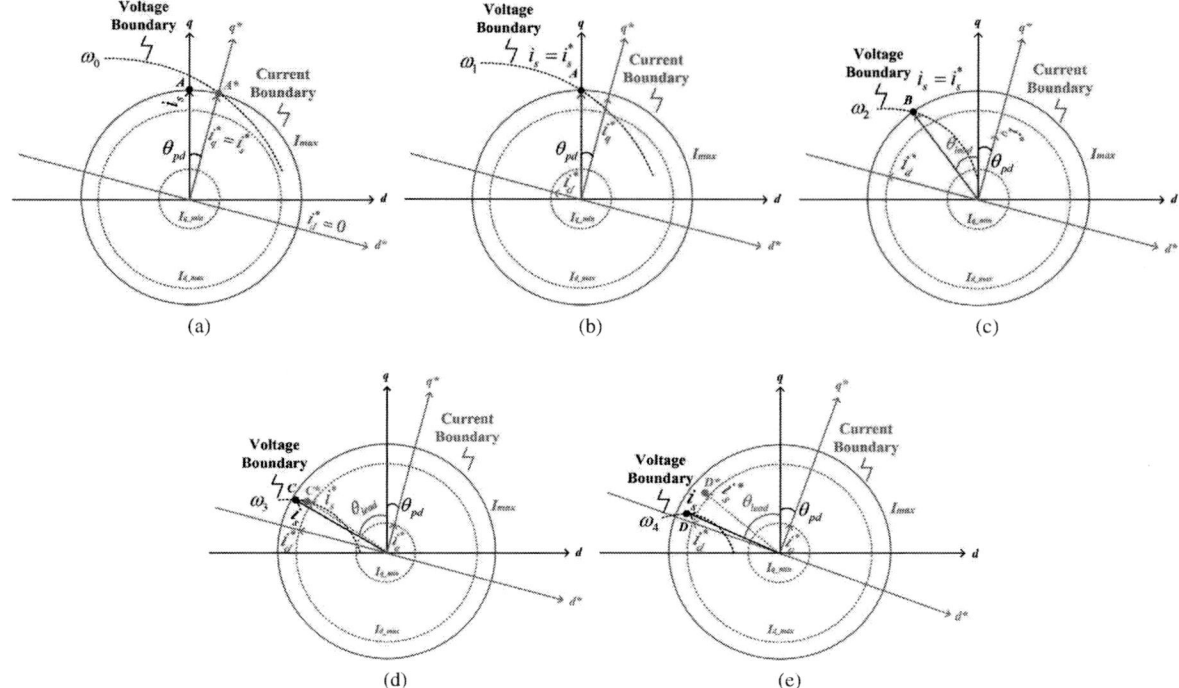

Fig. 7. Current and voltage constraint boundaries of the PMSM with different rotation speed in the motoring mode (the maximum available current is assumed as I_{max}), (a) Case 0: $\omega_r \leq \omega_0$, (b) Case 1: $\omega_0 < \omega_r \leq \omega_1$, (c) Case 2: $\omega_1 < \omega_r \leq \omega_2$, (d) Case 3: $\omega_2 < \omega_r \leq \omega_3$, (e) Case 4: $\omega_3 < \omega_r$.

that the current component of i_s^* on the q axis is less than that of the i_s, which reflects the reduced torque capacity of the motor.

Case 4 ($\omega_3 < \omega_r$): The phase angle (θ^*) reaches θ_{max} once ω_r achieves ω_3. It is obvious that the current component of i_s^* on q^* axis will be restricted as I_{q_min} and the current component on d^* axis will be limited as $-I_{d\_max}$ with continuous increase of rotation speed of the motor. As a consequence, operation point of the motor at ω_4 ($\omega_4 > \omega_3$) moves from D to D^* (see Fig. 7 e). It is noted that the point D^* is out of the voltage constraint boundary, which implies that the motor will lose control.

TABLE I. Impact of position measurement delay angle on performance of PMSM drives in a wide speed range

Speed Ranges	$\omega_r \leq \omega_0$	$\omega_0 < \omega_r \leq \omega_2$	$\omega_2 < \omega_r \leq \omega_3$	$\omega_3 < \omega_r$
Impact	Slight Affection	No Affection	Reduced Torque	Lose Control

Impact of the position measurement delay angle on performance of the PMSM in a wide speed range is summarized in Table. I. It is clearly observed that the position measurement delay angle should be properly compensated to effectively avoid the motor being out of control in deep FW region. In the following section, the calibration and compensation of the position delay angle will be demonstrated.

IV. CALIBRATION AND COMPENSATION OF POSITION MEASUREMENT DELAY ANGLE

Fig. 8 shows the relationship between d^*-q^* and d-q reference frames when PMSM operates in counter clockwise (CCW) and clockwise (CW) directions. Assuming the motor

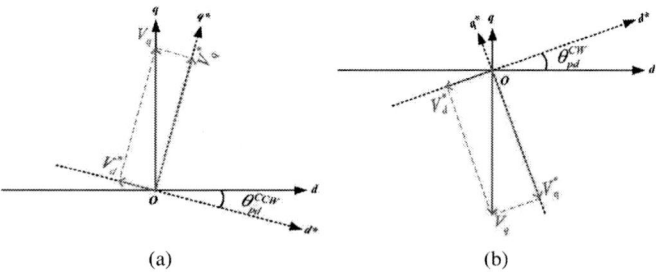

Fig. 8. Relationship between d^*-q^* and d-q reference frames when PMSM operates in CCW and CW directions, (a) CCW, (b) CW.

operates in none FW region and $i_d^r = i_q^r = 0$, the voltage on d-q axes (V_d and V_q) in steady state can be approximated as:

$$\begin{cases} V_d = 0 \\ V_q = \varphi_f \cdot p\omega_r \end{cases} \tag{10}$$

With the impact of the position delay angle (θ_{pd}), amplitude of the voltage on d^*-q^* axes (V_d^* and V_q^*) can be obtained as:

$$\begin{cases} |V_d^*| = |\varphi_f \cdot p\omega_r| \cdot \sin(\theta_{pd}) \\ |V_q^*| = |\varphi_f \cdot p\omega_r| \cdot \cos(\theta_{pd}) \end{cases} \tag{11}$$

As a result, the position delay angle can be determined as:

$$\begin{cases} \theta_{pd}^{CCW} = \arctan\left(\dfrac{-V_d^*}{V_q^*}\right) \\ \theta_{pd}^{CW} = \arctan\left(\dfrac{V_d^*}{V_q^*}\right) \end{cases} \qquad (12)$$

Fig. 9 shows the calibration setup of position measurement delay angle, which is similar with the one demonstrated in [21]. The dynamo motor provides various rotation speed in both CCW and CW directions, and the current reference of the PMSM under test is set as zero. Before the calibration, the zero-position offset angle of the PMSM should be aligned based on the methods proposed in [8], [10], [21]. Fig. 10 shows

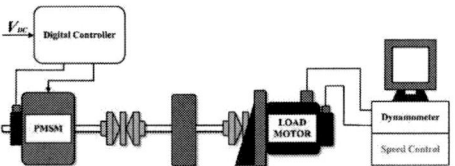

Fig. 9. Calibration setup of the position measurement delay angle for PMSM drives.

the measured V_d^* and V_q^* of the motor with various rotation speed in CCW and CW directions. Then, the calibrated position measurement delay angle can be acquired based on (12) accordingly.

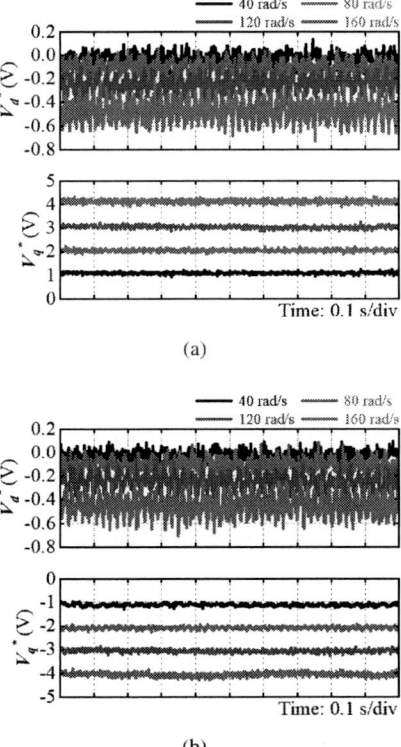

Fig. 10. V_d^* and V_q^* with various rotation speed in CCW and CW directions, (a) CCW, (b) CW.

Fig. 11 illustrates the calibrated position measurement delay angle with different rotation speed. Meanwhile, the relationship between compensation angle (θ_{comp}, unit: °) and rotation

speed (ω_r, unit: rad/s) can be obtained as (13) with the curve fitting method based on the mean value of the measured position delay angle ($\theta_{Ave} = \frac{\theta_{pd}^{CCW} + \theta_{pd}^{CW}}{2}$).

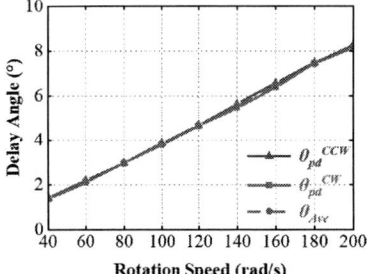

Fig. 11. Calibrated position measurement delay angle of the PMSM drives with various rotation speed in CCW and CW directions.

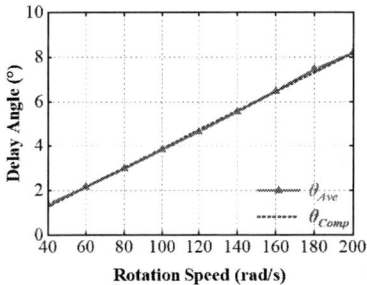

Fig. 12. The calibrated position measurement delay angle in average and the compensation angle with various rotation speed.

$$\theta_{Comp}(\omega_r) = \begin{cases} 0.0431 \cdot \omega_r - 0.4371 & \omega_r > 0 \\ 0 & \omega_r = 0 \\ 0.0431 \cdot \omega_r + 0.4371 & \omega_r < 0 \end{cases} \qquad (13)$$

Thus, θ_{Comp} can be adopted to compensate the position measurement delay angle in applications (see Fig. 12).

V. EXPERIMENT VERIFICATION

The experiment setup is given in Fig. 13, which is the same with the one applied in [22]. In this test, the dynamo motor provides the rotation speed of the motor accelerates from standstill to 3000 rpm (314 rad/s) within 0.5 s, and the current reference (i_s) of the PMSM is given as 80 A. In addition,

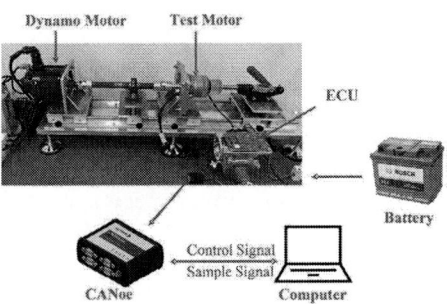

Fig. 13. Experiment setup.

979-8-3315-1612-3/25 $31.00 © 2025 IEEE

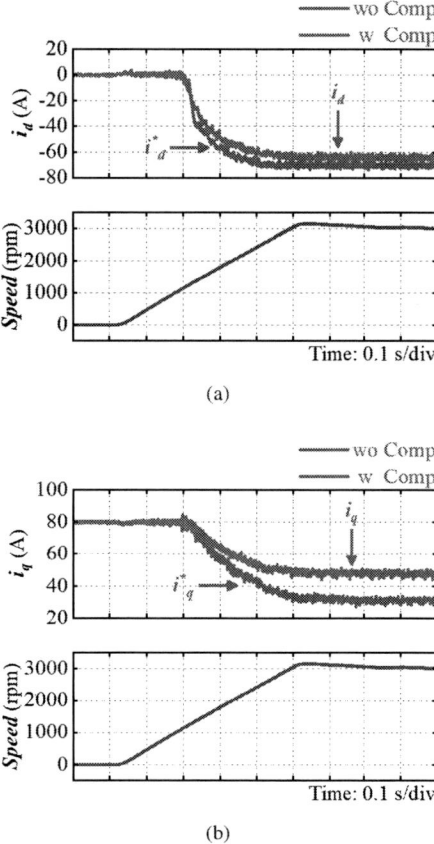

Fig. 14. d and q axes current with and without position measurement delay angle compensation when ω_r accelerates from standstill to 314 rad/s, (a) d axis current, (b) q axis current.

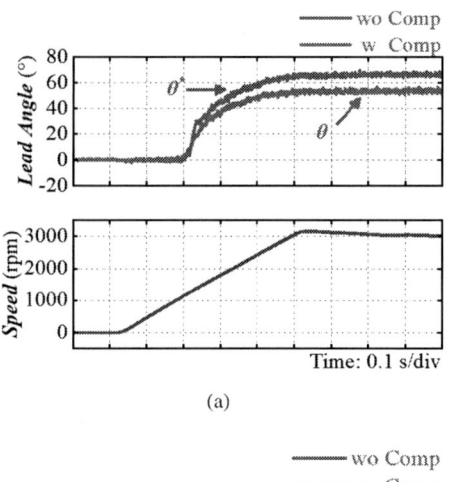

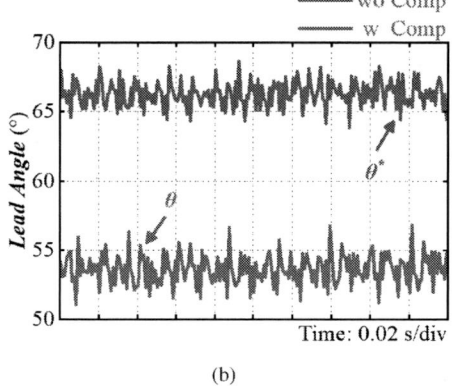

Fig. 15. The phase lead angles with and without position measurement delay angle compensation when ω_r varies from standstill to 314 rad/s, (a) The whole acceleration process, (b) The steady-state when ω_r= 314 rad/s.

T_c and T_{PWM} are set as 100 μs and 50 μs, respectively. Experiment results are shown in Figs. 14 and 15.

Fig. 14 illustrates the measured d and q axes current with (i_d and i_q) and without (i_d^* and i_q^*) position measurement delay angle compensation. Then, the phase angle θ (with compensation) and θ^* (without compensation) can be obtained based on (14). According to Fig. 15, it can be clearly observed that the average values of θ and θ^* are 53° and 66° when ω_r= 314 rad/s, respectively. As a result, the position delay angle ($\theta_{pd} = \theta^*$- θ, unit: °) can be acquired as 13°. It is noted that θ_{pd} is in good agreement with θ_{Comp} (13.09°) calculated based on (13) as ω_r= 314 rad/s, which verifies the cases 1 and 2 analyzed in Section III.

$$\begin{cases} \theta = \arctan\left(-\dfrac{i_d}{i_q}\right) \\ \theta^* = \arctan\left(-\dfrac{i_d^*}{i_q^*}\right) \end{cases} \tag{14}$$

It is also noted that the influence of θ_{pd} can be fully compensated with θ_{lead} regulated by FW strategy once θ^* is no more than θ_d.

$$\theta_d = \arctan\left(\frac{I_{d\_max}}{\sqrt{I_{max}^2 - I_{d\_max}^2}}\right) \tag{15}$$

Since the maximum available speed of the dynamo motor is 314 rad/s, verification in a higher speed range cannot be realized directly. In order to further verifies impact of position measurement delay angle in deep FW region, a battery with insufficient power capacity is applied. Therefore, a reduced dc-bus voltage shrinks the radium of voltage constraint boundary with the increased rotation speed, which can ensure the motor operating in deep FW region even though $\omega_r \leq 314$ rad/s. According to Fig. 16, it is interesting to observe that the motor can still operate stably with the delay angle compensation in a wide speed range. However, the motor will be out of control without the delay angle compensation once θ^* exceeds θ_{max}.

$$\theta_{max} = \arctan\left(\frac{I_{d\_max}}{I_{q\_min}}\right) \tag{16}$$

As a result, the case 4 analyzed in Section III has been well validated.

VI. CONCLUSION

In this paper, impact of position measurement delay angle on performance of PMSM drives for EPS in a wide speed range has been elaborated. According to theoretical analysis and relative experiment results, it can be figured out that the

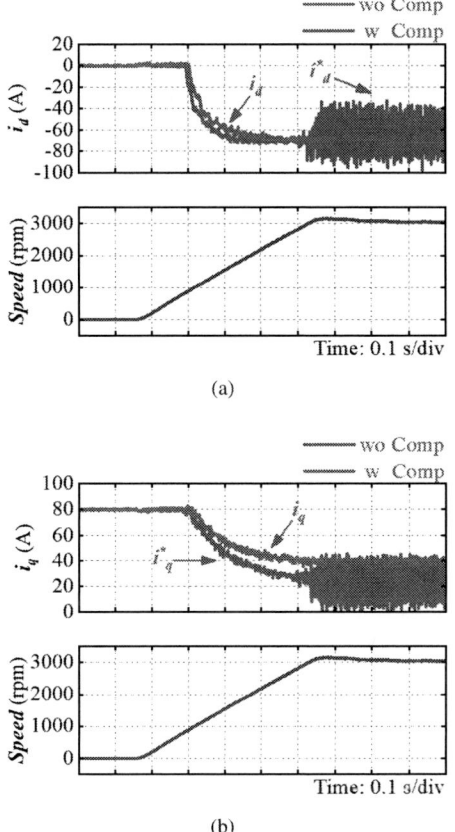

Fig. 16. d and q axes current waveforms measured based on a battery with insufficient power capacity when ω_r accelerates from standstill to 314 rad/s, (a) d axis current, (b) q axis current.

delay angle impacts the output torque of the motor slightly in low and medium rotation speed ranges. However, the reduced torque will be caused once the motor enters deep FW region. With continuous increase of the rotation speed in deep FW region, the delay angle will cause the motor lose control ultimately. In addition, the calibration and compensation of the position measurement delay angle have also been demonstrated, which can ensure the PMSM drives operating stably in the deep FW region.

References

[1] G. Liu, A. Kurnia, R. De Larminat, P. Desmond, and T. O'Gorman, "A low torque ripple PMSM drive for EPS applications," in *Proc. IEEE Applied Power Electronics Conference and Exposition (APEC)*, 2004, pp. 1130–1136.

[2] G. Lee, W. Choi, S. Kim, S. Kwon, and J. Hong, "Torque ripple minimization control of permanent magnet synchronous motors for EPS applications," *Int. J. Automot. Technol.*, vol. 12, no. 2, pp. 291–297, 2011.

[3] A. Zaremba, M. Liubakka, and R. Stuntz, "Control and steering feel issues in the design of an electric power steering system," in *Proc. American Control Conference (ACC)*, vol. 1, Philadelphia, Jun. 1998, pp. 36–40.

[4] D. Lee, K.-S. Kim, and S. Kim, "Controller design of an electric power steering system," *IEEE Trans. Control Syst. Technol.*, vol. 26, no. 2, pp. 748–755, 2018.

[5] K.-Y. Cho, Y.-K. Lee, H. Mok, H.-W. Kim, B.-H. Jun, and Y. Cho, "Torque ripple reduction of a PM synchronous motor for electric power steering using a low resolution position sensor," *J. Power Electron.*, vol. 10, no. 6, pp. 709–716, 2010.

[6] X. Li, J. Wu, Z. Zhang, and B. Zhuang, "Development of EPS rare-earth permanent-magnet brushless direct current motor," in *Proce. International Conference on Electrical Machines and Systems (ICEMS)*, vol. 3, 2005, pp. 2302–2305 Vol. 3.

[7] A. G. Garganeev, A. Ibrahim, D. I. Ulyanov, and A. A. Antropov, "Study and control of electric power steering system based on permanent magnet synchronous motor," in *Proc. IEEE International Conference of Young Professionals in Electron Devices and Materials (EDM)*, 2024, pp. 1170–1174.

[8] Y. Wu, H. Li, L. Wang, and S. Yin, "Zero-position offset calibration of PMSM based on I/f control strategy and slide mode observer for electric power steering system," *IEEE J. Emerg. Sel. Topics Power Electron.*, vol. 12, no. 2, pp. 2214–2225, 2024.

[9] S. S. Kuruppu and Y. Zou, "Post production PMSM position sensor offset error quantification via voltage estimation," in *Proc. IEEE Energy Conversion Congress and Exposition (ECCE)*, Detroit, Oct. 2020, pp. 3355–3361.

[10] D. Kim, J. Kim, H. Lim, J. Park, J. Han, and G. Lee, "A sudy on accurate initial rotor position offset detection for a permanent magnet synchronous motor under a no-Load condition," *IEEE Access*, vol. 9, pp. 73 662–73 670, 2021.

[11] G. Liu, A. Kurnia, R. De Larminat, and S. Rotter, "Position sensor error analysis for EPS motor drive," in *Proc. IEEE International Electric Machines and Drives Conference (IEMDC)*, Madison, Jun. 2003, pp. 249–254.

[12] X. Song, J. Fang, and B. Han, "High-precision rotor position detection for high-speed surface PMSM drive based on linear hall-effect sensors," *IEEE Trans. Power Electron.*, vol. 31, no. 7, pp. 4720–4731, 2016.

[13] D. Chen, J. Li, J. Chen, and R. Qu, "On-line compensation of resolver periodic error for PMSM drives," *IEEE Trans. Ind. Appl.*, vol. 55, no. 6, pp. 5990–6000, 2019.

[14] J. Bocker, S. Beineke, and A. Bahr, "On the control bandwidth of servo drives," in *Proc. European Conference on Power Electronics and Applications*, Barcelona, Sep. 2009, pp. 1–10.

[15] A. K. Singh, R. Raja, T. Sebastian, and A. Gebregergis, "Effect of position measurement delay on the performance of PMSM drive," in *Proc. IEEE Energy Conversion Congress and Exposition (ECCE)*, 2018, pp. 4622–4627.

[16] S. Kim, K. Park, D. Kang, and G. H. Lee, "High-performance permanent magnet synchronous motor control with electrical angle delayed component compensation," *IEEE Access*, vol. 11, pp. 129 467–129 478, 2023.

[17] C. Lian, F. Xiao, J. Liu, and S. Gao, "Analysis and compensation of the rotor position offset error and time delay in field-oriented-controlled PMSM drives," *IET Power Electron.*, vol. 13, no. 9, pp. 1911–1918, 2020.

[18] C. Cai, H. Zhang, J. Liu, and Y. Gao, "Modeling and simulation of BLDC motor in electric power steering," in *Proc. Asia-Pacific Power and Energy Engineering Conference (APPEEC)*, 2010, pp. 1–4.

[19] J.-J. Chen and K.-P. Chin, "Minimum copper loss flux-weakening control of surface mounted permanent magnet synchronous motors," *IEEE Trans. Power Electron.*, vol. 18, no. 4, pp. 929–936, 2003.

[20] J. Shi and Y.-S. Lu, "Field-weakening operation of cylindrical permanent-magnet motors," in *Proc. IEEE International Conference on Control Applications (ICCA)*, 1996, pp. 864–869.

[21] S. S. Kuruppu, "In-system calibration of position sensor offset in PMSM drives," in *Proc. IEEE International Electric Machines and Drives Conference (IEMDC)*, Hartford, May 2021, pp. 1–5.

[22] X. Liu, H. Li, Y. Wu, L. Wang, and S. Yin, "Dynamic dead-time compensation method based on switching characteristics of the MOSFET for PMSM drive system," *Electronics*, vol. 12, no. 23, 2023.

Physical Parameter Estimation for a Two-Level VSI Three-Phase PMSM Electric Drivetrain

Bernard Steyaert
Columbia University
bernard.steyaert@columbia.edu

Ananda Tjakra Adisurja
Columbia University
ananda.tjakra@columbia.edu

Matthias Preindl
Columbia University
matthias.preindl@columbia.edu

Abstract—**A methodology for computing the physical parameters of a two-level VSI three-phase PMSM drivetrain is described. The method uses a load test at constant speed, and lumps resistive voltage drop, core loss resistance, and inverter nonlinearity together. Motor inductance, flux offset, and drivetrain resistance are calculated for a group of setpoints. The method uses a constant speed load test with only current sensors and a position sensor, and overcomes the rank deficiency problem by including adjacent points in an overdetermined unconstrained least squares minimization problem. Results demonstrate satisfactory physical parameter estimation with a 15% increase in airgap torque estimation accuracy compared to nameplate while withstanding a robustness to ±20% speed errors and ±10% current space harmonics.**

Index Terms—**Electric Drives, parameter estimation, saturation, motor drives**

I. INTRODUCTION

The combination of the two-level Voltage Source Inverter (VSI) and three-phase Permanent Magnet Synchronous Machine (PMSM) has become a popular combination for electric drivetrains. High breakdown voltage switches such as SiC (600V-1200V) and GaN (400V-650V) are more efficient than previously-popular high voltage IGBT switches and power MOSFETs. PMSMs have replaced induction machines (IM) in many drivetrain sectors, such as automotive, due to their small form factor and high power density. Accurate parameterization of this common drivetrain configuration is key to maximizing performance.

Drivetrain parameter estimation methods include online and offline methods [1]. Some new online parameter estimation methods include Model Reference Adaptive Systems (MRAS) [2], and Optimization-based estimation [3] while more established methods are Extended Kalman Filters (EKF), and Recursive Least Squares (RLS) [4], [5]. Offline parameter estimation techniques precompute parameters and store them in look up tables [6], [7]. Popular methods include the finite element method (FEM) [8], and load tests [9]. Parameter variations from nameplate can lead to torque estimation errors, up to $\approx$ 15% as shown in this study. Most approaches model machine mutual inductances find them to be very close, but not exactly equal [5]. This is due to the rank-deficiency problem, in which there are two PMSM motor equations, and more than two parameters to estimate. In [4], [5], [9], self inductance and mutual inductance are combined into flux, and the differentiation between the two is lost. When using

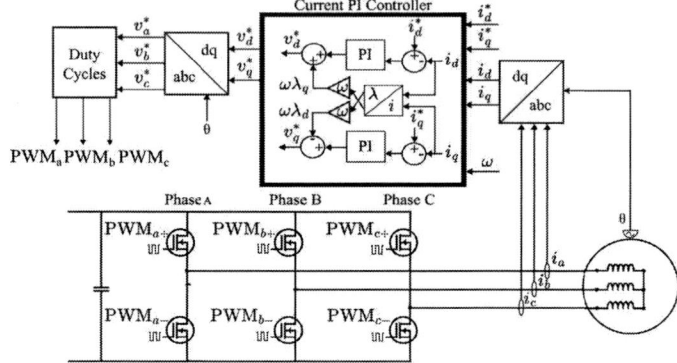

Fig. 1. Drivetrain Topology and Control Diagram

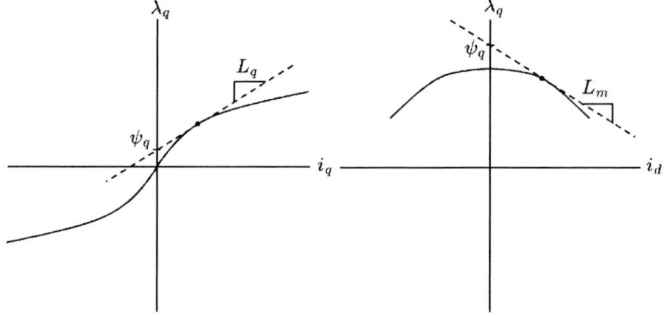

Fig. 2. Example current-flux relationship showing saturation of the q-axis flux

current control, the mutual inductance is neglected, so the aforementioned approach is acceptable. Alternatively, these terms can be included in a flux calculation and flux control can be used. Our approach uses a constant-speed load test under a set of steady-state current setpoints. The rank deficiency problem is solved by averaging the experimental data across adjacent setpoints, then using unconstrained least squares to solve for the parameters.

This study uses a lumped-parameter model to model inverter nonlinearity, which includes dead-time. The most important aspect of this method is ensuring physical motor parameters, that is to say positive self inductances, positive resistance, and equal mutual inductances. This has two major implications 1) it gives the inductance matrix desirable properties by positive semidefiniteness and 2) ensures stability properties of the

979-8-3315-1612-3/25 $31.00 © 2025 IEEE

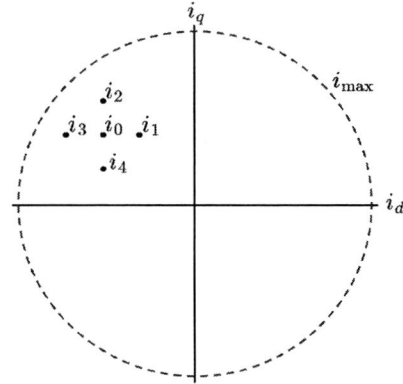

Fig. 3. Setpoint i_0 and adjacent setpoints $i_1 - i_4$

plant. The drivetrain model is presented in Section II, the parameter estimation algorithm is shown in Section III, the experimental setup and results are discussed in Section IV, and the study is concluded in Section V.

II. DRIVETRAIN MODEL

The drivetrain topology under study is shown in Fig. 1. The two-level three-phase VSI consists of three half bridges. The switches can be SiC or GaN, the only topological difference being SiC has reverse conducting diodes. Each of the three half bridges of the inverter electrifies one phase of a three-phase PMSM via the middle point of the two switches. The PMSM model is generalized to have a non-unity saliency ratio, i.e. the inductances of the d-axis and q-axis may be different, $L_d \neq L_q$, as is the case for the interior permanent magnet synchronous machine (IPMSM).

A. Motor Inductance and Dynamics

The inductances and flux offsets discussed in this study are strictly the differential inductances, not the secant inductances. Secant inductance is defined by the global ratio of flux and current

$$L_d(i_{dq})_{\text{sec}} = \frac{(\lambda_d - \psi_{pm})}{i_d}, \tag{1a}$$

$$L_q(i_{dq})_{\text{sec}} = \frac{\lambda_q}{i_q}, \tag{1b}$$

$$L_m(i_{dq})_{\text{sec}} = \frac{\lambda_q}{i_d} = \frac{(\lambda_d - \psi_{pm})}{i_q}, \tag{1c}$$

whereas differential inductance is defined by the localized ratio of flux and current

$$L_d(i_{dq})_{\text{diff}} = \frac{\Delta \lambda_d}{\Delta i_d}, \tag{2a}$$

$$L_q(i_{dq})_{\text{diff}} = \frac{\Delta \lambda_q}{\Delta i_q}, \tag{2b}$$

$$L_m(i_{dq})_{\text{diff}} = \frac{\Delta \lambda_q}{\Delta i_d} = \frac{\Delta \lambda_d}{\Delta i_q}. \tag{2c}$$

$$\tag{2d}$$

With each differential inductance there is an offset term that can be calculated as

$$\psi_d(i_{dq})_{\text{diff}} = \lambda_d - L_{d,\text{diff}} i_d - L_{m,\text{diff}} i_q, \tag{3a}$$

$$\psi_q(i_{dq})_{\text{diff}} = \lambda_q - L_{q,\text{diff}} i_q - L_{m,\text{diff}} i_d, \tag{3b}$$

where ψ_{pm} is lumped into $\psi_{d,\text{diff}}$. For the remainder of this paper all inductances are differential and the $\cdot_{\text{diff}}$ subscript is omitted.

An observation is that the differential inductances are the slopes of the first-order Taylor approximation of the flux map $\lambda_{dq} = f(i_{dq})$. This is shown as an example in Fig. 2 sliced on two axis

$$\begin{bmatrix} \lambda_d(i_d, i_q) \\ \lambda_q(i_d, i_q) \end{bmatrix} = \begin{bmatrix} L_d & L_m \\ L_m & L_q \end{bmatrix} \begin{bmatrix} i_d \\ i_q \end{bmatrix} + \begin{bmatrix} \psi_d \\ \psi_q \end{bmatrix}, \tag{4a}$$

$$\lambda_{dq} = \mathbf{L} i_{dq} + \psi. \tag{4b}$$

The relationship between current and flux at any given operating point can be modelled similarly to a transformer, with a coupling coefficient k relating the three inductances [10]. Just as in a transformer, the coupling coefficient k would model how magnetically coupled the machine's d-axis and q-axis are,

$$k = \frac{|L_m|}{\sqrt{L_d L_q}}. \tag{5}$$

It should also be noted that in this formulation the inductance matrix $\mathbf{L}$ is symmetric, furthermore the energy in the system cannot be negative, therefore the matrix is positive semidefinite

$$i_{dq}^T \mathbf{L} i_{dq} \geq 0 \Rightarrow \mathbf{L} \succeq 0. \tag{6}$$

And finally it should be noted that the inductance matrix $\mathbf{L}$ is invertible as long as

$$\det(\mathbf{L}) \neq 0 \iff L_d L_q \neq L_m^2. \tag{7}$$

Invertibility of $\mathbf{L}$ is a useful property when computing the reverse map $i_{dq} = g(\lambda_{dq})$ and positive semidefiniteness is useful in matrix decomposition.

It is practically impossible for $\det(\mathbf{L}) = 0$, as that would imply a coupling coefficient $k = 1$, which would mean the d-axis and q-axis are perfectly coupled, which does not make any practical sense. Therefore $\mathbf{L}$ is always invertible.

The electric dynamic equation for the system in the dq reference frame can be written in flux as

$$\frac{d\lambda_d}{dt} = \omega \lambda_q + v_d - R_s i_d, \tag{8a}$$

$$\frac{d\lambda_q}{dt} = -\omega \lambda_d + v_q - R_s i_q. \tag{8b}$$

The flux at the origin ($i_{dq} = \mathbf{0}$) using (4) can be substituted yielding the dynamic equations in current as

$$\frac{d(L_d i_d + L_m i_q + \psi_d)}{dt} = \omega(L_q i_q + L_m i_d + \psi_q) + v_q - R_s i_d, \tag{9a}$$

$$\frac{d(L_q i_q + L_m i_d + \psi_q)}{dt} = -\omega(L_d i_d + L_m i_q + \psi_d) + v_d - R_s i_q. \tag{9b}$$

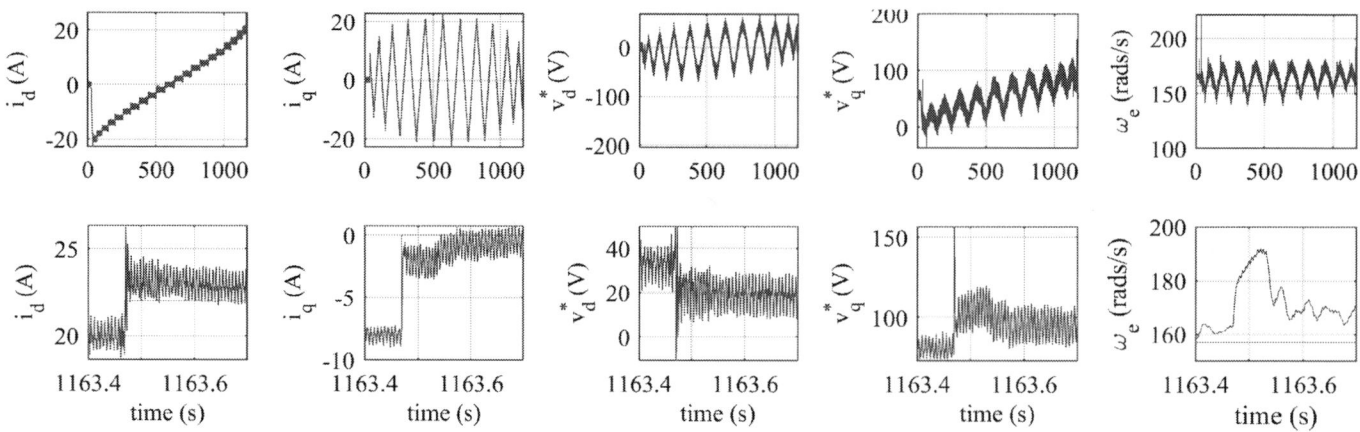

Fig. 4. Experimental data (Top) full sequence of 373 current setpoints (Bottom) zoomed in specific transition between setpoint $i_{dq} = [20A; -8A]$ to $i_{dq} = [22A; 0A]$, (left to right) i_d, i_q, v_d^*, v_q^*, ω_e

In (9), usually the cross saturation is assumed to be negligible ($L_m = 0$) and the flux offsets ψ_d and ψ_q are reduced to $\psi_d = \psi_{pm}$, $\psi_q = 0$, where ψ_{pm} is the flux of the permanent magnet. Saturation is typically neglected as well, fixing the values of L_q and L_d. This makes for a very simple equation at the sacrifice of not modelling saturation or cross saturation.

These assumptions are an approximation which do not hold when the machine saturates or has significant magnetic cross coupling. As shown in Fig. 2, the differential self inductance L_q and flux offset ψ_q change as the machine saturates at higher currents. As the flux increases more slowly, L_q decreases and ψ_q increases. The same occurs in the d-axis.

B. Control and Measurement

The only sensors required (not including the dyno drive) for the parameter estimation are current sensors and a position sensor. Assuming there is minimal leakage current and the system is balanced, only two current sensors are needed, as the third current can be calculated as $i_c = i_a - i_b$. The position sensor's angle measurement is used to compute the Park-Clarke transform to use dq-axis vector control.

The current controller used for the experiment is shown in Fig. 1, a constant current reference i_{dq}^* is compared against an experimental i_{dq}, which arises from the current sensor measurements i_{abc} and the Park-Clarke transform. Using equation (8), neglecting saturation and cross saturation, a simple PI controller for each axis is used. The input to the PI is current error, the output is voltage reference. Additionally, a feedforward term $\omega \mathbf{J} \lambda_{dq}$, $\mathbf{J}$ defined in (17), is introduced as well (although not needed for stability). This feedforward voltage term accounts for the cross-coupled back-emf feedback voltage of the drive system. With the feedforward term cancelling the cross coupling terms, setting $L_m = 0$ and fixing L_d, L_q, ψ_d, ψ_q constant one obtains the familiar system

$$L_d \frac{di_d}{dt} = v_d - R_s i_d, \quad L_q \frac{di_q}{dt} = v_q - R_s i_q, \quad (10)$$

which has the transfer functions

$$\frac{i_d(s)}{v_d(s)} = \frac{1}{R_s + sL_d}, \quad \frac{i_q(s)}{v_q(s)} = \frac{1}{R_s + sL_q}. \quad (11)$$

Using (11) as plant transfer functions one can easily tune two PI controllers (one for d-axis one for q-axis) to create closed-loop stability. Any offset in the feedforward term will be integrated out. An important note about (11) is that if either the self inductances L_d, L_q or resistance R_s is ever negative, the system will no longer be stable (pole in right half plane). Therefore it is imperative during parameter estimation that $L_d \geq 0$, $L_q \geq 0$, $R_s \geq 0$.

The output from the PI controllers is a voltage reference v_{dq}^*, which converts into a three phase reference v_{abc}^* via the inverse Park-Clarke transform, then to a duty cycle $d_{abc} = v_{abc}^*/v_{dc}$. A separate constant speed drive holds the electrical speed ω constant, which at steady state current and thus flux ($\frac{d\lambda_{dq}}{dt} = 0$) from (8) and (9) leads to the voltage references

$$v_d^* = -\omega(L_q i_q + L_m i_d + \psi_q) + \bar{R}_s i_d, \quad (12a)$$
$$v_q^* = \omega(L_d i_d + L_m i_q + \psi_d) + \bar{R}_s i_q. \quad (12b)$$

Additional terms about the drivetrain from any other nonlinearities will be coupled into this equation, creating a lumped parameter model, in the $\bar{R}_s$ parameter. There is also a thermal dependency on the drivetrain parameters [11], as well as spatial harmonics [12], [13] which create cogging torque. For this study the machine is assumed to be room temperature (25C) and the spatial harmonics will be averaged out in post processing.

In summary, the control variable is current i_{dq} via two PI controllers, and the measured variables are current i_{dq}, voltage v_q^*, and speed ω. The unknowns are inductances L_d, L_q, L_m, flux offsets ψ_d, ψ_q, and resistance $\bar{R}_s$.

C. Inverter Lumped Parameter Model

The $\bar{R}_s i_{dq}$ term in (12) differs from $R_s i_{dq}$ of (8) in that in addition to the resistive voltage drop, it also includes PM-induced core loss and inverter dead-time nonlinearity (among

979-8-3315-1612-3/25 $31.00 © 2025 IEEE

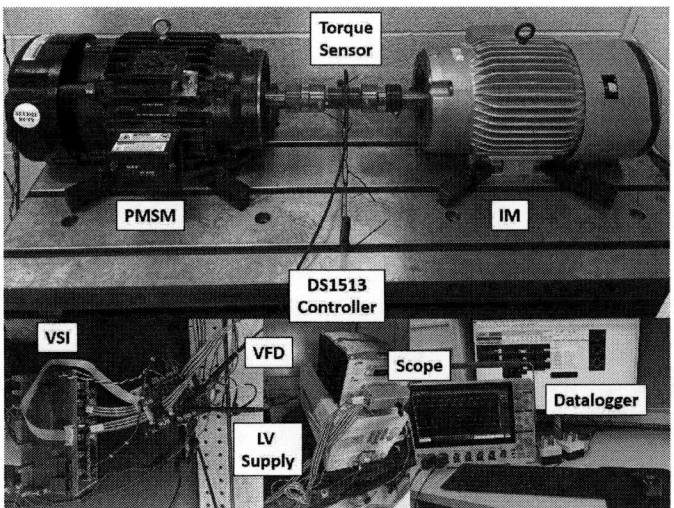

Fig. 5. Experimental Setup

other nonlinearities). The PM-induced core loss voltage drop is a function of all the motor parameters and speed, power conservation can be used to obtain the unwieldy equation [14]

$$v_{\text{pm,core,dq}} = \begin{bmatrix} R_c & 0 \\ 0 & R_c \end{bmatrix} \begin{bmatrix} i_d \\ i_q \end{bmatrix}, \tag{13a}$$

$$R_c|_{i_d=0} = \frac{\sqrt{v_d^2 + v_q^2} - \omega^2(\psi_{pm}^2 + (L_d - L_q)i_q^2)}{v_q i_q - \omega \psi_{\text{pm}} i_q - i_q^2 R_s}. \tag{13b}$$

The inverter nonlinearity has two primary components, the first is a resistive effect, the second is a voltage stepping effect [15], the former can be modelled as

$$v_{\text{dead,dq}} = \frac{1}{6}\left((\frac{t_{\text{off}} - t_{\text{on}} - t_{\text{dead}}}{t_s})V_{dc} - V_{ds} - V_{\text{diode}} \right) D_{dq}, \tag{14a}$$

$$D_{dq} = 2\begin{bmatrix} \cos(\theta) & \cos(\theta - \frac{2\pi}{3}) & \cos(\theta + \frac{2\pi}{3}) \\ -\sin(\theta) & -\sin(\theta - \frac{2\pi}{3}) & \sin(\theta + \frac{\pi}{3}) \end{bmatrix}\begin{bmatrix} \text{sign}(i_a) \\ \text{sign}(i_b) \\ \text{sign}(i_c) \end{bmatrix} \tag{14b}$$

where t_s, t_{off}, t_{on}, t_{dead} are the period time, turn off delay, turn on delay, and dead time; V_{ds} is the voltage drop across the drain and source of the transistor, and V_{diode} is the reverse conducting diode drop (in the case of SiC). These terms can be lumped into the resistance $\bar{R}_s$ as

$$\bar{R}_s i_d = R_s i_d + v_{\text{pm,core,d}} + v_{\text{dead,d}} \tag{15a}$$

$$\bar{R}_s i_q = R_s i_q + v_{\text{pm,core,q}} + v_{\text{dead,q}}, \tag{15b}$$

where R_s is the DC stator resistance.

D. Torque

The airgap torque of the motor is given by the cross product of the current and flux

$$T = \frac{3}{2}p i_{dq}^T \mathbf{J} \lambda_{dq}, \tag{16}$$

where p is the number of pole pairs and $\mathbf{J}$ is the cross-coupling matrix

$$\mathbf{J} = \begin{bmatrix} 0 & -1 \\ 1 & 0 \end{bmatrix}. \tag{17}$$

Substituting the localized differential flux from (4b) the following parameterized torque expression is derived

$$T_{\text{param}} = \frac{3}{2}p(i_d i_q(L_d - L_q) + L_m(i_q^2 - i_d^2) - i_d\psi_q + i_q\psi_d), \tag{18}$$

which depends on the operating point of the machine.

Using nameplate self inductances L_d and L_q, and the approximations $L_m = 0$, $\psi_d = \psi_{\text{pm}}$, $\psi_q = 0$, the nameplate torque is the familiar expression

$$T_{\text{name}} = \frac{3}{2}p i_q \left(\psi_m + (L_d - L_q)i_d\right). \tag{19}$$

The difference between parameterized torque (18) and nameplate torque (19) can be quite large in some instances.

III. PARAMETER ESTIMATION

Equation (12) can be re-arranged to obtain equation (20), in the form $\mathbf{A}x = b$, where $\mathbf{A}$ and b are measurements, and x are the parameters to estimate, or unknowns. Importantly, the mutual inductance (L_m) is one term.

1) Overcoming Rank Deficiency: One measurement results in the $\mathbf{A}$ matrix having two rows and being rank two. However there are six parameters to estimate. To overcome this rank deficiency, atleast three datapoints (six rows) are needed per measurement, in which two of the rows must have i_{dq} which are linearly independent, shown in equation (20).

In practice, for every i_{dq} setpoint, four adjacent points are added to the $\mathbf{A}$ matrix as two rows each, the $\mathbf{A}$ matrix will have ten rows, and the b vector will have ten elements. An example of adjacent points used to estimate the drivetrain parameters at a given point i_0 is shown in Fig. 3. Furthermore, to average out motor forward/reverse mechanical and electrical asymmetries, the measurements should be repeated for negative ω, making each equation setpoint have an $\mathbf{A}$ matrix of twenty rows.

$$\begin{bmatrix} i_d & 0 & -\omega i_q & -\omega i_d & 0 & -\omega \\ i_q & \omega i_d & 0 & \omega i_q & \omega & 0 \end{bmatrix}\begin{bmatrix} \bar{R}_s \\ L_d \\ L_q \\ L_m \\ \psi_d \\ \psi_q \end{bmatrix} = \begin{bmatrix} v_d^* \\ v_q^* \end{bmatrix}, \tag{20a}$$

$$\begin{bmatrix} i_{d0} & 0 & -\omega i_{q0} & -\omega i_{d0} & 0 & -\omega \\ i_{q0} & \omega i_{d0} & 0 & \omega i_{q0} & \omega & 0 \\ \vdots & & & & & \\ i_{dj} & 0 & -\omega i_{qj} & -\omega i_{dj} & 0 & -\omega \\ i_{qj} & \omega i_{dj} & 0 & \omega i_{qj} & \omega & 0 \end{bmatrix}\begin{bmatrix} \bar{R}_s \\ L_d \\ L_q \\ L_m \\ \psi_d \\ \psi_q \end{bmatrix} = \begin{bmatrix} v_{d0}^* \\ v_{q0}^* \\ \vdots \\ v_{dj}^* \\ v_{qj}^* \end{bmatrix}, \tag{20b}$$

$$\mathbf{A} \qquad\qquad x \quad = \quad b. \tag{20c}$$

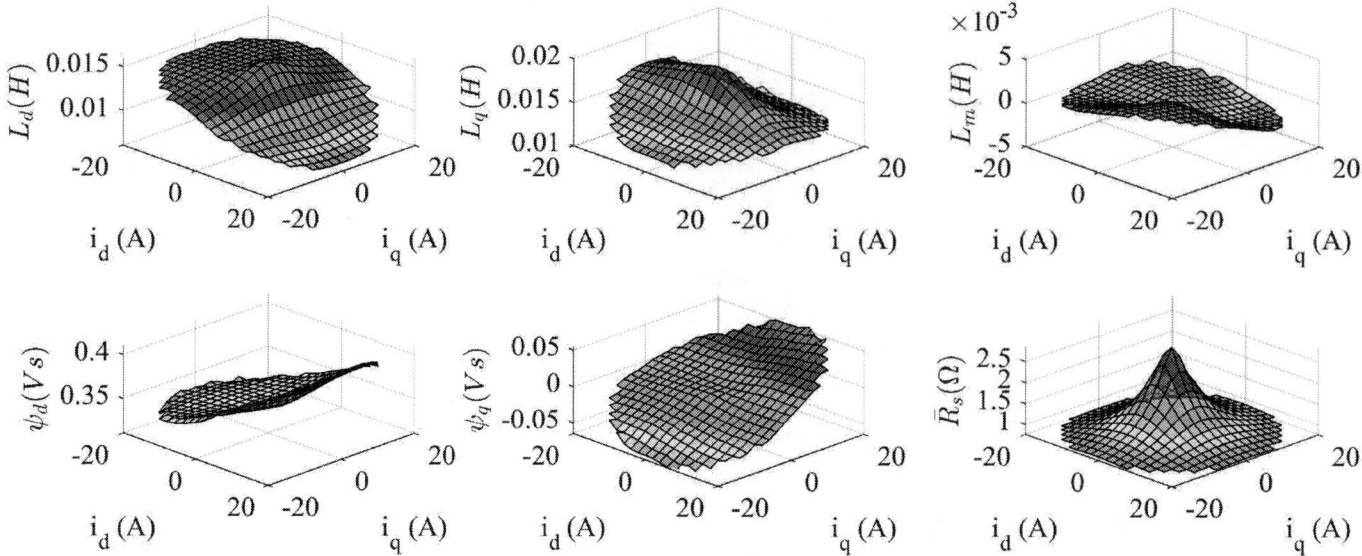

Fig. 6. Resulting drivetrain parameters for a two-level VSI three-phase PMSM (linearly interpolated)

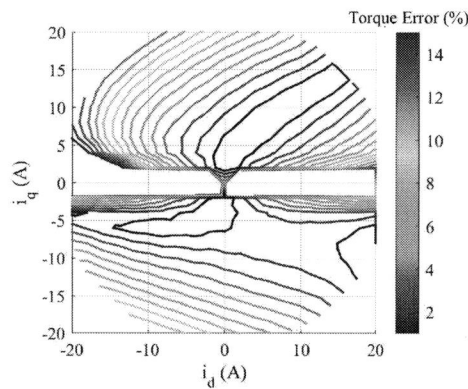

Fig. 7. Nameplate torque error $T_{\text{error}} = \frac{|T_{\text{name}} - T_{\text{param}}|}{|T_{\text{param}}|} \cdot 100$

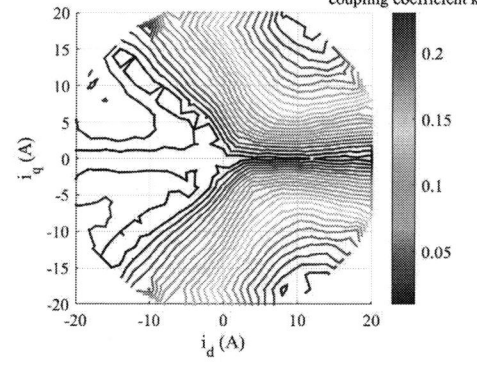

Fig. 8. PMSM Experimental Coupling Coefficient $k = L_m / \sqrt{L_d L_q}$

The estimated parameters can be solved by minimizing the residual of the equation $\mathbf{A}x - b = 0$, which for the overdetermined case has the known solution $x = \mathbf{A}^\dagger b$ using unconstrained least squares. If the problem is set up correctly, the self-inductances L_d and L_q as well lumped resistance $\bar{R}_s$ will be strictly positive. The mutual inductance L_m and flux offsets ψ_d, ψ_q may be positive or negative depending on the contour of the flux map (see Fig. 2).

The reference speed ω does not strictly need to be held constant (dyno may oscillate or have steady state offsets), and in practice may vary for each measurement, this will not change the rank of the matrix as measurements for each setpoint are averaged. The speed ω should be kept to a value low enough where $v_{\text{pm,core,dq}}$ is small and the machine will not exceed rated power across all i_{dq} setpoints, but speed should be large enough for the back-emf voltages ($\omega\lambda_{dq}$) to be large enough to be accurately measured. A speed of one third base speed is recommended [9].

IV. EXPERIMENTAL RESULTS

An internally-designed two-level three-phase VSI was used utilizing SiC switches. The DC bus voltage was 750V. The switching and control frequency were fixed to 20kHz. The switches are the Wolfspeed C2M0025120D, which are $1200V$ V_{ds} rated, and have an on-resistance of 25mΩ at room temperature.

TABLE I
PARAMETER ESTIMATION SUMMARY

Parameter	$i_{dq} = 0A$	$i_d = [20A, 0A]$	$i_q = [0A, 20A]$
$\bar{R}_s$ (Ω)	2.8335	0.8069	0.8617
L_d (mH)	15.4	7.0	12.3
L_q (mH)	19.1	13.4	12.0
L_m (mH)	≈ 0.0	≈ 0.0	≈ 0.0
ψ_d (Vs)	0.343	0.408	0.350
ψ_q (Vs)	-0.005	-0.004	≈ 0.0

The controller used was the DSPACE MicroAutoBox III DS1513 and datalogging was handled by DSPACE ControlDesk. The speed-controlled drive consists of an open-loop controlled VFD drive AE-V81 and Baldor CEM3770T induction machine. The open-loop speed control leads to steady state errors of $\pm 20 - 30\%$ off speed reference depending on the load torque. The setup is shown in Fig. 5.

The PMSM used is the SyMax 213TPFSA10096A, the drive has nameplate parameters $L_d = 15.4$mH, $L_q = 19.05$mH, $\bar{R}_s = 2.833\Omega$, $\psi_{pm} = 0.343$Vs, and 5 pole pairs. The machine's thermal current limit $\|i_{dq}\| \leq i_{max}$ is $i_{max} = 20$A. Within this current limit the setpoints were regularly gridded 2A apart for a total of 373 setpoints. Setpoints have atleast three adjacent points, most have four (see Fig. 3). The electrical transient time between setpoints was roughly 10ms, shown in Fig. 4. Each setpoint was run for 3s and the measurement averaged, averaging space harmonics. The reference speed ω was fixed to 157.1 rads/s (300 RPM) and the experiment was repeated for positive and negative speeds, eliminating machine asymmetries. As can be seen in Fig. 4 there is a $\pm 10\%$ variation in currents due to space (position-dependent) harmonics.

The machine parameters were solved for each setpoint using (20) and are shown in Fig. 6, and summarized in TABLE I. As expected, the self inductances L_d and L_q decrease as the machine reaches higher current in i_d and i_q respectively, due to saturation effects; the mutual inductance L_m is positive and negative but overall is very small, near zero. The flux offset ψ_d at the origin is the PM flux $\psi_d|_{i_{dq}=0} = \psi_{pm} = .343Vs$ and increases as i_d increases and decrease when i_d decreases. Flux offset ψ_q acts similarly along the i_q axis but with no offset. The lumped resistance term $\bar{R}_s$ near the origin ($i_{dq} = 0$) is large (2.83Ω) and dominated by the core loss resistance term in (15), because $v_{pm,core}$ is primarily dependent on ω and ψ_{pm} which are both constants and i_{dq} is very small. Then as the magnitude of current $\|i_{dq}\|$ increases, the $v_{pm,core}$ term subsequently shrinks, and resistive voltage drop term in (15) dominates, $\bar{R}_s$ drops to 0.81Ω.

The parameterized torque from (18) is calculated for each of the 373 setpoints and compared to the nameplate torque from (19), the error shown in Fig. 7. The error is calculated as

$$T_{error} = \frac{|T_{name} - T_{param}|}{|T_{param}|} \cdot 100, \qquad (21)$$

and ranges from near zero to $\approx 15\%$. The error is highest at large negative d-axis current and small q-axis current. The error is difficult to measure near $i_q \approx 0$ as the torque approaches zero. The coupling coefficient k from (5) is shown in Fig. 8, it varies from near zero to .23.

V. CONCLUSION AND FUTURE WORK

A methodology for computing physical parameters of a two-level VSI three-phase PMSM drivetrain is presented using a constant speed load test with only current sensors and a position sensor. The resulting parameters demonstrate satisfactory lumped parameter estimation for the inverter and physical inductances and resistances. The parameters were successfully estimated and physical, ensuring 1) the inductance matrix is positive semidefefinite, and 2) the plant is stable. The experiment was robust against $\pm 20 - 30\%$ speed errors, $\pm 10\%$ current spatial harmonics, and models torque up to 15% more accurately than just the nameplate value. Future work will include more terms to model nonlinearities, and motor drive topological changes.

REFERENCES

[1] M. S. Rafaq and J.-W. Jung, "A comprehensive review of state-of-the-art parameter estimation techniques for permanent magnet synchronous motors in wide speed range," *IEEE Transactions on Industrial Informatics*, vol. 16, DOI 10.1109/TII.2019.2944413, no. 7, pp. 4747–4758, 2020.

[2] G. Pei, J. Liu, L. Li, P. Du, L. Pei, and Y. Hu, "Mras based online parameter identification for pmsm considering vsi nonlinearity," in *2018 IEEE International Power Electronics and Application Conference and Exposition (PEAC)*, DOI 10.1109/PEAC.2018.8590394, pp. 1–7, 2018.

[3] L. Zhou, M. Eull, and M. Preindl, "Derivation and review of optimization-based estimation for motor drives and power electronics," *IEEE Transactions on Industry Applications*, vol. 60, DOI 10.1109/TIA.2024.3382639, no. 4, pp. 6593–6611, 2024.

[4] X. Li and R. Kennel, "Comparison of state-of-the-art estimators for electrical parameter identification of pmsm," in *2019 IEEE International Symposium on Predictive Control of Electrical Drives and Power Electronics (PRECEDE)*, DOI 10.1109/PRECEDE.2019.8753197, pp. 1–6, 2019.

[5] A. Brosch, S. Hanke, O. Wallscheid, and J. Böcker, "Data-driven recursive least squares estimation for model predictive current control of permanent magnet synchronous motors," *IEEE Transactions on Power Electronics*, vol. 36, DOI 10.1109/TPEL.2020.3006779, no. 2, pp. 2179–2190, 2021.

[6] B. Steyaert, E. Swint, W. W. Pennington, and M. Preindl, "Piecewise affine magnetic modeling of permanent-magnet synchronous machines for virtual-flux control," *Energies*, vol. 15, no. 19, 2022.

[7] B. W. Steyaert, E. Swint, W. W. Pennington, and M. Preindl, "Piecewise affine modeling of wound-rotor synchronous machines for real-time motor control," *IEEE Transactions on Industrial Electronics*, vol. 70, no. 6, pp. 5571–5580, 2023.

[8] G.-H. Kang, J.-P. Hong, G.-T. Kim, and J.-W. Park, "Improved parameter modeling of interior permanent magnet synchronous motor based on finite element analysis," *IEEE Transactions on Magnetics*, vol. 36, DOI 10.1109/20.877809, no. 4, pp. 1867–1870, 2000.

[9] E. Armando, R. I. Bojoi, P. Guglielmi, G. Pellegrino, and M. Pastorelli, "Experimental identification of the magnetic model of synchronous machines," *IEEE Transactions on Industry Applications*, vol. 49, DOI 10.1109/TIA.2013.2258876, no. 5, pp. 2116–2125, 2013.

[10] B. Hesterman, "Analysis and modeling of magnetic coupling," 01 2007.

[11] O. Wallscheid and J. Böcker, "Global identification of a low-order lumped-parameter thermal network for permanent magnet synchronous motors," *IEEE Transactions on Energy Conversion*, vol. 31, DOI 10.1109/TEC.2015.2473673, no. 1, pp. 354–365, 2016.

[12] K. Rahman and S. Hiti, "Identification of machine parameters of a synchronous motor," *IEEE Transactions on Industry Applications*, vol. 41, DOI 10.1109/TIA.2005.844379, no. 2, pp. 557–565, 2005.

[13] W. Liang, J. Wang, P. C.-K. Luk, W. Fang, and W. Fei, "Analytical modeling of current harmonic components in pmsm drive with voltage-source inverter by svpwm technique," *IEEE Transactions on Energy Conversion*, vol. 29, DOI 10.1109/TEC.2014.2317072, no. 3, pp. 673–680, 2014.

[14] I. Lar and M. M. Radulescu, "Equivalent core-loss resistance identification for interior permanent-magnet synchronous machines," in *2012 XXth International Conference on Electrical Machines*, DOI 10.1109/ICElMach.2012.6350104, pp. 1667–1671, 2012.

[15] J.-W. Choi and S.-K. Sul, "Inverter output voltage synthesis using novel dead time compensation," *IEEE Transactions on Power Electronics*, vol. 11, DOI 10.1109/63.486169, no. 2, pp. 221–227, 1996.

A Novel Two-Dimensional Random Switching Frequency PWM Method for Variable Frequency Drives

Mostafa Abarzadeh
Industrial Control Division (ICD)
Eaton Corporation
Menomonee Falls, WI, US
MostafaAbarzadehBayrami@eaton.com

Kevin Lee
Industrial Control Division (ICD)
Eaton Corporation
Menomonee Falls, WI, US
KevinLee@Eaton.com

Abstract— In this paper, a new two-dimensional random switching frequency pulse-width-modulation (2D-RSF-PWM) method is proposed for variable frequency drives (VFDs). The proposed 2D-RSF-PWM method is based on two newly defined parameters which are spread factor and number of repetitions. Hence, random switching frequency pattern with limited variation band of switching frequency is achieved by selecting the proper values of these two parameters. Accordingly, applying the proposed 2D-RSF-PWM method results in improved conductive EMI performance at high frequencies, enhanced switching noise reduction at low frequencies, and better total harmonic distortion (THD) in comparison to the conventional SPWM and RSF-PWM methods. Moreover, the input EMI filters are designed for both cases of applying the proposed 2D-RSF-PWM and the traditional SPWM methods to meet the IEC 61800-3 standard Category C2 requirements. Comparison between two designed input EMI filters to meet IEC 61800-3 standard demonstrates that the required values of CM and DM inductors of the input EMI filter are decreased by factor of 3 by applying the proposed 2D-RSF-PWM method in comparison to those values by applying the conventional SPWM method. The presented results and theoretical analysis verify the performance of the proposed 2D-RSF-PWM method.

Keywords— *EMI reduction, pulse width modulation, variable frequency drive, voltage source inverter, Random switching frequency pulse width modulation.*

I. INTRODUCTION

VFDs have been widely used in various applications such as industrial motor drives, automotive, marine, and electric aircraft propulsion systems. Increasing trends toward reduction of emanated electromagnetic interference (EMI) of VFDs at high frequencies to meet the relevant standards requirements, and reduction of switching noise at low frequencies to decrease the acoustic noise, whistling sound, and vibration of electric motors necessitate special attention and consideration for VFDs. The provided solutions to reduce emanated EMI of power electronic converters can be categorized into 1) filtering and suppression of the generated EMI by utilizing passive and active EMI filters and 2) reducing the generated EMI of converter by applying advanced modulation methods. Various passive, hybrid, and active EMI filters have been introduced in the literature for two-level and multilevel converters to filter and suppress the emanated EMI of converters. However, these filters usually require higher number of stages and higher volume, and they generate higher power loss and compel extra cost to overall converter architecture [1-5].

Reduction of emanated EMI noise at high frequencies by applying advanced modulation methods is an alternate and more effective approach in comparison to utilizing bulky, lossy, and costly passive input EMI filters and output filters to achieve the desired EMI performance at the input and the demanded operating performance of the electric motor at the output [1-5]. Various random switching frequency PWM (RSF-PWM) methods have been proposed to spread the spectrum of band of harmonics of switching frequency to reduce the peak value of amplitude of switching harmonic spectrum, and thus reduce the size of the required EMI filter [6-18]. In [6], the dithering technique with limited switching frequency range has been applied to reduce the emanated EMI of the AC drives. In [7, 8], various chaotic PWM methods have been described for EMI and mechanical resonance reduction in motor drives for electric vehicle (EV) applications. In [9], the EMI reduction technique by employing a periodic switching frequency modulation has been introduced. The presented EMI reduction in [9] utilizes the periodic pattern to change the switching frequency instead of randomizing the switching frequency. In [10], the effect of the modulation techniques on acoustic noise of inverters has been studied. In [11], the constant sampling frequency RSF-PWM technique has been introduced for DC-DC buck converter. In [12], the reduced switching frequency RSF-PWM method has been introduced for 2L inverters. In [14], the selective frequency cancellation method has been proposed for random space vector PWM method. In [15], the RSF-PWM method by utilizing different switching frequencies has been introduced for AC drives. In [16], the constant switching frequency random PWM method has been introduced for DC-DC buck converter. However, utilizing RSF-PWM methods deteriorates the output current THD, increases the switching noise at low frequencies, can cause mechanical resonance of the electric motor, can interfere with closed-loop control performance of the VFD, and can interfere and deteriorate the performance of the output filter of the VFD [6-18]. Hence, in this paper, the new 2D-RSF-PWM method is proposed to address and solve the above-mentioned issues with the introduced SPWM and RSF-PWM methods.

979-8-3315-1612-3/25 $31.00 © 2025 IEEE

II. THE PROPOSED 2D-RSF-PWM METHOD

The proposed 2D-RSF-PWM method is based on two newly defined parameters which are the spread factor (δ) and the number of repetitions (k). The spread factor provides limited variation band of the random switching frequency around main carrier frequency, and the number of repetitions provides the update rate of switching frequency variation in the defined limited variation band. Hence, in contrast with the introduced RSF-PWM methods in the literature, the proposed 2D-RSF-PWM method is a pseudo RSF-PWM with three different frequencies which are the main carrier frequency, limited variation band of the random switching frequency around the main carrier frequency defined by the spread factor, and the frequency of update rate of the carrier frequency variation defined by the number of repetitions. The theoretical analysis and proof of the proposed 2D-RSF-PWM method is presented as in the following. Generated random pulse train by applying the proposed 2D-RSF-PWM method is shown in Fig. 1.

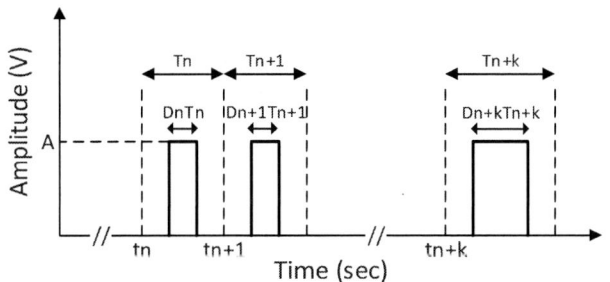

Fig. 1. Generated random pulse train by applying the proposed 2D-RSF-PWM method.

The modulated pulse train by applying pulse width modulation (PWM) technique with triangular carrier signal is expressed as follows,

$$g_n(t)$$
$$= \begin{cases} 0, & t \le t_n + \dfrac{T_n - D_n T_n}{2} \\ A, & t_n + \dfrac{T_n - D_n T_n}{2} < t \le t_n + \dfrac{T_n + D_n T_n}{2} \\ 0, & t > t_n + \dfrac{T_n + D_n T_n}{2} \end{cases} \quad (1)$$

$$g(t) = \lim_{N \to \infty} \sum_{n=1}^{N} g_n(t) \quad (2)$$

Where A is the amplitude of the pulse train in ON state, D_n is pulse width of the modulated signal, $T_n = 1/f_n$ is the time period of nth switching signal and t_n is the start point of T_n time period. Accordingly, by applying Wiener–Khintchine theorem to calculate the power spectral density (PSD) of $g(t)$ and in order to minimize or completely eliminate the certain frequency component from the output voltage of the converter, the PSD of pulse train of $g(t)$ at the specific target frequency of f_{tgt} should satisfy the following equation,

$$\begin{aligned} G(\omega_{tgt}) &= \int_{-\infty}^{+\infty} g(t) e^{-j\omega_{tgt}t} \\ &= \int_{-\infty}^{+\infty} g(t) \cos(\omega_{tgt}t)\, dt \\ &\quad - j \int_{-\infty}^{+\infty} g(t) \sin(\omega_{tgt}t)\, dt = 0 \end{aligned} \quad (3)$$

Hence, in the proposed 2D-RSF-PWM method, in order to minimize the emanated noise at the specific target frequency of f_{tgt}, the following generalized condition should be satisfied for the defined $c(f_{tgt})$ as defined in (4) and (5).

Hence, considering the fact that the variation of reference signal is negligible in comparison to carrier frequency, the switching frequencies of f_m should be selected in a way that the nth term of the first summation of (5) and $(n + k)^{th}$ term of the second summation of (5) are cancelled out. Accordingly, for each n, condition (6) should be satisfied. The required condition to satisfy (6) is expressed in (7) and t_{n+k} is defined as follows,

$$c(f_{tgt}) = \int_{-\infty}^{+\infty} g(t) \sin(2\pi f_{tgt}t + \varphi)\, dt = \lim_{N \to \infty} \sum_{m=1}^{N} \left(\int_{t_m + \frac{1/f_m - D_m/f_m}{2}}^{t_m + \frac{1/f_m + D_m/f_m}{2}} A \sin(2\pi f_{tgt}t + \varphi)\, dt \right) = 0 \quad (4)$$

$$\to c(f_{tgt}) = \frac{A}{\omega_{tgt}} \sum_{m=1}^{\infty} \cos\left[\omega_{tgt}\left(t_m + \frac{1/f_m - D_m/f_m}{2} \right) + \varphi \right] - \frac{A}{\omega_{tgt}} \sum_{m=1}^{\infty} \cos\left[\omega_{tgt}\left(t_m + \frac{1/f_m + D_m/f_m}{2} \right) + \varphi \right] = 0 \quad (5)$$

$$\frac{A}{\omega_{tgt}} \cos\left[\omega_{tgt}\left(t_n + \frac{1/f_n - D_n/f_n}{2} \right) + \varphi \right] - \frac{A}{\omega_{tgt}} \cos\left[\omega_{tgt}\left(t_{n+k} + \frac{1/f_{n+k} + D_{n+k}/f_{n+k}}{2} \right) + \varphi \right] = 0 \quad (6)$$

$$2\pi f_{tgt}\left(t_n + \frac{1/f_n - D_n/f_n}{2} \right) + \varphi + 2\kappa\pi = 2\pi f_{tgt}\left(t_{n+k} + \frac{1/f_{n+k} + D_{n+k}/f_{n+k}}{2} \right) + \varphi \quad (7)$$

$$t_{n+k} = t_n + \sum_{m=n}^{n+k-1} {}^1\!/_{f_m} \tag{8}$$

Hence, by substituting (8) into (7), the corresponding switching frequency of f_{n+k} to minimize or completely remove the PSD at f_{tgt} is calculated as follows,

$$f_{n+k} = \frac{f_{tgt}(1 + D_{n+k})}{2\kappa + {}^{f_{tgt}}\!/_{f_n}(1 - D_n) - 2\sum_{m=n}^{n+k-1} {}^{f_{tgt}}\!/_{f_m}} \tag{9}$$

In (9), k represents the number of repetitions of carrier frequency and κ is an integer number which is inversely proportional to the value of the switching frequency (f_{n+k}) and it is proportional to the range of carrier frequency variation (f_{min}, f_{max}). Therefore, by increasing the value of κ, the value of switching frequency is decreased whereas the range of carrier frequency variation (f_{min}, f_{max}) is increased. The impact of the value of κ on the switching frequency and the range of carrier frequency variation is shown in (10) by setting the number of repetitions to $k = 1$.

$$f_{n+1} = \frac{f_{tgt}(1 + D_{n+1})}{2\kappa - {}^{f_{tgt}}\!/_{f_n}(1 + D_n)} \tag{10}$$

Hence, (9) proves that the proposed 2D-RSF-PWM method can be achieved by changing the values of the number of repetitions of carrier frequency (k) and the factor of κ. Accordingly, considering (9), the following two new parameters are defined in the proposed 2D-RSF-PWM method to represent the impact of k and κ.

$$\delta = \frac{\Delta f_C}{f_C}$$
$$f_{rep} = {}^{f_C}\!/_k : \text{The frequency of change of carrier} \tag{11}$$
frequency variation

where f_C is the main carrier frequency, and Δf_C is the range of carrier frequency variation (f_{min}, f_{max}). As expressed in (11), the newly defined parameter δ represents the value of spread factor and the parameter k represents the values of the number of repetitions of carrier frequency in the proposed 2D-RSF-PWM method. Accordingly, the spread energy band (B) of the output voltage around the main carrier frequency by applying the proposed 2D-RSF-PWM method is calculated as,

$$B = \delta \cdot f_C + {}^{f_C}\!/_k \tag{12}$$

Moreover, the spread energy band (B_h) of the output voltage around h^{th} harmonic order of the carrier frequency by applying the proposed 2D-RSF-PWM method is calculated as,

$$B_h = \delta \cdot f_C \cdot h + {}^{f_C}\!/_k \tag{13}$$

As presented in (12) and (13), employing the spread factor of $\delta < 1$ and $k > 1$ in the proposed 2D-RSF-PWM method results in two degrees of freedom selective harmonic spectrum amplitude reduction at various frequencies. Accordingly, at lower frequencies and considering smaller value of h, the number of repetitions of carrier frequency (k) has more impact

on the spread energy band. On the other hand, at higher frequencies and by increasing the value of h, the impact of spread factor (δ) value is more dominant in the spread energy band. Therefore, by proper selection of values of δ and k, both conducted EMI reduction at the range of 150 kHz-30 MHz and harmonic spectrum amplitude reduction at the frequencies lower than 150 kHz can be achieved. The spectrums of spread bands of harmonics by applying the SPWM, the conventional RSF-PWM, and the proposed 2D-RSF-PWM methods are presented in Fig. 2.

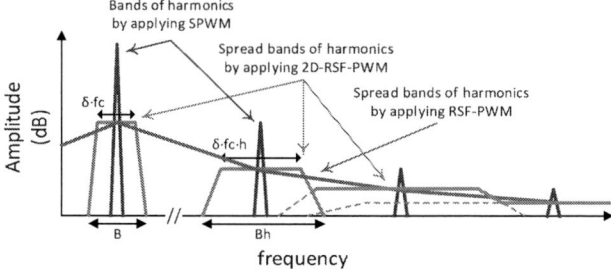

Fig. 2. Spectrums of spread bands of harmonics by applying the SPWM (Blue), the conventional RSF-PWM (Red), and the proposed 2D-RSF-PWM (Green) methods.

The major novelties and improvements of the proposed 2D-RSF-PWM method are as follows. 1) The proposed 2D-RSF-PWM method is a tunable pseudo RSF-PWM method based on two newly defined parameters. Hence, by proper selection of the value of spread factor and the number of repetitions, a random switching frequency pattern with limited variation band of switching frequency is achieved. 2) By proper selection of the spread factor and the number of repetitions in the proposed 2D-RSF-PWM method, improved conductive EMI performance at high frequencies, enhanced switching noise reduction at low frequencies, and better total harmonic distortion (THD) are achieved in comparison to the conventional SPWM and RSF-PWM methods.

III. RESULTS

The schematic diagram of the VFD with the input EMI filter and LISN network, and the implemented experimental setup are presented in Figs. 3 and 4, respectively. In Fig. 3, C_{par} represents the lumped parasitic capacitors between the outputs and DC link rails to the ground. The parameters of the implemented VFD are presented in Table I. Fig. 5 presents the frequency spectrum of the output phase voltage of the VFD at the range of 1 MHz by applying the SPWM, RSF-PWM, and the proposed 2D-RSF-PWM methods. As presented in Fig. 5 and Table II, the maximum amplitude of switching noise of the output voltage is decreased by 10 dB by applying the proposed 2D-RSF-PWM method in comparison to that value by applying the SPWM method. In addition, the maximum amplitude of switching noise of the output voltage is at 11.94 kHz, 12 kHz, and 9 kHz by applying the proposed 2D-RSF-PWM, the SPWM, and the RSF-PWM methods, respectively. Therefore, applying the proposed 2D-RSF-PWM method does not deviate the maximum amplitude of switching noise from the main carrier frequency, whereas applying the RSF-PWM method leads to at least 25% deviation in that value in comparison to the proposed

2D-RSF-PWM and the SPWM methods. Hence, by employing the proposed 2D-RSF-PWM method, the VFD is much less prone to unwanted resonances with the motor and the passive filter network in comparison to applying the RSF-PWM method. Moreover, frequency spectrum of the output phase voltage of the VFD at low frequency range of <150 kHz by applying above-mentioned three methods is shown in Fig. 6. As presented in Fig. 6, the switching noise at low frequency range and between the multiples of the main carrier frequency is decreased by up to 46 dB by applying the proposed 2D-RSF-PWM in comparison to that value by employing the RSF-PWM. Therefore, the achieved switching noise reduction at low frequency range by applying the proposed 2D-RSF-PWM leads to significant reduction of overlapping the mechanical resonant frequency of the motor and motor vibration. Furthermore, the load current THD values by applying the SPWM, the RSF-PWM, and the proposed 2D-RSF-PWM methods are presented in Table II. As presented in Table II, the load current THD values are 6.46%, 9.34%, and 6.60% by applying the SPWM, the RSF-PWM, and the proposed 2D-RSF-PWM methods, respectively. Hence, by applying the proposed 2D-RSF-PWM method, the current THD value is reduced by 2.74% in comparison to that value by applying the RSF-PWM method. In addition, as presented in Table II, by applying the proposed 2D-RSF-PWM method, the switching noise at 150 kHz is decreased by 9 dBμV in comparison to that value by applying the SPWM method. Hence, the required attenuation of the input EMI filter to meet the IEC 61800-3 standard Category C2 is 9 dB less than that value by applying the SPWM method. The values of components of the designed input EMI filters for both cases of applying the proposed 2D-RSF-PWM and the SPWM methods are presented in Table III. As can be seen in Table III, the values of CM and DM inductors are decreased by factor of 3 by applying the proposed 2D-RSF-PWM method in comparison to those values by applying the conventional SPWM method. Fig. 7 shows the conducted EMI of the VFD by utilizing the input EMI filter Design#1 and by applying the SPWM, the RSF-PWM, and the proposed 2D-RSF-PWM methods. As presented in Fig. 7, utilizing the input EMI filter Design#1 and applying the proposed 2D-RSF-PWM method fulfills the IEC 61800-3 standard Category C2 requirements.

TABLE I: PARAMETERS OF THE IMPLEMENTED VFD

VFD and PWM Methods		
Parameter		**Value(s)**
DC-link Voltage		680 VDC
DC-link Capacitance		1100 μF
Fundamental frequency		60 Hz
The Proposed 2D-RSF-PWM	Main carrier frequency	12 kHz
	δ	0.2
	k	5
SPWM carrier frequency		12 kHz
RSF-PWM Max. frequency		12 kHz

TABLE II: COMPARISON BETWEEN SPWM, RSF-PWM, AND THE PROPOSED 2D-RSF-PWM METHODS

Parameter	PWM Method	Value(s)
Switching Noise at the Output Voltage @ 150 kHz	SPWM	138 dBμV
	RSF-PWM	129 dBμV
	Proposed 2D-RSF-PWM	129 dBμV
Output Current THD	SPWM	6.46%
	RSF-PWM	9.34%
	Proposed 2D-RSF-PWM	6.60%

TABLE III: INPUT EMI FILTER PARAMETERS FOR THE PROPOSED 2D-RSF-PWM AND THE SPWM METHODS

	Parameter	Value(s)
Design#1: for the proposed 2D-RSF-PWM method	L_{CM}	1.5 mII
	L_{DM}	15 μH
	C_Y	3.3 nF
	C_X	1 μF
Design#2: for the SPWM method	L_{CM}	4.5 mH
	L_{DM}	45 μH
	C_Y	3.3 nF
	C_X	1 μF

IV. CONCLUSIONS

The new 2D-RSF-PWM method was proposed for VFD applications. The proposed 2D-RSF-PWM method is a tunable pseudo RSF-PWM method with two newly defined parameters which are the spread factor and the number of repetitions. Therefore, improved conductive EMI performance at high frequencies, enhanced switching noise reduction at low frequencies, and better total harmonic distortion (THD) were achieved by applying the proposed 2D-RSF-PWM method in comparison to the conventional SPWM and RSF-PWM methods. Moreover, the input EMI filters were designed for both cases of applying the proposed 2D-RSF-PWM and the traditional SPWM methods to meet the IEC 61800-3 standard Category C2 requirements. Accordingly, the required values of CM and DM inductors of the input EMI filter were decreased by factor of 3 by applying the proposed 2D-RSF-PWM method in comparison to those values by applying the conventional SPWM method. The presented results and theoretical analysis verified the performance of the proposed 2D-RSF-PWM method.

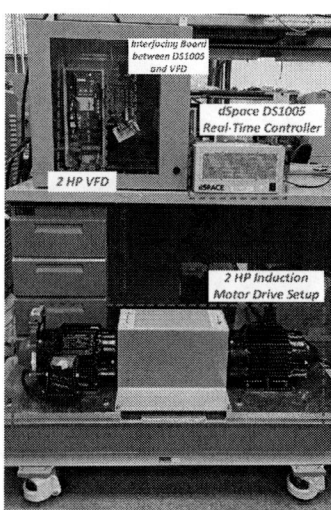

Fig. 4. The implemented experimental setup.

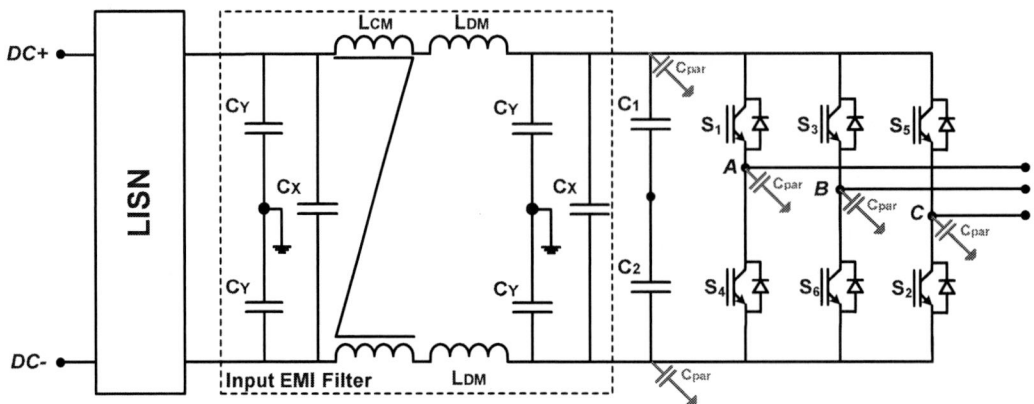

Fig. 3. The schematic diagram of the VFD with the input EMI filter.

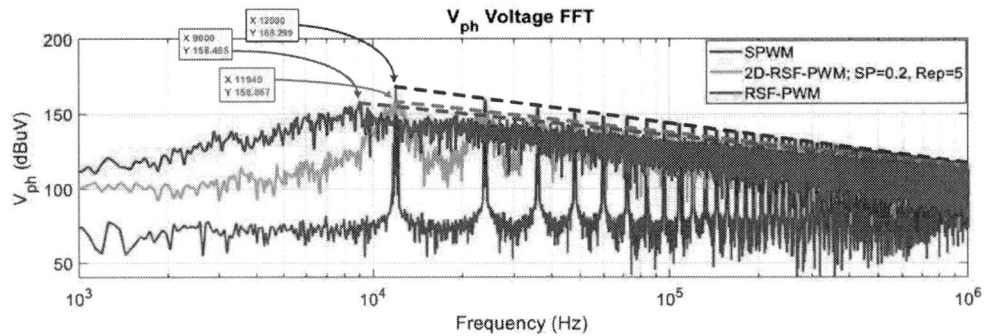

Fig. 5. Comparison between frequency spectrum of the output phase voltage of the VFD at the range of 1 MHz by applying the SPWM (Blue), the RSF-PWM (Red), and the proposed 2D-RSF-PWM (Green) methods.

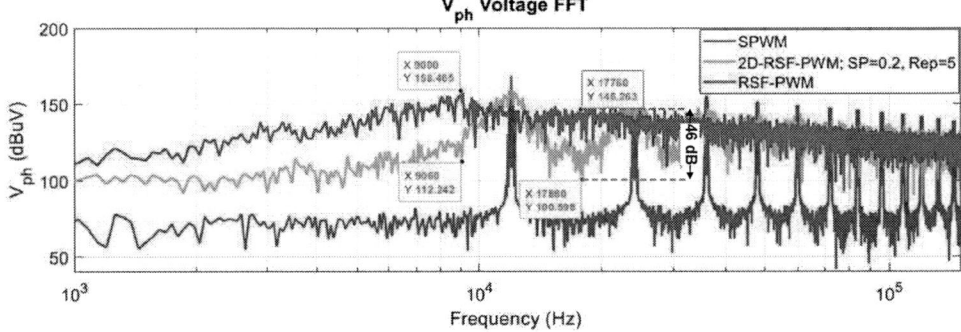

Fig. 6. Comparison between frequency spectrum of the output phase voltage of the VFD at low frequency range of <150 kHz by applying the SPWM (Blue), the RSF-PWM (Red), and the proposed 2D-RSF-PWM (Green) methods.

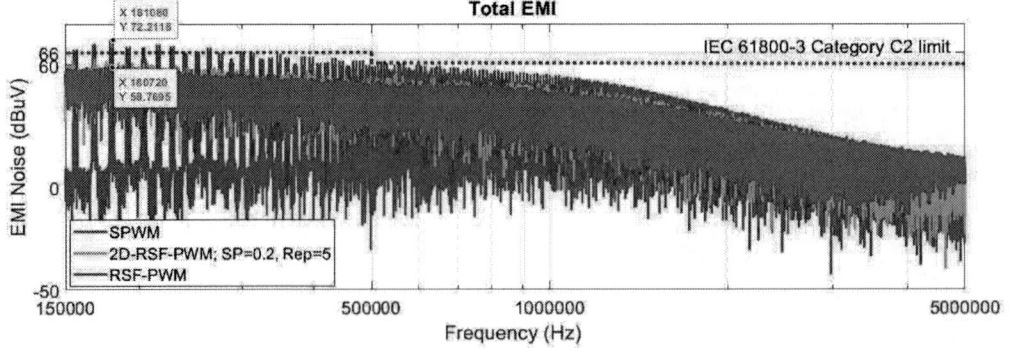

Fig. 7. The conducted EMI of the VFD by utilizing the input EMI filter Design#1 and by applying the SPWM (Blue), the RSF-PWM (Red), and the proposed 2D-RSF-PWM (Green) methods.

979-8-3315-1612-3/25 $31.00 © 2025 IEEE   1265

REFERENCES

[1] K. Lee, G. Shen, W. Yao and Z. Lu, "Performance Characterization of Random Pulse Width Modulation Algorithms in Industrial and Commercial Adjustable-Speed Drives," IEEE Transactions on Industry Applications, vol. 53, no. 2, pp. 1078-1087, March-April 2017.

[2] D. Han, W. Lee, S. Li and B. Sarlioglu, "New Method for Common Mode Voltage Cancellation in Motor Drives: Concept, Realization, and Asymmetry Influence," IEEE Transactions on Power Electronics, vol. 33, no. 2, pp. 1188-1201, Feb. 2018.

[3] C. T. Morris, D. Han and B. Sarlioglu, "Reduction of Common Mode Voltage and Conducted EMI Through Three-Phase Inverter Topology," IEEE Transactions on Power Electronics, vol. 32, no. 3, pp. 1720-1724, March 2017.

[4] G. Ala, G. C. Giaconia, G. Giglia, M. C. Di Piazza and G. Vitale, "Design and Performance Evaluation of a High Power-Density EMI Filter for PWM Inverter-Fed Induction-Motor Drives," IEEE Transactions on Industry Applications, vol. 52, no. 3, pp. 2397-2404, May-June 2016.

[5] Z. Zhang, L. Wei, P. Yi, Y. Cui, P. S. Murthy and A. M. Bazzi, "Conducted Emissions Suppression of Active Front End (AFE) Drive Based on Random Switching Frequency PWM," IEEE Transactions on Industry Applications, vol. 56, no. 6, pp. 6598-6607, Nov.-Dec. 2020.

[6] A. A. Fardoun and E. H. Ismail, "Reduction of EMI in AC Drives Through Dithering Within Limited Switching Frequency Range," IEEE Transactions on Power Electronics, vol. 24, no. 3, pp. 804-811, March 2009.

[7] Z. Zhang, T. W. Ching, C. Liu and C. H. T. Lee, "Comparison of chaotic PWM algorithms for electric vehicle motor drives," IECON 2012 - 38th Annual Conference on IEEE Industrial Electronics Society, Montreal, QC, Canada, 2012, pp. 4087-4092.

[8] Zheng Wang, K. T. Chau and M. Cheng, "A chaotic PWM motor drive for electric propulsion," 2008 IEEE Vehicle Power and Propulsion Conference, Harbin, China, 2008, pp. 1-6.

[9] D. Gonzalez et al., "Conducted EMI Reduction in Power Converters by Means of Periodic Switching Frequency Modulation," IEEE Transactions on Power Electronics, vol. 22, no. 6, pp. 2271-2281, Nov. 2007.

[10] I. P. Tsoumas and H. Tischmacher, "Influence of the inverter's modulation technique on the audible noise of electric motors," 2012 XXth International Conference on Electrical Machines, Marseille, France, 2012, pp. 2981-2987.

[11] Y. -S. Lai, Y. -T. Chang and B. -Y. Chen, "Novel Random-Switching PWM Technique With Constant Sampling Frequency and Constant Inductor Average Current for Digitally Controlled Converter," IEEE Transactions on Industrial Electronics, vol. 60, no. 8, pp. 3126-3135, Aug. 2013.

[12] S. Bhattacharya, D. Mascarella, G. Joos and G. Moschopoulos, "Reduced switching random PWM technique for two-level inverters," 2015 IEEE Energy Conversion Congress and Exposition (ECCE), Montreal, QC, Canada, 2015, pp. 695-702.

[13] Y. Shi, Y. Zhang, L. Wang and H. Li, "Reduction of EMI Noise Due to Nonideal Interleaving in a 100 kW SiC PV Converter," IEEE Transactions on Power Electronics, vol. 34, no. 1, pp. 13-19, Jan. 2019.

[14] A. Peyghambari, A. Dastfan and A. Ahmadyfard, "Selective Voltage Noise Cancellation in Three-Phase Inverter Using Random SVPWM," IEEE Transactions on Power Electronics, vol. 31, no. 6, pp. 4604-4610, June 2016, doi: 10.1109/TPEL.2015.2473001.

[15] S. A. Jawdeh, P. Li and A. Bazzi, "A Novel Random PWM Technique for Inverters and AC Drives," 2024 IEEE Transportation Electrification Conference and Expo (ITEC), Chicago, IL, USA, 2024, pp. 1-5.

[16] T. Morales-Leal, A. Moreno-Munoz, M. A. Ortiz-López, S. R. Geninatti and F. J. Quiles-Latorre, "New Random PWM Method at Constant Switching Frequency and Maximum Harmonic Reduction Created With a Flexible FPGA-Based Test Bench," IEEE Access, vol. 11, pp. 19385-19394, 2023.

[17] P. Zhang, S. Wang and Y. Li, "Generalized N-State Random Pulse Position Discontinuous PWM for High-Frequency Harmonics Reduction and Performance Improvement at High Modulation Ratios," IEEE Transactions on Power Electronics, Early Access, doi: 10.1109/TPEL.2024.3426564.

[18] W. Zhao, J. Feng, T. Tao, C. Wang and S. Liu, "High-Frequency Harmonics and Vibration Reduction for Dual Three-Phase PMSM Using Multiple Randomized SVPWM Strategy," IEEE Transactions on Power Electronics, vol. 39, no. 9, pp. 11455-11467, Sept. 2024.

Optimized Maximum Torque and Minimum Loss Fault-Tolerant Control Schemes for Dual Three-Phase PMSM

Maaz Syed Mohammad and Dong-Choon Lee
Department of Electrical Engineering, Yeungnam University
Gyeongsan, South Korea
Email: mohammadmaaz789@yu.ac.kr, dclee@yu.ac.kr

Abstract— **This paper proposes a fault-tolerant control (FTC) scheme for dual three-phase (DTP) permanent magnet synchronous motors (PMSMs), which is based on an optimization of loss function, utilizing a weighted sum approach to achieve both maximum torque (MT) and minimum loss (ML) control during open-switch faults (OSF) in DTP inverters. The MT control achieves a torque of 0.87 p.u., whereas the ML control further reduces losses by 20 % compared to conventional ML, simultaneously maintaining a torque of 0.76 p.u. during fault conditions. To simplify the hardware implementation, a phase shift based approach is proposed, which reduces the complexity in handling faults. Instead of requiring 12 different sets of equations to address all possible fault scenarios, this method requires only two. Experimental results have validated the effectiveness of the proposed methods and comparative analysis are provided.**

Keywords—Optimization, Fault tolerant control, Dual three phase PMSM.

I. INTRODUCTION

Multiphase drive systems offer significant advantages for modern industrial and transportation applications due to their fault tolerance, enabling continued operation even with phase or switch failures, though with reduced performance. Dual three-phase permanent magnet synchronous machines (DTP-PMSMs), with their two sets of windings arranged at a 30° phase shift, are particularly valued for reducing torque ripple and enhancing performance without sacrificing fault tolerance [1], [2].

Consequently, advancements in fault-tolerant control strategies have focused on leveraging phase redundancy, incorporating topology reconfiguration and robust closed-loop methods to sustain performance under fault conditions, while optimal post-fault current references help minimize losses and torque ripple, maximizing average torque [2], [3]. Optimal post-fault current references can be divided into two distinct categories under the vector space decomposition modelling (VSD): cases where $i_d=0$ and where $i_d\neq0$. The existing literature offers numerous optimization techniques where the current in synchronous-frame current i_d is set to zero.

In methods maintaining $i_d=0$ under fault conditions, the simplest approach is to shut down the faulty three-phase set and operate only the healthy set, yielding 0.5 p.u. torque with twice the losses ($\Sigma i^2 R_s = 6.0 R_s$) for the same torque output [4]. This leads to the development of two distinct methods to address each issue separately: maximum torque (MT) control and minimum loss (ML) control [5]. ML control provides about 0.55 p.u. torque with lower losses about 4.5 R_s, whereas MT control achieves a higher torque of 0.57 p.u. but incurs greater losses at approximately 6.0 R_s. To mitigate the

problems regarding the high losses of MT, a combined approach known as full range minimum loss (FRML) control was developed [6]. Other methodologies to achieve FRML have also been proposed in literature using online-equation-based techniques [7]. Another method was to achieve multiple switch FTC with an online MPC approach for a 5-phase PMSM drive [8].

The challenge with DTP-PMSM control lies in the numerous control adjustments required due to the six-phase, 12-switch configuration, which introduces many control variable combinations for harmonic axis management. To address this, generalized and adaptive harmonic current references have been introduced to handle ML, MT, and MT1, where MT1 achieves 0.67 p.u. torque [9]. Another approach to universal control focuses on applying both MT and ML strategies across all six faulty-phase scenarios [10]. Alternatively, optimal DC current injection into a specific faulty phase in a nine-phase induction motor has been shown to achieve an impressive 0.9 p.u. torque. For dual three-phase systems, it is suggested that this method could lead to a maximum torque increase of 0.11 p.u. compared to conventional MT for DTP-PMSM [11].

All the aforementioned techniques offer straightforward control by maintaining the i_d constraint. To achieve even greater torque with reduced losses, relaxing the constraint $i_d=0$ becomes necessary. Low copper loss methodologies have been proposed which substantially increase the torque range to 0.76 p.u. with considerably lower losses of 3.61 R_s [12]. Other approaches focus solely on maximizing torque, achieving levels of 0.66 p.u. or 0.73 p.u. [13], [14]. All these methods yield non-conventional, non-sinusoidal current references, which pose challenges for implementation in online systems. In addition, none of the existing approaches provide a generalization for these non-sinusoidal solutions.

This paper proposes an optimization methodology that employs a weighted sum approach to address open-switch faults in a two-level inverter for DTP-PMSM drives. The weighted sum approach enhances the existing loss function by adjusting the contribution of each phase, facilitating the implementation of various methods like ML and MT through this optimization process. The proposed ML method achieves approximately 0.76 p.u. torque with losses of 3.61 R_s, in contrast, the MT method yields 0.86 p.u. torque with losses of 3.94 R_s. In addition, to generalize the 12 cases for non-sinusoidal solutions, a phase shift-based generalization method is introduced. The proposed method can also produce low torque with minimal ripple and lower losses, even under single open-circuit fault conditions, ensuring reliability in critical applications such as aerospace actuation.

979-8-3315-1612-3/25 $31.00 © 2025 IEEE

II. PROPOSED OPTIMIZATION FOR ML AND MT DURING OPEN-SWITCH FAULTS.

A. VSD-Based Modelling of DTP-PMSM

For modeling the DTP-PMSM using VSD, the corresponding transformation matrix can be expressed as:

$$[VSD] = \frac{1}{3}\begin{bmatrix} 1 & -\frac{1}{2} & -\frac{1}{2} & \frac{\sqrt{3}}{2} & -\frac{\sqrt{3}}{2} & 0 \\ 0 & \frac{\sqrt{3}}{2} & -\frac{\sqrt{3}}{2} & \frac{1}{2} & \frac{1}{2} & -1 \\ 1 & -\frac{1}{2} & -\frac{1}{2} & -\frac{\sqrt{3}}{2} & \frac{\sqrt{3}}{2} & 0 \\ 0 & -\frac{\sqrt{3}}{2} & \frac{\sqrt{3}}{2} & \frac{1}{2} & \frac{1}{2} & -1 \\ 1 & 1 & 1 & 0 & 0 & 0 \\ 0 & 0 & 0 & 1 & 1 & 1 \end{bmatrix}. \tag{1}$$

The machine current and voltage vectors in the original dual three-phase axis (a,b,c,x,y,z) can be converted to three orthogonal stationary reference frames, $\alpha\text{-}\beta$, $xst\text{-}yst$ and $z_1\text{-}z_2$ by using

$$\begin{bmatrix} f_\alpha \\ f_\beta \\ f_{xst} \\ f_{yst} \\ f_{z1} \\ f_{z2} \end{bmatrix} = [VSD]\begin{bmatrix} f_a \\ f_b \\ f_c \\ f_x \\ f_y \\ f_z \end{bmatrix}, \tag{2}$$

where symbol f represents machine variables, such as current, voltage, and flux. According to VSD modelling, $\alpha\text{-}\beta$ subspace contributes to generating the rotating MMF of the machine, thereby enabling electromechanical energy conversion. On the contrary, the $xst\text{-}yst$ subspace consists of harmonics with the order of $6k \pm 1$, where $k = 1, 3, 5, \cdots$. The z_1 and z_2 subspace are for the zero sequence components.

Park transformation is used to convert the variables ($\alpha\text{-}\beta$) in the stationary reference frame to the synchronous reference frame:

$$\begin{bmatrix} f_d \\ f_q \end{bmatrix} = \underbrace{\begin{bmatrix} \cos(\theta_e) & \sin(\theta_e) \\ -\sin(\theta_e) & \cos(\theta_e) \end{bmatrix}}_{[PT]}\begin{bmatrix} f_\alpha \\ f_\beta \end{bmatrix}, \tag{3}$$

where f_d and f_q represents variables in the synchronous reference frame and θ_e does the rotor electrical angle.

Utilizing (1) - (3), the voltage equations for the DTP-PMSM can be expressed as

$$\begin{bmatrix} v_d \\ v_q \\ v_{xst} \\ v_{yst} \end{bmatrix} = \begin{bmatrix} R_s & -\omega_e L_q & 0 & 0 \\ -\omega_e L_q & R_s & 0 & 0 \\ 0 & 0 & R_s & 0 \\ 0 & 0 & 0 & R_s \end{bmatrix}\begin{bmatrix} i_d \\ i_q \\ i_{xst} \\ i_{yst} \end{bmatrix}$$
$$+ \frac{d}{dt}\begin{bmatrix} L_d & 0 & 0 & 0 \\ 0 & L_q & 0 & 0 \\ 0 & 0 & L_{ls} & 0 \\ 0 & 0 & 0 & L_{ls} \end{bmatrix}\begin{bmatrix} i_d \\ i_q \\ i_{xst} \\ i_{yst} \end{bmatrix} + \begin{bmatrix} 0 \\ \omega_e\psi_f \\ 0 \\ 0 \end{bmatrix}, \tag{4}$$

where v_d, v_q, v_{xst} and v_{yst} are the d, q, xst and yst axis voltages. i_d, i_q, i_{xst} and i_{yst} are the d, q, xst and yst axis currents. R_s is stator

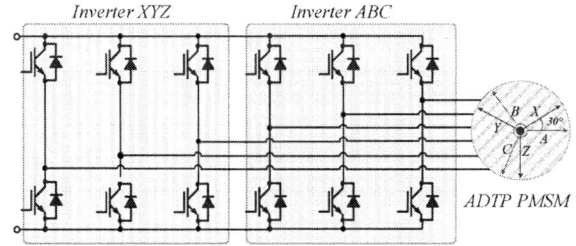

Fig. 1. DTP inverter and DTP-PMSM drive system.

resistance, ω_e is electrical angular speed of the rotor, ψ_f is rotor flux, L_d, L_q, and L_{ls} are d-axis, q-axis, and leakage inductances, respectively. The torque equation can be expressed as

$$T_e = 3P_p\psi_f i_q + 3P_p(L_d - L_q)i_d i_q, \tag{5}$$

where P_p stands for the number of pole pairs. The reluctance torque is considered zero due to the construction of the surface mount PMSM, therefore (5) can be rewritten as

$$T_e = 3P_p\psi_f i_q. \tag{6}$$

The DTP-PMSM drive system is shown in Fig.1.

B. Current Equations Under Faulty Conditions

An open-switch fault in a two-level inverter may occur in either the top or bottom switches, thereby restricting the current in the positive or negative direction, respectively. In consequence, any phase experiencing a single switch fault will limit the capacity to produce the affected portion of the waveform under fault-tolerant conditions.

The DTP-PMSM phase currents using VSD under healthy operating conditions can be calculated using (2) and (3) as follows:

$$\begin{bmatrix} i_a \\ i_b \\ i_c \\ i_x \\ i_y \\ i_z \end{bmatrix} = [VSD]^{-1}[PT]^{-1}\begin{bmatrix} i_d \\ i_q \\ i_{xst} \\ i_{yst} \end{bmatrix}. \tag{7}$$

For simplicity, phase a is considered to be in a faulty state, and the simplified phase current equation from (7) can be expressed as:

$$i_a = \begin{cases} i_{xst} + i_d\cos(\theta_e) - i_q\sin(\theta_e) \leq 0, top\,switch\,fault \\ i_{xst} + i_d\cos(\theta_e) - i_q\sin(\theta_e) \geq 0, bottom\,switch\,fault \end{cases} \tag{8}$$

As highlighted earlier, a constraint regarding the current polarity is introduced in (8), depending on whether the top or bottom switch is in operation. Similarly, using (7), the remaining healthy phases can be determined as follows:

$$\left.\begin{aligned} i_b &= \frac{\sqrt{3}}{2}(i_q\cos(\theta_e)+i_d\sin(\theta_e)-i_{yst})-\frac{1}{2}(i_{xst}+i_d\cos(\theta_e)-i_q\sin(\theta_e)), \\ i_c &= \frac{\sqrt{3}}{2}(-i_q\cos(\theta_e)-i_d\sin(\theta_e)+i_{yst})-\frac{1}{2}(i_{xst}+i_d\cos(\theta_e)-i_q\sin(\theta_e)), \\ i_x &= \frac{\sqrt{3}}{2}(-i_{xst}+i_d\cos(\theta_e)-i_q\sin(\theta_e))+\frac{1}{2}(i_{yst}+i_q\cos(\theta_e)+i_d\sin(\theta_e)), \\ i_y &= \frac{\sqrt{3}}{2}(i_{xst}-i_d\cos(\theta_e)+i_q\sin(\theta_e))+\frac{1}{2}(i_{yst}+i_q\cos(\theta_e)+i_d\sin(\theta_e)), \\ i_z &= -i_{yst}-i_q\cos(\theta_e)-i_d\sin(\theta_e), \end{aligned}\right\} \tag{9}$$

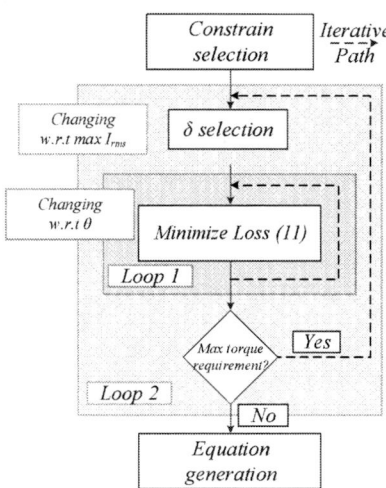

Fig. 2. Optimization process with loop 1 (minimization of 11) and loop 2 (weight optimization).

It can be observed in (9) that, in the proposed optimization algorithm, the i_{xst}-i_{yst} subspace remains in the stationary reference frame to minimize the need for multiple Park transformations.

C. Proposed Loss Function for Optimization

The main objective of fault-tolerant control is to keep the PMSM within thermal stress limits, making it essential to maintain copper losses below the rated amount. In consequence, the losses per phase in a faulty condition should remain within the rated copper loss limit to prevent additional permanent damage to the DTP-PMSM drive system. The losses per phase can be calculated by

$$i_{fault}^2 R_s \le i_{rated}^2 R_s, \tag{10}$$

where i_{fault} is the rms value of an arbitrary phase during fault, and i_{rated} is the rated rms for the DTP-PMSM in normal condition.

Conventionally, the loss minimization objective considered in the literature has been the summation of all phase losses. In this paper, instead of the conventional method, a weighted sum loss function is introduce as

$$Loss = \delta_1(i_a^2 R_s) + \delta_2(i_b^2 R_s) + \delta_3(i_c^2 R_s) + \delta_4(i_x^2 R_s) + \delta_5(i_y^2 R_s)$$
$$+ \delta_6(i_z^2 R_s), \tag{11}$$

where $\delta_1, \delta_2, ... \delta_6$ are the weighting factors assigned to each individual loss contributor ranging from 0 to 1. The weighting factors represents the relative significance or priority assigned to minimizing losses in each phase. For example, δ_1 is assigned to the average loss in phase a, thereby influencing the rms current in phase a as well. The sensitivity of the objective function can be discussed in terms of the impact of each weighting factor on the optimization process. Considering $\delta_1 = 0.5$ to be the normal condition, a higher priority (ranging from 0.5 to 1) would indicate a significant reduction in the loss contribution of phase a, consequently leading to a reduction in the rms of i_a. Conversely, a lower priority of δ_1 either means exclusion from reduction ($\delta_1=0$) or a lower decrement in the phase current rms (ranging from 0 to 0.5). Changes in the

priority can dictate the influence of that particular phase on the total losses.

In the conventional VSD approach, the entire DTP-PMSM is treated as a combined system, making it difficult to control the contribution of each three-phase winding. By employing (11), the loss contributions in individual phases can be adjusted, which can be utilized to achieve a maximum torque control strategy.

D. Proposed Offline Optimization Algorithm for ML and MT

The final loss function can be established by substituting (9) into (11), with (8) used as the constraint equation, depending on the fault condition. Initially, to maintain the losses as low as possible in the DTP-PMSM, the weighting factors are selected to be equal, i.e. $\delta_1, \delta_2, ... \delta_6 = 0.5$. This approach provides a solution in which all phase currents are minimized, regardless of the location of the fault. As mentioned in the introduction, the difference between the proposed method and the conventional approaches, aside from (11), also differs in the constraint equation of i_d. To achieve zero torque ripple, i_d is no longer constrained to be zero; instead, the average of i_d is constrained to zero in this algorithm.

Newton's method is employed to calculate the minimum values of i_d, i_{xst} and i_{yst} with i_q considered to be a constant. This method iteratively utilizes the curvature information of the surface, or the second order derivative, to converge towards a minimized solution. The Newton step can be computed as follows [15]:

$$(i_d, i_{xst}, i_{yst})_{k+1} = (i_d, i_{xst}, i_{yst})_k$$
$$- \nabla^2 f((i_d, i_{xst}, i_{yst})_k)^{-1} \nabla f((i_d, i_{xst}, i_{yst})_k), \tag{12}$$

where k is the current position , ∇f is the gradient and $\nabla^2 f$ is the hessian of the matrix. After minimization, this approach will yield the minimized values for the selected weights, i.e. 0.5. This strategy is categorized as ML control.

Therefore, by simply adjusting the six weighting factors, a different solution can be obtained by varying loss contributions. To achieve MT control, these weights can be iteratively selected to maximize the torque output. To achieve maximum torque in faulty condition T_{fault} is set as

$$T_{fault} = \frac{(i_{rms}^{max})_{fault}}{(i_{rms}^{max})_{normal}} \times T_{normal}. \tag{13}$$

For MT, the fault-tolerant control scheme should minimize i_{rms}^{max} of the DTP-PMSM. This means the factor $(i_{rms}^{max})_{fault}/(i_{rms}^{max})_{normal}$ should approach unity to achieve higher torque. It is proposed to optimize the weights considering (13) using gradient descent defined by [15]

$$(\delta_1, \delta_2 + ... + \delta_6)_{n+1} = (\delta_1, \delta_2 + ... + \delta_6)_n - \gamma \cdot \nabla I_{rms}. \tag{14}$$

In (14), the subscripts n, γ, and ∇I_{rms} represent the current iteration, rate of descent and gradient of the currents in the direction of the weights, respectively.

Fig. 2 consolidates the whole optimization process. Both the iterative paths are highlighted as loop 1 and loop 2. Loop 2 is only in effect if the requirement of maximum torque is

979-8-3315-1612-3/25 $31.00 © 2025 IEEE   1269

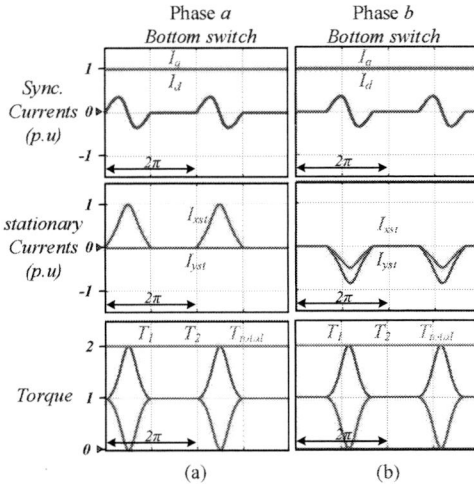

Fig. 3. ML-based optimized current references generated with constant torque (bottom-switch open). (a) Phase *a*. (b) Phase *b*.

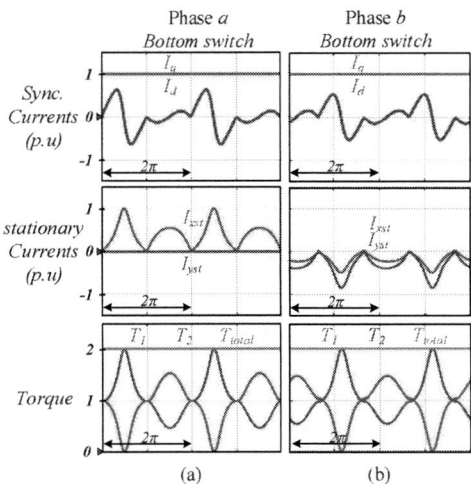

Fig. 4. MT-based optimized current references generated with constant torque (bottom-switch open). (a) Phase *a*. (b) Phase *b*.

enabled. This method is generally applicable across various torque and speed points; however, operation in the field-weakening region would require additional constraints to account for torque degradation.

III. OPTIMIZATION RESULTS FOR ML AND MT, GENERALIZATION AND COMPARISON TO EXISTING METHODOLOGIES

A. Optimized Minimum Loss (ML) Strategy

The primary objectives of this strategy is to minimize the copper losses in the DTP system with an open-switch fault and maintain a ripple-free total average torque. Fig. 3 illustrates the optimized current references when either phase *a* or *b* experiences an open-switch fault.

In Fig. 3(a), the faulty case of the bottom switch in phase *a* is shown, which limits the ability of phase *a* to produce negative current. During ML, a half of the current cycle can still be maintained using conventional healthy state control, with i_d, i_{xst}, i_{yst} all equal to zero. It can be seen in Fig. 3(a) that i_d and i_q in the synchronous reference frame as well as i_{xst} and i_{yst} in the stationary reference frame are zero for a half of the control period, whereas the other half is optimized to keep the total torque (T_{total}) constant. Similarly, the optimized waveforms for the faulty bottom-switch in phase *b* are depicted in Fig. 3(b). The proposed ML technique results in a loss value of *3.61 R_s* and achieves a maximum torque of 0.76 p.u.

B. Optimized Maximum Torque (MT) Strategy

Following the loop 2 optimization process for the open-switch fault in phase *a*, the current references shown in Fig. 4 are obtained. As compared to the ML strategy, the references contain additional harmonics, and the portion from the healthy period mentioned in section III A is no longer applicable. To achieve maximized torque with no torque ripple, the harmonic components are added extensively to the three axes apart from i_q evident by Fig. 4(a) and (b). The total torque achievable in the proposed MT strategy is about 0.87 p.u. with total 6-phase copper losses of about *3.94 R_s*.

C. Reference Current Generalization Using Phase Shift

The currents shown in Fig. 3 and 4 represent the new current references for a bottom open-switch fault in either phase *a* or *b*. The curve fitting toolbox in MATLAB has been used to fit these curves. The curve-fitted equation for MT under bottom-switch fault of phase *a* is as follows:

$$
\begin{aligned}
i_d &- i_q \cdot \{0.6375\sin(2.177\theta - 0.2786) + 0.1441\sin(4.892\theta - 4.543)\} \\
i_{xst} &= i_q \cdot \{0.7937\sin(1.17\theta - 0.267) + 0.1722\sin(3.928\theta - 1.638)\} \\
i_{yst} &= i_q \cdot \{0\}
\end{aligned}
$$
$$\underbrace{\qquad\qquad\qquad}_{for\ 0<\theta_A<\pi,} \qquad (15)$$

$$
\begin{aligned}
i_d &= i_q \cdot \{0.1333\sin(2.025\theta - 3.024) + 0.01088\sin(3.679\theta + 4.635)\} \\
i_{xst} &= i_q \cdot \{0.5901\sin(1.033\theta + 2.985) + 0.0521\sin(2.612\theta - 1.31)\} \\
i_{yst} &= i_q \cdot \{0\}
\end{aligned}
$$
$$\underbrace{\qquad\qquad\qquad}_{for\ \pi<\theta_A<2\pi}$$

Similarly, for bottom-switch fault of phase *b*, it can be expressed as:

$$
\begin{aligned}
i_d &= i_q \cdot \{0.5346\sin(2.12\theta_B - 0.1906) + 0.1092\sin(4.689\theta_B - 4.223)\} \\
i_{xst} &= i_q \cdot \{0.4079\sin(1.121\theta_B - 3.271) + 0.07974\sin(3.737\theta_B - 1.086)\} \\
i_{yst} &= i_q \cdot \{0.7064\sin(1.124\theta_B + 2.924) + 0.1373\sin(3.754\theta_B - 1.209)\}
\end{aligned}
$$
$$\underbrace{\qquad\qquad\qquad}_{for\ 0<\theta_B<\pi,} \qquad (16)$$

$$
\begin{aligned}
i_d &= i_q \cdot \{0.1427\sin(1.925\theta_B + 3.403) + 0.008432\sin(4.178\theta_B + 2.475)\} \\
i_{xst} &= i_q \cdot \{0.2626\sin(1.036\theta_B - 0.2735) + 0.03161\sin(2.462\theta_B - 3.871)\} \\
i_{yst} &= i_q \cdot \{0.4584\sin(1.043\theta_B - 0.2127) + 0.05796\sin(2.408\theta_B - 3.532)\}
\end{aligned}
$$
$$\underbrace{\qquad\qquad\qquad}_{for\ \pi<\theta_B<2\pi}$$

It can be observed that the equations use different θ values, namely θ_a and θ_b, corresponding to the generalization process of the curve-fitted equations.

Evident by the phase current equations, (8) and (9), a generalization pattern can be seen in terms of optimization constrain depending on the phase and switch being faulty. For selecting the optimization equation phase *a* and phase *z* currents lead to similar constrain equations:

979-8-3315-1612-3/25 $31.00 © 2025 IEEE

TABLE I
GENERALIZED PHASE SHIFT AND EQUATIONS

Phase	Top switch	Bottom Switch	Top switch	Bottom Switch
A	$\theta_a = \theta_e - \pi$	$\theta_a = \theta_e$	$i_{xst} = -i_{xst\_a}$ $i_{yst} = i_{yst\_a}$	$i_{xst} = i_{xst\_a}$ $i_{yst} = i_{yst\_a}$
B	$\theta_b = \theta_e + \dfrac{\pi}{3}$	$\theta_b = \theta_e - \dfrac{2\pi}{3}$	$i_{xst} = -i_{xst\_b}$ $i_{yst} = -i_{yst\_b}$	$i_{xst} = i_{xst\_b}$ $i_{yst} = i_{yst\_b}$
C	$\theta_c = \theta_e - \dfrac{\pi}{3}$	$\theta_c = \theta_e - \dfrac{4\pi}{3}$	$i_{xst} = -i_{xst\_b}$ $i_{yst} = i_{yst\_b}$	$i_{xst} = i_{xst\_b}$ $i_{yst} = -i_{yst\_b}$
X	$\theta_x = \theta_e - \dfrac{7\pi}{6}$	$\theta_x = \theta_e - \dfrac{\pi}{6}$	$i_{xst} = -i_{yst\_b}$ $i_{yst} = i_{xst\_b}$	$i_{xst} = i_{yst\_b}$ $i_{yst} = -i_{xst\_b}$
Y	$\theta_y = \theta_e - \dfrac{5\pi}{6}$	$\theta_y = \theta_c + \dfrac{\pi}{6}$	$i_{xst} = -i_{yst\_b}$ $i_{yst} = -i_{xst\_b}$	$i_{xst} = i_{yst\_b}$ $i_{yst} = i_{xst\_b}$
Z	$\theta_z = \theta_e - \dfrac{\pi}{2}$	$\theta_z = \theta_e + \dfrac{\pi}{2}$	$i_{xst} = i_{yst\_a}$ $i_{yst} = i_{xst\_a}$	$i_{xst} = i_{yst\_a}$ $i_{yst} = -i_{xst\_a}$

$$i_a = i_{xst} + i_d \cos(\theta_e) - i_q \sin(\theta_e) \leq 0 \ \ or \geq 0,$$
$$i_z = -i_{yst} - i_q \cos(\theta_e) - i_d \sin(\theta_e) \leq 0 \ \ or \geq 0, \qquad (17)$$

The differences lie in the phase shift and dependency changes, xst-yst axis polarity. The constraint equation for i_a depends on i_{xst} and is independent of i_{yst}, but the opposite is true for i_z. Therefore, the equations for phase a can be applied to phase z, provided that the necessary phase shift and dependency adjustments are made. Similarly, the other four phases (i_b, i_c, i_x and i_y) according to (9), share similar constraint equations for the optimizer and can therefore use the same equations as (16).

Table I highlights the phase shift required for each switch fault and the corresponding adjustments to dependencies. The highlighted cells indicate the standards, with the bottom open-switch fault for phase a taken as the reference and aligned to the rotor electrical angle. The angle shifts in the other phases are based on the half-cycle missing due to the open-switch fault. In conclusion, rather than requiring 12 different sets of equations for ML or MT, the proposed generalization method needs only two sets of equations per fault method.

D. Comparison to Existing Methodologies

A comparison of the proposed FTC strategies with existing methods from the literature is presented in Table II. The table provides a thorough analysis of the i_d constraint, maximum achievable torque, maximum losses, and generalization aspects of each method, as discussed in the literature. Conventional ML (CML) and maximum torque (CMT) strategies yield the least amount of torque and highest amount of losses. As shown by the generalization in [3], $i_d=0$ constraint is simpler to implement in terms of final deployment. A key benefit of the proposed optimization strategy is that all objectives can be achieved by simply modifying the constraint equation, unlike [10], which requires an additional DC expression for offset injection. The proposed ML (PML) and MT (PMT) techniques yield optimal

TABLE II
COMPARISON TO EXISTING METHODS

Method	$i_d \neq 0$	Max torque(p.u.)	Losses (p.u) (Σ I²)	Generalized approach
Normal	✗	1	3.00	✗
CML[5]-[6]	✗	0.55	4.50	✗
CMT[5]-[6]	✗	0.57	6.00	✗
MT1[9]	✗	0.67	3.75	✓
MT2[11]	✗	0.67	3.75	✗
ML[12]	✓	0.77	3.61	✗
MT3[13]	✓	0.66	4.63	✗
MT4[14]	✓	0.73	4.57	✗
PML	✓	**0.76**	**3.61**	✓
PMT	✓	**0.87**	**3.94**	✓

solutions, in addition to providing generalization for conditions where $i_d \neq 0$.

IV. DSP IMPLEMENTATION AND EXPERIMENTAL RESULTS

A. Implementation in DSP

The curve-fitted equations (15) and (16), along with the necessary phase shift, polarity, and dependencies (Table I), can be stored in the DSP and retrieved following any fault detection and diagnosis method. Fig. 5 shows the complete FTC diagram using VSD. The fault mode can be set to 0, 1, or 2, corresponding to the healthy state, ML control, and MT control, respectively. Once a fault is diagnosed, the fault mode shifts from 0 to the specified user mode. The relevant equation is then retrieved, along with the phase shift and dependency values from Table I, to initiate the FTC with minimal reconfiguration.

The experimental setup with the DTP inverter and DTP-PMSM is shown in Fig. 6. The parameters used in the experiments for the ML and MT methods are listed in Table III. The operating conditions for the DTP-PMSM were set at a speed of 1000 rpm and a load torque of 6.4 Nm, representing

979-8-3315-1612-3/25 $31.00 © 2025 IEEE

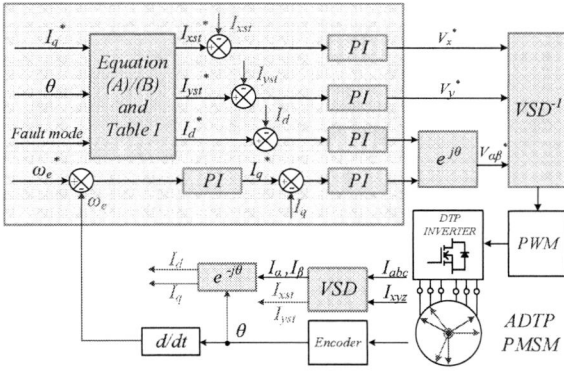

Fig. 5. Block diagram for implementation of FTC.

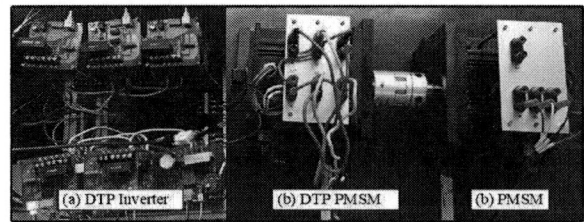

Fig. 6. Experimental setup. (a)DTP inverter. (b)DTP-PMSM. (c)Load PMSM.

TABLE III.
PARAMETERS OF EXPERIMENT

V_{dc}	160 V
P_p (Number of pole pairs)	5
Rated Speed	2000 rpm
Rated Torque	14 Nm
Stator resistance (R_s)	0.21 Ω
Stator inductance $(L_d = L_q)$	0.65 mH
Flux of PM (ψ_f)	0.063 wb
Moment of inertia (J)	0.00698 kgm^2
Switching frequency (f_{sw})	8 kHz

de-rated conditions to ensure safe operation under fault scenarios.

B. Experimental results

Fig. 7 and 8 show the experimental results for the optimized ML and MT techniques applied to the DTP-PMSM. For bottom-switch fault in phase *b*, Fig. 7(a) and (b) show the i_d and i_q in the synchronous reference frame and the i_{xst} and i_{yst} in the stationary reference frame, respectively. The phase currents depicted in Fig. 7(c) and (d) demonstrate the non-sinusoidal nature of the phase current due to additional harmonics. Despite this, Fig. 7(e) confirms that the total torque remains constant, as intended by the optimization.

To highlight the generalization capability of the proposed methodology, the waveforms shown in Fig. 7, in combination with the phase shift and polarity adjustments outlined in Table I, were utilized to generate Fig. 8. This demonstrates that (16) can be adapted for various fault conditions across different phases. By applying the phase shift and polarity changes outlined in Table I, the optimized current references and torque characteristics shown in Fig. 7 can be extended to phase *c*, as illustrated in Fig. 8.

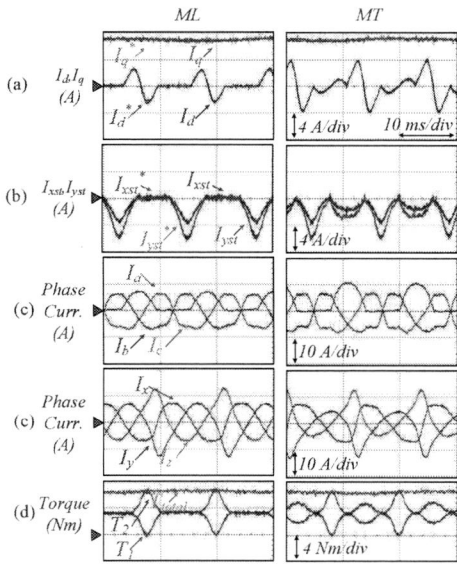

Fig. 7. FTC for bottom switch in phase *b*. (a) I_d and I_q. (b) I_{xst} and I_{yst}. (c) Phase currents (I_{abc}). (d) Phase currents (I_{xyz}). (e) Individual and total torques (T_1, T_2, and T_{total}).

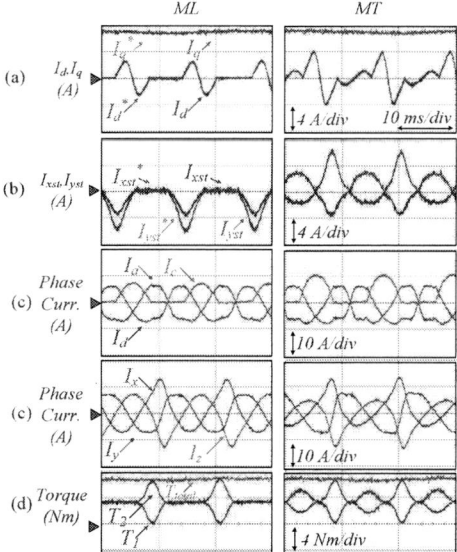

Fig. 8. FTC for bottom switch in phase *c*. (a) I_d and I_q. (b) I_{xst} and I_{yst}. (c) Phase currents (I_{abc}). (d) Phase currents (I_{xyz}). (e) Individual and total torques (T_1, T_2, and T_{total}).

V. CONCLUSIONS

In this paper, an optimization methodology for FTC of DTP-PMSMs has been studied. The proposed approach utilized a weighted sum method to address open-switch faults, offering a flexible optimization of torque and loss profiles. The results have demonstrated that the proposed MT and ML techniques significantly outperformed conventional methods, achieving 0.86 p.u. and 0.76 p.u. torque, respectively, with lower losses. Moreover, a phase-shift-based generalization method was introduced to handle the 12 fault scenarios, providing a manageable approach for non-sinusoidal current references. Experimental results have validated the effectiveness of the proposed strategies, showing minimal torque ripple and confirming the viability of the generalized control approach for multiple fault conditions.

REFERENCES

[1] Z. Zhu, S. Wang, B. Shao, L. Yan, P. Xu, "Advances in dual-three-phase permanent magnet synchronous machines and control techniques," *Energies*, vol. 14, no. 22, p. 7508, 2021, doi: 10.3390/en14227508.

[2] A. G. Yepes, O. Lopez, I. Gonzalez-Prieto, M. J. Duran, and J. Doval-Gandoy, "A comprehensive survey on fault tolerance in multiphase AC drives, part 2: phase and switch open-circuit faults," *Machines*, vol. 10, no. 3, pp. 1–78, 2022, doi: 10.3390/machines10030208.

[3] A. G. Yepes, O. Lopez, I. Gonzalez-Prieto, M. J. Duran, and J. Doval-Gandoy, "A comprehensive survey on fault tolerance in multiphase AC drives, part 1: general overview considering multiple fault types," *Machines*, vol. 10, no. 3, pp. 1–134, 2022, doi: 10.3390/machines10030208.

[4] M. A. Shamsi-Nejad, B. Nahid-Mobarakeh, S. Pierfederici, and F. Meibody-Tabar, "Fault tolerant and minimum loss control of double-star synchronous machines under open phase conditions," *IEEE Trans. Ind. Electron.*, vol. 55, no. 5, pp. 1956–1965, 2008, doi: 10.1109/TIE.2008.918485.

[5] H. S. Che, M. J. Duran, E. Levi, M. Jones, W. P. Hew, and N. A. Rahim, "Postfault operation of an asymmetrical six-phase induction machine with single and two isolated neutral points," *IEEE Trans. Power Electron.*, vol. 29, no. 10, pp. 5406–5416, 2014, doi: 10.1109/TPEL.2013.2293195.

[6] F. Baneira, J. Doval-Gandoy, A. G. Yepes, Ó. Lopez, and D. Pérez-Estévez, "Control strategy for multiphase drives with minimum losses in the full torque operation range under single open-phase fault," *IEEE Trans. Power Electron.*, vol. 32, no. 8, pp. 6275–6285, 2017, doi: 10.1109/TPEL.2016.2620426.

[7] G. Yang, H. Hussain, S. Li, J. Zhang, and J. Yang, "A unified fault-tolerant strategy for multiphase machine with minimum losses in full torque operation range based on closed-form expressions," *IEEE Trans.*

Power Electron., vol. 37, no. 10, pp. 12463–12473, 2022, doi: 10.1109/TPEL.2022.3172389.

[8] W. Huang, X. Zhu, H. Zhang, and W. Hua, "Generalized fault-tolerant model predictive control of five-phase PMSM drives under single/two open-switch faults," *IEEE Trans. Ind. Electron.*, vol. 70, no. 8, pp. 7569–7579, 2023, doi: 10.1109/TIE.2023.3239853.

[9] P. Shi, X. Wang, X. Meng, M. He, Y. Mao, and Z. Wang, "Adaptive fault-tolerant control for open-circuit faults in dual three-phase PMSM drives," *IEEE Trans. Power Electron.*, vol. 38, no. 3, pp. 3676–3688, 2023, doi: 10.1109/TPEL.2022.3223411.

[10] K. Yu, Z. Wang, X. Wang, M. Gu, J. Hang, and S. Ding, "Post-fault strategy of universal control for dual three-phase PMSM under single open-phase fault considering current amplitude," *IEEE Trans. Transp. Electrif.*, vol. PP, p. 1, 2023, doi: 10.1109/TTE.2023.3340871.

[11] J. Sun, Z. Zheng, C. Li, K. Wang, and Y. Li, "Optimal fault-tolerant control of multiphase drives under open-phase/open-switch faults based on DC current injection," *IEEE Trans. Power Electron.*, vol. 37, no. 5, pp. 5928–5936, 2021, doi: 10.1109/tpel.2021.3135280.

[12] X. Wang, Z. Wang, M. He, Q. Zhou, X. Liu, and X. Meng, "Fault-tolerant control of dual three-phase PMSM drives with minimized copper loss," *IEEE Trans. Power Electron.*, vol. 36, no. 11, pp. 12938–12953, 2021, doi: 10.1109/TPEL.2021.3076509.

[13] W. Wang, J. Zhang, M. Cheng, and S. Li, "Fault-tolerant control of dual three-phase permanent-magnet synchronous machine drives under open-phase faults," *IEEE Trans. Power Electron.*, vol. 32, no. 3, pp. 2052–2063, 2017, doi: 10.1109/TPEL.2016.2559498.

[14] L. Jin, Y. Mao, X. Wang, P. Shi, L. Lu, and Z. Wang, "Optimization-based maximum-torque fault-tolerant control of dual three-phase PMSM drives under open-phase fault," *IEEE Trans. Power Electron.*, vol. 38, no. 3, pp. 3653–3663, 2023, doi: 10.1109/TPEL.2022.3222224.

[15] N. K. Vishnoi, *Algorithms for Convex Optimization*, 1st ed. Cambridge: Cambridge University Press, 2021. doi: 10.1017/9781108699211.

AUTHOR INDEX

Abarzadeh, Mostafa 1261
Abbas, Asad 2973
Abotaleb, Youssef 1850
Abrams, Kerry J. 1781
Abramson, Rose A. 291, 2805
Abu-Rub, Omar 3071
Abu-Zaher, Mustafa 2327
Acero, Jesús 2468
Addin, Ali Sharaf 2960
Adeli, Mohammad Hassan 1489
Ademane, Harsha 3133
Adisurja, Ananda Tjakra 1255
Adragna, Claudio 958
Afrasiabi, Seyedeh Nazanin 1279
Afridi, Khurram K. 1640, 1646
Agarwal, Anant 2986
Ahammed, Md Tanvir 2220
Ahmad, Faheem 175
Aider, Youssef 1026
Aiello, Natale 738
Ajmal, Aidha Muhammad 3024
Akamatsu, Keiji 1728
Akter, Tanzila 1844, 2407
Akuta, Hector 761
Alam, Md Didarul 1746
Alam, Muhammad Muneeb 1051, 2569
Alassi, Abdulrahman 3071
Alathamneh, Mohammad 1953
Al-Durra, Ahmed 622, 2871, 3064
Alenezi, Ali 1217
Alexander, Mark 2162
Aleyasin, Seyed Hossein 1408
Ali, Abdelrahman 429
Ali, Jana A. Sheikh 3071
Ali, Kawsar 1529
Alkhatib, Mohamed 1940
Allen, Mark G. 1791
Allgeier, Jan 1919
Allioua, Abdelmoumin 2125
Alou, Pedro 1197
Al-Smadi, Mohammad K. 2779, 2840
Altin, Necmi 1489
Álvarez, Ignacio 3109
Aly, Mokhtar 746, 895, 2290, 2327
Alzahrani, Ahmad 1230
Alzate, Cesar 401
Amano, Yoshiki 3096
Amarathunga, Supun 3030

Amirabadi, Mahshid 1465, 1983
Amitkumar, K. S. 1279
Amler, Adrian 1759, 1767
Amor, Yacine Ayachi 1781
An, Jongchan 3000
Anand, Aniket 1096, 3147
Anand, Sandeep 69
Anantha, Neeraj 1121
Andapally, Bharadwaj Reddy 3119
Andersen, Michael A. E. 246
Anderson, Blake 1850
Ando, Masato 2681
Anekal, Latha 1224
Anjum, Waseah 1148
Antoszczuk, Pablo Daniel 479
Anurag, Anup 9, 442, 1318
Ao, Chengkang 171
Arai, Takamasa 2821
Araki, Hideo 3077
Aravind, G. 1610, 2785
Arduini, Douglas 1159
Asadi, Peyman 2162
Asel, Thaddeus J. 2419
Ashikaga, Toru 2284
Asllani, Besar 2051
Atkinson, Joshua 401
Attanasio, Rosario 3133, 3304
Attukadavil, Jenson Joseph C. 1481
Atwimah, Samuel K. 185, 207
Aunsborg, Thore Stig 175
Avenas, Yvan 1396, 2562
Aygun, Deniz 195
Azzopardi, Stéphane 2718
Bader, Samuel 1681
Bae, Jung-Soo 2228
Bae, Youngmin 3000
Baek, Jaeil 491
Bagci, F. Selin 880
Bahrami-Fard, Milad 930
Bak, Yeongsu 1734
Bakhshai, Alireza 3083
Balakrishnan, Manu 1286
Balamurali, Aiswarya 1096
Balda, Juan C. 27
Balen, Gleisson 1935
Balutto, Mattia 479
Banaie, Amin 1184
Banerjee, Arijit 3089

Bansal, Divyanshu	1610, 2785
Bantemits, Georgios	479
Bao, Mingjun	2628, 2968
Bao, Xiaokun	1143
Barbosa, Peter M.	2002
Barbosa, Peter	9, 442, 1318, 2082, 2296
Barik, Tapas	1184
Barros, Stayner Nóbrega	689
Barzegarkhoo, Reza	90
Basu, Arka	3253
Basu, Shibaji	464
Batard, Christophe	1076
Bau, Plinio	195
Bauer, Pavol	609
Bavi, Danial	385
Bazzi, Ali	2332, 2510
Beckemeyer, Randy	2082
Beig, Abdul R.	2647
Beinarys, Rytis	2009
Belanger, Matthew	2833
Belikov, Juri	1622
Belkhode, Satish	164, 3334
Benson, Mikayla	2413
Bergveld, H.J.	1451
Bertolini, Alessandro	2640
Beura, Kalpana	1940
Beushausen, Steffen	2589
Bezerra, Pedro A.M.	2361
Bhagat, Chinmay	3285, 3291
Bhambay, Rajul	2920
Bhattacharya, Subhashish	370, 552, 1347, 1866
Bhuse, Tejas	1, 54
Biadene, D.	2014
Bian, Fengwei	3312
Bien, Franklin	1629
Blaabjerg, Frede	696, 912, 1501
Blanco, Cristian	1935
Blaquière, Jean-Marc	2718
Blij, Nils Hans Van Der	479
Boby, Mathews	1279
Boisseau, Sébastien	828
Boisson, Guillaume Piquet	1396
Bojoi, Radu	472, 1408
Bolaños, Robert E.	1190
Boles, Jessica D.	1012
Bonanno, Giovanni	1666
Borowy, Bogdan S.	1326
Boroyevich, Dushan	2228
Bosch, Michael	2387
Boutet, Jérôme	828
Bracken, Christopher	544, 3119
Bradford, Paul	2393

Brandão, Danilo I.	1355
Brandão, Dener A. de L.	1355
Briz, P.	147
Briz, Pablo	3109
Brown, Alyssa	231
Brown III, Buck F.	1153
Brückner, Thomas	2960
Bruyere, Paul	2562
Bu, Jiankang	854
Bugade, Vikas	821
Burdío, José-Miguel	2468
Burgos, Rolando	111, 409, 1495, 1551, 2992
Burnett, Hunter	401
Burt, Graeme	3167
Buttay, Cyril	2051
Cairnie, Mark	2228, 2616
Calabretta, Michele	1070
Cammarata, Federica	252
Campbell, Steven	2797
Cao, Hanqing	1810
Cao, Hui	27
Cao, Yue	3036, 3048
Carretero, Claudio	2468
Castro, Alejandro	1427
Catanoso, Matthew	1791
Cattani, Alberto	2640
Cazzaniga, Daniele	958
Cazzitti, Sacha J.	1512
Cerutti, Stefano	738
Cervera, Pedro Alou	788
Cervone, Andrea	1305
Chae, Jongyoon	1899
Chagas, Rafael Bogo Portal	2446
Chakkalakkal, Sreejith	3147
Chakraborty, Shiladri	69
Chambon, Clément	828
Chamorro, Luis Ruiz	788
Chandrasekhar, Nurani	2889
Chang, Che-Wei	1495, 1551, 1564
Chang, Chuan-En	664
Chang, Jun-Yang	16
Chang, Yi-Chun	2143
Chareyron, Mathilde	828
Chatterjee, Bhaskar	1919
Chatterjee, Kallol	821
Chaturvedi, Shivam	2624, 3155
Chaturvedi, V.	1451
Chaudhary, Jai Aditya	3304
Chavarria, Jose	491
Chavez, Fredo	385
Cheema, Muhammad Ali Masood	768
Chellamuthu, Anand	1286

Chen, Cai	2343, 2369, 2426
Chen, Ching-Jan	664, 2131, 2143, 2725, 2735, 2741
Chen, Chun-Yen	938
Chen, Eric	1274
Chen, Guozhu	2059
Chen, Hao	906
Chen, Hongyu	2968
Chen, Hua	518
Chen, Hung-Chi	2687
Chen, Jiahong	1114
Chen, Jiann-Fuh	2043
Chen, Kai-Hui	887
Chen, Kevin J.	1047
Chen, Minjie	139, 349, 510, 566, 1274, 1693, 1882, 2438
Chen, Qiling	2375
Chen, Shih-Gang	938
Chen, Tianxiao	2361
Chen, Ting	2846
Chen, Wanjun	192
Chen, Xi	2628, 2968
Chen, Xingyu	1537, 1741
Chen, Yilun	2535
Cheng, Eric Ka-Wai	3227
Cheng, Jinpeng	1832
Cheng, Kuang-Yao	2157
Cheng, Lin	966, 1687
Cheng, Qi	2236
Cheng, Tzu-Ping	2900
Cheng, Yan	1047
Cheng, Yun-Keng	887
Cheshire, Audrey	682
Chetri, Chandan	3114
Chiu, Huang-Jen	389
Chiu, Jui-Yang	16, 900
Cho, Jaeyong	3187
Choi, Beomseok	491
Choi, Dongho	2311
Choi, Dongmin	1899
Choi, Jinsoo	3187
Choi, Jungwon	1874
Choi, Seokwon	2268
Choi, Seungdeog	943, 951, 1026, 1420, 1858
Choi, Sunghyuk	1659
Choi, Sungjin	3006
Choksi, Kushan	2582
Choo, Vin Loong	1576
Choong, Yin Quen	505
Chowdhury, Vikram Roy	645, 761, 1465, 3059
Chuang, Cheng-Ta	664, 2725
Chung, Henry Shu-Hung	98, 1507, 1582
Ciabattoni, Matteo	1646
Ciardo, S. Yuri	252

Clark, Landon	919
Cobos, Álvaro	1427
Cobos, José A.	1427
Coday, Samantha	971, 2249
Collings, William M.	185
Cong, Yizhou	2986
Contreras-Barrios, René	629
Coomans, Bart	195
Corradini, Luca	1, 54, 334, 2764
Costa, Levy F.	1334, 1341
Costinett, Daniel	3253, 3267
Cox, James	538
Cronin, Jared	2865
Croston, José Andrés Aguilar	2051
Crovetti, Paolo Stefano	738
Cruz, Alfonso	860
Cruz, Mario F.	670
Cui, Hongchang	202
Cui, Wen Tao	1108
Cui, Yujia	2932
Curbow, Austin	1167
D'Amato, Davide	689
Da-Cunha-Alves, Wendell	429
Dai, Hang	3174
Dang, Yongliang	278, 2482
Dannehl, Kai	1774
Dardeer, Mostafa	906
Darvish, Peyman	2453
Das, Shuvangkar Chandra	1184
Datta, Kishalay	1715
Datta, Promit	586
Davari, S. Alireza	2290
De, Vivek	518, 1681
Deboi, Brian	1167
Deboy, Gerald	1444, 2260
Defaz, Samuel	2576, 2582
Dekka, Apparao	2647
Delmar, Aria	1242
Deneke, Niklas	848
Deng, Jianting	3312
Deniz, Erkan	1489
Deppe, Conner	2393
Derbey, Alexis	2562
Desai, Nachiket	1681
Descamps, Anne-Sophie	1076
Deshpande, Ankit Vivek	1459
Dev, Archit	2920
DeVoto, Douglas	1824
Diao, Naizhe	757
Dieckerhoff, Sibylle	2361
DiMarino, Christina	586, 1836, 2228, 2616
Ding, Peiyang	2375

Ding, Wenlong 2713
Divan, Deepak 164, 3334
Do, Huong 491, 1681
Dobakhshari, Sina Salehi 1673
Dominguez, Miguel Alvarez 1640
Dong, Minhai 2075
Dong, ... 111, 1495, 1551, 1564, 2992
Driesen, Johan 3124
Driussi, Francesco 479
Dryden, Daniel M. 2419
Du, Bangli 436, 2752
Duan, Bin 2713
Dujic, Drazen 266, 1063, 1305
Dutta, Soham 711
Dworakowski, Piotr 2051
Eguchi, Shinichiro 2828
Ekuewa, Oluwaseun Isaiah 2973
Elasser, Youssef 510, 566, 2438
Elezab, Ahmed 670
El-Fouly, Tarek H.M. 622
Ellis, Nathan M. 2276
Ellis, Philip 1781
El-Refaie, Ayman M. 1551
El-Refaie, Ayman 1230, 1495
El-Saadany, Ehab F. 622
Elsanabary, Ahmed 746
Elshaer, Mohamed 3155
Emadi, Ali 670, 3147
Endo, Shun 2681
Eni, Emanuel 2746
Enjeti, Prasad 727, 1217, 1459, 3054
Enomoto, Jun 3194
Enslin, Johan 1153
Espinar, Alberto 3100
Espinoza, Angel 214
Estrin, Julia 132
Etta, Dheeraj 1640, 1646
Evzelman, Michael 594
Expósito, Alberto Delgado 788, 1803
Fahimi, Babak 930, 3160
Fahmy, Youssef A. 272
Falkenberg, Niklas 2772
Fan, Junchong 1203
Fan, Yucheng 2981
Farantos, Evangelos 1184
Farivar, Glen G. 1927
Fassi, Youssof 828
Fein, Martin 2348
Feng, Hao 1832
Feng, Kaiyuan 2894
Feng, Wenda 3174
Fernandes, Arnold 1311

Fernandes, Baylon G. 1481
Ferrari, Maximiliano 637
Figueroa, Alejandro 1427
Filho, Braz de J.C. 1355, 1615
Fiore, Michele 1070
Flannery, John 285
Flaten, Paul 682
Forouzesh, Mojtaba 1673, 1892
Forsyth, Andrew J. 1512
Foster, Geoffrey M. 207
Fox, Aidan P. 185
Fox, Matthew 1791
Francés, Airán 868, 3298
Francois, Thomas W. 1311
Frank, Simon 2348
Freeman, Andrew 1242
Fu, Minfan 809, 2846, 3206
Fu, Pengyu 1203, 2986
Fujisaki, Keisuke 1797
Fujita, Jun 1383
Funaki, Tsuyoshi 2813
Funatani, Kenji 2654
Funatsu, Shohei 1237
Furukawa, Akihiko 1383
Gaafar, Mahmoud A. 775, 906, 2327
Gajare, Siddhesh 214
Galamb, Andrew 2527
Gallage, Nirashi Polwaththa 874
Gangadhar, Pratheesh 2920
Gao, Alex 2149
Gao, Ju 171, 225
Gao, Mingze 2992
Gao, Xiang 2846
Gao, Xiaoguang 2070
Gao, Yuan 524, 1034
Gao, Yuntian 278, 2482
Garcia, Enrique 538
García, Pablo 1935
Garcia, Ricardo 214
García, Sofía 3298
García-Espinosa, Antoni 1774
Garza-Arias, Enrique 1459
Gasparini, Alessandro 2640
Gato, Jose 3119
Gautam, Sushanta 185
Gauthier, Jean-Yves 2051
Gauttam, Gaureej 3316
Geboers, Tim 436
Gennaro, Francesco 738
Georgescu, Sorin 180
Georgiev, Daniel G. 185, 207
Gessner, Joerg 1889

Ghanayem, Haneen	1953
Ghartemani, Masoud Karimi	943
Ghitelman, Kolman Puterman	2101
Ghosh, Mohendro Kumar	1326
Ghosh, Prosenjit	2541
Ghosh, Subarto Kumar	1420
Giardine, Francesca	151
Gil, Pablo M.	1701
Ginot, Nicolas	1076
Giuffrida, Simone	472
Gockel, Hendrik	2125
Goetz, Stefan M.	1754, 2846, 3206
Goicoechea, Javier	1427
Gomez-Rivera, Luis F.	1774
Gong, Jiakun	219
Gong, Minxiang	518
Gong, Taehyeon	3006
Gong, Xiaowu	1114
Gonzalez, Reynaldo S.	1190
Gonzalez-Castaño, Catalina	629
Goodrich, Dakota	719
Goto, Akiko	1569
Gouy, Louison	1076
Graber, Lukas	860, 3321
Grainger, Brandon	544, 1326
Green, Andrew J.	2419
Griepentrog, Gerd	2125
Grigoryan, Davit	566, 1882, 2438
Groon, Fabian	90
Guan, Quanxue	895
Guenther, Robert	1203
Guichon, Jean-Michel	2562
Guillod, Thomas	1816
Gunawardena, Pasan	3030
Guo, Heng	2713
Guo, Jiacheng	2375
Guo, Weisheng	3181
Guo, Xiaoqiang	757
Guo, Zhengchen	2703
Guo, Zhongyin	2070
Gurudiwan, Shubhangi	719, 2194
Guthrie, Travis	2162
Gutierrez, Harold	1159
Ha, Jung-Ik	457, 1659, 2268, 2937
Habibolahi, Zahra Sadat	2202
Haddadi, Aboutaleb	1184
Hajisadeghian, Hossein	1666
Halawa, Ali	1473
Hamani, Rachid	1889
Hameed, Aamna Nasir	1673
Hameed, Asad	1972
Han, Yi	103

Hanhart, Michael	2757
Hanna, Rachelle	1396
Hansen, Frederik Lillebæk	2380
Hanson, Alex J.	231, 1121, 2521
Hanson, Alex	77, 2857
Hao, Weijia	2109
Harbi, Ibrahim	895, 2290
Haryani, Nidhi	442
Hasan, Abu Shahir Md Khalid	1844, 2407
Hasan, Md Zakir	1026
Hasan, Syed Imam	1294, 2698
Hassan, Alaaeldien	2327
Hassan, Najam Ul	834
Hassan, Nazmul	1746
Hata, Katsuhiro	1084, 1102, 2284, 2551
Hayashi, Tetsuya	423
He, Bill	3129
He, Binghui	1673
He, JiangBiao	919, 1368
He, Jiayin	171, 225
He, Junlei	3129
He, Xinlong	2066
Heckel, Thomas	1759
Hedenik, Marina	1519
Hedeshi, Hamid Montazeri	2202
Hegde, Anantha	1728
Heinen, Stefan	2757
Heiries, Vincent	828
Heldwein, Marcelo Lobo	2446
Hemming, Samuel	670
Heo, Go Woon	1723
Herbert, Edward	2495
Hernandez, Arturo Sanchez	530
Herzer, Stefan	1286
Higashiyama, Koji	1728
Hiller, Marc	1919, 2348
Hiraki, Eiji	321, 2654
Hiraoka, Toshio	285
Hirase, Yuko	1946
Hisamochi, Hirofumi	1414
Hobart, Karl D.	185, 207
Hoene, Eckart	2361
Hokmabad, Hossein Nourollahi	1622
Hong, Kang	2096
Hontz, Micheal R.	207
Horibe, Masahiro	2821
Hornbuckle, Malachi	363, 2241
Horowitz, Logan	151, 2276
Hosani, Khalifa Al	1940, 2871, 3064
Hossain, Md Maksudul	2407
Hossain, Mohammad Safayet	1184
Hou, Ting	2375

Hou, Zhengming	1913, 2851
Houska, Brad	3334
Howell, Brandon	2162
Hsieh, Chun-Yu	2735
Hsieh, Hsin-Che	815
Hsu, Jun-Ming	938
Hu, Borong	1439, 2597
Hu, Changsheng	2894
Hu, Changyu	3129
Hu, Jhih-Cheng	2692
Hu, Jiangang	2932
Hu, Shoudong	2764
Huang, Alex Q.	1786
Huang, Cheng	505, 1946
Huang, Hao-Ran	664, 2131
Huang, Ming-Shi	938, 2692
Huang, PengHao	1217
Huang, Peng-Hao	727
Huang, Qinghui	1173, 2603
Huber, J.	2014
Huber, Jonas	1318
Huber, Laszlo	442
Hudgins, Jerry L.	2877
Hudgins, Jerry	1850
Huh, Kum-Kang	3174
Hui, Shu Yuen Ron	3275
Hung, Chien-Chih	16, 900
Hung, Yu-Ting	2735
Huo, Zhenguo	2660
Husain, Iqbal	1746
Husev, Oleksandr	1622, 2173
Hussain, Amir	1990
Hwang, Yun Seong	733
Iannuzzo, Francesco	738, 1070
Ibáñez-Muñoz, Esteban	629
Ibrahim, Ahmed	775
Ibrahim, Eltaib Abdeen D.	775
Ibrahim, Hasan	727, 3054
Ibrahim, Mohamed	670
Ide, Tomoya	1946
Ikriannikov, Alexandr	2149
Iliæ, Milan	2764
Ilka, Reza	1368
Imaeda, Yuta	2431
Imaoka, Jun	2431
Imperiali, Luc	1318
Inokuchi, Seiichiro	2356
Inoue, Shuntaro	782
Irie, Yusuke	2828
Ishido, Ryosuke	1797
Ishihara, Masataka	321, 2654
Ishikura, Yuki	3285, 3291
Ishizuka, Yoichi	2828
Ishraq, Naveed	34, 1135
Islam, Md Khurshedul	943, 951
Islam, Md Majharul	2407
Islam, Nasherul	2059
Islam, Sarwar	1824
Ismail, Ahmed H.	2453
Isobe, Takanori	1946
Ito, Yuki	3248
Itoh, Jun-Ichi	21, 48, 2913
Ivimey, Arjun	464
Iwabuchi, Akio	2828
Iwamoto, Motomitsu	1108
Iyer, Rahul K.	157
Iyer, Vignesh	2764
Jacobs, Alan G.	207
Jafarian, Yousefreza	3083
Jahns, Thomas	3174
Jain, Akshat	658
Jain, Praveen	464, 616, 2953, 3083
Jalakas, Tanel	1622
Jalalabadi, Esmaeil	416
Janabi, Ameer	2597
Jayalath, Sampath	3212
Jeong, Seogyong	834
Jeong, Won Hyo	2937
Jerez, Raiphy	2249
Jha, Kunal	1519
Ji, Shengchang	278, 2482
Ji, Shiqi	2857
Ji, Yichao	966, 1687
Ji, Yingfeng	2889
Jia, Xiaoting	1564
Jiang, C.Q.	795, 3181
Jiang, W.L.	1451
Jiang, Wei	2343, 2426
Jiang, Xi	1114
Jiang, Yang	978
Jiang, Yongbin	3220
Jiao, Dong	1913, 2851
Jiao, Yang	416
Jin, Feng	429
Jin, Liyang	1020, 1564
Jin, Sicong	3018
Jin, Zhiyang	860
Jing, Mengmeng	2713
Jo, Hyeonu	3200
Jo, Hyunkyeong	1629
Jochmans, Thomas	258
Johnson, Brian	711
Johnson, Ken	2510
Jørgensen, Asger Bjørn	357, 1034

Jørgensen, Jannick Kjær	175, 357
Joshi, Kishor	943
Juds, Mark A.	1326, 3119
Jung, Jee-Hoon	834
Jung, Jun-Hyung	689
Jurkov, Alexander	124, 132
Kabashima, Takamune	1728
Kachura, Avram	449, 1905
Kai, Toshihiro	423
Kalathy, Abirami	616, 2953
Kallfass, Ingmar	2387, 3241
Kamalapur, Aakash	2228
Kamran,	252
Kanakri, Haitham	2029
Kanathipan, Kajanan	768
Kandeel, Youssef	285
Kang, Byeong-Woo	2948
Kang, Doug	180
Kang, Eunjin	3012
Kang, Gyeong-Gu	566, 2438
Kang, Seung Hyun	733
Kang, Yong	2066, 2343, 2369, 2426
Kano, Yuko	782
Kanungo, Gautam Dey	821
Kar, Narayan C.	1096
Karanth, Shashank	2746
Karimi-Ghartemani, Masoud	1858
Kataoka, Soya	1237
Katsura, Kenshiro	1299
Kaufmann, Maik	1286
Kawahara, Chihiro	2356
Kawamoto, Keisuke	1569
Kawano, Akihiro	1977
Kelkar, Kapil	1519
Kennel, Ralph	895
Kerekes, Tamás	738, 3042
Khaburi, Davood Arab	895
Khadka, Purushottam	1040, 2400
Khalid, Saad	2569
Khalife, Khalil	479
Khan, Faisal	1824
Khan, N.	1451
Khan, Nisar Ahmed	2569
Khan, Shahid Aziz	2624, 3155
Khandelwal, Sourabh	385
Khandla, Dhaval	2920
Khanna, Mudit	854
Khanna, Raghav	185, 207
Khatua, Mausamjeet	1681
Khorasani, Ramin Rahimzadeh	2101
Kim, Byeong-Il	1734
Kim, Chae-Lyn	3200

Kim, Daehyun	3187
Kim, Dong Hwan	1723
Kim, Dongmin	1899
Kim, Han-Gyu	951
Kim, Hongrae	1746
Kim, Hyeon Soo	733
Kim, Jae-Seong	925
Kim, Jaewon	727, 3054
Kim, Jeonghun	761
Kim, Jonghoon	2973, 3000, 3006, 3012
Kim, Jong-Hun	834
Kim, Joon-Seok	1734
Kim, Jungho	1629
Kim, Katherine A.	880
Kim, Minhyeok	3012
Kim, Min-Sik	834
Kim, Myeong-Ho	834
Kim, Namwon	703
Kim, Sung-Oh	2943
Kim, Yura	3006
Kimball, Jonathan W.	1311
Kimpara, Renata	703
Kirtley, James L.	2474
Kisacikoglu, Mithat John	1602
Kishikawa, Ryoko	2821
Kishimoto, Sumiaki	285
Kitano, Junichi	3194
Klidbari, Mohammadreza Khodaparast	2202
Klymenko, Mariia	1590
Knapp, Jeffrey	854
Knappstein, Lukas	2772
Knoll, Jack	2228, 2616
Ko, Bomyeong	3006
Kobayashi, Takumi	3248
Koch, Dominik	2387
Koehler, Andrew D.	185, 207
Koga, Shunsaku	3194
Koga, Takahiro	2828
Kokkonda, Raj Kumar	1347
Kolar, J.W.	2014
Kolar, Johann W.	1318
Kolli, Nithin	1347
Komiyama, Yutaro	3248
Komo, Hideo	2356
Kondo, Hiroki	1102
Kondo, Ryota	2813
Kong, Jiaze	2167
Kong, Jie	2380
Kong, Rui	696
Konishi, Akihiro	3248
Koppolu, Manoj	2920
Korrani, Majid Ghasemi	930, 3160

Kosaka, Takashi	1237, 3096
Koseoglou, Sokratis	479
Kotani, Junichi	579
Kouro, Samir	775, 2327
Kozak, Joseph P.	1211
Kozielski, Kyle	3147
Kragl, Robert	1051
Krishnamoorthy, Harish S.	3316
Krishnamurthy, Harish	1681
Krishnamurthy, Karthik	3129
Krishnan, Sahana	151, 291, 2805
Kritprajun, Paychuda	1184
Ku, Han	900
Kubulus, Pawel Piotr	1034
Kularatna, Nihal	378, 874
Kularatna-Abeywardana, Dulsha	874
Kulasekaran, Siddharth	491
Kumar, Misha	2002, 2082
Kumar, Pavan	530, 3312
Kusaka, Keisuke	3261
Kusunoki, Shigeru	2551
Kutrolli, Uiliam	2332
Kwak, Jin Woong	2541
Kwon, Hyukjae	566, 2438
Kwon, Man Jae	733
Ladhar, Manraj Singh	2322
Laha, Arpan	616, 2953
Lahuerta, Óscar	2468
Lai, Jih-Sheng	815, 1058, 1913, 2851
Lai, Rixin	2138
Lai, Yanwen	1173, 2603
Laird, Ian	2088
Lam, John	768, 2022
Lamar, Diego G.	1701, 1959
Lawniczak, Celine	1129
Lawson, Wayne	1403, 1781
Lazzarin, Telles Brunelli	342
Le, Duc Dung	3155
Le, DucDung	2624
Le, Hoang	2647
Le, Thanh-Long	2718
Leary, Alex M.	2516
Lee, Bonyoung	1629
Lee, Byoung Kuk	733, 1723, 3200
Lee, Byunghun	834
Lee, Chen-Chan	1058
Lee, Dongcheol	3000
Lee, Dong-Choon	1267
Lee, Dongsu	457
Lee, Eun Woo	2311
Lee, Hoi	2236
Lee, Jaea	3012

Lee, Jaehyeong	3006
Lee, Ju-A	3200
Lee, Jun Young	2311
Lee, June-Seok	1734, 2311
Lee, Justin	2138
Lee, Juwon	457
Lee, Kahyun	2937
Lee, Kangbeen	2413
Lee, Kevin	327, 1261, 2907
Lee, Kyo-Beum	925, 2317, 2943, 2948
Lee, Kyungmin	2547, 3281
Lee, Miyoung	3000
Lee, Po-Chang	900
Lee, Seongkyu	3012
Lee, Seunghyun	3012
Lee, Sungjun	3006
Lee, Taewoo	1659
Lee, Ting-Lun	2143
Lee, Wen-Hsuan	2043
Lee, Woongkul	1473, 2413
Lee, Yun-Jin	2317
Lehman, Brad	761, 1465, 1983
Lehmeier, Thomas	1767
Lei, Weihao	1143
Lei, Yiming	3312
Leslie, Alec	401
Leyrer, Thomas	2920
Li, Bing	2932
Li, Chun-I	2741
Li, Duo	307
Li, Haoran	510, 566, 1882, 2438
Li, Heyuan	809, 2846, 3206
Li, Hui	1248, 2075
Li, Jiajun	1590
Li, Lingyun	524
Li, Peidong	3036, 3048
Li, Pengwei	2332
Li, Qiang	202, 299, 429, 498, 1433, 1537, 1557, 1741, 2228, 2488
Li, Ruqi	1159
Li, Sichao	3129
Li, Tien-Sheng	111
Li, Xiang	3312
Li, Xiaoling	1824
Li, Xindong	3212
Li, Xinze	1143
Li, Xuewen	751
Li, Yang	2576
Li, Yanqiao	1590
Li, Yaohua	3220
Li, Yi	1153
Li, Yilei	2035

Li, Yiming 1173
Li, Yuan 761, 1465
Li, Yunwei 3030
Li, Zehui 485
Li, Zhenchao 1305
Lian, Zhina 1090
Liang, Gaowen 1927
Liang, Jingyuan 1108
Liang, Katherine 363, 2241
Liang, Tsorng-Juu 887
Liang, Yaogan 1084
Liao, Hong-Xuan 2692
Liao, Hsuan 2043
Liao, Kuo Fu 2043
Liao, Mian 139, 349, 1882
Libbos, Elie 3089
Lim, Gyu Cheol 2937
Lim, Je-Yeong 1723
Lim, Jong-Hun 1723
Lin, Fanfan 1143
Lin, Jesse 1211
Lin, Jinshu 2075
Lin, Lei 2535
Lin, Qing 409
Lin, River 1159
Lin, Wei-Ren 258
Linares, Daniel Ríos 1375
Liserre, Marco 90, 118, 689, 1148
Liske, Andreas 2348
Liu, Baihan 2343, 2426
Liu, Caifeng 2066
Liu, Chen 3042
Liu, Chien-Lung 2692
Liu, Ching-Yao 1058
Liu, Christopher 1403
Liu, Chun-Hung 1026
Liu, Gao 357, 1034
Liu, Hanbing 3232
Liu, Haoyang 2361
Liu, Hong 2634
Liu, Hongru 2675
Liu, Hualong 1363, 1597
Liu, Jia 751
Liu, Jiahong 3042
Liu, Jiaxin 2343, 2369, 2426
Liu, Jinjie 1114
Liu, Jinjun 751
Liu, Kevin 1274
Liu, Liming 1746
Liu, Ming 2521
Liu, Sijia 2369, 2426
Liu, Wen-Chin B. 315

Liu, Wentao 1090
Liu, Xiaosen 1544, 2556, 2675
Liu, Xiaoshan 429
Liu, Y. 1451
Liu, Yan-Fei 1673, 1892
Liu, Yang 2675, 3321
Liu, Yifu 3129
Liu, Yongjie 1501, 3042
Liu, Yu-Chen 2179
Liu, Yunting 1179
Liu, Zeguo 1687
Liu, Zengyang 2488
Liu, Zhan 2521
Liu, Zhanlei 278, 2482
Liu, Ziheng 171, 225
Locher, Fabrice 2495
Locke, William 3141
Lodge, Finlay 3167
Logi, Sean 880
Long, Haihong 651, 2981
Long, Teng 1439, 2597
Loparo, Kenneth A. 2698
Lope, Ignacio 2468
López, Abraham 1959
Lopez-Torres, Carlos 1774
Lu, Che-Yu 1967, 2900
Lu, Fengwang 98
Lu, Guo-Quan 586, 2228
Lu, Lucas 416
Lu, Mowei 1754, 2846, 3206
Lu, Wei 2117
Luan, Shaokang 1034
Lucía, O. 147
Lucía, Óscar 3109
Luckett, Benjamin 919
Luise, Claudio 2640
Lukic, Srdjan 2527
Lumod, Phen 1159
Luo, Fang 2576, 2582
Luo, Tianming 2035
Lv, Jianwei 2343, 2369, 2426
Ma, D. Brian 2541
Ma, Dingkun 2375
Ma, Guangji 2070
Ma, Hangxiao 978
Ma, Tianlu 3181
Ma, Zhedong 1173, 2603
Ma, Zhiyuan 1786
Ma, Zhuxuan 27
Maaz, Syed Mohammad 1267
Mabuchi, Yuuichi 2681
MacFadyen, Martin 3167

Madadi, Mehrnaz 370
Maddela, Avinash 1715
Maekawa, Sari 2924
Mahbub, S. Tahmid 157
Maheshwari, Anuj 3089
Maji, Sounak 1640
Major, Joshua 1824
Mak, Pui-In 978
Maksimoviæ, Dragan 1, 54, 334, 682, 2764
Malannino, Claudia 252
Mallik, Ayan 34, 1135
Mallik, Ranajay 658
Mandrile, Fabio 472
Manjrekar, Madhav 2883
Mannan, Tahmid Ibne 1420, 1858
Manos, Konstantinos 1274, 1693
Mansour, Mahmoud 719
Mantooth, H. Alan 1844, 2407
Manzoni, Stefano 958
Marcault, Emmanuel 1396
Marellapudi, Aniruddh 164, 3334
Marianne, Julien 828
Marin, Brandon 491
Marquardt, Rainer 2960
Martin, Alexander 1211
Martin, Sébastien 828
Martin, Trent 1, 54
Martinez, Wilmar 238, 258, 436, 2167, 2752, 3124
Martinez-Limia, Alberto 1051
Martins, João R.R.O. 1889
Martins, Rui P. 978
März, Martin 1759, 1767
Mather, Barry 645, 3059
Mathieu, Frédéric 2495
Mathúna, Cian Ó 285
Matiushkin, Oleksandr 1622, 2173
Matsumori, Hiroaki 1237, 3096
Matsumoto, Hirokazu 579
Matsumoto, Yohei 2681
Matsuo, Takayoshi 2932
Mattavelli, P. 2014
Mattavelli, Paolo 2667
Maureira-Riquelme, Ángel 629
Mauromicale, Giuseppe 1070
Mavencamp, Dan 2157
Mazariegos, Pablo 1427
Mazzer, Simone 1444, 2254, 2260
McDonald, Brent 42
McGrew, Tyler 1557
Mekhilef, Saad 746
Mendes, Arthur 2992
Mercier, Patrick P. 315

Metwly, Mohamed Y. 919
Meyer, Stefan 1034
Miao, Honglei 2059
Michelis, Stefano 479
Milivojeviæ, Nikola 1, 54
Min, Hao 1090
Min, Hyungki 1629
Min, Run 2109
Minato, Yuichiro 2913
Mirafzal, Behrooz 2461
Mirkoviæ, Nikola 788
Mishima, Taichi 3248
Mishra, Santanu K. 2213
Mitcheson, Paul D. 1653
Mitrovic, Vladimir 409
Mitsui, Koji 1299
Miyamae, Masaki 2681
Miyanjou, Kazuki 1977
Miyazaki, Tatsuya 1797
Mo, Liping 795
Mo, Xianghao 1375
Mohammad, Mostak 1635
Mohammadi, Sajjad 2474
Mohseni, Parham 2173
Moniruzzaman, Md 943, 951, 1420
Montejano, Misael 637
Monticone, Francesco 1646
Montoya-Acevedo, Diego 629
Moon, Gun-Woo 1899
Moon, Jinyeong 1473, 2413
Moorthy, Radha Sree Krishna 2797
Morris, Lauryn 1311
Moschopoulos, Gerry 1972
Motoori, Shuichiro 1977
Motto, Eric 2356
Mou, Di 2857
Mou, Shin 2419
Mounesi, Reza 1791
Moursi, Mohamed Shawky El 2871, 3064
Mousavi, Mahdi S. 2290
Mu, Qiang 1388, 2790, 3328
Mu, Wei 2597
Mu, Xuchu 978
Muduli, Utkal Ranjan 1940, 2871, 3064
Mueller, Lukas 538
Muenz, Ulrich 1184
Mühlethaler, Jonas 2495
Mujica, Gabriel 868, 3298
Mukhopadhyay, Anwesha 3267
Mukunoki, Yasushige 2356
Müller, Kilian 3241
Mulumudi, Guru Abhilash 1135

Munk-Nielsen, Stig	175, 357, 1034
Murakami, Haruhiko	1569
Muravleva, Ekaterina	1850
Murillo-Yarce, Duberney	1959
Murray, Samantha K.	1905
Murukesan, Karthick	180
Muscat, Isaac	449
Musolino, Francesco	738
Mustakin, Zaheen	1388, 2790
Na, Woonki	2973, 3000, 3006, 3012
Nabila, Kashfia Tajmim	2877
Nabizadah, Ahmad	3160
Nag, Kumar Joy	990, 997
Nagahara, Teruaki	1569
Nagai, Yoshiyuki	423
Nagano, Masanori	285
Nagar, Anshul	2973
Nagasawa, Shinobu	2610
Nagayoshi, Kenichi	1102, 3096
Nakagaki, Akito	2654
Nakagawa, Shigeki	1797
Nakamura, Hirokazu	1728
Nakamura, Keisuke	1237
Nakano, Satoshi	2551
Nakashima, Junichi	2356
Nakata, Yosuke	1383
Nakata, Yuki	21, 2913
Nam, David	2992
Namadmalan, Alireza	2474
Namba, Akira	1797
Namburi, Krishna	1294
Naradhipa, Adhistira M.	498
Narasimhan, Sneha	1866
Narumanchi, Sreekant	1824
Nasiri, Adel	1489, 1791
Nassaji, Abolfazl	2290
Nassar, Rajaie	586, 2228
Nations, Mark	552
Naval, Sourav	1012
Navarro-Rodríguez, Ángel	1935
Neal, Adam T.	2419
Nelms, R.M.	1953, 2703
Nelson, Blake	395, 1167
Nelson, Tolen M.	207
Nelson, Tolen	185
Ng, Wai Tung	983, 1108
Ngo, Khai D. T.	2228
Ngo, Khai	586
Ngo, Minh	111
Nguyen, Allen T.	840
Nguyen, Calvin	1274
Nguyen, Duy T.	231, 1121

Nguyen, Hien	1, 54
Nguyen, Kien	3248
Nguyen, Tung-Tan	389
Ni, Chuan	2117
Nielsen, Morten Rahr	357
Nikmaram, Behnam	2290
Ning, Guangdong	809
Ning, Guangfu	2096
Ning, Shangxian	2660
Nishijima, Kimihiro	1977
Nishimura, Keigo	3096
Nishio, Haruhiko	1108
Nishizawa, Shin-Ichi	2551
Nitta, Honami	1797
Noesges, Brenton A.	2419
Noguchi, Koichiro	1569
Noh, Young-Seok	518
Norman, Patrick	3167
Notake, Koki	1299, 1414
Núñez, Guillermo	1197
Nuzzo, Jeremy	2387
O'Driscoll, Seamus	285, 2009
Oberdieck, Karl	1051, 2589
Oboreh-Snapps, Oroghene	1311
Ochiai, Yuki	3261
Ohi, Toshi	2821
Ohno, Takashi	21
Ohodnicki, Paul R.	544, 2516, 3119
Ohodnicki, Paul	370, 1326
Okamoto, Takahiro	321
Olalla, David	3100
Olimmah, Marshal	395
Onar, Omer C.	1635
Onishi, Hiroyuki	2431
Onuma, Naoto	2681
Opificius, Julian	401
Orabi, Mohamed	775, 906, 2327
Orlando, Tailan	342
Orr, Allison	1211
Oruganti, V.S.R.Varaprasad	801
Ota, Hiroaki	3248
Ou, Shuyu	1501, 3042
Ouyang, Ziwei	246, 252, 1810
Pahlevani, Majid	616, 2953
Pakala, Sriharsh	505
Palani, Praveenkumar	62
Pallantla, Manikanta	2708
Palmal, Manas	1874
Pan, Ci	192
Pan, Qishan	2207
Panja, Pijush Kanti	821
Paplham, Tyler W.	2516

Parashar, Sanket	1347
Paredes-Camacho, Alejandro	1774
Park, Junhyeong	3187
Park, Sung-Bum	3187
Parkhideh, Babak	1388, 2790
Parreiras, Thiago M.	1355, 1615
Pasupuleti, Sai Sushma	3316
Patle, Nagesh	2805
Paul, Sayan	1, 54, 334
Paulino, Glaucio H.	1274
Pavone, Mario Giuseppe	738
Peña-Alzola, Rafael	1935, 3167
Peng, Fang Z.	761
Peng, Hongjie	171, 225
Peng, Xiaochuan	1090
Penof, David	1519
Pereira, Joao	637
Pereira, Lucas	1388, 2790
Pereira, Thiago Antonio	118
Peretz, Mor Mordechai	594
Pérez, Fernando	868, 3298
Pérez, Sara	1197
Perez-Farre, Quirc	1774
Perreault, David J.	132
Perreault, David	124
Petriæ, Ivan Z.	157
Petriæ, Ivan	2764
Petrillo, Gaia	266
Petucco, Andrea	2667
Pfost, Martin	573, 1129, 1576, 2772
Philippe, Antoine	1396
Phukan, Ripun	2082, 2296
Phung, Thanh Hai	195
Picot-Digoix, Mathis	2718
Piel, Joshua J.	2419
Pietrini, Giorgio	670
Pigott, J.	1451
Pilawa-Podgurski, Robert C. N.	151, 157, 291, 558, 2276, 2805
Pillonnet, Gaël	315
Pirson, Nicolas	258
Pizzuto, Matteo	1096
Plum, Thomas	1919
Pong, Man-Hay	389
Pool-Mazun, Erick	1459
Popoviæ, Zoya	682
Porras, David A.	27
Porter, Matthew	1020, 1564
Pou, Josep	1927
Pourjafar, Saeid	2173
Prabhakar, Siva	69
Pradhan, Rachit	670

Prakash, Surya	1940
Preindl, Matthias	272, 1255
Prodiæ, Aleksandar	307, 990, 997
Punjabi, Shobhana	1159
Qahouq, Jaber A. Abu	2779, 2840
Qi, Nianzun	357, 1034
Qian, Ting	2117
Qian, Yijie	524
Qiblawey, Yazan	3071
Qin, Yuan	1564
Qin, Zian	609
Qiu, Tian	1040, 2400
Queiroz, Samuel S.	1334, 1341
Quenette, Vincent	1889
Rabenold, Elizabeth	2249
Radhakrishnan, Kaladhar	491, 1681
Radici, Christian	1403, 1512, 1781
Rafiq, Aamir	395
Rahman, Md Rashedur	943, 951
Rahman, Mohammad Dehan	1844, 2407
Rahouma, Ahmed	27
Rajagopal, Narayanan	1836
Rajpurohit, Chirayu	2764
Raju, Soniya	378
Rallabandi, Vandana	1635, 3174
Ram, Achala	2920
Ramasubramanian, Deepak	1184
Ramirez, Juan	1211
Ramkumar, S.	2708
Ramos, Gabriel V.	1355, 1615
Ramos, Regina	1197, 1375
Ran, Li	1832
Rana, Dilip	1040
Rana, Mandeep S.	2213
Rao, Yifan	1274
Rashid, Syed Saeed	1640, 1646
Rathore, Vikas Kumar	594
Raval, Vishwam	727, 3054
Ravichandran, Krishnan	1681
Rawat, Shubham	1347
Raychowdhury, Arijit	518
Reddy, Narsimha	3054
Redondo, Alejandro	868, 3298
Reinotas, Jurgis	1754
Ren, Linhao	2343
Ren, Sheng	3181
Ren, Xufu	1439
Restrepo, Carlos	629
Rettner, Cornelius	1759
Richardeau, Frédéric	2718
Rikiishi, Yasuhiro	2284
Ripamonti, Giacomo	479

Ristic-Smith, Aleksandar 1529
Rivas-Davila, Juan 363, 2241
Rizkalla, Maher 2029
Rizzolatti, Roberto 1444, 2254, 2260
Roberts, Gianluca 307
Rodgers, Aidan 1242
Rodriguez, Ezequiel Ramos 1927
Rodriguez, Fernando 3100
Rodriguez, José 746, 895, 2290, 2327
Rodríguez, Juan 1701, 1959
Rogers, Daniel 1529
Rogers, Michael 1569, 2356
Ronanki, Deepak 2647
Rong, Mingzhe 3220
Rong, Zhenshuai 1439
Rosa, Bruno M.G. 1653
Round, Simon 1803
Roy, Soham 1121
Rubinic, Jaksa 416
Rueß, Manuel 3241
Ruiz, Juan M. 2002
Ruiz, Juan 442, 2082, 2296
Ruoff, Dominik Alexander 2589
Ruppert, Daniel 1759
Russo, Andrea 252
Ruszczyk, Adam 1803
Sa, Satyam 103, 449
Saberi, Sajad 2840
Sadasivan, Arya 2461
Sadilek, Tomas 401
Sado, Kerry 2833
Saeedifard, Maryam 2051, 3071, 3321
Saelens, Jonathan 1311
Saggini, Stefano 479, 1444, 2260
Saha, Subrata 1237
Saha, Tarak 3174
Sahoo, Subham 696, 912, 1501
Sai, Ranajit 285
Saiga, Kazuma 2551
Saito, Shoji 1569
Saito, Wataru 2551
Sakai, Hiroto 3077
Salari, Omid 3083
Salehi, Maryam 2883
Samanta, Akash 3141
Sambo, Haifah B. 291
Sandoval, Rolando 1459
Sangwongwanich, Ariya 738, 1501, 3042
Sanjakdar, Omar 1396, 2562
Santi, Enrico 2833, 2865
Santos, Ion Leandro Dos 342
Santos Jr., Euzeli Cipriano Dos 2029

Sanusi, Bima Nugraha 246, 1810, 2035
Saraf, Pushkar 77
Sarajian, Ali 895
Sarda, Radhika 62, 1927
Sarlioglu, Bulent 3174
Sarnago, H. 147
Sarnago, Héctor 3109
Sarofim, Seif 449
Sati, Shraf Eldin 622
Sato, Yuji 1383
Sato, Yuki 579
Satterlee, Ryan 3133
Satyamsetti, Vijayakrishna 1403
Sauter, Bailey 2764
Sayed-Ahmed, Ahmed 2932
Sba, Baher Abu 2453
Sbabo, Paolo 2667
Schanen, Jean-Luc 2562
Scheideler, William 1590
Scherer, Yohannes Amilcar Tekle 342
Sebastián, Javier 1959
Sebata, Kohei 1977
Sekiya, Hiroo 3248
Selvarasu, Uthandi 761, 1465
Sen, Paresh C. 1892
Sen, Tanuj 139, 349, 1882
Sengstock, Jonathan 1242
Sengupta, Arkadeb 90, 118, 1148
Seo, Gab-Su 602, 645, 3059
Seo, Seoktae 1629
Sethupandi, Abishek 62
Seugnet, Léo 2718
Shadmand, Mohammad B. 3071
Shafei, Ahmad El 1326
Shah, Shreyas B. 670
Shahbazi, Reza 1179
Shahsavar, Tala Hemmati 1622
Shang, Shuye 3227
Shao, Hang 2138
Shao, Linbo 1020, 1564
Sharma, Mohit 3141
Shen, Andy 3129
Shen, Xiaobing 436, 2167
Shi, Guannan 1564
Shillaber, Luke 2597
Shimada, Takae 2681
Shimosako, Shumei 3077
Shin, Se-Un 834
Shivdikar, Saumil 2400
Shoji, Tomokazu 2821
Shrestha, Niranjan 801
Shu, Wenze 524

Siddiquee, Ashraf 1294, 1602
Silveira, Hector Bessa 342
Sim, Dong Hyeon 3200
Sim, Si Yuan 505
Singh, Anurag 1, 54, 334
Singh, Prashant 1026
Singla, Rishabh 3054
Siraj, Ahmed 1040, 2400
Sitta, Alessandro 1070
Smith, John 637
Solecki, Alex 1242
Solomentsev, Michael 77
Son, Gibong 1741
Song, Chen 2075
Song, Keqi 1507
Song, Minwoo 3012
Song, Qihao 202
Song, Xiaoqing 1844, 2407
Song, Yubo 696
Song, Zhihao 327
Sönmez, Ertuðrul 1051
Soundararajan, Soundhariya G. 238
Souri, Naser 2202
Sowers, Elizabeth A. 2419
Sozer, Yilmaz 1294, 1602, 2698
Spiazzi, Giorgio 2667
Spieler, Matthias 1495, 1551
Sridhar, Sundaramoorthy 299
Sriram, Vaisambhayana B. 62, 1927
Srivastava, Shubham 2213
Starke, Michael 637, 703
Stauth, Jason T. 1590, 1715
Steiner, Mark 2356
Stella, Fausto 1408
Steyaert, Bernard 1255
Steyn-Ross, Alistair 378, 874
Stillwell, Andrew 1242
Stokowski, Nicole 1242
Strache, Sebastian 2569
Strathman, Sophia A. 1311
Streit, Jochen 1051
Strezelecki, Ryszard 2173
Stricula, Justin 401
Sturdivant, Maurice 544
Su, Gui-Jia 1635
Su, Mei 2096
Sugie, Hisashi 2730
Sui, Qingcheng 436, 2752
Sukita, Yohei 1102
Sullivan, Charles R. 840, 1816
Sun, Bosheng 1990
Sun, Kai 2488

Sun, Lingwei 1108
Sun, Peiyuan 2375
Sun, Ruize 192
Sun, Weifeng 524
Sun, Xiuhu 3328
Sun, Zhen 3275
Sun, Ziang 2981
Sund, Jade 971
Sune, Joseph Benzaquen 164, 3334
Suntharalingam, Piranavan 670
Suzuki, Asamira 1728
Swaminathan, Madhavan 2101
Sweet, Mark 3167
Swoboda, Philipp 2348
Syed, Hadiuzzaman 1051
Szczublewski, Austin M. 185
Tadakuma, Toshiya 2610
Taguchi, Koichi 2356
Taha, Wesam 3147
Tajima, Shin 782
Takahashi, Keita 2610
Takahashi, Yoshiaki 1414
Takamiya, Makoto 1084, 1102, 2551
Takamura, Yota 1797
Takayama, Naoki 2681
Takeda, Ryo 2821
Takeuchi, Kosuke 21
Takeuchi, Toshiro 2828
Takishima, Kenta 423
Takizawa, Sota 2924
Tan, Matthew 1882
Tan, Siew-Chong 3227
Tanaka, Kenichiro 579
Tanaka, Ryota 423
Tanaka, Shinsaku 2284
Tanaka, Toshiyuki 2828
Tang, Ho-Tin 1507, 1582
Tang, Wenyuan 1363, 1597
Tang, Yi 3220
Tant, Mike 1824
Tariquzzaman, Md. 3036, 3048
Tarutani, Masayoshi 1383
Tatetsu, Riku 1977
Tayebi, Milad 854
Teng, Fei 2527
Teng, Yiyina 757
Terauchi, Naoya 285
Terzija, Vladimir 757
Thacker, Thimothy 2992
Then, Han Wui 1681
Thevar, Madasamy Palavesha 62
Thike, Rajendra 1279

Thirumoorthi, Sathya Rupan............................1866
Thurlbeck, Alastair P.....................................1602
Tian, Fanghao 2167, 3124
Tian, Jiachen...2327
Tian, Xiaoyang..1754
Tingbari, Vincent Masabiar............................2973
Tomey, Hala..1211
Tomioka, Shohei...579
Tong, Junhong...1786
Tong, Qiaoling 2109
Torres, Javier..1197
Torres, Renato Amorim...................................1495
Touhami, Mustapha 1012
Tran, Ngoc Ho 2569
Trescases, O. 1451
Trescases, Olivier.................. 103, 449, 676, 1905
Tripathi, Anshuman........................... 62, 1927
Tsai, Chieh-Ju 664, 2131, 2143, 2725, 2735, 2741
Tschanz, James 1681
Tseng, Chien-Hao...2725
Tsou, Ming-Chang..887
Tsuchida, Takayuki...285
Tuzizila, Jeremie..401
Uchida, Yasuo...48
Uddarraju, Praneeth 1311
Uddin, Muhammad Fasih 2453
Uegaki, Shin 1383
Uematsu, Takeshi...3248
Ulrich, Burkhard 1707, 2303, 2502
Umanand, L. 1610, 2785
Umar, Jamil...2973
Umar, Muhammad F. 3071
Umetani, Kazuhiro 321, 2654
Ursino, Mario........................... 1444, 2254, 2260
Uzum, Alper........................... 1294, 2698
Vagnon, Eric...2562
Vanderwegen, Wout...238
Varadarajan, Kamal...180
Vasiæ, Miroslav 788, 1375, 1803
Vedula, Inder... 1, 54
Vergès, Gaël..1905
Vico, Enrico..1408
Vines, Peter...................... 1403, 1512, 1781
Vinnac, Sébastien...2718
Vitale, Gianni 3133, 3304
Vohl, Kenny..2757
Wagner, Tomas..3100
Walters, Andrew..951
Wang, Cheng Feng....................... 103, 449
Wang, Daming 2369
Wang, Haiyan 3312
Wang, Haoyu 485, 983, 1544, 2207, 2556, 2675, 2857

Wang, Hongjie..719, 2393
Wang, Huai......................................912, 2380, 3042
Wang, Jin.......................................1203, 2986
Wang, Jinyan...171, 225
Wang, Jun.......................................538, 1850, 2088
Wang, Kaiyuan...3227
Wang, Kejia 505
Wang, Kun..2088
Wang, Kunrong...2162
Wang, Laili..2375
Wang, Lei...2162
Wang, Liang 983
Wang, Libing...3174
Wang, Lichong 757
Wang, Linguo...2070
Wang, Lisheng..1248
Wang, Liwei...1153
Wang, Maojun 225
Wang, Meng...3312
Wang, Mengqi................................2624, 3155
Wang, Pinhe...................................246, 2035
Wang, Qiong 498
Wang, Rudy...9, 1318
Wang, Rui...1063
Wang, Shaozhe...1459
Wang, Shukai...................................566, 1882, 2438
Wang, Shumeng...761, 1465
Wang, Shuo 1173, 2603
Wang, Sunqing...3018
Wang, Wei...2692
Wang, Xiao 219
Wang, Xiaohua..3220
Wang, Xiaosheng 795
Wang, Xiaoting..3030
Wang, Xiaoyu 416
Wang, Xinlin 809
Wang, Xiongfei..1615
Wang, Xuan 1791
Wang, Xuliang..................................1544, 2556, 2675
Wang, Yan 1544, 2556, 2675
Wang, Yao 3275
Wang, Yibo....................................795, 3181
Wang, Yicheng..3147
Wang, Yiju...1368
Wang, Yulei 219
Wang, Yunxin..2675
Wang, Zijian...1537
Wang, Ziyao 485
Wang, Zuoshuai..3018
Watabe, Kiyoto..2551
Watanabe, Hiroki...21, 48
Watanabe, Kenichi 1102, 3096

Wehr, Erik	2757
Wei, Anran	1983
Wei, Bo	327
Wei, Jinxiao	1832
Wei, Xing	3042
Wei, Xuanjing	1403
Wei, Yuxin	2713
Weihs, Leon	2757
Weiser, Mathias C.J.	2387, 3241
Weng, Sheldon	1681
Wens, Mike	195
Wheeler, Patrick	895
Wicht, Bernhard	848
Wick, Lukas	2380
Williamson, Sheldon	801, 1224, 2322, 3114, 3141
Wilson, Marcus	378
Winkler, Joseph	848
Wipprecht, Lukas	2303
Wojewoda, Leigh	491
Wong, Andy	2833
Wouters, Hans	238, 258, 2167
Wright, Jason	401
Wu, Alan	307
Wu, Chih-Chiang	2687
Wu, Hsiang-Kai	2687
Wu, Shang-Syun	2179
Wu, Taotao	1090
Wu, Teng	2634, 2660
Wu, Tsai-Fu	16, 900
Wu, Xin	651, 2981
Wu, Xinke	1995
Wu, Yang	139, 349
Wu, Yanqing	1995
Wu, Yingzhe	1248
Wu, Yue	1114, 3220
Wu, Yuxuan	2582
Wunderlich, Andrew	2865
Wunderlich, Ralf	2757
Xi, Zichen	1020
Xia, Xiaoyi	2022
Xiang, Zhangwei	429
Xiao, Junjie	609
Xie, Biyun	919
Xu, Dehong	651, 2894, 2981
Xu, Guo	2096
Xu, Haoran	2109
Xu, Huangsheng	2207
Xu, Limei	2075
Xu, Shen	524
Xu, Wentao	1012
Xu, Wenzhe	2426
Xu, Xinmiao	1433

Xu, Yun	1114
Xu, Ziyang	2521
Xue, Hui	2117
Xue, Yuxiang	1248
Yabuta, Shigenori	2821
Yagielski, John	3174
Yamaguchi, Koji	1299, 1414
Yamamoto, Keisuke	3194
Yamamoto, Masayoshi	2431
Yamanaka, Kimito	1797
Yan, Decheng	3334
Yan, Yiyang	2343, 2426
Yan, Zhaoheng	1114
Yan, Zhixing	357, 1034
Yang, Bowen	602
Yang, Garam	3000
Yang, Hélène T.W. Ma	983
Yang, Juchen	1203
Yang, Liu	2369
Yang, Qichen	860
Yang, Qiuzhe	1537
Yang, Robert	180
Yang, Xin	1020, 1564
Yang, Xingyu	1953
Yang, Xinliang	409
Yang, Yirui	1173, 2603
Yang, Yongheng	3024
Yang, Yun	3227, 3275
Yang, Zineng	1020
Yao, Wenxi	327
Yao, Yuzhou	1203
Yasko, Mohamed	3124
Yato, Shinji	3077
Ye, Liang	285
Ye, Zhengyu	2117
Yeo, Howe Li	62
Yeo, Sungku	2547, 3281
Yi, Lifang	2413
Yi, Zheyuan	2488
Yin, Shan	1248, 2075
Yin, Tianxiang	2535
Yoneyama, Rei	2356
Yoshimoto, Kantaro	423
Yoshimura, Yuto	2654
You, Longxiang	3018
Youssef, Mohamed Z.	3083
Yu, Hao	2343, 2426
Yu, Jingshu	1681
Yu, Ruiyang	854
Yu, Sheng-Han	2131
Yu, Sheng-Yang	42, 1990
Yu, Wensong	2220

Yu, Xiang	1892	Zhang, Xiong	2343, 2426
Yuan, Hao	2117	Zhang, Yi	2380
Yuan, Huan	3220	Zhang, Yichi	2380
Yuan, Jiaqi	670	Zhang, Yifan	2343, 2369, 2426
Yuan, Jingyi	966	Zhang, Yifu	2746
Yuan, Song	1114	Zhang, Yingjie	2603
Yuan, Tianlong	429	Zhang, Yuanxin	2857
Yuan, Tianshu	2375	Zhang, Yuhao	202, 1020, 1564
Yun, Dam	1659, 2268	Zhang, Yuxin	2973
Zaabi, Omar Al	1940	Zhang, Zhe	2907
Zade, Aditya	719, 1004, 2194	Zhang, Zhenbin	2846, 3206
Zaitsu, Toshiyuki	1977, 2654	Zhang, Zheyu	1040, 1153, 2400
Zaizen, Shohei	2551	Zhang, Zhi Jin	3321
Zaman, Mohammad Shawkat	676	Zhang, Zhining	1203
Zan, Xin	3232	Zhang, Zichen	1850
Zane, Regan	719, 1004, 2194	Zhao, Delin	3220
Zeineldin, Hatem H.	622	Zhao, Fangzhou	1615
Zekorn, Tobias	2757	Zhao, Hongbo	357, 1034
Zeng, Hank	2138	Zhao, Shuang	2369, 3227
Zeng, Jia-En	1967	Zhao, Shuofeng	1824
Zeng, Wenliang	510	Zhao, Tiefu	1388, 2790, 3328
Zeng, Zheng	219	Zhao, Tuo	1274
Zhan, Cao	278, 2482	Zhao, Wending	1995
Zhang, Ben	3181	Zhao, Yifan	2521, 2846, 3206
Zhang, Bing	2070	Zhao, Yue	27, 2453
Zhang, Bo	192	Zheng, Zexiang	2343, 2369
Zhang, Bohua	573	Zhou, Daniel H.	1693
Zhang, Boran	2675	Zhou, Daniel	566, 1274, 2438
Zhang, Boyi	1850, 2296	Zhou, Dao	2380
Zhang, Cheng	1512, 3212	Zhou, Fei	2541
Zhang, Chenghui	2713	Zhou, Feng	2624
Zhang, Chi	9	Zhou, Jiale	1388, 2790, 3328
Zhang, Desheng	2109	Zhou, Kunxiao	809
Zhang, Fuxing	2059	Zhou, Lufan	1803
Zhang, Haijin	3312	Zhou, Mingde	2207
Zhang, Hely	1286	Zhou, Wenqi	2589
Zhang, Heng	2369	Zhou, Xigen	2157
Zhang, Hengbin	1248	Zhou, Yan	1767
Zhang, Hong	2375	Zhou, Yi	651
Zhang, Honglang	2075	Zhou, Yuan	2535
Zhang, Jiazheng	2628, 2968	Zhou, Yuebin	2369
Zhang, Jincheng	978	Zhou, Zongjie	1047
Zhang, Jinfeng	1439	Zhu, Jiaqi	3312
Zhang, Li	2535	Zhu, Jinli	761, 1465
Zhang, Qingzheng	2894	Zhu, Junjie	2070
Zhang, Renjie	3220	Zhu, Lingyu	278, 2482
Zhang, Shengke	214	Zhu, Liyan	1020
Zhang, Shiqi	757	Zhu, Yicheng	558, 2276
Zhang, Tianyi	2066	Zhu, Zhenhai	1995
Zhang, Weihang	978	Zhuo, Fang	2327
Zhang, Xiangrong	3275	Zolfi, Pouya	1230
Zhang, Xin	505, 1143, 3018	Zou, Huanghaohe	1786

Zou, Jiaao..2066
Zou, Jiarui.............................. 558, 2276, 2805
Zou, Mingrui...219
Zou, Xudong...2066
Zou, Xuecheng..2109
Zufferli, Kevin 1444, 2260
Zuo, Yu 258, 436, 2167, 2752
Zuo, Zhiling..2070
Zynger-Capaverde, Betina..............................2562

IEEE
445 Hoes Lane
Piscataway, NJ 08854-4141

ISBN 979-8-3315-1612-3